THE NEW PALGRAVE
DICTIONARY OF ECONOMICS
SECOND EDITION

THE NEW PALGRAVE
DICTIONARY OF ECONOMICS
SECOND EDITION

Edited by Steven N. Durlauf and Lawrence E. Blume

Volume 7   **real balances – stochastic volatility models**

palgrave
macmillan

This edition published 2008 by
PALGRAVE MACMILLAN
Houndmills, Basingstoke, Hampshire RG21 6XS and
175 Fifth Avenue, New York, N.Y. 10010
Companies and representatives throughout the world

PALGRAVE MACMILLAN is the global academic imprint of the Palgrave Macmillan division of St. Martin's Press, LLC and of Palgrave Macmillan Ltd. Macmillan® is a registered trademark in the United States, United Kingdom and other countries. Palgrave is a registered trademark in the European Union and other countries.

ISBN-10 0-230-22643-4   Volume 7
ISBN-13 978-0-230-22643-2   Volume 7

ISBN-10 0-333-78676-9   8-Volume set
ISBN-13 978-0-333-78676-5   8-Volume set

This book is printed on paper suitable for recycling and made from fully managed and sustained forest sources. Logging, pulping and manufacturing processes are expected to conform to the environmental regulations of the country of origin.

A catalogue record for this book is available from the British Library.

Library of Congress Cataloging-in-Publication Data

The new Palgrave dictionary of economics / edited by Steven N. Durlauf and Lawrence E. Blume. – 2nd ed.
      v. cm.
   Rev. ed. of: The New Palgrave : a dictionary of economics. 1987.
   Includes bibliographical references.
   ISBN 978-0-333-78676-5 (alk. paper)

   1. Economics – Dictionaries.   I. Durlauf, Steven N.   II. Blume, Lawrence.   III. New Palgrave.

HB61.N49 2008
330.03–dc22
                                                    2007047205

10   9   8   7   6   5   4   3   2   1
17   16   15   14   13   12   11   10   09   08

Typesetting and XML coding by Macmillan Publishing Solutions, Bangalore, India
Printed in China

Contents

Publishing history

First edition of *Dictionary of Political Economy*,
edited by Robert Harry Inglis Palgrave, in three volumes:

Volume I, printed 1894.
Reprinted pages 1–256 with corrections, 1901, 1909.
Reprinted with corrections, 1915, 1919.

Volume II, printed 1896.
Reprinted 1900.
Reprinted with corrections, 1910, 1915.

Volume III, printed 1899.
Reprinted 1901.
Corrected with appendix, 1908.
Reprinted with corrections, 1910, 1913.
Reprinted, 1918.

New edition, retitled *Palgrave's Dictionary of Political Economy*,
edited by Henry Higgs, in three volumes:

Volume I, printed 1925.
Reprinted 1926.

Volume II, printed 1923.
Reprinted 1925, 1926.

Volume III, printed February 1926.
Reprinted May 1926.

The New Palgrave: A Dictionary of Economics,
edited by John Eatwell, Murray Milgate and Peter Newman.
Published in four volumes.

First published 1987.
Reprinted 1988 (twice).
Reprinted with corrections 1991.
Reprinted 1994, 1996.

First published in paperback 1998.
Reprinted 1999, 2003, 2004.

The New Palgrave Dictionary of Economics 2nd edition,
edited by Steven N. Durlauf and Lawrence E. Blume.
Published in eight volumes

List of entries A–Z

R
(CONTINUED)

real balances

By the term 'real balances' is meant the real value of the money balances held by an individual or by the economy as a whole, as the case may be. The emphasis on real, as distinct from nominal, reflects the basic assumption that individuals are free of 'money illusion'. It is a corresponding property of any well-specified demand function for money that its dependent variable is real balances. Indeed, Keynes in his *Treatise on Money* (1930, vol. 1, p. 222) designated the variation on the Cambridge equation that he had presented in his *A Tract on Monetary Reform* (1923, ch. 3: 1) as 'The "Real-Balances" Quantity Equation'.

Implicit – and sometimes explicit – in the quantity-theory analysis of the effect of (say) an increase in the quantity of money is the assumption that the mechanism by which such an increase ultimately causes a proportionate increase in prices is through its initial effect in increasing the real value of money balances held by individuals and consequently increasing their respective demands for goods: that is, through what is now known as the 'real-balance effect'. This effect, however, was not assigned a role in the general-equilibrium system of equations with which writers of the interwar period attempted to describe the workings of a money economy. In particular, these writers mistakenly assumed that in order for their commodity demand functions to be free of money illusion, they had to fulfil the so-called 'homogeneity postulate', which stated that these functions depended only on relative prices, and so were not affected by a change in the absolute price level generated by an equi-proportionate change in all money prices (Leontief, 1936, p. 192). Thus they failed to take account of the effect of such a change on the real value of money balances and hence on commodity demands. This in turn led them to contend that there existed a dichotomy of the pricing process, with equilibrium relative prices being determined in the 'real sector' of the economy (as represented by the excess-demand equations for commodities), while the equilibrium absolute price level was determined in the 'monetary sector' (as represented by the excess-demand equation for money): (Modigliani, 1944, sec. 13). This, however, is an invalid dichotomy, for it leads to contradictory implications about the determinacy or, alternatively, stability of the absolute price level (Patinkin, 1965, ch. 8).

Nor was the real-balance effect taken account of in Keynes's *General Theory* and in the subsequent Hicks (1937)–Modigliani (1944) IS–LM exposition of this theory, which rapidly became the standard one of macroeconomic textbooks. According to this exposition, the only way in which a decline in wages and prices can increase employment is by its effect in increasing the real value of money balances, hence reducing the rate of interest, and hence (through its stimulating effect on investment) increasing the aggregate demand for goods and hence employment. A further and basic tenet of this exposition was that there was a minimum below which the rate of interest could not fall. So if the wage decline were to bring about a lowering of the rate of interest to this minimum before full employment were reached, any further decline in the wage rate would be to no avail. In brief, the economy would then be caught in the 'liquidity trap'. And even though Keynes had stated in the *General Theory*, 'whilst this limiting case might become practically important in the future, I know of no example of it hitherto' (p. 207), the Keynesian theory of employment was for many years interpreted in terms of this 'trap'.

It was against this background that Pigou (1943, 1947) pointed out that the increase in the real value of money holdings generated by the wage and price decline increased the aggregate demand for goods directly, and not only indirectly through its downward effect on the rate of interest. Pigou's rationale was that individuals saved in order to accumulate a certain amount of wealth relative to their income, and that indeed the savings function depended inversely on the ratio of wealth to income. Correspondingly, as wages and prices declined, the real value of the monetary component of wealth increased and with it the ratio of wealth to income, causing a decrease in savings, which means an increase in the aggregate demand for consumption goods. Pigou's argument (which was formulated for a stationary state) thus had the far-reaching theoretical implication that even if the economy were caught in the 'liquidity trap', there existed a low enough wage rate that would generate a full-employment level of aggregate demand. In this way Pigou (1943, p. 351) reaffirmed the 'essential thesis of the classicals' that 'if wage-earners follow a competitive wage policy, the economic system must move ultimately to a full-employment stationary state'.

In his exposition and elaboration of Pigou's argument (which inter alia brought out the significance of the argument for dynamic stability analysis), Patinkin (1948) labelled the direct effect on consumption of an increase in the real value of money balances as the 'Pigou effect'. However, in subsequent recognition of the fact that this effect is actually an integral part of the quantity theory – as well as the fact that Pigou had been anticipated in drawing the implications of this effect for the Keynesian system by Haberler (1941, pp. 242, 389, and 403) and Scitovsky (1941, pp. 71–2) – Patinkin (1956, 1965) relabelled it the 'real-balance effect' and presented it as a component of the wealth effect.

In an immediate comment on Pigou's article, Kalecki (1944) pointed out that the definition of 'money' relevant for the real-balance effect is not the usual one of currency *plus* demand-deposits: for example, in the case of a price decline, the increase in the real value of the demand deposits has an offset in the corresponding increase in the real burden on borrowers of the loans they had received from the banking system. Thus (emphasized Kalecki) the monetary concept relevant for the real-balance effect in a gold-standard economy is only the gold reserve of the monetary system.

More generally, the relevant concept is 'outside money' (equivalent to the monetary base, sometimes also referred to as 'high-powered money'), which is part of the net wealth of the economy, as distinct from 'inside money', which consists of the demand deposits created by the banking system as a result of its lending operations and which accordingly is not part of net wealth (Gurley and Shaw, 1960). This distinction was subsequently challenged by Pesek and Saving (1967), who contended that banks regard only a small fraction of their deposits as debt, so that these deposits too should be included in net wealth. In criticism of this view, Patinkin (1969, 1972a) showed that if perfect competition prevails in the banking system, the present value of the costs of maintaining its demand deposits equals the value of these deposits, so that the latter cannot be considered as a component of net wealth. This is also the case if imperfect competition with free entry prevails in the system. On the other hand, if – because of restricted entry – the banking sector enjoys abnormal profits, then the present value of these profits should be included in net wealth for the purpose of measuring the real-balance effect.

There remains the question of whether – for the purpose of measuring the real-balance effect – one should include government interest-bearing debt, as contrasted with the non-interest-bearing debt (viz., government fiat money) which is a component of the monetary base. Clearly, in a world of infinitely lived individuals with perfect foresight, the former does not constitute net wealth and hence is not a component of the real-balance effect: for the discounted value of the tax payments which the representative individual must make in order to service and repay the debt obviously equals the discounted value of the payments on account of interest and principal that he will receive. Nor is the assumption of infinitely lived individuals an operationally meaningless one: for as Barro (1974) has elegantly shown, if in making his own consumption plans, the representative individual with perfect foresight is sufficiently concerned with the welfare of the next generation to the extent of leaving a bequest for it, he is acting as if he were infinitely lived.

More specifically, Barro's argument is as follows: assume that an individual of the present generation achieves his optimum position by consuming C_o during his lifetime and leaving a positive bequest of B_o for the next generation. Clearly, such an individual could have increased his consumption to $C_0 + \Delta C_0$ and reduced his bequest to $B_0 - \Delta C_0$ – but preferred not to do so. Assume now that the individual also holds government bonds payable by the next generation, and let the real value of these bonds increase as the result of a decline in the price level, expected to be permanent. The revealed preference of the present generation for the consumption-bequest combination C_o, B_o implies that this increase in the real value of its holdings of government interest-bearing debt will not cause it to increase its consumption at the expense of the next generation. In brief, government debt in this case is effectively not a component of wealth and hence of the real-balance effect.

Needless to say, the absence of perfect foresight, and the fact that individuals might not leave bequests (as is indeed assumed by the life-cycle theory of consumption) means that government interest-bearing debt should to a certain extent be taken account of in measuring the real-balance effect – or what in this context is more appropriately labelled the 'net-real-financial-asset effect' (Patinkin, 1965, pp. 288–94).

If we assume consumption to be a function of permanent income, and if we assume that the rate of interest which the individual uses to compute the permanent income flowing from his wealth is 10 per cent and the marginal propensity to consume out of permanent income is 0.80, then the marginal propensity to consume out of wealth (and out of real balances in particular) is the product of these two figures, or 0.08. However, in the case of consumers' durables (in the very broad sense that includes – besides household appliances – automobiles, housing, and the like), the operation of the acceleration principle implies an additional real-balance effect in the short run. In particular, assume that when the individual decides on the optimum composition of the portfolio of assets in which to hold his real wealth, W, he also considers the proportion, q, of these assets that he wishes to hold in the form of consumers durables, K_d, so that his demand for the *stock* of consumer-durable goods is $K_d = qW$. Assume now that wealth increases solely as a result of an increase in real balances, M/p. This leaves the representative individual with more money balances in relation to his other assets than he considers optimal. As a result he will attempt to shift out of money and into these other assets until he once again achieves an optimum portfolio. In the case of consumers' durables, this means that in addition to his preceding demand for new consumer-durable goods, he has a demand for

$$C_d = \Delta K_d = q[\Delta(M/p)] = q[(M/p)_t - (M/p)_{t-1}]$$

units, where $(M/p)_t$ represents real balances at time t. In general, the individual will plan to spread this additional demand over a few periods. In any event, once an optimally composed portfolio is again achieved, this additional effect disappears, so that the demand for new consumers' durables (which in the case of a stationary state is solely a replacement demand) will once again depend only on the ordinary real-balance effect as described at the beginning of this paragraph (Patinkin, 1967, pp. 156–62).

It is, of course, true that the process of portfolio adjustment generated by the monetary increase will cause a reduction in the respective rates of return on the other assets in the portfolio, so that the initial wealth effect of the monetary increase will be followed by substitution effects. Now, Keynes limited his analysis in the *General Theory* to portfolios consisting only of money and securities; hence (as indicated above) an increase in the quantity of money could increase the demand for goods only indirectly through the substitution effect created by the downward pressure on the rate of interest. But once one takes account of the broader spectrum of assets held by individuals, one must also take account of the direct real-balance effect on the purchase of these other assets as well.

Various empirical studies have shown that the real-balance effect as here defined (viz., as part of the wealth effect) is statistically significant (Patinkin, 1965, note M; Tanner, 1970). Other studies have demonstrated the statistical significance of yet another definition of this effect: namely, as the effect on the demand for commodities of an excess supply of money, defined as the excess of the existing stock of money over its 'desired' or 'long-run' level (Jonson, 1976; Laidler and Bentley, 1983; cf. also Mishan, 1958). It seems to me, however, that such a demand function is improperly specified: for though (as indicated above) the excess supply of money has a role to play in the consumption function (and particularly in that for consumers' durables), the complete exclusion of the real-balance effect *cum* wealth effect from the aforementioned demand function implies that in equilibrium there is no real-balance effect – an implication that is contradicted by the form of demand functions as derived from utility maximization subject to the budget constraint (Patinkin, 1965, pp. 433–8, 457–60; Fischer, 1981).

Granted the statistical significance of the real-balance effect, the question remains as to whether it is strong enough to offset the adverse expectations generated by a price decline – including those generated by the wave of bankruptcies that might well be caused by a severe decline. In brief, the question remains as to whether the real-balance effect is strong enough to assure the stability of the system: to assure that automatic market forces will restore the economy to a full-employment equilibrium position after an initial shock of a decrease in aggregate demand (Patinkin, 1948, part II; 1965, ch. 14: 1). On the assumption of adaptive expectations, Tobin (1975) has presented a Keynesian model with the real-balance effect which under certain circumstances is unstable. On the other hand, McCallum (1983) has shown that under the assumption of rational expectations, the model is generally stable.

In any event, no one has ever advocated dealing with the problem of unemployment by waiting for wages and prices to decline and thereby generate a positive real-balance effect that will increase aggregate demand. In particular, Pigou himself concluded his 1947 article with the statement that such a proposal had 'very little chance of ever being posed on the chequer board of actual life'.

Thus the significance of the real-balance effect is in the realm of macroeconomic theory and not policy.

Correspondingly, recognition of the real-balance effect in no way controverts the central message of Keynes's *General Theory*. For this message – as expressed in the climax of that book, Chapter 19 – is that the only way a general decline in money wages can increase employment is through its effect in increasing the real quantity of money, hence reducing the rate of interest, and hence stimulating investment expenditures; but that even if wages were downwardly flexible in the face of unemployment, this effect would be largely offset by the adverse expectations and bankruptcies generated by declining money wages and prices, so that the level of aggregate expenditures and hence employment would not increase within an acceptable period of time. In Keynes's words: 'the economic system cannot be made self-adjusting along these lines' (ibid., p. 267). And there is no reason to believe that Keynes would have modified this conclusion if he had also taken account of the real-balance effect of a price decline (Patinkin, 1948, part III; 1976, pp. 110–11).

The above discussion has considered only the real-balance effect on the demand for goods. In principle, this effect also operates on the supply of labour: for the greater the real balances and hence wealth of the individual, the greater his demand for leisure as well, which means the smaller his supply of labour. This influence, however, has received relatively little attention in the literature (but see Patinkin, 1965, p. 204; Phelps, 1972; Barro and Grossman, 1976, pp. 14–16).

Another limitation of the discussion is that it deals only with a closed economy. In the analysis of an open economy, the real-balance effect plays an important role in some of the formulations of the monetary approach to the balance of payments.

DON PATINKIN

See also **money illusion; quantity theory of money.**

Bibliography

American Economic Association. 1951. *Readings in Monetary Theory*. Philadelphia: Blakiston.

Barro, R.J. 1974. Are government bonds net wealth? *Journal of Political Economy* 82(November–December), 1095–117.

Barro, R.J. and Grossman, H.I. 1976. *Money, Employment and Inflation*. Cambridge: Cambridge University Press.

Fischer, S. 1981. Is there a real-balance effect in equilibrium? *Journal of Monetary Economics* 8(July), 25–39.

Gurley, J.G. and Shaw, E.S. 1960. *Money in a Theory of Finance*. Washington, DC: Brookings Institution.

Haberler, G. von. 1941. *Prosperity and Depression: A Theoretical Analysis of Cyclical Movements*, 3rd edn. Geneva: League of Nations.

Hicks, J.R. 1937. Mr Keynes and the 'classics': a suggested interpretation. *Econometrica* 5(April), 147–59. Reprinted in *Readings in the Theory of Income Distribution*,

Philadelphia: Blakiston for the American Economic Association, 1946, 461–76.

Jonson, P.D. 1976. Money and economic activity in the open economy: the United Kingdom, 1880–1970. *Journal of Political Economy* 84(October), 979–1012.

Kalecki, M. 1944. Professor Pigou on 'The classical stationary state': a comment. *Economic Journal* 54(April), 131–2.

Keynes, J.M. 1923. *A Tract on Monetary Reform*. London: Macmillan.

Keynes, J.M. 1930. *A Treatise on Money*. Vol. I: *The Pure Theory of Money*. London: Macmillan.

Keynes, J.M. 1936. *The General Theory of Employment, Interest and Money*. London: Macmillan.

Laidler, D. and Bentley, B. 1983. A small macro-model of the post-war United States. *Manchester School* 51(December), 317–40.

Leontief, W. 1936. The fundamental assumption of Mr Keynes' monetary theory of unemployment. *Quarterly Journal of Economics* 51(November), 192–7.

McCallum, B.T. 1983. The liquidity trap and the Pigou Effect: a dynamic analysis with rational expectations. *Economica* 50(November), 395–405.

Mishan, E.J. 1958. A fallacy in the interpretation of the cash balance effect. *Economica* 25(May), 106–18.

Modigliani, F. 1944. Liquidity preference and the theory of interest and money. *Econometrica* 12(January), 45–88. Reprinted in American Economic Association (1951), 186–240.

Patinkin, D. 1948. Price flexibility and full employment. *American Economic Review* 38(September), 543–64. Reprinted with revisions and additions in American Economic Association (1951), 252–83.

Patinkin, D. 1956. *Money, Interest, and Prices*. Evanston, Ill.: Row, Peterson.

Patinkin, D. 1965. *Money, Interest, and Prices*, 2nd edn. New York: Harper & Row.

Patinkin, D. 1967. *On the Nature of the Monetary Mechanism*. Stockholm: Almqvist and Wicksell. Reprinted in Patinkin (1972b), 143–67.

Patinkin, D. 1969. Money and wealth: a review article. *Journal of Economic Literature* 7(December), 1140–60.

Patinkin, D. 1972a. Money and wealth. In Patinkin (1972b), 168–94.

Patinkin, D. 1972b. *Studies in Monetary Economics*. New York: Harper & Row.

Patinkin, D. 1976. *Keynes' Monetary Thought: A Study of Its Development*. Durham, North Carolina: Duke University Press.

Pesek, B.P. and Saving, T.R. 1967. *Money, Wealth and Economic Theory*. New York: Macmillan.

Phelps, E.S. 1972. Money, public expenditure and labor supply. *Journal of Economic Theory* 5(August), 69–78.

Pigou, A.C. 1943. The classical stationary state. *Economic Journal* 53(December), 343–51.

Pigou, A.C. 1947. Economic progress in a stable environment. *Economic* 14(August), 180–88. Reprinted in American Economic Association (1951), 241–51.

Scitovsky, T. 1941. Capital accumulation, employment and price rigidity. *Review of Economic Studies* 8(February), 69–88.

Tanner, J.E. 1970. Empirical evidence on the short-run real balance effect in Canada. *Journal of Money, Credit and Banking* 2(November), 473–85.

Tobin, J. 1975. Keynesian models of recession and depression. *American Economic Review* 65(May), 195–202.

real bills doctrine

The 'real bills doctrine' has its origin in banking developments of the 17th and 18th centuries. It received its first authoritative exposition in Adam Smith's *Wealth of Nations*, was then repudiated by Thornton and Ricardo in the famous Bullionist Controversy, and was finally rehabilitated as the 'law of reflux' by Tooke and Fullarton in the currency–banking debate of the mid-19th century. Even now, echoes of the real bills doctrine reverberate in modern monetary theory.

The central proposition is that bank notes which are lent in exchange for 'real bills', that is, titles to real value or value in the process of creation, cannot be issued in excess; and that, since the requirements of the non-bank public are given and finite, any superfluous notes would return automatically to the issuer, at least in the long run. The grounds for rejecting the real bills doctrine have been many and varied. The main counter-argument is that overissue is not merely possible but inevitable in the absence of any external principle of limitation; in this view, commercial wants are insatiable and excess notes would not return to the issuer but undergo depreciation in the exact proportion to their excess.

By the time the real bills doctrine appeared in the economic literature, fractional reserve banking was already well established, releasing unproductive hoards for trade and investment. This did not satisfy John Law, that 'reckless, and unbalanced but most fascinating genius' (Marshall, 1923, p. 41n.). He outlined a primitive real bills doctrine in the course of his proposal for a land bank, which would issue paper money on 'good security'. He imagined, however, that the need for a metallic reserve was superseded by the abolition of legal convertibility, and that *economic* convertibility would *always* be maintained by conformity with the real bills doctrine (Law, 1705, p. 89).

The problem was that, as a mercantilist, Law identified money with capital; he believed that creating paper money was equivalent to increasing wealth. It was his attempt to 'break through' the metallic barrier that gave him 'the pleasant character mixture of swindler and prophet' (Marx, 1894, p. 441). The spectacular collapse of Law's 'System' set off a negative reaction against financial innovation, which was reflected in Cantillon's 'anti-System' (Rist, 1940, p. 73) and in Hume's opposition to what he called 'counterfeit money' (1752, p. 168). A more positive effect was a shift in the focus of political economy itself to the production process. This shift was led by the Physiocrats and by Adam Smith, whose 'original and profound' (Marx, 1859, p. 168) analysis of money and banking was developed in the context of classical value theory.

A decade before the *Wealth of Nations*, Sir James Steuart had attempted to revive Law's ideas from a 'neomercantilist' viewpoint (1767, book IV, pt. 2). For Smith, by contrast, the role of bank credit was to increase not the quantity of capital but its *turnover* (1776, pp. 245–6; also Ricardo, *Works*, III, pp. 286–7). Output was fixed by the level of accumulation, which for all the classical economists included the speed of its turnover. Credit had the effect both of reducing the magnitude of reserve funds which economic agents needed to hold and of allowing the money material itself – treated as an element of circulating capital and an unproductive portion of the social wealth – to be displaced by paper, thus providing 'a sort of wagon-way through the air'.

Smith followed Law and Steuart, however, in arguing that an overissue of bank notes could not take place if they were advanced upon 'real' bills of exchange, that is, those 'drawn by a real creditor upon a real debtor', as opposed to 'fictitious' bills, that is, those 'for which there was properly no real creditor but the bank which discounted it, nor any real debtor but the projector who made use of the money' (1776, p. 239; also p. 231). When a banker discounted fictitious bills, the borrowers were clearly 'trading, not with any capital of their own, but with the capital which he advances to them'. When, on the other hand, real bills were discounted, bank notes were merely substituted for a substantial proportion of the gold and silver which would otherwise have been idle, and therefore available for circulation (p. 231). The quantity of notes was thus equivalent to the maximum value of the monetary metals that would circulate in their absence at a given level of economic activity (p. 227).

This development of the classical law of circulation applied to credit and fiduciary money alike, with the difference that in the latter case overissue in the 'short run' might result in a permanent depreciation of the paper. By contrast, credit-money, that is, banknotes which were exchanged for real bills, could never be in long-run excess:

> The coffers of the bank ... resemble a water-pond, from which, though a stream is continually running out, yet another is continually running in, fully equal to that which runs out; so that, without any further care or attention, the pond keeps always equally, or very near equally full. (p. 231)

Only what Smith called 'over-trading' would upset this balance, by promoting excessive credit expansion and an accompanying drain of bullion.

Although the real bills doctrine was accepted by the Bank of England Directors as a guide to monetary management, it was challenged in the bullion controversy following the suspension of cash payments in 1797 as 'the source of all the errors of these practical men' (Ricardo,

Works, III, p. 362; also Thornton, 1802, p. 244 and *passim*). In the view of the 'bullionists',

> The refusal to discount any bills but those for *bona fide* transactions would be as little effectual in limiting the circulation; because, though the Directors should have the means of distinguishing such bills, which can by no means be allowed, a greater proportion of paper currency might be called into circulation, not than the wants of commerce could employ but greater than what could remain in the channel of currency without depreciation. (Ricardo, *Works*, III, p. 219)

Indeed, there was no other limit to the depreciation, and corresponding rise in the price level, 'than the will of the issuers' (*Works*, III, p. 226).

Nevertheless, the bullionist argument itself was open to challenge, because it confused money with credit. The inconvertible paper of the Bank Restriction was issued not as forced currency but on loan; it was therefore responsible not for increasing the money supply but simply altering its *composition*, by substituting one financial asset for another in the hands of the public. Only when cash payments were restored, however, was any further attempt made to rehabilitate the real bills doctrine, this time as the 'law of reflux': provided notes were lent on sufficient security, 'the reflux and the issue will, in the long run, always balance each other' (Fullarton, 1844, p. 64; Tooke, 1844, p. 60). The 'Banking School' called this law 'the great regulating principle of the internal currency' (Fullarton, 1844, p. 68). Their opponents, the 'Currency School' orthodoxy, 'never achieved better than this average measure of security'; and, after all, the average 'is not to be despised' (Marx, 1973, p. 131). The real bills doctrine made its next appearance in the Federal Reserve Act of 1913. In banking at least, discretion has always been the better part of valour.

ROY GREEN

See also **Banking School, Currency School, Free Banking School.**

Bibliography

Cantillon, R. 1755. *Essai sur la nature du commerce en general.* trans. H. Higgs. London: Macmillan, 1931.

Fullarton, J. 1844. *On the Regulation of Currencies.* London: John Murray.

Hume, D. 1752. *Essays, Literary, Moral and Political.* London: Ward, Lock & Co., n.d.

Law, J. 1705. *Money and Trade Considered.* Edinburgh: Anderson.

Marshall, A. 1923. *Money, Credit and Commerce.* London: Macmillan.

Marx, K. 1859. *A Contribution to the Critique of Political Economy.* Moscow: Progress Publishers, 1970.

Marx, K. 1894. *Capital.* vol. 3, Moscow: Progress Publishers, 1971.

Marx, K. 1973. *Grundrisse*. Harmondsworth: Penguin.

Ricardo, D. 1951–73. *The Works and Correspondence of David Ricardo*, ed. P. Sraffa. Cambridge: Cambridge University Press.

Rist, C. 1940. *History of Monetary and Credit Theory from John Law to the Present Day*. London: Allen & Unwin.

Smith, A. 1776. *An Inquiry into the Nature and Causes of the Wealth of Nations*. London: Routledge, 1890.

Steuart, J. 1767. *An Inquiry into the Principles of Political Oeconomy*. Edinburgh: Oliver & Boyd, 1966.

Thornton, H. 1802. *An Enquiry into the Nature and Effects of the Paper Credit of Great Britain*. London: LSE Reprint Series, 1939.

Tooke, T. 1844. *An Inquiry into the Currency Principle*. London: LSE Reprint Series, 1959.

real bills doctrine versus the quantity theory

Drawing on two very different hypotheses about the link between nominal money and economic activity, the real bills doctrine and the quantity theory of money represent sharply divergent advice on the conduct of monetary policy. The quantity theory has many prominent advocates, but the real bills doctrine has had a dominant influence in the history and practice of central banking. Further, the real bills doctrine was at the core of the Congressional act creating the US Federal Reserve System so that its importance echoes down to the current day.

The real bills doctrine views money as playing a decidedly passive role, calling for monetary expansion in line with economic activity. According to this view, economic activity is linked to business trade credit and the issuance of short-term debt instruments. Banks should freely purchase these 'real bills' with banknote issue, where the modifier 'real' refers to short-term debt instruments used to finance productive activity as opposed to speculation. The doctrine dates to at least 1705 with the publication of *Money and Trade Considered* by John Law, who suggested that banknote issue should be secured by and thus linked to the nominal value of land. The most famous statement of the doctrine is by Adam Smith, whose linkage of note issue to bills of exchange gave the doctrine its name:

> When a bank discounts to a merchant a real bill of exchange drawn by a real creditor upon a real debtor, and which, as soon as it becomes due, is really paid by that debtor; it only advances to him a part of the value which he would otherwise be obliged to keep by him unemployed, and in ready money for answering occasional demands. The payment of the bill, when it becomes due, replaces to the bank the value of what it had advanced, together with the interest. The coffers of the bank, so far as its dealings are confined to such customers, resemble a water pond, from which, though a stream is continually running out, yet another is continually running in, fully equal to that which runs

out; so that, without any further care or attention, the pond keeps equally, or very nearly full. (1776, p. 304)

Smith's water-pond metaphor illustrates the real-bills view that note issue would be self-regulating when tied to economic activity, that is, money issue could never be excessive when issued against short-term commercial bills.

The fundamental criticism of the real bills doctrine is that the value of commercial bills (or, in Law's case, the value of land) is tied proportionately to the price level. A commercial bill necessarily includes the dollar value of the goods or services to which it is linked. Thus, under the real bills doctrine, nominal note issue is tied to the nominal price level. If the price level is influenced by the money supply, then we have a circularity problem: nominal prices determine note issue, and note issue affects prices. Henry Thornton first noted the danger of this inflationary circle in his 1802 *An Enquiry into the Nature and Effects of the Paper Credit of Great Britain*. (David Ricardo was also a prominent opponent of the doctrine.) The thrust of Thornton's criticism was that the real bills doctrine provided no limit on banknote issue. Smith seems to have avoided Thornton's criticism because in Smith's system the gold standard provided an overall restraint on note issue. An excessive banknote issue would result in a bank losing its gold holdings, and see a drain on its 'coffers'. (See Laidler, 1981; 1984, for a defence of Smith.) But in a world with an inconvertible paper currency Thornton's inflationary critique is devastating.

Humphrey (2001) provides an algebraic description of the real bills doctrine. Suppose that the needs for trade credit are proportional to nominal production, PY, where P denotes the price level and Y denotes real production. The real bills doctrine would imply that banknote issue and thus the money supply (M) should be proportionally linked to the needs of trade credit so that we have:

$$M = kPY$$

where k is the constant of proportionality between trade credit and nominal production. The Thornton inflationary critique is now obvious: even with an exogenous level of output (Y), there is no way of determining the two endogenous variables, the money supply (M) and price level (P). A real bills counter-argument would be that the price level is exogenous to money, that is, the money supply has no direct effect on prices. As discussed below, the quantity theory makes the exact opposite claim.

The real bills doctrine and the Great Depression

Remarkably, the real bills doctrine survived Thornton and Ricardo's withering 19th century criticism to find a central place in 20th century US monetary history. In a fascinating account, Meltzer (2003) and Humphrey (2001) trace the flowering of the real bills doctrine into the US Federal Reserve Act of 1913. US Federal Reserve Banks existed for the purpose of 'accommodating commerce and business' and were supposed to discount only 'eligible paper', which the Act defined as 'notes, drafts, and bills of exchange arising out of actual commercial transactions'. Although, like Adam Smith, the Act presumed the existence of the gold standard, the real bills doctrine was deemed sufficient even in the absence of a specie constraint. For example, in the *Tenth Annual Report* (1924) of the Board of Governors of the Federal Reserve System, it is noted that 'there is little danger that the credit created and distributed by the Federal Reserve Banks will be in excessive volume if restricted to productive issues' (1924, p. 28). The Report further suggested no link between money and prices: 'The interrelationship of prices and credit is too complex to admit of any simple statement' (1924, p. 32). Adolph Miller, a founding member of the Federal Reserve Board and co-author of the Report, rejected the notion that 'changes in the level of prices are caused by changes in the volume of credit and currency...or that changes in the volume of credit and currency are caused by Federal Reserve policy' (quoted in Meltzer, 2003, pp. 187–8).

Meltzer (2003) convincingly argues that it was this belief in the self-regulating nature of the real bills doctrine that led the Federal Reserve to stand idly by as the US economy spiralled into the Great Depression in the early 1930s. From a real-bills perspective, monetary policy was very loose during these years because Reserve Banks stood ready to discount bills at historically low nominal rates of interest. Meltzer (2003, p. 321) concludes that

> the real bills doctrine implied that the correct policy was a passive one. Most [Federal Reserve] governors had always held these views ... The economies of the United States and much of the rest of the world became victims of the Federal Reserve's adherence to an inappropriate theory and the absence of basic economic understanding such as that developed by [Henry] Thornton and [Irving] Fisher.

The quantity theory

In sharp contrast to the real bills doctrine, the quantity theory held as its fundamental principle that the quantity of nominal money (M) is largely exogenous and is the principal force determining the endogenous price level (P). This argument was first articulated by David Hume (1752). An immediate corollary is that changes in the price level, that is, inflation, are primarily determined by movements in the supply of money. In the words of the celebrated quantity theorist Milton Friedman (1956, pp. 20–1):

> there is perhaps no other empirical relation in economics that has been observed to recur so uniformly under so wide a variety of circumstances as the relation between substantial changes over short periods in the stock of money and in prices; the one is invariably linked with the other and is in the same direction; this uniformity is, I suspect, of the same order as many of the uniformities that form the basis of the physical sciences.

The quantity theory's causal link between M and P included the concept of long-run monetary neutrality: exogenous changes in M would eventually be exactly matched by proportional changes in P. This inference is grounded on the stability of real money demand. In the words of Friedman: 'The quantity theory is in the first instance a theory of the demand for money' (1956, p. 4); 'The quantity theorist accepts the empirical hypothesis that the demand for money is highly stable – more stable than functions such as the consumption function that are offered as alternative key relations' (1956, p. 16). If we let $L(R,Y)$ denote real money demand as a function of the nominal interest rate (R) and the level of real production (Y), we have a money market equilibrium condition given by:

$$L(R, Y) = \frac{M}{P}.$$

The proportionality hypothesis is then quite clear: for a stable level of L, exogenous changes in M must be matched by changes in P of the exact same magnitude.

The quantity theory also included the concept of short-run non-neutrality. In the words of Hume (1752, p. 38):

> When any quantity of money is imported into a nation, it is not at first disposed into many hands but is confined to the coffers of a few persons, who immediately seek to employ it to advantage ... It is easy to trace the money in its progress through the whole commonwealth, where we shall find that it must first quicken the diligence of every individual before it increase the price of labour.

'There is always an interval before matters be adjusted to their new situation' (1752, p. 40). Quantity theorists would argue that increases in M are initially met by increases in production (Y) and declines in interest rates (R), but that in the long run R and Y would return to their original levels and that P would thus fully reflect the new higher level of M.

The quantity theory is closely associated with the quantity equation which can be derived as follows. The previous money demand relationship can be re-written as

$$M \frac{Y}{L(R, Y)} = PY.$$

If we define the velocity of money as

$$V \equiv \frac{Y}{L(R, Y)}$$

then we can write this relationship as the celebrated quantity equation:

$$MV = PY.$$

This is Pigou's (1927) variant of Irving Fisher's (1922) classic equation of exchange. The quantity equation is a useful device for expositing the two central tenets of the quantity theory of money: (*a*) in the long run, output (*Y*) and velocity (*V*) are exogenous to money, so that exogenous movements in the money supply (*M*) are met by proportional movements in prices (*P*), and (*b*) in the short run, movements in the money supply are met by some combination of movements in velocity, prices and output, so that changes in *M* have non-neutral effects on output. The quantity equation can also be used to illustrate Thornton's inflationary critique of the real bills doctrine. For a given level of the nominal rate and an exogenous level of production, velocity is determined by the money demand function, but there is no restriction on the size of *M* or the size of *P*.

The contemporary policy debate

From the vantage point of the outset of the 21st century, there is a sense in which the quantity theory has won numerous intellectual battles but lost the war. Most economists subscribe to the principles of long-run monetary neutrality and short-run non-neutrality. Most would also agree that the quantity equation can be a useful intellectual organizing device. Finally, a standard result in any monetary theory course is the nominal indeterminacy that arises under an exogenous interest-rate operating procedure (for example, Sargent, 1987, ch. 4). This result is just the modern statement of Thornton's 1802 criticism of the real bills doctrine. Hence, it would appear that the quantity theory is in the ascendant.

But remnants of the real bills doctrine are pervasive in both monetary policy implementation and theoretical work. In terms of policy, essentially all central banks in the industrialized world typically ignore or downplay movements in monetary aggregates and instead conduct monetary policy according to an interest rate operating procedure, a close descendant of a real-bills policy. The rationale for such a policy choice is the assertion that the demand for money and thus velocity are unstable. Such a policy implies seasonal movements in monetary aggregates to accommodate movements in real activity, a passive money supply movement that is directly out of a real-bills playbook.

From a theoretical perspective, there have been two prominent recent contributions in favour of interest rate policy. First, Sargent and Wallace (1982) provide something of a rehabilitation of the real bills doctrine by developing a model in which fluctuating nominal interest rates are harmful, and in which a policy of pegging the nominal interest rate at zero is Pareto efficient. Second, Woodford (2003) has pioneered an effort to conduct monetary policy analysis in 'cashless' models – models in which the price level is well defined even though there is no money in the model and the central bank follows an interest-rate operating procedure. We review each of these contributions in turn.

Sargent and Wallace (1982) consider a two-period-lived overlapping-generations model in which fiat money

is held even though nominal interest rates are positive because of a legal restriction on private real lending. There are three types of agents: poor savers, rich savers, and borrowers. Using their logarithmic preference specification, the two classes of savers have a constant desired level of savings, say, S^P for the poor and $S^R \gg S^P$ for the rich. The borrowers have a demand for loans given by

$$D^L = \frac{D}{1 + r}$$

where r is the real interest rate, and $D > S^R$. (Sargent and Wallace, 1982, consider the case in which the demand for loans fluctuates deterministically, but this is unimportant for their basic result.) The legal restriction is that borrowers cannot issue small-denomination notes. Hence, poor savers cannot lend directly to the borrowers, but can only save by accumulating fiat money. The equilibrium conditions for the money and credit markets are given by:

$$\text{Money market}: \quad S^P = \frac{M_t}{P_t}$$

$$\text{Credit market}: \quad S^R = \frac{D}{1 + r_t}$$

where M_t and P_t denote the time-t money supply and price level, and r_t is the real rate of interest. Under what Sargent–Wallace call a 'quantity-theory' regime, the central bank keeps the money supply fixed at some $M_t = M$. In this case, the price level and the real interest rate are constant and calculated from the above equilibrium conditions. This equilibrium is clearly not Pareto optimal as agents do not face the same inter-temporal rate of return – that is, rich savers earn a return of $r > 0$, while poor savers earn a zero real return on currency holdings.

Under a 'real-bills' regime the central bank stands ready to lend cash at a zero nominal rate of interest so that

$$(1 + r_t) = \frac{P_t}{P_{t+1}}.$$

In particular, the central bank purchases the 'real bills' issued by the borrowers. To finance these purchases the central bank creates the new fiat money denoted by N_t. The borrowers can then use this cash to purchase goods from the poor savers. By purchasing the borrowers' bonds with fiat money, the central bank is effectively opening up an avenue by which poor savers can lend to borrowers. Without this central bank intervention, the positive nominal rates in the credit market are symptoms of a problem – the inability of a fixed money stock to promote proper credit allocation. The real bills equilibrium conditions are given by:

$$\text{Money market}: \quad S^P = \frac{M_t + N_t}{P_t}$$

Credit market : $\dfrac{N_t}{P_t} + S^R = \dfrac{D}{(P_t/P_{t+1})}.$

Combining, we have that an equilibrium under the real-bills regime is defined by a price sequence that satisfies:

$$S^P + S^R = \frac{M_t}{P_t} + \frac{D}{(P_t/P_{t+1})}.$$

Solving, we have:

$$P_t = \left(\frac{D}{S^P + S^R}\right)P_{t+1} + \left(\frac{1}{S^P + S^R}\right)M_t.$$

Assuming $D < (S^P + S^R)$, the set of stationary equilibria are given by

$$P_t = \left(\frac{1}{S^P + S^R}\right)\sum_{j=0}^{\infty}\left(\frac{D}{S^P + S^R}\right)^j M_{t+j}$$

where the path of the money supply is free. In the special case in which the money supply grows at a constant rate g we have

$$P_t = \left(\frac{1}{S^P + S^R - D(1+g)}\right)M_t.$$

Note that, if g becomes large enough, the monetary equilibrium disappears.

Sargent and Wallace restrict the analysis to a particular equilibrium in which the beginning-of-period money supply is held fixed, $M_t = M$ for $t = 0, 1, 2, 3 \ldots$. However, the money supply grows and contracts *within* each period as the central bank accommodates the supply of one-period bonds issued by the borrowers ('real bills') with the passive expansion of N_t. In this equilibrium the price level is constant and the real return on savings is zero. This equilibrium is Pareto efficient, in contrast to the Pareto inefficiency of the quantity-theory regime. This is an argument in favour of the real bills doctrine and represents Sargent and Wallace's rehabilitation of the doctrine.

There are difficulties with this conclusion. First, the real-bills equilibrium selected by Sargent and Wallace does not Pareto-dominate the quantity-theory regime (rich savers are worse off under the real-bills regime). Second, there is an infinite number of other real-bills equilibria, all defined by the behaviour of the money stock, and not all of these are Pareto efficient. For example, if the money supply grows at a constant rate $g > 0$ the real-bills equilibrium is not Pareto efficient. Finally, Thornton's inflationary critique of the real-bills regime endures: since the money supply is entirely free, there are no restrictions on the short-term and long-term price level.

The second body of recent theoretical work that has a real-bills flavour is provided by Woodford (2003). The title of Woodford's treatise is *Interest and Prices*, a title that makes clear a principal assertion in the work: the money supply is largely irrelevant to price-level determination. The key relationship in the work is the Fisher equation linking nominal rates (i_t) to inflation rates and real rates (r_t):

$$i_t = r_t + p_{t+1} - p_t$$

where p_t is the log of the price level. For simplicity let us suppose that the real rate is exogenous. If the central bank conducts policy according to an exogenous nominal interest rate policy, then the Fisher equation uniquely determines the growth rate of prices (the inflation rate), but not the level of prices. This is, again, the Thornton critique of the real bills doctrine. But Woodford assumes that the central bank follows an endogenous interest rate policy in which the nominal rate responds to movements in prices:

$$i_t = \alpha p_t.$$

Assuming that $\alpha > 0$, the unique stationary equilibrium is given by:

$$p_t = \sum_{j=0}^{\infty}\left(\frac{1}{1+\alpha}\right)^{j+1} r_{t+j}.$$

From a quantity-theory perspective this is a remarkable conclusion: the price level is determined without any mention being made of the money supply. Where is the money demand curve? Either it does not matter (as the money supply moves passively to hit the interest rate target) or it does not even exist (a 'cashless' world). Woodford's (2003) analysis thus rejects the quantity theory as a useful guide for policy, and at the same time provides a 21st-century response to Thornton's 19th-century critique of the real bills doctrine: the money supply should be adjusted passively to hit the interest-rate target (as under a real-bills policy), but the interest-rate target should be moved endogenously to ensure price-level stability.

In the intellectual clash of ideas there are typically no clear winners or losers, but instead a synthesis of the combatants. This is surely true of the debate between the real-bills doctrine and the quantity theory of money. Current monetary policy practice and theory has a notable real-bills flavour in the near-universal use of interest rates as the operating target. To repeat, the advantage of such a policy is that it allows the money supply to respond automatically to and thus accommodate natural movements in real economic activity. But Thornton and the quantity theorists provide a cautionary critique: under an exogenous interest rate policy, there is no way of limiting the inflationary circle between note issue and the price level. To respond to this quantity-theory critique, Woodford (2003) and others have proposed an endogenous interest-rate policy of the form outlined above. This is just one manifestation of the synthesis of the two combatants in this intellectual debate.

TIMOTHY S. FUERST

See also **monetarism; quantity theory of money; real bills doctrine.**

Bibliography

Board of Governors of the Federal Reserve System. 1924. *Tenth Annual Report of the Federal Reserve Board: Covering Operations for the Year 1923.* Washington: Government Printing Office.

Fisher, I. 1922. *The Purchasing Power of Money,* Reprinted, 2nd edn. New York: August M. Kelley, 1963.

Friedman, M. 1956. The quantity theory of money: a restatement. In *Studies in the Quantity Theory of Money.* Chicago: University of Chicago Press.

Hume, D. 1752. Of interest; of money. In *Writings on Economics,* ed. E. Rotwein. Madison: University of Wisconsin Press, 1970.

Humphrey, T.M. 1974. The quantity theory of money: its historical evolution and role in policy debates. *Federal Reserve Bank of Richmond Economic Review* 1974(May/June), 2–19.

Humphrey, T.M. 1982. The real bills doctrine. *Federal Reserve Bank of Richmond Economic Review* 1982(September/October), 3–13.

Humphrey, T.M. 2001. Monetary policy frameworks and indicators for the federal reserve in the 1920s. *Federal Reserve Bank of Richmond Economic Quarterly* 87(1), 65–92.

Laidler, D. 1981. Adam Smith as a monetary economist. *Canadian Journal of Economics* 14, 185–200.

Laidler, D. 1984. Misconceptions about the real bills doctrine: a comment on Sargent and Wallace. *Journal of Political Economy* 92, 149–55.

Law, J. 1705. *Money and Trade Considered.* Edinburgh: Anderson.

Meltzer, A.H. 2003. *A History of the Federal Reserve, Volume I: 1913–1951.* Chicago: University of Chicago Press.

Pigou, A.C. 1917. The value of money. *Quarterly Journal of Economics* 32(November), 38–65.

Pigou, A.C. 1927. *Industrial Fluctuations.* London: Macmillan.

Sargent, T.J. 1987. *Macroeconomic Theory,* 2nd edn. Orlando, FL: Academic Press.

Sargent, T.J. and Wallace, N. 1982. The real-bills doctrine vs. the quantity theory: a reconsideration. *Journal of Political Economy* 90, 1212–36.

Smith, A. 1776. *An Inquiry into the Nature and Causes of the Wealth of Nations.* Indianapolis: Liberty Press, 1976.

Thornton, H. 1802. *An Enquiry into the Nature and Effects of the Paper Credit of Great Britain.* London: LSE Reprint Series, 1939.

Woodford, M. 2003. *Interest and Prices.* Princeton: Princeton University Press.

real business cycles

Real business cycles are recurrent fluctuations in an economy's incomes, products, and factor inputs – especially labour – that are due to non-monetary sources. Long and Plosser (1983) coined the term 'real business cycles' and used it to describe cycles generated by random changes in technology. Other real sources of fluctuations that have been studied include changes in tax rates and government spending, tastes, government regulation, terms of trade and energy prices.

Kydland and Prescott (1982), who studied the *quantitative* predictions of a stochastic growth model with shocks to technology, found that covariances between model series and autocorrelations of model output were consistent with corresponding statistics for US data. These findings were viewed as surprising, for two reasons. First, the findings ran counter to the idea that monetary shocks are the driving force behind business cycle fluctuations. Second, the policy implication for Kydland and Prescott's model was that stabilization policies are counterproductive. Fluctuations arise when households optimally respond to changes in technology.

The methodology that Kydland and Prescott (1982) used in their study of business cycles transformed the way in which applied research in macroeconomics is done. For this reason, the term 'real business cycles' is often associated with a methodology rather than Kydland and Prescott's original findings. Indeed, the methods of their 1982 paper have been used to study many different sources of business cycles, including monetary shocks.

Most real business cycle (RBC) models are variants or extensions of a neoclassical growth model. One such prototype is introduced. It is then shown how RBC theorists, following Kydland and Prescott (1982), use theory to make predictions about actual time series. Extensions of the prototype model are discussed. Current issues and open questions follow.

Prototype real business cycle model

Households choose sequences of consumption and leisure to maximize expected discounted utility. When aggregated, preferences are defined for a stand-in household that maximizes the expected value of

$$\sum \beta^t u(c_t, 1 - h_t)N_t, \tag{1}$$

where u is the utility function, c_t is per capita consumption at date t, $1-h_t$ is per capita leisure at date t, N_t is the population at date t which grows at rate η, and β is a discount factor.

The technology available in period t is $z_t F_t(K_t, H_t)$, where $z_t F_t$ is the output produced at date t with K_t units of capital and H_t hours. The function F_t has constant returns to scale so that doubling the inputs doubles the output. The variable z_t is a stochastic technology shock assumed to follow a Markov process. The variation in z modelled here is variation in the effectiveness of factor inputs, capital and labour, to produce final goods and services or *total factor productivity* (TFP). Fluctuations in TFP arise from many possible sources. For example, improvements in TFP can arise from new inventions or innovations in existing production processes. Reductions in TFP can arise from increased regulation on producers.

Households are endowed with time each period, normalized without loss of generality to 1, which they can allocate to work or to leisure. They can invest x_t (per capita) in new capital goods. Doing so yields

$$N_{t+1}k_{t+1} = N_t[(1-\delta)k_t + x_t], \qquad (2)$$

where k_t is per capita beginning-of-period t capital, k_{t+1} is per capita end-of-period t capital, and δ is the rate of per period depreciation.

Households face taxes on purchases of consumption and investment and on incomes to capital and labour. With taxation, the household budget constraint in period t is

$$(1+\tau_{ct})c_t + (1+\tau_{xt})x_t = r_t k_t - \tau_{kt}(r_t - \delta)k_t$$
$$+ (1-\tau_{ht})w_t h_t + \psi_t. \qquad (3)$$

Variables r_t and w_t are pre-tax payments to capital and labour, respectively. Variables τ_{ct}, τ_{xt}, τ_{kt}, and τ_{ht} are tax rates on consumption, investment, capital, and labour, respectively. These tax rates are assumed to be stochastic and follow a Markov process. Variable ψ_t is the per capita transfer payment at date t made by the government to each household. Total transfer payments are equal to tax revenues less total spending by the government. The per capita spending of the government at date t is g_t.

To derive explicit predictions about the behaviour of these households, it is necessary to first define and then compute an equilibrium for the economy. In doing so, it is convenient to de-trend any variables that grow over time and deal only with stationary processes. To be precise, assume that there is a constant rate of improvement in production processes over time so that $F_t(K_t, H_t) \equiv F(K_t, (1+\gamma)^t H_t)$ with F homogeneous of degree 1. If the per capita capital stock grows at rate γ and z_t and h_t are stationary, then output grows at rate γ. Certain assumptions on utility and the process for government spending also ensure that components of output grow at rate γ. Denote by $\tilde{v}_t$ the de-trended level of variable v_t, that is, $\tilde{v}_t = v_t/(1+\gamma)^t$.

A competitive equilibrium is defined as household policy functions for consumption $c(\tilde{k}, \tilde{K}, s)$, investment $x(\tilde{k}, \tilde{K}, s)$, and hours $h(\tilde{k}, \tilde{K}, s)$, where k is the (de-trended) stock of capital for the household, $\tilde{K}$ is the (de-trended) aggregate stock of capital, and $s = (\log z, \tau_c, \tau_x, \tau_k, \tau_h, \log \tilde{g})$; pricing functions $w(\tilde{K}, s)$ and $r(\tilde{K}, s)$; a function governing the evolution of the aggregate capital stock $\tilde{K}' = \Psi(\tilde{K}, s)$ that maps the current state into the capital stock next period $(\tilde{K}')$, and a function $\Phi(s', s)$ governing the transition of the stochastic shocks from s to s' such that (a) households maximize the expected value of (1) subject to (2) and (3) with the initial capital stock $\tilde{k}_0$ and functions for prices, aggregate capital, and the transition of s taken as given; (b) productive factors are paid their marginal products; (c) expectations are rational so

that $\tilde{k} = \tilde{K}$ and

$$\Psi(\tilde{k}, s) = [(1-\delta)\tilde{k} + x(\tilde{k}, s)]/[(1+\eta)(1+\gamma)];$$

and (d) markets clear:

$$c(\tilde{k}, \tilde{k}, s) + x(\tilde{k}, \tilde{k}, s) + g(s) = z(s)F(\tilde{k}, h).$$

Note that, in forming expectations about the future, households take processes for prices, tax rates and transfers as given. If households behave competitively, they assume that their own choice of capital next period does not affect the economy-wide level of capital. Therefore, in computing optimal decision functions for the household, it is necessary to distinguish the household's holdings of capital and the aggregate holdings of capital.

Comparing model predictions with data

Given equilibrium functions, properties of the model time series can be compared with data in a straightforward way. Starting with initial conditions on the state, the evolution of the state is determined by functions Ψ and Φ, resulting in sequences $\{\tilde{k}, s\}_{t=0}^{\infty}$ for the state. Equilibrium price and decision functions are then used with these sequences for the state to determine sequences of all prices and allocations.

A standard assumption for the transition $\Phi(s', s)$ is the vector autoregression

$$s_{t+1} = P_0 + Ps_t + Q\varepsilon_{t+1},$$

where each element of ε_t is a normally distributed random variable, independent of the other elements of ε and across time, with mean equal to zero and variance equal to 1. Allowing non zero off-diagonals in the matrices P and Q allows for correlations in the elements of the vector s. For example, a standard assumption is that tax rates and spending are positively correlated.

If the elements of the matrix QQ' are not large, the equilibrium evolution of the capital stock is well approximated by the following function:

$$\log \tilde{k}_{t+1} = A_0 + A_k \log \tilde{k}_t + B_k s_t,$$

which is linear in the log of the de-trended, per capita capital stock and the stochastic states. Similarly, the logarithms of consumption, investment, output and hours of work can be well approximated as linear functions of $\log \tilde{k}_t$ and s_t. (See Marimon and Scott, 1999, for an introduction to log-linear methods and nonlinear methods.)

Stacking the results in matrix form yields a system of equations

$$X_{t+1} = AX_t + B\varepsilon_{t+1}$$
$$Y_t = CX_t + \omega_t,$$

where X contains all variables of interest, some of which may not be observable, and Y is a vector of observables. This system can be easily simulated and lends itself nicely to standard methods of estimating model parameters. (See Anderson et al., 1996.)

An important feature of the analysis in Kydland and Prescott (1982) was the construction of the same statistics for the model and for the US data. Employing this methodology requires two necessary steps. The first concerns measurement: data series must be consistent with model series. For example, consumer durable expenditures are investments much like expenditures on new housing. National accountants treat expenditures on durables and housing differently, but the prototype model does not. Thus, revising the national accounts to include services, rents and depreciation of durables is necessary for data and model series to be consistent. The second step of Kydland and Prescott's (1982) methodology concerns reporting: the same statistics should be computed for the model and the revised data. Such comparisons are useful in highlighting similarities and deviations, which are both necessary ingredients to further the development of good theory.

Applying the two methodological tenets to the prototype model and US data yields a number of interesting results. Both the theory and the US data display pro-cyclical movements in consumption and investment, with the movements in investment being far greater in percentage terms. With tax rates and government spending fixed at mean US levels, the theory predicts fluctuations in per capita hours that are too smooth relative to US hours, and a correlation between hours worked and productivity that is too high relative to the correlation in US data. When fiscal shocks consistent with US policy are introduced, the theory predicts movements in per capita hours and a correlation between hours worked and productivity that are in line with the data.

Extensions of the prototype

During the 1980s and 1990s, business cycle research was exploratory but methodologically rooted. Researchers investigated the effects of many different shocks, the mechanisms that propagate them, and the welfare implications – in a consistent way that made clear what factors were important and why. A brief history is provided here, but interested readers are referred to the volume edited by Cooley (1995) and to a summary of more recent work in King and Rebelo (1999) and Rebelo (2005).

Kydland and Prescott (1982) and Long and Plosser (1983) emphasize technology shocks as an important source of fluctuations. Greenwood, Hercowitz and Huffman (1988) also explore the role of technology shocks for the business cycle but restrict attention to technological changes affecting the productivity of new capital goods and allow for accelerated depreciation of old capital. Mendoza (1995) includes shocks to the terms of trade in an international business cycle model and shows that responses of real exchange rates to productivity shocks and terms-of-trade shocks are quite different, both qualitatively and

quantitatively. Braun (1994), Christiano and Eichenbaum (1992), and McGrattan (1994) add fiscal shocks which are important for movement in hours and labour productivity, as noted above. Kim and Loungani (1992) add shocks to energy prices and show that the addition has only a modest impact on the variability of output and hours. Cooley and Hansen (1989) include monetary shocks and a cash-in-advance constraint and show that these additions have negligible effects on business cycle predictions.

The original technology-driven business cycle models under-predicted fluctuations in observed hours and over-predicted the correlation between hours and productivity, leading to further investigations of the model of the labour market and alternative mechanisms for propagating shocks. High – possibly infinite – elasticities were required in the original RBC models to generate fluctuations in aggregate hours comparable to the data. Rogerson (1988) motivates an infinite aggregate elasticity of labour supply in a world with variation in the fraction of people working: individuals work a standard workweek or not at all. This idea is implemented in an RBC model by Hansen (1985), who finds a significant increase in hours fluctuations relative to Kydland and Prescott (1982).

Another factor affecting the labour market is explored by Benhabib, Rogerson and Wright (1991) and Greenwood and Hercowitz (1991) who introduce home production. These researchers show that business cycle predictions depend crucially on the willingness and opportunity of households to substitute time in home work and market work. Under plausible parameterizations, the models do in fact generate greater variability of hours and lower correlations between hours and productivity.

The empirical performance of the RBC model is also improved when labour-market search frictions are introduced, as in Andolfatto (1996) and Merz (1995). Labour-market search models have also been used to study movements in unemployment and vacancies.

Current research and open questions

RBC research has evolved beyond the study of business cycles. The methodology that Kydland and Prescott (1982) introduced is now being applied to central questions in labour, finance, public finance, history, industrial organization, international macroeconomics and trade.

Within business cycle research, some open questions remain. What is the source of large cyclical movements in TFP? This question is especially interesting in the case of the US Great Depression, when TFP declined significantly (Cole and Ohanian, 2004). Are movements in TFP primarily due to new inventions and processes that are, by the nature of research and development, stochastically discovered? Or are movements in TFP primarily due to changing government regulations that may alter the efficiency of production? Are they due to unmeasured investments that fluctuate over time? The answers matter for policymakers, and they matter for economists who calculate the welfare costs or gains of changing policies.

ELLEN R. MCGRATTAN

See also **business cycle measurement; international real business cycles; monetary business cycles models; political business cycles; welfare costs of business cycles.**

I wish to thank Gary Hansen, Lee Ohanian and Ed Prescott for their comments on an earlier draft. The views expressed herein are those of the author and not necessarily those of the Federal Reserve Bank of Minneapolis or the Federal Reserve System.

Bibliography

Anderson, E.W., Hansen, L.P., McGrattan, E.R. and Sargent, T.J. 1996. Mechanics of forming and estimating dynamic linear economies. In *Handbook of Computational Economics*, vol. 1, ed. H. Amman, D. Kendrick, and J. Rust. Amsterdam: North-Holland.

Andolfatto, D. 1996. Business cycles and labor-market search. *American Economic Review* 86, 112–32.

Benhabib, J., Rogerson, R. and Wright, R. 1991. Homework in macroeconomics: household production and aggregate fluctuations. *Journal of Political Economy* 99, 1166–87.

Braun, R.A. 1994. Tax disturbances and real economic activity in the postwar United States. *Journal of Monetary Economics* 33, 441–62.

Christiano, L. and Eichenbaum, M. 1992. Current real-business-cycle theories and aggregate labor-market fluctuations. *American Economic Review* 82, 430–50.

Cole, H.L. and Ohanian, L.E. 2004. New Deal policies and the persistence of the Great Depression: a general equilibrium analysis. *Journal of Political Economy* 112, 779–816.

Cooley, T.F. 1995. *Frontiers of Business Cycle Research*. Princeton: Princeton University Press.

Cooley, T.F. and Hansen, G.D. 1989. The inflation tax in a real business cycle model. *American Economic Review* 79, 733–48.

Greenwood, J. and Hercowitz, Z. 1991. The allocation of capital and time over the business cycle. *Journal of Political Economy* 99, 1188–214.

Greenwood, J., Hercowitz, Z. and Huffman, G.W. 1988. Investment, capacity utilization, and the real business cycle. *American Economic Review* 78, 402–17.

Hansen, G.D. 1985. Indivisible labor and the business cycle. *Journal of Monetary Economics* 16, 309–27.

Kim, I.-M. and Loungani, P. 1992. The role of energy in real business cycle models. *Journal of Monetary Economics* 29, 173–89.

King, R.G. and Rebelo, S. 1999. Resuscitating real business cycles. In *Handbook of Macroeconomics*, vol. 1B, ed. J. Taylor and M. Woodford. Amsterdam: North-Holland.

Kydland, F.E. and Prescott, E.C. 1982. Time to build and aggregate fluctuations. *Econometrica* 50, 1345–70.

Long, Jr., J.B. and Plosser, C.I. 1983. Real business cycles. *Journal of Political Economy* 91, 39–69.

Marimon, R. and Scott, A., eds. 1999. *Computational Methods for the Study of Dynamic Economies*. New York: Oxford University Press.

McGrattan, E.R. 1994. The macroeconomic effects of distortionary taxation. *Journal of Monetary Economics* 33, 573–601.

Mendoza, E.G. 1995. The terms of trade, the real exchange rate, and economic fluctuations. *International Economic Review* 36, 101–37.

Merz, M. 1995. Search in the labor market and the real business cycle. *Journal of Monetary Economics* 36, 269–300.

Rebelo, S. 2005. Real business cycle models: past, present, and future. *Scandinavian Journal of Economics* 107, 217–38.

Rogerson, R. 1988. Indivisible labor, lotteries and equilibrium. *Journal of Monetary Economics* 21, 3–16.

real cost doctrine

Real cost doctrine is the doctrine that the supply price of a good is the price required to overcome the disutility involved in producing it. The worker, in other words, produces output up to the point at which his (decreasing) marginal utility of income equals his (increasing) marginal disutility of labour. The real cost doctrine can be seen as a half-way house inhabited by economists who had adopted a subjective theory of value but stopped short of the 'alternative cost' doctrine whereby the supply price of a resource is equal to its potential earning in its next most productive use. Much of the discussion which took place between English and Austrian economists concerned whether, and to what extent, the two doctrines logically came to the same thing.

Jevons (1871) formulated the real cost doctrine in terms of the diagram in Figure 1. Jevons assumes here (no such assumption is strictly necessary) that the worker at the start of the day not only enjoys his work but that, for a while, his enjoyment increases as he warms up to it. But, as the hours pass, the fatigue and boredom come to predominate over pleasure at an ever-increasing rate. The worker will maximize his surplus of utility over disutility by stopping at point X ($ab = bc$.)

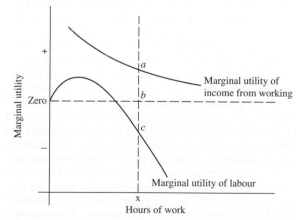

Figure 1

The idea that subjective disutility of labour is central in determining output and price is, perhaps, Jevons's most unquestionably original idea. Not only is it absent from the work of Walras and Menger, but its prefigurations in the classical period are rare and rudimentary when compared with the pre-1871 analyses of marginal utility theory. (Jennings, 1855, points out that marginal disutility of labour increases as the working day progresses but fails to build anything upon it.)

Marshall's theory of price determination, unveiled in his *Principles of Economics* (1890), differs little from Jevons's. Yet what looked radical in Jevons appears almost backward-looking in Marshall. This has something to do with the extension and dissemination of neoclassical principles in the intervening 20 years. But it also stems from a difference of presentation grounded in the contrast between Jevons's impatience with and Marshall's deference towards the Ricardian tradition. Much of Marshall's frequent praise for the English classical economists deftly sidesteps the question of how far they had actually been right. In the *Principles*, however, not only are cost and utility considerations given equal importance when determining price, but the fact that Marshall's conception of cost is ultimately a Jevonian 'subjective disutility' one is played down. It receives the strongest emphasis when Marshall argues that the capitalist as well as the worker undergoes real costs in the productive process, the capitalist's cost being that of 'waiting' rather than consuming his wealth immediately. (Nassau Senior had invoked Marx's sarcasm by speaking of capitalist 'abstinence' in the same context: Marshall tried both to circumvent the ridicule by renaming abstinence 'waiting' and to defend Senior from a neoclassical perspective, pointing out that *at the margin* of *aggregate* saving, considerable immediate sacrifice was undoubtedly involved.)

The rival doctrine, that of alternative cost, was espoused principally by the Austrians Wieser and Böhm-Bawerk and advertised in Britain by Wicksteed. All three denied the existence of any such thing as a supply curve, 'supply' simply being reverse demand. Böhm-Bawerk cited a horse fair: the buyer's utility from acquiring a horse and the seller's utility from keeping his horse played not just an equal but an identical role in determining price. Hence only a demand curve need be drawn; at the equilibrium price, it crosses the vertical line representing the fixed stock of horses. Both Marshallian and Austrian analysis predict the same price.

But, of course, the fixed stock of horses makes this a very simple case: we are ignoring the cost of producing them. Such considerations, however, were no problem to the Austrians, who proclaimed that the costs of factors of production and raw materials ultimately depended on utilities from alternative uses forgone. Thus, as regards the labour market, the wage in a particular industry was governed by the demand for labour in other industries. Each worker had to be paid enough to keep him out of his next best paid available job. The Jevonian notion of disutility of labour dropped out of the picture, Böhm-Bawerk (1894) arguing against it on the *empirical* ground that few

workers had the chance to make fine adjustments to the length of their working day. To this Edgeworth retorted that the Austrian doctrine implied that individuals made the choice to work or not to work once and for all at the beginning of their careers – it could not handle variations in labour supply due to variations in the wage rate.

The debate as a whole thus seemed to imply that the choice between real cost and alternative cost depended on whether *flexible* labour supply at the *individual* level (assumed by Jevons) or *inflexible* labour supply at the *aggregate* level (implied by the Austrians) was the more objectionable violation of reality. Yet logically the two theories come to exactly the same thing, and are seen to do so as long as the two 'sides' make one clarification apiece.

Austrians must make it clear that 'forgone utility' includes not only forgone income but also forgone leisure (when you work at all) and forgone non-pecuniary benefits (when you choose a less pleasant but better-paid job in preference to a more pleasant but worse-paid one). Böhm-Bawerk (1894) did spell this out.

Real cost theorists must make it clear that when a baker ponders whether to work another hour, what matters is not the disutility of the work as compared with doing nothing, but the disutility of work as compared with what he would choose to do (it might still be nothing!) if he were not baking. Equally it is not the 'gross' marginal utility of income which matters but the marginal utility of the *additional* income gained from spending another hour at the bakery rather than doing something else (other paid work, some leisure activity, or nothing). Edgeworth (1894) *failed* to spell this out; had he done so, a number of economists might have realized sooner than they actually did that both theories ultimately come to the same thing. (See Hobson, 1926, for an example of confusion persisting well into the 20th century.)

J. MALONEY

See also **Marshall, Alfred; opportunity cost.**

Bibliography

Blaug, M. 1985. *Economic Theory in Retrospect*. 4th edn. Cambridge: Cambridge University Press.

Böhm-Bawerk, E. von. 1894. One word more on the ultimate standard of value. *Economic Journal* 4, 719–24.

Edgeworth, F.Y. 1894. Professor Böhm-Bawerk on the ultimate standard of value. *Economic Journal* 4, 518–21.

Hobson, J.A. 1926. *Free Thought in the Social Sciences*. London: G. Allen & Unwin.

Jennings, R. 1855. *Natural Elements of Political Economy*. London: Longman, Brown, Green, Longmans.

Jevons, W.S. 1871. *Theory of Political Economy*, ed. and with an introduction by R.D. Collison Black. Harmondsworth: Penguin. 1970.

Marshall, A. 1890. *Principles of Economics*. London: Macmillan.

Wicksteed, P.H. 1910. *The Common Sense of Political Economy*. London: Macmillan.

real exchange rates

The real exchange rate plays a crucial role in models of the open economy. How the real exchange rate should be defined, how it behaves over time, and what determines it at various time horizons are all questions that have been posed over the years. They have taken on heightened importance in recent years, as the scope of international transactions has expanded and more and more economic activity is either directly or indirectly affected by economic activity in other countries.

The most common definition of the real rate is the nominal exchange rate adjusted by price levels,

$$q_t \equiv s_t - p_t + p_t^* \qquad (1)$$

where s is the log exchange rate defined in units of home currency per unit of foreign, and p and p^* are log price levels. If purchasing power parity (PPP) holds, then q is always unity (or a constant if price indices are used). One should expect PPP to hold in a world where transportation and transactions costs were negligible, consumption baskets were identical, and no arbitrage profits existed. Absent these conditions, the real exchange rate will vary.

One way of thinking about the determinants of movements in the real exchange rate is to appeal to a decomposition. Suppose the price index is a geometric average of traded and non-traded good prices:

$$p_t = \alpha p_t^N + (1 - \alpha) p_t^T \qquad (2)$$

where the lower-case letters denote logged values. Then substituting (2) into (1) yields:

$$q_t \equiv (s_t - p_t^T + p_t^{T*}) + [-\alpha(p_t^N - p_t^T) + \alpha^*(p_t^{N*} - p_t^{T*})] \qquad (3)$$

$$q_t \equiv q_t^T + [\omega_t] \qquad (3')$$

Equation (3) indicates that the real exchange rate can be expressed as the sum of two components: (i) the relative price of tradables q^T, (ii) the intercountry relative price of non-tradables in terms of tradables in the home country ω.

The determinants of the real exchange rate

If PPP holds only for tradable goods, then only the second term in eq. (3') can be non-zero, and the relative tradables–non-tradables price is the determining factor in the value of the real exchange rate. Another possibility is that all goods are tradable but not perfectly substitutable; then the imperfect substitutes model results, and q^T is equivalent to q. More generally, both terms on the right hand side of eq. (3') can take on non-zero values. In either of these cases, there are a large number of variables that could influence each relative price. And of course, there is nothing to rule out both relative price channels as being operative. In popular discussion, all three definitions of 'the real exchange rate' are used, sometimes leading to considerable confusion.

Most models of the real exchange rate can be categorized according to which specific relative price serves as the object of focus. If the relative price of non-tradables is key, then the resulting models – in a small country context – have been termed 'dependent economy' (Salter, 1959; Swan, 1960) or 'Scandinavian' model. In the former case, demand-side factors drive shifts in the relative price of non-tradables. In the latter, productivity levels and the nominal exchange rate determine the nominal wage rate and hence the price level, and thence the relative price of non-tradables. In this latter context, the real exchange rate is a function of productivity (Krueger, 1983, p. 157). Consequently, the two sets of models both focus on the relative non-tradables price but differ in their focus on the source of shifts in this relative price. Since the home economy is small relative to the world economy (hence, one is working with a one-country model), the tradable price is pinned down by the rest-of-the-world supply of traded goods. Hence, the 'real exchange rate' in this case is $(p^N - p^T)$.

The relative price of tradables definition is most appropriate when considering the relative price that achieves external balance in trade in goods and services. This variable is also what macroeconomic policymakers refer to as 'price competitiveness'; hence, anything that affects the markup of price over cost – including both the level of demand, input costs, and market structure – can determine the real exchange rate.

Notice the dichotomy between the relative price of tradables and the relative price of non-tradables breaks down when countries specialize in the production of goods. Then the real exchange rate is the same as the terms of trade; purchasing power parity would occur only if the two goods were perfect substitutes (see Lucas, 1982; Stockman, 1980).

Empirical modelling and results

Real exchange rate dynamics

In one special case, there is no need to model the real exchange rate. If relative PPP is *assumed* to hold, then q is a constant. Empirically, this is clearly not true in the short run but could be in the long run. Consequently, tremendous effort has been invested in investigating whether q is trend stationary, even though trend stationarity is not the same as purchasing power parity holding (the stronger condition of mean stationarity is required). Numerous studies have evaluated the trend stationarity of q directly by application of unit root tests, or indirectly by assessing whether the component series of q exhibit common long-term trends. Broadly speaking, the conclusions in this literature are mixed. Generally, panel methods, long time samples, and the use of producer or wholesale price indices provide more evidence in favour of a trend stationarity q than do pure time series methods, short samples, and the use of consumer price indices (see Rogoff, 1996; Taylor and Taylor, 2004). These results leave open the possibility that economic variables affect the movement of exchange rates over the short as well as the long run.

Modelling real exchange rate movements as a function of economic variables

The modelling of the real exchange rate determinants can be divided into two main categories. The first category includes models of the nominal exchange rate which, by virtue of the assumption of sticky prices, become models of the real exchange rate. First and foremost among these are sticky price monetary models that incorporate exchange rate overshooting, such as Dornbusch (1976) and Frankel (1979). In the long run, purchasing power parity holds, so that these models are only short-run models.

The second category includes models that focus on the determinants of the long-run real exchange rate. By far dominant in this category are those that centre on the relative price of non-tradables. These include the specifications based on the approaches of Balassa (1964) and Samuelson (1964) that model the relative price of non-tradables as a function of sectoral productivity differentials, including Hsieh (1982), Canzoneri, Cumby and Diba (1999) and Chinn (1999; 2000). They also include those models that search more broadly and include demand-side determinants of the relative price, such as DeGregorio and Wolf (1994). Engel (1999) has cast doubt upon the relevance of the relative non-tradables price. He demonstrates that for the G-7 economies, the variability of q^T as proxied by the tradable components of the CPI is comparable to the variability of ω even at horizons of 15 years.

More recently, some version of the portfolio balance model has been resurrected. Lane and Milesi-Ferretti (2002) have forwarded a model wherein the real rate depends upon net foreign assets. Early panel evidence in favour of the importance of this factor is to be found in Gagnon (1996).

Some methodological approaches do not fall neatly into one or the other category. The analysis by Mark and Choi (1997) is one instance. They compare the usefulness of monetary and real factors in predicting exchange rate changes over long horizons, and find – surprisingly – that monetary factors have persistent effects on the real exchange rate. Using a different methodology, namely, a structural vector autoregression, Clarida and Gali (1995) find that monetary and demand-side factors dominate in the determination of exchange rates. Also relying upon a structural (permanent-transitory) decomposition involving the real exchange rate and the current account, Lee and Chinn (2006) find that positive permanent shocks (interpreted as productivity innovations) tend to appreciate the currency and (at least for the United States)

have an impact comparable in magnitude to those of temporary shocks.

MENZIE D. CHINN

See also **cointegration; exchange rate dynamics; monetary business cycle models (sticky prices and wages); nominal exchange rates; purchasing power parity; real exchange rates; terms of trade; tradable and non-tradable commodities; unit roots.**

Bibliography

Balassa, B. 1964. The purchasing power parity doctrine: a reappraisal. *Journal of Political Economy* 72, 584–96.

Canzoneri, M.B., Cumby, R.E. and Diba, B. 1999. Relative labor productivity and the real exchange rate in the long run: evidence for a panel of OECD countries. *Journal of International Economics* 47, 245–66.

Chinn, M. 1999. Productivity, government spending and the real exchange rate: evidence for OECD countries. In *Equilibrium Exchange Rates*, ed. R. MacDonald and J. Stein. Boston: Kluwer.

Chinn, M. 2000. The usual suspects: productivity and demand shocks and Asia-Pacific real exchange rates. *Review of International Economics* 8, 20–43.

Clarida, R. and Gali, J. 1995. Sources of real exchange rate movements: how important are nominal shocks? *Carnegie-Rochester Conference Series on Public Policy* 41, 9–66.

DeGregorio, J. and Wolf, H. 1994. Terms of trade, productivity, and the real exchange rate. Working Paper No. 4807. Cambridge, MA: NBER.

Dornbusch, R. 1976. Expectations and exchange rate dynamics. *Journal of Political Economy* 84, 1161–76.

Engel, C. 1999. Accounting for US real exchange rate changes. *Journal of Political Economy* 107, 507–38.

Frankel, J. 1979. On the mark: a theory of floating exchange rates based on real interest differentials. *American Economic Review* 69, 610–22.

Gagnon, J. 1996. Net foreign assets and equilibrium exchange rates: panel evidence. International Finance Discussion Papers No. 574. Washington, DC: Board of Governors of the Federal Reserve System.

Hsieh, D. 1982. The determination of the real exchange rate: the productivity approach. *Journal of International Economics* 12, 355–62.

Krueger, A.O. 1983. *Exchange-Rate Determination.* Cambridge: Cambridge University Press.

Lane, P.R. and Milesi-Ferretti, G.M. 2002. External wealth, the trade balance, and the real exchange rate. *European Economic Review* 46, 1049–71.

Lee, J. and Chinn, M. 2006. Current account and real exchange rate dynamics in the G-7 countries. *Journal of International Money and Finance* 25, 257–74.

Lucas, R. 1982. Interest rates and currency prices in a two-country world. *Journal of Monetary Economics* 10, 335–59.

Mark, N. and Choi, D.Y. 1997. Real exchange rate prediction over long horizons. *Journal of International Economics* 43, 29–60.

Rogoff, K. 1996. The purchasing power parity puzzle. *Journal of Economic Literature* 34, 647–68.

Salter, W.A. 1959. Internal and external balance: the role of price and expenditure effects. *Economic Record* 35, 226–38.

Samuelson, P. 1964. Theoretical notes on trade problems. *Review of Economics and Statistics* 46, 145–54.

Stockman, A. 1980. A theory of exchange rate determination. *Journal of Political Economy* 88, 673–98.

Swan, T. 1960. Economic control in a dependent economy. *Economic Record* 36, 51–66.

Taylor, A.M. and Taylor, M.P. 2004. The purchasing power parity debate. *Journal of Economic Perspectives* 18(4), 135–58.

Williamson, J. 1994. *Estimating Equilibrium Exchange Rates*. Washington, DC: Institute for International Economics.

real rigidities

'Real rigidities' is the name given to a large class of business cycle propagation mechanisms. Real rigidities appear essential to any successful explanation of business cycles.

The definition of real rigidities

Although the term 'real rigidities' appears vague, it in fact refers to a precise concept. Consider an economy of symmetric price-setting firms that is at its flexible-price equilibrium, and suppose that the money supply increases with prices unchanged, so that aggregate output increases. Now ask by how much a representative firm would want to increase its price if it faced no barriers to nominal price adjustment. By definition, the smaller the amount the firm would want to increase its price in response to a given increase in aggregate output, the greater the degree of real rigidity.

One can see the meaning of 'real rigidity' more formally by observing that, in the experiment described above, the profits of the firm that is considering changing its price, neglecting any costs of price adjustment, typically can be written in the form $V(p_i - p, y)$, where p_i is the firm's price, p is the aggregate price level, and y is the departure of output from its flexible-price level (all in logs). In most models of this type, $V(\cdot)$ is a smooth function. The first-order condition for the profit-maximizing price is $V(p_i^* - p, y) = 0$ (subscripts denote partial derivatives). At the flexible-price equilibrium, $p_i^* = p$ and $y = 0$. Starting from that equilibrium, the derivative of the representative firm's desired relative price, $p_i^* - p$, with respect to y is thus

$$\left. \frac{d(p_i^* - p)}{dy} \right|_{p_i^* - p = 0,\, y = 0} = \frac{V_{12}(0,0)}{-V_{11}(0,0)}$$

$$\equiv \phi.$$

For the flexible-price equilibrium to be stable, ϕ must be positive. By definition, a lower value of ϕ corresponds to greater real rigidity. Note that real rigidity is defined entirely in terms of relations among real variables: it refers to the (lack of) responsiveness of desired real prices to aggregate real output.

The definition of real rigidity in models without symmetric price-setting firms is analogous: any force that reduces the amount that price setters would change their relative prices in response to movements in aggregate output that are the result of changes in aggregate demand is a real rigidity.

Real rigidities and business cycles

Real rigidities are crucial to business cycles. At a general level, real rigidities make firms less inclined to take actions that dampen movements in aggregate output. As a result, they increase the responsiveness of output to disturbances.

The importance of real rigidities is easiest to see in a static model where firms face fixed costs of changing prices. Consider the model sketched above, with two extensions. First, replace the profit function with a second-order approximation around the flexible-price equilibrium. This implies that the representative firm's loss in profits from failing to charge its profit-maximizing price (neglecting costs of price adjustment) is $K(p_i - p_i^*)^2$, where $K \equiv -V_{11}$. It also implies that the representative firm's profit-maximizing price is given by $p_i^* - p = \phi y$, where ϕ is as defined before. Second, assume that each firm faces a fixed cost $Z > 0$ of changing its nominal price.

The economy begins at its flexible-price equilibrium. We want to know by how much output can change in response to a change in aggregate demand before firms change their prices. Non-adjustment is an equilibrium as long as the representative firm's losses from failing to adjust are less than Z. Prior to the shock, the representative firm's price equals the aggregate price level, p. If the firm adjusts its price, it sets it to the new profit-maximizing level, $p + \phi y$. Thus the condition for non-adjustment to be an equilibrium is $K[p - (p + \phi y)]^2 < Z$, or $|y| < 1/(\phi\sqrt{Z/K})$. Thus, when ϕ is lower – that is, when real rigidity is greater – the range over which aggregate demand shocks affect real activity is greater.

Real rigidities are not just important to models with nominal rigidity, however. Consider, for example, the following minimalist real business cycle model. The markets for labour and goods are perfectly competitive, and the representative firm's production function is $y_i = a + n_i$ (y is output, a is productivity, and n is labour input, again all in logs). Labour supply is $n = \gamma w$, $\gamma > 0$, where w is the (log) real wage. In this model, the elasticity of profit-maximizing relative prices to demand-driven output fluctuations (that is, to variations in y with a fixed, which must come from variations in n) equals the elasticity of the real wage with respect to y, which is $1/\gamma$. Thus a larger value of γ corresponds to greater real rigidity.

The production function implies that labour demand is perfectly elastic at $w = a$. Labour-market equilibrium therefore requires that $n = \gamma a$. A larger value of γ therefore implies that productivity shocks have a larger impact on employment, and thus that the output effects of the shocks are magnified to a greater extent. Thus, even in this purely Walrasian model of fluctuations, real rigidity acts as a propagation mechanism.

Real rigidities also act as amplification mechanisms in dynamic models of price adjustment. Consider an economy with barriers to price adjustment where output is above its flexible-price level, and suppose that some firms have an opportunity to change their prices, and that their new prices will be in effect for more than one period. Greater real rigidity increases the persistence of the departure of output from its flexible-price level, for three reasons. First, as in the static model of price adjustment, it reduces the benefits of price adjustment, and so makes firms more inclined not to adjust at all. Second, the fact that only some firms have the opportunity to adjust means that output will continue to be above its flexible-price level. Greater real rigidity then implies that the firms that adjust will respond by less, thus drawing out the period of above-normal output. Third, the fact that other firms will be in the same situation when they adjust their prices means that they will adjust by less, which in turn dampens the adjustments of the firms that adjust immediately.

There is a close link between real rigidity and strategic complementarity in profit-maximizing prices. If we assume the stylized aggregate demand curve $y = m - p$ (where m reflects factors that shift aggregate demand), then the expression for the representative firm's profit-maximizing relative price, $p_i^* - p = \phi y$, implies $p_i^* = \phi m + (1 - \phi)p$. Thus greater real rigidity corresponds to greater strategic complementarity in desired prices: when ϕ is lower, each firm wants its price to move more closely with other prices.

Real rigidity and strategic complementarity in desired prices are not identical, however. To see this, suppose the aggregate demand equation is instead $y = \beta(m - p)$, $\beta > 0$. Then $p_i^* = \phi\beta m + (1 - \phi\beta)p$. Thus β affects strategic complementarity but not real rigidity. And it is real rigidity that is key to cyclical fluctuations. Nonetheless, because of the close link between the two concepts, and because many business cycle models assume $y = m - p$, the terms real rigidity and strategic complementarity in prices are often used interchangeably.

Types of real rigidities

Since any force that reduces the responsiveness of profit-maximizing relative prices to demand-driven output fluctuations is a real rigidity, there are many possible real rigidities. Some might not be commonly thought of as 'rigidities'. For example, as the simple real business cycle model shows, more elastic labour supply is a type of real rigidity.

Such neoclassical sources of real rigidity, however, are almost surely not strong enough to generate output fluctuations of the size and nature that we observe. In Walrasian models, the real wage is likely to rise sharply with the quantity of labour. For this not to occur, either the long-run elasticity of labour supply must be high or the intertemporal elasticity of substitution in labour supply must be high and short-run aggregate fluctuations must have a large transitory component. Neither of these conditions appears to hold in practice. In models of nominal disturbances and barriers to nominal price adjustment, the result is large incentives for price changes, and thus little nominal rigidity. In productivity-driven real business cycle models, the result is small movements in labour input, so that the dynamics of aggregate output largely mimic the dynamics of the underlying productivity shocks. Researchers have therefore turned their attention to non-Walrasian sources of real rigidity.

It appears difficult to understand substantial employment fluctuations without non-Walrasian real rigidities in the labour market. At a general level, what is needed is for some force causing workers to be off their labour supply curves, at least in the short run, so that the cyclical behaviour of the real wage is not governed by the elasticity of labour supply. For example, if there is equilibrium unemployment because of efficiency wages, the cyclical behaviour of the real wage depends on how the efficiency wage varies with aggregate output. As a result, the real wage can be (though it need not be) less procyclical than in a Walrasian labour market, with the result that fluctuations in employment and output are greater.

A more subtle real rigidity in the labour market arises if labour is imperfectly mobile in the short run (because of search frictions, for example), so that each firm faces an upward-sloping labour supply curve rather than perfectly elastic supply at the economy-wide wage. Consider, for example, a firm contemplating cutting its price and increasing production in a recession. With imperfectly mobile labour, this requires paying a higher real wage. Thus the amount the firm wants to reduce its price is smaller – that is, real rigidity is greater.

There can also be important real rigidities in other markets. In the goods market, forces making desired markups countercyclical act as real rigidities. When desired markups are more countercyclical, then, for a given degree of procyclicality of real marginal costs, desired movements in relative prices are smaller. Countercyclical desired markups can stem from a variety of sources. One simple but potentially important possibility is that, when economic activity is greater, firms' incentives to disseminate information and consumers' incentives to acquire it are greater, and so demand is more elastic.

Another feature of goods markets that can act as a real rigidity is input-output links among firms. If the prices charged by intermediate suppliers are sticky, the costs that firms face for intermediate inputs tend to rise by less than the suppliers' costs in a boom, thereby reducing the amount that firms would raise their prices if they were free to do so.

Capital-market imperfections can also create real rigidities. Capital-market imperfections can cause financing costs to be countercyclical, as higher output increases cash flow (and hence firms' ability to use internal finance) and raises asset values (and hence increases collateral and reduces the cost of external finance). With one component of costs countercyclical, desired prices are less procyclical. To give another example, financial difficulties in recessions can increase the importance of short-term profits to firms relative to expanding their customer base, and so can make desired markups countercyclical.

There is an important distinction between two broad categories of real rigidities. One category consists of forces, such as limited short-run labour mobility among firms, that increase real rigidity by affecting what happens when one firm changes its prices and others do not. The other consists of forces, such as factors that reduce the procyclicality of the real wage, that increase real rigidity by affecting what happens when all firms' output moves together. In terms of the definition of real rigidity as $V_{12}(0,0)/[-V_{11}(0,0)]$, the first category consists of forces raising $-V_{11}(0,0)$, and the second consists of forces reducing $V_{12}(0,0)$.

The distinction between these two categories is important for two reasons. First, real rigidities that result from forces that affect what happens when one firm changes its price with other firms' prices fixed are not relevant to the properties of business cycle models with identical firms and fully flexible prices. Second, the two types of real rigidities have different microeconomic implications. Most importantly, factors that increase real rigidity by affecting what happens when one firm changes its price and others do not increase the costs of departures from the profit-maximizing price; as a result, they typically predict smaller movements in firms' relative prices in response to many types of shocks.

Selected literature
Ball and Romer (1990) establish that in a static setting, imperfect competition and barriers to nominal adjustment alone are unlikely to generate substantial nominal rigidity. They show the general importance of real rigidities to static menu-cost models and stress that forces making desired real wages relatively unresponsive to output fluctuations are likely to be essential to generating substantial nominal rigidity (see also Blanchard and Fischer, 1989, ch. 8). Earlier work by Akerlof and Yellen (1985) incorporates substantial real rigidity in a model of price stickiness, although it does not explicitly analyse the importance of real rigidity to the results. Haltiwanger and Waldman (1989) show that strategic complementarity magnifies the impact of non-responders on equilibrium outcomes, a result that is closely related to the finding that real rigidities magnify the effects of barriers to price adjustment.

Kimball (1995) establishes the central role of real rigidities to the persistence of output fluctuations in models of staggered price adjustment, and stresses the importance of the distinction between forces that affect what happens when all firms' outputs move together and forces that affect what happens when one firm changes its price with other firms' prices fixed (see also Blanchard, 1987). Klenow and Willis (2006) show the differing microeconomic implications of the two categories of real rigidities. Romer (2006, ch. 6) provides a general discussion of the importance of real rigidities to static and dynamic models of nominal rigidity, catalogues many specific real rigidities and provides numerous references.

Real rigidities pervade modern business cycle models. In real business cycle models, for example, such common features as indivisible labour supply, variable capital utilization and labour hoarding, and learning-by-doing (see, for example, Rogerson, 1988; Burnside and Eichenbaum, 1996; and Chang, Gomes and Schorfheide, 2002) magnify the effects of disturbances precisely because they are real rigidities. In models with price stickiness, some important recent analyses where real rigidities play a central role include Mankiw and Reis (2002), Gertler and Leahy (2006) and Carvalho (2006).

DAVID ROMER

See also **cyclical markups; monetary business cycle models (sticky prices and wages); new Keynesian macroeconomics; real business cycles; sticky wages and staggered wage setting.**

Bibliography
Akerlof, G.A. and Yellen, J.L. 1985. A near-rational model of the business cycle, with wage and price inertia. *Quarterly Journal of Economics* 100, 823–38.

Ball, L. and Romer, D. 1990. Real rigidities and the non-neutrality of money. *Review of Economic Studies* 57, 183–203.

Blanchard, O.J. 1987. Aggregate and individual price adjustment. *Brookings Papers on Economic Activity* 1987(1), 57–109.

Blanchard, O.J. and Fischer, S. 1989. *Lectures on Macroeconomics*. Cambridge. MA: MIT Press.

Burnside, C. and Eichenbaum, M. 1996. Factor-hoarding and the propagation of business-cycle shocks. *American Economic Review* 86, 1154–74.

Carvalho, C. 2006. Heterogeneity in price stickiness and the real effects of monetary shocks. *Frontiers of Macroeconomics* 2(1), Article 1.

Chang, Y., Gomes, J.F. and Schorfheide, F. 2002. Learning-by-doing as a propagation mechanism. *American Economic Review* 92, 1498–520.

Gertler, M. and Leahy, J. 2006. A Phillips curve with an Ss foundation. Working Paper No. 11971. Cambridge, MA: NBER.

Haltiwanger, J. and Waldman, M. 1989. Limited rationality and strategic complements: the implications for macroeconomics. *Quarterly Journal of Economics* 104, 463–83.

Kimball, M.S. 1995. The quantitative analytics of the basic neomonetarist model. *Journal of Money, Credit, and Banking* 27, 1241–77.

Klenow, P.J. and Willis, J.L. 2006. Real rigidities and nominal price changes. Working Paper No. 06-03. Kansas City, MO: Federal Reserve Bank of Kansas City.

Mankiw, N.G. and Reis, R. 2002. Sticky information versus sticky prices: a proposal to replace the new Keynesian Phillips curve. *Quarterly Journal of Economics* 117, 1295–328.

Rogerson, R. 1988. Indivisible labor, lotteries and equilibrium. *Journal of Monetary Economics* 21, 3–16.

Romer, D. 2006. *Advanced Macroeconomics*, 3rd edn. New York: McGraw-Hill.

real wage rates (historical trends)

The real wage indicates the purchasing power of a worker's income. The real wage is the ratio of the nominal wage (what someone is actually paid) to a price or price index. Sometimes that price is the price of the product of a competitive firm, in which case the real wage is the marginal product of labour and has a productivity interpretation. In the more common case, however, and the one this article deals with, the price deflator is a consumer price index. In this case, the real wage measures the standard of living of the worker. Since that bears on central questions of economic growth and distribution, real wages have been an important tool for measuring and interpreting economic growth and stagnation over the last millennium.

Measuring real wages raises practical problems that are particularly acute in historical investigations. First, one needs information about wages, prices and spending shares to perform any calculations. Data sources for these have to be developed, which ultimately involves extensive archival work. There are conceptual problems as well since many people in the past received some income in kind as well as cash and many were also employed on piece rates that must be converted to earnings before an assessment of their purchasing power can be made. Second, an index number must be chosen to compute the consumer price index. While theorists have advanced many useful arguments about why some formulae are better than others, data limitations often force compromises. One of the most extreme was the once common practice of deflating wages by the price of grain. Other products were ignored, as well as the inconvenient fact that most people in the West ate bread, not grain. Most recent studies have avoided this practice. Third, new products and improvements in the quality of old products bedevil historical studies as they do modern ones. Although product innovation was less common in the past, the creation of the global economy led to the introduction into Europe or mass availability of maize, potatoes, tomatoes, chilli peppers, sugar, tobacco, cotton cloth, tea, coffee and porcelain. Also, comparisons of real wages between continents with radically different diets raise the question of new products in a cross-sectional context. How do you compare the standard of living of an English worker eating bread, beef and beer with a Chinese worker eating fish and rice?

Real wages and economic growth in developed countries

Economic theorists have divided the history of the world into two phases. Before the onset of modern economic growth around 1800, income per head grew very slowly, if at all. Increases in productivity simply resulted in more people. The real wage moved inversely with the population and remained constant in the long run. This was the Malthusian phase of history.

The second phase began in about 1800. Technology improved steadily raising income per head. Population growth was restrained, so an increase in the labour force did not swamp the increase in labour demand. Consequently, the real wage rose in step with productivity. This has been called the Solow phase in view of Solow's (1956) growth model. While these models can be nuanced, as we will see, they provide a starting point for real wage history: is it consistent with these models?

We can measure the real wage over the past 800 years thanks to the accumulated research of historians who have written 'price histories' of cities since the mid-19th century. The price historian finds an institution like a college or hospital that has existed for centuries and examines its accounts to abstract the prices of the things it bought and the wages it paid. Oxford and Cambridge colleges were the first to be studied (Rogers, 1866–92) and since then many European cities have been investigated. Phelps-Brown and Hopkins (1955; 1956) were the first to take advantage of this material and construct a real wage index for English building workers from 1264 to 1954. More recently, Allen (2001) and Clark (2005) have reworked this material and added new evidence to compute new real wage series. While there are differences among these authors on issues like real wage change in the Industrial Revolution, the broad outlines of the real wage story are the same (Figure 1).

The Malthusian and Solow phases stand out in this figure. Before the 19th century, England followed the Malthusian pattern. There was no long-run trend in the real wage, although there were fluctuations. These coincided with population swings; in particular, the real wage rose after the Black Death in 1348–9, which killed about one-third of the population, and fell in the 16th century as the population started to rebound. Real wages only rose above the pre-industrial peak once the Industrial Revolution was well under way or completed. The rise since then has been spectacular by comparison, and today the real wage is ten times its level in the pre-industrial world. This is the Solow phase of economic history.

The period of the Industrial Revolution, roughly 1770–1860, is something of a problem. Were real wages rising then or falling? The classical economists, who were writing in the early 19th century, were pessimistic about the prospects of workers. While they agreed that capitalism was likely to cause output per capita to rise, they also believed that real wages would remain constant at 'subsistence'. For Malthus, Ricardo and other mainstream economists, the reason was demographic: wages were the income of the bulk of the population, and a higher wage would lead them to have more children and live longer. The result would be an increase in the workforce that would push wages back to subsistence. While radicals like Marx and Engels rejected the demographic model, they agreed that wages would not rise under capitalism. Their explanation, however, turned on the demand for labour rather than its supply. Marx and Engels believed that technological progress would be so rapid and so labour-saving that the demand for labour would always fall short of the supply – again forcing wages back to subsistence. Only collective action or state interference would prevent this. None of the classical economists, in other words, expected the 'Malthusian economy' to transmute into the 'Solow economy'.

By the 20th century, it was clear that these arguments were wrong, for living standards were far higher than they had been 100 years before, as Figure 1 shows. Kuznets (1955) raised the possibility that inequality went through an 'invented U' patterned during economic development. In his model, this worked through the wage structure itself. At the outset, workers were in low productivity, low-wage sectors. As the modern sector grew, it attracted workers by offering higher wages. Inequality increased as workers moved to that sector since those employed there were earning more than their counterparts in agriculture. Inequality in wages declined as the process of labour reallocation was completed, for all workers were then earning the higher wage paid in the modern sector.

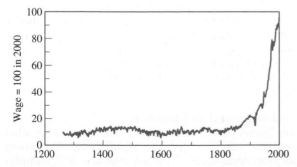

Figure 1   Real wage in London, 1200–2000. *Sources*: Data before 1914: Allen (2001). Later wage and price indices: Phelps Brown and Hopkins (1955), Mitchell (1998), ILO (various years).

The problem of economic development in poor countries provoked Lewis (1954) to propose a model of growth and distribution that was more classical in spirit and that emphasized the differential movements of output per head and the real wage. Lewis divided the economy into two sectors. In the traditional sector, consisting of peasant agriculture and the urban 'informal' economy, the main inputs were land and labour, and the latter was in surplus. In the modern, industrial sector, output was produced with capital and labour, and the former was scarce. Growth occurred as capital was accumulated and the modern sector expanded. Its labour force was drawn from the traditional sector. Surplus labour in that sector meant that the marginal product of labour was low, perhaps even zero, and income was shared and at a subsistence level. An elastic supply of labour kept the real wage in the modern sector at subsistence even though output per worker was rising. This process of rising inequality would continue, in Lewis's view, until the modern sector had expanded to absorb all the surplus labour. Only then would the real wage rise in step with output per worker.

How well do these theories fare in practice? The question has been extensively researched and vigorously debated in the case of the British Industrial Revolution. Lindert and Williamson (1983) were the first to apply modern economic methods to the question. They computed economy-wide average earnings and a consumer price index founded on budget surveys and corresponding prices. Their conclusion was guardedly 'optimistic' in that they found the average real wage rose sharply after 1815. This conclusion was not universally accepted. Feinstein (1998) computed an alternative price index that significantly reduced the rate of real wage growth leading to his title 'pessimism perpetuated'. Clark (2005), on the other hand, proposed yet another price index that tilted the conclusions back in a Lindert–Williamson direction. Most recently, Allen (2007a) has plumped for 'pessimism preserved'.

These disagreements reflect the limitations of the data, which are only a poor reflection of the ideal information discussed above. There were no comprehensive and representative samples of consumer spending, and even the annual series of individual prices are problematic. Quality change, in particular the growing use of cotton rather than wool in clothing, is only imperfectly grasped with the available information. There is considerable scope for contradictory – yet plausible – readings of the evidence.

The impact of economic development on wage rates has been pursued for many other countries with mixed results. Over the long term, real wage change in Western Europe has followed a pattern like that for England shown in Figure 1 (Scholliers and Zamagni, 1995). The United States has been repeatedly studied, and revisions to price and wage indices have been as thoroughgoing for the USA, as they have been for Britain. For instance, Douglas's (1930) conclusion that real wages only rose by eight per cent during the boom from 1890 to 1914 was overturned by Rees (1961), who found that real wages increased by 40 per cent. Over the long term, of course, real wages have risen dramatically in America, but the real wage lagged behind GDP per head during early industrialization, according to Margo's (2000) study of the period 1800–60.

Outside the advanced Organisation for Economic Co-operation and Development (OECD) countries, the link between economic growth and real wage advance is much weaker. The economic boom experience by Tsarist Russia, for instance, was not reflected in urban or rural real wages (Allen, 2003a). Latin America enjoyed a substantial rise in GDP per head in the 20th century, with only an elusive impact on real wages. In Mexico, which has been studied more than most countries, there were periods when real wages surged and others when real wages collapsed. The declines look about as big as the gains, but the uncertainty arising from the introduction of new goods makes definitive conclusions hazardous (Bortz and Aguila, 2006).

Real wages and the great divergence

The difference in real wages can be computed between two places as well as between two times, and the geographical dimension has allowed real wages to play an important role in the 'great divergence' debate. Since 1800, incomes have grown most rapidly in the most prosperous countries, so their lead over the poor countries has increased. Charting this divergence is the first step in explaining it. Real GDP per head may be the best indicator, and economists have tried to extrapolate it far into the past. Errors, however, accumulate, and the estimates become increasingly problematic the further back one goes. Real wages provide an alternative, and simpler, approach to the problem. The real wage is individual in its focus – what could a particular worker buy with his or her income? – and so avoids the economy-wide assumptions of national income accounting. The real wage also requires fewer data, and it can be computed directly for dates in the past without the need to extrapolate backwards.

Indeed, the classical economists, who established a long-standing view on the subject, expressed Europe's lead in terms of real wages. Adam Smith (1776, pp. 74–5, 91, 187, 206) saw the world in terms of a wage ladder: workers in England and the Netherlands had a higher standard of living than workers in France or elsewhere on the European continent. Workers in Asia lagged behind Europeans. 'The real price of labour, the real quantity of the necessaries of life which is given to the labourer…is lower both in China and Indostand than it is through the greater part of Europe.'

This view has recently been challenged by scholars of Asia who have argued that Asia was as prosperous as Europe at the time Smith wrote. According to Pomeranz (2000, p. 49), 'It seems likely that average incomes in Japan, China, and parts of South-East Asia were comparable to (or higher than) those in western Europe even in the late eighteenth century.'

How does Smith's wage ladder stand up in terms of modern evidence? The price histories of European cities provide a start, for they allow us to compute real wage differences across Europe from the late Middle Ages to the 19th century (van Zanden, 1999; Allen, 2001). While today real wages are similar across Western Europe, the calculations show that the last time this was even approximately true was in the late 15th century. Between 1500 and 1750, real wages in Amsterdam and London, the booming maritime cities of north-western Europe, were trendless, while they fell sharply in other parts of Europe under the impact of population growth not offset by economic expansion (Allen, 2003b). Incomes had diverged in Europe, in the manner described by Smith, before the Industrial Revolution. Indeed, it was decades, if not a century, before modern economic growth was perceptible in the real wage data. So far as Europe was concerned, the great divergence preceded the Industrial Revolution rather than being its sequel.

What about Europe and Asia? It is only very recently that comparisons have been made across the continents. Parthasarathi's (1998) study of England and India supported the revision view, but a broader collection of data supports Smith's assessment (Allen, 2007b). Comparisons with Japan and China also show that real wages there were like those of the backward parts of Europe. Even the Yangtze Delta, the most advanced region in China, had real wages on a par with those in Milan, not London or Amsterdam (Allen et al., 2007). While the Ottoman Empire has not received as much attention as east Asia in the revisionist historiography, Özmucur and Pamuk (2002) have shown that the real wage in Istanbul was also like that in Italy.

The worldwide conclusions require comparisons across regions with radically different diets. The comparisons are made by computing the cost of Smith's 'quantity of necessaries of life which is given to the labourer'. Figure 2 shows full-time, full-year earnings for a labourer deflated by the cost of maintaining a family on a mainly carbohydrate diet yielding 1,920 calories per adult male equivalent. In each region, the cheapest available carbohydrate is used for the calculation. A value of 1 indicates that the labourer's wage equalled this 'bare bones' subsistence, and, indeed, that was the case in the 18th century in much of Europe and Asia. Living standards were higher, however, in Amsterdam and London.

Real wages and globalization

The history of the global economy has attracted attention and real wages have played an important role in exposing its properties. Research on the 19th and 20th centuries has aimed to establish trends in real wages across countries as well as over time (Allen, 1994; Williamson, 1995). O'Rourke and Williamson (1999) argued that

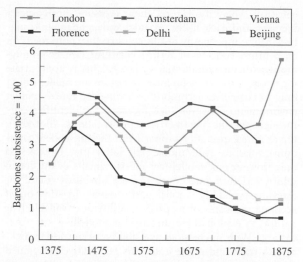

| —■— | London | —■— | Amsterdam | —■— | Vienna |
| —■— | Florence | —■— | Delhi | —■— | Beijing |

Figure 2 Subsistence ratio for labourers, various world cities, 1375–1875. Income/cost of subsistence basket. *Sources*: Allen (2001; 2007b); Allen et al. (2007).

international trade and migration tightly bound economies and determined their relative factor prices. Trends in real wages, in other words, were determined by the evolution of the global economy rather than by the internal forces of capital accumulation and technical change that most previous theories have emphasized. In a study of the British economy, O'Rourke and Williamson (2005) argued that international factors determined factor prices from 1850 onwards and perhaps from as early as 1750. The relative importance of internal and external factors in determining the real wage is a lively area of current research.

ROBERT C. ALLEN

See also **economic growth in the very long run; Industrial Revolution.**

Bibliography

Allen, R.C. 1994. Real incomes in the English-speaking world, 1879–1913. In *Labour Market Evolution: The Economic History of Market Integration, Wage Flexibility and the Employment Relation*, ed. G. Grantham and M. MacKinnon. London: Routledge.

Allen, R.C. 2001. The great divergence in European wages and prices from the middle ages to the First World War. *Explorations in Economic History* 38, 411–47.

Allen, R.C. 2003a. *Farm to Factory: A Reassessment of the Soviet Industrial Revolution*. Princeton: Princeton University Press.

Allen, R.C. 2003b. Poverty and progress in early modern Europe. *Economic History Review* 56, 403–43.

Allen, R.C. 2007a. Pessimism preserved: real wages in the British industrial revolution. Working Paper No. 314. Department of Economics, Oxford University.

Allen, R.C. 2007b. India in the great divergence. In *The New Comparative Economic History: Essays in Honor of Jeffery G. Williamson*, ed. T.J. Hatton, K.H. O'Rourke and A.M. Taylor. Cambridge, MA: MIT Press.

Allen, R.C., Bassino, J.-P., Ma, D., Moll-Murata, C. and van Zanden, J.-L. 2007. Wages, prices, and living standards in China, Japan, and Europe, 1738–1925. Working Paper No. 316. Department of Economics, Oxford University.

Beveridge, L. 1965. *Prices and Wages in England from the Twelfth to the Nineteenth Century*. London: Frank Cass.

Bortz, J. and Aguila, M. 2006. Earning a living: a history of real wage studies in twentieth-century Mexico. *Latin America History Review* 41, 112–38.

Boskin, M.J., Dulberger, E., Gordon, R., Griliches, Z. and Jorgenson, D. 1996. *Toward a More Accurate Measure of the Cost of Living*. Final Report to the Senate Finance Committee from the Advisory Commission to Study the Consumer Price Index. Washington, DC: Government Printing Office.

Clark, G. 2005. The condition of the working class in England, 1209–2004. *Journal of Political Economy* 113, 1307–40.

Diewert, W.E. 1976. Exact and superlative index numbers. *Journal of Econometrics* 4, 115–45.

Douglas, P.H. 1930. *Real Wages in the United States: 1890–1926*. Boston: Houghton Mifflin.

Feinstein, C.H. 1998. Pessimism perpetuated: real wages and the standard of living in Britain during and after the Industrial Revolution. *Journal of Economic History* 58, 625–58.

ILO (International Labour Organization) Various years. *Yearbook of Labour Statistics*. Geneva: ILO.

Kuznets, S. 1955. Economic growth and income inequality. *American Economic Review* 45, 1–28.

Lewis, W.A. 1954. Economic development with unlimited supplies of labour. *Manchester School of Economics and Social Studies* 22, 139–91.

Lindert, P.H. and Williamson, J.G. 1983. English workers' living standard during the Industrial Revolution: a new look. *Economic History Review* 36, 1–25.

Margo, R.A. 2000. *Wages and Labor Markets in the United States 1820–1860*. Chicago: University of Chicago Press.

Mitchell, B.R. 1998. *International Historical Statistics: Europe, 1750–1993*. London: Macmillan Reference.

O'Rourke, K.H. and Williamson, J.G. 1999. *Globalization and History*. Cambridge, MA: MIT Press.

O'Rourke, K.H. and Williamson, J.G. 2005. From Malthus to Ohlin: trade, industrialisation and distribution since 1500. *Journal of Economic Growth* 10, 5–34.

Özmucur, S. and Pamuk, S. 2002. Real wages and standards of living in the Ottoman Empire, 1489–1914. *Journal of Economic History* 62, 293–321.

Parthasarathi, P. 1998. Rethinking wages and competitiveness in the eighteenth century: Britain and south India. *Past & Present* 158, 79–109.

Phelps Brown, E.H. and Hopkins, S.V. 1955. Seven centuries of building wages. *Economica* NS 22, 195–206.

Phelps Brown, E.H. and Hopkins, S.V. 1956. Seven centuries of the prices of consumables, compared with builders' wage rates. *Economica* NS 23, 296–314.

Pomeranz, K. 2000. *The Great Divergence: China, Europe, and the Making of the Modern World*. Princeton: Princeton University Press.

Rees, A. 1961. *Real Wages in Manufacturing, 1890–1914*. Princeton: Princeton University Press.

Rogers, J.E.T. 1866–92. *A History of Agriculture and Prices in England*, vols. 7. Oxford: Clarendon Press.

Scholliers, P. and Zamagni, V. 1995. *Labour's Reward: Real Wages and Economic Change in 19th- and 20th-Century Europe*. Aldershot: Edward Elgar.

Smith, A. 1776. *An Inquiry into the Nature and Causes of the Wealth of Nations*, ed. E. Cannan. New York: The Modern Library, 1937.

Solow, R.M. 1956. A contribution to the theory of economic growth. *Quarterly Journal of Economics* 70, 65–94.

van Zanden, J.L. 1999. Wages and the standard of living in Europe, 1500–1800. *European Review of Economic History* 3, 175–97.

Williamson, J.G. 1995. The evolution of global labor markets since 1830: background evidence and hypotheses. *Explorations in Economic History* 32, 141–96.

realized volatility

Return volatility is critical for a range of issues in financial economics. In theory, an asset price reflects its return covariation with economy-wide risk factors, often captured through its covariance with returns of factor-replicating financial portfolios, including the broad (stock) market. Hence, assessments of asset pricing, fund performance and portfolio allocation are all directly linked to expectations of the future volatility and covariability of financial assets. Likewise, individual asset volatilities are key inputs to derivatives pricing and risk management. Finally, in recent years volatility *realizations* have become the object of direct contracting as the payoff on so-called volatility swaps is determined by the future value attained by a specified measure of return volatility over the contract horizon.

As a consequence, return volatility has been studied extensively in the literature. Until recently, there were two dominant paradigms. One uses parametric time series models within the GARCH (Engle, 1982; Bollerslev, 1986) or (genuine) stochastic volatility (Shephard, 2005) class to obtain conditional return variance estimates and forecasts. A second exploits market prices of volatility sensitive contracts, such as options, to back out the (implied) expected future volatility. Although the latter approach also conditions on a specific pricing model, it embodies a wider information set as market prices reflect the views of market participants, not just historical returns. However, derivative prices carry premiums for bearing volatility risk and thus provide a less direct measure of expected (physical) volatility. Hence, these measures are complementary and each likely provides independent information. More importantly in this context, they both focus on an a priori concept of return volatility, largely synonymous with the conditional variance. This is appropriate for many purposes as financial decisions are made subject only to current expectations regarding the future market environment. However, such measures are identified only through specific parametric representations. Moreover, the *ex ante* expectations differ from the subsequent (random) volatility realizations. The latter may be assessed only from *ex post* model-free measurements of the actual return variation. Such measures are obviously useful for assessing (volatility) model performance. In addition, if accurate measures of realized volatility are available it is natural to exploit these directly for modelling and forecasting. With increasing availability of intra-day tick-by-tick trade and quote data, this perspective has gained in popularity and a voluminous literature is evolving on the approach. This article presents a brief overview of these developments and associated empirical applications.

Historical volatility

The concept of realized volatility refines and extends the *historical volatility* measure which has a fairly long precedent in the literature. To make the argument transparent, we initially consider an extremely simplified environment. Assume we are given the daily closing logarithmic asset price, denoted p_t. The associated daily continuously compounded return is then $r_t = p_t - p_{t-1}$. Moreover, assume the returns are conditionally mean zero with i.i.d. standardized residuals, that is,

$$r_t = \sigma_t \cdot z_t, \quad \text{with} \quad z_t \sim iid(0,1) \quad \text{and}$$
$$Var(z_t^2) = \omega.$$
(1)

The goal of realized volatility measurement is to provide a model-free estimator for the return variation or volatility, σ_t^2, given only the concurrent return observations. Obviously, if volatility is time-varying, this is problematic. We have, at the daily level, conditional on current volatility,

$$E[r_t^2] = Var(r_t) = \sigma_t^2, \quad Var(r_t^2) = \omega \cdot \sigma_t^4, \quad \text{and}$$
$$E[r_t^2]/[Var(r_t^2)]^{1/2} = \omega^{-1/2}.$$
(2)

Hence, the concurrent squared return is an unbiased estimator of the underlying return variance. Unfortunately, it is extraordinarily noisy. The signal-to-noise ratio, defined as the mean of the estimator relative to the standard deviation, equals $\omega^{-1/2}$. Invariably, $\omega > 1$ at daily (and lower) frequencies, so the standard deviation of (estimated) realized volatility exceeds the expected value. Matters improve if we assume constant volatility, at σ_t^2, over the month representing, say, K daily returns. Letting $r_{t:K} = r_t + \ldots + r_{t+K}$ denote the monthly return, we may exploit the historical volatility indicator, $[r_t^2 + \ldots + (r_{t+K})^2]$ rather than simply $(r_{t:K})^2$. We then have,

$$Var(r_{t:K}) = E[r_t^2 + \ldots + (r_{t+K})^2] = K \cdot \sigma_t^2, \quad \text{and}$$
$$Var[r_t^2 + \ldots + (r_{t+K})^2] = K \cdot \omega \cdot \sigma_t^4.$$
$$(3)$$

The signal-to-noise ratio for the monthly realized volatility is $(K/\omega)^{1/2}$ or a factor $K^{1/2}$ larger than for eq. (2). Equation (3) may also readily be converted into an estimator for daily volatility based on the sample mean of the daily squared return over the month. This estimator is consistent with convergence rate $K^{1/2}$. However, given the simplifying assumptions, this measure is best viewed as an informal gauge of the underlying level of volatility. It was applied for computation of annual realized volatility from monthly data by Officer (1973) and monthly volatility series from daily data by, amongst others, Merton (1980), and French, Schwert and Stambaugh (1987).

The two equivalent interpretations of eq. (3) have distinct properties within a more realistic time-varying volatility setting. Since it is untenable to assume constant volatility for a month, or even one day, estimation of daily volatility from a surrounding set of daily returns covering, say, one month is inherently problematic. Nonetheless, the monthly realized volatility measure is robust to variation in volatility as explained more formally below. Of course, given the rapidly evolving markets, we would often need to assess the time variation in realized volatility at a daily level. The above reasoning suggests this will require access to high-frequency intraday data. This is the starting point for the modern realized volatility literature.

Realized volatility as an *ex post* return variability measure

Given the round-the-clock activity on financial markets, volatility is naturally seen as evolving stochastically in continuous time. A complete characterization of the volatility realization then consists of a full specification of the actual sample path. However, at most one price is observed at each point in time so *instantaneous* volatility cannot be assessed without relying on adjacent observations, which is justified only under auxiliary assumptions – just as daily volatility cannot be estimated from a single return. In contrast, realized volatility seeks to measure the temporally cumulated (instantaneous) volatility so the target is the (average) realization of volatility over a non-negligible interval, allowing for a feasible and robust estimator with desirable properties.

As a benchmark for analysis, let the logarithmic asset price, $p(t)$, be a continuous time stochastic process observed in a frictionless market. For brevity and clarity, the formal exposition is cast in a univariate setting, but all results for realized volatility generalize readily to the multivariate case. To avoid arbitrage, and subject only to weak auxiliary conditions, the price process constitutes a (special) semi-martingale (Harrison and Kreps, 1978; Back, 1991). The price process may then quite generally be represented as follows,

$$dp(t) = \mu(t)dt + \sigma(t)dB(t) + j(t)dJ(t), \quad t \in [0, T],$$
$$(4)$$

where $\mu(t)$ is a predictable, continuous process with bounded variation, the volatility process $\sigma(t)$ is strictly positive, $B(t)$ denotes a standard Brownian motion, $J(t)$ is a jump indicator taking the values zero (no jump) or unity (jump) and, finally, the $j(t)$ indicates the jump in the return process if a jump occurs at time t and $j(t) \equiv 0$ otherwise. We assume the jump intensity, denoted $\lambda(t)$, to be bounded so there is a finite number of jump in the price path per time period. This is standard in the asset pricing literature although it does rule out some valid Lévy representations.

Equation (4) implies an instantaneous expected return of $\{\mu(t) + \lambda(t)E[j(t)]\}\, dt$, which is an order smaller than the instantaneous innovation, $\sigma(t)(dt)^{1/2} + j(t)$ namely order dt versus $dt^{1/2}$. Hence, for short horizons, the return variability is dominated by the unpredictable martingale component and the mean return is negligible. These features are captured formally by the notion of return quadratic variation defined below.

We denote the discretely observed continuously compounded return at time t, based on price observations at times t and $t-h$, for $h > 0$, by $r(t, h) = p(t) - p(t - h)$. From equation (4), the h-period return then has the representation,

$$r(t, h) = p(t) - p(t - h) = \int_{t-h}^{t} \mu(\tau)d\tau$$
$$+ \int_{t-h}^{t} \sigma(\tau)dB(\tau) + \sum_{t-h \leq \tau < t} j(\tau).$$
$$(5)$$

Formally, the sample path variation of the return process over $[t-h, t]$ is given by the *quadratic variation* of the logarithmic price process,

$$QV(t, h) \equiv \int_{t-h}^{t} \sigma^2(s)ds + \sum_{t-h < s \leq t} j^2(s)$$
$$\equiv IV(t, h) + \sum_{t-h < s \leq t} j^2(s).$$
$$(6)$$

Equation (6) attributes the return variation to the diffusive volatility and the cumulative squared jumps. The first term is denoted the *integrated variance*. This quantity is often the focus of the broader realized volatility literature as many studies ignore jumps. The integrated variance is also of direct relevance for option pricing under stochastic volatility (Hull and White, 1987). This is in part due to the following result for the special case with neither jumps nor correlation between the return and volatility processes, that is, $B(t)$ is independent of $\sigma(s)$ for all $0 \leq s, \ t \leq T$. Conditional on the mean component and integrated variance, we then have,

$$r(t,h)/\{\mu(t,h), IV(t,h)\} \sim N(\mu(t,h), IV(t,h)),$$
$$(7)$$

where $\mu(t,h) = \int_{t-h}^{t} \mu(\tau)d\tau$. Since (innovations to) the mean component is of smaller order than the integrated variance for low values of h, the dominant feature is the time-varying second moment given by the realizations of integrated variance. Hence, the return distribution is a normal mixture governed by the integrated variance process. If return and volatility innovations are correlated, the distribution is no longer mixed normal, but the interpretation of the integrated variance as a return variability measure is maintained.

Of course, short of having a continuum of price observations available, the relevant quantities in eqs (6) or (7) are not directly observable. However, in theory, the quadratic variation can be approximated closely by the corresponding cumulative squared return process, motivating the following definition. Let $[t-h, t]$ be split into $M = h/\Delta$ sub-intervals of length Δ, with $0 < \Delta \ll h$, and define the *realized volatility* constructed from the equally spaced Δ-period returns as

$$RV(t, h; \Delta) = \sum_{m=1}^{M} r^2(t - h + m \cdot \Delta, \Delta).$$
$$(8)$$

The basic theory for semi-martingales ensures that realized volatility is consistent for quadratic variation in the sense that, for finer and finer sampling of intra-day returns, eq. (8) will, in the limit, provide a perfect measure of the realizations of the latent quadratic variation, that is,

$$RV(t, h; \Delta) \rightarrow QV(t, h), \quad \text{as } \Delta \rightarrow 0$$
$$\text{(and } M = h/\Delta \rightarrow \infty).$$
$$(9)$$

This provides a formal basis for *ex post* measurement of realized volatility without parametric assumptions. It is a model-free measure of actual realizations while standard approaches provide parametric model forecasts of future volatility. We have the following approximate relationship between parametric forecasts and realized volatility, for small h,

$$Var[r(t,h)|\mathscr{F}_{t-h}; \theta] \simeq E[RV(t, h; \Delta)|\mathscr{F}_{t-h}]$$
$$\simeq E[QV(t, h)|\mathscr{F}_{t-h}],$$
$$(10)$$

where the left most expression denotes the conditional variance over $[t-h, t]$ conditional on the available information at time $t-h$, $\mathscr{F}_{t-h}$, and the (true) model parameter vector, θ. Andersen, Bollerslev and Diebold (ABD) (2008) provide an in-depth discussion of the approximation behind eq. (10). The relation shows that realized volatility is *the* natural benchmark for assessing volatility forecast performance.

Realized volatility can in principle be used to estimate the instantaneous volatility of a pure diffusion consistently. Ruling out jumps in the price process, but allowing volatility to be caglad (left-continuous, right limit sample path) and thus having potential discontinuities, we have

$$QV(t, h) \rightarrow \sigma^2(t), \quad \text{as } h \rightarrow 0. \quad (11)$$

This insight is certainly not new. Merton (1976; 1980) discusses the result explicitly, and Foster and Nelson (1996) develop asymptotic results. However, this limiting operation is impractical. Equation (10) merely requires that $M = h/\Delta \rightarrow \infty$. For eq. (11) to hold, with quadratic variation replaced by a feasible realized volatility estimator, Δ must converge to zero at a faster rate than h. This requires a double limiting procedure with ever more data sampled within an ever shrinking neighbourhood of t. Intensive sampling over short intervals magnifies the microstructure effects stemming from price discreteness, bid-ask bounce, temporary order-driven dependencies and other institutional features affecting returns at the highest frequencies. The issue of how to deal with such complications has inspired an extensive literature, summarized succinctly in Hansen and Lunde (2006). The practical complications induced by microstructure noise can be illustrated through the asymptotic theory for the realized volatility estimator.

For a purely diffusive price process, it follows from Jacod and Protter (1998) and Barndorff-Nielsen and Shephard (BNS) (2002a; 2002b; 2004a), that asymptotically, as $\Delta \rightarrow 0$,

$$\sqrt{h/\Delta}[RV(t, h; \Delta) - IV(t, h)] \sim N(0, 2 \cdot IQ(t, h)),$$
$$(12)$$

so the ratio of the integrated quarticity, $IQ(t, h) = \int_{t-h}^{t} \sigma^4(s)ds$, to the number of intra-day returns, $M = h/\Delta$, determines the precision of the realized volatility estimator, generalizing and improving the result discussed below eq. (3). A simple way to convey the implications of microstructure noise is to impose an exogenous bound, say Δ^*, below which no useful information from sampling is available as the semi-martingale assumption is blatantly violated at the highest frequencies. Hence, over an interval of length h, only $M^* = h/\Delta^*$ observations can be exploited for inference, and h must be of a certain size for realized volatility measures to possess meaningful precision. Equation (11) requires Δ to vanish at a rapid rate, so the bound Δ^* is quickly binding. If only a handful of returns is available within a few minutes of t, the sampling scheme behind eq. (11) breaks down. In contrast, for Δ^* fixed at one or five minutes, it is feasible to estimate the quadratic variation with reasonable accuracy for h equal to one trading day.

Alternative return variation measures

For a pure diffusion in a frictionless market, the basic realized volatility estimator exploiting all available observations is optimal for the quadratic variation. However, once price jumps and microstructure noise are introduced, matters become more complex. A number of issues may be addressed through alternative return variation measures, including the ability to disentangle the effect of jumps from the diffusive volatility, to estimate quantities needed for feasible inference about realized volatility, and to develop more robust-to-noise measures of integrated variance.

First, absent jumps and under appropriate regularity, one may extend the theory for the integrated variance to include the integrated variation of arbitrary powers, that is, for $\Delta \to 0$ the *realized power variation of order p* consistently estimates the p'th order integrated power variation ($p \le 1/2$),

$$(E|Z|^p)^{-1}(h/\Delta)^{1-p/2}\sum_{m=1}^{M}|r(t-h+m\cdot\Delta,\Delta)|^p$$

$$\to \int_{t-h}^{t}\sigma^p(s)ds$$

$$(13)$$

where Z denotes a standard normal variable. For $p = 4$, this result provides a simple estimator for the integrated quarticity which may be plugged into eq. (12) to yield a feasible distribution theory for realized volatility. An asymptotic distribution theory akin to eq. (12) is available for realized power variation (see Barndorff-Nielsen et al., 2006b). Another extension, the *realized k-skip bipower variation*, is also consistent for the integrated variance,

$$BV(t,h;k;\Delta) \equiv \frac{\pi}{2}\sum_{m=k+1}^{M}|r(t-h+m\cdot\Delta,\Delta)|$$

$$\times |r(t-h+(m-k)\cdot\Delta,\Delta)|$$

$$\to IV(t,h)$$

$$\equiv \int_{t-h}^{t}\sigma^2(s)ds.$$

$$(14)$$

For $k = 1$, $BV(t,h;1,\Delta) \equiv BV(t,h;\Delta)$, is termed the *realized bipower variation*. These measures have convenient robustness properties. First, eq. (14) remains valid even if the return process follows a jump-diffusion. Hence, the bipower measures annihilate the jumps asymptotically and thus provide simple consistent estimators for the integrated variance. This allows for separation of the jump and diffusive contributions to the realized return variation, as

$$RV(t,h;\Delta) - BV(t,h;\Delta) \to QV(t,h) - IV(t,h)$$

$$= \sum_{t-h<s\le t} j^2(s), \quad as\ \Delta \to 0.$$

$$(15)$$

Combined with the asymptotic theory for realized power and bipower variation, the result renders formal statistical tests for the presence and impact of jumps feasible. This is applied for separate analysis of the diffusive and jump components (see BNS, 2004b; 2006; Huang and Tauchen, 2005; ABD, 2007). Alternative jump tests have recently been developed by Jiang and Oomen (2005), Andersen, Bollerslev and Dobrev (2007), and Lee and Mykland (2006).

The bipower variation measures also display robustness against microstructure noise. To first order, when not sampling at the very highest frequencies, the impact of noise may be mimicked by adding an i.i.d. process to the 'efficient' prices to generate noisy observations. This noise component induces negative return correlation which inflates the realized volatility estimator, resulting in a potentially strong upward bias. This may be alleviated by sampling more sparsely although this uses less data and reduces efficiency, pointing towards a bias-variance trade-off: one should sample sparsely enough to avoid a significant bias but frequently enough that efficiency is not compromised. An informal bias diagnostic is to apply the realized volatility estimator for different underlying frequencies over the full sample, that is, h is on the order of years or a decade. The long horizon minimizes sampling variability so that, absent the microstructure bias, all the measures should centre closely on the sample realized return volatility. A *volatility signature plot* depicts those realized volatility measures against the underlying sampling frequency with Δ ranging from seconds to a full day. For liquid financial markets, signature plots typically indicate inflated values at the highest frequencies which then decay quite smoothly to a stable level for sampling between 5 and 40 minutes. Andersen, Bollerslev, Diebold and Labys (ABDL) (2000a) suggest that the shortest sampling intervals not displaying a significant bias may be desirable choices. Of course, this criterion rewards unbiasedness over efficiency. Bandi and Russell (2005a) explicitly trade off the microstructure bias with the efficiency gains from more data. An alternative is to apply skip-k bipower variation measures. Huang and Tauchen (2005) document that these tend to work well when applied to noisy observations from jump diffusions. Andersen, Bollerslev, Frederiksen and Nielsen (ABFN) (2006a) extend the volatility signature plots to include both power variation and skip-k bipower variation measures. Such generalized plots provide insights into the robustness of the realized quarticity and other quantities used for jump tests and may guide the choice of an adequate sampling frequency for analysis of a range of different issues in the presence of market frictions.

Finally, procedures have been developed to correct for microstructure bias while utilizing more of the available data. The proposals include the subsampling idea alluded to in Zhou (1996), and extended and formalized by Zhang, Mykland and Aït-Sahalia (2005) and Aït-Sahalia, Mykland and Zhang (2006) as well as the kernel based methods of Barndorff-Nielsen, Hansen, Lunde and Shephard (2006), and numerous filtering approaches discussed in Hansen and Lunde (2006).

Empirical applications

Hsieh (1991) is perhaps the first to apply intra-day returns for historical volatility measurement of the daily return variation. Closely related work appears in publications by the Olsen & Associates group. This is surveyed in Dacorogna et al. (2001). Zhou (1996) offers the first systematic study of the realized volatility estimator combining theoretical and empirical issues. Interestingly, he discusses contamination by market microstructure noise as well as ideas for a variety of feasible corrections. Comte and Renault (1998) also comment on estimating diffusive spot volatility through the empirical counterpart to quadratic variation.

In parallel work, Andersen and Bollerslev (AB) (1997; 1998b) explore the dynamics of high-frequency return volatility, documenting the striking effectiveness of cumulative absolute and squared returns as daily volatility measures. These findings inspired theoretical inquiries and are followed by the statement of consistency of realized volatility for the quadratic variation for a general multivariate jump-diffusion setting in ABDL (2001). In concurrent work, BNS (2001; 2002a) and Meddahi (2002) provide initial asymptotic theory for the realized volatility estimator, with the diffusive multivariate case treated in BNS (2004a).

Empirical work almost invariably operates with h equal to one trading day (or more). This is due to the pronounced intra-day volatility pattern which induces systematic shifts in the quadratic return variation over different segments of the trading day. As noted in AB (1997; 1998b) this type of largely deterministic effects are alleviated at the daily frequency, rendering this the natural basis for analysis. The list of empirical applications is growing rapidly. A brief overview of the topics explored along with selective, but not exhaustive, references are provided below.

The most common use of realized volatility is as a basis for volatility forecasting and evaluation. AB (1998a) document the potential of realized volatility for assessment of standard volatility forecasts, a theme further explored in Andersen, Bollerslev and Lange (1999) and rationalized more formally in Andersen, Bollerslev and Meddahi (2004) using powerful analytical techniques

developed in Meddahi (2001). An integrated approach to measurement, modelling and forecasting of realized volatility is developed in ABDL (2003). Ghysels, Santa-Clara and Valkanov (2006) show that a combination of realized volatility measures for different frequencies and horizons may enhance forecast performance. Engle and Gallo (2006) follow a similar strategy but with a different modelling approach. ABD (2007) improve performance by separating the jump and diffusive volatility components in the forecasting procedure. Other studies on the presence and importance of price jumps using realized volatility related jump statistics are ABFN (2006b), Tauchen and Zhou (2006), and Fleming and Paye (2006), Andersen, Bollerslev and Dobrev (2007), Jiang and Oomen (2005), and Lee and Mykland (2006). In the same spirit, Liu and Maheu (2005) find that more jump-robust power variation measures are preferred to realized volatility for forecasting purposes. Earlier studies of forecast performance include Blair, Poon and Taylor (2001) and Martens (2002). Some initial studies of the role of microstructure noise and discretization error for forecasting and forecast evaluation is provided by Aït-Sahalia and Mancini (2006), ABM (2005; 2006), and Ghysels and Sinko (2006). The issue of how to include (noisy) overnight return information into the volatility measures and forecasts is addressed by Fleming, Kirby and Ostdiek (2003) and Hansen and Lunde (2005).

The evidence for long-range persistence in volatility is particularly striking when analysed via realized volatility rather than via daily return observations. AB (1998a) demonstrate that return series spanning only a couple of years are sufficient to identify a distinct hyperbolic decay in the realized power variation series. Moreover, the implied degree of fractional integration appears to be stable across subsamples, at around 0.35–0.45, implying a stationary volatility series. This finding is confirmed by virtually all subsequent studies of realized volatility exploring the issue, including ABDL (2001, 2003), Areal and Taylor (2002), Martens (2002), Zumbach (2004), Deo, Hsieh and Hurvich (2005), and Deo, Hurvich and Lu (2006). Moreover, related early work by the Olsen & Associates group also note the presence of scaling laws in volatility measures obtained from high frequency returns, see, for example, Müller et al. (1990) and the review in Dacorogna et al. (2001). This inspired an extensive amount of empirical work in the 'econophysics' area on volatility scaling laws which is summarized in Mantegna and Stanley (2000). These robust empirical results suggest that the long-memory property is not driven by occasional structural breaks but is present in the data generating process at high frequencies. As such, it sheds new light on a contentious issue which is not readily resolved without an effective volatility measure that improves the signal-to-noise ratio, allowing for more effective inference. Explicit estimation of the long memory features may be circumvented through a model with multiple volatility components, each governed by an autoregressive process, as such structures approximate the long-memory dependencies very well. This approach has been applied for realized volatility series by ABD (2007), BNS (2001), and Corsi (2003), among others.

The no-arbitrage implications for the return dynamics expressed in eqs (4) and (5), are necessarily quite weak and general, but do nonetheless have distributional implications which are potentially testable. As highlighted by eq. (7), auxiliary assumptions produce a mixture of normals result akin to the mixture-of-distributions theory, originating from work by Clark (1973) and Tauchen and Pitts (1983). The novelty of eq. (7) is the potential *ex post* observability of the mixing variable, the integrated variance, which enables direct inference regarding the distributional implications without any parametric assumptions. In practice there will be some discretization error and microstructure distortions, but the properties of the realized volatility estimator should facilitate powerful tests. ABDL (2000b) confirm that returns standardized by realized volatility are much closer to normal than the standardized residuals from the usual volatility models based on daily data, even if the normality is not exact. Thomakos and Wang (2003) reach the identical conclusion for a different set of assets. Of course, eq. (7) is not valid in the presence of price jumps or dependence between the volatility and return innovation processes. The former issue may be addressed through jump identification procedures which seek to annihilate the impact of the jumps. The latter issue is accommodated by sampling the return process in 'financial time', consisting of calendar periods representing equal increments to the integrated variance process, as noted by Peters and de Vilder (2006) for a diffusive return process. This approach is extended to a jump-diffusive setting by Andersen, Bollerslev and Dobrev (2007) who also explore practical implementation issues for the associated distributional tests in detail. More extensive data sets and alternative jump identification techniques are considered in ABFN (2006b). They quite generally obtain jump-adjusted, financial-time sampled returns that are indistinguishable from i.i.d. standard Gaussian variates through realized volatility based empirical procedures. In sum, the results corroborate the general framework, and the tools developed for jump identification and measurement of the quadratic variation deliver empirically meaningful series of jumps and quadratic variation which are fully consistent with the theoretical underpinnings. In the process, direct evidence of the importance of jumps and the asymmetric return-volatility relationship is provided. Jumps are present for all asset classes and constitute a non-negligible fraction of overall return variation. For equities, negative returns tend to induce higher volatility than corresponding positive returns, a feature broadly recognized in the prior literature. Interestingly, there are also signs of significant asymmetric relationships for other asset classes, although both magnitude and sign may change over time. Similar issues are studied by Fleming and Paye (2006) and closely related topics are explored by Maheu and McCurdy (2002).

Multivariate applications of the realized volatility estimator are in principle straightforward. They are used to study the broad correlation patterns among individual stocks in ABDE (2001), for volatility timing in portfolio allocation in Fleming, Kirby and Ostdiek (2003), for estimation of systematic market risk exposure (time-varying market betas) in Andersen et al. (2005), for assessment of a broader set of risk loading coefficients in Bollerslev and Zhang (2003), and for dynamic portfolio choice in Bandi, Russell and Zhou (2008). In spite of these initial explorations, it is clear that the multivariate setting introduces additional practical complications as there is evidence of significant delays in the reaction of one security price to movements in another related asset. Sheppard (2006) explores these issues in some depth even if general prescriptions for practice do not directly follow. The favoured approach in current work is to include temporal cross-correlation patterns through measurement of the relations between the return of one asset with lead and lag returns for the other asset – in the spirit of the corrections for non-trading effects in the estimation of betas from daily data in Scholes and Williams (1977). Hayashi and Yoshida (2005; 2006) and Griffin and Oomen (2006) study such realized covariance and correlation estimators. Bandi and Russell (2005b) and Zhang (2006) seek to trade off bias and efficiency optimally. An alternative approach is proposed in Bauer and Vorkink (2006).

A natural comparison for realized volatility based forecasts is with forecasts implied by traded financial contracts such as options. In fact, an intriguing parallel exists between expected future realized volatility and the pricing of volatility swaps, with the latter reflecting the expected future integrated variance under the risk-neutral (pricing) measure if jumps in the price path are absent. Hence, systematic differences between realized volatility and implied volatility measures reflect the market prices of volatility risk (see, for example, Britten-Jones and Neuberger, 2000; Carr and Madan, 1998; Carr and Wu, 2005). Moreover, Bondarenko (2004) shows that the implied volatility result remains approximately valid for jump diffusions and returns sampled at discrete intervals only. Recent empirical papers on the performance of realized volatility forecasts versus implied volatility measures include, for example, Andersen, Frederiksen and Staal (2006), Bollerslev, Gibson and Zhou (2005), Bollerslev and Zhou (2006), Busch, Christensen and Nielsen (2006), Chan, Kalimipulli and Sivakumar (2006) and Pong et al. (2004). The findings confirm that realized volatility forecasts contain information for future return variability over-and-beyond implied volatility forecasts, while standard volatility forecasts obtained from models utilizing only daily data are fully encompassed by the market based measures.

Future directions for research

Most of the empirical work associated with the realized volatility concept has focused directly on the measurement precision and forecast performance. In order for the approach to enter routinely in more mainstream applications within asset pricing, risk management and portfolio allocation the empirical studies must broaden in scope. As reviewed above, this has begun to happen, but much work remains. One recent example of using the concept for model specification testing is Andersen and Benzoni (2005). The study documents a serious deficiency in the ability of affine term structure models to accommodate the observed dynamics of realized yield volatility for the US Treasury market. Theoretically, the concurrent yield curve should span yield volatility, both *ex post* and *ex ante*, but this property is systematically violated as the yield variation at every maturity displays genuine stochastic features not associated with simultaneous directional shifts in the yield curve. The result extends earlier findings based on *ex ante* volatility measures at the monthly frequency. For generalizations of term structure models operating within the popular and tractable affine setting, the realized volatility measures promise to be valuable diagnostic tools.

The applications of realized volatility measures will surely continue to grow and broaden as the advantages of the enhanced precision in the measurement of the return variability are much too large to be ignored and many important questions await thorough analysis from this perspective. Of particular interest is the development of practical approaches to generate reliable measures for the high-dimensional case involving a large set of assets. On the theoretical front, work remains in terms of understanding the relative advantages of the different robust alternatives to the basic realized volatility measure. One potentially promising avenue is to further develop the locally constant volatility technique developed by Mykland (2006), seeking to provide a formal, yet simple and powerful, tool for asymptotic theory while allowing for time-varying price dynamics.

TORBEN G. ANDERSEN

See also **ARCH models; continuous and discrete time models; kernel estimators in econometrics; law(s) of large numbers; long memory models; martingales; mean-variance analysis; measurement error models; mixture models; options; stochastic volatility models; Wiener process.**

Bibliography

Aït-Sahalia, Y. and Mancini, L. 2006. Out of sample forecasts of quadratic variation. Manuscript, Princeton University and University of Zürich.

Aït-Sahalia, Y., Mykland, P.A. and Zhang, L. 2006. Ultra high frequency volatility estimation with dependent microstructure noise. Manuscript, Princeton University.

Andersen, T.G. and Benzoni, L. 2005. Can bonds hedge volatility risk in the U.S. Treasury market? A specification test for affine term structure models. Manuscript, Kellogg School, Northwestern University and University of Minnesota.

Andersen, T.G. and Bollerslev, T. 1997. Heterogeneous information arrivals and return volatility dynamics: uncovering the long-run in high frequency returns. *Journal of Finance* 52, 975–1005.

Andersen, T.G. and Bollerslev, T. 1998a. Answering the skeptics: yes, standard volatility models do provide accurate forecasts. *International Economic Review* 39, 885–905.

Andersen, T.G. and Bollerslev, T. 1998b. Deutschemark-Dollar volatility: intraday activity patterns, macroeconomic announcements, and longer run dependencies. *Journal of Finance* 53, 219–65.

Andersen, T.G., Bollerslev, T. and Diebold, F.X. 2007. Roughing it up: including jump components in the measurement, modeling and forecasting of return volatility. *Review of Economics and Statistics* 89.

Andersen, T.G., Bollerslev, T. and Diebold, F.X. 2008. Nonparametric volatility measurement. In *Handbook of Financial Econometrics*, ed. L.P. Hansen and Y. Aït-Sahalia. Amsterdam: North-Holland.

Andersen, T.G., Bollerslev, T., Diebold, F.X. and Ebens, H. 2001. The distribution of realized stock return volatility. *Journal of Financial Economics* 61, 43–76.

Andersen, T.G., Bollerslev, T., Diebold, F.X. and Labys, P. 2000a. Great realizations. *Risk* 13, 105–8.

Andersen, T.G., Bollerslev, T., Diebold, F.X. and Labys, P. 2000b. Exchange rate returns standardized by realized volatility are (nearly) Gaussian. *Multinational Finance Journal* 4, 159–79.

Andersen, T.G., Bollerslev, T., Diebold, F.X. and Labys, P. 2001. The distribution of realized exchange rate volatility. *Journal of the American Statistical Association* 96, 42–55.

Andersen, T.G., Bollerslev, T., Diebold, F.X. and Labys, P. 2003. Modeling and forecasting realized volatility. *Econometrica* 71, 579–625.

Andersen, T.G., Bollerslev, T., Diebold, F.X. and Vega, C. 2003. Micro effects of macro announcements: real-time price discovery in foreign exchange. *American Economic Review* 93, 38–62.

Andersen, T.G., Bollerslev, T., Diebold, F.X. and Wu, G. 2005. A framework for exploring the macroeconomic determinants of systematic risk. *American Economic Review* 95, 398–404.

Andersen, T.G., Bollerslev, T. and Dobrev, D. 2007. No-arbitrage semi-martingale restrictions for continuous-time volatility models subject to leverage effects, jumps and i.i.d. noise: theory and testable distributional implications. *Journal of Econometrics* 138, 125–80.

Andersen, T.G., Bollerslev, T., Frederiksen, P.H. and Nielsen, M.Ø. 2006a. Comment on realized variance and market microstructure noise. *Journal of Business & Economic Statistics* 24, 173–9.

Andersen, T.G., Bollerslev, T., Frederiksen, P.H. and Nielsen, M.Ø. 2006b. Continuous-time models, realized volatilities, and testable distributional implications for daily stock returns. Manuscript, Kellogg School, Northwestern University.

Andersen, T.G., Bollerslev, T. and Lange, S. 1999. Forecasting financial market volatility: sample frequency vis-à-vis forecast horizon. *Journal of Empirical Finance* 6, 457–77.

Andersen, T.G., Bollerslev, T. and Meddahi, N. 2004. Analytic evaluation of volatility forecasts. *International Economic Review* 45, 1079–110.

Andersen, T.G., Bollerslev, T. and Meddahi, N. 2005. Correcting the errors: volatility forecast evaluation using high-frequency data and realized volatilities. *Econometrica* 73, 279–96.

Andersen, T.G., Bollerslev, T. and Meddahi, N. 2006. Market microstructure noise and realized volatility forecasting. Manuscript, Kellogg School, Northwestern University.

Andersen, T.G., Frederiksen, P.H. and Staal, A.D. 2006. The information content of realized volatility forecasts. Manuscript, Kellogg School, Northwestern University.

Areal, N.M.P.C. and Taylor, S.J. 2002. The realized volatility of FTSE-100 futures prices. *Journal of Futures Markets* 22, 627–48.

Back, K. 1991. Asset prices for general processes. *Journal of Mathematical Economics* 20, 371–95.

Bandi, F. and Russell, J.R. 2005a. Microstructure noise, realized volatility, and optimal sampling. Manuscript, Graduate School of Business, University of Chicago.

Bandi, F. and Russell, J.R. 2005b. Realized covariation, realized beta and microstructure noise. Manuscript, Graduate School of Business, University of Chicago.

Bandi, F., Russell, J.R. and Zhou, Y. 2008. Using high-frequency data in dynamic portfolio choice. *Econometric Reviews*, forthcoming.

Barndorff-Nielsen, O.E., Graversen, S.E., Jacod, J. and Shephard, N. 2006a. Limit theorems for bipower variation in econometrics. *Econometric Theory* 22, 677–720.

Barndorff-Nielsen, O.E., Hansen, P.R., Lunde, A. and Shephard, N. 2006b. Designing realised kernels to measure the ex-post variation of equity prices in the presence of noise. Manuscript, Nuffield College, Oxford.

Barndorff-Nielsen, O.E. and Shephard, N. 2001. Non-Gaussian Ornstein–Uhlenbeck-based models and some of their uses in financial economics. *Journal of the Royal Statistical Society, Series B* 63, 167–241.

Barndorff-Nielsen, O.E. and Shephard, N. 2002a. Econometric analysis of realised volatility and its use in estimating stochastic volatility models. *Journal of the Royal Statistical Society, Series B* 64, 253–80.

Barndorff-Nielsen, O.E. and Shephard, N. 2002b. Estimating quadratic variation using realised variance. *Journal of Applied Econometrics* 17, 457–77.

Barndorff-Nielsen, O.E. and Shephard, N. 2003. Realized power variation and stochastic volatility models. *Bernoulli* 9, 243–65.

Barndorff-Nielsen, O.E. and Shephard, N. 2004a. Econometric analysis of realised covariation: high-frequency covariance, regression and correlation in financial econometrics. *Econometrica* 72, 885–925.

Barndorff-Nielsen, O.E. and Shephard, N. 2004b. Power and bipower variation with stochastic volatility and jumps. *Journal of Financial Econometrics* 2, 1–37.

Barndorff-Nielsen, O.E. and Shephard, N. 2006. Econometrics of testing for jumps in financial economics using bipower variation. *Journal of Financial Econometrics* 4, 1–30.

Bauer, G.H. and Vorkink, K. 2006. Multivariate realized stock market volatility. Manuscript, Bank of Canada and Sloan School, Massachusetts Institute of Technology.

Blair, B.J., Poon, S.-H. and Taylor, S.J. 2001. Forecasting S&P 100 volatility: the incremental information content of implied volatility and high-frequency index returns. *Journal of Econometrics* 105, 5–26.

Bollerslev, T. 1986. Generalized autoregressive conditional heteroskedasticity. *Journal of Econometrics* 31, 307–27.

Bollerslev, T., Gibson, M. and Zhou, H. 2005. Dynamic estimation of volatility risk premia and investor risk aversion from option-implied and realized volatilities. Manuscript, Department of Economics, Duke University.

Bollerslev, T. and Mikkelsen, H.O. 1996. Modeling and pricing long memory in stock market volatility. *Journal of Econometrics* 73, 151–84.

Bollerslev, T. and Zhang, B.Y.B. 2003. Measuring and modeling systematic risk in factor pricing models using high-frequency data. *Journal of Empirical Finance* 10, 533–58.

Bollerslev, T. and Zhou, H. 2006. Volatility puzzles: a simple framework for gauging return-volatility regressions. *Journal of Econometrics* 131, 123–50.

Bondarenko, O. 2004. Market price of variance risk and performance of hedge funds. Manuscript, University of Illinois at Chicago.

Britten-Jones, M. and Neuberger, A. 2000. Option prices, implied price processes, and stochastic volatility. *Journal of Finance* 55, 839–66.

Busch, T., Christensen, B.J. and Nielsen, M.Ø. 2006. The role of implied volatility in forecasting realized volatility and jumps in foreign exchange, stock and bond markets. Manuscript, University of Aarhus and Cornell University.

Carr, P. and Madan, D. 1998. Towards a theory of volatility trading. In *Volatility: New Estimation Techniques for Pricing Derivatives*, ed. R. Jarrow. London: Risk Books.

Carr, P. and Wu, L. 2005. Variance risk premia. Manuscript, Courant Institute, New York University.

Chan, W.H., Kalimipulli, M. and Sivakumar, R. 2006. The economic value of using realized volatility in the index options market. Manuscript, Wilfrid Laurier University and University of Waterloo.

Clark, P.K. 1973. A subordinated stochastic process model with finite variance for speculative prices. *Econometrica* 41, 135–55.

Comte, F. and Renault, E. 1998. Long memory in continuous time stochastic volatility models. *Mathematical Finance* 8, 291–323.

Corsi, F. 2003. A simple long-memory model of realized volatility. Manuscript, University of Southern Switzerland.

Dacorogna, M.M., Gencay, R., Müller, U., Olsen, R.B. and Pictet, O.V. 2001. *An Introduction to High-Frequency Finance*. San Diego, CA: Academic Press.

Deo, R., Hsieh, M. and Hurvich, C.M. 2005. Tracing the source of long memory in volatility. Manuscript, Stern School, New York University.

Deo, R., Hurvich, C.M. and Lu, Y. 2006. Forecasting realized volatility using a long-memory stochastic volatility model: estimation, prediction and seasonal adjustment. *Journal of Econometrics* 131, 29–58.

Engle, R.F. 1982. Autoregressive conditional heteroskedasticity with estimates of the variance of U.K. inflation. *Econometrica* 50, 987–1008.

Engle, R.F. and Gallo, G.M. 2006. A multiple indicators model for volatility using intra-daily data. *Journal of Econometrics* 131, 3–27.

Fleming, J., Kirby, C. and Ostdiek, B. 2003. The economic value of volatility timing using realized volatility. *Journal of Financial Economics* 67, 473–509.

Fleming, J. and Paye, B.S. 2006. High-frequency returns, jumps and the mixture-of-normals hypothesis. Manuscript, Jones School, Rice University.

Foster, D. and Nelson, D.B. 1996. Continuous record asymptotics for rolling sample estimators. *Econometrica* 64, 139–74.

French, K.R., Schwert, G.W. and Stambaugh, R.F. 1987. Expected stock returns and volatility. *Journal of Financial Economics* 19, 3–29.

Ghysels, E., Santa-Clara, P. and Valkanov, R. 2006. Predicting volatility: how to get the most out of returns data sampled at different frequencies. *Journal of Econometrics* 131, 59–95.

Ghysels, E. and Sinko, A. 2006. Volatility forecasting and microstructure noise. Manuscript, University of North Carolina, Chapel Hill.

Griffin, J.E. and Oomen, R.C.A. 2006. Covariance measurement in the presence of non-synchronous trading and market microstructure noise. Manuscript, University of Warwick.

Hansen, P.R. and Lunde, A. 2005. A realized variance for the whole day based on intermittent high-frequency data. *Journal of Financial Econometrics* 3, 525–54.

Hansen, P.R. and Lunde, A. 2006. Realized variance and market microstructure noise. *Journal of Business & Economic Statistics* 24, 127–61.

Harrison, J.M. and Kreps, D. 1978. Martingales and arbitrage in multiperiod securities markets. *Journal of Economic Theory* 20, 381–408.

Hayashi, T. and Yoshida, N. 2005. On covariance estimation of non-synchronously observed diffusion processes. *Bernoulli* 11, 359–79.

Hayashi, T. and Yoshida, N. 2006. Estimating correlations with nonsynchronous observations in continuous diffusion models. Manuscript, Columbia University and University of Tokyo.

Hsieh, D.A. 1991. Chaos and nonlinear dynamics: application to financial markets. *Journal of Finance* 46, 1839–77.

Huang, X. and Tauchen, G. 2005. The relative contribution of jumps to total price variation. *Journal of Financial Econometrics* 3, 456–99.

Hull, J. and White, A. 1987. The pricing of options on assets with stochastic volatilities. *Journal of Finance* 42, 281–300.

Jacod, J. and Protter, P. 1998. Asymptotic error distributions for the Euler method for stochastic differential equations. *Annals of Probability* 26, 267–307.

Jiang, G.J and Oomen, R.C.C. 2005. A new test for jumps in asset prices. Manuscript, Eller College of Management, University of Arizona and Warwick Business School.

Lee, S.S and Mykland, P.A. 2006. Jumps in real-time financial markets: a new nonparametric test and jump dynamics. Manuscript, University of Chicago.

Liu, C. and Maheu, J.M. 2005. Modeling and forecasting realized volatility: the role of power variation. Manuscript, University of Toronto.

Maheu, J.M. and McCurdy, T.H. 2002. Nonlinear features of realized FX volatility. *Review of Economics and Statistics* 84, 668–81.

Mantegna, R.N. and Stanley, H.E. 2000. *An Introduction to Econophysics – Correlations and Complexity in Finance*. Cambridge, MA: Cambridge University Press.

Martens, M. 2002. Measuring and forecasting S&P500 index-futures volatility using high-frequency data. *Journal of Futures Market* 22, 497–518.

Meddahi, N. 2001. An Eigen function approach for volatility modeling. Manuscript, Department of Economics, University of Montreal and CIRANO.

Meddahi, N. 2002. A theoretical comparison between integrated and realized volatilities. *Journal of Applied Econometrics* 17, 479–508.

Merton, R.C. 1976. Option pricing when underlying stock returns are discontinuous. *Journal of Financial Economics* 3, 125–44.

Merton, R.C. 1980. On estimating the expected return on the market: an exploratory investigation. *Journal of Financial Economics* 8, 323–61.

Müller, U.A., Dacorogna, M.M., Olsen, R.B., Pictet, O.V., Schwarz, M. and Morgenegg, C. 1990. Statistical study of foreign exchange rates, empirical evidence of a price change scaling law, and intraday analysis. *Journal of Banking and Finance* 14, 1189–208.

Mykland, P.A. 2006. A Gaussian calculus for inference from high frequency data. Manuscript, University of Chicago.

Officer, R.R. 1973. The variability of the market factor of New York stock exchange. *Journal of Business* 46, 434–53.

Peters, R.T. and de Vilder, R.G. 2006. Testing the continuous semimartingale hypothesis for the S&P 500. *Journal of Business & Economic Statistics* 24, 444–54.

Pong, S., Shackleton, M.B. Taylor, S.J. and Xu, X. 2004. Forecasting currency volatility: a comparison of implied volatilities and ARFIMA models. *Journal of Banking & Finance* 28, 2541–63.

Protter, P. 1992. *Stochastic Integration and Differential Equations: A New Approach*, 2nd edn. New York: Springer.

Scholes, M. and Williams, J. 1977. Estimating betas from nonsynchronous data. *Journal of Financial Economics* 5, 309–27.

Shephard, N. 2005. *Stochastic Volatility. Selected Readings.* Oxford: Oxford University Press.

Sheppard, K. 2006. Realized covariance and scrambling. Manuscript, University of Oxford.

Tauchen, G.E. and Pitts, M. 1983. The price variability-volume relationship on speculative markets. *Econometrica* 51, 485–505.

Tauchen, G.E. and Zhou, H. 2006. Identifying realized jumps on financial markets. Manuscript, Duke University and Federal Reserve Board of Governors.

Taylor, S.J. 1986. *Modeling Financial Time Series.* Chichester, UK: Wiley.

Thomakos, D.D. and Wang, T. 2003. Realized volatility in the futures markets. *Journal of Empirical Finance* 10, 321–53.

Zhang, L. 2006. Estimating covariation: Epps effect, microstructure noise. Manuscript, University of Chicago at Illinois.

Zhang, L., Mykland, P.A. and Aït-Sahalia, Y. 2005. A tale of two time scales: determining integrated volatility with noisy high-frequency data. *Journal of the American Statistical Association* 100, 1394–411.

Zhou, B. 1996. High-frequency data and volatility in foreign exchange rates. *Journal of Business and Economic Statistics* 14, 45–52.

Zumbach, G. 2004. Volatility processes and volatility forecasts with long memory. *Quantitative Finance* 4, 70–86.

reciprocity and collective action

Advancing the common interest of a group sometimes requires its members to sacrifice their private interests. Such situations, in which individual incentives are not properly aligned with shared goals, are called collective action problems. They arise frequently in economic and social life, for instance in the context of political mobilization, electoral turnout, pollution abatement, common property management and the provision of public goods. They can involve relatively small groups such as families, teams, or business partnerships, or very large groups that cut across national boundaries.

In his classic work on collective action, Mancur Olson (1965) conjectured that individuals would be unable to overcome such problems unless their behaviour was constrained by rules that were externally imposed and enforced. Along similar lines, Garret Hardin (1968) argued in an influential paper that, left to their own devices, individuals would face a 'tragedy of the commons' which could be overcome only by 'mutual coercion, mutually agreed upon'. This view continues to have considerable currency in economics in the form of the *free-rider hypothesis*, which maintains that voluntary contributions that are socially beneficial but privately costly will not generally be observed (Bergstrom, Blume and Varian, 1986).

Despite the compelling logic underlying the free-rider hypothesis, there are numerous instances of groups having overcome collective action problems without external pressure, sometimes by designing and abiding by their own set of rules, and sometimes on the basis of less formal arrangements codified in social norms. The success of OPEC in constraining production to maintain price levels is based on a mutually beneficial agreement among member countries that has been sustained despite strong incentives for some producers to free-ride on the restraint practised by others. On a smaller scale, many examples of successful collective action in the management of local fisheries, forests, and other renewable resources have been documented (Bromley, 1992; Ostrom, 1990). Such resources are often held as common property, and the maintenance of sustainable stocks requires restraint in individual extraction levels. Restraint is typically enforced by formal or informal sanctions, and participation in such punishment mechanisms is itself a form of collective action. There also exist examples of collective action in the absence of any sanctioning mechanism. For instance, voter turnout is often substantial in large elections, contrary to the predictions of the free-rider hypothesis.

It has been argued that many instances of successful collective action arise in small and stable groups whose members interact with each other repeatedly. Under such circumstances, pro-social behaviour can be fully consistent with the standard economic hypotheses of rationality and self-interest. When interactions are repeated, self-interested cooperation can arise if one believes that non-cooperative actions will be punished in future periods. Moreover, such threats of punishment can be credible if abstaining from punishment is itself punished. Formally, cooperative behaviour can be sustained in subgame perfect equilibrium if interactions are infinitely (or indefinitely) repeated (Fudenberg and Maskin, 1986). Hence the tension between individual and common interest is less severe and collective action more likely to arise in small and stable groups.

While the threat of future punishment or the promise of future reward might motivate collective action in some instances, there are many situations in which individual actions are unobservable or repetition too infrequent for such considerations to be decisive. Voter turnout, for instance, or private donations to charity are not easily explained as self-interested responses to material incentives. Similarly, sacrifices involving risks to life and limb, as in the case of battlefield heroism or spontaneous collective violence, are unlikely to be driven by a calculated response to future costs and benefits. What, then, could account for such phenomena?

There is now a considerable body of experimental evidence to suggest that many individuals are willing to take actions that further the common interest provided that they are reasonably sure that other group members will also take such actions. Furthermore, they are willing to sanction the opportunistic behaviour of others even at some cost to themselves (Fehr and Gächter, 2000). The widespread prevalence of such preferences for reciprocity suggests that collective action can sometimes be viewed as a coordination problem: if the members of a group confidently expect others to further the common good, such expectations can be self-fulfilling. On the other hand, expectations of widespread free-riding can also be self-fulfilling, so building confidence in the behaviour of others is a critical ingredient of successful collective action. Communication among group members can help coordinate expectations, and it is therefore not surprising that allowing for communication among experimental subjects can result in dramatically increased levels of cooperation. This is the case even if communication takes the form of 'cheap talk', with neither threats nor promises being enforceable (Ostrom, Walker and Gardner, 1992).

If preferences for reciprocity are indeed part of the explanation for successful collective action, this raises the question of how such preferences have come to be widespread in human populations in the first place. The existence of a willingness to sacrifice one's own material interest for the common good poses an evolutionary puzzle. In order to survive and spread in human populations, the possession of such preferences must confer on an individual some advantage relative to those who are entirely self-interested. One intriguing possibility is that, despite being disadvantageous to individuals within groups, traits that are advantageous for the group itself may survive because of competition among groups:

> There can be no doubt that a tribe including many members who, from possessing in a high degree the spirit of patriotism, fidelity, obedience, courage, and sympathy, were always ready to give aid to each other and to sacrifice themselves for the common good, would be victorious over other tribes; and this would be natural selection. (Darwin, 1871, p. 166)

In order to be effective, however, this mechanism requires variability across groups to be sustained while variability within groups is suppressed (Sober and Wilson, 1998). Whether or not the conditions for this are empirically plausible remains an open question.

There exist other channels through which a preference for reciprocity can be materially advantageous to individuals. One is assortative interaction: if individuals with preferences for reciprocity are more likely to interact with each other than with opportunists, the former can end up with higher material payoffs than the latter. Such assortation arises naturally in structured populations with local interaction. Even in unstructured populations with random matching, a propensity to reciprocate or to sanction opportunistic behaviour can

confer an advantage provided that such preferences are observable to others. The visible possession of such propensities can alter the behaviour of those with whom one is interacting in such a manner as to be materially rewarding. Even opportunistic individuals might be induced to behave cooperatively in interactions with those who have a credible reputation for reciprocity. Such considerations can provide the basis for an evolutionary theory of reciprocity (Sethi and Somanathan, 2001).

Reciprocity is a key feature of successful collective action, both in repeated interactions and in more spontaneous settings. The willingness to further the common good even at considerable personal cost is widespread in human populations, but is often contingent on the willingness of others to do the same. This perspective suggests that collective action problems are not insurmountable, but that communication and coordination are critical in overcoming them.

RAJIV SETHI

See also **collective action; common property resources; cooperation; coordination problems and communication; public goods; social norms; social preferences.**

Bibliography

Bergstrom, T.C., Blume, L. and Varian, H. 1986. On the private provision of public goods. *Journal of Public Economics* 29, 25–49.

Bromley, D.W. 1992. *Making the Commons Work: Theory, Practice and Policy.* San Francisco: Institute for Contemporary Studies.

Darwin, C. 1871. *The Descent of Man and Selection in Relation to Sex.* London: Murray.

Fehr, E. and Gächter, S. 2000. Cooperation and punishment in public goods experiments. *American Economic Review* 90, 980–94.

Fudenberg, D. and Maskin, E. 1986. The folk theorem in repeated games with discounting or with incomplete information. *Econometrica* 54, 533–54.

Hardin, G. 1968. The tragedy of the commons. *Science* 162, 1243–8.

Olson, M. 1965. *The Logic of Collective Action: Public Goods and the Theory of Groups.* Cambridge, MA: Harvard University Press.

Ostrom, E. 1990. *Governing the Commons: The Evolution of Institutions for Collective Action.* Cambridge: Cambridge University Press.

Ostrom, E., Walker, J. and Gardner, R. 1992. Covenants with and without a sword: self-governance is possible. *American Political Science Review* 86, 404–17.

Sethi, R. and Somanathan, E. 2001. Preference evolution and reciprocity. *Journal of Economic Theory* 97, 273–97.

Sober, E. and Wilson, D.S. 1998. *Unto Others: The Evolution and Psychology of Unselfish Behavior.* Cambridge, MA: Harvard University Press.

recursive competitive equilibrium

The underlying structure of most dynamic business-cycle and consumption-based asset-pricing models is a variant of the neoclassical stochastic growth model. Such models have been analysed by, among others, Cass (1965), Brock and Mirman (1972), and Donaldson and Mehra (1983). They focus on how an omniscient central planner seeking to maximize the present value of expected utility of a representative agent optimally allocates resources over the infinite time horizon.

Production is limited by an aggregate production function subject to technological (total factor productivity) shocks. The solution to the planning problem is characterized by time-invariant decision rules, which determine optimal consumption and investment each period. These decision rules have as arguments the economy's period aggregate capital stock and the shock to technology.

Business cycles, however, are not predicated on the actions of a central planner, but arise from interactions among economic agents in competitive markets. Given the desirable features of the stochastic growth paradigm – the solution methods are well known and the model generates well-defined proxies for all the major macro aggregates: consumption, investment, output, and so on – it is natural to ask if the allocations arising in that model can be viewed as competitive equilibria. That is, do price sequences exist such that economic agents, optimizing at these prices and interacting through competitive markets, achieve the allocations in question as competitive equilibria? This is the essential question of dynamic-decentralization theory.

Alternative approaches to dynamic decentralization: valuation equilibrium

One way of modelling uncertain dynamic economic phenomena is to use Arrow–Debreu general equilibrium structures and to search for optimal actions conditional on the sequence of realizations of all past and present random variables or shocks. The commodities traded are contingent claim contracts. These contracts deliver goods (for example, consumption and capital goods) at a future date, contingent on a particular sequential realization of uncertainty. Markets are assumed to be complete, so that, for any possible future realization of uncertainty (sequence of technology shocks) up to and including some future period, a market exists for contracts that will deliver each good at that date contingent on that realization (event). This requires a very rich set of markets. All trading occurs in the first period: consumers contract to receive consumption and investment goods and to deliver capital goods in all future periods contingent on future states so as to maximize the expected present value of their utility of consumption over their infinite lifetimes. Firms choose their production plans so as to maximize the present value of discounted profits. Given current prices, they contract to deliver consumption and investment goods to, and to receive capital goods from, the consumer-investors. Under standard preference structures, these contingent choices never need to be revised. That is, if markets reopen, no new trades will occur.

In its most general formulation, a valuation equilibrium is characterized simply as a continuous linear functional that assigns a value to each bundle of contingent commodities. Only under more restrictive assumptions can this function be represented as a price sequence (Bewley, 1972; Prescott and Lucas, 1972; Mehra, 1988). The basic result is that for any solution to the planner's problem – that is, sequences of consumption, investment and capital goods – a set of state-contingent prices exists such that these sequences coincide with the contracted quantities in the valuation equilibrium.

This decentralization concept is quite broad and applies to central-planning formulations much more general than the neoclassical growth paradigm. It reminds us that the financial structure underlying the stochastic growth paradigm is fundamentally one of complete contingent commodity markets. Nevertheless, it is a somewhat unnatural perspective for macroeconomists (all macro policies must be announced at time zero), and it presumes a set of markets much richer than any observed. These shortcomings led to the development of the concept of a recursive competitive equilibrium.

Recursive competitive theory

An alternative approach that has proved very useful in developing testable theories is to replace the attempt to locate equilibrium sequences of contingent functions with the search for time-invariant equilibrium decision rules. These decision rules specify current actions as a function of a limited number of 'state variables' which fully summarize the effects of past decisions and current information. Knowledge of these state variables provides the economic agents with a full description of the economy's current state. Their actions, together with the realization of the exogenous uncertainty, determines the values of the state variables in the next sequential time period. This is what is meant by a recursive structure. In order to apply standard time-series methods to any testable implications, these equilibrium decision rules must be time-invariant.

Recursive competitive theory was first developed by Mehra and Prescott (1977) and further refined in Prescott and Mehra (1980). These papers also establish the existence of a recursive competitive equilibrium and the supportability of the Pareto optimal through the recursive price functions. Excellent textbook treatments are contained in Harris (1987), Stokey, Lucas and Prescott (1989) and Ljungqvist and Sargent (2004). Since its introduction, it has been widely used in exploring a vide variety of economic issues including business-cycle fluctuations, monetary and fiscal policy, trade-related phenomena, and regularities in asset price co-movements. (See, for example, Kydland and Prescott, 1982; Long and Plosser, 1983; Mehra and Prescott, 1985.)

The recursive equilibrium abstraction postulates a continuum of identical economic agents indexed on the unit interval (again with preferences identical to those of the representative agent in the planning formulation), and a finite number of firms. As in the valuation equilibrium approach, consumers undertake all consumption and saving decisions. Firms, which have equal access to a single constant-returns-to-scale technology, maximize their profits each period, and are assumed to produce two goods: a consumption good and a capital good. Unlike in the valuation equilibrium approach, trading between agents and firms occurs every period. (This is in contrast to markets in an Arrow–Debreu setting where, as mentioned earlier, no trade would occur if markets were to reopen.) At the start of each period, firms observe the technological shock to productivity and purchase capital and labor services, which are supplied inelastically at competitive prices. The capital and labour are used to produce the capital and consumption goods. At the close of the period, individuals, acting competitively, use their wages and the proceeds from the sale of capital to buy the consumption and capital goods produced by the firms. Consumers then retain the capital good into the next period, when it again becomes available to firms and the process repeats itself. Note that firms are liquidated at the end of each period (retaining no capital assets while technology is freely available), and that no trades between firms and consumer-investors extend over more than one time period. Capital goods carried over from one period to the next are the only link between periods, and period prices depend only on the state variables in that period.

Formally, a *recursive competitive equilibrium* (RCE) is characterized by *time invariant* functions of a limited number of 'state variables', which summarize the effects of past decisions and current information. These functions (decision rules) include (*a*) a pricing function, (*b*) a value function, (*c*) a period allocation policy specifying the individual's decision, (*d*) a period allocation policy specifying the decision of each firm and (*e*) a function specifying the law of motion of the capital stock.

While the restrictive structure of markets and trades makes this concept less general than the valuation equilibrium approach, it provides an interpretation of decentralization that is better suited to macro-analysis. More recently, the recursive equilibrium concept has been generalized to admit an infinitely lived firm which maximizes its value. When an RCE is Pareto optimal, its allocation coincides with that of the associated planning problem. The solution to the central-planning stochastic-growth problem may then be regarded as the aggregate investment and consumption functions that would arise from a decentralized, recursive homogeneous consumer economy. We illustrate this with the help of an example below, which considers an economy with a single capital good.

The reader is referred to Prescott and Mehra (1980) for the more general case with multiple capital types.

An example

Consider the simplest central planning stochastic growth paradigm

$$w(k_0, \lambda_0) = \max E\left\{ \sum_{t=0}^{\infty} \beta^t u(c_t) \right\} \qquad \text{(P1)}$$

subject to

$$c_t + k_{t+1} \leq \lambda_t f(k_t, l_t), \quad \lambda_0, k_0 \ \text{given}, \ l_t = 1 \ \forall t.$$

In this formulation, $u(\cdot)$ is the period utility function of a representative consumer defined over his period t consumption c_t; k_t denotes capital available for production in period t and l_t denotes period t labour supply which is inelastically supplied by the consumer-investor at $l_t = 1$, for all t. The expression $f(k_t, l_t)$ represents the period technology (production function) which is shocked by the bounded stationary stochastic factor λ_t. (It is assumed that λ_t is subject to a stationary Markov process with a bounded ergodic set.) E denotes the expectations operator and the central planner is assumed to have rational expectations; that is, he uses all available information to rationally anticipate future variables. In particular he knows the conditional distribution of future technology shocks $F(\lambda_{t+1}; \lambda_t)$. For the purposes of this example we restrict preferences to be logarithmic and assume a Cobb–Douglas technology (to the best of my knowledge, this parameterization is the simplest example known to result in closed form solutions): $u(c_t) = \ln c_t$ and $f(k_t, l_t) = k_t^a l_t^{1-a}$. We also assume that $\alpha, \ \beta < 1$ and that capital fully depreciates each period.

These conditions are sufficient to guarantee a closed form solution to the planning problem:

$$c_t = (1 - \alpha\beta)k_t^\alpha \lambda_t, \ \text{and}$$

$$k_{t+1} = i_t = \alpha\beta k_t^\alpha \lambda_t$$

where we identify as investment, i_t, the capital stock held over for production in period $t + 1$. These allocations are Pareto optimal.

We will show that the investment and consumption policy functions arising as a solution to this problem may be regarded as the aggregate investment and consumption functions arising from a *decentralized* homogenous consumer economy.

We first qualitatively describe the RCE underlying this model, and then demonstrate the relevant equilibrium price and quantity functions explicitly. The one capital good is assumed to produce two goods – a consumer good and an investment (capital) good. At the beginning of each period, firms observe the shock to productivity (λ_t) and purchase capital and labour from individuals at competitively determined rates. Both capital and labour are used to produce the two output goods. Individuals use their proceeds from the sale of capital and labour services to buy the consumption good (c_t) and the investment good (i_t) at the end of the period. This investment good is used as capital (k_{t+1}) available for sale to the firm next period and the process continues recursively.

To cast this problem formally as a recursive competitive equilibrium, we introduce some additional notation. Let k_t denote the capital holdings of a particular (measure zero) individual at time t, and $\underline{k}_t$ the distribution of capital amongst other individuals in the economy. This latter distinction allows us to make formal the competitive assumption: all the economic participants will assume that $\underline{k}_t$ is exogenous to them and that the price functions depend solely on this aggregate (in addition to the technology shock). Clearly, in equilibrium, $k_t = \underline{k}_t$ for our homogeneous consumer economy. In addition, let p_i, p_c and p_k be the price of the investment, consumption and capital goods respectively and p_l be the wage rate. These prices are presumed to be functions of the economy-wide state variables exclusively and all participants take these prices as given for their own decision making purposes. The 'state variables' characterizing the economy are $(\underline{k}, \lambda)$ and the individual are $(k, \underline{k}, \lambda)$.

We use the symbols (c, i, k, l) to denote points in the 'commodity space' for the firm and the consumer. The c in the commodity point of the firm is a function specifying the consumption good supplied by the firm and is written as $c^s(\underline{k}_t, \lambda_t)$. Similarly, the c in the commodity point of the individual is the amount of the consumption good demanded by the individual and is written as $c^d(k_t, \underline{k}_t, \lambda_t)$. In equilibrium (as mentioned earlier, in equilibrium $k_t = \underline{k}_t$), since the market clears, of course $c^s = c^d$. The same comments apply to the other elements of the commodity point.

In the decentralized version of this economy, the problem facing a typical household is

$$v(k_0, \underline{k}_0, \lambda_0) = \max \; E \left\{ \sum_{t=0}^{\infty} \beta^t \ln \; c^d(k_t, \underline{k}_t, \lambda_t) \right\}$$

(P2)

subject to

$$p_c(\underline{k}_t, \lambda_t) \; c^d(k_t, \underline{k}_t, \lambda_t) + p_i(\underline{k}_t, \lambda_t) \; i^d(k_t, \underline{k}_t, \lambda_t)$$
$$\leq p_k(\underline{k}_t, \lambda_t) \; k^s(k_t, \underline{k}_t, \lambda_t) + p_l(\underline{k}_t, \lambda_t) \; l^s(k_t, \underline{k}_t, \lambda_t)$$

$$k_{t+1} \equiv k^s(k_{t+1}, \underline{k}_{t+1}, \lambda_{t+1}) = i^d(k_t, \underline{k}_t, \lambda_t),$$
$$l^s(k_t, \underline{k}_t, \lambda_t) \leq 1$$
and
$$\underline{k}_{t+1} = \psi(\underline{k}_t, \lambda_t)$$

is the law of motion of the aggregate capital stock.

With capital and labour priced competitively each period, the firm's objective function is especially simple – maximize period profits. The firm's problem then is

$$\max \; \{ p_c(\underline{k}_t, \lambda_t) \; c^s(\underline{k}_t, \lambda_t) + p_i(\underline{k}_t, \lambda_t) \; i^s(\underline{k}_t, \lambda_t)$$
$$- p_k(\underline{k}_t, \lambda_t) \; k^d(\underline{k}_t, \lambda_t) - p_l(\underline{k}_t, \lambda_t) \; l^d(\underline{k}_t, \lambda_t) \}$$

subject to

$$c_t^s + i_t^s \leq \lambda_t (k_t^d)^\alpha (l_t^d)^{1-\alpha}.$$

Via Bellman's principle of optimality, the recursive representation of the individual's problem P2 is

$$v(k_t, \underline{k}_t, \lambda_t) = \max_{\{c^d, i^d, l^s, k^d\}} \{ \ln \; (c^d(k_t, \underline{k}_t, \lambda_t))$$
$$+ \beta \int v(i^d(k_t, \underline{k}_t, \lambda_t), \psi(\underline{k}_t, \lambda_t), \lambda_{t+1}) dF(\lambda_{t+1}|\lambda_t) \}$$

subject to

$$p_c(\underline{k}_t, \lambda_t) \; c^d(k_t, \underline{k}_t, \lambda_t) + p_i(\underline{k}_t, \lambda_t) \; i^d(k_t, \underline{k}_t, \lambda_t)$$
$$\leq p_k(\underline{k}_t, \lambda_t) \; k^s(k_t, \underline{k}_t, \lambda_t) + p_l(\underline{k}_t, \lambda_t) \; l^s(k_t, \underline{k}_t, \lambda_t)$$

$$k_{t+1} \equiv k^s(k_{t+1}, \underline{k}_{t+1}, \lambda_{t+1}) = i^d(k_t, \underline{k}_t, \lambda_t),$$
$$l^s(k_t, \underline{k}_t, \lambda_t) \leq 1$$
and
$$\underline{k}_{t+1} = \psi(\underline{k}_t, \lambda_t)$$

is the law of motion of the aggregate capital stock.

The firm of course, simply maximizes its period profits and hence does not have a multiperiod problem.

The following functions that are a solution to the individual and firm maximization problem above satisfy the definition of recursive competitive equilibrium:

1. A value function $v(k_0, \underline{k}_0, \lambda_0) = E \left\{ \sum_{t=0}^{\infty} \beta^t \ln \left[(1 - \alpha\beta) \lambda_t \underline{k}_t^{\alpha-1} \{ \alpha(k_t - \underline{k}_t) + \underline{k}_t \} \right] \right\}$. It can be shown that $v(\underline{k}_0, \underline{k}_0, \lambda_0) = A + B \ln \underline{k}_0 + C \ln \lambda_0$ where A, B and C are constants which are functions of the preference and technology parameters.

2. A continuous pricing function $p(\underline{k}_t, \lambda_t) = \{ p_c(\underline{k}_t, \lambda_t), p_i(\underline{k}_t, \lambda_t), p_k(\underline{k}_t, \lambda_t), p_l(\underline{k}_t, \lambda_t) \}$ that has the same dimensionality as the commodity point, where

$$p_c(\underline{k}_t, \lambda_t) = p_i(\underline{k}_t, \lambda_t) = 1$$

(We have chosen the consumption good to be the numeraire.)

$$p_k(k_t, \lambda_t) = \alpha \lambda_t k_t^{\alpha-1}$$
$$p_l(k_t, \lambda_t) = (1 - \alpha) \lambda_t k_t^{\alpha-1}.$$

3. Consumption and investment functions for the individual that are a function of the current state of the individual $(k, \underline{k}, \lambda)$

$$c^d(k_t, \underline{k}_t, \lambda_t) = (1 - \alpha\beta)\lambda_t \underline{k}_t^{\alpha-1}\{\alpha(k_t - \underline{k}_t) + \underline{k}_t\}$$

$$l^s(k_t, \underline{k}_t, \lambda_t) = 1$$

$$i^d(k_t, \underline{k}_t, \lambda_t) = \alpha\beta\lambda_t \underline{k}_t^{\alpha-1}\{\alpha(k_t - \underline{k}_t) + \underline{k}_t\}$$

$$k^s(k_{t+1}, \underline{k}_{t+1}, \lambda_{t+1}) = i^d(k_t, \underline{k}_t, \lambda_t).$$

4. Decision rules for the firm that are contingent on the state of the economy $(\underline{k}, \lambda)$

$$c^s(\underline{k}_t, \lambda_t) = (1 - \alpha\beta)\lambda_t \underline{k}_t^{\alpha},$$

$$l^d(\underline{k}_t, \lambda_t) = 1,$$

$$i^s(\underline{k}_t, \lambda_t) = \alpha\beta\lambda_t \underline{k}_t^{\alpha},$$

$$k^d(\underline{k}_{t+1}, \lambda_{t+1}) = i^s(\underline{k}_t, \lambda_t).$$

5. The law of motion for the capital stock specifying the next period capital stock as a function of the current state of the economy $(\underline{k}_t, \lambda_t)$

$$\underline{k}_{t+1} = \psi(\underline{k}_t, \lambda_t) = \alpha\beta\lambda_t \underline{k}_t^{\alpha}.$$

6. The consumption and investment decisions of the individual $c^s(k, \underline{k}, \lambda)$, $l^s(k, \underline{k}, \lambda)$ and $i^s(k, \underline{k}, \lambda)$ maximize the expected utility subject to the budget constraint. So that

$$v(k_t, \underline{k}_t, \lambda_t) = \ln\left((1 - \alpha\beta)\lambda_t \underline{k}_t^{\alpha-1}\right.$$
$$\times (\alpha(k_t - \underline{k}_t) + \underline{k}_t))$$
$$+ \beta \int v(\alpha\beta\lambda_t \underline{k}_t^{\alpha-1}(\alpha(k_t - \underline{k}_t) + \underline{k}_t),$$
$$\times \alpha\beta\lambda_t \underline{k}_t^{\alpha})dF(\lambda_{t+1}|\lambda_t).$$

7. The decision rules of the firm $c^d(\underline{k}_t, \lambda_t)$, $l^d(\underline{k}_t, \lambda_t)$, $i^d(\underline{k}_t, \lambda_t)$ maximize firm profit.

Demand equals supply

$$c^d(k_{t+1}, \underline{k}_{t+1}, \lambda_{t+1}) = c^s(\underline{k}_t, \lambda_t),$$

$$l^s(k_{t+1}, \underline{k}_{t+1}, \lambda_{t+1}) = l^d(\underline{k}_t, \lambda_t)$$

$$\text{and } i^s(k_{t+1}, \underline{k}_{t+1}, \lambda_{t+1}) = i^d(\underline{k}_t, \lambda_t).$$

The law of motion of the representative consumers capital stock is consistent with the maximizing behaviour of agents $\psi(\underline{k}_t, \lambda_t) = i^d(\underline{k}_t, \underline{k}_t, \lambda_t)$. It is readily demonstrated that since $v(\underline{k}_0, \underline{k}_0, \lambda_0) = w(\underline{k}_0, \lambda_0)$, the competitive allocation is Pareto optimal. See eqs (P1) and (P2).

Having formulated expressions for the prices of the various assets and their laws of motion, it is a relatively simple matter to calculate rates of return (price ratios) and study their dynamics. For an application to risk premia, see Donaldson and Mehra (1984).

Some researchers have formulated models that can be cast in this same recursive setting, yet whose equilibria are not Pareto-optimal. As a consequence, the model's equilibrium can no longer be obtained as the solution to a central-planning-optimum formulation. These models incorporate various features of monetary phenomena, distortionary taxes, non-competitive labour market arrangements, externalities, or borrowing-lending constraints. Besides increasing general model realism, such features enable the models not only to better replicate the stylized facts of the business cycle, but also to provide a rationale for interventionist government policies. Monetary models of this class include those of Lucas and Stokey (1987, a monetary exchange model) and Coleman (1996, a monetary production model). Bizer and Judd (1989) and Coleman (1991) present models in which non-optimality is induced by tax distortions, while Danthine and Donaldson (1990) present a model in which non-optimality results from efficiency-wage considerations. In these models, equilibrium is characterized as an aggregate-consumption and an aggregate-investment function which jointly solves a system of first-order optimality equations on which market-clearing conditions have been imposed. Coleman (1991) provides a widely applicable set of conditions under which these suboptimal equilibrium functions exist. As already noted, however, these optimality conditions cannot, in general, characterize the solution to an optimum problem.

RAJNISH MEHRA

See also **Arrow–Debreu model of general equilibrium; decentralization; neoclassical growth theory; real business cycles.**

Bibliography

Bewley, T. 1972. Existence of equilibria in economies with infinitely many commodities. *Journal of Economic Theory* 4, 514–40.

Bizer, D. and Judd, K. 1989. Taxation and uncertainty. *American Economic Review Papers and Proceedings* 19, 331–6.

Brock, W.A. and Mirman, L.J. 1972. Optimal economic growth and uncertainty: the discounted case. *Journal of Economic Theory* 4, 497–513.

Cass, D. 1965. Optimal growth in an aggregative model of capital accumulation. *Review of Economic Studies* 32, 233–40.

Coleman, W.J. 1991. Equilibrium in a production economy with an income tax. *Econometrica* 59, 1091–104.

Coleman, W.J. 1996. Money and output: a test of reverse causation. *American Economic Review* 86, 90–111.

Danthine, J.P. and Donaldson, J.B. 1990. Efficiency wages and the business cycle puzzle. *European Economic Review* 34, 1275–301.

Donaldson, J.B. and Mehra, R. 1983. Stochastic growth with correlated production shock. *Journal of Economic Theory* 29, 282–312.

Donaldson, J.B. and Mehra, R. 1984. Comparative dynamics of an equilibrium intertemporal asset pricing model. *Review of Economic Studies* 51, 491–508.

Harris, M. 1987. *Dynamic Economic Analysis*. New York: Oxford University Press.

Kydland, F.E. and Prescott, E.C. 1982. Time to build and aggregate fluctuations. *Econometrica* 50, 1345–71.

Ljungqvist, L. and Sargent, T.J. 2004. *Recursive Macroeconomic Theory*, 2nd edn. Cambridge, MA: MIT Press.

Long, J.B., Jr. and Plosser, C.I. 1983. Real business cycles. *Journal of Political Economy* 91, 39–69.

Lucas, R.E., Jr. and Stokey, N. 1987. Money and interest in a cash advance economy. *Econometrica* 55, 491–513.

Mehra, R. 1988. On the existence and representation of equilibrium in an economy with growth and non-stationary consumption. *International Economic Review* 29, 131–5.

Mehra, R. and Prescott, E.C. 1977. Recursive competitive equilibria and capital asset pricing. In R. Mehra, Essays in financial economics. Doctoral dissertation, Carnegie Mellon University.

Mehra, R. and Prescott, E.C. 1985. The equity premium: a puzzle. *Journal of Monetary Economics* 15, 145–62.

Prescott, E.C. and Lucas, R.E., Jr. 1972. A note on price systems in infinite dimensional space. *International Economic Review* 13, 416–22.

Prescott, E.C. and Mehra, R. 1980. Recursive competitive equilibria: the case of homogeneous households. *Econometrica* 48, 1365–79.

Stokey, N., Lucas, R.E. and Prescott, E.C. 1989. *Recursive Methods in Economic Dynamics*. Cambridge, MA: Harvard University Press.

recursive contracts

In contract theory it is standard to introduce a participation constraint (PC) insuring that the contract offered to the agent delivers a utility higher than the best outside option. In a dynamic set-up agents may abandon the contract at any point in time, even after the contract has been in place for a while. For example, workers can leave a labour contract at almost no cost, or a borrower can stop repaying the loan if he or she declares bankruptcy. The possibility that the agent does not continue with the plan of the contract is usually called 'default'. Hence, in a dynamic context, it is natural to require that the PC is satisfied in all periods, in order to avoid default.

It turns out that, if a PC in all periods and realizations is introduced in the design of the optimal contract, standard dynamic programming does not apply, the Bellman equation does not hold, and the solution is not guaranteed to be a time-invariant function of the usual state variables. This complicates enormously the solution of these models.

To discuss this in a simple risk-sharing model, consider two agents $i = 1,2$ with utility function $E_0 \sum_{t=0}^{\infty} \beta^t u(c_t^i)$, where $\beta \in (0,1)$ is the discount factor and u the instantaneous utility. Each agent receives a stochastic endowment w_t^i and the realization of endowments is known both to the agents and the principal. The principal has full commitment, and will stick to his announced plan. Endowments provide the only supply of consumption good so that the following feasibility condition holds

$$c_t^1 + c_t^2 = w_t^1 + w_t^2 \tag{1}$$

A Pareto-optimal risk-sharing contract (implemented by a competitive equilibrium under complete markets) would set $\frac{u'(c_t^1)}{u'(c_t^2)}$ constant for all periods, so that agents would share all idiosyncratic risks. This allocation would be chosen as the optimal contract if agents would commit to never leave the risk-sharing arrangement. We refer to this allocation as the first best. The optimum satisfies the usual recursive structure in dynamic models, namely, that $c_t = F(w_t)$ where F is a time-invariant function and $w_t = (w_t^1, w_t^2)$.

Assume now agents cannot commit to staying in the contract for ever. An agent can leave the contract and consume for ever his individual endowment, so that a contract can only be implement if it satisfies

$$E_t \sum_{j=0}^{\infty} \beta^j u(c_{t+j}^i) \geq V_i^a(w_t)$$

at all periods and realizations, where $V_i^a(w_t) \equiv E_t \sum_{j=0}^{\infty} \beta^j u(w_{t+j}^i)$ is the utility of consuming in autarky for ever after t.

It is clear that the above PC is likely to be violated by the first best allocation. In periods when w_t^i is high, the right side of the PC is high, but the agent has to surrender a large part of his endowment in the first best and the left side of the PC is too low. Therefore, PCs are often binding and they make the first best unfeasible.

A Pareto-optimal risk-sharing contract with PCs can now be found by maximizing the weighted utility of the two agents $E_0 \sum_{t=0}^{\infty} \beta^t [\lambda u(c_t^1 c) + (1 - \lambda) u(c_t^2)]$ subject to the above PC for all periods and realizations and for both agents. The parameter λ indexes all such Pareto-optimal allocations. The result is an optimal contract under full commitment by the principal and partial commitment by the agents.

The Bellman equation does not give the solution to this problem. A key feature of standard dynamic programming is that the set of feasible actions must depend only on variables that were determined last period and the current shock. But it is possible to evaluate if a certain consumption level $\bar{c}_t^i$ satisfies the PC at time t only if future plans for consumption are known.

Intuitively, a promise of higher consumption in the future makes a lower consumption today compatible with the PC. But in order to implement this plan the principal has to 'remember' all the promises for higher consumption that were made in the past. Therefore, the optimal solution is unlikely to be a function of only today's endowment, the principal also needs to recall if, say, ten periods ago, the PC of one of the agents was binding.

As argued by Kydland and Prescott (1977), the same problem arises in models of optimal policy. The future restricts today's actions through the first order conditions of optimality of the agents, this causes the Bellman equation to fail and, in their language, the solution was time inconsistent. We find the same difficulty in contracting models of private information with incentive constraints, where some relevant piece of information is hidden from the principal, and more generally, in game theoretical models where an agent optimizes subject to the plans for the future of another agent.

If the Bellman equation fails, the solution could depend on all past shocks, and solving for the variables as a function of all past shocks would be very difficult. Too many variables would appear as arguments of the decision function. To overcome this difficulty the 'recursive contracts' literature provides several alternatives. The general idea is to recover a recursive formulation by adding a co-state variable.

One approach builds on the paper of Abreu, Pierce and Stachetti (1990; hereafter APS). To show how this can be applied in the above risk-sharing model with PCs, consider the case where w_t is i.i.d. and has two possible realizations $\overline{w}$ and $\overline{\overline{w}}$ with probabilities π and $(1 - \pi)$. Denote the utility of agent i for the whole future at t if $w_t = \overline{w}$ by $\overline{V}_t^i \equiv E_t$ $(\sum_{j=0}^{\infty} \beta^j u(c_{t+j}^i)|w_t = \overline{w})$, and let $\overline{\overline{V}}_t^i$ be the analogue for realization $\overline{\overline{w}}$. The above PC can be reformulated as

$$V_t^i = u(c_t^i) + \beta\left(\pi\overline{V}_{t+1}^i + (1 - \pi)\overline{\overline{V}}_{t+1}^i\right)$$

(2)

$$\overline{V}_{t+1}^i \geq V_i^a(\overline{w}), \quad \overline{\overline{V}}_{t+1}^i \geq V_i^a(\overline{w}) \text{ for all } t > 0,$$

where V_t^i is the actual realized utility. The first equation insures that V_t^i is the expected discounted utility, the second guarantees that the PC holds.

We can view the planner's choice at t as choosing the promised utilities $\overline{V}_{t+1}^i$, $\overline{\overline{V}}_{t+1}^i$ and consumption c_t^i, while V_t^i is given by past choices. It is clear that, in the APS approach, today's choice variables $x_t = (\overline{V}_{t+1}^i, \overline{\overline{V}}_{t+1}^i, c_t^i)$ are restricted by yesterday's promised utilities only, and the Bellman equation delivers the optimal contract after the realized V_t^i is included in the list of state variables. The promised utility V_t^i plays the same role as capital in a standard growth model, and (2) plays the role of the transition equation. Therefore, the optimal solution for the choices can be described recursively by a time-invariant function $x_t = F(w_t, V_t)$ for all $t > 0$.

A crucial caveat is that (2) is not sufficient to insure that the PCs are satisfied. The principal could choose arbitrarily high consumption and have ever higher Vs to satisfy (2), in a sort of Ponzi scheme for utility. The

promised utilities have to be further restricted to belong to a feasible set. Let us call $\overline{S} \subset R$ the feasible set of utilities such that, for each element $v \in \overline{S}$, there is a sequence of consumptions $\{c_{t+j}^i\}$ that satisfy (1) and the PCs such that $v = E_t(\sum_{j=0}^{\infty} \beta^j u(c_{t+j}^i)|w_t = \overline{w})$. Results in APS insure that this set is convex. Since in this case $\overline{S} \subset R$, this set is an interval and there exist bounds $\overline{V}_L^i$ and $\overline{V}_U^i$ such that adding the constraints

$$\overline{V}_L^i \leq \overline{V}_{t+1}^i \leq \overline{V}_U^i$$

(and similarly for $\overline{\overline{V}}_t^i$) to (2) is enough to insure feasibility. These bounds can be easily introduced in the Bellman equation and this guarantees that the chosen consumption sequences satisfies the PC. The only complication is that upper bound $\overline{V}_U$ needs to be computed separately, as it is not a datum of the problem ($\overline{V}_L^i$ is trivially equal to $V_i^a(\overline{w})$).

Another difference with standard dynamic programming is that the initial utility V_0^1 is an outcome of the solution and it is not fixed beforehand. This feature shows how time inconsistency arises in this model, since the choice for V in period zero is not given, but in future periods it is given from the past.

Promised utilities as co-states have been used extensively in models with incentive or participation constraints. Among others, Phelan and Townsend (1991) studied a model of risk-sharing with incentive constraints, Kocherlakota (1996) analysed the risk-sharing model with the PC described above, Hopenhayn and Nicolini (1997) a model of unemployment insurance and Alvarez and Jermann (2000) a decentralized version of the above risk-sharing model with debt constraints. In models of Ramsey equilibria it has been used by Golosov, Kocherlakota and Tsyvinski (2003) to study optimal taxation under private information and Chang (1998) in a model of optimal monetary policy.

The main problem with this approach is the computation of the set of feasible utilities $\overline{S}$. In the specific model described above this is not too costly, because it involves finding only two numbers, namely, the upper bounds $\overline{V}_U$, $\overline{\overline{V}}_U$. But the difficulties multiply when more than one co-state variable is needed. For example, if a third agent is included in the above risk-sharing model, the co-state variables would be (V_t^1, V_t^2). Results in APS guarantee that the set of feasible utilities $\overline{S} \subset R^2$ is convex, but now it is a generic set, not an interval. Computing a set is much harder than computing two numbers. Some papers overcome these difficulties; for example, Abraham and Pavoni (2005), who show how to find such a set in a model of saving under private information, or the paper of Judd, Yeltekin and Conklin (2003). But the difficulties increase very fast with the dimensionality of the promised utilities.

Furthermore, in some models, the set of feasible promised utilities changes every period. If a 'traditional' state variable (say, capital stock) appears in the problem, the set of feasible utilities is different depending on the level of capital, so that the feasible set is now given by a correspondence $\overline{S}(k)$. The researcher now needs to solve for a mapping from capital stock to sets. Phelan and Stacchetti (2001) compute in this way the optimal fiscal policy in a model with capital.

An alternative to APS is the Lagrangean approach described in Marcet and Marimon (1998). The Lagrangean for the optimal risk-sharing problem with PC is

$$L = E_0 \sum_{t=0}^{\infty} \beta^t \left[\lambda u(c_t^1) + (1 - \lambda)u(c_t^2) \right.$$
$$\left. + \sum_{i=1,2} \gamma_t^i \left(E_t \sum_{j=0}^{\infty} \beta^j u(c_{t+j}^i) - V_i^a(w_t) \right) \right]$$

where $\gamma_t^i \geq 0$ is the Lagrange multiplier of the PC. This can be rewritten as

$$L = E_0 \sum_{t=0}^{\infty} \beta^t [(\lambda + \mu_t^1)u(c_t^1) + (1 - \lambda + \mu_t^2)u(c_t^2)]$$
$$\text{s.t. } \mu_t^i = \mu_{t-1}^i + \gamma_t^i, \quad \gamma_t^i \geq 0$$
$$\mu_{-1}^i = 0$$

In this formulation, only current and past variables enter in the objective and in the constraints of this Lagrangean, and a proper initial condition for μ is given. In this approach, μ_t plays the role of the co-state variable instead of the promised utility in the APS approach. A saddle point functional equation (analogous but not equal to the Bellman equation) is satisfied, insuring that the optimal solution satisfies $(c_t, \gamma_t) = G(\mu_{t-1}, w_t)$ with $\mu_{-1}^i = 0$ for a time invariant function G.

The equilibrium satisfies $\frac{u'(c_t^1)}{u'(c_t^2)} = \frac{1 - \lambda + \mu_t^2}{\lambda + \mu_t^1}$. If the PC for agent i is binding, the corresponding γ_t^i is strictly positive, the weight μ_t^i goes up and so does c_t^i. The increase in μ is permanent (at least until another PC is binding). In this way the principal avoids default by spreading the reward over time in order to enhance smoothing of consumption.

Note that the initial value of μ is given and equal to zero, while in future periods μ_{t-1} needs to be set according to past Lagrange multipliers. Therefore, the initial value of the co-state does not need to be found separately as in APS. It is clear that, if the principal could re-optimize ignoring past commitments at sometime t, he or she would ignore the past co-state and reset $\mu = 0$. This is how time inconsistency is reflected in this formulation.

In the Lagrangean approach there is no need to find the set of feasible utilities. The only constraint on the co-states is the non-negativity constraint on γs. Application to models with capital accumulation and several co-states is much easier; for example, Marcet and Marimon (1992) solve a risk-sharing growth model with PC as described above and capital accumulation, Aiyagari et al. (2002) in a Ramsey equilibrium for fiscal policy under incomplete markets, where debt is a state variable, Attanasio and Ríos-Rull (2000) risk-sharing in small villages, Scott (2007) a model of optimal taxes with capital, Kehoe and Perri (2002) international capital flows with capital accumulation under PC, King, Kahn and Wolman (2003) optimal monetary policy, Cooley, Marimon and Quadrini (2004) a model of investment under private information, Abraham and Carceles-Poveda (2006) discuss how to decentralize a model with participation constraints, and Ferrero and Marcet (2004) and Scholl (2004) a model of temporary exclusion in the case of default. The drawback of the Lagrangean approach is that, at this writing, the theory for the non-convex case and for the private information case is still incomplete.

ALBERT MARCET

See also **agency problems; Bellman equation; dynamic programming; income taxation and optimal policies; optimal fiscal and monetary policy (with commitment); optimal taxation; risk sharing; time consistency of monetary and fiscal policy.**

Bibliography

Abraham, A. and Carceles-Poveda, E. 2006. Endogenous incomplete markets, enforcement constraints, and intermediation. *Theoretical Economics* 1, 439–59.

Abraham, A. and Pavoni, N. 2005. The efficient allocation of consumption under moral hazard and hidden access to the credit market. *Journal of the European Economic Association* 3, 370–81.

Abreu, D., Pierce, D. and Stacchetti, E. 1990. Towards a theory of discounted repeated games with imperfect monitoring. *Econometrica* 58, 1041–63.

Aiyagari, R., Marcet, A., Sargent, T.J., Seppälä, J. 2002. Optimal Taxation without State-Contingent Debt, *Journal of Political Economy* 110, 1220–54.

Alvarez, F. and Jermann, U.J. 2000. Efficiency, equilibrium, and asset pricing with risk of default. *Econometrica* 68, 775–98.

Attanasio, O. and Ríos-Rull, J.V. 2000. Consumption smoothing in island economies: can public insurance reduce welfare? *European Economic Review* 44, 1225–58.

Chang, R. 1998. Credible monetary policy with long-lived agents: recursive approaches. *Journal of Economic Theory* 81, 431–61.

Cooley, T.F., Marimon, R. and Quadrini, V. 2004. Aggregate consequences of limited contract enforceability. *Journal of Political Economy* 112, 817–47.

Ferrero, G. and Marcet, A. 2004. Limited commitment and temporary exclusion. Mimeo, Institut d'Analisi Economica, CSIC.

Golosov, M., Kocherlakota, N.R. and Tsyvinski, A. 2003. Optimal indirect and capital taxation. *Review of Economic Studies* 70, 569–87.

Hopenhayn, H.A. and Nicolini, J.P. 1997. Optimal unemployment insurance. *Journal of Political Economy* 105, 412–38.

Judd, K.L., Yeltekin, S. and Conklin, J. 2003. Computing supergame equilibria. *Econometrica* 71, 1239–54.

Kehoe, P.J. and Perri, F. 2002. International business cycles with endogenous incomplete markets. *Econometrica* 70, 907–28.

King, R.G., Kahn, A. and Wolman, A.L. 2003. Optimal monetary policy. *Review of Economic Studies* 70, 825–60.

Kocherlakota, N.R. 1996. Implications of efficient risk sharing without commitment. *Review of Economic Studies* 63, 595–609.

Kydland, F.E. and Prescott, E.C. 1977. Rules rather than discretion: the inconsistency of optimal plans. *Journal of Political Economy* 85, 473–92.

Marcet, A. and Marimon, R. 1992. Communication, commitment and growth. *Journal of Economic Theory* 58, 219–49.

Marcet, A. and Marimon, R. 1998. Recursive contracts. Working paper, Universitat Pompeu Fabra.

Phelan, C. and Stacchetti, E., 2001. Sequential equilibria in a Ramsey tax model. *Econometrica* 69, 1491–518.

Phelan, C. and Townsend, R.M. 1991. Computing multi-period, information-constrained optima. *Review of Economic Studies* 58, 853–81.

Scholl, A. 2004. Do endogenous incomplete markets explain cross-country consumption correlations and the dynamics of the terms of trade? Mimeo, Humboldt University, Berlin.

Scott, A. 2007. Optimal taxation and OECD labor taxes. *Journal of Monetary Economics* 54, 925–44.

recursive preferences

1 Introduction

Recursive methods have become a standard tool for studying economic behaviour in dynamic stochastic environments. In this chapter, we characterize the class of preferences that is the natural complement to this framework, namely *recursive preferences*.

Why model preferences rather than behaviour? Preferences play two critical roles in economic models. First, preferences provide, in principle, an unchanging feature of a model in which agents can be confronted with a wide range of different environments, institutions, or policies. For each environment, we derive behaviour (decision rules) from the same preferences. If we modelled behaviour directly, we would also have to model how it adjusted to changing circumstances. The second role played by preferences is to allow us to evaluate the welfare effects of changing policies or circumstances. Without the ranking of opportunities that a model of preferences provides, it's not clear how we should distinguish good policies from bad.

Why recursive preferences? Recursive preferences focus on the trade-off between current-period utility and the utility to be derived from all future periods. Since an agent's actions today can affect the evolution of opportunities in the future, summarizing the future consequences of these actions with a single index, that is, future utility, allows multi-period decision problems to be reduced to a series of two-period problems, and in the case of a stationary infinite-horizon problem, a single, time-invariant two-period decision problem. As we will see, this logic applies equally well to environments in which current actions affect the values of random events for all future periods. In this case, the two-period trade-off is between current utility and a *certainty equivalent* of random future utility. This recursive approach not only allows complicated dynamic optimization problems to be characterized as much simpler and more intuitive two-period problems, it also lends itself to straightforward computational methods. Since many computational algorithms for solving stochastic dynamic models themselves rely on recursive methods, numerical versions of recursive utility models can be solved and simulated using standard algorithms.

2 The stationary recursive utility function

Assume time is discrete, with dates $t = 0, 1, 2, \ldots$. At each $t > 0$, an event z_t is drawn from a finite set $\mathscr{Z}$, following an initial event z_0. The t-period history of events is denoted by $z^t = (z_0, z_1, \ldots, z_t)$ and the set of possible t-histories by $\mathscr{Z}^t$. Environments like this, involving time and uncertainty, are the starting point for much of modern economics. A typical agent in such a setting has preferences over payoffs $c(z^t)$ for each possible history. A general set of preferences might be represented by a utility function $U(\{c(z^t)\})$. In what follows, we will think of consumption as a scalar. This is purely for exposition since the extension to a vector of consumption at each point in time is straightforward.

Consider the structure of preferences in this dynamic stochastic environment. We define the class of stationary recursive preferences by

$$U_t = V[c_t, \mu_t(U_{t+1})], \qquad (1)$$

where U_t is short-hand for utility starting at some date-t history z^t, U_{t+1} refers to utilities for histories $z^{t+1} = (z^t, z_{t+1})$ stemming from z^t, V is a time aggregator, and μ_t is a certainty-equivalent function based on the conditional probabilities $p(z_{t+1}|z^t)$. As with other utility functions, increasing functions of U, with suitable adjustment of μ, imply the same preferences. This structure of preferences leads naturally to recursive solutions of economic problems, with (1) providing the core of a Bellman equation.

In general, the properties of U_t depend on both the properties of the time aggregator and the certainty equivalent. Since the certainty equivalent will be scaled such that $\mu(x) = x$ when x is a perfect certainty, the time aggregator V is all that matters in deterministic settings. Similarly, for a purely static problem with uncertainty, the certainty-equivalent function μ is all that matters. We consider the specification of each of these components in turn.

It is important to note that the utility functions presented in this article are not ad hoc but rather have clear axiomatic foundations, and can be derived from more primitive assumptions on preference orderings. Since utility functions are the typical starting point for applied research, we skip this step and refer the interested reader to the axiomatic characterizations of recursive preferences in the papers cited at the end of this article.

3 The time aggregator

Time preference is a natural starting point. Suppose there is no risk and c_t is one-dimensional. Preferences might then be characterized by a general utility function $U(\{c_t\})$. A common measure of time preference in this setting is the marginal rate of substitution between consumption at two consecutive dates (c_t and c_{t+1}, say) along a constant consumption path ($c_t = c$ for all t). If the marginal rate of substitution is

$$\text{MRS}_{t,t+1} = \frac{\partial U / \partial c_{t+1}}{\partial U / \partial c_t},$$

then time preference is captured by the discount factor

$$\beta(c) \equiv \text{MRS}_{t,t+1}(c).$$

(Picture the slope, $-1/\beta$, of an indifference curve along the '45-degree line'.) If $\beta(c)$ is less than one, the agent is said to be impatient: along a constant consumption path (that is, in the absence of diminishing marginal utility considerations), the agent requires more than one unit of consumption at $t+1$ to induce a sacrifice of one unit at t. For the traditional time-additive utility function,

$$U(\{c_t\}) = \sum_{t=0}^{\infty} \beta^t u(c_t), \tag{2}$$

$\beta(c) = \beta < 1$ regardless of the value of c, so impatience is built in and constant. A popular and useful special case of this utility function implies a constant elasticity of intertemporal substitution by assuming $u(c) = c^\rho / \rho$ for $\rho < 1$. Note that we can define the utility function in (2) recursively:

$$U_t = u(c_t) + \beta U_{t+1}, \tag{3}$$

for $t = 1, 2, \ldots$. The constant elasticity version can be expressed

$$U_t = [(1 - \beta)c_t^\rho + \beta U_{t+1}^\rho]^{1/\rho}, \tag{4}$$

where $\rho < 1$ and $\sigma = 1/(1 - \rho)$ is the intertemporal elasticity of substitution. (To put this in additive form, use the

transformation $\hat{U} = U^\rho / \rho$.) Note that U_t is homothetic and that the scaling we have chosen measures utility on the same scale as consumption:

$$U(c, c, c, \ldots) = c.$$

More generally, impatience summarized by the discount factor, $\beta(c)$, could vary with the level of consumption. Koopmans (1960) derives a class of stationary recursive preferences by imposing conditions on a general utility function U for a multi-dimensional consumption vector c. In the Koopmans class of preferences, time preference is a property of the time aggregator V. Consider our measure of time preference:

$$U_t = V(c_t, U_{t+1}) = V[c_t, V(c_{t+1}, U_{t+2})].$$

The marginal rate of substitution between c_t and c_{t+1} is therefore

$$\text{MRS}_{t,t+1} = \frac{V_2(c_t, U_{t+1}) V_1(c_{t+1}, U_{t+2})}{V_1(c_t, U_{t+1})}.$$

A constant consumption path at c is defined by $U = V(c, U)$, implying $U = g(c) = V[c, g(c)]$ for some function g.

In modern applications, we typically work in reverse order: we specify a time aggregator V and use it to characterize the overall utility function U. Any U constructed this way defines preferences that are stationary and dynamically consistent. In contrast to time-additive preferences, discounting depends on the level of consumption c.

The most common example of Koopmans's structure in applications is a generalization of eq. (3):

$$V(c, U) = u(c) + \beta(c)U,$$

where there is no particular relationship between the functions u and β. For this example, the intertemporal trade-off is given by

$$\text{MRS}_{t,t+1} = \beta(c_t) \left[\frac{u'(c_{t+1}) + \beta'(c_{t+1})U_{t+2}}{u'(c_t) + \beta'(c_t)U_{t+1}} \right].$$

When $\beta'(c) \neq 0$, optimal consumption plans will depend on the level of future utility. And along a constant consumption path, discounting is decreasing (increasing) in consumption when $\beta'(c) < 0$ ($\beta'(c) > 0$). Also note that U_t in this example is not homothetic.

4 The risk aggregator

Turn now to the specification of risk preferences, which we consider initially in a static setting. Choices have risky consequences or payoffs, and agents have preferences defined over those consequences and their probabilities. To be specific, let us say that the state z is drawn with probability $p(z)$ from the finite set $\mathcal{Z} = \{1, 2, \ldots, Z\}$. Consequences ($c$, say) depend on the state and the agent's preferences are represented by a utility function of state-contingent consequences ('consumption'):

$$U(\{c(z)\}) = U[c(1), c(2), \ldots, c(Z)].$$

At this level of generality there is no mention of probabilities, although we can well imagine that the probabilities of the various states will show up somehow in U. We regard the probabilities as known, which you might think of as an assumption of 'rational expectations'.

We prefer to work with a different (but equivalent) representation of preferences. Suppose, for the time being, that c is a scalar; very little of the theory depends on this, but it streamlines the presentation. We define the *certainty equivalent* of a set of consequences as a certain consequence μ that gives the same level of utility:

$$U(\mu, \mu, \ldots, \mu) = U[c(1), c(2), \ldots, c(Z)].$$

If U is increasing in all its arguments, we can solve this for the certainty-equivalent function $\mu(\{c(z)\})$. Clearly μ represents the same preferences as U, but we find its form particularly useful. For one thing, it expresses utility in payoff ('consumption') units. For another, it summarizes behaviour towards risk directly: since the certainty equivalent of a sure thing is itself, the impact of risk is simply the difference between the certainty equivalent and expected consumption.

The traditional approach to preferences in this setting is expected utility, which takes the form

$$U(\{c(z)\}) = \sum_z p(s)u[c(z)] = Eu(c),$$

or

$$\mu(\{c(z)\}) = u^{-1}\left(\sum_z p(z)u[c(z)]\right) = u^{-1}[Eu(c)].$$

Preferences of this form have been used in virtually all economic theory. The utility function of Kreps and Porteus employs a general time aggregator and an expected utility certainty equivalent. Following Epstein and Zin, many recent applications, particularly in dynamic asset pricing models, use the homothetic version of this utility function which combines the constant elasticity time aggregator in (4) with a linear homogeneous (constant relative risk aversion) expected utility certainty equivalent.

Empirical research both in the laboratory and in the field has documented a variety of difficulties with the predictions of expected utility models. In particular, people seem more averse to bad outcomes than implied by expected utility. In response to this evidence, there is a growing body of work that looks at decision making under uncertainty outside of the traditional expected utility framework. Without surveying all of these extensions, we demonstrate the basic mechanics of recursive utility with non-expected utility certainty equivalents by studying one particular analytically convenient class of preferences in detail, the Chew–Dekel class. Notable among the alternatives to this class are recursive and dynamic extensions of the Gilboa and Schmeidler 'max-min' preferences.

The Chew–Dekel certainty equivalent function μ for a set of payoffs and probabilities $\{c(z), p(z)\}$ is defined implicitly by a *risk aggregator* M satisfying

$$\mu = \sum_z p(z)M[c(z), \mu]. \tag{5}$$

Such preferences satisfy a weaker condition than the notorious independence axiom that underlies expected utility, yet like expected utility, they lead to first-order conditions in decision problems that are linear in probabilities, hence easily solved and amenable to econometric analysis. We assume M has the following properties: (i) $M(m, m) = m$ (sure things are their own certainty equivalents), (ii) M is increasing in its first argument (first-order stochastic dominance), (iii) M is concave in its first argument (risk aversion), and (iv) $M(kc, km) = kM(c, m)$ for $k > 0$ (linear homogeneity). Most of the analytical convenience of the Chew–Dekel class follows from the linearity of eq. (5) in probabilities. (Note that this implies that indifference curves on the probability simplex are linear, but not necessarily parallel.)

Examples of tractable members of the Chew–Dekel class include the following:

1. *Expected utility.* A version with constant relative risk aversion (that is, linear homogeneity) is implied by

$$M(c, m) = c^\alpha m^{1-\alpha}/\alpha + m(1 - 1/\alpha).$$

If $\alpha \leq 1$, M satisfies the conditions outlined above. Applying (5), we find

$$\mu = \left(\sum_z p(z)c(z)^\alpha\right)^{1/\alpha},$$

the usual expected utility with a power utility function.

2. *Weighted utility.* A relatively easy way to generalize expected utility given (5): weight the probabilities by a function of outcomes. A constant-elasticity version follows from

$$M(c, m) = (c/m)^\gamma c^\alpha m^{1-\alpha}/\alpha + m[1 - (c/m)^\gamma/\alpha].$$

For M to be increasing and concave in c in a neighbourhood of m, the parameters must satisfy either (a) $0 < \gamma < 1$ and $\alpha + \gamma < 0$ or (b) $\gamma < 0$ and $0 < \alpha + \gamma < 1$. Note that (a) implies $\alpha < 0$, (b) implies $\alpha > 0$, and both imply $\alpha + 2\gamma < 1$. The associated certainty equivalent function is

$$\mu^\alpha = \frac{\sum_z p(z)c(z)^{\gamma+\alpha}}{\sum_x p(x)c(x)^\gamma} = \sum_z \hat{p}(z)c(z)^\alpha,$$

where

$$\hat{p}(z) = \frac{p(z)c(z)^{\gamma}}{\sum_x p(x)c(x)^{\gamma}}.$$

This version highlights the impact of bad outcomes: they get greater weight than with expected utility if $\gamma < 0$, less weight otherwise.

3. *Disappointment aversion.* Another model that increases sensitivity to bad events 'disappointments') is defined by the risk aggregator

$$M(c,m) = \begin{cases} c^{\alpha}m^{1-\alpha}/\alpha + m(1-1/\alpha) & c \geq m \\ c^{\alpha}m^{1-\alpha}/\alpha + m(1-1/\alpha) \\ \quad + \delta(c^{\alpha}m^{1-\alpha} - m)/\alpha & c < m \end{cases}$$

with $\delta \geq 0$. When $\delta = 0$ this reduces to expected utility. Otherwise, disappointment aversion places additional weight on outcomes worse than the certainty equivalent. The certainty equivalent function satisfies

$$\mu^{\alpha} = \sum_z p(z)c(z)^{\alpha} + \delta \sum_z p(z)I[c(z) < \mu]$$

$$[c(z)^{\alpha} - \mu^{\alpha}] = \sum_z \hat{p}(z)c(z)^{\alpha},$$

where $I(x)$ is an indicator function that equals one if x is true and zero otherwise, and

$$\hat{p}(z) = \left(\frac{1 + \delta I[c(z) < \mu]}{1 + \delta \sum_x p(x)I[c(x) < \mu]}\right)p(z).$$

It differs from weighted utility in scaling up the probabilities of all bad events by the same factor, and scaling down the probabilities of good events by a complementary factor, with good and bad defined as better and worse than the certainty equivalent. (This implies a 'kink' in state-space indifference curves at certainty, which is referred to as 'first-order' risk aversion.) All three expressions highlight the *recursive* nature of the risk aggregator M: we need to know the certainty equivalent to know which states are bad so that we can compute the certainty equivalent (and so on).

5 Optimization and the Bellman equation

For an illustrative application of recursive utility, we turn to the classic Merton–Samuelson consumption/portfolio-choice problem. Consider a stationary Markov environment with states z and conditional probabilities $p(z'|z)$. Preferences are represented by a constant-discounting/constant-elasticity aggregator and a general linear homogeneous certainty equivalent. A dynamic consumption/portfolio problem for this environment is characterized by the Bellman equation which implicitly defines the value function:

$$J(a,z) = \max_{c,w}\{(1-\beta)c^{\rho} + \beta\mu[J(a',z')]^{\rho}\}^{1/\rho},$$

subject to the wealth constraint, $a' = (a-c)\sum_i w_i r_i$ $(z,z') = (a-c)\sum_i w_i r_i' = (a-c)r_p'$, where a denotes wealth, r_p is the return on the portfolio $(w_1, w_2, \ldots, w_{N-1}, 1 - \sum_{i=1}^{N-1} w_i)$, of assets with risky returns $(r_1, r_2, \ldots, r_N)$. The budget constraint and linear homogeneity of the time and risk aggregators imply linear homogeneity of the value function: $J(a, z) = aL(z)$ for some scaled value function L. The scaled Bellman equation is

$$L(z) = \max_{b,w}\{(1-\beta)b^{\rho} + \beta(1-b)^{\rho}$$

$$\mu[L(z')r_p(z,z')]^{\rho}\}^{1/\rho},$$

where $b \equiv c/a$. Note that $L(z)$ is the marginal utility of wealth in state z.

This problem divides into separate portfolio and consumption decisions. The portfolio decision solves: choose $\{w_i\}$ to maximize $\mu[L(z')r_p(z,z')]$. The portfolio first-order conditions are

$$\sum_{z'} p(z'|z)M_1[L(z')r_p(z,z'),\mu]$$

$$\times L(z')[r_i(z,z') - r_j(z,z')] = 0 \qquad (6)$$

for any two assets i and j.

Given a maximized μ, the consumption decision solves: choose b to maximize L. The intertemporal first-order condition is

$$(1-\beta)b^{\rho-1} = \beta(1-b)^{\rho-1}\mu^{\rho}. \qquad (7)$$

If we solve for μ and substitute into the (scaled) Bellman equation, we find

$$\mu = [(1-\beta)/\beta]^{1/\rho}[b/(1-b)]^{(\rho-1)/\rho}$$

$$L = (1-\beta)^{1/\rho}b^{(\rho-1)/\rho}. \qquad (8)$$

The first-order condition (7) and value function (8) allow us to express the relation between consumption and returns in a familiar form. Since μ is linear homogeneous, the first-order condition implies $\mu(x'r_p') = 1$ for

$$x' = L'/\mu = \left[\beta(c'/c)^{\rho-1}(r_p')^{1-\rho}\right]^{1/\rho}.$$

The last equality follows from $(c'/c) = (b'/b)(1-b)r_p'$, a consequence of the budget constraint and the definition of b. The intertemporal first-order condition can therefore be expressed

$$\mu(x'r_p') = \mu\left(\left[\beta(c'/c)^{\rho-1}r_p'\right]^{1/\rho}\right) = 1,$$

(9)

a generalization of the tangency condition for an optimum (set the marginal rate of substitution equal to the price ratio). Similar logic leads us to express the portfolio first-order conditions (6) as

$$E\left[M_1(x'r_p', 1)x'(r_i' - r_j')\right] = 0.$$

If we multiply by the portfolio weight w_j and sum over j we find

$$E\left[M_1(x'r_p', 1)x'r_i'\right] = E\left[M_1(x'r_p', 1)x'r_p'\right].$$

(10)

Euler's theorem for homogeneous functions allows us to express the right side as

$$E\left[M_1(x'r_p', 1)x'r_p'\right] = 1 - EM_2(x'r_p', 1).$$

Whether this expression is helpful depends on the precise form of M. For example, with disappointment aversion, (10) is

$$E\left[z^{\alpha-1}(1 + \delta I[z<1])\frac{r_i'}{r_p'}\right] = 1 + \delta EI[z<1],$$

where $z = [\beta(c'/c)^{\rho-1}r_p']^{1/\rho}$. This reduces to the Kreps–Porteus model when $\delta = 10$, and to the time-additive expected utility model when, in addition, $\rho = \alpha$.

6 Conclusion

A recursive utility function can be constructed from two components: (a) a time aggregator that completely characterizes preferences in the absence of uncertainty and (b) a risk aggregator that defines the certainty equivalent function that characterizes preferences over static gambles and is used to aggregate the risk associated with future utility. We looked at natural candidates for each of these components and gave an example of how Bellman's equation can be used to characterize optimal plans in a dynamic stochastic environment when agents have recursive preferences.

7 Further reading

For more on this subject, see Backus, Routledge and Zin (2004) and the references cited there. Much of the material in this chapter builds from Epstein and Zin (1989), who extend the preferences in Kreps and Porteus (1978) to allow for a stationary infinite-horizon model and for non-expected utility certainty equivalents. They also derive the consumption/portfolio-choice results of Section (5). For more on time aggregators, see Koopmans (1960), Uzawa (1968), Epstein and Hynes (1983), Lucas and Stokey (1984), and Shi (1994). Common departures from expected utility are documented in Kreps (1988, ch. 14) and Starmer (2000). Epstein and Schneider (2003) and Hansen and Sargent (2004) propose different dynamic and recursive extensions of the max-min risk preference of Gilboa and Schmeidler (1993). The Chew–Dekel risk aggregator was proposed by Chew (1983; 1989) and Dekel (1986). Examples within this class: weighted utility (Chew, 1983), disappointment aversion (Gul, 1991), semi-weighted utility (Epstein and Zin, 2001), and generalized disappointment aversion (Routledge and Zin, 2003).

DAVID K. BACKUS, BRYAN R. ROUTLEDGE, AND STANLEY E. ZIN

See also **Bellman equation; time preference.**

Bibliography

Backus, D.K., Routledge, B.R. and Zin, S.E. 2004. Exotic preferences for macroeconomists. In *NBER Macroeconomics Annual 2004*, vol. 19, ed. M. Gertler and K. Rogoff. Cambridge, MA: MIT Press.

Chew, S.H. 1983. A generalization of the quasi-linear mean with applications to the measurement of inequality and decision theory resolving the Allais paradox. *Econometrica* 51, 1065–92.

Chew, S.H. 1989. Axiomatic utility theories with the betweenness property. *Annals of Operations Research* 19, 273–98.

Dekel, E. 1986. An axiomatic characterization of preferences under uncertainty: weakening the independence axiom. *Journal of Economic Theory* 40, 304–18.

Epstein, L.G. and Hynes, J.A. 1983. The rate of time preference and dynamic economic analysis. *Journal of Political Economy* 91, 611–35.

Epstein, L.G. and Schneider, M. 2003. Recursive multiple-priors. *Journal of Economic Theory* 113, 1–31.

Epstein, L.G. and Zin, S.E. 1989. Substitution, risk aversion, and the temporal behavior of consumption and asset returns: a theoretical framework. *Econometrica* 57, 937–69.

Epstein, L.G. and Zin, S.E. 2001. The independence axiom and asset returns. *Journal of Empirical Finance* 8, 537–72.

Gilboa, I. and Schmeidler, D. 1993. Updating ambiguous beliefs. *Journal of Economic Theory* 59, 33–49.

Gul, F. 1991. A theory of disappointment aversion. *Econometrica* 59, 667–86.

Hansen, L.P. and Sargent, T.J. 2004. Misspecification in recursive macroeconomic theory. Manuscript, University of Chicago, January.

Koopmans, T.C. 1960. Stationary ordinal utility and impatience. *Econometrica* 28, 287–309.

Kreps, D.M. 1988. *Notes on the Theory of Choice*. Boulder, CO: Westview Press.

Kreps, D.M. and Porteus, E.L. 1978. Temporal resolution of uncertainty and dynamic choice theory. *Econometrica* 46, 185–200.

Lucas, R.E. and Stokey, N.L. 1984. Optimal growth with many consumers. *Journal of Economic Theory* 32, 139–71.

Routledge, B.R. and Zin, S.E. 2003. Generalized disappointment aversion and asset prices. Working Paper No. 10107. Cambridge, MA: NBER.

Shi, S. 1994. Weakly nonseparable preferences and distortionary taxes in a small open economy. *International Economic Review* 35, 411–28.

Starmer, C. 2000. Developments in non-expected utility theory. *Journal of Economic Literature* 38, 332–82.

Uzawa, H. 1968. Time preference, the consumption function, and optimum asset holdings. In *Value, Capital, and Growth: Papers in Honour of Sir John Hicks*, ed. J.N. Wolfe. Chicago: Aldine.

redistribution of income and wealth

The topic of redistribution is sometimes interpreted narrowly in rather dry terms: as the description and quantification of the simple fact of change in an income or wealth distribution. This can apply both to an actual change that takes place through time and also to the apparent alteration of the distribution at a point in time by taxes and transfers, and principally involves problems of measurement that are common to other fields of applied economics. However, redistribution can also be seen as a specific goal for economic policymakers: as such it is a subject of special interest in its own right. Sections 1–4 below concentrate primarily on this second interpretation; some issues arising under the first interpretation are considered in Section 5.

1 The reason for wanting to redistribute

Perhaps the simplest and most direct reason for wishing to see a redistribution of income, consumption or wealth in the community is simple fellow feeling on the part of the citizens of the community. This can be incorporated into the utilitarian approach to welfare judgements within the tradition of Bentham and Mill, in two ways. One might suppose that judgements about distribution are made in a state of primordial ignorance about one's own position in the distribution: social aversion to inequality is thus rationalized as individual aversion to risk (Harsanyi, 1955). Secondly, it might be supposed that the poor are made to feel worse off in their plight by the very knowledge that the well-to-do are well-to-do, and the rich are made to feel uncomfortable by the low living standards of the poor – see Hochman and Rodgers (1969). Thus the problems of inequality are rationalized within individual utilities as 'externalities' in a manner similar to health hazards from pollution. A weakness of this approach is that it puts a heavy burden on the

particular configuration of individual preferences that happen to be present within a given community at a given moment: should one *really* only redistribute if enough citizens happen to feel upset by it? And what if some citizens *like* knowing that the very poor are very poor?

An alternative approach is to take the motivation for redistribution as a direct moral imperative – see Tawney (1965), Rawls (1971); improvement in the well-being of the disadvantaged is perceived as a social objective in its own right, along with other apparently desirable goals such as civil liberties and growth in national income.

2 The objectives of redistribution

Whatever the precise reasons for wishing to redistribute income or wealth may be, in *broad* terms the principal goals of redistribution policy can be stated very simply: the primary objectives are usually some goal of greater equity and of 'social insurance'; and as a secondary, though important, desideratum, one is usually also concerned with economic efficiency.

In order to examine these objectives in more detail two concepts need to be carefully distinguished: redistribution '*ex ante*' – the rearrangement of the structure of *income-earning opportunities* – and redistribution '*ex post*' – the reallocation of income or wealth that results from the economic processes of production and exchange, whatever individual opportunities may have been. In practice the two concepts may be difficult to disentangle since a policy measure that apparently just rearranges the prizes (such as an income-tax scheme) may also have repercussions on some people's *ex ante* opportunities (by, for example, affecting market wages); but both are relevant to a discussion of the relationship between equity and other goals.

In a very simplified model of the distribution of income, 'equity' can be expressed fairly easily: if one considers that the cake has been cut very unequally, then one sets about trying to even up the slices. But in a dynamic view of the economy where people make economic choices which affect their future incomes, the slices-of-a-fixed-size-cake analogy can be misleading, and the position may be further complicated when those choices have to be made in the face of uncertainty. Obviously the size of the national cake to be 'shared out' is not, in practice, fixed: individual incomes (and hence the total income in the community) are determined by the choices people make as to how much they work, save, and take entrepreneurial risks, and again the total stock of wealth obviously also depends on the rate at which people save. So the elementary equity question of who ought to get what cannot be divorced in practice from the issue of how individual incomes and wealth holdings are generated: efficiency considerations have to be taken into account in the pursuit of greater equity. There is a second, more subtle, difficulty: because of incompetence, ignorance or plain 'bad luck' people who may have looked alike in terms of their original economic opportunities turn out to be very dissimilar in terms of outcomes once a few rounds have been played of the great economic game that determines how much everybody actually gets. Hence there is a good case for a government concerned with distributional equity to pay attention to both the *ex ante* and the *ex post* concepts of redistribution (Hammond, 1981).

For this reason an interest in social insurance is often taken to be a natural counterpart of a concern for equity. The public provision of protection against the slings and arrows of outrageous fortune is particularly important for those events for which conventional insurance markets are likely to give inadequate coverage, such as unemployment or ill health, for example – see Atkinson (1987). By filling such gaps social insurance may actually improve the efficient working of the economy. Besides this, social insurance can also apply to *ex post* redistribution that is intended to circumvent the otherwise unsatisfactory workings of some markets. For example, the markets for private insurance and savings might, under ideal circumstances, allow people to look after themselves effectively; but in practice problems such as imperfect information and the consequent rationing of insurance of credit to those people who are perceived to be good risks will mean that coverage is far from complete (Arrow, 1985). Hence the provision of state pensions as a means of cushioning the possibly unfortunate effects of restrictions on savings by people of modest means.

3 What should be redistributed?

Whether it is *income* (the flow of spending power during a given period) or *wealth* (the command over resources that a person may possess at a given point in time) that is to be redistributed depends to some extent on the precise definition of these terms (in particular the relevant period over which income is measured and the range of assets to be counted in as personal wealth) and also on the degree of importance that one attaches to *ex ante* or *ex post* concepts of redistribution. For example, some components of wealth (land, financial assets) may be regarded as part of the range of economic opportunities which results in the flow of spendable income. Again weekly money income might be more relevant than broader concepts of wealth or long-term income if one's primary concern is for redistribution to alleviate short-term need rather than to alter the structure of economic opportunities (Atkinson, 1983, ch. 3).

However, the issue of what one ought to use in order to achieve the objectives of redistribution cited above raises further questions. One of the most important of these is whether one ought to redistribute income itself (which yields purchasing power over consumption goods) or rather the consumption goods directly. The standard answer provided by economists is that cash is unquestionably more effective, since it allows individuals to be the judges of what is best for their welfare and to make substitutions between different goods under varying market conditions in pursuit of that welfare: money to buy soup is supposedly more effective than the provision of soup kitchens. However, this conclusion is strictly relevant only if one imposes a number of stringent conditions, for example, the assumption that everyone has access to perfect market opportunities and accurate information on which to base his judgement in the market. It is invalid in the presence of multiple market equilibria (Foldes, 1967). It ignores pressing requirements of crises such as war and famine: extreme circumstances may require direct intervention to act more swiftly and reliably to maintain living standards than the often capricious and sluggish movements of the 'invisible hand'.

4 The available instruments

Among the more obvious instruments available for *ex post* redistribution are taxes on income, wealth and the transfer of wealth via gifts and bequests, and transfer payments such as pensions and social-security benefits. However, it is not easy to draw a hard-and-fast line around the range of instruments that might be taken to be redistributive tools, particularly if one is concerned with description rather than prescription. There appears to be a good case in practice for including 'indirect' taxes (such as value added tax), subsidies and also those benefits 'in kind' which are bestowed on *particular* households or persons, since the impact of these items on personal spending power is usually fairly clear. This may, for example, be extended to include such goods as state-provided education. However the precise distributional impact of publicly provided goods that are really consumed *jointly* by the community (in which category we might include items such as public sanitation, the police services, or even national defence) is less easy to determine, but should not be assumed to be negligible.

As an alternative to raising taxes and the public provision of goods and services, a government wishing to redistribute real spending power may choose to intervene directly in the market mechanism. The most obvious example of this policy is price control. This term applies not only to rationing and the regulation of prices paid by consumers for goods – which can be an effective method of intervention to achieve redistribution – in emergencies such as wartime, but also to the control of prices that individuals receive for services that they may supply (for example, minimum wage legislation) or assets that they possess (control of house rents).

The instruments available for the purposes of *ex ante* redistribution (that is, the means of reorganizing the *opportunities* for creating income and accumulating wealth) are more disparate. One has the immediate problem that the range of policies considered to be available is strongly influenced by the economic philosophy which one considers to be relevant to the analysis and by the political and social system within the community. Take a prime example of this: education. There are many opinions on the potential for using this as a redistributional tool, some of which may be crudely summarized by the following three views: (*a*) it is a passport to higher positions on a ladder of economic opportunity whose rungs are pretty rigidly fixed, so that greater equality can be achieved simply by changing the method of issuing the passports; (*b*) it forms part of a complex of personal or family investment decisions, whereby intervention in the provision of education might upset the efficient allocation of the market mechanism without doing anything to alleviate the inequality of economic opportunity; (*c*) even if effective redistribution *could* be achieved in principle, substantial reorganization of educational opportunities is bound to be limited by what are seen as fundamental freedoms of choice. Note that the divergence of view concerns both economic role of education and the extent to which one is free to use it as an instrument of public policy (Le Grand, 1982, ch. 4).

5 The effectiveness of the policy

Any attempt to quantify the effectiveness of redistribution policy has to surmount a number of extremely troublesome obstacles.

In the first place one has to confront the problem of 'unequal inequalities', which essentially arises from an attempt to compare intrinsically complex social states. Even if one puts this in elementary terms, whereby every person's welfare is accurately measured by his or her income, a fundamental difficulty remains: apart from special circumstances – for example, a comparison involving a hypothetical state of perfect equality – the question of which of two distributions is the more unequal does not generally have a clear-cut answer. In practice, even a very successful redistribution policy will have diminished rather than completely eliminated real income differences, so that an assessment of the policy's impact necessarily involves a comparison of the apparent change in inequality that has been achieved relative to the degree of inequality that would have obtained otherwise. There is no single method for measuring such inequality changes that commands universal support, and hence no generally agreed measuring rod to ascertain the extent of redistribution under all circumstances (Cowell, 1977; Foster, 1985). One of the practical difficulties to which this gives rise is that it is difficult to be dogmatic about labelling policy instruments in terms of degrees of 'progressivity' (Lambert, 1985). Moreover, in many cases redistribution may involve not just a narrowing (or indeed expansion) of income differentials, but also a *re-ranking* of income receivers within the pecking order so that, to quantify redistribution effectively, more is required than a simple measurement of the change in overall dispersion (Cowell, 1985).

The second problem follows directly from this: who is to say what *would* have happened otherwise and, therefore, what change in inequality has actually been achieved? If one is merely concerned with the documentation of trends in the perceived inequities of income distribution through time, this may not be too difficult. But if at any moment of history one attempts to draw the inference that 'according to our chosen inequality index, the inequality of disposable income would have been 20 per cent higher than it is now but for the high marginal tax rates on upper income groups', then one is making a much bolder assertion about how the underlying economic mechanisms are supposed to work. For the very presence of the instruments of redistribution policy will have influenced the choices people make about their jobs, business enterprises and savings, which in turn, can be expected to affect the resulting income distribution. The 'distribution before tax' – obtainable from a statistical office's published figures – cannot automatically be taken to be the same thing as the distribution *without* the tax – the income distribution that one might expect to see if the relevant redistribution instrument were to be abolished.

Some allowance for this problem is usually possible in the case of *ex post* redistribution instruments – for example, it is possible to estimate the likely repercussion on the supply of different types of labour that will arise because of the supplementation of some people's incomes by public transfers and the reduction in other people's incomes through taxation (Hausman, 1985; Killingsworth, 1983), or the impact on private savings of the presence of state-provided pensions and social insurance schemes (Danziger, Haveman and Plotnick, 1981; Kotlikoff, 1984). However, the allowance to be made for these feedback effects is usually quite sensitive to the particular model of household behaviour that is applied.

Despite these reservations, some broad conclusions are possible. Very narrowly based measures run the danger of the 'demarcation trap': for example, subsidizing particular commodities or taxing only certain forms of income or wealth may present some people with an incentive to change their behaviour, or even misrepresent their true status, so as to profit by the artificial distinctions drawn by the selective tax or subsidy scheme. The effectiveness of the measure may thereby be reduced and, even if this does not happen, the discrimination of the scheme may itself create substantial inequities by treating essentially similar people in different ways. On the other hand, very broadly based measures may scatter their shot so widely that much of it misses the target: blanket allowances or exceptions within income- or wealth-tax laws, and some broadly defined educational subsidies are often found to be regressive in their actual *ex post* impact on income and wealth. Finally it is usually the case that *taxes*, taken as a whole, turn out not to be very progressive in terms of their *ex post* impact whereas *transfers* usually are.

F.A. COWELL

See also **progressive and regressive taxation; social insurance.**

Bibliography

Arrow, K.J. 1985. The economics of agency. In *Principals and Agents: The Structure of Business*, ed. J. Pratt and R. Zeckhauser. Cambridge, MA: Harvard Business School Press.

Atkinson, A.B. 1983. *The Economics of Inequality*, 2nd edn. Oxford: Oxford University Press.

Atkinson, A.B. 1987. Income maintenance and social insurance: a survey. In *Handbook of Public Economics*, vol. 2, ed. A.J. Auerbach and M.S. Feldstein. Amsterdam: North-Holland.

Cowell, F.A. 1977. *Measuring Inequality*. Oxford: Philip Allan.

Cowell, F.A. 1985. Measures of distributional change: an axiomatic approach. *Review of Economic Studies* 52, 135–51.

Danziger, S., Haveman, R.H. and Plotnick, R. 1981. How income transfer programs affect work, savings and the income distribution: a critical review. *Journal of Economic Literature* 19, 975–1028.

Foldes, L.P. 1967. Income redistribution in money and in kind. *Economica* 34, 30–41.

Foster, J. 1985. Inequality measurement. In *Fair Allocation*, ed. H.P. Young. Providence, RI: American Mathematical Society.

Hammond, P.J. 1981. *Ex-ante* and *ex-post* welfare economics. *Economica* 48, 235–50.

Harsanyi, J.C. 1955. Cardinal welfare, individualist ethics and interpersonal comparisons of utility. *Journal of Political Economy* 63, 309–21.

Hausman, J.A. 1985. Taxation and labour supply. In *Handbook of Public Economics*, vol. 1, ed. A.J. Auerbach and M.S. Feldstein. Amsterdam: North-Holland.

Hochman, J.M. and Rodgers, J.D. 1969. Pareto optimal redistribution. *American Economic Review* 59, 542–57.

Killingsworth, M.R. 1983. *Labour Supply*. Cambridge: Cambridge University Press.

Kotlikoff, L.J. 1984. Taxation and savings: a neoclassical perspective. *Journal of Economic Literature* 22, 1576–629.

Lambert, P.J. 1985. Tax-progressivity: a survey of the literature. Working Paper. London: Institute for Fiscal Studies.

Le Grand, J. 1982. *The Strategy of Equality*. London: Allen & Unwin.

Rawls, J. 1971. *A Theory of Justice*. Cambridge, MA: Harvard University Press.

Tawney, R.H. 1965. *Equality*. London: Allen & Unwin.

reduced rank regression

Reduced rank regression is an explicit estimation method in multivariate regression that takes into account the reduced rank restriction on the coefficient matrix.

Reduced rank regression model: We consider the multivariate regression of Y on X and Z of dimension p, q, and k, respectively: $Y_t = \Pi X_t + \Gamma Z_t + \varepsilon_t, t = 1, \ldots, T$. The hypothesis that Π has reduced rank less than or equal to r is expressed as $\Pi = \alpha\beta'$, where α is $p \times r$, and β is $q \times r$, where $r < \min(p,q)$, and gives the reduced rank model

$$Y_t = \alpha\beta'X_t + \Gamma Z_t + \varepsilon_t, t = 1, \ldots, T. \quad (1)$$

Reduced rank regression algorithm: In order to describe the algorithm, which we call $RRR(Y, X|Z)$, we introduce the notation for product moments $S_{yx} = T^{-1}\sum_{t=1}^{T} Y_t X_t'$, $S_{yx.z} = S_{yx} - S_{yz}S_{zz}^{-1}S_{zx}$, and so on. The algorithm consists of the following steps:

1. First, regress Y and X on Z and form residuals $(Y|Z)_t = Y_t - S_{yz}S_{zz}^{-1}Z_t$, $(X|Z)_t = X_t - S_{xz}S_{zz}^{-1}Z_t$ and product moments

$$S_{yx.z} = T^{-1}\sum_{t=1}^{T} (Y|Z)_t(X|Z)_t' = S_{yx} - S_{yz}S_{zz}^{-1}S_{zx},$$

and so on.
2. Next, solve the eigenvalue problem

$$\left|\lambda S_{xx.z} - S_{xy.z}S_{yy.z}^{-1}S_{yx.z}\right| = 0, \quad (2)$$

where $|.|$ denotes determinant. The ordered eigenvalues are $\Lambda = \text{diag}(\lambda_1, \ldots, \lambda_q)$ and the eigenvectors are $V = (v_1, \ldots, v_q)$, so that $S_{xx.z}V\Lambda = S_{xy.z}S_{yy.z}^{-1}S_{yx.z}V$, and V is normalized so that $V'S_{xx.z}V = I_p$ and $V'S_{yx.z}S_{xx.z}^{-1}S_{xy.z}V = \Lambda$. The singular value decomposition provides an efficient way of implementing this procedure; see Doornik and O'Brien (2002).
3. Finally, define the estimators

$$\hat{\beta} = (v_1, \ldots, v_r)$$

together with $\hat{\alpha} = S_{yx.z}\hat{\beta}$, and $\hat{\Omega} = S_{yy.z} - S_{yx.z}\hat{\beta}$ $(\hat{\beta}'S_{xx.z}\hat{\beta})^{-1}\hat{\beta}'S_{xy.z}$. Equivalently, once $\hat{\beta}$ has been determined, $\hat{\alpha}$ and $\hat{\Gamma}$ are determined by regression.

The technique of reduced rank regression was introduced by Anderson and Rubin (1949) in connection with the analysis of limited information maximum likelihood and generalized to the reduced rank regression model (1) by Anderson (1951). An excellent source of information is the monograph by Reinsel and Velu (1998), which contains a comprehensive survey of the theory and history of reduced rank regression and its many applications.

Note the difference between the unrestricted estimate $\hat{\Pi}_{OLS} = S_{yx.z}S_{xx.z}^{-1}$ and the reduced rank regression estimate $\hat{\Pi}_{RRR} = S_{yx.z}\hat{\beta}(\hat{\beta}'S_{xx.z}\hat{\beta})^{-1}\hat{\beta}'$ of the coefficient matrix to X.

Applications of the reduced rank model and algorithm

The reduced rank model (1) has many interpretations depending on the context. It is obviously a way of achieving fewer parameters in the possibly large $p \times q$ coefficient matrix Π. Another interpretation is that, although X is needed to explain the variation of Y, in practice only a few, r, factors are needed as given by the linear combinations $\beta'X$ in (1).

Restrictions on Π: Anderson (1951) formulated the problem of estimating Π under $p-r$ unknown restrictions $\ell'\Pi = 0$. In (1) these are given by the matrix $\ell = \alpha_\perp$, that is, a $p \times (p - r)$ matrix of full rank for which $\alpha'_\perp \alpha = 0$. The matrix $\alpha_\perp$ is estimated by solving the dual eigenvalue problem $|\lambda S_{yy.z} - S_{yx.z}S_{xx.z}^{-1}S_{xy.z}| = 0$, which has eigenvalues Λ and eigenvectors W, and the estimate is $\hat{\alpha}_\perp = (w_{r+1}, \ldots, w_p)$. If $p=q$, we can choose $W = S_{yy.z}^{-1}S_{yx.z}V\Lambda^{-1/2}$.

Canonical correlations: Reduced rank regression is related to canonical correlations (Hotelling, 1936). This is most easily expressed if $p = q$, where we find

$$\begin{pmatrix} W & 0 \\ 0 & V \end{pmatrix}' \begin{pmatrix} S_{yy.z} & S_{yx.z} \\ S_{xy.z} & S_{xx.z} \end{pmatrix} \begin{pmatrix} W & 0 \\ 0 & V \end{pmatrix}$$
$$= \begin{pmatrix} I_p & \Lambda^{1/2} \\ \Lambda^{1/2} & I_q \end{pmatrix}.$$

This shows that the variables $W'Y$ and $V'X$ are the empirical canonical variates.

Instrumental variable estimation: Let the variables U, V, and X be of dimension p, q, and k respectively with $k \geq q$. Assume that they are jointly Gaussian with mean zero and variance Σ, and that $E(U - \gamma'V)X' = 0$, so that X is an instrument for estimating γ. This means that $\Sigma_{ux} = \gamma'\Sigma_{vx}$, so that

$$E\left(\begin{pmatrix} U \\ V \end{pmatrix} \Big| X\right) = \begin{pmatrix} \gamma'\Sigma_{vx} \\ \Sigma_{vx} \end{pmatrix} \Sigma_{xx}^{-1}X$$
$$= \begin{pmatrix} \gamma' \\ I_q \end{pmatrix} \Sigma_{vx}\Sigma_{xx}^{-1}X = \alpha\beta'.$$

It follows that the $(p + q) \times k$ coefficient matrix in a regression of $Y = (U', V')'$ on X has rank q. Thus a reduced rank regression of Y on X is an algorithm for estimating the parameter of interest γ using the instruments X. This is the idea in Anderson and Rubin (1949) for the limited information maximum likelihood estimation.

Non-stationary time series: The model

$$\Delta Y_t = \alpha\beta'Y_{t-1} + \Gamma\Delta Y_{t-1} + \varepsilon_t, t = 1, \ldots, T \tag{3}$$

determines a multivariate times series Y_t, and the reduced rank of $\alpha\beta'$ implies non-stationarity of the time series. Under suitable conditions (see Johansen, 1996) Y_t is

non-stationary and ΔY_t and $\beta'Y_t$ are stationary. Thus Y_t is a cointegrated time series; see Engle and Granger (1987).

Common features: Engle and Kozicki (1993) used model (3) and assumed reduced rank of the matrix $(\alpha, \Gamma) = \xi\eta'$, so that $\Delta Y_t = \xi\eta'(Y'_{t-1}\beta, \Delta Y'_{t-1})' + \varepsilon_t$. In this case $\xi'_\perp\Delta Y_t = \xi'_\perp\varepsilon_t$ determines a random walk, where the common cyclic features have been eliminated.

Prediction: Box and Tiao (1977) analysed the model $Y_t = \Pi Y_{t-1} + \Gamma Y_{t-2} + \varepsilon_t$, and asked which linear combinations of the current values, $v'Y_t$, are best predicted by a linear combination of the past (Y_{t-1}, Y_{t-2}), and hence introduced the analysis of canonical variates in the context of prediction of times series.

The Gaussian likelihood analysis

If the errors ε_t in model (1) are i.i.d. Gaussian $N_p(0,\Omega)$, and independent of $\{X_s, Z_s, s \leq t\}$, the (conditional or partial) Gaussian likelihood is

$$-\frac{T}{2}\log|\Omega| - \frac{1}{2}\sum_{t=1}^{T}(Y_t - \alpha\beta'X_t - \Gamma Z_t)'$$
$$\times \Omega^{-1}(Y_t - \alpha\beta'X_t - \Gamma Z_t).$$

Anderson (1951) introduced the *RRR* algorithm as a calculation of the maximum likelihood estimator of $\alpha\beta'$. The Frisch–Waugh theorem shows that one can partial out the parameter Γ by regression, as in the first step of the algorithm. We next regress $(Y|Z)$ on $(\beta'X|Z)$ and find estimates of α and Ω, and the maximized likelihood function as functions of β:

$$\hat{\alpha}(\beta) = S_{yx.z}\beta(\beta'S_{xx.z}\beta)^{-1}, \tag{4}$$

$$\hat{\Omega}(\beta) = S_{yy.z} - S_{yx.z}\beta(\beta'S_{xx.z}\beta)^{-1}\beta'S_{xy.z},$$

$$L_{\max}^{-2/T}(\beta) = |\hat{\Omega}(\beta)|.$$

The identity

$$\left| \begin{pmatrix} S_{yy.z} & S_{yx.z}\beta \\ \beta'S_{xy.z} & \beta'S_{xx.z}\beta \end{pmatrix} \right| = |S_{yy.z}|$$
$$\times |\beta'S_{xx.z}\beta - \beta'S_{xy.z}S_{yy.z}^{-1}S_{yx.z}\beta|$$
$$= |\beta'S_{xx.z}\beta||S_{yy.z} - S_{yx.z}\beta(\beta'S_{xx.z}\beta)^{-1}\beta'S_{xy.z}|$$
$$= |\beta'S_{xx.z}\beta||\hat{\Omega}(\beta)|$$

shows that $L_{\max}^{-2/T}(\beta) = |S_{yy.z}||\beta'(S_{xx.z} - S_{xy.z}S_{yy.z}^{-1}S_{yx.z})\beta|/|\beta'S_{xx.z}\beta|$ so that β has to be chosen to minimize this. Differentiating with respect to β we find that β has to satisfy the relation $S_{xx.z}\beta = S_{xy.z}S_{yy.z}^{-1}S_{yx.z}\beta\xi$ for some $r \times r$ matrix ξ. This shows (see Johansen, 1996) that the space spanned by the columns of β, is spanned by r of the eigenvectors of (2), and hence, that choosing the largest λ_i

gives the smallest value of $L_{max}^{-1/T}(\beta)$, so that

$$\hat{\beta} = (v_1, \ldots, v_r), L_{max}^{-2/T}(\hat{\beta}) = |S_{yy.z}| \prod_{i=1}^{r} (1 - \lambda_i).$$

Hypothesis testing: The test statistic for rank of Π can be calculated from the eigenvalues because the eigenvalue problem solves the maximization of the likelihood for all values of r simultaneously. The likelihood ratio test statistic for the hypothesis rank$(\Pi) \leq r$, as derived by Anderson (1951), is

$$-2\log LR(\text{rank}(\Pi) \leq r) = T \sum_{i=r+1}^{\min(p,q)} \log(1 - \lambda_i).$$

$$(5)$$

Bartlett (1938) suggested using this statistic to test that r canonical correlations between Y and X were zero and hence that Π had reduced rank.

The simplest hypothesis to test on β is $\beta = H\phi$. We can estimate β under this restriction by $RRR(Y, H'X|Z)$ and therefore calculate the likelihood ratio statistic using reduced rank regression. If, on the other hand, we have restrictions on the individual vectors $\beta = (H_1\phi_1, \ldots, H_r\phi_r)$, then reduced rank regression does not provide a maximum likelihood estimator, but we can switch between reduced rank regressions as follows. For fixed $\beta_1, \ldots, \beta_{i-1}, \beta_{i+1}, \ldots, \beta_r$, we can find an estimator for ϕ_i and hence $\beta_i = H_i\phi_i$ by

$$RRR(Y, H_i'X|Z, (\beta_1, \ldots \beta_{i-1}, \beta_{i+1}, \ldots, \beta_r)'X).$$

By switching between the vectors in β, we have an algorithm which is useful in practice and which maximizes the likelihood in each step.

A switching algorithm: Another algorithm, that is useful for this model, is to consider the first order condition for β, when Γ has been eliminated, which has solution

$$\hat{\beta}(\alpha, \Omega) = S_{xx.z}^{-1} S_{xy.z} \Omega^{-1} \alpha (\alpha' \Omega^{-1} \alpha)^{-1}.$$

Combining this with (4) suggests a switching algorithm, as follows.

First choose some initial estimator $\hat{\beta}_0$, then switch between estimating α and Ω for fixed β by least squares, and estimating β for fixed α and Ω by generalized least squares.

This switching algorithm maximizes the likelihood function in each step and any limit point will be a stationary point. It seems to work well in practice. There are natural hypotheses one can test in the reduced rank model, like general linear restrictions on β, which are not solved by the reduced rank regression algorithm, whereas the above algorithm can be modified to give a solution.

Asymptotic distributions in the stationary case: The asymptotic distributions of the estimators and test statistics can be described under the assumption that the process (Y_t, X_t, Z_t) is stationary with finite second moments. It can be shown that estimators are asymptotically Gaussian and test statistics for hypotheses both for rank and for β are asymptotically χ^2; see Robinson (1973).

Asymptotic distributions in the non-stationary case: If the processes are non-stationary a different type of asymptotics is needed. As an example consider model (3) for $I(1)$ variables. When discussing the asymptotic distribution of the estimators, the normalization by $\hat{\beta}' S_{xx.z} \hat{\beta} = I_r$ is not convenient, and it is necessary to identify the vectors differently.

One can then prove (see Johansen, 1996), that the estimates of α, Γ and Ω are asymptotically Gaussian and have the same limit distribution as if β were known: that is, the asymptotic distribution they have in the regression of ΔY_t on the stationary variables $\beta' Y_{t-1}$ and ΔY_{t-1}.

The asymptotic distribution of $\hat{\beta}$ is mixed Gaussian, where the mixing parameter is the (random) limit of the observed information. Therefore, by normalizing on the observed information, we obtain asymptotic χ^2 inference for hypotheses on β.

The limit distribution of the likelihood ratio test statistic for rank, see (5), is given by a generalization of the Dickey–Fuller distribution:

$$DF_{p-r} = \text{tr}\left\{ \int_0^1 (dW)W' \left(\int_0^1 WW'du \right)^{-1} \right.$$
$$\left. \times \int_0^1 W(dW)' \right\},$$

where W is a standard Brownian motion in $p-r$ dimensions. The quantiles of this distribution can at present only be calculated by simulation if $p - r > 1$. The limit distribution has to be modified if deterministic terms are included in the model.

SØREN JOHANSEN

See also **instrumental variables; maximum likelihood.**

Bibliography

Anderson, T.W. 1951. Estimating linear restrictions on regression coefficients for multivariate normal distributions. *Annals of Mathematical Statistics* 22, 327–51.

Anderson, T.W. and Rubin, H. 1949. Estimation of the parameters of a single equation in a complete system of stochastic equations. *Annals of Mathematical Statistics* 20, 46–63.

Bartlett, M.S. 1938. Further aspects of the theory of multiple regression. *Proceedings of the Cambridge Philosophical Society* 34, 33–40.

Box, G.E.P. and Tiao, G.C. 1977. A canonical analysis of multiple time series. *Biometrika* 64, 355–65.

Doornik, J.A. and O'Brien, R.J. 2002. Numerically stable cointegration analysis. *Computational Statistics and Data Analysis* 41, 185–93.

Engle, R.F. and Granger, C.W.J. 1987. Co-integration and error correction: representation, estimation and testing. *Econometrica* 55, 251–76.

Engle, R.F. and Kozicki, S. 1993. Testing for common factors (with comments). *Journal of Business Economics and Statistics* 11, 369–78.

Hotelling, H. 1936. Relations between two sets of variables. *Biometrika* 28, 321–77.

Johansen, S. 1996. *Likelihood Based Inference on Cointegration in the Vector Autoregressive Model*. Oxford: Oxford University Press.

Reinsel, G.C. and Velu, R.P. 1998. *Multivariate Reduced Rank Regression*. Lecture Notes in Statistics. New York: Springer.

Robinson, P.M. 1973. Generalized canonical analysis for time series. *Journal of Multivariate Analysis* 3, 141–60.

regime switching models

Many economic time series occasionally exhibit dramatic breaks in their behaviour, associated with events such as financial crises (Jeanne and Masson, 2000; Cerra and Saxena, 2005; Hamilton, 2005) or abrupt changes in government policy (Hamilton, 1988; Sims and Zha, 2006; Davig, 2004). Of particular interest to economists is the apparent tendency of many economic variables to behave quite differently during economic downturns, when underutilization of factors of production rather than their long-run tendency to grow governs economic dynamics (Hamilton, 1989; Chauvet and Hamilton, 2006). Abrupt changes are also a prevalent feature of financial data, and the approach described below is quite amenable to theoretical calculations for how such abrupt changes in fundamentals should show up in asset prices (Ang and Bekaert, 2002a; 2000b; Garcia, Luger and Renault, 2003; Dai, Singleton and Yang, 2003).

Consider how we might describe the consequences of a dramatic change in the behaviour of a single variable y_t. Suppose that the typical historical behaviour could be described with a first-order autoregression,

$$y_t = c_1 + \phi y_{t-1} + \varepsilon_t, \qquad (1)$$

with $\varepsilon_t \sim N(0, \sigma^2)$, which seemed to adequately describe the observed data for $t = 1, 2, \ldots, t_0$. Suppose that at date t_0 there was a significant change in the average level of the series, so that we would instead wish to describe the data according to

$$y_t = c_2 + \phi y_{t-1} + \varepsilon_t \qquad (2)$$

for $t = t_0 + 1, t_0 + 2, \ldots$ This fix of changing the value of the intercept from c_1 to c_2 might help the model to get back on track with better forecasts, but it is rather unsatisfactory as a probability law that could have generated the data. We surely would not want to maintain that the change from c_1 to c_2 at date t_0 was a deterministic event that anyone would have been able to predict with

certainty looking ahead from date $t = 1$. Instead, there must have been some imperfectly predictable forces that produced the change. Hence, rather than claim that expression (1) governed the data up to date t_0 and (2) after that date, what we must have in mind is that there is some larger model encompassing them both,

$$y_t = c_{s_t} + \phi y_{t-1} + \varepsilon_t, \qquad (3)$$

where s_t is a random variable that, as a result of institutional changes, happened in our sample to assume the value $s_t = 1$ for $t = 1, 2, \ldots, t_0$ and $s_t = 2$ for $t = t_0 + 1, t_0 + 2, \ldots$. A complete description of the probability law governing the observed data would then require a probabilistic model of what caused the change from $s_t = 1$ to $s_t = 2$. The simplest such specification is that s_t is the realization of a two-state Markov chain with

$$\Pr(s_t = j | s_{t-1} = i, s_{t-2} = k, \ldots, y_{t-1}, y_{t-2}, \ldots)$$
$$= \Pr(s_t = j | s_{t-1} = i) = p_{ij}. \qquad (4)$$

On the assumption that we do not observe s_t directly, but only infer its operation through the observed behavior of y_t, the parameters necessary to fully describe the probability law governing y_t are then the variance of the Gaussian innovation σ^2, the autoregressive coefficient ϕ, the two intercepts c_1 and c_2, and the two state transition probabilities, p_{11} and p_{22}.

The specification in (4) assumes that the probability of a change in regime depends on the past only through the value of the most recent regime, though, as noted below, nothing in the approach described below precludes looking at more general probabilistic specifications. But the simple time-invariant Markov chain (4) seems the natural starting point and is clearly preferable to acting as if the shift from c_1 to c_2 was a deterministic event. Permanence of the shift would be represented by $p_{22} = 1$, though the Markov formulation invites the more general possibility that $p_{22} < 1$. Certainly in the case of business cycles or financial crises, we know that the situation, though dramatic, is not permanent. Furthermore, if the regime change reflects a fundamental change in monetary or fiscal policy, the prudent assumption would seem to be to allow the possibility of it changing back again, suggesting that $p_{22} < 1$ is often a more natural formulation for thinking about changes in regime than $p_{22} = 1$.

A model of the form of (3)–(4) with no autoregressive elements ($\phi = 0$) appears to have been first analysed by Lindgren (1978) and Baum et al. (1980). Specifications that incorporate autoregressive elements date back in the speech recognition literature to Poritz (1982), Juang and Rabiner (1985), and Rabiner (1989), who described such processes as 'hidden Markov models'. Markov-switching regressions were introduced in econometrics by Goldfeld and Quandt (1973), the likelihood function for which was first correctly calculated by Cosslett and Lee (1985). The formulation of the problem described here, in which all objects of interest are calculated as a by-product of an iterative algorithm similar in spirit to a Kalman filter, is due to Hamilton (1989; 1994). General characterizations of moment and stationarity conditions for such processes can be found in Tjøstheim (1986), Yang (2000), Timmermann (2000), and Francq and Zakoïan (2001).

Econometric inference

Suppose that the econometrician observes y_t directly but can only make an inference about the value of s_t based on what we see happening with y_t. This inference will take the form of two probabilities

$$\xi_{jt} = \Pr(s_t = j|\Omega_t; \theta) \tag{5}$$

for $j = 1, 2$, where these two probabilities sum to unity by construction. Here $\Omega_t = \{y_t, y_{t-1}, \ldots, y_1, y_0\}$ denotes the set of observations obtained as of date t, and θ is a vector of population parameters, which for the above example would be $\theta = (\sigma, \phi, c_1, c_2, p_{11}, p_{22})'$, and which for now we presume to be known with certainty. The inference is performed iteratively for $t = 1, 2, \ldots, T$, with step t accepting as input the values

$$\xi_{i,t-1} = \Pr(s_{t-1} = i|\Omega_{t-1}; \theta) \tag{6}$$

for $i = 1, 2$ and producing as output (5). The key magnitudes one needs in order to perform this iteration are the densities under the two regimes,

$$\begin{aligned} \eta_{jt} &= f(y_t|s_t = j, \Omega_{t-1}; \theta) \\ &= \frac{1}{\sqrt{2\pi}\sigma} \exp\left[-\frac{(y_t - c_j - \phi y_{t-1})^2}{2\sigma^2}\right], \end{aligned} \tag{7}$$

for $j = 1, 2$. Specifically, given the input (6) we can calculate the conditional density of the tth observation from

$$f(y_t|\Omega_{t-1}; \theta) = \sum_{i=1}^{2} \sum_{j=1}^{2} p_{ij}\xi_{i,t-1}\eta_{jt} \tag{8}$$

and the desired output is then

$$\xi_{jt} = \frac{\sum_{i=1}^{2} p_{ij}\xi_{i,t-1}\eta_{jt}}{f(y_t|\Omega_{t-1}; \theta)}. \tag{9}$$

As a result of executing this iteration, we will have succeeded in evaluating the sample conditional log likelihood of the observed data

$$\log f(y_1, y_2, \ldots, y_T|y_0; \theta) = \sum_{t=1}^{T} \log f(y_t|\Omega_{t-1}; \theta) \tag{10}$$

for the specified value of θ. An estimate of the value of θ can then be obtained by maximizing (10) by numerical optimization.

Several options are available for the value ξ_{i0} to use to start these iterations. If the Markov chain is presumed to be ergodic, one can use the unconditional probabilities

$$\xi_{i0} = \Pr(s_0 = i) = \frac{1 - p_{jj}}{2 - p_{ii} - p_{jj}}.$$

Other alternatives are simply to set $\xi_{i0} = 1/2$ or estimate ξ_{i0} itself by maximum likelihood.

The calculations do not increase in complexity if we consider an $(r \times 1)$ vector of observations $\mathbf{y}_t$ whose density depends on N separate regimes. Let $\Omega_t = \{\mathbf{y}_t, \mathbf{y}_{t-1}, \ldots, \mathbf{y}_1\}$ be the observations through date t, $\mathbf{P}$ be an $(N \times N)$ matrix whose row j, column i element is the transition probability p_{ij}, $\boldsymbol{\eta}_t$ be an $(N \times 1)$ vector whose jth element $f(\mathbf{y}_t|s_t = j, \Omega_{t-1}; \theta)$ is the density in regime j, and $\hat{\boldsymbol{\xi}}_{t|t}$ an $(N \times 1)$ vector whose jth element is $\Pr(s_t = j|\Omega_t, \theta)$ Then (8) and (9) generalize to

$$f(\mathbf{y}_t|\Omega_{t-1}; \theta) = \mathbf{1}'(\mathbf{P}\hat{\boldsymbol{\xi}}_{t-t|t-1} \odot \boldsymbol{\eta}_t) \tag{11}$$

$$\hat{\boldsymbol{\xi}}_{t|t} = \frac{\mathbf{P}\hat{\boldsymbol{\xi}}_{t-t|t-1} \odot \boldsymbol{\eta}_t}{f(\mathbf{y}_t|\Omega_{t-1}; \theta)} \tag{12}$$

where $\mathbf{1}$ denotes an $(N \times 1)$ vector all of whose elements are unity and $\odot$ denotes element-by-element multiplication. Markov-switching vector autoregressions are discussed in detail in Krolzig (1997). Vector applications include describing the co-movements between stock prices and economic output (Hamilton and Lin, 1996) and the tendency for some series to move into recession before others (Hamilton and Perez-Quiros, 1996). There further is no requirement that the elements of $\boldsymbol{\eta}_t$ be Gaussian densities or even from the same family of densities. For example, Dueker (1997) studied a model in which the degrees of freedom of a Student t distribution change depending on the economic regime.

One is also often interested in forming an inference about what regime the economy was in at date t based on observations obtained through a later date T, denoted $\hat{\boldsymbol{\xi}}_{t|T}$. These are referred to as 'smoothed' probabilities, an efficient algorithm for whose calculation was developed by Kim (1994).

Extensions

The calculations in (11) and (12) remain valid when the probabilities in $\mathbf{P}$ depend on lagged values of $\mathbf{y}_t$ or strictly exogenous explanatory variables, as in Diebold, Lee and Weinbach (1994), Filardo (1994) and Peria (2002). However, often there are relatively few transitions among regimes, making it difficult to estimate such parameters accurately, and most applications have assumed a time-invariant Markov chain. For the same reason, most applications assume only $N = 2$ or 3 different regimes, though there is considerable promise in models with a much larger number of regimes, either by tightly parameterizing the relation between the regimes (Calvet and Fisher, 2004), or with prior Bayesian information (Sims and Zha, 2006).

In the Bayesian approach, both the parameters θ and the values of the states $\mathbf{s} = (s_1, s_2, \ldots, s_T)'$ are viewed as random variables. Bayesian inference turns out to be greatly facilitated by Monte Carlo Markov chain methods, specifically, the Gibbs sampler. This is achieved by sequentially (for $k = 1, 2, \ldots$) generating a realization $\theta^{(k)}$ from the distribution of $\theta|\mathbf{s}^{(k-1)}, \Omega_T$ followed by a realization $\mathbf{s}^{(k)}$ from the distribution of $\mathbf{s}|\theta^{(k)}, \Omega_T$. The first distribution, $\theta|\mathbf{s}^{(k-1)}, \Omega_T$, treats the historical regimes generated at the previous iteration, $s_1^{(k-1)}, s_2^{(k-1)}, \ldots, s_T^{(k-1)}$, as if fixed known numbers. Often this conditional distribution takes the form of a standard Bayesian inference problem whose solution is known analytically using natural conjugate priors. For example, the posterior distribution of ϕ given other parameters is a known function of easily calculated OLS coefficients. An algorithm for generating a draw from the second distribution, $\mathbf{s}|\theta^{(k)}, \Omega_T$, was developed by Albert and Chib (1993). The Gibbs sampler turns out also to be a natural device for handling transition probabilities that are functions of observable variables, as in Filardo and Gordon (1998).

It is natural to want to test the null hypothesis that there are N regimes against the alternative of $N + 1$, for example when $N = 1$, to test whether there are any changes in regime at all. Unfortunately, the likelihood ratio test of this hypothesis fails to satisfy the usual regularity conditions because, under the null hypothesis, some of the parameters of the model would be unidentified. For example, if there is really only one regime, the maximum likelihood estimate $\hat{p}_{11}$ does not converge to a well-defined population magnitude, meaning that the likelihood ratio test does not have the usual χ^2 limiting distribution. To interpret a likelihood ratio statistic, one instead needs to appeal to the methods of Hansen (1992) or Garcia (1998). An alternative is to rely on generic tests of the hypothesis that an N-regime model accurately describes the data (Hamilton, 1996), though these tests are not designed for optimal power against the specific alternative hypothesis of $N + 1$ regimes. A test recently proposed by Carrasco, Hu and Ploberger (2004) that is easy to compute but not based on the likelihood ratio statistic seems particularly promising. Other alternatives are to use Bayesian methods to calculate the value of N implying the largest value for the marginal likelihood (Chib, 1998) or the highest Bayes factor (Koop and Potter, 1999), or to compare models on the basis of their ability to forecast (Hamilton and Susmel, 1994).

A specification where the density depends on a finite number of previous regimes, $f(\mathbf{y}_t|s_t, s_{t-1}, \ldots, s_{t-m}, \Omega_{t-1}; \theta)$, can be recast in the above form by a suitable redefinition of regime. For example, if s_t follows a 2-state Markov chain with transition probabilities $\Pr(s_t = j|s_{t-1} = i)$ and $m = 1$, one can define a new regime variable s_t^* such

that $f(\mathbf{y}_t|s_t^*, \Omega_{t-1}; \theta) = f(\mathbf{y}_t|s_t, s_{t-1}, \ldots, s_{t-m}, \Omega_{t-1}; \theta)$ as follows:

$$s_t^* = \begin{cases} 1 & \text{when } s_t = 1 \text{ and } s_{t-1} = 1 \\ 2 & \text{when } s_t = 2 \text{ and } s_{t-1} = 1 \\ 3 & \text{when } s_t = 1 \text{ and } s_{t-1} = 2 \\ 4 & \text{when } s_t = 2 \text{ and } s_{t-1} = 2 \end{cases}.$$

Then s_t^* itself follows a 4-state Markov chain with transition matrix

$$\mathbf{P}^* = \begin{bmatrix} p_{11} & 0 & p_{11} & 0 \\ p_{12} & 0 & p_{12} & 0 \\ 0 & p_{21} & 0 & p_{21} \\ 0 & p_{22} & 0 & p_{22} \end{bmatrix}.$$

More problematic are cases in which the order of dependence m grows with the date of the observation t. Such a situation often arises in models whose recursive structure causes the density of y_t given Ω_{t-1} to depend on the entire history $y_{t-1}, y_{t-2}, \ldots, y_1$ as is the case in ARMA, GARCH or state-space models. Consider for illustration a GARCH(1,1) specification in which the coefficients are subject to changes in regime, $y_t = h_t v_t$, where $v_t \sim N(0, 1)$ and

$$h_t^2 = \gamma_{s_t} + \alpha_{s_t} y_{t-1}^2 + \beta_{s_t} h_{t-1}^2. \tag{13}$$

Solving (13) recursively reveals that the conditional standard deviation h_t depends on the full history $\{y_{t-1}, y_{t-2}, \ldots, y_0, s_t, s_{t-1}, \ldots, s_1\}$. One way to avoid this problem was proposed by Gray (1996), who postulated that, instead of being generated by (13), the conditional variance is characterized by

$$h_t^2 = \gamma_{s_t} + \alpha_{s_t} y_{t-1}^2 + \beta_{s_t} \tilde{h}_{t-1}^2 \tag{14}$$

where

$$\tilde{h}_{t-1}^2 = \sum_{i=1}^{N} \hat{\xi}_{i,t-1|t-2}(\gamma_i + \alpha_i y_{t-2}^2 + \beta_i \tilde{h}_{t-2}^2).$$

In Gray's model, h_t in (14) depends only on s_t since $\tilde{h}_{t-1}^2$ is a function of data Ω_{t-1} only. An alternative solution, due to Haas, Mittnik and Paolella (2004), is to hypothesize N separate GARCH processes whose values h_{it} all exist as latent variables at date t,

$$h_{it}^2 = \gamma_i + \alpha_i y_{t-1}^2 + \beta_i h_{i,t-1}^2 \tag{15}$$

and then simply pose the model as $y_t = h_{s_t} v_t$. Again, the feature that makes this work is the fact that h_{it} in (15) is a function solely of the data Ω_{t-1} rather than the states $\{s_{t-1}, s_{t-2}, \ldots, s_1\}$.

A related problem arises in Markov-switching state-space models, which posit an unobserved state vector z_t characterized by

$$z_t = F_{s_t} z_{t-1} + Q_{s_t} v_t$$

with $v_t \sim N(0, I_n)$, with observed vectors y_t and x_t governed by

$$y_t = H'_{s_t} z_t + A'_{s_t} x_t + R_{s_t} w_t$$

for $w_t \sim N(0, I_r)$. Again, the model as formulated implies that the density of y_t depends on the full history $\{s_t, s_{t-1}, \ldots, s_1\}$. Kim (1994) proposed a modification of the Kalman filter equations similar in spirit to the modification in (14) that can be used to approximate the log likelihood. A more common practice recently has been to estimate such models with numerical Bayesian methods, as in Kim and Nelson (1999).

JAMES D. HAMILTON

See also **Markov chain Monte Carlo methods; Markov processes; maximum likelihood; mixture models; nonlinear time series analysis; numerical optimization methods in economics; structural change.**

Bibliography

Albert, J. and Chib, 1993. Bayes inference via Gibbs sampling of autoregressive time series subject to Markov mean and variance shifts. *Journal of Business and Economic Statistics* 11, 1–15.

Ang, A. and Bekaert, G. 2002a. International asset allocation with regime shifts. *Review of Financial Studies* 15, 1137–87.

Ang, A. and Bekaert, G. 2002b. Regime switches in interest rates. *Journal of Business and Economic Statistics* 20, 163–82.

Baum, L., Petrie, E., Soules, G. and Weiss, N. 1980. A maximization technique occurring in the statistical analysis of probabilistic functions of Markov chains. *Annals of Mathematical Statistics* 41, 164–71.

Calvet, L. and Fisher, A. 2004. How to forecast long-run volatility: regime-switching and the estimation of multifractal processes. *Journal of Financial Econometrics* 2, 49–83.

Carrasco, M., Hu, L. and Ploberger, W. 2004. Optimal test for Markov switching. Working paper. University of Rochester.

Cerra, V. and Saxena, S. 2005. Did output recover from the Asian crisis? *IMF Staff Papers* 52, 1–23.

Chauvet, M. and Hamilton, J. 2006. Dating business cycle turning points. In *Nonlinear Time Series Analysis of Business Cycles*, ed. C. Milas, P. Rothman and D. van Dijk. Amsterdam: Elsevier.

Chib, S. 1998. Estimation and comparison of multiple change-point models. *Journal of Econometrics* 86, 221–41.

Cosslett, S. and Lee, L.-F. 1985. Serial correlation in discrete variable models. *Journal of Econometrics* 27, 79–97.

Dai, Q., Singleton, K. and Yang, W. 2003. Regime shifts in a dynamic term structure model of U.S. Treasury bonds. Working paper, Stanford University.

Davig, T. 2004. Regime-switching debt and taxation. *Journal of Monetary Economics* 51, 837–59.

Diebold, F., Lee, J.-H. and Weinbach, G. 1994. Regime switching with time-varying transition probabilities. In *Nonstationary Time Series Analysis and Cointegration*, ed. C. Hargreaves. Oxford: Oxford University Press.

Dueker, M. 1997. Markov switching in GARCH processes and mean-reverting stock-market volatility. *Journal of Business and Economic Statistics* 15, 26–34.

Filardo, A. 1994. Business cycle phases and their transitional dynamics. *Journal of Business and Economic Statistics* 12, 299–308.

Filardo, A. and Gordon, S. 1998. Business cycle durations. *Journal of Econometrics* 85, 99–123.

Francq, C. and Zakoïan, J.-M. 2001. Stationarity of multivariate Markov-switching ARMA models. *Journal of Econometrics* 102, 339–64.

Garcia, R. 1998. Asymptotic null distribution of the likelihood ratio test in Markov switching models. *International Economic Review* 39, 763–88.

Garcia, R., Luger, R. and Renault, E. 2003. Empirical assessment of an intertemporal option pricing model with latent variables. *Journal of Econometrics* 116, 49–83.

Goldfeld, S. and Quandt, R. 1973. A Markov model for switching regressions. *Journal of Econometrics* 1, 3–16.

Gray, S. 1996. Modeling the conditional distribution of interest rates as a regime-switching process. *Journal of Financial Economics* 42, 27–62.

Haas, M., Mittnik, S. and Paolella, M. 2004. A new approach to Markov-switching GARCH models. *Journal of Financial Econometrics* 2, 493–530.

Hamilton, J. 1988. Rational-expectations econometric analysis of changes in regime: an investigation of the term structure of interest rates. *Journal of Economic Dynamics and Control* 12, 385–423.

Hamilton, J. 1989. A new approach to the economic analysis of nonstationary time series and the business cycle. *Econometrica* 57, 357–84.

Hamilton, J. 1994. *Time Series Analysis*. Princeton, NJ: Princeton University Press.

Hamilton, J. 1996. Specification testing in Markov-switching time-series models. *Journal of Econometrics* 70, 127–57.

Hamilton, J. 2005. What's real about the business cycle? *Federal Reserve Bank of St. Louis Review* 87, 435–52.

Hamilton, J. and Lin, G. 1996. Stock market volatility and the business cycle. *Journal of Applied Econometrics* 11, 573–93.

Hamilton, J. and Perez-Quiros, G. 1996. What do the leading indicators lead? *Journal of Business* 69, 27–49.

Hamilton, J. and Susmel, R. 1994. Autoregressive conditional heteroskedasticity and changes in regime. *Journal of Econometrics* 64, 307–33.

Hansen, B. 1992. The likelihood ratio test under non-standard conditions. *Journal of Applied Econometrics* 7, S61–S82. Erratum, 11(1996), 195–8.

Jeanne, O. and Masson, P. 2000. Currency crises, sunspots, and Markov-switching regimes. *Journal of International Economics* 50, 327–50.

Juang, B.-H. and Rabiner, L. 1985. Mixture autoregressive hidden Markov models for speech signals. *IEEE Transactions on Acoustics, Speech, and Signal Processing* 30, 1404–13.

Kim, C. 1994. Dynamic linear models with Markov-switching. *Journal of Econometrics* 60, 1–22.

Kim, C. and Nelson, C. 1999. *State-Space Models with Regime Switching*. Cambridge, MA: MIT Press.

Koop, G. and Potter, S. 1999. Bayes factors and nonlinearity: evidence from economic time series. *Journal of Econometrics* 88, 251–81.

Krolzig, H.-M. 1997. *Markov-Switching Vector Autoregressions: Modelling, Statistical Inference, and Application to Business Cycle Analysis*. Berlin: Springer.

Lindgren, G. 1978. Markov regime models for mixed distributions and switching regressions. *Scandinavian Journal of Statistics* 5, 81–91.

Peria, M. 2002. A regime-switching approach to the study of speculative attacks: a focus on EMS crises. In *Advances in Markov-Switching Models*, ed. J. Hamilton and B. Raj. Heidelberg: Physica Verlag.

Poritz, A. 1982. Linear predictive hidden Markov models and the speech signal. *Acoustics, Speech and Signal Processing, IEEE Conference on ICASSP '82*, vol. 7, 1291–4.

Rabiner, L. 1989. A tutorial on hidden Markov models and selected applications in speech recognition. *Proceedings of the IEEE* 77, 257–86.

Sims, C. and Zha, T. 2006. Were there switches in U.S. monetary policy? *American Economic Review* 96, 54–81.

Timmermann, A. 2000. Moments of Markov switching models. *Journal of Econometrics* 96, 75–111.

Tjøstheim, D. 1986. Some doubly stochastic time series models. *Journal of Time Series Analysis* 7, 51–72.

Yang, M. 2000. Some properties of vector autoregressive processes with Markov-switching coefficients. *Econometric Theory* 16, 23–43.

regional and preferential trade agreements

Strongly influenced by the perception that restricted commerce and preferences in trade relations had contributed to the Great Depression of the 1930s and the subsequent outbreak of war, the discussions leading to the General Agreement on Tariffs and Trade (GATT) in 1947 were driven by the desire to create an international economic order based on a liberal and non-discriminatory multilateral trade system. Enshrined in Article I of the GATT, the principle of non-discrimination (commonly referred to as the most-favoured-nation or MFN clause) precludes member countries from discriminating against imports based upon the country of origin. However, in an important exception this central prescript, the GATT, through its Article XXIV, permits its members to enter into preferential trade agreements (PTAs), provided these preferences are complete. In so doing, it sanctions the formation of free trade areas (FTAs), whose members are obligated to eliminate internal import barriers, and customs unions (CUs), whose members additionally agree on a common external tariff against imports from non-members. Additional derogations to the principle of non-discrimination now include the Enabling Clause, which allows tariff preferences to be granted to developing countries (in accordance with the Generalized System of Preferences) and permits preferential trade agreements among developing countries in goods trade. Among the more prominent existing PTAs are the North American Free Trade Agreement (NAFTA), the European Economic Community (EEC) and the European Free Trade Association (EFTA), all formed under Article XXIV, and the Mercosur (the CU between the Argentine Republic, Brazil, Paraguay, and Uruguay) and the ASEAN (Association of South East Asian Nations) Free Trade Area (AFTA), both formed under the Enabling Clause.

Static welfare analysis

Motivated by ongoing discussions concerning optimal trade arrangements in the post-war period, especially over the possibility of a European customs union, Jacob Viner (1950, pp. 41–50) developed a seminal analysis of the economics of preferential trade. Viner's analysis undermines the presumption that cutting tariffs is necessarily welfare improving. On the one hand, because of discriminatory liberalization, there will be commodities that a member country may 'newly import from the other but which it formerly did not import at all because the price of the protected domestic good was lower than the price of any foreign source plus the duty'. Viner calls this shift from a high-cost to a lower-cost point 'trade creation' and associates it with welfare-improvement for the importing country. He also argues that, on the other hand, 'there may be other commodities, which one of the members will now newly import from the other', whereas before the PTA it 'imported them from a third country, because that was the cheapest possible source of supply even after the payment of duty'. He calls this shift in imports from a low-cost third country to a higher-cost member country 'trade diversion,' associating it with an increase in the cost of imports and, thus, welfare losses for the importing country.

The demonstration that preferential trade liberalization may be welfare decreasing stimulated a substantial theoretical literature on the 'static' welfare effects of PTAs. Post-Vinerian analysis of the welfare effects of preferential trade include Meade's (1955) more explicit and comprehensive formulation of the problem in a three-country three-good setting. Meade argues that not only the magnitudes of trade creation and trade diversion but also the extent of cost reductions (in the former) and increases (in the latter) were necessary to arrive at a welfare evaluation. Subsequent analysis also developed examples of both welfare improving trade diversion and welfare-decreasing trade creation in general equilibrium contexts broader than those considered by Viner (see, for instance, Gehrels, 1956–7, Lipsey, 1957, and Bhagwati, 1971). However, the intuitive appeal of the concepts of trade creation and trade diversion has ensured their continued use in the economic analysis of preferential trade agreements, especially in policy analysis (see Panagariya, 2000, for a comprehensive survey).

The effects of preferences on intra-union and extra-union terms-of-trade are analysed by Mundell (1964), who argues that a country granting tariff preferences moves intra-union terms of trade against it and in favour of its partner by increasing its demand for imports from its partner. Extra-union terms of trade are improved for the partner, but change ambiguously for the preference-granting country. Thus, tariff preferences have asymmetric effects on the preference-granting and preference-receiving country. More sharply, Panagariya (1997a) shows how, even with fixed extra-union terms of trade, if a preference-granting country continues to import from the rest of the world, intra-union terms-of-trade losses (manifesting themselves as intra-union tariff revenue transfers as also seen in Berglas, 1979, and Riezman, 1979) unambiguously worsen its welfare, while its preference-receiving partner unambiguously gains.

Wonnacott and Lutz (1987), Krugman (1991) and Summers (1991) propose geographic proximity between partner countries as important in ensuring that preferential liberalization improves welfare. Specifically, they suggest that countries entering into preferential arrangements with geographically proximate countries are likely to do better than in agreements with distant countries, because the former are more likely than the latter to be trade creating, leading to a larger improvement in welfare. Proximate countries are thus argued to be 'natural' partners for preferential trade. Bhagwati and Panagariya (1996) and Panagariya (1997b), however, provide a number of examples in which, between two otherwise identical potential partners, a country achieves a superior outcome by granting trade preferences to the distant partner. For instance, it may be that a preference granted to a distant partner leads to a smaller transfer (loss) of tariff revenue with a closer country, since, with an initial non-discriminatory tariff, the liberalizing country imports less from the more distant partner. Thus, the 'natural trading partners' hypothesis is shown to lack general theoretical validity.

Numerous studies have attempted to evaluate quantitatively the trade creation and trade diversion effects of PTAs. Recently, focusing on the effects of PTAs on excluded countries, Chang and Winters (2002) show how Mercosur was associated with significant declines in the prices of non-members' exports to the region. Yeats (1998) shows how under Mercosur the greatest increases of intra-union trade flows were in goods in which the member countries had the least comparative advantage, confirming the trade diversionary effects of preferential liberalization.

Srinivasan, Whalley and Wooton (1993) note that the econometric frameworks used in most *ex post* studies of trade flows generally lack microeconomic underpinnings, making an evaluation of the associated welfare consequences difficult. Krishna (2003) develops an econometric framework for the analysis of PTAs with a strong welfare-theoretic foundation, so that the estimated parameters relating to trade creation and trade diversion effects fit directly into theoretically derived welfare expressions. His application of this framework to the evaluation of the natural trading partners hypothesis does not find any support in US data.

Necessarily welfare-improving preferential trade areas

The generally ambiguous welfare results provided by the theoretical literature raised an important question relating to the *design* of necessarily welfare-improving PTAs. A classic result stated independently by Kemp (1964) and Vanek (1965) and proved subsequently by Ohyama (1972) and Kemp and Wan (1976) provides a welfare-improving solution for the case of CUs. Starting from a situation with an arbitrary structure of trade barriers, if two or more countries freeze their net external trade vector with the rest of the world through a set of common external tariffs and eliminate the barriers to internal trade (which implies the formation of a CU), the welfare of the union as a whole necessarily improves (weakly) and that of the rest of the world does not fall. A Pareto-improving preferential trade agreement may thus be achieved. The logic behind the Kemp–Wan theorem is as follows. By fixing the combined, net extra-union trade vector of member countries at its pre-union level, non-member countries are guaranteed their original level of welfare. Moreover, if we take the extra-union trade vector as an endowment, the joint welfare of the union is maximized by allowing free trade of goods internally (thus equating the marginal rate of substitution and marginal rate of transformation for each pair of commodities to each other and across all agents in the union). The PTA thus constructed has a common internal price vector, implying further a common external tariff for member countries. This customs union is (weakly) welfare improving; the rest of the world is no worse off and the welfare of member countries is jointly improved (weakly). Welfare improvement is achieved even if additional 'non-economic' objectives (such as maintaining the output of a sector or its employment of a factor) are introduced, as Krishna and Bhagwati (1997) show. The

Kemp–Wan–Ohyama design, by freezing the external trade vector and thus eliminating trade diversion, offers a way to sidestep the complexities and ambiguities inherent in the analysis of PTAs. It has played an important role in shaping the way that economists think about issues relating to the design and implementation of PTAs.

The Kemp–Wan–Ohyama analysis of welfare improving CUs does not extend easily to FTAs, however. In the case of an FTA, member-specific tariff vectors imply that the domestic-price vectors differ across member countries and the FTA generally fails to equalize marginal rates of substitution across its members. Without a common internal price vector, however, the Kemp–Wan–Ohyama methodology lacks application. Nevertheless, Panagariya and Krishna (2002) have provided a corresponding construction of necessarily welfare-improving FTAs. The Panagariya–Krishna FTA, in complete analogy with the Kemp–Wan CU, freezes the external trade vector of the area, with the essential difference that the trade vector of each member country with the rest of the world is frozen at the pre-FTA level. Since, in FTAs, different member countries impose different external tariffs, it is necessary to specify a set of rules of origin to prevent a subversion of FTA tariffs by importing through the lower-tariff member country and directly trans-shipping goods to the higher-tariff country (which, if allowed, would bring the FTA arbitrarily close to a CU). The Panagariya–Krishna solution requires that all goods for which *any* value is added within the FTA are to be traded freely. Importantly, the proportion of domestic value added in final goods does not enter as a criterion in the rules of origin.

Theory thus suggests that ensuring welfare improvement requires that, along with elimination of internal barriers, external tariff vectors should eliminate trade diversion – member countries should continue to import the same amounts from the rest of the world as they did initially. There have, however, been significant departures in practice. While Article XXIV of the GATT stipulates that internal restrictions be eliminated on 'substantially all trade', the qualifier 'substantially' is vulnerable to abuse. Numerous goods are typically exempt from internal liberalization by member countries. Furthermore, restrictive rules of origin also serve to ensure a level of protection from both intra- and extra-union imports, as Krueger (1999) notes. On external tariffs, Article XXIV requires that external barriers not be more restrictive than initially. For FTAs, since countries retain individual tariff vectors, this could be taken to imply that no tariff is to rise. For CUs, since a common external tariff is to be chosen and initial tariffs on the same good likely vary across countries, the tariff vector would necessarily change for each country. The expectation is that that the 'general incidence' of trade barriers should not be higher or more restrictive than before. As Bhagwati (1993) notes, it is clear that Article XXIV's ambiguity in this regard leaves plenty of room for protectionist behaviour by member countries. The 1994 'Understanding on

the Interpretation' of Article XXIV issued by the GATT provides greater clarity on the issue of measurement and choice of the common external tariff – indicating that the GATT secretariat would compute weighted average tariff rates and duties collected in accordance with the methodology used in the assessment of tariff offers in the Uruguay Round of trade negotiations and examine trade flow and other data to arrive at suitable measures of non-tariff barriers. Nevertheless, it may be observed that leaving external barriers at their initial level and removing internal barriers do not eliminate trade diversion. Indeed, with this configuration, some trade diversion is practically guaranteed.

Preferential trade agreements and multilateral free trade

Recent analysis in the literature has focused on issues concerning the expansion of trade blocs, the endogenous determination of policy (relating to trade preferences within a PTA and extra-union trade), and the effects of preferential agreements on the multilateral trade system (that is, whether trade blocs will serve as 'building blocs' or 'stumbling blocs' in the path to multilateral free trade, in Bhagwati's, 1993, phrasing).

Krugman (1993) analyses the welfare consequences of exogenously formed and expanded trade blocs. Considering a fully symmetric structure of countries, each specialized in production in a differentiated product variety, Krugman asks how world welfare is affected by the expansion of trade blocs if member countries liberalize fully their mutual trade but apply optimal tariffs against non-members. As the (symmetric) trade blocs increase in size, their market power increases and so do the (optimal) tariffs they impose on non-members. On the other hand, increasing the number of countries within a bloc increases the volume of goods that is traded freely. Krugman finds that the net effect on world welfare is non-monotonic in bloc size. Specifically, world welfare (which is maximized with global free trade) falls as the world is divided up into trade blocs but rises again as bloc sizes decrease and the trade diversion losses (relative to trade creation gains) fall. Bond and Syropoulos (1996) show how generalizing the assumptions of Krugman's model relating to consumption preferences and the pattern of production and trade may alter the relationship between optimal tariffs and market size so that optimal tariffs fall as bloc size increases. More severely, Deardorff and Stern (1994) and Srinivasan (1993) question the robustness of Krugman's conclusions concerning non-monotonicity of welfare itself, demonstrating a substantial divergence in results when Krugman's assumptions regarding the structure of endowments and comparative advantage are changed.

A different strand of the literature has examined endogenously determined trade blocs and the internal political and economic incentives (if any) for their successive expansion. Taking the 'interest-group' approach to trade policy determination, Grossman and Helpman (1995) and Krishna (1998) both model the influence of powerful producers in considering entry into a PTA. While the models and analytic frameworks differ in detail, they come to a similar and striking conclusion, namely, that PTAs that divert trade are more likely to win internal political support. This is so because governments must respond to conflicting pressures from their exporting sectors, which gain from lower trade barriers in the partner, and from their import-competing sectors, which suffer from lower trade barriers at home, when deciding on whether to form or enter a PTA.

As Krishna (1998) argues, trade diversion effectively shifts the burden of the gain to member-country exporters from member-country import-competing sectors and onto non-member producers, who have little political clout inside the member countries. Krishna (1998) also argues that such PTAs will lower the within-union incentives for any subsequent multilateral liberalization – producers in trade-diverting PTAs may oppose multilateral reform since this would take away the gains from benefits of preferential access that they enjoyed in the PTA that diverted trade to them. Under some circumstances, the within-union incentives for further multilateral liberalization are completely eliminated.

Levy (1997) models trade policy as being determined by majority voting. Countries are assumed to differ in their endowments of factors (labour and capital). Countries are also assumed to produce different varieties of goods – so that trade reform will result in gains to individuals due to the greater number of varieties that are available for consumption. However, it should also be clear that any changes in trade policy result in changes in the distribution of income (by altering the relative rewards to the different factors of production). The arguments that emerge out of this framework are as follows. First, preferential trade integration with partners with similar relative factor endowments (that is, with similar capital–labour ratios) is more likely to receive majority support – since this results in minimal income redistribution and still provides variety gains from trade. Second, bilateral agreements could render infeasible multilateral liberalization (which, even if it brings greater variety gains, would involve trade with countries with quite different relative endowments of capital and labour and could therefore result in much more drastic income redistribution).

McLaren (2002) provides an analysis of the role of sunk costs and trade policy determination. He argues, roughly speaking, that the expectation of a preferential trading agreement could induce agents in the economy to undertake costly and irreversible investments that makes the members within the bloc more specialized towards each other and less so towards the rest of the world. In other words, they increase dependence on each other and lower it towards the rest of the world, and thus reduce their desire to liberalize trade with other countries. Exploring the first-mover advantage that member countries gain having invested in sunk costs, Freund (2000a) finds that with preferential trade member countries gain and that non-members lose relative to multilateral outcomes, with the former dominating the latter in magnitude.

A parallel literature has raised the question of what external tariffs will be chosen by member countries, examining, in particular, whether external tariffs can be expected to rise or fall following a PTA. No clear answers to this question emerge. Panagariya and Findlay (1996) find that external tariffs rise after tariff preferences are granted, as political lobbying for protection is directed away from imports from the partner country to imports from the rest of the world. Emphasizing tariff revenue competition between FTA members, Richardson (1995) find that external tariffs may fall in an FTA as welfare-minded member countries competitively reduce tariffs (so as to retain the source of extra-union imports and earn tariff revenue). In a general equilibrium context, with political lobbying over tariffs, Cadot, de Melo and Olarreaga (1999) reach a similar conclusion for FTAs, while finding that CUs are likely to raise their external tariffs. Cadot, de Melo and Olarreaga (2002) confirm these results for the case of quantity restrictions, where the protective effect that a quantity restriction imposed by a member country has in partner country markets proves central to the analysis. However, when collective action problems over lobbying for external trade policy dominate, FTAs may choose higher external tariffs than CUs, as Richardson (1994) argues. Finally, in a symmetric three-country oligopoly model, Ornelas (2005) finds that a PTA's endogenously determined external tariff may be lower than the pre-union MFN tariff. Cho and Krishna (2006) demonstrate that the opposite may obtain with asymmetric costs across partner countries, the likelihood of the external tariff being higher than the pre-union MFN tariff increasing with the inefficiency of the partner country relative to non-members.

Empirical analysis has offered mixed results as well. Bohara, Gawande and Sanguinetti (2004) report evidence of lowered external tariffs following Mercosur, while Cho (2006) finds tariffs in Mexico to be higher on average following NAFTA and systematically higher in goods in which its trading partners were inefficient suppliers (as proxied by pre-FTA export levels relative to the rest of the world). Limão (2006) examines data on US trade barriers and finds that those imported goods on which any partners received preferences were subject to smaller (subsequent) multilateral liberalization than others, suggesting a negative effect of preferences on multilateral reforms.

The economic incentives of non-member countries are considered by Baldwin (1995), who argues that PTA expansion could have 'domino' effects – increasing the size of a bloc increases the incentive for others to join it (as they then gain preferential access to increasingly large markets). On the assumption of open-membership rules (that is, insiders do not oppose the entry of new members who abide by the same rules as the members), the successive expansion of the PTA could then lead to multilateral free trade – a conclusion that is also reached by the work of Yi (1996), which develops a model of endogenous coalition formation to addresses this question.

Aghion, Antras and Helpman (2004) analyse the links between bilateral and multilateral negotiations over trade policy as a dynamic bargaining game in which a leading country endogenously decides whether to sequentially negotiate free trade agreements with subsets of countries or engage in simultaneous multilateral bargaining with all countries at once. They show that, if a coalition formed between the leading country and a follower generates a negative effect on outsiders (that is, there are negative coalitional externalities), the leader prefers sequential bargaining to multilateral bargaining. Conversely, positive coalition externalities imply that multilateral bargaining is preferred. Importantly, while political economy pressures may cause bilateral agreements to impede multilateral agreement, as in Levy (1997) and Krishna (1998), examples where bilateral agreements enable multilateral agreement are also found.

Self-enforcing trade agreements (which work by balancing any benefits that member countries may achieve by deviating from the agreement with the future losses they suffer due to punishments for the deviation) have been variously analysed in the international trade literature. Since bilateral (multilateral) agreements may alter both the benefits of deviating from an existing multilateral (bilateral) agreement and the future punishment costs of this deviation, self-enforcing agreements provide a context in which the links between preferential trade agreements and multilateralism may be studied. Bagwell and Staiger (1997a; 1997b) consider the impact of FTAs and CUs on multilateral tariff cooperation during a transition period when the exogenously agreed-upon lowering of tariffs within FTAs and CUs is implemented. Saggi (2006) shows how exogenously specified FTAs and CUs may undermine self-enforced multilateral tariff cooperation, the former by lowering the cooperation incentives of non-member countries and the latter by lowering the cooperation incentives of members. Freund (2000b) finds that exogenous multilateral liberalization may encourage and help sustain self-enforcing PTAs, thus explaining the recent trend towards bilateralism as a causal response to multilateralism. A similar causal link is explored by Cadot, de Melo and Olarreaga (2001), which views bilateral agreements as an endogenous (protective) response to the competitive pressures that domestic producers face with multilateral liberalization.

Conclusions

A half-century of research has advanced significantly our understanding of the implications of trade discrimination even if the frequently equivocal theoretical and empirical results have established among economists and policy-makers an ambivalent attitude towards preferential trade agreements. However, concerns regarding the fragmentation of the world trade system have grown with the rapid proliferation of preferential trade in recent years. Several hundred PTAs are currently in existence. Indeed, many countries belong to multiple PTAs – resulting in a confusing criss-crossing of trade preferences that Bhagwati (1995) has aptly described as 'spaghetti-bowl' regionalism. Several more preferential agreements are in process. With this inexorable erosion of non-discriminatory disciplines within the trade system, research on preferential trade is certain to remain central to the field of international trade policy for many years to come.

PRAVIN KRISHNA

See also **Mercosur; North American Free Trade Agreement (NAFTA); trade policy, political economy of.**

Bibliography

Aghion, P., Antras, P. and Helpman, E. 2004. Negotiating free trade. Working Paper No. 10721. Cambridge, MA: NBER.

Bagwell, K. and Staiger, R. 1997a. Multilateral cooperation during the formation of free trade areas. *International Economic Review* 38, 291–319.

Bagwell, K. and Staiger, R. 1997b. Multilateral cooperation during the formation of customs unions. *Journal of International Economics* 42, 91–123.

Baldwin, R. 1995. A domino theory of regionalism. In *Expanding European Regionalism: The EU's New Members*, ed. R. Baldwin, P. Haaparanta and J. Kiander. Cambridge: Cambridge University Press.

Berglas, E. 1979. Preferential trading: the n commodity case. *Journal of Political Economy* 87, 315–31.

Bhagwati, J. 1971. Trade-diverting customs unions and welfare improvement: a clarification. *Economic Journal* 81, 580–7.

Bhagwati, J. 1993. Regionalism and multilateralism: an overview. In *New Dimensions in Regional Integration*, ed. J. de Melo and A. Panagariya. Cambridge: Cambridge University Press.

Bhagwati, J. 1995. U.S. trade policy: the infatuation with free trade areas. In *The Dangerous Drift to Preferential Trade Agreements*, ed. J. Bhagwati and A. Krueger. Washington, DC: AEI Press.

Bhagwati, J. and Panagariya, A. 1996. *Free Trade Areas or Free Trade? The Economics of Preferential Trade Areas*. Washington, DC: AEI Press.

Bohara, A., Gawande, K. and Sanguinetti, P. 2004. Trade diversion and declining tariffs: evidence from Mercosur. *Journal of International Economics* 64, 65–88.

Bond, E. and Syropoulos, C. 1996. The size of trading blocs, market power and world welfare effects. *Journal of International Economics* 40, 411–37.

Cadot, O., de Melo, J. and Olarreaga, M. 1999. Regional integration and lobbying for tarriffs against non-members. *International Economic Review* 40, 635–58.

Cadot, O., de Melo, J. and Olarreaga, M. 2001. Can bilateralism ease the pains of multilateral trade liberalization? *European Economic Review* 45, 27–44.

Cadot, O., de Melo, J. and Olarreaga, M. 2002. Harmonizing external quotas in an FTA: a step backward? *Economics and Politics* 14, 259–82.

Chang, W. and Winters, A. 2002. How regional blocs affect excluded countries: the price effects of Mercosur. *American Economic Review* 92, 889–904.

Cho, M.-J. 2006. On the external trade barriers of preferential trade agreements: empirical evidence from Mexican manufacturing industries. Mimeo, Brown University.

Cho, M.-J. and Krishna, P. 2006. On the external trade barriers of preferential trade agreements. Mimeo, Brown University.

Deardorff, A. and Stern, R. 1994. Multilateral trade negotiations and preferential trading arrangements. In *Analytical and Negotiating Issues in the Global Trading System*, ed. A. Deardorff and R. Stern. Ann Arbor: University of Michigan Press.

Freund, C. 2000a. Different paths to free trade: the gains from regionalism. *Quarterly Journal of Economics* 115, 1317–41.

Freund, C. 2000b. Multilateralism and the endogenous formation of preferential trade agreements. *Journal of International Economics* 52, 359–76.

Gehrels, F. 1956–57. Customs union from a single–country viewpoint. *Review of Economic Studies* 24, 61–4.

Grossman, G. and Helpman, E. 1995. The politics of free trade arrangements. *American Economic Review* 85, 667–90.

Kemp, M. 1964. *The Pure Theory of International Trade.* Englewood Cliffs, NJ: Prentice-Hall.

Kemp, M. and Wan, H. 1976. An elementary proposition concerning the formation of customs unions. *Journal of International Economics* 6, 95–8.

Krishna, P. 1998. Regionalism and multilateralism: a political economy approach. *Quarterly Journal of Economics* 113, 227–50.

Krishna, P. 2003. Are regional trading partners 'natural'? *Journal of Political Economy* 111, 202–31.

Krishna, P. and Bhagwati, J. 1997. Necessarily welfare-improving customs unions with industrialization constraints: the Cooper–Massell–Johnson–Bhagwati proposition. *Japan and the World Economy* 154, 169–87.

Krueger, A. 1999. Free trade agreements as protectionist devices: rules of origin. In *Trade, Theory and Econometrics: Essays in Honor of John S. Chipman*, ed. J. Melvin, J. Moore and R. Riezman. Studies in the Modern World Economy, vol. 15. London: Routledge.

Krugman, P. 1991. The move to free trade zones. In *Policy Implications of Trade and Currency Zones*, Federal Reserve Bank of Kansas City. Kansas City: Federal Reserve Bank of Kansas City.

Krugman, P. 1993. Regionalism versus multilateralism: analytical notes. In *New Dimensions in Regional Integration*, ed. J. de Melo and A. Panagariya. Cambridge: Cambridge University Press.

Levy, P. 1997. A political economic analysis of free trade arrangements. *American Economic Review* 87, 506–19.

Limão, N. 2006. Preferential trade agreements as stumbling blocks for multilateral trade liberalization: evidence for the United States. *American Economic Review* 96, 896–914.

Lipsey, R. 1957. The theory of customs unions: trade diversion and welfare. *Economica* 24, 40–6.

McLaren, J. 2002. A theory of insidious regionalism. *Quarterly Journal of Economics* 117, 571–608.

Meade, J. 1955. *The Theory of Customs Unions.* Amsterdam: North-Holland.

Mundell, R. 1964. Tariff preferences and the terms of trade. *Manchester School of Economic Studies* 32, 1–13.

Ohyama, M. 1972. Trade and welfare in general equilibrium. *Keio Economic Studies* 9, 37–73.

Ornelas, E. 2005. Endogenous Free trade agreements and the multilateral trading system. *Journal of International Economics* 67, 471–97.

Panagariya, A. 1997a. The Meade model of preferential trading: history, analytics and policy implications. In *International Trade and Finance: New Frontiers for Research: Essays in Honor of Peter Kenen*, ed. B. Cohen. Cambridge: Cambridge University Press.

Panagariya, A. 1997b. Preferential trading and the myth of natural trading partners. *Japan and the World Economy* 9, 471–89.

Panagariya, A. 2000. Preferential trade liberalization: the traditional theory and new developments. *Journal of Economic Literature* 161, 316–60.

Panagariya, A. and Findlay, R. 1996. A political economy analysis of free trade areas and customs unions. In *The Political Economy of Trade Reform: Essays in Honor of Jagdish Bhagwati*, ed. R. Feenstra, D. Irwin and G. Grossman. Cambridge, MA: MIT Press.

Panagariya, A. and Krishna, P. 2002. On necessarily welfare-enhancing free trade areas. *Journal of International Economics* 57, 353–67.

Richardson, M. 1994. Why a free trade area? The tariff also rises. *Economics & Politics* 6, 79–95.

Richardson, M. 1995. Tariff revenue competition in a free trade area. *European Economic Review* 39, 1429–37.

Riezman, R. 1979. A 3 × 3 model of customs unions. *Journal of International Economics* 9, 341–54.

Saggi, K. 2006. Preferential trade agreements and multilateral tariff cooperation. *International Economic Review* 47, 29–57.

Srinivasan, T. 1993. Discussion on regionalism vs multilateralism: analytical notes. In *New Dimensions in Regional Integration*, ed. A. Panagariya and J. de Melo. Washington, DC: World Bank.

Srinivasan, T., Whalley, J. and Wooton, I. 1993. Measuring the effects of regionalism on trade and welfare. In

Regional Integration and the Global Trading System, ed. K. Anderson and R. Blackhurst. New York: St Martin's Press.

Summers, L. 1991. Regionalism and the world trading system. In *Policy Implications of Trade and Currency Zones*, Federal Reserve Bank of Kansas City. Kansas City: Federal Reserve Bank of Kansas City.

Vanek, J. 1965. *General Equilibrium of International Discrimination*. Cambridge, MA: Harvard University Press.

Viner, J. 1950. *The Customs Unions Issue*. New York: Carnegie Endowment for International Peace.

Wonnacott, P. and Lutz, M. 1987. Is there a case for free trade areas? In *Free Trade Areas and US Trade Policy*, ed. J. Schott. Washington, DC: Institute for International Economics.

Yi, S.-S. 1996. Endogenous formation of customs unions under imperfect competition: open regionalism is good. *Journal of International Economics* 41, 153–77.

Yeats, A. 1998. Does Mercosur's trade performance raise concerns about the effects of regional trade arrangements? *World Bank Economic Review* 12, 1–28.

regional development, geography of

Differences in economic activity across regions have interested economists since Adam Smith, who argued that high overland transport costs in the interior of Africa and Asia 'seem in all ages' to have had hindered economic development. However, economists' attraction to the study of spatial variations in economic activity has fluctuated over time. Standard trade theory based on comparative advantage helps to explain how the location of economic activity is affected by the spatial distribution of primary resources (such as land, labour and water), but standard trade theory says little about the interdependence of location decisions by economic agents, nor does it consider in any depth the more detailed aspects of physical geography (climate, soils, topography, disease epidemiology).

Neoclassical growth models focus on the accumulation of physical, human and technological capital, which individually or together complement raw labour and land as factors of production, but only recent theory (particularly in the work dubbed the 'new economic geography') has begun to grapple with location choices and the spatial concentration of industry (Henderson, 1988; Krugman, 1991; Fujita, Krugman and Venables, 1999). While these newer theories have contributed importantly to our understanding of why some regions develop more than others, and why cities arise and where they are located, they rarely incorporate Smith's observation that spatial differences in economic activity are also related to variations in physical geography, which intrinsically make some places more productive than others at particular points in time. Nor do they yet go into depth on regional development policy, that is, the use of economic

incentives to attract industry to one location or another. A full theory of regional development will integrate theories of agglomeration economies with physical geography and with public economics.

Theories of agglomeration

Economic activity and population around the globe are concentrated in highly dense metropolitan areas, which suggest that there is an important economic benefit of economic agglomeration (spatial co-location of economic agents). Alfred Marshall (1920) suggested that spatial concentration happens because of knowledge spillovers, larger markets for specialized skills, and backward and forward linkages associated with large local markets.

The initial literature to tackle the intractability of modelling economic geography grew from the von Thünen model (1826), which begins with the existence of a city and derives characteristics about land rents and land use surrounding the city; the resulting unplanned, efficient outcome is a concentric ring pattern of production referred to as 'von Thünen cones'. The model doesn't, however, attempt to explain the *raison d'être* of the city itself.

Later models aimed to explain why population and economic activity tend to agglomerate in the first place. Spatial concentration occurs because production is cheaper due to the large amount of nearby economic activity in agglomeration economies. These increasing returns to scale exist for several reasons: larger markets support more highly specialized products; efficiency increases as a large number of producers and consumers allows for less idle time (a source of increasing returns called demand smoothing); economies of scale of intermediate inputs make production cheaper even for sectors without increasing returns; externalities diffuse learning and expertise, as people can see each others' products and work methods; and search costs are lowered when the search process is spatially concentrated. Florida (1995) pioneered the concept of the 'learning region': to minimize transport costs and maximize learning, firms benefit from spatially concentrating their activities, and thus firms looking to augment their capabilities have strong incentive to locate in these learning regions.

New economic geography

The 'new economic geography' of recent decades grew from the Dixit and Stiglitz (1977) model of monopolistic competition under increasing returns to scale. Though admittedly a special case, this model became a workhorse in many fields, and a foundation for the new economic geography. The theoretical backbone of new economic geography is the core–periphery model in Krugman (1991), which looks at three effects: the 'market-access effect' (monopolistic firms locate in big markets and export to small markets), the 'cost-of-living effect' (cost of living is cheaper where there are more firms, due to low transport costs), and the 'market-crowding effect' (imperfectly competitive firms look to locate in regions with few competitors). The model was an important step forward in understanding spatial dynamics, but has the downside of being difficult to manipulate analytically and requires numerical simulations (instead of explicit expressions) to derive results.

Another important concept in the location of economic activity is that of clusters, especially in the work of Porter (1990; 1995; 1998a; 1998b). A cluster is a group of interconnected companies and institutions in a particular location (perhaps a city, or a state, or even a group of neighbouring countries). Companies in a cluster benefit from important complementarities, spillovers and a relationship with public institutions, which improve productivity and productivity growth, and stimulate new business formation. The important contribution of this literature is that a firm's comparative advantage (or 'competitive advantage' in the business phrase) can include characteristics outside the firm itself; often geography and location have important implications on how firms or industries can compete in the market.

One of the striking implications of the new economic geography is that spatial concentration arises in a homogeneous region, where there is no fundamental geographical advantage to locating in one place or another. The precise location of firms is accidental. Early advantages in agglomeration can lead to a snowball effect. First movers in regional development can achieve a lasting competitive advantage by attracting other mobile workers and investors. Growth proceeds with 'preferential attachment' to the places that get an early start.

The role of physical geography

In addition to the new economic geography models of agglomeration, a second basic approach seeking to shed light on growth poles and regional development is based on intrinsic geographical advantages. The assumption of homogeneous space is abandoned, and the role of coasts, hinterlands, rivers, mountains and a vast array of other geographical variables is brought to the fore. Adam Smith himself asserted that the division of labour is limited by the extent of the market, so that coastal regions, by virtue of their ability to engage in sea-based trade, enjoy a wider scope of the market than interior regions. More recently, climatic conditions have been found to have pervasive effects on regional development through disease ecology, agricultural productivity, transport costs, vulnerability to natural hazards, water stress and other factors that may affect economic performance.

Several studies (Gallup, Sachs and Mellinger, 1999; Bloom and Sachs, 1998) have noted that tropical areas are consistently poorer than temperate-zone areas, and hypothesize that this may be related to the effects of tropical ecology on human health and agricultural productivity.

Tropical infectious diseases, for example, impose very high burdens on human health that in turn may lead to shortfalls in economic performance much larger than their direct short-run effects on health. Another study (Gallup and Sachs, 2000) found that, after purchased inputs such as capital, labour and fertilizers are controlled for, the average productivity of tropical food production falls short of the productivity of temperate-zone food production. In the course of economic development, this poor performance in food productivity may have had serious adverse effects on nutrition levels, with adverse consequences for human capital accumulation, labour productivity and susceptibility to infectious disease. These geographical penalties can often be compensated by other kinds of interventions (such as malaria control or improved agronomic practices), but, since those interventions require added resources, affected regions may persistently lag behind more fortuitously located regions.

Geographical advantages can trigger subsequent agglomeration based on increasing returns to scale. The agglomeration is then self-reinforcing, even after the initial spatial advantage loses some of its importance. For example, Chicago's port is not as important as when it was the main driver of the city's growth in the middle of the 19th century. Glaeser (2005) illustrates that New York's rise in the 19th century was due to a technological change that moved ocean shipping from a point-to-point system to a hub and spoke system, and the city's geography made it the natural hub. Today, however, New York's pre-eminence is based not mainly on the port, but on the legacies of the earlier success: finance, business, remarkable infrastructure and the benefits of agglomeration.

Changing dimensions of geography

It is important to stress the changing nature of a region's geographic advantage as technology changes. In early civilizations, when transport and communications were too costly to support much interregional and international trade, regional advantage came from agricultural productivity and local transport rather than from access to oceans. As a result, early civilizations almost invariably emerged in highly fertile river valleys such as those around the Nile, Indus, Tigris, Euphrates, Yellow and Yangtze rivers. These civilizations produced high-density populations that in later eras were often disadvantaged by their remoteness from international trade. As the advantages of overland trade between Europe and Asia gave way to oceanic commerce in the 16th century and later, and as the trade routes to the Americas were discovered, economic advantage shifted from the Middle East and eastern Mediterranean to the North Atlantic. In the 19th century, the high costs of transporting coal for steam power meant that industrialization almost invariably depended on proximity to coal fields.

In the late 20th century, air transport and telecommunications have reduced the advantages of coastlines relative to hinterlands. The telecommunications sector, in particular, is deeply affecting the global division of labour and the nature of agglomeration economies. The disadvantages of interior and distant regions may well be eased or eliminated by the advances in telecommunications which allow for more disbursed production and new growth poles far from traditional trade routes. It is notable that Bangalore has become a booming centre of global information technology, despite being an inland city in southern India, and despite the weakness of India's roads and ports at the time of Bangalore's ascendancy. The examples of Bangalore and of course California's Silicon Valley show that today's competitive advantage has to do much more with the location of excellent universities and an attractive living environment for highly skilled and mobile information workers, much like the 'learning regions' described by Florida (1995).

Regional policy design

The presence of agglomeration economies, increasing returns and clusters suggests that countries can identify areas of potential growth poles and use policy tools and public investment to trigger these processes. Special policy instruments such as export-processing zones and special tax promotion schemes have helped developing countries to establish clusters in textiles and apparel, electronics, consumer appliances, software and automotive components, to name just a few industries where active industrial policy has played a hand. In the case of growth poles in the knowledge economy (such as Silicon Valley and Bangalore), the importance of government support for higher education and R&D and for the creation of science parks is especially apparent. Spillovers from military technology may play a role as well.

It is clear, however, that the successful development strategies of some countries cannot produce the same salubrious results when implemented in very different settings. When China opened some coastal pockets for foreign direct investment, these Special Economic Zones quickly blossomed into vibrant export platforms and created backward linkages with the immediate hinterland. When landlocked Mongolia turned the entire country into a free trade and investment zone in the late 1990s, however, the inflow of foreign capital was a trickle compared with China's experience, and was based mainly on primary commodities (such as copper). Even within China, the coastal provinces in the east have boomed relative to the interior provinces of western China. Physical geography therefore continues to condition economic development. Geographical determinism should be avoided, however; special geographical hindrances may well call for special compensating investments (in roads, disease control, telecommunications, and so on), or for promotion of a judicious choice of industries (those that can be sustained in the face of high transport costs, for example).

Empirical studies

Empirical evidence supports the idea that economies of scale, agglomeration forces (Davis and Weinstein, 1998; 1999; Midelfart-Knarvik et al., 2000; Overman and Puga, 2002; Hanson, 2005), and backward and forward linkages (Midelfart-Knarvik and Steen, 1999) help explain why economic activity clusters together, and that the von Thünen model helps explain economic dynamics near cities (Fafchamps and Shilpi, 2003). The traditional core–periphery model has considerable empirical support, given that the core regions of the global economy (particularly North America, Western Europe, and Japan), enjoy ever-increasing levels in productivity. At a smaller scale, studies of wages in the United States and in developing countries show that *ceteris paribus* workers earn much more in urban areas than rural areas, reflecting their higher productivity (Glaeser and Mare, 1994; Bairoch, 1988).

While looking for the presence of increasing returns to scale yields insights, it does not address the constraints physical geography may place upon economic growth. For example, Adam Smith's observations on the role of access to navigable water still hold. Cross-country empirical research affirms that the level and growth rate of per capita income continue to be strongly positively correlated with geographic variables such as climate and coastal proximity (Gallup, Sachs and Mellinger, 1999; Mellinger, Sachs and Gallup, 2000), while within-country differences in growth rates in India and China are clearly related to geography as well (Demurger et al., 2002; Sachs, Bajpai and Ramiah, 2002). Smith's observations also implicitly underscore the highly favourable economic geography enjoyed by the nations of Western Europe. Extensive ocean shorelines host a succession of natural harbours, and numerous navigable rivers penetrate deep into the interior. In addition, despite the large landmass of the United States, 57 per cent of income was generated in counties within 80 km from the coast, though these counties account for only 13 per cent of land mass (Rappaport and Sachs, 2003).

Future theoretical and empirical work in understanding regional development should aim to disentangle the forces of differential geography and self-organizing agglomeration economies. Policy studies should examine in depth how regional development policy has been used in the past, and which instruments are particularly important. Economists and business specialists should aim to provide new tools to help specific regions identify appropriate instruments for regional development, including which kinds of industries are likely to flourish in which kinds of spatial settings.

JEFFREY D. SACHS AND GORDON MCCORD

See also **location theory; Marshall, Alfred; new economic geography; spatial economics; Thünen, Johann Heinrich von; urban agglomeration.**

Bibliography

Bairoch, P. 1988. *Cities and Economics Development: From the Dawn of History to the Present.* Chicago: University of Chicago Press.

Baldwin, R., Forslid, R., Martin, P., Ottaviano, G. and Robert-Nicoud, F. 2003. *Economic Geography and Public Policy.* Princeton: Princeton University Press.

Bloom, D. and Sachs, J.D. 1998. Geography, demography, and economic growth in Africa. *Brookings Papers on Economic Activity* 1998(2), 207–95.

Clark, G., Feldman, M.P. and Gertler, M. 2000. *The Oxford Handbook of Economic Geography.* Oxford: Oxford University Press.

Davis, D.R. and Weinstein, D.E. 1998. Market access, economic geography, and comparative advantage: an empirical assessment. Working Paper No. 6787. Cambridge, MA: NBER.

Davis, D. and Weinstein, D.E. 1999. Economic geography and regional production structure: an empirical investigation. *European Economic Review* 43, 379–407.

Demurger, S., Sachs, J.D., Wing, T.-W., Bao, S., Chang, G. and Mellinger, A. 2002. Geography, economic policy, and regional development in China. *Asian Economic Papers* 1, 146–97.

Dixit, A.K. and Stiglitz, J.E. 1977. Monopolistic competition and optimum product diversity. *American Economic Review* 67, 297–308.

Dunning, J.H. (ed.) 2002. *Regions, Globalization, and the Knowledge-Based Economy.* New York: Oxford University Press.

Fafchamps, M. and Shilpi, F. 2003. The spatial division of labor in Nepal. *Journal of Development Studies* 39(6), 23–66.

Florida, R. 1995. Towards the learning region. *Futures* 27, 527–36.

Fujita, M., Krugman, P. and Venables, A.J. 1999. *The Spatial Economy: Cities, Regions, and International Trade.* Cambridge, MA: MIT Press.

Gallup, J.L. and Sachs, J.D. 2000. Agriculture, climate, and technology: why are the tropics falling behind? *American Journal of Agricultural Economics* 82, 731–77.

Gallup, J.L., Sachs, J.D. and Mellinger, A.D. 1999. Geography and economic development. *International Regional Science Review* 22, 179–232.

Glaeser, E.L. 2005. Urban colossus: Why is New York America's largest city? Working Paper No. 11398. Washington, DC: NBER.

Glaeser, E. and Mare, D. 1994. Cities and skills. Working Paper No. E94-11, Hoover Institution.

Hanson, G.H. 2005. Market potential, increasing returns, and geographic concentration. *Journal of International Economics* 67, 1–24.

Henderson, J.V. 1988. *Urban development: Theory, Fact, and Illusion.* Oxford: Oxford University Press.

Henderson, J.V. and Thisse, J.-F. (eds.) 2004. *Handbook of Regional and Urban Economics*, vol. 4. Amsterdam: North-Holland.

Krugman, P. 1991. Increasing returns and economic geography. *Journal of Political Economy* 99, 483–99.

Marshall, A. 1920. *Principles of Economics,* 8th edn. London: Macmillan.

Mellinger, A.D., Sachs, J.D. and Gallup, J.L. 2000. Climate, coastal proximity, and development. In *Oxford Handbook of Economic Geography*, ed. G.L. Clark, M.P. Feldman and M.S. Gertler. Oxford: Oxford University Press.

Midelfart-Knarvik, K.H., Overnman, H.G., Redding, S.J. and Venables, A.J. 2000. The location of European industry. Economic Papers No. 142. European Commission Directorate-General for Economic and Financial Affairs.

Midelfart-Knarvik, K. and Steen, F. 1999. Self-reinforcing agglomerations? An empirical industry study. *Scandinavian Journal of Economics* 101, 515–32.

O'Flaherty, B. 2005. *City Economics.* Cambridge, MA: Harvard University Press.

Overman, H.G. and Puga, D. 2002. Unemployment clusters across European regions and countries. *Economic Policy* 34, 115–47.

Porter, M. 1990. *The Competitive Advantage of Nations.* New York: Free Press.

Porter, M. 1995. The competitive advantage of the inner city. *Harvard Business Review* 73, 55–71.

Porter, M. 1998a. Clusters and competition: new agendas for companies, governments and institutions. In *On Competition.* Boston: Harvard Business School Press.

Porter, M. 1998b. The microeconomic foundations of economic development. In World Economic Forum, *Global Competitiveness Report 1998.* Geneva: WEF.

Rappaport, J. and Sachs, J.D. 2003. The United States as a coastal nation. *Journal of Economic Growth* 8, 5–46.

Sachs, J.D., Bajpai, N. and Ramiah, A. 2002. Understanding regional economic growth in India. *Asian Economic Papers* 1(3), 32–62.

Saxenian, A.L. 2006. *The New Argonauts: Regional Advantages in a Global Economy.* Cambridge, MA: Harvard University Press.

von Thünen, J.H. 1826. *Der Isolierte Staat in Beziehung auf Landtschaft und Nationalökonomie.* Hamburg. Trans V.M. Wartenberg as *Von Thünen's Isolated State.* Oxford: Pergamon Press, 1966.

regression-discontinuity analysis

The regression discontinuity (RD) data design is a quasi-experimental evaluation design first introduced by Thistlethwaite and Campbell (1960) as an alternative approach to evaluating social programmes. The design is characterized by a treatment assignment or selection rule which involves the use of a known *cut-off* point with respect to a continuous variable, generating a discontinuity in the probability of treatment receipt at that point. Under certain comparability conditions, a comparison of average outcomes for observations just left and right of the cut-off can be used to estimate a meaningful causal impact. While interest in the design had previously been mainly limited to evaluation research methodologists (Cook and Campbell, 1979; Trochim, 1984), the design is currently experiencing a renaissance among econometricians and empirical economists (Hahn, Todd and van der Klaauw, 1999; 2001; Angrist and Krueger, 1999; Porter, 2003). Among the main econometric contributions have been the formal derivation of identification conditions for causal inference and the introduction of semiparametric estimation procedures for the design. At the same time, a large and rapidly growing number of empirical applications are providing new insights into the applicability of the design, which have led to the development of several sensitivity and validity tests.

The popularity of the RD design in applied economic research can be linked to several of its features. First, the assignment rules in many existing programmes and procedures for allocating social resources, frequently lend themselves to RD evaluations. In many cases, programme resources are allocated based on some type of formula that has a cut-off structure. One area of economic research where the design has proven especially fruitful in recent years has been the evaluation of educational interventions. Education programmes are frequently assigned to schools or students who score below a cut-off on some scale (student performance, poverty), and school and programme funding decisions are often based on allocation formulas containing discontinuities. Similarly, the design has proven useful in evaluating the socioeconomic impacts of a diverse set of government programmes and laws, many of which use eligibility cutoffs or funding formulas involving thresholds in allocating scarce resources to those potential recipients who need or deserve them most (see van der Klaauw, 2007a). A second attractive feature of the design is that it is intuitive and its results can be easily communicated, often with a visual portrayal of sharp changes in both treatment assignment and average outcomes around the cut-off value of the assignment variable (Bloom et al., 2005). Third, a researcher can choose from among several different methods to estimate effects that have credible causal interpretations (Hahn, Todd and van der Klaauw, 2001).

Consider the general problem of evaluating the impact of a binary treatment on an outcome variable, using a random sample of individuals where for each individual i we observe an outcome measure y_i and a binary treatment indicator t_i, equal to one if treatment was received and zero otherwise. The evaluation problem that arises in determining the effect of t on y, is due to the fact that each individual either receives or does not receive treatment and is never observed in both states. Let $y_i(1)$ be the outcome given treatment, and $y_i(0)$ the outcome in absence of treatment. Then the actual outcome we observe equals $y_i = t_i y_i(1) + (1 - t_i) y_i(0)$. A common regression model representation for the observed outcome can then be written as

$$y_i = \beta + \alpha_i t_i + u_i \qquad (1)$$

where $\alpha_i = y_i(1) - y_i(0)$ and $y_i(0) = E[y_i(0)] + u_i = \beta + u_i$. Non-random assignment or selection into treatment implies that a comparison of average outcomes of treatment recipients and non-recipients $(E[y_i(1)|t_i = 1]$ and $E[y_i(0)|t_i = 0])$ would generally not provide us with a valid treatment effect estimate.

Hahn, Todd and van der Klaauw (HTV) (2001) analysed the conditions under which a discontinuity in the treatment assignment or selection rule can be exploited to solve the selection bias problem and to identify a meaningful causal effect. Following Trochim (1984) they distinguish between two different forms of the design, depending on whether the treatment assignment is related to the assignment variable by a deterministic function (*sharp design*) or a stochastic one (*fuzzy design*). In the case of a sharp RD design, individuals are assigned to or selected for treatment solely on the basis of a cut-off score on an observed continuous variable x. This variable, alternatively called the assignment, selection, running or ratings variable, could represent a single characteristic or a composite variable constructed using several characteristics. Those who fall below some distinct cutoff point $\bar{x}$ are placed in the control group $(t_i = 0)$, while those on or above that point are placed in the treatment group $(t_i = 1)$ (or vice versa). Thus, assignment occurs through a known and measured deterministic decision rule: $t_i = t(x_i) = 1\{x_i \geq \bar{x}\}$ where $1\{.\}$ is the indicator function. As the assignment variable itself may be correlated with the outcome variable, the assignment mechanism is clearly not random.

However, if we have reason to believe that persons close to the threshold with very similar x values are comparable, then we may view the design as almost experimental near $\bar{x}$, suggesting that we could evaluate the causal impact of treatment by comparing the average outcome for those with ratings just above to those with ratings just below the cutoff. More formally, consider the following *local continuity* (*LC*) assumption:

$E[u_i|x]$ and $E[\alpha_i|x]$ are continuous in x at $\bar{x}$, or equivalently, $E[y(1)|x]$ and $E[y(0)|x]$ are continuous at $\bar{x}$,

then on the assumption that the density of x is positive in a neighbourhood containing $\bar{x}$,

$$\lim_{x \downarrow \bar{x}} E[y_i|x] - \lim_{x \uparrow \bar{x}} E[y_i|x] = \lim_{x \downarrow \bar{x}} E[\alpha_i t_i|x]$$
$$- \lim_{x \uparrow \bar{x}} E[\alpha_i t_i|x] + \lim_{x \downarrow \bar{x}} E[u_i|x]$$
$$- \lim_{x \uparrow \bar{x}} E[u_i|x] = E[\alpha_i|x = \bar{x}].$$

$$(2)$$

The RD approach therefore identifies the average treatment effect for individuals close to the discontinuity point. Note that the continuity assumption formalizes the idea that individuals just above and below the cut-off need to be 'comparable', requiring them to have similar average potential outcomes when receiving and when not receiving treatment. While in the absence of additional assumptions (such as a common effect assumption) one could learn about treatment effects only for a sub-population of persons near the discontinuity point, as pointed out by HTV this local effect is highly relevant to policymakers who are contemplating less restrictive eligibility rules and marginal expansions of programmes via a change in the cut-off.

The continuity assumption required for identification is not innocuous. Even if treatment receipt is determined solely on the basis of a cut-off score on the assignment variable, this is not a sufficient condition for the identification of a meaningful causal effect. The continuity assumption rules out coincidental functional discontinuities in the $x - y$ relationship such as those caused by other programmes employing assignment mechanisms based on the exact same assignment variable and cut-off. In addition, the continuity restriction generally rules out certain types of behaviour both on the part of potential treatment recipients who exercise control over their value of x and programme administrators in choosing the assignment variable and cut-off point. Lee (2007) analyses the conditions under which an ability to manipulate the assignment variable may invalidate the RD identification assumptions. He shows in the context of a sharp RD design that as long as individuals do not have *perfect* control over the position of the assignment variable relative to the cut-off score, the continuity assumption will be satisfied.

While in the sharp RD design treatment assignment is known to depend on the selection variable x in a deterministic way, in the case of a fuzzy design (Campbell, 1969), treatment assignment depends on x in a stochastic manner but in such a way that the propensity score function $\Pr(t = 1|x)$ is again known to have a discontinuity at $\bar{x}$. Instead of a $0 - 1$ step function, the selection probability as a function of x would now contain a jump smaller than 1 at $\bar{x}$. The fuzzy design can occur in case of misassignment relative to the cut-off value in a sharp design, with values of x near the cut-off appearing in both treatment and control groups. This situation is analogous to having no-shows (treatment group members who do not receive treatment) and/or crossovers (control group member who do receive the treatment) in a randomized experiment. This could occur if in addition to the position of the individual's score relative to the cut-off value, assignment is based on additional variables observed by the administrator, but unobserved by the evaluator.

A comparison of average outcomes of recipients and non-recipients, even if near the cut-off, would not generally lead to correct inferences regarding an average treatment effect. However, as shown by HTV, one can again exploit the discontinuity in the selection rule to identify a causal impact of interest by noting that under the LC assumption and with a locally constant treatment

effect ($\alpha_i = \alpha$ in a neighbourhood around $\bar{x}$), the treatment effect α is identified by

$$\frac{\lim_{x \downarrow \bar{x}} E[y_i|x] - \lim_{x \uparrow \bar{x}} E[y_i|x]}{\lim_{x \downarrow \bar{x}} E[t_i|x] - \lim_{x \uparrow \bar{x}} E[t_i|x]}, \qquad (3)$$

where the denominator is always non-zero because of the known discontinuity of $E[t|x]$ at $\bar{x}$.

In the case of varying treatment effects, HTV show that under the local continuity assumption, and a local conditional independence assumption requiring t_i to be independent of α_i conditional on x near $\bar{x}$, the ratio above identifies $E[\alpha_i|x = \bar{x}]$, the average treatment effect for cases with values of x close to $\bar{x}$. The conditional independence assumption is a strong assumption which may be violated if individuals self-select into or are selected for treatment on the basis of expected gains from treatment. HTV show that, under a weaker local monotonicity assumption similar to that assumed by Imbens and Angrist (1994), the ratio (3) will instead identify a local average treatment effect (LATE) at the cut-off point, which represents the average treatment effect of the 'compliers', that is, the subgroup of individuals whose treatment status would switch from non-recipient to recipient if their score x crossed the cut-off. More recently Battistin and Rettore (2003) considered the special case where an eligibility rule divides the population into eligibles and non-eligibles according to a sharp RD design, and with eligible individuals self-selecting into treatment. In this case the LC assumption alone is sufficient for the ratio to identify $E[\alpha_i|t_i = 1, x = \bar{x}]$, the average treatment effect on the treated, for those near the cut-off.

As indicated by these identification results, estimation of treatment effects in an RD design involves estimating boundary points of conditional expectation functions. The most common empirical strategy in the literature has been to adopt parametric specifications for the conditional expectations functions. Consider the following alternative representation of outcome eq. (1) in case of a sharp RD design:

$$y_i = m(x_i) + \delta t_i + e_i, \qquad (4)$$

where $e_i = y_i - E[y_i|t_i, x_i]$, $t_i = 1\{x_i \geq \bar{x}\}$, $m(x) = E[u_i|x] + (E[\alpha_i|x] - E[\alpha_i|\bar{x}])1\{x \geq \bar{x}\}$. Then under the local continuity assumption $m(x)$ will be a continuous function of x at $\bar{x}$, and $\delta = E[\alpha_i|\bar{x}]$ (the average treatment effect at $\bar{x}$) will measure the discontinuity in the average outcome at the cut-off. This suggests that if the correct specification of $m(x)$ were known, and was included in the regression, we could consistently estimate the treatment effect for the sharp RD design. This idea of including a specification of $m(x)$ in the regression of y on t in order to correct for selection bias caused by selection on observables, is in the econometrics literature known as the *control function approach* (Heckman and Robb, 1985). A popular choice among empirical researchers has been to use global polynomials or to use splines (piecewise polynomials) which, even though globally continuous, have a knot at the cut-off (Trochim, 1984; van der Klaauw, 2002; McCrary, 2007).

In the case of a fuzzy RD design, when assuming local independence of t_i and α_i conditional on x, then in a neighbourhood of $\bar{x}$,

$$y_i = m(x_i) + \delta E[t_i|x_i] + w_i, \qquad (5)$$

where $w_i = y_i - E[y_i|x_i]$ and $m(x) = E[u_i|x] + (E[\alpha_i|x] - E[\alpha_i|\bar{x}])E[t|x]$. With the local continuity assumption again implying that $m(x)$ will be continuous at the cut-off, and with $E[t_i|x_i]$ being discontinuous at $\bar{x}$, δ in this regression will measure the ratio in (3), which in this case equals the average local treatment effect $E[\alpha_i|\bar{x}]$. Similarly, δ can be interpreted as a local average treatment effect if we replaced the local independence assumption with the local monotonicity condition of Imbens and Angrist (1994).

This naturally leads to the two-stage procedure adopted by van der Klaauw (2002), where in the first stage we estimate the propensity score function specified as $t_i = E[t_i|x_i] + v_i = f(x_i) + \gamma 1\{x_i \geq \bar{x}\} + v_i$ where $f(\cdot)$ is continuous in x at $\bar{x}$ and γ measures the discontinuity in the propensity score function at $\bar{x}$. In the second stage the control function-augmented outcome equation is then estimated with t_i replaced by the first-stage estimate of $E[t_i|x_i] = \Pr[t_i = 1|x_i]$ as in Maddala and Lee (1976). With correctly specified $f(x)$ and $m(x)$ functions, this two-stage procedure yields a consistent estimate of the treatment effect. The approach is similar in spirit to those proposed earlier in the RD evaluation literature by Spiegelman (1979) and Trochim and Spiegelman (1980). Note that in case of a parametric approach, if we assume the same functional form for $m(x)$ and $f(x)$, then the two-stage estimation procedure described here will be equivalent to two-stage least squares (in case of linear-in-parameter specifications) with $1\{x_i \geq \bar{x}\}$ and the terms in $m(x)$ serving as instruments. Because of the popularity of this particular parametrization, the RD approach is often interpreted as being equivalent to an instrumental variable approach, as it implicitly imposes an exclusion restriction by excluding $1\{x_i \geq \bar{x}\}$ as a variable in the outcome equation.

Valid parametric inference for the estimation of the treatment effect requires a correct specification of the control function $m(x)$ and of $f(x)$ in the treatment equation. To mitigate the potential for misspecification bias, several semiparametric estimation procedures have been proposed for estimating $m(x)$ and $f(x)$, or equivalently for estimating the limits $\lim_{x\downarrow\bar{x}} E[z|x]$ and $\lim_{x\uparrow\bar{x}} E[z|x]$ in (3) semiparametrically. These methods rely on less-restrictive smoothness conditions away from the discontinuity, with estimates based mainly on data in a neighbourhood on either side of the cut-off point. Asymptotically this neighbourhood needs to shrink, as with usual nonparametric estimation, implying that we should expect a slower than parametric rate of convergence in estimating treatment impacts. HTV considered the use of kernel and local linear regression estimators, while Porter (2003) proposed estimating the limits using local polynomial regression and partially linear model estimation. RD estimators based on local polynomial regression and partially linear model estimation have better boundary behaviour than the kernel-based estimator and as shown by Porter, achieve the optimal rate of convergence. This result is based on a known degree of smoothness of the conditional expectation functions. Sun (2005) proposed an adaptive estimator to first estimate the degree of smoothness in the data prior to implementing either estimator.

The internal validity of the RD approach relies on the local continuity of conditional expectations of potential outcomes around the discontinuity point. While this assumption is fundamentally untestable, a number of validity tests have been developed to bolster the credibility of the RD design. First, economic behaviour may lead to sorting of individuals around the cut-off point, where those below the cut-off may differ on average from those just above the cut-off. Such precise sorting around the cut-off would generally be accompanied by a discontinuous jump in the density of the assignment variable at the cutoff. Several approaches have been used for assessing this possibility (McCrary, 2007; Lee, 2007; Chen and van der Klaauw, 2007; Lemieux and Milligan, 2004). Second, one can test for evidence that individuals on either side of the cut-off are observationally similar by directly comparing average characteristics (McEwan and Urquiola, 2005) or by repeating the RD analysis treating the characteristics as outcome variables (van der Klaauw, 2007b). Alternatively, one can test for an imbalance of relevant characteristics by assessing the sensitivity of RD estimates to the inclusion of observed characteristics as controls (van der Klaauw, 2002; Lee, 2007). Third, in some applications data are available from a baseline period in which the programme did not yet exist, or for a group of individuals that was not eligible for treatment. In such a case the credibility of the design can be significantly enhanced by repeating the RD analysis with such data. Finding a zero treatment effect in such a falsification test would suggest that a non-zero post-programme effect was not an artifact of the specific RD model specification, estimation approach chosen or caused by another programme using the same cut-off and assignment variable.

Finally, while this exposition has focused on the binary treatment case with a selection rule containing a single discontinuity at a known cut-off, the approach can be readily extended to one where there are multiple treatment dose levels and multiple cut-off or 'cut-off ranges' within which the treatment dose varies continuously (van der Klaauw, 2007a). Similarly, the approach can be modified to cover cases where the assignment or selection variable is discrete instead of continuous (Lee and Card, 2006).

WILBERT VAN DER KLAAUW

See also **causality in economics and econometrics; natural experiments and quasi-natural experiments; propensity score; selection bias and self-selection; semiparametric estimation; treatment effect.**

Bibliography

Angrist, J.D. and Krueger, A.B. 1999. Empirical strategies in labor economics. In *Handbook of Labor Economics*, vol. 3, ed. O. Ashenfelter and D. Card. Amsterdam: North-Holland.

Battistin, E. and Rettore, E. 2003. Another look at the regression discontinuity design. Working Paper No. CWP01/03, CeMMAP, Institute for Fiscal Studies.

Bloom, H.S., Kemple, J., Gamse, B. and Jacob, R. 2005. Using regression discontinuity analysis to measure the impacts of reading first. Paper presented at the Annual Conference of the American Educational Research Association, Montreal, Canada, April.

Campbell, D.T. 1969. Reforms as experiments. *American Psychologist* 24, 409–29.

Chen, S. and van der Klaauw, W. 2007. The work disincentive effects of the disability insurance program in the 1990s. *Journal of Econometrics*.

Cook, T.D. and Campbell, D.T. 1979. *Quasi-Experimentation: Design and Analysis Issues for Field Settings*. Boston: Houghton-Mifflin.

Hahn, J., Todd, P. and van der Klaauw, W. 1999. Evaluating the effect of an antidiscrimination law using a regression-discontinuity design. Working Paper No. 7131. Cambridge, MA: NBER.

Hahn, J., Todd, P. and van der Klaauw, W. 2001. Identification and estimation of treatment effects with a regression-discontinuity design. *Econometrica* 69, 201–09.

Heckman, J.J. and Robb, R. 1985. Alternative methods for evaluating the impact of interventions. In *Longitudinal Analysis of Labor Market Data*, ed. J. Heckman and B. Singer. New York: Cambridge University Press.

Imbens, G.W. and Angrist, J. 1994. Identification and estimation of local average treatment effects. *Econometrica* 62, 467–76.

Lee, D.S. 2007. Randomized experiments from non-random selection in U.S. house elections. *Journal of Econometrics*. (forthcoming).

Lee, D.S. and Card, D. 2006. Regression discontinuity inference with specification error. Technical Working Paper No. 322. Cambridge, MA: NBER.

Lemieux, T. and Milligan, K. 2004. Incentive effects of social assistance: a regression discontinuity approach. Working Paper No. 10541. Cambridge, MA: NBER.

Maddala, G.S. and Lee, L. 1976. Recursive models with qualitative endogenous variables. *Annals of Economic and Social Measurement* 5, 525–45.

McCrary, J. 2007. Testing for manipulation of the running variable in the regression discontinuity design. *Journal of Econometrics*.

McEwan, P.J. and Urquiola, M. 2005. Economic behavior and the regression-discontinuity design: evidence from class size reduction. Working paper, Columbia University.

Porter, J. 2003. Estimation in the regression discontinuity model. Unpublished manuscript, Harvard University.

Spiegelman, C.H. 1979. Estimating the effect of a large scale pretest posttest social program. *Proceedings of the Social Statistics Section, American Statistical Association*, pp. 370–3.

Sun, Y. 2005. Adaptive estimation of the regression discontinuity model. Working paper, University of California, San Diego.

Thistlethwaite, D. and Campbell, D. 1960. Regression-discontinuity analysis: an alternative to the ex post facto experiment. *Journal of Educational Psychology* 51, 309–17.

Trochim, W.K. 1984. *Research Design for Program Evaluation: The Regression-Discontinuity Approach*. Beverly Hills, CA: Sage.

Trochim, W. and Spiegelman, C.H. 1980. The relative assignment variable approach to selection bias in pretest-posttest group designs. *Proceedings of the Survey Research Section, American Statistical Association*, pp. 376–80.

van der Klaauw, W. 2002. Estimating the effect of financial aid offers on college enrollment: a regression-discontinuity approach. *International Economic Review* 43, 1249–87.

van der Klaauw, W. 2007a. Regression-discontinuity analysis: a survey of recent developments in economics. Unpublished manuscript, Federal Reserve Bank of New York.

van der Klaauw, W. 2007b. Breaking the link between poverty and low student achievement: an evaluation of title I. *Journal of Econometrics*.

regular economies

General equilibrium theory describes those states of an economy in which the individual plans of many agents with partially conflicting interests are compatible with each other. Such a state is called an equilibrium. The concept of an equilibrium simply being based on a consistency requirement lends itself to the study of specific questions of quite different character. Indeed, equilibrium theory provides a unifying framework for the analysis of questions arising in various branches of economic theory. In our opinion it is fruitful to view equilibrium theory as a method of thinking applicable to a variety of problems of different origin.

Ideally one would like to have general principles which ensure that equilibria exist, that they are unique, and that, therefore, the equilibria resulting from different policy measures can unequivocally be compared. Moreover, one would like to know whether equilibria have some desirable properties when no single agent can exert an essential influence on the global outcome to his personal advantage. These welfare questions are particularly interesting because the concept of an equilibrium itself is not based on the well-being of the economic agents. Finally, although the concept of an equilibrium as described above is static in nature, one would like to have a dynamic theory according to which some equilibrium is approached in the course of time.

These and related questions such as the computability of equilibria have been studied in the past with different degrees of success. There are general principles which yield the existence of an equilibrium in an astonishingly large variety of cases. Furthermore, the welfare properties of equilibria are well understood. However, it is easy to construct examples of economies with an infinite number of equilibria and it appears to be very difficult to provide conditions which lead, without being artificial or and hoc, to the uniqueness of an equilibrium. As a consequence, comparative statics does not have a basis which makes it generally a well-defined problem. Also, the difficulties encountered when studying the uniqueness issue present severe obstacles for the development of a dynamic theory.

The theory of regular economies may be viewed as an effort to advance general equilibrium theory in the absence of a satisfactory uniqueness result. The seminal paper is Debreu (1970). Debreu explicitly allows for the multiplicity of equilibria. However, he requires each equilibrium to be *locally* unique. Each equilibrium is well determined and robust in the sense that it is not destroyed by a small change in the parameters.

A regular economy is an economy with a certain, finite number of equilibria, all of which respond continuously to small parameter changes. Hence each of these equilibria can be traced for some while during a parameter change. Thus there is a basis for doing comparative statics locally, that is to say as long as the equilibrium under consideration stays robust. If, at a certain point, it ceases to be robust, a drastic change is to be expected, the size and direction of which are probably hardly predictable. The focus of the theory of regular equilibria is more on the continuous behaviour of robust equilibria than on drastic changes.

It is most remarkable that Debreu (1970), by using concepts and techniques developed in the mathematical field of differential topology, has introduced a new kind of thought into economic analysis. In the meantime this way of thinking has penetrated many areas of economic theory at different levels. One of the first applications has occurred in the technically advanced area of core theory, where the continuous dependence of the set of price equilibria on the characteristics of the agents, which is guaranteed in a regular economy, plays an important role. An application on a purely conceptual level in oligopoly theory is incorporated in the notion of a demand function which an oligopolist faces in the Cournot–Nash context. The graph of this function is considered as given by the equilibria of an exchange economy with initial endowments as varying parameters.

The dependence of the equilibria on initial endowments will be discussed in detail in the next section because this case is particularly suited to illustrate basic ideas of the theory of regular economies.

Debreu's theorem on regular equilibria

The purpose of this section is to describe the kind of reasoning typical for the theory of regular economies in a prototypical situation. It is desirable to deal with parameter variations taking place in some Euclidean space because the mathematical structures to be used are most familiar in this case. We shall study exchange economies which differ by the allocation of initial endowments.

There are l commodities and m consumers. Individual initial endowments are supposed to be positive in each component. If we denote the strictly positive orthant in $\mathbb{R}^l$ by P, then an initial allocation is a vector $(e_1, \ldots, e_m) \in P^m$. Since the demand function f_i of each consumer i is considered as fixed, an economy E is fully specified by $(e_1, \ldots, e_m)$. The space of all economies under consideration can thus be identified with P^m, an extremely simple subset of a Euclidean space. We want to examine how the exchange equilibria of an economy–there may be several such equilibria–depend on the particular economy $E \in P^m$.

We assume that all goods are desired so that attention may be restricted to strictly positive relative prices. Price systems are normalized; to be specific we consider price systems in

$$S = \left\{ p = (p_1 \ldots p_2) \gg 0 \,\big|\, \|p\| = \left(\sum_{k=1}^{l} p_k^2 \right)^{1/2} = 1 \right\}.$$

If consumer i initially possesses the commodity bundle e_i, his demand at the price system p is $f_i(p, p \cdot e_i) \in \mathbb{R}_+^l$, where $p \cdot e_i = w_i > 0$ is i's wealth. Hence the aggregate excess demand of the economy E, given by the initial allocation $(e_1, \ldots, e_m) \in P^m$, at p is

$$Z_E(p) = \sum_{i=1}^{m} \left[f_i(p, p \cdot e_i) - e_i \right].$$

We assume Walras's Law which states that the value $p \cdot Z_E(p)$ of the excess demand is identically equal to zero. Furthermore, every f_i is supposed to be continuous.

The desirability of all commodities will be captured in the following condition, which is always satisfied when consumers have strictly monotone preferences.

(D) If the price of at least one good approaches zero and the wealth $w_i > 0$ of every agent stays away from zero, then

$$\sum_{i=1}^{m} \|f_i(p, w_i)\|$$

tends to infinity.

An *equilibrium price system* of E is a price system $p \in S$ at which the consumption plans $f_i(p, p \cdot e_i)$ of all agents are consistent, i.e. a zero of the excess demand function Z_E. It is not difficult to show the following consequence of the desirability assumption (D) by a fixed point argument:

Every economy $E \in P^m$ has at least one equilibrium if (D) *holds.*

We would like to know how the equilibrium prices vary when the initial allocation is modified. Therefore we look at the graph Γ of the correspondence ('multi-valued function') Π which assigns to every economy $E \in P^m$ its set $\{p \in S + Z_E(p) = 0\}$ of equilibrium price systems. Defining $Z : P^m \times S \to \mathbb{R}^l$ by $Z(E, p) = Z_E(p)$ we get

$$\text{graph } (\Pi) = \Gamma = Z^{-1}(0).$$

Since Z is a continuous function, Γ is a closed set. It is well known that, in the case of a (single-valued) function, the closedness of the graph is intimately related to the continuity of the function. Here, where Π is a correspondence rather than a function, we obtain the following continuity result: *the graph Γ of the equilibrium price correspondence Π is upper hemi-continuous and compact-valued, if* (D) *holds.*

This is tantamount to the following explicit statement. If (E_n) is a sequence of economies in P^m converging to $E \in P^m$ and if $p_n \in \Pi(E_v)$ is an equilibrium price system to E_n for all n, then the sequence (p_n) has a subsequence which converges to an equilibrium price system of the limit economy E, provided (D) holds.

To improve our understanding of Γ, we assume that the demand functions f_i are continuously differentiable (C^1 for short) and we invoke the implicit function theorem in the following manner. Walras's Law allows us to disregard one market, say the lth, and to concentrate on

$$\hat{Z} : P^m \times S \to \mathbb{R}^{l-1}$$

which is obtained from Z by deleting the last component. Let p be an equilibrium price system of E, i.e. $\hat{Z}(E, p) = 0$. A simple calculation yields that the derivative $d\hat{Z}(E, p)$ has maximal rank at (E, p). Therefore, the graph Γ is given by a smooth surface of dimension lm. That is to say each point in Γ has a neighbourhood in Γ which can be mapped onto an open subset of $\mathbb{R}^{lm}$ by a C^1 diffeomorphism, i.e. a C^1 map with a C^1 inverse. Such a locally Euclidean space is called a C^1 manifold: see Figure 1.

We have seen that *the graph Γ of the equilibrium price correspondence Π is a C^1 manifold*, but Figure 1 suggests more. In Figure 1, Γ is not only locally Euclidean, there is even a global diffeomorphism between Γ and $\mathbb{R}^{lm}$. Indeed, one can show that this global equivalence holds (see Balasko, 1975).

The equilibrium price correspondence is continuous except at two points, E_1 and E_2. If a parameter variation leads through E_1 (or E_2) the equilibrium may be forced to jump. The equilibrium reached after the jump, however, is robust in the sense that no sudden change must occur

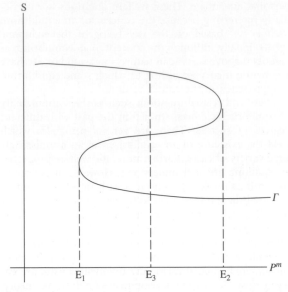

Figure 1

when one passes through E_1 (or E_2) again. One can imagine a situation such as in E_1 takes place when a slight reduction in the supply of an important raw material leads to a drastic increase in its price. If later on the supply begins to increase again prices perhaps vary but stay at their high level. A reversion of this phenomenon may occur when the supply has reached the much higher level corresponding to E_2.

In Figure 2 we have drawn a two-dimensional parameter space. The following remarkable phenomenon may happen here.

There are two paths, A and B, in the parameter space which have their starting point and endpoint in common. Following either path there is no need for the equilibrium to jump. However, the two equilibria reached at the end are quite different equilibria of the same economy. In other words, if two or more policy variables are at one's disposal one must be aware of the possibility that the final outcome depends very well on the order in which the variables are utilized.

The economies E_1 and E_2 in Figure 1 are characterized by the fact that the graph Γ has a vertical tangent above E_1 and above E_2. Similarly, in Figure 2, Γ has vertical tangents above all points on the cusp drawn in the bottom plane, which represents P^m. Apparently qualitative changes of the equilibrium price set at an economy E are associated with vertical tangents of Γ above E. This motivates the following definitions. A *critical point* of the projection pr: $\Gamma \to P^m$ is a point in Γ at which the derivative of pr has rank less than dim $P^m = lm$. A *critical value* of pr: $\Gamma \to P^m$ is the image of a critical point. A *regular value* of pr: $\Gamma \to P^m$ is a point in P^m which is not a critical value. Figures 1 and 2 suggest that almost all

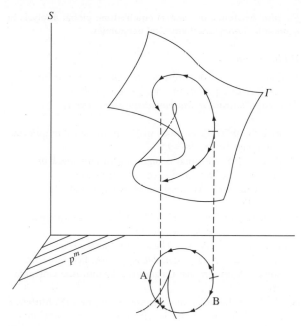

Figure 2

points in P^m are regular values. Indeed, the concepts introduced above are defined in differential topology in a quite general context and Sard's theorem, an analytical tool of great importance, asserts that the critical values of a sufficiently differentiable mapping are rare. More precisely, *Sard's theorem* applied to our particular problem yields that the set of critical values of pr: $\Gamma \to P^m$ is a (Lebesgue) null set.

Null sets are small in a probabilistic sense. At this point we make essential use of the space of economies P^m being part of a Euclidean space. If, for instance, consumers' demand functions or preferences are allowed to vary instead of consumers' endowments, it is not clear how null sets are to be defined. However, one can express quite easily when two demand functions or preference orderings are close to each other. That is to say metric structures are very often naturally given when there is no obvious way to define null sets. A set can then be defined to be small in a topological sense if its closure is nowhere dense.

Furthermore, if the concepts of smallness in the probabilistic and in the topological sense are both well-defined, as they are in the case of variable initial endowments, one has to be aware of the fact that the two variants of the intuitive notion of smallness apply to quite different sets. Defining a *critical economy* $E \in P^m$ as a critical value of pr: $\Gamma \to P^m$ and *regular economy* as a regular value of pr we ask ourselves whether the null set of critical economies has a null closure. We know already that the desirability assumption (D) implies that the equilibrium price correspondence is upper hemi-continuous and compact-valued or, in more intuitive terms, that Γ has

only finitely many layers above some compact ball B of economies in P^m. Hence the points in Γ which lie above B and have a vertical tangent form a compact set. Projecting this set down to B yields a compact set, the set of critical economies in B. Since this set is also null by Sard's theorem, it is nowhere dense. We obtain:

The set of critical economies in P^m is a closed null set if (D) holds.

Let $E \in P^m$ be a regular economy. Then E has a finite number of equilibria and this number is locally constant. If E has r equilibria, then there is a neighbourhood U of E and there are rC^1 functions $g_1, \ldots, g_r$ such that the set $\Pi(E')$ of equilibrium price systems of any economy $E' \in U$ is given by $\{g_1(E'), \ldots, g_r(E')\}$. In particular, *the equilibrium price correspondence Π is continuous in a neighbourhood of a regular economy.*

These results, with minor differences, have been obtained by G. Debreu (1970), whose proof, however, differs from the exposition given here.

Extensions

When one wants to extend the theory of regular equilibria to more general spaces of economies, it is often useful to employ a definition of a regular economy which focuses on the given economy and does not refer to the graph Γ or to the parameter space. To motivate the following definition we contrast Figure 3, in which the excess demand of a critical economy such as E_1 or E_2 in Figure 1 is drawn, with Figure 4, which shows the graph of a regular economy such as E_3. It is assumed that there are two goods so that it suffices, according to Walras's Law, to look at the excess demand ζ_1 for good 1.

In Figure 3 there is one equilibrium at which $d\zeta_1/dp_1$ vanishes. Shifting the graph of ζ_1 a little upwards destroys this equilibrium. In Figure 4, however, $d\zeta_1/dp_1$ does not vanish at any equilibrium and all equilibria are robust.

Let the excess demand function $\zeta : S \to \mathbb{R}^l$ of an economy E be C^1. A price system $p \in S$ is called a *regular equilibrium price system* if $\zeta(p) = 0$ and the matrix

$$\left[\frac{\partial \zeta_h}{\partial p_k}(p) \right]_{h,k=1,\ldots,l-1}$$

is regular. A *regular economy* is an economy all equilibrium price systems of which are regular. One can show that this definition, introduced by E. and H. Dierker (1972), is independent of the way in which goods are indexed and that it is consistent with the definition given above.

The results on regular economies obtained in various frameworks are quite similar to those established in the previous section. It is shown that almost all economies, in an appropriate sense, are regular. Every regular equilibrium is locally unique and can be traced along its path when the economy varies gradually, as long as it stays regular. Economic models in which results of this kind

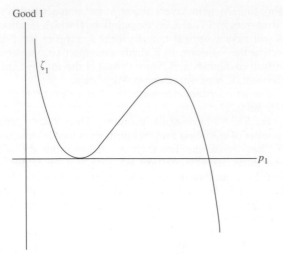

Good 1

ζ_1

p_1

Figure 3

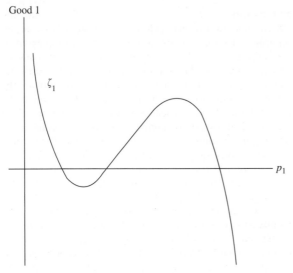

Good 1

ζ_1

p_1

Figure 4

have been precisely formulated and verified deal with variations in consumption and production (see, in particular, Smale, 1974). Also the case of many consumers, that is to say of consumption sectors described by the distribution of consumers' characteristics, has been treated. The basic mathematical tool is always some variant of Sard's Theorem. References can be found in my survey article (Dierker, 1982).

The study of regular equilibria has led to a revival of the differentiable viewpoint in general equilibrium theory and related areas. Readers interested in this modern development are referred to the excellent book by A. Mas-Colell (1985), which also contains an extensive, systematic presentation of the theory of regular equilibria.

EGBERT DIERKER

See also **existence of general equilibrium; global analysis in economic theory; mathematical economics.**

Bibliography

Balasko, Y. 1975. Some results on uniqueness and on stability of equilibrium in general equilibrium theory. *Journal of Mathematical Economics* 2, 95–118.

Debreu, G. 1970. Economies with a finite set of equilibria. *Econometrica* 38, 387–92.

Dierker, E. 1982. Regular economies. In *Handbook of Mathematical Economics*, ed. K. Arrow and M. Intriligator. Amsterdam: North-Holland, ch. 17, 795–830.

Dierker, E. and Dierker, H. 1972. The local uniqueness of equilibria. *Econometrica* 40, 867–81.

Mas-Colell, A. 1985. *The Theory of General Economic Equilibrium: A Differentiable Approach.* Econometric Society Monographs, Cambridge: Cambridge University Press.

Smale, S. 1974. Global analysis and economics IV: finiteness and stability of equilibria with general consumption sets and production. *Journal of Mathematical Economics* 1, 119–27.

regulation

Since the 1980s, the term '*régulation*' has suggested state intervention in the name of economic management though its opposite, 'dérégulation', has been more widely used. In the area of economic policy and in accordance with Keynesian precepts, regulation indicates the adjustment of macroeconomic activity by means of budgetary or monetary contra-cyclical interventions. In the area of public management, a complete body of literature, under the name of regulation theory, has investigated the methods for organizing the decentralization of the supply of various public utilities.

This term is also used in physics and biology. In mechanics, a regulator is a means to stabilize the rotary speed of a machine. In biology, regulation corresponds to the reproduction of substances such as DNA. In general terms, the theory of systems involves the study of the role of a set of negative and positive feedback loops in relation to the stability of a complex network of interactions.

Here, a third and different, but not totally unrelated, meaning of the term will be developed. Theories of régulation constitute an area of research which has focused on analysing long-term transformations in capitalist economies. Initially, it focused on American and French capitalisms (Aglietta, 1982; Benassy, Boyer and Gelpi, 1979) but it was progressively extended first to major OECD economies (Mazier, Basle and Vidal, 1999) then to Latin American countries (Hausmann, 1981; Ominami, 1985) and ultimately Asian countries (Bertoldi, 1989; Boyer, 1994). A general presentation of the present state of the theory is to be found in Boyer and Saillard (2002) and a large sample of national case studies in Jessop (2001). Basically, the theory of régulation combines Marxian intuitions and Kaleckian macroeconomics with institutionalist and historicist studies, mobilizing most of the tools of modern economic analysis.

At a primary level, a form of régulation denotes any dynamic process of adaptation of production and social demand resulting from a conjunction of economic adjustments linked to a given configuration of social relations, forms of organization and productive structures (Boyer, 1990).

Most economic theories emphasize the general invariables of eminently abstract systems, in which history serves merely as a confirmation or, failing that, as a perturbation. Neoclassical theory studies the shift of a stable equilibrium after an external shock, Keynesian economists stress the role of effective demand and fine tuning whatever the context and the period. Even Marxists tend to extrapolate, as general laws, the quite specific evolutions observed in the early phases of capitalism. In contrast, the régulation approach seeks a broader interaction between history and theory, social structures, institutions and economic regularities (de Vroey, 1984).

The starting point is the hypothesis that accumulation has a central role and is the driving force of capitalist societies. This necessitates a clarification of factors that reduce or delay the conflicts and disequilibria inherent in the formation of capital, and which allow for an understanding of the possibility of periods of sustained growth (Boyer and Mistral, 1978). These factors are associated with particular regimes of accumulation, namely, the form of articulation between the dynamics of the productive system and social demand, the distribution of income between wages and profits on the one hand, and on the other hand the division between consumption and investment. It is then useful to explain the organizational principles which allow for mediation between such contradictions as the extension of productive capacity under the stimulus of competition on product markets, for labour and finance. The notion of institutional form – defined as a set of fundamental social relations (Aglietta, 1982) – enables the transition between constraints associated with an accumulation regime and collective strategies; between economic dynamics and individual behaviour. A small number of key institutional forms, which are the result of past social struggles and the imperatives of the material reproduction of society, frame and channel a multitude of partial strategies which are decentralized and limited in terms of their temporal horizon. Five main institutional forms do shape accumulation regimes.

The forms of competition describe by what mechanisms the compatibility of a set of decentralized decisions by firms and individuals is ensured. They are competitive while the *ex post* adjustment of prices and quantities ensures a balance; they are monopolist if the *ex ante*

socialization of revenue is such that production and social demand evolve together (Lipietz, 1979). The type of monetary constraint explains the interrelations between credit and money creation: credit is narrowly limited in terms of movement of reserves when money is predominantly metallic; the causality is reversed when on the contrary the dynamics of credit conditions the money supply in systems where the external parity represents the only constraint weighing upon the national monetary system (Benassy, Boyer and Gelpi, 1979). The nature of institutionalized compromises defines different configurations of relations between the state and the economy (André and Delorme, 1983; Jessop, 1990): the state-as-referee when only general conditions of commercial exchange are guaranteed; as the interfering state when a network of régulation and budgetary interventions codifies the rights of different social groups. Modes of support for the international regime are also derived from a set of rules which organize relations between the nation state and the rest of the world in terms of commodity exchange, migration, capital movements and monetary settlements. History goes beyond the traditional contrast between an open and a closed economy, free trade and protectionism; it makes apparent a variety of configurations (Mistral, 1986; Lipietz, 1986a). Finally, forms of wage relations indicate different historical configurations of the relationship between capital and labour, that is, the organization of the means of production, the nature of the social division of labour and work techniques, type of employment and the system of determination of wages, and finally, workers' way of life including the welfare state. If, in the first stages of industrialization, wage-earners are defined first of all as producers, during the second stage they are simultaneously producers and consumers.

At this point appears the notion of régulation, as a conjunction of mechanisms and principles of adjustment associated with a configuration of wage relations, competition, state interventions and hierarchization of the international economy. Finally, a distinction between 'small' and 'big' crises is called for (Boyer, 1990). The former, which are of a rather cyclical nature, are the very expression of régulation in reaction to the recurrent imbalances of accumulation. The latter are of a structural nature: the very process of accumulation throws into doubt the stability of institutional forms and the régulation which sustains it because the profit does not recover by contrast with conventional business cycles.

Thus, in long-term dynamics as well as in short-term development, institutions are important. Historical research confirms that sometimes institutional forms make an impression on the system in operation; at other times they register major changes in direction. At the end of a period which can be counted in decades, the very mode of development – that is, the conjunction of the mode of régulation and the accumulation regime – is affected: there will be changes in the tendencies of long-term growth and eventually in inflation, specificities of cyclical processes (Mazier, Basle and Vidal, 1999).

So a periodization of advanced capitalist economies emerges which is not part of the traditional Marxist theory. Despite the rise in monopoly, the interwar period is still marked by competitive regulation. After the Second World War an accumulation regime without precedent is instituted – that of intensive accumulation centered on mass consumption (Bertrand, 1983) – known as Fordist and channelled through monopolist-type regulation.

In fact, the alteration in wage relations – in particular the transition to Fordism, that is, the synchronization of mass production and wage-earners' access to the 'American way of life' – and in monetary management, that is, transition to internally accepted credit money – seems to have played a greater role than the change in modes of competition or conjunctural fine tuning à la Keynes (Aglietta, 1982; Aglietta and Orlean, 1982; Boyer, 1988).

Since the 1960s, many economies have been experiencing a big crisis without historical precedent: stagflation, absence of cumulative depression, breaking-down of most previous economic regularities, length of the period of technological and institutional restructuring (Boyer and Mistral, 1978; Lipietz, 1985). In consequence, it is logical that former economic policies lose their efficacy (Boyer, 1990). First, because the crisis is not cyclical but structural; this invalidates the policy of fine-tuning; second, because the structural changes which permitted the 1929 crisis to be overcome have become blocked and cannot be repeated (Lipietz, 1986b).

Since the formative years, the research programme has been developing both extensively and intensively. The collapse of the Soviet bloc economies has pointed to the need to investigate the necessary and sufficient institutions required for a viable capitalist economy (Emergo, 1995; Hollingsworth and Boyer, 1997): economic viability depends on the compatibility of a complete set of institutional forms. In the epoch of financialization (Aglietta, 1998), information and communication technologies diffusion (Boyer, 2004), rise of services (Petit, 1986) and strengthening of foreign competition (Lipietz, 1986a), no clear follower to Fordism has yet emerged and diffused. Nevertheless, since the mid-1970s a series of trials and errors concerning the reform of the monetary regime, the tax and welfare system, competition and wage relations has finally delineated a new institutional architecture, quite complex to analyse. Conversion, layering and recomposition of existing institutional forms have replaced the strong synchronization associated with major crises and world wars (Boyer, 2005b).

The large number of international comparisons has systematically exhibited the persisting diversity of various brands of capitalism. Within industrialized countries: market dominated, corporate-led, state governed and social democratic, versions with some possible sub-variants, coexist (Amable, 2004). An equivalent but different variety is observed for Latin American countries (Quémia, 2001). Consequently, the financial crises experienced by Mexico, Brazil and Argentina are quite different, even if they all point out the destabilizing role of global finance upon contrasted domestic accumulation regimes (Boyer and Neffa, 2004).

These numerous structural changes call for new directions for the research agenda of régulation theory. Can the concepts of complementarity, hierarchy, isomorphism and coevolution explain how various mixes of institutions can cohere and define a coherent accumulation regime (Boyer, 2005a; Socio-Economic Review, 2005)? What kind of political economy analysis can explain the emergence and restructuring of institutional forms, especially the choice of monetary regime, the configuration of the welfare state or the nature of insertion into the world economy (Palombarini, 1999)? How to analyse multilevel régulation modes, especially in order to understand the complex process of European integration (Boyer and Dehove, 2001)? Finally, is not the anthropogenetic model, based on the production of humankind by education, health care and culture, a possible follower of the Fordist regime (Boyer, 2004)?

ROBERT BOYER

See also **Beveridge, William Henry; Braudel, Fernand; capitalism; collective bargaining; competition; contemporary capitalism; development economics; growth and institutions; institutional economics; Kalecki, Michal; Marx, Karl Heinrich; profit and profit theory.**

Bibliography

Aglietta, M. 1982. *Regulation and Crisis of Capitalism*. New York: Monthly Review Press.
Aglietta, M. 1998. Le capitalisme de demain. Note de la fondation Saint-Simon (101), November.
Aglietta, M. and Orlean, A. 1982. *La violence de la monnaie*. Paris: PUF.
Amable, B. 2004. *The Diversity of Modern Capitalisms*. Oxford: Oxford University Press.
André, Ch. and Delorme, R. 1983. *L'état et l'économie*. Paris: Seuil.
Benassy, J.P., Boyer, R. and Gelpi, R.M. 1979. Régulation des économies capitalistes et inflation. *Revue économique* 30(3), 397–441.
Bertoldi, M. 1989. The growth of Taiwanese economy 1949–1989: success and open problems of a model of growth. *Review of Currency Law and International Economics* 39, 245–88.
Bertrand, H. 1983. Accumulation, régulation, crise: un modèle sectionnel théorique et appliqué. *Revue économique* 34, 305–43.
Boyer, R. 1979. Wage formation in historical perspective: the French experience. *Cambridge Journal of Economics* 3, 99–118.
Boyer, R. 1988. Technical change and the theory of 'regulation'. In *Technical Change and Economic Theory:*

The Global Process of Development, ed. G. Dosi et al. London: Pinter Publishers.

Boyer, R. 1990. *The Regulation School*. New York: Columbia University Press.

Boyer, R. 1994. Do labour institutions matter for economic development? In *Workers, Institutions and Economic Growth in Asia*, ed. G. Rodgers. Geneva: ILO/ILLS.

Boyer, R. 2004. *The Future of Economic Growth*. Cheltenham: Edward Elgar.

Boyer, R. 2005a. Coherence, diversity and the evolution of capitalisms. *Evolutionary and Institutional Economic Review* 2(1), 43–80.

Boyer, R. 2005b. How and why capitalisms differ. *Economy and Society* 34, 509–57.

Boyer, R. and Dehove, M. 2001. Théories de l'intégration européenne. *La Lettre de la Régulation*, No. 38, 1–3.

Boyer, R. and Mistral, J. 1978. *Accumulation, inflation, crises*. Paris: PUF.

Boyer, R. and Neffa, J., eds. 2004. *La crisis Argentina (1976– 2001)*. Madrid, Buenos Aires: Editorial Mino y Davila.

Boyer, R. and Saillard, Y., eds. 2002. *Regulation Theory: The State of the Art*. London: Routledge.

de Vroey, M. 1984. A regulation approach interpretation of the contemporary crisis. *Capital and Class* 23 (Summer), 45–66.

Emergo: Journal of Transforming Economies and Societies. 1995. Special issue, vol. 2(4).

Hausmann, R. 1981. State landed property, oil rent and accumulation in Venezuela. Ph.D. thesis, Cornell University.

Hollingsworth, R.J. and Boyer, R., eds. 1997. *Contemporary Capitalism*. Cambridge: Cambridge University Press.

Jessop, B. 1990. *State Theory. Putting Capitalist States in their Places*. Oxford: Polity Press.

Jessop, B., ed. 2001. *Regulation Theory and the Crisis of Capitalism*, 5 vols. Cheltenham: Edward Elgar.

Lipietz, A. 1979. *Crise et inflation, pourquoi?* Paris: Maspéro.

Lipietz, A. 1985. *The Magic World: From Value to Inflation*. London: Verso.

Lipietz, A. 1986a. New tendencies in the international division of labor. In *Production, Work, Territory*, ed. A.J. Scott and M. Storper. London: Allen & Unwin.

Lipietz, A. 1986b. Behind the crisis: the exhaustion of a regime of accumulation. *Review of Radical Political Economics* 18(1–2), 13–32.

Mazier, J., Basle, M. and Vidal, J.F. 1999. *When Economic Crises Endure*. London: M.E. Sharpe.

Mistral, J. 1986. Régime international et trajectoires nationales. In *Capitalisme fin de siècle*, ed. R. Boyer. Paris: PUF.

Ominami, C. 1985. *Les transformations dans la crise des rapports nord– sud*. Paris: La Découverte.

Palombarini, S. 1999. *Vers une théorie régulationniste de la politique économique*. *L'Année de la Régulation 1999*, vol. 3, Paris: La Découverte.

Petit, P. 1986. *Slow Growth and the Service Economy*. London: Frances Pinter.

Quémia, M. 2001. *Théorie de la régulation et développement: trajectoires latino-américaines. L'Année de la régulation*, vol. 5, Paris: Presses de Sciences-Po.

Socio-Economic Review. 2005. A dialogue on institutional complementarity. Vol. 2(1), 43–80.

More information on régulation theory is available from the Association Recherche et Régulation. Online. Available at http://web.upmf-grenoble.fr/regulation, accessed 19 October 2006.

Reid, Margaret Gilpin (1896–1991)

Margaret Reid, a leading scholar in analysis of the economics of consumer behaviour, was made a distinguished fellow of the American Economic Association in 1980. She was Professor of Economics at Iowa State College (1930–43), the University of Illinois (1948–51), and the University of Chicago (1951–61).

A realistic theorist, Reid always looked behind data to processes that generate structural relationships. Her 1934 book on household production anticipated by three decades analyses built on the allocation of time, and she was the first (1947) to use wage-equivalent time measures of household work.

Already in Iowa she had questioned attempts to improve resource allocations by farm women that disregarded the nature of income effects. She went on to criticize assessments of the war-time cost-of-living index that neglected effects of changing incomes on the quality of goods traded, and she became the 'directing' member of the technical committee responsible for a report to the President's Commission on the Cost of Living (1945). Later on she challenged conventional treatments of income elasticities of consumption in general and of housing expenditures in particular (1952; 1962).

The concepts of 'permanent' and 'transitory' income were early a part of Reid's thinking (1952; 1953). Friedman drew on Reid in his 1957 application of the permanent income hypothesis to short-term shifts in consumption and saving, and Modigliani built on her work in his treatment of 'life stages' (Modigliani and Ando, 1960 and subsequently). In Reid's hands the concepts of 'permanent' and 'transitory' income evolved subtly and progressively in multiple facets of the analysis of consumer behaviour. After her retirement she probed interactions between health and income both over life cycles and across cohorts.

MARY JEAN BOWMAN

Selected works

1934. *Economics of Household Production*. New York: Wiley.

1943. *Food for People*. New York: Wiley.

1945. (With associates.) Appendix IV. Prices and the cost of living in wartime – an appraisal of the Bureau of Labor

Statistics Index of the Cost of Living in 1941–44. In *Report of the President's Committee on the Cost of Living*. Washington, DC: Government Printing Office.

1947. The economic contribution of homemakers. In *Women's Opportunities and Responsibilities*, ed. L.M. Young. Philadelphia: Annals of the American Academy of Political and Social Science.

1952. Effect of income concept upon expenditure curves of farm families. In *Conference on Research in Income and Wealth, Studies*, vol. 15, part IV. Philadelphia: NBER.

1953. Savings by family units in consecutive periods. In *Savings in the Modern Economy*, ed. W.W. Heller, F.M. Boddy and C.L. Nelson. Minneapolis: University of Minnesota Press.

1955. Food, liquor and tobacco. In *America's Needs and Resources: A New Survey*, ed. J.F. Dewhurst. New York: Twentieth Century Fund.

1960. Comments on J. Crockett and I. Friend, 'A Complete Set of Consumer Demand Relationships'. In *Proceedings of the Conference on Consumption and Savings*, vol. I, ed. I. Friend and R. Jones. Philadelphia: Wharton School of Finance and Commerce, University of Pennsylvania.

1962. *Housing and Income*. Chicago: University of Chicago Press.

Bibliography

Friedman, M. 1957. *A Theory of the Consumption Function*. Princeton: Princeton University Press.

Modigliani, F. and Ando, A. 1960. The 'permanent income' and the 'life cycle' hypothesis of saving behavior: comparisons and tests. In *Proceedings of the Conference on Consumption and Savings*, ed. I. Friend and R. Jones. Philadelphia: Wharton School of Finance and Commerce, University of Pennsylvania.

religion and economic development

The number of micro-level social anthropological studies is continually growing. Many of these concentrate on what to the economist may appear odd aspects of society such as ritual and religion ... and to which he pays little or no attention. For instance, an understanding of the complex of Hindu religious beliefs as they operate at village level ... is directly relevant to the problem of developing India's economy. This is but one of numerous examples which can be quoted to support the claim that development economists work in the dark unless they acquaint themselves with the relevant socio-political literature. (Epstein, 1973, p. 6)

How times have changed since Scarlett Epstein first lamented economists' general neglect of the role of religion in the study of economic development. She need not have been quite so fearful: contemporary economics has

seen the light, as it were, increasingly demanding a perspective on religion in order better to understand how it interacts with economic decision making. The increasing resilience of religion in both developed and developing countries, influencing globally both political will and popular debate, has been observed by scholars investigating the economics of religion (Iannaccone, 1998; Stark and Finke, 2001; Glaeser, 2005). Recent studies have investigated how religion affects growth (Guiso, Sapienza and Zingales, 2003; North and Gwin, 2004; Noland, 2005; Barro and McCleary, 2003; Glahe and Vorhies, 1989) with emphasis on particular religious traditions such as Islam, Hinduism or Catholicism (Kuran, 2004; Sen, 2004; Fields, 2003). Other studies have focused on the impact of religion on fertility (Lehrer, 2004; McQuillan, 2004). Still others examine the impact of religion on political outcomes (Glaeser, Ponzetto and Shapiro, 2005) and the role of religious organizations as insurance (Dehejia, DeLeire and Luttmer, 2005). Other studies examine how the causality may run the other way, from economic development to religion (Berman, 2000; Botticini and Eckstein, 2005; Goody, 2003).

Several theories have been advanced to account for the links between religion and development. First, there are theories that typify the 'rational choice' approach to religion and development. This approach considers the resilience of religion as a rational economic response to changes in the political, ecological and economic environments in which religions operate. In addition, a range of other structural theories encompass family socialization, social networks and a belief in other-worldly or supernatural elements. However, regardless of the scholastic tradition from which one approaches the study of religion, examining the interactions between religion and development poses significant challenges: first, to understand the endogenous interactions between religion and economic growth; second, to examine the techniques and methods needed to quantify these interactions; and third, to evaluate the impact of religion on development policy more widely.

Early writings

The economic concern with religion and development is not new, nor is it restricted to scholars of the 21st century. The writings of Thomas Aquinas, notably the *De Regno (De Regimine Principum) ad Regem Cypri*, written in 1267, dealt extensively with religion and public finance. Indeed, some scholars have considered the ideas in this work, as in Aquinas's *Summa Theologica* (1265–72), strikingly relevant for poverty reduction today; their themes of the 'universal common good' and 'global civil society' have implications for current debates about globalization and human development (Linden, 2003). The links between religion and development also feature in Joseph Schumpeter's *History of Economic Analysis* (1954). Jacques Le Goff authored *La Naissance du Purgatoire* (1981), which argued that purgatory was a necessary religious innovation for medieval capitalist development. However, it was in 1904 that Max Weber put forward his famous theory of the Protestant ethic and the spirit of capitalism, arguing that economic development in northern Europe could be explained by developments that were associated with Protestantism – the concern with savings, entrepreneurial activity, the frugality which Puritanism demanded, and the literacy needed to read the scriptures. The essence of Weber's thesis was that nascent capitalism emerged in the 16th century in Europe on account of the Protestant ethic which arose from the Reformation. Ascetic Protestantism encouraged diligence, discipline, self-denial and thrift. Both Lutheran and Calvinist doctrines urged adherents robustly to undertake their 'calling'. Spiritual grace from religion was attained by demonstrating temporal success in one's calling. The Protestant ethic thus involved the diligent undertaking of one's calling as a religious obligation, which promoted a work ethic that increased savings, capital accumulation, entrepreneurial activity, and investment, all of which in turn fostered economic development. Many scholars have criticized Weber's thesis, typified in the writings of Tawney (1926) and Gorski (2005). Tawney was concerned with reverse causality: how religion affected development, and in turn how economic and social changes themselves acted on religious beliefs. In his words, '"The capitalist spirit" is as old as history, and was not, as has sometimes been said, the offspring of Puritanism' (1926, p. 225). Tawney argued that Puritanism both helped mould the social order and in turn was moulded by it. Gorski (2005) focuses more on whether Weber's thesis stands up to closer historical scrutiny, highlighting other aspects of the Reformation that contributed to economic development such as Protestant migration, reforms to landholding, fewer religious holidays, and insurgencies, all of which influenced labour supply and the actions of government in Protestant countries.

The economic view of religion

Against this backdrop, recent academic interest linking religion and development has centred on the economics of religion. Studies in the economics of religion have focused on applying the tools of modern economic analysis to the analysis of religious institutions, faith-based welfare programmes and the economic regulation of the church (Oslington, 2003). Three principal themes emerge: first, identifying what determines religion and religiosity; second, examining how religion and religiosity may be described as social capital; and third, understanding the micro and macro consequences of religiosity.

Adam Smith (1776) made reference to the church in the *Wealth of Nations*; and recent work by economists such as Becker and Iannaccone has been very important for the development of this field. The broadly socio-economic view of religion, which expounds the rational choice

approach, is set out in the work of Azzi and Ehrenberg (1975), Iannaccone (1998), Stark, Iannaccone and Finke (1996), and Stark and Finke (2000). The focus here has been both on the supply side (the structures of religious organizations) and on the demand side (the preferences of consumers in religious economies). The micro view explains religious activity as the outcome of rational choice, with utility derived both in the individual's lifetime and in the afterlife. For example, if we think of religion as a club good, then many practices are used by religions to screen potential free riders and to ensure better monitoring of the existing faithful (Iannaccone, 1992). Religion also influences individual welfare through the externalities occasioned by social behaviour (Becker and Murphy, 2000). Religious forces are important as they change the environment in which individuals operate, directly affecting individuals' choices and behaviour by changing the utilities of goods. Moreover, greater trust fostered by the religious environment can encourage repeated interactions, leading to more cooperative behaviour within networks.

It is in this way that the second theme – religion as social capital – becomes important. Three aspects are emphasized here: social networks, social norms, and sanctions to penalize deviations from norms. Corresponding to this emphasis, economists of religion have been examining 'spiritual capital' – or religious capital – which embodies the norms, networks and sanctions exercised by groups that are organized on the basis of religion and religious networks.

Finally, the macro and micro consequences of religiosity have been examined. For example, there are a number of channels through which religious capital might affect economic growth. Religious capital affects output by changing the manner in which technology and human capital are used. Religious capital exerts a positive impact on human capital by increasing education. For example, particularly in many less developed countries, religious networks are important not only for the religious services they provide but also for their non-religious services, specifically with respect to health and education. Moreover, as religious institutions provide this insurance function, these networks determine the extent to which education is taken up (Borooah and Iyer, 2005). In developed countries, too, this would have implications for religious market structure and the growth of residential neighbourhoods that may be based upon faith-based activities (Gruber, 2005). So understanding the economic consequences of religion is of central concern.

The empirics of religion and development

Most empirical economic studies of religion and development attempt to solve classic decompositions of the form $Y_i - Y_j = \sum \beta (X_i - X_j)$ where the idea is to examine the various factors (X) that affect measures of religious attendance or behaviour (Y) across individuals (i, j), or more widely across countries, or alternatively in varied historical time periods, thence to arrive at conclusions based on the effects suggested by the parameters (β) estimated.

Empirical studies of religion and development across countries have investigated religious movements, examining particularly sect behaviour, with an emphasis on contrasting the 'European experience of religious monopoly' with the 'American case of religious cacophony' (Warner, 1993, p. 1081), drawing implications for the issue of whether regulation of religious organizations is necessary. This concern manifests itself in a plethora of research projects, especially on religion in the United States (Marty, 1986–96; Finke and Stark, 1988; Warner, 1993). In cross-country studies, economists have also revisited Weber's hypothesis. Barro and McCleary (2003) assess the effect of religious participation and beliefs on a country's rate of economic progress. Using international survey data for 59 countries drawn from the World Values Survey and the International Social Sciences Program conducted between 1981 and 1999, these authors find that greater diversity of religions is associated with higher church attendance and stronger religious beliefs. For a given level of church attendance, increases in some religious beliefs – notably belief in heaven, hell and an afterlife – tends to increase economic growth.

Other studies have focused more on particular religions in varied historical time periods. For example, very useful insights have been gained by focusing on Islam and on Judaism. For Islam, there have been detailed investigations into financial systems in the Middle East including *zakat* (alms for charity) and the manner in which Islamic banks have been using a financing method equivalent to the rate of interest to overcome adverse selection and information problems. There has also been more detailed investigation into Islamic law and financial activity historically with implications for poverty reduction in the Middle East (Kuran, 2004). There is research that has examined Jewish occupational selection using historical data from the eighth and ninth centuries onward to explain the selection of Jews into urban, skilled occupations prompted by educational and religious reform in earlier centuries (Botticini and Eckstein, 2005). Data are also being used to elucidate the role of religion in explaining historical differences in education among Hindus and Muslims in India (Borooah and Iyer, 2005).

A primary focus of current studies of religion and development is on explaining differences across individuals. For example, using data from the General Social Survey and the US Census, Gruber (2005) investigates religious market structure by estimating the effects of religious participation on economic measures of well-being, and concluded that residing in an area with more co-religionists improves well-being through the impact of increased religious participation. This particular study is also valuable from the methodological point of view, as it addresses a common problem in empirical studies of religion and development – the persistent endogeneity of religion to economic measures of well-being – and consequently the common econometric problem of how best to identify religion effects. While this particular study successfully uses ethnic heritage to provide an exogenous source of variation, and is thereby able to draw out cleanly the effects of religious participation on the variables of interest, econometrically the potential endogeneity of most religion variables is possibly the single most significant limitation of incorporating religion into empirical work in economics. This is mirrored in the many efforts to identify the effects of religion which generally have not been able to deal with self-selection issues easily.

To this end, fields such as economic demography have much to offer the study of religion and development. For example, recent research in economics has made a start towards examining the religious and economic reasons behind fertility differences between religious groups, especially in developing countries (Iyer, 2002). The economics of religion has also elucidated the study of politics, both local and international: Glaeser (2005) presents an economic model of religious group behaviour and the so-called 'political economy of hatred'. The economic approach to religion has been evaluating whether religion and politics are mutually exclusive. Glaeser, Ponzetto and Shapiro (2005) link religion with strategic extremism – the issues and platforms espoused by political parties, and the manner in which private information matters for this. Other studies have focused on terrorism and display a more general preoccupation with understanding views and attitudes in the Muslim world (Gentzkow and Shapiro, 2004).

Drawing a perspective from all these classes of studies, it strikes one that emerging economies are experiencing appreciable modern economic growth, yet this is coterminous with the increasing resilience of religious institutions. And it is this dichotomy between the sacred and the secular which epitomises the puzzle of the relationship between religion and economic development. It seems reasonable to address this puzzle by combining quantitative analysis of sample data with nuanced qualitative evaluations of the textual theology of religion, linking these to the manner in which individuals and institutions interpret religion at a local level. As well, an appreciation of the approach of the interdisciplinary economist would permit a more informed understanding of all these concerns. Economists will enthusiastically study religion and economic development in the future, and they will do so with ascetic assiduity – researching data with all the intensity of religious fervour in order to provide thoughtful prophecy for development policy.

SRIYA IYER

See also **Islamic economic institutions; religion, economics of; social capital; social interactions (empirics); Weber, Max.**

Bibliography

Aquinas, St Thomas. 1265–72. *Summa Theologica*. Online. Available at http://www.ccel.org/ccel/aquinas/summa.html, accessed 22 June 2006.

Aquinas, St Thomas. 1267. *De Regno (De Regimine Principum) ad Regem Cypri*. In *Aquinas: Political Writings*, ed. and trans. R.W. Dyson, Cambridge: Cambridge University Press, 2002.

Azzi, C. and Ehrenberg, R. 1975. Household allocation of time and church attendance. *Journal of Political Economy* 83, 27–56.

Barro, R.J. and McCleary, R. 2003. Religion and economic growth across countries. *American Sociological Review* 68, 760–81.

Becker, G. and Murphy, K. 2000. *Social Economics: Market Behavior in a Social Environment*. Cambridge, MA: Harvard University Press.

Berman, E. 2000. Sect, subsidy and sacrifice: an economist's view of ultra-orthodox Jews. *Quarterly Journal of Economics* 115, 905–53.

Borooah, V. and Iyer, S. 2005. *Vidya, veda* and *varna*: the influence of religion and caste on education in rural India. *Journal of Development Studies* 41, 1369–404.

Botticini, M. and Eckstein, Z. 2005. Jewish occupational selection: education, restrictions, or minorities? *Journal of Economic History* 65, 922–48.

Dehejia, R., DeLeire, T. and Luttmer, E. 2005. Insuring consumption and happiness through religious organizations. Working Paper No. 11576. Cambridge, MA: NBER.

Epstein, T. 1973. *South India: Yesterday, Today and Tomorrow*. London: Macmillan.

Fields, B. 2003. *The Catholic Ethic and Global Capitalism*. Aldershot, UK and Burlington, VT: Ashgate.

Finke, R. and Stark, R. 1988. Religious economies and sacred canopies: religious mobilization in American cities, 1906. *American Sociological Review* 53, 41–9.

Gentzkow, M. and Shapiro, J. 2004. Media, education and anti-Americanism in the Muslim world. *Journal of Economic Perspectives* 18(3), 117–33.

Glaeser, E. 2005. The political economy of hatred. *Quarterly Journal of Economics* 120, 45–86.

Glaeser, E., Ponzetto, G. and Shapiro, J. 2005. Strategic extremism: why Republicans and Democrats divide on religious values. *Quarterly Journal of Economics* 120, 1283–330.

Glahe, F. and Vorhies, F. 1989. Religion, liberty and economic development: an empirical investigation. *Public Choice* 62, 201–15.

Goody, J. 2003. Religion and development: some comparative considerations. *Development* 46(4), 64–7.

Gorski, P. 2005. The little divergence: the Protestant Reformation and economic hegemony in early modern Europe. In *The Protestant Ethic Turns 100: Essays on the Centenary of the Weber Thesis*, ed. W. Swatos and L. Kaelber. Boulder and London: Paradigm Publishers.

Gruber, J. 2005. Religious market structure, religious participation, and outcomes: is religion good for you? *Advances in Economic Analysis & Policy* 5(1), article 5.

Guiso, L., Sapienza, P. and Zingales, L. 2003. People's opium? Religion and economic attitudes. *Journal of Monetary Economics* 50, 225–82.

Iannaccone, L. 1992. Sacrifice and stigma: reducing free-riding in cults, communes, and other collectives. *Journal of Political Economy* 100, 271–91.

Iannaccone, L. 1998. Introduction to the economics of religion. *Journal of Economic Literature* 36, 1465–95.

Iannaccone, L., Finke, R. and Stark, R. 1997. Deregulating religion: the economics of church and state. *Economic Inquiry* 35, 350–64.

Iyer, S. 2002. *Demography and Religion in India*. Delhi: Oxford University Press.

Kuran, T. 2004. Why the Middle East is economically underdeveloped: historical mechanisms of institutional stagnation. *Journal of Economic Perspectives* 18(3), 71–90.

Lehrer, E. 2004. Religion as a determinant of economic and demographic behaviour in the United States. *Population and Development Review* 30, 707–26.

Le Goff, J. 1981. *La Naissance du Purgatoire*. Paris: Gallimard.

Linden, I. 2003. *A New Map of the World*. London: Darton, Longman and Todd.

Marty, M. 1986–96. *Modern American Religion*, 3 vols. Chicago: University of Chicago Press.

McQuillan, K. 2004. When does religion influence fertility? *Population and Development Review* 30, 25–56.

Noland, M. 2005. Religion and economic performance. *World Development* 33, 1215–32.

North, C. and Gwin, C. 2004. Religious freedom and the unintended consequences of state religion. *Southern Economic Journal* 71, 103–17.

Oslington, P., ed. 2003. *Economics and Religion*, vols. 1 and 2. The International Library of Critical Writings in Economics 167. Cheltenham: Edward Elgar.

Schumpeter, J. 1954. *History of Economic Analysis*. London: Oxford University Press.

Sen, A. 2004. Democracy and secularism in India. In *India's Emerging Economy: Performance and Prospects in the 1990s and Beyond*, ed. K. Basu. Cambridge, MA and London: MIT Press.

Smith, A. 1776. *An Inquiry Into the Nature and Causes of the Wealth of Nations*, 5th edn, ed. E. Cannan. London: Methuen and Co. Ltd., 1904.

Stark, R. and Finke, R. 2000. *Acts of Faith: Explaining the Human Side of Religion*. Berkeley: University of California Press.

Stark, R. and Finke, R. 2001. Beyond church and sect: dynamics and stability in religious economies. In *Sacred Markets and Sacred Canopies: Essays on Religious Markets and Religious Pluralism*, ed. T. Jelen. Lanham: Rowman and Littlefield.

Stark, R., Iannaccone, L. and Finke, R. 1996. Religion, science, and rationality. *American Economic Review* 86, 433–7.

Tawney, R. 1926. *Religion and the Rise of Capitalism*. London: Penguin Books, 1990.

Warner, R. 1993. Work in progress toward a new paradigm for the sociological study of religion in the United States. *American Journal of Sociology* 98, 1044–93.

Weber, M. 1904. *The Protestant Ethic and the Spirit of Capitalism*, trans. T. Parsons. London: Routledge, 1992.

religion, economics of

Adam Smith laid the foundations for the economic study of religion in *The Wealth of Nations* (1776, pp. 788–814). He argued that self-interest motivates the clergy; that market forces constrain churches just as they constrain secular firms; that competition improves the quality of religious services provided; and that government regulation distorts the provision of religion, reducing quality and promoting conflict. He also outlined a theory of sectarianism, a theory of religious violence and civility, and a general theory of Church and State.

After this inspired start the economics of religion lay dormant and nearly dead for two centuries. It is now enjoying a rebirth, animated by new data, methods and theory. Economists and other social scientists have harnessed rational choice models and modern empirical tools to study secularization, pluralism, church growth, religious extremism, conversion, fertility, Church–State relations, and more. The field now claims hundreds of papers, scores of contributors, an annual conference and international association (the Association for the Study of Religion, Economics, and Culture), university research centres, and even an AEA subject code (Z12). (New university centres are at Harvard, George Mason University, the University of Southern California and in Canberra, Australia.)

Current research on religion and economics falls into three related subfields: economic theories of religion, studies of religion's economic consequences, and religious assessments of economic policy. Adam Smith's critique of state-supported religion in the *Wealth of Nations* exemplifies the first subfield; Max Weber's 'Protestant ethic' conjecture the second. Together these two subfields constitute *the economics of religion* – the subject of this article. Our goal is to introduce readers to the distinctive economic ideas and models that have enhanced the social-scientific study of religious beliefs, behaviour, and institutions. (For a more complete review of the literature prior to 1998, see Iannaccone, 1998.)

This article makes no attempt to survey the field of religious economics, both because the latter tends to be religion-specific and because it is far from the mainstream of economic research. Religious economics seeks to evaluate economic behaviour and institutions in the light of sacred precepts. Mahmoud El-Gamal's recent book, *Islamic Finance* (2006), is a good example, examining whether current practices in banks that follow Islamic law actually serve the objectives of those laws. The literature on religious economics is large, diverse, and as old as religion itself – including, for example, the many biblical injunctions concerning property, slavery, wages, tithing, interest, wealth and poverty. With the help of economists and philosophers, contemporary clerics continue to debate the merits of income inequality, tax laws, private property, deficit spending, monetary policy, income redistribution, workers rights, interest rates, banking laws, entrepreneurship, government regulation, international trade, debt relief, unionization, entitlement programmes and much more. For representative readings in religious economics, see Oslington (2003) and the *Journal of Markets and Morality* (published by the Acton Institute for the Study of Religion and Liberty).

Social scientists once viewed religion as a dying vestige of our primitive and pre-scientific past. Modern research and contemporary events have destroyed this simplistic view of human history. The rise of radical Islam, the revival of religious practice in much of the former Soviet Union, the explosive growth of Protestantism in Latin America and sub-Saharan Africa, and the contribution of religion to identity (and often to conflict) all over the world testify to the continuing vitality of religion. And although religious belief and activity have declined in many economically advanced countries since the 1960s, the corresponding US data display remarkable stability, whether one focuses on rates of attendance, contributions, membership, or belief. Indeed, religiosity has become one of the strongest predictors of voting patterns and political orientation in America (Glaeser et al., 2006).

We cannot say why economics ignored religion for so long. The brilliant and iconoclastic economist Kenneth Boulding discussed economic features of religion long before the modern revival, but his insights seem to have gone largely unnoticed by economists or sociologists. (Boulding's essays on religion and economics from the 1950s appear in Boulding, 1970.) The other social sciences have subfields dedicated to the study of religion and most have sought to understand the connection between religious and economic trends – the most famous and influential generalization being Max Weber's 'Protestant ethic' thesis. (Weber studied economic history and was well-acquainted with Smith's *Wealth of Nations*. His essay, 'The Protestant Sects and the Spirit of Capitalism', describes how denominational membership enhanced the reputation and business prospects of Americans around 1900: see Weber, 1920. The essay appears to have been inspired by Smith's theory about the ways in which sect membership benefits a poor person: see Smith, 1776.) It seems likely, however, that most economists saw religion as too far removed from the realm of rational choice and market behaviour. We encourage the reader to revisit the issue of religion and rationality after reading this essay.

1 Economics, sociology, and rational choice

Nearly all economic theories rely on the twin assumptions of *rational choice* and *stable preferences*. In the realm of religion, this means choosing which religion, if any, to accept and how extensively to participate in it. These optimal choices need not be permanent. Economic models do a good job explaining differences in religious activity, both over time and across individuals. In keeping with the assumption of stable preference, however, these explanations rarely invoke varied norms, tastes or beliefs. A good economic story explains behaviour in terms of optimal responses to varying circumstances – such as prices, incomes, skills, experiences, technologies or resource constraints.

Although the previous paragraph merely extends modern economic orthodoxy to the realm of religion, it borders on sociological heresy. The commitment of economists to rational choice and stable preferences must be understood as relative, not absolute. Since the late 1970s, economists have devoted a great deal of attention to modelling preference formation. Formal models of religious capital formation (Iannaccone, 1984; 1990) are, in fact, directly linked to Becker's (1996) subsequent work on rational addiction and taste change. Recent work in the fields of behavioural, experimental and evolutionary economics underscores the extent to which choice systematically deviates from rationality; and social norms, social networks and imperfect information constrain choices further still. But it would be wrong to conclude that economists and sociologists now embrace a common 'world view' – as is readily apparent when one contrasts the papers presented by economists and sociologists at the joint annual meetings of the *Association for the Study of Religion, Economics, and Culture* and the *Society for the [Social] Scientific Study of Religion*. Most sociologists remain very sceptical about the value of formal models, rational choice theory and methodological individualism – a legacy passed down from the founders of the field, who promoted sociology as a corrective to errors and omissions of economics (Swedberg, 1990). Add the influence of Weber (1920; 1963), who made 'rationality' central to his analysis of religion while using the word in ways foreign to most contemporary economists, and 'doctrinal' debate is unavoidable. But the overall response to economic forays into religion has been surprisingly ecumenical, with several leading sociologists of religion going so far as to characterize economic theory and market models as 'the new paradigm' for religious research (Stark and Finke, 2000; Warner, 1993; Young, 1997).

2 Households and consumer choice

Economists finally returned to the study of religion in the 1970s, inspired by Gary Becker's path-breaking work on economics of the family. The first papers modelled church attendance and religious contributions as a special form of household production – one that involved trade-offs between time and money inputs, secular versus religious outputs, and present versus afterlife utility (Azzi and Ehrenberg, 1975; Ehrenberg,

1977). Formally, households maximize an intertemporal utility function $U = U(Z_1, \ldots, Z_n, A)$, where Z_t denotes secular consumption activities in period t, and A is consumption activity in an afterlife (of possibly infinite duration). In each period (of this life) households can spend their time, T, and goods, X, on either secular consumption or religious activities, $Z_t = Z(T_{Z_t}, X_{Z_t})$, $R_t = R(T_{R_t}, X_{R_t})$. Religious activities over a lifetime create afterlife consumption, $A = A(R_1, \ldots, R_n)$. Combined with a standard lifetime budget constraint, and on the assumption that the marginal product of religious activity does not decrease with age, the Azzi–Ehrenberg model predicts that religious activity increases with age, *ceteris paribus*. The model also predicts that households with high value of time (high wages) will substitute goods for time in producing religious activity.

As Azzi and Ehrenberg (and many others) have shown, religious activity does tend to increase with age. But it is not at all clear that the Azzi–Ehrenberg model captures the principal cause of this age effect. Ulbrich and Wallace (1983) found that activity increases with age even among those who do *not* believe in the afterlife. And Iannaccone (1984) showed that even in the absence of afterlife expectations, the rational accumulation of religious human capital (that is, rational religious 'addiction') could simultaneously account for the observed age effect as well as observed patterns of religious conversion, intermarriage, and marital stability (see also Lehrer and Chiswick, 1993; Neuman, 1986).

Predictions concerning religious substitution are on much stronger ground. Substitution of goods for time is observed across individuals, households and denominations. Although we cannot directly observe most religious commodities, we can observe the inputs used to produce them. The principal time and money inputs – attendance and contributions – are routinely measured in surveys. More specialized studies provide detailed information on time (such as time devoted to religious services, private prayer and worship, religious charity, and many other religious activities) and money (such as expenditures for special attire, transport, religious books and paraphernalia, sacrificial offerings, and contributions used to finance staff, services and charitable activities of religious organizations). Several studies, including Ehrenberg (1977), Iannaccone (1990), Hungerman (2005) and Gruber (2004), have found that attendance and donations are substitutes – and the recent work demonstrates that substitution remains strong even after one controls for endogeneity bias.

Both in theory and in fact, substitution induces different methods of religious organization and worship across different socio-economic strata. High-income congregations tend to hold shorter services, make heavy use of professional staff and inhabit more elaborate facilities. Longer services, volunteer workers, rented meeting halls and pot-luck dinners are typical of poorer congregations. We observe these differences within denominations and even within congregations (as members improve their socio-economic status), but the differences are especially stark *across* the denominations of a religious tradition, such as Reform Judaism versus Orthodox Judaism or Episcopalians versus Southern Baptists. Many Episcopalian or Presbyterian congregations have plenty of money to cover salaries and operating expenses but remain hard-pressed to recruit volunteers for their choirs, youth programmes, committees and other traditional programmes. For such denominations prosperity has proved a mixed blessing.

Economic trends forced adaptation and none more so than the growth of women's wages and workforce participation. As women have moved into the labour force and overall family earnings have grown, congregations have had to purchase many services formerly supplied by volunteers. The pattern is illustrated by Luidens and Nemeth's (1994) study of expenditure trends in Presbyterian denominations, which found that their (fourfold) increase in real per-capita giving from the 1940s to the 1990s was spent primarily on local congregational services previously supplied by volunteers.

3 Religion, magic and uncertainty

Contemporary theories of rational religious belief begin with just a few assumptions about human nature and the human condition – in essence scarcity, rationality, and the capacity to conceive of supernatural beings or forces (Iannacone and Berman, 2006; cf. Stark and Bainbridge, 1987). From these, they derive a universal demand for supernaturalism and a universal distinction between magic (emphasizing control of impersonal supernatural forces) and religion (emphasizing interaction with supernatural beings). Specialized suppliers arise naturally in both realms, but markets for magic and religion operate quite differently. It is relatively easy to test (and disprove) a magician's ability to control supernatural forces, but much harder to falsify a priest's claims concerning God. In practice, only religion can sustain long-term relationships, high levels of commitment, and moral communities. As Emile Durkheim (1915, p. 42) famously observed, 'there is no church of magic'.

As we shall see in Section 6, a strong religion can induce its members to foreswear all other suppliers of supernatural goods and services. But exclusivity is not a 'natural' outcome. Given the tremendous uncertainty that surrounds the supernatural, rational consumers are inclined to patronize many different suppliers – investing, so to speak, in diversified *portfolios* of supernatural commodities (Iannaccone, 1995). Diversification over different supernatural products and suppliers is pervasive in the (non-communal) market for magic, including the so-called 'New Age' movement. It also prevails in most polytheistic settings, including in the Greco-Roman world, and it remains common in Asian religious traditions. Judaism, Christianity and Islam display a much greater capacity to sustain exclusivity, but (as we

shall discuss in Section 6) only within communal settings that promote collective action, strong social ties, and large investments in religious capital.

4 Religious capital

James Coleman's (1988, p. 97) concept of 'social capital' helps connect rational choice theory to sociological analysis. Iannaccone's (1984; 1990) concept of 'religious capital' offers an analogous bridge from rational choice theory to the sociology of religion. Both concepts are inspired by human capital theory (Becker, 1964; Schultz, 1961), and both emphasize relationships rather than purely individual capacities.

Let S_{R_t} denote the stock of relationships, sensitivities and skills that alter a person's real or perceived benefits from religious activity at time t. Religious commodity production thus depends on current inputs of time and money and the current stock of religious capital, $R_t = R(T_{R_t}, X_{R_t}, S_{R_t})$. The SR variable can encompass a range of concepts, including religious habits, spiritual capital and social capital. Indeed, the mathematical models and empirical analyses remain essentially the same whether one frames the model in terms of the formation of religious 'preferences' or the accumulation of (unobservable) religious 'capital'. In either case, however, religious experience has two key features. First, past experience alters the value of current religious activities and thereby affects rates of religious participation: $S_{R_t} = F(T_{R_t}, X_{R_t}, S_{R_{t-1}})$. Second, most religious experience is 'context specific', yielding maximal benefits within the context of specific relationships, congregations, denominations and traditions. Religious capital remains a distinctive form of social and human capital because religions claim to promote relationships with supernatural beings. This enables religious institutions to maintain exceptionally high levels of commitment, but not without collective production, exclusivity and sacrifice. The bundle appeals less to people with better secular opportunities; hence we observe a 'church-to-sect' *spectrum* of denominations within most religious traditions. For details, see Iannaccone (1995) and Iannaccone and Berman (2006).

Capital models yield predictions that are well supported by evidence, including: (*a*) children tend to choose the same or similar religious denominations as did their parents; (*b*) conversion (like career choice) tends to occur early in adulthood, leaving time to accumulate religion-specific capital; (*c*) interfaith marriage is less likely when religious capital accumulation is high; (*d*) shared-faith marriages lead to higher rates of religious participation (due to complementarities in household production, and not mere sorting of more religious partners into shared-faith marriages); and (*e*) shared-faith marriages have lower rates of divorce and higher rates of fertility (Iannaccone, 1990; Lehrer and Chiswick, 1993; Waite and Lehrer, 2003).

Religion also contributes to extended relationships, social networks and shared norms. Indeed, Coleman's

(1988) seminal article on social capital concerned the impact of (Catholic) religious schools. Empirical studies find that nearly half of all associational memberships, personal philanthropy and volunteering in the United States is church-related leading Putnam (2000) to conclude that '[f]aith communities ... are arguably the single most important repository of social capital in America'. Yet social capital research has yet to give much attention to religion – see, for example, the literature review by Sobel (2002). There remain tremendous opportunities for policy-relevant research on religion's contribution to cooperation (Sosis and Ruffle, 2003), social multipliers (Becker and Murphy, 2000), threshold effects (Granovetter, 1978), public preferences (Kuran, 1995), and much more.

5 Measuring the effects of religious capital

Numerous empirical studies suggest that religious belief and participation yield a wide range of benefits, including mental and physical health, longevity, reduced substance abuse and marital stability (see Koenig, McCullough and Larson, 2001, for an extensive review of the relevant research). The statistical results must, however, be viewed with caution. We lack good instruments for religion on both the supply and demand sides, and most research examines only contemporary American data. Problems of spurious correlation and unobserved heterogeneity may afflict many published studies, as Heaton (2006) notes in his re-analysis of data on religion and crime. On the other hand, the positive association between religion and health has held up despite many different efforts to root out spuriousness, and Freeman's (1986) careful data analysis provides compelling evidence that church attendance really does lead to higher employment rates, higher school attendance, less crime and lower alcohol consumption and drug use among Black males in the United States.

There are many plausible reasons why religiosity might promote beneficial outcomes. As Adam Smith emphasized in his *Theory of Moral Sentiments* (1759), faith in an omniscient deity can solve otherwise intractable problems of self- and social control. Since religion is the quintessential credence good, religious institutions tend to be relatively efficient producers of moral restraint. And there can be no doubt that communities of faith do provide many concrete services while seeking to instil faith in the young, maintain faith among adults and constrain deviant behaviour. The potential benefits from these mechanisms are underscored by the (not undisputed) evidence that religious constraints on sexual conduct have reduced or limited AIDS among Muslims in central Africa and Christians in Uganda (Green, 2003).

6 Club models of religion

Club models have made major contributions to our understanding of 'cults', 'sects' and religious extremism. They also account for characteristics of religion that seem inconsistent with rational choice and risk-aversion – including the success of groups that demand exclusivity, sacrifice and stigma.

Club models start with the fundamental fact that religious 'commodities' are more compelling and gratifying when they are produced and consumed *in groups*. Effective congregations require highly committed members, not mere customers. In this respect, effective congregations are more like families than firms. This suggests that models in which the ith member's religious satisfaction has the form $R_i = F(T_{R_i}, X_{R_i}, Q)$, where Q is an index of the religious inputs of all the *other* group members.

As economists well know, shirking and free-riding constantly threaten collective action, especially in large groups. Paying people to attend church, accept church doctrine or support fellow members fails to solve the problem because a member's commitment and inputs to the group are difficult to observe, and payment rewards the wrong motivations. But the problems *can* be mitigated by seemingly gratuitous costs – the sacrifice and stigma characteristic of deviant religious. Sacrifice and stigmas enhance utility by screening out people who lack commitment and boosting involvement among those who remain in the group. Such groups manifest many distinctive characteristics that empirical researchers have long associated with 'sectarian' religions, including distinctive diet, dress or sexual conduct; physical separation from mainstream society; painful or costly rites; rules that limit social contact with non-members; and prohibitions restricting normal economic or recreational activities. (For more on the modern theory of church and sect, see Iannaccone, 1988; 1991.)

Sect theory also accounts for people's willingness to forgo religious diversification despite the obvious risk associated with most religious assurances. Sectarian religions can maintain levels of commitment and involvement that compensate for the increased risk associated with exclusivity. Corresponding constraints can almost never be sustained in standard, secular markets (nor in the impersonal market for magic) because exclusivity does not enhance the production non-collective goods and services (Iannaccone, 1995).

The club model has received wide acceptance, in part because it fits the data so well. Both cross-sectional surveys and case studies find substantially higher levels of mutual aid and social cohesion in more sectarian religious communities. Iannaccone's work on (mostly Christian) sects has been extended to radical religious Jews (Berman, 2000) and Muslims (Berman and Stepanyan, 2003; Chen, 2004). Despite some lingering debate over the extent of free-rider problems in mainline churches or the actual level of costs imposed by contemporary conservative churches, the basic model remains the natural starting point for studies of high-cost groups. The club model works well not only for religious groups routinely called sects, cults and fundamentalists, but also for communes, gangs, radical militias and terrorist organizations. The basic insight is that an organization designed to exclude free-riders and limit free-riding will be well equipped to exclude defectors and limit defection, the Achilles' heel of militias and terrorists. Thus religious sects prove to be especially effective at terrorism, militia activity and suicide terrorism (Berman, 2003; Berman and Laitin, 2005; Berman and Stepanyan, 2003).

7 Churches as firms

Many religious organizations are legally designated as firms, and many more look surprisingly firm-like. Around the time economists became interested in religious households, several sociologists of religion began thinking of churches as firms, re-examining old data sources with new theories of rational exchange, entrepreneurship and market competition (Finke and Stark, 1988; Stark and Bainbridge, 1985). Finke and Stark (1992) trace the explosive growth of Methodist and Baptist churches in 19th century America to superior marketing, organization and clergy incentives. By the 1990s, these economic and sociological streams of scholarship together included studies of sectarianism, denominational vitality, 'franchising' of religious brands, religious extremism, doctrinal innovation, Church and State, religious markets, non-Western faiths, religious history, and more. Ekelund et al. (1996) analyse numerous features of medieval Catholicism in terms of its monopoly status. Drawing upon standard theories of monopoly, rent seeking and transaction costs, they offer economic explanations for interest rate restrictions, marriage laws, the Crusades, the organization of monasteries, indulgences, and the doctrines of heaven, hell and purgatory (see also Ekleund, Hébert, and Tollison, 2006). Work on churches as firms continues to grow rapidly, in part because firms are easier to model than clubs, but also because the theory of the firm is so rich in predictions and data.

8 Religious markets and government intervention

Whether we think of them as clubs or as firms, individual denominations collectively constitute a *religious market* as long as they provide services that are substitutes. The theories of religion described above predict the existence of different market segments: exclusive 'sects' that operate like clubs, inclusive 'churches' sustained by a core of professionals which are more firm-like, and markets for 'magic' organized around simple exchanges between practitioners and clients.

Almost all economists and sociologists of religion accept the notion that religion in America constitutes a vast competitive market, overflowing with 'products' that range from New Age paraphernalia to orthodox liturgies. Scholars likewise accept that market success requires

entrepreneurship, innovation, and sensitivity to the demands of consumers. As a result, themes that rarely surfaced prior to Finke and Stark's *Churching of America* (1992) now parade as common sense. Even the harshest critics of rational choice theory (such as Bruce, 1999) emphasize the centrality of religious choice in today's world.

The most informative studies closely study how markets actually work. Market-oriented research must carefully address numerous issues, including product attributes, marketing strategies, incentive structures, exchange relationships, consumer characteristics, and Church–State relationships. Andrew Chesnut's (2003) study of rapidly growing religious movements in Latin America illustrates this point by showing how specific religions offer distinctive products that directly address the health- and family-oriented concerns of poor and middle-class women. Anthony Gill (1998) shows that Catholic bishops are much more likely to side with the poor in Latin American countries where Protestant growth threatens the Church's historic monopoly.

Adam Smith (1776, pp. 788–814) argued that established religions face the same incentive problems that plague other state-sponsored monopolies: lack of competition generates a low quality product.

> The teachers of [religion] …, in the same manner as other teachers, may either depend altogether for their subsistence upon the voluntary contributions of their hearers; or they may derive it from some other fund to which the law of their country many entitle them…. Their exertion, their zeal and industry, are likely to be much greater in the former situation than the latter. In this respect the teachers of new religions have always had a considerable advantage in attacking those ancient and established systems of which the clergy, reposing themselves upon their benefices, had neglected to keep up the fervour of the faith and devotion in the great body of the people.

Iannaccone tested Smith's conjecture with modern data. Figure 1 illustrates that within predominantly Protestant countries, church attendance declines sharply as the religious market becomes more concentrated. (The Herfindahl-style 'Protestant Concentration Index' proxies state support for particular religions and has the form $H = \sum_i S_i^2$, where S_i is the population share of the ith Protestant denomination.) All other surveyed measures of religiosity, including belief in God, fall with concentration as well. The data, and Smith's theory, strongly suggest that America's 'religious exceptionalism' is largely a product of religious laissez-faire. North and Gwin (2004) report similar results using a much larger number of countries and more direct measures of Church–State relationships.

Several studies have found positive correlations between local levels of religious diversity and religious activity *within* the USA, including an especially well-crafted

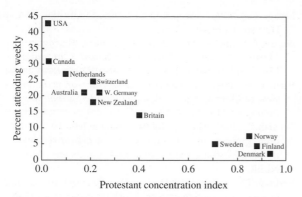

Figure 1 Market concentration and church attendance. *Source*: Gallup polls. See Iannaccone (1991) for details.

study by Finke, Guest, and Stark (1996). But much of this work suffers from specification problems that inevitably arise when rates of religious *membership* (as opposed to more direct measures of belief and participation) are regressed onto (membership-based) measures of religious diversity (Voas, Olson and Crockett, 2002). Nor is it clear that concentration *should* signal the presence of inefficient religious 'monopoly' across cities or states given the nation's minimal barriers to religious entry and innovation. There is, however, strong historical evidence that religious competition raises religious participation (note especially the work of Finke and Stark, 1992, and Olds, 1994, who show that post-colonial disestablishment led to the rapid growth in overall church membership rates, clergy demand, and the – primarily Baptist and Methodist – non-established denominations, while the major denominations that had enjoyed the support of colonial governments – primarily Episcopal, Presbyterian, and Congregational) rapidly *lost* market share).

9 Macroeconomic consequences of religion
Weber's *The Protestant Ethic and the Spirit of Capitalism* (1920) famously claimed that the Calvinist doctrine of predestination triggered a mental revolution within Protestantism that gave rise to modern capitalism. This remains the most influential single conjecture on the macroeconomic effects of religion, which is unfortunate since nearly all subsequent empirical research has shown it to be *false*.

Almost all the capitalist institutions that Weber emphasized actually preceded the Reformation (Stark, 2005; Tawney, 1998). Across and within European countries economic development was uncorrelated with religion (Samuelsson, 1993; Delacroix and Nielsen, 2001). The second country to industrialize was Belgium, which is Catholic. Although Germany and the Netherlands were early developers and majority Protestant, the fastest growth within those countries was among Catholic families of the Rhineland and Amsterdam, which were majority Catholic.

Despite much work by historians and sociologists, there is no consensus concerning the macroeconomic impact of Protestantism, Christianity, monotheism, or religiosity in general. Economists have recently entered this field of enquiry with cross-national studies of survey and census data. Barro and McCleary's (2003) cross-national analysis of survey and census data suggests that belief in hell boosts economic growth whereas frequency of church attendance retards growth (perhaps because the former induces honest and industry whereas the latter reduced time spent working). Using cross-national data from the World Value Surveys, Guiso, Sapienza and Zingales (2003) find that religious beliefs in general, and Christian beliefs in particular, are positively associated with economic attitudes conducive to higher per capita income and growth. Many other economists have begun doing similar studies, but data problems abound. In addition to standard econometric difficulties, there are scarcely any cross-national religious surveys that predate the 1980s; we cannot validate most responses concerning religion; and the meaning of religious participation and belief varies dramatically across cultures. There is better evidence of links between religious and socio-economic variables at the level of individuals and groups than for countries or cultures. For example, average family size and socio-economic status differ quite substantially across different religious groups in both rich and poor nations (Chiswick, 1983; Iyer, 2002). Historical studies do suggest strong relationships between religious and economic institutions, most notably those of medieval Europe. Ekelund, Hébert and Tollison (1996) interpret many distinctive features of medieval Catholicism as forms of rent seeking, and there is no doubt that the Church was by far the most important economic institution in medieval Europe. Richardson (2005) offers strong evidence that the doctrine of purgatory gained rapid acceptance because it served to link religious and economic activities (within guilds) in a way that solved commitment problems that arose because of the social disruptions induced by the Black Death. Timur Kuran (2004) makes a compelling case that specific Islamic legal institutions contributed significantly to the economic decline in Muslim countries relative to those of Europe over the past 500 years.

Recent attempts to promote development in poor and post-Communist countries affirm the importance of ethical norms and moral precepts, many of which have religious foundations (Hayek, 1988, pp. 135–40). Communism may be the most striking example of an economically and socially destructive religion, albeit a religion without traditional deities. In this sense, the strongest evidence for Weberian-style theory may be negative: some powerful systems of belief *do* retard economic progress.

10 Religious militancy
American economists have tended to ignore religion as a subject of public policy, in part because the 'establishment'

and 'free exercise' clauses of the First Amendment radically limit the religious role of government. Within this constitutionally mandated environment of religious laissez-faire (which initially constrained the federal government, but later extended to the states), Americans have maintained extraordinarily high rates of religious activity, diversity and tolerance. But elsewhere, religion remains a major factor in wars, civil unrest and ethnic conflict.

Adam Smith recognized that a detached and lazy clergy was just one cost associated with the marriage of Church and State. When government favours a particular religion in return for its support of the state, the favoured group inevitably demands the suppression of its competitors, and all other groups resist suppression and fight to capture favoured status. It is no coincidence that the USA has remained remarkably free of religious partisanship and militancy while other nations burn with religious conflict. Policies analogous to those embodied in the First Amendment's free exercise and establishment clauses may be key components of the so-called 'war on terror' (Iannaccone and Berman, 2006).

Conclusion
The economics of religion has animated research on secularization, pluralism, church growth, religious extremism, religious markets, the consequences of religion, and more. Forecasting the future of the field is a task best left to prophets. Yet promising areas include the study of non-Western religions, religious militancy, religion and demography, the relationship between religious decline and the growth of the welfare state, and the role of religion in the formation of preferences and social capital. Insights from experimental economics, behavioural economics, game theory, industrial organization, and the economics of information and uncertainty have scarcely been explored. And if the past is any indication of the future, economists still have much to learn from religious historians, sociologists, anthropologists, and other scholars after 200 years of wandering in the secular wilderness.

LAURENCE R. IANNACCONE AND ELI BERMAN

See also **Smith, Adam; social capital.**

Bibliography

Azzi, C. and Ehrenberg, R. 1975. Household allocation of time and church attendance. *Journal of Political Economy* 84, 27–56.

Barro, R.J. and McCleary, R.M. 2003. Religion and economic growth across countries. *American Sociological Review* 68, 760.

Becker, G.S. 1964. *Human Capital: A Theoretical and Empirical Analysis*, 1st edn. New York: Columbia University Press for the NBER.

Becker, G.S. 1996. *Accounting for Tastes*. Cambridge, MA: Harvard University Press.

Becker, G.S. and Murphy, K.M. 2000. *Social Economics: Market Behavior in a Social Environment*. Cambridge, MA; London: Belknap Press of Harvard University Press.

Berman, E. 2000. Sect, subsidy and sacrifice: an economist's view of Ultra-Orthodox Jews. *Quarterly Journal of Economics* 115, 905–53.

Berman, E. 2003. Hamas, Taliban and the Jewish underground: an economist's view of radical religious militias. Working paper No. 10004, Cambridge, MA: NBER.

Berman, E. and Laitin, D.D. 2005. Hard targets: theory and evidence on suicide attacks. Working Paper No. 11740. Cambridge, MA: NBER.

Berman, E. and Stepanyan, A. 2003. Fertility and education in radical Islamic sects: evidence from Asia and Africa. Mimeo, UC San Diego.

Boulding, K.E. 1970. *Beyond Economics: Essays on Society, Religion, and Economics*. Ann Arbor, MI: Ann Arbor Paperbacks.

Bruce, S. 1999. *Choice and Religion: A Critique of Rational Choice Theory*. Oxford and New York: Oxford University Press.

Chen, D. 2004. Club goods and group identity: evidence from Islamic resurgence during the Indonesian financial crisis. Working paper, University of Chicago.

Chesnut, R.A. 2003. *Competitive Spirits: Latin America's New Religious Economy*. New York: Oxford University Press.

Chiswick, B.R. 1983. The earnings and human capital of American Jews. *Journal of Human Resources* 18, 313–35.

Coleman, J.S. 1988. Social capital in the creation of human capital. *American Journal of Sociology* 94(Supplement), S95–S120.

Delacroix, J. and Nielsen, F. 2001. The beloved myth: Protestantism and the rise of capitalism in nineteenth century Europe. *Social Forces* 80, 509–53.

Durkheim, E. 1915. *The Elementary Forms of the Religious Life*. Trans. K.E. Fields. New York: The Free Press, 1995.

Ehrenberg, R.G. 1977. Household allocation of time and religiosity: replication and extension. *Journal of Political Economy* 85, 415–23.

Ekelund, R.B., Hébert, R.F. and Tollison, R.D. 2006. *The Marketplace of Christianity*. Cambridge, MA: MIT Press.

Ekelund, R.B., Hébert, R.F., Tollison, R.D., Anderson, G.M. and Davidson, A.B. 1996. *Sacred Trust: The Medieval Church as an Economic Firm*. New York: Oxford University Press.

El-Gamal, M.A. 2006. *Islamic Finance: Law, Economics and Practice*. Cambridge: Cambridge University Press.

Finke, R., Guest, A.M. and Stark, R. 1996. Mobilizing religious markets: pluralism and religious participation in the empire state, 1850–1865. *American Sociological Review* 61, 203–18.

Finke, R. and Stark, R. 1988. Religious economies and sacred canopies: religious mobilization in American cities, 1906. *American Sociological Review* 53, 41–9.

Finke, R. and Stark, R. 1992. *The Churching of America, 1776–1990: Winners and Losers in Our Religious Economy.* New Brunswick, NJ: Rutgers University Press.

Freeman, R.B. 1986. Who escapes? The relation of church-going and other background factors to the socio-economic performance of black male youths from inner-city poverty tracts. In *The Black Youth Employment Crisis*, ed. R.B. Freeman and H.J. Holzer. Chicago: University of Chicago Press.

Gill, A. 1998. *Rendering Unto Caesar: The Catholic Church and the State in Latin America.* Chicago: University of Chicago Press.

Glaeser, E.L., Ward, B.A., Glaeser, E.L. and Ward, B.A. 2006. Myths and realities of American political geography. *Journal of Economic Perspectives* 20(2), 119–44.

Granovetter, M. 1978. Threshold models of collective behavior. *American Journal of Sociology* 83, 1420–43.

Green, E.C. 2003. *Rethinking Aids Prevention: Learning from Successes in Developing Countries.* Westport, CT: Praeger.

Gruber, J. 2004. Pay or pray? The impact of charitable subsidies on religiosity. Working Paper No. 10374. Cambridge, MA: NBER.

Guiso, L., Sapienza, P. and Zingales, L. 2003. People's opium? Religion and economic attitudes. *Journal of Monetary Economics* 50, 225–82.

Hayek, F.A. 1988. *The Fatal Conceit: The Errors of Socialism.* Chicago: University of Chicago Press.

Heaton, P. 2006. Does religion really reduce crime? *Journal of Law and Economics* 49, 147–72.

Hungerman, D. 2005. Are church and state substitutes? Evidence from the 1996 Welfare Reform. *Journal of Public Economics* 89, 2245–67.

Iannaccone, L.R. 1984. Consumption capital and habit formation with an application to religious participation. Ph.D. thesis, University of Chicago.

Iannaccone, L.R. 1988. A formal model of church and sect. *American Journal of Sociology* 9 (Supplement), 241–68.

Iannaccone, L.R. 1990. Religious participation: a human capital approach. *Journal for the Scientific Study of Religion* 29, 297–314.

Iannaccone, L.R. 1991. The consequences of religious market regulation: Adam Smith and the economics of religion. *Rationality and Society* 3, 156–77.

Iannaccone, L.R. 1992. Sacrifice and stigma: reducing free-riding in cults, communes, and other collectives. *Journal of Political Economy* 100, 271–92.

Iannaccone, L.R. 1995. Risk, rationality, and religious portfolios. *Economic Inquiry* 38, 285–95.

Iannaccone, L.R. 1998. An introduction to the economics of religion. *Journal of Economic Literature* 36, 1465–95.

Iannaccone, L.R. and Berman, E. 2006. Religious extremism: the good, the bad, and the deadly. *Public Choice* 128, 109–29.

Iyer, S. 2002. *Demography and Religion in India.* Oxford and New York: Oxford University Press.

Koenig, H.G., McCullough, M.E. and Larson, D.B. 2001. *Handbook of Religion and Health.* Oxford and New York: Oxford University Press.

Kuran, T. 1995. *Private Truths, Public Lies: The Social Consequences of Preference Falsification.* Cambridge, MA: Harvard University Press.

Kuran, T. 2004. *Islam and Mammon: The Economic Predicaments of Islamism.* Princeton: Princeton University Press.

Lehrer, E.L. and Chiswick, C.U. 1993. Religion as a determinant of marital stability. *Demography* 30, 385–404.

Luidens, D. and Nemeth, R. 1994. Congregational vs. denominational giving: an analysis of giving patterns in the Presbyterian church in the United States and the reformed church in America. *Review of Religious Research* 36, 111–22.

Neuman, S. 1986. Religious observance within a human capital framework: theory and application. *Applied Economics* 18, 1193–202.

North, C.M. and Gwin, C.R. 2004. Religious freedom and the unintended consequences of the establishment of religion. *Southern Economic Journal* 71, 103–17.

Olds, K. 1994. Privatizing the church: disestablishment in Connecticut and Massachusetts. *Journal of Political Economy* 102, 277–97.

Oslington, P., ed. 2003. *Economics and Religion.* Cheltenham: Edward Elgar.

Putnam, R.D. 2000. *Bowling Alone: The Collapse and Revival of American Community.* New York: Simon and Schuster.

Richardson, G. 2005. Craft guilds and Christianity in late-medieval England: a rational-choice analysis. *Rationality and Society* 17, 139–89.

Samuelsson, K. 1993. *Religion and Economic Action: The Protestant Ethic, the Rise of Capitalism, and the Abuses of Scholarship.* Toronto: University of Toronto Press.

Schultz, T.W. 1961. Investment in human capital. *American Economic Review* 51, 1–17.

Smith, A. 1759. *The Theory of Moral Sentiments.* Ed. D.D. Raphael and A.L. Macfie, Indianapolis: Liberty Fund, 1984.

Smith, A. 1776. *An Inquiry into the Nature and Causes of the Wealth of Nations.* Indianapolis: Liberty Classics, 1981.

Sobel, J. 2002. Can we trust social capital? *Journal of Economic Literature* 40, 139–54.

Sosis, R.H. and Ruffle, B.J. 2003. Religious ritual and cooperation: testing for a relationship on Israeli religious and secular kibbutzim. *Current Anthropology* 44, 713–22.

Stark, R. 2005. *The Victory of Reason: How Christianity Led to Freedom, Capitalism, and Western Success.* New York: Random House.

Stark, R. and Bainbridge, W.S. 1985. *The Future of Religion.* Berkeley and Los Angeles: University of California Press.

Stark, R. and Bainbridge, W.S. 1987. *A Theory of Religion.* Bern: Peter Lang Publishing.

Stark, R. and Finke, R. 2000. *Acts of Faith: Explaining the Human Side of Religion.* Berkeley and Los Angeles: University of California Press.

Swedberg, R. 1990. *Economics and Sociology: Redefining their Boundaries: Conversations with Economists and Sociologists*. Princeton: Princeton University Press.

Tawney, R.H. 1926. *Religion and the Rise of Capitalism*. Repr., with new introduction by Adam B. Seligman. New Brunswick, NJ, and London: Transaction, 1998.

Ulbrich, H. and Wallace, M. 1983. Church attendance, age, and belief in the afterlife: some additional evidence. *Atlantic Economic Journal* 11, 44–51.

Voas, D., Olson, D.V.A. and Crockett, A. 2002. Religious pluralism and participation: why previous research is wrong. *American Sociological Review* 67, 212–30.

Waite, L.J. and Lehrer, E.L. 2003. The benefits from marriage and religion in the United States: a comparative analysis. *Population and Development Review* 29, 255–75.

Warner, R.S. 1993. Work in progress toward a new paradigm in the sociology of religion. *American Journal of Sociology* 98, 1044–93.

Weber, M. 1920. *The Protestant Ethic and the Spirit of Capitalism*. Trans. S. Kalberg, 3rd edn. Los Angeles: Roxbury, 2002.

Weber, M. 1963. *The Sociology of Religion*, trans. E. Fischoff. Boston: Beacon Press.

Young, L.A. 1997. *Rational Choice Theory and Religion: Summary and Assessment*. New York: Routledge.

rent

'Rent' is the payment for use of a resource, whether it be land, labour, equipment, ideas, or even money. Typically the rent for labour is called 'wages'; the payment for land and equipment is often called 'rent'; the payment for use of an idea is called a 'royalty'; and the payment for use of money is called 'interest'. In economic theory, the payment for a resource where the availability of the resource is insensitive to the size of the payment received for its use is named 'economic rent' or 'quasi-rent' depending on whether the insensitivity to price is permanent or temporary.

To early economists, 'rent' meant payments for use of land; Ricardo, in particular, called it the payment for the 'uses of the original and indestructible powers of the soil' (Ricardo, 1821, p. 33). Subsequently, in recognition that a distinctive feature of what was called 'land' was its presumed indestructibility (i.e. insensitivity of amount supplied to its price), the adjective 'economic' was applied to the word 'rent' for any resource the supply of which is indestructible (maintainable for ever at no cost) and non-augmentable, and hence invariant to its price. In the jargon of economics, the quantity of present and future available supply is completely inelastic with respect to price, a situation graphically represented by a vertical supply line in the usual 'Marshallian' price-quantity graphs.

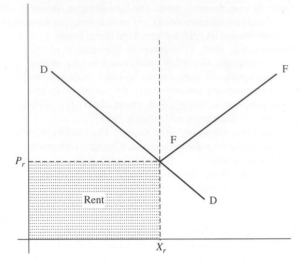

Figure 1

Economic rent

The concept of 'economic rent' is graphically depicted by the standard demand and supply lines in Figure 1 with a vertical supply curve (quantity supplied invariant to price) at the amount X_r. At all prices the supply is constant. The entire return to the resource is an 'economic rent'. If the aggregate quantity of such resources may in the future be increased by production of more indestructible units of the resource in response to a higher price (but the amount available at any moment is fixed regardless of the rent for its services), the supply line at the current moment is vertical. The supply curve for future amounts slopes upward from the existing amount, as depicted by the line FF in Figure 1. The long run rent would be P_r and the equilibrium stock would be X_r: at that equilibrium stock the 'market supply' (in Marshall's terminology) would be a vertical line. Thus, the supply of indestructible units would have depended on past anticipated prices about the present prices, but the supply of current units would be insensitive to the current price or rent. The return could be called 'economic rent', except that no convention has been developed with respect to the terminology for this situation of indestructible but augmentable resources.

Quasi-rent

Closely related to 'economic rent' is 'quasi-rent', a term apparently initiated by Alfred Marshall (Marshall, 1920, pp. 74, 424–6). Because virtually every existing resource is unresponsive to a change in price for at least some very small length of time, the return to every resource is like an 'economic rent' for at least a short interval of time. In time, the supplied amount will be altered, either by production or non-replacement of current items. Yet, the fact that the amount available is not instantly affected by price led to the term 'quasi-rent', which denotes a return, variations in which do not affect the current amount supplied to the demander but do affect the supply in the future.

If a rental (payments) stream to an existing resource is not sufficient to recover the costs incurred in its production the durability of that existing resource will nevertheless enable the resource to continue to provide services, at least for some limited time. In other words, because of the resource's durability it will continue for some interval to yield services even at a rent insufficient to recover its cost of production, but sufficient for current costs of use including interest on its salvage value (which is its highest value in some other use). Any excess over those current costs is a 'quasi-rent'.

Quasi-rent resembles an 'economic rent' in that it exceeds the amount required for its current use, albeit temporarily – except that a flow of rents that did not cover all 'quasi-rent' would preserve it for only a finite future interval, after which the resource would be diminished until not worth more than its salvage value. If the resource received a payment exceeding all the initially anticipated and the realized costs of production and operation, it will have achieved a profit, that is, more than pure interest on the resource's investment cost. The question exists as to whether 'quasi-rent' means just that portion of the rent in excess of the minimum operating costs over the remaining life of the asset, or all the excess, including profits, if any. Convention seems still to be missing. Marshall seems to have excluded interest on the investment as well as any profits from what he called quasi-rents, so that any excess over variable costs of operation were partitioned into quasi-rents, interest on investment and profits (Marshall, 1920, pp. 412, 421, 622).

Composite quasi-rent

'Composite quasi-rent' was another important, but subsequently ignored, concept coined by Marshall (Marshall, 1920, p. 626). When two separately owned resources are so specific to each other that their joint rent exceeds the sum of what each could receive if not used together, then that joint rent to the pair was called 'composite quasi-rent'. The two resources presumably already had been made specific to each other (worth more together than separately) by some specializing interrelated investments. Marshall cited the example of a mill and a water power site, presumably a mill built next to a dam to serve the mill, each possibly separately owned. One or both of the parties could attempt to hold up or extract a portion of the other party's expropriable quasi-rent. It is interesting to quote Marshall about this situation:

> The mill would probably not be put up till an agreement had been made for the supply of water power for a term of years; but at the end of that term similar difficulties would arise as to the division of the aggregate producer's surplus afforded by the water power

and the site with the mill on it. For instance, at Pittsburg when manufacturers had just put up furnaces to be worked by natural gas instead of coal, the price of the gas was suddenly doubled. And the history of mines affords many instances of difficulties of this kind with neighbouring landowners as to rights of way, etc., and with the owners of neighbouring cottages, railways and docks (Marshall, 1920, p. 454).

A reason for attributing importance to the concept of 'composite quasi-rent' is now apparent. If it arises with resources that have been made specific to each other in the sense that the service value of each depends on the other's presence, the joint value of composite quasi-rent might become the object of attempted expropriation by one of the parties, especially by the one owning the resource with controllable flow of high alternative use value. To avoid or reduce the possibility of this behaviour, a variety of preventative arrangements, contractual or otherwise, can be used prior to making the investments in resources of which at least one will become specific to the other. These include, among a host of possibilities: joint ownership, creation of a firm to own both, hostages and bonding, reciprocal dealing, governmental regulation, and use of insurers to monitor uses of interspecific assets. This is not the place to discuss these arrangements, beyond asserting that without the concept of 'quasi-rent' and especially 'expropriable quasi-rent' – which Marshall called 'composite quasi-rent' – a vast variety of institutional arrangements would otherwise be inexplicable as a means of increasing the effectiveness of economic activity.

Though Marshall briefly mentioned similar problems between employers and employees, I have not found any subsequent exposition by him about the precautionary contractual arrangements and institutions that attempt to avoid this problem, which has become a focus of substantial important research on what is called, variously, 'opportunism, shirking, expropriable quasi-rents, principal–agent conflicts, monitoring, problems of measuring performance, asymmetric information, etc.'.

Ricardian rent

The rents accruing to different units of some otherwise homogeneous resource may differ and result in differences of rent over the next most valued use, differences that are called 'Ricardian rents'. This occurs where the individual units, all regarded as of the same 'type' in other uses, are actually different with respect to some significant factor for its use *here*, though this factor, which is pertinent *here*, is irrelevant in any other uses. Examples of such factors can be location, special fertility, or talent that is disregarded in the other potential uses. For some questions, the inaccurate 'homogenization' can be a convenient simplification, but for explaining each unit's actual rents, it can lead to confusion and misunderstanding. The service value, hence rents, for the use of the services here may differ, though equal in every relevant respect elsewhere. Whether the specific use uniqueness is created by natural talent or sheer accident, the special differences in use value here imply differences in payments, often called 'Ricardian rents' to distinguish them from differences in rents (prices) obtained because of monopolizing or unnatural restrictions on any potential competitors, which may lead to higher rents, called 'monopoly rents' for the protected resources.

Differential rents

'Differential rents' are another category representing rent differences in a sort of reverse homogeneity. Units of resource that are equal with respect to their value in use *here* differ among themselves in their values of use elsewhere. This can be represented graphically as in Figure 2. The differential rents of successive units are represented by the differences between the price line and the curve RR, which arrays the units from those with the lowest alternative use values to the highest, a curve labelled RR. The arrayed units are not homogeneous for uses elsewhere, so even if identical for use here, calling them successive units of the same good is misleading. They are not totally homogeneous; if they were, each unit would have the same as any other unit's use value and rent elsewhere. A curve like RR is equivalent to Marshall's particular expenses curve, which arrayed units according to each individual unit's cost of production, or use value elsewhere, from lowest to highest (Marshall, 1920, p. 810n). The difference between price or rent here and the value on the RR curve is called 'producers' surplus' or 'differential rent'. In sum, 'Ricardian rents' indicate differences in rents to units that are equal in their best alternative use values, but different in their rent value here, while 'differential

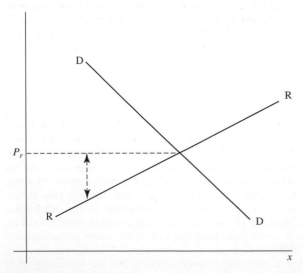

Figure 2

rents' are the premia to units that are the same value here but different in their best alternative use values.

It is worth digressing to note that an upward rising true supply curve, which reflects increasing marginal costs of production, is different from the RR curve. In the true supply curve the area between the supply curve and the price line does not represent any of the above mentioned rents nor 'producers' surplus' (as it does with the RR curve). It is the portion of earnings of the supplier that exceed the variable costs and are applicable to cover the costs (possibly past investment costs) that are invariant to the rate of output. That area does not represent any excess of rental or sale value of units produced over their full costs, since only the variable costs are under the marginal cost curve. It represents the classic distribution of income to capital, if, for example, labour is presumed to be a variable input and capital a fixed input.

High rents a result, not cause, of high prices

An earlier unfortunate analytic confusion occurred in the common misimpression that high rents of land made its products more expensive. Thus the high rent of land in New York was and is still often believed to make the cost of living, or the cost of doing business, higher in New York. Or higher rent for some agricultural land is believed to increase the cost of growing corn on that land. Proper attention to the meaning of 'demand' and 'costs' would have helped avoid that confusion. Demand *here* for some unit for resource is the highest value use of that resource if used *here*. The cost of using it *here* is the highest valued forsaken alternative act elsewhere. For any resource the cost of its use *here* is its best value elsewhere, that is, its demand elsewhere. Land rent is high for 'this' use because the land's value in some other use is high. The reason the rent is high *here* and can be paid is that its use value *here* is bid by competitors for its use *here* into the offered rent and exceeds the value in some other use. The product of the land can get higher price *here*; that is why the rent is bid up so high, even though the particular winning bidder then believes a high price of the products must be obtained because the rent was high, rather than the reverse. As with every marketable resource, its highest value use *here* determines its rent, rather than the reverse. It was the implication of this kind of analysis that Marshall attempted to summarize in the famous aphorism, which he attributed to Ricardo (1817): 'Rent does not enter into [Money] Cost of production' (Marshall, 1890, p. 482).

Probably the source of the confusion in believing that high rents of land caused high prices for products produced on expensive land is that an individual user of that expensive resource has to be able to charge a higher price for the product, if the rent is to be covered. Bidders for that land compete for the right to the land that can yield a service worth so much – though to any individual successful bidder that rent has to be paid regardless of how well the successful bidder may be at actually achieving the highest valued use of the land. Hence it may appear to an individual bidder that the rent determines the price that must be charged, rather than, as is the correct interpretation, the achievable high valued use enables the high bid for the land for the person best able to detect and achieve that highest valued use.

Function of rent

Some people were aware of this bidding for the 'land' and concluded that the rent served no social purpose, since the land would exist anyway. But the high receipt resulting from competitive bidding for its uses serves a useful purpose. It reveals which uses are the highest valued and directs the land to that use. In principle, a 100 per cent tax on the land rent would not alter its supply (assuming initially that 'land' is the name of whatever has a fixed indestructible supply). This would be correct if in this case the 'owner' of the land had any incentive left to heed the highest bidder where the highest bid determines the rent. The assertion assumes that somehow the highest valued use can be known and that amount of tax be levied without genuine bona fide competitive bids for its use, a dubious if not plainly false proposition.

Monopoly rent

Let the word 'monopoly' denote any seller whose wealth potential is increased by restrictions on other potential competitors, restrictions that are artificial or contrived in not being naturally inevitable. Laws prohibiting others from selling white wine, or opening restaurants, or engaging in legal practice are examples. It should be immediately emphasized that this does not imply nor is it to be inferred that all such restrictions are demonstrably undesirable. Nevertheless, the increased wealth potential is a 'monopoly rent'. Whether it is realized by the monopolist as an increase in wealth depends upon the costs of competing for the imposition of such restrictions. Competition for 'monopoly rents' may transfer them to, for example, politicians who impose the restrictions, and in turn may be dissipated by competition among politicians seeking to be in a position to grant such favours. The 'monopoly rents' may be dissipated (by what is often called 'rent-seeking' competition for such monopoly status of rights to grant it) into competitive payments for resources that enable people to achieve status to grant such restrictions. Those who initially successfully and cheaply obtained such 'monopoly' status may obtain a wealth increase, just as successful innovators obtain a profit stream before it is eliminated by competition from would-be imitators.

ARMEN A. ALCHIAN

See also **Malthus, Thomas Robert; Marshall, Alfred; Ricardo, David; Thünen, Johann Heinrich von; West, Edward.**

Bibliography

Marshall, A. 1890. *Principles of Economics.* 1st edn, London: Macmillan.
Marshall, A. 1920. *Principles of Economics.* 8th edn, London: Macmillan; reprinted, 1946.
Ricardo, D. 1821. *Principles of Political Economy and Taxation.* 3rd edn, London: Dent Dutton, 1965.

rent control

Rent controls of one kind or another affect roughly 40 per cent of the world's urban dwellers. Rent control is usually thought of as a policy applied to private markets, but publicly provided housing (for example, much urban housing in Russia and in China) is also subject to controls. In addition to regulations governing rents, controls often address additional contract features such as security of tenure and required maintenance. Actual rent control regimes vary enormously in their design and in their effects.

History

Rent controls are often instituted in response to a major economic or political shock which limits the responsiveness of the housing market. Controls were introduced in the Second World War in Europe, North America, and, under European colonial influence, much of the developing world as well. Most jurisdictions in the United States and Canada removed controls in the post-war years; however, controls of varying degrees of stringency were maintained in much of Europe and the developing world. Poorer countries tend to have more stringent regimes, though enforcement patterns vary at least as much as *de jure* codes.

Exactly why controls exist, or at least are retained after wartime or similar emergencies are clearly over, is still debated. An obvious point of political economy is that there are more tenants than landlords; but there is little correlation between the fraction of a country's population renting and the stringency of controls, according to Malpezzi and Ball (1991). On the other hand the relatively small number of US cities with rent control tend to have large renter populations, notably New York. Fischel (2001) presents several interesting conjectures about the political economy of controls, notably that homeowners might ally with landlords to oppose controls because they fear negative spillovers from reduced maintenance of stringently controlled buildings, as well as shifting property tax burdens. The strong opposition to relaxation of controls in New York, while nearby uncontrolled jurisdictions see little agitation for imposition, might be analysed in Kahneman and Tversky's (1979) loss aversion framework. A clear understanding of the political economy of controls awaits future research.

Features

One key feature is whether regulations set the level of rents, or control increases in rent. Others include how controlled rents are adjusted for changes in costs (with cost pass-through provisions, or adjustments for inflation); how close the adjustment is to changes in market conditions; how it is applied to different classes of units; or whether rents are effectively frozen over time. Other key provisions which vary from place to place include breadth of coverage, how initial rent levels are set, treatment of new construction, whether rents are reset for new tenants, and tenure security provisions. Rent control's effects can vary markedly depending on these specifics, and on market conditions, as well as enforcement practices.

Theory

Rent control can be analysed as an implicit tax on housing capital. In the simplest case, where imposition of controls reduces the price of an existing stock of rental housing, the tax is borne by landlords for the benefit of tenants. Over time, as the market adjusts to controls, the incidence of the 'tax' becomes more complicated.

Much of the debate in the literature about the efficacy of controls stems from maintained assumptions about the nature of the housing market, and the regulator, in turn. The first question is: is the housing market best modelled as a competitive market, or one where landlords have market power, for example from information asymmetries? If the former, then clearly rent control reduces the efficiency of the rental market, although the magnitude of such effects can be debated, and distributional arguments remain. If the latter, a second question readily follows: does the regulator have sufficient knowledge, and an appropriately designed set of regulations, to improve on the market outcome? Arnott (1995) ably reviews the contrast between competitive and 'market power' theoretical approaches, and also discusses why it is so difficult to resolve these issues empirically.

Whatever one's priors about market power, there are many alternative adjustment mechanisms which can arise in a notionally controlled market. Four of the adjustments can be embodied in rent control laws: (*a*) indexing (keeping real rents constant), (*b*) reassessment for new tenants, (*c*) differential pricing of new and existing units, and (*d*) differential pricing for upgraded units. Three are market responses which many would generally consider undesirable outcomes, namely, (*e*) outright evasion, (*f*) side payments such as key money, (*g*) adoption by tenants of maintenance expenditures, and (*h*) accelerated depreciation and abandonment, (*i*) distortions in consumption, not only in the composite housing services but also crowding, length of stay, mobility and tenure choice.

Key questions are: What are the efficiency losses from controls? Are the benefits to some tenants worth the costs? Do they redistribute income as intended? Several broad approaches have been taken in the empirical literature to answer these questions.

Static analysis

One of the first published studies of the costs and benefits of rent control is Olsen's. Using data from New York City in 1968, Olsen (1972) found the average controlled rent for an apartment was $999 a year (for comparison, the average income was $6,229). Olsen first estimated how much the controlled units would rent for in the absence of controls. The average estimated uncontrolled rent for controlled units was $1,405, implying a subsidy (static cost to landlords) of $406. Olsen next estimated how much households in controlled units would spend in the uncontrolled market, given their income and family size. The average estimated market expenditure for the controlled households was $1,470, indicating that they consumed slightly less housing than they would have in the free market. Olsen then computed the economic benefit of rent control to each surveyed controlled tenant using a simple consumer surplus model. Olsen's estimate of the average net benefit is $213, little more than half the gross subsidy of $406.

Examining the distribution of these benefits among controlled households, Olsen found the annual benefit decreased by about one cent for every dollar of additional income, $9 a year of head's age, and $69 per additional household member. Rent control in New York City in 1968 appears to redistribute income, but very weakly, and in no way proportional to its cost.

A number of other studies have been carried out along these lines (Malpezzi and Ball, 1991, review several). For example, in Cairo, Egypt, monthly rents for a typical unit are less than 40 per cent of estimated market rents. But 'key money' (illegal upfront payments to landlords) and other side payments make up about a third of the difference.

Dynamic analysis

Murray et al. (1991) is an early study of rent control dynamics. A simulation model was used to predict the time path of rents and the quantity of housing services given alternative control regimes. The magnitude of the effects varied substantially with details of the regime. In general, Murray et al. find that dynamic losses can be substantial; in fact they outweigh static consumer's surplus losses by as much as a factor of 18. Generally tenant benefits are were substantially less than landlord costs; the transfer efficiency in three representative cases ranged from 65 per cent to 83 per cent.

Another potential dynamic effect of controls, with possible spillovers to labour markets, is reduced household mobility. Several studies, such as Munch and Svarer (2002) and Simmons-Moseley and Malpezzi (2006), find that household mobility is inversely related to the estimated net benefits received from a control regime.

Given their potential importance, dynamic effects of controls are understudied. For example, no one has yet credibly analysed the effects of controls on the aggregate supply of housing. Reviews of the theoretical literature by Arnott (1995), and of the empirical literature by Turner and Malpezzi (2003), point out that empirical work lags theory in this area. Malpezzi and Ball (1991) did find that countries with stricter rent control regimes invested less in housing, in the aggregate; but while the analysis accounted for income and demographics on the demand side, other potential constraints on housing supply (for example, land use constraints, financial constraints) were not well specified. Since these may well be correlated with the strength of controls, these results cannot be viewed as the final word. Given the myriad ways real world regimes work, and the variety of possible ways around controls (legal and illegal) the size of the net aggregate effect on supply remains unknown.

Distributional issues

Such evidence as exists casts doubt on controls' effectiveness as income transfer mechanisms. In Kumasi and Rio, benefits were found to be somewhat 'progressive' in the common sense of the term (larger benefits to poorer households). On the other hand, in Cairo and Bangalore, no relationship was found between the benefits gained from reduced rent and household income, because rent control is not well targeted to low-income groups (Malpezzi and Ball, 1991). In fact, research on New York controls by Glaeser and Luttmer (2003) suggests that previous research largely *underestimates* the misallocation of housing under controls, and that, because of excess demand for controlled units, benefits are more or less randomly distributed.

Another questionable assumption behind redistribution as a rationale for controls is the notion that landlords are rich and tenants are poor. In Cairo, Kumasi and Bangalore, the income of tenants and landlords was compared; and, while the landlords' median income was higher in all three, there was significant overlap. In Cairo, for example, about 25 per cent of tenants had incomes that were higher than the landlord median, and about 25 per cent of landlords had incomes lower than the tenant median. There is no guarantee the transfers will occur only from high-income landlords to low-income tenants.

Most careful empirical studies find that at least some tenants are, on balance, worse off under controls because of constraints on housing consumption. And in markets with significant uncontrolled sectors, rent controls can drive up the price of uncontrolled housing, an important unintended consequence further complicating the incidence of its costs.

STEPHEN MALPEZZI

See also **housing policy in the United States; housing supply.**

Bibliography

Arnott, R. 1995. Time for revisionism on rent control. *Journal of Economic Perspectives* 9(1), 99–120.
Fischel, W.A. 2001. *The Homevoter Hypothesis: How Home Values Influence Local Government Taxation, School Finance, and Land-Use Policies*. Cambridge, MA: Harvard University Press.
Glaeser, E.L. and Luttmer, E.F.P. 2003. The misallocation of housing under rent control. *American Economic Review* 93, 1027–46.
Kahneman, D. and Tversky, A. 1979. Prospect theory: an analysis of decision under risk. *Econometrica* 47, 263–91.
Malpezzi, S. and Ball, G. 1991. Rent control in developing countries. Discussion Paper No. 129. Washington, DC: World Bank.
Munch, J.R. and Svarer, M. 2002. Rent control and tenancy duration. *Journal of Urban Economics* 52, 542–60.
Murray, M.P., Rydell, C.P., Barnett, C.L., Hillstead, C.E. and Neels, K. 1991. Analyzing rent control: the case of Los Angeles. *Economic Inquiry* 29, 601–25.
Olsen, E.O. 1972. An econometric analysis of rent control. *Journal of Political Economy* 80, 1081–100.
Simmons-Mosley, T.X. and Malpezzi, S. 2006. Household mobility in New York City's regulated rental housing market. *Journal of Housing Economics* 15, 38–62.
Turner, B. and Malpezzi, S. 2003. A review of empirical evidence on the costs and benefits of rent control. *Swedish Economic Policy Review* 10, 11–56.

rent seeking

The term 'rent-seeking' was introduced by Ann O. Krueger (1974), but the relevant theory had already been developed by Gordon Tullock (1967). The basic and very simple idea is best explained by reference to Figure 1. On the horizontal axis we have as usual the quantity of some commodity sold, on the vertical axis its price. Under competitive conditions the cost would be the line labelled PP and that would also be its price. Given a demand curve,

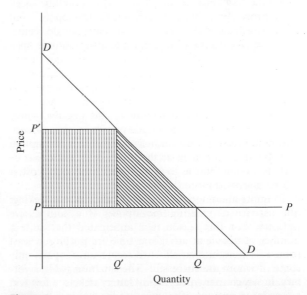

Figure 1

DD, quantity Q would be sold at that price. If a monopoly were organized, it would sell Q' units at a price of P'.

The traditional theory of monopoly argued that the net loss to society is shown by the shaded triangle, which represents the consumer surplus that would have been derived from the purchase of those units between Q' and Q, that are now neither purchased nor produced. The dotted rectangle, on the other hand, has traditionally been regarded simply as a transfer from the consumers to the monopolist. Since they are all members of the same society, there is no net social loss from this transfer.

This argument tends to annoy students of elementary economics (because they don't like monopolists), but until the development of the work on rent seeking it was nevertheless thought to be correct by most economists. Its basic problem, however, is that it assumes that the monopoly is created in a costless manner, perhaps by an act of God, whereas in fact real resources are used to create monopolies.

Most discussion of rent seeking has tended to concentrate on those monopolies that are government created or protected, probably because these are observed to be the commonest and strongest. It should be kept in mind, however, that purely private monopolies are possible – indeed, some actually exist. Concentration on government-created monopolies (or restrictions of various sorts that increase certain peoples' income) is probably reasonable, granted the contemporary frequency of such activities. Nevertheless, as we point out below there are certain significant areas where private rent seeking causes net social loss.

In the initial work both of Tullock and Krueger it was assumed that profit-seeking businessmen would be willing to use resources in an effort to obtain a monopoly, whether it was privately or government sponsored, up to the point where the last dollar so invested exactly counterbalanced the improved probability of obtaining the monopoly. From this it was deduced that the entire dotted rectangle (Figure 1) would be exhausted. Although this assumption is open to question (see Tullock, 1980), for the time being we will continue to assume that in effect there is no transfer from purchasers to the monopolist, but simply a social loss which comes from the fact that resources have been invested in unproductive activity, i.e. the negatively productive activity of creating a trade restriction of some sort. Theoretical reasons exist for believing that this assumption probably does not fit perfectly anywhere, but it is just as likely to overestimate as to underestimate the social cost; it will be discussed more thoroughly below.

To quote an aphorism frequently used in rent seeking: 'the activity of creating monopolies is a competitive industry.' For this reason it is anticipated that quite a number of people at any given time are putting at least some resources into an effort to secure a monopoly, only some of whom are successful. The situation is like a lottery, in which many people buy lottery tickets, a few win a very large amount of money and the rest lose, perhaps large or small amounts, depending on how much they have committed. In almost all existing lotteries, of course, the total investment of resources by the gamblers is considerably greater than the total payoff, whereas here it is still assumed that total resources committed to rent-seeking equal the total monopoly profits.

Thus the activity of creating monopolies could both absorb very large resources, particularly those resources that take the form of exceptionally talented individuals who devote their attention to this difficult and highly rewarded activity, and lead to considerable redistribution of wealth in the community. Suppose that ten different lobbyists go to Washington representing ten different associations, and each spends one million dollars over the course of a couple of years in the hope of influencing Congress to provide them with a monopoly. Only one of the lobbyists is successful and the monopoly turns out to have a present discounted value of ten million dollars. There is a substantial redistribution of resources from the unsuccessful lobbyists to the successful.

This substantial redistribution has occurred simultaneously with a considerable waste of resources in general, both because these highly intelligent people could otherwise be doing something of higher productivity and because the economy's use of resources has been further distorted by the creation of the monopoly. Further, although so far the discussion has been primarily about monopoly, actually very many possible interventions in the market process raise the same problem. A simple maximum or minimum price may have very large redistributive effects and the people who thus benefit may put considerable resources into receiving them. Of course there are many situations in which one lobbyist is pushing for a particular restriction and another lobbyist is pushing against it. The second activity is sometimes called 'rent avoidance', but it is costly and of course would not exist if there were not also rent seeking activity.

Another area is simple direct transfers. A tax on A for the purpose of paying B will lead to lobbying activity for the tax on the part of B and against it on the part of A. The total of these two lobbying activities could very well equal the total amount transferred (or prevented from being transferred), although one or other of these entrepreneurs will of course gain if his lobby is successful. Assume that A puts in $50 for lobbying to get $100 from B and B puts in $50 lobbying against that. Regardless of the outcome, one party will gain $50 from his lobbying. Society has lost $100.

Of course it is not true that everyone in society is in an equally good position to seek rents. Some kinds of interest are more readily organized than others and we would anticipate that they would win. There are however very many such interests and anyone who spends any time in Washington quickly realizes that there is a major industry engaged in just this kind of activity.

Actual social cost however is clearly very much greater than the mere cost of the various lobbying organizations in Washington. In particular it is normally necessary for the rent-seeking group to undertake directly productive activities in a way that is markedly inefficient, because it is necessary to introduce a certain element of deception into the process. In 1937, when the US Civil Aeronautics Board was organized, it would not have been politically feasible to put a direct tax on purchasers of airline tickets and use it to pay off the stockholders of the airline companies. Regulation, which has a similar effect but at a very much higher cost to the users of airlines per dollar of profit to the owners, was however, politically possible. The necessity of using inefficient methods of transferring funds to the potential beneficiary, because the efficient methods would be just too open and above board, is often one of the major costs of rent seeking. The rent avoidance lobbyist would have had too easy a time if the proposal had been a tax on uses of airlines for the benefit of the stockholders.

Note that in this case the argument against rent seeking turns out also to be an argument against political corruption. Suppose you are in a society which has an exchange control system and that it is possible to buy foreign currency by bribing an official in the exchange control office. This is the kind of situation dealt with by Krueger (1974), who was able to obtain a measure of the total social cost in Turkey and India where the amounts of the necessary bribes were well known; the cost varied from 7–15 per cent of the total volume of transactions.

Traditionally economists have tended to view this kind of bribery as in itself desirable, because it gets around an undesirable regulation. However, it leads to rent seeking. In this case the rent seeking does not come from the users of the permits but from the competition to get into the position where you can receive the bribe. Throughout the underdeveloped world, large numbers of people take fairly elaborate educational programmes which have no real practical value for their future life and engage in long periods of complicated political manoeuvring in hope that they will be appointed, let us say, a customs inspect in Bombay. Since these young men have a free career choice presumably the expected returns from this career are the same as in any other. The difference is that a doctor, say, begins earning money immediately on completing medical school whereas the young man who has studied economics and is now trying to obtain appointment as customs inspector will have a considerable period of time in which he is not appointed at all. Indeed, there will probably be enough such candidates that he has only perhaps one chance in five of being so appointed. The total cost of the rent seeking is the inappropriate education and the political manoeuvring of the five people of whom only one is appointed.

So far we have assumed that the total cost of rent seeking is the present discounted value of the income stream represented by the dotted rectangle in Figure 1. This assumes a special form for the function which 'produces' the monopoly or other privilege. It must be linear, with each dollar invested having exactly the same payoff in probability of achieving the monopoly as the previous dollar (Tullock, 1980). Most functions do not have this form, instead they are either increasing or decreasing cost functions.

If the organizing of private monopolies, or of influencing the government into giving you public monopolies, is subject to diseconomies of scale, then total investment in rent seeking will be less than the total value of the rents derived even if we assumed a completely competitive market with completely free entry. When there are economies of scale the situation is even more unusual. Either there is no equilibrium at all or there is a pseudo-equilibrium, in which total investment to obtain the rents is greater than the rents themselves. This is called a pseudo-equilibrium, because although it meets all the mathematical requirements for an equilibrium, it is obviously absurd to assume that people would, to take a single example, pay $75.00 for a 50–50 chance of $100.

Obviously, what is needed is empirical research, and an effort to measure the production functions appropriate to rent seeking. So far, however, no one has been able to develop a very good way of making such measurements. It seems likely that it would be easier to measure the costs of generating political influence than of private monopolies, if only because many of the expenditures used to influence the government appear in accounts in various places. The costs of private monopolies on the other hand, tend to be much more readily concealed. This does not mean that they do not exist.

The reader has no doubt been wondering what is wrong with rents and why we concern ourselves deeply with rent seeking. The answer to this is that the term itself is an unfortunate one. Obviously, we have nothing against rents when they are generated by, let us say, discovering a cure for cancer and then patenting it. Nor do we object to popular entertainers like Michael Jackson earning immense rents on a rather unusual collection of natural attributes together with a lot of effort on his part to build up his human capital. On the other hand, we do object to the manufacturer of automobiles increasing the rent on his property, and his employees increasing the rent on their union memberships, by organizing a quota against imported cars. All of these things are economic rents, but strictly speaking the term 'rent seeking' applies only to the latter. Its meaning might be expanded to seeking rents from activities which are themselves detrimental. The man seeking a cure for cancer is engaged in an activity which clearly is not detrimental to society. Thus we may observe immediately that activities aimed at deriving rents cover a continuum, but that the term 'rent seeking' is only used for part of that continuum.

The analysis of 'rent seeking' has been one of the most stimulating fields of economic theory in recent years. The realization that the explanation of the social cost of monopoly which was contained in almost every elementary text in economics was wrong, or at the very least seriously incomplete, came as quite a surprise. Revision of a very large part of economic theory in order to take this error into account is necessary. And history also needs to be revised. That J.P. Morgan was an organizer of cartels and monopolies during most of his life is well known, as is the fact that he received very large fees for this, fees which were part of the rent seeking cost of generating these monopolies. It is possible to argue that as a stabilizing factor in the banking system, Morgan more than repaid to the United States the social cost of his monopolistic activities in industry. But that there was a very large rent seeking cost is obvious. This cost is in addition to the deadweight cost of the monopolies.

To date, research on rent seeking has to a considerable extent changed our way of looking at things. We now talk of a great deal of government activity as rent seeking on the part of somebody or other. It was known that special interest existed, but we have traditionally tended to underestimate its cost greatly because we looked only at the deadweight costs of the distortion introduced into the economy. The realization that the actual cost is much greater socially, that the large-scale lobbying industry is truthfully a major social cost, is new although presumably, at all times, anyone who thought about the matter must have realized that these highly talented people could produce more in some other activity.

GORDON TULLOCK

See also **bribery; directly unproductive profit-seeking (DUP) activities.**

Bibliography

Buchanan, J., Tollison, R. and Tullock, G., eds. 1980. *Toward a Theory of the Rent-Seeking Society.* College Station, Texas: Texas A and M University Press.
Krueger, A.O. 1974. The political economy of the rent-seeking society. *American Economic Review* 64, 291–303.
Tullock, G. 1967. The welfare cost of tariffs, monopolies, and theft. *Western Economic Journal (now Economic Inquiry)* 5, 224–32.
Tullock, G. 1980. Efficient rent seeking. In Buchanan, Tollison and Tullock, eds., 91–112.

repeated games

Repeated games provide a formal and quite general framework to examine why self-interested agents manage to cooperate in a long-term relationship.

Formally, repeated games refer to a class of models where the same set of agents repeatedly play the same game, called the 'stage game', over a long (typically, infinite) time horizon. In contrast to the situation where agents interact only once, *any* mutually beneficial outcome can be sustained as an equilibrium when agents interact repeatedly and frequently. A formal statement of this fact is known as the folk theorem.

Repeated games and the general theories of efficiency

Thanks to the developments since the mid-1970s, economics now recognizes three general ways to achieve efficiency: (*a*) competition; (*b*) contracts; and (*c*) long-term relationships. For standardized goods and services, with a large number of potential buyers and sellers, promoting market competition is an effective way to achieve efficiency. This is formulated as the classic First and Second Welfare Theorems in general equilibrium theory. There are, however, other important resource allocation problems which do not involve standardized goods and services. Resource allocation within a firm or an organization is a prime example, as pointed out by Ronald Coase (1937), and examples abound in social and political interactions. In such cases, aligning individual incentives with social goals is essential for efficiency, and this can be achieved by means of *incentive schemes* (penalties or rewards). The incentive schemes, in turn, can be provided in two distinct ways: by a formal contract or by a long-term relationship. The penalties and rewards specified by a formal contract are enforced by the court, while in a long-term relationship the value of future interaction serves as the reward and penalty to discipline the agents' current behaviour. The theory of contracts and mechanism design concern the former case, and the theory of repeated games deals with the latter. These theories provide general methods to achieve efficiency, and have become important building blocks of modern economic theory.

An example: collusion of gas stations and the trigger strategy

Consider two gas stations located right next to each other. They have identical and constant marginal cost c (the wholesale price of gasoline) and compete by publicly posting their prices. Suppose their joint profit is maximized when they both charge $p = 10$, whereby each receives a large profit π. Although this is the best outcome for them, they have an incentive to deviate. By slightly undercutting its price, each can steal all the customers from its opponent, and its profit (almost) doubles. The only price free from such profitable deviation is $p = c$, where their profit is equal to zero. In other words, the only Nash equilibrium in the price competition game is an *inefficient* (for the gas stations) outcome where both charge $p = c$. This situation is the rule rather than the exception: the Nash equilibrium in the stage game, the only outcome that agents can credibly achieve in a one-shot interaction, is quite often inefficient for them. This is because agents seek only their private benefits, ignoring the benefits or costs of their actions for their rivals.

In reality, however, gas stations enjoy positive profits, even when there is another station nearby. An important reason may well be that their interaction is not one-shot. Formally, the situation is captured by a *repeated game*, where the two gas stations play the price competition game (the stage game) over an infinite time horizon $t = 0, 1, 2, \ldots$. Consider the following repeated game strategy:

1. Start with the optimal price $p = 10$.
2. Stick to $p = 10$ as long as no player (including oneself) has ever deviated from $p = 10$.
3. Once anyone (including oneself) deviated, charge $p = c$ for ever.

This can be interpreted as an explicit or implicit agreement of the gas stations: charge the monopoly price $p = 10$, and any deviation triggers cut-throat price competition ($p = c$ with zero profit). Let us now check whether each player has any incentive to deviate from this strategy. Note that, if neither station deviates, each enjoys profit π every day. As we saw above, a player can (almost) double its stage payoff by slightly undercutting the agreed price $p = 10$. Hence the short-term gain from deviation is at most π. If one deviates, however, its future payoff is reduced from π to zero in each and every period in the future. Now assume that the players discount future profits by the *discount factor* $\delta \in (0, 1)$. The number δ measures the value of a dollar in the next period. The discounted future loss is $\delta\pi + \delta^2\pi + \cdots = \frac{\delta}{1-\delta}\pi$. If this is larger than the short-term gain from defection (π), no one wants to deviate from the collusive price $p = 10$. The condition is $\pi \leq \delta/(1 - \delta)\pi$, or equivalently, $1/2 \leq \delta$.

Next let us check whether the players have an incentive to carry out the threat (the cut-throat price competition $p = c$). Since $p = c$ is the Nash equilibrium of the stage game, charging $p = c$ in each period is a best reply if the opponent always does so. Hence, the players are choosing mutual best replies. In this sense, the threat of $p = c$ is credible or self-enforcing.

In summary, under the strategy defined above, players are choosing mutual best replies *after any history*, as long as $1/2 \leq \delta$. In other words, the strategy constitutes a *subgame perfect equilibrium* in the repeated game. Similarly, in a general game, any outcome which Pareto dominates the Nash equilibrium can be sustained by a strategy which reverts to the Nash equilibrium after a deviation. Such a strategy is called a *trigger strategy*.

Three remarks: multiple equilibria, credibility of threat and renegotiation, and finite versus infinite horizon

A couple of remarks are in order about the example. First, the trigger strategy profile is not the only equilibrium of the repeated game. The repetition of the stage game Nash equilibrium ($p = c$ for ever) is also a subgame perfect equilibrium. Are there any other equilibria? Can we

characterize *all* equilibria in a repeated game? The latter question appears to be formidable at first sight, because there are an infinite number of repeated game strategies, and they can potentially be quite complex. We do have, however, some complete characterizations of all equilibria of a repeated game, such as folk theorems and self-generation conditions as will be discussed subsequently.

Second, one may question the credibility of the threat ($p = c$ for ever). In the above example, credibility was formalized as the subgame perfect equilibrium condition. According to this criterion, the threat $p = c$ is credible because a *unilateral* deviation by a *single* player is never profitable. The threat $p = c$, however, may be upset by *renegotiation*. When players are called upon to carry out this grim threat after a deviation, they may well get together and agree to 'let bygones be bygones'. After all, when there is a better equilibrium in the repeated game (for example, the trigger strategy equilibrium), why do we expect the players to stick to the inefficient one ($p = c$)? This is the problem of *renegotiation proofness* in repeated games. The problem is trickier than it appears, however, and economists have not yet agreed on what is the right notion of renegotiation proofness for repeated games. The reader may get a sense of difficulty from the following observation. Suppose the players have successfully renegotiated away $p = c$ to play the trigger strategy equilibrium again. This is self-defeating, however, because the players now have an incentive to deviate, as they may well anticipate that the threat $p = c$ will be again subject to renegotiation and will not be carried out. For a comprehensive discussion of this topic (and also of a number of major technical results on repeated games), see an excellent survey by D. Pearce (1990).

Third, let me comment on the assumption of an *infinite* time horizon. Suppose that the gas stations are to be closed by the end of next year (due to a new zoning plan, for example). This situation can be formulated as a *finitely* repeated game. On the last day of their business, the gas stations just play the stage game, and therefore they have no other choice but to play the stage game equilibrium $p = c$. In the penultimate day, they rationally anticipate that they will play $p = c$ *irrespective* of their current action. Hence they are effectively playing the stage game in the penultimate day, and again they choose $p = c$. By induction, the *only* equilibrium of the finitely repeated price competition is to charge $p = c$ in *every* period. The impossibility of cooperation holds no matter how long the time horizon is, and it is in sharp contrast to the infinite horizon case.

Although one may argue that players do not really live infinitely long (so that the finite horizon case is more realistic), there are some good reasons to consider the infinite horizon models. First, even though the time horizon is finite, if players do not know in advance exactly *when* the game ends, the situation can be formulated as an infinitely repeated game. Suppose that, with probability $r > 0$, the game ends at the end of any given period. This implies that, *with probability 1*, the game ends in a finite horizon. Note, however, that the expected discounted profit is equal to $\pi(0) + (1 - r)$ $\delta\pi(1) + (1 - r)^2\delta^2\pi(2) + \cdots$, where $\pi(t)$ is the stage payoff in period t. This is identical to the payoff in an infinitely repeated game with discount factor $\delta' = (1 - r)\delta$. Second, the drastic 'discontinuity' between the finite and infinite horizon cases in the price competition example hinges on the uniqueness of equilibrium in the stage game. Benoit and Krishna (1985) show that, if each player has multiple equilibrium payoffs in the stage game, the long but finite horizon case enjoys the same scope for cooperation as the infinite horizon case (the folk theorem, discussed below, approximately holds for T-period repeated game, when $T \to \infty$).

The repeated game model

Now let me present a general formulation of a repeated game. Consider an infinitely repeated game, where players $i = 1, 2, \ldots, N$ repeatedly play the same stage game over an infinite time horizon $t = 0, 1, 2, \ldots$. In each period, player i takes some action $a_i \in A_i$, and her payoff in that period is given by a stage game payoff function $g_i(a)$, where $a = (a_1, \ldots, a_N)$ is the action profile in that period. The repeated game payoff is given by

$$\Pi_i = \sum_{t=0}^{\infty} g_i(a(t))\delta^t,$$

where $a(t)$ denotes the action profile in period t and $\delta \in (0, 1)$ is the discount factor. It is often quite useful to look at the *average payoff* of the repeated game, which is defined to be $(1 - \delta)\Pi_i$. Note that, if one receives the same payoff x in each period, the repeated game payoff is $\Pi_i = x + \delta x + \delta^2 x + \cdots = x/(1 - \delta)$. This example helps to understand the definition of average payoff: in this case $(1 - \delta)\Pi_i$ is indeed equal to x, the payoff per period.

A *history* up to time t is the sequence of realized action profiles before t: $h^t = (a(0), a(1), \ldots, a(t - 1))$. A *repeated game strategy* for player i, denoted by s_i, is a complete contingent action plan, which specifies a current action after any history: $a_i(t) = s_i(h^t)$ (a minor note: to determine $a_i(0)$, we introduce a dummy history h^0 such that $a_i(0) = s_i(h^0)$). A repeated game strategy profile $s = (s_1, \ldots, s_N)$ is a *subgame perfect equilibrium* if it specifies mutual best replies after any history.

The folk theorem

Despite the fact that a repeated game has an infinite number of strategies, which can be arbitrarily complicated, we do have a *complete* characterization of equilibrium payoffs. The folk theorem shows exactly which payoff points can be achieved in a repeated game.

Before stating the theorem, we need to introduce a couple of concepts. First, let us determine the set of physically achievable average payoffs in a repeated game.

Note that, by alternating between two pure strategy outcomes, say u and v, one may achieve any point between u and v as the average payoff profile. Hence, an average payoff profile can be a weighted average (in other words, a convex combination) of pure strategy payoff profiles in the stage game. Let us denote the set of all such points by V. Formally, the set of *feasible average payoff profiles* V is the smallest convex set that contains the pure strategy payoff profiles of the stage game.

Second, let us determine the points in V that cannot possibly be an equilibrium outcome. For example, if a player has an option to stay out to enjoy zero profit in each period, it is a priori clear that her equilibrium average payoff cannot be less than zero. In general, there is a payoff level that a player can guarantee herself in any equilibrium, and this is formulated as the *minimax* payoff. Formally, the minimax payoff for player i is defined as $\underline{v}_i = \min_{\alpha_{-i}} \max_{\alpha_i} g_i(\alpha)$, where $\alpha = (\alpha_1, \ldots, \alpha_N)$ is a mixed action profile (α_i is a probability distribution over player i's pure actions) and $g_i(\alpha)$ is the associated expected payoff. To understand why min and max are taken in that particular order, consider the situation where player i always *correctly anticipates what others do*. If player i knows that others choose $\alpha_{-i} = (\alpha_1, \ldots \alpha_{i-1}, \alpha_{i+1}, \ldots, \alpha_N)$, he can play a best reply against α_{-i} to obtain $\max_{\alpha_i} g_i(\alpha)$. Note well that $\max_{\alpha_i} g_i(\alpha)$ is a function of α_{-i}. In the worst case, where others take the most damaging actions α_{-i}, player i obtains the minimax payoff (this is exactly what the definition says). From this definition it is clear that, in any equilibrium of the repeated game, *the average payoff to each player is at least her minimax payoff*. In any equilibrium, each player correctly anticipates what others do, and simply by playing the stage game best reply in each period, any player can make sure that her average payoff is more than her minimax payoff. (A comment: we consider mixed strategies in the definition of the minimax payoff because in many games the minimax payoff is smaller when we consider mixed strategies.)

From what we saw, now it is clear that the set of equilibrium average payoff profiles of a repeated game is *at most* $V^* = \{v \in V | \forall i \ v_i > \underline{v}_i\}$. (The points with $v_i = \underline{v}_i$ are excluded to avoid minor technical complications.) The set V^* is called the *feasible and individually rational payoff set*. This is the set of physically achievable average payoff profiles in the repeated game where each player receives more than her minimax payoff. The folk theorem shows that any point in this 'maximum possible region' can indeed be an equilibrium outcome of the repeated game. (Throughout this article, I maintain a minor technical assumption that each player has a finite number of actions in the stage game.)

Folk theorem In an N-player infinitely repeated game, any feasible and individually rational payoff profile $v \in V^*$ can be achieved as the average payoff profile of a subgame perfect equilibrium when the discount factor δ is close enough to 1, provided that either N = 2, or N $\geq$ 3 and no two players have identical interests.

Formally, no two players have identical interests if there are *no* players i and j $(i \neq j)$ whose payoffs satisfy $g_i(a) = bg_j(a) + c$, $b > 0$ (that is, no two players have the same preferences over the stage game outcomes). This is a 'generic' condition that is almost always satisfied: the case where players have identical interests is very special in the sense that the equality $g_i(a) = bg_j(a) + c$ fails by even a slight change of the payoff functions. Hence, the folk theorem provides a general theory of efficiency: it shows that, for virtually any game, any mutually beneficial outcome can be achieved in a long term relationship, if the discount factor is close to 1. *Although game-theoretic predictions quite often depend on the fine details of the model, this result is a notable exception for its generality.*

The crucial condition in the folk theorem is a high discount factor. The discount factor δ may measure the (subjective) patience of a player, or, it may be equal to $1/(1 + r)$, where r is the interest rate per period. Although the discount factor may not be directly observable (in particular, in the former case), it should be high when one period is short. Hence, an empirically testable implication is that players who have daily interaction (such as the gas stations in our example) have a better scope for cooperation than those who interact only once a year. An important message of the folk theorem is that a high *frequency of interaction* is essential for the success of a long term relationship.

The name 'folk theorem' comes from the fact that game theorists had anticipated that something like it should be true long before it was precisely formulated and proved. In this sense, the assertion had been folklore in the game theorist community. The proof is, however, by no means obvious, and there is a body of literature to prove the theorem in various degrees of generality. Early contributions include Aumann (1959), Friedman (1971) and Rubinstein (1979). The statement above is based on Fudenberg and Maskin (1986) and its generalization by Abreu, Dutta and Smith (1994). The proof is constructive: a clever strategy, which has a rather simple structure, is constructed to support any point in V^*.

Repeated games versus formal contracts

To discuss the scope of applications, I now compare a long-term relationship (repeated game) and a formal contract as a means to enforce efficient outcomes. As our gas station example shows, quite often an agent has an incentive to deviate from an efficient outcome, because it increases her private returns at the expense of the social benefit. Such a deviation can be deterred if we impose a sufficiently high penalty so that the *incentive constraint*

gain from deviation $\leq$ penalty

is satisfied. This is the basic and common feature of repeated games and contracts. A formal contract explicitly specifies the penalty and it is enforced by the court. In repeated games, the penalty is indirectly imposed through future interaction. In this sense the theory of repeated games can be regarded as the theory of *informal or relational contracts*.

When is a long-term relationship a better way to achieve cooperation than a formal contract? First, a long-term relationship is useful when a formal contract is too costly or impractical. For example, it is often quite costly for a third party (the court) to verify whether there was any deviation from an agreement, while defections may be directly observed by the players themselves. In practice, what constitutes 'cooperation' is often so fuzzy or complicated that it is hard to write it down explicitly, although the players have a common and good understanding about what it is. 'Pulling enough weight' in a joint research project may be a good example. In those situations, a long-term relationship is a more practical way to achieve cooperation than a formal contract. In fact, a classic study by Macaulay (1963) indicates that the vast majority of business transactions are executed without writing formal contracts. Second, there are some cases where a court powerful enough to enforce formal contracts simply does not exist. For example, in many problems in development economics and economic history, the legal system is highly imperfect. Even for developed countries in the modern age, there are no legal institutions which have enough binding power to enforce international agreements. Hence, repeated games provide a useful framework to address such problems as the organization of medieval trade, informal mutual insurance in developing countries, international policy coordination, and measures against global warming. Lastly, there is no legal system to enforce cartels or collusion, because the existing legal system refuses to enforce any contract that violates antitrust laws. Hence a long-term relationship is the only way to enforce a cartel or collusive agreement.

Is the folk theorem a negative result?

The theory of repeated games based on the folk theorem is often criticized because it does not, as the criticism goes, have any predictive power. The folk theorem basically says that anything can be an equilibrium in a repeated game. One could argue, however, that this criticism is misplaced if we regard the theory of repeated games as a theory of informal contracts. Just as anything can be enforced when the party agrees to sign a binding contract, in repeated games any (feasible and individually rational) outcome is sustained if the players agree on an equilibrium. Enforceability of a wide range of outcomes is the essential property of effective contracts, formal or informal. The folk theorem correctly captures this essential feature.

This criticism is valid, however, in the sense that the theory of repeated games does not provide a widely accepted criterion for equilibrium selection. When we regard a repeated game as an informal contract, where the players explicitly try to agree on which equilibrium to play, the problem of equilibrium selection boils down to the problem of bargaining. In such a context, it is natural to assume that an efficient point (in the set of equilibria) is played. In the vast majority of applied works of repeated games with symmetric stage games (such as the gas stations example), it is common to look at the best symmetric equilibrium. In contrast, when players try to find an equilibrium through trial and error, the theory of repeated games is rather silent about which equilibrium is likely to be selected. A large body of computer simulation literature on the evolution of cooperation, pioneered by Axelrod (1984), may be regarded as an attempt to address this issue.

Imperfect monitoring

So far we assumed that players can perfectly observe each other's actions. In reality, however, long term relationships are often plagued by *imperfect monitoring*. For example, a country may not verify exactly how much CO_2 is emitted by neighbouring countries. Workers in a joint project may not directly observe each others' effort. Electronic appliance shops often offer secret discounts for their customers, and each shop may not know exactly how much is charged by its rivals. In such situations, however, there are usually some pieces of information, or *signals*, which imperfectly reveal what actions have been taken. Published meteorological data indicates the amount of CO_2 emission, the success of the project is more likely with higher effort, and a shop's sales level is related (although not perfectly) to its rivals' prices.

According to the nature of the signals, repeated games with imperfect monitoring are classified into two categories: the case of *public monitoring*, where players commonly observe a public signal, and the case of *private monitoring*, where each player observes a signal that is not observable to others. Hence, the CO_2 emission game and the joint-project game are examples with imperfect public monitoring (published meteorological data and the success of the project are publicly observed), while the secret price-cutting game by electronic shops is a good example with imperfect private monitoring (one's sales level is private information).

This difference may appear to be a minor one, but, somewhat surprisingly, it is not. The imperfect *public* monitoring case shares many features with the *perfect* monitoring case, and we now have a good understanding of how it works. In contrast, the imperfect private monitoring case is not fully understood, and we have only some partial characterizations of equilibria. In what follows, I sketch the main results in the imperfect public and private monitoring cases.

Imperfect public monitoring

At first sight, this case might look much more complicated than the perfect monitoring case, but those two cases are similar in the sense that they share a *recursive structure*. Consider the set W^* of all average payoff profiles associated with the subgame perfect equilibria of a perfect monitoring repeated game. Any point $w \in W^*$ is a weighted average of the current payoff g and the continuation payoff $w' : (1 - \delta)g + \delta w'$. The continuation payoff typically changes when a player deviates from g, in such a way that the short-term gain from deviation is wiped out. Subgame perfection requires that all continuation payoffs are chosen from the equilibrium set W^*. In this sense, W^* is generated by itself, and this stationary or recursive structure turns out to be quite useful in characterizing the set of equilibria.

The set of equilibria in an imperfect public monitoring game also shares the same structure. Consider the equilibria where the public signal determines which continuation equilibrium to play. When a player deviates from the current equilibrium action, it affects both her current payoff and (through the public signal) her continuation payoff. The equilibrium action should be enforceable in the sense that any gain in the former should be wiped out in the latter, and this is easier when the continuation payoff admits large variations. Formally, given the range of continuation payoffs W, we can determine the set $B(W)$ of enforceable average payoffs. The larger the set W is, the more actions can be enforced in the current period (and therefore the larger the set $B(W)$ is). As in the perfect monitoring case, the equilibrium payoff set $W = W^*$ generates itself: it satisfies the *self-generation condition* of Abreu, Pearce and Stacchetti (1990) $W \subseteq B(W)$. W^* is the largest (bounded) set satisfying this condition, and the condition is in fact satisfied with equality. Conversely, it is easy to show that any (bounded) set satisfying the self-generation condition is contained in the equilibrium payoff set W^*.

This provides a simple and powerful characterization of equilibria, which is an essential tool to prove the folk theorem in the imperfect public monitoring case. The folk theorem shows that, despite the imperfection of monitoring, we can achieve any feasible and individually rational payoff profile under a certain set of conditions.

Before presenting a formal statement, let me sketch the basic ideas behind the folk theorem. When monitoring is imperfect, players have to be punished when a 'bad' signal outcome ω is observed, and this may happen with a positive probability even if no one defects. For example, in the joint project game, the project may fail even though everyone works hard. A crucial difference between the perfect and imperfect monitoring cases is that, in the latter, punishment occurs *on the equilibrium path*. The resulting welfare loss, however, can be negligible under certain conditions.

Consider a two-player game, where the probability distribution of the signal $\omega \in \Omega = \{\omega^1, \ldots, \omega^K\}$, when no one defects, is given by $P^* = (p^*(\omega^1), \ldots, p^*(\omega^K))$ in Figure 1. Suppose that each player's defection changes the probability distribution to exactly the same point P'. Then, there is absolutely no way to tell which player deviates, so that the only way to deter a defection is to *punish all players simultaneously*, when a 'bad' outcome emerges. This means that surplus is thrown away, and we are bound to have substantial welfare loss. Now consider a case where different players' actions affect the signal asymmetrically: player 1's defection leads to point P', while the defection by player 2 leads to P''. In this asymmetric case, one can *transfer* future payoff from player 1 to 2 when player 1's defection is suspected. Under such a transfer, surplus is never thrown away, and this enables us to achieve efficiency.

More precisely, consider the normal vector x of the hyperplane separating P' and P'' in the figure, and let $w_1 = x$ and $w_2 = -x$ be the continuation payoffs of player 1 and player 2 respectively. Figure 1 indicates that player 1's expected continuation payoff $P \cdot w_1 = P \cdot x$ is reduced by her own defection ($P' \cdot x < P^* \cdot x$). Similarly, player 2's defection reduces her expected continuation payoffs ($P^* \cdot (-x) > P''(-x)$). Note that this asymmetric punishment scheme does not reduce the joint payoff, because by construction $w_1 + w_2$ is identically equal to 0. This is an essential idea behind the folk theorem under imperfect public monitoring: *When different players' deviations are statistically distinguished, asymmetric punishment deters defections without welfare loss.*

When can we say that different players' deviations are statistically *distinguished*? Note well that the above construction is impossible when P'' is exactly in between P^* and P' (that is, when P'' is a convex combination of P^* and P'). Such a case can be avoided if P^*, P' and P'' are linearly independent. The linear independence of the equilibrium signal distribution (P^*) and the distributions associated with the players' unilateral deviations (P' and P''), is a precise formulation of what it means when the signal 'statistically distinguishes different players' deviations'.

Let us now generalize this observation. Given an action profile (for simplicity of exposition, assume it is pure) to be sustained, there is an associated signal distribution P^*. Consider any pair of players i and j, and let $|A_k|$ be the number of player k's actions ($k = i, j$) in the stage game. Since each player $k = i, j$ has $|A_k| - 1$ ways to deviate, we have $|A_i| + |A_j| - 2$ signal distributions associated with their unilateral deviations. If those distributions and the equilibrium distribution P^*, altogether $|A_i| + |A_j| - 1$ vectors, are linearly independent, we say that the signal can discriminate between deviations by i and deviations by j. This is called the *pairwise full rank condition*. This holds only when the dimension of the signal space ($|\Omega|$, the number of signal outcomes) is larger than the number of those vectors (that is, $|\Omega| \geq |A_i| + |A_j| - 1$). Conversely, if this inequality is satisfied, the pairwise full rank condition holds 'generically' (that is, it holds unless the signal distributions have a very special structure, such as exact symmetry). This leads us to the folk theorem under imperfect public monitoring (this is a restatement of Fudenberg, Levine and Maskin, 1994, in terms of genericity):

Folk theorem under imperfect public monitoring
Suppose that the signal space is large enough in the sense that $|\Omega| \geq |A_i| + |A_j| - 1$ holds for each pair of players i and j. Then, for a generic choice of the signal distributions and the stage game, any feasible and individually rational payoff profile $v \in V^*$ can be asymptotically achieved by a sequential equilibrium as the discount factor δ tends to 1.

In contrast to the perfect monitoring case, the proof is non-constructive. Rather than explicitly constructing equilibrium strategies, the theorem is proved by showing that any smooth subset of V^* is self-generating. In fact, the exact structure of the equilibrium strategy profile to sustain, for example, an efficient point is not so well understood. Sannikov (2005) shows that detailed structure of equilibrium strategies can be obtained if the model is formulated in continuous time.

Imperfect private monitoring

Now consider the case where all players receive a private signal about their opponents' actions. Although this has a number of important applications (a leading example is the secret price cutting model), this part of research is still in its infancy. Hence, rather than just summarizing definitive results as in the previous subsections, I explain in somewhat more technical detail the source of difficulties and the nature of existing approaches.

The difficulties come from a subtle but crucial difference from the perfect or public monitoring case. I explain below the difference from three viewpoints, in the increasing order of technicality.

1. In the perfect or public monitoring case, players share a mutual understanding about when and whom to

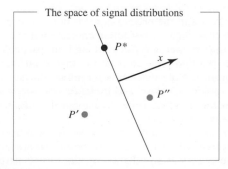

The space of signal distributions

P^*

x

P''

P'

Figure 1

punish. They can cooperate to implement a specific punishment, and, more importantly, they can mutually provide the incentives to carry out the punishment. This convenient feature is lost when players have diverse private information about each others' actions.

2. In the perfect or public monitoring case, public information directly tells the opponents' future action plans. In the private monitoring case, however, each player has to draw statistical inferences about the history of the opponents' private signals to estimate what they are going to do. The inferences quickly become complicated over time, even if players adopt relatively simple strategies.

3. In the perfect or public monitoring case, the set of equilibria has a recursive structure, in the sense that a Nash equilibrium of the repeated game is always played after any history. Now consider a Nash equilibrium of, for example, the repeated Prisoner's Dilemma with imperfect private monitoring. After the equilibrium actions in the first period, say (C,C), players condition their action plans on their private signals ω_1 and ω_2. Hence the continuation play is a *correlated equilibrium*, where it is common knowledge that the probability distribution of the correlation device (ω_1, ω_2) is given by $p(\omega_1, \omega_2 | C,C)$. When player 1 deviates to D in the first period, however, the distribution of correlation device is *not* common knowledge: player 1 knows that it is $p(\omega_1, \omega_2 | D,C)$, while player 2 keeps the equilibrium expectation $p(\omega_1, \omega_2 | C,C)$. Hence, after a deviation, the continuation play is no longer a correlated equilibrium in the usual sense. In addition, the space of the correlation device (the history of private signals) becomes increasingly rich over time. Therefore, the equilibria in the private monitoring case do not have a compact recursive structure; a continuation play is chosen from a different set, depending on the history.

One way to get around these problems is to allow communication (Compte, 1998; Kandori and Matsushima, 1998). In their equilibrium, players truthfully communicate their private signal outcomes in each period. The equilibrium is constructed in such a way that each player's report of her signal is utilized to discipline *other* players and does *not affect one's own continuation payoff*. This implies that each player is indifferent about what to report, and therefore truth telling is *a* best reply. Such an equilibrium, which depends on the history of publicly observable messages, works in much the same way as the equilibria in the public monitoring case. Hence, with communication, the folk theorem is obtained in the private monitoring case.

The remaining issue is to characterize the equilibria in the private monitoring case without communication. From the viewpoint of potential applications, this is important, because collusion or cartel enforcement is a major applied area of repeated games, where communication is explicitly prohibited by the antitrust law.

One may expect that, when players' private information admits sufficient positive correlation, an equilibrium can be constructed in a similar way to the public monitoring case. Sekiguchi (1997) is the first to construct a non-trivial (and nearly efficient) equilibrium in the private monitoring game without communication, and his construction is basically built on such an idea. Strong correlation of private information is, however, not assumed in his model but is derived endogenously. He assumes that private signals provide nearly perfect observability and considered *mixed* strategies. In such a situation, the privately observed random variables, the action-signal pairs, are strongly correlated (because a player's random action is strongly correlated with another player's signal under nearly perfect observability). Mailath and Morris (2002) show that, in general, there is 'continuity' between the public and private but sufficiently correlated monitoring cases, in the sense that any strategy with a *finite memory* works in either case.

Those papers are examples of the *belief-based approach*, which directly addresses the statistical inference problem (see point 2. above). Some other papers follow this approach, and they provide judiciously constructed strategies in rather specific examples, where the inference problem becomes tractable. Aside from the case with near perfect correlation, however, we are yet to have generally applicable results or techniques from this approach.

More successful has been the *belief-free approach*, where an equilibrium is constructed in such a way that the inference problem becomes *irrelevant*. As a leading example, here I explain Ely and Valimaki's work (2002) on the repeated Prisoner's Dilemma with imperfect private monitoring. Each player's strategy is a Markov chain with two states, *R* (reward) and *P* (punishment). A specific action is played in each state (*C* in *R*, and *D* in *P*), and the transition probabilities between the states depend on the realization of the player's private signal. Choose those transition probabilities in such a way that the *opponent* is always indifferent between C and D *no matter which state the player is in*. This requirement can be expressed as a simple system of dynamic programming equations, which has a solution when the discount factor is close to 1 and the private signal is not too uninformative. By construction, any action choice is optimal against this strategy after any history, and in particular this strategy is a best reply to itself (so that it constitutes an equilibrium). Note that one's incentives do not depend on the opponent's state, and therefore one does not have to draw the statistical inferences about the history of the opponent's private signals.

There are certain difficulties, however, in obtaining the folk theorem with such a class of equilibria. First, players may be punished simultaneously in this construction, and our discussion about the public monitoring case shows that some welfare loss is inevitable (unless monitoring is nearly perfect). Second, even if we restrict our attention to the nearly perfect monitoring case, there is a certain set of restrictions imposed on the action profiles that can be sustained by such a belief-free equilibrium.

Those difficulties can be resolved when we consider *block strategies*. Block strategies treat the stage games in T consecutive periods as if they were a single stage game, or a block stage game, and applies the belief-free approach with respect to those block stage games. It is now known that, by using the block strategies, the folk theorem under private monitoring holds in the nearly perfect monitoring case (Hörner and Olszewski, 2006) and for some two-player games where monitoring is far from perfect (Matsushima, 2004). In the former, the block structure of the stage game helps to satisfy the restrictions imposed on the sustainable actions in belief-free equilibria. In the latter, an equilibrium is constructed where players choose constant actions in each block. This means that players have T samples of private signals for the constant actions, so that the observability practically becomes nearly perfect when T is large. With this increased observability and some restrictions on payoff functions, the folk theorem is obtained. For this construction to be feasible, the signals have to satisfy certain strong conditions, such as independence (across players).

The general folk theorem, or a general characterization of equilibria, for the private monitoring case is yet to be obtained, and it remains an important open question in economic theory. A comprehensive technical exposition of the perfect monitoring, imperfect public monitoring, and private monitoring cases can be found in Mailath and Samuelson (2006).

KANDORI MICHIHIRO

See also **cartels; cooperation; reputation; social norms.**

Bibliography

Abreu, D., Dutta, P. and Smith, L. 1994. The folk theorem for repeated games: a NEU condition. *Econometrica* 62, 939–48.

Abreu, D., Pearce, D. and Stacchetti, E. 1990. Towards a theory of discounted repeated games with imperfect monitoring. *Econometrica* 58, 1041–64.

Aumann, R. 1959. Acceptable points in general cooperative N-person games. In *Contributions to the Theory of Games*, vol. 4, ed. R.D. Luce and A.W. Tucker. Princeton: Princeton University Press.

Axelrod, R. 1984. *Evolution of Cooperation*. New York: Basic Books.

Benoit, J.P. and Krishna, V. 1985. Finitely repeated games. *Econometrica* 53, 905–22.

Coase, R. 1937. The nature of the firm. *Economica* n.s. 4, 386–405.

Compte, O. 1998. Communication in repeated games with imperfect private monitoring. *Econometrica* 66, 597–626.

Ely, J. and Valimaki, J. 2002. A robust folk theorem for the Prisoner's Dilemma. *Journal of Economic Theory* 102, 84–105.

Friedman, J. 1971. A non-cooperative equilibrium for supergames. *Review of Economic Studies* 38, 1–12.

Fudenberg, D., Levine, D. and Maskin, E. 1994. The folk theorem with imperfect public information. *Econometrica* 62, 997–1040.

Fudenberg, D. and Maskin, E. 1986. The folk theorem in repeated games with discounting or with incomplete information. *Econometrica* 54, 533–54.

Hörner, J. and Olszewski, W. 2006. The folk theorem for games with private almost-perfect monitoring. *Econometrica* 74, 1499–544.

Kandori, M. and Matsushima, H. 1998. Private observation, communication and collusion. *Econometrica* 66, 627–52.

Macaulay, S. 1963. Non-contractual relations in business: a preliminary study. *American Sociological Review* 28, 55–67.

Mailath, G. and Morris, S. 2002. Repeated games with imperfect private monitoring: notes on a coordination perspective. *Journal of Economic Theory* 102, 189–228.

Mailath, G. and Samuelson, L. 2006. *Repeated Games and Reputations: Long-Run Relationships*. Oxford: Oxford University Press.

Matsushima, H. 2004. Repeated games with private monitoring: two players, *Econometrica* 72, 823–52.

Pearce, D. 1990. Repeated games: cooperation and rationality. In *Advances in Economic Theory*, ed. J. Laffont. Cambridge: Cambridge University Press.

Rubinstein, A. 1979. Equilibrium in supergames with overtaking criterion. *Journal of Economic Theory* 21, 1–9.

Sannikov, Y. 2005. Games with imperfectly observable actions in continuous time. Mimeo, University of California, Berkeley.

Sekiguchi, T. 1997. Efficiency in repeated Prisoner's Dilemma with private monitoring. *Journal of Economic Theory* 76, 345–61.

reputation

In a dynamic setting signals sent now may affect the current and future behaviour of other players; thus, signals can have effects unrelated to their current costs and benefits. It is the interplay between signals and their long-run consequences that is studied in the literature on reputation.

The literature on reputation has two main themes. The first is that introducing a small amount of incomplete information in a dynamic game can dramatically change the set of equilibrium payoffs: introducing something to signal can have big implications in a dynamic model. These kinds of result can also be interpreted as providing a robustness check. Dynamic and repeated games typically have many equilibria, and reputation results allow us to determine which equilibria continue to be played when a game is 'close' to complete information. The second theme of the literature on reputations is that introducing incomplete information in a dynamic game may introduce new and important signalling dynamics in the players' strategies. Thus reputation effects tell us something about behaviour. This theme is particularly important in applications to macroeconomics and to industrial organization, for example. For either of these themes to be relevant it is necessary to have a dynamic game with incomplete information, so work on reputation has been influenced by, and influences, the larger literature on repeated and dynamic games of incomplete information. An excellent detailed treatment of reputation can be found in Mailath and Samuelson (2006).

An example

Most of the results below will be described in the context of a simple infinitely repeated trading game. The row player is a seller who can produce high or low quality. The column player is a buyer. Producing high quality is always expensive for the seller, so she would rather produce low quality; the buyer, however, wants to buy only a high-quality product. The only non-standard element is that the buyer regrets not buying a high-quality product. The trading game (Figure 1) has a unique equilibrium (L, N).

Let us record some facts about this game. The set

$$V \equiv \{(x, y) : x > 0, y > -1/3, y \leq x \text{ and }$$
$$y \leq 3 - 2x\} \subset \mathbb{R}^2,$$

illustrated in Figure 2, is the set of feasible and strictly individually rational payoffs for the trading game. The axes are drawn through the minmax payoffs to make V clear. If the seller could commit to a pure strategy, she would prefer to choose H as the buyer's best response to this is B. However, she could do even better by committing to a mixed strategy; playing $(3/4, 1/4)$ for example would also ensure the buyer played B and give the seller a bigger payoff. Reputation arguments can provide ways for these commitment payoffs to be achieved by sellers who are not actually committed to anything.

The trading game is played in each of the periods $t = 1, 2, \ldots$ with perfect monitoring; at the end of the period the players get to observe all payoffs and the pure action taken by their opponent. If both players' discount factors, $\delta < 1$, were sufficiently large, any point in V could be sustained as an equilibrium payoff. If the seller is long lived but faces an infinite sequence of buyers who each

live one period, then any point on the line segment joining $(0,0)$ to $(1,1)$ is an equilibrium payoff. (No seller payoff above 1 is achievable if mixed actions are not observable; see Fudenberg, Kreps and Maskin, 1990.)

The stage is now set. To understand how reputation works we will need to introduce something for the seller to signal. Its commitment to high quality? Its low cost of high quality? Its commitment to always ripping off customers…? At this stage it is unnecessary to be specific, and we will concentrate on the general issues of learning. There are two types of sellers, 'strong' and 'normal', that the buyer may face in a game. The seller is told their type by nature at time $t=0$. The buyer, however, is unaware of nature's selection and spends the rest of the game looking at the seller's behaviour and trying to figure out what type she is. The normal seller plays action $a \in \{H, L\}$ with probability $\tilde{\sigma}^t(a)$ at time t, and the strong seller plays a with probability $\hat{\sigma}^t(a)$ at time t. Everything we say in the section below applies to the case where normal and strong sellers follow history-dependent strategies. (These behaviour strategies do depend on the – public – history of play before time t, but let us keep this out of our notation.) An initially uninformed buyer attaches probability p^t to the strong type and $1 - p^t$ to the normal type at time t; again this depends on the observed history. Our buyer expects the seller to play a with probability $\bar{\sigma}^t(a) = p^t \tilde{\sigma}^t(a) + (1 - p^t)\hat{\sigma}^t(a)$, and as time passes the buyers observe the outcomes of this strategy and revise their prior accordingly.

Tricks with Bayes's rule and martingales

Now we generate three properties of learning that are extensively used in the reputations literature. We will call them the 'merging' property, the 'right ballpark' property and the 'finite surprises' property. These properties are based on some simple facts about how Bayesian agents revise their beliefs, that is, how uncertainty about the seller's type is processed by the buyers or any other observer of its behaviour. A more advanced treatment of

		Buy (B)	Not Buy (N)
High Quality	(H)	(1, 1)	(−1, −1)
Low Quality	(L)	(2, −1)	(0, 0)

Figure 1 A trading game

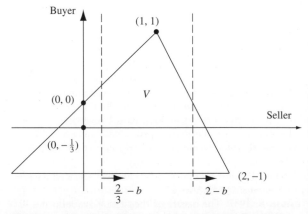

Figure 2 Sets of equilibrium payoffs and reputation bounds

these results can be found in Sorin (1999). We defer any derivation of reputation results to the next section, so a reader could skip this section.

How does the buyer revise his or her beliefs in the light of an observed action a^t? A plain application of Bayes' rule tells us

$$p^{t+1} = \frac{Pr(a^t \cap \text{Strong})}{Pr(a^t)} = \frac{p^t \hat{\sigma}^t(a^t)}{\bar{\sigma}^t(a^t)}.$$

Or, in terms of the change in the beliefs

$$p^{t+1} - p^t = \frac{p^t[\hat{\sigma}^t(a^t) - \bar{\sigma}^t(a^t)]}{\bar{\sigma}^t(a^t)}$$

$$= \frac{p^t(1 - p^t)[\hat{\sigma}^t(a^t) - \tilde{\sigma}^t(a^t)]}{\bar{\sigma}^t(a^t)}.$$

These equalities are powerful tools when combined with the properties of the priors.

Merging property. This tells us exactly how the long-run behaviour of the sellers is related to the buyer's long run beliefs. Either $p^t(1 - p^t) \to 0$ and the buyer eventually learns the type of the seller and can perfectly predict their actions, or all types of the seller end up behaving in the same way $\hat{\sigma}^t(a^t) - \tilde{\sigma}^t(a^t) \to 0$ and again the buyer can perfectly predict their actions. Nothing else can happen!

The stochastic process $\{p^t\}$ is a martingale on $[0,1]$ with respect to public histories. To see this there is a simple calculation we can do.

$$E(p^{t+1}|h_t) = \sum_{a^t} Pr(a^t)p^{t+1}$$

$$= \sum_{a^t} \bar{\sigma}^t(a^t)\frac{p^t \hat{\sigma}^t(a^t)}{\bar{\sigma}^t(a^t)} = p^t.$$

(The expectation $E(\cdot)$ is taken with respect to the buyer's beliefs about future play.) Bounded martingales converge almost surely (see Williams, 1991, for example), which implies $|p^{t+1} - p^t| \to 0$ almost surely. Applying this to the second equality above (noting that $|\bar{\sigma}^t(a^t)| \leq 1$), we get

$$p^t(1 - p^t)|\hat{\sigma}^t(a^t) - \tilde{\sigma}^t(a^t)| \to 0, \qquad \text{(Merging)}$$

almost surely. This kind of result is extensively used in Hart (1985) and the literature that stems from his work.

Right ballpark property. The strong seller knows that the future will evolve according to the strategy $\hat{\sigma}$ (we use $\hat{Pr}(\cdot)$ and $\hat{E}(\cdot)$ to denote her probability measure and its expectation). This seller might ask, as she plays out an equilibrium, how little probability the buyers can attach to the strong seller, or how low p^t could get when she plays $\hat{\sigma}$. Of course, when the seller is in fact the strong type it is very unlikely that p^t becomes low – beliefs must stay in the right ballpark. (For example, if $\hat{\sigma}$ was actually

a pure strategy the strong seller cannot ever believe p^t will decrease. As she plays $\hat{\sigma}$ there will be periods in which the normal type of seller could have done something different, so observing the actions of $\hat{\sigma}$ will cause buyers to revise p^t upwards.)

From the perspective of the strong seller, the likelihood ratio is a martingale:

$$\hat{E}\left(\frac{1 - p^{t+1}}{p^{t+1}}\bigg|h_t\right) = \frac{1 - p^t}{p^t}.$$

(The calculation is just like the earlier one for p^t, where we use $\hat{Pr}(a^t) = \hat{\sigma}(a^t)$.) Let τ be the first time, s say, that $p_s \leq v$ and let C^t be the event that $\tau \leq t$. That is, sometime in the first t periods $p_s < v$. Then the martingale property combined with the optional stopping theorem (for example, Williams, 1991) implies

$$\frac{1 - p^0}{p^0} = \hat{E}\left(\frac{1 - p^{t+1}}{p^{t+1}}\right) \geq \hat{Pr}(C^t)E\left(\frac{1 - p^\tau}{p^\tau}\bigg|C^t\right)$$

$$\geq \hat{Pr}(C^t)\frac{1 - v}{v}.$$

The above gives an upper bound on $\hat{Pr}(C^t)$ that is independent of t. Thus it also bounds the probability that p_t is ever below v:

$$\hat{Pr}(\exists t \text{ s.t. } p_t < v) \leq \frac{v}{p^0}. \qquad \text{(Right Ballpark)}$$

Hence, the strong seller knows that it is very unlikely that the buyer's posterior will ever be close to certain she is actually the normal seller.

Finite surprises property. The strong seller might also ask how many times (as she plays $\hat{\sigma}$) the uninformed buyers will make a big mistake in predicting her strategy, that is, how many periods does $||\hat{\sigma}^t - \bar{\sigma}^t|| > v$ occur when the seller actually plays $\hat{\sigma}$. Here we are helped by the fact that our seller has only two actions, so the variation distance between the mixed actions is just twice the difference in probability of the realized action $||\hat{\sigma}^t - \bar{\sigma}^t|| = 2|\hat{\sigma}^t(a^t) - \bar{\sigma}^t(a^t)|$. Let M_N be the event that there are more than N *mistakes*, $||\hat{\sigma}^t - \bar{\sigma}^t|| > v$, before time T. The finite surprises property is that independently of the equilibrium $\hat{Pr}(M_N) \to 0$ as $T, N \to \infty$. Thus, it is very unlikely that there are many periods in which the buyers do not think the seller will play as the strong type if the seller is indeed this type.

Jensen's inequality applied to the likelihood ratio above implies that the prior is a submartingale, that is, $\hat{E}(p^{t+1}|h_t) \geq p^t$. There is a second property of martingales we can now use: they cannot move around very much: $\sum_{t=1}^{T} \hat{E}((p^{t+1} - p^t)^2) \leq 1$. (A proof of this fact follows from $\hat{E}(p^{t+1} - p^t)^2 \leq \hat{E}((p^{t+1})^2 - (p^t)^2)$.) A substitution from the first Bayes' rule equality above then tells us

$$1 \geq \sum_{t=1}^{T} \hat{E}((p^t[\hat{\sigma}^t(a^t) - \bar{\sigma}^t(a^t)])^2).$$

It is obvious that only a few of the (non-negative) terms in the sum above can be much above zero, otherwise the upper bound will be violated. The right ballpark property tells us it is very unlikely that $p^t < v$. On the event $\{p^t \geq v \forall t\} \cap M_N$, the p^t in the above expectation is greater than v and there are at least N differences that are bigger than $v/2$, so the sum is at least $Nv(v/2)^2$, hence

$$1 \geq \sum_{t=1}^{T} \hat{E}(p^t[\hat{\sigma}^t(a^t) - \bar{\sigma}^t(a^t)])^2$$
$$\geq \hat{Pr}d(\{p^t \geq v \forall t\} \cap M_N) \frac{Nv^3}{4}.$$

Using the fact that $Pr(A \cap B) \geq Pr(B) - Pr(A^c)$ we now have an upper bound on $\hat{Pr}(M_N)$.

$$\frac{4}{Nv^3} + \hat{Pr}(\exists t \text{ s.t. } p_t < v) \geq \hat{Pr}(M_N).$$

The right ballpark property gives us

$$\hat{Pr}(M_N) \leq \frac{v}{p^0} + \frac{4}{Nv^3}. \quad \text{(Finite Surprises)}$$

As the size of the surprises becomes small $v \to 0$ and the number of surprises becomes large $Nv^3 \to \infty$, the strong seller must attach smaller and smaller probability to M_N. Fudenberg and Levine (1989; 1992), for example, invoke this property.

Basic reputation results: behaviour

The three tools above are sufficient to establish most well-known reputation results. The arguments below are entirely general, and are widely applied, but we use them only in the trading game. To make things simple, suppose that for some reason the strong seller is committed to playing $(b, 1-b)$, that is, in every period t the strong buyer provides high quality with probability b. We reserve the discussion of more complicated types of reputations for a later section.

From the perspective of the buyer, any equilibrium will consist of two phases: an initial phase when there is learning and signalling about the seller's type (this is sometimes called reputation building, although often reputation destruction is what occurs), and a terminal phase when the learning has virtually settled down. It is the merging property that tells us there must be this latter phase. The play in the game moves into this second phase either because the buyer is almost sure he knows the type of the seller (reputation considerations have vanished) or because the sellers are playing in the same

way. Thus the equilibria of dynamic signalling games are inherently non-stationary, which is in contrast to much of the work on repeated games. Of course, Markovian equilibria can be calculated but these too will exhibit the two phases of play. The initial learning, when reputation builds or is destroyed, depends on the particular equilibrium and the game being studied. This phase may last only one period (if a once and for all revealing action is taken by a seller) but frequently it is long and has a random duration (if both types of seller randomize, for example).

Let us first examine reputation destruction in the case where $b \approx 1$, so the strong seller is committed to high quality and only very occasionally slips up. There is an equilibrium of this game where the normal type of seller will offer low quality more often than the strong type, and thereby gradually reveal her type (destroy her reputation for being good). Nevertheless, as this occurs she will enjoy heightened payoffs. The trade-offs our normal seller experiences in this game are what drive the reputation destruction. A seller offering low quality today enjoys the benefit of a higher payoff now, but the observation of low quality typically leads the buyers to revise downwards their probability of the strong seller and buy less in the future, whereas a seller offering high quality will lead the buyer's posterior on the strong seller to be revised upwards and an increased likelihood of buying in the future. Exactly how the normal seller chooses to trade off long-run benefits and short-run costs is unclear. It is possible that pooling dominates and that future buying is so strong that the normal seller prefers to offer high quality today even if it costs something in the short run. However, in this equilibrium the normal seller perceives the long-run benefits to be relatively small and prefers to offer low quality today. The normal seller can be thought of as exploiting, or cashing in, the value of her accumulated reputation. We also know, from the finite surprises property, that there will be finite opportunities for the normal seller to do this. Relatively soon there will come a time where the buyers know the seller is normal and purchase accordingly.

Reputation building (as opposed to destruction) is more likely in a world where there is the possibility that one is thought to be bad, for example, if the strong type is committed to ripping customers off and only occasionally produces a good product ($b \approx 0$). In such a world the normal seller wants to tell buyers she is not this type, because by playing as the strong type she is doomed to never trade. She is building a reputation for *not* being the strong type. To do this the normal type will have to incur the cost of repeatedly offering high quality, even if the buyer is not buying. This is expensive and will drag down the normal seller's equilibrium payoff. But, as above, it will increase the likelihood of future buying by decreasing the likelihood of a strong seller. In contrast to the reputation destruction case, there are short-run costs borne by the normal type to achieve long-run gains.

Again, the nature of these costs and benefits rely on the buyers' uncertainty about the seller's type.

Basic reputation results: payoffs

Reputation issues can have an extreme effect on payoffs, and this is what first came to the attention of economists. The general question of how the presence of something to signal in the repeated game affects the equilibrium payoffs could be answered in a number of ways. One way would be to calculate equilibria explicitly. This is usually difficult and would not establish results that hold for *all* equilibria.

Instead, a different approach is taken that is described in the following recipe:

1. If the seller is strong, then in finite time the buyers will believe they face a seller who plays arbitrarily close to $(b, 1-b)$ for ever.
2. Figure out what the buyers will do when the seller is strong.
3. Use step 2 to evaluate the normal seller's payoff if she pretends to be strong for ever.
4. At a Nash equilibrium the answer to step 3 is a lower bound on the normal seller's equilibrium payoff.

Step 1 is independent of the model and is a result of our earlier calculations. The right ballpark property tells us that p^t does not tend to zero when the seller is strong. The merging property then implies either $p^t \to 1$, or eventually all remaining normal types of buyer are also playing arbitrarily close to $(b, 1-b)$. In either of these cases, at a large but finite time the buyers believe that they face a seller who will always play $(b, 1-b)$.

Before proceeding to apply this recipe, we illustrate its power with the remarkable results we expect to get. Let us first consider a world where buyers are short run. We will show that introducing an arbitrarily small probability that there is a strong seller places a lower bound on the normal seller's equilibrium payoffs of $2-b$ (when $b > 1/3$). Thus for b close to $1/3$ the equilibrium payoffs in the complete information game (the segment joining $(0,0)$ and $(1,1)$) and the incomplete information game are disjoint! Moreover, the normal seller can get almost his maximum feasible payoff at every equilibrium. In the second case, where buyers are also long run, we will get less strong conclusions; nevertheless, we will show that the normal type of seller must get at least $2/3 - b$ when $b > 1/3$. These payoffs are illustrated in Figure 2.

The really difficult part of our recipe is step 2, because we have to understand how the buyers will behave in equilibrium. We therefore need to consider as separate cases what happens if buyers are short run or long run. Also, the amount of discounting that the sellers do affects the answer to step 3, so we need to consider different arguments for different amounts of discounting. The following catalog moves from simple to more elaborate arguments and from stronger to weaker reputation effects.

Reputation without discounting: short-term buyers

When a buyer lives only one period he plays a best response to the seller's current action. By step 1 in the very long run this will be B if $b > 1/3$ and N if $b > 1/3$. Step 3 is simple; by playing $\hat{\sigma}$ for ever the normal seller knows that in a large but finite time she can ensure the buyer will behave as above and so she will receive a stage game payoff approximately $R^*(b)$, where

$$R^*(b) := \begin{cases} 2 - b & b > 1/3, \\ -b & b < 1/3. \end{cases}$$

If there is no discounting, and limits are correctly taken, $R^*(b)$ will equal the normal type's payoff from playing $\hat{\sigma}$ for ever. Thus, at step 4, at any Nash equilibrium the normal type must get at least $R^*(b)$.

In a general game R^* is equal to the seller's payoff from playing the strong type's stage game strategy when the buyer plays his or her unique best response. (If the best response is not unique this is not correct.)

Reputation with discounting: short-term buyers

Step 2 is as above – we still have short-term buyers. When the normal seller discounts payoffs, however, playing $\hat{\sigma}$ and eventually getting $R^*(b)$ every period does not tell us what her payoff discounted to time zero will be. There is an order of limits issue; as the discounting of the seller becomes weaker ($\delta \to 1$), it could be that the equilibria change and there are more and more periods where the seller is not getting $R^*(b)$. It is now that the finite surprises property plays an important role. First notice that, when v is chosen appropriately and $\|\bar{\sigma} - \hat{\sigma}\| < v$, then playing a best response to $\bar{\sigma}$ is the same as playing a best response to $\hat{\sigma}$. Hence, it is only when a surprise occurs that the normal seller is not getting $R^*(b)$ from playing $\hat{\sigma}$. But the probability of more than N surprises can be made very small *independently of the discounting*. So, as the discounting becomes weak and N periods have a small effect on total discounted payoff, there is a small probability of the normal seller of getting anything less than $R^*(b)$ when she plays $\hat{\sigma}$. Any Nash equilibrium, therefore, gives the normal seller at least $R^*(b)$. This is the kind of argument first made in specific cases by Kreps and Wilson (1982) and Milgrom and Roberts (1982), and generalized in Fudenberg and Levine (1989; 1992).

Reputation without discounting: one long-run buyer

If the buyer lives for many periods, he will not necessarily play a short-run best response to $(b, 1-b)$ even if he expects it to be played for ever. We can, however, use some weaker information. At an equilibrium the buyer must on average get at least $-1/3$ (his minmax payoff) against $(b, 1-b)$. This implies that the buyer has to buy with at least probability $1/3$ when $b > 1/3$ and buy with at most probability 1/3 when $b < 1/3$. There are, consequently, some bounds on the normal seller's payoff when she has played $\hat{\sigma}$ for a sufficiently long time. While playing $(b, 1-b)$ she gets $2-b$ when the buyer buys and $-b$ if not; thus, if the buyer buys with probability greater than 1/3, she expects to receive a payoff of at least $2/3 - b$. If the buyer buys with at most probability 1/3, she expects to get at least $-b$. The seller is not discounting, so what she gets in the long run from playing $\hat{\sigma}$ is also what she expects to get at time zero. Our answer to step 3, therefore, is

$$ R^{\dagger}(b) := \begin{cases} (2/3) - b & b > 1/3, \\ -b & b < 1/3; \end{cases} $$

and we have a weaker lower bound on the normal type's payoff.

In an arbitrary game $R^{\dagger}$ is equal to the seller's worst payoff from playing as the strong type when the buyer plays a response that gives him more than his minmax payoff. In certain cases this can be a very strong restriction – for example, if the seller has a pure strategy that minmaxed the buyer and there is a unique response for the buyer that ensured he received his minmax payoff. Certain games, known as games of conflicting interests, have the property that the best action for the seller to commit to is pure and minmaxes the buyer. $R^{\dagger}$ is a very tight bound for such games.

Reputation with discounting: one long-run buyer

This final case combines most of the above issues. If the seller discounts the future much less than the buyer, then in the long run the seller must get $R^{\dagger}(b)$ from playing $\hat{\sigma}$. If a normal seller pretends to be strong, the buyers think there are at most N periods when the strong strategy is not played. Imagine now we have a buyer who cares only about what happens in the next t' periods. Such a buyer can think there are at most $t'N$ periods in which $\hat{\sigma}$ is not played for the next t' periods. (This kind of argument is due to Schmidt, 1993.) As the seller becomes very patient Nt' periods become of vanishing importance and the normal seller's payoff is bounded below by $R^{\dagger}(b)$. If the seller and buyer discount equally, however, reputation effects cannot be found except in some very special cases.

Imperfect monitoring: temporary and bad reputation

The analysis of reputation given above presupposes perfect monitoring by the buyers and sellers of each others' actions. In many dynamic and repeated games this is not likely. To what extent do the above results continue to hold when the players are not able to see exactly what their opponent did in any one period? Perhaps reputations are harder to establish if the observed behaviour is noisy? On the other hand, perhaps deviations from the strong type's action are harder to detect and so reputations last longer and are more valuable.

The merging, right ballpark and finite surprises properties all hold true under imperfect monitoring, with a suitable redefinition, provided there is enough statistical information for the buyer to eventually identify the seller's behaviour. (This is a full-rank condition on the players' signals.) As a result, the bounds on payoffs given in the previous section continue to hold.

Under imperfect monitoring with adequate statistical information there is one new behavioural feature of these games – reputation is almost always temporary, that is, the buyer will eventually get to know the seller's type. To see why this is so, let us amend the game in Figure 1 by restricting the buyer to imperfectly observe the seller's action. With probability $1 - \varepsilon$ the buyer observes the seller's true action in the current period, but with probability $\backslash \varepsilon$ he observes the reverse action. (We must also assume the buyer does not see his own payoffs, otherwise he can deduce the seller's action from his payoff.) Consider a game where the seller always provides high quality $(b = 0)$ and suppose that reputation is permanent in such a game. Then p would, at least some of the time, converge to a number that is not zero or one. (Remember beliefs have to converge.) The merging property tells us that, in this case where the limit of beliefs is between zero and one, the buyer will be certain the normal seller is always providing high quality. Such buyers will ignore the occasional low-quality product as just unlucky outcomes, and there will be no loss of seller reputation if the buyer ever receives low quality. The normal type of seller can, therefore, deviate from always providing high quality, gain one unit of profit, and not face any costs in terms of loss of reputation. This cannot be an equilibrium. The initial claim that reputation is permanent has to be false as a result of this contradiction. The details of this argument can be found in Cripps, Mailath and Samuelson (2004).

When the monitoring is not statistically informative, 'bad reputation', due to Ely and Valimaki (2003), is a possibility. Uninformative monitoring is a particular problem in repeated extensive form games, because players do not get to see the actions their opponent would have taken on other branches of the game tree. Bad reputation may arise in our example if the buyer could take an action (such as not to buy) that stopped the seller being able to signal her type. Then, the normal seller might find herself permanently stuck in a situation where she cannot sell. This is not particularly surprising if the buyers were strongly convinced they faced a strong seller that almost always provided low quality. However, in certain circumstances this problem is much more severe: even if the buyers were almost certain the seller were normal, every equilibrium has trade ending in a bounded and finite time. Thus, it is possible that introducing something for the seller to signal has huge negative costs for her equilibrium payoffs. To illustrate this, suppose the seller were a restaurant with imperfect control over quality, although it does have a strategy (for example, doubling the butter and salt content!) that makes it more likely the buyer will think the meal he received is good – but is actually damaging to the buyer. When play has reached the position where just one more bad meal will lead the buyer to permanently avoid the restaurant, then the restaurant will choose to use this unhealthy strategy. Knowing this, the buyer will choose to go elsewhere for his last but one meal too, and there is an unravelling of the putative equilibrium. Buyers eat at the restaurant only if they get very few bad meals, because they know they are in for clogged arteries and high blood pressure after that. Bad reputation arises because the seller cannot resist the temptation of taking actions that are actually unfavourable to the buyer in an effort to regain his good opinion. They actually have the reverse effect of ultimately driving the buyers away.

Reputation for what?

In our discussion we consider a strong type of seller who is committed to playing a particular fixed (random) action in each period. Is this form of uncertainty the only relevant one, or are there other potential types of strong seller that may do even better for our normal seller? There are two alternatives to consider: the strong seller is committed to playing a history-dependent strategy, or the strong player is equipped with a payoff function and her strategy is determined by an equilibrium.

If the seller faces a sequence of short-term buyers, then committing to a fixed stage game action is the best she could ever do, because each buyer's optimization focuses on what the seller does in the current period – the future is irrelevant. Even when the buyers are long lived, there are circumstances where committing to play a fixed action imparts a strategic advantage in repeated play, for example in most coordination or common interest games. However, there are other repeated games, such as the Prisoner's Dilemma, and dynamic games where committing to a fixed stage action is worthless. What the seller would like to do is to commit to a strategy, such as tit-for-tat, which would persuade a sufficiently patient buyer to cooperate with the strong type. Provided some rather strong conditions are satisfied, this is possible.

Our recipe for reputation results will break down when we consider strong sellers with payoffs rather than actions; nevertheless, reputation results are possible. For example, if the strong seller had payoffs of 2 for high quality and zero for low quality he would be strategically identical to a seller who always provided high quality.

Many players: social reputation and other considerations

Thus far we have resolutely stuck to a model of two players, but it is clear that reputation is a pervasive social and competitive phenomenon. Here we sketch some of the issues in many-player reputation. The literature on this area is in its infancy; very little can be said with much certainty now.

The easiest case to deal with is what happens as the number of uninformed players (the buyers in our example) increases. Here the benefit to the seller of building a reputation for high quality increases, as providing a good product today means the seller is more likely to trade with many buyers tomorrow. In a way, increasing the number of buyers is like making the seller more patient, and so we would expect the seller to be more inclined to build a reputation in this case.

A second case would be where there are very large numbers of informed buyers trying to acquire reputations for individual or group characteristics. Models of career concerns are similar to reputation models and have many workers trying to acquire reputations for individual characteristics. Also, there are models of group reputation, such as Tirole (1996), where a particular class of individuals behaves in a particular way to perpetuate the 'group's' reputation. In both these types of model the large numbers assumption allows one individual's reputation decision to be treated as virtually independent of others. Thus they can be analysed using quite simple tools.

A final case is where a few informed agents are in competition or collusion with each other. Collusion in team reputation obviously introduces a public goods issue. If one player contributes to the good name of the group, he or she does not get to enjoy the full benefits of the contribution. Typically, therefore, reputations for such teams are harder to establish. One might conjecture that competition appears to drive a player towards excessive investment in reputation, but there are many effects at work that we do not completely understand. For example, competitors may also act to undermine their rival's reputation and to interfere with its development. This is a fertile region for applied and theoretical investigations.

MARTIN W. CRIPPS

See also **repeated games; signalling and screening.**

Bibliography

Cripps, M.W., Mailath, G.J. and Samuelson, L. 2004. Imperfect monitoring and impermanent reputations. *Econometrica* 72, 407–32.

Ely, J. and Valimaki, J. 2003. Bad reputation. *Quarterly Journal of Economics* 118, 785–814.

Fudenberg, D., Kreps, D. and Maskin, E. 1990. Repeated games with long-run and short-run players. *Review of Economic Studies* 57, 555–74.

Fudenberg, D. and Levine, D.K. 1989. Reputation and equilibrium selection in games with a patient player. *Econometrica* 57, 759–78.

Fudenberg, D. and Levine, D.K. 1992. Maintaining a reputation when strategies are imperfectly observed. *Review of Economic Studies* 59, 561–79.

Hart, S. 1985. Nonzero-sum two-person repeated games with incomplete information. *Mathematics of Operations Research* 10, 117–53.

Kreps, D. and Wilson, R. 1982. Reputation and imperfect information. *Journal of Economic Theory* 27, 253–79.

Mailath, G.J. and Samuelson, L. 2006. *Repeated Games and Reputations: Long-Run Relationships*. Oxford: Oxford University Press.

Milgrom, P. and Roberts, J. 1982. Predation, reputation and entry deterrence. *Journal of Economic Theory* 27, 280–312.

Schmidt, K. 1993. Reputation and equilibrium characterization in repeated games of conflicting interests. *Econometrica* 61, 325–51.

Sorin, S. 1999. Merging, reputation, and repeated games with incomplete information. *Games and Economic Behavior* 29, 274–308.

Tirole, J. 1996. A theory of collective reputations (with applications to the persistence of corruption and to firm quality). *Review of Economic Studies* 63, 1–22.

Williams, D. 1991. *Probability with Martingales*. Cambridge: Cambridge University Press.

resale markets

Resale markets seem necessary to correct misallocations of assets, where misallocations may be the result of mistakes in initial purchasing decisions, or more generally of changes in the state of the economy. For the sake of illustration, a car owner may after a while find it desirable to buy a new car, and he may be willing to resell his old car on the second-hand market. A manager of a firm holding a Universal Mobile Telephone System (UMTS) licence may be willing to resell her licence to another firm if she realizes that the firm is unable to cover its cost (generated by the licence acquisition). A homeowner may need to resell his house if he has to move to another country or jurisdiction.

A question of primary interest is whether such resale markets are good for the economy. Or, to put it differently, whether, when and how should such resale markets be regulated? This article starts with the laissez-faire viewpoint on this issue; it then proceeds to show how asymmetric information and commitment issues mitigate that viewpoint.

The laissez-faire viewpoint

The classical neoliberal viewpoint as represented by the Chicago School would favour laissez-faire. Within the present context, this would imply that resale markets should not be regulated. The premise of this line of thought is that resale markets give the right flexibility so that assets can be allocated to the right agents at any point in time. This view has important consequences for the theory of mechanism and market design. Indeed, it implies that the initial allocation of property rights is irrelevant, as resale markets should be able to correct any misallocations (this is one version of the so-called Coase Theorem – Coase, 1960). Thus, according to this view, a government interested in maximizing economic efficiency should worry neither about the method of privatization nor about how to allocate licences for operating public services. It should simply allow for well-functioning resale markets.

Of course, very few economists truly believe that real resale markets can achieve such a fantastic job of always allocating assets to the right agents at the right time. On the academic side, Akerlof (1970) provides an early theoretical example of market failure in the context of the market for used cars (more on this below). Coase himself argues that transaction costs which are numerous are likely to invalidate the above angelic view about resale markets. On the 'real world' side, it seems implausible that the method of privatization or the allocation of licences for the use of public services is irrelevant for economic efficiency. In fact, recent years have seen a rapid growth of auction methods to allocate licences or privatize publicly owned firms, suggesting an interest on the part of practitioners in market design. It is worth pointing out that, in the case of licence auctions, most governments have chosen not to allow for resale markets, suggesting some distrust towards their functioning.

In the tradition of Coase, the words 'transaction costs' will be interpreted to mean any reason why inefficiencies may arise in transactions. Of course, some of the reasons need not be related to the intuitive notion of transaction costs, and one could alternatively use the more neutral terminology of 'market imperfections'. The rest of this article will review how theoretical insights from the mechanism design literature and the bargaining literature help identify significant sources of transaction costs. The review will abstract from transferability issues, which is a legitimate idealization for transactions that are not too big for the financial capabilities of the parties. The theoretical insights will then be used to shed some light on whether and how to regulate resale markets.

The role of private information

It is relatively intuitive to see why private information may be a source of inefficiency in transactions. A seller who privately knows her valuation for the object for sale has an incentive to pretend that she values the object more than she really does, in the hope that this will lead the buyer to increase his purchasing price. Similarly, a buyer has an incentive to pretend that he values the good less than he really does, in the hope that he will obtain a lower selling price. But such distortions inevitably induce inefficiencies whenever the gains of trade are not large enough. This intuition has been formalized in the work of Myerson and Satterthwaite (1983), who show that, if the distributions of valuations are independently distributed between a seller and a buyer, and if it is not known who values the good more, inefficiencies must arise in any bargaining game in which no outside money is given to the bargaining parties. One of the strengths of Myerson and Satterthwaite's work is that it applies to any bargaining game, including protocols in which a broker could help improve the bargaining outcome and protocols allowing for several stages of bargaining. The result is obtained by relying on the so-called revelation principle, which allows for the derivation of constraints that should be satisfied in any Nash–Bayes equilibrium of any game (whether static or dynamic): these constraints are the so-called incentive constraints – an agent with valuation v should find his own strategy no worse than the strategy of the same agent with valuation v' – and the participation constraints – an agent should get at least what he could get by staying outside the game. Myerson and Satterthwaite then proceed to show that these constraints together with the constraint that the bargaining parties receive no outside money cannot be simultaneously satisfied unless there are inefficiencies (see Milgrom, 2004, for an exposition of this and other impossibility results).

The above buyer–seller set-up assumes that agents know how valuable the good is to them. This is referred to as a 'private values set-up'. Akerlof (1970) identifies another source of bargaining inefficiency in set-ups in which the value to the buyer is a function of the information held by the seller – this is sometimes called an informational externality and referred to as an 'interdependent values set-up'. For example, a seller of a used car may know the quality of his or her car, and the quality of the car obviously affects the valuation of both the seller and the buyer. In an elegant example, in which the buyer is known to value the good α times as much as the seller with $2 > \alpha > 1$ and the quality (identified here with the valuation of the seller) is distributed uniformly on $[0,1]$, Akerlof shows that there can be no trade. The no-trade result arises because a selling price of p would be acceptable to the seller only if the quality is below p, resulting in an average quality of $p/2$. But such an average quality does not justify buying the good at price p for the buyer, as $\alpha p/2 - p < 0$. One of the beauties in Akerlof's example is that it illustrates that, even in situations in which it is common knowledge that the buyer values the good more than the seller, there is no trade in equilibrium. Even though Akerlof restricts his analysis to special trading mechanisms, the inefficiency he identifies can be shown to arise in any equilibrium of any bargaining game, with the use of the same mechanism design techniques as those of Myerson and Satterthwaite. It also extends (even though not in the extreme form of no trade) to other classes of problems with interdependent values (see Samuelson, 1984).

In the above bargaining set-ups, a specific form of property rights was assumed. Within the same examples, other efficiency conclusions would arise with alternative property right structures, thereby illustrating how the initial allocation of property rights may affect efficiency in the presence of informational asymmetries. Obviously, in Akerlof's interdependent values example, if the person valuing the good more is initially the owner of the good there is no inefficiency, which thereby offers a simple illustration of this idea. (See Jehiel and Pauzner, 2006, for further elaboration.) In the private values situation considered by Myerson and Satterthwaite, if the two parties are *ex ante* symmetric and initially own 50 per cent shares of the object, a double auction (in which the party quoting the highest price would buy the 50 per cent shares of the other party at a selling price in between the two quoted prices) would result in an efficient allocation of property rights. Cramton, Gibbons and Klemperer (1987) generalize the latter insight by showing that mixed ownership is economically superior in partnership dissolution problems with private values.

The above bargaining inefficiencies implicitly assume that no outside money can be introduced on to the bargaining table. Otherwise, with large enough subsidies, efficiency could be obtained in the above bargaining set-ups, thereby suggesting that an appropriate public intervention may eliminate the inefficiency due to asymmetric information. However, in interdependent values situations in which agents hold multidimensional signals that are independently distributed across agents, Jehiel and Moldovanu (2001) show that the sole incentive constraints make it generically impossible to achieve the first-best allocation no matter how much money is introduced on to the bargaining table. This result is especially relevant in transactions involving several items because then private information is naturally multidimensional. The result then implies that no public intervention can eliminate the bargaining inefficiencies. (A similar conclusion arises even with one-dimensional private information if the single crossing condition is violated; see Maskin, 1992.)

The above results assume that there is no correlation in the private information held by the various agents. Whenever there are correlations, incentive constraints are less severe because the report made by agent i can be used to deter misreports by agent j. The works of Crémer and McLean (1985; 1988) and Johnson, Pratt and Zeckhauser (1990) (see also Myerson, 1981) suggest that inefficiencies can be totally eliminated even under moderate correlations if agents are risk neutral and transfers can be arbitrarily large. However, limited liability and risk aversion (which seem plausible, especially if very large transfers are involved) ensure that the qualitative insights obtained for the case without correlation continue to hold with moderate correlation (see Robert, 1991). Hence, inefficiencies due to asymmetric information continue to hold even in the correlated case, as long as correlation is not too large. (See also Compte and Jehiel, 2006, who argue within Myerson and Satterthwaite's private values set-up that inefficiencies may arise even with large correlation whenever agents have the option to leave the bargaining table at any time, thereby obtaining their reservation utility.)

As already mentioned, the above inefficiencies hold even if multiple stages of bargaining are allowed, as long as the only inferences of the players come from the equilibrium play of the other parties and not from the release of new hard information (either in an exogenous manner or through endogenous information acquisition). If new information becomes available, the situation is different. Obviously, if the private information held by the various agents become public at some stage, then at this stage bargaining parties with full commitment abilities should be able to implement an efficient agreement. This is because, if inefficiencies were to arise at that stage, a party could propose a Pareto improvement with no further move, keeping the generated surplus for herself: this can be viewed as an application of the Coase Theorem. But, even if one adopts the view that eventually private information becomes publicly available, a critical issue is about how long this takes. If it takes very long, inefficiencies are still likely to be significant because the transitory phase is long. If it does not take long and full commitments are possible, efficiency can be expected.

The role of commitment

The above reported results assume full commitment abilities on the part of the bargaining parties. Another major source of inefficiencies is the limited commitment abilities of the agents. From the viewpoint of mechanism design, the relaxation of commitment abilities of the proposing party (sometimes called Principal) is generally thought of as a bad thing. But one should be cautious here about the criterion used to assess what 'good' or 'bad' means. Clearly, from the viewpoint of the Principal limited commitment ability is a bad thing because it puts additional constraints on the Principal's maximization exercise. However, from the viewpoint of society (as measured by social welfare), the conclusion is far from clear. For example, Coase's conjecture suggests that a monopolist with no commitment ability may end up pricing his good efficiently if consumers are forward-looking (they anticipate the distribution of future prices correctly) and patient enough. In a similar vein, the commitment ability of an auctioneer may allow him to use inefficient reserve prices, which he might be unable to exploit under weaker commitment scenarios. (See McAfee and Vincent, 1997, for a formal approach, and Zheng, 2002, for an optimal auction model in which, even though the seller can commit not to lower his reservation price if there is no interested buyer, buyers can resell the object if they wish.) Clearly, more work is required to understand the pros and cons of commitment from a mechanism design perspective with non-benevolent principals.

In a number of transactions, the transacting parties impose a cost or benefit on third parties: think of the sale of pollution rights or the sale of technologies through patents in imperfectly competitive markets. From the viewpoint of the transaction, this corresponds to an externality in the sense that the trade between a subset of agents affects the payoffs of other agents (see Jehiel, Moldovanu and Stacchetti, 1996). Abstracting from informational asymmetries, Jehiel and Moldovanu (1999) in a one-object environment and Gomes and Jehiel (2005) in a general multi-object environment study resale markets in such set-ups with allocative externalities. They establish that the lack of commitment ability may induce long-run inefficiencies in resale markets whenever there are allocative externalities and agents are patient and forward-looking. Furthermore, if we take as given the legal constraints governing how goods can be exchanged, the initial allocation of property rights is shown to have no effect on the long-run properties of the equilibrium pattern of sales in such markets, as long as parties are forward-looking and patient enough. Thus, in such a complete information world, the lack of commitment ability induces inefficiencies in the presence of allocative externalities and at the same time makes it irrelevant how the initial property rights are allocated.

Practical implications

What are the lessons to be drawn from these theoretical observations? What do these results imply for the desirability of resale markets?

A first category of problems concerns those situations in which private information is persistent. Then the above inefficiency results show that in most scenarios, no matter how exchanges are organized, no matter whether or not resales are permitted, and no matter how well resale markets work, inefficiencies are inevitable. In interdependent value situations with multidimensional signals, even subsidies may not be enough to eliminate the inefficiencies.

Full commitments including controls over resales would seem desirable from a mechanism design viewpoint, as long as the proposing parties seek to maximize total welfare. However, with non-benevolent agents there is no reason in general to expect the full commitment scenario to be preferable to weaker commitment scenarios whenever private information is persistent.

A second category of problems concerns those situations with vanishing private information that will be identified with complete information. Then resale markets permit an efficient allocation of goods whenever agents care solely about their own allocation (that is, when there are no externalities). However, when there are allocative externalities in the sense that the allocation of agent i directly influences the well-being of agent j, resale markets do not allow parties with limited commitment abilities to reach an efficient state of the economy. Yet, even when there are allocative externalities, the efficiency of the economy is unaffected by the initial allocation of property rights, suggesting that in such situations the only role for government interventions is through the legal framework, not the allocation of property rights. For example, it may be desirable from this perspective to require by law that the transacting parties compensate those agents suffering from the transaction.

In complete information situations, it would seem that full commitments including controls over resales should improve efficiency. However, that view ignores the reality of a changing environment, which is one of the basic rationales for the existence of resale markets. Because the economy is changing, resale markets are necessary. The complete contracting scenario implicitly assumed by the full commitment idea is impractical in that it might involve agents that are not even present in the economy (think of a future homeowner who may not yet be born and whose future possession already exists). From a practical viewpoint, the main issue is about understanding the effect of the legal framework that governs resale markets on the overall efficiency of the economy. Some insights about how the legal framework might improve the economic performance of resale markets have been suggested above (see the idea of compensating those agents who suffer from the transaction). Admittedly, more work on both the theoretical and empirical sides is required to understand this as well as the additional effect of persistent private information on resale markets.

PHILIPPE JEHIEL

See also **bargaining; Coase theorem; efficient allocation; incentive compatibility; market failure; mechanism design.**

Bibliography

Akerlof, G. 1970. The market for 'lemons': quality uncertainty and the market mechanism. *Quarterly Journal of Economics* 84, 488–500.
Coase, R. 1960. The problem of social cost. *Journal of Law and Economics* 3, 1–44.
Compte, O. and Jehiel, P. 2006. Veto constraint in mechanism design: inefficiency with correlated types. Mimeo. Paris-Jourdan Sciences Economiques and University College London.
Cramton, P., Gibbons, R. and Klemperer, P. 1987. Dissolving a partnership efficiently. *Econometrica* 55, 615–32.
Crémer, J. and McLean, R. 1985. Optimal selling strategies under uncertainty for a discriminating monopolist when demands are interdependent. *Econometrica* 53, 345–62.
Crémer, J. and McLean, R. 1988. Full extraction of the surplus in Bayesian and dominant strategy auctions. *Econometrica* 56, 1247–57.
Gomes, A. and Jehiel, P. 2005. Dynamic processes of social and economic interactions: on the persistence of inefficiencies. *Journal of Political Economy* 113, 626–67.
Jehiel, P. and Moldovanu, B. 1999. Resale markets and the assignment of property rights. *Review of Economic Studies* 66, 971–91.
Jehiel, P. and Moldovanu, B. 2001. Efficient design with interdependent valuations. *Econometrica* 69, 1237–59.
Jehiel, P., Moldovanu, B. and Stacchetti, E. 1996. How (not) to sell nuclear weapons. *American Economic Review* 86, 814–29.
Jehiel, P. and Pauzner, A. 2006. Partnership dissolution with interdependent values. *RAND Journal of Economics* 37, 1–22.
Johnson, S., Pratt, J. and Zeckhauser, R. 1990. Efficiency despite mutually payoff–relevant private information: the finite case. *Econometrica* 58, 873–900.
McAfee, P. and Vincent, D. 1997. Sequentially optimal auctions. *Games and Economic Behavior* 18, 246–76.
Maskin, E. 1992. Auctions and privatization. In *Privatization*, ed. H. Siebert. Kiel: Institut für Weltwirtschaften der Universität Kiel.
Milgrom, P. 2004. *Putting Auction Theory to Work.* Cambridge: Cambridge University Press.
Myerson, R. 1981. Optimal auction design. *Mathematics of Operations Research* 6, 58–73.
Myerson, R. and Satterthwaite, M. 1983. Efficient mechanisms for bilateral trading. *Journal of Economic Theory* 28, 265–81.
Robert, J. 1991. Continuity in auction design. *Journal of Economic Theory* 55, 169–79.
Samuelson, W. 1984. Bargaining under asymmetric information. *Econometrica* 52, 995–1005.
Zheng, C. 2002. Optimal auction with resale. *Econometrica* 70, 2197–224.

research joint ventures

A *research joint venture* (RJV) is an agreement between two or more partners to perform research and development (R&D), where each partner has an active role in the generation of new knowledge and technology. As such, a RJV is distinct from the *ex ante* or *ex post* agreement to acquire knowledge or technology as in *R&D contracting* or the *licensing* of technology respectively. Many times RJV and *R&D cooperation* are used as synonyms in the literature.

Two features distinguish R&D from ordinary capital investments. First, R&D is a public good (Arrow, 1962). The use by one firm of the information produced by its R&D investments does not diminish the amount of information available to other firms. Second, and related to its public good nature, R&D investment is plagued by an *externality* problem. Firms investing in R&D typically cannot fully appropriate the returns to their own R&D investments. This tends to reduce the incentive to invest in R&D when firms act non-cooperatively (Spence, 1984; d'Aspremont and Jacquemin, 1988).

Both of these characteristics of R&D investments have a profound impact on the optimal way of organizing R&D, as they affect the incentives to invest in R&D.

From a welfare perspective the optimal economy-wide organization would involve the free distribution of the knowledge produced by these R&D investments. However, such a policy would provide little incentive for private investment in R&D in the first place. RJVs provide a mechanism to bridge this divide between public policy and private incentive.

Incentives to form RJVs

Given the public-good nature of R&D, firms do have an incentive to jointly develop technology and share the costs and risk of these projects. Mariti and Smiley (1983) provide evidence for the importance of cost and risk-sharing for the success of R&D cooperation. Developing new technology from scratch implies incurring a high (fixed) cost. Transferring and sharing knowledge that is already developed has a low (marginal) cost. Therefore, firms with complementary products (Röller, Tombak and Siebert, 1997) or complementary knowledge (Sakakibara, 1997) have an incentive to form RJVs to share knowledge for the development of new products. Furthermore, from a *transaction costs* perspective R&D collaboration allows access to specialized and complementary know-how, while at the same time allowing for a transfer of technology at lower transaction costs than with arm's length arrangements. As a result the total cost of developing new knowledge through a RJV is reduced (Pisano, 1990; Oxley, 1997).

While knowledge transfer and cost sharing provide the most common and trivial incentive for the formation of RJVs, the industrial organization literature emphasizes competitive motives for engaging in R&D cooperation and RJVs. R&D is imperfectly appropriable and R&D results, therefore, leak out involuntarily to rival firms. These models concentrate on horizontal R&D cooperation among rival companies as a mechanism to internalize these *spillovers*. The R&D process is represented as a two-stage, non-tournament model where in a first stage firms make R&D investments that (strategically) affect second-stage output market decisions through either a cost-reducing or a demand-enhancing effect. Firms can cooperate – form an RJV – in the R&D stage, but may continue to compete in the product market (for example, Katz, 1986; d'Aspremont and Jacquemin, 1988; Kamien, Müller and Zang, 1992; Suzumura, 1992; Leahy and Neary, 1997). From this literature we discern three important issues conditioning the interrelation between the profitability of RJVs and spillovers: *coordination*, *free-riding* and *information sharing*.

Coordination

Cooperation in these models is typically industry-wide and takes the form of firms coordinating R&D choices in order to maximize joint profits. As a result investment in R&D in an RJV is increasing in the level of the spillover as the firms internalize the positive effect these spillovers have on their partners. In addition, when spillovers are high enough – that is, above a critical level – coordination in R&D will result in higher R&D investment than in non-coordinating firms. At the critical spillover level, the profitability of cooperative and non-cooperative R&D strategies coincides. (When goods are substitutes, the level of product differentiation and the number of rivals are important parameters that determine the critical spillover level; de Bondt, Slaets and Cassiman, 1992.)

Coordination through joint profit maximization without incurring any explicit costs to R&D cooperation increases the firms' profitability in these models. But, more importantly, spillovers increase the profitability of cooperation in R&D. Furthermore, for spillovers above the critical level, firms have an increasing incentive to engage in R&D coordination (De Bondt and Veugelers, 1991). This means that, when spillovers are high enough, firms have an increasing incentive to engage in R&D coordination. Such cooperation would furthermore enhance welfare as R&D investment and market output increase.

Free riding

Most models focus on the welfare and profitability of R&D cooperation, ignoring the stability of such cooperation. The stability of RJVs can be threatened by free riding of non-participating companies on the output of the venture, or by free riding by partners who may conceal their technological expertise while trying to absorb as much as possible of the partner's knowledge (Shapiro and Willig, 1990). Kesteloot and Veugelers (1994) find that cooperative agreements that are profitable, and at the same time also stable, require involuntary – outgoing – spillover levels that are not too high. (Using a repeated game, cheating can be prevented by grim-trigger strategies specifying an eternal dissolution of an industry-wide venture. An alternative approach to solve the internal stability problem is through the organizational design of the venture. Perez-Castrillo and Sandonis, 1996, characterize incentive compatible and individually rational contracts that lead to disclosure of knowledge and, hence, the formation of profitable research joint ventures.) Hence, although higher spillover levels increase the profits from cooperation through coordination, they also increase the profits from cheating by a partner and from free riding by an outsider to the cooperative agreement. Therefore, cooperative ventures become more profitable the more able firms are to restrict outgoing spillovers by protecting their information while selectively sharing information with partners.

Information sharing

Some models take into account the fact that firms can indeed manage spillovers by voluntarily increasing the spillovers among cooperating partners. Such information sharing is found to further increase the profitability of cooperation in R&D. In addition, information sharing not only increases the profitability of R&D cooperation; it also makes such agreements more stable. Eaton and Eswaran (1997) show that, when technology trading cartels are not necessarily industry-wide, information sharing is an even stronger stabilizing force. In this case a much stronger punishment can be specified, namely, the ejection of the cheating firm from a technology-trading coalition, followed by the continuation of information sharing by the non-cheating members. Similarly, De Bondt and Wu (1997) find that information sharing produces larger coalition sizes that are both internally and externally stable.

Katsoulacos and Ulph (1998) explicitly model the choice of spillovers by cooperating and non-cooperating firms, and find that RJVs will always share at least as much information as non-cooperating firms because the former maximize joint profits. When firms act non-cooperatively, however, one would expect that the aim is to minimize the creation of spillovers – the *outgoing* spillovers – through the use of effective legal and strategic protection measures while at the same time to maximize the *incoming* spillovers. Kamien and Zang (2000) show that firms that coordinate their R&D expenditures maximize information flows – their *incoming* spillovers – through the choice of very broad research directions for the RJV. If the firms cannot coordinate their R&D expenditures, they are more concerned about managing their *outgoing* spillovers by choosing a more narrow research approach. This result emphasizes a potential dual role of spillovers: outgoing spillovers which might jeopardize the cooperative agreement, and incoming spillovers which increase the attractiveness of the cooperative agreement. In an empirical paper Cassiman and Veugelers (2002) indeed show that *incoming* spillovers and appropriability have important and separately identifiable effects: firms with higher incoming spillovers and better appropriation have a higher probability of cooperating in R&D.

RJVs and social welfare

When firms are allowed to form RJVs, R&D investments increase with the level of spillovers, exceeding the non-cooperative investment level when the spillovers are substantial (d'Aspremont and Jacquemin, 1988). Competing firms that cooperate in R&D might thus increase not only profits but also welfare when the spillovers are substantial. Policywise, a case can then be made for allowing RJVs to form when spillovers are high. However, when spillovers are low firms acting non-cooperatively with respect to R&D bring about higher welfare than when allowed to form an RJV (Suzumura, 1992). The only effect of a RJV in this case is to reduce R&D competition, which in turn decreases welfare (Katz, 1986). (It has often been suggested that RJVs might also facilitate collusion in the output market. A necessary condition for a RJV to be welfare improving in this case is that total R&D investments increase. Martin, 1997, analyses the increased potential for tacit collusion in RJVs, while Yi, 1995, looks at the welfare effects of product market collusion by an industry-wide RJV. Greenlee and Cassiman, 1999, discuss the effects of collusion in the output market on RJV formation.) This theoretical finding has fuelled the debate on the issue of relaxing antitrust regulation with respect to RJVs. In evaluating cooperative R&D, regulators often use the same 'rule of reason' as in the case of mergers. Given the dynamic nature of R&D, insensitive

application of static merger guidelines may lead to undesirable outcomes (Ordover and Willig, 1985). Appropriate standards for evaluating RJVs should be developed. Jorde and Teece (1990) propose the creation of an administrative procedure for evaluating and possibly certifying cooperative R&D agreements in order to establish a safe harbour from antitrust litigation. But Shapiro and Willig (1990) argue that this would provide too much protection to RJVs, especially because the regulator needs a great deal of information to evaluate a RJV, and much of this information might be proprietary.

Policymakers have attempted to address these issues. In the USA firms can register their RJVs under the National Cooperative Research Act (NCRA). By registering under the NCRA, firms become exempt from treble damages under antitrust regulation. However, cooperative R&D ventures need not register under the NCRA. In that case, they are liable under the usual antitrust regulation. Scott (1988) actually notes that cooperative research registered under the NCRA predominantly falls into industries without severe appropriability problems, while supposedly welfare-enhancing RJVs do not seem to register, leading to a suspicion of adverse selection of RJVs under the NCRA (Cassiman, 2000).

In Europe the 1986 Single European Act amendments to the Treaty of Rome gave the Community specific responsibility for strengthening 'the scientific and technological basis of European industry'. In addition to the EEC block exemption of Article 85(1) of the EC treaty for cooperative ventures in R&D, a variety of programmes were initiated, many of which explicitly fostered inter-firm cooperation tied to Community funding for part of the R&D costs of the proposed projects (Martin, 1996). Nevertheless, the debate on the exact implementation of these policies is still ongoing and has initiated a broader debate on the interaction between innovation policy and competition policy.

Conclusion

While the industrial organization models of RJVs have focused on imperfect appropriation among competitors, several areas for research on RJVs remain thoroughly unexplored. First, empirical work has indicated that most RJVs are formed with customers, suppliers or research organizations rather than with competitors. Recent empirical work has started to tackle the issue of different types of partners for the RJVs, but little theoretical work has followed (Fritsch and Lukas, 2001; Belderbos, Carree and Lokshin, 2004; Veugelers and Cassiman, 2005).

Second, and related, we still know very little about the actual effect of engaging in RJVs on firm (innovation) performance. Brandstetter and Sakakibara (1998) find some evidence of the formation of RJVs on research productivity, and Belderbos, Carree and Lokshin (2004) show that cooperation in R&D leads firms to generate more sales from products that are new to the market. But most empirical studies interpret R&D cooperation as an indirect indication of RJV's profitability. To really uncover the incentives to engage in RJVs, we need to understand how RJVs improve the innovation performance of firms relative to alternative organizational forms.

Finally, little progress has been made yet in understanding the organization of RJVs from a theory of the firm perspective. Why would firms make joint investments in R&D and share property rights and decision rights over the outcome of future research outcomes? When is this efficient or when does it enhance the competitiveness of firms?

BRUNO CASSIMAN

See also **externalities; information sharing among firms.**

Bibliography

Arrow, K. 1962. Economic welfare and the allocation of resources for invention. In *The Rate and Direction of Inventive Activity: Economic and Social Factors*, ed. R. Nelson. Princeton: Princeton University Press.

Belderbos, R., Carree, M. and Lokshin, B. 2004. Cooperative R&D and firm performance. *Research Policy* 33, 1477–92.

Brandstetter, L. and Sakakibara, M. 1998. Japanese research consortia: a microeconomic analysis of industrial policy. *Journal of Industrial Economics* 46, 207–33.

Cassiman, B. 2000. Research joint ventures and optimal R&D policy with asymmetric information. *International Journal of Industrial Organization* 18, 283–314.

Cassiman, B. and Veugelers, R. 2002. R&D cooperation and spillovers: some empirical evidence from Belgium. *American Economic Review* 92, 1169–84.

d'Aspremont, C. and Jacquemin, A. 1988. Cooperative and non-cooperative R&D in duopoly with spillovers. *American Economic Review* 78, 1133–7.

De Bondt, R., Slaets, P. and Cassiman, B. 1992. Spillovers and the number of rivals for maximum effective R&D. *International Journal of Industrial Organization* 10, 35–54.

De Bondt, R. and Veugelers, R. 1991. Strategic investment with spillovers. *European Journal of Political Economy* 7, 345–66.

De Bondt, R. and Wu, C. 1997. Research joint venture cartels and welfare. In *R&D Cooperation: Theory and Practice*, ed. J. Poyago-Theotoky. London: Macmillan.

Eaton, B. and Eswaran, M. 1997. Technology trading coalitions in supergames. *RAND Journal of Economics* 28, 135–49.

Fritsch, M. and Lukas, R. 2001. Who cooperates on R&D? *Research Policy* 30, 297–312.

Greenlee, P. and Cassiman, B. 1999. Product market objectives and the formation of research joint ventures. *Managerial and Decision Economics* 20, 115–30.

Jorde, T. and Teece, D. 1990. Innovation and cooperation: implications for competition and antitrust. *Journal of Economic Perspectives* 4(3), 75–96.

Kamien, , Müller, E. and Zang, I. 1992. Research joint ventures and R&D cartels. *American Economic Review* 82, 1293–306.

Kamien, M. and Zang, I. 2000. Meet me halfway: research joint ventures and absorptive capacity. *International Journal of Industrial Organization* 18, 995–1012.

Katsoulacos, Y. and Ulph, D. 1998. Endogenous spillovers and the performance of research joint ventures. *Journal of Industrial Economics* 46, 333–58.

Katz, M. 1986. An analysis of co-operative research and development. *RAND Journal of Economics* 17, 527–43.

Kesteloot, K. and Veugelers, R. 1994. Stable R&D co-operation with spillovers. *Journal of Economics and Management Strategy* 4, 651–72.

Leahy, D. and Neary, P. 1997. Public policy towards R&D in oligopolistic industries. *American Economic Review* 87, 642–62.

Mariti, P. and Smiley, R. 1983. Co-operative agreements and the organisation of industry. *Journal of Industrial Economics* 38, 183–98.

Martin, S. 1997. Public policy toward cooperation in research and development the European Community Japan the United States. In *Competition Policy in the Global Economy*, ed. L. Waverman, W. Comanor and A. Goto. London: Routledge.

Martin, S. 1996. Protection, promotion and cooperation in the European semiconductor industry. *Review of Industrial Organization* 11, 721–35.

Ordover, J. and Willig, R. 1985. Antitrust for high-technology industries: assessing research joint ventures and mergers. *Journal of Law and Economics* 28, 311–33.

Oxley, J.E. 1997. Appropriability hazards and governance in strategic alliances: a transaction costs approach. *Journal of Law, Economics and Organization* 13, 387–409.

Perez-Castillo, D. and Sandonis, J. 1996. Disclosure of know-how in research joint ventures. *International Journal of Industrial Organization* 15, 51–75.

Pisano, G. 1990. The R&D boundaries of the firm: an empirical analysis. *Administrative Science Quarterly* 35, 153–76.

Röller, L., Tombak, M. and Siebert, R. 1997. Why firms form research joint ventures: theory and evidence. Discussion Paper No. 1654, CEPR.

Sakakibara, M. 1997. Heterogeneity of firm capabilities and cooperative research and development: an empirical examination of motives. *Strategic Management Journal* 18, 134–64.

Scott, J.T. 1988. Diversification versus cooperation in R&D investment. *Managerial and Decision Economics* 9, 173–86.

Shapiro, C. and Willig, R. 1990. On the antitrust treatment of production joint ventures. *Journal of Economic Perspectives* 4(3), 113–30.

Spence, M. 1984. Cost reduction, competition and industry performance. *Econometrica* 52, 101–21.

Suzumura, K. 1992. Cooperative and noncooperative R&D in an oligopoly with strategic commitments. *American Economic Review* 82, 1307–20.

Veugelers, R. and Cassiman, B. 2005. R&D cooperation between firms and universities: some empirical evidence from Belgian manufacturing. *International Journal of Industrial Organization* 23, 355–79.

Yi, S. 1995. R&D cooperation, product-market collusion and welfare. Working paper, Department of Economics, Dartmouth College.

reservation price and reservation demand

The simplest example of a reservation price is that price below which an owner will refuse to sell a particular object in an auction. Since the owner could always, in principle, enforce such a price by outbidding everyone else, this leads immediately to the more general concept of a reservation price as that price at which the owner of a fixed stock will choose to *retain* some given amount from that stock, rather than supply more, and of the amount retained as the owner's 'reservation demand' at the price in question. Considering alternative hypothetical prices, one sees that the owner's supply curve of the commodity can equally well be described as an 'own (reservation) demand' curve, where 'supply' and 'own demand' sum identically to the given stock. The same is naturally true of the market supply curve. Thus consider the standard example of the determination of the price of first-edition copies of a certain old book. A demand curve may be drawn up for those who at present own no copies. Taking account of each present owner's reservation price (or prices for those who possess more than one copy), we may also draw up a supply curve. (Of course 'supply' by present owners may be negative at low prices.) Confrontation of the demand and supply curves will then show the market-clearing price. Equally, however, we could have drawn up the 'reservation demand' curve of present owners, summed it with the demand curve of non-owners and then confronted the 'total' demand curve with the given stock. Since 'supply' and 'reservation demand' sum identically to total stock, at every price, the alternative diagram inevitably shows the same market-clearing price as does the first; it does not show the number of books traded, however.

It will be clear that an agent's reservation price for any type of commodity can be expected to depend on one or more of the following considerations: the scope for direct 'own use' of the commodity; the agent's present need for liquidity; the agent's other resources; the perishability of the commodity and thus the various elements of storage costs (including interest costs); expectations about future prices, there being always a speculative element in the reservation price of any commodity which is not immediately perishable. These considerations all emerge in theories of 'factor supply', for example in the theory of household labour supply. Since 'labour time' is instantly perishable, there is no strictly speculative element to take into account (although someone seeking work may refuse a particular job offer because the wage offered is below a 'reservation wage' based on expectations as to the wage that can be obtained after further job searching). The conventional theory is, however, firmly based on viewing labour supply in terms of the 'reservation demand' for time not spent in market employment, and it is this that leads to the familiar argument that the income effect of a 'wage' change can both be large and contrary to the substitution effect, with the result that labour supply may be either positively or negatively related to the level of the 'wage'. Analogous arguments bear on the supply of land services by landowners who have an 'own use' for their land, on the supply of agricultural products, and so on. The reservation price concept is also useful in the context of privately owned natural resources, a context which introduces two further determinants of reservation price. The lowest price at which a natural resource owner will be prepared to extract the resource will naturally depend on extraction costs, both the present extraction costs and those expected in the future; it will also depend on the expected growth rate, if any, of the resource. It is to be noted that the 'neoclassical rule of free goods' would never have to be applied to primary inputs for which (a) there was a positive price below which supply would be zero, and (b) demand at a zero price would be positive (both conditions holding for all prices of other commodities).

It was noted above, in connection with the market for first-edition copies of a book, that the 'total' demand curve diagram gives the same information with respect to price, and less information with respect to quantity, than does the more conventional supply and demand diagram. How then could P.H. Wicksteed – whose name is so strongly associated with the concept of a supply curve being merely a 'reversed demand curve' – have been so insistent that the former diagram is actually *superior* to the conventional one? (See Wicksteed, 1910, Book II, Ch. IV, and 1914.) Because the 'total' demand curve diagram emphasizes the idea that essentially the *same* kind of forces underlie the conventional supply curve as underlie the usual demand curve, thus breaking down the idea that there is an asymmetry in market forces, with subjective factors being dominant on the 'demand side' and objective ones on the 'supply side'. The diagram in which a single demand curve (inclusive of reservation demand) confronts a fixed supply is at once congenial to any author both seeking to stress the subjective elements of the economic process and upholding the opportunity cost doctrine as against the real cost doctrine. While acknowledging that the demand and supply curves diagram illuminates the process through which the market clearing price is *discovered*, therefore, Wicksteed insisted that the other diagram brings out far more clearly the fundamental *determinants* of that price, namely, subjective marginal valuations and given supplies. With reference to continuously produced commodities, as opposed to first-edition copies, maintenance of this viewpoint would presumably require that the 'given stocks' referred to should be those of primary inputs. Here it may be noted that, even in the course of denouncing the conventional supply curve, Wicksteed admitted that 'as we recede from the market and deal with long periods … cases may arise in which something like a "supply curve" seems legitimate' and that nature does not have 'reserve prices in which she expresses her own demand!' (1914, p. 16, n.1).

IAN STEEDMAN

See also **Wicksteed, Philip Henry.**

Bibliography

Wicksteed, P.H. 1910. *The Common Sense of Political Economy, Including a Study of the Human Basis of Economic Law*. London: Macmillan.
Wicksteed, P.H. 1914. The scope and method of political economy in the light of the 'marginal' theory of value and distribution. *Economic Journal* 24(March), 1–23.

residential real estate and finance

Residential real estate is in any definition a major asset class. The average Swedish household invests three-quarters of its net wealth in its own home. Yet it was not until after 1990 that central questions in finance were asked about real estate. Is the market for real estate informationally efficient? What is the optimal fraction of real estate in a household portfolio? What role do financial constraints play in the pricing of real estate? These are particularly challenging questions in view of the special nature of residential real estate assets: properties are heterogeneous, transactions are infrequent, the trading parties are typically amateurs, and the market is best characterized as a search market where identical properties may trade at quite different prices. For all these reasons the data problems are of a different order of magnitude from those in the core areas of finance. Naturally, progress has been slow and we should not expect answers ever to be as sharp as for assets like stocks and bonds.

This article is organized around the questions posed above. Other areas, in particular the important field of mortgages and mortgage-backed securities, are not discussed.

Market efficiency

Standard theories of portfolio choice and asset pricing presume that markets are informationally efficient in the sense that it is impossible to make profits from trading strategies based on publicly available information, such as past returns. There is ample evidence indicating that real estate markets are not efficient in this sense. Time series studies of real estate returns typically find a strong pattern of positive autocorrelation on quarterly or yearly data (Case and Shiller, 1989; Englund and Ioannides, 1996). Such a pattern could in principle reflect time-varying risk premia, but this interpretation appears implausible. A problem with most studies of housing returns is that they measure only the time variation in the capital-gains part of returns and ignore the value of housing services (the implicit rent). An exception is Meese and Wallace (1994), which is based on micro evidence on unregulated rents. They confirm, for the San Francisco Bay Area, that returns on owner-occupied homes are indeed predictable

based on past returns, but they also show that the profits involved are not sufficient to cover realistic transaction costs for a round-trip trade. There is no money to be made by shifting between renting and owning, with housing consumption fixed, but it may be profitable to time moves according to predicted returns. A general conclusion is that transaction costs in a broad sense are important in understanding real estate markets.

Portfolio choice

Research on portfolio choice has been hampered by a lack of reliable high-frequency data. Much recent research has been stimulated by the repeat-sales indexes for US metropolitan areas developed and analysed by Case and Shiller (1989). Goetzmann (1993) uses the Case–Shiller indexes to compute optimal portfolios (efficient frontiers) in mean-variance space, taking into account the idiosyncratic component of housing return, that is, the added risk of an individual home above the general return risk captured by a price index. He finds optimum housing shares to be on the order of 10–50 per cent of household net wealth depending on risk attitudes. It is well known from portfolio analysis of other assets that calculated portfolio shares are quite sensitive to input data, particularly expected return, and hence should be treated with caution. Nevertheless, later studies using data for European countries (using hedonic indexes not available in the United States) have obtained similar results (see, for example, Englund, Hwang and Quigley, 2002, for Stockholm; le Blanc and Lagarenne, 2004, for Paris; and Iacoviello and Ortalo-Magné, 2004, for London). The discrepancy between computed optimal portfolio shares and real world numbers, often in the order of several hundred per cent, is striking and has provided a challenge for further research.

The standard mean-variance analysis is obviously oversimplified in several ways. First, it is static. Grossman and Laroque (1990) consider lifetime portfolio choice when utility is derived from a durable good (housing), which can only be traded at a cost (proportional to house value). Housing trades are determined in analogy with Ss-models from inventory theory, and optimal portfolios are shown to be mean-variance efficient like in the static case.

Second, the standard analysis does not account for housing services as a consumption good separate from non-durable goods. Flavin and Yamashita (2002) analyse a two-good version of the Grossman and Laroque model with a stochastic relative price of housing. Based on correlations calculated from Case–Shiller indexes, their model indicates that the optimal fraction of financial assets going into stocks is inversely related to the fraction invested in housing, and hence should increase with age, consistent with empirical observations. More recently some authors have analysed models with finite lifetimes, using numerical solution techniques. A key factor in determining the attractiveness of investing in housing is the correlation between labour income and the returns to housing: the stronger the correlation, the smaller is the optimal housing portfolio share.

Third, we have so far assumed housing to be consumed by owning, disregarding the alternative of renting. The issue of tenure choice is a classic one in the housing literature; see Henderson and Ioannides (1983) for a two-period model that brings out some of the basic features. Only rarely have issues of risk been included in the analysis. Among the exceptions are Rosen, Rosen and Holtz-Eakin (1984) and Turner (2003), who find that volatile house prices deter young households from entering into owner-occupancy. These studies do not explicitly measure the relative risks of owning compared with renting. More recently, Sinai and Souleles (2005) have emphasized that owning one's home is a way of hedging the risk associated with stochastic variations in the cost of renting. This is a particularly important aspect for households with a long expected stay in the same dwelling or the same housing market. Empirically, Sinai and Souleles confirm, for US households, that the probability of homeownership is indeed an increasing function of rent risk.

For most households, net wealth falls far short of the value of the house they demand for consumption purposes. Hence, any portfolio study that includes housing has to take a stand on the availability of borrowing. In fact, most households are constrained in their access to borrowing, at least when young, and financial constraints exert an important influence on savings and housing choices over the life cycle; see King (1980) for an early study emphasizing borrowing constraints. Integrating down-payment constraints into models of dynamic portfolio and tenure choice remains an important topic for future study.

Asset pricing and financing constraints

The standard approach to real estate price determination (as in Poterba, 1984) is explicitly couched in asset pricing terms: the price is the discounted value of the housing services generated by the property net of operating and maintenance costs. In principle, housing services could be valued based on market rents for comparable dwellings. In applying this approach, lip service is often paid to risk-adjusting the discount rate. It is fair to say, however, that there is no established theory or pragmatic consensus on the choice of discount rate. In recent years there has been a surge of interest in integrating housing into the standard asset-pricing paradigm. So far, however, interest has focused on the impact on financial asset prices of introducing housing collateral rather than on pricing real estate assets.

More attention has been paid to the direct impact of financial constraints on pricing. If the representative homebuyer is constrained by borrowing opportunities rather than by lifetime resources, then wealth shocks have a direct impact on housing demand. This implies that a shock to the demand and supply of housing services will be reinforced through its impact on financing constraints. An income shock, for example, will increase demand and housing prices, thereby releasing borrowing constraints. This will in turn give an extra boost to demand and prices. There will be a 'financial multiplier': the more important financial constraints are, the more sensitive prices will be to shocks to underlying fundamentals. This view of real estate pricing was formulated by Stein (1995) and has been inserted into an overlapping-generations framework with demographic fluctuations by Ortalo-Magné and Rady (2006). Its empirical validity has been investigated in some studies. As an example, Lamont and Stein (1999) show that variations in the sensitivity of house prices to income shocks across US states can be explained by differences in loan-to-value ratios. Financial constraints may also explain the strong impact of variations in house prices on consumption observed in many studies; see, for example, Case, Quigley and Shiller (2005) for the United States and internationally.

Historically, mortgage lending has been further restricted by regulations in virtually all countries. Dismantling these regulations has in many cases caused price booms. But borrowing constraints remain important facts of life even in unregulated market environments, and there are large differences across countries even today, reflecting history and legal institutions. Chiuri and Jappelli (2003) show that average downpayment ratios vary from close to 50 per cent in Italy to a little above 10 per cent in Sweden and United Kingdom. They find that these differences, which they largely ascribe to legal tradition – relating to foreclosure, for example – explain differences in homeownership rates across countries, in particular the age when young households enter into owner occupancy.

The future

Not only has the area of real estate economics been lagging in its adoption of new analytical frameworks from finance, markets have also been slow in adopting new financial instruments and contracts to handle better the important risks many household confront in relation to their housing investment. While households have access to a wide variety of mortgage instruments, markets remain seriously incomplete and fail to offer flexible and liquid contracts related to housing price risks. As Robert Shiller (2003) has forcefully argued, this is one of the macro risks in society that remain uninsurable despite their fundamental importance for individual welfare. Options or futures on relevant housing price indexes could go a long way towards providing such insurance. It remains to be seen how long it will take to develop liquid markets in such instruments.

PETER ENGLUND

See also **capital asset pricing model; efficient markets hypothesis; household portfolios; housing supply.**

Bibliography

Case, K., Quigley, J. and Shiller, R. 2005. Comparing wealth effects: the stock market versus the housing market. *Advances in Macroeconomics* 5, 1–32.

Case, K. and Shiller, R. 1989. The efficiency of the market for single-family homes. *American Economic Review* 79, 125–37.

Chiuri, M. and Jappelli, T. 2003. Financial market imperfections and homeownership: a comparative study. *European Economic Review* 47, 857–75.

Englund, P., Hwang, M. and Quigley, J. 2002. Hedging housing risk. *Journal of Real Estate Finance and Economics* 24, 167–200.

Englund, P. and Ioannides, Y. 1996. House price dynamics: an international empirical perspective. *Journal of Housing Economics* 6, 119–36.

Flavin, M. and Yamashita, T. 2002. Owner-occupied housing and the composition of the household portfolio. *American Economic Review* 92, 345–62.

Goetzmann, W. 1993. The single family home in the investment portfolio. *Journal of Real Estate Finance and Economics* 6, 201–22.

Grossman, S. and Laroque, G. 1990. Asset pricing and optimal portfolio choice in the presence of illiquid durable consumption goods. *Econometrica* 58, 25–51.

Henderson, J. and Ioannides, Y. 1983. A model of housing tenure choice. *American Economic Review* 73, 98–113.

Iacoviello, M. and Ortalo-Magné, F. 2004. Hedging housing risk in London. *Journal of Real Estate Finance and Economics* 27, 191–209.

King, M. 1980. An econometric model of tenure choice and demand for housing as a joint decision. *Journal of Public Economics* 53, 137–59.

Lamont, O. and Stein, J. 1999. Leverage and house-price dynamics in U.S. cities. *RAND Journal of Economics* 30, 498–514.

le Blanc, D. and Lagarenne, C. 2004. Owner-Occupied housing and the composition of the household portfolio: the case of France. *Journal of Real Estate Finance and Economics* 29, 259–75.

Meese, R. and Wallace, N. 1994. Testing the present value relation for housing prices: should I leave my house in San Francisco. *Journal of Urban Economics* 35, 245–66.

Ortalo-Magné, F. and Rady, S. 2006. Housing market dynamics: on the contribution of income shocks and credit constraints. *Review of Economic Studies* 73, 459–85.

Poterba, J. 1984. Tax subsidies to owner-occupied housing: an asset-market approach. *Quarterly Journal of Economics* 99, 729–52.

Rosen, H., Rosen, K. and Holtz-Eakin, D. 1984. Housing Tenure, uncertainty, and taxation. *Review of Economics and Statistics* 66, 405–16.

Shiller, R. 2003. *The New Financial Order: Risk in the 21st Century.* Princeton: Princeton University Press.

Sinai, T. and Souleles, N. 2005. Owner-occupied housing as a hedge against rent risk. *Quarterly Journal of Economics* 120, 763–89.

Stein, J. 1995. Prices and trading volume in the housing market: a model with downpayment constraints. *Quarterly Journal of Economics* 110, 379–406.

Turner, T. 2003. Does investment risk affect the housing decisions of families? *Economic Inquiry* 41, 675–91.

residential segregation

The term 'residential segregation' describes a housing market equilibrium marked by systematic disparities in the physical location of households belonging to different racial, ethnic, socio-economic, or other social groups.

While history is replete with examples of groups forced to live in complete isolation from the remainder of society, residential segregation is not inherently a dichotomous phenomenon. Rather, housing markets may exhibit varying degrees of segregation; social scientists have endeavoured to quantify this variation for the better part of a century. The term 'ghetto' is often ascribed to social groups experiencing segregation that exceeds a loosely defined threshold.

Residential segregation may be the outcome of a past residential sorting process wherein centralized authorities restricted some agents' location choices. Very simple economic theory, and an increasing amount of empirical evidence, however, point to the conclusion that modern-day residential segregation is driven primarily by the operation of decentralized market forces.

Even if residential segregation is a pure market phenomenon, many observers harbour concerns that segregated housing market equilibria are suboptimal from a social welfare perspective. Some debate exists as to whether segregated housing markets are inefficient. Arguments hinge on whether households are fully informed at the time they make location decisions, or whether they face borrowing constraints. It is a less controversial observation that residential segregation has important implications for distributional equity. In segregated equilibria, for example, wealthy households have the opportunity to avoid subsidizing the local public good consumption of poorer households. Over the past several decades, there have been many attempts to estimate the relationship between residential segregation and inequality between groups, in both the short term and the long term.

Here, basic evidence is provided on the existence and magnitude of contemporary residential segregation. This evidence draws heavily on the experience of racial and immigrant groups in the United States; the measurement of segregation in other nations is limited in scope and often confined to very recent observations. As discussed below, this is more a reflection of data limitations than any genuine lack of interest. The basic economic theory of why segregation exists is then outlined, and empirical evidence that has been brought to bear on the issue discussed. The concluding discussion considers the potential implications of segregation on socio-economic outcomes and human capital investment.

Measuring segregation

There are many ways to measure segregation (Massey and Denton, 1988). The metrics most commonly used in sociology and economics require the existence of neighbourhood-level data on the distribution of groups in a city or region. Some measures, including the spectral segregation index (Echenique and Fryer, 2005), require additional data on the physical location of these neighbourhoods and, in some cases, their land area. A central challenge to the systematic measurement of segregation is the lack of comparable neighbourhood-level data across nations, or even cities within nations, and over time. The United States, for example, has collected data on the race of its inhabitants since 1790, but did not report race at a consistently defined neighbourhood level until 1940. The United Kingdom did not systematically collect information on the ethnic identity of its inhabitants until the 1991 Census.

Given the existence of required data, the most commonly used segregation indices classify the residential separation of any particular group between two extremes: perfect segregation, where group members never share a neighbourhood with individuals not belonging to the group, and perfect integration, where group members form an equal share of the population in all neighbourhoods. The dissimilarity index (Duncan and Duncan, 1955), records groups on the scale from 0 (perfectly integrated) to 1 (perfectly segregated) using the following formula:

$$D = \frac{1}{2} \sum_i \left| \frac{A_i}{A} - \frac{B_i}{B} \right| \qquad (1)$$

where i indexes neighborhoods, A_i and B_i represent the number of group members and others in neighborhood i, respectively, and A and B represent the total population of group members and others in the city or region. The dissimilarity index has a relatively intuitive interpretation: it is the share of group members, or others, who would have to switch neighbourhoods in order to achieve perfect integration. While many demographers state a preference for other indices based on various criteria, the dissimilarity index is most commonly used in existing literature.

Stylized facts

The absence of neighbourhood-level data makes it difficult to gauge the contemporary level of segregation in

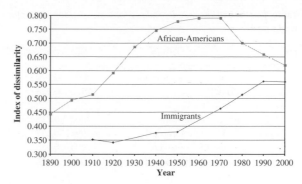

Figure 1   Dissimilarity of African-Americans and immigrants in the United States, 1890–2000

many cities, let alone historical levels. The most comprehensive historical data pertain to American cities. Cutler, Glaeser and Vigdor (1999; 2005) use these data to compute long-term trends in dissimilarity for African-Americans and foreign-born individuals, respectively. Figure 1 plots weighted averages of these measures across cities, with weights equal to the population of the group in question in each city. Immigrant segregation is computed separately for each country-of-origin group in each city; the immigrant time series represents the weighted average of these data.

As the relatively low initial levels of black–white dissimilarity indicate, urban ghetto neighbourhoods were relatively uncommon in the United States at the turn of the 20th century. The birth of the African-American ghetto coincided with the so-called Great Migration of blacks from the southern part of the United States to northern cities between the First World War and 1965. Segregation reached its peak at the end of this period of migration. In 1970 dissimilarity levels in some areas, chiefly large industrial cities of the north-eastern and mid-western United States, were at or near 0.90. Since that time, segregation has fallen pervasively throughout the nation but most acutely in rapidly growing cities in the southern and western parts of the country.

Immigrant segregation, quite strikingly, displays the opposite trend to racial segregation in the United States. Immigrant dissimilarity remained stable at relatively low levels through the first half of the century, then rose steadily. Thus, even as racial ghettos have declined over the past few decades, immigrant segregation has risen. Cutler, Glaeser and Vigdor (2005) present further data indicating that the rise in average segregation can be attributed primarily to the growth of groups that have always experienced high segregation, rather than to the increasing segregation of individual groups. The growing, highly segregated groups generally originate in less developed countries and gravitate toward the largest cities in the United States. The limited amount of data available from other nations supports the general trend found in the United States: individual racial and ethnic group are

Table 1 *Recent dissimilarity indices for various groups in major world cities*

City	Group	Year	Dissimilarity
Barcelona	Latin American immigrants	2001	0.290
Cape Town	Blacks*	1996	0.928
Chicago	African-Americans	2000	0.778
Cologne	Turkish immigrants**	1994	0.337
Lima	High SES households	1993	0.440
London	Blacks (Caribbean, African and Other)*	2001	0.468
London	South Asians*	2001	0.544
Los Angeles	Mexican immigrants	2000	0.446
Mexico City	High SES households	2000	0.380
New York	African-Americans	2000	0.670
Santiago	High SES households	1992	0.490
Tokyo	Individuals over 65	1995	0.147

Note: dissimilarity indices measure the separation of each group from the remainder of the population, except as indicated.
*Dissimilarity from whites   **Dissimilarity from Germans   SES: socio-economic status
Sources: Barcelona: Martori i Cañas and Hoberg (2004); Cape Town: Rospabe and Selod (2003); Chicago and New York: Glaeser and Vigdor (2002); Cologne: Friedrichs (1998); Lima, Mexico City and Santiago: Arriagada Luco and Vignoli (2003); London: Burgess, Wilson and Lupton (2005); Los Angeles: Cutler, Glaeser and Vigdor (2005); Tokyo: Nakagawa (2003).

experiencing stable or declining segregation in most parts of the world.

Socio-economic segregation, or the degree to which households in poverty tend to cluster together in neighbourhoods, increased in the United States in the 1970s and 1980s, but showed some evidence of lessening in the 1990s.

Table 1 presents some representative dissimilarity index values for groups in some of the world's largest cities, using recently available data.

Why are groups segregated?

Theories of racial segregation can be classified into two types. The first type permits some form of discrimination in housing markets. The second models segregation as the equilibrium outcome of a fully competitive market. A potential third class of models explains one form of segregation as the direct consequence of a second form – for example, it explains racial segregation as a consequence of economic segregation. This third class is of less interest to attempts to explain segregation more generally.

In a discrimination-based model, location choices are constrained for members of one group, defined by race, ethnicity, or other observable characteristic. The constraints

on location choice might include explicit legal barriers or implicit patterns of 'steering' households towards certain locations. Historical examples of explicit legal barriers abound. In a few cases, governments have attempted to restrict location choices as a matter of public law, have enforced contracts between private parties restricting racial ownership or occupancy of property, or have adopted policies that had the effect of limiting the residential options of certain groups. In the United States, federal legislation had made most explicit forms of housing market discrimination illegal by the end of the 1960s.

While few observers would argue that explicit racial or ethnic barriers to location choice persist in the developed world, the existence and prevalence of implicit discriminatory patterns is a subject of continuing debate. Government policies such as zoning laws, which local governments use to regulate the density and nature of residential development within their borders, may implicitly perpetuate segregation. Housing audit studies provide evidence of discriminatory behaviour among real estate agents, mortgage brokers, or landlords. In these studies, auditors of different races present carefully matched, fictionalized credentials to housing market agents. The behaviour of these agents is then analysed to uncover any systematic differences in treatment by race. Recent studies, such as Ondrich, Ross and Yinger (2003), find evidence of significant racial disparities in treatment in the United States.

While disparities in treatment of housing market auditors can be interpreted as evidence of continued racism, such behaviour can also be consistent with unbiased, profit-maximizing motives. As in models of statistical discrimination in labour markets or other settings, agent behaviour could be motivated by accurate perceptions of differences in average preferences across racial groups.

Such an interpretation is consistent with the second type of racial segregation theory, which posits that segregated housing market equilibria are fully consistent with decentralized, unconstrained household choices. Preference-based theories of segregation owe some intellectual debt to Tiebout's (1956) vision of residential sorting, but evolve most clearly from Schelling's (1978) simulation of residential sorting in the presence of very slight preferences for neighbours of one's own group. Schelling's simulations show that a small initial concentration of same-group neighbours can rapidly evolve into a vast enclave community. This process of 'tipping' is driven by group members' heightened willingness to pay for locations in close proximity to the initial cluster. As the enclave grows in size, it becomes disproportionately more attractive to group members than to others. So long as the segregated group in question maintains a steady population share in the entire market, the enclave is very unlikely to dissipate.

Much anecdotal evidence supports the Schelling model. The neighbourhood integration that has taken place in the United States since 1970, for example, has left most African-American enclaves untouched. Rather, integration has occurred either in newly developed neighbourhoods on the fringe of urban areas or in locations marked by significant demolition and redevelopment.

While intuitively appealing and supported by anecdotal observation, true empirical tests of preference-based theories are rendered difficult by the unobservability of household preferences. Econometric models associated with the measurement of willingness to pay, such as discrete choice models, often assume away the existence of housing market discrimination (for example, Bayer, McMillan and Rueben, 2004). Survey-based methods of eliciting preferences are valid only to the extent that respondents can accurately separate their valuation of neighbourhood racial composition from all other attributes, and truthfully reveal this valuation. What survey evidence that exists supports the notion that groups harbour preferences for same-group neighbours (Vigdor, 2003).

Why might individuals care about the racial or ethnic composition of their neighbourhoods? Group members may prefer to congregate in enclaves in order to take advantage of scale economies enabling the supply of group-specific community institutions or consumer goods. Individuals may also seek to limit exposure to other groups on the basis of stereotyped perceptions of inferiority, greater criminality, or other characteristics. It is also possible that individuals care, not about the race of their neighbors directly, but about characteristics correlated with race, such as socio-economic status. These varying hypotheses have dramatically different implications for the social value of segregation. Unfortunately, these various explanations are observationally equivalent. Each predicts that segregation occurs in equilibrium because willingness to pay for housing in a group enclave is relatively higher among group members.

While there is currently no consensus on the importance of housing market discrimination in perpetuating segregation, Cutler, Glaeser and Vigdor (1999) present evidence that any such importance has declined. In 1940, at a time when many forms of housing market discrimination were legal – and in some cases practised by government itself – restrictions on African-American location choice had the impact of increasing equilibrium prices in segregated areas. By 1970 that premium had disappeared, suggesting that these artificial barriers to mobility had been removed.

Does segregation influence economic outcomes?

A number of hypothesized causal mechanisms link segregation to socio-economic outcomes. The 'spatial mismatch' hypothesis contends that segregation reduces the average income of certain groups to the extent that their residential enclaves are located at some distance from growing employment centres (Kain, 1968). Segregation may also lead to differences in education quality across racial or ethnic groups, to the extent that schooling is tied to residential location. Finally, there may be other localized factors that differ across neighbourhoods and have the net impact of leading to different human capital investment trajectories. For example, children growing up in different neighbourhoods may develop different consumption or investment preferences by being exposed to different types of role models.

Numerous attempts have been made to empirically estimate the impact of segregation on outcomes, whether operating immediately through spatial mismatch-type mechanisms or developmentally. Much of this empirical literature is plagued by a fundamental endogeneity problem: since individuals choose their own neighbourhoods, any correlation between neighbourhood characteristics and individual outcomes might reflect selection rather than any causal effect of the former on the latter. Researchers have implemented three strategies for circumventing these selection problems. The first is to focus on the outcomes of young adults, whose location choices are presumably determined by their parents rather than themselves. Vigdor (2002) points out that the strategy of studying young adults is suspect in the presence of inter-generational transmission of economic outcomes.

A second basic strategy for identifying the impact of segregation on outcomes in the presence of selective migration is to model location choice and socio-economic outcomes simultaneously. Some research in this vein makes use of individual data-sets with detailed geographic identification, recently made available by the US Census Bureau. A simultaneous equation model can uncover the true causal impact of segregation on outcomes if it employs an instrumental variable – a factor than affects location choice but otherwise bears no correlation to individual outcomes. In practice, identifying a valid instrumental variable is very difficult.

Recently, researchers have addressed selective migration concerns by turning their attention to randomized mobility experiments, in which a 'treatment' group is offered a voucher redeemable for housing only in certain neighbourhoods, while a 'control' group is offered no such aid. While these experiments generally do not permit examination of the causal impact of segregation per se, they do allow a more general study of the potential importance of neighbourhood characteristics in determining outcomes. In general, studies find little impact of neighbourhood factors on the socio-economic outcomes of adults. There is more evidence in favour of developmental impacts on youth. Orr et al. (2003) present an overview of research results stemming from one such randomized mobility experiment, the Moving to Opportunity demonstration programme.

JACOB L. VIGDOR

See also **finance; ghettoes; housing policy in the United States; immigration and the city; neighbours and neighbourhoods; spatial mismatch hypothesis; symmetry breaking; urban economics; urban housing demand.**

Bibliography

Arriagada Luco, C. and Vignoli, J. 2003. Segregación residencial en áreas metropolitanas de América Latina: magnitud, caracteristicas, evolución e implicaciones de politica. Series Población y Desarrollo No. 47. Santiago de Chile: UN Comisión Económica para América Latina y El Caribe.

Bayer, P., McMillan, R. and Rueben, K. 2004. Residential segregation in general equilibrium. Working Paper no. 11095. Cambridge. MA: NBER.

Burgess, S., Wilson, D. and Lupton, R. 2005. Parallel Lives? Ethnic segregation across schools and neighborhoods. *Urban Studies* 42, 1027–56.

Cutler, D., Glaeser, E. and Vigdor, J. 1999. The rise and decline of the American ghetto. *Journal of Political Economy* 107, 455–506.

Cutler, D., Glaeser, E. and Vigdor, J. 2005. Is the melting pot still hot? Explaining the resurgence of immigrant segregation. Working Paper No. 11295. Cambridge, MA: NBER.

Duncan, O. and Duncan, B. 1955. A methodological analysis of segregation indexes. *American Sociological Review* 20, 210–17.

Echenique, F. and Fryer, R.G., Jr. 2005. On the measurement of segregation. Working Paper No. 11258. Cambridge, MA: NBER.

Friedrichs, J. 1998. Ethnic segregation in Cologne, Germany, 1984–1994. *Urban Studies* 35, 1745–63.

Glaeser, E. and Vigdor, J. 2002. Residential segregation: promising news. In *Redefining Urban & Suburban America: Evidence from Census 2000*, vol. 1, ed. B. Katz and R. Lang. Washington, DC: Brookings Institution Press.

Kain, J. 1968. Housing segregation, negro employment, and metropolitan decentralization. *Quarterly Journal of Economics* 82, 175–97.

Martori i Cañas, J. and Hoberg, K. 2004. Indicadores cuantativos de segregación residencial. El caso de la población inmigrante en Barcelona. *Scripta Nova* 8 (169). Online. Available at http://www.ub.es/geocrit/sn/sn-169.htm, accessed 1 August 2005.

Massey, D. and Denton, N. 1988. The dimensions of racial segregation. *Social Forces* 67, 281–315.

Nakagawa, M. 2003. Analysis of age discrimination in the rental housing Market in Japan: an approach using a fair housing audit. Discussion paper No. 577. Osaka: Institute of Social and Economic Research, Osaka University.

Ondrich, J., Ross, S. and Yinger, J. 2003. Now you see it, now you don't: why do real estate agents withhold available houses from black customers? *Review of Economics and Statistics* 85, 854–73.

Orr, L., Feins, J., Jacob, R., Beecroft, E., Sanbonmatsu, L., Katz, L., Liebman, J. and Kling, J. 2003. *Moving to Opportunity: Interim Impacts Evaluation*. Washington, DC: Office of Policy Development and Research, US Department of Housing and Urban Development.

Rospabe, S. and Selod, H. 2003. Does city structure cause unemployment? The case study of Cape Town. Unpublished manuscript. Paris: Laboratoire d'Economie Appliquée.

Schelling, T. 1978. *Micromotives and Macrobehavior*. New York: W.W. Norton.

Tiebout, C. 1956. A pure theory of local expenditures. *Journal of Political Economy* 64, 416–24.

Vigdor, J. 2002. Locations, outcomes, and selective migration. *Review of Economics and Statistics* 84, 751–55.

Vigdor, J. 2003. Residential segregation and preference misalignment. *Journal of Urban Economics* 54, 587–609.

reswitching of technique

Reswitching of technique refers to the virtual adoption of production techniques, either by the individual producer or by the economic system as a whole. Standard economic theory treats technical adoption on the assumption that there is a multiplicity of techniques for producing any given good, and that the producer, as a rational decision maker, will switch from one technique to another according to a certain hypothetical sequence as the prices of productive factors are changed. This sequence would depend on the ranking of techniques in terms of capital per man or 'capital intensity', so that a lower rate of interest (which is equal to the rate of profit in equilibrium) would be associated with the 'adoption' of a technique characterized by higher capital per man. This process is known as capital deepening.

The development of discrete production models in the 1950s led to the discovery that this view of 'rational' technical adoption is not necessarily well founded. David Champernowne (1953) and Joan Robinson (1956) pointed out that a movement of the rate of interest in a given direction might make it optimal once again to use techniques that had been previously excluded. This phenomenon is known as *reswitching of technique*.

The original discovery was associated with the belief that reswitching was nothing more than a 'curiosum', which could not be left out on grounds of pure logic but was nevertheless unlikely to happen. The discussion of this phenomenon by Piero Sraffa (1960) showed that reswitching is the normal outcome of a situation in which the various production processes are characterized by different proportions between 'direct' labour and the quantity of 'past' labour. (This latter is the quantity of labour that is indirectly required in a production process, being required in producing its intermediate inputs.) Sraffa's analysis also provides a clear insight into the reasons for technical reswitching along the hypothetical sequence associated with changes in the rate of profit. It is worthwhile considering his example in some detail.

The 'pure products' case

It is useful to start with the consideration of a special category of commodities, which we might call of the *pure product* type. These are commodities that are never used as productive inputs, so that their price reflects production cost, but cost is never influenced by the variation of their particular prices.

Let a and b be commodities of that type, and let them be produced with different proportions of direct labour to past labour. (This structure of labour requirements is representative of the differences in the proportions between labour and intermediate inputs in the production processes of the two commodities.)

Let a require more labour than b if we consider labour applied eight years before the year in which the product is ready, whereas b requires more labour than a in the cases of labour applied in the current year and 25 years earlier. This situation may be represented as follows (n is the date at which labour is applied):

(i) $n = 8$

$$l_{a(8)} = v + 20$$
$$l_{b(8)} = v$$

(ii) $n = 0$

$$l_{a(0)} = x$$
$$l_{b(0)} = x + 19$$

(iii) $n = 25$

$$l_{a(25)} = y$$
$$l_{b(25)} = y + 1$$

We are now in a position to examine in which way the cost difference between the two products may vary if the rate of profit is raised from 0 to a maximum value of 25 per cent. (An increase of the rate of profit is equivalent to a change in the weight of the different labour terms in each cost equation.)

The cost difference is expressed by the following equation:

$$p_a - p_b = 20w(1 + r)^8 - \left[19w + w(1 + r)^{25}\right]. \tag{1}$$

On the assumption that the wage rate (w) is inversely related to the rate of profit according to the following expression:

$$w = 1 - \frac{r}{25\%},$$

the cost difference equation will be represented by the curve in Figure 1.

The cost of a rises relatively to b as r increases between zero and nine per cent. The reason for this is that the change of r leaves the value of current labour unaffected, whereas the 'excess labour' of date 8 is much greater than the excess labour of date 25. The increase in the value of $l_{b(25)}$ is more than offset by the increase in the value of $l_{a(8)}$ and the compound effect of these two variations is an increase in the cost difference. Beyond $r = 9\%$, the increasing weight of remote labour terms brings the cost

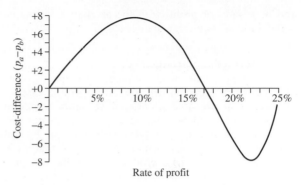

Figure 1

difference down. This reduction stops at $r = 22\%$, since at this particular level of the rate of profit the decline of the wage rate starts offsetting the increase in the value of remote labour terms due to a higher r.

The above argument has straightforward implications for technical choice in the case of commodities of the pure product type. For in this case we can take for granted that the price of each commodity reflects its cost of production, whereas this price has no influence at all on the cost. Under such conditions, eq. (1) permits us to examine in which way the relative profitability of two techniques is varied as r goes from 0 to r (max). In fact, we may take eq. (1) to illustrate the difference between the unit costs of production of the same commodity produced with two alternative techniques. (For reasons of symmetry with the previous argument we call such alternative techniques a and b respectively.) Figure 1 can be applied to this particular case. An immediate short-coming would be that a change in the price of direct to 'dated' labour, as reflected in an increasing r, is associated with a positive excess of unit cost p_a over unit cost p_b until the curve intersects the horizontal axis for the first time. This involves that, over this interval, technique b is more profitable than technique a. A further increase of r (until r (max)) is associated with a negative difference $(p_a - p_b)$, so that technique a is more profitable than technique b. However, the same figure shows that the *reduction* of the cost difference stops at $r = 22\%$. For any r such that $22\% < r < r(\text{max})$, the cost difference is increasing once again. This increase stops at $r = 25\%$, when techniques a and b become equally profitable.

The movement of the cost difference when r is increasing shows that the relative profitability of techniques a and b is subject to fluctuations which depend on the particular interval within which r is changed. The relative profitability of technique a with respect to technique b is initially decreasing, then increasing, finally decreasing again. These fluctuations show that the 'unevenness' of the input structure may bring about multiple switches between the two techniques as we consider a steadily increasing r: the same technique might be adopted at low and high rates of profit, with the alternative technique being adopted at intermediate levels of r.

The 'intermediate products' case

It might appear that the above picture gets greatly complicated when we consider the more general case of products that are used as productive inputs either of themselves or of other commodities. For in this new situation the price of a commodity reflects its production cost, but this cost might itself be influenced by that price. (Directly in the case of a product used in its own production, indirectly in the case of a product that is, at some stage, a necessary means of production for at least one of its inputs.)

An immediate consequence of the consideration of interdependence between production processes is that inspection of the cost difference equation is no longer sufficient in order to assess the relative profitability of alternative techniques. The mutual influence between prices and production costs brings about the need of comparing systems of interrelated techniques (production technologies) rather than individual techniques. This requires consideration of the price system that will be associated with each technology at any given distribution of income between wages and profits.

The analysis of the 'intermediate products' case can be carried out by examining a simple model with two alternative two-good technologies A and B, in which all products are used as inputs of themselves and of the other commodity. We shall also assume that the two technologies differ only in the technique used to produce commodity 1.The two price systems may be written as follows:

$$(a_{11}p_1 + a_{21}p_2)(1 + r) + l_1(a)w = p_1$$

$$(a_{12}p_1 + a_{22}p_2)(1 + r) + l_2(a)w = p_2,$$

$$(2.1)$$

$$(b_{11}p_1 + b_{21}p_2)(1 + r) + l_1(b)w = p_1$$

$$(b_{12}p_1 + b_{22}p_2)(1 + r) + l_2(b)w = p_2,$$

$$(2.2)$$

where $a_{ij}(i, j = 1, 2)$ and $b_{ij}(i, j = 1, 2)$ are the quantities of commodity i required to produce one unit of commodity j with technologies A and B respectively, $l_i(a)$ and $l_i(b)$ are the quantities of labour entering one unit of commodity i with technologies A and B respectively, $p_i(i = 1, 2)$ is the price of product i, w is the unit wage and r the rate of profit. The quantities a_{ij}, b_{ij}, $l_i(a)$ and $l_i(b)$ are known, whereas r, w, p_i are unknown.

Either product is common to both systems. We may thus choose either commodity 1 or 2 as the common standard of prices (*numéraire*) in both systems. If we put the price of commodity 1 equal to unity, commodity 1 becomes the common *numéraire* of price systems (2.1) and (2.2). At this stage, it is found convenient to assess

the relative profitability of alternative technologies by considering the functional relationship between r and w for each technology.

The systems of eqs. (2.1) and (2.2) would each be associated with a particular relation between the rate of profit and the unit wage. The wage–profit relationships for the two systems would respectively be given by the following expressions:

$$w_A = \frac{1 - (a_{22} + a_{11})(1 + r) + (a_{11}a_{22} - a_{21}a_{12})(1 + r)^2}{(1 + r)[a_{21}l_2(a) - a_{22}l_1(a)] + l_1(a)}$$

$$(3.1)$$

$$w_B = \frac{1 - (b_{22} + b_{11})(1 + r) + (b_{11}b_{22} - b_{21}b_{12})(1 + r)^2}{(1 + r)[b_{21}l_2(b) - b_{22}l_1(b)] + l_1(b)}$$

$$(3.2)$$

It may be immediately noted that w is always a decreasing function of r, independently of the sign of the second order derivative (see also Morishima, 1966, p. 521). We may also note that the unit wage is expressed in terms of the same numéraire in (3.1) and (3.2). This suggests that the relationships between r and w (also known as factor-price frontiers) can be plotted as negatively sloped curves on the same diagram.

The intersections between the two curves occur at those levels of the rate of profit which are associated with the same unit wage in both technologies. The number of intersections can be obtained by equating w in eqs. (3.1) and (3.2) and solving for r. The resulting equation will generally have more than one positive solution (Bruno, Burmeister and Sheshinski, 1966, p. 534). In the case of technologies such that each product is a necessary input for all commodities including itself (all products are *basic commodities*), the maximum number of intersections is given by the number of distinct commodities in the two alternative systems of production (Bharadwaj, 1970). This implies that, in the two-good technologies of our example, there will be at most two intersections. Figure 2 represents a case in which there are two intersections in the positive quadrant.

Technologies A and B can now be compared, on grounds of profitability, by considering which technology yields the higher rate of profit for any given wage. (Or, alternatively, which technology yields the higher wage rate for any given rate of profit.)

Figure 2 makes clear that the relative profitability of the two technologies is subject to fluctuation as r increases from 0 to $r^*(B)$ (the maximum rate of profit with technology B). At a low level of the rate of profit $(r < r_1)$, technology A is more profitable ('cheaper') than B. At $r = r_1 = r_2$, A and B are equally profitable. At levels of r between r_1 and r_2, B is more profitable than A. But at any rate of profit higher than r_2, A is again more profitable than B.

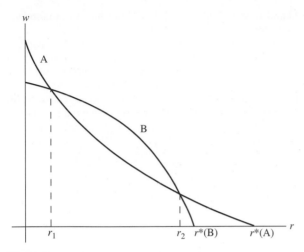

Figure 2

Reswitching of technique may be shown to be possible between complete production systems as well as between individual techniques. Shortly after the identification of the reswitching possibility by Champernowne (1953) and Robinson (1956), and its subsequent analysis by Sraffa (1960), Morishima (1964) and Hicks (1965), David Levhari (1966) proposed the argument that reswitching between production systems is possible only in the case of a 'reducible' or 'decomposable' technology matrix, so 'that reswitching would not occur with technologies producing only basic commodities ('irreducible' or 'in decomposable' technologies). Levhari's argument was disproved by Pasinetti and others (Pasinetti, 1966; Morishima, 1966; Garegnani, 1966). It was also acknowledged to be false by Levhari and Samuelson (Levhari and Samuelson, 1966; Samuelson, 1966). Conditions excluding reswitching were then discovered by Bruno, Burmeister and Sheshinski (1966) and other authors (Starrett, 1969). Their outstanding feature is the introduction of technological assumptions that eliminate those 'complicated patterns of price-movement with several ups and downs' (Sraffa, 1960, p. 37) on which the very possibility of reswitching is founded.

As shown above, the possibility of reswitching in the comparison between alternative states of the economy is associated with differences in the proportions between labour and intermediate inputs for any given pair of production techniques.

This implies that reswitching can be observed only if the economic system is represented in such a way as to bring in view the 'ups and downs' of relative prices. This property was implicitly recognized by John Hicks (1973), when he noted that the possibility of reswitching arises when techniques are 'no longer capable of being distinguished by a single parameter', so that any switch along the technological frontier 'will be a matter of balance between advantages and disadvantages, a balance which

itself is affected by prices' (Hicks, 1973, pp. 44–5). This consideration is at the basis of Hicks's 'simple profile', in which any given technique is described by a single parameter (the ratio of construction labour to utilization labour) and reswitching is excluded (Hicks, 1973, pp. 44–6). Joseph Stiglitz took a different view when he distinguished between reswitching as a possibility relative to the comparison among steady states, and 'recurrence of techniques' as an outcome for an economy 'on its optimal development trajectory' (Stiglitz, 1973, p. 138). In particular, Stiglitz noted that 'recurrence of techniques may occur in technologies which do not allow reswitching', and that 'in technologies in which there is reswitching there may be no recurrences' (Stiglitz, 1973, p. 139). John Wright (1975, p. 22) examined a related issue showing that reswitching can be avoided if one assumes an appropriate 'rate of fall of discount rate through time'. A few years later, Edwin Burmeister and Peter Hammond (1977) explored a related issue, and suggested that reswitching can be excluded as soon as we allow economies to 'jump' over intermediate states (techniques) along a given optimal adjustment path. Recent literature has examined the likelihood of reswitching from the computational or the empirical point of view. In particular, Stefano Zambelli (2004) has shown that a discrete production model is significantly likely to (computationally) generate a reswitching economy, whereas Zonghie Han and Bertram Schefold (2006) have found the empirical likelihood of reswitching to be significant but not very high. Another strand of literature investigated the relationship between the possibility of reswitching and the stability of optimal paths. In this connection, John Barkley Rosser further explored the problem set-up examined in Burmeister and Hammond (1977) and noted the existence of a trade off between the observability of reswitching and the smoothness of optimal adjustment paths (Rosser, 1983; 2000). More recently, Michael Mandler (2005) and Bertram Schefold (2005) have discussed alternative conditions under which reswitching may or may not be associated with unstable economic dynamics.

Synthesis and appraisal

The capital controversy of the 1960s has conclusively shown that the logical possibility of reswitching is of a general nature. Disagreement about the implications of reswitching for economic theory as a whole does not conceal the fact that a crucial discovery in the theory of technical choice was made. In particular it was shown that choice of technique is related to income distribution in a much more complex way than it was once thought to be, and that the rate of interest (or the rate of profit) cannot provide an unambiguous ranking of different technical alternatives as the distribution of income is varied.

The discussion of reswitching called attention to a paradox that had long been overlooked. This is that rational choice, in its classical formulation, presupposes not only agents capable to rank alternatives in a consistent way, but also objective states of the world making such a consistent ranking feasible (see Urmson, 1950, pp. 154–9; Scazzieri, 1982). The reswitching debate has shown that a 'granular' representation of production techniques leads to a complex pattern of interaction such that any given technique may be associated with two or more different positions on the profitability ranking of techniques (see above). This discovery was made possible by the consideration of price movements in a capital-using economy (see above). Its most immediate implication has been to cast doubt upon the representation of capital structure in terms of simple aggregate parables. However, reswitching also called attention to another, perhaps more fundamental, feature of technical choice. This is the dual nature of the grading procedure associated with choice. For grading situations express not only the agent's ability to rank states of the world in a consistent way, but also the possibility to rank those states in terms of 'objective' characteristics independent of the agent's preferences and choices. The reswitching debate has proved that the latter prerequisite may be a will-o'-the-wisp as soon as we consider the complex interactions that take place in a production economy.

R. SCAZZIERI

See also **capital theory; capital theory (paradoxes); preference reversals; reverse capital deepening; technical change.**

Bibliography

Bharadwaj, K. 1970. On the maximum number of switches between two production systems. *Schweizerische Zeitschrift für Volkswirtschaft und Statistik* 106, 409–29.

Bruno, M., Burmeister, E. and Sheshinski, E. 1966. The nature and implications of the reswitching of techniques. *Quarterly Journal of Economics* 80, 526–53.

Burmeister, E. and Hammond, P. 1977. Maximin paths of heterogeneous capital accumulation and the instability of paradoxical steady states. *Econometrica* 45, 853–70.

Champernowne, D. 1953. The production function and the theory of capital: a comment. *Review of Economic Studies* 21, 112–35.

Garegnani, P. 1966. Switching of technique. *Quarterly Journal of Economics* 80, 554–67.

Han, Z. and Schefold, B. 2006. An empirical investigation of paradoxes: reswitching and reverse capital deepening in capital theory. *Cambridge Journal of Economics* 30, 737–65.

Hicks, J. 1965. *Capital and Growth*. Oxford: Clarendon Press.

Hicks, J. 1973. *Capital and Time: A Neo-Austrian Theory*. Oxford: Clarendon Press.

Levhari, D. 1966. A nonsubstitution theorem and switching of techniques. *Quarterly Journal of Economics* 79, 98–105.

Levhari, D. and Samuelson, P. 1966. The nonswitching theorem is false. *Quarterly Journal of Economics* 80, 518–19.

Mandler, M.A. 2005. Well-behaved production economies. *Metroeconomica* 56, 477–94.

Morishima, M. 1964. *Equilibrium, Stability and Growth: A Multi-Sectoral Analysis*. Oxford: Clarendon Press.

Morishima, M. 1966. Refutation of the nonswitching theorem. *Quarterly Journal of Economics* 80, 520–25.

Pasinetti, L. 1966. Changes in the rate of profit and switches of techniques. *Quarterly Journal of Economics* 80, 503–17.

Robinson, J. 1956. *The Accumulation of Capital*. London: Macmillan.

Rosser, J.B., Jr. 1983. Reswitching as a cusp catastrophe. *Journal of Economic Theory* 31, 182–93.

Rosser, J.B., Jr. 2000. *From Catastrophe to Chaos: A General Theory of Economic Discontinuities*. 2nd edn, vol. 1: *Mathematics, Microeconomics, Macroeconomics, and Finance*. Boston, Dordrecht and London: Kluwer Academic Publishers.

Samuelson, P.A. 1966. A summing up. *Quarterly Journal of Economics* 80, 568–83.

Scazzieri, R. 1982. Scale and efficiency in models of production. In *Advances in Economic Theory*, ed. M. Baranzini. Oxford: Basil Blackwell.

Schefold, B. 2005. Reswitching as a case of instability of intertemporal equilibria. *Metroeconomica* 56, 438–76.

Sraffa, P. 1960. *Production of Commodities by Means of Commodities*. Cambridge: Cambridge University Press.

Starrett, D. 1969. Switching and reswitching in a general production model. *Quarterly Journal of Economics* 80, 673–87.

Stiglitz, J.E. 1973. Recurrence of techniques in a dynamic economy. In *Models of Economic Growth*, ed. J.A. Mirrlees and N.H. Stern. London and Basingstoke: Macmillan.

Urmson, J.O. 1950. On grading. *Mind* 59, 145–69.

Wright, J.F. 1975. The dynamics of reswitching. *Oxford Economic Papers*, n.s. 27, 21–46.

Zambelli, S. 2004. The 40% neoclassical aggregate theory of production. *Cambridge Journal of Economics* 28, 99–120.

retirement

The common-sense definition of retirement is leaving employment of a substantial nature by a worker in his or her fifties, sixties or older with no intention of returning to work. However, this definition has no empirical counterpart because we do not observe intentions in the data. Rather, empirical work typically measures retirement in one of two ways. First, a worker is said to retire when he or she leaves the labour force in his or her fifties or older for a 'considerable' period of time. The 'considerable' period may be limited by the length of the observation period in panel data, but it is meant to distinguish retirement from normal job change by workers in their fifties or sixties. The second definition is an affirmation by the worker that he is retired. This definition aims to address right-censoring in panel data by using the individual's own assessment of retirement status. Because many workers state that they are retired after they have left a career job yet continue to work, this definition often adds the requirement of departure from the labour force. Which definition should be preferred will depend on the empirical analysis and the objective of the research. For some research questions, the definition can make a substantial difference, for example in the study of 'unretirement'. In this article I think of retirement as the transition from being in the labour force to not being in the labour force by people in their fifties or older.

Historical trends in labour force participation

In 1957 the labour force participation rate of men aged 60–64 in the United States was about 83 per cent; by 1987 it had fallen to 55 per cent, and since then has risen to about 58 per cent. The participation rate of women aged 60–64 rose over this time period because of the historical increase in the participation rate of women of younger ages: an increasing rate of retirement by older women was offset by an increasing number of women reaching age 60 and still in the labour force. Although the levels and rates of decline are somewhat different, participation rates of older men fell sharply in nine European countries and Canada. What caused these very large declines? In the United States and in many European countries the generosity of the public pension system increased sharply in the late 1960s and 1970s. For example, a good measure of the generosity of the system in the United States is the monthly Social Security benefit for men were they to retire at age 65. The average of those Social Security benefits was $307 in 1957 and $649 in 1987 (both in 1987 dollars), for an annual growth rate of 2.5 per cent. Since 1987 the real growth rate has been just 1.1 per cent and that growth has been due to wage growth, not to changes in the programme rules which have been stable. The coincidence of the decline in labour force participation with the increase in Social Security benefits suggests that Social Security was at least partly responsible for the decline, but there were changes in other determinants of labour supply as well. The private pension system expanded, and real household income increased both because of a rise in earnings and an increase in dual earner households. One objective of research on retirement has been to quantify the contributions of these and other sources to the decline in labour force participation, to predict the future course of labour force participation of the older population, and to understand the response to policy change such as alterations in the structure and generosity of Social Security.

The leading edge of the baby-boom generation will begin to retire in substantial numbers in about 2008, leading to a worsening of the financial health of the Social Security and Medicare systems in the United States. For example, the ratio of the population 65 or over to the working age population (ages 20–64) is a commonly used measure of demographic ageing. In 2000 this ratio was 0.21; it is forecast to increase to 0.36 by 2030, an increase of 72 per cent. The retirement of the baby-boom generation will affect the Social Security and Medicare trust funds, requiring adjustments to those programmes. What will be the effect of those changes on retirement? In particular could policy delay retirement without unduly harming workers while improving fiscal balance? To make a good assessment of the effects of policy requires a model of retirement behaviour.

Data

Since the 1970s the most important advance in our ability to study retirement behaviour has been the development of the Health and Retirement Study (HRS). The HRS is a longitudinal data collection on about 20,000 people aged 51 or over in the United States. The HRS was fielded in 1992 with the express purpose of providing data with which to study retirement and health, and their interactions. As such it contains data on all the relevant economic variables that affect retirement, many health variables and many other non-economic variables that have additional effects. The HRS is a biennial longitudinal survey, and as of 2006 it had fielded eight waves. The original cohort was initially 51–61, so that by wave 8 it was 65–75 and had mostly retired. New cohorts aged 51–61 were added in 1998 and in 2004, and they were re-interviewed in successive waves. Based on the success of the HRS, the English Longitudinal Study of Ageing was modelled on HRS and was fielded in England in 2002. It is also a biennial panel. The Survey of Health, Ageing and Retirement in Europe was fielded in 2004 in 11 European countries, and a second wave with an expanded roster of countries followed in 2006. It is modelled on the HRS and ELSA with the aim of providing data that will permit international comparative studies.

Economic models of retirement

Retirement is an aspect of labour supply, and so the same general framework applies. However, in a number of ways it is easier to study retirement than hours worked: most of the retirement incentives are well measured; typically (although not always) retirement can be freely chosen whereas the choice of hours may be constrained by the demands of employers. As a consequence the response of retirement to incentives is substantial whereas the response of hours to the wage rate, at least among males, is small. Although some models of retirement are very complex, many of the ideas can be illustrated with a simplified version of a retirement model which, nonetheless, incorporates most of the important aspects of economic model of retirement.

Retirement must be placed in a life-cycle context because the gain from additional work is an addition to lifetime economic resources, and its value depends on life expectancy. Consider a worker who will live another N years and who is contemplating whether to retire. Should he work another year he would lose a year of leisure which has utility of U and which initially I assume is constant no matter what the age of the worker. He would gain a year's income which he could add to his stock of wealth. The increase in utility from the income is $V' \times wage$: the marginal utility of wealth multiplied by the annual wage. To maximize utility the worker should not work when $U > V' \times wage$. Under the universal assumption that the marginal utility of consumption declines in consumption, V' will be smaller at older ages with wealth held constant: at older ages the fixed amount of wealth would have to be consumed over fewer years so that per period consumption would be greater than at younger ages. Greater consumption would cause the marginal utility of consumption to be lower and therefore the marginal utility of wealth to be lower. At some age V' declines enough that $V' \times wage < U$, and at that age the worker would leave the labour force.

In a complete life-cycle model, consumption and, therefore, saving would be chosen by the worker as well as retirement. Yet we would like to think of a 'wealth' effect on retirement. Variation in wealth across workers and the accompanying variation in retirement ages can be generated by variation in wages to which the worker reacts both in the choice of consumption and the retirement age. In this example, wealth is endogenous; but we might think of some variation in wealth that is exogenous to the model. Examples would be variation in initial wealth or through inheritances, variation in rates of return on assets or variation in required expenditures during the working life such as the number of children. Having in mind some exogenous variation in wealth across individuals, I will speak of a 'wealth effect' on retirement, but it should be understood that its estimation is difficult because it is endogenous in a complete life-cycle model of retirement behaviour.

These ideas are illustrated in a model of retirement choice. In this model the worker's problem is to choose the retirement age R and consumption level c to maximize lifetime utility

$$\max_{R,c} \left\{ \int_0^R (u(c) + L)dt + \int_R^N (u(c) + L + U)dt \right\}$$

where c is consumption which is assumed to be constant. (In this model with fixed lifespan, consumption will be constant if the interest rate and the subjective time rate of discount are the same.) L is baseline utility from leisure and U is the additional utility from leisure that someone gets when retired. The lifetime budget constraint is $Nc = Rw$ where w is the fixed wage. A corner solution is possible when someone places little value on leisure: he will work his entire life. But in the more usual case of retirement before age N, the solution satisfies the lifetime budget constraint and the first-order condition

$$u'(c) = U/w$$

where u' is marginal utility. Then some manipulation will show that $dR/dU < 0$, and $dc/dU < 0$: an increased value of leisure in retirement will reduce the retirement age and the budget constraint will require a reduction in consumption. Also, $dR/dN > 0$: increases in life expectancy will increase the retirement age. Because those in good health have greater life expectancy, healthy people will work longer, independent of any healthy effect on productivity or on the disutility of work.

An increase in the wage will increase consumption: $dc/dw > 0$. The effect of w on R is indeterminate because of the income and substitution effects whose relative magnitudes depend on the utility function. For example, if utility is constant relative risk aversion so that $u'(c) = c^{-\gamma}$, then $dR/dw > 0$ if $\gamma < 1$ and $dR/dw < 0$ if $\gamma > 1$.

In the context of this model and other models that allow consumption to be chosen, wealth will be an object of choice. We can observe co-variation in wealth and R across individuals due to variation in w, but that co-variation will not show how R would change were we to add additional wealth to someone's wealth holdings. In this model such an addition would reduce R whereas the observed variation in assets at retirement associated with variation in R could either be positive or negative depending on the details of the utility function.

Because u' is constant in age, the model predicts that once retired, no one will 'unretire'.

The retirement hazard

A common object of study in retirement research is the retirement hazard: the probability of retirement at age t given working at t. The retirement hazard can be found from the simple retirement model by considering only the part of the population still working at age t. Among those workers find those who will chose $R = t + 1$; the ratio of the number of those workers to the number of workers at t is the retirement hazard.

Estimation of a hazard model requires panel data where the hazard would be expressed in discrete time: the probability of retirement between t and $t + 1$ conditional on working at t. Retirement hazard models are a rather natural way to think about the retirement process particularly in the context of time-varying covariates such as the wage rate or health – just as in the simple model, an increase in wealth will increase the retirement hazard because of the reduction in V'. The model does not make a prediction about the variation in the retirement hazard with age, which will depend on the rate at which V' declines with wealth.

Other predictions depend on whether we are thinking of long-run comparisons across individuals, or short-run reactions by an individual to a change in the environment. For example, if we compare the retirement behaviour of two individuals, one who has a high wage rate and one who has a low wage rate, we cannot predict who will retire earlier because their saving rates would have been different and so their wealth levels would be different: in the comparison of U with $u'w$, u' and w move in opposite directions.

A good deal of the work on retirement comes from extensions of this simple model to take account of complexities in the budget set, changes in U with age and uncertainty about the future. A leading example is the study of the effects of private pensions on retirement. Private pensions are either defined contribution (DC) pensions or defined benefit (DB) pensions. In a DC plan, the employer and/or the employee puts money as specified by the plan into an investment account usually at each pay period. The amount is a small fraction of pay (say, six per cent) which implicitly increases the wage by a small per cent. The account grows at the rate determined by the portfolio held in the account. At retirement the funds are available to the retired worker typically to spend as he wishes. Thus the plan is defined by the contribution rules (hence, DC). What the worker actually receives will depend on the performance of the portfolio.

In a DB plan the worker will receive a pension or annuity at retirement which is based on the years of service with the employer, on the age at retirement and on a measure of earnings in the last few years of employment. Thus the pension plan is defined by the benefit that a worker will receive on retirement. Most DB plans have the curious feature that the benefit will depend on the age of retirement in a highly nonlinear way. If PV_a is the expected present value of lifetime pension benefits (pension wealth) conditional on retiring at a, $PV_{a+1} - PV_a$ is the addition to pension wealth (additional compensation) from working from a to $a+1$. DB plans often have a critical age, say A, at which a full or unreduced pension benefit is paid to a worker who retires at that age. Workers who retire before A may have their pension benefit reduced substantially. Then the apparent compensation from working from $A-1$ to A is the wage plus $PV_A - PV_{A-1}$. It is not hard to find examples where the total compensation is more than twice the wage. Said differently, the pension is reduced sufficiently for early retirement so that the gain in pension wealth from a year's work exceeds the wage. Furthermore, often pensions are not adjusted upward if a worker retires past A even though for a given pension level the expected present value declines with age. For example, for a single male aged 60 under the assumption that the pension is not indexed and that the nominal interest rate is five per cent, the pension should increase by about eight per cent per year of delayed retirement after age 60 to keep pension wealth constant. If the pension is not adjusted upward at all, the implicit wealth loss from delaying retirement for a year would be about eight per cent of the expected present value of the pension. Assuming the pension replaces 50 per cent of the pre-retirement wage and assessing pension wealth at 12.2 times annual pension income (which is PV_A in this example) means that the loss in pension wealth from delaying retirement is

about 50 per cent of the wage. Said differently, the worker would be working for just half of the apparent wage.

The large gain before age 60 in pension wealth creates a large gain in compensation for working from 59 to 60, and the large loss reduces compensation substantially should the worker not retire at 60. These changes in DB pension wealth modify the money wage to produce net compensation. Net compensation for working from age 59 to 60 would be large and net compensation for working from 60 to 61 would be small. It is likely that a worker aged 59 would calculate that $U < V' \times wage$ because $wage$, which is understood to be the net wage, would be large. A worker aged 60 would calculate $U > V' \times wage$ because the net wage would be small. Thus we would observe many retirements at age 60. More generally spikes in compensation induced by DB plans cause correspondingly large spikes in the retirement hazard at critical ages such as A. An important part of research on retirement is to obtain data on the details of DB plans so as to relate retirement to these spikes.

DC plans matter, but mainly as an addition to wealth. Typically DC plans do not have special ages at which the implicit compensation is very large or small: rather they add a small percentage to the implicit wage. (However, DC plans can have early withdrawal penalties which may affect retirement among those who have no private savings.)

This simple model of retirement has a number of advantages: estimation would show the effect of changing the net wage or changing wealth on the retirement hazard, and the estimations follow in a straightforward way from what is directly observable in data. It is clear what variation in the data produces the results. The model has considerable flexibility: the retirement hazards can be age specific so that the wage and wealth effects are different for each age. Because the estimation only requires the net wage and wealth at t and the retirement outcome at $t+1$, just two waves of panel data are needed. Data far out of sample are not needed: for example, to study the retirement hazard of 59 year-olds, one does not need data on what their wage would be should they work until, say, 65.

However, the simple model has a number of disadvantages. Sometimes DB pensions can induce nonconvexities in the lifetime budget constraint as shown in Figure 1. In that figure the vertical axis is lifetime

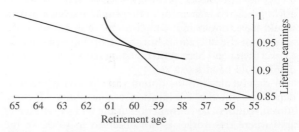

Figure 1   Lifetime earnings and stylized indifference curve

earnings on an arbitrary scale, and the horizontal axis is age at retirement, inverted to show increasing years of retirement. The lifetime budget constraint has a slope equal to the net wage. In our example the large implicit net wage from working while 59 causes the slope of the budget constraint to steepen, and the small implicit net wage from working while 60 causes it to flatten. We would expect that normal shaped indifference curves would cause large numbers to retire at age 60, and, indeed, this is what is observed in data when we have good information about DB plans. The simple model would replicate this clustering at 60. But the simple model would not replicate the prediction that very few would retire at 58: the apparent gain from working while 57 is about the same as the gain from working at 55, 56 and 58, so the simple model would predict that about the same number would retire at each of those ages. But workers can foresee that if they work until they are 59 they will have the option of working from age 59 to age 60, resulting in considerable financial gain. Of course the reason the simple model would not replicate the data is that it makes only a two period comparison: retirement at t compared with retirement at $t+1$. It does not make global utility comparisons.

The simple model does not take account of uncertainty. For example, if a possible decline into bad health will require considerable health care expenditures at some future date, a worker may consider retiring later to build up precautionary saving. In this simple model any such tendency would show up in other estimated parameters.

A second type of model is designed to handle non-convexities in the budget set. It is the option value model. In a simplified form it specifies a utility function in which utility depends on years of leisure and on lifetime earnings (Stock and Wise, 1990). A worker will chose the retirement that maximizes utility which is shown at age 60 in the figure. A worker aged 58 would observe that the gain from working another year would be relatively modest, but that the gain from working two more years would be substantial. He would work another year so as to have the option of working the year after that.

The main advantage of the option value model over the simple model is that it can account for non-convexities in the budget set. If properly specified and estimated, it can simulate a greater range of policy options than the simple model. For example, an expansion of the budget set at age 63 by the introduction of a work bonus could only locally affect predicted retirement in the simple model: a worker would have to remain in employment until age 63 in the absence of the alteration. But in the option value model workers who had been contemplating retirement at 60, 61 or 62 could be affected.

A disadvantage of the option value model is that it is dependent on the specification of the utility function. Also it requires the construction of the budget set even at ages where the worker is not observed to work. In the extreme, it requires the construction of the budget set for all future ages. For example, if a worker continues to work at age 57, it may be that he has strong tastes for work or it may be that he has a DB plan with a large incentive to work until age 60. To study his retirement behaviour at age 57 we have to construct the budget set that he perceives at age 57.

In the same way as the simple model, the option value model does not account for uncertainty. This creates some tension because the model assumes that at age t the worker uses information about the environment at t and has expectations about what the environment will be at $t+1$, at $t+2$, at $t+3$ and so forth. Based on this information he may decide to retire at, say, $t+4$. At $t+1$ he will use information about the environment at $t+1$ which will usually be different from the information that was used at t about the environment at $t+1$. This new information along with new expectations about the future environment may cause an alteration in the intended retirement age. Yet the model does not allow the decision at t to be influenced by the knowledge that new information will be arriving and that the (tentative) decision could be changed.

Social Security and retirement

Social Security, the public pension system in the United States, is a DB pension but it differs from private (employer provided) pensions in at least two ways. First, it is almost universal so that its empirical effects on retirement are difficult to study due to the lack of programme variation across individuals. Second, at critical retirement ages Social Security is approximately actuarially fair; that is, $PV_{a+1} - PV_a \approx 0$ for most workers so that it does not generate the strong retirement incentives of private DB plans. Nonetheless, it is clear that Social Security has an important influence on retirement. First, the retirement hazard is much greater at 62 than at 61 or at 63, as illustrated in Figure 2. Age 62 is the age at which a worker can first claim Social Security benefits, and there is no other explanation for the elevated hazard. Until year 2003, 65 was the normal retirement age under Social Security, and, in addition, the age of entitlement to Medicare, the health care insurance plan for the elderly. Second, this pattern is found in international comparisons where there is programme variation that can help identify programme effects (Gruber and Wise, 1999).

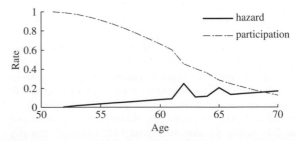

Figure 2   Stylized labour force participation and retirement hazard rates

Despite the empirical evidence of the effect of Social Security on retirement, its influence is difficult to explain in an economic model: why should a worker retire at 62 and claim Social Security benefits rather than at some other age, when there are no economic gains from doing so? One possible explanation is based on a liquidity constraint: low-wage workers have been forced by the Social Security system to save more than they would desire and so they do not engage in any private saving. Thus they reach their early sixties with greater-than-desired retirement resources but access to those resources is conditional on retirement. As a consequence, they retire at 62 when they are first able to access them. A difficulty with this explanation is that many workers who have wealth (demonstrating that they were not forced to oversave by the Social Security system) also retire at 62. A second possible explanation concerns the rate of return (about three per cent real) used in the calculation of $PV_{a+1} - PV_a$. If some individuals believe they can obtain a higher rate on their investments, they can increase their lifetime resources by taking Social Security payments and investing them at a higher rate of return than the implicit rate associated with delayed claiming. It is difficult to evaluate this explanation because we do not know the rates of return people expect. A third explanation is that most people believe their life expectancies to be less than average; in that case they are better off taking Social Security benefits early because they may die before they have received substantial benefits. While individual variation in subjective survival may explain the desire of some to retire and claim Social Security benefits at 62, the average subjective survival as measured by subjective survival in the HRS is close to life-table survival rates. Thus, these factors may explain some (small) part of the excessive retirements at age 62, but they are inadequate for explaining the major part of the excess.

We expect that greater wealth will lead to earlier retirement, and, indeed a contributory factor to the decline in labour force participation in the older population is probably increases in wealth. However, a wealth effect is difficult to show empirically, for several reasons. Wealth is measured with substantial error, and this tends to reduce any estimated effects. Taste variation for retirement can mean that observed wealth is not causative for retirement but rather the result of a desired retirement age. For example, those who place little value on leisure will want to retire late in life and so will save at a low rate. Thus when they reach normal retirement age they will be observed to have low wealth and not to retire. But the delay in retirement is not the result of the lack of wealth: rather the lack of wealth is the result of wanting to retire late. In this model it is necessary to find an instrumental variable to correct for the endogeneity and observation error of wealth. As always this is difficult.

The wage rate measures the price of leisure. In the simple model which says that retirement will occur when $U > V' \times wage$ an increase in the wage would lead to delayed retirement. It is, of course, necessary to control for wealth in an estimation aimed at finding a wage effect: those with high wages in the past will have accumulated more wealth, reducing V', the marginal utility of wealth. If we do not account for wealth, we may observe little relationship between the wage and retirement in data.

Health

The HRS as well as the international data gathering efforts collected many non-economic variables that are likely to influence retirement. The leading class of additional variables measures health. In the evaluation of whether $U > V' \times wage$, U is understood as the utility of leisure relative to the utility of working. It is likely that worsening health increases U because it reduces the utility of working more than it reduces the utility from leisure. Thus a first-order effect of poor health is earlier retirement, both in cross-person comparisons (comparing those in poor health with those in good health) and within person comparisons (comparing those suffering a decline in health with those having stable health). If health declines on average with age, the retirement hazard will tend to increase in age. However, health also influences V', the marginal utility of wealth, although the direction of that influence is not solidly established. If the institutional setting exposes individuals to considerable health spending risk, V' will be higher among those in poor health because of the high productivity of (private) spending on health. If individuals are fully insured against health spending risk, V' may be lower among those in worse health: health could prevent individuals from fully enjoying what money can purchase. Of course, in simple cross-person comparisons other influences on retirement vary systematically with health and they must be controlled. For example, those in worse health have less wealth, causing V' to be larger; they have reduced life expectancy, which reduces V'; and they have lower wages. Thus the relationship between health and $V' \times wage$ is ambiguous. However, as an empirical matter, we observe in panel data that those in worse health retire earlier than those in better health.

The response to an unexpected decline in health (a health shock) is easier to understand because some factors that vary across persons are constant. If exposure to health care expenditure risk is relatively small, V' would decline because of reduced life expectancy and (possibly) because of a reduced ability to spend wealth. If, in addition, the health shock caused a decline in the wage, $V' \times wage$ would decline leading to an increased likelihood of retirement. Indeed, empirically health shocks such as a heart attack are associated with elevated retirement hazard rates.

The availability of health insurance on the job and of employer-sponsored retiree health insurance should affect retirement before the age of 65 because it will change both the expected value of out-of-pocket health care costs and the variance in those costs (Blau and Gilleskie, 2001; 2006). For couples the situation is more complex because in retirement one spouse can be covered by the health insurance of the other. This variation in the provision of health spending insurance provides opportunities for the identification of an insurance effect on retirement.

Accounting for uncertainty

The effects of uncertainty are put in sharpest perspective under the assumption that it is costly to return to work once retired. If it is not costly, an individual can simply return to work as new information arrives in the future, buffering the effects of any negative shocks. This means the decision to 'retire' has less consequence. It is undoubtedly true that it is costly to return to work once retired, although the magnitude of the cost varies substantially across persons because of differences in specific human capital by occupation.

A worker contemplating retirement should be thinking about uncertainties that he would face should he continue to work and uncertainties should he retire. The first type would include wage growth, the likelihood of job displacement, the likelihood of a health event that would limit work, the evolution of his pension entitlement and other job characteristics. Remaining on the job gives an option to experience these outcomes both positive and negative. Retiring means both forgoing these options and forgoing the option of continuing to work both in the coming year and in subsequent years. The second type of uncertainty, that associated with the state of retirement, includes the rates of return on assets, health expenditures in the health insurance environment associated with retirement, survival or life expectancy, uncertainty about one's own utility function especially about one's ability to enjoy wealth, and the utility associated with full-time, uninterrupted leisure.

It is obvious that decision-making under these kinds of uncertainty is difficult. For example, economic resources have to last for many years on average; yet a typical survival curve shows that there are significant chances of dying shortly following retirement and significant chances of dying many years after retirement. In 2003 a 65-year-old man had a life expectancy of 16.8 years, or, stated differently, he could expect to die at age 81.8. However, he had an 11 per cent chance of dying before age 70 and an 11 per cent chance of surviving to age 93. To find the optimal or even satisfactory consumption path is difficult: on the one side he would need to guard against running out of resources should he survive to 93; on the other side excessively low consumption is likely to lead to his dying with considerable wealth, which, if he has no bequest motive, is wasted.

The market solution to this problem is annuities. However, the private purchase of annuities is minimal: among 65–69-year-olds just three per cent receive any income from privately purchased annuities; among those aged 70–74, six per cent receive income from annuities. There are a number of possible explanations for the lack of privately purchased annuities. The rate of return is low because of the profit of the sellers. The price of annuities is actuarially unfair to most people because the typical purchaser lives longer than average, which increases the seller's break-even price. Finally, some people may have a bequest motive: complete annuitization would eliminate any bequests.

In my view, none of these explanations is adequate to explain the lack of annuitization. Profits are not unusually high compared with profit margins in other financial products. While annuities are actuarially unfair for many people, they are not unfair for people who expect to be long lived; yet, annuity purchase is low among such people. Even people with a bequest motive should find it advantageous to annuitize partially. A possible explanation that has been little explored concerns the actual insurance that annuities provide. Privately purchased annuities are not indexed, so people will be concerned about the real value of consumption an annuity would be able to finance at advanced old age. Even a fairly moderate level of inflation will reduce substantially the real value of an annuity over 25 years. Also, the appropriate time horizon is 25–30 years: what is the probability the annuity provider will still be in business?

Estimation of the effects of uncertainty on retirement is conducted in the context of a dynamic programming (DP) model (Rust and Phelan, 1997). The model will specify all possible future states of the world and assign utilities to them conditional on economic resources. Then by well-established backward solution methods the algorithms will find the expected utility associated with continued work and associated with retiring, where the expectation is taken with respect to the joint distribution of stochastic elements in the model. Thus, the analyst will supply the probability distribution of the age of death, the probability distributions of rates of return on assets, the probability distribution of health shocks and associated spending, and so forth. The model predicts continued work when the expected utility from work is greater than the expected utility from retiring, and it predicts retirement at the age when the reverse becomes true. For the reasons discussed in the context of the certainty retirement model, expected utility associated with continued work declines with age and expected utility associated with retirement increases with age, so a worker will eventually be predicted to retire. The model is adjusted with respect to parameters and specification until the predicted retirement ages match most closely those observed in the data.

The data requirements of such a DP model are immense: the analyst needs to assign probabilities to all future exogenous outcomes such as mortality, asset returns and so forth, but the probabilities are not observable. The appropriate probabilities are subjective probabilities: those used by the respondent when making the retirement decision under uncertainty. In particular, the probabilities need not be the same as any observable probabilities of the corresponding events. For example, a population life table displays an estimate of the population mortality risk at each age. Even if the population subjective survival probabilities match those in the life table, individuals should have subjective survival probabilities that deviate from the life table because each person has risk factors that will alter the objective survival probabilities. People with above-average health will survive longer than people with below-average health; people with less education die earlier than people with more education. However, the subjective survival probabilities of people with those characteristics need not correspond even to the objective survival probabilities conditional on those observable characteristics. First, people undoubtedly have private information about their true survival probabilities. Second, they may well have biased subjective survival probabilities in the sense that the probabilities on which people base their retirement decisions are not good predictors of their actual survival. In a similar manner, people have subjective probability distributions over rates of return on assets that may not correspond at all to historic market rates of return or to rates of return predicted by any model based on rational expectations. In this situation, there are no objective data from which the analyst can find the probabilities of the stochastic events required by the dynamic programming model.

Because subjective probabilities are so important in the study of intertemporal decision making, including retirement, the HRS asks respondents to state their probabilistic beliefs about important stochastic events such as survival. Although the model requires survival probabilities to each future age, knowing even the subjective survival probability to a single age is a considerable improvement over using life tables: based on a model of the relationship between subjective survival, actual survival and a life table, one can estimate individualized subjective survival curves (Gan, Hurd and McFadden, 2005).

The advantages of DP models are that they incorporate uncertainty in a formal manner and in principle they can provide an estimate of the effects of uncertainty on retirement. For example, they could predict the response to a mean-preserving spread in survival such as a 20 per cent change of dying before age 70 and a 20 per cent change of surviving to age 93. DP models produce estimates of utility function parameters and so they are capable of out-of-sample prediction. For example, they could predict retirement patterns were the normal retirement age under Social Security increased to age 70.

The disadvantages of dynamic programming models of retirement include the data requirements. Because of the complexity and data requirements, dynamic programming models of retirement are able to account for only a limited number of stochastic events. A significant problem is that the data are not subject to validation: thus a model failure could be due to an incorrectly specified model or to invalid data, particularly the probability distributions of the stochastic events. A second disadvantage is that it is difficult to understand what in the data is causing observed model outcomes because of the complexity of the model.

Joint retirement

Most people are married when they reach retirement age, and, because of increases in the labour force participation of wives, often both spouses work. Among working husbands aged 55–59 in 2004, 74 per cent of their wives also work. On average, husbands are about three years older than their wives, so that, when the husbands reach age 62, their wives will be about 59. A husband may be influenced by his Social Security benefits to retire but it may be disadvantageous for the wife to retire at 59. Nonetheless, we observe in data some coordination of retirement dates; that is, the probability a wife will retire given the retirement of the husband is greater than the unconditional probability that a wife will retire, and similarly for the conditional probability a husband will retire (Blau, 1998; Gustman and Steinmeier, 2000). A way to model this is to assume a household utility function in which the value of leisure of one spouse is increased by the leisure of the other spouse. That is, their leisures are complements. In a reduced form, the retirement of the husband will be influenced by the incentives he faces such as his wage rate and pension provisions, but also by the incentives his wife faces. Notice that these effects are in addition to any operating through the lifetime budget constraint: if their leisures are not complements, we should observe the early retirement of the husband balanced by the late retirement of the wife in compensation for the loss of earnings of the husband.

Joint retirement offers an arena for the study of household decision-making when there may be conflict between husband and wife as in a collective model of household utility. A typical empirical implementation of the collective model studies the demand for various purchased goods; but it is difficult to know which spouse benefits from a particular purchase. For example, suppose we observe that in households where the husband earns substantially more than the wife the household spends relatively little on clothing. Is this evidence that the power allocation in the household is related to relative earnings? The answer would depend on the assumption that wives benefit more from clothing purchases than husbands. In the case of early retirement, however, we observe who the primary beneficiary of the leisure is, providing sharper identification of the collective model versus the unitary model.

We should anticipate that models of joint retirement will become more important and useful. First, an increasing number of wives reach traditional retirement ages while working, so the quantitative importance of joint retirement will increase. Second, the strong influence of DB pension plans on retirement incentives has made the cost of coordinating retirement high when there are conflicts between optimal retirement dates induced by the plans. With the shift to DC plans, these conflicts will become quantitatively less important in the population because DC plans do not have the sharp retirement incentives of DB plans. We should expect an increase in the amount of coordinated retirements.

Behavioural economics and retirement

The discussion of the determination of retirement has assumed rational decision-making in the context of the life-cycle model: individuals and couples are assumed to maximize expected lifetime utility conditional on beliefs about the probabilities of future events. In that set-up we are still very far from testing important aspects of the model because of our limited measures of those beliefs: an apparent failure of the model could be due to its being an incorrect characterization of decision-making but it could also be due to our lack of valid measures of those beliefs. Nonetheless, there is evidence at least in saving behaviour that the forward-looking model does not apply to all people. Some people strongly prefer the present to the future; they lack the ability to process relevant information; they apparently use rules of thumb; they are heavily influenced by defaults; they do not take actions that would result in fairly large financial gains even though the cost of those actions seems to be small. These examples are about saving behaviour and portfolio choice. It is more difficult to find evidence of non-optimizing behaviour in retirement choice, possibly because it is more difficult to understand what the optimal decision is. For example, job characteristics including distaste for work, perceived discrimination, health and how it interacts with job characteristics could all have large influences on retirement; yet they are mostly unobserved.

In view of these difficulties it is worthwhile asking what would constitute evidence against the rational retirement model. One type of evidence would be an empirically important 'normative' retirement age: a high rate of retirement at an age that may be economically disadvantageous (or at least not economically advantageous) where that age is determined by social norms or convention. Thus, a substantial number of individuals who retire at that age would have better economic outcomes had they retired at some other age. However, to find this empirically faces considerable difficulty. At the population level in the United States excessive retirements occur at 62 and 65, ages of importance under Social Security and Medicare. To be certain that these retirements are due to convention rather than to rational economic choice, we need a great deal of information about expectations and personal tastes, and we need to

have confidence that our models are complete. In my view we are not in a position to assert that individuals are making a mistake when they retire at those ages.

A somewhat different category of evidence concerns economic preparation for retirement. Do we observe large numbers of workers retiring with inadequate resources to finance their consumption in retirement? If the answer is 'yes', retirement is suboptimal in the sense that the marginal utility of wealth is too high for the retirement age chosen. While there is controversy in the literature about the empirical facts concerning preparation for retirement, the main assertion in the literature about a lack of maximizing behaviour has been that saving is suboptimal, not that retirement is suboptimal. Abstracting from spending on health shocks, this argument has considerable validity because saving is under the control of the individual whereas the individual has somewhat less control over retirement. For example, unemployment often leads to retirement because of the difficulty for older workers to find re-employment. Or unmeasured job characteristics such as discrimination against older workers may make continued employment uncomfortable.

Future course of research

The economic models of retirement that have been discussed here and estimated in the literature assume that the decision to work at a given wage is made by the worker. Yet, there is a large literature on the desire of employers to shed older workers; furthermore, there is unemployment at older ages, implying that sometimes retirement is not completely voluntary. Although there are laws against age discrimination in employment, it is likely that employers are able to put some pressure on older workers to retire. However, this observation about the demand side of the labour market for older workers has been formed during an era of increasing number of workers in their forties and fifties when firms may, indeed, have felt they had too many older workers. But as the baby-boom generation begins to retire, the attitude of firms towards older workers should change: firms will want to retain them. Thus, an important research question concerns the evolution of the demand side. How will employers accommodate older workers who may not want to work full time or with the intensity of younger workers? Connected to this question is the long-run productivity of older workers. Apparently firms have wanted to shed older workers because their productivity relative to their costs (including the cost of health care) declined with age. With changing technological requirements on the job, will this unfavourable age-related decline in productivity worsen or improve over time?

We have witnessed a long-run improvement in the health of the older population both in terms of life expectancy and in terms of disability. According to economic theory, increased longevity should cause an increase in the retirement age because of the necessity of financing increased years in retirement. A decline in disabilities will allow more workers to remain in the workforce, and better health is likely to lead to greater productivity at any given age. An important research objective is to quantify these effects and to forecast how changes in retirement will affect important policy concerns. For example, in the United States working past age 65 reduces Medicare expenditures because employer-provided health insurance pays for health care before Medicare. Should enough workers remain in the labour force, the financing difficulties with Medicare will be partially solved, requiring less vigorous policy intervention.

Research on the interaction between health and retirement will likely increase in importance. For example, the sanguine scenario of later retirement because of a reduction in the rate of disability depends on continuing improvements in health. Yet there is considerable uncertainty in forecasts of health, particularly because of the high levels of obesity in the working-age population. We do not have a good understanding of how obesity leads to disability.

Although we have made some progress in our understanding of intertemporal decision-making, a great deal remains to be done. We need to understand how people make such decisions, what information they use, what expectations or probability distributions they have and how they form them, how those expectations evolve as they approach retirement, and so forth. These investigations would be helped were we to have methods of estimating at the individual level preference parameters such as risk aversion and the subjective time rate of discount independently of actual choices. A good example is portfolio choice, where we would like to estimate risk aversion. Lacking data on expected rates of return, we have to make assumptions about those expectations so that risk aversion is conditional on those assumptions. If we knew something about risk aversion we could estimate beliefs about expected rates of return.

These objectives will likely lead to a greater use of subjective data combined with objective data. For example, rather than just studying the determinants of actual retirement, we can study the determinants of the subjective probability of retirement at some target age, say, 62. In panel data, this method can control for unobserved heterogeneity in a straightforward way because we can observe the change in the subjective probability of retirement as the environment changes (Chan and Stevens, 2004).

Because of the continued evolution in survey methods and the ongoing data collection by the HRS, we will have greater sample sizes with which to estimate retirement models. We should be able to observe the effects of natural experiments that can help identify our models. For example, the HRS was in the field in 2002 for the beginning of the natural experiment of the increase in the normal retirement age under Social Security. We will be able to find directly any movement in the retirement spike associated with that age. The continued data collection, natural experiments, innovations in survey design, and the greater use of subjective data should lead to considerable progress in modelling and retirement decision and in quantifying the determinants of retirement.

MICHAEL HURD

See also **labour supply; pensions; Social Security in the United States.**

Financial support from the National Institute on Ageing is gratefully acknowledged. Many thanks to Susann Rohwedder for her valuable advice and suggestions.

Bibliography

Blau, D.M. 1998. Labor force dynamics of older married couples. *Journal of Labor Economics* 16, 595–629.

Blau, D. and Gilleskie, D. 2001. Retiree health insurance and the labor force behavior of older men in the 1990's. *Review of Economics and Statistics* 83, 64–80.

Blau, D. and Gilleskie, D. 2006. Health insurance and retirement of married couples. *Journal of Applied Econometrics* 21, 935–53.

Chan, S. and Stevens, A.H. 2004. Do changes in pension incentives affect retirement? A longitudinal study of subjective retirement expectations. *Journal of Public Economics* 88, 1307–33.

Gan, L., Hurd, M. and McFadden, D. 2005. Individual subjective survival curves. In *Analyses in the Economics of Aging*, ed. D. Wise. Chicago: University of Chicago Press.

Gruber, J. and Wise, D. 1999. *Social Security and Retirement Around the World*. Chicago: University of Chicago Press.

Gustman, A.L. and Steinmeier, T.L. 2000. Retirement in a family context: a structural model for husbands and wives. *Journal of Labor Economics* 18, 503–45.

Rust, J. and Phelan, C. 1997. How Social Security and Medicare affect retirement behavior in a world of incomplete markets. *Econometrica* 65, 781–831.

Stock, J. and Wise, D. 1990. Pensions, the option value of work and retirement. *Econometrica* 58, 1151–80.

returns to scale

The technique of production of a commodity y may be characterized as a function of the required inputs x_i:

$$y = f(x_1, x_2, \ldots x_n)$$

If all inputs are multiplied by a positive scalar, t, and the consequent output represented as $t^s y$, then the value of s may be said to indicate the magnitude of returns to scale.

If $s = 1$, then there are constant returns to scale: any proportionate change in all input results in an equiproportionate change in output. If $s > 1$, there are increasing returns to scale. If $s < 1$ (though not less than zero, given the possibility of free disposal) then there are decreasing returns to scale.

These mathematical definitions suggest a symmetry between the three classifications of returns to scale. This appearance of symmetry is entirely spurious.

The original arguments from which is derived the economic rationale underlying the various categories of returns to scale are to be found in the works of the classical economists. Yet there, as Sraffa (1925) pointed out, each category is derived from quite different economic phenomena. Increasing returns derived from the process of accumulation and technological change, associated as they were with the division of labour attendant upon the extension of the market. Decreasing returns were held to derive from the limited availability of land, and were an important component of the theory of income distribution, being the foundation of the theory of rent.

Yet it was from these disparate origins that Marshall (1890) attempted to formulate a unified, symmetric, analysis of returns to scale which would provide the rationale for the construction of the supply curve of a competitive industry, derived in turn from the equilibria of the firms within the industry. Marshall himself recognized the incompatibility of the assumption of competition and presence of increasing returns (1890, Appendix H). Piero Sraffa (1925; 1926) exposed the entire exercise as ill-founded by demonstrating that neither increasing nor decreasing returns to scale are compatible with the assumption of perfect competition in the theory of the firm or of the partial-equilibrium industry supply curve – a result which, although prominently published and debated, has apparently escaped the notice of those who still draw that bogus U-shaped cost curve whilst purporting to analyse the equilibrium of the competitive firm.

The difficulties identified by Sraffa rest upon the *economic* rationales for variable returns to scale.

The idea of constant returns to scale derives essentially from the proposition that a given set of production conditions may be replicated so long as all the requisite inputs may be varied in the same proportion. Indivisibilities in the production process may limit exact replication to particular levels of output. But the concept, though less precise, is not in any way diminished by the presence of indivisibilities, particularly if the optimal scale of operation of a given technique is small relative to the overall level of output.

The presence of decreasing returns *to scale* would suggest that replication is, for some reason, impossible. Yet if all inputs are correctly enumerated and all increased in the same proportion, then, barring indivisibilities, there can be no barrier to replication. Decreasing returns can derive only from a fixed input (or an input which cannot be increased in the same proportion as others) which prevents replication. In other words, there is no such thing as decreasing returns *to scale.* Decreasing returns derives from *substitution,* from the necessity of changing input proportions.

Whilst decreasing returns to scale do not exist, increasing returns to scale are typically based on propositions so general as to defy precise clarification.

There are some examples in which outputs are an increasing function of inputs for purely technical reasons. The capacity of a pipeline, for example, is defined by the area of its cross-section, πr^2 whereas the circumference of that cross-section is equal to $2\pi r$. If it were possible to increase capacity merely by increasing the circumference (if the walls of the pipe did not require strengthening), then a quadrupling of capacity could be achieved simply by doubling the material inputs.

There is one odd symmetry in this 'technical' case of increasing returns. Whereas decreasing returns can derive only from substitution and not from scale, increasing returns can derive only from scale, not from substitution! Choice of optimal proportions of inputs (with free disposal and no indivisibilities) will always ensure *at least* constant returns.

Such technical examples are not, however, the examples which typically come to mind in the discussion of increasing returns to scale. More typical are examples of mass production, of production lines, or, today, of production integrated by means of sophisticated information systems. Yet these examples, which are akin to Adam Smith's analysis of increasing returns, are associated more with technological change, and with the possibilities for change inherent in a larger, or more rapidly growing, market, than with a simple increase in the scale of identical inputs. Generalization of the concept to 'dynamic increasing returns' (Young, 1928; Kaldor, 1966) increasing returns associated with growth of output further distances the idea of increasing returns from the formal characteristics of scale.

These arguments suggest that the concept of 'returns to scale' is not merely a very limited means of characterizing technology, but it is also a very *limiting* concept. None of the interesting characteristics of the relationship between scale of production and method of production are captured by the idea of returns to scale. Indeed, the only really satisfactory formal characterization of returns to scale is that of constant returns – and this only because replication is formally a precise notion, however empty empirically.

JOHN EATWELL

Bibliography

Kaldor, N. 1966. *Causes of the Slow Rate of Economic Growth in the United Kingdom.* Cambridge: Cambridge University Press.

Marshall, A. 1890. *Principles of Economics.* 9th (Variorum) edn, London: Macmillan, 1961.

Smith, A. 1776. *An Inquiry into the Nature and Causes of the Wealth of Nations.* London: Methuen, 1961.

Sraffa, P. 1925. Sulla relazioni fra costo e quantità prodotta. *Annali di Economia* 2, 277–328.

Sraffa, P. 1926. The laws of returns under competitive conditions. *Economic Journal* 36, 535–50.

Young, A.A. 1928. Increasing returns and economic progress. *Economic Journal* 38, 527–42.

returns to scale measurement

Knowing the degree of returns to scale (RTS) in a firm, or its average in an industry or economy, is important for a variety of economic questions. First, it is important for assessing the plausibility of models of endogenous growth, which typically require at least constant returns to *reproducible* inputs, and thus increasing returns overall. Second, the size of scale economies is an important determinant of the gains from trade. Third, knowing the RTS is important for assessing the plausibility of certain business-cycle models, which often rely on the existence of substantial increasing returns to scale (IRS). Fourth, the RTS is a lower bound on the size of the markup of price over marginal cost, which is a quantity of great interest in industrial organization. Finally, a basic tenet of productive efficiency requires that the value marginal product of an input be equalized across the uses to which that input could be devoted. Knowing the RTS in each use is important for checking that this condition holds.

Assume that firms have a production function for gross output:

$$Y = F(\tilde{K}, \tilde{L}, M, T). \qquad (1)$$

Firms use capital services $\tilde{K}$, labour services $\tilde{L}$, and intermediate inputs of materials and energy, M. T indexes 'technology', not directly observed, which is defined to include any inputs that affect firm-level production but are not compensated by the firm (including, for example, Marshallian externalities as well as exogenous technical change). All variables are functions of time.

Assume $\tilde{L} = EHN$, where the number of employees, N, and hours worked per employee, H, are observed, but the effort of each employee, E, is unobserved. Capital services are the product of the observed capital stock, K, and its unobserved utilization rate, Z (for example, the number of shifts the machine is operated): $\tilde{K} = ZK$. The capital stock and the number of employees may be quasi-fixed (costly to adjust). The adjustment cost can be modelled explicitly in the production function (see Berndt, 1986).

F is (locally) homogeneous of arbitrary degree γ in the priced inputs. Constant returns implies $\gamma = 1$. RTS equals the sum of output elasticities:

$$\gamma = \frac{F_1 \tilde{K}}{Y} + \frac{F_2 \tilde{L}}{Y} + \frac{F_3 M}{Y}, \qquad (2)$$

where F_J is the marginal product of input J. Assuming firms minimize cost, we can denote the firm's cost function by $C(Y)$. γ also equals the inverse of the elasticity of cost with respect to output:

$$\gamma(Y) = \frac{C(Y)}{Y C'(Y)} = \frac{C(Y)/Y}{C'(Y)} = \frac{AC}{MC}, \qquad (3)$$

where AC equals average cost and MC equals marginal cost. IRS may reflect overhead (fixed) costs or decreasing marginal cost; both imply that average cost exceeds marginal cost. If increasing returns take the form of overhead costs, then $\gamma(Y)$ is not a constant structural parameter, but depends on the level of output the firm produces. To make this point more clear, consider a special case of (1):

$$Y = G(K, L, M, T) - \Phi, \qquad (4)$$

where Φ is a *flow* (per-period) fixed cost and G is homogeneous of degree ρ in K, L and M. In this case, $\gamma = \rho(Y + \Phi)/Y$. Thus, RTS, γ, may strictly exceed ρ. Some papers use empirical estimates of γ to calibrate ρ. Since this procedure is not generally correct, some of the results in these papers (for example, the existence of sunspot equilibria), do not follow from the existence of IRS per se. Indeed, IRS is compatible with increasing marginal costs ($\rho < 1$), as in the standard Chamberlinian model of imperfect competition.

Even if firms are identical, the RTS of the *aggregate* production function (either of an industry or an economy) is not necessarily the γ of every firm; it also depends on the dynamics of firm entry and exit. Suppose that in the long run all changes in aggregate output are accommodated by changes in the number of firms, with firm-level output remaining constant. Then the aggregate production function has constant returns to scale in the long run, but increasing returns when the number of firms is fixed in the short run. (However, if the new firms produce new varieties of goods, then the aggregate function may exhibit a form of increasing returns through a 'love of variety' in production, as in Ethier, 1982.)

Firms may charge a price P with a markup, μ, where $\mu \equiv P/MC$. RTS is a technical property of the production function, whereas the markup is a behavioural parameter. However, from (3), the two are linked:

$$\gamma = \frac{C(Y)}{Y C'(Y)} = \frac{P}{C'(Y)} \frac{C(Y)}{PY} = \mu(1 - s_\pi), \qquad (5)$$

where s_π is the share of pure economic profit in gross revenue. As long as pure economic profits are small, as most estimates suggest, eq. (5) shows that μ approximately equals γ. Large markups thus require large increasing returns. Since most studies estimate low profit rates, and since $\mu \geq 1$, eq. (5) shows that firm-level RTS must either be approximately constant or increasing.

Internal IRS also *requires* that firms charge a markup, to avoid losses.

One can estimate RTS from either the production function or the cost function, using the implications of eq. (2) or eq. (3) (see Berndt, 1986, for an exposition of the cost approach). The two literatures have developed to have different aims. The cost-function literature typically takes second-order approximations to the underlying production function, which allows it to estimate elasticities of substitution between inputs, but pays little attention to the issue that observed factor prices may not be allocative, especially at high frequencies. The production-function literature takes first-order approximations, but devotes more attention to correcting biases from unobserved right-hand side variables (the quantity analogue of unobserved true factor prices). Neither literature has found a good solution for dealing with issues of endogenous regressors (for example, the presence of output in the firm cost function when one allows for non-constant returns to scale).

Taking logs of both sides of (1) and differentiating with respect to time gives:

$$dy = \frac{F_1 K}{Y} d\tilde{k} + \frac{F_2 L}{Y} d\tilde{l} + \frac{F_3 M}{Y} dm + dt.$$

$$(6)$$

Small letters denote growth rates (so dy, for example, equals $\dot{Y}/Y$); the output elasticity with respect to technology is normalized to one.

Cost minimization puts additional structure on (6). (The advantage of the cost minimization framework is that it is unnecessary to specify the potentially very complicated, dynamic profit maximization problem that gives rise to P or μ.) Suppose that firms take the price of all J inputs, P_J, as given by competitive markets. The first-order conditions for cost-minimization then imply that

$$PF_J = \mu P_J. \qquad (7)$$

If firms make pure economic profits, these appear in the data as factor payments (most often to capital, sometimes to labour). In order for (7) to hold, the prices of capital and labour must be defined as the rental price (or shadow rental price) of capital and the competitive wage rate for labour. The relationship still holds if some factors are quasi-fixed (costly to adjust), as long as we define the input price of the quasi-fixed factors as the appropriate *shadow* prices, or implicit rental rates.

Using eqs (5) and (7), we can write each output elasticity as the product of RTS multiplied by total expenditure on each input divided by total *cost* (not revenue). Thus, for example,

$$\frac{F_1 ZK}{Y} = \gamma \frac{P_K K}{\sum P_J J} \equiv \gamma c_K. \qquad (8)$$

c_J are the *cost shares* of each type of input, and sum to 1.

Substitute these output elasticities into (6) and use the definition of input services:

$$\begin{aligned} dy &= \gamma \Big[c_K d\tilde{k} + c_L d\tilde{l} + c_M dm \Big] + dt \\ &= \gamma [c_K (dk + dz) + c_L (dn + dh + de) + c_M dm] + dt \\ &= \gamma [c_K dk + c_L (dn + dh) + c_M dm] \\ &\quad + \gamma [c_K dz + c_L de] + dt \\ &\equiv \gamma dx + \gamma du + dt \end{aligned} \qquad (9)$$

Defining dx as a share-weighted average of conventional (observed) input growth, and du as a weighted average of unobserved variation in capital utilization and effort, we obtain our basic estimating equation for γ, the last line of (9). Note that to create the cost shares c_J one needs to construct an estimate of pure profits, as in Hall (1990). Alternatively, one can assume zero economic profit on average and use the observed revenue shares.

Regarding (9) as an estimating equation, one immediately faces three issues.

First, the econometrician usually does not observe utilization du directly. In this case, the regression suffers from measurement error. Unlike classical measurement error, variations in utilization du are likely to be (positively) correlated with changes in the measured inputs dx, leading to an upward bias in the estimated γ.

Second, should one take the output elasticities as constant (appropriate for a Cobb–Douglas production function or for a first-order log-linear approximation), or time-varying? That is, should one allow γ and the share-weights in (9) to change over time? If the elasticities are not truly constant over time, then treating them as constant may introduce bias.

Third, even if the output elasticities are constant and all inputs are observable, one faces the 'transmission problem': The technical change term, dt, is likely to be correlated with a firm's input choices, leading to biased OLS estimates of γ. In principle, one can solve this problem by instrumenting the right-hand-side variables, or by using a proxy for dt, following Olley and Pakes (1996).

Approaches to controlling for du also involve the use of proxies. One method builds on the intuition that firms view all inputs (whether observed by the econometrician or not) identically. For example, a firm should equate the marginal cost of obtaining more services from the observed intensive margin (for example, working current workers longer hours) and from the unobserved intensive margin (working them harder each hour). If the costs of increasing hours and effort are convex, firms will choose to use both margins. Thus, changes in an observed input – for example, hours per worker – provide a measure of unobserved changes in the intensity of work. This suggests a regression of the form:

$$dy = \gamma dx + \kappa dh + dt, \qquad (10)$$

where dh is the growth rate of hours per worker. Basu and Fernald (2001) summarize research showing that regression (10) controls for variable effort. In addition, if the cost of varying the workweek of capital takes the form of a shift premium – for example, one needs to pay workers more to work at night – then this regression corrects for variations in utilization of capital as well as labour. (If the cost of varying capital's workweek is 'wear and tear' – that is, capital depreciates in use – then the regression is somewhat more complicated, but theory still suggests appropriate proxies.)

In principle, allowing for time-varying factor shares in an estimating equation like (10) is always preferable to having constant shares, since using time-varying shares approximates the true function to a second order. However, attempting to estimate the time-varying shares requires observing (or estimating) the true shadow cost of inputs at each point in time. If observed factor payments at each point in time do not correspond to the factor's true cost each period – for example, if firms smooth wage payments by offering workers insurance through an implicit contract – then treating the observed prices as allocative may introduce larger biases (see also Carlton, 1983, on intermediate goods prices).

Since one is unlikely to observe allocative factor prices period-by-period, one probably should take a first-order approximation and assume constant, not time-varying, elasticities. For estimating the RTS a first-order approximation may suffice, and it will be accurate as long as the true average factor price is the mean of the observed prices over the sample period.

So far, the discussion has concerned the estimation of internal returns to scale. However, a number of interesting models assume the existence of spillovers between competitive firms with internal constant returns, leading to *external* increasing returns in the aggregate production function. The empirical literature searching for such spillovers follows two sharply divergent tracks. The search for high-frequency spillovers is usually atheoretical, and amounts to augmenting disaggregated estimating equations like (10) with measures of aggregate activity. However, since most such exercises do not attempt to control for unobserved changes in utilization (omitting, for example, the κdh term in (10)), they are vulnerable to the charge that the putative externalities are actually proxies for unobserved changes in internal inputs. Furthermore, Basu (1995) presents a model where apparent external effects are actually driven by a different economic mechanism, and shows that his model can be distinguished from true technological spillovers by examining gross-output data, as opposed to the commonly used value-added data. Performing the test, apparent externalities are found in value-added but not in gross output, suggesting they are not true spillovers.

However, the search for long-run external effects is based firmly on the economic insight that knowledge creation has built-in increasing returns, since knowledge is non-rival. Thus, there is a long tradition of searching for externalities to R&D, summarized by Griliches (1998). R&D spillovers appear to be a fact, but their exact magnitude is still an issue subject to debate. And there is no consensus at all on whether the magnitude of the spillover is large enough to permit fully endogenous long-run growth.

So far, the discussion has been couched in terms of firm-level output, or aggregation over identical firms. For some applications, one wants to know the RTS for an industry or a sector but allow – plausibly – for the possibility that firms have heterogeneous characteristics, including different γ's. It turns out that, in this realistic scenario, there is not even an unambiguous definition of increasing returns to scale. Basu and Fernald (1997) show that industry output growth equals:

$$d\bar{y} = \bar{\gamma}d\bar{x} + d\bar{u} + R + d\bar{t}. \qquad (11)$$

$\bar{\gamma}$ is the average RTS across firms; $d\bar{y}, d\bar{x}$ and $d\bar{u}$ are appropriately – weighted averages of firm-level output and input growths; R represents various reallocation (or aggregation) effects; and $d\bar{t}$ is an appropriately – weighted average of firm-level technology.

The intuition for 'R' is that γ need not be the same across firms within an industry (or the economy). Output growth therefore depends on the *distribution* of input growth as well as on its mean: if inputs grow faster in firms where they have above-average marginal products (γ is higher), industry output grows more rapidly as well. Thus, aggregate productivity growth is not just firm-level productivity growth writ large; comparing eqs (9) and (11) shows that there are qualitatively new effects at the aggregate level. Is the RTS of an industry just the average of firm-level RTS, $\bar{\gamma}$, or does it include the aggregation effects, R, which are also the result of deviations from constant returns and perfect competition? The answer will depend on the economic question being asked (see Basu and Fernald, 1997, section V), but empirically the magnitudes are often quite different.

SUSANTO BASU

See also **capital utilization; cyclical markups; external economies; multiple equilibria in macroeconomics; production functions; returns to scale; technical change.**

Bibliography

Basu, S. 1995. Intermediate goods and business cycles: implications for productivity and welfare. *American Economic Review* 85, 512–31.

Basu, S. and Fernald, J.G. 1997. Returns to scale in U.S. production: estimates and implications. *Journal of Political Economy* 105, 249–83.

Basu, S. and Fernald, J.G. 2001. Why is productivity procyclical? Why do we care? In *New Developments in Productivity Analysis*, ed. C. Hulten, E. Dean and M. Harper. Cambridge, MA: NBER.

Berndt, E. 1986. *The Practice of Econometrics: Classic and Contemporary.* New York: Addison-Wesley.

Carlton, D.W. 1983. Equilibrium fluctuations when price and delivery lags clear the market. *Bell Journal of Economics* 14, 562–72.

Ethier, W.J. 1982. National and international returns to scale in the modern theory of international trade. *American Economic Review* 72, 389–405.

Griliches, Z. 1998. *R&D and Productivity.* Chicago: University of Chicago Press.

Hall, R.E. 1990. Invariance properties of Solow's productivity residual. In *Growth/Productivity/Unemployment: Essays to Celebrate Bob Solow's Birthday,* ed. P. Diamond. Cambridge, MA: MIT Press.

Olley, G.S. and Pakes, A. 1996. The dynamics of productivity in the telecommunications equipment industry. *Econometrica* 64, 1263–97.

returns to schooling

1 Introduction

The return to schooling is the internal rate of return on an additional year of schooling: the discount rate at which the present value of the gains associated with the investment equals the costs. The notion of treating education as a capital investment – and calculating the return accordingly – was proposed by Walsh (1935) in an aptly titled article, 'Capital Concept Applied to Man'. Subsequent contributions (Mincer, 1958; 1974; Becker, 1962; 1964; 1967) have elaborated the theoretical underpinnings of this exercise, while advances in data availability and econometric methods have led to refinements in the empirical procedures used to calculate the return to schooling (see Griliches, 1977; Card, 2001; Harmon, Oosterbeek and Walker, 2003, for surveys).

Following Mincer (1974), the term return to schooling also refers to the coefficient of years of schooling in a linear regression of log earnings on years of schooling and controls for labour market experience. Under certain simplifying assumptions this coefficient is approximately equal to the internal rate of return to an additional year of schooling (see Section 2.1 below). More generally, however, applied economists use the term return to schooling to denote the causal effect of additional schooling on log earnings, holding constant experience (or in some cases age). In this sense, which I will adopt below, the return to schooling is a structural parameter that may vary with the level of schooling, personal characteristics, and the economic environment. Moreover, the observed *ex post* returns to schooling can differ from the *ex ante* returns that were anticipated when the schooling decision was made (Cunha and Heckman, 2006).

2 Theoretical framework

2.1 The internal rate of return and equalizing differences

The internal rate of return is an accounting concept that can be implemented without reference to a particular theory of wages and schooling. As was recognized by Walsh (1935), however, if there is free entry into different schooling options, and if increases in the supply of workers with a given schooling level reduce relative wages for the group, internal rates of return to different choices will be driven down to a common level.

(Walsh, 1935, p. 284, wrote: 'Investment in training … tends to be made as long as the returns promise to cover the cost of that training with an ordinary commercial profit. And this of course is the fundamental characteristic of the competitive, equalizing market …')

Using this 'equalizing differences' framework, Mincer (1958) showed that the equilibrium wage differential between two occupations requiring differing amounts of schooling will equal the difference in years of schooling multiplied by the discount rate.

Willis (1986) considers the choice of an optimal schooling level S (measured in units of time) under four assumptions: (1) individuals maximize the discounted present value of earnings using a common interest rate r; (2) earnings are zero while in school, and equal to $f(S) g(t - S)$ at age t (where age is measured in units of time since the completion of compulsory schooling); (3) the duration of work life is independent of S; (4) the only cost of schooling is the opportunity cost of forgone earnings. Under these assumptions, the internal rate of return for a marginal increase in schooling from an initial level S_0 is $f'(S_0)/f(S_0)$ – that is, the proportional earnings differential per year of education between people with schooling S_0 and those with a little more (or less), holding constant work experience. (Under the assumptions specified, the internal rate of return r equates $V(S_0,r)$ and $V(S_0+\varepsilon, r)$, where $V(S,r) = \int_S^{S+n} f(S)g(t - S)e^{-rt}dt = f(S)e^{-rS} \int_0^n g(x)e^{-rx}dx$. Equality implies that $e^{r\varepsilon} = f(S_0 + \varepsilon)/f(S_0) \approx 1 + \varepsilon f'(S_0)/f(S_0)$ and taking the limit as $\varepsilon \to 0$ gives $r = f'(S_0)/f(S_0)$.)

If people can choose freely between schooling opportunities, in equilibrium log earnings will be a linear function of years of schooling (with slope r), and the internal rate of return for any schooling choice will equal r. Consistent with this insight, one of the most important regularities in labour economics is that a regression of log earnings on years of schooling and controls for experience yields a coefficient that is comparable to a discount rate for a risky investment – of the order of 5–15 per cent per year. Though the precise magnitude of such a coefficient varies over time and across labour markets, the predictability of the magnitude of the estimated return to schooling is unmatched in any other area of empirical microeconomics.

2.2 An extended model

While a simple equalizing differences framework provides a useful starting point for understanding the relationship between earnings and schooling, it does not explain why different people choose different levels of schooling. In fact, children's education choices are strongly correlated with their parent's schooling and socio-economic status, and with their own test scores in early grades. (See Card, 1999, and Solon, 1999.)

These correlations raise a fundamental question: to what extent do people with more education have other attributes – like ability or privileged family background – that would cause them to earn more even in the absence of extra schooling? In the literature this possibility is known as ability bias. A closely related question is whether people who acquire additional schooling have higher returns than those who do not – a sorting or self-selection bias of the type identified by Roy (1951).

Becker (1967) presented a simple model of earnings and schooling determination that can be used to address these issues. In this model, an individual faces a market opportunity locus $y(S)$ that gives the level of earnings y associated with different schooling choices S, and chooses a level of schooling by equating the marginal benefit of schooling with the marginal cost. Following Card (1995a), it is convenient to assume the individual chooses S to maximize a utility function $U(S, y) = \log y - h(s)$, where h is an increasing convex function. An optimal schooling choice satisfies the first-order condition

$$h'(S) = y'(S)/y(S).$$

Note that, because the objective function is linear in $\log y$, the optimal choice of schooling is independent of factors that generate a parallel shift in the $\log y(S)$ function. Griliches (1977) presented a more general model of preferences with the feature that a uniform upward shift in log earnings for all levels of schooling leads to a lower schooling choice.

Individual heterogeneity in the optimal schooling outcome arises from two sources: differences in the costs of (or tastes for) schooling, represented by heterogeneity in $h(S)$; and differences in the economic benefits of schooling, represented by heterogeneity in the marginal return $y'(S)/y(S)$. A tractable assumption is that both functions are linear in S, with additive heterogeneity components:

$$y'(s)/y(S) = b_i - k_1 s,$$
$$h'(S) = r_i + k_2 S.$$

Here b_i and r_i are random variables with means $\bar{b}$ and $\bar{r}$ and some joint distribution across individuals (indexed by i), and k_1 and k_2 are non-negative constants. This specification implies that the optimal schooling choice is linear in the individual-specific heterogeneity terms:

$$S_i^* = (b_i - r_i)/k,$$

where $k = k_1 + k_2$.

The assumed model for the marginal returns to schooling implies that log earnings are generated by a model of the form

$$\log y_i = \alpha_i + b_i S_i - \frac{1}{2} k_1 S_i^2,$$

where α_i is a person-specific constant of integration. This is a generalization of the semi-logarithmic functional form adopted in Mincer (1974) and hundreds of subsequent studies. In particular, individual heterogeneity potentially affects both the *intercept* of the earnings equation (via α_i) and the *slope* of the earnings-schooling relation (via b_i). In general the optimal schooling choice will be positively correlated with b_i, leading to a 'self-selection bias' that arises because people with higher returns to schooling acquire more schooling. If α_i is also positively correlated with S_i (via a positive correlation with b_i or a negative correlation with r_i) the relationship between earnings and schooling will also include an 'ability bias', that is, a bias that arises because people with a higher level of earnings for each level of schooling have characteristics that lead them to acquire more schooling.

A particularly simple version of this model has only two schooling choices (Willis and Rosen, 1979). In this case the model reduces to a discrete choice model for the longer schooling option, and an earnings equation with a random intercept and random coefficient on a dummy representing the longer schooling option. A more general version arises if one relaxes the linearity assumptions for the marginal costs and marginal returns, but maintains additive heterogeneity: that is, $y'(S)/y(S) = b_i + \lambda(S)$, $h'(S) = r_i + \mu(S)$. In this case, Rau-Binder (2006) shows the optimal schooling is $S = \theta^{-1}(b_i - r_i)$, where $\theta(S) = \mu(S) - \lambda(S)$, and log earnings are generated by a model of the form $\log y_i = \alpha_i + \phi(S) + b_i S_i$, where $\phi'(S) = \lambda(S)$.

What does this class of models imply about the return to schooling? For individual i, the marginal return to the last unit of schooling is:

$$\beta_i = b_i - k_1 S_i^* = b_i(1 - k_1/k) + r_i k_1/k,$$

which varies across people unless one of two conditions is satisfied: either $b_i = \bar{b}$ for all i and $k_1 = 0$ (so each additional unit of schooling has the same proportional effect on earnings for everyone); or $r_i = \bar{r}$ for all i and $k_2 = 0$ (so everyone uses the same discount rate and invests in schooling until the return on their last unit of schooling is driven down to $\bar{r}$). Even if one of these conditions is satisfied and β_i is constant across the population, it is not necessarily true that one can obtain an unbiased estimate of the average marginal return to schooling $\bar{\beta} = E[\beta_i]$ from observational data on earnings and schooling. In the first case (homogeneous returns) the implied earnings model is

$$\log y_i = \alpha_i + \bar{b} S_i.$$

Only if α_i and S_i are uncorrelated will an ordinary least squares (OLS) regression yield a consistent estimate of $\bar{b}$. In the second case (homogeneous interest rates) the implied earnings model is:

$$\log y_i = \alpha_i + \bar{r} S_i + \tfrac{1}{2} k_1 S_i^2.$$

Since people with higher values of b_i invest in more schooling, the implied relationship between earnings and schooling is convex, leading to an upward bias in the OLS estimator relative to the true marginal return to schooling, $\bar{r}$ (Mincer, 1997). Any correlation between α_i and S_i will confound the situation even further.

For the general case where marginal returns vary across the population, Card (1999) shows that an OLS regression of earnings on schooling yields a coefficient b_{ols} that has probability limit

$$\text{plim } b_{ols} = \bar{\beta} + \lambda + \psi \bar{S},$$

where $\lambda = \text{cov}[\alpha_i, S_i]/\text{var}[S_i]$ represents an ability bias term and $\psi = \text{cov}[b_i, S_i]/\text{var}[S_i]$ represents a self-selection or sorting bias term. (This expression assumes that the heterogeneity terms have symmetric distributions – see Card, 1999.) Since people with higher returns at each level of schooling will tend to acquire more schooling, the sorting bias term should be positive, although the magnitude may be small. The sign of the ability bias term is less clear: several studies – including the seminal paper by Willis and Rosen (1979) – have obtained negative estimates of λ. In any case, observed pay differences between people with different levels of education may imply rates of return that are above or below $\bar{\beta}$, the average marginal return to education in the population.

The Mincer–Willis equalizing differences model is a long-run general equilibrium model in which wage differentials across education groups are determined by a free entry condition on the supply side. Becker's (1967) model, in contrast, is a partial equilibrium model describing the schooling decisions made by different individuals in a given cohort, taking the earnings generating function as given. Once these decisions are made, shifts in the demand and supply for different education groups can lead to realized returns that are higher or lower than were originally anticipated *ex ante*. Moreover, the fraction of a cohort that acquires higher education can affect their *ex post* returns – a general equilibrium effect. In the mid-1970s for example, the college-high school premium in the United States was relatively low, and analysts described an 'oversupply' of college-educated labour (Freeman, 1976). Within 15 years, however, the premium bounced back, and it now appears that cohorts born in the 1950s have enjoyed higher returns to education than they expected *ex ante*.

2.3 Dynamic models of schooling

A more realistic alternative to the static Becker (1967) model is one in which young people make a series of decisions about whether to enrol in school (for example, Keane and Wolpin, 1997; 2001). If they do, their education increments by an amount which may depend on effort and ability, and they then become eligible to enter a higher level of schooling the next period. Individuals also choose a level of savings or borrowing which can depend on tuition costs, earnings, family transfers, and access to loans and grants. This class of models sheds light on a number of features that are inconsistent with (or simply ignored by) a static framework. For example, a dynamic model can be used to formally address the question of how students learn about their potential returns to different levels of schooling (Arcidiacono, 2004), and how schooling choices are affected by risk aversion and access to credit markets (Keane and Wolpin, 2001).

A dynamic framework is also helpful for understanding the distribution of observed education choices in the presence of 'sheepskin' or 'degree' effects that create non-concavities in the earnings–schooling relationship. In the United States, for example, people with three years of college education have about the same earnings as those with only two years of college (Park, 1994). (Likewise, people with three years of high school earn about the same as those with only two years of high school; see Hungerford and Solon, 1987.) From a static modelling perspective it is unclear why anyone would ever plan to leave college after three years. From a dynamic perspective, however, the outcome of three years of college can be explained by noting that the true return to the third year of college is the option value of entering the fourth year (Altonji, 1993). Students begin their third year of college knowing it is a necessary step to graduation, but may receive some information – for example, about their ability to complete the programme – that causes them to re-evaluate the costs and benefits of enrolment and drop out without graduating.

A dynamic perspective suggests that one should calculate the distribution of final education outcomes conditional on starting a specific education programme, and use this distribution, in combination with the estimated costs and earnings for each outcome, to measure the *ex ante* return to programme entry. (In fact, such a calculation is explicitly built into dynamic optimization models like the ones estimated by Keane and Wolpin, 1997 and Eckstein and Wolpin, 1999.) An interesting case in point is entry to a junior college, which has three main outcome possibilities: early dropout, completion of an Associates (AA) degree, or entry (with two years of college credit) to a four-year college programme. The third node creates an option value to entering an academic programme at junior college that is ignored in simple *ex post* comparisons of earnings between those who are observed holding an AA degree and those with only high school education.

3 Evidence on the returns to schooling

3.1 Mincerian studies

Most of the existing evidence on the returns to schooling is based on Mincer's (1974) 'human capital earnings function': an OLS regression of log earnings on years of completed schooling and a polynomial of post-schooling experience (that is, current age minus an estimate of age at the completion of schooling). As noted in Section 2.1, under certain simplifying assumptions the coefficient of schooling can be interpreted as an estimate of the internal rate of return to alternative education choices. Though the empirical validity of these assumptions varies from application to application, Mincer's model has fitted in hundreds of studies of earnings determination around the world.

Several issues arise in the specification of the human capital earnings function (HCEF) that affect the magnitude of the estimated returns to schooling. One is the choice of earnings measure. Since better-educated people tend to work more hours per week and weeks per year than those with less education, the estimated returns to schooling are usually larger for annual earnings than for weekly or hourly earnings (Card, 1999).

Arguably, earnings should also include the cash value of work-related benefits like health insurance and pensions, leading to an additional source of 'returns' to schooling. A related issue is the treatment of taxes and transfer income during periods of nonwork, for example, from unemployment insurance and welfare programmes. From the perspective of an individual investor, the return to a given schooling choice presumably depends on the expected net incomes associated with the choice (that is, taking into account expected taxes and transfers). Interestingly, the earnings measures available in conventional surveys for many European countries (for example, France and Spain) are net of social security and income taxes, whereas the earnings measures available for other countries (in particular the United States) exclude taxes. Thus, there is some adjustment for taxes built-in to conventional human capital earnings functions estimated for many European countries, but not for the United States. Finally, schooling may affect longevity or health, leading to another indirect effect on earnings.

A second issue is functional form. Mincer's equalizing differences framework implies that log earnings are related to the opportunity cost of a given schooling choice, measured in years of forgone earnings, plus an additive term in years of post-schooling experience. Mincer (1974) assumed a linear path for on-the-job investments in human capital after the completion of schooling and showed that earnings would then depend on a *quadratic* function of years of post-schooling experience. Unless the assumptions underlying this derivation are correct, however, the conditional expectation of earnings, given education and age, will differ from this highly restrictive functional form. Empirically, the model adopted by

Mincer (1974) is probably too restrictive (Lemieux, 2006). For example, Murphy and Welch (1990) conclude that a model with a third- or fourth-order polynomial in experience provides a significant improvement in fit.

Researchers have generalized the HCEF by including dummies for degrees (or a complete set of dummies for all possible schooling choices), by including interactions between schooling and experience (or cohort), and by including interactions between schooling and characteristics like gender, cognitive ability, family background, and school quality. Estimation results from such models can be used to calculate 'returns' to schooling that vary by the level or type of schooling and by individual characteristics. Although the resulting estimates cannot be strictly interpreted as internal rates of return, it is conventional to refer to the implied marginal effects as *returns to schooling*.

Related to the issue of functional form is the question of whether post schooling choices – like occupation or industry – should be added as controls to HCEF. From the perspective of calculating the returns to alternative schooling choices, the answer is 'no', since some of the return to additional schooling is the increased chance of working in a more highly paid occupation or industry (Becker, 1964). A more subtle issue is region or urban location, since some part of the wage differential associated with these choices is caused by differences in the cost of living (which presumably should be subtracted from earnings to calculate the return to schooling).

Recent surveys of the returns to schooling based on the Mincerian HCEF (Psacharopoulos, 1994; Psacharopoulos and Patrinos, 2004; Harmon, Walker and Westergaard-Nielson, 2001; 2003) suggest that returns are in the range of 5–15 per cent for most OECD countries, and somewhat higher in developing countries, on average. In Europe, returns appear to be relatively low in Scandinavia (around 5 per cent) and relatively high in the United Kingdom and Ireland (10 per cent or more). Estimated returns in the United States are comparable to those in the United Kingdom, with evidence of a positive trend in both countries over the 1980–2000 period (Katz and Murphy, 1992; Card and Lemieux, 2001; Gosling and Lemieux, 2003). Using meta-analytic techniques, Harmon, Walker and Westergaard-Nielson (2001) conclude that estimated returns are on average 1–2 points lower when the sample is limited to the public sector, when the earnings model includes controls for occupation or 'ability' measures, and when allowances are made for taxes. They also conclude that returns are slightly higher for women than men. In the United States, estimated returns for the mid-1990s from a conventional HCEF based on hourly earnings were about 10 per cent for men and 11 per cent for women (Card, 1999, Table 1).

3.2 Causal studies

In his pioneering study Walsh noted: 'No doubt the students who go on from high school to college are, on average, richer in natural endowments than those who are left behind. They are a selected lot …' (Walsh, 1935, pp. 272–3). Two main methods have been developed to control for the potential selection biases that confound simple earnings comparisons between people with different levels of schooling: (1) comparisons of siblings or twins; (2) comparisons based on interventions or exogenous factors that affected the education choices of one group relative to another. Detailed discussions of these methods are presented in Card (1995a; 1999; 2001), Krueger and Lindahl (2001), Harmon, Oosterbeek and Walker (2003), and Blundell, Dearden and Sianesi (2004). This section presents a brief overview of some of the main methodological issues – and some of the associated findings – without attempting a comprehensive review.

Gorseline (1932) first proposed the use of sibling comparisons to control for selection biases between different education groups. The basic idea can be illustrated using a variant of the 'homogeneous returns' model discussed in Section 2.2. Letting y_{ij} and S_i denote the earnings and schooling of sibling j ($j = 1,2$) from family i, the homogeneous returns model posits:

$$\log\ y_{ij} = \alpha_{ij} + \bar{b}S_{ij},$$

where α_{ij} represents the level of earnings that sibling j would receive in the absence of schooling. One possible assumption is that $\alpha_{i1} = \alpha_{i2}$: that is, that the two siblings have equal 'ability'. In this case, one can obtain an unbiased estimate of the true return to schooling from a within-family regression, since

$$\log\ y_{i1} - \log\ y_{i2} = \bar{b}(S_{i1} - S_{i2}).$$

Chamberlain and Griliches (1975) re-analysed Gorseline's sample of Indiana brothers and obtained a within-family estimate of $\bar{b}$ equal to 0.080 – only slightly below the estimate of 0.082 obtained from a conventional earnings model estimated by OLS on the same data. (Chamberlain and Griliches, 1975, also included the sibling's differences in age and age-squared as added regressors.) Of course siblings may not have identical abilities, and if they don't, a within-family estimator b_w can be worse (that is, more biased) than the corresponding OLS estimator b_{ols}. Assuming a homogenous returns model is correct, the bias in the OLS estimate is

$$\text{plim}\ b_{ols} - \bar{b} = \text{cov}[\alpha_{ij}, S_{ij}]/\text{var}[S_{ij}],$$

while the bias in the within-family estimator is

$$\text{plim}\ b_w - \bar{b} = \text{cov}[\alpha_{ij} - \alpha_{i2}, S_{i1} - S_{i2}]/\text{var}[S_{i1} - S_{i2}].$$

Although differencing eliminates the shared component of α_{i1} and α_{i2}, it is possible that the remaining within-family difference in ability is large relative to the within-family variance of schooling, implying a larger bias in b_w

than b_{ols}. (A similar analysis can be conducted when both the slope and intercept of the earnings function have person-specific components; see Card, 1999.)

One approach to the concern over ability differences between siblings is to focus on identical (monozygotic) twins. Unfortunately, schooling differences are small among identical twins, and even a little measurement error in reported education can lead to large attenuation bias in the within-twin estimate of the return to schooling. Ashenfelter and Krueger (1994) proposed an innovative solution based on asking each twin about its own and its sibling's education. Their method has been widely adapted in the literature (for example, Ashenfelter and Rouse, 1998; Miller, Mulvey and Martin, 1995; Bonjour et al., 2003) and leads to estimated returns within twins that are comparable to the corresponding OLS estimates, or only slightly smaller. (As noted by Bound and Solon, 1999, if the measurement error in schooling is mean-reverting, the Ashenfelter–Krueger approach 'over-corrects' and leads to an upward bias in the resulting estimator.)

Despite the intuitive appeal of identical twins to some researchers, others (for example, Bound and Solon, 1999) have questioned whether twins who choose different schooling levels are really 'identical' or whether the small differences in upbringing and experience that lead them to choose different schooling also contribute to their different earnings. Fundamentally, the problem is that the source of the differential schooling choices is unobserved, so different observers can argue that the choice was driven by factors that are either correlated or uncorrelated with earnings. A similar problem arises in 'matching' estimates of the return to schooling (see Blundell, Dearden and Sianesi, 2004), which attempt to compare earnings between people who are very similar in all dimensions except their choice of schooling. Indeed, a perfect matching algorithm applied to a sample of twins would presumably match twins to each other, leading to a within-family estimate of the return to schooling.

A second approach to the issue of selection bias is the use of instrumental variables (IV) methods. Specifically, the researcher posits the existence of a variable Z that is exogenous to individuals but affects their schooling choices. As shown by Heckman (1978) in a different context, the equation relating S to Z need not represent a well-specified model, only a linear projection. For example, assume:

$$S_i = Z_i\pi + \xi_i.$$

If earnings are generated by the homogeneous returns model:

$$\log\ y_i = \alpha_i + \bar{b}S_i,$$

and Z_i is orthogonal to α_i, then a consistent estimate of $\bar{b}$ can be obtained by IV, using Z_i as an instrument for S_i. Individual-level instruments that have been proposed include quarter of birth (Angrist and Krueger, 1991), the sex composition of one's siblings (Butcher and Case, 1994), and distance to the nearest college (Card, 1995b). Other IV studies use school system reforms such as changes in the minimum school-leaving age (Harmon and Walker, 1995; Oreopoulos, 2006; Meghir and Palme, 2005), changes in tuition at local state colleges (Kane and Rouse, 1993; Fortin, 2006), and expansions in local infrastructure (Duflo, 2001).

Many IV studies yield estimated returns to schooling that are as large as or slightly larger than the corresponding IV estimates (see for example, Card, 2001; Harmon, Oosterbeek and Walker, 2003). Since the IV approach was motivated by the concern that OLS leads to an *overestimate* of the returns to schooling, this is potentially puzzling, and three explanations have been offered. First, OLS estimates are downward-biased by measurement error in education, and the measurement error bias may offset any upward selectivity bias (Griliches, 1977). Second, the search for IV designs that yield statistically significant estimates may create a 'publication bias' in favour of samples and specifications with relatively large IV coefficients (Ashenfelter, Harmon and Oosterbeek, 1999). Third, if the returns to education vary across the population, certain instruments may identify returns for subgroups with relatively high marginal returns to schooling (Card, 1995a).

The third explanation can be most easily understood in the context of a social experiment with a randomly assigned intervention (indexed by $Z_i = 1$). Let (S_{i0}, y_{i0}) represent the schooling and earnings outcomes for person i if he or she were assigned to the control group, and (S_{i1}, y_{i1}) denote the outcomes if he or she was assigned to treatment (note that only one of these pairs is observed). The treatment effect on schooling for person i is $\Delta S_i = S_{i1} - S_{i0}$, while the effect on log earnings is $\Delta \log y_i = \log y_{i1} - \log y_{i0}$. Assuming that individual i's marginal return to schooling in the absence of the intervention is β_i, and that the intervention only affects earnings through its effect on schooling, $\Delta \log y_i = \beta_i \Delta S_i$. An IV estimate of the return to schooling based on assignment status is numerically equal to the difference in mean log earnings between the treatment and control groups, divided by the corresponding difference in their average schooling, and has probability limit

$$\text{plim } b_{IV} = \frac{E[\log y_i | Z_i = 1] - E[\log y_i | Z_i = 0]}{E[S_i | Z_i = 1] - E[S_i | Z_i = 0]}$$
$$= \frac{E[\beta_i \Delta S_i]}{E[\Delta S_i]}.$$

If $E[\beta_i \Delta S_i] = E[\beta_i] E[\Delta S_i]$, then the IV estimator gives a consistent estimate of the average marginal return to education $\bar{\beta} = E[\beta_i]$. This will be true if the intervention induces the same change in schooling for everyone, or more generally if $E[\Delta S_i | \beta_i]$ is independent of β_i. Otherwise, provided that $\Delta S_i \geq 0$ for all i (that is, no one reduces schooling because of the intervention) the IV estimate is a weighted average of the β_i's, with the weight for person i equal to $\Delta S_i / E[\Delta S_i]$.

An intervention that induces larger gains in schooling for people with high values of β_i can lead to an IV estimate that overstates $\bar{\beta}$. Card (1995a) argued that this might be true for interventions – like the increases in the minimum school leaving age studied by Harmon and Walker (1995) and Oreopoulos (2006) – that mainly affect children from disadvantaged family backgrounds who stop school early because of high marginal costs rather than because of low marginal benefits. An alternative explanation for the finding that an IV estimate exceeds the corresponding OLS estimate is that the assumptions underlying the particular instrumental variable are invalid. In particular, in the absence of a true experiment, one can never 'prove' that the instrument is as good as randomly assigned. Even in an experimental setting, it is also possible that an intervention has an independent causal effect on earnings, confounding the interpretation of the IV estimate. (For example, Willis and Rosen, 1979, and Heckman and Li, 2004, use parental education as instruments for schooling, though others have argued that parental education has an independent effect on earnings.)

A generalization of IV that is useful when a researcher believes there may be random payoffs to schooling is a control function approach, first used in the schooling context by Garen (1984). (Other recent applications include Conneely and Uusitalo, 1997, Blundell, Dearden and Sianesi, 2004, and Rau-Binder, 2006.) This method relies on assumptions about the relationship between the error component in the equation relating schooling to the instrument(s) Z, and the random slope and intercept in the earnings equation. Assuming these are satisfied, a control function approach can recover unbiased estimates of the average marginal return to schooling, as well as useful information on how the returns to education vary with the unobserved factors driving the choice of schooling (Rau-Binder, 2006).

4 Summary

The idea of treating schooling as an investment that yields internal rates of return comparable to other investments in the economy has proven extremely useful, and has led to an unusually coherent body of research that combines theoretical modelling and detailed empirical analysis. Much of the existing empirical work is conducted in the framework of Mincer's (1974) human capital earnings function, which relates the logarithm of earnings to completed schooling – measured in years to reflect the opportunity cost of the investment – and a control for post-schooling experience. In a strict equalizing differences framework the coefficient of schooling is the internal rate of return to schooling. In a more general framework that recognizes the endogenous nature of the schooling decision, and the importance of ability differences that partially determine the choice of schooling, observed differences in earnings across different education groups will not necessarily reveal the rate of return to schooling for any one person, or for the population as a whole. Nevertheless, existing evidence from studies of siblings and twins, and from studies that focus on arguably exogenous sources of variation in education choices, suggest that the return to schooling is in the range of 5–15 per cent, and not too different from the value implied by the simple Mincerian approach.

DAVID CARD

See also **Becker, Gary S.; control functions; Griliches, Zvi; human capital; Mincer, Jacob; Rosen, Sherwin; Schultz, T.W.; selection bias and self-selection.**

Bibliography

Altonji, J.J. 1993. The demand for and return to education when education outcomes are uncertain. *Journal of Labor Economics* 11, 48–83.

Angrist, J.D. and Krueger, A.B. 1991. Does compulsory school attendance affect schooling and earnings? *Quarterly Journal of Economics* 106, 979–1014.

Arcidiacono, P. 2004. Ability sorting and the returns to college major. *Journal of Econometrics* 121, 343–75.

Ashenfelter, O., Harmon, C. and Oosterbeek, H. 1999. A review of estimates of the schooling/earnings relationship, with tests for publication bias. *Labour Economics* 6, 453–70.

Ashenfelter, O. and Krueger, A.B. 1994. Estimates of the economic return to schooling from a new sample of twins. *American Economic Review* 84, 1157–73.

Ashenfelter, O. and Rouse, C. 1998. Income, schooling, and ability: evidence from a new sample of twins. *Quarterly Journal of Economics* 113, 869–95.

Becker, G.S. 1962. Investment in human capital: a theoretical analysis. *Journal of Political Economy* 70, 9–49.

Becker, G.S. 1964. *Human Capital: A Theoretical and Empirical Analysis, with Special Reference to Education.* New York: Columbia University Press.

Becker, G.S. 1967. *Human Capital and the Personal Distribution of Income.* Ann Arbor, MI: University of Michigan Press.

Blundell, R., Dearden, L. and Sianesi, B. 2004. Evaluating the impact of education on earnings in the U.K.: models, methods, and results from the NCDS. Working Paper No. 03/20, Institute for Fiscal Studies.

Bonjour, D., Cherkas, L., Haskel, J., Hawkes, D. and Spector, T. 2003. Returns to education: evidence from UK twins. *American Economic Review* 93, 1799–812.

Bound, J. and Solon, G. 1999. Double trouble: on the value of twins-based estimation of the returns to education. *Economics of Education Review* 18, 169–82.

Butcher, K.F. and Case, A. 1994. The effect of sibling composition on women's education and earnings. *Quarterly Journal of Economics* 109, 531–63.

Card, D. 1995a. Earnings, schooling, and ability revisited. In *Research in Labor Economics*, vol. 14, ed. S. Polachek. Greenwich, CT: JAI Press.

Card, D. 1995b. Using geographic variation in college proximity to estimate the return to schooling. In *Aspects of Labour Market Behaviour: Essays in Honour of John Vanderkamp*, ed. L.N. Christofides, E.K. Grant and R. Swidinsky. Toronto: University of Toronto Press.

Card, D. 1999. The causal effect of education on earnings. In *Handbook of Labor Economics*, vol. 3A, ed. O. Ashenfelter and D. Card. Amsterdam: North-Holland.

Card, D. 2001. Estimating the return to schooling: progress on some persistent econometric problems. *Econometrica* 69, 1127–60.

Card, D. and Lemieux, T. 2001. Can falling supply explain the rising return to college for younger men? A cohort-based analysis. *Quarterly Journal of Economics* 116, 705–46.

Chamberlain, G. and Griliches, Z. 1975. Unobservables with a variance–covariance structure: ability, schooling, and the economic success of brothers. *International Economic Review* 16, 422–49.

Conneely, K. and Uusitalo, R. 1997. Estimating heterogeneous treatment effects in the Becker schooling model. Discussion Paper, Industrial Relations Section, Princeton University.

Cunha, F. and Heckman, J.J. 2006. Identifying and estimating the distributions of *ex post* and *ex ante* returns to schooling: a survey of recent developments. Working Paper, University of Chicago.

Duflo, E. 2001. Schooling and labor market consequences of school construction in Indonesia: evidence from an unusual policy experiment. *American Economic Review* 91, 795–813.

Eckstein, Z. and Wolpin, K.I. 1999. Why youths drop out of high school: the impact of preferences, opportunities and abilities. *Econometrica* 67, 1295–339.

Fortin, N. 2006. Higher education policies and the college premium: cross-state evidence from the 1990s. *American Economic Review* 96, 959–87.

Freeman, R.B. 1976. *The Over-Educated American.* New York: Academic Press.

Garen, J. 1984. The returns to schooling: a selectivity bias approach with a continuous choice variable. *Econometrica* 52, 1199–218.

Gorseline, D.E. 1932. *The Effect of Schooling Upon Income.* Bloomington: University of Indiana Press.

Gosling, A. and Lemieux, T. 2003. Labor market reforms and changes in wage inequality in the United Kingdom and the United States. In *Seeking a Premier Economy: The Economic Effects of British Economic Reforms, 1980–2000*, ed. D. Card, R. Blundell and R.B. Freeman. Chicago: University of Chicago Press.

Griliches, Z. 1977. Estimating the returns to schooling: some econometric problems. *Econometrica* 45, 1–22.

Harmon, C. and Walker, I. 1995. Estimates of the economic return to schooling for the United Kingdom. *American Economic Review* 85, 1278–86.

Harmon, C., Oosterbeek, H. and Walker, I. 2003. The returns to education: microeconomics. *Journal of Economic Surveys* 17, 115–56.

Harmon, C., Walker, I. and Westergaard-Nielson, N. 2001. Introduction. In *Education and Earnings in Europe*, ed. C. Harmon, I. Walker and N. Westergaard-Nielson. Aldershot: Edward Elgar.

Harmon, C., Walker, I. and Westergaard-Nielson, N. 2003. The returns to education: microeconomics. *Journal of Economic Surveys* 17, 115–56.

Heckman, J.J. 1978. Dummy endogenous variables in a simultaneous equation system. *Econometrica* 46, 931–59.

Heckman, J.J. and Li, X. 2004. Selection bias, comparative advantage, and heterogeneous returns to education: evidence from China in 2000. *Pacific Economic Review* 9, 155–71.

Hungerford, T. and Solon, G. 1987. Sheepskin effects in the return to education. *Review of Economics and Statistics* 69, 175–7.

Kane, T.J. and Rouse, C.E. 1993. Labor market returns to two- and four-year colleges: is a credit a credit and do degrees matter? Working Paper No. 4268. Cambridge, MA: NBER.

Katz, L.F. and Murphy, K.M. 1992. Changes in relative wages, 1963–1987: supply and demand factors. *Quarterly Journal of Economics* 107, 35–78.

Keane, M.P. and Wolpin, K.I. 1997. The career decisions of young men. *Journal of Political Economy* 105, 473–521.

Keane, M.P. and Wolpin, K.I. 2001. The effect of parental transfers and borrowing constraints on educational attainment. *International Economic Review* 42, 1051–103.

Krueger, A.B. and Lindahl, M. 2001. Education for growth: why and for whom? *Journal of Economic Literature* 39, 1101–36.

Lemieux, T. 2006. The 'Mincer equation' thirty years after *Schooling, Experience, and Earnings*. In *Jacob Mincer: A Pioneer of Modern Labor Economics*, ed. S. Grossbard-Shechtman. Heidelberg: Springer.

Meghir, C. and Palme, M. 2005. Educational reform, ability and parental background. *American Economic Review* 95, 414–24.

Miller, P.W., Mulvey, C. and Martin, N. 1995. What do twins studies reveal about the economic returns to education? A comparison of Australian and US findings. *American Economic Review* 85, 586–99.

Mincer, J. 1958. Investment in human capital and personal income distribution. *Journal of Political Economy* 66, 281–302.

Mincer, J. 1974. *Schooling, Experience, and Earnings*. New York: NBER (distributed by Columbia University Press).

Mincer, J. 1997. Changes in wage inequality, 1970–1990. *Research in Labor Economics* 16, 1–18.

Murphy, K.M. and Welch, F. 1990. Empirical age–earnings profiles. *Journal of Labor Economics* 8, 202–29.

Oreopoulos, P. 2006. Estimating average and local average treatment effects of education when compulsory schooling laws really matter. *American Economic Review* 96, 152–75.

Park, J.H. 1994. Returns to schooling: a peculiar deviation from linearity. Working Paper No. 335, Industrial Relations Section, Princeton University.

Psacharopoulos, G. 1994. Returns to investment in education: a global update. *World Development* 22, 1325–43.

Psacharopoulos, G. and Patrinos, H.A. 2004. Returns to investment in education: a further update. *Education Economics* 12, 111–34.

Rau-Binder, T. 2006. Semiparametric estimation of microeconometric models with endogenous regressors and sorting. Manuscript, Department of Economics, University of California Berkeley.

Roy, A.D. 1951. Some thoughts on the distribution of earnings. *Oxford Economic Papers* 3, 135–46.

Solon, G. 1999. Intergenerational mobility in the labor market. In *Handbook of Labor Economics*, vol. 3A, ed. O.C. Ashenfelter, D. Card and R. Layard. Amsterdam: North-Holland.

Walsh, J.R. 1935. Capital concept applied to man. *Quarterly Journal of Economics* 49, 255–85.

Willis, R. 1986. Wage determination: a survey and reinterpretation of human capital earnings functions. In *Handbook of Labor Economics*, vol. 1, ed. O. Ashenfelter and R. Layard. Amsterdam: North-Holland.

Willis, R. and Rosen, S. 1979. Education and self-selection. *Journal of Political Economy* 87, S7–S36.

revealed preference theory

Economists do not observe preferences. They may, however, observe demand behaviour – the choices made by consumers. Is there a way for economists to tell whether the observed behaviour is generated through the maximization of a preference relation or utility function? Since most economic theories are ultimately based on a consumer who maximizes a preference or utility, the question is clearly important for developing and testing theories.

Revealed preference theory answers this question by characterizing choice behaviour that is generated by preference or utility maximization. Relating choice behaviour and preference maximization is also a goal of integrability theory. What distinguishes the theories from each other, and from the other parts of rationality theory, is the special nature of their tools: integrability theory uses mathematical integration in its proofs, and usually states its hypotheses in differential form; revealed preference theory uses a variety of mathematical tools for its proofs, and its hypotheses are usually in a discrete 'revelation' form. The distinctions are not always sharp, however, and we shall see areas in which the theories overlap.

Samuelson invented revealed preference theory in 1938. The basic idea, much of the terminology, and some of the axioms are due to him. In the following outline, a useful paradigm is the one that guided the first three decades: a consumer with a finite-dimensional euclidean commodity space, facing 'competitive' budgets determined by fixed positive prices, and satisfying a budget equality constraint.

1. The problem of rationality

From the economist's point of view, unobservable preferences generate observable choices. Since many preference relations may generate the same choice correspondence, the map from preferences to choice correspondences is many-one: We cannot hope to find *the* preference generating choices, but only *some* preferences – a set of 'equivalent' preferences. For example:

It is well known in preference theory that a lexicographic preference on the plane does not admit a real-valued utility function (Debreu, 1954). A hasty conclusion might be that there is no hope of representing a lexicographic-maximizing consumer as a utility-maximizing consumer. Too hasty! For her behaviour clearly maximizes this function g on the non-negative plane (for positive prices):

$$g(x_1, x_2) = x_1 \qquad (1)$$

Even if her 'intention' is to maximize a lexicographic preference, she acts *as if* her intention were to maximize g. In fact, even if the choices were made by a committee, a machine, or any other mindless decision maker, we can still say the actions are *as if* the intent were g-maximization.

This example shows the distinction between a typical question in utility and preference theory ('Does *this* preference have a utility function?'), and the basic question in revealed preference and integrability theories ('Is this demand generated by *some* preference?'). It also demonstrates the need for precise definitions. (Our notation will follow the glossaries of Richter, 1966, 1971.)

To describe choices, the theory requires an underlying set X and a family $\mathscr{B}$ of subsets $B \subset X$. (Often X is the non-negative orthant of n-space and each B is a 'competitive' budget determined by positive prices and income.) We call any $B \in \mathscr{B}$ a *budget*. A *choice* or *demand correspondence* h is a function assigning to each $B \in \mathscr{B}$ a subset $h(B) \subset B$ interpreted as the set of elements chosen from B. And any binary relation on X is called a *preference*. Rationality theory relates choices h on $(X, \mathscr{B})$ to preferences R on X in two ways.

(i) If we start with a preference relation R we can ask what kind of choice it generates. There are two obvious senses in which it could generate a choice h. First, we might have, for all $B \in \mathscr{B}$

$$h(B) = \{x \in B : \forall_{y \in B} xRy\}, \qquad (2)$$

i.e., the set of elements chosen from B is the set of R-most preferred elements in B. Then we say that R *rationalizes* h (Richter, 1971).

Alternatively, we might have, for all $B \in \mathscr{B}$

$$h(B) = \{x \in B : \forall_{y \in B} yRx\}, \qquad (3)$$

i.e., the set of elements chosen from B is the set of elements in B for which nothing in B is R-more preferred. Then we say that R *motivates* h (Kim and Richter, 1986).

Definition (2) is appropriate if we think of R as a 'weak' (i.e. reflexive) relation, while (3) is appropriate if we think of R as a 'strict' (asymmetric) relation.

(ii) Conversely, if we start with a choice h we can ask whether any preference R generates, or 'explains' h. If there exists some R generating h in the sense of (2) (Richter, 1966, 1971), then we say that h is *rational*. Often we are interested in *reflexive*-rationality (rationalization by a reflexive preference), *transitive*-rationality (rationalization by a transitive preference), *regular*-rationality (rationalization by a reflexive, transitive, and total preference), etc. For example, *utility–rationality* requires the existence of a function $f : X \rightarrow R^1$ satisfying

$$h(B) = \{x \in B : \forall_{y \in B} f(x) \geqq f(y)\} \qquad (4)$$

for all $B \in \mathscr{B}$ – i.e., the set of elements chosen from B is the set of those elements in B with the highest utility.

If there exists some R generating h in the sense of (3), then we say that h is *motivated* (Kim and Richter, 1986). Again, we are often interested in *asymmetric*-motivation (motivation by an asymmetric preference), etc. In fact, h is rational if and only if it is motivated (Clark, 1985; Kim and Richter, 1986). Of course, the example (1) makes it clear that such a rationalizing or motivating R will not usually be unique: there will be a whole equivalence class of such relations generating the same choice (Kim and Richter, 1986).

It is important to note that rationality and motivation have been defined as properties of demand, not of preference. We do not say, for example, that a particular preference is rational or irrational. Instead, the definitions relate demand and preferences.

An economist who derives comparative statics results from preference maximization is answering questions of type (i). The issue arises – for both theoretical development and empirical testing – whether any further results can be derived, or whether all the (independent) consequences of preference maximization have been found. This is usually a much more difficult problem. A major task of both revealed preference and integrability theory is to address this issue, by answering questions of type (ii). The two questions are parts, then, of the fundamental Problem of Rationality: give necessary (i) and sufficient (ii) conditions for a demand to be rational (of a particular type), or motivated (of a particular type). Revealed preference theory solves the problem through axioms with a unique flavour.

2. Revealed preference solutions

It is important to distinguish revealed preference definitions from revealed preference axioms, and these in turn from revealed preference theorems.

2.1. Revelation definitions

If consumer (i.e. choice) h selects alternative $x \in B$ from B – i.e., if $x \in h(B)$ – when alternative y could have been selected – i.e., if $y \in B$ – then we write xVy. And it is natural to say that x is *revealed as good as* y. If also $x \neq y$ then we write xSy, and it is natural to say that x is *revealed preferred to* y. This terminology of Samuelson's is very suggestive, because if $\succeq$ is any rationalization, then xVy implies $xx \succeq y$ as does xSy. In fact, if $x \in h(B)$ & $y \in (B)h(B)$ and if $\succeq$ is regular, then its asymmetric part $\succ$ also satisfies $x \succ y$. So an observer of h can deduce properties common to all rationalizations. But beware: xSy is a statement about *choice*, not about a particular preference.

Unlike the psychologist, who may be able to present an individual with binary choices, and thereby uncover a total ordering, the economist will typically observe S as only a partial ordering. This is one of the challenging features of revealed preference theory. It is why, mathematically, revealed preference theory is a study of partial orders, in contrast to the classical theory of preference, which is a theory of total orders. It is also why there is generally more than one preference in the equivalence class of preferences that rationalize or motivate a given choice.

2.2. Revelation axioms

We describe four revealed preference axioms. Samuelson proposed the *asymmetry* of S as a basic axiom of consumer theory: for all $x, y \in X$

$$xSy \Rightarrow ySx. \qquad (5)$$

In other words, if x is revealed preferred to y (under some budget), then y is never (under any budget) revealed preferred to x. As Samuelson noted, this is a property of any single-valued demand function maximizing a regular preference. This is now called the *Weak Axiom of Revealed Preference*.

Houthakker noted other necessary consequences of regular-rationality, for single-valued demand functions: there can be no cycles of the form

$$xSy_1 Sy_2 S \ldots Sy_k Sx. \qquad (6)$$

In other words, x is never, even indirectly, revealed preferred to itself. Houthakker proposed this as a new axiom, now called the *Strong Axiom of Revealed Preference*. If we define xHy to mean that xSy or $xSv_1 Sv_2 S \ldots Sv_k Sy$, then we can rephrase Houthakker's axiom as saying that H is asymmetric. In other words, if x is (even indirectly) revealed preferred to y, then y is never (even indirectly) revealed preferred to x.

Richter noted still another consequence of regular-rationality. For this it is convenient to define xWy

to mean either xVy or $xVu_1 V \ldots Vu_k Vy$. Clearly regular-rationality implies: for all $x, y \in X$ & $B \in \mathscr{B}$

$$x \in h(B) \& y \in B \& yWx \Rightarrow y \in h(B). \qquad (7)$$

In other words, if x is chosen from B, and if y is also available in B and is revealed (even directly) as good as x, then y is also chosen from B. This is the *Congruence Axiom of Revealed Preference*. He also noted a behavioural consequence of any rationality: for all $x, y \in X$ & $B \in \mathscr{B}$

$$x \in B \forall_{y \in B} xVy \Rightarrow x \in h(B). \qquad (8)$$

In other words, if x is in B and is revealed as good as everything in B, then x is chosen from B. This is the *V-Axiom*.

We will use these axioms to discuss the main solutions to the Problem of Rationality.

2.3. Revelation theorems

(a) *Weak Axiom.* Samuelson proposed the Weak Axiom in 1938, as a foundation for all consumer theory (1938 a,b). He did not name it, but he suggested that (for single-valued demand functions) it followed from maximizing a utility function (cf. also Samuelson (1955), pp. 110–11). In the opposite direction, his idea of founding consumer theory on it was implicitly a conjecture that it implied utility-rationality, or at least regular-rationality. Indeed, after preliminary work by I.M.D. Little, Samuelson succeeded in showing that, for two commodities and Lipschitz-continuous demand functions, the Weak Axiom implied regular-rationality (Samuelson, 1948).

(b) *Strong Axiom.* Then in 1950 Houthakker (1950) proposed the Strong Axiom (by a different name) as a basis for consumer theory, and showed that, for any number of commodities, it implied utility-rationality for Lipschitz-continuous demand functions. Samuelson (1950) then gave the Weak and Strong Axioms their modern names.

In 1959, Uzawa (1960, 1971) developed a more precise analogue of Houthakker's result, showing that the Strong Axiom and a Lipschitzian hypothesis on the demand implied irreflexive-transitive-monotone-convex-lower semi continuous-motivation. His proof was along the lines of the Samuelson–Little–Samuelson–Houthakker analytic methods.

Although the Strong Axiom implied the Weak, it was still not clear whether the Weak implied the Strong. Indeed, Rose (1958) showed that the Weak Axiom does imply the Strong Axiom, when there are only two commodities and prices are positive (needed!). Then Gale (1960) constructed an example with three commodities, showing that the Weak Axiom did not imply the Strong. And Kihlstrom, Mas-Colell and Sonnenschein (1976) showed how to obtain very easily many examples, for any number of commodities greater than two. And Shafer (1977b), affirming a conjecture of Samuelson (1953), showed that the full strength of the Strong Axiom is needed: even for three goods, there is no upper bound on the length of S-cycles that must be ruled out. In the opposite direction, several authors have discussed special conditions under which the Weak Axiom does imply the Strong (Arrow, 1959; Uzawa, 1960, 1971).

Richter (1966) used set-theoretic methods – very different from the analytic methods of Samuelson, Little, Houthakker and Uzawa – to simplify the proofs, eliminate extraneous assumptions, and strengthen the rationality results. In a framework of abstract budget spaces, and without the technical assumptions required by the earlier analytical approaches, he showed that the Strong Axiom is equivalent to regular-rationality for demand functions. Thus the Strong Axiom completely exhausts the theory of demand functions maximizing a regular preference.

Richter (1966) also showed that, if a competitive demand satisfies the Strong Axiom, then it is utility-rational if its range is well behaved, but it may not be utility-rational otherwise (Richter, 1971).

Extensions

There have been many extensions. Richter (1966) showed that the V-Axiom characterized rationality, and the Congruence Axiom characterized regular-rationality, for demand correspondences. (Hansson (1968) gave an alternative criterion for regular-rationality.) Other extensions have obtained stronger properties of the rationalization under special hypotheses (Hurwicz and Richter, 1971; Mas-Colell, 1978, Theorem 1; Richter, 1986; Matzkin and Richter, 1986); uniqueness of the rationalization within certain classes (Mas-Colell, 1977); revealed preference axioms characterizing more general rationality types (Richter, 1971; Kim and Richter, 1986; Kim, 1987); dual axioms (Sakai, 1977; Richter, 1979); and axioms for stochastic rationality (McFadden and Richter, 1970).

Applications

Several applications have supported Samuelson's original idea that revealed preference could provide an alternative to preference theory as a foundation for consumer theory. Revealed preference techniques have been applied to prove the existence of competitive equilibrium (Wald, 1936, 1951); to prove the stability of competitive equilibrium (Arrow and Hurwicz, 1958, 1960); to prove the Hicks Composite Commodity Theorem (Richter, 1970; Calsamiglia, 1978); to analyse and characterize aggregate excess demand functions (Debreu, 1974; McFadden et al., 1974); to prove aggregation properties for correspondences (Shafer, 1977a); to prove properties of measurable demand correspondences (Yamazaki, 1984); to prove theorems about social choice functions (Plott, 1973); etc.

3. Revealed preference and integrability

With the same rationality goal as revealed preference theory, integrability theory uses axioms on the Slutsky or Antonelli matrices to characterize rational choice (cf. Hurwicz, 1971; see also INTEGRABILITY OF DEMAND). Under some smoothness assumptions on the demand function, the basic theorems state that symmetry and negative semidefiniteness of these matrices is necessary and sufficient for (upper-semicontinuous-) regular-rationality.

Samuelson established a link between revealed preference theory and integrability theory by showing that his Weak Axiom implied negative semidefiniteness of the matrices (Samuelson, 1938b, 1955, pp. 111–14). Later Kihlstrom, Mass-Colell, and Sonnenschein (1976) demonstrated that negative semidefiniteness was equivalent to a Weak Weak Axiom.

This left open the question of finding a revealed preference axiom equivalent to the symmetry. The Strong Axiom was clearly too strong, since it already implied regular-rationality, and therefore both symmetry and negative-semidefiniteness. Then Hurwicz and Richter (1979a,b) showed that a differential axiom of Ville (1946, 1951) provided the exact strength needed. Although it does not even imply the Weak Axiom, it is similar in spirit to the Strong Axiom and can be given a revealed preference interpretation. It thus serves, like Kihlstrom, Mas-Colell and Sonnenschein's Weak Weak Axiom, as a bridge between the Revealed Preference and Integrability approaches to consumer rationality. Richter (1979) discussed these bridges from the viewpoint of duality.

4. Other notions of rationality

Many economists have used notions of rationality different from Richter's notion (2).

Sometimes the term 'rational' has been applied to preference, rather than demand. (In such applications it is often a synonym for 'transitive'.) In Uzawa (1957) and Arrow (1959), on the other hand, it was applied to demand, but only in terms of axioms on demand behaviour. By contrast, (2) is applied to demand, but in terms that relate both demand and preferences.

Some economists have used weaker notions of rationality than (2), requiring only: for all $x, y \in X$ & $B \in \mathscr{B}$

$$h(B) \subset \{x \in B : \forall_{y \in B} xRy\}. \tag{9}$$

In other words, every element chosen from B is R-most preferred in B, but B may contain other R-most preferred elements that are not chosen. We will call this *subsemi-rationality*, although it has often been referred to as rationality.

A drawback of this concept is its loose linkage of preference and demand. Any constant function, for example, satisfies (9). On the other hand, if one interprets $h(B)$ as a set of incomplete observations, then one might wonder whether, with more observations of choices from B, the set $h(B)$ of chosen elements might grow. Then one might want to find a preference R

satisfying just (9), rather than insisting (as does (2)) that *R* explain *precisely* the observed set *h*(*B*).

Afriat (1967) gave conditions on a demand function, over a finite set of budgets, that are necessary and sufficient for it to be subsemi-rationalized by a continuous monotone concave function. His work was clarified by Diewert (1973) who gave a criterion for continuous-monotone-concave-subsemirationality in terms of a linear programming problem. Varian (1983) restated Afriat's finite-budgets result in terms of a Generalized Axiom of Revealed Preference – weaker than the Strong Axiom.

Matzkin and Richter (1986) obtained full rationality by replacing the Generalized Axiom with the Strong Axiom, which they proved was necessary and sufficient for continuous-monotone-strictly   concave-utility-rationality in the finite case. No revealed preference criterion for concave-regular-rationality is known for the not-necessarily finite case.

MARCEL K. RICHTER

See also **demand theory; integrability of demand; Samuelson, Paul Anthony.**

Bibliography

Afriat, S.N. 1967. The construction of utility functions from expenditure data. *International Economic Review* 8, 67–77.

Arrow, K.J. and Hurwicz, L. 1958. On the stability of competitive equilibrium, I. *Econometrica* 26, 522–52.

Arrow, K.J. 1959. Rational choice functions and orderings. *Economica* N.S. 26, 121–7.

Arrow, K.J. and Hurwicz, L. 1960. Some remarks on the equilibria of economic systems. *Econometrica* 28, 640–46.

Calsamiglia, X. 1978. Composite goods and revealed preference. *International Economic Review* 19, 395–404.

Clark, S.A. 1985. A complementary approach to the strong and weak axioms of revealed preference. *Econometrica* 53, 1459–63.

Debreu, G. 1954. Representation of a preference ordering by a numbering function. In *Decision Processes*, ed. R.M. Thrall, C.H. Coombs and R.L. Davis. New York: Wiley, 159–65.

Debreu, G. 1974. Excess demand functions. *Journal of Mathematical Economics* 1, 15–21.

Diewert, W.E. 1973. Afriat and revealed preference theory. *Review of Economic Studies* 40, 419–25.

Gale, D. 1960. A note on revealed preference. *Economica* . NS 27, 348–54.

Hansson, B. 1968. Choice structures and preference relations. *Synthese* 18, 443–58.

Houthakker, H.S. 1950. Revealed preference and the utility function. *Economica* NS 17, 159–74.

Hurwicz, L. 1971. On the problem of integrability of demand functions. In *Preferences, Utility and Demand*, ed. J.S. Chipman, L. Hurwicz, M.K. Richter, and H.F. Sonnenschein. New York: Harcourt, Brace, Jovanovich, ch. 9.

Hurwicz, L. and Richter, M.K. 1971. Revealed preference without demand continuity assumptions. In *Preferences, Utility and Demand*, ed. J.S. Chipman, L. Hurwicz, M.K. Richter, and H.F. Sonnenschein. New York: Harcourt, Brace, Jovanovich, ch. 3.

Hurwicz, L. and Richter, M.K. 1979a. An integrability condition with applications to utility theory and thermodynamics. *Journal of Mathematical Economics* 6, 7–14.

Hurwicz, L. and Richter, M.K. 1979b. Ville axioms and consumer theory. *Econometrica* 47, 603–19.

Kihlstrom, R., Mas-Colell, A. and Sonnenschein, H. 1976. The demand theory of the weak axiom of revealed preference. *Econometrica* 44, 971–8.

Kim, T. and Richter, M. 1986. Nontransitive-nontotal consumer theory. *Journal of Economic Theory* 38, 324–63.

Kim, T. 1987. Intransitive indifference and revealed preference. *Econometrica*.

Mas-Colell, A. 1977. The recoverability of consumers' preferences from market demand behavior. *Econometrica* 45, 1409–30.

Mas-Colell, A. 1978. On revealed preference analysis. *Review of Economic Studies* 45, 121–31.

Matzkin, R. and Richter, M.K. 1986. Testing concave rationality. Department of Economics, University of Minnesota, Minneapolis.

McFadden, D. and Richter, M.K. 1970. Stochastic rationality and revealed stochastic preference. Presented to the 1970 Winter Meetings of the Econometric Society.

McFadden, D. and Richter, M.K. 1988. Stochastic rationality and revealed stochastic preference. In *Uncertainty, Preferences and Optimality. Essays in Honor of Leonid Hurwicz*, ed. J.S. Chipman, D. McFadden, and M.K. Richter, New York.

McFadden, D., Mas-Colell, A., Mantel, R. and Richter, M.K. 1974. A characterization of community excess demand functions. *Journal of Economic Theory* 9, 361–74.

Plott, C.R. 1973. Path independence, rationality, and social choice. *Econometrica* 41, 1075–91.

Richter, M.K. 1966. Revealed preference theory. *Econometrica* 34, 635–45.

Richter, M.K. 1971. Rational Choice. In *Preferences, Utility, and Demand*, ed. J.S. Chipman, L. Hurwicz, M.K. Richter and H.F. Sonnenschein. New York: Harcourt, Brace, Jovanovich, ch. 2.

Richter, M.K. 1979. Duality and rationality. *Journal of Economic Theory* 20, 131–81.

Richter, M.K. 1986. Continuous demand functions. Department of Economics, University of Minnesota, Minneapolis.

Rose, H. 1958. Consistency of preference: the two-commodity case. *Review of Economic Studies* 25, 124–5.

Sakai, Y. 1977. Revealed favorability, indirect utility, and direct utility. *Journal of Economic Theory* 14, 113–29.

Samuelson, P.A. 1938a. A note on the pure theory of consumer's behaviour. *Economica* NS 5, 61–71.

Samuelson, P.A. 1938b. A note on the pure theory of consumer's behaviour: an addendum. *Economica* NS 5, 353–4.

Samuelson, P.A. 1948. Consumption theory in terms of revealed preference. *Economica* NS 15, 243–53.

Samuelson, P.A. 1950. The problem of integrability in utility theory. *Economica* NS 17, 355–85.

Samuelson, P.A. 1953. Consumption theorems in terms of overcompensation rather than indifference comparisons. *Economica* NS 20, 1–9.

Samuelson, P.A. 1947. *Foundations of Economic Analysis.* Cambridge, Mass.: Harvard University Press.

Shafer, W.J. 1977a. Revealed preference and aggregation. *Econometrica* 45, 1173–82.

Shafer, W.J. 1977b. Revealed preference cycles and the Slutsky matrix. *Journal of Economic Theory* 16, 293–309.

Uzawa, H. 1957. Note on preference and axioms of choice. *Annals of the Institute of Statistical Mathematics* 8, 35–40.

Uzawa, H. 1960. Preference and rational choice in the theory of consumption. In *Mathematical Methods in the Social Sciences, 1959*, ed. K.J. Arrow, S. Karlin, and P. Suppes. Stanford: Stanford University Press, ch. 9.

Uzawa, H. 1971. Preference and rational choice in the theory of consumption. In *Preferences, Utility, and Demand*, ed. J.S. Chipman, L. Hurwicz, M.K. Richter and H.F. Sonnenschein. New York: Harcourt, Brace, Jovanovich, ch. 1.

Varian, H.R. 1983. Non-parametric tests of consumer behaviour. *Review of Economic Studies* 50, 99–110.

Ville, J. 1946. Sur les conditions d'existence d'une ophélimité totale et d'un indice du niveau des prix. *Annales de l'Université de Lyon* 9, Sec. A(3), 32–9.

Ville, J. 1951. The existence conditions of a total utility function. *Review of Economic Studies* 19, 123–8.

Wald, A. 1936. Über einige Gleichungssyteme der mathematischen ökonomie. *Zeitschrift für Nationalökonomie* 7, 637–70.

Wald, A. 1951. On some systems of equations of mathematical economics. *Econometrica* 19, 368–403.

Yamazaki, A. 1984. The critical set of a demand correspondence in the price space and the weak axiom of revealed preference. *Hitotsubashi Journal of Economics* 25, 137–44.

revelation principle

Communication is central to the economic problem (Hayek, 1945). Opportunities for mutually beneficial transactions cannot be found unless individuals share information about their preferences and endowments. Markets and other economic institutions should be understood as mechanisms for facilitating communication. However, people cannot be expected to reveal information when it is against their interests; for example, a seller may conceal his willingness to sell at a lower price. Rational behaviour in any specific communication mechanism can be analysed using game-theoretic equilibrium concepts, but efficient institutions can be identified only by comparison with all possible communication mechanisms. The revelation principle is a technical insight that allows us, in any given economic situation, to make general statements about all possible communication mechanisms.

The problem of making statements about all possible communication systems might seem intractably complex. Reports and messages may be expressed in rich languages with unbounded vocabulary. Communication systems can include both public announcements and private communication among smaller groups. Communication channels can have noise that randomly distorts messages. A communication mechanism may also specify how contractually enforceable transactions will depend on agents' reports and messages. So a general communication mechanism for any given set of agents may specify (*a*) a set of possible reports that each agent can send, (*b*) a set of possible messages that each agent can receive from the communication system, and (*c*) a probabilistic rule for determining the messages received and the enforceable transactions as a function of the reports sent by the agents. However, the revelation principle tells us that, for many economic purposes, it is sufficient for us to consider only a special class of mechanisms, called 'incentive-compatible direct-revelation mechanisms'.

In these mechanisms, every economic agent is assumed to communicate only with a central mediator. This mediator may be thought of as a trustworthy person or as a computer at the centre of a telephone network. In a direct-revelation mechanism, each individual is asked to report all of his private information confidentially to the mediator. After receiving these reports, the mediator then specifies all contractually enforceable transactions, as a function of these reports. If any individual controls private actions that are not contractually enforceable (such as efforts that others cannot observe), then the mediator also confidentially recommends an action to the individual. A direct-revelation mechanism is any rule for specifying how the mediator determines these contractual transactions and privately recommended actions, as a function of the private-information reports that the mediator receives.

A direct-revelation mechanism is said to be 'incentive compatible' if, when each individual expects that the others will be honest and obedient to the mediator, then no individual could ever expect to do better (given the information available to him) by reporting dishonestly to the mediator or by disobeying the mediator's recommendations. That is, the mechanism is incentive compatible if honesty and obedience is an equilibrium of the resulting communication game. The set of incentive-compatible direct-revelation mechanisms has good mathematical properties that often make it easy to analyse because it can be defined by a collection of linear inequalities, called 'incentive constraints'. Each of these incentive constraints expresses a requirement that an individual's expected utility from using a dishonest or disobedient strategy should not be greater than the individual's expected utility from being honest and obedient, when it is anticipated that everyone else will be honest and obedient.

The analysis of such incentive-compatible direct-revelation mechanisms might seem to be of limited interest, because real institutions rarely use such fully centralized mediation and often generate incentives for dishonesty or disobedience. For any equilibrium of any general communication mechanism, however, there exists an incentive-compatible direct-revelation mechanism that is essentially equivalent. This proposition is the revelation principle. Thus, the revelation principle tells us that, by analysing the set of incentive-compatible direct-revelation mechanisms, we can derive general properties of all equilibria of all coordination mechanisms.

The terms 'honesty' and 'obedience' here indicate two fundamental aspects of the general economic problem of communication. In a general communication system, an individual may send out messages or reports to share information that he knows privately, and he may also receive messages or recommendations to guide actions that he controls privately. The problem of motivating individuals to report their private information honestly is called 'adverse selection', and the problem of motivating individuals to implement their recommended actions obediently is called 'moral hazard'. To describe the intuition behind the revelation principle, let us consider first the special cases where only one or the other of these problems exists.

Pure adverse selection

First, let us formulate the revelation principle for the case of pure adverse selection, as developed in Bayesian social choice theory. In this case we are given a set of individuals, each of whom has some initial private information that may be called the individual's 'type', and there is a planning question of how a social allocation of resources should depend on the individuals' types. Each individual's payoff can depend on the resource allocation and on the types of all individuals according to some given utility function, and each type of each individual has some given probabilistic beliefs about the types of all other individuals. A general communication system would allow each individual i to send a message m_i in some rich language, and then the chosen resource allocation would depend on all these messages according to some rule $\gamma(m_1, \ldots m_n)$. In any equilibrium of the game defined by this communication system, each individual i must have some strategy σ_i for choosing his message as a function of his type t_i, so that $m_i = \sigma_i(t_i)$.

For the given equilibrium $(\sigma_1, \ldots, \sigma_n)$ of the given social-choice rule γ, the revelation principle is satisfied by a mediation plan in which each individual is asked to confidentially report his type t_i to a central mediator, who then implements the social choice

$$\mu(t_1, \ldots t_n) = \gamma(\sigma_1(t_1), \ldots, \sigma_n(t_n)).$$

So the mediator computes what message would be sent by the reported type of each individual i under his or her strategy σ_i, and then the mediator implements the resource allocation that would result from these messages under the rule γ. It is easy to see that honesty is an equilibrium under this mediation plan μ. If any individual could gain by lying to this mediator, when all others are expected to be honest, then this individual could have also gained by lying to himself when implementing his equilibrium strategy σ_i under the given mechanism γ, which would contradict the optimality condition that defines an equilibrium. So μ is an incentive-compatible direct-revelation mechanism that is equivalent to the given general mechanism γ with the given equilibrium $(\sigma_1, \ldots, \sigma_n)$.

In this case of pure adverse selection, the revelation principle was introduced by Gibbard (1973), but for a narrower solution concept (dominant strategies, instead of Bayesian equilibrium). The revelation principle for the broader solution concept of Bayesian equilibrium was recognized by Dasgupta, Hammond and Maskin (1979), Harris and Townsend (1981), Holmstrom (1977), Myerson (1979), and Rosenthal (1978).

Pure moral hazard

Next let us formulate the revelation principle for the case of pure moral hazard, as developed in Aumann's (1974) theory of correlated equilibrium. In this case we are given a set of individuals, each of whom controls some actions, and each individual's payoff can depend on the actions $(c_1, \ldots, c_n)$ that are chosen by all individuals, according to some given utility function $u_i(c_1, \ldots c_n)$. That is, we are given a game in strategic form. In this case of pure moral hazard, nobody has any private information initially, but a communication process could give individuals different information before they choose their actions. In a general communication system, each individual i could get some message m_i in some rich language, with these messages $(m_1, \ldots, m_n)$ being randomly drawn from some joint probability distribution ρ. In any equilibrium of the game generated by adding this communication system, each individual i has some strategy σ_i for choosing his action c_i as a function of his message m_i, so that $c_i = \sigma_i(m_i)$.

For the given equilibrium $(\sigma_1, \ldots, \sigma_n)$ of the game with the given communication system ρ, the revelation principle is satisfied by a mediation plan in which the mediator randomly generates recommended actions in such a way that the probability of recommending actions $(c_1, \ldots, c_n)$ is the same as the probability of the given communication system ρ yielding messages $(m_1, \ldots, m_n)$ that would induce the players to choose $(c_1, \ldots, c_n)$ in the σ equilibrium. That is, the probability $\mu(c_1, \ldots, c_n)$ of the mediator recommending $(c_1, \ldots, c_n)$ is

$$\mu(c_1, \ldots, c_n) = \rho(\{(m_1, \ldots, m_n) | \sigma_1(m_1) = c_1, \ldots, \sigma_n(m_n) = c_n\}).$$

Then the mediator confidentially tells each individual i only which action c_i is recommended for him. Obedience is an equilibrium under this mediation plan μ because, if any individual could gain by disobeying this mediator when all others are expected to be obedient, then this individual could have also gained by disobeying himself in implementing his equilibrium strategy σ_i in the given game with communication system ρ. So μ is an incentive-compatible direct-revelation mechanism that is equivalent to the given mechanism ρ with the given equilibrium $(\sigma_1, \ldots, \sigma_n)$.

General formulations

Problems of adverse selection and moral hazard can be combined in the framework of Harsanyi's (1967) Bayesian games, where players have both types and actions. The revelation principle for general Bayesian games was formulated by Myerson (1982; 1985). A further generalization of the revelation principle to multistage games was formulated by Myerson (1986). In each case, the basic idea is that any equilibrium of any general communication system can be simulated by a maximally centralized communication system in which, at every stage, each individual confidentially reports all his private information to a central mediator, and then the mediator confidentially recommends an action to each individual, and the mediator's rule for generating recommendations from reports is designed so that honesty and obedience form an equilibrium of the mediated communication game.

The basic assumption here is that, although the motivations of all economic agents are problematic, we can find a mediator who is completely trustworthy and has no costs of processing information. Asking agents to reveal all relevant information to the trustworthy mediator maximizes the mediator's ability to implement any coordination plan. But telling any other agent more than is necessary to guide his choice of action would only increase the agent's ability to find ways of profitably deviating from the coordination plan.

For honesty and obedience to be an equilibrium, the mediation plan must satisfy incentive constraints which say that no individual could ever expect to gain by deviating to a strategy that involves lying to the mediator or disobeying a recommendation from the mediator. In a dynamic context, we must consider that an individual's most profitable deviation from honesty and obedience could be followed by further deviations in the future. So, to verify that an individual could never gain by lying, we must consider all possible deviation strategies in which the individual may thereafter choose actions that can depend disobediently on the mediator's recommendations (which may convey information about others' types and actions).

When we use sequential equilibrium as the solution concept for dynamic games with communication, the set of actions that can be recommended in a sequentially incentive-compatible mechanism must be restricted somewhat. In a Bayesian game, if some action d_i could never be optimal for individual i to use when his type is t_i, no matter what information he obtained about others' types and actions, then obedience could not be sequentially rational in any mechanism where the mediator might ever recommend this action d_i to i after he reports type t_i. Myerson (1986) identified a larger set of *co-dominated actions* that can never be recommended in any sequentially incentive-compatible mechanism. Suppose that, if any individual observed a zero-probability event, then he could attribute this surprise to a mistake by the trembling hand of the mediator. Under this assumption, Myerson (1986) showed that the effect of requiring sequential rationality in games with communication is completely characterized by the requirement that no individuals should ever be expected to choose any co-dominated actions. (See Gerardi and Myerson, 2007.)

Limitations

The revelation principle says that each equilibrium of any communication mechanism is equivalent to the honest-obedient equilibrium of an incentive-compatible direct-revelation mechanism. But this direct-revelation mechanism may have other dishonest equilibria, which might not correspond to equilibria of the original mechanism. So the revelation principle cannot help us when we are concerned about the whole set of equilibria of a communication mechanism. Similarly, a given communication mechanism may have equilibria that change in some desirable way as we change the players' given beliefs about each others' types, but these different equilibria would correspond to different incentive-compatible mechanisms, and so this desirable property of the given mechanism could not be recognized with the revelation principle.

The assumption that perfectly trustworthy mediators are available is essential to the mathematical simplicity of the incentive-compatible set. Otherwise, if individuals can communicate only by making public statements that are immediately heard by everybody, then the set of equilibria may be smaller and harder to compute.

In principal-agent analysis we often apply the revelation principle to find the incentive-compatible mechanism that is optimal for the principal. If the principal would be tempted to use revealed information opportunistically, then there could be loss of generality in assuming that the agents reveal all their private information to the principal. But we should not confuse the principal with the mediator. The revelation principle can still be applied if the principal can get a trustworthy mediator to take the agents' reports and use them according to any specified mechanism.

There are often questions about whether the allocation selected by a mechanism could be modified by subsequent exchanges among the individuals. An individual's right to offer his possessions for sale at some future date could be accommodated in mechanism design by additional moral-hazard constraints.

For example, suppose the principal can sell an object each day, on days 1 and 2. The only buyer's value for such objects is either low $1 or high $3, low having probability 0.25. To maximize the principal's expected revenue with the buyer participating honestly, an optimal mechanism would sell both objects for $3 if the buyer's type is high, but would sell neither if the buyer is low. But if no sale is recommended then the principal could infer that the buyer is low and would prefer to sell for $1. Suppose now that the principal cannot be prevented from offering to sell for $1 on either day. With these additional moral-hazard constraints, an optimal mechanism uses randomization by the mediator to conceal information from the principal. If the buyer reports low then the mediator recommends no sale on day 1 and selling for $1 on day 2. If the buyer reports high, then with probability 1/3 the mediator recommends no sale on day 1 and selling for $3 on day 2, but with probability 2/3 recommends selling for $1.50 on both days. A no-sale recommendation on day 1 implies probability 0.5 of low, so that obedience yields the same expected revenue $0.5 \times (0 + 1) + 0.5 \times (0 + 3)$ as deviating to sell for $1 on both days.

A proliferation of such moral-hazard constraints may greatly complicate the analysis, however. So in practice we often apply the revelation principle with an understanding that we may be overestimating the size of the feasible set, by assuming away some problems of mediator imperfection or moral hazard. When we use the revelation principle to show that a seemingly wasteful mechanism is actually efficient when incentive constraints are recognized, such overestimation of the incentive-feasible set would not weaken the impact of our results (as long as this mechanism remains feasible).

Centralized mediation is emphasized by the revelation principle as a convenient way of characterizing what people can achieve with communication, but this analytical convenience does not imply that centralization is necessarily the best way to coordinate an economy. For fundamental questions about socialist centralization versus free-market decentralization, we should be sceptical about an assumption that centralized control over national resources could not corrupt any mediator. The power of the revelation principle for such questions is instead its ability to provide a common analytical framework that applies equally to socialism and capitalism. For example, a standard result of revelation-principle analysis is that, if only one producer knows the production cost of a good, then efficient incentive-compatible mechanisms must allow this monopolistic producer to take positive informational rents or profits (Baron and Myerson, 1982). Thus the revelation principle can actually be used to support arguments for decentralized multi-source production, by showing that problems of profit-taking by an informational monopolist can be just as serious under socialism as under capitalism.

Nash (1951) advocated a different methodology for analysing communication in games. In Nash's approach, all opportunities for communication should be represented by moves in our extensive model of the dynamic game. Adding such communication moves may greatly increase the number of possible strategies for a player, because each strategy is a complete plan for choosing the player's moves throughout the game. But if all communication will occur in the implementation of these strategies, then the players' initial choices of their strategies must be independent. Thus, Nash argued, any dynamic game can be normalized to a static strategic-form game, where players choose strategies simultaneously and independently, and Nash equilibrium is the general solution for such games.

With the revelation principle, however, communication opportunities are omitted from the game model and are instead taken into account by using incentive-compatible mechanisms as our solution concept. Characterizing the set of all incentive-compatible mechanisms is often easier than computing the Nash equilibria of a game with communication. Thus, by applying the revelation principle, we can get both a simpler model and a simpler solution concept for games with communication. But, when we use the revelation principle, strategic-form games are no longer sufficient for representing general dynamic games, because normalizing a game model to strategic form would suppress implicit opportunities for communicating during the game (see Myerson, 1986). So the revelation principle should be understood as a methodological alternative to Nash's strategic-form analysis.

ROGER MYERSON

See also **mechanism design.**

Bibliography

Aumann, R. 1974. Subjectivity and correlation in randomized strategies. *Journal of Mathematical Economics* 1, 67–96.

Baron, D. and Myerson, R. 1982. Regulating a monopolist with unknown costs. *Econometrica* 50, 911–30.

Dasgupta, P., Hammond, P. and Maskin, E. 1979. The implementation of social choice rules: some results on incentive compatibility. *Review of Economic Studies* 46, 185–216.

Gerardi, D. and Myerson, R. 2007. Sequential equilibrium in Bayesian games with communication. *Games and Economic Behavior* 60, 104–34.

Gibbard, A. 1973. Manipulation of voting schemes: a general result. *Econometrica* 41, 587–601.

Harris, M. and Townsend, R. 1981. Resource allocation under asymmetric information. *Econometrica* 49, 1477–99.

Harsanyi, J. 1967. Games with incomplete information played by Bayesian players. *Management Science* 14, 159–82, 320–34, 486–502.

Hayek, F. 1945. The use of knowledge in society. *American Economic Review* 35, 519–30.

Holmstrom, B. 1977. On incentives and control in organizations. Ph.D. thesis, Stanford University.

Myerson, R. 1979. Incentive-compatibility and the bargaining problem. *Econometrica* 47, 61–73.

Myerson, R. 1982. Optimal coordination mechanisms in generalized principal-agent problems. *Journal of Mathematical Economics* 10, 67–81.

Myerson, R. 1985. Bayesian equilibrium and incentive compatibility: an introduction. In *Social Goals and Social Organization*, ed. L. Hurwicz, D. Schmeidler and H. Sonnenschein. Cambridge: Cambridge University Press.

Myerson, R. 1986. Multistage games with communication. *Econometrica* 54, 323–58.

Nash, J. 1951. Non-cooperative games. *Annals of Mathematics* 54, 286–95.

Rosenthal, R. 1978. Arbitration of two-party disputes under uncertainty. *Review of Economic Studies* 45, 595–604.

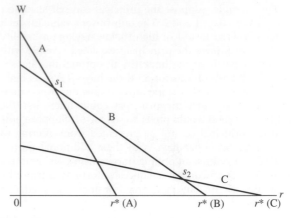

Figure 1

reverse capital deepening

It has long been taken for granted that there is an inverse monotonic relationship between the rate of interest (or the rate of profit) and the quantity of capital per worker. This belief was founded on the principle of substitution, whereby 'cheaper' is substituted for 'more expensive' as the relative price of two inputs is changed.

In the field of capital theory, the principle of substitution persuaded many economists, such as E. von Böhm-Bawerk (1889), J.B. Clark (1899) and F.A. von Hayek (1941), that a lower rate of interest (which is equal to the rate of profit in equilibrium) is associated with the use of more 'capital intensive' techniques, and thus with the substitution of capital for other productive factors, such as labour or land. This process is called capital deepening.

Recent discussions have shown that this is not necessarily true, since a lower rate of interest might be associated with *lower*, rather than higher, capital per worker. This phenomenon is called *reverse capital deepening*.

This discovery was made at the same time as it was realized that it is not generally possible to order 'efficient' techniques in such a way that technical choice becomes a monotonic function of the rate of interest (and of the rate of profit).

It can be shown that both reverse capital deepening and reswitching of technique are related to the same fundamental property of the economic system: the possibility (in fact, the near generality) of nonlinear wage–profit relationships. To illustrate this proposition, it is useful to begin by considering the hypothetical case of linear wage–profit relationships (see Figure 1).

The linearity of the three wage–profit relationships makes reswitching impossible as r increases between 0 and $r^*(C)$ (which is the maximum rate of profit with technology C). The reason is that no wage–profit line can ever be crossed more than once by another wage–profit line. In this special case, there is an inverse monotonic relationship between the rate of profit and the quantity of capital per worker. This may be shown as follows. We can read the net final output per worker on the w-axis of Figure 1

at the point at which $r = 0$. (At that point the net final output per worker coincides with the maximum wage.) The net final output per worker associated with technology A is higher than the net final output per worker associated with technology B. The net final output per worker associated with technology B is higher than the net final output per worker associated with technology C. At switchpoints s_1 and s_2 the wage is the same for both technologies between which substitution takes place. It follows that, at switchpoint s_1, profit per worker is higher with technology A than with technology B. Similarly, at switchpoint s_2, profit per worker is higher with technology B than with technology C. Assuming that the rate of profit is uniform across technologies, we find that, at s_1, A is associated with higher capital per worker than B. A higher rate of profit (or rate of interest) is thus associated with substitution of 'less capital' for 'more capital'. In this particular case, the traditional approach to capital theory would seem to be well founded.

However, these properties disappear altogether once we drop the assumption of linear wage–profit relationships. (It might be interesting to inquire into the economic meaning of straight wage–profit relationships, which are possible only in the case of a technology characterized by a uniform proportion between labour and intermediate inputs in all production processes: only in this case a change in the rate of profit leaves relative prices unaffected.)

But in general wage–profit relationships are of the nonlinear type, which means that the proportion between labour and intermediate inputs is generally different from one production process to another. This feature of the wage–profit frontier makes it possible for wage–profit curves to intersect more than once, thus bringing about the possibility of multiple switching. Under the same circumstances it can be shown that the relationship between the rate of profit and capital per worker is no longer of the inverse monotonic type. This can be seen in the reswitching case (Figure 2), but it can

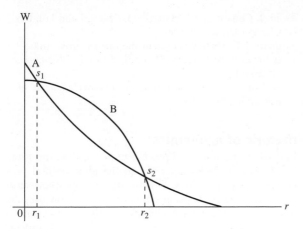

Figure 2

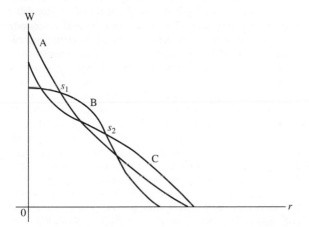

Figure 3

higher capital per worker (C) for a technology with lower capital per worker (B).

Complementarity in production is often at the root of apparently perverse price behaviour (see Broome, 1978). One reason for this had been noted by John Hicks, when he wrote that the 'net effect' of a change in the price of productive factor x upon the price of complementary factor y 'is ... compounded out of two contrary tendencies, a direct effect tending to raise it, and indirect effect tending to reduce it; either may be dominant' (Hicks, 1946, p. 107). Reverse capital deepening is an especially important instance of a widespread phenomenon associated with indirect effects in a production economy. This possibility is associated with other phenomena which are not compatible with traditional beliefs about capital and capital accumulation. Simple inspection of Figure 2 or 3 shows that at a switchpoint associated with reverse capital deepening (s_2 in either figure) a technology with higher capital per worker and higher net final product per worker is substituted for a technology with lower capital per worker and lower net final product per worker. At such switchpoints a higher rate of profit (and rate of interest) could be associated with a higher ratio of capital per worker to net final product per worker, that is, with a higher capital–output ratio.

Figures 2 and 3 also alert us as to the possibility that a technology adopted at a high rate of interest is associated with higher maximum consumption per head than a technology adopted at a lower rate of interest. In addition, transition to a lower rate of interest may involve the switch to a lower maximum consumption per head. (This can be seen at switchpoint s_2 in either figure, where maximum consumption per head can be read on the w-axis at point $r = 0$.) This behaviour of consumption per head in relation to the rate of interest is clearly incompatible with the view that a higher rate of interest brings about a special type of exchange, in which less consumption in the current period is substituted against higher consumption in the future. Reverse capital deepening alerts us as to the possibility that a higher rate of interest might be associated with greater current consumption per head than the consumption per head feasible with the technology adopted at a lower rate of interest (see Bruno, Burmeister and Sheshinski, 1966; Samuelson, 1966).

The relevance of reverse capital deepening is that the foundation of the traditional view that technical choice is a monotonic function of the rate of interest is seriously questioned. Similarly, the widespread policy implications of the traditional view are also questioned. However, there is as yet no full agreement as to the main consequences of this result. For example, Christopher Bliss (1975, p. 279) has noted that reverse capital deepening makes it impossible to see the accumulation of capital as a process associated with 'a continuous increase in consumption per capita,...a continuous decline in the rate of interest and ... a continuous increase in the real wage rate'. He also called attention to the fact that the 'extended accumulation

also be seen in the case in which the wage–profit curves never intersect more than once on the efficiency frontier (Figure 3). (See also Pasinetti, 1966.)

In Figure 2, reswitching is associated with reverse capital deepening. Technology A is the more profitable at levels of the rate of profit lower than r_1, it is overtaken by technology B at rates of profit between r_1 and r_2, it becomes again the more profitable at rates of profit higher than r_2. At the same time, switchpoint s_1 is associated with the substitution of the technology with lower value of capital per worker (B) for the technology with the higher value of capital per worker (A), whereas at switchpoint s_2 the opposite happens: the technology with higher capital per worker (A) is substituted for the technology with lower capital per worker (B), in spite of the fact that the rate of profit is higher (reverse capital deepening).

In Figure 3, there is no reswitching but we still have a reverse capital deepening. For no wage–profit curves cross one another more than once on the efficiency frontier, but at switchpoint s_2 an increasing rate of profit is associated with the substitution of a technology with history' of an economic system moving through real time is normally different from the hypothetical history we can tell by 'travelling' across steady states (1975, pp. 194, 280–1). In particular, he emphasized that the statement that the rate of interest may be expected to fall as capital deepening takes place 'cannot be interpreted' in the case of extended accumulation history, as we would have, in that case, 'a whole structure of interest rates ... not a single rate of interest' (1975, p. 294). A different point of view has been expressed by Pierangelo Garegnani, who has maintained that capital paradoxes in general, and reverse capital deepening in particular, by making traditional beliefs untenable, suggest a 'correction' to traditional theory, which would make it reasonable to expect 'instabilities or tendencies to zero of wages, or of net returns on capital' (Garegnani, 1990, p. 76). This author's view is that, rather than introducing such a correction, one should drop any idea of a causal connection from marginal products to the distribution of the social product, and further develop the conjecture that distribution is brought about by 'more complex economic and social forces like those envisaged by the old classical economists' (Garegnani, 1990, pp. 76–7). As it is common in theoretical sciences (see Kuhn, 1970; 2000), the discovery of an apparent anomaly has induced economists to look for a more general theory, or to switch to an altogether different framework.

R. SCAZZIERI

See also **capital theory; capital theory (paradoxes); production functions; reswitching of technique.**

Bibliography

Bliss, C.J. 1975. *Capital Theory and the Distribution of Income.* Amsterdam, Oxford, New York: North-Holland Publishing Company and American Elsevier Publishing Company.
Böhm-Bawerk, E. von. 1889. *Positive Theorie des Kapitales.* Zweite Ableitung: *Kapital und Kapitalzins.* Trans. as *The Positive Theory of Capital*, London: Macmillan, 1891.
Broome, J. 1978. Perverse prices. *Economic Journal* 88, 778–87.
Bruno, M., Burmeister, E. and Sheshinski, E. 1966. The nature and implications of the reswitching of techniques. *Quarterly Journal of Economics* 80, 526–53.
Clark, J.B. 1899. *The Distribution of Wealth.* New York: Macmillan.
Garegnani, P. 1990. Quantity of capital. In *The New Palgrave: Capital Theory*, ed. J. Eatwell et al. London and Basingstoke: Macmillan.
Hayek, F.A. von. 1941. *The Pure Theory of Capital.* London: Routledge.
Hicks, J. 1946. *Value and Capital*, 2nd edn. Oxford: Clarendon Press.
Kuhn, T.S. 1970. *The Structure of Scientific Revolutions*, 2nd edn enlarged. Chicago: University of Chicago Press.
Kuhn, T.S. 2000. *The Road Since Structure: Philosophical Essays, 1970– 1993*, with an autobiographical interview, ed. J. Conant and J. Haugeland. Chicago and London: Chicago University Press.
Pasinetti, L.L. 1966. Changes in the rate of profit and switches of techniques. *Quarterly Journal of Economics* 80, 503–17.
Samuelson, P. 1966. A summing up. *Quarterly Journal of Economics* 80, 568–83.

rhetoric of economics

Rhetoric is the study and practice of persuasive expression, an alternative since the Greeks to the philosophical programme of epistemology. The rhetoric of economics examines how economists persuade – not how they say they do, or how their official methodologies say they do, but how in fact they persuade colleagues and politicians and students to accept one economic assertion and reject another.

Some of their devices arise from bad motives, and bad rhetoric is what most people have in mind when they call a piece of writing 'rhetorical'. An irrelevant and inaccurate attack on Milton Friedman's politics while criticizing his economics would be an example, as would a pointless and confusing use of mathematics while arguing a point in labour economics. The badness does not reside in the techniques themselves (political commentary or mathematical argument) but in the person using them, since all techniques can be abused. Aristotle noted that 'if it be objected that one who uses such power of speech unjustly might do great harm, *that* is a charge which may be made in common against all good things except virtue itself'. Cato the Elder demanded that the user of analogy (or in our time the user of regression) be *vir bonus dicendi peritus*, the *good* man skilled at speaking. The protection against bad science is good scientists, not good methodology.

Rhetoric, then, can be good, offering good reasons for believing that the elasticity of substitution between capital and labour in American manufacturing, say, is about 1.0. The good reasons are not confined by syllogism and number. They include good analogy (production is *just like* a mathematical function), good authority (Knut Wicksell and Paul Douglas thought this way, too), good symmetry (if mining can be treated as a production function, so should manufacturing). Furthermore, the reasonings of syllogism and number are themselves rhetorical, that is, persuasive acts of human speech. An econometric test will depend on how apt is an analogy of the error term with drawings from an urn. A mathematical proof will depend on how convincing is an appeal to the authority of the Bourbaki style. 'The facts' and 'the logic' matter, of course; but they are part of the rhetoric, depending themselves on the giving of good reasons.

Consider, for example, the sentence in economics, 'The demand curve slopes down'. The official rhetoric says that economists believe this because of statistical evidence – negative coefficients in demand curves for pig iron or negative diagonal items in matrices of complete systems of demand – accumulating steadily in journal articles. These are the tests 'consistent with the hypothesis'. Yet most belief in the hypothesis comes from other sources: from introspection (what would I do?); from thought experiments (what would they do?); from uncontrolled cases in point (such as the oil crisis); from authority (Alfred Marshall believed it); from symmetry (a law of demand if there is a law of supply); from definition (a higher price leaves less for expenditure, including this one); and above all, from analogy (if the demand curve slopes down for chewing gum, why not for housing and love too?). As may be seen in the classroom and seminar, the range of argument in economics is wider than the official rhetoric allows.

The rhetoric of economics brings the traditions of rhetoric to the study of economic texts, whether mathematical or verbal texts. It is a literary criticism of economics, or a jurisprudence, and from literary critics like Wayne Booth (1974) and lawyers such as Chaim Perelman (Perelman and Olbrechts-Tyteca, 1958) much can be learned. Although its precursors in economics are methodological criticisms of the field (such as Frank Knight, 1940), censorious joking (such as Stigler, 1977), and finger-wagging presidential addresses (such as Leontief, 1971, or Mayer, 1980), the main focus of the work has been the analysis of how economists seek to persuade, whether good or bad (Klamer, 1984; Henderson 1982; Kornai, 1983; McCloskey, 1986). Econometrics has its own rhetorical prehistory, more self-conscious than the rest (Leamer, 1978), reaching back to the founders of decision theory and Bayesian statistics.

The movement has parallels in other fields. Imre Lakatos (1976), Davis and Hersh (1981), and others have uncovered a rhetoric in mathematics; Rorty (1982), Toulmin (1958), and Rosen (1980) in technical philosophy; and numbers of scientists in their own fields (Polanyi, 1962; Medawar, 1964). Historians and sociologists of science have since the 1960s accumulated much evidence that science is a conversation rather than a mechanical procedure (Kuhn, 1977; Collins, 1985). The analysis of conversation from scholars in communication and literary studies (Scott, 1967) has provided ways of rereading various fields (a sampling of these is contained in Nelson, Megill and McCloskey, 1987).

A rhetoric of economics questions the division between scientific and humanistic reasoning, not to attack quantification or to introduce irrationality into science, but to make the scientific conversation more aware of itself. It is a programme of greater, not less, rigour and relevance, of higher, not lower, standards in the conversations of mankind.

DONALD N. MCCLOSKEY

See also **philosophy and economics**

Bibliography

Booth, W. 1974. *Modern Dogma and the Rhetoric of Assent.* Chicago: University of Chicago Press.

Collins, H.M. 1985. *Changing Order: Replication and Induction in Scientific Practice.* London: Sage.

Davis, P.J. and Hersh, R. 1981. *The Mathematical Experience.* Boston: Houghton Mifflin.

Henderson, W. 1982. Metaphors in economics. *Economics* 18(4), 147–53.

Klamer, A. 1984. *Conversations with Economists: New Classical Economists and Opponents Speak Out on the Current Controversy in Macroeconomics.* Totowa, NJ: Rowman and Allanheld.

Knight, F. 1940. 'What is truth' in economics? *Journal of Political Economy* 48, 1–32.

Kornai, J. 1983. The health of nations: reflections on the analogy between medical science and economics. *Kyklos* 36(2), 191–212.

Kuhn, T. 1977. *The Essential Tension: Selected Studies in Scientific Tradition and Change.* Chicago: University of Chicago Press.

Lakatos, I. 1976. *Proofs and Refutations: the Logic of Mathematical Discovery.* Cambridge: Cambridge University Press.

Leamer, E. 1978. *Specification Searches: Ad Hoc Inferences with Nonexperimental Data.* New York: Wiley.

Leontief, W. 1971. Theoretical assumptions and nonobserved facts. *American Economic Review* 61, 1–7.

Mayer, T. 1980. Economics as a hard science: realistic goal or wishful thinking? *Economic Inquiry* 18, 165–78.

McCloskey, D.N. 1986. *The Rhetoric of Economics.* Madison: University of Wisconsin Press.

Medawar, P. 1964. Is the scientific paper fraudulent? *Saturday Review*, 1 August.

Nelson, J., Megill, A. and McCloskey, D.N., eds. 1987. *The Rhetoric of the Human Sciences: Papers and Proceedings of the Iowa Conference.* Madison: University of Wisconsin Press.

Perelman, C. and Olbrechts-Tyteca, L. 1958. *The New Rhetoric: A Treatise on Argumentation.* Notre Dame, IN: University of Notre Dame Press.

Polanyi, M. 1962. *Personal Knowledge: Towards a Post-critical Philosophy.* Chicago: University of Chicago Press.

Rorty, R. 1982. *The Consequences of Pragmatism: Essays.* Minneapolis: University of Minnesota Press.

Rosen, S. 1980. *The Limits of Analysis.* New York: Basic Books.

Scott, R. 1967. On viewing rhetoric as epistemic. *Central States Speech Journal* 18(1), 9–17.

Stigler, G.J. 1977. The conference handbook. *Journal of Political Economy* 85, 441–3.

Toulmin, S. 1958. *The Uses of Argument.* Cambridge: Cambridge University Press.

Ricardo, David (1772–1823)

David Ricardo was one of the most outstanding political economists of the 19th century and one of the most influential of all time. Born on 18 April 1772 at 36 Broad Street Buildings in the City of London, he was the third of

15 surviving children of Abraham Israel Ricardo (1733–1812) and his wife, Abigail Delvalle (1753–1801). Abraham's family were Sephardic Jews who had emigrated from Portugal to Holland at the end of the 16th century. His father (David's grandfather) is described as 'a non-official broker in funds and stocks' on the Amsterdam exchange (Heertje, 2004, p. 283). Abraham also became a stockbroker, first in Amsterdam and then in London, where he moved in 1760. He married Abigail (from an established London Sephardic family of tobacco and snuff merchants) in 1769 and was granted citizenship in 1771. As related by David's brother, Moses Ricardo, their father

> was a man of good intellect, but uncultivated. His prejudices were exceedingly strong; and they induced him to take the opinions of his forefathers in points of religion, politics, education &c., upon faith, and without investigation. Not only did he adopt this rule for himself, but he insisted on its being followed by his children; his son [David], however, never yielded his assent on any important subject, until after he had thoroughly investigated it. It was perhaps in opposing these strong prejudices that he was first led to that freedom and independence of thought for which he was so remarkable. (Ricardo, 1951–55, *Works*, X, p. 5; hereafter '*Works*')

According to Moses Ricardo, the young David was allotted a 'common-school education', typical of 'those who are destined for a mercantile line of life' (*Works* X, p. 3), the emphasis being on reading, writing and arithmetic. Less typically, at the age of 11 David was sent to Amsterdam for two years. Details of the visit are sketchy. It has been suggested that he attended the famous Talmud Tora, attached to the Portuguese Synagogue in Amsterdam, although recent scholarship has favoured a more mundane account in which he was privately tutored in the 'common-school' subjects with the addition of French and Spanish (Heertje, 2004). Following his return to London, David's full-time education continued only until he reached the age of 14, when he began working for his father as a clerk and messenger on the London Stock Exchange, although he was allowed 'any masters for private instruction whom he chose to have' during his spare time (Moses Ricardo, *Works* X, p. 3). He was later to complain bitterly of years of neglected education 'at the most essential period of life' (*Works* VII, p. 305), to which he frequently attributed his difficulties in written composition.

In 1792 the Ricardo family moved to Bow, close to the house of Edward Wilkinson, a Quaker and surgeon. Before long, David was romantically involved with Wilkinson's daughter, Priscilla Ann (1768–1829), whom he married on 20 December 1793. The young couple was promptly disowned by both sets of parents. Although the breach with the Wilkinson's was short-lived, it is said that David neither spoke to nor saw his mother again. He was also disinherited and removed from his father's business (he was reconciled with his father after his mother's

death). The marriage was the occasion of his breach with the Jewish faith, which, according to Sraffa, was 'the culmination of a gradual estrangement [with Judaism] … in progress for some time before' (Sraffa, *Works* X, pp. 38–9). He was subsequently to attend the meetings of the non-conformist Unitarians.

The break with his father was not to prove an insurmountable obstacle to David's financial prospects. With the assistance of City friends, he embarked on his spectacularly successful career as a jobber on the stock market and a loan contractor for government stock. Before long he had amassed a considerable fortune, and in 1815, the year in which he made his single most profitable transaction, he began a gradual retirement from business. The total value of his estate at death has been estimated at between £675,000 and £775,000, roughly equivalent to more than £500 million ($950,000,000) at 2006 prices.

As Ricardo's wealth grew, so too did his family and social standing. Eight children were born between 1795 and 1810. From 1812, the family's prestigious London address was 56 Upper Brook Street, Grosvenor Square. To this was added in 1814 Ricardo's country seat of Gatcombe Park near the small village of Minchinhampton, where he is reported to have financed almshouses, endowed a school, provided an infirmary and started a savings bank. His petition for his own coat of arms was also granted in 1814. Having entered the squirearchy (he was High Sheriff of Gloucestershire in 1818), acquired his reputation as a leading intellectual and political economist and become a prominent Member of Parliament (he took his seat in 1818), his company was sought increasingly by luminaries of the aristocracy, the political classes and the intelligentsia. Ricardo was highly gratified by his success, no more so than as a recognized authority on his favourite subject, namely, as he described it, 'political economy'.

It is said that Ricardo's interest in political economy was stimulated by chancing upon a copy of Adam Smith's *Wealth of Nations* in a travelling library while on a visit to Bath. Prior to this time, we are told by his brother that a predilection for subjects of an abstract and general nature had led to a leisurely interest in science, including mathematics, chemistry, geology and mineralogy (in 1807 he was a founding member of the London Institution for the Advancement of Literature and the Diffusion of Useful Knowledge, which was charged with promoting science, as well as literature and the arts, and in 1808 he joined the Geological Society of London). Yet, although his interest was awakened, to the point where he become an avid reader of the early articles on political economy in the *Edinburgh Review*, he was for several years too preoccupied with furthering his financial career to treat political economy as anything more that 'an agreeable subject for half an hour's chat' (as he later reminisced to his old friend Hutches Trower, *Works* VII, p. 246). The turning point came in 1809.

The free convertibility of paper currency into gold had been suspended under the Bank Restriction Act of 1797, following the run on the Bank that had been provoked by fears of a French invasion (in the context of the French wars of 1792–1815). In the aftermath of restriction, the market price of gold (in terms of the now unconvertible paper currency) had risen above the (fixed) mint price and the 'exchanges' had become 'unfavourable', so that premiums were now to be paid on bills of exchange drawn on overseas banks for the purposes of settling international debts. This gave rise to the first phase of the 'Bullion Controversy' (c.1797–1801), with contributions from writers including Henry Thornton and Lord Peter King. The controversy was concerned with the reasons for the depreciation of 'paper' relative to gold and the deterioration in the exchanges, with the 'bullionists' (as represented by King) arguing that the fault lay with the Bank of England for overissuing paper, while the 'anti-bullionists' (including Thornton) emphasized the role of special government payments overseas and poor domestic harvests, independently of the Bank's issues (this was the debate that dominated the early entries on political economy in the *Edinburgh Review*). After 1801, however, the price of gold fell back towards the mint price, the exchanges improved and the controversy duly subsided, to be revived in 1808 by a further marked depreciation of paper relative to gold (of around 20 per cent) and an accompanying fall in the exchanges. It was to this second phase of the controversy that Ricardo's first publication was directed, taking the form of an anonymous letter to the *Morning Chronicle* newspaper, published 29 August 1809.

With uncharacteristic rhetorical excess (Ricardo intoned gravely about the 'present evil' of the depreciation of paper and the 'disastrous consequences' and 'future ruin' that might follow) the argument in the letter followed the standard bullionist position that the root cause of the 'ills' was the 'over-issues of the Bank [of England]', the remedy being (in a remark that anticipates his later 'Ingot Plan') that 'the Bank be enjoined by Parliament gradually to withdraw … notes from circulation, without obliging them, in the first instance, to pay in specie' (*Works* III, p. 21). The argument was developed in further letters to the *Morning Chronicle*, and in his (signed) pamphlets, *The High Price of Bullion* (1810–11) and the *Reply to Mr. Bosenquet* (1811). The *Bullion* essay was a straightforward elaboration of Ricardo's position (in which he acknowledged that he 'can add but little to the arguments which have been so ably urged by Lord King', *Works* III, pp. 51–2), with an appendix to the fourth edition in which he developed his plan to resume convertibility by requiring the Bank of England to pay on demand in bullion ingots (not specie) bank notes of the value of at least £20, the alleged advantages being that this would reduce the supply of domestic paper, prevent an excessive demand on the Bank for gold (since the demand for bullion ingots would be less than the demand for specie), and prevent the withdrawal of low face-value bank notes (although Ricardo was later to acknowledge that a secondary market in bullion could facilitate the exchange of small notes pro rata). The second pamphlet was a reply

to criticisms of the parliamentary Bullion Committee Report (1810), with which Ricardo broadly agreed, and of Ricardo's own currency writings.

The contributions to the Bullion Controversy brought Ricardo to the attention of political and intellectual figures including Thomas Robert Malthus and James Mill (Ricardo was also in correspondence with Spencer Perceval, the Tory Prime Minister, and the opposition Whig leader, George Tierney). Both Malthus and Mill were to play critical roles in the development of Ricardo's subsequent career, although their influences were profoundly different. At the time of Ricardo's entry on the public stage, Malthus was a seasoned writer, the author of the *Essay on Population* and, arguably, the leading political economist of the day. Although he and Ricardo became, and remained, close friends, their relationship was marked by disagreement over many areas within the new 'science' of political economy. While Ricardo borrowed from some of Malthus's writings (including population theory and the theory of differential rent) other aspects of his work evolved dialectically from epistolary skirmishes with his contemporary.

Malthus introduced himself to Ricardo by letter in June 1811, by which time it seems that Mill was already considered a close friend and an ally on the bullion question. Mill had been an early contributor to the *Edinburgh Review*, but by this time his attention had turned to writing his *A History of British India* and it seems unlikely that he had much influence over the content of Ricardo's political economy with the exception of the 'law of markets' (in short, the doctrine that 'supply creates its own demand'). But that is not to detract from Mill's influence on Ricardo in other respects. Mill advised, encouraged, cajoled and even (only semi-humorously) bullied the ever-reticent Ricardo, who almost certainly would not have completed his major work without Mill's incessant prodding (see J.S. Mill, 1873, p. 42). It was also James Mill, as associate and disciple of Jeremy Bentham (with whom Ricardo also became personally acquainted), who was to coach the initially sceptical Ricardo in political utilitarianism and persuade him to enter parliament.

Currency issues dominated the early Ricardo–Malthus correspondence, but by late summer of 1813 their attention began to turn to a new subject, the forces governing movements in the general rate of profit. Up to this point Ricardo had robustly endorsed Adam Smith's 'competition of capitals thesis', according to which the general rate of profit is regulated by the intensity of competition in the labour market (determining movements in wage rates) and in the output market (determining prices). Exactly how Ricardo himself interpreted Smith's doctrine is not clear, nor is it possible to pinpoint the reason for the change of focus in the Ricardo–Malthus correspondence. However, it is certain that Ricardo was newly emphasizing the conditions of producing food as an influence on profits. His position was developed in lost 'papers on the profits of Capital', following which, in response to a clamour by the landed aristocracy for a revision of the old Corn Law, and the report in May 1814 of the Parliamentary Committee on the Corn Trade, his deliberations become more narrowly centred on the effects on profits of restrictions on the free importation of corn. The outcome was his *Essay on The Influence of a Low Price of Corn on the Profits of Stock; Shewing the Inexpediency of Restrictions on Importation* (1815; reprinted in *Works* IV).

The central argument of the *Essay* may be given as follows. On the assumption of an economy closed to the importation of foreign corn (the principal subsistence commodity and wage good), the increasing demand for corn from a growing population must be met either by the more intensive cultivation of land or by cultivating land that is less fertile or more disadvantageously situated relative to the final market. Either way, the expansion of output will encounter diminishing returns which in turn lead to a higher corn price, higher money wages and, therefore, a lower agricultural rate of profit. Only the landlords benefit, because they receive more *differential* rent: following Malthus, rent is the difference in return from the 'best' and the 'worst' land on the assumption that the return from the 'worst' land is sufficient only to give farmers the general rate of profit; ergo, landlords benefit as that difference increases. To complete the argument, the reduction in profitability is transmitted to capitalists generally by means of higher money wages. As for labourers, the argument appears to be that they might also suffer in consequence of a ('temporary') fall in labour demand, itself the result of lower profitability. Hence Ricardo's provocative conclusion that 'the interest of the landlord is always opposed to the interest of every other class in the community' (*Works* IV, p. 21).

The *Essay* was a transitional work in which Ricardo repudiated some of the fundamental tenets of the prevailing orthodoxy, as derived from Adam Smith and upheld by Malthus, but failed to supply a fully convincing logical alternative. It was James Mill who persuaded Ricardo to develop his ideas in the form of a major treatise. Two years later Mill's exhortations were rewarded with the publication of *On the Principles of Political Economy and Taxation* (1817; reprinted in *Works* I). However, before Ricardo could begin serious work on the *Principles* he had to fulfil his commitment to Pascoe Grenfell M.P., who had enlisted Ricardo's support for an assault on the Bank of England. Ricardo was more than happy to oblige ('I always enjoy any attack upon the Bank', *Works* VI, pp. 268–9), and the result was his *Proposals for an Economical and Secure Currency; with Observations on the Profits of the Bank of England, as they regard the Public and the Proprietors of Bank Stock* (1816; reprinted in *Works* IV).

In language suggested by James Mill, Ricardo lamented that 'a great and opulent body like the Bank of England' should 'wish to augment their hoards by undue gains wrested from the hands of an overburthened people' (*Works* IV, p. 93). It was intolerable that a mere 'company of merchants' should make vast profits by overcharging on the management of the public debt and other public business, through their 'seignorage' on the issue of paper money and by reducing their unprofitable stock of bullion (as they were enabled to do by the Restriction Act). For the longer term (after the expiry of the Bank Charter in 1833), Ricardo's preferred solution was to strip the Bank of its management of the money supply, which he would entrust to 'commissioners responsible to parliament only, the state' (*Works* IV, p. 114) (this plan was developed in the *Plan for the Establishment of a National Bank* [1824], drafted by Ricardo in 1823, reprinted in *Works* IV). For the shorter term, he suggested that the government should seek more favourable terms for the management of the debt. Above all, however, he again advocated a swift return to a fully convertible paper currency, to be achieved by his Ingot Plan. The result would indeed be an 'economical and secure currency' which, along with its other advantages (of cheapness in comparison with a fully metallic currency and of stability by constraining movements in the market price of gold) would facilitate short-term, compensating changes in the money supply in response to fluctuations in the availability of credit.

With the *Proposals* dispatched to the printers, the way was open for Ricardo to commence work on his *Principles*; or, be more precise, it was *almost* so, for he still had to contend with a hectic social life, recurring bouts of lethargy and defeatism, continuing business interests, the demands of a large family and a 'temptation of being out in the air in fine weather' (*Works* VI, p. 263). Fortunately for posterity, the summer of 1816 offered very few outdoor temptations and Ricardo dedicated himself to his task. The *Principles* was published on 19 April, 1817. It was the result of little more than six or seven months' sustained activity on Ricardo's part.

The 'principal problem in Political Economy' is defined in the *Principles* as the determination of the 'laws' which regulate 'the natural course of rent, profit, and wages' over time. These issues had been addressed in the *Essay* and, indeed, the *Principles* was initially conceived as an *Essay* writ large. In the process of writing the later work, however, its scope was enlarged in previously unforeseen ways as Ricardo developed his ideas. The result was a volume comprising 31 chapters, covering not only the 'laws' governing rent, profit and wages, but also a labour theory of value, a theory of international comparative advantage, monetary theory, several chapters devoted to 'the influence of taxation on different classes of the community', and strictures on the writings of predecessors and contemporaries. The 'core' theoretical analysis as it relates to 'the natural course' may be summarized as follows.

In terms of the newly adopted 'pure' labour theory of value, (changes in) the exchangeable value of competitively produced, freely reproducible commodities are determined exclusively by (changes in) the *quantities* of labour expended on their production, where the relevant quantity of labour is the *greatest* quantity expended per unit of output sold. The theory applies only when commodities exchange at their *natural prices*, defined by uniform wage and profit rates (rent is excluded as a component of price, as explained shortly). In addition, it is assumed that one domestically produced commodity, gold (not to be confused with its real-world namesake), serves as the 'invariable standard' (the *numéraire*) in terms of which all prices are expressed, its 'invariability' defined in terms of a given and unchanging labour input per unit of its output. It follows that any change in a commodity's gold-denominated natural price is an exact reflection of a corresponding change in the labour expended on *its* production. This theory of value was used by Ricardo beyond the first chapter of the *Principles*.

Next, there is the theory of differential rent, derived from Malthus. As Ricardo explained, the relevance of the theory is not confined to agriculture but applies whenever units of the same (homogeneous) class of commodity are produced by different quantities of labour. If all units sell at the same natural price, determined by the *greatest* labour input per unit; and, if the rate of profit from the sale of the unit requiring the greatest labour input is equal to the general (uniform) rate; then, an additional surplus revenue will be earned on units requiring a lower labour input, and it is this additional surplus that constitutes (differential) *rent*. (If we assume with Ricardo that capitalist producers, in agriculture and elsewhere, are profit-maximizers, they will *always* extend production to the point where revenue from the sale of the incremental output, requiring the greatest labour input, is sufficient *only* to yield the general rate of profit; moreover, this greatest labour input *must* determine price, otherwise the general rate of profit could not be received and the output would never be produced.)

On wages, Ricardo introduces the distinction between the market price of labour (or market wage) and the natural price (or natural wage), the latter defined (in money terms) as 'that price which is necessary to enable the labourers, one with another, to subsist and to perpetuate their race, without either increase or diminution' (*Works* I, p. 93). The price that *is* necessary depends on the 'real' natural wage: on 'the quantity of food, necessaries, and conveniences [which] become essential ... from habit' (*Works* I, p. 93). Habits may change over time (perhaps under the influence of education, as Ricardo hoped) but, for analytical purposes, the natural (real) wage is a datum. In the event that the market wage is above or below the natural wage, a Malthusian-style population mechanism is triggered: population expands (or contracts) and the market wage returns to the natural level.

To turn to profits, the eponymous chapter in the *Principles* is chiefly a revision of the central argument from the *Essay*, although it does contain the ingredients for a more general theory of the rate of profit. Now in terms of

the labour theory of value, the attempt to expand the output of corn in a closed economy encounters diminishing returns in the form of a greater labour input per unit of output; hence, the (natural) price of corn rises proportionately. This in turn increases money wages, because corn enters the given real wage. The rate of profit (calculated with reference to the output produced by the greatest labour input) must therefore fall (since the rise in the natural price of corn is proportionate only to the increase in the *quantity* of labour expended on its production and does not reflect the increased *cost* of that labour) and, by the reasoning explained above, differential rent increases. Natural prices outside agriculture are either unchanged or, if corn is required as a material input, rise only to reflect the increase in the labour expended on their production; hence, the fall in the agricultural rate is communicated to other sectors by an increase in money wages. Perforce, the 'natural course or rent, profit, and wages' is for rent to increase, the rate of profit to fall and (money) wages to rise, although it must be stressed that this 'prediction' is *entirely* contingent on a host of assumptions and should not be taken as evidence of a gloomy or pessimistic attitude by Ricardo to Britain's economic prospects (such an inference, although commonly made, could not be further from the truth).

Finally, the more general theory from Ricardo's analysis is that (changes in) the rate of profit depend exclusively on (changes in) the labour expended on the production of the given real wage where, to borrow J.S. Mill's distinction, the 'labour expended' covers the 'direct' labour and the 'indirect' labour expended on the production of the non-labour inputs to the production of wage-goods. Provided one grants him his assumptions (of a labour theory of value, a given real wage and known labour conditions of production), Ricardo had thus produced a strong candidate for the first logically coherent theory of the determination of the general rate of profit in the history of economic thought.

There was a great deal riding on the success of the *Principles*, not just Ricardo's growing reputation as a political economist. Mill had suggested in 1815 that Ricardo should enter Parliament, a suggestion from which the latter had recoiled with horror. One year later he was becoming more amenable to Mill's plan, writing to his friend: 'If my book succeeds ... perhaps my ambition may be awakened, and I may aspire to rank with senators' (*Works* VII, p. 113). Much to Ricardo's relief, the book did succeed to an extent far surpassing his self-deprecatory expectations.

Ricardo entered Parliament on 26 February 1819 as the independent member for the rotten borough of Portarlington in Ireland: a constituency which he never visited, with 12 or so electors in the 'pocket' of Lord Portarlington to whom Ricardo had advanced £25,000 as a loan on the mortgage of his estates.

Ricardo availed himself of every opportunity to educate the House of Commons in the 'true principles of political economy'. These principles dictated the gradual repeal of trade restrictions generally and the of the Corn Law in particular; the gradual repeal of the Poor Laws; the repayment of the National Debt (his heroic proposal to replay the debt over two or three years by the imposition of a property tax was met with widespread incredulity); minimal taxation and a balanced budget; and a return to a convertible currency. With the signal exception of convertibility (Peel's Bill of 1819 for the Resumption of Cash Payments owed much to his proposals), Ricardo mostly found himself on the losing side, but that did nothing to shake his convictions. His parliamentary contributions are testimony to his belief in political economy as a subject of direct empirical relevance (the view of Ricardo as a pure theorist is a travesty). They also mark him as a zealous advocate of a free-market capitalist system with minimal government interference, who believed that Great Britain 'would be the happiest country in the world, and its progress in prosperity would be beyond the power of imagination to conceive, if we got rid of two great evils – the national debt and the corn laws' (*Works* V, p. 55). Additionally, he spoke out on a range of 'liberal' issues including religious tolerance, slavery, freedom of speech and the right to petition. He also aligned himself with the 'radical' cause for the reform of parliament.

The contention that 'good government' would not be achieved without a reform of parliament had been put to Ricardo by James Mill in 1815 but was at that time rejected on the grounds that Mill exaggerated the 'sinister interest' of politicians in pursuing their own selfish interests and undervalued the corrective influence of enlightened public opinion. Three years later Ricardo's position had changed. Partly as a result of Mill's bombardment of Ricardo with 'radical' messages, partly because of his growing conviction that the Tory government was failing to pursue 'right measures', and after reading Jeremy Bentham's *Plan of Parliamentary Reform*, Ricardo was won over to the 'radical' cause. As he came to argue, 'good government' – government 'administered for the happiness of the *many*, and not for the benefit of the *few*' (*Works* VII, p. 299) – required that politicians should 'legislate for the public benefit only, and not ... attend to the interests of any particular class' (*Works* VIII, p. 275); yet, under present arrangements, politicians fell prey to the interests of particular classes, particularly the landed class; hence the necessity for reform. However, Ricardo's proposals fell some way short of those of his 'radical' contemporaries. The introduction of the secret ballot was, for him, an almost sufficient basis for securing good government under existing circumstances, although he did make a case for triennial parliaments and a modest extension of the franchise to include householders. He might therefore be described as a moderate reformer in the utilitarian tradition of Bentham and Mill.

The infamous proposal for the speedy repayment of the national debt was also presented to the public in an invited article on the *Funding System* (1820), written in autumn 1819 for publication in the *Supplement to the Encyclopaedia Britannica*. This article is noteworthy for its exposition of what has come to be known, misleadingly, as the 'Ricardian equivalence theorem'. To take Ricardo's own argument, suppose that a war involves the expenditure of £20 million. This can be financed either by raising £20 m in taxes or by borrowing £20 m and repaying by taxes £1 m per annum in perpetuity (at an assumed annual interest rate of five per cent), or by borrowing the £20 m and (for example) repaying by taxes £1.2 m per annum, which would clear the interest (at five per cent) and the initial £20 m over a period of 45 years. 'In point of economy', as Ricardo stated, 'there is no real difference in either of the modes', because the present value of £1 m per annum in perpetuity or £1.2 m over 45 years, both at five per cent annually, *is* £20 m, hence the idea of *equivalence*. But, he continued, 'the people who pay the taxes never so estimate them, and therefore do not manage their private affairs accordingly': the different modes are *not* equivalent because, according to him, individuals are prone to undervalue the true cost of repaying a loan over time. That being so, Ricardo's proposal was for the pay-as-you-spend mode of financing which, he believed, would make people 'less disposed wantonly to engage in an expensive contest [namely war], and if engaged in it ... be sooner disposed to get out of it' (*Works* IV, p. 186).

While Ricardo was writing his *Funding System*, his friend Malthus was putting the finishing touches to his own *Principles*, published in April 1820, which contained an unsparing critique of Ricardo's central doctrines (Malthus's *Principles* together with Ricardo's comments are reprinted in *Works* II). Of all Malthus's criticisms, those levelled at Ricardo's treatment of value were the most acute, thus prompting Ricardo to a major revision of his first chapter for the third edition of his *Principles* (1821) (a lightly revised second edition of the *Principles* had been published in 1819). In addition to the defence against Malthus, the third edition is distinguished by a new chapter 'On Machinery' in which Ricardo, stimulated by the work of John Barton, famously declared that 'the opinion entertained by the labouring class, that the employment of machinery is frequently detrimental to their interests, is not founded on prejudice and error, but is conformable to the correct principles of political economy' (*Works* I, p. 392). To avoid misunderstanding, although there may be a very distant family resemblance between Ricardo's analysis and the standard 'neoclassical' case of factor substitution in response to changes in factor prices within a timeless framework, a principal difference is that Ricardo was describing a process *over time*, in which the substitution of machinery for labour was likely only to apply to new ventures. Nor did his analysis end there, since the capitalists were expected to expand accumulation in consequence of their higher profits, so tending to reverse the fall in the demand for labour (and wages).

Following the third edition of the *Principles*, Ricardo's next and last publication within his own lifetime was *On Protection to Agriculture* (1822; reprinted in *Works* IV): a veritable tour de force, written in little more than three weeks.

A new Corn Law had been passed in 1815 which prohibited the importation of foreign corn until the price had been at least 80 shillings per quarter for six weeks, by which time the ports could be opened to duty-free importation. Imports had been triggered in 1817–19, with prices first rising too 111 shillings in June 1817 and then (under the pressure of importation, followed by good domestic harvests) falling steadily to 55 shillings in the second half of 1820. After more bumper harvests, 1822 then witnessed the lowest average corn prices since 1792, with a fall to 34 shillings in November. The 'agricultural distress' was widespread and severe, and the powerful landed interest turned to Parliament for assistance. A parliamentary committee was established to investigate the causes and possible remedies for the distress, with Ricardo as one of its members.

Ricardo was not optimistic that Parliament would, or could, shift its position towards a free trade in corn; as he wrote in correspondence, 'I have no hope of good measures being adopted, the landlords are too powerful in the House of Commons to give us any hope that they will relinquish the tax which they have in fact contrived to impose on the rest of the community' (*Works* IX, p. 158). He was proved right. *On Protection* was his response to the protectionist report of the committee, in which he maintained his position that free trade was the only long-term solution while proposing a revised version of the 1815 Act (with measures for dampening price fluctuations). The pamphlet is also distinguished by sharp restatements of Ricardo's central doctrines, a wealth of detailed empirical analysis and a pungent defence of his own position with regard to Peel's Bill of 1819 for the resumption of cash payments. *On Protection* shows him at the peak of his career, a true master of his subject and a political economist in the most rounded sense.

In the summer of 1822 Ricardo embarked with his family on a four-month grand tour of Continental Europe. Upon his return he resumed his hectic life as an active parliamentarian, attended meetings of the Political Economy Club (which he had co-founded in 1821), drafted his plan for an independent national bank, and continued with his deliberations on 'value'. He was also looking forward to hosting a visit to Gatcombe from his old friend Hutches Trower, to whom he wrote: 'we shall walk and ride, we will converse on politics, on Political Economy, and on Moral Philosophy, and neither of us will be the worse for the exercise of our colloquial powers (*Works* IX, p. 377). But it was not to be. On 11 September 1823 Ricardo died from the effects of an abscess in the middle ear. He was buried at Hardenhuish Park, Wiltshire, on the estate of his daughter, Henrietta, and her husband, Thomas Clutterbuck.

The newspaper obituaries of the time were lavish in their praise of Ricardo's achievements, both as a political economist and as a 'Senator' (see Peach, 2003). By his friends, he was applauded for having virtually revolutionized economic theory, not merely for its own sake but as means of guiding government policy and thus promoting the 'general happiness' of society. Of course, his critics were to asses his contributions less kindly, but in producing what was arguably the first coherent supply-side analysis of value, distribution and growth – never mind anything else – his place in doctrinal history was assured.

The following sections consider in more details various aspects of Ricardo's work and a selection of the main interpretative disputes that continue to surround it.

Monetary contributions, the law of markets and comparative advantage

As Peake (1978, p. 31) rightly observed, 'Ricardo's total productive output was dominated by monetary questions', and it was in this area that he had the greatest practical influence in his own lifetime.

A simple approximation to Ricardo's 'model' includes the following assumptions: the domestic currency consists entirely of paper money ('paper') issued by a central bank; money prices are a function of the supply of paper (*ceteris paribus*); the bank allows the free convertibility of paper into gold on demand at a permanently fixed mint price, initially equal to the globally determined market price in terms of paper; and all profit-seeking economic agents have virtually perfect market information. Now suppose that the bank increases its supply of paper. Domestic commodity prices rise, as does the market price of gold in terms of paper. (Ricardo treated gold as just another commodity, a view that later ensnared him in the position that the exchangeable value of gold is determined by the labour expended on its production even though its value is *incessantly* fluctuating in response to changes in the volume and pattern of world trade.) Gold has become relatively cheaper (or 'redundant') because, by assumption, it may be purchased at the lower mint price. Profit-seeking agents therefore exchange paper for gold which, because of its new relative cheapness compared with other domestic commodities, is now exported in preference to those commodities in exchange for foreign produce. Hence, the supply of domestic paper contracts, domestic prices fall, the market price of gold (in paper) returns to the mint price and the *status quo ante* is restored.

Now suppose that paper is no longer freely convertible into gold at the mint price. As before, the supply of paper is increased, but the 'stimulus which a redundant currency gives to the exportation of the coin [namely gold] … cannot, as formerly, relieve itself' (*Works* III, p. 78), so the market price of gold *remains* above the mint price. In addition, the paper cost of bills of exchange drawn on foreign banks will increase to reflect the fall in paper relative to gold (the foreign exchange becomes 'unfavourable'). Hence (following Lord King) Ricardo's 'two unerring tests' of a depreciation in Bank-notes, 'the rate of exchange and the price of bullion' (*Works* III, p. 75). (Ricardo flatly rejected the measurement of depreciation either in terms of changes in the exchangeable value of gold for domestic commodities – because commodities 'are continually varying in value' among themselves – or in terms of subjectively perceived 'enjoyment', 'because two persons may derive very different degrees of enjoyment from the possession of the same commodity'; *Works* IV, pp. 59, 61.)

As to the consequences of 'depreciation', the picture is mixed. Ricardo stressed that the rate of interest is determined by the rate of profit in the 'real' economy (by the 'competition of capitals' or by the conditions of producing wage goods, in the earlier and later writings respectively); and that the 'trifling' effect on the rate of profit (hence on output) of an increase in paper is confined to an interval 'of momentary duration' before money wages adjust to restore the (assumed) given real wage (*Works* III, pp. 91–2, 318–19, 329; *Works* V, p. 446; *Works* VI, pp. 16–17). This dominant position supports a (mostly) neutral money interpretation, but it also raises the question of why Ricardo became so exercised by depreciation if its real effects were insignificant. The answer is possibly to be found in his concern with the effects of rising prices on recipients of fixed money incomes, especially 'monied men' (see *Works* III, pp. 21, 95–6; *Works* VI, p. 68), regarded by Sayers (1953, p. 65) as a 'shattering' inconsistency with the view 'that long-run effects come quickly and easily'. In addition, the later Ricardo was also to remark on the danger of an 'easy' inconvertible paper-money regime in facilitating speculation and over-trading (*Works* V, pp. 397, 446).

Whatever the economic costs of depreciation, Ricardo campaigned tirelessly for a resumption of convertibility at the pre-restriction mint price of gold. In his evidence before the Parliamentary Committees on Cash Payments (1819) he argued, with heroic simplicity, that, in the prevailing circumstances of a four per cent premium in the market over the mint price of gold, a reduction in paper currency of about four per cent would be sufficient to restore parity, with a consequent fall of domestic prices generally also of around four per cent (*Works* V, pp. 416–17). This objective could be achieved 'in a few months' (*Works* V, p. 396). However, 'by a consideration of the fears which I think many people very unreasonably entertained', he was 'reconciled' to a plan for the phased return to the old mint price over one or two years (*Works* V, p. 451). The logic, if not the detail, of Ricardo's argument was accepted by the committees, leading to Peel's Bill (1819) with its provision for a staged return to convertibility at the old par over a period from February 1820 to May 1823. Payments were to be made only in bullion ingots in the first two stages, in line with Ricardo's recommendation. Contrary to his proposals,

however, the third stage gave the Bank the *option* of making payments in specie, while the fourth and final stage saw a return to *full* convertibility at par.

Ricardo regretted that Parliament had not adopted his plan in its totality, but was otherwise supportive of the bill. Certainly, he did not foresee the events that were to follow which led, on his 1822 estimation, to a ten per cent depreciation of paper, thus making 'the reverting to a fixed currency as difficult a task to the country as possible' (*Works* IX, pp. 140, 152). The fault lay not with his analysis, however, but with the Bank of England, who had (in his opinion) needlessly purchased large quantities of gold in anticipation of resumption, thus raising its market price *independently* of note issues.

Ricardo may have been justified in blaming the Bank, but he must also stand accused of reasoning *as if* his simple model had captured all relevant aspects of reality. He was aware, on one level, that nominal inflation or deflation was *not* determined exclusively by changes in the Bank's supply of paper, and that the market price of gold could differ from the mint price *independently* of changes in the domestic money supply. He had noted in his early monetary writings that the 'regulator of prices' must include not only the quantity of money, but also 'the rapidity of its circulation' and 'the mass of commodities' (*Works* III, p. 311); later (1815–16), in response to post-war economic conditions, he allowed that the quantity of money was also determined by the independent behaviour (in context, the bankruptcy) of the country banks, and he conceded that 'bullionists', himself included, had underrated the effects on the market price of gold from changes in world demand for the metal (*Works* VI, p. 344; *Works* IV, p. 62); finally, under hostile questioning from some members of the Parliamentary Committees on Resumption, he admitted the further qualification that changes in the general state of confidence could affect domestic prices by influencing the availability of credit, itself a substitute for currency (*Works* V, p. 419). Yet, for the most part (as with his 'four per cent' calculation, noted above), he argued as if these counteracting influences were nugatory to the point that they could be ignored completely. This was a treacherous foundation on which to build economy policy.

It was also Ricardo's habitual presumption that real-world economic actors behaved 'rationally' and it is for this reason that he was highly critical of the Bank for purchasing gold when (on his analysis) it was not in their interest to do so. The same presumption was at the heart of his version of the law of markets, described by Keynes as the (flawed) doctrine that 'supply creates its own demand in the sense that the aggregate demand price is equal to the aggregate supply price for all levels of output and employment' (Keynes, 1936, 21–2).

The 'law' is commonly attributed to Jean-Baptiste Say although its roots extend back to Adam Smith's *Wealth of Nations*. It seems likely, however, that it was derived by Ricardo from James Mill, who had sketched out the argument in his 1808 review of William Spence's *Britain Independent of Commerce* in the *Edinburgh Review*. It was first used by Ricardo in his early monetary writings to argue that foreign markets will never be so 'glutted' by British produce as to constrain further British exports when money becomes comparatively dearer (that is, the opposite of 'redundancy'). Later, it was used to bolster the argument that the *only* cause of a permanent reduction in general profitability is an increase in the labour expended on the production of wage goods. It was also invoked by Ricardo in the distressed aftermath of the French wars to support his unshakably optimistic view that recovery was *always* imminent.

Ricardo's version of the 'law' may be reduced to the following propositions: first, commodities will continue to be produced only if they return at least the going general rate of profit; second, capitalist producers (and *only* capitalist producers) are not 'for any length of time … ill-informed of the commodities which [they] can most advantageously produce' (*Works* I, p. 290); third, the desire to consume *something* is 'implanted in every man's breast; nothing is required but the means' (*Works* I, p. 292); fourth, all money income is spent, either by the direct recipients or by those to whom the recipients lend (all) their unspent money income (there is no hoarding). If, to take the extreme case, commodities *always* exchange at natural prices (which implies that the producers earn precisely the going general rate of profit) and all income is *always* spent (on the same output), we have 'Say's identity' version of the law, defined by an excess demand for money of zero at all times. This is the version that Keynes attributed to Ricardo. Yet, although Ricardo's emphasis was always on equilibrium or long-period tendencies, he did (as he was forced to by external events) allow for strictly 'temporary' periods of capital misallocation in which capitalists produce the 'wrong' commodities and demand money (to satisfy creditors) in excess of revenue. The 'Say's equality' version of the 'law', allowing for 'temporary' disequilibria of excess demand for money, would therefore better describe his position. Above all, however, what is most striking is his belief that the 'law' encapsulated real-world tendencies, to the extent that he condemned out of hand all proposals for relief works (because only the capitalists knew best how to allocate capital) and, ultimately, was left totally bemused by the scale and duration of the post-war distress (see *Works* VIII, p. 277).

Economists have been considerably more impressed by his statement of comparative advantage in the chapter 'On Foreign Trade' in the *Principles*. Following Ricardo's own example, assume two commodity bundles of cloth (x_1) and wine (x_2), both of which could in principle be produced in England or Portugal ($[x_1, x_2]$ and $[x_1^*, x_2^*]$ respectively). To produce the bundles in England would require 100 labourers for cloth (a_1) and 120 labourers for wine (a_2); and to produce them in Portugal would require 90 labourers for cloth (a_1^*) and 80 labourers for

wine (a_2^*). Portugal therefore has an absolute advantage in the production of *both* bundles. Labour (alias 'capital') is immobile internationally, trade initially takes place by way of real barter and, implicitly, bundles are produced under constant returns to scale in both countries.

As Ricardo states, it will be advantageous for England to specialize in making cloth and Portugal to specialize in wine, because both countries thereby obtain more of the other commodity per unit of their domestic labour than if they attempted to make it themselves. For example, if Portugal used 80 labourers to make cloth she could obtain only $0.89x_1$, but if she can exchange $1x_2^*$ (also the produce of 80 labourers) for $1x_1$, as Ricardo supposes, then it would be 'advantageous for her to export wine in exchange for cloth' (*Works* I, p. 135). Similarly, if England used 100 labourers to produce wine she could obtain only $0.833x_2$, so she also benefits by exchanging $1x_1$ (the produce of 100 labourers) for $1x_2^*$.

Ricardo's example implies that the pattern of specialization is dictated by the 'four magic numbers' (Samuelson, 1972, p. 378), namely, $a_1/a_2 < a_1^*/a_2^*$. However, the principal purpose of the analysis was not so much to illustrate comparative advantage per se, but to show that the 'same [labour theory] rule which regulates the relative value of commodities in one country, does not regulate the relative value of the commodities exchanged between two or more countries' (*Works* I, p. 133). If, as Ricardo supposes, there is a rate of exchange of $1x_1$ for $1x_2$, 'England would give the produce of the labour of 100 men, for the produce of the labour of 80': something that 'could not take place between the individuals of the same country' (*Works* I, p. 135).

It was also Ricardo's purpose to show that the introduction of money (gold) would leave the analysis unaffected: gold will be 'distributed in such proportions amongst the different countries … as to accommodate [itself] to the natural traffic which would take place if no such metal existed, and the trade between countries were purely a trade of barter' (*Works* I, p. 137). To give the flavour of the argument, suppose England and Portugal each produce both commodities and that the initial gold prices are $px_1 = px_1^*$ which, given the 'magic numbers', implies $p_2 > p_2^*$. Wine is therefore exported from Portugal to England and is paid for by gold. But, on Ricardo's quantity theory reasoning, the influx of gold to Portugal, and its efflux from England, will raise prices in the former country and reduce them in the latter. Hence, the specie-flow mechanism ensures that $px_1 = px_1^*$ is unsustainable (the same would apply to $px_2 = px_2^*$, by similar reasoning) and that the price of Portuguese cloth must exceed the price of English cloth (just as the price of wine must be higher in England than in Portugal), thus leading to complete specialization. Contrary to Ricardo, however, 'the natural traffic which would take place if no such metal existed', hence relative world prices, are not unique (as he implied), with the range of possible outcomes defined by the condition of $a_1/a_2 \leq p_1/p_2 \leq a_1^*/a_2^*$.

Debate continues as to whether Ricardo was the true originator of the comparative advantage doctrine, or whether that accolade should be awarded to his contemporary, Robert Torrens (see Ruffin, 2002; 2005, for a recent case in Ricardo's favour). But, regardless of who may have crossed the line first, it is with Ricardo's name that comparative advantage has become indelibly linked.

Early writings on profit (1813–15) and the 'corn model' interpretation

In the introduction to his masterful edition of Ricardo's *Collected Works* it was suggested by Piero Sraffa that the early Ricardo had devised a model in which corn is the sole agricultural input and output, thus supplying a 'rational foundation' for the 'principle of the determining role of the profits of agriculture', putatively articulated by Ricardo in 1814 with the words 'it is the profits of the farmer which regulate the profits of all other trades' (*Works* VI, p. 104). By implication, when Ricardo wrote that agricultural profits 'regulate' other profits, he had intended a statement of *unique determination* in full awareness of the logically required assumptions. Sraffa revealed, however, that the corn model (or 'corn ratio theory of profits') was 'never stated by Ricardo in any of his extant letters and papers' although, on the basis of indirect textual evidence, he claimed that Ricardo 'must have formulated it' either in lost papers or conversation (*Works* I, p. xxxi). Later, with the publication of Sraffa (1960), it transpired that the corn model had additional significance as a simple precursor of Sraffa's own 'Standard system' in which corn is the sole 'basic commodity'; and Sraffa also disclosed that the interpretation was the outcome of his *own* theoretical work: 'it was only when the Standard system and the distinction between basics and non-basics had emerged in the course of the present investigation that the ['corn model'] interpretation of Ricardo's theory suggested itself as a natural consequence' (Sraffa, 1960, p. 93).

The corn model interpretation was widely embraced. With a beguiling pedagogical simplicity, it could 'explain' Ricardo's regulatory statements and his later development of the pure labour theory, with Malthus entering the story to remind Ricardo that agricultural capital does *not* consist entirely of corn, thus necessitating a new (labour) theory of value. However, beginning in the early 1970s with the work Samuel Hollander, doubts have increasingly been aired about the textual basis for the interpretation. What follows is the view of one such critic. (For a sample of critical interpretations, see Faccarello, 1982; Hollander, 1973; 1975; 1979; and Peach, 1984; 1993; 2001. The case for the defence has been made by Eatwell, 1975, and Garegnani 1982, among others.)

If Ricardo's writings are sifted for confirmation for the interpretation – in other words, if the corn model is *presumed* – then it is easy enough to find 'evidence' in its favour. Without that presumption, the picture is rather different. Thus, Ricardo's assertion in correspondence that the 'rate of profits and of interest must depend on the proportion of production to the consumption necessary to such production' (*Works* VI, p. 108) is said by Sraffa to be the 'nearest that Ricardo comes to an explicit statement on these [corn model] lines' (*Works* I, p. xxxii). Yet, although it is *possible* to conceive of such a 'proportion' in material terms, the expression itself provides no evidence of the way it was conceived by Ricardo; moreover, very similar expressions had been used by him in his earlier monetary writings in which there is no question of him having adopted corn model assumptions. As for the regulatory statements, while it is *possible* to impose a corn model rationalization, the problem is in establishing that the same rationalization was applied by Ricardo. Here, too, the evidence is disobliging. The *Essay* (1815) is replete with such statements (for example, 'The general profits of stock depend wholly on the profits of the last portion of capital employed on the land'; *Works* IV, p. 21), but we can be sure they were not thought *by Ricardo* to depend on the corn model because he *explicitly* assumed heterogeneous inputs to agriculture (including 'buildings, implements, &c.'; *Works* IV, p. 10). Indeed, an arresting feature of the *Essay* is that its specious corn model appearance derives from the use of corn (alias wheat) to *value* the physically heterogeneous agricultural capital. It was Ricardo's 'failure' to *revalue* this capital as corn became more difficult to produce (in the initial agricultural phase of the argument) that drew Malthus's criticism and led, ultimately, to Ricardo's adoption of the labour theory, not an assumption that agricultural capital *comprises* of corn alone.

A question for those who reject the corn model is how the pre-*Essay* Ricardo could arrive at his 'regulatory' position if, as he believed at the time, a rise in the price of corn would be followed by a rise in prices generally. Samuel Hollander has conjectured that Ricardo may have invoked a monetary constraint, so that an agriculturally induced rise in money wages would not be passed on in higher prices. Alternatively, it may have been that his view of pricing was *integral* to the analysis: with price rises common to output *and* the heterogeneous inputs to agriculture, Ricardo might have reasoned that an increase in the capital–output ratio *must* reduce profitability. Admittedly, these alternative interpretations do not have the simple elegance and logical consistency of the corn model but, then again, the period 1813–15 was one in which Ricardo was struggling to establish new ideas within an inherited theoretical framework, much of which was later to be discarded. The existence of contradictions and unresolved theoretical issues during this period is unsurprising.

The labour theory of value

The unmodified or 'pure' labour theory of value (PLTV) was adopted by Ricardo in early 1816, on the basis of which he drafted material that would form the first seven chapters of the *Principles* (up to and including the chapter 'On Foreign Trade'). But then he discovered a source of modification to the PLTV resulting from differences in capital structure between production processes. At first, the discovery impeded his progress, but then it seems he had the inspiration to turn it to his advantage (so he thought) in the form of the 'curious effect': the iconoclastic demonstration that prices *fall* following a general rise in wages and consequent reduction in the rate of profit. What he did not do, however, was provide any justification for using the PLTV in the light of the 'curious effect' analysis.

A simple 'dated labour' framework may serve to illustrate Ricardo's position. Assume three commodities (x_1, x_2, x_3), each produced by ten units of homogeneous labour (L) applied over two discrete production periods ($t-1$, t), with the following conditions of production: $10L_t \rightarrow x_1$; $5L_t + 5L_{t-1} \rightarrow x_2$; $10L_{t-1} \rightarrow x_3$. If we denote the uniform wage and profit rates as w and r, each taking period t values, the natural price equations for the commodities are:

$$px_1 = 10L_t.w(1 + r) \qquad (1)$$

$$px_2 = 5L_t.w(1 + r) + 5L_{t-1}.w(1 + r)^2 \qquad (2)$$

$$px_3 = 10L_{t-1}(1 + r)^2. \qquad (3)$$

A PLTV requires $px_1 - px_2 = px_3$, because each commodity is produced with the same quantity of labour. However, it is evident (with $r > 0$) that $px_3 > px_2 > px_1$. Moreover, if distribution (between w and r) changes, there will be price fluctuations *even though* labour inputs are unchanged. Thus, assume that x_1 is the *numéraire* commodity (so that $px_1 \equiv 1$); in principle, on the basis of (1) a new (lower) r can be calculated for a given rise in w, and these numbers may be entered into (2) and (3) to obtain the new natural prices of x_2 and x_3. With the 'compounding' (or magnification) of the effect of the lower r on px_2 and, even more so, on px_3, the result will be a fall of both prices (expressed in terms of x_1) with px_3 falling to the greater extent.

To relate the above to the chapter 'On Value' in the first two editions of the *Principles*, the differences in production conditions (or capital structure) are, in Ricardo's terms, a reflection of differences in (a) the durability of fixed capital; (b) the ratios of fixed to circulating capital; (c) the durability of circulating capital (added in the second edition), where the fixed–circulating capital distinction depends, essentially, on the time required to repay a capital expenditure (the longer the time, the more 'fixed' the expenditure). In the case given by Ricardo, the two extremes (corresponding to x_1 and x_3) are a commodity produced by unassisted labour in one year and a commodity produced by unassisted machinery that lasts 100 years. Then, taking the former commodity as his

'invariable standard' (alias *numéraire*), he calculates that a fall of seven per cent in the rate of profit would reduce the price of the latter by 68 per cent: a vivid illustration of the 'curious effect' (*Works* I, p. 60).

The effect implies, by Ricardo's own testimony, that the PLTV is subject to a (truly) 'considerable modification' from differences in capital structure, but the really curious feature of the first two editions of the *Principles* is that the PLTV (used beyond the first chapter) had been undermined by its own author. This does, indeed, deserve the obloquy of a shattering inconsistency, and one not lost on Malthus, who, in his *Principles* (1820, *Works* II), employed the ingenious tactic of using Ricardo's own analysis to demonstrate the untenability of the PLTV. His criticisms hit home.

Ricardo comprehensively rewrote the chapter 'On Value' for the third edition of the *Principles*, newly adopting two strategies for the defence of the PLTV. First, he ruthlessly extirpated all the numerical examples that *had* suggested a 'considerable modification' to the PLTV and replaced them by others, according to which the 'greatest effects which could be produced on the relative prices of … goods from a rise of wages, could not exceed 6 or 7 per cent' (*Works* I, p. 36). Second, he introduced a new section 'On an invariable measure of value', where he indicated his desire to find a 'perfect measure of value' in terms of which prices would change *only* to reflect changes in the quantities of labour expended on production. This was tantamount to claiming that the discovery of the 'perfect' standard would itself sanction a PLTV, his problem being, however, that any commodity standard must be produced with *some* capital structure and, as he had demonstrated, the 'unwanted' price fluctuations are inescapable if capital structures differ. Hence his second-best solution of assuming that the standard is produced using an (unweighted) 'average' capital structure (cf. x_2, above), allegedly characteristic of 'most commodities': at least for *them* a PLTV would apply. There would still be a 'curious effect' with a fall in profitability, just as some commodities (such as our x_1) would now *rise* in price, but this was announced *sotto voce* (*Works* I, p. 46) and the effect was nowhere near as 'curious', at least in its magnitude, as it had been before.

Through this process of 'double indemnification' Ricardo had, *for the first time*, justified his use of the PLTV in explicit acknowledgement of the problems caused by differences in capital structures: either the differences are small and can be ignored, or (really a variation on the same theme) all the relevant commodities, including the standard, are part of a 'general mass' with the same capital structure. The 'exceptions' to the PLTV may therefore be ignored.

If we leave aside the dubious merit of the defence, its very inclusion is evidence that Ricardo was not retreating in his advocacy of the PLTV, contrary to claims by earlier commentators including J. Hollander (1904) and Cannan (1929, p. 177). That view was laid to rest by Sraffa, who opined that 'the theory of edition 3 appears to be the same, in essence and in emphasis, as that of edition 1' (*Works* I, p. xxxviii): a view that may itself be criticized for undervaluing the scale and significance of the changes (cf. Hollander, 1979, p. 217). But why was the theory so important to him?

One attraction is that it provided him with (in its own terms) a logically coherent framework for establishing his central theoretical propositions, particularly of the dependency of the general rate of profit on the conditions of producing wage goods.

A second possibility is that the theory appealed because of its (supposed) empirical relevance; hence Stigler's attribution to Ricardo of 'an *empirical* labour theory of value, that is, a theory that the relative quantities of labour … are the dominant determinants of relative value' (Stigler, 1958, p. 60; emphasis in original). However, although Ricardo did make empirical claims on behalf of theory (for which, it must be said, no evidence was adduced), those claims were arguably more a reflection of his commitment than its basis.

There was also an increasing tendency on Ricardo's part to identify the very *essence* of value with expended labour time. This 'value', referred to him at different times as 'natural value', 'positive value' and 'real value', was conceived as an attribute of *individual* commodities; hence the criticism, *fully accepted by Ricardo*, that he had moved beyond a purely relative usage of 'value' (as in Stigler's interpretation) and had turned it into something absolute (*Works* IX, p. 38). From this perspective, the role of the 'perfect' standard was to harmonize the 'labour values' with (relative) cost-of-production 'values' (or natural prices), since (changes in) the latter would become an exclusive reflection of (changes in) the former. As Ricardo forlornly conceded, however, 'perfection' is ruled out by unequal capital structures: the point he develops in *Absolute Value and Exchangeable Value* (*Works* IV, pp. 361–412), poignantly truncated by his final illness.

The analytical and 'philosophical' attractions are therefore central to understanding Ricardo's PLTV commitment. Of course, with the benefit of nearly two centuries of hindsight, it could be (and has been) argued that the labour theory can be jettisoned, to be replaced (say) with Sraffa's physically specified input–output equations. That argument may be formally correct, but we would no longer have *Ricardo's* theoretical and conceptual system. For him, the labour theory of value was both fundamental and indispensable.

The 'new view'

The inappropriately styled 'new view' (anticipated by Cannan, 1893, pp. 247–53, 350, with modern restatements by Casarosa, 1978, Hicks and Hollander, 1977, and Hollander, 1990; 2001; 2002, among others) can be treated either as a stand-alone interpretation of Ricardo's

treatment of wages or as part of a more far-reaching attempt to assimilate Ricardo's work to 'neoclassical' economics.

In the second great 'rehabilitation' of Ricardo (the first was J.S. Mill's attempt to have him reinstated as 'the greatest political economist': Mill, 1848, p. 397), Alfred Marshall applied his principle of 'generous interpretation' to distance Ricardo from the labour theory of value (by that time with its Marxian connotations) and absorb him within the mainstream intellectual tradition. Thus he averred that Ricardo had been 'feeling his way' towards a subjective utility analysis and that, despite appearances, he had attributed coordinate importance to supply *and* demand in the determination of natural prices (Marshall, 1920, Appendix I). Interestingly, however, Marshall's generosity deserted him when it came to Ricardo's treatment of wages, which he regarded as indefensible (Pigou, 1925, p. 413).

Most subsequent commentators have considered Marshall's' interpretation as far too generous, the prominent exception being Samuel Hollander, who goes even further in claiming to find a 'fundamentally important core of general-equilibrium economics' in Ricardo's work, implying a 'strong continuity of doctrine' between Ricardo's and later 'neoclassical' analysis (Hollander, 1987, pp. 6–7; cf. Morishima, 1989). As part of this 'general equilibrium' analysis, Ricardo had (allegedly) treated the wage rate as an *endogenous variable*, and it is this feature that is emphasized by the 'new view' interpretation.

The 'non-wage' aspects of the 'neoclassical' Ricardo may be dealt with briefly. First, with regard to utility (in the sense of subjective satisfaction), there is no question that it was treated as a *precondition* for exchangeable value and, in circumstances of fixed supplies, it was also conjectured by Ricardo that prices would be *proportionate* to 'utilities' (*Works* II, pp. 24–5; *Works* VIII, 276–7). However, there was no attempt by him to develop an analysis of *diminishing* marginal utility and, for the purpose of explaining exchangeable values (at natural prices), his emphasis was on *objective* determination by quantities of labour time. As to the 'coordinate' influence of supply and demand, this confuses the *process* by which market prices tend to their natural levels (which does involve output variations and, therefore, a pre-'neoclassical' species of supply-and-demand reasoning) with the *determination* of the natural price levels: for Ricardo, the latter is *independent* of supply and demand, thus effectively denying any *theoretical* relationship between output and (labour) conditions of production (see, for example, *Works* VIII, p. 207). Finally, on 'general equilibrium', there is not a single developed instance of such an analysis in the entire corpus of Ricardo's writings. It is an interpretation obtained only by reconstructing his work and, in the process, obliterating his own hallmark emphasis on unidirectional relationships.

With the 'new view', however, there is at least a textual basis. The most compelling evidence is from three paragraphs in the chapter 'On Wages'. The first paragraph opens thus: 'In the natural advance of society, the wages of labour will have a tendency to fall, as far as they are regulated by supply and demand', the assumption being that 'the supply of labourers will continue to increase as the same rate, while the demand for them will increase at a slower rate' (*Works* I, p. 101). Wages are therefore falling continuously in 'the natural advance' and only reach their 'natural' level (defined for a stationary population) in the terminal stationary state. However, 'we must not forget, that wages are also regulated by the prices of the commodities on which they are expended' (*Works* I, p. 101) and, particularly, by the rising price of corn (from diminishing returns on the land). Money wages therefore *rise* in the 'natural advance' but not by so much as to fully compensate the labourers for the rising corn price, so that real wages secularly decline as before. The effect of diminishing agricultural returns is in this way 'shared' between capitalists and labourers and has come to be known as the 'shared incidence principle'.

There is no doubt that the new view passages exist, and there are also muted refrains of the analysis elsewhere in the *Principles* (*Works* I, pp. 215, 220). At the same time, the natural wage analysis – with the natural wage, defined for a stationary population, as the active centre of gravity for market wages in *all* stages of society – is by far the dominant analysis in the *Principles*; and, unlike the new view, it is the only one consistent with the repeated claim that real-wage variations are of only 'temporary' significance, particularly with regard to movements in the general rate of profit. Based solely on the *Principles*, the proposition that the new view represents Ricardo's *true* position is difficult to sustain.

Malthus, for one, did not recognize Ricardo as a (kindred) new view theorist; hence the trenchant criticisms of Ricardo's natural wage analysis in his own *Principles* (Malthus, 1820. *Works* II, pp. 256–64). As Samuel Hollander (2007) has emphasized, however, Ricardo protested that he maintained 'no other doctrine than that which has been well explained by Mr. Malthus' (*Works* II, p. 288). Yet he also reaffirmed his *own* definition of the natural wage (*Works* II, pp. 227–8), which is inexplicable if he truly agreed with Malthus (for whom Ricardo's natural wage would be irrelevant outside the stationary state).

While it cannot be denied, then, that Ricardo was on some level sympathetic to the new view analysis, he was at no time an unequivocal exponent of that doctrine. Even in his later writings (such as *On Protection to Agriculture*, *Works* IV), in which the natural wage is not mentioned explicitly, the real wage is treated as a given and fixed entity, without a trace of the new view. It is also significant that Ricardo was never to criticize the writings of contemporaries, including their avowed representations of his own position, for the (universal) failure to include the new view (Peach, 2007). His own credentials as a new view theorist must therefore remain in considerable doubt.

Conclusion: Ricardo as a 'classical' economist?

Ricardo was to achieve great fame as a political economist during the tragically short period in which he wrote on the subject, although his ideas, especially his policy proposals, were often bitterly contested by critics of differing political and theoretical persuasions (see Peach, 2003). Following his death, his name, if not always his own doctrines, lived on through the writings of his 'New School' disciples, notably James Mill, Thomas De Quincey and the indefatigable J.R. McCulloch, and through the efforts of J.S. Mill. With the advent of 'neo-classical' economics, however, Ricardo's stock began to plummet, with Marshall's attempted rehabilitation to no avail. By the time Sraffa's edition of the *Collected Works* appeared in the 1950s, Ricardo's positive contribution was not uncommonly reduced to anaemic generalities such as the development of a 'professional frame of mind' or an 'abstract deductive approach', the onward progress of economic science having established the 'inadequacy' of much of his substantive work.

The *Collected Works* prompted a flurry of new scholarly interest in Ricardo, but it was only after the publication of Sraffa (1960) that he was subjected to his third major 'rehabilitation', this time not as a 'mainstream' economist (as with J.S. Mill and Marshall) but as a precursor of *Sraffa's* economics or, as related by Sraffa's followers, as a founder of the 'classical' (or 'surplus') tradition that Sraffa (1960) had revived.

The defining characteristics of 'classical' economics are alluded to in the Preface of Sraffa (1960) and amount to the assumption of *given* outputs and methods of production. The distribution between wages and profits may then be 'solved' by taking one distributive variable as given and calculating the other as a residual (or 'surplus'). But how well does this apply to Ricardo's approach? At one level – the calculation of profit as a 'surplus' – it does so well enough. Where the problems arise is with the other attribution of given outputs.

According to Sraffa, 'The "principal problem in Political Economy" was in [Ricardo's] view the division of the national product between classes and in the course of that investigation he was troubled by the fact that the size of this product appears to change when the division changes' (*Works* I, p. xlviii). Hence, 'the problem of value which interested Ricardo was how to find a measure of value which would be invariant to changes in the division of the product' (*Works* I, p. xlviii); and, as Sraffa remarks parenthetically, Ricardo may have come close to solving his 'problem' with the 'average' standard adopted in the third edition of the *Principles*: 'If measured in such a standard, the average price of all commodities, and their aggregate value, would remain unaffected by a rise or fall of wages' (*Works* I, pp. xliv–xlv).

Several objections can be made against Sraffa's interpretation. First and foremost, it implies that Ricardo's 'principal problem' was with purely 'notional' redistributions of a *given national product* (that is, with given and unchanging outputs). However, Ricardo's own 'principal problem', as he defined it himself, was with the 'natural course of rent, profit, and wages' *over time*, and for the purpose of *his* investigation there will be at least one output, that of corn, that *cannot* be treated as given and unchanging in terms of its conditions of production. Second, as Ricardo clarified, *his* analysis of distribution was to be framed at the level of the individual firm, or farm, not in terms of social or 'national' aggregates. Third, there is no evidence that *he* envisaged a 'price-balancing' function for his 'average' standard (that is, to ensure constancy in the total value of national output); indeed, Ricardo's opinion was that *all* distributions-induced price changes are evidence of a 'defect' in the standard.

Ricardo was not a full-fledged 'classical' economist in the Sraffa mould. To describe him more loosely as a 'surplus theorist' is unexceptionable, although by focusing on only one (albeit important) area of Ricardo's writings it is also a 'thin' characterization. Ricardo was a towering intellectual force whose work ranged over all the main areas of political economy. Forcing him into classificatory boxes of a later construction is a disservice to him and a hindrance to those who would seek to understand the full richness and extent of his historical significance.

TERRY PEACH

See also **classical economics; Sraffian economics.**

Bibliography

Cannan, E. 1893. *A History of the Theories of Production and Distribution in English Political Economy from 1776 to 1848.* London: Percival.

Cannan, E. 1929. *A Review of Economic Theory.* London: P.S. King & Son.

Casarosa, C. 1978. A new formulation of the Ricardian system. *Oxford Economic Papers* 1, 38–63.

Eatwell, J.L. 1975. The interpretation of Ricardo's *Essay on Profits. Economica* 42, 182–7.

Faccarello, G. 1982. Sraffa versus Ricardo: the historical irrelevance of the 'corn profit' model. *Economy and Society* 11, 122–37.

Garegnani, P. 1982. On Hollander's interpretation of Ricardo's early theory of profits. *Cambridge Journal of Economics* 6, 65–77.

Heertje, A. 2004. The Dutch and Portuguese background of David Ricardo. *European Journal of the History of Economic Thought* 11, 281–94.

Hicks, J. and Hollander, S. 1977. Mr. Ricardo and the moderns. *Quarterly Journal of Economics* 91, 351–69.

Hollander, J.H. 1904. The development of Ricardo's theory of value. *Quarterly Journal of Economics* 19, 455–91.

Hollander, S. 1973. Ricardo's analysis of the profit rate, 1813–15. *Economica* 40, 260–82.

Hollander, S. 1975. Ricardo and the Corn Profit Model: Reply to Eatwell. *Economica* 52, 188–202.

Hollander, S. 1979. *The Economics of David Ricardo.* London: Heinemann.

Hollander, S. 1987. *Classical Economics.* London: Blackwell.

Hollander, S. 1990. Ricardian growth theory: a resolution of some problems in textual interpretation. *Oxford Economic Papers* 42, 730–50.

Hollander, S. 2001. Classical economics: a reification wrapped in an anachronism? In *Reflections on the Classical Canon in Economics*, ed. E.L. Forget and S. Peart. London: Routledge.

Hollander, S. 2002. The canonical classical growth model: content, adherence and priority. In *Competing Economic Theories*, ed. S. Nisticò and D. Tosato. London: Routledge.

Hollander, S. 2007. Ricardo as a 'classical economist': the 'new view re-examined: A reply to Dr Peach. *History of Political Economy* 39, 307–12.

Keynes, J.M. 1936. *The General Theory of Employment, Interest and Money.* London: Macmillan.

Marshall, A. 1920. *Principles of Economics.* 8th edn. London: Macmillan, 1979.

Mill, J. 1808. Review of Britain Independent of Commerce, by W. Spence. *Edinburgh Review* 11, 429–49.

Mill, J.S. 1848. *Principles of Political Economy.* New York: Kelley, 1987.

Mill, J.S. 1873. *Autobiography.* London: Penguin, 1989.

Morishima, M. 1989. *Ricardo's Economics: A General Equilibrium Theory of Distribution and Growth.* Cambridge: Cambridge University Press.

Peach, T. 1984. David Ricardo's early treatment of profitability: a new interpretation. *Economic Journal* 94, 733–51.

Peach, T. 1993. *Interpreting Ricardo.* Cambridge: Cambridge University Press.

Peach, T. 2001. Hollander de Vivo and the 'further evidence' for the corn model interpretation of Ricardo: a conspiracy of silence? *Cambridge Journal of Economics* 25, 685–692.

Peach, T. 2003. *David Ricardo: Critical Responses*, 4 vols. London: Routledge.

Peach, T. 2007. Ricardo as a 'classical economist': the 'new view' re-examined. *History of Political Economy* 39, 293–306.

Peake, C.F. 1978. Henry Thornton and the development of Ricardo's economic thought. *History of Political Economy* 10, 193–212. In *David Ricardo: Critical Assessments* (second series), vol. 5, ed. J. Cunningham Wood. London: Routledge, 1994.

Pigou, A.C. 1925. *Memorials of Alfred Marshall.* London: Macmillan.

Ricardo, D. 1951–55. *The Works and Correspondence of David Ricardo*, ed. P. Sraffa with the collaboration of M.H. Dobb, vols. I–X, Cambridge: Cambridge University Press.

Ruffin, R.J. 2002. David Ricardo's discovery of comparative advantage. *History of Political Economy* 34, 725–48.

Ruffin, R.J. 2005. Debunking a myth: Torrens on comparative advantage. *History of Political Economy* 37, 711–22.

Samuelson, P.A. 1972. The way of an economist. In *The Collected Papers of Paul A. Samuelson*, ed. R.C. Merton. Cambridge, MA: MIT Press.

Sayers, R.S. 1953. Ricardo's views on monetary questions. *Quarterly Journal of Economics* 67, 30–49. In *David Ricardo: Critical Assessments*, vol. 4, ed. J. Cunningham Wood. Beckenham: Croom Helm, 1985.

Sraffa, P. 1960. *Production of Commodities by Means of Commodities: Prelude to a Critique of Economic Theory.* Cambridge: Cambridge University Press.

Stigler, G.J. 1958. Ricardo and the 93% labour theory of value. *American Economic Review* 48, 357–67. In *David Ricardo: Critical Assessments*, vol. 2, ed. J. Cunningham Wood. Beckenham: Croom Helm, 1985.

Ricardian equivalence theorem

The Ricardian equivalence theorem is the proposition that the method of financing any particular path of government expenditure is irrelevant. More precisely, whether government purchases are financed by levying lump-sum taxes or by issuing government bonds does not affect the consumption of any household, nor does it affect capital formation. In this sense, financing government purchases by lump-sum taxes is *equivalent* to financing these purchases by issuing bonds. The fundamental logic underlying this proposition was presented by David Ricardo in Chapter XVII ('Taxes on Other Commodities than Raw Produce') of *The Principles of Political Economy and Taxation*. Although Ricardo clearly explained why government borrowing and taxes could be equivalent, he warned against accepting the argument on its face: 'From what I have said, it must not be inferred that I consider the system of borrowing as the best calculated to defray the extraordinary expenses of the state. It is a system which tends to make us less thrifty – to blind us to our real situation' (1821, pp. 162–3).

The question of whether lump-sum taxes and government debt are equivalent arises in the specification of the consumption function. The aggregate consumption function plays an important role in models of national income determination, and aggregate consumption is often specified to depend on contemporaneous aggregate disposable income and on aggregate wealth. The question is whether the public's holding of bonds issued by the government should be treated as part of aggregate net wealth. Indeed, this is the eponymous question of Barro's (1974) classic article on Ricardian equivalence. If consumers recognize that government bonds, in the aggregate, represent future tax liabilities, then these bonds would not be part of aggregate wealth. If, on the other hand, consumers do not recognize, or for some reason do not care about, the implied future tax liabilities associated with these bonds, then they should be counted as part of aggregate wealth in an aggregate consumption function. Patinkin (1965, p. 289), citing Carl Christ and

Christ's discussions with Milton Friedman, recognized this question and specified that a fraction k of the stock of outstanding government bonds is to be treated as wealth. Under the Ricardian equivalence view, k would equal zero; under the view that consumers ignore all future tax liabilities, k would equal 1. Bailey (1971) also examined the question of whether future tax liabilities affect aggregate consumption in a model of national income determination, though his formulation of the aggregate consumption function does not explicitly include aggregate wealth.

The question of whether government bonds are net wealth and the question of the effects of alternative means of financing a given amount of government expenditure are, in many contexts, basically the same question. For purposes of exposition, it is clearest to focus on one particular formulation of the question. The discussion here will focus on the question of the choice between current taxation and debt finance.

The underlying logic of the Ricardian equivalence theorem is quite simple and can be displayed by considering a reduction in current lump-sum taxes of 100 dollars per capita. This reduction in government tax revenue is financed by the sale of government bonds on the open market in the amount of 100 dollars per capita. For simplicity, suppose that the bonds are one-year bonds with an interest rate of five per cent per year. In addition, suppose that the population of taxpayers is constant over time. In the year following the tax cut, the bonds are redeemed by the government. In order to pay the principal and interest on the bonds, taxes must be increased by 105 dollars per taxpayer in the second year.

Now consider the response of households to this intertemporal rearrangement of their tax liabilities. Households can afford to maintain their originally planned current and future consumption by increasing their current saving by 100 dollars. In fact, the additional 100 dollars of private saving could be held in the form of newly issued government bonds. In the second year, when the government increases taxes by 105 dollars to redeem the bonds, households pay the extra tax using the principal and interest on the bonds. Thus, the originally planned path of consumption continues to be feasible after the tax change. In addition, since the originally planned path of consumption was chosen by the consumer before the tax change, it would continue to be chosen after the tax change since all relative prices remain unchanged. Therefore, household behaviour is invariant to the switch between tax finance and debt finance for a given amount of government spending.

In the basic example, the tax cut in the current year is financed by the issuance of one-year government bonds. However, the invariance result continues to hold if the current tax cut is financed by the issuance of N-year bonds. The argument is that once again each consumer uses the extra 100 dollars of disposable income in the first year to purchase 100 dollars of newly issued government bonds. If these government bonds pay interest in years before the bond is redeemed, then the government must increase lump-sum taxes in those years to service the bonds. Consumers who are holding the bonds and receive interest use the interest on their bonds to pay the increased taxes. Then, when the bonds mature after N years, each consumer uses the principal and final interest on these bonds to pay the higher taxes that are levied to redeem the debt. Once again, consumers can afford to maintain the originally planned path of current and future consumption and find it optimal to do so.

Having seen that the Ricardian equivalence theorem holds even if long-term bonds are issued to pay for the current tax cut, it is natural to ask whether the invariance result continues to hold even if some or all of the currently living consumers die before the bonds are redeemed. The first answer to this question would appear to be that consumers who are alive during the tax cut, but who die before the newly issued bonds are retired, would have a reduction in the present value of their taxes and thus an increase in the present value of their disposable income. Equivalently, such consumers could afford to increase their current and future consumption. It is not necessary for these consumers to hold the extra bond that is issued in the first year because they will not have to use the bonds to pay for the future tax increase needed to redeem the bonds. Therefore, these consumers would tend to increase their current and future consumption, *ceteris paribus*.

A self-interested consumer who receives a tax cut financed by government bonds will increase his consumption if he knows with certainty that he will die before future taxes are collected to fully repay the newly issued bonds. But if the consumer is uncertain about when he will die, the situation involves some additional considerations. I begin by ignoring survival-contingent assets such as annuities and life insurance, and I will assume that all consumers have positive net financial assets so that I can put aside issues related to borrowing costs for consumers who may die before repaying their loans. To keep the argument simple, suppose that lump-sum taxes are reduced in the current year by 100 dollars per taxpayer, and the government finances the tax cut by issuing 20-year zero-coupon bonds. Twenty years in the future the government will increase lump-sum taxes to pay off these bonds. The present value of the future tax increase is 100 dollars per current taxpayer. If the number of taxpayers 20 years in the future is the same as in the current year, then tax increase in the future will be $100(1 + r)^{20}$ per taxpayer, where r is the annual interest rate on the government bonds. In this case, the current tax cut will not affect the current consumption of any taxpayer. A current taxpayer could use the 100 dollars from the current tax cut to buy 100 dollars of government bonds, and simply plan to hold on to the bonds for 20 years. In the event that the consumer is still alive and consuming 20 years in the future, he can use his bonds,

which will have grown in value to $100(1 + r)^{20}$ to pay the additional lump-sum taxes in that year, without changing consumption in that year (or in the current year). In the event that the consumer dies before 20 years elapse, he will, of course, consume the same level, namely zero, as in the absence of the tax cut. Thus, buying and holding 100 dollars of government bonds in the current year just allows the consumer to maintain consumption unchanged at all ages and in all states (that is, the state in which the consumer is alive in 20 years and the state in which he is not alive in 20 years). Therefore, Ricardian equivalence holds in this case.

The example in the preceding paragraph illustrates that Ricardian equivalence can hold even when selfish consumers receive a bond-financed tax cut and, with some unpredictability, die before the taxes are levied to fully repay the bonds. A crucial step in the argument is the assumption that the number of future taxpayers is the same as the number of current taxpayers. But with some taxpayers dying over time, the only way to maintain a constant number of taxpayers is for new taxpayers to arrive – through birth or immigration – at the same rate at which taxpayers are dying. In an economy with a growing population, that is, in an economy in which the sum of the birth rate and net immigration rate exceeds the death rate, the increase in future lump-sum taxes *per taxpayer* will be smaller than $100(1 + r)^{20}$, because the cost of paying off government bonds is spread among a larger number of taxpayers. Therefore, a consumer who receives a tax cut of 100 dollars in the current year will face a future tax increase that has a present value smaller than 100 dollars, and so will increase consumption in the current period (and in the future period, if he is alive). Alternatively, in an economy with a shrinking population, a lump-sum tax increase will have the opposite effect and will reduce current consumption. (An analytic version of this example is in Abel, 1989.)

Ricardian equivalence is often illustrated in the context of perfect markets. If consumers face uncertainty about the length of their lives, and if they do not have bequest motives of any sort, they will want to hold annuities, which are assets that pay off if the owner of the annuity is alive, but pay zero if the owner is not alive. If all consumers face the same publicly known probability, p, of dying each year, then the actuarially fair annual gross rate of return on annuities will be $(1 + r)/(1 - p)$. If all consumers invest one dollar in an annuity that pays a lump sum in 20 years, then in 20 years each survivor will receive $[(1 + r)/(1 - p)]^{20}$ dollars. Whether consumers who receive a 100 dollar lump-sum tax cut in the current year will change their current consumption depends on the amount of the tax increase *per taxpayer* 20 years in the future when the bonds used to finance the tax cut are paid off. If the birth rate and the net immigration rate are both zero, then the population of taxpayers in 20 years will be a fraction $(1 - p)^{20}$ of the population in the current year. Thus, to repay the principal and interest on the

100 dollars of bonds issued per current taxpayer, the lump-sum tax will have to increase by $[(1 + r)/(1 - p)]^{20}$ dollars per current taxpayer. Thus, a current taxpayer could use the 100 dollar tax cut in the current period to purchase 100 dollars of annuities in the current period. If the consumer survives for 20 years, the payoff of the annuity, $[(1 + r)/(1 - p)]^{20}$, will be just sufficient to pay the increased lump-sum tax in that year. Thus, Ricardian equivalence holds in this case with perfect annuities and a zero birth rate and zero net immigration rate. Ricardian equivalence will fail to hold, however, if the birth rate is positive, because the tax burden in 20 years will be spread among a group of taxpayers consisting of surviving taxpayers from the current period plus additional taxpayers. In this case, the tax increase per future taxpayer will be smaller than $[(1 + r)/(1 - p)]^{20}$. Thus, recipients of the tax cut in the current period would be able to increase consumption in the current period somewhat and use the remainder of the tax cut to buy enough annuities to pay the increased lump-sum tax and to increase consumption in 20 years.

The examples with uncertain longevity illustrate that, as emphasized by Weil (1989), the departure from Ricardian equivalence does not result solely from the chance of dying before the future tax increase. In the examples in which the tax cut in the current year induces an increase in current consumption, the effect results from the fact that future increase in taxes per future taxpayer is smaller than the current tax cut per taxpayer. In the case of perfect annuities, this effect is made possible by a positive birth rate, and in the case without annuities the effect was made possible by growing population resulting from a death rate lower than the birth rate plus the net immigration rate.

Altruistic consumers

If consumers are entirely self-interested, then escaping future taxes through death can lead to departures from the Ricardian equivalence theorem, as discussed above. However, Robert Barro (1974) presented an ingenious argument that extends the Ricardian equivalence theorem to situations in which consumers die before future taxes are increased to repay the bonds that are issued to finance the current tax cut. Before discussing the substantive content of Barro's argument, it is interesting to observe that the term 'Ricardian equivalence theorem' apparently was first used by James Buchanan (1976) in a published comment on Barro's paper. Buchanan's comment begins by pointing out Barro's failure to credit Ricardo with the idea that debt and taxes may be equivalent and, indeed, the comment is titled, 'Barro on the Ricardian Equivalence Theorem'. Previously, Buchanan had referred to this result as the 'equivalence hypothesis' (1958, p. 118).

Barro postulated that consumers have bequest motives of a particular form that has been labelled 'altruistic'.

An altruistic consumer obtains utility from his own consumption as well as from the utility of his children. Therefore, a consumer who is altruistic toward all of his children cares not only about his own consumption but also indirectly about the consumption of all his children. Furthermore, if all of the altruistic consumer's children are also altruistic and care about the utility of all of their children, then the altruistic consumer cares indirectly about the consumption of all of his grandchildren. Provided that all consumers are altruistic, the argument can be extended ad infinitum with the important implication that an altruistic consumer cares, at least indirectly, about the entire path of current and future consumption of himself and all of his descendants.

Barro's insight that an intergenerationally altruistic consumer cares about the entire path of his family's consumption defuses the argument that consumers who know they will escape future taxes through death will increase consumption in response to a current tax cut. For altruistic consumers, it does not matter whether they themselves or their descendants pay the higher taxes necessary to pay the principal and interest on the newly issued bonds. In response to a 100 dollar tax cut in the current year, an altruistic consumer will not change his consumption but will hold an additional 100 dollars of government bonds. If the bonds are not redeemed until after the consumer dies, he will bequeath them to his children who can then use the bonds to pay the higher taxes in the year in which the bonds are redeemed, or else bequeath the bonds to their children if the bonds are not redeemed during their lifetimes.

The fact that a consumer leaves a bequest is not prima facie evidence that he is altruistic in the sense defined above. Bequests may arise as the accidental outcome of an untimely death or they may arise for motives other than altruism. For instance, if the utility that a consumer obtains from leaving a bequest depends only on the size of the bequest, then he will not care about tax increases that may be levied on his children or his children's children. In this case Ricardian equivalence would not hold.

The argument that each current and future consumer in a family of intergenerationally altruistic consumers cares about his own consumption as well as the consumption of all of his descendants for ever raises the question of whether the government must ever pay off the newly issued government bonds. If the government could roll over the principal and interest on this debt for ever, so that it would never be necessary to increase future taxes, it would seem that a current tax cut financed by issuance of government bonds would reduce the present value of the taxes paid by the current and future members of the family and hence would lead to an increase in the family's consumption. If the government attempted to roll over its debt each year by issuing new bonds, the quantity of these bonds would grow in perpetuity at the rate of interest. If the rate of interest exceeds the economy's growth rate, then these bonds would not willingly be held in private portfolios. Alternatively, if the rate of interest falls short of the economy's growth rate – a condition that signals an inefficient over-accumulation to capital – then, as pointed out by Feldstein (1976), it is possible for the government to roll over the debt permanently. Carmichael (1982) has shown that in this case the altruistic bequest motive will not be operative (that is, the non-negativity constraint on bequests will bind) but that an altruistic gift motive from children to parents (which specifies that a consumer's utility depends on his own consumption and the utility of his parents) may be operative. If the gift motive is operative, then Carmichael argues that Ricardian equivalence will hold, despite the fact that government bonds may be regarded as net wealth.

Departures from Ricardian equivalence

Now that we have described a fairly general set of conditions under which Ricardian equivalence holds, it is useful to discuss several of the conditions that might lead to a violation of Ricardian equivalence. A clear overview of reasons why the Ricardian equivalence theorem may not provide an accurate description of the actual effects of debt finance vs. tax finance is provided by Tobin (1980).

The basic argument underlying the Ricardian equivalence theorem is that it makes no difference whether the government issues debt in the amount of 100 dollars per capita or whether it collects taxes of 100 dollars per capita since in the latter case consumers can borrow 100 dollars per capita to pay the higher taxes. In the former case, public borrowing is increased by 100 dollars per capita, and in the latter case private borrowing is increased by 100 dollars per capita. Under the appropriate conditions it makes no difference whether the borrowing is by the public sector or by the private sector. In order for the choice between debt finance and tax finance to have an effect, it must be the case that any changes in government borrowing cannot be fully offset by changes in private sector behaviour. Equivalently, there must be something that the government can do in credit markets that the private sector cannot do.

The government can borrow by issuing bonds, but in some situations consumers may not be able to borrow. For instance, a young consumer with a high prospective income might like to borrow to increase his consumption when young with the intention of repaying the loan when his income is higher in the future. However, for a variety of reasons it may simply not be possible for the young consumer to borrow the desired amount; if this is the case, the consumer is described as 'liquidity-constrained'. A liquidity-constrained consumer who receives a tax cut in the current period may choose to consume some, or even all, of the tax cut rather than save the entire tax cut. In effect, the current tax cut allows the consumer to

borrow in order to increase current consumption, which is what the liquidity-constrained consumer wanted to do anyway. The current tax cut financed by an issue of government bonds can be viewed as the government borrowing on behalf of the consumer. Although this example makes it seem clear that a liquidity-constrained consumer would increase his current consumption in response to a current tax cut, some caution is required in interpreting this result. Unless the reason for the liquidity constraint is specified, one cannot determine what will be the effect of the tax cut. For example, suppose that a consumer is able to borrow some funds, but is liquidity-constrained in the sense that he would like to borrow even more funds. If his creditors determine how much they are willing to lend by looking at his ability to repay the loan, then, in response to the prospective tax increase accompanying the current tax cut, his lenders may reduce the amount they are willing to lend by the amount of the tax cut. In this case, Ricardian equivalence would hold.

The Ricardian equivalence theorem requires not only that consumers be intergenerationally altruistic, but that their bequest motives be operative in the sense that consumers can bequeath whatever amount they choose subject to their budget constraint. To be more precise, it is possible that an altruistic consumer may like to leave a negative bequest to his children, but he is constrained from leaving a bequest less than zero. The fact that a consumer may want to leave a negative bequest does not necessarily violate the assumption that the consumer is altruistic. It may be that the consumer's children will all be so much wealthier than the consumer that, even though the consumer cares about the utility of his children, he could achieve higher utility by taking some of his children's resources and consuming them himself. Formal conditions that imply that altruistic consumers would like to leave negative bequests have been presented by Drazen (1978) and Weil (1987). Under these conditions, if the consumer is constrained from leaving a negative bequest, he will instead leave a zero bequest. In such cases, a tax cut that is followed by a tax increase after the consumer's death will reduce the present value of the taxes paid by the consumer and he will increase his consumption. In effect, the current tax cut helps the consumer achieve the desired negative bequest by giving him current resources and taking resources away from his descendants.

Another reason for departure from the Ricardian equivalence theorem is that policy may redistribute resources among families that have different marginal propensities to consume. For instance, suppose that one half of the consumers receive a 200 dollar tax cut in the current year and the other half of the consumers have unchanged taxes in the current year. The government finances the tax cut by issuing bonds in the amount of 100 dollars per capita, and in the following period it redeems the bonds and pays the interest. For simplicity, suppose that the population is constant and that the interest rate on government bonds is five per cent per year. Then in the year following the tax cut there is a tax increase of 105 dollars per consumer. Finally, suppose that this tax increase is levied on all consumers equally. In this case, the tax cut in the current year redistributes resources from the consumers whose taxes are unaffected to the consumers whose taxes are reduced in the current year. The recipients of the transfer will increase their consumption and the other consumers will reduce their consumption. The reallocation of consumption across consumers may be viewed as a violation of Ricardian equivalence. Whether aggregate consumption rises or falls depends on the marginal propensities to consume of the recipients of the transfer compared with the marginal propensities to consume of the other consumers. If all consumers have equal marginal propensities to consume, then there will be no effect on aggregate consumption or capital accumulation. However, if, for instance, the recipients of the transfers have a higher marginal propensity to consume than the other consumers, then aggregate consumption would increase. It should be pointed out that, in some sense, this example does not represent a violation of the Ricardian equivalence theorem, because it ignores the possibility that there might exist an insurance market for individual tax liabilities. If there were such a market, then consumers could have insured themselves against the redistribution of taxes. Such markets do not generally exist, but whether the Ricardian equivalence theorem holds may depend on the reason why these markets do not exist.

To see the role of insurance markets in a different context, consider consumers who each contribute 100 dollars to a social security fund during their working life. Suppose that at the end of their working lives some of the consumers die and the others survive and live in retirement. Although the number of consumers who die at retirement may be predictable, the identities of those who will die are not predictable. The surviving retired consumers each receive an equal share of the social security fund (with accrued interest) to which they contributed while they were working. Each survivor's social security income is greater than the 100 dollars (plus interest) which he contributed, because the fund contains the contributions plus interest of his peers who died at the end of the working life.

Does the introduction of this type of social security system affect consumption and capital accumulation or does Ricardian equivalence imply that consumption and capital accumulation will be unaffected? To answer this question, it is useful to observe that this stylized social security system has the characteristics of an actuarially fair annuity. That is, consumers pay a premium when young (the social security tax) and receive a payment if, and only if, they survive to old age. Furthermore, if all consumers face the same probability of dying, the

rate of return to the survivors is equal to the actuarially fair rate of return. If there were a competitive annuity market offering the actuarially fair rate of return, the social security system would have no effect on consumption or capital accumulation. Workers who are taxed 100 dollars are essentially forced to hold 100 dollars of the publicly provided actuarially fair annuity called social security; however, these consumers can afford, and will choose, to maintain their originally planned consumption and bequests by reducing their holdings of privately supplied annuities by 100 dollars. This reduction in the holding of private annuities allows consumers to re-establish their initial portfolios of annuities and other assets while maintaining consumption unchanged. Thus, the Ricardian equivalence theorem holds in this example, provided that consumers each originally planned to hold at least 100 dollars of private annuities.

If the probability of surviving until retirement differs across consumers, and if individual consumers are better informed about their own survival probabilities than are insurance companies, then the funded social security system described above will affect consumption. The reason is that, if an insurance company offered annuities at a price that would be actuarially fair to the average consumer, it would suffer from what is known as 'adverse selection'. As a simple example, suppose that insurance companies know the average mortality probability but have no additional information about the mortality probabilities of individual consumers. If an insurance company offered annuities at a price that would be actuarially fair to the average consumer, then consumers who believe they are healthier than average would view these annuities as a bargain; consumers who believe they are less healthy (or engage in more dangerous activities) than average would view these annuities as overpriced because these consumers have a smaller chance of living to reap the rewards. As the healthy consumers would buy a disproportionately large share of annuities, they would, on average, inflict losses on the sellers of these annuities and would induce these sellers to charge a higher price for annuities. However, the social security system can supply its annuities at the actuarially fair price for the average consumer because a compulsory social security system is immune to adverse selection. That is, because the government can determine the amount of the publicly provided annuity held by each individual, it does not have to worry that a disproportionately large share of annuities are held by healthy consumers. Therefore, as shown in Abel (1986), the annuity offered by the social security system would yield a higher rate of return than private annuities, or, equivalently, would be made available at a lower price to consumers. Because of the difference in the prices of the publicly provided and privately supplied annuities, consumers could not exactly offset the effects of social security by transacting in private annuity markets.

The example in which adverse selection leads to violation of the Ricardian equivalence theorem was constructed to obey the strict set of rules demanded by strong adherents to the view that the choice between debt finance and tax finance is irrelevant. In particular, the following assumptions were maintained: (a) consumers are forward-looking and understand that a bond-financed tax cut implies an increase in future taxes; (b) consumers have operative altruistic bequest motives so that they care about taxes after their death; (c) there is a complete set of competitive markets; and (d) only lump-sum taxes are changed. However, actual economies display several important departures from each of these assumptions. Violations of these assumptions are discussed below.

First, despite the widespread appeal of rational expectations in modern economics, it may simply be that consumers do not fully appreciate the link between a current tax cut and an increase in future taxes. If consumers did not understand this link at all, then a current tax cut would tend to increase current consumption.

Second, consumers may not have a bequest motive, either because they have no children or because they do not care about the welfare of anyone else. Even if consumers do have a bequest motive, it may not be operative as discussed above. Even if the bequest motive is operative, it may not be of the appropriate form for the Ricardian equivalence theorem to hold. If a consumer's utility depends directly on the size of the bequest he leaves rather than on the utility of his heirs, then a current tax cut followed by a tax increase on his heirs, would tend to raise the current consumption of the consumer. The reason is that he does not care about his heirs' utility per se. His bequest yields utility directly just as any other consumption good. As a result of the decrease in taxes he must pay over his lifetime, the consumer will have a higher level of lifetime income and can increase his own consumption and the bequest he leaves. If his own consumption and the bequest are both normal goods in his utility function, then he will choose to increase both.

Even if all consumers have operative altruistic bequest motives, a tax cut may increase current consumption. If all consumers have several children, but if each consumer cares about the utility of only one of his children, then there will be consumers in future generations whose utility is ignored by all current consumers. To the extent that future taxes are levied on these consumers, some part of future tax liabilities associated with a current tax cut will be ignored by current consumers. In this case, a tax cut would increase contemporaneous aggregate consumption.

Bernheim and Bagwell (1988) have challenged the plausibility of the assumption of intergenerational altruism by showing that this assumption leads to some untenable conclusions. If consumers A and B are unrelated to each other and both are altruistic toward consumer C, then A and B are effectively linked to each other, if both consumers A and B both plan to give positive transfers (bequests) to consumer C. For example, unrelated grandparents (consumers A and B) who plan to make positive transfers to their common grandchildren (consumers C) are effectively linked to each other. If the government transfers a dollar from consumer A to consumer B, these consumers can, and will choose to, undo this transfer and maintain their originally chosen patterns of consumption. The mechanism for undoing the government transfer is for consumer A to reduce his transfer to consumer C by one dollar and for consumer B to increase his transfer to consumer C by one dollar. Bernheim and Bagwell have argued that, if one takes intergenerational altruism seriously, such linkages are so widespread that all consumers are effectively linked to each other. In this case, all government transfers among consumers, including a transfer from future taxpayers to current taxpayers in the form of a bond-financed tax cut, would have no effect. Bernheim and Bagwell go on to show that even non-lump-sum taxes, and indeed prices themselves, would not affect consumption. Rather than conclude that all taxes and prices are irrelevant, Bernheim and Bagwell conclude that their findings cast doubt on the policy conclusions, including Ricardian equivalence, that are based on the assumption of intergenerational altruism.

Third, various types of insurance markets may be absent or, as described above, may suffer from adverse selection. Chan (1983) and Barsky, Mankiw and Zeldes (1986) have argued that, if there are no markets for insuring against unpredictable fluctuations in after-tax income, then a current tax cut could increase current consumption. The argument, which was outlined by Barro (1974, p. 1115) and Tobin (1980, p. 60), is that to the extent that individual tax liabilities are proportional to income the tax system provides partial insurance against fluctuations in individual disposable income. Therefore, the increase in tax rates that follows a lump-sum tax cut in the current year will reduce the variability of future disposable income. The reduction in the riskiness of future disposable income reduces current precautionary saving that consumers undertake to guard against low future consumption. The counterpart of the reduction in precautionary saving is an increase in current consumption.

Fourth, most taxes are not lump-sum taxes. Generally, taxes are levied on economic activities, and changes in these taxes provide incentives to alter the levels of these activities. Although the existence of distortionary taxes does not in all cases imply that Ricardian equivalence is violated when applied to lump-sum tax changes, it does strain the interpretation of empirical tests of Ricardian equivalence that examine historical data on deficits and consumption.

As discussed above, there are many potential sources of departure from the Ricardian equivalence theorem, and ultimately the importance of these departures is an empirical question. The existing literature that attempts to test empirically whether Ricardian equivalence holds has produced mixed results, some claiming to show that it holds, and others the opposite. In judging the empirical relevance of the Ricardian equivalance theorem, however, the important question from the viewpoint of fiscal policy formulation is not whether the theorem holds exactly but whether there are departures from it that are quantitatively substantial. Existing empirical work has not yet produced a consensus on this question.

ANDREW B. ABEL

See also **government budget constraint; public debt; public finance.**

Bibliography

Abel, A. 1986. Capital accumulation and uncertain lifetimes with adverse selection. *Econometrica* 54, 1079–97.

Abel, A. 1989. Birth, death and taxes. *Journal of Public Economics* 39, 1–15.

Bailey, M. 1971. *National Income and the Price Level*, 2nd edn. New York: McGraw-Hill.

Barro, R. 1974. Are government bonds net wealth? *Journal of Political Economy* 82, 1095–117.

Barsky, R., Mankiw, G. and Zeldes, S. 1986. Ricardian consumers with Keynesian propensities. *American Economic Review* 76, 676–91.

Bernheim, B. and Bagwell, K. 1988. Is everything neutral? *Journal of Political Economy* 96, 308–38.

Buchanan, J. 1958. *Public Principles of Public Debt*. Homewood, IL: Richard D. Irwin.

Buchanan, J. 1976. Barro on the Ricardian equivalence theorem. *Journal of Political Economy* 84, 337–42.

Carmichael, J. 1982. On Barro's theorem and debt neutrality: the irrelevance of net wealth. *American Economic Review* 72, 202–13.

Chan, L. 1983. Uncertainty and the neutrality of government financing policy. *Journal of Monetary Economics* 11, 351–72.

Drazen, A. 1978. Government debt, human capital and bequests in a lifecycle model. *Journal of Political Economy* 86, 337–42.

Feldstein, M. 1976. Perceived wealth in bonds and social security: a comment. *Journal of Political Economy* 84, 331–6.

Patinkin, D. 1965. *Money, Interest and Price*, 2nd edn. New York: Harper and Row.

Ricardo, D. 1821. *The Principles of Political Economy and Taxation*. London: M. Dent and Sons, 1911.

Tobin, J. 1980. *Asset Accumulation and Economic Activity*. Chicago: University of Chicago Press.

Weil, P. 1987. 'Love thy children': reflections on the Barro debt neutrality theorem. *Journal of Monetary Economics* 19, 377–91.

Weil, P. 1989. Overlapping families of infinitely-lived agents. *Journal of Public Economics* 38, 183–98.

Ricardian trade theory

Ricardian trade theory takes cross-country technology differences as the basis of trade. By abstracting from the roles of cross-country factor endowment differences and cross-industry factor intensity differences, which are the primary concerns of factor proportions theory (such as Heckscher–Ohlin and Specific Factor models), Ricardian trade theory offers a simple and yet powerful framework within which to address many positive and normative issues of international trade. It is particularly well-equipped to examine the effects of country sizes, of technology changes and transfers, and income distributions. Furthermore, its simple production structure makes it relatively easy to allow for many tradable goods and many countries, and hence capable of generating valuable insights, which are lost in the standard two-country, two-goods model of international trade.

Let us start with the Ricardian model with a continuum of tradable goods, adopted from Dornbusch, Fischer and Samuelson (DFS) (1977). The world consists of two countries, Home and Foreign. There is a continuum of competitive industries, indexed by $z \in [0, 1]$, each producing a homogenous tradable good, also indexed by z. There is only one non-tradable factor of production, called labour. (Or, if there are many non-tradable factors, they can be aggregated into a single composite factor.) Let $a(z)$ and $a^*(z)$ be the Home and Foreign unit labour requirements of good z, that is, labour input required to produce one unit of output z at Home and Foreign. Without loss of generality, we can index z so that Home's relative efficiency, $A(z) \equiv a^*(z)/a(z)$, is non-increasing in z. In Figure 1, it is strictly decreasing. In short, Home (Foreign) has a *comparative advantage* in low-indexed (high-indexed) goods.

Let w and w^* denote the wage rates at Home and Foreign. Then, the prices in autarky are given by $p(z) = wa(z)$ at Home and $p^*(z) = w^*a^*(z)$ at Foreign. Under free trade (and in the absence of any trade costs), the price of each good is equalized across the two countries and is given by $p(z) = p^*(z) = \min\{wa(z), w^*a^*(z)\}$. Then, for a given relative wage rate or a given level of the *factoral terms of trade*, w/w^*, there is a marginal good, m, defined by

$$\frac{w}{w^*} = A(m), \qquad (1)$$

such that Home produces only goods in $[0,m]$, and Foreign produces only goods in $[m,1]$, and the prices become

$$p(z) = p^*(z) = wa(z), \quad z \in [0, m];$$
$$p(z) = p^*(z) = w^*a^*(z), \quad z \in [m, 1]. \qquad (2)$$

To pin down the relative wage rate, we must specify the demand conditions. To keep it simple, let us assume that there are $L(L^*)$ households at Home (Foreign), each supplying one unit of labour, and that every house-hold shares the symmetric Cobb–Douglas preferences defined over $z \in [0, 1]$, as $U = \int_0^1 \log[c(z)]dz$ and $U^* = \int_0^1 \log[c^*(z)]dz$. Then, the world income (and the world total expenditure), $wL + w^*L^*$, is also equal to the world expenditure on each good. Since Home produces the goods in $[0,m]$, the total expenditure on the Home goods is $m(wL + w^*L^*)$, which must be equal to the Home income, wL, in equilibrium. This condition yields

$$\frac{w}{w^*} = \frac{m}{1-m}\left[\frac{L^*}{L}\right] \qquad (3)$$

which is depicted by the upward sloping curve in Figure 1. It is upward-sloping because a higher m means that a larger fraction of the world expenditure goes to the goods produced by the Home labour, hence its relative wage goes up. As shown by the intersection of the two curves in Figure 1, eq. (1) and eq. (3) jointly determine the equilibrium relative wage rate, w/w^*, and the equilibrium patterns of trade and specialization, m, as Home exports and Foreign imports goods in $[0,m)$ and Home imports and Foreign exports goods in $(m,1]$. In short, *the patterns of trade follow the patterns of comparative advantage.*

The standard two-country, two-goods Ricardian model, found in many college textbooks, may be recovered as a special case of this model, where $A(z) = A_1$ for $z \in [0, \alpha]$ and $A(z) = A_2$ for $z \in (\alpha, 1]$, with $A_1 > A_2$, as shown in Figure 2. By aggregating all the goods in $[0, \alpha]$ as a composite good, called Good 1, and all the goods in $(\alpha, 1]$ as another composite good, called Good 2, the model becomes a two-sector model, where the households have the preferences, $U = \alpha\log(C_1) + (1-\alpha)\log(C_2)$ and $U^* = \alpha\log(C_1^*) + (1-\alpha)\log(C_2^*)$. Viewed this way, the model highlights the restrictive feature of the two-good assumption in the textbook Ricardian model.

Gains from trade and country size effects

The Home and Foreign welfares are measured by $U = \int_0^1 \log[w/p(z)]dz$ and $U^* = \int_0^1 \log[w^*/p^*(z)]dz$, respectively. In autarky, they are equal to

$$U_A = -\int_0^1 \log[a(z)]dz; \quad U_A^* = -\int_0^1 \log[a^*(z)]dz; \qquad (4)$$

and, under free trade, they are equal to

$$U_T = -\int_0^m \log[a(z)]dz + \int_m^1 \log\left[\frac{w}{w^*a^*(z)}\right]dz;$$

$$U_T^* = \int_0^m \log\left[\frac{w^*}{wa(z)}\right]dz - \int_m^1 \log[a^*(z)]dz. \qquad (5)$$

Subtracting (4) from (5) and using (1) show that the welfare changes from autarky to free trade by:

$$\Delta U = U_T - U_A = \int_m^1 \log\left[\frac{A(m)}{A(z)}\right]dz;$$

$$\Delta U^* = U_T^* - U_A^* = \int_0^m \log\left[\frac{A(z)}{A(m)}\right]dz, \qquad (6)$$

both of which are strictly positive in the case depicted in Figure 1. More generally, both countries gain from trade, as long as $0 < m < 1$ and $A(0) > A(m) > A(1)$. Note that this condition could hold even when one country, say Home, has *absolute* advantage over the other, say Foreign, that is, $A(z) > 1$ for $z \in [0, 1]$. Clearly, such absolute advantage allows Home households to enjoy higher wage income and hence a higher standard of living than Foreign households, $w > w^*$, and $U_T - U_T^* = \log(w) - \log(w^*) > 0$. Yet both countries gain from trade as long as trade allows them to

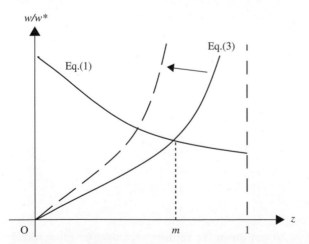

Figure 1 The equilibrium factor terms of trade and patterns of specialization

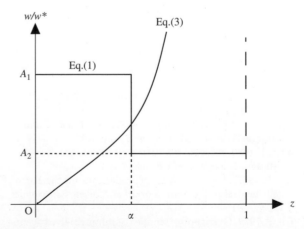

Figure 2 The two-goods case

specialize in the goods that they are relatively good at producing.

In the two-goods case, shown in Figure 2, both countries gain from trade only when $m=\alpha$ and $A_1>w/w^*>A_2$, which requires that $A_2(1-\alpha)/\alpha<L^*/L<A_1(1-\alpha)/\alpha$. Home does not gain from trade if $w/w^*=A_2$, which occurs when $L^*/L\leq A_2(1-\alpha)/\alpha$, and Foreign does not gain from trade if $w/w^*=A_1$, which occurs when $L^*/L\geq A_1(1-\alpha)/\alpha$. This result of the two-sector model is often interpreted as saying that a large country cannot gain from trade, as it remains *incompletely* specialized, or that a country must *fully* specialize in order to gain from trade, but this is due to an artificial feature of the two-goods model which restricts the cross-country differences in technology.

What is generally true is that, as one country becomes large (small) relative to the rest of the world, its gains from trade become smaller (large). As shown in Figure 1, an increase in L^*/L, which shifts the upward-sloping curve to the left, leads to a higher w/w^* and a lower m, which implies a higher ΔU and a lower ΔU^*, as seen from (6). For example, a faster population growth in the South (Foreign) allows the North (Home) to specialize further, which improves its factoral terms of trade and its standard of living at the expense of the South. (The same phenomenon might be described by the protectionist in the North as saying, 'because of the cheap labour in the South, the North loses its competitive advantage and industries move from the North to the South'.)

This also suggests that a country with a small population could enjoy higher per capita income even with limited technological superiority. In Figure 3, Home's technologies are inferior in almost all the goods, yet, thanks to its relatively small population, Home enjoys higher per capita income. This may explain why countries like Norway and Switzerland enjoy a high standard of living even though their geography and climates are not particularly suitable to most economic activities. With smaller populations, they can maintain a high standard of living by specializing in a narrower range of activities that they are particularly good at. This effect is difficult to see within a two-goods model.

These results suggest diminishing returns to scale (DRS) at the country level, even though technologies satisfy the constant returns to scale (CRS) property. This is because the endogeneity of the terms of trade, w/w^*, introduces *de facto* diminishing returns. To see some macroeconomic implications, let us reinterpret the model in the following way. Home and Foreign produce their GDPs, Y and Y^*, by the CRS aggregate production function, $\log(Y)=\int_0^1\log[c(z)]dz$ and $\log(Y^*)=\int_0^1\log[c^*(z)]dz$, where $c(z)$ and $c^*(z)$ denote the inputs of the tradeable intermediate goods, $z\in[0,1]$. The representative household at Home and at Foreign supplies L and L^* units of the composite of the primary factors, which may include not only labour but also physical and human capital. Then, the expressions analogous to (5) become

$$\log\left[\frac{Y}{L}\right]=\int_m^1\log\left[\frac{w}{w^*a^*(z)}\right]dz-\int_0^m\log[a(z)]dz;$$

$$\log\left[\frac{Y^*}{L^*}\right]=\int_0^m\log\left[\frac{w^*}{wa(z)}\right]dz-\int_m^1\log[a^*(z)]dz.$$

The effect of a change in L and L^* can be seen by totally differentiating the above expressions, which yields the following growth accounting:

$$\frac{dY}{Y}=\frac{dL}{L}+(1-m)\left[\frac{dw}{w}-\frac{dw^*}{w^*}\right];$$
$$\frac{dY^*}{Y^*}=\frac{dL^*}{L^*}+m\left[\frac{dw^*}{w^*}-\frac{dw}{w}\right].$$

If L and L^* grow at the same rate, w/w^* remains constant, and hence both Y and Y^* also grow at the same rate. However, if L grows faster than L^*, then Y grows slower than L and Y^* grows faster than L^* through the terms of trade effect. This example also suggests that, even when there are increasing returns to scale (IRS) in the aggregate production technologies, naive cross-country growth regression exercises which do not take into account interdependence among countries might fail to uncover economies of scale. See also Acemoglu and Ventura (2002), which studies how such a terms-of-trade mechanism generates stable cross-country distribution of income in the world even when different countries accumulate factors at different rates.

Technology changes and transfers

Because it takes cross-country technology differences as the basis of trade, the Ricardian model is well-suited to study the effects of technology changes. Let

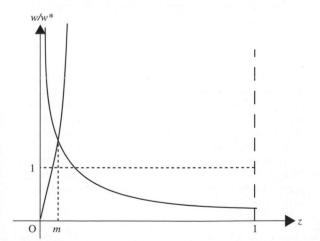

Figure 3   Gains from trade and country size effects

$g(z) \equiv -\{d\log[a(z)]$ and $g^*(z) \equiv -d\log[a^*(z)]$ denote the rate of productivity change in industry z at Home and Foreign. By totally differentiating (5), the Home welfare changes can be expressed as

$$dU = \int_0^m g(z)dz + \int_m^1 g^*(z)dz$$
$$+ \left[\frac{1-m}{1+\xi(m)m(1-m)}\right][g(m) - g^*(m)]$$

where $\xi(z) \equiv -d\log[A(z)]/dz$. For example, it is easy to see that Home always gains from productivity growth in its export sectors, $g(z) > 0$ for $z \in [0, m]$. Does Home gain from productivity growth abroad, as well? If Foreign experiences a *uniform* productivity growth in its export sectors (that is, $g^*(z) = g^* > 0$ for all $z \in [m, 1]$), then the answer is yes because

$$dU = \left[\frac{m(1-m)^2\xi(m)}{1+\xi(m)m(1-m)}\right]g^* > 0,$$

unless $\xi(m) = 0$, in which case productivity gains in the Foreign export sectors are entirely offset by the terms of trade change.

On the other hand, Home may lose from Foreign productivity growth if it is concentrated around the marginal sector, m. The following example, taken from Jones (1979), illustrates this possibility. Let $A(z) = A_1 = a_1^*/a_1$ for $z \in [0, \alpha_1]$; $A(z) = A_2 = a_2^*/a_2$ for $z \in (\alpha_1, \alpha_1 + \alpha_2]$, and $A(z) = A_3 = a_3^*/a_3$ for $z \in (\alpha_1 + \alpha_2, 1]$, with $A_1 > A_2 > A_3$. Again, we may view this example as a three-sector model, by aggregating all the goods within each segment, as Goods 1, 2, and 3. When the upward-sloping curve intersects with $A(z)$ at its middle segment, $w/w^* = A_2$. Then, the Home and Foreign welfares are given by

$$U_T = (1 - \alpha_1 + \alpha_2)\log(a_2^*/a_3^*);$$
$$U_T^* = -(\alpha_1 + \alpha_2)\log(a_2^*) - (1 - \alpha_1 + \alpha_2)\log(a_3^*),$$

where we normalize $a_1 = a_2 = a_3 = 1$ without loss of generality to simplify the expression. The Home welfare declines (and the Foreign welfare improves) unambiguously when Foreign productivity growth takes the form of a reduction in a_2^*, depicted by the arrow in Figure 4. The reason for the Home loss is that the Home purchasing power measured in Good 2 remains unchanged as Foreign productivity growth is completely offset by the increase in the Foreign wage, while the Home purchasing power measured in Good 3 declines, as the increase in the Foreign wage makes it more expensive.

The above example again demonstrates the restrictiveness of the two-goods assumption. In particular, it suggests that the widely used distinction between 'export-biased' and 'import-biased' technology changes in two-sector models, first introduced by Hicks (1953), is of very

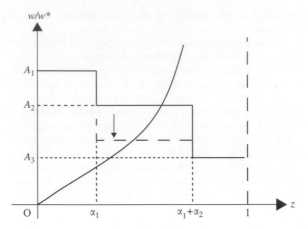

Figure 4   The three-goods case

limited value, as it can be applied only when these changes are uniform across all the export (or import) sectors. Indeed, it can be misleading because the effects of technology changes that take place in *some* export sectors may be very different from those of 'export-biased technological changes' in the two-sector model. (A similar point can be made for the analysis of trade policies. In the standard competitive two-sector model of trade, the government cannot improve its national welfare by providing export subsides to its export sector. This should be interpreted as saying that export subsidies provided *uniformly* to all of its export sectors cannot improve its national welfare. Indeed, using the Ricardian models similar to the one above, Itoh and Kiyono, 1987, showed that *selective* export subsidies that *target* sectors around the marginal sector can improve the national welfare.)

The same mechanism could operate when the technologically lagging country succeeds in catching up with the technologically leading country. Suppose $a(z) = 1$ for all $z \in [0,1]$ and $A(z) = a^*(z) = \Lambda^{1-z}$, where $\Lambda > 1$ is a parameter, representing the extent to which Foreign 'lags behind' Home technologically. For example, each tradable good is produced by performing two tasks by x_1 and x_2, with the Cobb–Douglas production function, $F^z(x_1, x_2) = [x_1/z]^z[x_2/(1-z)]^{1-z}$, and that the unit labour requirement for task 1 is equal to one everywhere, while the unit labour requirement for task 2 is one at Home and $\Lambda > 1$ at Foreign. One may think that Λ reflects the technology gap, which affects when performing task 2, but not task 1. Then, Home has absolute advantage in all the goods, but Foreign has comparative advantage in the high-indexed goods, which can be produced mostly by performing the simple task 1. (Krugman, 1986, offered another story behind a similar parameterization of $A(z)$.) With this parameterization, we have

$$U_T = \frac{(1-m)^2}{2}\log\Lambda, \qquad U_T^* = -\frac{1-m^2}{2}\log\Lambda,$$

where m is determined by the condition, $mL^*/(1-m)L = \Lambda^{1-m}$. Home could lose if Foreign succeeds in narrowing the gap (that is, a reduction in Λ). The reason is easy to understand. As the gap narrows, Foreign becomes more similar to Home, and Home gains little from trading with a country similar to itself. Indeed, if Foreign catches up completely, $\Lambda = 1$, Home loses all the gains from trading with Foreign because the two countries become identical. Note that the underlying mechanism in this example is the same as in the three-goods example. When Foreign narrows the gap, their productivity growth is not uniform across its export sectors. It is larger in the sectors in which Home has bigger absolute advantages. However, it is false to say that Home suffers because Foreign productivity growth is 'import-biased'. The Home loss is caused by Foreign productivity growth around the marginal sector, not at the lower end of the spectrum.

Nontraded goods, trade costs, and effects of globalization

We have been examining the effects of trade by comparing the two extreme cases: autarky, where no goods are traded, and 'free' trade, where all goods are costlessly tradeable. Let us now introduce some trade costs and examine the effects of (partial) trade liberalization by reducing the trade costs. Matsuyama (2007c) conducts such exercises by following DFS (1977), which proposed two alternative ways of introducing trade costs in their model: traded–nontraded dichotomy and uniform iceberg costs.

Traded–nontraded dichotomy

Suppose that only the goods in $[0,k]$ are tradable at zero cost and $A(z) \equiv a^*(z)/a(z)$ is continuous, and strictly decreasing in z, within this range. On the other hand, trade costs are so high for the goods in $(k,1]$ that they need to be produced locally. At this point, we do not have to specify the production technologies for these nontradables.

Given the marginal good, $m \in [0,k]$, defined by $A(m) = w/w^*$, Home produces all the goods in $[0,m]$ for both countries and Foreign produces all the goods in $(m, k]$. In addition, each country produces all the goods in $(k,1]$ locally. Therefore, the total expenditure on the goods produced at Home is equal to $m(wL + w^*L^*) + (1-k)wL$, which must be equal to the Home income, wL, in equilibrium. This condition yields

$$\frac{w}{w^*} = \frac{m}{k-m}\left[\frac{L^*}{L}\right]. \tag{7}$$

The equilibrium is determined jointly by eq. (1) and eq. (7), as shown in Figure 5. DFS (1977) used this extension to study the classical transfer problem. In the presence of the nontraded goods, the German households spend a larger share of their income on the goods produced in Germany than the households abroad. Because of this 'home bias' in demand, an exogenous income

transfer from Germany to the Allies (the war reparations after the Treaty of Versailles of 1919) shifts demand away from the German goods, which leads to a deterioration of the German terms of trade, imposing the additional burden on the German economy.

Let us use this model to study the effects of a globalization. Imagine that some nontradables become tradable. For example, the governments might decide to lift the bans on trading some goods that can be traded costlessly. Or advances in information and communication technologies might open up the possibility of trading some labour services at zero cost. The effects, of course, depend on the relative efficiency of the two countries in producing these newly tradables. Consider the case where $A(z) = A$ for all $z \in (k,1]$, and that a fraction g of these goods becomes newly tradable at zero cost. Then, if $w/w^* > A$, all of the newly tradeables are produced at Foreign. Therefore, given the marginal good, $m \in [0,k]$, Home produces all the goods in $[0,m]$ for both countries and $(1-g)(1-k)$ fraction of the goods (those which remain nontradable) locally. Hence, in equilibrium, $wL = m(wL + w^*L^*) + (1-g)(1-k)wL$. On the other hand, if $w/w^* < A$, all of the newly tradables are produced at Home. Therefore, Home produces $m + g(1-k)$ fraction of the goods for both countries and $(1-g)(1-k)$ fraction of the goods locally. Hence, in equilibrium, $wL = [m + g(1-k)](wL + w^*L^*) + (1-g)(1-k)wL$. Thus, we have, instead of (7),

$$\frac{w}{w^*} = \frac{m+g(1-k)}{k-m}\left[\frac{L^*}{L}\right] \text{ for } \frac{w}{w^*} < A;$$

$$\frac{w}{w^*} = \frac{m}{k+g(1-k)-m}\left[\frac{L^*}{L}\right] \text{ for } \frac{w}{w^*} > A; \tag{8}$$

and $w/w^* = A$, otherwise. Note that setting $g = 0$ in eq. (8) recovers eq. (7). A higher g shifts the graph to the right above $w/w^* = A$ and to the left below $w/w^* = A$. For each value of g, eq. (1) and eq. (8) jointly determines the marginal good and the Home relative wage, which we denote by $m(g)$ and $A(m(g))$.

Suppose that, before globalization, $g = 0$, the equilibrium Home relative wage, $A(m(0))$, is higher than A, as shown in Figure 5. The arrow indicates the shift caused by an increase in g, which is small enough to keep the Home relative wage higher than A. When some nontraded sectors are opened up, Home abandons the production of these new tradeables, and instead starts producing and exporting the goods in $(m(0), m(g))$, which Home imported previously. The Home relative wage declines as a result, from $A(m(0))$ to $A(m(g))$. The Home and Foreign welfares may be evaluated by

$$U(g) = \int_{m(g)}^{k} \log\left[\frac{A(m(g))}{A(z)}\right] dz$$

$$+ g(1-k)\log\left[\frac{A(m(g))}{A}\right];$$

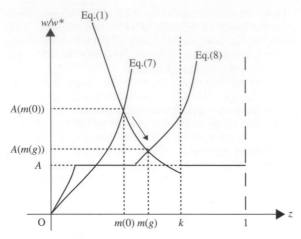

Figure 5   Non-uniform globalization

$$U^*(g) = m(g)\log\left[\frac{A}{A(m(g))}\right]$$
$$+ \int_{m(g)}^{k} \log\frac{A}{A(z)}\,dz - \log A,$$

where we use the normalization, $A(z) = a^*(z)/a(z) = a^*(z)$ for all $z \in [0,1]$, to simplify the expressions. A globalization (an increase in g) affects the Home welfare through two effects that operate in the opposite directions. On one hand, it allows Home to reallocate its labour to the sectors where they have higher relative efficiency, that is, from A to $A(m(g))$ or higher. On the other hand, its relative wage rate, or the factoral terms of trade, $w/w^* = A(m(g))$, deteriorates. The overall effect is generally ambiguous. However, if an increase in g brings down its relative wage rate $A(m(g))$ sufficiently close to A, the positive reallocation effect is dominated by the negative terms of trade effect, so that a further globalization harms the Home welfare. In contrast, a globalization unambiguously improves the Foreign welfare, because both effects operate positively.

The possibility that Home could lose when a globalization takes this form should not be too surprising, because it can be viewed as a form of non-uniform technological changes. Indeed, this mechanism may capture some of the widely held concerns that high-wage countries might lose from 'outsourcing' simple tasks to low-wage countries.

Uniform iceberg cost
Suppose now that all the goods, $z \in [0,1]$, are tradeable but subject to the iceberg cost. Each good, when shipped abroad, melts away in transit and only a fraction $g < 1$ arrives at the destination. Thus, in order to supply one unit of each good, the exporter must produce $1/g > 1$ units of the good. Then, Home exports to Foreign only

when $a(z)w/g < a^*(z)w^*$, or $w/w^*g < A(z)$, and Foreign exports to Home only when $a(z)w > a^*(z)w^*/g$, or $wg/w^* > A(z)$. Thus, there are two marginal goods, defined by

$$\frac{w}{w^*g} = A(m^-) > A(m^+) = \frac{wg}{w^*}, \qquad (9)$$

such that Home produces all the goods in $[0, m^-)$ for both countries; Foreign produces all the goods in $(m^+, 1]$ for both countries, and each produces the goods in $[m^-, m^+]$, which becomes (endogenously) nontraded goods. The demand condition now becomes

$$\frac{w}{w^*} = \frac{m^-}{1 - m^+}\left[\frac{L^*}{L}\right]. \qquad (10)$$

Eqs (9) and (10) jointly determine three endogenous variables, m^-, m^+, and w/w^*.

One could proceed to examine the effects of a reduction in the trade cost, by increasing g. This is left as an exercise for interested readers.

Multiple countries
The two-country assumption is clearly restrictive for certain purposes, such as analysing the income distribution across countries, studying the patterns of bilateral trade flows, let alone the issues related to the regional integration, such as NAFTA. It is relatively straightforward to extend the two-goods Ricardian model for an arbitrary number of countries; see, for example, Becker (1952) for a finite number of countries and Matsuyama (1996) and Yanagawa (1996) for a continuum of countries. It has been a challenge to allow for an arbitrary number of goods *and* countries in a tractable way. For example, Wilson (1980) extended the DFS model in many dimensions, including a finite number of countries, but it does not permit more than a local perturbation analysis. Acemoglu and Ventura (2002), in their analysis of the cross-country income distribution, assumed the extreme form of technological heterogeneity by adopting the Armington assumption, which prevents the patterns of specialization from changing endogenously.

Eaton and Kortum (2002) developed a parsimonious representation of the Ricardian model with a continuum of goods, which allows for an arbitrary number of countries with the iceberg costs that are uniform across sectors but vary across country pairs. Their key idea is to view the technology heterogeneity across countries as a realization from the Frechet distributions, instead of trying to index the goods in a particular order. This yields simple expressions relating the bilateral trade volumes to technology and geographical barriers, and they use these expressions to estimate the parameters needed to quantify the effects of various policy experiments. For further development, see Alvarez and Lucas (2004).

Multi-stage trade and vertical specialization

Sanyal (1983) proposed a reinterpretation of the DFS model, according to which the final good is produced through many stages of production, $z \in [0,1]$, in order to analyse trade in intermediate inputs and vertical specialization. If the order in which these inputs need to be produced in the vertical chain of production perfectly coincides with the pattern of comparative advantage, as Sanyal assumed, these inputs are traded only once, as one country specializes in the earlier stages of production and the other specializes in the later stages. Under more general patterns of comparative advantage, however, these inputs may be traded back and forth many times. In such a setting, even a small reduction in trade costs could cause a large and nonlinear increase in the volume of trade, as documented by Hummels, Ishii and Yi (2001).

More general preferences

With the Cobb–Douglas specification, each good receives a constant share of the expenditure regardless of the prices. Its homotheticity implies that the rich and the poor consume all the goods in the same proportions (when they all face the identical prices), so that the demand compositions are independent of the income distribution within each country. These features, while greatly simplifying the analysis, are too restrictive for addressing many important issues related to growth and development.

Consider, for example, the Fisher–Clark–Kuznets thesis, that is, the changing patterns of sectoral compositions, with the decline of agriculture, the rise and the fall of manufacturing, and the rise of the service sectors. To understand such patterns of structural change in the context of a global economy, Matsuyama (2007b) relaxed the Cobb-Douglas assumption to allow for non-unitary price and non-unitary income elasticities in the three-goods (two tradable and one nontradable) Ricardian model.

Non-homothetic preferences also play the key roles in many models of North–South trade. Flam and Helpman (1987), Stokey (1991), and Matsuyama (2000) all built two-country (North and South) Ricardian models with a continuum of goods, with the open-ended goods space, $z \in [0, \infty)$, and considered non-homothetic preferences with the property that, as the household's income goes up, its demand compositions shift towards higher-indexed goods. When the South, the poorer country, has comparative advantages in lower-indexed goods, the demand has home biases (in spite of the absence of any trade costs). Furthermore, the asymmetry of demand generates many comparative statics results that are absent in the standard Ricardian model. For example, in Matsuyama (2000), immiserizing growth might occur; uniform productivity growth in the South might make the South worse off, as all the benefits go to the North. Or, as the South's population grows, some industries migrate from the North to the South, and new industries are born in the North, generating the patterns of product cycles. Flam and Helpman (1987) and Matsuyama (2000) also looked at the roles of income distributions within each country by endowing different households with different amounts of labour.

Endogenous technologies and increasing returns

So far, we have taken the cross-country differences in technology as exogenous and examined their effects on patterns of specialization and trade. However, the patterns of trade and specialization may also affect technologies. Many Ricardian models with endogenous technologies have been developed to examine such two-way causality between technology and trade. Endogenous technologies have also been used as a natural way of introducing increasing returns in production. Due to the space constraint, we cannot do justice to this vast literature, which contains many alternative approaches to endogenize technologies (static external economies of scale, dynamic increasing returns due to learning-by-doing with or without inter-industry spillovers, agglomeration economies with endogenous product varieties, R&D activities, and so on) with a wide range of results with different policy implications. The interested reader should start with a survey by Grossman and Helpman (1995).

Beyond technologies: policy-induced and institution-based comparative advantage

The Ricardian set-up has also been used to explain how the differences in national policies and institutions give rise to the patterns of comparative advantage, even in the absence of any inherent technology differences. In Copeland and Taylor (1994), the clean environment is a normal good, so that the rich North chooses a higher pollution tax than the poor South. As a result, the North (South) ends up having comparative advantages in less (more) polluting industries. In Matsuyama (2005), Costinot (2006), and Acemoglu, Antras and Helpman (2007), industries differ according to the severity of agency or contractual problems, and the country with a better (worse) institutional set-up to deal with these problems has comparative advantages in industries that are more (less) subject to these problems. One may view this line of research as an attempt to endogenize technology differences. Unlike the literature surveyed by Grossman and Helpman (1995), however, the main objective here is to look at the deeper or more fundamental causes of technology differences, rather than looking at the two-way causality between technology and trade.

Finally, because the Ricardian trade theory abstracts from the roles of factor endowment differences across countries and factor intensity differences across industries as the basis of trade, it is an ideal set-up in which to isolate the roles of factor endowments and intensity differences that are unrelated to the basis of trade. For example, Matsuyama (2007a) uses a two-country Ricardian model to examine how factor intensity affects the extent of globalization and how globalization affects factor prices when certain factors are used more intensively in international trade than in domestic trade. The model is Ricardian in the sense that the patterns of comparative advantage are determined entirely by the exogenous technological differences. The factor proportions matter, however, because they determine the extent of globalization, as the effective trade costs vary with the relative endowments of the factor used intensively in international trade.

KIMINORI MATSUYAMA

See also **comparative advantage; globalization; international trade theory; terms of trade.**

Bibliography

Acemoglu, D., Antras, P. and Helpman, E. 2007. Contracts and technology adoption. *American Economic Review* 97, 916–43.

Acemoglu, D. and Ventura, J. 2002. The world income distribution. *Quarterly Journal of Economics* 117, 659–94.

Alvarez, F. and Lucas, R.E. Jr. 2004. General equilibrium analysis of the Eaton–Kortum model of international trade. Working Paper, University of Chicago.

Becker, G. 1952. A note on multi-country trade. *American Economic Review* 42, 558–68.

Copeland, B. and Taylor, S. 1994. North–South trade and the environment. *Quarterly Journal of Economics* 109, 755–87.

Costinot, A. 2006. On the origins of comparative advantage. Working paper, UCSD.

Dornbusch, R., Fischer, S. and Samuelson, P.A. 1977. Comparative advantage, trade and payments in a Ricardian model with a continuum of goods. *American Economic Review* 67, 823–39.

Eaton, J. and Kortum, S. 2002. Technology, geography, and trade. *Econometrica* 70, 1741–79.

Flam, H. and Helpman, E. 1987. Vertical product differentiation and North–South trade. *American Economic Review* 77, 810–22.

Grossman, G. and Helpman, E. 1995. Technology and trade. In *Handbook of International Economics*, vol. 3, ed. G. Grossman and K. Rogoff. Amsterdam: North-Holland.

Hicks, J.R. 1953. An inaugural lecture. *Oxford Economic Papers* 5, 117–35.

Hummels, D., Ishii, J. and Yi, K.-M. 2001. The nature and growth of vertical integration in world trade. *Journal of International Economics* 54, 75–96.

Itoh, M. and Kiyono, K. 1987. Welfare enhancing export subsidies. *Journal of Political Economy* 95, 115–37.

Jones, R.W. 1979. Technical progress and real incomes in a Ricardian trade model. In *International Trade: Essays in Theory*, ed. R.W. Jones. Amsterdam: North-Holland.

Krugman, P. 1986. A 'technology-gap' model of international trade. In *Structural Adjustment in Developed Open Economies*, ed. K. Jungenfelt and D. Hague. London: Macmillan.

Matsuyama, K. 1996. Why are there rich and poor countries?: symmetry-breaking in the world economy. *Journal of the Japanese and International Economies* 10, 419–39.

Matsuyama, K. 2000. A Ricardian model with a continuum of goods under nonhomothetic preferences: demand complementarities, income distribution and north-south trade. *Journal of Political Economy* 108, 1093–120.

Matsuyama, K. 2005. Credit market imperfections and patterns of international trade and capital flows. *Journal of the European Economic Association* 3, 714–23.

Matsuyama, K. 2007a. Beyond icebergs: towards a theory of biased globalization. *Review of Economic Studies* 74, 237–53.

Matsuyama, K. 2007b. Productivity-based theory of manufacturing employment declines: a global perspective. Working paper, Department of Economics, Northwestern University.

Matsuyama, K. 2007c. Uniform versus non-uniform globalization. Working paper, Department of Economics, Northwestern University.

Sanyal, K.K. 1983. Vertical specialization in a Ricardian model with a continuum of stages of production. *Economica* 50, 71–8.

Stokey, N. 1991. The volume and composition of trade between rich and poor countries. *Review of Economic Studies* 58, 63–80.

Wilson, C.A. 1980. On the general structure of Ricardian models with a continuum of goods. *Econometrica* 48, 1675–702.

Yanagawa, N. 1996. Economic development in a Ricardian world with many countries. *Journal of Development Economics* 49, 271–88.

risk

The phenomenon of *risk* is one of the key determining factors in the formation of investment decisions, the operation of financial markets, and several other aspects of economic activity.

Risk versus uncertainty

The most fundamental distinction in this branch of economic theory, due to Knight (1921), is that of 'risk' versus 'uncertainty'. A situation is said to involve *risk* if the randomness facing an economic agent presents itself in the form of exogenously specified or scientifically calculable *objective probabilities*, as with gambles based on a roulette wheel or a pair of dice. A situation is said to involve *uncertainty* if the randomness presents itself in the form of alternative possible *events*, as with bets on a horse race, or decisions involving whether or not to buy earthquake insurance.

The standard approach to the modelling of preferences under uncertainty (as opposed to risk) has been the *state-preference approach* (for example, Arrow, 1964; Debreu, 1959, ch. 7; Hirshleifer, 1965; 1966; Karni, 1985; Yaari, 1969). Given the absence of exogenously specified objective probabilities, this approach represents the randomness facing the individual by a set of mutually exclusive and exhaustive *states of nature* or *states of the world* $\mathscr{S} = \{s_1, \ldots, s_n\}$. Depending upon the particular application, this partition of all conceivable futures may either be very coarse, as with the pair of states (it snows here tomorrow, it doesn't snow here tomorrow) or else very fine, so that the description of a single state might read 'it snows more than three inches here tomorrow *and* the temperature in Paris at noon is 73° *and* the price of gold in New York is over \$900.00/ounce'. The objects of choice in this framework consist of *state-payoff bundles* of the form $(c_1, \ldots, c_n)$, which specify the payoff that the individual will receive in each of the respective states. As with regular commodity bundles, individuals are assumed to have preferences over state-payoff bundles which can be represented by indifference curves in the *state-payoff space* $\{(c_1, \ldots, c_n)\}$.

Even though the state-preference approach has led to important advances in the analysis of choice under uncertainty (see, for example, the above citations), the advantages of being able to draw on modern probability theory has led economists to hypothesize that an individual's *beliefs* in such settings can nevertheless still be represented by so-called *personal probabilities* or *subjective probabilities*, which take the form of an additive *subjective probability measure* $\mu(\cdot)$ over the state space $\mathscr{S}$. In such a case, a given state-payoff bundle $(c_1, \ldots, c_n)$ will be viewed as yielding outcome c_i with probability $\mu(s_i)$, so that the individual would evaluate the bundle $(c_1, \ldots, c_n)$ in the same manner as he or she would evaluate a casino gamble which yielded the payoffs $(c_1, \ldots, c_n)$ with respective objective probabilities $(\mu(s_1), \ldots, \mu(s_n))$. The hypothesis that individuals have such probabilistic beliefs and evaluate state-payoff bundles in such a manner is termed the *hypothesis of probabilistic sophistication*, and permits a unified application of probability theory to the analysis of decisions under both objective risk and subjective uncertainty. The joint hypothesis of probabilistic sophistication and expected utility risk preferences has been axiomatized by Ramsey (1926), Savage (1954), Anscombe and Aumann (1963), Pratt, Raiffa and Schlaifer (1964) and Raiffa (1968, ch.5), and probabilistic sophistication without expected utility has been axiomatized by Machina and Schmeidler (1992).

Choice under risk: the expected utility model

For reasons of expositional ease, we consider a world with a single commodity (for example, wealth). An agent making a decision under either risk or probabilistic uncertainty can therefore be thought of as facing a choice set of alternative univariate probability distributions. In order to consider both discrete (for example, finite outcome) distributions as well as distributions with density functions, we represent each such probability distribution by means of its cumulative distribution function $F(\cdot)$, where $F(x) \equiv \text{prob}(\tilde{x} \geqslant x)$ for the random variable $\tilde{x}$.

In such a case we can model the agent's preferences over alternative probability distributions in a manner completely analogous to the approach of standard (that is, non-stochastic) consumer theory: he or she is assumed to possess a ranking $\geqslant$ over distributions which is complete, transitive and continuous (in an appropriate sense), and hence representable by a real-valued *preference* function $V(\cdot)$ over cumulative distribution functions, in the sense that $F^*(\cdot) \geqslant F(\cdot)$ (that is, the distribution $F^*(\cdot)$ is weakly preferred to $F(\cdot)$) if and only if $V(F^*) \geqslant V(F)$.

Of course, as in the non-stochastic case, the above set of assumptions implies nothing about the functional form of the preference functional $V(\cdot)$. For reasons of both normative appeal and analytic convenience, economists typically assume that $V(\cdot)$ is a *linear functional* of the distribution $F(\cdot)$, and hence takes the form

$$V(F) \equiv \int U(x)\, dF(x) \qquad (1)$$

for some function $U(\cdot)$ over wealth levels x, where $U(\cdot)$ is referred to as the individual's *von Neumann–Morgenstern utility function*. (For readers unfamiliar with the *Riemann–Stieltjes integral* $\int U(x)dF(x)$, it represents nothing more than the expected value of $U(\tilde{x})$ when $\tilde{x}$ possesses the cumulative distribution function $F(\cdot)$. Thus if $\tilde{x}$ took the values $x_1, \ldots, x_n$ with probabilities $p_1, \ldots, p_n$ then $\int U(x)dF(x)$ would equal $\sum U(x_i)p_i$, and if $\tilde{x}$ possessed the density function $f(\cdot) = F'(\cdot)$ then $\int U(x)dF(x)$ would equal $\int U(x)f(x)dx$.)

Since the right side of (1) may be thought of as the mathematical expectation of $U(\tilde{x})$, this specification is known as the expected utility model of preferences over random prospects (for a more complete statement of this model, see expected utility hypothesis). Within this framework, an individual's attitudes towards risk are reflected in the shape of his or her utility function $U(\tilde{x})$. Thus, for example, an individual would always prefer shifting probability mass from lower to higher outcome levels if and only if $U(x)$ were an increasing function of x, a condition which we shall henceforth always assume. Such a shift of probability mass is known as a *first order stochastically dominating shift*.

Risk aversion

The representation of an individual's preferences over distributions by the shape of his or her von Neumann–Morgenstern utility function provides the first step in the

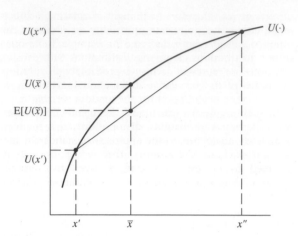

Figure 1 Von Neumann–Morgenstern utility function of a risk-averse individual

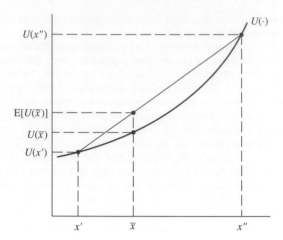

Figure 2 Von Neumann–Morgenstern utility function of a risk-loving individual

modern economic characterization of risk. After all, whatever the notion of 'riskier' means, it is clear that bearing a random wealth $\tilde{x}$ is riskier than receiving a certain payment of $\bar{x} = E[\tilde{x}]$ (the expected value of the random variable $\tilde{x}$). We therefore have from Jensen's inequality that an individual would be *risk averse*, that is, would always prefer a payment of $E[\tilde{x}]$ (and obtaining utility $U(E[\tilde{x}])$) to bearing the risk $\tilde{x}$ (and obtaining expected utility $E[U(\tilde{x})]$) if and only if his or her utility function were concave. This condition is illustrated in Figure 1, where the random variable $\tilde{x}$ is assumed to take on the values x' and x'' with respective probabilities 2/3 and 1/3.

Of course, not all individuals need be risk averse in the sense of the previous paragraph. Another type of individual is a *risk lover*. Such an individual would have a *convex* utility function, and would accordingly prefer receiving a random wealth $\tilde{x}$ to receiving its mean $E[\tilde{x}]$ with certainty. An example of such a utility function is given in Figure 2.

Standard deviation as a measure of risk

While the above characterizations of risk aversion and risk preference allow for the derivation of many results in the theory of choice under risk, they say nothing about which of a pair of non-degenerate random variables $\tilde{x}$ and $\tilde{y}$ is the more risky. Since real-world choices are almost never between risky and riskless situations but rather over alternative risky situations, such a means of comparison is necessary.

The earliest and best-known univariate measure of the riskiness of a random variable $\tilde{x}$ is its *variance* $\sigma^2 = E[(\tilde{x} - E[\tilde{x}])^2]$ or alternatively its *standard deviation* $\sigma = E[(\tilde{x} - E[\tilde{x}])^2]^{1/2}$. The tractability of these measures, as well as their well-known statistical properties, led to the widespread use of mean-standard deviation analysis in the 1950s and 1960s, and in particular to the

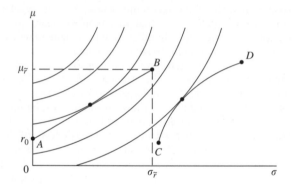

Figure 3 Portfolio analysis in the mean-standard deviation model

development of modern portfolio theory by Markowitz (1952; 1959), Tobin (1958) and others. As an example of this, consider Figure 3. Points A and B correspond to the distributions of a riskless asset with (per dollar) gross return r_0 and a risky asset with random return $\tilde{r}$ with mean $\mu_{\tilde{r}}$ and standard deviation $\sigma_{\tilde{r}}$. An investor dividing a dollar between the two assets in proportions $\alpha{:}(1-\alpha)$ will possess a portfolio whose return has a mean of $\alpha \cdot r_0 + (1 - \alpha) \cdot \mu_{\tilde{r}}$ and standard deviation $(1 - \alpha) \cdot \sigma_{\tilde{r}}$, so that the set of attainable (μ, σ) combinations consists of the line segment connecting the points A and B in the figure. It is straightforward to show that, if the individual were also allowed to *borrow* at rate r_0 in order to finance purchase of the risky asset (that is, could sell the riskless asset short), then the set of attainable (μ, σ) combinations would be the ray emanating from A and passing through B and beyond.

If we then represent the individual's risk preferences by means of indifference curves in this diagram, we obtain his or her optimal portfolio (the example in the figure implies an equal division of funds between the two assets). In the more general case of choice between a pair of risky assets, the set of (μ, σ) combinations generated by alternative divisions of wealth between them will trace out a possibly nonlinear locus such as the one between points C and D in the diagram, with the curvature of this locus determined by the degree of statistical dependence (that is, covariance) between the two random returns.

As mentioned, the representation and analysis of risk and risk-taking by means of the variance or standard deviation of a distribution proved tremendously useful in the theory of finance, culminating in the mean-standard deviation-based *capital asset pricing model* of Sharpe (1964), Lintner (1965), Mossin (1966) and Treynor (1999). However, by the late 1960s the mean-standard deviation approach was under attack for two reasons.

The first reason (known since the 1950s) was the fact that an expected utility maximizer would evaluate all distributions solely on the basis of his or her means and standard deviations if and only if their von Neumann–Morgenstern utility function took the quadratic form $U(x) \equiv ax + bx^2$ for $b \Leftarrow 0$. The sufficiency of this condition is established by noting that $E[U(\tilde{x})] = E[a\tilde{x} + b\tilde{x}^2] = a \cdot E[\tilde{x}] + b \cdot (E[\tilde{x}]^2 + \sigma^2)$. To prove necessity, note that the distributions that yield a 2/3:1/3 chance of the outcomes $(x - \delta) : (x + 2\delta)$ and a 1/3:2/3 chance of the outcomes $(x - 2\delta) : (x + \delta)$ both possess the same mean and variance for each x and δ, so that $(2/3) \cdot U(x - \delta) + (1/3) \cdot U(x + 2\delta) \equiv (1/3) \cdot U(x - 2\delta) + (1/3) \cdot U(x + \delta)$ for all x and δ. Differentiating with respect to δ and simplifying yields $U'(x + 2\delta) + U'(x - 2\delta) \equiv U'(x + \delta) + U'(x - \delta)$ for all x and δ. This implies that $U'(\cdot)$ must be linear and hence that $U(\cdot)$ must be quadratic.

The assumption of quadratic utility is objectionable. If an individual with such a utility function is risk averse (that is, if $b < 0$), then (a) utility will decrease as wealth increases beyond $1/(2b)$, and (b) the individual will be more averse to constant additive risks about high wealth levels than about low wealth levels – in contrast to the observation that those with greater wealth take greater risks (see for example Hicks, 1962, or Pratt, 1964).

Borch (1969) struck the second and strongest blow to the mean-standard deviation approach. He showed that, for any two points (μ_1, σ_1) and (μ_2, σ_2) in the (μ, σ) plane which a mean-standard deviation preference ordering would rank as indifferent, it is possible to find random variables $\tilde{x}_1$ and $\tilde{x}_2$ which possess these respective (μ, σ) values and where $\tilde{x}_2$ first order stochastically dominates $\tilde{x}_1$. However, *any* person with an increasing von Neumann–Morgenstern utility function would strictly prefer $\tilde{x}_2$ to $\tilde{x}_1$. In response to these arguments and the additional criticisms of Feldstein (1969), Samuelson (1967) and others, the use of mean-standard deviation analysis in economic theory waned. See, however, the work of Meyer (1987) for a partial rehabilitation of such two-moment models of preferences.

Besides the variance or standard deviation of a distribution, several other univariate measures of risk have been proposed. Examples include the *mean absolute deviation* $E[|\tilde{x} - E[\tilde{x}]|]$, the *interquartile range* $F^{-1}(.75) - F^{-1}(.25)$, and the classical statistical measures of *entropy* $\sum \ln(p_i) \cdot p_i$ or $\int \ln(f(x)) \cdot f(x) \cdot dx$. Although they provide the convenience of a single numerical index, each of these measures is subject to problems of the sort encountered with the variance or standard deviation. In particular, the entropy measure is based exclusively on the *probability levels* of a random variable, and is particularly unresponsive to its *outcome values* – for example, the 50:50 gambles over the values \$49:\$51 and \$0:\$100 possess identical entropy levels.

Increasing risk

By the late 1960s, the failure to find a satisfactory univariate measure of risk led to another approach to this problem. Working independently, several researchers (Hadar and Russell, 1969; Hanoch and Levy, 1969; and Rothschild and Stiglitz, 1970; 1971) developed an alternative characterization of increasing risk. The appeal of this approach is twofold. First, it formalizes three different intuitive notions of increasing risk. Second, it allows for the straightforward derivation of comparative statics results in a wide variety of economic situations. Unlike the univariate measures described above, however, this approach provides only a partial ordering of random variables. In other words, not all pairs of random variables can be compared with respect to their riskiness.

We now state three alternative formalizations of the notion that a cumulative distribution function $F^*(\cdot)$ is riskier than another distribution $F(\cdot)$ with the same mean. In the following, all distributions are assumed to be over the outcome interval $[0, M]$ unless otherwise indicated.

The first definition of increasing risk captures the notion that 'risk is what all risk averters hate'. Thus an increase in risk must lower the expected utility of all risk averters. Formally:

(A) $F^*(\cdot)$ and $F(\cdot)$ have the same mean and $\int U(x)dF^*(x) \leq \int U(x)dF(x)$ for every concave utility function $U(\cdot)$.

Note that this relationship will *not* be satisfied by every pair of distributions with the same mean. That is to say, there exist pairs $F(\cdot)$ and $F^*(\cdot)$, with the same mean, but such that some risk-averse utility functions prefer $F(\cdot)$ to $F^*(\cdot)$ but other risk-averse utility functions prefer $F^*(\cdot)$ to $F(\cdot)$. This reflects the above-stated fact that comparative risk is a *partial* rather than a *complete* order over the family of probability distributions, even over families of distributions with a common mean. (Although comparative risk is not a complete order, it is a *transitive* order, in the sense that, if the pair $F^*(\cdot)$ and $F(\cdot)$ satisfy condition (A), and the pair $F^{**}(\cdot)$ and $F^*(\cdot)$ satisfy condition (A), then the pair $F^{**}(\cdot)$ and $F(\cdot)$ will also satisfy condition (A).)

The second characterization of the notion that a random variable $\tilde{y}$ with distribution $F^*(\cdot)$ is riskier than a variable $\tilde{x}$ with distribution $F(\cdot)$ is that $\tilde{y}$ consists of the variable $\tilde{x}$ plus an additional zero-mean noise term $\tilde{\varepsilon}$. One possible specification of this is that $\tilde{\varepsilon}$ statistically independent of $\tilde{x}$. However, this condition is too strong in the sense that it does not allow the variance of $\tilde{\varepsilon}$ to depend upon the magnitude of $\tilde{x}$, as in the case of heteroskedastic noise. Instead, Rothschild and Stiglitz (1970) modelled the addition of noise by the condition:

(B) $F(\cdot)$ and $F^*(\cdot)$ are the respective cumulative distribution functions of the random variables $\tilde{x}$ and $\tilde{x} + \tilde{\varepsilon}$, where $E[\tilde{\varepsilon}|x] \equiv 0$ for all values of x.

The third notion of increasing risk involves the concept, due to Rothschild and Stiglitz (1970), of a *mean preserving spread*. Intuitively, such a spread consists of moving probability mass from some region in the centre of a probability distribution out to its tails in a manner that preserves the expected value of the distribution, as seen in the top panels of Figures 4 and 5. In the discrete case of Figure 4, probability mass is moved from the pair of outcome values b and c out to the outcome values a and d. In the continuous density case of Figure 5, probability mass is moved from the interval (b, c) out to the intervals (a, b) and (c, d). We can unify, generalize and formalize this condition by saying that $F^*(\cdot)$ differs from $F(\cdot)$ by a 'mean preserving spread' if they have the same mean and there exists a single crossing point x_0 such that $F^*(x) \geq F(x)$ for all $x \leq x_0$ and $F^*(x) \leq F(x)$ for all $x \geq x_0$ (see the middle panels of Figures 4 and 5). Since it is clear that *sequences* of such spreads will also lead to riskier distributions, the third characterization of increasing risk is:

(C) $F^*(\cdot)$ may be obtained from $F(\cdot)$ by a finite sequence, or as the limit of an infinite sequence, of mean preserving spread.

Although the single crossing property of the previous paragraph serves to characterize cumulative distribution functions that differ by a single mean preserving spread, distributions that differ by a sequence of such spreads will typically not satisfy the single crossing condition. However, if we consider the integrals of these cumulative distribution functions, we see from the bottom panels of Figures 4 and 5 that a mean preserving spread will always serve to raise or preserve the value of this integral for each x, and (since $F^*(\cdot)$ and $F(\cdot)$ have the same mean) will exactly preserve it for $x=M$. In contrast to the single crossing property, this so-called 'integral condition' *will* continue to be satisfied by distributions which differ by a sequence of one or more mean preserving spreads. Accordingly, we may rewrite condition (C) above by the analytically more convenient:

(C′) The integral $\int_0^x [F^*(\xi) - F(\xi)] \cdot d\xi$ is non-negative for all $x > 0$, and is equal to 0 at $x=M$.

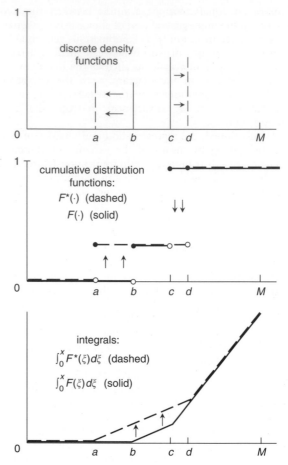

Figure 4   Mean preserving spread of a discrete distribution

Rothschild and Stiglitz (1970) showed that these three concepts of increasing risk are the same by proving that conditions (A), (B) and (C/C′) are equivalent. Thus, a single partial ordering of distribution functions corresponds simultaneously to the notion that risk is what risk averters hate, to the notion that adding noise to a random variable increases its risk, and to the notion that moving probability mass from the centre of a probability distribution to its tails increases the riskiness of the distribution. The original Rothschild–Stiglitz formulation and proofs have since been further strengthened and extended by Machina and Pratt (1997).

This characterization permits the derivation of general and powerful comparative statics theorems concerning economic agents' responses to increases in risk. The general framework for these results is that of an individual with a von Neumann–Morgenstern utility function $U(x, \alpha)$ which depends upon both the outcome of some random variable $\tilde{x}$ as well as a *control variable* α which the individual chooses so as to maximize expected utility $\int U(x, \alpha) dF(x; r)$, where the distribution function $F(\cdot; r)$ depends upon

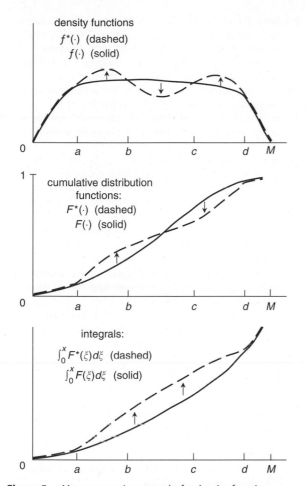

density functions
$f^*(\cdot)$ (dashed)
$f(\cdot)$ (solid)

cumulative distribution functions:
$F^*(\cdot)$ (dashed)
$F(\cdot)$ (solid)

integrals:
$\int_0^x F^*(\xi)d\xi$ (dashed)
$\int_0^x F(\xi)d\xi$ (solid)

Figure 5 Mean preserving spread of a density function

some exogenous parameter r (x for example might be the return on a risky asset, and α the amount invested in it). For convenience, we assume that $F(0;r) \equiv prob(\tilde{x} \leq 0) \geq 0$ for all r. The first order condition for this problem is then:

$$\int U_\alpha(x,\alpha)dF(x;r) = 0 \qquad (2)$$

where $U_\alpha(x,\alpha) = \partial U(x,\alpha)/\partial\alpha$, and we assume that the second derivative $U_{\alpha\alpha}(x,\alpha) = \partial^2 U(x,\alpha)/\partial\alpha^2$ is always negative to insure we have a maximum. Implicit differentiation of (2) then yields the comparative statics derivative:

$$d\alpha/dr = -\int U_\alpha(x,\alpha)dF_r(x;r) \bigg/ \int U_{\alpha\alpha}(x,\alpha)dF(x;r) \qquad (3)$$

where $F_r(x;r)=\partial F(x;r)/\partial r$. Since the denominator of this expression is negative by assumption, the sign of $d\alpha/dr$ is

given by the sign of the numerator $\int U_\alpha(x,\alpha)dF_r(x;r)$. Integrating by parts twice yields:

$$\int U_\alpha(x,\alpha)dF_r(x;r)$$
$$= \int U_{xx\alpha}(x,\alpha) \cdot \left[\int_0^x F_r(\xi;r)d\xi\right]dx$$
$$= \int U_{xx\alpha}(x,\alpha) \cdot \left[\frac{d}{dr}\int_0^x F(\xi;r)d\xi\right]dx \qquad (4)$$

Thus, if increases in the parameter r imply increases in the riskiness of the distribution $F(\cdot,r)$, it follows from condition (C′) that the signs of the square-bracketed terms in (4) will be non-negative, so that the effect of r upon α depends upon the sign of $U_{xx\alpha}(x,\alpha) = \partial^3 U(x,\alpha)/\partial x^2\partial\alpha$. Thus, if $U_{xx\alpha}(x,\alpha)$ is uniformly negative a mean preserving increase in risk in the distribution of x will lead to a fall in the optimal value of the control variable α, and vice versa. Another way to see this is to note that if $U_\alpha(x,\alpha)$ is concave in x then a mean preserving increase in risk will lower the left side of the first order condition (2), which (since $U_{\alpha\alpha}(x,\alpha) \leq 0$) will require a drop in α to re-establish the equality. Economists, mathematicians and scientists routinely use this technique when analysing models involving risk; see for example Rothschild and Stiglitz (1971), Dionne, Eeckhoudt and Gollier (1993), Eeckhoudt, Gollier and Schlesinger (1996), Jewitt (1987), Ormiston (1992), Tzeng (2001), Nowak (2004), Chateauneuf, Cohen and Meilijson (2004), Baker (2006), and Beladi, de la Vina and Firoozi (2006).

Related topics
The characterization of risk outlined in the previous section has been extended along several lines. Diamond and Stiglitz (1974), for example, have replaced the notion of a mean preserving spread with that of a mean *utility* preserving spread to obtain a general characterization of a *compensated increase in risk*. They relate this notion to the well known Arrow–Pratt characterization of comparative risk aversion (see EXPECTED UTILITY HYPOTHESIS).

In addition, researchers such as Ekern (1980), Fishburn (1982), Fishburn and Vickson (1978), Hansen, Holt, and Peled (1978), Tesfatsion (1976), and Whitmore (1970) have extended the above work to the development of a general theory of *stochastic dominance*, which provides a whole sequence of similarly characterized partial orders on distributions, each presenting a corresponding set of equivalent conditions involving algebraic conditions on the distributions, types of spreads, and classes of utility functions which prefer (or are averse to) such spreads. The comparative statics analysis presented above may be similarly extended to such characterizations. An extensive bibliography of the stochastic dominance literature is given in Bawa (1982). Finally, various extensions of the notions of increasing risk and stochastic dominance to the case of multivariate distributions may be found in Epstein and Tanny (1980), Fishburn and Vickson (1978), Huang, Kira and Vertinsky (1978), Lehmann (1955), Levhari, Parousch and Peleg (1975), Levy and Parousch (1974), Russell and Seo (1978), Sherman (1951), and Strassen (1965); see also the mathematical results in Marshall and Okun (1979).

MARK J. MACHINA AND MICHAEL ROTHSCHILD

See also **expected utility hypothesis; risk aversion; uncertainty.**

Bibliography

Anscombe, F. and Aumann, R. 1963. A definition of subjective probability. *Annals of Mathematical Statistics* 34, 199–205.

Arrow, K. 1964. The role of securities in the optimal allocation of risk-bearing. *Review of Economic Studies* 31, 91–6.

Baker, E. 2006. Increasing risk and increasing informativeness: equivalence theorems. *Operations Research* 54, 26–36.

Bawa, V. 1982. Stochastic dominance: a research bibliography. *Management Science* 28, 698–712.

Beladi, H., de la Vina, L. and Firoozi, F. 2006. On information value and mean-preserving transformations. *Applied Mathematics Letters* 19, 843–8.

Borch, K. 1969. A note on uncertainty and indifference curves. *Review of Economic Studies* 36, 1–4.

Chateauneuf, A., Cohen, M. and Meilijson, I. 2004. Four notions of mean-preserving increase in risk, risk attitudes and applications to the rank-dependent expected utility model. *Journal of Mathematical Economics* 40, 547–71.

Debreu, G. 1959. *Theory of Value: An Axiomatic Analysis of General Equilibrium*. New Haven: Yale University Press.

Diamond, P. and Rothschild, M., eds. 1989. *Uncertainty in Economics: Readings and Exercises*, 2nd edn. New York: Academic Press.

Diamond, P. and Stiglitz, J. 1974. Increases in risk and in risk aversion. *Journal of Economic Theory* 8, 337–60.

Dionne, G., Eeckhoudt, L. and Gollier, C. 1993. Increases in risk and linear payoffs. *International Economic Review* 34, 309–19.

Eeckhoudt, L., Gollier, C. and Schlesinger, H. 1996. Changes in background risk and risk taking behavior. *Econometrica* 64, 683–9.

Ekern, S. 1980. Increasing nth degree risk. *Economics Letters* 6, 329–33.

Epstein, L. and Tanny, S. 1980. Increasing generalized correlation: a definition and some economic consequences. *Canadian Journal of Economics* 13, 16–34.

Feldstein, M. 1969. Mean-variance analysis in the theory of liquidity preference and portfolio selection. *Review of Economic Studies* 36, 5–12.

Fishburn, P. 1982. Simplest cases of *n*'th degree stochastic dominance. *Operations Research Letters* 1, 89–90.

Fishburn, P. and Vickson, R. 1978. Theoretical foundations of stochastic dominance, in Whitmore and Findlay (1978).

Gärdenfors, P. and Sahlin, N.-E., eds. 1988. *Decision, Probability, and Utility: Selected Readings*. Cambridge: Cambridge University Press.

Hadar, J. and Russell, W. 1969. Rules for ordering uncertain prospects. *American Economic Review* 59, 25–34.

Hanoch, G. and Levy, H. 1969. The efficiency analysis of choices involving risk. *Review of Economic Studies* 36, 335–46.

Hansen, L., Holt, C. and Peled, D. 1978. A note on first degree stochastic dominance. *Economics Letters* 1, 315–9.

Hey, J., ed. 1997. *The Economics of Uncertainty, Vol. II: Uncertainty and Dynamics*. Cheltenham: Edward Elgar.

Hicks, J. 1962. Liquidity. *Economic Journal* 72, 787–802.

Hirshleifer, J. 1965. Investment decision under uncertainty: choice-theoretic approaches. *Quarterly Journal of Economics* 79, 509–36.

Hirshleifer, J. 1966. Investment decision under uncertainty: applications of the state-preference approach. *Quarterly Journal of Economics* 80, 252–77.

Huang, C., Kira, D. and Vertinsky, I. 1978. Stochastic dominance for multi-attribute utility functions. *Review of Economic Studies* 45, 611–16.

Jewitt, I. 1987. Risk aversion and the choice between risky prospects: the preservation of comparative statics results. *Review of Economic Studies* 54, 73–85.

Karni, E. 1985. *Decision Making Under Uncertainty: The Case of State Dependent Preferences*. Cambridge, MA: Harvard University Press.

Knight, F. 1921. *Risk, Uncertainty and Profit*. Boston: Houghton Mifflin.

Kyberg, H. and Smokler, H., eds. 1964. *Studies in Subjective Probability*. New York: Wiley.

Lehmann, E. 1955. Ordered families of distributions. *Annals of Mathematical Statistics* 26, 399–419.

Levhari, D., Paroush, J. and Peleg, B. 1975. Efficiency analysis for multivariate distributions. *Review of Economic Studies* 42, 87–91.

Levy, H. and Paroush, J. 1974. Toward multivariate efficiency criteria. *Journal of Economic Theory* 7, 129–42.

Lintner, J. 1965. The valuation of risk assets and the selection of risky investments in stock portfolios and capital budgets. *Review of Economics and Statistics* 44, 243–69.

Machina, M. and Pratt, J. 1997. Increasing risk: some direct constructions. *Journal of Risk and Uncertainty* 14, 103–27.

Machina, M. and Schmeidler, D. 1992. A more robust definition of subjective probability. *Econometrica* 60, 745–80. Reprinted in Hey (1997).

Markowitz, H. 1952. Portfolio selection. *Journal of Finance* 7, 77–91.

Markowitz, H. 1959. *Portfolio Selection: Efficient Diversification of Investment*. New Haven, CT: Yale University Press.

Marshall, A. and Okun, I. 1979. *Inequalities: Theory of Majorization and Its Applications*. New York: Academic Press.

Meyer, J. 1987. Two-moment decision models and expected utility maximization. *American Economic Review* 77, 421–30.

Mossin, J. 1966. Equilibrium in a capital asset market. *Econometrica* 34, 768–83.

Nowak, M. 2004. Interactive approach in multicriteria analysis based on stochastic dominance. *Control and Cybernetics* 33, 463–76.

Ormiston, M. 1992. First and second degree transformations and comparative statics under uncertainty. *International Economic Review* 33, 33–44.

Pratt, J. 1964. Risk aversion in the small and in the large. *Econometrica* 32, 122–36. Reprinted in Diamond and Rothschild (1989).

Pratt, J., Raiffa, H. and Schlaifer, R. 1964. The foundations of decision under uncertainty: an elementary exposition. *Journal of the American Statistical Association* 59, 353–75.

Raiffa, H. 1968. *Decision Analysis: Introductory Lectures on Choice Under Uncertainty*. Reading, MA: Addison Wesley.

Ramsey, F. 1926. Truth and probability. In *The Foundations of Mathematics and other Logical Essays*, ed. R.B. Braithwaite. London: Routledge & Kegan Paul, 1931. Reprinted in Kyberg and Smokler (1964), Ramsey (1978), and Gärdenfors and Sahlin (1988).

Ramsey, F. 1978. In *Foundations: Essays in Philosophy, Mathematics and Economics*, ed. D.H. Mellor. London: Routledge & Kegan Paul.

Rothschild, M. and Stiglitz, J. 1970. Increasing risk: I. A definition. *Journal of Economic Theory* 2, 225–43. Reprinted in Diamond and Rothschild (1989).

Rothschild, M. and Stiglitz, J. 1971. Increasing risk: II. its economic consequences. *Journal of Economic Theory* 3, 66–84.

Rothschild, M. and Stiglitz, J. 1972. Addendum to 'increasing risk: I. a definition'. *Journal of Economic Theory* 5, 306.

Russell, W. and Seo, T. 1978. Ordering uncertain prospects: the multivariate utility functions case. *Review of Economic Studies* 45, 605–11.

Samuelson, P. 1967. General proof that diversification pays. *Journal of Financial and Quantitative Analysis* 2, 1–13.

Savage, L. 1954. *The Foundations of Statistics*. New York: John Wiley and Sons. Rev. edn, New York: Dover Publications, 1972.

Sharpe, W. 1964. Capital asset prices: a theory of market equilibrium under conditions of risk. *Journal of Finance* 19, 425–42.

Sherman, S. 1951. On a theorem of Hardy, Littlewood, Polya, and Blackwell. *Proceedings of the National Academy of Science* 37, 826–31. (Errata. *Proc. Nat. Ac. Sci.* 38, 382.)

Strassen, V. 1965. The existence of probability measures with given marginals. *Annals of Mathematical Statistics* 36, 423–39.

Tesfatsion, L. 1976. Stochastic dominance and the maximization of expected utility. *Review of Economic Studies* 43, 301–15.

Tobin, J. 1958. Liquidity preference as behavior toward risk. *Review of Economic Studies* 25, 65–86.

Treynor, J. 1999. Toward a theory of market value of risky assets. In *Asset Pricing and Portfolio Performance*, ed. R. Korajczyk. London: Risk Books. In Korajczyk (1999).

Tzeng, L. 2001. Increase in risk and weaker marginal-payoff-weighted risk dominance. *Journal of Risk and Insurance* 68, 329–37.

Whitmore, G. 1970. Third-degree stochastic dominance. *American Economic Review* 60, 457–9.

Whitmore, G. and Findlay, M. 1978. *Stochastic Dominance: An Approach to Decision Making Under Risk*. Lexington, MA: Heath.

Yaari, M. 1969. Some remarks on measures of risk aversion and on their uses. *Journal of Economic Theory* 1, 315–29. Reprinted in Diamond and Rothschild (1989).

risk aversion

Arrow–Pratt theory of risk aversion

The classical theory of risk aversion, due to Pratt (1964) and Arrow (1965), is rooted in the expected utility theory of decision making. An agent's preferences are assumed to have an expected utility representation. The objects of choice are real valued random variables defined either on a finite or infinite set of states of the world with probabilities of states that may be either objective or subjective. The intended interpretation of a random variable is as an agent's risky wealth.

An agent whose expected utility representation of preferences is written $E[u(\tilde{x})]$, where u is the von Neumann–Morgenstern utility function and E denotes the expectation (expected value), is *risk averse* if

$$E[u(\tilde{x})] \leq u(E(\tilde{x})) \qquad (1)$$

for every risky wealth $\tilde{x}$. If (1) holds with strict inequality for every non-deterministic $\tilde{x}$, the agent is *strictly risk averse*. Jensen's inequality implies that, if utility function u is concave, the agent is risk averse. The converse is also true. Thus, the concavity of u is a necessary and sufficient condition for risk aversion. Moreover, strict concavity of u is a necessary and sufficient condition for strict risk aversion. Examples of strictly concave von Neumann–Morgenstern utility functions, commonly used in applied work, include the negative exponential utility $u(w) = e^{-\alpha w}$ with $\alpha > 0$, the logarithmic utility $u(w) = \ln(w)$, and the power utility $u(w) = \frac{1}{1-\alpha} w^{1-\alpha}$ with $\alpha > 0$, $\alpha \neq 1$.

It is useful to have a measure of the intensity of risk aversion. The most natural measure is *risk compensation*. It is by definition the amount $\rho(w, \tilde{z})$ of deterministic wealth one could extract from an agent in exchange for relieving her of zero-expectation risk $\tilde{z}$ at an initial deterministic wealth w,

$$E[u(w + \tilde{z})] = u(w - \rho(w, \tilde{z})). \qquad (2)$$

A risk-averse agent has non-negative risk compensation for every zero-expectation risk, at every level of initial wealth. Risk compensation makes possible interpersonal comparisons of risk aversion and, for any agent, comparisons of risk aversion at different levels of her initial wealth. If risk compensation $\rho_1(w, \tilde{z})$ of an agent with von Neumann–Morgenstern utility function u_1 is greater than or equal to risk compensation $\rho_2(w, \tilde{z})$ of another agent with utility function u_2, for every deterministic wealth w and risk $\tilde{z}$ with $E(\tilde{z}) = 0$, then the agent with u_1 is said to be *more risk averse* than the one with u_2. An agent has increasing, decreasing or constant risk aversion if, for every zero-expectation risk, her risk compensation is increasing, decreasing or constant in w, for every $\tilde{z}$ with $E(\tilde{z}) = 0$.

Another measure of risk aversion is *certainty equivalent*. It is by definition the amount $c(\tilde{x})$ of deterministic wealth such that an agent is indifferent between this deterministic wealth and risky wealth $\tilde{x}$,

$$E[u(\tilde{x})] = u(c(\tilde{x})). \qquad (3)$$

For a risk-averse agent, certainty equivalent is lower than the expectation of risky wealth. Since certainty equivalent and risk compensation are related by $c(w + \tilde{x}) = w - \rho(w, \tilde{z})$, these two measures can be interchangeably used in the Arrow–Pratt theory of risk aversion.

Although with considerable intuitive appeal, risk compensation is not all that practical. It is only implicitly defined in (2). The basic insight of Arrow and Pratt is that there is a simple measure of risk aversion which is in a certain sense equivalent to risk compensation, namely, the *Arrow–Pratt measure of absolute risk aversion*

$$A(w) = -\frac{u''(w)}{u'(w)}. \qquad (4)$$

The negative of the second derivative $u''(w)$ is a mathematical measure of the degree of concavity of u. It is rescaled by the first derivative (assumed non-zero) which makes the measure invariant under any affine transformation of u. For 'small' risk $\tilde{z}$ with $E(\tilde{z}) = 0$, risk compensation $\rho(w, \tilde{z})$ equals approximately half the product of the variance $\sigma^2(\tilde{z})$ and the Arrow–Pratt measure at w,

$$\rho(w, \tilde{z}) \cong \frac{1}{2} A(w) \sigma^2(\tilde{z}) \qquad (5)$$

as can be demonstrated using quadratic approximation of expected utility $E[u(w + \tilde{z})]$.

The important theorem of Pratt establishes an equivalence of the two measures as criteria for comparative risk aversion. For that theorem, let u_1 and u_2 be two strictly increasing von Neumann–Morgenstern utility functions, twice differentiable with continuous second derivatives.

Theorem (Pratt, 1964): The following conditions are equivalent:

(a) $A_1(w) \geq A_2(w)$ *for every* w.
(b) $\rho_1(w, \tilde{z}) \geq \rho_2(w, \tilde{z})$ *for every* w *and every random variable* $\tilde{z}$ *with* $E(\tilde{z}) = 0$,
(c) u_1 *is a concave transformation of* u_2*; that is,* $u_1 = f \circ u_2$ *for* f *concave and strictly increasing.*

There is a version of the Pratt's theorem which has equalities in (a) and (b) and an affine transformation in (c). This version implies that the Arrow–Pratt measure identifies the von Neumann–Morgenstern utility function up to an affine transformation. For example, the negative exponential utility is (up to an affine transformation) the only strictly concave utility with constant absolute risk aversion. There is also a strict version of the Pratt's theorem with strict inequalities in (a) and (b) and a strictly concave transformation in (c).

A corollary to the Pratt's Theorem says that an agent has increasing, decreasing or constant risk aversion if and only if her Arrow–Pratt measure of risk aversion is increasing, decreasing or constant. One needs only to consider utility functions $u_1(w) = u(w)$ and $u_2(w) = u(w + \Delta w)$ for arbitrary $\Delta w > 0$.

Arrow (1965) and Pratt (1964) extended their theory of measurement of risk aversion to relative risk, that is, risk per dollar of an agent's wealth. Risk compensation $\rho_r(w, \tilde{\zeta})$ for relative risk $\tilde{\zeta}$ with $E(\tilde{\zeta}) = 0$ at initial wealth w is defined by

$$E[u(w + w\tilde{\zeta})] = u(w - w\rho_r(w, \tilde{\zeta})). \qquad (6)$$

The Arrow–Pratt measure of relative risk aversion is $R(w) = -\frac{u''(w)}{u'(w)} w$. The measures ρ_r and R are related in the same way that their counterparts for absolute risk are related. Power and logarithmic utility functions have constant relative risk aversion.

Risk compensation is defined when the agent's initial position is risk-free. The approximation (5) of risk compensation by the Arrow–Pratt measure holds at a risk-free position, too. Measures of risk aversion that can be used when the initial position is risky have been developed by Ross (1981) and Machina and Neilson (1987).

When random variables are vector-valued (multivariate), the Arrow–Pratt theory can be applied to risk in one coordinate (for example, consumption of one good) when values of other coordinates are deterministic. Alternatively, multivariate risk aversion can be defined by requiring condition (1) to hold for every multivariate random variable $\tilde{x}$. Multivariate random variables arise when objects of choice are consumption plans of multiple goods or consumption plans over multiple time periods. Multivariate risk aversion is equivalent to concavity of the von Neumann–Morgenstern utility function and implies that the induced ordinal preferences over multiple goods under certainty are convex. The theory of comparative risk aversion has been extended to the multivariate case by Kihlstrom and Mirman (1974) under the restriction that utility functions induce the same ordinal preferences.

Rabin and Thaler (2001) have pointed out a peculiar feature of risk aversion under expected utility. If an agent rejects a small actuarially favourable gamble at every level of wealth (or at a big enough range of wealth), then she will reject a gamble with a modest loss and an arbitrarily large gain. They presented a calibration exercise which shows that any risk-averse agent who rejects an even-chance gamble of losing $10 or winning $11 will turn down an even-chance gamble of losing $1,000 or winning any sum of money. The significance of Rabin and Thaler's observation is a subject of current debate.

Risk aversion without expected utility

A representation of preferences which is closely related to but more general than the expected utility is the state-dependent expected utility. For a finite set S of states of the world, it is written $\sum_{s \in S} \pi_s u_s(x_s)$, where π_s is the probability of state s and u_s is the state-dependent utility. Werner (2005a) has shown that an agent with state-dependent expected utility is risk averse in the sense of having a preference for deterministic outcomes over risky outcomes with equal expectations if and only if the utility functions u_s are state independent and concave. When utility functions u_s are state dependent, a risk-free wealth may have risky utility and, more importantly, risky marginal utility. Karni (1985) has developed a theory of aversion to risk in marginal utility defined by an agent being unwilling to take an actuarially fair gamble when starting from a position of risk-free marginal utility of wealth. Measures of risk aversion analogous to the measures introduced by Arrow and Pratt can be defined, and an extension of the Pratt's theorem obtains for utility functions that have the same set of wealth profiles with risk-free marginal utility. State-dependent utilities arise in instances of behaviour under health risk.

The Arrow–Pratt theory of risk aversion is based on the simple notion that every risky outcome is more risky than the deterministic outcome with equal expectation. For preferences that do not have an expected utility representation (state-independent, or not), this concept of 'more risky' is too restrictive to deliver a meaningful notion of risk aversion. A weaker concept of 'more risky' has been introduced by Rothschild and Stiglitz (1970). It is a partial ordering of random variables according to the integrals of their cumulative distribution functions. For two random variables $\tilde{x}$ and $\tilde{y}$ with the same expectation, $\tilde{x}$ is *more risky* than $\tilde{y}$ if

$$\int_{-\infty}^{w} F_y(t) \, dt \leq \int_{-\infty}^{w} F_x(t) \, dt, \qquad \forall w, \tag{7}$$

where F_x is the cumulative distribution function assigning to each w the probability $Prob\{\tilde{x} \leq w\}$.

An agent whose utility function on random variables is monotone decreasing with respect to the ordering of more risky is said to be *strongly risk averse* (see Cohen, 1995). It follows that a strongly risk-averse agent, when starting from a risky position $\tilde{x}$, is unwilling to take a gamble $\tilde{z}$ with zero expectation conditional on each possible realization x of $\tilde{x}$, that is, $E(\tilde{z}|\tilde{x}=x)=0$ for every x. The ordering (7) has been known in mathematics as the second-order stochastic dominance and the strongly risk-averse functions have been known as the Schur concave functions (see Marshall and Olkin, 1979). Chew, Karni and Safra (1987) derived necessary and sufficient conditions for strong risk aversion of two types of utility functions: the rank-dependent expected utility of Quiggin (1982), and the dual utility of Yaari (1987). Characterization results for general utility functions can be found in Machina (1982), Chew and Mao (1995), and Dana (2005). For an expected utility, strong risk aversion and risk aversion in the sense of (1) are equivalent. Concavity of the von Neumann–Morgenstern utility function is necessary and sufficient for strong risk aversion of expected utility (see Rothschild and Stiglitz, 1970).

An important representation of preferences under uncertainty, more general than the expected utility, is the maxmin (or multiple-prior) expected utility (see Gilboa and Schmeidler, 1989). Under the maxmin expected utility representation an agent has a set $\|$ of probability measures (priors) on states instead of a single probability measure, and a von Neumann–Morgenstern utility function u. The set $\|$ is assumed to be closed and convex. An agent's utility of risky wealth $\tilde{x}$ is

$$\min_{P \in P} E_P[u(\tilde{x})], \tag{8}$$

where $E_P[u(\tilde{x})]$ is the expectation of $u(\tilde{x})$ with respect to probability measure P. Multiplicity of probability measures reflects the agent's ambiguous information about states of the world. Taking the minimum reflects the concern with the 'worst case' scenario. If the set $\|$ consists of all probability measures, then the maxmin expected utility (8) equals $\min_s u(x_s)$ meaning that the agent follows the Wald's criterion of choice. Maxmin expected utility may exhibit risk aversion with respect to some probability measure in the set of priors. If the von Neumann–Morgenstern utility function u is concave, the agent prefers deterministic wealth in the amount of $E_P(\tilde{x})$ over risky wealth $\tilde{x}$ for every probability measure P in her set of priors. Thus the agent is risk averse in the Arrow–Pratt sense with respect to every P in the set $\|$. Wald's minimum utility is also strongly risk averse with respect to every probability measure. Many maxmin expected utility functions are not distribution invariant under any probability measure on states, rendering the question of strong risk aversion meaningless for these functions. Werner (2005b) proposes a concept of more risky, stronger than (7), such that adding a gamble with zero conditional expectation makes an initial risky position more risky, but without identifying random variables with their probability distributions. For many (but not all) sets of priors, maxmin expected utility with concave von Neumann–Morgenstern utility function is risk averse in that sense under some probability measure from the set of priors.

Some implications of risk aversion

The choice of insurance coverage provides a good illustration of implications of risk aversion on agents' behaviour. Suppose that an expected-utility maximizing individual with initial wealth w faces a risk of losing L with probability π or not losing it with probability $1 - \pi$. She is offered insurance against the loss at price p per dollar of coverage. A strictly risk-averse individual will choose full coverage giving her risk-free wealth $w - pL$, if the insurance is priced actuarially fair, that is $p = \pi$. If it is priced above the fair value, that is $p > \pi$, then the optimal coverage will be partial. Schlesinger (1997) shows that these results continue to hold under risk aversion without expected utility.

A risk-averse investor who invests her initial risk-free wealth among many risky assets and a risk-free asset will choose an optimal portfolio with risky payoff only if the expected return on that portfolio exceeds the risk-free return. For a strictly risk-averse investor, the expected return on the optimal portfolio must strictly exceed the risk-free return if the payoff is risky. This is the fundamental *risk–return trade-off* in asset markets and it is a consequence of risk aversion. It continues to hold when the investor's initial position includes an endowment portfolio of assets. In an equilibrium in competitive asset markets where many strictly risk-averse investors trade their endowment portfolios, the market portfolio (that is, the outstanding supply of assets) must have expected return that exceeds the risk-free return. This is so because the return on the market portfolio is a weighted sum of the returns on investors' optimal portfolios, with weights equal to the respective shares of total wealth. Expected returns on optimal portfolios exceed the risk-free return, with some exceeding it strictly, if the payoff of the market portfolio is risky. Thus, risk aversion provides a qualitative explanation of the expected return in equity markets exceeding the risk-free return. Attempts to give a quantitative explanation of the observed high excess return on equities over risk-free bonds have led to the *equity premium puzzle* (see Mehra and Prescott, 1985).

JAN WERNER

See also **expected utility hypothesis; non-expected utility theory; risk; stochastic dominance.**

Bibliography

Arrow, K. 1965. *Aspect of the Theory of Risk Bearing.* Helsinki: Yrjö Jahnsson Foundation.

Chew, S., Karni, E. and Safra, Z. 1987. Risk aversion in the theory of expected utility with rank dependent preferences. *Journal of Economic Theory* 42, 370–81.

Chew, S. and Mao, M. 1995. A Schur concave characterization of risk aversion for non-expected utility preferences. *Journal of Economic Theory* 67, 402–35.

Cohen, M. 1995. Risk-aversion concepts in expected- and non-expected-utility models. *Geneva Papers on Risk and Insurance Theory* 20, 73–91.

Dana, R.-A. 2005. A representation result for concave Schur-concave functions. *Mathematical Finance* 15, 613–34.

Gilboa, I. and Schmeidler, D. 1989. Maxmin expected utility with nonunique prior. *Journal of Mathematical Economics* 18, 141–53.

Karni, E. 1985. *Decision Making under Uncertainty.* Cambridge, MA: Harvard University Press.

Kihlstrom, R. and Mirman, L. 1974. Risk aversion with many commodities. *Journal of Economic Theory* 8, 361–88.

Machina, M. 1982. Expected utility analysis without the independence axiom. *Econometrica* 50, 277–323.

Machina, M. and Neilson, W. 1987. The Ross characterization of risk aversion: strengthening and extension. *Econometrica* 55, 1139–49.

Marshall, A. and Olkin, I. 1979. *Inequalities: Theory of Majorization and its Applications.* New York: Academic Press.

Mehra, R. and Prescott, E. 1985. The equity premium: a puzzle. *Journal of Monetary Economics* 15, 145–61.

Pratt, J. 1964. Risk aversion in the small and in the large. *Econometrica* 32, 132–6.

Quiggin, J. 1982. A theory of anticipated utility. *Journal of Economic Behavior and Organization* 3, 323–43.

Rabin, M. and Thaler, R. 2001. Anomalies: risk aversion. *Journal of Economic Perspectives* 15(1), 219–232.

Ross, S. 1981. Some stronger measures of risk aversion in the small and in the large with applications. *Econometrica* 49, 621–38.

Rothschild, M. and Stiglitz, J. 1970. Increasing risk. I: a definition. *Journal of Economic Theory* 2, 225–43.

Schlesinger, H. 1997. Insurance demand without the expected-utility paradigm. *Journal of Risk and Insurance* 64(1), 19–39.

Werner, J. 2005a. A simple axiomatization of risk-averse expected utility. *Economics Letters* 88, 73–7.

Werner, J. 2005b. Risk and risk aversion when states of nature matter. Mimeo.

Yaari, M. 1987. The dual theory of choice under risk. *Econometrica* 55, 95–115.

risk-coping strategies

Risk is pervasive, not least in developing countries. A high dependence on agriculture for income and employment means that people's livelihoods are strongly affected by climatic vagaries. Poor health care and immunization means that illness is common, affecting labour supply. Poor infrastructure and market institutions result in limited market integration, leading to a high sensitivity of prices to local shocks. Economic instability, conflict and political instability add further to a high-risk environment in many developing countries.

While risk is widespread, in developing countries insurance markets are typically missing or incomplete. This causes potentially serious welfare losses, especially since government-led alternatives such as social security are largely missing, compounded by imperfections in credit markets that limit the extent to which credit can be used as to substitute for insurance. Economic agents are typically risk averse and, even in the poorest settings, the evidence suggests that they do not just passively undergo the consequences of risk. Instead, given risk aversion, they try to make activity and asset portfolio choices to balance their need to make a living, but without exposing themselves to too much risk. The strategies used to achieve this are often referred to as 'risk management and coping strategies'. The economic analysis of these strategies has been one of the areas in which research based on some of the poorest high-risk rural settings in developing countries has made a substantial contribution to the economic literature in general. Our focus in this overview is on the empirical literature, often drawing on the evidence generated from the sample based on six Indian villages for which the International Crops Research Institute for the Semi-Arid Tropics (ICRISAT) collected exceptionally detailed panel data over ten years in the high-risk environment of the semi-arid tropics of India (Morduch, 2004).

Self-insurance via savings

It is possible to distinguish a number of commonly observed different risk strategies. First, we can consider strategies that aim to cope with the consequences of shocks. Risk aversion is sufficient to induce households to try to smooth consumption or nutrition, and indeed standard models of consumption smoothing when insurance and credit markets are imperfect can shed light on this type of strategy (Deaton, 1992). When shocks occur, households may decide to curb their consumption loss through the sacrifice of existing assets. Households may pre-empt this by trying to self-insure against risk through precautionary savings. A precautionary motive for savings would be sufficient for savings to increase in response to increased risk, so that a buffer stock is built to deplete when shocks occur. Even though formal credit markets in high-risk settings are typically underdeveloped, informal credit transactions may also be used for smoothing purposes.

There is a large literature testing whether households in developing countries smooth consumption, building on standard models of permanent income and often using shocks to identify the relevant effects (for example,

Paxson, 1992). Furthermore, which assets tend to be used for this purpose in the face of shocks has occasionally been investigated. (Finding positive evidence of using assets to smooth is not sufficient to show that any savings were built up for precautionary motives to start with. This would require evidence that greater risk indeed increased savings, which is harder to test.) Nevertheless, the evidence suggests that in some settings productive assets are sold off for smoothing, for example in work by Rosenzweig and Wolpin (1993) using the ICRISAT data, while in other settings (for example, in Burkina Faso, as in Fafchamps, Udry and Czukas, 1998) livestock were not sold off despite serious income shocks. Furthermore, there is much anecdotal evidence that, during famines in mixed farming environments, livestock is not being sold despite serious human nutritional losses and risks.

Squaring these findings with basic theoretical models remains a challenge, and different suggestions can be made. For example, different technologies may underlie the pattern of returns to different assets, so that optimal portfolio mixes would suggest that assets are depleted at different rates when shocks occur. Another possibility is that asset returns and prices are risky as well, so that the reliability of the buffer is limited. For example, if incomes, asset prices and returns have a high positive covariance, then selling assets when incomes are low may not be the optimal strategy, since the future gains from holding on to assets is actually high. Alternatively, behavioural theories based on experiments, such as that risk-loving rather than risk-averse behaviour may be displayed in the face of losses that are potentially very large (as in Kahneman and Tversky, 1979), may provide some insight to these puzzles.

A related strategy to self-insurance observed in poor settings involves the key asset available to the poor, namely, labour. In response to shocks, labour supply is adjusted to increase involvement in productive activities, including off-farm or temporary migration (for example, Kochar, 1995). Furthermore, children may be taken out of school and into work in response to income shocks (Jacoby and Skoufias, 1997).

Risk-sharing through mutual support

A second common strategy to cope with the consequences of risk involves non-market institutions based on risk sharing, whereby households or other economic agents use transfers to smooth outcomes across a group of people when shocks occur. Conceptually, this is the cross-sectional equivalent of standard permanent income models: it is concerned with smoothing over space rather than over time. Unless risk preferences differ between economic agents, unlike self-insurance strategies this strategy is relevant only for idiosyncratic shocks, not covariate risk. Townsend (1994) is the seminal paper on high-risk developing-country environments. The basic prediction of Townsend's model, under specific assumptions, is that household consumption is dependent on average village resources but not on individual income realizations. This provides a clear basis for empirical testing of the perfect risk-sharing hypothesis: do idiosyncratic shocks to income affect consumption or nutrition outcomes within a well-defined setting? Using the ICRISAT data from India, he finds that perfect risk-sharing is rejected within the village, even though substantial risk-pooling takes place. Other studies (including some on the same data) suggest that some risk-sharing typically occurs in villages, but the evidence is typically not consistent with perfect risk-sharing.

The failure of perfect risk-sharing in village settings has attracted much attention in terms of theoretical work. Work has focused on accounting for information asymmetries, private savings and the role of enforceability problems related to these informal risk-sharing contracts *ex post* in repeated game contexts (for example, Ligon, Thomas and Worrall, 2002). The work on enforceability has found that constrained efficient contracts will contain updating rules offering higher weights in the risk-sharing contracts in particular states of the world to those facing stronger incentives to leave, and that over time, the weights have memory. The consequence is that these risk-sharing contracts take on features more common in credit contracts, leading to their description as 'quasi-credit' arrangements. There is some empirical evidence consistent with these models.

The lack of perfect risk-sharing within villages has led to further work investigating whether risk-sharing occurs in other settings, for example within households or extended families, or in other social groupings. Partial risk-sharing has been documented across ethnic groups as well as within families. More recently, risk-sharing across networks has been explored as well. It is relatively straightforward to analyse risk-sharing within networks beyond specifically exogenously defined groupings, such as caste, but for most group or network links network membership has to be endogenously modelled. Theoretical work has emerged analysing the formation and stability of insurance networks, including in the face of group deviations (for example, Genicot and Ray, 2003). Better data-sets focusing on linkages between households in communities has also allowed further evidence to emerge of the extent and nature of risk-sharing within networks and groups. Integrating the endogeneity of network formation for insurance purposes in empirical analysis nevertheless remains a challenge.

The literatures on intertemporal consumption smoothing and risk-sharing in developing countries appear to converge at least in terms of diagnosis: consumption is relatively smooth in the face of income shocks, but not perfectly. Nevertheless, it is often not clear in the tests whether consumption smoothing in practice occurs through transfers or through self-insurance or credit. Even in the ICRISAT data on India, as used in the Townsend data, this has remained an issue of contention

(Townsend, 1995; Morduch, 2004). One strategy is to specifically study the actual responses to shocks (such as dissavings, credit transactions or transfers) and their contribution to smoothing. Deaton and Paxson (1994) provide an alternative test to distinguish whether insurance or credit is responsible for observed smoothing by looking at the changing distribution over time of consumption for a particular cohort in a number of countries, including Taiwan. If consumption smoothing is present due to formal or informal insurance, then inequality can be expected to remain constant over time. If smoothing consumption occurs through credit market transactions, then inequality can be expected to increase over time due to changes in permanent income. Their results suggest that credit markets are more important than insurance for the observed patterns of smoothing.

Income smoothing

Households do not just use strategies that aim to cope with the consequences of shocks; they may try to reduce or mitigate the risk they face, not least given the limits to risk-sharing and self-insurance. To put it more directly, they may aim to smooth income (Morduch, 1995). In rural settings, this strategy has been a central force in shaping farming systems and institutions. The most straightforward strategy is income diversification, whereby income sources are combined and the resulting portfolio faces reduced risk even if the underlying income processes are equally risky when taken separately (as long as they are not perfectly covariate). In some cases, risk management may imply diversifying into (or even specializing in) specific low-risk technologies or activities, such as growing drought-resistant crops or gathering firewood for sale. Social institutions have also developed to include means of reducing exposure to risk. Geographically dispersed marriage patterns, such as those observed by Rosenzweig and Stark (1989) in villages in southern India, can be interpreted as linked to risk diversification within extended families. Local institutions to manage commons and natural resources or land tenure arrangements appear to include risk management as part of their rationale.

Testing the specific role of risk in observed diversification patterns in activities and assets is nevertheless not straightforward. Given the multiple market imperfections faced by the poor, for example in labour markets, the fact of observed multiple activities is not sufficient to sustain the conclusion that risk is its cause. Furthermore, the opportunity to shape risk faced by households also implies econometric problems in standard tests of consumption smoothing and risk sharing, requiring exogenous sources of variation in risk faced. While rainfall variation may provide a useful instrument for common risk, finding sources of exogenous variation for the identification of idiosyncratic shocks is more difficult. These problems suggest that further exploration of risk management strategies will remain methodologically challenging. Furthermore, most investigations of risk strategies in developing countries have been in rural settings, with a focus on agricultural households in relatively stationary environments. This was mainly due to the lack of suitable panel data-sets in urban settings. With more long-term data-sets becoming available, more attention can be paid to the changes in risk strategies following increased integration of local economies and the rise in migration opportunities.

Risk strategies and persistent poverty

All the above hints at an important consequence. While risk strategies contribute to avoiding serious consumption fluctuations, they are not without consequences for welfare, investment and poverty. More specifically, households tend to trade risk and smooth consumption in the short run for lower mean welfare outcomes in the long run. Precautionary motives for saving and credit constraints may induce asset portfolios to focus more on liquidity than on returns. Sales of productive assets for smoothing or taking children out of school to increase labour supply will reduce permanent income. Income smoothing strategies will involve leaving aside profitable opportunities for activity and asset portfolios with a lower mean return. Evidence from villages in the ICRISAT sample in India suggests that these effects may well be substantial, with those households with limited protection against risk (identified by rainfall variability) opting for portfolios with lower returns (Rosenzweig and Binswanger, 1993). More specifically, they find that an increase in rainfall variability by one standard deviation would reduce returns to assets by 35 per cent for the poorest wealth tercile of farmers but have no effect on the richest tercile, which is likely to be better protected against risk through its assets or access to credit. If anything, this type of evidence suggests that risk and the lack of appropriate insurance or protection may well be one of the factors that keep poor people poor.

Risk strategies and policy responses

The perceived failure to keep consumption smooth in the face of risk and the long-run costs attached to existing risk strategies has also stimulated an increasing interest in finding appropriate policy responses, not least since insured risk appears to affect the ability of many poor people to grow out of poverty. Standard transfer schemes, such as food aid or cash transfers, may all provide protection against shocks. However, how existing risk strategies could be strengthened remains less explored (Dercon, 2004). For example, it is clear that self-insurance through savings could offer substantial benefits in terms of protecting against the consequences of risk. Even if insurance and credit markets were not functioning well, improving the availability of better savings product could assist poor people to improve their

risk management. Similarly, informal insurance schemes could be strengthened, for example by linking micro-insurance initiatives to indigenous group-based systems. A number of insurance-related initiatives have been taken in this respect by governments and NGOs. For example, weather-indexed insurance contracts that trigger payments if rainfall falls below a predetermined level are being piloted in a number of countries. Health insurance schemes, often based at local health facilities, have also become more widespread. Much work is still needed on developing these initiatives, not least in terms of evaluating their effectiveness with appropriate techniques, for example randomized interventions.

STEFAN DERCON

See also **famines; Kahneman, Daniel; micro-credit; risk; risk aversion; risk sharing.**

Bibliography

Deaton, A. 1992. *Understanding Consumption.* Oxford: Clarendon Press.

Deaton, A. and Paxson, C. 1994. Intertemporal Choice and inequality. *Journal of Political Economy* 102, 437–67.

Dercon, S., ed. 2004. *Insurance against Poverty.* Oxford: Clarendon Press.

Fafchamps, M. and Lund, S. 2003. Risk-sharing networks in rural Philippines. *Journal of Development Economics* 71, 261–87.

Fafchamps, M., Udry, C. and Czukas, K. 1998. Drought and saving in West Africa: are livestock a buffer stock? *Journal of Development Economics* 55, 273–305.

Genicot, G. and Ray, D. 2003. Group formation in risk-sharing arrangements. *Review of Economic Studies* 70, 87–113.

Jacoby, H. and Skoufias, E. 1997. Risk, financial markets and human capital in a developing country. *Review of Economic Studies* 64, 311–35.

Kahneman, D. and Tversky, A. 1979. Prospect theory: an analysis of decision under risk. *Econometrica* 47, 263–92.

Kochar, A. 1995. Explaining household vulnerability to idiosyncratic income shocks. *American Economic Review* 85, 159–64.

Ligon, E., Thomas, J. and Worrall, T. 2002. Informal insurance with limited commitment: theory and evidence from village economies. *Review of Economic Studies* 69, 209–44.

Morduch, J. 1995. Income smoothing and consumption smoothing. *Journal of Economic Perspectives* 9(3), 103–14.

Morduch, J. 2004. Consumption smoothing across space. In *Insurance against Poverty*, ed. S. Dercon. Oxford: Clarendon Press.

Paxson, C. 1992. Using weather variability to estimate the responses of savings to transitory income in Thailand. *American Economic Review* 82, 15–33.

Rosenzweig, M. and Binswanger, H. 1993. Wealth, weather risk and the composition and profitability of agricultural investments. *Economic Journal* 103, 56–78.

Rosenzweig, M. and Stark, O. 1989. Consumption smoothing, migration, and marriage: evidence from rural India. *Journal of Political Economy* 97, 905–26.

Rosenzweig, M. and Wolpin, K. 1993. Credit market constraints, consumption smoothing, and the accumulation of durable production assets in low-income countries: investment in bullocks in India. *Journal of Political Economy* 101, 223–44.

Skees, J., Varangis, P., Larson, D. and Siegel, P. 2004. Can financial markets be tapped to help poor people cope with weather risks. In *Insurance against Poverty*, ed. S. Dercon. Oxford: Clarendon Press.

Townsend, R. 1994. Risk and insurance in village India. *Econometrica* 62, 539–91.

Townsend, R. 1995. Consumption insurance: an evaluation of risk-bearing systems in low-income economies. *Journal of Economic Perspectives* 9(3), 83–102.

risk sharing

Any two agents may be said to share risk if they employ state-contingent transfers to increase the expected utility of both by reducing the risk of at least one. A very wide variety of human institutions seem to play an important role in risk sharing, including insurance, credit, financial markets, and sharecropping in developing countries.

To be precise, consider a set of agents indexed by $i = 1, \ldots, n$ each with von Neumann–Morgenstern utility function U_i and a finite set of possible states of the world $s = 1, \ldots, S$, each of which occurs with probability $p(s)$. For simplicity, suppose that each agent i receives a quantity of a single consumption good $x_i(s)$ in state s, thus receiving expected utility

$$EU_i(x_i) = \sum_{s=1}^{S} p(s) U_i(x_i(s)),$$

where x_i denotes the random variable, $x_i(s)$ denotes its realizations, and E is the expectation operator. We assume that U_i is strictly increasing, weakly concave, and continuously differentiable for all $i = 1, \ldots, n$, so that all agents are at least weakly risk averse. Define the risk faced by agent i to be a quantity

$$R_i(x_i) = EU_i(x_i) - EU(x_i).$$

This cardinal measure orders probability distributions in the same manner as Rothschild and Stiglitz (1970). We say that i faces *idiosyncratic risk* if $R_i(x_i) > 0$ and $\mathrm{corr}(U_i'(x_i), U_j'(x_j)) < 1$ for some j, where U_j' denotes j's marginal utility. If any agent i bears idiosyncratic risk, then there exists a set of state-contingent transfers of the consumption good between i and j, $\{\tau_i^j(s)\}$ which will strictly increase the expected utility of each, while strictly decreasing the risk of at least one of i and j. Implementing such transfers is risk sharing.

Full risk sharing

What might be termed *full risk sharing* (Allen and Gale, 1988; Rosenzweig, 1988) is a situation in which all idiosyncratic risk is eliminated. While agents may still face risk, this risk is shared, so that marginal utilities of consumption are perfectly correlated across all agents. Full risk sharing is a hallmark of any Pareto-efficient allocation in an Arrow–Debreu economy, provided that agents have von Neumann–Morgenstern preferences, no one is risk-seeking, and at least one agent is strictly risk averse.

Let us establish the necessity of full risk sharing for any interior Pareto-efficient allocation in a simple multi-period endowment economy. The environment is similar to that described above, but agents consume in several periods indexed by $t = 1, \ldots, T$, with agent i discounting future expected utility using a discount factor β_i. Different states of the world are realized in each period, with the probability of state $s_t \in \{1, \ldots, S\}$ being realized in period t allowed to depend on the period, and so given by $p_t(s_t)$. Then consider the problem facing a social planner, who assigns state-contingent consumption allocations to solve

$$\max_{\{(c_{it}(s))\}} \sum_{i=1}^{n} \lambda_i \sum_{t=1}^{T} \beta_i^{t-1} \sum_{s_t=1}^{S} p_t(s_t) U_i(c_{it}(s_t))$$

subject to the resource constraints

$$\sum_{i=1}^{n} c_{it}(s_t) \leq \sum_{i=1}^{n} x_{it}(s_t),$$

which must be satisfied at every period t and state s_t; the planner takes as given the initial state s_0 and a set of positive weights $\{\lambda_i\}$. By varying these weights one can compute the entire set of interior Pareto-efficient allocations (Townsend, 1987).

If we let $\mu_t(s_t)$ denote the Lagrange multiplier associated with the resource constraint for period t in state s_t, then the first order conditions for the social planner's problem are

$$\lambda_i \beta_i^{t-1} p_t(s_t) U_i'(c_{it}(s_t)) = \mu_t(s_t). \tag{1}$$

Since this condition must be satisfied in all periods and states for every agent, it follows that

$$U_i'(c_{it}(s_t)) = \frac{\lambda_j}{\lambda_i} \left(\frac{\beta_j}{\beta_i} \right)^{t-1} U_j'(c_{it}(s_t))$$

for any period t, any pair of agents (i, j) and any state s_t, so that $\mathrm{corr}(U_i'(c_{it}), U_j'(c_{jt})) = 1$, and we have full risk sharing.

Thus far, we've considered risk sharing only in the context of an endowment economy. However, the thrust of the claims advanced above holds much more generally. If we were, for example, to add production and some kind of intertemporal technology (for example, storage), the first order conditions of the planner's problem with respect to state-contingent consumptions (1) would remain unchanged – the effect of these changes would be that the Lagrange multipliers $\{\mu_t(s_t)\}$ would change. This is an illustration of what is sometimes called 'separability' between production and consumption, which typically prevails only when there is full risk sharing (see, for example, Benjamin, 1992).

Risk sharing can also be thought of as a means to smooth consumption across possible states of the world. This suggests a connection to the permanent-income hypothesis, which at its core involves agents smoothing consumption across periods. And indeed, it's easy to show that full risk sharing in every period implies the kind of smoothing across periods implied by the consumption Euler equation. However, the consumption Euler equation doesn't imply full risk sharing.

Tests of full risk sharing
The insight that Pareto-efficient allocation among risk-averse agents implies full risk sharing has led to tests of versions of (1). The usual strategy involves adopting a convenient parameterization of U_i, and then calculating the logarithm of both sides of (1). For example, if $U_i(c) = \frac{c^{1-\gamma}}{1-\gamma}$, with $\gamma > 0$, then this yields the relationship

$$\gamma \log c_{it}(s_t) = \log \frac{\mu_t(s_t)}{p_t(s_t)} - \log \frac{\lambda_i}{\beta_i} - t \log \beta_i.$$

$$(2)$$

This is a simple consumption function, which we would expect to be consistent with any efficient allocation. The quantity $\frac{\mu_t(s_t)}{p_t(s_t)}$ is related to the aggregate supply of the consumption good. Note that this is the only determinant of consumption which depends on the random state. This reflects the fact that the only risk borne by agents in an efficient allocation will be aggregate risk. The second term varies with neither the state nor the date, and can be thought of as depending on the levels of consumption that agent i can expect (in a decentralization of this endowment economy, λ_i could be interpreted as a measure of i's time zero wealth). The final term has to do with differences in agents' patience.

Now, suppose one has panel data on realized consumption for a sample of agents over some period of time. If we let $\tilde{c}_{it}$ denote observed consumption for agent i in period t, (2) implies the estimating equation

$$\log \tilde{c}_{it} = \eta_t + \alpha_i + \delta_i t + \varepsilon_{it},   (3)$$

where $\eta_t = \log \frac{\mu_t(s_t)}{\gamma p_t(s_t)}$, $\alpha_i = -\log \frac{\lambda_i}{\gamma \beta_i}$, $\delta_i = -\log \beta_i$, and ε_{it} is some disturbance term. Because this final disturbance term isn't implied by the model, it's typically motivated by assuming that it's related either to measurement error in consumption or to some preference shock.

The reduced form consumption eq. (3) can be straightforwardly estimated by using ordinary least squares, but this doesn't constitute a test of full risk sharing. To construct such a test, one typically uses data on other time-varying, idiosyncratic variables which would plausibly influence consumption under some alternative model which predicts less than full risk sharing. Perhaps the most obvious candidate for such a variable is some measure of income, for example the observed endowment realizations $\tilde{x}_{it}$ referred to in the model above. Then one can add (the logarithm of) this variable to reduced form as an additional regressor, yielding an estimating equation of the form

$$\log \tilde{c}_{it} = \eta_t + \alpha_i + \delta_i t + \phi \log \tilde{x}_{it} + \varepsilon_{it}   (4)$$

(Mace, 1991; Cochrane, 1991; Deaton, 1992; Townsend, 1994). Then full risk sharing and an auxiliary assumption that ε_{it} is mean independent of the regressors implies the exclusion restriction $\phi = 0$, which can be easily tested.

Partial risk sharing
Restrictions along the lines of (4) have been used to test for full risk sharing in a wide variety of settings, including within-dynasty risk sharing (Hayashi, Altonji and Kotlikoff, 1996) in the United States, risk sharing across countries (Obstfeld, 1994), risk sharing within networks in the Philippines (Fafchamps and Lund, 2003), and risk sharing across households in India (Townsend, 1994), Africa, or the United States (Mace, 1991). A typical finding is that the estimated response of consumption to income shocks is small but significant, leading one to reject the null hypothesis of full risk sharing. In this case it is tempting to interpret the estimated relationship as determining the response of consumption to income. However, this is generally a mistake. By rejecting the hypothesis of full risk sharing one also rejects the model which generated the hypothesis, so that theory no longer supports the interpretation of (4) as a consumption function.

Given this kind of evidence against full risk sharing, scholars have been led to devise and test alternative models in which some kind of friction leads to agents bearing some idiosyncratic income risk. Two promising frictions are private information and limited commitment. In the case of private information, realized or announced incomes may provide a useful signal regarding hidden actions or information, and thus an agent's consumption will optimally depend on this signal, leading to a balance between risk sharing and incentives (Holmström, 1979); Ligon (1998) tests this weaker risk-sharing hypothesis in three Indian villages, and is unable to reject it. In the case of limited commitment, an agent who receives an unusually large endowment realization may be tempted to renege on a pre-existing risk-sharing arrangement unless she receives a larger share of resources (Kocherlakota, 1996); a test of this model in the same three Indian villages by Ligon, Thomas and Worrall (2002) finds that this model predicts a response of consumption to income of just the right magnitude. Still, the construction, estimation, and testing of well-specified models which predict only partial risk sharing remains in its infancy.

ETHAN LIGON

See also **Euler equations; permanent-income hypothesis.**

Bibliography

Allen, F. and Gale, D. 1988. Optimal security design. *Review of Financial Studies* 1, 229–63.

Benjamin, D. 1992. Household composition, labor markets, and labor demand: testing for separation in agricultural household models. *Econometrica* 1, 287–322.

Cochrane, J. 1991. A simple test of consumption insurance. *Journal of Political Economy* 99, 957–76.

Deaton, A. 1992. *Understanding Consumption*. Oxford: Clarendon Press.

Fafchamps, M. and Lund, S. 2003. Risk-sharing networks in rural Philippines. *Journal of Development Economics* 71, 261–87.

Hayashi, F., Altonji, J. and Kotlikoff, L. 1996. Risk-sharing between and within families. *Econometrica* 64, 261–94.

Holmström, B. 1979. Moral hazard and observability. *Bell Journal of Economics* 10, 74–91.

Kocherlakota, N. 1996. Implications of efficient risk sharing without commitment. *Review of Economic Studies* 63, 595–610.

Ligon, E. 1998. Risk-sharing and information in village economies. *Review of Economic Studies* 65, 847–64.

Ligon, E., Thomas, J. and Worrall, T. 2002. Informal insurance arrangements with limited commitment: theory and evidence from village economies. *Review of Economic Studies* 69, 209–44.

Mace, B. 1991. Full insurance in the presence of aggregate uncertainty. *Journal of Political Economy* 99, 928–56.

Obstfeld, M. 1994. Risk-taking, global diversification, and growth. *American Economic Review* 84, 1310–29.

Rosenzweig, M. 1988. Risk, implicit contracts and the family in rural areas of low-income countries. *Economic Journal* 98, 1148–170.

Rothschild, M. and Stiglitz, J. 1970. Increasing risk: I. A definition. *Journal of Economic Theory* 2, 225–43.

Townsend, R. 1987. Microfoundations of macroeconomics. In *Advances in Economic Theory: Fifth World Congress*, ed. T. Bewley. Cambridge: Cambridge University Press.

Townsend, R. 1994. Risk and insurance in village India. *Econometrica* 62, 539–91.

Robbins, Lionel Charles (1898–1984)

Lionel Robbins, who in 1961 became Baron Robbins of Clare Market, was one of the major academic economists of the interwar period. He remained active after the Second World War but never really regained the centre of the stage that he had occupied. He was also a great public servant for his country, serving it well and loyally in many aspects of social, political and cultural life. He was truly a 'Renaissance man'.

Robbins was born in 1898 in Middlesex, the son of Rowland Richard Robbins – for many years President of the National Farmers' Union – and Rosa Marion Robbins. He spent one year reading for an Arts degree at University College London and then volunteered for war service with the Royal Artillery. He saw active service on the Western Front, was wounded and invalided back to England in 1918. He was an undergraduate at the London School of Economics and Political Science from 1920 to 1923, from which he graduated with a BSc (Econ.) degree, choosing political ideas as his major field of study, and having had the left-wing Harold Laski as his tutor. Beveridge employed him as a research assistant for a year, after which Robbins was a tutor in economics at New College, Oxford. He returned to teach economics at LSE from 1925 to 1927, then back to New College as Fellow (1928–9) and finally, at the incredibly young age of 31, back to the Senior Professorship in Economics at LSE to succeed Allyn Young.

Apart from government service during the Second World War, Robbins remained at LSE as Head of the Economics Department until 1960 when, on accepting the Chairmanship of the *Financial Times,* the University of London forced him to resign his professorship – a move than brought Robbins great personal distress, although he retained his connection with LSE and taught courses there until a year or so of his death in 1984.

Outside academic and government advisory activity, Robbins had a distinguished record in arts administration, being connected with both the National Gallery and the Royal Opera House, but he may perhaps be best remembered, in such 'outside' activities, for his contribution to the structure of higher education in the United Kingdom. He chaired the committee – commonly referred to as the Robbins Committee on Higher Education – that proposed the criterion that all qualified applicants should receive a place, and financial support, to read for a degree. The acceptance of the 'Robbins Principle' led to a vast expansion of degree courses, especially in the social sciences in the 1960s and early 1970s in the UK.

Robbins's contributions to economics may be considered under four headings; economic theory, methodology and philosophy of economics, the theory of economic policy, and the history of economic thought.

Those who only knew Robbins later in his life often forget that he made his initial mark in economics as a theorist. Three contributions here are worth noting; he launched a sustained attack on Marshall's concept of the Representative Firm which was apparently so successful that it drove the concept out of the pages of microeconomic texts. Robbins basically argued that the understanding of the equilibrium neither of the firm nor of the industry was aided by introducing the Representative Firm, hence it should be eliminated from analysis. More recent work has shown a greater sympathy towards Marshall's construct and it seems clear now that Robbins failed to understand the exact dynamic problem that Marshall was trying to cope with and why the Representative Firm was an important contribution to this problem.

Robbins also pioneered the micro-analysis of the labour supply function. Although he did not explicitly use the division of a wage change into an income and substitution effect, he showed clearly why the sign on the response of hours to a real wage rate change would be ambiguous.

In macroeconomics Robbins was a firm exponent of the Austrian theory of the trade cycle and here he was greatly influenced by Frederick von Hayek, whom he brought to LSE from Vienna in 1928. The central feature of the Austrian analysis was that depression was due primarily to under-saving (or excess consumption) and these views, which Robbins expounded as an explanation of the 1930s depression in his book *The Great Depression*, led to a head-on collision between the senior LSE economists and the Cambridge School centred around Keynes. This rift was not finally healed until the wartime collaboration in Whitehall between Robbins and Keynes. After the war in the Marshall Lectures for 1946, published as *The Economic Problem in Peace and War*, Robbins announced his conversion to full employment policies via control of aggregate demand, although it is not clear that he became a Keynesian.

The second area where Robbins made a major contribution and where he wrote what is probably his best known work in economics was that of the methodology and philosophy of economics. His *Nature and Significance of Economic Science* was one of the most cited, if not most read, books on the subject in the period 1932–60, and it influenced greatly economists' views about the nature of their discipline. There are several strands to the book, none original in themselves, but Robbins put them together in beautifully clear prose and in a very persuasive manner. The major themes were; first, that economic science could be clearly demarcated from those discussions of economic issues that involved value judgements – by which latter term Robbins meant evaluative statements of the form 'better or worse' where interpersonal comparisons of utility were involved. He also argued that there was a clear demarcation between economic science and other branches of social enquiry such as social psychology, sociology, politics and so on.

The second major theme was that the subject matter of economic science was not a particular activity (for example, Cannan's view that economics was the science of wealth), but rather an aspect of all human conduct. This aspect was the 'fact' of economic scarcity – a manifestation of unlimited ends on the part of individuals and society and means of satisfying those ends that were limited in supply. In

words so often quoted in economics texts Robbins defined economic science as 'that science that studies the relationship between ends and means that have alternative uses' – a definition that is more than reminiscent of Menger's exposition of the economizing process.

These two aspects of the *Nature and Significance* were widely accepted by the world of academic economists and are still propagated. But they have always had their critics; in particular, the view that there is a body of scientific economics 'free from value' is much disputed.

The third aspect of the book – Robbins's views on the procedures for checking the validity of economic theory – was less fortunate in its effect on the development of the subject. Robbins appeared to argue that the central propositions of economics were derived from very basic, and obvious, assumptions and a process of logical deduction from these assumptions. Moreover, these deductions gave essentially qualitative predictions. Robbins expressed great scepticism about the feasibility and meaningfulness of quantitative work in economics, and by the implication of his message inhibited the development of econometric testing in economics.

Robbins's contributions to discussions of economic policy were basically consistent throughout his career, although the purity of his earlier thoughts was muddled as he grew older. His major policy theme was his advocacy of, what may be loosely termed, economic liberalism. Robbins decreasingly argued this on the grounds of some alleged theoretical or a priori superiority of market solutions over collectivist or interventionist plans, but rather as an empirical point that the liberal solution seemed best to combine liberty and efficiency. In his earlier writings, for example *The Economic Causes of War* (1939a) and *The Economic Basis of Class Conflict* (1939b) he adopted an extreme free trade position and it was this stance as much as macro-theory debate that led to his conflict with Keynes in the 1930s. His later work revealed a much greater readiness to allow ad hoc exceptions to strict economic liberalism – he espoused, among other measures, the Beveridge plan, grants for higher education, subsidies for the arts, control of the exports of works of art, overall macro-control for full employment. Probably the most rounded statement of his policy beliefs is to be found in his *Economic Problem in Peace and War*.

Finally, mention must be made of Lionel Robbins's contribution to the teaching and study of the history of economic thought. He, together with one or two other scholars of his generation – like his great friend, Jacob Viner – kept interest in the subject alive and flourishing when many economists regarded it, as they still do, as irrelevant to their studies. Much of his influence came via his masterly teaching of the subject and via the important theses that were produced under his supervision, as much as from his own specific contributions. He also aided the production of important series in the history of economic thought such as the LSE reprints and the collected works of Bentham and J.S. Mill.

Of his specific contributions, two are minor classics, his *Theory of Economic Policy in Classical Political Economy* (1952) and *Robert Torrens and the Evolution of Classical Economics* (1958). In the former work, Robbins argued very persuasively, if not entirely convincingly, that the British classical economists did not adhere to the Continental laissez-faire dogma but rather argued for freedom in economic relationships as a general principle with many ad hoc exceptions. He further tried to clear them of any anti-working class bias.

The book on Torrens is a perfect example of how to survey the collected works of a writer who, though not of the first rank of classical economists, is nonetheless a useful writer by whom to assess the general achievement of the classical school.

B.A. CORRY

See also **value judgements.**

Selected works

1932. *An Essay on the Nature and Significance of Economic Science.* London: Macmillan.

1934. *The Great Depression.* London: Macmillan.

1937. *Economic Planning and International Order.* London: Macmillan.

1939a. *The Economic Causes of War.* London: Jonathan Cape.

1939b. *The Economic Basis of Class Conflict.* London: Macmillan.

1947. *The Economic Problem in Peace and War.* London: Macmillan.

1952. *The Theory of Economic Policy in English Classical Political Economy.* London: Macmillan.

1958. *Robert Torrens and the Evolution of Classical Economics.* London: Macmillan.

1971. *Autobiography of an Economist.* London: Macmillan.

Robertson, Dennis (1890–1963)

Dennis Robertson was born in 1890, the son of a clergyman and schoolmaster, and was educated at Eton and Trinity College, Cambridge. After taking a Part I in Classics and Part II in Economics he was elected a Fellow of Trinity College in 1914 and in 1930 became a Reader in the University of Cambridge. He left Cambridge in 1938 to become a Professor in the University of London but during most of his time in that post he was seconded to the Treasury on war-related work. Elected in 1944 to succeed Pigou in the Chair of Political Economy, he returned to the University of Cambridge, holding that position until his retirement in 1957. He died in Cambridge in 1963.

Economics in Cambridge when Robertson commenced working at it was dominated by Marshall. Not by the man himself (although still alive he had retired in

1908) but by his analytical methods and by his *Principles of Economics*. It was quite natural that the topic selected by Robertson for his fellowship dissertation should involve a 'Marshallian' approach to a subject on which Marshall himself had written relatively little: the nature and causes of fluctuations in the general level of economic activity. As Robertson recorded in the introduction to the published version of this dissertation:

> In some of the more abstract portions of this essay I shall make use, without further explanation or apology, of the processes and terminology in common use among the school of economic thought associated in this country chiefly with the name of Dr Marshall. My reason is that after a study of many facts and theories I am deliberately of the opinion that one cause of the obscurity which still surrounds this problem is that in the attack upon it full and systematic use has never hitherto been made of the weapons supplied by this particular intellectual armoury. (1915, p. 11)

Although Robertson did not suspect it then, the refinement and further development of the ideas about cycles and growth in economic activity presented in this study were to occupy him for the next 20 years. Two different sorts of factors led him in this direction. The first was the need to develop a framework for designing an organized policy response to the large-scale dislocation of economic life which had followed the end of the First World War, while the second was a more specific, personal, influence. In the early 1920s Keynes commissioned him to write an introductory textbook (in the Cambridge Economic Handbook series) to be entitled, simply, 'Money'. The difficulties he encountered in attempting to provide an elementary account of monetary theory made Robertson particularly aware that, even in its more sophisticated variants, existing theoretical work provided an inadequate basis for dealing with the economic problems of the 1920s. The combined influence of these two resulted in a prolonged period of reflection and research, yielding a series of loosely related publications which recorded the development of a fairly comprehensive analytical scheme interrelating the problems of money, the trade cycle, economic growth, and the role of the state in promoting economic progress.

Robertson's approach to this analysis involved the development of successively more complicated, more 'realistic' models of economies each of which constituted a different, abstract, 'type'. All 'types' shared the characteristics that production was undertaken on the basis of 'rational' decision-making by competing producer 'groups', each making different products with a fixed labour force and a productive process involving fixed capital. Now although each type of economy was both a *production* and an *exchange* economy it was the possibility that these activities could be 'organized' in different ways that distinguished the different types. Production could be organized in two ways, cooperatively or non-cooperatively, while exchange could also be organized in two ways, direct exchange or monetary exchange. In total there were, then, four types of economies. The distinction between the two types of productive organization turned on the decision-making functions of the members of the groups: in a cooperative group decisions were *made* and *carried out* by the group members acting together, while in a non-cooperative group 'entrepreneurs' made decisions and 'workers' carried them out. In respect of the organization of exchange it was on the existence and use of money that the distinction rested, in one case exchange was carried out by 'direct barter', while in the other, money supplied through a (potentially) government-controlled banking system provided the means of exchange.

Robertson's basic analytical building block was the 'cooperative non-monetary economy', an economy where each competing industrial group made its employment and, thus, output decision cooperatively, and exchanged its output without the use of money. Although in such an economy no distinction was made between the members of the group, a distinction was made between two different categories of producer groups, those providing consumer goods and those producing capital goods. The first group, consumer goods producers exchanged some of their output with the second group for capital goods, thereby providing consumption goods for capital goods producers. Now an economy of this type, Robertson argued, would experience cyclical fluctuations in aggregate output deriving from the effect of gestation lags on the time pattern of the supply of capital goods and of the durability of capital goods on the time-pattern of demand for their replacement.

A non-cooperative non-monetary economy would, though, experience fluctuations of even greater severity than those felt in an otherwise identical cooperative economy. This proposition derived directly from the fact that in a non-cooperative economy production decisions were taken by entrepreneurs who hired workers to carry them out, and workers and entrepreneurs had differing 'interests'. These divergent interests were reflected most importantly in the different utility attached to leisure by the two classes. An entrepreneur, for example, would wish to expand output further in the boom and contract it further in the slump than the workers in his group; and since entrepreneurs were in control, their interests prevailed. Although the degree of fluctuation in the non-cooperative economy was more pronounced than in the cooperative, Robertson adopted it as the benchmark which defined the 'appropriate' degree of fluctuation to be aimed at by policymakers concerned with stabilization. He did so because he maintained that the failure to recognize that production was, in practice, organized non-cooperatively could lead to an attempt to reduce fluctuations too much. Such attempts, by altering the structure of incentives, could damage the longer-run growth possibilities of the economy.

The cooperative monetary economy construct was built directly on to the foundations provided by the cooperative non-monetary economy and this type of economy exhibited, therefore, a cyclical pattern in the production of fixed capital which generated cyclical fluctuations in output as a whole. Now the introduction of money also required a slight change of focus, since in the monetary case Robertson concentrated not on fixed capital but on the demand for circulating capital, essentially on the demand for consumption goods which were consumed by those engaged in the process of production. This concern with circulating capital was necessarily associated with the analysis of monetary economies because Robertson made the assumption (reflecting British banking practice) that it was with the finance of the acquisition of circulating capital that the banking system was concerned. His analysis then described the policies which, if adopted by the banking system, would lead to fluctuations being of no greater amplitude than in the corresponding non-monetary economy. A failure to implement such policies would lead to fluctuations in the price level, and thus in output, of greater magnitude than was 'appropriate'.

The difference made by the substitution of cooperation in the monetary type turned principally on the effect on decision-making of changes in income distribution. In particular, it was assumed that only entrepreneurial incomes adjusted quickly to changes in the price level, so that variations in the price level over the cycle were an additional source of influence on production decisions. The nature of this influence led entrepreneurs to expand their activities further in the boom (as rising prices increased their profits) and contract them further in the slump (as falling prices reduced their profits) than would have been the case in the corresponding cooperative economy. But these changes in income distribution were not permanent. In the course of the boom workers managed to restore real wages to pre-recovery levels, the expansion of output would be slowed, and in the slump, as entrepreneurs restored profits to their pre-depression levels, the contraction of output would be slowed. The end of the boom and the slump, though, if an 'appropriate' monetary policy were adopted, would be dictated by the behaviour of the underlying non-cooperative non-monetary economy. So non-cooperation in the monetary case had additional effects only on the amplitude of cyclical fluctuations.

Robertson also developed a set of tools to analyse the process of cyclical change in monetary economies. He divided time up into a sequence of market periods (during each of which the supply of goods was fixed) and then focused on the dynamics of the transfer of resources from current consumption by those already in employment to those newly employed to increase output during the expansion phase of the cycle. The mechanism generating this transfer was a price-level increase as the newly employed (whose wages had been borrowed from the banking system) outbid the existing employed on the goods market. Robertson's aim was to show how the magnitude of the price-level increase was determined and the nature of the monetary policies which could be adopted in order to minimize it. The rate of inflation was shown to depend upon the relationship between the rate at which the banking system made new loans to producers and the rate at which this new money was absorbed into the money-holdings of the existing employed. The faster the new money was absorbed, that is, the faster that the existing employed gave up their claims on current output, the smaller the rise in the price-level accompanying the transfer of resources from the public to the expanding producers. To the extent that this money was not immediately absorbed, the existing employed were *forced* to share current output with those producers by price-level changes. By minimizing these changes, then, the monetary authority through its control of the banking system would also be able to minimize the amount of 'forced' saving which accompanied the recovery. A similar approach was also applied to the non-cooperative case, but here policy design was more difficult because the inflation led to changes in the distribution of income between workers and entrepreneurs. Even so, monetary policy could play a useful role in reducing fluctuations to their 'irreducible' non-monetary amplitude.

The central concern of Robertson's analytical work was to provide an explanation of fluctuations in aggregate activity which was closely linked to a broader concern, that of remedying the adverse effects of such fluctuations. The identification of the use of capitalistic (though not necessarily capitalist) production methods as the source of fluctuations, though, left with a rather ambivalent attitude to possible remedies: capitalistic production methods always produced cycles, but also brought with them the possibility of economic progress. And he thought that there was a trade-off between these two, greater stability being associated with slower growth, less stability with faster growth:

> From some points of view the whole cycle of industrial change presents the appearance of a perpetual immolation of the present upon the altar of the future. During the boom sacrifices are made out of all proportion to the enjoyment over which they will ultimately give command: during the depression enjoyment is denied lest it should debar the possibility of making fresh sacrifices. Out of the welter of industrial dislocation the great permanent riches of the future are generated. (1926, p. 254)

He concluded that the choice between these two conflicting goals was ultimately a question of: 'ethics, rather than economics'.

The theoretical framework sketched above had emerged by the early 1930s. But its further development was interrupted by the publication in 1936 of Keynes's *General Theory of Employment, Interest and Money.*

Robertson's response to this book was to examine how the *General Theory* might affect his vision of how the world worked. The central issue for Robertson was whether Keynes had provided a more satisfactory explanation than he had himself of the forces which determined the behaviour of the trend rate of growth of economic activity. The distinguishing feature of Keynes's approach identified by Robertson was in the treatment of the theory of the rate of interest. He interpreted as Keynes's central proposition the contention that there was an inherent tendency for the rate of interest to remain above the level consistent with the maintenance of full employment. And although Robertson was prepared to accept that an argument could, in principle, be made out along such lines he did not accept that Keynes had succeeded in doing so. In particular he maintained that while 'liquidity preference' might make the interest rate 'sticky' in the short period, with its downward movement resistant to monetary expansion, he rejected such an approach to the long-period theory of interest rate determination, summarizing the argument in the following way:

> ... the rate of interest is what it is because it is expected to become other than it is; if it is not expected to become other than it is, there is nothing left to tell us why it is what it is. The organ which secretes it has been amputated, and yet it somehow still exists – a grin without a cat. ('Mr Keynes and the Rate of Interest' in *Essays in Monetary Theory*, 1940, p. 36)

Keynes's theoretical argument was, therefore, flawed. And the associated case for stabilizing the economy at a level other than that identified in Robertson's own analysis as 'appropriate' was consequently not proven.

The first repercussion of this reaction to the *General Theory* was an estrangement between Robertson and Keynes, virtually ending a close friendship which had lasted for more than 20 years (Robertson having been a student of Keynes, then a fellow teacher and collaborator in research). It then motivated Robertson's decision to leave Cambridge for London in 1938. Moreover, even after Keynes's death in 1946, strained and difficult relations with Keynes's disciples in Cambridge left him a somewhat isolated figure. The impact of Keynes's *General Theory* on Robertson's professional life was no less significant, the whole terrain of the area in which he worked was changed: from being on the creative frontier of the subject he felt himself forced into the role of commentator and critic. In the years after 1936 he wrote almost nothing new in what had been his specialist field. An explanation was provided in a letter to a friendly reviewer of one of his collections of essays who had called upon Robertson to prepare a monograph combining and extending his earlier analytical work, and to whom he wrote:

> ... I'm afraid there is no chance of my responding to your challenge and trying to produce a full length synthetic Theory of Money or Fluctuations or What-you-will. I'm

too old and too lazy! But even if I were younger and less lazy, I think history had made it impossible. I believe that once Keynes had made up his mind to go the way he did it was my particular function to … [elucidate and criticise the details of his work] … and to go on pegging away at them (as is still necessary). It will not be easy for *anyone* for another twenty years to produce a positive and constructive work which is not in large measure a commentary on Keynes, – that is the measure of his triumph. For me, it would now be psychologically impossible, and the attempt is not worth making. (Private letter of D.H. Robertson to T.J. Wilson, 31 October 1953)

M. ANYADIKE DANES

Selected works

1915. *A Study of Industrial Fluctuation.* London: P.S. King & Son. Reprinted with a new introduction, in *Reprints of Scarce Works on Political Economy*, London: London School of Economics and Political Science, 1948.

1922. *Money.* Cambridge Economic Handbook. London: Nisbet & Co. Revised edn, 1924; new edn, 1948.

1923. *The Control of Industry.* Cambridge Economic Handbook. London: Nisbet & Co. Rev. edn, 1928; new edn (with S.R. Dennison), 1960.

1926. *Banking Policy and the Price Level.* London: P.S. King & Son. Reprinted, 1926; reprinted with revisions, 1932; reprinted in the USA with a new preface, New York: Augustus M. Kelley, 1949.

1931a. *Economic Fragments.* London: P.S. King & Son.

1931b. (With A.C. Pigou.) *Economic Essays and Addresses.* London: P.S. King & Son.

1940. *Essays in Monetary Theory.* London: P.S. King & Son.

1950. *Utility and All That.* London: George Allen & Unwin.

1956. *Economic Commentaries.* London: Staples Press.

1957–9. *Lectures on Economic Principles.* 3 vols. London: Staples Press. Paperback edn in one volume, London: Fontana, 1963.

1960. *Growth, Wages, Money.* The Marshall Lectures at the University of Cambridge. London: Cambridge University Press.

Bibliographic addendum

Fletcher, G. 2007. *Dennis Robertson: Essays on his Life and Work.* Basingstoke: Palgrave Macmillan.

Robinson Crusoe

Written by Daniel Defoe, *Robinson Crusoe* was first published in 1719–20. By the end of the 19th century there were many references made to a Crusoe economy to illustrate the principles of supply and demand economic theory. Crusoe thus became a representative rational economic individual, allocating his available resources to obtain maximum satisfaction in the present or future.

The figure of Crusoe as the personification of supply and demand economic theory can be found in W.S. Jevons's *Theory* (1871), C. Menger's *Principles* (1871), P. Wicksteed's *Alphabet of Economic Science* (1888), E. Böhm-Bawerk's *Theory of Capital* (1890), A. Marshall's *Principles* (1891) and K. Wicksell's *Value, Capital and Rent* (1893). The principal uses of the device were to show how an isolated individual would allocate consumption items so as to maximize utility in a marginalist fashion and distribute labour effort between producing items for consumption or investment (creating 'capital'). Calculations were made according to the relative amounts of pleasure and pain immediately or ultimately involved in the various activities. Marshall also used Crusoe to illustrate producer and consumer surplus, while F.Y. Edgeworth's *Mathematical Psychics* (1881) introduced 'the black', Friday, when discussing issues in the theory of commodity exchange.

The role of a Crusoe economy was not simply to illustrate various components of supply and demand theory. It was also utilized to support the claim that the principles of rational behaviour, as defined by that theory, could be applied to any type of economy – from the isolated individual to 'modern civilization'. This point was made particularly clear in J.B. Clark's *The Distribution of Wealth* (1899). Similar references to a Crusoe economy can be found in textbooks today.

Two general characteristics of an economic Crusoe's actions are important to note. First, he must be able to calculate in a precise fashion making fine decisions between whether to work or rest, to consume or save/invest. Second, he has no resources other than those available in the island environment. Both characteristics mean the economic Crusoe bears no relation to the Crusoe in Defoe's novel. Defoe's Crusoe wastes time because he cannot calculate in a marginalist fashion; he cannot rationally allocate labour time because labour is as useful in one pursuit as another; and he would not have survived without items salvaged from the shipwreck. Other decisions, such as whether to consume or save, also preclude calculation on the basis of relative pleasure and pain (White, 1982). Moreover, the relation between Crusoe and Friday is not based on voluntary reciprocal exchanges, as in the supply and demand parable, but rather on power and violence (Hymer, 1980). The economic Crusoe could not, therefore, have been produced by relying on the letter of Defoe's text.

It is possible to find some references to Crusoe by English political economists during the 1830s, but these were sporadic and no attempts were made to utilize Crusoe in a systematic fashion. An economic Crusoe thus appears only after mid-century with references in F. Bastiat's *Economic Harmonies* (1850) and H. Gossen's *Entwickelung* (Gossen, 1854). These references owed a good deal to the rewriting of Defoe's text within the literary genre of the 'Robinsonade'.

The Robinsonade literature dates from the early 18th century (Gove, 1941) and includes voyage or shipwreck narratives, imaginary voyages to 'isolated lands' and more general discussions of colonial settlements which depicted various stages of societal development. The last group of Robinsonade texts bears some resemblance to the four-stage theory of societies produced during the Scottish Enlightenment, remnants of which can be found in the work of the classical political economists (Meek, 1976). One such remnant was the illustrative device, used by A. Smith and D. Ricardo, of hunters exchanging commodities according to the labour embodied in them. While Marx was critical of this device, he noted it made sense in the context of the previous century's Robinsonades. However he considered the later discussion of Crusoe by Bastiat for example, was 'twaddle' because it depicted an individual 'outside society' (Marx, 1857–8, pp. 83–5).

Bastiat's Crusoe relied on a different type of Robinsonade literature, particularly J.H. Campe's *Robinson the Younger* (1779/80). Campe rewrote Defoe's tale to show Crusoe's survival on the island was not dependent on the shipwreck items. Gossen also appealed to Campe's novel to illustrate his marginalist explanation of work and consumption decisions. By the mid-19th century, then, the 'individualist' Robinsonade was utilized by those theorists who conceptualized the economy as a series of voluntary exchanges, where the principles of economic activity were those of the individual writ large.

English supply and demand economists could also draw upon a discernible shift in the readings of Defoe's text by literary commentators after 1850. Earlier commentary had stressed the novel was useful for showing, especially to children and the 'working classes', the virtue of work and the need to accept the given social organization ordained by Divine Providence. Commentary after mid-century represented Crusoe more as an individual calculating costs and benefits in the manner of an English shop keeper. It was even argued Crusoe represented a 'national ideology' in that regard. The remarkable similarity between this Crusoe and the illustrative device of English supply and demand economic theory suggests the latter was able to appropriate the former as a recognizable referent.

The economic Crusoe served, in effect, as a useful defensive device against 'historical' critics of economic theory such as T.E. Cliffe Leslie and J.K. Ingram. Writing between the mid-1860s and early 1880s, the critics argued that there were no universal laws of economic behaviour since behaviour could change according to the type of society being considered. Supply and demand theorists, such as Jevons, rejected that criticism, claiming historical studies could only confirm the 'universal' laws of behaviour assumed in the theory (Jevons, 1876, pp. 196–7). In this context, the economic Crusoe provided an apparently tangible reference point when supply and demand theory began its analysis with the actions of an 'isolated' or representative individual. Indeed, Gossen had used Campe's Crusoe in precisely that fashion when criticizing

the German 'National Economists' in 1854 (Gossen, 1854, pp. 45–7). The role of an economic Crusoe, as both illustrative and defensive device for supply and demand theory, was thus inscribed from its inception.

M.V. WHITE

See also **economic man; rational behaviour.**

Bibliography

Gossen, H.H. 1854. *Entwickelung der Gesetze des menschlichen Verkehrs und der daraus fliessenden Reglen für menschliches Handeln*. Brunswick: Vieweg.
Gove, P.B. 1941. *The Imaginary Voyage in Prose Fiction*. New York: Columbia University Press.
Hymer, S. 1980. Robinson Crusoe and the secret of primitive accumulation. In *Growth, Profits and Property*, ed. E.J. Nell. New York: Cambridge University Press.
Jevons, W.S. 1876. The future of political economy. In *The Principles of Economics and Other Papers*, ed. H. Higgs. London: Macmillan. 1905.
Marx, K. 1857–8. *Grundrisse*. Harmondsworth: Penguin, 1973.
Meek, R.L. 1976. *Social Science and the Ignoble Savage*. Cambridge: Cambridge University Press.
White, M.V. 1982. Reading and rewriting. The production of an economic Robinson Crusoe. *Southern Review* 15(2), 115–42.

Robinson, Edward Austin Gossage (1897–1993)

Austin Robinson was educated at Marlborough College and Christ's College, Cambridge. During the First World War he served as a pilot in the RNAS and the RAF. After finishing his studies at Cambridge he became a Fellow of Corpus Christi College, from 1923 to 1926. In 1926 he married Joan, daughter of Major-General Sir Frederick Maurice and later to become the eminent economist. From 1926 to 1928 Robinson was tutor to the Maharaja of Gwalior. He returned to Cambridge as a university lecturer in economics in 1929, and from then on was an important figure on the Cambridge economics scene. He became Professor of Economics in 1950. He retired in 1965 (and it so happened that Joan Robinson was appointed to his chair). After his retirement, he continued to play a prominent role in Cambridge economics, as well as on the national and international scene.

Austin Robinson's first book, *The Structure of Competitive Industry* (1931), established his reputation as an economist. This seminal work drew on Alfred Marshall's writings on industry, and considered in detail the problems involved in determining the optimum size of firm. But although it emphasized the importance of scale, and inspired much of the later empirical work in this area, it also recognized that low British productivity in manufacturing industry was not primarily the consequence of scale, but of attitudes towards work and competition. All subsequent writing on this subject owed a considerable debt to Robinson. He followed up his work on competitive industry with a book on *Monopoly* (1941), as well as with a number of articles, including work on Africa. He was a member of the group surrounding Keynes when he was formulating the *General Theory,* and wrote a review of it in *The Economist,* insisting on signing it (against the traditions of the paper) because of the exceptionally controversial nature of the subject.

Robinson's long association with the *Economic Journal* began in 1934, as Assistant Editor to Keynes, and was later to be followed by much editorial work. Robinson did distinguished service during the war. He was a member of the Economic Section, War Cabinet Office, from 1939 to 1942, and from 1942 to 1945 was Economic Adviser and Head of Programmes Division, Ministry of Production. This was followed by a period as Economic Adviser to the Board of Trade. He returned to Cambridge in 1946, but served a further period in government on the Economic Planning Staff from 1945 to 1947. He was joint editor of the *Economic Journal* from 1944 to 1970, and played a leading role in the profession in other ways, holding a number of important posts, including that of managing editor of the Royal Economic Society's edition of Keynes's works. He was much involved in the work of the new International Economic Association: he was President from 1959 to 1962 and editor of its publications for many years. A good deal of his subsequent writing and editorial work, much of it on the problems of developing countries, was carried out in the context of the work of the IEA.

Austin Robinson's career was a remarkable one. He combined writing, teaching, editorial work and administration with advising governments in both the developed and developing world. He played a leading role in the economics profession for an exceptionally long period, internationally as well as in Britain, and did so throughout with much distinction.

Z.A. SILBERSTON

See also **Royal Economic Society.**

Selected works

1931. *The Structure of Competitive Industry.* London: Nisbet & Co.; Cambridge: Cambridge University Press.
1941. *Monopoly.* London: Nisbet & Co.; Cambridge: Cambridge University Press.

Robinson, Joan Violet (1903–1983)

Joan Robinson (née Maurice) was born at Camberley, Surrey, on 31 October 1903. She died in Cambridge on 5 August 1983.

She is the only woman (with the possible, but controversial, exception of Rosa Luxemburg) among the great economists. In 1975, which was proclaimed Woman's Year, most economists in the United States expected that she would naturally be chosen for the Nobel Memorial Prize in Economics for that year. She had received triumphant acclaim, as a Special Ely Lecturer, at the American Economic Association annual meeting three years earlier, in spite of the harsh hostility that her theories had always met in the United States. The American magazine *Business Week,* after sounding out the American economics profession, felt so sure of the choice as to anticipate the event by publishing a long article on her, presenting her explicitly as being 'on everyone's list for this year's Nobel Prize in Economics'. But the Swedish Royal Academy missed that opportunity (and alas, never regained it). Ever since, in shop-talk among economists all over the world, Joan Robinson has become the greatest Nobel Prize winner that never was.

Basic biography

Joan Robinson was the daughter of Major General Sir Frederick Maurice and of Helen Marsh (who was herself the daughter of a Professor of Surgery and Master of Downing College, Cambridge). Sir Frederick pursued a brilliant career in the British Army, but in 1918 he found himself at the centre of a public debate, and he gave up his army career on a point of principle. This was very much in the family tradition. Sir Frederick's grandfather – Joan Robinson's great-grandfather – was Frederick Denison Maurice, the Christian Socialist who lost his chair of theology at King's College London, for his refusal to believe in eternal damnation.

Joan Robinson certainly had many of these traits: toughness and endurance of character, nonconformism and unorthodoxy of views, the absence of any reverential feeling or timidity, even in the face of the world's celebrities, a passionate longing for the new and the unknown.

She was educated at St Paul's Girls' School in London. (Curiously enough, Richard Kahn was educated in the boys' section of the same school.) In October 1922, she was admitted to the University of Cambridge, going up to Girton College, where she read economics at a time when the dominant figures in Cambridge were Marshall and Pigou. Marshall had retired (he died in 1924) but was extremely influential not only in Cambridge but in the whole of the British Isles. Pigou, his favourite pupil and chosen successor, was the Professor of Political Economy, at whose lectures Cambridge students absorbed the official *verbum* of Marshallian economics. Keynes was a sort of outsider, part-time in Cambridge and part-time in London, always involved with government policies, either at the Treasury or in public opposition. In those days he lectured on strictly orthodox monetary theory and policies. His lectures were not given regularly but were well attended.

The intellectual environment must have appeared solidly traditional. Joan graduated in 1925, as a good girl would: with second class honours.

In the following year (1926), she married E.A.G. Robinson (later Professor Sir Austin Robinson), who was six years her senior and at the time a junior Fellow of Corpus Christi College. Together they left Cambridge and set off for India, where they stayed for two years. Austin Robinson served as tutor of the Maharajah of Gwalior. Joan was there as Austin's wife but did some teaching at the local school. When they returned, after their two-year Indian engagement, Austin Robinson took a permanent post as Lecturer in Economics at Cambridge, where they settled for life. They had two daughters.

It was on the return to Cambridge from India (summer 1928) that Joan Robinson began to do some College supervision of undergraduates, and then to do economics research in earnest. The Cambridge intellectual environment had changed dramatically. After Edgeworth's death (1926), Keynes became the sole editor of the *Economic Journal* and was engaged on his *Treatise on Money* (Keynes, 1930). Most of all, he had brought to Cambridge Piero Sraffa, the young Italian economist who had dared to launch a scathing attack on Marshallian economics (Sraffa, 1926). Moreover, some new stars were rising in the firmament of Keynes's entourage – Frank Ramsey, the brilliant mathematician; Ludwig Wittgenstein, the Austrian philosopher whom Keynes persuaded to come to Cambridge; and Richard Kahn, Keynes's favourite pupil. It was with Richard Kahn that Joan Robinson began an intense intellectual partnership that lasted for her whole life.

On a strictly academic level, Joan Robinson slowly ascended the academic ladder: Junior Assistant Lecturer in 1931, Full Lecturer in 1937, Reader in 1949. It was suggested in Cambridge that the fact that her husband was in the same faculty kept her back at all stages of her academic career. She became full professor only on Austin Robinson's retirement, in 1965. Her association with the Cambridge colleges was more irregular. But she was, in succession, a Fellow of Girton College and of Newnham College. Yet whatever the formal position in the faculty or in the Cambridge colleges, she was for years one of the major attractions in Cambridge for many generations of undergraduates, not only in economics. In the post-war period, she was certainly the best-known member of the Cambridge economics faculty abroad. An indefatigable traveller, she did not limit her foreign visits to universities; she also wanted to know local customs and local conditions of life, even far away from urban centres. Her strong constitution and temperamental toughness helped her enormously. A friend from Makerere University, who took her, when she was already in her seventies, on a month's travel in tribal Africa was amazed at how much she could endure in terms of living in most primitive conditions with raw food, lack of facilities and exposure to harsh tropical weather, day and night.

It would be impossible to list here all the places she visited or the talks, seminars and public lectures she gave all over the world. She rarely stayed in Cambridge during the summer or term vacations or during her sabbatical years, though punctually and punctiliously returning there on the eve of the terms of her teaching. Asia was her favourite continent (especially India and China). But hundreds of students in North and South America, Australia, Africa and Europe also knew her at first hand.

In Cambridge she rarely missed her classes, lectures and seminars and she was a regular attendant of other people's seminars, especially visitors', never avoiding discussion and confrontation. Professor Pigou – a well-known misogynist – had included her in his category of 'honorary men'.

She was extremely popular with the students – a clear, brilliant, stimulating teacher. She was a person who inspired strong feelings – of love and hate. Her opponents were frightened by her, and her friends really admired, almost worshipped her. Her nonconformism in everyday life and even in her clothing (most of which she bought in India) was renowned.

She retired from her professorship in Cambridge on 30 September 1971. On retirement she did not agree to continue lecturing in Cambridge. (Later on, in the late 1970s, she gave in partially, giving a course of lectures on 'the Cambridge tradition'.) But her writing and lecturing abroad, at the invitation of economics faculties and students all over the world, continued unabated.

When, in the late 1970s, King's College (Keynes's college) finally dropped the traditional anachronistic ban on women and became co-educational, Joan Robinson, upon an enthusiastic and unanimous proposal by all economists of the college, became the first woman to be made an Honorary Fellow of King's College. (She had earlier become an Honorary Fellow of Girton College and of Newnham College.)

Towards the end of her life, she became very concerned and disappointed with the direction in which economic theory had turned and with the ease with which the younger economists could bend their elegant models to suit the new conservative moods and the selfish economic policies of politicians and governments. Her friends also noticed a sort of stiffening rigidity in her views that had not appeared before. This was unfortunate, as it contributed to increasing the hostility of her opponents towards her.

She suffered a stroke in early February 1983, from which she never recovered. She lay for a few months in a Cambridge hospital, and died peacefully six months later.

Distinctive traits of her intellectual personality

In order to understand better the nature of Joan Robinson's contributions to economic theory, it may be helpful to begin by considering explicitly a few characteristic traits of her intellectual personality.

Joan Robinson had a remarkable analytical ability. Since she did not normally use mathematics, this remarkable intellectual ability was of a nature that defies conventional description. In her early works she made use of geometrical representations, backed up by calculus (normally provided by Richard Kahn). In her mature works, her way of reasoning took up a more personal feature. Her style is difficult to imitate (as when she invites the readers to follow her in the construction of economic exercises) but very effective. The results are always impressive. Those who used to argue with her knew that she could grasp and keep in the back of her mind (to be brought out at the appropriate moment) a whole series of chain effects and interdependences which her interlocutors could hardly imagine.

She was not the type of person who could go on thinking in isolation. The way she could best express herself was by having somebody in constant confrontation. She could put her views best either in opposition or in support of somebody else's position. This made her extraordinarily open to concepts and contributions coming from the people she encountered. The accurate historian of economic ideas will probably find in her works traces of almost every person she met. It is therefore important, in considering Joan Robinson's contributions, to keep in mind at least the most important economists who influenced her. These include her teachers (Marshall through Pigou, Keynes, Shove), her contemporaries (Sraffa, Kaldor, and Kalecki, through whom she went back to Marx, but especially Richard Kahn, who read, criticized and improved every single one of her works) and also a whole series of other (younger) people – pupils and students.

This raises the question of her originality. The prefaces to her books are packed with acknowledgements, sometimes heavy acknowledgements – consider, for example, the following excerpt from the *Economics of Imperfect Competition*:

> … this book contains some matter which I believe to be new. Of not all the new ideas, however, can I definitely say that 'this is my own invention'. I particularly have had the constant assistance of Mr R.F. Kahn … many of the major problems … were solved as much by him as by me. (Robinson, 1933, p. v)

But one must remember what has been said above. In fact, Joan Robinson was a highly original thinker, but of a particular type. Besides the contributions to economic theory that are distinctly hers she had her own highly original way, even in small details, of presenting other authors' views, which she always did through a distinctly personal re-elaboration. Sometimes the re-elaboration is so personal as to sound parochial. But this trait is not exclusive to Joan Robinson. Cambridge parochialism is shared by almost all purely Cambridge-bred economists since Marshall (Keynes included). It sometimes creates unnecessary difficulties of communication with economists outside Cambridge (that is, with the overwhelming majority!) or introduces a few odd notes into an otherwise impeccable performance.

One can clearly detect an evolution in Joan Robinson's approach to economics that with age strengthened her innovative tendencies. It looks as if she was very cautious in her early years, preoccupied at first with building up solid analytical foundations. But as soon as she felt sure of her analytical equipment, she began to venture more and more into the exciting field of innovation. In her mature works her typical style became established. A sort of mixture of educational, temperamental and intellectual factors made her one of the leading unorthodox economists of the 20th century. Always impatient with dogmas, constantly fighting for new unorthodox ideas, relentlessly attacking established beliefs, she acquired a sort of vocation to economic heresies (see Robinson, 1971). Her attitude reminds one of a dictum by Pietro Pomponazzi, the Italian Renaissance philosopher: 'It is better to be a heretic if one wishes to find the truth.'

Strongly related to this attitude is the social message that comes from her writings. Her 'box of tools' and her logical chain of arguments were not proposed for their own sake; they were always aimed at practical action, with a view to the world's most pressing problems – unemployment before the war, underdevelopment and the struggle of ex-colonial nations after the war (very noticeable is her special concern for Asia and her enthusiasm, at points rather naive, for Communist China). Consistently, she has been among the strongest assertors – second perhaps only to Gunnar Myrdal – of the non-neutrality of economic science and of the necessity of stating explicitly one's convictions and beliefs.

And yet, in spite of her bold attacks and her satirical mood, her literary style is surprisingly feminine – rich with fable-like parables, with down-to-earth examples from everyday life ('the price of a cup of tea …') and with similes from scenes and examples taken from nature (the *Accumulation of Capital* begins with the economic life of the robin). Her sparkling prose and her entertaining asides make Joan Robinson one of the most brilliant writers among economists and certainly one of the most enjoyable and delightful to read.

Her scientific achievements

Joan Robinson wrote numerous books and an enormous number of articles, most of which have been collected in her *Collected Economic Papers* (1951–79).

They fall neatly into three broad groups, corresponding to the three basic phases of her intellectual development. A first group belongs to the phase of her by now classic *Economics of Imperfect Competition* (1933). A second group belongs to the phase of explanation, propagation and defence of Keynes's *General Theory*. Finally, a third group of writings grew around the major work of her maturity, *The Accumulation of Capital*

(1956). Other books and articles have originated from miscellaneous or wider interests or from the desire to provide students with economics exercises or with a non-orthodox economics textbook (Robinson and Eatwell, 1973c). Altogether, they make an impressive list. Even neglecting her articles (most of which are reprinted in the books), her bibliography contains no fewer than 24 books.

The most widely known of Joan Robinson's works is still the first, *The Economics of Imperfect Competition* (1933). It was the book of her youth, which placed her immediately in the forefront of the development of economic theory. It is a work conceived in Cambridge, at the end of a decade characterized by an intense controversy on cost curves and the laws of returns (see Sraffa, 1926, and the Symposium on the 'laws of returns' by Robertson, Sraffa and Shove, 1930). With this controversy in the background, Joan Robinson's book emerges in 1933 as a masterpiece in the traditional sense of the word. The restrictive conditions of perfect competition on which Marshall's theory was constructed are abandoned, and perfect competition is shown to be a very special case of what in general is a monopolistic situation. A whole new analysis of market behaviour is carried out on new, more general, assumptions; and yet the whole method of analysis, the whole approach – though refined and perfected – is still the traditional Marshallian one. Sraffa's criticism of the master is accepted, but is incorporated into the traditional fold by a generalization of Marshall's own theoretical framework. The outcome is extremely elegant and impressive. The whole matter of market competition is clarified. Marshall's ambiguities are eliminated, the various market conditions are rigorously defined, a whole technical apparatus (a 'box of analytical tools') is developed to deal with market situations in the general case (from demand and supply curves to marginal cost and marginal revenue curves). In a sense, therefore, rather than a radical critique, the *Economics of Imperfect Competition* might well be regarded as the completion and coronation of Marshallian analysis. This may help to explain why Joan Robinson herself came to like that book less and less, as her thought later developed on different lines. In 1969 she came to the point of writing a harsh eight-page criticism of it. Very courageously she published it, on the occasion of a reprint of the book, as a Preface to the second edition!

The book had appeared almost simultaneously with the *Theory of Monopolistic Competition* by Edward Chamberlin (1933); and the two books are normally bracketed together as indicating the decisive breakaway of economic theory from the assumptions of perfect competition. Chamberlin always complained about this association. For although the two books represent the simultaneous discovery of basically the same thing, made quite independently by two different authors, they are in fact substantively different.

It may also be added that, looked at in retrospect, these two books do not appear so conclusive a contribution to the theory of the firm as they appeared to be in the 1930s. The behaviour of firms in oligopolistic markets and the policies of the large corporations have turned out to require more complicated analysis. At the same time, the assumption of perfect competition, far from being completely dead, has recently come back in different guises in the works of many theoretical economists. Yet there is no doubt that the two books remain there to represent a definite turning-point in the development of the theory of the firm – so much so as to be referred to as representing the 'monopolistic competition revolution' (Samuelson, 1967). Very characteristically, Edward Chamberlin, after writing the *Theory of Monopolistic Competition,* spent the whole of his life in refining, completing and adding appendices to his masterpiece (no fewer than eight editions!). For Joan Robinson, the *Economics of Imperfect Competition* was only the first step on a very long way to a series of works in quite different and varied fields of economic theory.

It should be added that the *Economics of Imperfect Competition* was not Joan Robinson's only contribution to microeconomic theory in the 1930s. Her name appears again and again on the pages of the avant-garde economic journals of the time. From among her papers, explicit mention must be made at least of her remarkably lucid article on 'rising supply price' and of her contribution to clarifying the meaning of Euler's theorem as applied to marginal productivities, in the traditional theory of production (see her *Collected Papers*, vol. 1).

But something of extraordinary importance was happening in Cambridge in the 1930s. Keynes was in the process of producing his revolutionary work (Keynes, 1936). Joan Robinson abandoned the theory of the firm and threw herself selflessly and entirely into the new paths opened up by him. This was a really brave decision, if one thinks that her first book had gained her great reputation in the economic profession. Very rarely do we find someone who, after striking success and becoming a leading figure in a certain field, pulls out of it and puts himself or herself into the shadow of someone else, be this someone else even of the stature of Keynes. Joan Robinson did precisely that. She was one of the members – actually an important member, as is revealed by the recent publication of her correspondence with Keynes (see Keynes, 1973; 1979) – of that group of young economists known as the 'Cambridge Circus' (and including Kahn, Sraffa, Harrod, Meade, besides Austin and Joan Robinson) who regularly met for discussion, and played a crucial role in the evolving drafts of Keynes's *General Theory.*

It must be said that the new Keynes's ways were more congenial to her temperament. They were a break with tradition and this suited her nonconformist attitude; they dealt with the deep social problems of unemployment and this appealed to her social conscience. It is in this vein that she published her *Essays in the Theory of Employment* (1937a) and her *Introduction to the Theory of*

Employment (1937b). These twin books were simply meant to be a help to the readers of Keynes's *General Theory*. In fact, they turned out to be much more than that. In particular, Joan Robinson contributes to the clarification of a major piece of Keynesian theory – the process through which investments determine savings – which had remained rather obscure from the *General Theory*. For her, this appeared important because it broke a crucial link in traditional theory, which presented the rate of interest as a compensation for the 'sacrifice' of supplying capital (that is, for saving). Joan Robinson stresses the role of investment as an independent variable, while total saving is shown as being determined passively by investment through the operation of the multiplier; the conclusion being that the rate of interest cannot be remunerating anybody's 'sacrifice'. Even more so in depression times, when thrift – a 'private virtue' – becomes a 'public vice'. Other concepts, introduced by Joan Robinson at the time, that were to remain permanently in the following economic literature on the theory of employment are those concerning what she called 'beggar-my-neighbour' policies, 'disguised unemployment' and the generalization of the Marshall–Lerner conditions on international trade, in terms of 'the four elasticities'.

Towards the end of the 1930s, Joan Robinson met Kalecki, and discovered that quite independently of, and in fact earlier than, Keynes he had come to the same conclusions. Kalecki had started from a Marxist background, against which Keynes was prejudiced. This led her to re-reading Marx and to re-thinking her own position vis-à-vis Marxian theory (Robinson, 1942).

Joan Robinson's flirtation with Marx is very curious. It has all the charm of a meeting and all the clamour of a clash. She is no doubt attracted by Marx's general conception of society. She finds in Marx much that she approves of. But she finds his scientific nucleus embedded in, and in need of being liberated from, ideology. To obtain this, she says, one must work hard. Her writings on Marx are specifically aimed at 'separating the wheat of science from the chaff of ideology'. Needless to say, this has caused her a lot of trouble with the Marxists. It should be kept in mind that in Continental Europe discussions on Marx have a long and complex tradition of philological heaviness and ideological passion. Joan Robinson's discussion is short and simple. She is always looking at Marx as 'a serious economist'. Accordingly, she always tries to go straight to what she thinks is his economic analysis. Her insistence on the necessity of rescuing Marx, as a scholar and a first-rate analytical mind, has recently been vindicated, especially after the publication of Sraffa's book (1960; see also, for example, Samuelson, 1971).

But the post-war period was opening up new vistas. With Keynes's *General Theory* in the background, Joan Robinson saw a formidable task ahead, consisting in nothing less than a reconstruction of economic theory.

This led, after a decade of intense work, to the publication of her second major contribution to economic theory – *The Accumulation of Capital* (1956), the work of her maturity and the one that expresses Joan Robinson's genius at her best. Here she has chosen to move on new and controversial ground. While in her first book the direction – once established – was clear and she had to fill in the details, here the direction itself is not entirely clear and has to be continually adjusted. The details acquire less importance and may well be abandoned altogether and replaced with others at a second attempt. As a consequence, a lot of re-writing had to be done.

The Rate of Interest and Other Essays (1952), with its central essay devoted to a 'Generalization of the *General Theory*', turned out to be a sort of preparation. *The Accumulation of Capital* represents the central nucleus of what she perceived as a new framework for economic theory. Then the *Exercises in Economic Analysis* (1960a), the *Essays in the Theory of Economic Growth* (1962a) and a series of other articles fill in the gaps, clarify obscurities, and take the arguments further.

The 'Generalization of the *General Theory*' represents Joan Robinson's response to an interchange with Harrod, following Harrod's *Towards a Dynamic Economics* (1948) and also his earlier review of her *Essays in the Theory of Employment* (1937a). Joan Robinson breaks away from the limitations of the short run, but has not yet defined clearly her direction. Yet, once the process of 'generalization', that is, 'dynamization' of the *General Theory* is started, the author is compelled to recast the Keynesian arguments in terms of the more fundamental categories of capital accumulation, labour supply, technical progress and natural resources. Through this recasting, it became inevitable that she should go to the earlier methodological approach (common to Ricardo and Marx) of stating the problems in terms of social aggregates. The evidence of her intense searching may be found at the end of the book in a chapter of 'acknowledgements and disclaimers', where she describes in succession the way she has been influenced by, or has reacted to, Marx, Marshall, Rosa Luxemburg, Kalecki and Harrod.

The years of transition from the *Rate of Interest and Other Essays* (1952) to the *Accumulation of Capital* (1956) had been marked by a series of intense discussions in Cambridge, especially with Kahn, Sraffa, Kaldor and Champernowne. In the end, Joan Robinson emerged centring her attention on the problem of capital accumulation as the basic process in the development of a capitalist economy. She began with a scathing attack on the traditional concept of 'production function' (in a well-known article, now in her *Collected Papers*, vol. 2, which elicited a chain of angry responses: see, for example, Solow, 1955–6, and Swan, 1956). Then she patiently proceeded to a reconstruction. A crucial step was her own way of rediscovering the Swedish economist Knut Wicksell.

The Accumulation of Capital (1956) bears the same title as Rosa Luxemburg's book, to whose translation into

English Joan Robinson wrote an introduction (Luxemburg, 1951). This was a great tribute to another woman economist. But we should not be misled. Joan Robinson's book belongs to an entirely different age and takes an entirely different approach. Set into a Keynesian framework extended to the long run, it takes its origin from a welding together of Harrod's economic dynamics and of Wicksell's capital theory. The main question Joan Robinson poses to herself is by now a typically classical one: what are the conditions for the achievement of a cumulative long-term growth of income and capital (what she characteristically christened a 'golden age'); and what is the outcome of this process, in terms of growth of gross and net output and of the distribution of income between wages and profits, given a certain evolution through time of the labour force and of technology? To answer these question Joan Robinson builds up a two-sector dynamic model with a finite number of techniques; and goes on to show the interactions of the relations between wages and profits, the stock of capital and the techniques of production, entrepreneurial expectations and the degree of competition in the economy, bringing in the effects of higher degrees of mechanization and both 'neutral' and 'biased' technical progress. The basic model and the basic answers are all worked out very quickly in the book. The rest is then devoted to relaxing the simplifying assumptions. The whole analysis is carried out *without* the use of mathematics. This is remarkable. Joan Robinson squeezes out of the model, one by one, all the answers that are needed. The non-use of mathematics has certain obvious disadvantages. Though the analysis need not necessarily be any less rigorous, in many passages it is not so easy to follow. It has, however, some advantages, which Joan Robinson is very ready and able to exploit. She succeeds, for example, in freeing herself from the symmetry that a mathematically formulated model normally imposes. In Joan Robinson's model, certain results are always more likely to happen than their symmetrical counterpart. Symmetry and formal elegance play no part; only relevance does, or at least it does in the way perceived by the author.

The overall result is, again, impressive. The oversimplified dynamic model of Harrod is enormously enriched by the introduction of the choice among a finite number of alternative techniques. At the same time the Wicksellian analysis of accumulation at a given technology is completed by the new analysis of a constant flow of inventions of various types. And this marriage of Harrod's model to Wicksellian analysis is made to fructify in a number of directions. So many and so rich are in fact these directions that Joan Robinson herself did not pursue all of them, as became evident from the abundant literature that followed.

To this literature, Joan Robinson contributed a whole series of essays and books (see her *Collected Papers*, vols 2–5, and J. Robinson, 1960b; 1962a), which represent clarifications and further elaborations. They also represent her way of recasting and adjusting her arguments in response to opposition from her critics and to comments, remarks and stimuli of any sort from her friends, as well as her way of coming to grips with results – not always or not entirely compatible with hers – coming from the works of other scholars, colleagues and pupils, who were broadly working on similar problems and with the same aims.

Meanwhile, proceeding on parallel lines, many other separate strands of thinking were emerging from her remarkable intellectual activity. At least a few must briefly be mentioned here.

First, a whole series of concepts and ideas were coming to fruition, which – though not belonging to her major fields of interest – came to complete her overall coverage of economic theory: writings on the theory of international trade (including her professorial inaugural lecture at Cambridge on *The New Mercantilism*, 1966a), on Marxian economics (at various stages in her career), and on the theory of economic development and planning, reproducing her lectures delivered during her world travels or coming from calm reflection, once she had returned home (see her *Collected Papers*, and also J. Robinson, 1970b; 1979b).

Second, her deeply felt concern with economics students and economics teaching in general gave origin to books, such as Joan Robinson (1966b; 1971) and especially (with Eatwell) (1973c), which contributed to giving substance to, and disseminating all over the world, her strongly felt conviction that an overall approach to economic reality, alternative to that of traditional economics, does exist and is viable.

Third, the ideas, reflections, rationalizations, accumulated in the course of her life took the form of books such as *Economic Philosophy* (1962b) and *Freedom and Necessity* (1970a), which were concerned with wider issues than economics itself, attempting to give an overall conception of the world and a whole philosophy of life. These writings contribute, not marginally, to place Joan Robinson among the influential thinkers of this century. At the same time, they may well be enjoyed, by the general reader, even more than her masterpieces. From a purely literary point of view, they make delightful reading.

It should be added that there are, moreover, many themes which, while not being exclusively connected with any specific work of Joan Robinson's, recur time and again in her writings, so as to have become characteristically associated with her approach. Here are a few: (*a*) the concept of 'entrepreneurs' animal spirits' – an expression picked up from Keynes and developed as an important element contributing to explain investment in capitalist economies; (*b*) the conviction that Marshall's notions of prices and rate of profit, with reference to industry, are much more akin to Ricardo's notions than to Walras's; (*c*) a sharp distinction between 'logical' time and 'historical' time, both of which have a place in economic analysis but with different roles. On this point Joan Robinson's characterization of the evolution of an economy in historical time as concerning decisions to be taken between 'an irrevocable past and an uncertain future' is well known; (*d*) an equally sharp distinction between *comparisons* of equilibrium–growth positions and *movements* from one equilibrium–growth position to another, in dynamic analysis; (*e*) a tendency, especially in the later part of her life, to shift nearer and nearer to the positions of Kalecki, as opposed to those of Keynes, in interpreting the overall working of the institutions of capitalist economies, especially with reference to what she found as a more satisfactory integration in Kalecki of the concept of effective demand with the process of price formation.

Finally, one must mention specifically an issue which may well continue to give rise to controversial evaluations. This concerns the role that may be assigned to Joan Robinson in the well-known controversy on capital theory that flared up between the two Cambridges in the 1960s (see Pasinetti et al., 1966). One view on this issue is that Joan Robinson had the merit of anticipating the controversy by her (already mentioned) attacks on the neoclassical production function in the mid-1950s (see Harcourt, 1972). Another view is that Joan Robinson, herself a victim of her emotional temperament, started her attacks on the traditional concepts too early and misplaced the whole criticism, by neglecting the really basic point (the phenomenon of reswitching of techniques; see Sraffa, 1960) that in the end delivered the fatal blow to the neoclassical notion of production function. What one can say for certain is that a hint at the reswitching phenomenon does appear in the *Accumulation of Capital*, but is relegated to the role of a curiosum, in an entirely secondary section. Perhaps the phenomenon had been pointed out to her but she grossly underestimated its importance. What is curious is that she continued to underestimate it even after it was brought to the foreground (see her 'Unimportance of Re-switching' in *Collected Papers*, vol. 5).

But at this point the works of Joan Robinson merge into those of that remarkable group of Cambridge economists – notably, Piero Sraffa, Nicholas Kaldor and Richard Kahn, among others, besides Joan Robinson herself (on this, see the Preface to Pasinetti, 1981) – who happened to be concentrated in Cambridge in the post-war period and who took up, continued and expanded the challenge that Keynes had launched on orthodox economic theory. This remarkable group of economists started a stream of economic thought which is obviously far from complete. Its basic features, however, are clear enough; they embody a determined effort to shift the whole focus of economic theorizing away from the problems of optimum allocation of given resources, where it had remained for almost a century, and move it towards the fundamental factors responsible for the dynamics of industrial societies. This shift of focus inevitably brings into the foreground the once central themes of capital accumulation, population growth, production expansion, income distribution, and thus technical progress and structural change.

It is perhaps too early to try to evaluate the relative role played by Joan Robinson as a member of this remarkable group of economists. The single components of the group have made contributions which are sometimes complementary, at other times overlapping, and at yet other times even partly contradictory. To mention only one major problem, Piero Sraffa's book appeared too late for Joan Robinson to be able to incorporate it into her theoretical framework; and the brave efforts she later made to this effect are not always convincing. They actually reveal here and there a sort of ambivalent attitude. At the same time, her *Accumulation of Capital* ventures into fields of economic dynamics which Sraffa does not touch at all. Quite obviously, the common fundamental thrust behind post-Keynesian analysis does not presuppose complete identity of views or complete harmony of approach.

Future developments will clarify issues and will reveal which of the lines of approach proposed are the most useful, fruitful or fecund. There can be little doubt, however, that if this theoretical movement is going to prove successful, quite a lot of rewriting will have to be done in economic theory. If, and when, this rewriting occurs, Joan Robinson's contributions are going to take a major place.

LUIGI L. PASINETTI

Selected works

1933. *The Economics of Imperfect Competition*. London: Macmillan. 2nd edn, 1969.

1937a. *Essays in the Theory of Employment*. London: Macmillan.

1937b. *Introduction to the Theory of Employment*. London: Macmillan.

1942. *An Essay on Marxian Economics*. London: Macmillan.

1951. *Collected Economic Papers*, vol. 1. Oxford: Basil Blackwell. (Vol. 2, 1960b; vol. 3, 1965; vol. 4, 1973a; vol. 5, 1979a.)

1952. *The Rate of Interest and Other Essays*. London: Macmillan.

1956. *The Accumulation of Capital*. London: Macmillan.

1960a. *Exercises in Economic Analysis*. London: Macmillan.

1960b. *Collected Economic Papers*, vol. 2. Oxford: Basil Blackwell.

1962a. *Essays in the Theory of Economic Growth*. London: Macmillan.

1962b. *Economic Philosophy*. London: C.A. Watts.

1965. *Collected Economic Papers*, vol. 3. Oxford: Basil Blackwell.

1966a. *The New Mercantilism – An Inaugural Lecture*. Cambridge: Cambridge University Press.

1966b. *Economics – An Awkward Corner*. London: Allen & Unwin.

1970a. *Freedom and Necessity.* London: Allen & Unwin.

1970b. *The Cultural Revolution in China.* London: Penguin Books.

1971. *Economic Heresies: Some Old-fashioned Questions in Economic Theory.* London: Macmillan.

1973a. *Collected Economic Papers*, vol. 4. Oxford: Basil Blackwell.

1973b. ed. *After Keynes.* Papers presented to Section F (economics) of the 1972 annual meeting of the British Association for Advancement of Science. Oxford: Basil Blackwell.

1973c. (With J. Eatwell.) *An Introduction to Modern Economics.* New York: McGraw-Hill.

1978. *Contributions to Modern Economics.* Oxford: Basil Blackwell.

1979a. *Collected Economic Papers*, vol. 5. Oxford: Basil Blackwell.

1979b. *Aspects of Development and Underdevelopment.* Cambridge: Cambridge University Press.

1980. *Further Contributions to Modern Economics.* Oxford: Basil Blackwell.

Bibliography

Chamberlin, E. 1933. *The Theory of Monopolistic Competition.* Cambridge, MA: Harvard University Press.

Harcourt, G. 1972. *Some Cambridge Controversies in the Theory of Capital.* Cambridge: Cambridge University Press.

Harrod, R.F. 1948. *Towards a Dynamic Economics.* London: Macmillan.

Keynes, J.M. 1930. *A Treatise on Money.* 2 vols. London: Macmillan.

Keynes, J.M. 1936. *The General Theory of Employment, Interest and Money.* London: Macmillan.

Keynes, J.M. 1973, 1979. *The Collected Writings of John Maynard Keynes*, vols 13 and 14 (1973) and 29 (1979), ed. D.C. Moggridge. London: Macmillan for the Royal Economic Society.

Luxemburg, R. 1913. *The Accumulation of Capital.* Trans. A. Schwarzschild, with an Introduction by J. Robinson, London: Routledge & Kegan Paul, 1951.

Pasinetti, L.L. 1974. *Growth and Income Distribution: Essays in Economic Theory.* Cambridge: Cambridge University Press.

Pasinetti, L.L. 1981. *Structural Change and Economic Growth: A Theoretical Essay on the Dynamics of the Wealth of Nations.* Cambridge: Cambridge University Press.

Pasinetti, L.L., Levhari, D., Samuelson, P.A., Bruno, M., Burmeister, E., Sheshinski, E., Morishima, M. and Garegnani, P. 1966. Contributions to 'Paradoxes in Capital Theory – A Symposium'. *Quarterly Journal of Economics* 80, 503–83.

Robertson, D.H., Sraffa, P. and Shove, G. 1930. Increasing returns and the representative firm: a symposium. *Economic Journal* 40, 76–116.

Samuelson, P.A. 1967. The monopolistic competition revolution. In *Monopolistic Competition: Studies in Impact*, ed. R.M. Kuenne. New York: Wiley & Sons.

Samuelson, P.A. 1971. Understanding the Marxian notion of exploitation: a summary of the so-called transformation problem between Marxian values and competitive prices. *Journal of Economic Literature* 9, 339–431.

Solow, R.M. 1955–6. The production function and the theory of capital. *Review of Economic Studies* 23(2), 101–8.

Sraffa, P. 1926. The laws of returns under competitive conditions. *Economic Journal* 36, 535–50.

Sraffa, P. 1960. *Production of Commodities by Means of Commodities: Prelude to a Critique of Economic Theory.* Cambridge: Cambridge University Press.

Swan, T.W. 1956. Economic growth and capital accumulation. *Economic Record* 32, 334–61.

Wicksell, K. 1901. *Lectures on Political Economy.* Vol. 1, ed. L. Robbins, London: Routledge & Kegan Paul, 1934.

Bibliographic addendum

Essay collections on Robinson's contributions to economics include B. Gibson, ed., *Joan Robinson's Economics: A Centennial Celebration*, Cheltenham: Edward Elgar, 2005, and G.C. Harcourt, ed., *Joan Robinson: Critical Assessments of Leading Economists*, London: Routledge, 2002. A combination of biography and bibliography of Robinson's work is J. Cicarelli and J. Cicarelli, *Joan Robinson: A Bio-Bibliography*, Westport, CT: Greenwood Press, 1996.

robust control

Robust control considers the design of decision or control rules that fare well across a range of alternative models. Thus robust control is inherently about model uncertainty, particularly focusing on the implications of model uncertainty for decisions. Robust control originated in the 1980s in the control theory branch of the engineering and applied mathematics literature, and it is now perhaps the dominant approach in control theory. Robust control gained a foothold in economics in the late 1990s and has seen increasing numbers of economic applications in the past few years. (For related surveys see Hansen and Sargent, 2001; and Backus, Routledge and Zin, 2005. For a more comprehensive view of the leading approach to robust control in economics, see Hansen and Sargent, 2008.)

The basic issues in robust control arise from adding more detail to the opening sentence above – that a decision rule performs well across alternative models. To begin, define a model as a specification of a probability distribution over outcomes of interest to the decision maker, which is influenced by a decision or control variable. Then model uncertainty simply means that the decision maker faces subjective uncertainty about the specification of this probability distribution. A first key issue in robust control, then, is to specify the class of alternative models which the decision maker entertains. As we discuss below, there are many approaches to doing so, with the most common cases taking a benchmark *nominal model* as a starting point and considering perturbations of this model. How to specify and measure the magnitude of the perturbations are key practical considerations.

With the model set specified, the next issue is how to choose a decision rule and thus what it means for a rule to 'perform well' across models. In Bayesian analysis, the decision maker forms a prior over models and proceeds as usual to maximize expected utility (or minimize expected loss). Just as we defined a model as a probability distribution, a Bayesian views model uncertainty as simply a hierarchical probability distribution with one layer consisting of shocks and variables to be integrated over, and another layer averaging over models. In contrast, most robust control applications focus on minimizing the worst case loss over the set of possible models (a minimax problem in terms of losses, or max-min expected utility). Stochastic robust control problems thus distinguish sharply between shocks which are averaged over and models which are not. The robust control approach thus presumes that decision makers are either unable or unwilling to form a prior over the forms of model misspecification. Of course, decision makers must be able to specify the set of models as discussed above, but typically this involves bounding the set of possibilities in some way rather than fully specifying each alternative. Finally, there are some approaches which seek a middle ground between the average case and the worst case, for example by maximizing expected utility subject to a bound on the worst case loss. These have been less prominent both in control theory (Limebeer, Anderson and Hendel, 1994, is one example) and in economics (Tornell, 2003, is one exception), and thus will not be discussed further. For the remainder of the article robust control will mean a minimax approach.

Robustness and worst case analysis

Broadly speaking, the control theory literature has adopted the worst-case philosophy out of concerns for stability. A basic desideratum for robust control in practice is that the system remain stable in the face of perturbations, and since instability may be equated with infinite loss, minimizing the worst case outcomes will insure stability (when possible). Moreover, many engineering applications have specific performance objectives which must be maintained, and a cost function penalizing deviations is not clearly specified. However, in dealing with economic agents rather than controlled machines, decision theoretic criteria naturally come into

play. In this sphere, robust control is closely related to the notions of Knightian uncertainty, ambiguity and uncertainty aversion, which are all roughly equivalent (although sometimes differing in formalization).

Starting with the observations of the classic Ellsberg (1961) paradox – that (some) decision makers prefer environments with known odds to those with uncertain probabilities – there has been a broad literature in decision theory which has weakened the Savage axioms to incorporate preferences which display such aversion to uncertainty or ambiguity. The most widely used characterization is due to Gilboa and Schmeidler (1989), who axiomatized ambiguity preferences with multiple priors. Decision-making with multiple priors can be represented as max-min expected utility: maximizing the utility with respect to the least favourable prior from a convex set of priors. More recently, Epstein and Schneider (2003) have extended the static environment of Gilboa and Schmeidler to a dynamic context, where the set of priors is updated over time. Hansen et al. (2006) formally established the links between robust control and ambiguity aversion, showing that the model set of robust control as discussed above can be thought of as a particular specification of Gilboa and Schmeidler's set of priors. Moreover, although the ambiguity preferences are characterized by posing particular counterfactuals which require multiple priors, once the least favourable prior is chosen, behaviour could be rationalized as Bayesian with that prior. Thus from a Bayesian viewpoint Sims (2001) views robust control as a means of generating priors, which then naturally leads to questioning whether the worst case prior accurately reflects actual beliefs and preferences. (See also Svensson, 2001. Hansen et al., 2006, show how to back out the Bayesian prior which rationalizes robust decision-making.)

Finally, in many cases robust or ambiguity-averse preferences are similar to enhanced risk aversion, and in some cases they are observationally equivalent. This insight dates to Jacobson (1973) and Whittle (1981) in the control theory literature, and the relations between robust control and a particular specification of Kreps and Porteus (1978), Epstein and Zin (1989) and Duffie and Epstein (1992) recursive utility with enhanced risk aversion have been shown by Anderson, Hansen and Sargent (2003), Hansen et al. (2006) and Skiadas (2003).

Control theory background

Since many of the ideas and inspiration for robust control in economics come from control theory, we give here just a broad outline of its development. More detail and different perspectives can be found in the books by Zhou, Doyle and Glover (1996), Başar and Bernhard (1995), and Burl (1999). Throughout the late 1960s and early 1970s optimal control came into its own, largely through the work of Kalman on linear quadratic (LQ) control and filtering. While this approach remains widely used today throughout economics, starting in the late 1970s and early 1980s the control theory literature started to change as theory and practice showed some of the limitations of the LQ approach. Although LQ control with full observation (the so-called linear quadratic regulator or LQR) was known to be robust to some types of model perturbations, Doyle (1978) showed that there are no such assurances in the case of partial observation (the so-called linear-quadratic-Gaussian or LQG case, which is an LQR control matched with a Kalman filter). Doyle's paper title and abstract are classic in the literature – title: 'Guaranteed Margins for LQG Regulators', abstract: 'There are none'.

Spurred by this and related work, control theorists started to move away from LQ control to look for a more robust approach. Zames (1981) was influential in the development of H_∞ control as a more robust alternative to LQ control. Loosely speaking, in LQ control the quadratic cost means that performance is measured with a 2-norm across frequencies. By contrast, H_∞ uses an ∞ − norm that looks at the peak of the losses across frequencies. It is also interpretable as the maximal magnification of the disturbances to outputs of interest. While the early robust control literature used a frequency domain approach, in the late 1980s Doyle and others developed state space formulations (see Doyle et al., 1989, for example) which gave explicit solutions and allowed for alterative formalizations. For example, the H_∞ approach was given alternative justifications in terms of penalizing disturbances from the nominal model, which can be implemented as a dynamic game between a decision maker seeking to minimize losses and a malevolent agent seeking to maximize loss. (See Başer and Bernhard, 1995, for a development of this approach.) Finally, the uncertainty sets in the H_∞ approach are unstructured – they represent perturbations of the model which are bounded but have no particular form. The implications of structured perturbations have been studied more recently. Some examples include parametric perturbations, unmodelled dynamics, or uncertainty only about particular channels or connections in a model. Applications with structured uncertainty use the structured singular value (also known as μ) rather than the H_∞ norm as a measure of performance. Although there are some important stability and performance criteria, in general constructing control rules is a more daunting task, and the theory is not as fully developed as the unstructured case.

The Hansen–Sargent approach

In the economics literature, the most prominent and influential approach to robust control is due to Hansen and Sargent (and their co-authors), which is summarized in their monograph Hansen and Sargent (2008). This approach starts with a nominal model and uses entropy as a distance measure to calibrate the model uncertainty set. More specifically, the model set consists of those models whose relative entropy or Kullback–Leibler distance from the nominal model is bounded by a specified value. Note that this puts no structure on the uncertainty, but only restricts the alternative models to those which are difficult to distinguish statistically from the nominal model. In practice, a Lagrange multiplier theorem is typically used to convert the entropy constraint into a penalty on perturbations from the model. Then the solution of the control problem is found via a dynamic game implementation: the agent maximizes utility by his choice of control, while an evil agent minimizes utility by his choice of perturbation, while being penalized by the entropy of the deviations. Relative to the control theory literature such as Başar and Bernhard (1995), the main differences are that all models are stochastic, while control theory largely uses deterministic models. One exception is Petersen, James, and Dupuis (2000) who use a similar approach to consider uncertain stochastic systems. In addition, discounting is not typically considered in control theory, while it is natural in economics. In full information problems discounting has relatively little effect, but it raises important issues in problems with partial information (see Hansen and Sargent, 2005a; 2005b). Finally, the Hansen–Sargent approach naturally extends beyond the LQ setting laid out in Hansen, Sargent and Tallarini (1999), with some examples in Anderson, Hansen and Sargent (2003), Cagetti et al. (2002) and Maenhout (2004).

To be more concrete, consider an LQ example where x_t is the state, i_t is the agent's control, and ε_t is an i.i.d. Gaussian shock. The nominal model is:

$$x_{t+1} = Ax_t + Bi_t + C\varepsilon_{t+1}, \qquad (1)$$

and the agent's intertemporal preferences are:

$$E_0 \sum_{t=0}^{\infty} \beta^t \left(x_t' Q x_t + i_t' R i_t \right) \qquad (2)$$

where $0 < \beta < 1$ and Q and R are negative definite matrices. The approach of Hansen and Sargent perturbs the nominal model with an additional 'misspecfication shock' w_{t+1} which is allowed to be correlated with the state x_t:

$$x_{t+1} = Ax_t + Bi_t + C(\varepsilon_{t+1} + w_{t+1}). \qquad (3)$$

The shock w_{t+1} is used to represent alternative models. These models are made to be close to the nominal model in an entropy sense by imposing the bound:

$$E_0 \sum_{t=0}^{\infty} \beta^t w_{t+1}' w_{t+1} \leq \eta \qquad (4)$$

for some constant $\eta \geq 0$. The agent then maximizes (2) with respect to the worst case perturbed model (3) from the set (4). Using a Lagrange multiplier theorem, the constraint set can be converted to a penalty and the decision problem can by solved recursively by solving the Bellman equation for a two-player zero sum game:

$$V(x) = \max_i \min_w \{\{x'Qx + i'Ri + \beta\theta w'w$$
$$+ \beta E[V(Ax + Bi + C(\varepsilon + w))|x]\}\} \quad (5)$$

where $\theta > 0$ is a Lagrange multiplier on the constraint (4) and the expectation is over the Gaussian shock ε. Often this multiplier formulation is taken as the starting point, for example Maccheroni, Marinacci and Rustichini (2006) characterize preferences of this form, with θ governing the degree of robustness. As $\theta \to \infty$ the penalization becomes so great that only the nominal model remains (thus $\eta \to 0$), and the decision rule is less robust. Conversely, there is typically a minimal value of θ beyond which the value is $V(x) = -\infty$. This gives the most robust decision rules, allowing for the largest uncertainty set.

Adding structure to the uncertainty set

The approach discussed above uses unstructured uncertainty, and has been well developed and extended in different dimensions. We now discuss some alterative approaches which put more structure on the uncertainty set. There are many reasons to do so. It may be that some of the models that are close to the nominal model in a statistical sense may not be plausible economically. Alternatively, the decision makers may have a discrete set of models in mind, and bounding them all in one uncertainty set may include extraneous implausible models. Perhaps most substantively, the decision maker may be more confident some aspects of the model relative to others. Some examples of this include knowing the model up to the values of parameters, or being more certain about the dynamics of certain variables in the model. Not taking into account the particular structure may give a misleading impression of the actual uncertainty the decision makers face.

There are many ways of building in structured uncertainty, and the distinctions between cases are not always clear. For example, consider the same nominal model (1) as above, but suppose that instead of the unstructured perturbations (3) the uncertainty is instead solely in the values of the parameters A and B. Thus we can represent the parametric perturbed models as:

$$x_{t+1} = (A + \hat{A})x_t + (B + \hat{B})i_t + C\varepsilon_{t+1}$$
$$(6)$$

for some matrices $\hat{A}$ and $\hat{B}$. Of course it's possible to rewrite (6) as a version of (3) with:

$$w_{t+1} = \hat{A}x_t + \hat{B}i_t, \quad (7)$$

so in principle parametric perturbations are just a special case of the unstructured uncertainty. However what makes a substantive difference is how uncertainty is measured, that is whether we restrict w_{t+1} as in (4) or whether we restrict the parameters $\hat{A}$ and $\hat{B}$, say by bounding them in a confidence ellipsoid around the nominal model. Moreover, as (7) makes clear the differences between the uncertainty measurements will depend on the actual control rule i_t in place. Onatski and Williams (2003) provide an example of a simple estimated model where the uncertainty specifications matter dramatically for outcomes. In particular, the optimal policy for the largest possible unstructured uncertainty set (that is for the minimal value of θ) leads to instability for relatively small parametric perturbations. Thus the particular structure and measurement of uncertainty can have important implications for decisions. (Peterson, James and Dupuis, 2000, modify the unstructured approach described above to deal with structured uncertainty by separating the entropy penalty for unstructured perturbations from a different penalization for structured perturbations.)

Some economic applications with structured uncertainty include the following:

1. The simplest cases are uncertainty sets with discrete possible models. Some examples include: Levin and Williams (2003), who consider both Bayesian and minimax approaches; Cogley and Sargent (2005); and Svensson and Williams (2006), who focus on a Bayesian approach, and the recent work of Hansen and Sargent (2006), who have built this type of structure into their robust approach.

2. Another common form is parameter uncertainty within a fully specified model. Brainard (1967) is the classic reference from a Bayesian perspective with many references in this line, while Giannoni (2002) and Chamberlain (2000) consider minimax approaches.

3. Somewhat more broad are cases with different parametric model specifications. For example this includes uncertainty about dynamics (lags and leads), variables which may enter, uncertainty about data quality, and other features which are built into parametric extensions of the nominal model. Examples include the model error modelling approach of Onatski and Williams (2003) and the empirical specifications of Brock, Durlauf and West (2003).

4. Finally, the model sets may be nonparametric but structured in particular ways. For example, Onatski and Stock (2002) consider different structured types of uncertainty such as linear time-invariant perturbations, nonlinear time-varying perturbations, and perturbations which only enter particular parts of the model. Other examples include nonparametric specifications of uncertainty which differs across frequencies as in Onatski and Williams (2003) and Brock and Durlauf (2005).

NOAH WILLIAMS

See also **ambiguity and ambiguity aversion; model uncertainty; stochastic optimal control; uncertainty.**

Bibliography

Anderson, E., Hansen, L.P. and Sargent, T. 2003. A quartet of semi-groups for model specification, robustness, prices of risk, and model detection. *Journal of the European Economic Association* 1, 68–123.

Backus, D.K., Routledge, B.R. and Zin, S.E. 2005. Exotic preferences for macroeconomists. In *NBER Macroeconomics Annual 2004*, ed. M. Gertler and K. Rogoff. Cambridge, MA: MIT Press.

Başar, T. and Bernhard, P. 1995. H_∞-*Optimal Control and Related Minimax Design Problems*. Boston: Birkhauser.

Brainard, W. 1967. Uncertainty and the effectiveness of policy. *American Economic Review* 57, 411–25.

Brock, W.A. and Durlauf, S.N. 2005. Local robustness analysis: theory and application. *Journal of Economic Dynamics and Control* 29, 2067–92.

Brock, W.A., Durlauf, S.N. and West, K.D. 2003. Policy evaluation in uncertain economic environments. *Brookings Papers on Economic Activity* 2003(1), 235–301.

Burl, J.B. 1999. *Linear Optimal Control: $\mathcal{H}_2$ and $\mathcal{H}_\infty$ Methods*. Menlo Park, CA: Addison-Wesley.

Cagetti, M., Hansen, L.P., Sargent, T.J. and Williams, N. 2002. Robustness and pricing with uncertain growth. *Review of Financial Studies* 15, 363–404.

Chamberlain, G. 2000. Econometric applications of maxmin expected utility theory. *Journal of Applied Econometrics* 15, 625–44.

Cogley, T. and Sargent, T.J. 2005. The conquest of U.S. inflation: learning and robustness to model uncertainty. *Review of Economic Dynamics* 8, 528–63.

Doyle, J.C. 1978. Guaranteed margins for LQG regulators. *IEEE Transactions on Automatic Control* 23, 756–7.

Doyle, J.C., Glover, K., Khargonekar, P.P. and Francis, B.A. 1989. State-space solutions to standard $\mathcal{H}_2$ and $\mathcal{H}_\infty$ control problems. *IEEE Transactions on Automatic Control* 34, 831–47.

Duffie, D. and Epstein, L.G. 1992. Stochastic differential utility. *Econometrica* 60, 353–94.

Ellsberg, D. 1961. Risk, ambiguity and the Savage axioms. *Quarterly Journal of Economics* 75, 643–69.

Epstein, L. and Schneider, M. 2003. Recursive multiple-priors. *Journal of Economic Theory* 113, 1–31.

Epstein, L. and Zin, S. 1989. Substitution, risk aversion and the temporal behavior of consumption and asset returns: a theoretical framework. *Econometrica* 57, 937–69.

Giannoni, M. 2002. Does model uncertainty justify caution? Robust optimal monetary policy in a forward-looking model. *Macroeconomic Dynamics* 6, 111–44.

Gilboa, I. and Schmeidler, D. 1989. Maxmin expected utility with non-unique prior. *Journal of Mathematical Economics* 18, 141–53.

Hansen, L.P. and Sargent, T.J. 2001. Acknowledging misspecification in macroeconomic theory. *Review of Economic Dynamics* 4, 519–35.

Hansen, L.P. and Sargent, T.J. 2005a. Robust estimation and control under commitment. *Journal of Economic Theory* 124, 258–301.

Hansen, L.P. and Sargent, T.J. 2005b. Robust estimation and control without commitment. Working paper, Department of Economics, University of Chicago.

Hansen, L.P. and Sargent, T.J. 2006. Fragile beliefs and the price of model uncertainty. Working paper, Department of Economics, University of Chicago.

Hansen, L.P. and Sargent, T.J. 2008. *Robustness*. Princeton, NJ: Princeton University Press.

Hansen, L.P., Sargent, T. and Tallarini, T. 1999. Robust permanent income and pricing. *Review of Economic Studies* 66, 873–907.

Hansen, L.P., Sargent, T.J., Turmuhambetova, G.A. and Williams, N. 2006. Robust control and model misspecification. *Journal of Economic Theory* 128, 45–90.

Jacobson, D.H. 1973. Optimal stochastic linear systems with exponential performance criteria and their relation to deterministic differential games. *IEEE Transactions on Automatic Control* AC 18, 1124–31.

Kreps, D.M. and Porteus, E.L. 1978. Temporal resolution of uncertainty and dynamic choice. *Econometrica* 46, 185–200.

Levin, A.T. and Williams, J. 2003. Robust monetary policy with competing reference models. *Journal of Monetary Economics* 50, 945–75.

Limebeer, D.J.N., Anderson, B.D.O. and Hendel, B. 1994. A Nash game approach to mixed H_2/H_∞ control. *IEEE Transactions on Automatic Control* 39, 69–82.

Maccheroni, F., Marinacci, M. and Rustichini, A. 2006. Dynamic variational preferences. *Journal of Economic Theory* 128, 4–44.

Maenhout, P.J. 2004. Robust portfolio rules and asset pricing. *Review of Financial Studies* 17, 951–83.

Onatski, A. and Stock, J.H. 2002. Robust monetary policy under model uncertainty in a small model of the US economy. *Macroeconomic Dynamics* 6, 85–110.

Onatski, A. and Williams, N. 2003. Modeling model uncertainty. *Journal of the European Economic Association* 1, 1087–122.

Petersen, I.R., James, M.R. and Dupuis, P. 2000. Minimax optimal control of stochastic uncertain systems with relative entropy constraints. *IEEE Transactions on Automatic Control* 45, 398–412.

Sims, C.A. 2001. Pitfalls of a minimax approach to model uncertainty. *American Economic Review* 91, 51–4.

Skiadas, C. 2003. Robust control and recursive utility. *Finance and Stochastics* 7, 475–89.

Svensson, L.E. 2001. Robust control made simple. Working paper, Department of Economics, Princeton University.

Svensson, L.E.O. and Williams, N. 2006. Monetary policy with model uncertainty: distribution forecast targeting.

Working paper, Department of Economics, Princeton University.

Tornell, A. 2003. Exchange rate anomalies under model misspecification: a mixed optimal/robust approach. Working paper, Department of Economics, UCLA.

Whittle, P. 1981. Risk sensitive linear quadratic Gaussian control. *Advances in Applied Probability* 13, 764–77.

Zames, G. 1981. Feedback and optimal sensitivity: model reference transformations, multiplicative seminorms, and approximate inverses. *IEEE Transactions on Automatic Control* 26, 301–20.

Zhou, K., Doyle, J. and Glover, K. 1996. *Robust and Optimal Control*. Upper Saddle River, New Jersey: Prentice Hall.

robust estimators in econometrics

Econometrics often deals with data under, from the statistical point of view, non-standard conditions such as heteroscedasticity or measurement errors, and the estimation methods thus need either to be adapted to such conditions or to be at least insensitive to them. Methods insensitive to violation of certain assumptions – for example, insensitive to the presence of heteroscedasticity – are in a broad sense referred to as 'robust' (for example, robust to heteroscedasticity). On the other hand, there is also a more specific meaning of the word 'robust', which stems from the field of robust statistics. This latter notion defines robustness rigorously in terms of the behaviour of an estimator both at the assumed (parametric) model and in its neighbourhood in the space of probability distributions. Even though the methods of robust statistics have been used only in the simplest settings, such as estimation of location, scale or linear regression for a long time, they have motivated a range of new econometric methods, which we focus on in this article.

The concepts and measures of robustness are introduced first (Section 1), followed by the most common types of estimation methods and their properties (Section 2). Various econometric methods based on these common estimators are discussed in Section 3, covering tasks from time-series regression over GMM estimation to simulation-based methods.

1 Measures of robustness

Robustness properties can be formulated within two frameworks: qualitative and quantitative robustness. *Qualitative robustness* is concerned with the situation in which the shape of the underlying (true) data distribution deviates slightly from the assumed model. It focuses on questions like stability and performance loss over a family of such slightly deviating distributions. *Quantitative robustness* is involved when the sensitivity of estimators to a proportion of aberrant observations is studied.

A simple example can make this clear. Suppose one has collected a sample on an individual's income (after say ten years of schooling) and one is interested in estimating the mean income. If $\{x_i\}_{i=1}^n$ denotes the logarithm of this data and we suppose that they have a cumulative distribution function (cdf) F, assumed to be $N(\mu,\sigma^2)$, the maximum likelihood estimator (MLE) is $\bar{x} = \int u dF_n(u) = T(F_n)$, where $F_n(u) = n^{-1}\sum_{i=1}^n I(x_i \leq u)$, and $\mu = \int u dF(u) = T(F)$. Qualitative robustness asks here the question: how well will μ be estimated if the true distribution is in some neighbourhood of F? Quantitative robustness would concentrate on the question: will $T(F_n)$ be bounded if some observations $x_i \to \infty$? In fact, the latter question is easy to answer: if $x_i \to \infty$ for some i, $T(F_n) = \bar{x} \to \infty$ as well. So we can say here in a loose sense that $\bar{x}$ is not quantitatively robust.

1.1 Formalities

In the following we present a mathematical set-up that allows us to formalize our thoughts on robustness.

The notion of the sensitivity of an estimator T is put into theory by considering a model characterized by a cdf F and its neighbourhood $\mathscr{F}_{\varepsilon,G}$: distributions $(1 - \varepsilon)F + \varepsilon G$, where $\varepsilon \in (0,1/2)$ and G is an arbitrary probability distribution, which represents data contamination. Hence, not all data necessarily follow the pre-specified distribution, but the ε-part of data can come from a different distribution G. If $H \in \mathscr{F}_{\varepsilon,G}$, the estimation method T is then judged by how sensitive or robust the estimates $T(H)$ are to the size of $\mathscr{F}_{\varepsilon,G}$, or alternatively, to the distance from the assumed cdf F. Two main concepts for robust measures analyse the sensitivity of an estimator to infinitesimal deviations, $\varepsilon \to 0$, and to finite (large) deviations, $\varepsilon > 0$, respectively. Despite generality of the concept, easy interpretation and technical difficulties often limit our choice to point-mass distributions (Dirac measures) $G = \delta_x, x \in \mathbb{R}$, which simply represents an (erroneous) observation at point $x \in \mathbb{R}$. This simplification is also used in the following text.

1.2 Qualitative robustness

The influence of infinitesimal contamination on an estimator is characterized by the *influence function*, which measures the relative change in estimates caused by an infinitesimally small amount ε of contamination at x (Hampel et al., 1986). More formally,

$$IF(x; T, F) = \lim_{\varepsilon \to 0} \frac{T\{(1 - \varepsilon)F + \varepsilon\delta_x\} - T(F)}{\varepsilon}. \tag{1}$$

For each point x, the influence function reveals the rate at which the estimator T changes if a wrong observation appears at x. In the case of sample mean $\bar{x} = T(F_n)$ for

$\{x_i\}_{i=1}^n$, we obtain

$$IF(x; T, F_n) = \lim_{\varepsilon \to 0} \left[(1 - \varepsilon) \int u dF_n(u) \right.$$

$$+ \varepsilon \int u d\delta_x(u) - \int u dF_n(u)]/\varepsilon$$

$$= \lim_{\varepsilon \to 0} \left[-\int u dF_n(u) + \int u d\delta_x(u) \right]$$

$$= x - \bar{x}.$$

The influence function allows us to define various desirable properties of an estimation method. First, the largest influence of contamination on estimates can be formalized by the *gross-error sensitivity*,

$$\gamma(T, F) = \sup_{x \in \mathbb{R}} IF(x; T, F), \qquad (2)$$

which under robustness considerations is finite and small. Even though such a measure can depend on F in general, the qualitative results (for example, $\gamma(T, F)$ being bounded) are typically independent of F. Second, the sensitivity to small changes in data, for example moving an observation from x to $y \in \mathbb{R}$, can be measured by the *local-shift sensitivity*

$$\lambda(T, F) = \sup_{x \neq y} \frac{||IF(x; T, F) - IF(y; T, F)||}{||x - y||}. $$

$$(3)$$

Also, this quantity should be relatively small since we generally do not expect that small changes in data cause extreme changes in values or sensitivity of estimates. Third, as unlikely large or distant observations may represent data errors, their influence on estimates should become zero. Such a property is characterized by the *rejection point*,

$$\rho(T, F) = \inf_{r > 0} \{ r : IF(x; T, F) = 0, ||x|| \geq r \},$$

$$(4)$$

which indicates the non-influence of large observations.

1.3 Quantitative robustness
Alternatively, the behaviour of the estimator T can be studied for any finite amount ε of contamination. The most common property looked at in this context is the estimator's bias $b(T, H) = E_H\{T(H)\} - E_F\{T(F)\}$, which measures a distance between the estimates for clean data, $T(F)$, and contaminated data, $T(H), H \in \mathscr{F}_{\varepsilon, G}$. The corresponding *maximum-bias* curve measures the maximum bias of T on $\mathscr{F}_{\varepsilon, G}$ at any ε:

$$B(\varepsilon, T) = \sup_{x \in \mathbb{R}} b\{T, (1 - \varepsilon)F + \varepsilon\delta_x\}. \qquad (5)$$

Although the computation of this curve is rather complex, Berrendero and Zamar (2001) provide general methodology for its computation in the context of linear regression.

The maximum-bias curve is not only useful on its own, but allows us to define further scalar measures of robustness. The most prominent is the *breakdown point* (Hampel, 1971), which is defined as the smallest amount ε of contamination that can cause an infinite bias:

$$\varepsilon^*(T) = \inf_{\varepsilon \geq 0} \{ \varepsilon : B(\varepsilon, T) = \infty \}. \qquad (6)$$

The intuitive aim of this definition specifies the breakdown point $\varepsilon^*(T)$ as the smallest amount of contamination that makes the estimator T useless. Note that in most cases $\varepsilon^*(T) \leq 0.5$ (He and Simpson, 1993). This definition and the upper bound, however, apply only in simple cases, such as location or linear regression estimation (Davies and Gather, 2005). The most general definition of breakdown point formalizes the idea of 'useless' estimates in the following way: an estimator is said to break down if, under contamination, it is not random anymore, or, more precisely, it can achieve only a finite set of values (Genton and Lucas, 2003). This definition is based on the fact that estimates are functions of observed random samples and are thus random quantities themselves unless they fail. Although the latter definition includes the first one, the latter one may generally depend on the underlying model F, for example in time-series context.

2 Estimation approaches
Denote by F_n an empirical distribution function (edf) corresponding to a sample $\{x_i\}_{i=1}^n \in \mathbb{R}$ drawn from a model based on probability distribution F. Most estimation methods can be defined as an extremum problem, minimizing a contrast $\int h(z, \theta)dF(z)$ over θ in a parameter space, or as a solution of an equation, $\int g(z, \theta)dF(z) = 0$ in θ. The estimation for a given sample utilizes finite-sample equivalents of these integrals, $\int h(z, \theta)dF_n(z)$ and $\int g(z, \theta)dF_n(z)$, respectively.

Consider the pure location model $X_i = \mu + \sigma\varepsilon_i, i = 1, \ldots, n$, with a known scale σ and $\varepsilon \sim F$. The cdf of X is then $F\{(x - \mu)/\sigma\}$. With a quadratic contrast function $h(x, \theta) = (x - \theta)^2$, the estimation problem is to minimize $\int (x - \theta)^2 dF\{(x - \mu)/\sigma\}$ with respect to θ. For known F, this leads to $\theta = \mu$ and one sees that, without loss of generality, one can assume $\mu = 0$ and $\sigma = 1$. For the sample $\{x_i\}_{i=1}^n$ characterized by edf F_n, the location parameter μ is estimated by

$$\hat{\mu} = \arg \min_{\theta} \int (x - \theta)^2 dF_n(x) = n^{-1} \sum_{i=1}^n x_i = \bar{x}.$$

Note that for $g(x, \theta) = x - \theta$, the parameter μ is the solution to $\int g(x, \theta)dF(x) = 0$. The estimator may therefore be alternatively defined through $\mu = T(F) = \int u dF(u)$.

As indicated in the introduction, this standard estimator of location performs unfortunately rather poorly under the sketched contamination model. Estimating a population mean by the least squares (LS) or sample mean $\bar{x} = T(F_n)$ has the following properties. First, the influence function (1)

$$IF(x; T, F) = \lim_{\varepsilon \to 0} \frac{T\{(1 - \varepsilon)F + \varepsilon \delta_x\} - T(F)}{\varepsilon}$$

$$= \lim_{\varepsilon \to 0} \frac{\{(1 - \varepsilon) \int u dF(u) + \varepsilon x\} - \int u dF(u)}{\varepsilon}$$

$$= \lim_{\varepsilon \to 0} \varepsilon^{-1} \left\{ -\varepsilon \int u dF(u) + \varepsilon x \right\}$$

$$= x - \int u dF(u) = x - T(F).$$

Hence, the gross-error sensitivity (2) $\gamma(T, F) = \infty$, the local-shift sensitivity (3) $\lambda(T, F) = 0$, and the rejection point (4) $\rho(T, F) = \infty$. Second, the maximum-bias (5) is infinite for any $\varepsilon > 0$ since

$$\sup_{x \in \mathbb{R}} \| T\{(1 - \varepsilon)F + \varepsilon \delta_x\} - T(F) \|$$

$$= \sup_{x \in \mathbb{R}} \| - \varepsilon T(F) + \varepsilon x \| = \infty.$$

Consequently, the breakdown point (6) of the sample mean $\bar{x} = T(F_n)$ is zero, $\varepsilon^*(T) = 0$.

Thus, none of robustness measures characterizing the change of T under contamination of data (even infinitesimally small) is finite. This behaviour, typical for LS-based methods, motivated alternative estimators that have the desirable robust properties. In this section, the M-estimators, S-estimators and τ-estimators are discussed as well as some extensions and combination of these approaches. Even though there is a much wider range of robust estimation principles, we focus on those already studied and adopted in various areas of econometrics.

2.1 M-estimators

To achieve more flexibility in accommodating requirements on robustness, Huber (1964) proposed the *M-estimator* by considering a general extremum estimator based on $\int \rho(z, \theta) dF(z)$, thus minimizing $\int \rho(z, \theta) dF_n(z)$ in finite samples. Providing that the first derivative $\psi(z, \theta) = \partial \rho(z, \theta)/\partial \theta$ exists, an M-estimator can be also defined by an implicit equation $\int \psi(z, \theta) dF_n(z) = 0$.

This extremely general definition is usually adapted to a specific estimation problem such as location, scale or regression estimation. In a univariate location model, $F(z)$ can be parameterized as $F(z - \theta)$ and hence one limits $\rho(z, \theta)$ and $\psi(z, \theta)$ to $\rho(z - \theta)$ and $\psi(z - \theta)$. In the case of scale estimation, $F(z) = F(z/\theta)$ and consequently $\rho(z, \theta) = \rho(z/\theta)$ and $\psi(z, \theta) = \psi(z/\theta)$. In linear regression, $z = (x, y)$ and a zero-mean error term $\varepsilon = y - x^\top \theta$. Analogously to the location case, one can then consider

Table 1   *Examples of ρ and ψ functions used with M-estimators*

	$\rho(t)$	$\psi(t)$
Least squares	t_2	$2t$
Least absolute deviation	$|t|$	Sign (t)
Quantile estimation	$\{\tau - I(x<0)\}x$	$\tau - I(x<0)$
Huber: for $|t| \leq c$	t^2	$2t$
for $c < |t|$	$c|t|$	c sign(t)
Hample: for $|t| \leq c$	t^2	$2t$
for $a < |t| \leq b$	$a|t|$	a sign(t)
for $b < |t| \leq c$	$\frac{ac}{c-b}t - \frac{a}{c-b}t^2$sign$(t)$	$a(c - |t|)/(c - b)$
for $c < |t|$	$a|t|$	0
Biweight (Tukey)	$-(c^2 - t^2)^3 I(|t| \leq c)/6$	$t(c^2 - t^2)^2 I(|t| \leq c)/6$
Sine (Andrews)	$-c \cos(x/c)I(|t| \leq \pi c)$	$\sin(x/c)I(|t| \leq \pi c)$

$\rho(z, \theta) = \rho(y - x^\top \theta)$ and $\psi(z, \theta) = \psi(y - x^\top \theta)x$, or more generally, $\rho(z, \theta) = \rho(y - x^\top \theta, x)$ and $\psi(z, \theta) = \psi(y - x^\top \theta, x)$ (GM-estimators). Generally, we can express $\rho(z, \theta)$ as $\rho\{\eta(z, \theta)\}$, $\psi\{\eta(z, \theta)\}$, where $\eta(z, \theta) \sim F$.

Some well-known choices of univariate objective functions ρ and ψ are given in Table 1; functions $\rho(t)$ are usually assumed to be non-constant, non-negative, even and continuously increasing in $|t|$. This documents flexibility of the concept of M-estimators, which include LS and quantile regression as special cases.

On the other hand, many of the ρ and ψ functions in Table 1 depend on one or more constants $a, b, c \in \mathbb{R}$. If an estimator T is to be invariant to the scale of data, one can apply the estimator to rescaled data, that is, to minimize $\int \rho\{(z - \theta)/s\} dF_n(z)$ or to solve $\int \psi\{(z - \theta)/s\} dF_n(z) = 0$ for a scale estimate s like the median absolute deviation (MAD). Alternatively, one may also estimate parameters θ and scale s simultaneously by considering $\rho(z, \{\theta, s\}) = \rho\{(z - \theta)/s\}$ or

$$\psi(z, \{\theta, s\}) = \{\psi_1(z, \{\theta, s\}), \psi_s(z, \{\theta, s\})\}.$$

Let us now turn to the question how the choice of functions ρ and ψ determines the robust properties of M-estimators. First, the influence function of an M-estimator can generally depend on several quantities such as its asymptotic variance or the position of explanatory variables in the regression case, but the influence function is always proportional to function $\psi(z, b)$. Thus, the finite gross-error sensitivity, $\gamma(T, F) < \infty$, requires bounded $\psi(t)$ (which is not the case with LS). Similarly, the finite rejection point, $\rho(T, F) < \infty$, leads to $\psi(t)$ being zero for all sufficiently large t (the M-estimators defined by such a ψ-function are called redescending). Hampel et al. (1986) shows how, for a given bound on $\gamma(T, F)$, one can determine the most efficient choice of ψ function (for example, the skipped median, $\psi(t) = sign(t)I(|t| < K), K > 0$, in the location case).

More formally, the optimality of M-estimators in the context of qualitative robustness can be studied by the *asymptotic relative efficiency* (ARE) of an estimator $\hat{\theta}^1$

relative to another estimator $\hat{\theta}^2$:

$$ARE(\hat{\theta}^1, \hat{\theta}^2) = \frac{\text{as. } \text{var}(\hat{\theta}^1)}{\text{as. } \text{var}(\hat{\theta}^2)}. \qquad (7)$$

For example, at the normal distribution with $\hat{\theta}^1$ and $\hat{\theta}^2$ being the least absolute deviation (LAD) and LS estimators, ARE equals $2/\pi \approx 0.64$. Under the Student cdf t_5, the ARE of the two estimators climbs up to ≈ 0.96. For Huber's M-estimator, we see that its limit cases are the median for $c \to 0$ and the mean for $c \to \infty$. At the normal distribution and for $c = 1.345$, we have ARE of about 0.95. This means that this M-estimator is almost as efficient as MLE, but does not lose so drastically in performance as the standard mean under contamination because of the bounded influence function.

Whereas the influence function of M-estimators is closely related to the choice of its objective function, the global robustness of M-estimators is in a certain sense independent of this choice. Maronna, Bustos and Yohai (1979) showed in linear regression that the breakdown point of M-estimators is bounded by 1/p, where p is the number of estimated parameters. As a remedy, several authors proposed one-step M-estimators that are defined, for example, as the first step of the iterative Newton– Raphson procedure, used to minimize $\int \rho(z, \theta) dF(z)$, starting from initial robust estimators $\hat{\theta}^0$ of parameters and $\hat{s}^0$ of scale (see Welsh and Ronchetti, 2002, for an overview). Possible initial estimators can be those discussed in subsections 2.2 and 3.3. For example for an M-estimator of location $\hat{\theta}$ defined by a function $\psi(x, \theta) = \psi(x - \theta)$, its one-step counterpart can be defined at sample $\{x_i\}_{i=1}^n$ by

$$\hat{\theta} = \hat{\theta}^0 + \hat{s}^0 \sum_{i=1}^n \psi\left(\frac{x_i - \hat{\theta}^0}{\hat{s}^0}\right) \Big/ \sum_{i=1}^n \psi'\left(\frac{x_i - \hat{\theta}^0}{\hat{s}^0}\right),$$

where $\hat{\theta}^0$ and $\hat{s}^0$ represent initial robust estimators of location and scale like the median and MAD, respectively. Such one-step estimators, under certain conditions on the initial estimators, preserve the breakdown point of the initial estimators, and at the same time have the same first-order asymptotic distribution as the original M-estimator (Simpson, Ruppert and Carroll, 1992; Welsh and Ronchetti, 2002). Further development of such ideas includes an adaptive choice of parameters of function ψ in the iterative step (Gervini and Yohai, 2002).

2.2 S-estimators

An alternative approach to M-estimators achieving high breakdown point (HBP) was proposed by Rousseeuw and Yohai (1984). The S-estimators are defined by minimization of a scale statistics $s^2(z, b) = s\{\eta(z, b)\}$ defined as the M-estimate of scale,

$$\int \rho[\eta(z, b)/s\{\eta(z, b)\}] dF_n(z) = K = \int \rho(t) dF(t),$$

at the model distribution F; the functions ρ and η are those defining M-estimators in subsection 2.1. More generally, one can define S-estimators by means of any scale-equivariant statistics s^2, that is, $s\{c\eta(z, b)\} = |c|s\{\eta(z, b)\}$. Under this more general definition, S-estimators include as special cases LS and LAD estimators. Further, they encompass several well-known robust methods including least median of squares (LMS) and least trimmed squares (LTS): whereas the first defines the scale statistics $s^2\{\eta(z,b)\}$ as the median of squared residuals $\eta(z,b)$, the latter uses the scale defined by the sum of h smallest residuals $\eta(z,b)$. In order to appreciate the difference to M-estimators, it is worth pausing for a moment and to present LMS, the most prominent representative of S-estimators, in the location case:

$$\arg\min_{\theta} \text{ med } \{(x_1 - \theta)^2, \ldots, (x_n - \theta)^2\}.$$

Due to its definition, the S-estimators have the same influence function as the M-estimator constructed from the same function ρ. Contrary to M-estimators, they can achieve the highest possible breakdown point $\varepsilon^* = 0.5$. For example, this is the case of LMS and LTS. For Gaussian data, the most efficient (in the sense of ARE (7)) among the S-estimators with $\varepsilon^* = 0.5$ is, however, the one corresponding to $K = 1.548$ and ρ being the Tukey biweight function (see Table 1). Given the HBP of S-estimators, their maximum-bias behaviour is of interest too. Although it depends on the function ρ and constant K (Berrendero and Zamar, 2001), Yohai and Zamar (1993) proved that LMS minimizes maximum bias among a large class of (residual admissible) estimators, which includes most robust methods.

An important shortcoming of HBP S-estimation is, however, its low ARE: under Gaussian data, efficiency relative to LS varies from zero per cent to 27 per cent. Thus, S-estimators are often used as initial estimators for other, more efficient methods. Nevertheless, if an S-estimator is not applied directly to sample observations, but rather to the set of all pairwise differences of sample observations, the resulting generalized S-estimator exhibits higher relative efficiency for Gaussian data, while preserving its robust properties (Croux, Rousseeuw and Hossjer, 1994; Stromberg, Hossjer and Hawkins, 2000).

2.3 τ-estimators

The S-estimators improve upon M-estimators in terms of their breakdown-point properties, but at the cost of low Gaussian efficiency. Although one-step M-estimators based on an initial S-estimate can remedy this deficiency to a large extent, their exact breakdown properties are not known. One alternative approach, proposed by Yohai and Zamar (1988), extends the principle of S-estimation in the following way. Assuming that ρ_1 and ρ_2 are nonnegative, even, and continuous functions, the M-estimate $s^2(z, \theta) = s^2\{\eta(z, \theta)\}$ of scale can be defined as in the case of S-estimation,

$$\int \rho_1[\eta(z,\theta)/s\{\eta(z,\theta)\}]dF_n(z) = K$$

$$= \int \rho_1(t)dF(t).$$

Next, the τ-estimate of scale is defined by

$$\tau^2(z,\theta) = s^2\{\eta(z,\theta)\}$$

$$\int \rho_2[\eta(z,\theta)/s\{\eta(z,\theta)\}]dF_n(z)$$

and the corresponding τ-*estimator* of parameters θ is then defined by minimizing the τ-estimate of scale, $\tau^2(z,\theta)$.

As a generalization of S-estimation, the τ-estimators include S-estimators as a special case for $\rho_1 = \rho_2$ because then $\tau^2(x,\theta) = \theta s^2(z,\theta)$. On the other hand, if $\rho_2(t) = t^2$, $\tau^2(z,\theta) = \int \eta^2(z,\theta)dF_n(z)$ is just the standard deviation of model residuals. Compared with S-estimators, the class of τ estimators can improve in terms of relative Gaussian efficiency because its breakdown point depends only on function ρ_1, whereas its asymptotic variance is a function of both ρ_1 and ρ_2. Thus, ρ_1 can be defined to achieve the breakdown point equal to 0.5 and ρ_2 consequently adjusted to reach a pre-specified relative efficiency for Gaussian data (for example, 95 per cent).

3 Methods of robust econometrics

The concepts and methods of robust estimation discussed in Sections 1 and 2 are typically proposed in the context of a simple location or linear regression models, on the assumption of independent, continuous and identically distributed random variables. This, however, rarely corresponds to assumptions typical for most econometric models. In this section, we therefore present an overview of developments and extensions of robust methods to various econometric models. As the M-estimators are closest to the commonly used LS and MLE, most of the extensions employ M-estimation. The HBP techniques are not that frequently found in the economics literature (Zaman, Rousseeuw and Orhan, 2001; Sapra, 2003) and are mostly applied only as a diagnostic tool.

In the rest of this section, robust estimation is first discussed in the context of models with discrete explanatory variables, models with time-dependent observations, and models involving multiple equations. Later, robust alternatives to general estimation principles, such as MLE and generalized method of moments (GMM), are discussed. Before doing so, let us mention that dangers of data contamination are not studied only from the theoretical point of view. There is a number of studies that check the presence of outliers in real data and their influence on estimation methods. For example, there is evidence of data contamination and its adverse effects on

LS and MLE in the case of macroeconomic time series (Balke and Fomby, 1994; Atkinson, Koopman and Shephard, 1997), in financial time series (Sakata and White, 1998; Franses, van Dijk and Lucas, 2004), marketing data (Franses, Kloek and Lucas, 1999), and many other areas. These adverse effects include biased estimates, masking of structural changes, and creating seemingly nonlinear structures, for instance.

3.1 Discrete variables

To achieve a HBP, many robust methods such as LMS often eliminate a large portion of observations from the calculation of their objective function. This can cause non-identification of parameters associated with categorical variables. For example, having data on income $\{y_i\}_{i=1}^n$ of men and women, where gender is indicated by $\{d_i\}_{i=1}^n \in \{0,1\}$, one can estimate the mean income of men and women by a simple regression model $y_i = a + bd_i$. If a HBP method such as LMS or LTS is used to estimate the model and it eliminates a large portion of observations from the calculation (for example, one half of them), the remaining data could easily contain only income of men or only income of women, and consequently the mean income of one of the groups could not be then identified. Even though this seems unlikely in our simple example, it becomes more pronounced as the number of discrete variables grows (see Hubert and Rousseeuw, 1997, for an example).

A common strategy employs a robust estimator with a HBP for a model with only continuous variables, and using this initial estimate, the model with all variables is estimated by an M-estimator. Such a combined procedure preserves the breakdown point of the HBP estimator: even though a misclassified values of categorical explanatory variables can bias the estimates, this bias will be bounded in common models as the categorical variables are bounded as well. See Hubert and Rousseeuw (1997) and Maronna and Yohai (2000), who combine an initial S-estimator with an M-estimator.

3.2 Time series

In time series, there are several issues not addressed by the standard theory of robust estimation because of time-dependency of observations. First, the asymptotic behaviour of various robust methods has to be established; see Koenker and Machado (1999) and Koenker and Xiao (2002) for L_1 regression; Künsch (1984) and Bai and Wu (1997) for M-estimators; and Sakata and White (2001), Zinde-Walsh (2002) and Čížek (2006) for various S-type estimators. In these cases, the results are usually established for general nonlinear models.

Second, the effects of data contamination are more complex and widespread due to time dependency: an error in one observation is transferred, by means of a model, to others close in time. The possible effects of outliers in time series are elaborated by Chen and Liu (1993) and Tsay, Pena and Pankratz (2000), for instance.

The first work also offers a sequential identification of outliers (an alternative procedure based on τ-estimators is offered by Bianco et al., 2001). Consequently, the robust properties in time series differ from those experienced in cross-sectional data. For example, the breakdown point is asymptotically zero in the case of M-estimators (Sakata and White, 1995) and can be much below 0.5 for various S-estimators (Genton and Lucas, 2003).

A further issue specific to time series is testing for stationarity of a series. Effects of outliers are in this respect similar to those of neglected structural changes. To differentiate between random outliers and real structural changes, robust tests for change-point detection have been proposed by Gagliardini, Trojani and Urga (2005) and Fiteni (2002; 2004); the last of these papers uses τ-estimation. The asymptotics of M-estimators under unit-root assumption and the corresponding tests have been established, for example, by Lucas (1995), Koenker and Xiao (2004), and Haldrup, Montans and Sanso (2005). An early reference is Franke, Härdle and Martin (1984).

3.3 Multivariate regression

An important application of robust methods in economics concerns the multivariate regression case. This is relatively straightforward with exogenous explanatory variables only, see Koenker and Portnoy (1990), Bilodeau and Duchesne (2000), and Lopuhaä (1992) for the M-, S- and τ-estimation, respectively. Estimating general simultaneous equations models has to mimic either three-stage LS or full information MLE (Marrona and Yohai, 1997). Whereas Koenker and Portnoy (1990) follow with the weighted LAD the first approach, Krishnakumar and Ronchetti (1997) use M-estimation together with the second strategy.

3.4 General estimation principles

There are naturally many more model classes for which one can construct robust estimation procedures. Since most econometric models can be estimated by means of MLE or GMM, it is however easier to concentrate on robust counterparts of these two estimation principles. There are other estimation concepts, such as nonparametric smoothing, that can employ robust estimation (Härdle, 1982), but they go beyond the scope of this article.

First, recent contributions to robust MLE can be split to two groups. One simply defines a weighted maximum likelihood, where weights are computed from an initial robust fit (Windham, 1995; Markartou, Basu and Lindsay, 1997). Alternatively, some erroneous observations can be excluded completely from the likelihood function (Clarke, 2000; Marazzi and Yohai, 2004). This approach requires existence of an initial robust estimate, and thus it is not useful for models for which there are no robust methods available. The second approach is motivated by the S-estimation, namely, LTS, and defines the maximum trimmed likelihood as an estimator maximizing the product of the h largest likelihood contribution, that is,

those corresponding only to h most likely observations (Hadi and Luceño, 1997). This estimator has been studied mainly in the context of generalized linear models (Müller and Neykov, 2003), but its consistency is established in a much wider class of models (Čížek, 2004).

Second, more widely used GMM has also attracted attention from its robustness point of view. A special case – instrumental variable estimation – has been studied, for example, by Wagenvoor and Waldman (2002) and Kim and Muller (2007). See also Chernozhukov and Hansen (2006) for instrumental variable quantile regression. More generally, Ronchetti and Trojani (2001) have proposed an M-estimation-based generalization of GMM, studied its robust properties, and designed corresponding tests. This work became a starting point for others, who have extended the methodology of Ronchetti and Trojani (2001) to robustify simulation-based methods of moments (Genton and Ronchetti, 2003; Ortelli and Trojani, 2005).

P. ČÍŽEK AND W. HÄRDLE

See also **adaptive estimation; categorical data; computational methods in econometrics; generalized method of moments estimation; maximum likelihood; measurement error models; time series analysis; two-stage least squares and the *k*-class estimator.**

This work was supported by the Deutsche Forschungsgemeinschaft through the SFB 649 'Economic Risk'.

Bibliography

Atkinson, A., Koopman, S. and Shephard, N. 1997. Detecting shocks: outliers and breaks in time series. *Journal of Econometrics* 80, 387–422.

Bai, Z.D. and Wu, Y. 1997. General M-Estimation. *Journal of Multivariate Analysis* 63, 119–35.

Balke, N. and Fomby, T. 1994. Large shocks, small shocks, and economic fluctuations: outliers in macroeconomic time series. *Journal of Applied Econometrics* 9, 181–200.

Berrendero, J. and Zamar, R. 2001. Maximum bias curves for robust regression with non-elliptical regressors. *Annals of Statistics* 29, 224–51.

Bianco, A., Ben, M., Martínez, E. and Yohai, V. 2001. Regression models with ARIMA errors. *Journal of Forecasting* 20, 565–79.

Bilodeau, M. and Duchesne, P. 2000. Robust estimation of the SUR model. *Canadian Journal of Statistics* 28, 277–88.

Chen, C. and Liu, L.-M. 1993. Joint estimation of model parameters and outlier effects in time series. *Journal of the American Statistical Association* 88, 284–97.

Chernozhukov, V. and Hansen, C. 2006. Instrumental quantile regression inference for structural and treatment effect models. *Journal of Econometrics* 132, 491–525.

Čížek, P. 2004. General trimmed estimation: robust approach to nonlinear and limited dependent variable models. Discussion Paper No. 2004/130, CentER, Tilburg University.

Čížek, P. 2006. Least trimmed squares in nonlinear regression under dependence. *Journal of Statistical Planning and Inference* 136, 3967–88.

Clarke, B. 2000. An adaptive method of estimation and outlier detection in regression applicable for small to moderate sample sizes. *Probability and Statistics* 20, 25–50.

Croux, C., Rousseeuw, P. and Hossjer, O. 1994. Generalized S-estimators. *Journal of the American Statistical Association* 89, 1271–81.

Davies, L. and Gather, U. 2005. Breakdown and groups. *Annals of Statistics* 33, 988–93.

Fiteni, I. 2002. Robust estimation of structural break points. *Econometric Theory* 18, 349–86.

Fiteni, I. 2004. τ-estimators of regression models with structural change of unknown location. *Journal of Econometrics* 119, 19–44.

Franses, P., Kloek, T. and Lucas, A. 1999. Outlier robust analysis of longrun marketing effects for weekly scanning data. *Journal of Econometrics* 89, 293–315.

Franses, P., van Dijk, D. and Lucas, A. 2004. Short patches of outliers, ARCH and volatility modelling. *Applied Financial Economics* 14, 221–31.

Franke, J., Härdle, W. and Martin, R. 1984. *Robust and Nonlinear Time Series Analysis*. Berlin: Springer.

Gagliardini, P., Trojani, F. and Urga, G. 2005. Robust GMM tests for structural breaks. *Journal of Econometrics* 129, 139–82.

Genton, M. and Lucas, A. 2003. Comprehensive definitions of breakdown-points for independent and dependent observations. *Journal of the Royal Statistical Society, Series B* 65, 81–94.

Genton, M. and Ronchetti, E. 2003. Robust indirect inference. *Journal of the American Statistical Association* 98, 67–76.

Gervini, D. and Yohai, V. 2002. A class of robust and fully efficient regression estimators. *Annals of Statistics* 30, 583–616.

Hadi, A. and Luceño, A. 1997. Maximum trimmed likelihood estimators: a unified approach, examples, and algorithms. *Computational Statistics & Data Analysis* 25, 251–72.

Haldrup, N., Montans, A. and Sanso, A. 2005. Measurement errors and outliers in seasonal unit root testing. *Journal of Econometrics* 127, 103–28.

Hampel, F. 1971. A general qualitative definition of robustness. *Annals of Mathematical Statistics* 42, 1887–96.

Hampel, F., Ronchetti, E., Rousseeuw, P. and Stahel, W. 1986. *Robust Statistics: The Approach Based on Influence Function*. New York: Wiley.

Härdle, W. 1982. Robust regression function estimation. *Journal of Multivariate Analysis* 14, 169–80.

He, X. and Simpson, D. 1993. Lower bounds for contamination bias: globally minimax versus locally linear estimation. *Annals of Statistics* 21, 314–37.

Huber, P. 1964. Robust estimation of a location parameter. *Annals of Mathematical Statistics* 35, 73–101.

Hubert, M. and Rousseeuw, P. 1997. Robust regression with both continuous and binary regressors. *Journal of Statistical Planning and Inference* 57, 153–63.

Kim, T.-H. and Muller, C. 2006. Two-stage Huber estimation. *Journal of Statistical Planning and Inference* 137, 405–18.

Koenker, R. and Machado, J. 1999. Goodness of fit and related inference processes for quantile regression. *Journal of the American Statistical Association* 94, 1296–310.

Koenker, R. and Portnoy, S. 1990. M-estimation of multivariate regressions. *Journal of the American Statistical Association* 85, 1060–8.

Koenker, R. and Xiao, Z. 2002. Inference on the quantile regression process. *Econometrica* 70, 1583–612.

Koenker, R. and Xiao, Z. 2004. Unit root quantile autoregression inference. *Journal of the American Statistical Association* 99, 775–87.

Krishnakumar, J. and Ronchetti, E. 1997. Robust estimators for simultaneous equations models. *Journal of Econometrics* 78, 295–314.

Kunsch, H. 1984. Infinitesimal robustness for autoregressive processes. *Annals of Statistics* 12, 843–63.

Lopuhaä, H. 1992. Multivariate τ-estimators. *Canadian Journal of Statistics* 19, 307–21.

Lucas, A. 1995. An outlier robust unit root test with an application to the extended Nelson–Plosser data. *Journal of Econometrics* 66, 153–73.

Marazzi, A. and Yohai, V. 2004. Adaptively truncated maximum likelihood regression with asymmetric errors. *Journal of Statistical Planning and Inference* 122, 271–91.

Markatou, M., Basu, A. and Lindsay, B. 1997. Weighted likelihood estimating equations: the discrete case with applications to logistic regression. *Journal of Statistical Planning and Inference* 57, 215–32.

Maronna, R., Bustos, O. and Yohai, V. 1979. Bias- and efficiency-robustness of general M-estimators for regression with random carriers. In *Smoothing Techniques for Curve Estimation*, ed. T. Gasser and M. Rossenblatt. Berlin: Springer.

Maronna, R. and Yohai, V. 1997. Robust estimation in simultaneous equations models. *Journal of Statistical Planning and Inference* 57, 233–44.

Maronna, R. and Yohai, V. 2000. Robust regression with both continuous and categorical predictors. *Journal of Statistical Planning and Inference* 89, 197–214.

Müller, C. and Neykov, N. 2003. Breakdown points of trimmed likelihood estimators and related estimators in generalized linear models. *Journal of Statistical Planning and Inference* 116, 503–19.

Ortelli, C. and Trojani, F. 2005. Robust efficient method of moments. *Journal of Econometrics* 128, 69–97.

Ronchetti, E. and Trojani, F. 2001. Robust inference with GMM estimators. *Journal of Econometrics* 101, 37–69.

Rousseeuw, P. and Yohai, V. 1984. Robust regression by means of S-estimators. In *Robust and Nonlinear Time*

Series Analysis, ed. J. Franke, W. Härdle and R. Martin. Heidelberg: Springer.

Sakata, S. and White, H. 1995. An alternative definition of finite-sample breakdown point with application to regression model estimators. *Journal of the American Statistical Association* 90, 1099–106.

Sakata, S. and White, H. 1998. High breakdown point conditional dispersion estimation with application to S&P 500 daily returns volatility. *Econometrica* 66, 529–67.

Sakata, S. and White, H. 2001. S-estimation of nonlinear regression models with dependent and heterogeneous observations. *Journal of Econometrics* 103, 5–72.

Sapra, S. 2003. High-breakdown point estimation of some regression models. *Applied Economics Letters* 10, 875–78.

Simpson, D., Ruppert, D. and Carroll, R. 1992. On one-step GM estimates and stability of inferences in linear regression. *Journal of the American Statistical Association* 87, 439–50.

Stromberg, A., Hossjer, O. and Hawkins, D. 2000. The least trimmed differences regression estimator and alternatives. *Journal of the American Statistical Association* 95, 853–64.

Tsay, R., Pena, D. and Pankratz, A. 2000. Outliers in multivariate time series. *Biometrika* 87, 789–804.

Wagenvoor, R. and Waldman, R. 2002. On B-robust instrumental variable estimation of the linear model with panel data. *Journal of Econometrics* 106, 297–324.

Welsh, A. and Ronchetti, E. 2002. A journey in single steps: robust one-step M-estimation in linear regression. *Journal of Statistical Planning and Inference* 103, 287–310.

Windham, M. 1995. Robustifying model fitting. *Journal of the Royal Statistical Society, Series B* 57, 599–609.

Yohai, V. and Zamar, R. 1988. High breakdown-point estimates of regression by means of the minimization of an efficient scale. *Journal of the American Statistical Association* 83, 406–13.

Yohai, V. and Zamar, R. 1993. A minimax-bias property of the least a-quantile estimates. *Annals of Statistics* 21, 1824–42.

Zaman, A., Rousseeuw, P. and Orhan, M. 2001. Econometric applications of high-breakdown robust regression techniques. *Economics Letters* 71, 1–8.

Zinde-Walsh, V. 2002. Asymptotic theory for some high breakdown point estimators. *Econometric Theory* 18, 1172–96.

Rogers, James Edwin Thorold (1823–1890)

Rogers was educated at King's College London, and Magdalen College, Oxford. From 1859 until his death he held the first Tooke Professorship of Statistics and Economic Science at King's College London. In 1862 he was elected Drummond Professor of Political Economy in the University of Oxford, a post he lost in 1868 largely because of his outspoken radical views, but to which he was re-elected in 1888. He was ordained but abandoned the clerical profession. From 1880 to 1886 he served as a rather inconspicuous member of the House of Commons.

His chief work is his monumental *History of Agriculture and Prices*, where he did much to turn economic history into the field of distribution and attempted to use more exact methods in economic historical investigations on a large scale. His work is marred by his casual deductions. He argued for a high standard of living of the English labourer during the Middle Ages and explained the subsequent deterioration by legislative interference by the landowners controlling the government.

Politically, he was greatly influenced by his friend and brother-in-law Richard Cobden. He was firmly opposed to extensive government intervention. He did however support trade unions as providing the remedy for nearly all social ills. His advocacy of laissez-faire separates him from the rest of the English Historical School, his allies in attacking theoretical economics in looking to economic history as a realistic foundation for the proper understanding and solution of contemporary social and economic problems.

O. KURER

Selected works

1884. *Six Centuries of Work and Wages: The History of English Labour.* London: Swan Sonnenschein.

1886–1902. *A History of Agriculture and Prices in England. From the Year After the Oxford Parliament (1259) to the Commencement of the Continental War (1793)*, 7 vols. Oxford: Clarendon.

1888. *The Economic Interpretation of History.* New York: Putnam.

1892. *The Industrial and Commercial History of England.* Ed. A.G.L. Rogers. New York: Putnam. Published posthumously.

Bibliography

Ashley, W.J. 1889. James E. Thorold Rogers. *Political Science Quarterly* 4, 381–407.

De Marchi, N.B. 1976. On the early dangers of being too political an economist: Thorold Rogers and the 1868 election to the Drummond Professorship. *Oxford Economic Papers* 28, 364–80.

Hewins, W.A.S. 1897. James Edwin Thorold Rogers. In *Dictionary of National Biography.* Oxford: Oxford University Press.

Wood, J.C. 1983. *British Economists and the Empire, 1860–1914.* Beckenham, Kent: Croom Helm.

Roos, Charles Frederick (1901–1958)

Born on 18 May 1901, in New Orleans, Roos completed his Ph.D. in mathematics at Rice Institute in 1926. Influenced directly by his supervisor Evans (1922; 1924; 1930) and indirectly by Volterra, his main interests in graduate work were the calculus of variations, integral equations, and applications of those areas of mathematics to problems in dynamic economics.

Although he published several brilliant articles (Roos, 1925; 1927a; 1927b; 1927c; 1928; 1930), Roos found no journal which would readily accept manuscripts in which he combined economics, mathematics and sometimes statistics at suitably advanced levels (cf. Roos, 1934, p. xiii). Spurred by similar frustrations, Frisch and Roos jointly took the initiative which led to creation of the Econometric Society in 1930 (of which Roos became President in 1948) and publication of its journal, *Econometrica*, from 1933 on.

In 1930 Roos set out to write a treatise on dynamic economics; he published an important book under that title in 1934. It was reviewed enthusiastically by Tintner (1936) and uncomprehendingly by Freeman (1935). *Dynamic Economics* (1934) is a brilliant combination of mathematical economic theory and applied econometrics. Roos's mathematical approach inspired Tintner to write a dozen articles on dynamic economic theory (for example, Tintner, 1938).

Roos held a series of administrative positions during 1931–7 and published a major book on *NRA Economic Planning* (1937). In 1938 he founded an econometric consulting firm in New York and directed it until his death. Examples of his later work are Roos and von Szeliski (1939a; 1939b) and Roos (1955; 1957). He died in Milwaukee on 7 January 1958.

Hotelling (1958) describes Roos as 'a unique and outstanding figure', while Davis (1958) presents a complete list of his writings.

KARL A. FOX

Selected works

1925. A mathematical theory of competition. *American Journal of Mathematics* 47(July), 163–75.

1927a. Dynamical economics. *Proceedings of the National Academy of Sciences* 13, 145–50.

1927b. A dynamical theory of economic equilibrium. *Proceedings of the National Academy of Sciences* 13, 280–85.

1927c. A dynamical theory of economics. *Journal of Political Economy* 35, 632–56.

1928. A mathematical theory of depreciation and replacement. *American Journal of Mathematics* 50, 147–57.

1930. A mathematical theory of price and production fluctuations and economic crises. *Journal of Political Economy* 38, 501–22.

1934. *Dynamic Economics: theoretical and statistical studies of demand, production and prices*. Cowles Commission Monograph No. 1. Bloomington, IN: Principia Press.

1937. *NRA Economic Planning*. Cowles Commission Monograph No. 2. Bloomington, IN: Principia Press.

1939a. (With V. von Szeliski.) *The Dynamics of Automobile Demand*. Detroit: General Motors Corporation.

1939b. (With V. von Szeliski.) The concept of demand and price elasticity; the dynamics of automobile demand. *Journal of the American Statistical Association* 34, 652–66.

1955. Survey of economic forecasting techniques. *Econometrica* 23, 363–95.

1957. *Dynamics of Economic Growth: the American Economy, 1957–1975*. New York: Econometric Institute.

Bibliography

Davis, H.T. 1958. Charles Frederick Roos. *Econometrica* 26(4), 580–89. Contains a complete bibliography of Roos's writings (91 items).

Evans, G.C. 1922. A simple theory of competition. *American Mathematical Monthly* 29(10), 371–80.

Evans, G.C. 1924. The dynamics of monopoly. *American Mathematical Monthly* 31(February), 77–83.

Evans, G.C. 1930. *Mathematical Introduction to Economics*. New York: McGraw-Hill.

Freeman, H.A. 1935. Review of C.F. Roos, *Dynamic Economics*. *American Economic Review* 25, 520.

Hotelling, H. 1958. C.F. Roos, econometrician and mathematician. *Science* 128, 1194–5.

Tintner, G. 1936. Review of Dynamic Economics. *Journal of Political Economy* 44, 404–9.

Tintner, G. 1938. The theoretical derivation of dynamic demand curves. *Econometrica* 6, 375–80.

Roscher, Wilhelm Georg Friedrich (1817–1894)

Roscher was born in Hannover into a well-established civil service family. He studied history and political science in Göttingen and Berlin. In 1840 he became lecturer in both subjects at Göttingen, in 1843 he was appointed extraordinary professor of political economy, and in the next year was promoted professor. In 1848 he transferred to Leipzig, where he taught for the rest of his life. Roscher had a Protestant background and was deeply religious, adhering to a rather 'primitive form of religious belief' (Max Weber, 1903–6).

Roscher may be considered as one of the most important German economists of his time. He was one of the founders and the leading exponent of the German 'older' Historical School. He did not develop any new theory: his main contribution to political economy lay in the field of method. He adhered to what he called the 'historical-physiological method', as opposed to the 'idealistic method' (1842; 1854–94, vol. 1, pp. 43–56). This inductive method intended to provide a description of the actual course of economic development and of real economic life. Thus, Roscher tried to analyse laws of economic development by comparing the history of different people and nations and showing analogies in stages of their development. The emphasis was on historical relativism: economic behaviour depended to a large extent on the specific national and historic conditions of the different people and nations. This implied that a nation had to be regarded as a whole, as an 'organic unity', and not as the mere sum of individuals.

This was opposed to what Roscher called the 'idealistic method', which intended to provide an ideal picture, logically derived from abstract principles, of the functioning of the economic system. An example of this was the classical economists' deduction of economic laws from a system of hypotheses. Although Roscher emphasized that in economic analysis there existed generally no definite causal relationships but reciprocal ones, he did not reject the existence of 'laws of motion' within economic life. However, these laws were distinct from laws of natural science in that they dealt with free human beings gifted with reason and hence with changing motives for action (1854–94, vol. 1, pp. 26–9). Roscher was closer to the theoretical system of the classics than the exponents of the 'younger' historical school. He tended to regard it as the appropriate system of analysis of the current stage of economic development. He only modified and supplemented it with a careful historical analysis, but he may still be regarded as being in the classical tradition.

The first volume of Roscher's main work, *System der Volkswirtschaft* (1854–94: 1854) still looked very much like a traditional textbook. It analysed essentially the same topics as the classical economists – production, distribution and prices. Roscher was already strongly influenced by supply and demand approaches, but still determined the exchange value of a commodity by its cost of production. His theory of rent was Ricardian and his thinking about population development followed Malthusian patterns. Differing from classical textbooks, Roscher supplemented the theoretical analysis with a historical description – the reader finds the history of rent, interest and wages, of population development, of the prices of necessary and luxury commodities, and of luxury in general. Roscher accepted the classical notion of individual self-interest as a central axiom of modern economic behaviour, but he did not follow the classical patterns in deriving his economic principles from this assumption. As a result of his religious beliefs, he included human conscience as a regulating mechanism into his analysis of the role of self-interest (1854–94, vol. 1, pp. 20–3).

The other four volumes of the *System der Volkswirtschaft* (1859; 1881; 1886; 1894), which may be perceived as his main contribution to applied economics, were even more historically oriented and focused on agriculture, trade and industry, public finance, social policy and poor relief.

Roscher classified economic development into stages of maturity. The economic factors that govern the development of nations were land, labour and capital which subsequently dominated the different stages (1861, ch. 1). Later, Roscher presented a more detailed analysis of stages of political and societal development (1892) on the

basis of a classification of the different forms of government during history: early patriarchal kingdom, aristocracy of knights and priests, absolute monarchy, democracy. The latter then degenerated into a plutocracy, which is followed by a military dictatorship Roscher called 'Caesarismus'. Roscher did not systematically attempt an integration of his theory of political development and the stages of economic evolution.

He wrote several contributions on the history of economic thought. His compendium on the history of political economy in Germany (1874) was his most outstanding work and has remained important. Roscher may be regarded as the most eminent historian of cameralism and early German political economy. His treatise on economic problems of the location of large towns (1871) was an original contribution to economic theory.

Roscher supported German imperialism. In order to secure raw materials and markets for German goods, as well as to relieve the national labour market and prevent social unrest, he advocated an expansive German colonial policy, especially towards Eastern Asia, where he saw Germany's colonial future (1885). He was a conservative but he remained all his life unaffiliated to any political party or group.

B. SCHEFOLD

See also **Historical School, German.**

Selected works

1842. *Leben, Werk und Zeitalter des Thukydides.* Göttingen.
1854–94. *System der Volkswirtschaft.* Stuttgart: Cotta. Vol. 1: *Die Grundlagen der Nationalökonomie*, 1854. Trans. from 13th edn by J.J. Lalor as *Principles of Political Economy*, 2 vols, New York, 1878. Vol. 2: *Nationalökonomik des Ackerbaues und der verwandten Urproduktionen*, 1859. Vol. 3: *Nationalökonomik des Handels und Gewerbefleißes*, 1881. Vol. 4: *System der Finanzwissenschaft*, 1886. Vol. 5: *System der Armenpflege und der Armenpolitik*, 1894.
1861. *Ansichten der Volkswirtschaft aus dem geschichtlichen Standpunkt.* Leipzig and Heidelberg: Winter.
1871. *Betrachtungen über die geographische Lage der grossen Städte.* Leipzig.
1874. *Geschichte der Nationalökonomik in Deutschland.* Munich: Oldenbourg.
1885. *Kolonien, Kolonialpolitik und Auswanderung.* Leipzig: Winter Part II, ch. 1. Translated from 3rd edn by E.H. Baldwin and E.G. Bourne as *The Spanish Colonial System*, New York, 1904.
1892. *Politik: Geschichtliche Naturlehre der Monarchie, Aristokratie und Demokratie.* Stuttgart.

Bibliography

Cunningham, W. 1894–5. Why had Roscher so little influence in England? *Annals of the American Academy of Political and Social Sciences* 5, 317–34.

Weber, M. 1903–6. *Roscher und Knies und die logischen Probleme der Historischen Nationalökonomie.* In *Gesammelte Aufsätze zur Wissenschaftslehre.* Tübingen: Mohr, 1922.

Rosen, Sherwin (1938–2001)

Rosen was born in Chicago on 29 September 1938. He died in Chicago on 17 March 2001. He earned his BS in economics from Purdue University in 1960. He obtained his graduate economics degrees from the University of Chicago: his MA in 1962 and his Ph.D. in 1966. His first appointment was as assistant professor of economics at the University of Rochester in 1964. Promoted to associate professor in 1968 and full professor in 1970, he became the Kenan Professor of Economics in 1975. Rosen returned to the University of Chicago in 1977, and became the Bergman Professor of Economics in 1983. From 1992 until his death he served as the Edwin A. and Betty L. Bergman Distinguished Service Professor. In addition, he served as department chairman during 1988–94. He was a Senior Research Associate of the National Bureau of Economic Research from 1968 and a Senior Research Fellow of the Hoover Institution during 1983–96 before becoming a Senior Fellow in 1997.

Rosen was elected a fellow of the Econometric Society in 1976, and a fellow of the American Academy of Arts and Sciences in 1984. He became a member of the National Academy of Sciences in 1998. He was President of the Midwest Economic Association during 1996–7, President of the Society of Labor Economists in 2000, and President of the American Economics Association for 2001.

Rosen was a prolific scholar and one of the leading economists of his generation. His contributions spanned many fields, including equilibrium theory, human capital theory, income distribution theory and investment theory. His research provided the theoretical underpinnings of labour economics, urban economics and health economics. A unifying aim of his research is to explain differential market outcomes. Price differences of goods can be explained by their differential amounts of characteristics. These price differences could be driven by differences in preferences arising from wealth differences or differences in technologies available to firms. For example, cars sell for different prices because they contain different attributes, and workers earn different amounts across jobs because the jobs have different characteristics. Earnings may differ if workers differ in their human capital. Life-cycle earnings are explained from the characterization of human capital accumulation beyond formal schooling. Returns to higher education are best modelled as arising from revealed preference of workers, both college-educated and non-college educated. Earnings may differ between identical workers because they are in different job classifications. Small differences in worker productivity can manifest themselves in large earnings differences if there are production scale economies (creating superstars, for example), strong complementarities, or increasing returns in skill use. Finally, differences in returns and investments can arise from predictable future demand shifts or unpredicted contemporaneous demand shocks. Rational investment cycles are likely if investment is small relative to the stock of capital, and if the seed capital is a large proportion of the stock of capital.

Rosen was the author of two books, *A Disequilibrium Model of Demand for Factors of Production* (Nadiri and Rosen, 1973) and *Markets and Diversity* (2005), and editor of three collections: *Studies in Labor Markets* (1981), *Organizations and Institutions: Sociological and Economic Approaches to the Analysis of Social Structure* (Rosen and Winship, 1988), and *Implicit Contract Theory* (1994).

Equilibrium theory

Rosen's 1974 article 'Hedonic prices and implicit markets: product differentiation in pure competition' is the quintessential example of his work in equilibrium theory. Consider the following labour market application. Rosen's analysis allows for a job to be characterized by N dimensions, but for clarity we focus on only two, its wage and its dirtiness. Some jobs are dirtier than others. They provide meaner working conditions including unheated and/or non-air-conditioned workplaces, less pleasant coworkers, few or no promotion possibilities, high unemployment risk, large variability in hours demanded by the employer, inflexible hours of work, fewer vacations, worse fringe benefits like poor or no health insurance or disability insurance, poor pensions, and so on. Consider aggregating all of these features into a single measure called 'dirt'. A worker likes wages and dislikes dirt. Employers can offer any combination of wages and dirt as a package to prospective workers. Assume that (*a*) worker preferences are convex; hence a worker's dislike for dirt increases with the level of dirt on the job, and he or she requires ever larger increases in wages to accept an additional unit of dirt as the level of dirt increases; and (*b*) firm production technologies are convex; firms require greater wage reductions for a unit reduction in dirt as the job becomes less dirty. There are three extreme cases. First, if all workers are identical in wealth and preferences, and all employers have access to the same technology of production, then there is a single equilibrium point. This occurs at the tangency of the representative worker's iso-utility curve and the representative employer's iso-profit curve. With free entry, competition drives the equilibrium to the zero profit iso-profit curve. In the second case, suppose all workers are identical in wealth and preferences, but employers have different technologies. For example, mining firms find it more costly than software design firms to provide cleaner work environments. In equilibrium the economist will observe a locus of points, which traces out the representative worker's iso-utility curve, and each observed increase in dirt is associated exactly with the worker's compensating differential to accept the increased dirtiness. Finally, suppose workers have different preferences, say because of wealth differences. Assume that dirt is an inferior good. Let all firms have access to a single technology. In equilibrium the economist observes a locus of points, which traces out the representative firm's zero iso-profit curve. The second example identifies preferences, the third example identifies technology. Of course, the world is not so stark or clean for an economist. Preferences are heterogeneous, workers have differing skill levels, firms have different technologies. Thus, econometrically the problem is one of finding controls that allow for identification (see Ekeland, Heckman and Nasheim, 2004). One important application to the labour market is Murphy (a Rosen student) and Topel (1987).

Rosen's (1974) paper serves as the benchmark for thinking about how markets link customers of multiple characteristic goods and services with the suppliers of these complex goods and services. One important application of this model is by Roback (1982), a Rosen student. Her model examines the compensating differentials in worker wages and land rents arising from differences in location-specific amenities, say, climate or population density. Another application of this hedonic approach is the examination of the increased wages that firms pay to workers in order to induce them to accept greater risks to their health or, in particular, their lives. Rosen and Thaler (1976) allow variation in earnings due to variation in on-the-job risks to life, controlling for productivity (schooling and experience) and in other job characteristics to identify the reservation price of mortality risk for the typical worker. This allows for the calculation of the economic value of a life. Rosen (1988) revisits this arena by examining the valuation placed on increasing longevity. These two papers served as inspiration for an entire sub-field of health economics, highlighted by Murphy and Topel (2003).

Rosen (1978) examines the assignment solution of workers to tasks within an organization, in a world with a fixed number of inputs and many worker types. Rosen shows that the division of labour corresponding to the optimum assignment determines the marginal rates of substitution between worker types or between job categories. Thus the division of labour determines the extent of product and factor market substitutions in the economy. This paper provides an application of economics to the optimal determination of job types, or the efficient bundling of activities into a job. Rosen (1982b) extends the analysis. It is further generalized in Tamura (1992); with a continuum of intermediate tasks, and N different worker types, each of measure 1, output can be shown to come from the following reduced form:

$$Y = \left\{ \sum_{i=1}^{N} h_i^{\rho} \right\}^{\frac{1}{\rho}} \tag{1}$$

where type i workers have h_i units of human capital and $0 < \rho < 1$. With each individual a set of measure zero, each worker is paid the marginal product of his or her human capital and, given the constant returns to scale in the distribution of human capital, output is completely exhausted. However, since $\rho < 1$, there is an agglomeration economy in participation. Earnings for an individual of type j are the product of the marginal product of human capital of type j workers, w_j, and the amount of human capital of type j workers, h_j, or:

$$y_j = w_j h_j = \left\{ \sum_{i=1}^{N} h_i^{\rho} \right\}^{\frac{1-\rho}{\rho}} h_j^{\rho-1} h_j = \left\{ \sum_{i=1}^{N} h_i^{\rho} \right\}^{\frac{1-\rho}{\rho}} h_j^{\rho} \tag{2}$$

Assume that the human capital of worker type j grows at rate λ_j. Suppose workers of type j have more human capital than workers of type i. If the growth rates of human capital differ across type, then the relative earnings of these two worker types will change. In particular, notice:

$$\frac{y_{jt+1}}{y_{it+1}} = \left(\frac{\lambda_j h_{jt}}{\lambda_i h_{it}} \right)^{\rho} = \left(\frac{\lambda_j}{\lambda_i} \right)^{\rho} \frac{y_{jt}}{y_{it}} \tag{3}$$

Thus earnings become more (less) unequal if $\lambda_j > (<) \lambda_i$. Nothing about differences in firm investments in technical change is required, merely differences in the abilities of workers to continue to accumulate human capital. As Rosen (1972a; 1972b) notes, higher education can help individuals become better learners. Thus rising wage inequality can be the result of rising task specialization of the more skilled. Hence the works of Acemoglu (1998; 2002) can be thought of as arising from underlying primitives of differential worker abilities to learn.

Rosen also made fundamental contributions with Li, Mussa and Suen. Mussa and Rosen (1978) provide an equilibrium analysis of the product quality choice of monopolists. Li and Rosen (1998) examine the effect of breaches of contracts, unravelling, on optimal assignments of workers to firms when worker quality is uncertain. Li, Rosen and Suen (2001) examine the role of committees in information aggregation. If individuals have idiosyncratic information, committees help to aggregate the information. However, if committee members have conflicting preferences, the only equilibrium truth-telling rules are binary: yes or no, promote or do not promote, hire or not hire, keep or fire.

Human capital theory

Rosen applied his characteristics approach in order to make fundamental contributions to human capital theory. Rosen (1972a) models jobs as producing both output and learning opportunities for workers. Jobs differ in their learning opportunities. These opportunities are costly to firms; they produce less market output in return for producing more skills for workers in the future. With identical workers and many firms in equilibrium, young workers seek out the firms with the best learning opportunities. Workers accept lower earnings to pay the firm for the learning opportunities associated with their job. As they gain experience and skill, but have fewer years of work remaining, they switch to jobs offering less rapid learning and greater production. The theory produces occupational switching and the typical age-earnings profile. With heterogeneity in ability, the model is capable of producing a distribution of outcomes by age. The most able learners choose jobs with the most rapid learning possibilities, while less able learners choose to forgo those jobs entirely. Rosen (1972b) displays an early grasp of dynamic programming. He formulates the optimal accumulation of knowledge from learning by producing by analysing the excess valuation of production over and above current profits for the acquisition of higher future profits. He formulates a model of optimal knowledge accumulation as an explanation for technological progress. Curiously, Rosen notes that a stationary solution to an infinite horizon problem is not possible. However modern endogenous growth models in fact do take his first functional form in the paper. As long as output grows at a constant rate, knowledge growth will continue at a constant rate. Rosen presages Romer (1986) and Lucas (1988) by arguing that knowledge creation is likely to have important spillovers across workers and industries.

In a contribution to a Feschrift volume for his advisor H. Gregg Lewis, on the occasion of his retirement from the University of Chicago, Rosen (1976) applies a novel twist on the problem of life-cycle earnings. He considers the standard formulation of time t wealth value of human capital:

$$W(t) = \int_{t}^{N} y(s)e^{-r(s-t)}ds \qquad (4)$$

where N is an exogenously specified retirement age, $y(s)$ is the earnings at age s, and the individual faces a constant interest rate, r. Differentiating (4) with respect to t and rearranging produces:

$$rW(t) - \dot{W}(t) = y(t) \qquad (5)$$

The standard interpretation is to consider the first term as the potential earnings, the second term as the dollar cost of human capital investment and the right-hand side as observed earnings. In his words (1976, p. S46), 'The method adopted here is to go behind the scenes of (5) and use the theory to parameterize $y(t)$ directly in the form of restrictions on the unobservable $W(t)$.'

Rosen considers the accumulation of human capital as a self-directed process as in Ben-Porath (1967); $y = Rk(1-s)$, where s is the proportion of knowledge

spent on learning. Rosen considers two possible tractable formulations on the earnings generating function: (a) the learning function depends on total capital resources spent on accumulation, $\dot{k} = h(sk)$, which produces $y = Rk - h^{-1}(\dot{k})$, and (b) the learning function is linear in the stock of knowledge, $\dot{k} = h(s)k$, which produces $y = Rk[1 - h^{-1}(\frac{\dot{k}}{k})]$. Rosen chooses the latter functional form. The reader familiar with endogenous growth models will immediately see that his preferred specification is the Ak model of Jones and Manuelli (1990), Rebelo (1991) and Lucas (1988). Rosen also assumes that children are born with a fixed fraction of their parents' capital, such that at age 0, $k^{t+1}(0) = \gamma k^{t}(0) > k^{t}(0)$; thus he formulates Lucas (1988) without the human capital externalities, but with perpetual growth! Despite the difficulty imposed by finite time horizon models, Rosen derives closed form solutions for the optimal rate of human capital accumulation, $s(t)$, as well as for $k(t)$ and $y(t)$. Unfortunately, economists appear to have misgivings about working with hyperbolic sines, cosines and cotangents! Rosen (1976) produces the standard life-cycle shapes of observed earnings, potential earnings and human capital investments. From his explicit analytic solutions, Rosen is able to estimate the structural parameters of this model using census data. His estimates are broadly consistent with empirical results on rates of returns to schooling. More interestingly, he conducts counterfactual experiments about the nature of college. Schooling could be purely vocational in substance, increasing the knowledge of the future worker. It could also make the individual permanently more productive at future learning. Rosen posits different pairs of learning efficiencies and initial knowledge immediately after college completion that make the college graduate indifferent to college or work after high school graduation. He conducts the same counterfactual for high school graduates. Clearly, this thought experiment is one that foreshadows his seminal work with Robert Willis.

Willis and Rosen (1979) present a version of the Roy (1951) model for educational choice. Individuals can choose between stopping after high school graduation and continuing on to college. In their model there exists comparative advantage. Revealed preference implies that those workers that stopped after high school chose optimally to ignore college because their own rate of return to college education would be less than their cost of funds. Revealed preferences of college graduates imply the opposite. Now, if in addition high school graduates have an absolute advantage in high school occupations relative to what college graduates could earn in those jobs as high school graduates, estimated rates of return to college would be biased. The estimated rate of return to college would be below the true return to the college graduate, but more than the prospective return to the high school graduate. This revealed preference of educational-occupational selection indicates that there might not exist much ability bias in estimated rates of return to college. After nearly 30 years of empirical work, this is the dominant view in the economics profession (see Ashenfelter and Krueger, 1994; Ashenfelter and Zimmerman, 1997; Ashenfelter and Rouse, 1998).

Rosen (1983) identifies an increasing returns feature to human capital. The key point is that the marginal cost of creating human capital is independent of the intensity of use of human capital. That is to say, human capital investment is like a sunk cost. Once acquired, the marginal cost of using human capital is zero. The more intensely an individual uses his or her human capital, the greater the return to the human capital. Identical workers have an incentive to specialize their human capital or to endogenously choose their comparative advantage. This endogenous comparative advantage is a further extension of Willis and Rosen (1979). Thus, more specialized workers spend their careers in large markets that more fully utilize their skills. The largest metropolitan areas will be home to the most specialized human capital, in any field. Medical specialists will agglomerate in large metro areas and not smaller cities or rural areas; see Baumgardner (1988a; 1988b), a Rosen student. The increasing returns to utilization and endogenous comparative advantage model he envisions are explored in Tamura (1992; 1996; 2002; 2006).

Income distribution theory

Rosen made seminal contributions to understanding the functional distribution of earnings in the economy. Underlying his work is a search for the answer to the fundamental question: 'Why are earnings so skewed?' Furthermore, his work operates under the constraint that the answer should arise from a minimal amount of heterogeneity in underlying individual talent. Ideally, *ex ante* identical individuals would produce the observed skewed earnings distribution. One can view Rosen's work in human capital theory specifically as producing answers to this question with close to this ideal assumption of identical initial human capital endowments. In addition to his human capital programme, his research in this category includes Lazear and Rosen (1981) and his solo authored works (1981; 1982a; 1986a; 1997a). In Lazear and Rosen (1981) and Rosen (1986a), workers are paid as a result of internal relative comparisons. Assume that worker effort is not observable. If individual worker productivity is measured with noise, but a large proportion of that noise is common for all workers of the firm or for workers at similar levels within the firm, the use of relative productivity in order to determine compensation is efficient. This is because, by using relative comparisons, the effects of the unmeasured noise tend to be eliminated or greatly mitigated. For all workers, the wage bill must equal the value produced by the workers. However, workers are paid in relation to their place in the tournament, and hence paid in line with their job title. Increasing the spread between job levels or ranks raises the effort level of workers in the

tournament. The larger the total wage bill, with the number of workers at the firm held constant, the greater is the average effort level, and the greater the average ability of workers. In noisier environments, the spread between winning and losing workers must be greater than the earnings spread in more predictable environments. This larger spread is required because increasing noise dissipates the return to worker effort. Thus, in order to elicit the same level of effort, noisier industries must have greater earnings disparity.

Lazear and Rosen (1981) deals with a single contest. However, internal hierarchies are tournaments with many rounds. As a worker successfully progresses up the job ladder, there are fewer and fewer rounds left to play. In order to maintain the efficient level of effort, the prize gap must increase. Hence the gap between the CEO and the second in command of the firm must be larger than the gap between the second in command and his or her direct subordinates. Even if the CEO is only marginally better than the second in command, this larger prize must be given in order to provide the correct incentives throughout the organization. The pay gap serves to motivate not only the CEO but all workers in the internal hierarchy, and especially those close to the CEO in rank. As Lazear (2003, p. 13) notes, '(t)he theory helps explain why there is a larger spread in earnings between the top and bottom in new industries than in old ones.' As a consequence newer industries often pay workers in stocks or stock options in order to enlist greater effort levels. When the winners of these new companies in new industries are anointed, the stock options induce huge pay differentials within and across firms in this industry. Rosen (1986a) also shows that the single elimination tournament among players with heterogeneous talents is more likely to be the efficient tournament design than a round robin format. It promotes survival of the fittest at a more rapid rate. Rosen (1986a) shows this in an environment of 'symmetric ignorance', or the 'veil of ignorance', in which all players know only the common distribution from which all players' talents are drawn, including their own. Through Bayesian updating, survival in each round provides information about the ability of the contestant. These papers were among the first to apply game theory to labour economics. Furthermore, in the conclusion, Rosen (1986a) identifies the interesting area of further research, namely, the effects of player optimism or pessimism. His conjectures again presage the seminal works of Benabou and Tirole (2002; 2003; 2004) on micro models of behavioural attitudes.

On the question of skewed income distribution from small initial differences, Rosen (1981) shows that markets where costs of reproduction are trivial overwhelmingly choose to reward the individual who is perceived to be the best, even when the best is only trivially better than the second-best performer. Hence the entertainment industry with low-cost reproduction of movie prints greatly increases the skewness of the earnings distributions of actors, producers and directors in comparison with the earnings distributions of these same labour inputs in the days of the travelling show, or the Broadway theatre. Adding books, LPs, video tapes, CDs, DVDs, and so on continues to lower the cost of 'owning' a performance. Hence, an individual perceived by the market to be the best will harvest the overwhelming bulk of the demand. The individual performance is captured or recorded once, and then can be replicated at near zero marginal cost. An additional example is the falling costs of journal publication, producing rising skewness in the earnings distribution in academics.

This research leads directly to Rosen (1982a). The CEO can supply the same effort level working for a family firm with $1 million of revenue or a publicly traded firm with $100 billion of revenue. The marginal return to talent, however, greatly varies between the two. Hence, those workers with the lowest disutility of effort, or the greatest productivity of effort, will be more valuable working for organizations with greater sales. Essentially, managers are distributing their efforts across a greater scope of inputs, just as superstar performers spread their efforts to ever larger groups of customers. Once again, marginally better managers will earn significantly more than slightly less able managers because they work with a much larger scale of complementary inputs.

Rosen's (1997a) presidential address to the Society of Labor Economists shows that there is an endogenous reason for income inequality among *ex ante* identical workers. His model relies on non-convexity of preferences. These can arise from a variety of primitives: Friedman (1953) provides one; Bergstrom (1986) utilizes state dependent utility functions; Becker, Murphy and Werning (2005) use status; Becker, Murphy and Tamura (1990) and Tamura (1994) produce one with non-convexities in human capital. With these assumptions, Rosen demonstrates that the equilibrium and efficient outcome includes occupation lotteries or specialized investment in order to convexify the non-convex portion of utility. The winners get to enjoy higher utility, and the losers enter a lower level of utility. *Ex ante*, individuals are better off for entering into the lottery.

Investment theory

Another area that receives considerable attention from Rosen is investment theory. Rosen applies his customary analytical insights to understanding the dynamics of investment, particularly in areas where the 'time to build' aspect is significant (see, for example, Kydland and Prescott, 1982). This occurs in Rosen (1983), where the costs of acquiring human capital are separable from the rate of intensity of use. Furthermore, Rosen (1987) focuses on the role that investment in anticipation of future demand plays in price and quantity dynamics. These issues are explored in more detail in Rosen and Topel (1988), who produce a rational explanation for boom–bust cycles,

observed in the hog market and elsewhere. With a rising supply price, rational individuals build in anticipation of demand. When investment is a small fraction of the stock of the durable good, anticipated future demand shocks produce contemporaneous price changes. If an anticipated large permanent increase in demand will occur five years into the future, then investment will occur today. The immediate rise in investment and the slow shifting out of supply leads to a reduction in current rentals to the service flow. Investment continues rising because the value of the durable good continues to rise as the number of periods before the permanent demand shift shrinks. Until the demand shock appears, rental rates continue to fall; this is the bust phase of the cycle. When the demand shift arrives, rental rates jump, but less than they would have with no anticipatory investment; this is the boom phase of the cycle. Rosen and Topel produce a bust-boom cycle that is created because of the anticipated nature of the demand shock. Unanticipated shocks would produce even more dramatic changes in the rental rate of the durable, but no boom–bust characteristics. These insights are evident in Rosen (1992) and, from two dissertations he supervised, Siow (1984) and Zarkin (1985).

One might ask when known future demand shocks would arise. Two examples are the baby boom and Disney World. The baby boom, starting in 1946 and continuing through 1964, produced above-trend rates of fertility in American women. It is known that by age six a child must be enrolled in school. Hence, with generally little uncertainty, college students in the mid-1940s would have foreseen an increase in the demand for primary school teachers starting in 1951, for secondary school teachers in 1959, and for college faculty in the 1960s. In the second example, Walt Disney announced the construction of Disney World in Orlando, Florida, in the mid-1960s. It opened to the public only in 1970; but the model predicts increased construction of hotels, housing, schools, shopping areas, and so on in anticipation of the future increase in population. Examples of unexpected shocks would be the space race induced by Sputnik, the Soviet satellite, in the late 1950s, the space-science bust of the late 1960s, and the unexpected end of the baby boom.

Murphy, Rosen and Scheinkman (1994) apply the dynamic model of investment to the cattle industry. The long gestation cycle of cows (eight months) and their relatively short reproductive life (eight to ten years) implies that the breeding stock is likely to be a very large portion of the overall cattle herd. Thus, demand shocks are likely to greatly affect the breeding stock and hence the industry's ability to respond to future demand shocks. The authors show that their model does an excellent job of fitting the data from 1875 to 1990, despite the change in technology arising from corn feeding as opposed to range feeding, introduced in the 1930s and 1940s, which halved the time of the beef production cycle.

Rosen (1999) re-examines the Irish potato famine. He disproves the idea that potatoes were a Giffen good. As in the cattle industry, seed potatoes are a large proportion of the crop. Rosen argues that rational expectations of Irish potato farmers, who assumed that the potato blight was a temporary and not a permanent productivity shock, sealed their doom, since they did not consume their seed stock. When the blight turned out to be permanent, their exposure to imminent starvation was *ex post* predictable and tragic.

Conclusion

A measure of Rosen's influence on the economics profession can be seen by the number of published academic tributes to him (see, for example, Hartog, 2002; Lazear, 2003; Sanderson, 2001). In addition to his fertile research, Rosen possessed a great talent for synthesis, not only in his own work, as testified by Lazear (2003), but in entire fields. This is evident by his seminal contributions in this regard for human capital theory in 'Human capital: a survey of empirical research' (1977), 'Implicit contracts: a survey' (1985), 'The theory of equalizing differences' (1986b), 'Public employment taxes and the welfare state in Sweden' (1997b) and 'Theories of the distribution of earnings' (Neal and Rosen, 2000). Rosen was influential in much of Lazear's work (1995) on personnel economics.

Sherwin Rosen married Sharon Girsburg from Chicago. They were the embodiment of the marriage covenant, a beacon to all who knew them. They shared their love for 40 years, and had two daughters, Jennifer and Adria. Sherwin Rosen was a beloved professor at the University of Rochester and the University of Chicago. He was treasured by his colleagues and affectionately admired by graduate students. His concern for the success of his junior colleagues and of graduate students was legendary. His keen insight lit the seminars and classes, and his infectious laughter filled the hearts of his colleagues and graduate students. His work continues to illuminate the way for the economics profession, and his memory inspires and warms his former colleagues and students.

ROBERT TAMURA

See also **compensating differentials; hedonic prices; human capital; personnel economics; Roy model; superstars, economics of; value of life.**

Selected works

1972a. Learning and experience in the labour market. *Journal of Human Resources* 7, 326–42.

1972b. Learning by experience as joint production. *Quarterly Journal of Economics* 86, 366–82.

1973. (With M. Nadiri.) *A Disequilibrium Model of Demand for Factors of Production*. New York: NBER, Columbia University Press.

1974. Hedonic prices and implicit markets: product differentiation in pure competition. *Journal of Political Economy* 82, 34–55.

1976. A theory of life earnings. *Journal of Political Economy* 84, S45–68.

1976. (With R. Thaler.) The value of saving a life: evidence from the labor market. In *Household Production and Consumption: NBER Studies in Income and Wealth*, vol. 40, ed. N. Terleckyj. New York: Columbia University Press.

1977. Human capital: a survey of empirical research. In *Research in Labor Economics*, vol. 1, ed. R. Ehrenberg. Amsterdam: North-Holland.

1978. Substitution and the division of labor. *Economica* 45, 235–50.

1978. (With M. Mussa.) Monopoly and product quality. *Journal of Economic Theory* 18, 301–17.

1979. (With R. Willis.) Education and self selection. *Journal of Political Economy* 87, S7–36.

1981. The economics of superstars. *American Economic Review* 71, 845–58.

1981. ed. *Studies in Labor Markets*. Chicago: NBER, University of Chicago Press.

1981. (With E. Lazear.) Rank-order tournaments as optimal labour contracts. *Journal of Political Economy* 89, 841–64.

1982a. Authority, control and the distribution of earnings. *Bell Journal of Economics* 13, 311–23.

1982b. Further notes on the division of labour and the extent of the market. Working paper. Chicago: University of Chicago.

1983. Specialization and human capital. *Journal of Labor Economics* 1, 43–9.

1985. Implicit contracts: a survey. *Journal of Economic Literature* 23, 1144–75.

1986a. Prizes and incentives in elimination tournaments. *American Economic Review* 76, 701–15.

1986b. The theory of equalizing differences. In *Handbook of Labor Economics*, vol. 1, ed. O. Ashenfelter and R. Layard. Amsterdam: North-Holland.

1987. Dynamic animal economics. *American Journal of Agricultural Economics* 69, 547–57.

1988. The value of changes in life expectancy. *Journal of Risk and Uncertainty* 1, 285–304.

1988. (With R. Topel.) Housing investment in the United States. *Journal of Political Economy* 96, 718–40.

1988. (With C. Winship, eds.) Organizations and institutions: sociological and economic approaches to the analysis of social structure. Special issue of *American Journal of Sociology* 94.

1992. The market for lawyers. *Journal of Law and Economics* 35, 215–46.

1994. ed. *Implicit Contract Theory*. London: Edward Elgar.

1994. (With K. Murphy and J. Scheinkman.) Cattle cycles. *Journal of Political Economy* 102, 468–92.

1997a. Manufactured inequality. *Journal of Labor Economics* 15, 189–96.

1997b. Public employment taxes and the welfare state in Sweden. In *The Welfare State in Transition,* ed. R. Freeman, B. Swedenborg and R. Topel. Chicago: NBER, University of Chicago Press.

1998. (With H. Li.) Unraveling in matching markets. *American Economic Review* 88, 371–87.

1999. Potato paradoxes. *Journal of Political Economy* 107, S294–313.

2000. (With D. Neal.) Theories of the distribution of earnings. In *Handbook of Income Distribution,* vol. 1, ed. A. Atkinson, and F. Bourguignon. Amsterdam: North-Holland.

2001. (With H. Li and W. Suen.) Conflicts and common interests in committees. *American Economic Review* 91, 1478–97.

2002. Markets and diversity. *American Economic Review* 92, 1–15.

2005. *Markets and Diversity.* Cambridge, MA: Harvard University Press.

Bibliography

Acemoglu, D. 1998. Why do new technologies complement skills? Directed technical change and wage inequality. *Quarterly Journal of Economics* 113, 1055–89.

Acemoglu, D. 2002. Directed technical change. *Review of Economic Studies* 69, 781–810.

Ashenfelter, O. and Krueger, A. 1994. Estimates of the economic return to schooling from a new sample of twins. *American Economic Review* 84, 1157–73.

Ashenfelter, O. and Rouse, C. 1998. Income, schooling and ability: evidence from a new sample of identical twins. *Quarterly Journal of Economics* 113, 253–84.

Ashenfelter, O. and Zimmerman, D. 1997. Estimates of the returns to schooling from sibling data: fathers, sons, brothers. *Review of Economic Statistics* 79, 1–9.

Baumgardner, J. 1988a. The division of labor, local markets, and worker organizations. *Journal of Political Economy* 96, 509–27.

Baumgardner, J. 1988b. Physicians' services and the division of labor across local markets. *Journal of Political Economy* 96, 948–82.

Becker, G., Murphy, K. and Tamura, R. 1990. Human capital, fertility and economic growth. *Journal of Political Economy* 98, S12–37.

Becker, G., Murphy, K. and Werning, I. 2005. The equilibrium distribution of income and the market for status. *Journal of Political Economy* 113, 282–310.

Benabou, R. and Tirole, J. 2002. Self-confidence and personal motivation. *Quarterly Journal of Economics* 117, 871–915.

Benabou, R. and Tirole, J. 2003. Intrinsic and extrinsic motivation. *Review of Economic Studies* 70, 489–520.

Benabou, R. and Tirole, J. 2004. Willpower and personal rules. *Journal of Political Economy* 112, 848–87.

Ben-Porath, D. 1967. The production of human capital and the life cycle in earnings. *Journal of Political Economy* 75, 352–65.

Bergstrom, T. 1986. Soldiers of fortune. In *Essays in Honor of Kenneth J. Arrow,* vol. 2, ed. W. Heller, R. Starr and D. Starrett. New York: Cambridge University Press.

Ekeland, I., Heckman, J. and Nesheim, L. 2004. Identification and estimation of hedonic models. *Journal of Political Economy* 112, S60–109.

Friedman, M. 1953. Choice, chance and the personal distribution of income. *Journal of Political Economy* 61, 277–90.

Hartog, J. 2002. Desperately seeking structure: Sherwin Rosen (1938–2001). *Economic Journal* 112, 519–31.

Jones, L. and Manuelli, R. 1990. A convex model of equilibrium growth: theory and policy implications. *Journal of Political Economy* 98, 1008–38.

Kydland, F. and Prescott, E. 1982. Time to build and aggregate fluctuations. *Econometrica* 50, 1345–70.

Lazear, E. 1995. *Personnel Economics.* Cambridge: MIT Press.

Lazear, E. 2003. Sherwin Rosen. In *Biographical Memoirs of the National Academy of Sciences,* vol. 83. Washington, DC: National Academies Press. Online. Available at http://www.nap.edu/readingroom/books/biomems/srosen.pdf, accessed 10 August 2005.

Lucas, R., Jr. 1988. On the mechanics of economic development. *Journal of Monetary Economics* 22, 3–42.

Murphy, K. and Topel, R. 1987. Unemployment, risk, and earnings: testing for equalizing wage differences in the labor market. In *Unemployment and the Structure of Labor Markets,* ed. K. Lang and J. Leonard. New York: Basil Blackwell.

Murphy, K. and Topel, R., eds. 2003. *Measuring the Gains from Medical Research: An Economic Approach.* Chicago: University of Chicago Press.

Rebelo, S. 1991. Long-run policy analysis and long-run growth. *Journal of Political Economy* 99, 500–21.

Roback, J. 1982. Wages, rent and the quality of life. *Journal of Political Economy* 90, 1257–78.

Romer, P. 1986. Increasing returns and long-run growth. *Journal of Political Economy* 94, 1002–37.

Roy, A. 1951. Some thoughts on the distribution of earnings. *Oxford Economic Papers* 3, 135–46.

Sanderson, A. 2001. Sherwin Rosen, 1938–2001. *Journal of Sports Economics* 2, 211–12.

Siow, A. 1984. Occupational choice under uncertainty. *Econometrica* 52, 631–45.

Tamura, R. 1992. Efficient equilibrium convergence: heterogeneity and growth. *Journal of Economic Theory* 58, 355–76.

Tamura, R. 1994. Fertility, human capital and the 'wealth of families'. *Economic Theory* 4, 593–603.

Tamura, R. 1996. Regional economies and market integration. *Journal of Economic Dynamics and Control* 20, 825–45.

Tamura, R. 2002. Human capital and the switch from agriculture to industry. *Journal of Economic Dynamics and Control* 27, 207–42.

Tamura, R. 2006. Human capital and economic development. *Journal of Development Economics* 79, 26–72.

Zarkin, G. 1985. Occupational choice: an application to the market for public school teachers. *Quarterly Journal of Economics* 100, 409–46.

Rosenstein-Rodan, Paul Narcyz (1902–1985)

Rodan was one of the founders and first leaders of the field of development economics. His formative intellectual years were in the Austrian School of economics at the University of Vienna. He moved to the Department of Political Economy at University College London, in 1931.

Rodan's early essays in economics show a preoccupation with themes which reappeared throughout his professional career: the interaction and complementarity of economic processes (1933) and their temporal patterns (1934). Rodan's seminal article on developing countries (1943) argued that complementarities and externalities in demand and production created a need for the programming of investment. The arguments were subsequently extended to justify the need for an across-the-board 'big push' for a successful start to the development process (1963). He was among the first to apply the concept of 'disguised unemployment', described by Joan Robinson (1936), to developing countries as a persisting rather than cyclical problem.

Rodan first became actively engaged in development policy during his tenure at the World Bank from 1947 to 1954. In 1954 he moved to the Department of Economics at the Massachusetts Institute of Technology, where he produced an influential article (1961) which demonstrated that feasible levels of assistance to developing countries would substantially improve their growth performance. After retirement from MIT in 1968 he moved to the University of Texas and then to Boston University in 1972, where he established and worked in the Center for Latin American Development Studies until his death. Rodan was an active policy adviser to international agencies and governments of many countries and served on the Panel of Experts, the 'Nine Wise Men' of the Alliance for Progress, from 1961 to 1966.

RICHARD S. ECKAUS

Selected works

1933. La complementarita, prime delle tre fase del progresso della teoria economica pura. *Riforma Sociale* 44, 257–308.
1934. The role of time in economic theory. *Economica* NS 1, 77–97.
1943. Problems of industrialization of eastern and southeastern Europe. *Economic Journal* 53, 202–11.
1957. Disguised unemployment and underemployment in agriculture. *Monthly Bulletin of Agricultural Economics and Statistics* 6(7–8), 1–6.
1961. International aid for underdeveloped countries. *Review of Economics and Statistics* 43, 107–38.

1963. Notes on the theory of the 'big push' in economic development. In *Proceedings of a Conference of the International Economics Association,* ed. H.S. Ellis. London: Macmillan.

Bibliography

Robinson, J. 1936. Disguised unemployment. *Economic Journal* 46, 225–37.

Rosenthal, Robert W. (1944–2002)

Robert W. Rosenthal (1944–2002) was an economic theorist whose thoughtful papers inspired a wide range of new ideas. As Radner and Ray (2003) point out, Rosenthal (1978) gives one of the first formal statements of the revelation principle, a result noted in Myerson's first paper (1979) on the subject. Rosenthal (1979) initiated the study of repeated games with varying opponents, a modelling device used by Milgrom, North and Weingast (1990), Kandori (1992), and others to study social norms and other issues. He also wrote influential papers on pricing (Rosenthal, 1980; 1982), multi-unit auctions (Krishna and Rosenthal, 1996), purification of mixed strategy equilibria (Radner and Rosenthal, 1982; Aumann et al., 1983), sovereign debt (Fernandez and Rosenthal, 1990), analysis of experimental data (Brown and Rosenthal, 1990), and many other topics.

He is arguably best-known for his 1981 *Journal of Economic Theory* paper in which he discussed what Binmore (1987) named the 'centipede game'. Like its older cousin, the Prisoner's Dilemma, the centipede game beautifully summarizes a fundamental and intriguing strategic problem. Like the game which inspired but was overshadowed by it, Selten's (1978) chain store paradox, the centipede calls into question one of the most basic principles of game theory, namely, backward induction.

Consider the game shown in Figure 1. In this game, backward induction predicts that 1 plays *A* and 2 plays *D*. The reasoning seems very compelling. If 2 is rational, then, faced with a choice between a payoff of 3 and a payoff of 4, he obviously chooses 4. Hence 2 will play *D*. If 1 knows that 2 is rational, 1 knows that 2 will play *D*. Hence, if 1 is rational, he will choose *A* to get 4 instead of playing *B* which would yield 3. Thus the hypothesis that each player is rational and knows the other is rational seems to predict the backward induction solution. In longer games, there will be longer chains of reasoning, of course. However, the reasoning above has led many to conclude that backward induction is the implication of rationality and common knowledge of rationality.

A version of Rosenthal's centipede is shown in Figure 2. Here backward induction predicts that 1 chooses *d* at his first choice node, ending the game right away. Now the reasoning seems more suspect. If 2 is rational, he should choose *D* at the end rather than *A*. If 1 anticipates this, he

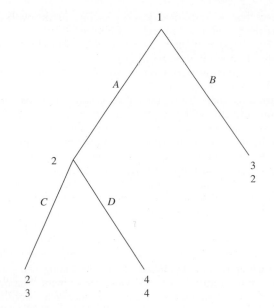

Figure 1

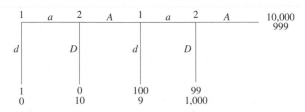

Figure 2

related critique of the use of iterated dominance in implementation theory.

Interestingly, McKelvey and Palfrey's (1992) experiments with the centipede game led them to develop the notion of quantal response equilibrium (McKelvey and Palfrey, 1995), an idea which echoes Rosenthal's own analysis. Rosenthal suggested that players might make 'mistakes' where these mistake probabilities would be decreasing in the cost of the mistakes, an idea he explored further in Rosenthal (1989). Quantal response equilibrium is another formulation of this idea.

BARTON L. LIPMAN

See also **behavioural economics and game theory; epistemic game theory: an overview; quantal response equilibria; rationality, bounded.**

I thank Hseuh Ling Huynh, Andy Postlewaite and Debraj Ray for comments.

Selected works

1978. Arbitration of two-party disputes under uncertainty. *Review of Economic Studies* 45, 595–604.

1979. Sequences of games with varying opponents. *Econometrica* 47, 1353–66.

1980. A model in which an increase in the number of sellers leads to a higher price. *Econometrica* 48, 1575–9.

1981. Games of perfect information, predatory pricing and the chain-store paradox. *Journal of Economic Theory* 25, 92–100.

1982. A dynamic model of duopoly with customer loyalties. *Journal of Economic Theory* 27, 69–76.

1982. (With R. Radner.) Private information and pure-strategy equilibria. *Mathematics of Operations Research* 7, 401–9.

1983. (With R. Aumann, Y. Katznelson and R. Radner.) Approximate purification of mixed strategies. *Mathematics of Operations Research* 8, 327–41.

1989. A bounded-rationality approach to the study of noncooperative games. *International Journal of Game Theory* 18, 273–92.

1990. (With J. Brown.) Testing the minimax hypothesis: a re-examination of O'Neill's game experiment. *Econometrica* 58, 1065–81.

1990. (With R. Fernandez.) Strategic models of sovereign-debt renegotiations. *Review of Economic Studies* 57, 331–49.

1992. (With J. Glazer.) A note on Abreu–Matsushima mechanisms. *Econometrica* 60, 1435–8.

1996. (With V. Krishna.) Simultaneous auctions with synergies. *Games and Economic Behavior* 17, 1–31.

Bibliography

Aumann, R. 1995. Backward induction and common knowledge of rationality. *Games and Economic Behavior* 8, 6–19.

should choose *d* at his last decision node. Similar reasoning shows that 2 should choose *D* at his first decision node and that 1 should choose *d* at his first node. Yet it is clear that each player must be virtually certain about his opponent's choice at the next move to justify choosing down rather than across, a certainty that seems extremely implausible in practice.

Many writers have argued that Rosenthal's centipede shows the paradoxical nature of backward induction. Consider player 1 at his second decision node. Here he is supposed to be certain that player 2 will choose *D* at the following node, justifying his own choice of *d*. Yet he also knows that 2 should have chosen *D* at the previous node and did not. If 2 failed to be rational in the past, why should 1 remain confident that he will be rational in the future? If 1 does have doubts, perhaps he should play *a* — a move which would make 2 glad to have deviated from 'rationality' at the preceding move!

Rosenthal's work led to a major debate on the question of backward induction. See, for example, Aumann (1995; 1998), Binmore (1987; 1996) and Reny (1993). See also Glazer and Rosenthal (1992) for a conceptually

Aumann, R. 1998. On the centipede game. *Games and Economic Behavior* 23, 97–105.

Binmore, K. 1987. Modeling rationality players, I. *Economics and Philosophy* 3, 9–55.

Binmore, K. 1996. A note on backward induction. *Games and Economic Behavior* 17, 135–7.

Kandori, M. 1992. Social norms and community enforcement. *Review of Economic Studies* 59, 81–92.

McKelvey, R. and Palfrey, T. 1992. An experimental study of the centipede game. *Econometrica* 60, 803–36.

McKelvey, R. and Palfrey, T. 1995. Quantal response equilibria in normal form games. *Games and Economic Behavior* 10, 6–38.

Milgrom, P., North, D. and Weingast, B. 1990. The role of institutions in the revival of trade: the law merchant, private judges, and the champagne fairs. *Economics and Politics* 2, 1–23.

Myerson, R. 1979. Incentive compatibility and the bargaining problem. *Econometrica* 47, 61–73.

Radner, R. and Ray, D. 2003. Robert W. Rosenthal. *Journal of Economic Theory* 112, 365–8.

Reny, P. 1993. Common belief and the theory of games with perfect information. *Journal of Economic Theory* 59, 257–74.

Selten, R. 1978. The chain-store paradox. *Theory and Decision* 9, 127–59.

Rostow, Walt Whitman (1916–2003)

Walt Whitman Rostow, economic historian, historian of economic thought, pioneer of modern development economics, and social scientist with interests in demography, politics, sociology and cultural aspects of development, was born in 1916. A professor of economics and history at the University of Oxford in 1946–7, Cambridge University, 1949–50, Massachusetts Institute of Technology 1950–61 and University of Texas at Austin 1968–2003, he is best known for his *Stages of Economic Growth: A Noncommunist Manifesto* (1960) and for his service as National Security Advisor to US President Lyndon B. Johnson during the Vietnam War. He led an active intellectual life engaged in public policy issues up to his death in 2003.

Several themes developed in his first publication, 'Investment and the Great Depression' (1938), recur in his first book, *Essays on the British Economy in the Nineteenth Century* (1948), and his *Process of Economic Development* (1953). His book co-authored with A.D. Gayer and Anna J. Schwartz, *The Growth and Fluctuation of the British Economy* (1952), was considered a classic study, and his work co-authored with Max Millikan, *A Proposal: Key to an Effective Foreign Policy* (1957), made his reputation in the field of foreign policy.

These books established Rostow as one of the world's foremost economic historians of his age.

His *Stages of Economic Growth* was a blockbuster. It stepped on many toes, assuring his reputation as the one of the most controversial economists of the last half of the 20th century. At the time, his model clashed with that of Harrod–Domar. They modelled steady-state (equilibrium) growth, with no historical context, and focused on two variables: saving and output–capital ratios. Naturally, an economic historian would ask how an economy got there in the first place. Rostow though he saw a pattern in how countries got there. Development proceeded through five stages: traditional society, preconditions for take-off, take-off to sustained growth, drive to maturity and age of high mass consumption. His critics saw these stages as 'empty boxes', not empirically verifiable and devoid of predictive power. Most were especially critical of the take-off stage. He had not demonstrated empirically the necessity of a significant rise in the saving and output–capital ratios. His critics were not convinced about his dating of stages for the seven countries he studied. Besides, he had not heeded Marshall's dictum, *Natura non facit saltum* – nature does nothing in jumps. Controversy swirled over his discontinuous, disequilibrium approach to economic growth.

His work was so upsetting to many of the world's most distinguished researchers in the field of development that the International Economic Association convened a conference in Konstanz in 1960 devoted exclusively to Rostow's work. This exclusivity was a first for the Association, and is indicative of the importance placed on his work. If Rostow did not convince his critics, or they him, the conference gave him worldwide notoriety, and his ideas were embraced by many economists in developing countries. Twenty years later the controversy continued.

Seizing on earlier criticism, Rostow published a massive volume, *The World Economy: History and Prospect* (1978). It examines world economic history from 1790 to 1976 in terms of population dynamics, long-term trends, cyclical fluctuations in production, prices and international trade. It extends the work of *Stages* with later data, and expands coverage to 20 countries.

In 1982 Charles P. Kindleberger and Guido Di Tella edited a three-volume Festschrift in Rostow's honour. A reviewer, Mancur Olson (1985), noted a paradox: many of the contributors were critics, and in a Festschrift! He pondered over an interesting question: how can so many distinguished critics also be admirers? Henry Rosovsky (1965) probably had the right explanation in an earlier comment: 'I invariably learn more by disagreeing with Professor Rostow than I do by agreeing with most other writers.'

Among economists with roots in the 1960s, Rostow's visible positions in the US government made him the most influential. He helped to form the Alliance for Progress and was President John F. Kennedy's representative on it. As an architect of the Vietnam War and

President Johnson's National Security Advisor, he became controversial in the political arena. He knew many of the world's leaders and was known by most of them. Through his public service, he became the Keynes of his day.

He continued to write books on important issues such as East–West relations, verification of nuclear arsenals, foreign aid and world population problems. He died in 2003 at age 87 just before his final book, *Concept and Controversy* (2003), was published. For many years, Rostow's ideas energized the field of economic development. That, even alone, is a major contribution to economics.

DOUGLAS CALVIN DACY

See also **growth and cycles; Kondratieff cycles.**

Selected works

1938. Investment and the Great Depression. *Economic History Review* 8, 136–58.

1948. *Essays on the British Economy of the Nineteenth Century.* London: Oxford University Press.

1952. (With A. Gayer and A. Schwartz.) *The Growth and Fluctuation of the British Economy.* Oxford: Oxford University Press.

1953. *The Process of Economic Development.* New York: W.W. Norton.

1957. (With M. Millikan.) *A Proposal: Key to an Effective Foreign Policy.* New York: Harper.

1960. *The Stages of Economic Growth: A Noncommunist Manifesto.* Cambridge: Cambridge University Press.

1960. *The United States in the World Arena.* New York: Harper.

1963. *The Economics of Take-Off into Sustained Growth.* London: Macmillan.

1975. *How It All Began: Origins of the Modern Economy.* New York: McGraw-Hill.

1978. *The World Economy: History and Prospect.* Austin: University of Texas Press.

1990. *Theorists of Economic Growth from David Hume to the Present.* New York: Oxford University Press.

1998. *The Great Population Spike and After.* New York: Oxford University Press.

2003. *Concept and Controversy: Sixty Years of Taking Ideas to Market.* Austin: University of Texas Press.

Bibliography

Kindleberger, C. and di Tella, G. 1982. *Economics in the Long View: Essays in Honour of W.W. Rostow.* London: Macmillan.

Olson, M. 1985. Review of C. Kindleberger and G. di Tella, eds., *Economics in the Long View: Essays in Honor of W.W. Rostow. Journal of Economic Literature* 23, 622–5.

Rosovsky, H. 1965. The take-off into sustained controversy. *Journal of Economic History* 25, 271–5.

rotating saving and credit associations (ROSCAs)

Rotating saving and credit associations (roscas) are the simplest form of collective financial institution. A rosca is a group of individuals who meet at regular intervals, each of whom contributes at each meeting a pre-determined amount to a collective 'pot' which is then given to one member. The latter is then excluded from receiving the pot in future meetings, while still being obliged to contribute to the pot. The meeting process repeats itself until each member has received the pot, thereby completing a cycle. Then the rosca can start a new cycle. From this description, the main virtues of roscas are clear: they do not require storage of funds, accounting and durations of obligations are transparent, and there are no complicated interest payments or debt management. Roscas are very popular in developing countries. For instance, average membership rates in Indonesia have been estimated at 40 per cent of the population (Armendariz de Aghion and Morduch, 2005), 20 per cent in Taiwan (Levenson and Besley, 1996) and 40 per cent in a Kenyan slum (Anderson and Baland, 2002). Although roscas do exist alongside more formal financial institutions, they are often the sole saving and credit institution in many rural areas.

Roscas vary widely in terms of the size of the contributions, the number of members and the frequency of meetings. Also, the process by which the pot is allocated can be a lottery (random roscas), or follow a fixed order imposed, for instance, by the leaders in the group (fixed roscas), or be determined by a bidding process (bidding roscas).

The literature identified four differing motives for individuals to save through roscas. First, roscas allow individuals to purchase indivisible goods earlier in expected terms than through the accumulation of individual savings (Besley, Coate and Loury, 1993). Roscas thus provide an implicit positive interest rate to those receiving the pot early. Second, as emphasized by Anderson and Baland (2002), roscas may be used by married women as a way to commit the household to higher saving rates than what can be done at home. Given the presence of social sanctions, husbands are then forced to comply with the saving rate imposed by the rosca. Relatedly, Gugerty (2006) argues that people facing intertemporal inconsistency join roscas to bind themselves to a particular saving pattern (see also Ardener, 1964; Ambec and Treich, 2007). Lastly, bidding roscas provide some insurance to their members against short-term income shocks by providing implicit short-run credit to those willing to pay the highest bid.

The most common problem of roscas has to do with enforcement. Indeed, the first members to receive the pot are de facto borrowers from the other members and, absent social sanctions, are better off not repaying their debts. Given that the size of the pot is fixed, they can always replicate (and also do better than) the best that the rosca can offer them by saving on their own. Social sanctions are then necessary to discipline members. Also, as argued in Anderson, Baland and Moene (2003), the rule for allocating the pot can be chosen to partially address this issue. Selection of members is another issue faced by roscas, particularly bidding roscas, where higher bidders may be exposed to more intrinsic risks than others (see Eeckhoudt and Munshi, 2005; Klonner and Rai, 2006).

JEAN-MARIE BALAND

See also **micro-credit**.

Bibliography

Ambec, S. and Treich, N. 2007. Roscas as financial agreements to cope with self-control problems. *Journal of Development Economics* 82, 120–37.

Anderson, S. and Baland, J.-M. 2002. The economics of ROSCAs and intrahousehold allocation. *Quarterly Journal of Economics* 117, 963–95.

Anderson, S., Baland, J.-M. and Moene, K. 2003. Sustainability and organizational design in Roscas: some evidence from Kenya. Mimeo, Center for Research in the Economics of Development (CRED), University of Namur.

Ardener, S. 1964. The comparative study of rotating credit associations. *Journal of the Royal Anthropological Institute of Great Britain and Ireland* 44, 202–29.

Armendariz de Aghion, B. and Morduch, J. 2005. *The Economics of Microfinance*. Cambridge, MA: MIT Press.

Besley, T., Coate, S. and Loury, G. 1993. The economics of rotating savings and credit associations. *American Economic Review* 83, 792–810.

Eeckhoudt, J. and Munshi, K. 2005. Mitigating regulatory inefficiency: the non-market response to financial regulation in India. Mimeo, Brown University.

Gugerty, M. 2006. You can't save alone: commitment and rotating savings and credit associations in Kenya. Working Paper No. 2006-08, Evans School of Public Affairs, University of Washington.

Klonner, S. and Rai, A. 2006. Adverse selection in credit markets: evidence from bidding roscas. Mimeo, Cornell University.

Levenson, A. and Besley, T. 1996. The anatomy of an informal financial market: rosca participation in Taiwan. *Journal of Development Economics* 51, 45–68.

Rothbard, Murray N. (1926–1995)

Murray Rothbard was influential in continuing the tradition of the Austrian school of economics in America. In more than two dozen books and hundreds of articles, his work spanned economics, history, philosophy and political science. He earned his Ph.D. from Columbia University, but was influenced mainly by Ludwig Von Mises's seminar at New York University. Rothbard was a strong believer in *apriorism*, the idea that economic laws could be discovered using logical reasoning (as opposed to empirical testing), and he attempted to build on and extend the economic logic of Mises and others in that tradition. Rothbard's treatise, *Man, Economy, and State* (1962), analysed the economics of market exchange, while his follow-up volume, *Power and Market* (1970), analysed the economics of government intervention. An underlying theme of his work is that the market is the realm of mutually beneficial exchange, whereas the government is the realm of coercion where some gain at the expense of others.

Rothbard considered economics to be a value-free science, but he believed economic reasoning can be used to determine whether normative views are internally consistent. He was strongly critical of government intervention in the economy, arguing against those who believe that government policies can be Pareto-superior and make all people better off. For example, Rothbard was one of the only economists writing in the 1950s and 1960s to argue against all antitrust laws. He thought that perfect competition was an unattainable ideal, and he said that monopolies or cartels do not pose problems on the free market. He believed that the only monopolies that warrant concern are those sanctioned by government. Rothbard was critical of arguments about market failure in general, insisting that mainstream notions of economic efficiency were unrealistic. He criticized the welfare economics of his day on the grounds that it rests on unscientific interpersonal comparisons of utility.

In addition, Rothbard wrote a great deal on economic history, often documenting government getting in the way of markets. For example, his 1963 book *America's Great Depression* argued that government caused and lengthened the Great Depression through distortionary monetary and regulatory policies. Rothbard also devoted much of his writing to political philosophy, and here too he was unabashedly libertarian. Rothbard's contribution is particularly noteworthy because he was one of the first economists to argue that markets do not depend on the existence of government. Before him, even the most free-market theorists, such as Ludwig von Mises, Henry Hazlitt, Ayn Rand, and Friedrich Hayek, had simply assumed that services like law enforcement must be provided collectively by the state. But in *Power and Market* and *For a New Liberty* (1973) Rothbard maintained that public goods such as law enforcement must be analysed in terms of marginal units and, as with other goods, those marginal units can be provided privately. He pointed to historical examples of private law enforcement and speculated how a purely private system might function. Rothbard's ideas advancing private property anarchism were radical, but they influenced many economists who now write about alternatives to government law (Stringham, 2006). Rothbard's thorough libertarian views pushed free-market thinking to become more free market.

BENJAMIN POWELL AND EDWARD STRINGHAM

See also **Austrian economics; Hayek, Friedrich August von; libertarianism; Menger, Carl; Mises, Ludwig Edler von.**

Selected works

1962. *Man, Economy, and State: A Treatise on Economic Principles*. Princeton: Van Nostrand.
1963. *America's Great Depression*. Princeton: Van Nostrand.
1970. *Power and Market*. Kansas City: Sheed Andrews and McMeel.
1973. *For a New Liberty: The Libertarian Manifesto*. New York: Macmillan.
Many of Rothbard's writings are freely available online at http://www.rothbard.org, accessed 26 September 2006.

Bibliography

Raimondo, J. 2000. *An Enemy of the State: The Life of Murray N. Rothbard*. Amherst, NY: Prometheus Books.
Stringham, E., ed. 2006. *Anarchy and the Law: The Political Economy of Choice*. Somerset, NJ: Transaction Publishers.

rotten kid theorem

The 'rotten kid theorem' was proposed (and named) by Gary Becker in an influential 1974 article, and further discussed in his book, *A Treatise on the Family* (1981). The theorem claims the following. If a family has a caring household head who gives money to all other household members, then each member, no matter how selfish, will maximize his or her own utility by taking actions that lead to maximization of total family income. Thus, all family members will act harmoniously in the family interest – at least if they know what is good for them. (Becker also applies this result to two-adult families where one adult has a much higher income than the other.)

Here is an informal argument in support of the theorem. If, in equilibrium, the household head makes gifts to all kids, then expenditure on each kid's consumption is ultimately decided by the head. The post-gift distribution of consumption among family members will maximize the head's utility subject to the constraint that family expenditure not exceed family income. If consumption of every family member is a 'normal good', then it must be that, taking account of gifts from the head, a selfish child can maximize his own consumption only by taking those actions that maximize total family income.

This argument for the rotten kid theorem can be made rigorous with sufficiently strong assumptions about tastes and technology. Consider a family with n selfish kids and a household head who cares about the happiness of each kid. Assume that there is just one consumption good. Let x_0 denote the amount of the

good consumed by the household head and x_i the amount consumed by kid i. The utility function of each kid i is $u_i(x_i) = x_i$, and the utility function of the household head, $u(x_0, \ldots, x_n)$, is strictly increasing in all the x_i's. Every household member earns some personal income, the amount of which depends on her own actions a_i, but possibly also on the actions of other household members. Let a be the vector of actions chosen by household members, let $m_i(a)$ be i's personal income, and let $m(a) = \sum m_i(a)$. Feasible allocations must satisfy the household budget constraint, $\sum_i x_i = m(a)$. For any income y, define $(x_0(y), \ldots x_n(y))$ as the allocation that maximizes $u(x_0, \ldots, x_n)$ subject to $\sum_i x_i = y$. Assume that consumption for each i is a *normal good* so that $x_i(y)$ is a strictly increasing function of y. Finally, assume that the household head has personal income large enough so that in equilibrium he chooses to donate money to all other persons in the household. This means that, for all feasible a and for each kid, i, $x_i(m(a)) > m_i(a)$. Consider the following two-stage game. In the first stage, household members choose their actions and thus determine total family income $m(a)$. In the second stage the household head finds the allocation $x(m(a))$ that maximizes $u(x_1, \ldots, x_n)$ subject to $\sum_i x_i = m(a)$ and donates $x_i(m(a)) - m_i(a)$ to kid i. In the first stage of the game, each kid realizes that, after the head has redistributed income, her own consumption will be $x_i(m(a))$. The normal goods assumption implies that $x_i(m(a))$ is an increasing function of $m(a)$. Therefore, the self-interest of each kid coincides with maximizing total family income, $m(a)$. (To ensure that a maximum exists, assume that each m_i is continuous and that each a_i must be chosen from a closed bounded set.)

The trouble with the rotten kid theorem is that it fails to hold in models that make slight concessions toward realism. Bergstrom (1989) shows that, in general, the rotten kid theorem fails if kids care about their activities as well as about consumption. For example, if leisure is a complement to consumption, a child can manipulate the parents' transfer in his or her favour by taking too much leisure. Lindbeck and Weibull (1988) and Bruce and Waldman (1990) show that the rotten kid theorem fails when individuals can choose between current and future consumption. Lundberg and Pollak (2003) show a dramatic failure of the rotten kid theorem when families choose between discrete options like whether to move house or whether to have a child.

Bergstrom (1989) explored the most general conditions under which a rotten kid theorem can be proved. He showed that, in general, a necessary and sufficient condition for the conclusion of the rotten kid theorem to be satisfied is that there is 'conditional transferable utility'. This means that the utility possibility sets corresponding to all possible activity choices are nested and are bounded above by parallel straight line segments. For example, there is conditional transferable utility if kids care only about their consumption, so that $u_i(x_i, a) = x_i$, and if total family income is $m(a)$. Then the utility possibility frontier conditional on a is the simplex $\{(u_1, \ldots, u_n)|\sum_1^n u_i \leq m(a)$ and $u_i \geq 0$ for all $i\}$. In general, however, if the kids' utilities depend on their actions, kids will be able to influence the 'slope' of the utility possibility frontier by their choice of actions, a. For example, a selfish kid may benefit by choosing an action that reduces family income but makes it 'cheaper' for the parent to invest in her utility rather than that of her sibling. Bergstrom shows that the most general class of environments for which there is conditional transferable utility requires that each kid i has a utility function of the form $u(x_i, a) = A(a)x_i + B_i(a)$ where x_i is i's expenditure on consumer goods and a is the vector of family members' activities. This allows the possibility that activities a_i generate externalities in consumption as well as in income-earning. (Bergstrom and Cornes, 1983, show that in a public goods economy the efficient quantity of public goods is independent of income distribution if and only if preferences can be represented in this form, which is dual to the Gorman polar form for public goods.) Then, for any a, the upper boundary of the utility possibility set is $\{u|\sum u_i = A(a)m(a) + \sum B_i(a)\}$. If utilities of kids are normal goods for the head, then each kid will maximize her utility by maximizing $F(a) = A(a)m(a) + \sum B_i(a)$. Thus selfish kids would act in the family interest, as the rotten kid theorem asserts.

An interesting debate in evolutionary biology parallels the economists' rotten kid theorem. Alexander (1974) maintained that natural selection favours genetic lines in which offspring act so as to maximize family reproductive success. Dawkins (1976) disputed Alexander's argument, citing Hamilton's theory of kin selection (1964), which implies that in sexual diploid species offspring value the reproductive success of their siblings at only half of their own. Alexander (1979) conceded Dawkins's point, but offered an additional reason that offspring would act in the interest of their parents, namely, that 'the parent is bigger and stronger than the offspring, hence in a better position to pose its will'. Bergstrom and Bergstrom (1999) propose an evolutionary model that could support the Becker–Alexander conclusion that children will act in the family interest. They construct a two-locus genetic model, where a gene at one locus controls an animal's behaviour when the animal is a juvenile and a gene at the other controls its behaviour when it is a parent. Then the frequency of recombination between genes at these two loci determines the evolutionary outcome of parent–offspring conflict. If recombination between these genes is rare, offspring will tend to act in the genetic interest of their parent. If recombination is frequent, there can be an equilibrium where some offspring successfully 'blackmail' their parents into giving them more resources than is optimal for the family's reproduction.

THEODORE C. BERGSTROM

See also **Becker, Gary S; family economics.**

Bibliography

Alexander, R. 1979. *Darwinism and Human Affairs*. Seattle: University of Washington Press.

Alexander, R. 1974. The evolution of social behavior. *Annual Review of Ecological Systems* 5, 25–83.

Becker, G. 1974. A theory of social interactions. *Journal of Political Economy* 82, 1063–93.

Becker, G. 1981. *A Treatise on the Family*. Cambridge, MA: Harvard University Press.

Bergstrom, T. and Cornes, R. 1983. Independence of allocative efficiency from distribution in the theory of public goods. *Econometrica* 51, 1753–65.

Bergstrom, T. 1989. A fresh look at the rotten kid theorem. *Journal of Political Economy* 97, 1138–59.

Bergstrom, C. and Bergstrom, T. 1999. Does Mother Nature punish rotten kids? *Journal of Bioeconomics* 1, 47–72.

Bruce, N. and Waldman, M. 1990. The rotten-kid theorem meets the Samaritan's Dilemma. *Quarterly Journal of Economics* 105, 155–65.

Dawkins, R. 1976. *The Selfish Gene*. Oxford: Oxford University Press.

Hamilton, W. 1964. The genetical evolution of social behavior, I and II. *Journal of Theoretical Biology* 7, 1–52.

Lindbeck, A. and Weibull, J. 1988. Altruism and efficiency: the economics of fait accompli. *Journal of Political Economy* 96, 1165–82.

Lundberg, S. and Pollak, R. 2003. Efficiency in marriage. *Review of the Economics of the Household* 1, 153–67.

Roy model

The Roy (1951) model of self-selection on outcomes is one of the most important models in economics. It is a framework for analysing comparative advantage. The original model analysed occupational choice with heterogeneous skill levels and has subsequently been applied in many other contexts. We first discuss the model. We then summarize what is known about identification of the model. We end by describing some applications based on the model and its extensions.

Basic models

In the original Roy (1951) model, agents can pursue one of two possible occupations: hunting and fishing. They cannot pursue both at the same time. There is no interaction among agents so the choice of one agent does not affect the choice of another agent either through prices or through external effects. Let π_f and π_r be the price of fish and rabbits respectively in the village. Let F_i denote the number of fish that individual i would catch if he chooses to fish. Similarly let R_i denote the number rabbits he would catch. Then individual i's wage is

$$w_{fi} = \pi_f F_i$$

if he fishes and

$$w_{ri} = \pi_r R_i$$

if he hunts. The income that worker i receives for working in sector j is thus proportional to π_j (where $j \in \{r, f\}$). If workers are pure income maximizers, they will choose the occupation with higher income. Thus a worker chooses to fish if $w_{fi} > w_{ri}$. If F_i and R_i are continuous random variables, $\Pr(\pi_r R_i = \pi_f F_i) = 0$, so the indifference set is negligible. A fundamental aspect of the Roy model is that it allows for heterogeneity in (F_i, R_i). This heterogeneity can arise from inherent ability differences or human capital investment.

An important issue is self-selection. Under what conditions will the best workers self-select into an occupation? Will people who self-select be above average? For example, for fishing, under what conditions is the average productivity of people working in the fishing sector above the population mean productivity:

$$E[\log(F_i) | \pi_f F_i \geq \pi_r R_i] > E[\log(F_i)]?$$

Assume, as did Roy (1951), that log skills are jointly normally distributed

$$\begin{bmatrix} \log(F_i) \\ \log(R_i) \end{bmatrix} \sim N\left(\begin{bmatrix} \mu_f \\ \mu_r \end{bmatrix}, \begin{bmatrix} \sigma_{ff} & \sigma_{fr} \\ \sigma_{fr} & \sigma_{rr} \end{bmatrix} \right).$$

Then it is straightforward to show that

$$\begin{aligned} &E[\log(F_i) | \pi_f F_i \geq \pi_r R_i] \\ &= \mu_f + \underbrace{\frac{(\sigma_{ff} - \sigma_{fr})}{\sigma} \lambda \left(\frac{\log(\pi_f) - \log(\pi_r) + \mu_f - \mu_r}{\sigma} \right)}_{\text{selection effect}} \end{aligned}$$

where σ^2 is the variance of $\log(F_i/R_i)$ and $\lambda(\cdot)$ is the inverse Mills ratio. (See SELECTION BIAS AND SELF-SELECTION.)

The function λ is positive but decreasing in its arguments with $\lim_{\pi_f \to \infty} \lambda(\cdot) = 0$. The selection effect is the second term on the right-hand side of this expression. There is a parallel expression for $E(\log R_i | \pi_F F_i \leq \pi_R R_i)$ with the subscripts f and r interchanged.

Recall that $E[\log(F)] = \mu_f$ and that λ and σ must both be positive. Therefore, the question of whether there is positive selection into fishing depends only upon the sign of $\sigma_{ff} - \sigma_{fr}$. It does not depend on skill prices. Moreover, since

$$\sigma^2 = (\sigma_{ff} - \sigma_{fr}) + (\sigma_{rr} - \sigma_{fr}) > 0,$$

at least one of $(\sigma_{ff} - \sigma_{fr})$ and $(\sigma_{rr} - \sigma_{fr})$ must be positive. Thus, there must be positive selection into one of the occupations, and there can be positive selection into both.

If, however, there is positive selection into only one occupation, the question arises as to which occupation is most likely to have positive selection. Roy argues that relatively simple tasks (setting traps for rabbits in his case) can be described by a small standard deviation of skill. For more difficult skills (fishing in his example) the standard deviation will be relatively higher as there is a bigger difference between the most skilled and the least skilled. Thus, if fishing is the more difficult task, $\sigma_{ff} > \sigma_{rr}$, there must be positive selection into fishing (that is, $E(\log(F_i) | \pi_f F_i \geq \pi_r R_i) > E(\log(F_i))$).

Whether there is positive selection into hunting depends on the value of σ_{fr} relative to σ_{rr}. When $\sigma_{fr} < 0$, we will see positive selection into hunting. At the other extreme, if hunting and fishing are perfectly correlated, then σ_{fr} must be larger than σ_{rr}, and there is negative selection into hunting. Intuitively, since F and R are perfectly positively correlated, and F is more dispersed, persons with low values of F can avoid low incomes by using their value of R. Persons with high values of F (and R) should fish because the upper tail of F is more dispersed. For cases in between, either positive or negative selection is possible depending on the sign of $\sigma_{rr} - \sigma_{fr}$. Heckman and Honoré (1990) generalize this result to a broader class of distribution functions.

This model has been generalized in a number of ways. There can be more than two occupational choices. Following Heckman and Sedlacek (1985), one can assume that individuals possess a vector of skills S_i and that different tasks use the different skills according to the function $T_j(S_i)$. We still let π_j denote task prices so that we can write an individual's wage at task j as

$$w_{ji} = \pi_j T_j(S_i).$$

Another extension of the model allows individuals to care about aspects of the job other than just their wages (see Heckman and Sedlacek, 1985). Let $U_{ji}(w)$ be the utility that individual i would receive from performing task j under wage level w. This allows for some tasks (such as playing basketball) to be generally preferred to more unpleasant tasks (such as cleaning bathrooms). Individuals then choose the occupation that yields the highest level of utility for them $U_{ji}(w_{ij})$. This is the generalized Roy model in which the generalization comes in the agent decision rules.

The generalized Roy model can be trivially extended to a model of labour force participation by allowing non-market work to be one of the tasks. To see this, let $j = 0$ denote the home sector as in Gronau (1974) and Heckman (1974). Of course, in general, there will not be a market price for home-produced goods, but one can interpret $T_0(S_i)$ as the value of goods produced at home. One could also assume that staying at home is pure leisure in which $T_0(S_i) = 0$, but people enjoy staying at home $U_{0i}(0) > U_{ji}(0)$ for $j > 0$. The Roy model has been generalized to allow for uncertainty in agent decision making in Cunha, Heckman and Navarro (2005). See the reviews in Heckman, Lochner and Todd (2006) and Cunha and Heckman (2007).

Identification

The economics of these models is simple, but identification and estimation are considerably more difficult. Heckman and Honoré (1990) consider identification of the basic Roy model with two occupations and income maximization. They consider two different cases: (a) the standard Roy model, in which the two occupations represent two different sectors of the economy and the econometrician has data on wages in both sectors; and (b) a case motivated by labour supply in which the econometrician has wage data from one sector (the market sector) but not from the other (the home sector). It is important to keep in mind that the comparative advantage decision at the heart of the Roy Model is just one factor that can lead to selection bias. SELECTION BIAS AND SELF-SELECTION discusses the more general framework for thinking about sample selection and also discusses in some detail how the Roy model fits into this framework.

Heckman and Honoré (1990) consider identification from a single cross section. When one can observe wages in both sectors, under log normality, the Roy model is identified even without any regressors in the model. However, when one relaxes the log normality assumption, without regressors in the outcome equation the model is no longer identified. This is true despite the strong assumption of agent income maximization.

Heckman and Honoré (1990) provide conditions under which one can identify these models using variation across markets, or by using variation in observables within a market. To see the intuition behind the latter case, consider the model in which

$$\log(F_i) = g_f(Z_{fi}, X_i) + \varepsilon_{fi}$$
$$\log(R_i) = g_r(Z_{ri}, X_i) + \varepsilon_{ri},$$

and prices are normalized to 1. In this context, it is helpful for identification to have an exclusion restriction – that is, a variable Z_{fi} that varies separately from (X_i, Z_{ri}) and a variable Z_{ri} that varies separately from (X_i, Z_{fi}). As long as there is sufficient variation in the excluded variables, Heckman and Honoré (1990) show that with a location normalization the full model is identified provided that $(\varepsilon_{fi}, \varepsilon_{ri})$ are independent of (Z_{fi}, Z_{ri}, X_i), that is, they identify g_f, g_r and the joint distribution of $(\varepsilon_{fi}, \varepsilon_{ri})$. (They also establish identification when only one sector's output is observed.)

To see the intuition for why the model is identified, consider an 'identification at infinity' argument. For

convenience, take the location normalization to be

$$E(\varepsilon_{fi}) = E(\varepsilon_{ri}) = 0.$$

Suppose that g_r is such that for any x, say x_0,

$$\lim_{z_r \to -\infty} g_r(z_r, x_0) = -\infty.$$

Let $J_i \in \{f, r\}$ be an indicator of the occupation that was chosen by individual i. Then

$$\lim_{z_r \to -\infty} E[\log(F_i)|J_i = f, X_i = x, Z_{fi} = z_f, Z_{ri} = z_r]$$
$$= g_f(z_f, x) + \lim_{z_r \to -\infty} E(\varepsilon_{fi}|g_f(z_f, x)$$
$$+ \varepsilon_{fi} > g_r(z_r, x) + \varepsilon_{ri}, X_i = x,$$
$$Z_{fi} = z_f, Z_{ri} = z_r)$$
$$= g_f(z_f, x).$$

By varying (z_f, x) one can trace out g_f. This occurs because conditioning on the event

$$g_f(z_f, x) + \varepsilon_f > g_r(z_r, x) + \varepsilon_r$$

becomes irrelevant as z_r becomes arbitrarily small. Identification of g_r is analogous using variation in z_f.

To identify the joint distribution of $(\varepsilon_{fi}, \varepsilon_{ri})$ note that from the data one can observe

$$\Pr(J_i = f, \log(F_i) < s|X_i = x, Z_{fi} = z_f, Z_{ri} = z_r)$$
$$= \Pr(g_f(z_f, x) + \varepsilon_{fi} > g_r(z_r, x)$$
$$+ \varepsilon_{ri}, g_f(z_f, x) + \varepsilon_{fi} < s|X_i$$
$$= x, Z_{fi} = z_f, Z_{ri} = z_r)$$
$$= \Pr(\varepsilon_{fi} - \varepsilon_{ri} < g_f(z_f, x) - g_r(z_r, x),$$
$$\varepsilon_{fi} < s - g_f(z_f, x)|X_i$$
$$= x, Z_{fi} = z_f, Z_{ri} = z_r)$$

which is the cumulative distribution function of $(\varepsilon_{fi} - \varepsilon_{ri}, \varepsilon_{fi})$ evaluated at the point $(g_f(z_f, x) - g_r(z_r, x), s - g_f(z_f, x))$. By varying the point of evaluation one can identify the joint distribution of $(\varepsilon_{fi} - \varepsilon_{ri}, \varepsilon_{fi})$ from which one can derive the joint distribution of $(\varepsilon_{fi}, \varepsilon_{ri})$. Thus the model is identified. Heckman and Honoré (1989) also present conditions for identification of a competing risk version of a Roy model when there are no exclusion restrictions ($Z_r = X = Z_f$) but g_f and g_r can be independently varied. Buera (2006) makes stronger differentiability assumptions and relaxes the separability assumption in the choice equation. He also identifies a Roy model without exclusion restrictions.

Identifying the more general model where individuals choose fishing when

$$U_{fi}(w_{fi}) > U_{ri}(w_{ri})$$

is possible under a variety of assumptions. Consider the separable case in which

$$U_{fi}(w_{fi}) - U_{ri}(w_{ri}) = h(Q_i, Z_{fi}, Z_{ri}, X_i) + v_i$$

where Q_i is an additional variable that might affect the relative utilities of the two options. The function h is identified up to a normalization (see, for example, Matzkin, 1992).

Identification of parts of the model follows from the preceding reasoning. If there is a variable that affects sectoral choice, but not wages as a fisherman, we can identify g_r. Note that this exclusion restriction could be in the form of either Q_i or Z_{ri}. We can then identify the joint distribution of (v_i, ε_{ri}) using an argument analogous to the above. Using the same argument we can identify g_f and the joint distribution of (v_i, ε_{fi}). A formalization of this argument can be found in Heckman (1990) for the case in which h is linear and is extended in Heckman and Smith (1998), Carneiro, Hansen and Heckman (2003) and Heckman and Navarro (2007). One cannot, without further assumptions, identify the joint distribution of $(v_i, \varepsilon_{ri}, \varepsilon_{fi})$. (Abbring and Heckman, 2007, present conditions for identification of the joint distribution by restricting dependence relations. See also Aakvik, Heckman and Vytlacil, 2005.)

If one is interested in evaluating policies in which wages can change, this reduced form model is not enough since there is no separation of wage effects from non-wage effects in the choice model. Assume further that we can write

$$
\begin{aligned}
h(Q_i, Z_{fi}, Z_{ri}, X_i) + v_i &= \alpha_1 F_i - \alpha_2 R_i \\
&\quad + h^*(Q_i, Z_{fi}, Z_{ri}, X_i) + v_i^* \\
&= \alpha_1 g_f(Z_{fi}, X_i) - \alpha_2 g_r(Z_{ri}, X_i) \\
&\quad + h^*(Q_i, Z_{fi}, Z_{ri}, X_i) \\
&\quad + \alpha_1 \varepsilon_{fi} - \alpha_2 \varepsilon_{ri} + v_i^*.
\end{aligned}
$$

Identification of this model is possible if there are exclusion restrictions in Z_{fi} and Z_{ri}, that is, if there are components of Z_{fi} and Z_{ri} that do not affect h^*. Under sufficient variation of these variables and imposing a normalization, the model is identified. An interesting special case of the model is when $\alpha_1 = \alpha_2$. In this case one needs a somewhat weaker exclusion restriction in that one could use variation in X_i. That is, we could use a variable that affects labour market outcomes, but not sectoral choice directly.

Empirical models
There are many examples that build on the Roy model, but in labour economics three stand out. The earliest empirical application of this model is to the labour supply decision (Heckman, 1974; Gronau, 1974). We refer interested readers to LABOUR SUPPLY rather than discuss

these models explicitly. The second application is to occupational choice, which is most closely linked to the original Roy model. The third, and perhaps most well known, application is to education.

We start by describing the empirical applications of the model to education. Willis and Rosen (1979) consider a model in which students decide whether to attend college. Students may have a comparative advantage in either the college sector or the high school sector. Their model assumes that decisions about schooling are made in an environment of perfect certainty on the principle of income maximization. They assume access to outcome measures in two periods. The decision to attend college depends on interest rates which are not observable to the econometrician. (One could reinterpret their model as a generalized Roy model if one interprets the interest rate as representing utility differences rather than interest rates.) Semiparametric identification requires two types of exclusion restrictions: a variable that influences the decision to attend college but not directly wages, and a variable that influences wages but not the decision to attend college directly. For the former type of exclusion restriction, Willis and Rosen (1979) use family background variables, arguing that they will be correlated with interest rates but uncorrelated with wages. For the latter type they use test scores, arguing that they are related to skill as in the Roy model, but unrelated to the interest rate.

Although they discuss comparative advantage in the labour market, as did Roy, they do not present direct empirical evidence on this question because they cannot estimate the joint distribution of schooling outcomes across both choices. They present some indirect evidence on the importance of comparative advantage in the labour market because they can identify the counterfactual means of what college students would earn if they had been high school students and what high schools students would earn had they been college students.

There are many extensions of this model, including Taber (2000), Cameron and Taber (2004) and Heckman, Lochner and Todd (2006). Cunha, Heckman and Navarro (2005) and Cunha and Heckman (2007) extend the model to allow for uncertainty, to identify agent information and to directly test for comparative advantage in the labour market by identifying the joint distribution of outcomes for the two counterfactual states (college and high school).

In a series of papers, Heckman, Lochner and Taber (1998a; 1998b; 1998c) estimate a general equilibrium version of this model. That is, they allow the skill prices π_j to be endogenous. They show that accounting for equilibrium effects is essential for estimating the impact of policy on earnings inequality. In particular Heckman, Lochner and Taber (1998b) show that ignoring equilibrium effects overstates the impact of a tuition subsidy on college enrolment by an order of magnitude. They also decompose the policy effect on earnings inequality into its various components.

Other papers estimate a Roy model of occupational choice. Most notably, Heckman and Sedlacek (1985; 1990) estimate models in which workers choose between industrial sectors. In some cases they allow for non-market work. They show how to estimate the model, but reject a pure Roy model. They show instead that a more general model with utility maximization and non-participation can fit the data well. Gould (2002) extends this framework to address the changing wage structure. He shows that workers choose sectors to maximize their comparative advantage and that this activity tends to decrease earnings inequality. However, he shows that the importance of this effect decreases over time as sectors increasingly value more similar skill sets.

Keane and Wolpin (1997) and Eckstein and Wolpin (1999) estimate dynamic discrete choice models of occupational and educational choice that extends the Roy model to a dynamic setting with uncertainty with serially independent shocks. Agents in their model make labour supply, education and occupational choice simultaneously. Heckman and Navarro (2007) present a nonparametric identification analysis of a dynamic discrete choice model with serially correlated shocks. Abbring and Heckman (2007) survey the dynamic discrete choice literature, including these papers.

JAMES HECKMAN AND CHRISTOPHER TABER

See also **labour economics; labour supply; returns to schooling; selection bias and self-selection.**

Bibliography

Aakvik, A., Heckman, J.J. and Vytlacil, E.J. 2005. Estimating treatment effects for discrete outcomes when responses to treatment vary: an application to Norwegian vocational rehabilitation programs. *Journal of Econometrics* 125, 15–51.

Abbring, J.H. and Heckman, J.J. 2007. Econometric evaluation of social programs, part III: distributional treatment effects, dynamic treatment effects, dynamic discrete choice, and general equilibrium policy evaluation. In *Handbook of Econometrics*, vol. 6, ed. J. Heckman and E. Leamer. Amsterdam: North-Holland.

Buera, F.J. 2006. Non-parametric identification and testable implications of the Roy model. Unpublished manuscript, Department of Economics, Northwestern University.

Cameron, S.V. and Taber, C. 2004. Estimation of educational borrowing constraints using returns to schooling. *Journal of Political Economy* 112, 132–82.

Carneiro, P., Hansen, K. and Heckman, J.J. 2003. Estimating distributions of treatment effects with an application to the returns to schooling and measurement of the effects of uncertainty on college choice. 2001 Lawrence R. Klein Lecture. *International Economic Review* 44, 361–422.

Cunha, F. and Heckman, J.J. 2007. Identifying and estimating the distributions of ex post and ex ante

returns to schooling: a survey of recent developments. *Labour Economics* 14, 870–93.

Cunha, F., Heckman, J.J. and Navarro, S. 2005. Separating uncertainty from heterogeneity in life cycle earnings. The 2004 Hicks Lecture. *Oxford Economic Papers* 57, 191–261.

Eckstein, Z. and Wolpin, K.I. 1999. Why youths drop out of high school: the impact of preferences, opportunities, and abilities. *Econometrica* 67, 1295–339.

Gould, E.D. 2002. Rising wage inequality, comparative advantage, and the growing importance of general skills in the United States. *Journal of Labor Economics* 20, 105–47.

Gronau, R. 1974. Wage comparisons – a selectivity bias. *Journal of Political Economy* 82, 1119–43.

Heckman, J.J. 1974. Shadow prices, market wages, and labor supply. *Econometrica* 42, 679–94.

Heckman, J.J. 1990. Varieties of selection bias. *American Economic Review* 80, 313–8.

Heckman, J.J. and Honoré, B.E. 1989. The identifiability of the competing risks model. *Biometrika* 76, 325–30.

Heckman, J.J. and Honoré, B.E. 1990. The empirical content of the Roy model. *Econometrica* 58, 1121–49.

Heckman, J.J., Lochner, L.J. and Taber, C. 1998a. Explaining rising wage inequality: explorations with a dynamic general equilibrium model of labor earnings with heterogeneous agents. *Review of Economic Dynamics* 1, 1–58.

Heckman, J.J., Lochner, L.J. and Taber, C. 1998b. General-equilibrium treatment effects: a study of tuition policy. *American Economic Review* 88, 381–6.

Heckman, J.J., Lochner, L.J. and Taber, C. 1998c. Tax policy and human-capital formation. *American Economic Review* 88, 293–7.

Heckman, J.J., Lochner, L.J. and Todd, P.E. 2006. Earnings equations and rates of return: the Mincer equation and beyond. In *Handbook of the Economics of Education*, ed. E.A. Hanushek and F. Welch. Amsterdam: North-Holland.

Heckman, J.J. and Navarro, S. 2007. Dynamic discrete choice and dynamic treatment effects. *Journal of Econometrics* 136, 341–96.

Heckman, J.J. and Sedlacek, G.L. 1985. Heterogeneity, aggregation, and market wage functions: an empirical model of self-selection in the labor market. *Journal of Political Economy* 93, 1077–125.

Heckman, J.J. and Sedlacek, G.L. 1990. Self-selection and the distribution of hourly wages. *Journal of Labor Economics* 8(1, Part 2), S329–63.

Heckman, J.J. and Smith, J.A. 1998. Evaluating the welfare state. In *Econometrics and Economic Theory in the Twentieth Century: The Ragnar Frisch Centennial Symposium*, ed. S. Strom. New York: Cambridge University Press.

Keane, M.P. and Wolpin, K.I. 1997. The career decisions of young men. *Journal of Political Economy* 105, 473–522.

Matzkin, R.L. 1992. Nonparametric and distribution-free estimation of the binary threshold crossing and the binary choice models. *Econometrica* 60, 239–70.

Roy, A. 1951. Some thoughts on the distribution of earnings. *Oxford Economic Papers* 3, 135–46.

Taber, C.R. 2000. Semiparametric identification and heterogeneity in discrete choice dynamic programming models. *Journal of Econometrics* 96, 201–29.

Willis, R.J. and Rosen, S. 1979. Education and self-selection. *Journal of Political Economy* 87(5, Part 2), S7–S36.

Roy, René François Joseph (1894–1977)

René Roy was born in Paris on 21 May 1894. He entered the Ecole Polytechnique in 1914, and joined the army on 15 August 1914. He was seriously wounded on 14 April 1917 at the Chemin des Dames, as a result of which he was blinded at the early age of 23. This tragedy, which meant the collapse of all his youthful hope and dreams, brought him to the slough of despond, and exceptional spiritual strength alone enabled him eventually to accept the unacceptable with serenity and to undertake a double career as an engineer and economist that was to last 60 years.

He studied at the Ecole Polytechnique (from 1918 to 1920) graduating first in his year, and then at the Ecole Nationale des Ponts et Chaussées (1920–2). He entered the Ministry of Public Works and Transport as a state engineer in 1922, specializing in problems of local railway networks and urban transport until his retirement in 1964. He died in Paris in 1977.

In parallel with this activity, he became Professor of General Political Economy and Social Economy at the Ecole des Ponts et Chaussées in 1929, and Professor of Econometrics at the Statistical Institute of the University of Paris in 1931. In 1949 he taught econometrics at the Ecole d'Application de l'Institut National de la Statistique et des Etudes Economiques (School of Instruction of the National Institute of Statistics and Economic Studies). From 1947 he was in charge of an Econometrics Seminar at the National Centre of Scientific Research.

He was elected President of the Paris Statistical Society in 1949 and of the International Econometrics Society in 1953. He was also a fellow of the International Statistical Society (1949), a member of the Academy of Moral and Political Science (1951), and an honorary fellow of the Royal Statistical Society (1957). He received the degree of Doctor Honoris Causa from the University of Geneva in 1964.

René Roy's research was focused mainly on transport, demand functions, economic indices, fields of choice and their respective relationships. His main published works are *Le régime économique des voies ferrées d'intérêt local* (1925), his doctoral thesis; *La demande de biens de consommation directe* (1935); *De l'utilité – contribution à la théorie des choix* (1942); 'Les nombres indices' (*Journal de la Société de Statistique de Paris*, 1949); and *Eléments d'économétrie* (1970). In addition, in collaboration with François Divisia and Jean Dupin, he published in 1953–4 *A la recherche du franc perdu*, whose three volumes cover the movement of prices, production and wealth respectively in France from 1914 to 1950. Roy's analysis of the basic relationships of demand functions and price and quantity index numbers are contained in his 1949 publication.

René Roy's ability to analyse very difficult questions and constantly stay abreast of the main publications of his era was a truly remarkable achievement for a totally sightless person. He showed that accomplishment is possible in the face of an irremediable adversity by dint of unremitting energy associated with remarkable intelligence. His book *Vers la lumière* (1930) gives us his message as a blind man.

MAURICE ALLAIS

See also **indirect utility function.**

Selected works

1925. *Le régime économique des voies ferrées d'intérêt local.* Paris.

1930. *Vers la lumière.* Paris: Bibliotheque-Charpentier.

1935. *Le demande de biens de consommation directe.* Paris: Hermann.

1942. *De l'utilité – contribution à la théorie des choix.* Paris: Hermann.

1949. Les nombres indices. *Journal de la Société de Statistique de Paris* 90(1–2), 15–34.

1953–4. (With F. Divisia and J. Dupin.) *A la recherche du franc perdu.* Paris: Société d'Editions Hommes et Mondes.

1970. *Eléments d'économétrie.* Paris: Presses Universitaires de France.

Royal Economic Society

Originally founded in 1890 as the British Economic Association (BEA), the Royal Economic Society (RES) assumed its current title in 1902 when it obtained a Privy Council charter and royal patronage. The RES is now unquestionably the leading organization of professional economists in Britain, with its flagship publication the *Economic Journal* (*EJ*), a world-class general journal for theoretical and applied research (having in 2004 an International Statistical Institute, ISI, journal citation ranking of 15/172 and 1.723 impact factor). Such dominance, however, has not always been the case and was not easily achieved, with the RES's fortunes, like those of many other long-established economics associations, subject to a changing complex of pressures, including at times competitors.

The RES was the eventual institutional result of the long process whereby political economy was transformed into economics as the 'sciences of the social' were dissolved and reconstructed into the modern social sciences in the second half of the 19th century. The establishment of the BEA followed a long period of consultations over whether a new learned society was required to propagate what has come to be known in the modern literature as Marshall's mission for the professionalization of economics (Middleton, 1998), or whether the Royal Statistical Society (established 1834) and/or Section F (Economics and Statistics, established 1835) of the British Association for the Advancement of Science, were sufficient vehicles.

The year 1890 witnessed also the publication of Marshall's *Principles of Economics* and the completion of the first instalment of the first Palgrave dictionary, which was published the following year, as was the first issue of the *EJ*. For Keynes (1940, p. 409), these happy concurrences made this the beginnings of the 'modern age of British economics'. This now looks somewhat questionable: the BEA and *EJ* were part of a pre-professionalization trend towards 'clearer demarcation and definition of the field of scholarly endeavour' (Kadish and Freeman, 1990, p. 23) and the achievements of the professionalization agenda were as yet limited. That this was the case is apparent from what was one of the central issues of the debate preceding the BEA's formation: would it be a closed society of professional economists (at this time, it was still not true that a majority of these consisted of academics) or would it be open to all, however imperfect their claim to be called an economist? Mindful of the contemporaneous example of the American Economic Association (AEA), where many of its leading figures had been deeply involved in methodological and policy disputations which had been unhelpful to the professionalization agenda, the BEA's founders resolved to follow a more cautious and restricted policy than its transatlantic counterpart: membership was not open but dependent upon a candidate's approval by Council (with the chosen designated as Fellows from 1902 to 1964 under the Royal Charter); the BEA was to refrain from organizing discussions and conferences, partly for fear of exposing the substantial differences of opinion within their ranks; and, in its early years, the Council routinely chose a prominent public figure rather than an academic as its president, beginning with the then Chancellor of the Exchequer, G. J. Goschen.

While the early years of the BEA were often difficult, the *EJ* was an undoubted success, and this notwithstanding the rival Oxford publication, the *Economic Review*, which had been launched a few months earlier (and survived until 1914). The *EJ* was not conceived as a specialist publication for an exclusively academic or quasi-academic economics audience, but instead as a 'means of disseminating economic truth amongst readers from all walks of life, while setting new standards of economic investigation' (Kadish and Freeman, 1990, pp. 36–7). This general informational role was to endure strongly until the 1970s and was still be present in the 1990s when, as the journal became increasingly internationalized, it almost exclusively focused on scholarly papers and its policy forum, with less emphasis on book reviews and with reports on its learned society activities hived off to the website (http://www.res.org.uk).

The *EJ*'s initial editor was F. Y. Edgeworth, followed by Keynes (singularly and jointly, 1912–45) and then R. F. Harrod (1945–61), during which time the *EJ* consolidated its status as the leading British economic journal, despite the appearance of several rivals. Cambridge, Oxford and London economists initially dominated, but from the 1970s onwards, as the balance of professional influence and authority shifted away from the older centres, the Council had to respond to pressures to make the RES a more democratic organization. Concurrently, the *EJ* editors found that the ever-increasing scale, scope and technical nature of the discipline necessitated increased personnel to provide expert opinion on journal submissions. An editorial board was established in 1971 and has since evolved; most importantly, it now has more foreign (largely, but not exclusively, US) than British economists.

RES membership is now over 3,300 individuals, of whom 60 per cent are not British residents, and there are a further 2,400 institutional subscribers to the *EJ* and the *Econometrics Journal* (established 1998). Notwithstanding the internationalization of the *EJ*, the RES mission statement remains essentially British: to be the 'professional association which promotes the encouragement of the study of economic science in academic life, government service, banking, industry and public affairs'. Increasingly an umbrella organization for a large number of activities, mainly but not exclusively to do with university economics, the RES maintains also its activities as a major publisher of scholarly editions, of which the 30-volume collected writings of Keynes is one of its major achievements, even if it did nearly bankrupt the Society. It has also finally resolved one of the issues of disagreement at its creation: since 1990 the RES has operated an annual conference, selected papers from which appear in a special issue of the *EJ*, itself now enlarged for its second century from a quarterly to a bimonthly publication.

ROGER MIDDLETON

See also **Keynes, John Maynard; Marshall, Alfred.**

Bibliography

Coats, A.W. 1968. The origins and early development of the Royal Economic Society. *Economic Journal* 78, 349–71.

Kadish, A. and Freeman, R.D. 1990. Foundation and early years. In *A Century of Economics: 100 Years of the Royal Economic Society and the Economic Journal*, ed. J.D. Hey and D. Winch. Oxford: Basil Blackwell.

Keynes, J.M. 1940. The Society's jubilee, 1890–1940. *Economic Journal* 50, 401–9.

Middleton, R. 1998. *Charlatans or Saviours? Economists and the British Economy from Marshall to Meade.* Cheltenham: Edward Elgar.

Rubin causal model

The Rubin Causal Model (RCM) is a formal mathematical framework for causal inference, first given that name by Holland (1986) for a series of previous articles developing the perspective (Rubin, 1974; 1975; 1976; 1977; 1978; 1979; 1980). There are two essential parts to the RCM, and a third optional one. The first part is the use of 'potential outcomes' to define causal effects in all situations – this part defines 'the science', which is the object of inference, and it requires the explicit consideration of the manipulations that define the treatments whose causal effects we wish to estimate. The second part is an explicit probabilistic model for the assignment of 'treatments' to 'units' as a function of all quantities that could be observed, including all potential outcomes; this model is called the 'assignment mechanism', and defines the structure of experiments designed to learn about the science from observed data or the acts of nature that lead to the observed data. The third possible part of the RCM framework is an optional distribution on the quantities being conditioned on in the assignment mechanism, including the potential outcomes, thereby allowing model-based Bayesian 'posterior predictive' (causal) inference. This part of the RCM focuses on the model-based analysis of observed data to draw inferences for causal effects, where the observed data are revealed by applying the assignment mechanism to the science. A full-length text that discusses estimation and inference for causal effects from this perspective is Imbens and Rubin (2006).

Implications of the RCM for research design

Before defining each of these three parts of the RCM, it is helpful to consider the implications of this structure for applied research about causal effects. The first part implies that we should always start by carefully defining all causal estimands (quantities to be estimated) in terms of potential outcomes, which are all values that could be observed in some real or hypothetical experiment that compares the results under an active treatment with the results under a control treatment. That is, causal effects are defined by a comparison of (*a*) the values that would be observed if the active treatment were applied and (*b*) the values that would be observed if, instead, the control treatment were applied. This step contrasts with the common practice of defining causal effects in terms of parameters in some model, where the manipulations defining the active versus control treatments are often left implicit and ill-defined, with the resulting causal inferences correspondingly weak and ill-defined. This first

part can be completely abstract and can take place before any data are observed or even collected. In the RCM, however, there is 'no causation without manipulation' (Rubin, 1975, p. 238), where the manipulation (that is, the treatment) could be real or hypothetical. The collection of potential outcomes with and without this manipulation defines the scientific objective of causal inference in all studies, whether randomized, observational or entirely hypothetical.

The second part of the RCM, the assignment mechanism, implies that, given the defined science, we should continue by explicating the design of the real or hypothetical study being used to estimate that science. The assignment mechanism describes why some study units will be (or were) exposed to the active treatment and why other study units will be (or were) exposed to the control treatment, and the reasons are formalized by the mathematical statement of the assignment mechanism. When the study is a true experiment, the assignment mechanism may involve the consideration of background (that is, pretreatment assignment) variables for the purpose of creating strata of similar units to be randomized into treatment and control, thereby improving the balance of treatment and control groups with respect to these background variables (that is, covariates). A true experiment automatically cannot use any outcome (post-treatment) variables to influence design because they are not yet observed. If the observed data were not generated by a true experiment, but rather by non-randomized observational data, there still should be an explicit design phase. That is, in an observational study, the same guidelines as in an experiment should be followed.

More explicitly, the design step in the analysis of an observational data set for causal inference should structure the data to approximate (or reconstruct or replicate) a true randomized experiment as closely as possible. In this design step, the researcher never uses or even examines any outcome data but rather identifies subsets of units such that the treatments can be thought of as being randomly assigned within the subsets. This assumed randomness of treatment assignment is assessed by examining, within these subsets of units, the similarity of the distributions of the covariates in the treatment group and in the control group. Because this design step is focused on creating these subsets of units with balanced distributions of covariates between treatment and control groups, and never uses outcome data, the researcher cannot select a design to produce a desired answer, even unconsciously.

The third part of the RCM is optional; it derives inferences for causal effects from the observed data by conceptualizing the problem as one of imputing the missing potential outcomes. That is, once outcome data are available (that is, observations of the potential outcomes corresponding to the treatments actually received by the various units), then the modelling of the outcome data given the covariates should be structured to derive predictions of those potential outcomes that would have been observed if the treatment assignments had been different. This modelling will generate stochastic predictions (that is, imputations) for all missing potential outcomes in the study, which, when combined with the actually observed potential outcomes, will allow the calculation of any causal-effect estimand. Because the imputations of the missing potential outcomes are stochastic, repeating the process results in different values for the causal-effect estimand. This variation across the multiple imputations (Rubin, 1987; 2004a) generates interval estimates and tests for the causal estimands. Typically, in practice this third part is implemented using simulation-based methods, such as Markov chain Monte Carlo computation applied to Bayesian models.

The conceptual clarity in the first two steps of the RCM often allows previously difficult causal inference situations to be easily formulated and handled. The optional third part often extends this success by relying on modern computational power to handle analytically intractable problems. With this overview in place, we consider features of the RCM in more detail.

Potential outcomes and causal effects

For defining causal effects, there are three basic primitives – concepts that are fundamental and on which we must build: units, treatments and potential outcomes. A unit is a physical object, for example a person, at a particular point in time. A treatment is an action that can be applied or withheld from a unit. We focus on the case of two treatments, although the extension to more than two treatments is simple in principle although not necessarily so with real data. Associated with each unit are two potential outcomes: the value of an outcome variable Y at a future point in time if the active treatment is applied, and the value of Y at the *same* future point in time if instead the control treatment is applied. The objective is to learn about the causal effect of the application of the active treatment relative to the control on Y, where, by definition, the causal effect is a comparison of the two potential outcomes. For example, the unit could be a person 'now' without a job, the active treatment could be participating in a job training programme, and the control could be not participating. The outcome Y could be the total income over the next three years, with the two potential outcomes being the total income with and without job training; the causal effect of being trained versus not being trained is the comparison of the person's three-year total income with and without the training.

Notationally, let W indicate which treatment the unit receives: $W = 1$ the active treatment, and $W = 0$ the control treatment. Also let $Y(1)$ be the value of the potential outcome if the unit received the active version, and $Y(0)$ the value if the unit received the control version. The causal effect of the active treatment relative to the control is the comparison of $Y(1)$ and $Y(0)$ – typically the difference, $Y(1) - Y(0)$, or perhaps the difference in logs, $\log[Y(1)] - \log[Y(0)]$, or some other comparison, possibly the ratio. We can observe only one or the other of $Y(1)$ and $Y(0)$ as indicated by $W : Y_{obs} = WY(1) + (1 - W)Y(0)$. The 'fundamental problem facing inference for causal effects' (Rubin, 1978, p. 38) is that, for any individual unit, we observe the value of the potential outcome for this unit under only one of the possible treatments, namely, the treatment actually assigned, and the potential outcome under the other treatment is missing. Thus, inference for causal effects is a missing-data problem – the 'other' value is missing, so the nature of causal inference is that at least 50 per cent of the values of the potential outcomes are missing. Covariates have values that are unaffected by the treatments, such as age or sex of the unit in the job training example, and are denoted by X. Even when X represents a lagged Y, such as total income last year, $Y(1) - X$ is not the causal effect of training unless $Y(0) = X$, but rather a change of income across time.

To clarify the RCM set-up with potential outcomes, consider a specific difficult case: what is the causal effect of race on hiring practices? To consider this explicitly causal question in the RCM, we must consider the manipulations that define the active and control treatments. Literally changing one's race is presumably impossible given current medical technology, but one can conceptualize experiments that can plausibly capture what researchers want to know, that is, how employers react to race when all else is constant. For example, suppose that résumés are submitted by mail to groups of employers, where the treatment to be applied to each résumé (that is, each unit) is the name attached to it (see, for example, Bertrand and Mullainathan, 2004). Here, the active treatment is the use of a distinctive African-American name on the résumé, and the control treatment is the use of a traditional name. In this case, the explication of what is meant by 'the causal effect race' is through the description of the manipulations, and the causal effect to be estimated is thereby well-defined: the causal effect of having a résumé with an African-American name compared with a traditional name on the resultant hiring outcome. Whether that effect corresponds to what the investigator wants to estimate or to what others believe is relevant to policy is another issue, but the causal nature of the comparison is clear. If it is not the desired quantity estimand or is deemed not relevant, then other more appropriate manipulations should be described.

Suppose, now that there are N units rather than only one. To make the representation with only two potential outcomes for each unit adequate, must accept an assumption, the stable unit treatment value assumption (SUTVA; Rubin, 1980), which rules out interference between units (Cox, 1958) and rules out different versions of the treatments for the units (for example, no 'technical errors'; Neyman, 1935; Rubin, 1990b). SUTVA can be weakened, but still some such assumption regarding the full set of potential outcomes is required. Often, in practice, SUTVA is made more plausible by aggregating the units. For example, training some of the unemployed in a local labour market may affect job opportunities for others in that local market. Therefore changing the unit of analysis to be the local labour market in a study with many geographically separated local labour markets may make it more plausible that there is no effect of the exposure of one unit to the treatment on other units.

Under SUTVA, all causal estimands (quantities to be estimated) can be defined from the matrix of values with i^{th} row: $(X_i, Y_i(0), Y_i(1))$, $i = 1, \ldots, N$. A causal estimand involves a comparison of $Y_i(0)$ and $Y_i(1)$ on all N units, or on a common subset of units; for example, the average causal effect across all units that are female as indicated by their X_i, or the median $Y_i(1)$ minus the median $Y_i(0)$ for the set of units with X_i indicating male and $Y_i(0)$ indicating no income. By definition, all relevant scientific information that is recorded is encoded in this matrix, and so the labelling of its rows is a random permutation of $1, \ldots, N$; that is, the N-row matrix $\{X, Y(0), Y(1)\}$ is row exchangeable. For convenience, we refer to this array of values as the 'science', functions of which we wish to estimate.

Brief history of potential outcomes to define causal effects

The basic idea that causal effects are the comparisons of potential outcomes on a common set of units seems so direct that it must have ancient roots, and we can find elements of this definition of causal effects among both philosophers (for example, Mill, 1843, p. 327) and experimenters (for example, Fisher, 1918, p. 214). But apparently there was no formal notation for potential outcomes until Neyman (1923), which appears to have been the first place where a mathematical analysis is written for a randomized experiment. This notation became standard for work in randomized experiments with randomization-based inference, and was a major advance. Independently and nearly simultaneously, Fisher (1925) recommended physically randomizing treatments to units in experiments, as well as a different, but compatible, method of randomization-based inference, although Fisher apparently never used the potential outcomes notation. But despite the almost immediate acceptance in the late 1920s of Fisher's proposal for randomized experiments, and of Neyman's notation for potential outcomes in randomized experiments, and of both men's proposals for randomization-based inference, this potential outcome notation was not used for causal inference more generally for a half century thereafter, apparently not until introduced by Rubin (1974). As a result, the insights into causal inference that accompanied the use of the potential outcomes notation were entirely limited to the relatively simple setting of randomization-based inference in randomized experiments.

The approach used in nonrandomized settings, during the half-century following the introduction of Neyman's seminal notation for randomized experiments, was based on mathematical models (for example, regression models) relating the observed value of the outcome variable $Y_{obs,i}$ to X_i and W_i, and then defining causal effects as parameters (for example, regression coefficients) of these models. This was the standard approach in medical and social science, including economics, and led to substantial confusion – the role of randomization cannot even be directly stated mathematically using the observed outcome notation. Of course, there were seeds of this first part of the RCM in social science before 1974, in particular in economics, in Tinbergen (1930), Haavelmo (1944) and Hurwicz (1962), but we can find no previous use of explicit notation like Neyman's to define causal effects. The use of the idea of potential outcomes certainly did appear in discussions in economic theory, for example, in the context of supply and demand functions (for example, Haavelmo, 1944) or the Roy (1951) model, but these discussions did not lead to inference in terms of potential outcomes. Instead, inference took place in terms of the specification of simultaneous equations using observed quantities and distributional properties of error terms (for example, Heckman and Robb, 1984, in the context of program evaluation models).

Nevertheless, the potential outcome part of the RCM framework for defining causal effects, namely, a generalization of Neyman's notation to allow non-randomized data, seems to have been basically accepted and adopted by most researchers by the end of the 20th century; compare, for example, Imbens and Angrist (1994) and Heckman, Ichimura and Todd (1998) with the earlier formulation in Heckman and Robb (1984). An article exploring whether the full potential outcomes framework can be avoided when conducting causal inference is Dawid (2000), which included discussion by others that was largely supportive of the propriety of potential outcomes for causal inference.

The assignment mechanism and assignment-based causal inference

The second part of the RCM framework is the specification of an 'assignment mechanism': a probabilistic model for how some units received the active treatment and how other units received the control – how we conceptualize the design for how some potential outcomes were revealed and others remained hidden (that is, missing). The assignment mechanism is fundamental to causal inference. It specifies the conditional probability of each vector of assignments $W = (W_1, \ldots, W_N)^T$ given the matrix of all covariates and potential outcomes:

$$\Pr(W | X, Y(0), Y(1)). \tag{1}$$

It appears that Rubin (1975) was the first place that expressed the possible dependence of the assignment vector on the potential outcomes in this direct way, which allows the statement of what makes randomized experiments special, and more generally, generates a classification of assignment mechanisms. Again, economic theory sometimes implied a specific assignment mechanism, but this theory was never explicitly stated as in the general form of (1). For example, individuals may choose the occupation that maximizes their earnings, as in the Roy model, which would lead to $W_i = \text{argmax}_w(Y_i(W))$, or more generally individuals may optimize an objective function that involves expectations over the unknown components of the potential outcomes. Imbens and Rubin (2006) provide details of such examples.

Randomized experiments are special in that they have 'unconfounded' and 'probabilistic' assignment mechanisms. Unconfounded assignment mechanisms (Rubin, 1990a) are free of dependence on either $Y(0)$ or $Y(1)$:

$$\Pr(W|X, Y(0), Y(1)) = \Pr(W|X). \quad (2)$$

Assignment mechanisms are 'probabilistic' (or 'probability' as in Rubin, 1990a) if each unit has a positive probability of receiving either treatment:

$$0 < \Pr(W_i = 1|X, Y(0), Y(1)) < 1. \quad (3)$$

'Strongly ignorable' assignment mechanisms (Rosenbaum and Rubin, 1983a) satisfy (2) and (3), and thus have unit level probabilities, or 'propensity scores', $\Pr(W_i = 1|X_i)$, that are strictly between 0 and 1, and are free of all potential outcomes.

Ignorable assignment mechanisms (Rubin, 1978), are free from dependence on missing potential outcomes but may depend on observed potential outcomes $Y_{\text{obs}} = \{Y_{\text{obs},i}\}$:

$$\Pr(W|X, Y(0), Y(1)) = \Pr(W|X, Y_{\text{obs}}). \quad (4)$$

Ignorable but confounded assignment mechanisms arise in practice, especially in sequential experiments. All strongly ignorable assignment mechanisms are unconfounded, and all unconfounded assignment mechanisms are ignorable, but not the other way. Strongly ignorable assignment mechanisms allow particularly straightforward estimation of causal effects, and are the basic template for the analysis of observational studies. More generally, observational studies have possibly confounded, non-ignorable, assignment mechanisms. A confounded assignment mechanism is one that depends on the potential outcomes, and so does not satisfy (2); a non-ignorable assignment mechanism does not even satisfy (4), and thus allows treatment assignment (or, to use common economics terminology, 'selection') to depend on unobserved values, that is, the missing potential outcomes, $Y_{\text{mis}} = \{Y_{\text{mis},i}\}$, $Y = \{Y_{\text{obs}}, Y_{\text{mis}}\}$.

When the assignment is strongly ignorable, it can generally be represented as a 'regular' assignment mechanism, which is proportional to the product of the propensity scores:

$$\Pr(W|X, Y(0), Y(1)) \propto \prod_{i=1}^{N} \Pr(W_i = 1|X_i). \quad (5)$$

Regular assignment mechanisms are the basic template in the RCM for the analysis of observational data, because two units with the same propensity score but different treatments are essentially randomized into the two treatment conditions. Therefore, with regular assignment mechanisms, matching on the propensity score (for example, as in Rosenbaum and Rubin, 1984), or subclassifying on it (for example, as in Rosenbaum and Rubin, 1985), restores the assumed underlying experimental design, and inference is straightforward based only on the assignment mechanism. These assignment-based methods of inference are due to Neyman (1923) and Fisher (1925), and they involve the calculation of large-sample confidence intervals and exact significance tests for null hypotheses, respectively; both are discussed in Rubin (1990a; 1990b; 1991). For the validity of either Fisher's or Neyman's approach, the analysis must formally be defined a priori, as part of the design. But the existence of these assignment-based modes of inference helps justify the view in the RCM that the model for the assignment mechanism is more fundamental for causal inference than the model for the science, which is not needed for randomization-based inference.

Thus, in the RCM an observational study should be designed as if its data arose from a 'broken' randomized experiment, where the unknown propensity scores must be reconstructed on the basis of the covariates X prior to ever observing any potential outcomes. In such settings, it is often quite advantageous to use estimated propensity scores (for example, as in Rosenbaum and Rubin, 1984; Rubin and Thomas, 1992a; 1992b; 1996; 2000; Hirano, Imbens and Ridder, 2003). When estimated propensity scores for some units are so low that they have essentially no chance of being treated, then those units should be discarded from further consideration when estimating the treatment effect in the treated (see, for example, Peters, 1941; Belson, 1956; Cochran and Rubin, 1973; Rubin, 1973a; 1973b; 1977; Rosenbaum and Rubin, 1985; Dehijia and Wahba, 1999; Crump et al., 2005). The result of the design phase should be treatment and control groups with very similar distributions of observed Xs, because of either matching or subclassification. If a dataset does not permit similar X distributions to be constructed in treatment and control groups, it cannot be used to support causal inferences without extraneous assumptions justifying extrapolations. Rubin (2002) offers an example of such matching and subclassification in the context of the US tobacco litigation, and Rubin (2006a) is a book devoted to matched sampling.

A striking example of the applied success of the above approach to inference in observational studies is Dehijia and Wahba (1999), which reanalysed the classic Lalonde (1986) data on job-training experiments but using the assignment-based approach of the RCM. In contrast to the wild variety of contradictory, but highly significant, answers found by the traditional econometric methods, Dehijia and Wahba used matching on the propensity score to arrive at inferences that tracked those from the underlying randomized experiment in the overall sample and in a variety of subsamples (see also Abadie and Imbens, 2006).

Posterior predictive, or model-based, causal inference

The third part of the RCM involves an optional distribution on the N-row array of science, $\Pr(X, Y(0), Y(1))$, thereby allowing Bayesian, or model-based, inference as well as assignment-based inference. An important virtue of the RCM framework is that it distinctly separates the science – its definition in the first part (and a possible model for it in the third part) from the design of what is revealed about the science – the assignment mechanism in the second part, which can also involve some scientific insights as when it is assumed to be generated by equilibrium conditions, as in supply and demand models, or by optimizing behaviour, and so on.

Bayesian inference for causal effects directly and explicitly confronts the missing potential outcomes, Y_{mis}, by using the specification for the assignment mechanism and the specification for the underlying data to derive the posterior predictive distribution of Y_{mis}, that is, the distribution of Y_{mis} given all observed values:

$$\Pr(Y_{\text{mis}}|X, Y_{\text{obs}}, W).$$

This distribution is posterior because it conditions on all observed values (X, Y_{obs}, W) and is predictive because it predicts (stochastically) the missing potential outcomes. From this distribution and all of the observed values (the observed potential outcomes, Y_{obs}; the observed assignments, W; and observed covariates, X), the posterior distribution of any causal effect can, in principle, be calculated. This conclusion is immediate if we view the posterior predictive distribution as specifying how to take a random draw of Y_{mis}. Once a value of Y_{mis} is drawn, any causal effect can be directly calculated from the drawn value of Y_{mis} and the observed values of X and Y_{obs}, for example, the median causal effect for males: $\text{med}\{Y_i(1) - Y_i(0)|X_i \text{ indicate males}\}$. Repeatedly drawing values of Y_{mis} and calculating the causal effect for each draw generates the posterior distribution of the desired causal effect. Thus, we can view causal inference

entirely as a missing data problem, where we multiply-impute (Rubin, 1987; 2004a) the missing potential outcomes to generate a posterior distribution for the causal effects.

For example, the treated units have $Y_i(1)$ observed and $Y_i(0)$ missing. Under ignorability, the regression of $Y_i(0)$ on X_i among treated units, for which there is no direct evidence, can be shown to be the same as the regression of $Y_i(0)$ on X_i among controls, for which we have data. Thus, this third part of the RCM tells us to build a realistic model of $Y_i(0)$ given X_i among control subjects, and use it to impute the missing $Y_i(0)$ among the treated from their X_i values, while being wary of issues of extrapolation beyond the observed range of X_i control values. Analogously, build a model of $Y_i(1)$ given X_i among the treated, and use it to impute the missing $Y_i(1)$ among controls. The general structure is outlined in Rubin (1978), and is developed in detail in Imbens and Rubin (2006); a chapter-length summary appears in Rubin (2006).

Advantages of the RCM

Because of the flexibility in the RCM for (a) formulating causal estimands, and (b) positing assignment mechanisms, it can handle difficult cases in principled ways. With observational studies, estimated propensity scores play a key role, because the initial analysis proceeds as if the assignment mechanism were unconfounded. To assess the consequences of this assumption, sensitivity analyses can be conducted under various hypothetical situations, typically with fully missing covariates, U, such that treatment assignment is unconfounded given U but not given the observed data. Assumed relationships (given X) between U and W, and between U, $Y(0)$ and $Y(1)$, are then varied, for example, as in Rosenbaum and Rubin (1983b), utilizing the third part of the RCM. Ideally, this speculation occurs at the design stage. Extreme versions of sensitivity analyses lead to large-sample bounds (for example, Manski, 2003).

A complication, common when the units are people, is non-compliance with assigned treatment. Early work related to this issue can be found in economics using the terminology of instrumental variables, and the bridge from this terminology to the basic RCM is developed in Imbens and Angrist (1994) and in Angrist, Imbens and Rubin (1996), and the connection to the full RCM approach is presented in Imbens and Rubin (1997) and in Hirano et al. (2000). Another complication is censoring due to death, where units may 'die' before the final outcome can be measured. This problem is formulated from the RCM perspective in Rubin (2006b), with bounds given in Zhang and Rubin (2003); see Zhang, Rubin and Mealli (2007) for application to the evaluation of job-training programmes. This topic is also related to 'direct' and 'indirect' causal effects (Rubin, 2004b; 2005). Combinations of such complications are considered in Barnard et al. (2003) in the context of a school choice example, as well as in Mealli and Rubin (2002; 2003), Jin and Rubin (2007) and Frangakis and Rubin (1999; 2001) in other contexts. The above examples can all be viewed as special cases of 'principal stratification' (Frangakis and Rubin, 2002).

The references in the preceding paragraph are clearly idiosyncratic in the sense of their being specific applications of the RCM in which the authors of this article have been participants, and are not representative, but we hope they provide indications of the breadth of recent applications of the RCM.

GUIDO W. IMBENS AND DONALD B. RUBIN

See also **Bayesian econometrics; Bayesian statistics; econometrics; matching; matching estimators; treatment effect.**

Bibliography

Abadie, A. and Imbens, G.W. 2006. Large sample properties of matching estimators for average treatment effects. *Econometrica* 74, 235–67.

Angrist, J.D., Imbens, G.W. and Rubin, D.B. 1996. Identification of causal effects using instrumental variables. *Journal of the American Statistical Association* 91, 444–72 (an Applications Invited Discussion Article with discussion and rejoinder).

Barnard, J., Hill, J., Frangakis, C. and Rubin, D. 2003. A principal stratification approach to broken randomized experiments: a case study of vouchers in New York City. *Journal of the American Statistical Association* 98, 299–323 (with discussion and rejoinder).

Bertrand, M. and Mullainathan, S. 2004. Are Emily and Greg more employable than Lakisha and Jamal? A field experiment on labor market discrimination. *American Economic Review* 94, 991–1013.

Belson, W.A. 1956. A technique for studying the effect of a television broadcast. *Applied Statistics* 5, 195–202.

Cochran, W.G. and Rubin, D.B. 1973. Controlling bias in observational studies: a review. *Sankhya*, A 35, 417–46.

Cox, D.R. 1958. *The Planning of Experiments*. New York: Wiley.

Crump, R., Hotz, J., Imbens, G. and Mitnik, O. 2005. Moving the goalposts: addressing limited overlap in estimation of average treatment effects by changing the estimand. Unpublished manuscript, Department of Economics, University of California, Berkeley.

Dawid, A.P. 2000. Causal inference without counterfactuals. *Journal of the American Statistical Association* 95, 407–24 (with discussion).

Dehijia, R. and Wahba, S. 1999. Causal effects in non-experimental studies: re-evaluating the evaluations of training programs. *Journal of the American Statistical Association* 94, 1053–62.

Fisher, R.A. 1918. The causes of human variability. *Eugenics Review* 10, 213–20.

Fisher, R.A. 1925. *Statistical Methods for Research Workers*, 1st edn. Edinburgh: Oliver and Boyd.

Frangakis, C. and Rubin, D.B. 1999. Addressing complications of intention-to-treat analysis in the combined presence of all-or-none treatment-noncompliance and subsequent missing outcomes. *Biometrika* 86, 366–79.

Frangakis, C.E. and Rubin, D.B. 2001. Addressing an idiosyncrasy in estimating survival curves using double sampling in the presence of self-selected right censoring. *Biometrics* 57, 333–42 (with discussion and rejoinder, 343–53).

Frangakis, C.E. and Rubin, D.B. 2002. Principal stratification in causal inference. *Biometrics* 58, 21–9.

Haavelmo, T. 1944. The probability approach in econometrics. *Econometrica* 15, 413–19.

Heckman, J., Ichimura, H. and Todd, P. 1998. Matching as an econometric evaluation estimator. *Review of Economic Studies* 65, 261–94.

Heckman, J. and Robb, R. 1984. Alternative methods for evaluating the impact of interventions. In *Longitudinal Analysis of Labor Market Data*, ed. J. Heckman and B. Singer. Cambridge: Cambridge University Press.

Hirano, K., Imbens, G. and Ridder, G. 2003. Efficient estimation of average treatment effects using the estimated propensity score. *Econometrica* 71, 1161–89.

Hirano, K., Imbens, G., Rubin, D.B. and Zhou, X. 2000. Estimating the effect of an influenza vaccine in an encouragement design. *Biostatistics* 1, 69–88.

Holland, P.W. 1986. Statistics and causal inference. *Journal of the American Statistical Association* 81, 945–70.

Hurwicz, L. 1962. On the structural form of interdependent systems. In *Logic, Methodology, and Philosophy of Science, Proceedings of the 1960 International Congress*, ed. E. Nagel, P. Suppes and A. Tarski. Stanford, CA: Stanford University Press.

Imbens, G.W. and Angrist, J. 1994. Identification and estimation of local average treatment effects. *Econometrica* 62, 467–76.

Imbens, G.W. and Rubin, D.B. 1997. Bayesian inference for causal effects in randomized experiments with noncompliance. *Annals of Statistics* 25, 305–27.

Imbens, G.W. and Rubin, D.B. 2006. *Causal Inference in Statistics and the Medical and Social Sciences*. Cambridge: Cambridge University Press.

Jin, H. and Rubin, D.B. 2007. Principal stratification for causal inference with extended partial compliance: application to Efron–Feldman data. *Journal of the American Statistical Association*.

Lalonde, R. 1986. Evaluating the econometric evaluations of training programs. *American Economic Review* 76, 604–20.

Manski, C.F. 2003. *Partial Identification of Probability Distributions*. New York: Springer-Verlag.

Mealli, F. and Rubin, D.B. 2002. Assumptions when analyzing randomized experiments with noncompliance and missing outcomes. *Health Services Outcome Research Methodology* 3, 225–32.

Mealli, F. and Rubin, D.B. 2003. Assumptions allowing the estimation of direct causal effects: discussion of 'Healthy, wealthy, and wise? Tests for direct causal paths between health and socioeconomic status' by Adams et al. *Journal of Econometrics* 112, 79–87.

Mill, J.S. 1843. A System of Logic. In *Collected Works of John Stuart Mill*, vol. 7, ed. J.M. Robson. Toronto: University of Toronto Press, 1973.

Neyman, J. 1923. On the application of probability theory to agricultural experiments: essay on principles, section 9. Translated in *Statistical Science* 5(1990), 465–80.

Neyman, J. 1935. Statistical problems in agricultural experimentation. Supplement to *Journal of the Royal Statistical Society, B* 2, 107–8 (with discussion). (With cooperation of K. Kwaskiewicz and St. Kolodziejczyk.)

Peters, C.C. 1941. A method of matching groups for experiments with no loss of population. *Journal of Educational Research* 34, 606–12.

Rosenbaum, P.R. and Rubin, D.B. 1983a. The central role of the propensity score in observational studies for causal effects. *Biometrika* 70, 41–55.

Rosenbaum, P.R. and Rubin, D.B. 1983b. Assessing sensitivity to an unobserved binary covariate in an observational study with binary outcome. *Journal of the Royal Statistical Society, B* 45, 212–18.

Rosenbaum, P.R. and Rubin, D.B. 1984. Reducing bias in observational studies using subclassification on the propensity score. *Journal of the American Statistical Association* 79, 516–24.

Rosenbaum, P.R. and Rubin, D.B. 1985. Constructing a control group using multivariate matched sampling incorporating the propensity score. *American Statistician* 39, 33–8.

Roy, A.D. 1951. Some thoughts on the distribution of earnings. *Oxford Economic Papers* 3, 135–46.

Rubin, D.B. 1973a. Matching to remove bias in observational studies. *Biometrics* 29, 159–83. Correction note: 1974. *Biometrics* 30, 728.

Rubin, D.B. 1973b. The use of matched sampling and regression adjustment to remove bias in observational studies. *Biometrics* 29, 185–203.

Rubin, D.B. 1974. Estimating causal effects of treatments in randomized and nonrandomized studies. *Journal of Educational Psychology* 66, 688–701.

Rubin, D.B. 1975. Bayesian inference for causality: the importance of randomization. *Proceedings of the Social Statistics Section of the American Statistical Association*, 233–9.

Rubin, D.B. 1976. Inference and missing data. *Biometrika* 63, 581–92.

Rubin, D.B. 1977. Assignment of treatment group on the basis of a covariate. *Journal of Educational Statistics* 2, 1–26.

Rubin, D.B. 1978. Bayesian inference for causal effects: the role of randomization. *Annals of Statistics* 6, 34–58.

Rubin, D.B. 1979. Discussion of 'Conditional independence in statistical theory' by A.P. Dawid. *Journal of the Royal Statistical Society*, Series B 41, 27–8.

Rubin, D.B. 1980. Discussion of 'Randomization analysis of experimental data in the Fisher randomization test' by Basu. *Journal of the American Statistical Association* 75, 591–3.

Rubin, D.B. 1987. *Multiple Imputation for Nonresponse in Surveys*. New York: Wiley.

Rubin, D.B. 1990a. Formal modes of statistical inference for causal effects. *Journal of Statistical Planning and Inference* 25, 279–92.

Rubin, D.B. 1990b. Neyman (1923) and causal inference in experiments and observational studies. *Statistical Science* 5, 472–80.

Rubin, D.B. 1991. Practical implications of modes of statistical inference for causal effects. *Biometrics* 47, 1213–34.

Rubin, D.B. 2000. The utility of counterfactuals for causal inference. Comment on A.P. Dawid, 'Causal inference without counterfactuals'. *Journal of the American Statistical Association* 95, 435–8.

Rubin, D.B. 2002. Using propensity scores to help design observational studies: application to the tobacco litigation. *Health Services and Outcomes Research Methodology* 2, 169–88.

Rubin, D.B. 2004a. *Multiple Imputation for Nonresponse in Surveys*. New York: Wiley. Reprinted with new appendices as a Wiley Classic.

Rubin, D.B. 2004b. Direct and indirect causal effects via potential outcomes. *Scandinavian Journal of Statistics* 31, 161–70 (with discussion and reply, 196–8).

Rubin, D.B. 2005. Causal inference using potential outcomes: design, modeling, decisions. 2004 Fisher Lecture. *Journal of the American Statistical Association* 100, 322–31.

Rubin, D.B. 2006a. *Matched Sampling for Causal Effects*. Cambridge: Cambridge University Press.

Rubin, D.B. 2006b. Causal inference through potential outcomes and principal stratification: applications to studies with 'censoring' due to death. *Statistical Science* 21, 299–321.

Rubin, D.B. 2007. Statistical inference for causal effects, with emphasis on applications in epidemiology and medical statistics. In *Handbook of Statistics: Epidemiology and Medical Statistics*, ed. C.R. Rao, J.P. Miller and D.C. Rao. Amsterdam: North-Holland.

Rubin, D.B. and Thomas, N. 1992a. Affinely invariant matching methods with ellipsoidal distributions. *Annals of Statistics* 20, 1079–93.

Rubin, D.B. and Thomas, N. 1992b. Characterizing the effect of matching using linear propensity score methods with normal covariates. *Biometrika* 79, 797–809.

Rubin, D.B. and Thomas, N. 1996. Matching using estimated propensity scores: relating theory to practice. *Biometrics* 52, 249–64.

Rubin, D.B. and Thomas, N. 2000. Combining propensity score matching with additional adjustments for prognostic covariates. *Journal of the American Statistical Association* 95, 573–85.

Tinbergen, J. 1930. Determination and interpretation of supply curves: an example. *Zeitschrift fur Nationalokonomie*. Reprinted in *The Foundations of Econometric Analysis*, ed. D.F. Hendry and M.S. Morgan. Cambridge: Cambridge University Press, 1997.

Zhang, J. and Rubin, D.B. 2003. Estimation of causal effects via principal stratification when some outcomes are truncated by 'death'. *Journal of Educational and Behavioral Statistics* 28, 353–68.

Zhang, J., Rubin, D. and Mealli, F. 2007. Evaluating the effects of job training programs on wages through principal stratification. *Advances in Econometrics* 21.

Ruggles, Richard (1916–2001)

Richard Ruggles and his wife, Nancy Ruggles, with whom he co-authored almost all of his work, did pioneering work in the field of national economic accounting. Mr Ruggles attended Harvard for both undergraduate and graduate study, earning his BA in 1939, an MA in 1941 and his Ph.D. in 1942. After earning his doctorate, Richard Ruggles joined the Office of Strategic Services as an economist during the Second World War. He worked for the office in London, where he estimated the production rates of tanks at German factories using photographs of the serial numbers from captured or destroyed tanks. In 1945–6 he was with the US Strategic Bombing Survey in Tokyo and Washington. Mr Ruggles returned briefly to Harvard as an instructor in 1946 before joining the Yale faculty a year later as an assistant professor of economics. He was named an associate professor in 1949 and a full professor in 1954. He was appointed the Stanley Resor Professor of Economics in 1954. He chaired the department of economics from 1959 to 1962. He also conducted research for numerous government agencies and bodies, including the United Nations, the Organization of American States, the Federal Reserve Board, the Bureau of the Census and the National Bureau of Economic Research, as well as the Ford Foundation.

Three principal themes emerge in the work of the Ruggles. The first is the reconciliation of macrodata with microdata. National accounts were developed during the 1930s and 1940s by Simon Kuznets, Richard Stone and Richard Ruggles, among others. The 1950s saw the development of microdata such as the Current Population Survey (CPS) in the United States. In principle, the data contained in microdata sources such as income should be consistent with the corresponding entries in the national accounts. In practice, however, this was seldom the case. Several requirements are put forward by Nancy and Richard Ruggles to fully integrate the two sources. First, the definition of sectors should be the same. For example, while household microdata include only households, the macro 'household' accounts often include non-profits and group quarter residents, such as those living in college dormitories, nursing homes and prisons. Second, definitions and imputations, such as the treatment of pensions and imputed rent, should be consistent between the two sources. Third, alignment of macro and microdata should not rely exclusively on macro totals. For example, in national accounts personal interest is computed as a residual whereas in microdata the household provides a direct estimate.

The second theme is the synthesis of microdata from several sources. Since household surveys can ask only a limited number of questions, different surveys concentrate on different characteristics. The CPS focuses on demographics and income, while the Consumer Expenditure Survey is very strong on expenditures and the Survey of Consumer Finances concentrates on assets and liabilities. Another problem is that different microdata focus on different parts of the income distribution. The CPS focuses mainly on the middle classes but its income data are weak for the lower and upper tails while the Internal Revenue Service Tax Model, a sample of tax returns, contains detailed income data on the upper tail but limited information on the bottom tail since these families do not file tax returns.

The solution proposed by the Ruggles is a statistical match of microdata. The idea is to merge microdata files which are complementary in terms of the variables they contain or the parts of the distribution that they cover. One such successful match described in their work was between the 1970 Census of Population and the 1969 Tax Model.

A third theme is the importance of institutional sectoring for the analysis of economic behaviour. In several papers, the Ruggles focus on the measurement of savings. Though most theories of savings, such as the life-cycle model, implicitly assume that all savings is done by households, Nancy and Richard Ruggles argue that savings is done by different institutions. In their accounting scheme, they develop separate current and capital accounts for the household, enterprise, and government sectors. They find that the household and the enterprise sectors are each self-financing. On net, the household sector channels almost no financial savings to the enterprise sector, and almost all investment done by enterprises is financed through enterprise savings. These results have wide-ranging implications for theories of savings and investment.

EDWARD N. WOLFF

Selected works

1947. *An Introduction to National Income and Income Analysis*. New York: McGraw-Hill.

1993. Accounting for saving and capital formation in the United States, 1947–1991. *Journal of Economic Perspectives* 7(2), 3–17.

1970. (With N.D. Ruggles.) *The Design of Economic Accounts.* New York: NBER.

1975. (With N.D. Ruggles.) The role of microdata in the national economic accounts. *Review of Income and Wealth* 21, 203–16.

1986. (With N.D. Ruggles.) The integration of macro- and micro data for the household sector. *Review of Income and Wealth* 32, 245–76.

1992. (With N.D. Ruggles.) Household and enterprise saving and capital formation in the United States, 1947–1991: market transactions view. *Review of Income and Wealth* 38, 119–64.

1977. (With N.D. Ruggles and E.N. Wolff.) Merging microdata: rationale, practice and testing. *Annals of Economic and Social Measurement* 6, 407–28.

S

Saint-Simon, Claude-Henri (1760–1825)

Born in 1760 into a noble family, Saint-Simon spent the first 40 years of his life as a soldier and speculator before devoting himself to the study of science and society. Commissioned in 1778, he served with the French forces in the Caribbean and in America, taking part in the Battle of Yorktown (1781). In 1787 he left the army and became associated with a Spanish project for a canal linking Madrid to the Atlantic, in which project he intended to direct the workforce of 6,000 men. The outbreak of the French Revolution prompted his return to France, where he became President of the Municipal Assembly in Falvy, near Péronne. His ambitions for social improvement led him into the purchase of aristocratic and church property from the government, and a financial partnership he formed to this end met with great success. A period of imprisonment in 1793–4 ended with the fall of Robespierre, and in the ensuing period his business interests expanded rapidly. On the proceeds of his financial successes he founded a salon and became a patron of the sciences. During the peace of 1801–2 he travelled to England, and then to Geneva to visit Madame de Staël. While in Geneva he published his first text of any significance on the reform of society, *Lettre d'un habitant de Genève à l'humanité* (1802). Returning to Paris via Germany, he published further pieces on social reorganization, though his writing was interrupted by the collapse of his personal fortune in 1806. With the support of a former servant, Saint-Simon found time for full-time study and in 1807 was able to publish an introduction to the scientific tasks of the 19th century. In 1814 he was joined in his work by the historian Augustin Thierry, and with his aid assumed the role of a leading publicist for liberal interests. This found its most direct expression in the editing of a series of journals: *L'Industrie* (1816–18); *Le Politique* (1819); and *L'Organisateur* (1819–20). This last brought him some recognition in France and abroad, and led him to the publication of a series of pieces in 1821 under the title *Du Système industriel*, one of his most important works. Despite this success, he felt that his work failed to gain appropriate recognition, and he attempted suicide in 1823. Nursed back to health by a loyal band of followers, he continued writing and studying, his final years being marked increasingly by his interest in religious sentiment as a means of social change. He died in May 1825.

The work of Saint-Simon has been variously described as corporatist, totalitarian and even anarchist. Some care is needed in such characterizations, for the work of Saint-Simon must be distinguished from that of his followers like Comte or Halévy who contributed to the formation of 'Saint-Simonianism'; and it is also necessary to treat with care the accounts of Saint-Simon to be found in the work of those heavily influenced by him, such as Proudhon or Durkheim. Saint-Simon's own writing was not presented in a systematic manner, nor did he pay much regard to scholarly niceties when developing his ideas. Nevertheless, his name is associated with a number of important ideas which have marked the development of social thought.

Saint-Simon believed that the study of society should be conducted on a scientific basis; that a positive, empirical science of society was both necessary and possible. Society, he argued, was like an organism governed by natural laws; and a 'healthy' society was one which is well-organized. Proper recognition of this fact would make possible the reconstruction of society on sound foundations – utopia would become constructible through the application of science to society.

Future society would be *industrial* society, in which 'general directors' would ensure that useful work was unhindered and government would therefore administer things, not people. Politics would become the 'science of production' – the link to 19th-century socialist thought is here quite evident. 'Industry' embraced all kinds of productive activity, and so 'industrial society' is one of productive activity in general, not a vision of a technological or manufacturing future. Accordingly, society is primarily arranged into 'industrious' and 'idle' classes. The future society was not to be a classless one: differences between groups would continue to exist, but would not be a source of social antagonism.

In later life the teachings of Saint-Simon assumed an ever more spiritual cast, a development that was promoted after his death by the group of followers that he had gathered. In the early 1830s the sect had a large number of influential adherents, although a formal organization in the shape of a 'Church' soon collapsed. Napoleon III was an open admirer of Saint-Simon's ideas, and the creation of the Crédit Mobilier investment bank, a model commercial bank involved in economic and financial developments, owed much to Saint-Simon's ideas.

K. TRIBE

Selected works

1964. *Social Organization, The Science of Man and other writings*, ed. F.M.H. Markham. New York: Harper & Row.
1966. *Oeuvres de Claude-Henri de Saint-Simon*. 6 vols, Paris: Editions Anthropos.
1975. *Henri Saint-Simon (1760– 1825), Selected Writings on Science, Industry and Social Organisation*, ed. K. Taylor, London: Croom Helm.

Samuelson, Paul Anthony (born 1915)

Paul Anthony Samuelson (born in Gary, Indiana, in 1915) has made fundamental contributions to nearly all branches of economic theory. Besides the specific analytic contributions, Samuelson more than anyone else brought economics from its pre-1930s verbal and diagrammatic mode of analysis to the quantitative mathematical style and methods of reasoning that have dominated for the last three decades. Beyond that, his *Economics* (McGraw Hill, first edition, 1948, now in its twelfth edition, the first with a co-author, William D. Nordhaus) has educated millions of students, teaching that economics, however dismal, need not be dull.

Ten eminent economists describe and evaluate his work in their respective fields in Brown and Solow (1983). Arrow (1967) and Lindbeck (1970) provide useful overall reviews. (See also the papers in Feiwel, 1982.)

Samuelson's work consists of *Foundations of Economic Analysis* (1947, reprinted in an enlarged edition in 1983), *Economics, Linear Programming and Economic Analysis* (1958, joint with Robert Dorfman and Robert M. Solow) and his *Collected Scientific Papers* (Volumes I and II, 1966; Volume III, 1972, Volume IV, 1977, and Volume V, 1986). The five volumes of the *Collected Papers* include 388 articles, most of them indeed scientific.

Bliss in his 1967 review of the first two volumes of the *Collected Scientific Papers* comments on the impossibility for anyone other than Samuelson of reviewing his work. The task has not been made any easier by the publication of another two volumes of collected papers, and by the 144-page summary of developments in economic theory since the *Foundations* in the 1983 enlarged edition. Rather than try to be comprehensive, I will describe the major analytic contributions in several areas, ending with macroeconomics where I also discuss Samuelson's views and advice on economic policy. I conclude with a description of his role at and through MIT.

Although the topic-by-topic approach is unavoidable, the man and the economist is more than the sum of his contributions in several areas. The verve and sparkle of his style, the breadth of his economic and general knowledge, the mastery of the historical setting and the generosity of his hyphenated freight-train allusions to predecessors, are unique. Samuelson's presidential address to the American Economic Association (1961: II, ch. 113) is a good sampler. (References to the *Collected Scientific Papers* (*CSP*) will give year of publication of the original article where needed, followed by volume number, and chapter and/or page number as needed.)

I Background

Samuelson has provided fragments of his autobiography in 'Economics in a Golden Age: A Personal Memoir' (1972: IV, ch. 278), and in biographical articles on contemporaries and teachers. He attended 14 schools, in Gary, Indiana, on the North Side of Chicago, in Florida, and then at Hyde Park High in Chicago. From Hyde Park High he entered the University of Chicago in January 1932, taking his first economics course from Aaron Director. 'It was as if I was made for economics' (1972: IV, p. 885). Milton Friedman and George Stigler were Chicago graduate students at the time. Jacob Viner's famous course

in economic theory provided the sound non-mathematical microeconomics that any economist needs to truly understand the field (1972: IV, ch. 282; see also Bronfenbrenner, 1982).

In 1935 he moved to graduate school at Harvard, propelled by a fellowship that required him to leave Chicago and attracted, he claims, by the ivy and the monopolistic competition revolution. Samuelson spent five years at Harvard, the last three as a Junior Fellow. It was the time of both the Keynesian and monopolistic competition revolutions, and 'Harvard was precisely the right place to be' (1972: IV, p. 889). The teachers he mentions most are Hansen, Leontief, Schumpeter and E.B. Wilson, the mathematical physicist and mathematical economist.

His fellow students make up the larger part of the honour roll of early post-Second World War United States economics (1972: IV, p. 889). Among them was his wife of 40 years, Marion Crawford (1915–78, author of a well-known 1939 article on the tariff. Abram Bergson (particularly his 1938 article on the social welfare function) and Lloyd Metzler are most mentioned among his other fellow students. Samuelson was the dominant presence among the students: Cary Brown in conversation describes the excitement as his papers were analysed and absorbed by the graduate students; James Tobin in correspondence says that the students loved the seminars, where Samuelson could be counted on to put down his seniors with brash brilliance.

The Keynesian revolution and Alvin Hansen had a greater impact on Samuelson's work and attitudes than the monopolistic competition revolution and Chamberlin. Chamberlin is barely mentioned in his reminiscences of Harvard and his only monopolistic competition article appeared in 1967 in the Chamberlin *festschrift* (III, ch. 131). Much of Samuelson's work assumes perfect competition, but none of his macroeconomics or his policy advice gives any credence to the view that the macro-economy is better left alone than treated by active policy (except perhaps his views on flexible exchange rates).

His first published article 'A Note on the Measurement of Utility' (1937: I, ch. 20) appeared when he was a 21-year-old graduate student. By 1938 the flow was up to five articles a year, a rate of production that has been maintained with perturbations for half a century. And of course, since 1948 he has produced a new edition of *Economics* almost every three years.

Samuelson moved to MIT as an Assistant Professor in 1940 and has remained there since. Harvard's failure to match MIT's offer at that time has been the subject of much speculation. Samuelson himself has been eager to find excuses for Harvard (1972: IV, ch. 278, footnote 11, p. 896). His is not the best position from which to judge or to write freely; he has noted that academic life, and by implication the chairman of the Economics Department, Burbank, one of the few of whom Samuelson speaks harshly in print, were not innocent of anti-Semitism in that pre-Second World War era. Burbank was a political power in the Department and University. His attitude to mathematical economics can be gauged by the fact that indifference curves were outlawed in the introductory course he supervised.

It is hard to believe that even the Harvard of 1940 would have been unable to find room for an economist of Samuelson's already recognized stature unless a non-academic reason or reasons stood in the way. Among those reasons were anti-Semitism, his then brashness, and his brilliance: indeed Schumpeter is rumoured to have told his colleagues that it would have been easier to forgive their vote if it had been based on anti-Semitism rather than on the fact that Samuelson was smarter than they were.

Samuelson has been at MIT since 1940, virtually without a break. Except for a few months away on a Guggenheim, he has taken time off only in Cambridge, Mass. He proudly claims that he has never been in Washington for as long as a week – though he was a major adviser to President Kennedy. His only departure from academic economics came in 1944–5, when he worked at MIT's Radiation Laboratory. He became one of 12 MIT Institute Professors in 1966.

He has gathered all the honours the profession can offer: the first John Bates Clark medal (1947); the second Nobel Memorial Prize in Economics (1970); he has been President of the American Economic Association (1961), the Econometric Society (1951), and the International Economic Association (1965–8); and he has been awarded numerous other prizes and honorary degrees.

Although many graduate students have passed through his classes and been profoundly affected by him, there is no Samuelson school of economics, no overarching grand design for either economics or the world that is uniquely his. It is for that reason that his contributions have to be discussed field by field. The nearest that he has come to proclaiming a vision is in the *Foundations*.

II Foundations of Economic Analysis

Foundations, published in 1947, is based on Samuelson's 1941 David Wells prize-winning dissertation, *Foundations of Analytic Economics*, subtitled 'The Observational Significance of Economic Theory'. Its themes are partially described by the subtitle and by the motto from J. Willard Gibbs, 'Mathematics is a language'. The thesis, dated 1940, is very close in content to the *Foundations*.

The *Foundations* in places claims to be an attempt to derive empirically meaningful comparative equilibrium results from two general principles, that of maximization, and Samuelson's *correspondence principle*. The correspondence principle states that the hypothesis of dynamic stability of a system yields restrictions that make it possible to answer comparative equilibrium questions.

The maximizing theme recurs in Samuelson's 1970 Nobel Prize lecture, 'Maximum Principles in Analytic Economics' (III, ch. 130). The point is not the now common view that only models in which everyone is relentlessly maximizing are worth considering. Rather it is that the properties of the maximum (for instance, second-order conditions) usually imply the comparative static properties of the system. Samuelson also invokes the generalized Le Chatelier principle, which loosely interpreted states that elasticities are larger the fewer constraints are imposed on a system. Analogies from physics (and biology) figure prominently in Samuelson's analytic methods and explanations of his results.

The correspondence principle was intended to do for market or macroeconomic comparative statics what maximization did for the comparative statics of the individual or firm. The principle can be useful when the analyst knows something about the dynamic behaviour of a system, but as noted by Tobin (1983), is ambiguous in that different dynamics may be consistent with the same steady-state behaviour.

The simplest example of the ambiguity can be seen in a demand–supply diagram where the supply curve is negatively sloped. Whether a tax on the good will increase or reduce price depends on which curve is more steeply sloped. Whether the market is stable or not depends on the same fact and whether quantity or price rises in response to excess demand – that is, whether dynamics are Marshallian or Walrasian.

In the introduction to the 1983 enlarged edition, Samuelson records correctly that the *Foundations* was better off for not sticking to its narrow themes. Substance keeps breaking in on the methodology. The treatment of the theory of the consumer and firm, developed in detail, does not differ in essence from that of Hicks in *Value and Capital*. But where Hicks hides the mathematics in appendices, Samuelson flaunts his in the text. Nonetheless Samuelson takes pains to provide economic insight, including interpretations of Lagrange multipliers as shadow prices. These portions of the *Foundations* apparently existed in 1937–8 and were written independently of *Value and Capital* (Bronfenbrenner, 1982, p. 349), though not of course of Hicks and Allen (I, ch. 1, p. 4).

The theory of revealed preference (see below) receives prominence, as do two chapters on welfare economics, and in Part II chapters on the stability of general equilibrium. A few pages on money in the utility function (pp. 117–24) remain authoritative. The mathematical appendices on maximization and difference equations have been useful despite an elliptical style that leaves many steps to be filled in by the user.

Samuelson's thesis is dated 1940: *Foundations* is the work of a 25-year-old. There are signs of youth in the eagerness to proselytize for the new mathematical faith and its overreaching in trying to impose an entirely coherent theme on the material. But the book bears the unmistakable mark of the master, in command of the economics of his material, at home with technique, and most remarkably for a young man in a hurry, thoroughly familiar and patient with the literature. It is, as Schumpeter no doubt remarked, a remarkable performance.

III Consumer theory and welfare economics

Samuelson's first published paper (1937: I, ch. 20) set up a finite horizon continuous time intertemporal optimization model of a consumer with additively separable utility function and exponential discounting, and derived the result that the profile of consumption is determined by the relation between the interest rate and rate of time preference. The focus is however the measurability of utility.

The theory of revealed preference, his major achievement in consumer theory, made its unnamed appearance in 1938 in 'A note on the pure theory of consumer's behaviour' in *Economica* (I, ch. 1; see also Houthakker, 1983, and Mas-Colell, 1982, for exceptionally lucid accounts). The purpose was to develop the entire theory of the consumer free of 'any vestigial traces of the utility concept' (I, p. 13). Rather than postulate a utility function or, as Hicks and Allen had done, a preference ordering, Samuelson imposed conditions directly on the choices made by individuals – their preferences as revealed by their choices. The key condition was the weak axiom of revealed preference, applying to choices made in two situations, say zero and one. With prices and quantities of goods j, $j = 1, \ldots, n$ in situation i given by p^i_j and x^i_j, the axiom is

$$\sum_j p^0_j \left(x^1_j - x^0_j \right) \leq 0 \text{ implies } \sum_j p^1_j \left(x^1_j - x^0_j \right) \leq 0$$

In words, if the individual chooses consumption bundle zero when he could have chosen the bundle one, he will not choose one when zero is available.

This minimal condition of consistency is shown to imply most of the conditions on demand implied by utility theory. But the symmetry and negative definitiveness of the Slutsky matrix could not be established using the weak axiom. Equivalently, the issue was the so-called integrability of demand functions, with the question being whether the preference map could be recovered given enough observations on the individual's choices. Houthakker (1950) solved the problem, by proposing the strong axiom of revealed preference, namely that for any finite string of choices in which B is revealed preferred to A, C is revealed preferred to B, ..., and Z is revealed preferred to Y, then A is *not* revealed preferred to Z. In this case, and given appropriate continuity conditions on demand, the demand functions are integrable and an entire preference map, satisfying the Slutsky conditions, can be recovered from the individual's choices.

The full equivalence between the properties of the demand functions of an individual and the preference ordering is the leading example of Samuelson's definition of the operational or observational significance of economic theory. Samuelson regards a theory as meaningful if it is potentially refutable by data. A single consumer could make a succession of choices that contradict the strong axiom. But the theory is not operational in the sense that a modern econometrician would want it to be: it does not apply to aggregate data, nor, in the form in which Samuelson left it, does it apply to choices that are made in chronological time.

Revealed preference links the theory of demand, index numbers, and parts of welfare economics. The link between demand and index number theory comes in the *Foundations'* (pp. 147–8) recognition that the fundamental index number problem is to deduce from price and quantity information alone whether an individual is better off. Using the weak axiom, Samuelson demonstrates the conditions under which, in a comparison of two situations, it is possible to say whether an individual is better off in one (*Foundations*, pp. 156–63). He argues that index numbers add no information on the essential question and indeed may be positively misleading in tempting the observer to attach significance to the numerical scale of measurement.

A similar concern no doubt motivates Samuelson's long-standing hostility to the use of consumer surplus measures. He has frequently argued that there is no need for the concept. He asserts in *Foundations* (p. 197) that there is no need for consumer's surplus in answering, for example, the question of whether Robinson Crusoe, a socialist state, or a capitalist one, should build a particular bridge. That view may have been moderated over the decades: the 1985 Samuelson–Nordhaus *Economics* (p. 418) states that the concept 'is extremely useful in making many decisions about public goods – it has been employed in decisions about airports, roads, dams, subways, and parks' (bridges are conspicuously absent).

The revealed preference axiom comes into play too in Samuelson's 'Evaluation of Real National Income' (1950: II, ch. 77), a largely negative report on the then new welfare economics that attempted to deduce from aggregate data criteria that would make it possible to say whether society was better off in one situation than another. Taking, as he has since 1938, the viewpoint that a Bergsonian social welfare function is the best way of understanding social welfare issues, Samuelson showed that no index-number type national income comparison between situations A and B could reveal whether society's feasible utility possibility frontier (a useful Samuelson innovation, apparently simultaneously invented by Allais) in A lies uniformly outside that of B. And, he argued, we could claim situation A is better than B only if that is the case.

In the *Foundations* (chapter 8) Samuelson draws extensively on the Bergsonian social welfare function to elucidate definitively the notion of Pareto optimality and the 'germ of truth in Adam Smith's doctrine of the Invisible Hand' (*Foundations*, 1983 edn, p. xxiv).

Arrow (1967) is critical of Samuelson's failure to look behind the social welfare function, and of his failure to link it to actual policy decisions. Similar sentiments are conveyed along with a more complete evaluation of Samuelson's welfare economics in Arrow (1983). Samuelson (1967: III, ch. 167) asserts that the Bergson Social Welfare Function and the Arrow Constitution Function are distinct concepts, though the argument is difficult to follow.

The expected utility theorem shows Samuelson wrestling for decades with his doubts over the independence axiom (I: ch. 12, 1950; ch. 13, 1952; ch. 14, 1952; *Foundations*, 1983, pp. 503–18). Despite his tentative 1983 acceptance of the expected utility formulation, he notes with approval Machina's (1982) development of expected utility without the independence axiom. Of course, these doubts have not kept him from making creative use of the expected utility approach in models of portfolio choice and finance.

IV Capital theory

The theory of capital and growth sections of the first four volumes of *CSP* account for 38 papers, the largest single category. Although capital theory is the branch of economics most vulnerable to Samuelson's comparative technical advantage, and although both his earliest papers are placed in that category in *CSP* (1937: I, ch. 17, ch. 20) the output in this area is concentrated in *CSP* III, covering the years from 1965 to 1970. Solow (1983) provides a fine review of this part of Samuelson's research, some of which he co-authored.

Among the early papers, the 1943 Schumpeter *festschrift* contribution 'Dynamics, Statics, and the Stationary State' (I, ch. 19) discusses the economics of the steady state and the possibility of a zero interest rate. Samuelson argues that a steady state with a zero real interest rate is possible if the rate of time preference of the infinitely lived individuals is zero; he has in mind a situation in which the marginal product of capital can be driven to zero. In this article (I, p. 210), as in his first paper (I, p. 216), Samuelson makes highly favourable reference to Ramsey, in contrast to the famous unflattering 1946 remark (II, p. 1528). The well-known argument that a zero rate of interest is impossible because income generating assets would have an infinite value is rejected, on the ground that an infinite value is not a problem since assets could trade against each other at finite price ratios. Some second thoughts are presented in a 1971 paper (IV, ch. 217); curiously, Samuelson discusses the Schumpeter issue entirely in a model with infinite horizon maximizers rather than in an overlapping generations framework.

The modern contributions in *CSP* I include the famous 1958 consumption loans model, which will be examined in the macroeconomics section, and the surrogate production function (1962: I, ch. 28). As Solow (1983) notes, much of the capital theory in *CSP* is related to developments in Dorfman, Samuelson and Solow (1958), which itself grew out of a 1949 Samuelson three-part memorandum for the Rand Corporation.

Notable among the contributions is a variety of turnpike theorems. A turnpike theorem is conjectured in the 1949 Samuelson memorandum, and fully worked out in the 1958 volume. The theorem states that for any accumulation programme, starting from an initial vector of capital goods, and with specified terminal conditions, as the horizon lengthens the optimal programme spends an increasing proportion of its time near the von Neumann ray; more generally in problems with intermediate consumption, the economy spends time near the modified golden rule. Several of the papers in the capital and growth section of *CSP* III contain turnpike theorems. A periodic turnpike result is reported in 1976 (IV, ch. 224).

The surrogate production function was an attempt to justify the aggregate production function as being consistent with an underlying model with heterogeneous capital goods and production techniques, and one type of labour. The article names and uses the factor price frontier, noting that it had been used earlier by others, including himself (in 1957: I, ch. 29). Samuelson shows that a downward sloping factor price frontier is traced out in a competitive multi-capital goods multi-technique economy, with higher steady state wages accompanying a lower steady state interest rate. Further, this frontier has the same properties as in the one-sector model, with the slope of the factor price frontier equal to the capital labour ratio. The theorem is correct, but as noted by Solow (1983), the conditions for it to obtain are special.

Under more general conditions, the famous reswitching result may occur in which a given technique of production that had been used at a low interest rate comes back into use again at a high interest rate (see the November 1966 *Quarterly Journal of Economics*). Reswitching implies that the one-sector neoclassical production function cannot be viewed as a general 'as if' construct that describes the behaviour of economies with several techniques of production. Cambridge (England) critics of neoclassical capital theory viewed reswitching as a confirmation of the view that marginal productivity had nothing to do with distribution, since the same techniques of production might be used with two (or many) different distributions of income. Various criticisms are offered by Robinson (1975) and responded to with forebearance in *CSP* (1975: IV, ch. 216).

Samuelson started the surrogate production function article by denying the need for any concept of aggregate capital. That position would be strengthened by the reswitching result. However, as with so many useful constructs in economics, the concept of aggregate capital has survived the demonstration that its validity may be limited. Neither Samuelson nor other neoclassics have been constrained by reswitching from using one-sector production functions or marginal productivity factor-pricing conditions.

The property that the slope of the factor price frontier is equal to the capital–labour ratio is one example of the duality between price and quantity that Samuelson began to emphasize in the *Foundations* and has used repeatedly since. *Foundations* (p. 68) contains the Roy's Identity envelope condition that the derivative of the minimized cost function of the firm with respect to the wage of factor *i* is the demand for factor *i*. It also provides shadow price interpretations of Lagrange multipliers. Samuelson has used duality in optimal growth and linear programming problems ('Market mechanisms and maximization', 1949: I, ch. 33, is a gem) and in *CSP* (1965: III, ch. 134).

V Dynamics and general equilibrium

Chapters IX through XI of *Foundations* cover stability analysis and dynamics, in both individual markets and the economy at large. The basic assumption of this dynamics is the 'law of supply and demand' that price rises in response to excess demand.

The impetus for the multi-market analysis came partly from Hicks's *Value and Capital* (1939) discussion of stability, in which there is no explicit dynamical system. The Samuelson approach is general equilibrium, though it does not start from the primitives of endowments. As Hahn (1983) notes, the underlying microeconomics is not specified. Samuelson nonetheless set the agenda of the next 15 years for the study of dynamics in a more explicitly general equilibrium framework, and most important, in a framework in which the issue of stability is precisely posed.

Explicit use of the law of supply and demand in theoretical work has fallen out of favour, though the Phillips curve can be interpreted as using that approach. The monopolistic competition wing of macroeconomics prefers to model price setting by firms and workers explicitly rather than rely on an auctioneer, and the equilibrium approach assumes prices are continuously at market-clearing levels. The older approach is used in disequilibrium macroeconomics, but is typically regarded as suspect.

Samuelson has not been a general equilibrium theorist in the sense of one striving for maximum generality. He has been general equilibrium in the sense opposed to partial equilibrium: he frequently works with models of the whole economy, in growth and capital theory, in trade and macroeconomics, and in his excursions into the history of thought.

The most micro-oriented of these general equilibrium contributions are the non-substitution theorem (1951: I, ch. 36) and factor–price equalization. The non-substitution theorem was presented at a 1949 conference, and was obtained independently by Samuelson and Georgescu-Roegen (I, p. 521). Consider an economy where labour is the only primary factor, and where goods are used either for consumption or as input into the production of other goods. Suppose the production function for each good is neoclassical, permitting substitution among factors of production, but there is no joint production.

The theorem is that relative prices in this economy are independent of demand, that is, are determined on the supply side alone. There is a single least cost way of producing each good, where cost is determined by direct and indirect labour requirements. Hahn (1983) provides a clear account of the theorem, and generalizations to dynamic systems with capital (1961: I, ch. 37). The question in the system with capital is whether, given the interest rate, the relative price structure is unique. Conditions for uniqueness are discussed in Hahn. The link with the surrogate production function, published at about the same time, is clear. The nonsubstitution theorem is used also in Samuelson's discussions of Ricardo (1959: I, chs 31, 32).

VI International trade

'Our subject puts its best foot forward when it speaks out on international trade' (1969: III, p. 683), and some of Samuelson's best-known contributions are undoubtedly in this field. Jone's 1983 article describes Samuelson's considerable impact on trade theory: on the gains from trade; the transfer problem; the Ricardian model; the Heckscher–Ohlin–Samuelson model; and the Viner–Ricardo model.

Earliest among the well-known contributions is the 1941 Stolper–Samuelson result (II, ch. 66), which uses the two-sector, two-country Heckscher–Ohlin model with identical production functions in the two countries to analyse the effects of the opening of trade, or the imposition of a tariff, on the wage. The result is that protection will benefit the factor that is relatively (to the other country) scarce. Or, the opening of trade benefits the relatively plentiful factor. But the paper contains more than that result. As Jones (1983) notes, it introduces the basic elements of Heckscher–Ohlin theory for small-scale trade models – and those models were the analytic core of real trade theory for decades.

Stolper–Samuelson flags the issue of factor–price equalization, the question of whether trade in goods alone can produce the factor price equalization that would obtain if factors were freely mobile. Ohlin claimed that trade would cause a necessarily incomplete tendency to equalization. Samuelson (1948 and 1949: II, ch. 67; 68) showed in the Heckscher–Ohlin context conditions under which equalization would be complete: identical production functions in the two countries, no factor-intensity reversals, and similarity of the ratio of endowments (so that countries are not specialized in production). The paper was remarkable and surprising, and did not suffer from the happy coincidence that a 1933 Abba Lerner contribution rediscovered by Lionel Robbins had independently reached the same conclusions in a similar model.

Factor price equalization in more generality is considered in the famous 1953 paper 'Prices of factors and goods in general equilibrium' (II, ch. 70), which caused a substantial literature including Gale–Nikaido (1965). It is striking that many of Samuelson's famous papers led to prolonged discussion of the exact conditions needed for his particular results to obtain: he opened more doors in economics than he closed.

The transfer problem is an old issue in the literature that arose in the 1920s, after Second World War, and arises again in contemplation of the world debt crisis. Samuelson's 1952 and 1954 papers (II, chs 74, 75) are classics in this extensive literature, on the issue of whether a transfer from one country to another (such as German reparations) is likely also to worsen the terms of trade of the country making the transfer, which Samuelson describes as the orthodox presumption. In the modern context the orthodox presumption would be that the developing countries will have to suffer a terms of trade loss to run current account surpluses to reduce their indebtedness. Samuelson typically argues that there is no presumption about the terms of trade shift, though the orthodox presumption is more likely to hold where there are non-traded goods or impediments to trade (1971: III, ch. 163).

Samuelson's contributions to trade theory are classics: the contributions are basic, the models are tractable and fecund, the problems come from the real world as well as the literature, the articles continue to reward the reader. And they continue to be read.

VII Finance

Despite his long-time personal interest in capital markets, Samuelson's contributions to finance theory started only as he turned 50. These papers are concentrated in *CSP* III and IV; the earlier ones are self-reviewed in 'Mathematics of speculative price' (1972: IV, ch. 240). Merton (1983) describes and evaluates six of Samuelson's favourite papers in finance, broadly defined to include his 1952 paper on expected utility and the independence axiom (I, ch. 14).

The two most important papers are 'Proof that properly anticipated prices fluctuate randomly' (1965: III, ch. 198) and 'Rational theory of warrant pricing' (1965: III, ch. 199). 'Proof …' provides a first precise formulation of the consequences for speculative prices of market efficiency. The theorem describes the behaviour of the current price of a commodity for delivery at a given future date, for example June 1990 wheat. Assuming that speculators do not have to put up any money to enter the contract, the result is that the market price should be the expectation at each date of the June 1990 wheat price. Given rational expectations, there is no serial correlation in the changes in price. Hence 'properly anticipated prices fluctuate randomly'.

Samuelson says of this theorem, which is now entirely basic: 'This theorem is so general that I must confess to having oscillated over the years between regarding it as trivially obvious (and almost trivially vacuous) and regarding it as remarkably sweeping. Such perhaps is characteristic of basic results' (III, p. 186).

Note what the theorem does not say, using the exchange rate as the example. The theorem is not that the exchange rate fluctuates randomly; predictable inflation or predictable business cycle fluctuations can cause predictable movements in the exchange rate. Rather it is the current price of foreign exchange at a *given* future date that fluctuates randomly. The notion that efficiency produces random motion is itself fascinating. But far more important is the restriction on empirical behaviour implied by efficiency that Samuelson derives in a well-defined context. Testing for efficiency of speculative markets has become a major industry.

'Rational theory of warrant pricing' missed its target, but it is as Merton (1983) remarks, a near miss. Samuelson had pursued option pricing for well over a decade. He supervised Kruizenga's 1956 MIT dissertation on the topic, and was familiar with Bachelier's 1900 continuous time stochastic calculus calculation of rational option prices. Samuelson derived a partial differential equation for the option price that depends, among other variables, on the expected return on the stock and the required return on the option. The remarkable feature of the Black–Scholes solution to the problem is that the rational price of the warrant does not depend on the expected return on the stock, but rather on the risk-free rate. Nonetheless, the Samuelson differential equation can be specialized to the correct Black–Scholes equation.

Other contributions to finance theory include papers on diversification (1967: III, ch. 201), and on conditions under which mean-variance analysis can be justified (1970: III, ch. 203) – with continuous time models providing the best argument for the procedure.

VIII Macroeconomics

All the Samuelson contributions described to this point are firmly neoclassical. His work in macroeconomics presents a more mixed picture. I take up in turn the early multiplier–accelerator model, which is not at all price-theory oriented, the neoclassical synthesis, Samuelson the policy adviser and commentator, and the entirely neoclassical consumption loans model.

The multiplier–accelerator model

In a 1959 note (II, ch. 84) on the multiplier–accelerator model, Samuelson describes his contribution as being the algebraic generalization of a numerical example of Alvin Hansen's. The model (1939: II, chs 82, 83) is a simple one in which current consumption is proportional to lagged output and investment is determined by the difference between current and lagged consumption (the accelerator).

This implies a second-order difference equation, which can generate asymptotic or oscillatory damped approaches to equilibrium, or oscillatory or non-oscillatory explosive paths for output. Although Frisch and Slutsky had already written on the ability of stochastic difference equations to mimic cycle-like behaviour, Samuelson does not – except for a quotation from J.M. Clark that receives little emphasis – link his second-order equation with a stochastic forcing term.

Samuelson (1939: II, p. 1111), while emphasizing the simplicity of the algebraic analysis, argues for the empirical importance of the accelerator. This judgment has held up over time as flexible accelerator effects continue to feature strongly in modern estimated investment functions. From the theoretical point of view, the multiplier–accelerator model is interesting for the lack of concern over microfoundations. Where a 1980s macroeconomist might agonize over the microfoundations of the consumption function, over the accelerator, or over the impact of rational expectations of future output on investment, Samuelson proceeds constructively with a simple implicitly fix-price model. The famous 45-degree diagram popularized in *Economics* – and for several editions on the cover – forcefully emphasizes Samuelson's view that aggregate demand is the key determinant of output.

> In the 1940 'Theory of pump-priming reexamined' (II, ch. 85) he stipulates the basic features of the private economy forming the environment within which governmental action must take place … – (1) The economic system is not perfect and frictionless so that there exists the possibility of unemployment and under-utilization of productive resources …

This view pervades Samuelson's macroeconomics. Indeed, when asked recently his view of the causes of wage and price stickiness, he replied that he decided 40 years ago that wages and prices were sticky, that he could understand the behaviour of the economy and give policy advice on that basis, that he had seen nothing since then to lead him to change his view on the issue – and that he had not seen a payoff to researching the question.

He was of course aware of the issues. An abstract of a paper presented at the 1940 meetings of the Econometric Society (II, ch. 88) describes a totally modern discussion of the question of whether general involuntary unemployment is impossible in a world of price flexibility. His penetrating 1941 review of Pigou's *Employment and Equilibrium* (II, ch. 89) outlines a simple classical model in which price flexibility through its effects on aggregate demand produces full employment even with a constant real wage. This is not however Pigou's model; according to Samuelson, Pigou adopts a model in which money wage flexibility is an alternative to active monetary policy. Samuelson never regarded the Pigou effect as being of real world significance (1963: II, ch. 115).

The neoclassical synthesis

Tobin (1983, p. 197) describes the neoclassical synthesis as Samuelson's greatest contribution to macroeconomics. The synthesis is outlined in articles in the early 1950s (1951: II, ch. 98; 1953: II, ch. 99; 1955: II, ch. 100) and developed in successive editions of *Economics*. It argues that monetary and fiscal policy can be used to keep the economy close to full employment, and the monetary–fiscal mix can be used to determine the rate of investment.

The synthesis represents the views of mainstream macroeconomics in the 1950s and 1960s, and perhaps in the 1970s and even the 1980s. Its activist spirit was evident in the Kennedy administration. Its acceptance must have been helped by the widespread use of Samuelson's *Economics* and by the many clones that preached its message.

Perhaps the most notorious component of the neoclassical synthesis is the 1960 Samuelson–Solow 'Analytic aspects of anti-inflation policy' (II, ch. 102), which presents a United States Phillips curve. This article is frequently cited as containing the view that the Phillips curve presents society's long-run tradeoffs between inflation and unemployment.

It does not. The paper starts by discussing the difficulties of distinguishing cost–push from demand–pull inflation. Samuelson and Solow then plot the scatter of percentage changes in average hourly earnings in manufacturing against the unemployment rate (the years plotted are not specified, but include the 1930s). The discussion that follows considers alternative points on the Phillips curve as policy choices for the next few years. But the authors warn explicitly that the discussion is short-term, and that it would be wrong to think that the menu of choices represented by the Phillips curve 'will maintain its same shape in the longer run. … [I]t might be that … low-pressure demand would so act upon wage and other expectations as to shift the curve downward in the longer run …' (II, p. 1352). This is though hardly a clear demonstration of the vertical long-run Phillips curve – for Samuelson–Solow suggest that low demand might also cause the Phillips curve to shift up (a notion that many in Europe now find entirely believable) – but it is clear evidence that the authors were not guilty of believing the Phillips curve would stay put no matter what. In conversation, Samuelson has said that he was always the Kennedy administration pessimist about the long-run Phillips curve tradeoff.

The policy adviser and commentator

Samuelson has long taken an active part in economic policy debates, through Congressional testimony, as consultant to the Treasury and the Fed, in his *Newsweek* column that ran every three weeks from 1966 to 1981, in other newspaper columns, public addresses, advice to candidates and Presidents, and in contributions at academic conferences and in symposia.

His views reflect the neoclassical synthesis, a disdain for rules rather than discretion in determining policy, and an almost shameless eclecticism. He knows the macroeconomic numbers and can speak the language of policy discussions. He is a cautious forecaster, rarely committing numbers to print, preferring to decide on which side of the consensus to place his bets. His 1941 consumption function remains his only econometric work (II, ch. 87); he has said that the major disappointment in economics in the last 40 years has been the failure of econometric evidence to settle disputes.

Macroeconomics is Samuelson's primary applied economics field, with finance the second. He keeps up with the current state of the macroeconomy, drawing on forecasts and empirical work of others. He is sceptical of individual forecasts though a law of averages permits him to put some trust in the mean or median forecast. His eclecticism makes his policy views less exciting than those of economists with a strong view of the way the world works – but he has never sought to be interesting rather than right. (This despite his 1962: II, ch. 113, p. 1509) comment on John Stuart Mill: 'It is almost fatal to be flexible, eclectic, and prolific if you want your name to go down in the history books ….')

Nonetheless, Samuelson's implied attitude to the applications of economics gives pause. As Arrow (1967) notes, his work reveals ambivalence about the relevance of neoclassical price theory. He shows no great faith that his microeconomics can be applied to the real world. No doubt comparative advantage plays a role in that attitude. But the theoretical sophistication he brings from microeconomics does not distinguish his macroeconomic policy advice and forecasting from that of the pack; his neoclassical training is not seriously used in Samuelson's applied macroeconomics. *Economics* may be evidence however that he values simple microeconomics.

The consumption loans model

In the classroom Samuelson has confessed that among his many offspring the consumption loans model (I, ch. 21) is his favourite. The affection is amply rewarded: within macroeconomics the two-period lived overlapping generations structure of the model has been used in countless papers in which a tractable framework with an explicit time structure is needed. The original consumption loans model examined the role of money or bonds as institutions for making Pareto-improving trades feasible; the structure has been used subsequently to examine the dynamics of capital accumulation, the burden of the debt, Ricardian equivalence, social security, the role of money, the effects of open market operations, intertemporal substitution of leisure, labour contracts, government financial intermediation, and more.

The set-up for the original model is one in which people live two periods, with utility functions defined over consumption in the two periods. Each young person receives an endowment of one nonstorable chocolate in

period 1. In the absence of trade each person could consume only in period 1. Trades are possible in which the current young give part of their chocolate to the current old in return for chocolate to be received next period from the then young. But there is no double coincidence of wants, no direct way of making the bilateral trades.

Now comes the ostensible point of the model: the social contrivance of money makes trade possible, and its introduction is a Pareto-improving change given the pattern of endowments. The consumption loans model has been much criticized as a model of money, because it implies the velocity of circulation is one per generation. Equivalently, the criticism is that the model describes money as effectuating integenerational transactions whereas in practice other assets, such as bonds, serve that role. (Patinkin 1983 discusses the consumption loans model as a basis for monetary theory and also Samuelson's excursions into the history of monetary thought.)

This is certainly correct. But the significance of the consumption loans model is not its rationale for the existence of money. Rather the model has been so influential and popular because it provides a simple *tractable* general equilibrium structure for modelling intertemporal problems with life-cycle maximizing individuals. The earlier examples prove how easily the general structure can be adapted. It can also be adapted to more periods of life (in the original article Samuelson extended lifetimes to three periods), with 50 period lifetime models being easily solvable on computers. Its strength lies in the elegance and robustness with which it captures the essential point that finite lived individuals exist in an infinitely lived economy (we are each but not all dead in the long run).

Nearly 30 years after the consumption loans model was first published, Malinvaud (1987) drew attention to the little-noticed earlier discovery and extended development of the overlapping generations model by Allais (1947). No doubt Samuelson's eminence and location in the United States, as well as publication in *Journal of Political Economy* rather than a book, had much to do with his independent discovery providing the impetus for the exploitation of this extraordinarily useful model – though even so, it took several years before the overlapping generations structure found its way into common use.

IX Samuelson and MIT

MIT had famous economists before Samuelson: Francis A. Walker, third president of MIT (1881–99) and first president of the American Economic Association (1886–92), and Davis R. Dewey, president of the AEA (1909) and editor of the *American Economic Review* (1911–40). But the modern era, in which the Department of Economics has risen to worldwide prominence within the profession begins with the arrival of Samuelson in

1940. Brown and Solow (1983) describe the MIT Department of the 1930s, and the transformation that nearly began in 1941 after Samuelson arrived and the first Ph.D. class, including Lawrence Klein and George P. Shultz, was about to get under way. Second World War intervened, and it was only in the late 1940s and early 1950s that the faculty and the Ph.D. programme reached full strength.

The MIT department and Ph.D. programme have been consistently among the best in the world since the early 1960s. The names of the faculty members are well known. Equally remarkable is the collection of eminent economists who are MIT Ph.Ds, whose names are legion but whom it would only be invidious to begin to list.

Samuelson's role in this success was pivotal but not domineering. His research habits (including sheer hard work), the open-door policy for students (a lesser burden for someone of whom the students were in awe than for others) and fellow faculty, his absolute refusal to use authority instead of reason in faculty meetings, his zest for conversation about economics, economists, and all else, made him a role model for a department where cooperation and friendliness have been extraordinary. He helped shape the department but he did not dictate its shape; he told one of his young co-authors that as a young man he decided that at age 40 he would stop taking initiatives in the department, at 50 he would venture an opinion only when asked, and at 60 would stop attending faculty meetings. Within the margin of error allowed to economists, he held to that resolution.

Samuelson the teacher played a lesser role. His worldwide fame (and that of other faculty members) doubtless was a major reason many of the outstanding students were there. But, at least in the last two decades, he supervised relatively few theses. His method of supervision was ideally suited to better students, for he would ask broad questions and give general guidance rather than involve himself in details.

His classroom lectures in the period 1966–9 when I heard them were not a model of organization. His advanced theory lectures were given in the first class of the day and it was always possible to tell whether the traffic had been bad that day by whether his hand-written mimeographed lecture notes were available at the beginning of the lecture or only later. The time until the notes arrived was taken up by stories setting the historical background for the problem, and anecdotes about the protagonists. The day he lectured en route to deliver his contribution to the Irving Fisher *festschrift* (1967: III, ch. 184) was especially memorable, though word filtered back from New Haven that his Yale audience was less than enchanted by the stories. His students were not surprised to find in his Nobel lecture (1970: III, ch. 130) both that he had been warned that the lecture was to be serious, and that he started a less than serious story with that warning.

His lectures were simply not designed for the novice. But they were superb for those with some background. He explained finer points, threw out open questions, made unexpected connections between topics, and communicated the zest with which he approaches economics.

X Concluding comments

Among the missing from this list of Samuelson's contributions are his work in the history of thought (where he has been more interested in clarifying analysis than in evaluating contributions), his methodological articles, the famous public goods theorem, the recent work on mathematical biology, the informative and entertaining biographies of contemporaries, the frank self-evaluations, and *Economics*.

The extraordinary success of *Economics* is something of a mystery, for the book is not easy – as witness the fact that simpler texts that follow Samuelson's structure have found a large market. *Economics* is a multi-level book that in its appendices, footnotes and allusions goes far beyond elementary economics. Depending on what students retain from their economics courses, *Economics* may have done much to raise the level of public discourse about economic policy.

Samuelson's self-evaluations, as in 'Economics in a Golden Age' (1972: IV, ch. 278), must have shocked many readers. The typical self-effacing scientist does not include stories of Newton and Gauss in his intellectual autobiography. Reflection leads to a different perspective: it would have been easy for Samuelson not to 'tell the truth and shame the devil' (1972: IV, p. 881). But how much more interesting it is to have the account of how Samuelson views his own achievements.

Samuelson was described in 1967 as 'knocking on the door … of the pantheon of the greats …' (Seligman, p. 160). He may have been let in by now. But the final word has to be left to Franco Modigliani, who, after the speeches at the 1983 party at which Samuelson was presented with the Brown–Solow *festschrift*, walked over to the seated Samuelson, wagged his finger at him, and said 'You', and after a pause 'You have enriched our lives.'

STANLEY FISCHER

Selected works

1947. *Foundations of Economic Analysis*. Cambridge, Mass.: Harvard University Press. Enlarged edn, 1983.

1948. *Economics, an Introductory Analysis*. 12th edn, co-author William Nordhaus, New York: McGraw-Hill, 1985.

1958. (With R. Dorfman and R.M. Solow.) *Linear Programming and Economic Analysis*. New York: McGraw-Hill.

1966–86. *The Collected Scientific Papers of Paul A. Samuelson*, vols I–V. Vols I and II, ed. J.E. Stiglitz, Cambridge, Mass.: MIT Press, 1966; vol. III, ed. R.C. Merton, Cambridge, Mass. and London: MIT Press, 1970; vol. IV, ed. H. Nagatani and K. Crowley, Cambridge, Mass. and London: MIT Press, 1977; vol. V. ed. K. Crowley, Cambridge, Mass. and London: MIT Press, 1986.

Bibliography

Allais, M. 1947. *Economie et intérêt*. Paris: Imprimerie Nationale.

Arrow, K.J. 1967. Samuelson collected. *Journal of Political Economy* 75(5), October, 730–37.

Arrow, K.J. 1983. Contributions to welfare economics. In Brown and Solow (1983).

Bachelier, L. 1900. Théorie de la speculation, trans. A.J. Boness and ed. P. Cootner, as *The Random Character of Stock Market-Prices,* Cambridge, Mass.: MIT Press, 1967.

Bliss, C.J. 1967. Review of *Collected Scientific Papers of Paul A. Samuelson,* 2 vols. *Economic Journal* 77(306), June, 338–45.

Bronfenbrenner, M. 1982. On the superlative in Samuelson. In Feiwel (1982).

Brown, E.C. and Solow, R.M. (eds) 1983. *Paul Samuelson and Modern Economic Theory.* New York: McGraw-Hill.

Feiwel, G., ed. 1982. *Samuelson and Neoclassical Economics.* Boston: Kluwer-Nijhoff.

Gale, D. and Nikaido, H. 1965. The Jacobian matrix and the global univalence of mappings. *Mathematische Annalen* 159(2), 81–93.

Hahn, F.H. 1983. On general equilibrium and stability. In Brown and Solow (1983).

Hicks, J.R. 1939. *Value and Capital.* Oxford: Oxford University Press.

Houthakker, H. 1950. Revealed preference and the utility function. *Economica* 17, May,159–74.

Houthakker, H. 1983. On consumption theory. In Brown and Solow (1983).

Jones, R.W. 1983. International trade theory. In Brown and Solow (1983).

Lindbeck, A. 1970. Paul Samuelson's contribution to economics. *Swedish Journal of Economics* 72, 341–54.

Machina, M.J. 1982. 'Expected utility' analysis without the independence axiom. *Econometrica* 50(2 March), 277–324.

Malinvaud, E. 1987. The overlapping generations model in 1947. *Journal of Economic Literature*, 25, 103–5.

Mas-Colell, A. 1982. Revealed preference after Samuelson. In Feiwel (1982).

Merton, R.C. 1983. Financial economics. In Brown and Solow (1983).

Patinkin, D. 1983. Monetary economics. In Brown and Solow (1983).

Robinson, J. 1975. The unimportance of reswitching. *Quarterly Journal of Economics* 89(1), February, 32–9.

Samuelson, M.C. 1939. The Australian case for protection re-examined. *Quarterly Journal of Economics* 54, November, 143–9.

Seligman, B.B. 1967. On the question of operationalism. *American Economic Review* 57(1), March, 146–61.

Solow R.M. 1983. Modern capital theory. In Brown and Solow (1983).

Tobin, J. 1983. Macroeconomics and fiscal policy. In Brown and Solow (1983).

Sargan, John Denis 1924–1996

J.D. Sargan was born on 23 August 1924, in Doncaster, Yorkshire, where he spent his childhood. He was Emeritus Professor of Econometrics at the London School of Economics when he died at his home in Theydon Bois, Essex, on Saturday 13 April 1996. He received his secondary education at Doncaster Grammar School and, at the age of 17, gained a State Scholarship for entrance to St Johns College, Cambridge, where he took a first in mathematics and was Senior Wrangler.

Immediately after obtaining his degree, he was drafted into war work as a junior scientific officer attached to the RAF in Haverfordwest, where he provided basic statistical advice on the testing of new weapons systems. Like many of his generation Sargan's enthusiasm for economics was aroused by Keynes's *General Theory of Employment, Interest and Money*, and he decided to use his knowledge of mathematics and statistics to help tackle some of the pressing economic problems that faced society in the post-war years. Accordingly, in 1946 he returned to Cambridge to do more statistics, particularly time series, and to read economics, taking advantage of regulations that enabled him to complete a BA degree in economics in a year. More detailed biographical information is given in Hendry and Phillips (2003), on which much of the following discussion draws.

Starting his professional career as a lecturer in economics at Leeds University in 1948, Sargan went on to become the leading British econometrician of his generation, playing a central role in establishing the technical basis for modern time-series econometric analysis. In a distinguished career spanning more than 40 years as a teacher, researcher and practitioner, particularly during the period that he was Professor of Econometrics at the LSE, Sargan transformed both the role of econometrics in the analysis of macroeconomic time series and the teaching of econometrics. His influence on British econometrics was profound and continues today in the traditions he established.

Early research

Much of Sargan's research in the first decade of his career at Leeds University over 1948–58 was devoted to economic issues associated with the distribution of wealth, duopoly, production and growth. His paper (1957a) on the distribution of wealth is recognized to this day as the most general analytic treatment of the determination of the wealth distribution. His work (1958a; 1961a) on the instability of Leontieff's dynamic input–output model also attracted attention, showing that the Leontief model is not well adapted to explaining the behaviour of a decentralized economy. In addition to these researches on economic issues, he also published an early paper (1953a) on subjective probability and Bayesian thinking in economics, and another paper (1953b) on some of the statistical properties of the covariogram and periodogram.

Sargan's first foray into econometric methodology began with his paper (1957b) on 'The Dangers of Oversimplification', a discussion of the path-breaking analysis of the Oxford Savings Survey by Malcolm Fisher. Sargan's comments revealed a concern with three issues that recurred in his later research on econometric methodology: the abstract and constrained form of economic-theory models relative to the complexities of the data under analysis; the oversimplified nature of many estimated regression equations, excluding effects that were likely to be important in practice; and the problems of interpreting tests of large numbers of hypotheses. The first two concerns may have led to his subsequent interest in estimating relatively general and unrestricted models, and the third to his ideas about 'data mining' and model selection which became manifest in later research that was published posthumously in Sargan (2001a; 2001b). This discourse on oversimplification was closely followed by two major papers in econometrics that developed a theory of instrumental variable (IV) estimation, published in *Econometrica* in 1958 and the *Journal of the Royal Statistical Society* (JRSS) in 1959.

Instrumental variables
The two IV papers broke new ground, taking several large steps forward in the analytical treatment of simultaneous equations and the econometric methodology of estimation and inference. They quickly established Sargan as a technically accomplished new thinker in the econometrics arena, and they remain of lasting significance. The *Econometrica* paper laid out the methodology of IV estimation as we presently know it, provided asymptotics, related the approach to canonical correlation analysis and limited information maximum likelihood, gave tests for overidentification and under-identification, developed significance tests and confidence intervals, suggested instruments for use in practical work, and discussed the accuracy of the asymptotic theory. In the course of the latter discussion, Sargan pointed out that biases in estimation are likely to be large when the structural equation is almost unidentified, thereby foreshadowing some concerns over the effects of weak instrumentation that have occupied professional interest in recent years, following their explicit treatment in P. Phillips (1989), Nelson and Startz (1990) and Staiger and Stock (1997).

The *JRSS* paper advanced the analysis of IV estimation by considering the more general case where the structural coefficients $a = a(\theta)$ satisfied some analytic constraints and could be functionalized on a vector of fundamental parameters θ, applying the theory to the case of structural models with autoregressive errors, constructing statistical tests and confidence intervals, and again looking at over-identified and unidentified cases. The framework of this paper made it possible to consider problems of dynamic specification in a rigorous manner by means of statistical testing in a nonlinear in parameters context, thereby laying a foundation for much subsequent research in econometric methodology including Sargan's own later work (1980) on dynamic simplification.

Focus on econometric theory
Sargan took up a Fulbright scholarship in the United States for two years from 1958, spending the first academic year at the University of Minnesota, teaching summer school at Stanford University, moving to the University of Chicago for 1959–60 and visiting the Cowles Foundation at Yale in 1960. These visits firmly focused his growing interest on the econometric theory of estimating structural economic models from time-series data and, together with the publication of his two papers on IV estimation, brought his work in econometrics to the attention of the North American academic community. From this point forward, Sargan's career fell under the spell of a deep fascination with the design of statistical methods suitable for studying empirical economic problems and the intellectual problems involved in working out their finite sample and asymptotic properties.

In July 1960, Sargan returned to Leeds University to take up a readership, and his growing reputation for insightful, rigorous and powerful analyses led to his election to a fellowship of the Econometric Society in 1963. In the same year, he was recruited by the London School of Economics as a Reader in Statistics, in the same department as Jim Durbin, before joining A.W.H. (Bill) Phillips (already famous for the Phillips machine and the Phillips curve) in the economics department as Professor of Econometrics in 1965.

The period of the early 1960s saw the publication of some of Sargan's most influential papers and the formation of fundamental ideas that would play a major role in his later research. Two articles in *Econometrica* (1961b; 1964b) studied maximum likelihood estimation of structural systems. The first set up a framework that enabled structural estimation in the presence of autoregressive errors, thereby accomplishing a marriage of two earlier theories. The second elegantly established the asymptotic equivalence of full information maximum likelihood (FIML) and three-stage least squares (3SLS), thereby confirming the asymptotic efficiency of the latter. A third paper, presented at the Copenhagen meetings of the Econometric Society in 1963 and later abstracted in *Econometrica* (1964b), conceived the notion of approximating the distributions of econometric estimators by means of Edgeworth expansions. This paper was never published, but it gradually evolved into a major research programme concerned with the theory and application of Edgeworth expansions, formally beginning nearly a decade later with the publication of Sargan and Mikhail (1971).

The 'Colston paper'
A fourth paper was prepared for the Colston Society conference on National Economic Planning held at

Bristol University in 1963 and was published in 1964. This 'Colston paper' (1964a), as it has become known, is possibly Sargan's most famous paper and is certainly his most important contribution to empirical econometric methodology. The paper laid out the conceptual basis of the so-called 'LSE approach' to econometric modelling, so Sargan is justly credited with the foundation of that approach. The main characteristics of the 'LSE approach to econometric modelling' (which in fact draws on work from many other institutions) are blending prior economic theory ideas with thorough data analysis to develop empirical models consistent with both sources of information, but with neither having precedence. In the context of time series, this led to an emphasis on commencing empirical modelling from relatively general dynamic equations capable of capturing the properties of the data while representing the relevant economic theories, rather than estimating stochastic implementations of theory models. Few papers can have contained so many novel ideas, each of which really deserved a separate article.

The paper is characteristically self-effacing and modest about its many practical contributions, though technically brilliant and economical in its execution. First, as a framework for constructing models, Sargan considered the use of 'long-run' economic analysis to specify the equilibrium of a model and introduced 'equilibrium-correction' mechanisms as a behavioural dynamic, following some earlier work (particularly A. Phillips, 1954) on trade cycle adjustment mechanisms. In doing so, Sargan established what is now perhaps the most widely used form of time-series econometric equation in empirical work. Second, Sargan viewed the presence of autoregressive errors in time-series models as a simplification (by virtue of the implicit factorization) of more general system dynamic reactions, and he constructed mis-specification tests that were valid even after estimating dynamic equations. This work translated into Sargan's later concern (1980) for direct tests of dynamic specification and simplification strategies in inference. To address another practical problem of empirical research, the Colston paper formulated a procedure for comparing linear against logarithmic specifications, and investigated the impact of data transformations on the selection of models. The paper further proposed a nonlinear in parameters IV estimator for models where the data were subject to measurement errors, devised and implemented operational computer programs for the new econometric methods, and included a proof that the required iterative computations would converge. Finally, the paper illustrated the methodology by matching the econometric theory to the specific, topical and difficult empirical problem of wage and price determination in the United Kingdom. Previous models had related the changes in the variables, namely, wage inflation and price inflation. Such formulations precluded any relationship between the levels of wages and prices, which could therefore drift apart over time. Sargan argued that economic agents are concerned about the level of real wages, not just price inflation, so he formulated a model with a long-run equilibrium and incorporated real wages in the wage equation, thereby distinguishing the equation from many other models, including the Phillips curve. The paper also included a data-based proxy for 'inflation expectations', which was called 'an extrapolation of past price movements into the future' and the disequilibrium of real wages from its target depended on unemployment, productivity and political factors. In modern parlance, the levels variables were integrated whereas the differences and the equilibrium errors were not, so the equation implicitly required cointegration between the levels. Sargan's analysis highlighted the role of real-wage resistance in wage bargains, interpreting the equilibrium correction – the deviation of real wages from a productivity trend – as a 'catch-up' mechanism for recouping losses incurred from unanticipated inflation. As the 1960s proceeded, this real-wage resistance proved to be an insuperable barrier to the successful implementation of incomes policy in the UK. The Colston paper also included a policy discussion in which permanent and transitory effects were distinguished to ascertain which changes would persist and which fade out (such as devaluations).

Prior to Sargan's Colston paper, it was common in empirical econometric practice to test for residual autocorrelation (for example, by the Durbin–Watson statistic), and if it were shown to be present, estimate a 'generalized' model that allowed for an autoregressive-error process. Sargan reversed this convention, interpreting autoregressive errors as a restriction on the dynamic specification of a model that, when valid, permitted the adoption of a more parsimonious representation. He also stressed that empirical specifications should be stringently evaluated, and formulated tests for the validity of the instrumental variables used in estimation and for higher-order autoregressive errors based on the residuals from the estimated equations. Despite the existence of this test, which was valid in dynamic equations, the Durbin–Watson test continued to be widely and invalidly used for many years in dynamic systems. Regarding computation, the paper carefully addressed the logic of the calculations both to embed all of the estimators in a common framework and to ensure as efficient an iterative procedure as possible, including good selections of the initial values and step lengths, and checks for multiple optima. Sargan's demonstration that the step-wise iterative computations converged to a local optimum was the first of its kind in econometrics and reflected his keen interest in numerical analysis.

Thus, in matters of econometric theory, empirical methodology, numerical computation, empirical application and the integration of economic ideas and econometric technique, the Colston paper was a watershed of new ideas and stands as one of the classic works of econometrics.

Many new avenues of research were opened up, leading through equilibrium correction to cointegration analysis, encompassing, general-to-specific modelling, and a greater emphasis on model evaluation (Hendry, 1995, provides an overview).

Advanced theory

While the Colston paper constituted Sargan's most influential work from the perspective of empirical practice, the challenges that fired his intellectual passion principally lay elsewhere – in advanced theory. His greatest theoretical interest was in developing a finite sample distribution theory of estimation and inference, perhaps the most technically demanding field of econometric analysis. His main contributions in this area began with the publication of Sargan and Mikhail (1971), continued throughout the 1970s and 1980s, and covered all approaches – exact analytical derivation, asymptotic series approximations of both distributions and moments, and simulation-based methods. Despite the near-intractable nature of the manifold problems in this field, Sargan devoted a huge effort to produce solutions, pushing the frontiers of knowledge forward in remarkable ways in each of the main approaches.

Since economic systems are typically dynamic and/or simultaneous, the finite sample distributions of most econometric estimators and tests are extremely complicated, and the exact derivation of these distributions is a technically daunting task in all but the most trivial cases. Even when an exact theory is developed, the final results are often of limited applicability, rely on strong distributional assumptions and do not extend to dynamic settings because of formidable mathematical complexity. Sargan (1976a, Appendix A) provided the first general exact results for the distribution of the IV estimator in a structural equation, but he was able to resolve the distribution in closed form (as distinct from integral form) only in the just identified case. The general overidentified case was resolved subsequently in P. Phillips (1980).

Even in cases where the exact distribution itself is unattainable, certain interesting features of the distribution may be established, such as the existence or non-existence of moments. In this regard, Sargan (1970) gave an elegant demonstration of the fact that structural form FIML estimators, for instance, have no finite integral-order moments (mean, variance, and so on), thereby establishing that that these distributions typically have thick tails. By contrast, IV estimators generally have finite moments up to an order that is determined by the degree of over-identification of the structural equation and, on this topic, Sargan (1978) provided a definitive analysis of moment existence for the 3SLS estimator. In related work that was eventually published in his collected papers, Sargan (1975b) examined the tail behaviour of reduced form estimators and here showed that FIML estimators have finite moments to a certain order (determined by the

sample size) whereas IV estimators like 3SLS typically have no finite reduced form moments in overidentified cases. These exact results reveal that FIML procedures can offer some advantages, in terms of reduced outlier activity, when the fitted reduced form is used, for example in prediction.

More general results can be obtained using series expansion and other approximations. Indeed, Sargan hoped that general approximation formulae using Edgeworth asymptotic series could be developed and incorporated into regression software, possibly with the use of computerized algebra, and then used to adjust critical values and improve inference. That goal has not yet been realized and, even with the advent of more recent bootstrap technology, continues to be elusive, partly because available approximations are rarely accurate enough and partly because major difficulties are encountered with all approaches in time-series models as the zone of non-stationarity is approached.

In terms of computer-intensive methods, Sargan helped at an early stage in the development and implementation of ideas (such as the use of antithetic and control variates) that made simulation methods viable and computationally efficient (Sargan, 1976a, Appendix D). He also made important headway in validating approaches based on moment approximations (Sargan, 1974a), and even considered the complex case where Monte Carlo estimates and moment approximations are developed in cases where the actual moments fail to exist (Sargan, 1982), so that the approximations characterize pseudo-moments (or moments of suitable approximating distributions). Pseudo-moment expansions of this type provide an intriguing way of interpreting the descriptive moment statistics conventionally reported in Monte Carlo experiments. When the moments of the underlying distribution are infinite, Sargan's results reveal that such simulation-based moment statistics can be validly interpreted as estimates of the actual moments of the Edgeworth approximating distributions up to a certain order, depending on the sample size and the number of replications. This work resolved a major potential concern in the reporting of simulation-based research, since many simulation experiments are conducted in settings where the existence of moments has not been established.

Sargan's work on asymptotic expansions of the finite sample distributions of econometric estimators and test statistics was extraordinary in its coverage and its generality, dealing with IV estimators (Sargan and Mikhail, 1971), full information maximum likelihood (Sargan, 1970), k-class estimators (Sargan, 1975a), asymptotic chi-squared criteria (Sargan, 1980), and the theory of validity of the expansions in econometric contexts (Sargan and Satchell, 1986) together with formulae and algorithms for implementation of the approach (Sargan, 1976a). The final reference in this list is Sargan's famous Walras–Bowley lecture, which was presented at the 1974

San Francisco meetings of the Econometric Society and contained a multifaceted analytical development of the subject complete with four long technical appendices dealing with different approaches and detailing formulae that must have been obtained over many years of research. In a lucid discussion of density expansions in a general setting, this paper gave explicit formulae for the components of the Edgeworth expansion for a general form of econometric statistic and revealed the dependence of the correction terms on the form of the statistic and the cumulants of the sample moments on which the statistic depended. Importantly, given subsequent research, the paper also supplemented the idea of analytic expansions with a simulation-based approach (originally due to George Barnard) that is now recognized as a version of the modern parametric bootstrap.

Sargan's Walras–Bowley lecture and several of his other papers in this demanding field are filled with technical innovations and show little sign of aging even after decades of subsequent research. Although asymptotic expansions have been found an unreliable means of improving inferential accuracy, Sargan's theoretical contributions helped blaze the trail of finite-sample theory in the 1970s and early 1980s, and they furnish a substantial body of results that have improved our understanding of the properties of econometric estimators and tests. Edgeworth expansions of the sort Sargan sought to validate and implement are now routinely used (for example, Hall, 1992; Horowitz, 2001) to validate the improvements delivered by bootstrap methods in practical econometric applications.

Other contributions

In addition to the main themes of his research outlined above, Sargan made several other intriguing contributions to econometric theory. His work (1975a) on 'large models', for instance, still stands as a lone pioneering piece of technical analysis of the consequences of having a system whose size is large relative to the available database, and was strangely unlike any of the other papers published in the symposium on large macroeconometric models in which it appeared. Instead, as Robinson (2003) has argued, Sargan's ideas on large simultaneous systems are more relevant to the semiparametric methods that are now commonplace in econometrics.

Sargan's (1974a) work on continuous time stochastic models represents another major contribution, again in a very different field. This paper provided the first formal asymptotic study of the effects of approximating open-loop linear differential equation systems with discrete time simultaneous equations. Such discrete approximations were in use in practical work, and later have become more popular through the use of Euler approximations of differential equations in financial econometrics. Sargan's work built on an earlier study by Bergstrom (1966) and analysed the order of magnitude, in terms of

the sampling interval, of the inconsistency in various IV and FIML econometric estimates of the coefficients in the continuous system. The paper also examined more applied issues such as the impact of this bias on forecasting.

Research on identification

In other important research, Sargan (1980) addressed the thorny issue of the effects of near non-identification on modelling and inference. Early researchers on simultaneous equations methodology had recognized the importance of, but practical difficulties in assessing, identification. Tests for under-identification (such as those in Sargan, 1958b) were a manifestation of this concern. In practical work, however, these tests are seldom used, and most empirical research proceeds by assuming an equation is identified by order conditions. Sargan recognized that, in the event of near lack of identification, the asymptotic properties of econometric estimators and tests would be affected – in fact, in an early discussion, Sargan (1959, section 6) hints at some of the possibilities, including slower rates of convergence than the usual $\sqrt{n}$ rate for a sample of size n. Subsequently, Sargan (1975b) explored the relationship between identification and consistent estimability in systems of simultaneous stochastic equations. Then, in his presidential address to the Econometric Society in 1980, he considered nonlinear-in-parameter models that were 'nearly unidentified', in the sense that the first-order rank condition for local identification failed, but higher-order defining shape conditions held so that there was still identification. In singular cases like these, which followed the earlier discussion in the 1959 paper, Sargan found that the conventional asymptotic theory for IV estimation broke down, with slower rates of convergence and a non-normal limit theory applying. Sargan (1983b) later showed that similar problems of singularity occur in dynamic models with autoregressive errors. This work on near lack of identification anticipated future research, and its arena of application has proved to be far wider than may originally have been envisaged. It is especially relevant, for instance, in microeconometric applications where the relevance condition is weak and the IVs are sometimes barely correlated with the regressors. A prominent example in this field has been the study of the impact of schooling on earnings, where intrinsic ability affects both, is unmeasured and therefore contaminates the equation error. In such cases, the search for an instrumental variable that satisfies orthogonality with the error can lead to some arcane choices that end up being only weakly correlated with the regressors they service (Angrist and Krueger, 1991). The impact of such weak (or nearly irrelevant) instruments in applied econometric work is now an intensive area of research – see Andrews and Stock (2005) for an overview.

More time-series analysis

While Sargan retired before unit root and cointegration theory revolutionized time-series econometrics in the 1980s, he had studied Gaussian random walks, presenting an early paper on the subject at the UK econometric study group held at LSE in 1973, some results of which later appeared in joint work with Bhargava (1983c) in *Econometrica*. In further work, Sargan and Bhargava (1983d) showed that in regression models with moving average errors where there is a root on or near the unity circle, the likelihood function can have a local maximum at unity with reasonably high probability and that the limit theory is nonnormal in the unit root case, invalidating conventional tests. Accordingly, their paper argued against the empirical practice of checking for overdifferencing and in support of a most powerful invariant test of independence based on the Berenblut–Webb (1973) statistic. This approach has subsequently received much attention in the unit root testing context.

This brief summary of Sargan's theoretical contributions to econometrics shows the enormous range of his research interests. While almost all of the econometric theory he developed related to time series models fitted by time-domain methods, he also worked on adapting frequency-domain methods to simultaneous equations models in econometrics (Espasa and Sargan, 1977), missing data (Sargan and Drettakis, 1974b) and took some interest in panel data problems (Sargan and Bhargava, 1983e). By the time he retired in 1984, he had worked on most of the important problems and research areas in econometrics of his generation.

Overview

Sargan's appointment at the LSE in 1965 took its econometrics group to the technical forefront in research. He can be credited with the creation of a generation of econometricians in the UK who were trained to high technical levels in all aspects of quantitative economics but who were especially strong in econometric theory and methodology. His devotion to teaching and research training was exemplary. In total he supervised 36 successful doctorates in a host of fields covering much of the discipline of econometric method and many of its applications. Sargan held a 'modern' view of dissertation research as a process by which students learnt the practice of research by means of intimate involvement on the part of a supervisor. In this regard, his generosity to his students and colleagues was famous at the LSE and beyond, and undoubtedly played a major role in attracting doctoral students in econometrics. A full listing of his graduate students and their dissertation titles is contained in Sargan (1988a).

Sargan's contributions earned him international distinction and honours. In 1980 he served as President of the Econometric Society, presiding over the World Congress of the Society at Aix-en-Provence. He was made a Fellow of the British Academy in 1981 and assumed the Tooke Professorship of Economic Science and Statistics at the LSE in 1982. On retirement in 1984, he became Emeritus Professor of Economic Science and Statistics at the University of London, and an international conference was held in his honour at Oxford University. He became an honorary foreign member of the American Academy of Arts and Sciences in 1987, was awarded a fellowship of the LSE in 1990, and received an honorary doctorate from the University of Carlos III, Madrid in 1993, where a further celebratory conference was held for him.

The wide range of Sargan's research is celebrated by the topics addressed in the volume edited by Hendry and Wallis (1984), which commemorated his 60th birthday. He was interviewed for the journal *Econometric Theory* by P. Phillips (1985). Maasoumi edited his collected works, published as Sargan (1988a), which, together with his advanced econometrics lecture notes edited by Desai (Sargan, 1988b) well illustrate his analytic rigour and intellect. Three issues of econometrics journals have appeared in his memory: one in *Journal of Applied Econometrics*, 2001, which was on empirical macroeconometrics; another in *Econometric Reviews*, 2001, gave a biographical history of Sargan's career and printed several of his still unpublished papers; a third, in *Econometric Theory*, 2003, brought together two of Denis Sargan's essays on econometrics published for the first time, a laudation by Antoni Espasa, and three memorial essays offering an intellectual overview of his work.

Sargan had an enormous intellectual influence within the UK, both on the training of econometric theorists and on econometric practice. Outside the UK, his influence has not been as strong and, particularly in North America, different traditions and interests have prevailed. The Colston volume was an obscure source for economists and this undoubtedly limited the impact of his work on econometric methodology; and his choice of problems in econometric theory also did not always relate well to the immediate concerns of empirical researchers or other econometricians – he had his own vision of what the subject needed, and he pursued that vision with determination. Yet, when the history of econometrics in the second half of the 20th century is written, Sargan's place among the leaders of the econometric profession in that era is assured. The research agenda that he initiated has proved to be of tremendous scope, affecting almost every major area of the discipline. His scientific works show a remarkable durability, some of them (like the Colston paper and Walras–Bowley lecture) having the status of enduring classics. The world of econometric theory and its applications has moved on, but the themes of Sargan's research program persist in ongoing work, and his technical accomplishments remain part of the edifice of theory, technique and methodology that we collectively call econometrics.

DAVID F. HENDRY AND PETER C.B. PHILLIPS

See also **cointegration; computational methods in econometrics; data mining; distributed lags; econometrics; Edgeworth, Francis Ysidro; finite sample econometrics; full and limited information methods; inflation dynamics; instrumental variables; maximum likelihood; measurement error models; Phillips, Alban William Housego; serial correlation and serial dependence; simultaneous equations models; specification problems in econometrics; two-stage least squares and the *k*-class estimator; vector autoregressions.**

We thank many individuals for their information and help in writing this biography. In particular, Mary Sargan kindly provided details of Sargan's early life, and we have drawn on reviews, obituaries and memoirs written with, or by, Lord Meghnad Desai, Neil Ericsson, Toni Espasa, Esfandiar Maasoumi, Grayham Mizon, Hashem Pesaran, Peter Robinson and Kenneth Wallis, as well as our own memoir (Hendry and Phillips, 2003).

Selected works

1953a. Subjective probability and the economist. *Yorkshire Bulletin of Economic and Social Research* 5, 53–64.

1953b. An approximate treatment of the properties of correlogram and periodogram. *Journal of the Royal Statistical Society, Series B* 15, 140–52.

1957a. The distribution of wealth. *Econometrica* 25, 568–90.

1957b. The dangers of oversimplification. *Bulletin of the Oxford Institute of Statistics* 19, 171–8.

1958a. The instability of the Leontief dynamic model. *Econometrica* 26, 381–92.

1958b. The estimation of economic relationships using instrumental variables. *Econometrica* 26, 393–415.

1959. The estimation of relationships with autocorrelated residuals by the use of the instrumental variables. *Journal of the Royal Statistical Society, Series B* 21, 91–105.

1961a. Lags and the stability of dynamic systems: a reply. *Econometrica* 29, 670–3.

1961b. The maximum likelihood estimation of economic relationships with autoregressive residuals. *Econometrica* 29, 414–26.

1964a. Wages and prices in the United Kingdom: a study in econometric methodology. In *Econometric Analysis for National Economic Planning*, vol. 16 of Colston Papers, ed. P. Hart, G. Mills and J. Whitaker. London: Butterworth Co.

1964b. 3SLS and FIML estimates. *Econometrica* 32, 77–81.

1970. The finite sample distribution of FIML estimators. In J. Sargan, *Contributions to Econometrics*, vol. 1, ed. E. Maasoumi. Cambridge: Cambridge University Press, 1988.

1971. (With W. Mikhail.) A general approximation to the distribution of instrumental variables estimates. *Econometrica* 39, 131–69.

1974a. Some discrete approximations to continuous time stochastic models. *Journal of the Royal Statistical Society, Series B* 36, 74–90.

1974b. (With E. Drettakis.) Missing data in an autoregressive model. *International Economic Review* 15, 39–58.

1975a. Asymptotic theory and large models. *International Economic Review* 16, 75–91.

1975b. The identification and estimation of sets of simultaneous stochastic equations. In J. Sargan, *Contributions to Econometrics*, vol. 1, ed. E. Maasoumi. Cambridge: Cambridge University Press, 1988.

1976a. Econometric estimators and the Edgeworth approximation. *Econometrica* 44, 421–48.

1976b. The existence of moments of estimated reduced form coefficients. In J. Sargan, *Contributions to Econometrics*, vol. 2, ed. E. Maasoumi. Cambridge: Cambridge University Press, 1988.

1977. (With A. Espasa.) The spectral estimation of simultaneous equation systems with lagged endogeneous variables. *International Economic Review* 18, 583–605.

1978. The existence of moments of 3SLS estimators. *Econometrica* 46, 1329–50.

1980. Some tests for dynamic specification for a single equation. *Econometrica* 48, 879–97.

1982. On Monte Carlo estimates of moments that are infinite. In *Advances in Econometrics*, ed. R. Basmann and G. Rhodes, Jr. Greenwich, CT: JAI Press.

1983a. Identification and lack of identification. *Econometrica* 51, 1605–33.

1983b. Identification of models with autoregressive errors. In *Studies in Econometrics, Time Series and Multivariate Statistics*, ed. S. Karlin, T. Amemiya and L. Goodman. New York: Academic Press.

1983c. (With A. Bhargava.) Testing residuals from least squares regression for being generated by the Gaussian random walk. *Econometrica* 51, 153–74.

1983d. (With A. Bhargava.) Maximum likelihood estimation of regression models with first order moving average errors when the root lies on the unit circle. *Econometrica* 51, 799–820.

1983e. (With A. Bhargava.) Estimating dynamic random effects models from panel data covering short time periods. *Econometrica* 51, 1635–59.

1986. (With S. Satchell.) A theory of validity for Edgeworth expansions. *Econometrica*, 54, 189–213.

1988a. *Contributions to Econometrics*, 2 vols., ed. E. Maasoumi. Cambridge: Cambridge University Press.

1988b. *Lectures on Advanced Econometrics*. New York: Basil Blackwell.

2001a. Model building and data mining. *Econometric Reviews* 20, 159–70.

2001b. The choice between sets of regressors. *Econometric Reviews* 20, 171–86.

Bibliography

Andrews, D. and Stock, J. 2005. Inference with weak instruments. Discussion Paper No. 1530. Cowles Foundation, Yale University.

Angrist, J. and Krueger, A. 1991. Does compulsory school attendance affect schooling and earnings? *Quarterly Journal of Economics* 106, 979–1014.

Berenblutt, I. and Webb, G. 1973. A new test for autocorrelated errors in the linear regression model. *Journal of the Royal Statistical Society, Series B* 35, 33–50.

Bergstrom, A. 1966. Nonrecursive models as discrete approximations to systems of stochastic differential equations. *Econometrica* 34, 173–82.

Hall, P. 1992. *The Bootstrap and Edgeworth Approximation.* New York: Springer Verlag.

Hendry, D. 1995. *Dynamic Econometrics.* Oxford: Oxford University Press.

Hendry, D. and Phillips, P. 2003. John Denis Sargan 1924–1996. *Proceedings of the British Academy* 120, 385–409.

Hendry, D. and Wallis, K. 1984. *Econometrics and Quantitative Economics.* New York: Basil Blackwell.

Horowitz, J. 2001. The bootstrap. In *Handbook of Econometrics*, vol. 5, ed. J. Heckman and E. Leamer. Amsterdam: North-Holland.

Nelson, C. and Startz, R. 1990. Some further results on the small sample properties of the instrumental variable estimator. *Econometrica* 58, 967–76.

Phillips, A. 1954. Stabilization policy in a closed economy. *Economic Journal* 64, 290–313.

Phillips, A. 1959. The estimation of parameters in systems of stochastic differential equations. *Biometrika* 46, 67–76.

Phillips, P. 1980. The exact finite sample density of instrumental variable estimators in an equation with $n+1$ endogenous variables. *Econometrica* 48, 861–78.

Phillips, P. 1985. The ET interview: John Denis Sargan. *Econometric Theory* 1, 119–39.

Phillips, P. 1989. Partially identified econometric models. *Econometric Theory* 5, 181–240.

Robinson, P. 2003. Denis Sargan: some perspectives. *Econometric Theory* 19, 481–94.

Staiger, D. and Stock, J. 1997. Instrumental variables regression with weak instruments. *Econometrica* 65, 557–86.

satisficing

A decision maker who chooses the best available alternative according to some criterion is said to optimize; one who chooses an alternative that meets or exceeds specified criteria, but that is not guaranteed to be either unique or in any sense the best, is said to satisfice. The term 'satisfice', which appears in the *Oxford English Dictionary* as a Northumbrian synonym for 'satisfy', was borrowed for this new use by H.A. Simon (1956), in 'Rational Choice and the Structure of the Environment'.

Optimization and its problems

In the literature of economics and statistical decision theory, rationality has usually been defined in such a way as to imply some form of optimization, for example, maximization of utility subject to budget constraints. In simple situations, like the illustrative examples used in economics textbooks, computing a maximum may be a simple process, requiring, perhaps, nothing more onerous than taking a first derivative and setting it equal to zero. Even in much more complex situations, involving thousands of linear equalities and inequalities but also a linear criterion function, the powerful methods of linear programming often permit optimal choices to be found with tolerable amounts of computing effort.

In many (most?) real-world situations, however, genuine optima (maxima or minima) are simply not computable within feasible limits of effort (*see* RATIONALITY, BOUNDED). This is especially true when decisions must be made without benefit of computer, but it is frequently true even when powerful computing facilities are available. The complexity of the world is not limited to thousands or even tens of thousands of variables and constraints, nor does it always preserve the linearities and convexities that facilitate computation.

The satisficing alternative

Faced with a choice situation where it is impossible to optimize, or where the computational cost of doing so seems burdensome, the decision maker may look for a satisfactory, rather than an optimal alternative. Frequently, a course of action satisfying a number of constraints, even a sizeable number, is far easier to discover than a course of action maximizing some function.

The example has been given of searching for a needle in a haystack. Given a probability density distribution of needles of varying degrees of sharpness throughout the haystack, searching for the sharpest needle may require effort proportional to the size of the haystack. The task of searching for a needle sharp enough to sew with requires an effort that depends only on the density of needles of the requisite sharpness, and not at all on the size of the stack. The attractiveness of the satisficing criterion derives from this independence of search cost from the size and complexity of the choice situation.

In a formal sense, a process of satisficing could always be converted into a process of optimizing by taking into account the cost of search, and only searching up to the point where the expected gain derivable from another minute of search is just equal to the opportunity cost of that minute (Simon, 1955; Stigler, 1961). However, this conversion imposes a new, possibly heavy, informational and computational burden upon the chooser: the burden of estimating the expected marginal return of search and the opportunity cost. Solving these estimation problems may be as difficult as making the original choice, or even more difficult. An alternative is to search until a satisfactory alternative is found.

Conversely, most of the so-called optimization models of operations research and management science can more profitably be viewed as satisficing models. In the application of OR optimization techniques, some highly simplified approximation to a real-world situation is reduced to a formal model (for example, a linear programming or integer programming model), and an optimum is then calculated for this approximation with respect to a similarly approximate criterion function. The resulting 'optimal' decision will often provide a satisfactory decision for the real-world situation, but without guarantee that it will be better than a decision arrived at by some alternative satisficing technique.

How may the satisficer set the level of the criteria that define 'satisfactory'? Psychology proposes the mechanism of aspiration levels: if it turns out to be very easy to find alternatives that meet the criteria, the standards are gradually raised; if search continues for a long while without finding satisfactory alternatives, the standards are gradually lowered. Thus, by a kind of feedback mechanism, or 'tâtonnement', the decision maker converges toward a set of criteria that are attainable, but not without effort. The difference between the aspiration level mechanism and the optimization procedure is that the former calls for much simpler computations than the latter. It is somewhat analogous to the difference between adaptive and rational expectations, respectively, in the theory of choice under uncertainty.

Incommensurability of goals and outcomes

Satisficing can also provide another kind of computational advantage over optimizing. Human decision makers often find it very difficult to make trade-offs among aspects or dimensions of value that seem to them incommensurable. Of course, it is the function of the utility function to insure commensurability in all cases, but we may not wish to postulate in advance that such a function exists, or may not know how to characterize it. Three classes of situations into which incommensurability is especially likely to intrude are: (1) cases of uncertainty, where, for each alternative, a bad outcome under one contingency must be balanced against a good outcome under another; (2) cases of multiperson choice, where one person's gain is another's loss; and (3) cases where each choice involves gain along one dimension of value and loss along another very different one.

It has been observed empirically that in circumstances of these kinds, and especially when each outcome entails unpleasant as well as pleasant consequences, decision makers do not proceed promptly to a choice, but instead seek to avoid the necessity for comparison. One common reaction, for example, is to refuse to choose among the given set of alternatives, and instead, to initiate a search for a new alternative that: in case (1), will ensure at least a minimally satisfactory outcome under all contingencies; in case (2), will ensure all participants a satisfactory outcome; and in case (3), will ensure an outcome that is at least minimally satisfactory along all dimensions. The acceptance of such alternatives comes within our definition of satisficing (Hogarth, 1980).

Consequences for economic theory

It is easier to reconcile a satisficing than an optimizing theory of economic decision making with what is known empirically of actual choice behaviour and of the computational limits of the human mind. On the other hand, a great deal is given up by the substitution of the former for the latter – given up in terms of the strength and variety of theorems that can be derived from the postulate of rationality in the two cases. To make predictions about behaviour on the basis of a satisficing theory requires much more empirical data about, for example, aspiration levels and their adaptivity, than does prediction on the basis of the optimizing theory. The magnitude of the difference becomes less, however, when we recognize that the optimizing theory says nothing about the shape or content of the utility function. It simply postulates a consistency of behaviour over time that may not be found if the decision maker is satisficing instead of optimizing.

In the last analysis, a choice between the two kinds of postulates will have to be made in terms of their relative effectiveness and accuracy in predicting and explaining economic behaviour, at both micro and macro levels. There is still little consensus in the economics profession as to the circumstances under which one postulate or the other will be the more advantageous.

HERBERT A. SIMON

See also **behavioural economics and game theory; economic man; rationality, bounded.**

Bibliography

Hogarth, R.M. 1980. Judgment and Choice: The Psychology of Decision. New York: Wiley.
Radner, R. 1975. Satisficing. *Journal of Mathematical Economics* 2, 253–62.
Simon, H.A. 1955. A behavioral model of rational choice. *Quarterly Journal of Economics* 69, 99–118. Reprinted in Simon (1982), ch. 7.2.
Simon, H.A. 1956. Rational choice and the structure of the environment. *Psychological Review* 63, 129–38. Reprinted in Simon (1982), ch. 7.3.
Simon, H.A. 1982. *Models of Bounded Rationality*. Cambridge, Mass.: MIT Press.
Stigler, G.J. 1961. The economics of information. *Journal of Political Economy* 69, 213–25.
Winter, S.G. 1971. Satisficing, selection and the innovating remnant. *Quarterly Journal of Economics* 85, 237–61.

Sauvy, Alfred (1898–1990)

Sauvy was born 31 October 1898, in Villeneuve-de-la-Raho, France. He graduated from the Ecole Polytechnique, and is known as a demographer, economist, statistician and sociologist. He was an adviser to Jean

Monnet and Paul Reynaud 1938–1940, a member of the Population Commission of the United Nations from 1947 to 1974, and he occupied a chair in social demography at the College of France from 1959 to 1969. He founded the Institut National d'Etudes Démographiques, one of the world's leading centres of demographic research, which he directed from 1945 to 1962. He was president of the International Union for the Scientific Study of Population from 1961 to 1963. A prolific author, his works include 45 books and many articles, on a broad range of topics from French economic history since the First World War to the effect of technological change on employment and the history of thought in demography. But he is best known among demographers and economists for his two-volume treatise *Théorie générale de la population* (1966), published in English in 1969 as *The General Theory of Population*.

The General Theory of Population attempts both theoretical and substantive generality and is largely independent of the English-language literature on the subject. Thus one finds no references to or integration of the English literature on the economics of fertility or the consequences of population change or optimum population theory. Rather, the treatise presents a highly individual view of the subject. The book is full of briefly presented but penetrating insights on a wide variety of topics, often illustrated with descriptive data. For an economist, however, the main interest of the book is in its more rigorous development and extension of the concepts of economic–demographic equilibrium, and of maximum, minimum and optimum population. The analysis is entirely comparative statics, based on the assumption of first increasing, then decreasing returns to labour and population. At the minimum population and again at the maximum, the average product of labour equals the subsistence level; in between these population sizes there remains an economic surplus after subsistence needs are met. Per capita product is of course maximized when the marginal product equals the average; more interestingly, total surplus is maximized when the marginal product of labour equals subsistence, at what is termed the 'power optimum' population – with the notion that a costly collective social goal can best be met at this size. The military optimum will be at a point between the power optimum and the maximum, since it requires both soldiers and surplus output. The implications of inequalities in income distribution are also discussed; the labour intensity of the consumption goods demanded by the rich will influence the size of the equilibrium population. Sauvy's theories, particularly of the determinants of the equilibrium population, have had an important influence on the thinking of economic demographers and social historians on the subject of homeostasis in human populations.

RONALD LEE

Selected works

1966. *Théorie générale de la population*. Paris: PUF. Trans. C. Campos as *The General Theory of Population*. New York: Basic Books, 1969.

Savage, Leonard J. (Jimmie) (1917–1971)

L.J. (Jimmie) Savage, né Leonard Ogashevitz, was born in Detroit on 20 November 1917 and died in New Haven on 1 November 1971. His interests were encyclopedic: as a youth he immersed himself in the *Book of Knowledge,* and at the time of his death he was preparing for the Peabody Museum a demonstration-exhibit on animal odorants. The dominant theme of Savage's professional work was the mathematical analysis of normative behaviour.

He received a BS (1938) and Ph.D. (1941) from the University of Michigan. In the early 1940s he obtained a broad postdoctoral exposure to pure and applied mathematics at the Institute for Advanced Study in Princeton, at Cornell, Brown, the Statistical Research Group of Columbia, the Courant Institute at New York University, and at Woods Hole Marine Biological Laboratory. From 1946 to 1960 he was at the University of Chicago, where he was central to the development of the statistics programme. Subsequently, he held professorships at Michigan and at Yale. Always, he was intellectually generous with students, colleagues, visitors and correspondents.

Savage's basic views and results on normative behaviour appear in his *Foundations* (1954). His essential theme, still being elaborated, is the relation between a person's probability for an event and his utility for the event. In particular, his probabilities and utilities must be consistent with the principle of maximizing his expected utility. These results flow from compelling axioms of coherent behaviour and they recommend specific strategies for applied statistics, such as the use of Bayesian statistics and the likelihood principle.

Savage's axioms imply that all probabilities reflect individual experience so that there is no reason for two people to have the same probability for a particular event. His theory conflicts with traditional views that hold probabilities to be basic constants of nature. At first Savage thought this conflict would not be significant in applying statistical theory but he remarked in the preface to the second edition of the *Foundations* (1972) that he was not successful in bringing the theories of statistics together at the applied level. He recognized the long process from elegant theory to serious applications. His paper on elicitation of probabilities (1971) develops methods to implement his theory of personal probability. And Savage (1977) warns against holding theoretical foundations as adequate to cover all aspects of applied statistics.

Savage's work on the foundations of statistics had major antecedents in Frank Ramsey and B. De Finetti, whose work was developed, polished and taught to a generation of scholars by Savage himself. Hewitt and Savage (1955) is both elegant mathematics and an extension of a basic result of de Finetti in the foundations of statistics. Dubins and Savage (1965) stems from a normative problem and bears mathematical fruit. Exposition of the basic ideas of applied Bayesian statistics combined with the new theory of stable estimation appears in Edwards, Lindman and Savage (1963). Additional biographical and critical analysis as well as most of Savage's published papers appear in the selection prepared by Ericson et al. (1981).

I. RICHARD SAVAGE

See also **Bayesian statistics; expected utility hypothesis.**

Selected works

1948. (With M. Friedman.) The utility analysis of choices involving risk. *Journal of Political Economy* 56, 279–304.

1954. *The Foundations of Statistics.* New York: John Wiley & Sons. 2nd rev. edn, New York: Dover Publications, 1972.

1955. (With E. Hewitt.) Symmetric measures on Cartesian products. *Transactions of the American Mathematical Society* 80, 470–501.

1963. (With W. Edwards and H. Lindman.) Bayesian statistical inference for psychological research. *Psychological Review* 70, 192–242.

1965. (With L.E. Dubins.) *How to Gamble If You Must: Inequalities for Stochastic Processes.* New York: McGraw-Hill. Reprinted with a new Bibliographic Supplement and Preface by L.E. Dubins as *Inequalities for Stochastic Processes (How to Gamble If You Must)*, New York: Dover Publications, 1976.

1971. Elicitation of personal probabilities and expectations. *Journal of the American Statistical Association* 66, 783–801.

1977. The shifting foundations of statistics. In *Logic, Laws and Life: Some Philosophical Complications*, ed. R.G. Colodny. Pittsburgh: University of Pittsburgh Press.

1981. *The Writings of Leonard Jimmie Savage – A Memorial Selection.* Prepared by a Committee (W.H. DuMouchel, W.A. Ericson (chair), B. Margolin, R.A. Olshen, H.V. Roberts, I.R. Savage and A. Zellner) for the American Statistical Association and the Institute of Mathematical Statistics, Washington, DC.

Savage's subjective expected utility model

In his seminal book *The Foundations of Statistics*, Savage (1954) advanced a theory of decision making under uncertainty and used that theory to define choice-based subjective probabilities. He intended these probabilities to express the decision maker's beliefs, thereby furnishing Bayesian statistics with a behavioural foundations.

The interpretation of probability as a numerical expression of beliefs is as old as the idea of probability

itself. According to Hacking (1984), the notion of probability emerged in the 1650s with a dual meaning: (*a*) the relative frequency of a random outcome in repeated trials and (*b*) a measure of a decision maker's degree of belief in the truth of propositions or the likely realization of events. Both the 'objective' and the 'subjective' probabilities, as these interpretations are now called, played important roles in the developments that lead to the formulation of Savage's subjective utility model.

In the early stages of their respective evolutions, the notion of utility was predicated on the existence of objective probabilities, and the notion of subjective probabilities presumed the existence of some form of utility. The ideas of utility and expected utility-maximizing behaviour were originally introduced by Bernoulli (1738). Bernoulli's preoccupation with resolving the famous St Petersburg paradox justifies his taking for granted the existence of probabilities in the sense of relative frequencies. In the same vein, von Neumann and Morgenstern's (1944) axiomatic characterization of expected utility-maximizing players facing opponents who may employ a randomizing device to determine the choice of a pure strategy assumes that probabilities of these strategies are relative frequencies.

In the late 1920s and 1930s, Ramsey and de Finetti formalized the concept of choice-based subjective probability assuming that individuals seek to maximize expected utility when betting on the truth of propositions. In the behaviourist tradition, they explored the possibility of inferring the degree of confidence a decision maker has in the truth of a proposition from his betting behaviour and quantifying the degree of confidence, or belief, by probability. Invoking the axiomatic approach and taking the existence of utilities as given, Ramsey (1931) sketched a proof of the existence of subjective probabilities. De Finetti (1937) proposed a definition of subjective probabilities assuming linear utility and no arbitrage opportunities.

These developments culminated in the work of Savage. While synthesizing the ideas of de Finetti and von Neumann and Morgenstern, Savage introduced a new analytical framework and conditions that are necessary and sufficient for the existence and joint uniqueness of utility and probability, and the characterization of individual choice as expected utility-maximizing behaviour.

Savage's analytical framework

Decision making under uncertainty pertains to situations in which a choice of a course of action, by itself, does not determine a unique outcome. To formalize this notion Savage (1954) introduced an analytical framework consisting of a set S, whose elements are *states of the world* (or states, for brevity); an arbitrary set C, of *consequences*; and the set F, of *acts* (that is, functions from the set of states to the set of consequences). Acts correspond to courses of action, consequences describe anything that may happen to a person, and states are possible resolutions of uncertainty, that is, 'a description of the world so complete that, if true and known, the consequences of every action would be known' (Arrow, 1971, p. 45). Implicit in this definition is the notion that there is a unique true state. Events are subsets of the set of states. An event is said to obtain if it includes the true state.

A decision maker is characterized by a preference relation, $\succcurlyeq$, on F. The statement $f \succcurlyeq f'$ has the interpretation 'the course of action f is at least as desirable as the course of action f''. Given $\succcurlyeq$, the strict preference relation $\succ$ and the indifference relation $\sim$ are defined as follows: $f \succ f'$ if $f \succcurlyeq f'$ and not $f' \succcurlyeq f$; $f \sim f'$ if $f \succcurlyeq f'$ and $f' \succcurlyeq f$.

The preference structure

The evaluation of a course of action in the face of uncertainty involves the decision maker's taste for the possible consequences and his beliefs regarding their likely realization. Savage's subjective expected utility theory postulates a preference structure, depicted axiomatically, permitting the numerical expression of the decision maker's valuation of the consequences by a utility function, that of his beliefs by a (subjective) probability measure on the set of all events, and the evaluation of acts by the mathematical expectations of the utility with respect to the subjective probability.

To state Savage's postulates, I employ the following notation and definitions. Given an event E and acts f and h, let $f_E h$ be the act such that $(f_E h)(s) = f(s)$ if $s \in E$, and $(f_E h)(s) = h(s)$ otherwise. An event E is *null* if $f_E h \sim f'_E h$ for all acts f and f', otherwise it is *non-null*. A constant act is an act that assigns the same consequence to all the states. To simplify the exposition, I denote the constant acts by their values (that is, if $f(s) = x$ for all s, I denote the act f by x).

The first postulate asserts that the preference relation is transitive and that all acts are comparable.

P.1 A preference relation is a transitive and complete binary relation on F.

The second postulate, also known as the sure thing principle, requires that the preference between acts depend solely on the consequences in states in which the payoffs of the two acts being compared are distinct. This implies that the valuation of the consequences of an act in one event is independent of the payoffs of the same act in the complementary event.

P.2 For all acts, f, f', h, h' and every event E, $f_E h \succcurlyeq f'_E h$ if and only if $f_E h' \succcurlyeq f'_E h'$.

The sure thing principle makes it possible to define conditional preferences as follows. For every event E, $f \succcurlyeq_E f'$ iff $f \succcurlyeq f'$ and for every s not in E, $f(s) = f'(s)$.

The third postulate asserts that the ordinal ranking of consequences is independent of the event and the act that yield them.

P.3 For every non-null event E and all constant acts, x and y, $x \succcurlyeq y$ if and only if $x_E f \succcurlyeq y_E f$ for every act f.

In view of P.3, it is natural to refer to an act that assigns to an event E a consequence that ranks higher than the consequence it assigns to the complement of E as a *bet* on E. Ramsey (1931) was the first to suggest that a decision maker's belief that an event E is at least as likely to obtain as another event E' should manifest itself through preference for a bet on E over the same bet on E'.

The fourth postulate, which requires that the betting preferences be independent of the specific consequences that define the bets, formalizes this idea.

P.4 For all events E and E' and constant acts x, y, x' and y' such that $x \succ y$ and $x' \succ y'$, $x_E y \succcurlyeq x_{E'} y$ if and only if $x'_E y' \succcurlyeq x'_{E'} y'$.

Postulates P.1–P.4 imply the existence of a transitive and complete relation on the set of events that has the interpretation 'at least as likely to obtain as' representing the decision maker's beliefs as qualitative probabilities. They also imply that the decision maker's risk attitudes are event-independent.

The fifth postulate renders the decision making problem and the qualitative probabilities non-trivial by ruling out that the decision maker is indifferent among all acts.

P.5 For some constant acts x and x', $x \succ x'$.

The sixth postulate introduces a form of continuity of the preference relation. It asserts that no consequence is either infinitely better or infinitely worse than any other consequence. Put differently, the next postulate requires that there be no consequence that, if it were to replace the payoff of an act on a non-null even, no matter how unlikely, will reverse a strict preference ordering of two acts.

P.6 For all acts f, g, and h satisfying $f \succ g$, there is a finite partition $(E_i)_{i=1}^{n}$ of the set of states such that, for all i, $f \succ g_{E_i} h$ and $h_{E_i} f \succ g$.

A probability measure is non-atomic if every non-null event may be partitioned into two non-null sub-events. Formally, π is a non-atomic probability measure on the set of states if for every event E and number $0 < \alpha < 1$ there is an event $E' \subset E$ such that $\pi(E') = \alpha \pi(E)$. Postulate P.6 implies that there are infinitely many states of the world and that, if there exists a probability measure representing the decision maker's beliefs, it must be non-atomic. Moreover, the probability measure is defined on the set of all events, hence it is finitely additive (that is, for every event E, $0 \leq \pi(E) \leq 1$, $\pi(S) = 1$ and for any two disjoint events, E and E', $\pi(E \cup E') = \pi(E) + \pi(E')$.

The seventh postulate is a monotonicity requirement asserting that, if the decision maker considers an act strictly better (worse) than each of the payoffs of another act on a given non-null event, then the former act is conditionally strictly preferred (less preferred) than the latter.

P.7 For every event E and all acts f and f', if $f \succ_E f'(s)$ for all s in E then $f \succcurlyeq_E f'$ and if $f'(s) \succ_E f$ for all s in E then $f' \succcurlyeq_E f$.

Representation

Savage's theorem establishes an equivalence between a preference relation having the properties described by the seven postulates and a preference relation induced by the maximization of the expectations of a utility function on the set of consequences with respect to a probability measure on the set of all events. The utility function is unique up to a positive affine transformation and the probability measure is unique.

Savage's theorem: Let $\succcurlyeq$ be a preference relation on F. Then the following two conditions are equivalent:

(a) $\succcurlyeq$ satisfies postulates P.1–P.7.
(b) There exists a unique, non-atomic, finitely additive, probability measure π on S such that $\pi(E) = 0$ if and only if E is null, and a bounded, unique up to a positive affine transformation, real-valued function u on C such that, for all acts f and g, $f \succcurlyeq f'$ if and only if $\int_S u(f(s))\, d\pi(s) \geq \int_S u(f'(s))\, d\pi(s)$.

Interpretation and criticism

In Savage's theory, consequences are assigned utilities that are independent of the underlying state of the world, and events are assigned probabilities that are independent of acts. These assignments, however, are not implied by the postulates. This observation merits elaboration.

The structure of the preference relation, in particular postulates P.3 and P.4, implies that the preference relation is state independent. In other words, the ordinal rankings of both consequences and bets are independent of the underlying events. This implies event-independent risk attitudes but does not, by itself, rule out the possibility that the states affect the decision maker's well-being, or that the utility of the consequences is state dependent. Put differently, Savage's model implies state-independent preferences but not a state-independent utility function. The utility and probability that figure in the representation of the preferences in Savage's theorem are unique as a pair, that is, the probability is unique given the utility and the utility is unique (up to a positive affine transformation) given the probability. It is possible, therefore,

to define new probability measures and state-dependent utility functions – and thereby to obtain a new subjective expected utility representation – without violating any of Savage's postulates. For instance, let γ be a bounded, positive, non-constant, real-valued function on S, and let $\Gamma = \int_S \gamma(s)\, d\pi(s)$. For every event E, define $\hat{\pi}(E) = \int_E \gamma(s) d\pi(s)/\Gamma$ and let $\hat{u}(x, s) = \Gamma u(x)/\gamma(s)$ for all s in S and x in C. Then, for every act f, $\int_S u(f(s))\, d\pi(s) = \int_S \hat{u}(f(s), s)\, d\hat{\pi}(s)$. Because γ is arbitrary and non-constant, $\pi \neq \hat{\pi}$. This shows that the uniqueness of the probability in Savage's theory is predicated on the convention that the utility function is state independent (that is, constant acts are constant utility acts). This convention is not implied by the postulates, it has no choice manifestation, and its validity is not subject to refutation in the context of Savage's analytical framework. Moreover, the employment of this convention renders the definition of probability in Savage's model arbitrary and the claim that it represents the decision maker's beliefs scientifically untenable. That said, it is noteworthy that, in so far as the theory of decision making under uncertainty is concerned, because all the representations obtained using the procedure outlined above are equivalent, the failure to correctly quantify the decision maker's beliefs is not critical. In so far as providing choice-based foundations of Bayesian statistics is concerned, however, this failure is fatal.

A somewhat related aspect of Savage's model that is similarly unsatisfactory concerns the interpretation of null events. Ideally, an event should be designated as null and be ascribed zero probability if and only if the decision maker believes it to be impossible. In Savage's model an event is null if the decision maker displays indifference among all acts that agree on the payoff on the complement of the said event. However, this definition does not make a distinction between an event that the decision maker perceives as impossible and one whose possible outcomes he perceives as equally desirable. It is possible, therefore, that events that the decision maker believes possible, or even likely, are defined as null and assigned zero probability. Consider this example: a passenger who is indifferent to the size of his estate in the event that he dies is about to board a flight. For such a passenger, a plane crash is a null event and is assigned zero probability, even though he may believe that the plane could crash. This problem renders the representation of beliefs by subjective probabilities dependent on the implicit and unverifiable assumption that in every event some outcomes are strictly more desirable than others. If this assumption is not warranted, the procedure may result in a misrepresentation of beliefs.

The requirement that the preferences be state independent imposes significant limitations on the range of applications of Savage's theory. Choosing a disability insurance policy, for example, is an act whose consequences – the indemnities – depend on the realization of the decision maker's state of health. In addition to affecting the decision maker's well-being, it is conceivable that alternative states of disability influence his risk attitudes. Disability may also alter the decision maker's ordinal ranking of the consequences, which is a violation of P.3. For instance, a leg injury may reverse a decision maker's preferences between going hiking and attending a concert. Similar observations apply to the choice of life and health insurance policies.

Savage presented his seven postulates as principles that a rational individual ought to follow rather than an hypothesis describing how individuals actually choose among courses of action in the face of uncertainty. Indeed, almost from the moment of it inception, the descriptive validity of Savage's model – in particular, the sure thing principle, which is responsible for the specific functional form of the representation and the separability and linearity in the probabilities – has been questioned. It has repeatedly been shown in experimental settings that the theory fails systematically to predict subjects' choice. The most severe and remarkable criticism in this regard is due to Ellsberg (1961), who demonstrated using simple thought experiments that individuals display choice patterns that are inconsistent with the existence of beliefs representable by a probability measure.

EDI KARNI

See also **Bernoulli, Daniel; de Finetti, Bruno; Ramsey, Frank Plumpton; Savage, Leonard J. (Jimmie); state-dependent preferences; utility; von Neumann, John.**

Bibliography

Arrow, K. 1971. *Essays in the Theory of Risk Bearing.* Chicago: Markham Publishing Co.

Bernoulli, D. 1738. Specimen theoriae novae de mensura sortis. *Commentarii Academiae Scientiatatum Imperalas Petropolitanae* 5, 175–92. Translated as 'Exposition of a new theory on the measurement of risk', *Econometrica* 22 (1954), 23–6.

de Finetti, B. 1937. La prévision: ses lois logiques, ses sources subjectives. *Annals de l'Institute Henri Poincaré* 7, 1–68. Trans. H. Kyburg in H. Kyburg and H. Smokler, eds., *Studies in Subjective Probabilities.* New York: John Wiley and Sons, 1964.

Ellsberg, D. 1961. Risk, ambiguity, and the Savage axioms. *Quarterly Journal of Economics* 75, 643–59.

Hacking, I. 1984. *The Emergence of Probabilities.* Cambridge: Cambridge University Press.

Ramsey, F. 1931. Truth and probability. In *The Foundations of Mathematics and Other Logical Essays,* ed. R. Braithwaite and F. Plumpton. London: K. Paul, Trench, Truber and Co.

Savage, L. 1954. *The Foundations of Statistics.* New York: John Wiley.

von Neumann, J. and Morgenstern, O. 1944. *Theory of Games and Economic Behavior.* Princeton, NJ: Princeton University Press.

Sax, Emil (1845–1927)

Sax was born in Jauernig (then, Austrian-Silesia; today, Javornik in Czechoslovakia). He studied in Vienna, where he became a university lecturer. After some practical activity (among other things in the railway organization), he became, in 1879, professor of political economy at the University of Prague. From 1879 to 1885 he was a member of the Vienna Chamber of Deputies. In 1893 he abandoned his Prague professorship and retired to Abbazia, Istria (then, Italy; today, Opatija in Croatia), where he died in 1927.

Sax holds a peculiar place within the older Austrian School. He shared its basic idea according to which 'value' virtually opens the way to the explanation of all economic problems. However, in methodological matters, in the conception of economics, and in the interpretation of the value phenomenon itself, he went his own way. For Sax 'value' is not the rationally perceived significance of a commodity for the welfare of a person, but an emotional relationship between the person and the world of goods. His conception of economics is based on the distinction between individualism and collectivism. Behaviour is individualistic if it results (self-determined) from the individual personality; behaviour is collectivistic if the individual is motivated only as a member of a (larger and stable) group and in relation to this group. According to Sax, these two fundamental forces shape all economic and social phenomena in a characteristic way: the simple feeling of value becomes, individualistically, the exchangeable value; collectivistically, the complicated determination of value within a group. In relation to the social environment, these two fundamental forces can appear egoistic, altruistic and (as a mixture of both) mutualistic. Sax believed he had overcome the psychological one-sidedness of the Classical School, and therefore he considered the findings of his theoretical work as exact results of inductive research.

According to Sax, the distinction between individualism and collectivism corresponds to the (value based) theories of private and public economy. The absolute tax level (today called tax–GNP ratio) is determined by evaluations of such individuals in whom the fundamental force of collectivism is effective; they take care of the fact that the levels of the private and the public sphere are balanced. The relative tax level (today called tax apportionment to individuals) is deduced by Sax from the 'equivalence of value' of the individual tax liabilities; this results in the application of the equal sacrifice theory. For Sax, compulsion in taxation merely substitutes for a lack of correct insight. The 'statemindedness of those governing' and the 'resistance of those governed' prevent the abuse of power.

Besides fiscal theory problems, Sax paid particular attention to the then young science of transport economics. His statements on transport policy (1878–9) are founded on a theoretically analysed historical experience, although the theoretical sections do not reach up to the abstract (mathematical) level such as Launhardt's treatment of transport problems. Despite the fact that much of Sax's writings on transport economics have since become obsolete, the book remains an excellent piece of applied economics.

K. SCHMIDT

Selected works

1878–9. *Die Verkehrsmittel in Volks- und Staatswirthschaft.* 2nd edn, 3 vols, Berlin: Springer, 1918–22.

1884. *Das Wesen und die Aufgaben der Nationalökonomie. Ein Beitrag zu den Grundproblemen dieser Wissenschaft.* Vienna: Hölder.

1887. *Grundlegung der theoretischen Staatswirthschaft.* Vienna: Hölder.

1924. *Die Wertungstheorie der Steuer. Zeitschrift für Volkswirtschaft und Sozialpolitik, NS 4.* Vienna and Leipzig: Deuticke.

Bibliography

Beckerath, E. von. 1930. Emil Sax, ein Nachruf. *Zeitschrift für Nationalökonomie 1.* Vienna: Springer.

Beckerath, E. von. 1956. Sax, Emil. *Handwörterbuch der Sozialwissenschaften*, vol. 9., Stuttgart: Fischer; Tübingen: Mohr; Göttingen: Vandenhoeck & Ruprecht.

Say, (Jean-Baptiste) Léon (1826–1896)

French statesman, financier and economist, born in Paris in 1826; died there on 22 April 1896. He was the son of Horace Emile Say, the grandson of Jean-Baptiste Say, the nephew of Louis Auguste Say and Charles Comte. Léon Say became one of the most prominent statesmen of the French Third Republic. He served as Finance Minister from 1872 to 1879, and again in 1882, overseeing the largest financial operation of the century – payment of war reparations in Germany. His financial policies were directed towards a decrease in public expenditures and the removal of barriers to internal trade. A brilliant speaker and debater, he railed against socialism from the left and protectionism from the right. With Gambetta and Freycinet, he launched the ambitious programme of public works that bears the latter's name. Upon leaving the Cabinet, Say returned to his seat in parliament, assuming the leadership of the free trade party. He was at one time considered for the presidency of the republic, but was gradually set apart from his constituency by a rising tide of radicalism.

As an economist, Say's talents fall somewhere between the modest gifts of his father and the more imposing skills of his grandfather. He left no large work nor did he create any school of thought. Like his father, he was faithful to the doctrines of his namesake, and was a competent editor of his grandfather's works. In his youth he associated briefly with Léon Walras in a scheme to promote cooperative associations of production. He later

became a frequent contributor to the *Journal des économistes,* mostly on economic policy, and a lecturer at the Ecole des Sciences Politiques, which was the prototype of the London School of Economics and Political Science. Say had a broad knowledge of history and theory, and he was capable of sustained exposition at a high level. As an example, his *Solutions dómocratiques de la question des impôts'* (1886) was directed against the idea of using taxation as a means of social equalization. He argued, instead, that the basis of taxation should always be real (based on property), never personal.

A curious parallel exists in the careers of Say and Turgot, whose name Say declared he could not even pronounce without emotion. They shared a body of ideas and a similar destiny. Both achieved eminence as finance ministers in the French government, only to be turned out upon losing public favour. Say however, helped to immortalize his predecessor by writing one of the earliest biographies of Turgot.

R.F. HÉBERT

Selected works

1886. *Les solutions démocratiques de la question des impôts.* 2 vols, Paris: Guillaumin.

1888. *Turgot,* trans. M.B. Anderson. Chicago: A.C. McClurg & Co.

1891. *Economie sociale.* Paris: Guillaumin.

1891–2. *Nouveau dictionaire d'économie politique.* 2 vols. Paris: Guillaumin.

1896. *Finances publiques, liberté du commerce.* Paris: Guillaumin.

1898–1901. *Les Finances de la France sous la Troisième République.* 4 vols. Paris: C. Levy.

Bibliography

Michel, G. 1900. *Léon Say, sa vie, ses oeuvres.* 2nd edn. Paris: Calmann Levy.

Picot, G.M.R. 1907. Léon Say. In *Etudes d'histoire contemporaine. Notices historiques*, vol. 2. Paris: Hachette.

Say, Jean-Baptiste (1767–1832)

Although Jean-Baptiste Say is remembered primarily for Say's Law, one of the cornerstones of classical economics, he was also an early proponent of the utility theory of value, and was therefore very much at odds with his classical contemporaries, to whom labour was the source of value. Say's best-known work, his *Traité d'économie politique* (published in five editions, from 1803 to 1826) was intended as a shorter and more systematic presentation of economics than Adam Smith's *Wealth of Nations.* The success of this book made Say the best-known expositor of Smith in Europe and America, and he became in 1815 France's first professor of political economy. Translations of the *Traité* were used as textbooks at universities on both sides of the Atlantic.

Say was not, however, a mere uncritical expositor of Smith. The central importance of labour in Smith's discussions of value was replaced by Say's concern to show utility as the ultimate foundation of value. Production itself was defined as the production of utility, not of physical output. He noted in the first chapter of his *Traité* that this was subjective utility, which the economist must take as given data, however much moralists might attempt to change people's valuations. Businessmen also played a much more important and honourable role in Say than in Smith – Say having been a businessman himself and descended from a mercantile family. The *Traité d'économie politique* also went beyond Smith in developing what Say called 'one of the most important truths of political economy' – that supply creates its own demand, the doctrine ultimately named Say's Law.

Much controversy has surrounded the question of Say's originality in developing this principle, or rather, related series of principles. Claims have been made for James Mill as the real author of Say's Law. However, Mill's earliest published discussion of issues involving aggregate supply and demand came in an 1804 review in *The Literary Journal* of a book by Lauderdale – one year after the first edition of Say's *Traité* was published. While the chapter ('Des Débouchés') in which Say's Law was first set forth was very brief in the first edition, there was further discussion of the same principle in that same edition, notably in Chapter 5 of the second volume. Later editions brought these scattered discussions together in an enlarged chapter on aggregate supply and demand, but much of the substance was there from the beginning. Mill was only the first of many to reformulate and elaborate what Say had done.

Say was concerned with the methodology as well as the substantive propositions of economics. He advocated systematic analysis rather than naive empiricism, but was also highly critical of the abstract deductive method of Ricardo and his followers. According to Say, Ricardo 'pushes his reasonings to their remotest consequences, without comparing their results with those of actual experience'. During a friendly correspondence with Ricardo, Say pointed out that facts 'are the masters of us all'. In the introduction to his *Traité d'économie politique*, Say also expressed his fear of 'our always being misled in political economy, whenever we have subjected its phenomena to mathematical calculation'.

Say was in touch with the leading economists of his day, by mail and in person. He never resolved his differences with Ricardo as to whether value was based on labour or utility, but in attempting to clarify his position in 1822, Say spoke of 'the last quantity of useful things' as being crucial – a suggestion of the missing *marginal* concept essential to the utility theory of value. In his correspondence with Sismondi and Malthus, he came ultimately to reconcile Say's Law with their theories

of aggregate disequilibrium. The fifth edition of his *Traité d'économie politique* in 1826 incorporated some of Sismondi's reasoning at the end of his chapter on Say's Law of markets (unfortunately, the English translation is from the previous edition) and called this to Malthus's attention as an admitted 'restriction' on this doctrine. A later textbook by Say, *Cours complet d'économie politique*, published in 1828–9, followed the chapter on Say's Law with one entitled 'Limits of Production', a phrase from Sismondi along with Sismondian analysis in the chapter.

Say was a policy-oriented economist rather than a model-builder like Ricardo. In his introduction to the new restrictions added to his chapter on the law of markets, Say remarked: 'Now, we are studying practical political economy here.' To Malthus he wrote: 'It is better to stick to facts and their consequences than to syllogisms.'

THOMAS SOWELL

Say's Law

Say's Law, the apparently simple proposition that supply creates its own demand, has had many different meanings, and many sets of reasoning underlying each meaning – not all of these by Jean-Baptiste Say. Historically, Say's Law emerged in the wake of the Industrial Revolution, when the two striking new economic phenomena of vastly increased output and the economy's cyclical inability to maintain sales and employment led some to fear that there was some inherent limit to the growth of production – some point beyond which there would be no means of purchasing it all. At the very least, some feared, there would not naturally or automatically be generated sufficient purchasing power to absorb the ever-growing output of the industrial economy, unless special policy arrangements were made to insure that income would be large enough to purchase output. Robert Owen and Karl Rodbertus exemplified these views, which were not those of any school of economists.

Say's Law attempted to answer such concerns by pointing out that the production of output tends of itself to generate purchasing power equal to the value of that output: supply creates its own demand. But Say's Law did not spring forth, full blown, like Minerva from the head of Zeus. It emerged piecemeal over a span of years, enveloped in controversies that ultimately involved nearly every noted economist of the early 19th century, and as its elaboration proceeded its definitions shifted under polemical stress. Moreover, the basic terms of discourse in economics were themselves in a process of evolution. The words 'supply' and 'demand' had different meanings for those economists like Sismondi and Malthus, groping towards the schedule or functional meanings of today, from those in the writings of David Ricardo or John Stuart Mill, who rigidly defined the terms as quantity supplied and quantity demanded. They repeatedly argued past each other.

The central meaning of Say's Law was implied by J.B. Say's rhetorical question, 'how could it be possible that there should now be bought and sold in France five or six times as many commodities as in the miserable reign of Charles VI?' This dramatized his main point, that there was no long-run limit to the growth of output, or of the demand for it. This did not deny that a *short run* derangement of the economy could take place, but Say described this as 'an evil which can only be passing'. Moreover, he attributed these short-run phenomena to a wrong mixture of output, compared to consumer demand, rather than to aggregate overproduction. This was an ad hoc addendum, not logically implied by the principle of supply creating its own demand. Say thus created a subsidiary meaning of Say's Law – a denial that there was such a thing, even in the short run, as aggregate overproduction – which is to say, that there was no such thing as an equilibrium aggregate output. There could be a 'partial glut' of particular commodities produced in excess of demand but there could be no 'general glut' of commodities in the aggregate. It was this vulnerable subsidiary argument which Sismondi and Malthus attacked in the 19th century and Keynes in the 20th.

The first edition of Say's *Traité d'économie politique* in 1803 contained the crucial propositions of Say's Law, though not all in his chapter on markets ('Des Débouchés'). The quantity of products demanded was 'without a doubt' determined by the quantity of products created, according to Say. 'The demand for products in general is therefore always equal to the sum of the products', he said. The distinction between secular stagnation and short-run downturns, and between partial and general gluts, was also present from the first edition of Say's *Traité*. These ideas all reappeared in James Mill's writings shortly afterward, but Say's priority is clear, both from the dates of the publications and from Mill's citation of Say in his own early writings, depriving him of even subjective originality.

The elder Mill did, however, make a significant contribution to the evolution of Say's Law. Where Say had asserted the inherent sufficiency of demand to purchase supply in terms of half the goods being essentially bartered for the other half, or in terms of saved money being spent as investment, Mill added the behavioural theory that people produced only because of, and only to the extent of, their demand for other goods. Each individual's supply equalled his demand *ex ante;* therefore society's supply must equal society's demand *ex ante*. Unfortunately, Mill also cited the *ex post* identity of supply and demand as evidence, as did J.R. McCulloch, Robert Torrens and John Stuart Mill.

While Say's priority over James Mill is readily established, the notion that in the long run aggregate demand 'has no known limits' was stated by the Physiocrat Mercier de la Rivière in the year of Say's birth. Nor was

this a passing remark. His book, *L'Ordre naturel et essential des sociétés politique* (1767), especially Chapter 36, contained both the concept of a circular flow of money and of goods, and discussions of the conditions under which the existing level of aggregate output would be reproduced in subsequent time periods, as well as the conditions in which it would fall because receipts failed to cover the supply price of inputs. Yet it was not through Mercier de la Rivière but through Say that Say's Law entered the mainstream of economics. Ironically, both Say and Mill attacked Mercier de la Rivière. While his statement that aggregate demand is unlimited anticipated both of them, his concept of aggregate equilibrium and disequilibrium was anathema to the subsidiary version of Say's Law that was an integral part of the doctrine during the early 19th century.

Adam Smith also anticipated Say when he asserted in *The Wealth of Nations* (p. 407) that 'a particular merchant' could have a glut of goods but that a whole nation could not. Moreover, Smith's doctrine that savings rather than consumption promoted growth provided yet another dimension to Say's Law and the controversies surrounding it. One possible meaning of Smith's statement was that a shift in the savings function – a willingness to save and invest more at a given rate of return – would tend to increase future output. But another interpretation, equally permissible in the absence of functional concepts, was that an increased quantity saved would promote future growth. Lord Lauderdale interpreted Smith in the latter sense, and attacked this proposition on grounds that there was some equilibrium level of savings and investment which, if exceeded, would reduce rates of return to a point that would cause the existing levels of savings to decline in subsequent time periods. In short, Lauderdale argued that there could be a general glut of capital, just one step from saying that there could be a general glut of aggregate output.

The leading critics of Say's Law during the classical era – Sismondi, Malthus and Lauderdale – all asserted short-run disequilibrium, not long-run stagnation, but the long-run comparative-statics approach of the Ricardians made it especially difficult for them to understand what the critics were saying in short-run dynamic terms. While Say himself ultimately came to understand – and reproduce in his later writings – the aggregate equilibrium theories which he now reconciled with the central meaning of Say's Law, more than 20 years later John Stuart Mill was still representing Lauderdale, Sismondi and Malthus as stagnationists. However, after completely misinterpreting their positions, J.S. Mill also set forth the most sophisticated analysis of the issues in classical economics in the second of his *Essays on Some Unsettled Questions in Political Economy* (1844).

After denouncing the 'mistakes', the 'completely erroneous' ideas and 'palpable absurdities' of those who emphasized the need for adequate aggregate demand, J.S. Mill nevertheless conceded that there could be 'general excess' in the sense that when money was not immediately respent, a seller 'does not therefore necessarily add to the *immediate* demand for one commodity when he adds to the supply of another'. Thus there may be 'a superabundance of all commodities relative to money'. Both the previous classical economists and such critics as Sismondi and Malthus had analysed the issue of aggregate output in essentially barter terms, despite incidental references to money. The theory of equilibrium aggregate output in Sismondi and Malthus was based on a balance of the utility of output and the disutility of the efforts required to produce it – a balance which could be temporarily unbalanced in their short-run dynamic models, though not in the long-run comparative statics model of James Mill.

John Stuart Mill's model, in which the role of money was important, was a different dimension, though not unique – Robert Torrens having expressed similar ideas more than two decades earlier. However, on the crucial issue of an aggregate equilibrium output in a non-monetary model, J.S. Mill remained adamant that there could be only internal disproportionality, not aggregate overproduction. Output, 'if distributed without miscalculation among all kinds of produce in the proportion which private interest would dictate, creates, or rather constitutes, its own demand'. The issue remained to J.S. Mill one of internal 'proportions', not aggregate amounts.

Discussions of Say's Law virtually disappeared from economics for at least a generation after John Stuart Mill wrote on it in the 1840s. Even the sweeping challenges of neoclassical economics to classical orthodoxy, beginning in the 1870s, largely bypassed the issue of Say's Law. Isolated criticisms came from beyond the pale – from Marx, Hobson and assorted cranks. Within the economics profession, Say's Law was one of those things simply assumed and ignored. Early in the 20th century, Knut Wicksell explored the relationship between the quantity theory of money and Say's Law. The classical assumption that money is demanded only for transactions during the current period is incompatible with the price level being determined by the quantity of money, for a change in the price level requires that, at some point in the process, there must be either an excess or deficient money demand for goods in the aggregate, causing the general price level to go up or down in response. However, Wicksell's own belated recognition by the English-speaking world meant that Say's Law did not become a major concern again until the appearance of John Maynard Keynes's *General Theory of Employment, Interest and Money* in 1936.

Modern, and especially Post Keynesian, discussions of Say's Law have revealed it to be not one, but a number of related, propositions. The most general of these propositions is that the aggregate value of goods supplied (including money) equals the aggregate value of goods demanded (including money). Thus an excess supply of goods is the same as an excess demand for money. This proposition has been christened 'Walras' Law'. Where there is assumed to be no excess demand for money, as in James Mill, for example, then aggregate supply is identically equal to aggregate demand. This proposition has been christened 'Say's Identity'. When the equality of aggregate supply and demand is stated as an equilibrium condition – a sense in which equilibrium output theorists like Sismondi and Malthus could subscribe to it – then it merely states that both equilibrium and disequilibrium levels of output may exist. This proposition has been christened 'Say's Equality'. This Ricardo, James Mill, and initially Say, all denied.

The Keynesian Revolution not only produced a more sophisticated theory of aggregate equilibrium, but also contributed to the distortion of Say's Law, which Keynes reduced to Say's Identity. According to Keynes, Say's Law 'is equivalent to the proposition that there is no obstacle to full employment'. Only the cruder statements of the Ricardians said that.

Say's Law has been an important proposition in many ways. By indicating that the possibility of purchasing output from the income generated during its production is ultimately not limited by the mere size of output, Say's Law exposed the fallacy of recurrent popular fears that economic growth must collide with some impassable limit. The modern Post Keynesian delineations of the different senses of Say's Law (Walras' Law, Say's Identity, Say's Equality) more precisely specify the conditions of aggregate equilibrium and disequilibrium, and indicate the theories of economic behaviour behind them. Finally, the long history of controversies over Say's Law sheds light on the enormous difficulties involved when even intelligent thinkers with honesty and goodwill try to understand each other's theories without clearly defined terms and without a clear sense of the conceptual framework of the opposing views. In short, its implications reach beyond economics to intellectual history in general.

THOMAS SOWELL

Bibliography

Becker, G. and Baumol, W.J. 1960. The classical monetary theory: the outcome of the discussion. In *Essays in Economic Thought*, ed. J.J. Spengler and W.R. Allen. Chicago: Rand McNally.

Malthus, T.R. 1820. *Principles of Political Economy*. New York: Augustus M. Kelley, 1951.

Mill, J. 1803. *Commerce Defended*, ch. 6. London: C. and R. Baldwin.

Say, J.-B. 1803. *Traité d'économie politique*, vol. 1, ch. 15. Paris: Chez Rapilly, 1826.

Sismondi, J.C.L. de. 1827. *Nouveaux principes d'économie politique*, vol. 1, books 1, 4; vol. 2, Appendix. Paris: Delaunay Libraire.

Sowell, T. 1972. *Say's Law: An Historical Analysis*. Princeton: Princeton University Press.

Sowell, T. 1974. *Classical Economics Reconsidered*. Princeton: Princeton University Press.

Spengler, J.J. 1960. The Physiocrats and Say's law of markets. In *Essays in Economic Thought*, ed. J.J. Spengler and W.R. Allen. Chicago: Rand McNally.

Scandinavia, economics in

Denmark–Norway before 1814

Ludvig Holberg (1694–1754) was the first to forcefully advocate political economy as a science and academic discipline in the dual monarchy of Denmark–Norway. Today he is recognized as the Molière of northern Europe, but few know about his contribution to economics. Strongly influenced by the natural law philosopher Samuel Pufendorf (1632–94), but also by English and French Enlightenment philosophy, he studied economic phenomena and problems from the perspective of moral philosophy. His contributions cover a wide spectrum and include his literary authorship, his achievements at the University of Copenhagen, and his efforts in re-establishing Sorö Academy as a modern centre of higher education. The latter was made possible in 1747 when Holberg bequeathed his estates to Sorö Academy.

The Academy placed modern sciences on in its curriculum, and quickly became an alternative to the University of Copenhagen, which was marked by a strong theological influence. At the Academy students were taught 'Political Economy, Commerce and Cameral Sciences'. This proved a great success. In the second half of the 18th century the Academy functioned as the academic home for social scientists and social critics. At Sorö Jens Schelderup Sneedorff (1724–64) was appointed the first professor of political economy and public law in 1751. He and his successor Andreas Schytte (1726–77) were influenced both by German cameralism and, like Holberg, by English and French philosophers. Schytte wrote the first textbooks in political economy in the local language.

At the university Ole Stockfleth Pihl (1729–65) was appointed the first professor in political economy in 1761. Before that he had been the editor and publisher of the monthly *Oeconomisk Journal*. As the position had no salary he resigned after two years. The next professor, Johan Christian Fabricius (1745–1808), was appointed in 1772. His salary was so small that after four years he accepted a chair in natural history, political economy and cameral sciences at University of Kiel, the second university in the dual monarchy. Here he created 'an economic garden' and wrote a widely used textbook in political economics.

Another important factor in the development of political economy as a science was the establishment of *Danmark og Norges Oeconomiske Magazin* in 1756. Its initiator and editor, the bishop of Bergen Erik Pontoppidan (1698–1764), was an enlightened mercantilist, whom the king had called on to carry out reforms at the university. His reforms were not successful, but the *Economic Magazin* became a sanctuary and a workshop for those who were engaged with economic questions in the middle of the 18th century. Otto Diderich Lütken (1713–88), its most important contributor, is considered the most original of economic thinkers in Denmark–Norway in the 17th century. He published several essays discussing theoretical as well as practical economic issues. In one article he claims, as Malthus did 40 years later, that there is a connection between population and available food, and that population would increase until a shortage of food put an end to further growth.

The influence of the *Economic Magazin* was considerable. People connected with this periodical and with Sorö Academy gained influence from the mid-18th century and far into the 19th century. Their thinking found expression in the agricultural and social reforms carried out towards the end of the 19th century, and is credited with giving impetus to a translation of Adam Smith's *Wealth of Nations* in 1779, initiated by Norwegian tradesmen.

Denmark after 1814

Characteristics of Danish development

Unlike in many other countries, social development in Denmark has been characterized by its continuity. No political changes have been sufficiently radical to create a break in academic traditions. In 1849 the country changed from being an absolute monarchy to being a democracy, but A.F. Bergsøe was professor in political economy during the whole period 1845–54. In 1901, after decades of political struggle, a right-wing government was replaced by parliamentarism and a left-wing government, but H.W. Scharling, Minister of Finance in the right-wing government, was professor in political economy without interruption from 1869 to 1911. Furthermore, the economic systems in Denmark have not been subjected to disturbances great enough to leave traces in the science of economics. The Danish economy has been marked by relatively stable growth for 200 years.

Apart from continuity, a dominant fact is the small size of the country. Until 1936, the University of Copenhagen was the only institution offering university-level teaching in economics. Courses in economics were started in 1848, and from that time there were two chairs in political economy, one of them a chair in cameralism and public economics dating back to 1762. In 1886, Harald Westergaard was given a personal chair, bringing the number of professors to three. This number did not increase again until after the First World War, and not dramatically until after the Second World War. In the 1960s and 1970s, the number of professors begins to shoot up, with five universities and 14 full professors in economics in 1960, and eight universities and 34 chairs in 1995. These figures actually underestimate the growth; before 1960, there were scarcely any teachers who were not full professors, whereas the number of assistant and associated professors today is much greater than the number of professors.

International contacts

Perhaps the tiny domestic research environment made international contacts even more necessary than would have been the case in larger countries. An example of this can be seen in the marginal revolution at the beginning of the 1870s. When this revolution was started by Jevons, Menger and Walras, they had no contact with each other. Jevons died in 1882 without having heard of Menger (Howey, 1972). In Denmark, the publication of Walras's first volume was reviewed by *Nationaløkonomisk Tidsskrift* in 1875, and the reviewer (an institutional economist, A. Petersen) found, 'surprisingly', that Walras did not know Jevons's work (Kærgård, 1996).

Danish economists corresponded in French with Walras and in English with Jevons at a time when Danish economics was largely German-oriented. Contact with other Scandinavian economists was particularly close. From 1863 onwards, there were regular Scandinavian conferences on political economy, and a 'Marstrand Meeting' for Scandinavian economic researchers was arranged for most years between 1936 and 1985.

All this has changed in recent decades, when Danish economists have become an ordinary part of the general international research community. At the same time, the trend has been to move from publishing in books towards publishing in international journals, and from publishing in Danish to publishing in English.

Danish contributions

If we consider the contributions made by Danish economists, we find none who have come even close to an established position in the history of economics. Hutchinson's standard text *A Review of Economic Doctrine 1870–1929* mentions 359 names; these include six Swedes, one Norwegian and no Danes (Boserup, 1980).

Thus the history of Danish economics cannot until the most recent decades show internationally known names, but it is filled with overlooked precursors and with discoveries described in Danish, never known outside the domestic border. Mentioned can be Otto Ditlev Lütken, who wrote on population growth and scarcity of food before Malthus (Sæther, 1993); Bing and Julius Petersen, who wrote on neoclassical distribution theory in 1873 (Whitaker, 1982); Westergaard, who in the 1870s was the first to use mathematical maximization theory in economics (Creedy, 1980; Davidsen, 1986; and Kærgård and Davidsen, 1998); Mackeprang, whose thesis of 1906 was the first econometric analysis (Kærgård, 1984); Warming's description of the identification problem from 1906

(Kærgård, 1984, and Kærgård, Andersen and Topp, 1998); Wulf and Warming's development of the multiplier theory from 1896 to 1932 (Boserup, 1969; Topp, 1981); Frederik Zeuthen's discussion of monopolistic competition in the late 1920s (Brems, 1976); Jørgen Pedersen's description of fiscal policy in 1937 (Topp, 1988) and Gelting's derivation of the balanced budget multiplier in 1941 (Hansen, 1975). None of these discoveries were published in English, and were not made known internationally until after the theories had become widely known.

There are several possible explanations for the high number of unknown Danish contributions to economic theory. Brems (1986) suggests two barriers to the dissemination of economic theories: a linguistic barrier (Anglo-Saxons do not speak German and French) and a mathematical barrier. Unlike the economists in the larger European countries, those from the small-language communities such as Denmark understood all major languages, and were therefore better acquainted with all the international schools and could combine their ideas. However, economists from smaller countries often wrote in their own language, and consequently their work never became widely known. With so few professors of economics in small countries, it was furthermore necessary for them to be very versatile, and they therefore tended to move from one subject to another, and a more persistent approach is necessary to be established as a pioneer (one might recall Walras's battle over years to achieve recognition).

Danish economists: generalists not specialists

We can see from the above that the relationship between Danish economists and the various international schools has been like that between butterflies and flowers. They have generally fluttered from school to school, taking from each what they felt useable; very few Danish economists have been orthodox disciples of one of the recognized schools. A couple of examples can be mentioned. At the time of the great methodological battle in Germany between the neoclassical school and the Christian-Social-Historical School, Westergaard was in close contact with both schools. In the 1870s, he corresponded with Jevons concerning mathematical-economic problems, and at the same time he was the leading representative in Denmark of the Christian-Social-Historical school (Kærgård, 1995). During the debate among neoclassicists, monetarists and Keynesians in the 1960s and 1970s, Anders Ølgaard played a central role in economic debates in Denmark as Chairman of the Danish Board of Economic Advisors, arguing from a typical Keynesian viewpoint, at the same time as he was writing a substantial treatise on neoclassical growth theory (Ølgaard, 1966).

Danish economists were almost always more than just theorists. Between 1870 and 1970 there were a total of 15 people who held chairs in political economy at the

University of Copenhagen; of these, six were members of parliament, and of those six three held posts as members of the government. Another was member of the Copenhagen municipal government. Among the remaining eight were Bertil Ohlin, who was in Copenhagen only for a short time and later became politically active in Sweden; Erik Hoffmeyer, who was the first director of the Central Bank of Denmark for more than 30 years; Harald Westergaard, who was a leading church politician and social reformer; and Carl Iversen and Anders Ølgaard, two chairmen of the Board of Economic Advisors. All 15 were active in the public debate, writing numerous newspaper articles. Jointly, they held almost innumerable positions in commercial life, councils and commissions. Recent decades have revealed a completely different type of university economist, with purely academic and theoretical interests.

Norway after 1814

In 1814 Christen Smith (1785–1816) was appointed professor of botany and political economy at the first Norwegian university in Oslo. Botany and political economy would be considered a strange combination of subjects nowadays, but at that time the logic of such an arrangement was clear. The wealth of nature would create prosperity for the people. Unfortunately, before he could take up his position Smith died during a British-led botanic expedition to Congo.

Breakthrough of political economy

His successor, Gregers Fougner Lund (1786–1836), was not appointed until 1822. He wanted to move political economy away from mercantilism towards economic liberalism. His views were supported by Jacob Aall (1773–1844), ironmaster and member of parliament, who, with his essays on economic problems, became highly influential.

Anton Schweigaard (1808–70), who took over the chair in law, political economy and statistics in 1836, dominated economic thinking in Norway for almost half a century. He supported the liberal economic policy recommended by the classical economists. However, he did not follow them blindly, since in his opinion they sometimes carried their policies too far. In spite of being a spokesman for free trade he rejected the doctrine of laissez-faire. On many questions he was closer to the Continental economists, especially Say in France and Hermann and Rau in Germany. With Schweigaard political economy had gained a firm foothold as a science.

Tradition and renewal

When Schweigaard died, his former student, Torkel H. Aschehoug (1822–1909), professor of law, took over his teaching responsibilities in economics and statistics. Until his death he dominated political economy

within the academic world and beyond. He was behind, or strongly supported, several important events: the creation of the Statistical Bureau of Census in 1876 and the establishment in 1883 of *Statsøkonomisk forening*, an association for Norwegian economists, which he chaired for 20 years. The latter became a forum where the enlightened elite of bureaucrats, government ministers, parliamentarians and academics discussed economic issues. From 1887 it published an economic journal, *Statsøkonomisk Tidsskrift* (renamed *Norsk Økonomisk Tidsskrift* in 1997), in which economists discussed both theoretical and practical issues.

A second chair in 'pure economics' was created in 1877, and an independent study programme in 'political economy' was established in 1905. Aschehough wanted to give an account of economic science in the Norwegian language. The first edition of his *Socialøkonomik*, completed in 1891, dealt with the theories of the classics, and the moderns Böhm-Bauwerk, Jevons, Menger, Schmoller and Walras. Later editions, however, were strongly influenced by Marshall's *Principles of Economics* and in particular his theory of value.

Professional build-up

Oskar Jæger (1863–1933), Peder Thorvald Aarum (1867–1926) and Ingvar Wedervang (1891–1961) were the central persons in the Norwegian economic profession between Aschehoug and Ragnar Frisch (1895–1973).

Jæger's contributions span from treatises on methodology to thoughts on public finance, including an active, although disputed, participation in economic politics. His historical lectures in political economy were concerned with the development of 'modern' analysis from an Austrian point of view. He mentions Marshall, of course, but his mainstay is Böhm-Bawerk.

Aarum's university career was relatively short, but his textbooks in theoretical and practical economics gave him considerable influence. He followed Marshall, and claimed that the interactions of demand and supply in the market simultaneously determined price and quantity. Market equilibrium became a key concept. He also introduced the extensive use of diagrammatic exposition in his lectures and books. Aarum was regarded as 'the modern' among Norwegian economists.

Wedervang is considered one of the great profession builders in Norwegian economics. He was behind the parliament's decision to create a new chair in economics at the university in 1931 for Ragnar Frisch. Together with Frisch, he was behind the establishment of the Rockerfeller supported economic institute in 1932. He and Frisch were its first directors. Furthermore, he was in the forefront when a new five-year study programme in economics was adopted in 1934, and when the parliament decided to establish the Norwegian School of Economics and Business Administration in Bergen in 1936.

The Oslo School

In 1919 Ragnar Frisch graduated from the study programme in political economy. After studies in France, Germany, England, the USA and Italy he became Aarum's research assistant in 1925. After defending his doctoral dissertation '*Sur un problème d'économie pure*' in 1926 he again went to the United States, but returned when the university made its offer in 1931. During the 1930s Frisch participated actively in international economic activities and conferences. He was among the small group of initiators who, in 1931, established the Econometric Society. In 1933 he became the first editor of *Econometrica*, a position he held for more than 20 years. When the Prize in Economic Sciences in Memory of Alfred Nobel was created in 1969, Frisch together with Jan Tinbergen received it for their development and application of dynamic models for the analysis of economic problems.

In the beginning of the 1930s not only Norway's economy was at a low ebb, so too was the status of economics as a science. Frisch started his grand project of bringing economics as a science out of 'the fog'. He believed that economic theory should be based on mathematical models and quantitative analysis. The new economics should be shaped in a precise mathematical language. It was only with mathematical models that it would be possible to carry out complicated analysis and reasoning. He promoted this with enthusiasm, genius and force. All opposition was brushed aside. The study programme in economics was changed into a programme with a strong emphasis on mathematical analysis and economic research. His best students were attached to the institute as research assistants.

Frisch created a revolution, but change did not come without conflict. He was applauded, but also met with opposition from his colleagues. There was, however, no organized opposition against him. When Wedervang left in 1937 to become the first rector of the new business school, his and other positions were filled with Frisch's students. On the strength of the new study programme and his new staff, Frisch succeeded, in a short time, in creating his own school within economic research. This Oslo School, which to this day influences Norwegian economics, particularly at the University of Oslo, broke with tradition by introducing quantitative methods into economic research and teaching. The development of national accounts, national budgets and economic planning were given top priority. This work was strengthened by Leif Johansen (1930–82), who became Frisch's assistant in 1952 and took over Frisch's chair when he retired in 1965. Among Johansen's most important contribution was his doctoral dissertation, 'A Multi-Sectoral Study of Economic Growth', which became the basis for the long-term economic planning by the Ministry of Finance. Macroeconomic planning, research and policy became the alfa and omega in the Norwegian post-war economy.

Trygve Haavelmo (1911–99) joined Frisch as a research assistant in 1933. In 1938 he was visiting professor at University of Aarhus and in 1939 a research fellow at Harvard University. Caught in the United States by the war, he worked for Nortraship, an organization set up by the Norwegian government in exile to administer the war effort of the Norwegian merchant marine. After the war he stayed a year with Cowles Commission in Chicago, where, according to Schumpeter (1954, p. 1163), he 'exerted an influence that would credit to the lifetime work of a professor'. On returning to Norway he was appointed professor of economics in 1948, a position he held until his retirement in 1979. With his research contributions, teaching, generosity and gentle personality, he had a decisive influence on the development of economics. He won the Nobel Prize in 1989 for his fundamental contributions to econometrics.

During the economic depression of the inter-war period Frisch developed a deep mistrust of the market economy and the working of the price mechanism. National economic planning administered and managed by well-trained economists was, in his opinion, clearly superior to the shifting bustles of the market. As a consequence Frisch, as well as Johansen, who was a member of the Communist Party, were great admirers of the Soviet economic planning system, and claimed it was superior to the market economies of the Western world. They were therefore not easily attuned to other ideas.

Challenges to the Oslo School

Karl H. Borch (1919–86) was in 1959 recruited to the Norwegian School of Economics and Business Administration (NHH), first as a university fellow and from 1963 as a professor of insurance. This institution was at the time not so strongly focused on research. However, Borch stood out as an eminent researcher and a spiritual leader for the younger researchers. With his international network he strongly urged his students to pursue doctoral studies abroad and particularly in North America.

The new competence-building and international recognition achieved by Borch and his colleagues, together with the economic developments in the 1970s, slowly broke the monopoly and the influence of the Oslo School in Norwegian economics and politics. Economic planning in the Frisch–Johansen tradition was from the mid-1980s no longer central. More emphasis was placed on the functioning of competitive markets under uncertainty. Two of Borch's students should in particular be mentioned: Jan Mossin (1936–87) and Agnar Sandmo (b. 1938). Mossin was among a group of international researchers who independently contributed to the development of the modern theory for financial markets, the capital asset pricing model. Sandmo's research, to a large extent focused on the theory of taxation, is based on the assumption that we live in a world where we must deal with uncertainty and where there are limited opportunities for action. Markets and social institutions do not function in an ideal way. We must accept compromises and second-best solutions. Another prominent economist at

NHH, Victor D. Norman (b. 1946), who earned his Ph.D. from MIT in 1972, has made significant contributions to international trade theory.

This work had a marked influence on Norwegian monetary and fiscal policies and also laid the basis for increased independence of the central bank. This line of research was also pursued by the Norwegian, Finn E. Kydland (b. 1943). from NHH, who, in 2004, together with Edward C. Prescott (b. 1940) was awarded the Nobel Prize for their contribution to dynamic macroeconomics, notably the time consistency of economic policy and the driving forces behind business cycles.

Sweden

Institutional evolution

As an academic discipline, political economy in Sweden can be dated back to 1741, when the first chair was created at Uppsala University. Official policy in the mid-1700s aimed at promoting economic growth, and economic debate was flourishing. The creation of three more chairs in political economy before 1760 was an element in this effort. However, because of changed priorities and loss of territories, a decline soon set in, and during most of the 19th century political economy at Swedish universities was quite weak.

At the end of the 19th century and beginning of the 20th century, culture and science in Sweden were especially influenced by the German-language area. Study tours were mainly directed to Germany. Doctoral theses in economics were written in either Swedish or German; the first in English was published in 1929, while virtually all have been in English in the 2000s. Half of the books on economics acquired by university libraries in 1903–7 were published in Germany or Austria, and only a fourth in the UK or the USA. Fifty years later the proportions were almost reversed. The two world wars and Nazi oppression help explain the transition from German to Anglo-Saxon influence, as do faster growth of American population and academic research, less importance of geographical proximity and possibly a lingering dominance of the Historical School in Germany (Sandelin, 2001).

In the absence of a better measure, the growth and specialization of academic economics may be described by the number and scope of chairs. At the beginning of the 20th century, there were only two chairs of economics in Sweden, one in Uppsala (David Davidson) and one in Lund (occupied in 1901 by Knut Wicksell). Both were located in the faculty of law, and both also included fiscal law. In 1903 a chair in economics and sociology was created in Gothenburg (Gustaf Steffen), and in 1904 one in economics and public finance was created in Stockholm (Gustav Cassel). By comparison, in 1996, a few years before the principles for appointing professors were radically changed, there were 57 chairs in Sweden, of which 45 were directed towards a special field

within economics, and only one formally included more than economics (Sandelin, 1998, p. 2; 2000; Sandelin, Sarafoglou and Veiderpass, 2000, p. 46).

Early Swedish economists such as Wicksell, Cassel, Heckscher, and the Stockholm School economists had a common forum in the journal *Ekonomisk Tidskrift*, founded by David Davidson in 1899. Its name was changed in 1965 to *The Swedish Journal of Economics* — then in 1976 to *The Scandinavian Journal of Economics*, when the circle of contributors and editors was widened. Those changes left room for a Swedish-language journal directed to a broader audience and dedicated to practical economic problems; so *Ekonomisk Debatt* was born in 1973, published by *Nationalekonomiska föreningen*, the economists' association, founded in 1877.

Before the neoclassical breakthrough

International currents are visible in early Swedish thought. Some early authors have been labelled mercantilists, the most influential probably Anders Berch, who became the first professor of political economy appointed at a Swedish university (Uppsala, in 1741), and who published the first textbook, *Inledningen til almänna hushålningen* (1747), which then enjoyed a monopoly in academic teaching for more than 80 years. Opponents of mercantilist ideas arose, among them the clergyman Anders Chydenius, who published liberal booklets in the mid-1760s that have resulted in him being called a Swedish physiocrat.

Despite this beginning, political economy at Swedish universities was not strong at the beginning of the 19th century, though there were a few representatives of classical economic thought. Lars Rabenius – who appreciated Adam Smith's ideas, though with reservations – published a textbook in 1829 which finally replaced Berch's old book. Carl Adolph Agardh, who had attended and been influenced by Say's lectures in Paris in the 1820s, thought that the classical economists gave the state too modest a role, so his ideas were more akin to the Historical School.

Gustaf Steffen, who was professor of economics and sociology in Gothenburg during 1903–29, was the last Swedish professor who can be classified with the Historical School. The majority of university economists during this period were turning towards neoclassical ideas (Lönnroth, 1991, 1998; Magnusson, 1987).

The early modern generation

David Davidson, Knut Wicksell and Gustav Cassel introduced modern economics into Sweden around 1900; Eli Heckscher may also be included in this group, although his main works were published later.

David Davidson (1854–1942) was Wicksell's teacher and an important adviser on domestic monetary and fiscal policy, though he did not address an international audience. The editor of the *Ekonomisk Tidskrift* for 40 years and one of its main contributors, he also influenced policy directly as a member of several government committees on taxation (Uhr, 1991).

Knut Wicksell (1851–1926), an ardent participant in public discussions on social matters of all kinds, was the most important introducer of neoclassical economic thought into Sweden. His book *Value, Capital and Rent* (1893) was permeated with derivatives and marginalist concepts. The original German edition has the subtitle *nach den neueren nationalökonomischen Theorien* ('according to the new economic theories') and it is the theories of Walras, Jevons, Menger, and — especially concerning capital — Böhm-Bawerk that he tries to bring together.

Wicksell's analysis of just taxation from the perspective of the benefit principle in his next book, *Finanztheoretische Untersuchungen* (1896), has become an unavoidable point of reference. Likewise, his idea of a cumulative process of inflation, expounded in *Interest and Prices* (1898), is still referred to. Wicksell's *Lectures on Political Economy* (vol. 1, 1901; vol. 2, 1906) is not simply a textbook version of the ideas developed in his earlier books, but contains refined and modified approaches to questions raised earlier.

Gustav Cassel (1866–1945) began as a mathematician but turned later to economics. He studied with Wagner and Schmoller in Berlin, and around 1900 evinced doubt about the benefit of unlimited competition; later he became more sceptical of government intervention. His basic economic thought, expounded in his *Theoretische Sozialökonomie* (1918), was evidently much influenced by Walras, although he did not give him proper credit in that book. During the 1920s Cassel worked in various positions with international monetary problems, and during many decades he was, like Wicksell, Heckscher, Ohlin, and others, a persistent participant in public discussions, publishing several hundred newspaper articles (Magnusson, 1991; Carlson, 1994).

Especially because of his book *Mercantilism* (1931), the youngest in the group, Eli F. Heckscher (1879–1952), is internationally known mainly as an economic historian. As such, he pleaded for the integration of historical and neoclassical analysis. His lasting contribution to economic theory is an article in the *Ekonomisk Tidskrift* in 1919, which provided the basis of the so-called Heckscher–Ohlin theory in international trade (Henriksson, 1991).

As noted, the early modern generation was extensively involved in public debate; Wicksell considered it his 'foremost duty to educate the Swedish people' (Jonung, Hedlund-Nyström and Jonung, 2001, p. 19). This attitude was taken over by several of the Stockholm School economists.

The Stockholm School

Both Cassel and Heckscher were advocates of traditional economic liberalism, and sceptical of major government intervention. Around 1930, when nobody could overlook the problem of unemployment, this scepticism was challenged by a group of young economists, some of whom were disciples of Cassel and Heckscher. Dag Hammarskjöld (1905–61), Alf Johansson (1901–81), Karin Kock (1891–1976), Erik Lindahl (1891–1960), Erik Lundberg (1907–87), Gunnar Myrdal (1898–1987), Bertil Ohlin (1899–1979), and Ingvar Svennilsson (1908–72) were members of this group, called 'the Stockholm School' by Ohlin in an article in the *Economic Journal* in 1937. Ohlin believed that the Stockholm School had developed a theory of employment and had demonstrated how employment can be stimulated by economic policy, before Keynes did so.

Although several individual members are well-known, sometimes for other contributions, such as Myrdal's institutional analyses and Ohlin's contributions to the theory of international trade, the Stockholm School did not live on in the way Keynes's thinking did; it was hardly more than a national phenomenon, for several reasons. The few university positions in Sweden could not absorb all of them. They wrote mainly in Swedish, often in the form of government reports. They emphasized the dynamics of economic problems, which were difficult to present pedagogically. They also tended to analyse special rather than general cases, in the belief that useful general conclusions were difficult to draw. And their approach in some ways conflicted with techniques coming into vogue after the Second World War (Siven, 1985; Jonung, 1987).

The first post-war decades

Although it can be considered as a national phenomenon in the sense that it had little influence outside Sweden, the Stockholm School itself was not devoid of influences from abroad. It was mainly a theoretical school, and theory is more international than empirical knowledge.

Swedish economics was hardly more internationalized in the 1940s and 1950s than it was during the preceding decades. As before, most economic research was performed when students wrote their *magnum opus*, the doctoral dissertation. (A new system of graduate education, similar to the American, was introduced in 1969.) And most of the dissertations were more empirical than theoretical, focusing on the Swedish economy.

Outside the university world, the trade union economists Rudolf Meidner and Gösta Rehn developed ideas on the relationship between inflation and employment, and recommended a general deflationary policy, combined with selective measures directed towards those parts of the economy that would suffer from it. The latter part of the recommendation – selective means – was politically accepted and characterized the 'Swedish model' of actual economic policy for many years.

Disappearance of national traits

The closer we come to our own time, the less reason there is to classify economic thought geographically.

National traits have become less evident as communications have improved. In an evaluation of Swedish economic research, Dixit, Honkapohja and Solow (1992, p. 129) concluded that

> over the past three or four decades the literature of analytical economics has become almost completely homogenous worldwide. Mainstream economists in all countries now contribute to a single international literature as part of a single intellectual community. ... One can easily imagine a new idea or technique arising anywhere in the world of mainstream economics, and being pursued at first by its originator and his or her graduate students, but one cannot easily imagine a distinctively national school arising within the mainstream. Good ideas circulate much too rapidly.

Nevertheless, we may point to a couple of Swedish characteristics. Persson, Stern and Gunnarsson (1992, p. 118) found that non-mainstream economists like neo-Ricardians and Post Keynesians were much less cited by Swedish economists than by the world's economists on average, and this probably remains true. Similarly, as found by Dixit, Honkapohja and Solow (1992, p. 139), the application of advanced econometric techniques seems still to prevail over the creation of new ones.

Stockholm University's Institute for International Economic Studies has remained the most successful research unit in Sweden for several decades. As in other small European countries, many Swedish university economists have traditionally been involved in government committees and commissions. A change may have occurred in recent years, however, partly as a consequence of faster growth in the supply of qualified economists than in the demand for people willing to accept such side-commissions, and partly because young economists may be giving pure research higher priority than before.

NIELS KÆRGÅRD, BO SANDELIN AND ARILD SÆTHER

See also **Cassel, Gustav; Davidson, David; Frisch, Ragnar Anton Kittel; Haavelmo, Trygve; Heckscher, Eli Filip; Johansen, Leif; Lindahl, Erik Robert; Lundberg, Erik Filip; Myrdal, Gunnar; Ohlin, Bertil Gotthard; Stockholm School; Wicksell, Johan Gustav Knut; Zeuthen, Frederik Ludvig Bang.**

Bibliography

Aarum, T. 1924–8. *Læren om Samfundets Økonomi I and II.* Kristiania (Oslo): Olaf Norlis Forlag.

Andersen, P. 1983. On rent of fishing grounds: a translation of Jens Warming's 1911 article, with an introduction. *History of Political Economy* 15, 391–6.

Aschehoug, T. 1908. *Socialøkonomik.* Kristiania (Oslo): Aschehoug og Nygaard forlag.

Bergh, T. and Hanisch, T. 1984. *Vitenskap og politikk. Linjer i norsk sosialøkonomi gjennom 150 år.* Oslo: H. Aschehoug & Co. (W. Nygaard).

Bisgaard, H. 1902. *Den danske nationaløkonomien idet 18. århundre* [Danish Economic Thought in the 18th century]. Copenhagen: Hagerups forlag.

Boserup, M. 1969. A note on the prehistory of the Kahn multiplier. *Economic Journal* 79, 667–9.

Boserup, M. 1980. The international transmission of ideas: a small-country case study. *History of Political Economy* 12, 420–33.

Brems, H. 1976. From the years of high theory: Frederik Zeuthen (1888–1959). *History of Political Economy* 8, 400–11.

Brems, H. 1986. *Pioneering Economic Theory 1630–1980: A Mathematical Restatement.* Baltimore: Johns Hopkins University Press.

Carlson, B. 1994. *The State as a Monster: Gustav Cassel and Eli Heckscher on the Role and Growth of the State.* Lanham, NY and London: University Press of America.

Cassel, G. 1918. *The Theory of Social Economy.* Revised English translation, New York: Kelley, 1967.

Creedy, J. 1980. The early use of Lagrange multipliers in economics. *Economic Journal* 90, 371–6.

Danmark og Norges Oeconomiske Magazin (1757–64), 1–8. Copenhagen.

Davidsen, T. 1986. Westergaard, Edgeworth and the use of Lagrange multipliers in economics. *Economic Journal* 96, 808–11.

Dixit, A., Honkapohja, S. and Solow, R. 1992. Swedish economics in the 1980s. In *Economics in Sweden: An Evaluation of Swedish Research in Economics*, ed. L. Engwall. London and New York: Routledge.

Frisch, R. 1947. *Notater til økonomisk teori*, 2nd, 3rd and 4th edns. Oslo: Socialøkonomisk Institutt, University of Oslo.

Hansen, B. 1975. Introduction to Jørgen Gelting's 'Some alternations on the Financing of Public Activity'. *History of Political Economy* 7, 32–5.

Heckscher, E. 1919. Utrikeshandelns inverkan på inkomstfördelningen. *Ekonomisk Tidskrift* 21, del II, 1–32.

Heckscher, E. 1931. *Mercantilism.* English translation London: Allen and Unwin, 1935.

Henriksson, R. 1991. Eli F. Heckscher: the economic historian as economist. In *The History of Swedish Economic Thought*, ed. B. Sandelin. London and New York: Routledge.

Howey, R. 1972. The origins of marginalism. *History of Political Economy* 4, 281–302.

Jæger, O. 1922. *Teoretisk Socialøkonomik.* Kristiania (Oslo): Centrum Avskrivnings byraa.

Jonung, L. 1987. Stockholmsskolan – vart tog den vägen? *Ekonomisk Debatt* 15, 318–26.

Jonung, L., Hedlund-Nyström, T. and Jonung, C., eds. 2001. *Att uppfostra svenska folket. Knut Wicksells opublicerade manuskript.* Stockholm: SNS Förlag.

Kærgård, N. 1984. The earliest history of econometrics: some neglected Danish contributions. *History of Political Economy* 16, 437–44.

Kærgård, N. 1995. Cooperation not opposition: marginalism and socialism in Denmark 1871–1924. In *Socialism and Marginalism in Economics 1879– 1930*, ed. I. Steedman. London and New York: Routledge.

Kærgård, N. 1996. Denmark and the marginal revolution. In *Research in the History of Economic Thought and Methodology*, vol. 14, ed. W. Samuels and J. Biddle. Greenwich and London: JAI Press Inc.

Kærgård, N., Andersen, P. and Topp, N.-H. 1998. Jens Warming – an odd genius. In *European Economists of the Early 20th Century*, vol. 1, ed. W. Samuels. Cheltenham: Edward Elgar.

Kærgård, N. and Davidsen, Th. 1998. Harald Westergaard: from young pioneer to established authority. In *European Economists of the Early 20th Century*, vol. 1, ed. W. Samuels. Cheltenham: Edward Elgar.

Lönnroth, J. 1991. Before economics. In *The History of Swedish Economic Thought*, ed. B. Sandelin. London and New York: Routledge.

Lönnroth, J. 1998. Gustaf Steffen. In *European Economists of the Early 20th Century. Volume 1: Studies of Neglected Thinkers of Belgium, France, The Netherlands and Scandinavia*, general ed. W. Samuels. Cheltenham and Northampton: Edward Elgar.

Magnusson, L. 1987. Mercantilism and 'reform' mercantilism: the rise of economic discourse in Sweden during the eighteenth century. *History of Political Economy* 19, 415–33.

Magnussson, L. 1991. Gustav Cassel, popularizer and enigmatic Walrasian. In *The History of Swedish Economic Thought*, ed. B. Sandelin. London and New York: Routledge.

Munthe, P. 1992. *Norske økonomer. Sveip og Portretter*. Oslo: Oslo University Press.

Nielsen, A. 1948. *Det Statsvidenskabelige Studium i Danmark før 1848* [Studies in Political Economy in Denmark before 1848]. Copenhagen: Nordisk Forlag.

Ohlin, B. 1937. Some notes on the Stockholm theory of savings and investment. *Economic Journal* 47, 53–69, 221–40.

Ølgaard, A. 1966. *Growth, Productivity and Relative Prices*. Amsterdam: North-Holland.

Oxenbøll, E. 1977. *Dansk økonomisk tænkning 1700– 1770* [Danish Economic Thought 1700–1770]. Copenhagen: Copenhagen University Press.

Persson, O., Stern, P. and Gunnarsson, E. 1992. Swedish economics on the international scene. In *Economics in Sweden: An Evaluation of Swedish Research in Economics*, ed. L. Engwall. London and New York: Routledge.

Sandelin, B. 1998. Introduction: Swedish economics 1900–1960. In *Swedish Economics, Volume I: Introduction and Selected Essays*, ed. B. Sandelin. London and New York: Routledge.

Sandelin, B. 2000. Nationalekonomin i Sverige under 100 år. *Ekonomisk Debatt* 28, 59–69.

Sandelin, B. 2001. The de-Germanization of Swedish economics. *History of Political Economy* 33, 517–39.

Sandelin, B., Sarafoglou, N. and Veiderpass, A. 2000. The post-1945 development of economics and economists in Sweden. In *The Development of Economics in Western Europe since 1945*, ed. A. Coats. London and New York: Routledge.

Sæther, A. 1981. *Den økonomiske tenkning i Danmark-Norge på 1700 tallet* [Economic Thought in Denmark–Norway in the 18th Century]. Kristiansand: Agder State College.

Sæther, A. 1993. Otto Diederich Lütken – 40 years before Malthus? *Population Studies* 47, 511–17.

Schumpeter, J. 1954. *History of Economic Analysis*. New York: Oxford University Press.

Siven, C.-H. 1985. The end of the Stockholm School. *Scandinavian Journal of Economics* 87, 577–93.

Statsøkonomisk tidsskrift 1890– 1936. Kristiania (Oslo): Aschehoug og Nygaard.

Topp, N.-H. 1981. A nineteenth-century multiplier and its fate: Julius Wulff and the multiplier theory in Denmark, 1896–1932. *History of Political Economy* 13, 824–45.

Topp, N.-H. 1988. Fiscal Policy in Denmark 1930–45. *European Economic Review* 32, 512–18.

Uhr, C. 1991. David Davidson: the transition to neoclassical economics. In *The History of Swedish Economic Thought*, ed. B. Sandelin. London and New York: Routledge.

Whitaker, J. 1982. A neglected classic in the theory of distribution. *Journal of Political Economics* 90, 333–55.

Wicksell, K. 1893. *Value, Capital and Rent*. English translation, London: George Allen and Unwin, 1954.

Wicksell, K. 1896. *Finanztheoretische Untersuchungen nebst Darstellung und Kritik des Steuerwesens Schwedens*. Jena: Verlag von Gustav Fischer.

Wicksell, K. 1898. *Interest and Prices: A Study of the Causes Regulating the Value of Money*. English translation, London: Macmillan, 1936.

Wicksell, K. 1901. *Lectures on Political Economy. Vol. 1, General Theory*. English translation, London: Routledge, 1934.

Wicksell, K. 1906. *Lectures on Political Economy. Vol. 2, Money*. English translation, London: Routledge, 1935.

Schelling, Thomas C. (born 1921)

Thomas Schelling was born in 1921 in Oakland, California. He received a BA from the University of California at Berkeley in 1944 and a Ph.D. in economics from Harvard University in 1951. Between 1948 and 1953 he worked in the Paris Marshall Plan Headquarters, in the White House, and in the Executive Office of the president, on negotiations relating to foreign aid, the European Payments Union, and NATO. He taught at Yale University (1953–8), then at Harvard University (1958–90), and finally at the University of Maryland at College Park (1990–2003). He also had a long association with the RAND Corporation, with appointments as an adjunct fellow for most of his career (1956–2002) and as a full-time researcher in 1958. He was awarded the Nobel Prize in economics in 2005.

Schelling is known for works that use the tools of economics to illustrate major social phenomena while also making foundational theoretical advances. His publications illuminate patterns and paradoxes concerning military strategy and arms control, nuclear proliferation, conflict and bargaining, coordination and conventions, tipping points and critical mass, racial segregation and integration, addiction, health policy, and business ethics. In terms of style, he has avoided the formalization that now characterizes most economic research. Using mathematics sparsely, he has managed to convey intricate theoretical arguments mainly through eloquent prose and penetrating examples.

In the 1950s, when Schelling's academic career took off, policymakers around the globe were consumed by the rivalry between the two nuclear-armed superpowers, the United States and the Soviet Union. His first classic book, *The Strategy of Conflict* (1960), framed the challenge facing the two sides as coordinating on a commonly expected and mutually acceptable outcome. Avoiding nuclear confrontation required the negotiators to focus on a particular set of concessions. The underlying logic, Schelling proceeded to show, applies to a very broad set of problems in which communication is incomplete, if not impossible. People who have a mutual interest in coordinating their behaviours will look for a 'focal point' capable of generating a common expectation as to what is feasible.

One of his famous examples involves two strangers who are instructed to meet each other on a particular day in New York City. Unable to communicate, they look for an obvious place to meet, so obvious that each will know that it is obvious to both of them. At the time that Schelling was writing, the information desk at Grand Central Station provided just such a focal point. In this case, people achieved coordination not by speculating on what the other would do but by identifying a common course of action with the understanding that the other party was trying to do the same. Typically, the solution entailed a set of actions that stood out among its numerous alternatives. Hence, there was no uniquely 'correct' answer. What made an alternative 'correct' was simply that enough people thought so.

In a class of contexts, the required focal point lies in the structure of the game being played. For instance, in certain games with multiple equilibria it consists of a Pareto-dominant equilibrium – an outcome that no one can improve upon without harming at least one other player. Schelling's key insight, which the Grand Central Station experiment encapsulates, is that in a wide range of other contexts the focal point emerges from factors not captured by a formal representation of the game. It may depend on such factors as analogy, precedent, accidental arrangement, symmetry, aesthetic configuration, even on what the parties know about each other.

The Strategy of Conflict laid the foundations for a version of applied game theory that focuses not on zero-sum games in which players have diametrically opposed interests but on positive-sum games in which the players have both common and conflicting interests. The nuclear arms race offered the paradigmatic case: each superpower wanted to avoid touching off a mutually destructive nuclear showdown but also to dominate the other. In identifying lessons for policymakers, Schelling played a pioneering role in the development of various concepts included in the basic toolkit of modern game theory: commitment, credibility, threats, and brinkmanship. He also introduced, albeit informally, the concept now known as subgame perfection, which is a generalization of backward induction. His insight that mutual nuclear deterrence requires the threats of adversaries to be credible was an early illustration of an idea now central to thinking about strategic interactions in economics, political science, sociology, and beyond.

Another of Schelling's classic books, *Micromotives and Macrobehavior* (1978), deals with settings in which a group's aggregate behaviour is more than the sum of the behaviours of its members. What unites the members is that in acting and reacting to their environment they fail to perceive, and usually do not care, how their own choices combine with those of others to produce unintended and unanticipated consequences for the whole group. One of his influential applications of this insight concerns racial segregation. A popular explanation for racial segregation in American cities was deep-seated racism. Schelling showed that racial segregation could arise even in places where racism was not particularly acute. It could emerge, in fact, even in a community whose members all wanted to live in a racially mixed neighbourhood.

To see the underlying logic, consider a population whose members are either white or black. Regardless of colour, everyone prefers to live in a racially diverse neighbourhood. By the same token, each member of the population is also averse to being in the minority, though to varying degrees. Suppose now that a certain neighbourhood is 52 per cent white and 48 per cent black. Because whites are in the majority, the neighbourhood attracts disproportionately many whites. The accentuated imbalance is acceptable to the white majority. Its members do not mind if their share rises, say, to 65 per cent. But black residents who are most sensitive to being in the minority begin to move out, which reduces the proportion of blacks even further. That reduction then triggers further exits. The upshot is that the neighbourhood becomes fully white even though that was not the intention of anyone in either the majority or the minority.

In developing this analysis, Schelling familiarized the social sciences with concepts that have since gained broad applications. One of these is 'critical mass,' a shorthand for the minimum level of activity, often defined as a number or ratio, required to make that activity self-sustaining. If a neighbourhood will experience black flight once the black proportion falls to 40 per cent, that percentage marks the critical mass of blacks required to keep the neighbourhood integrated. A related Schelling concept is 'tipping', which entails a cumulative process dependent on differences in critical mass across individuals. In the presence of such differences, a behaviour can feed on itself. A few black departures can induce further departures by making the share of blacks dip below more individual thresholds, and the process can repeat itself until none remain.

Schelling's analysis points to the dangers of inferring individual characteristics from observations of collective outcomes, and of jumping to conclusions about aggregate behaviour from what one knows about individual preferences. That individually rational behaviours may generate persistent inefficiencies was already understood. The 'Prisoner's Dilemma' offers a case in point. Schelling's work on the tensions between 'micromotives' and 'macrobehaviour' helped to show that such tensions are much more pervasive than had been appreciated. In the case of racial sorting, people who would rather live in a more or less balanced neighbourhood than in a racially homogeneous one end up with the less preferred outcome. Moreover, once racial segregation has run its course, it is difficult to reverse, because few will voluntarily move into a neighbourhood in which they would form a tiny minority.

Identifying analogous tensions in a rich variety of other settings, Schelling demonstrated that market activity can produce lasting inefficiencies when an expectation-driven interactive process shapes the choices of participants. His examples included the custom of sending holiday cards. People feel obliged to send cards, he observed, to people from whom they expect to receive them, even when they sense that they will receive them only because the senders expect to receive cards themselves. Accordingly, two acquaintances may send each other cards for years on end, even though both find the custom burdensome and each is capable of ending the habit unilaterally. Individuals may keep sending cards to people they have not seen for decades for no other reason than the suspicion that cessation could signal something undesirable. Society as a whole would be better off, Schelling infers, with a 'bankruptcy proceeding' through which all holiday-card lists are obliterated to allow people to start over, motivated only by the holiday spirit, without accumulated obligations.

Schelling's insight into the possibilities of disharmony between 'micromotives' and 'macrobehaviour' has stimulated a wide range of other studies based on the observation that people fail to account for the externalities of their decisions. Refined or extended versions of his framework have been used to explain, among other phenomena, the emergence and disappearance of clothing fashions, the unpredictability of revolutions, the dynamics of jury decisions, the significance of momentum in political campaigns, the broadening of ethnic cleavages, norms against price competition, and economically dysfunctional behaviours inimical to development. The micro–macro interactions characteristic of Schelling's works appear also in various network models in which local interactions can trigger unintended and disproportionate global consequences.

Where *The Strategy of Conflict* and *Micromotives and Macrobehavior* address problems involving interactions among people, *Choice and Consequence* (1984), Schelling's third classic book, focuses on conflicts within individuals. Economics has traditionally treated the individual as a unified and internally consistent utility maximizer. Yet difficulties with reconciling conflicting impulses are central to the human experience. The problem of addiction, Schelling suggests, entails a failure to manage inner conflicts. It arises from inadequate self-control. The key insight of *Choice and Consequence* is that problems of self-control are commonly dampened, if not resolved fully, through tricks of the mind, social institutions, and public policies.

Many American taxpayers understate the number of their dependants to the Internal Revenue Service, the tax collection agency of the American government, to have an excessive portion of their income withheld, thus ensuring a hefty refund at the end of the year. Because excess withholdings yield no interest, this practice amounts to an expensive method of saving. If people find it convenient, the reason is that they see themselves as lacking the self-control to allow savings accumulate in a bank. By placing savings in the custody of the IRS, a person's responsible and forward-looking self gains control over the impulsive self that coexists with it.

Addictions stem from self-control difficulties, which anti-addiction treatments are designed to address. Obese people who cannot lose weight on their own check into 'fat farms' that regulate their diet. Likewise, alcoholics enrol in programmes that either directly limit their access to alcohol or boost their self-control through support networks. In either case, the addict effectively conspires with society to favour his responsible self over his impulsive self. In studying various self-control problems for which social remedies have been devised, Schelling finds that society does not always side with the responsible and forward-looking self. With regard to terminally ill people who are in so much pain that they want to die immediately, yet cannot bring themselves to take the final step, in most countries both the law and public morality side with the self that wants to keep living. Thus, people are prohibited from writing legally binding contracts to get assistance with dying.

The fundamental contribution of Schelling's writings on personal self-control has been in spreading awareness of a broad category of human problems that call for social intervention informed by a combination of economics and psychology. Education campaigns aimed at making people understand the consequences of their behaviours will not be effective, at least not by themselves. Most alcoholics know well that their addiction makes them unproductive, unhealthy and socially disconnected. Their troubles stem not from ignorance but from an inability to mediate conflicting demands within themselves. The period since Schelling published these insights has only raised their significance. With obesity now an acute social problem in developed and underdeveloped countries alike, and other addictions spreading as a result of rising prosperity, problems of self-control now lie at the centre of vast research programmes.

Over and beyond his penetrating and often pioneering insights into specific economic, social and political problems, Thomas Schelling has made an enduring contribution to economics and its sister disciplines by showing that a deep understanding of social systems requires attention to cultural context, social dynamics, strategic interactions, and human complexities. To make sense of why a person would deliberately place an alarm clock far away from his bed, one must realize that he may be tormented by warring impulses. Although his problem can be modelled as one of utility maximization, only by taking account of the competition within him can one understand why one possible solution is likely to work better than another. Only by considering the cultural context can one understand why strangers asked to meet each other are more likely to succeed in certain cities and times than in others. No approach limited to the formal properties of payoff matrices, or the strangers' formal decision problems, will suffice. And only by understanding the mechanics of strategic decision-making, and the factors that make threats credible, can one appreciate how the massive nuclear stockpiles of the United States and the Soviet Union allowed human civilization to survive the Cold War.

TIMUR KURAN

See also **behavioural economics and game theory; preference reversals; residential segregation; self-confirming equilibria; signalling and screening; social norms; threshold models.**

Selected works

1960. *The Strategy of Conflict*. Cambridge, MA: Harvard University Press.
1966. *Arms and Influence*. New Haven, CT: Yale University Press.
1971. Dynamic models of segregation. *Journal of Mathematical Sociology* 1, 143–86.
1978. *Micromotives and Macrobehavior*. New York: W.W. Norton.
1984. *Choice and Consequence: Perspectives of an Errant Economist*. Cambridge, MA: Harvard University Press.

Bibliography

Ayson, R. 2004. *Thomas Schelling and the Nuclear Age*. London: Frank Cass.
Swedberg, R. 1990. Thomas C. Schelling. In *Economics and Sociology*. Princeton: Princeton University Press.
Zeckhauser, R. 1989. Distinguished Fellow: reflections on Thomas Schelling. *Journal of Economic Perspectives* 3(2), 153–64.

Schlesinger, Karl (1889–1938)

Karl Schlesinger was born in Budapest; in 1919, after Béla Kun's communist revolution in Hungary, he moved to Vienna, where he committed suicide when Hitler occupied Austria in March 1938. As early as 1914 he had published his important work on monetary theory, *Theorie der Geld- und Kreditwirtschaft*, which went, however, more or less unnoticed at that time because it used mathematical tools and was written in German – a forbidding combination at a time when the only German-speaking economists interested in theory, the Austrians, were rather averse to mathematical economics. Schlesinger was also an exceptional figure in so far as he was not a university teacher but a banker and influential member of the financial community. Nevertheless, he became a respected member of the Vienna Economic Society and, in the 1930s, one of the most active participants in Karl Menger's mathematical colloquium.

As an economic theorist, Schlesinger was a Walrasian, in fact the only Walrasian (with the exception, perhaps, of Wicksell) who significantly advanced Walras's theory of the demand for money balances and of equilibrium in the money market. In his 1914 book, Schlesinger clearly distinguished between payments the magnitudes and future dates of which are fixed, and those whose time profile is subject to uncertainty. While the first type of payment streams offers no choice but generates a money 'demand' equal to the maximum cumulative payments deficit for a given period (though Schlesinger correctly points to the possibility of modifying the payment stream by investing and disinvesting temporary cash surpluses), the second type, which lies at the centre of Schlesinger's analysis, gives rise to a choice between higher and lower cash reserves held as an insurance against illiquidity losses. Schlesinger determines the individual demand for these precautionary balances from the equality between the marginal utilities respectively of interest income lost due to holding a cash reserve and of the insurance service provided by this reserve. He also demonstrated the economies of scale from an increase in the number (but not in the nominal magnitudes) of transactions. Finally, Schlesinger derives an aggregate money demand function, additively separable in transactions and precautionary demand, virtually identical to the one set up by Keynes much later, and determines the partial equilibrium money rate of interest as that which equalizes aggregate demand and stock supply of money. Schlesinger also addressed himself to problems of international monetary economics: in a publication of 1916 he advocated and gave a clear exposition of the purchasing power parity theory. In 1931, in the context of a book review, Schlesinger developed a rigorous and detailed mathematical analysis of money creation on the level of individual banks and for the financial system as a whole.

Apart from his writings on money, Schlesinger made another original and remarkable contribution to economic theory, viz. to the mathematical theory of

Walrasian general economic equilibrium described by n zero-profit conditions (equating commodity prices which are given functions of quantities produced with the respective sums of products of factor input coefficients and factor prices) and m factor market equilibrium conditions (equating factor supplies with the respective sums of products of factor input coefficients and quantities of goods produced). At the beginning of the 1930s it came to be recognized that such a system of equations need not have a solution, at least not an economically meaningful solution (in non-negative output and factor prices). In 1934 Schlesinger suggested, in Menger's colloquium, to introduce non-negative slack variables on the demand side of the factor market equations and to enlarge the system of $n+m$ equations by additional m equations setting the respective products of slack variables and factor prices equal to zero. Schlesinger had arrived at this ingenious idea independently of an identical proposal made by Zeuthen in 1932. Going definitely beyond Zeuthen, however, he also raised the conjecture that this procedure would solve the existence problem. Schlesinger's conjecture was proved to hold true by A. Wald, with whom Schlesinger had taken lessons in mathematics.

G. SCHWÖDIAUER

Selected works

1914. *Theorie der Geld- und Kreditwirtschaft*. Munich: Duncker & Humblot. Partial English translation as 'Basic principles of the money economy', *International Economic Papers* 9 (1959), 20–38.

1916. *Die Veränderungen des Geldwertes im Kriege*. Vienna: Manz.

1931. Bankpolitik, von F. Somary [book review]. *Archiv für Sozialwissenschaft und Sozialpolitik* 66, 1–35.

1934. Über die Produktionsgleichungen der ökonomischen Wertlehre. *Ergebnisse eines mathematischen Kolloquiums* 6, 10–11.

Bibliography

Menger, K. 1973. Austrian marginalism and mathematical economics. In *Carl Menger and the Austrian School of Economics*, ed. J.R. Hicks and W. Weber. Oxford: Oxford University Press.

Morgenstern, O. 1968. Schlesinger, Karl. In *International Encyclopedia of the Social Sciences*, vol. 14. New York: Macmillan and Free Press.

Nagatani, K. 1978. *Monetary Theory*. Amsterdam: North-Holland.

Patinkin, D. 1965. *Money, Interest and Prices*. Supplementary Note D, 573–80. 2nd edn. New York: Harper & Row.

Schmoller, Gustav von (1838–1917)

Schmoller was born in Heilbronn, the son of a Württemberg civil servant. He studied Staatswissenschaften (a combination of economics, history and administrative science) in Tübingen. After a short period in the financial department of the Württemberg civil service, which he had to quit because of his pro-Prussian views, be became Professor in Halle (1864–72), Strassburg (1872–82), and Berlin (1882–1913).

Schmoller was the leading economist of Imperial Germany. He was the leader of the 'Kathedersozialisten' (socialists of the chair), and founder and long-time chairman of the Verein für Socialpolitik. He was editor or co-editor of several publications such as *Staats- und sozialwissenschaftliche Forschung* and *Jahrbuch für Gesetzgebung, Verwaltung und Volkswirtschaft im Deutschen Reich* – later known simply as *Schmollers Jahrbuch*; he was named official historian of Brandenburg and Prussia, and supervised the publication of the *Acta Borussica* and the *Forschungen zur brandenburgischen und preussischen Geschichte*. Thus Schmoller was one of the major organizers of research in the social sciences. He is said to have controlled almost every important academic appointment in economics in the German Reich.

As the outspoken leader of the 'younger' Historical School, Schmoller was against the abstract axiomatic–deductive approach of the classicals and neoclassicals (1893; 1900, pp. 1–124). When Menger, the Austrian marginal utility theorist, attacked Schmoller's point of view and asserted the necessity of applying the exact methods of natural sciences and abstract logical reasoning to political economy, the Methodenstreit (struggle over methods) began, which was by and large a dispute between the inductive and the deductive method. It occupied two generations of German-speaking economists, produced a vast literature and was perceived essentially as 'a history of wasted energies' (Schumpeter) by theoretical economists of the next generation. However, it may also be viewed as the expression of the endeavour to preserve seminal insights into the historical and changing nature of economic and social phenomena against simplified and mechanistic views of the laws of 'rational' behaviour, and as such it had important consequences for the development of neighbouring disciplines, especially sociology.

Although Schmoller put the emphasis on the inductive method, he was not excluding deduction from economic reasoning. In his opinion, it was of the utmost importance for the application of deductive methods and for economic theory formation in general to be based on the knowledge of sufficient historical facts and material. He advocated a somewhat interdisciplinary approach that would also take into account the psychological, sociological and philosophical aspects of the problems. Through detailed and monographic historical research he intended to free political economy from 'false abstractions' (1904, p. vi), and to put it on a solid empirical foundation. His most important historical studies were his works on the history of the weavers guild of Strassburg (1879), on the guilds in 17th- and 18th-century Brandenburg and Prussia, on the Prussian silk industry in the 18th century, on the history of Prussian financial policy (1898a) and on the history of German towns in general and Strassburg in particular (1922). He was also interested in the history and formation of social classes and the historical development of class struggle (1900, Book II; 1904, pp. 496–577; 1918). He further made some important contributions to the study of mercantilism (1898a, ch. 1; 1904, pp. 580–605), which he regarded essentially as the process of the formation of national state and the national economy. The adoption of mercantilist policies was of special significance for Germany, whose backwardness in the 17th and 18th century Schmoller ascribed to the absence of a centralized national state and the consequent domination of particularist regional and local interests. Schmoller perceived the enlightened and despotic sovereigns – especially the Prussian kings – as the only power that would implement a policy aimed at the breaking-up of these particularist tendencies and at the establishment of large and unified economic territories. An important step in that direction was the abolition of town autonomy after 1713 under Friedrich Wilhelm I (1922, pp. 231–428).

This glorification of the Prussian state and its rulers was probably the most characteristic feature of Schmoller's work. He regarded the Prussian monarchy with its corps of loyal civil servants, which he perceived as standing above the social classes and their egoist interests, as the central achievement of German history. Only this type of government had been able to overcome the earlier feudal corporate state and the class rule of the Junker (1898a, p. 302), and was at present capable of implementing social reforms.

Social reform and social justice were central to Schmoller's thinking. We may regard him as a conservative in the specific German or, better, Prussian sense of the word. He rejected Marxism, Manchester Liberalism and also reactionary, anti-reformist views such as those of the historian Heinrich von Treitschke, with whom he had a famous polemic on the notion of social reform (1874–5; Small, 1924–5).

Schmoller advocated a paternalistic social policy to raise the material and cultural standard of the working classes as the only means to prevent revolution, integrate the workers into the monarchic state, and keep the traditions of Prussia alive. He even envisaged an alliance between the monarchy and the working classes (1918, p. 648).

It is nowadays generally agreed that Schmoller's influence on the development of the economic sciences in Germany was rather unfortunate: it contributed to the neglect of economic theory in Germany for a full half century. Neither Schmoller nor his pupils achieved their goal of building a new theory based on the historical material they collected, however valuable it was. Schmoller's main work, the *Grundrisse* (1900; 1904) remained rather traditional in its theoretical part – the treatment of value and prices was not too far away from mainstream neoclassical economics – and constituted all in all a rather incoherent analysis. Perhaps this was the main reason why Schmoller's work and with it the whole Historical School was to fall into oblivion in Germany soon after his death.

B. SCHEFOLD

See also **Historical School, German.**

Selected works

1870. *Zur Geschichte der deutschen Kleingewerbe im 19. Jahrhundert: Statistische und nationalökonomische Untersuchungen.* Halle: Waisenhaus.

1879. *Die Strassburger Tucher- und Weberzunft; Urkunden und Darstellungen nebst Regesten und Glossar. Ein Beitrag zur Geschichte der deutschen Weberei und des deutschen Gewerberechts vom XIII.–XVII. Jahrhundert.* Strassburg: Trübner.

1888. *Zur Litteraturgeschichte der Staats- und Sozialwissenschaften.* Leipzig: Duncker & Humblot.

1898a. *Umrisse und Untersuchungen zur Verfassungs-, Verwaltungs-und Wirtschaftsgeschichte besonders des preussischen Staates im 17. und 18. Jahrhundert.* Leipzig: Duncker & Humblot.

1898b. *Über einige Grundfragen der Sozialpolitik und der Volkswirtschaftslehre.* Leipzig: Duncker & Humblot. It contains the works *Über einige Grundfragen des Rechts und der Volkswirtschaft* (1874–5), *Die Volkswirtschaft, die Volkswirtschaftslehre und ihre Methode* (1893), and *Wechselnde Theorien und feststehende Wahrheiten im Gebiete der Staats- und Sozialwissenschaften und die heutige deutsche Volkswirtschaftslehre* (1897).

1900–1904. *Grundriss der allgemeinen Volkswirtschaftslehre.* Vol. 1, 1900; vol. 2, 1904. Munich and Leipzig: Duncker & Humblot.

1918. *Die soziale Frage:* In *Klassenbildung, Arbeiterfrage, Klassenkampf,* ed. L. Schmoller. Munich und Leipzig: Duncker & Humblot.

1922. *Deutsches Städtewesen in älterer Zeit.* Bonner Staatswissenschaftliche Untersuchungen, Bonn and Leipzig: Schröder.

Bibliography

Brinkmann, C. 1937. *Gustav Schmoller und die Volkswirtschaftslehre.* Stuttgart: Kohlhammer.

Small, A.W. 1924–5. The Schmoller–Treitschke controversy. *American Journal of Sociology* 30, 49–86.

Schmookler, Jacob (1918–1967)

Schmookler was born in Woodstown, New Jersey, in 1918 and died in Minneapolis, Minnesota, in 1967. In 1951 he received his Ph.D. in economics from the University of Pennsylvania, where he was a student of Simon Kuznets. He subsequently held teaching positions at Michigan State University and the University of Minnesota.

Schmookler's work helped to establish the importance of technological change as a contributor to economic growth. His article 'The Changing Efficiency of the American Economy, 1869 to 1938' appeared in 1952 (Schmookler, 1972, pp. 3–36), several years before the seminal papers of Abramovitz and Solow. However, in his later work he also analysed the specific economic mechanisms that determined the allocation of resources among different categories of invention. By an extremely careful and original use of patent data, Schmookler demonstrated the decisive role played by changes in demand in shaping the pattern of inventive activity. In his most important work, *Invention and Economic Growth* (1966), he showed how changes in demand have accounted for variations in inventive activity in a specific industry (such as railroad equipment, petroleum refining and building) over time, as well as different rates of inventive activity among different industries at a given moment of time.

Schmookler's writing showed that technological change need not be treated as an exogenous variable. On the contrary, he showed that the changing direction of inventive activity could be accounted for by readily identifiable economic variables, most especially changes in the pattern of demand that determine the size of the prospective market, and hence potential profitability, for particular classes of inventions. Thus, inventions are significant not only because they influence the growth rate of an economy but also because they constitute forms of economic activity in their own right.

N. ROSENBERG

Selected works

1966. *Invention and Economic Growth*. Cambridge, MA: Harvard University Press.

1972. *Patents, Invention and Economic Change*, ed. Z. Griliches and L. Hurwicz. Cambridge, MA: Harvard University Press. (This volume contains the patent data which formed the basis for Schmookler's research. It also contains a complete bibliography of his writings.)

scholastic economics

The tradition in economic thought known as scholastic economics was part of the system of scholastic moral theology and philosophy taught in the schools and universities of western Europe in the Middle Ages. Its purpose was the construction of viable norms of commercial behaviour in a Christian world. The literary authorities on which it was based include the Bible and the writings of the Church fathers, and papal and conciliar decrees from the time of the early Church to the Carolingian period. In the 12th century the theologian Peter Lombard in his *Sentences* and the canonist Gratian in his *Decretum* codified much of this material along with later decrees. In the 13th century Gratian's work was augmented

with the *Decretals* of Pope Gregory IX to form the *Corpus Juris Canonici*. The major names in scholastic economics were theologians writing with a leaning towards canon law. A majority of them were mendicant friars. The canonists observed a distinction between the norms that apply in the internal forum of conscience and the more liberal ones that found expression in the external forum of the ecclesiastical courts. The latter were partly adapted from Roman law expressed in the *Code* and *Digest* of the *Corpus Juris Civilis*. The translations of Aristotle's *Ethics* and *Politics* gave access to some other bits of ancient texts that invited economic reasoning. The literary vehicles of scholastic economics were commentaries on the Bible and the *Sentences* and on Aristotle's works, theological *summae*, records of academic disputations, sermons, handbooks for confessors, as well as a few treatises dealing specifically with some given economic subject.

The schoolmen were primarily concerned with external economics, that is, with buying and selling and other forms of exchange of goods and services. This is so because of the ethical problem encountered in the exchange situation. The key word is consent. Mutual consent is required for an economic contract to be valid, but traders may be tempted by avarice to obtain terms of exchange to which the other party does not truly consent. In the case of buying and selling, the schoolmen envisaged a positive and a negative approach to consent. Because no one suffers injustice voluntarily, the former approach consisted in estimating the JUST PRICE. The latter consisted in identifying factors incompatible with consent. A just price could be estimated on the basis of cost or market factors, the main thesis being that a competitive market price is a just price and that cost must adapt to the market. According to Aristotle, justice requires equivalence between goods given and received in exchange or between goods and money. It used to be argued that this is meaningless because there would be no motivation for anyone to exchange if what he got was of equal value to what he gave. This objection rests on a confusion of exchange value with utility, a confusion of which the schoolmen were not guilty. In their discussions of justice, equality referred to value in exchange, though not to an exact point; in the words of Thomas Aquinas (d. 1274), the just price was rather an estimate that was not completely precise. According to John Duns Scotus (d. 1308) it permitted of a certain latitude. Scotus also noted that the parties to a sale would often disagree about the just price. Both might then yield a little in order to reach an agreement, and, if they were then satisfied, an element of gift was involved in the contract. Consent, and hence justice, was still preserved.

Consent and justice are precluded in cases of fraud and coercion. In reply to a Roman law maxim stating that a thing is simply worth the amount at which it can be sold, a number of schoolmen from William of Auxerre (d. 1231) and Roland of Cremona (d. 1259) protested that this is not true if the thing can be sold at that amount only by means of fraud. Fraud is wilful misrepresentation of the object or conditions of exchange. The buyer must know what he buys. But knowledge also presumes a mental capacity to receive and digest information. Peter Olivi (d. 1298) states that terms of exchange are invalid if their acceptance by the buyer is due to his mental backwardness, ignorance or inexperience. A confusion regarding fraud arose because Roman law permitted the parties to a sale to outwit and 'deceive' one another in the course of the haggling and bargaining about price. The law accepted agreements thus obtained unless the result of the bargaining involved a deviation of more than one half of the just price. The theologians were inclined to reject this one-half idea and insisted on a just price within rather narrower limits. In principle, 'deception' as an element of the bargaining process must be distinguished from deliberate and blatant fraud, such as telling lies about the nature and quality of merchandise, hiding defects, using false weights and measures, soaking wool or diluting wine, and the like, in order to obtain an unjust price. No law condones this. Thomas Aquinas's catalogue of fraudulent practices in the *Summa Theologiae* is representative of the schoolmen's position.

Besides fraud and deception, consent is incompatible with practices considered by the schoolmen to be coercive. Monopoly is mentioned in Aristotle's *Politics* but merely in a context that did not invite disapproval. Monopoly and collusion among sellers to set minimum prices are prohibited in the *Code* of Justinian. The issue was brought into the commentary tradition on the *Decretals* by the canonist Hostiensis (d. 1271) and through him to a number of authors of works for the internal forum. Among the more important ones, mention may be made of John of Freiburg (d. 1314), Astesanus of Asti (d. 1330), Raniero of Pisa (d. c. 1348), Antonino of Florence (d. 1459), and Angelo Carletti (d. 1495). According to these thinkers, merchants agree among themselves that only one of them shall trade in a certain necessary commodity or that all shall sell at the same excessive price. A related tradition can be traced back to Alexander of Hales (d. 1240). Alexander, in his *Summa Theologica* (1951–7, III, 490, p. 724) berates those who 'take over the whole marketplace of commodities' in order to raise prices. Later authors, from Astesanus to Angelo Carletti, paraphrase Alexander and rebuke those who disturb the run of the market by such means. John Duns Scotus has a name for them: they are 'regraters'. Angelo vividly describes the merchants standing at the city gates buying up all the new grain, preventing its reaching the marketplace to be sold there. Battista Trovamala (d. c. 1495) explains how the price of victuals can be raised by impeding or detaining ships bringing home such goods or by giving earnest money for the purchase of all the spice in the city in order to sell it later at a profit.

From Carolingian price regulation, transmitted via canon law, the schoolmen received two ordinances known by their incipits (opening words) as *Placuit* and *Quicumque*. *Placuit* forbids price discrimination against travellers in local markets. Such people should pay the market price or (in an alternative version, which comes to much the same thing) pay what the residents pay. This injunction was cited by John of Freiburg and Astesanus and numerous followers in the Dominican and Franciscan penitential traditions. The great preacher Bernardino of Siena (d. 1444) in his vernacular sermons in the cities of Tuscany repeatedly impressed on his crowds of listeners the simple message that all should pay the same price. It was argued that residents ganging up against foreigners is not morally very different from merchants conspiring against local customers.

Quicumque lacks this conspiracy element. Originally a capitulary of Charlemagne included in the *Decretum*, it censures the malpractice of buying up hoards of victuals or wine cheaply at the time of harvest or vintage in order later to sell them at a much higher price (practices known as 'engrossing' or 'forestalling'). Large profits on such operations are possible because 'scarcity is created'– literally: 'dearth is induced' (*caristia inducatur*). This phrase originated in canon law literature; its earliest known appearance is in a gloss by Laurence of Spain (d. 1248). It proved to possess a remarkable literary appeal. Besides later canonists, it was copied by Albert the Great (d. 1280), Ulrich of Strasbourg (d. 1277), Thomas Aquinas and Antonino of Florence in theological works, by John of Freiburg, Bartolomeo of San Concordio (d. 1347), Durand of Champagne (d. c. 1350), Angelo Carletti and Battista Trovamala in their penitential summas, and by Henry of Langenstein (d. 1397), Matthew of Cracow (d. 1410), and John Gerson (d. 1429) in their treatises on economic contracts, as well as by numerous other schoolmen. Many of these authors state that buyers faced with 'created scarcity' are forced to pay excessive prices. A person paying more than usual because of scarcity due to crop failure or other natural misadventures can of course also be said to be forced to do so. Unlike much of modern economics, however, scholastic economics maintained a distinction between personal and impersonal compelling agents, because the former could be held morally responsible for their acts.

The reason why the schoolmen came down so hard on creating scarcity is that scarcity is one of the basic phenomena of all economic life. Economic activity is, in a word, a common struggle against scarcity. Using a terminology adopted from Peter Olivi by Bernardino f Siena, value rests on *raritas* (scarcity), *virtuositas* (usefulness, in an objective sense), and *complacibilitas* (desirability, pleasingness). These factors vary from person to person, from time to time, and from place to place, and determine the outcome of all voluntary transactions. In a beautiful piece of analysis, Richard of Middleton (d. 1307) shows that, just as a profit can be made in long-distance trade because just prices differ with plenty and scarcity in different locations, thus, also, both parties can profit justly in local exchange because they value what they receive more highly than what they give – an argument that readily can be restated in Olivi's terms. The idea that both parties to an exchange are better off because they would not otherwise have exchanged is not a major insight recently discovered. Moreover, it is a truism. The schoolmen's point is that both parties profit by *just* exchange voluntarily agreed upon in the absence of fraud and duress.

The schoolmen held no comprehensive theory of wages. To some extent, arguments about the just price could be applied by analogy to the just wage. Possession of qualified labour skills and professional competence, when these are in demand, also represents economic power. The recognition of this fact by the *Ethics* commentator Gerald Odonis (d. 1349) and some of those who copied his cases should not be permitted to obscure the fact that power in medieval labour relations was heavily concentrated on the part of the employer rather than on the part of the employee. Thomas Aquinas emphasized that manual labourers should be paid a decent wage and that it should be paid promptly because they lived from hand to mouth. This admonition was repeated by numerous authors of penitential manuals, including some well-known names like Henry of Langenstein, John Gerson and Antonino of Florence. Because the wage was in large part fixed by custom or by institutions like the guilds, their emphasis was on paying promptly rather than on paying justly. Antonino's main contribution in this area was his attack on abuse of the truck system, whereby workers were paid in goods rather than in money, goods that they had to sell at a loss. Battista Trovamala notes that a double fraud was sometimes involved in truck practices. Merchants who barter goods rather than buy and sell them may do so justly but ostensibly at more than the just price. If a merchant has workers in his employ, he might claim that the goods paid them in lieu of money were bought, rather than bartered, at these inflated prices.

Scholastic monetary theory was largely Aristotelian. Scholars do not agree on whether Aristotle held a metallist theory of money. The schoolmen were definitely metallists. Money is a medium of exchange, a measure of value, and a store of wealth. In order to serve well in these capacities, certain properties are required. A number of commentaries on the *Ethics* and the *Politics* contain lists of such properties. That by Henry of Friemar (d. 1340) has five items: small size, to prevent subtraction (clipping and so forth); a stamp of the sovereign to prevent falsification; a determined weight so that prices can be fixed exactly; durability to ensure uncorrupted future validity; a content of precious material, like gold and silver, for easy and prompt valuation. John Buridan (d. c. 1360) adds a sixth requirement: coins of small denomination for the petty purchases of the poor. In exchange between merchants of different regions, money-changing (*campsoria*) was necessary. Henry of Ghent

(d. 1293) points out that the changer may grant himself a fee for his labour spent counting and guarding money. It is true that money is non-vendible, says Giles of Lessines (d. 1304), but money-changing is not properly a sale but a form of barter. Guido Terreni (d. 1342) makes a different distinction. Sometimes two kinds of money are exchanged in a place where both are current, one being domestic currency and the other foreign currency permitted at a certain rate of exchange. If this rate underestimates the foreign currency, it may be bought and sold by weight. If exchange is made in a place where neither or only one kind of money is current, non-current money becomes a commodity. It can be bought or sold by weight at a just price and taken to a location where it is current and worth more.

Whereas tampering with coins for a quick profit might be tempting for the individual citizen, debasement of the currency in order to replenish the treasury proved irresistible to medieval monarchs. In early 14th-century France debasement had reached ruinous proportions. Some schoolmen spoke out against it. Drawing on Guido Terreni and others, John Buridan argues as follows. The primary measure of goods in exchange is human need. It is by that yardstick that money is a measure as well. Granted that we do not often need gold and silver, wealthy people desire them for display, and so we find that they are worth the same, or nearly the same, in specie as in bullion. Only if money is thus measured in proportion to need can goods be measured in proportion to money. For whatever proportion they have to need, the same proportion they will have to money proportioned to need. If a certain coin is in circulation and the king issues another coin, it is true that he can stipulate its value in relation to the former, for instance declaring that one new penny is to be given in exchange for three of the old. If it is not worth that much, however, or very near it, according to the relation of its material to human need, the king sins and profits unjustly at the cost of the common people. Much of the credit for the scholastic theory of money has been granted to Nicole Oresme (d. 1382), whose merit is a detailed and knowledgeable description of the various ways in which the currency can be debased and of the social consequences of debasement.

The gravest specific economic sin was usury, unjust profit on a loan, usually a loan of money. Usury could also occur in credit sales or other sales involving time. Such arrangements can be interpreted as combinations of a cash sale and a loan. The schoolmen considered usury to be condemned in Luke 6:35: 'Lend, hoping for nothing again.' In addition, several arguments from natural reason were put forward. According to Peter the Chanter (d. 1197), the usurer profits merely because time passes, but time belongs to God. Thomas Aquinas argued that money is consumed in use by being spent, just as natural fungibles like grain and wine are consumed when we eat and drink them; money, therefore, has no use separate from its substance, and to charge a profit for lending it is to make the borrower pay twice for the same thing or to pay for something that does not exist. Gerardo of Siena (d. 1336) tied in with Aquinas by declaring that money is sterile, not because it is 'barren metal' (an idea wrongly attributed to Aristotle) but because it is a fungible. Contrary to fruit-bearing things, all fungibles are barren because they are always of the same value in terms of themselves. This intriguing argument has proved difficult to refute. An argument from coercion was originally proposed by William of Auxerre. The borrower pays usury because he is forced by need to do so. Adopting a broad definition of need, Thomas accepted this argument and proceeded to demonstrate the analogy between exploiting a needy borrower and exploiting a needy buyer, thus combining the two main areas of scholastic economics by a common formula. Whereas usury was universally condemned, the schoolmen recognized the lender's right to compensation for all damage incurred owing to default in repayment. Some extrinsic titles to interest within the period of the loan were also hesitantly recognized. After all, in the words of Gerald Odonis, two persons cannot use the same money at the same time. The full development of extrinsic titles was mainly the work of later canonists.

The decline of scholastic economics is a complex process impossible to date. In one sense part of it is still present in the European moral consciousness. In another sense, of a continuously developing system, it did not last much longer than to the latest authors cited in this article, that is, to the end of the 15th century. The so-called 'second scholasticism' associated with the Counter-Reformation School of Salamanca broke with medieval scholasticism to an extent that makes treatment under the same heading unadvisable. The change is well illustrated by the shift of focus from the attack on usury to the defence of titles to interest. In a larger perspective this and other changes are expressions of the fading-out of the natural law philosophy of Aquinas in favour of the natural rights philosophy on which most of modern economic thought is based.

ODD LANGHOLM

See also **Albert the Great, Saint Albertus Magnus; Aquinas, St Thomas; just price.**

Bibliography

Alexander of Hales. 1951–7. *Summa Theologica*, 4 vols., Quaracchi: Collegii S. Bonaventurae.

Aquinas, St T. 1961–65. *Summa Theologiae*. 5 vols., Madrid: Biblioteca de autores Cristianos.

Corpus Iuris Canonici. 1879. 2 vols. Leipzig: Taucnitz.

Corpus Iuris Civilis Iustiniani cum commenttariis. 1627. 8 vols. Osnabrück: Zeller, 1965–6.

de Roover, R. 1967. *San Bernardino of Siena and Sant'Antonino of Florence: The Two Great Economic Thinkers of the Middle Ages.* Boston: Baker Library, Harvard Graduate School of Business Administration.

Friedberger, W. 1967. *Der Reichtumserwerb im Urteil des Hl. Thomas von Aquin und der Theologen im Zeitalter des Frühkapitalismus.* Passau: Passavia.

Gilchrist, J. 1969. *The Church and Economic Activity in the Middle Ages.* London: Macmillan.

Gordon, B. 1975. *Economic Analysis before Adam Smith.* London: Macmillan.

Langholm, O. 1983. *Wealth and Money in the Aristotelian Tradition.* Oslo: Universitetsforlaget.

Langholm, O. 1992. *Economics in the Medieval Schools.* Leiden: Brill.

Langholm, O. 1998. *The Legacy of Scholasticism in Economic Thought.* Cambridge: Cambridge University Press.

Langholm, O. 2003. *The Merchant in the Confessional.* Leiden: Brill.

Little, L.K. 1978. *Religious Poverty and the Profit Economy in Medieval Europe.* Ithaca: Cornell University Press.

Noonan, Jr. J.T., 1957. *The Scholastic Analysis of Usury.* Cambridge, MA: Harvard University Press.

Peter Lombard. 1971, 1981. *Sententiae*, 2 vols., Rome: Collegii S. Bonaventurae.

Todeschini, G. 1994. *Il Prezzo della Salvezza: Lessici medievali del pensiero economico.* Rome: Studi superiori NIS.

Viner, J. 1983. Religious thought and economic society. *History of Political Economy* 10, 9–189.

Wood, D. 2002. *Medieval Economic Thought.* Cambridge: Cambridge University Press.

Scholes, Myron (born 1941)

Best known for his Nobel prize-winning work on derivatives pricing, Myron Scholes made significant contributions to a wide range of topics in financial economics, from asset pricing to dividend policy and tax incentives. Born 1 July 1941 in Timmins, Ontario, Canada, he earned a Bachelor's degree in Economics from McMaster University in Hamilton, Ontario in 1962. He then entered the MBA course at the University of Chicago, transferring to the Ph.D. course after his second year. He earned his MBA in 1964 and his Ph.D. in 1968, writing his dissertation on the effects of information and signalling on the shape of the demand curves for traded securities. Upon finishing graduate studies he took a position as Assistant Professor of Finance at the Sloan School of Management at the Massachusetts Institute of Technology. After five years he moved to the Graduate School of Business at the University of Chicago. He was first Visiting Professor, and then in 1974 he accepted a permanent position. He stayed there until 1981, when he became Visiting Professor at Stanford University for two years, becoming a permanent faculty member of the university's Business School and Law Schools in 1983, remaining there until his retirement in 1996. He is currently the Frank E. Buck Emeritus Professor of Finance at Stanford, as well as a Managing Partner of Oak Hill Capital Management and the Chairman of Oak Hill

Platinum Partners. He serves on the board of directors of several corporations.

It was during his time at MIT that Scholes worked with Fischer Black and Robert Merton to develop the Black–Scholes formula for option pricing. An option is the right but not the obligation to buy or sell a security on or before a certain date at a certain price. Options are members of a family of securities known as derivatives, because their value is derived from the value of another security. Scholes and Merton earned a Nobel Prize in economics in 1996 for 'a new method to determine the value of derivatives'. Almost every MBA and many undergraduates across the world learn the famous Black–Scholes formula, which is widely used by derivatives traders; indeed, traders often quote option 'prices' not in dollars and cents but in terms of one of the inputs into the Black–Scholes formula – the volatility of the return on the stock on which the option is written, or 'vol'. The formula has had ramifications far beyond the area of derivatives, offering new ways of thinking about related areas, such as corporate finance, capital budgeting, and financial markets and institutions.

Black (1989) and Bernstein (1992) provide detailed histories of the development of the formula, which can be summarized as follows. Between 1968 and 1969 Fischer Black derived a partial differential equation that described the value of an option as a function of the maturity of the option and the underlying stock price. He did so by applying the capital asset pricing model (CAPM) to options, rather than stocks, at each instant in time. Unable to solve the equation, Black teamed up with Scholes in 1969.

Scholes's contribution to the formula was the logic used to solve the equation – logic that is still taught to students studying finance. Black and Scholes noted that the differential equation did not involve the expected return on the stock and that, therefore, any expected stock return is consistent with the option price, including the risk-free rate of interest. They then turned to a much earlier work, Sprenkle (1961), which contains a formula for the expected future value of an option as a function of the stock return. Using the risk-free rate as the stock return in this formula gave them the expected future value of the option. Using the risk-free rate to discount this future value back to its present value, Black and Scholes came up with what is now known as the Black–Scholes formula, which did indeed solve Black's differential equation. The published version of this work appeared in the *Journal of Political Economy* in 1973. It did not, however, contain Black's derivation of the formula, but one suggested by Robert Merton, who pointed out that it is possible to replicate option payoffs with those from a portfolio of stocks and bonds. Hedging the option with this portfolio produces a risk-free position, and equating the return on this risk-free position with the risk-free rate of interest produces Black's differential equation.

Scholes's contributions to financial economics are not limited to his research on option pricing. For example, Black, Jensen and Scholes (1972) performs one of the earliest and most prominent tests of the CAPM, finding that the systematic risk of a stock explains less of the stock's return than the amount predicted by the CAPM. In related work, Scholes and Williams (1977) devise a method for testing asset pricing models with high frequency data, even when stocks are traded asynchronously. This research was an outgrowth of Scholes's work to develop a database on daily stock prices for the Center for Research in Security Prices at the University of Chicago. Another notable paper is Bulow and Scholes (1983), which explores seeming anomalous features of pension plans such as early retirement benefits or ad hoc increases in benefits. These features allow employees to have claims, in excess of vested benefits, on the assets of defined-benefit pension plans. They explain this anomaly as a product of employee human capital, which ends up being compensated via the pension plan in the optimal contract with stockholders.

Although important, these papers were diversions from Scholes's main line of research: the effects of taxation on financial decisions and financial markets. His most prominent work in this area examines the effects of dividend taxation on security prices. Black and Scholes (1974) was the lead article in the first issue of the *Journal of Financial Economics*. The paper tests the idea in Modigliani and Miller (1961) that changes in dividend policy can affect firm value if capital gains are taxed at a lower rate than dividends, because the firm can easily increase its value by cutting dividends and instead distribute cash to shareholders via stock repurchases. Black and Scholes found no evidence that dividend policy affects returns, despite the existence of differential taxation. Their explanation of the result appealed to the idea of tax clienteles. Firms adjust the supply of shares at any particular dividend yield to meet the demands of different groups of investors with different tax preferences. Market equilibrium then implies that no firm can increase its value by changing its dividend policy. Perhaps motivated by these results, Miller and Scholes (1978) present sufficient conditions for dividend policy to be irrelevant, even in the presence of taxes. Miller and Scholes (1982a) again test the relationship between returns and dividends, taking care to purge any reaction of returns to dividend of signalling or information effects. In contrast to Black and Scholes (1974), they find significant tax effects.

Scholes's interest in taxes extended beyond dividend policy. He also worked on tax issues relating to the cost of capital, capital structure, property values and the optimal liquidation of assets. His research with Mark Wolfson examined corporate tax planning under uncertainty and information asymmetry. Much of this work is summarized in their 1992 book, *Taxes and Business Strategy: A Planning Approach*. Scholes's interest in taxation also intersected with his interest in the use of stock to compensate employees. For example, Miller and Scholes (1982b) argue that deferred compensation programmes structured to optimize employee incentives are observationally equivalent to those structured to optimize taxes. In his 1990 presidential address to the American Finance Association (Scholes, 1991), he outlined the interplay between taxes and incentives in deferring employee compensation. When an employee accepts deferred compensation, he essentially agrees to allow the firm to invest in its own stock on his account. The tax advantage to deferred compensation therefore depends on the after-tax rates of return that can be earned by the corporation compared to the return an employee could earn investing on his own. The employee stock ownership implicit in deferred compensation, however, aligns incentives of employees with those of shareholders. Optimal compensation schemes trade off these benefits with any possible tax costs.

In the latter part of his career Scholes once again became interested in derivatives, this time from a more practical and institutional angle. He became the managing director and co-head of the fixed-income derivatives group at Salomon Brothers. In 1994 he and Robert Merton joined several colleagues from Salomon to start a firm called Long-Term Capital Management (LTCM). Their basic strategy was to take a long position in one instrument and to offset the long position by a short position in a similar instrument or its derivative. Most of LTCM's bets were variations on the same theme, convergence between short positions in liquid Treasuries and long positions in theoretically underpriced instruments that commanded a credit or liquidity premium. While profitable for several years, this strategy unravelled in 1998 when Russia defaulted on its rouble and domestic dollar debt. Money fled into high quality instruments such as Treasuries, and LTCM's short positions in these instruments plummeted in value. Although LTCM failed, it presaged the enormous growth in hedge funds that we see today, in 2007.

TONI M. WHITED

See also **Black, Fischer; Merton, Robert C.; Miller, Merton; options.**

Bibliography

Bernstein, P.L. 1992. *Capital Ideas: The Improbable Origins of Modern Wall Street*. New York: Free Press.

Black, F. 1989. How we came up with the option formula. *Journal of Portfolio Management* 15, 4–8.

Black, F., Jensen, M.C. and Scholes, M.S. 1972. The capital asset pricing model: some empirical tests. In *Studies in the Theory of Capital Markets*, ed. M.C. Jensen. New York: Praeger.

Black, F. and Scholes, M.S. 1973. The pricing of options and corporate liabilities. *Journal of Political Economy* 81, 637–54.

Black, F. and Scholes, M.S. 1974. The effects of dividend yield and dividend policy on common stock prices and returns. *Journal of Financial Economics* 1, 1–22.

Bulow, J. and Scholes, M. 1983. Who owns the assets in a defined-benefit pension plan? In *Financial Aspects of the U.S. Pension System*, ed. Z. Bodie and J. Shoven. Chicago: University of Chicago Press.

Miller, M. and Scholes, M.S. 1978. Dividends and taxes. *Journal of Financial Economics* 6, 333–64.

Miller, M. and Scholes, M.S. 1982a. Dividends and taxes: some empirical results. *Journal of Political Economy* 90, 1118–41.

Miller, M. and Scholes, M.S. 1982b. Executive compensation taxes and incentives. In *Financial Economics: Essays in Honor of Paul Cootner*, ed. K. Cootner and W. Sharpe. Englewood Cliffs, NJ: Prentice-Hall.

Modigliani, F. and Miller, M.H. 1961. Dividend policy, growth and the valuation of shares. *Journal of Business* 34, 411–33.

Scholes, M.S. 1991. Stock and compensation. *Journal of Finance* 46, 803–23.

Scholes, M.S. and Williams, J. 1977. Estimating betas from nonsynchronous data. *Journal of Financial Economics* 5, 309–27.

Scholes, M.S. and Wolfson, M.A. 1992. *Taxes and Business Strategy: A Planning Approach*. Englewood Cliffs, NJ: Prentice-Hall.

Sprenkle, C.M. 1961. Warrant prices as indications of expectations. *Yale Economic Essays* 1, 179–232.

school choice and competition

The economic idea of school choice and competition was introduced by Milton Friedman in two related essays, both entitled 'The Role of Government in Education' (1955; 1962). His proposals remain the core of the school choice idea, but myriad economists and non-economists have refined it since. The essence of the idea is that students should be free to choose among schools that compete for them on a level playing field. Students should carry the resources for their education with them in the form of a tax-financed fee. Schools that attract students should be able to expand, possibly by creating branded spin-offs. Schools that fail to attract students should close. Families should play the lynchpin role by making choices that govern students' experiences and schools' sustainability. The government, in contrast, should create structures that make schools compete on the basis of the value they add to students, where parents are the ultimate judges of value. For instance, the government should raise taxes that support fees, ensure that schools provide accurate information to parents, prevent financial malfeasance, enforce contracts and health and safety standards, and provide mechanisms that clear students' preference lists for schools and clear the market in school buildings and equipment.

The logic of school choice

Friedman's first insight was that externalities associated with primary and secondary education provide substantial motivation for the government to finance schools, but much less motivation for the government to *run* them. A more modern view would suggest that the government does not merely play a financing role but also plays the market referee role that it adopts in a variety of other industries with public goods dimensions. The government might also engage in limited regulation. For instance, if society decides that all students should learn civics in order to be responsible citizens, then a plausible regulation might impose a test of civics knowledge. So long as the government does not attempt to manage schools, but confines itself to setting parameters and auditing outcomes, the essence of the school choice idea is retained.

Friedman's second insight was that the typical state-run school has a virtual monopoly over certain students and is therefore likely to be x-inefficient, taking rents in forms such as overemployment and a quiet life for managers (for instance, overpaying for school supplies or failing to fire problem teachers). The effect of school choice on x-efficiency is paramount: most supporters envision it to be the main benefit of school choice. Some hypothesize that school choice might also deliver benefits by allowing students to match themselves to pedagogical methods or peers especially conducive to their learning. However, neither of these matching benefits is essential to the logic of school choice, and it is fairly easy to describe situations in which allowing students to choose their peers helps some to learn and hinders others (Epple and Romano, 1998).

School choice can be and has been implemented in numerous forms. The schools involved may be private ('school vouchers'), state schools that are independently managed ('charter schools'), or state schools run by fully autonomous local governments. In practice, it is the supply side that distinguishes true school choice. Beware of programmes that adopt the choice nomenclature but exhibit supply side inflexibility (schools do not readily open, expand, shrink or close) or that structure fees in such a way that schools do not compete on a level playing field. The essence of the school choice idea is that incentives flow from the consequences of competition among schools on the basis of their value-added (as judged by parents). If competition is inconsequential or biased in favour of certain schools, then the school choice idea is not being implemented. Most magnet school programmes, controlled choice programmes, and open enrolment programmes fail to be school choice – largely because competition is inconsequential. On the other hand, some phenomena not explicitly labelled as school choice fulfil the idea to some extent. Examples include tuition tax credits (see Lips and Feinberg, 2006) and competition among autonomous local school districts, a phenomenon called Tiebout choice (Tiebout, 1956).

Evaluations of school choice programmes

Modern work on school choice takes three main forms: empirical evaluations of school choice programmes, design work (theoretical analysis, simulation and experiments) on demand-side problems, and design work on supply side problems.

The primary task for an evaluation is identifying the general equilibrium effects of school choice on schools' x-efficiency and on students' peer groups. (It is far more important to identify the effects on x-efficiency than the effects on sorting. This is because it appears – as explained below – that sorting is susceptible to management.) To identify the general equilibrium effects, a researcher needs to find fairly random circumstances that cause some geographic areas to have more pressure from school choice than other areas do. By comparing areas randomly subjected to more and less school choice, general equilibrium effects are revealed. So far, the most convincingly random circumstances have come from the prevalence of natural boundaries (more natural boundaries translate into more local jurisdictions and thus more Tiebout choice: see Hoxby, 2000) and laws that translate small differences in state schools' performance into large differences in the school choice they face. For instance, Chakrabarti (2006c), using regression discontinuity combined with the implementation of a school choice policy, compares Florida schools ranked just above and below the state's threshold for failure. Those ranked just below were exposed to school choice pressure starting in a certain year; those just above were exposed to none. Chakrabarti finds statistically significant improvements in x-efficiency in the traditional state schools exposed to pressure. Unfortunately for the sake of research, the amount of pressure on these schools was small and discontinued after a short time because the programme in question ended.

Some researchers have attempted to rely solely on the implementation of school choice programmes to identify the general equilibrium effects of school choice. However, simple before–after comparisons rarely generate convincing estimates because societies do not implement school choice randomly. They usually do so when concerned about current conditions and after policy debates that often result in the simultaneous introduction of other educational policies. A counterfactual constructed from pre-school choice data is therefore rarely convincing. Even worse, because the policy they are studying is often uniform across the state or country in question, authors of before–after studies are often tempted to use *endogenous* geographic variation in the take-up of the choice programme. This is highly problematic because students who take up the programme are far less likely to be satisfied with their traditional state school than are students who have the same opportunity to take it up but refuse to do so. Hsieh and Urquiola (2006) is one example of such problematic studies. Of course, more credence can be given to studies that combine before–after variation with plausibly exogenous geographic variation in the *availability* (not take-up) of the choice programme

(Akabayashi, 2006; Ahlin, 2003; Chakrabarti, 2006a; Hoxby, 2004).

If researchers are to have a serious chance of identifying the general equilibrium effects of school choice, they must study an instance in which the supply of choice schools is fairly ample and can expand with demand. After all, x-efficiency effects cannot be expected if students cannot leave traditional state schools because there are no places available at choice schools. Also, the student sorting consequences of a tiny programme (such as one that affects only a few per cent of students) are unlikely to be distinguishable. Ironically, the very conditions that impede general equilibrium analysis facilitate analysis of the partial effect of particular choice schools on the students who attend them. If choice schools are inadequately supplied so that they enrol only a tiny share of students and are routinely oversubscribed, then researchers are in a fairly good position to identify the change in achievement caused by attending a particular choice school – not including the effects that general equilibrium might have on the school.

Why is this? Because students select to participate in choice programmes, selection biases plague studies that attempt to estimate the partial effect of voucher-taking schools, charter schools, and other choice schools on the students who attend them. Researchers have attacked the selection problem with a variety of methods, but the only one that enjoys widespread approbation is randomizing applicants so that some are offered a place and others are not. Randomization often occurs naturally when a choice school is oversubscribed. It allows researchers to compare lotteried-in and lotteried-out students, all of whom have the motivation and other qualities that made them self-select into applying. Formally, one estimates a local average treatment-on-the-treated effect by instrumenting for attendance at a choice school with an indicator for being lotteried into a choice school (see Angrist, Imbens, and Rubin, 1996 for a description of the method). Most randomization-based studies have found positive and significant effects of attending a choice school on outcomes such as test scores, educational attainment, parental satisfaction, altruism, and some other social outcomes (Angrist et al., 2001; Angrist, Bettinger and Kremer, 2006; Bettinger and Slonim, 2006; Howell and Peterson, 2002; Howell et al., 2002; Hoxby and Rockoff, 2005; Hoxby and Murarka, 2007a; 2007b; Peterson and Howell, 2004; Peterson et al., 1999; Rouse, 1998; West, Peterson and Campbell, 2001; Wolf, Howell and Peterson, 2000; Wolf, Peterson and West, 2001; Wolf and Hoople, 2006). Randomization-based studies have non-trivial limitations, however. Most importantly, it is not obvious that the partial effect matters much. If a choice school created incentives or sorting such that students' achievement rose equally in all schools (including the school itself), we would surely consider the school to be a success. Yet the school's partial effect would be nil. In addition, the results of randomized methods cannot be extrapolated to students who differ substantially from the applicants, and one cannot evaluate choice schools that are undersubscribed.

In addition to randomization, researchers have tried a variety of methods – instrumental variables, regression discontinuity, matching – to estimate an unbiased partial effect of choice schools. Although such methods could in theory be useful, they have in practice been substantially less convincing than randomization. Researchers have also often tried panel methods, but these are really never convincing owing to problems that are familiar to anyone who has studied the extensive programme evaluation literature on self-selection into training schemes. If a student attends a traditional state school for some years and then applies to, say, a charter school, his application is not a random event but deliberate selection into treatment. (Charter schools are state schools that families are fully free to choose and that have autonomous management and finances. The fee each child brings with him is tax-financed, applicants are admitted via a lottery, and schools can lose their charter if they do not produce positive results.) The student's pre-application achievement trajectory, no matter how well estimated, is an unreliable predictor of how he would perform post-application in the counterfactual world in which he was forced to stay in his traditional state school. His decision to apply to a choice school is an indication of concern about his current trajectory and of an intention to change it in some way. Therefore, in the counterfactual world, his trajectory might reasonably be expected to fall (if the choice school was the only treatment likely to help him) or rise (if he were to undertake other remedies in the absence of a choice school's being available): see Hoxby and Murarka (2007b) for further discussion of this point.

The final major problem that affects empirical evaluations is the substantial divergence between most actual school choice programmes and the school choice idea. For instance, many programmes have fees that are only a fraction of the per-pupil amounts enjoyed by traditional state schools, thereby creating a playing field that is obviously not level. Most school choice programmes have strict limits on the number of students and schools who can participate so that the programmes generate no pressure for x-efficiency once the limits are reached, as they often quickly are (see Lips and Feinberg, 2006). If such divergences between the idea and implementation of school choice were the inevitable result of applying an idea to a world of imperfect information, then researchers could interpret their estimates as the effects of *practical* school choice. The divergences are not inevitable, however; they result from legislators bowing to opponents of school choice. They thus leave researchers with an awkward choice between (i) abstaining from interpreting their results as a test of school choice and (ii) extrapolating their results beyond the data in order to form a meaningful test. Good studies do some of both, combining careful extrapolation with warnings not to interpret their results as definitive tests of the school choice idea (for example, Chakrabarti, 2006b). Unfortunately, numerous studies do neither.

Design work on demand-side problems in school choice

The second major form of modern work on school choice is design work (simulation, theoretical analysis and actual experiments) on demand-side problems. Prominent demand-side problems are how to ensure that families use accurate information about schools (especially schools' value-added when making decisions; how to ensure that students get their most preferred school (given the competing demands of other students) in a world where students submit a preference ordering; and how to incorporate social priorities such as desires to have schools that are socio-economically integrated or neighbourhood-focused. Progress on demand-side problems is occurring in part because such problems are susceptible to remedy, though perhaps not solution, with established tools. Also, districts that run school choice programmes have demonstrated a willingness to provide data for simulations and even to run experiments. For instance, although no information on schools' value-added is perfect or complete, econometrics and data systems have been focused on providing such information more often and more accurately. Districts like Charlotte-Mecklenberg, North Carolina, have engaged in experiments, giving more complete and digestible information to randomly selected families. Such experiments suggest that even modest improvements in the information received by families cause them to revise their choices (Hastings and Van Weelden, 2007).

Just as important, economists have proposed mechanisms for processing students' preference lists in order to guarantee that students get their most preferred choice, given other students' preferences. A good mechanism is one that is strategy-proof and that produces a stable match. A strategy-proof mechanism is one in which a student can do no better than submit his true preferences. A stable match is one in which there is no student and school pair where the student prefers the school over his assignment and the school gives the student a higher priority than a student who has been assigned to the school. For example, Random Serial Dictatorship is strategy-proof and stable. It takes the preference lists submitted by students, assigns each student a random number, assigns the first student his top choice, the second student his top choice among available schools, and so on. Not only theoretical proofs but also evidence based on actual experiments – that is, districts switching their mechanisms – have shown that such mechanisms successfully clear the market (Abdulkadiroğlu, Pathak and Roth, 2006; Abdulkadiroğlu et al., 2006; Pathak, 2006). For instance, after New York City switched to a stable, strategy-proof mechanism, families and schools engaged

in substantially less strategic behaviour and families were more satisfied with the schools their children ended up attending.

Stable, strategy-proof mechanisms can incorporate some social priorities. In practice, priorities are most often given to siblings, economically disadvantaged children, children who attend schools classified as failing, disabled children, and children with limited English. (Race- and ethnicity-based priorities are being phased out by courts.) With social priorities, not only students but also schools submit ordered lists. Schools' lists, however, are based on the priorities, not on the personal preferences of school staff. The deferred-acceptance mechanism deals with the lists as follows.

> (Step 1) Each student 'proposes' to his first choice. Each school tentatively assigns its seats to its proposers one at a time following only their priority order. Any proposers who remain when all seats are assigned are rejected.
> (Each subsequent step) Each student who was rejected in the previous step 'proposes' to his or her next choice. Each school considers the students it has been holding together with its new proposers and tentatively assigns its seats to these students one at a time following their priority order. Any proposers who remain when all seats as assigned are rejected.
> (Termination) The mechanism stops when no student proposal is rejected. Each student is assigned to his or her final tentative assignment.

The above description is borrowed from Pathak (2006). Another mechanism that can be used in this situation is top trading cycles (Abdulkadiroğlu and Sönmez, 2003).

The availability of mechanisms that incorporate social priorities implies that the general equilibrium effects of school choice on student sorting can be *managed*. By *managed*, one means that unwelcome or extreme outcomes such as segregation can be avoided. One does not mean that a robust school choice programme would have no effects on student sorting: it is hard to imagine a programme in which students' choices are meaningful and yet so micro-managed that sorting looked the same before and after programme implementation.

Design work on supply side problems in school choice

The third major form of modern work on school choice is design work on supply side problems. This work attempts to address the questions about which programme designs are most likely to ensure a level playing field for schools in competition with one another. Most importantly, what should the structure of fees be? What should vouchers or charter school fees be? How much should they be topped-up for students who are disabled, limited-English proficient, economically disadvantaged or otherwise expensive to educate? There is also a series of questions about growing and shrinking schools. For instance, should a school that is very oversubscribed be given funds to expand its school or move to a larger one? Should shrinking schools be given incentives to cede their space to growing schools? Unlike the demand-side work, supply side work is limited and may remain so for some time. This is partly because actual experiments are rare: governments choose the supply side policies of their school choice programme after considerable political wrangling and then stick with them. Moreover, supply side work that uses theory or simulation has so far not produced general truths. Instead, it generates results that are sensitive to assumptions about peer effects and the effects of school spending on achievement (see Epple and Romano, 1998; Nechyba, 2003). Yet there is no way, at present, to ground such assumptions in agreed facts. Suppose one asks how much additional money is needed by a school if it is to achieve the same value-added with a child whose parents are poor secondary school drop-outs as it would with a child whose parents are affluent college graduates. Estimates vary enormously, are very imprecise, and are sometimes even 'wrong-signed' (Hanushek, 2002). Yet, an estimate is what one needs to model the effect of raising the amount by which the basic fee is topped up for disadvantaged students.

At present, as of 2007, evidence-based design work on the supply side is fairly indirect. Authors occasionally compare a programme before and after its design has been changed (Chakrabarti, 2006a; Hoxby, 2004). Comparing programmes across states has also been tried (Chakrabarti, 2006b; Hoxby, 2006), but this is inherently difficult because states differ along many lines, not just a single dimension of school choice design. Because, for a test of the school choice idea, it is crucial that schools compete on an even playing field, design work on supply side problems is important. It is most likely to be advanced by deliberate experimentation within school choice programmes – for instance, varying the fee structure in interesting ways across different metropolitan areas within a state.

CAROLINE HOXBY

See also **education production functions; educational finance; local public finance; matching and market design; mechanism design; Tiebout hypothesis.**

Bibliography

Abdulkadiroğlu, A., Pathak, P. and Roth, A. 2006. Strategy-proofness versus efficiency: redesigning the New York City High School match. Working paper, New York University and Columbia University.

Abdulkadiroğlu, A., Pathak, P., Roth, A. and Sönmez, T. 2006. Changing the Boston school choice mechanism. Working Paper No. 11965. Cambridge, MA: NBER.

Abdulkadiroğlu, A. and Sönmez, T. 2003. School choice: a mechanism design approach. *American Economic Review* 93, 729–47.

Ahlin, Å. 2003. Does school competition matter? Effects of a large-scale school choice reform on student performance. Working paper, Nationalekonomiska institutionen, Uppsala University.

Akabayashi, H. 2006. Average effects of school choice on educational attainment: evidence from Japanese high school attendance zones. Working paper, Keio University.

Angrist, J., Bettinger, E. and Kremer, M. 2006. Long-term consequences of secondary school vouchers: evidence from administrative records in Colombia. *American Economic Review* 96, 847–62.

Angrist, J., Bettinger, E., Bloom, E., King, E. and Kremer, M. 2001. Vouchers for private schooling in Colombia: evidence from a randomized natural experiment. Working Paper No. 8343. Cambridge, MA: NBER.

Angrist, J., Imbens, G. and Rubin, D. 1996. Identification of causal effects esing instrumental variables. *Journal of the American Statistical Association* 91, 444–55.

Bettinger, E. and Slonim, R. 2006. Using experimental economics to measure the effects of a natural educational experiment on altruism. *Journal of Public Economics* 90, 1625–48.

Chakrabarti, R. 2006a. Can increasing private school participation and monetary loss in a voucher program affect public school performance? Evidence from Milwaukee. Working paper, Federal Reserve Bank of New York.

Chakrabarti, R. 2006b. Impact of voucher design on public school performance: evidence from Florida and Milwaukee voucher programs. Working paper, Federal Reserve Bank of New York.

Chakrabarti, R. 2006c. Vouchers, public school response and the role of incentives: evidence from Florida. Working paper, Federal Reserve Bank of New York.

Epple, D. and Romano, R. 1998. Competition between private and public schools, vouchers, and peer-group effects. *American Economic Review* 88, 33–62.

Friedman, M. 1955. The role of government in education. In *Economics and the Public Interest*, ed. R. Solo. New Brunswick, NJ: Rutgers University Press.

Friedman, M. 1962. The role of government in education. In *Capitalism and Freedom*. Chicago: University of Chicago Press.

Hanushek, E. 2002. Publicly provided education. In *Handbook of Public Economics*, ed. A. Auerbach and M. Feldstein. Amsterdam: North-Holland.

Hastings, J. and Van Weelden, R. 2007. Preferences, information, and parental choice behavior in public school choice. Working Paper No. 12995. Cambridge, MA: NBER.

Howell, W. and Peterson, P. 2002. *The Education Gap: Vouchers and Urban Schools*. Washington, DC: Brookings Institution.

Howell, W., Wolf, P., Campbell, D. and Peterson, P. 2002. School vouchers and academic performance: results from three randomized field trials. *Journal of Policy Analysis and Management* 21, 191–218.

Hoxby, C. 2000. Does competition among public schools benefit students and taxpayers? *American Economic Review* 90, 1209–38.

Hoxby, C. 2004. School choice and school competition: evidence from the United States. *Swedish Economic Policy Review* 10, 9–65.

Hoxby, C. 2006. The supply of charter schools. In *Charter Schools against the Odds*, ed. P. Hill. Stanford: Hoover Institution Press.

Hoxby, C. and Murarka, S. 2007a. How New York City's charter schools affect student achievement: results of the first year of evaluation. Working Paper No. 13255. Cambridge, MA: NBER.

Hoxby, C. and Murarka, S. 2007b. Methods of assessing the achievement of students in charter schools. In *Charter Schools: Their Growth and Outcomes*, ed. M. Berends. Mahwah, NJ: Lawrence Erlbaum.

Hoxby, C. and Rockoff, J. 2005. The impact of charter schools on student achievement. Working paper, Harvard University.

Hsieh, C. and Urquiola, M. 2006. The effects of generalized school choice on achievement and stratification: evidence from Chile's voucher program. *Journal of Public Economics* 90, 1477–503.

Lips, D. and Feinberg, E. 2006. School choice: 2006 progress report. Washington, DC: Heritage Foundation. Online. Available at http://www.heritage.org/Research/Education/bg1970.cfm, accessed 26 June 2007.

Nechyba, T. 2003. What can be (and what has been) learned from general equilibrium simulation models of school finance? *National Tax Journal* 56, 387–414.

Pathak, P. 2006. Lotteries in student assignment. Working paper, Harvard University.

Peterson, P. and Howell, W. 2004. Voucher research controversy: new looks at the New York City evaluation. *Education Next* 4, 73–8.

Peterson, P., Myers, D., Howell, W. and Mayer, D. 1999. School choice in New York City. In *Earning and Learning: How Schools Matter*, ed. P. Peterson and S. Mayer. Washington, DC: Brookings Institution.

Rouse, C. 1998. Private school vouchers and student achievement: an evaluation of the Milwaukee parental choice program. *Quarterly Journal of Economics* 113, 553–602.

Tiebout, C. 1956. A pure theory of local expenditures. *Journal of Political Economy* 64, 416–24.

West, M., Peterson, P. and Campbell, D. 2001. School choice in Dayton, Ohio after two years: an evaluation of the parents advancing choice in education scholarship program. Working Paper No. RWP02-021. Cambridge, MA: Kennedy School of Government.

Wolf, P. and Hoople, D. 2006. Looking inside the black box: what school factors explain voucher gains in Washington, DC? *Peabody Journal of Education* 81, 7–26.

Wolf, P., Howell, W. and Peterson, P. 2000. School choice in Washington, D.C.: an evaluation after one year. Working Paper No. PEPG/00–08. Program on Education Policy and Governance, Harvard University.

Wolf, P., Peterson, P. and West, M. 2001. Results of a school voucher experiment: the Case of Washington, D.C. after two years. Working Paper No. PEPG/01–05. Program on Education Policy and Governance, Harvard University.

Schultz, Henry (1893–1938)

Schultz was one of a small group of pioneering econometricians who, in the 1920s and 1930s, laid the groundwork for the phenomenal development of mathematical economics and econometrics that occurred after the Second World War. His graduate courses in mathematical economics and statistics inspired students to address themselves to economic problems in quantitative terms, to reformulate economic theory in empirically testable form, and to test theories by means of diligent search for relevant statistical information and careful application of appropriate statistical analysis. His own research, culminating in his magnum opus, *The Theory and Measurement of Demand* (1938), could well serve as a model for a proper approach to economic analysis today. Schultz devoted all his professional life to the integration of pure economic theory with empirical analysis. Unlike considerable econometric work today, which is often empirical without much grounding in economic theory, his statistical analysis is solidly based on mathematical economic theory, as well as on the statistical theory of correlation and curve-fitting. Elegant summaries of both fields, based in large part upon his lecture notes and his research work, appear in the book, along with the empirical studies of demand for a large number of agricultural commodities for which the theories served as the basic foundation. The student wishing to get a good introduction to mathematical economics as formulated by Cournot, Walras and Pareto, and to the fundamentals of Gaussian curve-fitting analysis, will find clear and lucid presentations of these subjects in Schultz's book. At the same time he will not fail to be impressed by the extraordinary concern for statistical accuracy and precision demonstrated in the empirical analysis throughout the book.

Schultz, who was born in Russian Poland on 4 September 1893, was brought by his parents to the United States in 1907, as part of the large wave of migration of Russian Jews after the Russo–Japanese War of 1905. Despite the family's poverty, Schultz's drive and determination enabled him to enter college and even to pursue a graduate education – no small feat in those days for the oldest child of an immigrant family. After receiving his AB from City College in 1916, Schultz entered Columbia University, where he came under the lasting influence of one of the world's leading econometricians, Professor Henry L. Moore. Schultz felt so indebted to him that in 1938, when he was himself internationally recognized as an outstanding authority in econometrics, he dedicated his major work to 'Professor Henry Ludwell Moore, trail blazer in the statistical study of demand'.

Schultz's studies were interrupted by the First World War, during which he was wounded in the Meuse–Argonne offensive. In the spring and summer of 1919, an Army scholarship enabled him to study at the London School of Economics and at the Galton Laboratory of University College, London, under two leading statisticians, A.L. Bowley and Karl Pearson. After returning to the United States in 1920, he served with several agencies of the United States government and became Director of Statistical Research of the Children's Bureau of the Department of Labor. At the same time he continued his academic work, receiving the Ph.D. degree from Columbia University in 1925. His dissertation on 'The Statistical Law of Demand as Illustrated by the Demand for Sugar' was published the same year in the *Journal of Political Economy*.

The following year Schultz received his appointment at the University of Chicago. His courses in mathematical economics and statistics were highly organized and presented his students with a clear and systematic exposition both of the classic texts and of the most important up-to-date journal articles from the English, French, Italian and Russian literature. He was a voracious reader of economic and statistical literature, and enriched his courses by examination of related material in the fields of biology, physics and psychology.

It was the research laboratory, however, that absorbed most of his energies. Almost from the time he came to the University of Chicago, he embarked on an ambitious research project intended to harness theory and empirical analysis for the purpose of filling some of the 'empty boxes' in the theory of exchange. At first his interest focused on determining the coefficients of elasticity of demand as well as the magnitudes of the average shifts in demand over time. At a later stage be became interested in testing the consistency of demand coefficients for a set of commodities whose demands were interrelated. The work was so thoroughly organized, documented and proof-checked, and the calculations were so systematically set out, with automatic sum-checks at every appropriate point, that research assistants picking up a worksheet prepared by others even a decade earlier had no difficulty tracing the sources and checking the accuracy of every figure on the sheet.

Shortly after his book was published, Schultz was killed, together with his wife and two daughters, on 26 November 1938, in a car accident in California, where he had gone to teach on sabbatical leave from the University of Chicago. Paul Douglas wrote of Schultz:

> All in all, he was about the finest man it has ever been my privilege to know in academic life. The world of scholarship and of science (for I may so use the term in

connection with him) was the richer for his fruitful life. We are much the poorer for his death at the full height of his powers.

JACOB L. MOSAK

Selected works

1925. The statistical law of demand as illustrated by the demand for sugar. *Journal of Political Economy* 33, Part I, 481–504; Part II, 577–637.
1938. *The Theory and Measurement of Demand*. Chicago: University of Chicago Press.

Schultz, T.W. (1902–1998)

Theodore Schultz shared the Nobel Prize in 1979 with Sir Arthur Lewis 'for pioneering research into economic development research with particular consideration of the problems of developing countries'.

Early in his career, Schultz studied agricultural organization and production, which evolved into studies of economic growth and finally into studies of economic development. His training and early contributions were in agricultural economics, but it is too limiting to label him an agricultural economist (as did the press release announcing his Nobel Prize award). He made seminal contributions in agriculture but also was a 20th-century pioneer advocating the importance of human capital and particularly education for understanding economic growth in developed and developing countries. In this brief sketch I review his intellectual contributions to economics.

Biographical details

T.W. Schultz was born on 30 April 1902 in South Dakota and died on 26 February 1998. He received his undergraduate degree in 1927 from South Dakota State University. He was a student of John R. Commons and received his Ph.D. in Agricultural Economics from the University Wisconsin in 1930. Besides the Nobel Prize he received the Francis A. Walker Medal from the American Economic Association (1972), and the Leonard Elmhurst Medal from the International Agricultural Economic Association (1976). He was President of the American Economic Association in 1960, a member of the National Academy of Science (1974), and a fellow of the American Farm Economic Association (1957), the American Academy of Arts and Sciences (1958), the American Philosophical Society (1962), a guest of the Soviet Academy of Science (1960) and a Founding member of the National Academy of Education (1965). Also, he received about half a dozen honorary degrees from universities in the United States and abroad. Nerlove (1999) offers a longer and more detailed appraisal of Schultz's contributions and provides a complete listing of Schultz's prolific writings; see Johnson (1998) and Nerlove (1999) for a commentary on Schultz's personality. They give strong testimony to Schultz's role as mentor and colleague.

Style

Three features of Schultz's style characterize his writings. First, his wit and humility are much on display. Schultz travelled the world and served as an advisor to many governments and international organizations but he retained his humble Midwestern perspective and values. His work is serious and deeply considered, but presented with an irreverence to self and others that makes for easy and enjoyable reading. (A hint of this irreverence is evident in his 1979 Nobel Banquet Speech; Schultz, 1980.)

Second, his analysis is thoroughly modern; he thought in terms of supply and demand, and gained much analytical depth through careful assessment of opportunities, incentives and information. His analyses considered a broad sweep of potential factors such as political economy and institutional constraints, besides the more obvious economic factors all economists are trained to recognize. Indeed, one characterization of Schultz's approach is that he merged the analytical insight of Irving Fisher with the breath and style of argumentation of (his mentor) John R. Commons. Schultz applied these tools with equal force to problems in agriculture, labour, public economics, macroeconomics and development. He was not bound by traditional field–subfield definitions, but, rather, his writing is problem-focused; for example, on understanding the economics of being poor. In his Presidential Address to the American Economics Association (AEA), for instance, Schultz (1961) applied ideas of human capital to age–earnings profiles, population flows from rural to urban areas and deciphering the mysteries of economic growth for developed and developing countries. And, as another example, in *Transforming Traditional Agriculture* (1964), analyses of the US agricultural market are seamlessly interwoven with analyses of agricultural markets in developing countries. This uniformity of approach is all the more astonishing when compared to his contemporaries who viewed farmers as 'different' from non-farmers and especially saw differences between farmers in the United States and those in developing countries. Schultz had enormous faith in people's common-sense response to incentives they face. He believed people were knowledgeable about economic opportunities and responded to those opportunities. He tacitly assumed preferences were constant and sought explanation and prediction in responses to differences in opportunities or abilities. This perspective also made him critical of many government-sponsored agricultural and anti-poverty programmes. Schultz recognized that programmes could distort individual incentives in unanticipated ways or would fail because they did not remove barriers permitting individuals to act on existing economic incentives.

Schultz wrote frequently about 'disequilibrium' and its importance. However, I think the appropriate modern term is 'dynamics', not disequilibrium. Academic economists perceive a clinical Walrasian auctioneer at work to obtain equilibrium. Schultz saw a myriad of 'equilibrium distortions' that create profit opportunities, which individuals perceive and arbitrage. Schultz was interested in how farmers, students and entrepreneurs reallocate resources in response to new information on economic costs and returns. In 'The Value of the Ability to Deal with Disequilibrium' (1975) Schultz argues that the more educated are more able and thus quicker to process information on the economic environment. Their early arbitrage activity provides, Schultz argued, another return to education.

Third, even more impressive than Schultz's broad conceptual approach to economics was his tireless work in linking economic theory with economic measurement. In his empirical studies, understanding the economic phenomenon takes centre stage, not statistical technique or theoretical elegance. Empirical measures are fully described. He considered possible biases and other deficiencies in each. Finally, Schultz assesses the likely quantitative magnitude of each bias and the consequences for empirical analysis. Every effort is made to get the closest connection between available (generally aggregate or tabular) measures and the theoretical concepts. The honesty and the fullness of the presentation makes the analysis compelling.

Agricultural economics

In his 1979 Nobel Prize lecture Schultz summarized the motivation for his research thus: 'Most of the people in the world are poor, so if we knew the economics of being poor, we would know much of the economics that really matters. Most of the world's poor people earn their living from agriculture, so if we knew the economics of agriculture, we would know much of the economics of being poor' (Schultz, 1992). His seminal works on agriculture include *Agriculture in an Unstable Economy* (1945) and *Production and Welfare of Agriculture* (1949) and *Transforming Traditional Agriculture* (1964).

According to Schultz (1993, p. 2) understanding the economics of being poor starts with the hypothesis that there are relatively few significant economic inefficiencies in established communities where most people are poor. In these traditional agricultural settings, farmers have used the same technology and the same factor inputs for generations. Consequently, they have acquired significant experience in their abilities and the means of production available to them. They live and operate within a stationary environment in which there has been no significant change in technology for generations. Thus, contrary to the claims of Schultz's contemporaries, it is not that these households saved and invested too little and did not respond to normal economic incentives. Rather, it was that they did not have profitable opportunities. Schultz pushed the frontier on technological change in agriculture as the key factor for transforming traditional agriculture by creating profit opportunities for investment. Schultz recognized that the Green Revolution created new high-yielding varieties, which were more responsive to fertilizers and other modern inputs that helped offset the diminishing marginal utility of land and created profit opportunities that led to investment and economic growth. Indeed, it is the availability of new technology and high return investment opportunities that characterizes modern agriculture. Schultz encouraged his student Zvi Griliches to investigate the diffusion of the new hybrids (Griliches, 1957). *Transforming Traditional Agriculture* (1964) is Schultz's forceful summary of the transition. Thomas Balogh's (1964) vitriolic review of *Transforming Traditional Agriculture* is the best evidence of its revolutionary nature.

Human capital

Schultz's research in agricultural production evolved into a study of economic growth. In the 1930s, Schultz began to see that new fertilizers expanded the productivity capacity of land. Yet he quickly realized that technological advances could not explain all the gains in productivity; a search was on for a more complete explanation. In the 1940s he came to see 'acquired ability of labour' as a major source of the unexplained gains in productivity. Scarce resources produced these augmented abilities; the stage was set for a formal study of the investment in man.

To study the investment in man, a new concept of capital was needed. Schultz recognized that many eminent economists before him (notably Adam Smith, Irving Fisher and Frank Knight) considered human abilities as capital. In his writings in the 1950s and 1960s, Schultz was critical of Alfred Marshall's perspective on capital that included only physical equipment. (Schultz, 1993, p. 4, presents a revised, more generous assessment of Marshall's perspective on human capital. From interactions with Marshall's defenders, Schultz accepted that analytically Marshall saw human investments as a form of capital but considered human capital as impractical because it was divorced from the marketplace.) Schultz argued that Marshall's narrow definition of capital, among other deficiencies, led to the popular perception that economics studied only material things. And more perniciously the restricted definition led to the notion that productivity of labour is homogenous and independent of capital, so only the number of hours of work matter. (Marshall's view remained popular among leading theorists outside Chicago and Columbia. The role of physical capital was emphasized by the World Bank, for example, throughout the 1960s and 1970s. Human capital made its way into aggregate growth models only in the 1980s. It is telling that the 1979 Nobel Committee press release considers 'Schultz on the Human Factor' and makes only one reference to human capital, and then only in quotes. Some 13 years later human capital was

more accepted and used without quote marks in the press release announcing Gary Becker's Nobel Award.)

By the mid-1950s, while a Fellow at the Center for Advanced Study, Schultz's research in the economics of education took shape:

> During the year at the Center, I began to see that the productive essences of what I was identifying as capital and labour were not constant but were being improved over time and that these improvements were being left out in what I was measuring as capital and labour. It became clear to me also that in the United States many people are investing in man are having a pervasive influence upon economic growth, and that the key investment in human capital is education. (Schultz, 1963, p. viii)

Schultz spent the last 40 years of his career understanding investments 'in man'.

Human capital to Schultz was the acquisition of all useful skills and knowledge that is part of deliberate investment (Schultz, 1961, p. 1). Rather than offer formal definitions, Schultz defined human capital by example:

> Much of what we call consumption constitutes investment in human capital. Direct expenditures on education, health, and internal migration to take advantage of better job opportunities are clear examples. Earnings foregone by mature students attending school and by workers acquiring on-the-job training are equally clear examples. Yet, nowhere do these enter our national accounts. The use of leisure time to improve skills and knowledge is widespread and it too is unrecorded. In these and similar ways the *quality* of human effort can be greatly improved and its productivity enhanced. I shall contend that such investments in human capital accounts for most of the impressive rise in real earnings per worker. (Schultz, 1961, p. 1)

Schultz's research on human capital sought to clarify the investment process and the incentives to invest in human capital. He studied mainly formal education and organized research. The application of the investment approach in the intervening 40 years expanded to consider an array of different forms in many vistas.

It was Schultz's overarching view of human capital that made him a 20th-century pioneer in human capital theory. Jacob Mincer and Gary Becker were pioneers as well, but they focused on the effect of human capital on the level and distribution of earnings: it was Schultz who pushed the profession to see human capital investments in their totality – education, training, work experience, migration and health.

This totality of vision made Schultz a leader in the re-emergence of economic demography. He organized two conferences, in 1972 and 1973, whose proceedings were originally published in the March 1973 and March 1974 in issues of the *Journal of Political Economy* and subsequently published in book form as *Economics of the*

Family: Marriage, Children and Human Capital (1974). The conferences generated several seminal papers, including Willis on fertility, Becker and Lewis on the quality–quantity dimensions of children, and Mincer and Polacheck on human capital and women's earnings. In 'Fertility and Economic Values' (1974) Schultz recognized the large advances published in these studies but pushed the field to extend the static models to consider richer models of life-cycle behaviour and to collect panel data necessary to support their estimation – this was a prophetic vision, as it neatly summarized work in the economics of fertility over the next 30 years.

Envoi

As I reread several of Schultz's papers, I lament the cost of the short half-life of ideas in economics; of what we miss by not reading some of the original formulations. Textbook treatments offer modern notation and language with a well-honed presentation of ideas that speeds initial learning. Yet the vibrancy and freshness (and, yes, sometimes confusion over issues that get sorted out later) of the original frequently is lost. We have lost sight of Schultz's original contributions and it is our loss. Researchers interested in migration, economic demography, development and economic growth would be well served to read some of T.W. Schultz classics. His broad view of economics continues to be fresh and makes for fruitful reading.

JAMES R. WALKER

See also **agricultural economics; agriculture and economic development; education in developing countries; human capital; poverty.**

The National Science Foundation and the National Institute of Child Health and Human Development provided research support. I thank Glen Cain, T. Paul Schultz and Kenneth I. Wolpin for comments.

Selected works

1945. *Agriculture in an Unstable Economy.* New York: McGraw-Hill.

1949. *Production and Welfare of Agriculture.* New York: Macmillan.

1953. *The Economic Organization of Agriculture.* New York: McGraw-Hill.

1960. Capital formation by education. *Journal of Political Economy* 68, 571–83.

1961. Investment in human capital. *American Economic Review* 51, 1–17.

1962. Reflections on investment in man. *Journal of Political Economy* 70, S1–S8.

1963. *The Economic Value of Education.* New York: Columbia University Press.

1964. *Transforming Traditional Agriculture.* New Haven, CT: Yale University Press.

1974. Fertility and economic values. In *Economics of the Family, Marriage, Children and Human Capital.* Chicago: University of Chicago Press.

1975. The value of the ability to deal with disequilibrium. *Journal of Economics Literature* 13, 827–46.

1980. Banquet speech, Nobel Banquet, 10 December 1979. In *Les Prix Nobel. The Nobel Prizes 1979*, ed. W. Odelberg. Stockholm: Nobel Foundation. Online Available at http://nobelprize.org/nobel_prizes/economics/laureates/1979/schultz-speech.html, accessed 4 June 2007.

1992. The economics of being poor. In *Nobel Lectures, Economics 1969–80*, ed. A. Lindbeck. Singapore: World Scientific Publishing Co. Online. Available at http://nobelprize.org/nobel_prizes/economics/laureates/1979/schultz-lecture.html, accessed 10 June 2007.

1993. *The Economics of Being Poor.* New York: Basil Blackwell.

Bibliography

Balogh, T. 1964. Review of Transforming Traditional Agriculture. *Economic Journal* 70, 996–9.

Griliches, Z. 1957. Hybrid corn: an exploration in the economics of technological change. *Econometrica* 4, 501–22.

Johnson, D.G. 1998. In memoriam: Theodore W. Schultz. *Economic Development and Cultural Change* 47, 29–213.

Nerlove, M. 1999. Transforming economics: Theodore W. Schultz, 1902–1998: in memoriam. *Economic Journal* 109, F726–48.

Schumpeter, Joseph Alois (1883–1950)

Economist and social scientist of Austrian origin, Schumpeter was born in Triesch, Moravia, on 8 February 1883, the son of the owner of a textile factory, and died in Taconic, Connecticut, on 8 January 1950. Schumpeter attended an academic high school in Vienna and studied law and economics at the University of Vienna. In 1908 he published an important book on the nature and content of economic theory, which established his fame as the ablest among the younger group of Austrian economists. Schumpeter had F. von Wieser and E. von Böhm-Bawerk as his teachers. After being nominated at the University of Czernowitz, he became professor of economics at the University of Graz in 1911. His famous book on *The Theory of Economic Development* was published in 1912. Much of his later work on business cycles and the evolution of capitalism into socialism is to a certain extent an elaboration and improvement of the ideas and analysis presented in his book on economic development.

From 1925 to 1932 Schumpeter was a professor at the University of Bonn and in 1932 he became a permanent professor of economics at the Harvard University, a post he held until 1950. His impressive work on *Business Cycles* appeared in 1939, and in 1942 he published *Capitalism, Socialism and Democracy,* in which he predicted the gradual decay of capitalism. On the basis of his numerous works, one would be inclined to think that

Schumpeter devoted his whole life to teaching, writing and theorizing. In fact his life has been even more colourful. From spring 1919 up to October of the same year he held the position of Austrian Minister of Finance in Renner's cabinet. During his term he was in favour of sound finance and a capital levy; he even started as a strong defender of massive socialization. He changed his mind, however, partly because he acknowledged the necessity to import capital and gradually his attitude caused tensions with his socialist colleagues in the cabinet, which finally led to his downfall. Between his professorships in Graz and Bonn, Schumpeter accepted the presidency of a Viennese private bank, the Biedermann Bank. Around 1926 the bank went bankrupt and Schumpeter was left with huge debts, probably due to speculations with borrowed money. Schumpeter was married three times, for the first time in 1907 to an Englishwoman, a marriage that ended in divorce in 1920. In 1925 he married a Viennese, 21 years his junior, who died in 1926 in childbirth. In 1937 Schumpeter married Elizabeth Boody, who, after his death in 1950, edited his monumental work, the *History of Economic Analysis* (Stolper, 1994).

In Schumpeter's interpretation of capitalism, the entrepreneur, who applies new combinations of factors of production, plays a central role. He is the innovator, and the agent of economic change and development. Centred around the role of the Schumpeterian entrepreneur is the rise and decay of capitalism. The gifted few, pioneering in the field of new technologies, new products and new markets, carry out innovations and, joined some time later by many imitators, they are at the heart of the short and long cycles observed in economic life. The importance Schumpeter assigns to the creation of money and overexpansion of credit in his early work foreshadows his later work. He argues that since in static analysis there is no room for (*a*) new combinations of factors of production; (*b*) the entrepreneur (in the Schumpeterian sense); and (*c*) profit, there is no need for the further creation of money. He also questions whether in a static context one can speak of economic development at all. Here he is not questioning that economic phenomena and magnitudes change, but he is suggesting that the causes of change may not lie in economic factors. In modern economic theory, this would be expressed by asking whether economic development is due to endogenous or exogenous factors. In the case of exogenous factors, Schumpeter would not speak of economic development at all, for he regards factors such as population growth, consumer preferences, technical development and social organization as non-economic factors. On the other hand, he argues that changes in human nature and social organization can in fact be attributed to economic causes, which can then be regarded as endogenous factors.

The somewhat ambiguous terminology Schumpeter uses in order to describe and explain the development of the economic process is repeated in his book on economic development. Saving is no longer regarded as a factor leading to economic development in the sense of entrepreneurial innovations. Capital formation and the increase of population determine the growth rate in the stationary economy. The terms 'development' and 'economic development' are not used to describe the actual course of economic events but to distinguish between changes caused in the economic process by endogenous factors and other changes.

Not all endogenous factors lead to economic development, and Schumpeter explicitly excludes continuous endogenous changes. His theory of economic development is reduced to the treatment of spontaneous and discontinuous changes in the economic cycle. The endogenous changes Schumpeter has in mind are not found on the demand side of the economic process, but on the supply side. Economic development consists of the discontinuous introduction of new combinations of products and means of production. The five examples mentioned by Schumpeter show that the term 'new combination' must be taken in a very broad sense; it comprises a new product, a new method of production, the opening-up of a new market, the utilization of new raw materials, and the reorganization of sectors of the economy.

He restricts the meaning of the word 'enterprise' to the creation of new combinations, and the meaning of the word 'entrepreneurs' to those economic figures who introduce new combinations. Schumpeter's entrepreneur operates in an uncertain world, has the courage to start up new ventures, and must be strong enough to swim against the tide of society. In Schumpeter's view, new combinations can be financed only if a successful appeal can be made to the banking system to create money.

According to Schumpeter, ups and downs in economic development can be explained quite simply by the fact that new combinations or innovations appear, if at all, discontinuously in groups or swarms. The appearance of entrepreneurs in 'bursts' is due exclusively to the fact that the appearance of one or a few entrepreneurs facilitates the appearance of others. This is the only reason for an upswing in the business cycle. The downturn sets in as a result of smaller profit margins due to imitation and a new equilibrium will be reached after the diffusion process is completed.

In his book on business cycles, Schumpeter sharpened his analytical tools. He introduced the concept of the production function, which – in his words – tells us all we need to know about the technological aspects of production. Schumpeter regards the setting up of a new production function as the introduction of new combinations. The changes caused by innovations are no longer regarded as economic development but as economic evolution. He uses the term 'technical development' only for innovations that involve the introduction of new methods of production. Innovations must be distinguished from inventions. The application of new combinations by entrepreneurs is possible without inventions, while inventions as such need not necessarily lead to innovations and need not have any economic consequences. Innovation itself is the independent endogenous factor that causes economic life to go through a number of cycles.

Innovations lead to cyclical fluctuations whose length is determined by both the character and the period of implementation of the innovations. Schumpeter applies this general explanation to the 40-month Kitchin, the ten-year Juglar and the 60-year Kondratieff cycles. The combination of the use of the innovations, overinvestment and of credit expansion going too far, brings the upswing to an end. The recession, which in Schumpeter's view is a healthy phase of restructuring, sets in, paving the way for a new burst of future innovations. Schumpeter's prediction of the decay of capitalism is based on the vision that it is not economic failure but rather the economic success of capitalism that causes the march into socialism. Social rather than economic factors are according to Schumpeter responsible for the structural change in the organization of society (Schumpeter, 1939).

Typical of this economic scene is the process of creative destruction. In Schumpeter's view it is the essential fact about capitalism. The process of creative destruction concerns the implementation of new combinations that incessantly modifies the economic structure from within. The competitive character of capitalism is mainly determined by creative destruction and far less by the case of textbook competition, in which prices play such a dominant role. This process of creative destruction must be judged by its long-term results. Schumpeter has the highest possible opinion of the dynamic character and productive capability of capitalism. In weighing up the static optimal allocation of resources in case of perfect competition, and dynamic efficiency of monopolistic structures, in particular with regard to innovative activities, he has an outspoken preference for monopoly and oligopoly and a disdain for free competition. Schumpeter did not adhere to the theory that vanishing investment opportunities and a slowdown of technical change would lead to stagnation and in the end to a breakdown of capitalism (Metcalfe, 1998).

The rate of growth of production is not reduced because the technical possibilities are exhausted, but because capitalism suffers from a change in the behaviour of entrepreneurs. On the one hand, it is now easier than before to do things that lie outside familiar routine – innovation itself is being reduced to routine. Technological progress is increasingly becoming the business of teams of trained specialists who make technical change a predictable process. On the other hand, characteristics such as personality, willpower and a dynamic attitude count less in environments which have become accustomed to economic and social change.

Economic development thus becomes more and more impersonal and mechanical. The success of the capitalist mode of production makes capitalism itself redundant: capitalism undermines the social institutions which protect it. These institutions are the remnants of the feudal system and the existence of many small businesses and farmers. The disappearance of these social forms weakens the political position of the bourgeoisie. The elimination of the socioeconomic function of the entrepreneur, especially in large corporations where technical change is a matter of routine and management is bureaucratized, reinforced by the growing influence of the public sector, further undermines the bourgeoisie. Above all, however, capitalism produces an army of critical and frustrated intellectuals who by their negative attitude contribute to the decline of capitalism and help to establish an atmosphere in which private property and bourgeois values are daily subjected to heavy criticism (for example, in newspapers). In short, capitalism's economic success leads to its political failure (Schumpeter, 1942).

Schumpeter's thoughts on the effects of monopolistic and oligopolistic markets on technical change and on the influence of technology on the emergence of big business have given a great impetus to the study of the relationship between technical change and market structure. His sharp distinction between innovations and inventions has triggered off several reactions, and both theoretical analysis and empirical research have been directed to the question whether innovations are really as independent of inventions as Schumpeter supposed. A critical evaluation of Schumpeter's theory has led to a discussion of the nature of technical change, the time-lag between invention and application, the significance of patents and the diffusion process of technical change. Schumpeter's predictions about the decay of capitalism have not come about (Heertje, 1981). He seems to have underestimated the dynamic character of capitalism, the importance of which he himself emphasized so eloquently. He did not realize that a new generation of entrepreneurs might come to the fore, prepared to apply new technology and start a new wave of innovations. There is much more room for Schumpeterian entrepreneurial activity, especially on a small-scale basis, than Schumpeter foresaw.

It can be admitted that in large firms in particular the emergence of new technical knowledge is to a certain extent mechanized and that the decision-making process about its application, in the sphere of both production and marketing, is often hampered by bureaucratic features, threatening the static and dynamic efficiency of the firm and therefore the level of output. But these decisions still take place in a world of uncertainty and financial risk. These latter aspects of the decision-making process come more to the fore as the size of the firm becomes smaller. Within these firms, individuals who take final and major decisions and bear the responsibility of profits and losses do still exist. If, due to a lack of such individuals, Schumpeterian developments arise, the result will often be a new dynamic leadership. Many small firms still exist or come into being in order to gain a place in the market. Their managers take initiatives and often have to overcome resistance of consumers to achieve their market goals. It may be true that nowadays nearly everybody is confronted with change and new developments at a very high rate, but this does not imply that everybody accepts this state of affairs. So, on the whole, there are traces of a Schumpeterian development in Western economies, but mechanization and routinization of the entrepreneurial function are by no means the general picture. Economic life is still a melting pot of conflicting tendencies, ups and downs, major risks and minor certainties. In short, it is the dynamic, ever-changing scene for entrepreneurs who have to be innovative and sensitive to new opportunities. Those entrepreneurs who follow the Schumpeterian line will be punished by the market, that is, by competitors who are prepared to take risks and to bear losses.

On balance, it seems that innovation as a process of development, application and diffusion of new technical possibilities has not been reduced to routine. Nor have people become so reconciled to change that personal characteristics such as willpower and perseverance are no longer needed to break traditional patterns. However, even this kind of modification of Schumpeter's views does not mean that over some much longer run he may still not turn out to have been right. It therefore seems appropriate to look more carefully at the empirical evidence about the entrepreneurial function and about the acceptability of the level of output as the sole yardstick for economic performance. This may enable us to look into the future of capitalism, as well (Heertje, 2006).

Is the essence of the entrepreneurial function really the exploitation of new technical possibilities and of new opportunities in general? On the basis of Schumpeter's distinction between inventions and innovations, it is natural to identify entrepreneurial activity with innovative activity. But in the no man's land between invention and innovation there is a missing link; a link which according to Kirzner comprises three essential entrepreneurial components in human action: (*a*) the alertness to information; (*b*) the awareness of new existing opportunities, waiting to be noticed; and (*c*) the response to possibilities offered by the market system (Kirzner, 1986). Although in many cases those who are on the lookout for new opportunities are the same men and women who exploit them, Kirzner's refinement of Schumpeter's characterization of entrepreneurship is of the utmost importance – particularly if this view is combined with the idea of the market as an ongoing process of creative discovery. It may be argued that certain stages of the application and execution of new ideas can be routinized. However, the property of being alert to marketable applications of what already exists, but is currently overlooked by others, is still an individualistic characteristic, not one for being automatized. Even if we comprehend in the definition of

the entrepreneurial function the application and implementation of new combinations, it is still true that first of all one has to be alert to what may be applied. In this sense entrepreneurship never was and never will be a matter of routine (Arena and Romani, 2002).

With regard to the level of output as a yardstick for economic performance, it may be observed that other aspects of welfare are taken into account as well. The operational meaning of welfare, in the broad sense of the level of the satisfaction of wants in so far as this depends on the allocation of scarce resources, comes more and more to the fore. The level of output is no longer the only thing that matters; the quality of growth, job satisfaction and other immaterial aspects of welfare are becoming more and more important. They are often the source for new methods of production and products, which give new impetus to capitalism. As an Austrian, Schumpeter would certainly welcome the broadening of the economic dimension to the formal and subjective concept of welfare (Hennipman, 1995).

Furthermore, socialism as an alternative to capitalism is on a worldwide scale less and less attractive as a system able to take care of aspirations and heterogeneous preferences of individuals and of technical change. Since the beginning of the 1980s the economic recovery inspired the reinforcement of the market. Against this background a revival of Schumpeter's ideas in economic theory took place. While Keynes dominated economic thinking during a large part of the 20th century, Schumpeter is a major source of inspiration ever since. His concern with permanent and endogenous changes in economic life show up the limited significance of static equilibrium theory and paved the way for a neo-Austrian emphasis on the importance of the market mechanism as a vehicle for the discovery of new products and new methods of production, in short of the market as a dynamic institution. The dynamics of the market mechanism combined with the economics of research and development, new methods of production and products and of the diffusion of knowledge and of the application of new technology have led in modern economic theory to a Schumpeterian-inspired approach as part of endogenous growth theory (Aghion and Howitt, 1998). It comes as no surprise that in 1986 the International J.A. Schumpeter Society was founded in Augsburg, Germany. Since then every two years an international conference is organized.

Within the realm of social theory, Schumpeter's distinction between political and methodological individualism should also be mentioned. In particular, the concept of methodological individualism has proved to be very important for the analysis of social phenomena, outside the sphere of the market mechanism. As a method of analysis, methodological individualism prescribes starting from the economic behaviour of the individual in order to build a theory about the structure and working of the political process and about the behaviour of groups. In this sense, Schumpeter's thinking is the opposite of Marx's analysis in terms of the class struggle. The modern theory of public choice, in which the maximization of individual welfare of politicians and bureaucrats plays an essential role, in order to describe their social behaviour as part of the government, is a direct application of methodological individualism. Related to this development is the economic theory of democracy, of which Schumpeter is a forerunner. In his view, the democratic method is that institutional arrangement for arriving at political decision in which individuals acquire the power to decide by means of a competitive struggle for the people's vote. In other words, Schumpeter introduces the idea that democracy is a type of horizontal coordination in the public sector that can be compared to the role of the market mechanism in the private sector of the economy. The political process is regarded as a market process in which the voters are the demanders and politicians and bureaucrats are the suppliers.

This idea appears to be very fruitful in both theory and practice and contributes to Schumpeter's fame as a social and economic thinker of lasting significance.

Schumpeter's view of society is based on the integration of historical facts, philosophical considerations and sociological visions. While Marx predicted the breakdown of capitalism as an inevitable consequence of the objective inner structure of the system, determined in its development by technical change, Schumpeter, although also pointing out structural changes from within, leaves room for the role of individuals, who by their behaviour can turn the tide. His message on the march into socialism is not meant to be defeatist; it would only be so if all differences between individuals disappeared and, in particular, leadership vanished (Shionoya, 1995).

Schumpeter will always be referred to for his impressive contributions to the history of economic thought. His *Economic Doctrine and Method*, originally published in German in 1914, is an early expression of his interest. His book *Ten Great Economists* provides further evidence, and his monumental work, the *History of Economic Analysis*, is a lasting culmination. It illustrates his detailed knowledge of the vast literature on economic theory since the days of the Greeks and Romans.

On the whole his discussion of the theoretical contributions of numerous economists is fair, generous and well-balanced. There are, however, a few notable exceptions. He ranks Cournot higher than Ricardo, and although we find both economists in modern economic theory, it seems fair to conclude that Ricardo's contributions to economics leave a broader scope and impact. Furthermore, he considers Walras the greatest economist of all time, greater than, for example, Marshall. His great appreciation for two mathematical economists is in contrast with his own non-mathematical treatment of economics, although he even became a founder and first president of the Econometric Society.

Reading Schumpeter, one realizes that his lasting significance stems from historical description and non-mathematical theoretical analysis. His inability to put his ideas about the development of economic life into a mathematical form does change our assessment of him. But whatever the final evaluation of Schumpeter may be, it cannot be denied that he gave new direction to the development of economic science by posing some entirely new questions. Schumpeter's preoccupation with the dynamics of economic life broke the spell of the static approach to economic problems.

Throughout his life Schumpeter was an *enfant terrible*, who was always ready to take extreme positions for the sake of argument, and often seized the chance to irritate people. But he was also a giant on whose shoulders many later scholars contributing to economic science stood. As a political economist he is no longer in the shadow of Keynes, but in the centre of the economic scene, both in the theoretical and empirical sense.

ARNOLD HEERTJE

See also **creative destruction; creative destruction (Schumpeterian conception); endogenous growth theory.**

Selected works

1912. *The Theory of Economic Development*. Leipzig: Duncker & Humblot. Trans. R. Opie. Cambridge, MA: Harvard University Press, 1934. Reprinted, New York: Oxford University Press, 1961.
1914. *Economic Doctrine and Method*. Trans. R. Aris. London: George Allen & Unwin, 1954.
1939. *Business Cycles*. 2 vols. New York: McGraw-Hill.
1942. *Capitalism, Socialism and Democracy*. New York: Harper & Brothers. London: George Allen & Unwin, 1943. 5th edn, London: George Allen & Unwin, 1976.
1951. *Ten Great Economists*. New York: Oxford University Press.
1954. *History of Economic Analysis*, ed. E. Boody. New York: Oxford University Press; London: George Allen & Unwin. Repr. with a new introduction by M. Perlman, London: Routledge, 1994.
1986. *Aufsätze zur Wirtschaftspolitik*. Tübingen: J.C.B. Mohr.

Bibliography

Aghion, P. and Howitt, P. 1998. *Endogenous Growth Theory*. London: MIT Press.
Allen, R.L. 1991. *Opening Doors: The Life and Work of Joseph Schumpeter*. London: Transaction.
Arena, R. and Romani, P.-M. 2002. Schumpeter on entrepreneurship. In *The Contribution of Joseph Schumpeter to Economics*, ed. R. Arena and C. Dangel-Hagnauer. London: Routledge.
Bös, D. and Stolper, W., eds. 1984. *Schumpeter oder Keynes*. Berlin: Springe-Verlag.
Frisch, H., ed. 1982. *Schumpeterian Economics*. New York: Praeger.

Harris, S.E., ed. 1951. *Schumpeter, Social Scientist.* Cambridge, MA: Harvard University Press.

Heertje, A., ed. 1981. *Schumpeter's Vision.* New York: Praeger.

Heertje, A. 2006. *Schumpeter on the Economics of Innovation and the Development of Capitalism.* Cheltenham: Edward Elgar.

Hennipman, P. 1995. *Welfare Economics and the Theory of Economic Policy*, ed. D.A. Walker and J. van den Doel. Aldershot: Edward Elgar.

Kirzner, I.M. 1986. *Discovery and the Capitalist Process.* Chicago: University of Chicago Press.

Metcalfe, J.S. 1998. *Evolutionary Economics and Creative Destruction.* London: Routledge.

Reisman, D. 2004. *Schumpeter's Market.* Cheltenham: Edward Elgar.

Shionoya, Y. 1995. *Schumpeter and the Idea of Social Science.* Cambridge: Cambridge University Press.

Stolper, W.F. 1994. *Joseph Alois Schumpeter: The Public Life of a Private Man.* Princeton: Princeton University Press.

Swedberg, R. 1991. *J.A. Schumpeter, His Life and Work.* Cambridge: Polity Press.

Schumpeterian growth and growth policy design

Three main ideas underlie the Schumpeterian growth paradigm: (*a*) growth is primarily driven by technological innovations; (*b*) innovations are produced by entrepreneurs who seek monopoly rents from them; (*c*) new technologies drive out old technologies.

The Schumpeterian growth model (Aghion and Howitt, 1992; 1998) grew out of modern industrial organization theory (Tirole, 1988). It focuses on quality-improving innovations that render old products obsolete, and hence involves the force that Schumpeter called 'creative destruction'. In this article we argue that the Schumpeterian paradigm holds the best promise of delivering a systematic, integrated, and yet operational framework for analysing and developing context-dependent growth policies.

Schumpeterian theory begins with a production function specified at the industry level:

$$Y_{it} = A_{it}^{1-\alpha} K_{it}^{\alpha}, \quad 0 < \alpha < 1 \tag{1}$$

where A_{it} is a productivity parameter attached to the most recent technology used in industry i at time t. In this equation, K_{it} represents the flow of a unique intermediate product used in this sector, each unit of which is produced one-for-one by capital. Aggregate output is just the sum of the industry-specific outputs Y_{it}. (Although the theory focuses on individual industries and explicitly analyses the microeconomics of industrial competition, the assumption that all industries are *ex ante* identical gives it a simple aggregate structure. In particular, it is easily shown that aggregate output depends on the aggregate capital stock K_t according to the Cobb–Douglas aggregate per-worker production function: $Y_t = A_t^{1-\alpha} K_t^{\alpha}$ where the labour-augmenting productivity factor A_t is just the unweighted sum of the sector-specific A_{it}'s. As in neoclassical theory, the economy's long-run growth rate is given by the growth rate of A_t, which here depends endogenously on the economy-wide rate of innovation.)

Each intermediate product is produced and sold exclusively by the most recent innovator. A successful innovator in sector i improves the technology parameter A_{it} and is thus able to displace the previous innovator as the incumbent intermediate monopolist in that sector, until displaced by the next innovator.

First implication: *faster growth generally implies a higher rate of firm turnover, because the process of creative destruction generates entry of new innovators and exit of former innovators.*

There are two main inputs to innovation, namely, the private expenditures made by the prospective innovator, and the stock of innovations that have already been made by past innovators. The latter input constitutes the publicly available stock of knowledge to which current innovators are hoping to add. The theory is quite flexible in modelling the contribution of past innovations. It encompasses the case of an innovation that leapfrogs the best technology available before the innovation, resulting in a new technology parameter A_{it} in the innovating sector i, which is some multiple γ of its pre-existing value. And it also encompasses the case of an innovation that catches up to a global technology frontier $\bar{A}_t$ which we typically take to represent the stock of global technological knowledge available to innovators in all sectors of all countries. In the former case the country is making a leading-edge innovation that builds on and improves the leading-edge technology in its industry. In the latter case the innovation is just implementing technologies that have been developed elsewhere.

For example, consider a country in which in any sector leading-edge innovations take place at the frequency μ_n and implementation innovations take place at the frequency μ_m. Then the change in the economy's aggregate productivity parameter A_t will be:

$$A_{t+1} - A_t = \mu_n(\gamma - 1)A_t + \mu_m(\bar{A}_t - A_t) \tag{2}$$

and hence the growth rate will be:

$$g_t = \frac{A_{t+1} - A_t}{A_t} = \mu_n(\gamma - 1) + \mu_m(a_t^{-1} - 1) \tag{3}$$

where:

$$a_t = A_t / \bar{A}_t$$

is an inverse measure of 'distance to the frontier'. We then obtain a second important implication of the paradigm.

Second implication: *by taking into account the fact that innovations can interact with each other in different ways in countries or sectors at various distances from the frontier, Schumpeterian theory provides a framework in which to analyse how a country's growth performance will vary with its proximity to the technological frontier a_t, to what extent the country will tend to converge to that frontier, and what kinds of policy changes are needed to sustain convergence as the country approaches the frontier.*

We could take as given the critical innovation frequencies μ_m and μ_n that determine a country's growth path as given, just as neoclassical theory often takes the critical saving rate s as given. However, Schumpeterian theory goes deeper by deriving these innovation frequencies endogenously from the profit-maximization problem facing a prospective innovator, just as the Ramsey model endogenizes s by deriving it from household utility maximization. This maximization problem and its solution will typically depend upon institutional characteristics of the economy such as property rights protection and the financial system, and also upon government policy.

Equation (3) incorporates Gerschenkron's (1962) 'advantage of backwardness', in the sense that the further the country is behind the global technology frontier (that is, the smaller a_t is) the faster it will grow, given the frequency of implementation innovations. As in Gerschenkron's analysis, the advantage arises from the fact that implementation innovations allow the country to make larger quality improvements the further it has fallen behind the frontier. As we shall see below, this is just one of the ways in which distance to the frontier can affect a country's growth performance.

In addition, as stressed by Acemoglu, Aghion and Zilibotti (2006) (AAZ), growth equations like (3) make it quite natural to capture Gerschenkron's idea of 'appropriate institutions'. Suppose indeed that the institutions that favour implementation innovations (that is, that lead to firms emphasizing μ_m at the expense of μ_n) are not the same as those that favour leading-edge innovations (that is, that encourage firms to focus on μ_n); we then obtain the following:

Third implication: *far from the frontier a country or sector will maximize growth by favouring institutions that facilitate implementation; however, as it catches up with the technological frontier, to sustain a high growth rate the country will have to shift from implementation-enhancing institutions to innovation-enhancing institutions as the relative importance of leading-edge innovations for growth is also increasing.*

As formally shown in AAZ, failure to operate such a shift can prevent a country from catching up with the frontier level of per capita GDP, and Sapir et al. (2003) argued that this failure largely explains why Europe stopped catching up with US per capita GDP from the mid-1970s. More specifically, suppose that the global frontier (the United States) grows at some rate $\bar{g}$. Then eq. (3) implies that in the long run a country that engages in implementation investments (with $\mu_m > 0$) will ultimately converge to the same growth rate as the world technology frontier. That is, the relative gap a_t that separates this economy from the technology frontier will converge asymptotically to the steady-state value:

$$\hat{a} = \frac{\mu_m}{\bar{g} + \mu_m - \mu_n(\gamma - 1)} \qquad (4)$$

which is an increasing function of the domestic innovation rates and a decreasing function of the global productivity growth rate. An insufficient emphasis on innovation (μ_n) in Europe will reduce $\hat{a}$, that is, the long-run level of European per capita GDP compared with that of the United States.

The model can also explain why, since the mid-1990s, the EU has been growing at a lower rate than the United States. A plausible story, which comes out naturally from the above discussion, is that the European economy caught up technologically to the United States following the Second World War, but then its growth began to slow down before the gap with the United States had been closed, because its policies and institutions were not designed to optimize growth when close to the frontier. That by itself would have resulted in a growth rate that fell to that of the United States but no further. But then what happened was that the information technology revolution resulted in a revival of $\bar{g}$ in the late 1980s and early 1990s. Since Europe was as not as well placed as the United States to benefit from this technological revolution, the result was a reversal of Europe's approach to the frontier, which accords with the Schumpeterian steady-state condition (4); and the fact that Europe is not adjusting its institutions and policies in order to produce the growth-maximizing innovation policy acts as a force delaying growth convergence with the United States. (Endogenizing μ_m can also generate divergence in growth rates. For example, human capital constraints as in Howitt and Mayer-Foulkes, 2005, or credit constraints as in Aghion, Howitt and Mayer-Foulkes, 2005, make the equilibrium value of μ_m increasing in a, which turns the growth equation (3) into a nonlinear equation. That μ_m be increasing in a follows in turn from the assumption that the cost of innovating is proportional to the frontier technology level that is put in place by the innovation – Ha and Howitt, 2007, provide empirical support for this proportionality assumption – whereas the firm's investment is constrained to be proportional to current local productivity. Then, countries very far from the frontier and/or with very low degrees of financial development or of human capital will tend to grow in the long run at a rate which is strictly lower than the frontier growth rate $\bar{g}$. However, our empirical analysis in this paper shows that this source of divergence does not apply to EU countries.)

In the next section we concentrate on a particular policy area, namely, entry and exit.

Entry and exit

Is it always growth-enhancing to liberalize entry and to facilitate exit? That exit could be growth-enhancing follows immediately from the fact that creative destruction is about better technologies or inputs replacing old and increasing obsolete technologies. Now, what about entry? Is it unambiguously good for innovation and growth by incumbent firms? As above, the answer may depend upon firms' distance from the technological frontier. In particular, suppose that firms can improve their technologies only in a gradual (or step-by-step) fashion, and that new potential entrants are endowed with the current frontier technology. Then, incumbent firms that are initially close to the frontier can match or even leapfrog a potential entrant's technology, and therefore can deter entry by innovating. In contrast, firms that are initially far below the frontier cannot prevent entry by innovating as they can never match an entrant's technology. What does this imply for how these firms will react to increased entry threat? The answer is simple: a greater entry threat will induce firms that are close to the frontier to invest more in innovation in order to protect their monopoly rents, whereas it will discourage firms far below the frontier from investing in innovation as such investment is less likely to be of any use the greater is the probability of entry.

In short, exit can be growth-enhancing and, regarding the effects of entry, the closer a firm or sector is to the frontier, the more positively (or the less negatively) it will react to increased entry threat. These predictions have been corroborated by a variety of empirical findings. First, Aghion et al. (2005a) investigate the effects of entry threat on total factor productivity (TFP) growth of UK manufacturing establishments, using panel data with more than 32,000 annual observations of firms in 166 different four-digit industries over the 1980–93 period. They estimate the equation:

$$Y_{ijt} = \alpha + \beta E_{jt} + \gamma E_{jt} \cdot D_{jt} + \eta_i + \tau_t + \varepsilon_{ijt}$$

$$(5)$$

where Y_{ijt} is TFP growth in firm i, industry j, year t, η and τ are fixed establishment and year effects, and E_{jt} is the industry entry rate, measured by the change in the share of UK industry employment in foreign-owned plants. In order to verify that this effect of entry on incumbent productivity growth is a result of increased incumbent

innovation rather than technology spillover from, or copying of, the superior technologies brought in by the entrants, Aghion et al. (2005a) also estimate eq. (5) using a patent count rather than productivity growth as the dependent variable. They provide direct evidence that the escape competition is stronger for industries that are closer to the frontier. Specifically, the interaction coefficient γ is highly significantly negative in all estimations. A one-standard deviation increase in the entry variable above its sample mean would reduce the estimated number of patents by 10.8 per cent in an industry far from the frontier (at the 90th percentile of D_{jt}) and would increase the estimated number by 42.6 per cent in an industry near the frontier (at the tenth percentile). Thus it seems that the positive effect of entry threat on incumbent productivity growth in Europe is indeed much larger now than it was immediately after the Second World War, and that the relative neglect of entry implications of competition policy is having an increasingly detrimental effect on European productivity growth.

On exit and growth, in ongoing research Aghion, Antras and Prantl combine UK establishment-level panel data with the input–output table to estimate the effect on TFP growth arising from growth in high-quality input in upstream industries, and also from exit of obsolete input-producing firms in upstream industries. Specifically, we take a panel of 23,886 annual observations of more than 5,000 plants in 180 four-digit industries between 1987 and 1993, together with the 1984 UK input–output table, to estimate an equation of the form:

$$g_{ijt} = \alpha + \beta \cdot q_{jt-1} + \gamma \cdot x_{jt-1} + \delta \cdot Z_{ijt-1}$$
$$+ \eta_i + \phi_j + \tau_t + \varepsilon_{ijt}$$

$$(6)$$

where g_{ijt} is the TFP growth rate of firm i in industry j, q_{jt-1} is a measure of upstream quality improvement, and x_{jt-1} is a measure of exit of obsolete upstream input-producing firms. Establishment, industry and year effects are included, along with the other controls in Z_{ijt-1}, including a measure of the plant's market share.

The result of this estimation is a significant positive effect of both upstream quality improvement and upstream input-production exit. These results are robust to taking potential endogeneity into account by applying an instrumental variable approach, using instruments similar to those of Aghion et al. (2005a) described above. The effects are particularly strong for plants that use more intermediate inputs, that is, plants with a share of intermediate product use above the sample median. Altogether, the results we find are consistent with the view that quality-improving innovation is an important source of growth. The results are, however, not consistent with the horizontal innovation model, in which there should be nothing special about the entry of foreign firms, and according to which the exit of upstream firms should if anything reduce growth by reducing the variety of inputs being used in the industry.

Conclusion and comparison with alternative endogenous growth models

What have we learned from our discussion so far? First, that Schumpeterian growth theory features quality-improving innovations that displace previous technologies, and that such innovations are motivated by the prospect of monopoly rents. Second, that, due to the natural conflict between new and old technologies, a higher rate of growth is likely to be associated with a higher rate of firm turnover (entry and exit).

In particular, exit can have a positive effect on productivity growth in downstream industries because it replaces less efficient input producers by more efficient ones. Third, that quality improvements can be generated, either by imitating current frontier technologies or by innovating upon previous local technologies, and that the relative importance of either type of innovation depends upon a sector's or a country's initial distance from the corresponding technological frontier. Fourth, that the same policy will tend to have contrasting effects on sectors or countries at different distances from the frontier, and that therefore growth policy must be adapted to the particular context of a sector or country. Fifth, that entry and delicensing have a more positive effect on growth in sectors or countries that are closer to the technological frontier, but have a less positive effect on sectors or countries that lie far below the frontier. This suggests that, although disregarding entry was of no great concern during the 30 years immediately after the Second World War, when Europe was still far behind the United States and catching up with it, nevertheless, now that Europe has come close to the world technology frontier this relative neglect of entry considerations is having an increasingly depressing effect on European growth. It also suggests a role for complementary policies aimed at reallocating resources and workers from laggard to more frontier sectors and activities in order to maximize the positive effects of competition and entry on productivity growth.

Finally, one may want to contrast the Schumpeterian growth paradigm with the two alternative models of endogeneous growth. The first version of endogenous growth theory was the so-called AK theory (see Frankel, 1962; Romer, 1986; Lucas, 1988), whereby knowledge accumulation is a serendipitous by-product of capital accumulation by the various firms in the economy. Here thrift and the resulting capital accumulation are the keys to growth, not novelty and innovation. The second model of endogenous growth theory is by Romer (1990), according to which aggregate productivity is a function of the degree of product variety. Innovation causes productivity growth in the product-variety paradigm by creating new, but not necessarily improved, varieties of products. The driving force of long-run growth in the product-variety paradigm is innovation, as in the Schumpeterian paradigm. In this case, however, innovations do not generate better intermediate products, just more of them. Also as in the Schumpeterian model, the equilibrium R&D investment and innovation rate results from a research arbitrage equation that equates the expected marginal payoff from engaging in R&D with the marginal opportunity cost of R&D. But the fact that there is just one kind of innovation, which always results in the same kind of new product, means that the product-variety model is limited in its ability to generate context-dependent growth, and is therefore of limited use for policymakers.

In particular, the theory makes it very difficult to talk about the notion of technology frontier and of a country's distance from the frontier. Consequently, it has little to say about how the kinds of policy appropriate for promoting growth in countries near the world's technology frontier may differ from those appropriate for technological laggards, and thus little to say by way of explaining why Asia is growing fast with policies that depart from the Washington consensus, or why Europe grew faster than the United States during the first three decades after the Second World War but not thereafter. In addition, nothing in this model implies an important role for exit and turnover of firms and workers; indeed increased exit in this model can do nothing but reduce the economy's GDP, by reducing the variety variable that uniquely determines aggregate productivity. As we just argued above, these latter implications of the product-variety model are inconsistent with an increasing number of recent studies demonstrating that labour and product market mobility are key elements of a growth-enhancing policy near the technological frontier.

PHILIPPE AGHION

See also **creative destruction; endogenous growth theory; information technology and the world economy; Schumpeter, Joseph Alois.**

Bibliography

Acemoglu, D., Aghion, P. and Zilibotti, F. 2006. Distance to frontier, selection, and economic growth. *Journal of the European Economic Association* 4, 37–74.

Aghion, P., Bloom, N., Blundell, R., Griffith, R. and Howitt, P. 2005b. Competition and innovation: an inverted-U relationship. *Quarterly Journal of Economics* 120, 701–28.

Aghion, P., Blundell, R., Griffith, R., Howitt, P. and Prantl, S. 2004. Entry and productivity growth: evidence from micro-level panel data. *Journal of the European Economic Association* 2, 265–76.

Aghion, P., Blundell, R., Griffith, R., Howitt, P. and Prantl, S. 2005a. The effects of entry on incumbent innovation and productivity. Working Paper No. 12027. Cambridge, MA: NBER.

Aghion, P., Burgess, R., Redding, S. and Zilibotti, F. 2005. Entry liberalization and inequality in industrial performance. *Journal of the European Economic Association* 3, 291–302.

Aghion, P. and Howitt, P. 1992. A model of growth through creative destruction. *Econometrica* 60, 323–51.

Aghion, P. and Howitt, P. 1998. *Endogenous Growth Theory.* Cambridge, MA: MIT Press.

Aghion, P., Howitt, P. and Mayer-Foulkes, D. 2005. The effect of financial development on convergence: theory and evidence. *Quarterly Journal of Economics* 120, 173–222.

Frankel, M. 1962. The production function in allocation and growth: a synthesis. *American Economic Review* 52, 995–1022.

Gerschenkron, A. 1962. *Economic Backwardness in Historical Perspective.* Cambridge, MA: Harvard University Press.

Ha, J. and Howitt, P. 2007. Accounting for trends in productivity and R&D: a Schumpeterian critique of semi-endogenous growth theory. *Journal of Money, Credit, and Banking* 39, 733–74.

Helpman, E. 1993. Innovation, imitation, and intellectual property rights. *Econometrica* 61, 1247–80.

Howitt, P. and Mayer-Foulkes, D. 2005. R&D, implementation and stagnation: A Schumpeterian theory of convergence clubs. *Journal of Money, Credit and Banking* 37, 147–77.

Lucas, R. 1988. On the mechanics of economic development. *Journal of Monetary Economics* 22, 3–42.

Nickell, S. 1996. Competition and corporate performance. *Journal of Political Economy* 104, 724–46.

Nicoletti, G. and Scarpetta, S. 2003. Regulation, productivity and growth. *Economic Policy* 36, 11–72.

Romer, P. 1986. Increasing returns and long-run growth. *Journal of Political Economy* 94, 1002–37.

Romer, P. 1990. Endogenous technical change. *Journal of Political Economy* 98, 71–102.

Sapir, A. et al. 2003. *An Agenda for a Growing Europe.* Oxford: Oxford University Press.

Tirole, J. 1988. *The Theory of Industrial Organization.* Cambridge, MA: MIT Press.

science, economics of

The economics of science aims to understand the impact of science on the advance of technology, to explain the behaviour of scientists, and to understand the efficiency or inefficiency of scientific institutions.

The first economics of science may have been Adam Smith's idealistic, but sadly untrue, discussion in the *Theory of Moral Sentiments* (1759, p. 124) of Newton having been motivated purely by curiosity rather than a desire to achieve fame and fortune. If Smith's account was the beginning of a positive economics of science, then Charles Babbage's argument (1830) for the reform

of British scientific institutions may count as one of the earliest instances of a mainly normative economics of science. Also usually mentioned as an early founder of the normative economics of science is the American philosopher C.S. Peirce (1839–1914), who advocated the application of economic tools of analysis to decide which scientific projects to adopt (see Wible, 1998).

The 'modern' economics of science grew out of three main topics The first topic addressed how the advance of science contributed to the advance of technology, and hence productivity and growth. The second topic, which overlaps with concerns in the history and philosophy of science, addressed how science advances. A third topic is the empirical data collection and econometric analysis of the supply, demand, compensation and productivity of scientists.

Diamond (1996) and Stephan (1996) both provide broad surveys of the economics of science, with Diamond perhaps devoting more space to interdisciplinary and policy issues. Special attention has been paid to the contributions to the economics of science of three of the field's founders: Mansfield, Griliches and Stigler (Diamond, 2003; 2004; 2005). Several of the more important papers in the economics of science through 1998 are included in the two-volume collection edited by Stephan and Audretsch (2000).

In what follows, we first consider the literature that most makes the case for why the economics of science should be a priority for our attention: the literature on science as a contributor to economic productivity and growth. We proceed to briefly summarize some economic discussion of some of the 'deep' issues in the economics of science, which sometimes overlap with issues in philosophy of science, such as the objectives of scientists (truth, fame, fortune?), and the constraints that are most relevant to their choices about which projects to pursue and which theories to adopt. Next we look at some of the studies that have attempted to model and measure a variety of aspects of the market for science and scientists. Finally, we give examples of some of the studies in normative economics of science that argue for changes in funding or for institutional reform.

Impact of science on technology, productivity, and growth

The importance of technology as a driver of economic growth and well-being has been appreciated since Adam Smith's *Wealth of Nations* (1776), and emphasized most notably by Schumpeter (1942). If technology is the main driver of economic growth, the next question is: what is the main driver of technology?

Rosenberg (1982) made the credible point that most economists, for most of the history of the profession, had viewed the process by which new technologies are developed and adopted as a 'black box'. In the years since, partly led by Rosenberg himself, economists have increasingly attempted to say more about what goes on inside the box, especially concerning the role of science in advancing technology.

Several economic historians have examined the role of science in the advance of technology and economic growth over the broad sweep of history. Mokyr (1990; 2002), Rosenberg (Mowery and Rosenberg, 1989; Rosenberg and Birdzell, 1986), and Landes (1998), broadly agree that the advance of science is a necessary but not sufficient condition for substantial and rapid advance in technology and economic growth.

Nelson (1959) catalogued many examples of how science had contributed to the advance of technology. More recently, Rosenberg (Mowery and Rosenberg, 1989, pp. 11–14) has claimed that the distinction between science and technology is often hard to make, providing several examples of how scientific advances have resulted from the pursuit of 'practical' results. Although most economists adopt Nelson's view that mainly new science enables the advance of new technology, it is not hard to find examples where the advance of science was enabled by new instruments provided by advanced technology (Ackermann, 1985).

Mokyr (1990), in his broad economic history of the advance of technology over the ages, generally finds the advance to be slow and fitful until the industrial revolution. Until the mid-1800s, the relationship between science and technology was loose (Mansfield, 1968; Mokyr, 1990, pp. 167–70). Those who advanced science and technology shared an attitude of optimism about the prospects of understanding and controlling nature (Landes, 1969). But, beginning in the mid-19th century, and especially with the development of commercial laboratories toward the end of the 19th century, the relationship between science and technology became closer, with advances in science more often and more clearly being a necessary condition for technological advances.

Beginning with Nelson's taxonomic paper (1959), evidence for this latter claim has been provided by economists in a variety of forms. Griliches's main contribution, in a pair of papers (1957; 1958), was to measure the return to scientific research on hybrid corn and to measure and explain varying rates of adoption (see Diamond, 2004). Surveys of research managers by Nelson (1986) and by Mansfield (1991; 1992) provided evidence that science is sometimes important for technical change, although the importance varies considerably with the industry and with the sub-field of science.

The development by Romer (1986; 1990) and others of the 'new growth theory' attracted further attention to science as a driver of technology, because such models include a more prominent role for knowledge ('recipes') than earlier models. Such models implied the possibility of increasing returns to investments in knowledge if various spillover effects were large enough. Stimulated partly by such models and partly by independent research by economists such as Griliches (see Diamond, 2004), considerable empirical work has been undertaken measuring the spillover effects of scientific knowledge; for example, Jaffe (1989), Adams (1990) and Jaffe, Trajtenberg and Henderson (1993).

Deep understanding of science

Stigler and Becker (1977) have argued that everyone has the same utility function. This contrasts with Adam Smith's suggestion in his *Theory of Moral Sentiments* (1759, p. 124) that scientists, or at least the best scientists, were more purely motivated by curiosity. Gordon Tullock (1966, pp. 34–6), and Kenneth Arrow (2004) suggest that scientists have a range of motives, from those who fit the Smith ideal to those motivated by fame and fortune (Levy, 1988). However, in their research on the behaviour of scientists, economists usually follow Stigler and Becker in assuming that scientists mainly value income and prestige. Scientists valuing both income and prestige might explain Stern's finding (2004) that industrial scientists are willing to give up some income in exchange for greater ability to publish their results.

The process of theory choice among scientists has been explained using economic tools (Diamond, 1988b; Hull, 1988; Goldman and Shaked, 1991). The explanations have been criticized by Hands (1997). Brock and Durlauf (1999) build on the work of Kitcher (1993) in their construction of a dynamic model of scientists' adoption of new theories. A key assumption of their set-up is that one source of a scientist's utility is the 'conformity' of a scientist's views with those of other scientists. The model allows the possibility that science is progressive, even if social considerations have some weight in scientists' utility functions, partially answering many of those in the social studies of science field who believe that the admission of social considerations undermines the special cognitive status of science.

Stigler authored several papers that presented evidence on hypotheses about the determinants of successful science. He provided evidence and arguments on questions such as: does a scientist's biography help us understand the scientist's contributions (1976), and how efficiently is error weeded out in science (1978)? Stigler generally framed his studies as seeking just to understand, not to reform, although the results sometimes stimulated thoughts of reform in others. A fuller account of Stigler's contributions to the economics of science appears in Diamond (2005).

The market for science and scientists

Michael Polanyi (1962) optimistically portrayed science as an efficient marketplace of ideas. Much research in the economics of science in the last few decades shares Polanyi's research programme of explaining the behaviour of scientists and scientific institutions on the basis of rational optimization within an efficient marketplace.

Many studies in the economics of science fall within the domain of labour economics, and assume that scientists are rational maximizers of income, and sometimes of prestige. Within labour economics, a significant theoretical and empirical literature has developed that examines the economics of higher education. Since many scientists, and especially most scientists who are credited with major scientific discoveries, have been associated with universities, this literature is relevant to the economics of science, even when the examples or data are not drawn explicitly from science.

Some of the earliest economics of science studies collected data to analyse the supply of and demand for scientists. Early examples of this genre were Blank and Stigler (1957) and Arrow and Capron (1959). Richard Freeman's 'cobweb' model (1975) of the labour market for physicists had an unstable equilibrium because students' occupational choices in the model are based on systematically biased forecasts of the future demand for physicists. Siow later (1984) showed that professional labour markets are better characterized by assuming the students' forecasts are based on rational expectations.

Human capital theory and measurement have been used to estimate earnings functions for scientists, including as independent variables measures of productivity, such as articles produced, citations received, and teaching evaluations. Many of the studies also include one or more variables intended to measure hypotheses of discrimination, such as gender and race variables (for example, McDowell, 1982). Yet other studies include variables intended to measure what has recently been called 'social capital' and has previously been identified with Robert K. Merton's (1968) 'Matthew effect' (the rich get richer) or with 'old-boy' networks.

One goal of many of the earnings regression studies has been to learn how much of the variation in academic salaries can be explained on the basis of variation in measures of academic productivity. Lovell's (1973) paper was one of the first to include measures of research productivity in an academic earnings regression. Early studies tended to focus on number of articles published as the measure of research productivity. A pair of papers by Stigler and Friedland (1975; 1979) helped establish the credibility of citations as a measure of academic productivity. One of the first to include citations in an earnings function as a measure of productivity may have been Laband (1986). Subsequent studies using the citation measure include Hamermesh, Johnson and Weisbrod (1982), Diamond (1986b), Sauer (1988), and Kenny and Studley (1995). A review of the literature on bibliometric measures of productivity can be found in Diamond (2000).

The simplest models of compensation assume that workers are paid the value of their current productive output (their 'marginal revenue product'). To account for observed anomalies with this hypothesis, especially in professional labour markets, a literature has developed

supposing that there are long-term implicit labour contracts. For example, universities may provide scientists with insurance for variability of research output by paying the scientist more than the value of their output in low-output years and less than the value of output in high-output years. If scientists are uncertain at the beginning of their careers whether they will become high- or low-productivity scientists, they may also demand insurance against the possibility of their being low-productivity scientists (S. Freeman, 1977). This might explain the observed greater variability in measures of scientists' productivity than in measures of their salaries.

An alternative explanation is to make use of a compensating differentials argument (Frank, 1984). The assumption is that scientists receive utility from being paid more than other scientists. So the top scientist would be paid less than the value of her productivity, because she is receiving a compensating differential in the form of being at the top of the pecking order.

Implicit contract models have also been developed to try to explain important scientific labour market practices, such as academic tenure. For example, Carmichael (1988) argues that tenure is an institutional device to reduce the costs to incumbent faculty of correctly identifying promising new faculty, while Waldman (1990) claims that faculty value tenure because it serves as a signal to outside institutions of the faculty members' quality, and hence increases outside higher salary offers. Siow (1998) claims that specializing is risky, since subfields of specialization may suddenly become obsolete; so without tenure as a form of insurance, faculty would under-specialize.

Implicit contract models are often clever and sometimes plausible. But as clever, plausible models multiply that explain the same stylized facts (for example, the existence of academic tenure), the credibility of the exercise may suffer. It may also be worth mentioning that, ceteris paribus, economists will be more popular with their peers if they create models justifying tenure, and other academic institutions, than if they create models showing tenure is inefficient.

Other mainly empirical studies have examined the mobility of academic scientists between university positions (Rees, 1993), and the mobility of industrial scientists between technical and managerial jobs (Biddle and Roberts, 1994). Another extensive, mainly empirical, literature makes use of standard theory on the optimal allocation of time over the life cycle to motivate analysis of scientific productivity over the life cycle. Life-cycle investment models (for example, Diamond, 1984) often suggest that it makes sense to invest in human capital early in the life cycle. These models often imply concave age–productivity profiles. Empirical evidence confirms this generalization (Diamond, 1986a; Stephan and Levin, 1992), but with very different peak productivity ages for different fields of science. Age-related differences in the rate of acceptance of new theories have also been examined (Hull, Tessner and Diamond, 1978; Diamond, 1980; 1988a; 1988b; Levin, Stephan and Walker, 1995).

Efficiency and reform of scientific institutions

In the previous section we discussed research that for the most part argues that the institutions for rewarding and allocating resources to scientists can be explained as efficient aspects of a well-functioning market of ideas. Bartley (1990) has argued that while a 'marketplace of ideas' is an appropriate goal and standard, it is an inaccurate description of the current institutions of science. In one of his headings (1990, p. ix), he claims that our current institutions for academic science are 'where consumers do not buy, producers do not sell, and owners do not control'.

In an early study, Stigler (1965) provided evidence that, when economics underwent a transition from a science done by amateurs to one done by professionals, the discipline became much more theoretical and mathematical, and much less applied and policy-oriented. Stigler's friend and colleague, Milton Friedman, argued (1981) that the funding of the National Science Foundation (NSF) had had a similar effect, and argued further that this effect had slowed the advance of knowledge. The debate was renewed 13 years later (Friedman, 1994; Griliches, 1994). Edward Lazear (1997) has developed a model implying more modest advice for the NSF: the foundation should give fewer but larger awards.

Other economists have studied the funding of science. Arrow (1962), Johnson (1972) and others have argued that science is a public good that will be under-provided by the private sector.

Dasgupta and David (1994) accept the public goods argument of the 'old' economics of science of Arrow (1962), but want to add to it findings of some sociologists on the secrecy that sometimes results from the competition for priority, in order to develop a 'new' economics of science. Their 'new' economics of science argues for greater government funding of science, accompanied by increased incentives for scientists to share their findings sooner with other scientists and with those seeking to apply the findings to new technologies.

Romer (2001) argues that, if roughly half a million more scientists and engineers were supplied and appropriately deployed, the US economy could sustain a half a per cent greater rate of growth in GDP. He suggests that major changes would be required in academic institutions and government policies to achieve this goal, but he believes the resulting implications for the economy of success would be 'staggering' (2001, p. 227).

Kealey (1996) and Martino (1992) explicitly dispute the traditional public goods argument for government support of science on the grounds that private industry often has both the incentives and the ability to do substantial high-quality scientific research. Hanson (1995) supports an alternative private form of science funding when he argues that greater scientific innovation would occur if more of the funding for science came from a betting market, where those who predict accurately the ultimate outcome of currently debated scientific questions receive more resources.

Although not opposing the science-as-public-good theory as strongly as Kealey and Martino, Rosenberg (1990) has emphasized the incentives that private firms have to invest in science. He examined firms that hired Ph.D. scientists and that allowed the scientists considerable leeway in the allocation of their time and in the publication of their results. He argues that this was in the firm's interest because of the value of such scientists as a resource in keeping up with and explaining scientific advances relevant to the firm's product development efforts. Besides Rosenberg's paper, there is a considerable literature measuring returns to firm investment in R&D. Some of these studies might be considered part of the economics of science to the extent that they study 'basic research', a label that is sometimes used interchangeably with 'science'. Examples of this literature are surveyed in Audretsch et al. (2002).

Several scholars have attempted to measure the extent to which public expenditures on science add to the total funding of science, or the extent to which they simply crowd out private funding on science. Diamond (1999), for example, using highly aggregated time-series measures of government and industry investment in science, found no evidence of crowding out. An evaluative survey of this literature has been published by David, Hall and Toole (2000).

Some economists have explained the behaviour of some scientists and the structure of some scientific institutions in terms of rent-seeking behaviour. Rent seeking is a zero-sum process in which an agent invests resources to obtain an uncompensated transfer from another agent (Tullock, 1967). In one example, McKenzie (1979) suggested that there is a fixed fund for department salaries, and that department members can increase their share of the fund either by being more productive themselves or by sabotaging the productivity of others, perhaps, for example, by the calling of unnecessary meetings. Other rent-seeking accounts of academic institutions have been provided by Brennan and Tollison (1980) and Grubel and Boland (1986).

We mentioned in the previous section economic models of academic tenure that attempt to explain the institution as an efficient response to features of the academic labour market. Others (for example, Rogge and Goodrich, 1973) have followed Alchian (1959) in presenting a basically rent-seeking account of tenure as an inefficient institution that exists because it is in the interests of a sufficiently powerful special interest group.

In an account highly complementary to the rent-seeking hypothesis, Goolsbee (1998) has studied data on federal funding of science and found that it largely results in windfalls for scientists. Goolsbee's results call into question the extent to which federal funding actually increases the amount of science produced.

ARTHUR M. DIAMOND, JR.

See also **Arrow, Kenneth Joseph; crowding out; Griliches, Zvi; implicit contracts; public goods; rent seeking; Stigler, George Joseph; technology.**

Bibliography

Ackermann, R. 1985. *Data, Instruments and Theory*. Princeton, NJ: Princeton University Press.

Adams, J. 1990. Fundamental stocks of knowledge and productivity growth. *Journal of Political Economy* 98, 673–702.

Alchian, A. 1959. Private property and the relative cost of tenure. In *The Public Stake in Union Power*, ed. P. Bradley. Charlottesville, VA: University Press of Virginia.

Arrow, K. 1962. Economic welfare and the allocation of resources for inventions. In *The Rate and Direction of Inventive Activity: Economic and Social Factors*, ed. R. Nelson. Princeton: Princeton University Press.

Arrow, K. 2004. Foreword. In *Reflections of Eminent Economists*, ed. M. Szenberg, and L. Ramrattan. Cheltenham, UK: Edward Elgar.

Arrow, K. and Capron, W. 1959. Dynamic shortages and price rises: the engineer–scientist case. *Quarterly Journal of Economics* 73, 292–308.

Audretsch, D., Bozeman, B., Combs, K., Feldman, M., Link, A., Siegel, D., Stephan, P., Tassey, G. and Wessner, C. 2002. The economics of science and technology. *Journal of Technology Transfer* 27, 155–203.

Babbage, C. 1830. *Reflections on the Decline of Science in England, and on Some of Its Causes*. London: B. Fellowes.

Bartley, W., III 1990. *Unfathomed Knowledge, Unmeasured Wealth: On Universities and the Wealth of Nations*. LaSalle, IL: Open Court.

Biddle, J. and Roberts, K. 1994. Private sector scientists and engineers and the transition to management. *Journal of Human Resources* 29, 82–107.

Blank, D. and Stigler, G. 1957. *The Demand and Supply of Scientific Personnel*. New York: National Bureau of Economic Research.

Brennan, H. and Tollison, R. 1980. Rent seeking in academia. In *Toward a Theory of the Rent-Seeking Society*, ed. J. Buchanan, R. Tollison and G. Tullock. College Station, TX: Texas A & M University Press.

Brock, W. and Durlauf, S. 1999. A formal model of theory choice in science. *Economic Theory* 14, 113–30.

Carmichael, H. 1988. Incentives in academics: why is there tenure? *Journal of Political Economy* 96, 453–72.

Dasgupta, P. and David, P.A. 1994. Toward a new economics of science. *Research Policy* 23, 487–521.

David, P., Hall, B. and Toole, A. 2000. Is public R&D a complement or substitute for private R&D? A review of the econometric evidence. *Research Policy* 29, 497–529.

Diamond, A., Jr. 1980. Age and the acceptance of cliometrics. *Journal of Economic History* 40, 838–41.

Diamond, A., Jr. 1984. An economic model of the life-cycle research productivity of scientists. *Scientometrics* 6, 189–96.

Diamond, A., Jr. 1986a. The life-cycle research productivity of mathematicians and scientists. *Journal of Gerontology* 41, 520–25.

Diamond, A., Jr. 1986b. What is a citation worth? *Journal of Human Resources* 21, 200–15.

Diamond, A., Jr. 1988a. The polywater episode and the appraisal of theories. In *Scrutinizing Science: Empirical Studies of Scientific Change*, ed. A. Donovan, L. Laudan and R. Laudan. Dordrecht: Kluwer Academic Publishers.

Diamond, A., Jr. 1988b. Science as a rational enterprise. *Theory and Decision* 24, 147–67.

Diamond, A., Jr. 1996. The economics of science. *Knowledge and Policy* 9(2, 3), 6–49.

Diamond, A., Jr. 1999. Does federal funding 'crowd in' private funding of science? *Contemporary Economic Policy* 17, 423–31.

Diamond, A., Jr. 2000. The complementarity of scientometrics and economics. In *The Web of Knowledge: A Festschrift in Honor of Eugene Garfield*, ed. B. Cronin and H. Adkins. Medford, NJ: Information Today, Inc.

Diamond, A., Jr. 2003. Edwin Mansfield's contributions to the economics of technology. *Research Policy* 32, 1607–17.

Diamond, A., Jr. 2004. Zvi Griliches's contributions to the economics of technology and growth. *Economics of Innovation and New Technology* 13, 365–97.

Diamond, A., Jr. 2005. Measurement, incentives, and constraints in Stigler's economics of science. *European Journal of the History of Economic Thought* 12, 637–63.

Frank, R. 1984. Are workers paid their marginal products? *American Economic Review* 74, 549–71.

Freeman, R. 1975. Supply and salary adjustments to the changing science manpower market: Physics, 1948–1973. *American Economic Review* 65, 27–39.

Freeman, S. 1977. Wage trends as performance displays productive potential: a model and application to academic early retirement. *Bell Journal of Economics* 8, 419–43.

Friedman, M. 1981. An open letter on grants. *Newsweek*, 18 May, 99.

Friedman, M. 1994. National science foundation grants for economics: correspondence. *Journal of Economic Perspectives* 8(1), 199–200.

Goldman, A. and Shaked, M. 1991. An economic model of scientific activity and truth acquisition. *Philosophical Studies* 63, 31–55.

Goolsbee, A. 1998. Does government R&D policy mainly benefit scientists and engineers? *American Economic Review* 88, 298–302.

Griliches, Z. 1957. Hybrid corn: an exploration in the economics of technological change. *Econometrica* 25, 501–22.

Griliches, Z. 1958. Research cost and social returns: hybrid corn and related innovations. *Journal of Political Economy* 66, 419–31.

Griliches, Z. 1994. National science foundation grants for economics: response. *Journal of Economic Perspectives* 8(1), 203–5.

Grubel, H. and Boland, L. 1986. On the efficient use of mathematics in economics: some theory, facts and results of an opinion survey. *Kyklos* 39, 419–42.

Hamermesh, D., Johnson, G. and Weisbrod, B. 1982. Scholarship, citations and salaries: economic rewards in economics. *Southern Economic Journal* 49, 472–81.

Hands, D. 1997. Caveat emptor: economics and contemporary philosophy of science. *Philosophy of Science* 64(4), S107–S116.

Hanson, R. 1995. Could gambling save science? Encouraging an honest consensus. *Social Epistemology* 9, 3–33.

Hull, D. 1988. *Science as a Process: An Evolutionary Account of the Social and Conceptual Development of Science.* Chicago: University of Chicago Press.

Hull, D., Tessner, P. and Diamond, A., Jr. 1978. Planck's principle: do younger scientists accept new scientific ideas with greater alacrity than older scientists?. *Science* 202, 717–23.

Jaffe, A. 1989. Real effects of academic research. *American Economic Review* 79, 957–69.

Jaffe, A., Trajtenberg, M. and Henderson, R. 1993. Geographic localization of knowledge spillovers as evidenced by patent citations. *Quarterly Journal of Economics* 108, 577–98.

Johnson, H. 1972. Some economic aspects of science. *Minerva* 10, 10–18.

Kealey, T. 1996. *The Economic Laws of Scientific Research.* New York: St Martin's Press.

Kenny, L. and Studley, R. 1995. Economists' salaries and lifetime productivity. *Southern Economic Journal* 62, 382–93.

Kitcher, P. 1993. *The Advancement of Science.* Oxford: Oxford University Press.

Laband, D. 1986. Article popularity. *Economic Inquiry* 24, 173–80.

Landes, D. 1969. *The Unbound Prometheus: Technological Change 1750 to the Present.* Cambridge: Cambridge University Press.

Landes, D. 1998. *The Wealth and Poverty of Nations.* New York: W.W. Norton.

Lazear, E. 1997. Incentives in basic research. *Journal of Labor Economics* 15, S167–S197.

Levin, S., Stephan, P. and Walker, M. 1995. Planck's principle revisited: a note. *Social Studies of Science* 25, 275–83.

Levy, D. 1988. The market for fame and fortune. *History of Political Economy* 20, 615–25.

Lovell, M. 1973. The production of economic literature: an interpretation. *Journal of Economic Literature* 11, 27–55.

Mansfield, E. 1968. *The Economics of Technological Change.* New York: W.W. Norton.

Mansfield, E. 1991. Academic research and industrial innovation. *Research Policy* 20, 1–12.

Mansfield, E. 1992. Academic research and industrial innovation: a further note. *Research Policy* 21, 295–96.

Martino, J. 1992. *Science Funding.* New Brunswick, NJ: Transaction Publishers.

McDowell, J. 1982. Obsolescence of knowledge and career publication profiles: some evidence of differences among fields in costs of interrupted careers. *American Economic Review* 72, 752–68.

McKenzie, R. 1979. The economic basis of departmental discord in academe. *Social Science Quarterly* 59, 653–64.

Merton, R. 1968. The Matthew effect in science. *Science* 159(3810), 56–63.

Mokyr, J. 1990. *The Lever of Riches: Technological Creativity and Economic Progress.* Oxford: Oxford University Press.

Mokyr, J. 2002. *The Gifts of Athena: Historical Origins of the Knowledge Economy.* Princeton, NJ: Princeton University Press.

Mowery, D. and Rosenberg, N. 1989. *Technology and the Pursuit of Economic Growth.* Cambridge: Cambridge University Press.

Nelson, R. 1959. The simple economics of basic scientific research. *Journal of Political Economy* 67, 297–306.

Nelson, R. 1986. Institutions supporting technical advance in industry. *American Economic Review, Papers and Proceedings* 76, 186–9.

Polanyi, M. 1962. The republic of science: its political and economic theory. *Minerva* 1, 54–73.

Rees, A. 1993. The salaries of Ph.D.'s in academe and elsewhere. *Journal of Economic Perspectives* 7(1), 151–8.

Rogge, B. and Goodrich, P. 1973. Education in a free society. In *Education in a Free Society*, ed. A. Burleigh. Indianapolis: Liberty Fund.

Romer, P. 1986. Increasing returns and long-run growth. *Journal of Political Economy* 94, 1002–37.

Romer, P. 1990. Endogenous technological change. *Journal of Political Economy* 98(5), S71–102.

Romer, P. 2001. Should the government subsidize supply or demand in the market for scientists and engineers? In *Innovation Policy and the Economy*, vol. 1, ed. A. Jaffe, J. Lerner and S. Stern. Cambridge, MA: MIT Press.

Rosenberg, N. 1982. *Inside the Black Box: Technology and Economics.* Cambridge: Cambridge University Press.

Rosenberg, N. 1990. Why do firms do basic research (with their own money)? *Research Policy* 19, 165–74.

Rosenberg, N. and Birdzell, L., Jr. 1986. *How the West Grew Rich.* New York: Basic Books.

Sauer, R. 1988. Estimates of the returns to quality and coauthorship in economic academia. *Journal of Political Economy* 96, 855–66.

Schumpeter, J. 1942. *Capitalism, Socialism, and Democracy.* New York: Harper and Row.

Siow, A. 1984. Occupational choice under uncertainty. *Econometrica* 52, 631–45.

Siow, A. 1998. Tenure and other unusual personnel practices in academia. *Journal of Law, Economics and Organization* 14, 152–73.

Smith, A. 1759. *The Theory of Moral Sentiments.* Oxford: Clarendon Press, 1976.

Smith, A. 1776. *An Inquiry Into the Nature and Causes of the Wealth of Nations*. Indianapolis: Liberty Press, 1976.

Stephan, P. 1996. The economics of science. *Journal of Economic Literature* 34, 1199–235.

Stephan, P. and Audretsch, D., eds. 2000. *The Economics of Science and Innovation*. 2 vols. Cheltenham: Edward Elgar.

Stephan, P. and Levin, S. 1992. *Striking the Mother Lode in Science: The Importance of Age, Place, and Time*. Oxford: Oxford University Press.

Stern, S. 2004. Do scientists pay to be scientists? *Management Science* 50, 835–53.

Stigler, G. 1965. Statistical studies in the history of economic thought. In *Essays in the History of Economics*, Chicago: The University of Chicago Press.

Stigler, G. 1976. The scientific uses of scientific biography, with special reference to J.S. Mill. In *Papers of the Centenary Conference: James and John Stuart Mill*, ed. J. Robson and M. Laine. Toronto: University of Toronto Press.

Stigler, G. 1978. The literature of economics: the case of the kinked oligopoly demand curve. *Economic Inquiry* 16(2), 185–204.

Stigler, G. and Becker, G. 1977. De gustibus non est disputandum. *American Economic Review* 67(2), 76–90.

Stigler, G. and Friedland, C. 1975. The citation practices of doctorates in economics. *Journal of Political Economy* 83, 477–507.

Stigler, G. and Friedland, C. 1979. The pattern of citation practices in economics. *History of Political Economy* 11, 1–20.

Tullock, G. 1966. *The Organization of Inquiry*. Durham, NC: Duke University Press.

Tullock, G. 1967. The welfare costs of tariffs, monopolies, and theft. *Western Economic Journal* 5, 224–32.

Waldman, M. 1990. Up-or-out contracts: a signaling perspective. *Journal of Labor Economics* 8, 230–50.

Wible, J. 1998. *The Economics of Science: Methodology and Epistemology as If Economics Really Mattered*. London: Routledge.

scientific realism and ontology

Economists customarily talk about the 'realism' of economic models and of their assumptions and make descriptive and prescriptive judgements about them: this model has more realism in it than that model, the realism of assumptions does not matter, and so on. This is not the way philosophers mostly use the term 'realism'; thus there is a major terminological discontinuity between the two disciplines. The following remarks organize and critically elaborate some of the philosophical usages of the term and show some of the ways in which they relate to economists' concerns. In the philosophy of science, scientific realism is the mainstream position – or rather a heterogeneous collection of positions – that includes

ideas about the nature of scientific theories and how they are related to the real world and about the goals and achievements of scientific inquiry. However, most of what philosophers have contributed around these ideas is not designed to deal with the peculiarities of economics, so some important adjustments are needed to make scientific realism an interesting position for economists.

Economists and their critics, as well as the consumers of economic research, are often intrigued by two kinds of issue that connect with realism: Is this model about something real, something that really exists (or instead about some imaginary fiction or social construct perhaps)? Is this model true (partly true, approximately true) about something real (instead of just being useful or convenient or persuasive)? These are the general questions discussed here – not by answering them, but by way of clarifying the conceptual prerequisites of trying to answer them, and in particular by way of examining what it is to be a realist about them. Misuses of 'realism' are abundant enough to warrant this exercise. The focus will be on ontology, and on how realism relates to it, so we will be mainly talking about ontological realism.

What exists?

Ontological realism is a philosophical thesis that deals with two questions: What exists? What is existence? Consider the first question. Slightly different ways of putting it include:

What is there in the world?
What is the furniture of the world?
What is the world made of?
What is its structure?
What is the case?
What is the way the world is?
What is the way the world works?

To such questions one expects answers of the form, 'X exists' or 'Z is the case' or 'WWW is the way the world works'. But isn't it the task of science to provide answers of this form to such questions? What is the role of distinct ontological reflection? The quick response is in two parts: the answers provided by ordinary scientific practice ('normal science') are often implicit and only presupposed, while ontological reflection seeks to make them explicit; and the explicit answers that daily scientific practice supplies are mostly more specific and concrete that those offered by focused ontological scrutiny. There is an overlap and continuity between 'normal' science and ontological reflection regarding the kinds of answers they seek to offer to those questions. Consistency between the two is desirable, even though occasional (and often fruitful) conflicts arise.

Scientists may claim that there are neural brain states and processes that cause human behaviour, or that there are beliefs and wants that causally produce such behaviour. Scientists may make claims about there being causal connections between certain aggregate variables, such as the money stock and inflation, or between inflation and unemployment. Such claims have ontological presuppositions, but practising scientists seldom engage themselves in explicating and elucidating them. Are there mental states such as beliefs and wants in addition to brain states? In referring to preferences and expectations in their explanations, are economists presupposing that they exist? Are there macroeconomic aggregates in addition to individuals and their attributes? Do economists commit themselves to their existence when invoking them in explaining economic phenomena? And what is causation all about? Is there just one kind of causation out there, or are there perhaps different kinds of causal connections and other causal facts, depending, among other things, on whether we talk about connections between brains, minds, individual actions, institutions, or economic aggregates?

Explicitly or implicitly, economists hold presuppositional beliefs about such ontological matters. They give, or imply, answers to the sorts of questions above, provoked by developments and debates in economics, such as those around neuroeconomics and microfoundations. What does this have to do with realism? Not much, as such. Various alternative answers to those questions are compatible with ontological realism. One can be a realist about brains or minds, or both, and one can be a realist about human individuals or social institutions, or both. There is no single privileged 'realism' that would determine the contents of our ontology – there is no such thing as 'the realist ontology'. It indicates a misunderstanding of 'realism' to suggest that there is one general 'realist ontology' that has specific contents concerning issues such as law and causation, or the relations between human individuals and social structures. Realism as such does not imply specific answers to questions 'What exists?' or 'What is the case?' and the like.

Ontological realism is always Realism-about-X, so we get a variety of realisms depending on the value of X. So one can choose to be, or not to be, a realist about electrons, molecules, cells, minds, human individuals, and social organizations, as well as more generally about relations and causal processes, natures and necessities, numbers and sets, parts and wholes, material states and moral values. One can coherently be a realist about some such things while being an anti-realist about others; one can be a realist about molecules without being a realist about morality, or vice versa. And, again, choosing to be a realist does not as such determine what you are a realist about. You choose Realism-about-X as a package, so it is not your choice of realism that implies your choice of some specific X.

What is existence?

It also works the other way around. Your choice of X as the kind of thing that you think exists does not yet make

you a realist. In order to qualify as a realist it is not enough to hold that the hardware of the brain exists or that social structures are causally powerful, or what have you. The general form of the thesis of ontological realism is 'X exists' or 'Z is the case' or some such. It is not enough to be specific about the X and the Z; one also needs to say more about the meanings of 'exists' and 'is the case'. No answer to 'what exists' – no list of things that are claimed to exist or to be the case – is alone sufficient for ontological realism. We also need to carefully answer the question, 'What is existence?'

Irreducible existence

Two requirements are needed in order to come up with an appropriate idea of existence. First, realism requires that existence claims be understood literally. Thus, one is a realist about X if one takes X to exist, and by this one means, literally, that X exists rather than ultimately meaning that something else exists. One is a realist about ions and institutions just in case one takes ions and institutions to exist, period. On the other hand, one is not a realist about X if one holds that X exists in the sense that Y exists by virtue of the fact that X is ultimately Y. In such an anti-realist manoeuvre one substitutes a reductionist reading for a literal reading of existence claims. This table at which I write exists in the sense that a certain bundle of atoms exists – tables are ultimately nothing but bundles of atoms. Or to say that the table exists is another way of saying that a certain collection of sense data exists – middle-sized material objects are collections of sense data, literally speaking. Minds and mental states such as expectations exist, but their existence is a matter of human brains and neurons existing. A business corporation exists, but to say so is just a convenient way of speaking: it exists more precisely in the sense that an organized collective of human individuals exists – social collectives are sets of individual people after all. This way of using existence claims amounts to an ontologically reductionist strategy that allows for existence claims that are not supposed to be taken literally: an appropriately literal reading is a post-reduction reading. In contrast, an ontological realist about X reads 'X exists' literally and insists on its irreducibility to 'Y exists'.

Consider causation. In order to qualify as a realist about causation it is not sufficient to hold the view that there are causal facts in the world. It is also required that causation be viewed as an irreducible notion, one that cannot be analysed fully in terms of other, non-causal notions. For example, an ontological empiricist about causation might say that causation exists, and then add that it exists in the sense that empirical regularities or constant conjunctions of observable events exist – simply because this is what causation is, in the end. Here causation will be reduced to, or analysed into, non-causal facts. In philosophers' jargon, 'causal realism' is a name for a position that denies such a reduction and instead requires claims about causation to be taken at face value. For a realist, causation is, literally, a matter of causing, producing, bringing about, propagating, enabling, inhibiting and so on. A causal realist analyses causation in causal terms, whereas a causal anti-realist analyses causation in non-causal terms and thereby analyses causation away.

So conceived, causal realism is an ontological position. This is compatible with epistemologies of causation that employ non-causal terms. David Hume seems to have been a causal realist – causation in the world is a matter of causes producing their effects – while at the same time his epistemological scepticism suggested that we only have epistemic access to constant conjunctions (see, for example, Strawson, 1989). Or consider Granger-causality, defined partly in terms of predictability of effects. Predictability is an epistemic notion, so a causal realist should not include it in his concept of causation in the world. But Granger-causality does not require such inclusion, so as such it does not rule out the possibility of causal realism. The general idea behind it, that of providing an 'operational definition' of causation, does not require Humean scepticism either: Granger-causality can be thought of as a (fallible) element in the economists' imperfect epistemic endeavours to discover irreducibly causal relations in the world.

Independent existence

The second requirement for a position to qualify as realist is that it must hold that whatever exists has to exist in some suitable independent way. On this issue the peculiar features of society and the social sciences impose special requirements on the appropriate conception of realism. Ontological realism about some X – in its answer to the question, 'what is existence?' – claims that X exists independently of some Z. Further versions of this idea depend on how 'independent' and 'Z' are specified.

The usual manner of defining ontological realism is in terms of mind-independence: some X exists independently of the human mind. But even though this idea would seem to apply to many sorts of things, it is problematic in the social sciences. While it may be plausible to claim that galaxies and quarks exist mind-independently, this does not seem a good idea in the case of, say, people's preferences and expectations or a society's institutions and organizations. The conventional existence test that tends to appeal to most people's intuitions is to imagine a situation without minds: take away human minds, will galaxies survive? Yes, they will. Take away human minds, will social institutions survive? No way. Social objects are mind-dependent. This does not require, of course, that whatever there is in society is intentionally created by people, as intended results of their purposeful action. Much of what there is in society is being produced as unintended consequences of the actions of people with minds.

Not only are social objects mind-dependent in general, they are also representation-dependent. They are dependent on representations in general for the trivial reason that people are animals active in producing and using representations as an integral feature of human action. Moreover, and more strongly, many – but surely not all – social things are dependent on representations *about them*: the existence of X is dependent on representations of X – on being represented *as* X. Many contracts between economic actors only exist when linguistically represented. Take away the representation, and the contract ceases to exist. Likewise, the euro, the currency in use in most members of the European Union, is partly constituted by representations, among them certain treaties of the European Council. In contrast, the existence of DNA molecules is not dependent on any particular representations of them: supposing they exist at all, they existed both before and after Watson and Crick's double helix representation, and independently of it.

Social reality is variously shaped by, and dependent on, people's beliefs and expectations, goals and wants, plans and impulses, emotions and reasonings, speech acts such as promises and persuasions, agreements and disagreements, collaborations and rivalries, meanings and their interpretations, customs and conventions, and so on. None of this is mind-independent in certain obvious senses.

Science-independence

Thus ontological realism about society and social sciences requires some other idea of independent existence. Here it is advisable to start being more precise about the issue: we are concerned with the ontology of *scientific* realism. This suggests that we need some notion of science-independence (an idea that has been largely ignored by philosophers debating scientific realism). A scientific realist takes galaxies and quarks to be independent of astronomy and physics, of the theories and explanations and procedures in these sciences. But consider social facts. Much of what there is in society is increasingly dependent on science, natural science included. This dependence works through various powerful influences of science on the world views and technologies prevalent in society. It is through these channels that people's beliefs and aspirations, and society's norms and institutions, are shaped by science. These constitute the stuff of which societies are made, so social matters are not science-independent in an obvious sense.

What about the economy and the science of economics, how do they relate to one another? We can repeat some of the things we said above, and we can add an idea of a stronger dependence. The Lucas critique provides one expression to the old and obvious idea that the economy is not fully economics-independent. Many economic facts seem to be dependent on theories and procedures in the science of economics. Among the more striking cases, just think of the dependence of certain practices of finance in the real world on certain theories of finance – such as the Black–Scholes–Merton formula for option pricing (see MacKenzie, 2006). Economic theories and research results shape people's beliefs and world views, and policy advice based on economic theories and research results shapes economic policies, and these in turn shape the economy. Economic theories, people's beliefs and economic facts are furthermore often connected through mechanisms of self-fulfilment and self-defeat. The economy is thus variously dependent on economics. One might suspect that the notion of science-independence does not serve scientific realism well in the case of economics.

In order to see why all this does not undermine scientific realism about economics and the economy, we need to pay attention to the notion of independence itself in 'exists independently of'. Of the many kinds of independence, it will be sufficient to distinguish two general categories (Mäki, 2002).

Consider again the idea above about society being science-dependent just because science shapes the world views and technologies in society, and these in turn shape people's behaviour and social institutions. What kind of claim is this? I take shaping (or whatever similar terms one may want to use) to be a matter of causal influence. Thus, a causal claim is being made: social matters are causally dependent on science. This is a relief to a realist about society and social sciences since such causal dependence can easily be accommodated by scientific realism.

Scientific realism does not need to deny or despair over the causal dependence of the economy on economics either. The causal influences of economics on the economy travel through obvious channels. Policymakers and others (such as students, investors, entrepreneurs, workers, consumers) learn, directly or indirectly, about economic theories, explanations and predictions, and are inspired by them enough to modify their beliefs. These modified beliefs make a difference for the behaviour of these actors, and this has consequences for the economy. The connections, and hence the dependencies, are causal.

The same holds for cases in which the connections from theory to the world change the latter so as to make the theory more closely correspond to the world – often characterized as the 'performativity' of theory. If it is the case that students of economics act more than other students in accordance with the conventional behavioural assumptions of economic theory, this might be because their image of appropriate human behaviour and thereby their actual behaviour is influenced by what they are taught in class. If certain practices in real world finance are in line with the Black–Scholes–Merton formula for option pricing, this may be because the theoretical formula has managed to travel from academic research to economic practice so as to shape the latter. In such cases, the connections are causal.

Rather than constituting a threat to realism, such causal connections between economic theory and the economy pose a constructive challenge to realism.

The economy is organized into causal structures and processes, and investigated by economists. Adding economic theory to the picture in some cases means adding another causal set of connections to our image of social reality – science is not done outside of society, it is very much part of social reality, after all. This invites further scientific inquiry into the detailed features of these causal connections. Good social science illuminates the roles that ideas play in the social world. Good social science should also illuminate the roles that scientific ideas play in the social world.

A realist is comfortable with causal connections in general, and so is not disturbed by causal connections from theories to the world either. To see why, it is important to be precise about the nature of this dependence. Indeed, literally speaking, there is no causation flowing from theories to the world here. Economic theories do not shape the economy. People do. People, in various roles as economic actors, are inspired, directly or indirectly, by the contents of economic theories, this shapes their beliefs, their beliefs shape their motives, and those motives drive them in action that shapes the economy. Phrases such as 'self-fulfilling' and 'self-defeating' theories and predictions are therefore not to be taken literally: theories do not fulfil or defeat themselves (Mäki, 2002; 2005a). If they had the capacity to do so, scientific realism would be defeated. Likewise, the idea of 'performativity' of economic theories is not to be understood literally: theories or their utterances alone do not serve performative functions in the sense of themselves giving rise to what those theories are about.

Social construction

So scientific realism is comfortable with a lot of science-dependence of society. What sort of science-dependence does scientific realism rule out? As an analogy, consider any of the numerous economist jokes (such as 'No reality, please. We're economists.' Blaug, 2002, p. 36). There is no way the joke could be funny without being perceived as such. The joke's being funny is constituted by some people – not necessarily economists – finding it funny. Being funny and being perceived as funny are inseparable, not because the funniness of the joke (causally) makes people laugh, but because their laughter (conceptually) makes it funny. The connection is not causal – as in funny jokes causing people to find them funny – but rather conceptual or constitutive – being perceived as funny is part of the definition of 'funny,' or partly constitutes what it is to be funny.

Economic facts in social reality would have the same status as funny jokes if they were facts only if perceived to be so by economists. The economy being in state S and some of its developments being governed by mechanism M are so just because economists hold theories and other beliefs that say so. This is what the realist will deny. A realist will grant there is a formal contract between two economic actors only provided these actors believe it is there, and they – perhaps together with third parties – have agreed to represent the contract. But a scientific realist about the economics of contracts takes those contracts to be science-independent in the sense of not being created by acts of economic theorizing. The contracts exist and have the properties they have independently of being believed or claimed to be so by economists. While facts about contracts are constituted by the beliefs and representations by contracting parties, they are not constituted by acts of scientific theorizing about them. Creating a theory of X is not a matter of creating X. Creating a theory of (the causation of) business cycles does not, just by that token, create (that very causation of) business cycles. Saying so does not make it so.

This idea of non-causal science-independence suggests how to identify some of the opponents of scientific realism. They are those who do not subscribe to the non-causal science-independence of matters of fact in the world, including social and economic facts. Of contemporary (both academic and broader cultural) relevance are forms of social constructivism (see Hacking, 1999). This is a doctrine that comes in an obscure variety of forms that are characteristically not distinguished from one another. Modest versions claim that beliefs about the world are outcomes of social construction (including education, negotiation, persuasion, testimony, imitation, indoctrination, herd behaviour, group pressure, cognitive path dependency, and other social processes). More radical versions claim that the world itself is socially constructed. More weakly, one may argue that beliefs and myths about gender, race and schizophrenia are socially constructed – or that scientific theories about atomic structure and evolution are so constructed. More strongly, one may argue that gender and race, as well as quarks and evolution, are products of social processes of cognitive construction.

Realism is comfortable with weaker forms of social constructivism. Generally held beliefs about the world are socially constructed – simply because cognition is essentially social activity, with numerous cognitive agents interacting with one another under changing institutional constraints. Moreover, parts of the world itself are socially constructed in obvious ways. Indeed, society itself is socially constructed: social objects, properties, states, and processes are outcomes of social processes. Since this is a general fact about social reality, ontological realism about society grants it.

So what does realism rule out when it comes to society and social sciences? It is incompatible with positions that deny the non-causal science-independence of social matters. Claiming that social facts are socially constructed by social scientists just by way of constructing concepts and theories about society is beyond the scientific realist. A special version of this is the idea that social facts – certain ordinary kinds of facts about society outside of academia – emerge as an outcome of rhetorical

persuasion by economists of their audiences: certain facts about, say, the causation of inflation emerge as soon as masses of economists and others are persuaded by monetarist economists about the causal link between money and inflation. What precisely is the problem with this thought that bothers a realist? First consider what realism accepts. What the realist will accept is that the beliefs held by economists about the causation of inflation are socially constructed by ordinary social processes of academic (and possibly non-academic) persuasion. The realist also accepts that the beliefs held by people other than academic economists – such as politicians, journalists, union leaders and consumers – are outcomes of persuasion. Finally, realism accepts that such beliefs may motivate behaviour that has consequences for the causal connection between money and inflation. Such a connection between beliefs and behaviour is part of the causal structure giving rise to actual inflation rates. So what does realism deny? It denies the radical social constructivist idea that, when an economist puts forth a theory about the causation of inflation and manages to get it generally accepted, this alone will give rise to the causal facts described by the theory. This is to deny that agreement without action is sufficient for making or changing the world.

For a realist, theory construction does not amount to world construction. Even theory dissemination alone is not sufficient for world construction (where 'world' refers to what the theory is about). Of course, it is a social fact that masses of people are persuaded about whatever, and this indeed is an outcome of rhetorical practices. What a realist should deny is that what those people are persuaded about is itself constructed just by those rhetorical practices. Again, it is a different idea that people so persuaded may take action and this action may make a difference for some social facts. This requires that the beliefs adopted in consequence of persuasion become causally efficacious in relation to actual behaviour. But this gives us what I called causal construction, and it is compatible with scientific realism.

Model worlds

Much of social constructivism may be inspired by the observation that scientists do not have direct access to the details and complexities of the real world. Scientists do not directly investigate the messy concreteness and complexity of the world, but rather engage in various manoeuvres that seek to 'prepare' that world for closer inspection. Economists will easily recognize this as part of their model building practices. Faced with the immense complexity of the world, economists are forced to simplify their images of it by way of using various procedures of omission, idealization and abstraction. The models they build and employ isolate small slices of the world for detailed scrutiny while leaving most details out of the picture. Such procedures of isolation are among those manoeuvres of preparation, and they result in models that appear to describe simple imaginary model worlds rather than the real world.

One may then say that models and model worlds are constructed rather than, say, discovered, and that they are socially constructed just because scientific work is intrinsically a social activity. The realist grants this much, but then goes on to insist that, even if economic models of the social world are socially constructed by economists, the social world itself is not socially constructed by the modelling practices of economists.

At this point it will be helpful to have an account of models that is able to deal with this set of issues (Mäki, 2005b). I take models to be what can be called substitute systems or surrogate systems that serve as direct subjects of inquiry. They are substitute systems in that they are examined instead of the real systems that they replace and stand for. The real systems 'out there' may be too complex (such as social systems), too far away in time or space (such as the origin of the species) to be capable of being directly investigated, or it might be unethical to examine the real system directly (such as using human subjects for medical experimentation). Much of scientists' activity is a matter of building and manipulating such substitute systems in order to learn about their properties and behaviour. A realist about models and the real world would add that the properties of models are directly examined by scientists in order to indirectly learn about the properties of the real world.

A radical social constructivist will not buy this image. In a radical constructivist framework, there is no distinction between model systems and real systems, nor is there room for an idea of indirect acquisition of information about the latter by way of examining the former. There are only socially constructed model systems, and it is a matter of power and persuasion which models will be taken to provide the facts. This is one way of illuminating the strong social constructivist ontology in a nutshell.

Truth

This takes us to the issue of truth. Economists appear to have great difficulties with applying the concept of truth. They characteristically believe that, since models do not reproduce the whole complexity of their subject matter, they are necessarily false, or that, because models involve false assumptions, models themselves are false. In order to see why these beliefs about the falsehood of models are false, we should take a realist perspective to the ontology of truth. This will help further clarify the very concept of realism. Here the focus is on what truth is (rather than what is true) (see Mäki, 2004).

Truth is a property of truth bearers. A truth bearer is true just in case there is a truth maker that makes the truth bearer true. The sentence, or statement, 'The cat is on a mat' is a truth bearer that is made true by its truth maker in the world, namely, the cat being on a mat. 'Education is a major cause of economic growth' is a truth bearer that is made true by its truth maker, namely, education being causally responsible for at least 15 per cent of the growth in the output of an economy (supposing this is how we have defined 'major'). Now all models of the connection between education and growth omit lots of things in the world and make false idealizing assumptions about others. But whether one takes a model to convey true information about the world depends on what one takes the relevant truth bearers to be. In the case that the intended truth bearer is, say, 'Education is responsible for about one quarter of economic growth (in some spatio-temporally specified location)', then its truth makers include those situations in the real world in which education is causally responsible for 23–27 per cent of growth (supposing this interval is what we meant by 'about one quarter'). In the case that the intended truth bearer is something different, namely, a description of the causal mechanism by way of which education contributes to growth, then its truth maker would be that mechanism in the real world (such as some skill mechanism or a signalling mechanism, depending on what we seek to describe with the model).

No model offers the whole truth of its domain, and no model, if spelled out in full, is devoid of false elements (false if considered as potential truth bearers). Understanding that models can be true, or can be used for making true claims, requires understanding the intended relevant truth bearers. These intuitions provide the basis for resisting the popular idea that models cannot possibly be true.

The realist ontology of truth so conceived is rather straightforward. Realism requires that the relevant truth makers be real, that they exist in a suitably independent manner. Truth makers are, after all, just those things the existence of which ontological realism deals with. Truth makers must exist without being non-causally science-dependent. Creating a model of a signalling mechanism does not create that signalling mechanism: the mechanism exists (if it does) independently of any economic models about it – it exists even though there were no models about it.

What does realism say about truth making, the way in which truth makers make truth bearers true? Again, a suitable independence requirement is imposed. The way in which truth makers make true statements or models true must be independent of our ways of coming to hold beliefs about their truth. In particular, epistemic matters do not enter truth by way of, say, making truth making dependent on evidential considerations or our capacity to recognize truth. So we do not make true models true by coming to recognize them as true in virtue of having collected lots of supportive evidence, or having become persuaded by the most prestigious economists, or the like. While we make truth bearers, we do not make them true. Facts do.

Possible existence and possible truth

Here is an important further qualification of what it takes to be a realist about the world and models about it. Above, I have said that it is not sufficient to hold that some entity X exists and that theory T is true about X in order to count as a realist about X and T (it is not enough because one has to add further restrictions in terms of irreducibility, non-causal science-independence, and independence of epistemic attitudes, for example). It is now time to weaken this by saying that holding those ideas is not necessary either. It is enough for a realist to hold the view that X might exist and T might be true about X, that it is possible for X to be the case and T to be true. Such a view will reveal a realist attitude: there is a fact of the matter concerning whether or not X exists and whether or not T is true. It is an attitude that will give real existence and objective truth a chance, but one that at the same time is prepared for concluding that X does not exist or T is not true, after all. This is an advisable attitude for epistemically insecure and self-critical scientific practice based on consistent fallibilism.

Ontological realism about X is primarily an account of what it is for X to exist (if it does). To this, one may add the weaker claim that X might exist, or the stronger claim that X does exist. Likewise with truth. Realism about truth is primarily an account of what it is for a truth bearer T to be true. One may then add the weaker claim that T might be true, or the stronger claim that T indeed is true. Naturally each such claim requires different sorts of supportive argument.

The bite of ontology

Does ontology make a difference for scientific practice and its evaluation? That ontology may have some bite is evident even in the case of the weak version of realism put in terms of possible existence. Some economists criticize the model of perfect competition on the grounds that perfect competition is not even conceivable: its perfection has been taken so far that such a competitive market has become an impossibility (Richardson, 1960). If this is correct, then ontological realism would be an inappropriate attitude regarding perfect competition as depicted by the model.

In general, economists hold, or entertain, ontological convictions, explicitly or implicitly, and these may have consequences regarding the preferred ways of theorizing and explaining. Whatever mismatch there may be between ontology and economic theory, sometimes it plays an important role in driving intellectual progress, sometimes not. When there is a clash between the ontological convictions of an economist and the apparent ontological implications of a theory held by this economist, there may be a healthy pressure to revise the theory so as to realign ontology and theory. Such a clash may, in suitable circumstances, serve as a driving tension that creates pressure to modify theories or invent new

theories such that the economist will be able to take the theory with an ontologically realist attitude. An economist may, deep down in his convictions, view certain facts – such as increasing returns, bounded rationality, and institutional structure – as causally powerful features of the economy. At the same time, the modelling conventions and techniques of the discipline may be such that the economist will be unable to incorporate such things in his models. There is a tension that needs to be resolved, and the resolution will be sought in the spirit of ontological realism.

Yet, in general, one's favourite ontology does not determine the contents of theory or model used, nor is it determined by these. The impact of ontology is rather a matter of constraint. In many cases even this much impact is too much. An economist may correctly say things that are quite appropriate in their respective contexts even though those things may clash with deeper ontological convictions. An economist may appropriately say 'this country is applying monetarist tenets in its economic policies' even though the deeper conviction may be in line with ontological individualism claiming that no such collective entities as countries exist – only individuals do. Or an economist may say that 'individuals have preferences' while a philosopher of mind or a neuroeconomist may endorse the ontological claim to the effect that 'preferences do not exist' simply because she believes nothing mental does: preferences really are just configurations of neural states. To claim that individuals have preferences as part of economic discourse does not contradict the eliminativist materialism endorsed by the philosopher – just as the claim 'there are 15 chairs in the seminar room' may be correct even though one's philosophical ontology may imply that chairs do not exist, while bundles of atoms do. In such cases, there is no conflict between the two claims in each pair due to the idea that one of the claims, and only one of the claims, informs us, after all, how things are 'at bottom' or 'ultimately'.

A parallel idea is that ontology does not determine methodology, but rather serves as a constraint on it. Thus ontological individualism (the belief that only individuals exist) does not imply a commitment to methodological individualism – the obligation to spell out the microfoundations of economic theories and explanations. The economist may legitimately decide that the things he or she wishes to know can be established using aggregative models not grounded in individual behaviour. For example, it may be that certain facts depend on distributions across individuals and not on individuals themselves. Similarly, the conventional procedure of theoretical isolation (isolating certain phenomena that are to be analysed) by building models of small and simple hypothetical worlds does not imply the ontological conviction that the actual systems in the real world are equally simple and isolated from other aspects of reality. What economists do believe is that, in order to study complex systems, simple models are needed, because the real world is too complicated and because the limited questions asked about it require focusing attention on a limited set of mechanisms (while ontological realism may be used to require that these mechanisms exist also in the complex real world).

Methodology is under-determined by ontology. It depends on ontology but it also depends on other things such as the cognitive interests of the inquirers and their limited cognitive capacities. But even this much is enough for concluding that ontology matters.

USKALI MÄKI

See also **assumptions controversy; causality in economics and econometrics; Lucas critique; methodological individualism; methodology of economics; microfoundations; models; philosophy and economics; rhetoric of economics.**

Bibliography

Blaug, M. 2002. Ugly currents in modern economics. In *Fact and Fiction in Economics. Realism, Models, and Social Construction*, ed. U. Mäki. Cambridge: Cambridge University Press.

Hacking, I. 1999. *The Social Construction of What?* Cambridge, MA: Harvard University Press.

MacKenzie, D. 2006. *An Engine, Not a Camera: How Financial Models Shape Markets*. Cambridge, MA: MIT Press.

Mäki, U. 2002. Some non-reasons for non-realism about economics. In *Fact and Fiction in Economics. Realism, Models, and Social Construction*, ed. U. Mäki. Cambridge: Cambridge University Press.

Mäki, U. 2004. Some truths about truth for economists, their critics and clients. In *Economic Policy-Making under Uncertainty: The Role of Truth and Accountability in Policy Advice*, ed. P. Mooslechner, H. Schuberth and M. Schurtz. Cheltenham: Edward Elgar.

Mäki, U. 2005a. Reglobalising realism by going local, or (how) should our formulations of scientific realism be informed about the sciences. *Erkenntnis* 63, 231–51.

Mäki, U. 2005b. Models are experiments, experiments are models. *Journal of Economic Methodology* 12, 303–15.

Richardson, G. 1960. *Information and Investment*. Oxford: Clarendon.

Strawson, G. 1989. *The Secret Connexion: Causation, Realism, and David Hume*. Oxford: Clarendon.

Scitovsky, Tibor (1910–2002)

Tibor Scitovsky was born in Budapest. He received a degree in law from the University of Budapest and a degree in economics from the London School of Economics. He migrated to the United States in 1939 and served in the US Army during the Second World War. He taught at Stanford, the University of California at Berkeley, Yale, Harvard, and the London School of Economics. He also worked at the OECD from 1966 to 1968.

His writings are brilliant, original, succinct, lucid and full of subtlety. They always enlighten and move the debate forward. He made fundamental and lasting contributions to a large number of subjects: welfare economics, international trade, economic development and microeconomics. One can discern a unifying theme to these varied contributions; it is to indicate ways in which neoclassical equilibrium analysis fails to capture important aspects of economic reality and, therefore, leads to misleading policy implications from efficiency, stability or welfare points of view. He stresses dynamics, and interdependence among the utilities of consumers and decision-outcomes of producers, as the major sources of divergence of social optima from market equilibria of perfectly competitive economies.

His work in welfare economics points to the impossibility of purging policy analysis from value judgements concerning the optimality of the initial and final distributions of income. This is true whether the initial and final situations are efficient or inefficient, in static terms, or whether or not the possibility for compensation of losers by winners exists. (Compensation restores the original distribution and therefore implies the judgement that the original distribution was optimal.) Rather than abandon the possibilities for policy recommendations entirely, economists should make explicit the value judgements that underlie their policy advice.

His contributions to economic development make the capturing of external economies the cornerstone of development strategy. His classic paper 'Two Concepts of External Economies' (1954) distinguishes between technological and pecuniary external economies. In developing countries, the existence of pecuniary externalities argues for the planning of coordinated investment decisions since market prices provide imperfect signals when those decisions are interdependent. He also argues for economic integration and export-led growth in economies too small to secure the advantages of both economies of large-scale production and pecuniary economies of balanced growth.

In trade theory, his 'Reconsideration of the Theory of Tariffs' (1942) points to the parallelism between tariffs and the monopolist's markup (or monopsonist's markdown) for exploiting his trading partners and argues that market forces could never approximate free trade, which would have to be imposed and enforced by international agreement or by a dominant large power against each nation's selfish preferences.

His most controversial but most original book, *The Joyless Economy* (1976), tries to introduce into consumption theory the psychologist's classification of satisfactions into comfort, stimulation and pleasure, with emphasis on the psychological trade-off between them. The second part of the book explores the implications of the consumer's ignorance of that psychological trade-off on the rationality of his choice behavior, using American lifestyles as an illustration.

Reading Scitovsky's writings, one is made painfully aware of how much has been lost by the modern trend to mathematize and computerize. By comparison, modern economics appears mechanical and myopic, lacking in subtlety and sweep. Many of the themes raised in his writings appear as fresh today as they were when they were first written. And there are many points still worth following up decades after they were first made. His later work on the integration of microeconomics with macroeconomics, contained in his analysis of the real side of inflation, which builds on the price-maker price-taker distinction first introduced by him in his book on *Welfare and Competition* (1952), is a case in point.

IRMA ADELMAN

Selected works

1942. Reconsideration of the theory of tariffs. *Review of Economic Studies* 9(2), 89–110.

1952. *Welfare and Competition: The Economics of a Fully Employed Society*. London: Allen & Unwin.

1954. Two concepts of external economies. *Journal of Political Economy* 62, 70–82.

1958. *Economic Theory and Western European Integration*. London: Allen & Unwin.

1964. *Papers on Welfare and Growth*. London: Allen & Unwin.

1969. *Money and the Balance of Payments*. Chicago: Rand McNally.

1970. (With I.M.D. Little and M.F. Scott.) *Industry and Trade in Some Developing Countries*. Oxford: Oxford University Press.

1976. *The Joyless Economy*. Oxford: Oxford University Press.

1986. *Human Desire and Economic Satisfaction*. Brighton: Wheatsheaf; New York: New York University Press.

Scott, William Robert (1868–1940)

W.R. Scott ('the chief' as A.L. Macfie used to describe him) was born in Armagh on 31 August 1868, the eldest son of Charles and Margaret Scott. He was educated at Canon Stewart's Preparatory School and then St Columba's College, Rathfarnham. Scott went to Trinity College Dublin in 1885, graduating BA in 1889 and MA two years later. 1891 saw his first major publication, *An Introduction to Cudworth's Treatise Concerning Eternal and Immutable Morality*. Scott's philosophical interests were also marked by his *Simple History of Ancient Philosophy* (1894).

In 1896 Scott took up the post of assistant to the professor of moral philosophy in the University of St. Andrews, and three years later became the University's first lecturer in political economy. Scott was responsible for planning the teaching of economics until 1915 when he was translated to the Adam Smith Chair of

Political Economy in Glasgow, in succession to William Smart. Scott died in Glasgow on 3 April 1940, after a brief illness.

Scott was extremely active as examiner, teacher, researcher, and adviser to government with a marked interest in contemporary, as well as historical, issues. There are three identifiable strands to his work.

The first was through his interest in contemporary economic and social problems, encouraged by his chairmanship for many years of his family's firm of millers in Tyrone. It led to an active involvement in public affairs, in days when economists were less consulted than subsequently. Apart from his membership and chairmanship of committees, national and regional, especially in the 1920s, Scott's name became attached to several departmental and other reports, even when he was not the sole or main author.

He was appointed by the Secretary for Scotland to examine home industries in the North. The ensuing report was published in 1914. He subsequently addressed the *Economic Problems of Peace after War* (1917; 1918), and later worked (with James Cunnison) on *Industries of the Clyde Valley during the War* (1924a) as part of a Carnegie Series.

A second stream is represented by his pioneering work in economic history. Scott published an edition of *The Records of a Scottish Cloth Manufactory at New Mills, Haddingtonshire 1681–1703* (1905). This was followed by one of his best known studies, the definitive, three volume, *Constitution and Finance of English, Scottish and Irish Joint Stock Companies to 1720* (1910–12). He was president of the Economic History Society from 1928 until his death.

A third contribution is represented by Scott's work as an historian of ideas. His book on *Francis Hutcheson* (1900) is the work of a philosopher well versed in economics and is still a classic. Scott also dramatically advanced contemporary knowledge of Hutcheson's most famous pupil with the publication of *Adam Smith as Student and Professor* (1937) which featured his discovery of important Smith papers in the Buccleuch MSS. Scott's ability to find records was impressive. Some criticism of his handling of them is possible but many of the difficulties which impeded his work have been reduced for later scholars by the discovery and cataloguing of many relevant records of economic and intellectual history which he helped to bring about.

At the time of his death Scott was working on a bibliography, which was published under the auspices of the British Academy, and edited by his successor in Glasgow, Alec Lawrence Macfie.

As his obituarist noted in the *Glasgow Herald* (4 April 1940),

> If one today were to seek the model of his famous and beloved predecessor (Adam Smith) it would be impossible to find a closer re-incarnation than William

Robert Scott. He had the same temper, controlled, wide-eyed impartiality of mind allied with an absorbing fire of enthusiasm for reasoned practice and disinterested policy. This is the tribute he would most have appreciated, and it is one which all his friends and students would endorse as the most appropriate.

The British Academy published an appreciation by Sir John Clapham, *William Robert Scott 1868–1940* (London, 1940).

ANDREW SKINNER

Selected works

1891. *An Introduction to Cudworth's Treatise concerning Eternal and Immutable Morality, with Life of Cudworth and a few Critical Notes.* London: Longmans.

1893. *Geography of Ptolemy Elucidated*, ed. T.G. Rylands. Dublin: Ponsonby & Weldrick.

1894. *A Simple History of Ancient Philosophy.* London: E. Stock.

1900. *Frances Hutcheson: His Life, Teaching, and Position in the History of Philosophy.* Cambridge: Cambridge University Press.

1903. *Free Trade in Relation to the future of Britain and the Colonies: A Plea for an Imperial Policy: a lecture.* St Andrews: W.C. Henderson & Son.

1903–4. The fiscal policy of Scotland before the Union. *Scottish Historical Review* 1, 173–90.

1903–5. Scottish industrial undertakings before the Union. *Scottish Historical Review* 1(1903–4), 407–15; 2 (1905), 53–60, 287–97, 406–11; 3 (1906), 71–6.

1905. *Records of a Scottish Cloth Manufactory at New Mills, Haddingtonshire, 1681–1703.* Edinburgh: Scottish History Society.

1907. Scottish industry before the Union; Scottish industry after the Union. In *The Union of 1707*, ed. P. Hume Brown. Glasgow.

1910–12. *The Constitution and Finance of English, Scottish and Irish Joint Stock Companies to 1720.* 3 vols. Cambridge: Cambridge University Press.

1911a. Is increasing utility possible? In *Celebration of the Five-Hundredth Anniversary of the Foundation of St. Andrews.* Glasgow.

1911b. *Scottish Economic Literature: A List of Authorities.* Glasgow: Scottish Exhibition of History, Art and Industry.

1913. The trade of Orkney at the end of the eighteenth century. *Scottish Historical Review* 10, 360–8.

1917. Mercantile shipping in the Napoleonic wars. *Scottish Historical Review* 14, 272–5.

1917–18. *Economic Problems of Peace after War.* The W. Stanley Jevons Lectures at University College London. Cambridge: Cambridge University Press.

1920. William Cunningham, 1849–1919. *Proceedings of the British Academy* 9, 465–74.

1921. ed. Adam Smith. *An Inquiry into the Nature and Causes of the Wealth of Nations.* 2 vols. London: Bohn's Standard Library.

1923. Books as links of empire: *The Wealth of Nations. Empire Review.*

1924a. (With J. Cunnison) *The Industries of the Clyde Valley during the War.* Oxford: Clarendon Press.

1924b. Adam Smith. *Proceedings of the British Academy.* London.

1924c. Adam Smith and the City of Glasgow. *Proceedings of the Royal Philosophical Society of Glasgow* 52.

1926a. Alfred Marshall, 1842–1924. *Proceedings of the British Academy.* London.

1926b. Scottish land settlement. In *Scotland During the War, in Economic and Social History of the World War*, ed. R.D. Jones et al. London: Humphrey Milford.

1928. Joseph Shield Nicholson, 1850–1927. *Proceedings of the British Academy.* London.

1929–30. Economic resiliency. *Economic History Review* 2, 291–9.

1931. The manuscript of Adam Smith's Glasgow lectures. *Economic History Review* 3, 91–2.

1934. Adam Smith and the Glasgow merchants. *Economic Journal* 44, 506–8.

1935a. Adam Smith at Downing Street, 1766–67. *Economic History Review* 6, 79–89.

1935b. The manuscript of an early draft of part of *The Wealth of Nations. Economic Journal* 45, 427–38.

1936. New light on Adam Smith. *Economic Journal* 46, 401–11.

1937. *Adam Smith as Student and Professor, with Unpublished Documents.* Glasgow: Jackson.

1938a. A manuscript criticism of the *Wealth of Nations* in 1776 by Hugh Blair. *Economic History* 4, 47–53.

1938b. *Adam Smith: An Oration.* Glasgow: Jackson.

1939. *Greek Influence on Adam Smith.* Athens: Pyros Press.

1940. Studies relating to Adam Smith during the last fifty years. In *Proceedings of the British Academy*, ed. A.L. Macfie. London.

Scrope, George Poulett (1797–1876)

George Poulett Scrope was one of the most prolific contributors to the literature of political economy in the mid-19th century. He was also one of the more able critics of features of the Ricardian economic orthodoxy which came to dominate that literature. Scrope is at his best as an economist in the series of articles he contributed when chief economics reviewer for the *Quarterly Review* (1831–3). His *Principles of Political Economy* (1833c) is disappointing by comparison.

After education at Harrow, Scrope entered Pembroke College, Oxford in 1815. During the following year he moved to St John's College, Cambridge where he graduated in 1821. He married the heiress Emma Phipps Scrope, altering his surname (which had been 'Thomson') to that of his wife, and establishing himself as one of the leading gentlemen of the county of

Wiltshire. Scrope was appointed a magistrate of the county in 1823.

Research in geology was an early and enduring involvement. Scrope's distinguished work in this field led to his election to the Geological Society (1824). Two years later, he became a Fellow of the Royal Society, and he continued to publish papers on geological subjects until shortly before his death.

Scrope entered Parliament, and he remained Member for Stroud from 1833 to 1867. During his first 20 years in the Commons he spoke frequently and was a member of numerous parliamentary committees. His contributions in the legislature distinguish him as a 'man with a philosophic and inquiring mind, trying to explain the upheavals in economic relations, and also to guide policy on behalf of interests that went beyond his own personal gain' (Fetter, 1980). In debate, Scrope found himself in alliance at times with doctrinaire Ricardians such as Joseph Hume. Scrope supported repeal of the Corn Laws and the Navigation Acts. He was also in favour of parliamentary reform. On a variety of issues, however, he was decidedly 'unorthodox'. Those issues included: the nature of the currency and the structure of the banking system; the public funding of education; the maintenance of outdoor relief for the unemployed; and closer government regulation of working conditions in factories. The problems of Ireland were among his special concerns, and he was a leading advocate of the extension of poor law provisions to that country.

The interventions of Scrope within Parliament were supplemented by his publication of numerous pamphlets on current policy issues. In addition, he contributed periodical articles in which he assailed the economic theories of the followers of David Ricardo, and the population doctrine of Thomas Robert Malthus. Scrope rejects a theory of value based on cost of production and he argues for recognition of a relationship between value-in-use and value-in-exchange. He finds the reasoning of Richard Jones on rent much superior to that of Ricardo. Noting the absence of any account of a basis for profit in Ricardian theory, he constructs an abstinence theory of interest and allies this to a risk–effort theory of profit which incorporates the concept of quasi-rent. Scrope is also concerned with the possibilities of over-saving and of a general glut of markets. In this latter respect, his ideas are akin to those of Malthus. However, Scrope is a persistent and incisive critic of Malthusian population theory and the policy implications which his contemporaries drew from it.

BARRY GORDON

Selected works

Scrope's bibliography (Sturges, 1984) contains 175 items on economic, geological, and local history topics. The following is a list of some of the more notable publications dealing with economic issues.

1830a. *The Currency Question Freed from Mystery*. London.
1830b. *On Credit – Currency and its Superiority to Coin*. London.
1831a. The political economists. *Quarterly Review* 44(January), 1–51.
1831b. Malthus and Sadler – population and emigration. *Quarterly Review* 45(April), 97–245.
1832. The rights of industry and the banking system. *Quarterly Review* 47(July), 407–55.
1833a. Martineau's monthly novels. *Quarterly Review* 49(April), 136–51.
1833b. *An Examination of the Bank Charter Question*. London: John Murray.
1833c. *Principles of Political Economy*. London: Longman. Reprinted, New York: Kelley, 1969. Italian trans., 1855. 2nd edn, published as *Political Economy for Plain People*, 1873.
1848. Irish clearances and improvement of waste lands. *Westminster and Foreign Quarterly Review* 50(October), 163–87.

Bibliography

Fetter, F.W. 1958. The economic articles in the *Quarterly Review* and their authors, 1809–1852. *Journal of Political Economy* 66, Pt I, 47–64; Pt II, 154–70.
Fetter, F.W. 1980. *The Economist in Parliament: 1780–1868*. Durham, NC: Duke University Press.
Gordon, B. 1965. Say's Law, effective demand and the contemporary British periodicals, 1820–1850. *Economica* 32, 438–46.
Gordon, B. 1969. Criticism of Ricardian views on value and distribution in the British periodicals, 1820–1850. *History of Political Economy* 1, 390–87.
Opie, R. 1928. A neglected English economist: George Poulett Scrope. *Quarterly Journal of Economics* 44, November, 101–37.
Rashid, S. 1981. Political economy and geology in the early nineteenth century: similarities and contrasts. *History of Political Economy* 13, 726–44.
Rudwick, M.S. 1974. Poulett Scrope on the volcanoes of Auvergne: Lyellian time and political economy. *British Journal for the History of Science* 7, 205–42.
Sturges, P. 1984. *A Bibliography of George Poulett Scrope: Geologist, Economist and Local Historian*. Kress Library Publication No. 24. Boston, MA: Harvard Business School.

search-and-matching models of monetary exchange

In this article we review a class of equilibrium search (matching) models that can be used to study the trading process, and in particular to develop a formal theory of *money as a medium of exchange*. Developing such a theory is one of the longest-standing issues in economics, but it met with at best limited success prior to the development of search-based models, which provide a natural framework in which to formalize venerable stories about money helping to facilitate exchange.

These stories, going back to Smith, Jevons, Menger, Wicksell, and others – many of which are reprinted in Starr (1990) – concern a *double coincidence of wants problem* in bilateral exchange, as discussed below. Overlapping generations models (for example, Wallace, 1980) provide an alternative approach. Ostroy and Starr (1990) survey earlier attempts to develop microfoundations for monetary theory, including Jones (1976), which is similar in spirit if not detail to modern search models. There is not the space here to discuss the pros and cons of the various approaches, but it seems fair to say search and matching models now dominate the area.

Background

Diamond (1982) introduced a framework that, although it cannot be used directly, can be extended naturally to build microfoundations for monetary economics. In his model, a [0,1] continuum of infinitely lived agents interact in an economy where activity takes place in two distinct sectors: one for production and one for exchange. In the first sector, agents encounter potential production opportunities randomly over time according to a Poisson process with arrival rate α. Each opportunity yields a unit of output at cost $c \geq 0$, where c is random with CDF $F(c)$. Since c is observed before a production decision is made, given an opportunity, there is a reservation cost k such that agents produce if $c \leq k$. For now, these goods are indivisible, and agents can store at most one at a time.

All goods yield utility of consumption $u > 0$, except by assumption agents cannot consume their own output; hence they must trade. Traders with goods meet bilaterally in the exchange sector according to a Poisson process with arrival rate γ. Upon meeting they trade, consume, and return to production. Since all goods are the same, and indivisible, every meeting yields trade, and every trade is a one-for-one swap. Generally, $\gamma = \gamma(N)$ depends on the measure of agents in the exchange sector N. This is based on a matching technology that gives the number of agents who meet a partner per unit time as $m(N)$, with $m'(N) > 0$, implying $\gamma(N) = m(N)/N$ for all $N > 0$.

Let V_0 and V_1 be the value functions for producers and traders. The flow Bellman equation for a producer is

$$rV_0 = \alpha E \max\{V_1 - V_0 - c, 0\}$$
$$= \alpha \int_0^k (k - c)dF(c),$$

where $k = V_1 - V_0$. Similarly, for a trader

$$rV_1 = \gamma(N)(u + V_0 - V_1) = \gamma(N)(u - k).$$

(We focus on steady states; for dynamics, see for example Diamond and Fudenberg, 1989.)

In words, the flow value rV_0 equals the arrival rate of opportunities times the expected option value of switching from production to exchange, while rV_1 equals the arrival rate of meetings times the gain from trading and switching back. Combining these equations,

$$rk = \gamma(N)(u - k) - \alpha \int_0^k (k - c)dF(c).$$

Given N this has a unique solution for k. Given k, in steady state, the flow of agents from production to exchange must equal the flow back,

$$(1 - N)\alpha F(k) = m(N).$$

An equilibrium is a pair (N, k) satisfying these last two equations. It is simple to derive results concerning existence, comparative statics, and so on. As Diamond emphasizes, under increasing returns in $m(N)$, if any non-degenerate equilibrium (one with production) exists then multiple such equilibria exist. Under constant returns, a unique non-degenerate equilibrium exists if parameters fall in a certain range – for example, u is not too low, r not too high, and so on. To complete our review of this basic model, notice that exchange is trivial, even though it is restricted to bilateral trade, because there is only one good (or, all goods are the same). To make money interesting we need to generalize this. (Diamond, 1984, took a short cut to getting money into the model with a cash-in-advance constraint. By changing the environment as we do below, we see this is not only uninteresting, it is unnecessary.)

To ease the presentation we first simplify the production process. Assume everyone is *always* in the exchange sector, and, instead of carrying goods around, they can produce whenever they meet someone, at deterministic cost $C \geq 0$. Now, following Kiyotaki and Wright (1991; 1993), assume goods come in varieties, say colours. Each agent produces a particular colour, but different agents like to consume different colours. The simplest specification assumes agents get $u = U > C$ from any good in some set, and $u = 0$ from other goods, and x is the probability output in the relevant set (that is, an agent wants what the other agent can produce) in any random meeting. Also, since agents can produce whenever they want, to simplify things we assume goods are non-storable.

When goods *are* storable, Kiyotaki and Wright (1989) determine endogenously which objects serve as media of exchange, potentially including commodity plus fiat money. That model illustrates the trade-off between fundamental properties like storability and equilibrium properties like acceptability. It has many implications – for example, there can be multiple equilibria with different monies, objects with bad fundamental properties may end up as money, and so on. Generalizations and applications of the model include Marimon, McGrattan and Sargent (1990); Aiyagari and Wallace (1991; 1992);

Kehoe, Kiyotaki and Wright (1993); Wright (1995); and Duffy and Ochs (1999). Here, by making goods nonstorable, we focus on determining how an economy operates when there is a single candidate medium of exchange, namely, fiat money.

On the assumption that exchange requires mutual agreement, which occurs when I want to consume your good and you want to consume mine, trade now occurs only in a meeting with probability x^2, at least if the event that I want your good is independent of the event that you want mine (see below). This captures nicely the famous *double coincidence problem* with direct barter: trade requires meeting someone who produces something you like – which would be a coincidence – and also likes what you produce – a double coincidence. Payoffs are given by $rV_B = \gamma x^2(U - C)$, where the subscript on V_B stands for 'barter'. If x is small, which is the case if there is a lot of specialization, double coincidence meetings are rare and V_B is very low.

But is it really necessarily the case that trade occurs iff both parties want to consume what the other produces? Following ideas in Kocherlakota (1998), suppose agents get together at the start of time and discuss when to trade. Clearly, they agree that whenever *either* agent wants what the other produces he should get it, since this maximizes *ex ante* welfare

$$rV_C = \gamma[x^2(U - C) + x(1 - x)U \\ - (1 - x)xC] = \gamma x(U - C),$$

where the subscript on V_C stands for 'cooperation' (or perhaps 'commitment' or 'credit'). As long as $x < 1$, $V_C > V_B$. However, suppose agents cannot commit now to do things when they meet later that are not in their interest at that time. Then trades must satisfy incentive compatibility (IC), the binding condition being that you should be willing to produce in meetings where you do not consume.

If we can keep a public record of all agents' behaviour, we can try to use *trigger strategies* to support cooperative trade as follows: instruct agents to cooperate as long as everyone else does; but if anyone deviates, trigger to … 'something bad'. One can argue the worst trigger is 'autarky' which yields $V_A = 0$; or it may be 'barter' which yields V_B. In the former case the relevant IC condition is $-C + V_C \geq V_A$, which simplifies to $rC \leq \gamma x(U - C)$; in the latter case it is $-C + V_C \geq V_B$, which simplifies to $rC \leq \gamma x(1 - x)(U - C)$. In either case, if r is small we can sustain cooperative trade. Moreover, one can prove formally that money has no role here (Kocherlakota, 1998; Wallace, 2001); instead of proving this here we move to models where money does have a role.

First-generation search models of money

Suppose it is difficult to use triggers because, say, there is incomplete monitoring or record keeping, or, to take the simplest situation, suppose agents have no memory – they just cannot recall what happened in previous meetings! Kocherlakota (1998), Kocherlakota and Wallace (1998), Wallace (2001), Corbae, Temzilides and Wright (2003), Araujo (2004), and Aliprantis, Camera and Puzzello (2007) explore less extreme variations, but our assumption allows us to make the point more easily. In our 'memoryless' world, your continuation payoff V_M cannot depend on what you do in a given meeting. Hence, the relevant constraint to get you to produce without consuming is $-C + V_M \geq V_M$, which is violated for any $C > 0$. There is no scope for using threats to sustain cooperation without memory (generally, there is limited scope when memory is imperfect, which is what we need; we use the starkest case merely for tractability).

Suppose we introduce into this world a new object called fiat *money*. By definition, a medium of exchange is an object that is accepted in trade not to be used for consumption – or production – but to be traded again later for something else. When an object serving as a medium of exchange is for some people at some times a consumption good, it is called commodity money. When an object with no consumption value serves as a medium of exchange it is fiat money. At the start of time, we endow a fraction M of the population each with $m = 1$ unit of money and the rest with $m = 0$. Initially, those with $m = 0$ can produce; after this, agents can produce after they consume but not before. This implies agents with money cannot produce, and at any point in time everyone either has $m = 1$ or $m = 0$. Now, even without memory, agents have an option other than pure barter: offer money for goods. Let Π be the probability a random producer accepts such an offer, and let π be your best response.

If V_m and V_p are the value functions of agents with and without money,

$$rV_p = \gamma(1 - M)x^2(U - C) \\ + \gamma M x \pi(V_m - V_p - C) \\ rV_m = \gamma(1 - M)x\Pi(U + V_p - V_m).$$

For example, rV_p equals the arrival rate of agents with goods $\gamma(1 - M)$, times the double coincidence probability x^2, times the gain from barter $U - C$; plus the arrival rate of agents with money γM, times the probability of trade $x\pi$, times the gain $V_m - V_p - C$. We restrict attention to pure strategies (mixed strategy equilibria are not robust here; see Shevchenko and Wright, 2004). Then the best response condition is $\pi = 0$ if $V_m - V_p < C$ and $\pi = 1$ if $V_m - V_p > C$. It is easy to see $\pi = 0$ is always an equilibrium, and $\pi = 1$ is an equilibrium iff

$$rC \leq \gamma(1 - M)x(1 - x)(U - C).$$

Naturally, $\pi = 0$ is an equilibrium – if no one else accepts money, why would you? It is more interesting that $\pi = 1$ can be an equilibrium, since then intrinsically worthless money is valued, as a medium of exchange. Given M, one can check $\pi = 1$ yields higher payoffs than $\pi = 0$. Alternatively, if we choose M to maximize welfare, one can check $M > 0$ iff x is not too big. Hence, introducing money can improve welfare, even given the assumption that money holders cannot produce. The convention of money as a medium of exchange is good because it eases trade. Now, $\pi = 1$ is only an equilibrium when r is not too high, and one can check the cutoff for r here is more stringent (and payoffs lower) than when we had memory and triggers – that is, money is not a perfect mechanism. One reason money is not as good as memory is the *random* nature of matching. The problem is that you might, for example, have two meetings in a row where you want a good from someone who does not want your good, and in the second one you will have run out of money (this can also happens with positive probability when we relax the upper bound of unity on money holdings). However, in an *endogenous* (rather than random) matching model this never happens – when you have no money you do not go to someone whose good you like, but to someone who likes your good; see Corbae, Temzilides and Wright (2003). Still, money can do pretty well here, and if we cannot use triggers it is the only way to improve on pure barter.

The model is obviously crude, yet it gets at the essence of money. To recap, the results assume the following explicit frictions: (*a*) a double coincidence problem (generated here by random bilateral matching, although there are other devices in the related literature); (*b*) imperfect commitment; and (*c*) imperfect memory (or anything else that makes it difficult to use triggers). These frictions are severe – but no one said it was going to be easy to get money into economic theory in an interesting way. There are many extensions and applications of this model (some of which are surveyed in Rupert et al., 2000), but in the interest of space, we now move on to models where prices are endogenous. We mention one extension to endogenous specialization in Kiyotaki and Wright, 1993, based on ideas in Adam Smith. Consider the case where the probability that someone accepts your good x is a *choice variable*: if you want a large fraction of the population to like your output, you cannot specialize too much, which reduces productivity. Thus, the arrival rate in the production sector – to return to Diamond's two sector set-up – is $\alpha(x)$, with $\alpha' < 0$. When choosing x, you take the average X as given, and in equilibrium $x = X$. Two results follow. First, monetary equilibria have lower x than non-monetary equilibria, so the use of money enhances specialization and productivity. Second, $x \to 0$ as $\gamma \to \infty$, so when frictions vanish, agents specialize completely, and since the double coincidence probability is x^2, barter completely disappears.

Second-generation models

Suppose that goods are no longer indivisible, but can be consumed and produced in any amount $q \geq 0$, which yields utility $U(q)$ and disutility $-C(q)$, respectively. These functions have all the usual properties, plus $C(0) = U(0) = 0$. We maintain for now the assumptions that money is indivisible, money holders cannot produce, and everyone holds $m \in \{0,1\}$. But we relax the assumption of independence in generating the double coincidence problem: the probability that I like your good is x, but now the probability that I like your good and you like mine is y, and not necessarily x^2, in general. (Consider N goods and N types, where type n produces good n, but likes good $n + 1$ modulo N. If $N = 2$ then $x = y = 1/2$ (if I like your good you must like mine), while if $N \geq 3$ then $x = 1/N$ and $y = 0$ (if I like your good you cannot like mine). It is only under independence, which does not hold in these examples, that we necessarily have $y = x^2$.)

Conditional on money being accepted ($\pi = 1$), we have

$$rV_p = \gamma(1 - M)y[U(\hat{q}) - C\hat{q}] + \gamma M x[V_m - V_p - C(\bar{q})]$$
$$rV_m = \gamma(1 - M)x[U(\bar{q}) + V_p - V_m],$$

where $\hat{q}$ is the amount traded in barter and $\bar{q}$ the amount traded for money. It facilitates the presentation to start with the case $y = 0$ and then give general results. Now, to determine the equilibrium value of money, as in Shi (1995) or Trejos and Wright (1995), we say the following: when I meet you and want your good, if I have $m = 1$ while you have $m = 0$, we *bargain* over the q you produce for my money, taking as given $\bar{q}$ in all other meetings. Equilibrium is a fixed point, $q = \bar{q}$, and the price level is $p = 1/q$.

One can use any bargaining solution, including generalized Nash

$$\max [U(q) + V_p - V_m]^\theta [V_m - V_p - C(q)]^{1-\theta},$$

for any $\theta \in (0,1)$; we use $\theta = 1$ because it is so easy. (See Rupert et al., 2001, for $\theta \in (0,1)$ and other generalizations. Alternatives to bargaining studied in versions of this model include posting – for example, Curtis and Wright, 2004 – and auctions – Julien, Kennes and King, 2007. Or, instead of imposing a particular pricing mechanism, one can study the entire set of incentive-feasible trades – Wallace, 2001.) When $\theta = 1$, agents with $m = 0$ get no gains from trade since they have no bargaining power (and $y = 0$). Hence $V_p = 0$, $V_m = C(q)$, and the Bellman equation for V_m reduces to

$$rC(q) = \gamma(1 - M)x[U(q) - C(q)].$$

This is 1 equation in q, with two solutions: 0 and a unique $q > 0$. Again, we get equilibrium where an intrinsically worthless object is valued as a medium of exchange. It is easy to do comparative statics, welfare analysis, and so on in this model (for example, it is immediate that q falls and p rises when M or r increase).

Once we reintroduce some barter – that is, once we allow $y > 0$ – one can show that, in addition to $q = 0$, generically there either exist two equilibria with $q > 0$ or no equilibrium with $q > 0$. If y is small then equilibrium with $q > 0$ always exists. It is not much harder to analyse the general case $\theta \in (0,1)$. This model has a large number of variations, extensions and applications – too many to review here (again see Rupert et al., 2000). Suffice it to say that the basic results of first-generation models more or less go through, with additional insights concerning prices.

Third-generation models

The approach sketched above provides a compelling microfoundation for monetary theory: it is based on sound economic thinking going back to some very famous economists, brought up to date with modern and rigorous methods and ideas. Still, obviously those first- and second-generation models are quite abstract and quite special. In particular, the assumption that agents hold $m \in \{0,1\}$ is severe, and precludes using the models for much quantitative and policy analysis. The difficult part of relaxing this and allowing any $m \geq 0$ is that we need to keep track of the distribution of m across agents, which is complicated by the random nature of matching and the endogenous amount of money spent in each match. There are several ways to deal with this problem. Some analytic results are available in Green and Zhou (1998) and Camera and Corbae (1999) for example, while computation methods are used by Molico (2006).

Another approach is to amend the environment to get around this problem while hopefully maintaining the spirit and essence of the matching models outlined above. There are two main ways to do this, following either Shi (1997) or Lagos and Wright (2005). The Shi model assumes the fundamental decision makers are not individuals, but families, each with a large number of members. If the individual members experience independent random meetings, when they return to the household at the end of each period the total amount of money in the family is pinned down by the law of large numbers. Hence, each household starts the next period with the same (deterministic) amount of money. There are many extensions and applications of this framework (see Shi, 2006, for some references).

The Lagos–Wright model alternatively assumes that at the end of each round of decentralized trade agents go to a centralized market where they can (among other things) rebalance their money holdings. On the assumption of quasi-linear utility, all agents choose the same m for next period, independent of the amount with which they start. Again, agents enter each round of decentralized trade with the same m here, just as in the family model (although there are several interesting differences between the approaches). Versions of either model are easily used for quantitative and policy analysis. These models are perhaps still special, since they use 'tricks' to harness the distribution of m, but this is merely for technical convenience in deriving analytic results. If one is willing to use a computer, most of the special assumptions can be avoided (see, for example, Chiu and Molico, 2006).

Conclusion

We have reviewed several generations of search-and-matching models of the exchange process that can be used to provide microfoundations for monetary economics. While the literature is big, and growing fast, it is to be hoped that this article conveys some of the main ideas and models in an accessible fashion.

RANDALL WRIGHT

See also **bargaining; barter; commodity money; fiat money; inside and outside money; money; money, classical theory of; money and general equilibrium; search theory; search theory (new perspectives).**

Bibliography

Aiyagari, S.R. and Wallace, N. 1991. Existence of steady states with positive consumption in the Kiyotaki Wright model. *Review of Economic Studies* 58, 901–16.

Aiyagari, S.R. and Wallace, N. 1992. Fiat money in the Kiyotaki Wright model. *Economic Theory* 2, 447–64.

Aliprantis, C.D., Camera, G. and Puzzello, D. 2007. Anonymous markets and monetary trading. *Journal of Monetary Economics* 54, 1905–28.

Araujo, L. 2004. Social norms and money. *Journal of Monetary Economics* 51, 241–56.

Camera, G. and Corbae, D. 1999. Monetary patterns of exchange with search. *International Economic Review* 40, 985–1008.

Chiu, J. and Molico, M. 2006. Liquidity, redistribution and the welfare cost of inflation. Mimeo, Bank of Canada.

Corbae, D., Temzilides, T. and Wright, R. 2003. Directed matching and monetary exchange. *Econometrica* 71, 731–56.

Curtis, E. and Wright, R. 2004. Price setting, price dispersion and the value of money, or the law of two prices. *Journal of Monetary Economics* 51, 1599–621.

Diamond, P.A. 1982. Aggregate demand management in search equilibrium. *Journal of Political Economy* 90, 881–94.

Diamond, P.A. 1984. Money in search equilibrium. *Econometrica* 52, 1–20.

Diamond, P.A. and Fudenberg, D. 1989. Rational expectations business cycles in search equilibrium. *Journal of Political Economy* 97, 606–19.

Duffy, J. and Ochs, J. 1999. Emergence of money as a medium of exchange: an experimental study. *American Economic Review* 89, 847–77.

Green, E. and Zhou, R. 1998. A rudimentary model of search with divisible money and prices. *Journal of Economic Theory* 81, 252–71.

Jones, R. 1976. The origin and development of media exchange. *Journal of Political Economy* 84, 757–75.

Julien, B., Kennes, J. and King, I. 2007. Bidding for money. *Journal of Economic Theory*.

Kehoe, T.J., Kiyotaki, N. and Wright, R. 1993. More on money as a medium of exchange. *Economic Theory* 3, 297–314.

Kiyotaki, N. and Wright, R. 1989. On money as a medium of exchange. *Journal of Political Economy* 97, 927–54.

Kiyotaki, N. and Wright, R. 1991. A contribution to the pure theory of money. *Journal of Economic Theory* 53, 215–35.

Kiyotaki, N. and Wright, R. 1993. A search theoretic approach to monetary economics. *American Economic Review* 83, 63–77.

Kocherlakota, N. 1998. Money is memory. *Journal of Economic Theory* 81, 232–51.

Kocherlakota, N. and Wallace, N. 1998. Optimal allocations with incomplete record keeping and no commitment. *Journal of Economic Theory* 81, 272–89.

Lagos, R. and Wright, R. 2005. A unified framework for monetary theory and policy analysis. *Journal of Political Economy* 113, 463–84.

Marimon, R., McGrattan, E.R. and Sargent, T.J. 1990. Money as a medium of exchange in an economy with artificially intelligent agents. *Journal of Economic Dynamics and Control* 14, 329–73.

Molico, M. 2006. The distribution of money and prices in search equilibrium. *International Economic Review* 47, 701–22.

Ostroy, J.M. and Starr, R.M. 1990. The transaction role of money. In *Handbook of Monetary Economics*, ed. B. Friedman and F. Hahn. Amsterdam: North-Holland.

Rupert, P., Schindler, M., Shevchenko, A. and Wright, R. 2000. The search-theoretic approach to monetary economics: a primer. *Federal Reserve Bank of Cleveland Review* 36, 10–28.

Rupert, P., Schindler, M. and Wright, R. 2001. Generalized search theoretic model of monetary exchange. *Journal of Monetary Economics* 48, 605–22.

Shevchenko, A. and Wright, R. 2004. A simple model of money with heterogeneous agents and partial acceptability. *Economic Theory* 24, 877–85.

Shi, S. 1995. Money and prices: a model of search and bargaining. *Journal of Economic Theory* 67, 467–96.

Shi, S. 1997. A divisible search model of fiat money. *Econometrica* 65, 75–102.

Shi, S. 2006. Viewpoint: a microfoundation of monetary economics. *Canadian Journal of Economics* 39, 643–88.

Starr, R.M. 1990. *General Equilibrium Models of Monetary Economics*. New York: Academic Press.

Trejos, A. and Wright, R. 1995. Search, bargaining, money and prices. *Journal of Political Economy* 103, 118–41.

Wallace, N. 1980. The overlapping generations model of fiat money. In *Models of Monetary Economies*, ed. J.H. Kareken and N. Wallace. Minneapolis: Federal Reserve Bank of Minneapolis.

Wallace, N. 2001. Whither monetary economics. *International Economic Review* 42, 847–70.

Wright, R. 1995. Search, evolution, and money. *Journal of Economic Dynamics and Control* 19, 181–206.

search models of unemployment

To the average person (perhaps even the average economist), unemployment is often equated with a state of involuntary idleness. This commonly held view, however, is inconsistent with the way in which unemployment is in fact defined and measured. According to International Labor Organization (ILO) conventions, which are followed by most national labour force surveys, unemployment relates to those individuals who are not employed but who are *actively searching* for work (over the reference period of the survey – for example, the previous four weeks). Non-employed (jobless) individuals who have not engaged in active job search are classified as non-participants. This latter category includes the set of so-called 'discouraged workers'.

Since active job search consumes time and other resources, relating unemployment to a state of idleness seems wrong. Furthermore, since individuals are free to allocate their time across many competing activities, with time devoted to search yielding at least the prospect of a future payoff, the notion of unemployment as an involuntary state is potentially misleading. To understand unemployment – or, at least, unemployment *data* (as opposed to some other pet notion of unemployment) – one must therefore entertain a theory that explains the circumstances under which individuals find it in their interest to remain non-employed while searching for work. Search models of unemployment are designed to do just this.

Search theory

A hallmark of conventional labour market theory (the neoclassical model and its sticky-wage variants) is the assumption that labour is exchanged in a centralized marketplace. Among other things, a centralized market embodies the idea that there is perfect information concerning the location of all jobs and workers. In any such environment, devoting precious time to an activity like search literally makes no sense (whether or not the wage is market-clearing); that is, individuals either have a worthwhile job opportunity to exploit or they do not. In the former case, they become employed; in the latter, they become non-employed (and would be labelled as non-participants by standard labour force surveys).

The starting point for any search model of unemployment then is to dispense with the notion of a centralized marketplace for labour. Instead, the labour market is viewed as a set of decentralized locations, where firms and workers can potentially meet to form mutually beneficial relationships. In a decentralized market, meetings are to some extent determined by search effort and to some extent by chance. In many ways, the labour market resembles a matching market for couples; that is, one is generally aware that the opposite side of the market consists of better and worse matches (we seldom take the view that there are *no* potential matches). The exact location of the better matches is unknown, but may be discovered with some effort. In the meantime, it may make sense to refrain from matching with 'substandard' opportunities that are currently available. But since search is costly, it will generally not be optimal to wait for one's 'soulmate' to come along. Furthermore, since relationships are not perfectly durable, there is no reason to expect the stock of singles to converge to zero over time. Nor is it clear, given the technology of match-formation, that having everyone matched at all times (irrespective of match quality) is in any way desirable – even if it is feasible; and even if people generally desire to be matched.

In the context of the labour market then, the key friction that potentially rationalizes job search (and, hence, unemployment) is imperfect information over the location of one's best job opportunity. In such an environment, job search constitutes a form of investment in the acquisition of information. The idea of job search as an information-gathering activity has been in place for some time; see, for example, Stigler (1962) as well as the several papers and references contained in Phelps (1970). Perhaps the most influential early formalization of the theory of job search is provided by McCall (1970); see also Sargent (1987, ch. 2). In the next section, I present the basic idea of this classic literature by way of a simple model.

A simple model

I begin by describing a simple 'one-sided' search model; for example, the case in which an individual searches for a job opportunity that pays a given wage depending on match quality. Each time period $t = 1, 2, \ldots, \infty$ is divided into two subperiods that I call stage 1 and stage 2. There will be no intertemporal aspect to individual decision-making, so the model's dynamic equilibrium can be thought of as a sequence of static equilibria.

Each person enters stage 1 endowed with one unit of (indivisible) time, a job opportunity that pays a real wage $w \in [0, \overline{w}]$, and a leisure opportunity that generates a return $0 < b < \overline{w}$. One can interpret b as either an unemployment insurance benefit or, more generally, as the consumption available when non-employed (for example, home production). The real wage w can be interpreted as the quality of the job–worker match associated with any given job opportunity. Let us assume that individuals do not know the precise location of their best job match (that is, the match that would yield them $\overline{w}$). However, it is assumed that individuals know the

distribution of match qualities associated with the given set of available jobs; let $F(w)$ denote the fraction of job matches that yield a wage no greater than w. The act of job search is modelled as a random draw from $F(w)$.

At the beginning of stage 1, a choice has to be made: should I search or not? The act of not searching means that time must be allocated during stage 1 across one of two activities: work or leisure. The utility payoff to work is $u(w)$, while the utility payoff to leisure is $u(b)$, where $u'' < 0 < u'$. On the assumption that I choose not to search, optimal behaviour entails choosing to work if $w \geq b$. Alternatively, I could choose to search instead. In this case, my time endowment is 'transported' to stage 2. Here, the cost of job search is given by the assumption of *no recall*; that is, I must abandon my current opportunity w to exercise my search option. Let $w' \sim F(w')$ denote the wage associated with a new job opportunity I find as a result of my search. At this stage, I now have the opportunity to either work or enjoy leisure; that is, I will only choose to work at this new job if $w' \geq b$. Hence, the expected utility payoff from search is given by:

$$S(b) \equiv F(b)u(b) + \int_b^{\overline{w}} u(w') \, \mathrm{d}F(w').$$

Implicit in this discussion is the assumption that consumption levels during stage 1 and 2 are viewed as perfect substitutes. I am also assuming that consumption is non-storable and the absence of insurance/loan markets.

The problem has been set up here in such a way that it will never be optimal for a person to choose leisure during stage 1. This is because there is always a zero-cost option to consume leisure during stage 2. Hence, the relevant choice during stage 1 is whether to work or search. The optimal strategy is to set a *reservation wage* $w_R(b)$, such that one should choose work if $w \geq w_R(b)$ and choose search if $w < w_R(b)$. This reservation wage must satisfy:

$$u(w_R) = S(b). \tag{1}$$

That is, a person's reservation wage is defined to be the wage that would make a person just indifferent between working at that wage and abandoning it in favour of another activity (in this case, job search). In other words, the reservation wage is a measure of an individual's 'choosiness' over job opportunities. Note that as $S'(b) = F(b) u'(b) > 0$, the reservation wage is an increasing function of the option value associated with leisure. Intuitively, a person with a good outside option (b) can afford to be more discriminating with respect to the quality of any given job opportunity w.

Thus far I have described the choice problem of an individual beginning with an endowed wage/leisure opportunity (w, b). I would like to now describe how this behaviour translates into the evolution of aggregates over time. To render static decision-making optimal in this environment, assume that all individuals begin each period with a match quality w drawn from the distribution $G(w)$. The interpretation here is that economy experiences a 'structural' shock every period that 'shuffles' individual match qualities, but otherwise leaves aggregate production possibilities unchanged. Following this structural shock, individuals behave in the manner described above. For the sake of simplicity, assume that all individuals share a common value for b that remains constant over time.

It is an easy matter now to characterize the equilibrium unemployment rate. At the beginning of each period, individuals are randomly assigned a match quality w, whose distribution is $G(w)$. The fraction $[1 - G(w_R)]$ will choose to exploit their employment opportunity and the fraction $G(w_R)$ will find it optimal to abandon their opportunity in favour of search. Job searchers find new job opportunities with associated match qualities w' drawn from the distribution $F(w')$. Those who find a job that pays $w' \geq b$ choose to work; those that find a job that pays $w' < b$ choose leisure instead. Thus, $F(b)$ denotes the fraction of job-searchers who are 'unsuccessful' in their search. On the assumption that the reference period for a labour force survey is the length of a period (and that the survey is performed at the end of the period), the equilibrium unemployment rate is given by:

$$U = G(w_R)F(b).$$

In other words, the unemployed are those who: (a) performed no work during the reference period; and (b) were actively searching for work during the reference period.

Associated with the 'steady-state' unemployment rate U are steady-state flows of workers making transitions from employment to unemployment (job destruction) and transitions from unemployment to employment (job creation); these flows are given by:

$$JC = [1 - G(w_R)F(b)]U;$$
$$JD = G(w_R)F(b)(1 - U).$$

Any given individual may, of course, experience a variety of employment/unemployment histories.

The model developed above constitutes an example of the modern approach to the theory of unemployment. Note that unemployment here is interpreted as an equilibrium phenomenon; in particular, it is not the product of irrational behaviour or markets that fail to clear. Instead, unemployment is the natural by-product of an economy subject to ongoing structural disturbances that depreciate the value of existing employer–employee matches, and where match formation takes time owing to the imperfect information pertaining to the location of new job opportunities.

One by-product of the modern theory of unemployment is that renders obsolete much of the traditional language that was used to describe the phenomenon (see Rogerson, 1997). Consider, for example, the traditional classification of unemployment into its 'voluntary' and 'involuntary' components. In the model developed above, there is a sense in which unemployment is both voluntary and involuntary. It is voluntary in the sense that the model people (as with real people) can choose not to search, and instead allocate time to inferior job opportunities or home production activities. On the other hand, it is also involuntary in the sense that the economic circumstances that compel individuals to become unemployed are often beyond their immediate control. By the same token, however, the same classification might be made for employed workers (for example, the existence of those who cannot afford not to work, or the so-called working poor). Whether these traditional labels have any substantive meaning, however, is questionable. In particular, note that the well-being of these model people (and presumably, people in reality) does not depend on how we (as theorists) choose to label them.

Welfare

A classic question in the theory of the labour market is whether an equilibrium level of unemployment corresponds in any way to an efficient allocation. Not surprisingly, the answer to this question depends critically on the nature of the economic environment; see Diamond (1982), Mortensen (1982), Hosios (1990) and Moen (1997). In the present context, the optimal level of unemployment can be characterized as the solution to the following planning problem:

$$\max_{c, w_R} \left\{ u(c) : c \leq \int_{w_R}^{w} w \, dF(w) + F(w_R) \left[F(b)b + \int_{b}^{\overline{w}} w' \, dF(w') \right] \right\}.$$

The socially optimal reservation wage is given by:

$$w_R^* = F(b)b + \int_{b}^{\overline{w}} w' \, dF(w').$$

The optimal unemployment rate corresponds to the equilibrium unemployment rate if $w_R^* = w_R$, where w_R is the solution to (1). As it turns out, this will be the case if $u'' = 0$ (risk-neutral individuals). However, with risk-averse individuals, one can easily establish that $w_R < w_R^*$, which implies here that the equilibrium unemployment rate is too *low*. The suboptimality of equilibrium in this particular model has nothing to do with search per se; rather, it is due to the assumed lack of insurance. The intuition is straightforward: in the absence of a well-functioning insurance market, individuals are too willing to hold on to marginal jobs rather than risk an even worse outcome in the event of an unsuccessful job search.

One striking implication of the modern theory of unemployment is that the unemployment rate bears no obvious relation to any sensible measure of social welfare. Consider, for example, two economies A and B, identical in every respect except that $b_A > b_B$ (so that the residents of economy A are in some sense 'wealthier' than those of economy B). In this case, the unemployment rate in economy A will be higher than in economy B. Nevertheless, given the choice, individuals would rather live in the high-unemployment economy. On the other hand, suppose that the two economies differ only in terms of F, with F_A stochastically dominating F_B. In this case, economy A will have lower unemployment and higher welfare. The basic lesson here is that 'economic performance' should not be measured in terms of how individuals choose to allocate their time across competing activities, but rather should be measured in terms of the level and distribution (or stochastic properties) of broadly defined consumption.

Unemployment insurance

Governments in many countries operate a fiscal policy known as unemployment insurance (UI). UI systems are characterized by transfers of income to those who are unemployed (or otherwise meet certain eligibility requirements) that are typically financed via a payroll tax (or out of general tax revenue). Presumably, the motivation for such transfers rests on the belief that: (1) private insurance markets are unavailable (or work poorly) and (2) self-insurance (via precautionary saving) is either grossly inefficient, or perhaps beyond the means of many workers (a more cynical view interprets UI as one of many government transfer schemes designed to benefit various special interests at the expense of the general taxpayer).

Search models of unemployment have been applied in both positive and normative investigations concerning the effects of UI programmes. For example, in the context of the model developed above, let b denote a UI benefit that is financed via a payroll tax τ. For given programme parameters (b, τ), the reservation wage now satisfies:

$$u((1 - \tau)w_R) = F(b)u(b)$$
$$+ \int_b^{\overline{w}} u((1 - \tau)w')\, dF(w'), \qquad (2)$$

which implicitly characterizes $w_R(b, \tau)$. In equilibrium, the tax and benefit level is related by the government budget constraint:

$$\tau\left[\int_{w_R}^{\overline{w}} w\, dF(w) + F(w_R)\int_b^{\overline{w}} w'\, dF(w')\right]$$
$$= F(w_R)F(b)b. \qquad (3)$$

For a given UI policy parameter (say, an exogenous level of b), eqs (2) and (3) characterize an equilibrium $(\hat{w}_R, \hat{\tau})$. The positive and normative implications of the theory can then be deduced by varying the policy parameter b. Andolfatto and Gomme (1996) consider a similar such experiment, albeit in a considerably richer environment that (among other things) models the UI system in greater detail (in particular, UI programmes are typically characterized by several programme parameters, including eligibility requirements, replacement rates and benefit duration parameters).

Several papers have recently examined the issue of how an optimal UI system should be designed. This question becomes particularly interesting when one reasonably assumes that workers have private information concerning the nature of their job opportunities and/or the intensity with which they search. In a classic paper, Shavell and Weiss (1979) demonstrate that an optimal UI programme should 'front load' UI payments, with benefit levels declining monotonically over the unemployment spell. The high initial benefit level provides the desired insurance, while the declining benefit profile mitigates adverse incentive effects by stimulating (unobserved) search intensity. Wang and Williamson (1996) and Hopenhayn and Nicolini (1997) flesh out other properties of optimal UI systems in the context of dynamic moral hazard environments. Among other things, these authors report that optimally designed programmes can deliver potentially large welfare benefits relative to existing systems.

Business cycles

Search models of unemployment have also been used to interpret various aspects of the business cycle. Naturally, a key set of questions deal with the cyclical properties of unemployment itself. However, researchers have also investigated the extent to which 'search frictions' in the labour market shape the pattern of the business cycle more generally (see Mortensen and Pissarides, 1994; Merz, 1995; Andolfatto, 1996).

An empirical relationship that has drawn considerable attention in the literature is the so-called Beveridge curve; that is, the tendency for unemployment and vacancies to move in opposite directions over the business cycle (for example, see Blanchard and Diamond, 1989). The level of job vacancies, as measured by the help-wanted index, is highly volatile and tends to lead unemployment over the cycle. A natural interpretation of these facts is that as job opportunities become more plentiful, unemployed workers are able to find jobs more easily, with search frictions preventing instantaneous adjustment. Vacancies themselves can be interpreted as the business sector equivalent of unemployed workers. Since finding suitable workers is costly and time-consuming, and since employment relationships once formed are durable, recruiting intensity constitutes a form of capital spending that presumably reacts to actual and expected shocks in much the same way as other forms of capital spending (for example, see Howitt, 1988).

A seminal paper in this area is Pissarides (1985), who appeals to the concept of an aggregate matching technology to model (in a reduced-form manner) the outcome of uncoordinated search in the labour market (see also Pissarides, 2000; Petrongolo and Pissarides, 2001). According to this approach, match formation (job creation) is the product of the aggregate search intensity of workers (unemployment) and firms (vacancies), both of which serve as complementary inputs into an aggregate matching function, that is, $m_t = M(v_t, u_t)$. Ignoring individual differences, m/v and m/u can be interpreted as the probabilities that vacant jobs and unemployed workers contact each other. If M displays constant returns to scale, then these probabilities depend only on the labour-market tightness variable $\theta \equiv v/u$. In particular, $m/v = q(\theta)$ and $m/u = p(\theta)$, where $q(\theta) \equiv M(\theta, 1)$ and $p(\theta) \equiv M(1, \theta^{-1})$. Note that $q'(\theta) < 1$ and $p'(\theta) > 1$. In other words, firms find it more difficult to find workers in a tighter labour market; while the converse is true for workers.

With a fixed labour force L, the stock of employed workers at date t is given by $L - u_t$. Again note that the stock of employment constitutes a form of capital. Employment capital depreciates over time as matches dissolve owing to idiosyncratic shocks that affect the viability of individual relationships. Let $0 < \sigma < 1$ denote the fraction of employment relationships that are terminated at the end of each period. In this case, the job destruction flow is given by $\sigma(L - u_t)$. For simplicity, assume that σ is exogenous. At the same time, the stock of employment is replenished by the flow of job creation, $p(\theta_t)u_t$, so that the stock of unemployment evolves over time according to:

$$u_{t+1} = u_t - \sigma(L - u_t) + p(\theta_t)u_t. \qquad (4)$$

In a steady state, the flow of job creation just offsets the flow of job destruction, so that the 'natural' rate of unemployment satisfies (normalizing $L = 1$):

$$u = \left(\frac{\sigma}{\sigma + p(\theta)} \right). \qquad (5)$$

Equations (4) and (5) relate the dynamics of the unemployment rate to the labour-market tightness variable θ, as well as to parameters describing the structure of the matching market. In the simple version considered here, the unemployed search passively (workers have one unit of time that they allocate either to work or search at zero utility cost). Consequently, the equilibrium $\theta_t = v_t/u_t$ is determined entirely by vacancy creation, which is endogenized as follows.

Assume that each firm has a single job that, together with a worker, produces $y_t > 0$ units of output. Assume

that y_t follows a first-order Markov process with $G(y', y) = \Pr[y_{t+1} \leq y' | y_t = y]$. For simplicity, assume that match quality is identical across all firm–worker pairs and that wages are determined by an exogenous bargaining process that divides output in some manner between firm and worker; that is, let $0 < \xi < 1$ denote the fraction of output that accrues to the firm (alternatively, one could model wage determination by imposing a Nash bargaining solution). Let J denote the capital value of a firm currently matched with a worker; this value must satisfy the following Bellman equation:

$$J(y) = \xi y + \beta(1 - \sigma) \int J(y')G(\mathrm{d}y', y),$$

$$(6)$$

where $0 < \beta < 1$ denotes a discount parameter. Implicit in this formulation is that the value of the firm falls to zero in the event that the match is dissolved. Note that if match productivity exhibits positive persistence, then $\int J(y')G(\mathrm{d}y', y)$ is increasing in y. In other words, a positive productivity shock (increase in y) is associated with information that leads firms to revise upward their estimate of the returns to match formation.

Finally, consider the cost–benefit analysis associated with vacancy creation. Assume that creating (and maintaining) a vacancy entails the flow cost $\kappa > 0$, measured in units of output. A vacant job potentially meets an unemployed worker with probability $q(\theta)$, with the match producing a flow of output beginning in the subsequent period. Let Q denote the capital value of a vacant job; this value must satisfy the following Bellman equation:

$$Q = -\kappa + \beta \left[q(\theta) \int J(y')G(\mathrm{d}y', y) + (1 - q(\theta))Q' \right].$$

$$(7)$$

Assuming free entry into vacancy creation, the level of v will expand as long as $Q > 0$. But note that for a fixed level of unemployment u, any expansion in the number of vacancies increases labour-market tightness θ and therefore reduces the success probability $q(\theta)$. In equilibrium, θ adjusts to a point that renders further vacancy creation unprofitable; that is, $Q = Q' = 0$. Consequently, the equilibrium labour-market tightness variable $\theta(y)$ is determined by:

$$q(\theta)\beta \int J(y')G(\mathrm{d}y', y) = \kappa. \qquad (8)$$

The economic mechanism at work here can be described as follows. Consider a positive productivity shock. Since the shock is persistent, the return to job creation (contemporaneous recruiting activities that augment the future stock of employment) increases (see (6)). Naturally, in response to these bright prospects, firms increase their recruiting activities and create new vacancies. However, as the competition for new workers intensifies, the probability of success falls to the point where further expansion becomes uneconomical (see (8)). On impact then, the effect of the shock leaves current unemployment unchanged, but increases the demand for labour (supply of vacancies). The dynamic effects of the shock can then be traced out by appealing to eq. (4). In particular, note that since the job-finding probability for workers $p(\theta)$ is increasing in θ, the effect of the shock is to lower the future rate of unemployment. These effects continue to be propagated forwards in time through the search mechanism.

Current issues

Several authors have observed that the unemployment rate appears to exhibit a type of cyclical asymmetry, rising sharply at the onset of a recession, but declining only slowly over the course of a subsequent recovery (for example, see Neftçi, 1984; Hussey, 1992). In an influential study, Davis and Haltiwanger (1992) investigate US manufacturing data and report that job destruction appears to be significantly more volatile than job creation over the cycle. The natural conclusion that follows from this body of work is that recessions are attributable to shocks that lead to brief, but sharp, increases in job losses followed by relatively dampened, but prolonged, periods of job creation as the business sector slowly rebuilds its employment capital (see, for example, Hall, 1995).

Shimer (2005a) has recently cast doubt on this 'conventional' view of the cycle. In his detailed examination of Current Population Survey data, Shimer concludes that almost all the cyclical variability in the unemployment rate is attributable to fluctuations in the job-finding probability (and in particular, the job-finding rate associated with transitions from unemployment to employment, rather than from non-employment to employment). Surprisingly (relative to received wisdom), the separation rate appears to be very nearly acyclical and relatively stable. In addition to these patterns, Shimer reports a very high degree of correlation between the job-finding probability and the vacancy–unemployment ratio.

Taken together, this new body of evidence suggests that, at least for the purpose of understanding cyclical behaviour, one may reasonably begin by organizing ideas around a simple Pissarides-style search model along the lines described above; that is, with a constant labour force L, an exogenous separation probability σ, and a job-finding probability $p(\theta)$ that varies with labour-market tightness. Whether a suitably calibrated version of this model can account for observation, however, remains a topic of current debate (for example, see Shimer, 2005b; Hagedorn and Manovskii, 2005; Hall, 2005; Mortensen, 2005; as well as Hornstein, Krusell and Violante, 2005 for a nice summary).

Apart from cyclical phenomena, several outstanding issues remain unresolved in terms of understanding

secular and cross-country measures of labour-market activity. At the forefront of these phenomena is the dramatic rise in European unemployment rates relative to the United States since the early 1970s (for example, Ljungqvist and Sargent, 1998; Blanchard and Wolfers, 2000). But unemployment is just one of the three major classifications of time allocation, along with employment and non-participation (the latter two of which contain many more people than the former). Rogerson (2001) documents several interesting facts concerning the cross-country differences and low-frequency movements in employment-to-population ratios. Among other things, cross-country differences are large and persistent over time, with considerable movement within the distribution of employment-to-population ratios across countries. Furthermore, while employment and unemployment closely mirror each other at business cycle frequencies, the same is not true at lower frequencies. This latter observation suggests that the commonly invoked short-cut of abstracting from participation decisions for business cycle analysis may be inappropriate when investigating the causes of secular movements and cross-country differences in time allocation.

On the theoretical front, several issues remain the subject of ongoing research. Paramount among these are investigating the microeconomic foundations of the matching technology and the process of wage determination in decentralized markets. On these and related matters, the interested reader may refer to Rogerson, Shimer and Wright (2005) – a recent survey that contains 167 references while claiming to only 'scratch the surface' of this interesting and rapidly expanding body of research.

DAVID ANDOLFATTO

See also **involuntary unemployment; matching; matching and market design; search theory; unemployment insurance.**

Bibliography

Andolfatto, D. 1996. Business cycles and labor-market search. *American Economic Review* 86, 112–32.

Andolfatto, D. and Gomme, P. 1996. Unemployment insurance and labor market activity in Canada. *Carnegie-Rochester Series on Public Policy* 44, 47–82.

Blanchard, O. and Diamond, P. 1989. The Beveridge curve. *Brookings Papers on Economic Activity* 1, 1–60.

Blanchard, O. and Wolfers, J. 2000. The role of shocks and institutions in the rise of European unemployment. *Economic Journal* 110, 1–33.

Davis, S. and Haltiwanger, J. 1992. Gross job creation, gross job destruction, and employment reallocation. *Quarterly Journal of Economics* 107, 818–63.

Diamond, P. 1982. Wage determination and efficiency in search equilibrium. *Review of Economic Studies* 49, 217–27.

Hagedorn, M. and Manovskii, I. 2005. The cyclical behavior of equilibrium unemployment and vacancies revisited. Working paper. University of Pennsylvania.

Hall, R.E. 1995. Lost jobs. *Brookings Papers on Economic Activity* 1, 221–56.

Hall, R.E. 2005. Employment fluctuations with equilibrium wage stickiness. *American Economic Review* 95, 50–65.

Hopenhayn, H. and Nicolini, J.P. 1997. Optimal unemployment insurance. *Journal of Political Economy* 105, 412–38.

Hornstein, A., Krusell, P. and Violante, G. 2005. Unemployment and vacancy fluctuations in the matching model: inspecting the mechanism. *Federal Reserve Bank of Richmond Economic Quarterly* 91, 19–51.

Hosios, A.J. 1990. On the efficiency of matching and related models of search and unemployment. *Review of Economic Studies* 57, 79–98.

Howitt, P. 1988. Business cycles with costly search and recruiting. *Quarterly Journal of Economics* 103, 147–65.

Hussey, R. 1992. Nonparametric evidence on asymmetry in business cycles using aggregate employment time-series. *Journal of Econometrics* 51, 217–31.

Ljungqvist, L. and Sargent, T. 1998. The European unemployment dilemma. *Journal of Political Economy* 106, 514–50.

McCall, J.J. 1970. Economics of information and job search. *Quarterly Journal of Economics* 84, 113–26.

Merz, M. 1995. Search in the labor market and the real business cycle. *Journal of Monetary Economics* 36, 269–300.

Moen, E. 1997. Competitive search equilibrium. *Journal of Political Economy* 105, 385–411.

Mortensen, D.T. 1982. Property rights and efficiency in mating, racing, and related games. *American Economic Review* 72, 968–79.

Mortensen, D.T. 2005. More on unemployment and vacancy fluctuations. Working paper. Northwestern University.

Mortensen, D.T. and Pissarides, C. 1994. Job creation and job destruction in the theory of unemployment. *Review of Economic Studies* 61, 397–415.

Neftçi, S. 1984. Are economic time-series asymmetric over the cycle? *Journal of Political Economy* 92, 307–28.

Petrongolo, B. and Pissarides, C. 2001. Looking into the black box: a survey of the matching function. *Journal of Economic Literature* 39, 390–431.

Phelps, E.S. 1970. *Microeconomic Foundations of Employment and Inflation Theory*. New York: W.W. Norton.

Pissarides. 1985. Short-run equilibrium dynamics of unemployment, vacancies, and real wages. *American Economic Review* 75, 676–90.

Pissarides. 2000. *Equilibrium Unemployment Theory*, 2nd edn. Cambridge, MA: MIT Press.

Rogerson, R. 1997. Theory ahead of language in the economics of unemployment. *Journal of Economic Perspectives* 11(1), 73–92.

Rogerson, R. 2001. The employment of nations: a primer. *Federal Reserve Bank of Cleveland Economic Review* 4, 27–50.

Rogerson, R., Shimer, R. and Wright, R. 2005. Search-theoretic models of the labor market: a survey. *Journal of Economic Literature* 43, 959–88.

Sargent, T.J. 1987. *Dynamic Macroeconomic Theory*. Cambridge, MA: Harvard University Press.

Shavell, S. and Weiss, L. 1979. The optimal payment of unemployment insurance benefits over time. *Journal of Political Economy* 87, 1347–62.

Shimer, R. 2005a. Reassessing the ins and outs of unemployment. Working paper. University of Chicago.

Shimer, R. 2005b. The cyclical behavior of equilibrium unemployment and vacancies. *American Economic Review* 95, 25–49.

Stigler, G.J. 1962. Information in the labor market. *Journal of Political Economy* 70, 94–105.

Wang, C. and Williamson, S. 1996. Unemployment insurance with moral hazard in a dynamic economy. *Carnegie-Rochester Conference on Public Policy* 44, 1–41.

search theory

Walrasian analysis presumes that resource allocation can be adequately modelled using the assumption of instantaneous and costless coordination of trade. In contrast, search theory is the analysis of resource allocation with specified, imperfect technologies for informing agents of their trading opportunities and for bringing together potential traders. The modelling advantages of assuming a frictionless coordination mechanism, plus years of hard work, permit Walrasian analysis to work with very general specifications of individual preferences and production technologies. In contrast, search theorists have explored a variety of special allocation mechanisms together with very simple preferences and production technologies. Lacking more general theories, we examine the catalogue of analyses that have been completed.

Paralleling the Walrasian framework, we first examine individual choice and then equilibrium. There are a large number of variations on the basic search-theoretic choice problem. We explore one set-up in detail, while mentioning some of the variations that have been developed. Coordination of trade involves two separate steps: information gathering about opportunities, and arrangement of individual trades. One simple case is where information gathering is limited to visiting stores sequentially, combining the costs of collecting goods and of gathering information. Alternatively, there can be an information gathering mechanism which is independent of the process of ordering and receiving the good. We begin with models where the only information gathering is associated with visiting stores, and then look at the changes that come from additional devices for information spread.

Once two potential traders have met there are several ways of determining whether they trade and the terms of trade if they do. Among these are price setting on a take-it-or-leave-it basis, idealized negotiations where any mutually advantageous trade occurs at a price satisfying some bargaining solution, and more realistic negotiation processes that recognize the time and cost of negotiation, the possibility of a negotiating impasse, and the possible arrival of alternatives for one or the other of the trading partners. We explore the first two mechanisms.

One final distinction in the literature is between one-time purchases of commodities and ongoing trade relations. Infrequently purchased consumer goods are the classic example of the former, while the employment relationship is the classic example of the latter. Introducing on-going relationships permits the exploration of delayed learning of the quality of a match and associated rearrangements through quits and firings. Intermediate between these two cases is a situation such as that of frequently purchased consumer goods, where past trades facilitate future trades but do not bring about the closeness of an employment relationship. We discuss mainly the one-shot purchase. The discussion of individual choice and partial equilibrium will be given in terms of a consumer purchase. The parallel discussion of labour markets is only briefly mentioned.

I. Individual choice

Consider a consumer in a store who is deciding whether to make a purchase or to visit another store with an unknown price. Denote by $U(p, 1)$ the utility that the consumer receives (net of purchase costs) if the purchase is made in the first store at a price equal to p. This assumes an ability to purchase the optimal number of units at a constant per unit price of p. If the purchase is made at the second store at price p, utility is $U(p, 2)$. This utility is less than $U(p, 1)$ because of the cost and the time delay from visiting a second store. We assume that the entire purchase is made at a single store, that it is impossible to return to the first store, and that there are no other stores that can be visited. Ignoring the possibility of making no purchase and no further searches, the alternative to purchasing in store 1 at price p, is a single visit to store 2 where the price will be drawn from a (known) distribution which we denote $F(p)$. The purchase should be made in store 1 if the utility of purchase there is at least as large as the expected utility of purchase in store 2:

$$U(p, 1) \geq \int U(p, 2) \mathrm{d}F(p). \qquad (I-1)$$

As long as the consumer views the distribution of prices in store 2 as independent of the price in the first store, the rule in (I–1) yields a cut-off price, p^*, given by (I–2):

$$U(p^*, 1) = \int U(p, 2) \mathrm{d}F(p). \qquad (I-2)$$

For prices above p^*, optimal behaviour calls for visiting the second store, while for prices below p^*, optimal behaviour calls for making a purchase in the first store. Thus p^* is the cut-off price. Implicit in this formulation is the assumption that it is not desirable to make some purchase in store 1 and the remaining purchase in store 2. While this assumption is true for many consumer goods it is certainly not true for all of them. Without this assumption the decision resembles portfolio choice and has not been explored in the literature. A similar analysis applies to the search for high quality.

If the consumer does not know with certainty the distribution of prices in the second store, the consumer's beliefs about those prices may depend upon the price observed in the first store. We write the subjective distribution of prices in the second store, conditional on an observed price of p_1 in the first store as $F(p; p_1)$. The purchase should be made in the first store if p_1 satisfies the inequality:

$$U(p_1, 1) \geq \int U(p, 2) \mathrm{d}F(p; p_1). \qquad (I-3)$$

With no restriction on the beliefs of the consumer as to the structure of prices found in both stores, the set of prices resulting in a purchase in store 1 does not necessarily satisfy a cut-off price rule. For example, if either a high or a low observed price implies the same price in both stores, while an intermediate price in store 1 implies a low price in store 2, then the consumer should purchase in store 1 at the high and low prices but not at an intermediate price. Thus, the intermediate price might signal a price war. If the information content of the price found in store 1 is a greater likelihood of similar prices in store 2, the optimality of a cut-off price rule is restored (Rothschild, 1974). For the remainder of this entry we restrict analysis to the case of known distributions. The caveats implicit in this counter example should be kept in mind.

Returning to the set-up with a known distribution, we can increase the options of the shopper by adding the possibility of returning to the first store after observing the price in the second store. Denote by $U(p, 3)$ the utility that is realized if this option is followed. The utility function $U(p, 3)$ is less than $U(p, 2)$, which, in turn, is less than $U(p, 1)$. Once in the second store, the choice is between buying there and returning to the first store with both prices known. Therefore it pays to purchase in the first store in the first period if the price there, p_1, satisfies the inequality:

$$U(p_1, 1) \geq \int \max[U(p, 2), U(p_1, 3)] \mathrm{d}F(p).$$

$$(I-4)$$

That is, the purchase should be made in store 1 if utility there is higher than expected utility with optimal behaviour in choosing between the second store and returning to the first store. This is a particularly simple example of the backwards induction that can be used to solve the finite horizon sequential shopping problem. Behaviour in the first store, (I–4), again satisfies a cut-off price rule if the utility function has constant search costs and discount rate.

We now specialize the example by assuming additive, constant search costs c and utility discounting with a discount factor R. That is, $U(p, 2)$ equals $RU(p, 1) - c$, with $R \leq 1$. Returning to the choice problem without a return to store 1, we denote by $V(p_1)$ expected utility on observing the price p_1 in store 1, given optimal behaviour:

$$V(p_1) = \max\left[U(p_1), -c + R \int U(p)\mathrm{d}F(p)\right].$$
$$(I-5)$$

The value of being in a store that has price p_1 is the larger of (i) the utility from making the optimal purchase at that price, and (ii) the expected utility if the search cost c is paid and the purchase is made in the second store. Using this function V, we can describe choice in the first period of a new three-period search problem with no return to previous stores. The optimal rule is to purchase if

$$U(p_1) \geq -c + R \int V(p)\mathrm{d}F(p). \qquad (I-6)$$

That is, purchase is made in the first period if the achievable utility there is at least as large as that achievable with optimal behaviour, beginning with a visit to a randomly selected second store. The latter utility is the discounted expected optimized utility minus the search costs of the visit, recognizing that the second period choice is again a choice between a purchase and a search in the following period. By having $F(p)$ independent of p_1, we are sampling with replacement rather than sampling without replacement from the known distribution of prices. The choice rule given in (I–6) again shows cut-off price behaviour for period one choice. However, the cut-off price is higher in the second period than in the first because of the reduction in options as the end of the search process comes closer. Denoting the cut-off prices in the two periods by p_1^* and p_2^*, they satisfy the two equations:

$$U(p_1^*) = -c + R \int V(p)\mathrm{d}F(p) \qquad (I-7a)$$

$$U(p_2^*) = -c + R \int U(p)\mathrm{d}F(p) \qquad (I-7b)$$

$V(p)$ is at least as large as $U(p)$ since it represents the choice between purchase and searching again. Thus $p_1^* \leq p_2^*$, with a strict inequality in problems where the

search cost and discount rate are not so large as to always imply a purchase in the current store.

There are additional reasons for cut-off prices to rise over time or equivalently, in a job search setting, for reservation wages to fall over time. In many settings, search costs rise over time. The utility of a purchase or of finding a job can fall over time. In the job setting, these can arise from declining wealth being used to finance consumption while searching for a job and from the shortening period over which any job might be held.

A known finite horizon for the end of search is incorrect in many settings. In addition, with many periods, the backwards induction optimization process is a cumbersome description of individual choice. Fortunately, the infinite horizon stationary case is easy to analyse. In this setting, a parallel analysis to that in (I–7) is a straightforward application of dynamic programming principles. With the assumption of a stationary environment the cut-off price is the same period after period. Denote by p^* the cut-off price and by V the optimized expected value of utility after paying the search cost to enter a store but before observing its price. Then V equals the utility of purchase if a purchase is made plus the probability of not making a purchase times the discounted optimized utility from facing the same problem one period later after paying search cost c:

$$V = \int_0^{p^*} U(p)\mathrm{d}F(p) + [1 - F(p^*)][-c + RV].$$
$$(I-8)$$

Solving (I–8) for V we have:

$$V = \frac{\int_0^{p^*} U(p)\mathrm{d}F(p) - c[1 - F(p^*)]}{1 - R[1 - F(p^*)]}.$$
$$(I-9)$$

The optimal p^* maximizes V and can be calculated by differentiation. More intuitively, we note that a purchase just worth making will give the same utility as will waiting to search again:

$$U(p^*) = -c + RV. \qquad (I-10)$$

Rearranging terms, we can write the implicit equation for p^*:

$$(1 - R)U(p^*) = -c + R \int_0^{p^*} [U(p) - U(p^*)]\mathrm{d}F(p).$$
$$(I-11)$$

Using this first order condition we can analyse the comparative statics of optimal search behaviour. Naturally, the cut-off price increases if the search cost increases or if the discount factor becomes smaller. Interestingly, an increase in the riskiness of the distribution of prices (holding constant mean utility from a randomly selected price, $\int U(p)\mathrm{d}F$) makes search more valuable and so lowers the cut-off price. This result follows from the structure of optimal choice – decreases in low prices make search more attractive while increases in high prices are irrelevant since no purchase is made at high prices. Analysis of the relationship between the expected number of searches and the distribution of prices is complicated since it depends on the shape of that distribution.

Thus far we have assumed that all stores are *ex ante* identical; that is, that a choice to search is a choice to draw from the distribution $F(p)$. In many problems one can choose where to search. In that case, one is choosing which distribution $F(p)$ to sample from or, if there are limited draws allowed from a particular distribution, the sequence of distributions from which prices should be sampled. Interestingly, the reservation prices which tell whether to purchase or to sample again from a given distribution also serve to rank distributions.

In the choice problem analysed so far we have used discrete time, with the arrival of one offer in each time period. There are two straightforward generalizations. First, one might have the opportunity to receive more than one offer in any period, with the number of offers received being a function of the chosen level of search costs. In this way one can model the choice of search intensity. Second, the process of attempting to locate stores might have a stochastic rather than a determinate time structure. The simplest such model has the arrival of purchase opportunities satisfying the Poisson distribution law. That is, at any moment of time there is a constant flow probability of an offer arriving, any such offer being an independent draw from the distribution of available prices. Let us denote by a, the arrival rate of these offers; and by c, the constant search cost from being available to receive these offers. Utility is discounted at the constant (exponential) rate r. One can derive the optimal cut-off price and the optimized level of expected utility by analysing the discrete time process as above and passing to the limit. As a more intuitive alternative, let us think of the opportunity to purchase as an asset, where V now represents the value of that asset. The utility discount rate plays the role of an interest rate in asset theory. The asset is priced properly when the rate of discount times the value of the asset equals the expected flow of benefits from holding that asset. The expected flow of benefits is the gain that will come from making a purchase at a price below the cut-off price rather than continuing to search, adjusted for the probability of such an event, less search costs. Thus asset value satisfies

$$rV = a \int_0^{p^*} [U(p) - V]\mathrm{d}F(p) - c.$$

$$(I-12)$$

It is worthwhile to make any purchase with a higher utility than that from continued search. Thus the cut-off price satisfies

$$rU(p^*) = a \int_0^{p^*} [U(p) - U(p^*)]\mathrm{d}F(p) - c.$$

$$(I-13)$$

Again, one can introduce search intensity by having the Poisson arrival rate be a function of the search cost. In the equilibrium discussions below we will use the choice problem in the form (I–13).

So far we have ignored events after a purchase. In the labour setting this is equivalent to the assumption that taking a job is the end of search. In practice individuals frequently shift from job to job with no intervening period of unemployment. One can model job choice recognizing the possibility of continued search while working. Such an analysis must consider the rules that cover compensation between the parties in the event of a quit or firing, with no compensation and compensatory and liquidated damages being the situations analysed in the literature. The search for a better job is only one aspect of turnover. Also, one can model learning about the quality of match in a particular job as a function of the time on the job and the stochastic realization of experience. With a shadow value for quitting to search for a new job, one then has a second aspect of the theory of turnover.

The formulation of job taking given above has been combined with data on individual experience to examine empirically the determinants of the distribution of spells of unemployment. Since this essay focuses on equilibrium and the empirical literature has not examined the determinants of the distribution of opportunities, we do not explore this sizeable and interesting literature, nor the estimates of the effect of unemployment compensation on the distribution of unemployment spells. For an example, see Kiefer and Neumann (1979).

In the model above we have assumed that no additional information is received during the search process. In practice, individuals are simultaneously searching for many different consumer goods and often for jobs and investment opportunities as well. The relations among search processes coming from the arrival of information and the random positions with simultaneous search for many different goods have not been explored in the literature. Focusing on search for a single good, we have added several new factors to the theory of demand, particularly the cost of attempting to purchase elsewhere and the knowledge and beliefs of shoppers about opportunities elsewhere. In practice, these are important determinants of demand.

II. Equilibrium with bargaining

The theory of choice above is a simple version of the complex problem people face when making decisions about information gathering and purchases over time.

That simplicity yields a choice theory that can be embedded in an equilibrium model. To complete an equilibrium model, we need to model the determination of two endogenous variables: the arrival rate of purchase opportunities and the distribution of their prices. In this section, we consider prices that satisfy the bargaining condition of equal division of the gains from trade. In the next section, we consider take-it-or-leave-it prices set by suppliers. In both cases we assume that there are no reputations either of soft bargaining or low price setting that affect the arrival rate of potential customers. We begin by assuming that all buyers are identical and all sellers are identical. This case brings out the role of search in determining the level of prices. Below we consider determinants of the distribution of prices.

Axiomatic bargaining theory relates the terms of trade to the threat points of the two bargainers and the shapes of their utility functions. To avoid complications from the latter, we assume that a single unit is purchased and utility from purchase equals a constant, u_d, minus the price paid. We also assume that each seller has a single unit to sell. The utility from a sale is the price received less the cost of the good, u_s. One might think of this as a homogeneous used car market. To divide equally the gains from trade, the differences between the utility position with the trade and the utility position without it are equalized for the two parties. The value of purchasing at price p is $u_d - p$; expected utility without a trade is V_d, the optimized expected utility from continued search. We restrict ourselves to an economic environment where all trades take place at the same price. With a degenerate distribution of prices, we can rewrite the value equation (I–12) as

$$rV_d = a_d(u_d - p - V_d) - c_d. \qquad \text{(II–1)}$$

For suppliers, the utility from a sale is $p - u_s$. The gain from selling now rather than later is $p - u_s$, less the value of having a car for sale, V_s. The carrying cost of having a car available for sale can be incorporated in the search cost. The value equation for suppliers is

$$rV_s = a_s(p - u_s - V_s) - c_s. \qquad \text{(II–2)}$$

We ignore the sufficient conditions for search to be worthwhile ($V_d, V_s \geq 0$).

Equal division of the gains from trade implies

$$
\begin{aligned}
u_d - p - V_d &= \frac{r(u_d - p) + c_d}{r + a_d} \\
&= \frac{r(p - u_s) + c_s}{r + a_s} = p - u_s - V_s.
\end{aligned}
$$
$$\text{(II–3)}$$

We have assumed the same utility discount rate for both parties. Thus we have a relationship between the equilibrium price, the arrival rates of trading

opportunities, the search costs, and the utility from ownership. Solving (II–3) for the equilibrium price, we have:

$$p = \frac{(r + a_s)(ru_d + c_d) + (r + a_d)(ru_s - c_s)}{r(2r + a_s + a_d)}. $$
$$\text{(II–4)}$$

Without direct search costs ($c_d = c_s = 0$), the position of the price between the seller's reservation price of u_s and the demander's reservation price u_d depends on the relative ease of finding alternative trading partners. As it becomes very easy to find buyers (a_s becomes infinite), the price goes to u_d. Alternatively, as it becomes very easy to find suppliers (a_d becomes infinite), the price goes to u_s. Furthermore, an increase in one's search cost pushes the price in an unfavourable direction. In this extremely simplified setting, (II–4) brings out the new element that search theory brings to equilibrium analysis, namely, the dependence of equilibrium prices on the abilities of traders to find alternatives. Implicit in Walrasian theory is the idea that a perfectly substitutable trade can be found costlessly and instantaneously. In this restricted sense, there is no consumer surplus in a Walrasian equilibrium.

To complete the theory we need to determine the two endogenous arrival rates of trading partners. Assuming a search process without history, these depend on the underlying technology for bringing together buyers and sellers and the stocks of buyers (N_d) and sellers (N_s). We write the arrival rates as $a_d(N_d, N_s)$ and $a_s(N_d, N_s)$. The two arrival rate functions satisfy the accounting identity between the numbers of purchases and of sales:

$$a_d(N_d, N_s)N_d = a_s(N_d, N_s)N_s. \qquad \text{(II–5)}$$

Next, we must examine the determinants of the stocks of buyers and sellers. This theory can be based on the stocks of traders or the flows of new traders. One extreme example is that the steady state stocks of buyers and sellers are exogenous. One then inserts the functions a_d and a_s in the price equation (II–4).

An alternative extreme to perfect inelasticity is the assumption of perfectly elastic supplies of buyers and sellers at given reservation values for search, $\bar{V}_s$ and $\bar{V}_d$. Assuming reservation values that are consistent with the existence of equilibrium with positive trade, the equality of gains from trade (II–3) implies

$$p = \frac{u_d + u_s + \bar{V}_s - \bar{V}_d}{2}. \qquad \text{(II–6)}$$

The numbers of traders actively searching adapts to give this simple formula. Substituting from (II–6) in (II–1) and (II–2) we have the necessary values of a_d and a_s and so two equations for N_s and N_d.

For a market with professional suppliers one can consider the case of inelastic demand ($\bar{N}_d$) and a perfectly elastic supply ($\bar{V}_s$). If we assume further that demanders visit suppliers at a rate and cost independent of the number of suppliers, then a_d and c_d are parameters. Solving (II–3) for p in terms of the exogenous variables we now have

$$p = \frac{(r + a_d)(u_s + \bar{V}_s) + r u_d + c_d}{2r + a_d}.$$

$$(II-7)$$

In this case the response of price to an increase in the cost of the good or the reservation utility of suppliers is $(r + a_d)/(2r + a_d)$, which is less than one. The speed of the search process relative to the interest rate determines the extent to which search equilibrium is different from Walrasian equilibrium. In a labour setting, an analogue to (II–7) shows how unemployment compensation affects wages by changing search costs.

Efficiency

There are two decisions implicit in the model above – whether to enter the search market and whether to accept a particular trade opportunity. The decision to enter a search market, like the choice of search intensity, affects the ease of trade of others. There is nothing in the process that determines prices which reflects the externalities arising from the impact of changed numbers on the opportunities to trade. Thus, in general, equilibrium will not be efficient and one has the possibility of both too much entry and too little entry.

In order to explore the efficiency of the choice of acceptable trades, we need a reason for waiting for a better deal in the future. This can be done by introducing differences in traders or differences in matches between preferences of demanders and goods on sale. However formulated, we have the proposition that the marginally acceptable trade generates no surplus to the two agents making that trade, yet the marginal trade changes the search environment of others. This involves externalities of the same kind as the entry decision already discussed. Again, in general, equilibrium is not efficient.

Individual differences

There are many patterns of differences among demanders in their evaluations of different goods. We explore two simple cases which have been dubbed quality differences and variety differences. With quality differences, all demanders have the same utility evaluation of goods. One asks how the price of a good varies with the quality of the good. With variety differences, all demanders have the same distribution of utility evaluations of the set of goods in the market, but demanders disagree as to which is better. There is then an issue of 'matching' preferences with goods. One asks how the price in a transaction varies with the quality of the match.

We use q as the index of universally agreed on quality, and denote by $p(q)$ the price paid in a transaction for a good of quality q. By suppressing all other differences, we have the same price in all the purchases of a good of any quality. We denote by $V_s(q)$ the optimized net value to a supplier of having a unit of quality q for sale. Paralleling (II–2), we can calculate the net gain to a supplier of selling his unit. This gain, $p(q) - u_s(q) - V_s(q)$, satisfies

$$p(q) - u_s(q) - V_s(q) = \frac{r[p(q) - u_s(q)] + c_s}{r + a_s}.$$

$$(II-8)$$

For the demander, we denote by V_d the value of entering the search market to make a purchase, and by $u_d(q)$ the utility, gross of purchase price, of purchasing a unit of quality q. Paralleling (I–12), the utility discount rate times the value of being a demander is equal to the net flow of gains from search. The gross flow of gains equals the arrival rate of purchase opportunities times the expected gain from a purchase. The expected gain is the utility of buying the good less the price that has to be paid for the good less the shadow value of being a searcher. Denoting the distribution of qualities in a randomly selected trade encounter by $F(q)$, the value of being a demander satisfies

$$r V_d = a_d \int [u_d(q) - p(q) - V_d] dF(q) - c_d.$$

$$(II-9)$$

A full equilibrium analysis of this model would require determination of the distribution $F(q)$ as well as the arrival rates a_d and a_s. $V_s(q)$ would play an important role in determining $F(q)$. Such a model could consider investment in human capital with a search labour market. We will not carry out such an analysis, but focus merely on the relative prices $p(q)$, given a non-degenerate distribution $F(q)$. This problem is kept simple by the uniformity of product evaluations, which results in consumers' purchasing the first unit encountered, just as in the homogeneous case above. In any trade, the gains are shared equally between buyer and seller. Using (II–8) and (II–9) to eliminate V_d and $V_s(q)$ in the equal gain condition (II–3), we have the equilibrium price function

$$(2r + a_s)p(q) = (r + a_s)u_d(q) + r u_s(q) - c_s + (r + a_s)$$

$$\times \left\{ c_d - a_d \int [u_d(z) - p(z)] dF(z) \right\} \Big/$$

$$\times (r + a_d). \qquad (II-10)$$

This generalization of the homogeneous case, (II–4), shows a price that rises with quality assuming that cost does.

$$p'(q) = \frac{(r + a_s)u_d'(q) + r u_s'(q)}{2r + a_s}. \qquad (II-11)$$

The speed of search relative to the interest rate determines the magnitude of deviation from the Walrasian result that with identical demanders all transactions give the same utility level $[p'(q) = u'_d(q)]$.

With pure quality differences, all consumers have the same expected utility from search, while suppliers have expected utilities which vary with the quality of goods for sale. In a symmetric variety model, both demanders and suppliers have the same expected utility from search. The variable q now represents the quality determined by the particular match of demander and good. We view the distribution of these qualities, $F(q)$, as given and the same for all demanders and all goods. Implicitly we are assuming random matching between demanders and different goods. It is now the case that a sufficiently poor match will not result in a trade. We denote by $u_d(q)$ the utility evaluation, gross of purchase price, of buying a good, by $u_s(q)$ the cost of supplying a good, and by $p(q)$ the price when the quality of a match is q. The value of search for a supplier satisfies

$$rV_s = a_s \int_{q_1}^{q_2} [p(q) - u_s(q) - V_s] dF(q) - c_s$$

$$(\text{II-12})$$

where q_1 is the lower bound of match qualities at which it is mutually advantageous to carry out a trade. At the lowest acceptable quality, q_1, $p(q_1)$ is equal to $u_s(q_1) + V_s$. The value of search for a demander continues to satisfy (II-9). The assumption that all mutually advantageous trades are taken implies that q_1 also equates the gain from a purchase $u_d(q_1) - p(q_1)$ with the utility from search V_d. Equating the gains from trade for buyer and seller and solving for the price we have

$$2p(q) = u_d(q) + u_s(q) - V_d + V_s.$$

$$(\text{II-13})$$

Price increases with match quality to reflect the changed cost of supply, $u'_s(q)$, plus half the change in surplus, $[u'_d(q) - u'_s(q)]/2$.

Recapitulating our analysis of search equilibrium with bargaining, we have seen two themes. The first is how the search for trading partners introduces an additional element in the determination of trading prices: namely, the relative ease of the two potential trading partners in finding alternative trades. Secondly, the presence of a costly trade coordination mechanism is naturally replete with externalities as the availability of traders affects the trading opportunities of others.

In the model used in this section negotiation is instantaneous while search is slow. A fascinating literature explores equilibrium in models where the negotiation process is an explicit game of exchanging bids that can be interrupted by the arrival of an alternative trading partner (cf. Rubinstein and Wolinsky, 1985).

III. Equilibrium with price setting

In contrast to the bargaining theory used above, we now assume that prices are set on a take-it-or-leave-it basis by suppliers. This rule of (not) bargaining over prices gives the supplier a potential for monopoly power. The search for alternatives limits this monopoly power. The fundamental question is how much. We begin with the assumption that the only source of price information is visiting randomly chosen suppliers sequentially one at a time. We assume many identical suppliers, implying equal profitability of different pricing strategies used in equilibrium. If all buyers have identical positive search costs and identical demand curves that yield a unique profit maximizing price, then the unique equilibrium is the price that would be set by a monopolist. This result assumes a sufficient number of suppliers that buyers will not search for a single low price. This extreme result comes from the uniformity of trading opportunities. The best a buyer can do is wait to make exactly the same deal in the future. Therefore a buyer is always willing to pay a little bit more today than he has to pay in the future. Thus the demand curve for an individual seller coincides with the underlying demand curve in the neighbourhood of the equilibrium price. Even though this result is limited to unrealistic cases, it is interesting that the price is independent of the cost and speed of search, as long as search is not costless and instantaneous.

Given the pervasive reality of price distributions in retail markets, it has been natural for the literature to concentrate on generating equilibria without uniform prices. With differences in demanders, either from differences in underlying characteristics or from differences in their history of past purchases, the equilibrium can involve a distribution of prices and the structure of that distribution will depend upon search costs. In this case, consumers care about the characteristics of other consumers since these characteristics affect price setting behaviour. Similarly, with differences among suppliers the equilibrium price distribution varies with search costs.

Information gathering

When visiting a store is the only way to learn its price, price quotations are gathered one at a time. Separating the gathering of price information from going to stores does not necessarily change the model. If price quotations are still gathered one at a time, the cost of going to purchase the good can be deducted from the utility of acquiring it, leaving the model unchanged. However, the separation of the gathering of information from the collection of goods opens up the possibility of sometimes receiving price quotations one at a time and sometimes two or more at a time. This possibility destroys the single price equilibrium in the model of identical buyers and sellers. To see this result, note that profit per sale is continuous in price but, with uniform prices, the number of sales is discontinuous in price since a slight decrease in price wins all sales when a firm's price is one of two that are learned simultaneously. With positive profit made on each sale it would always pay to decrease price slightly below the uniform price of all other suppliers. With constant costs the competitive price is not a possible equilibrium either since a price increase gains profits when one is the only price quote while losing zero profit sales when one is not the only price quote. Thus there is necessarily a distribution of prices in equilibrium. Without price reputations, a store can choose any price it wants without affecting the flow of information about that store. Therefore, with identical firms the equilibrium will satisfy an equal profit condition. There will be low-price high-volume stores and high-price low-volume stores. One way to complete this model is to allow purchasers a choice of intensity of search which stochastically generates varying numbers of price quotations per period. We examine three additional models – price guides, advertising, and word of mouth.

Price guide

In this extension of the model we continue to have consumers seek price information one price at a time. In addition, consumers can purchase a guide to lowest cost shopping, with the purchase cost varying across consumers. A consumer who purchases such a guide is directed to one of the lowest price stores; a consumer who does not follows the search procedure described above. Assuming free entry of identical firms with U-shaped costs and an equilibrium where some consumers purchase the price guide and some do not but otherwise consumers are identical, we have a two-price equilibrium. Some of the stores set the price at the competitive equilibrium level. These stores sell to all consumers who purchase price information and those sequential shoppers who are lucky enough to find one of these stores on their first shopping visit. The remaining stores have higher prices, equal to the cut-off price for searching consumers or the profit maximizing price for selling to such a consumer, whichever is lower. The fraction of stores of the two kinds and the aggregate quantity of stores per consumer are determined by the zero profit condition for the two pricing strategies. When more consumers purchase the price guide, there will be more stores setting the competitive price and a drop in the cut-off price of searching consumers. This external benefit to searching consumers implies the inefficiency of the original equilibrium. A very slight subsidization of the cost of the price guide involves a second-order efficiency cost to the purchase of guides, no effect on firms (which have zero profits), and a first-order gain to searching consumers.

Advertising

It is obviously counterfactual to have all the information flows resulting from actions by shoppers. Advertising is a pervasive modern phenomenon. We continue to assume that stores have no price reputations. If the form of advertising is direct communication of prices to individual consumers, we can construct a model that again results in a distribution of prices. Stochastic communication from stores to consumers naturally generates a distribution of the number of price quotes that consumers receive. Any specific model of the stochastic structure of attempted communication will generate a distribution of numbers of price quotes learned by consumers. Free entry then implies a particular equilibrium distribution of prices provided some consumers receive a single price quotation and others receive more than one.

Word of mouth

It is natural to model both the seeking of price information and the spreading of price information as costly activities. However, some price information passes between consumers as a costless activity, part of the pleasure of discussing life. The presence of word-of-mouth communication in addition to sequential shopping alters equilibrium. The natural way to model word-of-mouth price communication brings price reputations into the model, since the prices set in one period affect communications about stores in future periods when their prices might be different. In order to isolate the effect of word of mouth we consider a very artificial model. Stores set prices which must hold for two periods. Consumers shop in the first or second period but are otherwise identical. In the first period, there is only sequential search, visiting stores one at a time as modelled above. Between the first and second periods there are random contacts between first-period shoppers and second-period shoppers. In this way, each second-period shopper receives information about the price in some positive number of stores. We assume that some people hear of only one store, while others hear of at least two. Then there will be a distribution of prices, with the structure of the distribution depending on the details of the word-of-mouth process. This analysis can be extended by having shoppers tell not only of the prices they paid, but also of prices they have heard from others. Both types of communication require a model of memory. The density of stores has different effects on equilibrium prices for different models of memory. This approach has been used in a setting of search for quality rather than low price to argue that doctors' fees can be higher where there are more doctors per capita (Satterthwaite, 1979).

Recapitulating the analysis of search equilibrium with price setting but not price reputation, we have seen two themes. One is the tendency for even low-cost search to generate sizeable amounts of monopoly power because of similar incentives for all suppliers. The second is a tendency for equilibrium to have a distribution of prices. Since price distributions are a widespread phenomenon in decentralized economies, it is reassuring that the theory produces such distributions.

IV. Additional issues

We have considered the search analogue to competitive equilibrium. It was assumed that there were many small firms, whose behaviour was adequately approximated while ignoring their impacts on certain aspects of equilibrium. Search theory has also examined equilibria with small numbers of firms. It may pay a monopolist to have a distribution of prices across his outlets rather than a single price as a method of discriminating among consumers with different search costs, even though the need to search for a low price adds to the cost of purchase of the good (Salop, 1977). In a duopoly or oligopoly setting, it is natural to consider randomized pricing strategies which again give rise to a distribution of prices (Shilony, 1977). This may be one of the many factors that go into the empirical fact of sales by retail outlets.

The technology of shopping in the models above is extremely simple. Little has been done to marry the underlying search issues with some of the realities of the geographic distributions of consumers and firms and the normal travels of shoppers. Similarly, little has been done to model the search basis for the role of intermediaries.

Price reputations

All the models mentioned above omit or severely limit the inter-temporal links in profitability that arise from price reputations. This is a major hole in the existing literature. Probably significant progress in this area will have to await the discrimination of cases in which optimal strategies (whether determinate or stochastic) are stationary, from those in which optimal strategies involve building up a reputation which is then run down. In such a setting analysis will be very sensitive to the assumptions made about consumer knowledge both of existing prices and of price strategies followed by firms. It would be nice to have both an empirical evaluation of the level of consumer ignorance about opportunities, and a theoretical structure capable of examining the relationship between equilibrium and the extent to which consumers are accurately informed.

Conclusion

Walrasian theory assumes that consumers are perfectly informed about the prices of all commodities in the economy. This assumption is central for the law of one price, that a homogeneous commodity sells at the same price in all transactions in a given market. This assumption is also central for a variety of inequalities on prices, limiting price differences to be less than transportation costs. These inequalities are consequences of the absence of opportunities for arbitrage profits. In order to make a rigorous arbitrage argument, there must be simultaneous purchase and sale of the same commodity at different prices net of transportation costs. If the purchase and sale are at different times, there is likely to be risk for the would-be arbitrageur. Similarly, a proper arbitrage argument requires homogeneous commodities. It is improper to apply arbitrage arguments to labour markets for example, although migration arguments may lead to similar conclusions. In search theory with a known distribution of prices, there is a cost to finding any trading partner and possibly a large cost to finding one willing to trade at some particular price. This idea captures one aspect of the limitations on the extent of arbitrage arguments.

Realistically, one must recognize that infrequent traders are often ill-informed about the distribution of prices in the market. This introduces two important changes in the basic theory. One is that gathering information changes beliefs about the distribution of prices, as well as revealing the location of possible transactions. The second is the incentive created for sellers to find consumers whose beliefs make them willing to transact at high prices. The differences between the search for suckers and the hunt for the highest value use of resources has not been clearly drawn in the literature, yet this distinction is valid and important for evaluating the functioning of some markets. Search-based theory and empirical work have a long way to go until we have satisfactory answers to a number of allocation questions that are totally ignored in a Walrasian setting. Nevertheless, the theory has already shown how informational realities can seriously alter the conclusions of Walrasian theory.

It would have been highly duplicative to have reviewed search theory of the labour market as well as that of the retail market. For a survey of labour search theory and a partial guide to the literature, see Mortensen (1984). Individual patterns of unemployment spells are the key empirical fact requiring revision of the Walrasian paradigm.

The failure of the profession, thus far, to produce a satisfactory integration of micro and macroeconomics based on the Walrasian paradigm (with or without price stickiness) raises the thought that such an integration might come out of search theory. For a presentation of this view and discussion of some applications of search ideas to macro unemployment issues, see Diamond (1984).

P. DIAMOND

See also **exchange; search theory (new prespectives).**

Bibliography

Diamond, P. 1984. *A Search-Equilibrium Approach to the Micro Foundations of Macroeconomics.* Cambridge, Mass.: MIT Press.

Kiefer, N. and Neumann, G. 1979. An empirical job search model with a test of the constant reservation wage hypothesis. *Journal of Political Economy* 87, 69–82.

Mortensen, D. 1984. Job search and labor market analysis. In *Handbook of Labour Economics*, ed. R. Layard and O. Ashenfelter. Amsterdam: North-Holland.

Rothschild, M. 1974. Searching for the lowest price when the distribution is not known. *Journal of Political Economy* 82, 689–711.

Rubinstein, A. and Wolinsky, A. 1985. Equilibrium in a market with sequential bargaining. *Econometrica* 53, 1133–50.

Salop, S. 1977. The noisy monopolist: imperfect information, price dispersion and price discrimination. *Review of Economic Studies* 44, 393–406.

Satterthwaite, M. 1979. Consumer information, equilibrium industry price, and the number of sellers. *Bell Journal of Economics* 10(2), 483–502.

Shilony, Y. 1977. Mixed pricing in oligopoly. *Journal of Economic Theory* 14, 373–88.

search theory (new perspectives)

Search theory is an analysis of resource allocation in economic environments with trading frictions. These frictions include the difficulty of bringing potential traders together, coordinating agents' decisions, informing agents of trading opportunities and keeping records of agents' trading histories. In the market, trading frictions appear in various forms of transactions cost and they generate important regularities in quantities and prices. For example, there are unemployed workers, underutilized capital, and unsold goods in inventory, which indicate that markets are unable to exhaust all potentially desirable trades. Also, the law of one price predicted for a frictionless economy is at odds with the dispersion of prices often observed for similar goods.

In his classic article on SEARCH THEORY in the 1987 edition of *The New Palgrave* (reproduced in this edition), Diamond surveyed the early development of the theory. He formulated two types of search models. One is sequential search models, in which agents on one side of the market post prices. Agents on the other side receive price quotes at an exogenous rate and decide sequentially whether to accept the quotes. The other type is random-matching models, in which agents determine prices through bargaining after they are matched. Diamond explained how these models can generate price dispersion in the equilibrium. He also showed that the market fails to allocate resources efficiently in these models. Since 1987, the literature has extensively applied the two models to analyse economic issues such as price (wage) inequality, unemployment, business cycles, marriages, and investments in physical and human capital.

Search theory has also experienced developments on the theoretical side. One development is the exploration of the mechanisms which can improve efficiency of the market. A particular mechanism is directed search or competitive search, as pioneered by Peters (1991). This exploration has led to the formulations of search as a strategic game. Another development uses search to construct a microfoundation for monetary theory. This article will focus on directed search, with only a brief description of the search theory of money. For reference, we call the models in Diamond's SEARCH THEORY random (or undirected) search models.

The main difference between directed and undirected search lies in whether individual agents are able to use pricing mechanisms to change their matching frequency. Random search models specify pricing and matching as two independent processes. In particular, the matching frequency is either a parameter or a function of aggregate variables only. Because prices play a limited role, the market cannot internalize the externalities in the search process. By contrast, directed search models allow individual agents to use pricing mechanisms to directly affect the matching frequency. The explicit trade-off between the matching frequency and price improves efficiency.

Because efficiency is a fundamental issue in economics, it is the primary motivation for studying directed search. Another motivation is simply the fact that directed search is a realistic description of many markets. Sellers often advertise prices and buyers know many price offers before they visit stores (firms). Incorporating directed search may lead to a robust explanation for phenomena such as persistent unemployment – however, we will not pursue this empirical agenda here.

Instead, we will start with random search models and illustrate the inefficiency of the equilibria. Then, we will describe three models of directed search and related issues. Our conclusion will follow the brief description of monetary search theory. To simplify the language, we will treat the market as a labour market and let the time horizon be one period. The models are applicable to the goods market with straightforward modifications and they can be extended to infinite horizon (see Cao and Shi, 2000).

Random search and inefficiency

Consider a labour market that contains a large number of workers and firms. The number of workers searching for jobs is a fixed number u. All workers are the same and they are risk neutral. When employed, a worker produces goods whose value is $y > 0$. When unemployed, a worker enjoys leisure, the utility of which is normalized to 0. The number of vacancies is v, which is determined by competitive entry of firms. A potential firm can incur a cost c to create a vacancy, where $0 < c < y$. The technology of production has constant returns to scale so that a firm treats each vacancy separately. Normalize the production cost to 0.

Let us use a matching function, $M(u,v)$, to describe the total number of matches in the period. Let $\theta = u/v$ denote the 'tightness' of the market. The matching probability for a worker is $p(\theta) = M(u, v)/u$ and the matching probability for a vacancy is $q(\theta) = M(u, v)/v$. Assume that M is increasing, concave and differentiable in each argument for all θ such that $p, q \in (0,1)$. Moreover, the function has constant returns to scale. Thus, $p(\theta)$ is decreasing in θ, $q(\theta)$ is increasing in θ, and $q(\theta) = \theta p(\theta)$. Moreover, assume that $q(\theta)$ is concave, $q(0) = 0$, and $q(\infty) = 1$. The matching share of workers is defined as

$$s(\theta) = \frac{u}{M}\frac{\partial M(u,v)}{\partial u}.$$

Then, $s(\theta) = 1 + \theta p'(\theta)/p$. The matching share of firms is $(1-s)$.

Once a worker and a firm are matched, the two choose the wage for the worker, w. Assume that this is done with Nash bargaining, which maximizes the geometrically weighted surplus of the two sides of the match: $w^\sigma(y-w)^{1-\sigma}$. Here, the worker's bargaining weight is $\sigma \in [0,1]$. The solution for the wage share is $w/y = \sigma$.

The value of a vacancy is $J = q(\theta)(y - w)$ and the value of a worker's search is $V = p(\theta)w$. With competitive entry of firms, a firm's net profit is zero; that is, $J = c$. This equilibrium condition can be rewritten as $(1 - \sigma)q(\theta) = c/y$. A unique solution for θ exists if $0 < c < (1 - \sigma)y$.

In the equilibrium, some workers are unemployed and some jobs are vacant. However, the existence of unemployment alone is not a sufficient indication of inefficiency. With the matching technology, not all resources can be fully utilized. The appropriate notion of efficiency must respect the constraint of this technology.

Let us measure efficiency with a social welfare function. Define social welfare as the weighted sum of expected values of agents in the economy, where all agents are given the same weight. This measure is also equal to the expected utility of an agent who is ignorant of whether he or she is a worker or a firm. Because firms earn zero net profit, the welfare level is equal to the sum of workers' values, uV. Using the condition of competitive entry to substitute w, we can express the welfare level as aggregate output minus the vacancy cost, that is, $u[p(\theta)y - c/\theta]$. Because u is exogenous, the level of θ that maximizes welfare satisfies $-\theta^2 p'(\theta) = c/y$. If we compare this efficient outcome with the equilibrium outcome, we can see that the equilibrium is efficient if and only if:

$$\sigma = s(\theta).$$

This condition is the Hosios condition (see Hosios, 1990). It is required for efficiency for the following reason. The social value created by a marginal firm is $y[\partial M(u,v)/\partial v] = y(1-s)q$. In contrast, the firm's value in the equilibrium is $q(y-w)$. For the equilibrium to be efficient, the firm's value in the equilibrium must be equal to its social value. This requirement is met if and only if the wage share is equal to the matching share of workers, as the Hosios condition requires.

More specifically, a firm's entry into the market creates two externalities. One is positive – the presence of an additional firm increases the matching frequency of workers. The other externality is negative – an additional firm reduces the matching frequency of other firms. The Hosios condition ensures that the two externalities cancel out with each other. If $\sigma > s(\theta)$, a firm is under-compensated for its entry cost and the amount of entry is deficient; if $\sigma < s(\theta)$, a firm is over-compensated and the amount of entry is excessive.

With random matching, the equilibrium cannot satisfy the Hosios condition generically, because both sides of the condition involve exogenous elements of the model. In particular, when the matching function is Cobb–Douglas, the matching share s is a constant which is unrelated to the workers' wage share.

The inefficiency is not specific to the random-matching model. Instead, it is common to all undirected search models. For example, if a sequential search model is used instead of a random-matching model, then firms post wages rather than bargain over wages. There can be a non-degenerate distribution of wage shares in the equilibrium (see Burdett and Mortensen, 1998). However, the matching share will still be independent of the wage share and the independence leads to the violation of the Hosios condition.

Directed search and efficiency

Directed search links the wage share to the matching share by explicitly modelling an agent's trade-off between the wage and the matching frequency. To capture this trade-off, suppose that all agents expect each wage level to be associated with a market tightness by a function $\theta(w)$. Search is 'directed' in the sense that, by posting a particular wage, a firm expects to change the matching probability by affecting workers' applications. For a firm posting wage w, the matching probability is $q(\theta(w))$; for a worker who applies for to wage w, the matching probability is $p(\theta(w))$. The functions $p(\theta)$ and $q(\theta)$ have the properties assumed above. Given the tightness function, each firm chooses a wage to post to maximize the expected value $J = q(\theta(w))(y - w)$, and each worker chooses to apply for a wage that maximizes the expected value $V = p(\theta(w))w$. The equilibrium tightness must be consistent with competitive entry and workers' application decisions.

Without restricting the function $\theta(\cdot)$, there can be many equilibria. For example, take an arbitrary wage $w_0 \in (0, y)$, and let θ_0 satisfy: $q(\theta_0)(y - w_0) = c$. Suppose that workers believe that all firms will post only wage w_0. With this belief, workers will apply only for wage w_0. But if no worker applies to other wages, then all firms will indeed post only wage w_0. That is, the pair (w_0, θ_0), together with the particular belief, is an equilibrium. In this equilibrium, $\theta(w)$ is not well-defined for $w \neq w_0$, because there is no firm or worker at such wages.

One way to avoid this problem is to introduce a small measure of non-optimizing firms that post every feasible wage and to analyse the limit of the equilibrium when this measure approaches zero. Another way is to impose restrictions on the beliefs out of the equilibrium, as we do here. Let E be the set of equilibrium wages. For $w^* \in E$, denote the expected value of applying to w^* as $V^* = p(\theta(w^*))w^*$. We require that, for every $w^* \in E$, the function $\theta(\cdot)$ must satisfy $p(\theta(w))w = V^*$ for w in a neighbourhood of w^*. That is, a firm believes that workers will apply for a deviating wage to such an extent that they will be indifferent between the deviating wage and the equilibrium wage.

The restriction implies the following features of the trade-off between the wage level and the matching probability. First, because $p(\theta)$ is a decreasing function, a worker who applies for to a wage higher than an equilibrium wage expects to face a tighter market and, hence, a lower matching probability. Similarly, a firm that posts a wage higher than an equilibrium wage expects to increase its matching probability. Second, because $p(\theta)$ is differentiable, the restriction implies that the function $\theta(\cdot)$ is differentiable. Thus, the trade-off between the wage level and the matching probability is smooth.

To characterize the equilibrium, suppose $w^* \in E$, with $V^* = p(\theta(w^*))w^*$. Each firm takes V^* as given and chooses w to solve the following problem:

$$\max \ q(\theta(w))(y - w) \ \text{s.t.} \ p(\theta(w))w = V^*.$$

Under the earlier assumptions on the function $q(\theta)$, the problem above is a concave problem and the solution is interior for all $V^* \in (0, y)$. Using the relationship $q(\theta) = \theta p(\theta)$, we can derive the first-order condition of the problem as $w^*/y = s(\theta)$. The equilibrium satisfies the Hosios condition!

As before, we can determine the tightness in the equilibrium by the entry condition, $J = c$. Then, the worker's indifference condition recovers V^*. It is easy to see that the market tightness is identical to the efficient one. Thus, the equilibrium is efficient.

The reason why the equilibrium is efficient can be related to hedonic pricing. With directed search, the market functions as if there is a price (in terms of wage) for every level of tightness. The inverse of the function $\theta(\cdot)$ serves as such a pricing function. Given this function, each worker chooses to apply for a wage level that maximizes his or her expected utility and each firm posts a wage to maximize expected profit. In the equilibrium, the market prices the tightness efficiently. That is, the increase in wage that a firm is willing to give for a marginal increase in the tightness is equal to the increase in wage that a worker asks for to compensate for a tighter market. As a result, the equilibrium internalizes search externalities. Because of this link to hedonic pricing, directed search is also called competitive search (see Moen, 1997).

Directed search can also induce the efficient amount of investment. Suppose that each firm chooses the level of capital before entering the labour market. Anticipating that the equilibrium wage will divide the match surplus efficiently, firms will choose the efficient level of capital.

Strategic formulation of directed search

In the above analysis, the matching function is a black box – it is specified exogenously as in models of undirected search. Because the matching function is important for the analysis of efficiency, it is important to derive a matching function from agents' strategic behaviour. Peters (1991) and Burdett, Shi and Wright (2001) formulate such a strategic game of directed search. The formulation also justifies the restriction above on the beliefs out of equilibrium. Let us describe the game where both u and v are fixed numbers greater than or equal to two. Competitive entry can be introduced in the same way as above.

The one-period game is as follows. First, all firms post wages simultaneously. Each worker observes all firms' posted wages. (The essence of the model is the same if each worker can observe only two wages that are randomly drawn from posted wages.) Then, all workers choose the firms to which they apply. Assume that a worker can apply for only one job in the period, but the worker can use mixed strategy in the application. After receiving applicants, a firm randomly chooses one to be employed. Production takes place immediately and an employed worker is paid the posted wage. Then, the game ends.

There are many equilibria of this game that are asymmetric in the sense that identical agents do not use the same strategy. When $u = v = 2$, for example, one asymmetric equilibrium is that one worker applies only to one firm and the other worker applies only to the other firm, while the two firms post zero wage. In this equilibrium, there is no unemployment – unemployment is eliminated by implicit coordination between the two workers. That is, a worker believes that the other worker will not apply for the same job as he or she does. Other asymmetric equilibria involve trigger strategies that also feature implicit coordination. Such coordination is unlikely to be attainable when there are many agents in the market.

To emphasize the lack of coordination, we focus on the symmetric equilibrium, where all (identical) workers use the same mixed strategy to apply for the jobs. In this equilibrium, it is probable that two or more workers will apply for the same job, in which case some workers will be unemployed.

To characterize the equilibrium, consider a particular firm, called firm A. Suppose that other firms post wage w, but firm A posts wage x. If x is close to w, some workers will apply to firm A: if no other worker applied to firm A, a lone applicant to firm A would be employed with certainty, which would generate higher expected utility than applying to w. In fact, workers will increase the probability of applying to firm A until the expected utility from this application is the same as that from applying to other firms. Let a be the probability with which a worker applies to firm A. Then, firm A will receive one or more workers with probability $[1-(1-a)^u]$, and the expected number of applicants received by the firm will be ua. A worker who applies to firm A will be employed with probability $[1-(1-a)^u]/(ua)$. Because a worker's application probabilities across the firms must add up to one, a worker applies to each firm other than firm A with probability $\pi(a)=(1-a)/(v-1)$. The probability of employment in such a firm is $[1-(1-\pi(a))^u]/(u\pi(a))$. For a worker to be indifferent between firm A and other firms, the expected payoff must be the same from these firms. That is,

$$\frac{1-(1-a)^u}{ua}x = \frac{1-[1-\pi(a)]^u}{u\pi(a)}w.$$

This equation defines a smooth function $a = f(x,w)$. This function serves the same role as the tightness function did in the above formulation of directed search – it describes how a firm's wage offer will affect workers' application, given other firms' wage offers. Note that f is an increasing function of x. Taking other firms' wage offers as given, firm A chooses x to solve:

$$\max \ (y-x)[1-(1-a)^u] \quad \text{s.t.} \quad a = f(x,w).$$

Denote the solution to this problem as $x = g(w)$.

A symmetric equilibrium is a wage level w such that $w = g(w)$. In this equilibrium, $a = \pi(a) = 1/v$. The first-order condition of the above maximization problem, evaluated in the equilibrium, yields:

$$w = y\left[\frac{(1-1/v)^{-u}-1}{u/v} - \frac{1}{v-1}\right]^{-1}.$$

The formulation above reveals two features of a market with a finite number of agents. First, the number of matches generated in the equilibrium is $v[1-(1-1/v)^u]$. This matching technology exhibits decreasing returns to scale. The reason is that when the number of agents increases, the coordination failure becomes more severe, and so the number of matches per agent falls. Second, when a firm chooses its wage offer, it cannot take as given the payoff which a potential applicant can get by applying elsewhere. We have made this interdependence explicit with the notation $\pi(a)$. That is, when firm A raises the offer, it will attract all workers to apply for it with a higher probability, which will increase the probability of employment at other firms. For any given offer by other firms, a worker's payoff of applying to those firms will increase as a result of the wage increase by firm A. These two features complicate the analysis.

Fortunately, the complexity disappears in the limit when the market becomes infinitely large. Suppose that u and v approach infinity, with a fixed ratio $\theta = u/v$. Then, the matching probability is $(1-e^{-\theta})$ for a firm and $(1-e^{-\theta})/\theta$ for a worker. These matching probabilities have all the properties assumed above and, in particular, they are independent of the scale. Moreover, $u\pi(a) \to \theta$, which is independent of an individual firm's offer, x. The payoff to a worker who applies to a firm other than firm A is $w(1-e^{-\theta})/\theta$, which is also independent of x. In the limit economy, the equilibrium satisfies the Hosios condition and it is efficient.

Other pricing mechanisms and price dispersion

Price-posting is not the only mechanism to direct search. There are other mechanisms of directed search that can generate efficiency, such as auction (for example, Julien, Kennes and King, 2000). In contrast to price-posting, auction induces price dispersion. Thus, efficiency does not necessarily require identical workers to be paid the same wage in an economy with risk-neutral agents.

Consider the following game with first-price auctions. Each firm announces a reserve wage and the following scheme. If two or more workers participate in the firm's auction, the participants bid on the wage and the lowest bidder is employed at the bid wage; if two or more workers have the lowest bid, one of them is chosen randomly by the firm; if only one worker participates, the worker is paid the reserve wage. After observing all firms' announcements, workers choose the auction in which they will participate. A worker can participate in only one firm's auction, although the choice can be a mixed strategy.

Choose an arbitrary firm and call it firm A. Let x be the reserve wage announced by firm A and a the probability with which a worker participates in this firm's auction. For a worker who participates in firm A's auction, there are two possible outcomes. The first is that the worker is the only participating worker, in which case the worker gets wage x. This outcome occurs with probability $(1-a)^{u-1}$. The other possibility is that the firm receives one or more other participants. In this case, the participants bid the wage down to 0. Thus, by participating in firm A's auction, a worker expects to obtain a value $(1-a)^{u-1}x$. For firm A, there are also two cases. If only one worker participates in the firm's auction, profit is $(y-x)$. This case occurs with probability $ua(1-a)^{u-1}$. If two or more workers participate, profit is y. The probability for this case is $[1-(1-a)^u-ua(1-a)^{u-1}]$. Thus, the expected value (profit) of firm A is:

$$ua(1-a)^{u-1}(y-x) + [1-(1-a)^u$$
$$- ua(1-a)^{u-1}]y.$$

For a firm other than firm A, let r be the reserve wage announced by the firm, $\pi(a)$ the probability of a worker's participation in the firm's auction, and $V(r,a)$ the expected value for a worker from such participation.

In order for a worker to be indifferent between firm A's auction and other firms' auctions, the expected value must be the same; that is, $(1-a)^{u-1}x = V(r,a)$. Taking this condition as a constraint and taking other firms' auctions as given, firm A chooses x to maximize the expected profit above. Let $x = g(r)$ be the optimal choice. Then, a symmetric equilibrium is a reserve wage r such that $r = g(r)$.

As in the case of wage-posting, the characterization of the equilibrium is simplified in the limit economy where $u \to \infty$ and $\theta = u/v \in (0, \infty)$. In such a limit, we have $\pi(a)v \to 1$. Hence, $\pi(a)$ and $V(r,a)$ are independent of a. Solving the above maximization yields $r = y$. Thus, in contrast to directed search with wage posting, auction generates a wage differential. Some employed workers are paid their productivity but others are paid their reservation wage, 0.

Despite the dispersion of wages, the equilibrium is efficient. With risk-neutral agents, it is expected wage, rather than the actual wage, that is important for efficiency. With auction, the expected payoff is $ye^{-\theta}$ to a worker and $y[1-(1+\theta)e^{-\theta}]$ to a firm. These expected payoffs are the same as those in directed search with wage posting.

Related issues

Risk aversion and asymmetric information: when workers are risk averse, different mechanisms of directed search can differ in efficiency. For example, price-posting generates lower risks in workers' income than auction. If the insurance market is imperfect, then wage-posting may give higher expected utility to workers than auction. Moreover, unemployment insurance, financed by lump-sum taxes, can improve welfare in this case (see Acemoglu and Shimer, 1999). On the other hand, price-posting is unlikely to be efficient when there is asymmetric information about the quality of goods or workers' productivity. Auction can allocate resources more efficiently in the presence of private information.

Heterogeneity and assortative matching: directed search models can be extended to allow workers to be heterogeneous. To achieve efficiency in such an extension, firms must rank different types of workers in addition to announcing wages. Heterogeneity can also appear on both sides of the market. In this case, an interesting question is whether the matching pattern is assortative, that is, whether similar attributes are matched with each other. In a frictionless economy, the efficient matching pattern is positively assortative, provided that the attributes on the two sides of the market are complementary. Moreover, the competitive equilibrium can implement the efficient matching outcome. When search frictions are introduced through undirected search, the equilibrium pattern of matches is neither assortative nor efficient (for example, Sattinger, 1995; Shimer and Smith, 2000). Introducing directed search can restore efficiency. However, the efficient matching pattern may be non-assortative when utility is transferable. There is a trade-off between the matching quality and the matching rate (for example, Shi, 2001; 2002).

Multiple applications: most search models assume that an agent on one side of the market can visit only one agent on the other side of the market in a given period; for example, a worker can apply only for one job at a time. This assumption may not be realistic in some markets. When workers can apply for multiple jobs simultaneously, there is a new source of the failure of coordination among firms: two firms may select the same worker and one of them will fail to obtain the worker. If the left-out firm has no recourse to other applicants it received, then the equilibrium is inefficient even with directed search. However, there are rules of selection, such as the one described by Gale and Shapley (1962), that can eliminate this difficulty of coordination and restore efficiency.

On-the-job search: search on the job is rarely examined in directed search models; sequential (undirected) search models have dominated the analysis on this topic (see Mortensen, 2003; 2007; LABOUR MARKET SEARCH). These models are constructed typically in continuous time. They assume that each worker, whether employed or not, receives a wage offer according to a Poisson process. While an unemployed worker receives one offer at a time, an employed worker effectively receives two offers – his current wage and the new offer from another firm. In this environment, the equilibrium must have a continuous distribution of wages with no mass point in the interior of the support; otherwise, a firm's payoff function would be discontinuous on the right side of the mass point. These models yield strong predictions on the shape of the wage distribution, some of which are counter-factual. Allowing on-the-job search to be directed can make the predictions more realistic and generate limited wage mobility endogenously, as shown by Delacroix and Shi (2006).

Search as a microfoundation for monetary theory

A surprising development of search theory is its use in monetary theory. For monetary economics, a fundamental question is why intrinsically useless objects, such as fiat money, can have a positive value in the equilibrium (see Wallace, 2007: FIAT MONEY). A familiar but informal answer is that such objects relieve the difficulty of exchange by acting as media of exchange. To capture this role of money, traditional monetary theory has used shortcuts while keeping the assumption of frictionless (Walrasian) markets. Examples include the requirement that agents must hold cash in advance of purchases and the assumption that money yields direct utility which cannot be generated by other assets. These short cuts seem incompatible with the Walrasian markets in the model and they are unable to explain why different media of exchange can have different values. To formalize the difficulty of exchange, Kiyotaki and Wright (1993) abandoned the short cuts and replaced the Walrasian exchange with random bilateral matches. The resulting model is a value theory of money, which gives money a role in improving efficiency of the market. Shi (1995) and Trejos and Wright (1995) integrate sequential bargaining into monetary search models to determine prices: see Shi (2006), Wright (2007) and MATCHING for a survey.

Monetary theory has gone one step further to analyse optimal trading mechanisms. Using the method of mechanism design, the theory characterizes the set of allocations that are compatible with agents' incentives in the presence of search frictions. Next, the theory examines the efficient allocations and asks whether the implementation of these allocations entails particular types of trade, such as the use of money, banking, or a payments system – for example, Green and Zhou (2005). This analysis has clarified the relationship between optimal trading mechanisms and different components of search frictions, such as the difficulty for agents to meet, the difficulty for the society to keep record of agents' transactions, and the difficulty of enforcing trades.

Conclusions

Search theory was initially formulated to understand price dispersion and unemployment. Recent research has shifted the focus to the pricing mechanism and efficiency in frictional economies. Directed search is formulated to allow agents to explicitly make a trade-off between prices and matching frequency. The main finding is that directed search can restore efficiency that failed in earlier search models. However, even the efficient allocation cannot fully utilize all resources, because of the constraint of the matching technology. Moreover, the efficient allocation may not have the assortative pattern that emerges in a frictionless economy. The literature has explored different pricing mechanisms of directed search and used search to develop a microfoundation for monetary theory.

By focusing on pricing mechanisms and efficiency, the research has brought search theory close to the task of analysing the interactions between trades inside economic organizations and outside in the market. These interactions are important for explaining the observed forms of contracts and trading institutions. Monetary search theory has already taken up this task by using the approach of mechanism design. Other fields can also benefit from incorporating search frictions. An example is the literature on optimal dynamic contracts. By introducing search, one can endogenize the duration in which an agent stays with a particular contract. In general, the integration of search theory and contract theory awaits future research.

Another direction of future research is to explore the empirical implications of directed search. Most empirical work on wage distribution and business cycles has employed random search models (see Mortensen, 2003;

Andolfatto, 2007; SEARCH MODELS OF UNEMPLOYMENT). It is known that wage dispersion in these models is sensitive to the assumption of undirected search. Moreover, these models need a large amount of heterogeneity among workers and firms to explain the observed distribution of wages, which seems to diminish the spirit of search. Directed search may offer a useful alternative approach in the empirical investigation.

SHOUYONG SHI

See also **fiat money; labour market search; matching; search models of unemployment; search theory.**

Bibliography

Acemoglu, D. and Shimer, R. 1999. Efficient unemployment insurance. *Journal of Political Economy* 107, 893–928.

Andolfatto, D. 2007. Search models of unemployment. In *The New Palgrave Dictionary of Economics*, 2nd edn. ed. L. Cornell and S. Durlauf. London: Macmillan.

Burdett, K. and Mortensen, D. 1998. Wage differentials, employer size, and unemployment. *International Economic Review* 39, 257–73.

Burdett, K., Shi, S. and Wright, R. 2001. Pricing and matching with frictions. *Journal of Political Economy* 109, 1060–85.

Cao, M. and Shi, S. 2000. Coordination, matching, and wages. *Canadian Journal of Economics* 33, 1009–33.

Delacroix, A. and Shi, S. 2006. Directed search on the job and the wage ladder. *International Economic Review* 47, 651–99.

Diamond, P. 1987. Search theory. In *The New Palgrave Dictionary of Economics*, ed. J. Eatwell, M. Milgate and P. Newman. London: Macmillan.

Gale, D. and Shapley, L. 1962. College admissions and the stability of marriage. *American Mathematical Monthly* 69, 9–15.

Green, E. and Zhou, R. 2005. Money as a mechanism in a Bewley economy. *International Economic Review* 46, 351–71.

Hosios, A. 1990. On the efficiency of matching and related models of search and unemployment. *Review of Economic Studies* 57, 279–98.

Julien, B., Kennes, J. and King, I. 2000. Bidding for labor. *Review of Economic Dynamics* 3, 619–49.

Kiyotaki, N. and Wright, R. 1993. A search-theoretic approach to monetary economics. *American Economic Review* 83, 63–77.

Moen, E.R. 1997. Competitive search equilibrium. *Journal of Political Economy* 105, 385–411.

Mortensen, D. 2003. *Wage Dispersion: Why Are Similar People Paid Differently*. Cambridge: MIT Press.

Mortensen, D. 2007. Labour market search. In *The New Palgrave Dictionary of Economics*, 2nd edn. ed. L. Cornell and S. Durlauf. London: Macmillan.

Peters, M. 1991. Ex ante price offers in matching games: non-steady state. *Econometrica* 59, 1425–54.

Sattinger, M. 1995. Search and the efficient assignment of workers to jobs. *International Economic Review* 36, 283–302.

Shi, S. 1995. Money and prices: a model of search and bargaining. *Journal of Economic Theory* 67, 467–96.

Shi, S. 2001. Frictional assignment. 1. efficiency. *Journal of Economic Theory* 98, 232–60.

Shi, S. 2002. A directed search model of inequality with heterogeneous skills and skill-biased technology. *Review of Economic Studies* 69, 467–91.

Shi, S. 2006. A microfoundation of monetary economics. *Canadian Journal of Economics* 39, 643–88.

Shimer, R. and Smith, L. 2000. Assortative matching and search. *Econometrica* 68, 343–70.

Trejos, A. and Wright, R. 1995. Search, bargaining, money and prices. *Journal of Political Economy* 103, 118–41.

Wallace, N. 2007. Fiat money. In *The New Palgrave Dictionary of Economics*, 2nd edn. ed. L. Cornell and S. Durlauf. London: Macmillan.

Wright, R. 2007. Matching models. In *The New Palgrave Dictionary of Economics*, 2nd edn. ed. L. Cornell and S. Durlauf. London: Macmillan.

seasonal adjustment

Seasonal adjustment of economic time series dates back to the 19th century and it is based on an attitude properly expressed by Jevons, who wrote:

> Every kind of periodic fluctuation, whether daily, weekly, monthly, quarterly, or yearly, must be detected and exhibited not only as a subject of a study in itself, but because we must ascertain and eliminate such periodic variations before we can correctly exhibit those which are irregular or non-periodic, and probably of more interest and importance. (1884, p. 4)

The most popular model behind seasonal adjustment in the beginning of the 20th century was either the so-called additive unobserved components (UC) model

$$X_t = T_t + C_t + S_t + I_t,$$
$$t = 1, 2, \ldots, n \tag{1}$$

where the observed series X_t is divided into a trend component, T_t, a business cycle component, C_t, a seasonal component, S_t, and an irregular component, I_t, or the multiplicative UC model

$$X_t = T_t * C_t * S_t * I_t,$$
$$t = 1, 2, \ldots, n \tag{2}$$

which is applied if the series is positive and the oscillations increase with the level of the series.

The definitions of the individual components could vary, but Mills (1924, p. 357) defined the trend component, T_t, as the smoothed, regular, long-term movement of the series X_t, while the seasonal component, S_t, contains fluctuations that are definitely periodic in character with a period of one year, that is, 12 months or 4 quarters. The business cycle component, C_t, is less markedly periodic, but characterized by a considerable degree of regularity with a period of more than one year, while the irregular component, I_t, has no periodicity. A detailed description of the historical development is given in Hylleberg (1986).

The rationale behind seasonal adjustments is that the unobserved components model is useful, that the components are independent, and that the components of interest are the trend and cycle components.

The assumption of independence is highly questionable, as the actual economic time series is a result of economic agents' reaction to some exogenous seasonally varying explanatory factors such as the climate, the timing of religious festivals and business practices. For typical economic agents, decisions designed to smooth seasonal fluctuations will naturally interact with non-seasonal fluctuations, since the costs of such smoothing will necessarily be interrelated through budget constraints and so forth. Therefore, not only is the independence assumption economically unreasonable, but seasonal patterns may be expected to change if economic agents change their behavioural rules.

Hylleberg (1992, p. 4) defines seasonality as

> A systematic, although not necessarily regular, intra-year movement caused by the changes of the weather, the calendar, and timing of decisions, directly or indirectly through the production and consumption decisions made by agents of the economy. These decisions are influenced by endowments, the expectations and preferences of the agents, and the production techniques available in the economy.

Such a view of seasonality is somewhat different from the views expressed by most statistical data-producing agencies. The views of the statistical offices are well represented by the arguments put forward by OECD (1999, p. vii), where the implied definition of seasonality stresses the fixed timing of certain events during the year. Likewise, they indicate that the reason for changes in the seasonal pattern is 'the trading day effect', that is, the changing number of working days in a month, the changing number of Saturdays, and movable feasts such as Easter, Pentecost, Chinese New Year and Korean Full Moon Day. Obviously, such factors do influence the seasonal pattern in economic time series, but in the longer run technical progress and economic considerations based on these will imply changes in the seasonal pattern as well. In addition, the seasonal economic time series may constitute an invaluable and plentiful source of data for testing theories about economic behaviour, as the seasonal pattern is a recurrent though changing event where the pattern, despite the changes, is somewhat easier to forecast than many other economic phenomena. (For a general discussion of seasonality and the literature, see Hylleberg, 1986. For a presentation and discussion of the results since then, see Hylleberg, 1992; Franses, 1996; Ghysels and Osborn, 2001; Brendstrup et al., 2004.)

Seasonal adjustment and treatment of the seasonal components may in practice be undertaken in two ways: simply applying the seasonally adjusted data produced by the statistical agencies, or integrating the modelling and adjustment into the econometric analysis undertaken.

Officially applied seasonal adjustment programmes

Several different methods for seasonal adjustment are in actual use, but the most popular programme is the X-12-ARIMA seasonal adjustment programme (see Findley et al., 1998) which is a further development of the popular X-11 seasonal adjustment programme (see Shiskin, Young and Musgrave, 1967; Hylleberg, 1986). Another popular seasonal adjustment programme is the TRAMO/SEATS programme developed in Gomez and Maravall (1996).

X-12-ARIMA seasonal adjustment programme

The main characteristics of the X-11 seasonal adjustment method for the monthly multiplicative model (see (2)),

$$X_t = TC_t * S_t * TD_t * H_t * I_t, \qquad (3)$$

where TC_t is the combined trend-cycle component, while TD_t is the trading day component, and H_t the holiday component, is the repeated application of selected moving averages such as a 12-month centred moving average to estimate TC_t followed by an actual extraction of the estimated trend-cycle component. The extraction by the moving average filters takes place after a prior adjustment for trading days and certain holidays, and a varying seasonal pattern is taken care of by applying so-called Henderson moving averages with a 9, 13 or 23 number of terms. The Henderson trend filters are used in preference to simpler moving averages because they can reproduce polynomials of up to degree 3, thereby capturing trend-turning points.

In addition, treatment for so-called extreme observations was possible, and a refined asymmetric moving averages filter is used at the ends of the series.

In order to robustify the initial seasonally adjusted series against data revisions, the X-11 seasonal adjustment method was improved by extending the series by forecasts and backcasts from an ARIMA model before seasonally adjusting the series (see Dagum, 1980).

The X-12-ARIMA seasonal adjustment programme described in Findley et al. (1998) extends the facilities of X-11-ARIMA by adding a modelling module denoted RegARIMA, which not only facilitates modelling the processes in order to forecast and backcast the time

series, but also facilitates modelling of trading day and holiday effects, detection of outlier effects, dealing with missing data, detection of sudden level changes, and detection of changes in the seasonal pattern, trading day effects and so forth. The second major improvement on the earlier programmes is the inclusion of a module for diagnostics which contains many helpful 'tests'. The third improvement is a user-friendly interface.

Although X-12 is a major improvement to X-11, it has its critics. For example, Wallis (1998) doubts that the trend estimation procedure taken over from X-11 is still the best available despite the results obtained since the mid-1970s, and he stresses the need to give the user of the adjusted numbers an indication of their susceptibility to revision.

TRAMO/SEATS seasonal adjustment programme

The main difference between the X-12 programme and the TRAMO/SEATS programme is that the former uses signal-to-noise ratios to choose between the different moving average filters available while SEATS uses signal extraction with filters derived from a time series (ARIMA) model.

The programme also contains a preadjustment programme, TRAMO, which basically performs tasks similar to RegARIMA in X-12.

The signal extraction is based on an additive model such as (1) or

$$Y_t = \mu_t + \gamma_t + \varepsilon_t, \qquad (4)$$

where μ_t is the trend-cycle component, γ_t the seasonal component, and ε_t is the irregular component. It is then assumed that the μ_t and γ_t can be modelled as two distinct ARIMA processes

$$A_C(L)\,(1-L)^d\mu_t = B_C(L)v_t \quad \text{and}$$
$$A_S(L)\,(1-L^s)^D\gamma_t = B_S(L)w_t \qquad (5)$$

where the processes v_t, w_t and ε_t are independent, serially uncorrelated processes with zero means and variances σ_u^2, σ_w^2 and σ_ε^2, and d and D are integers, while L is the lag operator. This class of model is also called the unobserved components autoregressive integrated moving-average model (UCARIMA) by Engle (1978).

Hence, the TRAMO/SEATS programme requires the estimation of the UCARIMA parameters for each specific series, which in principle should allow computation of the correct number of degrees of freedom. This is not possible in X-12 due to the adjustments undertaken within the programme based on the characteristics of the individual series.

A discussion of the merits and drawbacks of X-12 and TRAMO/SEATS may be found in Ghysels and Osborn (2001), Hood, Ashley and Findley (2004), and several working papers from EUROSTAT (see Mazzi and Savio, 2005), which find that X-12 is slightly preferable to

TRAMO/SEATS when applied to short time series – a result to be expected as the model-based approach requires more data. In fact, the main differences between the two leading competitors reflect the difference between the model-based approach of TRAMO/SEATS, which tailors a seasonal filter to each series, and the uniform filter applied by X-12 (see below). However, the model-based approach relies on a very restrictive set of models, and the uniform filter approach is not really applying the same filter, as individual characteristics like outliers, smoothness, and so on, have an influence on the filter.

Seasonal adjustment as an integrated part of the analysis

The main objective of the production of seasonally adjusted time series is to give the policy analyst or adviser easy access to a common time series data-set that has been purged of what is considered noise contaminating the series. Obviously, the application of the seasonally adjusted data may be more or less formal and meticulous, ranging from eyeball analysis to thorough econometric analysis.

However, although the application of officially seasonally adjusted data may have the advantage of saving costs, it also implies that the user runs a severe risk of not making the most effective use of the information available and – perhaps more serious – of applying a dataset distorted by the applied seasonal filter.

The possible reasons for these shortcomings are:

- the seasonal component is a noise component but

  – the wrong seasonal adjustment filter has been applied, or
  – the data have been seasonally adjusted individually without consideration being given to the fact that they are often used as input to a multivariate analysis;

- the seasonal components of different time series may be closely connected and contain valuable information across series.

Filtering the data before they are applied may of course distort the outcome of the analysis if the wrong filter is used, but even if the filter is 'correct' as seen from the individual series, the filtering may produce biased estimates of the parameters in certain cases where, for instance, a regression model is applied (see Hylleberg, 1986, p. 3). However, this result is complicated by the application of other transformations to the original series. Which filter to apply may in fact depend on the order of the applied transformations, as shown by Ghysels (1997).

Hence, in order to optimally model many economic phenomena, seasonality may need to be treated as an integrated part of an econometric analysis based on unadjusted quarterly, monthly, weekly and daily time series or panel data observations. This may be done in many different ways depending on the specific context and the set of reasonable assumptions one can make within that context.

As both X-12 and TRAMO/SEATS seasonal adjustment programmes are available to the individual researcher, they may both be applied as part of an integrated approach and their use somewhat adapted to the specific analysis. In what follows we discuss some alternative methods, of three kinds: pure noise models, time series models and economic models.

Pure noise models

The first group comprises seasonal adjustment methods which are based on the assumption that the seasonal component is noise. Thus, the group also contains the officially applied seasonal adjustment programmes presented earlier. The seasonal adjustment methods in this group are distinguished by their ability to take care of a changing seasonal component.

Seasonal dummies

The use of seasonal dummy variables to filter quarterly and monthly times series data is a very simple, straightforward and therefore popular method in econometric applications. The dummy variable method is designed to take care of a constant, stable seasonal component. The popularity of the seasonal dummy variable method is partly due to its simplicity and the flexible way it can be used either as a pre-filtering device whereby the series are regressed on a set of seasonal dummy variables and the residuals used in the final regression, or within the regression as an extension of the set of regressors with seasonal dummy variables (see Frisch and Waugh, 1933; Lovell, 1963).

Band spectrum regression and band pass filters

A natural and quite flexible way to analyse time series with a strong and somewhat varying periodic component is to perform the analysis in the frequency domain, where the time series is represented as a weighted sum of cosine and sine waves. Hence, the time series are Fourier-transformed and the seasonal filtering of the time series may take place by removing specific frequency components from the Fourier-transformed data series.

Application of such filters dates back a long time (see Hannan, 1960). Band spectrum regression is further developed and analysed by Engle (1974; 1980) and Hylleberg (1977; 1986). The so-called real business cycle literature has since named it 'band pass filtering' (see Baxter and King, 1999).

Let us assume that we have data series with T observations in a vector y and a matrix X related by $y = X\beta + \varepsilon$, where ε is the disturbance term and β a coefficient vector. Band spectrum regression is then performed as a regression in the transformed model $A\Psi y = A\Psi X\beta + A\Psi \varepsilon$, where the transformation by the matrix Ψ is a finite Fourier transformation of the data. The transformation by the diagonal matrix A with zeros and ones on the diagonal, symmetric around the south-west north-east diagonal, is a filtering which removes the frequency components corresponding to the elements with the zeros. Hence, by an appropriate choice of zeros in the main diagonal of A the exact seasonal frequencies and possibly a band around them may be filtered from the series.

An obvious advantage of the band spectrum regression representation is that the model $A\Psi y = A\Psi X\beta + A\Psi \varepsilon$ lends itself directly to a test for the appropriate filtering, as argued in Engle (1974). In fact the test is just the well-known so-called Chow test applied to a stacked model with the null hypothesis that the parameters are the same over the different frequencies. A drawback of band spectrum regression is that the temporal relations between series may be affected in a complicated way by the two-sided filter (see Engle, 1980; Bunzel and Hylleberg, 1982).

Seasonal integration and seasonal fractional integration
A simple filter often applied in empirical econometric work is the seasonal difference filter $(1 - L^s)^d$, where s is the number of observations per year and d the number of times the filter should be applied to render the series stationary at the long run and seasonal frequencies (see Box and Jenkins, 1970).

In the unit root literature a time series is said to be integrated of order d if its d'th difference has a stationary and invertible ARMA representation. Hylleberg et al. (1990) generalize this to seasonal integration and denote, for instance, a quarterly series y_t, $t = 1, 2, \ldots, T$ represented by the model $(1 - L^4)y_t = \varepsilon_t$, $\varepsilon_t \sim$ i.i.d. $(0, \sigma^2)$ as integrated of order 1 at frequency θ since $(1 - L^4) = (1 - L)(1 + L)(1 + L^2)$ has real roots at the unit circle at the frequencies $\theta = \{0, \frac{1}{2}, [\frac{1}{4}, \frac{3}{4}]\}$, where θ is given as the share of a total circle of 2π.

Many empirical studies have applied the so-called Hylleberg, Engle, Granger, and Yoo (HEGY) test for seasonal unit roots developed by Hylleberg et al. (1990) and Engle et al. (1993) for quarterly data, extended to monthly data by Beaulieu and Miron (1993), and to daily data integrated at a period of one week by Kunst (1997). These tests are extensions of the well-known Dickey–Fuller test for a unit root at the long-run zero frequency (Dickey and Fuller, 1979) and at the seasonal frequencies (Dickey, Hasza and Fuller, 1984).

The HEGY test is the simplest and most easily applied test for seasonal unit roots. In the quarterly case the test is based on an autoregressive model $\phi(L)y_t = \varepsilon_t$, $\varepsilon_t \sim iid$ $(0, \sigma^2)$ where $\phi(L)$ is a lag polynomial with possible unit roots at frequencies $\theta = \{0, \frac{1}{2}, [\frac{1}{4}, \frac{3}{4}]\}$. A rewritten linear regression model where the possible unit roots are isolated in specific terms is

$$\phi^*(L)y_{4t} = \pi_1 y_{1t-1} + \pi_2 y_{2t-1} + \pi_3 y_{3t-2} + \pi_4 y_{3t-1} + \varepsilon_t$$
$$y_{1t} = (1 + L + L^2 + L^3)y_t$$
$$y_{2t} = -(1 - L + L^2 - L^3)y_t$$
$$y_{3t} = -(1 - L^2)y_t$$
$$y_{4t} = (1 - L^4)y_t. \qquad (6)$$

The lag polynomial $\phi^*(L)$ is a stationary and finite polynomial by assumption. Denoting integration of order d at frequency θ by $I_\theta(d)$ we thus have $y_{1t} \sim I_0(1), y_{2t} \sim I_{\frac{1}{2}}(1)$, and $y_{3t} \sim I_{[1/4,3/4]}(1)$, while $y_{1t} \sim I_{\frac{1}{2},[1/4,3/4]}(0)$, $y_{2t} \sim I_{0,[1/4,3/4]}(0)y_{3t} \sim I_{0,\frac{1}{2}}(0)$, and $y_{4t} \sim I_{0,\frac{1}{2},[1/4,3/4]}(0)$, provided $y_t \sim I_{0,\frac{1}{2},[1/4,3/4]}(1)$.

The HEGY tests of the null hypothesis of a unit root are conducted by 't-value' tests on π_1 for the long-run unit root, π_2 for the semi-annual unit root, and 'F-value tests' on π_3 and π_4 for the annual unit roots. In fact, the 't-value' tests on π_1 is just the unit root test of Dickey and Fuller with a special augmentation applied. As in the Dickey–Fuller cases the statistics are not t or F distributed but have non-standard distributions, which for the 't' are tabulated in Fuller (1976) while critical values for the 'F' test are tabulated in Hylleberg et al. (1990).

As in the Dickey–Fuller case the correct lag-augmentation in the auxiliary regression (6) is crucial. The errors need to be rendered white noise in order for the size to be close to the stipulated significance level, but the use of too many lag coefficients reduces the power of the tests.

Obviously, if the data-generating process (DGP) contains a moving average component, the augmentation of the autoregressive part may require long lags (see Hylleberg, 1995) and the HEGY test may be seriously affected by autocorrelation in the errors, moving average terms with roots close to the unit circle, so-called structural breaks, and noisy data with outliers.

The existence of seasonal unit roots in the DGP implies a varying seasonal pattern where 'summer may become winter'. In most cases such an extreme situation is not logically possible, and the findings of seasonal unit roots should be taken as an indication of a varying seasonal pattern and the unit root model as a parsimonious approximation to the DGP.

Another test where the null is no unit root at the zero frequency is suggested by Kwiatkowski et al. (1992) and extended to the seasonal frequencies by Canova and Hansen (1995), and further developed by Busetti and Harvey (2003). See Hylleberg (1995) for a comparison of the Canova–Hansen test and the HEGY test. See also Taylor (2005) for a variance ratio test.

Arteche (2000) and Arteche and Robinson (2000) have extended the analysis to include non-integer values of d in the definition of a seasonally integrated process. In case d is a number between 0 and 1 the process is called fractionally seasonally integrated. The fractionally integrated seasonal process is said to have strong dependence or long memory at a frequency ω since the autocorrelations at that frequency die out at a hyperbolic rate, in contrast to the much faster exponential rate in the weak dependence case where $d = 0$. In the integrated case where $d = 1$ the autocorrelations never die out.

The difficulty with the fractional model is estimation of the parameter d. Even in the quarterly case there are three possible d parameters, and the testing procedure may become very elaborate, requiring, for instance, a sequence of clustered tests as in Gil-Alana and Robinson (1997).

Time series models

The time series models may be univariate models such as the Box–Jenkins model, unobserved components models, time varying parameter models or evolving seasonal models, or multivariate models with seasonal cointegration or periodic cointegration, or models with seasonal common features.

Univariate seasonal models

The Box–Jenkins model. In the traditional analysis of Box and Jenkins (see Box and Jenkins, 1970), the time series where s is the number of quarters, months, and so on, in the year were made stationary by application of the filters $(1 - L)$ *and/or* $(1 - L^s) = (1 - L)S(L)$, where $S(L) = (1 + L + L^2 + L^3 + \ldots \ldots L^{s-1})$, as many times as was deemed necessary from the form of the resulting autocorrelation function. After stationarity had been obtained, the filtered series were modelled as an autoregressive moving average model (ARMA) model. Both the AR and the MA part could be modelled as consisting of a non-seasonal and seasonal lag polynomial. Hence, the so-called seasonal ARIMA model has the form

$$\phi(L)\phi_s(L^s)\,(1 - L^s)^D (1 - L)^d y_t = \theta(L)\theta_s(L^s)\varepsilon_t$$

$$(7)$$

where $\phi(L)$ and $\theta(L)$ are invertible lag polynomials in L, while $\phi_s(L^s)$ and $\theta_s(L^s)$ are invertible lag polynomials in L^s, and D and d integers.

In light of the results mentioned in the section on seasonal unit roots, the modelling strategy of Box and Jenkins may easily be refined to allow for situations were the non-stationarity exists only at some of the seasonal frequencies.

The 'structural' or unobserved components model. When modelling processes with seasonal characteristics, one must apply complicated and high-ordered polynomials in the ARMA representation. As an alternative to this, the unobserved components model (UC) discussed earlier was proposed. It is easily seen that the UCARIMA model is a general ARIMA model with restrictions on the parameters. Alternatively, the UC model may be specified as a so-called structural model following Harvey (1993).

The structural model is based on a very simple and quite restrictive modelling of the components of interest

such as trends, seasonals and cycles. The model is often specified as (4). The trend μ_t is normally assumed to be stationary only in first or second differences, whereas the seasonal component γ_t is stationary when multiplied by the seasonal summation operator $S(L)$. In the basic structural model (BSM) the trend is specified as

$$\mu_t = \mu_{t-1} + \beta_{t-1} + \eta_t$$
$$\beta_t = \beta_{t-1} + \zeta_t \qquad (8)$$

where each of the error terms is independently distributed. (If $\sigma_\zeta^2 = 0$ this collapses to a random walk plus drift. If $\sigma_\eta^2 = 0$ as well it corresponds to a model with a linear trend.) The seasonal component is specified as

$$S(L)\gamma_t = \sum_{j=0}^{n-1} \gamma_{t-j} = \omega_t \qquad (9)$$

where s is the number of periods per year and where $\omega_t \sim N(0, \sigma_\omega^2)$. (This specification is known as the dummy variable form, since it reduces to a standard deterministic seasonal component if $\sigma_\omega^2 = 0$. Specifying the seasonal component this way makes it slowly changing by a mechanism that ensures that the sum of the seasonal components over any s consecutive time periods has an expected value of zero and a variance that remains constant over time.) The BSM model can also be written as

$$y_t = \frac{\xi t}{\Delta^2} + \frac{\omega_t}{S(L)} + \varepsilon_t, \qquad (10)$$

where $\xi_t = \eta_t - \eta_{t-1} + \xi_{t-1}$ is equivalent to an MA(1) process. Expressing the model in the form (10) makes the connection to the UCARIMA model in (4) clear.

Estimation of the general UC model is treated in Hylleberg (1986) and estimation of the structural model is treated in Harvey, Koopman and Shephard (2004).

In the structural approach the problems of specifying the ARMA models for the components is thus avoided by a priori restrictions. Harvey and Scott (1994) argue that the type of model above, which has a seasonal component evolving relatively slowly over time, can fit most economic time series, irrespective of the apparently strong assumptions of a trend component with a unit root and a seasonal component with all possible seasonal unit roots present.

Periodic models and other time varying parameter models. The periodic model extends the non-periodic time series models by allowing the parameters to vary with the seasons. The so-called periodic autoregressive model of order h (PAR(h)) assumes that the observations in each of the seasons can be described using different autoregressive models (see Franses, 1996).

Consider a quarterly times series y_t which is observed for N years. The stationary PAR(h) quarterly model can be written as

$$y_t = \sum_{s=1}^{4} \mu_s D_{s,t} + \sum_{s=1}^{4} \phi_{1s} D_{s,t} y_{t-1}$$

$$+ \ldots + \sum_{s=1}^{4} \phi_{hs} D_{s,t} y_{t-h} + \varepsilon_t \qquad (11)$$

with $s = 1, 2, 3, 4$, $t = 1, 2, \ldots, T = 4N$, and where $D_{s,t}$ are seasonal dummies, or as $y_t = \mu_s + \phi_{1s} y_{t-1} + \cdots + \phi_{ps} y_{t-h} + \varepsilon_t$.

It has been shown that any PAR model can be described by a non-periodic ARMA model (Osborn, 1991). In general, however, the orders will be higher than in the PAR model. For example, a PAR(1) corresponds to a non-periodic ARMA(4,3) model. Furthermore, it has been shown that estimating a non-periodic model when the true DGP is a PAR can result in a lack of ability to reject the false non-periodic model (Franses, 1996). Fitting a PAR model does not prevent the finding of a non-periodic AR process, if the latter is in fact the DGP. In practice it is thus recommended that one starts by selecting a PAR(h) model and then tests whether the autoregressive parameters are periodically varying using the method described above.

A major weakness of the periodic model is that the available sample for estimation $N = n/s$ often is too small. Furthermore, the identification of a periodic time series model is not as easy as it is for non-periodic models.

Now, let us rewrite the series y_t, $t = 1, 2, 3, \ldots T$, as $y_{s,\tau}$, where $s = 1, 2, 3, 4$ indicating the quarter, and $\tau = 1, 2, \ldots, n$ indicating the year. The PAR(1) process can then be written as

$$y_{s,\tau} - \phi_s y_{s-1,\tau} + \varepsilon_{s,\tau}, \ s - 1, 2, 3, 4; \ \tau - 1, 2, \ldots, n$$
$$(12)$$

where $y_{0,\tau} = y_{4,\tau-1}$, or in vector notation

$$\Phi(L) Y_\tau = U_\tau \qquad (13)$$

where

$$\Phi(L) = \begin{bmatrix} 1 & 0 & 0 & 0 \\ -\phi_2 & 1 & 0 & 0 \\ 0 & -\phi_3 & 1 & 0 \\ 0 & 0 & -\phi_4 & 1 \end{bmatrix}$$

$$+ \begin{bmatrix} 0 & 0 & 0 & -\phi_1 \\ 0 & 0 & 0 & 0 \\ 0 & 0 & 0 & 0 \\ 0 & 0 & 0 & 0 \end{bmatrix} L$$

$$Y_\tau' = [y_{1,\tau}, y_{2,\tau}, y_{3,\tau}, y_{4,\tau}]$$
$$U_\tau' = [\varepsilon_{1,\tau}, \varepsilon_{2,\tau}, \varepsilon_{3,\tau}, \varepsilon_{4,\tau}]$$

with L operating on the seasons, that is, $L\, y_{s,\tau} = y_{s-1,\tau}$ and especially $L\, y_{1,\tau} = y_{0,\tau} = y_{4,\tau-1}$. The PAR(1) process in (13) is stationary provided $|\Phi(z)| = 0$ has all its roots outside the unit circle, which is the case if and only if $|\phi_1 \phi_2 \phi_3 \phi_4| < 1$.

The model may be estimated by maximum likelihood or OLS. Testing for periodicity in (11) amounts to testing the hypothesis $H_0 : \phi_{is} = \phi_i$ for $s = 1, 2, 3, 4$ and $i = 1, 2, \ldots, p$, and this can be done with a likelihood ratio test, which is asymptotically χ^2_{3p} under the null, irrespective of any unit roots in y_t (see Boswijk and Franses, 1995).

The vector representation of the PAR model forms an effective vehicle for generating estimation and testing procedures directly from the general result for stationary vector autoregression (VAR) models, but it also creates an effective way to handle the non-stationary case and compare the periodic models with the models with seasonal integration.

In the non-stationary case, a periodically integrated process of order 1(PI(1)) is defined as a process, where there exists a quasi-difference

$$D_s y_{s,\tau} = 1 - \alpha_s y_{s-1,\tau}$$

$$\alpha_1 \alpha_2 \alpha_3 \alpha_4 = 1$$

$$\text{not all } \alpha_s = 1, s = 1, 2, 3, 4. \qquad (14)$$

such that $D_s y_{s,\tau}$ has a stationary and invertible representation. Notice that the $PI(1)$ process is neither an integrated $I_0(1)$ process nor a seasonally integrated $I_{0,\frac{1}{2}[1/4,3/4]}(1)$ process as shown in Ghysels and Osborn (2001).

The periodic models can be considered special cases of what are referred to as the time-varying parameter models (see Hylleberg, 1986). These are regression models of the form

$$Y_t = X_t' \beta_t + u_t$$
$$B(L) \ (\beta_t - \bar{\beta}) = A\gamma_t + \xi_t \qquad (15)$$

which can be written in state-space form and estimated using the Kalman filter. However, the number of parameters is often greater than the number of observations, and in practice one may be forced to restrict the parameter space. Gersovitz and MacKinnon (1978), applying Bayesian techniques, adopt the sensible assumption that the parameters vary smoothly over the seasons.

The evolving seasonals model. The evolving seasonals model was promulgated by Hannan, Terrell and Tuckwell (1970). The model has been revitalized by Hylleberg and Pagan (1997), who show that the evolving seasonals model produces an excellent vehicle for analysing different commonly applied seasonal models as it nests many of them. The model has been used by Koop and Dijk (2000) to analyse seasonal models from a Bayesian perspective.

The evolving seasonals model for a quarterly time series is based on a representation like

$$
\begin{aligned}
y_t &= \alpha_{1t}\cos(\lambda_1 t) + \alpha_{2t}\cos(\lambda_2 t) + 2\alpha_{3t}\cos(\lambda_3 t) \\
&\quad + 2\alpha_{4t}\sin(\lambda_3 t), \\
&= \alpha_{1t} + \alpha_{2t}\cos(\pi t) + 2\alpha_{3t}\cos(\pi t/2) \\
&\quad + 2\alpha_{4t}\sin(\pi t/2), \\
&= \alpha_{1t}(1)^t + \alpha_{2t}(-1)^t \\
&\quad + \alpha_{3t}[i^t + (-i)^t] + \alpha_{4t}[i^{t-1} + (-i)^{t-1}],
\end{aligned}
\tag{16}
$$

where $\lambda_1 = 0$, $\lambda_2 = \pi$, $\lambda_3 = \pi/2$, $\cos(\pi t) = (-1)^t$, $2\cos(\pi t/2) = [i^t + (-i)^t]$, $2\sin(\pi t/2) = [i^{t-1} + (-i)^{t-1}]$, $i^2 = -1$, while $\alpha_{jt}, j = 1, 2, 3, 4$; is a linear function of its own past and a stochastic term e_{jt}, $j = 1, 2, 3, 4$: For instance,

$$
\begin{aligned}
\alpha_{1t} &= \rho_1\alpha_{1,t-1} + e_{1t}, \\
\alpha_{2t} &= \rho_1\alpha_{2,t-1} + e_{2t}, \\
\alpha_{3t} &= \rho_3\alpha_{3,t-2} + e_{3t}, \\
\alpha_{4t} &= \rho_4\alpha_{4,t-3} + e_{4t}.
\end{aligned}
\tag{17}
$$

In such a model, $\alpha_{1t}(1)^t = \alpha_{1t}$ represents the trend component with the unit root at the zero frequency, $\alpha_{2t}(-1)^t$ represents the semi-annual component with the root -1, while $\alpha_{3t}[i^t + (-1)^t]\, \alpha_{4t}[i^{t-1} + (-1)^{t-1}]$ represents the annual component with the complex conjugate roots $\pm i$. In Hylleberg and Pagan (1997) it is shown that the HEGY auxiliary regression in (6) has an evolving seasonals model representation, and also that the Canova–Hansen test and the PAR(h) model may be presented in the framework of the evolving seasonals model.

Multivariate seasonal time series models

The idea that the seasonal components of a set of economic time series are driven by a smaller set of common seasonal features seems a natural extension of the idea that the trend components of a set of economic time series are driven by common trends.

If the seasonal components are seasonally integrated, the idea immediately leads to the concept of seasonal cointegration, introduced in Engle, Granger and Hallman (1989), Hylleberg et al. (1990), and Engle et al. (1993). In case the seasonal components are stationary, the idea leads to the concept of seasonal common features (see Engle and Hylleberg, 1996), while so-called periodic cointegration considers cointegration season by season (see Birchenhal et al., 1989; Ghysels and Osborn, 2001).

Seasonal cointegration. Seasonal cointegration exists at a particular seasonal frequency if at least one linear combination of series which are seasonally integrated at the particular frequency is integrated of a lower order.

Consider the quarterly case where y_t and x_t are both integrated of order 1 at the zero and at the seasonal frequencies, that is, the transformations corresponding to 6 are $\{y_{1t}, x_{1t}\} \sim I_0(1)$, $\{y_{2t}, x_{2t}\} \sim I_1(1)$ and $\{y_{3t}, x_{3t}\} \sim I_{[1/4,3/4]}(1)$. Cointegration at the frequency $\theta = 0$ then exists if $y_{1t} - k_1 x_{1t} \sim I_0(0)$ for some non-zero k_1, cointegration at the frequency $\theta = \frac{1}{2}$ exists if $y_{2t} - k_2 x_{2t} \sim I_{\frac{1}{2}}(0)$ for some non-zero k_2, while cointegration at the frequency $\theta = [1/4, 3/4]$ exists if $y_{2t} - k_3 x_{2t} - k_4 x_{2,t-1} \sim I_{[1/4,3/4]}(0)$ for some non-zero pair $\{k_3, k_4\}$. The complex unit roots at the annual frequency $[1/4,3/4]$ lead to the concept of polynomial cointegration, where cointegration exists if one can find at least one linear combination including a lag of the seasonally integrated series which is stationary.

In Hylleberg et al. (1990) and Engle et al. (1993), seasonal cointegration is analysed along the path set up in Engle and Granger (1987).

The well-known drawbacks of this method, especially when the number of variables included exceeds two, is partly overcome by Lee (1992), who extends the maximum likelihood (ML)-based methods of Johansen (1995) for cointegration at the long-run frequency, to cointegration at the semi-annual frequency $\theta = \frac{1}{2}$.

To adopt the ML-based cointegration analysis at the annual frequency $\theta = [1/4, 3/4]$ with the complex pair of unit roots $\pm i$ is somewhat more complicated, however. The general results may be found in Johansen and Schaumburg (1999), and Cubadda (2001) applies the results of Brillinger (1981) on the canonical correlation analysis of complex variables to obtain tests for cointegration at all the frequencies of interest, that is, at the frequencies 0 and π with the real unit roots ± 1 and at the frequency $\theta = [1/4, 3/4]$ with the complex roots $\pm i$.

Periodic cointegration. Periodic cointegration extends the notion of seasonal cointegration by allowing the coefficients in the cointegration relations to be periodic (see Ghysels and Osborn, 2001).

Consider the example given above with two quarterly time series y_t and x_t, $t = 1, 2, \ldots, T$ which are integrated of order 1 at the zero and seasonal frequencies implying that a transformation by the fourth difference $1 - L^4$ will make the two series stationary. Such series are called seasonally integrated series. Let us rewrite the series as $y_{s,\tau}$ and $x_{s,\tau}$ with $s = 1, 2, 3, 4$ indicating the quarter, and $\tau = 1, 2, \ldots, n$ indicating the year. Hence, the eight yearly series $y_{s,\tau}$, $x_{s,\tau}$ $s = 1, 2, 3, 4$ are all integrated of order 1 at the zero frequency.

Hence, full periodic cointegration exists (see Boswijk and Franses, 1995) if $y_{\tau t} - k_s x_{\tau r} \sim I_0(0)$ for some non-zero k_s, $s = 1, 2, 3, 4$, $\tau = 1, 2, 3$. In case stationarity is only obtained for some $s = 1, 2, 3, 4$, partially periodic cointegration exists.

Several interesting and useful results reviewed in Ghysels and Osborn (2001) follow:

1. Two seasonally integrated series may fully or partially periodically cointegrate.

2. Two $I_0(1)$ processes cannot be periodically cointe-grated. They are either non-periodically cointegrated or not cointegrated at all.

3. If two $PI(1)$ processes cointegrate in one quarter. they cointegrate in all four quarters.

Periodic cointegration is a promising but currently not fully exploited area of research, which has the inherent problem that it requires a large sample. It is therefore not surprising that the recent advances in this area happen when data are plentiful (daily) and it is possible to restrict the model appropriately (Haldrup et al., 2007).

Common seasonal features. Although economic time series often exhibit non-stationary behaviour, stationary economic variables exist as well, especially when conditioned on some deterministic pattern such as linear trends, seasonal dummies, breaks and so on. However, a set of stationary economic times series may also exhibit common behaviour and, for instance, share a common seasonal pattern. The technique for finding such patterns, known as common seasonal features (see Engle and Hylleberg, 1996; Cubadda, 1999) is based on earlier contributions defining common features by Engle and Kozicki (1993) and Vahid and Engle (1993).

Consider a multivariate autoregression written in error correction form as

$$\Delta Y_t = \sum_{j=1}^{p} B_j \Delta Y_{t-j} + \Pi \nu_{t-1} + \Gamma z_t + \varepsilon_t, t$$
$$= 1, 2, \ldots, T,$$

(18)

where Y_t is $k \times 1$ vector of observations on the series of interest in period t and the error correction term is $\Pi \nu_{t-1}$. The vector ν_t contains the cointegrating relations at the zero frequency, and the number of cointegrating relations is equal to the rank of Π. If Π has full rank equal to k the series are stationary. In the quarterly case the vector z_t is a vector of trigonometric seasonal dummies, such as $\{\cos(2\pi ht/4 + 2\pi j/T), \quad h = 1, 2; \quad j \in (-\delta T \leqslant j \leqslant \delta T), \quad \sin(2\pi h4 + 2\pi j/T), \quad h = \quad 1, 2; \quad j \in (-\delta T \leqslant \leqslant \delta T), j \neq 0,$ when $h = 2\}$. The use of trigonometric dummy variables facilitates the 'modelling' of a varying seasonal pattern, since a proper choice of δ takes care of the neighboring frequencies to the exact seasonal frequencies. If δ is 0, the filter is equivalent to the usual seasonal dummy filter.

The implication of a full rank of the $k \times m$ matrix Γ, equal to $\min[k, m]$, is that different linear combinations of the seasonal dummies in z_t are needed in order to explain the seasonal behaviour of the variables in Y_t. However, if there are common seasonal features in these variables we do not need all the different linear combinations, and the rank of Π is not full. Thus, a test of the number of common seasonal features can be based on the rank of Π (see Engle and Kozicki, 1993).

The test is based on a reduced rank regression similar to the test for cointegration by Johansen (1995). Hence, the hypotheses are tested using a canonical correlation analysis between of z_t and ΔY_t, where both sets of variables are purged of the effect from the other variables in (18).

This kind of analysis has proved useful in some situations, but it is difficult to apply in cases where the number of variables is large, and the results are sensitive to the lag-augmentation as in the case of cointegration. In addition, the somewhat arbitrary nature of the choice of z_t poses difficulties.

Economic models of seasonality

Many economic time series have a strong seasonal component, and obviously economic agents must react to that. Hence, the seasonal variation in economic time series must be an integrated part of the optimizing behaviour of economic agents, and the seasonal variation in economic time series must be a result of the optimizing behaviour of economic agents, reacting to exogenous factors such as the weather, the timing of holidays, and so on.

The fact that economic agents react and adjust to seasonal movements on one hand and influence them on the other, implies that the application of seasonal data in economic analysis may widen the possibilities for testing theories about economic behaviour. The relative ease with which the agents may forecast at least some of the causes of the seasonality may be quite helpful in setting up testable models for production smoothing, for instance.

Apart from what is caused by the easiness of forecasting exogenous factors, the type of optimizing behaviour and the agents' reactions to a seasonal phenomenon may be expected not to differ fundamentally from what is happening in a non-seasonal context. However, the recurrent characteristic of seasonality may be exploited.

The economic treatment of seasonal fluctuation has been discussed in the real business cycle (RBC) literature (for example, Chatterjee and Ravikumar, 1992; Braun and Evans, 1995), working with a utility optimizing consumer faced with some feasibility constraint. However, in most of this RBC branch seasonality arises from deterministic shifts in tastes and technology. A few other papers incorporate seasonality through stochastic productivity shocks (see for example Wells, 1997; Cubadda, Savio and Zelli, 2002).

Another area is the production smoothing literature (for instance, Ghysels, 1988; Miron and Zeldes, 1988; Miron, 1996) and habit persistence (for example, Osborn, 1988), where a model for seasonality and habit persistence is presented in a life-cycle consumption model.

SVEND HYLLEBERG

See also **data filters.**

The author is grateful for helpful comments from Niels Haldrup and Steven Durlauf.

Bibliography

Arteche, J. 2000. Gaussian semiparametric estimation in seasonal/cyclical long memory time series. *Kybernetika* 36, 279–310.

Arteche, J. and Robinson, P. 2000. Semiparametric inference in seasonal and cyclical long memory processes. *Journal of Time Series Analysis* 21, 1–25.

Baxter, M. and King, R. 1999. Measuring business cycles: approximate band-pass filters for economic time series. *Review of Economics and Statistics* 81, 575–93.

Beaulieu, J. and Miron, J. 1993. Seasonal unit roots in aggregate US data. *Journal of Econometrics* 55, 305–28.

Birchenhal, C., Bladen-Howell, R., Chui, A., Osborn, D. and Smith, J. 1989. A seasonal model of consumption. *Economic Journal* 99, 837–43.

Boswijk, H. and Franses, P. 1995. Periodic cointegration: representation and inference. *Review of Economics and Statistics* 77, 436–54.

Box, G. and Jenkins, G. 1970. *Time Series Analysis, Forecasting, and Control.* San Francisco: Holden-Day.

Braun, R. and Evans, C. 1995. Seasonality and equilibrium business cycle theories. *Journal of Economic Dynamics and Control* 19, 503–31.

Brendstrup, B., Hylleberg, S., Nielsen, M., Skipper, L. and Stentoft, L. 2004. Seasonality in economic models. *Macroeconomic Dynamics* 8, 362–94.

Brillinger, D. 1981. *Time Series: Data Analysis and Theory.* San Francisco: Holden Day.

Bunzel, H. and Hylleberg, S. 1982. Seasonality in dynamic regression models: a comparative study of finite sample properties of various regression estimators including band spectrum regression. *Journal of Econometrics* 19, 345–66.

Busetti, F. and Harvey, A. 2003. Seasonality tests. *Journal of Business and Economic Statistics* 21, 421–36.

Canova, F. and Hansen, B. 1995. Are seasonal patterns constant over time? A test for seasonal stability. *Journal of Business and Economic Statistics* 13, 237–52.

Chatterjee, S. and Ravikumar, B. 1992. A neoclassical model of seasonal fluctuations. *Journal of Monetary Economics* 29, 59–86.

Cubadda, G. 1999. Common cycles in seasonal non-stationary time series. *Journal of Applied Econometrics* 14, 273–91.

Cubadda, G. 2001. Complex reduced rank models for seasonally cointegrated time series. *Oxford Bulletin of Economics and Statistics* 63, 497–511.

Cubadda, G., Savio, G. and Zelli, R. 2002. Seasonality, productivity shocks, and sectoral comovements in a real business cycle model for Italy. *Macroeconomic Dynamics* 6, 337–56.

Dagum, E. 1980. The X-11-ARIMA seasonally adjustment method. Technical Report 12-564E. Ottawa: Statistics Canada.

Dickey, D. and Fuller, W. 1979. Distribution of the estimators for autoregressive time series with a unit root. *Journal of the American Statistical Association* 74, 427–31.

Dickey, D., Hasza, D. and Fuller, W. 1984. Testing for unit roots in seasonal time series. *Journal of the American Statistical Association* 79, 355–67.

Engle, R. 1974. Band spectrum regression. *International Economic Review* 15, 1–11.

Engle, R. 1978. Estimating structural models of seasonality. In *Seasonal Analysis of Economic Time Series*, ed. A. Zellner. Washington, DC: US Census Bureau.

Engle, R. 1980. Exact maximum likelihood methods for dynamic regressions and band spectrum regressions. *International Economic Review* 21, 391–407.

Engle, R. and Granger, C. 1987. Co-integration and error correction: representation, estimation and testing. *Econometrica* 55, 251–76.

Engle, R., Granger, C. and Hallman, J. 1989. Merging short and long run forecasts: an application of seasonal cointegration to monthly electricity sales forecasting. *Journal of Econometrics* 40, 45–62.

Engle, R., Granger, C., Hylleberg, S. and Lee, H. 1993. Seasonal cointegration – the Japanese consumption function. *Journal of Econometrics* 55, 275–98.

Engle, R. and Hylleberg, S. 1996. Common seasonal features: global unemployment. *Oxford Bulletin of Economics and Statistics* 58, 615–30.

Engle, R. and Kozicki, S. 1993. Testing for common features. *Journal of Business and Economic Statistics* 11, 369–80.

Findley, D., Monsell, B., Bell, W., Otto, M. and Chen, B. 1998. New capabilities and methods of the X-12-ARIMA seasonal adjustment program. *Journal of Business and Economic Statistics* 16, 127–76.

Franses, P. 1996. *Periodicity and Stochastic Trends in Economic Time Series.* Oxford: Oxford University Press.

Frisch, R. and Waugh, F. 1933. Partial time regressions as compared with individual trends. *Econometrica* 1, 387–401.

Fuller, W. 1976. *Introduction to Statistical Time Series.* New York: John Wiley and Sons.

Gersovitz, M. and MacKinnon, J. 1978. Seasonality in regression: an application of smoothness priors. *Journal of the American Statistical Association* 73, 264–73.

Ghysels, E. 1988. A study towards a dynamic theory of seasonality for economics time series. *Journal of the American Statistical Association* 83, 68–72.

Ghysels, E. 1997. Seasonal adjustments and other data transformations. *Journal of Business and Economic Statistics* 15, 410–18.

Ghysels, E. and Osborn, D. 2001. *The Econometric Analysis of Seasonal Time Series.* Cambridge: Cambridge University Press.

Gil-Alana, L. and Robinson, P. 1997. Testing of unit root and other non-stationary hypotheses in macroeconomic time series. *Journal of Econometrics* 80, 241–68.

Gomez, V. and Maravall, A. 1996. *Programs TRAMO and SEATS.* Madrid: Banco de Espana.

Haldrup, N., Hylleberg, S., Pons, G. and Sansó, A. 2007. Common periodic correlation features and the interaction of stocks and flows in daily airport data. *Journal of Business and Economic Statistics* 25, 21–32.

Hannan, E. 1960. *Time Series Analysis*. London: Methuen.

Hannan, E., Terrell, R. and Tuckwell, N. 1970. The seasonal adjustment of economic time series. *International Economic Review* 11, 24–52.

Harvey, A. 1993. *Time Series Models*. London: Prentice Hall/ Harvester Wheatsheaf.

Harvey, A. and Scott, A. 1994. Seasonality in dynamic regression models. *Economic Journal* 104, 1324–45.

Harvey, A., Koopman, S. and Shephard, N. eds. 2004. *State Space and Unobserved Component Models: Theory and Applications*. Cambridge: Cambridge University Press.

Hood, C., Ashley, J. and Findley, D. 2004. An empirical evaluation of the performance of TRAMo/SEATS on simulated series. Technical report. Washington, DC: US Census Bureau.

Hylleberg, S. 1977. A comparative study of finite sample properties of band spectrum regression estimators. *Journal of Econometrics* 5, 167–82.

Hylleberg, S. 1986. *Seasonality in Regression*. Orlando, FL: Academic Press.

Hylleberg, S. ed., 1992. *Modelling Seasonality*. Oxford: Oxford University Press.

Hylleberg, S. 1995. Tests for seasonal unit roots: general to specific or specific to general. *Journal of Econometrics* 69, 5–25.

Hylleberg, S., Engle, R., Granger, C. and Yoo, S. 1990. Seasonal integration and cointegration. *Journal of Econometrics* 44, 215–38.

Hylleberg, S. and Pagan, A. 1997. Seasonal integration and the evolving seasonals model. *International Journal of Forecasting* 13, 329–40.

Jevons, W. 1884. *Investigations in Currency and Finances*. London: Macmillan.

Johansen, S. 1995. *Likelihood-Based Inference in Cointegrated Vector Autoregressive Models*. Oxford: Oxford University Press.

Johansen, S. and Schaumburg, E. 1999. Likelihood analysis of seasonal cointegration. *Journal of Econometrics* 88, 301–39.

Koop, G. and Dijk, H. 2000. Testing for integration using evolving trend and seasonals models: a Bayesian approach. *Journal of Econometrics* 97, 261–91.

Kunst, R. 1997. Testing for cyclical non-stationarity in autoregressive processes. *Journal of Time Series Analysis* 18, 123–35.

Kwiatkowski, D., Phillips, P., Schmidt, P. and Shin, Y. 1992. Testing the null hypothesis of stationarity against the alternative of a unit root – how sure are we that economic time series have a unit root? *Journal of Econometrics* 54, 159–78.

Lee, H. 1992. Maximum likelihood inference on cointegration and seasonal cointegration. *Journal of Econometrics* 54, 1–47.

Lovell, M. 1963. Seasonal adjustment of economic time series. *Journal of the American Statistical Association* 58, 993–1010.

Mazzi, G.L. and Savio, G. 2005. The seasonal adjustment of short time series. Technical Report KS-DT-05-002. EUROSTAT.

Mills, F. 1924. *Statistical Methods*. London: Pitman.

Miron, J. 1996. *The Economics of Seasonal Cycles*. Cambridge, MA: MIT Press.

Miron, J. and Zeldes, S. 1988. Seasonality, cost shocks and the production smoothing model of inventories. *Econometrica* 56, 877–908.

OECD. 1999. Feature article: seasonal adjustment. *Main Economic Indicators*. November. Paris: OECD.

Osborn, D. 1988. Seasonality and habit persistence in a life cycle model of consumption. *Journal of Applied Econometrics* 3, 255–66.

Osborn, D. 1991. The implications of periodically varying coefficients for seasonal time-series processes. *Journal of Econometrics* 48, 373–84.

Shiskin, J., Young, A. and Musgrave, J. 1967. The X-11 variant of the census method II seasonal adjustment program. Technical Paper No. 15. Washington, DC: US Census Bureau.

Taylor, A. 2005. Variance ratio tests of the seasonal unit root hypothesis. *Journal of Econometrics* 124, 33–54.

Vahid, F. and Engle, R. 1993. Common trends and common cycles. *Journal of Applied Econometrics* 8, 341–60.

Wallis, K. 1998. Comment. *Journal of Business and Economic Statistics* 16, 164–5.

Wells, J. 1997. Business cycles, seasonal cycles, and common trends. *Journal of Macroeconomics* 19, 443–69.

second best

One of the passages most often quoted in the literature on economic policy is the following from a seminal paper by R.G. Lipsey and K. Lancaster (1956, p. 11):

> The general theorem for the second best optimum states that if there is introduced into a general equilibrium system a constraint which prevents the attainment of one of the Paretian conditions, the other Paretian conditions, although still attainable, are, in general, no longer desirable.

The implication of this theorem was that most of the simple and general guidelines for policy provided by welfare economics – for example, the 'Paretian conditions' stating that price should equal marginal cost – would not be relevant for real-world economies which are likely to be subject to constraints on policy. The Lipsey–Lancaster article seems to have come as a shock to economists in general and has since had a significant impact on the theory, and practice, of economic policy. Apparently, until the publication of this article, the conventional wisdom was that it was desirable to pursue a 'piecemeal policy', here and there fulfilling the 'Paretian conditions' – which, if applied everywhere, would lead to a Pareto optimum – regardless of whether these conditions actually were attained elsewhere.

This state of affairs in 1956 was somewhat puzzling considering that the Lipsey–Lancaster conclusion was not entirely novel. As early as 1909, V. Pareto himself had argued that free trade (which in modern terminology may be interpreted as fulfillment, as far as possible, of the 'Paretian conditions') may not be preferable to protection and that individuals may not end up in a better position if one of several distortions to resource allocation were eliminated. Both of these arguments are in line with the general theory of second best. Even closer to the Lipsey–Lancaster result was the statement by Paul Samuelson in his *Foundations* (1947, p. 252) that a 'given divergence in a subset of the optimum conditions necessitates alterations in the remaining ones'.

Second best reasoning had also been prevalent in various areas of applied welfare economics. (For reviews using different perspectives, see Lipsey and Lancaster, 1956; Negishi, 1972; and McKee and West, 1981.) Thus, concerning *optimal pricing* in the presence of monopolies and other forms of imperfect competition, Hicks (1940) had argued that price equal to marginal cost in an industry A is not compatible with efficiency if other industries B are not competitive. As marginal inputs in A at the expense of B are then worth more than is reflected by input prices, the marginal cost confronting A is less than the true social marginal cost and therefore unsuitable as a benchmark for the price in this industry.

Early 'second best results' had also been obtained in *public finance*. For example, given that leisure is untaxable, ordinary income taxation cannot be considered more efficient than indirect taxation of one commodity, as in both cases at least one 'Paretian condition' (that between commodities and leisure vs. that between the commodity subject to an indirect tax and the other commodities, respectively) is violated. Hence, it is an open question which tax system is the better of these two imperfect alternatives (Little, 1951). Somewhat later, it was shown that a 'second best optimal' way of raising a given amount of government revenue, barring the use of lump-sum taxes, was a set of unequal indirect taxes with low tax rates on commodities which are substitutes for leisure and high tax rates on commodities complementary to leisure (Corlett and Hague, 1953). In fact, a similar result had already been obtained by Ramsey in 1927. These cases clearly illustrate that when all 'Paretian conditions' cannot be met, it may not be efficient to fulfil some of them.

The field of *trade policy* had been especially rich in providing examples of second best reasoning. Viner (1950) showed that in a world of trade protection, a reduction of some trade barriers or introduction of a customs union for some of the trading countries – both of which constitute steps towards free trade and the fulfillment of some of the 'Paretian conditions' – will not necessarily increase efficiency in world production. In the customs union case, the explanation is that the positive welfare effect of trade creation within the union may be outweighed by the negative welfare effect of trade diversion between member countries and the rest of the world.

Meade (1955a; 1955b) dealt with a number of trade policy problems as well as some domestic policy problems where it would not be possible to reach a Pareto optimum – or Utopian optimum, as he tellingly called it. Assuming the existence of several market imperfections and efficiency-distorting policy interventions, he analysed the effect of reducing or eliminating one of them and tried to determine what policy rule would perform better. Meade coined the term 'theory of second best' for this type of policy analyses whose real-world relevance and basic similarities were elucidated especially in his *Trade and Welfare* (1955a).

In retrospect, it may be argued that the catalogue of separate but similar policy issues in distorted economies provided by Meade goes as far as the theory of second best has reached. But Lipsey and Lancaster were the ones who put second best problems on the map of the average economist. This was accomplished by their attempt to present a general theory of second best, covering the main characteristics of the particular cases dealt with by Meade and others up to that point. Although their 1956 article contained a number of comments on these particular cases as well as reservations on their general theory, it was their concise version of this theory that gained most of the attention.

The Lipsey–Lancaster general theorem of the second best departs from 'the typical choice situation in economic analysis' where an objective function $F(x_1, \ldots, x_n)$ is to be maximized or minimized subject to a constraint $\Phi(x_1, \ldots, x_n) = 0$. Lipsey and Lancaster called the solution to this problem the Paretian optimum. To make the problem explicitly harmonize with what is generally meant by a Paretian optimum, we may interpret $x_1, \ldots, x_n$ as the elements of the consumption vectors of all individuals in the economy in some given order (x_1 being Alpha's consumption of commodity I, x_2 Alpha's consumption of commodity II, and so on, up to x_n, the last individual's consumption of the last commodity). As in most of the literature on the subject, we assume for simplicity that the objective function reveals an interest in efficiency alone (that is, attaining any Pareto optimum) and not in distribution (that is, one particular Pareto optimum).

Optimizing $F(\cdot)$ subject to $\Phi(\cdot) = 0$, where Φ can be seen as the transformation function specifying the constraint given by available technology and initial resources, we get the following necessary optimum conditions (assuming that all functions are well behaved):

$$F_i/F_n = \Phi_i/\Phi_n \quad i = 1, \ldots, n-1. \tag{1}$$

These 'Paretian conditions' – or first best Pareto optimum conditions as they are now commonly called – may be interpreted as requiring equality between the marginal rates of substitution and the marginal rates of transformation. The purpose of deriving these conditions in the present context, it should be stated explicitly, is (a) to check whether they are fulfilled in a particular situation and, if they are not, (b) to provide guidelines for policy.

Lipsey and Lancaster then tried to formulate an additional constraint which would cover most of what the literature had observed as obstacles to achieving a first best Pareto optimum. They attempted to accomplish this about as generally as when the function Φ is taken to represent the production constraint of the economy. They argued that, if for some reason monopoly elements, externalities or other so-called imperfections were 'out of bounds' for policy intervention, one of the conditions (1) could not be fulfilled due to a constraint

$$F_1/F_n = k\Phi_1/\Phi_n \qquad (2)$$

with $k \neq 1$ and k – 'for simplicity' – assumed to be constant. The resulting problem amounts to the optimization of the Lagrangean function

$$F - \lambda\Phi - \mu(F_1/F_n - k\Phi_1/\Phi_n) \qquad (3)$$

where λ and μ are Lagrangean multipliers. A solution to this problem, it should be noted, is also a Pareto optimum as it does not allow anyone to become better off without making someone else worse off, given the two constraints now in force. The necessary optimum conditions can be written

$$\frac{F_i}{F_n} = \frac{\Phi_i + (\mu/\lambda)(Q_i - kR_i)}{\Phi_n + (\mu/\lambda)(Q_n - kR_n)}, \quad i = 2,\dots,n-1 \qquad (4)$$

where

$$Q_i = (F_n F_{1i} - F_1 F_{ni})/F_n^2 \quad \text{and}$$
$$R_i = (\Phi_n \Phi_{1i} - \Phi_1 \Phi_{ni})/\Phi_n^2.$$

Aside from some special cases (see below), this means that, in second best optimum, $F_i/F_n \neq \Phi_i/\Phi_n$, from which follows the theorem quoted in the introduction.

These second best optimum conditions are obviously quite complicated – in fact, so complicated that in many cases a great deal of detailed information would be required even to determine the signs of the second derivatives F_{ni}, Φ_{ni}, etc. Hence, it would often be impossible to know whether in second best optimum $F_i/F_n > \Phi_i/\Phi_n$ or the opposite. Moreover, it is no longer possible to translate the second best optimum conditions into intuitively simple relationships between price and marginal cost, which was true for (some of) the first best optimum conditions (eq. 1).

The great impact of this result on economists in general must be attributed to the simplicity of the theorem itself, given that, in essence, the same thing had been said on earlier occasions. A large part of the ensuing debate concerned whether this simplicity was warranted by real-world conditions. In particular, (a) the origin and form of the additional constraint were questioned. A second dominating issue in the debate concerned (b) the complexity of the rules for second best policy and attempts at identifying important cases where simple first best optimum conditions are still valid in second best situations. We deal with these two issues in turn.

The nature of the additional constraint

In regard to the alleged generality of the formulation of the Lipsey–Lancaster theory, the question was asked: What exactly does this constraint (2) stand for? Clearly, the optimization problem in the 'second best literature' refers to a national government *or* an independent unit of government (such as a public monopoly) which operates as if it tried to optimize function F. Obviously, the government-unit perspective (represented for example, by, the work of Davis and Whinston, 1965; 1967, and developed by McFadden, 1969) could allow a number of constraints like (2) concerning variables that are out of reach for this unit. But what about additional constraints imposed on the overall allocation problem confronting the *national government*? Here, two interpretations have been considered in the literature.

First, for an initial state of the economy in which one of the conditions (1) is violated, it may be *technically* impossible for the government to have this condition fulfilled along with all the others. For example, markets for certain commodities such as specific kinds of insurance may not be possible to introduce or may be too costly to administer. Or it may be impossible or prohibitively expensive to correct for certain externalities or instances of imperfect competition. Likewise, when the government, in an attempt to attain a feasible Pareto optimum, needs to raise money for subsidies or for the production of public goods, there may not be any non-distortive taxation scheme available (note, for example, the 'impossibility' of taxing leisure assumed above).

It should be noted that constraints of this technical type are irremovable, in the same way that the constraint imposed by the transformation function is irremovable. Thus, if a market economy does not by itself reach a first best Pareto optimum, the government could not reach one either when policy constraints of this type are strictly binding. Then, a Lipsey–Lancaster Paretian optimum does not exist and the only optimum conceivable is in fact what has here been called a second best optimum (see, for example, McKee and West, 1981).

Second, the constraint (2) can be interpreted as a *behavioural* constraint on policy, reflecting the fact that certain measures, although technically speaking feasible and in principle capable of removing the restraint, are not at the government's disposal or just not believed to be so. For example, the law may prohibit or delimit the use of a specific policy instrument. Or the government may have other and hierarchically higher goals than Pareto optimality: it may for example, simply dislike nationalization of certain industries. Or the government may want to avoid the use of a policy for 'political reasons', believing, for example, that it would lose the next election if this policy was used. Traditions, idiosyncrasies, and so on could play a similar role. Economists who have paid attention to the origin of constraints like (2) seem to have adhered primarily to this 'behavioural' interpretation.

Obviously, the 'behavioural' type of constraint need not be such that all policies with the same effects on the objective function are restrained to the same extent. (For a different perspective which holds that constraints are or should be, in some narrow sense, 'rational', see Faith and Thompson, 1981.) For example, assume that there are two policy instruments, say, a tax and a regulation, each of which, if unconstrained, would have attained the first best, as k could then be made equal to 1, but that policy now is constrained so that just one of them is ruled out. If so, a constraint of type (2) would not exist. This means that, in general, it is not possible to presume what constraints on policy instruments imply in terms of the relations between endogenous variables such as the marginal rates of transformation and substitution in (2). (Actually, the literature has not been able to present any great number of cases where policy constraints yield an expression like (2) with k constant and not equal to one.) Instead, to obtain a solution to the allocation problem, it must be specified exactly what the actual constraints on policy instruments are, that is, to what domain or what combinations with other instruments or variables each policy instrument is restricted.

Specifying the policy constraints in this way has some important implications for a general theory of second best (McManus, 1959; 1967; Bohm, 1967).

First, it cannot be known beforehand whether policy constraints prevent the attainment of a first best optimum or not.

Second, the actual impact of policy constraints, including now also the technical ones mentioned earlier, will depend on the actual behavioural properties of the agents operating in the market economy. Thus, actual rules of market behaviour would, at least in principle, have to be added as constraints on the objective function along with the constraints on policy.

Third, the many possible forms of the policy constraints imply that there cannot be any *general* second best optimum conditions in the sense that there exist general first best optimum conditions (when no policy constraints exist) in terms of a specific relationship between marginal rates of substitution and marginal rates of transformation. In fact, constraints on policy instruments will require that the optimum-feasible solution be derived directly in terms of optimum-feasible values for the policy instruments. Thus, there would not even be any role for policy guidelines in the form of second best conditions such as (4). This, of course, does not preclude the existence of special cases, which *ex post* turn out to coincide with the Lipsey–Lancaster formulation. These are the cases where the constraint on policy happened to affect only the relationship between one marginal rate of substitution and one marginal rate of transformation exactly in the way specified by (2), with no impact whatsoever on the use of policy instruments that could influence other such rates in the economy.

Given that additional constraints on the allocation problem can have any shape – with (2) being a possible *ex post* formulation of one of many special cases – the use of the term 'second best' becomes somewhat unclear. Should a *second best problem* be defined as an allocation problem with constraints on policy regardless of whether a first best optimum will turn out to be impossible? Or should it be reserved for such problems where analysis will eventually show that a first best optimum cannot be reached? Although the second alternative is in line with the intended problem formulation in Lipsey–Lancaster, it is obviously inconvenient as it cannot be used until after the problem has been solved. The term *second best optimum*, on the other hand, is predominantly used for a constrained optimum not equal to a first best optimum and is not likely to cause much of a problem. Hence, a practical, and nowadays probably the dominant, terminology is to distinguish between first best and second best problems according to the first-mentioned definition, where a second best problem may have a first best *or* a second best optimum solution.

First best rules for second best problems

To deal with the second issue prominent in the literature on second best, we return to the problem as it was formulated by Lipsey and Lancaster. They had argued that second best conditions, in contrast to first best ones, were so complicated and required so much information that, on this account, the conditions could not be used for practical policy. This spurred a number of economists to undertake a rescue operation, in which they tried to show that in many instances the simple first best conditions would still be relevant for the controllable part of the economy. To the extent this was true, it would restore at least part of the relevance of piecemeal policy, that is, policy guided by principles which are unaffected by the exact nature of the circumstances in the uncontrollable part of the economy.

First, it has been pointed out that first best rules may be optimal even with the particular Lipsey–Lancaster formulation of the second best problem, for example when $Q_i - kR_i = Q_j - kR_j = 0$; $i, j = 2, \ldots, n$; $i \neq j$ (see Santoni and Church, 1972; Dusansky and Walsh, 1976; Rapanos,

1980). Similar special cases may be found for other forms of additional constraints on the objective function (Mishan, 1962).

Second, and of more general interest for practical policy, it was pointed out that if the additional constraints affect only a limited set of markets in the economy, the relation between this sector and the remaining 'perfectly controllable' sector may be such that first best conditions are optimal in the latter sector even when they turn out to be unattainable in the former sector. This must be true, of course, if the two sectors are completely independent of each other with different primary inputs, no input deliveries between the sectors and different final consumers. Approximately the same results hold if interdependence between sectors is negligible, for example due to one of the sectors being relatively small (Mishan, 1962).

Other attempts at identifying similar cases of separability have been made without much success in terms of general and easily applicable principles for ascertaining when the first best optimum conditions that are still attainable should in fact be attained. Moreover, the very idea of identifying two sectors, one of which is imperfectly controllable, has appeared to be less and less attractive as a description of the real world. Instead, in most countries, income taxes, distorting institutional rules and regulations, and so on emerge as irremovable constraints affecting the economy as a whole. Moreover, as the typical allocation problems confronting real-world governments are beset with a multitude of policy constraints – technical as well as behavioural – it is only by pure chance that optimum conditions for the irrelevant first best problem can be found to be a priori relevant. This does not mean, of course, that first best optimum conditions never will *turn out* to be valid in second best optimum or that information may not be so inadequate that these conditions appear to be acceptable as a rule of thumb (see Ng, 1977).

Thus, the outcome of the literature on general second best theory up to this point is disillusioning in several respects. There do not seem to be any *general* second best problems of the type formalized by Lipsey and Lancaster, much less any general second best optimum *conditions*. Granted that there remains some disagreement on the purpose of second best theory, what has emerged from the literature by way of a general description of second best problems can be summarized as follows:

All economies – even real-world centrally planned economies – have at least some (most often, a very large number of) given behaviour functions which contribute to determining the outcome of any policy 'intervention'. Hence, these functions must be observed in the formulation of the allocation problem; or, which is the same thing, they must be included as constraints on the optimization of the objective function. 'At the same time, the authorities have at their command a set of policy instruments which enter these functions as arguments.

The optimum is then found by [optimizing the objective function] subject to all the constraints over the domain of these instruments. Indeed the whole problem has little practical interest without some such explicit policy formulation' (McManus, 1967, p. 321). The fact that this is a highly demanding analytical task in actual practice requires simplifications and approximations of the models to be used, but it cannot justify an oversimplification of the actual problem to be tackled.

This in effect may seem to take us back to the case-by-case approach of applied welfare economics that was used by Meade and others in the beginning of the 1950s (represented in later and technically more elaborate studies by, for example, Boiteux, 1956; Rees, 1968; and Guesnerie, 1975). However, matters have changed in one important respect since then. Much more empirical knowledge is now available concerning behaviour of individual markets which in itself improves the outlook for practical second best policy.

This is not to say that attempts to construct a general theory of second best have not made a significant contribution to economic theory and policy. Should one result be highlighted, it may quite likely be the general theorem of second best as quoted in the introduction. If nothing else, this theorem has probably made economists more careful when providing governments with policy prescriptions.

PETER BOHM

See also **marginal and average cost pricing; optimal tariffs; Pareto efficiency.**

Bibliography

Allingham, M. and Archibald, G.C. 1975. Second best and decentralisation. *Journal of Economic Theory* 10, 157–73.
Boadway, T.J. and Harris, R. 1977. A characterisation of piecemeal second best policy. *Journal of Public Economics* 8(2), 169–90.
Bohm, P. 1967. On the theory of 'second best'. *Review of Economic Studies* 34, 301–14.
Boiteux, M. 1956. Sur le gestion des monopoles publics astreints à l'équilibre budgetaire. *Econometrica*.
Translated into English as: On the management of public monopolies subject to budgetary constraints, *Journal of Economic Theory* 3(3) (1971), 219–40.
Corlett, W.J. and Hague, D.C. 1953. Complementarity and the excess burden of taxation. *Review of Economic Studies* 21, 21–30.
Davis, O.A. and Whinston, A.B. 1965. Welfare economics and the theory of second best. *Review of Economic Studies* 32, 1–14.
Davis, O.A. and Whinston, A.B. 1967. Piecemeal policy in the theory of second best. *Review of Economic Studies* 34, 323–31.
Dusansky, R. and Walsh, J. 1976. Separability, welfare economics, and the theory of second best. *Review of Economic Studies* 43, 49–51.

Faith, R. and Thompson, E. 1981. A paradox in the theory of second best. *Economic Enquiry* 19, 235–44.

Guesnerie, R. 1975. Production of the public sector and taxation in a simple second best model. *Journal of Economic Theory* 10, 127–56.

Hatta, T. 1977. A theory of piecemeal policy recommendations. *Review of Economic Studies* 44, 1–21.

Hicks, J.R. 1940. The rehabilitation of consumers' surplus. *Review of Economic Studies* 8, 108–16.

Kawamata, K. 1977. Price distortion and the second best optimum. *Review of Economic Studies* 44, 23–29.

Lipsey, R.G. and Lancaster, K. 1956. The general theory of second best. *Review of Economic Studies* 24(1), 11–32.

Little, I.M.D. 1951. Direct versus indirect taxes. *Economic Journal* 61, 577–84.

McFadden, D. 1969. A simple remark on the second best Pareto optimality of market equilibria. *Journal of Economic Theory* 1, 26–38.

McKee, M. and West, E.G. 1981. The theory of second best: a solution in search of a problem. *Economic Inquiry* 19, 436–48.

McManus, M. 1959. Comments on the general theory of second best. *Review of Economic Studies* 26, 209–24.

McManus, M. 1967. Private and social costs in the theory of second best. *Review of Economic Studies* 34, 317–21.

Meade, J.E. 1955a. *Trade and Welfare*, [*including the*] *Mathematical Supplement*. London, New York: Oxford University Press.

Meade, J.E. 1955b. *The Theory of Customs Unions*. Amsterdam: North-Holland.

Mishan, E.J. 1962. Second thoughts on second best. *Oxford Economic Papers* 14, 205–17.

Negishi, T. 1972. *General Equilibrium and International Trade*. Amsterdam: North-Holland.

Ng, Y.K. 1977. Towards a theory of third-best. *Public Finance* 32(1), 1–15.

Pareto, V. 1909. *Manuel d'Economie Politique*. Paris: Girard et Brière.

Ramsey, F.P. 1927. A contribution to the theory of taxation. *Economic Journal* 37, 47–61.

Rapanos, V.T. 1980. A comment on the theory of second best. *Review of Economic Studies* 47, 817–19.

Rees, R. 1968. Second-best rules for public enterprise pricing. *Economica* 35, 260–73.

Samuelson, P.A. 1947. *Foundations of Economic Analysis*. Cambridge, MA: Harvard University Press.

Santoni, G. and Church, A. 1972. A comment on the general theorem of second best. *Review of Economic Studies* 39, 527–30.

Sontheimer, K.C. 1971. An existence theorem for the second best. *Journal of Economic Theory* 3, 1–22.

Viner, J. 1950. *The Customs Union Issue*. New York: Carnegie Endowment for International Peace; London: Stevens & Sons.

second economy (unofficial economy)

The second economy in the Soviet-type command economies (STEs) was defined by Grossman (1977) as all economic activities that are either undertaken directly for private gain or are knowingly illegal in some substantial way. Both legal and illegal economic activities fall within this definition. The second economy served as a precursor of the unofficial sector in the economies in transition. Accordingly, the institution of unofficial economy in transition is properly understood as an heir to the second economy in the STEs.

The official primary coordination mechanism in the STEs was central planning. The well-known difficulties with implementing this mechanism created both the incentives and the opportunity for the existence of a large second economy. The scope of the legal part of the second economy depended on the laws of a given STE, but it usually included private agriculture, small-scale construction, certain professional services (for example, lawyers working privately), and so on. While the legal second economy was important, it was usually smaller than the illegal sector, also referred to as unofficial, underground, shadow economy, or black or parallel markets. Moreover, the concept of legal second economy is not applicable to market economies. Accordingly, this article focuses on the unofficial economic activities both in the STEs and during transition to markets, with less attention devoted to the legal second economy in the STEs. For the economies in transition, the unofficial economy is defined here as all value-added market-based economic activities that either evade taxation or are not registered, or both.

The most common illegal economic activities performed by individuals in the former USSR and other STEs were the theft of state property, corruption, and 'speculation' defined as resale of goods by individuals for profit (Grossman, 1979). While the first two types of activities are present in all modern economies, the dominant role of the state and state-owned enterprises made these activities particularly widespread in the command economies. The general illegality of speculation was unique to the STEs. Other significant illegal economic activities included unofficial production by individuals and small teams, much of it taking place at state-owned enterprises on company time and using state-owned equipment.

The unofficial economic activities among individuals in the STEs were mainly caused by pervasive price controls, prohibition of many private economic activities that are common in market economies, high taxes on the permitted activities, ubiquitous and poorly protected public property, enormous discretionary power of the bureaucracy, and social attitudes that accommodated economic crimes that did not directly harm specific individuals. None of these causes was unique to the STEs, but their rarely found confluence provided a hospitable environment for the unofficial economy, so that the entire system could be called a quasi-market economy

(Leitzel, 1995). In many STEs the second economy was a normal part of everyday life for a typical consumer (see Grossman, 1977; 1979; 1989, for fascinating descriptions related to the former USSR).

A large 'shadow economy' existed also among socialist enterprises. Much of it had to do with the informal or semi-formal exchanges of goods between state-owned enterprises performed without the prior direction by the planners. Such exchanges, even when legal by themselves, were often accompanied by illegal side payments and other informal inducements. In addition, enterprises sometimes hired labour or acquired other inputs from individuals in the unofficial economy.

Measuring the unofficial economy

The more or less precise dimensions of the illegal economy are all but impossible to ascertain. The approaches to measuring it in various countries are reviewed by Schneider and Enste (2000). With respect to STEs and the economies in transition, the two most commonly used methods have been the surveys of either consumers or firms, and the electricity consumption method. Both approaches have certain advantages and disadvantages. Unless a survey is conducted among emigrants, as was usually the case prior to the collapse of the Communist regimes, the survey questions about unofficial economy are often indirect. For example, the consumers may be asked about their expenditures and official incomes and savings. If reported expenditures exceed legitimate incomes, the difference can serve as a basis for estimating unofficial income. In surveys of firms, the managers may be asked about hidden sales in their industry rather than in their specific firm. The reliability of the survey method depends crucially on the sample selection and on the veracity of the respondents' answers.

The electricity consumption approach assumes a close connection between electricity consumption and the true size of economic activity in a country. In the simplest case, the elasticity of electricity consumption with respect to the combined official and unofficial GDP is taken to be one, implying that the dynamics of electricity output coincides with the dynamics of total GDP. Assuming some initial size of the unofficial economy and knowing the change of the official GDP would then permit calculating the change in the size of the unofficial sector. The two main disadvantages of this method are the need to know the initial size of the unofficial economy and the elasticity of electricity consumption by GDP. The latter information is particularly difficult to infer for the economies in transition that are undergoing structural reforms and experiencing large changes of relative prices. However, an important advantage of this approach is its ease of implementation in a uniform fashion across time and countries. Moreover, it can be used to extrapolate a point estimate of the unofficial economy both into the future and into the past.

Among the major STEs, survey-based estimates of the second economy have been published only for the former USSR. The estimates based on the surveys of Soviet emigrants relate to the second half of 1970s and range approximately between ten per cent and 30 per cent of incomes of urban households. (The lower estimate is due to Ofer and Vinokur, 1992. The higher estimate is from Grossman, 1991.) An alternative set of estimates based on the Soviet-era official family budget survey data puts the second economy at around 23 per cent of household income (Kim, 2003). Using these estimates to obtain the shares of second economy in GDP is difficult, however, particularly because no estimates of the shadow economy among the socialist enterprises have been published. The size of the second economy also varied greatly across the former Soviet republics. In the Caucuses and in central Asian republics both legal and illegal private economic activities were widespread, while in the Baltic republics the second economy was relatively insignificant (Grossman, 1991; Kim, 2003; Alexeev and Pyle, 2003).

Kaufman and Kaliberda (1996) cite a number of micro studies of the unofficial economy in east European countries in the early 1990s and extrapolate these estimates backwards to 1989 using the electricity consumption method. The results range from six per cent of GDP in Czechoslovakia to over 20 per cent of GDP in Bulgaria, Romania and Hungary.

The unofficial sectors in the economies in transition have been measured by the above methods as well as by currency demand method and the latent variable estimation. The former approach estimates a currency demand function for a country and assumes that all 'excess' demand for cash relative to this demand function is due to the growth of the unofficial economy. The latter approach uses latent variable estimation techniques to infer the size of the unofficial economy as an unobserved variable influenced by a number of different factors and affecting several observed economic indicators. Each of these methods has its strengths and weaknesses that are discussed in Schneider and Enste (2000) and Schneider (2005). The latter work combines the two approaches to estimate the unofficial economy as a share of GDP in 110 countries in year 2000 and some earlier years. The 2000 estimates for the economies in transition range from 13.1 per cent in China to over 60 per cent in Georgia and Azerbaijan.

The quantitative importance of the second economy raises the issue of the nature of its impact on welfare, on the 'first' economies, and on the attempts at partial reform. More important for the economics of transition is the transformation of the relationship between the official and unofficial economies and the dynamics of the latter as a result of radical market-oriented reforms.

The second economy's impact on the STEs

The second economy appears to be necessary for the survival of an STE. The marginal effect of the second economy's expansion is less clear. A growing second economy increases the influence of market forces on a STE, obviating various types of non-price rationing and correcting some resource misallocations. For instance, rationing was undermined even for such a seemingly easy-to-control good as residential housing (Alexeev, 1988). Also, certain theoretical concepts applied to a STE such as 'forced savings' or 'monetary overhang' become questionable in the presence of a developed second economy (Alexeev, Gaddy and Leitzel, 1991). On the other hand, second economy growth may have significant negative effects, as is discussed later.

Ericson (1983) provides the first formal analysis of the second economy's impact on the official sector. In his model, enterprise manager i, $i = 1, \ldots, n$, receives a planned output target, $y^i \in R_+$, and the input allocations given by $\omega^i \in R_+^{m+1}$, where the first component, ω_0, represents the amount of official funds and the last m components denote the amounts of material inputs. The production function is described by $f^i : X^i \to R_+$, where $X^i \subset R^m$ is the set of all feasible combinations of material inputs. Given f^i, the plan is not necessarily feasible. The manager's preferences are represented by a utility function monotonic in $x_0^i \in R_+$, and $y^i = f^i(x^i)$. If the initial inputs are misallocated, the managers have incentives to trade legally at the official prices $p = (1, p_1, \ldots, p_m)$. However, these fixed prices are not likely to equilibrate the secondary market. The disequilibrium generates incentives for the managers to induce other managers to trade by offering informal side-payments from the manager's hidden slush fund of loose cash, the initial amount of which is c^i. The flexible prices in this informal market are denoted by $a = (0, a_1, \ldots, a_m) \in R^{m+1}$. It is also assumed that a fixed proportion, α, of this cash 'sticks' to the palms of individuals who handle the transactions. Given this leakage, the manager faces the following problem (index i is suppressed):

$$\text{Max}_z \{u(x_0, f(x))\} \quad \text{s.t.} \quad pz \leq 0 \text{ and}$$
$$az \leq c - \alpha(a_+ z_- + a_- z_+),$$

where $z_j = x_j - \omega_j$, $j = 0, 1, \ldots, m$, represent trades of inputs among managers and cash prices of material inputs may be either positive or negative, depending on whether payment is necessary to induce a purchase or sale. In addition, $z_+ \equiv \text{max}(0, z)$ and $z_- \equiv \text{max}(0, -z)$, with a_+ and a_- defined similarly. Note that $a_0 = 0$, reflecting the assumption that the official funds cannot buy cash and there is no need for financial inducements to transfer official funds. (As is discussed below, the wall between the official funds and loose cash had seriously eroded by the late 1980s as a result of partial reforms in the USSR and other STEs. This erosion was in part responsible for the eventual collapse of the command economic system.)

Under certain technical conditions, this exchange economy has a suitably defined equilibrium, and any trades that occur with the inducement of loose cash represent a Pareto improvement. Given the managers' preferences, the second economy in this model always enhances plan fulfillment. (This equilibrium, however, is not in general constrained Pareto-optimal – where the 'constraint' is given by $pz \leq 0$ – because the leakage might stop trading before Pareto optimal allocation is reached.)

Of course, the first economy and the overall economic welfare may suffer if the official plans are inefficient. Also, the second economy always enhances plan fulfilment in Ericson's framework only because enterprise managers' preferences do not exhibit a trade-off between plan fulfillment and personal gain from unofficial dealings. Ericson's model takes no account of the potential negative externalities that the second economy transactions can create in an otherwise distorted economy. Some of these externalities are examined by Stahl and Alexeev (1985) in a model of a queue-rationed exchange economy. Here, N consumers face inelastic supply of goods $X \in R_{++}^n$ with a fixed official price $p_1 \in R_{++}^n$. If a shortage arises, the demand is rationed by queues with a deterministic waiting time $\tau \in R_+^n$. Consumers are endowed with money, M^i, and leisure, L^i. Consumer i has well-behaved preferences $u^i(x^i, l^i)$ with respect to consumption of goods, $x^i \in R_+^n$, and leisure, $l^i \in R_+$. In the absence of black markets, each consumer solves (superscript i is suppressed):

$$\text{Max}_{x, l} u(x, l) \quad \text{s.t.} \quad p_1 x \leq M \text{ and } \tau x + 1 \leq L$$

An appropriately defined equilibrium that requires no excess demand always exists.

Black markets are introduced by allowing consumers to resell goods they have purchased in the official markets. Denote the black market trades by $y^i \in R^n$, with $y_j^i > 0$ if the consumer buys the corresponding good and $y_j^i < 0$ if he sells. Black market prices, $p_2 \in R_+^n$, are flexible. The new budget constraints become (superscript i is suppressed):

$$p_1(x - y) + p_2 y \leq M$$
$$\tau(x - y) + 1 \leq L$$

A necessary condition for no excess demand is $p_2 = p_1 + w\tau$, where w is interpreted as an implicit wage for standing in line. Accordingly, the additional requirements for an equilibrium include $p_2 = p_1 + w^* \tau^*$ and $\sum y^* = 0$. Equilibrium exists if $\sum M \leq p_1 X$. Given the aggregate queuing time and p_1, the equilibrium with black markets is efficient while this is not usually true in the absence of black markets.

The main insight, however, is that the introduction of black markets does not always result in a Pareto improvement over pure queue rationing. This happens because black markets create wealth effects that may result in longer queues. Nonetheless, in this framework the poor typically prefer queuing with black markets to pure queue rationing (Polterovich, 1993). The comparisons for the rich are ambiguous.

In addition to the above effects, the unofficial economic activities both in the consumption sector and among socialist enterprises reduce the feedback to the policymakers from their actions by covering up policy consequences and resource misallocations, and rendering the official statistics, including those on consumer incomes, savings, consumption and employment, less relevant. Also, the efficiency of unofficial transactions is reduced by the need for secrecy, which impedes information flows within black markets and limits the range of available contract enforcement mechanisms, resulting in greater uncertainty and a suboptimally small scale of operations, among other things. Moreover, the potential profitability of illegal transactions breeds corruption and weakens the official institutions.

On balance, the effect of the second economy on the official economy and on overall welfare is theoretically ambiguous on the margin.

The second economy's impact on the pre-reform crises and on reforms within an STE

As a command economy matures and becomes increasingly complex, the functioning of the traditional administrative coordination mechanism deteriorates. Many STEs responded to their worsening economic situation by introducing reforms aimed at liberalizing parts of the economy, particularly labour markets and consumer-oriented sectors. Such partial reforms had generally failed to improve overall economic performance, largely because of the second-best considerations. Black markets, however, may alleviate some of the adverse consequences of partial reforms. For example, Boycko (1992) and Osband (1992) demonstrate that wage increases caused by partial liberalization of labour markets in an STE lead to a cycle of greater shortages, longer queues, and, therefore, lower output if prices of consumer goods remain fixed. This 'pre-reform crisis' is moderated when black markets are brought into the picture, because the additional queuing is done mainly by lower-productivity agents who can then resell the goods in black markets while without the unofficial economy everybody has to queue up more (Alexeev and Sabyr, 2004). In another example, Leitzel (1998) shows beneficial effects of black markets by introducing them into a simple model of repressed inflation developed by Lipton and Sachs (1990). In Leitzel's model, retail trade employees divert goods in short supply to black markets, thereby reducing welfare losses due to rising repressed inflation and pre-empting the adverse distributional consequences of full price liberalization.

The presence of black markets may also worsen the effects of partial reforms. Consider, for example, the so-called 'dual track' partial reform mechanism (Lau,

Qian and Roland, 2000). It attempts to conduct reforms in a Pareto-improving fashion by preserving the output and supply quotas that exist in an STE while liberalizing the markets for the above quota output. If the government can enforce the pre-reform quotas, this mechanism avoids the potential severe misallocation of resources due to partial liberalization emphasized by Murphy, Shleifer and Vishny (1992) and, at the same time, prevents excessive redistribution that may result from full liberalization. However, the violation of pre-reform quotas via the unofficial trades can generate large rents. If such violations are allowed to take place, resource misallocation may easily occur.

The growth of the second economy by itself represents an implicit market reform of an STE, so that partial reforms simply provide an official permission for certain unofficial activities that have been already widespread (Leitzel, 1995). In part, the second economy growth pre-empts the potential negative effects of full liberalization, making radical market reforms politically more acceptable. In addition, the second economy familiarizes the general population with the workings of the market, develops trust and entrepreneurship, thereby facilitating popular acceptance of the market mechanism (Grossman, 1989). However, to the extent the second economy improves the functioning of the overall economy, it also reduces the potential benefits of full liberalization relative to the unreformed or partially reformed economy and, therefore, may postpone radical reforms.

Whatever the impact of the second economy on the performance of partial reforms, it is clear that historically such reforms spurred an explosive growth of the unofficial sector in the STEs. Leitzel (2003) argues that this growth was unavoidable given partial relaxation of state controls that made evasion harder to detect while leaving most of the formal distortions in place. In the USSR, for example, the permission of crypto-private firms and transactions between them and state-owned enterprises greatly reduced the ability of central planners to monitor state-owned enterprises. Enterprise managers then employed transfer pricing schemes and self-dealing to channel state-provided resources to private firms related to the managers. Tax collections plummeted while subsidies increased, undermining the state budget and the entire state sector.

The unofficial economy during transition to markets

Radical market reforms of the STEs seek to turn these economies, including most of their unofficial sectors, into well-functioning legal markets. Due to general economic liberalization, particularly the removal of such a major cause of black markets as price controls, the unofficial economies can be expected to diminish. This indeed seems to have occurred in some countries, but not in others. Using the electricity consumption approach,

Johnson, Kaufmann and Shleifer (1997) estimate that between 1989 and 1995 the unofficial economy grew significantly in Russia, Ukraine and some other former Soviet republics while it remained approximately the same in Hungary, Slovakia and Estonia, and declined noticeably in Poland. (Note that the electricity consumption approach attributes the entire change – decline – in measured output relative to electricity consumption to the unofficial economy. Alternatively, the dynamics of GDP-to-electricity consumption ratios in these economies could have been influenced by relative prices and availability of electricity, government policies with respect to energy conservation, and the changing structure of the economy.)

Also, according to Johnson, McMillan and Woodruff (2000), private manufacturing firms in Russia and Ukraine tend to hide their activities from tax authorities on a dramatically greater scale than do their counterparts in eastern Europe. The firms in Russia and Ukraine under-reported around 30–40 per cent of their sales and 25–40 per cent of their employee salaries, as compared with five to ten per cent in Poland, Slovakia and Romania.

On the assumption that the measurements are approximately correct, what may explain such different dynamics? One answer relies on path dependency that arises from the possibility of multiple equilibria in the relationship between the official and unofficial economies. Consider, for example, a labour allocation model by Johnson, Kaufmann and Shleifer (1997). The aggregate amount of labour normalized to one is allocated between the official economy (subscript F) and the unofficial sector (subscripted by I), $L_F + L_I = 1$. The agents in each sector $J = F, I$ maximize their utilities, $U_J = 1(1 - t_J)Q_J$, where t_J is the tax rate and Q_J denotes the quantity of the output-enhancing public good in each sector by, respectively, the government and the mafia. (Q_J can be used only by sector J agents.) Let $Q_J = (T_J)^\beta$ where $T_J = t_J Q_J L_J$ is aggregate tax revenue collected in sector J. This model has an unstable interior equilibrium and two stable corner equilibria: in one $L_F = L$, $L_I = 0$, and in the other $L_F = 0$, $L_I = L$. Therefore, depending on the initial conditions, the economy may end up either entirely above or below ground. If parameter β is greater in the official economy than in the unofficial sector, the latter equilibrium is inefficient.

An important assumption in this model is that, while the government is hurt by labour escaping underground, it does not attempt to either stop the escape or fight the mafia that facilitates underground activities. Roland and Verdier (2003) show that one reason for government passivity may be the lack of resources for law-enforcement activities. Suppose economic agents choose between becoming producers and predators. Producers pay taxes to fund law enforcement to fight off predators. The more agents become producers, the greater are the resources for law enforcement. That is, similarly

to Johnson, Kaufmann and Shleifer (1997), a fiscal externality is present. Assume also that law enforcement has fixed costs in order to become effective and that a 'compliance externality' exists, so that the probability of a predator being punished is inversely related to the number of predators. Under certain assumptions on the utilities of producers and predators, this model again has multiple equilibria, in some of which the government lacks the resources to fight the predators. (The government may also refrain from fighting the mafia that provides public goods underground because the presence of predatory mafia actually benefits predatory government. Alexeev, Janeba and Osborne (2004) demonstrate that, when public goods are expensive to provide, the state has higher revenues in the economy with the mafia than without it. This is because, when public goods provision is difficult and few of them are provided, the main effect of the mafia is to increases the costs of underground economic activities. This makes it possible for the state to increase its tax rate without having too many entrepreneurs escape underground.)

On the assumption that the above models reflect some essential features of the transition processes, an important policy issue is how a country can avoid a 'bad' equilibrium or escape one if it finds itself in it. Roland and Verdier suggest that the promise of EU accession can be one mechanism that solves coordination problems in their model described above. Obviously, this solution is not available to all economies in transition. Some argue that the collapse of government institutions and the ensuing flight of taxpayers underground can be avoided by conducting reforms gradually. China is often cited as a successful example of a gradual approach. However, such strategy is risky both because it may not avoid the deterioration of government institutions (for example, despite Ukraine's gradual reforms, the data mentioned in the beginning of this section testify to the similarities between Ukraine's and Russia's unofficial economies) and because it may postpone genuine reforms indefinitely, as appears to have happened in Belarus and Uzbekistan. Friedman et al.'s (2000) findings based on a cross-section of countries suggest a rather pessimistic conclusion that the size of the unofficial economy depends largely on the quality of institutions, which in turn is determined by exogenous factors such as geography, religion, linguistic fractionalization, and the origin of the legal system, while the policy parameters that are more easily controlled by governments such as tax rates do not matter very much. These empirical results imply that over-regulation, corruption and a weak rule of law are associated with larger unofficial economy. (The positive relationship between corruption and unofficial economy is less intuitive than the other correlations. This is because the economic agents may either use the unofficial economy to escape bribe requests or bribe officials to evade taxation and regulations.)

The high tax burden can, of course, also serve as a reason for shifting resources into the unofficial economy.

However, high tax rates can also generate revenues necessary to increase the output of productivity-enhancing public goods, making the official economy more attractive. The net effect of tax rates on the unofficial economy is ambiguous. We note also that the impact of taxation presumably depends on the effective tax rates that are often determined by how the tax system is administered rather than on the statutory tax rates as long as the latter are within a reasonable range.

Despite the above considerations, if the government of an economy in transition is truly intent on legalizing a significant part of the unofficial economy, the rationalization of the tax system can serve as a reasonable first step, as Alexeev, Conrad and Hay (2004) argue and as Russia's tax reforms of 2001–3 demonstrate. Ivanova, Keen and Klemm (2005) and Sinelnikov-Mourylev et al. (2003) show that a dramatic reduction of the highest marginal tax rates on personal income in Russia induced many taxpayers to legalize their incomes, leading to an increase in government revenues. Of course, tax reform alone is not sufficient. At the very least, it needs to be accompanied by reform of government regulations, administration and the courts. Unfortunately, these administrative reforms may be difficult to implement in those economies in transition that need them most.

While many economies in transition may find it difficult to reduce the unofficial sector, its size appears to be smaller than in a number of developing countries at a comparable level of development such as Mexico, South Korea and Chile. (See Schneider and Enste, 2000. Campos, 2000, argues, however, that the nature of the unofficial sectors in the economies in transition and the developing economies may be different. In the latter, the unofficial economy appears to consist mostly of small labour-intensive businesses trying to escape excessive government regulation and taxation. In the former, large firms hiding their activities account for a larger part of the unofficial economy. If true, these differences would testify to a greater degree of government capture by large firms in the economies in transition, implying that reforms aimed at reducing unofficial economies there may be more difficult to implement. This argument may be plausible, but it is hard to test given the available data and the structural differences between major economies in transition and developing economies. For example, greater capital intensity of the unofficial sector in the economies in transition can be simply a consequence of greater capital intensity of the overall economy.)

More important, the impact of the unofficial economy on welfare in any real-world institutional environment, and particularly in the rather distorted economies in transition, is ambiguous, even on the margin. Among other things, this is because the possibility of going underground often provides an effective check on the power of governments to over-regulate and overtax.

MICHAEL ALEXEEV

See also **bribery; command economy; informal economy; rationing; tax compliance and tax evasion; transition and institutions.**

Bibliography

Alexeev, M. 1988. Market vs. rationing: the case of soviet housing. *Review of Economics and Statistics* 70, 414–20.

Alexeev, M., Conrad, R. and Hay, J. 2004. Nalogooblozhenie i pravovaia reforma v perekhodnoi ekonomike: predvaritel'nyi analiz [Taxation and legal reform in an economy in transition: a preliminary analysis]. In *Pravovye reformy i ekonomicheskii rost* [Legal Reform and Economic Growth], ed. E. Berglof and S. Shishkin. Moscow: CEFIR.

Alexeev, M., Gaddy, C. and Leitzel, J. 1991. An economic analysis of the Ruble overhang. *Communist Economies and Economic Transformation* 3, 467–79.

Alexeev, M., Janeba, E. and Osborne, S. 2004. Taxation and evasion in the presence of extortion by organized crime. *Journal of Comparative Economics* 32, 375–88.

Alexeev, M. and Pyle, W. 2003. A note on measuring the unofficial economy in the former Soviet republics. *Economics of Transition* 11, 153–75.

Alexeev, M. and Sabyr, L. 2004. Black markets and pre-reform crises in former socialist economies. *Economic Systems* 28, 1–12.

Boycko, M. 1992. When higher incomes reduce welfare: queues, labor supply, and macro equilibrium in socialist economies. *Quarterly Journal of Economics* 107, 907–20.

Campos, N. 2000. Never at noon: on the nature and causes of the transition shadow, Discussion paper. Prague: CERGE-EI.

Ericson, R. 1983. On an allocative role of the Soviet second economy. In *Marxism, Central Planning, and the Soviet Economy: Economic Essays in Honor of Alexander Erlich*, ed. P. Desai. Boston: MIT Press.

Friedman, E., Johnson, S., Kaufmann, D. and Zoido-Lobaton, P. 2000. Dodging the grabbing hand: the determinants of unofficial activity in 69 countries. *Journal of Public Economics* 76, 459–93.

Grossman, G. 1977. The 'second economy' of the USSR. *Problems of Communism* 26, 25–40.

Grossman, G. 1979. Notes on the illegal private economy and corruption. In *Soviet Economy in a Time of Change*, vol. 1, ed. Joint Economic Committee of Congress. Washington: US Government Printing Office.

Grossman, G. 1989. The second economy: boon or bane for the reform of the first economy. In *Economic Reforms in the Socialist World*, ed. S. Gomulka, Y. Ha and C. Kim. Armonk, NY: M.E. Sharpe.

Grossman, G. 1991. Wealth estimates based on the Berkeley–Duke émigré questionnaire: a statistical compilation. Berkeley–Duke Occasional Papers on the Second Economy in the USSR, No. 27.

Ivanova, A., Keen, M. and Klemm, A. 2005. The Russian flat tax reform. *Economic Policy* 20, 397–444.

Johnson, S., Kaufmann, D. and Shleifer, A. 1997. The unofficial economy in transition. *Brookings Papers on Economic Activity* 1997(2), 159–239.

Johnson, S., McMillan, J. and Woodruff, C. 2000. Why do firms hide? Bribes and unofficial activity after communism. *Journal of Public Economics* 76, 495–520.

Kaufmann, D. and Kaliberda, A. 1996. Integrating the unofficial economy into the dynamics of post-socialist economies: a framework for analysis and evidence. In *Economic Transition in Russia and the New States of Eurasia*, ed. B. Kaminski. London: M.E. Sharpe.

Kim, B. 2003. Informal economy activities of Soviet households: size and dynamics. *Journal of Comparative Economics* 31, 532–51.

Lau, L., Qian, Y. and Roland, G. 2000. Reform without losers: an interpretation of China's dual-track approach to reforms. *Journal of Political Economy* 108, 120–63.

Leitzel, J. 1995. *Russian Economic Reform*. London: Routledge.

Leitzel, J. 1998. Goods diversion and repressed inflation: notes on the political economy of price liberalization. *Public Choice* 94, 255–66.

Leitzel, J. 2003. *The Political Economy of Rule Evasion and Policy Reform*. London: Routledge.

Lipton, D. and Sachs, J. 1990. Creating market economy in eastern Europe: the case of Poland. *Brookings Papers on Economic Activity* 1990(1), 75–133.

Murphy, K., Shleifer, A. and Vishny, R. 1992. The transition to a market economy: pitfalls of partial reform. *Quarterly Journal of Economics* 107, 889–906.

Ofer, G. and Vinokur, A. 1992. *The Soviet Household Under the Old Regime: Economic Conditions and Behavior in the 1970s*. Cambridge, MA: Cambridge University Press.

Osband, K. 1992. Economic crisis in a shortage economy. *Journal of Political Economy* 100, 673–90.

Polterovich, V. 1993. Rationing, queues, and black markets. *Econometrica* 61, 1–28.

Roland, G. and Verdier, T. 2003. Law enforcement and transition. *European Economic Review* 47, 669–85.

Schneider, F. 2005. Shadow economies around the world: what do we really know? *European Journal of Political Economy* 21, 598–642.

Schneider, F. and Enste, D. 2000. Shadow economies: size, causes, and consequences. *Journal of Economic Literature* 38, 77–114.

Sinelnikov-Mourylev, S., Batkibekov, S., Kadochnikov, P. and Nekipelov, D. 2003. Otsenka rezul'tatov reformy podokhodnogo naloga v Rossiiskoi Federatsii [Evaluation of personal income tax reform in the Russian Federation]. Working Paper No. 52. Moscow: Institute for Economy in Transition (IET).

Stahl, D. and Alexeev, M. 1985. The influence of black markets on queue-rationed CPE. *Journal of Economic Theory* 35, 234–50.

seemingly unrelated regressions

A seemingly unrelated regression (SUR) system comprises several individual relationships that are linked by the fact that their disturbances are correlated. Such models have found many applications. For example, demand functions can be estimated for different households (or household types) for a given commodity. The correlation among the equation disturbances could come from several sources such as correlated shocks to household income. Alternatively, one could model the demand of a household for different commodities, but adding up constraints leads to restrictions on the parameters of different equations in this case. On the other hand, equations explaining some phenomenon in different cities, states, countries, firms or industries provide a natural application as these various entities are likely to be subject to spillovers from economy-wide or worldwide shocks.

There are two main motivations for use of SUR. The first one is to gain efficiency in estimation by combining information on different equations. The second motivation is to impose and/or test restrictions that involve parameters in different equations. Zellner (1962) provided the seminal work in this area, and a thorough treatment is available in the book by Srivastava and Giles (1987). A recent survey can be found in Fiebig (2001). This article selectively overviews the SUR model, some of the estimators used in such systems and their properties, and several extensions of the basic SUR model. We adopt a classical perspective, although much Bayesian analysis has been done with this model (including Zellner's contributions).

Basic linear SUR model

Suppose that y_{it} is a dependent variable, $x_{it} = (1, x_{it,1}, x_{it,2}, \ldots, x_{it,K_i-1})'$ is a K_i-vector of explanatory variables for observational unit i, and u_{it} is an unobservable error term, where the double index it denotes the tth observation of the ith equation in the system. Often t denotes time and we will refer to this as the time dimension, but in some applications t could have other interpretations, for example as a location in space. A classical linear SUR model is a system of linear regression equations,

$$y_{1t} = \beta_1' x_{1t} + u_{1t}$$
$$\vdots$$
$$y_{Nt} = \beta_N' x_{Nt} + u_{Nt}$$

where $i = 1, \ldots, N$, and $t = 1, \ldots, T$. Denote $L = K_1 + \cdots + K_N$. Further simplification in notation can be accomplished by stacking the observations either in the t dimension or for each i. For example, if we stack for each observation t, let $Y_t = [y_{1t}, \ldots, y_{Nt}]'$, $\tilde{X}_t = \mathrm{diag}(x_{1t}, x_{2t}, \ldots, x_{Nt})$, a block-diagonal matrix with $x_{1t}, \ldots, x_{Nt}$ on its diagonal, $U_t = [u_{1t}, \ldots u_{Nt}]'$, and $\beta = [\beta_1', \ldots, \beta_N']'$. Then,

$$Y_t = \tilde{X}_t'\beta + U_t. \tag{1}$$

Another way to present the SUR model is to write it in a form of a multivariate regression with parameter restrictions. For this, define $X_t = [x_{1t}', x_{2t}', \ldots, x_{Nt}']'$ and $A(\beta) = \text{diag}(\beta_1, \ldots, \beta_N)$ to be a $(L \times N)$ block diagonal coefficient matrix. Then, the SUR model in (1) can be rewritten as

$$Y_t = A(\beta)'X_t + U_t, \tag{2}$$

and the coefficient $A(\beta)$ satisfies

$$\text{vec}(A(\beta)) = G\beta, \tag{3}$$

for some $(NL \times L)$ full rank matrix G. In the special case where $K_1 = \cdots = K_N = K$, we have $G = \text{diag}(i_1, \ldots, i_N) \otimes I_K$, where i_j denotes the jth column of the $N \times N$ identity matrix I_N.

Assumption
In the classical linear SUR model, we assume that for each $i = 1, \ldots, N$, $x_i = [x_{i1}, \ldots, x_{iT}]'$ is of full rank K_i, and that conditional on all the regressors $X' = [X_1, \ldots, X_T]$, the errors U_t are i.i.d. over time with mean zero and homoskedastic variance $\Sigma = E(U_t U_t'|X)$. Furthermore, we assume that Σ is positive definite and denote by σ_{ij} the (i,j)th element of Σ: that is, $\sigma_{ij} = E(u_{it} u_{jt}|X)$.

Under this assumption, the covariance matrix of the entire vector of disturbances $U' = [U_1, \ldots, U_T]$ is given by $E[\text{vec}(U)(\text{vec}(U))'] = \Sigma \otimes I_T$.

Estimation of β
In this section we summarize four estimators of β that have been widely used in applications of the classical linear SUR. Other estimators (such as Bayes, empirical Bayes or shrinkage estimators) have also been proposed. Interested readers should refer to Srivastava and Giles (1987) and Fiebig (2001).

1. Ordinary least squares (OLS) estimator. The first estimator of β is the ordinary least squares (OLS) estimator of Y_t on regressor $\tilde{X}_t$,

$$\hat{\beta}_{OLS} = \left(\sum_{t=1}^{T} \tilde{X}_t \tilde{X}_t' \right)^{-1} \sum_{t=1}^{T} \tilde{X}_t Y_t.$$

This is just the vector that stacks the equation-by-equation OLS estimators, $\hat{\beta}_{OLS} = (\hat{\beta}_{1,OLS}', \ldots, \hat{\beta}_{N,OLS}')'$, where $\hat{\beta}_{i,OLS} = (\sum_{t=1}^{T} x_{it} x_{it}')^{-1} \sum_{t=1}^{T} x_{it} y_{it}$.

2. Generalized least squares (GLS) and feasible GLS (FGLS) estimator. When the system covariance matrix Σ is known, the GLS estimator of β is

$$\hat{\beta}_{GLS} = \left(\sum_{t=1}^{T} \tilde{X}_t \Sigma^{-1} \tilde{X}_t' \right)^{-1} \sum_{t=1}^{T} \tilde{X}_t \Sigma^{-1} Y_t.$$

When the covariance matrix Σ is unknown, a feasible GLS (FGLS) estimator is defined by replacing the unknown Σ with a consistent estimate. A widely used estimator of Σ is

$$\hat{\Sigma} = (\hat{\sigma}_{ij}),$$

where $\hat{\sigma}_{ij} = \frac{1}{T}\sum_{t=1}^{T} \hat{e}_{it}\hat{e}_{jt}$ and $\hat{e}_{kt}$ is the OLS residuals of the kth equation: that is, $\hat{e}_{kt} = y_{kt} - \hat{\beta}_{k,OLS} x_{kt}$, $k = i, j$. Then

$$\hat{\beta}_{FGLS} = \left(\sum_{t=1}^{T} \tilde{X}_t \hat{\Sigma}^{-1} \tilde{X}_t' \right)^{-1} \sum_{t=1}^{T} \tilde{X}_t \hat{\Sigma}^{-1} Y_t.$$

The FGLS estimator is a two-step estimator where OLS is used in the first step to obtain residuals $\hat{e}_{kt}$ and an estimator of Σ. The second step computes $\hat{\beta}_{FGLS}$ based on the estimated Σ in the first step. This estimator is sometimes referred to as the restricted estimator as opposed to the unrestricted estimator proposed by Zellner that uses the residuals from an OLS regression of (2) without imposing the coefficient restrictions (3), that is, from regressing each regressand on all distinct regressors in the system.

3. Gaussian quasi-maximum likelihood estimator (QMLE). The Gaussian log-likelihood function is

$$L(\beta, \Sigma) = \text{const} + \frac{T}{2}\det\Sigma - \frac{1}{2}\sum_{t=1}^{T}$$
$$\times \left(Y_t - \tilde{X}_t'\beta \right)'\Sigma^{-1}\left(Y_t - \tilde{X}_t'\beta \right),$$

or equivalently,

$$L(\beta, \Sigma) = \text{const} + \frac{T}{2}\det\Sigma - \frac{1}{2}\sum_{t=1}^{T}$$
$$\times \left(Y_t - A(\beta)'X_t \right)'\Sigma^{-1}\left(Y_t - A(\beta)'X_t \right),$$

where $A(\beta)$ denotes the coefficient A in (2) with the linear restriction of (3), and the QMLE $(\hat{\beta}_{QMLE}, \hat{\Sigma}_{QMLE})$ maximizes $L(\beta, \Sigma)$. When the vector U_t has a normal distribution, this estimator is the maximum likelihood estimator.

4. Minimum distance (MD) estimator. The idea of the MD estimator is to obtain an estimator of the unrestricted coefficient A in (2), $\hat{A}$, and then, obtain an estimator of β by minimizing the distance between $\hat{A}$ and β in (3). For this, assume that $T > L$ and that the whole regressor matrix X has full rank L. When $\hat{A}$ is the OLS estimator of $A(\beta)$, that is $\hat{A} = (\sum_{t=1}^{T} X_t X_t')^{-1} \sum_{t=1}^{T} X_t Y_t'$, the optimal MD estimator $\hat{\beta}_{MD}$ minimizes the optimal

MD objective function

$$Q_{MD}(\beta) = \left(\text{vec}(\hat{A}) - G\beta\right)' \left(\hat{\Sigma}^{-1} \otimes \sum_{t=1}^{T} X_t X_t'\right)$$
$$\times \left(\text{vec}(\hat{A}) - G\beta\right).$$

In this case, we have

$$\hat{\beta}_{MD} = \left(G'\left(\hat{\Sigma}^{-1} \otimes \sum_{t=1}^{T} X_t X_t'\right)G\right)^{-1}$$
$$\times \left(G'\left(\hat{\Sigma}^{-1} \otimes \sum_{t=1}^{T} X_t X_t'\right)\text{vec}(\hat{A})\right).$$

Relationship between the estimators
Some of the above estimators are tightly linked. For example, *if we use the same consistent estimator $\hat{\Sigma}$, the FGLS and the MD estimators above are identical: that is, $\hat{\beta}_{FGLS} = \hat{\beta}_{MD}$.* Also, if we use the QMLE estimator of Σ, $\hat{\Sigma}_{QMLE}$ in place of $\hat{\Sigma}$, $\hat{\beta}_{QMLE}$ is identical to $\hat{\beta}_{FGLS}$ (and to $\hat{\beta}_{MD}$). By the Gauss–Markov theorem, the GLS estimator $\hat{\beta}_{GLS}$ is more efficient than the OLS estimator $\hat{\beta}_{OLS}$ when the system errors are correlated across equations. However, this efficiency gain disappears in some special cases described in Kruskal's theorem (Kruskal, 1968). A well-known special case of this theorem is when the regressors in each equation are the same. For other cases, readers can refer to Greene (2003, ch. 14) and Davidson and MacKinnon (1993, pp. 294–5). The efficiency gain relative to OLS tends to be larger when the correlation across equations is larger and when the correlation among regressors in different equations is smaller.

Note also that efficient estimators propagate misspecification and inconsistencies across equations. For example, if any equation is misspecified (for example some relevant variable has been omitted), then the entire vector β will be inconsistently estimated by the efficient methods. In this sense, equation-by-equation OLS provides some degree of robustness since it is not affected by misspecification in other equations in the system.

Distribution of the estimators
In the literature on the classical linear SUR, the FGLS estimator $\hat{\beta}_{FGLS}$ is often called the SUR estimator (SURE). The usual asymptotic analysis of the SURE is carried out when the dimension of index t, T, increases to infinity with the dimension of index i, N, kept fixed. For asymptotic theories for large N, T, one can refer to Phillips and Moon (1999). Under regularity conditions, the asymptotic distributions as $T \to \infty$ of the aforementioned

estimators are

$$\sqrt{T}\left(\hat{\beta}_{OLS} - \beta\right)$$
$$\Rightarrow N\left(0, \left[E\left(\tilde{X}_t \tilde{X}_t'\right)\right]^{-1} E\left(\tilde{X}_t \Sigma \tilde{X}_t'\right)\left[E\left(\tilde{X}_t \tilde{X}_t'\right)\right]^{-1}\right)$$

and

$$\sqrt{T}\left(\hat{\beta}_{GLS} - \beta\right), \quad \sqrt{T}\left(\hat{\beta}_{FGLS} - \beta\right),$$
$$\sqrt{T}\left(\hat{\beta}_{MD} - \beta\right) \Rightarrow N\left(0, \left[E\left(\tilde{X}_t \Sigma^{-1} \tilde{X}_t'\right)\right]^{-1}\right)$$
$$\equiv N\left(\left(G'(\Sigma^{-1} \otimes E(X_t X_t'))G\right)^{-1}\right).$$

It is straightforward to show that the SUR estimator using the information in the system is more efficient (has a smaller variance) than the estimator of the individual equations. By using the above distributional results, it is straightforward to construct statistics to test general nonlinear hypotheses.

Finite sample properties of SURE have been studied extensively either analytically in some restrictive cases (Zellner, 1963; 1972; Kakwani, 1967), by asymptotic expansions (Phillips, 1977; Srivastava and Maekawa, 1995) or by simulation (Kmenta and Gilbert, 1968). Most work has focused on the two-equation case. The above approximations appear to be good descriptions of the finite-sample behaviour of the estimators analysed when the number of observations, T, is large relative to the number of equations, N. In particular, efficient methods provide an efficiency gain in cases where the correlation among disturbances across equations is high and when correlation among regressors across equations is low. Non-normality of disturbances has also been found to deteriorate the quality of the above approximations. Bootstrap methods have also been proposed to remedy these documented departures from normality and improve the size of tests.

Extensions
In this section we discuss several extensions of the classical linear SUR model where the assumption on the error terms is no longer satisfied.

Autocorrelation and heteroskedasticity
As in standard univariate models, non-spherical disturbances can be accommodated by either modelling the residuals or computing robust covariance matrices. In addition to standard dynamic effects, serial correlation can arise in this environment due to the presence of individual effects (see Baltagi, 1980). One could define the equivalent of White (in the case of heteroskedasticity) or HAC (in the case of serial correlation) standard errors to conduct inference with the OLS estimator as in the single-equation framework.

For efficiency in estimation some parametric assumption on the disturbance process is often imposed (see Greene, 2003). For example, in the case of heteroskedasticity, Hodgson, Linton, and Vorkink (2002) propose an adaptive estimator that is efficient under the assumption that the errors follow an elliptical symmetric distribution that includes the normal as a special case. An intermediate approach is to use a restricted (or parametric) covariance matrix to try to capture some efficiency gains in estimation, and then use a nonparametric heteroskedasticity and autocorrelation (HAC) consistent estimator of the covariance matrix to do inference. This two-tier approach (dubbed quasi-FGLS) has been suggested by Creel and Farell (1996).

Endogenous regressors

When the regressor X_t in the SUR model is correlated with the error term U_t, one needs instrumental variables (IVs), say, $Z_t = [z'_{1t}, \ldots, z'_{Nt}]'$ to estimate β. We suppose that the IVs satisfy the usual rank condition. The generalized method of moments (GMM) estimator (or the IV estimator), then, utilizes the moment condition

$$E\big[\operatorname{vec}(Z_t U'_t)\big] = 0.$$

The optimal GMM estimator $\hat{\beta}_{GMM}$ is derived by minimizing the GMM objective function with the optimal choice of weighting matrix given by $(\hat{\Sigma} \otimes \big(\sum_{t=1}^{T} Z_t Z'_t\big))^{-1}$,

$$
Q_{GMM}(\beta) = \left[\sum_{t=1}^{T} \operatorname{vec}\Big\{Z_t\big(Y_t - A(\beta)'X_t\big)'\Big\}\right]'
$$
$$
\times \left(\hat{\Sigma} \otimes \sum_{t=1}^{T} Z_t Z'_t\right)^{-1}
$$
$$
\times \left[\sum_{t=1}^{T} \operatorname{vec}\Big\{Z_t\big(Y_t - A(\beta)'X_t\big)'\Big\}\right].
$$

Then, we have

$$
\hat{\beta}_{GMM} = \left\{ G'\left(\hat{\Sigma} \otimes \left(\sum_{t=1}^{T} X_t Z'_t\right)\left(\sum_{t=1}^{T} Z_t Z'_t\right)^{-1}\right. \right.
$$
$$
\left. \times \left(\sum_{t=1}^{T} Z_t X'_t\right)\right)^{-1} G\right\}^{-1} \times G'\left\{\hat{\Sigma}^{-1} \otimes \left(\sum_{t=1}^{T} X_t Z'_t\right)\right.
$$
$$
\left. \times \left(\sum_{t=1}^{T} Z_t Z'_t\right)^{-1}\left(\sum_{t=1}^{T} Z_t X'_t\right)\right\} \operatorname{vec}(\hat{A}_{2SLS}),
$$

where

$$
\hat{A}_{2SLS} = \left\{\left(\sum_{t=1}^{T} X_t Z'_t\right)\left(\sum_{t=1}^{T} Z_t Z'_t\right)^{-1}\left(\sum_{t=1}^{T} Z_t X'_t\right)\right\}^{-1}
$$
$$
\times \left(\sum_{t=1}^{T} X_t Z'_t\right)\left(\sum_{t=1}^{T} Z_t Z'_t\right)^{-1}\left(\sum_{t=1}^{T} Z_t Y'_t\right)
$$

is the two-stage least squares estimator of $A(\beta)$. When X_t is exogenous, so that $X_t = Z_t$, the GMM objective function $Q_{GMM}(\beta)$ and minimum distance objective function $Q_{MD}(\beta)$ are identical, and in this case $\hat{\beta}_{GMM} = \hat{\beta}_{MD}$.

Vector autoregressions

When the index t in the SUR model denotes time and the regressors x_{it} include lagged dependent variables, the classical linear SUR model becomes a vector autoregression model (VAR) with exclusion restrictions. In this case, the regressors X are no longer strictly exogenous, and the assumption in the previous section is violated. A special case is when the order of the lagged dependent variables is one. In this case, for $\{y_{it}\}_t$ to be stationary, it is necessary that the absolute value of the coefficient of y_{it-1} is less than one. If the coefficient of y_{it-1} is one, $\{y_{it}\}_t$ is non-stationary. Non-stationary SUR VAR models have been used in developing tests for unit roots and cointegration in panels with cross-sectional dependence: see for example Chang (2004), Groen and Kleibergen (2003) and Larsson, Lyhagen, and Lothgren (2001).

Seemingly unrelated cointegration regressions

When the non-constant regressors in X_t are integrated non stationary variables but the errors in U_t are stationary, we call model (1) (or equivalently (2)) a seemingly unrelated cointegration regression model; see Park and Ogaki (1991), Moon (1999), Mark, Ogaki and Sul (2005), and Moon and Perron (2004). These papers showed that for efficient estimation of β, an estimator of the long-run variance of U_t, not of the spontaneous covariance Σ as in the previous section, should be used in FGLS. In addition, some modification of the regression is necessary when the integrated regressors and the stationary errors are correlated. Empirical applications in the main references include tests for purchasing power parity, the relation between national saving and investment, and tests of the forward rate unbiasedness hypothesis.

Nonlinear SUR (NSUR)

An NSUR model assumes that the conditional mean of y_{it} given x_{it} is nonlinear, say $h_i(\beta, x_{it})$, that is, $y_{it} = h_i(\beta, x_{it}) + u_{it}$. Defining $H(\beta, X_t) = (h_1(\beta, x_{1t}), \ldots, h_N(\beta, x_{Nt}))$, we write the NSUR model in a multivariate nonlinear regression form,

$$Y_t = H(\beta, X_t) + U_t.$$

In this case, we may estimate β using (quasi) MLE assuming that Y_t are Gaussian conditioned on X_t or

GMM utilizing the moment condition that $E[g(X_t)U_t'] = 0$ for any measurable transformation g of X_t.

HYUNGSIK ROGER MOON AND BENOIT PERRON

See also **Bayesian econometrics; bootstrap; cointegration; generalized method of moments estimation; heteroskedasticity and autocorrelation corrections; linear models; serial correlation and serial dependence; spatial econometrics; statistical inference; vector autoregressions.**

Bibliography

Baltagi, B. 1980. On seemingly unrelated regressions with error components. *Econometrica* 48, 1547–52.

Chang, Y. 2004. Bootstrap unit root tests in panels with cross-sectional dependency. *Journal of Econometrics* 120, 263–93.

Creel, M. and Farell, M. 1996. SUR estimation of multiple time-series models with heteroskedasticity and serial correlation of unknown form. *Economic Letters* 53, 239–45.

Davidson, R. and MacKinnon, J. 1993. *Estimation and Inference in Econometrics.* Oxford: Oxford University Press.

Fiebig, D.G. 2001. Seemingly unrelated regression. In *A Companion to Theoretical Econometrics*, ed. B. Baltagi. Oxford: Blackwell.

Greene, W. 2003. *Econometric Analysis*, 5th edn. Englewood Cliffs, NJ: Prentice-Hall.

Groen, J. and Kleibergen, F. 2003. Likelihood-based cointegration analysis in panels of vector error correction models. *Journal of Business and Economic Statistics* 21, 295–318.

Hodgson, D., Linton, O. and Vorkink, K. 2002. Testing the capital asset pricing model efficiently under elliptical symmetry: a semiparametric approach. *Journal of Applied Econometrics* 17, 619–39.

Larsson, R., Lyhagen, J. and Lothgren, M. 2001. Likelihood-based cointegration tests in heterogeneous panels. *Econometrics Journal* 4, 109–42.

Kakwani, N.C. 1967. The unbiasedness of Zellner's seemingly unrelated regression equations estimator. *Journal of the American Statistical Association* 62, 141–2.

Kmenta, J. and Gilbert, R.F. 1968. Small sample properties of alternative estimators for seemingly unrelated regressions. *Journal of the American Statistical Association* 63, 1180–200.

Kruskal, W. 1968. When are Gauss–Markov and least squares estimators identical? *Annals of Mathematical Statistics* 39, 70–5.

Mark, N., Ogaki, M. and Sul, D. 2005. Dynamic seemingly unrelated cointegrating regressions. *Review of Economic Studies* 72, 797–820.

Moon, H.R. 1999. A note on fully-modified estimation of seemingly unrelated regression models with integrated regressors. *Economics Letters* 65, 25–31.

Moon, H.R. and Perron, P. 2004. Efficient estimation of SUR cointegration regression model and testing for purchasing power parity. *Econometric Reviews* 23, 293–323.

Park, J. and Ogaki, M. 1991. Seemingly unrelated canonical cointegrating regressions. Working Paper No. 280, University of Rochester.

Phillips, P.C.B. 1977. Finite sample distribution of Zellner's SURE. *Journal of Econometrics* 6, 147–64.

Phillips, P.C.B. and Moon, H.R. 1999. Linear regression limit theory for nonstationary panel data. *Econometrica* 67, 1057–111.

Srivastava, V.K. and Giles, D.E.A. 1987. *Seemingly Unrelated Regression Equations Models.* New York: Dekker.

Srivastava, V.K. and Maekawa, K. 1995. Efficiency properties of feasible generalized least squares estimators in SURE models under non-normal disturbances. *Journal of Econometrics* 66, 99–121.

Zellner, A. 1962. An efficient method of estimating seemingly unrelated regression equations and tests of aggregation bias. *Journal of the American Statistical Association* 57, 500–9.

Zellner, A. 1963. Estimators for seemingly unrelated regression equations: some finite sample results. *Journal of the American Statistical Association* 58, 977–92.

Zellner, A. 1972. Corrigenda. *Journal of the American Statistical Association* 67, 255.

selection bias and self-selection

The problem of selection bias in economic and social statistics arises when a rule other than simple random sampling is used to sample the underlying population that is the object of interest. The distorted representation of a true population as a consequence of a sampling rule is the essence of the selection problem. Distorting selection rules may be the outcome of decisions of sample survey statisticians, self-selection decisions by the agents being studied, or both.

A random sample of a population produces a description of the population distribution of characteristics that has many desirable properties. One attractive feature of a random sample generated by the *known rule* that all individuals are equally likely to be sampled is that it produces a description of the population distribution of characteristics that becomes increasingly accurate as sample size expands.

A sample selected by any rule not equivalent to random sampling produces a description of the population distribution of characteristics that does not accurately describe the true population distribution of characteristics no matter how big the sample size. Unless the rule by which the sample is selected is known or can be recovered from the data, the selected sample cannot be used to produce an accurate description of the underlying population. For certain sampling rules, even knowledge of the rule generating the sample does not suffice to recover the population distribution from the sampled distribution.

This entry defines the problem of selection bias and presents conditions required to solve the problem.

Examples of various types of commonly encountered sampling frames are given and specific economic selection mechanisms are presented. Assumptions required to use selected samples to determine features of the population distribution are discussed.

The analytical framework developed to understand the inferential problems raised by selection bias is also fruitful in understanding the economics of self-selection. The prototypical choice theoretic model of self-selection is that of Roy (1951). In his model, agents choose among a variety of discrete 'occupational' opportunities. Agents can pursue only one 'occupation' at a time. While every person can, in principle, do the work in each 'occupation', at least at some level of competence, self-interest drives individuals to choose that 'occupation' which produces the highest income (utility) for them. As in the statistical selection bias problem, there is a latent population (of skills). Observed (utilized) skill distributions are the outcome of a selection rule by agents. The relationship between observed and latent skill distributions is of considerable interest and underlies recent work on worker hierarchies (see Willis and Rosen, 1979). The 'occupations' can be: (a) market work or non-market work (b) unemployed and searching or working at the offered wage (c) working in one province or working in another, or (d) any choice among a set of mutually exclusive opportunities.

Because the insights in the Roy model underlie much recent research, we present a brief exposition of it and demonstrate how it can be or has been fruitfully extended to a variety of settings. An important issue, closely linked to the problem of identifying population parameters from selected sample distributions, is the empirical content of economic models of self-selection and worker hierarchies. Are they artefacts of distributional assumptions for unobservable skills or are they genuine behavioural hypotheses?

1. A definition and some examples of selection bias

Any selection bias model can be described by the following set-up. Let $\mathbf{Y}$ be a vector of outcomes of interest and let $\mathbf{X}$ be a vector of 'control' or 'explanatory' variables. The population distribution of $(\mathbf{Y}, \mathbf{X})$ is $F(\mathbf{y}, \mathbf{x})$. To simplify the exposition we assume that the density is well defined and write it as $f(\mathbf{y}, \mathbf{x})$.

Any sampling rule can be interpreted as producing a non-negative weighting function $\omega(\mathbf{y}, \mathbf{x})$ that alters the population density. Let $(\mathbf{Y}^*, \mathbf{X}^*)$ denote the sampled random variables. The density of the sampled data $g(\mathbf{y}^*, \mathbf{x}^*)$ may be written as

$$g(\mathbf{y}^*, \mathbf{x}^*) = \omega(\mathbf{y}^*, \mathbf{x}^*) f(\mathbf{y}^*, \mathbf{x}^*) / \int \omega(\mathbf{y}^*, \mathbf{x}^*) f(\mathbf{y}^*, \mathbf{x}^*) d\mathbf{y}^* d\mathbf{x}^*$$

(1.1)

where the denominator of the expression is introduced to make the density $g(\mathbf{y}^*, \mathbf{x}^*)$ integrate to one as is required for proper densities.

Alternatively, the weight may be defined as

$$\omega^*(\mathbf{y}^*, \mathbf{x}^*) = \frac{\omega(\mathbf{y}^*, \mathbf{x}^*)}{\int \omega(\mathbf{y}^*, \mathbf{x}^*) f(\mathbf{y}^*, \mathbf{x}^*) d\mathbf{y}^* d\mathbf{x}^*}$$

so that

$$g(\mathbf{y}^*, \mathbf{x}^*) = \omega^*(\mathbf{y}^*, \mathbf{x}^*) f(\mathbf{y}^*, \mathbf{x}^*). \qquad (1.2)$$

Sampling schemes for which $\omega(\mathbf{y}, \mathbf{x}) = 0$ for some values of $(\mathbf{Y}, \mathbf{X})$ create special problems. For such schemes, not all values of $(\mathbf{Y}, \mathbf{X})$ are sampled. Let indicator variable $i(\mathbf{x}, \mathbf{y}) = 0$ if a potential observation at values $\mathbf{y}, \mathbf{x}$ cannot be sampled and let $i(\mathbf{y}, \mathbf{x}) = 1$ otherwise. Let $\Delta = 1$ record the occurrence of the event 'a potential observation is sampled, i.e. the value of $\mathbf{y}, \mathbf{x}$ is observed' and let $\Delta = 0$ if it is not. In the population, the proportion that is sampled is

$$\Pr(\Delta = 1) = \int i(\mathbf{y}, \mathbf{x}) f(\mathbf{y}, \mathbf{x}) d\mathbf{y} \, d\mathbf{x}. \quad (1.3)$$

while

$$\Pr(\Delta = 0) = 1 - \Pr(\Delta = 1).$$

For samples in which $\omega(\mathbf{y}, \mathbf{x}) = 0$ for a non-negligible proportion of the population $(\Pr(\Delta = 0) > 0)$ it is clarifying to consider two cases. A *truncated sample* is one for which $\Pr(\Delta = 1)$ is not known and cannot be consistently estimated. For such a sample, (1.1) is the density of all of the sampled $\mathbf{Y}$ and $\mathbf{X}$ values. A *censored sample* is one for which $\Pr(\Delta = 1)$ is known or can be consistently estimated. The sampling rule in this case is such that values of $\mathbf{y}, \mathbf{x}$ for which $\omega(\mathbf{y}, \mathbf{x}) = 0$ are not known but it is known whether or not $i(\mathbf{y}, \mathbf{x}) = 0$ for all values of $\mathbf{Y}, \mathbf{X}$. In this case it is notationally convenient to define $(\mathbf{Y}^*, \mathbf{X}^*) = (\mathbf{0}, \mathbf{0})$ for values of $\mathbf{y}, \mathbf{x}$ such that $\omega(\mathbf{y}, \mathbf{x}) = i(\mathbf{y}, \mathbf{x}) = 0$. Such a definition is innocuous provided that in the population there is no point mass (concentration of probability mass) at $(\mathbf{0}, \mathbf{0})$. (Any value other than $(\mathbf{0}, \mathbf{0})$ can be selected provided that there is no point mass at that value.) Given $\Delta = 0$ the distribution of $\mathbf{Y}^*, \mathbf{X}^*$ is

$$G(\mathbf{y}^*, \mathbf{x}^*) = 1 \quad \text{for} \quad \Delta = 0$$

at

$$\mathbf{Y}^* = \mathbf{0} \quad \text{and} \quad \mathbf{X}^* = \mathbf{0}.$$

The joint density of $\mathbf{Y}^*, \mathbf{X}^*, \Delta$ for the case of a censored sample is obtained by combining (1.1) and

(1.3). Thus

$$g(\mathbf{y}^*, \mathbf{x}^*, \delta) = \left[\frac{\omega(\mathbf{y}^*, \mathbf{x}^*)f(\mathbf{y}^*, \mathbf{x}^*)}{\int \omega(\mathbf{y}^*, \mathbf{x}^*)f(\mathbf{y}^*, \mathbf{x}^*)\mathrm{d}\mathbf{y}^* \, \mathrm{d}\mathbf{x}^*}\right]^\delta$$

$$\times \left[\int i(\mathbf{y}, \mathbf{x})f(\mathbf{y}, \mathbf{x})\mathrm{d}\mathbf{y} \, \mathrm{d}\mathbf{x}\right]^\delta$$

$$\times [1]^{1-\delta}\left[\int (1 - i(\mathbf{y}, \mathbf{x}))f(\mathbf{y}, \mathbf{x})\mathrm{d}\mathbf{y} \, \mathrm{d}\mathbf{x}\right]^{1-\delta}.$$

$$(1.4)$$

The first term on the right-hand side of (1.4) is the conditional density of $\mathbf{Y}^*, \mathbf{X}^*$ given $\Delta = 1$. The second term is the probability that $\Delta = 1$. The third term is the conditional density of $\mathbf{Y}^*, \mathbf{X}^*$ given $\Delta = 0$. This density assigns unit mass to $\mathbf{y}^* = 0$, $\mathbf{x}^* = \mathbf{0}$ when $\Delta = 0$. The fourth term is the probability that $\Delta = 0$. Notice that in the case in which $\omega(\mathbf{y}, \mathbf{x}) > 0$ for all $\mathbf{y}, \mathbf{x}$, $\Delta = 1$ and (1.4) is identical to (1.1).

In a random sample $\omega(\mathbf{y}^*, \mathbf{x}^*) = 1$ (and so $\omega^*(\mathbf{y}^*, \mathbf{x}^*) = 1$). In a selected sample, the sampling rule weights the data differently. Values of $(\mathbf{Y}, \mathbf{X})$ are over-sampled or under-sampled relative to their occurrence in the population. In the case of truncated samples, the weight is zero for certain values of the outcome.

In many problems in economics, attention focuses on $f(\mathbf{y}|\mathbf{x})$, the conditional density of $\mathbf{Y}$ given $\mathbf{X} = \mathbf{x}$. In such problems knowledge of the population distribution of $\mathbf{X}$ is of no direct interest. If samples are selected solely on the $\mathbf{x}$ variables ('selection on the exogenous variables'), $\omega(\mathbf{y}, \mathbf{x}) = \omega(\mathbf{x})$ and there is no problem about using selected samples to make valid inference about the population conditional density. This is so because in the case of selection on the exogenous variables

$$g(\mathbf{y}^*, \mathbf{x}^*) = f(\mathbf{y}^*|\mathbf{x}^*)\frac{\omega(\mathbf{x}^*)f(\mathbf{x}^*)}{\int \omega(\mathbf{x}^*)f(\mathbf{x}^*)\mathrm{d}\mathbf{x}}$$

and

$$g(\mathbf{x}^*)\frac{\omega(\mathbf{x}^*)f(\mathbf{x}^*)}{\int \omega(\mathbf{x}^*)f(\mathbf{x}^*)\mathrm{d}\mathbf{x}^*}.$$

Thus

$$g(\mathbf{y}^*|\mathbf{x}^*) = \frac{g(\mathbf{y}^*, \mathbf{x}^*)}{g(\mathbf{x}^*)} = f(\mathbf{y}^*|\mathbf{x}^*).$$

For such problems, sample selection distorts inference only if selection occurs on $\mathbf{y}$ (or $\mathbf{y}$ and $\mathbf{x}$). Sampling on both $\mathbf{y}$ and $\mathbf{x}$ is termed *general stratified sampling*.

From a sample of data, it is not possible to recover the true density $f(\mathbf{y}, \mathbf{x})$ without knowledge of the weighting rule. On the other hand, if the weighting rule is known $(\omega(\mathbf{y}^*, \mathbf{x}^*))$ the density of the sampled data is known $(g(\mathbf{y}^*, \mathbf{x}^*))$, the support of $(\mathbf{y}, \mathbf{x})$ is known and $\omega(\mathbf{y}, \mathbf{x})$

is non-zero, then $f(\mathbf{y}, \mathbf{x})$ can always be recovered because

$$\frac{g(\mathbf{y}^*, \mathbf{x}^*)}{\omega(\mathbf{y}^*, \mathbf{x}^*)} = \frac{f(\mathbf{y}^*, \mathbf{x}^*)}{\int \omega(\mathbf{y}^*, \mathbf{x}^*)f(\mathbf{y}^*, \mathbf{x}^*)\mathrm{d}\mathbf{y}^* \, \mathrm{d}\mathbf{x}^*}$$

$$(1.5)$$

and by hypothesis both the numerator and denominator of the left-hand side are known. From the requirement that $(\mathbf{y}^*, \mathbf{x}^*)$ has a well defined density

$$\int f(\mathbf{y}^*, \mathbf{x}^*)\mathrm{d}\mathbf{y}^* \, \mathrm{d}\mathbf{x}^* = 1.$$

Integrating the left-hand side of (1.5) it is possible to determine $\int \omega(\mathbf{y}^*, \mathbf{x}^*)f(\mathbf{y}^*, \mathbf{x}^*)\mathrm{d}\mathbf{y}^* \, \mathrm{d}\mathbf{x}^*$ and hence to use (1.5) to recover the population density of the data.

The requirements that (a) the support of $(\mathbf{y}, \mathbf{x})$ is known and (b) $\omega(\mathbf{y}, \mathbf{x})$ is nonzero are not innocuous. In many important problems in economics requirement (b) is not satisfied: the sampling rule excludes observations for certain values of $\mathbf{y}, \mathbf{x}$ and hence it is impossible without invoking further assumptions to determine the population distribution of $(\mathbf{Y}, \mathbf{X})$ at those values. If neither the support nor the weight is known, it is impossible, without invoking strong assumptions, to determine whether the fact that data are missing at certain $\mathbf{y}, \mathbf{x}$ values is due to the sampling plan or that the population density has no support at those values. We now turn to some specific sampling plans of interest in economics.

Example 1

Data are collected on incomes of individuals whose income Y exceeds a certain value c (for cut-off value). The rule is to observe Y if $Y > c$. Thus $\omega(y) = 1$ if $y > c$ and $\omega(y) = 0$ if $y \leq c$. Because the weight is zero for some values of y, we know that knowledge of the sampling rule does *not* suffice to recover the population distribution. From a random sample of the entire population, the social scientist knows or can consistently estimate (a) the sample distribution of Y above c and (b) the proportion of the original random sample with income below c ($F(c)$ where F is the distribution function of Y). The social scientist does not observe values of Y below c.

In this example, observed income is a *truncated random variable*. The point of truncation is c. The *sample* of observed income is said to be *censored*. If the proportion of the original random sample with income below c is not known and cannot be consistently estimated, the *sample* is truncated. In a truncated sample, nothing is known about the proportion of the underlying population that can appear in the sample. A sample is truncated only if $\omega(\mathbf{y}) = 0$ for some intervals of $\mathbf{y}$ (for $\mathbf{y}$ continuous) or if $\omega(\mathbf{y}) = 0$ at values of $\mathbf{y}$ at which there is finite probability mass. In a censored sample, the proportion of the underlying population that can appear in the sample is known, at least to an arbitrarily high degree of approximation, as sample size increases.

Let $Y^* = Y$ if $Y > c$. Define $Y^* = 0$ otherwise (the choice of the value for Y^* when Y is not observed is inessential and any value can be used in place of 0 provided that the true distribution places no mass at the selected value). Define an indicator variable $\Delta = 1$ if $Y > c$. $\Delta = 0$ otherwise. Then the distribution of Y^* is

$$G(y^*|Y > 0) = F(y^*|Y > c) = F(y^*|\delta = 1)$$
$$= \frac{F(y^*)}{1 - F(c)}, y^* > c. \qquad (1.6a)$$

$$G(y^*|Y^* > 0) = 1 \quad \text{for} \quad Y^* = 0(\Delta = 0). \qquad (1.6b)$$

Observe that (1.6a) is obtained from (1.1) by setting $\omega(y^*) = 1$ if $y > c$, and $\omega(y^*) = 0$ otherwise, and integrating up with respect to y^*. The distribution of Δ is

$$\text{pr}(\Delta) = [1 - F(c)]^\delta [F(c)]^{1-\delta}.$$

The joint distribution of (Y^*, Δ) is

$$F(y^*, \delta) = F(y^*|\delta)\text{Pr}(\delta)$$
$$= \left\{\frac{F(y^*)}{(1 - F(c))}\right\}^\delta$$
$$\times [1 - F(c)]^\delta (1)^{1-\delta} [F(c)]^{1-\delta}$$
$$= [F(y^*)]^\delta [F(c)]^{1-\delta}. \qquad (1.7)$$

Note that (1.7) is obtained from (1.4) by setting $\omega(y) = 0, y < c, \omega(y) = 1$ otherwise, by setting $\iota(y) = \omega(y)$ and by integrating up with respect to y^*. For normally distributed Y, (1.7) is the 'Tobit' distribution.

The difference between the information in a truncated sample and the information in a censored sample is encapsulated in the contrast between (1.6a) and (1.7). Clearly there is more information in a censored sample than in a truncated sample because one can obtain (1.6a) from (1.7) (by conditioning on $\Delta = 1$) but not vice versa.

Inferences about the population distribution based on assuming that $F(y^*|Y > c)$ closely approximates $F(y)$ are potentially very misleading. A description of population income inequality based on a subsample of high income people may convey no information about the true population distribution.

Without further information about F and its support, it is not possible to recover F from $G(y^*)$ from either a censored or a truncated sample. Access to a censored sample enables the analyst to recover $F(y)$ for $y > c$ but obviously does not provide any information on the shape of the true distribution for values of $y \leq c$.

This problem is routinely 'solved' by assuming that F is of a known functional form. This solution strategy does not always work. If F is normal, then it can be recovered from a censored or truncated sample (Pearson, 1901). If F is Pareto, F cannot be recovered from either a truncated or a censored sample (see Flinn and Heckman, 1982). If F is real analytic (i.e. possesses derivatives of all order) and the support of Y is known, then F can be recovered (Heckman and Singer, 1985).

Example 2

Expand the discussion in the previous example to a linear regression setting. Let

$$Y = X\beta + U \qquad (1.8)$$

be the population earnings function where Y is earnings, X is a regressor vector assumed to be distributed independently of mean zero disturbance U. 'β' is a suitably dimensioned parameter vector. Conventional assumptions are invoked to ensure that ordinary least squares applied to a random sample of earnings data consistently estimates β.

Data are collected on incomes of persons for whom Y exceeds c. Again the weight depends solely on y, i.e. $\omega(y, x) = 0, y \leq c, \omega(y, x) = 1, y > c$. The social scientist knows or can consistently estimate (a) the sample distribution of Y above c (b) the sample distribution of the X for Y above c and (c) the proportion of the original random sample with income below c. The social scientist does not observe values of Y below c.

As before, let $Y^* = Y$ if $Y > c$. Define $Y^* = 0$ otherwise. $\Delta = 1$ if $Y > c$, $\Delta = 0$ otherwise. The probability of the event $\Delta - 1$ given $X - x$ is

$$\text{Pr}(\Delta = 1|X = x) = \text{Pr}(Y > c|X = x)$$
$$= \text{Pr}(Y > c - x\beta|X = x).$$

Invoking independence between U and X and letting F_u denote the distribution of U,

$$\text{Pr}(\Delta = 1|X = x) = 1 - F_u(c - x\beta) \qquad (1.9a)$$

and

$$\text{Pr}(\Delta = 0|X = x) = F_u(c - x\beta). \qquad (1.9b)$$

The distribution of Y^* conditional on X is

$$G(y^*|Y > 0, X = x) = F(y^*|X = x, Y > c)$$
$$= F(y^*|X = x, \Delta = 1)$$
$$= \frac{F_u(y^* - x\beta)}{1 - F_u(c - x\beta)}, \quad y^* > c. \qquad (1.10a)$$

$$G(y^*|Y \leq 0) = 1 \quad \text{for} \quad Y^* = 0(\Delta = 0). \qquad (1.10b)$$

The joint distribution of (Y^*, Δ) given $\mathbf{X} = \mathbf{x}$ is

$$F(y^*, \delta | \mathbf{X} = \mathbf{x}) = F(y^* | \delta, \mathbf{x}) \Pr(\delta | \mathbf{x}).$$
$$= \{F_u(y^* - \mathbf{x}\boldsymbol{\beta})\}^\delta \{F_u(c - \mathbf{x}\boldsymbol{\beta})\}^{1-\delta}.$$
$$(1.11)$$

In particular,

$$E(Y^* | \mathbf{X} = \mathbf{x}, \Delta = 1) = \mathbf{x}\boldsymbol{\beta} + E(U | \mathbf{X} = \mathbf{x}, \delta = 1)$$
$$= \mathbf{x}\boldsymbol{\beta} + \int_{c - \mathbf{x}\boldsymbol{\beta}}^{\infty} \frac{z \, dF_u(z)}{(1 - F_u(c - \mathbf{x}\boldsymbol{\beta}))}$$
$$(1.12)$$

where z is a dummy variable of integration. In contrast, the population mean regression function is

$$E(Y | \mathbf{X} = \mathbf{x}) = \mathbf{x}\boldsymbol{\beta}. \qquad (1.13)$$

The contrast between (1.12) and (1.13) is illuminating. Many behavioural theories in social science produce empirical counterparts of (1.8) with population conditional expectations like (1.13). Such theories sometimes restrict the signs, permissible values and other relationships among the coefficients in $\boldsymbol{\beta}$. When the theoretical model is estimated on a selected sample ($\Delta = 1$) the true conditional expectation is (1.12) not (1.13). The conditional mean of U depends on $\mathbf{x}$. In terms of conventional omitted variable analysis, $E(U | \mathbf{X} = \mathbf{x}, \Delta = 1)$ is omitted from the regression. Since this term is a function of $\mathbf{x}$ it is likely to be correlated with $\mathbf{x}$. Least squares estimates of $\boldsymbol{\beta}$ obtained on selected samples which do not account for selection are biased and inconsistent.

To illustrate the nature of the bias, it is useful to draw on the work of Cain and Watts (1973). Suppose that X is a scalar random variable (e.g. education) and that its associated coefficient is positive ($\beta > 0$). Under conventional assumptions about U (e.g. mean zero, independently and identically distributed and distributed

independently of X), the population regression of Y on X is a straight line. The scatter about the regression line and the regression line are given in Figure 1. When $Y > c$ is imposed as a sample inclusion requirement, lower population values of U are excluded from the sample in a way that systematically depends on x ($Y > c$ or $U > c - x\beta$). As x increases, the conditional mean of $U[E(U | X = x, \Delta = 1)]$ decreases. Regression estimates of β that do not correct for sample selection (that is, include $E(U | X = x, \Delta = 1)$ as a regressor) are downward biased because of the negative correlation between x and $E(U | X = x, \Delta = 1)$. See the flattened regression line for the selected sample in Figure 1.

In models with more than one regressor, no sharp result on the sign of the bias in the regression estimate that results from ignoring the selected nature of the sample is available except when the $\mathbf{X}$ variables are from certain distributions (e.g. normal, see Goldberger, 1983). None the less, the key result – that conventional least squares estimates of $\boldsymbol{\beta}$ obtained from selected samples are biased and inconsistent – remains true.

As in example 1, it is fruitful to distinguish between the case of a truncated sample and the case of a censored sample. In the truncated sample case, no information is available about the fraction of the population that would be allocated to the truncated sample [$\Pr(\Delta = 1)$]. In the censored sample case, this fraction is known or can be consistently estimated. In the censored sample case it is fruitful to distinguish two further cases: (a) the case in which $\mathbf{X}$ is not observed when $\Delta = 0$ and (b) the case in which it is. Case (b) is the one most fully developed in the literature (Heckman and MaCurdy, 1981).

Note that the conditional mean $E(U | \mathbf{X} = \mathbf{x}, \Delta = 1)$ is a function of $c - \mathbf{x}\boldsymbol{\beta}$ solely through $\Pr(\Delta = 1 | \mathbf{x})$. Since $\Pr(\Delta = 1 | \mathbf{x})$ is monotonic in $c - \mathbf{x}\boldsymbol{\beta}$ the conditional mean depends solely on $\Pr(\Delta = 1 | \mathbf{x})$ and the parameters F_u i.e. since

$$F_u^{-1}(1 - \Pr(\Delta = 1 | \mathbf{x})) = c - \mathbf{x}\boldsymbol{\beta},$$
$$E(U | \mathbf{X} = \mathbf{x}, \Delta = 1)$$
$$= \int_{F_u^{-1}[1 - \Pr(\Delta = 1 | \mathbf{x})]}^{\infty} \frac{z \, dF_u(z)}{\Pr(\Delta = 1 | \mathbf{x})}.$$

This relationship demonstrates that the conditional mean is a function of the probability of selection. As the probability of selection goes to 1, the conditional mean goes to zero. For samples chosen so that the values of $\mathbf{x}$ are such that the observations are certain to be included in the sample, there is no problem in using ordinary least squares on selected samples to estimate $\boldsymbol{\beta}$. Thus in Figure 1, ordinary least squares regressions fit on samples selected to have large $\mathbf{x}$ values closely approximate the true regression function and become arbitrarily close as $\mathbf{x}$ becomes large. The condition mean in (1.12) is a surrogate for $\Pr(\Delta = 1 | \mathbf{x})$. As this probability goes to one,

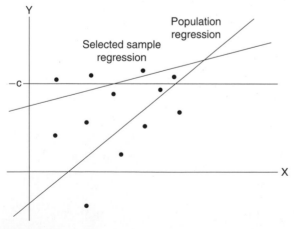

Figure 1

the problem of sample selection in regression analysis becomes negligibly small.

Heckman (1976) demonstrates that $\boldsymbol{\beta}$ and F_u are identified if U is normally distributed and standard conditions invoked in regression analysis are satisfied. Gallant and Nychka (1984) and Cosslett (1984) establish conditions for identification for non-normal U. In their analyses, F_u is consistently non-parametrically estimated.

Example 3

The next example considers *censored random variables*. This concept extends the notion of a truncated random variable by letting a more general rule than truncation on the outcome of interest generate the selected sample. Because the sample generating rule may be different from a simple truncation of the outcome being studied, the concept of a censored random variable in general requires at least two distinct random variables.

Let Y_1 be the outcome of interest. Let Y_2 be another random variable. Denote observed Y_1 by Y_1^*. If $Y_2 < c$, Y_1 is observed. Otherwise Y_1 is not observed and we can set $Y_1^* = 0$ or any other convenient value (assuming that Y_1 has no point mass at $Y_1 = 0$ or at the alternative convenient value). In terms of the weighting function ω, $\omega(y_1, y_2) = 0$ if $y_2 > c$, $\omega(y_1, y_2) = 1$ if $y_2 \le c$.

Selection rule $Y_2 < c$ does not necessarily restrict the range of Y_1. Thus Y_1^* is not in general a truncated random variable. Define $\Delta = 1$ if $Y_2 < c$; $\Delta = 0$ otherwise. If $F(y_1, y_2)$ is the population distribution of (Y_1, Y_2), the distribution of Δ is

$$\Pr(\Delta = \delta) = [1 - F_2(c)]^{1-\delta}[F_2(c)]^{\delta}, \quad \delta = 0, 1,$$

where F_2 is the marginal distribution of Y_2. The distribution of Y_1^* is

$$G(y_1^*) = F(y_1^* | \delta = 1) = \frac{F(y_1^*, c)}{F_2(c)}, \quad \Delta = 1,$$

$$(1.14a)$$

$$G(y_1^* = 0) = 1, \quad \Delta = 0. \qquad (1.14b)$$

Note that (1.14a) is the distribution function corresponding to the density in (1.1) when $\omega(y_1, y_2) = 1$ if $y_2 \le c$ and $\omega(y_1, y_2) = 0$ otherwise.

The joint distribution of (Y_1^*, Δ) is

$$G(y_1^*, \delta) = [F(y_1^*, c)]^{\delta}[1 - F_2(c)]^{1-\delta}.$$

$$(1.15)$$

This is the distribution function corresponding to density (1.4) for the special weighting rule of this example. In a censored sample, under general conditions it is possible to consistently estimate $\Pr(\Delta = \delta)$ and $G(y_1^*)$. In a truncated sample, only conditional distribution (1.14a) can be estimated. A degenerate version of this model has $Y_1 \equiv Y_2$. In

that case, censored random variable Y_1 is also a truncated random variable. Note that a censored random variable may be defined for a truncated or censored sample.

Example 3 and variants of it have wide applicability in economics. Let Y_1 be the wage of a woman. Wages of women are observed only if women work. Let Y_2 be an index of a woman's propensity to work. In Gronau (1974) and Heckman (1974), Y_2 is postulated as the difference between reservation wages (the value of time at home determined from household preference functions) and potential market wages Y_1. Then if $Y_2 < 0$, the woman works. Otherwise, she does not. $Y_1^* = Y_1$ if $Y_2 < 0$ is the observed wage.

If Y_1 is the offered wage of an unemployed worker, and Y_2 is the difference between reservation wages (the return to searching) and offered market wages, $Y_1^* = Y_1$ if $Y_2 < 0$ is the accepted wage for an unemployed worker (see Flinn and Heckman, 1982). If Y_1 is the potential output of a firm and Y_2 is its profitability, $Y_1^* = Y_1$ if $Y_2 > 0$. If Y_1 is the potential income in occupation one and Y_2 is the potential income in occupation two, $Y_1^* = Y_1$ if $Y_1 - Y_2 < 0$ while $Y_2^* = Y_2$ if $Y_1 - Y_2 \ge 0$. We develop this example at length in section 2 where we consider explicit economic models of self-selection. There we discuss the identifiability of this model.

Example 4

This example builds on example 3 by introducing regressors. This produces the *censored regression model* (Heckman, 1976; 1979). In example 3 set

$$Y_1 = \mathbf{X}_1 \boldsymbol{\beta}_1 + U_1 \qquad (1.16a)$$

$$Y_2 = \mathbf{X}_2 \boldsymbol{\beta}_2 + U_2 \qquad (1.16b)$$

where $(\mathbf{X}_1, \mathbf{X}_2)$ are distributed independently of (U_1, U_2), a mean zero, finite variance random vector. Conventional assumptions are invoked to ensure that if Y_1 and Y_2 can be observed, least squares applied to a random sample of data on $(Y_1, Y_2, \mathbf{X}_1, \mathbf{X}_2)$ would consistently estimate $\boldsymbol{\beta}_1$ and $\boldsymbol{\beta}_2$. $Y_1^* = Y_1$ if $Y_2 < 0$. If $Y_2 < 0$, $\Delta = 1$. Then the regression function for the selected sample is

$$E(Y_1^* | \mathbf{X}_1 = \mathbf{x}_1, Y_2 < 0)$$
$$= E(Y_1^* | \mathbf{X}_1 = \mathbf{x}_1, \Delta = 1)$$
$$= \mathbf{X}_1 \boldsymbol{\beta}_1 + E(U_1 | \mathbf{X}_1 = \mathbf{x}_1, \Delta = 1)$$

$$(1.17)$$

and the regression function for the population is

$$E(Y_1 | \mathbf{X}_1 = \mathbf{x}_1) = \mathbf{X}_1 \boldsymbol{\beta}_1. \qquad (1.18)$$

As in the regression analysis of truncated random variables, there is an illuminating contrast between the conditional expectation for the selected sample (1.17) and the population regression function (1.18). The two functions differ by the conditional mean of

$U_1[E(U_1|\mathbf{X}_1 = \mathbf{x}_1, \Delta = 1)]$. In the regression analysis of truncated random variables, ordinary least squares estimates of $\boldsymbol{\beta}$ (in equation (1.14)) are biased and inconsistent because the conditional mean is improperly omitted from the selected sample regression. The same analysis applies to the regression analysis of censored random variables. The conditional mean is a surrogate for the probability of selection $[\Pr(\Delta = 1|\mathbf{x}_2)]$. As $\Pr(\Delta = 1|\mathbf{x}_2)$ goes to one, the problem of sample selection bias becomes negligible. However, in the censored regression case, a new phenomenon appears. If there are variables in $\mathbf{X}_2$ not in $\mathbf{X}_1$, such variables may appear to be statistically important determinants of Y_1 when ordinary least squares is applied to data generated from censored samples.

As an example, suppose that survey statisticians use some extraneous (to $\mathbf{X}_1$) variables to determine sample enrolment. Such variables may appear to be important determinants of Y_1 when in fact they are not. They are important determinants of Y_1^*. In an analysis of self-selection, let Y_1 be the wage that a potential worker could earn were he to accept a market offer. Let Y_2 be the difference between the best non-market opportunity available to the potential worker and Y_1. If $Y_2 < 0$, the agent works. The conditional expectation of observed wages ($Y_1^* = Y$ if $Y_2 < 0$) given $\mathbf{x}_1$ and $\mathbf{x}_2$ will be a non-trivial function of $\mathbf{x}_2$. Thus variables determining non-market opportunities will determine Y_1^*, even though they do not determine Y_1. For example, the number of children less than six may appear to be significant determinants of Y_1 when inadequate account is taken of sample selection, even though the market does not place any value or penalty on small children in generating wage offers for potential workers.

Heckman (1976) develops the analysis of this model when (U_1, U_2) is normally distributed. Gallant and Nychka (1984) and Cosslett (1984) demonstrate that under mild restrictions on $F(u_1, u_2)$, if there is one continuous valued variable in $\mathbf{X}_2$ not in $\mathbf{X}_1$ (so that there is no exact linear dependence between $\mathbf{X}_2$ and $\mathbf{X}_1$), $\boldsymbol{\beta}_1, \boldsymbol{\beta}_2$ and $F(u_1, u_2)$ can be consistently non-parametrically estimated. Heckman and MaCurdy (1986) develop this class of models at length.

Example 5

This example demonstrates how self-selection bias affects the interpretation placed on estimated consumer demand functions when there is self-selection. We postulate a population of consumers with a quasi-concave utility function $U(\mathbf{Z}, E)$ which depends on the consumption of goods and preference shock E which represents heterogeneity in preferences among consumers. The support of E is $\mathbf{E}$. For price vector $\mathbf{P}$ and endowment income M, the consumer's problem is to

$$\text{Max } U(\mathbf{Z}, E) \quad \text{subject to} \quad \mathbf{P}'\mathbf{Z} \le M.$$

In the population $\mathbf{P}$ and M are distributed independently of E. First-order conditions for this problem are

$$\frac{\partial U(\mathbf{Z}, E)}{\partial \mathbf{Z}} \le \lambda \mathbf{P}, \tag{1.19}$$

where λ is the Lagrange multiplier associated with the budget constraint. Focusing on the demand for the first good, Z_1, none of it is purchased if at zero consumption of Z_1

$$\left. \frac{\partial U(\mathbf{Z}, E)}{\partial Z_1} \right|_{Z_1=0} \le \lambda P_1. \tag{1.20}$$

that is, marginal valuation is less than marginal cost in utility terms. Conventional interior solution demand functions for Z_1 are defined for a given $\mathbf{P}$, M only for values of E such that

$$\left. \frac{\partial U(\mathbf{Z}, E)}{\partial Z_1} \right|_{Z_1=0} \ge \lambda P_1. \tag{1.21}$$

Let the set of E for which conventional interior solution consumer demand functions for Z_1 are defined be denoted by $\underline{\underline{E}}$. Then

$$\underline{\underline{E}} = \left\{ E \text{ such that } \left. \frac{\partial U(\mathbf{Z}, E)}{\partial Z_1} \right|_{Z_1=0} \right.$$
$$\left. \ge \lambda P_1 \text{ for given } \mathbf{P}, M \right\}.$$

Let $\Delta_1 = 0$ if the consumer does not purchase Z_1. Let $\Delta_1 = 1$ otherwise. If $F(\varepsilon)$ is the population distribution of E, the proportion purchasing none of good Z_1 given $\mathbf{P}, M$ is

$$\Pr(\Delta_1 = 0|\mathbf{P}, M) = 1 - \int_{\underline{\underline{E}}} dF(\varepsilon).$$

Provided inequality (1.21) is satisfied, $\Delta_1 = 1$ and interior solution demand function

$$Z_1 = Z_1(\mathbf{P}, M, E) \tag{1.22}$$

is well defined and $Z_1 = Z_1^*$. When $\Delta_1 = 0$, observed $Z_1 = Z_1^* = 0$.

Equation (1.22) is the conventional object of interest in consumer theory. Partial derivatives of that function *holding E and the other arguments constant* have well defined economic interpretations. Suppose that some non-negligible proportion of the population buys none of good Z_1. Regression estimates of the parameters of (1.22) using Z_1^* approximate the conditional expectation

$$E(Z_1|\Delta_1 = 1, \mathbf{P}, M) = \int_{\underline{\underline{E}}} Z_1(\mathbf{P}, M, \varepsilon) dF(\varepsilon). \tag{1.23}$$

The derivatives of (1.23) are different from the derivatives of (1.22). In order to define these derivatives, it is helpful to define $I_{\underline{\underline{E}}}(E)$ as an indicator function for set $\underline{\underline{E}}$ which equals one if $E \in \underline{\underline{E}}$ and equals zero otherwise. When prices or income change, the set of values of E that satisfy inequality (1.21) changes. Let $\underline{\underline{E}} + \Delta \underline{\underline{E}}_{\mathbf{P}}$ be the set of E values that satisfy (1.21) when there is a finite price change $\Delta \mathbf{P}$. $I_{\underline{\underline{E}} + \Delta \underline{\underline{E}}_{\mathbf{P}}}(E)$ is an indicator function which equals one when $E \in \underline{\underline{E}} + \Delta \underline{\underline{E}}_{\mathbf{P}}$. Then the derivatives of (1.23) are, for the jth price

$$
\frac{\partial E(Z_1 | \Delta = 1, \mathbf{P}, M)}{\partial P_j} = \int_{\underline{\underline{E}}} \frac{\partial Z_1(\mathbf{P}, M, \varepsilon)}{\partial P_j} dF(\varepsilon)
$$

$$
+ \lim_{\Delta P_j \to 0} \int_{\underline{\underline{E}}} \frac{\left[\left(I_{\underline{\underline{E}} + \Delta \underline{\underline{E}}_{P_j}}(\varepsilon) - I_{\underline{\underline{E}}}(\varepsilon) \right) Z(\mathbf{P}, M, \varepsilon) \right]}{\Delta P_j} dF(\varepsilon).
$$

$$(1.24)$$

When the limit in the second term does not exist, the derivative does not exist. We assume for expositional convenience that the limit is well defined.

The first expression on the right-hand side of (1.24) is the average effect of price change on commodity demand. The second term on the right-hand side of (1.24) arises from the change in sample composition of E as the proportion of non-purchasers changes in response to price change. This term generates the selection bias.

Neither term is the same as the price derivative of (1.22) for an arbitrary value of $E = \varepsilon$ although the first term on the right-hand side of (1.24) approximates the price derivative of (1.22) for some value of $E = \varepsilon$.

A similar decomposition of the derivatives of the conditional demand function can be performed if it is defined solely for a sample of non-zero purchasers (see Heckman and MaCurdy, 1981; 1986).

Just as in the statistical sample selection bias problem, there is a population of interest. In this case, the population parameters of interest are the distribution of E and the parameters of $U(\mathbf{Z}, E)$. Those who buy Z_1 are a self-selected sample of the population. Estimates of population parameters estimated on self-selected samples are biased and inconsistent. There is a population distribution of $Z_1(\mathbf{P}, M, E)$ generated by the distribution of E. Observations of Z_1 are obtained only if $E \in \underline{\underline{E}}(\omega(E) = 1$ if $E \in \underline{\underline{E}}, \omega(E) = 0$ otherwise). Alternatively one can express the inclusion criteria in terms of the latent population distribution of Z_1 induced by E (given $\mathbf{P}$ and M) and write $\omega(z_1) = 1$ if $z_1 > 0$, $\omega(z_1) = 0$ if $z_1 \leq 0$.

Heckman (1974) and Heckman and MaCurdy (1981) provide further discussion of this type of model which is widely used in applied economics and consider issues of identifiability for such models.

Example 6. Length biased sampling

Let T be the duration of an event such as a completed unemployment spell or a completed duration of a job with an employer. The population distribution of T is $F(t)$ with density $f(t)$. The sampling rule is such that *individuals* are sampled at random. Data are recorded on a completed spell *provided that at the time of the interview the individual is experiencing the event*. Such sampling rules are in wide use in many national surveys of employment and unemployment.

In order to have a sampled completed spell, a person must be in the state at the time of the interview. Let '0' be the date of the survey. Decompose any completed spell T into a component that occurs before the survey T_b and a component that occurs after the survey T_a. Then $T = T_a + T_b$. For a person to be sampled, $T_b > 0$. The density of T given $T_b = t_b$ is

$$
f(t|t_b) = \frac{f(t)}{1 - F(t_b)}, \quad t \geq t_b. \tag{1.25}
$$

Suppose that the environment is stationary. The population entry rate into the state at each instant of time is k. From each vintage of entrants into the state distinguished by their distance from the survey date t_b, only $1 - F(t_b) = Pr(T > t_b)$ survive. Aggregating over all cohorts of entrants, the population proportion in the state at the date of the interview is P where

$$
P = \int_0^\infty k(1 - F(t_b)) dt_b \tag{1.26}
$$

which is assumed to exist. The density of T_b^*, sampled pre-survey duration, is

$$
g(t_b^* | t_b^* > 0) = \frac{k(1 - F(t_b^*))}{P}. \tag{1.27}
$$

The density of sampled completed durations is thus

$$
\begin{aligned}
g(t^*) &= \int_0^{t^*} f(t^* | t_b^*) g(t_b^* | t_b^* > 0) dt_b^* \\
&= k \frac{f(t^*)}{1 - F(t_b^*)} \frac{1 - F(t_b^*)}{P} \int_0^{t^*} dt_b^* \\
&= k \frac{t^* f(t^*)}{P}.
\end{aligned}
$$

Observe from (1.26) that by a standard integration by parts argument

$$
P = k \int_0^\infty (1 - F(z)) dz = k \int_0^\infty z \, dF(z) = kE(T).
$$

Note that

$$
g(t^*) = \frac{t^* f(t^*)}{E(T)}. \tag{1.28}
$$

In this form (1.28) is equivalent to (1.1) with $\omega(t) = t$. Hence the term 'length biased sampling'. Intuitively, longer spells are oversampled when the requirement is imposed that a spell be in progress at the time the survey is conducted ($T_b > 0$). Suppose, instead, that individuals are randomly sampled and data are recorded on the *next* spell of the event (after the survey date). As long as successive spells are independent, such a sampling frame does not distort the sampled distribution because no requirement is imposed that the sampled spell be in progress at the date of the interview. It is important to notice that the source of the bias is the requirement that $T_b > 0$, not that only a fraction of the population experiences the event ($P < 1$).

The simple length weight ($\omega(t) = t$) that produces (1.28) is an artefact of the stationarity assumption. Heckman and Singer (1985) consider the consequences of non-stationarity and un-observables when there is selection on the event that a person be in the state at the time of the interview. They also demonstrate the bias that results from estimating parametric models on samples generated by length biased sampling rules when inadequate account is taken of the sampling plan. Vardi (1983, 1985) and Gill and Wellner (1985) consider nonparametric identification and estimation of models with densities of the form (1.28).

It is unfortunate that the lessons of length biased sampling are not adequately appreciated in economics. Two widely cited studies by Clark and Summers (1979) and Hall (1982) use length biased data to prove, respectively, that unemployment and employment spells are 'surprisingly long'. Whether their findings are artefacts of sampling plans remains to be determined.

Example 7. Choice based sampling

Let D be a discrete valued random variable which assumes a finite number of values I. $D = i$, $i = 1, \ldots, I$ corresponds to the occurrence of state i. States are mutually exclusive. In the literature the states may be modes of transportation choice for commuters (Domencich and McFadden, 1975), occupations, migration destinations, financial solvency status of firms, schooling choices of students, etc. Interest centres on estimating a population choice model

$$\Pr(D = i | \mathbf{X} = \mathbf{x}), \quad i = 1, \ldots, I. \quad (1.29)$$

The population density of $(D, \mathbf{X})$ is

$$f(d, \mathbf{x}) = \Pr(D = d | \mathbf{X} = \mathbf{x}) h(\mathbf{x}) \quad (1.30)$$

where $h(\mathbf{x})$ is the density of the data.

In many problems, plentiful data are available on certain outcomes while data are scarce for other outcomes. For example, interviews about transportation preferences conducted at train stations tend to over-sample train riders and under-sample bus riders. Interviews about

occupational choice preferences conducted at leading universities over-sample those who select professional occupations.

In choice based sampling, selection occurs solely on the D coordinate of $(D, \mathbf{X})$. In terms of (1.1) (extended to allow for discrete random variables), $\omega(d, \mathbf{X}) = \omega(d)$. Then sampled $(D^*, \mathbf{X}^*)$ has density

$$g(d^*, \mathbf{x}^*) = \frac{\omega(d^*) f(d^*, \mathbf{x}^*)}{\sum_{i=1}^{I} \int \omega(i) f(i, x^*) dx^*} \quad (1.31)$$

Notice that the denominator can be simplified to

$$\sum_{i=1}^{I} \omega(i) f(i)$$

where $f(d^*)$ is the marginal distribution of D^* so that

$$g(d^*, \mathbf{x}^*) = \frac{\omega(d^*) f(d^*, \mathbf{x}^*)}{\sum_{i=1}^{I} \omega(i) f(i)} \quad (1.32)$$

Also, integrating (1.31) with respect to $\mathbf{x}$ using (1.32) we obtain

$$g(d^*) = \frac{\omega(d^*) f(d^*)}{\sum_{i=1}^{I} \omega(i) f(i)} \quad (1.33)$$

which makes transparent how the sampling rule causes the sampled proportions to deviate from the population proportions. Note further that as a consequence of sampling only on D, the population conditional density

$$h(\mathbf{x}^* | d^*) = \frac{f(d^*, \mathbf{x}^*)}{f(d^*)} \quad (1.34)$$

can be recovered from the choice-based sample. The density of $\mathbf{x}$ in the sample is thus

$$g(\mathbf{x}^*) = \sum_{i=1}^{I} h(\mathbf{x}^* | i) g(i). \quad (1.35)$$

Then using (1.32)–(1.35) we reach

$$g(d^* | \mathbf{x}^*) = f(d^* | \mathbf{x}^*)$$
$$\times \left\{ \left[\frac{\omega(d^*)}{\sum_{i=1}^{I} \omega(i) f(i)} \right] \left[\frac{1}{\sum_{i=1}^{I} f(i | \mathbf{x}^*) \frac{g(i)}{f(i)}} \right] \right\}. \quad (1.36)$$

The bias that results from using choice based samples to make inference about $f(d^* | \mathbf{x}^*)$ is a consequence of neglecting the terms in braces on the right-hand side of (1.36). Notice that if the data are generated by a random sampling rule, $\omega(d^*) = 1$, $g(d^*) = f(d^*)$ and the term in braces is one.

Manski and Lerman (1977), Manski and McFadden (1981) and Cosslett (1981) provide illuminating discussions of choice based sampling.

Example 8. Size biased sampling

Let N be the number of children in a family. $f(N)$ is the density of discrete random variable N. Suppose that family size is recorded only when at least one child is interviewed. Suppose further that each child has an independent and identical chance β of being interviewed. The probability of sampled family size of $N^* = n^*$ is

$$g(n^*) = \frac{\omega(n^*)f(n^*)}{E[\omega(N^*)]} \qquad (1.37)$$

where $\omega(n^*) = 1 - (1 - \beta)^{n^*}$ (the probability that at least one child from a family of size n^* will be sampled) and

$$E[\omega(N^*)] = \sum_{n^*}\left(1 - (1 - \beta)^{n^*}\right)f(n^*)$$

is the probability of observing a family. In a large population $\beta \to 0$ with increasing population size. Using l'Hospital's rule, and assuming that passage to the limit under the summation sign is valid

$$\lim_{\beta \to 0} g(n^*) = \frac{n^*f(n^*)}{E(N^*)} \qquad (1.38)$$

Thus the limit form of (1.37) is identical to (1.28). Larger families tend to be oversampled and hence a misleading estimate of family size will be produced from such samples. Since the model is formally equivalent to the length biased sampling model, all references and statements about identification given in example 6 apply with full force to this example. See the discussion in Rao (1965).

2. Economic models of self-selection

We begin our analysis by expositing the Roy model of self-selection for workers with heterogeneous skills. The statistical framework for this model has been outlined in examples 3 and 4. Following Roy, we assume that there are two market sectors in which income-maximizing agents can work. Agents are free to enter the sector that gives them the highest income. However, they can work in only one sector at a time.

Each sector requires a unique sector-specific task. Each agent has two skills, T_1 and T_2, which he cannot use simultaneously. The model is short run in that aggregate skill distributions are assumed to be given. There are no costs of changing sectors, and investment is ignored.

Because of this assumption, the model presented here applies to environments with certain or uncertain prices for sector-specific tasks. For simplicity and without any loss of generality (given the preceding assumptions), we assume an environment of perfect certainty.

Let T_i be the amount of sector i specific task a worker can perform. The price of task i is π_i. An agent works in sector 1 if his income is higher there, that is

$$\pi_1 T_1 > \pi_2 T_2 \qquad (2.1)$$

Indifference between sectors is a negligible probability event if the $T_i = 1, 2$ are assumed to be continuous nondegenerate random variables. Throughout we assume that prices are positive $(\pi_i > 0)$.

The log wage in task i of an individual with endowment T_i is

$$\ln W_i = \ln \pi_i + \ln T_i \qquad (2.2)$$

The proportion of the population working at task i is the proportion of the population for whom

$$T_1 > \frac{\pi_2}{\pi_1} T_2.$$

Roy assumes that $(\ln T_1, \ln T_2)$ is normally distributed with mean (μ_1, μ_2) and covariance matrix Σ. Letting (U_1, U_2) be a mean zero normal vector, agents in the Roy model choose between two possible wages:

$$\ln W_1 = \ln \pi_1 + \mu_1 + U_1$$

or

$$\ln W_2 = \ln \pi_2 + \mu_2 + U_2.$$

Workers enter sector 1 if $\ln W_1 > \ln W_2$. Otherwise they enter sector 2.

Letting

$$\sigma^* = \sqrt{\mathrm{var}(U_1 - U_2)}$$

and

$$c_i = (\ln(\pi_i/\pi_j) + \mu_i - \mu_j)/\sigma^*, \quad i \neq j,$$
$$\Pr(i) = P(\ln W_i > \ln W_j) = \Phi(c_i),$$
$$i \neq j, \quad i, j = 1, 2$$

where $\Phi(\ )$ is the cumulative distribution function of a standard normal variable. When standard sample selection bias formulae are used (see, e.g. Heckman, 1976), the mean of log wages observed in sector i is

$$E(\ln W_i | \ln W_i > \ln W_j)$$
$$= \ln \pi_i + \mu_i + \frac{\sigma_{ii} - \sigma_{ij}}{\sigma^*} \lambda(c_i),$$
$$i, j = 1, 2, \quad i \neq j,$$

$$(2.3)$$

where

$$\lambda(c) = \frac{\frac{1}{\sqrt{2\pi}}\exp\left(-\frac{1}{2}c^2\right)}{\Phi(c)}$$

is a convex monotone decreasing function of c with $\lambda(c) \geq 0$ and

$$\lim_{c\to\infty}\lambda(c) = 0, \qquad \lim_{c\to-\infty}\lambda(c) = \infty.$$

Convexity is proved in Heckman and Honoré (1986).

The variance of log wages observed in sector i

$$\text{var}(\ln W_i|\ln W_i > \ln W_j) =$$
$$\sigma_{ii}\{\rho_i^2[1 - c_i\lambda(c_i) - \lambda^2(c_i)] + (1 - \rho_i^2)\}, \quad i\neq j$$

$$(2.4)$$

where $\rho_i = \text{correl}(U_i, U_i - U_j), i\neq j = 1, 2$. The variance of the log of observed wages never exceeds σ_{ii}, the population variance, because the term in braces in (2.4) is never greater than unity. In general, sectoral variances decrease with increased selection. For example, if ρ_1 and ρ_2 do not equal zero, as π_1 increases with π_2 held fixed so that people shift from sector 2 to sector 1, the variance in the log of wages in sector 1 increases while the variance in the log of wages in sector 2 decreases.

Using the fact that $W_i = \pi_i T_i$, we may use (2.3) to write

$$E(\ln T_1|\ln W_1 > \ln W_2) = \mu_1 + \frac{\sigma_{11} - \sigma_{12}}{\sigma^*}\lambda(c_1),$$

$$(2.5a)$$

$$E(\ln T_2|\ln W_1 > \ln W_2) = \mu_2 + \frac{\sigma_{22} - \sigma_{12}}{\sigma^*}\lambda(c_2).$$

$$(2.5b)$$

Focusing on (2.5a) and noting that λ is positive for all values of c_1 (except $c_1 = \infty$), the mean of log task 1 used in sector 1 exceeds, equals, or falls short of the population mean endowment of log task 1 as $\sigma_{11} - \sigma_{12}$ is greater than, equal to, or less than zero. If endowments of tasks are uncorrelated ($\sigma_{12} = 0$) self-selection always causes the mean of $\ln T_1$ employed in sector 1 to be above the population mean μ_1. The opposite case occurs when $\sigma_{11} - \sigma_{12}$ is negative. This case can arise only when values of $\ln T_1$ and $\ln T_2$ are sufficiently positively correlated. If this occurs, the mean of log task 1 used in sector 1 falls below the population mean μ_1. Since covariance matrices must be positive semi-definite, $\sigma_{11} + \sigma_{22} - 2\sigma_{12} \geq 0$. Thus if $\sigma_{11} - \sigma_{12} < 0, \sigma_{22} - \sigma_{12} > 0$ so the mean of log task 2 employed in sector 2 necessarily lies above the population mean μ_2. In the Roy model the unusual case can arise in at most one sector. Notice from (2.5) that only if $\sigma_{11} - \sigma_{12} = 0$ (so $\rho_1^2 = 0$) is the variance of log task 1 employed in sector 1 identical to the variance of log task 1 in the population. Otherwise, the sectoral

variance of observed log task 1 is less than the population variance of log task 1.

To gain further insight into the effect of self-selection on the distribution of earnings for workers in sector 1, it is helpful to draw on some results from normal regression theory. The regression equation for $\ln T_2$ conditional on $\ln T_1$ is

$$\ln T_2 = \mu_2 + \frac{\sigma_{12}}{\sigma_{11}}(\ln T_1 - \mu_1) + \varepsilon_2, \quad (2.6)$$

where $E(\varepsilon_2) = 0$ and $\text{var}(\varepsilon_2) = \sigma_{22}[1 - (\sigma_{12}^2/\sigma_{11}\sigma_{22})]$.

Figure 2 plots regression function (2.6) for the case $\sigma_{12} = \sigma_{11}$ and $\mu_2 > \mu_1 > 0$. For each value of $\ln T_1$, the population values of $\ln T_2$ are normally distributed around the regression line. Individuals with high values of $\ln T_1$ also tend to have a high value of $\ln T_2$. Assuming $\pi_1 = \pi_2$, individuals with $(\ln T_1, \ln T_2)$ endowments above the 45° line of equal income shown in Figure 1 choose to work in sector 2, while those individuals with endowments below this line work in sector 1. Because $\sigma_{12} = \sigma_{11}$, the regression function is parallel to the line of equal income.

The distribution of ε_2 about the regression line is the same for all values of $\ln T_1$. When individuals are classified on the basis of their $\ln T_1$ values the same proportion of individuals work in sector 1 at all values of $\ln T_1$. For this reason the distribution of $\ln T_1$ employed in sector 1 is the same as the latent population distribution. If π_1 is raised (or π_2 is lowered) so that the 45° equal income line is shifted upward, the same proportion of people enter sector 1 at each value of $T_1 = t_1$. Figure 3

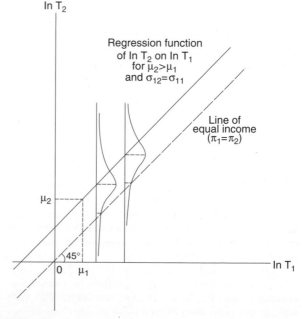

Figure 2

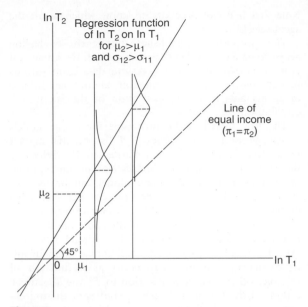

Figure 3

plots regression function (2.6) for the case $\sigma_{12} > \sigma_{11}$ and $\mu_2 > \mu_1 > 0$.

As before we set $\pi_1 = \pi_2$. Individuals with endowments above the 45° line choose to work in sector 2, while those with endowments below this line work in sector 1. When individuals are classified on the basis of their T_1 values, the fraction of people working in sector 1 decreases the higher the value of T_1. Self-selection causes the mean of log task 1 employed in sector 1 to be less than the mean of log task 1 in the total population. People with high values of T_1 are under-represented in sector 1 and low T_1 values are over-represented. In the extreme, when ln T_1 and ln T_2 are perfectly positively correlated, all high-income individuals are in sector 2, while all the low-income individuals are in sector 1. The highest-paid sector 1 worker earns the same as the lowest-paid sector 2 worker (Roy, 1951; Willis and Rosen, 1979). In this case there is really only one skill dimension and individuals can be unambiguously ranked along this scale.

If π_1 is raised (or π_2 is lowered) so that the line of equal income is shifted upward, the mean of ln T_1 employed in sector 1 must rise. The only place left to get T_1 is from the high end of the T_1 distribution. Unlike the case of $\sigma_{12} = \sigma_{11}$ in which a 10 per cent increase in π_1 results in a 10 per cent increase in measured average earnings in sector 1, when $\sigma_{12} > \sigma_{11}$ a 10 per cent increase in π_1 results in a greater than 10 per cent increase in the measured average earnings in sector 1 as the average quality of the sector 1 workforce increases. The variance of log wages in sector 1 increases.

If $\sigma_{11} < \sigma_{12}$ then $\sigma_{12} < \sigma_{22}$ in order for Σ to be a covariance matrix. In the population, log task 2 must have greater variability than log task 1. Individuals with high T_1 values tend to have high T_2 values. But the

population distribution of log task 2 has more mass in the tails. The higher an agent's value of T_1, the more likely it is that he will be able to get higher income in sector 2. At the lower end of the distribution, the process works in reverse: lower T_1 individuals on average have poor T_2 values. Self-selection causes the ln T_1 distribution in sector 1 to have an evacuated right tail, an exaggerated left tail, and a lower mean than the population mean of ln T_1.

If $\sigma_{12} < \sigma_{11}$ (a case not depicted graphically), the proportion of each T_1 group working in sector 1 increases, the higher the value of T_1. The mean of the log task employed in sector 1 exceeds μ_1. A 10 per cent increase in π_1 produces an increase of less than 10 per cent in the average earnings of workers in sector 1 as the mean of ln T_1 employed in sector 1 declines. In fact if $\sigma_{12} > \sigma_{22}$ it is possible for an increase in π_1 to cause measured sector 1 wages to decline. Thus through a selection phenomenon it is possible for the average wage of people working in sector 1 to decline even though the price per unit skill increases there.

How robust are these conclusions if the normality assumption is relaxed? Heckman and Sedlacek (1985) show that many propositions derived from assumed normality of skills do not hold up for more general distributions. For example, increasing selection need not decrease sectoral variances. The effects of selection on mean employed skill levels are ambiguous. Heckman and Honoré (1986) demonstrate that in a single cross-section of data, it is possible to identify all of the parameters of the model from the data if the normality assumption is invoked. However, in a single cross-section many other models can explain the data equally well. In particular, intuitive notions about the degree of correlation or dependence among skills have no empirical content and so models of skill 'hierarchies' based on the extent of such dependence have no content for single cross-sections of data with all individuals facing common prices.

To show this, write the density of skills as $f(t_1, t_2)$. Let

$$Z_1 = \begin{cases} T_1 & \text{if } T_1 > T_2 \\ 0 & \text{otherwise} \end{cases}$$

$$Z_2 = \begin{cases} T_2 & \text{if } T_2 > T_1 \\ 0 & \text{otherwise} \end{cases}$$

Prices are normalized to unity ($\pi_1 = \pi_2 = 1$). Then the density of Z_1 is

$$Q_1'(z_1) = \int_{\{t_2 | t_2 < z_1\}} f(z_1, t_2)\, dt_2$$
$$= \int_0^{z_1} f(z_1, t_2)\, dt_2.$$

The density of Z_2 is

$$Q_2'(z_2) = \int_0^{z_2} f(t_1, z_2)\, dt_1.$$

Note that $Q'_1(n)$ and $Q'_2(n)$ summarize all of the available data on observed earnings.

Now if T_1, T_2 are independent with cdf's F^*_1 and F^*_2 respectively

$$Q'_1(n) = f^*_1(n)F^*_2(n)$$
$$Q'_2(n) = F^*_1(n)f^*_2(n).$$

Define

$$\bar{Q}(n) = \int_0^n [Q'_1(l) + Q'_2(l)] \, dl$$
$$= F^*_1(n)F^*_2(n).$$

Then

$$\int_\phi^\infty \frac{Q'_1(n)}{\bar{Q}(n)} \, dn = \int_\phi^\infty \frac{f^*_1(n)}{F^*_1(n)} \, dn$$
$$= -\ln F^*_1(\phi).$$

Thus we can write

$$F^*_i(\phi) = \exp - \left(\int_\phi^\infty \left[\frac{Q'_i(n)}{\bar{Q}(n)} \right] dn \right) \quad i = 1, 2$$

so that we can always rationalize the data on wages in a single cross-section by a model of skill independence, and economic models of skill hierarchies have no empirical content for a single cross-section of data.

Suppose, however, that the observing economist has access to data on skill distributions in different market settings i.e. settings in which relative skill prices vary. To take an extreme case, suppose that we observe a continuum of values of π_1/π_2 ranging from zero to infinity. Then it is possible to identify $F(t_1, t_2)$ and it is possible to give empirical content to models based on the degrees of dependence among latent skills.

This point is made most simply in a situation in which Z is observed but the analyst does not know Z_1 or Z_2 (i.e. which occupation is chosen). When $\pi_1/\pi_2 = 0$ everyone works in occupation two. Thus we can observe the marginal density of t_2. When $\pi_1/\pi_2 = \infty$ everyone works in occupation one. As π_1/π_2 pivots from zero to infinity it is thus possible to trace out the full joint distribution of (T_1, T_2).

To establish the general result, set $\sigma = \pi_2/\pi_1$. Let $F(t_1, t_2)$ be the distribution function of T_1, T_2. Then

$$\Pr(Z \leq n) = \Pr(\max(T_1, \sigma T_2) \leq n)$$
$$= \Pr\left(T_1 \leq n, T_2 \leq \frac{1}{\sigma} n \right)$$
$$= F\left(n, \frac{n}{\sigma} \right).$$

As σ varies between 0 and ∞ the entire distribution can be recovered since N is observed for all values in $(0, \infty)$.

Note that it is not necessary to know which sector the agent selects.

This proposition establishes the benefit of having access to data from more than one market. Heckman and Honoré (1986) show how access to data from various market settings and information about the choices of agents aids in the identification of the latent skill distributions.

The Roy model is the prototype for many models of self-selection in economics. If T_1 is potential market productivity and T_2 is non-market productivity (or the reservation wage) for housewives or unemployed individuals, precisely the same model can be used to explore the effects of self-selection on measured productivity. In such a model, T_2 is never observed. This creates certain problems of identification discussed in Heckman and Honoré (1986). The model has been extended to allow for more general choice mechanisms. In particular, selection may occur as a function of variables other than or in addition to T_1 and T_2. Applications of the Roy model include studies of the union–non-union wage differential (Lee, 1978), the returns to schooling (Willis and Rosen, 1979), and the returns to training (Bjorklund and Moffitt, 1986, and Heckman and Robb, 1985). Amemiya (1984) and Heckman and Honoré (1986) present comprehensive surveys of empirical studies based on the Roy model and its extensions.

JAMES J. HECKMAN

Bibliography

Amemiya, T. 1984. Tobit models: a survey. *Journal of Econometrics* 24, 3–61.

Bjorklund, A. and Moffitt, R. 1986. Estimation of wage gains and welfare gains from self selection models. *Review of Economics and Statistics* 24, 1–63.

Cain, G. and Watts, H. 1973. Toward a summary and synthesis of the evidence. In *Income Maintenance and Labor Supply*, ed. G. Cain and H. Watts. Madison: University of Wisconsin Press.

Clark, K. and Summers, L. 1979. Labor market dynamics and unemployment: a reconsideration. *Brookings Papers on Economic Activity*, 13–60.

Cosslett, S. 1981. Maximum likelihood estimation from choice based samples. *Econometrica*.

Cosslett, S. 1984. Distribution free estimator of regression model with sample selectivity. Unpublished manuscript, University of Florida.

Domencich, T. and McFadden, D. 1975. *Urban Travel Demand*. Amsterdam: North-Holland.

Flinn, C. and Heckman, J. 1982. New methods for analyzing structural models of labor force dynamics. *Journal of Econometrics* 18, 5–168.

Gallant, R. and Nychka, R. 1984. Consistent estimation of the censored regression model. Unpublished manuscript, North Carolina State.

Gill, R. and Wellner, J. 1985. Large sample theory of empirical distributions in biased sampling models. Unpublished manuscript, University of Washington.

Goldberger, A. 1983. Abnormal selection bias. In *Studies in Econometrics, Time Series and Multivariate Statistics*, ed. S. Karlin, T. Amemiya and L. Goodman. Wiley, NY.

Gronau, R. 1974. Wage comparisons – a selectivity bias. *Journal of Political Economy* 82(6), 1119–1144.

Hall, R. 1982. The importance of lifetime jobs in the US economy. *American Economic Review* 72, September, 716–724.

Heckman, J. 1974. Shadow prices, market wages and labor supply. *Econometrica* 42(4), 679–94.

Heckman, J. 1976. The common structure of statistical models of truncation, sample selection and limited dependent variables and a simple estimator for such models. *Annals of Economic and Social Measurement* 5(4), 475–92.

Heckman, J. 1979. Sample selection bias as a specification error. *Econometrica* 47(1), 153–62.

Heckman, J. and Honoré, B. 1986. The empirical content of the Roy model. Unpublished manuscript, University of Chicago.

Heckman, J. and MaCurdy, T. 1981. New methods for estimating labor supply functions. In *Research in Labor Economics*, Vol. 4, ed. R. Ehrenberg. Greenwich, Conn.: JAI Press.

Heckman, J. and MaCurdy, T. 1986. Labor econometrics. In *Handbook of Econometrics*, vol. 3, ed. Z. Griliches and M. Intriligator. Amsterdam: North-Holland.

Heckman, J. and Robb, R. 1985. Alternative methods for evaluating the effect of training on earnings. In *Longitudinal Analysis of Labor Market Data*, ed. J. Heckman and B. Singer. Cambridge: Cambridge University Press.

Heckman, J. and Sedlacek, G. 1985. Heterogeneity, aggregation and market wage functions. *Journal of Political Economy* 93, December, 1077–125.

Heckman, J. and Singer, B. 1985. Econometric analysis of longitudinal data. In *Handbook of Econometrics*, Vol. III, ed. Z. Griliches and M. Intriligator. Amsterdam: North-Holland.

Lee, L.F. 1978. Unionism and wage rates: a simultaneous equations model with qualitative and limited dependent variables. *International Economic Review* 19, 415–33.

Manski, C. and Lerman, S. 1977. The estimation of choice probabilities from choice based samples. *Econometrica* 45, 1977–88.

Manski, C. and McFadden, D. 1981. Alternative estimates and sample designs for discrete choice analysis. In *Structural Analysis of Discrete Data with Econometric Applications*, ed. C. Manski and D. McFadden. Cambridge: MIT Press.

Pearson, K. 1901. Mathematical contributions to the theory of evolution. *Philosophical Transactions*, 195, 1–47.

Rao, C.R. 1965. On discrete distributions arising out of methods of ascertainment. In *Classical and Contagious Distributions*, ed. G. Patil. Calcutta: Pergamon Press.

Roy, A.D. 1951. Some thoughts on the distribution of earnings. *Oxford Economic Papers*, 3, 135–46.

Vardi, Y. 1983. Nonparametric estimation in the presence of length bias. *Annals of Statistics* 10, 616–20.

Vardi, Y. 1985. Empirical distributions in selection bias models. *Annals of Statistics*, 13, 178–203.

Willis, R. and Rosen, S. 1979. Education and self selection. *Journal of Political Economy* 87, S7–S36.

self-confirming equilibria

A self-confirming equilibrium is the answer to the following question: what are the possible limiting outcomes of purposeful interactions among a collection of adaptive agents, each of whom averages past data to approximate moments of the conditional probability distributions of interest? If outcomes converge, a law of large numbers implies that agents' beliefs about conditional moments become correct on events that are observed sufficiently often. Beliefs are not necessarily correct about events that are infrequently observed. Where beliefs are correct, a self-confirming equilibrium is like a rational expectations equilibrium. But there can be interesting gaps between self-confirming and rational expectations equilibria where beliefs of some important decision makers are incorrect.

Self-confirming equilibria interest macroeconomists because they connect to an influential 1970s argument made by Christopher Sims that advocated rational expectations as a sensible equilibrium concept. This argument defended rational expectations equilibria against the criticism that they require that agents 'know too much' by showing that we do not have to assume that agents start out 'knowing the model'. If agents simply average past data, perhaps conditioning by grouping observations, their forecasts eventually become unimprovable.

Research on adaptive learning has shown that that the glass is 'half full' and 'half empty' for this clever 1970s argument. On the one hand, the argument is correct when applied to competitive or infinitesimal agents: by using naive adaptive learning schemes (various versions of recursive least squares), agents can learn every conditional distribution that they require to play best responses within an equilibrium. On the other hand, large agents (for example, governments in macro models) who can influence the market outcome cannot expect to learn everything that they need to know to make good decisions: in a self-confirming equilibrium, large agents may base their decisions on conjectures about off-equilibrium-path behaviours that turn out to be incorrect. Thus, a rational expectations equilibrium is a self-confirming equilibrium, but not vice versa.

While agents' beliefs can be incorrect off the equilibrium path, the self-confirming equilibrium path still restricts them in interesting ways. For macroeconomic applications, the government's model must be such that its off-equilibrium path beliefs rationalize the decisions (its Ramsey policy or Phelps policy, in the language of

Sargent, 1999) that are revealed along the equilibrium path. The restrictions on government beliefs required to sustain self-confirming equilibria have only begun to be explored in macroeconomics, mainly in the context of some examples like those in Sargent (1999). Analogous restrictions have been more thoroughly analysed in the context of games (Fudenberg and Levine, 1993).

The freedom to specify beliefs off the equilibrium path makes the set of self-confirming equilibria generally be larger than the set of Nash equilibria, which often admit unintuitive outcomes in extensive form games (Kreps, 1998). A widely used idea of refining self-confirming equilibrium is to embed the decision making problem within a learning process in which decision makers estimate unknown parameters through repeated interactions, and then to identify a stable stationary point of the learning dynamics (for example, Fudenberg and Kreps, 1993; 1995; Evans and Honkapohja, 2001).

The gap between a self-confirming equilibrium and a rational expectations equilibrium can be vital for a government designing a Ramsey plan, for example, because its calculations necessarily involve projecting outcomes of counterfactual experiments. For macroeconomists, an especially interesting feature of self-confirming equilibria is that, because a government can have a model that is wrong off the equilibrium path, a policy that it thinks is optimal can very well be far from optimal. Even if a policy model fits the historical data correctly and is unimprovable, one cannot conclude that the policy is optimal. As a result, it requires an entirely a priori theoretical argument to diminish the influence of a good fitting macroeconomic model on public policy (Sargent, 1999).

Formal definitions

An agent i is endowed with strategy space A_i and state space X_i. Generic elements of A_i and X_i are called a strategy and a state, respectively. A probability distribution $\mathscr{P}_i$ over $A_i \times X_i$ describes how actions and states are related. A utility function is $u_i : A_i \times X_i \to \mathbb{R}$. Let $\mu_i(\cdot : a_i)$ be a probability distribution over X_i, which represents i's belief about the state conditioned on action a_i. Agent i's decision problem is to solve

$$\max_{a_i \in A_i} \int_{X_i} u_i(a_i, x_i) d\mu_i(x_i : a_i). \qquad (1)$$

Single person decision problems
Here a self-confirming equilibrium is a simply a pair (a_i^*, μ_i^*) satisfying

$$a_i^* \in \arg\max_{a_i \in A_i} \int_{X_i} u_i(a_i, x_i) d\mu_i^*(x_i : a_i) \qquad (2)$$

$$\mu_i^*(x_i : a_i^*) = \mathscr{P}_i(x_i : a_i^*). \qquad (3)$$

(2) implies that the choice must be optimal given his subjective belief μ_i^*, while (3) says that the belief must be confirmed, conditioned on his equilibrium action a_i^*. Self-confirming equilibrium has the two key ingredients of rational expectations equilibrium: optimization and self-fulfilling property. The key difference is (3), which imposes a self-confirming property conditioned only on equilibrium action a_i^*. The decision maker can entertain $\mu_i(\cdot : a_i) \neq \mathscr{P}(\cdot : a_i)$, conditioned on $a_i \neq a_i^*$. In this sense, the agent can have multiple beliefs about the state conditioned on his own action.

If we strengthen things to require

$$\mu_i(\cdot : a_i) = \mathscr{P}(\cdot : a_i) \qquad \forall a_i, \qquad (4)$$

then we attain a rational expectations equilibrium. As will be shown later, (4) is called the unitary belief condition (Fudenberg and Levine, 1993), which is one of the three key features that distinguishes a self-confirming equilibrium from a rational expectations equilibrium or Nash equilibrium.

Multi-person decision problems
If we interpret the state space as the set of the strategies of the other players $X_i = A_{-i}$, we can naturally extend the basic definition to the situation where more than one person is making a decision. A self-confirming equilibrium is a profile of actions and beliefs, $\{(a_1^*, \mu_1^*), \ldots, (a_n^*, \mu_n^*)\}$ such that (2) and (3) hold for every $i = 1, \ldots, n$. As we move from single person to multi-person decision problems, however, (3) differs three ways from a Nash equilibrium, in addition to the unitary belief condition (4). (1) If there are more than two players, the belief of player $i \neq j$, k about player k's strategy can be different from player j's belief about player k's strategy (failure of consistency). (2) Similarly, player i can entertain the possibility that player j and player k correlate their strategies according to an un-modelled randomization mechanism, leading to correlated beliefs. (3) If we require that a self-confirming equilibrium should admit unitary and consistent beliefs, while excluding correlated beliefs, then the self-confirming equilibrium is a Nash equilibrium (Fudenberg and Levine, 1993).

Dynamic decision problems
Suppose that player i solves (1) repeatedly. The first step to embed self-confirming equilibria in dynamic contexts is to spell out learning rules that specify how beliefs respond to new observations. We define a learning rule as a mapping that updates belief μ_i into a new belief when new data arrive. Define $Z_i \subset X_i$ as a subspace of X_i that is observed by a decision maker. Let $\mathscr{M}_i$ be the set of probability distributions over $Z_i \times A_i$. These represent player i's belief about the state, that is, the model entertained by player i. A learning rule is defined as

$$\mathscr{T}_i : \mathscr{M}_i \times Z_i \to \mathscr{M}_i.$$

A belief $\mu_i^* \in \mathcal{M}_i$ is a steady state of the learning dynamics if

$$a_i^* \in \arg\max_{a_i \in A_i} \int_{x_i} u_i(a_i, x_i) d\mu_i^*(x_i : a_i) \quad (5)$$

$$\mu_i^* = \mathcal{T}_i(\mu_i^*, z_i) \quad (6)$$

for every z_i in the support of $\mathcal{P}_i(x_i : a_i^*)$.

The steady state of learning dynamics is a self-confirming equilibrium for a broad class of recursive learning dynamics including Bayesian (Fudenberg and Levine, 1993), and least square learning algorithms (Sargent, 1999; Evans and Honkapohja, 2001) are self-confirming equilibria.

Refinements

We can study the salience of the self-confirming equilibrium by examining the stability of the associated steady states. The stability property provides a useful foundation for selecting a sensible self-confirming equilibrium (Bray, 1982; Marcet and Sargent, 1989; Woodford, 1990; Evans and Honkapohja, 2001). With a possible exception of the Bayesian learning algorithm, most ad hoc learning rules are motivated by the simplicity of some updating scheme as well as its ability to support sufficiently sophisticated behaviour in the limit. By exploiting the convergence properties of learning dynamics, we can often devise a recursive algorithm to *calculate* a self-confirming equilibrium, that is, a fixed point of the $\mathcal{T}_i$. This approach to computing equilibria has occasionally proved fruitful to compute equilibria in macroeconomics (for example, Aiyagari et al., 2002).

In principle, a player need not know the other player's payoff in order to play a self-confirming equilibrium. Self-confirming equilibria allow a player to entertain any belief conditioned on actions not used in the equilibrium. This is one of the main sources of multiplicity. In the game theoretic context, a player can delineate the set of possible actions of the other players, even if he does not have perfect foresight. If each player knows the payoff of the other players, and if it is common knowledge that every player is rational, then a player can eliminate the actions of the other players that cannot be rationalized. By exploiting the idea of sophisticated learning of Milgrom and Roberts, (1990; 1991), Dekel, Fudenberg and Levine (1999) restricted the set of possible beliefs off the equilibrium path to eliminate evidently unreasonable self-confirming equilibrium.

Applications

Self-confirming equilibria and recursive learning algorithms are powerful tools to investigate a number of important dynamic economic problems such as (1) the limiting behaviour of learning systems (Evans and Honkapohja, 2001; Fudenberg and Levine, 1998); (2) the selection of plausible equilibria in games and dynamic macroeconomic models (Marcet and Sargent, 1989; Kreps, 1998); (3) the incidence and distribution of rare events that occasionally arise as large deviations from self-confirming equilibria (Cho, Williams and Sargent, 2002; Sargent, 1999); and (4) formulating plausible models of how agents respond to model uncertainty (Cho and Kasa, 2006).

Remarkably, related mathematics tie together all of these applications. The mean dynamics that propel the learning algorithms to self-confirming equilibria (in item (1)) are described by ordinary differential equations (ODE) derived through an elegant stochastic approximation algorithm (for example, Fudenberg and Levine, 1998; Woodford, 1990; Marcet and Sargent, 1989). Because the stationary point of the ODE is a self-confirming equilibrium, the stability of the ODE determines the selection criterion used to make statements about item (2) (Fudenberg and Levine, 1998; Evans and Honkapohja, 2001). Remarkably, by adding an adverse deterministic shock to that same ODE, we obtain a key object that appears in a *deterministic* control problem that identifies the large-deviation excursions in item (3) *away* from a self-confirming equilibrium (Cho, Williams and Sargent, 2002). Finally, that same large deviations mathematics is associated with robust control ideas that use entropy to model how agents cope with model uncertainty (Pandit and Meyn, 2006).

IN-KOO CHO AND THOMAS J. SARGENT

Financial support from the National Science Foundation (ECS-0523620) and Deutsche Bank through the Institute for Advanced Study at Princeton is gratefully acknowledged. Any opinions, findings, and conclusions or recommendations expressed in this material are those of the authors and do not necessarily reflect the views of the National Science Foundation, Deutsche Bank, or the Institute for Advanced Study.

Bibliography

Aiyagari, R., Marcet, A., Sargent, T.J. and Seppälä, J. 2002. Optimal taxation without state-contingent debt. *Journal of Political Economy* 110, 1220–54.

Bray, M.M. 1982. Learning, estimation and the stability of rational expectations. *Journal of Economic Theory* 26, 318–39.

Cho, I.-K. and Kasa, K. 2006. Learning and model validation. Working paper, University of Illinois and Simon Fraser University.

Cho, I.-K., Williams, N. and Sargent, T.J. 2002. Escaping Nash inflation. *Review of Economic Studies* 69, 1–40.

Dekel, E., Fudenberg, D. and Levine, D.K. 1999. Payoff information and self-confirming equilibrium. *Journal of Economic Theory* 89, 165–85.

Evans, G.W. and Honkapohja, S. 2001. *Learning and Expectations in Macroeconomics.* Princeton: Princeton University Press.

Fudenberg, D. and Kreps, D.M. 1993. Learning mixed equilibria. *Games and Economic Behavior* 5, 320–67.

Fudenberg, D. and Kreps, D.M. 1995. Learning in extensive games, I: self-confirming and Nash equilibrium. *Games and Economic Behavior* 8, 20–55.

Fudenberg, D. and Levine, D.K. 1993. Self-confirming equilibrium. *Econometrica* 61, 523–45.

Fudenberg, D. and Levine, D. 1998. *Learning in Games*. Cambridge, MA: MIT Press.

Kreps, D.M. 1998. Anticipated utility and dynamic choice. In *Frontiers of Research in Economic Theory: The Nancy L. Schwartz Memorial Lectures, 1983– 1997*, ed. D. P. Jacobs, E. Kalai and M.I. Kamien. Cambridge: Cambridge University Press.

Marcet, A. and Sargent, T.J. 1989. Convergence of least squares learning mechanisms in self referential linear stochastic models. *Journal of Economic Theory* 48, 337–68.

Milgrom, P. and Roberts, D.J. 1990. Rationalizability, learning and equilibrium in games with strategic complementarities. *Econometrica* 58, 1255–77.

Milgrom, P. and Roberts, D.J. 1991. Adaptive and sophisticated learning in normal form games. *Games and Economic Behavior* 3, 82–100.

Pandit, C. and Meyn, S.P. 2006. Worst-case large-deviations with application to queueing and information theory. *Stochastic Process and their Applications* 116, 724–56.

Sargent, T.J. 1999. *The Conquest of American Inflation*. Princeton: Princeton University Press.

Woodford, M.D. 1990. Learning to believe in sunspots. *Econometrica* 58, 277–307.

Seligman, Edwin Robert Anderson (1861–1939)

Seligman was born in New York City on 25 April 1861 and died on 18 July 1939 at Lake Placid, New York. An economist of unusual erudition, energy and wide-ranging interests, Seligman successfully combined a life of distinguished scholarship with philanthropy and active participation and leadership in a variety of reform causes. Raised in a talented and wealthy New York Jewish business family, Seligman studied privately under Horatio Alger Jr. and at Columbia Grammar School before graduating from Columbia in 1879. After three years' study in Berlin, Heidelberg (under Karl Knies), Geneva and Paris he returned to Columbia obtaining MA and LLB degrees in 1884, the Ph.D. cum laude in 1885 and a full professorship in political economy at age 30, a post he held until retirement in 1931. Dignified, wise and balanced in outlook, Seligman personified the best in late 19th-century efforts to blend orthodox classical and German historical economics. His original studies of neglected British and American economists, and his compilation of perhaps the world's greatest library of economic works, reveal his lifetime devotion to doctrinal history, while his widely read and durable *Economic Interpretation of History* (1902) testifies to the breadth and sensitivity of his historical knowledge. Like Henry Carter Adams, with whom he created the field of public finance in America, Seligman was influenced by Adolph Wagner. But he was more of a theorist than Adams, and his concepts of 'faculty' or ability to pay, and benefit, were the first systematic modern efforts to develop theoretical criteria of taxation. A severe critic of Henry George, Seligman nevertheless favoured taxes on land values and progressive inheritance taxes, and advocated proportional income taxes as early as 1894. Sympathetic to labour unions, federal railroad legislation, effective central banking measures and other moderate reform proposals, including deficit finance and public works during the depression of the 1930s, Seligman also advocated US aid to Europe after 1918 and the cancellation of their debts.

Seligman served on innumerable public bodies as a taxation and financial specialist, and as a social reformer, for example as Chairman of the Bureau of Municipal Research and with the National Civic Federation. A founder, first Treasurer and later President (1902–4) of the American Economic Association, he was an outstanding champion of academic freedom and co-founder of the American Association of University Professors, of which he was President 1919–20. His success as fund-raiser and Editor in Chief of the *Encyclopaedia of the Social Sciences* 1927–35 was a fitting culmination of an outstanding career of scholarly and public service.

A.W. COATS

Selected works

1892. *On the Shifting and Incidence of Taxation*. Baltimore: American Economic Association. 5th edn, revised, New York: Columbia University Press, 1927.

1894. *Progressive Taxation in Theory and Practice*. Baltimore: American Economic Association. 2nd edn, revised and enlarged, Princeton: American Economic Association, 1908.

1895. *Essays in Taxation*. New York and London: Macmillan. 10th edn, 1925.

1902. *The Economic Interpretation of History*. New York: Columbia University Press. 2nd edn, revised, 1924.

1905. *The Principles of Economics, with Special Reference to American Conditions*. New York: Longmans Green & Co. 12th edn, revised, 1929.

1911. *The Income Tax: A Study of the History, Theory, and Practice of Income Taxation at Home and Abroad*. New York: Macmillan. 2nd edn, revised and enlarged, 1914.

1925a. *Essays in Economics*. New York: Macmillan.

1925b. *Studies in Public Finance*. New York: Macmillan.

Selten, Reinhard (born 1930)

Reinhard Selten was born in Breslau, then in Germany, and now called Wroclaw, in Poland. His father, who was of Jewish origin, died of illness when he was 12 years old, and the family had a difficult time during the war. After having been refugees for a couple of years, they settled in Hessia, where Selten completed high school in 1951. In that same year, he had his first contact with game theory through a popular article in *Fortune*. From 1951 to 1957, Selten studied mathematics in Frankfurt, where, under the guidance of Ewald Burger, he wrote his Master's thesis on cooperative game theory, aiming to axiomatize a value for extensive form games. He continued to work for his Ph.D. in Frankfurt, in the institute of Professor Heinz Sauermann, where he soon became involved in laboratory experiments. After receiving his Ph.D. in mathematics in 1961, he remained in Frankfurt until 1969, when he moved to the Free University of Berlin. In 1972, he moved to Bielefeld, to the newly established Institute of Mathematical Economics. In 1984, he moved to the University of Bonn, where he set up the Bonn Experimental Laboratory, to which he is still affiliated. In 1994, he received the Nobel Prize in Economics, which he shared with John Nash and John Harsanyi. For more personal details, see his autobiography for the prize (Selten, 1995).

Selten describes himself as a 'methodic dualist': throughout his career, he has investigated both the consequences of ideal rationality in game situations, as well as the limitations of this concept to describe actual observed behaviour, and he has proposed several more descriptive theories. For Selten's perspective on his own work and how it hangs together, see Selten (1993). A selection of his path-breaking contributions has been collected in Selten (1999).

Perfect equilibria

John Nash has proposed the fundamental solution concept for non-cooperative games, that is, games in which the players cannot make binding agreements outside of the formal rules. A *Nash equilibrium* has a natural stability property: once having arrived at it, either through introspection, learning or evolution, no player has an incentive to unilaterally deviate. Nash showed that each game has at least one equilibrium, and that a non-self-destroying, single-valued theory of rationality has to prescribe playing such an equilibrium. See STRATEGIC AND EXTENSIVE FORM GAMES for more details for formal definitions of other concepts discussed in this article.

Applying John van Neumann's insight that any extensive form (that is, dynamic) game can be reduced to an equivalent one in which the players move just once and simultaneously, Nash focused mainly on the latter. In his early experiments, Selten was, however, working with dynamic oligopoly games. When calculating the Nash equilibria, to provide a benchmark to compare the actual outcomes with, he soon discovered that there typically were many, including those that could not be considered compatible with rationality. To eliminate these, he proposed the refinement of subgame perfection. The idea underlying this concept is that, once a subgame (a part that constitutes a game in itself) is reached, everything outside of it (the things that could have happened but did not) has become irrelevant, so that the logic underlying the equilibrium should be applied to the subgame as well.

By requiring subgame perfection, one eliminates equilibria that rely on non-credible threats or promises. As an example, suppose that two players have to divide $4. One player, the proposer, P, may offer 3, 2 or 1 to the responder, R, who can only accept or reject; if he accepts, the division is implemented, otherwise each player gets zero. Assume it is commonly known that each player cares only about his own monetary payoff and prefers more money to less. The game has a Nash equilibrium in which P offers 3 and R rejects any outcome in which he is offered less. The threat to reject any amount, however, is not credible: confronted with offer m, by assumption R prefers m to 0, hence, he should accept it. In the unique subgame perfect equilibrium, P offers 1, which R accepts.

Later, Selten discovered that subgame perfection is not sufficient to rule out all 'non-rational' equilibria. In Selten (1975), he introduced a further refinement: 'trembling-hand perfection', a concept that insists that equilibria be robust with respect to small perturbations of the strategies. Formally, it is assumed that, whenever a player has to move, with a small probability he makes a mistake, and an equilibrium is trembling-hand perfect if, under these circumstances, each player is still willing to play it. Ideal rationality is thus viewed as a limiting case of rationality with small errors.

Selten was the first to refine Nash's concept for analysing dynamic games. In the beginning of the 1970s, he was also one of the first to successfully apply these refined concepts to dynamic game models of imperfect competition, thus pioneering the use of game theory in industrial organization. Subgame perfection is now routinely applied in situations of strategic interaction involving a time element. The robustness approach proposed in Selten (1975) also turned out to be very fruitful. On the one hand, it gave rise to solution concepts, such as sequential equilibrium, that were easy to use in applications; on the other hand, it induced a search for further refinements and stability concepts that uncovered the deeper mathematical structure of ideal rationality; see Kohlberg and Mertens (1986). We refer to NASH EQUILIBRIUM, REFINEMENTS OF for further details.

Equilibrium selection and the cooperation with John Harsanyi

In 1965, during a workshop in Jerusalem, Selten met John Harsanyi, and they started to cooperate. During the academic year 1967–8, Selten visited Harsanyi in Berkeley to work on two-person bargaining under incomplete information. They decided to take the rationality assumption in situations of strategic interaction to its logical limit, and started a project that would be finished only 20 years later with Harsanyi and Selten (1988). The motivation for this work is that the traditional justification of the Nash equilibrium concept is incomplete: it relies on the assumption that each game has a unique solution, but a game typically has multiple Nash equilibria. The natural question thus is, whether, by strengthening the rationality requirements, a unique selection can be obtained.

An essential ingredient of their selection theory is the notion of risk dominance. The concept tries to assess whether, in a situation in which players are a priori uncertain about which equilibrium should be played, they can still coordinate on a single equilibrium and, if so, which one. The stag hunt game may illustrate the concept and can show that there may be a conflict between payoff dominance and risk dominance. Assume two players that each can choose between a safe action, S, and a risky one, R. S yields 1 for sure, while R gives payoff X, with $X > 1$, if the other player also chooses R, but it yields 0 otherwise. Both (R,R) and (S,S) are equilibria and the former Pareto dominates the latter. If X is not too large, however, one can gain only little by playing R and the downward risk is considerable. In Harsanyi and Selten (1988), risk dominance is formalized by means of the tracing procedure, a theoretical model of the thinking process that converts any mixed strategy profile into an equilibrium of the game. In our example, (S,S) is the risk dominant equilibrium if $X < 2$ and there is a conflict between risk dominance and payoff dominance if $1 < X < 2$. The concept has found applications also in other domains of game theory. For example, in an evolutionary setting, where players repeatedly play the game in a myopic fashion, adjusting strategies through time, it was found that risk dominance is related to the long-run stochastic stability of the equilibrium (see LEARNING AND EVOLUTION IN GAMES: AN OVERVIEW).

For extensive form games, Harsanyi and Selten propose subgame consistency, a natural extension of subgame perfection, as an important selection principle: if g is a subgame of G, then in g the solution of G should prescribe the solution of g. Again the idea is that, once g is reached, everything else has become irrelevant. To illustrate, let g be the stag hunt game with $X = 1.8$, and let G be the game in which player 1 first chooses whether to take up an outside option yielding both players 1.5 or to play g. A selection theory incorporating subgame consistency and risk dominance implies that player 1 should take up the outside option. On the other hand, the strategy 'go to g and play S' is a dominated strategy in the overall game, so that repeated elimination of dominated strategies, or forward induction, produces the outcome (R,R). The example shows that several desirable

properties easily conflict. The conclusion is that it is possible to construct a single valued theory of rationality in non-cooperative games, but that an ideal theory does not exist.

Bounded rationality and experiments

When, after finishing his Master's thesis in 1957, Selten started to work in the institute of Professor Heinz Sauermann on a project on the 'Theory of the Firm', under the influence of Herbert Simon's ideas he very quickly became convinced that it was necessary to model economic behaviour as boundedly rational. (Also see RATIONALITY, BOUNDED.) In Selten's view, bounded rationality (defined as the rationality exhibited by actual human behaviour) differs fundamentally from the ideal rationality that we have been discussing in this article thus far. Bounded rationality, hence, is not viewed as an approximation to full rationality; it simply is something different. Furthermore, since actual behaviour cannot be invented in the armchair, the development of theories of bounded rationality needs an empirical basis. Consequently, laboratory experimentation becomes an important source of empirical evidence, also because of the possibility of gathering data under controlled circumstances. See also EXPERIMENTAL ECONOMICS. Selten's first experimental paper, co-authored with Sauermann, had already appeared in 1959. Given the sharp distinction drawn between ideal rationality and bounded rationality, it is not too surprising that the work was not directed at testing a theory; rather, it was an explorative piece, trying to uncover regularities. Ever since that first paper, Selten's work in this area has tried to uncover the structure of boundedly rational decision making, with an emphasis on individual data.

Selten distinguishes between cognitive bounds on rationality and motivational bounds. The former arise from the limited human ability to think and compute. Motivational bounds are different: even when the standard rational solution appears obvious, and is fully understood, as in the finitely repeated Prisoner's Dilemma game, a player may not have the incentive to implement it. In his paper on the chain store game, Selten argued that we lack a 'behavioural trust' in abstract subgame perfection arguments, and he proposed a theory of decision making in which decisions can arise at three different levels: the routine level, using past experience and analogies; the imagination level, involving various scenarios; and the reasoning level in which the individual makes a conscious effort to analyse the situation in rational, logical way. Since different levels involve different decision costs, not all levels may be activated, and even if they are, then, in Selten's view, there is no reason why the decision at the higher level should be selected. This brief description may make clear why boundedly rational behaviour may indeed be very different from normative rationality.

An important tool in the uncovering of the structure of boundedly rational decision making has been the use of the strategy method: in experiments, subjects are asked not just to make decisions as the game goes along, but also to specify or programme strategies for the entire game. In this way more of the strategic reasoning of the subject is revealed. Selten (1998) provides a partial overview of some of the results that have been achieved so far and of how different boundedly rational behaviour is from standard textbook rationality. For example, the analysis starts with a superficial analysis of the situation, and, because of the cost of decision effort, may not proceed beyond that level; fairness and reciprocity considerations are important as are issues of *ex post* rationality; there is no optimization or formation of quantitative expectations; in dynamic interactions, the level of memory depth is low; and so on.

Conclusion

While we cannot do full justice to several other important contributions, it cannot go unmentioned that Selten was an early contributor to the application of game theory to biology; see GAME THEORY AND BIOLOGY. Originally, the fundamental solution concept in evolutionary game theory, that of evolutionarily stable strategies (ESS), had been defined only for symmetric static games; Selten extended it to dynamic games and to asymmetric games, thus expanding the range of applicability. It is important to note that evolutionary game theory is not normative but descriptive, with equilibrium being thought of as the result of a dynamic process. Even in games arising in economics and social situations, equilibrium is increasingly viewed in this way.

Selten has often worked in areas away from the mainstream. As he has written: 'Since I am slow, I have to try to be early.' In the areas that he pioneered in the first part of his career, many others have meanwhile joined him.

ERIC VAN DAMME

See also **experimental economics; game theory and biology; learning and evolution in games: an overview; Nash equilibrium, refinements of; rationality, bounded; strategic and extensive form games.**

Selected works

1959. (With H. Sauermann.) Ein Oligopolexperiment. *Zeitschrift für die gesamte Staatswissenschaft* 115, 427–71.
1965. Spieltheoretische Behandlung eines Oligopolmodells mit Nachfrageträgheit. *Zeitschrift für die gesamte Staatswissenschaft* 121, 301–24 and 667–89.
1975. Reexamination of the perfectness concept for equilibrium points in extensive games. *International Journal of Game Theory* 4, 25–55.
1978. The chain store paradox. *Theory and Decision* 9, 127–59.

1988. (With J.C. Harsanyi.) *A General Theory of Equilibrium Selection in Games.* Cambridge, MA: MIT Press.
1993. In search of a better understanding of economic behavior. In *Makers of Modern Economics*, ed. A. Heertje. New York: Simon and Schuster.
1995. Autobiography. *Les Prix Nobel. The Nobel Prizes* 1994, ed. T. Frängsmyr. Stockholm: Nobel Foundation. Online. Available at http://nobelprize.org/nobel_prizes/economics/laureates/1994/selten-autobio.html, accessed 23 March 2007.
1998. Features of experimentally observed bounded rationality. *European Economic Review* 42, 413–36.
1999. *Game Theory and Economic Behavior. Selected Essays*, 2.vols. Cheltenham-Northampton: Edward Elgar.

Bibliography

Kohlberg, E. and Mertens, J.-F. 1986. On the strategic stability of equilibria, *Econometrica* 54, 1003–38.

semiparametric estimation

Introduction

Semiparametric estimation methods are used to obtain estimators of the parameters of interest – typically the coefficients of an underlying regression function – in an econometric model, without a complete parametric specification of the conditional distribution of the dependent variable given the explanatory variables (regressors). A structural econometric model relates an observable dependent variable y to some observable regressors x; some unknown parameters β, and some unobservable 'error term' ε, through some functional form $y = g(x, \beta, \varepsilon)$; in this context, a semiparametric estimation problem does not restrict the distribution of ε (given the regressors) to belong to a parametric family determined by a finite number of unknown parameters, but instead imposes only broad restrictions on the distribution of ε (for example, independence of ε and x, or symmetry of ε about zero given x) to obtain identification of β and construct consistent estimators of it.

Thus the term 'semiparametric estimation' is something of a misnomer; the same estimator can be considered a parametric, semiparametric or nonparametric estimator depending upon the restrictions imposed upon the economic model. For example, if a random sample of dependent variables $\{y_i\}$ and regressors $\{x_i\}$ are assumed to satisfy a linear regression model $y_i = x_i'\beta + \varepsilon_i$, the classical least squares estimator can be considered a 'parametric' estimator of the regression coefficient vector β if the error terms $\{\varepsilon_i\}$ are assumed to be normally distributed and independent of $\{x_i\}$. It could alternatively be considered a 'nonparametric estimator' of the best linear predictor coefficients $\beta = \left[E(x_i x_i')\right]^{-1} E(x_i, y_i)$ if only the weak condition $E(x_i \varepsilon_i)$ is imposed (implying that

β is a unique function of the joint distribution of the observations). And the least squares estimator would be 'semiparametric' under the intermediate restriction $E(\varepsilon_i|x_i) = 0$, which imposes a parametric (linear) form for the conditional mean $E(y_i|x_i) = x_i'\beta$ of the dependent variable but imposes no further restrictions on the conditional distribution. So the term 'semiparametric' is a more suitable adjective for models which are partly (but not completely) parametrically specified than it is for the estimators of those parameters.

Nevertheless, while most econometric estimation methods that do not explicitly specify the likelihood function of the observable data (for example, least squares, instrumental variables, and generalized method-of-moments estimators) could be considered semiparametric estimators, 'semiparametric' is sometimes used to refer to estimators of a finite number of parameters of interest (here, β) that involve explicit nonparametric estimators of unknown nuisance functions (for example, features of the distribution of the errors ε). Such 'semiparametric estimators' use nonparametric estimators of density or regression functions as inputs to second-stage estimators of regression coefficients or similar parameters. Occasionally terms like 'semi-nonparametric', 'distribution-free' and even 'nonparametric' have been used to describe such estimation methods, with the latter terms referring to the treatment of the error terms in an otherwise-parametric structural model.

The primary objective of semiparametric methods is to identify and consistently estimate the unknown parameter of interest β by determining which combinations of structural functions $g(x,\beta,\varepsilon)$ and weak restrictions on the distribution of the errors ε permit this. Given identification and consistent estimation, the next step in the statistical theory is determination of the speed with which the estimator $\hat{\beta}$ converges to its probability limit β. The rate of convergence for estimators for standard parametric problems is the square root of the sample size n, while nonparametric estimators of unknown density and regression functions (with continuously distributed regressors) generically converge at a slower rate; if a semiparametric estimator can be shown to converge at the parametric rate, that is, if it is 'root-n consistent', then its relative efficiency to a parametric estimator (for a correctly specified parametric model) will not tend to zero as n increases. For inference, it is also useful to demonstrate the asymptotic (that is, approximate) normality of the distribution of $\hat{\beta}$ in large samples, so that asymptotic confidence regions and hypothesis tests can be constructed using normal sampling theory. Finally, for problems where existence of root-n consistent, asymptotically normal semiparametric estimators can be shown, the question of efficient estimation arises. The solution to this question has two parts – determination of the efficiency bound for the semiparametric estimation problem and construction of a feasible estimator that attains that bound.

Econometric applications

In econometrics, most of the attention to semiparametric methods dates from the late 1970s and early 1980s, which saw the development of parametric models for discrete and *limited dependent variable* (*LDV*) models. Unlike the linear regression model, those models are not additive in the underlying error terms, so the validity (specifically, the consistency) of maximum likelihood and related estimation methods depends crucially on the assumed parametric form of the error distribution. As shown for particular examples by Arabmazar and Schmidt (1981; 1982) and Goldberger (1983), failure of the standard assumption of normally distributed error terms makes the corresponding likelihood-based estimators inconsistent. This is in contrast to the linear regression model, where the maximum likelihood (classical least squares) estimator is consistent under much weaker assumptions than normally (and identically) distributed errors.

Much of the early literature on semiparametric estimation concentrated on a particular limited dependent variable model, the *binary response model*, which arguably presents the most challenging setting for identification and estimation of the underlying regression coefficients. Early examples of semiparametric identification assumptions and estimation methods for this model give a flavour of the approaches used for other econometric models, among them the *censored regression* and *sample selection* models. The discussion here treats only selected assumptions and estimators for these models, and not their numerous variants; more complete surveys of semiparametric models and estimation methods are given by Manski (1989), Powell (1994), Newey (1994), and Pagan and Ullah (1999).

Semiparametric binary response models

The earliest semiparametric estimation methods in the econometrics literature on LDV models concerned the *binary response model*, in which the dependent variable y assumed the values zero or 1 depending upon the sign of some underlying latent (unobservable) dependent variable y^* which satisfies a linear regression model $y^* = x'\beta + \varepsilon$; that is,

$$y_i = 1\{x_i'\beta - \varepsilon_i > 0\},$$

where '$1\{A\}$' denotes the indicator function of the event A; that is, it is 1 if A occurs and is zero otherwise. For a parametric model, in which the errors ε_i are assumed to be independent of x_i and distributed with a known marginal cumulative distribution function $F(\varepsilon)$, the average log-likelihood function takes the form

$$L_n(\beta) = \frac{1}{n}\sum_{i=1}^{n}\left[y_i\ln F(x_i'\beta) + (1 - y_i)\ln(1 - F(x_i'\beta))\right]$$

for a random sample of size n, and consistency of the corresponding the maximum likelihood estimator $\hat{\beta}_{ML}$

requires correct specification of F unless the regressors satisfy certain restrictions (as discussed by Ruud, 1986). When F is unknown, a scale normalization on β is required, and a constant (intercept) term will not be identified no normalization on the location of ε is imposed.

Manski (1975; 1985) proposed a semiparametric alternative, termed the 'maximum score' estimator, which defined the estimator to maximize the number of correct matches of the value of y_i with an indicator function $1\{x_i'\beta > 0\}$ of the positivity of the regression function. That is, the maximum score estimator $\hat{\beta}_{MS}$ maximizes the average 'score' function

$$S_n(\beta) = \frac{1}{n}\sum_{i=1}^{n}[y_i \cdot 1\{x_i'\beta > 0\} + (1 - y_i) \cdot 1\{x_i'\beta \leq 0\}]$$

over β. Unlike the maximum likelihood estimator $\hat{\beta}_{ML}$, consistency of $\hat{\beta}_{MS}$ requires only that the median of the error terms was zero given the regressors, that is, the conditional cumulative $F(\varepsilon|x)$ of given $x_i = x$ had $F(\lambda|x) > 1/2$ when $\lambda > 0$, and $F(\lambda|x) < 1/2$ when $\lambda < 0$. However, the estimation approach is generally not root-n consistent (as shown by Chamberlain, 1986). A variant of the maximum score estimator, proposed by Horowitz (1992), essentially 'smoothed' the indicator functions for positivity of $x_i'\beta$ in the minimand $S_n(\beta)$ using a continuous approximation to it, similar to the smoothing used in nonparametric kernel estimators of regression and density functions. The rate of convergence of the resulting 'smoothed maximum score' estimator can be made arbitrarily close to the root-n rate if the distribution of the regressors is sufficiently smooth.

To obtain root-n consistent estimators of the unknown β, the assumption on the error term ε can be strengthened to independence of ε and x. Han (1987) proposed an alternative to the maximum score estimator, termed the 'maximum rank correlation' estimator, which compared the sign of the difference $y_i - y_j$ of the dependent variable to the corresponding difference $(x_i - x_j)'\beta$ in the regression functions across all distinct pairs of observations i and j. The estimator $\hat{\beta}_{MRC}$ maximizes

$$M_n(\beta) = \binom{n}{2}^{-1}\sum_{i=1}^{n-1}\sum_{j=i+1}^{n} sgn(y_i - y_j)$$
$$\cdot\, sgn((x_i - x_j)'\beta)$$

⇔ over β, where $sgn(u) \equiv 1\{u > 0\} - 1\{u < 0\}$. The rationale for this estimator is based upon the monotonicity of $\Pr\{y_i = 1|x\} = F(x_i'\beta)$ in $x_i'\beta$ so that, given $y_i \neq y_j$, $\Pr\{y_i > y_j|x_i, x_j\}$ exceeds $1/2$ when $x_i'\beta > x_j'\beta$. Han's article gave conditions under which $\hat{\beta}_{MRC}$ was shown to be consistent, and Sherman (1993) showed that this estimator was root-n consistent and asymptotically normal.

An alternative estimation approach for β under the assumption of independence of u and x combines estimation of the parameter vector β with nonparametric estimation of the unknown distribution function F. Cosslett (1983) proposed a 'nonparametric maximum likelihood' estimator $\hat{\beta}_{NPML}$ obtained by simultaneously maximizing the likelihood function $L_n(\beta) = L_n(\beta; F)$ over both β and F, where the latter function is restricted to be nondecreasing with values in the unit interval. While consistency of this estimator could be established, its rate of convergence could not. An alternative estimation method, proposed by Klein and Spady (1993), used kernel regression methods to estimate the unknown distribution function F in the likelihood function. The resulting estimator was shown to be root-n consistent and asymptotically normally distributed under additional regularity conditions; furthermore, the estimator was shown to achieve the semiparametric efficiency bound for this problem, that is, its asymptotic covariance matrix is the smallest possible among regular estimators of β which impose only the independence restriction between x and u.

Still other estimators for β when u and x are independent exploit the single index regression structure of this model, since the conditional expectation of y_i given x_i only depends upon the 'single index' $x_i'\beta$:

$$E[y_i|x_i] \equiv g(x_i) = F(x_i'\beta).$$

If the vector of regressors x_i is continuously distributed with joint density function $f_X(x)$ which is continuous for all x, Stoker (1986) noted that the vector of slope parameters β is proportional to the expectation of the derivative of $g(x)$,

$$E\left[\frac{\partial g(x_i)}{\partial x}\right] = E[F'(x_i'\beta)] \cdot \beta.$$

Using integration-by-parts, this 'average derivative' can in turn be expressed as the expected value of the product of $-y_i$ and the derivative of the logarithm of the density f_X of the regressors,

$$E\left[\frac{\partial g(x_i)}{\partial x}\right] = -E\left[y_i\frac{\partial \log[f_X(x_i)]}{\partial x}\right].$$

Härdle and Stoker (1989) proposed a semiparametric estimator of this representation of β (up to scale) using nonparametric (kernel) estimators of f_X and its gradient, while Powell, Stock and Stoker (1989) constructed a similar estimator of the 'density-weighted average derivative'

$$E\left[f_X(x_i)\frac{\partial g(x_i)}{\partial x}\right] = E[f_X(x_i)F'(x_i'\beta)] \cdot \beta$$
$$= -2E\left[y_i\frac{\partial f_X(x_i)}{\partial x}\right],$$

which is also proportional to β under the single index restriction.

Though the motivation given here was based upon the binary response model under independence of the errors and regressors, the average derivative and weighted average derivative estimators apply to other models with a single index structure, for example, any *transformation model* with

$$y_i = T(x_i'\beta + \varepsilon_i),$$

for T a nondegenerate function (possibly unknown) and with ε_i continuously distributed and independent of x_i. The same is true for the 'single index regression' estimator proposed by Ichimura (1993), defined to minimize

$$R_n(\beta) = \frac{1}{n} \sum_{i=1}^{n} \left(y_i - \hat{F}(x_i'\beta; \beta) \right)^2 t_n(x_i);$$

in this expression, $\hat{F}(u; \beta)$ represents a nonparametric regression estimator of $E[y_i | x_i'\beta = u]$ and $t_n(x_i)$ represents a 'trimming' term which is zero whenever x_i lies outside a set for which F is sufficiently precisely estimated. Unlike the average derivative $\hat{\beta}_{AD}$ and weighted average derivative $\hat{\beta}_{WAD}$ estimators, which require the regressors to be jointly continuously distributed, root-n consistency and asymptotic normality of the single index regression estimator $\hat{\beta}_{SIR}$ require only that $x_i'\beta$ has a continuous distribution, so that some of the regressors can be discrete. The criterion function $R_n(\beta)$ is the nonlinear least squares analogue of the maximand for the Klein and Spady (1993) estimator (which also involved a similar trimming term $t_n(x_i)$). The asymptotic covariance matrices for both estimators have the same general form as the corresponding nonlinear least squares and maximum likelihood estimators with F known, except for the replacement of the cross product of the regressors x_i with the cross product of $x_i - E[x_i | x_i'\beta]$, adjusting the asymptotic covariance matrices upward to account for the nonparametric estimation of the unknown function F.

The problem of consistent estimation of β in binary response models is compounded for *panel data* models with *fixed effects* (that is, individual-specific intercept terms), written as

$$y_{it} = 1\{x_{it}'\beta + \alpha_i - \varepsilon_{it} > 0\}$$

for individuals i ranging from 1 to n and time periods t from 1 to T. For this model, even if the distribution function F of the error terms ε_{it} is known, the maximum likelihood estimators of β and the fixed effects $\{\alpha_i\}$ will generally be inconsistent if the number of time periods T is fixed as N increases. A consistent semiparametric estimation strategy using a variant of the maximum rank correlation estimator was proposed by Manski (1987); for the special case $T=2$ (that is, two time periods), the estimator $\hat{\beta}_{BPD}$ can be defined as the

maximizer of the criterion

$$P_{(n)}(\beta) = \frac{1}{n} \sum_{i=1}^{n} sgn(y_{i2} - y_{i1}) \cdot sgn\left((x_{i2} - x_{i1})'\beta\right),$$

which is analogous to $M_n(\beta)$, except that the differencing is across time periods rather than across individuals. While consistency of $\hat{\beta}_{BPD}$ was established under weak conditions on the error terms, it is not possible to obtain a root-n consistent estimator unless the errors are logistic (Chamberlain, 1993) or other restrictive assumptions (for example, independence of the fixed effect α_i and the regressors x_{it}, or the conditions in Honoré and Lewbel, 2002; Lee, 1999) are imposed.

Other semiparametric econometric models

Many of the identifying assumptions imposed on semiparametric binary response models give identification and yield consistent estimators for other limited dependent variable models, though these models can sometimes be identified and consistently estimated using assumptions that are uninformative for binary response. Consider, for example, the censored regression ('Tobit') model, in which the dependent variable y_i satisfies a linear regression model if it is nonnegative, and is zero otherwise:

$$y_i = \max\{0, x_i'\beta + \varepsilon_i\}.$$

For this model, as for the binary response model, the dependent variable is a monotonic function of the error term ε_i; since monotone transformations by definition preserve orderings, the median (or any other percentile) of this monotonic transformation of ε_i is the monotonic transformation evaluated at the median. Thus the assumption that the errors ε_i have conditional median zero given x_i implies that the conditional median of y_i given x_i takes the form $\max\{0, x_i'\beta\}$, depending only on the unknown coefficients β and not on the shape of the distribution of ε_i. Using this fact, and the characterization of medians as minimizers of a least absolute deviations criterion, Powell (1984) proposed estimation of the unknown β vector by the minimizer $\hat{\beta}_{CLAD}$ of

$$Q_n(\beta) = \frac{1}{n} \sum_{i=1}^{n} |y_i - \max\{0, x_i'\beta\}|$$

for this model; it is analogous to the maximum score estimator $\hat{\beta}_{MS}$ for the binary response model, which can be defined as the minimizer of the sample average absolute deviation of y_i from its conditional median function $1\{x_i'\beta > 0\}$ for binary response with median zero errors. (The maximum rank correlation estimator $\hat{\beta}_{MRC}$ and binary panel data estimator $\hat{\beta}_{BPD}$ can also be expressed as solutions to least absolute deviations problems.) Unlike $\hat{\beta}_{MS}$, though, the censored median estimator $\hat{\beta}_{CLAD}$ is root-n consistent and asymptotically normally distributed under weak regularity conditions, without need for a scale normalization. An alternative estimator for this model, which involved a nonparametric estimator of the probability that y_i equals zero given x_i, was proposed by Buchinsky and Hahn (1998).

A stronger restriction on the error distribution is conditional symmetry about zero given the regressors; while this restriction is no more informative than the implied zero median restriction for binary response, it yields different identification approaches for censored regression. Specifically, the 'symmetrically censored' residual

$$u_i(\beta) \equiv \min\{y_i - x_i'\beta, x_i'\beta\}$$
$$= \min\{\max\{-x_i'\beta, \varepsilon_i\}, x_i'\beta\}$$

is an even function of ε_i when the regression function $x_i'\beta$ is positive, and thus is itself conditionally symmetric about zero. This implies a population moment restriction

$$0 = E\big[1\{x_i'\beta > 0\}\psi(\tilde{u}_i(\beta)) \cdot x_i\big],$$

for $\psi(u) = -\psi(-u)$ an odd function of its argument. Powell (1986) proposed a 'symmetrically censored least squares' estimator of β based upon this restriction with $\psi(u) = u$; like the censored median estimator $\hat{\beta}_{CLAD}$ – which exploits the same moment condition with $\psi(u) = sgn(u)$ – the estimator $\hat{\beta}_{SCLS}$ is root-n consistent and asymptotically normally distributed under weak assumptions. Neither estimator involves explicit nonparametric estimation of the error distribution, a feature shared by the maximum score estimator $\hat{\beta}_{MS}$ and its relatives $\hat{\beta}_{MRC}$ and $\hat{\beta}_{BPD}$ for binary response.

As for the binary response model or most limited dependent variable models, consistent estimation of slope coefficients using panel data with fixed effects is challenging, with maximum likelihood estimators for β being inconsistent when the number of time periods is fixed and the number estimated fixed effects increases. For the special case $T = 2$, writing

$$y_{it} = \max\{0, x_{it}'\beta + \alpha_i + \varepsilon_{it}\},$$

Honoré (1992) noted that the difference in 'identically trimmed' residuals

$$\tilde{u}_i(\beta) = \max\{-x_{i1}'\beta, y_{i2} - x_{i2}'\beta\}$$
$$\quad - \max\{-x_{i2}'\beta, y_{i1} - x_{i1}'\beta\}$$
$$= \max\{-x_{i1}'\beta, -x_{i2}'\beta, \alpha_i + \varepsilon_{i2}\}$$
$$\quad - \max\{-x_{i1}'\beta, -x_{i2}'\beta, \alpha_i + \varepsilon_{i2}\}$$

would be symmetrically distributed about zero if the error terms ε_{i1} and ε_{i2} were identically distributed given x_{i1} and x_{i2} and value of the fixed effect α_i. This implies population moment conditions of the form

$$0 = E[\psi(\tilde{u}_i(\beta)) \cdot (x_{2i} - x_{i1})],$$

again with $\psi(u)$ an odd function of its argument. Setting $\psi(u) = sgn(u)$ and $\psi(u) = u$ yields root-n consistent and asymptotically normal estimators which are similar to the censored least absolute deviations estimator $\hat{\beta}_{CLAD}$ and symmetrically-censored least squares estimator $\hat{\beta}_{SCLS}$, respectively.

Other estimation approaches for censored regression involve explicit nonparametric estimation of features of the distribution of the error terms, which is common for other semiparametric econometric models. One such model is the *semiparametric regression* (or *semilinear regression*) model, for which some regressors enter linearly while others enter nonparametrically. The model can be written algebraically as

$$y_i = x_i'\beta + \lambda(w_i) + \varepsilon_i$$
$$\equiv x_i'\beta + u_i,$$

where the error terms ε_i are restricted to satisfy $E[\varepsilon_i|x_i, w_i] = 0$, or, equivalently, $E[u_i|x_i, w_i] = E[u_i|w_i] \equiv \lambda(w_i)$; the regressors x_i and w_i thus enter parametrically (linearly) or nonparametrically in the conditional mean of y_i. Robinson (1988) exploited the fact that

$$y_i - E[y_i|w_i] = (x_i - E[x_i|w_i])'\beta + \varepsilon_i$$

to construct a root-n consistent, asymptotically-normal estimator of β by applying least squares estimation to this equation, replacing the unknown quantities $E[y_i|w_i]$ and $E[x_i|w_i]$ by nonparametric (kernel) estimators. For the parameters β to be identified for this model, the covariance matrix of the 'residual regressors' must be nonsingular, ruling out functional dependence of x_i on w_i.

Though the semilinear regression model is not itself a limited dependent variable model, it arises as a consequence of 'selectivity bias' in a bivariate limited dependent variable model, the *censored selection model*, in which a linear latent 'outcome' variable $y_i^* = x_i'\beta + \varepsilon_i$ is observed only if some related binary 'selection' variable d_i equals 1:

$$d_i = 1\{w_i'\delta - \eta_i > 0\},$$
$$y_i = d_i \cdot (x_i'\beta + \varepsilon_i),$$

where the regressors x_i and w_i are observed and the unobserved error terms η_i and ε_i need not be mutually independent. Heckman (1979) showed that, for the uncensored ($d_i = 1$) subsample from this model, the dependent variable satisfied a semilinear regression model, since

$$E[y_i|d_i = 1, x_i, w_i] = x_i'\beta + \lambda(w_i'\delta);$$

when the errors are jointly normal, as Heckman assumed, the function $\lambda(u)$ has a known parametric form, but is nonparametric if the error distribution is not in a parametric family. Cosslett (1991) developed a consistent two-step estimator for the regression parameters β in the outcome equation, computing a binary nonparametric maximum likelihood estimator $\hat{\delta}_{NPML}$ of δ in the first step and using a step-function approximation to $\lambda(u)$ in a least squares fit of the outcome equation for the uncensored observations. Ahn and Powell (1993) proposed a root-n consistent two-step estimator of β for the related semilinear model

$$E[y_i|d_i = 1, x_i, w_i] = x_i'\beta + \lambda^*(p(w_i)),$$

where the 'propensity score' $p(w_i) \equiv E[d_i|w_i]$ is first estimated by a nonparametric regression method; this semilinear model is implied by a generalization of the original censored selection model, replacing the linear form of the regression function $w_i'\delta$ in the selection equation with an unknown function of the regressors w_i.

Some variations of the censored selection model admit other semiparametric identification strategies. For example, if the selection equation is censored rather than binary, for example, if

$$d_i = \max\{0, w'\delta + \eta_i\},$$
$$y_i = 1\{d_i > 0\} \cdot (x_i'\beta + \varepsilon_i),$$

then Honoré, Kyriazidou and Udry (1997) construct a root-n consistent two-step estimator of β under the assumption that the errors η_i and ε_i are jointly symmetric about zero given the regressors w_i and x_i, using a symmetrically censored least squares estimator of δ in the first step and exploiting the symmetry of $y_i - x_i'\beta$ about zero given that $0 < d_i < 2w_i'\delta$ in the second step. In contrast, estimation of censored selection models for panel data with fixed effects is no less challenging than for binary panel data models; Kyriazidou (1997) proposes a consistent (but not root-n consistent) two-step estimator for the panel data selection model

$$d_{it} = 1\{w_{it}'\delta + v_i - \eta_{it} > 0\},$$
$$ty_{it} = d_{it} \cdot (x_{it}'\beta + \alpha_i + \varepsilon_{it}),$$

using Manski's (1987) binary panel data estimator to estimate δ in the first step and a semilinear regression estimator similar to the Ahn and Powell (1993) approach in the second step.

JAMES L. POWELL

See also **nonlinear panel data models; nonparametric structural models; quantile regression; robust estimators in econometrics; selection bias and self-selection; Tobit model.**

Bibliography

Ahn, H. and Powell, J.L. 1993. Semiparametric estimation of censored selection models with a nonparametric selection mechanism. *Journal of Econometrics* 58, 3–29.

Arabmazar, A. and Schmidt, P. 1981. Further evidence on the robustness of the Tobit estimator to heteroscedasticity. *Journal of Econometrics* 17, 253–8.

Arabmazar, A. and Schmidt, P. 1982. An investigation of the robustness of the Tobit estimator to non-normality. *Econometrica* 50, 1055–63.

Buchinsky, M. and Hahn, J. 1998. An alternative estimator for the censored quantile regression model. *Econometrica* 66, 653–72.

Chamberlain, G. 1986. Asymptotic efficiency in semiparametric models with censoring. *Journal of Econometrics* 32, 189–218.

Chamberlain, G. 1993. Feedback in panel data models. Unpublished manuscript, Department of Economics, Harvard University.

Cosslett, S.R. 1983. Distribution-free maximum likelihood estimator of the binary choice model. *Econometrica* 51, 765–82.

Cosslett, S.R. 1991. Distribution-free estimator of a regression model with sample selectivity. In *Nonparametric and Semiparametric Methods in Econometrics and Statistics*, ed. W.A. Barnett, J.L. Powell and G. Tauchen. Cambridge: Cambridge University Press.

Goldberger, A.S. 1983. Abnormal selection bias. In *Studies in Econometrics, Time Series, and Multivariate Statistics*, ed. S. Karlin, T. Amemiya and L. Goodman. New York: Academic Press.

Han, A.K. 1987. Non-parametric analysis of a generalized regression model: the maximum rank correlation estimator. *Journal of Econometrics* 35, 303–16.

Härdle, W. and Stoker, T.M. 1989. Investigating smooth multiple regression by the method of average derivatives. *Journal of the American Statistical Association* 84, 986–95.

Heckman, J.J. 1979. Sample selection bias as a specification error. *Econometrica* 47, 153–62.

Honoré, B.E. 1992. Trimmed LAD and least squares estimation of truncated and censored regression models with fixed effects. *Econometrica* 60, 533–65.

Honoré, B.E., Kyriazidou, E. and Udry, C. 1997. Estimation of type 3 Tobit models using symmetric trimming and pairwise comparisons. *Journal of Econometrics* 76, 107–28.

Honoré, B.E. and Lewbel, A. 2002. Semiparametric binary choice panel data models without strictly exogenous regressors. *Econometrica* 70, 2053–63.

Horowitz, J.L. 1992. A smoothed maximum score estimator for the binary response model. *Econometrica* 60, 505–31.

Ichimura, H. 1993. Semiparametric least squares (SLS) and weighted SLS estimation of single index models. *Journal of Econometrics* 58, 71–120.

Klein, R.W. and Spady, R.H. 1993. An efficient semiparametric estimator for discrete choice models. *Econometrica* 61, 387–421.

Kyriazidou, E. 1997. Estimation of a panel data sample selection model. *Econometrica* 65, 1335–64.

Lee, M-J. 1999. A root-n consistent semiparametric estimator for related effect binary response panel data. *Econometrica* 60, 533–65.

Manski, C.F. 1975. Maximum score estimation of the stochastic utility model of choice. *Journal of Econometrics* 3, 205–28.

Manski, C.F. 1985. Semiparametric analysis of discrete response, asymptotic properties of the maximum score estimator. *Journal of Econometrics* 27, 205–28.

Manski, C.F. 1987. Semiparametric analysis of random effects linear models from binary panel data. *Econometrica* 55, 357–62.

Manski, C.F. 1989. *Analog Estimation Methods in Econometrics*. New York: Chapman & Hall.

Newey, W.K. 1994. The asymptotic variance of semiparametric estimators. *Econometrica* 62, 1349–82.

Pagan, A.P. and Ullah, A. 1999. *Nonparametric Econometrics*. Cambridge: Cambridge University Press.

Powell, J.L. 1984. Least absolute deviations estimation for the censored regression model. *Journal of Econometrics* 25, 303–25.

Powell, J.L. 1986. Symmetrically trimmed least squares estimation of Tobit models. *Econometrica* 54, 1435–60.

Powell, J.L. 1994. Estimation of semiparametric models. In *Handbook of Econometrics*, vol. 4, ed. R.F. Engle, and D.L. McFadden. Amsterdam: North-Holland.

Powell, J.L., Stock, J.H. and Stoker, T.M. 1989. Semiparametric estimation of weighted average derivatives. *Econometrica* 57, 1403–30.

Robinson, P. 1988. Root-N-consistent semiparametric regression. *Econometrica* 56, 931–54.

Ruud, P. 1986. Consistent estimation of limited dependent variable models despite misspecification of distribution. *Journal of Econometrics* 32, 157–87.

Sherman, R.P. 1993. The limiting distribution of the maximum rank correlation estimator. *Econometrica* 61, 123–37.

Stoker, T.M. 1986. Consistent estimation of scaled coefficients. *Econometrica* 54, 1461–81.

Sen, Amartya (born 1933)

Amartya Sen was born in Santiniketan, Bengal in 1933 on the campus of Rabindranath Tagore's Viswa Bharati (both a school and a college) where his maternal grandfather taught Sanskrit and where both his mother and he had been students. At Santiniketan he concentrated on Sanskrit, mathematics and physics, before studying economics (with mathematics minor) at Presidency College, Calcutta and then at Trinity College, Cambridge. His Ph.D. thesis at Cambridge University on the 'choice of techniques' was the basis of his election to a competitive Prize Fellowship at Trinity, and this allowed him time to pursue his interests in logic, epistemology, and moral and political philosophy – in addition to economics. Some of his later work in social choice theory and welfare economics would draw on the intersection of these interests, but he has also made major contributions to many of these fields separately.

Sen has taught at several universities: Trinity College, Cambridge; Jadavpur University in Calcutta; the Delhi School of Economics; the London School of Economics; Oxford, where he was Drummond Professor of Political Economy and Fellow of All Souls College; and Harvard, where he is Lamont University Professor, Professor of Economics and Philosophy, and Senior Fellow of the Society of Fellows. He has also held visiting appointments at MIT, Stanford, Berkeley and Cornell. During 1998–2003 Sen was back at Cambridge – as Master of Trinity College. (For further details of his personal and academic life, see Sen, 1999a.) He is widely regarded as one of the finest teachers anywhere, with his lectures and classes always overflowing or oversubscribed. Amartya Sen has been president of the Econometric Society, the American Economic Association, the Indian Economic Association, and the International Economic Association – and he holds more than 90 honorary doctorates. He has won innumerable awards including the Agnelli International Prize in Ethics, the Feinstein World Hunger Award, the Nobel Memorial Prize in Economics, the Bharat Ratna (highest civilian award in India), the Eisenhower Medal, Companion of Honour, UK, and the George C. Marshall Award. His research output is prodigious, and as of 2007 he had published more than 25 books (his works have been translated into more than 30 languages) and about 400 articles. Details of his publications (as well as honorary fellowships and awards) may be found in his CV, at http://www.economics.harvard.edu/faculty/sen/cv.pdf. A list of his publications up to 1998 is available in the *Scandinavian Journal of Economics* (1999).

Sen has made fundamental contributions to several areas in economics including social choice theory, economic measurement, and welfare economics. He has also worked in many other areas, such as axiomatic choice theory, rationality and economic behaviour, development economics, gender and feminist economics, famines and hunger, economic methodology, project evaluation and cost–benefit analysis, and so on. In this brief review of his contributions to economics, I will concentrate on the first three areas, and will also have to overlook his major work in philosophy, history, and on India's economy, society, culture and politics (his CV lists 20 separate areas, of which this review considers just *three*).

As one reviewer put it, Sen is as comfortable writing in the *Journal of Philosophy* or *Philosophy and Public Affairs* as he is in *Econometrica* or the *Economic Journal*. His writings outside economics address some of the most important issues and theories in ethics, justice, and legal and moral philosophy.

In the space available for this review, some areas within economics where Sen's research has had a major impact can only be flagged. These, in general, are more accessible and have been the subject of several other accounts of his work. Thus, in development economics Sen's research began with his Ph.D. thesis, published as *Choice of Techniques* (1968, 3rd edition). It made rigorous a series of arguments that had often been advanced casually concerning the need for 'appropriate' (labour-intensive) technology in poor and populous countries. Further work developed the idea of 'labour surplus' in a framework that brought the division of labour within the household to economists' attention. The game-theoretic concepts used in this work were also applied to explain low rates of saving in developing countries. He has also studied important questions of policy, with research on cost–benefit analysis and accounting prices, including 'shadow wages'. A summary of his contributions to employment and technological choice in different institutional and labour market contexts may be found in his monograph *Employment, Technology and Development* (1975). A more extensive collection of his writings on development economics, covering many other topics, is available in *Resources, Values and Development* (1984).

Another major contribution was his work on the understanding of famine, drawn together in his book *Poverty and Famines: An Essay on Entitlement and Deprivation* (1981a). He introduced the notion that famines may be due to a decline not in the overall supply of food but in the purchasing power, more generally 'entitlements', of vulnerable people – in particular, the poor. The real incomes of a section of the population can decline radically for a variety of reasons (unemployment, relative price changes, production failure, and so on) which can lead to a collapse of their command over food even without any change in *overall* food production and supply. This analysis can provide both a better explanation of famines as well as a more effective approach to the remedying of starvation and hunger – through the regeneration of 'food entitlements' (see Drèze and Sen, 1989).

In the area of gender and family economics, Sen (1990) has highlighted the deprivation suffered by females both inside and outside the household. In traditionally unequal societies, gender bias in nutrition, health care and medical attention – as well as women often having to work harder than men – has led Sen to broaden the information base in assessing relative female disadvantage. He has used mortality and survival information to draw attention to the coarsest aspects of gender-related inequality, and has estimated the number of 'missing women' by comparing the female–male ratio with what would be expected in the absence of gender bias in mortality (including female foeticide) (see Sen, 1992b; 2003).

Sen has been instrumental in the recharacterization of development away from the metric of gross national product per capita and its growth. In his book *Development as Freedom* (1999b), he argues that development can be seen as a process of expanding the real freedoms that people enjoy. Freedoms depend not only on individual incomes but on other determinants such as social and economic arrangements, and political and civil

rights. He argues that substantive freedoms are not only the primary ends of development but also among its principal means. His related work on 'capabilities' has directly influenced the assessment of development by the United Nations Development Programme in its *Human Development Reports*.

Sen's work in economics combines foundational and theoretical originality, and the willingness to reconsider basic assumptions. His approach is unusual for its breadth of concern coupled with an uncompromising rigour of analysis. It is important to stress the range of his scholarship in an age when the profession is becoming increasingly specialized. For all its diversity there is considerable harmony in his work, with research in one area informing that in another. In this academic biography I shall review his contributions in the areas cited by the Royal Swedish Academy of Sciences in its press release on 14 October 1998 for his award of the 1998 Bank of Sweden Prize in Economic Sciences in Memory of Alfred Nobel. Under the general heading of 'welfare economics', the citation lists three specific areas: social choice, welfare distributions, and poverty.

Contributions to social choice theory

Sen's book *Collective Choice and Social Welfare* (1970a) is a classic work in the theory of social choice. It is a magisterial study that contains profound and original theorems, including the famous 'liberal paradox' that has spawned a secondary literature of hundreds of articles in learned journals. His analysis of the causes of the various paradoxes of collective choice, including voting and other decision-making procedures, has been central in the economics, philosophy and political theory literatures. His deep insights into Arrow's impossibility theorem, particularly concerning its informational base, have greatly enhanced our understanding of the entire subject of social choice theory. Sen's seminal papers in social choice, including many written after 1970, are reprinted in his book *Choice, Welfare and Measurement* (1982), and later papers are reprinted in his collection *Rationality and Freedom* (2002).

As Arrow (1999, p. 163) states,

> ...we cannot do justice to Sen's work in social welfare on the basis of one or two seminal papers, although there have been several such, as I will point out. Rather it is the work as a whole and the way the various parts interplay that must be understood to see the importance of Sen's contribution. His exploration of the notions of social welfare takes place at every level of analysis, formal-mathematical, conceptual, and empirical. It is by far the most comprehensive study of its kind, drawing on profound understanding of both economics and moral philosophy.

Some formal notation is needed to help understand Sen's contributions in social choice theory. Let X be the set of social states, and n the number of individuals in the society. Each individual i ($i=1, 2, \ldots, n$) has a weak preference relation R_i over the set X, where R_i is assumed to be an *ordering*, that is, reflexive, complete and transitive. P_i denotes the corresponding strict preference relation. The social weak preference relation is denoted by R, with P its asymmetric factor.

An Arrow *social welfare function* (SWF) is a functional relation that specifies one social ordering for any given n-tuple of individual orderings $\{R_i\}$, one ordering for each person:

$$R = f(\{R_i\}) = f(R_1, R_2, \ldots, R_n).$$

Arrow was concerned with the case in which the value of the function $f(\cdot)$, viz. R, is required to be an ordering, that is, reflexive, complete and transitive.

Four conditions are imposed on the function $f(\cdot)$.

(U) (*unrestricted or universal domain*). The domain of $f(\cdot)$ includes all possible n-tuples of individual orderings of X.
(P) (*weak Pareto principle*). For any pair of social states $\{x, y\}$, if for all i: $xP_i y$, then xPy.
(I) (*pairwise independence of irrelevant alternatives*). The social preference between two alternatives x and y depends only on the individual preferences between x and y (and not on individual preferences over other, 'irrelevant', alternatives).
(ND) (*non-dictatorship*). There is no individual i such that for all preference n-tuples in the domain of $f(\cdot)$, for each ordered pair (x, y),

$$xP_i y \rightarrow xPy.$$

Denoting the set of individuals in the society as H, Arrow's impossibility theorem can be stated as follows.

Theorem (Arrow) If H is finite and the number of alternatives is at least three, there is no SWF satisfying conditions (U), (P), (I) and (ND).

In other words, a SWF satisfying conditions (U), (P) and (I) must be dictatorial.

This result has turned out to be very robust, and nobody has done more to enhance our understanding of it than Sen. He suggested weakening the requirement of transitivity of the social preference relation R in Arrow's theorem as a way out of the dilemma (Sen, 1969, theorem V). If we demand only *quasi-transitivity* of R, that is, transitivity of the strict preference relation P, it turns out to generate not dictatorship but (still) an oligarchy. Combined with conditions (U), (P) and (I), a reflexive, complete and quasi-transitive social preference relation yields an oligarchy (Sen, 1970a, attributes this result to Allan Gibbard in an unpublished paper). An oligarchy is a group of individuals G who are together decisive (whenever $xP_i y$ for all i in G then xPy) and every member of G has a veto (for any i, $xP_i y$ implies *not* yPx). Thus the replacement of transitivity by quasi-transitivity translates the possibility of dictatorship to an oligarchy with veto powers (that is, a group that is decisive with each person in the group having a veto).

One extreme case of an oligarchy is a dictator (one-person oligarchy). The other extreme makes the oligarchy group include every individual in the society. In that case, the fact that all individuals taken together happen to be decisive is not remarkable; it follows immediately from the Pareto principle. But it also gives every member of the society 'veto' power in the sense that if anyone prefers any x to any y, this precludes the possibility of y being socially preferred to x.

The oligarchy group will consist of everyone in the society if the condition of anonymity is imposed, that is, the social ranking does not depend on who holds which preferences.

(A) (*anonymity*). If $\{R_i\}$ is a permutation of $\{R_i'\}$, then $f(\{R_i\}) = f(\{R_i'\})$.

Anonymity rules out any proper subset of H being oligarchic (including dictatorship), and leaves the whole of H as the oligarchy group. Condition (A) together with (U), (P) and (I) are necessary and sufficient for a quasi-transitive R to be the 'weak Pareto-extension rule' where x is socially preferred to y if and only if everyone prefers x to y (Sen, 1970a, theorem 5*3). Hence x and y are socially indifferent if they are weak Pareto-incomparable, which covers a much wider variety of cases than their being Pareto-indifferent. The Pareto-extension rule thus gives everyone a 'veto'.

The consequences of further weakening of quasi-transitivity of the social preference relation R to *acyclicity*, admitting no cycle of strict preference, have also been explored as a way out of the Arrow problem. Acyclicity is sufficient if we are interested simply in choosing the best element(s) of a subset S of X – called the choice set $C(S)$. Sen shows (1970a, lemma 1*1) that if R is reflexive and complete, and X is finite, then a necessary and sufficient condition for $C(S)$ to be non-empty for every non-empty subset S of X is that R be *acyclical*.

A *social decision function* (SDF) is defined as a function $f(\cdot)$ that maps n-tuples of individual orderings $\{R_i\}$, one ordering for each person, into reflexive, complete and *acyclic* social preference relations R. It turns out that, even without quasi-transitivity, the 'veto' result continues to be obtained for an SDF – by supplementing the weaker demand of acyclicity with some other conditions (a variety of such results is discussed in Sen, 1977a). While the existence of vetoers may be less unattractive than that of a dictator, it is, according to Sen (1986, p. 1085), 'unappetizing enough not to provide a grand resolution of the Arrow problem'.

As Sen (1993, p. 507) states,

> ... these and other weakenings [of the 'consistency' requirement of the social preference relation R] cannot avoid the 'spirit' of Arrow's impossibility theorem. The impossibility can be regenerated through balancing the weakening of 'social preference' by corresponding strengthenings, which are plausible enough, of other conditions, in particular, the non-dictatorship requirement (avoiding not just a dictator, but also an oligarchy, or a vetoer, or a partial vetoer, and so on).

There is a detailed review and assessment of such theorems in Sen (1986).

The discussion so far has imposed some sort of consistency or rationality condition on the social preference relation R, for example, transitivity, quasi-transitivity or acyclicity. However, as Sen (1993; 1995) notes, the idea of consistency or rationality applied to 'social preferences' is more difficult to defend than in the case of an individual's preferences. The question thus arises whether dropping the requirement that social choice be based on a (transitive) binary preference relation negates Arrow's theorem reformulated in 'choice-functional' terms. A *functional collective choice rule* (FCCR) makes the value of the function $f(\{R_i\})$ not a social preference relation R but a *choice function* $C(S)$, which specifies for every non-empty set S of social states a non-empty subset $C(S)$ of states chosen from S:

$$C(S) = f(\{R_i\}).$$

For a FCCR the Arrow conditions (U), (P), (I) and (ND) can be translated in different ways, but the typical translation has taken the form of restricting choices over pairs only. The consistency conditions for the choice function $C(\cdot)$, which are analogous to rationality conditions for the social preference relation R, can be classified or factorized into requirements of two essentially different types – *contraction* consistency and *expansion* consistency. (Sen, 1971, has shown that standard contraction consistency – Property α – and standard expansion consistency – Property γ – are necessary and sufficient for the choice function to be binary – in that its informational content can be exactly captured by a binary relation defined on X.) A series of choice-functional impossibility theorems has been established using alternative forms of such consistency conditions, together with (U), (P), (I) and (ND) translated into choice-functional terms. As in the social-relational case, the conditions imposed on the FCCR (apart from non-dictatorship) can lead it to be dictatorial, oligarchic, or have a vetoer (the results are reviewed in Sen, 1977a; 1986).

In the choice-functional context the robustness of Arrow's impossibility is demonstrated in a remarkable theorem by Sen (1993). He shows (1993, theorem 3) that, without imposing *any* inter-menu consistency condition on the choice function, the Arrow result survives in the form of a dictator who is decisive in rejecting any dispreferred alternative from a given set S, irrespective of the preferences of others. The condition (P) is translated in the choice context as the rejection of Pareto-inferior states; (U) as unrestricted domain applying to the FCCR; (I) as the rejection decisiveness of a group of individuals over any ordered pair (x, y) in S not being affected by individual preferences over pairs other than (x, y), that is, over 'irrelevant' alternatives; and (ND) as there being no individual who is rejection-decisive over a given set S of social states.

Since Sen's proof invokes only one set (or menu) of social states S, no inter-menu consistency condition for social choice is considered. By the same token, the dictator that emerges from combining the modified conditions (U), (P) and (I) with the FCCR is a rejection dictator for the given set S – and not necessarily for another set of social states, which could have a different individual as rejection dictator. The non-dictatorship condition in Sen's theorem is thus stronger than requiring the absence of a single individual who is a rejection dictator across all sets of social states. His theorem identifies exactly how far it is possible to go down the route of weakening and finally eliminating any consistency requirement for the social choice mechanism without escaping Arrow's problem. With no consistency condition on social choice, it is still not possible to avoid an individual who dictates the rejection of every state in a given set S, no matter what the other individuals prefer. Sen's theorem demonstrates just how robust is Arrow's impossibility. The source of Arrow's problem seems to lie elsewhere than in the consistency conditions imposed on the choice function $C(\cdot)$ or the social preference relation R (see above).

Interpersonal comparisons and social welfare functionals
Sen has argued that the real source of the Arrow impossibility problem is 'the tension between the informational eschewal implicitly imposed by Arrow's set of axioms and the demands of discriminating social choice also entailed by the same axioms. The positive possibilities lie, therefore, in making more room for use of information (both utility and non-utility information) in social choice' (1993, p. 514, n. 38). Thus, in the original (social-relational) form of Arrow's theorem, the domain of the social welfare function (SWF) consists of n-tuples of individual *orderings* $\{R_i\}$ only, with no information on valuation of intensity of preferences, or interpersonal comparisons of utility. Sen (1970a) formulated a framework to enrich the informational base to allow different assumptions of measurability and comparability of individual utilities. This has opened up an entire new field with rich and important results.

The informational base of the Arrovian approach can be improved by making the social preference relation R not a function of the n-tuple of individual orderings $\{R_i\}$, but the n-tuple of individual utility functions $\{u_i(\cdot)\}$.

For example, the classical utilitarian characterization of social welfare is a special case of such a form. However, the difficulty with this way of formalizing the functional relations arises from the fact that, given the measurability and comparability assumptions of individual utilities, the utility function has to be represented not by *one* n-tuple of individual utilities, but by a *set* of n-tuples of individual utilities which are *informationally identical* (for the given assumptions of measurability and interpersonal comparability). For example, utility functions can be nominally varied through alternative representations without involving any 'real' change (for instance, doubling all the utility numbers), and such informationally equivalent representations should lead to the same social ordering R. This problem is met in Sen's (1970a) approach of *social welfare functionals* through imposing a class of invariance requirements, which demand the same outcome for each of the n-tuples of utility functions in the class that could reflect the same underlying reality – given the chosen characterization of measurability and comparability of utilities.

A social welfare functional (SWFL) specifies exactly one social ordering R over the set X of social states for any given n-tuple $\{u_i(\cdot)\}$ of individual utility functions defined over X, that is, $R=F(\{u_i\})$. The invariance requirement takes the general form of specifying that, for any two n-tuples of utility functions $\{u_i\}$, $\{u_i'\}$ that are related in a particular way (reflecting the assumptions that are made of measurability and interpersonal comparability of individual utilities), the social ordering generated must be the same, that is, $F(\{u_i\})=F(\{u_i'\})$. Thus Arrow's SWF corresponds to a SWFL where the invariance class permits monotonic increasing transformations of the individual utility functions u_i; these transformations can be *different* for each i – thus incorporating ordinal non-comparability of the utility functions $\{u_i\}$ which represent the orderings $\{R_i\}$ in Arrow's SWF.

Using this SWFL framework, Sen (1970a, theorem 7*1) showed that utilitarianism requires cardinal 'unit' but not 'level' comparability (that is, comparability of utility differences but not of levels). He also showed that cardinality of individual utilities without interpersonal comparability does not avoid Arrow's impossibility (1970a, theorem 8*2). He noted that Rawls's maximin criterion only required ordinal interpersonal comparisons 'to discover who is the worst-off person', and then the 'minimal element in the set of individual welfares is maximized' (Sen, 1970a, pp. 136–7). Many other results were presented by him using the SWFL framework, including several concerning 'partial comparability' (Sen, 1970c). The field he opened up has generated a great number of important results, including axiomatic characterizations of utilitarianism, and the maximin (or its lexicographic version, leximin) criterion as a positional dictatorship rule which requires only 'level' comparability of ordinal utility functions. Hammond (1976), d'Aspremont and Gevers (1977), Maskin (1978) and

Roberts (1980a), among others, have derived many of the results in this area (individual references for other contributions can be found in Sen, 1986). Roberts (1980b) has presented a comprehensive characterization of the welfare functions that result from assuming different degrees of measurability and comparability of individual utilities, including cardinal full comparability and ratio-scale full comparability.

The impossibility of the Paretian liberal

Apart from Arrow's impossibility theorem, perhaps the most well-known result in social choice theory is Sen's (1970b) impossibility of a Paretian liberal. Sen (1979, p. 539) has argued that the Arrow impossibility can be 'seen as resulting from combining a version of welfarism ruling out the use of non-utility information with making the utility information remarkably poor (particularly in ruling out interpersonal utility comparisons)'. Welfarism entails that two social states are ranked exclusively on the basis of the individual utilities in the respective states, with no regard to the non-utility features of the states.

It turns out that the impossibility of the Paretian liberal is closely related to the difficulties with welfarism. Paretianism can be seen as essentially a weak form of welfarism, which makes non-utility information redundant in the special case where everyone's utility rankings of two social states coincide. Considerations of liberty require the use of non-utility information, viz. the specification of an individual's 'protected sphere' with the social ranking respecting the individual's ranking of states that fall within it. Sen (1970a, theorem 6*1) shows that this use of non-utility information goes not only against welfarism but it can go even against Paretianism.

Sen introduces a condition called 'minimal liberalism' as follows.

(L) (*minimal liberalism*). There are at least two individuals such that for each such individual i there is a personal domain with at least one pair of social states $\{x, y\}$ such that:

$$xP_i y \rightarrow xPy, \quad \text{and} \quad yP_i x \rightarrow yPx.$$

Theorem (Sen) There is no social decision function $f(\cdot)$ that satisfies (U), (P) and (L).

Arrow (1999) was greatly impressed by this result, stating that it 'brilliantly combines simplicity and depth'. He added that he found it

> very surprising [because] both the Pareto judgment and the idea that each individual has some private domain of choice, even if others would make different choices over that domain, are hard to deny; and independence, which on the whole is central to most variations of the Impossibility Theorem, is not assumed here. The paradox arises because 'nosy' preferences of others about choices that are in an individual's domain of private

choice enter into the Pareto judgment. The result is not only surprising analytically but also addresses profound ethical questions on the relation between even the vestigial remnant of utilitarianism contained in the Pareto principle and the existence of individual 'rights', a scope (however small) over which the individual has complete control. (Arrow, 1999, pp. 165–6)

Sen's 'impossibility of the Paretian liberal' is rightly seen as a seminal result in social choice theory. (The choice-functional version of the 'impossibility of the Paretian liberal' was originally presented in Sen, 1970a, pp. 81–2, with conditions (U), (P) and (L) translated for a FCCR and with *no* consistency condition imposed on the choice function. A formal proof is also given in Sen, 1993, theorem 2.)

Contributions to economic measurement
Inequality and welfare
The contributions of Sen to the literature on inequality and welfare measurement go back to his classic monograph *On Economic Inequality* (1973a) (*OEI-1973*). This book has been re-issued after a quarter century with a substantial annexe written jointly with James Foster. In the annexe there is a review of the themes in Sen (1973a) that have motivated a great deal of subsequent work in the area and also an acknowledgement that a 'significant part of *OEI-1973* was, in fact, Atkinson-inspired' (Sen and Foster, 1997, p. 114). Indeed, the literature on inequality measurement has been much influenced by Atkinson's celebrated paper on the ranking of income distributions (see also Kolm, 1969). Atkinson (1970) demonstrated the equivalence of three different rankings of income distributions – according to Lorenz dominance, the principle of transfers, and social welfare for all additively separable concave welfare functions.

Formally, let the vector $\mathbf{y} = (y_1, y_2, \ldots, y_n)$ denote an income distribution among n individuals with person i receiving income y_i, for $i = 1, 2, \ldots, n$. Let the vector $\mathbf{x} = (x_1, x_2, \ldots, x_n)$ denote a different income distribution among the n individuals with the same total income. Atkinson's theorem then shows that the following statements are equivalent:

(i) $\mathbf{x}$ Lorenz dominates $\mathbf{y}$;
(ii) $\mathbf{x}$ can be obtained from $\mathbf{y}$ by a sequence of income transfers from richer to poorer individuals;
(iii) $\sum_i U(x_i) \geq \sum_i U(y_i)$ for any non-decreasing concave function $U(\cdot)$, that is $\mathbf{x}$ yields as much welfare as $\mathbf{y}$ for any additively separable, symmetric, non-decreasing, concave social welfare function $(\sum_i U(y_i))$.

In fact, a slightly stronger theorem than this can be proved by adopting a weaker criterion for the welfare ranking (iii). The welfare function $\sum_i U(y_i)$ can be replaced by a symmetric, non-decreasing, quasi-concave function of individual incomes $W(y_1, y_2, \ldots, y_n)$ (Dasgupta, Sen and

Starrett, 1973). The quasi-concavity restriction on the welfare function can be weakened still further. For the theorem to go through, it is clear that the *weakest* requirement on the function is that welfare does *not decrease* by a transfer of income from a richer to a poorer individual. Such a function may be called egalitarian or, as in the literature, S-concave – a property that is weaker than quasi-concavity (Rothschild and Stiglitz, 1973). Thus we consider two further welfare rankings of the income distributions $\mathbf{x}$ and $\mathbf{y}$:

(iv) $\mathbf{x}$ yields as much welfare as $\mathbf{y}$ for all symmetric, non-decreasing, quasi-concave welfare functions;
(v) $\mathbf{x}$ yields as much welfare as $\mathbf{y}$ for all symmetric, non-decreasing, egalitarian or S-concave welfare functions.

As shown in Anand (1983, p. 338), the stronger theorem establishing equivalence of (iii) with (iv) and (v) follows as an immediate corollary to Atkinson's theorem. Since the class of additively separable concave functions is contained in the class of quasi-concave functions, which in turn is contained in the class of egalitarian or S-concave functions, it follows that (v) implies (iv), and (iv) implies (iii). But from Atkinson's theorem, (iii) implies (ii), and by the very definition of an egalitarian welfare function, (ii) implies (v); therefore, (iii) implies (v). Hence the chain of implications is complete, with $(v) \rightarrow (iv) \rightarrow (iii) \rightarrow (v)$, and there is no information loss in ranking by welfare functions from the more restrictive additively separable class. It is enough to check that distribution $\mathbf{x}$ yields as much welfare as distribution $\mathbf{y}$ for all members of the class of additively separable concave functions, and it will automatically do so for all members of the more general class of S-concave functions too. Hence, Atkinson's theorem in fact establishes the equivalence of (i), (ii) and (v).

Sen's own wide-ranging contributions to the measurement of inequality and welfare begin with *On Economic Inequality* (1973a) and are followed inter alia by Sen (1974; 1976a; 1978; 1992a). In *OEI-1973* he argues for inequality to be seen as a 'quasi-ordering' (or partial ordering) without insisting that it must be a complete ordering; he has in fact defended such assertive incompleteness in many different evaluative contexts. Sen (1973a, pp. 72–4) shows that when there are multiple criteria each of which yields a complete ordering (for example, welfare functions or inequality indices in a given class), then their intersection generates a partial ordering. Thus the Lorenz ranking of distributions $\mathbf{x}$ and $\mathbf{y}$ (with the same mean income) is their intersection quasi-ordering generated by all welfare functions in the class of additively separable concave functions. Sen initially used the intersection (or unanimity) quasi-ordering approach to measure relative inequality in a positive or descriptive sense by employing different statistical measures of inequality, which avoids 'exclusive reliance on any particular measure and on the complete ordering generated by it' (1973a, p. 72). Consider the class of inequality indices that satisfy mean and population-size independence, and the Pigou–Dalton condition (viz. a transfer of income from a richer to a poorer person reduces the value of the inequality index). Then all indices from this class will show less inequality for a Lorenz dominant distribution (Anand, 1983, pp. 339–40), and the intersection quasi-ordering generated by all measures in this class is the Lorenz partial ordering (Sen and Foster, 1997, pp. 142–5).

When Lorenz curves cross, further structure has to be imposed on our social values (or relative inequality indices) – in terms of agreement about sub-classes of welfare functions (inequality indices) or choice of a specific one. Sen provides an axiomatic derivation for a welfare function by drawing on his definition of the Gini coefficient as an affine transformation of the rank-order weighted sum of individual income levels. With individuals labelled in non-descending order of income so that $y_1 \leq y_2 \leq \cdots \leq y_n$, and μ their mean income, Sen (1973a, p. 31) expresses the Gini coefficient as:

$$
\begin{aligned}
G &= (n+1)/n - (2/n^2\mu) \\
&\quad \times [ny_1 + (n-1)y_2 + \cdots + 2y_{n-1} + y_n] \\
&= (n+1)/n - (2/n^2\mu)\sum_i (n+1-i)y_i
\end{aligned}
$$

so that the poorest person receives a weight of n, the ith poorest person a weight of $(n+1-i)$, and the richest (or nth poorest) person a weight of unity. Hence the weights on incomes are based on the *ranking* of individual welfare (assumed to be monotonic in income) levels, which leads to a social welfare ordering of distributions of the same total income by the negative of the Gini coefficient.

The social welfare function behind the Gini coefficient has been used by Sen in different contexts, and has been defended by him in that it 'makes much use of [ordinal] level comparability without bringing in interpersonally comparable cardinal welfare functions, but at the same time shuns Rawlsian extremism' (Sen, 1974, p. 398). The rank-order weighting scheme leads him to a distributionally-adjusted measure of real national income (Sen, 1976a), which is the country's mean income μ multiplied by $(1 - G)$, where G is the Gini coefficient. The real income, or 'welfare' in the space of income, measure $\mu(1 - G)$, was the motivation for the development of the generalized Lorenz curve, and Sen's measure is simply twice the area under the generalized Lorenz curve – just as $(1 - G)$ is twice the area under the ordinary Lorenz curve.

Poverty measurement

Expressing dissatisfaction with the earlier literature on poverty measurement, Sen (1976b) proposed a new index of poverty which he derived axiomatically. Two types of indices have been used to measure the extent of poverty once the poverty line has been chosen. The most common index is the headcount ratio H, which simply counts the proportion of people below the poverty line. The other index is the income-gap ratio I, which measures the proportionate average income shortfall of the poor from the poverty line. The former index ignores the amounts by which the incomes of the poor fall below the poverty line, while the latter is independent of the number actually in poverty. Both, moreover, are insensitive to a transfer of income from the poor to the very poor. In other words, neither measure is sensitive to the distribution of income among the poor. Sen's measure of poverty incorporates all three of these concerns into a single index.

The index is derived axiomatically after the general form for the poverty measure is taken to be a 'normalized weighted sum of the income gaps of the poor'. Two axioms then suffice to derive the index. The first specifies the weights on the income gaps of the poor, and the second the normalization of the index. Sen argues for weights based on the rank order of the poor below the poverty line, and normalizes his index such that when all the poor have the same income the index is simply $H \cdot I$. The weight on the income gap of the ith poorest person is the number of poor people with incomes at least as large as that of person i. From the definition of the Gini coefficient above, this weighting scheme will yield the Gini coefficient of the income distribution among the poor, G_p.

For an ordered income distribution $\mathbf{y}$ (with $y_1 \leq y_2 \leq \cdots \leq y_n$), let z be the poverty line, q the number of people with income less than or equal to z, and v the mean income of the poor. Then $H = (q/n)$, $I = (z-v)/z$, and Sen's poverty measure

$$
\begin{aligned}
P &= (q/n) \cdot [z - v + \{q/(q+1)\}vG_p]/z \\
&= (q/n) \cdot [z - v(1 - G_p)]/z \text{ for large } q \\
&= H \cdot [I + (1-I)G_p].
\end{aligned}
$$

The effect of the weighting scheme is to augment the average income gap $(z - v)$ by the Gini coefficient G_p times the mean income v of the poor. Thus, the equivalent of an additional income loss arises when inequality among the poor is taken into account. The correction for this loss involves deflating the mean income v of the poor by $(1 - G_p)$, which yields the familiar equally distributed equivalent income corresponding to the rank-order welfare function. Hence the weighted income gap is calculated by taking the difference not between the poverty line and the mean income of the poor, but between the poverty line and the equally distributed equivalent income of the poor.

Weighting schemes based on welfare functions different from the rank-order function will produce different expressions for equally distributed equivalent income and, by the same token, different measures of inequality.

Hence, Sen's approach will generate a different index of poverty for each different welfare function.

Sen's pioneering article has led to an enormous theoretical and empirical literature on poverty measurement. There have been different normalizations of the Sen index, and authors have used different weights on the income gaps of the poor, for example, those that depend on the *size* of a poor person's income gap rather than on the *number* of poor people with higher incomes (as with rank-order weights). The huge theoretical literature that has been generated by the original article is surveyed in Sen and Foster (1997).

Contributions to welfare economics
Utility, income and capability
Sen has criticized utilitarianism from many points of view – among them that it is welfarist and concerned solely with the *sum-total* of individual utilities, not with the interpersonal *distribution* of that sum. Consider the pure distribution problem for two people in which person 1 derives exactly twice as much utility as person 2 from any given level of income, say, because person 2 has some handicap (for example, being a cripple). In this case the utilitarian solution is to give person 1 a higher income than person 2. Even if income were equally divided, person 1 would have enjoyed more utility than person 2 (twice as much); but instead of reducing this inequality the utilitarian rule compounds it – by giving more income to person 1 who is already better off.

The perverse consequences of utilitarianism on inequality in both income and utility spaces led Sen (1973a) to formulate the Weak Equity Axiom (WEA) – which, according to Arrow (1999, p. 167), is another 'example of Sen's seminal role in the area of formal theories of social choice'. WEA states that, in distributing a given total income, an individual getting less utility from any level of income should get a higher income. This axiom is in general inconsistent with utilitarianism, and its consequences have been much explored in the social choice literature; for example, similar equity axioms over the set of social states X have been used to characterize the leximin rule (the results are discussed in Sen, 1977b; 1986). WEA highlights interest in equality as a criterion independent of utility or income. But equality of what?

What is missing in the utility or income framework is some notion of 'basic capabilities': a person being able to do certain basic things. The ability to move about is relevant for a cripple, but there are others such as 'the ability to meet one's nutritional requirements, the wherewithal to be clothed and sheltered, the power to participate in the social life of the community' (Sen, 1980, p. 218).

Sen is critical of the use of both income (or commodity possession) and utility as measures of well-being,

arguing that they constitute the wrong space in which to make such assessments. He argues that 'well-being' has to do with *being* well, which must take into account the capability to live long, be well-nourished, be literate, and so on. As he puts it, the 'value of the living standard lies in the living, and not in the possessing of commodities, which has derivative and varying relevance' (Sen, 1987, p. 25). What is valued intrinsically are people's achievements – their 'beings' and 'doings' – or their 'capabilities to function'. Income can have importance as an instrument for expanding capabilities, while utility may provide evidence of achievement. But Sen's argument is that the space in which well-being should be assessed has to be more directly linked to what matters most – not its instrumental antecedents (income) nor its evidential correlates (utility).

Poverty as capability deprivation
The conversion of income into achieved well-being is subject to great variation. The mapping of income into basic capabilities is significantly affected by personal heterogeneities (such as age, disability and illness), gender and social roles, environmental and epidemiological conditions, among many other factors. There is thus an important need to go beyond income information in poverty analysis, in particular to see poverty as capability deprivation.

Apart from the basic capabilities involved in leading a minimally acceptable life, such as avoiding hunger or undernourishment or preventable morbidity, there are elementary social functionings, such as 'appearing in public without shame' or 'taking part in the life of the community'. The commodity or income requirements of such capability fulfillment will vary between communities within a country, and between countries. Indeed, as Sen (1983, p. 161) has argued convincingly, 'poverty is an absolute notion in the space of capabilities' but it often takes 'a relative form in the space of commodities'. For example, to avoid shame from the inability to meet the demands of custom or convention in a society, the relative income requirements will typically be higher in richer countries. These requirements vary with what *others* in the community standardly have (for example, in 18th century Europe, according to Adam Smith, 'a creditable day-labourer would be ashamed to appear in public without a linen shirt' – cited in Sen, 1983, p. 161). *Relative* deprivation in the space of incomes can thus lead to *absolute* deprivation in the space of capabilities (Sen, 1992a, p. 115).

As Atkinson (1999, p. 186) notes, '[i]n this way, Sen exposes the popular confusion that an absolute approach implies a constant real income standard when measuring poverty. Rising living standards in the community as a whole may lead to a poverty line that increases in real terms, even when our concern is limited to absolute deprivation in the space of capabilities.'

A concluding remark

The importance of the subject of welfare economics cannot be overemphasized. Welfare judgements lie at the heart of economic policy analysis and prescription. Yet there has been a tendency to avoid examining welfare statements critically or the assumptions underlying them. Sen's work in welfare economics is of far-reaching significance. His critique of utilitarianism and the Pareto principle, and his insistence on embodying distributional judgements and non-utility information (for example, personal liberty, rights and capabilities), have dramatically shifted the focus of the subject. Sen has done more than anyone else to investigate, scrutinize and develop the foundations of welfare economics and social choice theory. He has used formal analysis, conceptual elucidation, measurement theory, philosophical reasoning, and empirical work to advance the subject – as no one has done before him. The range and reach of Sen's contributions to welfare economics are truly formidable. Their totality essentially defines the present subject matter of this critical branch of economics.

SUDHIR ANAND

See also **Arrow, Kenneth Joseph; Arrow's theorem; development economics; Gini ratio; inequality (measurement); poverty; social choice; welfare economics.**

Selected works

1968. *Choice of Techniques*, 3rd edn. Oxford: Basil Blackwell and Delhi: Oxford University Press.

1969. Quasi-transitivity, rational choice and collective decisions. *Review of Economic Studies* 36, 381–93.

1970a. *Collective Choice and Social Welfare*. San Francisco: Holden-Day.

1970b. The impossibility of a Paretian liberal. *Journal of Political Economy* 78, 152–7.

1970c. Interpersonal aggregation and partial comparability. *Econometrica* 38, 393–409.

1971. Choice functions and revealed preference. *Review of Economic Studies* 38, 307–17.

1973a. *On Economic Inequality*. Oxford: Clarendon Press.

1973b. Behaviour and the concept of preference. *Economica* 40, 241–59.

1973. (With P. Dasgupta and D. Starrett.) Notes on the measurement of inequality. *Journal of Economic Theory* 6, 180–7.

1974. Informational basis of alternative welfare approaches: aggregation and income distribution. *Journal of Public Economics* 3, 387–403.

1975. *Employment, Technology and Development*. Oxford: Clarendon Press and New York: Oxford University Press.

1976a. Real national income. *Review of Economic Studies* 43, 19–39.

1976b. Poverty: an ordinal approach to measurement. *Econometrica* 44, 219–31.

1977a. Social choice theory: a re-examination. *Econometrica* 45, 53–89.

1977b. On weights and measures: informational constraints in social welfare analysis. *Econometrica* 45, 1539–72.

1977c. Rational fools: a critique of the behavioural foundations of economic theory. *Philosophy and Public Affairs* 6, 317–44.

1978. Ethical measurement of inequality: some difficulties. In *Personal Income Distribution*, ed. W. Krelle and A.F. Shorrocks. Amsterdam: North-Holland.

1979. Personal utilities and public judgements: or what's wrong with welfare economics? *Economic Journal* 89, 537–58.

1980. Equality of what? In *The Tanner Lectures on Human Values*, Vol. 1, ed. S.M. McMurrin. Salt Lake City: University of Utah Press and Cambridge: Cambridge University Press.

1981a. *Poverty and Famines: An Essay on Entitlement and Deprivation*. Oxford: Clarendon Press and New York: Oxford University Press.

1981b. Ingredients of famine analysis: availability and entitlements. *Quarterly Journal of Economics* 95, 433–64.

1982. *Choice, Welfare and Measurement*. Oxford: Basil Blackwell.

1983. Poor, relatively speaking. *Oxford Economic Papers* 35, 153–69.

1984. *Resources, Values and Development*. Oxford: Basil Blackwell and Cambridge, MA: Harvard University Press.

1985. *Commodities and Capabilities*. Amsterdam: North-Holland.

1986. Social choice theory. In *Handbook of Mathematical Economics*, Vol. 3, ed. K.J. Arrow and M.D. Intriligator. Amsterdam: North-Holland.

1987. *The Standard of Living*, Tanner Lectures with rejoinders by Bernard Williams and others, ed. G. Hawthorne. Cambridge: Cambridge University Press.

1989. (With J. Drèze.) *Hunger and Public Action*. Oxford: Clarendon Press.

1990. Gender and cooperative conflicts. In *Persistent Inequalities*, ed. I. Tinker. New York: Oxford University Press.

1992a. *Inequality Reexamined*. Oxford: Clarendon Press, New York: Russell Sage Foundation and Cambridge, MA: Harvard University Press.

1992b. Missing women. *British Medical Journal* 304 (March), 587–8.

1993. Internal consistency of choice. *Econometrica* 61, 495–521.

1995. Rationality and social choice. *American Economic Review* 85, 1–24.

1997. Maximization and the act of choice. *Econometrica* 65, 745–79.

1997. (Annexe with J.E. Foster.) *On Economic Inequality*. Expanded edition with an annexe 'On Economic Inequality after a Quarter Century'. Oxford: Clarendon Press and New York: Oxford University Press.

1999a. Autobiography. In *Les Prix Nobel. The Nobel Prizes 1998*, ed. T. Frängsmyr. Stockholm: Nobel Foundation, 1999. Available at http://nobelprize.org/nobel_prizes/economics/laureates/1998/sen-autobio.html, accessed 3 June 2007.

1999b. *Development as Freedom*. New York: Alfred Knopf and Oxford: Oxford University Press.

1999c. The possibility of social choice. *American Economic Review*. 89, 349–78.

2002. *Rationality and Freedom*. Cambridge, MA and London, England: Harvard University Press.

2003. Missing women—revisited. *British Medical Journal*. 327 (December), 1297–8.

Bibliography

Anand, S. 1983. *Inequality and Poverty in Malaysia: Measurement and Decomposition*. New York: Oxford University Press.

Arrow, K.J. 1999. Amartya K. Sen's contributions to the study of social welfare. *Scandinavian Journal of Economics* 101, 163–72.

Atkinson, A.B. 1970. On the measurement of inequality. *Journal of Economic Theory* 2, 244–63.

Atkinson, A.B. 1999. The contributions of Amartya Sen to welfare economics. *Scandinavian Journal of Economics* 101, 173–90.

d'Aspremont, C. and Gevers, L. 1977. Equity and the informational basis of collective choice. *Review of Economic Studies* 44, 199–209.

Hammond, P.J. 1976. Equity, Arrow's conditions, and Rawls' difference principle. *Econometrica* 44, 793–804.

Kolm, S.-C. 1969. The optimal production of social justice. In *Public Economics*, ed. J. Margolis and H. Guitton. London: Macmillan.

Maskin, E.S. 1978. A theorem on utilitarianism. *Review of Economic Studies* 45, 93–6.

Roberts, K.W.S. 1980a. Possibility theorems with interpersonally comparable welfare levels. *Review of Economic Studies* 47, 409–20.

Roberts, K.W.S. 1980b. Interpersonal comparability and social choice theory. *Review of Economic Studies* 47, 421–39.

Rothschild, M. and Stiglitz, J.E. 1973. Some further results on the measurement of inequality. *Journal of Economic Theory* 6, 188–204.

Scandinavian Journal of Economics. 1999. Bibliography of A.K. Sen's publications, 1957–1998. Vol. 101, 191–203.

Senior, Nassau William (1790–1864)

Born at Compton Beauchamp in Berkshire, the eldest son of John Raven Senior, Vicar of Durnford, Nassau Senior studied for the Bar in London, was the first Drummond Professor of Political Economy at Oxford, 1825–30, and was elected to a second term, 1847–52. In 1831 he was appointed Professor of Political Economy at King's College, London, but was soon forced to resign over his controversial recommendation that some of the revenues of the established Church in Ireland be turned over to the Roman Catholics. Senior became a respected adviser: he served on the Commission for inquiring into the Administration and Operation of the Poor Laws (1832–4), being mainly responsible for the writing of its report, and was consulted by Lord John Russell on Irish Poor Law Reform in 1836. In 1841 Senior drew up the Report of the Commission on the condition of the Unemployed Hand-loom Weavers (on this see Stigler, 1949, Lecture 3). Two years after it was founded in 1821 Senior was elected to membership of the Political Economy Club and remained a member, except for the years 1848–53, until his death.

Senior first came to the notice of those conversant with political economy through an article on the Corn Laws in the *Quarterly Review* (1821), and he was a regular contributor to the *Edinburgh Review* from 1841 to 1855. He was at home in both literary and political circles in London and cultivated an interest in Continental affairs via frequent travels and the company of men of influence, among whom were Guizot and De Tocqueville. Conversations with such men in France and Italy and in Ireland were assiduously recorded (and checked for accuracy) and, together with a traveller's observations on these and other countries, filled many journals. Several of these were published, including conversations with De Tocqueville spanning the years 1834–59.

It was Senior's intention to publish a systematic account of political economy, collecting the ideas that were largely scattered in periodicals, lectures, official reports and pamphlets into a major treatise. The plan was not fulfilled and his main printed legacy is his 1836 *Outline of the Science of Political Economy*, plus a collection of pamphlets and public letters published by Augustus Kelley, entitled *Selected Writings on Economics* (1966) and his *Three Lectures on the Rate of Wages* (1830a). S. Leon Levy undertook the work Senior never brought to fruition, in a volume entitled *Industrial Efficiency and Social Economy* (1928). This is a composite, organized by a plan of the editor's own making, and comprising selections from periodical articles, reports and – mainly – manuscript lectures from Senior's second term as Drummond Professor. The work is supposed to represent Senior's 'mature' thoughts, but the manner of its composition makes it of very limited value from a scholarly point of view.

Senior belonged to the band of eminent political economists of the second quarter of the 19th century who may be called respectful dissenters from Ricardo's doctrines. He did not dissent on methodological grounds, as did Whewell and Richard Jones (De Marchi and Sturges, 1973; Hollander, 1985, vol. 1, chs 1–3), but on value and distribution he followed Smith and Say more closely than Ricardo. Thus 'value' to Senior meant value in exchange rather than cost of production or labour cost (1836, pp. 13–14). And distribution he preferred to treat as a question of factor incomes rather than of functional shares. The expression 'high wages', for example, Senior used to stand for high nominal or real wages rather than for a large share of labour's product actually received by labour (1830a, pp. 2–3). This is not to say that he anticipated the marginal productivity theory of distribution, although at least as far as labour and capital are concerned he liked to think of the incomes attributable to each as payment for services rendered or, even more especially, as a 'reward … [for] conduct' (1836, p. 89).

Senior's approach to distribution displays tensions that are unavoidable in trying to combine a Ricardian concern with macro-issues such as capital and population growth and the time-path of wages with a Smithian predilection to treat value and distribution as the outcome of voluntary exchanges entered into by individuals. To illustrate: Senior retained the Ricardian theory of rent, whereby rent is an intra-marginal surplus. This surplus accrues to ownership. Where, however, there is competitive access to the powers of nature, the price of the product equals the sum of wages, a reward for labour services, and profit, a return for waiting or 'abstinence' (1836, p. 89). This falls short of an integrated theory of distribution, reflecting as it does the institutional fact of appropriation, on the one hand, and economic conduct by free agents, on the other. Similarly, in discussing the time-path of wages, Senior reverted to a wages-fund approach: given labour productivity, 'the rate of wages depends on the extent of the fund for the maintenance of labourers, compared with the number of labourers to be maintained' (1830a, p. xii). This refers to the average wage; but we are not told how individual contracts struck between worker and employer upon the value of labour services relate to this average. The problem is familiar to modern theorists wanting to make explicit the microfoundations of macroeconomics; though the major difficulty, changing relative wages, Senior neatly sidestepped by following Smith in the conviction that, once established, relative wage scales remained fixed (1830b, p. 15).

Senior's Ricardianism is perhaps most evident in his views on trade and on the international aspects of money. In Three Lectures on the Transmission of the Precious Metals, delivered at Oxford in June 1827 and printed the following year, on an issue he felt to be 'next to the Reformation, next to the question of free religion, the most momentous that has ever been submitted to human decision' (1828, p. 88), Senior employed Hume's doctrine (and Ricardo's) to show that no country can have a permanently favourable or unfavourable balance of trade or exchange rate, then used this against the mercantilist view. His main concern is the efficient allocation of labour, and this leads him to the Ricardian conclusion that, if a domestic tax on one industry hurts its international competitiveness, then a 'countervailing duty' on the competing import is 'not a departure from the

principles of free trade but an application of them' (1828, p. 70; compare Ricardo 1822, vol. 4, p. 217). In another set of three lectures, this time On the Cost of Obtaining Money (1830b), Senior addressed the question of international comparisons of wages and argued that the productivity of labour measured in the goods required to import precious metals (in a non-mining country) determines whether wages are high or low. In both sets of lectures Senior discusses paper money and reaches the basic Ricardian conclusion that variations in the amount of the currency, whether metal or paper, may cause sudden disturbances but these will be transitory.

It is worth stressing that the Stoic tradition so evident in Smith's writings – self-respect issuing from prudent behaviour, most notably self-restraint – also infuses Senior's discussions of distribution and related social issues. Senior takes it as given that men tend to be myopic and to prefer taking their ease to working (Senior, 1827, p. 8; 1836, pp. 26 ff.). He therefore considered the supply of goods to be a result of prudential exertions to overcome these obstacles (1836, pp. 15–16). There is no such exertion attached to mere ownership. Hence the income of a landlord is a transfer and categorically distinct from wages and profits, which are properly considered rewards. They are like the good which results from confronting and overcoming unavoidable evil (pain). Although Senior is properly extolled for having glimpsed marginal utility, his contribution on the side of supply and the overcoming of obstacles is at least as interesting.

Marx, it need hardly be added, saw Senior as 'a mere apologist of the existing order' (Marx, 1905–10, vol. 3, p. 353). This in part refers to Senior's view that profit is the reward of 'abstinence'. Senior's sacrifice–reward approach to the sharing out of the price of product at the margin, however, was adopted fully by the later Ricardians, John Stuart Mill (1848, p. 400) and John Elliott Cairnes (1874, p. 74).

Senior's views on method too are indistinguishable in all but two details from John Stuart Mill's. One reasons in political economy from true premises known by introspection (the desire for wealth, subject to a preference for least effort and present enjoyment) or from observation (the laws of return and of population). Being true these premises will, upon correct reasoning, yield true principles. In the science of political economy, therefore, there resides as much certainty as in any science outside axiomatic logic (Senior, 1827, p. 11). Nonetheless, the psychological drives for wealth, leisure and present satisfaction produce counteracting conduct, so that it is difficult to assign motives (causes) to or predict behaviour (1827, p. 9).

Mill tended to bundle the three psychological motives together and make the wealth motive do most of the work. He also chose to reason hypothetically, as if the desire for wealth were the sole motivation of an individual. This meant that his results could be true only in the abstract, or to the degree to which that assumption was in fact true. In later years Senior, despite his early warning that counteracting motives could not readily be disentangled (1827, p. 9), argued against Mill's simplifying approach. His Four Introductory Lectures from the second period of his tenure of the Drummond Professorship (1847–52) lamented the slow progress made in political economy. He found reason in the Millian manner of dealing with the science. Hypothetical reasoning, in Senior's view, rendered the subject unattractive, because unrealistic, and laid the reasoner open to error, either from forgetting some relevant additional cause or from forgetting that the reasoning itself was based on arbitrary assumptions and was not directly transferable to real world situations (Senior, 1852, pp. 63–5). In more recent discussion, T.W. Hutchison has kept Senior's objections alive, protesting in particular a tendency to confuse tautological with empirical propositions, and to leap straight from abstract models to policy conclusions (Hutchison, 1938). While Senior's concern is understandable, to the extent to which his earlier judgement was well-founded – 'that we are liable to the greatest mistakes when we endeavour to assign motives [causes] to … conduct' (1827, p. 9) – it is not clear that he gains anything by his later modified approach, and there is in that case less practical difference between his position and Mill's than some modern commentators have made out (Bowley, 1937, pp. 59–62).

Both Senior and Mill cautioned against applying the principles of political economy without the utmost care and a broad and detailed knowledge of the facts applying to any case in question. Senior, however, nonetheless confidently offered advice on the great issues of the 1830s and 1840s: Poor Law reform; the Factory Act (Ten Hours Bill); agricultural unrest, overpopulation and emigration; free trade and the role of banks in commercial crises. In few of the views he expressed is there anything very remarkable – an exception is noted below in connection with the Ten Hours Bills – but he had a striking ability to cut through to the heart of complex issues, and he had a persuasive pen. To illustrate, his overriding criterion for judging all schemes of relief to the able-bodied poor was whether they destroyed incentives by separating effort from reward (1830a, preface; 1834, pp. 126 ff.). In a particular argument against reducing factory hours, he held that if capital replenishment is, say, 11/12ths of gross turnover, then interest plus profits depend essentially on the last hour of a 12-hour day. A ten-hour day, therefore, would spell ruin (1837, pp. 12–13). This not only presupposed constant returns to hours worked, but confused stocks and flows in the calculation of returns (Johnson, 1969).

N. DE MARCHI

See also **Jevons, William Stanley; Marshall, Alfred; Mill, James; Mill, John Stuart; Ricardo, David; Smith, Adam.**

Selected works

1821. Report on the state of agriculture. *Quarterly Review* 25(50), July.
1827. *An Introductory Lecture on Political Economy.* In Senior (1966).
1828. *Three Lectures on the Transmission of the Precious Metals.* In Senior (1966).
1830a. *Three Lectures on the Rate of Wages.* London: Murray.
1830b. *Three Lectures on the Cost of Obtaining Money.* In Senior (1966).
1834. *Report from his Majesty's Commissioners on the Administration and Practical Operation of the Poor Laws.* London: British Parliamentary Papers.
1836. *An Outline of the Science of Political Economy.* New York: Kelley Reprint, 1965.
1837. *Two Letters on the Factory Acts.* In Senior (1966).
1852. *Four Introductory Lectures on Political Economy.* In Senior (1966).
1872. *Correspondence and Conversations of Alexis De Tocqueville with Nassau William Senior from 1834 to 1859.* 2 vols, ed. M.C.M. Simpson, 2nd edn, London: Henry S. King.
1966. *Selected Writings on Economics. A Volume of Pamphlets 1827– 1852.* New York: Kelley Reprint.

Bibliography

Bowley, M. 1937. *Nassau Senior and Classical Economics.* London: Allen & Unwin.
Cairnes, J.E. 1874. *Some Leading Principles of Political Economy Newly Expounded.* London: Macmillan.
De Marchi, N. and Sturges, R.P. 1973. Malthus' and Ricardo's inductivist critics: four letters to William Whewell. *Economica* 40, 379–93.
Hollander, S. 1985. *The Economics of John Stuart Mill*, 2 vols. Oxford: Blackwell.
Hutchison, T.W. 1938. *The Significance and Basic Postulates of Economic Theory.* London: Macmillan.
Johnson, O. 1969. The 'last hour' of Senior and Marx. *History of Political Economy* 1, 359–69.
Levy, S.L., ed. 1928. *Industrial Efficiency and Social Economy* by Nassau W. Senior, 2 vols. New York: Henry Holt & Co.
Marx, K. 1905–10. *Theories of Surplus Values*, 3 vols. London: Lawrence & Wishart, 1969; 1972.
Mill, J.S. 1848. *Principles of Political Economy with Some of their Applications to Social Philosophy.* In *Collected Works of John Stuart Mill*, vols 2 and 3, ed. J.M. Robson. Toronto: University of Toronto Press, 1965.
Ricardo, D. 1822. On protection to agriculture. In *The Works and Correspondence of David Ricardo*, vol. 4, ed. P. Sraffa with the collaboration of M.H. Dobb, Cambridge: Cambridge University Press for the Royal Economic Society, 1951.
Stigler, G.J. 1949. The classical economists: an alternative view. In *Five Lectures on Economic Problems.* London: Longmans, Green for the London School of Economics and Political Science.

separability

1 Introduction

Separability, as discussed here, refers to certain restrictions on functional representations of consumer (or social) preferences or producer technologies. These restrictions add structure to the decision-making tasks undertaken by economic agents. They also allow the economic researcher to study the behaviour of these agents in a more effective manner.

To keep things simple we will focus on consumers. Consumers must make choices among a large number of goods – both consumption and leisure, both present and future. The apparent complexity of their decision-making problems may lead consumers to engage in simplified budgeting practices. Are these procedures consistent with rational behaviour? In modelling consumer behaviour, economists – theoretical and empirical – employ models that consider only a subset or several subsets of the complete list of goods and services consumed. Can these practices be justified? And, if so, what kinds of restrictions do these rationalizations place on the preferences and the behaviour of the consumer? In what follows, we discuss several proposals that address this class of problems and present their solutions.

One way to think about reducing the complexity of the allocation decision is to imagine that the consumer receives a lump sum of money that he or she first allocates to broad classes of commodities, such as food, shelter and recreation. Detailed decisions about how to spend the money that has been allocated to the food budget are postponed until one is actually in the store buying specific food items.

More formally, if the correct amount of money to spend on food commodities, for example, has been allocated, under what circumstances is the consumer able to dispense the food budget among food commodities knowing only the food prices? If the consumer can arrive at an optimal pattern of food expenditures in this way, preferences are said to be *decentralizable*. It is fairly obvious that, if food commodities were separable from all other commodities, decentralization would be possible. What is somewhat surprising, however, is that separability is necessary as well as sufficient for this practice to be rationalized. We first present more formally the concepts of decentralization and separability and then discuss their equivalence.

Consumption bundles are N-tuples, $x = (x_1, \ldots, x_N)$, that are elements of a consumption space, Ω. We take Ω to be a closed and convex subset of $\mathbb{R}^N_+$. Thus we begin with a consumer with a well-defined, neoclassical utility function, $U : \Omega \to \mathbb{R}$. Partition the set of variable indices, $I = \{1, 2, \ldots, N\}$, into two subsets $\{I^1, I^2\}$. This divides the set of goods into two groups, one and two, and we can write the consumption space as $\Omega = \Omega^1 \times \Omega^2$ and the consumption vector as $x = (x^1, x^2)$. The consumer faces commodity prices given by $p = (p_1, \ldots, p_N)$ and allocates

income y among the N goods. The price vector can also be written as $p = (p^1, p^2)$.

The consumer's utility maximization problem,

$$\max_x \; U(x) \quad \text{subject to} \quad p \cdot x = y, \qquad (1)$$

can be rewritten as

$$\max_{x^1, x^2} U(x^1, x^2) \quad \text{subject to} \quad p^1 \cdot x^1 + p^2 \cdot x^2 = y. \tag{2}$$

The solution to (1) [or (2)] is denoted by

$$x^* = (\Phi^1(y, p), \Phi^2(y, p)). \qquad (3)$$

Expenditure on goods in group two is denoted by y_2. Group two expenditure is optimal if $y_2 = p^2 \cdot \Phi^2(y, p)$.

2 Decentralization and separability

The consumer's utility maximization problem may be decentralized if the consumer is able to optimally allocate expenditure on group-two commodities knowing only group-two optimal expenditure and group-two prices. More formally, the utility maximization problem is *decentralizable* (for group two) if there exists a function ϕ^2 such that

$$\phi^2(y_2, p^2) = \Phi^2(y, p) \quad \text{if} \quad y_2 = p^2 \cdot \Phi^2(y, p).$$

Now the question is: what restriction on the consumer's utility function allows for decentralizability of the utility maximization problem? As it turns out, decentralizability for group-two goods is possible if and only if group-two goods are separable from group-one goods. We turn to the formal definition of separability.

According to the original Leontief–Sono definition of separability, the goods in group two are *separable* from the goods in group one if

$$\frac{\partial}{\partial x_k} \left(\frac{\partial U(x)/\partial x_i}{\partial U(x)/\partial x_j} \right) \equiv 0, \qquad (4)$$

for all $i, j \in I^2$, $k \in I^1$ (see Leontief, 1947a; 1947b; Sono, 1945; 1961).

The condition (4) says that marginal rates of substitution between pairs of goods in group two are independent of quantities in group one and hence that an aggregator function exists for group-two goods. Under fairly mild conditions, this aggregator function may be defined by

$$U^2(x^2) := U(O^1, x^2)$$

for any choice of a fixed vector, $O^1 \in \Omega^1$; that is, $(O^1, x^2) \in \Omega$. Different choices of O^1 give rise to different aggregator functions that are ordinally equivalent. Having accomplished this, we also define a macro function, $\mathbb{U}$, that is defined by

$$U(x^1, x^2) = \mathbb{U}(x^1, U^2(x^2)). \qquad (5)$$

Example Let $I = \{1,2,3,4\}$, $I^1 = \{1,2\}$, and $I^2 = \{3,4\}$. Consider the utility function given by

$$U(x_1, x_2, x_3, x_4) = x_1^{1/3} x_3^{1/3} x_4^{1/3} + x_2^{1/2} x_3^{1/4} x_4^{1/4}.$$

By defining $u_2 = U^2(x_3, x_4) = x_3^{1/3} x_4^{1/3}$ and $\mathbb{U}(x_1, x_2, u_2) = x_1^{1/3} u_2 + x_2^{1/2} u_2^{3/4}$, we observe that

$$U(x_1, x_2, x_3, x_4) = \mathbb{U}(x_1, x_2, u_2),$$

and hence that group two is separable from group one. One may also confirm this fact using the Leontief–Sono conditions. The reader should notice that the choice of the aggregator and macro function is not unique. An alternative choice would be $u_2 = U^2(x_3, x_4) = x_3^{1/2} x_4^{1/2}$ and $\mathbb{U}(x_1, x_2, u_2) = x_1^{1/3} u_2^{2/3} + x_2^{1/2} u_2^{1/2}$. In this case, the aggregator function is chosen to be homogeneous of degree one. The reader should also notice that group one is *not* separable from group two; separability is not a symmetric concept.

The formal result on decentralizability can now be stated. The consumer's utility maximization problem is decentralizable for group two if and only if group two is separable from group one – that is, if and only the utility function can be written as

$$U(x^1, x^2) = \mathbb{U}(x^1, U^2(x^2)).$$

The implication of this result is that the consumer utility maximization problem may be broken up into two parts:

1. Solve

$$\max_{x^2} U^2(x^2) \quad \text{subject to} \quad p^2 x^2 \leq y_2 \qquad (6)$$

to get the conditional demand function, $\phi^2(y_2, p^2)$, for group two.
2. Solve

$$\max_{x^1, y_2} \mathbb{U}(x^1, U^2(\phi^2(y_2, p^2))) \qquad (7)$$
$$\text{subject to} \quad p^1 x^1 + y_2 \leq y$$

to get the optimal demands for group one and the optimal income allocation for group two.

While separability is inherently an asymmetric concept, we may want to consider the symmetric case. We may also want to extend the analysis to consider R groups of goods. To this end, let $\{I^1, \ldots, I^R\}$ be a partition of the original set of variable indices I. The goods vector and the price vector can be written as $x = (x^1, \ldots, x^R)$ and

$p = (p^1, \ldots, p^R)$, respectively. Then the utility function is separable in the partition $\{I^1, \ldots, I^R\}$ if and only if the utility function may be written as

$$U(x^1, \ldots, x^R) = \mathscr{U}(U^1(x^1), \ldots, U^R(x^R)). \qquad (8)$$

Expenditure on goods in group r is denoted by y_r, $r = 1, \ldots, R$. Group r expenditure is optimal if $y_r = p^r \cdot \Phi^r(y, p)$.

The consumer's utility maximization problem may be decentralized if the consumer is able to optimally allocate expenditure on group r commodities knowing only group r optimal expenditure and group r prices. More formally, the utility maximization problem is *decentralizable* for the partition $\{I^1, \ldots, I^R\}$ if there exist functions Φ^r, $r = 1, \ldots, R$, such that

$$\phi^r(y_r, p^r) = \Phi^r(y, p) \quad \text{if} \quad y_r = p^r \cdot \Phi^r(y, p),$$
$$r = 1, \ldots, R.$$

A useful device for explicating the necessary and sufficient conditions for decentralizability of the utility maximization problem is the conditional indirect utility function. For the R group case it is defined as

$$H(y_1, \ldots, y_R, p) = \max_x \{U(x) : p^r x^r \leq y^r,$$
$$r = 1, \ldots, R\}. \qquad (9)$$

This function yields the maximum utility conditional on an income allocation, $(y_1, \ldots, y_R)$, among the R groups. In a second stage one can solve for the optimal income allocation by solving

$$\max_{y_1, \ldots, y_R} H(y_1, \ldots, y_R, p) \quad \text{subject to} \quad \sum_{r=1}^{R} y^r \leq y. \qquad (10)$$

Denote the solution to (10) by the expenditure-allocation functions

$$y_r = \theta^r(y, p), r = 1, \ldots, R. \qquad (11)$$

A remarkable feature of the conditional indirect utility function is that

$$H(\theta^1(y, p), \ldots, \theta^R(y, p), p) = W(y, p)$$
$$= \max_x \{U(x) : p \cdot x \leq y\}.$$

Thus, the overall consumer utility maximization may always be performed in the two steps given in (9) and (10) even in the absence of separability restrictions on the utility function.

The results for decentralizability in the R-group symmetric case can be stated in two parts. First, the utility function is separable in the partition $\{I^1, \ldots, I^R\}$ as in (8) if and only if the conditional indirect utility function may be written as

$$H(y_1, \ldots, y_R, p) = U(v^1(y_1, p^1), \ldots, v^R(y_R, p^R))$$

where

$$v^r(y_r, p^r) = \max_{x^r}\{U^r(x^r) : p^r x^r \le y_r\},$$
$$r = 1, \ldots, R,$$

$$(12)$$

are the conditional indirect utility functions for the aggregator functions, $U^r(x^r)$, $r = 1, \ldots, R$. Second, the consumer's utility maximization problem is decentralizable for the partition $\{I^1, \ldots, I^R\}$ if and only if the utility function is separable in the partition $\{I^1, \ldots, I^R\}$.

Separability in the partition $\{I^1, \ldots, I^R\}$ allows the following two-step budgeting procedure. In the first step the consumer solves the problem in (12), which yields conditional demand functions $x^r = \phi^r(y_r, p^r)$, $r = 1, \ldots, R$. In the second step, the consumer solves (10), which yields the optimal income allocation. While the first step economizes the informational requirements – knowledge of only in-group prices are needed – the second step requires the entire vector of prices. However, additional restrictions on the utility function will reduce both the computational and informational burden of the second step; these restrictions will be discussed shortly.

Decentralization may be possible in more than one partition of the set of commodities. Moreover, the separable sectors might well overlap. Suppose, for example, that U is a social welfare function and x^r is the consumption vector of consumers in group r. As some policies affect only a subset of the population, one might want to analyse separately the social welfare of, say, two subgroups of the population: (1) a particular ethnic group and (2) those in a particular geographical region. Clearly, the two groups could overlap. A deep result of Gorman (1968) indicates that the existence of such overlapping separable groups has powerful implications for the structure of the welfare function.

3 Additive structures

Let I^r and I^c be the two (separable) groups of interest and suppose that $I^1 := I^r - I^s \ne \varnothing$, $I^2 := I^r \cap I^s \ne \varnothing$, and $I^3 := I^s - I^r \ne \varnothing$. Let I^4 be the complement of $I^r \cup I^c$ in In $\mathscr{I}$. *Gorman's overlapping theorem* indicates that I^1, I^2 and I^3 are also separable from their complements in I. One of the important aspects of Gorman's overlapping theorem is that this structure is equivalent to the following representation of U:

$$U(x) = U(U^1(x^1) + U^2(x^2) + U^3(x^3), x^4).$$

$$(13)$$

(Additional regularity conditions on the 'essentiality' of each group are required for this result; see Gorman, 1968; Blackorby, Primont and Russell, 1978; 1998.)

Note that what drives this result on (groupwise) additive structures is not simply separability of each of the sectors, I^1, I^2 and I^3 from their complements but also separability of arbitrary unions of these sectors. This observation suggests the general result on groupwise-additive structures (see Debreu, 1959; Gorman, 1968):

$$U(x) = \left(\sum_{r=1}^{R} U^r(x^r)\right), \quad R > 2, \qquad (14)$$

if and only if arbitrary unions of subsets of the partition $\{I^1, \ldots, I^R\}$ are separable from their complements. Note that a special case of (14) is when each element of the partition I is a singleton ($R = n$), in which case,

$$U(x) = \left(\sum_{i=1}^{N} U^i(x_i)\right), \quad N > 2. \qquad (15)$$

That is, u is additive in the variables themselves.

Another example in the context of social welfare functions illustrates the power of Gorman's overlapping theorem. Suppose that the social decision rule satisfies the anonymity condition: only the individuals' utilities and not their names should matter in social evaluation. This standard social choice assumption is equivalent to symmetry of the social welfare function: permuting the names of individuals does not affect the value of the function. Let $\{1,2\}$ be the subset of individuals who are affected by some set of policies, and suppose that these policies are judged solely by their effects on these two individuals. This implies that $\{1,2\}$ is separable from its complement in the set of citizens and hence that

$$W(u_1, \ldots, u_R) = J(F(u_1, u_2), u_3, \ldots, u_R).$$

$$(16)$$

The fact that W is symmetric means that it can be written as

$$J(F(u_1, u_2), u_3, \ldots, u_R) = J(F(u_1, u_3), u_2, u_4, \ldots, u_R).$$

$$(17)$$

This in turn implies that $\{1,3\}$ is separable from its complement in the set of citizens. However, these sets have a non-empty intersection, $\{2\}$, and the overlapping theorem implies that the social welfare function can be written as

$$W(u_1, \ldots, u_R) = \mathbb{J}(f^1(u_1) + f^2(u_2)$$
$$+ f^3(u_3), u_4, \ldots, u_R).$$

Proceeding by induction yields a strong implication: the social welfare function must be additive:

$$W(u_1, \ldots, u_R) = \mathbb{W}(f^1(u_1) + \ldots f^R(u_R)).$$

The Gorman overlapping argument establishing the groupwise additive structure (13) requires the existence of at least two overlapping separable sets resulting in at least three separable sets. Put differently, separability of arbitrary unions of non-overlapping sets in a binary partition $\{I^1, I^2\}$ adds no restriction to separability of the two sets from their complements. But two-group additivity arises in many contexts. One example is the typical additive utility function in overlapping-generations models where each generation's finite lifetime is divided into two periods (often a 'work' period and a 'retirement' period). A special case of two-group additivity is the quasi-linear utility function that is so critical to the analysis of public goods and incentive compatibility. The stronger conditions required for two-group additivity are based on Sono's (1945; 1961) independence condition. A set of variables, I^1, is said to be *Sono independent* of I^2 if there exist functions ψ^{ji}, $i, j \in I^1$, such that

$$\frac{\partial}{\partial x_i}\left(\ln \frac{\partial U(x)/\partial x_k}{\partial U(x)/\partial x_j} \right) = \frac{\partial}{\partial x_i}\left(\ln \frac{\partial U(x)/\partial x_\ell}{\partial U(x)/\partial x_j} \right) = \psi^{ji}(x^1)$$
$$\forall i, j \in I^1, \forall k, \ell \in I^2$$

That is, the effect of changing the consumption of a commodity in sector 1 on the marginal rates of substitution between pairs of variables, one in sector I^1 and the other in sector I^2, depends only on the values of consumption levels in sector 1.

The necessary and sufficient condition for two-group additivity is as follows: U can be written as

$$U(x) = \mathcal{U}(U^1(x^1) + U^2(x^2))$$

if and only if I^1 is independent of I^2 and is separable from I^2. (This version of the two-group additivity result is attributable to Blackorby, Primont and Russell, 1978. A somewhat weaker version was proved by Sono, 1945; 1961, who maintained separability of I^2 from I^1 as well.) Although this structure is commonly referred to as 'separable', it is clearly stronger than separability as the term has historically been used.

4 Two-stage budgeting

It was emphasized in Section 2 that the principal motivation for separability is that it rationalizes decentralizability of the (possibly complex) expenditure-constrained

maximization problem. But, since separability only guarantees that sectoral expenditure is optimally allocated within the sector, full rationalization requires in addition that the optimal amount of money be allocated to each sector. This is accomplished by solving the allocation problem (10).

But solving this problem is tantamount to solving the overall optimization problem and does not seem to reduce the informational requirements. For this reason, Strotz (1957) and Gorman (1959) emphasized the existence of sectoral price aggregates (indexes) that can be used to simplify the first stage of the two-stage optimization problem. Formally, we say that price aggregation is possible if the allocation functions in (11) can be written as

$$y_r = \Theta^r(y, \Pi^1(p^1), \ldots, \Pi^R(p^R)), \quad r = 1, \ldots, R$$
$$(18)$$

(where the price-index functions, Π^r, $r = 1, \ldots, R$, are *not* assumed to be homogeneous of degree one or even homothetic).

Price aggregation is equivalent to the following structure for the direct utility function (in an appropriate permutation of the sectoral indices, $1, \ldots, R$):

$$U(x) = F\left(\sum_{r=1}^{D} f^r(x^r) + G(f^{D+1}(x^{D+1}), \ldots, f^R(x^R)) \right),$$
$$(19)$$

where each f^r, $r = D+1, \ldots, R$, is homothetic and hence can be normalized to be homogeneous of degree one and the sectoral indirect utility functions dual to the first D aggregator functions, defined by

$$v^r(y_r, p^r) = \max_{x^r}\{U^r(x^r) : p^r x^r \leq y_r\},$$
$$r = 1, \ldots, R,$$
$$(20)$$

have the structure,

$$v^r(y_r, p^r) = v^r\left(\frac{y_r}{\Pi^r(p^r)} \right) + w^r(p^r), \quad r = 1, \ldots, D,$$
$$(21)$$

where each w^r is homogeneous of degree zero in p^r and Π^r is homogeneous of degree 1 in p^r. (Gorman, 1959, assuming away the troublesome two-group case, proved a restricted version of this result. Exploiting Sono independence and some newer results not available to Gorman in 1959, Blackorby and Russell, 1997, proved the general result, showing that the entire structure needed for two-stage budgeting is, in fact, imbedded in the two-group case.)

An interesting special case of (19) is obtained if $D = 0$:

$$U(x) = F(f^1(x^1), \ldots, f^R(x^R)). \quad (22)$$

In this case, the first stage of the budgeting algorithm can be expressed as choosing $X^1, \ldots, X^R$ to maximize $F(X^1, \ldots, X^R)$ subject to the budget constraint, $\sum_r \Pi^r(p^r)X^r = y$. This structure allows the two-stage budgeting to be accomplished using only price and quantity aggregates in the first stage.

5 Closing remarks

We have chosen consumer demand theory as a way to illustrate the use of separability in economic analysis. However, separability assumptions, either explicit or implicit, are found in numerous areas in economics. For references to the literature, the reader can consult Blackorby, Primont and Russell (1978; 1998).

CHARLES BLACKORBY, DANIEL PRIMONT AND R. ROBERT RUSSELL

See also **duality; Gorman, W.M. (Terence); indirect utility function; Strotz, Robert H.**

Bibliography

Blackorby, C., Primont, D. and Russell, R.R. 1978. *Duality, Separability, and Functional Structure: Theory and Economic Applications*. New York and Amsterdam: North-Holland.

Blackorby, C., Primont, D. and Russell, R.R. 1998. Separability: a survey. In *Handbook of Utility Theory*, ed. S. Barbara, P.J. Hammond and C. Seidl. Dordrecht: Kluwer.

Blackorby, C. and Russell, R.R. 1997. Two-stage budgeting: an extension of Gorman's theorem. *Economic Theory* 9, 185–93.

Debreu, G. 1959. Topological methods in cardinal utility theory. In *Mathematical Methods in the Social Sciences*, ed. K. Arrow, S. Karlin and P. Suppes. Stanford, CA: Stanford University Press.

Gorman, W.M. 1959. Separability and aggregation. *Econometrica* 27, 469–81.

Gorman, W.M. 1968. The structure of utility functions. *Review of Economic Studies* 32, 369–90.

Leontief, W.W. 1947a. A note on the interrelation of subsets of independent variables of a continuous function with continuous first derivatives. *Bulletin of the American Mathematical Society* 53, 343–50.

Leontief, W.W. 1947b. Introduction to a theory of the internal structure of functional relationships. *Econometrica* 15, 361–73.

Sono, M. 1945. The effect of price changes on the demand and supply of separable goods [in Japanese]. *Kokumin Keisai Zasshi* 74, 1–51.

Sono, M. 1961. The effect of price changes on the demand and supply of separable goods. *International Economic Review* 2, 239–71.

Strotz, R.H. 1957. The empirical implications of a utility tree. *Econometrica* 25, 269–80.

sequence economies

A 'sequence economy' is a general equilibrium model in discrete time including specific provision for the availability of markets at a sequence of dates (Hicks, 1939; Radner, 1972). Markets reopen over time, and at each date firms and households act so that plans and prospects for actions on markets available in the future significantly affect their current actions.

This model is in contrast to the Arrow–Debreu model with a (complete) full set of futures markets (Debreu, 1959). There, all exchanges for current and future goods (including contingent commodities, futures contracts contingent on the realization of uncertain events) are transacted on a market at a single point in time. In the Arrow–Debreu model, there is no need for markets to reopen in the future; economic activity in the future consists simply of the execution of the contracted plans. The Arrow–Debreu model with a full set of futures markets appears unsatisfactory in that it denies commonplace observation: futures markets for goods and Arrow securities (contingent contracts payable in money) are not generally available for most dates or a sufficiently varied array of uncertain events; markets do reopen over time. The sequence economy model is an alternative that allows formalization and explanation of these observations.

Several major classes of theoretical model are set in the sequence economy framework: overlapping generations (Balasko and Shell, 1980; 1981; Geanakoplos and Polemarchakis, 1991; Wallace, 1980); temporary equilibrium (Grandmont, 1977; Lucas, 1978), sunspot equilibrium (Chiappori and Guesnerie, 1991), incomplete markets (Geanakoplos, 1990; Magill and Quinzii, 1996). These are general equilibrium models over sequential time emphasizing monetary and financial structure. Each of these areas has a large literature of its own. Typically these models assume a given (incomplete) structure of active financial markets without a detailed foundation for how markets come to be incomplete. This contrasts with the statement of the sequence economy model presented below, which derives market activity and incompleteness endogenously as an equilibrium outcome reflecting transaction costs.

The sequence economy model is particularly suitable to provide a microeconomic foundation for the store-of-value function of money (Hahn, 1971; Starrett, 1973). It is precisely because markets reopen over time that agents may find it desirable to carry abstract purchasing power from one date to succeeding dates. Typically, this will take the form of transactions on spot markets at a succession of dates with money or other financial assets held over time to reflect the (net) excess value of prior sales over purchases. This may occur simply because the model does not provide for futures markets or because futures markets, though available in principle, are in practice inactive. Endogenously determined inactivity of futures markets is the result of transaction costs which tend to make the use of futures markets disproportionately costly compared with spot markets.

There are three principal reasons for the excess cost of futures markets:

1. The necessarily greater complexity of futures contracts may require use of more resources (for example, for record keeping or enforcement) than spot markets.
2. The transaction costs of a futures contract are incurred (partly) at the transaction date, those of an equivalent spot transaction are incurred in the future. The present discounted value of the spot transaction costs incurred in the distant future may be lower than the futures market transaction cost incurred in the present, simply because of time-discounting.
3. Use of a full set of futures markets under uncertainty implies that most contracts transacted become otiose and are left unfulfilled as their effective dates pass and the events on which they were contingent do not occur. There is a corresponding saving in transaction costs associated with reducing the number of transactions required by use of a single spot transaction instead of many contingent commodity contracts, though this reduction may imply a different and inferior allocation of risk-bearing.

We now present a formal pure exchange sequence economy model with transaction costs (Kurz, 1974; Heller and Starr, 1976).

Commodity i for delivery at date τ may be bought spot at date τ or futures at any date t, $1 \leq t < \tau$. The complete system of spot and futures markets is available at each date (although some markets may be inactive). The time horizon is date K; each of H households is alive at time 1 and cares nothing about consumption after K. There are n commodities deliverable at each date; in the monetary interpretation of the model spot money is one of the goods. At each date and for each commodity, the household has available the current spot market, and futures markets for deliveries at all future dates. Spot and futures markets will also be available at dates in the future and prices on the markets taking place in the future are currently known. Thus in making his purchase and sale decisions, the household considers without price uncertainty whether to transact on current markets or to postpone transactions to markets available at future dates. There is a sequence of budget constraints, one for the market at each date. That is, for every date, the household faces a budget constraint on the spot and futures transactions taking place at that date, (4) below. The value of its sales to the market at each date (including delivery of money) must balance its purchases at that date.

In addition to a budget constraint, the agent's actions are restricted by a transaction technology. This technology specifies for each complex of purchases and sales at date t, what resources will be consumed by the process of transaction. It is because transaction costs may differ between spot and futures markets for the same good

that we consider the reopening of markets allowed by the sequence economy model. Specific provision for transaction cost is introduced to allow an endogenous determination of the activity or inactivity of markets. In the special case where all transaction costs are nil, the model is unnecessarily complex; there is no need for the reopening of markets, and the equilibrium allocations are identical to those of the Arrow–Debreu model. Conversely, in the case where some futures markets are prohibitively costly to operate and others are costless, then there is an incomplete array of spot and futures markets and the model is an example of that of Radner (1972).

All of the n-dimensional vectors below are restricted to be non-negative.

$x_\tau^h(t) = $ vector of purchases for any purpose at date t by household h for delivery at date τ.
$y_\tau^h(t) = $ vector of sales analogously defined.
$z_\tau^h(t) = $ vector of inputs necessary to transactions undertaken at time t. The index τ again refers to the date at which these inputs are actually delivered.
$\omega^h(t) = $ vector of endowments at t for household h.
$s^h(t) = $ vector of goods coming out of storage at date t.
$r^h(t) = $ vector of goods put into storage at date t.
$p_\tau(t) = $ price vector on market at date t for goods deliverable at date τ.

With this notation, $p_{it}(t)$ is the (scalar) spot price of good i at date t, and $p_{i\tau}(t)$ for $\tau > t$ is the futures price (for delivery at τ) of good i at date t.

The (non-negative) consumption vector for household h is

$$c^h(t) = \omega^h(t) + \sum_{\tau=1}^{t} \left[x_t^h(\tau) - y_t^h(\tau) - z_t^h(\tau) \right]$$
$$+ s^h(t) - r^h(t) \geqq 0, \quad (t = 1, \ldots, K). \tag{1}$$

That is, consumption at date t is the sum of endowments plus all purchases past and present with delivery date t minus all sales for delivery at t minus transaction inputs with date t (including those previously committed) plus what comes out of storage at t minus what goes into storage. We suppose that households care only about consumption and not about which market consumption comes from.

The household is constrained by its transaction technology, $T^h(t)$, and by its storage technology, $S^h(t)$. $T^h(t)$ specifies the resources, for example, how much leisure time and shoeleather, must be used to carry out a transaction. Let $x^h(t)$ denote the vector of $x_\tau^h(t)$'s [and similarly for $y^h(t)$ and $z^h(t)$]. We insist

$$[x^h(t), y^h(t), z(t)] \in T^h(t), \quad (t = 1, \ldots, K). \tag{2}$$

Naturally, storage input and output vectors must be feasible, so

$$[r^h(t), s^h(t+1)] \in S^h(t), \qquad (t-1, \ldots, K-1).$$
(3)

The budget constraints for household h are then:

$$p(t) \cdot x^h(t) \leqq p(t) \cdot y^h(t), \qquad (t = 1, \ldots, K).$$
(4)

Households may transfer purchasing power forward in time by using futures markets and by storage of goods that will be valuable in the future. Purchasing power may be carried backward by using futures markets. But these may be very costly transactions. In a monetary interpretation of the model, where money and promissory notes are present, the household can either hold money as a store of wealth, or it can buy or sell notes.

Let household h's action at date t be denoted $a^h(t) \equiv [x^h(t) y^h(t), z^h(t), r^h(t), s^h(t)]$. Let a^h be a vector of the $a^h(t)$'s, and define x^h, y^h, z^h, r^h and s^h similarly. Define $B^h(p)$ as the set of a^hs which satisfy constraints (1)–(4). The household chooses $a^h(t)$ to maximize $U^h(c^h)$ over $B^h(p)$. Denote the demand correspondence (i.e. the set of maximizing a^hs) by $\gamma^h(p)$.

The model can be interpreted as monetary or non monetary. We think of money as simply a 0th good that does not enter household preferences. Futures contracts in money are discounted promissory notes. $x^h_{0t}(t)$ is h's monetary receipts at t, $x^h_{0\tau}(t)$ is h's note purchase at t due at τ. Money is not treated as numeraire – positivity of its value cannot be assumed – it has a price $p_{0t}(t)$.

The correspondences $\gamma^h_t(p)$ are always homogeneous of degree zero in $p(t)$, as is seen from the definition of $B^h(p)$. We can therefore restrict the price space to the simplex. Let S^t denote the unit simplex of dimensionality, $n(K-t+1)$. Let $P = X^K_{t=1} S^t$, where X denotes a Cartesian product.

An equilibrium of the economy is a price vector $p^* \in P$ and an allocation a^{h*}, for each h, so that $a^{h*} \in \gamma^h(p^*)$ for all h and

$$\sum_{h=1}^{H} x^{h*} \leqq \sum_{h=1}^{H} y^{h*}$$
(5)

(the inequality holds coordinate-wise), where for any good i, t, τ such that the strict inequality holds in (5) it follows that $p^*_{i\tau}(t) = 0$. The equilibrium of a monetary economy is said to be *non-trivial* (that is, the economy is really monetary) if $p^*_{0t}(t) \neq 0$ for all t. Sufficient conditions for existence of equilibrium are continuity and convexity requirements typical of an Arrow–Debreu model appropriately extended. Transaction costs are often thought to be non-convex, leading to approximate

equilibrium rather than full equilibrium results (Heller and Starr, 1976).

In the case of fiat (unbacked) money, existence of non-trivial monetary equilibrium requires additional structure designed to maintain positivity of the price of money (boundedness of the price level expressed in monetary terms). This may take a variety of forms: the model may arbitrarily require that fiat money be held or turned in at a finite horizon; households may expect fiat money to be valuable in the future sustaining its value in the present; there may be taxes payable in fiat money. Alternatively, the model may assume an infinite horizon (typical of the overlapping generations model) so that the lack of backing for fiat money need not be experienced (though a nil value of fiat money in equilibrium is still a logical possibility).

In contrast to the Arrow–Debreu economy, a sequence economy equilibrium allocation is not generally Pareto efficient. This is not due simply to the presence of transaction costs; transaction costs technically necessary to a reallocation must be incurred, and they represent no inefficiency. The Arrow–Debreu model, however, uses a lifetime budget constraint. The corresponding constraint here is the sequence of budget constraints in (4). Transfer of purchasing power intertemporally – costless in the Arrow–Debreu model – is here a resource using activity; it requires purchase and sale of assets with resultant transaction cost. But the intertemporal transfer of purchasing power, unlike reallocation of goods among households, is needed not to satisfy technical or consumption requirements but rather to satisfy the administrative requirements of sequential budget constraint embodied in (4). Hence technically feasible Pareto-improving reallocations may be prevented in equilibrium by prohibitive transaction costs which would have to be incurred to satisfy the purely administrative requirements of crediting and debiting agents' budgets intertemporally (Hahn, 1971). If trade in monetary instruments is costless, however, then an equilibrium allocation is Pareto efficient (Starrett, 1973). Thus the sequence economy model provides a value-theoretic foundation for the store-of-value role of money.

ROSS M. STARR

See also **general equilibrium.**

Bibliography

Balasko, Y. and Shell, K. 1980. The overlapping-generations model, I. The case of pure exchange without money. *Journal of Economic Theory* 23, 281–306.

Balasko, Y. and Shell, K. 1981. The overlapping-generations model. II. The case of pure exchange with money. *Journal of Economic Theory* 24, 112–42.

Chiappori, P.A. and Guesnerie, R. 1991. Sunspot equilibria in sequential markets models. In *Handbook of Mathematical Economics*, vol. 4, ed. W. Hildenbrand and H. Sonnenschein. Amsterdam: North-Holland.

Debreu, G. 1959. *Theory of Value*. New York: Wiley.

Geanakoplos, J. 1990. An introduction to general equilibrium with incomplete asset markets. *Journal of Mathematical Economics* 19, 1–38.

Geanakoplos, J.D. and Polemarchakis, H.M. 1991. Overlapping generations. In *Handbook of Mathematical Economics*, vol. 4, ed. W. Hildenbrand and H. Sonnenschein. Amsterdam: North-Holland.

Grandmont, J.M. 1977. Temporary general equilibrium theory. *Econometrica* 45, 535–72.

Hahn, F.H. 1971. Equilibrium with transaction costs. *Econometrica* 39, 417–39.

Heller, W.P. and Starr, R.M. 1976. Equilibrium with non-convex transactions costs: monetary and non-monetary economies. *Review of Economic Studies* 43, 195–215.

Hicks, J.R. 1939. *Value and capital*. Oxford: Oxford University Press.

Kurz, M. 1974. Equilibrium in a finite sequence of markets with transactions cost. *Econometrica* 42, 1–20.

Lucas, Jr., R.E. 1978. Asset prices in an exchange economy. *Econometrica* 46, 1429–45.

Magill, M. and Quinzii, M. 1996. *Theory of incomplete markets*. Cambridge, MA: MIT Press.

Radner, R. 1972. Existence of equilibrium of plans, prices, and price expectations in a sequence of markets. *Econometrica* 40, 289–303.

Starrett, D.A. 1973. Inefficiency and the demand for 'money' in a sequence economy. *Review of Economic Studies* 40, 437–48.

Wallace, N. 1980. The overlapping generations model of money. In *Models of Monetary Economics*, ed. J.H. Kareken and N. Wallace. Minneapolis: Federal Reserve Bank of Minneapolis.

sequential analysis

Statistical experiments are of either fixed sample or sequential design. A *fixed sample size* experiment is one in which the sample size taken for experimentation is predetermined, while a sequential experiment involves monitoring incoming data to help determine an appropriate time to stop experimentation.

To formalize these notions, suppose the data can be observed one-at-a-time; let X_1, X_2, ... denote this possible stream of data. Examples include a series of products coming off an assembly line, a series of missiles being tested for accuracy, and a series of patients participating in a clinical trial.

A key concept is that of a stopping rule, R, which is a description of the manner in which the data stream will be used to determine cessation of the experiment.

Example 1 Consider the stopping rule R_1: stop experimentation after n observations have been taken. This stopping rule effectively defines what we earlier called a fixed sample size experiment, since we will take precisely n observations.

Example 2 Consider the stopping rules (where $\overline{X}_j = \Sigma_{i=1}^{j} X_i / j$) R_2: stop experimentation if $\overline{X}_{50} > 0.62$ or (failing that) when $n = 100$; R_3: after each new observation, X_j, check whether or not $\overline{X}_j \geq 0.5 + 0.823/\sqrt{j}$ if so, stop experimentation and otherwise take the next observation. Note that R_2 allows experimentation to stop only after 50 or 100 observations have been taken (this is often called 'group sequential' analysis), while R_3 gives rise to the possibility of stopping after any observation.

To see why stopping rules such as R_2 and R_3 can be desirable, consider a clinical trial investigating a new treatment in which, for the jth participating patient, the observation is a Bernoulli (θ) random variable, X_j, which can assume the values 1 (denoting treatment success) or 0 (denoting treatment failure). Thus θ is the probability of the treatment being successful. Suppose that the standard (old) treatment is known to have a success probability of $\frac{1}{2}$, so it is desired to test the hypothesis (H_0) that $\theta \leq \frac{1}{2}$ (the old treatment is better) versus the hypothesis (H_a) that $\theta > \frac{1}{2}$ (the new treatment is better).

A typical fixed sample size test of these hypotheses would proceed by choosing a sample size, say $n = 100$, observing $X_1, \ldots, X_{100}$ from $n = 100$ independent patients, and then rejecting if $\overline{X}_{100} \geq 0.582$. This is an $\alpha = 0.05$ level test. (We make no judgement here concerning the appropriateness of formulating this problem as a statistical hypothesis test.)

Suppose now that the experimenters happen to look at the data after 50 patients have participated in the trial, and observe that, for all 50, the treatment proved successful. This would appear to be overwhelming evidence that the new treatment is better, and would lead reasonable people to stop the clinical trial and recommend adoption of the new treatment. It is a rather surprising fact that this conclusion would be *forbidden* by classical statistics, because the original design called for a sample of size 100. (Classical analyses do not allow deviation from original experimental protocol.) It would have been possible, however, to plan for such a possible eventuality by adopting a sequential design, whereby after every observation (or every few observations) the possibility of stopping is allowed. Indeed, R_2 and R_3 are two such stopping rules, and had either been employed, the above-mentioned clinical trial would certainly have stopped by the time the overwhelming evidence had accumulated.

As indicated in the above example, the advantage of a sequential experimental design is that it allows one to stop the experiment precisely when sufficient evidence has accumulated. The disadvantages of a sequential design are that it can be more expensive (often it is cheaper per observation if the data is collected all at once or in large batches), and that it is harder to analyse from the classical perspective. This last point has to do with the fact that the stopping rule can significantly affect classical statistical measures.

Example 2 (continued) Suppose the stopping rule R_2 had been employed in the clinical trial (that is, an *interim analysis* at the halfway point in the trial had been performed). Also, suppose that, if one did stop after 50 observations (that is, if $\overline{X}_{50} > 0.62$), then H_0 would be rejected, and that, if the trial lasted for all 100 observations (that is, if $\overline{X}_{50} < 0.62$, so that the experiment did not stop at the halfway point), then H_0 would be rejected when $\overline{X}_{100} > 0.582$. It can be shown that, for a *fixed* sample of 50 observations, rejecting H_0 when $\overline{X}_{50} > 0.62$ is an $\alpha = 0.05$ level test, as is rejecting H_0 if $\overline{X}_{100} > 0.582$ for a *fixed* sample size experiment with $n = 100$. For the experiment using R_2, however, it can be shown that the level is $\alpha = 0.095$. (One obtains an error probability larger than each of the separate $\alpha = 0.05$ because use of R_2 gives 'two chances' to reject H_0.) Thus, if R_1 had been used and $\overline{X}_{100} = 0.582$ had been observed, one could claim significant evidence against H_0 at the $\alpha = 0.05$ level, while if R_2 had been used, one could not claim significance at the $\alpha = 0.05$ level.

It should be mentioned that there is considerable controversy over the issue of whether use of stopping rules should affect statistical conclusions. When classical measures are used, there is no typically a substantial effect. But, interestingly, for certain other statistical measures, such as Bayesian measures, the stopping rule has *no* effect. Thus, employment of the Bayesian approach to statistics allows one to collect data without having to pre-specify a rigid initial stopping rule, greatly increasing the flexibility of experimentation. For discussion of this issue, and support for the Bayesian viewpoint, see Berger and Wolpert (1984) and Berger (1985).

The founder of sequential analysis is generally acknowledged to be Abraham Wald, with Milton Friedman and W. Allen Wallis providing substantial motivational and collaborative support. Early history of sequential analysis is given in Wald (1947), which developed the basic formulation of the problem in terms of stopping rules and analysed a number of basic situations, such as the sequential probability ratio test (for testing between two simple hypotheses). Most of the subsequent work in sequential analysis has focused on either (*a*) evaluating classical measures, such as error probabilities, for special stopping rules (see Siegmund, 1985), or (*b*) determining optimal stopping rules. This last problem is very difficult, and can be rephrased as the problem of deciding if enough information is already available to reach a decision, or if another (or several) observations should be taken. The mathematics of this problem is essentially that of dynamic programming. For general reviews of sequential analysis, see DeGroot (1970), Ghosh (1970), Ghosh, Sen and Mukhopadhyay (1997),

Govindarajulu (1981), Berger (1985), Lai (2001), Sen and Ghosh (1991), and Siegmund (1985).

JAMES O. BERGER

Bibliography

Berger, J. 1985. *Statistical Decision Theory and Bayesian Analysis*. New York: Springer-Verlag.

Berger, J. and Wolpert, R. 1984. *The Likelihood Principle*. Hayward, CA: Institute of Mathematical Statistics.

DeGroot, M.H. 1970. *Optimal Statistical Decisions*. New York: McGraw-Hill.

Ghosh, B.K. 1970. *Sequential Tests of Statistical Hypotheses*. Reading, MA: Addison-Wesley.

Ghosh, M., Sen, P.K. and Mukhopadhyay, N. 1997. *Sequential Estimation*. New York: Wiley.

Govindarajulu, Z. 1981. *The Sequential Statistical Analysis of Hypothesis Testing, Point and Interval Estimation, and Decision Theory*. Columbus: American Science Press.

Lai, T.L. 2001. Sequential analysis: some classical problems and new challenges (with discussion). *Statistica Sinica* 11, 303–408.

Sen, P. and Ghosh, B. 1991. *Handbook of Sequential Analysis*. London: Marcel Dekker.

Siegmund, D. 1985. *Sequential Analysis: Tests and Confidence Intervals*. New York: Springer-Verlag.

Wald, A. 1947. *Sequential Analysis*. New York: Wiley.

serial correlation and serial dependence

1 Introduction

Serial correlation and serial dependence have been central to time series econometrics. The existence of serial correlation complicates statistical inference of econometric models; and in time series analysis, inference of serial correlation, or more generally, serial dependence, is crucial to characterize the dynamics of time series processes. Lack of serial correlation is also an important implication of many economic theories and economic hypotheses. For example, the efficient market hypothesis implies that asset returns are an martingale difference sequence (*m.d.s.*), and so are serially uncorrelated. More generally, rational expectations theory implies that the expectational errors of the economic agent are serially uncorrelated. In this article we first discuss various tests for serial correlation, for both estimated model residuals and observed raw data, and we discuss their relationships. We then discuss serial dependence in a nonlinear time series context, introducing related measures and tests for serial dependence.

2 Testing for serial correlation

Consider a linear regression model

$$Y_t = X_t' \beta^0 + \varepsilon_t, \quad t = 1, \cdots, n, \qquad (2.1)$$

where Y_t is a dependent variable, X_t is a $k \times 1$ vector of explanatory variables, β^0 is an unknown $k \times 1$ parameter vector, and ε_t is an unobservable disturbance with $E(\varepsilon_t | X_t) = 0$. Suppose X_t is strictly exogenous such that $\text{cov}(X_t, \varepsilon_s) = 0$ for all t, s. Then (2.1) is called a static regression model. If X_t contains lagged dependent variables, (2.1) is called a dynamic regression model.

For a linear dynamic regression model, serial correlation in $\{\varepsilon_t\}$ will generally render inconsistent the OLS estimator. To see this, we consider an AR(1) model

$$Y_t = \beta_0^0 + \beta_1^0 Y_{t-1} + \varepsilon_t = X_t' \beta^0 + \varepsilon_t,$$

where $X_t = (1, Y_{t-1})'$. If $\{\varepsilon_t\}$ also follows an AR(1) process, we will have $E(X_t \varepsilon_t) \neq 0$, rendering inconsistent the OLS estimator for β^0. It is therefore important to check serial correlation for estimated model residuals, which serves as a misspecification test for a linear dynamic regression model. For a static linear regression model, it is also useful to check serial correlation. In particular, if there exists no serial correlation in $\{\varepsilon_t\}$ in a static regression model, then there is no need to use a long-run variance estimator of the OLS estimator $\hat{\beta}$ (for example, Andrews, 1991; Newey and West, 1987).

2.1 Durbin–Watson test

Testing for serial correlation has been a longstanding problem in time series econometrics. The most well known test for serial correlation in regression disturbances is the Durbin–Watson test, which is the first formal procedure developed for testing first order serial correlation

$$\varepsilon_t = \rho \varepsilon_{t-1} + u_t, \qquad \{u_t\} \sim i.i.d.(0, \sigma^2)$$

using the OLS residuals $\{e_t\}_{t=1}^n$ in a static linear regression model. Durbin and Watson (1950; 1951) propose a test statistic

$$d = \frac{\sum_{t=2}^n (e_t - e_{t-1})^2}{\sum_{t=1}^n e_t^2}.$$

Durbin and Watson present tables of bounds at the 0.05, 0.025 and 0.01 significance levels of the d statistic for static regressions with an intercept. Against the one-sided alternative that $\rho > 0$, if d is less than the lower bound d_L, the null hypothesis that $\rho = 0$ is rejected; if ρ is greater than the upper bound d_U, the null hypothesis is accepted. Otherwise, the test is equivocal. Against the one-sided alternative that $\rho < 0$, $4 - d$ can be used to replace d in the above procedure.

The Durbin–Watson test has been extended to test for lag 4 autocorrelation by Wallis (1972) and for autocorrelation at any lag by Vinod (1973).

2.2 Durbin's h test

The Durbin–Watson d test is not applicable to dynamic linear regression models, because parameter estimation uncertainty in the OLS estimator $\hat{\beta}$ will have nontrivial

impact on the distribution of d. Durbin (1970) developed the so-called h test for first-order autocorrelation in $\{\varepsilon_t\}$ that takes into account parameter estimation uncertainty in $\hat{\beta}$. Consider a simple dynamic linear regression model

$$Y_t = \beta_0^0 + \beta_1^0 Y_{t-1} + \beta_2^0 X_t + \varepsilon_t,$$

where X_t is strictly exogenous. Durbin's h statistic is defined as:

$$h = \hat{\rho} \sqrt{\frac{n}{1 - n \hat{\text{var}}(\hat{\beta}_1)}},$$

where $\hat{\text{var}}(\hat{\beta}_1)$ is an estimator for the asymptotic variance of $\hat{\beta}_1$, $\hat{\rho}$ is the OLS estimator from regressing e_t on e_{t-1} (in fact, $\hat{\rho} \approx 1 - d/2$). Durbin (1970) shows that $h \xrightarrow{d} N(0, 1)$ as $n \to \infty$ under null hypothesis that $\rho = 0$.

2.3 Breusch–Godfrey test

A more convenient and generally applicable test for serial correlation is the Lagrange multiplier test developed by Breusch (1978) and Godfrey (1978). Consider an auxiliary autoregression of order p:

$$\varepsilon_t = \sum_{j=1}^p \alpha_j \varepsilon_{t-j} + z_t, \qquad t = p + 1, \cdots, n.$$

$$(2.2)$$

The null hypothesis of no serial correlation implies $\alpha_j = 0$ for all $1 \leq j \leq p$. Under the null hypothesis, we have $n R_{uc}^2 \xrightarrow{d} \chi_p^2$, where R_{uc}^2 is the uncentred R^2 of (2.2). However, the autoregression (2.2) is infeasible because ε_t is unobservable. One can replace ε_t with the OLS residual e_t:

$$e_t = \sum_{j=1}^p \alpha_j e_{t-j} + v_t, \qquad t = p + 1, \cdots, n.$$

Such a replacement, however, may contaminate the asymptotic distribution of the test statistic because $e_t = \varepsilon_t - (\hat{\beta} - \beta)' X_t$ contains the estimation error $(\hat{\beta} - \beta)' X_t$ where X_t may have nonzero correlation with the regressors e_{t-j} for $1 \leq j \leq p$ in dynamic regression models. This correlation affects the asymptotic distribution of $n R_{uc}^2$ so that it will not be χ_p^2. To purge this impact of the asymptotic distribution of the test statistic, one can consider the augmented auxiliary regression

$$e_t = X_t' \gamma + \sum_{j=1}^p \alpha_j e_{t-j} + v_t, \qquad t = p + 1, \cdots, n.$$

$$(2.3)$$

The inclusion of X_t will capture the impact of estimation error $(\hat{\beta} - \beta)' X_t$. As a result, the test statistic $n R^2 \xrightarrow{d} \chi_p^2$ under the null hypothesis, where, assuming that

X_t contains an intercept, R^2 is the centred squared multi-correlation coefficient in (2.3). For a static linear regression model, it is not necessary to include X_t in the auxiliary regression, because $\{X_t\}$ and $\{\varepsilon_t\}$ are uncorrelated, but it does not harm the size of the test if X_t is included. Therefore, the nR^2 test is applicable to both static and dynamic regression models. We note that Durbin's h test is asymptotically equivalent to the nR^2 test of (2.3) with $p = 1$.

2.4 Box–Pierce–Ljung test
In time series ARMA modelling, Box and Pierce (1970) propose a portmanteau test as a diagnostic check for the adequacy of an ARMA model

$$Y_t = \psi_0 + \sum_{j=1}^{r} \psi_j Y_{t-j} + \sum_{j=1}^{q} \theta_j \varepsilon_{t-j} + \varepsilon_t,$$

$$\{\varepsilon_t\} \sim i.i.d.(0, \sigma^2).$$

$$(2.4)$$

Suppose e_t is an estimated residual obtained from a maximum likelihood estimator. One can define the residual sample autocorrelation function

$$\hat{p}(j) = \frac{\hat{\gamma}(j)}{\hat{\gamma}(0)}, \quad j = 0, \pm 1, \cdots, \pm(n-1),$$

where $\hat{\gamma}(j) = n^{-1} \sum_{t=|j|+1}^{n} e_t e_{t-|j|}$ is the residual sample autocovariance function.

Box and Pierce (1970) propose a portmanteau test

$$Q \equiv n \sum_{j=1}^{p} \hat{\rho}^2(j) \xrightarrow{d} \chi^2_{p-(r+q)},$$

where the asymptotic χ^2 distribution follows under the null hypothesis of no serial correlation, and the adjustment of degrees of freedom $r+q$ is due to the impact of parameter estimation uncertainty for the r autoregressive coefficients and q moving average coefficients in (2.4).

To improve small sample performance of the Q test, Ljung and Box (1978) propose a modified Q test statistic:

$$Q^* \equiv n(n+2) \sum_{j=1}^{p} (n-j)^{-1} \hat{\rho}^2(j) \xrightarrow{d} \chi^2_{p-(r+q)}.$$

The modification matches the first two moments of Q^* with those of the χ^2 distribution. This improves the size in small samples, although not the power of the test.

The Q test is applicable to test serial correlation in the OLS residuals $\{e_t\}$ of a linear static regression model, with $Q \xrightarrow{d} \chi^2_p$ under the null hypothesis. Unlike for ARMA models, there is no need to adjust the degrees of freedom for the χ^2 distribution because the estimation error $(\hat{\beta} - \beta)' X_t$ has no impact on it, due to the fact that $\text{cov}(X_t, \varepsilon_s) = 0$ for all t, s. In fact, it could be shown that the nR^2

and Q statistics are asymptotically equivalent under the null hypothesis. However, when applied to the estimated residual of a dynamic regression model which contains both endogenous and exogenous variables, the asymptotic distribution of the Q test is generally unknown (Breusch and Pagan, 1980). One solution is to modify the Q test statistic as follows:

$$\hat{Q} \equiv n\hat{\rho}'(I - \hat{\Phi})^{-1}\hat{\rho} \xrightarrow{d} \chi^2_p \text{ as } n \to \infty,$$

where $\hat{\rho} = [\hat{\rho}(1), \cdots, \hat{\rho}(p)]'$ and $\hat{\Phi}$ captures the impact caused by nonzero correlation between $\{X_t\}$ and $\{\varepsilon_s\}$. See Hayashi (2000, Section 2.10) for more discussion.

2.5 Spectral density-based test
Much criticism has been levelled at the possible low power of the Box–Pierce–Ljung portmanteau tests, which also applies to the nR^2 test, due to the asymptotic equivalence between the Q test and the nR^2 test for a static regression. Moreover, there is no theoretical guidance on the choice of p for these tests. A fixed lag order p will render inconsistent any test for serial correlation of unknown form.

To test serial correlation of unknown form in the estimated residuals of a linear regression model, which can be static or dynamic, Hong (1996) uses a kernel spectral density estimator

$$\hat{h}(\omega) = \frac{1}{2\pi} \sum_{j=1-n}^{n-1} k(j/p)\hat{\gamma}(j)e^{-ij\omega}, \qquad \omega \in |-\pi, \pi|,$$

and compares it with the flat spectrum implied by the null hypothesis of no serial correlation:

$$\hat{h}_0(\omega) = \frac{1}{2\pi} \hat{\gamma}(0), \qquad \omega \in [-\pi, \pi].$$

Under the null hypothesis, $\hat{h}(\omega)$ and $\hat{h}_0(\omega)$ are close. If $\hat{h}(\omega)$ is significantly different from $\hat{h}_0(\omega)$ there is evidence of serial correlation. A global measure of the divergence between $\hat{h}(\omega)$ and $\hat{h}_0(\omega)$ is the quadratic form

$$L^2(\hat{h}, \hat{h}_0) \equiv \int_{-\pi}^{\pi} \left[\hat{h}(\omega) - \hat{h}_0(\omega) \right]^2 d\omega$$

$$= \sum_{j=1}^{n-1} k^2(j/p)\hat{\gamma}^2(j).$$

The test statistic is a normalized version of the quadratic form:

$$M_o \equiv \left[n \sum_{j=1}^{n-1} k^2(j/p)\hat{\rho}^2(j) - \hat{C}_o(p) \right] \Big/$$

$$\sqrt{\hat{D}_o(p)} \xrightarrow{d} N(0, 1)$$

where the centring and scaling factors

$$\hat{C}_o(p) = \sum_{j=1}^{n-1}(1 - j/n)k^2(j/p),$$

$$\hat{D}_o(p) = 2\sum_{j=1}^{n-2}(1 - j/n)[1 - (j+1)/n]k^4(j/p).$$

This test can be viewed as a generalized version of Box and Pierce's (1970) portmanteau test, the latter being equivalent to using the truncated kernel $k(z)=\mathbf{1}(|z|\leq 1)$, which gives equal weighting to each of the first p lags. In this case, M_o is asymptotically equivalent to

$$M_T \equiv \frac{n\sum_{j=1}^{p}\hat{\rho}^2(j) - p}{\sqrt{2p}} \xrightarrow{d} \frac{\chi_p^2 - p}{\sqrt{2p}}$$

$$\sim N(0,1) \quad as \ p \to \infty.$$

However, uniform weighting to different lags may not be powerful when a large number of lags is employed. For any weakly stationary process, the autocovariance function $\gamma(j)$ typically decays to 0 as lag order j increases. Thus, it is more efficient to discount higher order lags. This can be achieved by using non-uniform kernels. Most commonly used kernels, such as the Bartlett, Pazren and quadratic-spectral kernels, discount higher order lags. Hong (1996) shows that the Daniell kernel $k(z) = \sin(\pi z)/(\pi z)$, $-\infty < z < \infty$, maximizes the power of the M test over a wide class of the kernel functions when $p \to \infty$. The optimal kernel for hypothesis testing differs from the optimal kernel for spectral density estimation.

It is important to note that the spectral density test M applies to both static and dynamic regression models, and no modification is needed when applied to a dynamic regression model. Intuitively, parameter estimation uncertainty causes some adjustment of degrees of freedom, which becomes asymptotically independent when the lag order $p \to \infty$ as $n \to \infty$. This differs from the case where p is fixed.

For similar spectral density-based tests for serial correlation, see Paparoditis (2000), Chen and Deo (2004), and Fan and Zhang (2004).

2.6 Heteroskedasticity–robust tests

All the aforementioned tests assume conditional homoskedasticity or even *i.i.d.* on $\{\varepsilon_t\}$. This rules out high frequency financial time series, which have been documented to have persistent volatility clustering. Some effort has been devoted to robustifying tests for serial correlation. Wooldridge (1990; 1991) proposes a two-stage procedure to robustify the nR^2 test for serial correlation in estimated residuals $\{e_t\}$ of a linear regression model (2.1): (i) regress $(e_{t-1},\ldots,e_{t-p})$ on X_t and save the estimated $p \times 1$ residual vector $\hat{v}_t$; (ii) regress 1 on $\hat{v}_t e_t$ and obtain SSR, the sum of squared residuals; (iii) compare the $n-SSR$ statistic with the asymptotic χ_p^2 distribution. The first auxiliary regression purges the impact of parameter

estimation uncertainty in the OLS estimator $\hat{\beta}$ and the second auxiliary regression delivers a test statistic robust to conditional heteroskedasticity of unknown form.

Whang (1998) also proposes a semiparametric test for serial correlation in estimated residuals of a possibly non-linear regression model. Assuming that $\varepsilon_t = \sigma[Z_t(\alpha)]z_t$, where $\{z_t\} \sim$ i.i.d.(0,1), $\mathrm{var}(\varepsilon_t|I_{t-1}) = \sigma^2[Z_t(\alpha)]$ depends on a random vector with fixed dimension (for example, $Z_t(\alpha) = (\varepsilon_{t-1}^2, \cdots, \varepsilon_{t-K}^2)'$ for a fixed K), but the functional form $\sigma^2(\cdot)$ is unknown. This covers a variety of conditionally heteroskedastic processes, although it rules out non-Markovian processes such as Bollerslev's (1986) GARCH model. Whang (1998) first estimates $\sigma^2[Z_t(\alpha)]$ using a kernel method, and then constructs a Box–Pierce type test for serial correlation in the estimated regression residuals standardized by the square root of the nonparametric variance estimator.

The assumption imposed on $\mathrm{var}(\varepsilon_t|I_{t-1})$ in Whang (1998) rules out GARCH models, and both Wooldridge (1991) and Whang (1998) test serial correlation up to a fixed lag order only. Hong and Lee (2007) have recently robustified Hong's (1996) spectral density-based consistent test for serial correlation of unknown form:

$$\hat{M} \equiv \left[n^{-1}\sum_{j=1}^{n-1}k^2(j/p)\hat{\gamma}^2(j) - \hat{C}(p)\right] \bigg/ \sqrt{\hat{D}(p)},$$

where the centring and scaling factors

$$\hat{C}(p) \equiv \hat{\gamma}^2(0)\sum_{j=1}^{n-1}(1 - j/n)k^2(j/p)$$

$$+ \sum_{j=1}^{n-1}k^2(j/p)\hat{\gamma}_{22}(j),$$

$$\hat{D}(p) \equiv 2\hat{\gamma}^4(0)\sum_{j=1}^{n-2}(1 - j/n)[1 - (j+1)/n]k^4(j/p)$$

$$+ 4\hat{\gamma}^2(0)\sum_{j=1}^{n-2}k^4(j/p)\hat{\gamma}_{22}(j)$$

$$+ 2\sum_{j=1}^{n-2}\sum_{l=1}^{n-2}k^2(j/p)k^2(l/p)\hat{C}^2(0,j,l),$$

with $\hat{\gamma}_{22}(j) \equiv n^{-1}\sum_{t=j+1}^{n-1}[e_t^2 - \hat{\gamma}(0)][e_{t-j}^2 - \hat{\gamma}(0)]$ and $\hat{C}(0,j,l) \equiv n^{-1}\sum_{t=\max(j,l)+1}^{n}[e_t^2 - \hat{\gamma}(0)]e_{t-j}e_{t-l}$. Intuitively, the centring and scaling factors have taken into account possible volatility clustering and asymmetric features of volatility dynamics, so the $\hat{M}$ test is robust to these effects. It allows for various volatility processes, including GARCH models, Nelson's (1991) EGARCH, and Glosten, Jagannathan and Runkle's (1993) Threshold GARCH models.

Martingale tests

Several tests for serial correlation are motivated for testing the *m.d.s.* property of an observed time series $\{Y_t\}$, say asset returns, rather than estimated residuals of a regression model. We now present a unified framework to view some martingale tests for observed data.

Extending an idea of Cochrane (1988), Lo and MacKinlay (1988) first rigorously present an asymptotic theory for a variance ratio test for the m.d.s. hypothesis of $\{Y_t\}$. Because the m.d.s. hypothesis implies $\gamma(j)=0$ for all $j>0$, one has

$$\frac{\text{var}\left(\sum_{j=1}^{p} Y_{t-j}\right)}{p \cdot \text{var}(Y_t)} = \frac{p\gamma(0) + 2p\sum_{j=1}^{p}(1-j/p)\gamma(j)}{p\gamma(0)} = 1.$$

This unity property of the variance ratio can be used to test the m.d.s. hypothesis because any departure from unity is evidence against the m.d.s. hypothesis.

The variance ratio test is essentially based on the statistic

$$\text{VR}_0 \equiv \sqrt{n/p} \sum_{j=1}^{p}(1-j/p)\hat{\rho}(j) = \frac{\pi}{2}\sqrt{n/p}\left[\hat{f}(0) - \frac{1}{2\pi}\right],$$

where $\hat{f}(0)$ is a kernel-based normalized spectral density estimator at frequency 0, with the Bartlett kernel $k(z) = (1-|z|)\,\mathbf{1}\,(|z|\leq 1)$ and a lag order p. In other words, VR_0 is based on a spectral density estimator of frequency 0, and because of this, it is particularly powerful against long memory processes, whose spectral density at frequency 0 is infinity (see Robinson, 1994, for an excellent survey).

Under the m.d.s. hypothesis with conditional homoskedasticity, Lo and MacKinlay (1988) show that for any fixed p,

$$\text{VR}_0 \xrightarrow{d} N[0, 2(2p-1)(p-1)/3p] \text{ as } n \to \infty.$$

Lo and MacKinlay (1988) also consider a heteroskedasticity consistent variance ratio test:

$$\text{VR} \equiv \sqrt{n/p} \sum_{j=1}^{p}(1-j/p)\hat{\gamma}(j)/\sqrt{\hat{\gamma}_2(j)},$$

where $\hat{\gamma}_2(j)$ is a consistent estimator for the asymptotic variance of $\hat{\gamma}(j)$ under conditional heteroskedasticity. Lo and MacKinlay (1988) assume a fourth order cumulant condition that

$$E[(Y_t - \mu)^2(Y_{t-l} - \mu)(Y_{t-1} - \mu)] = 0,$$
$$j, l > 0, j \neq l. \tag{2.5}$$

Intuitively, this condition ensures that the sample autocovariances at different lags are asymptotically uncorrelated; that is, $\text{cov}[\sqrt{n}\hat{\gamma}(j), \sqrt{n}\hat{\gamma}(l)] \to 0$ for all $j \neq l$. As a result, the heroskedasticity-consistent VR has the same asymptotic distribution as VR_0. However, the condition in (2.5) rules out many important volatility processes, such as EGARCH and Threshold GARCH models. Moreover, the variance ratio test only exploits the implication of the m.d.s. hypothesis on the spectral

density at frequency 0; it does not check the spectral density at nonzero frequencies. As a result, it is not consistent against serial correlation of unknown form. See Durlauf (1991) for more discussion.

Durlauf (1991) considers testing the m.d.s. hypothesis for observed raw data $\{Y_t\}$, using the spectral distribution function

$$H(\lambda) \equiv 2\int_0^{\pi\lambda} h(\omega)d\omega = \gamma(0)\lambda$$
$$+ \sqrt{2}\sum_{j=1}^{\infty}\gamma(j)\frac{\sqrt{2}\sin(j\pi\lambda)}{j\pi}, \quad \lambda \in [0,1],$$

where $h(\omega)$ is the spectral density of $\{Y_t\}$:

$$h(\omega) = \frac{1}{2\pi}\sum_{j=-\infty}^{\infty}\gamma(j)\cos(j\omega), \quad \omega \in [-\pi, \pi].$$

Under the m.d.s. hypothesis, $H(\lambda)$ becomes a straight line:

$$H_0(\lambda) = \gamma(0)\lambda, \quad \lambda \in [0,1].$$

An m.d.s. test can be obtained by comparing a consistent estimator for $H(\lambda)$ and $\hat{H}_0(\lambda) = \hat{\gamma}(0)\lambda$.

Although the periodogram (or sample spectral density function)

$$\hat{I}(\omega) \equiv \frac{1}{2\pi n}\left|\sum_{t=1}^{n}(Y_t - \bar{Y})e^{it\omega}\right|^2 - \frac{1}{2\pi}\sum_{j=1-n}^{n-1}\hat{\gamma}(j)e^{-ij\omega}$$

is not consistent for the spectral density $h(\omega)$, the integrated periodogram

$$\hat{H}(\lambda) \equiv 2\int_0^{\lambda\pi}\hat{I}(\omega)d\omega$$
$$= \hat{\gamma}(0)\lambda + \sqrt{2}\sum_{j=1}^{n-1}\hat{\gamma}(j)\frac{\sqrt{2}\sin(j\pi\lambda)}{j\pi}$$

is consistent for $H(\lambda)$, thanks to the smoothing provided by the integration. Among other things, Durlauf (1991) proposes a Cramer–von Mises type statistic

$$CVM \equiv \frac{1}{2}n\int_0^1 \left[\hat{H}(\lambda)/\hat{\gamma}(0) - \lambda\right]^2 d\lambda$$
$$= n\sum_{j=1}^{n-1}\hat{\rho}^2(j)/(j\pi)^2.$$

Under the m.d.s. hypothesis with conditional homoskedasticity, Durlauf (1991) shows

$$CVM \xrightarrow{d} \sum_{j=1}^{\infty}\chi_j^2(1)/(j\pi)^2,$$

where $\{\chi_j^2(1)\}_{j=1}^{\infty}$ is a sequence of i. i. d. χ^2 random variables with one degree of freedom. This asymptotic distribution is nonstandard, but it is distribution-free and can be easily tabulated or simulated. An appealing property of Durlauf's (1991) test is its consistency against serial correlation of unknown form, and there is no need to choose a lag order p.

Deo (2000) shows that under the $m.d.s.$ hypothesis with conditional heteroskedasticity, Durlauf's (1991) test statistic can be robustified as follows:

$$CVM = \sum_{j=1}^{n-1} \frac{\left[\hat{\gamma}^2(j)/\hat{\gamma}_2(j)\right]}{(j\pi)^2} \xrightarrow{d} \sum_{j=1}^{\infty} \chi_j^2(1)/(j\pi)^2.$$

where $\hat{\gamma}_2(j)$ is a consistent estimator for the asymptotic variance of $\hat{\gamma}(j)$ and the asymptotic distribution remains unchanged. Like Lo and MacKinlay (1988), Deo (2000) also imposes the crucial fourth order joint cumulant condition in (2.5).

3 Serial dependence in nonlinear models

The autocorrelation function $\gamma(j)$, or equivalently, the power spectrum $h(\omega)$, of a time series $\{Y_t\}$, is a measure for linear association. When $\{Y_t\}$ is a stationary Gaussian process, $\gamma(j)$ or $h(\omega)$ can completely determine the full dynamics of $\{Y_t\}$.

It has been well documented, however, that most economic and financial time series, particularly high-frequency economic and financial time series, are not Gaussian. For non-Gaussian processes, $\gamma(j)$ and $h(\omega)$ may not capture the full dynamics of $\{Y_t\}$. We consider two nonlinear process examples:

- Bilinear (BL) autoregressive process:

$$Y_t = \alpha\varepsilon_{t-1}Y_{t-2} + \varepsilon_t, \{\varepsilon_t\} \sim i.i.d.(0,\sigma^2).$$
$$(3.1)$$

- Nonlinear moving average (NMA) process:

$$Y_t = \alpha\varepsilon_{t-1}\varepsilon_{t-2} + \varepsilon_t, \{\varepsilon_t\} \sim i.i.d.(0,\sigma^2).$$
$$(3.2)$$

For these two processes, there exists nonlinearity in conditional mean: $E(Y_t|I_{t-1}) = \alpha\varepsilon_{t-1}Y_{t-2}$ under (3.1) and $E(Y_t|I_{t-1}) = \alpha\varepsilon_{t-1}Y_{t-2}$ under (3.2). However, both processes are serially uncorrelated. If $\{Y_t\}$ follows either a BL process in (3.1) or a NMA process in (3.2), $\{Y_t\}$ is not $m.d.s.$ but $\gamma(j)$ and $h(\omega)$ will miss it. Hong and Lee (2003a) document that indeed, for foreign currency markets, most foreign exchange changes are serially uncorrelated, but they are all not $m.d.s.$ There exist predictable nonlinear components in the conditional mean of foreign exchange markets.

Serial dependence may also exist only in higher order conditional moments. An example is Engle's (1982) first order autoregressive conditional heteroskedastic (ARCH

(1)) process:

$$\begin{cases} Y_t = \sigma_t\varepsilon_t, \\ \sigma_t^2 = \alpha_0 + \alpha_1 Y_{t-1}^2, \\ \{\varepsilon_t\} \sim i.i.d.(0,1). \end{cases} \quad (3.3)$$

For this process, the conditional mean $E(Y_t|I_{t-1}) = 0$; which implies $\gamma(j) = 0$ for all $j > 0$.

However, the conditional variance, $var(Y_t|I_{t-1}) = \alpha_0 + \alpha_1 Y_{t-1}^2$, depends on the previous volatility. Both $\gamma(j)$ and $h(\omega)$ will miss such higher order dependence.

In nonlinear time series modelling, it is important to measure serial dependence, that is, any departure from $i.i.d.$, rather than merely serial correlation. As Priestley (1988) points out, the main purpose of nonlinear time series analysis is to find a filter $h(\cdot)$ such that

$$h(Y_t, Y_{t-1}, \cdots.) = \varepsilon_t \sim i.i.d.(0,\sigma^2).$$

In other words, the filter $h(\cdot)$ can capture all serial dependence in $\{Y_t\}$ so that the 'residual' $\{\varepsilon_t\}$ becomes an $i.i.d.$ sequence. One example of $h(\cdot)$ in modelling the conditional probability distribution of Y_t given I_{t-1}, is the probability integral transform

$$Z_t(\beta) = \int_{-\infty}^{Y_t} f(y|I_{t-1},\beta)\,dy,$$

where $f(y|I_{t-1}, \beta)$ is a conditional density model for Y_t given and I_{t-1}, and β is an unknown parameter. When $f(y|I_{t-1}, \beta)$ is correctly specified for the conditional probability density of Y_t given I_{t-1}, that is, when the true conditional density coincides with $f(y|I_{t-1}, \beta^0)$ for some β^0, the probability integral transforms becomes

$$\{Z_t(\beta^0)\} \sim i.i.d.U[0,1]. \quad (3.4)$$

Thus, one can test whether $f(y|I_{t-1}, \beta)$ is correctly specified by checking the $i.i.d.U[0,1]$ for the probability integral transform series.

3.1 Bispectrum and higher-order spectra

Because the autocorrelation function $\gamma(j)$ and the spectral density $h(\omega)$ are rather limited in nonlinear time series analysis, various alternative tools have been proposed to capture nonlinear serial dependence (for example, Granger and Terasvirta, 1993; Tjøstheim, 1996). For example, one often uses the third-order cumulant function

$$C(j,k) \equiv E[(Y_t - \mu)(Y_{t-j} - \mu)(Y_{t-k} - \mu)],$$
$$j, k = 0, \pm 1, \cdots.$$

This is also called the biautocovariance function of $\{Y_t\}$. It can capture certain nonlinear time series, particularly those displaying asymmetric behaviours such as skewness. Hsieh (1989) proposes a test based on $C(j, k)$

for a given pair of (j, k) which can detect some predictable nonlinear components in asset returns.

The Fourier transform of $C(j, k)$,

$$b(\omega_1, \omega_2) \equiv \frac{1}{(2\pi)^2} \sum_{j=-\infty}^{\infty} \sum_{k=-\infty}^{\infty} C(j, k) e^{-ij\omega_1 - ik\omega_2},$$

$$\omega_1, \omega_2 \in [-\pi, \pi],$$

is called the bispectrum of $\{Y_t\}$. When $\{Y_t\}$ is $i.i.d.$, $b(\omega_1, \omega_2)$ becomes a flat bispectral surface:

$$b_0(\omega_1, \omega_2) \equiv \frac{E(Y_t^3)}{(2\pi)^2}, \quad \omega_1, \omega_2 \in [-\pi, \pi].$$

Any deviation from a flat bispectral surface will indicate the existence of serial dependence in $\{Y_t\}$. Moreover, $b(\omega_1, \omega_2)$ can be used to distinguish some linear time series processes from nonlinear time series processes. When $\{Y_t\}$ is a linear process with $i.i.d.$ innovations, that is, when

$$Y_t = \alpha_0 + \sum_{j=1}^{\infty} \alpha_j \varepsilon_{t-j} + \varepsilon_t, \quad \{\varepsilon_t\} \sim i.i.d.(0, \sigma^2),$$

the normalized bispectrum

$$|\tilde{b}(\omega_1, \omega_2)|^2 \equiv \frac{|b(\omega_1, \omega_2)|^2}{h(\omega_1)h(\omega_2)h(\omega_1 + \omega_2)} = \frac{[E(\varepsilon_t^3)]^2}{2\pi\sigma^6}$$

is a flat surface. Any departure from a flat normalized bispectral surface will indicate that $\{Y_t\}$ is not a linear time series with $i.i.d.$ innovations.

The bispectrum $b(\omega_1, \omega_2)$ can capture the BL and NMA processes in (3.1) and (3.2), because the third order cumulant $C(j, k)$ can distinguish them from an $i.i.d.$ process. However, it may still miss some important alternatives. For example, it will easily miss ARCH (1) with $i.i.d.$ $N(0,1)$ innovation $\{\varepsilon_t\}$. In this case, $b(\omega_1, \omega_2)$ becomes a flat bispectrum and cannot distinguish ARCH (1) from an $i.i.d.$ sequence. One could use higher order spectra or polyspectra (Brillinger and Rosenblatt, 1967a; 1967b), which are the Fourier transforms of higher order cumulants. However, higher-order spectra have met with some difficulty in practice: Their spectral shapes are difficult to interpret, and their estimation is not stable in finite samples, due to the assumption of the existence of higher order moments. Indeed, it is often a question whether economic and financial data, particularly high-frequency data, have finite higher order moments.

3.2 Nonparametric measures of serial dependence

Nonparametric measures for serial dependence have been proposed in the literature, which avoid assuming the existence of moments. Granger and Lin (1994) propose a nonparametric entropy measure for serial dependence to identify significant lags in nonlinear time series. Define

the Kullback–Leibler information criterion

$$I(j) = \int ln\left[\frac{f_j(x, y)}{g(x)g(y)}\right] f_j(x, y) dx dy, \quad j = 1, 2, \ldots.$$

where $f_j(x, y)$ is the joint probability density of Y_t and Y_{t-j}, and $g(x)$ is the marginal probability density of $\{Y_t\}$. The Granger–Lin normalized entropy measure is defined as follows:

$$e^2(j) = 1 - \exp[-2I(j)],$$

which enjoys some appealing features. For example, $e(j) = 0$ if and only if Y_t and Y_{t-j} are independent, and it is invariant to any monotonic continuous transformation. Because $f_j(x, y)$ and $g(x)$ are unknown, Granger and Lin (1994) use nonparametric kernel density estimators. They establish the consistency of their entropy estimator (say $\hat{I}(j)$) but do not derive its asymptotic distribution, which is important for confidence interval estimation and hypothesis testing.

In fact, Robinson (1991) has elegantly explained the difficulty of obtaining the asymptotic distribution of $\hat{I}(j)$ for serial dependence, namely it is a degenerate statistic so that the usual root-n normalization does not deliver a well-defined asymptotic distribution. Robinson (1991) considers a modified entropy estimator

$$\hat{I}_\gamma(j) = n^{-1} \sum_{t=j+1}^{n} C_t(\gamma) ln\left[\frac{\hat{f}_j(Y_t, Y_{t-j})}{\hat{g}(Y_t)\hat{g}(Y_{t-j})}\right],$$

where $\hat{f}_j(\cdot, \cdot)$ and $\hat{g}(\cdot)$ are nonparametric kernel density estimators, $C_t(\gamma) = 1 - \gamma$ if t is odd, $C_t(\gamma) = 1 + \gamma$ if t is even, and γ is a pre-specified parameter. The weighting device $C_t(\gamma)$ does not affect the consistency of $\hat{I}_\gamma(j)$ to $I(j)$ and affords a well-defined asymptotic $N(0,1)$ distribution under the $i.i.d.$ hypothesis.

Skaug and Tjøstheim (1993a; 1996) use a different weighting function to avoid the degeneracy of the entropy estimator for serial dependence:

$$\hat{I}_w(j) = n^{-1} \sum_{t=1}^{n} w(Y_t, Y_{t-j}) ln\left[\frac{\hat{f}_j(Y_t, Y_{t-j})}{\hat{g}(Y_t)\hat{g}(Y_{t-j})}\right],$$

where $w(Y_t, Y_{t-j})$ is a weighting function of observations X_t and X_{t-j}. Unlike using Robinson's (1991) weighting device, $\hat{I}_w(j)$ is not consistent for the population entropy $I(j)$, but it also delivers a well-defined asymptotic $N(0, 1)$ distribution after a root-n normalization.

Intuitively, the use of weighting devices slows down the convergence rate of the entropy estimators, giving a well-defined asymptotic $N(0,1)$ distribution after the usual root-n normalization. However, this is achieved at the cost of an efficiency loss, due to the slower convergence rate. Moreover, this approach breaks down when $\{Y_t\}$ is uniformly distributed, as in the case of the probability integral transforms of the conditional density in (3.4). Instead of using a weighting device, Hong and

White (2005) exploit the degeneracy of $\hat{I}(j)$ and use a degenerate U-statistic theory to establish its asymptotic normality. Specifically, Hong and White (2005) show

$$nh\hat{I}(j) + hd_n^0 \xrightarrow{d} N(0, V),$$

where $h = h(n)$ is the bandwidth, and d_n^0 and V are non-stochastic factors. This approach preserves the convergence rate of the unweighted entropy estimator, giving sharper confidence interval estimation and more powerful hypothesis tests. It is applicable when $\{Y_t\}$ is uniformly distributed.

Skaug and Tjøstheim (1993b) also use an Hoeffding measure to test serial dependence (see also Delgado, 1996; Hong, 1998; 2000). The empirical Hoeffding measures are based on the empirical distribution functions, which avoid smoothed nonparametric density estimation.

3.3 Generalized spectrum

Without assuming the existence of higher order moments, Hong (1999) proposes a generalized spectrum as an alternative analytic tool to the power spectrum and higher order spectra. The basic idea is to transform $\{Y_t\}$ via a complex-valued exponential function

$$Y_t \rightarrow \exp(iuY_t), \quad u \in (-\infty, \infty),$$

and then consider the spectrum of the transformed series. Let $\psi(u) \equiv E(e^{iuY_t})$ be the marginal characteristic function of $\{Y_t\}$ and let $\psi_j(u, v) \equiv E[e^{i(uY_t + vY_{t-|j|})}], j = 0, \pm 1, \cdots,$ be the pairwise joint characteristic function of $(Y_t, Y_{t-|j|})$. Define the covariance function between transformed variables e^{iuY_t} and $e^{ivY_{t-|j|}}$:

$$\sigma_j(u, v) \equiv \text{cov}(e^{iuY_t}, e^{ivY_{t-|j|}}), \quad j = 0, \pm 1, \ldots.$$

Straightforward algebra yields $\sigma_j(u, v) = \psi_j(u, v) - \psi(u)\psi(v)$, which is zero for all u, v if and only if Y_t and $Y_{t-|j|}$ are independent. Thus $\sigma_j(u, v)$ can capture any type of pairwise serial dependence over various lags, including those with zero autocorrelation. For example, $\sigma_j(u, v)$ can capture the BL, NMA and ARCH (1) processes in (3.1)–(3.3), all of which are serially uncorrelated.

The Fourier transform of the generalized covariance $\sigma_j(u, v)$:

$$f(\omega, u, v) \equiv \frac{1}{2\pi} \sum_{j=-\infty}^{\infty} \sigma_j(u, v)e^{-ij\omega}, \quad \omega \in [-\pi, \pi],$$

is called the 'generalized spectral density' of $\{Y_t\}$. Like $\sigma_j(u, v)$, $f(\omega, u, v)$ can capture any type of pairwise serial dependencies in $\{Y_t\}$ over various lags. Unlike the power spectrum and higher order spectra, $f(\omega, u, v)$ does not require any moment condition on $\{Y_t\}$. When $\text{var}(Y_t)$ exists, the power spectrum of $\{Y_t\}$ can be obtained by differentiating $f(\omega, u, v)$ with respect to (u, v) at $(0, 0)$:

$$h(\omega) \equiv \frac{1}{2\pi} \sum_{j=-\infty}^{\infty} \gamma(j)e^{-ij\omega}$$

$$= -\frac{\partial^2}{\partial u \partial v} f(\omega, u, v)|_{(u,v)=(0,0)}, \quad \omega \in [-\pi, \pi].$$

This is the reason why $f(\omega, u, v)$ is called the 'generalized spectral density' of $\{Y_t\}$.

When $\{Y_t\}$ is *i.i.d.*, $f(\omega, u, v)$ becomes a flat generalized spectrum as a function of ω:

$$f_0(\omega, u, v) = \frac{1}{2\pi} \sigma_0(u, v), \quad \omega \in [-\pi, \pi].$$

Any deviation of $f(\omega, u, v)$ from the flat generalized spectrum $f_0(\omega, u, v)$ is evidence of serial dependence. Thus, $f(\omega, u, v)$ is suitable to capture any departures from *i.i.d.* Hong and Lee (2003b) use the generalized spectrum to develop a test for the adequacy of nonlinear time series models by checking whether the standardized model residuals are *i.i.d.* Tests for *i.i.d.* are more suitable than tests for serial correlation in nonlinear contexts. Indeed, Hong and Lee (2003b) find that some popular EGARCH models are inadequate in capturing the full dynamics of stock returns, although the standardized model residuals are serially uncorrelated.

Insight into the ability of $f(\omega, u, v)$ can be gained by considering a Taylor series expansion

$$f(\omega, u, v) = \sum_{m=0}^{\infty} \sum_{l=0}^{\infty} \frac{(iu)^m (iv)^l}{m! l!}$$

$$\times \left[\frac{1}{2\pi} \sum_{j=-\infty}^{\infty} \text{cov}(X_t^m, X_{t-|j|}^l)e^{-ij\omega} \right].$$

Although $f(\omega, u, v)$ has no physical interpretation, it can be used to characterize cyclical movements caused by linear and nonlinear serial dependence. Examples of nonlinear cyclical movements include cyclical volatility clustering, and cyclical distributional tail clustering (for example, Engle and Manganelli's (2004) CAVaR model). Intuitively, the supremum function

$$s(\omega) = \sup_{-\infty < u, v < \infty} |f(\omega, u, v)|, \quad \omega \in [-\pi, \pi],$$

can measure the maximum dependence at frequency ω of $\{Y_t\}$. It can be viewed as an operational frequency domain analogue of Granger and Terasvirta's (1993) maximum correlation measure

$$mm\rho(j) = \max_{g(\cdot), h(\cdot)} |corr[g(Y_t), h(X_{t-j})]|.$$

Once generic serial dependence is detected using $f(\omega, u, v)$ or any other dependence measure, one may like to explore the nature and pattern of serial dependence. For example, one may be interested in the following questions:

- Is serial dependence operative primarily through the conditional mean or through conditional higher order moments?
- If serial dependence exists in conditional mean, is it linear or nonlinear?
- If serial dependence exists in conditional variance, does there exist linear or nonlinear and asymmetric ARCH?

Different types of serial dependence have different economic implications. For example, the efficient market hypothesis fails if and only if there is no serial dependence in conditional mean.

Just as the characteristic function can be differentiated to generate various moments, generalized spectral derivatives, when they exist, can capture various specific aspects of serial dependence, thus providing information on possible types of serial dependence. Suppose $E[(Y_t)^{2\max(m,l)}] < \infty$ for some nonnegative integers m, l. Then the following generalized spectral derivative exists:

$$f^{(0,m,l)}(\omega, u, v) = \frac{\partial^{m+l}}{\partial u^m \partial v^l} f(\omega, u, v)$$

$$= \frac{1}{2\pi} \sum_{j=-\infty}^{\infty} \sigma_j^{(m,l)}(u, v) e^{-ij\omega},$$

where $\sigma_j^{(m,l)}(u, v) \equiv \partial^{m+l} \sigma_j(u, v) / \partial u^m \partial v^l$. As an illustrative example, we consider the generalized spectral derivative of order $(m, l) = (1, 0)$:

$$f^{(0,1,0)}(\omega, u, v) = \frac{1}{2\pi} \sum_{j=-\infty}^{\infty} \sigma_j^{(1,0)}(u, v) e^{-ij\omega}.$$

Observe $\sigma_j^{(1,0)}(0, v) \equiv \text{cov}(iY_t, e^{ivY_{t-|j|}}) = 0$ for all $v \in (-\infty, \infty)$ if and only if $E(Y_t|Y_{t-|j|}) = E(Y_t)$ a.s. The function $E(Y_t|Y_{t-|j|})$ is called the autoregression function of $\{Y_t\}$ at lag j. It can capture a variety of linear and nonlinear dependencies in conditional mean, including the BL and NMA processes in (3.1) and (3.2). (The use of $\sigma_j^{(1,0)}(0, v)$, which can be easily estimated by a sample average, avoids smoothed nonparametric estimation of $E(Y_t|Y_{t-|j|})$.) Thus, the generalized spectral derivative $f^{(0,1,0)}(\omega, u, v)$ can be used to capture a wide range of serial dependence in conditional mean. In particular, the function

$$s(\omega) = \sup_{-\infty < v < +\infty} |f^{(0,1,0)}(\omega, 0, v)|$$

can be viewed as an operational frequency domain analogue of Granger and Terasvirta's (1993) maximum mean correlation measure

$$mm(j) = \max_{h(\cdot)} |\text{corr}(Y_t, h(Y_{t-j}))|.$$

See Hong and Lee (2005) for more discussion.

Suppose one has found evidence of serial dependence in conditional mean using $f^{(0,1,0)}(\omega, u, v)$ or any other suitable measure, one can go further to explore whether there exists linear serial dependence in mean. This can be done by using the (1,1)-th order generalized derivative

$$f^{(0,1,1)}(\omega, 0, 0) = -h(\omega),$$

which checks serial correlation. Moreover, one can further use $f^{(0,1,l)}(\omega, u, v)$ for $l \geq 2$ to reveal nonlinear serial dependence in mean. In particular, these higher-order derivatives can suggest that there exist: (i) an ARCH-in-mean effect (for example, Engle, Lilien and Robins, 1987) if $\text{cov}(Y_t, Y_{t-j}^2) \neq 0$, (ii) a skewness-in-mean effect (for example, Harvey and Siddique, 2000) if $\text{cov}(Y_t, Y_{t-j}^3) \neq 0$, and (iii) kurtosis-in-mean effect (for example, Brooks, Burke and Persand, 2005) if $\text{cov}(Y_t, Y_{t-j}^4) \neq 0$. These effects may arise from the existence of a time-varying risk premium, asymmetry of market behaviours, and inadequate account for large losses, respectively.

YONGMIAO HONG

See also **kernel estimators in econometrics; spectral analysis.**

I thank Steven Durlauf (editor) for suggesting this topic and comments on an earlier version, and Jing Liu for excellent research assistance and references. This research is supported by the Cheung Kong Scholarship of the Chinese Ministry of Education and Xiamen University. All remaining errors are solely mine.

Bibliography

Andrews, D.W.K. 1991. Heteroskedasticity and autocorrelation consistent covariance matrix estimation. *Econometrica* 59, 817–58.

Bollerslev, T. 1986. Generalized autoregressive conditional heteroskedastcity. *Journal of Econometrics* 31, 307–27.

Box, G.E.P. and Pierce, D.A. 1970. Distribution of residual autocorrelations in autoregressive moving average time series models. *Journal of the American Statistical Association* 65, 1509–26.

Breusch, T.S. 1978. Testing for autocorrelation in dynamic linear models. *Australian Economic Papers* 17, 334–55.

Breusch, T.S. and Pagan, A. 1980. The Lagrange multiplier test and its applications to model specification in econometrics. *Review of Economic Studies* 47, 239–53.

Brillinger, D.R. and Rosenblatt, M. 1967a. Asymptotic theory of estimates of Kth order spectra. In *Spectral Analysis of Time Series*, ed. B. Harris. New York: Wiley.

Brillinger, D.R. and Rosenblatt, M. 1967b. Asymptotic theory of estimates of kth order spectra. In *Spectral Analysis of Time Series*, ed. B. Harris. New York: Wiley.

Brooks, C., Burke, S. and Persand, G. 2005. Autoregressive conditional kurtosis. *Journal of Financial Econometrics* 3, 399–421.

Campbell, J.Y., Lo, A.W. and MacKinlay, A.C. 1997. *The Econometrics of Financial Markets*. Princeton, NJ: Princeton University Press.

Chen, W. and Deo, R. 2004. A generalized portmanteau goodness-of-fit test for time series models. *Econometric Theory* 20, 382–416.

Cochrane, J.H. 1988. How big is the random walk in GNP? *Journal of Political Economy* 96, 893–920.

Delgado, M.A. 1996. Testing serial independence using the sample distribution function. *Journal of Time Series Analysis* 17, 271–85.

Deo, R.S. 2000. Spectral tests of the martingale hypothesis under conditional heteroscedasticity. *Journal of Econometrics* 99, 291–315.

Durbin, J. 1970. Testing for serial correlation in least squares regression when some of the regressors are lagged dependent variables. *Econometrica* 38, 422–1.

Durbin, J. and Watson, G.S. 1950. Testing for serial correlation in least squares regression: I. *Biometrika* 37, 409–28.

Durbin, J. and Watson, G.S. 1951. Testing for serial correlation in least squares regression: II. *Biometrika* 38, 159–78.

Durlauf, S.N. 1991. Spectral based testing of the martingale hypothesis. *Journal of Econometrics* 50, 355–76.

Engle, R. 1982. Autoregressive conditional hetersokedasticity with estimates of the variance of United Kingdom inflation. *Econometrica* 50, 987–1008.

Engle, R., Lilien, D. and Robins, R.P. 1987. Estimating time varying risk premia in the term structure: the ARCH-M model. *Econometrica* 55, 391–407.

Engle, R. and Manganelli, S. 2004. CARViaR: conditional autoregressive value at risk by regression quantiles. *Journal of Business and Economic Statistics* 22, 367–91.

Fan, J. and Zhang, W. 2004. Generalized likelihood ratio tests for spectral density. *Biometrika* 91, 195–209.

Glosten, R., Jagannathan, R. and Runkle, D. 1993. On the relation between the expected value and the volatility of the nominal excess return on stocks. *Journal of Finance* 48, 1779–801.

Godfrey, L.G. 1978. Testing against general autoregressive and moving average error models when the regressors include lagged dependent variables. *Econometrica* 46, 1293–301.

Granger, C.W.J. and Lin, J.L. 1994. Using the mutual information coefficient to identify lags in nonlinear models. *Journal of Time Series Analysis* 15, 371–84.

Granger, C.J.W. and Terasvirta, T. 1993. *Modeling Nonlinear Economic Relationships*. Oxford: Oxford University Press.

Harvey, C.R. and Siddique, A. 2000. Conditional skewness in asset pricing tests. *Journal of Finance* 51, 1263–95.

Hayashi, F. 2000. *Econometrics*. Princeton: Princeton University Press.

Hong, Y. 1996. Consistent testing for serial correlation of unknown form. *Econometrica* 64, 837–64.

Hong, Y. 1998. Testing for pairwise serial independence via the empirical distribution function. *Journal of the Royal Statistical Society, Series B* 60, 429–53.

Hong, Y. 1999. Hypothesis testing in time series via the empirical characteristic function: a generalized spectral

density approach. *Journal of the American Statistical Association* 94, 1201–20.

Hong, Y. 2000. Generalized spectral tests for serial dependence. *Journal of the Royal Statistical Society, Series B* 62, 557–74.

Hong, Y. and Lee, T.H. 2003a. Inference on predictability of foreign exchange rates via generalized spectrum and nonlinear time series models. *Review of Economics and Statistics* 85, 1048–62.

Hong, Y. and Lee, T.H. 2003b. Diagnostic checking for the adequacy of nonlinear time series models. *Econometric Theory* 19, 1065–121.

Hong, Y. and Lee, Y.J. 2005. Generalized spectral testing for conditional mean models in time series with conditional heteroskedasticity of unknown form. *Review of Economic Studies* 72, 499–51.

Hong, Y. and Lee, Y.J. 2007. Consistent testing for serial correlation of unknown form under general conditional heteroskedasticity. Working paper, Department of Economics, Cornell University, and Department of Economics, Indiana University.

Hong, Y. and White, H. 2005. Asymptotic distribution theory for nonparametric entropy measures of serial dependence. *Econometrica* 73, 837–901.

Hsieh, D.A. 1989. Testing for nonlinear dependence in daily foreign exchange rates. *Journal of Business* 62, 339–68.

Ljung, G.M. and Box, G.E.P. 1978. On a measure of lack of fit in time series models. *Biometrika* 65, 297–303.

Lo, A.W. and MacKinlay, A.C. 1988. Stock market prices do not follow random walks: evidence from a simple specification test. *Review of Financial Studies* 1, 41–66.

Nelson, D. 1991. Conditional heteroskedasticity in asset returns: a new approach. *Econometrica* 59, 347–70.

Newey, W.K. and West, K.D. 1987. A simple, positive semi-definite, heteroscedasticity and autocorrelation consistent covariance matrix. *Econometrica* 55, 703–8.

Paparoditis, E. 2000. Spectral density based goodness-of-fit tests for time series models. *Scandinavian Journal of Statistics* 27, 143–76.

Priestley, M.B. 1988. *Non-Linear and Non-Stationary Time Series Analysis*. London: Academic Press.

Robinson, P.M. 1991. Consistent nonparametric entropy-based testing. *Review of Economic Studies* 58, 437–53.

Robinson, P.M. 1994. Time series with strong dependence. In *Advances in Econometrics, Sixth World Congress*, vol. 1, ed. C. Sims. Cambridge: Cambridge University Press.

Skaug, H.J. and Tjøstheim, D. 1993a. Nonparametric tests of serial independence. In *Developments in Time Series Analysis*, ed. S. Rao. London: Chapman and Hall.

Skaug, H.J. and Tjøstheim, D. 1993b. A nonparametric test of serial independence based on the empirical distribution function. *Biometrika* 80, 591–602.

Skaug, H.J. and Tjøstheim, D. 1996. Measures of distance between densities with application to testing for serial independence. In *Time Series Analysis in Memory of E.J. Hannan*, ed. P. Robinson and M. Rosenblatt. New York: Springer.

Tjøstheim, D. 1996. Measures and tests of independence: a survey. *Statistics* 28, 249–84.

Vinod, H.D. 1973. Generalization of the Durbin–Watson statistic for higher order autoregressive processes. *Communications in Statistics* 2, 115–44.

Wallis, K.F. 1972. Testing for fourth order autocorrelation in quarterly regression equations. *Econometrica* 40, 617–36.

Whang, Y.J. 1998. A test of autocorrelation in the presence of heteroskedasticity of unknown form. *Econometric Theory* 14, 87–122.

Wooldridge, J.M. 1990. An encompassing approach to conditional mean tests with applications to testing nonnested hypotheses. *Journal of Econometrics* 45, 331–50.

Wooldridge, J.M. 1991. On the application of robust, regression-based diagnostics to models of conditional means and conditional variances. *Journal of Econometrics* 47, 5–46.

Shackle, George Lennox Sharman (1903–1992)

Shackle was born in Cambridge. Financial circumstances compelled him to take an external degree while working first as a bank clerk and then as a schoolmaster; it was only in 1935 that he was able to study under Hayek at the London School of Economics. This was an exciting time to be starting out and later, in one of his best-loved books, *The Years of High Theory* (1967), Shackle was to look back at the problem-solving activities responsible for the interwar theoretical breakthroughs. Within two years, and very much influenced by the latest work of Myrdal and Keynes, he completed his first doctorate (published as Shackle, 1938). By 1940 he was employed in wartime official service, having completed a second thesis that drew on material from his work as assistant to E.H. Phelps Brown at Oxford. Despite the demands of official work, he produced a series of articles on uncertain, crucial choices, whose outcomes may define, for good or bad, the chooser's future possibilities (see especially Shackle, 1942; 1943). These were reworked into his (1949) book and he rose rapidly, after returning to academia as Reader in Economics at Leeds University in 1950, to become Brunner Professor of Economic Science in the University of Liverpool in 1951. His retirement from Liverpool in 1969 saw no easing in his industry or in his desire to see economists deal with knowledge problems as analytical rudiments rather than refinements (see Shackle, 1972).

Although Shackle's (1949) analysis of crucial choices attracted immediate attention, it won few adherents. In this book, as in many of his subsequent works, Shackle argued that probabilistic notions are questionable if choice experiments can destroy any possibility of their own replication. (Post-Keynesians have extended his view in criticizing the rational expectations hypothesis.)

Shackle suggested that, in such situations, choosers would come to focus on particularly attention-arresting pairs of *possibilities* (one pair for each scheme of action). A possibility is not something which a chooser would expect to happen, given enough tries, with a particular frequency, but something whose taking place looks *ex ante* surprising (unsurprising) because potentially fatal obstacles to it can (cannot) be envisaged. Shackle insisted that possibility is not in general distributive: thoughts about a possible outcome not previously imagined need not affect assignments of potential surprise to its rivals, since potential surprise ratings do not sum to any fixed, bounded value. Despite this, many theorists found his 'potential surprise curves' difficult to distinguish from inverted probability distributions; they also took issue with his view that it is not rational for choosers to weigh together values for possibilities that are mutually exclusive. Behaviouralists were ill-disposed to the large role played by indifference surfaces in his analysis of how ascendant gain/loss pairings were focused upon and then ranked; whereas orthodox theorists (for example, Ford, 1983) argued that it would be irrational for choosers to focus in the way he proposed, and that his selection device – the 'gambler preferences' map – produced the questionable result that an investor will choose a portfolio consisting of no more than two types of financial asset.

Much of his noteworthy retirement output (especially his 1974 book) tried to make economists recognize that the incompatibility of speculators' expectations and changes in the 'state of the news' will make the relative demands for durable assets prone to kaleidoscopic instability. To many orthodox model-builders, his kaleidic conception of economic systems had unacceptably nihilistic implications, but it led some Post Keynesians to examine how institutions and policies might be designed to constrain explosive and implosive forces whose precise timings and strengths may be impossible to anticipate.

PETER EARL

Selected works

1938. *Expectations, Investment and Income*. Oxford: Oxford University Press.

1942. A theory of investment-decisions. *Oxford Economic Papers*, NS6(April), 77–94.

1943. The expectational dynamics of the individual. *Economica*, NS10(May), 99–129.

1949. *Expectation in Economics*. Cambridge: Cambridge University Press.

1967. *The Years of High Theory: Invention and Tradition in Economic Thought 1926–1939*. Cambridge: Cambridge University Press.

1972. *Epistemics and Economics*. Cambridge: Cambridge University Press.

1974. *Keynesian Kaleidics*. Edinburgh: Edinburgh University Press.

Bibliography

Ford, J.L. 1983. *Choice, Expectation and Uncertainty.* Oxford: Martin Robertson.

Ford, J.L. 1994. *G.L.S. Shackle: The Dissenting Economist's Economist.* Aldershot: Edward Elgar.

shadow pricing

When a businessman evaluates a project, he does it with a view to calculating the prospective profit from it. These calculations can be seen as taking place in two steps. At the first step, all the physical consequences of relevance to the businessman – the inputs to and outputs from the project – are assessed. At the second stage, these inputs and outputs are converted into costs and revenues, using *market prices.* It is natural that a private businessman should use the ruling market prices for costing inputs and for valuing sales, since these are the prices at which transactions take place and hence profit generated.

Consider now the evaluation of a project by a government. Such evaluation will differ at each of the two steps referred to above. At the first step, the government will be interested in *all* of the repercussions of the project, however indirect. This is because it is the government rather than a private businessman concerned with his own narrowly defined activities. At the second step, the government will wish to use not the ruling market prices but prices which reflect social costs and social benefits, in order to calculate what might be termed social profit. These prices are referred to as *shadow prices,* or accounting prices (see Little and Mirrlees, 1974), and the name suggests that they are to be used in lieu of the actual market prices.

Market prices are what they are. But how are shadow prices to be calculated? Clearly they depend on the government's objective function and on the constraints it faces. The shadow prices should be such that the social profit from the project is positive if and only if the project increases the value of the government's objective function. In a general competitive equilibrium, if the government's objective is economic efficiency, then it can be argued that for a small project the shadow prices do in fact coincide with market prices. If the government's objective includes the pursuance of equity, but it has lump sum instruments to carry this out, then shadow prices still coincide with market prices. Basically the government should use redistributive lump sum taxation to pursue equity and the project to pursue increases in aggregate economic welfare.

But if the government does not have a sufficient range of instruments to pursue effective redistribution without distortion it may be the case that, even with a full competitive equilibrium, shadow prices may differ from market prices. In addition to this, if the economy is not in a full competitive equilibrium, then the case for using shadow prices different from market prices can be argued strongly.

In programming terms, shadow prices are simply dual to the changes in the government's objective function. One justification for their use is the benefits of decentralization: local project evaluators are better equipped to analyse the physical consequences of a project, and this localized knowledge should be used in conjunction with centrally determined shadow prices to evaluate the social profitability of projects. But the real difficulties arise in specifying the objectives of the government and in specifying its constraints, and this is in turn related to who is thought of as doing the project evaluation.

The standard assumption is one of a unitary government with a given social welfare function – a benevolent dictator. But the reality is one where either the project evaluator is part of a government which is a coalition of interests, or the project evaluation is being done by an international agency which faces a government made up of conflicting and competing objectives. The logical procedure for an international agency should be clear – in evaluating a project it should incorporate a model of the political process to clarify the responses of various government instruments to the project. Sen (1972) gives an illuminating discussion of a project which requires importing an input on which there is already a quota – so that the border price of the input is very different from its domestic scarcity value. The Little and Mirrlees (1974) method of using border prices is predicated on the assumption that it is these prices which represent the transformation possibilities for the economy as a whole. But if the assessment of the political realities is such that this quota will not be removed by the government – because of the overriding influence of interest groups that benefit from the rents generated by the quota – then the domestic scarcity value should be used in costing the input.

Similarly, any project which alters significantly the distribution of income will have repercussions on the political process – and there will be attempts by groups who are adversely affected to restore their standard of living. Project evaluation in general, and shadow pricing in particular, should take these into account. Consider, for example, the shadow cost of labour. If the labour used on the project comes from the agricultural sector, and if this labour is a constraint on output, then agricultural output will fall. If government revenue depends on taxation of this output, this will fall too. If, in turn, government expenditure is a major source of non-agricultural (urban) incomes, then at constant fiscal deficit urban incomes will fall. This change in the distribution of income will be an important element in the shadow cost of labour. But suppose now that the political processes are such as to not allow a decline in urban living standards. Rather, government expenditure remains constant and the fiscal deficit increases. Now it is the increased burden on future generations which has to be taken into account. Either way, it should be clear that a model of the political process is crucial in specifying shadow prices even if the project evaluator (be it an international agency or a project evaluation unit within the government) is clear about what the objectives are.

RAVI KANBUR

See also **cost–benefit analysis.**

Bibliography

Little, I.M.D. and Mirrlees, J.A. 1974. *Project Appraisal and Planning for Developing Countries.* London: Heinemann.
Sen, A.K. 1972. Control areas and accounting prices: an approach to economic evaluation. *Economic Journal* 82 (Supp.), 486–501.

Shapley value

The *value* of an uncertain outcome (a 'lottery') is an a priori measure, in the participant's utility scale, of what he expects to obtain (this is the subject of 'utility theory'). The question is, how would one evaluate the prospects of a player in a multi-person interaction, that is, in a game?

This question was originally addressed by Lloyd S. Shapley (1953a). The framework was that of n-person games in coalitional form with side-payments, which are given by a set N of 'players', say 1, 2, ..., n, together with a 'coalitional function' v that associates to every subset S of N ('coalition') a real number $v(S)$, the maximal total payoff the members of S can obtain (the 'worth' of S). An underlying assumption of this model is that there exists a medium of exchange ('money') that is freely transferable in unlimited amounts between the players, and moreover every player's utility is additive with respect to it (that is, a transfer of x units from one player to another decreases the first one's utility by x units and increases the second one's utility by x units; the total payoff of a coalition can thus be meaningfully defined as the sum of the payoffs of its members). This requirement is known as existence of 'side payments' or 'transferable utility'. In addition, the game is assumed to be adequately described by its coalitional function (that is, the worth $v(S)$ of each coalition S is well defined, and the abstraction from the extensive structure of the game to its coalitional function leads to no essential loss; such a game is called a 'c-game'). These assumptions may be interpreted in a broader and more abstract sense. For example, in a voting situation, a 'winning coalition' is assigned worth 1, and a 'losing' coalition, worth 0. The essential feature is that the prospects of each coalition may be summarized by one number.

The *Shapley value* associates to each player in each such game a unique payoff – his 'value'. The value is required to satisfy the following four axioms. (EFF) *Efficiency* or *Pareto optimality*: The sum of the values of all players equals $v(N)$, the worth of the grand coalition of all players (in a superadditive game $v(N)$ is the maximal amount that the players can jointly get); this axiom combines feasibility and efficiency. (SYM) *Symmetry* or *equal treatment*: If two players in a game are substitutes (that is, the worth of no coalition changes when replacing one of the two players by the other one), then their values are equal. (NUL) *Null* or *dummy player*: If a player in a game is such that the worth of every coalition remains the same when he joins it, then his value is zero. (ADD) *Additivity*: The value of the sum of two games is the sum of the values of the two games (equivalently, the value of a probabilistic combination of two games is the same as the probabilistic combination of the values of the two games; this is analogous to 'expected utility'). The surprising result of Shapley is that these four axioms *uniquely determine* the values in *all* games.

Remarkably, the Shapley value of a player in a game turns out to be exactly his *expected marginal contribution to a random coalition*. The marginal contribution of a player i to a coalition S (that does not contain i) is the change in the worth when i joins S, that is, $v(S \cup \{i\}) - v(S)$. To obtain a random coalition S not containing i, arrange the n players in a line (for example, 1, 2, ..., n) and put in S all those that precede i in that order; all $n!$ orders are assumed to be equally likely. The formula for the Shapley value is striking, first, since it is a consequence of very simple and basic axioms and, second, since the idea of marginal contribution is so fundamental in much of economic analysis.

It should be emphasized that the value of a game is an a priori measure, that is, an evaluation before the game is actually played. Unlike other solution concepts (for example, core, von Neumann–Morgenstern solution, bargaining set), it need not yield a 'stable' outcome (the probable final result when the game is actually played). These final stable outcomes are in general not well determined; the value – which is uniquely specified – may be thought of as their expectation or average. Another interpretation of the value axioms regards them as rules for 'fair' division, guiding an impartial 'referee' or 'arbitrator'. Also, as suggested above, the Shapley value may be understood as the utility of playing the game (Shapley, 1953a; Roth, 1977).

In view of both its strong intuitive appeal and its mathematical tractability, the Shapley value has been the focus of much research and many applications. We can only briefly mention some of these here (together with just a few representative references). The reader is referred to the survey of Aumann (1978) and, for more extensive coverage, to the *Handbook of Game Theory* (Aumann and Hart, vol 1: 1992 [HGT1], vol 2: 1994 [HGT2], vol 3: 2002 [HGT3]), especially Chapters 53–58, as well as parts of Chapters 32–34 and 37.

Variations

Following Shapley's pioneering approach, the concept of *value* has been extended, modified and generalized.

Weighted values

Assume that the players are of unequal 'size' (for example, a player may represent a 'group', a 'department', and so on), and this is expressed by given (relative) weights. This setup leads to 'weighted Shapley values' (Shapley, 1953b); in unanimity games, for example, the values of the players are no longer equal but, rather, proportional to their weights [HGT3, Ch. 54].

Semi-values

Abandoning the efficiency axiom (EFF) yields the class of 'semi-values' (Dubey, Neyman and Weber, 1981). An interesting semi-value is the *Banzhaf index* (Penrose, 1946; Banzhaf, 1965; Dubey and Shapley, 1979), originally proposed as a measure of power in voting games. Like the Shapley value, it is also an expected marginal contribution, but here all coalitions not containing player *i* are equally likely [HGT3, Ch. 54].

Other axiomatizations

There are alternative axiomatic systems that characterize the Shapley value. For instance, one may replace the additivity axiom (ADD) with a *marginality axiom* that requires the value of a player to depend only on his marginal contributions (Young, 1985). Another approach is based on the existence of a *potential* function together with efficiency (EFF) (Hart and Mas-Colell, 1989) [HGT3, Ch. 53].

Consistency

Given a solution concept which associates payoffs to games, assume that a group of players in a game have already agreed to it, are paid off accordingly, and leave the game; consider the 'reduced game' among the remaining players. If the solution of the reduced game is the same as that of the original game, then the solution is said to be *consistent*. It turns out that consistency, together with some elementary requirements for two-player games, characterizes the Shapley value (Hart and Mas-Colell, 1989) [HGT3, Ch. 53], [HGT1, Ch. 18].

Large games

Assume that the number of players increases and individuals become negligible. Such models are important in applications (such as competitive economies and voting), and there is a vast body of work on values of large games that has led to beautiful and important insights (for example, Aumann and Shapley, 1974) [HGT3, Ch. 56].

NTU games

These are games 'without side payments', or 'with non-transferable utility' (that is, the existence of a medium of utility exchange is no longer assumed). The simplest such games, two-person pure bargaining problems, were originally studied by Nash (1950). Values for general NTU games, which coincide with the Shapley value in the side payments case, and with the Nash bargaining solution in the two-person case, have been introduced by Harsanyi (1963), Shapley (1969), Maschler and Owen (1992) [HGT3, Ch. 55].

Non-cooperative foundations

Bargaining procedures whose non-cooperative equilibrium outcome is the Shapley value have been proposed by Gul (1989) (see Hart and Levy, 1999; Gul, 1999) and Winter (1994) for strictly convex games, and by Hart and Mas-Colell (1996) for general games [HGT3, Ch. 53].

Other extensions

This includes games with communication graphs (Myerson, 1977), coalition structures (Aumann and Drèze, 1974; Owen, 1977; Hart and Kurz, 1983), and others [HGT2, Ch. 37], [HGT3, Ch. 53].

Economic applications

Perfect competition

In the classical economic model of perfect competition, the commodity prices are determined by the requirement that total demand equals total supply; this yields a *competitive* (or *Walrasian*) *equilibrium*. A different approach in such setups looks at the cooperative 'market game' where the members of each coalition can freely exchange among themselves the commodities they own. A striking phenomenon occurs: various game-theoretic solutions of the market games yield precisely the competitive equilibria. In particular, in perfectly competitive economies every Shapley value allocation is competitive and, if the utilities are smooth, then every competitive allocation is also a value allocation. This result, called the *value equivalence principle*, is remarkable since it joins together two very different approaches: competitive prices arising from supply and demand on the one hand, and marginal contributions to trading coalitions on the other. The value equivalence principle has been studied in a wide range of models (for example, Shapley, 1964; Aumann, 1975). While it is undisputed in the TU case, its extension to the general NTU case seems less clear (it holds for the Shapley NTU value, but not necessarily for other NTU values) [HGT3, Ch. 57].

Cost allocation

Consider the problem of allocating joint costs in a 'fair' manner. Think of the various 'tasks' (or 'projects', 'departments', and so on) as players, and let $v(S)$ be the total cost of carrying out the set S of tasks (Shubik, 1962). It turns out that the axioms determining the Shapley value are easily translated into postulates appropriate for solving cost allocation problems (for example, the efficiency axiom becomes 'total-cost-sharing'). Two notable applications are airport landing fees (a task here is an aircraft landing; Littlechild and Owen, 1973) and telephone billing (each time unit of a phone call is a player; the resulting cost allocation scheme was put into actual use at Cornell University; Billera, Heath and Raanan, 1978) [HGT2, Ch. 34].

Other applications

The value has been applied to various economic models; for example, models of taxation where a political power structure is given in addition to the economic data (Aumann and Kurz, 1977). Further references to economic applications can be found in Aumann (1985) [HGT3, Ch. 58], [HGT2, Ch. 33].

Political applications

What is the 'power' of an individual or a group in a voting situation? A trivial observation – though not always remembered in practice – is that the political power need not be proportional to the number of votes (see Shapley, 1981, for some interesting examples). It is therefore important to find an objective method of measuring power in such situations. The Shapley value (known in this setup as the *Shapley–Shubik index*; Shapley and Shubik, 1954) is, by its very nature, a most appropriate candidate. Indeed, consider a simple political game, described by specifying whether each coalition is 'winning' or 'losing'. The Shapley value of a player *i* turns out to be the probability that *i* is the 'pivot' or 'key' player, namely, that in a random order of all players those preceding *i* are losing, whereas together with *i* they are winning. For example, in a 100-seat parliament with simple majority (that is, 51 votes are needed to win), assume there is one large party having 33 seats and the rest are divided among many small parties; the value of the large party is then close to 50%, considerably more than its voting weight (that is, its 33% share of the seats). In contrast, when there are two large parties each having 33 seats and a large number of small parties, the value of each large party is close to 25% – much less than its voting weight of 33%. To understand this, think of the competition between the two large parties to attract the small parties to form a winning coalition; in contrast, when there is only one large party, the competition is between the small parties (to join the large party).

The Shapley value has also been used in more complex models, where 'ideologies' and 'issues' are taken into account (thus, not all arrangements of the voters are equally likely; an 'extremist' party, for example, is less likely to be the pivot than a 'middle-of-the-road' one; Owen, 1971; Shapley, 1977).

References to political applications of the Shapley value may be found in Shapley (1981); these include various parliaments (USA, France, Israel), the United Nations Security Council, and others [HGT2, Ch. 32].

SERGIU HART

See also **game theory.**

Bibliography

Aumann, R.J. 1975. Values of markets with a continuum of traders. *Econometrica* 43, 611–46.

Aumann, R.J. 1978. Recent developments in the theory of the Shapley Value. *Proceedings of the International Congress of Mathematicians*, Helsinki.

Aumann, R.J. 1985. On the non-transferable utility value: a comment on the Roth–Shafer examples. *Econometrica* 53, 667–78.

Aumann, R.J. and Drèze, J.H. 1974. Cooperative games with coalition structures. *International Journal of Game Theory* 3, 217–37.

Aumann, R.J. and Hart, S., eds. 1992 [HGT1], 1994 [HGT2], 2002 [HGT3]. *Handbook of Game Theory, with Economic Applications*, vols 1–3. Amsterdam: North-Holland.

Aumann, R.J. and Kurz, M. 1977. Power and taxes. *Econometrica* 45, 1137–61.

Aumann, R.J. and Shapley, L.S. 1974. *Values of Non-atomic Games*. Princeton: Princeton University Press.

Banzhaf, J.F. 1965. Weighted voting doesn't work: a mathematical analysis. *Rutgers Law Review* 19, 317–43.

Billera, L.J., Heath, D.C. and Raanan, J. 1978. Internal telephone billing rates: a novel application of non-atomic game theory. *Operations Research* 26, 956–65.

Dubey, P., Neyman, A. and Weber, R.J. 1981. Value theory without efficiency. *Mathematics of Operations Research* 6, 122–8.

Dubey, P. and Shapley, L.S. 1979. Mathematical properties of the Banzhaf Power Index. *Mathematics of Operations Research* 4, 99–131.

Gul, F. 1989. Bargaining foundations of Shapley value. *Econometrica* 57, 81–95.

Gul, F. 1999. Efficiency and immediate agreement: a reply to Hart and Levy. *Econometrica* 67, 913–18.

Harsanyi, J.C. 1963. A simplified bargaining model for the *n*-person cooperative game. *International Economic Review* 4, 194–220.

Hart, S. and Kurz, M. 1983. Endogenous formation of coalitions. *Econometrica* 51, 1047–64.

Hart, S. and Levy, Z. 1999. Efficiency does not imply immediate agreement. *Econometrica* 67, 909–12.

Hart, S. and Mas-Colell, A. 1989. Potential, value and consistency. *Econometrica* 57, 589–614.

Hart, S. and Mas-Colell, A. 1996. Bargaining and value. *Econometrica* 64, 357–80.

Littlechild, S.C. and Owen, G. 1973. A simple expression for the Shapley value in a special case. *Management Science* 20, 370–72.

Maschler, M. and Owen, G. 1992. The consistent Shapley value for games without side payments. In *Rational Interaction: Essays in Honor of John Harsanyi*, ed. R. Selten. New York: Springer.

Myerson, R.B. 1977. Graphs and cooperation in games. *Mathematics of Operations Research* 2, 225–29.

Nash, J.F. 1950. The bargaining problem. *Econometrica* 18, 155–62.

Owen, G. 1971. Political games. *Naval Research Logistics Quarterly* 18, 345–55.

Owen, G. 1977. Values of games with a priori unions. In *Essays in Mathematical Economics and Game Theory*, ed. R. Henn and O. Moeschlin. New York: Springer.

Penrose, L.S. 1946. The elementary statistics of majority voting. *Journal of the Royal Statistical Society* 109, 53–7.

Roth, A.E. 1977. The Shapley value as a von Neumann–Morgenstern utility. *Econometrica* 45, 657–64.

Shapley, L.S. 1953a. A value for *n*-person games. In *Contributions to the Theory of Games, II*, ed. H.W. Kuhn and A.W. Tucker. Princeton: Princeton University Press.

Shapley, L.S. 1953b. Additive and non-additive set functions. Ph.D. thesis, Princeton University.

Shapley, L.S. 1964. Values of large games VII: a general exchange economy with money. Research Memorandum 4248-PR. Santa Monica, CA: RAND Corp.

Shapley, L.S. 1969. Utility comparison and the theory of games. In *La Décision: agrégation et dynamique des ordres de préférence*. Paris: Editions du CNRS.

Shapley, L.S. 1977. A comparison of power indices and a nonsymmetric generalization. Paper No. P–5872. Santa Monica, CA: RAND Corp.

Shapley, L.S. 1981. Measurement of power in political systems. *Game Theory and its Applications*, Proceedings of Symposia in Applied Mathematics, vol. 24. Providence, RI: American Mathematical Society.

Shapley, L.S. and Shubik, M. 1954. A method for evaluating the distribution of power in a committee system. *American Political Science Review* 48, 787–92.

Shubik, M. 1962. Incentives, decentralized control, the assignment of joint costs and internal pricing. *Management Science* 8, 325–43.

Winter, E. 1994. The demand commitment bargaining and snowballing cooperation. *Economic Theory* 4, 255–73.

Young, H.P. 1985. Monotonic solutions of cooperative games. *International Journal of Game Theory* 14, 65–72.

Shapley–Folkman theorem

The Shapley–Folkman theorem places an upper bound on the size of the non-convexities (loosely speaking, openings or holes) in a sum of non-convex sets in Euclidean N-dimensional space, R^N. The bound is based on the size of non-convexities in the sets summed and the dimension of the space. When the number of sets in the sum is large, the bound is independent of the number of sets summed, depending rather on N, the dimension of the space. Hence the size of the non-convexity in the sum becomes small as a proportion of the number of sets summed; the non-convexity per summand goes to zero as the number of summands becomes large. The

Shapley–Folkman theorem can be viewed as a discrete counterpart to the Lyapunov theorem on non-atomic measures (Grodal, 2002).

The theorem is used to demonstrate the following properties:

- existence of approximate competitive general equilibrium in large finite economies with non-convex preferences (increasing marginal rate of substitution) or non-convex technology (bounded increasing returns; the U-shaped cost curve case);
- convergence of the core to the set of competitive equilibria (Arrow and Hahn, 1972; Anderson, 1978).

It may also be used to characterize the solution of non-convex programming problems (Aubin and Ekeland, 1976).

For $S \subset R^N$, S compact, define rad(S), the radius of S, as a measure of the size of S. Define $r(S)$, the inner radius of S, and $\rho(S)$ inner distance of S, as measures of the non-convexity (size of holes) of S. Let con$S$ denote the closed convex hull of S (smallest closed convex set containing S as a subset).

$$\mathrm{rad}(S) \equiv \inf_{x \in R^N} \sup_{y \in S} |x - y|;$$

$$r(S) \equiv \sup_{x \in \mathrm{con}S} \inf_{\{T \subset S \mid T \,\mathrm{spans}\, x\}} \mathrm{rad}(T);$$

$$\rho(S) \equiv \sup_{x \in \mathrm{con}S} \inf_{y \in S} |x - y|.$$

rad(S) is the radius of the smallest closed ball centred in con$S$ containing S. A set of points T is said to span a point x, if x can be expressed as a convex combination (weighted average) of elements of T. $r(S)$ is the smallest radius of a ball centred in the convex hull of S, so that the ball is certain to contain a set of points of S that span the ball's centre. Hence $r(S)$ represents a measure of breadth of non-convexities in S. $\rho(S)$ is the maximum distance from a point in con$S$ to (the nearest point of) S. Hence it represents the smaller of breadth or depth of non-convexities of S.

Let $S_1, S_2, \ldots, S_m$ be a (finite) family of m compact subsets of R^N. The vector sum of $S_1, S_2, \ldots, S_m$, denoted W is a set composed of representative elements of $S_1, S_2, \ldots, S_m$ summed together. W is defined as

$$W \equiv \sum_{i=1}^{m} S_i \equiv \left\{ w \mid w = \sum_{i=1}^{m} x^i, x^i \in S^i \right\}$$

where the sum in the brackets is taken over one element of each S_i.

Theorem (Shapley–Folkman): Let $S_1, \ldots, S_m$ be a family of m compact subsets of R^N; $W = \Sigma_{i=1}^{m} S_i$. Let $L \geq \mathrm{rad}(S_i)$ for all S_i; let $n = \min(N, m)$. Then for any $x \in \mathrm{con}W$

(i) $x = \Sigma_{i=1}^{m} x^i$, where $x^i \in \mathrm{con}S_i$ and with at most N exceptions, $x^i \in S_i$;

(ii) there is $y \in W$ so that $|x - y| \leq L\sqrt{n}$.

Corollary (Starr): Let $S_1, \ldots, S_m$ be a finite family of compact subsets of R^N. $W = \Sigma_{i=1}^{m} S_i$. Let $L \geq r(S_i)$ for all S_i, $n = \min(m, N)$. Then for any $x \in \mathrm{con}W$ there is $y \in W$ so that

$$|x - y| \leq L\sqrt{n}.$$

Corollary (Heller): Let $S_1, \ldots, S_m$ be a finite family of compact subsets of R^N; $W = \Sigma_{i=1}^{m} S_i$. Let $L \geq \rho(S_i)$ for all S_i, $n = \min(m, N)$. Then for any $x \in \mathrm{con}W$ there is $y \in W$ so that

$$|x - y| \leq Ln.$$

Statements and proofs of the theorem and corollaries along with applications are available in Arrow and Hahn (1972) and Green and Heller (1981). Development of the theorem is due to L.S. Shapley and J.H. Folkman (private correspondence) with publication in Starr (1969). Extensions, alternative proofs, and applications appear in the other references.

ROSS M. STARR

See also **perfect competition.**

Bibliography

Anderson, R.M. 1978. An elementary core equivalence theorem. *Econometrica* 46, 1483–7.

Anderson, R.M. 1988. The Second Welfare Theorem with nonconvex preferences. *Econometrica* 56, 361–82.

Arrow, K.J. and Hahn, F.H. 1972. *General Competitive Analysis*. San Francisco: Holden-Day.

Artstein, Z. and Vitale, R.A. 1975. A strong law of large numbers for random compact sets. *Annals of Probability* 3, 879–82.

Artstein, Z. 1980. Discrete and continuous bang-bang and facial spaces or: look for the extreme points. *SIAM Review* 22, 172–85.

Aubin, J.-P. and Ekeland, I. 1976. Estimation of the duality gap in nonconvex optimization. *Mathematics of Operations Research* 1(3), 225–45.

Cassels, J.W.S. 1975. Measure of the non-convexity of sets and the Shapley–Folkman–Starr theorem. *Mathematical Proceedings of the Cambridge Philosophical Society* 78, 433–6.

Chambers, C.P. 2005. Multi-utilitarianism in two-agent quasilinear social choice. *International Journal of Game Theory* 33, 315–34.

Ekeland, I. and Temam, R. 1976. *Convex Analysis and Variational Problems*. Amsterdam: North-Holland.

Green, J. and Heller, W.P. 1981. Mathematical analysis and convexity with applications to economics. In *Handbook of Mathematical Economics*, vol. 1, ed. K.J. Arrow and M. Intriligator. Amsterdam: North-Holland.

Grodal, B. 2002. The equivalence principle. In *Optimization and Operation Research, Encyclopedia of Life Support Systems (EOLSS)*, ed. U. Derigs. Cologne. Online.

Available at http://www.econ.ku.dk/grodal/EOLSS-final.pdf, accessed 5 April 2007.

Hildenbrand, W., Schmeidler, D. and Zamir, S. 1973. Existence of approximate equilibria and cores. *Econometrica* 41, 1159–66.

Howe, R. 1979. On the tendency toward convexity of the vector sum of sets. Discussion Paper No. 538, Cowles Foundation, Yale University.

Manelli, A.M. 1991. Monotonic preferences and core equivalence. *Econometrica* 59, 123–38.

Mas-Colell, A. 1978. A note on the core equivalence theorem: how many blocking coalitions are there? *Journal of Mathematical Economics* 5, 207–16.

Proske, F.N. and Puri, M.L. 2002. Central limit theorem for Banach space valued fuzzy random variables. *Proceedings of the American Mathematical Society* 130, 1493–501.

Starr, R.M. 1969. Quasi-equilibria in markets with non-convex preferences. *Econometrica* 37, 25–38.

Starr, R.M. 1981. Approximation of points of the convex hull of a sum of sets by points of the sum: an elementary approach. *Journal of Economic Theory* 25, 314–17.

Tardella, F. 1990. A new proof of the Lyapunov Convexity Theorem. *Applied Mathematics* 28, 478–81.

Weil, W. 1982. An application of the central limit theorem for Banach space valued random variables to the theory of random sets. *Probability Theory and Related Fields* 60, 203–8.

Zhou, L. 1993. A simple proof of the Shapley–Folkman theorem. *Economic Theory* 3, 371–2.

sharecropping

Sharecropping is a form of land leasing contract in which the tenant shares the final product with the landlord as a partial or total payment of the rent. A landowner leasing his land to a tenant may use several forms of land renting contracts.

'Sharecropping' usually designates all particular forms of land tenancy contracts in which the landlord allows the tenant to cultivate his land in return for a stipulated fraction of the product (the 'share'), possibly combined with other side payments. This institutional contractual agreement prevailed in many parts of the world and many different periods in the history of agriculture, from antiquity (Egypt, Mesopotamia and Greece), the Middle Ages and Renaissance in Europe, through to contemporary economies. Sharecropping is currently most commonly found in less developed countries where agriculture and land rental markets are more active, but it also still exists in many developed countries.

The sharecropping relationship assumes a variety of forms and is sometimes linked to agreements involving not only land and labour transactions but also credit, lending, insurance or marketing agreements. Indeed, within sharecropping arrangements landlords may also determine the crops to be grown, may choose to monitor some of the key moments of the agricultural process, and may defray a greater or lesser share of the costs of some inputs (other than labour) with a pre-specified fraction that may not necessarily be equal to the fraction of output retained by the landlord in the payment rule.

Is sharecropping inefficient?

Since Adam Smith, economists have taken an interest in sharecropping because of its apparent inefficiency based on the simple observation that the sharecropper receives only a share of the marginal productivity of his labour but bears its full marginal cost. The persistence of such institutions has thus puzzled many economists. More recently, sharecropping has also constituted the typical example of the principal–agent model, the basic paradigm of contract theory. Similar economic relationships occur in both developed and less developed countries when some party (the principal) delegates the use of some capital to another party in exchange for compensation depending on the returns obtained by the other party (the agent). Examples abound in capital markets (stock markets) where investors may let others use their capital in return for a share of the profits, in vertical relationships between producers and retailers in many industries (food and other consumption goods, or rental services), and in some labour contracts within firms where wages may depend on some measure of performance.

Researchers working on the theory of rural organization have attempted to explain not only the persistence of sharecropping but also the particular features it exhibits.

Sharecropping as an efficient risk-sharing contract

Concerns about the efficiency of sharecropping relationships have gone through several stages in the history of economic thought. While it was first thought to be an inefficient institutional arrangement, since Stiglitz (1974) it has been understood as possibly representing an efficient risk-sharing mechanism in environments where production is risky and other forms of insurance are not available. Sharecropping has the advantage over fixed-rent land leasing contracts of relieving the tenant of some of the risk. By sharing the product, the landlord and tenant also share its fluctuations due to risks related to the weather, diseases and other unpredictable factors affecting agricultural production. Through the payment of a rent contingent on agricultural production, the risk associated with variations in prices of marketed commodities is also shared by both parties. However, if the landlord is less risk averse, he should further protect the risk-averse peasant by simply using wage contracts. Moreover, the same risk-sharing opportunities could be provided without sharecropping simply by having workers combine wage contracts and rental contracts.

However, the landlord's ability to monitor the tenant's labour has also been also called into question. In most

places where absentee landlords delegate the use of land to a tenant, it seemed implausible that a contract precisely specifying the labour to be applied could be enforceable. Stiglitz (1974) shows that sharecropping could be an institutional arrangement designed both to share risks and to provide incentives in a situation where monitoring effort (labour supply) is costly. Sharecropping results, then, from a trade-off between incentives and risk sharing. Fixed-rent contracts provide 'perfect' incentives by giving the full marginal product to the tenant but at the cost of shifting all the risk on to the tenant, while fixed-wage contracts protect the tenant against production risk but also remove direct incentives to provide effort.

However, the two necessary ingredients of this trade-off have been successively challenged. Cheung (1969) criticizes the need to provide incentives through contracted remuneration, arguing that contracts could simply specify the optimal level of labour that the tenant should provide. Wage contracts would not then imply any inefficiency in the provision of effort and would completely insure tenants against income fluctuations. However, this reasoning implicitly assumes that monitoring is not costly, and empirical tests have shown in some contexts that input provision was actually lower under share contracts than under fixed-rent contracts (see, among others, Shaban, 1987). Conversely, the other side of the trade-off was also challenged by the apparent paucity of evidence as to the effect of risk on contractual forms. In fact, until recently a great deal of empirical work had failed to present evidence as to the effect of risk on the contract incentives that would be consistent with the alleged trade-off. This was mostly due to the significance of other trade-offs determining the choice of contracts (which we examine below) but also to the failure to recognize that the choice of contracts had to be modelled within a more general understanding of how land rental markets function. Dubois (2002) shows that taking into account the endogeneity of the choice to delegate use of land was important in the empirical analysis of contractual choices in order to avoid problems of selection bias in econometric estimates. Ackerberg and Botticini (2002) show that not taking into account the endogenous matching of landlords and tenants could also lead to an apparent absence of correlation between the incentive power of contracts and the crop risk. Referring to direct evidence on risk sharing in village economies using consumption data linked to contract choices, Dubois (2000) demonstrates that the sharecropping institution could actually play a role in consumption risk sharing.

Thus, in order to reduce shirking, the landlord could either expend resources in monitoring the worker or prefer to resort to a sharecropping contract. The persistence of sharecropping can thus be explained by this argument together with a number of other observed features. For example, the landlord has an incentive to encourage the tenant to use inputs (such as fertilizer or manure) which raise the worker's marginal product, therefore resulting in higher worker effort. This explains why, as is often observed in sharecropping contracts, the landlord may be prepared to bear a fraction of the costs of inputs that exceeds the fraction of the product received. Obviously, to implement cost sharing, costs have to be observable and verifiable. Why then does the landlord not simply enforce a specific level of input provision by the tenant? This relates to the information structure as to the appropriate level of input, about which the tenant may have better knowledge given the conditions of production. Sharing costs thus remains a useful incentive.

Another argument against interpreting sharecropping as a risk-sharing contract is that the terms of the contracts should logically vary with the level of risk represented by the specific environment, the crops grown, and both parties' degree of risk aversion. Empirical observation, however, shows that in practice the terms of sharecropping contracts exhibit little variation, especially when it comes to the share of the product, which is often one-half and sometimes one-third or two-thirds. This could be seen to be the result of an approximation process of optimal contracts, but Allen (1985) provides an interesting explanation of this phenomenon. In a model where landlords initially screen tenants with heterogeneous abilities before entering into fixed-rent contracts with only the more able farmers, a sharecropping contract emerges endogenously in a state of equilibrium. Interestingly, the optimal share of production for the tenant in this case has to be a trade-off between the gains accrued from shirking and leaving the relationship after a given period and the gains from not shirking and being taken on again with a fixed-rent contract following a period of screening. This trade-off clearly depends on the farmers' time preferences. Allen (1985) shows that, given the usual interest rates that apply in less developed countries, the optimal share has to be close to one-half.

Another argument calling into question the aforementioned rationale for sharecropping is that the theory of contracts predicts that optimal incentive contracts should depend on output in general in a nonlinear way. However, nonlinear contracts imply that different tenants would face different marginal prices for their production, giving opportunities for arbitrage and incentives to collude among farmers. Linear contracts may then be a way to avoid this problem in environments where monitoring harvest and trade between tenants may be difficult. Moreover, the gains represented by the use of nonlinear contracts may not be worth the potential additional costs of implementing them.

In addition to such attempts to determine the rationale of sharecropping, the expected effect of this kind of rural organization on agricultural innovation and development has been investigated. The adoption of innovations in agriculture in developing countries in particular has been a major preoccupation. Whether the contractual form affects the adoption of innovations, and if so how, have been the subjects of much investigation.

Sharecropping within the principal–agent paradigm

Most theoretical analysis of sharecropping contracts has been cast within the principal–agent paradigm, in which the landlord is generally considered to be the principal having the bargaining power to make a take-it-or-leave-it offer to the agent (the tenant). This analysis does not explain the decision of the landlord to delegate the use of land and the way landlords and tenants meet in the land rental market. It is only recently that these decisions have been taken into account in the analysis of sharecropping contracts, from both empirical and theoretical points of view. In this light, let us consider a model where the agricultural production function is linearly homogenous in land area (as generally admitted: see Stiglitz, 1974, and Otsuka, Chuma and Hayami, 1992) because agriculture is a spatial activity and induces constant returns to scale in the cultivated area. For a fixed amount of land, denote y the agricultural output of the next crop period, e the tenant's work effort (which can be considered as a measure of efficient labour time), x a state variable as, for example, land fertility at the beginning of the agricultural period, and define an agricultural production function f such that $y = \varepsilon f(x, e)$ where ε is a multiplicative positive random variable with mean one representing weather uncertainty. The effort e can represent labour tasks or other agricultural inputs and can be multidimensional. Its cost is $C(e)$. Assume also that an investment function controls land fertility dynamics such that the next period land fertility is $g(x, e)$ (adding a multiplicative positive random variable with mean one to represent the influence of the weather or other externalities on land fertility or ground quality would not change the following results). According to the contract signed, the principal pays the agent $T(y) = \alpha y + \beta$. The contract parameters (α, β) allow the landlord to propose different kinds of contact, from a fixed-wage contract where $\alpha = 0$, $\beta = w$ with w the wage, to a fixed-rent contract where $\alpha = 1$, $\beta = -R$ with R the rent paid to the landlord, through sharecropping contracts where $0 < \alpha < 1$ and β can be zero or not. Concerning preferences, we define $U(T(y)) - C(e)$ and $y - T(y)$ the agent's and principal's utility functions. We assume that U is increasingly concave because of risk aversion, while we treat the principal as risk neutral.

Moral hazard in sharecropping

When the agent's actions are unobservable to the principal or monitoring costs are prohibitively high, a moral hazard problem arises leading to effort shirking by the agent. The worker chooses his effort level to maximize his expected utility, given the terms of the contract and his outside wage opportunities. Thus, for a crop season, the principal proposes a contract to maximize his welfare given the agent's incentive compatibility (IC) constraint and its individual rationality (IR) constraint guaranteeing him an exogenous reservation utility denoted U. The maximization programme of the landlord can thus be written as

$$\underset{\alpha,\beta}{Max}\, E[(1-\alpha)y - \beta]$$

subject to

$$e^* \in \arg\max_e \; EU(\alpha y + \beta) - C(e) \qquad (1)$$

$$EU(\alpha y + \beta) - C(e^*) \geq \overline{U} \qquad (2)$$

Denoting f_e and f_{ee} the first and second derivatives of the production function with respect to effort, we can show that the solution to this programme is such that the individual rationality constraint (IR) is binding and the optimal share of production α^* satisfies the following equation

$$\alpha^* = 1 + \left(1 - \frac{E\varepsilon U'}{EU'}\right)\frac{f[\alpha^* f_{ee} - C'']}{f_e^2}$$

where e_α is the derivative of effort with respect to the share of output α received by the tenant and that satisfies $e_\alpha = \frac{f_e}{\alpha^* f_{ee} - C''}$. Because of the concavity of U, $0 < \frac{E\varepsilon U'}{EU'} < 1$. Moreover, with concavity of production with effort and convexity of cost of effort, the optimal share α^* is strictly lower than 1, thus corresponding to a sharecropping contract.

The exact form of the contract depends on both the properties of U and f, and the magnitude of uncertainty. First, the greater the (compensated) labour supply elasticity, that is, the more sensitive the worker is to incentives, the greater is the optimal share α^*, that is, the closer is the optimal contract to a rental contract. Second, the trade-off between incentives and risk sharing depends on the riskiness of production through ε and on risk preferences through U. With some assumptions on the distribution of ε or on the shape of U, it can be shown that the optimal share is lower for more risk-averse agents or more risky environments (Stiglitz, 1974). At the limit, if the tenant is risk neutral, then $\alpha^* = 1$ and a pure rental contract will be used. Conversely, the greater the risk aversion and the greater the risk, the closer the optimal contract is to a pure wage contract ($\alpha^* = 0, \beta^* > 0$). These predictions constituted the focus of many empirical tests that involved examining the determinants of the choice between sharecropping and fixed-rent contracts. A large number of empirical tests (Braido, 2005) have failed to provide evidence that the risk-sharing trade-off could explain the choice between fixed rent and sharecropping.

Also, these optimal sharecropping contracts ($0 < \alpha^* < 1, \beta^*$) involve a fixed payment β^* (either to or from the landlord). In practice, many contractual relations may have an implicit or explicit provision calling for such fixed payments. For example, payments from the landlord to the worker to finance stipulated inputs, like fertilizer, can be interpreted in this manner. However, the empirical observation and measurement of such fixed transfers is generally difficult, which explains why they have not been used for empirical testing of the theory.

In some contexts, given the relative paucity of evidence that the risk-sharing trade-off could explain the determinants of contract choice between fixed rent and sharecropping, other explanations related to transaction costs inherent to landlord–tenant relationships have been offered. One of the most significant transaction costs that seem to plague land rental contracts is linked to the question of land quality maintenance and investment. The risk-sharing argument remains completely silent about investment and land-maintenance problems that were often raised in transaction cost approaches with respect to land-rental contracts. Moreover, as already pointed out by Johnson (1950) and even Adam Smith, the problem of land-fertility maintenance and land overuse can also be cited to explain the choice of contract by landlords. In fact, although observable, land fertility may not be contractible. It may also elude verification due to the complexity of specifying the agricultural tasks related to land-quality maintenance and to difficulties in objectively measuring land quality. A moral hazard problem in land maintenance may thus appear. Delegating farming may lead to land overuse if the landlord and tenant do not have the same opportunity cost of usage of land (Allen and Lueck, 1993). A share contract may then curb the farmer's incentive to exploit land attributes. One way to see this in the previous model is to take into account the value of the land in the landlord's objective. The landlord will then anticipate the consequences of delegating the use of land for the future returns obtained, since the tenant's actions may affect the future land value. The land value $v(z)$, an increasing function of the land fertility index z, can be seen as the result of the expected discounted sum of all future profits obtained by the landlord for a given plot of land of quality z. Then, if the objective of the landlord is now

$$\underset{\alpha,\beta}{Max}\, E[(1-\alpha)y - \beta + v(z)]$$

subject to (IC) and (IR), the optimal share of output between the landlord and tenant (Dubois, 2002) is the solution to

$$\alpha^* = 1 + v'(z)\frac{g_e}{f_e} - \left(1 - \frac{E\varepsilon U'}{EU'}\right)\frac{f}{e_\alpha f_e}$$

where $z = g(x, e)$. With risk neutrality of the tenant, the optimal share is thus below 1 if the effort of production reduces land fertility ($g_e < 0$) because

$$\alpha^* = 1 + v'(z)\frac{g_e}{f_e}.$$

The contract here shows low-powered incentives and generally corresponds to a sharecropping contract even if there is no risk-sharing issue.

Multitasks and contract repetition in sharecropping

But other features of agricultural activity and contractual relationships have been used to explain the observation of sharecropping. Several forms of multitask moral hazard models and dynamic considerations provide interesting insights into this form of contracting.

First, the multitask moral hazard model of Holmström and Milgrom (1987), applied to sharecropping for example by Luporini and Parigi (1996), shows that low-powered incentives can be obtained as a way to mitigate the substitution of effort across tasks that cannot be monitored by the landlord even without risk aversion of the tenant. Luporini and Parigi (1996) consider the two distinct production tasks of subsistence crops and cash crops. In another kind of multitask model, with limited liability of the tenant instead of risk aversion, Ghatak and Pandey (2000) show that a sharecropping contract can be optimal when there is joint moral hazard in effort and in risk factor for output. Other quite different models with multiple labour inputs explain sharecropping differently according to a number of features that can be observed from time to time. In Bardhan and Srinivasan (1971) or Eswaran and Kotwal (1985), both the landowner and the tenant provide labour input. In Roumasset and Uy (1987), a model with an investment task, a production task and two periods provides a study in the reduction of agency costs by monitoring. Bardhan (1989, ch. 7) and Braverman and Stiglitz (1986) have a sharecropping model with a fertilizer input and non-observable labour effort. They determine the efficient incentives on both separable inputs through production sharing and cost sharing.

The other significant dimension of the landlord–tenant relationship that may explain the form of contracts is the fact that these contracts are often repeated and may have variable duration. The repetition of relationships between a landlord and a tenant actually called into question the rationale of supposedly short-term contracts of sharecropping generally observed empirically. Bardhan (1989, ch. 8) uses a two-period model to show the trade-off between production incentives, enhanced in the initial period by the threat of dismissal by the landowner, and land improvement incentives that decrease with a more powered contract. Dutta, Ray and Sengupta (1989) and Bose (1993) study a number of long-term contracts between landowners and landless peasants where infinitely repeated relationships with threats of eviction are examined. Eviction threats can actually serve as an incentive device in repeated sharecropping contracts (Banerjee, Gertler and Ghatak, 2002; Banerjee and Ghatak, 2004). Moreover, in a repeated moral hazard relation, spot-contract sequences may allow the outcomes of long-term contracts, which are Pareto-superior to short-term agreements, to be implemented (Fudenberg, Holmström and Milgrom, 1990; Malcomson and Spinnewyn, 1988).

Sharecropping is also sometimes considered to be part of a 'tenancy ladder' in agriculture allowing landless wage workers to become farmers before they become landlords. The farmer's financial constraints and limited liability may then play a significant role in explaining access to land rental markets and contractual forms (sharecropping with variable share or fixed-rent contracts) proposed by the landlord (Shetty, 1988; Ray and Singh, 2001; Laffont and Matoussi, 1995). In Laffont and Matoussi (1995), risk-neutral farmers are offered a sharecropping contract with more or fewer incentives rather than a fixed-rent contract, due to financial constraints that restrict the amount of working capital the tenant can use as affected by the rent and the share of inputs to be paid at the beginning of the crop season.

Finally, many other forms of sharecropping contracts, including some side transfers or those interlinked with credit (Mitra, 1983; Braverman and Stiglitz, 1982) or involving state contingent informal gifts and transfers (Sadoulet, Fukui and de Janvry, 1994), exist and may sometimes completely change the efficiency properties of such contracts. Thus, considerable care and attention should be devoted to describing contractual agreements so as to study such organizations and possibly recommend policy reforms for land rental markets.

PIERRE DUBOIS

See also **access to land and development; contract theory; Laffont, Jean-Jacques; peasants; peasant economy; principal and agent (i); principal and agent (ii); risk sharing; Stiglitz, Joseph E.**

Bibliography

Ackerberg, D. and Botticini, M. 2002. Endogenous matching and the empirical determinants of contract form. *Journal of Political Economy* 110, 564–91.
Allen, F. 1985. On the fixed nature of sharecropping contracts. *Economic Journal* 95, 30–48.
Allen, D. and Lueck, D. 1993. Transaction costs and the design of cropshare contracts. *RAND Journal of Economics* 24, 78–100.
Bardhan, P. 1989. *The Theory of Agrarian Institutions*. Oxford: Clarendon Press.
Bardhan, P. and Srinivasan, T. 1971. Cropsharing tenancy in agriculture: a theoretical and empirical analysis. *American Economic Review* 61, 48–64.
Banerjee, A., Gertler, P. and Ghatak, M. 2002. Empowerment and efficiency: tenancy reform in West Bengal. *Journal of Political Economy* 110, 239–80.
Banerjee, A. and Ghatak, M. 2004. Eviction threats and investment incentives. *Journal of Development Economics* 74, 469–88.
Bose, G. 1993. Interlinked contracts and moral hazard in investments. *Journal of Development Economics* 41, 247–73.

Braido, L. 2005. Insurance and incentives in sharecropping. In *Insurance: Theoretical Analysis and Policy Implications*, ed. P. Chiappori and C. Gollier. Cambridge, MA: MIT Press.

Braverman, A. and Stiglitz, J. 1982. Sharecropping and the interlinking of agrarian markets. *American Economic Review* 72, 695–715.

Braverman, A. and Stiglitz, J. 1986. Cost-sharing arrangements under sharecropping: moral hazard, incentive flexibility, and risk. *American Journal of Agricultural Economics* 68, 642–52.

Cheung, S. 1969. *The Theory of Share Tenancy*. Chicago: University of Chicago Press.

Dubois, P. 2000. Assurance parfaite, hétérogénéité des préférences et métayage au Pakistan. *Annales d'Economie et de Statistique* 59, 1–36.

Dubois, P. 2002. Moral hazard, land fertility and sharecropping in a rural area of the Philippines. *Journal of Development Economics* 68, 35–64.

Dutta, B., Ray, D. and Sengupta, K. 1989. Contracts with eviction in infinitely repeated principal–agent relationships. In *The Theory of Agrarian Institutions*, ed. P. Bardhan. Oxford: Clarendon Press.

Eswaran, M. and Kotwal, A. 1985. A theory of contractual structure in agriculture. *American Economic Review* 75, 352–67.

Fudenberg, D., Holmström, B. and Milgrom, P. 1990. Short-term contracts and long-term agency relationships. *Journal of Economic Theory* 51, 1–31.

Ghatak, M. and Pandey, P. 2000. Contract choice in agriculture with joint moral hazard in effort and risk. *Journal of Development Economics* 63, 303–26.

Holmström, B. and Milgrom, P. 1987. Aggregation and linearity in the provision of intertemporal incentives. *Econometrica* 55, 303–28.

Johnson, D. 1950. Resource allocation under share contracts. *Journal of Political Economy* 58, 111–23.

Laffont, J.-J. and Matoussi, M. 1995. Moral hazard, financial constraints and sharecropping in El Oulja. *Review of Economic Studies* 62, 381–99.

Luporini, A. and Parigi, B. 1996. Multi-task sharecropping contracts: the Italian Mezzadria. *Economica* 63, 445–57.

Malcomson, J. and Spinnewyn, F. 1988. The multiperiod principal–agent problem. *Review of Economic Studies* 40, 391–408.

Mitra, P. 1983. A theory of interlinked rural transactions. *Journal of Public Economics* 20, 167–91.

Otsuka, K., Chuma, H. and Hayami, Y. 1992. Land and labor contracts in agrarian economies: theories and facts. *Journal of Economic Literature* 30, 1965–2018.

Ray, T. and Singh, N. 2001. Limited liability, contractual choice, and the tenancy ladder. *Journal of Development Economics* 66, 289–303.

Roumasset, J. and Uy, M. 1987. Agency costs and the agricultural firm. *Land Economics* 63, 290–302.

Sadoulet, E., Fukui, S. and de Janvry, A. 1994. Efficient share-tenancy contracts under risk: the case of two rice-growing villages in Thailand. *Journal of Development Economics* 45, 225–43.

Shaban, R. 1987. Testing between competing models of sharecropping. *Journal of Political Economy* 95, 893–920.

Shetty, S. 1988. Limited liability, wealth differences and tenancy contracts in agrarian economies. *Journal of Development Economics* 29, 1–22.

Stiglitz, J. 1974. Incentives and risk sharing in sharecropping. *Review of Economic Studies* 41, 219–55.

Williamson, O. 1989. Transaction cost economics. In *Handbook of Industrial Organization*, vol. 1, ed. R. Schmalensee and R. Willig. Amsterdam: North-Holland.

Sharpe, William F. (born 1934)

William F. Sharpe is one of the founders of the modern theory of finance. Born in Boston in 1934, Sharpe received his BA in economics from UCLA in 1955. After graduation and military service, Sharpe joined the Rand Corporation and simultaneously pursued his Ph.D. from UCLA, which he received in 1961. He joined the department of economics at the University of Washington–Seattle in 1961, remaining until 1968. After a two-year stint at UC Irvine, he joined the Stanford Business School, where he remained for the rest of his career. In addition to his scholarly pursuits, he has been an active consultant for financial firms as well as textbook writer.

Sharpe's most famous contribution to economics is his development of the capital asset pricing model (CAPM). This work developed in two stages (Varian, 1993, provides a very clear discussion of the evolution of Sharpe's work on CAPM). First, mentored by Harry Markowitz, who was also at Rand, Sharpe studied the question of the construction of efficient portfolios in the presence of a riskless asset. This led to Sharpe's study of what he dubbed 'single factor' models but which are now more often called 'single index' models, in which the holding return on a given asset is a linear function of the return on the market portfolio. The single factor approach provides important computational advantages in constructing optimal portfolios of the type studied by Markowitz. This doctoral dissertation research was subsequently published in *Management Science* in 1963 as 'A Simplified Model for Portfolio Analysis'.

The analysis of single factor asset return models as a short cut to efficient portfolio construction was followed by Sharpe's investigation of what equilibrium risk–return relationships will emerge in a market of rational agents with mean/variance preferences, leading to his celebrated 1964 *Journal of Finance* paper 'Capital Asset Prices – A Theory of Market Equilibrium Under Conditions of Risk'. The demonstration that the riskiness of an asset is determined not by the variance of its holding return but rather by the now celebrated 'beta' of the asset, defined as the covariance of that holding return with the holding return on the market portfolio as a whole divided by the standard deviation of the market portfolio, is now a canonical idea in economics and is the basis for much of the modern theory of asset pricing. The underlying economic ideas of the CAPM find modern analogs in the use of Euler equations to characterize equilibrium expected asset returns. The main difference in Euler equation approaches from Sharpe's original formulation in this later work is the relaxation of the assumption that market participants have mean variance preferences with respect to asset returns; in its place explicit consumption-based utility functions are used. It is interesting to note that Sharpe's single factor model embodied the CAPM risk–return relationship by construction. Sharpe's Nobel Prize acceptance speech, published as Sharpe (1991b), is interesting for encapsulating his assessment of the model. See CAPITAL ASSET PRICING MODEL for discussion of the model in detail as well as its intellectual history; as Sharpe notes in his Nobel autobiography, a number of researchers were working on similar ideas to his.

Sharpe's subsequent research has spanned a wide range of issues in finance. Perhaps most noteworthy, Sharpe's interest in understanding risk and return relationships led to his development, initially in the context of mutual fund evaluation, of what is now called the Sharpe ratio, first discussed in his 1966 *Journal of Business* paper 'Mutual Fund Performance'. (Sharpe called it the reward to variability ratio.) The ratio is measured by taking the difference between the expected return on an asset (or portfolio) and a benchmark security and dividing by the standard deviation of this difference. When the benchmark security is riskless, the Sharpe ratio provides a simple characterization of the return to risk. Sharpe (1994) gives a nice summary of the statistic and its interpretation.

These achievements led to Sharpe's receipt, with Merton Miller and Harry Markowitz, of the 1990 Nobel Memorial Prize in economics as one of the 'pioneers in the theory of financial economics and corporate finance'. Varian (1993) provides a lovely discussion of why this joint award was so merited.

STEVEN N. DURLAUF

See also **capital asset pricing model.**

Selected works

1963. A simplified model for portfolio analysis. *Management Science* 9, 277–93.
1964. Capital asset prices – a theory of market equilibrium under conditions of risk. *Journal of Finance* 19, 425–42.
1965. Risk-aversion in the stock market – some empirical evidence. *Journal of Finance* 20, 416–22.
1966. Mutual fund performance. *Journal of Business* 39(1), Part II, 119–38.
1974. Imputing expected returns from portfolio composition. *Journal of Financial and Quantitative Analysis*, June, 463–72.

1976. Corporate pension funding policy. *Journal of Financial Economics* 3, 183–93.
1977. The capital asset pricing model: a 'multi-beta' interpretation. In *Financial Decision Making Under Uncertainty*, ed. H. Levy and M. Sarnat. New York: Academic Press.
1978. *Investments*. Englewood Cliffs, NJ: Prentice-Hall.
1978. (With R. Lanstein.) Duration and security risk. *Journal of Financial and Quantitative Analysis*, November, 653–68.
1981. Decentralized investment management. *Journal of Finance* 36, 217–34.
1989. (With G. J. Alexander.) *Fundamentals of Investments*. Englewood Cliffs, NJ: Prentice-Hall.
1991a. Autobiography. In *Les Prix Nobel. The Nobel Prizes 1990*, ed. T. Frängsmyr. Stockholm: Nobel Foundation. Online. Available at http://nobelprize.org/nobel_prizes/economics/laureates/1990/sharpe-autobio.html, accessed 11 October 2006.
1991b. Capital asset prices with and without negative holdings. *Journal of Finance* 46, 489–509.
1994. The Sharpe Ratio. *Journal of Portfolio Management* 21(1), 49–58.
2006. *Investors and Markets: Portfolio Choices, Asset Prices, and Investment Advice*. Princeton: Princeton University Press.

Bibliography

Varian, H. 1993. A portfolio of Nobel Laureates: Markowitz, Miller, and Sharpe. *Journal of Economic Perspectives* 7(1), 159–69.

Shephard, Ronald William (1912–1982)

Shephard was born in Portland, Oregon, or 22 November 1912 and died in Berkeley, California, on 22 July 1982.

He received his BA in Mathematics and Economics in 1935 and his Ph.D. in Mathematics and Statistics in 1940 at the University of California at Berkeley.

During the years 1943–6 he was a statistical consultant at the Bell aircraft corporation. In the years 1949–51 he worked under the direction of Oskar Morgenstern of Princeton University, producing his path-breaking work, *Cost and Production Functions* (1953). During 1950–2, he was a senior economist at the RAND Corporation and during 1952–6 he was the manager of the systems analysis department at the Sandia Corporation. From 1957 to 1980 he was a Professor of Industrial Engineering and Operations Research at the University of California at Berkeley.

Shephard made several fundamental contributions to economics. He was the first to rigorously derive a duality between cost and production functions; that is, given a knowledge of either function, the other may be derived from it. He also introduced the distance function to the economics literature in the course of establishing his duality theorems; the distance function is used to define a theoretical index number concept due to Malmquist. Shephard was also the first to derive the derivative property of the cost function (or Shephard's Lemma) starting from the cost function (the derivations by Hicks and Samuelson started from the production or utility function). Shephard also appreciated the econometric implications of Shephard's Lemma.

Shephard also defined the concept of a homothetic production or utility function: a function is homothetic if it is a monotonic transform of a linearly homogeneous function. He also deduced the implications of a homothetic function for its dual cost function.

Shephard also realized the importance of the assumption of homogeneous weak separability for index number and aggregation theory.

Finally, Shephard postulated an ingenious system of axioms or properties for a production function and then was able to deduce the classical law of diminishing returns to a subset of the factors as a theorem.

W.E. DIEWERT

See also **cost functions.**

Selected works

1953. *Cost and Production Functions.* Princeton: Princeton University Press.

1970a. *Theory of Cost and Production Functions.* Princeton: Princeton University Press.

1970b. Proof of the law of diminishing returns. *Zeitschrift für Nationalökonomie* 30, 7–34.

shrinkage-biased estimation in econometrics

In economics much empirical research proceeds in the context of incomplete subject matter theories and data based on sampling designs not devised by or known to the researcher. This leads to economic or econometric models in the form of ill-posed inverse problems, and partial or incomplete data, and brings a range of uncertainty to the data-processing and information-recovery process in general and the estimation and inference process in particular. In practice, procedures such as preliminary test statistics, tuning parameters and perhaps a bit of magic are invoked to identify a particular econometric-statistical model on which to base estimation and inference process. Given the uncertainty surrounding the model-discovery and post-data estimation and inference tasks, one basis for coping with or reducing the entropy level is to focus on the statistical implications of shrinkage estimation and the possibilities for combining competing estimation problems. The objective is to demonstrate estimation and inference methods that are free of subjective choices and tuning parameters and that have superior risk performance. In the process we demonstrate simple estimators

that are uniformly and non-trivially superior over the unknown parameter space to conventional parametric and pretest estimators used by most applied econometrics researchers.

1. The conventional statistical model and estimator base

In econometrics the most widely used estimation and inference techniques are based on linear statistical models and maximum likelihood (ML) and least squares concepts. These estimation and inference methods are supported by a body of theory that dates back over two centuries to Gauss and Legendre and includes the often cited Gauss–Markov theorem with its best linear unbiased estimator conclusion – a conclusion that appears to be generally accepted by applied econometricians. However, for the economic researcher who is interested in parameter estimation this statistical property, which is right on average, may have limited usefulness and also negative statistical implications. It is to these questions that we now turn.

In econometrics, many multivariate estimation problems can be reduced to the canonical form where $\mathbf{b} = (b_1, b_2, \ldots, b_K)'$ is a K variate normal random vector with $\mathbf{b} \sim N_K(\boldsymbol{\beta}, \boldsymbol{\Sigma_b})$. The mean vector $\boldsymbol{\beta}$ is unknown and the covariance $\boldsymbol{\Sigma_b}$ is usually assumed to be known up to a constant of proportionality. One common problem that gives rise to the above is estimating the location vector $\boldsymbol{\beta}$ for the linear statistical model when we observe a vector $\mathbf{y}$ such that $\mathbf{y} = X\boldsymbol{\beta} + \mathbf{e}$, where $\mathbf{e} \sim N(\mathbf{0}, \sigma^2 I_T)$ and $\delta(\mathbf{b}) = (X'X)^{-1}X'\mathbf{y}$ is the maximum likelihood estimator (MLE) with covariance $\boldsymbol{\Sigma_b} = \sigma^2(X'X)^{-1}$. The objective is to estimate the unknown vector $\boldsymbol{\beta}$ by an estimator $\delta(\mathbf{b})$ under the sum of squares of error loss measure

$$L(\boldsymbol{\beta}, \delta(\mathbf{b})) = \|\boldsymbol{\beta} - \delta(\mathbf{b})\|^2 = \sum_{i=1}^{K} (\beta_i - \delta(b_i))^2$$

$$(1.1)$$

where the unknown loss is a function of both parameters and data. The usual evaluation of the estimator $\delta(\mathbf{b})$ makes use of the risk function.

$$\rho(\boldsymbol{\beta}, \delta(\mathbf{b})) = E_{\boldsymbol{\beta}}^{\mathbf{b}}[L(\boldsymbol{\beta}, \delta(\mathbf{b}))] = E_{\boldsymbol{\beta}}^{\mathbf{b}}[\|\boldsymbol{\beta} - \delta(\mathbf{b})\|^2]$$

$$(1.1a)$$

where superscripts denote the random quantity over which the expectation is to be taken for fixed $\boldsymbol{\beta}$. Under (1.1), the maximum likelihood estimator $\delta(\mathbf{b}) = \mathbf{b}$ is minimax and has constant risk $\rho(\boldsymbol{\beta}, \delta(\mathbf{b})) = \text{tr}\boldsymbol{\Sigma_b} = \sigma^2\text{tr}(X'X)^{-1}$.

1.1 The Stein alternative

The first hint of difficulty for the MLE $\delta(\mathbf{b})$ in estimating the multivariate normal mean under quadratic loss (1.1) was when Stein (1955) demonstrated for the orthonormal symmetric case, $\Sigma_b = \sigma^2(X'X)^{-1} = I_K$, that the conventional estimator $\delta(\mathbf{b})$ is inadmissible when $K \geq 3$. This means there exists another estimator, say $\delta^s(\mathbf{b})$, where $\rho(\delta^s(\mathbf{b}), \boldsymbol{\beta}) \leq \rho(\delta(\mathbf{b}), \boldsymbol{\beta}) = K$ for all $\boldsymbol{\beta}$ and $\rho(\delta^s(\mathbf{b}), \boldsymbol{\beta}) < \rho(\delta(\mathbf{b}), \boldsymbol{\beta}) = K$, for some $\boldsymbol{\beta}$. In other words, under the usual measure of statistical performance, there is a superior estimator. Stein's inadmissibility proof did not go on to suggest the components of a risk dominating alternative estimator.

1.1.1 The James and Stein shrinkage estimator

Given the inadmissibility result that suggested there may be under quadratic loss other estimators that risk dominate the MLE, $\mathbf{b}$, James and Stein (1961) demonstrated the estimator

$$\delta^s(\mathbf{b}) = \left(1 - \frac{K-2}{\|\mathbf{b}\|^2}\right)\mathbf{b} \qquad (1.2)$$

that has uniformly smaller risk than $\mathbf{b}$. This estimator makes the adjustment in the MLE, $\mathbf{b}$, a smooth and known function of the data. The mean of (1.2) is

$$E[\delta^s(\mathbf{b})] = \boldsymbol{\beta} - (K-2)E\left[1/\chi^2_{(K+2,\lambda)}\right]\boldsymbol{\beta} \qquad (1.2a)$$

with risk

$$\rho(\delta^s(\mathbf{b}), \boldsymbol{\beta}) = K - (K-2)^2 E\left[1/\chi^2_{(K,\lambda)}\right] \qquad (1.2b)$$

where $\chi^2_{(K|2,\lambda)}$ is a non-central chi square random variable with non centrality parameter $\lambda = \boldsymbol{\beta}'\boldsymbol{\beta}/2$ (see Judge and Bock, 1978). When $\boldsymbol{\beta} = 0$ the risk of (1.2) is 2 and increases to K, the risk of the MLE, $\mathbf{b}$, as $\boldsymbol{\beta}'\boldsymbol{\beta} \to \infty$. Consequently, for the values of $\boldsymbol{\beta}$ close to the origin the risk gain may be considerable.

The James and Stein estimator (1.2) combines the MLE, $\mathbf{b}$, and the restricted-fixed vector, $\boldsymbol{\beta} = 0$, and shrinks $\mathbf{b}$ toward the null mean vector, $\boldsymbol{\beta} = 0$. A more general formulation that introduces explicitly an arbitrary origin, considers a fixed mean vector $\mathbf{r}$, and an estimator of the form

$$\delta^s_r(\mathbf{b}) = \left[1 - \frac{(K-2)}{\|\mathbf{b} - \mathbf{r}\|^2}\right](\mathbf{b} - \mathbf{r}) + \mathbf{r} \qquad (1.3)$$

which shrinks $\mathbf{b}$ toward the target vector $\mathbf{r} \in R^K$. This estimator has bias and risk characteristics in line with the James and Stein estimator (1.2).

If σ^2 is unknown, the optimal James and Stein estimator may be written as

$$\delta^s(\mathbf{b}) = [1 - ((K-2)/(T-K+2))(s/\mathbf{b}'\mathbf{b})]\mathbf{b} \qquad (1.4)$$

where $s/\sigma^2 = (T-K)\hat{\sigma}^2/\sigma^2$ has a $\chi^2_{(T-K)}$ random variable distribution that is independent of $\mathbf{b}$. Since $\mathbf{b}'\mathbf{b}/\sigma^2$ is distributed as a χ^2_K, the optimal James and Stein estimators (1.4) may be rewritten as

$$\delta^s(\mathbf{b}) = [1 - (T-K)(K-2)/(T-K+2)(1/u)]\mathbf{b} \qquad (1.5)$$

where $u = \mathbf{b}'\mathbf{b}/(K\hat{\sigma}^2)$ is the likelihood ratio statistic which has an F distribution with K and $(T-K)$ degrees of freedom and non-centrality parameter $\lambda = \mathbf{b}'\mathbf{b}/2\sigma^2$. Thus the shrinkage of the MLE is determined by the data and the hypothesis vector, which in this case is, $\boldsymbol{\beta} = \mathbf{0}$. The larger the value of the F test statistic, the smaller is the adjustment made in the MLE. It will be useful to keep in mind this continuous likelihood ratio shrinkage estimator when we discuss pretest estimators in Section 2.

Some of the Stein-like shrinkage estimators have desirable properties from both sampling theory and Bayesian inference points of view. For examples, when σ^2 is unknown the empirical Bayes counterpart is

$$\delta^B(\mathbf{b}) = [1 - ((K-2)/(T-K))(s/\mathbf{b}'\mathbf{b})]\mathbf{b} \qquad (1.6)$$

This estimator is dominated by (1.4).

1.1.2 A positive Stein shrinkage rule

Although under quadratic loss Stein rules improve on the minimax MLE and are themselves minimax, they are not admissible and other superior shrinkage rules exist. One such estimator is the positive Stein rule estimator.

$$\delta^{+s}(\mathbf{b}) = \left(1 - \frac{c}{\|\mathbf{b}\|^2}\right)\mathbf{b}I_{[c,\infty)}\left(\|\mathbf{b}\|^2\right)$$

$$- \mathbf{b} - \left(1 \wedge \frac{c}{\|\mathbf{b}\|^2}\right)\mathbf{b} \qquad (1.7)$$

where $a \wedge b = \min(a, b)$. Under this formulation, Baranchik (1964) demonstrated that for $(K-2) \leq c < 2(K-2)$ the positive rule estimator (1.7) uniformly improves on the James and Stein estimator (1.2) and thus proves its inadmissibility. There is no one value of c that is optimal, but Efron and Morris (1973) have demonstrated that rules that restrict c to $[(K-2), 2(K-2)]$, dominate shrinkage rules of c in $[(0,(K-2)]$. Although the positive rule estimator (1.7) is minimax under quadratic loss, it too is inadmissible.

For the more general non-symmetric case where Σ_b is just some positive definite symmetric matrix, the class of pseudo Stein–Bayes rules, $\delta^s(\mathbf{b})$, having uniformly smaller risk than the MLE, $\delta(\mathbf{b})$, is very large. For example, if we let $(K-2)/(T-K+2) = a$ and $s = (T-K)\hat{\sigma}^2$, we have the following Stein-type estimator proposed by Judge and Bock (1978, p. 240):

$$\delta(\mathbf{b}, s) = \left(1 - \frac{as}{\mathbf{b}'X'X\mathbf{b}}\right)\mathbf{b} \tag{1.8}$$

which, under squared loss is minimax if $0 \leq a \leq 2\left(\text{tr}(X'X)^{-1}d_L^{-1} - 2\right)/(T - K + 2)$ and $\text{tr}(X'X)^{-1} > 2d_L$, where d_L is the smallest characteristic root of $X'X$. To prevent overshrinking, the positive rule estimator $\delta(\mathbf{b}, s)^+ = c^+\mathbf{b}$, which dominates $\delta(\mathbf{b}, s) = c\mathbf{b}$, where $c^+ = \max\{c, 0\}$, should be used (Judge and Bock, 1978, p. 246).

1.1.3 Implications

In the previous subsections we have considered the problem where the econometrician wishes to estimate the parameters of a K dimensional vector $\boldsymbol{\beta} = (\beta_1, \beta_2, \ldots, \beta_K)'$, where $\boldsymbol{\beta}$ is the mean of an independent normal random vector $\mathbf{b} \sim N(\boldsymbol{\beta}, I_K)$ and $\boldsymbol{\beta}$ is unknown. For this problem James and Stein (1961) have demonstrated for $K \geq 3$ that the estimator

$$\delta^s(\mathbf{b}) = \left(1 - (K-2)/\|\mathbf{b}\|^2\right)\mathbf{b} \tag{1.9}$$

is uniformly better than the MLE, $\mathbf{b}$, under quadratic loss. This result holds for a range of linear statistical models and corresponding ML estimators. Given the high esteem in which the MLE is held, this seems at first blush to be impossible. In this estimator the estimate of each β_i depends not only on b_i, but also on the other b_i whose distributions are apparently independent of b_i. The result is a risk improvement in the MLE regardless of the values of β_i. The reaction of statisticians and econometricians to this seeming magic has been less than overwhelming. However, after a half century, using this shrinkage idea as one way of dealing with model uncertainty seems to be firmly established and, as we see in the next section, has led to the more general idea of combining estimation problems (Efron and Morris, 1973).

1.2 Combining estimation problems

The variants of the James and Stein estimators discussed in section 1.1 achieve their risk advantages by shrinking to a fixed vector. Another alternative is shrinking to a random vector and in this context Lindley (1962) suggested the estimator

$$\delta^2(\mathbf{b}) = \bar{\mathbf{b}} + \left\{1 - (K-3)/\|\mathbf{b}\|^2\right\}(\mathbf{b} - \bar{\mathbf{b}}) \tag{1.10}$$

that shrinks toward the grand mean $\bar{\mathbf{b}}$. This estimator dominates the MLE competitor when $K \geq 4$ and does especially well risk-wise when the b_i are near each other. Shrinking toward a random vector suggests consideration

of two estimation problems and two corresponding estimators that may have different sampling characteristics. Moving in this direction, Green and Strawderman (1991), in the spirit of the James and Stein estimator, proposed an estimator that involves the best weighted linear combination of two estimation problems, where 'best' is defined in terms of quadratic loss. An important point in their formulation is that the quadratic loss criterion is introduced up front and not considered as a result of the estimation process.

Although the Stein-rules of Section 1.1 may be developed from an empirical Bayes base (Judge and Bock, 1978, pp. 173–5), how the shrinkage rule (1.2) came about is a mystery. To make the combining rule transparent we develop the weighted linear combination estimator in some detail. For expository purposes, we continue to consider the orthonormal statistical model

$$\mathbf{y} = X\boldsymbol{\beta} + \mathbf{e}; \mathbf{e} \sim \left(\mathbf{0}, \sigma^2 I_n\right) \quad \text{and} \quad X'X = I_K \tag{1.11}$$

and the following ML estimator $\mathbf{b} = \hat{\boldsymbol{\beta}} = (X'X)^{-1}X'\mathbf{y} \sim (\boldsymbol{\beta}, \sigma^2 I_K)$, and a biased competitive estimator, $\tilde{\boldsymbol{\beta}} \sim (\boldsymbol{\beta} + \boldsymbol{\delta}, \tau^2 I_K)$, where $\boldsymbol{\delta}$ is a bias vector and $\tau^2 < \sigma^2$. For convenience assume σ^2 and τ^2 are known. The objective is to determine the best linear combination of $\hat{\boldsymbol{\beta}}$ and $\tilde{\boldsymbol{\beta}}$ where performance is evaluated in terms of quadratic loss. Under quadratic loss the risk of $\hat{\boldsymbol{\beta}}$ is $\sigma^2 K$ and the risk of $\tilde{\boldsymbol{\beta}}$ is $\tau^2 K + \boldsymbol{\delta}'\boldsymbol{\delta}$. This suggests that over the parameter space $\boldsymbol{\beta}'\boldsymbol{\beta}$ the risk functions of the two estimators may cross.

Given this situation, the question is whether there exists a weighted linear combination of the two estimators that leads to a combined estimator

$$\gamma\left(\hat{\boldsymbol{\beta}}, \tilde{\boldsymbol{\beta}}, \boldsymbol{\beta}\right) = \alpha\hat{\boldsymbol{\beta}} + (1 - \alpha)\tilde{\boldsymbol{\beta}} \tag{1.12}$$

that risk dominates $\hat{\boldsymbol{\beta}}$. The risk of the linear combination of the two estimators $\gamma(\hat{\boldsymbol{\beta}}, \tilde{\boldsymbol{\beta}}, \boldsymbol{\beta})$ is

$$\rho\left(\hat{\boldsymbol{\beta}}, \tilde{\boldsymbol{\beta}}, \boldsymbol{\beta}\right) = \alpha^2\sigma^2 K + (1 - \alpha)^2\left(\tau^2 K + \boldsymbol{\delta}'\boldsymbol{\delta}\right) \tag{1.13}$$

and the value of the mixing parameter α that minimizes (1.13) must satisfy

$$d\rho\left(\hat{\boldsymbol{\beta}}, \tilde{\boldsymbol{\beta}}, \boldsymbol{\beta}\right)/d\alpha = -\alpha(\sigma^2 K + \tau K + \boldsymbol{\delta}'\boldsymbol{\delta}) + \left(\tau^2 K + \boldsymbol{\delta}'\boldsymbol{\delta}\right) = 0 \tag{1.14}$$

such that

$$\alpha = \frac{\tau^2 K + \boldsymbol{\delta}'\boldsymbol{\delta}}{\sigma^2 K + \tau^2 K + \boldsymbol{\delta}'\boldsymbol{\delta}} \tag{1.15}$$

Consequently, we may write the minimum risk combination of the two estimators in (1.12) as

$$\gamma\left(\hat{\beta}, \tilde{\beta}, \beta\right) = \frac{(\tau^2 K + \delta' \delta)}{(\tau^2 K + \delta' \delta + \sigma^2 K)} \hat{\beta}$$
$$+ \frac{\sigma^2 K}{\sigma^2 K + \tau^2 K + \delta' \delta} \tilde{\beta} \quad (1.16)$$

Since $E\left[\left\|\hat{\beta} - \tilde{\beta}\right\|^2\right] = \sigma^2 K + \tau^2 K + \delta' \delta$, if we substitute $\left\|\hat{\beta} - \tilde{\beta}\right\|^2$ for its expected value (Judge and Bock, 1978, p. 175), we may rewrite (1.16) as

$$\gamma\left(\hat{\beta}, \tilde{\beta}, \beta\right) = \tilde{\beta} + \left(1 - \sigma^2 K / \left\|\hat{\beta} - \tilde{\beta}\right\|^2\right)\left(\hat{\beta} - \tilde{\beta}\right)$$
$$(1.17)$$

which is in the form of a shrinkage estimator, where $\hat{\beta}$ is shrunk toward the biased estimator $\tilde{\beta}$, with K as an approximation for the best linear combination of $\hat{\beta}$ and $\tilde{\beta}$.

In the context of Section 2 it is straightforward to demonstrate that $\gamma(\hat{\beta}, \tilde{\beta}, \beta)$ risk dominates the MLE, $\hat{\beta}$, with $K-2$ as the uniformly best combination. Building on this work, Judge and Mittelhammer (2004) generalized the orthonormal independent estimator combining problem and developed natural adaptive semiparametric estimation and inference methods for dependent estimators. The resulting semiparametric estimators are free of parametric choices and tuning parameters, and have good asymptotic and finite sample properties and superior risk performance. Extending Stein-like estimation to include a random shrinkage vector greatly extends the applicability of combining type estimators for a range of problems in econometric theory and practice.

2. Traditional pretest estimation

When there is uncertainty about the econometric model and thus the appropriate restrictions or hypothesis to impose, a traditional way to proceed is by statistical hypothesis testing based on the data at hand. The econometric literature abounds with exact and approximate tests for identifying sins of omission or commission relative to a variety of possibly false models. Although the two-stage estimation rule that results is used routinely in applied work, the econometric literature is strangely silent as to the statistical properties of the resulting two-stage estimator, or its sequential counterpart. To see the possible statistical significance of this two-stage process, continue with the orthonormal linear statistical model and notation used in Section 1.2 and follow Judge and Bock (1978; 1983). Under the statistical model and a fixed vector $\mathbf{r}$ and the MLE, $\hat{\beta}$ we may use likelihood ratio procedures to test the null hypothesis $H_0 : \beta = \mathbf{r}$ against the hypothesis $\beta \neq \mathbf{r}$, by

using the test statistic

$$u = \left(\hat{\beta} - \mathbf{r}\right)'\left(\hat{\beta} - \mathbf{r}\right) / K\hat{\sigma}^2, \quad (2.1)$$

which, if the hypotheses (restrictions) are correct, is distributed as a central F random variable with K and $(T-K)$ degrees of freedom. Of course if the restrictions are incorrect $E\left[\hat{\beta} - \mathbf{r}\right] = (\beta - \mathbf{r}) = \delta \neq 0$, and u is distributed as a non-central F with non-centrality parameter $\lambda = (\beta - \mathbf{r})'(\beta - \mathbf{r})/2\sigma^2 = \delta'\delta/2\sigma^2$. As a test mechanism the null hypothesis is rejected if $u \geq F_{(K,T-K)}^\alpha = c$, where c is determined for a given level of the test α by $\int_c^\infty dF_{(K,t-K)} = P\left[F_{(K,T-K)} \geq c\right] = \alpha$. This means that by accepting the null hypothesis we use the restricted least squares estimator $\beta^* = \mathbf{r}$ as our estimate of β, and by rejecting the null hypothesis $\beta - \mathbf{r} = \delta = 0$ we use the unrestricted least squares estimator, $\hat{\beta}$. The estimate that results is dependent upon a preliminary test of significance and this means the estimator used by many applied workers is of the form

$$\hat{\beta}^* = \begin{bmatrix} \beta^* & \text{if} & u < c, \\ \hat{\beta} & \text{if} & u \geq c. \end{bmatrix} \quad (2.2)$$

Alternatively the estimator may be written as

$$\hat{\beta}^* = I_{(0,c)}(u)\beta^* + I_{[c,\infty)}(u)\hat{\beta} = \beta - I_{(0,c)}(u)\left(\hat{\beta} - \mathbf{r}\right)$$
$$(2.3)$$

where $I_A(u)$ is an indicator function that takes the value 1 when $u \in A$ and takes the value 0 otherwise. This specification means that in a repeated sampling context the data, the linear hypotheses, and the selected level of statistical significance determine the combination of the two estimators that is chosen. From (2.3) the mean of the pretest estimator is

$$E\left[\hat{\beta}^*\right] = \beta - E\left[I_{(0,c)}(u)\left(\hat{\beta} - \mathbf{r}\right)\right] \quad (2.4)$$

which by theorem 2.1 in Judge and Bock (1978, p. 71) may be expressed as

$$E\left[\hat{\beta}^*\right] = \beta - \delta P\left[\chi_{(K+2,\lambda)}^2 / \chi_{(T-K)}^2 \leq cK/(T-K)\right]$$
$$(2.5)$$

Consequently, if $\delta = 0$, the pretest estimator is unbiased. This fortunate outcome aside, the size of the bias is affected by the probability of a random variable with a non-central F distribution being less than a constant, which is determined by the level of the test, the number of hypotheses, and the degree of hypothesis error, δ or λ. Since the probability is always equal to or less than one, the bias of the pretest estimator is equal to or less than the bias of the restricted estimator β^*. Following Judge and Bock (1978, p. 70) and using the discontinuous estimator rule (2.3), we may express the risk function for the pretest estimator, $\hat{\boldsymbol{\beta}}^*$, as

$$
\begin{aligned}
\rho\left(\boldsymbol{\beta}, \hat{\boldsymbol{\beta}}^*\right) &= E\left[\left(\hat{\boldsymbol{\beta}}^* - \boldsymbol{\beta}\right)'\left(\hat{\boldsymbol{\beta}}^* - \boldsymbol{\beta}\right)\right] \\
&= \sigma^2 K + (2\boldsymbol{\delta}'\boldsymbol{\delta} - \sigma^2 K) P\left[\frac{\chi^2_{(K+2,\lambda)}}{\chi^2_{(T-K)}} \le \frac{cK}{T-K}\right] \\
&\quad - \boldsymbol{\delta}'\boldsymbol{\delta}\ P\left[\frac{\chi^2_{(K+4,\lambda)}}{\chi^2_{(T-K)}} \le \frac{cK}{T-K}\right].
\end{aligned}
$$

(2.6)

Defining the terms in brackets as L(2) and L(4), we may write (2.6) compactly as

$$
\rho\left(\boldsymbol{\beta}, \hat{\boldsymbol{\beta}}^*\right) = \sigma^2 K + (2\boldsymbol{\delta}'\boldsymbol{\delta} - \sigma^2 K)L(2) - \boldsymbol{\delta}'\boldsymbol{\delta}L(4),
$$

(2.7)

where $1 > L(2) > L(4) > 0$. From the risk function (2.6) the following characteristics of the pretest estimator $\hat{\boldsymbol{\beta}}^*$ emerge: (i) if the restrictions are correct and $\boldsymbol{\delta} = 0$, the pretest estimator has a smaller risk than the ML estimator $\hat{\boldsymbol{\beta}}$ at the origin, $\boldsymbol{\delta} = 0$, and the risk depends on the level of significance α and correspondingly the critical value of the test c; (ii) as the hypothesis error $\boldsymbol{\delta}$ or λ grows, the risk of the pretest estimator $\hat{\boldsymbol{\beta}}^*$ increases, obtains a maximum after exceeding the risk of the MLE, $\hat{\boldsymbol{\beta}}$, and then monotonically decreases to approach $\sigma^2 K$, the risk of the MLE; (iii) as the hypothesis error $\hat{\boldsymbol{\beta}} - \mathbf{r} = \boldsymbol{\delta}$ increases and approaches infinity, the risk of the pretest estimator approaches $\sigma^2 K$, the risk of the MLE, from above; and (iv) the risk of the pretest estimator, $\hat{\boldsymbol{\beta}}^*$, varies with α the chosen level of significance. Thus, if one is to use the estimator $\hat{\boldsymbol{\beta}}^*$, the question as to the optimal level of significance α remains.

Finally, we note that Sclove, Morris and Radhakrishman (1972) demonstrated that the Stein-rule estimator

$$
\delta_u^S\left(\hat{\boldsymbol{\beta}}\right) = I_{[c,\infty)}(u)(1 - c^*/u)\left(\hat{\boldsymbol{\beta}} - \mathbf{r}\right) + \mathbf{r}
$$

(2.8)

where $c^* = (T - K)(K - 2)K^{-1}(T - K + 2)^{-1}$ and u is the likelihood ratio statistic defined in (2.1), is under quadratic loss uniformly superior to the pretest estimator, $\hat{\boldsymbol{\beta}}^*$, thus proving its inadmissibility.

3. Concluding remarks

Post-data evaluation procedures constitute a rejection of the concept of a true econometric model for which econometric theory provides a basis for estimation and inference. In this context, if a researcher is willing to forgo the property of unbiasedness, the Stein family of nonlinear biased estimation rules provides, under quadratic loss and conditions normally found in practice, a uniformly superior alternative. These estimators that shrink maximum likelihood estimates towards zero or some predetermined coordinate enjoy good statistical properties from both sampling theory and Bayesian points of view. Extension of the Stein-rule idea to weighted linear combining estimation problems leads to estimators that can be recommended from the standpoint of simplicity, generalizability and efficiency, and robustness over distribution assumptions and loss functions. If, in an information-theoretic context, one wishes to leave the parametric Stein-rule family, minimum divergence estimators that involve reference and subject distributions and a choice of distance measures offer attractive semiparametric estimation and inference alternatives (Mittelhammer et al., 2005).

Estimators that evolve by a two-stage estimation and then hypothesis-testing process lead to estimation rules that are (i) risk inferior, over a large range of the parameter space, to the data-based maximum likelihood estimator, and (ii) uniformly risk-inferior to Stein-rule alternatives. Recognizing this unfortunate situation. one of my colleagues once remarked that pretest estimation (single or repeated hypothesis tests) is the biggest unreported scandal in inferential statistics. My only quibble with this statement is that the statistical implications of pretesting have been reported for over a half-century. Hypothesis testing is like an addictive drug and has led econometricians to produce a plethora of conditional test statistics, to higher consumption of old and new hypothesis tests, and to model-discovery processes that lead to pretest estimators with negative and unknown sampling performance. Perhaps it is time for econometricians to just say 'no'!

GEORGE G. JUDGE

See also **econometrics; linear models; maximum likelihood; semiparametric estimation; testing.**

Bibliography

Baranchik, A. 1964. *Multiple Regression and Estimation of the Mean of Multivariate Normal Distribution.* Technical Report 51. Stanford University, Department of Statistics.

Bock, M., Yancey, T. and Judge, G. 1973. The statistical consequences of preliminary test estimators in regression. *Journal of American Statistical Association* 68, 107–10.

Efron, B. and Morris, C. 1973. Combining possibly related estimation problems. *Journal of the Royal Statistical Society,* Series B, 35, 379–421.

Green, E. and Strawderman, W. 1991. James-Stein-type estimator for combining unbiased and possibly biased estimators. *Journal of the American Statistical Associations* 86, 1001–6.

James, W. and Stein, C. 1961. Estimation with quadratic loss. In *Proceedings of the Fourth Berkley Symposium on Mathematical Statistics and Probability,* vol. 1. Berkeley and Los Angeles: University of California Press.

Judge, G. and Bock, M. 1978. *The Statistical Implications of Pretest and Stein-Rule Estimators in Econometrics*. New York: North-Holland.

Judge, G. and Bock, M. 1983. Biased estimation. In *Handbook of Econometrics*, vol. 1, ed. Z. Griliches and M. Intriligator. Amsterdam: North-Holland.

Judge, G., Hill, R., Griffiths, W., Lutkepohl, H. and Lee, T. 1988. *Introduction to the Theory and Practice of Econometrics*, ch. 20. New York: John Wiley.

Judge, G. and Mittelhammer, R. 2004. A semiparametric basis for combining estimation problems under quadratic loss. *Journal of American Statistical Association* 49, 479–87.

Lindley, D. 1962. Discussion of Professor Stein's paper. *Journal of the Royal Statistical Society*, Series B, 24, 285–8.

Mittelhammer, R., Judge, G., Miller, D. and Cardell, S. 2005. Minimum divergence moment based binary response models: estimation and inference. Working Paper No. 998. CUDARE, University of California, Berkeley.

Sclove, S., Morris, C. and Radhakrishman, R. 1972. Non-optimality of preliminary-test estimators for the multinormal mean. *Annals of Mathematical Statistics* 43, 1481–90.

Stein, C. 1955. Inadmissibility of the usual estimator for the mean of a multivariate normal distribution. *Proceedings of the Third Berkeley Symposium on Mathematical Statistics and Probability*, vol. 1. Berkeley and Los Angeles: University of California Press.

Sidgwick, Henry (1838–1900)

Henry Sidgwick was a Victorian-era philosopher, ethicist, classicist, economist, political and legal theorist, parapsychologist, educational reformer, and literary critic who spent his entire working life at Cambridge University, becoming a central figure in the early Cambridge School of economics. Educated at Rugby and Cambridge, where he studied classics and mathematics and joined the secret discussion society known as the Apostles, he eventually, in 1883, achieved the status of Knightbridge Professor of Moral Philosophy. With his wife, Eleanor Mildred Sidgwick (née Balfour), he helped found both the Society for Psychical Research and Newnham College, Cambridge, one of England's first colleges for women.

Sidgwick published widely, but his major books during his lifetime were *The Methods of Ethics* (first edition 1874), *The Principles of Political Economy* (first edition 1883), and *The Elements of Politics* (first edition 1891). Best known as an ethical philosopher, he was also very influential in other areas, particularly economics. He worked extensively with the Cambridge and London Charity Organization Societies, and in 1885 he was elected president of the economics and statistics section of the British Association; he also contributed to the first *Dictionary of Political Economy*, edited by R. Palgrave, advised his university on economic policy, and appeared as an expert witness for various government committees on economic policy matters.

The Cambridge School

Sidgwick's importance as an economic thinker has often been underestimated, in part because of his stormy relationship with Alfred Marshall, who is generally regarded as the founder of the Cambridge School (Groenewegen, 1995). Marshall harshly opposed Sidgwick as a 'University politician' and criticized his economic work, but Sidgwick clearly played a vital role in shaping both Marshall and their Cambridge context (Backhouse, 2006; Schultz, 2004). He was, as much as Marshall, caught up in the so-called marginalist revolution:

> as Jevons had admirably explained, the variations in the relative market values of different articles express and correspond to variations in the comparative estimates formed by people in general, not of the *total* utilities of the amounts purchased of such articles, but of their *final* utilities; the utilities, that is, of the last portions purchased. (Sidgwick, 1901, p. 82)

But Sidgwick was more involved than Marshall in the methodological debates of the time, seeking to balance both deductive and inductive approaches, and he was also wary of the evolutionary metaphors and talk of 'economic biology' to which Marshall was given. Much the better philosopher, Sidgwick's reflective, qualified utilitarianism and analysis of public goods and the role of the state anticipated early 20th-century welfare economics. Indeed, the welfare economics of Marshall's designated successor, A.C. Pigou, was more in line with Sidgwick's views than with Marshall's (O'Donnell, 1979; Backhouse, 2006). As early as 1913, J.S. Nicholson noted that, not only did Pigou 'apply the same general principle of utility, but the main trend of the argument is the same' (Nicholson, 1913, p. 420). Baumol (1965) complained that Sidgwick's overall approach and 'penetrating discussion' of the 'Pigouvian problem' (the possible divergence between private and social benefits and costs) was 'largely unrecognized', and quoted from book III of the *Principles*, where Sidgwick argues that 'there is no general reason' for supposing that it will always be the case that 'the individual can always obtain through free exchange adequate remuneration for the services he is capable of rendering to society' since, among other things, 'there are some utilities which, from their nature, are practically incapable of being appropriated by those who produce them or who would otherwise be willing to purchase them' (for example, lighthouses, improvements in climate).

Yet even the accounts of Sidgwick's work that recognize his significance often tend to stress different accomplishments. Blaug (1985, p. 479) suggests that the *Principles* may have been the first work 'to question the traditional idea that technical change is necessarily capital-using'. Stigler (1982, p. 41), when discussing how Edgeworth, Sidgwick and Marshall gave currency to the work of Cournot and Dupuit on monopoly and oligopoly, remarked that Sidgwick's *Principles* 'has two chapters (bk II, ch. IX and X) which are among the best in the history of microeconomics, dealing with the theories of human capital and noncompetitive behavior'. Chapter IX concludes that:

> the possessors of capital, real and personal, as well as persons endowed with rare natural gifts, are likely to have – by reason of their limited numbers – important advantages in the competition that determines relative wages; in consequence of which the remuneration of such persons may – and in England often does – exceed the wages of ordinary labour by an amount considerably larger than is required to compensate them for additional outlay or other sacrifices; such excess tending to increase as the amount of capital owned by any individual increases, but in a ratio not precisely determinable by general considerations. (Sidgwick, 1901, p. 337)

Chapter X investigates how and when self-interested action will lead to combination, and it provocatively argues – beyond Mill and against 'any economist of repute' – that in many ordinary cases it is possible for workers to combine and win higher wages without such gain having 'any manifest tendency to be counterbalanced by future loss'. More generally, and anticipating current analyses of labour markets as involving bargaining processes, Sidgwick argues that, when it comes to wage controversies:

> Economic science cannot profess to determine the normal division of the difference remaining, when from the net produce available for wages and profits in any branch of production we subtract the minimum shares which it is the interest of employers and employed respectively to take rather than abandon the business and seek employment for their labour and capital elsewhere. (1901, p. 355)

Economics and ethics

This fine-grained (but qualitative) analysis, attentive to expert opinion and the line between description and prescription, but yielding conclusions that for the times were quite sceptical or subversive in their implications, is characteristic of Sidgwick's scholarly work in general. Even his work in ethics mostly sought to set aside preaching and polemics, proceeding instead by a painstaking comparative investigation of the rational grounds for the major systematic ethical positions. His celebrated *Methods* sought, with something akin to scientific open-mindedness, 'to consider simply what conclusions will be rationally reached if we start with certain ethical premises, and with what degree of certainty and

precision' (Sidgwick, 1907, p. viii). Sidgwick concluded that, although commonsense moral rules – do not lie, do not murder, do not break promises – could be largely derived from utilitarianism, there was a 'dualism of the practical reason' when it came to egoism (pursue one's own greatest good) and utilitarianism (pursue the greatest good in general), each of which was as 'rational' as the other. The dualism of practical reason was in key respects a philosophical representation of collective action problems, and the conception of ultimate good or happiness figuring in both egoism and utilitarianism was a subtle hedonistic account of pleasurable or desirable consciousness that made it clear how difficult yet unavoidable interpersonal comparisons of utility could be. This conception would form the basis for Edgeworth's attempted 'hedonometry', or quantification of hedonism, revived in recent years by Kahneman (for example, Edgeworth, 1877; Kahneman, 1999).

Sidgwick himself favoured utilitarianism, though he was haunted by his inability to defend it fully. His utilitarianism nonetheless clearly (and admittedly) informed his economic views, especially when it came to the 'art' of political economy, which concerned the normative considerations that came to be called welfare economics. When it came to such normative political theory, Sidgwick largely assumed the utilitarian principle as the normative bottom line (Sidgwick, 1901). Despite their differences, Marshall was willing to give Sidgwick credit for his 'art', calling the third book of Sidgwick's *Principles* by 'far the best thing of its kind in any language' (in Pigou, 1925, p. 7). He also expressed an admiration for a broader orientation they shared: 'that we are not at liberty to play chess games, or exercise ourselves upon subtleties that lead nowhere' (Whitaker, 1996). Neither Sidgwick nor Marshall was enthusiastic about formalization for formalization's sake.

Still, it is not clear precisely what Marshall admired about Sidgwick's welfare economics. Although both stressed the role of education in helping to overcome poverty and economic inequality, Sidgwick arguably went beyond Marshall (and Mill) in setting out the cases of market failure, the limits of laissez-faire, and the limitations of economic analysis in general, both descriptive and normative. Backhouse (2002, p. 271) urges that the truly 'fundamental part of Sidgwick's argument' was distinguishing between wealth as 'as the sum of goods produced, valued at market prices' and wealth as 'the sum of individuals' utilities – what we would now term welfare'. This distinction 'made it possible, arguably for the first time, to conceive of welfare economics as something distinct from economics in general', while greatly complicating comparisons of utilities. Analogous arguments are evident in Sidgwick's sceptical account of the possibilities for comparing wealth in cross-cultural or trans-historical contexts. And the notion of unpurchased utilities, not measured by exchange values, allowed that, as Backhouse puts it, 'if the marginal utility of a particular good were higher for one person than for another, total utility could be raised by redistributing goods to those who valued them most. This would leave wealth at market prices unchanged.' Such views, conjoined with a belief in declining marginal utility, suggest serious redistributivist possibilities that both Sidgwick and Marshall downplayed:

> Marshall is much more aware of the quantitative side of the problem than is Sidgwick ... but no nearer a way to thinking quantitatively about how to achieve the best use of resources. They share both a philosophical viewpoint that inclines them towards egalitarianism and a conservatism that will not risk any interference with incentives, lest output be reduced. (Backhouse, 2006, p. 33)

Economics and politics

Perhaps Sidgwick was overly impressed by the idea that incentives were crucial to production and that complete communism would lead to splendidly equal destitution. Still, he took more gradualist forms of (market) socialism very seriously – so much so that the libertarian Hayek (1960, p. 419) could complain that Sidgwick's *Elements of Politics* scarcely 'represents what must be regarded as the British liberal tradition and is already strongly tainted with that rationalist utilitarianism which led to socialism'. The *Elements* and the *Principles* take the laissez-faire principle of individualism – that 'what one sane adult is legally compelled to render to others should be merely the negative service of non-interference, except so far as he has voluntarily undertaken to render positive services' – only as a starting point, from which to run a very long course of qualifications and exceptions: education, child care, poor relief, disease control, public works or goods (the famous lighthouse, pure research, the environment and defence), monopoly, collective bargaining and others. Sidgwick emphasizes two cases that sharply point up the limitations of economic individualism – the 'humane treatment of lunatics, and the prevention of cruelty to the inferior animals' – because such restrictions hardly aim at securing the freedom of the lunatics or the animals, but are 'a one-sided restraint of the freedom of action of men with a view to the greatest happiness of the aggregate of sentient beings' (Sidgwick, 1919, p. 141). These are only the most conspicuous of the many difficulties with a principle that betrays a naive faith in 'the psychological proposition that every one can best take care of his own interest' and the 'sociological proposition that the common welfare is best attained by each pursuing exclusively his own welfare and that of his family in a thoroughly alert and intelligent manner'.

Cautious as he was, Sidgwick (1903) was ultimately persuaded that the growth of federalism and large-scale state organizations was likely to continue, though he doubted that the social sciences were anywhere near to discovering actual laws of historical development. Moreover, even if he was cautious about economic or political socialism, he was relatively enthusiastic about ethical socialism, that is, about the possibility of humanity growing more altruistic and compassionate, regarding their labour as their contribution to the common good. He was under no illusions whatsoever, not only about the market failing to reflect claims of desert or merit, but also about the limitations of that abstraction, 'economic man', since historical and cultural or national context could dramatically alter the possibilities for moving beyond economic individualism. Perhaps the most disturbing and problematic aspect of his economic and political work was the way in which it lent itself to the racist and imperialistic tendencies of the British Empire (Schultz, 2004). Although he was more sceptical about claims of inherent racial differences than many of his contemporaries, he too often countenanced the possibility of such differences, accepting stereotypes about the varying fitness for physical versus mental labour of different peoples (Schultz and Varouxakis, 2005). And although on the economic side he tended to follow Turgot in doubting the material benefits of colonies, he allowed that there were other forms of remuneration involved:

> there are sentimental satisfactions, derived from justifiable conquests, which must be taken into account, though they are very difficult to weigh against the material sacrifices and risks. Such are the justifiable pride which the cultivated members of a civilised community feel in the beneficent exercise of dominion, and in the performance by their nation of the noble task of spreading the highest kind of civilization. (1919, p. 313)

It was in this highly Eurocentric manner that Sidgwick embraced the global leadership of the 'civilised' nations in (supposedly) spreading international law and morality and limiting the possibility of war (Sidgwick, 1898).

BARTON SCHULTZ

See also **Edgeworth, Francis Ysidro; happiness, economics of; Marshall, Alfred; Mill, John Stuart; Pigou, Arthur Cecil; utilitarianism and economic theory; welfare economics.**

Selected works

1898. *Practical Ethics*. London: Swan Sonnenschein.
1901. *The Principles of Political Economy*, 3rd edn. London: Macmillan.
1903. *The Development of European Polity*. London: Macmillan.
1904. *Miscellaneous Essays and Addresses*. London: Macmillan.
1907. *The Methods of Ethics*, 7th edn. London: Macmillan.
1919. *The Elements of Politics*, 4th edn. London: Macmillan.
1997, 1999. *The Complete Works and Select Correspondence of Henry Sidgwick*, gen. ed. B. Schultz. Charlottesville, VA: InteLex Corp.

Bibliography

Backhouse, R. 2002. *The Ordinary Business of Life*. Princeton: Princeton University Press.

Backhouse, R. 2006. Sidgwick, Marshall, and the Cambridge School of economics. *History of Political Economy* 38, 15–44.

Baumol 1965. *Welfare Economics and the Theory of the State*, 2nd edn. Cambridge, MA: Harvard University Press.

Blaug, M. 1985. *Economic Theory in Retrospect*. Cambridge: Cambridge University Press.

Edgeworth, F. 1877. *Old and New Methods of Ethics*. Oxford and London: James Parker and Co.

Groenewegen, P. 1995. *A Soaring Eagle: Alfred Marshall, 1842–1924*. Aldershot, UK and Brookfield, US: Edward Elgar.

Hayek, F. 1960. *The Constitution of Liberty*. Chicago: Regnery.

Kahneman, D. 1999. Objective happiness. In *Well-Being: The Foundations of Hedonic Psychology*, ed. D. Kahneman, E. Diener and N. Schwarz. New York: Russell Sage Foundation.

Nicholson, J. 1913. The vagaries of recent political economy. *Quarterly Review* 219.

O'Donnell, M. 1979. Pigou: an extension of Sidgwickian thought. *History of Political Economy* 11, 588–605.

Pigou, A., ed. 1925. *Memorials of Alfred Marshall*. London: Macmillan.

Schultz, B. 2004. *Henry Sidgwick, Eye of the Universe*. New York: Cambridge University Press.

Schultz, B. and Varouxakis, G., eds. 2005. *Utilitarianism and Empire*. Lanham, MD: Lexington Books.

Stigler, G. 1982. *The Economist as Preacher*. Chicago: University of Chicago Press.

Whitaker, J. 1996. *The Correspondence of Alfred Marshall, Economist*, 3 vols. New York: Cambridge University Press.

Sidrauski, Miguel (1939–1968)

Sidrauski was born and educated in Buenos Aires. He entered the University of Chicago Ph.D. programme in 1963, completed his dissertation in 1966, and accepted an assistant professor appointment at MIT. He died of cancer in September 1968, leaving his wife, Martha, and two-month old daughter, Carmela.

Sidrauski is best known in economics for his article 'Rational Choice and Patterns of Growth in a Monetary Economy' (1967a), based on his dissertation written under the supervision of Hirofumi Uzawa and Milton Friedman. The model is one of an economy with a representative intertemporally maximizing household, which derives instantaneous utility from both consumption and the holding of real balances. The thesis contains a careful discussion of the device of putting money in the utility function. The household can hold capital as well as money. Sidrauski derives necessary conditions for a maximum, and then studies the dynamics and steady states of inflation and capital accumulation in the model.

The key result is that steady state capital intensity is invariant to the rate of monetary expansion, and thus that money is superneutral between steady states. Sidrauski indicates in his thesis, which extends the paper in several directions, that the superneutrality result changes if money is given a productive role in the economy. In the dynamic analysis he assumes that expectations of inflation are adaptive.

In a non-maximizing money-and-growth model (1967b) Sidrauski confirms the Tobin result that an increase in the growth rate of money increases steady state capital intensity. This is based on the effects of an increase in the growth rate of money in reducing consumption. Sidrauski shows also in this paper that, with adaptive expectations, an increase in the growth rate of money first causes capital accumulation to fall and only later to increase, taking capital intensity above its initial level.

He published four other articles, including one on exchange rate determination, and a book, *Monetary and Fiscal Policy in a Growing Economy*. The book, written jointly with Duncan Foley and published posthumously in 1970, develops a three-asset (money, bonds and capital) two-sector Tobinesque growth model. The two sector structure gives a key role in the investment process to the relative price of capital, p_k, which is Tobin's q. The authors succeed in making the model answer questions about the effects on growth and inflation of policy changes such as fiscal expansion, increases in the growth rate of all outside assets, and open market operations. The model has not been widely used despite its usefulness and versatility; this may be because the full employment assumption makes it unattractive for use as a cyclical model, and the absence of explicit maximizing assumptions makes it unattractive to many who study long-term growth.

In his two years at MIT Sidrauski established himself as an excellent teacher and adviser, and as an economist of outstanding promise. Milton Friedman's eulogy (1969) speaks not only of his technical skill and promise, but also of his personal warmth and generosity. It concludes:

> The death of this young man is a grievous loss to our profession and to the world. Here was a man who would have pushed out the frontiers of our subject, would have changed and added to economic analysis, would have enlightened and informed generations of students – struck down at the very beginning of his career, full of promise but as yet almost bereft of fulfillment.

STANLEY FISCHER

Selected works

1967a. Rational choice and patterns of growth in a monetary economy. *American Economic Review, Papers and Proceedings* 57, 534–44.

1967b. Inflation and economic growth. *Journal of Political Economy* 75, 796–810.

1970. (With D.K. Foley.) *Monetary and Fiscal Policy in a Growing Economy.* London: Macmillan.

Bibliography

Friedman, M. 1969. Miguel Sidrauski. *Journal of Money, Credit and Banking* 1, 129–30.

sieve extremum estimation

1 Introduction

In econometrics literature, a model is called *parametric* if all of its parameters belong to finite-dimensional parameter spaces, and a model is *semiparametric* if its parameters of interests belong to finite-dimensional spaces but its nuisance parameters are in infinite-dimensional spaces. For example, a parameterized joint distribution model (such as logit or probit) is a parametric one and can be estimated by the method of maximum likelihood (ML). A model that only parameterizes some moment restrictions is a semiparametric one and can be estimated by the generalized method of moments (GMM). Since economics problems are complicated, researchers often find parametric likelihood models too restrictive and sensitive to deviations from the parametric specification. Moment-based semiparametric models are less restrictive yet still subject to misspecification of parametric moments.

Due to growing availability of large economic datasets and computational advances, nonparametric and semi-nonparametric models have become increasingly popular in both theoretical and applied econometrics. A model is *nonparametric* if all of its parameters are in infinite-dimensional parameter spaces, and a model is *semi-nonparametric* if it contains both finite-dimensional and infinite-dimensional unknown parameters of interests. For example, estimation of a conditional mean function $E[Y|X, W]$ without specifying its functional form is a nonparametric problem. If $E[Y|X, W]$ is specified as a partially linear form, $\beta'X + h(W)$, with β being unknown finite-dimensional parameter and $h()$ being unknown function, then it becomes a semi-nonparametric problem. Nonparametric and semi-nonparametric models are more flexible and robust. However, they involve infinite-dimensional parameter spaces; hence it can be computationally difficult to estimate such models using finite data sets. Moreover, even if one could solve the problem of optimizing a sample criterion over an infinite-dimensional parameter space that may not be compact, the resulting estimator may have undesirable large sample properties such as inconsistency and/or a very slow rate of convergence.

To tackle the difficulties encountered in semi-nonparametric problems, the *method of sieves* (Grenander, 1981) optimizes an empirical criterion over a sequence of approximating parameter spaces (that is, *sieves*); the sieves are less complex, but their complexity grows with the sample size so as to be dense in the original space. The resulting sieve estimator is consistent under very mild regularity conditions, and can generally reach optimal rate of convergence by balancing the bias part (which diminishes as sieve complexity grows), and the standard derivation part (which grows with sieve complexity). Most commonly used sieves in economics are finite dimensional (compact) approximating parameter spaces; when such sieves are used to estimate a semi-nonparametric model, the computation is as easy as to estimate a parametric model. The sieve method is particularly convenient when unknown functions enter the criterion function (or moment condition) nonlinearly, and/or when semi-nonparametric models contain complicated endogeneity and latent heterogeneity. It can easily incorporate prior information and constraints, often derived from economic theory, such as monotonicity, convexity, additivity, multiplicity, exclusion and non-negativity. It can simultaneously estimate the parametric and nonparametric parts in semi-nonparametric models, typically with optimal convergence rates for both parts.

The method of sieves consists of two key ingredients: a criterion function and sieve parameter spaces. Both the criterion functions and the sieve spaces can be very flexible, and we shall present some examples in the next two sections.

2 Examples of sieve spaces

The infinite-dimensional unknown parameter in a semi-nonparametric model can often be viewed as a member of some function space with certain regularities (for example, having bounded derivatives, monotone, concave). Thus, many deterministic approximation results developed in mathematics can be used to suggest sieves that provide good and computable approximations to an unknown function. Here we present some commonly used sieves in economics applications. Additional ones can be found in Judd (1998), DeVore and Lorentz (1993) and Chen (2006).

2.1 Finite-dimensional linear sieves with bounded support

A sieve is called a '(finite-dimensional) linear sieve' if it is a linear span of finitely many known basis functions. Linear sieves, including power series, Fourier series, splines and wavelets, form a large class of sieves that have been widely used in econometrics and statistics. We now provide some commonly used linear sieves for univariate functions with support $\chi = [0,1]$.

Polynomials. $\mathrm{Pol}(J_n)$ is the space of polynomials on $[0,1]$ of degree J_n or less:

$$\mathrm{Pol}(J_n) = \left\{ \sum_{k=0}^{J_n} a_k x^k, x \in [0,1] : a_k \in \mathscr{R} \right\}.$$

Trigonometrics. $\mathrm{TriPol}(J_n)$ is the space of trigonometric polynomials on $[0,1]$ of degree J_n or less:

$$\mathrm{TriPol}(J_n) = \left\{ a_0 + \sum_{k=1}^{J_n} [a_k \cos(2k\pi x) \right.$$

$$\left. + b_k \sin(2k\pi x)], x \in [0,1] : a_k, b_k \in \mathscr{R} \right\}.$$

$\mathrm{CosPol}(J_n)$ is the space of cosine polynomials on $[0,1]$ of degree J_n or less:

$$\mathrm{CosPol}(J_n) = \left\{ a_0 + \sum_{k=1}^{J_n} a_k \cos(k\pi x), x \in [0,1] : a_k \in \mathscr{R} \right\}.$$

$\mathrm{SinPol}(J_n)$ is the space of sine polynomials on $[0,1]$ of degree J_n or less:

$$\mathrm{SinPol}(J_n) = \left\{ \sum_{k=1}^{J_n} a_k \sin(k\pi x), x \in [0,1] : a_k \in \mathscr{R} \right\}.$$

We note that the classical trigonometric sieve, $\mathrm{TriPol}(J_n)$, is well suited for approximating periodic functions on $[0,1]$, while the cosine sieve, $\mathrm{CosPol}(J_n)$, is well suited for approximating aperiodic functions on $[0,1]$ and the sine sieve, $\mathrm{SinPol}(J_n)$, can approximate functions vanishing at the boundary points (that is, when $h(0) = h(1) = 0$).

Splines. Let J_n be a positive integer, and let t_0, $t_1, \ldots, t_{J_n}$, t_{J_n+1} be real numbers with $0 = t_0 < t_1 < \ldots < t_{J_n} < t_{J_n+1} = 1$. Partition $[0,1]$ into J_n+1 subintervals $I_j = [t_j, t_{j+1})$, $j = 0, \ldots, J_n-1$, and $I_{J_n} = [t_{J_n}, t_{J_n+1}]$. We assume that the knots $t_1, \ldots, t_{J_n}$ have bounded mesh ratio:

$$\frac{\max_{0 \leq j \leq J_n}(t_{j+1} - t_j)}{\min_{0 \leq j \leq J_n}(t_{j+1} - t_j)} \leq c \text{ for some constant } c > 0.$$

$$(2.1)$$

Let $r \geq 1$ be an integer. A function on $[0,1]$ is a *spline of order r*, equivalently, *of degree $m \equiv r - 1$, with knots* $t_1, \ldots, t_{J_n}$ if the following hold: (i) it is a polynomial of degree m or less on each interval I_j, $j = 0, \ldots, J_n$; and (ii) (for $m \geq 1$) it is $(m-1)$-times continuously differentiable on $[0,1]$. Such spline functions constitute a linear space of dimension J_n+r. For detailed discussions of univariate splines, see de Boor (1978) and Schumaker (1981). For a fixed integer $r \geq 1$, we let $\mathrm{Spl}(r, J_n)$ denote the space of splines of order r (or of degree $m \equiv r - 1$)

with J_n knots satisfying (2.1). Since

$$\mathrm{Spl}(r, J_n) = \left\{ \sum_{k=0}^{r-1} a_k x^k + \sum_{j=1}^{J_n} b_j [\max\{x - t_j, 0\}]^{r-1}, \right.$$

$$\left. x \in [0,1] : a_k, b_j \in \mathscr{R} \right\},$$

we also call $\mathrm{Spl}(r, J_n)$ the polynomial spline sieve of degree $m \equiv r - 1$.

2.2 Finite-dimensional linear sieves with unbounded support

In semi-nonparametric econometric applications, sometimes the parameters of interest are functions with unbounded supports. Here we present three finite-dimensional linear sieves that can approximate functions with unbounded supports well. In the following we let $L_2(\chi, \omega)$ denote the space of real-valued functions h such that $\int_\chi |h(x)|^2 \omega(x) dx < \infty$ for a smooth weight function $\omega : \chi \mapsto (0, \infty)$.

Orthogonal wavelets. Let $m \geq 0$ be an integer. A real-valued function ϕ is called a 'father wavelet' of degree m if it satisfies the following: (i) $\int_\mathscr{R} \phi(x) dx = 1$; (ii) ϕ and all its derivatives up to order m decrease rapidly as $|x| \to \infty$; (iii) $\{\phi(x - k) : k \in \mathbb{Z}\}$ forms a Riesz basis for a closed subspace of $L_2(\mathscr{R}, leb)$. A real-valued function ψ is called a 'mother wavelet' of degree m if it satisfies the following: (i) $\int_\mathscr{R} x^k \psi(x) dx = 0$ for $0 \leq k \leq m$; (ii) ψ and all its derivatives up to order m decrease rapidly as $|x| \to \infty$; (iii) $\{2^{j/2} \psi(2^j x - k) : j, k \in \mathbb{Z}\}$ forms a Riesz basis of $L_2(\mathscr{R}, leb)$.

Given an integer $m \geq 0$, there exist a father wavelet ϕ of degree m and a mother wavelet ψ of degree m, both compactly supported, such that for any integer $j_0 \geq 0$, any function g in $L_2(\mathscr{R}, leb)$ has the following wavelet m-regular multiresolution expansion:

$$g(x) = \sum_{k=-\infty}^{\infty} a_{j_0 k} \phi_{j_0 k}(x) + \sum_{j=j_0}^{\infty} \sum_{k=-\infty}^{\infty} b_{jk} \psi_{jk}(x), \quad x \in \mathscr{R},$$

where

$$a_{jk} = \int_\mathscr{R} g(x) \phi_{jk}(x) \, dx, \quad \phi_{jk}(x) = 2^{j/2} \phi(2^j x - k),$$
$$x \in \mathscr{R},$$

$$b_{jk} = \int_\mathscr{R} g(x) \psi_{jk}(x) \, dx, \quad \psi_{jk}(x) = 2^{j/2} \psi(2^j x - k),$$
$$x \in \mathscr{R},$$

and $\{\phi_{j_0 k}, k \in \mathbb{Z}; \psi_{jk}, j \geq j_0, k \in \mathbb{Z}\}$ is an orthonormal basis of $L_2(\mathscr{R}, leb)$; see Meyer (1992, theorem 3.3). We consider the finite-dimensional linear space spanned by this wavelet basis. For an integer $J_n > j_0$, set

$$\text{Wav}(m, 2^{J_n}) = \left\{ \sum_{k=0}^{2^{j_0}-1} \alpha_{j_0 k} \phi_{j_0 k}(x) \right.$$
$$\left. + \sum_{j=j_0}^{J_n-1} \sum_{k=0}^{2^j-1} \beta_{jk} \psi_{jk}(x), x \in \mathcal{R} : \alpha_{j_0 k}, \beta_{jk} \in \mathcal{R} \right\}$$

or, equivalently,

$$\text{Wav}(m, 2^{J_n}) = \left\{ \sum_{k=0}^{2^{J_n}-1} \alpha_k \phi_{J_n k}(x), x \in \mathcal{R} : \alpha_k \in \mathcal{R} \right\}.$$

Hermite polynomials. Hermite polynomial series $\{H_k: k = 1, 2, \ldots \ldots\}$ is an orthonormal basis of $L_2(\mathcal{R}, \omega)$ with $\omega(x) = \exp\{-x^2\}$. It can be obtained by applying the Gram–Schmidt procedure to the polynomial series $\{x^{k-1}: k=1,2,\ldots \ldots\}$ under the inner product $\langle f, g \rangle_\omega = \int_{\mathcal{R}} f(x)g(x) \exp\{-x^2\} dx$. Let $\text{HPol}(J_n)$ denote the space of Hermite polynomials on $\mathcal{R}$ of degree J_n or less:

$$\text{HPol}(J_n) = \left\{ \sum_{k=1}^{J_n+1} a_k H_k(x) \exp\left\{-\frac{x^2}{2}\right\}, x \in \mathcal{R} : a_k \in \mathcal{R} \right\}.$$

Then any function in $L_2(\mathcal{R}, leb)$ can be approximated by the $\text{Pol}(J_n)$ sieve as $J_n \to \infty$.

Laguerre polynomials. The Laguerre polynomial series $\{L_k: k = 1, 2, \ldots \ldots\}$ is an orthonormal basis of $L_2([0, \infty), \omega)$ with $\omega(x) = \exp\{-x\}$. It can be obtained by applying the Gram–Schmidt procedure to the polynomial series $\{x^{k-1}: k = 1, 2, \ldots \ldots\}$ under the inner product $\langle f, g \rangle_\omega = \int_0^\infty f(x)g(x) \exp\{-x\} dx$. Let $\text{LPol}(J_n)$ denote the space of Laguerre polynomials on $[0, \infty)$ of degree J_n or less:

$$\text{LPol}(J_n) = \left\{ \sum_{k=1}^{J_n+1} a_k L_k(x) \exp\left\{-\frac{x}{2}\right\}, x \in [0, \infty) : a_k \in \mathcal{R} \right\}.$$

Then any function in $L_2([0, \infty), leb)$ can be approximated by the $\text{LPol}(J_n)$ sieve as $J_n \to \infty$.

2.3 Finite-dimensional nonlinear sieves

A popular class of nonlinear sieves in econometrics is single hidden layer feedforward artificial neural networks (ANN) (see for example, Barron, 1993; Hornik et al., 1994; Chen and White, 1999). A typical ANN sieve is the one with a sigmoid activation function:

$$\text{sANN}(k_n) = \left\{ \sum_{j=1}^{k_n} \alpha_j S(\gamma_j' x + \gamma_{0,j}) : \gamma_j \in \mathcal{R}^d, \alpha_j, \gamma_{0,j} \in \mathcal{R} \right\},$$

where $S: R \to R$ is a sigmoid activation function, that is, a bounded non-decreasing function such that $\lim_{u \to -\infty} S(u) = 0$ and $\lim_{u \to \infty} S(u) = 1$. Some popular sigmoid activation functions include

- heaviside $S(u) = 1\{u \geq 0\}$

- logistic $S(u) = 1/(1+\exp\{-u\})$
- hyperbolic tangent $S(u) = (\exp\{u\} - \exp\{-u\})/(\exp\{u\} + \exp\{-u\})$
- Gaussian sigmoid $S(u) = (2\pi)^{-1/2} \int_{-\infty}^u \exp(-y^2/2) dy$

Additional examples of nonlinear sieves include spline sieves with data-driven choices of knot locations (or free-knot splines), and wavelet sieves with thresholding (Donoho et al., 1995). Nonlinear sieves are more flexible and may enjoy better approximation properties than linear sieves; see for example, Chen and Shen (1998) for the comparison of linear vs. nonlinear sieves.

2.4 Shape-preserving sieves

There are many sieves that can preserve the shape, such as non-negativity, monotonicity and convexity, of the unknown function to be approximated. Here we mention one of such shape-preserving sieves.

Cardinal B-spline wavelets. The cardinal B-spline of order $r \geq 1$ is given by

$$B_r(x) = \frac{1}{(r-1)!} \sum_{j=0}^r (-1)^j \binom{r}{j} [\max(0, x - j)]^{r-1},$$

(2.2)

which has support $[0, r]$, is symmetric at $r/2$ and is a piecewise polynomial of highest degree $r-1$. It satisfies $B_r(x) \geq 0$, $\sum_{k=-\infty}^{+\infty} B_r(x - k) = 1$ for all $x \in \mathcal{R}$, which is crucial to preserve the shape of the unknown function to be approximated. See Chui (1992, ch. 4) for a recursive construction of cardinal B-splines and their properties. Denote

$$\text{SplWav}(r - 1, 2^{J_n}) = \left\{ \sum_{k=-\infty}^{\infty} \alpha_k 2^{J_n/2} B_r(2^{J_n} x - k), \right.$$
$$\left. x \subset \mathcal{R} : \alpha_k \in \mathcal{R} \right\}.$$

Any non-decreasing continuous function on $\mathcal{R}$ can be approximated well by the $\text{SplWav}(r - 1, 2^{J_n})$ sieve with non-decreasing sequence $\{\alpha_k\}$ (that is, $\alpha_k \leq \alpha_{k+1}$). See Anastassiou and Yu (1992) on shape-preserving wavelet sieves.

2.5 Infinite-dimensional (nonlinear) sieves

Most commonly used sieve spaces are finite-dimensional truncated series such as those listed above. However, the general theory on sieve extremum estimation can also allow for infinite-dimensional sieve spaces. For example, any function θ that belongs to a Hölder space $\Theta = \Lambda^p([0,1])$ with smoothness $p > 1/2$ can be expressed as an infinite Fourier series $\theta(x) = \sum_{k=1}^\infty [a_k \cos(kx) + b_k \sin(kx)]$, and its derivative with fractional power $\gamma \in (0, p]$ can also be defined in terms of Fourier series:

$$\theta^{(\gamma)}(x) = \sum_{k=1}^{\infty} k^{\gamma} \left[\left(a_k \cos \frac{\pi\gamma}{2} + b_k \sin \frac{\pi\gamma}{2} \right) \cos(kx) \right.$$
$$\left. + \left(b_k \cos \frac{\pi\gamma}{2} - a_k \sin \frac{\pi\gamma}{2} \right) \sin(kx) \right].$$

More generally, if the parameter space Θ is a typical function space such as a Hölder, Sobolev or Besov space, then any function $\theta \in \Theta$ and its fractional derivatives can be expressed as infinite series of some known Riesz basis $\{B_k(\cdot)\}_{k=1}^{\infty}$ such as splines and wavelets (see for example, Meyer, 1992). An infinite-dimensional sieve space could then take the form:

$$\Theta_n = \{\theta \in \Theta : \theta(\cdot) = \sum_{k=1}^{\infty} a_k B_k(\cdot), \ \text{pen}(\theta) \leq b_n\}$$

$$\text{with } b_n \to \infty \text{ slowly}, \qquad (2.3)$$

where $\text{pen}(\theta)$ is a smoothness (or roughness) penalty term, such as $\text{pen}(\theta) = (\int_{\chi} |\theta^{(p)}(x)|^q dx)^{1/q}$, with $p > 1/2$ the smoothness of the function θ, and some $q \geq 1$. For example, Wahba (1990) considered smoothing spline sieve with $q = 2$ to approximate conditional mean function; Koenker, Ng and Portnoy (1994) considered smoothing spline sieve with $q = 1$ to approximate conditional quantile function. See Shen (1997) and van de Geer (2000) for more expressions of $\text{pen}(\theta)$.

3 Sieve extremum estimation

Let Θ be an infinite dimensional parameter space endowed with a (pseudo-) metric d. A typical semi-nonparametric econometric model specifies that there is a population criterion function $Q : \Theta \to \mathscr{R}$, which is uniquely maximized at a (pseudo-) true parameter $\theta_o \in \Theta$. The choice of $Q(\cdot)$ and the existence of θ_o are suggested by the identification of an econometric model. The (pseudo-) true parameter $\theta_o \in \Theta$ is unknown but is related to a joint probability measure $P_o(z_1, \ldots, z_n)$, from which a sample of size n observations $\{Z_t\}_{t=1}^{n}$, $Z_t \in \mathscr{R}^{d_z}$, $1 \leq d_z < \infty$, is available. Let $\hat{Q}_n : \Theta \to \mathscr{R}$ be an empirical criterion, which is a measurable function of the data $\{Z_t\}_{t=1}^{n}$ for all $\theta \in \Theta$, and converges to Q in some sense as the sample size $n \to \infty$.

When Θ is infinite dimensional and possibly not compact with respect to the (pseudo-) metric d, maximizing $\hat{Q}_n$ over Θ may not be well-defined; or even if a maximizer $\text{argsup}_{\theta \in \Theta} \hat{Q}_n(\theta)$ exists, it is generally difficult to compute, and may have undesirable large sample properties such as inconsistency and/or a very slow rate of convergence.

The method of sieves provides one general approach to resolve the difficulties associated with maximizing $\hat{Q}_n$ over an infinite dimensional space Θ by maximizing $\hat{Q}_n$ over a sequence of approximating spaces Θ_n, called *sieves*

by Grenander (1981), which are less complex but are dense in Θ. Popular sieves are typically compact, non-decreasing ($\Theta_n \subseteq \Theta_{n+1} \subseteq \ldots \subseteq \Theta$) and are such that for any $\theta \in \Theta$ there exists an element $\pi_n \theta$ in Θ_n satisfying $d(\theta, \pi_n \theta) \to 0$ as $n \to \infty$, where the notation π_n can be regarded as a projection mapping from Θ to Θ_n.

An *approximate sieve extremum estimate*, denoted by $\hat{\theta}_n$, is defined as an approximate maximizer of $\hat{Q}_n(\theta)$ over the sieve space Θ_n, that is,

$$\hat{Q}_n(\hat{\theta}_n) \geq \sup_{\theta \in \Theta_n} \hat{Q}_n(\theta) - O_P(\eta_n),$$
$$\text{with } \eta_n \to 0 \text{ as } \quad n \to \infty. \qquad (3.1)$$

When $\eta_n = 0$, we call $\hat{\theta}_n$ in (3.1) the *exact* sieve extremum estimate. The sieve extremum estimation method clearly includes the standard extremum estimation method by setting $\Theta_n = \Theta$ for all n. Following White and Wooldridge (1991, theorem 2.2), one can show that $\hat{\theta}_n$ in (3.1) is well defined and measurable under mild sufficient conditions: (i) $\hat{Q}_n(\theta)$ is a measurable function of the data $\{Z_t\}_{t=1}^{n}$ for all $\theta \in \Theta_n$; (ii) for any data $\{Z_t\}_{t=1}^{n}$, $\hat{Q}_n(\theta)$ is upper semicontinuous on Θ_n under the metric $d(\cdot, \cdot)$; and (iii) the sieve space Θ_n is compact under the metric $d(\cdot, \cdot)$.

For a semi-nonparametric model, $\theta_o \in \Theta$ can be decomposed into two parts $\theta_o = (\beta_o', h_o')' \in B \times \mathscr{H}$, where B denotes a finite dimensional compact parameter space, and $\mathscr{H}$ an infinite dimensional parameter space. In this case, a natural sieve space will be $\Theta_n = B \times \mathscr{H}_n$ with $\mathscr{H}_n$ being a sieve for $\mathscr{H}$, and the resulting estimate $\hat{\theta}_n = (\hat{\beta}_n, \hat{h}_n)$ in (3.1) will sometimes be called a simultaneous (or joint) sieve extremum estimate. For a semi-nonparametric model, we can also estimate the parameters of interest (β_o, h_o) by the *approximate profile sieve extremum estimation* that consists of two steps:

Step 1: for an arbitrarily fixed value $\beta \in B$, compute $\hat{Q}_n(\beta, \tilde{h}(\beta)) \geq \sup_{h \in \mathscr{H}_n} \hat{Q}_n(\beta, h) - O_P(\eta_n)$ with $\eta_n = o(1)$; *Step 2:* estimate β_o by $\hat{\beta}_n$ solving $\hat{Q}_n(\hat{\beta}, \tilde{h}(\beta)) \geq \max_{\beta \in B} \hat{Q}_n(\beta, \tilde{h}(\beta)) - O_P(\eta_n)$, and then estimate h_o by $\hat{h}_n = \tilde{h}(\hat{\beta}_n)$.

3.1 Sieve M-estimation

When $\hat{Q}_n(\theta)$ can be expressed as a sample average form $\hat{Q}_n(\theta) = \frac{1}{n} \sum_{t=1}^{n} l(\theta, Z_t)$, with $l : \Theta \times \mathscr{R}^{d_z} \to \mathscr{R}$ being the criterion based on a single observation, we also call the $\hat{\theta}_n$ solving (3.1) as an approximate *sieve maximum-likelihood-like (M-)* estimate. This includes sieve maximum likelihood (ML), sieve least squares (LS), sieve generalized least squares (GLS) and sieve quantile regression as special cases.

Example 3.1 (*single spell duration models with unobserved heterogeneity*): Let $G(\tau|\beta, u, x)$ be a parametric structural distribution function of duration T

conditional on a scalar of unobserved heterogeneity $U = u$ and a vector of observed heterogeneity $X = x$.

The distribution of observed duration given $X = x$ is

$$F(\tau|\beta, h, x) = \int G(\tau|\beta, u, x)dh(u),$$

where the unobserved heterogeneity U is modelled as a random factor with unknown distribution function $h(\cdot)$. Let $\{T_i, X_i\}_{i=1}^n$ be an i.i.d. sample of observations. Heckman and Singer (1984) propose sieve ML estimation of this semi-nonparametric single spell duration model with unobserved heterogeneity. Denote $\theta_o = (\beta_o', h_o)' \in B \times \mathscr{H}$ as the unknown true parameters, where B is a compact subset in $\mathscr{R}^{d_\beta}$ and $\mathscr{H}$ is the space of distribution functions. Let $\mathscr{H}_n$ denote a sieve for $\mathscr{H}$ such as the first-order spline sieve basis used in Heckman and Singer (1984). Then the sieve ML estimate is given by

$$\hat{\theta}_n = (\hat{\beta}_n, \hat{h}_n) = \arg \max_{(\beta,h)\in B \times \mathscr{H}_n} \frac{1}{n}$$
$$\times \sum_{i=1}^n \log\left\{\int g(T_i|\beta, u, X_i)dh(u)\right\}.$$

$$(3.2)$$

Series estimation is a special case of sieve M-estimation with *concave* criterion functions $\hat{Q}_n(\theta) = \frac{1}{n}\sum_{t=1}^n l(\theta, Z_t)$ and *finite-dimensional linear* sieve spaces Θ_n (that is, the sieves Θ_n are linear spans of finitely many known basis functions).

Example 3.2 (*multivariate LS regression*): We consider the series estimation of an unknown multivariate conditional mean function $\theta_o(\cdot) = h_o(\cdot) = E(Y|X = \cdot)$. Here $Z = (Y, X)$, Y is a scalar, X has support X that is a bounded subset of $\mathscr{R}^d$, $d \geq 1$. Suppose $h_o \in \Theta$, where Θ is a linear subspace of the space of functions h with $E[h(X)^2] < \infty$. Let $l(h, Z) = -[Y - h(X)]^2$ and $Q(\theta) = -E\{[Y - h(X)]^2\}$; then both are concave in h and Q is strictly concave in $h \in \Theta$.

Let $\{p_j(X), j = 1, 2, \ldots\}$ denote a sequence of known basis functions that can approximate any real-valued square integrable functions of X well; see the previous section for specific examples of such basis functions. Then

$$\Theta_n = \mathscr{H}_n = \{h : \chi \to \mathscr{R}, h(x)$$
$$= \Sigma_{j=1}^{k_n} a_j p_j(x) : a_1, \ldots, a_{k_n} \in \mathscr{R}\},$$

$$(3.3)$$

with $\dim(\Theta_n) = k_n \to \infty$ slowly as $n \to \infty$, is a finite-dimensional linear sieve for Θ, and $\hat{h} = \text{argmax}_{h \in \mathscr{H}_n} \frac{-1}{n}\sum_{t=1}^n [Y_t - h(X_t)]^2$ is a series estimator of the conditional mean $h_o(\cdot) = E(Y|X = \cdot)$. Moreover, this series estimator $\hat{h}$ has a simple closed-form expression:

$$\hat{h}(x) = p^{k_n}(x)'(P'P)^{-}\sum_{i=1}^n p^{k_n}(X_i)Y_i, \quad x \in \chi,$$

$$(3.4)$$

with $p^{k_n}(X) = (p_1(X), \ldots, p_{k_n}(X))'$, $P = (p^{k_n}(X_1), \ldots, p^{k_n}(X_n))'$ and $(P'P)^{-}$ the Moore–Penrose generalized inverse. The estimator $\hat{h}$ given in (3.4) is called a *series LS estimator*.

Many popular semi-nonparametric regression models, such as the partially linear regression of Engle et al. (1986) and Robinson (1988), the additive regression model of Stone (1985) and Andrews and Whang (1990), can be easily estimated via the series LS estimation.

3.2 Sieve MD estimation of semi-nonparametric conditional moment models

Many economic models imply semi-nonparametric conditional moment restrictions of the form

$$E[\rho(Z_t; \theta_o)|X_t] = 0, \quad \theta_o \equiv (\beta_o', h_o')',$$

$$(3.5)$$

where $\rho(\cdot; \cdot)$ is a column vector of residual functions whose functional forms are known up to unknown parameters, $\theta = (\beta', h')'$, and $\{Z_t' = (Y_t', X_t')\}_{t=1}^n$ is the data where Y_t is a vector of endogenous variables and X_t is a vector of conditioning variables. Here $E[\rho(Z_t, \theta)|X_t]$ denotes the conditional expectation of $\rho(Z_t, \theta)$ given X_t, and the true conditional distribution of Y_t given X_t is unspecified (and is treated as a nuisance function). The parameters of interest $\theta_o \equiv (\beta_o', h_o')'$ contain a vector of finite dimensional unknown parameters β_o and a vector of infinite dimensional unknown functions $h_o(\cdot) = (h_{o1}(\cdot), \ldots, h_{oq}(\cdot))'$, where the arguments of $h_{oj}(\cdot)$ could depend on Y, X, known index function $\delta_j(Z, \beta_o)$ up to unknown β_o, other unknown function $h_{ok}(\cdot)$ for $k \neq j$, or could also depend on unobserved random variables. This class of models (3.5) includes many semi-nonparametric models with endogeneity and/or latent heterogeneity as special cases. A leading, yet difficult example is the purely nonparametric instrumental variables (IV) regression $E[Y_{1i} - h_o(Y_{2i})|X_i] = 0$ studied by Newey and Powell (2003), Darolles, Florens and Renault (2006), Blundell, Chen and Kristensen (2007), and Hall and Horowitz (2005). A more difficult example is the nonparametric IV quantile regression $E[1\{Y_{1i} \leq h_o(Y_{2i})\} - \gamma|X_i] = 0$ for some known $\gamma \in (0,1)$ considered by Chernozhukov, Imbens and Newey (2007), Horowitz and Lee (2007) and Chen and Pouzo (2007). Both examples belong to the so-called 'ill-posed inverse' problems. See for example, Blundell and Powell (2003), Florens (2003) and Carrasco, Florens and Renault (2006) for additional examples of ill-posed inverse semi-nonparametric models.

Newey and Powell (2003) and Ai and Chen (2003) propose to estimate the model (3.5) by the *sieve minimum distance* (MD) procedure:

$$\sup_{\theta \in \Theta_n} \hat{Q}_n(\theta) = \sup_{\theta \in \Theta_n} -\frac{1}{n} \sum_{t=1}^{n} \hat{m}(X_t, \theta)' \{\hat{\Sigma}(X_t)\}^{-1} \hat{m}(X_t, \theta),$$

$$(3.6)$$

where Θ_n is compact sieves, $\hat{m}(X_t, \theta)$ is any nonparametrically consistent estimate of the conditional mean function $m(X_t, \theta) = E[\rho(Z, \theta)|X = X_t]$, and $\hat{\Sigma}(X_t)$ is a possibly nonparametrically consistent estimate of a positive definite weighting matrix $\Sigma(X_t)$.

Example 3.3 (*nonparametric external habit-based consumption asset pricing models*): A consumption-based asset pricing model assumes that at time zero a representative agent maximizes the expected present value of the total utility function $E_0\{\sum_{t=0}^{\infty} \delta^t u(C_t)\}$, where δ is the time discount factor and $u(C_t)$ is period t's utility. The consumption-based asset pricing models imply that for any traded asset indexed by ℓ, with a gross return at time $t+1$ of $R_{\ell,t+1}$, the following Euler equation holds:

$$E(M_{t+1} R_{\ell,t+1}|\mathbf{w}_t) = 1, \quad \ell = 1, \ldots, N,$$

$$(3.7)$$

where M_{t+1} is the intertemporal marginal rate of substitution in consumption, and $E(\cdot|\mathbf{w}_t)$ denotes the conditional expectation given the information set at time t (which is the sigma-field generated by $\mathbf{w}_t$). Hansen and Singleton (1982) have assumed that the period t utility takes the power specification $u(C_t) = [(C_t)^{1-\gamma} - 1]/[1 - \gamma]$, where γ is the curvature parameter of the utility function at each period, which implies the Euler equation:

$$E\left(\delta_o \left(\frac{C_{t+1}}{C_t}\right)^{-\gamma_o} R_{\ell,t+1} - 1 \Big| \mathbf{w}_t\right) = 0,$$
$$\ell = 1, \ldots, N,$$

$$(3.8)$$

where the unknown scalar parameters δ_o, γ_o can be estimated by Hansen's (1982) GMM. However, this classical power utility-based asset pricing model (3.8) has been rejected empirically.

Chen and Ludvigson (2003) combine the power utility specification with a nonparametric internal habit formation: $E_0\{\sum_{t=0}^{\infty} \delta^t [(C_t - H_t)^{1-\gamma} - 1]/[1 - \gamma]\}$, where $H_t = H(C_t, C_{t-1}, \ldots, C_{t-L})$ is the period t habit level. Here $H(\cdot)$ is a homogeneous of degree one unknown function of current and past consumption, and can be rewritten as $H(C_t, C_{t-1}, \ldots, C_{t-L}) = C_t h_o\left(\frac{C_{t-1}}{C_t}, \ldots, \frac{C_{t-L}}{C_t}\right)$ with $h_o(\cdot)$ unknown. It is obvious that one needs to impose $0 \leq h_o(\cdot) < 1$ so that $0 \leq H_t < C_t$. The following external habit specification is a special case of their model:

$$E\left(\delta_o \left(\frac{C_{t+1}}{C_t}\right)^{-\gamma_o} \frac{\left(1 - h_o\left(\frac{C_t}{C_{t+1}}, \ldots, \frac{C_{t+1-L}}{C_{t+1}}\right)\right)^{-\gamma_o}}{\left(1 - h_o\left(\frac{C_{t-1}}{C_t}, \ldots, \frac{C_{t-L}}{C_t}\right)\right)^{-\gamma_o}} R_{\ell,t+1} - 1 \Big| \mathbf{w}_t\right) = 0,$$

$$(3.9)$$

for $\ell = 1, \ldots, N$, where $\gamma_o > 0$, $\delta_o > 0$ are unknown scalar preference parameters, $h_o(\cdot) \in [0,1)$ is an unknown function and $H_{t+1} = C_{t+1} h_o\left(\frac{C_t}{C_{t+1}}, \ldots, \frac{C_{t+1-L}}{C_{t+1}}\right)$ is the habit level at time $t+1$. Chen and Ludvigson (2003) have applied the sieve method to estimate this model and its generalization which allows for internal habit formation of unknown form. Their empirical findings, using quarterly data, are in favor of flexible nonlinear internal habit formation.

3.3 Large sample properties

For an infinite-dimensional, compact parameter space Θ, Gallant and Nychka (1987) derived the consistency of sieve M-estimates; Newey and Powell (2003) established the consistency of sieve MD estimates. For an infinite-dimensional, possibly non-compact parameter space Θ, Geman and Hwang (1982) obtained the consistency of sieve MLE with i.i.d. data; White and Wooldridge (1991) obtained the consistency of sieve M-estimates with dependent and heterogeneous data; Chen (2006) and Chen and Pouzo (2007) established the consistency of sieve MD estimates that is applicable to general ill-posed semi-nonparametric problems.

For general theory on convergence rate of nonparametric part, we refer to Wong and Shen (1995) for sieve MLE, Shen and Wong (1994), Birgé and Massart (1998) and van de Geer (2000) for sieve M-estimation with i.i.d. data, Chen and Shen (1998) for sieve M-estimation with time series data, Newey (1997) and Huang (1998) for series LS estimation, and Chen and Pouzo (2007) for sieve MD estimation with i.i.d. data and allowing for ill-posed inverse problems.

For general theory on semiparametric efficiency and root-n normality of sieve estimates of smooth functionals, we refer to Shen (1997) for sieve MLE, Chen and Shen (1998) for sieve M-estimation with time series data, van de Geer (2000) for sieve M-estimation, Ai and Chen (2003) and Chen and Pouzo (2007) for sieve MD estimation allowing for ill-posed inverse problems. See Ai and Chen (2007) for a related result on root-n normality of sieve MD estimates of smooth functionals when the semi-nonparametric conditional moment models could be misspecified.

Unfortunately, so far there is no general theory on limiting pointwise distribution of sieve extremum estimators yet. Nevertheless, such results are established for series estimators (that is, sieve M-estimators with finite-dimensional linear sieves) (see Andrews, 1991b; Newey, 1997; Huang, 2003).

There is also no general theory on data-driven choice of smoothing parameters ('complexity of sieves') at the time of writing. There are some results for sieve M-estimation (see, for example, Barron, Birgé and Massart, 1999; Shen and Ye, 2002). There are well developed results for series LS regression and series density estimation (see for example, Li, 1987; Hurvich, Simonoff and Tsai, 1998; Andrews, 1991a; Coppejans and Gallant, 2002; Donald and Newey, 2001). Also see Chen (2006) for a detailed survey of theories on large sample properties of sieve estimation.

4 Economics applications

We conclude this article by listing some applications of the sieve extremum estimation in econometrics; see Chen (2006) for a more detailed review. In microeconometrics, Elbadawi, Gallant and Souza (1983) studied Fourier series LS estimation of demand elasticity. Heckman and Singer (1984) considered sieve ML estimation of a duration model where the unknown error distribution is approximated by a first-order spline. Hausman and Newey (1995) considered power series and spline series LS estimation of consumer surplus. Hahn (1998) used power series and splines in the two-step efficient estimation of the average treatment effect models. Newey, Powell and Vella (1999) considered series estimation of a triangular system of simultaneous equations. Gallant and Nychka (1987) proposed the Hermite polynomial sieve ML estimation of semiparametric sample selection model. Blundell, Chen and Kristensen (2007) considered a profile sieve MD procedure to estimate shape-invariant Engel curves with nonparametric endogenous expenditure. Hirano, Imbens and Ridder (2003) proposed a sieve logistic regression to estimate propensity score for treatment effect models. Chen, Fan and Tsyrennikov (2006) studied sieve MLE of semi-nonparametric multivariate copula models. Chen, Hong and Tamer (2005) made use of spline sieves to estimate nonlinear non-classical measurement error models with an auxiliary sample. Bierens and Carvalho (2007) applied Legendre polynomial sieve MLE to estimate a competing risks model of recidivism.

In time series econometrics, Engle et al. (1986) forecast electricity demand using a partially linear spline regression. Engle and Gonzalez-Rivera (1991) applied sieve MLE to estimate ARCH models where the unknown density of the standardized innovation is approximated by a first order spline sieve. Gallant and Tauchen (1989) employed Hermite polynomial sieve MLE to study asset pricing and foreign exchange rates. Gallant and Tauchen (1996) have proposed the combinations of Hermite polynomial sieve and simulated method of moments to effectively solve many complicated asset pricing models with latent factors, and their methods have been widely applied in empirical finance. White (1990) and Granger and Terasvirta (1993) suggested nonparametric LS forecasting via sigmoid ANN sieve. Chen, Racine and

Swanson (2001) used partially linear ANN and ridgelet sieves to forecast US inflation. Shintani and Linton (2004) proposed a nonparametric test of chaos via ANN sieves. Chen and Ludvigson (2003) employed a sigmoid ANN sieve to estimate the unknown habit function in a consumption asset pricing model. Chen and Conley (2001) made use of the shape-preserving wavelet spline sieve to estimate a spatial temporal model with flexible conditional mean and conditional covariance. Phillips (1998) applied orthonormal basis to analyse spurious regressions.

XIAOHONG CHEN

See also **generalized method of moments estimation; spline functions; wavelets.**

Bibliography

Ai, C. and Chen, X. 2003. Efficient estimation of models with conditional moment restrictions containing unknown functions. *Econometrica* 71, 1795–843.

Ai, C. and Chen, X. 2007. Estimation of possibly misspecified semiparametric conditional moment restriction models with different conditioning variables. *Journal of Econometrics*, 141, 5–43.

Anastassiou, G. and Yu, X. 1992. Monotone and probabilistic wavelet approximation. *Stochastic Analysis and Applications* 10, 251–64.

Andrews, D. 1991a. Asymptotic optimality of generalized C_L, cross-validation, and generalized cross-validation in regression with heteroskedastic errors. *Journal of Econometrics* 47, 359–77.

Andrews, D. 1991b. Asymptotic normality of series estimators for nonparametric and semiparametric regression models. *Econometrica* 59, 307–45.

Andrews, D. and Whang, Y. 1990. Additive interactive regression models: circumvention of the curse of dimensionality. *Econometric Theory* 6, 466–79.

Barron, A.R. 1993. Universal approximation bounds for superpositions of a sigmoidal function. *IEEE Transactions on Information Theory* 39, 930–45.

Barron, A., Birgé, L. and Massart, P. 1999. Risk bounds for model selection via penalization. *Probability of Theory Related Fields* 113, 301–413.

Bierens, H. and Carvalho, J. 2007. Semi-nonparametric competing risks analysis of recidivism. *Journal of Applied Econometrics*, 22, 971–93.

Birgé, L. and Massart, P. 1998. Minimum contrast estimators on sieves: exponential bounds and rates of convergence. *Bernoulli* 4, 329–75.

Blundell, R., Chen, X. and Kristensen, D. 2007. Semi-nonparametric IV estimation of shape-invariant Engel curves. *Econometrica*, 75, 1613–69.

Blundell, R. and Powell, J. 2003. Endogeneity in nonparametric and semiparametric regression models. In *Advances in Economics and Econometrics: Theory and Applications*, ed. M. Dewatripont, L.P. Hansen and S. Turnovsky. Cambridge: Cambridge University Press.

Carrasco, M., Florens, J.-P. and Renault, E. 2006. Linear inverse problems in structural econometrics estimation based on spectral decomposition and regularization. In *The Handbook of Econometrics*, vol. 6, ed. J.J. Heckman and E.E. Leamer. Amsterdam: North-Holland.

Chen, X. 2006. Large sample sieve estimation of semi-nonparametric models. In *The Handbook of Econometrics*, vol. 6, ed. J.J. Heckman and E.E. Leamer. Amsterdam: North-Holland.

Chen, X. and Conley, T. 2001. A new semiparametric spatial model for panel time series. *Journal of Econometrics* 105, 59–83.

Chen, X., Fan, Y. and Tsyrennikov, V. 2006. Efficient estimation of semiparametric multivariate copula models. *Journal of the American Statistical Association* 101, 1228–40.

Chen, X., Hong, H. and Tamer, E. 2005. Measurement error models with auxiliary data. *Review of Economic Studies* 72, 343–66.

Chen, X. and Ludvigson, S. 2003. Land of addicts? An empirical investigation of habit-based asset pricing models. Manuscript, New York University.

Chen, X. and Pouzo, D. 2007. On estimation of semi-nonparametric conditional moment models with possibly nonsmooth moments. Manuscript, New York University.

Chen, X., Racine, J. and Swanson, N. 2001. Semiparametric ARX neural network models with an application to forecasting inflation. *IEEE Transactions on Neural Networks* 12, 674–83.

Chen, X. and Shen, X. 1998. sieve extremum estimates for weakly dependent data. *Econometrica* 66, 289–314.

Chen, X. and White, H. 1999. Improved rates and asymptotic normality for nonparametric neural network estimators. *IEEE Transactions on Information Theory* 45, 682–91.

Chernozhukov, V., Imbens, G. and Newey, W. 2007. Instrumental variable estimation of nonseparable models. *Journal of Econometrics*, 139, 4–14;.

Chui, C. 1992. *An Introduction to Wavelets*. San Diego: Academic Press.

Coppejans, M. and Gallant, A.R. 2002. Cross-validated SNP density estimates. *Journal of Econometrics* 110, 27–65.

Darolles, S., Florens, J.-P. and Renault, E. 2006. Nonparametric instrumental regression. Mimeo, GREMAQ, University of Toulouse.

de Boor, C. 1978. *A Practical Guide to Splines*. New York: Springer-Verlag.

DeVore, R.A. and Lorentz, G.G. 1993. *Constructive Approximation*. Berlin: Springer.

Donald, S. and Newey, W. 2001. Choosing the number of instruments. *Econometrica* 69, 1161–91.

Donoho, D.L., Johnstone, I.M., Kerkyacharian, G. and Picard, D. 1995. Wavelet shrinkage: asymptopia? *Journal of the Royal Statistical Society, Series B* 57, 301–69.

Elbadawi, I., Gallant, A.R. and Souza, G. 1983. An elasticity can be estimated consistently without a prior knowledge of functional form. *Econometrica* 51, 1731–51.

Engle, R. and Gonzalez-Rivera, G. 1991. Semiparametric ARCH models. *Journal of Business and Economic Statistics* 9, 345–59.

Engle, R., Granger, C., Rice, J. and Weiss, A. 1986. Semiparametric estimates of the relation between weather and electricity sales. *Journal of the American Statistical Association* 81, 310–20.

Florens, J.P. 2003. Inverse problems and structural econometrics: the example of instrumental variables. In *Advances in Economics and Econometrics: Theory and Applications*, ed. M. Dewatripont, L.P. Hansen and S. Turnovsky. Cambridge: Cambridge University Press.

Gallant, A.R. and Nychka, D. 1987. Semi-non-parametric maximum likelihood estimation. *Econometrica* 55, 363–90.

Gallant, A.R. and Tauchen, G. 1989. Semiparametric estimation of conditional constrained heterogenous processes: asset pricing applications. *Econometrica* 57, 1091–120.

Gallant, A.R. and Tauchen, G. 1996. Which moments to match? *Econometric Theory* 12, 657–81.

Geman, S. and Hwang, C. 1982. Nonparametric maximum likelihood estimation by the method of Sieves. *Annals of Statistics* 10, 401–14.

Granger, C.W.J. and Terasvirta, T. 1993. *Modelling nonlinear economic relationships*. New York: Oxford University Press.

Grenander, U. 1981. *Abstract Inference*. New York: Wiley.

Hahn, J. 1998. On the role of the propensity score in efficient semiparametric estimation of average treatment effects. *Econometrica* 66, 315–32.

Hall, P. and Horowitz, J. 2005. Nonparametric methods for inference in the presence of instrumental variables. *Annals of Statistics* 33, 2904–29.

Hansen, L.P. 1982. Large sample properties of generalized method of moments estimators. *Econometrica* 50, 1029–54.

Hansen, L.P. and Singleton, K. 1982. Generalized instrumental variables estimation of nonlinear rational expectations models. *Econometrica* 50, 1269–86.

Hausman, J. and Newey, W. 1995. Nonparametric estimation of exact consumer surplus and deadweight loss. *Econometrica* 63, 1445–67.

Heckman, J. and Singer, B. 1984. A method for minimizing the impact of distributional assumptions in econometric models for duration data. *Econometrica* 68, 839–74.

Hirano, K., Imbens, G. and Ridder, G. 2003. Efficient estimation of average treatment effects using the estimated propensity score. *Econometrica* 71, 1161–89.

Hornik, K., Stinchcombe, M., White, H. and Auer, P. 1994. Degree of approximation results for feedforward networks approximating unknown mappings and their derivatives. *Neural Computation* 6, 1262–75.

Horowitz, J. and Lee, S. 2007. Nonparametric instrumental variables estimation of a quantile regression model. *Econometrica*, 75, 1191–208.

Huang, J.Z. 1998. Projection estimation in multiple regression with application to functional ANOVA models. *Annals of Statistics* 26, 242–72.

Huang, J.Z. 2003. Local asymptotics for polynomial spline regression. *Annals of Statistics* 31, 1600–35.

Hurvich, C., Simonoff, J. and Tsai, C. 1998. Smoothing parameter selection in nonparametric regression using an improved Akaike information criterion. *Journal of the Royal Statistical Society, Series B* 60, 271–93.

Judd, K. 1998. *Numerical Method in Economics*. Cambridge, MA: MIT Press.

Koenker, R., Ng, P. and Portnoy, S. 1994. Quantile smoothing splines. *Biometrika* 81, 673–80.

Li, K. 1987. Asymptotic optimality for C_p, C_L, cross-validation, and generalized cross-validation: discrete index set. *Annals of Statistics* 15, 958–75.

Meyer, Y. 1992. *Ondelettes et operateurs I: Ondelettes*. Paris: Hermann.

Newey, W.K. 1997. Convergence rates and asymptotic normality for series estimators. *Journal of Econometrics* 79, 147–168.

Newey, W.K. and Powell, J.L 2003. Instrumental variable estimation of nonparametric models. *Econometrica* 71, 1565–78.

Newey, W.K., Powell, J.L. and Vella, F. 1999. Nonparametric estimation of triangular simultaneous equations models. *Econometrica* 67, 565–603.

Phillips, P.C.B. 1998. New tools for understanding spurious regressions. *Econometrica* 66, 1299–325.

Robinson, P. 1988. Root-n-consistent semiparametric regression. *Econometrica* 56, 931–54.

Schumaker, L. 1981. *Spline Functions: Basic Theory*. New York: Wiley.

Shen, X. 1997. On methods of sieves and penalization. *Annals of Statistics* 25, 2555–91.

Shen, X. and Wong, W. 1994. Convergence rate of sieve estimates. *Annals of Statistics* 22, 580–615.

Shen, X. and Ye, J. 2002. Adaptive model selection. *Journal of American Statistical Association* 97, 210–21.

Shintani, M. and Linton, O. 2004. Nonparametric neural network estimation of Lyapunov exponents and a direct test for chaos. *Journal of Econometrics* 129, 1–33.

Stone, C.J. 1985. Additive regression and other nonparametric models. *Annals of Statistics* 13, 689–705.

Van de Geer, S. 2000. *Empirical Processes in M-estimation*. Cambridge: Cambridge University Press.

Wahba, G. 1990. *Spline Models for Observational Data*. Philadelphia: CBMS-NSF Regional Conference Series.

White, H. 1990. Connectionist nonparametric regression: multilayer feedforward networks can learn arbitrary mappings. *Neural Networks* 3, 535–50.

White, H. and Wooldridge, J. 1991. Some results on sieve estimation with dependent observations. In *Nonparametric and Semi-parametric Methods in Econometrics*

and Statistics, ed. W.A. Barnett, J. Powell and G. Tauchen. Cambridge: Cambridge University Press.

Wong, W.H. and Shen, X. 1995. Probability inequalities for likelihood ratios and convergence rates for sieve MLE's. *Annals of Statistics* 23, 339–62.

signalling and screening

Mathematical formulation

We consider here the simplest game-theoretic version of Spence's (1973; 1974) model. A worker's productivity, or *type*, is either θ_H or θ_L, with $\theta_H > \theta_L > 0$. Productivity is private information. Firms share a common prior p, with $p = \Pr\{\theta = \theta_H\} \in (0, 1)$. Before entering the job market, the worker chooses a costly education level $e \geq 0$. Workers maximize $U(w, e; \theta) = w - c(e; \theta)$, where w is their wage and $c(e; \theta)$ is the cost of education. Assume that $c(0; \theta) = 0$, $c_e(e; \theta) > 0$, $c_{ee}(e; \theta) > 0$, $c_\theta(e; \theta) < 0$ for all $e > 0$, where $c_e(\cdot; \theta)$ and $c_\theta(e; \cdot)$ denote the derivatives of the cost with respect to education and types, respectively, and $c_{ee}(\cdot; \theta)$ is the second derivative of cost with respect to education. The key assumption made in the literature is that on the cross-derivative: $c_{e\theta}(e; \theta) < 0$. That is, the marginal cost is lower for a high-productivity worker. This *single-crossing* condition ensures that the indifference curves of a high and a low-productivity worker cross at most once, with the indifference curve of the high-productivity worker having a smaller slope where they do. See Figure 1.

To focus on signalling, assume that education does not affect productivity. If a firm assigns probability $p(e)$ to the high-productivity worker conditional on education e, the worker's expected productivity is $(1 - p(e))\theta_L + p(e)\theta_H$. If the worker accepts wage w, the firm's profit is then $(1 - p(e))\theta_L + p(e)\theta_H - w$.

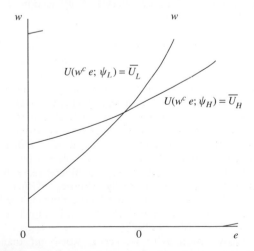

Figure 1

The basic signalling game

There is one worker and two firms. In the first stage, the worker chooses education. Education is observable. In the second stage, firms compete through wages in Bertrand-competition fashion. Following usual arguments, the wage $w(e)$ offered and accepted equals the expected productivity of the worker, given the observed education. A (perfect Bayesian) equilibrium specifies an education function $e^* : \{\theta_L, \theta_H\} \to R_+$, and a belief function $p^* : R_+ \to [0, 1]$, giving respectively the education chosen by each type and the probability assigned to a high-productivity worker conditional on each possible education level, so that the worker's choice is optimal given the wage determined by p^*, and the belief function is derived from this choice using Bayes's rule whenever possible.

Either the worker chooses distinct education levels depending on his productivity or he does not. An equilibrium is *separating* if $e^*(\theta_L) \neq e^*(\theta_H)$, and *pooling* otherwise. In the first case, education perfectly reveals productivity, so that the worker's wage equals his productivity: $w(e^*(\theta_i)) = \theta_i$, for $i = L, H$. In the second case, education reveals no information, and the wage is equal to his expected productivity given the prior p: $w(e^*(\theta_i)) = (1 - p)\theta_L + p\theta_H =: E(\theta)$, for $i = L, H$.

Observe that, in any separating equilibrium, the low-productivity worker gets the lowest possible wage. Since education is costly, this implies that he chooses no education: $e^*(\theta_L) = 0$. The high-productivity worker, on the other hand, gets the highest possible wage. Therefore, the corresponding education level $e^*(\theta_H)$ must be high enough to deter the low-productivity worker from pretending he has high productivity. That is, it must be that $e^*(\theta_H) \geq e'$, where e' solves $\theta_L - c(0; \theta_L) = \theta_H - c(e'; \theta_L)$. At the same time, the education level $e^*(\theta_H)$ cannot be too high, since the high-productivity worker must choose it. That it, it must be that $e^*(\theta_H) \leq e''$, where e'' solves $\theta_L - c(0; \theta_H) = \theta_H - c(e''; \theta_H)$. Single-crossing implies that the interval $[e', e'']$ is non-empty. Indeed, since the indifference curves $\{(e, w) : U(w, e; \theta_H) = U(\theta_L, 0; \theta_H)\}$ and $\{(e, w) : U(w, e; \theta_L) = U(\theta_L, 0; \theta_L)\}$ cross at $(0, \theta_L)$, the point (e'', θ_H) which lies along the first one must be to the right of the point (e', θ_H) which lies along the second. See Figure 2. Because a high-productivity worker is more willing to trade off an increase in education to induce an increase in wage, it is possible to find a suitable education level that is worth acquiring if and only if the worker's productivity is high.

Each education level $e^* \in [e', e'']$ can be supported in equilibrium, for suitable beliefs. For instance, by setting $p^*(e) = 0$ if $e < e^*$, and $p^*(e) = 1$ otherwise, a high-productivity worker optimally chooses $e^*(\theta_H) = e^*$. Observe that these equilibria are Pareto-ranked. The best equilibrium outcome, involving $e^* = e'$, is known as the *Riley outcome* (Riley, 1979).

The low-productivity worker is worse off than in the case in which signalling is not available. Without

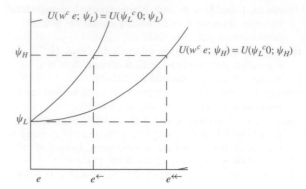

Figure 2

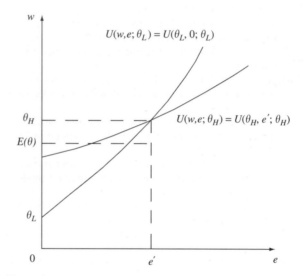

Figure 3

signalling, the worker would not acquire any education, independently of his type, and he would receive the wage $E(\theta)$. Here, instead, the low-productivity worker earns only θ_L. Surprisingly, the high-productivity worker may also be worse off. As no education is interpreted as evidence of low productivity, the outcome without signalling is no longer available to him in a separating equilibrium, his utility is at most $\theta_H - c(e'; \theta_H)$. Without signalling opportunities, his utility is $E(\theta) - c(0; \theta_H)$. While $E(\theta)$ tends to θ_H as p tends to one, e' is independent of p. Therefore, if p is large enough, the high-productivity is worse off. See Figure 3.

There is also a continuum of pooling outcomes. Let $\hat{e}$ solve $U(\theta_L, 0; \theta_L) = U(E(\theta), \hat{e}; \theta_L)$. Every level of education $e^* \in [0, \hat{e}]$ can be supported by the beliefs $p^*(e) = 0$ if $e < e^*$, $p^*(e) = p$ otherwise. Since education is costly, we need only check that the worker prefers $e(\theta_i) = e^*$ $(i = L, H)$ to no education at all. This is true for the low-productivity worker by definition of $\hat{e}$, and follows from single-crossing for the high-productivity worker. Here as well, the outcomes are Pareto-ranked, with the best equilibrium outcome, sometimes referred to as the *Hellwig outcome*, specifying $e^* = 0$ (Hellwig, 1987). In addition to these separating and pooling outcomes, there also exists a continuum of equilibria in mixed strategies.

The basic screening game

While it is standard in the literature to call signalling models those in which the informed party moves first, they are closely related to screening models, in which the uninformed parties take the lead. Classic references include Rothschild and Stiglitz (1976) and Wilson (1977) in the context of insurance markets. In these models, the two firms simultaneously announce a menu of pairs (e, w). Given these contracts, the worker chooses which contract to accept, if any. We sketch here the main results of this model. An equilibrium is *separating* if the worker accepts distinct contracts depending on his type, and *pooling* otherwise.

Observe that, in equilibrium, firms must just break even. Otherwise, if the worker of type $i = L, H$ accepts contract (e_i, w_i), a contract $(e_i, w_i + \varepsilon)$ for small $\varepsilon > 0$ would attract both types of worker, and the firm earning less than half the aggregate profits would gain by offering it.

Also, there can be no pooling equilibrium. Because a pooling contract (e^*, w^*) would have to break even, a firm whose rival offered this contract would gain by offering a contract (e, w) specifying a higher wage and education level, accepted only by the high-productivity worker. See Figure 4.

Finally, in any separating equilibrium, wages paid must equal the worker's productivity. In particular, the contract accepted by the low-productivity worker specifies education $e_L = 0$ and wage $w_L = \theta_L$. Indeed, if $(e_L, w_L) \neq (0, \theta_L)$, then the firm whose rival offered (e_L, w_L) would gain by offering a contract with either a slightly lower wage, or a slightly lower education, independently of the worker's type accepting it. Similarly, if the wage accepted by the high-productivity worker fell short of θ_H, a firm whose rival offered the contract accepted by the low-productivity worker would gain by offering a contract specifying a slightly higher wage and education than those specified by the contract accepted by the high-productivity worker. Since $w_i = \theta_i$, $i = L, H$, and $e_L = 0$, it follows that e_H solves $\theta_H - c(e_H; \theta_L) = \theta_L - c(0; \theta_L)$, that is, $e_H = e'$: if instead the low-productivity worker preferred his contract to (e_H, w_H), at least one firm would gain by offering a contract specifying a wage and education just below w_H and e_H.

However, such an equilibrium need not exist for large p. If $E(\theta) - c(0; \theta_H) > \theta_H - c(e'; \theta_H)$, the contract $(0, E(\theta) - \varepsilon)$ for small $\varepsilon > 0$ attracts both types of workers and makes profits. Thus, if the cost of sorting outweighs the gain, no equilibrium exists. As emphasized by Riley (2001), existence requires a strengthening of single-crossing, as marginal cost must be sufficiently lower for a high-productivity worker, given p. This is the same condition as earlier, under which the high-productivity worker prefers signalling to be unavailable. While equilibria in mixed strategies exist (Dasgupta and Maskin, 1986), they have not been characterized.

Extensions and refinements

To a large extent, the theoretical literature on signalling has focused on selecting among the equilibria, while the literature on screening has addressed the non-existence issue.

The early literature takes the view that the screening model ignores the dynamic adjustments between firms. To account for these, Wilson (1977) and Riley (1979) define equilibria differently. A set of contracts is a Wilson equilibrium if no firm has a profitable deviation that remains profitable once existing contracts that lose money after the deviation are withdrawn, while it is a Riley, or reactive equilibrium, if no firm has a profitable deviation that remains profitable once new contracts that make money after the deviation are added. Under either definition, equilibria exist. Wilson equilibria involve some pooling, while the unique Riley equilibrium is separating. Hellwig (1987) offers a game-theoretic treatment of Wilson by modelling a second stage in which firms may withdraw any contract offered previously. In the two-type case, the Hellwig outcome can be supported as an equilibrium.

Formal game-theoretic treatments of signalling appear in the 1980s. Many refinements have been applied to and inspired by the basic signalling game, shedding new light on the somewhat ad hoc selection procedures used previously. While sequential equilibrium (Kreps and

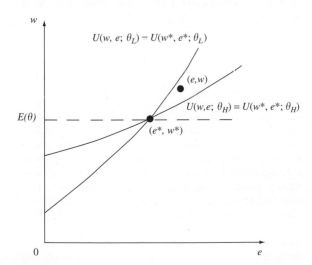

Figure 4

Wilson, 1982) does not reduce the multiplicity of equilibria, the *intuitive criterion* (Cho and Kreps, 1987) selects the Riley outcome in the basic signalling model. This result, as striking as it is, has several limitations. First, uniqueness does not obtain with more types. Second, the Riley outcome is not necessarily persuasive when the probability of the high-productivity worker is nearly one. As long as this probability is less than 1, the high-productivity worker acquires education $e' > 0$, independently of p. But if he is known to be of high productivity, we should expect him not to acquire education, as it serves no signalling purpose. Third, the motivation behind the intuitive criterion also underlies the more stringent *perfect sequential equilibrium* (Grossman and Perry, 1986). Yet such an equilibrium fails to exist in the situation described earlier, in which the basic screening game has no equilibrium. An alternative is offered by the concept of *undefeated equilibrium* (Mailath, Okuno-Fujiwara and Postlewaite, 1993), which selects the Riley outcome when it is also a perfect sequential equilibrium outcome, and the Hellwig outcome otherwise.

In settings with more types, stronger refinements are needed to select the Riley outcome. These include Banks and Sobel's (1987) *divinity* and *universal divinity*, Cho and Kreps's (1987) criterion *D1*, and Kohlberg and Mertens's (1986) *stability* concepts. See also Cho and Sobel (1990). It is worth pointing out that, with a continuum of types and under weak assumptions, the separating outcome is unique in the signalling model (Mailath, 1987), while no equilibrium exists in the screening model (Riley, 2001).

The single-crossing condition has been generalized to multidimensional signals by Engers (1987) for the case of screening and by Cho and Sobel (1990) and Ramey (1996) for the case of signalling. Quinzii and Rochet (1985) consider multidimensional types. Little is known about equilibria when single-crossing fails, as may occur in applications.

Maskin and Tirole (1992) enlarge the set of contracts. In the screening version, firms offer contracts that let the worker choose *ex post* among a set of pairs (e, w). In the signalling version, the worker offers such a contract to the firm. Under weak assumptions, the set of equilibrium outcomes coincides. In particular, only the Riley outcome obtains when the basic screening model has an equilibrium.

Acquiring education takes time, and there is no reason to expect firms to wait until graduation before drawing inferences. In Nöldeke and van Damme (1990) and Swinkels (1999), firms make offers before workers complete their education. Offers are public in Nöldeke and van Damme, and private in Swinkels. As the time between offers diminishes, only the Riley outcome satisfies Kohlberg and Mertens' never weak best response criterion when offers are public, while only the Hellwig outcome is a sequential equilibrium outcome when offers are private.

Following Spence's early suggestion, Nöldeke and Samuelson (1997) extend the basic signalling model by considering a dynamic model in which agents adjust their beliefs and actions to past market outcomes, and introduce perturbations into the process. The dynamic process admits at most two recurrent sets, closely related to the Riley and Hellwig outcomes. Several known refinements reappear in their characterization.

Application

Signalling has found many applications besides education, insurance and labour. Whenever possible, the reader is referred to surveys.

Industrial organization

Signalling helps explain limit (or predatory) pricing. Milgrom and Roberts (1982) show that low price may signal an incumbent's low cost. In Milgrom and Roberts (1986), advertising is a signal that a firm's experience good is of high quality. Bagwell and Riordan (1991) show that introductory pricing can serve the same purpose. In Gal-Or (1989), warranties signal product durability. See Tirole (1988) for a general survey, and Bagwell (2001) for a survey specific to advertising and pricing.

Finance

Myers and Majluf (1984) show that stock issues may signal the firm's value, so that the choice of financing (equity or debt) affects a firm's investment policy when the firm is better informed about returns than investors. In Bhattacharya (1979), dividends signal cash flows, as managers are better informed about those than investors. In Leland and Pyle (1977), the owner's stake signals the firm's underlying quality in an initial public offering, provided the owner is risk averse. These early applications have been extended in several directions, and their predictions empirically tested. See Allen and Morris (2001).

Political science

Signalling has been applied to electoral competition. Banks (1990a) shows how campaign platforms signal the candidates' future actions if elected, while Banks (1990b) argues that agenda-setting signals the bureaucrat's private information about the 'reversion level' if the proposal is turned down. See Banks (1991) for a survey. Lohmann (1993) shows how individuals engage in costly political actions to signal their private information, if politicians are responsive to turnout. Prat (2002) considers a signalling game in which an interest group has private information about candidates, based on which they can offer contributions that are used as campaign advertising.

Social norms

Bernheim (1994) develops a theory of social conformity based on signalling. Agents are motivated both by private tastes and status. When society censures extreme preferences, consumers with centrist preferences may choose to pool, while extremists refuse to conform. Fang (2001) provides a model of social culture rich enough to endogenously generate the single-crossing condition supporting the separating equilibrium. Austen-Smith and Fryer (2005) study signalling when workers have both a social and an economic type, and education affects both the wage and their perception by their peers. Peer pressure may induce educational underinvestment by accepted types.

Biology
Following Zahavi's (1975) 'handicap principle', asserting that animal signals are reliable because they are costly, a large literature on signalling has emerged in biology. Grafen (1990) provides a game-theoretic treatment. See Maynard Smith and Harper (2003) for a survey.

JOHANNES HÖRNER

See also **cheap talk.**

Bibliography
Allen, F. and Morris, S. 2001. Finance applications of game theory. In *Game Theory and Business Applications*, ed. K. Chatterjee and W.F. Samuelson. Boston, MA: Kluwer Academic Publishers.
Austen-Smith, D. and Fryer, R.G. 2005. An economic analysis of acting white. *Quarterly Journal of Economics* 119, 551–83.
Bagwell, K. 2001. Introduction. In *The Economics of Advertising*, ed. K. Bagwell. Cheltenham: Edward Elgar.
Bagwell, K. and Riordan, M.H. 1991. High and declining prices signal product quality. *American Economic Review* 81, 224–39.
Banks, J.S. 1990a. A model of electoral competition with incomplete information. *Journal of Economic Theory* 50, 309–25.
Banks, J.S. 1990b. Monopoly agenda control and asymmetric information. *Quarterly Journal of Economics* 104, 445–64.
Banks, J.S. 1991. *Signaling Games in Political Science*. New York: Harwood Academic.
Banks, J.S. and Sobel, J. 1987. Equilibrium selection in signaling games. *Econometrica* 55, 647–61.
Bernheim, B.D. 1994. A theory of conformity. *Journal of Political Economy* 102, 841–77.
Bhattacharya, S. 1979. Imperfect information, dividend policy and 'the bird in the hand' fallacy. *Bell Journal of Economics* 9(1), 259–70.
Cho, I.-K. and Kreps, D.M. 1987. Signaling games and stable equilibria. *Quarterly Journal of Economics* 102, 179–221.
Cho, I.-K. and Sobel, J. 1990. Strategic stability and uniqueness in signaling games. *Journal of Economic Theory* 50, 381–413.
Dasgupta, P. and Maskin, E. 1986. The existence of equilibria in discontinuous economic games, I: theory. *Review of Economic Studies* 53, 1–26.

Engers, M. 1987. Signalling with many signals. *Econometrica* 55, 663–74.
Fang, H. 2001. Social culture and economic performance. *American Economic Review* 91, 924–37.
Gal-Or, E. 1989. Warranties as a signal of quality. *Canadian Journal of Economics* 22, 50–61.
Grafen, A. 1990. Biological signals as handicaps. *Journal of Theoretical Biology* 144, 517–46.
Grossman, S. and Perry, M. 1986. Perfect sequential equilibrium. *Journal of Economic Theory* 39, 97–119.
Hellwig, M. 1987. Some recent developments in the theory of competition in markets with adverse selection. *European Economic Review* 31, 319–25.
Kohlberg, E. and Mertens, J.-F. 1986. On the strategic stability of equilibria. *Econometrica* 54, 1003–37.
Kreps, D.M. and Sobel, J. 1994. Signaling. In *Handbook of Game Theory*, vol. 2, ed. R.J. Aumann and S. Hart. Amsterdam: North-Holland.
Kreps, D.M. and Wilson, R. 1982. Sequential equilibria. *Econometrica* 50, 863–94.
Leland, H.E. and Pyle, D.H. 1977. Informational asymmetries, financial structure, and financial intermediation. *Journal of Finance* 32, 371–87.
Lohmann, S. 1993. A signaling model of informative and manipulative political action. *American Political Science Review* 87, 319–33.
Mailath, G.J. 1987. Incentive compatibility in signaling games with a continuum of types. *Econometrica* 55, 1349–65.
Mailath, G.J., Okuno-Fujiwara, M. and Postlewaite, A. 1993. Belief-based refinements in signalling games. *Journal of Economic Theory* 60, 241–76.
Maskin, E. and Tirole, J. 1992. The principal–agent relationship with an informed principal: the case of common values. *Econometrica* 60, 1–42.
Maynard, S.J. and Harper, D. 2003. *Animal Signals*. New York: Oxford University Press.
Milgrom, P. and Roberts, J. 1982. Limit pricing and entry under incomplete information: an equilibrium analysis. *Econometrica* 50, 443–59.
Milgrom, P. and Roberts, J. 1986. Price and advertising signals of product quality. *Journal of Political Economy* 94, 796–821.
Myers, S.C. and Majluf, N. 1984. Corporate financing and investment decisions when firms have information that investors do not have. *Journal of Financial Economics* 13, 187–221.
Nöldeke, G. and Samuelson, L. 1997. A dynamic model of equilibrium selection in signaling markets. *Journal of Economic Theory* 73, 118–56.
Nöldeke, G. and van Damme, E. 1990. Signalling in a dynamic labour market. *Review of Economic Studies* 57, 1–23.
Prat, A. 2002. Campaign advertising and voter welfare. *Review of Economic Studies* 69, 999–1017.
Quinzii, M. and Rochet, J.-C. 1985. Multidimensional signalling. *Journal of Mathematical Economics* 14, 261–84.

Ramey, G. 1996. D1 signaling equilibria with multiple signals and a continuum of types. *Journal of Economic Theory* 69, 508–31.

Riley, J. 1979. Informational equilibrium. *Econometrica* 47, 331–59.

Riley, J. 2001. Silver signals: twenty-five years of screening and signaling. *Journal of Economic Literature* 39, 432–78.

Rothschild, M. and Stiglitz, J. 1976. Equilibrium in competitive insurance markets: an essay on the economics of imperfect information. *Quarterly Journal of Economics* 80, 629–49.

Spence, A.M. 1973. Job market signalling. *Quarterly Journal of Economics* 90, 225–43.

Spence, A.M. 1974. *Market Signaling*. Cambridge, MA: Harvard University Press.

Swinkels, J. 1999. Education signalling with preemptive offers. *Review of Economic Studies* 66, 949–70.

Tirole, J. 1988. *The Theory of Industrial Organization*. Cambridge, MA: MIT Press.

Wilson, C.A. 1977. A model of insurance markets with incomplete information. *Journal of Economic Theory* 16, 167–207.

Zahavi, A. 1975. Mate selection – a selection for a handicap. *Journal of Theoretical Biology* 53, 205–14.

silver standard

The silver standard, the dominant monetary system for many centuries, lost much importance with the advent of the classical gold standard; and, due to US policy, residual monetary use of silver was virtually eliminated in the 1930s.

Definition of silver standard

A silver standard involves (*a*) a fixed silver content of the monetary unit, (*b*) 'free coinage' of silver, that is, privately owned silver in form other than domestic coin convertible into domestic silver coin at, or approximately at, the mint price (the inverse of the silver content of the monetary unit), (*c*) no restrictions on private parties (i) melting domestic coin into bullion, or (ii) importing or exporting silver in any form, and (*d*) full legal-tender status for domestic silver coin.

Other forms of money may exist, but silver is the primary money. Foreign silver coin may be given equal legal-tender status with domestic coin. Gold coin may be in circulation, but its value is in terms of the silver monetary unit and may fluctuate by weight, varying with the market gold–silver price ratio. Paper currency and deposits may exist, but, as liabilities of the issuer or bank, are payable in legal tender, that is, silver coin (or silver-convertible government or central-bank currency).

If silver (whether domestic or foreign coin, or both) constitutes the only money, then, even absent free coinage, the economy is clearly on a silver standard. This conclusion holds with gold coin circulating as well, providing it is circulating by weight or is a minor part of the money supply.

A silver standard might be effective even though the monetary system is legally bimetallic. If the coinage gold–silver price ratio is sufficiently below the market ratio, then gold, undervalued at the mint, will be sold on the world market (even in the form of melted domestic coin), while silver, overvalued, will be imported and coined. Ultimately, an effective silver standard may result.

Depreciation of the silver coinage involves an increased ratio of the legal (face) value of coins relative to silver content, usually by debasement (reducing the silver content, whether weight or fineness, of given-denomination coins) rather than by increasing the denomination of existing (given-weight-and-fineness) coins. In England, the penny (of sterling, 11/12th fineness) was steadily reduced in size from 24 grains in the eighth century to less than 1/3 that weight in 1601.

A silver standard, just as the gold standard, provides a constraint on the money stock. Depreciation of silver coinage was a way of escaping that constraint, even though the authority's objective typically was to increase government revenue (in the form of seigniorage) and/or to change the coinage ratio (under legal bimetallism).

Countries on silver standard to 1870

A silver standard first occurred in ancient Greece. Notwithstanding generally legal bimetallism, silver was everywhere the effective metallic standard – or at least the far-more-important coined metal in the money stock – well into the 18th century. Because of its relative scarcity and high density, gold was always much more valuable than silver on a per-ounce basis: coinage and market ratios were far above unity. So, with most transactions of low value compared with the unit of account, silver was better suited than gold to serve as a medium of exchange. In US history, 'one dollar' was both the smallest gold piece and the largest silver piece ever coined.

In England, from the Anglo-Saxon period until the late 13th century, the only coin in existence (with rare exceptions) was the silver penny, with 240 pence coined ideally from one pound of silver and later constituting one pound sterling (where 'sterling,' of course, denotes silver). This was a silver standard by default. With coinage of gold, in 1257, there was legal bimetallism; but the practice of denominating gold coins in (silver) shillings and pence was implicit recognition of an effective silver standard. Even the popular, consistently coined, (gold) guinea, first issued in 1663, was left to find its own market value in shillings and pence. However, by the turn of the 18th century, foreign gold–silver price ratios had been falling and, having been increased greatly in 1696, the British coinage ratio was not subsequently reduced enough to compensate. England went briefly on a bimetallic standard, and then on an effective gold standard, legalized in 1774 and 1816.

In the United States, since colonial times a silver standard was in effect, based on the Spanish dollar, the primary circulating silver coin, which varied much in weight and fineness. Yet the dollar was accepted everywhere at face value in terms of local (individual-state) pound–shilling–pence units of account. Gold coins were rated in dollars according to fine-metal content. The Coinage Act of 1792 placed the United States on a legal bimetallic standard; but the coinage ratio soon fell below the (increasing) world-market ratio. An effective silver standard resulted, until the coinage ratio was corrected in 1834.

In 1870, just before Germany united and established the gold standard (using as financing the French indemnity, emanating from the Franco–Prussian War), Netherlands, Denmark, Norway, Sweden, India, China, Straits Settlements, Hong Kong, Dutch East Indies, Mexico and some German states were on a silver standard. In the 1870s these European countries (and Dutch East Indies) abandoned silver in favour of gold. By 1885 almost all of western Europe – along with the United States, Britain, its dominions and various colonies – was on gold.

Asian abandonment of the silver standard prior to the First World War

Traditionally, Asian countries preferred silver to gold for both monetary and non-monetary use, and the low market ratios in the Far East reflected that fact. The silver standard continued after 1885 in the Asian countries listed above. Further, in the 1880s the Philippines and Japan went on de facto silver.

Until 1873, bimetallic France kept the world market gold–silver price ratio around a narrow band centred on the French coinage ratio of $15\frac{1}{2}$. When France ended bimetallism in 1873, the market ratio lost its anchor and escalated tremendously. The exchange rates between silver-standard and gold-standard currencies also lost their anchor. Following the market gold–silver price ratio, silver currencies depreciated greatly with respect to gold currencies. Exports were enhanced, imports were more expensive, debt and other obligations stated in terms of gold or gold currencies increased greatly in domestic currency, domestic inflation increased, and foreign investment was discouraged due to exchange-rate instability.

The problem of a depreciating currency was especially acute for India, which had the obligation of substantial recurring sterling-denominated 'home charges' to Britain (for debt service, pensions, military and other equipment, and so forth). In 1893 India abandoned the silver standard, and in 1898 went on the gold-exchange standard, pegging the (silver) rupee against the pound sterling.

In 1897 Japan switched from a de facto silver standard under legal bimetallism to a monometallic gold-coin standard, using as financing the indemnity received from defeated China in the Sino–Japanese War. In 1903 the Philippines adopted a gold-exchange standard, with the (silver) peso pegged to the US (gold) dollar. The impetus was transfer of the country from Spain to the United States, thanks to US victory in the Spanish–American War.

Mexico, a large silver producer, with both commodity exporters and silver producers in favour of a continued silver standard, finally adopted a gold-coin standard in 1905. At the beginning of the First World War, the silver standard encompassed only China, Hong Kong and a few minor countries.

Termination of the silver standard

The final blow to the silver standard was delivered by the United States, ironically after it left the gold standard. In December 1933, when the (fluctuating) market price of silver was 44 cents per ounce, President Roosevelt proclaimed that US mints should purchase all new domestically produced silver at a net price (to the depositor, or seller) of 64.65 cents per ounce (half the official, but inoperative, mint price of silver). In 1934 this policy was reinforced by the Silver Purchase Act, which directed the Treasury to purchase silver at home and abroad as long as (a) the Treasury stock of gold constituted less than one-quarter its total monetary stock, and (b) the market price did not exceed the US official mint price. Subsequently, the president ordered that all silver (with minor exceptions) then situated in the continental United States was to be delivered to US mints, at a net price of 50.01 cents per ounce. In 1935, in response to a higher foreign market price of silver (largely due to the US silver-purchase policy itself!), the president increased the net price for newly produced domestic silver to 71.11 cents.

The reason for the US silver-purchase policy was to provide a subsidy to the (politically powerful) domestic silver producers. Inadvertently, the policy effectively destroyed what remained of the silver standard. The last major country on the silver standard was China. As the gold-standard world suffered monetary and real deflation in 1929–30, the price of silver fell. The Chinese, silver-based, currency (yuan) therefore depreciated against the, gold-based, currencies of important trading partners (Britain, India, Japan). The enhanced competitiveness of export and import-competing industries, and resulting balance-of-payments surplus, prevented deflation. China lost some 'silver protection' in 1931, after Britain, India and Japan left the gold standard, as the yuan appreciated against the pound, rupee and yen; but the United States was still on the gold standard, and the yuan continued to fall, slightly, against the dollar. After the United States abandoned the gold standard, in 1933, the yuan appreciated against all four currencies.

While China had lost its 'silver protection' from the world depression, it nevertheless retained the silver standard and probably suffered less economically than its main trading partners. Disaster struck with the US silver policy of 1933–4. The huge increase in the US and market price of silver involved a corresponding appreciation of the yuan. Loss of competitiveness, balance-of-payments deficit, export of silver (and gold) to finance the deficit, and deflation followed. China had no choice but to leave the silver standard, effectively in 1934, and legally in 1935.

Other silver-standard, as well as silver-using, countries were also adversely affected by the US policy. Hong Kong followed China, and left the gold standard in 1935. Though not on the silver standard, various Latin American countries had a large silver coinage. These were token coins (face value higher than metallic-content value). Nevertheless, the high US price for silver encouraged the melting and export of these coins. The affected countries resorted to debasement and re-coining in order to retain their silver coinage.

Mexico was a special case. Silver coins constituted a high proportion of its money supply; but, as the world's largest producer of silver, Mexico benefited from a higher price for a major export. However, as other countries left the silver standard, the price of silver began to fall, and this advantage was reduced. Mexico prohibited melting or export of silver coins in 1935, and replaced the coins with paper money. Later, re-coinage occurred, and melting and export were again permitted. Yet the damage had been done, and Mexico was now on a 'managed paper standard', having lost the discipline provided by metallic money. In sum, in the 1930s, a US domestic-oriented policy reduced considerably such monetary use of silver as remained.

LAWRENCE H. OFFICER

See also **bimetallism; commodity money; gold standard.**

Bibliography

Bojanic, A.N. 1994. The silver standard in two late adherents: Mexico and India. Ph.D. thesis, Auburn University.

Brandt, L. and Sargent, T.J. 1989. Interpreting new evidence about China and U.S. silver purchases. *Journal of Monetary Economics* 23, 31–51.

der Eng, P.V. 1993. *The Silver Standard and Asia's Integration into the World Economy, 1850–1914.* Canberra: Australian National University.

Eichengreen, B. and Flandreau, M. 1996. The geography of the gold standard. In *Currency Convertibility: The Gold Standard and Beyond*, ed. J.B. de Macedo, B. Eichengreen and J. Reis. London and New York: Routledge.

Einzig, P. 1970. *The History of Foreign Exchange*, 2nd edn. London: Macmillan.

Feavearyear, Sir A. 1963. *The Pound Sterling: A History of English Money*, 2nd edn. Oxford: Clarendon Press.

Friedman, M. 1992. *Money Mischief: Episodes in Monetary History.* New York: Harcourt Brace Jovanovich.

Lai, C. and Gau, J.J. 2003. The Chinese silver standard economy and the 1929 Great Depression. *Australian Economic History Review* 43, 155–68.

Leavens, D.H. *Silver Money.* Bloomington, IN: Principia Press.

Officer, L.H. 1996. *Between the Dollar–Sterling Gold Points.* Cambridge: Cambridge University Press.

Redish, A. 2000. *Bimetallism: An Economic and Historical Analysis.* Cambridge: Cambridge University Press.

Wilson, T. 2000. *Battles for the Standard: Bimetallism and the Spread of the Gold Standard in the Nineteenth Century.* Aldershot: Ashgate.

Simon, Herbert A. (1916–2001)

Herbert A. Simon was born on 15 June 1916 in Milwaukee, Wisconsin, USA. He received his Ph.D. in political science from the University of Chicago in 1943, and taught at the Illinois Institute of Technology from 1942 to 1949 before going to Carnegie Mellon University in 1949, where he stayed until he died on 9 February 2001. Simon received major awards from many scientific communities, including the A.M. Turing Award (in 1975), the National Medal of Science (in 1986), and the Nobel Prize in Economics (in 1978). During his career, Simon also served on the Committee on Science and Public Policy and as a member of the President's Science Advisory Committee.

Simon made important contributions to economics, psychology, political science, sociology, administrative theory, public administration, organization theory, cognitive science, computer science and philosophy. His best known books include *Administrative Behavior* (1947), *Organizations* (1958, with James G. March), *The Sciences of the Artificial* (1969), *Human Problem Solving* (1972, with Allen Newell), and his autobiography, *Models of My Life* (1991). Although contributing to so many seemingly different domains and traditions, Simon's main research interest remained the same: understanding human decision making.

Early life

Simon spent his early years with his parents and his older brother on the West Side of Milwaukee in a middle-class neighbourhood. Attending public schools, Simon at first intended to study biology. However, after he went on a strawberry hunting trip, and discovered that he was colour-blind (he was unable to distinguish the strawberries from the plants), he changed his mind, thinking that colour blindness would be too big a handicap in biology. He then thought briefly about studying physics, but he gave up that idea after discovering that there weren't really any major advances left to be made in physics. 'They have all these great laws', he said in conversation. 'Newton had done it, no use messing around with it.' As a result, upon finishing high school in 1933 Simon enrolled instead at the University of Chicago, with an interest in making social science more mathematical, and an intention to major in economics. In keeping with his strong wish to be independent, Simon preferred reading alone to taking classes; and as he particularly refused to take the class in accounting, which was required for graduation in economics, he majored instead in political science.

Political science wasn't physics, of course, with all its 'great laws'. However, as a science it could encompass both theory and practice; and, being an empirical science, it had to take the data seriously. Furthermore, Simon found he was attracted to interdisciplinary thinking (in particularly psychology) in understanding political behaviour. The details of Simon's mature work differ, but the underlying ideas, interdisciplinary thinking and the necessity of bringing together theory and reality remain. Also present from the start was the essential idea of limited rationality, which would stay with Simon as he proceeded to translate his insights in political science and public administration into his work in economics, organization theory, psychology and artificial intelligence.

Early on Simon was invited by Clarence Ridley to participate as a research assistant in a project for the International City Managers' Association (Simon, 1991, p. 64). Together with Ridley, Simon published the results of this project in several articles as well as a book, *Measuring Municipal Activities* (1938). This brought an invitation to join the University of California's Bureau of Public Administration in order to study local government. While working in Berkeley on directing a study of the administration of state relief programmes, intended to demonstrate how quantitative empirical research could contribute to understanding and improving municipal government problems (1991, p. 82), Simon was also working on an early manuscript of his thesis, which became *Administrative Behavior* (1947), intended to reforming administrative theory. The first working title of *Administrative Behavior* was 'The Logical Structure of an Administrative Science'. Simon had intended the book to have a heavy philosophical component, in particular reflecting the influence of Rudolph Carnap. Furthermore, Simon introduced the importance of organizations to individual decision making; a theme later elaborated especially in *Organizations* (1958). 'Human rationality', he wrote, 'gets its higher goals and integrations from the institutional settings in which it operates and by which it is molded. ... [Therefore] ... [t]he rational individual is, and must be, an organized and institutionalized individual' (1947, pp. 101–2). Simon argued that organizations make it possible to

make decisions by constraining the set of alternatives to be taken into account and the considerations that are to be treated as relevant. Organizations can be improved by improving the ways in which those limits are defined and imposed. Finally, *Administrative Behavior* criticized existing administrative theory for being based on 'proverbs' (often contradictory common-sense principles), a perspective he wanted to replace with a more empirically oriented outlook investigating the nature of the decision processes in administrative organizations.

It was in *Administrative Behavior* that Simon first systematically examined the importance of limits to human rationality. 'The dissertation contains both the foundation and much of the superstructure of the theory of bounded rationality that has been my lodestar for nearly fifty years' (1991, p. 86). The core chapters of this book were intended to develop a theory of human decision making which was broad and realistic enough to accommodate both 'those rational aspects of choice that have been the principal concern of the economist, and those properties and limitations of the human decision making mechanisms that have attracted the attention of psychologists and practical decision makers' (1947, p. xi). Bringing together insights from economics and psychology, Simon laid the foundation for the later establishment of behavioural economics and for organization theory. In Simon's view, the significance of his early work was in replacing 'economic man' with 'administrative man' by bringing insights from psychology to bear on studying decision-making processes (1947, p. xxv).

While finishing his dissertation Simon moved to Illinois Institute of Technology, in an environment in Chicago in the early 1940s where most of his fellow researchers were believers in rational decision making, Simon remained a strong advocate of the idea of limited rationality. He began to discuss his ideas with prominent economists, in particularly those connected to the Cowles Commission, a group of mathematical economists doing pioneering research in econometrics, linear and dynamic programming, and decision theory, among other things. The economists connected to the Cowles Commission included such well-known names as Kenneth Arrow, Jacob Marshak, Tjalling Koopmans, Roy Radner and Gerard Debreu, and they held regular seminars to discuss their research. During the last years of Simon's stay in Chicago he began attending the Cowles Commission seminars; this became very important to Simon both because, as he noted in his autobiography, his interaction with Cowles almost made him 'a full time economist' (1991, p. 140) and because several members of the Cowles Commission would become good friends. Furthermore, Simon seems to have realized that among economists the possibilities for exploring the limits of rationality were themselves limited, for only later did he proceed to construct the broad behavioural programme upon which the foundation of the psychology of decision making in behavioural science could rest. 'In none of [the] early papers', wrote Simon, 'did I challenge the foundations of economic theory strongly' (1991, p. 270).

In 1949 Simon moved to Pittsburgh to join the newly established School of Industrial Administration at Carnegie Mellon University, an early engineering school trying to become a business school. Business education at that time wasn't oriented mostly towards research, but Simon and colleagues wanted to be different: they wanted to do research. They wanted their research to be relevant for business leaders, while at the same time emphasizing the tools of good science (Cooper, 2002). Early core courses in the programme included 'quantitative control and business' (consisting of basically accounting and statistics) taught by Bill Cooper, a sequence of micro and macroeconomics taught by Lee Bach, and organization theory taught by Simon. As a result of their early efforts to build up a research programme at Carnegie Mellon GSIA was picked by the Ford Foundation as one of the foremost places where the new science of behavioural economics could be developed. It became a pioneer in the establishment of business education in the United States, and must be seen as part of the Simon legacy, perhaps as important as his direct intellectual contributions.

Later work and career

Decision making was also the core of Simon's later work, and it became the basis of his other contributions to organization theory, economics, psychology, and computer science. Decision making, as Simon saw it, is purposeful yet not rational, because rational decision making would involve a complete specification of all possible outcomes conditional on possible actions in order to choose the single option that is best. In challenging neoclassical economics, Simon found that such complex calculation is not possible. As a result, Simon wanted to replace the assumption of global rationality in economics with an assumption that was more in correspondence with how humans make decisions, their computational limitations, and how they accessed information in their current environment (1955), thereby introducing the ideas of bounded rationality and satisficing. Satisficing is the idea that decision makers interpret outcomes as either satisfactory or unsatisfactory, with an aspiration level constituting the boundary between the two. Whereas decision makers in neoclassical rational choice theory would list all possible outcomes evaluated in terms of their expected utilities, and then chose the one that is rational and maximizes utility, decision makers in Simon's model face only two possible outcomes, and look for a satisfying solution, continuing to search only until they have found a solution which is good enough. The ideas of bounded rationality and satisficing became important for the subsequent development of economics.

The emphasis on bounded rationality introduced a more psychological and realistic assumption into the analysis. As Simon noted early on:

> [T]he first principle of bounded rationality is that the intended rationality of an actor requires him to construct a simplified model of the real situation in order to deal with it. He behaves rationally with respect to this model, and such behavior is not even approximately optimal with respect to the real world. To predict his behavior, we must understand the way in which this simplified model is constructed, and its construction will certainly be related to his psychological properties as a perceiving, thinking, and learning animal. (1957, p. 199)

Both satisficing and bounded rationality were introduced in 1955, when Herbert Simon published a paper that provided the foundation for a behavioural perspective on human decision making and introduced the ideas of satisficing and bounded rationality. The paper provided a critique of the assumption in economics of perfect information and unlimited computational capability, and replaced the assumption of global rationality with one that was more in correspondence with how humans (and other choosing organisms) made decisions, their computational limitations and how they accessed information in their current environments (1955, p. 99). In Simon's illustration of the problem, the influence of his early ideas outlined in *Administrative Behavior* is clear, echoing the view that decisions are reasoned and intendedly rational yet limited (1947). He first suggested a simple and very general model of behavioural choice which analyses choosing organisms (such as humans) in terms of basic properties to understand what is meant by rational behaviour. He introduced the simplifying assumptions (such as the choice alternatives, the payoff function, possible future states and the subset of choice alternatives which is considered, as well as the information about the probability that a particular outcome will lead to a particular choice) (1955, p. 102). But immediately afterwards he turned to the simplifications of this model, stressing that upon careful examination 'we see immediately what severe demands they make upon the choosing organism'. Whereas in models of rational choice the organism must be able to 'attach definite payoffs (or at least a definite range of payoffs) to each possible outcome', Simon suggested that 'there is a complete lack of evidence that, in actual human choice situations of any complexity, these computations can be, or are in fact, performed' (1955, p. 103). As a consequence of the lack of computational power, decision makers have to simplify the structure of their decisions (and thus satisfice), one of the most important lessons of bounded rationality.

In a companion paper 'Rational Choice and the Structure of the Environment' (1956), Simon introduced the idea that the environment influences decision making as much as do information processing abilities. He examined the influence of the structural environment on the problem of 'behaving approximately rationally, or adaptively' in particular environments (1956, p. 130). Simon would later elaborate these ideas in his book, *Sciences of the Artificial*, using the famous 'ant on the beach' metaphor to illustrate his idea (1969, pp. 51–3). The ant makes its way from one point to another, using a complex path, the complexity consisting of the patterns of the grains of sand along the way rather than internal constraints. Just so with human beings, Simon argues: 'Human beings, viewed as behaving systems, are quite simple. The apparent complexity of our behavior over time is largely a reflection of the complexity of the environment in which we find ourselves' (1969, p. 53).

Another early important paper (1951) concerned the nature of the employment relation. The paper began by emphasizing the traditional Simon view that models ought to correspond to the empirical realities that are neglected in most economic models of the employment contract (1951, p. 293). He then turned to a concept that was so central to him in *Administrative Behavior*, namely, the concept of authority. Central to the employment relation, Simon said, is the fact that the employer accepts a certain amount of authority of the employee for which he pays a wage and the employee accepts this authority within certain 'areas of acceptance' (1951, p. 294). His model applies the idea of satisfaction functions to the employment problem, yet it is still ripe for extension because it still is 'highly abstract and oversimplified, and leaves out of account numerous important aspects of the real situation' (1951, p. 302). The model appears to be considerably more realistic in the way it conceptualizes the nature of the employment relationship; yet it is still about 'hypothetically rational behavior in an area where institutional history and other nonrational elements are notoriously important' (1951, p. 302). The model suggests a way to reconcile administrative theory and economics through the economic nature of the employment relation; yet it is still limited by its 'assumptions of rational utility-maximization behavior incorporated in it' (1951, p. 305). Thus, Simon used the framework of economics (however limited it might be) to discuss an issue he had been interested in since his thesis, and he concluded his analysis by pointing out the limitations of a constrained model and the necessity of accounting also for non-rational elements.

Simon used this behavioural view of decision making to create a propositional inventory of organization theory, together with James March and Harold Guetzkow, which led to the book *Organizations* (1958). The book was intended to provide the inventory of knowledge of the (then almost non-existent) field of organization theory, and also a more proactive role in defining the field. Results and insights from studies of organizations in political science, sociology, economics and social psychology were summarized and codified. The book

expanded and elaborated ideas on behavioural decision making, search and aspiration levels, and the significance of organizations as social institutions. 'The basic features of organization structure and function', March and Simon wrote, 'derive from the characteristics of rational human choice. Because of the limits of human intellective capacities in comparison with the complexities of the problems that individuals and organizations face, rational behavior calls for simplified models that capture the main features of a problem without capturing al its complexities' (1958, p. 151). The book is now considered one of the classics and pioneering works in organization theory.

While Simon opposed some major developments in rational choice economics, he found value in the emerging field of operations research. Although Simon's marriage with operations research was neither entirely happy nor permanent, the fact that operations research was well suited to crossing disciplinary boundaries immediately appealed to him, in addition to its appeal in using computers for heuristic programming. Thus, Simon and Newell wrote (1958, p. 6):

> Even while operations research is solving well-structured problems, fundamental research is dissolving the mystery of how humans solve ill-structured problems. Moreover, we have begun to learn how to use computers to solve these problems, where we do not have systematic and efficient computational algorithms. And we now know, at least in a limited area, not only how to program computers to perform such problem-solving activities successfully; we know also how to program computers to learn to do these things.

Although most of the techniques used in operations research are techniques of constrained maximization, Simon found that they 'formed a natural continuity with my administrative measurement research (1991, p. 108). He found artificial intelligence to be the logical next step in operations research, something which would eventually bring Simon's insights into behavioural economics and organization theory to bear on management science, using empirical studies in decision making in organizations, constructing a mathematical model of the process under study, and then simulating it on a computer (1965).

Simon's interest in operations research is also evident in his work on the design of optimal production schedules, something which ultimately led to the book *Planning, Production, Inventories, and Work Force* (1960). Although initiated at the Cowles Commission, this worked was carried out at Carnegie Mellon University, which provided the context for most of Simon's academic life. It was also at Carnegie that it became clear that Simon was not 'just' another economist. Highly respected among most (if not all) distinguished economists of his time (see, for instance, Samuelson, 2004; Radner, 2004), Simon himself was much more than an economist. For instance, at Carnegie he quickly retooled himself as an organization theorist in order to carry out, with James G. March, a major Ford Foundation study on theories of decision making in organizations. Most important, at Carnegie Simon found both colleagues and an environment which could accommodate and appreciate his broad interests and honour his willingness to cross disciplinary boundaries in pursuing his vision. With the emerging emphasis on behavioural science at Carnegie came many contributions of a cross-disciplinary and interdisciplinary nature. The disciplinary boundary crossing that had previously been, if not difficult, then different from the mainstream became possible and more widespread with the behavioural research focus that Simon helped establish at Carnegie. Having found during his years at Chicago the limits of standard economic theory in dealing with limits to rationality, he turned his attention towards founding a research programme in behavioural economics to accommodate his vision.

Simon thus incorporated his early views on decision making and rationality into his contributions to psychology, computer science and artificial intelligence. For example, in his work with Allen Newell he attempted to develop a general theory of human problem solving which conceptualized both humans and computers as symbolic information-processing systems (1972). Their theory was built around the concept of an information processing system, defined by the existence of symbols, elements of which are connected by relations into structures of symbols. The book became as influential in cognitive science and artificial intelligence as Simon's earlier work had been in economics and organization theory.

During his amazingly productive intellectual life, Simon worked on many, sometimes different, things; yet he pursued really one vision (Augier and March, 2002). He contributed significantly to many scientific disciplines, yet found scientific boundaries themselves to be less important, even unimportant, vis-à-vis solving the questions he was working on. Even as Simon sought to develop the idea that you could simulate the psychological process of thinking, he tied his interest in economics and decision making closely to computer science and psychology. He used computer science to model human problem solving in a way that was consistent with his approach to rationality. He implemented his early ideas of bounded rationality and means–ends analysis into the heart of his work on artificial intelligence.

MIE AUGIER

See also **operations research.**

Selected works

1938. (With C. Ridley.) *Measuring Municipal Activities.* Chicago: International City Managers' Association.
1947. *Administrative Behavior.* New York: Free Press.

1951. A formal theory of the employment relationship. *Econometrica* 19, 293–305.
1955. A behavioral model of rational choice. *Quarterly Journal of Economics* 69, 99–118.
1956. Rational choice and the structure of the environment. *Psychological Review* 63, 129–38.
1957. *Models of Man.* New York: Wiley.
1958. (With A. Newell.) Heuristic problem solving: the next advance in operations research. *Operations Research* 6, 1–10.
1958. (With J. March.) *Organizations.* New York: Wiley.
1960. (With C. Holt, F. Modigliani and J. Muth.) *Planning, Production, Inventories, and Work Force.* Englewood Cliffs: Prentice Hall.
1963. Discussion: problems of methodology. *American Economic Review* 53, 229–31.
1965. *The Shape of Automation (for Men and Management).* New York: Harper and Row.
1969. *The Sciences of the Artificial.* Cambridge, MA: MIT Press.
1972. (With A. Newell.) *Human Problem Solving.* Englewood Cliffs: Prentice-Hall.
1991. *Models of My Life.* Cambridge, MA: MIT Press.

Bibliography

Augier, M. and March, J. 2002. A model scholar. *Journal of Economic Behavior and Organization* 49, 1–17.
Cooper, W. 2002. Auditing and accounting: impacts and aftermaths of R.M. Cyert's research in statistical sampling. In *The Economics of Choice, Change and Organization: Essays in Honor of Richard M. Cyert,* ed. M. Augier and J. March. Cheltenham, UK: Edward Elgar.
Mirowski, P. 2002. *Machine Dreams.* Cambridge: Cambridge University Press.
Radner, R. 2004. The best is the enemy of the good. In *Models of a Man: Essays in Memory of Herbert A. Simon,* ed. M. Augier and J. March. Cambridge, MA: MIT Press.
Samuelson, P. 2004. The Hawkins–Simon theorem revisited. In *Models of a Man: Essays in Memory of Herbert A. Simon,* ed. M. Augier and J. March. Cambridge, MA: MIT Press.

Simons, Henry Calvert (1899–1946)

Simons was born in Virden, Illinois, and died in Chicago. An economist at the University of Chicago from 1927 to 1946, he was the first professor of economics at the University of Chicago Law School. A leader of the 'Chicago School', he had an important influence on American thinking about economic policy.

Simons's central theme was stated in the title of his first writing to attract attention, a 1934 pamphlet: 'A Positive Program for Laissez Faire: Some Proposals for a Liberal Economic Policy'. The conjunction of the words 'positive' and 'laissez faire' set him apart from both the conventional conservatives of his time and the conventional liberals (in the American sense of interventionists). Simons visualized a division of labour between the government and the market. The market would determine what gets produced, how and for whom. The government would be responsible for maintaining overall stability, for keeping the market competitive and for avoiding extremes in the distribution of income. This system would preserve liberty by preventing concentration of power, and liberty is the primary virtue, followed closely by equality.

Simons's work was the response of a free society liberal – or, as he preferred, 'libertarian' – to the rise of totalitarianism in Europe, to the worldwide depression and to the attempt in the democracies, including the United States, to cope with the depression in ways that Simons regarded as threats to freedom. Simons's close friend, Professor Aaron Director, later said that Simons acted as if the end of the world was at hand. During the period of Simons's work, if not the end of the world at least the end of the free society could realistically be considered a serious possibility. Simons undertook to help to prevent that, by showing that the free society had not failed but that the government had failed to discharge its role in the free society.

The 1934 pamphlet contained the elements of a policy for a free economy that he was to restate and refine for the next 12 years, with some changes of emphasis. As he put it in 1934:

> The main elements in a sound liberal program may be defined in terms of five proposals or objectives (in a descending scale of relative importance):
>
> I. Elimination of private monopoly in all its forms … .
>
> II. Establishment of more definite and adequate 'rules of the game' with respect to money … .
>
> III. Drastic change in our whole tax system, with regard primarily for the effects of taxation upon the distribution of wealth and income … .
>
> IV. Gradual withdrawal of the enormous differential subsidies implicit in our present tariff system … .
>
> V. Limitation upon the squandering of our resources in advertising and selling activities. (1934, p. 57)

In later years the fifth of these items fell from the list.

The first proposal, 'elimination of private monopoly in all its forms', was substantially altered later. In 1934 Simons had said, 'The case for a liberal-conservative policy must stand or fall on the first proposal, abolition of private monopoly; for it is the *sine qua non* of any such policy' (p. 57). His measures for achieving this included limitation on the absolute size of corporations and on their relative size in their industries. He suggested, for example, that 'in major industries no ownership unit should produce or control more than 5 per cent of the total output' (p. 319).

By 1945 he was saying 'Industrial monopolies are not yet a serious evil' (1945, p. 35). Simons's concern about private monopoly had always been about its interaction with the state. He feared that government would support private monopolies and then have to become more powerful to control the warring monopolies it had created. The 1934 pamphlet was written at the time of Roosevelt's National Recovery Administration, which was promoting the universal cartelization of business under government aegis. But in 1945 all that was past and the political influence of business seemed too small to be a danger.

In 1945 what he had said about business monopoly he now said about labour unions. In 1934 Simons had expressed concern about labour monopolies, but in a rather subdued way. In the decade after the 1934 pamphlet, labour union membership quadrupled, and this growth showed no sign of diminishing. In his final credo (1945) Simons said 'the hard monopoly problem is labour organization'. For this problem he could offer no 'specific', only a rather uncertain question about 'the capability of democracy to protect the common interest' (1945, pp. 35–6).

As the Second World War drew to an end, the preservation of free international trade received more of Simons's attention. Peace was essential for all the goals he cherished. Even the fear of war would require a centralization of power in government that would be incompatible with personal freedom. Simons believed that economic nationalism would be the greatest threat to peace after the Second World War. Therefore, he devoted much of his work in the mid-1940s to arguing for a liberal international economic order.

While the emphasis of some points in Simons's initial policy agenda shifted, two items remained of major importance and constituted Simons's chief contribution aside from the general idea of conjoining 'positive program' with 'laissez faire'. These were the need for monetary certainty and stability and the need to finance government primarily by progressive taxation of 'income' defined in a comprehensive way.

His analysis of the monetary problem and proposals for its solution, already outlined in the 1934 pamphlet, were elaborated in a 1936 essay whose title defined the issue for years to come, 'Rules versus Authorities in Monetary Policy'. Simons believed that economic instability was due largely to the instability of the financial system. The system rested excessively on private debt, mainly short-term debt. Variations in the quantity and quality of this debt caused destabilizing variations in the quantity of money, in the quantity of 'near-money', and thereby of velocity, and in the financial requirements of business. The monetary authority, the central bank, was unreliable in discharging its responsibility to counter these devastating tendencies.

Simons's remedy for this condition was a radical reform of the financial structure and the establishment of a rule to govern the conduct of monetary policy. He regarded as an 'approximately ideal solution' one in which all property was held in equity form. Failing that, he would have preferred that all debt be in the form of perpetuities, or at least of very long maturities. He did not, however, expect to achieve even that much. But he was specific in recommending the insulation of the banking system and government finance from the malignancy of short-term debt. Banks would be required to hold reserves in currency and Federal Reserve deposits against 100 per cent of their deposits. The government would have only two kinds of debt: currency and consols.

This arrangement would give the government effective control of the quantity of money, a control that it would exercise by fiscal means – by altering the size of its own debt or the division of the debt between currency and controls. This control would be exercised 'under simple, definite rules laid down in legislation', to provide the private sector with the maximum certainty.

Simons wrestled continuously with the question of what the rules should be. His indecision appeared at the beginning, in 1934, when he referred to controlling 'the quantity, (or through quantity, the value) of effective money' (1934, p. 57). He debated with himself on this issue in the 'Rules versus Authorities' and elsewhere. He recognized that a rule aimed at the price level (or the value of money) would necessarily leave the authority with discretion to decide what quantity of money would achieve the goal. But he also feared that with the existing financial situation the velocity of money would be so variable that a quantity rule would yield great price-level instability. His solution to this dilemma was to opt for the price-level rule until reform of the financial system would reduce the quantity of near moneys and the instability of the debt structure, after which stabilizing the quantity of money would be the preferred rule.

Simons's only two books were on taxation. The first, *Personal Income Taxation* was his doctoral dissertation, written in the early 1930s and published in 1938. The second, *Federal Tax Reform*, was commissioned by the Committee for Economic Development, an organization of businessmen, mainly written in 1943 and published posthumously in 1950. A few main elements ran through all of his work on taxation. The nearly exclusive source of revenue should be taxation of personal income, meaning what has come to be called the Haig–Simons definition of income as the sum of the value of the taxpayers' consumption plus the addition to his net assets. (The reference is to Robert M. Haig, 'The Concept of Income', in *The Federal Income Tax*, ed. R.M. Haig, New York, 1921.) This definition should be applied as comprehensively as possible, for the sake of equity and economic efficiency. Simons fully explored the implication of that for the treatment of capital gains, gifts, income in kind and corporate profits. Finally, he emphasized the use of the progressive income tax as a means of reducing inequality both because reducing inequality was important and because the progressive income tax was a way of reducing inequality that was much more compatible with a free economy than other measures commonly proposed for that purpose.

Simons was a leading member of what became known in the 1930s as the 'Chicago School'. Other members at the time were Frank H. Knight, Lloyd W. Mints and Aaron Director; Jacob Viner shared many of their views but did not consider himself a member. Simons more than the others translated their general attitudes into specific policy proposals, which he advanced forcefully in his own writing and defended in a series of strong reviews of the writings of the opposition.

Simons's great attraction for his colleagues, students and sympathetic readers was a matter of personal and literary style as well as of substance. His writing was polished, ironical, free of technical jargon, statistics or mathematics, rising above 'mere' economic analysis to grand pronouncements on eternal subjects. It was not very difficult but difficult enough to leave the reader with a sense of accomplishment at having recognized its merits. He gave his readers and students a feeling of being initiated into a select club that had great insights that politicians, businessmen and most economists were intellectually, morally and ethically incapable of appreciating.

After the Second World War, and after his death in 1946, national discussion and, to some extent, policy turned in Simons's direction. There was no possibility of reverting to the negative conservatism of the pre-war years. But with a greatly enlarged federal budget and debt, and with the experience of inflation, the naive expansionism of Keynes's American disciples was no longer an acceptable policy. In this gap, Simons's ideas filled a need. A 'modern conservativism' emerged that accepted government responsibility for overall economic stability, was strongly anti-inflationary, sought a rule to govern stabilization policy, relied on tax changes rather than expenditure changes when positive fiscal measures were needed, opposed protectionism, sought to weaken the power of labour unions and accepted the progressive personal income tax as the main source of federal revenue. Simons's work contributed to this development. By the 1950s some of his principal concepts had become common currency in policy discussion – the combination of positive measures with laissez-faire, the rules-versus-authority issue and the Haig–Simons definition of the tax base. Many of his colleagues and students came into positions from which they could influence public opinion and policy.

By 1960 a new-generation Chicago School had come into prominence. Typified by Milton Friedman and George Stigler, they had been profoundly influenced by Simons as students but were departing substantially from his policy positions. Monetary history convinced them that Simons was wrong in opting for a price-level rule rather than a quantity-of-money rule for monetary policy. They concluded that antitrust activity, on which he had once placed so much emphasis, was on the whole destructive of competition. Whereas Simons never contemplated a peacetime federal budget exceeding 10 per cent of the national income, they were living with one exceeding 20 per cent, and that changed their views of many things. They came to doubt Simons's reliance on rational discussion as a way to improve government policy in a democracy; this led them, in the case of Friedman, to a search for constitutional amendments that would limit the political process or, in the case of Stigler, to concentrating on explaining rather than influencing the process. But still, they all retained the Simons vision of the good free society with a division of responsibility between the government and the market, and through them his voice was still heard 40 years after his death.

HERBERT STEIN

See also **Chicago School; taxation of income.**

Selected works

1934. A positive program for laissez-faire: some proposals for a liberal economic policy. First published as Public Policy Pamphlet No. 15, ed. H.D. Gideonse. Chicago: University of Chicago Press. Reprinted in Simons (1948).
1936. Rules versus authorities in monetary policy. *Journal of Political Economy* 54, 1–30. Reprinted in Simons (1948).
1938. *Personal Income Taxation.* Chicago: University of Chicago Press.
1945. Introduction: a political credo. First published in Simons (1948).
1948. *Economic Policy for a Free Society.* Chicago: University of Chicago Press. Contains 12 previously published and one previously unpublished article, a prefatory note by A. Director and a complete bibliography of Simons's writings.
1950. *Federal Tax Reform.* Chicago: University of Chicago Press. Contains a prefatory note by A. Director.

Simonsen, Mario Henrique (1935–1997)

Simonsen was born on 19 February 1935 in Rio de Janeiro, and died on 9 February 1997 in the same city. A talented mathematical economist, Simonsen played a prominent role in the development of economic research – especially the nature of chronic inflationary processes and their effects on the economic system – and the formulation of economic policy in Brazil from the mid-1960s to the mid-1990s. Simonsen graduated in civil engineering (Universidade do Brasil, 1957) and economics (Faculdade de Economia e Finanças do Rio de Janeiro, 1963), and in 1973 received his doctorate from the School of Graduate Studies in Economics (EPGE) at

Getulio Vargas Foundation (FGV, Rio) with a thesis about inflation in Brazil, published a few years earlier (Simonsen, 1970). However, he was essentially a self-taught man as far as economics is concerned. Indeed, he started teaching economics already in 1961, and in 1966 became director of EPGE, the first school of its kind in Brazil, founded a year earlier. He left Rio for Brasilia in 1974, when he assumed office as minister of finance (Fishlow, 1988; Simonsen, 1988d). He resumed his activities as director of EPGE in 1979, a position he held until 1993. From 1979 to 1995 Simonsen was also an outside member of the board of Citicorp. In 1981 he co-organized, with Rudiger Dornbusch, an international conference on indexation held in Rio; the conference volume is today a key reference in the field.

Cash-in-advance model

Simonsen wrote textbooks on microeconomics (1967–69) and macroeconomics (1974a; 1983a). One of the hallmarks of his microeconomics textbook was the application of the Kuhn–Tucker theorem to several problems in consumer and production theories. This is noteworthy in his early elaboration of what we now call cash-in-advance models of demand for money (Clower, 1967). Simonsen (1964; 1967–69) explicitly introduced the cash-in-advance constraint as an inequality in a nonlinear programming problem and provided a diagrammatic illustration of the interior and boundary solutions, which correspond to the money-in-utility function and cash-in-advance cases respectively. His micro book included application of general equilibrium analysis to discuss structural unemployment caused by institutional wage rigidities or by fixed coefficients in underdeveloped dual economies, and the argument that risk aversion provides a solution to the problem of the optimal size of the firm under perfect competition and constant returns.

Inertial inflation

Simonsen's 1974 macro textbook restated his 1970 inflation model. The model is remarkable for introducing into the literature the concept of 'inertial inflation'. As pointed out by Simonsen (1970), the dependence of the current rate of inflation on its past values means that cold turkey strategies of disinflation are costly. The inertial element – called the 'feedback component' – together with the 'autonomous component' (supply shocks) and the 'demand regulation component' (excess aggregate demand) decide the inflation rate in a given period of time according to a linear formula. The feedback component could be explained at the time either by contract indexation or by adaptive expectations of price changes. Simonsen chose the indexation assumption, because it reflected the institutional wage-setting mechanism of the Brazilian economy in the 1970s. Simonsen's 1970 feedback model differed from the accelerationist Phillips curve by explaining inflation acceleration as a result not of revised expectations but of a reduction in the adjustment interval, captured by changes in the feedback coefficient. Moreover, the model implied that, even if inflation expectations fell to zero, the feedback inertial mechanism would keep working due to wage staggering in the indexation process. On the assumption of zero excess aggregate demand and a less than unity feedback coefficient, the lower limit to the current rate of inflation was given by the autonomous component divided by 1 minus the feedback coefficient. In order to illustrate the argument, Simonsen distinguished between the peak and the average real wage concepts in a sawtooth pattern curve (called the 'Simonsen curve' in the Brazilian literature) with the real wage rate as a function of time for a given adjustment interval.

Indexation and incomes policy

Simonsen came back to indexation and wage staggering in the 1980s, when those topics became fashionable after Gray (1976) and Taylor (1979). In both his graduate macro textbook (1983a) and in his conference paper (1983b), he showed that an expanded Gray model under rational expectations and with perfect wage indexation – that is, with individual prices indexed to the current price index – supported Milton Friedman's argument in favour of escalator clauses as a way to ease the side effects of anti-inflationary policies. However, he also showed that, under the more realistic assumption of lagged indexation – with prices indexed to changes in the price index over some past interval – the result would be the opposite: the inflation rate predicted for the current period and later depends upon the inflation rate in the past period (which, by definition, cannot reflect new information about current or future monetary policy). Simonsen's demonstration that lagged indexation brings about a trade-off between inflation and unemployment confirmed his previous conclusion that cold-turkey disinflation is costly. This inflation inertia (or 'persistence') result has been obtained in other rational expectation models as well (Fuhrer and Moore, 1995). As shown by Simonsen (1986a), inflation inertia was not a feature of Taylor's 1979 wage staggering model, which could only generate price-level inertia. Simonsen argued for the de-indexation of the Brazilian economy as a crucial step to end chronic inflation, which was eventually achieved through the monetary reform that introduced a new currency in 1994, partly under the influence of his ideas (Andrade and Silva, 1996).

In a series of essays published between 1986 and 1989 and in his 1995 book, Simonsen used game theory to provide a model of inertial inflation and to bring out the role of incomes policies to reduce it. He generalized Townsend's (1978) argument about the correspondence between rational expectations macroeconomics and Nash equilibria, and suggested that the central weakness of the rational expectation hypothesis is the implicit assumption that rational participants in a non-cooperative game promptly move to a Nash equilibrium. From that perspective, inflation inertia is consequence of a coordination failure between wage and price setters in the economy after an observed change in macroeconomic policy. Incomes policy can be used to resolve this coordination failure, in the sense of providing information to speed up the location of Nash equilibria by economic agents.

International debt crisis

Another macroeconomic issue that attracted Simonsen's attention was the debt dynamics of developing countries. Simonsen (1983a, ch. 5; 1985) put forward an analytical framework to derive the conditions for a country's solvency. He advanced a differential equation that splits the rate of change of the country's net foreign debt into the interest payments and the resource gap, and expressed it in the form of the ratio to exports. Simonsen then showed that the condition for solvency is that the rate of growth of exports exceeds the interest rate, which he used to explain the breakdown of competitive recycling in the early 1980s (see also Krueger, 1987). Apart from his new approach to inertial inflation, Simonsen also applied game theory to study the non-cooperative behaviour of banks in providing competitive loans to highly indebted developing countries (1985) and to investigate dynamic bargaining problems between banks and developing countries (1989), partly based on the Rubinstein (1982) bargaining model (cf. Bulow and Rogoff, 1989). His model attempted to explain why creditors, organized as a cartel, preferred to deal with each country separately, and why debtors did not organize themselves as a cartel.

MAURO BOIANOVSKY

See also **Dornbusch, Rudiger; sovereign debt; wage indexation.**

Selected works

1964. A lei de Say e o efeito liquidez real. *Revista Brasileira de Economia* 18, 41–66.
1965. (With W. Baer.) Profit inflation and policy-making in an inflationary economy. *Oxford Economic Papers* 17, 279–90.
1967–69. *Teoria Microeconomica*, 4 vols. Rio: FGV.
1970. *Inflação: gradualismo vs. tratamento de choque*. Rio: APEC.
1974a. *Macroeconomia*, 2 vols. Rio: APEC.
1974b. The anti-inflationary policy. In *The New Brazilian Economy*, ed. M.H. Simonsen and R. Campos. Rio: Crown.
1983a. *Dinamica Macroeconomica*. São Paulo: McGraw-Hill.
1983b. Indexation: current theory and the Brazilian experience. In *Inflation, Debt, and Indexation*,

ed. M.H. Simonsen and R. Dornbusch. Cambridge, MA: MIT Press.

1985. The developing-country debt problem. In *International Debt and the Developing Countries*, ed. J. Cuddington and G. Smith. Washington, DC: World Bank.

1986a. Cinqüenta anos da *Teoria Geral do Emprego*. *Revista Brasileira de Economia* 40, 301–34.

1986b. Rational expectations, income policies and game theory. *Revista de Econometria* 6, 7–46.

1988a. Price stabilization and incomes policies: theory and the Brazilian case study. In *Inflation Stabilization – The Experiences of Israel, Argentina, Brazil and Mexico*, ed. M. Bruno et al. Cambridge, MA: MIT Press.

1988b. (With R. Dornbusch.) Inflation stabilization: the role of income policies and monetization. In *Exchange Rates and Inflation*, ed. R. Dornbusch. Cambridge, MA: MIT Press.

1988c. Rational expectations, game theory and inflationary inertia. In *The Economy as an Evolving Complex System*, vol. 5, ed. P.W. Anderson, K.J. Arrow and D. Pines. New York: Addison-Wesley.

1988d. Interview. In *An Oral History of Finance: 1967–1988*, compiled by the editors of *Institutional Investor*. New York: William Morrow.

1989. Macroeconomia e teoria dos jogos. *Revista Brasileira de Economia* 43, 315–71.

1995. *30 anos de indexação*. Rio: FGV.

Bibliography

Alberti, V., Sarmento, C. and Rocha, D. (eds.) 2002. *Mario Henrique Simonsen – um homem e seu tempo*. Rio: FGV.

Andrade, J. and Silva, M. 1996. Brazil's new currency: origin, development and perspectives of the Real. *Revista Brasileira de Economia* 50, 427–67.

Boianovsky, M. 2002. Simonsen and the early history of the cash-in-advance approach. *European Journal of the History of Economic Thought* 9, 57–71.

Bulow, J. and Rogoff, K. 1989. A constant recontracting model of sovereign debt. *Journal of Political Economy* 97, 155–78.

Clower, R. 1967. A reconsideration of the microfoundations of monetary theory. *Western Economic Journal* 6, 1–8.

Fishlow, A. 1988. Tale of two presidents. In *Democratizing Brazil: Problems of Transition and Consolidation*, ed. A. Stepan. Oxford: University Press.

Fuhrer, J. and Moore, G. 1995. Inflation persistence. *Quarterly Journal of Economics* 110, 127–59.

Gray, J. 1976. Wage indexation – a macroeconomic approach. *Journal of Monetary Economics* 2, 221–35.

Krueger, A. 1987. Origins of the developing countries' debt. *Journal of Development Economics* 27, 165–87.

McNelis, P. 1985. Review essay of Simonsen and Dornbusch 1983. *Journal of Money, Credit and Banking* 17, 274–80.

Rubinstein, A. 1982. Perfect equilibrium in a bargaining model. *Econometrica* 50, 97–110.

Taylor, J. 1979. Staggered wage setting in a macro model. *American Economic Review* 69, 108–13.

Townsend, R. 1978. Market anticipations, rational expectations and Bayesian analysis. *International Economic Review* 19, 481–94.

simplex method for solving linear programs

The data for the linear programming problem (LP) is stated below in standard form:

FIND Min $z, x_j \geq 0$:

$$a_{11}x_1 + \cdots + a_{1s}x_s + \cdots + a_{1n}x_n = b_1$$
$$\cdots\cdots\cdots\cdots\cdots\cdots\cdots\cdots\cdots\cdots$$
$$a_{r1}x_1 + \cdots + a_{rs}x_s + \cdots + a_{rn}x_n = b_r$$
$$\cdots\cdots\cdots\cdots\cdots\cdots\cdots\cdots\cdots\cdots$$
$$a_{m1}x_1 + \cdots + a_{ms}x_s + \cdots + a_{mn}x_n = b_m$$
$$\text{obj}: \quad c_1x_1 + \cdots + c_sx_s + \cdots + c_nx_n = z - b_0$$

$$(1)$$

Obj is the objective or cost equation defining z. In economic applications, the coefficient a_{ij}, depending on sign, is the input or output of item i per unit level of activity j and x_j is the level of activity j to be determined.

Any particular set of values $x^0 = (x_1^0, \ldots, x_n^0)$ that satisfies the first m equations of (1) is called a solution; if in addition $x_j^0 \geq 0$ for all j then x^0 is a feasible solution; if upon substitution into the obj equation of (1), x^0 yields a value of $z = \text{Min } z$, then x^0 is an optimal feasible solution.

The LP, however, could have been given in one of several other ways which, from a mathematical point of view, are all equivalent. Suppose the LP were originally stated as one of minimization of a linear form subject to a system of linear inequalities. It could then be easily converted to (1). For example, the relation $2u + 3v \leq 4$ can be replaced by the equation $2(x_1 - x_2) + 3(x_3 - x_4) + x_5 = 4$ where $x_j \geq 0$ by setting u and v each equal to the difference of two non-negative variables $u = x_1 - x_2$, $v = x_3 - x_4$, and introducing a non-negative *slack* variable x_5.

Commercial software for LP usually allows the user to specify which variables are unrestricted in sign and whether the relation is an equation or an inequality. The software program does not make the above substitutions but uses a modified form of the simplex algorithm designed to handle the mixed case of signed/unsigned variables and equation/inequality relations.

Pivoting defined

The simplex method consists of a sequence of $t = 0$, $1, 2 \ldots$ pivot steps (iterations) performed on system (1) which transforms it on each step to a new, mathematically equivalent, system of equations. Any solution of (1) is also a solution for the system of iteration t, and conversely. Thus feasible and optimal feasible solutions remain feasible and optimal after pivoting and so remain for all t. Since the generated systems all have the same solution set as (1), it is not necessary to store in the memory of the computer a record of all the intermediate steps.

We will use the same symbols a_{ij}, c_j, b_i to denote the updated system after pivoting as before. When necessary to distinguish as to which iteration t they pertain, we will use a superscript a_{ij}^t, c_j^t, b_i^t.

To perform a pivot step on system (1) iteration $t = 0$ or a subsequent iteration t, select any term $a_{rs}x_s$ where $a_{rs} \neq 0$, called the *pivot term*. Replace equation r by dividing it through by a_{rs}, then replace each equation $i \neq r$ by subtracting from it the new rth equation multiplied by a_{is}. Do the same thing with the objective equation by subtracting from it the rth equation multiplied by c_s. This eliminates the variable x_s from all equations except the rth. During a pivot step, the current solution x^t is also updated to a new solution x^{t+1} by some rule.

A number of variants of the simplex method based on pivoting from one iteration to the next are used to solve LPs. These include the dual simplex method, the primal-dual method, and the symmetric method. They differ only in the rules used for choosing the pivot term or the way the current solution is updated.

The simplex method to be described was first proposed in 1947; it can be stated in 20 or so instructions for a computer. Commercial codes based on the simplex method, however, usually involve thousands of instructions which are there to take advantage of sparsity (most coefficients of practical problems are zero), to make it easy to start from solutions to variants of the same problem, and to guarantee numerical accuracy of the solution for large-scale systems.

Outline of the procedure

The simplex method consists of phase I which finds a feasible solution if one exists, and phase II which finds an optimal one if one exists. Thus the method can terminate with (a) no feasible solution, (b) an optimal feasible solution, or (c) a class of feasible solutions whose values for the objective $z \to -\infty$. Each phase applies a special subroutine called the simplex algorithm to a related but different LP problem. We begin by describing this algorithm.

Simplex algorithm

This algorithm requires the system to be given in canonical form with the right-hand side constants $b_i \geq 0$. The system is said to be in canonical form if we can permute the order of the variables of the first m equations so that

coefficients of the first m variables form an identity matrix, i.e., a square array of all zeros except for a diagonal of all ones. We also require their corresponding terms in the obj equation be zero. We illustrate with an $m = 2$, $n = 5$ example.

FIND Min z, $(x_1, \ldots, x_5) \geq 0$:

$$2x_1 + 1x_3 + a_{14}x_4 + 1x_5 = 8$$
$$-3x_1 + 1x_2 - 7x_4 + 1x_5 = 6$$
$$\text{obj}: \quad 4x_1 + c_4x_4 + 1x_5 = z - 3. \quad (2)$$

By choosing the constants $c_4 = +5$ or -5 and $a_{14} = -2$ or $+1$, we have, in fact, four different examples. System (2) is in canonical form because we can reorder the variables so that x_3, x_2 come before the rest. When we do so the matrix of coefficients of x_3, x_2 in the first two equations is the 2×2 identity matrix:

$$\begin{bmatrix} 1 & 0 \\ 0 & 1 \end{bmatrix}. \quad (3)$$

The *ordered* set of m indices giving rise to the identity matrix, in the example $\{3, 2\}$, is called the basic set; its corresponding variables are called the basic variables; its set of coefficients is called the basis. Each iteration t will give rise to varying basic sets of m indices.

Termination

The simplex algorithm terminates with an optimal solution when a canonical system is generated on some iteration t with $c_j \geq 0$ for all j. This is the case in the example if $c_4 = +5$. Note $c_1 = 4$, $c_2 = c_3 = 0$, $c_4 = 5$, $c_5 = 1$. Phase I will always terminate in this way. Phase II can also terminate with a class of feasible solutions in which $z \to -\infty$. This happens when a canonical system is generated on some iteration t with some variable x_s whose $c_s < 0$ and all its other coefficients $a_{is} \leq 0$. In the example if $c_4 = -5$ and $a_{14} = -2$, then for variable x_4 this termination condition holds, namely: $c_4 = -5$, $a_{14} = -2$, $a_{24} = -7$.

Basic feasible solutions

The solution obtained by setting the values of all non-basic (independent) variables equal to zero and solving for the basic (dependent) variables is called a basic solution. Since the canonical form for each iteration t satisfies $b_i \geq 0$ for all i, the basic feasible solution is simply $x_j = 0$ for j non-basic and $x_{j_i} = b_i$ where $j_1, j_2, \ldots, j_m$ are the basic set of indices in the order that their coefficients form an identity matrix. In the example, $\{j_1, j_2\} = \{3, 2\}$; the basic feasible solution is $x_3 = 8$, $x_2 = 2$, $x_1 = x_4 = x_5 = 0$. Substituting this solution into the obj equation, we obtain $z = 3$.

Proof of optimality

To prove that the basic feasible solution yields Min $z = b_0$ when $c_j \geq 0$ for all j, we observe for our example with $c_4 = +5$ that the objective equation states that $z = 3 + 4x_1 + 5x_4 + x_5$. Therefore the value of $z \geq 3$ because $4x_1 + 5x_4 + x_5 \geq 0$ for all $x_j > 0$ and its lower bound $z = 3$ is attained for the basic feasible solution. Therefore $z = 3$ is minimum.

In general, the value of z for the basic feasible solution for iteration t is clearly $z = b_0$ and the obj equation can be rewritten $z = b_0 + \sum c_j x_j$. Therefore if $c_j \geq 0$ and $x_j \geq 0$ then $z \geq b_0$. Since the lower bound $z = b_0$ is attained for the basic feasible solution, this implies Min $z = b_0$.

Proof that $z \to -\infty$. We wish to show z has no lower bound when for some x_s, $c_s < 0$ and $a_{is} \leq 0$, for all i. In the example let $c_4 = -5$ and $a_{14} = -2$, then for x_4, $c_4 = -5$, $a_{14} = -2$, $a_{24} = -7$ which satisfies the termination condition. Setting all non-basic variables $= 0$ *except* x_4 and solving for the basic variables and z in terms of x_4, we have:

$$x_3 = 8 + 2x_4, \quad x_1 = x_5 = 0,$$
$$x_2 = 6 + 7x_4,$$
$$z = 3 - 5x_4. \tag{4}$$

As $x_4 \to +\infty$, a class of feasible solutions is generated in which $z \to -\infty$.

In general, setting all non-basic variables $x_j = 0$ except x_s, and solving for basic x_j and z in terms of x_s, we have:

$$x_j = b_i - a_{is} x_s \quad \text{for } j_i \text{ basic}$$
$$z = b_0 + c_s x_s. \tag{5}$$

Again we see for $a_{is} \leq 0$, $c_s < 0$, that a class of feasible solutions $x_j \geq 0$ is generated in which $z \to -\infty$.

Improving a basic feasible solution

Let s be such that $c_s = \text{Min } c_j$. If $c_s \geq 0$ the algorithm terminates with an optimal basic-feasible solution. If $c_s \leq 0$, then clearly setting all non-basic $x_j = 0$ except x_s and allowing x_s to increase causes $z = b_0 + c_s x_s$ to decrease towards $-\infty$; hence the more we can decrease x_s the better. However, the values of the basic variables in terms of $x_{j_i} = b_i - a_{is} x_s$ and therefore the maximum increase allowable for x_s in order to keep all x_j non-negative is $x_s = x_s^*$ where

$$x_s^* = \underset{i}{\text{Min}}(b_i / a_{is}) \tag{6}$$

where Min$_i$ is restricted to i such that $a_{is} > 0$. If there are no $a_{is} > 0$, then we have the termination case already discussed of $z \to -\infty$. Otherwise the minimum occurs at some $i = r$ and $x_s^* = b_r / a_{rs}$. When $x_s = x_s^*$, the rth basic variable assumes the value $x_{j_r} = b_r - a_{rs}(b_r / a_{rs}) = 0$. This suggests that the variable x_s replace x_{j_r} as rth basic variable by pivoting on $a_{rs} x_s$.

We illustrate this for our example with $c_4 = -5$ and $a_{14} = 1$. Since $c_4 = \text{Min } c_j$ and $c_4 < 0$, we have $s = 4$. Accordingly we set all non-basic $x_j = 0$ except x_4 and

solve for the values of basic variables in terms of x_4. Thus:

$$x_3 = 8 - 1x_4, \quad x_1 = x_5 = 0,$$
$$x_2 = 6 + 7x_4,$$
$$z = 3 - 5x_4. \tag{7}$$

We are blocked from increasing x_4 indefinitely because x_3 would become negative if $x_4 > 8$ and our class of generated solutions would no longer remain feasible. At $x_4 = 8$ we have two variables positive, $x_4 = 8$ and $x_2 = 62$, and all the rest $x_1 = x_3 = x_5 = 0$. Therefore we drop $j = 3$ from our basic set and replace it by $j = 4$ by pivoting on $a_{14} x_4$. Thus we have:

Iteration $t = 0$

$$2x_1 \qquad +1x_3 + \boxed{1x_4} + 1x_5 = 8$$
$$-3x_1 + 1x_2 \qquad -7x_4 + 1x_5 = 6 \qquad (8)$$
$$\text{obj:} \quad 4x_1 \qquad -5x_4 + 1x_5 = z - 3$$

Iteration $t = 1$ (after pivoting using $\boxed{1x_4}$ as pivot term).

$$2x_1 + 1x_3 + 1x_4 + 1x_5 = 8$$
$$11x_1 + x_2 + 7x_3 + 8x_4 = 62$$
$$\text{obj:} \quad 14x_1 + 5x_3 + 6x_5 = z + 37. \qquad (9)$$

We conclude that the basic feasible solution for iteration 1, namely $x_4 = 8$, $x_2 = 62$, $x_1 = x_3 = x_5 = 0$ and $z = -37$ is the optimal feasible solution. If the obj equation for iteration $t - 1$ had some $c_j < 0$, we would have continued the algorithm.

Phase 1

To initiate phase I, multiply by -1 all equations of (1) with $b_i < 0$, $i \neq 0$, so that (1) after modification $b_i \geq 0$. Next adjoin auxiliary variables, called artificials, $x_{n+1}, \ldots, x_{n+m}$ as shown below.

Phase I problem, Iteration $t = 0$.
FIND Min w, $x_j \geq 0$:

$$a_{11} x_1 + \cdots + a_{1s} x_s + \cdots + a_{1n} x_n + x_{n+1} = b_1$$
$$\cdots\cdots\cdots\cdots\cdots\cdots\cdots\cdots\cdots\cdots\cdots$$
$$a_{r1} x_1 + \cdots + a_{rs} x_s + \cdots + a_{rn} x_n + x_{n+r} = b_r$$
$$\cdots\cdots\cdots\cdots\cdots\cdots\cdots\cdots\cdots\cdots\cdots$$
$$a_{m1} x_1 + \cdots + a_{ms} x_s + \cdots + a_{mn} x_n + x_{n+m} = b_m$$
$$\text{obj:} \quad d_1 x_1 + \cdots + d_s x_s + \cdots + d_n x_n = w - d_0 \tag{10}$$

The obj equation has been replaced by a phase I obj defined by

$$d_j = -\sum_i a_{ij} \quad \text{and} \quad d_0 = +\sum_i b_i. \tag{11}$$

Note the system is in canonical form with $b_i \geq 0$ so that we are all set to apply the simplex algorithm.

Special rule

Once an artificial variable x_{n+i} is pivoted out of the set of basic variables on some iteration t and becomes non-basic, it is discarded (i.e., all terms involving x_{n+i} are dropped from the canonical form). Hence the pivot term $a_{rs}x_s$ will be one from among the first m rows and n columns of (10).

If we add the first m equations of (10) to the obj equation, we obtain by (11) that:

$$x_{n+1} + x_{n+2} + \cdots + x_{n+m} = w. \qquad (12)$$

Thus the phase I objective is equivalent to minimizing the sum of the artificial variables. If a feasible solution to (1) exists, then a feasible solution to (10) exists in which all $x_{n+i} = 0$ and therefore a feasible solution to (10) exists in which $w = 0$. Since $x_{n+i} \geq 0$, it follows for all feasible solutions to the phase I problem, $w \geq 0$ and Min $w \geq 0$. It is therefore impossible in phase I to find a class of solutions in which $w \to -\infty$. If the optimal solution yields Min $w > 0$, the simplex method is terminated with the statement that no feasible solution to (1) exists. If phase I terminates with $w = 0$, then we set up the phase II problem.

Transition to phase II

At the end of phase I if Min $w = 0$, then all artificial variables have value 0 in the basic solution. Usually there are no longer any artificial variables left among the basic ones in the canonical form. When this is the case, we replace the obj equation of phase I by the original one given as input data (1). Next we eliminate from the obj all terms $c_j x_j$ corresponding to $j = j_i$ in the basic set. This is done by subtracting from the obj equation the ith equation of the canonical form multiplied by c_{j_i}. The phase II problem is now in canonical form ready to apply the simplex algorithm to find Min z.

For example, suppose at the end of phase I we have: FIND Min z. $(x_1, \ldots, x_4) \geq 0$:

$$2x_1 + 1x_3 + a_{14}x_4 + 1x_5 = 8$$
$$- 3x_1 + 1x_2 - 7x_4 + 1x_5 = 6$$
$$\text{obj}: \quad c_1x_1 + c_2x_2 + c_3x_3 + c_4x_4 + c_5x_5 = z - 3.$$
$$(13)$$

The basic set is $\{j_1, j_2\} = \{3, 2\}$. Multiplying the first equation by c_3 and the second by c_2 and subtracting from obj, we eliminate the basic variables x_3, x_2 from the obj equation obtaining an obj equation of the form:

$$\text{obj}: \quad c_1x_1 + c_4x_4 + c_5x_5 = z - b_0. \quad (14)$$

It can happen, however, at the end of Phase I for some iteration t that Min $w = 0$ and some artificial variable,

say x_{n+r}, still is basic. Its basic solution value is $x_{n+r} = b_r = 0$. x_{n+r} is gotten rid of by pivoting on any term $a_{rs}x_s$ of the canonical form where $a_{rs} \neq 0$ and $s \leq n$. The new basic solution will have as rth basic variable $x_s = 0$ and x_{n+r}, now non-basic, is then discarded. This process is continued until all artificials are dropped.

There still remains the possibility that a pivot term for some r cannot be found because all $a_{rj} = 0$ for $j = 1, 2, \ldots, n$. In this case it is easy to prove that the rth equation of the original problem is *redundant* and the rth equation of the canonical form of the phase I problem can be discarded or, alternatively, x_{n+r} can be reclassified as belonging among the *true* variables – it will do no harm to include it because in all subsequent iterations its basic solution value will remain zero. Once the artificials are made non-basic and removed, we complete the transition as outlined in the paragraph above.

Upon termination of the simplex algorithm applied to the Phase II problem, the software program is directed to print out a statement about the type of termination. In the case of an optimum solution, this is followed by the list of indices of the obj, the basic variables, and alongside them the values of the corresponding b_0 and b_i.

In the case z is unbounded below, the list printed is

$$\begin{vmatrix} \text{obj} & b_0 & +c_s \\ j_i & b_i & -a_{is} \\ s & 0 & 1 \end{vmatrix} \quad \text{for } i = 1, \ldots, m. \quad (15)$$

This information permits one to generate z and a feasible solution for any choice of $x_s \geq 0$.

Proof of convergence

It is not difficult to show that if any basic set of indices were to be repeated in some subsequent iteration, the entire canonical form would be repeated including the value of z in the basic solution. In the example (8), the value of z in the basic solution is $z = 3$. After pivoting, see (9), the value of $z = -37$. We see that its value decreased from 3 to -37.

In general, if there is a *positive* decrease in the value of z in the basic solution from one iteration to the next, the canonical form cannot be repeated since the value of z is lower. On the other hand, the iterative process must stop sometime because there is only a finite number of canonical forms. But the only way it could have stopped is via one of the two termination conditions. Hence the iterative process is finite when there is a positive decrease on each iteration. This should not be interpreted, however, as meaning the algorithm is efficient because the number of ways to pick m objects out of n grows exponentially with increasing m and n.

Degeneracy

Should the pivot term occur on a row r whose $b_r = 0$, then the updated value of z in the basic solution is

$z = b_0 - b_r(c_s/a_{rs}) = b_0$, i.e., the change in value of z is zero and the proof of convergence given above is no longer applicable. In this case, one or more of the values of the basic variables in a basic solution are zero and the basic solution is said to be *degenerate*. There are examples of canonical forms with degenerate basic solutions, which after a number of pivots return to the original canonical form.

To avoid this possibility of cycling, special rules have been invented that are easy to implement but are not found in commercial codes. Almost all practical problems are degenerate. Failure to provide a rule has never (or almost never) caused the algorithm to cycle in practice. From a theoretical point of view, however, devices that prevent cycling are important because the simplex method is used as a powerful analytic tool for proving theorems like the duality theorem.

Economic interpretations

Feasible. In planning, a feasible solution is a plan or policy that is physically implementable. The plan may be feasible but not necessarily an optimal one.

Prices. Associated with a basic solution is a set of prices $(\pi_1, \pi_2, \ldots, \pi_m)$, also called *Lagrange Multipliers*, which are defined so that if we 'price out' the inputs and outputs of activities associated with basic variables, they *break even*. By this is meant for each j in the basic set:

$$c_j^0 - \sum_i \pi_i a_{ij}^0 = 0, \quad j = j_1, \ldots, j_m. \quad (16)$$

where a_{ij}^0, c_j^0 refer to a_{ij}, c_j of iteration $t = 0$.

The value of c_j of iteration t is denoted by c_j^t. It is easy to show that c_j^t can be obtained directly from the data of iteration $t = 0$ by 'pricing out' any activity j in terms of the prices associated with the current basis:

$$c_j^t = c_j^0 - \sum_i \pi_i a_{ij}^0, \quad j = 1, \ldots, n. \quad (17)$$

If $c_j^t < 0$ for any activity $j = j^*$, it pays to replace one of the activities of the basic set by activity j^*. The simplex method chooses among the activities j in the non-basic set $j^* = j_s$ that shows the greatest profitability per unit change of activity level x_j, namely s such that $c_s = \text{Min } c_j$ and $c_s < 0$.

Duality

The dual of the LP (1) iteration $t = 0$ is defined by:
FIND Max $\mathbf{z}$, $(\pi_1, \pi_2, \ldots, \pi_m)$, and $(y_1, \ldots, y_m)$:

$$y_j = c_j - \sum_i \pi_i a_{ij}, \quad j = 1, \ldots, n$$

$$\mathbf{z} = b_0 + \sum_i \pi_i b_i \quad (18)$$

It is easy to show that $\mathbf{z} \leq z$ for all feasible solutions to the original primal system (1) and feasible solutions to the dual system (18). This implies when feasible solutions to both the primal and dual systems exist that z has a finite lower bound and $\mathbf{z}$ has a finite upper bound. We have shown in this case that an optimal feasible solution exists to the primal system. Moreover for the optimal canonical form of some iteration t that π_i defined by (16), satisfies $c_j^t \geq 0$ in (17). Setting $y_j = c_j^t \geq 0$ for all j, we see that π_i and $y_i \geq 0$ satisfy (18). It is easy to show that z of this basic feasible solution satisfies Min $z = b_0^0 + \sum_i \pi_i b_i^0$ so that $\mathbf{z} = z$. It follows therefore that Max $\mathbf{z} = $ Min z. This is called the *strong* duality theorem; note that we have proved it using the properties of the simplex algorithm.

Computational experience

Since 1947 the simplex method and its variants have successfully solved each day thousands of large and small scale practical problems. New methods for solving LPs are constantly cropping up. Many LPs have special structures and special algorithms have been developed to solve them. For example there is considerable research on how to efficiently solve large-scale dynamic economic models under uncertainty. One approach makes use of parallel computers, random sampling, methods of decomposing the problem into many subproblems which are solved using the simplex method as a subroutine.

GEORGE B. DANTZIG

simulation-based estimation

Simulation-based estimation is an application of the general Monte Carlo principle to statistical estimation: any mathematical expectation, when unavailable in closed form, can be approximated to any desired level of accuracy through a generation of (pseudo-) random numbers. Pseudo-random numbers are generated on a computer by means of a deterministic method. (For convenience, we henceforth delete the qualification 'pseudo'.) Then a well-suited drawing of random numbers (or vectors) $Z_1, Z_2, \ldots, Z_H$ provides the Monte Carlo simulator $(1/H)\sum_{h=1}^H g(Z_h)$ of $E[g(Z)]$. Of course, one may also want to resort to many simulators improving upon this naive one in terms of variance reduction, increased smoothness and reduced computational cost. A detailed discussion of simulation techniques is beyond the scope of this article. Nor are we going to study Monte Carlo experiments, which complement a given statistical procedure by the observation of its properties on simulated data. Rather, our focus of interest is to show how Monte Carlo integration may directly help to compute estimators that would be unfeasible without resorting to simulators.

The article is organized as follows. Section 1 is devoted to the most natural use of Monte Carlo integration for estimation, which is finite sample bias correction. It encompasses the parametric bootstrap. More generally, we use throughout the framework of a fully parametric econometric model with possibly latent variables, as defined in Section 1. Section 2 emphasizes that the simulation tool actually provides at least an asymptotic bias correction in much more general settings, such as simultaneity bias, bias due to errors in variables or any kind of misspecification bias. The general approach is dubbed *simulation-based indirect inference* (SII). Instead of using SII only for bias correcting a poor initial estimator, we can actually take advantage of any instrumental piece of information, insofar as it (over)identifies the structural parameters of interest. Section 3 is devoted to the *simulated method of moments* (SMM) and its *simulated-score-matching* version. With an asymptotic point of view, it can be seen as a particular asymptotic case of SII when instrumental parameters are some well-chosen moments. Besides computation of moments to match, Monte Carlo integration can also be used for the direct assessment of the criterion to maximize for M-estimation, when it is not available in closed form. The objective of asymptotic efficiency in the context of a parametric model leads us to put forward the *simulated maximum likelihood* (SML) or a *simulated score* technique, both described in Section 4. Some alternative simulated M-estimators, convenient though inefficient, are also reviewed. Concluding remarks in Section 5 are mainly focused on the trade-off between efficiency and robustness to misspecification; the fact that the structural model is also providing a simulator raises new issues for this classical trade-off.

The exposition in this article of simulation-based estimation methods relevant for econometric applications is selective in several respects. We do not present Markov chain Monte Carlo methods and data augmentation techniques. These are especially popular in Bayesian statistics and econometrics, but also relevant for some applications in a classical inference setting. Generally speaking, any kind of random drawing in the parametric space is beyond the scope of this article. Finally, it should be borne in mind throughout that all the simulation-based estimation methods have a non-simulation-based counterpart. While it is well known that SMM and SML are the simulation-based counterparts of GMM (generalized method of moments) and MLE (maximum likelihood estimation) respectively, it may be less known that the approaches of bootstrap and indirect interface make sense even without simulations. The essential characteristic of these techniques is to insert a consistent estimator of the data generating process in a functional of the true data distribution. Simulations are only a tool to evaluate the resulting estimated functional which, more often than not, is not available in closed form. There are, however, interesting exceptions like linear indirect least squares. Moreover, even though we always refer to the general concept of Monte Carlo integration, it does not necessarily involve a large number of simulated paths. Asymptotic theory of simulation-based estimation techniques will be considered when the length of the observed sample path goes to infinity. Depending on the techniques, the number of simulated paths may or may not be constrained to tend to infinity to get consistent estimators.

1 General framework and simulation-based bias correction

Let us denote by θ a vector of p unknown parameters. We want to build an accurate estimator $\hat{\theta}_T$ of θ from an observed sample path of length T. Let us assume that we have at our disposal some initial estimator, denoted by $\hat{\beta}_T$. Note that we purposely use a letter β different from θ to stress that the estimator $\hat{\beta}_T$ may give a very inaccurate assessment of the true unknown θ^0 we want to estimate. In particular this estimator is potentially severely biased: its expectation $b_T(\theta^0)$ does not coincide with θ^0. The notation $b_T(\theta^0)$ refers to the so-called binding function (Gourieroux and Monfort, 1995) and depends on at least two things: not only on the true unknown value of the parameters of interest but also on the sample size. The bootstrap is a method for estimating the distribution of an estimator, and in particular its expectation, by re-sampling the data. We refer the reader to Hall (1992) and Horowitz (1997) for surveys from which we borrow here. Since the bootstrap estimate is built upon an estimator of the data distribution, it is always recommended to use a parametric estimator of it when available. This is why, since this article is about estimation in parametric models, we focus on the parametric bootstrap, which was first considered in econometrics for the linear regression model with Gaussian errors:

$$y_i = z_i'b + \sigma\varepsilon_i, \qquad \varepsilon_i \sim \mathrm{IIN}(0,1).$$

Of course, bootstrapping is not very useful in such a simple context but it may become relevant if for instance the dependent variable is replaced by its Box–Cox transformation with some unknown parameters (Horowitz, 1997). More generally we allow for any kind of non-linear transformation and also for dynamic models in reduced form, possibly including lagged endogenous variables among the explanatory variables (see Monfort and Van Dijk, 1995 for a thorough exposition of the general framework presented below):

$$y_t = r[z(1,t), y(0,t-1), \varepsilon_t; \theta], \qquad t = 1, \ldots, T \tag{1}$$

where (ε_t) is a white-noise process whose marginal distribution P_ε is known, (z_t) is a process which is independent of (ε_t), $z(1,t) = (z_\tau)_{1 \leq \tau \leq t}$, $y(0,t-1) = (y_\tau)_{1 \leq \tau \leq t-1}$ and $r(\cdot)$ is a known function. Model (0) defines the conditional pdf $f[y_t|x_t; \theta]$ where x_t includes the realization of all the predetermined variables $z(1,t)$ and $y(0,t-1)$. Then, by drawing independently from P_∂ it is possible to simulate values ε_t^h, $t=1, \ldots, T$ and

$h = 1, \ldots, H$ and to compute:

$$y_t^h(\theta) = r[z(1,t), y(0, t-1), \varepsilon_t^h; \theta],$$
$$t = 1, \ldots, T, \quad h = 1, \ldots, H.$$

The pdf of $y_t^h(\theta)$ is precisely $f[y_t | x_t; \theta]$. In other words, it is possible to perform *conditional simulations*, that is, to draw, for each t, from the conditional distribution whose pdf is $f[y_t | x_t; \theta]$ for any given value θ of the unknown parameters and for the observed value of x_t. Note that, in all simulation-based estimation methods considered below, the basic drawings ε_t^h will be kept fixed when θ changes.

For the purpose of SMM, it will actually be worthwhile making a distinction between such *conditional simulations* and (unconditional) *path simulations*, which may be the only ones feasible in the more general case of a non-linear state-space model defined as:

$$y_t = r_1[z(1,t), y(0, t-1), y^*(0, t), \varepsilon_{1t}; \theta], \quad t = 1, \ldots, T$$
$$y_t^* = r_2[z(1,t), y(0, t-1), y^*(0, t-1), \varepsilon_{2t}; \theta], \quad t = 1, \ldots, T$$

$$(2)$$

where $\varepsilon_t = (\varepsilon_{1t}', \varepsilon_{2t}')'$ is a white-noise process whose marginal distribution P_ε is known, (z_t) is independent of (ε_t), (y_t^*) is a process of latent variables and r_1 and r_2 are two known functions. The big difference between models (1) and (2) is that the latter only recursively defines the observed endogenous variables through a path of latent ones, making conditional simulation impossible. More precisely, from independent random draws ε_t^h, $t = 1, \ldots, T$ and $h - 1, \ldots H$ in P_ε we can now compute recursively:

$$y_t^{*h}(\theta) = r_2[z(1,t), y^h(0, t-1)(\theta), y^{*h}(0, t-1)(\theta), \varepsilon_{2t}^h; \theta],$$
$$t = 1, \ldots, T \; ; h = 1, \ldots H.$$

$$y_t^h(\theta) = r_1[z(1,t), y^h(0, t-1)(\theta), y^{*h}(0, t)(\theta), \varepsilon_{1t}^h; \theta],$$
$$t = 1, \ldots, T \; ; h = 1, \ldots H.$$

In other words, while each simulated path $y^h(0, T)(\theta)$, $h = 1, \ldots H$ has been correctly drawn from its distribution given the observed path $z(1, T)$ of exogenous variables for each possible value of θ, the draw of $y_t^h(\theta)$ at each given t is conditional to past *simulated* $y^h(0, t-1)(\theta)$ and not to past *observed* $y(0, t-1)$: hence the terminology *path simulations*. Note, however, that the model does not specify the probability distribution of exogenous variables and thus, all simulations are conditional to the observed path $z(1, T)$ of exogenous variables.

In both cases, model (1) or (2), since the spirit of bootstrap is re-sampling from a preliminary estimator, $\hat{\beta}_T$ gives rise to H bootstrap samples $y^h(0, T)(\hat{\beta}_T)$, $h = 1, \ldots H$. For each bootstrap sample, the same estimation procedure can be applied to get H estimators denoted as $\hat{\beta}_T^h(\hat{\beta}_T)$, $h = 1, \ldots H$. These H estimations characterize

the bootstrap distribution of $\hat{\beta}_T$ and allow us for instance to approximate the expectation $b_T(\theta^0)$ by $b_T(\hat{\beta}_T)$. Of course, $b_T(\hat{\beta}_T)$ is not known in general but may be approximated at any desired level of accuracy, from the Monte Carlo average $(1/H)\sum_{h=1}^{H} \hat{\beta}_T^h(\hat{\beta}_T)$, for H sufficiently large. As already mentioned, one may imagine non-simulation based versions of bootstrap when the binding function is available in closed form. In any case, the bias-corrected bootstrap estimator is then defined as:

$$\hat{\theta}_T = \hat{\beta}_T - [b_T(\hat{\beta}_T) - \hat{\beta}_T] \tag{3}$$

It is worth mentioning however that this parametric bootstrap procedure requires that we sufficiently trust the initial estimator $\hat{\beta}_T$ to consider that the estimated bias $[b_T(\hat{\beta}_T) - \hat{\beta}_T]$ gives a correct assessment of the true bias $[b_T(\theta^0) - \theta^0]$. This is the reason why Gourieroux, Renault and Touzi (2000) have rather proposed an iterative procedure which, at step j, improves upon an estimator $\hat{\theta}_T^j$ by computing $\hat{\theta}_T^{j+1}$ as:

$$\hat{\theta}_T^{j+1} = \hat{\theta}_T^j + \lambda[\hat{\beta}_T - b_T(\hat{\theta}_T^j)] \tag{4}$$

for some given updating parameter λ between 0 and 1. In other words, at each step, a new set of simulated paths $y^h(0, T)(\hat{\theta}_T^j)$, $h = 1, \ldots H$ is built and it provides a Monte Carlo assessment $b_T(\hat{\theta}_T^j)$ of the expectation of interest. It is worth noting that this does not involve new random draws of the noise ε. Note that (3) corresponds to the first iteration of (4) in the particular case $\lambda = 1$ with a starting value $\hat{\theta}_T^1 = \hat{\beta}_T$. While this preliminary estimator is indeed a natural starting value, the rationale for considering λ smaller than 1 is to increase the probability of convergence of the algorithm, possibly at the cost of slower convergence (if faster update would also work). If the algorithm converges, its limit will define an estimator $\hat{\theta}_T$ solution of:

$$b_T(\hat{\theta}_T) = \hat{\beta}_T \tag{5}$$

Gourieroux, Renault and Touzi (2000) study more generally the properties of the estimator (5), which is actually a particular case of SII estimators developed in the next section. The intuition is quite clear. Let us call $\hat{\beta}_T$ the naive estimator. Our preferred estimator $\hat{\theta}_T$ is the value of unknown parameters θ, which, if it had been the true one, would have generated a naive estimator which, in average, would have coincided with our actual naive estimation. In particular, if the bias function $[b_T(\theta) - \theta]$ is linear with respect to θ, we deduce $b_T[E(\hat{\theta}_T)] = E(\hat{\beta}_T) = b_T(\theta^0)$ and thus our estimator is unbiased. Otherwise, unbiasedness is only approximately true to the extent a linear approximation of the bias is reasonable. Since, in the context of stationary first order autoregressive processes, the negative bias of the OLS estimator of the correlation coefficient becomes more severely non-linear in the near unit root case, Andrews (1993) has put forward a median-unbiased estimator based on the principle (5) with median replacing expectation. The advantage of median is to be immune to nonlinear monotonic transformations. However, its generalization to multivariate parameters is problematic.

Another well-documented advantage of bootstrap is to provide asymptotic refinements when the initial procedure is not too bad. Gourieroux, Renault and Touzi (2000) have shown that SII does as well as bootstrap in this respect.

2 Simulation-based indirect inference

Let us start from the simple textbook example of a just-identified supply–demand system in equilibrium:

$$Q_t^s = \theta_1 p_t + \theta_2 z_{1t} + u_{1t}$$
$$Q_t^d = \theta_3 p_t + \theta_4 z_{2t} + u_{2t}$$
$$Q_t^s = Q_t^d = Q_t.$$

Then the reduced form can obviously be written as a bivariate regresssion of (Q_t, p_t) on (z_{1t}, z_{2t}) and the reduced form regression coefficients β are given as a function $\beta = b(\theta)$ of the structural parameters:

$$\theta = (\theta_1, \theta_2, \theta_3, \theta_4)$$
$$b(\theta) = (\theta_1 - \theta_3)^{-1}(-\theta_2\theta_3, \theta_1\theta_4, -\theta_2, \theta_4).$$

Under standard assumptions, the vector β of reduced form parameters can be consistently estimated by its OLS counterpart $\hat{\beta}_T$. Moreover, the binding function $\beta = b(\theta)$ relating the vector β of reduced form parameters to the vector θ of structural parameters is clearly one-to-one. Inverting the binding function is a straightforward exercise and suggests computing a consistent estimator $\hat{\theta}_T$ of the structural parameters as $\hat{\theta}_T = b^{-1}(\hat{\beta}_T)$. This estimator has been known since the early days of the simultaneous equations literature as the *indirect least squares* estimator. We conclude from this example that defining an indirect estimator $\hat{\theta}_T$ of the parameters of interest from an initial estimator $\hat{\beta}_T$ and a binding function $b_T(\cdot)$ by solving the equation:

$$b_T(\hat{\theta}_T) = \hat{\beta}_T \qquad (6)$$

may be worthwhile in many situations other than the bias-correction setting of Section 1. The vector β of the so-called *instrumental parameters* must identify the *structural parameters* θ but does not need to bear the same interpretation. However, the example of *indirect least squares* is too simple to display all the features of *indirect inference* as more generally devised by Smith (1993) and Gourieroux, Monfort and Renault (1993). Two complications may arise.

First, the binding function is not in general available in closed form and can be characterized only thanks to Monte Carlo integration. Moreover, by contrast with the simple linear example, the binding function does depend in general on the sample size T.

Second, most interesting examples allow for over-identification of the structural parameters, for instance through a bunch of instrumental variables in the simultaneous equation case. This is the reason why we refer henceforth to the auxiliary parameters β as instrumental parameters.

The key idea is that, as already explained in Section 1, our preliminary estimation procedure for instrumental parameters not only gives us an estimation $\hat{\beta}_T$ computed from the observed sample path but can also be applied to each simulated path $y^h(0, T)(\theta)$, $h = 1, \ldots H$, always associated to the observed path $z(1, T)$ of exogenous variables. Thus, we end up, for any fixed value of θ, with a set of H 'estimations' $\beta_T^{(h)}(\theta)$. Averaging them, we get a Monte Carlo binding function:

$$\beta_{T,H}(\theta) = (1/H) \sum_{h=1}^{H} \beta_T^{(h)}(\theta).$$

The exact generalization of what we did in Section 1 amounts to defining the binding function $b_T(\theta)$ as the probability limit (w.r.t. the random draw of the process ε) of the sequence $\beta_{T,H}(\theta)$ when H goes to infinity. However, for most non-linear models, the instrumental estimators $\beta_T^{(h)}(\theta)$ are not reliable for finite T but only for a sample size T going to infinity. It is then worth realizing that when T goes to infinity, for any given $h = 1, \ldots, H$, $\beta_T^{(h)}(\theta)$ should tend towards the so-called *asymptotic binding* function $b(\theta)$ which is also the limit of the *finite sample binding* function $b_T(\theta)$.

Therefore, as far as consistency of estimators when T goes to infinity is concerned, a large number H of simulations is not necessary and we will define more generally an *indirect estimator* $\hat{\theta}_T$ as solution of a minimum distance problem:

$$\text{Min}_\theta [\hat{\beta}_T - \beta_{T,H}(\theta)]' \Omega_T [\hat{\beta}_T - \beta_{T,H}(\theta)]$$

$$(7)$$

where Ω_T is a positive definite matrix converging towards a deterministic positive definite matrix Ω. In case of a completed Monte Carlo integration (H large) we end up with an approximation of the exact binding function-based estimation:

$$\text{Min}_\theta [\hat{\beta}_T - b_T(\theta)]' \Omega_T [\hat{\beta}_T - b_T(\theta)] \qquad (8)$$

which generalizes the bias-correction procedure of Section 1. As in Section 1, we may expect good finite sample properties of such an indirect estimator since, intuitively, the finite sample bias is similar in the two quantities, which are matched against each other and thus should cancel out in the matching process.

In terms of asymptotic theory, the main results under standard regularity conditions (see Gourieroux, Monfort and Renault, 1993) are:

(i) the indirect inference estimator $\hat{\theta}_T$ converges towards the true unknown value θ^0 insofar as the asymptotic binding function identifies it:

$$b(\theta) = b(\theta^0) \Rightarrow \theta = \theta^0;$$

(ii) the indirect inference estimator $\hat{\theta}_T$ is asymptotically normal insofar as the asymptotic binding function first order identifies the true value:

$$\frac{\partial b}{\partial \theta'}(\theta^0) \text{ is of full-column rank;}$$

(iii) we get an indirect inference estimator with a minimum asymptotic variance if and only if the limit-weighting matrix is proportional to the inverse of the asymptotic variance Σ_H of $\sqrt{T}[\hat{\beta}_T - \beta_{T,H}(\theta^0)]$;

(iv) the asymptotic variance of the efficient indirect inference estimator is the inverse of

$$\frac{\partial b'}{\partial \theta}(\theta^0)(\Sigma_H)^{-1}\frac{\partial b}{\partial \theta'}(\theta^0)$$

with

$$\Sigma_H = \left[1 + \frac{1}{H}\right]\Sigma_\infty.$$

An implication of these results is that, as far as asymptotic variance of the indirect inference estimator is concerned, the only role of a finite number H of simulations is to multiply the optimal variance (obtained with $H = \infty$) by a factor $(1+1/H)$. Actually, when computing the indirect inference estimator (7), one may be reluctant to use a very large H since it involves, for each value of θ along a minimization algorithm, computing H instrumental estimators $\beta_T^{(h)}(\theta)$, $h = 1, \ldots, H$. We will see in Section 3 several ways to replace these H computations by only one. However, this will come at a price, which is the probable loss of the nice finite sample properties of (7) and (8).

As a conclusion, let us stress that indirect inference is able, beyond finite sample biases, to correct for any kind of misspecification bias. The philosophy of this method is basically to estimate a simple model, possibly wrong, to get easily an instrumental estimator $\hat{\beta}_T$ while a direct estimation of structural parameters θ would have been a daunting task. Therefore what really matters is to use an instrumental parameter that captures the key features of the parameters of interest, while being much simpler to estimate. For instance, Pastorello, Renault and Touzi (2000) and Engle and Lee (1996) have proposed to first estimate a GARCH model as an instrumental model to indirectly recover an estimator of the structural model of interest, a stochastic volatility model much more difficult

to estimate directly. Other natural examples are models with latent variables such that an observed variable provides a convenient proxy. An estimator based on this proxy suffers from a misspecification bias, but we end up with a consistent estimator by applying the indirect inference matching. Examples of this approach are:

(i) Pastorello, Renault and Touzi (2000), who use Black and Scholes implied volatilities as a proxy of realizations of the latent spot volatility process.
(ii) Li (2006), who, following a suggestion of Renault (1997), uses observed bids in an auction market as a proxy of latent private values.

3 Simulated method of moments

SMM, as introduced by Ingram and Lee (1991) and Duffie and Singleton (1993), is the simulation-based counterpart of GMM to take advantage of the informational content of some conditional moment restrictions:

$$E\{K[y_t, z(1,t)]|z(1,t)\} = k[z(1,t); \theta^0].$$
$$(9)$$

The role of simulations in this context is to provide a Monte Carlo assessment of the population conditional moment function $k[z(1,t); \theta]$ when it is not easily available in closed form. Typically, with $y_t^h(\theta)$ drawn as above for $h = 1, \ldots, H$ (model (1) or (2)), a convenient Monte Carlo counterpart is: $(1/H)\sum_{h=1}^{H}K[y_t^h(\theta), z(1,t)]$. Even though we will mainly present SMM in this simple setting, two possible extensions are worth mentioning.

1. First, in dynamic settings, one may want to consider conditional moment restrictions given not only past and current exogenous variables but also past endogenous variables:

$$E\{K[y_t, z(1,t), y(0,t-1)]|z(1,t), y(0,t-1)\}$$
$$= k[z(1,t), y(0,t-1); \theta^0].$$

SMM can still be applied to this kind of dynamic moment insofar as one is able to draw simulated values $y_t^h(\theta)$ in the conditional probability distribution (corresponding to the value θ of parameters) of y_t given $x_t = [z(1,t), y(0,t-1)]$. In other words, we need conditional simulations and not only path simulations. As explained above, such conditional simulations are not feasible when the structural model is only defined through a recursive form (2). By contrast, either path simulations or conditional simulations work for static moment conditions (9).

2. Second, the introduction of latent variables paves the way for many other Monte Carlo assessments of the population expectation $E\{K[y_t, z(1,t)]|z(1,t)\}$. Let us assume to simplify that $y_t = r_1(y_t^*)$ with i.i.d. latent variables y_t^* endowed with a probability distribution with fixed support (independent of the unknown parameters

θ). Of course, the density function $f[y_t^*|z(1,t);\theta]$ does depend on θ but one may also pick, as a sampling tool called importance function, another distribution with a given density function $\varphi(u_t)$ on the same support. Then, instead of assessing the population expectation with its naive Monte Carlo counterpart

$$(1/H)\sum_{h=1}^{H} K[r_1(y_t^{*h}(\theta)), z(1,t)]$$

one may prefer to resort to

$$(1/H)\sum_{h=1}^{H} K[r_1(u_t^h), z(1,t)]\frac{f[u_t^h|z(1,t);\theta]}{\varphi(u_t^h)}$$

where the u_t^h, $h=1, \ldots, H$, are independently drawn from the distribution with density function $\varphi(u_t)$. This kind of *importance sampling* may be helpful, for instance, for removing some nasty lack of smoothness with respect to the unknown parameters.

As far as static conditional moment restrictions like (9) are concerned, the natural way to extend GMM with a Monte Carlo assessment of the population moment is to minimize with respect to the unknown parameters θ a norm of the sample mean of:

$$Z_t\left\{K[y_t, z(1,t) - (1/H)\sum_{h=1}^{H} K[y_t^h(\theta), z(1,t)]\right\}$$

where Z_t is a matrix of chosen instruments, that is a fixed matrix function of $z(1,t)$. It is then clear that the minimization programme which is considered is a particular case of (7) above with:

$$\hat{\beta}_T = (1/T)\sum_{t=1}^{T} Z_t K[y_t, z(1,t)]$$

and $\beta_{T,H}(\theta)$ defined accordingly. In other words, we reinterpret SMM as a particular case of *indirect inference*, when the instrumental parameters to match are simple moments rather than themselves defined through some structural interpretations. Note, however, that the moment conditions for SMM could be slightly more general since the function $K[y_t, z(1,t)]$ itself could depend on the unknown parameters θ. In any case, the general asymptotic theory sketched above for SII is still valid. It may be a little more involved when using importance sampling, since then the variance of the simulator no longer coincides with the variance of the initial moments, and then computing the asymptotic variance of the SMM estimator is no longer simply akin to multiplying standard formulas by a factor $[1+1/H]$. However, we still note that the number H of simulated paths does not need to be large for getting consistent and rather accurate estimators.

In contrast with general SII as presented above, an advantage of SMM is that the instrumental parameters to match, as simple moments, are in general easier to compute than estimated auxiliary parameters $\beta_T^{(h)}(\theta)$, $h=1, \ldots, H$, derived from some computationally demanding *extremum* estimation procedure. Gallant and Tauchen (1996) have taken advantage of this remark to propose a practical computational strategy for implementing indirect inference when the estimator $\hat{\beta}_T$ of the instrumental parameters is obtained as an M-estimator solution of:

$$\text{Max}_\beta(1/T)\sum_{t=1}^{T} q_t[y(0,t), z(1,t);\beta].$$

The key idea is then to define the moments to match through the (pseudo)-score vector of this M-estimator. Let us denote

$$K[y(0,t), z(1,t);\beta] = \frac{\partial q_t}{\partial \beta}[y(0,t), z(1,t);\beta]$$

and consider an SMM estimator of θ obtained as a minimizer of the norm of a sample mean of:

$$K[y(0,t), z(1,t);\hat{\beta}_T] - (1/H)$$
$$\sum_{h=1}^{H} K[y^h(0,t)(\theta), z(1,t);\hat{\beta}_T].$$

For a suitable GMM metric, such a minimization defines a so-called *simulated-score matching* estimator $\hat{\theta}_T$ of θ. In the spirit of Gallant and Tauchen (1996), the objective function that defines the initial estimator $\hat{\beta}_T$ is typically the log-likelihood of some auxiliary model. However, this feature is not needed for the validity of the asymptotic theory sketched below. Several remarks are in to order.

1. By contrast with a general SMM criterion, the minimization above does not involve the choice of any instrumental variable. Typically, over-identification will be achieved by choosing an auxiliary model with a large number of instrumental parameters β rather than by choosing instruments.

2. By definition of $\hat{\beta}_T$, the sample mean of $K[y(0,t), z(1,t);\beta]$ takes the value zero for $\beta = \hat{\beta}_T$. In other words, the minimization programme above amounts to:

$$\text{Min}_\theta\left\|(1/TH)\sum_{t=1}^{T}\sum_{h=1}^{H}\frac{\partial q_t}{\partial \beta}[y^h(0,t)(\theta), z(1,t);\hat{\beta}_T]\right\|\Omega_T$$

$$(10)$$

for a suitable GMM metric Ω_T.

3. It can be shown (see Gourieroux, Monfort and Renault, 1993) that under the same assumptions as for the asymptotic theory of SII, the *score-matching* estimator is consistent asymptotically normal. We get a score-matching estimator with a minimum asymptotic variance if and only if the limit-weighting matrix Ω is proportional to the inverse of the asymptotic conditional

variance of

$$\sqrt{T}\sum_{t=1}^{T}\frac{\partial q_t}{\partial \beta}[y(0,t),z(1,t);b(0^0)]$$

given the exogenous variables z. Then the resulting *efficient score-matching estimator* is asymptotically equivalent to the *efficient indirect inference estimator*.

4. Owing to this asymptotic equivalence, the score-matching estimator can be seen as an alternative to the efficient SII estimator characterized in Section 2. This alternative is often referred to as *efficient method of moments* (EMM) since, when $q_t[y(0,t),z(1,t);\beta]$ is the log-likelihood $f^a[y_t|z(1,t),y(0,t-1);\beta]$ of some auxiliary model, the estimator is as efficient as maximum likelihood under correct specification of the auxiliary model. More generally, the auxiliary model is designed to approximate the true data generating process as closely as possible and Gallant and Tauchen (1996) propose the *semi-nonparametric* (SNP) modelling to this end. These considerations and the terminology EMM should not lead us to believe that *score-matching* is more efficient than *indirect inference*. The two estimators are asymptotically equivalent even though the score-matching approach makes more transparent the required spanning property of the auxiliary model to reach the Cramer Rao efficiency bound of the structural model.

5. Another alleged advantage of the *score-matching* with respect to *parameter-matching* in SII is its low computational cost. The fact is that with a large number of instrumental parameters β, as will typically be the case with a SNP auxiliary model, it may be costly to maximize H times the log-likelihood of the auxiliary model (for each value of θ along an optimization algorithm) with respect to β to compute $\beta_T^{(h)}(\theta)$, $h=1,\ldots,H$. By contrast, the programme (10) minimizes only once the norm of a vector of derivatives with respect to β. One must realize, however, that not only is this cheaper computation likely to lose the expected nice finite sample properties of SII put forward in the previous section, but also that the point is not really about a choice between matching (instrumental) parameters β or matching the (instrumental) score $\sum_{t=1}^{T}\frac{\partial q_t}{\partial \beta}$. The key issue is rather the way to use H simulated paths of size T each as explained below.

6. It is worth realizing that the sum of TH terms considered in the definition (10) of the score-matching estimator is akin to consider only one simulated path $y^1(0,TH)(\theta)$ of size TH built from random draws as above (conditional simulations or path simulations) from a fictitious path $z^*(1,TH)$ of exogenous variables defined in the following way:

$$z_1^* = z_1,\ldots,z_T^* = z_T, z_{T+1}^* = z_1,\ldots,$$
$$z_{2T}^* = z_T, z_{2T+1}^* = z_1,\ldots,z_{TH}^* = z_T.$$

If for instance (z_t) is Markov of order 1, such a fictitious

path is a correct draw except possibly for H values, which is immaterial when T goes to infinity. From such a simulated path, estimation of instrumental parameters would have produced a vector $\beta_{TH}^{(1)}(\theta)$ that could have been used for *indirect inference*, that is to define an estimator $\hat{\theta}_T$ solution of:

$$\text{Min}_\theta[\hat{\beta}_T - \beta_{TH}^{(1)}(\theta)]'\Omega_T[\hat{\beta}_T - \beta_{TH}^{(1)}(\theta)]. \tag{11}$$

This *parameter-matching* estimator is not more computationally demanding than the corresponding *score-matching* estimator computed from the same simulated path as solution of:

$$\text{Min}_\theta\left\|(1/TH)\sum_{t=1}^{TH}\frac{\partial q_t}{\partial \beta}[y^1(0,t)(\theta),z^*(1,t);\hat{\beta}_T]\right\|\Omega_T. \tag{12}$$

Actually, (11) and (12) are numerically identical in the case of just-identification (dim θ = dim β). Then the choice of a GMM metric is immaterial and both estimators are basically the solution $\hat{\theta}_T$ of:

$$(1/TH)\sum_{t=1}^{TH}\frac{\partial q_t}{\partial \beta}[y^1(0,t)(\hat{\theta}_T),z^*(1,t);\hat{\beta}_T] = 0.$$

More generally, the four estimators (7), (10), (11) and (12) are asymptotically equivalent when T goes to infinity and the GMM weighting matrix is efficiently chosen accordingly. However, it is quite obvious that only (7) performs the right finite sample bias correction by matching instrumental parameters values estimated on both observed and simulated paths of lengths T. The trade-off is thus between giving up finite sample bias correction or paying the price for computing H estimated instrumental parameters.

4 Simulated M-estimators

Even though well-chosen moments to match may allow one to get accurate estimators, it is somewhat contradictory to resort to a fully parametric model to perform simulations needed for SMM while Hansen's (1982) GMM was semi-parametric in spirit. *Simulated maximum likelihood* (SML) methods aim at exploiting the whole parametric structure for efficient estimation. The key role of simulations would then be to provide an unbiased simulator of each conditional p.d.f. $f[y_t|z(1,t),y(0,t-1);\theta]$ (also denoted $f[y_t|x_t;\theta]$) because it is not available in closed form. This is typically the case in a model with latent variables, and then conditioning may provide a convenient simulator. The simplest example comes from model (2) when latent variables are exogenous. Let us write it without observed exogenous variables for sake of expositional simplicity:

$$y_t = r_1[y(0, t-1), y^*(0, t), \varepsilon_{1t}; \theta], \qquad t = 1, \ldots, T$$
$$y_t^* = r_2[y^*(0, t-1), \varepsilon_{2t}; \theta], \qquad t = 1, \ldots, T.$$

$$(13)$$

Then $f[y_t | x_t; \theta]$ is nothing but the expectation with respect to the probability distribution of $y^*(0, t)$ of $f[y_t | x_t, y^*(0, t); \theta]$. While the latter is easily deduced from the pdf of ε_{1t} through the measurement equation (the first equation above), the former is in general easy to compute from the evolution equation (the second equation above). Then Monte Carlo integration provides an unbiased estimate of $f[y_t | x_t; \theta]$ with $(1/H) \sum_{h=1}^{H} f[y_t | x_t, y^{*h}(0, t)(\theta); \theta]$ where $y^{*h}(0, t)(\theta)$, $h = 1, \ldots, H$, are independent draws obtained from independent draws in the known distribution of ε_{2t}. Of course, importance sampling must also be relevant in this context. Moreover, for a general model (2), when y does cause y^*, this naive approach may be very inefficient (in terms of speed of convergence of the variance towards zero) and one may want to refer to either accelerated versions of importance sampling (Danielsson and Richard, 1993), simulated expectation maximization algorithm (SEM, see for example Shephard, 1993) and other more sophisticated simulators, which are beyond the scope of this article.

Irrespective of the choice of the simulator, the main difficulty with SML is that the logarithm of an unbiased simulator of the likelihood is not an unbiased simulator of the log-likelihood. Therefore, in contrast with SMM, SML is consistent only when both H and T go to infinity. Note, however, that asymptotic bias corrections are possible (see Gourieroux and Monfort, 1996). Under standard regularity conditions, SML is asymptotically efficient insofar as T goes to infinity slower than H^2. Another version of the SML method is the *simulated score* method (see Hajivassiliou, 1993). This is basically a version of SMM where the latent score is used as a simulator for the score of the model on observables, while the later is not tractable analytically. This method should not be confused with *simulated score matching* of the former section, where the score to match was not something to simulate to get it in closed form.

Finally, it is worth mentioning that simulation-based estimation methods are not always to be recommended, even in some fully parametric modelling situations with latent variables making standard maximum likelihood unfeasible. Such situations typically occur with structural econometric models where equilibrium of a market or of a game implies a deterministic relationship between latent variables and observed ones. Then, not only is the above SML theory no longer valid but any other simulation based method will in general be highly inefficient because the dependence between the two blocks of unknown 'parameters', namely, structural parameters and latent variables, is sharp. In such contexts, some authors have, however, considered some simulated non-linear least squares methods (Laffont, Ossard and Vuong, 1995 for auction markets) or more generally simulated pseudo-likelihood methods (Laroque and Salanie, 1993). While the focus of interest of Laroque and Salanie (1993) on a disequilibrium model raised more involved issues due to non-differentiability, it is rather clear that some implied state GMM methods (Pan, 2002; Pastorello, Patilea and Renault, 2003) are more efficient than SMM in the contexts of smooth equilibrium relationships like those produced by option pricing theory. The key issue is to take advantage of the deterministic relationship between y_t and y_t^* (for a given value of the parameters θ) to track what would have been an efficient estimation if latent variables had been observed.

5 Concluding remarks

The econometrician's search for a well-specified parametric model ('quest for the Holy Grail' as stated by Monfort, 1996) and associated efficient estimators even remain popular when MLE becomes intractable due to highly non-linear dynamic structure including latent variables. The efficiency properties of SML, EMM and more generally of SMM and SII when the set of instrumental parameters to match is sufficiently large to span the likelihood scores are often advocated as if the likelihood score is well specified. However, the likely misspecification of the structural model requires a generalization of the theory of SII as recently proposed by Dridi, Guay and Renault (2007). As for MLE with misspecification (see White, 1982; Gourieroux, Monfort and Trognon, 1984) such a generalization entails two elements.

First, asymptotic variance formulas are complicated by the introduction of sandwich formulas. Ignoring this kind of correction is even more detrimental than for QMLE since two types of sandwich formulas must be taken into account, one for the data generating process (DGP) and one for the simulator (based either on model (1) or model (2)) which turns out to be different from the DGP in case of misspecification.

Secondly, and even more importantly, misspecification may imply that we consistently estimate a pseudo-true value, which is poorly related to the true unknown value of the parameters of interest. Dridi, Guay and Renault (2007) put forward the necessary (partial) encompassing property of the instrumental model (through instrumental parameters β) by the structural model (with parameters θ) needed to ensure consistency towards true values of (part of) the components of θ in spite of misspecification. The key issue is that, since structural parameters are recovered from instrumental ones by inverting a binding function $\beta = b(\theta)$, all components are interdependent. The requirement of encompassing typically means that, if one does not want to proceed under the maintained assumption that the structural model (1) or (2) is true, one must be parsimonious with respect to the number of moments to match or more generally to the scope of empirical evidence that is captured by the instrumental parameters β. In other words, robustness to misspecification requires an instrumental model choice strategy opposite to that commonly used for a structural model: the larger the instrumental model, the larger the risk of contamination of the estimated structural parameters of interest by what is wrong in the structural model. Of course, there is no such thing as a free lunch: robustness to misspecification through a parsimonious and well-focused instrumental model comes at the price of efficiency loss. Efficiency loss means not only lack of accuracy of estimators of structural parameters but also lack of power of specification tests.

ERIC RENAULT

See also **bootstrap; Markov chain Monte Carlo methods; simulation estimators in macroeconometrics; state space models; stochastic volatility models.**

Bibliography

Andrews, D. 1993. Exactly median unbiased estimation of first order autoregressive/unit root models. *Econometrica* 61, 139–65.

Danielsson, J. and Richard, J.F. 1993. Accelerated Gaussian importance sampler with application to dynamic latent variables models. *Journal of Applied Econometrics* 8, S153–73.

Dridi, R., Guay, A. and Renault, E. 2007. Indirect inference and calibration of dynamic stochastic general equilibrium models. *Journal of Econometrics* 136, 397–430.

Duffie, D. and Singleton, K. 1993. Simulated moments estimation of Markov models of asset prices. *Econometrica* 61, 929–52.

Engle, R.F. and Lee, G.G.J. 1996. Estimating diffusion models of stochastic volatility. In *Modeling Stock Market Volatility: Bridging the Gap to Continuous Time*, ed. P.E. Rossi. New York: Academic Press.

Gallant, A.R. and Tauchen, G. 1996. Which moments to match? *Econometric Theory* 12, 657–81.

Gourieroux, C. and Monfort, A. 1995. Testing, encompassing, and simulating dynamic econometric models. *Econometric Theory* 11, 195–228.

Gourieroux, C. and Monfort, A. 1996. Simulation-based econometric methods. *Core Lectures*. Oxford: Oxford University Press.

Gourieroux, C., Monfort, A. and Renault, E. 1993. Indirect inference. *Journal of Applied Econometrics* 8, S85–118.

Gourieroux, C., Monfort, A. and Trognon, A. 1984. Pseudo-maximum likelihood methods theory. *Econometrica* 52, 681–700.

Gourieroux, C., Renault, E. and Touzi, N. 2000. Calibration by simulation for small sample bias correction. In *Simulation-Based Inference in Econometrics*, ed. R. Mariano, T. Schuerman and M.J. Weeks. Cambridge: Cambridge University Press.

Hajivassiliou, V.A. 1993. Simulation estimation methods for limited dependent variable models. In *Handbook of Statistics*, vol. 11, ed. G.S. Maddala, C.R. Rao and H. Vinod. Amsterdam: North-Holland.

Hall, P. 1992. *The Bootstrap and Edgeworth Expansion*. New York: Springer.

Hansen, L.P. 1982. Large sample properties of generalized method of moments. *Econometrica* 50, 1029–54.

Horowitz, J.L. 1997. Bootstrap methods in econometrics: theory and numerical performance. In *Advances in Econometrics, seventh world congress*, ed. D. Kreps and K. Wallis. Cambridge: Cambridge University Press.

Ingram, B.F. and Lee, B.S. 1991. Estimation by simulation of time series models. *Journal of Econometrics* 47, 197–207.

Laffont, J.J., Ossard, H. and Vuong, Q. 1995. Econometrics of first-price auction. *Econometrica* 63, 953–80.

Laroque, G. and Salanie, B. 1993. Simulation based estimation of models with lagged latent variables. *Journal of Applied Econometrics* 8, S119–33.

Li, T. 2006. Indirect inference in structural econometric models. *Journal of Econometrics*, forthcoming.

Monfort, A. 1996. A reappraisal of misspecified econometric models. *Econometric Theory* 12, 597–619.

Monfort, A. and Van Dijk, H.K. 1995. Simulation-based econometrics. In *Econometric Inference Using Simulation Techniques*, ed. H.K. Van Dijk, A. Monfort and B.W. Brown. New York: Wiley.

Pan, J. 2002. The jump-risk premia implicit in options: evidence from an integrated time-series study. *Journal of Financial Economics* 63, 3–50.

Pastorello, S., Patilea, V. and Renault, E. 2003. Iterative and recursive estimation in structural non-adaptive models. Invited Lecture with discussion. *Journal of Business, Economics and Statistics* 21, 449–509.

Pastorello, S., Renault, E. and Touzi, N. 2000. Statistical inference for random-variance option pricing. *Journal of Business and Economic Statistics* 18, 358–67.

Renault, E. 1997. Econometric models of option pricing errors. In *Advances in Econometrics, Seventh World Congress*, ed. D. Kreps and K. Wallis. Cambridge: Cambridge University Press.

Shephard, N. 1993. Fitting nonlinear time series with applications to stochastic variance models. *Journal of Applied Econometrics* 8, S134–52.

Smith, A. 1993. Estimating nonlinear time series models using simulated vector autoregressions. *Journal of Applied Econometrics* 8, S63–84.

White, H. 1982. Maximum likelihood estimation of misspecified models. *Econometrica* 50, 1–25.

simulation estimators in macroeconometrics

The method of simulated moments (MSM) in macroeconometrics is a method for estimating the parameters of a dynamic, stochastic, general equilibrium (DSGE) model using simulated solution paths for the model's observable variables.

Regardless of the complexity of the DSGE model, implementation of MSM requires only simulated data from that model. Moments of the simulated model are then formally matched to moments of the observed data, and the parameter vector that minimizes the distance between the model and the empirical moments is the MSM estimator. MSM estimates typically have nice statistical properties, including consistency and asymptotic normality.

In general, the solution to a DSGE can be characterized as a vector stochastic process dependent upon $k \times 1$ parameter vector, θ, and $m \times 1$ fundamental shock vector $\{\varepsilon_n, n \geq 0\}$. Consider the simple stochastic growth model in which a representative agent chooses consumption, c_t, labour supply, h_t, and capital, k_t, to maximize expected utility subject to a resource constraint and endowment constraints:

$$\max_{\{c_n, k_n, h_n > 0\}} E_0 \sum_{n=1}^{\infty} \beta^n \ln(c_n)$$

$$s.t. \quad c_n + k_n = A_n k_{n-1}^{\alpha} h_n^{1-\alpha}$$

$$h_n \leq 1 \text{ and } k_0 \text{ given.}$$

Here, $\{A_n, n \geq 0\}$ is an exogenous stochastic process; for example:

$$A_n = \rho A_{n-1} + \varepsilon_n, \quad \varepsilon_n \sim N(0, \sigma^2). \quad (1.1)$$

In this example, the parameter vector is comprised of the four parameters in the model, $\theta = [\alpha \ \beta \ \rho \ \sigma^2]$, and the fundamental shock is $\{\varepsilon_n, n \geq 1\}$. The solution to the model is a collection of stochastic processes, $\{c_n, k_n, h_n, n \geq 1\}$ that depend on θ, k_0, and $\{\varepsilon_n, n \geq 1\}$.

It is assumed that a method exists for generating a realization of the stochastic process given a specific parameter vector, $\tilde{\theta}$, and a realization of the shocks, $\{\tilde{\varepsilon}_n\}_{n=1}^{N}$. In the example above, the decision rules for capital and consumption can be derived analytically (labour choice is trivial in this example):

$$k_n = \alpha\beta A_n k_{n-1}^{\alpha}$$
$$c_n = (1 - \alpha\beta)A_n k_{n-1}^{\alpha} \quad (1.2)$$

Given k_0, specific values for the parameters and a realization of the shock process (1.1), a finite realization of consumption and capital can be generated from (1.2). The goal of MSM is to use this simulated data from the model and an observed data set to obtain an estimate for θ. Denote by $\{y_n(\tilde{\theta}; \tilde{\varepsilon}_n)\}_{n=0}^{N}$ the $q \times 1$ vector of data simulated as a solution of a general DSGE model under parameter vector $\tilde{\theta}$ and shock realization $\{\tilde{\varepsilon}_n\}_{n=1}^{N}$; denote the observed data counterpart to these series by $\{x_t\}_{t=1}^{T}$, with N not necessarily equal to T. In the example, the simulated and observed data might consist of such variables as personal consumption expenditures, real GDP, and gross fixed investment.

Let $m(\bullet)$ represent an $r \times 1$ vector of functions. For instance,

$$m(\bullet) = [y_{1n}(y_{2n} - \bar{y}_2)^2 y_{3n}/y_{4n}]$$

If the model is a true description of the data-generating process for the observed data, then given the true parameter vector, θ_0, the following must hold:

$$E[m(y_n(\theta_0))] = E[m(x_t)] \quad (1.3)$$

That is, the moments implied by the model should be equal to the moments of the observed data when the model is evaluated at the parameter vector that generated the observed data.

In general, the theoretical moments in (1.3) cannot be evaluated analytically, so we employ the empirical counterpart to (1.3), given simulated data from the model and the observed data:

$$\frac{1}{N} \sum_{n=1}^{N} m(y_n(\theta_0)) \approx \frac{1}{T} \sum_{t=1}^{T} m(x_t) \quad (1.4)$$

Since we are using time averages to compute the expectations in (1.3), the equality in (1.4) can be satisfied only asymptotically. Hence, the estimation strategy involves choosing a parameter vector that minimizes the following quadratic form for some $r \times r$ symmetric weighting matrix W_T:

$$\Delta(\theta) = \left[\frac{1}{N} \sum_{n=1}^{N} m(y_n(\theta)) - \frac{1}{T} \sum_{t=1}^{T} m(x_t) \right]$$

$$\times W_T \left[\frac{1}{N} \sum_{n=1}^{N} m(y_n(\theta)) - \frac{1}{T} \sum_{t=1}^{T} m(x_t) \right]'$$

$$(1.5)$$

Hansen (1982) provides assumptions under which a similar estimator, the generalized method of moments (GMM) estimator, is both consistent and asymptotically normal. Lee and Ingram (1991) show that, under the conditions in Hansen (1982), the MSM estimator will have an asymptotic normal distribution with mean θ_0 and covariance matrix:

$$\Sigma = (B'WB)^{-1} B'W(1 + \tau)\Omega WB(B'WB)^{-1} \quad (1.6)$$

The constant τ is equal to T/N, and the matrices W, B and Ω are defined as follows:

$$W = \plim_{T \to 0} W_T,$$

$$B \equiv E[\partial m(y_n(\theta))/\partial \theta], \text{ and}$$

$$\Omega \equiv Var\left[\frac{1}{T} \sum_{t=1}^{T} m(x_t) \right].$$

Duffie and Singleton (1993) provide an alternative set of assumptions under which the MSM estimator is consistent. Two assumptions deserve particular attention. First, the two stochastic processes $\{y_n(\theta; \varepsilon_n), n \geq 0\}$ and $\{x_t, t \geq 0\}$ must be stationary and ergodic; this ensures that the time average of a moment converges asymptotically to its expectation. Second, eq. (1.3) must have a unique zero at θ_0 in order for the parameter vector to be exactly identified; if more than one value of θ satisfies eq. (1.3), the estimator will not necessarily converge to θ_0.

Clearly, the size of the asymptotic covariance matrix and, thus, the precision of the estimate of θ is determined by the choice of W, the length of the simulated series relative to the observed series, the matrix B, and the matrix Ω. We discuss each element in turn.

The only restriction on the choice of W_T is that it converge in probability to a constant matrix, W. An obvious choice for W_T might be the identity matrix. In that case,

$$\Sigma = (1 + \tau)(B'B)^{-1}B'\Omega B(B'B)^{-1}$$

When $W = I_r$, all moments are weighted equally, which may not necessarily be optimal from an efficiency standpoint. Although this is a straightforward choice for W, it does not produce the smallest asymptotic covariance matrix. With all else held constant, the smallest asymptotic covariance matrix is attained when W_T is chosen to equal $[(1 + \tau)\Omega]^{-1}$. In that case, $\Sigma = (1 + \tau)[B'\Omega^{-1}B]^{-1}$. The asymptotic covariance matrix can be further reduced by choosing N to be large relative to T (and τ close to zero). As the length of the simulated data series increases relative to the length of the observed series, the term $(1 + \tau)$ tends to 1.

The $r \times k$ matrix B is a measure of the sensitivity of the moments to changes in the parameter vector. Note that the number of moments must equal or exceed the number of parameters, $r \geq k$; if not, the $k \times k$ matrix $B'\Omega^{-1}B$ will not be invertible since the rank of this matrix can be no larger than the rank of its constituent matrices. If the partial derivatives of the moments with respect to a particular parameter are close to zero, then B will be close to non-invertible, as will $B'\Omega^{-1}B$, producing large standard errors. One solution is to choose a different set of moments. A smaller asymptotic covariance matrix is achieved when the moments chosen have larger derivatives (in absolute value) with respect to the parameters. Heuristically, the larger the derivative with respect to a parameter, the more informative is the moment about that parameter.

The matrix Ω must be estimated consistently, despite the likely presence of autocorrelation in the moments. Newey and West (1987) provide one such estimator that is consistent in the presence of both autocorrelation and heteroskedasticity. Many statistical packages provide algorithms for implementing heteroskedasticity and autocorrelation consistent (HAC) estimators for covariance matrices; Andrews (1991) provides a comparison of the properties of several such estimators.

To implement the procedure, the researcher must be able to generate realizations drawn from the stationary distribution of the stochastic process $\{y_n(\theta; \varepsilon_n), n \geq 0\}$. The finite realization of the stochastic process, however, depends on a set of starting values (for example, k_0 in the example above). The researcher must draw the starting values from the stationary distribution for the stochastic process; this, however, is problematic in practice. A practical solution is to generate simulations longer than are needed, and to drop a set of observations from the start of the sample.

If the number of moments exceeds the number of parameters to be estimated, $r > k$, this method also provides a test of fit for the model. Under the null hypothesis that the model provides a true description of the observed data, the product of the length of the observed series and the value of eq. (1.5) evaluated at the estimated value of θ is distributed as a chi-square random variable, $T \times \Delta(\hat{\theta}) \sim \chi^2(r - k)$. Of course, failure to reject the model only indicates that the model fits the data along the dimensions implied by the moments used in estimation.

The MSM is one of many approaches for estimating DSGE models. Calibration is similar in the sense that parameters are chosen to equate model and data moments; calibration, as normally implemented, lacks a statistical foundation. Like MSM, generalized method of moments (GMM) is a limited information method in that is uses a subset of the stochastic information implied by the model. GMM uses the conditional moments implied by the stochastic Euler equations produced by the DSGE's optimization problem. The advantage of GMM is that a complete solution to the DSGE need not be generated; however, the Euler equations may involve data series that do not have observable counterparts. In contrast, maximum likelihood estimation (MLE) is a full information method, but requires a complete characterization of the likelihood of the observed data implied by the DSGE. Since the DSGE is apt to be false along some dimension, the likelihood function will be mis-specified. Zhou (2001) and Ruge-Murcia (2003) provide a more detailed analysis and comparison of the statistical properties of MSM, GMM and MLE.

A final issue with MSM (and all classical estimation methods) is that it can yield parameter estimates that violate sensible restrictions on the model's parameter vector. For example, it is quite common to produce estimates of the rate of time discount, β, that exceed 1; Bayesian estimation methods (DeJong, Ingram and Whiteman, 2000) provide one approach to resolving this issue.

BETH F. INGRAM

See also **Bayesian methods in macroeconometrics; calibration; econometrics; simulation-based estimation.**

Bibliography

Andrews, D. 1991. Heteroskedasticity and autocorrelation consistent covariance matrix estimation. *Econometrica* 59, 817–58.

DeJong, D., Ingram, B. and Whiteman, C. 2000. A Bayesian approach to dynamic macroeconomics. *Journal of Econometrics* 98, 203–23.

Duffie, D. and Singleton, K. 1993. Simulated moments estimation of Markov models of asset prices. *Econometrica* 61, 929–952.

Hansen, L. 1982. Large sample properties of generalized method of moments estimators. *Econometrica* 50, 1029–54.

Lee, B. and Ingram, B. 1991. Simulation estimation of time-series models. *Journal of Econometrics* 47, 197–205.

Newey, W. and West, K. 1987. A simple, positive definite, heteroskedasticity and autocorrelation consistent covariance matrix. *Econometrica* 55, 703–8.

Ruge-Murcia, F. 2003. Methods to estimate dynamic stochastic general equilibrium models. Working Paper No. 17-2003, CIREQ, University of Montreal.

Zhou, H. 2001. Finite sample properties of EMM, GMM, QMLE, and MLE for a square-root interest rate diffusion model. *Journal of Computational Finance* 5, 89–122.

simultaneous equations models

Models that attempt to explain the workings of the economy typically are written as interdependent systems of equations describing some hypothesized technological and behavioural relationships among economic variables. Supply and demand models, Walrasian general equilibrium models, and Keynesian macromodels are common examples. A large part of econometrics is concerned with specifying, testing and estimating the parameters of such systems. Despite their common use, simultaneous equations models still generate controversy. In practice there is often considerable disagreement over their proper use and interpretation.

In building models economists distinguish between *endogenous* variables which are determined by the system being postulated and *exogenous* variables which are determined outside the system. Movements in the exogenous variables are viewed as autonomous, unexplained causes of movements in the endogenous variables. In the simplest systems, each of the endogenous variables is expressed as a function of the exogenous variables. These so-called 'reduced-form' equations are often interpreted as causal, stimulus–response relations. A hypothetical experimental is envisaged where conditions are set and an outcome occurs. As the conditions are varied, the outcome also varies. If the outcome is described by the scalar endogenous variable y and the conditions by the vector of exogenous variables x, then the rule describing the causal mechanism can be written as $y = f(x)$. If there are many outcomes of the experiment, y and f are interpreted as vectors; the rule describing how the ith outcome is determined can be written as $y_i = f_i(x)$.

Most equations arising in competitive equilibrium theory are motivated by hypothetical stimulus–response experiments. Demand curves, for example, represent the quantity people will purchase when put in a price-taking market situation. The conditions of the experiment are, in addition to price, all the other determinants of demand. In any given application, most of these determinants are viewed as fixed as the experiment is repeated; attention is directed at the handful of exogenous variables whose effects are being analysed. In an n good world, there are n such equations, each determining the demand for one of the goods as a function of the exogenous variables.

Reduced-form models where each equation contains only one endogenous variable are rather special. Typically, economists propose interdependent systems where at least some of the equations contain two or more endogenous variables. Such models have a more complex causal interpretation since each endogenous variable is determined not by a single equation but *simultaneously* by the entire system. Moreover, in the presence of simultaneity, the usual least-squares techniques for estimating parameters often turn out to have poor statistical properties.

Why simultaneity?

Given the obvious asymmetry between cause and effect, it would at first thought appear unnatural to specify a behavioural economic model as an interdependent, simultaneous system. Although equations with more than one endogenous variable can always be produced artificially by algebraic manipulation of a reduced-form system, such equations have no independent interpretation and are unlikely to be interesting. It turns out, however, that there are many situations where equations containing more than one endogenous variable arise quite naturally in the process of modelling economic behaviour. These so-called 'structural' equations have interesting causal interpretations and form the basis for policy analysis. Four general classes of examples can be distinguished.

1. Suppose two experiments are performed, the outcome of the first being one of the conditions of the second. This might be represented by the two equations $y1 = f_1(x)$ and $y_2 = f_2(x, y_1)$. In this two-step causal chain, both equations have simple stimulus–response interpretations. Implicit, of course, is the assumption that the experiment described by the first equation takes place before the experiment described by the second equation. Sequential models where, for example, people choose levels of schooling and later the market responds by offering a wage are of this type. Such *recursive* models are only trivially simultaneous and raise no conceptual problems although they may lead to estimation difficulties.

2. Nontrivial simultaneous equations systems commonly arise in multi-agent models where each individual equation represents a separate hypothetical stimulus–response relation for some group of agents, but the outcomes are constrained by equilibrium conditions. The simple competitive supply–demand model illustrates this case. Each consumer and producer behaves as though it has no influence over price or over the behaviour of other agents. Market demand is the sum of each consumer's demand and market supply is the sum of each producer's supply, with all agents facing the same price. Although the market supply and demand functions taken separately represent hypothetical stimulus–response situations where quantity is endogenous and price is exogenous, in the combined equilibrium model both price and quantity are endogenous and determined so that supply equals demand.

Most competitive equilibrium models can be viewed as (possibly very complicated) variants of this supply–demand example. The individual equations, when considered in isolation, have straightforward causal interpretations. Groups of agents respond to changes in their environment. Simultaneity results from market clearing equilibrium conditions that make the environments endogenous. Keynesian macromodels have a similar structure. The consumption function, for example, represents consumers' response to their (seemingly) exogenous wage income – income that in fact is determined by the condition that aggregate demand equal aggregate supply.

3. Models describing optimizing behaviour constitute a third class of examples. Suppose an economic agent is faced with the problem of choosing some vector y in order to maximize the function $F(y, x)$, where x is a vector of exogenous variables outside the agent's control. The optimum value, denoted by y^*, will depend on x. If there are G choice variables, the solution can be written as a system of G equations, $y^* = g(x)$. If F is differentiable and globally concave in y, the solution can be obtained from the first-order conditions $f(y^*, x) = 0$, where f is the G-dimensional vector of partial derivatives of F with respect to y. The two sets of equations are equivalent representations of the causal mechanism. The first is a reduced-form system with each endogenous variable expressed as a function of exogenous variables alone. The second representation consists of a system of simultaneous equations in the endogenous variables. These latter equations often have simple economic interpretations such as, for example, setting marginal product equal to real input price.

4. Models obtained by simplifying a larger reduced-form system are a fourth source of simultaneous equations. The Marshallian long-run supply curve, for example, is often thought of as the locus of price–quantity pairs that are consistent with the marginal firm having zero excess profit. Both price and quantity are outcomes of a complex dynamic process involving the entry and exit of firms in response to profitable opportunities. If, for the data at hand, entry and exit are in approximate balance, the reduced-form dynamic model may well be replaced by a static interdependent equilibrium model.

This last example suggests a possible reinterpretation of the equilibrium systems given earlier. It can be argued (see, for example, Wold, 1954) that multi-agent models are necessarily recursive rather than simultaneous because it takes time for agents to respond to their environments. From this point of view, the usual supply–demand model is a simplification of a considerably more complex dynamic process. Demand and supply in fact depend on *lagged* prices; hence current price need not actually clear markets. However, the existence of excess supply or demand will result in price movement which in turn results in a change in consumer and producer behaviour next period. When time is explicitly introduced into the model, simultaneity disappears and the equations have simple causal interpretations. But, if response time is short and the available data are averages over a long period, excess demand may be close to zero for the available data. The static model with its simultaneity may be viewed as a limiting case, approximating a considerably more complex dynamic world. This interpretation of simultaneity as a limiting approximation is implicit in much of the applied literature and is developed formally in Strotz (1960).

The need for structural analysis

These examples suggest that systems of simultaneous equations appear quite naturally when constructing economic models. Before discussing further their logic and interpretation, it will be useful to develop some notation. Let y be a vector of G endogenous variables describing the outcome of some economic process; let x be a vector of K 'predetermined' variables describing the conditions that determine those outcomes. (In dynamic models, lagged endogenous variables as well as the exogenous variables will be considered as conditions and included in x.) By a simultaneous equations model we mean a system of m equations relating y and x:

$$g_i(y, x) = 0 \qquad (i = 1, \ldots, m).$$

In the important special case where the functions are linear, the system can be written as the vector equation

$$By + \Gamma x = 0, \tag{1}$$

where B is an $m \times G$ matrix of coefficients, Γ is an $m \times K$ matrix of coefficients, and 0 is an m-dimensional vector of zeros. (The intercepts can be captured in the matrix Γ if we follow the convention that the first component of x is a 'variable' that always takes the value one.) A complete system occurs when $m = G$ and B is non-singular. Then the vector of outcome variables can be expressed as a linear function of the condition variables.

$$y = -B^{-1}\Gamma x = \Pi x. \qquad (2)$$

Although the logic of the analysis applies for arbitrary models, the main issues can most easily be illustrated in this case of a complete linear system.

If both sides of the vector equation (1) are pre-multiplied by any $G \times G$ nonsingular matrix F, a new representation of the model is obtained, say

$$B^*y + \Gamma^*x = 0 \qquad (3)$$

where $B^* = FB$ and $\Gamma^* = F\Gamma$. If F is not the identity matrix, the systems (1) and (3) are not identical. Yet if one representation is 'true' (that is, the real world observations satisfy the equation system) then the other is also 'true'. Which of the infinity of possible representations should we select? The obvious answer is that it does not matter. Any linear combination of equations is another valid equation. Any nonsingular transformation is as good as any other. For simplicity, one might as well choose the solved reduced form given by eq. (2).

In practice, however, we are not indifferent between the various equivalent representations. There are a number of reasons for that. First, it may be that some representations are easier to interpret or easier to estimate. The first-order conditions for a profit maximizing firm facing fixed prices may, depending on the production function, be much simpler than the reduced form. Secondly, if we contemplate using the model to help analyse a changed regime, it is useful to have a representation in which the postulated changes are easily described. This latter concern leads to the concept of the degree of *autonomy* of an equation.

In the supply–demand model, it is easy to contemplate changes in the behaviour of consumers that leave the supply curve unchanged. For example, a shift in tastes may modify demand elasticities but have no effect on the cost conditions of firms. In that case, the supply curve is said to be autonomous with respect to this intervention in the causal mechanism. Equations that combine supply and demand factors (like the reduced form relating quantity traded to the exogenous variables) are not autonomous. The analysis of policy change is greatly simplified in models where the equations possess considerable autonomy. If policy changes one equation and leaves the other equations unchanged, its effects on the endogenous variables are easily worked out. Comparative static analysis as elucidated, for example, by Samuelson (1947) is based on this idea. Indeed, the power of the general equilibrium approach to economic analysis lies largely in its separation of the behaviour of numerous economic agents into autonomous equations.

As emphasized by the pioneers in the development of econometrics, it is not enough to construct models that fit a given body of facts. The task of the economist is to find models that successfully predict how the facts will change under specified new conditions. This requires knowing which relationships will remain stable after the intervention and which will not. It requires the model builder to express for every equation postulated the class of situations under which it will remain valid. The concept of autonomy and its importance in econometric model construction is emphasized in the classic paper by Haavelmo (1944) and in the expository paper by Marschak (1953). Sadly, it seems often to be ignored in applied work.

The autonomy of the equations appearing in commonly proposed models is often questionable. Lucas (1976) raises some important issues in his critique of traditional Keynesian macromodels. These simultaneous equations systems typically contain distributed lag relations which are interpreted as proxies for expectations. Suppose, for example, consumption really depends on expected future income. If people forecast the future based on the past, the unobserved expectation variable can be replaced by some function of past incomes. However, since income is endogenous, the actual time path of income depends on all the equations of the model. Under rational expectations, any change in the behaviour of other agents or in technology that affects the time path of income will also change the way people forecast and hence the distributed lag. Thus it can be argued that the traditional consumption function is not an autonomous relation with respect to most interesting policy interventions. Sims (1980) pursues this type of argument further, finding other reasons for doubting the autonomy of the equations in traditional macroeconomic models and concluding that policy analysis based on such models is highly suspect. Although one may perhaps disagree with Sims's conclusion, the methodological questions he raises cannot be ignored.

Even if one accepts the view that structural equations actually proposed in practice often possess limited autonomy and are not likely to be invariant to many interesting interventions, it may still be the case that these equations are useful. A typical reduced-form equation contains all the predetermined variables in the system. Given the existence of feedback, it is hard to argue a priori about the numerical values of the various elements of Π. It may be much easier to think about orders of magnitude for the structural coefficients. Our intuition about the behaviour of individual sectors of the economy is likely to be considerably better than our intuition about the general equilibrium solution. As long as there are no substantial structural changes, specification in terms of structural equations may be appropriate even if some of the equations lack autonomy.

Some econometric issues

The discussion up to now has been in terms of exact relationships among economic variables. Of course, actual data do not lie on smooth monotonic curves of the type used in our theories. This is partially due to the fact that the experiments we have in the back of our mind when we postulate an economic model do not correspond exactly to any experiment actually performed in the world. Changing price, but holding everything else constant, is hypothetical and never observed in practice. Other factors, which we choose not to model, do in fact vary across our sample observations. Furthermore, we rarely pretend to know the true equations that relate the variables and instead postulate some approximate parametric family. In practice we work with models of the form

$$g_i(x, y, \theta, u_i) = 0 \quad (i = 1, \ldots, m)$$

where the g's are known functions, θ is a vector of unknown parameters, and the u's are unobserved error terms reflecting the omitted variables and the misspecification of functional form. In the special case where the functions are linear in x and y with an additive error, the system can be written as

$$By + \Gamma x = u, \tag{4}$$

where u is a m-dimensional vector of errors. If B is non-singular, the reduced form is also linear and can be written as

$$y = -B^{-1}\Gamma x + B^{-1}u = \Pi x + v. \tag{5}$$

Equation system (4) as it stands is empty of content since, for any value of x and y, there always exists a value of u producing equality. Some restrictions on the error term are needed to make the system interesting. One common approach is to treat the errors as though they were draws from a probability distribution centred at the origin and unrelated to the predetermined variables. Suppose we have T observations on each of the $G+K$ variables, say, from T time periods or from T firms. We postulate that, for each observation, the data satisfy eq. (4) where the parameters B and Γ are constant but the n error vectors $u_1, \ldots, u_T$ are independent random variables with zero means. Furthermore, we assume that the conditional distribution of u_t given the predetermined variables x_t for observation t is independent of x_t. A least-squares regression of each endogenous variable on the set of predetermined variables should then give good estimates of Π; if the sample size is large and there is sufficient variability in the regressors. However, unless the inverse of B contains blocks of zeros, eq. (5) implies that each of the endogenous variables is a function of all the components of u. In general, every element of y will be correlated with all the endogenous variables, if the correlation is small compared with the sample variation in those variables.

The conclusion that structural parameters in interdependent systems cannot be well estimated using least squares is widely believed by econometric theorists and widely ignored by empirical workers. There are probably two reasons for this discrepancy. First, although the logic of interdependent systems suggests that structural errors are likely to be correlated with all the endogenous variation, if the correlation is small compared with the sample variable in those variables, least squares bias will also be small. Given all the other problems facing the applied econometrician, this bias may be of little concern. Secondly, alternative estimation methods that have been developed often produce terrible estimates. Sometimes the only practical alternative to least squares is no estimate at all – a solution that is rarely chosen.

In some applications, the reduced form parameters Π are of primary concern. The structural parameters B and Γ are of interest only to the extent they help us learn about Π. For example, in a supply–demand model, we may wish to know the effect on price of changes in the weather. Price elasticities of supply and demand are not needed to answer that question. On the other hand, if we wish to know the effect of a sales tax on quantity produced, knowledge of the price elasticities might be essential. Although least squares is generally available for reduced-form estimation (at least if the sample is large), it is not obvious that, without further assumptions, good structural estimates are ever attainable. The key assumption of the model is that the G structural errors are uncorrelated with the K predetermined variables. These GK orthogonality assumptions are just enough to determine (say by equating sample moments to population moments) the GK parameters in Π. But the structural coefficient matrices B and Γ contain G^2+GK elements. Even with G normalization rules that set the units in which the parameters of each equation will be measured, there are more coefficients than orthogonality conditions. It turns out that structural estimation is possible only if additional assumptions (for example, that some elements of B and Γ are known a priori) are made. These considerations lead to three general classes of questions that have been addressed by theoretical econometricians: (1) When, in principle, can good structural estimates be found? (2) What are the best ways of actually estimating the structural parameters, given that it is possible? (3) Are there better ways of estimating the reduced-form parameters than by least squares?

These questions are studied in depth in standard econometrics textbooks and will not be examined here. The answers, however, do have a common thread. If each structural equation has more than K unknown parameters, structural estimation is generally impossible and least squares applied to the reduced form is optimal in large samples. If, on the other hand, each structural equation has fewer than K unknown coefficients, structural estimation generally is possible and least squares applied to the reduced form is no longer optimal. In this latter situation, various estimation procedures are available, some requiring little computational effort.

However, the sample size may need to be quite large before these procedures are likely to give good estimates.

The assumption that the errors are independent from trial to trial is obviously very strong and quite implausible in time-series analysis. If the nature of the error dependence can be modelled and if the lag structure of the dynamic behavioural equations is correctly specified, most of the estimation results that follow under independence carry over. Unfortunately, with small samples, it is usually necessary to make crude (and rather arbitrary) specifications that may result in very poor estimates. Despite the fact that simultaneous equations analysis in practice is mostly applied to time-series data, it can be argued that the statistical basis is much more convincing for cross-section analysis where samples are large and dependency across observations minimal. Even there, the assumption that the errors are unrelated to the predetermined variables must be justified before simultaneous equations estimation techniques can be applied.

The role of simultaneous equations

Many applied economists seem to view the simultaneous equations model as having limited applicability, appropriate only for a very small subset of the problems actually met in practice. This is probably unwise. Estimated regression coefficients are commonly used to explain how an intervention which changes one explanatory variable will affect the dependent variable. Except for very special cases, this interpretation requires us to believe that the proposed intervention will not affect any of the other explanatory variables and that, in the sample, the errors were unrelated to the variation in the regressors. That is, the mechanism that determines the explanatory variables must be unrelated to the causal mechanism described by the equation under consideration. Unless the explanatory variables were in fact set in a carefully designed controlled experiment, viewing the explanatory variables as endogenous and possibly determined simultaneously with the dependent variable is a natural way to start thinking about the plausibility of the required assumptions.

In a sense, the simultaneous equations model is an attempt by economists to come to grips with the old truism that correlation is not the same as causation. In complex processes involving many decision-makers and many decision variables, we wish to discover stable relations that will persist over time and in response to changes in economic policy. We need to distinguish those equations that are autonomous with respect to the interventions we have in mind and those that are not. The methodology of the simultaneous equations model forces us to think about the experimental conditions that are envisaged when we write down an equation. It will not necessarily lead us to good parameter estimates, but it may help us to avoid errors.

THOMAS J. ROTHENBERG

See also **econometrics; maximum likelihood.**

Bibliography

Haavelmo, T. 1944. The probability approach in econometrics. *Econometrica* 12, Supplement, July, 1–115.

Lucas, R. 1976. Econometric policy evaluation: a critique. In *The Phillips Curve and Labor Markets*, ed. K. Brunner and A. Meltzer. Carnegie-Rochester Conference Series on Public Policy No. 1. Amsterdam: North-Holland.

Marschak, J. 1953. Economic measurement for policy and prediction. In *Studies in Econometric Method*, ed. W. Hood and T. Koopmans. New York: Wiley.

Samuelson, P. 1947. *Foundations of Economic Analysis*. Cambridge, MA: Harvard University Press.

Sims, C. 1980. Macroeconomics and reality. *Econometrica* 48, 1–48.

Strotz, R. 1960. Interdependence as a specification error. *Econometrica* 28, 428–42.

Wold, H. 1954. Causality and econometrics. *Econometrica* 22, 162–77.

single tax

Single Tax is a generic label for the programme of Henry George and others to socialize land rent by substituting one heavy tax on land value for most other taxes. It is not an adequate descriptor but a slogan that caught on. 'Land value taxation' is more used today, especially for a limited tax by a local authority. In Scotland and England 'taxation of ground values' and 'site-value rating' are used.

Specifically, the 'Single Tax' slogan marked a shift in the movement after 1887 as George swung towards the Centre after purging the Marxists from his United Labour Party, losing Powderley and Gompers, and demurring to the quixotic demands of Fr. Edward McGlynn. He was losing Irish-American support from the hostility of Parnell and the Catholic hierarchy.

Thomas Shearman, a corporate lawyer, coined 'Single Tax' to differentiate George's free-market, pro-capitalist programme from those of others who had coalesced around him in the radical and protest awakening of 1879–87. Soon it also served to differentiate Georgism from Bellamy nationalism and Bryan inflationism. In Britain it distinguished Georgists from Wallace's land nationalizers, Hyndman's Marxists, Webb's Fabians, and Parnell's and Chamberlain's movements for peasant proprietorship.

A change of emphasis followed. George had originally striven for rent socialization, redistribution and augmented social spending. Critics on the Right saw too much taxation and levelling. The Single Tax slogan emphasized the counterpart benefits of relief from other taxes. Single Taxers would remove state and local property taxes from buildings and movable capital, and lower most transit and utility rates, often to zero, meeting deficits from the rent fund (anticipating the marginal-cost pricing policies elaborated by Hotelling and Vickrey). Critics on the Left now saw too limited a revenue; so did machine Democrats, militarists, public contractors, and of course landholders seeking developmental public works at the general expense.

The heavy national taxes in America were tariffs. With *Protection or Free Trade?* (1886) George attacked them, invoking Quesnay, Ricardo, Cobden and Bright. Two million copies were printed, equalling his earlier *Progress and Poverty*. He supported Grover Cleveland in 1888 and 1892, hoping that free trade would force Congress to turn to land for revenues. Single Taxers in Congress succeeded in having the Income Tax Act of 1894 include land rent and unearned increments in the base, even though that was likely to be the grounds for its being held unconstitutional, and was. Single Tax began to connote tax limitation.

In *Protection or Free Trade?* George had restated the Physiocratic doctrine of tax incidence, while broadening it to include urban land, which most Physiocrats (except Turgot) had oddly excluded. There is only so much taxable surplus to tap under any system, and most of it lodges in land rent. Single Tax was simply the way to tax this surplus without what is now called the excess burden of indirect taxation. This might refute the charge of inadequacy, but the point has not been widely understood, and some still question revenue adequacy.

Shearman invited more such questions when he went another step from the Left with his 'Single Tax Limited', the upper limit being two-thirds of economic rent. To some adherents Single Tax is a tax limitation device. George remained a 'Single Taxer, Unlimited'. He held that taxation is only a means to justice; justice means every infant has an equal right to the Earth, its use and its rents. He remained a populist who supported Bryan in 1896, even though cold to the free silver panacea.

In Scotland and England George was active and well received by the radical wing of the Liberal Party. But Single Tax continued to mean an extreme position which moderates shunned, even when the Liberal Party put a land tax plank in its platform after Gladstone retired in 1895. Liberals under Asquith and Lloyd George introduced a token land tax in their 1909 budget. The Single Tax bogey frightened the Conservative members of the House of Lords into an intransigent obstructionism that alienated the voters and was used to consolidate the power of the Commons, the Liberals and Lloyd George, who then temporized away the land tax. Labour reintroduced a land tax in 1931 under MacDonald and Snowden, but Neville Chamberlain scuttled it finally in 1934. Post-war Labour governments abandoned Single Tax as being too market-oriented.

Henry George died in 1897. Single Tax remained a power for another generation. Leaders like Shearman and Louis Post sought to professionalize the movement and reconcile it with middle-class values, a timely adaptation to the ethos of progress under scientific management. Somers, Pollock and Zangerle professionalized land assessment. Lawson Purdy helped found the National Tax Association. The Progressive and New Freedom movements absorbed many Single Taxers and reflected some of their ideals.

A century earlier at the court of Louis XV, François Quesnay and his Physiocrats had advanced the '*impôt unique*', an even more limited single tax restricted to farm land, and one-third the rent. Like Shearman, Quesnay argued efficiency and laissez-faire, not redistribution: it was the age of enlightened despotism, not of populism. They (and later their disciple Walras) called it 'co-proprietorship of land by the state'.

But there is a touch of class-levelling inherent in any proposal to tax land directly, however sugar-coated with the doctrine that landholders gain from removing other taxes, however limited by the safeguard of 'co-proprietorship'. Physiocracy could beguile a despot dreaming of energizing a decadent gentry, and liberating his people from tax farmers and a jumble of enervating excises that weakened France's economy. It was the fate of Quesnay, and of France, that the privileged gentry proved more sensitive to Physiocracy's threat than others were to its benefits.

Quesnay was closer than he knew to an age of populism; he might better have addressed the new constituency. Indirectly, he did through his influence on Jefferson, transmitted through his disciples Pierre Samuel Du Pont and Destutt de Tracy; and through his influence on Smith, Ricardo and Mill, whose special treatment of land rent set the stage for George's Single Tax.

MASON GAFFNEY

See also **George, Henry.**

Bibliography

Brown, H.G. 1932. *The Economic Basis of Tax Reform.* Columbia, MO: Lucas Bros.

Douglas, R. 1976. *Land, People and Politics.* London: Allison & Busby.

Geiger, G.R. 1936. *The Theory of the Land Question.* New York: Macmillan.

Groenewegen, P.D., ed. and trans. 1977. *The Economics of A.R.J. Turgot.* The Hague: Martinus Nijhoff.

Holland, D.M., ed. 1970. *The Assessment of Land Value.* Madison: University of Wisconsin Press.

Howe, F.C. 1925. *Confessions of a Reformer.* New York: Charles Scribner's Sons.

Lindholm, R.W. and Lynn, A.D., Jr., eds. 1981. *Land Value Taxation in Thought and Practice.* Madison: University of Wisconsin Press.

Lissner, W., ed. 1941. *American Journal of Economics and Sociology.* New York: Robert Schalkenbach Foundation.

Miller, J.D., ed. 1917. *Single Tax Yearbook.* New York: Single Tax Review Publishing Co.

Scott, A., ed. *Natural Resource Revenues: A Test of Federalism.* Vancouver: University of British Columbia Press.

Shearman, T. 1888. *Natural Taxation.* 2nd edn, New York: Doubleday, 1911.

Skouras, A. 1977. *Land and its Taxation in Recent Economic Theory.* Athens: Papazissis Publishers.

Young, A.N. 1916. *History of the Single Tax Movement in the United States.* Princeton: Princeton University Press.

Sismondi, Jean Charles Leonard Simonde de (1773–1842)

A number of concepts and theories that later became important in the history of economics first appeared in the writings of the Swiss economist J.C.L. Simonde de Sismondi. Whether or not these can be considered as his 'contributions' to economics is a question not unlike that as to whether a tree that falls in a deserted forest makes a sound. Sismondi developed the first aggregate equilibrium income theory and the first algebraic growth model. Yet both concepts had to be rediscovered and redeveloped by others before they entered the mainstream of economics, long after Sismondi's time. The fact that Sismondi wrote in French may have been part of the reason why his work made so little impact at a time when the development of classical economics was largely the work of British economists. However, the fame achieved by his French contemporary, Jean-Baptiste Say, suggests that language differences alone cannot explain the neglect of Sismondi. His economic writings were neglected in France and Switzerland as well.

When he was born in Geneva in 1773, his name was Jean Charles Leonard Simonde. After an exile in Italy, during which he determined that he was descended from a noble Italian family named Sismondi, he returned to Geneva in 1800 with his new surname, Simonde de Sismondi. However, he was sufficiently tentative about it to use his original name on his first book in economics, *De la richesse commerciale*, in 1803. Sismondi also wrote extensively on history, including a 16-volume history of Italy. All his writings were pervaded by considerations of public policy in general, and the interests of the less fortunate in particular.

Sismondi was born into a prosperous bourgeois family, which was despoiled of much of its wealth during Swiss political upheavals reflecting the contemporary revolution in France. Shifting political fortunes led not only to Sismondi's exile but to two imprisonments as well. After the turmoil subsided, Sismondi worked at a variety of occupations, including gentleman farmer and professor of philosophy.

Sismondi's first venture into economics, the two-volume *De la richesse commerciale*, was intended as a systematic exposition of the ideas of Adam Smith. Yet in it Sismondi also pointed out that he was presenting an 'absolutely new' way of looking at aggregate output changes. Crude arithmetic examples depicted output during a given year as a function of investment during a previous year, and showed how a closed economy differed from an economy with international trade, and how the latter differed when there was an export surplus and an import surplus. Algebraic formulas in his footnotes repeated the same arguments presented arithmetically in the text. But the book was little noticed, and so Sismondi's original efforts produced no contribution to the development of economics.

In the wake of the post-Napoleonic War depression, Sismondi turned his attention once more to economics and to issues of aggregate income equilibrium. In 1814, he produced a long article entitled 'Political Economy', written in English for the *Edinburgh Encyclopaedia*. In the midst of a summary presentation of classical economics appeared an early version of Sismondi's own theory of aggregate equilibrium income. This theory was elaborated in Sismondi's main economic work, the two-volume *Nouveaux principes d'économie politique* (1819). With this work he entered the controversy over Say's Law and general gluts.

According to Sismondi, the utility of output was balanced against the disutility of work, whether by Robinson Crusoe on an island or by a complex society. But, with different people doing partial balancing in isolation from one another in a complex economy, the aggregate balance was not always continuously assured. Whenever the disutility of labour exceeded the utility of output in a given time period, subsequent time periods would see a decline in aggregate output until the balance was restored. Conversely, when the utility of output exceeded that of labour, output would tend to rise.

The germ of this reasoning had already appeared in *L'Ordre naturel* by the Physiocrat Mercier de la Rivière in 1767. Sismondi elaborated it into a theory of equilibrium income, with which he challenged the reigning view, expressed in Say's Law, that there were no limits to production.

Say's Law, then as later, had many meanings. But one of the contemporary meanings was that there was no such thing as an equilibrium level of aggregate output. Whatever level of output was supplied could always find a demand, and where this did not happen, it was because the assortment of goods did not match consumer preferences, not because the total output was at an unsustainable level. Sismondi rejected this reasoning, arguing that the demand for leisure would at some point outweigh the demand for other goods, and that when production went beyond the point at which this happened, it would be unsaleable at cost-covering prices and so fail to be reproduced in subsequent time periods.

Sismondi understood the full implications of what he was saying and how it contradicted prevailing views. The balance of aggregate supply and demand he considered the most important question in economics, and especially so during the depression following the Napoleonic wars. J.B. Say and the Ricardians maintained that the unsaleability of some goods showed only that insufficient other goods had been produced to exchange with them – that output had the wrong internal proportions, not an excess in the aggregate, and that the proper proportions could be restored at a still higher level of aggregate production. In this view, there could be a partial glut of particular commodities but not a *general* glut of commodities.

Sismondi argued that there could be a general glut of commodities because one of the goods desired was leisure – that is, exemption from the production of commodities. He did not believe that this occurred in the normal course of free market competition but because government policy sometimes artificially fostered production at an unsustainable level. Like the orthodox economists of his time, Sismondi regarded money as an unessential factor, a 'veil' obscuring but not fundamentally changing the behaviour of economic aggregates. His disagreements with the classical economists had nothing to do with monetary controversies such as those in 20th-century macroeconomics.

When Sismondi's *Nouveaux principes* appeared in 1819, it was immediately attacked in the *Edinburgh Review* in October of that year, in the midst of a discussion of Robert Owen. Its reasoning was declared a 'fallacy' and once more 'proportions' – not aggregates – were declared to be the only prerequisites for markets to be cleared and increased output sustained. A glut was there defined as 'an increase in the supply of a particular class of commodities, unaccompanied by a corresponding increase in the supply of those other commodities which should serve as their equivalents'. In short, there were partial gluts but no general gluts and a still higher level of output was sustainable if properly proportioned internally. The basis of this reasoning was explicitly attributed to 'the celebrated M. Say', with a 'most clear and conclusive' treatment of the subject added by James Mill.

The appearance of T.R. Malthus's *Principles of Political Economy* the following year added fuel to the debate, for he too challenged Say's Law in the same way. Marx later characterized Malthus's book as simply the 'English translation' of Sismondi, but in fact it represented views which Malthus had long expressed in correspondence with Ricardo. Replies to both authors began to appear in both French and English publications during the 1820s, provoking rejoinders in books and articles. Their controversy over general gluts persisted for more than a decade, involving not only the leading economists of the time – Say, Ricardo, Malthus, Sismondi, Torrens, McCulloch and both Mills – but also Samuel Bailey, William Blake, Thomas Chalmers, and others either forgotten or little remembered in the history of economics. These published controversies were supplemented by correspondence between Sismondi and Say, Sismondi

and Ricardo, Malthus and Say, and Malthus and Ricardo. Only Say seems to have acknowledged that the theory of aggregate equilibrium income had relevance to one version of Say's Law that was current at the time. In the fifth edition of his *Traité d'économie politique* in 1826 he added three paragraphs to his chapter on the law of markets, discussing 'the limit to a growing production' and repeating (without citation) Sismondi's argument that when output's 'utility is not worth what it cost', such output is unsustainable. A year later he admitted in a letter to Malthus that his law of markets was 'subject to some restrictions' which he had included in the most recent edition of his book. (Unfortunately, the English translation of Say's *Traité* is from the previous edition.) Finally, in a textbook published in 1828–9, Say followed the chapter on his law of markets with a chapter entitled, 'Limits to Production' – a phrase from Sismondi.

No such impact or even acknowledgement occurred in British economic writings. There Sismondi and Malthus were answered as if they were arguing for secular stagnation instead of temporary aggregate disequilibrium. John Stuart Mill enshrined this misunderstanding of Sismondi and Malthus in his classic *Principles of Political Economy* in 1848. Thus things stood for nearly a century, until John Maynard Keynes resurrected Malthus, but not Sismondi, as his predecessor in aggregate equilibrium theory.

Sismondi's anticipations of later economic theory were not limited to aggregate income theory. In the course of dealing with that large topic he also proposed a theory of destabilizing responses to overproduction, which would initially take the economy further from equilibrium, though it would ultimately return to equilibrium 'after a frightful suffering'. He also dealt with the issue of the short-run shut down point of a firm, which he argued would produce even below cost-covering prices if much of its cost was fixed rather than variable. Sismondi also argued against the reigning Malthusian population theory, pointing out fatal ambiguities in the word 'tendency' as Malthus used it and using empirical evidence to show that the *historical* tendency was for food supply to grow faster than population.

In many ways Sismondi also anticipated Marx. Sismondi's emphasis on 'the proletarians', on an increasing concentration of capital, recurring business cycles, technological unemployment and economic dynamics in general all reappeared (without credit) in Marx's writings.

None of these pioneering efforts by Sismondi received either contemporary acknowledgement or later recognition by the profession. His loose and sometimes inconsistent writings and his emotional assertions made it easy to dismiss him and throw away his insights along with his errors. He left no disciples and his eclecticism provided no dogma around which a school could crystallize.

THOMAS SOWELL

Selected works

1814. *Political Economy.* New York: Augustus M. Kelley, 1966.

1819. *Nouveaux principes d'économie politique.* Paris: Delaunay, 1827.

Bibliography

Rappard, W.E. 1966. *Economistes Genèvois du XIXe siècle.* Geneva: Libraire Droz, 1966.

Salis, R. de 1932. *Sismondi, 1775–1842.* Paris: Libraire Ancienne Honoré Champion.

Sowell, T. 1968. Sismondi: a neglected pioneer. *History of Political Economy* 1, 62–88.

size of nations, economics approach to the

The number and size of sovereign states has been at the centre of human history for thousands of years, from the times of Sumerian city states to the post-cold war era. Plato in *The Laws* (360 BC, Book V) calculated the optimal size of a state as 5,040 heads of family. According to Aristotle in *The Politics* (350 BC, Book VII, ch. 4), experience had shown that a very populous state could rarely, if ever, be well governed (a view probably not shared by Aristotle's famous pupil, Alexander the Great). Montesquieu in *The Spirit of Laws* (1748, Book VIII, ch.16) wrote that 'in a small [republic], the interest of the public is more obvious, better understood, and more within the reach of every citizen'. A theory of optimal size was sketched by Beccaria (1764, p. 91), the Italian philosopher who inspired Bentham's utilitarian approach:

> The more society grows, the smaller part of the whole does each member become, and the republican sentiment diminishes proportionately if the laws neglect to reinforce it. Societies have, like human bodies, their circumscribed limits, increasing beyond which the economy is necessarily disturbed. It would seem that the size of a state ought to vary inversely with the sensitivity ['*sensibilità*'] of its constituency ... A republic grown too vast can escape despotism only by subdividing and then reuniting itself as a number of federated little republics.

These are selective quotations from an enormous philosophical, political and historical literature (see for example Dahl and Tufte, 1973). By contrast, for a long while economists have taken political borders as given. Only in recent years has a small but expanding economic literature started to address questions of country formation and break-up with the tools of economic analysis (for discussions, see Bolton, Roland and Spolaore, 1996; Alesina and Spolaore, 2003; Spolaore, 2006). This research is motivated by the fact that political borders are not a fixed part of the geographical landscape but human-made institutions, affected by the decisions and interactions of individuals and groups who pursue their objectives under constraints. The economics approach to the size of nations can be viewed as a natural extension of the research programme of modern political economics, whose aim is to 'endogenize' collective decisions and institutions.

This economic literature has addressed the determination and change of political borders in different political and economic environments and using various solution concepts. Friedman (1977) studies the formation of borders as set by rent-maximizing Leviathans. Findlay (1996) analyses the expansion of empires. Alesina and Spolaore (1997; 2003) derive and compare the number and size of countries under different solution concepts: efficient (that is, welfare-maximizing) borders, voting equilibria, equilibria under unilateral secessions, and equilibria in a world of rent-maximizing Leviathans. Bolton and Roland (1997) study the break-up of nations by direct majority vote, when income distributions differ across regions, and regional median voters have different preferences over redistribution. The relationship between economic integration and the size of countries has been analysed by Alesina and Spolaore (1997) and Alesina, Spolaore and Wacziarg (2000; 2005). Wittman (1991; 2000) focuses on welfare-maximizing solutions. Optimal secession rules have been studied by Bordignon and Brusco (2001). Le Breton and Weber (2003) analyse equilibria when groups of individuals can secede unilaterally, and compensation mechanisms are possible within countries. Contributions to the literature also include Casella and Feinstein (2002), Goyal and Staal (2003), and many others.

When one considers the size of nations from an economic perspective, a natural starting point is the trade-off between benefits and costs from a larger size. Important benefits of scale are associated with the provision of public goods, which are cheaper in per capita terms when more taxpayers pay for them (empirically, smaller countries do have larger governments). Larger countries can also better internalize cross-regional externalities, a point extensively studied in the literature on decentralization and fiscal federalism. Additional benefits from size come from insurance against natural and economic shocks through inter-regional transfers. But size also comes with costs. As countries become larger, congestion may overcome some of the above benefits. More importantly, an expansion of a country's borders is likely to bring about higher heterogeneity of preferences across different individuals. Being part of the same country implies sharing jointly supplied public goods and policies in ways that cannot satisfy everybody's preferences. Decentralization of some public goods and policies may offer a partial response to heterogeneity. However, many policies that characterize a sovereign state (basic characteristics of the legal system, foreign policy, defence policy) are indivisible and must be shared among the whole population. This induces a trade-off between economies of scale in the provision of public goods and heterogeneity of preferences. This trade-off has played a central role in the economic literature on the size of nations (see Alesina and Spolaore, 1997; 2003). The trade-off depends not only on the degree of heterogeneity of preferences but also on the political regime through which preferences are turned into policies. For example, rent-seeking Leviathans that are less concerned with the preferences of their subjects may pursue expansionary policies that lead to the formation of inefficiently large countries and empires. By contrast, democratization leads to secessions, and, in the absence of effective mechanisms to integrate populations with diverse preferences, self-determination can be associated with excessive fragmentation and costly break-up. Historically, successful societies are those that have managed to minimize the costs of heterogeneity while maximizing the benefits stemming from a diverse pool of preferences, skills and endowments.

Economic analyses of the size of nations have pointed out that the trade-off between benefits and costs of size is also a function of the degree of international economic integration (Alesina and Spolaore, 1997; Alesina, Spolaore and Wacziarg, 2000; 2005; Spolaore and Wacziarg, 2005). The size of the market may or may not coincide with the political size of a country as defined by its borders. Larger nations mean larger domestic markets when political borders imply barriers to international exchange. By contrast, market size and political size would be uncorrelated in a world of perfect free trade in which political borders imposed no costs on international transactions. Therefore, market size depends both on country size and on the trade regime. If market size matters, small countries can prosper in a world of free trade and high economic integration, while a large size is more important for economic success in a world of trade barriers and protectionism. In fact, empirically the effect of size on economic performance (income per capita, growth) tends to be higher for countries that are less open, and the effect of openness is much larger for smaller countries (Alesina, Spolaore and Wacziarg, 2000; 2005; Spolaore and Wacziarg, 2005). As economic integration increases, the benefits of a large political size are reduced, and political disintegration becomes less costly. Conversely, smaller countries tend to benefit from more openness. Hence, economic integration and political disintegration tend to go hand in hand.

The economic literature on the size of nations is connected to other established bodies of research, such as the literature on local public goods and clubs, pioneered by Tiebout (1956) and Buchanan (1965). But while local jurisdictions are not completely autonomous and people are free to move across them, the analysis of nations explicitly focuses on sovereign states that can impose direct barriers to economic exchange and mobility, and can use force in settling disputes with their neighbours. The study of nations is also connected to the literature on customs unions and trade blocs. In so far as modern nations tend to promote free trade within their own borders, countries can be seen as 'trade blocs' of regions. In general, free trade areas, customs unions, supranational organizations, confederations and sovereign states can be viewed as points on a continuum of increasing coordination and integration of political functions. A third body of connected research is the economic analysis of conflict, pioneered by Boulding (1962), Tullock (1974), Hirshleifer (1991; 1995), Grossman (1991) and others. Alesina and Spolaore (2005; 2006) and Spolaore (2004) explicitly link the economic literature on conflict with the economic approach to the size of nations, and analyse how the benefits of size tend to be larger in a world with wars and international conflict. The historical importance of military technology and violent resolution of conflict for the determination of political borders is likely to spur further work linking these variables to the other economic and political factors that affect the number and size of countries over time.

In summary, the economics approach to the size of nations, while in its infancy, builds on an extensive body of concepts and tools in order to explain the endogenous formation and break-up of sovereign states within the framework of modern political economy.

ENRICO SPOLAORE

See also **globalization; Tiebout hypothesis; war and economics.**

Bibliography

Alesina, A. and Spolaore, E. 1997. On the number and size of nations. *Quarterly Journal of Economics* 112, 1027–56.

Alesina, A. and Spolaore, E. 2003. *The Size of Nations.* Cambridge, MA: MIT Press.

Alesina, A. and Spolaore, E. 2005. War, peace, and the size of countries. *Journal of Public Economics* 89, 1333–54.

Alesina, A. and Spolaore, E. 2006. Conflict, defense spending, and the number of nations. *European Economic Review* 50, 90–120.

Alesina, A., Spolaore, E. and Wacziarg, R. 2000. Economic integration and political disintegration. *American Economic Review* 90, 1276–96.

Alesina, A., Spolaore, E. and Wacziarg, R. 2005. Trade, growth, and the size of countries. In *Handbook of Economic Growth*, ed. P. Aghion and S. Durlauf. Amsterdam: North-Holland.

Aristotle. 350 BC. *The Politics.* Online. Available at http://classics.mit.edu/Aristotle/politics.7.seven.html, accessed 22 August 2005.

Beccaria, C.B. di. 1764. *On Crimes and Punishments.* Indianapolis: Bobbs-Merrill, 1963.

Bolton, P. and Roland, G. 1997. The breakups of nations: a political economy analysis. *Quarterly Journal of Economics* 112, 1057–89.

Bolton, P., Roland, G. and Spolaore, E. 1996. Economic theories of the break-up and integration of nations. *European Economic Review* 40, 697–705.

Bordignon, M. and Brusco, S. 2001. Optimal secession rules. *European Economic Review* 45, 1811–34.

Boulding, K. 1962. *Conflict and Defense: A General Theory*. New York: Harper.

Buchanan, J. 1965. An economic theory of clubs. *Economica* 32, 1–14.

Casella, A. and Feinstein, J. 2002. Public goods in trade: on the formation of markets and political jurisdictions. *International Economic Review* 43, 437–62.

Dahl, R. and Tufte, E. 1973. *Size and Democracy*. Stanford: Stanford University Press.

Findlay, R. 1996. Towards a model of territorial expansion and the limits of empires. In *The Political Economy of Conflict and Appropriation*, ed. M. Garfinkel and S. Skaperdas. Cambridge: Cambridge University Press.

Friedman, D. 1977. A theory of the size and shape of nations. *Journal of Political Economy* 85, 59–77.

Goyal, S. and Staal, K. 2003. The political economy of regionalism. *European Economic Review* 48, 563–93.

Grossman, H. 1991. A general equilibrium model of insurrections. *American Economic Review* 81, 912–21.

Hirshleifer, J. 1991. The technology of conflict as an economic activity. *American Economic Review* 81, 130–34.

Hirshleifer, J. 1995. Anarchy and its breakdown. *Journal of Political Economy* 103, 26–52.

Le Breton, M. and Weber, S. 2003. The art of making everybody happy: how to prevent a secession? *IMF Staff Papers* 50, 403–35.

Montesqueu, Baron de. 1748. *The Spirit of Laws*. Online. Available at http://www.constitution.org/cm/sol_08.htm#016, accessed 22 August 2005.

Plato. 360 BC. *The Laws*. Online. Available at http://classics.mit.edu/Plato/laws.5.v.html, accessed 22 August 2005.

Spolaore, E. 2004. Economic integration, international conflict and political unions. *Rivista di Politica Economica* 94, 3–50.

Spolaore, E. 2006. The political economy of national borders. In *Oxford Handbook of Political Economy*, ed. B. Weingast and D. Wittman. Oxford: Oxford University Press.

Spolaore, E. and Wacziarg, R. 2005. Borders and growth. *Journal of Economic Growth* 10, 331–86.

Tiebout, C. 1956. A pure theory of local expenditures. *Journal of Political Economy* 64, 416–24.

Tullock, G. 1974. *The Social Dilemma: The Economics of War and Revolution*. Blacksburg, VA: University Publications.

Wittman, D. 1991. Nations and states: mergers and acquisitions; dissolution and divorce. *American Economic Review, Papers and Proceedings* 81, 126–9.

Wittman, D. 2000. The wealth and size of nations. *Journal of Conflict Resolution* 6, 885–95.

skill-biased technical change

Skill-biased technical change (SBTC) is a shift in the production technology that favours skilled (for example, more educated, more able, more experienced) labour over unskilled labour by increasing its relative productivity and, therefore, its relative demand. Ceteris paribus, SBTC induces a rise in the skill premium – the ratio of skilled to unskilled wages.

From factor-neutral to factor-biased technical change

Economic theory views production technology as a function describing how a collection of factor inputs can be transformed into output, and it defines technical change as a shift in the production function, that is, a change in output for given inputs. The traditional measure of economy-wide technological change, introduced by Solow (1957), is aggregate total factor productivity (TFP). Solow defines a TFP advancement as an increase in output that leaves marginal rates of transformations untouched for given inputs; thus, a change in TFP is a form of *factor-neutral* technical change.

For illustrative purposes, suppose that the aggregate production function is constant returns to scale and Cobb–Douglas in aggregate capital (K) and aggregate labor (L) services, that is, $Y = ZK^\alpha L^{1-\alpha}$, where Y is aggregate output, α is the elasticity of output to capital, and Z denotes precisely TFP. If output and input markets are competitive, then the share of income going to capital equals α. Solow's (1957) fundamental insight is that, armed with this estimate of α and measures of (Y, K, L) from national accounts, neutral technical change can be quantified 'residually'. This clever and parsimonious approach to growth accounting has dominated the literature for decades, creating an overwhelming consensus that neutral technological improvements are the primary source of growth in income per capita.

However, a key fact that recently emerged from the data highlights the limits of this conceptualization of technical change. Since the mid-1970s, the rental price of skilled labour has soared dramatically relatively to that of unskilled labour despite a major increase in the relative supply of skills: for example, the college wage premium – defined as the ratio between the wage of college graduates and the wage of high-school graduates – jumped from 1.45 in 1965 to 1.7 in 1995, while the relative supply of college skills tripled over the same period. Given the observed movements in the relative quantities, these price changes could not be generated by movements 'along the production function'. Neutral technical change is, by definition, silent on changes in relative prices. Therefore, to make sense of these recent developments one must introduce the concept of *factor-biased* technical change.

For this purpose, I now generalize the aggregate production function above by letting labour input, L, be a constant elasticity of substitution (CES) function of skilled and unskilled labour, L_s and L_u, with factor-specific productivities A_s and A_u:

$$L = [(A_s L_s)^\sigma + (A_u L_u)^\sigma]^{1/\sigma}, \sigma \le 1. \qquad (1)$$

At this point, it is not necessary to specify what makes a worker more skilled than another: it could be education, innate ability or experience. The (log of the) marginal rate of transformation (MRT) between the two labour inputs is

$$\ln(MRT_{s,u}) = \sigma \ln\left(\frac{A_s}{A_u}\right) + (1-\sigma)\ln\left(\frac{L_u}{L_s}\right). \qquad (2)$$

Note that the TFP term Z does not enter the above equation. A change in the ratio A_s/A_u is a form of factor-biased technical change since it modifies the marginal rates of transformation, at a given input ratio. In particular, under the empirically plausible parametric assumption $\sigma > 0$, technical change is *skill-biased* if A_s/A_u increases. With competitive input markets, the (log of the) skill premium can be read off the right-hand side of (2) as well. Therefore, skill-biased technical change induces an increase in the relative productivity of skilled labour that raises its relative demand and, ceteris paribus, the skill premium.

To take this logic one step further, given an estimate of the elasticity of substitution between types of labour $1/(1-\sigma)$, and time-series on relative wages and relative factor supplies, one can measure skill-biased technical change residually from (2). For example, with an elasticity of substitution of 1.4 (or $\sigma = 0.29$) between college graduates and the rest of the labour force, the dynamics of the US college premium and of the relative supply of college skills imply a growth of skill-biased technical change (that is, of the ratio A_s/A_u) in excess of ten per cent per year from 1963 to 1987 (Katz and Murphy, 1992).

The skill bias of information technologies

Recent shifts in technology have been skill-biased. But SBTC appears all but an unexplained residual very much like Solow TFP, a 'black box' that needs to be filled with economic content. What really accounts for this shift in the production process since the mid-1970s? The timing of the rise in the skill premium has coincided with the rapid diffusion of information and communication technologies in the work place. Thus, a natural candidate for this wave of SBTC is the 'information technology revolution'.

Expenditures in information processing equipment and software, as a share of US private non-residential fixed investment, rose from six per cent in 1960 to 40 per cent in 2000. At the heart of these dynamics there is a staggering improvement in the quality and

productivity of all those equipment goods relying heavily on semiconductors, like computers, software, and switching equipment underlying much of communication technology.

Ample microeconometric research and several case studies document a statistical correlation between the use of new technologies, like computers, and either the employment share of skilled workers (Bartel and Lichtenberg, 1987) or their wage share (Autor, Katz and Krueger, 1998) across industries. These studies firmly establish that the new technologies are deployed with better-qualified and better-paid labour, but they fail to explain *why*. This deeper question requires a quantitative theory built around an explicit economic mechanism.

Technology-skill complementarity

A large number of economic models in the literature provides a foundation for SBTC (for surveys, see Acemoglu, 2002; Aghion, 2002; Hornstein, Krusell and Violante, 2005). The central tenet of all these theories is technology-skill complementarity and takes three alternative formulations.

The first formulation is built on a defining feature of the post-war US economy: the sharp decline of the constant-quality relative price of equipment investment (Gordon, 1990; Greenwood, Hercowitz and Krusell, 1997), especially evident for information technologies whose prices fell at ten per cent per year. Krusell et al. (2000) argue that the substantial cheapening of equipment capital is the force behind SBTC. This decline in price led to an increased use of equipment capital in production. At least since Griliches (1969), various empirical papers support the idea that skilled labour is relatively more complementary to equipment capital than is unskilled labour. As a result of capital-skill complementarity in production, the faster growth of the equipment stock pushed up the relative demand for skilled labour and, in turn, the skill premium.

More explicitly, these authors generalize the aggregate production function to:

$$Y = K_s^\alpha \left[\gamma[\mu(K_e)^\rho + (1-\mu)(L_s)^\rho]^{\frac{\sigma}{\rho}} + (1-\lambda)(L_u)^\sigma \right]^{\frac{1-\alpha}{\sigma}}, \qquad (3)$$

where K_s denotes structures capital and K_e equipment capital. Profit-maximizing behaviour of price-taking firms implies that the skill premium (and the MRT between labor inputs) can be approximated as

$$\ln\left(\frac{w_s}{w_u}\right) \simeq \lambda \frac{\sigma - \rho}{\rho} \ln\left(\frac{K_e}{L_s}\right)^\rho + (1-\sigma)\ln\left(\frac{L_u}{L_s}\right). \qquad (4)$$

If $\sigma > \rho$, as estimated by Krusell et al., the relative demand for skills increases with the stock of equipment capital. Note the difference between eqs. (2) and (4): capital–skill complementarity gives economic content to the notion of SBTC by replacing an unobserved residual trend (A_s/A_u) with the actual upward trend in equipment–skilled labour ratio. This model replicates well the dynamics of the US skill premium since the mid-1960s. Moreover, historical evidence suggests that complementarity between skilled labour and capital characterized technological developments throughout the entire 20th century (Goldin and Katz, 1998).

The second formulation is inspired by the Nelson–Phelps view of human capital. In the words on Nelson and Phelps (1966, p. 70), 'educated people make good innovators, so that education speeds the process of technological diffusion'. In particular, they contend that more educated, able or experienced labour deals better with technological change. Skilled workers are less adversely affected by the turmoil created by major technological transformations since it is less costly for them to learn the additional knowledge needed to adopt a new technology. Therefore, rapid technological transitions, such as that witnessed since the mid-1970s, are skill-biased, as more able workers adapt better to change (Greenwood and Yorukoglu, 1997; Caselli, 1999; Galor and Moav, 2000).

Incidentally, this version of the technology–skill complementarity hypothesis, by emphasizing the importance of learning during episodes of drastic technical change, is consistent with the TFP slowdown experienced by most developed economies in the 1980s: upon the arrival of the new information technologies, aggregate productivity can fall temporarily as workers and firms learn how to deploy the new production methods at their best (Hornstein and Krusell, 1996; Aghion, 2002).

The Nelson–Phelps conjecture implies that the rise in the skill premium is transitory: it is only in the early adoption phase of a new technology that those who adapt more quickly can reap some benefits. As time goes by, there will be enough workers learning how to work with the new technology to offset the wage differential. Note the difference from the hypothesis set forth by Krusell et al. (2000), where the effect of capital deepening on the skill premium is permanent.

The third formalization of this hypothesis is based on Milgrom and Roberts (1990). These authors argue that information technologies reduce costs of data storage, communication, monitoring and supervision activities within the firm, which triggers a shift towards a new organizational design. In particular, the layers of the hierarchical structure can be reduced, so that the organization of the firm becomes 'flatter'. Workers no longer perform routinized, specialized tasks, but they are now responsible for a wide range of tasks within teams. Therefore, adaptable workers who have general skills and who are more versed at multi-tasking activities benefit from this transformation. In other words, the change in technology induces an organizational shift which is skill-biased. An elegant formalization of this hypothesis is contained in Garicano and Rossi-Hansberg (2004).

Microeconomic evidence consistent with all these formulations of the technology–skill complementarity hypothesis is offered by Autor, Levy and Murnane (2003). Based on data on the skill content and tasks of various occupations, they split job requirements into 'routine' and 'non-routine' tasks and document that, starting from the 1970s, the labour input of non-routine analytic and interactive tasks increased sharply relative to routine cognitive and manual tasks. This shift was concentrated in rapidly computerizing industries and it was pervasive at all educational levels. The authors interpret these findings as evidence that information technologies substituted for unskilled labour employed on simple and more repetitive tasks – more amenable to computerization – and complemented workers endowed with generalized problem-solving, complex communication, and analytical skills.

Endogenous direction of technical change

In the same vein as the endogenous growth literature developed in the 1990s, one could contend that not only the speed – as traditionally argued – but also the *direction* of technical change is endogenous. Profit incentives of innovators determine the amount of R&D activity directed towards different factors of production (Acemoglu, 1998). The main determinants of profit incentives are market size, relative prices and institutions. These forces can shed light on numerous episodes in the history of technology.

Under the assumption that the R&D is fixed, the market size of the innovation determines its revenues. The expansion of educated labour during the postwar period made it profitable to develop machines complementary to skilled workers. The vast rural–urban migration wave towards English cities during the late 18th century opened the way to the development of the factory system and, later, to the Tayloristic assembly line which quickly replaced skilled artisans' craft shops. Incidentally, this is a notable example of *unskill bias* which proves that, historically, the direction of technical progress has varied.

Profit maximization dictates that, ceteris paribus, innovation be directed towards those factors that are more intensely used in the production of highly priced goods. The recent burst of North–South trade increased the relative price of skill-intensive goods in the North, representing yet another force which pushed towards skill-biased innovations in the post-war period.

Labour institutions that keep wages high despite reductions in productivity induce firms to direct efforts towards labour-saving technologies. Such a fall in labour demand may explain the rise in European unemployment, after the upward wage push secured by the 'labour movement' in the 1970s. The hump-shaped dynamics of the European labour share between 1970 and 1990 validate this conjecture.

The theory of endogenous factor bias in technology is, potentially, far reaching. The main limit, at this early stage of development, is the lack of quantitative analysis of the proposed mechanisms. For example, is the acceleration in the growth of college skills of the 1970s large enough to generate the observed rise in the productivity of skilled labour and in the skill premium, under a plausible model calibration? Such questions remain unanswered to date.

To conclude: traditionally, in the growth literature technological progress is associated with productivity improvements that benefit all workers, and it is viewed as the chief long-run determinant of *average income* levels. The notion of 'skill bias' – and the literature that has recently blossomed around it – has introduced the theoretical possibility that technological progress benefits only a sub-group of workers, placing technical change also at the center of the *income distribution* debate.

GIOVANNI L. VIOLANTE

See also **biased and unbiased technological change; Griliches, Zvi; Industrial Revolution; information technology and the world economy; Solow, Robert; substitutes and complements; technical change; technology; total factor productivity.**

I am grateful to the CV Starr Center for research support.

Bibliography

Acemoglu, D. 1998. Why do new technologies complement skills? Directed technical change and wage inequality. *Quarterly Journal of Economics* 113, 1055–90.

Acemoglu, D. 2002. Technical change, inequality and the labor market. *Journal of Economic Literature* 40, 7–72.

Aghion, P. 2002. Schumpeterian growth theory and the dynamics of income inequality. *Econometrica* 70, 855–82.

Autor, D., Katz, L. and Krueger, A. 1998. Computing inequality: have computers changed the labor market? *Quarterly Journal of Economics* 113, 1169–213.

Autor, D., Levy, F. and Murnane, R. 2003. The skill content of recent technical change: an empirical exploration. *Quarterly Journal of Economics* 118, 1279–334.

Bartel, A. and Lichtenberg, F. 1987. The comparative advantage of educated workers in implementing new technology. *Review of Economics and Statistics* 69, 1–11.

Caselli, F. 1999. Technological revolutions. *American Economic Review* 89, 78–102.

Galor, O. and Moav, O. 2000. Ability biased technological transition, wage inequality within and across groups, and economic growth. *Quarterly Journal of Economics* 115, 469–97.

Garicano, L. and Rossi-Hansberg, E. 2004. Inequality and the organization of knowledge. *American Economic Review* 94, 197–202.

Goldin, C. and Katz, L. 1998. The origins of technology–skill complementarity. *Quarterly Journal of Economics* 113, 693–732.

Gordon, R. 1990. *The Measurement of Durable Goods Prices.* Chicago: University of Chicago Press.

Greenwood, J., Hercowitz, Z. and Krusell, P. 1997. Long-run implications of investment-specific technological change. *American Economic Review* 87, 342–62.

Greenwood, J. and Yorukoglu, M. 1997. 1974. *Carnegie–Rochester Conference Series on Public Policy* 46, 49–96.

Griliches, Z. 1969. Capital–skill complementarity. *Review of Economics and Statistics* 5, 465–68.

Hornstein, A. and Krusell, P. 1996. Can technology improvements cause productivity slowdowns? In *NBER Macroeconomics Annual*, vol. 11, ed. J. Rotemberg and B. Bernanke. Cambridge, MA: MIT Press.

Hornstein, A., Krusell, P. and Violante, G. 2005. The effects of technical change on labor market inequalities. In *Handbook of Economic Growth*, vol. 1B, ed. P. Aghion and S. Durlauf. Amsterdam: North-Holland.

Katz, L. and Murphy, K. 1992. Changes in relative wages, 1963–1987: supply and demand factors. *Quarterly Journal of Economics* 107, 35–78.

Krusell, P., Ohanian, L., Ríos-Rull, J. and Violante, G. 2000. Capital skill complementarity and inequality: a macroeconomic analysis. *Econometrica* 68, 1029–54.

Milgrom, P. and Roberts, J. 1990. The economics of modern manufacturing: technology, strategy, and organization. *American Economic Review* 80, 511–28.

Nelson, R. and Phelps, E. 1966. Investment in humans, technological diffusion, and economic growth. *American Economic Review* 56, 69–75.

Solow, R. 1957. Technical change and the aggregate production function. *Review of Economics and Statistics* 39, 312–20.

slavery

Slavery entails the ownership of one person by another. As a form of labour organization it has existed throughout history, in a large number of different societies. Indeed, while today we regard slavery as 'the peculiar institution', by historical standards it is wage-labour markets that are 'unusual'. Given slavery's long and varied history a precise definition is not always agreed upon, but there are certain general characteristics found in most societies. The status of slave is generally applied to outsiders – individuals not belonging to the dominant nation, religion or race – although the definition of exactly who is an outsider has varied; Orlando Patterson (1982) considers the basic characteristic of slavery to be 'social death', with a loss of honour as well as legal rights by the enslaved. So widespread and acceptable was slavery that in Europe and the Americas no movement developed to attack slavery as a system until the late 18th century. In Western thought slavery had long been regarded as a necessity or as a 'necessary evil'. The pre-19th-century discussions of slavery often pointed to the desirability (on grounds of religion and morality) of ameliorating the conditions of the enslaved or of facilitating the liberation of individual slaves, but it was only in the late 18th century that widespread thought was given to abolishing the institution itself (see Davis, 1966; 1975; 1984).

The economic basis of slavery

In some societies, particularly at low levels of income, slavery was entered into voluntarily as a means of obtaining a minimum level of living. Slavery generally played a quite different role in such cases than in the major slave societies of Greece, Rome, Brazil, the United States South and the West Indies (see Finley, 1980), where slave labour was used on large-scale agricultural units and in mines to produce goods to be sold in foreign or urban markets. These major slave societies were marked by an extensive external trade in slaves – war captives, kidnapped or otherwise acquired.

The economic basis of slavery seems quite straightforward (see, however, Pryor, 1977). While slavery existed in some parts of the world to provide prestige to owners or for purposes of lineage needs, slavery as an economic institution generally persisted because it provided slaveowners with an ability to capture a surplus of the value of production above the costs of the slave's subsistence. In 1900 the Dutch ethnographer H.J. Nieboer presented a detailed comparative study of slavery throughout the world. He argued that at relatively primitive stages of production, among hunters and fishers and in pastoral societies, slavery was generally non-existent. It was only with the development of settled agriculture in areas with productive land still available that slavery became widespread. The role of 'open resources' – free land – has been more formally presented by Evsey Domar (1970), who argues that only two of the following three conditions can hold simultaneously: free land, free peasants and non-working landowners. The basic argument is that if it were possible for workers to produce more than their subsistence, with unrestricted mobility they would move to freely available land; it is only with a form of labour coercion (serfdom or slavery) that landowners can obtain an income. The benefit of coerced labour to the slaveowner involves a redistribution of that part of the income above subsistence that would go to a free worker. Thus if everything (for example, crops grown, slave and free productivity) were equal, slavery would provide a means of redistributing the excess above subsistence from the labourer to the landed slaveowner.

While this model of forced redistribution points to the need for slave output to exceed subsistence to make the system desirable to slaveowners, it is incomplete as an explanation for most large-scale slave economies. The critical point has been that free labourers have avoided certain types of labour – producing certain crops, or working in certain locations, as well as limiting their labour force participation. Thus slavery expanded the available labour supply, and was particularly important to the production of certain outputs – from mines and in large-scale agricultural units – for which labour would have been available only at very high prices needed to offset non-pecuniary aspects of the labour process (Barzel, 1977). It is the existence of crops such as sugar, for which free labour cannot easily be attracted, that explains the development of New World slavery and accounts for the fact that where slavery has been economically important the slaves have performed functions quite different from those of free labourers.

The Domar–Nieboer argument, by itself, explains neither the existence of slavery as a form of coerced labour nor even the actual existence of coerced labour. The theoretical point is consistent with any form of coerced labour– slavery as well as serfdom – with the general pattern being that slavery dominated where it was necessary to move more labour into an area. The availability of free land by itself can have a quite opposite impact, as Adam Smith (1776) and others argued was the case for the northern states of the United States. There, free land (and labour mobility) meant a wider distribution of property and a more egalitarian society. The conditions for a successful cartel of slaveowners require methods of restricting the movement of the labour force, measures precluding direct bargaining between labourers and potential owners, and means of identifying and returning those slaves who attempt to leave the system. Why such restrictions would not be enforced against Europeans was no doubt the outcome of various cultural, religious and racial forces, so that the availability of free land provided quite different outcomes in the northern and in the southern states. Importantly, different technologies of crop production – the larger optimum scale for cotton, sugar, rice and tobacco, in comparison with the family farm for grains – meant that some form of coerced labour was needed to attract workers onto the southern plantations, while the prices of slaves became too high to permit large numbers of slave imports into the north.

Between the ending of Roman slavery and the origins of large-scale slavery in the New World settlements of the European powers, slavery persisted throughout Europe, but the relative absence of a large trading sector and the limited need to attract a larger population to new areas of settlement meant that slavery was generally economically unimportant and serfdom was a more frequent form of labour organization.

The transatlantic slave trade

The slave trade westward from Africa began with the movement of African slaves to the offshore islands by

Portuguese traders in the middle of the 15th century. From that year until the last slave was landed in Cuba in the late 1860s, more than ten million Africans landed in the New World. Allowing for death in the 'Middle Passage' about 12 million slaves left Africa (Curtin, 1969; Lovejoy, 1983). Higher numbers for the impact of the slave trade on Africa have been presented, allowing for deaths between enslavement and shipment, as well as for estimates of the deaths due to military actions which led to enslavement. While the transatlantic slave trade was most intense in terms of numbers carried per year, it has been estimated that the trans-Saharan slave trade to Arabia and the Middle East may have carried a comparable number over a longer time span. It appears, however, that except for some small areas, the slave trade did not lead to depopulation within Africa. And, although slavery had long existed in Africa, many believe that it was the European contact that transformed African slavery into a harsher institution.

Most countries of western Europe were involved in the Atlantic slave trade, and slaves were sent to all parts of the Americas. Although the high-risk nature meant that some voyages could be very profitable, recent work has cast doubt on the argument that the slave trade provided abnormal profits to European traders, given the African control of the inland traffic and competition among shippers. The first attacks upon slavery were aimed at restricting the transatlantic shipments of Africans. The British, after initial regulatory legislation beginning in 1788, and the United States, as a result of a constitutional compromise permitting its outlawing, ended the slave trade in 1808. Denmark, a minor carrier, had ended the slave trade in 1802. Due to British pressures other countries ended their slave trades, although the 'illegal' slave trade to Cuba and Brazil did not end until after mid-century (Eltis, 1987).

The slave trade was linked to European overseas expansion and played an important part in the settlement of the Caribbean colonies. Because of its early start and late ending Brazil was the largest of the New World recipients of slaves. Large numbers were sent to the British and the French West Indian colonies, whose populations soon became 80 to 90 per cent enslaved blacks. The United States, which was to become the largest slaveholding nation in the 19th century, received only a small part of the African slave trade, its large population being due to the unusually rapid rate of natural increase of the slave population. Cuba rose to dominance as a sugar producer in the 19th century but, unlike the other major sugar-producing Caribbean islands, its population was only about one-third slave, a ratio similar to that in Brazil and in the United States at that time.

The major use of slave labour during the period of the slave trade was in the production of sugar (see Deerr, 1949–50; Klein, 1986). In the case of mainland North America, although slavery existed in every colony, the major uses of slave labour in the 18th century were for tobacco production in the Chesapeake area and rice and indigo production in South Carolina. Cotton was grown in the West Indies and Brazil and, after the invention of the cotton gin in 1793, it became the most important crop produced by slave labour in the United States. These and other slave-grown crops in the Americas, such as coffee and cocoa, were grown on units larger than the family farm.

Slavery in the Americas

In the Americas, after some initial attempts at the enslavement of the native Americans, the condition of slave became limited to Africans and their descendants. The black–white ratio was generally highest in the Caribbean, declining as one moved north and south. There were pronounced differences in the demographic performance of slave populations in the Americas, as reflected in differences between the total number of slaves imported and the number of blacks in various areas. The most dramatic and widely noted differences were seen between the United States and the West Indies. In the West Indies, with few exceptions, the slave population was unable to maintain its numbers, and there was a continued need for new slaves brought from Africa. In the British Caribbean, for example, it is estimated that the approximately two million Africans received before 1808 left a population of only about 780,000 blacks at the time of emancipation (1834). The United States provided a quite different case – unique for a slave population and unusual for any population. There an estimated 600,000 slaves imported resulted in a black population of over 2.3 million in 1830, rising to about 4.4 million in 1860. The relatively high death rates during the period euphemistically known as 'seasoning' (the initial period of exposure to the new disease environment) accounts for some of the correlation between imports and mortality. Despite claims at the time there seems little evidence that planters systematically either worked slaves to death or deliberately engaged in the breeding of slaves.

There was both a considerably higher birth rate for slaves in the United States than in the West Indies (a rate about equal to that of United States whites, about 50 per cent above that in Europe and the West Indies), as well as a lower death rate. The higher birth rate was due in part to an earlier onset of menarche in the United States (due to better nutrition), a shorter child-spacing interval (reflecting differences in lactation practices) and a higher frequency of childbearers among adult women (due perhaps to differences in working conditions as well as in nutrition and health care) (see Steckel, 1985). The lower death rate reflected, in part, the differences between the location and work routines of tropical sugar plantations and the major uses of slave labour in the United States – tobacco and cotton (see Higman, 1984).

Recent studies of the New World slave economies indicate that, despite arguments of contemporaries and subsequent scholars, slavery was expanding throughout its period of existence and that there were no signs that slavery was becoming unprofitable and non-viable on economic grounds. Slave prices in Brazil, Cuba and the United States peaked around 1860 (Moreno Fraginals, Klein and Engerman, 1983). The United States Civil War ended slavery there, and while prices in Brazil and Cuba declined somewhat from peak levels, they remained higher than they had been before mid-century, until there were clear signals that the system was soon to be ended legislatively. The basic importance of labour coercion in sugar production can also be seen in the drive to import contract labour from India, China and Africa, in the West Indies after the ending of slavery – in some cases (most importantly Cuba) even while slavery still existed.

Whatever alternatives may be argued to have been ultimately in their self-interest, throughout the 19th century (and earlier) slavery was profitable to planters. Rather than facing economic difficulties, planters were benefiting from the increased European demands for sugar, cotton, coffee and other plantation commodities. Despite the theoretical logic of the arguments by Cairnes (1862) and others about the ultimate limits to slavery's profitability, for the major slave powers of the New World emancipation required political or military action to overcome a profitable slave economy in which few planters anticipated an immediate collapse in the productive value of their principal assets.

Emancipation and its economic effects

Several states of the northern United States ended slavery by judicial or legislative measures between the Revolutionary War and the early 19th century, as did several of the formerly Spanish colonies after their independence was achieved in the first part of the 19th century, but these were areas where slavery was relatively unimportant. The major sugar-producing area of Saint-Domingue (now Haiti) ended slavery, as the result of a major slave revolt, by the start of the 19th century, and after violently opposing attempts by its new leaders to reintroduce a plantation economy to produce sugar, it became an area devoted to small-scale peasant production.

The first major area to end slavery after the Haitian Revolution was the British West Indies in 1834. The legislation passed in 1833 in response to a decade of pressure by the antislavery movement, provided for (1) a cash payment to owners and slaves of £20 million based on (but less than one-half of) the 1823–30 market values; and (2) an 'apprenticeship' of from four to six years, depending upon the slave's occupation. It is estimated that the value of the monetary compensation, plus the labour dictated by apprenticeship, would be nearly equal to the average 1823–30 value of slaves although, as the

slaveowners pointed out, the loss in the value of land due to emancipation was not compensated (Fogel and Engerman, 1974b). The period of apprenticeship was terminated in 1838.

Metropolitan legislations ended slavery, with compensation, in the French West Indies and in the Danish West Indies in 1848, and in the Dutch colonies in 1863. The American Civil War provided an uncompensated end to United States slavery with the passage of the Thirteenth Amendment. The Moret Law of 1870 provided that all those born to slave mothers in Cuba and Puerto Rico (which then ended slavery, with compensation to masters, in 1873) were considered free, subject to a period of compelled labour. The Rio Branco Law of 1871 in Brazil included a similar provision – a 'law of the free womb' – with a period of controlled labour. In Cuba slavery was ended in 1880, subject to a proposed eight-year period of *patronato*, which was terminated in 1886. In Brazil slavery was ended without compensation in 1888.

The causes of this century-long process of emancipation have become a major historical controversy, with particular attention given to the movements in England and the United States. For England, the view that emancipation was the outcome of disinterested humanitarianism came under attack with the economic interpretation of Eric Williams (1944), which related the timing of the ending of the slave trade and of slave emancipation to the British Industrial Revolution. While the specific groups and mechanisms remain debated, the linkage now stresses the rise of individualism, the 'free labour ideology' and 'modernization', all of which meant that slavery was considered an unacceptable arrangement. Similarly for the United States the link between the 'free labour ideology' and anti-slavery has become central to the interpretation of various political issues of the 1850s which culminated in the Civil War.

The economic effects of slavery upon production can be seen clearly in the general patterns of output that developed after the emancipation of the slaves in most areas. With few exceptions, emancipation led to initial declines in the level of output, with particularly sharp declines in the production of the staple export commodities (Engerman, 1982). There was a movement of ex-slaves away from the plantation sector into small-scale agriculture. The ending of slavery thus demonstrated anew why most New World slavery had developed – people, given free choice, preferred not to work on plantations producing staple crops for exports, since landowners would (or could) not provide sufficient wages to provide an adequate voluntary plantation force from the local population.

There were, however, several notable exceptions. In Barbados the end of slavery did not end the plantation system nor did it lead to declines in sugar production. Rather the labour-to-land ratio was already so high that land for the ex-slaves to move to was unavailable. Barbados thus maintained its plantation sector, while serving as an area for labour outflow to other parts of the Caribbean through the 19th and 20th centuries. Another important exception was Cuba. By the time of emancipation the rise of the large central mill utilizing cane from smaller farms permitted an alternative there which offset the impact of declines in the plantation labour force, and so the output of sugar did not decline after slavery ended.

The two largest slave economies, the United States and Brazil, were nations where slave labour did not reach the proportionate dominance that it did in most of the Caribbean and they had rather different problems. United States sugar, rice, cotton and tobacco production declined with emancipation, with output recovering pre-Civil War levels subsequently, at speeds that were in inverse relation to the optimum scale of plantation production – tobacco and cotton recovering fastest, and sugar and rice least rapidly. Dramatic regional shifts occurred, with prolonged declines in the older tobacco areas of Virginia and the ultimate transfer of rice production from its antebellum base in South Carolina and Georgia to Louisiana. Cotton production expanded quickly after the post-emancipation decline, recovering antebellum peaks within a decade and the United States regained its dominance in world markets by 1880. Yet plantation production declined, and there emerged a system of small-scale farms, often sharecropped, which were less productive than were antebellum plantations; and while most blacks remained within the cotton sector, increased numbers of whites became involved in the cotton economy. Unlike most other ex-slave societies, with this expansion of cotton production the South was exporting a higher proportion of its agricultural output than before emancipation.

In Brazil, the emancipation of slaves had a sharp initial impact upon the expanding coffee industry. Recovery occurred with a decline in the importance of plantations and a shift in the nature of the labour force, with the attraction of immigrants from southern Europe (mainly Italy) to produce on small units. A move to smaller farms producing sugar for central mills also permitted recovery in the production of sugar, with limited numbers of ex-slaves remaining in sugar production.

This general pattern of ex-slave withdrawal from plantation work was a characteristic of the post-emancipation period throughout the Americas. Important in some parts of the Caribbean after slave emancipation was the attraction of a new labour force from overseas, through indentured labour transported under contract to work on plantations for specified periods of time. The areas of the British Caribbean expanding in the late slave period – Trinidad and British Guiana – regained pre-emancipation levels of sugar output within a plantation-based economy. The labour force on the plantations was not primarily ex-slave, but rather indentured labour brought in from Africa, Madeira, China and India, the latter being the predominant group. This system of contract labour had been employed initially in the British Indian Ocean colony of Mauritius. Called 'a new system of slavery', contract labour was a widely discussed and regulated form of labour movement by metropolitan powers as well as in areas of outflow and inflow until its abolition in 1917. The importance of the problem of maintaining a labour force on a continuous basis on sugar plantations is seen also in the expansion of contract labour from foreign areas to the newly emerging sugar-producing regions, such as Fiji, Hawaii, Natal and Australia, in the late 19th century. As late as 1880, the production of most cane sugar that entered export markets took place in areas where the predominant plantation labour force was based either on slavery or indentured labour (Engerman, 1983).

Slavery in economic thought

The consideration of slavery in the literature of economics has helped shape subsequent interpretations of the slave economy. Adam Smith (1776) has been the most frequently quoted economist against slavery, his arguments featuring in contemporary debates as well as historical writings. To Smith slavery was an inefficient system: the slave lacked incentives to work as well as to innovate in technological change. Smith explained the existence of slavery in the production of such crops as sugar and tobacco as indicative that these were so profitable that they could afford to utilize slave labour, something not possible with less profitable crops such as corn. Smith drew upon existing arguments, his proposition on relative incentives having a long history going back to the classical world; but Smith's reputation as a political economist served to make his arguments a central component in the emerging anti-slavery argument. Recent views on Smith stress less that he was presenting an empirical proposition than that his remarks on slavery should be regarded as a basic ideological statement.

Several of the classical economists, for example McCulloch (1825) and Mill (1848), agreed with Smith's contention as to the relative effectiveness of slave and free labour when both were undertaking the same type of work, but they stressed that slavery seemed essential for production in areas and in conditions where free labour could not be obtained, particularly in tropical areas for the production of plantation crops. These, however, they regarded either as special cases (where a different economics applied) or else a transient stage (of undefined duration) along the road to free labour.

The most systematic writer on the economics of slavery was John Elliot Cairnes, whose *The Slave Power* (1862) focused on the United States and combined a theoretical analysis of slave labour with a propagandistic attempt to influence British opinion during the Civil War. He argued that slave labour was inefficient – it is given reluctantly; it is unskilful; it is wanting in versatility – and that it precluded southern economic development because of its negative effects upon technology and upon the attitudes of the free white population towards labour. Cairnes did allow that slavery could survive under certain unusual or temporary conditions – the availability of new lands which would offset the retarding effects of the land exhausted by slave labour. This statement, one of theoretical tendencies, was consistent with arguments that expansion was economically necessary for the southern economy. The role of the increasing ratio of labour-to-land in ending slavery's profitability in the long run was also a theme of various American writers of the early 19th century, and the same point re-emerged in the historiography of United States slavery in the 1920s, with arguments about the natural limits of slavery's expansion, the imminent unprofitability of slavery and the 'needless' Civil War.

Discussions of the economic role of slavery and the causes of its ending were often presented in the 19th and 20th centuries. Yet because of the ideological views of many writers and the emotive implications of slavery it was often difficult to secure entirely accurate descriptions. To the Marxist, slavery is an inefficient economic system, incapable of high levels of productivity and of technical innovation – a system which, however necessary in its time, is incapable of generating sustained economic development. The decline of Roman slavery was thus attributed to its inability to innovate and adapt, and the re-emerged slavery of the modern world was similarly doomed to defeat in economic competition with the capitalistic order. (The relation of slavery and capitalism has itself become a debated subject, even among Marxists.) So, in the examination of the slave economies of the Americas, a perhaps surprising consistency of opinion between those coming from the classical, laissez-faire tradition and those coming from a Marxist perspective meant that a view of slavery as a backward, inefficient economic system had come to dominate the economic and historical literature on slavery.

Economic history and the economics of slavery

In the mid-20th century, work of a detailed empirical nature on slavery in the British Caribbean and in the United States was expanded. This has provided more detailed information as well as having led to new questions and issues being studied. Perhaps the dominant figure in the historiography of the British West Indies has been Eric Williams, the late Prime Minister of Trinidad and Tobago, whose most famous work *Capitalism and Slavery* (1944) dealt with three topics of importance: (1) the link of slavery and racism; (2) the role of slavery and the slave trade in the British Industrial Revolution; and (3) the impact of a declining West Indian economy upon the British abolition of the slave trade and the emancipation of slaves. Williams argued that it was the need for a cheap labour force that led to slavery, and that the justification for enslavement of Africans led to the development of racist beliefs about blacks – a view which

remains the subject of a major historical controversy. Using arguments provided by contemporaries justifying the slave trade, Williams traced an important role in financing and in providing markets for British industrial development to the slave economies of the West Indies and the slave trade with Africa. Williams's last two propositions have formed the basis of much of the recent work on slavery in West Indian economic history. Recent writings point to a more limited role for slavery and the slave trade in British industrialization than that advanced by Williams. Even more attention has been devoted to the question of the conditions for the abolition of slavery and its link to a possible decline of the economies of the British West Indies. The thesis of decline has come under strong attack, particularly by Seymour Drescher (1977), who argues that the politically mandated end of the slave trade led to declining West Indian economic fortunes, and not vice versa. The issue of the specifics of the movement to end the slave trade, and the relative contributions of ideological, class and economic forces has become a central historical issue (see Drescher, 1986; Solow and Engerman, 1987).

In the United States there has been a more specific concern by economists and economic historians with issues related to the economics of slavery. A key breakthrough in the new approach to the economics of slavery was an article by Conrad and Meyer (1958) which dealt with the profitability of antebellum slavery. This article was concerned with a major historical issue – was slavery economically unprofitable in the late antebellum period? – and was also intended more generally to demonstrate the value of an economic approach to historical problems. By framing the issue as one of the rate of return on investment, and using available data on interest rates, slave prices, life expectation, costs of consumption by slaves and of their supervision, and crop production and prices, they argued that a planter purchasing a slave at the market price in the late antebellum period would have earned a return equal to that upon alternative assets. The response to this article – both positive and negative, in regard to substance as well as method – has been enormous, and the economics of slavery became one of the most heated and widely discussed topics in American history. It was seen, however, that profitability, as measured by 'normal' profits on an existing asset, did not really adequately answer the question of the possible economic ending of slavery in the absence of the Civil War, since the comparison had been based on market price, and not on the cost of 'producing' a slave.

An analysis by Yasukichi Yasuba (1961) pointed out that, given the illegality of slave imports and the constraints on the demographic expansion of the slave population, the market price of slaves could exceed the costs of rearing slaves, yielding a rent to the slaveowning class (see also Evans, 1962). Yasuba showed that the surplus above rearing costs for slaves was rising in the late antebellum period, peaking just before the onset of the

Civil War. Thus, far from being on the verge of economic collapse, slavery was becoming more profitable to the slaveowning class, who did not foresee an economic end to their system in the immediate future. The linking of Easterlin's regional income estimates with Gallman's GNP estimates for 1840 to 1860 indicated that the South was growing about as rapidly in terms of per capita income as was the North, and in 1860 had reached a level of per capita income above that of the agrarian Midwest and most of the rest of the world (Fogel and Engerman, 1971; 1974a). While these estimates cover only a limited time span, they did help to provide a different view of the dynamics of the southern slave economy.

Questions of profitability, viability, and rates of growth of income were not seen, however, as of central historical interest by historians such as Eugene Genovese (1965), who argued that the important questions concerned rather the development and potential for industrialization in the slave economy, reflected particularly in the differences in economic structure in comparison with that of the northern states, as well as the political issues posed by the differing class structures of the two societies. It is argued that the antebellum expansion of the South was due to the demand for one major crop, cotton, leading to a less diversified and industrialized economy than in the North – a growth that could not be sustained. At debate remain the causes and consequences of the southern specialization in agriculture rather than the development of a larger manufacturing sector, and the implications of the limited industrialization, urbanization, and expansion of education (in comparison with the North).

An extensive debate on the efficiency of slavery in United States agriculture was generated by the application of total factor productivity estimates by Fogel and Engerman (1974a). The question was an old one – frequent comparisons of slave versus free labour had been made by contemporaries as part of the anti-slavery argument. In their analysis Fogel and Engerman used a sample of over 5,000 farms in cotton-producing counties, drawn by William Parker and Robert Gallman from the census manuscript schedules. The specific contention that in 1860 southern agriculture was more efficient than northern, in the sense of getting more output per unit of input, became widely discussed and criticized (see David et al., 1976; Wright, 1978). The extensive debate included discussion of alternative measures of factor inputs and adjustments for variations in crop-mix, as well as arguments about the emotive content of the term 'efficiency'. Nevertheless, this debate did lead to changes in depictions of the slave economy. More attention was paid to the flexibility of the economy, in terms of shifting patterns of production and location in response to economic stimuli as well as in the use of various mechanisms, in addition to the whip, to elicit work effort from the slaves. More attention was also given to the standard of living provided for the slaves and to their actual work experiences.

Much of the writing on United States slavery in the 1970s, coming from a variety of backgrounds and using different sources, also led to reinterpretations of the nature of slavery and the slave experience. Attention was given to the slave culture, affected, but not destroyed, by the controls of the master. While there remain disagreements about the frequency of the sales of slaves and the extent to which they separated spouses as well as young children, much work has established the strength of the slave family. Descriptions of slave religion, slave culture and the slave family all pointed to the capacity of the slaves to resist being reduced to Sambos – a point with obvious implications for the behaviour of masters as well. Slavery has come to be seen as a system which, with its initial imbalance of power, permitted a range of give-and-take between master and slave (see Genovese, 1974), with the power of the former not as complete, and the impact on the latter not as destructive, as earlier argued.

S.L. EGERMAN

Bibliography

Barzel, Y. 1977. An economic analysis of slavery. *Journal of Law and Economics* 20, 87–110.

Cairnes, J.E. 1862. *The Slave Power*. London: Parker, Son & Bourn.

Conrad, A.H. and Meyer, J.R. 1958. The economics of slavery in the antebellum south. *Journal of Political Economy* 66, 95–130.

Curtin, P.D. 1969. *The Atlantic Slave Trade*. Madison: University of Wisconsin Press.

David, P.A. et al. 1976. *Reckoning with Slavery*. New York: Oxford University Press.

Davis, D.B. 1966. *The Problem of Slavery in Western Culture*. Ithaca: Cornell University Press.

Davis, D.B. 1975. *The Problem of Slavery in the Age of Revolution 1770–1823*. Ithaca: Cornell University Press.

Davis, D.B. 1984. *Slavery and Human Progress*. New York: Oxford University Press.

Deerr, N. 1949–50. *The History of Sugar*. 2 vols. London: Chapman & Hall.

Domar, E. 1970. The causes of slavery or serfdom: a hypothesis. *Journal of Economic History* 30, 18–32.

Drescher, S. 1977. *Econocide*. Pittsburgh: University of Pittsburgh Press.

Drescher, S. 1986. *Capitalism and Antislavery*. London: Macmillan.

Eltis, D. 1987. *Economic Growth and the Ending of the Translatlantic Slave Trade*. New York: Oxford University Press.

Engerman, S.L. 1982. Economic adjustments to emancipation in the United States and British West Indies. *Journal of Interdisciplinary History* 13(2), 191–220.

Engerman, S.L. 1983. Contract labour, sugar, and technology in the nineteenth century. *Journal of Economic History* 43, 635–59.

Evans, R., Jr. 1962. The economics of American Negro slavery. In Universities National Bureau Committee for

Economic Research, *Aspects of Labor Economics*. Princeton: Princeton University Press.

Finley, M.I. 1980. *Ancient Slavery and Modern Ideology*. New York: Viking Press.

Fogel, R.W. and Engerman, S.L. 1971. The economics of slavery. In *The Reinterpretation of American Economic History*, ed. R.W. Fogel and S.L. Engerman. New York: Harper & Row.

Fogel, R.W. and Engerman, S.L. 1974a. *Time on the Cross*. 2 vols. 1: *The Economics of American Negro Slavery*; 2: *Evidence and Methods*. Boston: Little, Brown.

Fogel, R.W. and Engerman, S.L. 1974b. Philanthropy at bargain prices: notes on the economics of gradual emancipation. *Journal of Legal Studies* 3, 377–401.

Genovese, E.D. 1965. *The Political Economy of Slavery*. New York: Pantheon.

Genovese, E.D. 1974. *Roll, Jordan, Roll*. New York: Pantheon.

Higman, B.W. 1984. *Slave Populations of the British Caribbean, 1807–1834*. Baltimore: Johns Hopkins University Press.

Klein, H.S. 1986. *African Slavery in Latin America and the Caribbean*. New York: Oxford University Press.

Lovejoy, P.E. 1983. *Transformations in Slavery*. Cambridge: Cambridge University Press.

McCulloch, J.R. 1825. *The Principles of Political Economy*. Edinburgh: W. & C. Tait.

Mill, J.S. 1848. *Principles of Political Economy*. 2 vols. London: J.W. Parker.

Moreno Fraginals, M., Klein, H.S. and Engerman, S.L. 1983. The level and structure of slave prices on Cuban plantations in the mid-nineteenth century: some comparative perspectives. *American Historical Review* 88, 201–18.

Nieboer, H.J. 1900. *Slavery as an Industrial System*. The Hague: M. Nijhoff.

Patterson, H.O. 1982. *Slavery and Social Death*. Cambridge, MA: Harvard University Press.

Pryor, F.L. 1977. *The Origins of the Economy*. New York: Academic Press.

Solow, B.L. and Engerman, S.L., eds. 1987. *Caribbean Slavery and British Capitalism*. Cambridge: Cambridge University Press.

Smith, A. 1776. *An Inquiry into the Nature and Causes of the Wealth of Nations*. Ed. E. Cannan, New York: Modern Library, 1937.

Steckel, R.H. 1985. *The Economics of US Slave and Southern White Fertility*. New York: Garland Publishing.

Williams, E. 1944. *Capitalism and Slavery*. Chapel Hill: University of North Carolina Press.

Wright, G. 1978. *The Political Economy of the Cotton South*. New York: W.W. Norton.

Yasuba, Y. 1961. The profitability and viability of plantation slavery in the United States. *Economic Studies Quarterly* 12, September, 60–7. Repr. in *The Reinterpretation of American Economic History*, ed. R.W. Fogel and S.L. Engerman, New York: Harper & Row, 1971.

Slichter, Sumner Huber (1892–1959)

Slichter was both a wide-ranging general economist and a scholar in industrial relations, regarding the two disciplines as parts of a seamless whole. He wrote an introductory economics textbook (1931) and was probably the most widely read economist by the general public of his day. He was a highly respected economic forecaster, Paul Samuelson calling him 'our best economic forecaster for the period 1935–55', and served as president of the American Economic Association, 1940–1. In industrial relations his two large classics (1941 and 1960) grew out of extended field work.

Slichter took his undergraduate degree in 1913 at the University of Wisconsin (where his father was professor of mathematics), did graduate work there with John R. Commons and completed his doctorate (1918) at the University of Chicago with H.A. Millis. He taught at Cornell for a decade, moved to Harvard Graduate School of Business Administration in 1930 and joined the Department of Economics in 1935. In 1940 he was appointed the first university professor at Harvard.

Among the themes that Slichter stressed were that the American economy was not in danger of stagnation; that the Second World War would not be followed by a depression but rather a boom; that America was becoming a 'laboristic' economy in which value judgements of the community reflect those of employees; that the challenges that unions have presented to managements have created superior and better-balanced managements; and that a vigorous and healthy economy is associated with an upward creep in prices that results from strong demand for goods and services and a slow climb in labour costs.

JOHN T. DUNLOP

Selected works

1919. *The Turnover of Factory Labor*. New York: D. Appleton.

1931. *Modern Economic Society*. New York: Henry Holt.

1941. *Union Policies and Industrial Management*. Washington, DC: Brookings.

1943. *Present Savings and Postwar Markets*. New York: McGraw-Hill.

1947. *Trade Unions in a Free Society*. Cambridge, MA: Harvard University Press.

1948. *The American Economy: Its Problems and Prospects*. New York: Knopf.

1951. *What's Ahead for American Business?* Boston: Little, Brown.

1960. (With J.J. Healy and E.R. Livernash.) *The Impact of Collective Bargaining on Management*. Washington, DC: Brookings.

1961. *Potentials of the American Economy: Selected Essays of Sumner H. Slichter*, ed. J.T. Dunlop. Cambridge, MA: Harvard University Press. A full bibliography appears on pp. 435–56.

sliding scales (wages)

Theories of long-term employment contracts imply strong private incentives for employers and employees to link wages to product prices, quite apart from the external benefits emphasized by Weitzman (1984), whether or not wages are also indexed to other variables such as consumption goods' prices. Such arrangements have been extremely rare since the Second World War, but many schemes linking wage rates to product prices, referred to generally as 'sliding scales', were observed in Britain and the United States from the 1860s to the 1930s (Howard, 1920; Munro, 1885–6; Palgrave, 1896; Poole, 1938; US Industrial Commission, 1901, pp. 89–98, 135–6). Recent studies of historical sliding scales include Greenfield (1960), Treble (1987), South (1990), and Hanes (2007).

Most sliding scales were a result of negotiation or arbitration between unions and employers' associations in coal or iron-ore mining, or in the metals industries – iron, steel and tinplating. In the United States, sliding scales were also used in zinc, silver and copper mines, in glass manufacture, and in the textile mills of Fall River, Massachusetts, between 1905 and 1910. They were written agreements but not legally enforceable contracts, and were meant to hold for at most one year or for no fixed duration – either side could withdraw after some weeks' notice. Terms specified time- or piece-rate wages, but not employment levels. Nearly all sliding scales set minimum levels below which wages could not fall, no matter what happened to prices; some included maximums as well. Within these limits, wage adjustments took place at predetermined points in time, with intervals ranging from one week to six months, as a continuous or stepped-schedule function of prices observed in the previous interval. A number of scales based wages on the 'margin' between prices of products and material inputs. Agreements were frequently renegotiated in response to changes in external labour-market conditions or costs of non-labour inputs left unaccounted for in the scale's formula. For some sliding scales, prices were taken from press reports of open-market transactions. More frequently, prices were taken from employers' account books, examined by professional accountants approved by both parties or the arbitrator. Because firms did not want to reveal sales information to competitors, examiners typically reported only a summary statistic used in the wage formula.

In the United States, unions and their sliding scales were expelled from most iron and steel plants after the 1900s. Where unions survived they continued to negotiate sliding scales up to the 1920s (Daugherty, de Chazeau and Stratton, 1937, pp. 143–4). By the early 1940s, however, sliding scales had nearly disappeared, even in mining (US Bureau of Labor Statistics, 1940, p. 13). In Britain, sliding scales or similar 'proceeds-sharing' plans remained widespread within mining and the metals industries throughout the 1930s. These schemes were suspended at the beginning of the Second World War in response to the advent of price controls, and never revived (Burn, 1961, p. 27; Haynes, 1953, p. 26). Hanes (2007) argues that pre-1940s sliding scales were not examples of indexation as described by theoretical employment-contract models, but rather devices to reduce the frequency of costly strikes in the absence of contracts; that their scope was severely limited by the inability of rank-and-file union members to observe product prices, even in industries where price information was relatively plentiful; and that they disappeared after the 1930s because US unions gained the ability to enter long-term contracts, while British mining and metals industries remained under forms of government control that broke the link between product prices and labour demand.

CHRISTOPHER HANES

See also **wage indexation.**

Bibliography

Burn, D. 1961. *The Steel Industry 1939–1959: A Study in Competition and Planning.* Cambridge: Cambridge University Press.

Daugherty, C.R., de Chazeau, M.D. and Stratton, S.S. 1937. *The Economics of the Iron and Steel Industry.* New York: McGraw-Hill.

Greenfield, H.I. 1960. Sliding wage scales. Doctoral dissertation, University of Colombia.

Hanes, C. 2007. The rise and fall of the sliding scale, or why wages are no longer indexed to product prices. Working paper, Department of Economics, State University of New York at Binghamton.

Haynes, W.W. 1953. *Nationalization in Practice: The British Coal Industry.* Boston: Harvard University Graduate School of Business.

Howard, S. 1920. *The Movement of Wages in the Cotton Manufacturing Industry of New England since 1860.* Boston: National Council of American Cotton Manufacturers.

Munro, J.E.C. 1885–6. Sliding scales in the iron industry. *Transactions of the Manchester Statistical Society,* 1–43.

Munro, J.E.C. 1889–90. Sliding scales in the coal and iron industries from 1885 to 1889. *Transactions of the Manchester Statistical Society,* 119–71.

Palgrave, R.H.I. 1896. Sliding scale (wages). *Dictionary of Political Economy,* vol. 2, London: Macmillan.

Poole, A.G. 1938. *Wage Policy in Relation to Industrial Fluctuations.* London: Macmillan.

South, N.M. 1990. Price contingent wage contracts in British coal and iron and steel 1860–1913: theory and evidence. Ph.D. dissertation, University of Toronto.

Treble, J.G. 1987. Sliding scales and conciliation boards: risk-sharing in the late 19th century British coal industry. *Oxford Economic Papers* 39, 679–98.

US Bureau of Labor Statistics. 1940. Wage-adjustment provisions in union agreements. *Monthly Labor Review* 50, 6–15.

US Industrial Commission. 1901. *Report of the Industrial Commission on the Relations and Conditions of Capital and Labor.* Washington: GPO.

Weitzman, M.L. 1984. *The Share Economy.* Cambridge, MA: Harvard University Press.

Slutsky, Eugen (1880–1948)

Born in the Yaroslav province of Russia, Slutsky had troubled years as a student: he enrolled in the department of physics and mathematics at Kiev University, was expelled for taking part in student revolts, went abroad to the Munich Institute of Technology to study engineering and finally graduated in the department of law in 1911 back at Kiev University. He became a member of the faculty at Kiev Institute of Commerce in 1913 and full professor there in 1920. In 1926 he moved to Moscow as a staff member of the Conjuncture Institute; in 1934 he became a staff member of the Mathematical Institute of the University of Moscow and in 1936 a member of the Mathematical Institute of the Academy of Sciences, Moscow, a post which he held until his death.

Slutsky was a mathematician, statistician and economist. His fame as an economist rests mainly on one single contribution (1915), which went unnoticed until the 1930s, when it was discovered independently by Dominedò (1933, p. 790), Schultz (1935, pp. 439ff), and Allen (1936), and subsequently influenced the further development of consumer theory. Hicks – who, together with Allen (Hicks and Allen, 1934), had independently arrived at Slutsky's results – writes: 'The theory to be set out in this chapter and the two following is essentially Slutsky's … The present volume is the first systematic exploration of the territory which Slutsky opened up' (Hicks, 1939, p. 19). Building on earlier work by Pareto (who had already derived the formulae which express the change in the consumer's demand when any one of its arguments changes, but without seeing their implications), Slutsky showed that the effect of a price change on the quantity demanded can be divided into two effects. One is the effect of a *compensated variation* of price; if a price increases and the consumer is given an income increase so as to make possible the purchase of the same quantities of all the goods previously purchased, the individual – though being in the position to purchase the preceding bundle of goods – will no longer consider it preferable to any other, and there will take place some kind of *residual variation* of demand. This is called the *residual variability* by Slutsky (the substitution effect in Hicks's terminology). It should be noted that the compensated variation of price can also be defined in terms of the income change which leaves the consumer's

real income unchanged, that is which causes the consumer to remain on the *same* indifference curve (this is the concept used by Hicks, 1939, in the text, while in the mathematical appendix he gives the same definition as Slutsky). Although the two definitions are equivalent for infinitesimal changes (as was first shown by Mosak, 1942), Slutsky's is preferable from the operational point of view since it does not require knowledge of the consumer's indifference map. The other effect is the *income effect*, which gives the change in the consumer's purchases when his money income changes at unchanged prices. The two effects turn out to be independent and additive and their algebraic sum gives the *price effect*: this is, in Hicks' terminology, the 'Fundamental Equation of Value Theory', also called the Slutsky Equation.

Slutsky proved the complete properties of the various effects and of the demand curves. The income effect may be either normal (demand increases as income increases: 'relatively indispensable goods' in Slutsky's terminology) or abnormal ('relatively dispensable' goods). The 'own' substitution effect is always negative ('The residual variability of a good in the case of a compensated variation of its price, is always negative', [1915] 1953, p. 42) and the cross substitution effect is symmetric ('The residual variability of the j-th good in the case of a compensated variation of the price p_i is equal to the residual variability of the i-th good in the case of a compensated variation of the price p_j', [1915] 1953, p. 43). The 'own' price effect, therefore, is necessarily normal in the case of relatively indispensable goods. Slutsky also proved the relation which implies that the individual demand functions are homogeneous of degree zero. He gave a definition of complementary and competing goods, and made an important methodological point which is usually overlooked in his contribution: he stressed the need for *experiment* in order to obtain all the values of the relevant magnitudes (which cannot be obtained by observation of existing budgets) which enter into the definition. This emphasis on the need for experimental verification of economic laws, which concludes his contribution, is worthy of note and obviously arises from his statistical background.

Slutsky did no other noticeable work in economics but made important contributions to mathematical statistics and probability theory.

In (1914) he suggested the use of a χ^2 variate to test the goodness of fit of a regression line ('line' is taken in the broad sense, i.e. including both straight lines and curves); as a logical consequence, he introduced the concept of *minimum chi-square estimator* ('the most probable values of the coefficients will be those which bring our χ^2 to a minimum', 1914, p. 83) as a general method of fitting regressions. This paper was written several years before R.A. Fisher's work on the subject.

Slutsky was one of the originators of the theory of stochastic processes and time-series analysis. In his renowned (1927) paper he proved several important theorems. One is that the summation of random causes may be the source of cyclic or undulatory processes, and that these waves will show an approximate regularity in the sense that they can be approximated quite well by a relatively small number of terms (sine curves) of the Fourier series. Another is the sinusoidal limit theorem, which states that under certain conditions the summation of random causes will tend to give rise to a specific sine wave. For example, if one takes a moving average (of two terms) of a random series n times and then takes the mth difference of the result, and lets $n \to \infty$ so that m/n tends to a constant c between zero and one, it follows that the series will tend to a sine wave with wavelength arc cos $(1-c)/(1+c)$. A corollary of these theorems is the famous Slutsky–Yule Effect (so named because it was also independently discovered by Yule): if a moving average of a random series is taken (for example to determine trend), this may *generate* an oscillatory movement in the series where none existed in the original data.

Slutsky also worked in the theory of probability, where he studied the concept of asymptotic convergence in probability (e.g. 1925, 1928, 1929). He spent the last years of his life in preparing tables for the computation of the incomplete gamma-function and the chi-square probability distribution (1950).

GIANCARLO GANDOLFO

Selected works

A bibliography (1912–46) is contained in the memorial article (in Russian) by A.N. Kolmogorov, in *Uspekhi Matematicheskikh Nauk*, Vol. 3, No. 4, 1948. This bibliography is reproduced in Allen (1950). A collection of selected papers was published posthumously in Russian (1960). On Slutsky's life and works see also the memorial article (in Russian) by N. Smirnov, in *Izvestiia Akademiia Nauk SSSR*, Mathematical Series, Vol. 12, 1948, and Allen (1950).

1914. On the criterion of goodness of fit of the regression lines and on the best method of fitting them to the data. *Journal of the Royal Statistical Society* 77, Pt I, December, 78–84.
1915. Sulla teoria del bilancio del consumatore. *Giornale degli Economisti e Rivista di Statistica* 51(July), 1–26 Trans. as 'On the theory of the budget of the consumer' in *Readings in Price Theory*, ed. K.E. Boulding and G.J. Stigler, London: Allen & Unwin, 1953, 26–56.
1925. Über stochastische Asymptoten und Grenzwerte. *Metron* 5(December), 3–89.
1927. The summation of random causes as the source of cyclic processes. (In Russian.) *Problems of Economic Conditions*, The Conjuncture Institute, Moscow, Vol. 3, No. 1. Revised English version in *Econometrica* 5(April) 1937, 105–46.
1928. Sur un critérium de la convergence stochastique des ensembles de valeurs éventuelles. *Comptes rendu des séances de l'Académie des Sciences* 187, Paris, 17 July to 13 August.

1929. Quelques propositions sur les limites stochastiques éventuelles. *Comptes rendu des séances de l'Académie des Sciences* 189, Paris, 2 September.
1950. *Tables for the computation of the incomplete Gamma-function and the Chi-square probability distribution*. (In Russian.) Ed. A.N. Kolmogorov, Moscow: Akademiia Nauk SSSR (posthumous).
1960. *Selected Works*. (In Russian.) Moscow: Akademiia Nauk SSSR (posthumous).

Bibliography

Allen, R.G.D. 1936. Professor Slutsky's theory of consumers' choice. *Review of Economic Studies* 3(February), 120–29.
Allen, R.G.D. 1950. The work on Eugen Slutsky. *Econometrica* 18(July), 209–16.
Dominedò, V. 1933. Considerazioni intorno alla teoria della domanda, I-IIe principali premesse e caratteristiche delle curve statiche. *Giornale degli Economisti e Rivista di Statistica* 48(November), 765–807.
Hicks, J.R. 1939. *Value and Capital*. Oxford: Clarendon Press. 2nd edn, 1946.
Hicks, J.R. and Allen, R.G.D. 1934. A reconsideration of the theory of value. *Economica* 1, Pt I, February, 52–76; Pt II, May, 196–219.
Mosak, J. 1942. On the interpretation of the fundamental equation of value theory. In *Studies in Mathematical Economics and Econometrics in Memory of Henry Schultz*, ed. O. Lange, F. McIntyre and T.O. Yntema. Chicago: University of Chicago Press, 69–74.
Samuelson, P.A. 1947. *Foundations of Economic Analysis*. Cambridge, Mass.: Harvard University Press Enlarged edn, 1983.
Schultz, H. 1935. Interrelations of demand, price, and income. *Journal of Political Economy* 43(August), 433–81.

small-world networks

Small-world networks – so named on account of their resemblance to the 'small-world hypothesis' – exhibit short global path lengths in spite of considerable local structure. The small-world hypothesis, long an article of popular belief (Guare, 1990), was first investigated by Pool and Kochen (1978), who posed the following question: 'How many pairs of persons in the population can be joined by a single acquaintance, how many by a chain of two persons, how many by a chain of three, etc?' Pool and Kochen showed that when all individuals choose their acquaintances uniformly at random from the entire population, the number of pairs i, j connected by a path of length d increases exponentially fast in d. Thus under quite reasonable assumptions regarding the average number k of acquaintances per person, most pairs of individuals in even a very large population should be connected by paths only a few steps long. For example, for the population of the United States at the time – roughly 200 million – and assuming that individuals possessed an average of roughly 1,000 acquaintances, Pool and Kochen estimated that a randomly selected pair could be connected in only three or four steps.

This result, it turns out, is equivalent to a standard result from the theory of random graphs (Bollobas, 1985; Erdos and Renyi, 1959; Solomonoff and Rapoport, 1951), namely, that the average shortest path length separating two randomly chosen individuals should be proportional to $\log N$ (for $N \gg k \gg 1$), where N is the size of the population. (Strictly speaking, this result holds only when the network is connected, which is not guaranteed for small k; hence the second condition $k \gg 1$.) From this result, we can infer our first definition of what it means for the world to be 'small':

Definition 1: A network can be said to exhibit the small-world property when the average shortest path length of the network $L \propto \log N$.

Pool and Kochen noted, however, that social networks, far from being random, exhibit considerable structure; that is, individuals are more likely to know each other when they share certain traits, such as geographical proximity, socio-economic status or common interests. The interesting (but far less tractable) version of the small-world hypothesis is therefore not that path lengths can be short but that path lengths should remain short even when the network in question is far from random. Pool and Kochen also speculated that the increase in L resulting from the introduction of social structure to a random network would be a good measure of that structure. As we shall see, however, the validity of the small-world hypothesis actually implies the converse of this claim.

Six degrees of separation

Stimulated by Pool and Kochen's paper, the social psychologist Stanley Milgram, along with his graduate student Jeffrey Travers, decided to test the small-world hypothesis experimentally, using what they called the 'small-world method' (Milgram, 1967; Travers and Milgram, 1969):

1. Recruit one or more volunteers to be 'targets' of the experiment.
2. For each target, recruit some large number of initial 'senders'.
3. Give each sender a description of the target sufficient to identify him or her uniquely (for example, name, address, occupation and employer, wife's maiden name, and so on).
4. Instruct each sender to forward a message to the target in one of two ways:
   (a) If target is an acquaintance (that is, is known to sender on a first-name basis), forward directly to target.

(b) If not, forward message, target information, and instructions to an acquaintance who is 'closer' to target than sender.

5. Repeat steps 3 and 4 with additional senders until chain either *completes* (reaches target) or *terminates* (is not forwarded for any other reason). In Milgram's version, subsequent senders also received a list of previous senders, in order to avoid cycles; however, in practice these cycles are extremely unlikely, and this requirement has subsequently been dropped (Dodds, Muhamad and Watts, 2003).

Travers and Milgram implemented this protocol using a single target (a Boston stockbroker who was an acquaintance of Milgram's) and 296 senders: 100 from Boston, 100 blue-chip stockholders from Omaha, Nebraska, and 96 Omaha residents who were randomly selected from a list of people who had agreed to receive marketing literature. Famously, they found that the chains which reached the target were surprisingly short – approximately six steps on average, where, unsurprisingly, chains initiated by stockholders in Nebraska were shorter than those from randomly selected Nebraskans, and chains that began in Boston were shorter still. However, most chains (roughly 80 per cent) terminated before completion. Travers and Milgram also noted that a disproportionate fraction of completed chains were delivered at the last step by a small number of individuals (one man delivered 16 letters) whom Travers and Milgram dubbed 'sociometric stars'. Subsequently, small-world experiments have been conducted a number of times for different-sized populations (for reviews, see Garfield, 1979; Kleinfeld, 2002; Kochen, 1989). In most of these studies completion rates are low, but the chains that do complete are short (Dodds, Muhamad and Watts, 2003; Korte and Milgram, 1970).

If one assumes that chains terminate out of insufficient interest or motivation on the part of senders, and not because the underlying network is disconnected or non-navigable (this assumption has been disputed recently by Kleinfeld, 2002, but no internally self-consistent alternative has yet been proposed), then it is possible to compute the hypothetical distribution of chain lengths corresponding to zero attrition (Dodds, Muhamad and Watts, 2003; White, 1970). Because longer chains have more opportunities to terminate, this 'ideal' distribution necessarily yields higher estimates of chain length than estimates based solely on completed chains. Nevertheless, the revised estimates – typically between seven and nine steps – are still consistent with the small-world hypothesis, even for reasonably high attrition rates. Small-world experiments therefore suggest two surprising features of large social networks: (*a*) in spite of their considerable structure, any pair of randomly selected individuals is likely to be connected via some short path; and (*b*) these individuals can actually find such a path, given only local information about the network.

Small-world networks

Addressing the first property – that randomly selected individuals in a large network can be connected via a short chain of intermediaries – Watts and Strogatz (1998) analysed a network model that incorporated elements of both social structure and randomness. In the model 'social structure' was represented by a uniform one-dimensional lattice, where each node was connected to its k nearest neighbours on the lattice, and 'randomness' was characterized by a tunable parameter p, which specified the probability that a link in the lattice would be *randomly rewired* (see Figure 1). Defining the *clustering coefficient C* of a network as the average probability that two neighbours of a given node would themselves be neighbours, they showed that when $p = 0$ (completely ordered) the network is 'large' ($L(0) \sim N/2k$) and 'highly clustered' ($C(0) \simeq 3/4$), and when $p = 1$ (completely random) it is 'small' ($L(1) \sim \log N/\log k$) and 'poorly clustered' ($C(1) \sim k/N$), thus suggesting that path lengths are short only when clustering is low. In contrast with Pool and Kochen's prediction, however, they showed that the model exhibits a broad region of p values in which $C(p)$ is high relative to its random limit $C(1)$, yet $L(p)$ is roughly speaking as 'small' as possible (see Figure 2).

Watts and Strogatz coined the term 'small-world networks' to refer to networks in this class. Because the conditions required for any network to exhibit small-world properties (just a small fraction of long-range, random 'short cuts') were relatively weak, Watts and Strogatz predicted that many real-world networks – whether social networks or otherwise – would be small-world networks. They then checked this prediction by considering three network data-sets – the affiliation network of movie actors, the power transmission grid of the western United States, and the neural network of the nematode *C. elegans* – and found that all three examples satisfied the small-world criteria (Watts and Strogatz, 1998). Numerous authors have since investigated the properties of this model, and many large empirical networks have been found to exhibit small-world-like properties (see Newman, 2000; 2003; and Watts, 2004, for reviews of this literature, as well as Newman, Barabasi and Watts, 2006, for a collection of early papers).

Searchable small-world networks

The small-world network models of the kind proposed by Watts and Strogatz, however, do not satisfy the second striking feature of Travers and Milgram's results, namely, that individuals can locate short paths using only local information about the network structure. This point, first raised by Kleinberg (2000a; 2000b), led him to propose a more general class of partly ordered, partly random networks in which the spatial distribution of random links is permitted to vary according to the probability distribution $P_{ij} \propto r_{ij}^{-\gamma}$, where r_{ij} is the distance between nodes i and j measured on some underlying lattice of dimension D, and γ is a tunable parameter. Kleinberg then proved that only when $\gamma = D$ will the network be searchable in the sense that short paths not only exist but are also discoverable: when $\gamma < D$ short paths exist, but cannot be found using only local information; and when $\gamma > D$ the shortest paths, although discoverable, are not short. Subsequently, more realistic models of searchable small-world networks have been proposed, incorporating for example 'social' (Adamic and Adar, 2005; Watts, Dodds and Newman, 2002) as well as geographical (Liben-Nowell et al., 2005) distance, and also heterogeneity of degree (Simsek and Jensen, 2005).

What is a 'short' path anyway?

Although the observed path lengths in small-world experiments do indeed seem to be 'short', it is not clear that they are short in the sense of Definition 1 above; that is, proportional to $\log N$. Indeed, short of conducting controlled experiments in which N is varied systematically over several orders of magnitude, strict logarithmic scaling of chain lengths would be difficult to establish empirically. Furthermore, Chung and Lu (2002) have

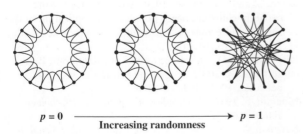

Figure 1 Schematic of the Watts–Strogatz model of partly ordered, partly random networks. *Note*: $p =$ the probability of random edge rewiring.

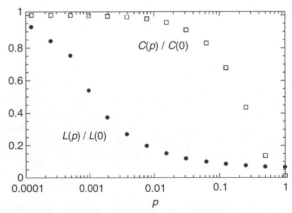

Figure 2 Normalized average path length $L(p)/L(0)$ and clustering coefficient $C(p)/C(0)$ versus p (note log scale)

demonstrated that so-called 'scale-free' random networks, whose degree distributions exhibit 'power law' tails (that is, $P[K>k] \propto k^{-\alpha}$ for $k \gg 1$, where $1<\alpha<2$), exhibit even shorter average path lengths than in Definition 1 (formally, $L \propto \log \log N$); whereas in contrast Kleinberg (2000b) has suggested that the small-world hypothesis is consistent with longer path lengths (formally, $L \propto (\log N)^{\mu}$, where $\mu \geq 1$). Ultimately, however, it may not matter, as for realistic network sizes all these definitions yield similar answers, and thus are unlikely to be empirically distinguishable. Also, in the presence of chain attrition, the absolute length of chains is more relevant than its scaling behaviour: from a practical perspective, if a chain doesn't complete, it doesn't matter whether or not it is as short as it could be; it is still too long. For these reasons, Watts, Dodds and Newman (2002) have proposed an alternative definition of 'short':

Definition 2: Given a message failure probability p and minimal required chain completion rate r, the small-world hypothesis requires (approximately) that average path length $L \leq \log r/\log(1 - p)$.

Sociometric stars

The final result of Travers and Milgram above – that chains tend to be completed by highly connected individuals, called variously 'stars', 'hubs', 'connectors' or 'funnels' – appears to be supported by recent theoretical models (Adamic et al., 2001; Adamic, Lukose and Huberman, 2003), which suggest that when otherwise random networks are characterized by extremely skewed degree distributions, the most connected individuals do indeed serve a critical role in directing decentralized searches. The hubs in these models, however, are analogous to airline network hubs, which tend to be connected directly to a large fraction of total airports in a given region, and, in particular, are connected to most other hubs. In social networks, by contrast, even the most gregarious individuals are estimated to know only on the order of 1,000 others (Bernard et al., 1991; Killworth et al., 1990), and thus are adjacent to a negligible fraction of the entire network. Hubs in social networks, therefore, cannot play the role that they do in so-called 'scale-free' networks. Furthermore, they do not need to: empirical work shows no evidence that 'special individuals' are required to resolve social search problems (Dodds, Muhamad and Watts, 2003), and recent theoretical models suggest that networks can be searchable even when all individuals have exactly the same acquaintance volume (Kleinberg, 2000a; 2000b; Watts, Dodds and Newman, 2002).

DUNCAN J. WATTS

See also **mathematics of networks; network formation; psychology of social networks; social networks in labour markets.**

Bibliography

Adamic, L. and Adar, E. 2005. How to search a social network. *Social Networks* 27, 187–203.

Adamic, L., Lukose, R. and Huberman, B. 2003. Local search in unstructured networks. In *Handbook of Graphs and Networks: From the Genome to the Internet*, ed. S. Bornholdt and H. Schuster. Weinheim: Wiley-VCH.

Adamic, L., Lukose, R., Puniyani, A. and Huberman, B. 2001. Search in power-law networks. *Physical Review E* 64, 46135–43.

Bernard, H., Johnsen, E., Killworth, P. and Robinson, S. 1991. Estimating the size of an average personal network and of an event population: some empirical results. *Social Science Research* 20, 109–21.

Bollobas, B. 1985. *Random Graphs*. New York: Academic Press.

Chung, F. and Lu, L. 2002. The average distances in random graphs with given expected degrees. *Proceedings of the National Academy of Sciences of the United States of America* 99, 15879–82.

Dodds, P., Muhamad, R. and Watts, D. 2003. An experimental study of search in global social networks. *Science* 301, 827–9.

Erdos, P. and Renyi, A. 1959. On random graphs. *Publicationes Mathematicae* 6, 290–7.

Garfield, E. 1979. It's a small world after all. In *Essays of an Information Scientist*, ed. Garfield. Philadelphia: ISI Press.

Guare, J. 1990. *Six Degrees of Separation: A Play*. New York: Vintage.

Killworth, P., Johnsen, E., Bernard, R., Shelley, G. and McCarty, C. 1990. Estimating the size of personal networks. *Social Networks* 12, 289–312.

Kleinberg, J. 2000a. Navigation in a small world. *Nature* 406–845.

Kleinberg, J. 2000b. The small-world phenomenon: an algorithmic perspective. In *Proceedings of the 32nd Annual ACM Symposium on Theory of Computing*. New York: Association of Computing Machinery.

Kleinfeld, J. 2002. The small world problem. *Society* 39, 61–6.

Kochen, M., ed. 1989. *The Small World*. Norwood, NJ: Ablex.

Korte, C. and Milgram, S. 1970. Acquaintance linking between white and negro populations: application of the small world problem. *Journal of Personality and Social Psychology* 15, 101–18.

Liben-Nowell, D., Novak, J., Kumar, R., Raghavan, P. and Tomkins, A. 2005. Geographic routing in social networks. *PNAS* 102, 11623–8.

Milgram, S. 1967. The small world problem. *Psychology Today* 2, 60–7.

Newman, M. 2000. Models of the small world. *Journal of Statistical Physics* 101, 819–41.

Newman, M. 2003. The structure and function of complex networks. *Siam Review* 45, 167–256.

Newman, M., Barabasi, A.-L. and Watts, D., eds. 2006. *The Structure and Dynamics of Networks*. Princeton, NJ: Princeton University Press.

Pool, I. and Kochen, M. 1978. Contacts and influence. *Social Networks* 1, 1–48.

Simsek, O. and Jensen, D. 2005. Decentralized search in networks using homophily and degree disparity. In *Proceedings of the Nineteenth International Joint Conference on Artificial Intelligence* (IJCAI 2005). Edinburgh.

Solomonoff, R. and Rapoport, A. 1951. Connectivity of random nets. *Bulletin of Mathematical Biophysics* 13, 107–17.

Travers, J. and Milgram, S. 1969. An experimental study of the small world problem. *Sociometry* 32, 425–43.

Watts, D. 2004. The 'new' science of networks. *Annual Review of Sociology* 30, 243–70.

Watts, D., Dodds, P. and Newman, M. 2002. Identity and search in social networks. *Science* 296, 1302–5.

Watts, D. and Strogatz, S. 1998. Collective dynamics of 'small-world' networks. *Nature* 393, 440–2.

White, H. 1970. Search parameters for small world problem. *Social Forces* 49, 259–64.

Smart, William (1853–1915)

William Smart was an entrepreneur turned academic economist. Born at Barrhead in Renfrewshire on 10 April 1853, he was educated at the University of Glasgow, where he was later to occupy the (newly created) Adam Smith Chair of Political Economy from 1896 until his death on 19 March 1915. However, his transition from student to professor was interrupted by a successful career in industry which began in the early 1870s and terminated in 1884 when the firm in which he was a partner was sold to the considerable financial advantage of its principals.

Smart's main contribution to economics probably remains his translations into English of the work of Böhm-Bawerk (1890; 1891a), and his edition of von Wieser (1893). In certain circles, Smart is felt to have been primarily responsible for introducing the work of the Austrian School to English readers. As well as making available the originals, Smart published in 1891 his own account of Austrian economics under the title *Introduction to the Theory of Value* – a book which went through three editions during Smart's lifetime. Smart's other work includes a book (1895) dealing principally with wages, consumption and currency. It seems that Smart's advocacy of a bimetallic standard had made his election to the Adam Smith Chair at Glasgow in 1896 more problematic than it might have been. He also wrote on the distribution of income (1899), the single tax (1900), and tariff reform (1904), the last two being essentially contributions to popular debates of the day.

As a young man, and while still a practising business-man, Smart was a Ruskinite. He was a member of the Guild of St George and his first publication was his inaugural address as president of the Ruskin Society of Glasgow (1880). Smart's own account of these intellectual influences on his early development can be found in his *Second Thoughts of an Economist* (1916).

Separate mention should be made of Smart's *Economic Annals of the Nineteenth Century* (1910–17), which he began as a result of the difficulties he had experienced in gathering information in his role as member of the Poor Law Commission in 1905. The simple rationale was to render more accessible official material related to actual economic conditions and debates of the period. Although Smart only saw through to completion two volumes of the *Annals* before he died (which cover less than one third of the 19th century), a glance at the material assembled in the extant volumes is sufficient to confirm their value.

MURRAY MILGATE

Selected works

1880. *John Ruskin: His Life and Work*. Manchester: A. Heywood & Sons.

1883. *A Disciple of Plato*. Glasgow: Wilson & McCormick.

1890. trans. E.v. Böhm-Bawerk, *Capital and Interest*. London and New York: Macmillan & Co.

1891a. trans. E.v. Böhm-Bawerk, *Positive Theory of Capital*. London and New York: Macmillan & Co.

1891b. *Introduction to the Theory of Value on the Lines of Menger, Wieser and Böhm-Bawerk*. London: Macmillan. 2nd edn, 1910; 3rd edn, 1914.

1893. ed. F.v. Wieser, *Natural Value*. Trans. C.A. Malloch. London and New York: Macmillan & Co.

1895. *Studies in Economics*. London: Macmillan & Co.

1899. *The Distribution of Income*. London: Macmillan & Co.

1900. *The Taxation of Land Values and the Single Tax*. Glasgow: J. MacLehose & Sons.

1904. *The Return to Protection*. London: Macmillan & Co.

1910–17. *Economic Annals of the Nineteenth Century*. 2 vols. London: Macmillan & Co.

1916. *Second Thoughts of an Economist*. London: Macmillan & Co.

Smith, Adam (1723–1790)

Biographical

Adam Smith was born in Kirkcaldy, on the east coast of Scotland, and baptized on 5 June 1723. He was the son of Adam Smith, Clerk to the Court Martial and Comptroller of Customs in the town (who died before his son was born) and of Margaret Douglas of Strathendry.

Smith attended the High School of Kirkcaldy, and then proceeded to Glasgow University. He first matriculated in

1737, at the not uncommon age of 14. At this time the university, or more strictly the college, was small. It housed only 12 professors who had in effect replaced the less specialized system of regents by 1727. Of the professoriate, Smith was most influenced by the 'never-to-be-forgotten' Francis Hutcheson (Corr., letter 274, dated 16 November 1787). Hutcheson had succeeded Gerschom Carmichael, the distinguished editor of Pufendorf's *De Officio Hominis et Civis* as Professor of Moral Philosophy.

Smith left Glasgow in 1740 as a Snell Exhibitioner at Balliol College to begin a stay of six years. The atmosphere of the college at this time was Jacobite and 'anti-Scotch'. Smith was also to complain: 'In the university of Oxford, the greater part of the publick professors have, for these many years, given up altogether even the pretence of teaching' (WN,V.i.f.8). But there were benefits, most notably ease of access to excellent libraries, which in turn enabled Smith to acquire an extensive knowledge of English and French literature, which was to prove invaluable, not least in terms of his knowledge of the sciences.

Smith left Oxford in 1746 and returned to Kirkcaldy without a fixed plan. But in 1748 he was invited to give a series of public lectures in Edinburgh, with the support of three men – the Lord Advocate, Henry Home; Lord Kames; and a childhood friend, James Oswald of Dunnikier.

The lectures, which are thought to have be *primarily* (not exclusively) concerned with rhetoric and belles letters, brought Smith £100 a year (Corr. letter 25, dated 8 June 1758). They also seem to have been wide-ranging.

Smith's reputation as a lecturer brought its reward. In 1751 he was elected to the Chair of Logic in Glasgow University, again with the support of Lord Kames. According to John Millar, Smith's most distinguished pupil, he devoted the bulk of his time to the delivery of a system of rhetoric and belles lettres, which was based on the conviction that the best way of:

> explaining and illustrating the various powers of the human mind, the most useful part of metaphysics, arises from an examination of the several ways of communicating our thoughts by speech, and from an attention to the principles of those literary compositions which contribute to persuasion or entertainment. (Stewart, I. 16)

Smith continued to teach the main part of his lecture course on logic after he had been translated to the Chair of Moral Philosophy in 1752. A set of lecture notes, discovered by J.M. Lothian in 1958, relate to the session 1762/3. The notes correspond closely to Millar's description of the course given more than a decade earlier, in that they are concerned with such problems as the development of language, style and the organization of forms of discourse which include the oratorical, narrative and didactical (scientific). Smith was primarily concerned

with the study of human nature and with the analysis of the means and forms of communication. He no doubt continued to lecture on these subjects to students of moral philosophy because he rightly believed them to be important (see J.M. Lothian, 1963: W.S. Howell, 1975).

Smith's lectures on language were published in expanded form as *Considerations Concerning the First Formation of Language*, in the *Philological Miscellany* for 1761. They were reprinted in the third edition of the *Theory of Moral Sentiments* in 1767.

Smith's teaching from the Chair of Moral Philosophy fell into four parts and in effect set the scene for the major published works which were to follow. Again on the authority of John Millar, it is known that Smith lectured on natural theology, ethics, jurisprudence and 'expediency' or economics, *in that order*. The lectures on natural theology (a sensitive subject at the time) have not yet been found. But Millar made it clear that the lectures on ethics form the basis for the *Theory of Moral Sentiments* and that the subjects covered in the last part of the course were to be further developed in the *Wealth of Nations* (Stewart, I. 20). As to the third part, on jurisprudence, Millar noted that:

> Upon this subject he followed the plan that seems to be suggested by Montesquieu; endeavouring to trace the gradual progress of jurisprudence, both public and private, from the rudest to the most refined ages, and to point out the effects of those arts which contribute to subsistence, and to the accumulation of property, in producing correspondent improvements or alterations in law and government. (Stewart, I. 19)

Illustration and confirmation of this claim proved impossible until 1896 when Edwin Cannan published an edition of the *Lectures on Jurisprudence*. The notes edited by Cannan are dated 1766, although they were taken in the session 1763/4. This was Smith's last session in Glasgow, so that these lectures, where 'public' (broadly constitutional law) precedes 'private' jurisprudence (concerning man's rights as a citizen), may reflect a preferred order. A second set of notes, this time relating to the previous session, were also found by J.M. Lothian as recently as 1958 and are here styled LJA.

Academically, the major event for Smith was the publication of the *Theory of Moral Sentiments* in 1759. The book was well received by both the public and Smith's friends. In a delightful letter Hume reminded Smith of the futility of fame and public approbation, and having encouraged him to be a philosopher in practice as well as profession, continued:

> Supposing therefore, that you have duly prepared yourself for the worst by these Reflections; I proceed to tell you the Melancholy News, that your Book has been most unfortunate: For the Public seem disposed to applaud It extremely. (Corr, letter 31, dated 12 April 1759)

The book was to establish Smith's reputation. There was a second revised edition in 1761 and further editions in 1767, 1774, 1781 and 1790.

Charles Townshend was among those to whom Hume had sent a copy of Smith's treatise. Townshend had married the widowed Countess of Dalkeith in 1755 and was sufficiently impressed by Smith's work as to arrange for his appointment as tutor to her son, the young Duke of Buccleuch. The position brought financial security (£300 sterling p.a. for the rest of his life), and Smith duly accepted, formally resigning his chair early in 1764.

Smith and his party left almost immediately for France to begin a sojourn of some two years. At the outset, the visit was unsuccessful, causing Smith to write to Hume, with some humour, that 'I have begun to write a book in order to pass away the time. You may believe I have very little to do' (Corr., letter 82, date 5 July 1764, Toulouse).

But matters improved with Smith's increasing familiarity with the language and the success of a series of short tours. In 1765 Smith, the Duke, and the Duke's younger brother Hew Scott, reached Geneva, giving Smith an opportunity to meet Voltaire, whom he greatly admired as 'the most universal genius perhaps which France has ever produced' (Letter, 17). The party arrived in Paris in mid-February 1766, where Smith's fame, together with the efforts of David Hume, secured him a ready entrée to the leading *salons* and, in turn, introductions to *philosophes* such as d'Alembert, Holbach and Helvetius.

During this period Smith met François Quesnay, the founder, with the Marquis de Mirabeau, of the Physiocratic School of economics (Meek, 1962). By the time Smith met Quesnay, the latter's model of the economic system as embodied in the *Tableau économique* (1757, trans. in Meek, 1962) had already been through a number of editions. Quesnay was then working on the *Analyse* (trans. in Meek, 1962), while it is also known that A.R.J. Turgot was currently engaged on his *Reflections on the Formation and Distribution of Riches* (trans. in Meek, 1973).

Smith, who had already developed an interest in political economy, had arrived in Paris at the very point in time that the French School had reached the zenith of its influence and output. The contents of Smith's library amply confirm his interest in this work (Mizuta, 2000).

Smith's stay in Paris had been enjoyable both socially and in academic terms. But it was marred by the developing quarrel between Hume and Rousseau and sadly terminated by the death of Hew Scott. Smith returned to London on 1 November 1766.

Smith spent the winter in London, where he was consulted by Townshend and engaged in corrections for the third edition of the *Theory of Moral Sentiments*. By the spring of 1767 (the year in which Sir James Steuart published his *Principles of Political Oeconomy*) Smith was back in Kirkcaldy to begin a study of some six years. It was during this period that he struggled with the *Wealth of Nations*. Correspondence of the time amply confirms the mental strain involved. But by 1773 Smith was ready to return to London, leaving his friends, notably David Hume, under the impression that completion was imminent. As matters turned out, it took Smith almost three more years to finish his book; a delay which may have been due to part to his increasing concern with the American War of Independence and with the wider issue of the relationship between the colonies and the 'mother country' (WN, IV. vii).

An Inquiry into the Nature and Causes of the Wealth of Nations was published by Strahan and Cadell on 9 March 1776, and elicited once more a warm response from Hume:

> Dear Mr. Smith: I am much pleas'd with your Performance, and the Perusal of it has taken me from a State of great Anxiety. It was a Work of so much Expectation, by yourself, by your Friends, and by the Public, that I trembled for its Appearance; but am now much relieved. Not but the Reading of it necessarily requires so much Attention, and the Public is disposed to give so little, that I shall still doubt for some time of its being at first very popular. (Corr., letter 150, dated 1 April 1776)

In fact, the book sold well, with subsequent editions in 1778, 1784, 1786 and 1789.

The year 1776 was marred for Smith by the death of David Hume, after a long illness, and by his concern over the future of the latter's *Dialogues Concerning Natural Religion*. This work, together with Hume's account of 'My Own Life' had been left in the care of William Strahan, to whom Smith wrote expressing the hope that the *Dialogue* should remain unpublished, although Hume himself had determined otherwise.

But Smith proposed to 'add to his life a very well authenticated account' of Hume's formidable courage during his last illness (Corr., letter 172, dated 5 September 1776). The letter was published in 1777, and as Smith wrote later to Andreas Holt, 'brought upon me ten times abuse than the very violent attack I had made upon the whole commercial system of Great Britain' (Corr., letter 208, dated October 1780).

In 1778 Smith was appointed Commissioner of Customs, due in part to the efforts of the Duke of Buccleuch. The office brought an income of £600, in addition to the pension of £300 which the Duke refused to discontinue (Corr., letter 208). Smith settled in Edinburgh, where he was joined by his mother and a cousin, Janet Douglas.

During 1778 Alexander Wedderburn sought Smith's advice on the future conduct of affairs in America. Smith's 'Thoughts on the State of the Contest with America' were written in the aftermath of the battle of Saratoga. The Memorandum was first published by G.H. Guttridge in the *American Historical Review* (vol. 38, 1932/3).

In this document, Smith rehearsed a number of arguments which he had already stated in WN (IV.vii.c). He advocated the extension of British taxes to Ireland and to America, provided that representatives from both countries were admitted to Parliament at Westminster in conformity with accepted constitutional practice. Smith noted that 'Without a union with Great Britain, the inhabitants of Ireland are not likely for many ages to consider themselves as one people' (WN,V.iii.89). With respect to America, he observed that her progress had been so rapid that 'in the course of little more than a century, perhaps, the produce of American might exceed that of British taxation. The seat of the empire would then naturally remove itself to that part of the empire which contributed most to the general defence and support of the whole' (WN, IV.vii.c.79).

But Smith also repeated a point already made in WN; namely, that the opportunity for union had been lost, and proceeded to review the bleak options, now all too familiar, which were actually open to the British government. Military victory was increasingly unlikely (WN, V.i.s.27) and military government, even in the event of victory, unworkable (Corr., letter 383). Voluntary withdrawal from the conflict was a rational but politically impracticable course, given the probable impact on domestic and world opinion (Corr., letter 383). The most likely outcome, in Smith's view, was the loss of the thirteen united colonies and the successful retention of Canada – the worst possible solution since it was also the most expensive in terms of defence (Corr., letter 385).

Smith worked hard as a Commissioner, and to an extent which, as he admitted, affected his literary pursuits (Corr., letter 208). But in this period he completed the third edition of WN (1784), incorporating major developments which were separately published as 'Additions and Corrections'. The third edition also features an index and a long concluding chapter to Book IV entitled 'Conclusion of the Mercantile System'.

After 1784 Smith must have devoted most of his attention to the revision of TMS. The sixth edition of 1790 features an entirely new Part VI which includes a further elaboration of the role of conscience, and the most complete statement which Smith offered as to the complex social psychology which lies behind man's broadly economic aspirations.

In addition to the essay on the 'Imitative Arts', which is mentioned in his letter to Andreas Holt (Corr., letter 208), Smith observed that 'I have likewise two other great works upon the anvil; the one is a sort of Philosophical History of all the different branches of Literature, of Philosophy, Poetry and Eloquence; the other is a sort of theory and History of Law and Government' (Corr., letter 248, dated 1 November 1785, addressed to the duc de la Rochefoucauld).

Smith's literary ambitions also feature in the Advertisement to the 1790 editions of TMS, where he drew attention to the concluding sentences of the first edition of 1759. In these passages Smith makes it clear that TMS and WN are parts of a single plan which he hoped to complete with a published account of 'the general principles of law and government, and of the different revolutions which they had undergone in the different ages and periods of society'. Smith's 'present occupations' and 'very advanced age' prevented him from completing this great work, although the approach is illustrated by LJA and LJB, and by those passages in WN which can now be recognized as being derived from them (most notably WN, III and V.i.a.b).

Smith died on 17 July 1790, having first instructed his executors, Joseph Hutton and James Block, to burn his papers, excepting those which were published in Essays on Philosophical Subjects (1795).

In what follows, Smith's system will be expounded in terms of the order of argument which he is known to have employed as a lecturer; namely, ethics, jurisprudence and economics. But it will be convenient to begin with his treatment and knowledge of the literature of science.

The literature of science

It should be recalled that each separate component of Smith's system represents scientific work in the style of Newton, contributing to a greater whole which was conceived in the same image. Smith's scientific aspirations were real, as was his consciousness of the methodological tensions which may arise in the course of such work.

Smith's interest in mathematics dates from his time as a student in Glasgow (Stewart, I. 7). He also appears to have maintained a general interest in the natural and biological sciences, facts which are attested by his purchases for the University Library (Scott, 1937, p. 182) and for his own collection (Mizuta, 2000). Smith's 'Letter to the Authors of the Edinburgh Review' (1756), where he warned against any undue preoccupation with Scottish literature, affords evidence of wide reading in the physical sciences, and also contains references to contemporary work in the French Encyclopédie as well as to the productions of Buffon, Daubenton and Reaumur. D.D. Raphael has argued that the Letter owes much to Hume (TMS, pp. 10, 11).

The essay on astronomy, which dates from the same period (it is known to have been written before 1758 and may well date from the Oxford period) indicates that Smith was familiar with classical as well as with more modern sources, such as Galileo, Kepler and Tycho Brahe, a salutary reminder that an 18th-century philosopher could work close to the frontiers of knowledge in a number of fields.

But Smith was also interested in science as a form of communication, arguing in the LRBL that the way in which this type of discourse is organized should reflect its purpose as well as a judgement as to the psychological characteristics of the audience to be addressed.

In a lecture delivered on 24 January 1763 Smith noted that didactic or scientific writing could have one of two aims: either to 'lay down a proposition and prove this, by the different arguments that lead to that conclusion' or to deliver a system in any science. In the latter case Smith advocated what he called the Newtonian method, whereby we 'lay down certain principles known or proved in the beginning, from whence we account for the several phenomena, connecting all together by the same Chain' (LRBL, ii. 133). Two points are to be noted. First, Smith makes it clear that Descartes rather than Newton was the first to use this method of exposition, even although the former was now perceived to be the author of 'one of the most entertaining Romances that have ever been wrote' (LRBL, ii. 134; see Letter 5). Secondly, his reference to the pleasure to be derived from the 'Newtonian method' (LRBL, ii. 134) draws attention to the problem of scientific motivation, a theme which was to be developed in the 'Astronomy', where Smith considered those principles 'which lead and direct philosophical enquiry'.

The 'Astronomy' takes as given certain results which had already been established in the lectures on language and in the *Considerations*; namely, that men have a capacity for acts of 'arrangement or classing, or comparison, and of abstraction' (LRBL, ii. 207; cf. Corr., letter 69, dated 7 February 1763).

But the essay on astronomy approaches the matter in hand in a different way by arguing that a mind thus equipped derives a certain pleasure from the contemplation of relation, similarity or order – or as Hume would have put it, from a certain association of ideas. Smith struck a more original note in arguing that when the mind confronts a new phenomenon which does not fit into an already established classification, or where we confront an unexpected association of ideas, we feel the sentiment of surprise, and then that of wonder (Astronomy, II. 9). This is typically followed by an attempt at explanation with a view to returning the 'imagination' to a state of tranquillity (Astronomy, II. 6).

Looked at in this way, the task of explanation is related to a perceived need, which can only be met if the account offered is coherent and conducted in terms which are capable of account for observed appearances in terms of 'familiar' principles. It was Smith's contention that the philosopher or scientist would react in the same way as the casual observer, and that nature as a whole 'seems to abound with events which appear solitary and incoherent', thus disturbing 'the easy movement of the imagination (Astronomy, II. 12). But he also observed that philosophers pursue scientific study 'for its own sake, as an original pleasure or good in itself' (Astronomy, III. 3).

The bulk of the essay is concerned to illustrate the extent to which the four great systems of thought which he identified were actually able to 'soothe the imagination', these being the systems of Concentric and Eccentric Spheres, together with the theories of Copernicus and Newton. But Smith added a further dimension to the argument by seeking to expose the dynamics of the process, arguing that each thought-system was subject to a process of modification as new observations were made. Smith suggested that each system was subjected to a process of development which eventually resulted in unacceptable degrees of complexity, thus paving the way for the generation of an alternative explanation of the same phenomena, but one which was better suited to meet the needs of the imagination by offering a simpler account (Astronomy, IV. 18, 28). In Smith's eyes, the work of Sir Isaac Newton thus marked the apparent culmination of a long historical process (Astronomy, IV. 76).

The argument as a whole also contains some radical conclusions. There is nothing in the analysis which suggests that the Newtonian (or Smithian) system embodies some final truth. At the same time, Smith seems to have given emphasis to what is now known as the problem of 'subjectivity' in science in arguing that scientific thought often represents a reaction to a perceived psychological need. He also likened the pleasure to be derived from great productions of the scientific intellect to that acquired when listening to a 'well composed concerto of instrumental music' (Imitative Arts, II. 30). Elsewhere he referred to a propensity, natural to all men, 'to account for all appearances from as few principles as possible' (TMS, VII.ii.2.14) and commented further on the ease with which the 'learned given up the evidence of their senses to preserve the coherence of the ideas of their imagination' (Astronomy, IV. 35). Smith also emphasized the role of the prejudices of sense and education in discussing the reception of new ideas (Astronomy, IV. 35).

He drew attention to the importance of analogy in suggesting that philosophers often attempt to explain the unusual by reference to knowledge gained in unrelated fields, noting that in some cases the analogy chosen could become not just a source of 'ingenious similitude' but the great hinge upon which everything turned' (Astronomy, II. 12).

Smith made extensive use of mechanistic analogies, sometimes derived from Newton, seeing in the universe 'a great machine' wherein we may observe 'means adjusted with the nicest artifice to the ends which they are intended to produce' (TMS, II.ii.3.5). In the same way he noted that 'Human society, when we contemplate it in a certain abstract and philosophical light, appears like a great, an immense machine' (TMS, VII.ii.1.2), a position which leads quite naturally to a distinction between efficient and final causes (TMS, II.ii.3.5), which is not inconsistent with the form of Deism associated with Newton himself. It is also striking that so sympathetic a thinker as Smith should have extended the mechanistic analogy to systems of thought.

> Systems in many respects resemble machines. A machine is a little system created to perform, as well as to connect together, in reality, those different

movements and effects which the artist has occasion for. A system is an imaginary machine invented to connect together in the fancy those different movements and effects which are already in reality performed. (Astronomy, IV. 19)

Each part of Smith's contribution is in effect an 'imaginary' machine which conforms closely to his own stated rules for the organization of scientific discourse. All disclose Smith's perception of the 'beauty of a systematical arrangement of different observations connected by a few common principles' (WN, V.i.f.25). The whole reveals much as to Smith's drives as a thinker, and throws an important light on his own marked (subjective) preference for system, coherence and order.

The Theory of Moral Sentiments

The *Theory of Moral Sentiments* shows clear evidence of a model, and of a form of argument which is in part designed to explain how so self-regarding a creature as man succeeds in erecting barriers against his own passions.

In Part VII of TMS, Smith reviewed different approaches to the questions confronting the philosopher in this field, basically as a means of differentiating his own contribution from them.

In Smith's view there were two main questions to be answered: 'First, wherein does virtue consist', and secondly, 'by what means does it come to pass, that the mind prefers one tenour of conduct to another'? (TMS, VII.i.2). In dealing with the first question, Smith described all classical and modern theories in terms of the emphasis given to the qualities of propriety, prudence and benevolence. In each case, he argued that the identification of a particular quality was appropriate, but rejected what he took to be undue emphasis on any one. He criticized those who found virtue in propriety, on the ground that this approach emphasized the importance of self-command at the expense of 'softer' virtues, such as sensibility. He rejected others who found virtue in prudence because of the emphasis given to qualities which are useful, thus echoing his criticism of David Hume in TMS, Part IV. In a similar way, while he admired benevolence, Smith argued that proponents of this approach (notably Francis Hutcheson) had neglected virtues such as prudence.

Smith's criticism of Hutcheson's teaching is remarkable for the emphasis which he gave to self-interest and his denial of Hutcheson's proposition that self-love 'was a principle which could never be virtuous in any degree or in any direction' (TMS, VII.ii.3.12). Smith also rejected the argument of Mandeville, whose fallacy it was 'to represent every passion as wholly vicious, which is so in any degree' (TMS, VII.ii.4.12). Smith contended that 'The conditions of human nature were peculiarly hard, if these affections, which, by the very nature of our being, ought frequently to influence our conduct, could upon no occasion appear virtuous, or deserve esteem and commendation from anybody' (TMS, VII.ii.3.18).

A further distinctive element in Smith's approach emerges in his treatment of the second question. He accepted Hutcheson's argument that the perception of right and wrong rests not upon reason but 'immediate sense and feeling' (TMS, VII.iii.2.9). But Smith rejected Hutcheson's emphasis on a special sense, the moral sense, which was treated as being analogous to 'external' senses, such as sight or touch. But in so doing Smith in effect elaborated on the argument of his teacher, who had already presented moral judgements as being disinterested and as based upon sympathy or fellow-feeling. Smith also enlarged on the role of the *spectator*, which had been a feature of the work done by Hutcheson and Hume.

Smith argued that the spectator may form a judgement with respect to the activities of another person by visualizing how he would have behaved or felt in similar circumstances. It is this capacity for acts of imaginative sympathy which permits the spectator to form a judgement as to the propriety or impropriety of the conduct observed, and as to the 'suitableness or unsuitableness, the proportion or disproportion which the affection seems to bear to the cause or object which excites it' (TMS, I.i.3.6).

Since we can 'enter into' the feelings of another person only to a limited degree, Smith was able to identify the 'amiable' virtue of sensibility with the quality of imagination, and that of self-command with a capacity to control expressions or feeling to such an extent as to permit the spectator to comprehend, and thus to 'sympathize', with them.

The argument was extended to take account of those actions which have consequences for other people, in suggesting that in such cases the spectator may seek to form a judgement as to the propriety of the *action* taken and of the *reaction* to it. The sense of *merit* is a compounded sentiment, made up of two distinct emotions; a direct sympathy with the sentiments of the agent, and an indirect sympathy with the gratitude of those who receive the benefit of his actions' (TMS, II.i.5.2). Conversely, a sense of *demerit* is compounded of 'antipathy to the affections and motives of the agent' and 'an indirect sympathy with the resentment of the sufferer' (TMS, II.i.5.4).

Smith further contended that 'Nature, when she formed man for society, endowed him with an original desire to please, and an original aversion to offend his brethren' (TMS, III.2.6).

But this general disposition is not of itself sufficient to ensure an adequate degree of control. The first problem which Smith confronted is that of *information*, a problem which arises from the fact that the actual spectator of the conduct of another person is unlikely to be familiar with his *motives*.

Smith solved this problem by arguing that we tend to judge our own conduct by trying to visualize the reaction of an imagined or 'ideal spectator' to it, that is, by seeking to visualize the reaction of a spectator, who is necessarily fully informed, with regard to our own motives. Smith gave more and more attention to the role of the ideal spectator in successive editions as an important source of control; that is, to the voice of 'reason, principle, conscience … the great judge and arbiter of our conduct' (TMS, III.3.4). Looked at in this way, the argument depends on man's desire not merely for praise, but praiseworthiness (TMS, III.2.32).

The second problem arises from the fact that Smith, following Hume, presents man as an active, self-regarding being, whose legitimate pursuit of the objects of ambition, notably wealth, can on some occasions have hurtful consequences for others. The difficulty here is that of *partiality* of view, even where we have the information which is needed to arrive at accurate judgements. When we are about to act, 'the eagerness of passion will seldom allow us to consider what we are doing with the candour of an indifferent person', while after we have acted, we often 'turn away our view from those circumstances which might render … judgement unfavourable' (TMS, III.4.3–4). The solution to this particular problem is found in man's capacity for generalization on the basis of particular experience:

> It is thus that the general rules of morality are formed. They are ultimately founded upon experience of what, in particular instances, our moral faculties, our natural sense of merit and propriety, approve, or disapprove of. … The general rule … is formed, by finding from experience, that all actions of a certain kind, or circumstanced in a certain manner, are approved or disapproved of. (TMS, III.4.8)

It is these rules that provide the yardstick against which man can judge his actions in all circumstances; rules which command respect by virtue of the desire to be praiseworthy and which are further supported by the fear of God (TMS, III.5.12).

Smith thus offered an explanation of the way in which men were fitted for society, arguing in effect that they typically erect a series of barriers to the exercise of their own (self-regarding) passions, which culminate in the emergence of generally accepted rules of behaviour.

The rules themselves vary in character. Those which relate to justice 'may be compared to the rules of grammar, the rules of the other virtues, to the rules which critics lay down for the attainment of what is sublime and elegant in composition. The one, are precise, accurate and indispensable. The other, are loose, vague, and indeterminate' (TMS, III.6.11).

But Smith was in no doubt that the rules of justice were indispensable. Justice 'is the main pillar that upholds the whole edifice' (TMS, II.ii.3.4). Smith added that the final precondition of social order was a system of positive law, embodying current conceptions of the rules of justice and administered by some system of magistracy:

> As the violation of justice is what men will never admit to from one another, the public magistrate is under the

necessity of employing the power of the commonwealth to enforce the practice of this virtue. Without his precaution, civil society would become a scene of bloodshed and disorder, every man revenging himself at his own hand whenever he fancied he was injured. (TMS, VII.iv.36)

Smith's ethical argument forms an integral part of his treatment of jurisprudence precisely because it is concerned to show how particular rules of behaviour emerge. In LJ the focus is narrower than in TMS, but it is still the spectator that is of critical importance whether Smith is discussing accepted standards of punishment or of law. Attention has also been drawn to the role of the magistrate in this connection (Bagolini, 1975) and of the Legislator (Haakonssen, 1981).

Smith's emphasis in TMS is interesting. He chose to concentrate on the means by which the mind forms judgements as to what is fit and proper to be done or to be avoided as distinct from trying to formulate specific rules of behaviour. He had recognized that while the processes of judgement might claim universal validity, specific judgements must be related to experience.

No one living in the age of Montesquieu could fail to be aware of variations in standards of accepted behaviour in different societies at the same point in time, and in the same societies over time. The point at issue seems to have been grasped by Edmund Burke in writing to Smith: 'A theory like yours founded on the Nature of man, which is always the same, will last, when those that are founded upon his opinions, which are always changing, will and must be forgotten' (Corr., letter 38, dated 10 September 1759).

But Smith did not deny that common elements could be found on the basis of experience. Although he did not complete his intended account of 'general principles' (TMS, VII.iv.37), Smith did provide an argument which related the discussion of private and public jurisprudence to four broad types of socio-economic *environment*, the stages of hunting, pasture, farming and commerce. The importance of the argument in the present context is that it was designed in part to explain the origin of government, thus solving a problem which was only noted in TMS. At the same time the historical dimension throws light on the causes of change in accepted rules of behaviour. As part of the same exercise, Smith supplied a successful account of the emergence of the state of commerce, the stage with which he, as an economist, was primarily concerned.

Emergence of the exchange economy

As we have seen in the last section, Smith's analysis of general rules of behaviour suggests that such rules are the result of man's capacity to form judgements as to what is fit and proper to be done or to be avoided. One implication of this argument is that men, at all times and places, form judgements by using the same mental processes. On the other hand it is clear that judgements formed on particular occasions will be related to experience and to the environment which happens to prevail. This in turn, means that accepted patterns of behaviour may vary between different societies at any point in time, but also that they may vary within a particular society over time. There is a comparative aspect, but also a concern with *change*. The point was caught by Dugald Stewart, Professor of Moral Philosophy in Edinburgh, and an acute commentator on Smith, when he noted that:

> When in such a period of society as that in which we live, we compare our intellectual acquirements, our opinions, manners and institutions with those which prevail among rude tribes, it cannot fail to occur to us as an interesting question, by what gradual steps the transition has been made from the first simple efforts of uncultivated nature, to a state of things so wonderfully artificial and complicated. (Stewart, II. 45)

Stewart appreciated that the problem of change applied to the sciences and the arts but also to the treatment of the 'astonishing fabric of ... political union' – our main concern at this point.

While Smith's interest in the history of civil society is illustrated very clearly in WN, there can be little doubt that his reputation was enhanced by the discovery of LJ. But at the same time, we should recall that interest in this area of study was widespread notably in Italy and France. The point is neatly caught by Voltaire, whom Smith greatly admired, when he observed that:

> My principal object is to know, as far as I can, the manners of peoples, and to study the human mind. I shall regard the order of succession of kings and chronology as my guides, but not as the objects of my work. (Quoted in Black, 1926, p. 4)

In Scotland, the theme was pursued by David Hume in his *History of England*, but also by Adam Ferguson (*History of Civil Society*, 1767), John Millar, William Robertson and Lord Kames, to name but a few (see Berry, 1997, chs 5 and 6).

At the same time there was a growing interest among both Scottish and French writers in the link between economic organizations (modes of subsistence) and patterns of behaviour in the fields of sociology and politics. The link that was established between the form of economic and the social and political structure was so explicit as to permit William Robertson, Histographer Royal and Principal of Edinburgh University, to state the main propositions with economy and accuracy. First, Robertson noted that:

> In every enquiry concerning the operations of men when united together in society, the first object of attention should be their mode of subsistence. According as that varies, their laws and policy must be different.

Secondly, Robertson drew attention to the relationship between property and power, noting for example that 'Upon discovering in what state property was at any particular period, we may determine with precision what was the degree of power possessed by the king or the nobility at that juncture', that is, by the government (Skinner, 1996, p. 99).

Smith managed to isolate four distinct modes of subsistence to which there corresponded different types of social and political structures, together with different patterns of 'manners', to use Hume's phrase. The thesis was common among Smith's Scottish contemporaries. The different modes of subsistence are represented by the stages of 'hunting, pasturage, farming and commerce' (LJB, p. 149). The most detailed treatment of the first two stages will be found in WN, Book V, where Smith considers the historical provision of defence and justice. The third and fourth stages are examined in WN, Book III.

Smith's historical sweep was wide ranging, starting as his discourse did from the record of early classical Greece before proceeding to the *Decline and Fall of the Roman Empire* in the West, and thus to the emergence of the modern state. It should be noted that while the perspective adopted in the third Book is European in its emphasis, the focus gradually narrows to the consideration of British and indeed English experience.

Early history

> When the German and Scythian nations over-ran the western provinces of the Roman empire, the confusions which followed so great a revolution lasted for several centuries. The rapine and violence which the barbarians exercised against the antient inhabitants, interrupted the commerce between the towns and the country. The towns were deserted, and the country was left uncultivated, and the western provinces of Europe, which had enjoyed a considerable degree of opulence under the Roman empire, sunk into the lowest state of poverty and barbarism (WN, III.ii.1).

At the same time, however, Smith argued that the domination of the barbarian nations had generated not only a desert but also an environment from which a particular form of European civilization was ultimately to emerge.

Smith's explanation of this general trend begins with the fact that the primitive tribes which overran the empire had already attained a relatively sophisticated form of the pasturage economy, with some idea of agriculture and of property in land. He argued that they would naturally use existing instructions in their new situation and that in particular their first act would be a division of the conquered territories.

> The chiefs and principal leaders of those nations, acquired or usurped to themselves the greater part of the lands ... A great part of them was uncultivated; but no part of them, whether cultivated or uncultivated,

was left without a proprietor. All of them were engrossed, and the greater part by a few great proprietors. (WN, III.ii.1)

In this way we move in effect from a developed version of one economic stage to a primitive version of another; from the state of pasture to that of 'agriculture'. Under the circumstances outlined, property in land became the source of power and distinction, with each estate assuming the form of a separate principality. As a result of this situation, Smith argued, a gradual change took place in the laws governing property, featuring the introduction of primogeniture and entails, designed to protect estates against division and to preserve a 'certain lineal succession'. The basic point emphasized was the 'The security of a landed estate … the protection which its owner could afford to those who dwelt on it, depended upon its greatness. To divide it was to ruin it, and to expose every part of it to be oppressed and swallowed up by the incursions of its neighbours' (WN, III.ii.3).

Such institutions as these quite obviously reflect a change in the mode of subsistence and in the form of property, thus presenting some important contrasts with the previous state of pasture. On the other hand, the great proprietor has still nothing on which to expend his surpluses other than the maintenance of dependants – and at the same time has a positive incentive to do so since they contribute to his military power and security. While Smith carefully distinguished between *retainers* and *cultivators* in this context, he took pains to emphasize that the latter group were in every respect as dependent on the proprietor as the first, and added that 'Even such of them as were not in a state of villanage, were tenants at will, who paid a rent in no respect equivalent to the subsistence which the land afforded them' (WN, III.iv.6).

In short, the period was marked by clear relations of power and dependence – but above all by disorder and conflict, and it was from this source that the first important changes in the outlines of the system were to come. As Smith put it by way of summary:

> In those disorderly times, every great landlord was a sort of petty prince. His tenants were his subjects. He was their judge, and in some respects their legislator in peace, and their leader in war. He made war according to his own discretion, frequently against his neighbours, and sometimes against his sovereign. (WN, III.ii.3)

It was this state of conflict, Smith suggests, which gave the proprietors some incentive to alter the pattern of landholding, in two quite different ways. First, Smith argued that the heavy demands which were inevitably made on their immediate tenants (as distinct from villains) for military service would inevitably change the quit-rent system in terms of which land was normally held. Smith argued in effect that the great lords would naturally begin to grant leases for a term of years, and

then in a form which gave security to the tenant's family and ultimately to his posterity. In this way, land came to be held in a *feudal* relationship, which was being designed to give both parties a benefit: the lord, in terms of the supply of military service, and the tenant, security in the use of land. Smith also noted certain consequential developments which reflected the basic purpose of the arrangement, in describing what he called the feudal *casualities*.

Secondly, Smith argued that the need for protection which had altered the relationship between the great lord and his tenants would also lead to patterns of alliance between members of the former group and, therefore, to arrangements which gave some guarantee of mutual service and support. It was for these reasons, Smith argued, that the lesser landowners gradually entered into *feudal* arrangements with those greater lords who could ensure their survival (thus enhancing their ability to do so), just as the great lords would be led to make similar arrangements amongst themselves and with the king. These changes took place about the ninth, tenth and eleventh centuries, and by imposing some shackles on the free enterprise of the proprietors contributed thereby to the emergence of a more orderly form of government.

However, while Smith did describe the *feudal* as a higher form of the agrarian economy than the *allodial*, he also took some pains to emphasize the limited possibilities for economic growth which it presented; limitations which were themselves the reflection of the political institutions now prevailing. He argued that the quit-rent system, so far as it survived, gave no incentive to industry, and that the institution of slavery ensured that it was in the interest of the ordinary individual to 'eat as much, and to labour as little as possible' (WN, III.ii.9). In the same way he also cited the disincentive effects of the arbitrary services and feudal taxes which were imposed. But, undoubtedly, Smith placed most emphasis on the continuing problem of political instability:

> The authority of government still continued to be, as before, too weak in the head and too strong in the inferior members, and the excessive strength of the inferior members was the cause of the weakness of the head. After the institution of feudal subordination, the king was as incapable of restraining the violence of the great lords as before. They still continued to make war according to their own discretion, almost continually upon one another, and very frequently upon the king; and the open country still continued to be a scene of violence, rapine, and disorder. (WN, III.iv.9)

Once again, a state of instability was to produce some change in the outlines of the social system, and once again the motive behind this change was *political* rather than *economic* – but now with the kings rather than the great lords as the main actors in the drama.

The exchange economy

The kind of economy which Smith described as appropriate to the agrarian state in its developed form is fundamentally a simple one. It consisted of a division between town and country, that is, between those who produce food and those who make the manufactured goods without which no large country could subsist – the critical point being, however, that such an economy was not wholly based on exchange. The cities which Smith described were small, and composed of those merchants, tradesmen and mechanics who were not bound to a particular place and who might find it in their (economic) interest to congregate together. Smith had in fact relatively little to say about the historical origins of such groupings, but he did emphasize that the inhabitants of the towns were in the same servile condition as the inhabitants of the country, and that the wealth which they did manage under such unfavourable conditions would be subject to the arbitrary exactions of both the king and those lords in whose territories they might happen to reside (WN, III.iii.2).

But evidently some developments must have been possible, for Smith examined the role of cities from the period in time when three distinctive features of royal policy with regard to them were already in evidence. First, Smith noted that cities had often been allowed to farm the taxes to which they were subject, the inhabitants thus becoming 'jointly and severally answerable' for the whole sum due (WN, III.iii.3). Second, he noted that in some cases these taxes, instead of being farmed for a given number of years, had been 'let in fee', that is 'forever, reserving a rent certain never afterwards to be augmented' (WN, III.iii.4). Third, Smith observed that the cities:

> were generally at the same time erected into a commonality or corporation, with the privilege of having magistrates and a town-council of their own, of making bye-laws for their own government, of building walls for their own defence, and of reducing all their inhabitants under a sort of military discipline, by obliging them to watch and ward … (WN, III.iii.6)

It was as a result of following these policies that some kings has achieved the apparently remarkable result of freezing the very revenues which were most likely to increase over time, and at the same time effectively curtailing their own power by erecting 'a sort of independent republicks in the heart of their own dominions' (WN, III.iii.7).

Smith advanced two main reasons to explain the apparent paradox. First, he argued that by encouraging the cities the king made it possible for a group of his subjects to defend themselves again the power of the great lords, when he personally was frequently unable to do so, and, secondly, that by imposing a limit on taxation 'he took away from those whom he wished to have for his friends, and, if one may so do, for his allies, all ground of jealousy and suspicion that he was ever afterwards to oppress them, either by raising the farm rent of their town, or by granting it to some other' (WN, III.iii.8).

The encouragement given to the cities represented in effect a tactical alliance which was beneficial to both parties. In speaking of the burghers, Smith remarked that 'Mutual interest … disposed them to support the king, and the king to support them against the lords. They were the enemies of his enemies, and it was his interest to render them as secure and independent of those enemies as he could' (WN, III.iii.8).

Smith also noted that this development was directly related to the weakness of kings, so that it was likely to be more significant in some countries than in others, and that in general the policy had been successful where employed. He also remarked that the granting of powers of self-government to the inhabitants of the cities had set in motion forces which were ultimately to weaken the authority of the kings through creating an environment within which the forces of economic development could, for the first time, be effectively released. In Smith's own words:

> Order and good government, and along with them the liberty and security of individuals, were, in this manner, established in cities at a time when the occupiers of land … were exposed to every sort of violence. But men in this defenceless state naturally content themselves with their necessary subsistence; because to acquire more might only tempt the injustice of their oppressors. On the contrary, when they are secure of enjoying the fruits of their industry, they naturally exert it to better their condition, and to acquire not only the necessaries, but the conveniencies and elegancies of life. (WN, III.iii.12)

The stimulus to economic growth and to further social change was thus seen to emanate from the cities; institutions which had themselves been developed and protected in an attempt to solve a political problem. From this point, Smith's attention shifted to the analysis of the process of economic growth in the manufacturing and trading sectors, before going on to examine its impact on the agrarian sector.

Smith clearly recognized that growth was limited by the size of the market and, since the agrarian sector was relatively backward, that the main stimulus to economic growth would have to come from foreign trade. He concluded that cities such as Venice, Genoa and Pisa, all of which enjoyed ready access to the sea, had provided the models for the process. In general Smith laid most emphasis on three sources of encouragement to the development of trade and manufactures. First, he argued that in many cases agrarian surpluses could be acquired by the merchants and used in exchange for foreign manufacturers, and suggested as a matter of fact that the early trade of Europe had largely consisted in the exchange 'of their own rude, for the manufactured products of more civilized nations'. Secondly, he suggested that over time the merchants would naturally seek to introduce similar manufactures at home (with a view to saving carriage). Such manufactures, it was suggested, would require the use of foreign *materials*, thus inducing an important change in the general pattern of trade. Thirdly, he argued that some manufactures would develop 'naturally', that is through the gradual refinement of the 'coarse and rude' products which were normally produced at home and which were, therefore, based on domestic materials. Smith suggested that such developments were normally found in those cities which were 'not indeed at a very great, but at a considerable distance from the sea coast, and sometimes even from all water carriage' (WN, III.iii.20). He suggested that manufacturers might well develop in areas to which artisans had been attracted by the cheapness of subsistence, thus allowing trade to develop within the locality:

> The manufacturers first supply the neighbourhood, and afterwards, as their work improves and refines, more distant markets. For though neither the rude produce, nor even the coarse manufacture, could, without the greatest difficulty, support the expence of a considerable land carriage, the refined and improved manufacture easily may. In a small bulk it frequently contains the price of a great quantity for rude produce. (WN, III.iii.20).

Smith cited the silk manufacture at Lyons and Spitalfields as examples of the first category of manufactures; those of Leeds, Halifax, Sheffield, Birmingham and Wolverhampton as examples of the second, the natural 'offspring of agriculture' (WN, III.iii.19, 20). He also added that manufacturers of the latter kind were generally posterior to those 'which were the offspring of foreign commerce' and that the process of development just outlined made it perfectly possible for the city within which economic development took place to grow up to a great wealth and splendour, while not only the country in its neighbourhood, but all those to which it traded, were in poverty and wretchedness' (WN, III.iii.13).

In the next stage of analysis, however, it was argued that the situation as outlined was unlikely to continue; that the development of manufacturers and trade within the cities was bound to impinge on the agrarian sector and, ultimately, to destroy the service relationships which still subsisted within it.

Essentially, this process may be seen to stem from the fact that the development of trade and manufacturers had given the proprietors a means of expending their wealth, other than in the maintenance of dependants. The development of commerce and manufacturers:

> gradually furnished the great proprietors with something for which they could exchange the whole surplus produce of their lands, and which they could consume themselves without sharing it either with tenants or

retainers. All for ourselves, and nothing for other people, seems, in every age of the world, to have been the vile maxim of the masters of mankind. (WN, III.iv.10)

This situation generated two results. First, since the proprietor's objective was now to increase his command over the means of exchange, it would be in his interest to reduce the number of retainers:

> till they were at last dismissed altogether. The same cause gradually led them to dismiss the unnecessary part of their tenants. Farms were enlarged, and the occupiers of land, notwithstanding the complaints of depopulation, reduced to the number necessary for cultivating it, according to the imperfect state of cultivation and improvement in those times. (WN, III.iv.13)

Secondly, since the purpose was now to maximize the disposable surplus, it would be in the proprietor's interest to change the forms of leasehold in order to encourage output and increase his returns. In this way, Smith traced the gradual change from the use of slave labour on the land, to the origin of the 'metayer' system where the tenant had limited property rights, until the whole process finally resulted in the appearance of farmers properly so called 'who cultivated the land with their own stock, paying a rent certain to the landlord' (WN, III.ii.14). Smith added that the same process would, over time, tend to lead to an improvement in the conditions of leases, until the tenants could be 'secured in their possession, for such a term of years as might give them time to recover with profit whatever they should lay out in the further improvement of land. The expensive vanity of the landlord made him willing to accept of this condition …' (WN, III.iv.13).

As a result of these two general trends, the great proprietors gradually lost their powers, both judicial and military, until a situation was reached where 'they became as insignificant as any substantial burgher or tradesman in a city. A regular government was established in the country as well as in the city, nobody having sufficient power to disturb its operations in the one, any more than in the other' (WN, III.iv.15).

Smith thus associated the decline in the feudal powers of the great proprietors with three general trends, all of which followed on the introduction of commerce and manufactures: the dissipation of their fortunes, the dismissal of their retainers, and the substitution of a cash relationship for the service relationships which had previously existed between the owner of land and those who cultivated it. He noted elsewhere that 'the gradual improvements of arts, manufactures, and commerce, the same causes which had destroyed the power of the great barons, destroyed in the same manner, through the greater part of Europe, the whole temporal power of the clergy' (WN, V.i.g.25).

As a result, an economic system was generated where the disincentives to 'industry' had been removed from the agrarian sector, and where both sectors were, for the first time, fully interdependent at the domestic level.

Smith argued in effect that the *quantitative* development of manufactures based on the cities had eventually produced an important *qualitative* change in creating the institutions of the exchange economy, that is, of the fourth economic stage. It is in this situation that the drive to better our condition, allied to the insatiable wants of man (referred to in the *Theory of Moral Sentiments*), provided the maximum possible stimulus to economic growth, and ensured that the gains accruing to town and country were eventually both mutual and reciprocal. As Smith put it:

> The great commerce of every civilized society, is that carried on between the inhabitants of the town and those of the country. It consists in the exchange of rude for manufactured produce, either immediately, or by the intervention of money, or of some sort of paper which represents money … The gains of both are mutual and reciprocal, and the division of labour is in this, as in all other cases, advantageous to all the different persons employed in the various occupations into which it is subdivided. (WN, III.i.1)

Such an economic and social structure effectively eliminated the direct dependence of the previous period, in that each productive service now commands a price. While the farmer, tradesman or merchant must depend upon his customers, yet 'Though in some measure obliged to them all … he is not absolutely dependent upon any one of them' (WN, III.iv.12). It will be noted that the whole process of historical change involved in the transition from the feudal to the commercial state depended on the activities of individuals who were unconscious of the ultimate end towards which such activities contributed. Or, as Smith put it in reviewing the actions of the proprietors and merchants during the latter stage of the historical process which we have outlined:

> A revolution of the greatest importance to the public happiness, was in this manner brought about by two different orders of people, who had not the least intention to serve the public. To gratify the most childish vanity was the sole motive of the great proprietors. The merchants and artificers, much less ridiculous, acted merely from a view to their own interest, and in pursuit of their own pedlar principle of turning a penny wherever a penny was to be got. Neither of them had either knowledge or foresight of that great revolution which the folly of the one, and the industry of the other, was gradually bringing about. (WN, III.iv.17)

Finally it should be noted that while Smith regarded the processes of history as inherently complex, he did nonetheless associate these processes with certain economic, social and constitutional trends. The growth of 'luxury and commerce' is represented as the inevitable outcome of normal human drives, and associated with the appearance of new sources of wealth. These new forms of wealth allied to the high degree of personal liberty appropriate to the new patterns of dependence also brought with them a new social and political order – a form of 'constitution' which was often cited as an explanation for, and in defence of, the English Revolution Settlement. In this way Whig principles could be put on a sound historical basis; a point which is neatly illustrated by John Ramsay of Ochtertyre's comment on Lord Kames's abandonment of his early Jacobite leanings. Ramsay expressed no surprise that Kames should have finally concluded that the Revolution was 'absolutely necessary' after 'studying history and conversing with first rate people' – no doubt including Smith. Rather similar sentiments were expressed by John Millar when he remarked that 'When we examine historically the extent of the tory, and of the whig principle, it seems evident, that from the progress of arts and commerce, the former has been continually diminishing, and the latter gaining ground in the same proportion' (quoted in Skinner, 1996, p. 91). There is little doubt that Smith shared such opinions, or that he rejoiced in a situation where the personal liberty of the subject had been confirmed at the expense of the absolutist pretensions of kings and the power of the old feudal aristocracy.

The latter theme had been elaborated in Hume's essay 'Of Refinement in the Arts'. Where 'luxury nourishes commerce and industry' Hume wrote, 'the peasants, by a proper cultivation of the land, become rich and independent; while the tradesmen and merchants acquire a share of the property, and draw authority and consideration to that middling rank of men, who are the best and firmest basis of public liberty' (1985, p. 277). Hume suggested that this development had brought about major constitutional changes, at least in England. The 'lower house is the support of our popular government; and all the world acknowledges, that it owed its chief influence and consideration to the increase of commerce, which threw such a balance of property into the hands of the commons. How inconsistent then is it to blame so violently a refinement in the arts, and to represent it as the bane of liberty and public spirit!' (1985, p. 278).

Hume's perception of the interconnection between economic growth and liberty moved Adam Smith to remark of 'the most illustrious philosopher and historian of the present age' (WN, V.i.g.3) that: 'Mr Hume is the only writer who, so far as I know, has hitherto taken notice of it' (WN, III.iv.4). This interesting but extraordinary statement was never corrected, perhaps as a tribute to Hume's originality.

The model

Smith's account of the origin of the exchange economy suggests that such an economic structure had to be regarded as a model with history. But he also recognized that this particular institutional structure must be associated with a particular set of 'customs and manners', to use Hume's phrase. The link here is with the analyses of the TMS and man's desire for approbation. It is a remarkable fact that the judgements offered with regard to the psychology of the 'economic man' are to be found in the TMS rather than in the WN.

For Smith, 'Power and riches appear ... then to be, what they are, enormous and operose machines contrived to produce a few trifling conveniences to the body, consisting of springs the most nice and delicate' (TMS, IV.i.8). But Smith continued to emphasize that the pursuit of wealth is related not only to the desire to acquire the means of purchasing 'utilities' but also to the need for status.

> From whence, then, arises that emulation which runs through all the different ranks of men, and what are the advantages which we propose by that great purpose of human life which we call bettering our condition? To be observed, to be attended to, to be taken notice of ... are all the advantages which we can propose to derive from it. (TMS, I.iii.2.1)

Smith also suggested that in the modern economy, men tend to admire not only those who have the capacity to enjoy the trappings of wealth, but also the qualities which contribute to that end.

Smith recognized that the pursuit of wealth and 'place' was a basic human drive which would involve sacrifices which are likely to be supported by the approval of the spectator. The 'habits of economy, industry, discretion, attention and application of thought, are generally supposed to be cultivated from self-interested motives, and at the same time are apprehended to be very praiseworthy qualities, which deserve the esteem and approbation of everybody' (TMS, IV.2.8). Smith developed this theme in a passage which was added to the TMS in 1790:

> In the steadiness of his industry and frugality, in his steadily sacrificing the ease and enjoyment of the present moment for the probable expectation of the still greater ease and enjoyment of a more distant but more lasting period of time the prudent man is always both supported and rewarded by the entire approbation of the impartial spectator. (TMS, VI.1.11)

The most polished accounts of the emergence of the exchange economy and of the psychology of the 'economic man' are to be found, respectively, in the third book of WN and in Part VI of TMS which was added in 1790. Yet both areas of analysis are old and their substance would have been communicated to Smith's students and understood by them to be what they might be seen to be: a preface to the treatment of political economy.

It is a subtle argument taken as a whole. Nicholas Phillipson had argued that Smith's ethical theory 'is redundant outside the context of a commercial society with a complex division of labour' (1983, pp. 179, 182). John Pocock concluded that:

> A crucial step in the emergence of Scottish social theory, is, of course, that elusive phenomenon, the advent of the four stages scheme of history. The progression from hunter to farmer, to merchant offered not only an account of increasing plenty, but a series of stages of increasing division of labour, bringing about in their turn an increasingly complex organisation of both society and personality. (1983, p. 242)

Early economic analysis
Hutcheson and the Lectures

The early analyses of questions relating to political economy are to be found primarily in two documents: the lectures delivered in 1762–63 and the text discovered by Cannan (1896, LJB). Cannan's discovery is the most significant in respect of both date and content. This version is the most complete and provides an invaluable record of Smith's teaching in this branch of his project in the last year of his professorship (1763–64).

The Cannan version yielded two important results.

First, Cannan was able to confirm Smith's debts to Francis Hutcheson. Hutcheson's economic analysis was not presented by him as a separate discourse, but rather woven into the broader fabric of his lectures on jurisprudence. Perhaps it was for this reason that historians of economic thought had rather neglected him. But the situation was transformed as a result of Cannan's work as he first noted that the *order* of Smith's lectures on 'expediency' followed that suggested by Hutcheson, albeit, significantly, in the form of a single discourse. The importance of the connection was noted by Cannan (1896, xxv–xxvi; 1904, xxxvi–xli).

Renewed interest in Hutcheson's *economic* analysis revealed that it had its own history. It is evident that he admired the work of his immediate predecessor in the Chair of Moral Philosophy, Gerschom Carmichael (1672–1729), and especially his translation of, and commentary on, the works of Pufendorf. In Hutcheson's address to the 'students of Universities', the *Introduction to Moral Philosophy* (1742) is described thus:

> The learned will at once discern how much of this compend is taken from the writings of others, from Cicero and Aristotle, and to name no other moderns, from Pufendorf's smaller work, *De Officio Hominis et Civis Juxta Legem Naturalem* which that worthy and ingenious man the late Professor Gerschom Carmichael of Glasgow, by far the best commentator on that book, has so supplied and corrected that the notes are of much more value than the text. (Taylor, 1965, p. 25)

It is to W.L. Taylor that we are indebted for the reminder that Carmichael and Pufendorf may have shaped

Hutcheson's economic *ideas*, thus indirectly influencing Smith. Taylor concluded that:

> The interesting point for the development of economic thought in all this is the very close parallelism between Pufendorf's *De Officio* and Hutcheson's *Introduction to Moral Philosophy*. Each man covered almost exactly the same field … The inescapable conclusion is that Francis Hutcheson to over almost in whole, from Carmichael, the economic ideas of Pufendorf. (1965, pp. 28–9)

Undoubtedly, both men followed a particular *order* of argument. Starting with the division of labour they sought to explain the manner in which disposable surpluses could be maximized, before going on to emphasize the importance of security of property and freedom of choice. This analysis led naturally to the problem of value and hence to the analysis of the role of money. What is distinctive about the analysis is the attention given to value *in exchange* where both writers emphasized the role of utility and disutility: perceived utility attaching to the commodities to be acquired, and perceived disutility embodied in the labour necessary to create the goods to be exchanged. The distinction between utility anticipated and realized is profoundly striking (Skinner, 1996, ch. 5). This tradition was continued by Smith both in LJ and WN, but with a change of emphasis towards the *measurement* of value – thus explaining Terence Hutchison's point that Smith retained *some* of his heritage (1988, p. 199; ch. 11).

Cannan's account revealed that in his lectures, Smith was concerned with a system featuring the activities of agriculture, manufacture and commerce (LJB, p. 210) where these activities are characterized by a division of labour (by sector and process) with the process of exchange facilitated by the use of money. The most polished area of analysis, as in the case of the WN, is that of price theory where Smith deployed his distinctions between 'market' and 'natural' price in a way which illuminated the processes by virtue of which 'equilibrium' positions tended to be attained. Examples such as these refer to particular ('partial') cases, but Smith may be said to have added further dimension to the argument by showing an understanding of the fact that the economic system can be seen under a more general aspect (Skinner, 1996, pp. 124–6).

This much is evident in his objection to particular regulations of 'police' (policy) on the ground that they distorted the use of resources by breaking what he called the 'natural balance of industry' while interfering with the 'natural connexion of all trades in the stock' (LJB, pp. 233–4). He concluded: 'Upon the whole, therefore, it is by far the best police to leave things to their natural course' (LJB, p. 235).

Smith's understanding of the interdependence of economic phenomena was quite as sophisticated as that of his master. Yet at the same time, it must be noted that his lecture notes do not confirm a clear distinction between factors of production (land, labour, capital) nor between those categories of return which correspond to them (rent, wages, profit). Nor is there any evidence of a macroeconomic model of the system as a whole: a model which Smith first met during his visit to Paris.

Paris, 1700: the Physiocrats

There was a great deal in Physiocratic writing that was to prove unattractive to some, most obviously, perhaps, the doctrine of legal despotism and a political philosophy which envisaged a constitutional monarch modelled upon the Emperor of China.

The attitudes of the disciplines to the teaching of the master, Quesnay, were also a source of aggravation, moving Hume to write to Morellet on the subject of his *Dictionnaire du Commerce*.

> I see that, in your prospectus, you care not to disoblige your economists … But I hope that in your work you will thunder them, and crush them, and pound them, and reduce them to dust and ashes! They are, indeed, the set of men the most chimerical and most arrogant that now exist. I wonder what could engaged our friend, M Turgot, to herd among them. (Hume, Corr, ii.205)

Ironically, Turgot himself was as deeply opposed to authority and received doctrine as Hume had been.

Hume's reaction also found an echo in France. Murray Rothbard has reminded us of an amusing passage in the works of Simon Nicolas Linguet (1736–1794), ridiculing the idea that the Physiocrats were not 'a cult or sect':

> Not a sect? You have a rallying cry, banners, a march, a trumpeter (Dupont), a uniform for your books, and a sign like freemasons. Not a sect? One cannot touch one of you but all rush to his aid. You laud and glorify each other, and attack and intimidate your opponents in unmeasured terms. (Rothbard, 1995, p. 377)

Smith himself objected to the fact that Quesnay's disciples followed 'implicitly, and without any sensible variation' the doctrine of the master such as there was 'upon this account little variety in the greater part of their works' (WN, IV.ix.38).

But Smith did recognize that the system:

> with all its imperfections, is, perhaps the nearest approximation to the truth that has yet been published upon the subject of political economy, and is upon that account well worth the consideration of every man who wishes to examine with attention the principles of that very important science. (IV.ix.38)

Quesnay's purpose was both practical and theoretical. As R.L. Meek has indicated, Quesnay announced his purpose in a letter to Mirabeau which accompanied the first edition of the *Tableau*.

> I have tried to construct a fundamental *Tableau* of the economic order for the purpose of displaying

expenditure and products in a way which is easy to grasp, and for the purpose of forming a clear opinion about the organisation which the government can bring about. (Meek, 1962, p. 108)

The statement is important in that it confirms the importance of government action in the context of a relatively underdeveloped economy which needed urgent support for the agrarian sector, a reform of the mercantile policies associated with Colbert, and in particular changes in the financial sector and in respect of fiscal policy. But Quesnay's statement also announced a clear understanding of the point that governments can act only on the basis of a knowledge of economic laws. Or, as Meek put it, with pardonable exaggeration:

> With the physiocrats, for the first time in the history of economic thought, we find a firm appreciation of the fact that areas of decision open to policy makers in the economic sphere have certain limits, and that a theoretical model of the economy is necessary to define these limits. (1962, p. 370).

The model in question seeks to explore the interrelationships between output, the generation of income, expenditure and consumption – or in Quesnay's words, a 'general system of expenditure, work, gain and consumption' (Meek, 1962, p. 374) which would expose the point that 'the whole magic of a well ordered society is that each man works for others, while believing that he is working for himself' (Meek, 1962, p. 70). Again, as Meek put it:

> In this circle of economic activity, production and consumption appeared as mutually interdependent variables, whose action and interaction in any economic *period*, proceeding according to certain socially determined laws, laid, the basis for a reputation of the process in the next economic period. (Meek, 1962, p. 19)

The model has a deliberately *abstract* quality but also a number of deficiencies. There is no clear analysis of the division of labour, as Smith understood the term, and no analysis of the problem of price determination and the allocation of resources. There is no formal allowance made for profit *at this point* and nor is there a division between capitalists and wage labour – to name but a few issues of importance.

But despite these criticisms, what Smith would have found in the 'Oeconomical Table' was a *model* of the economic process which represents the working of a macro-economic process as one which involves a series of withdrawals of commodities (consumption and investment goods) from the market, which is matched in turn by a process of continuous replacement, by virtue of production *in the same time period*, all in the *context of a capital-using system*. Smith could hardly fail to be struck by this model, or by the transformation effected by Turgot,

who, in effect, made good the bulk of the analytical deficiencies in Quesnay's account (Meek, 1973).

That Smith benefited from his examination of the French system taken as a whole was quickly noted by Cannan. In referring to the theories of distribution and to the macro-economic dimension, Cannan noted that:

> When we find that there is no trace of these theories in the *Lectures*, and that in the meantime Adam Smith had been to France … it is difficult to understand, why we should be asked, without any evidence, to refrain from believing that he came under physiocratic influence after and not before or during his Glasgow period. (1904, p. xxxi)

Economic analysis

A model of conceptualized reality

The concept of an economy involving a flow of goods and services, and the appreciation of the importance of inter-sectoral dependencies, were familiar in the 18th century. Such themes are dominant features of the work done, for example, by Sir James Steuart and David Hume. But what is distinctive about Smith's work, at least as compared to his *Scottish* contemporaries, is the emphasis given to the importance of *three distinct factors* of production (land, labour, capital) and to the three categories of return (rent, wages, profit) which correspond to them. What is distinctive to the modern eye is the way in which Smith deployed these concepts in providing an account of the flow of goods and services between the sectors involved and between the different socio-economic groups (proprietors of land, capitalists, and wage-labour). The approach is also of interest in that Smith, *following the lead of the French economists*, worked in terms of period analysis – the year was typically chosen, so that the working of the economy is examined within a significant time dimension as well as over a series of time periods. Both versions of the argument emphasize the importance of capital, fixed and circulating.

Smith can be seen to have addressed a series of problems which begin with an analysis of the division of labour, before proceeding to the discussion of value, price and allocation, and thence to the issue of distribution in any one time period and over time.

The analysis offered in the first book enabled Smith to proceed, in WN, Book II, to the discussion of both macro-statics and macro-dynamics, in the context of a model where all magnitudes are dated. What Smith had produced was a model of conceptualized reality which was essentially descriptive, and which was further illustrated by reference into an *analytical* system which, if on occasion subject to ambiguity, was none the less so organized as to meet the requirements of the Newtonian ideal. The intellectual system was intended to be *comprehensive*.

Value

Although Smith's model, in its post-Physiocratic form, has several distinct elements, the feature on which he continued to place more emphasis was the *division of labour*. In terms of the content of the model outlined in previous chapters, a division of labour is of course implied in the existence of distinct *sectors* or types of productive activity. But Smith also emphasizes the fact that there was specialization by types of employment, and even within each employment. To illustrate the basic point, Smith chose the celebrated example of the pin; a very 'trifling manufacture' which none the less required some 18 processes for its completion.

In Smith's hands, the argument was important for two main reasons. First, he was at some pains to point out that the division of labour (by process) helped to explain the relatively high productivity of labour in modern times; a phenomenon which he ascribed to:

1. The increase in 'dexterity' which inevitably results from making a single, relatively simple operation 'the sole employment of the labourer'.
2. The saving of time which would otherwise be lost in 'passing from one species of work to another'.
3. The associated use of machines which 'facilitate and abridge labour, and enable one man to do the work of many' (WN, I.i.5).

He further observed that the existence of specialization (by employment) necessarily involves a high degree of interdependence, in that each separate *manufacture* tends to rely on the output of other industries for different goods and services. It thus follows that the individual customer who purchases a single commodity must at the same time acquire, in effect, the separate outputs of a 'great variety of labour'. Smith added:

> If we examine ... all of these things, and consider what a variety of labour is employed about each of them, we shall be very sensible that without the assistance and co-operation of many thousands, the very meanest person in a civilised country could not be provided, even according to, what we very falsely imagine, the easy and simple manner in which he is commonly accommodated. (WN, I.ii.11)

However, the aspect of this discussion which is most immediately relevant is the light which it throws on the necessity of exchange. As Smith observed, once the division of labour is established, our own labour can supply us with only a very small part of our wants. He thus noted that even in the barter economy the individual can best satisfy the whole range of his needs by exchanging the surplus part of his own production, receiving in return the products of others. Where the division of labour is thoroughly established, it is to be expected that each individual is in a sense dependent on his fellows, and that 'Every man thus lives by exchanging, or becomes in some measure a merchant' (I.iv.1).

This observation brought Smith directly to the problem of value where he returned to an area which, interestingly, had become more a feature of Hutcheson's lectures than of his own. Here it is noteworthy that he employed the analytical (as distinct from the historical) device of the barter economy. However, despite the attempt to be 'perspicuous', these passages remain somewhat difficult largely because Smith uses a single term in handling two distinct but related problems.

> The word VALUE, it is to be observed, has two different meanings, and sometimes expresses the utility of some particular object, and sometimes the power of purchasing other goods which the possession of that object conveys. (WN, I.iv.13).

The first problem concerns the forces which determine the *rate* at which one good, or units of one good, may be exchanged for another; the second is concerned basically with the means by which we can measure the value of the total *stock* of goods created by an individual, and which is used in exchange for others. We may take these issues in turn.

As regards the *rate* of exchange, Smith isolated two relevant factors: the usefulness of the good *to be acquired*, and the 'cost' incurred in creating the commodity *to be given up*. The first of the relevant relationships is obviously that which exists between 'usefulness' and value. The elements of Smith's argument become apparent in his handling of the famous paradox, namely that:

> The things which have the greatest value in use have frequently little or no value in exchange; and, on the contrary, those which have the greatest value in exchange have frequently little or no value in use. Nothing is more useful than water: but it will scarce purchase anything. A diamond, on the contrary, has scarce any value in use; but a great quantity of other goods may frequently be had in exchange for it. (WN, I.iv.13)

The solution to this paradox can be stated in two stages, where the first involves an explanation as to why two such goods have *some* value, and the second an explanation as to why the two goods have *different* values.

Smith's handling of the first part of the problem is based on his recognition of the fact that both goods are considered to be 'useful' although noting that the 'utilities' of each are qualitatively different. In the former case (water) we place a value on the good because we can use it in a practical way, while in the latter (diamonds) we place a value on the good because it appeals to our 'senses', an appeal which, as Smith observed, constitutes a ground 'of preference', or 'source of pleasure'. He concluded:

> The demand for the precious stones arises altogether from their beauty. They are of no use, but ornaments. (WN, I.xi.c.32)

The utilities of the two goods thus emerge as being qualitatively different, although the significant point is seen to be that *both* have *some* value precisely because they represent sources of satisfaction to the individual.

Smith was then left with the second part of the initial problem, namely the explanation as to why the two goods have *different* values. Here again, the answer provided, while simple, is clear, embodying the argument that merit (value) is a function of scarcity. As Smith put it: 'the merit of an object which is in any degree useful or beautiful, is greatly enhanced by its scarcity' (WN, I.xi.c.31). Even more specifically he remarked:

Cheapness is in fact the same thing with plenty. It is only on account of the plenty of water that it is so cheap as to be got for the lifting, and on account of the scarcity of diamonds (for their real use seems not yet to be discovered) that they are so dear. (LJB, pp. 105–6)

Smith introduced the second major element in the problem by observing that the *rate* at which the individual will exchange one good for another must be affected not only by the utility of the good to be acquired, but also by the 'toil and trouble' involved in creating the good exchanged. In this connection he recognized that in acquiring the means of exchange (goods in the barter case), the individual must undergo the 'fatigues' of labour and thus 'lay down' a 'portion of his ease, his liberty, and his happiness' (WN, I.v.7).

In dealing with the *rate* of exchange, Smith may be seen to have placed most emphasis on the supply side of the problem, and explicitly argued that in the case of the barter economy 'the proportion between the quantities of labour necessary for acquiring different objects seems to be the only circumstance which can afford any rule for exchanging them for one another' (WN, I.vi.1). Thus he suggested that if it takes twice the labour to kill a beaver as it does to kill a deer then 'one beaver should naturally exchange for ... two deer'; an argument which may owe something to Hutcheson's emphasis on labour embodied. Smith left the analysis in this form although it will be apparent that the rate of exchange which he specified could only obtain where the perceived ratios of the utilities and disutilities are acceptable to the respective hunters. These are of course subjective judgements whose presence helps to confirm Cannan's opinion that the 'germ' of the WN is to be found in Hutcheson's treatment of value.

This is one way of looking at the problem of exchange value, which clearly shows a parallel with Hutcheson, but Smith seems to have treated it, not as an end in itself, but as a means of elucidating those factors which govern the value of *the whole stock of goods* which the individual creates, and which he proposes to use in exchange. It is of course the presence of this argument in the WN which helps to confirm Taylor's judgement to the effect that the treatment of value was dominated by a concern with the *measurement* of welfare.

Looking at the problem *in this way*, Smith went on to argue that:

The value of any commodity ... to the person who possesses it, and who means not to use or consume it himself, but to exchange it for other commodities, is equal to the quantity of labour which it enables him to purchase or command. Labour, therefore, is the real measure of the exchangeable value of all commodities. (WN, I.v.1)

Smith's meaning becomes clear when he remarks that the value of a *stock* of goods must always be in proportion to 'the quantity ... of other men's labour, *or what is the same thing, of the produce* of other men's labour, which it enables him to purchase or command. The exchangeable value of every thing must always be precisely equal to the extent of this power' (WN, I.v.3). In other words, Smith is here arguing that the real value of the goods which the workman has to dispose of (in effect his income) must be measured by the quantity of goods (expressed in terms of labour units) which he can command, and which he receives once the whole volume of (separate) exchanges has taken place.

As Smith observed, a clear difference between the barter and modern economies is to be found in the fact that, while in the former, goods are exchanged for goods, in the latter, goods are exchanged for a sum of money, which may then be expended in purchasing other goods. Under such circumstances the individual, as Smith saw, very naturally estimates the value of his receipts (received in return for undergoing the 'fatigues' of labour) in terms of money, rather than in terms of the quantity of goods he can acquire by virtue of his expenditure. However, Smith was at some pains to insist that the real measure of welfare (that is, our ability to satisfy our wants) was to be found in 'the money's worth' rather than the money, where the former is determined by the quantity of products (labour 'commanded') which either individuals or groups can purchase. On this basis, Smith went on to distinguish between the *nominal* and the *real* value of income, pointing out that if the three original sources of (monetary) revenue in modern times are wages, rent and profit, then the real value of each must ultimately be measured 'by the quantity of labour which they can, each of them, purchase or command' (WN, I.vi.9)

Price

Smith regarded rent, wages and profit as the types of return payable to the three 'great constituent orders' of society, and as the price paid for the use of the factors of production. The revenues which accrue to individuals and groups in society, and which permit them to purchase commodities, thus appear to be costs incurred by those who create commodities. These points were made quite explicitly by Smith when he remarked:

As the price or exchangeable value of every particular commodity, taken separately, resolves itself into some one or other or all of those three parts; so that of all the commodities which compose the whole annual produce of the labour of every country, taken complexly, must resolve itself into the same three parts, and be parcelled out among different inhabitants of the country, either as the wages of their labour, the profits of their stock, or the rent of their land. (WN, I.vi.17)

This argument obviously raises the problem of price and its determinants.

To begin with, Smith assumed the existence of given 'ordinary or average' rates of wages, profit and rent; rates which may be said to prevail within any given society or neighbourhood, during any given time period. This assumption is of considerable importance, for two main reasons. First, it indicates that in dealing with the problem of price, Smith was working in terms of a given (stable) level of aggregate demand for them. Secondly, the assumption of given rates of return is important in that these rates determine the supply price of commodities.

With these two points forming Smith's major premises, he proceeded to examine the determinants of price, and to produce a discussion which seems to involve two distinct, but related, problems. First, Smith set out to explain the forces which determine the prices of *particular* commodities. Secondly, he would appear to have used the above analysis as a means of explaining the phenomenon of *general* interdependence, and thus those forces which determine the manner in which a given stock of factors of production is allocated between different uses or employments.

In dealing with the first aspect of the problem, Smith implicitly examines the case of a single commodity manufactured by a number of sellers, opening the analysis by establishing a distinction between 'natural' and 'market' price. *Natural price* is now defined as that amount which is 'neither more nor less than what is sufficient to pay the rent of the land, the wages of labour, and the profits of the stock ... according to their natural rates' (WN, I.vii.4). In other words, where natural price prevails, the seller is just able to cover his costs of production, including a margin for 'ordinary or average' profit. By contrast, *market price* is defined as that price which may prevail at any given point in time, being regulated 'by the proportion between the quantity which is actually brought to market, and the demand of those who are willing to pay the natural price of the commodity', the 'effectual demanders' (WN I.vii.8). These two 'prices' are interrelated, the essential point being that while in the short run the market and natural prices may diverge, in the long run they will tend to coincide. If for example, the quantity offered by the sellers was less than that which the consumers were willing to take at a particular (natural) price, the consequence would be a competition among *consumers* to procure some of a limited stock. The price would then rise above the natural price, and the

rewards to factors (notably wages and profit in the short run) would diverge from the natural rates, leading to an influx of resources, and an expansion in the supply of the commodity, thus *tending* towards a return to a position of equilibrium. In making the latter point, Smith took note of the fact that in some cases demand could be postponed to another time period, while in others (for example, perishable necessaries) it could not.

In the second case, where supply exceeds demand, market prices will sink and with them rates of factor payment until factors leave the employment and the supply of commodities is thus reduced. Here again the competition between *suppliers* to rid themselves of excess stock will be affected by the nature of the commodity (durable or perishable). It will be noted that Smith makes allowance for interrelated adjustments in commodity and factor prices, that he makes due allowance for competition *between and among* buyers and sellers, while noting the distinction between durable and perishable goods.

Smith also observed that the result attained, namely that commodities in the long run are sold at their cost of production, can only hold good where there is perfect liberty (as distinct from perfect competition). The cost of production solution is, in short, only to be expected where free competition prevails.

The first stage of the discussion established that in the case of any one commodity, equilibrium will tend to be attained where the good is sold at its natural price, and where each of the relevant factors is paid for at its natural rate. Under these circumstances, equilibrium obtains precisely because there can be no tendency for resources to increase or decrease in this particular type of employment.

Now it is evident that if this process, and this result, holds good for all commodities taken separately, it must also apply to all commodities 'taken complexly', at least where a competitive situation prevails. That is, where the conditions which form the assumptions of the competitive case are satisfied *over the whole economy*, a position of equilibrium will tend to be attained where each different type of good is sold at its natural price, and where each factor in each employment is paid at its natural rate. The economy can then be said to be in a position of 'balance', since where the above conditions are satisfied there can be no tendency to move resources within or between employments. Where the necessary conditions are not satisfied (for example, as a result of changes in tastes) they will naturally tend to be re-established as a result of simultaneous adjustments in the factor *and* commodity markets. It will be observed that departure from, and re-attainment of, a position of equilibrium depends upon the essentially self-interested actions and reactions of consumers and producers. Smith's treatment of price and allocation thus provides one of the best examples of his emphasis on 'interdependence' and one of the most dramatic applications of his *analogy* of the Invisible Hand.

While Smith certainly conceived of 'balance' in terms of a situation where there was no tendency for resources to move between employments, he also recognized that a position of 'balance' need not involve an equality between monetary rates of return. The point follows directly from Smith's recognition of the fact that employments differ qualitatively and that such differences may serve to explain why, even in a position of 'balance', different money rates prevail. As Smith put it, 'certain circumstances in the employments themselves … either really, or at least in the imaginations of men, make up for a small pecuniary gain in some, and counterbalance a great one in others' (WN, I.x.a.2). Thus, for example, he noted that money wage rates would tend to vary between different types of employment according to the difficulty of learning the trade, the constancy of employment, and the degree of trust involved. In the same way he observed that both wages and profit would vary with the agreeableness of the work, the cost of training, and the probability of success in particular fields. In short, he was suggesting that money rates of return would tend to equality within employments of similar types, so that over the whole economy the relevant 'balance' would be one involving net advantages.

Before passing to the next stage of the argument it may be useful to make two points both of which refer to the TMS while at the same time bearing upon the present discussions of the allocative mechanism. The treatment of the doctrine of net advantages is connected to the argument advanced in TMS to the effect that men are motivated by the desire to be approved of. In the present context the arguments suggest that where a profession is widely admired, public approbation may become part of the reward. On the other hand, Smith's analysis suggests that men may only be induced to enter and to remain in particular professions if public disapprobation is compensated by an appropriate monetary reward (the trades of the butcher and the inn-keeper are described as 'odious').

Distribution

As we have seen, Smith's analysis of price and its determinants was built upon the assumption of given rates of factor payment of the kind that could prevail in a given time period – say one year. Briefly stated, the argument has three features: Smith attempts to explain *why* a particular form of return is paid to a particular factor of production (labour, capital and land); the nature of those forces which determine the rates of factor payment which prevail at a particular point in time, and finally, those forces which explain trends in factor payment over long periods of time.

Wages: Smith observed that payment for the factor labour is paid for by those classes which require the services involved. The process of wage determination in a *given time period* will then depend on the relative bargaining position of two groups (labour and

entrepreneurs) in a situation where the legal advantages typically lay with the 'masters'. Where labour is scarce compared with the demand for it, wage rates will tend to be relatively high; relatively low in the opposite case. Smith was thus able to argue that wage rates would be relatively high or low depending on the size of the working population and on the size of the capital stock destined for the employment of the factor (the wages fund). Wage rates may also be relatively high or low depending on the current definition of the subsistence wage, where the latter is defined as a level sufficient to sustain a constant level of population (I.viii.15). The argument leads on to the issue of *long-term adjustment*; Smith's position being that where the wage rate is below the subsistence level, population must contract, and where sustained above this level over a period of time, population must expand – the more typical case in the circumstances of, for example, Great Britain and the North American colonies.

In the context of the relatively short run it follows that wage rates may be equal to, above or below the subsistence rate and that the rates paid are related to the size of the working population and the size of the wages fund.

Profit: Smith did not consider that this form of return was payable for the work of 'inspection and direction' but rather as the reward accruing to those entrepreneurs who risked their capital in combining this factor with others, such as labour and land. The emphasis upon risk is to be noted.

At least as a broad generalization Smith felt able to argue that at *a given point* in time, the rate of profit prevailing would be determined by the capital stock available, taken in conjunction with the volume of business to be transacted by it. But Smith made an important qualification to this statement in arguing that even where the quantity of stock (capital) remains the same, the rate of profit will be related to the prevailing wage rate.

In the long run, however, Smith suggested that the rate of profit would tend to fall, thus establishing a proposition which was to have enduring vitality. Over time, Smith contended that the rate of profit would tend to decline, partly in consequence of an increase in the capital stock, and partly as a result of the increasing difficulty of finding 'a profitable method of employing any new capital'. He continued:

> When the stocks of many rich merchants are turned into the same trade, their mutual competition naturally tends to lower its profit, and when there is a like increase of stock in all the different trades carried on in the same society, the same competition must produce the same effect in them all. (I.xi.2)

It then follows that the 'diminution of profit is the natural effect of prosperity' (I.x.10).

Rent: is formally defined as the 'price paid for the use of land' (I.xi.a.1). Looked at this way, Smith made a point which is reminiscent of the French economists, in arguing that rent constitutes a surplus in the sense that it accrues to the owner of land independently of any effort made by him (I.xi.8) and that rent payments are generally the highest that can 'be got in the actual circumstances of the land' (I.xi. a.1). The reference to actual circumstances is important since Smith recognized that rent would vary both with the fertility and the situation of the land.

The analysis serves to suggest that at *any point in time*, or during any annual period, rent payments will be related to the stock of land actually in use where the latter is in turn related to the level of population. The argument also indicates that rent payments will be related not only to the fertility of the land and its situation, but also to the prevailing rates of profit and wages – another reminder of the interdependence of the different rates of return (I.xi.a.8). In the long run, Smith suggested that rent payments in the aggregate would tend to increase owing to the increased use of the available stock of land (I.xi.2). He added that the real value of the landlord's receipts would also increase over time since all 'those improvements in the productive powers of labour, which tend directly to reduce the real price of manufactures, tend indirectly to raise the real rent of land' (I.xi.4).

The argument just reviewed thus has a short- and long-run dimension. Smith was concerned with long-run trends in rates of return which, as in the short-run case, are interrelated. Thus Smith suggests that profits will decline as the size of the capital stock increases, that high rates of accumulation of capital will generate high market wage rates, leading to an increase in the level of population and in land use. But so far no explanation has been offered as to the source of the crucial increase in capital; a gap which was to be filled in the analyses of Book II.

Macroeconomics (I)

Smith's analysis of the 'circular flow' may be seen as a direct development of certain results already stated in connection with the theory of price. To begin with, it will be recalled that costs of production are incurred by those who create commodities, thus providing individuals with the means of exchange. It therefore follows that if the price of each good in a position of equilibrium comprehends payments made for rent, wages and profit, according to their natural rates, then 'it must be so with regard to all the commodities which compose the whole annual produce of the land and labour of every country, taken complexly' (WN, II.ii.2). On this basis, Smith concluded that 'The whole price or exchangeable value of that annual produce, must resolve itself into the same three parts, and be parcelled out among the different inhabitants' (WN, II.ii.2). If we ignore the problem of distribution (that is of a given level of income between rent, wages and profit), the result which Smith was endeavouring to establish may be stated to involve a relationship between aggregate output and aggregate income. In his own words, 'The gross revenue of all the inhabitants of a great country, comprehends the whole annual produce of their land and labour' (WN, II.ii.5).

It will be evident that a particular level of income, created by a particular level of aggregate output, represents that power to purchase goods which is available to all the members of 'a great society'. Smith then went on to observe that this level of purchasing power would be divided into two funds, consumption and saving. In fact, Smith offered no *formal* explanation of the forces which would determine the actual distribution of aggregate income or purchasing power between these two uses, at any particular point in time. He did, however, suggest that proprietors and labourers would tend to devote a high proportion of their income to consumption, the latter by virtue of the size of their receipts in relation to their basic needs, and the former by virtue of the habits of 'expence' associated with that class. The problem of balancing future against present enjoyments thus appeared to be mainly relevant for the entrepreneurial groups; groups whose functions and objectives dispose them to frugality, at least while actively engaged in the pursuit of fortune.

But Smith did clarify the problems here considered from the standpoint of expenditure. For example, he noted that the proportion of annual income for consumption, taking all groups 'complexly', would be used to purchase commodities which were either perishable or durable in character. He also noted that this type of expenditure could involve the purchase of services; services of kind which do not directly contribute to the annual output of commodities in physical terms and which thus cannot be said to contribute to the level of income associated with it. Smith formally described such labour as 'unproductive', but did not deny that such services were useful. With regard to *savings* Smith identified two sources and two uses. For example, he identified the agrarian, trading, and manufacturing interests as groups wherein 'the owners themselves employ their own capitals' (WN, II.iv.5), as distinct from the monied interest who may lend either for the purpose of consumption or of production.

Smith went on to argue that the undertaker or entrepreneur, engaged in agriculture, manufacture or trade could employ their own or borrowed resources for productive purposes, and divided their capitals into two categories both of which are reminiscent of Physiocratic teaching. *Fixed capital* was defined as that portion of savings used to purchase 'useful machines' or to improve, for example, the productive powers of land, the characteristic feature being that goods are created, and profits ultimately acquired, by using and retaining possession of the investment goods involved. *Circulating capital* was defined as that portion of savings used to purchase investment goods other than 'fixed implements', such as labour power or raw materials, the characteristic feature being that goods are produced through temporarily 'parting with' the funds so used. Smith made three points in the context of this discussion:

1. 'Every fixed capital is both originally derived from, and requires to be continually supported by, a circulating capital' (WN, II.i.24);
2. 'No fixed capital can yield any revenue but by means of a circulating capital (WN, II.i.25), while in addition
3. 'different occupations require very different proportions between the fixed and circulating capitals employed in them' (WN, II.i.6).

Macroeconomics (II)

While these points are important of themselves, they were to gain further significance when Smith moved to the next stage of his argument: the development of his version of the 'circular flow' where, again following the Physiocratic lead, he examined the functioning of the system in a given time period (such as a year). Taking the economic system as a whole, Smith suggested that the total stock of society could be divided into three categories:

There is, first, that part of the total stock which is reserved for immediate consumption, and which is held by *all* consumers (capitalists, labour and proprietors) reflecting purchases made in previous time periods. The characteristic feature of this part of the total stock is that it affords no revenue to its possessors since it consists in 'the stock of foods, clothes, household furniture, etc. which have been purchased by their proper consumers, but which are not yet entirely consumed' (WN, II.i.12).

Secondly, there is that part of the total stock which may be described as 'fixed capital' and which will be distributed between the various groups in society. This part of the stock, Smith suggested, is composed of the 'useful machines' purchased in preceding periods but currently held by the undertakers engaged in manufacture; the quantity of useful buildings and of 'improved land' in the possession of the 'capitalist' farmers and the proprietors, together with the 'acquired and useful abilities' of all the inhabitants (WN, II.i.13–17); that is, human capital.

Thirdly, there is that part of the total stock which may be described as 'circulating capital' and which again has several components, these being:

1. The quantity of money necessary to carry on the process of circulation. In this connection Smith observed that 'The sole use of money is to circulate consumable goods. By means of it, provisions, materials, and finished work, are bought and sold, and distributed to their proper consumers. The quantity of money, therefore, which can be annually employed in any country must be determined by the value of the consumable goods annually circulated within it' (WN, II.iii.23).
2. The stock of provisions and other agricultural products which are available for sale during the current period, but which are still in the hands of either the farmers or merchants.
3. The stock of raw materials and work in progress, which is held by merchants, undertakers, or those capitalists engaged in the agricultural sector (including mining, and so on).
4. The stock of manufactured goods (consumption and investment goods) created during the previous period, but which remain in the hands of undertakers and merchants at the beginning of the period examined (WN, II.i.19–22).

The logic of the process can be best represented by artificially separating the activities involved much in the manner of the Physiocratic model with which Smith was familiar. Let us suppose that at the beginning of the time period in question, the major capitalist groups possess the total net receipts earned from the sale of products in the previous period, and that the undertakers engaged in agriculture open by transmitting the total rent due to the proprietors of land, for the current use of that factor. The income thus provided will enable the proprietors to make the necessary purchases of consumption (and investment) goods in the current period, thus contributing to reduce the stocks of such goods with which the undertakers and merchants began the period. Secondly, let us assume that the undertakers engaged in both sectors, together with the merchant groups, transmit to wage labour the content of the wages fund, thus providing this socio-economic class with an income which can be used in the current period. It is worth noting in this connection that the capitalist groups transmit a fund to wage labour which formed a part of their *savings*, providing by this means an income which is available for current *consumption*. Thirdly, the undertakers engaged in agriculture and manufactures will make purchases of consumption and investment goods from each other, through the medium of retail and wholesale merchants, thus generating a series of expenditures linking the two major sectors. Finally, the process of circulation may be seen to be completed by the purchases made by individual undertakers within their own sectors. Once again these purchases will include consumption and investment goods, thus contributing still further to reduce the stocks of commodities which were available for sale when the period under examination began, and which formed part of the circulating capital of the society in question.

Given these points, we can represent the working of the system in terms of a series of flows whereby money income, accruing in the form of rent, wages and profit, is exchanged for commodities in such a way as to involve a series of withdrawals from the 'circulating' capital of society. As Smith pointed out, the consumption goods withdrawn from the existing stock may be entirely used up within the current period, used to *increase* the stock 'reserved for immediate consumption', or to *replace* the more durable goods, for example, furniture or clothes, which had reached the end of their lives in the course of the same period. Similarly, the undertakers, as a result of their purchases, may *add* to their stocks of raw materials and/or their fixed capital, or *replace* the machines which had finally worn out in the current period, together with the materials used up as a result of current productive activity. Looked at in this way, the 'circular flow' could be seen to involve purchases which take goods *from* the circulating capital of *society*, which is in turn matched by a continuous process of *replacement* by virtue of current production of materials and finished goods – where both types of production require the use of the fixed and circulating capitals of *individual* entrepreneurs. It is an essential part of Smith's argument that all available resources will normally be used:

> In all countries where there is tolerable security, every man of common understanding will endeavour to employ whatever stock he can command in procuring either present enjoyment or future profit. If it is employed in procuring present enjoyment, it is a stock reserved for immediate consumption. If it is employed in procuring future profit, it must procure this profit either by staying with him, or by going from him. In the one case it is a fixed, in the other it is a circulating capital. A man must be perfectly crazy who, where there is tolerable security, does not employ all the stock which he commands, whether it be his own or borrowed of other people, in some one or other of those three ways. (WN, II.i.30)

Smith elaborated on this argument in drawing attention to the point that the differing ways in which the *entrepreneurial* classes employ their capitals were interdependent. The point is reminiscent of Turgot:

> A capital may be employed in four different ways: either, first, in procuring the rude produce annually required for the use and consumption of the society; or, secondly, in manufacturing and preparing that rude produce for immediate use and consumption; or, thirdly, in transporting either the rude produce from the places where they abound to those where they are wanted; or, lastly, in dividing particular portions of either into such small parcels as suit the occasional demands of those who want them. (WN, II.v.1).

Macroeconomics (III): the sources of growth

In choosing to examine the working of the economy during a given time period such as a year, Smith gave his model a broadly short-run character although it is obviously one which included a time dimension. At the same time Smith did not seek to formulate *equilibrium* conditions (as Quesnay had done) for the model, at least in the sense that he did not try to develop an argument which used specified assumptions of a quantitative kind as a means of showing the conditions which must be satisfied before the following time period could open under conditions identical to those prevailing in the period actually examined.

Nor in dealing with the 'flow' did Smith suggest that the level of output attained during any given period would be exactly sufficient to replace the goods used up during its course. On the contrary, he argued that output levels attained in any year would be likely to exceed previous levels: an important reminder that Smith's predominant concern was with economic growth. In this connection, Smith noted that the 'annual produce of the land and labour of any nation can be increased in its value by no other means, but by increasing either the number of its productive labourers, or the productive power of those labourers who had before been employed' (WN, II.iii.32). Smith also observed that both the above sources of increased output required an 'additional capital' devoted either to increasing the size of the wages fund or to the purchase of 'machines and instruments which facilitate and abridge labour'; an additional capital which can only be acquired through net savings.

> By what a frugal man annually saves, he not only affords maintenance to an additional number of productive hands, for that or the ensuing year, but like the founder of a public workhouse, he establishes as it were a perpetual fund for the maintenance of an equal number in all times to come. (WN, II.iii.19)

It will be observed that net savings attained during the course of a single annual period will lead to higher output and income, where the latter becomes available during the course of the period examined. The argument can be extended from this point, in that higher levels of output and income attained in any one year make it possible to reach still higher levels of savings and investment in subsequent years, thus generating further increases in output and income. Once started, the process of capital accumulation and thus economic growth may be seen as self-generating, indicating that Smith's flow is to be regarded as spiral rather than as a circle of given dimensions. This indeed is the burden of Smith's argument in Book II; a fact which helps to explain some of its recurrent themes.

First, Smith frequently argued that net savings will always be possible during each annual period:

> Whatever a person saves from his revenue he adds to his capital, and either employs it himself in maintaining an additional number of productive hands, or enables some other person to do so, by lending it to him for an interest, that is, for a share of the profits. As the capital of an individual can be increased only by what he saves from his annual revenue or his annual gains, so the capital of a society, which is the same with that of all the individuals who compose it, can be increased only in the same manner. (WN, II.ii.15)

Secondly, Smith emphasized that:

> Parsimony, by increasing the fund which is destined for the maintenance of productive hands, tends to increase the number of those whose labour added to the value of the subject upon which it is bestowed. It tends therefore to increase the exchangeable value of the land and labour of the country. It puts into motion an additional quantity of industry, which given an added value to the annual produce. (WN, II, iii.18)

Smith's basic theme is that economic growth depends upon the accumulation of capital, and he went on from this point to draw attention to those factors which affect its *rate*.

In this connection he noted a number of issues:

1. The incidence of commercial failure since 'every injudicious project in agriculture, mines, fisheries, trade or manufactures, tends to diminish the funds destined for the maintenance of productive labour' (II.iii.26).
2. The cost of factors needed to maintain productive assets in a state of normal efficiency (II.ii.7).
3. The area of investment to which a specific injection of capital was applied – it being Smith's contention, for example, that agriculture would support a great quantity of productive labour even than manufactures (see. for example, WN, II.v).
4. The extent to which resources are devoted to the purchase of productive as distinct from unproductive labour. Productive labour for Smith involves the creation of commodities or 'fixed subjects or vendible commodities' which may be either investment or consumption goods and which contribute directly to the generation of income. Other forms of labour are described as unproductive although Smith did not deny that such services are useful. For example, he pointed out that the services of artists have a value to those who wish to pay for them. In the same way the services provided by governments are essential to the well-being of society. Yet all such services are by definition unproductive:

> The sovereign ... with all the officers both of justice and war who serve under him, the whole army and navy, are unproductive labourers. They are the servants of the public, and are maintained by a part of the annual produce of the industry of other people. (II.iii.2)

It therefore follows that the *rate of growth* will be affected especially by the size of the government sector. Smith concluded:

> According ... a smaller or great proportion ... is in any one year employed in maintaining unproductive hands, the more in the one case and the less in the other will remain for productive, and next years produce will be greater or smaller accordingly. (II.iii.3)

Policy

Smith's analytical apparatus, allied to his judgement with respect to the probable trends of the economy, led him to advance the claims of economic liberty; claims which had already featured in LJ and which date back to his days in Edinburgh (Stewart, IV.25). The argument is repeated in WN, where Smith called upon the sovereign to discharge himself from a duty:

> in the attempting to perform which he must always be exposed to innumerable delusions, and for the proper performance of which no human wisdom or knowledge could ever be sufficient; the duty of superintending the industry of private people, and of directing it towards the employments most suitable to the interests of the society. (WN, IV.ix.51)

The statement is familiar, yet conceals a point of great significance; namely, that while the institutions of the exchange economy are consistent with the emergence of personal freedom (for example, under the law), they are not of themselves sufficient to establish what Smith described as the 'system of natural liberty' (WN, IV.ix.51). In fact, one of the most important functions of government is that of *identifying* and *removing* impediments to the effective working of the economy. Smith drew attention, for example, to the adverse effects of the statute of apprenticeship, and of corporate privileges. Regulations of this kind were criticized on the ground that they were both impolitic and unjust: unjust in that controls over qualification for entry to a trade were a violation 'of this most sacred property which every man has in his own labour' (WN, I.x.c.12) and impolitic in that such regulations are not of themselves sufficient to guarantee competence. But Smith particularly emphasized that the regulations in question would adversely affect the working of the market mechanism. The 'statute of apprenticeship obstructs the free circulation of labour from one employment to another, even in the same place. The exclusive privileges of corporations obstruct it from one place to another, even in the same employment' (WN, I.x.c.42). He also commented on the problems presented by the Poor Laws and the Laws of Settlement (WN, IV.ii.42), which further restricted the free movement of labour from one geographical location to another.

Smith objected to positions of privilege, such as monopoly power, which he regarded as creations of the civil law. The institution was again represented as impolitic and unjust: unjust in that a monopoly position is one of privilege and advantage, and therefore 'contrary to that justice and equality of treatment which the sovereign owes to all the different orders of his subjects', impolitic in that the prices at which goods so controlled are sold are 'upon even occasion the highest that can be got' (WN, I.vii.27). He added that monopoly is 'a great enemy to good management' (WN, I.xi.b.5) and that the institution had the additional defect of restricting the flow of capital to the trades affected as a result of the legal barriers to entry which were involved.

It is useful to distinguish Smith's objection to monopoly from his criticism of one expression of it; namely, the mercantile system of regulation which he described as the 'modern system' of policy, best understood 'in our own country and in our own times' (WN, IV.2). Smith asserted that mercantile policy aimed to secure a positive balance of trade through the control of exports and imports, a policy whose 'logic' was best expressed in terms of the Regulating Acts of Trade and Navigation, which currently determined the pattern of trade between Great Britain and her colonies and which were designed to create in effect a self-sufficient Atlantic Economic Community.

Smith objected to current policies of the type described on the ground that they artificially restricted the market and thus damaged opportunities for economic growth. It was Smith's contention that such policies were liable to that general objection which may be made to all the different expedients of the mercantile system, 'the objection of forcing some part of the industry of the country into a channel less advantageous than that in which it would run of its own accord' (WN, IV.v.a.24). In WN Smith placed more emphasis on interference with the allocative mechanism than he had done in LJ, where greater attention had been given to the inconsistency which was involved in seeking a positive balance of trade, an argument which relied heavily on Hume's analysis of the specie flow.

While it is difficult to judge the extent to which the claim for economic liberty explains the contemporary reception of WN, it may have been a major factor, at least in Britain (Schumpeter, 1954, p. 185). There can be no doubt that later generations found Smith's argument (and rhetoric) attractive. The celebrations to mark the 50th anniversary of the book showed a wide and continuing acceptance of the doctrines of free trade. In 1876, at a dinner held by the Political Economy Club to mark the centenary of WN, one speaker identified free trade as the most important consequence of the work done by 'this simple Glasgow professor', and predicted that

> there will be what may be called a large negative development of Political Economy tending to produce an important beneficial effect; and that is, such a development of Political Economy as will reduce the functions of government within a smaller and smaller compass. (Black, 1976, p. 51)

This view still commands wide contemporary support.

There can be no argument with Jacob Viner's contention that 'Smith in general believed that there was, to say the least, a strong presumption against government activity' (Viner, 1928, p. 14). But as Viner also reminded his auditors during the course of the Chicago conference which celebrated the 150th anniversary of the publication of WN, 'Adam Smith was not a doctrinaire advocate of laissez-faire. He saw a wide and elastic range of activity for government' (1928, pp. 153–4). A number of examples, all identified by Viner in a classic article, may briefly be reviewed here.

First, Smith was prepared to justify specific policies to meet particular needs as these arose; the principle of intervention ad hoc. He defended the use of stamps on plate and linen as the most effectual guarantee of quality (WN, I.x.c.13), the compulsory regulation of mortgages (WN, V.ii.h.17), the legal enforcement of contracts (WN, I.ix.16) and government control of the coinage. In addition, he supported the granting of temporary monopolies to mercantile groups, to the inventors of new machines and, not surprisingly, to the authors of new books (WN, V.i.e.30). He further advised governments that where they were faced with taxes imposed by their competitors, retaliation could be in order, especially if such action had the effect of ensuring the 'repeal of the high duties or prohibitions complained of. The recovery of a great foreign market will generally more than compensate the transitory inconveniency of paying dearer during a short time for some sorts of goods' (WN, IV.ii.39).

Secondly, Smith advocated the use of taxation, not as a means of raising revenue but as a source of social reform, and as a means of compensating for what would now be described as a defective telescopic faculty. In the name of the public interest, Smith supported taxes on the retail sale of liquor in order to discourage the multiplication of alehouses (WN, V.ii.g.4) and differential rates on ale and spirits in order to reduce the sale of the latter (WN, V.ii.k.50). He advocated taxes on those proprietors of land who demanded rents in kind, and on those leases which prescribed a certain form of cultivation. In the same way, Smith argued that the practice of selling a future, for the sake of present, revenue should be discouraged on the ground that it reduced the working capital of the tenant and at the same time transferred a capital sum to those who would use it for the purposes of consumption (WN, V.ii.c. 12) rather than investment which would directly support productive labour.

Smith was well aware, to take a third example, that the modern version of the 'circular flow' depended on paper money and on credit; in effect, a system of 'dual circulation' involving a complex of transactions linking producers and merchants, and dealers and consumers (WN, II.ii.88). It is in this context that he advocated control over the rate of interest, to be set in such a way as to ensure that 'sober people are universally preferred, as borrowers, to prodigals and projectors' (WN, II.iv.15). He was also willing to regulate the small note issue in the interests of a stable banking system. To those who objected to such a proposal Smith replied that the interests of the community required it, and concluded that 'the obligation of building party walls, in order to prevent the communication of fire, is a violation of natural liberty, exactly of the same kind [as] the regulations of the banking trade which are here proposed' (WN,

II.ii.94). Although Smith's monetary analysis is not regarded as amongst the strongest of his contributions, it should be remembered that as a witness of the collapse of the Ayr Bank, he was acutely aware of the problems generated by a sophisticated credit structure, and that it was in this context that he articulated a very general principle; namely, that 'those exertions of the natural liberty of a few individuals, which might endanger the security of the whole society, are, and ought to be, restrained by the laws of all governments; of the most free, as well as of the most despotical' (WN, II.ii.94).

Fourthly, emphasis should be given to Smith's contention that a major responsibility of government must be the provision of certain public works and institutions for facilitating the commerce of the society which were 'of such a nature, that the profit could never repay the expense to any individual or small number of individuals, and which it, therefore, cannot be expected that any individual or small number of individuals should erect or maintain' (WN, V.i.c.1). The examples of public works which he provided include roads, bridges, canals and harbours – all thoroughly in keeping with the conditions of the time and with Smith's emphasis on the importance of transport as a contribution to the effective operation of the market and to the process of economic growth. But although the list is short by modern standards, the discussion is of interest for two main reasons.

First, Smith contended that public works or services should only be provided where market forces have failed to do so; secondly, he insisted that attention should be given to the requirements of efficiency and equity.

As Nathan Rosenberg (1960) has pointed out in an important article, Smith did not argue that governments should *directly* provide relevant services; rather, they should establish institutional arrangements so structured as to engage the motives and interests of those concerned. Smith tirelessly emphasized the point that in every trade and profession 'the exertion of the greater part of those who exercise it, is always in proportion to the necessity they are under of making that exertion' (WN, V.i.f.4); teachers, judges, professors, civil servants and administrators alike.

With regard to equity, Smith argued that public works, such as highways, bridges and canals should be paid for by those who use them and in proportion to the wear and tear occasioned – an expression of the general principle that the beneficiary should pay. He also defended direct payment on the ground of efficiency since only by this means would it be possible to ensure that necessary services would be provided where there was an identifiable need (WN, V.i.d.6).

Yet Smith recognized that it would not always be possible to fund or to maintain public services without recourse to general taxation. In this case he argued that 'local or provincial expenses of which the benefit is local or provincial' ought to be no burden on general taxation since 'It is unjust that the whole society should contribute towards an expense of which the benefit is confined to a part of society' (WN, V.i.i.3). However, he did agree that a general contribution would be appropriate in cases where public works benefit the whole society and cannot be maintained by the contribution 'of such particular members of the society as are most immediately benefited by them' (WN, V.i.i.6).

But here again, the main features of the system of liberty are relevant in that they affect the way in which taxation should be imposed. Smith pointed out on welfare grounds that taxes should be levied in accordance with the canons of equality, certainty, convenience and economy (WN, V.ii.b), and insisted that they should not be raised in ways which infringed the liberty of the subject – for example, through the odious visits and examinations of the tax-gatherer. Similarly, he argued that taxes ought not to interfere with the allocative mechanism (as, for example, taxes on necessities or particular employments) or constitute important disincentives to the individual effort on which the effective operation of the whole system depended (for example, taxes on profits or on the produce of land).

Ethics and history

The policy views which have just been considered are closely related to Smith's economic analysis. Others are only to be fully appreciated when seen against the background of his work on ethics and jurisprudence.

It will be recalled that for Smith moral judgement depends on a capacity for acts of imaginative sympathy, and that such acts can only take place within the context of some social group (TMS, III.i.3). However, Smith also observed that the mechanism of the impartial spectator might well break down in the context of the modern economy, due in part to the size of the manufacturing units and of the cities which housed them.

Smith observed that in the actual circumstances of modern society, the poor man could find himself in a situation where the 'mirror' of society (TMS, III.i.3) was ineffective. The 'man of rank and fortune is by his station the distinguished member of a great society, who attend to every part of his conduct, and who thereby oblige him to attend to every part of it himself'. But the 'man of low condition', while 'his conduct may be attended to' so long as he is a member of a country village, 'as soon as he comes into a great city, he is sunk, in obscurity and darkness. His conduct is observed and attended to by nobody, and he is therefore very likely to neglect it himself, and to abandon himself to every sort of low profligacy and vice' (WN, V.i.g.12).

In the modern context, Smith suggests that the individual thus placed would naturally seek some kind of compensation, often finding it not merely in religion but in religious sects; that is, small social groups within which he can acquire 'a degree of consideration which he never had before' (WN, V.i.g.12). Smith noted that the morals of such sects were often disagreeably 'rigorous and unsocial', recommending two policies to offset this.

The first of these is learning, on the ground that science is 'the great antidote to the poison of enthusiasm and superstition'. Smith suggested that government should institute 'some sort of probation, even in the higher and more difficult sciences, to be undergone by every person before he was permitted to exercise any liberal profession, or before he could be received as a candidate for any honourable office of trust or profit' (WN, V.i.g.14). The second remedy was through the encouragement given to those who might expose or dissipate the folly of sectarian bitterness by encouraging an interest in painting, music, dancing, drama – and satire (WN, V.i.g.15).

If the problems of solitude and isolation consequent on the growth of cities explain Smith's first group of points, a related trend in the shape of the division of labour helps to account for the second. In the earlier part of the argument, Smith had emphasized the gain to society at large which arose from improved productivity. But he noted later that this important source of economic benefit could also involve social costs:

> In the process of the division of labour, the employment of the far greater part of those who live by labour, that is, of the great body of the people, comes to be confined to a few very simple operations; frequently to one or two. But the understandings of the greater part of men are necessarily formed by their ordinary employments. The man whose life is spent in performing a few simple operations, of which the effects too are, perhaps, always the same, or very nearly the same, has no occasion to exert his understanding, or to exercise his invention in finding out expedients for removing difficulties which never occur. (WN, V.i.f.50)

Smith went on to point out that despite a dramatic increase in the level of *real income*, the modern worker could be relatively worse off than the poor savage, since in such primitive societies the varied occupations of all men – economic, political and military – preserve their minds from that 'drowsy stupidity, which, in a civilized society, seems to benumb the understanding of almost all the inferior ranks of people' (WN, V.i.f.51). It is the fact the 'labouring poor, that is the great body of the people' will fall into the state outlined that makes it necessary for government to intervene.

Smith's justification for intervention is, as before, market failure, in that the labouring poor, unlike those of rank and fortune, lack the leisure, means or (by virtue of their occupation) the inclination to provide education for their children (WN, V.i.f.53). In view of the nature of the problem, Smith's programme seems rather limited, based as it is on the premise that 'the common people cannot, in any civilized society, be so well instructed as people of some rank and fortune' (WN, V.i.f.54). However, he did argue that they could all be taught 'the most

essential parts of education … to read, write, and account' together with the 'elementary parts of geometry and mechanics' (WN, V.i.f.54, 55). Smith added:

> The publick can *impose* upon almost the whole body of the people the necessity of acquiring those most essential parts of education, by obliging every man to undergo an examination or probation in them before he can obtain the freedom in any corporation, or be allowed to set up any trade either in a village or town corporate. (WN, V.i.f.57; italics supplied)

Distinct from the above, although connected with it, is Smith's concern with the decline of martial spirit, which is the consequence of the nature of the fourth, or commercial, stage. He concluded that:

> Even though the martial spirit of the people were of no use towards the defence of society, yet to prevent that sort of mental mutilation, deformity and wretchedness, which cowardice necessarily involves in it, from spreading themselves through the great body of the people would still deserve the most serious attention of government. (WN, V.i.f.60)

Smith went on to liken the control of cowardice to the prevention of 'a leprosy or any other loathsome and offensive disease' – thus moving Jacob Viner to add public health to Smith's already lengthy list of governmental functions (Viner, 1928, p. 150). Such concerns have enabled Winch (1978) to find in Smith evidence of the *language* of an older, classical, concern with the problem of citizenship. Others (for example, see contributions in Hont and Ignatieff, 1983) have located Smith more firmly in the tradition of civic humanism.

The historical dimension of Smith's work also affects the treatment of policy, noting as he did that in every society subject to a process of transition, 'Laws frequently continue in force long after the circumstances, which first gave occasion to them, and which could also render them reasonable, are no more' (WN, III.ii.4). In such cases Smith suggested that arrangements which were once appropriate but are now no longer so should be removed, citing as examples the laws of succession and entail; laws which had been appropriate in the feudal period but which now had the effect of liming the sale and improvement of land. The continuous scrutiny of the relevance of particular laws is an important function of the 'legislator' (Haakonssen, 1981).

In a similar way, the treatment of justice and defence, both central services to be organized by the government, are clearly related to the discussion of the stages of history, an important part of the argument in the latter case being that a gradual change in the economic and social structure had necessitated the formal provision of an army (WN, V.i.a.b).

But perhaps the most striking and interesting features emerge when it is recalled that for Smith the fourth economic stage could be seen to be associated with a particular form of social and political structure which determines the outline of government and the context within which it must function. It may be recalled in this connection that Smith associated the fourth economic stage with the elimination of the relation of direct dependence which had been a characteristic of the feudal agrarian period. Politically, the significant and associated development appeared to be the diffusion of power consequent on the emergence of new forms of wealth which, *at least in the peculiar circumstances of England*, had been reflected in the increased significance ·of the House of Commons.

Smith recognized that in this context government was a complex instrument, that the pursuit of office was itself a 'dazzling object of ambition' – a competitive game with as its object the attainment of 'the great prizes which sometimes come from the wheel of the great state lottery of British politics' (WN, IV.vii.c.75).

Yet for Smith the most important point was that the same economic forces which had served to elevate the House of Commons to a superior degree of influence had also served to make it an important focal point for sectional interests – a development which could seriously affect the legislation which was passed and thus affect that extensive view of the common good which ought ideally to direct the activities of Parliament.

It is recognized in the *Wealth of Nations* that the landed, moneyed, manufacturing and mercantile groups all constitute special interests which could impinge on the working of government. Smith referred frequently to their 'clamourous importunity', and went so far as to suggest that the power possessed by employers generally could seriously disadvantage other classes in the society' (WN, I.x.c.61; cf. 1. viii.12,13).

Smith insisted that any legislative proposals emanating from this class:

> ought always to be listened to with great precaution, and ought never to be adopted till after having been long and carefully examined, not only with the most scrupulous, but with the most suspicious attention. It comes from an order of men, whose interest is never exactly the same with that of the public, who have generally an interest to deceive and even to oppress the public, and who accordingly have, upon many occasions, both deceived and oppressed it. (WN, I.xi.p.10)

He was also aware of the dangers of manipulation arising from deployment of the civil list (LJA, iv.175–6).

It is equally interesting to note how often Smith referred to the constraints presented by the 'confirmed habits and prejudices' of the people, and to the necessity of adjusting legislation to what 'the interests, prejudices, and temper of the times would admit of' (WN, IV.v.b.40, 53, and V.i.g; cf. TNS, VI.ii.2.16). Such passages add further meaning to the discussion of education. An educated people, Smith argued, would be more likely to see through the interested complaints of faction and

sedition. He added a warning and a promise in remarking that:

> In free countries, where the safety of government depends very much on the favourable judgment which the people may form of its conduct, it must surely be of the highest importance that they should not be disposed to judge rashly or capriciously concerning it. (WN, V.i.f.61)

Aftermath

J.S. Mill, the archetypal classical economist, of a later period, is known to have remarked that 'The *Wealth of Nations* is in many parts obsolete and in all, imperfect'. Writing in 1926, Edwin Cannan observed:

> Very little of Adam Smith's scheme of economics has been left standing by subsequent enquirers. No one now holds his theory of value, his account of capital is seen to be hopeless confused, and his theory of distribution is explained as an ill-assorted union between his own theory of prices and the Physiocrat fanciful Economic Table. (1926, p. 123)

In view of authoritative judgements such as these, it is perhaps appropriate to ask what elements in his story should command the attention of the modern historian or economist. A number of points might be suggested.

First, there is the issue of *scope*. As we have seen, Smith's approach to the study of political economy was through the examination of history and ethics. The historical analysis is important in that he set out to explain the origins of the commercial stage. The ethical analysis is important to the economist because it is here that Smith identifies the human values which are appropriate to the modern situation. It is here that we confront the emphasis on the desire for status (which is essentially Veblenesque) and the qualities of mind which are necessary to attain this end: industry, frugality, prudence.

But the TMS also reminds us that the pursuit of economic ends takes place with a social context, and the men maximize their chances of success by respecting the rights of others. In Smith's sense of the term, 'prudence' is essentially rational self-love. In a favourite passage from the TMS (II.ii.2.1) Smith noted, with regard to the competitive individual, that:

> In the race for wealth, and honours, and preferments, he may run as hard as he can, and strain every nerve and muscle, in order to outstrip all his competitors. But if he should justle, or throw down any of them, the indulgence of the spectators is entirely at an end. It is violation of fair play, which they cannot admit of.

Smith's emphasis upon the fact that self-interested actions take place within a social setting and that men are motivated (generally) by a desire to be approved of by their fellows, raises some interesting questions of continuing relevance. For example, in an argument which bears upon the analysis of the TMS, Smith noted in effect that the rational individual may be constrained in respect of economic activity or choices by the reaction of the spectator of his conduct – a much more complex case than that which more modern approaches may suggest. Smith made much of the point in his discussion of Mandeville's 'licentious system' which supported the view that private vices were public benefits, in suggesting that the gratification of desire should be consistent with observance of the rules of property – as defined by the spectator, that is, by an external agency. In an interesting variant on this theme, Etzioni has noted that we need to recognize 'at least two irreducible sources of valuation or utility: pleasure and morality' (1988, 21–4; cf. Oakley, 1994).

Secondly, there is a series of issues which arise from Smith's interest in political economy as a system. The idea of a single all-embracing conceptual system, whose parts should be mutually consistent, is not easily attainable in an age where the division of labour has increased the quantity of science through specialization. Smith was aware of the division of labour in different areas of sciences, and of the fact that specialization often led to systems of thought which were inconsistent with each other (Astronomy, IV, 35, 52, 67). But the division of labour within a *branch* of science, for example, economics, has led to a situation where sub-branches of a single subject may be inconsistent with one another.

To take a third point, it may be noted that one of the most significant features of Smith's vision of the economic process lies in the fact that it has a significant time dimension. For example, in dealing with the problem of value in exchange, Smith made due allowance for the fact that the process involves judgements with regard to the utility of the commodities to be acquired, and the disutility involved in creating the goods to be exchanged. In the manner of his predecessors (Hutcheson, Carmichael and Pufendorf), Smith was aware of the distinction between utility (and disutility) anticipated and realized, and, therefore, of the process of adjustment which would inevitable take place through time,

Smith's theory of price, which allows for a wide range of changes in taste, is also distinctive in that it allows for competition among and between buyers and sellers, while presenting the allocative mechanism as one which involves simultaneous and interrelated adjustments in both factor and commodity markets.

As befits a writer who was concerned to address the problems of change, and adjustments to change, Smith's position was also distinctive in that he was not directly concerned with the phenomenon of *equilibrium*. For Smith, the 'nature' (supply) price was, as it were:

> The central price, to which the prices of all commodities are continually gravitating … whatever may be the obstacles which hinder them from settling in this centre of response and continuance, they are constantly tending towards it. (WN, I.viii.15)

But perhaps the most intriguing feature of the *macro* model is to be found in the way in which it was linked to the analytics of Book I and in the way in which it was specified. As noted earlier, Smith argued that incomes were generated as a result of productive activity, thus making it possible for commodities to be withdrawn from the 'circulating' capital of society. As he pointed out, the consumption goods withdrawn from the existing stock may be used up in the present period, or added to the stock reserved for immediate consumption; or used to replace more durable goods which had reached the end of their lives in the current period. In a similar manner, entrepreneurs and merchants may also add to their stocks of materials, or to their holding of fixed capital, while replacing the plant which had reached the end of its operational life. It is equally obvious that entrepreneurs and merchants may add to, or reduce their *inventories* in ways which will reflect the changed patterns of demand for consumption and investment goods, and their past and current levels of production. Variation in the level of inventories has profound implications for the conventional theory of the allocative mechanism.

Smith's emphasis upon the point that different 'goods' have different life-cycles also means that the pattern of purchase and replacement may vary continuously as the economy moves through different time periods, and in ways which reflect the various age profiles of particular products as well as the pattern of demand for them. If Smith's model of the circular flow is to be seen as a spiral, rather than a circle, it soon becomes evident that this spiral is likely to expand (and possibly *contract*) through time at variable rates. This point does not seem to have attracted much attention.

It is perhaps this total vision of the complex working of the economy that led Mark Blaug to comment on Smith's distinctive and sophisticated grasp of the economic *process* and to distinguish this from his contribution to particular *areas* of economic analysis.

Blaug noted that:

> In appraising Adam Smith, or any other economist, we ought always to remember that brilliance in handling purely economic concepts is a very different thing from a firm grasp of the essential logic of economic relationships. Superior technique does not imply superior insight and vice-versa. Judged by standards of analytical competence, Smith is not the greatest of eighteenth century economists. But for an acute insight into the nature of the economic process, it would be difficult to find Smith's equal. (1985, p. 57)

Joseph Schumpeter, not always a warm critic of 'A. Smith', yet regarded WN as 'the peak success of (the) period:

> Though the *Wealth of Nations* contained no really novel ideas, and though it cannot rank with Newton's *Principia* or Darwin's *Origin* as an intellectual achievement, it

is a great performance all the same and fully deserved its success. (1954, p. 185)

Writing from a different, but related, point of view, A.L. Macfie noted that 'the Scottish method was more concerned with giving a broad, well balanced picture seen from different points of view than with logical rigour' (1967, 22–3).

It has been argued above that Smith's approach to the study of political economy has some distinctive features which deserve the attention of the modern student of the discipline, but which do not seem to loom large in modern teaching. But nor can it be said that the classical system which was to follow Smith did any better.

Richard Teichgraeber's research (1987) revealed that there 'is no evidence to show that many people exploited his arguments with great care before the first two decades of the nineteenth century'. He concluded:

> It would seem at the time of his death that Smith was widely known and admired as the author of the *Wealth of Nations*. Yet it should be noted that only a handful of his contemporaries had come to see his book as uniquely influential. (1987, p. 363)

But, as we have seen, there were commentators who understood Smith's analytical purpose, notably Thomas Pownall and Smith's biographer, Dugald Stewart. The latter indeed became an important channel of communication between Smith and a later generation of students – many of whom were to contribute to debates published by the *Edinburgh Review* in the early part of the new century (Winch, 1994, p. 91). Amongst the contributors we can number J.R. McCulloch (1789–1864), editor of the *Scotsman*, disciple of Ricardo and a writer who contributed nearly 80 articles to the *Review* between 1818 and 1837.

However, Black made a different point in his 'Historical Perspective' in observing that for Smith's early successors the *Wealth* was 'not so much a classical monument to be inspected, but as a structure to be examined and improved where necessary (1976, p. 44). It was thought that there were ambiguities in respect of Smith's treatment of value, interest, rent, population theory and the theory of economic growth. These ambiguities were reduced by the work of T.R. Malthus (1766–1834) whose *Essay on the Principle of Population* was first published in 1798; by the French economist, J.B. Say (1826–1896), *Traite d'economie politique* (1803) and especially by David Ricardo (1772–1823), *Principles* (1817). In this context we can name a number of writers who developed the short-run and dynamic themes associated with Say and Ricardo, such as James Mill (1773–1836), *Elements of Political Economy* (1821) and, of course, his son John Stuart Mill (1806–1873) whose *Principles* brought to close this early version of the classical system.

Among Say's contributions we must number a version of the classical short-run macroeconomic system which accepted the view that the supply of commodities generates income and purchasing power, and also the Smithian assumption that the only use of money was to circulate commodities while also accepting Smith's assumption that an act of savings would normally be matched by a decision to invest. In the relatively short run, the tendency of the economy was to full employment; a situation sustained by self-regulating mechanisms such as Smith's theory of price and allocation; a theory which was not seriously questioned. In Ricardo's case the emphasis was upon a generalized statement of Smith's labour embodied theory of value, a revised theory of rent, and a theory of growth which under the assumptions of constant technique and a closed economy, suggested that the normal progression of an economy was from an advancing state to a stationary state where no further growth was possible. Both models are formal, operating under specified assumptions. They are also essentially mathematical in character even if they clearly do owe much to Smith.

But there are important differences, arising not least from the fact that Smith's own approach was not narrowly 'mathematical' (Macfie, 1967, pp. 22–3), as compared, for example, with Ricardo (Baumol, 1962).

There was another difficulty arising from the fact that there was a tendency to assume that the basis of the subject of political economy dated from 1776, the year in which the *Wealth of Nations* first appeared. Donald Winch quotes an important passage from J. B. Say, Smith's committed, but by no means uncritical, disciple. Say wrote that:

> Whenever the inquiry into the *Wealth of Nations* is perused with the attention it so well merits, it will be perceived that until the epoch of its publication, the science of political economy did not exist. (Quoted in Winch, 1994, p. 103)

Terence Hutchison has argued that the 'losses and exclusions which ensued after 1776, with the subsequent transformation of the subject and the rise to dominance of the English classical orthodoxy were immense' (1988, p. 370). One such loss was the Physiocratic concept of the circular flow, to which Smith owed so much. Other losses occurred as a result of ignoring the contributions of Smith's close friend, David Hume, and the work of the latter's friend, Sir James Steuart (1713–1780), *Principles* (1767). The use of the historical method as applied to economic analysis and policy was one such loss and so too was the concern with structural unemployment, and the model of 'primitive' (pre-capitalist) accumulation. In addition the classical orthodoxy showed little interest in the problems presented by differential rates of growth in the context of international trade – hardly surprising in view of the fact that Smith largely ignored the problems identified by his two Scottish predecessors.

Ironically, Smith's own work did not always benefit from the work of those who 'inspected' the edifice. Here attention may be drawn to Smith's version of the 'circular flow' with its complex focus on period analysis and on the fact that all commodities have different life-cycles. Nor was much attention given to the *Smithian* use of time.

Ironically, the new orthodoxy also made it possible to think of political economy as a discipline which was quite separate from ethics and jurisprudence, thus obscuring Smith's true purpose. In referring to the way in which Smith organized his system of social science (ethics, jurisprudence, economics) Hutchison observed in a telling passage, that Smith was led as it by an Invisible Hand to promote an end which was no part of his original intention, that of 'establishing political economy as a separate, autonomous disciple' (1988, p. 355). A.L. Macfie made a related point in observing that 'it is a paradox of history that the analytics of Book I, in which Smith took his own line, should have eclipsed the philosophical and historical methods in which he so revelled, and which showed his Scots character' (1967, p. 21).

ANDREW SKINNER

See also **British classical economics.**

Selected works

Editions and abbreviations. An excellent edition of the *Lectures of Jurisprudence* was brought out by Edwin Cannan in 1896 (Oxford: Clarendon Press). Cannan also prepared a valuable edition of the *Wealth of Nations* in 1904 (London: Methuen). J.M. Lothian edited the *Lectures on Rhetoric* in 1963 (Edinburgh: Nelson).

Subsequent references are to the Glasgow edition of the *Works and Correspondence of Adam Smith* (Oxford, Clarendon Press, 1976–83) and follow the usages of that edition. The edition consists of:

I   *The Theory of Moral Sentiments* (TMS), ed. D.D. Raphael and A.L. Macfie, 1976.

II   *An Inquiry into the Nature and Causes of the Wealth of Nations* (WN), ed. R.H. Campbell, A.S. Skinner and W.B. Todd, 1976.

III   *Essays on Philosophical Subjects* (EPS), ed. D.D. Raphael and A.S. Skinner, 1980.
      This volume includes:

(i)   'The History of the Ancient Logics and Metaphysics' (Ancient Logics).

(ii) 'The History of the Ancient Physics' (Ancient Physics).

(iii) 'The History of Astronomy' (Astronomy).

(iv) 'Of the affinity between Certain English and Italian Verses' (English and Italian Verses).

(v)   'Of the External Senses' (External Senses).

(vi) 'Of the Nature of the Imitation which takes place in what are called the Imitative Arts' (Imitative Arts).

(vii) 'Of the Affinity between Music, Dancing and Poetry'. Items (i) to (vii), above, were prepared by W.P.D. Wightman.

(viii) 'Of the Affinity between Certain English and Italian Verses'.

(ix) Contributions to the *Edinburgh Review* (1755–6):
      (a) Review of Johnson's Dictionary.
      (b) A Letter to the authors of the *Edinburgh Review* (Letter).

(x)   Preface to William Hamilton's Poems on general Occasions. Items (viii) to (x), above, were prepared by J.C. Bryce.

(xi) Dugald Stewart, 'Account of the Life and Writings of Adam Smith LL.D' (Stewart), ed. I.S. Ross.

IV   *Lectures on Rhetoric and Belles Lettres* (LRBL), ed. J.C. Bryce; general editor, A.S. Skinner, 1983.
      This volume includes:
      'Considerations concerning the First Formation of Languages' (Considerations).

V   *Lectures on Jurisprudence* (LJ), ed. R.L. Meek, P.G. Stein and D.D. Raphael, 1978.
      This volume includes:

(i)   Student notes for the session 1762–3 (LJA).

(ii) Student notes for the session 1763–4 but dated 1766 (LJB).

(iii) The 'Early Draft' of the *Wealth of Nations* (ED).

(iv) Two Fragments on the Division of Labour, (FA and FB).

VI   *Correspondence of Adam Smith* (Corr.), ed. E.C. Mossner and I.S. Ross, 1977.
      This volume includes:

(i)   'A Letter from Governor Pownall to Adam Smith (1776)'.

(ii) 'Smith's thoughts on the State of the Contest with America, February 1778', ed. D. Stevens.

(iii) Jeremy Bentham's 'Letters' to Adam Smith (1787, 1790).

Associated volume

Essays on Adam Smith (EAS), ed. A.S. Skinner and T. Wilson. Oxford: Clarendon Press, 1975.

References to Corr. give letter number and date. References to LJ and LRBL give volume and page number from the MS. All other references provide section, chapter and paragraph number in order to facilitate the use of different editions. For example, Astronomy, II. 4='History of Astronomy', section II, para. 4. Stewart, I.12=Dugald Stewart, 'Account', section I, para. 12. TMS, I.i.5.5=TMS, Part I, section I, chapter 5, para. 5. WN, V.i.f.26=WN, Book V, chapter I, section 6, para. 26.

Bibliography

The literature on Smith is immense. But there are two useful surveys. The first is by Vivienne Brown, 'Mere Inventions of the Imagination: A Survey of Recent Literature on Adam Smith', *Economics and Philosophy* 13 (1997), 281–312. The second has been prepared by Anthony Brewer, 'Let Us Now Praise Famous Men: Assessments of Adam Smith's economics', *Adam Smith Review* 3 (2007).

Bagolini, L. 1975. The topicality of Adam Smith's notion of sympathy. In EAS.

Baumol, W.I. 1962. *Economic Dynamics*. London: Macmillan.

Berry, C.J. 1997. *Social Theory of the Scottish Enlightenment*. Edinburgh: Edinburgh University Press.

Black, J.B. 1926. *The Art of History*. New York: Methuen.

Black, R.D.C. 1976. Smith's contribution in historical perspective. In *Essays in Honour of Adam Smith*, ed. T. Wilson and A.S. Skinner. Oxford: Clarendon Press. Cited as EAS.

Blaug, M. 1985. *Economic Theory in Retrospect*. London: Heinemann.

Brown, M. 1988. *Adam Smith's Economics*. New York: Croom Helm.

Brown, V. 1994. *Adam Smith's Discourse*. London: Routledge.

Campbell, R.H. and Skinner, A.S. 1982. *Adam Smith*. London: Croom Helm.

Cannan, E. 1926. Adam Smith as an economist. *Economica* 6, 123–34.

Etzioni, A. 1988. *The Moral Dimension: Towards a New Economics*. London: Macmillan.

Fitzgibbons, A. 1995. *Adam Smith's System of Liberty, Wealth and Virtue*. Oxford: Clarendon Press.

Groenewegen, P. 1969. Turgot and Adam Smith. *Scottish Journal of Political Economy* 16, 271–87.

Haakonssen, K. 1981. *The Science of a Legislator*. Cambridge: Cambridge University Press.

Hollander, S. 1973. *The Economics of Adam Smith*. Toronto: Toronto University Press.

Hont, I. and Ignatieff, M., eds. 1983. *Wealth and Virtue: The Shaping of Political Economy in the Scottish Enlightenment*. Cambridge: Cambridge University Press.

Howell, W.S. 1975. Adam Smith's lectures on rhetoric: an historical assessment. In EAS.

Hume, D. 1985. *Essays Moral, Political and Literary*, ed. E.F. Miller. Indianapolis: Liberty Fund.

Hutchison, T. 1988. *Before Adam Smith*. Oxford: Blackwell.

Jeck, A. 1994. The macro structure of Adam Smith's theoretical system. *European Journal of the History of Economic Thought* 3, 551–76.

Jensen, H.E. 1984. Sources and contours: conceptualised reality. In *Adam Smith: Critical Assessments*, ed. J.C. Wood. London: Croom Helm.

Koebner, R. 1961. *Empire*. Cambridge: Cambridge University Press.

Lindgren, R. 1973. *The Social Philosophy of Adam Smith*. The Hague: Nijhoff.

Macfie, A.L. 1967. *The Individual in Society: Papers Relating to Adam Smith*. London: Allen and Unwin.

McNally, D. 1988. *Political Economy and the Rise of Capitalism*. Berkeley: University of California Press.

Meek, R.I. 1962. *The Economics of Physiocracy*. London: Allen and Unwin.

Meek, R.L. 1973. *Turgot on Progress, Sociology and Economics*. Cambridge: Cambridge University Press.

Meek, R.L. 1976. *Social Science and the Ignoble Savage*. Cambridge: Cambridge University Press.

Mizuta, H. 2000. *Adam Smith's Library: A Catalogue*. Oxford: Clarendon Press.

Oakley, A. 1994. *Classical Economic Man*. Aldershot: Edward Elgar.

O'Brien, D.P. 1975. *The Classical Economists*. Oxford: Clarendon Press.

Peacock, A.T. 1975. The treatment of the principles of public finance in the *Wealth of Nations*. In EAS.

Peil, J. 1999. *Adam Smith: Economic Science*. Cheltenham: Edward Elgar.

Phillipson, N. 1983. Adam Smith as a civic moralist. In Hont and Ignatieff (1983).

Pocock, J.G.A. 1983. Cambridge paradigms and Scotch philosophers. In Hont and Ignatieff (1983).

Rae, J. 1895. *Life of Adam Smith*. London: Macmillan.

Raphael, D.D. 1985. *Adam Smith*. Oxford: Oxford University Press.

Raphael, D.D. 2007. *The Impartial Spectator: Adam Smith's Moral Philosophy*. Oxford: Oxford University Press.

Robbins, I. 1953. *The Theory of Economic Policy in English Classical Political Economy*. London: Macmillan.

Rosenberg, N. 1960. Some institutional aspects of the Wealth of Nations. *Journal of Political Economy* 68, 557–70.

Ross, I. 1995. *Life of Adam Smith*. Oxford: Clarendon Press.

Rothbard, M. 1995. *An Austrian Perspective on the History of Economic Thought*. Aldershot: Edward Elgar.

Rothschild, E. 2001. *Economic Sentiments*. Cambridge, MA: Harvard University Press.

Schumpeter, J.A. 1954. *History of Economic Analysis*. London: Allen and Unwin.

Scott, W.R. 1900. *Francis Hutcheson: His Life, Teaching and Position in the History of Philosophy*. Cambridge: Cambridge University Press.

Scott, W.R. 1937. *Adam Smith as Student and Professor*. Glasgow: Jacksons.

Sen, A. 1987. *On Ethics and Economics*. Oxford: Blackwell.

Skinner, A.S. 1996. *A System of Social Science: Papers Relating to Adam Smith*, 2nd edn. Oxford: Clarendon Press.

Taylor, W.L. 1965. *Francis Hutcheson and David Hume as Predecessors of Adam Smith*. Durham, NC: University of North Carolina Press.

Teichgraeber, R. 1987. 'Less abused than I had reason to expect': the reception of the *Wealth of Nations* in Britain, 1776–1790. *Historical Journal* 30, 337–66.

Tribe, K. 1988. *Governing Economy: The Reformation of German Economic Discourse*. Cambridge: Cambridge University Press.

Viner, J. 1928. Adam Smith and laissez faire. In *Adam Smith, 1776–1926*. Chicago: Chicago University Press.

Vivenza, G. 2001. *Adam Smith and the Classics*. Oxford: Oxford University Press.

Werhane, P.H. 1991. *Adam Smith and his Legacy for Modern Capitalism*. Oxford: Oxford University Press.

Winch, D. 1978. *Adam Smith's Politics: An Essay in Historiographic Revision*. Cambridge: Cambridge University Press.

Winch, D. 1994. Nationalism and cosmopolitanism. In *Political Economy and National Realities*, ed. M. Albertone and A. Masoero. Turin: Fondazione Luigi Einaudi.

Young, J. 1977. *Economics as a Moral Science: The Political Economy of Adam Smith*. Cheltenham: Edward Elgar.

Smith, Bruce D. (1954–2002)

Bruce David Smith was born on 21 September 1954 in St. Paul, Minnesota, and died in Rochester, Minnesota, on 9 July 2002. Smith graduated with a BA in economics from the University of Minnesota in 1977 and obtained his Ph.D. in economics from the Massachusetts Institute of Technology in 1981. Smith's career began as an assistant professor at Boston College (1981–2). From there he moved to the research department at the Federal Reserve Bank of Minneapolis (1982–6), which was where his lifelong research interests in monetary history, monetary theory and financial intermediation blossomed among colleagues like Thomas Sargent, Neil Wallace, Ed Prescott, John Boyd and Warren Weber. After visits to Carnegie-Mellon and the University of California at Santa Barbara, he moved to the University of Western Ontario as an associate professor from 1987 to1990. Smith moved to Cornell University in 1990 and finally to the University of Texas at Austin in 1996, where he was the Fred Hofheinz Regent's Professor of Economics.

Smith's scholarly research consists of nearly one hundred published papers (a complete list of Smith's papers can be found in the *Federal Reserve Bank of Minneapolis Quarterly Review*, 2002). His research includes widely cited papers about monetary history, the causes and consequences of banking panics, the impact of financial market development on per-capita income and growth, the impact of inflation on financial market development, the macroeconomic effects of various types of credit market imperfections, and the optimal conduct of monetary and exchange rate policy.

The importance of financial intermediation for economic growth and stability underlies nearly all of Smith's research. In 'Taking intermediation seriously' (2003), his last single-authored paper, Smith argues that the information and spatial frictions which underlie a welfare-improving role for intermediation have important implications for the conduct of monetary policy. In particular, Smith lays out various theoretical models that show not only how excessive monetary expansion can lead to decreases in growth, but also how restrictive monetary policy (for instance, the Friedman rule) can be suboptimal.

One of Smith's most widely cited articles, 'Financial Intermediation and Endogenous Growth' (1991), co-authored with Valerie Bencivenga, provides a theoretical foundation for a large body of empirical research that finds a correlation between measures of financial market development and real output growth. Bencivenga and Smith provide a model with production externalities where liquidity provided by banks allows private agents to reduce the fraction of their savings held in the form of unproductive liquid assets and to increase their holdings of productive capital assets inducing growth.

Smith viewed economic history as a laboratory for understanding money and the theory of financial intermediation that he argued remains relevant for present-day policy. In 'Some colonial evidence on two theories of money: Maryland and the Carolinas', he argues that data from North and South Carolina is inconsistent with the quantity theory of money. For example, while the per capita stock of paper money more than tripled from 1755 to 1760, the price level increased by only seven per cent. On the other hand, data from Maryland, which had a unique method of backing the value of its currency with deliveries of sterling from the Bank of England, appears to be consistent with what Smith calls the Sargent–Wallace approach, where the value of money is determined in the same way that other privately issued assets are priced as the expected present discounted value of future cash flows (in this case, claims to sterling deliveries).

P. DEAN CORBAE

See also **financial intermediation; optimum quantity of money.**

Selected works

1985. Some colonial evidence on two theories of money: Maryland and the Carolinas. *Journal of Political Economy* 93, 1178–211.
1991. (With V. Bencivenga.) Financial intermediation and endogenous growth. *Review of Economic Studies* 58, 195–209.
2003. Taking intermediation seriously. *Journal of Money, Credit, and Banking* 35, 1319–57.

Bibliography

2002. The published work of Bruce D. Smith. *Federal Reserve Bank of Minneapolis Quarterly Review*, 26(4), 43–52. Online. Available at http://woodrow.mpls.frb.fed.us/research/qr/qr2644.pdf, accessed 10 March 2006.

Smith, Vernon (born 1927)

Vernon Lomax Smith is an American economist who shared the 2002 Nobel Prize in Economics (with the psychologist Daniel Kahneman) for his pioneering work on the methodology of laboratory experiments in economics. He is a remarkable scholar and a true pioneer in the quest to understand market institutions (such as auction mechanisms) and nonmarket institutions (such as bargaining rules), as well as the structure and motivation of individual behaviour. His methodological contributions helped overturn the traditional notion of economics as an inherently non-experimental science.

Smith was born on 1 January 1927 in Wichita, Kansas. He studied electrical engineering at the California Institute of Technology, receiving a bachelor's degree in 1949, then earned an MA in economics from the University of Kansas in 1952, and a Ph.D. from Harvard University in 1955. Smith's first teaching position was at the Krannert

School of Management at Purdue University (1955–67), where he began his work in experimental economics. After appointments at Stanford, the University of Massachusetts, the Center for Advanced Study in the Behavioral Sciences and Caltech, in 1975 he accepted a position at the University of Arizona, where he was to spend the next 26 years and build a body of research that would result in a Nobel Prize. Since 2001, Smith has been Professor of Economics and Law at George Mason University, where he is a research scholar in the Interdisciplinary Center for Economic Science, and a Fellow of the Mercatus Center.

Author or co-author of over 200 articles and books on capital theory, finance, natural resource economics and experimental economics, Smith has served on editorial boards of journals including (among many others) the *American Economic Review*. He has been influential in professional organizations, serving as president of the Public Choice Society, the Economic Science Association, and the Western Economic Association. His honours are many: he is a Fellow of the Econometric Society, the American Association for the Advancement of Science, and the American Academy of Arts and Sciences, and a Distinguished Fellow of the American Economic Association. He was elected a member of the National Academy of Sciences in 1995. Smith's collected papers were published by Cambridge University Press in 1991, with a second volume in 2000.

The press release issued by the Royal Swedish Academy of Sciences on the occasion of the 2002 Nobel Prize summarizes the contributions that earned Smith the award:

> Vernon Smith has laid the foundation for the field of experimental economics. He has developed an array of experimental methods, setting standards for what constitutes a reliable laboratory experiment in economics. In his own experimental work, he has demonstrated the importance of alternative market institutions, e.g., how the revenue expected by a seller depends on the choice of auction method. Smith has also spearheaded 'wind-tunnel tests', where trials of new, alternative market designs – e.g., when deregulating electricity markets – are carried out in the lab before being implemented in practice. His work has been instrumental in establishing experiments as an essential tool in empirical economic analysis. (Press Release: The Bank of Sweden Prize in Economic Sciences in Memory of Alfred Nobel, 9 October 2002)

Smith's early contributions focused on markets and how different ways of organizing exchange might lead to different outcomes. The next phase built upon the principles learned in the laboratory to design new institutions, as deregulation and privatization created unprecedented opportunities for the emergence of new markets around the world. Later work turned to the behaviour of agents in bilateral bargaining situations, involving pairs or small groups of participants, and exploring the nature and role of personal social exchange in decision-making in these settings. Recent research has explored the relationship between brain functions and decision-making (2001), the emergence of exchange systems in the absence of extant institutions that are typically imposed exogenously in experimental economies (2006a; 2006b), and the philosophy of science as it relates to the experimental method in economics. In every phase of his broad research agenda, he has produced important insights and critical methodological innovations.

Markets work, but not for the reasons we think

Vernon Smith tells the story of his early experiments (see Smith, 1991, pp. 154–8). As a graduate student at Harvard, he participated in the classroom experiment of Edward Chamberlin. As a teaching exercise, Chamberlin gave students seller costs or buyer values, then instructed them to circulate in the room, and negotiate prices with their counterparts. He then collected the prices, and displayed them, purportedly to illustrate that markets do not work as the perfectly competitive model suggests.

When Smith started teaching at Purdue, he adapted Chamberlin's experiment to his own classes, making a few modifications so that the market more closely resembled a stock market. He distributed costs and values to the participants, as Chamberlin had done, but had traders call out bids and offers. A pit boss recorded the bids and offers on the blackboard, and a trade occurred when a buyer accepted a seller's offer, or a seller accepted a buyer's bid. He also repeated the market, to allow students to learn the mechanics of the trading situation. He says, 'These two changes seemed to be the appropriate modifications to do a more credible job of rejecting competitive price theory' (1991, p. 155). To his surprise, the market converged in a couple of rounds to the predicted competitive equilibrium price and quantity. Thinking this might just be a fluke, he repeated it in another class, with the same result. Finally, imagining that the result might depend on the symmetric producer and consumer surpluses in his particular set-up, he ran another market with highly asymmetric surpluses, and once again 'the darned thing converged to competitive equilibrium'. In all cases, after a few rounds, all trades were taking place within a few cents of the same price. One can almost imagine him saying, 'Well look at that. Markets work!' These early 'double oral auction' (DOA, also referred to as an oral double auction) market experiments were published as Smith (1962), which also reports tests of the effects of shifts in supply and demand.

This exercise is such a strong illustration of the power of competitive markets that many economists use a classroom version of it to introduce their students to the supply and demand model. Participating in this experiment brings the model to life, and experiencing convergence to competitive equilibrium is an unforgettable lesson for students. By conducting his early experiments in the classroom, Smith inadvertently taught us how to teach economics.

These studies were followed by comparisons of different trading institutions. For example, compared with the posted offer institution found in most retail markets, where sellers post prices and buyers choose whether to buy, the DOA converges more quickly, and responds faster to changes in demand or supply. Smith termed it a 'disciplining' institution: the DOA is disciplining in the sense that buyers and sellers quickly discover whether they have left money on the table, and can modify their decisions immediately. In a posted offer market, this is less clear, as buyers who do not win the auction do not find out about other's values.

Another important contribution of Smith's early work is to discover the interactions between market structure and trading institutions in determining the outcome in the market. His work showed that market power is much harder to exercise in the DOA than in a posted offer market, where the monopolist can more easily sustain monopoly prices (Smith, 1981). This result stands outside standard economic theory, as research in industrial organization says little or nothing about the trading institution. A major contribution of experimental economics, and Smith's work in particular, is the insight that 'institutions matter'. This insight belongs in every Principles of Economics course, though it has not yet made its way into standard texts.

This new-found understanding of how markets and institutions work could then be harnessed to design new markets and institutions, using the laboratory as a 'wind tunnel'. The development of computer-assisted 'smart markets' expanded market design to a whole range of complex policy applications. Smith's work contributed to the design of many such markets, including airport landing slots, natural gas, wholesale electricity, as well as the design of the Arizona Stock Exchange, and the highly profitable auction by the US Federal Communications Commission of the portions of the microwave spectrum used for cellular phones, among other applications.

While Smith is known for his work showing how markets work, he also has explored situations when they might not work so well. This work showed that implementing the common knowledge assumption of perfectly competitive markets and game theory can be difficult, and this failure can lead to unexpected outcomes. For example, Smith and his colleagues have contributed to the growing field of behavioural finance by replicating the propensity of asset markets to bubble and crash. When valuations are based on subjects' beliefs about others' knowledge, beliefs and strategies, prices can deviate substantially from the underlying fundamentals of the asset (1988). In addition, he shows that more information is not necessarily better. When subjects are given information about others' payoffs, fulfilling one of the theoretical assumptions of competitive markets,

this can delay or distort convergence to equilibrium (Smith, 1976).

Dominated strategies are for playing

What happens where there is no competitive market to discipline trading behaviour? While people often act in accordance with the rational actor model in competitive situations, the negotiated outcomes of bargaining games allow scope for other motives to emerge. In the 1980s, laboratory tests of game-theoretic models began to produce results that revealed a penchant for fairness on the part of the participants. In a series of papers, Smith and his colleagues explored the importance of social distance between subjects. The greater the social distance, the more likely subjects were to behave selfishly in a dictator game (where one player decides the allocation of an endowment between himself and a counterpart) (1994; 1996a).

Subjects also take advantage of strategies that, according to theory, they should never play. Smith and his colleagues were the first to show that subjects will take advantage of an opportunity to punish a counterpart for unfair behaviour, even when that choice is costly for them (1996b). In this study and others, dominated strategies are chosen by subjects, when they are useful to punish bad behaviour by reducing a counterpart's payoff.

People are more rational than the agents in our models

Smith's Nobel lecture (Smith, 2003) surveys the territory that has been colonized by laboratory experimental economics. He organizes the research by distinguishing two types of rationality: constructivist, by which he means the sort of rationality built into the standard social science model of 'economic man'; and ecological, by which he refers to the use of reason to understand the emergent order and embodied intelligence of cultural rules, norms, and the institutions that result from regular human interactions, but that are not deliberately designed. Experimental economics taught Smith that our understanding of economic phenomena requires both constructivist and ecological rationality. From the discussion it is clear that his own research began with the constructivist model, but from his experience in the laboratory, he grew to appreciate that the richness of human behaviour required an augmented view. Smith's view of rationality grew out of a profound curiosity about and respect for human behaviour, and was shaped by his experience in the laboratory and his careful observation of human behaviour in a wide variety of settings.

CATHERINE C. ECKEL

See also **auctions (experiments); experimental economics; experimental methods in economics; experimental economics, history of.**

Selected works

1962. An experimental study of competitive market behavior. *Journal of Political Economy* 70, 111–37.

1976. Experimental economics: induced value theory. *American Economic Review* 66, 274–9.

1981. An empirical study of decentralized institutions of monopoly restraint. In *Essays in Contemporary Fields of Economics*, ed. G. Horwich and J.P. Quirk. West Lafayette: Purdue University Press.

1988. (With G. Suchanek and A. Williams.) Bubbles, crashes and endogenous expectations in experimental spot asset markets. *Econometrica* 56, 1119–51.

1991. *Papers in Experimental Economics*. New York: Cambridge University Press.

1994. (With E. Hoffman, K.A. McCabe and K. Shachat.) Preferences, property rights, and anonymity in bargaining games. *Games and Economic Behavior* 7, 346–80.

1996a. (With E. Hoffman and K.A. McCabe.) Social distance and other regarding behavior in dictator games. *American Economic Review* 86, 653–60.

1996b. (With K.A. McCabe and S.J. Rassenti.) Game theory and reciprocity in some extensive form experimental games. *Proceedings of the National Academy of Science* 93, 13421–8.

2000. *Bargaining and Market Behavior*. Cambridge: Cambridge University Press.

2001. (With K.A. McCabe, D. Houser, L. Ryan and T. Trouard.) A functional imaging study of cooperation in two-person reciprocal exchange. *Proceedings of the National Academy of Sciences* 98, 11832–5.

2003. Constructivist and ecological rationality in economics. *American Economic Review* 93, 465–508.

2006a. (With S. Crockett and B.J. Wilson.) Exchange and specialization as a discovery process. Working paper, George Mason University .

2006b. (With E. Kimbrough and B.J. Wilson.) Historical property rights, sociality, and the emergence of impersonal exchange in long-distance trade. Working paper, George Mason University.

SNP: nonparametric time series analysis

SNP is a method of multivariate nonparametric time series analysis. SNP is an abbreviation of 'seminonparametric' which was introduced by Gallant and Nychka (1987) to suggest the notion of a statistical inference methodology that lies halfway between parametric and nonparametric inference. The method employs an expansion in Hermite functions to approximate the conditional density of a multivariate process.

The leading term of this expansion can be chosen through selection of model parameters to be a Gaussian vector autoregression (VAR) model, a semi-parametric VAR model, a Gaussian ARCH model (Engle, 1982), a semiparametric ARCH model, a Gaussian GARCH model (Bollerslev, 1986), or a semiparametric GARCH model, either univariate or multivariate in each case. The unrestricted SNP expansion is more general than that of any of these models. The SNP model is fitted using maximum likelihood together with a model selection strategy that determines the appropriate order of expansion. Because the SNP model possesses a score, it is an ideal candidate for the auxiliary model in connection with efficient method of moment estimation (Gallant and Tauchen, 1996). Due to its leading term, the SNP approach does not suffer from the curse of dimensionality to the same extent as kernels and splines. In regions where data are sparse, the leading term helps to fill in smoothly between data points. Where data are plentiful, the higher-order terms accommodate deviations from the leading term. The method was first proposed by Gallant and Tauchen (1989) in connection with an asset pricing application. A C++ implementation of SNP is at http://econ.duke.edu/webfiles/arg/snp/, together with a User's Guide, which is an excellent tutorial introduction to the method.

Important adjuncts to SNP estimation are a rejection method for simulating from the SNP density developed in Gallant and Tauchen (1992), which can be used, for example, to set bootstrapped confidence intervals as in Gallant, Rossi and Tauchen (1992); nonlinear error shock analysis as described in Gallant, Rossi and Tauchen (1993), which develops the nonlinear analog of conventional error shock analysis for linear VAR models; and re-projection, which is a form of nonlinear Kalman filtering that can be used to forecast the unobservables of nonlinear latent variables models (Gallant and Tauchen, 1998).

As stated above, the SNP method is based on the notion that a Hermite expansion can be used as a general purpose approximation to a density function. Letting z denote an M–vector, we can write the Hermite density as $h(z) \propto [\mathscr{P}(z)]^2 \phi(z)$ where $\mathscr{P}(z)$ denotes a multivariate polynomial of degree K_z and $\phi(z)$ denotes the density function of the (multivariate) Gaussian distribution with mean zero and variance the identity matrix. Denote the coefficients of $\mathscr{P}(z)$ by a, which is a vector whose length depends on K_z and M. When we wish to call attention to the coefficients, we write $\mathscr{P}(z|a)$.

The constant of proportionality is $1/\int [\mathscr{P}(s)]^2 \phi(s) ds$ which makes $h(z)$ integrate to one. As seen from the expression that results, namely

$$h(z) = \frac{[\mathscr{P}(z)]^2 \phi(z)}{\int [\mathscr{P}(s)]^2 \phi(s) ds},$$

we are effectively expanding the square root of the density in Hermite functions of the form $\mathscr{P}(z)\sqrt{\phi(z)}$. Because the square root of a density is always square integrable and because the Hermite functions of the form $\mathscr{P}(z)\sqrt{\phi(z)}$ are dense for the collection of square integrable functions (Fenton and Gallant, 1996), every density has such an expansion. Because $[\mathscr{P}(z)]^2/\int [\mathscr{P}(s)]^2\phi(s)ds$ is a homogeneous function of the coefficients of the polynomial $\mathscr{P}(z)$, the coefficients can only be determined to within a scalar multiple. To achieve a unique representation, the constant term of the polynomial part is put to 1. Customarily the Hermite density is written with its terms orthogonalized and the C++ code is written in the orthogonalized form for numerical efficiency. But reflecting that here would lead to cluttered notation and add nothing to the ideas.

A change of variables using the location-scale transformation $y = Rz + \mu$, where R is an upper triangular matrix and μ is an M-vector, gives

$$f(y|\theta) \propto \{\mathscr{P}[R^{-1}(y - \mu)]\}^2$$
$$\times \{\phi[R^{-1}(y - \mu)]/|\det(R)|\}$$

The constant of proportionality is the same as above, $1/\int[\mathscr{P}(s)]^2\phi(s)ds$. Because $\{\phi[R^{-1}(y - \mu)]/|\det(R)|\}$ is the density function of the M-dimensional, multivariate, Gaussian distribution with mean μ and variance-covariance matrix $\Sigma = RR'$, and because the leading term of the polynomial part is 1, the leading term of the entire expansion is proportional to the multivariate, Gaussian density function. Denote the Gaussian density of dimension M with mean vector μ and variance matrix Σ by $n_M(y|\mu, \Sigma)$ and write

$$f(y|\theta) \propto [\mathscr{P}(z)]^2 n_M(y|\mu, \Sigma)$$

where $z = R^{-1}(y - \mu)$ for the density above.

When K_z is put to zero, one gets $f(y|\theta) = n_M(y|\mu, \Sigma)$ exactly. When K_z is positive, one gets a Gaussian density whose shape is modified due to multiplication by a polynomial in $z = R^{-1}(y - \mu)$. The shape modifications thus achieved are rich enough to accurately approximate densities from a large class that includes densities with fat, t-like tails, densities with tails that are thinner than Gaussian, and skewed densities (Gallant and Nychka, 1987).

The parameters θ of $f(y|\theta)$ are made up of the coefficients a of the polynomial $\mathscr{P}(z)$ plus μ and R and are estimated by maximum likelihood which is accomplished by minimizing $s_n(\theta) = (-1/n)\sum_{t=1}^n \log[f(y_t|\theta)]$. As mentioned above, if the number of parameters p_θ grows with the sample size n, the true density and various features of it such as derivatives and moments are estimated consistently (Gallant and Nychka, 1987).

This basic approach can be adapted to the estimation of the conditional density of a multiple time series $\{y_t\}$ that has a Markovian structure. Here, the term 'Markovian structure' is taken to mean that the conditional density of the M-vector y_t given the entire past $y_{t-1}, y_{t-2}, \ldots$ depends only on L lags from the past. For convenience, we will presume that the data are from a

process with a Markovian structure, but one should be aware that, if L is sufficiently large, then non-Markovian data can be well approximated by an SNP density (Gallant and Long, 1997). Collect these lags together as $x_{t-1} = (y_{t-1}, y_{t-2}, \ldots, y_{t-L})$, where L exceeds all lags in the following discussion.

To approximate the conditional density of $\{y_t\}$ using the ideas above, begin with a sequence of innovations $\{z_t\}$. First consider the case of homogeneous innovations; that is, the distribution of z_t does not depend on x_{t-1}. Then, as above, the density of z_t can be approximated by $h(z) \propto [\mathscr{P}(z)]^2 \phi(z)$ where $\mathscr{P}(z)$ is a polynomial of degree K_z. Follow with the location-scale transformation $y_t = Rz_t + \mu_x$ where μ_x is a linear function that depends on L_u lags

$$\mu_x = b_0 + Bx_{t-1}.$$

(If $L_u < L$, then some elements of B are zero.) The density that results is

$$f(y|x, \theta) \propto [\mathscr{P}(z)]^2 n_M(y|\mu_x, \Sigma)$$

where $z = R^{-1}(y - \mu_x)$. The constant of proportionality is as above, $1/\int[\mathscr{P}(s)]^2\phi(s)ds$. The leading term of the expansion is $n_M(y|\mu_x, \Sigma)$ which is a Gaussian vector autoregression or Gaussian VAR. When K_z is put to zero, one gets $n_M(y|\mu_x, \Sigma)$ exactly. When K_z is positive, one gets a semiparametric VAR density.

To approximate conditionally heterogeneous processes, proceed as above but let each coefficient of the polynomial $\mathscr{P}(z)$ be a polynomial of degree K_x in x. A polynomial in z of degree K_z whose coefficients are polynomials of degree K_x in x is, of course, a polynomial in (z, x) of degree $K_z + K_x$. Denote this polynomial by $\mathscr{P}(z, x)$. Denote the mapping from x to the coefficients a of $\mathscr{P}(z)$ such that $\mathscr{P}(z|a_x) = \mathscr{P}(z, x)$ by a_x and the number of lags on which it depends by L_p. The form of the density with this modification is

$$f(y|x, \theta) \propto [\mathscr{P}(z, x)]^2 n_M(y|\mu_x, \Sigma)$$

where $z = R^{-1}(y - \mu_x)$. The constant of proportionality is $1/\int[\mathscr{P}(s, x)]^2\phi(s)ds$. When K_x is zero, the density reverts to the density above. When K_x is positive, the shape of the density will depend upon x. Thus, all moments can depend upon x and the density can, in principal, approximate any form of conditional heterogeneity (Gallant and Tauchen, 1989).

In practice the second moment can exhibit marked dependence upon x. In an attempt to track the second moment, K_x can get quite large. To keep K_x small when data are markedly conditionally heteroskedastic, the leading term $n_M(y|\mu_x, \Sigma)$ of the expansion can be put to a Gaussian GARCH rather than a Gaussian VAR. SNP uses a modified BEKK expression as described in Engle and Kroner (1995); the modifications are to add leverage and level effects.

$$\Sigma_{x_{t-1}} = R_0 R_0'$$

$$+ \sum_{i=1}^{L_g} Q_i \Sigma_{x_{t-1-i}} Q_i'$$

$$+ \sum_{i=1}^{L_r} P_i (y_{t-i} - \mu_{x_{t-1-i}})(y_{t-i} - \mu_{x_{t-1-i}})' P_i'$$

$$+ \sum_{i=1}^{L_v} \max[0, V_i(y_{t-i} - \mu_{x_{t-1-i}})]$$

$$\max[0, V_i(y_{t-i} - \mu_{x_{t-1-i}})]'$$

$$+ \sum_{i=1}^{L_w} W_i x_{(1),t-1} x_{(1),t-i}' W_i'.$$

Above, R_0 is an upper triangular matrix. The matrices P_i, Q_i, V_i, and W_i can be scalar, diagonal, or full M by M matrices. The notation $x_{(1),t-i}$ indicates that only the first column of x_{t-i} enters the computation. The $\max(0, x)$ function is applied elementwise. Because $\Sigma_{x_{t-1}}$ must be differentiable with respect to the parameters of $\mu_{x_{t-2-i}}$, the $\max(0,x)$ function is approximated by a twice continuously differentiable cubic spline. Defining $R_{x_{t-1}}$ by the factorization $\Sigma_{x_{t-1}} = R_{x_{t-1}} R_{x_{t-1}}'$ and writing x for x_{t-1}, the SNP density becomes

$$f(y|x,\theta) \propto [\mathscr{P}(z,x)]^2 n_M(y|\mu_x, \Sigma_x)$$

where $z = R_x^{-1}(y - \mu_x)$. The constant of proportionality is $1/\int[\mathscr{P}(s,x)]^2 \phi(s) ds$. The leading term $n_M(y|\mu_x, \Sigma_x)$ is Gaussian ARCH if $L_g = 0$ and $L_r > 0$ and Gaussian GARCH if both $L_g > 0$ and $L_r > 0$ (leaving aside the implications of L_v and L_w).

A. RONALD GALLANT

See also **ARCH models; computational methods in econometrics; impulse response function; nonlinear time series analysis; nonparametric structural models.**

Research for this article was supported by the National Science Foundation.

Bibliography

Bollerslev, T. 1986. Generalized autoregressive conditional heteroskedasticity. *Journal of Econometrics* 31, 307–27.

Engle, R.F. 1982. Autoregressive conditional heteroskedasticity with estimates of the variance of United Kingdom inflation. *Econometrica* 50, 987–1007.

Engle, R.F. and Kroner, K.F. 1995. Multivariate simultaneous generalized ARCH. *Econometric Theory* 11, 122–50.

Fenton, V.M. and Gallant, A.R. 1996. Qualitative and asymptotic performance of SNP density estimators. *Journal of Econometrics* 74, 77–118.

Gallant, A.R., Hsieh, D. and Tauchen, G. 1997. Estimation of stochastic volatility models with diagnostics. *Journal of Econometrics* 81, 159–92.

Gallant, A.R. and Long, J.R. 1997. Estimating stochastic differential equations efficiently by minimum chi-square. *Biometrika* 84, 125–41.

Gallant, A.R. and Nychka, D.W. 1987. Seminonparametric maximum likelihood estimation. *Econometrica* 55, 363–90.

Gallant, A.R., Rossi, P.E and Tauchen, G. 1992. Stock prices and volume. *Review of Financial Studies* 5, 199–242.

Gallant, A.R., Rossi, P.E. and Tauchen, G. 1993. Nonlinear dynamic structures. *Econometrica* 61, 871–907.

Gallant, A.R. and Tauchen, G. 1989. Seminonparametric estimation of conditionally constrained heterogeneous processes: asset pricing applications. *Econometrica* 57, 1091–120.

Gallant, A.R. and Tauchen, G. 1992. A nonparametric approach to nonlinear time series analysis: estimation and simulation. In *New Directions in Time Series Analysis, Part II.*, ed. D. Brillinger et al. New York: Springer-Verlag.

Gallant, A.R. and Tauchen, G. 1996. Which moments to match? *Econometric Theory* 12, 657–81.

Gallant, A.R. and Tauchen, G. 1998. Reprojecting partially observed systems with application to interest rate diffusions. *Journal of the American Statistical Association* 93, 10–24.

social capital

Definitions?

The idea of *social capital* sits awkwardly in contemporary economic thinking. Although it has a powerful, intuitive appeal, the object has proven hard to track as an economic good. One can argue (Arrow, 2000) that it is misleading to use the term 'capital' to refer to whatever it is that 'social capital' happens to be, because capital is usually identified with tangible, durable and alienable objects (for example, buildings and machines), whose accumulation can be estimated and whose worth can be assessed. There is much to agree with this. But in regard to both heterogeneity and intangibility, social capital would seem to resemble knowledge and skills. So one can also argue that, since economists have not shied away from regarding knowledge and skills as forms of capital, we should not shy away in this case either.

In an early definition, social capital was identified with those 'features of social organization, such as trust, norms, and networks that can improve the efficiency of society by facilitating coordinated actions' (Putnam, Leonardi and Nanetti, 1993, p. 167). The characterization suffers from a weakness: it encourages us to amalgamate strikingly different objects, namely (and in that order), beliefs, behavioural rules, and such forms of capital assets as interpersonal links (or 'networks'), without establishing reasons why such an inclusive definition would prove useful in gaining an understanding of our social world. Subsequently, Putnam (2000, p. 19) suggested a redefinition: 'social capital refers to connections among individuals – social networks and the norms of reciprocity and trustworthiness that arise from them.' Since then authors have defined social capital even more inclusively, where attitudes towards others make their appearance as well: 'Social capital generally refers to trust, concern for one's associates, a willingness to live by the norms of one's community and to punish those who do not' (Bowles and Gintis, 2002, p. F419).

These definitions tell us that 'social capital' is an ingredient in the workings of *civil society* (Putnam, 1993; 2000). In a parallel development, the theory and empirics of common-property resources in poor countries (for example, coastal fisheries, village tanks, local forests, pasture lands, and threshing grounds) have revealed the character of those local institutions that enable mutually beneficial courses of action to be undertaken within communities (Dasgupta and Heal, 1979; Jodha, 1986; Ostrom, 1990; Dasgupta and Mäler, 1991; Bromley, 1992; Baland and Platteau, 1996). Development economists have also studied rotating savings and credit associations, irrigation management systems, and credit and insurance arrangements in poor countries (Ostrom, 1990; Udry, 1990; Besley, Coate and Loury, 1992; Grootaert and van Bastelaer, 2002). These studies suggest that social capital is a measure of the worth of *communitarian institutions*.

Where the state is weak or indifferent or rapacious, where markets do not work well or are even non-existent, communities enable people to survive, even if they do not enable them to live well. That may be why scholars writing on social capital have frequently imbued the notion with a warm glow. But there is a dark side to communities, often involving hierarchical social structures (for example, the Hindu caste system), rent-seeking groups, the Mafia, and street gangs. Ominously, the theory of repeated games (Fudenberg and Maskin, 1986) cautions us that communitarian relationships can involve allocations where some of the parties are *worse off* than they would have been if they had not been locked into those relationships; even though no overt coercion is visible, such relationships can be exploitative (see Dasgupta, 2000; 2005).

Why do people keep promises?

In order not to prejudge the character of communities, it is best not to worry about defining social capital, but to ask instead a fundamental question facing *any* group of people who have agreed on a joint course of action: *under what contexts can the members trust one another to try to carry out their terms of the agreement?*

Four points come to mind:

(i) *Mutual affection.* Consider the situation where the people involved care about one another. The household is

the most obvious example of an institution based on affection. To break a promise we have made to someone we care about is to feel bad. So we try not to do so.

(ii) *Pro-social disposition.* Another situation is where people are trustworthy, or where they reciprocate if others have behaved well towards them. Evolutionary psychologists have suggested that we are adapted to have a general disposition to reciprocate. Development psychologists have found that pro-social disposition can be formed by communal living, role modelling, education, and receiving rewards and punishments (be it here or in the afterlife).

We do not have to choose between the two viewpoints; they are not mutually exclusive. Our capacity to have such feelings as shame, guilt, fear, affection, anger, elation, reciprocity, benevolence, jealousy, and our sense of fairness and justice have emerged under selection pressure. Culture helps to shape preferences, expectations, and our notion of what constitutes fairness. Such notions in turn influence behaviour, which is known to differ among societies. But cultural coordinates enable us to identify the situations *in* which shame, guilt, fear, affection, anger, elation, reciprocity, benevolence, and jealousy arise; they do not displace the centrality of those feelings in the human make-up. By internalizing norms of behaviour, people enable the springs of their actions to include them. In short, they have a disposition to obey the norm, be it personal or social. When they do violate a norm, neither guilt nor shame would typically be absent, but frequently they will have rationalized the act. Making a promise is a commitment for such people; and it is essential for them that others recognize it to be so.

People are trustworthy to varying degrees. So, although pro-social disposition is not foreign to human nature, no society could rely on it exclusively; for how is one to tell to what extent someone is trustworthy?

Societies everywhere have therefore tried to establish institutions where people have the incentive to do business with one another. The incentives differ in their details, but they have one thing in common: *those who break agreements without acceptable cause are punished.* Broadly speaking, there are two ways in which the right incentives are created.

(iii) *External enforcement.* It could be that the agreement is translated into an explicit contract and enforced by an established structure of power and authority; that is, an *external enforcer.*

By an external enforcer we imagine here, for simplicity, the state. (There can, of course, be other external enforcement agencies; for example, tribal chieftains, warlords and so forth.) Consider that the rules governing transactions in the formal marketplace involve legal contracts backed by an external enforcer, namely, the state. So it is because you and the supermarket owner are confident that the state has the ability and willingness to enforce contracts that you and the owner of the supermarket are willing to transact when you go there to purchase goods.

What is the basis of that confidence? Simply to invoke an external enforcer for solving the credibility problem will not do; for why should the parties trust the state to carry out its tasks in an honest manner? A possible answer is that the government worries about its reputation. So, for example, a free and inquisitive press in a democracy, aided by a demanding civil society, helps to sober the government into believing that incompetence or malfeasance would mean an end to its rule, come the next election. Knowing that the government worries, the parties trust it to enforce agreements.

The above argument involves a system of interlocking beliefs about one another's abilities and intentions, one that supports an equilibrium in which the agreement is kept. Unfortunately, non-cooperation can also be held together by its own bootstraps. At a non-cooperative equilibrium the parties do not trust one another to keep their promises, because the external enforcer cannot be trusted to enforce agreements. To ask whether cooperation or non-cooperation would prevail is to ask which system of beliefs has been adopted by the parties about one another's intentions. Social systems have multiple equilibria.

(iv) *Mutual enforcement in long-term relationships.* Suppose the group of parties in question expect to face similar transaction opportunities in each period over an indefinite future. Imagine, too, that the parties cannot depend on the law of contracts because the nearest courts are far from their residence. There may even be no lawyers in sight. In rural parts of sub-Saharan Africa, for example, much economic life is shaped outside a formal legal system. But even though no external enforcer may be available, people there do transact. Credit involves saying, 'I lend to you now with your promise that you will repay me'; and so on. But why should the parties be sanguine that the agreements will not turn sour on account of malfeasance?

They would be sanguine if agreements were mutually enforced. The basic idea is this: *a credible threat by members of a community that stiff sanctions would be imposed on anyone who broke an agreement could deter everyone from breaking it.* The problem then is to make the threat credible. As the theory of repeated games has shown, the solution to the credibility problem in this case is achieved by recourse to social norms of behaviour.

By a *social norm* we mean a rule of behaviour (or strategy) that is followed by members of a community. For a rule of behaviour to be a social norm, it must be in the interest of everyone to act in accordance with the rule if all others were to act in accordance with it. Social norms are equilibrium rules of behaviour. The theory of repeated games has shown that, if people discount the future benefits from cooperation at a low enough rate, there are social norms that support cooperation.

As with the case of external enforcement, even when cooperation is a possible equilibrium under mutual enforcement, non-cooperation is an equilibrium too. If each party were to believe that all others would break the agreement from the start, each party would break the agreement at that stage. Failure to cooperate could be due simply to a collection of unfortunate, self-confirming beliefs.

Social capital as interpersonal networks

In common parlance, we reserve the term 'society' to denote a collective that has managed to equilibrate at a mutually beneficial outcome. Underlying each of the four contexts I have alluded to in which people trust one another to cooperate is a system of mutual beliefs. Because such a system of beliefs is likely to arise only if the parties know one another (at least indirectly), I believe it is best to regard *social capital* as *interpersonal networks*. The advantage of such a lean notion is that it does not prejudge the asset's quality. Just as a building can remain unused and a wetland can be misused, so a network can remain inactive or be put to use in socially destructive ways. There is nothing good or bad about interpersonal networks: other things being equal, it is the *use* to which a network is put by members that determines its quality.

Interpersonal networks are systems of communication channels linking people to one another. Networks include as tightly woven a unit as a nuclear family or a kinship group, and one as extensive as a voluntary organization, such as Amnesty International. We are born into certain networks and enter new ones. Personal relationships, whether or not they are long term, are emergent features within networks, and involve systems of mutual beliefs. For example, Seabright (1997) has suggested that civic engagements and communal activities heighten the disposition to cooperate (context (ii) above). The idea is that trust begets trust and that this gives rise to a positive feedback between civic and communal activities and a disposition to be so engaged. That positive feedback is, however, tempered by the cost of additional engagements (time), which, typically, rises with increasing engagements.

Networks and human capital

How did people who now interact with one another get to connect in the first place? In village economies in poor countries the answer is simple: mostly they have known one another from birth. People engaged in long-term relationships based on social norms (contexts (ii) and (iv) above) – or *communities*, for short – have to know one another, at least indirectly through people they know personally. Communities are *personal* and *exclusive*. Members have names, personalities and attributes. An outsider's word is not as good as an insider's. Markets, in contrast, are impersonal and inclusive.

In his pioneering work, Coleman (1988) saw social capital as an *input* in the production of human capital. In modern, mobile societies, people have to invest resources trying to meet people. Some of the investment is pleasurable, some not. Even so, just as academics are paid for what they mostly like doing anyway (as a return on investment in their education), networking would be expected to pay dividends even when maintaining networks is a pleasurable activity.

Burt (1992) has found among business firms in the United States, controlling for age, education and experience, that employees enjoying strategic positions in networks are more highly compensated than those who are not. His findings confirm that some of the returns from investment in network creation are captured by the investor. However, because of network externalities, not all the returns can be captured by the investor: when *A* and *B* establish a channel linking them, the investment improves both *A*'s and *B*'s earnings, but it also improves the earnings of *C*, who was already connected to *B*. The findings of Burt and his colleagues imply that membership in networks is a component of someone's 'human capital'. If firms pay employees on the basis of what they contribute to profitability, they would look not only at the conventional human capital employees bring with them (for example, health, education, experience, personality), but also the personal contacts they possess. It would be informative to untangle networks from the rest of human capital. This could reveal the extent to which returns from network investment are captured by the investor. But measurement problems abound. They may be insurmountable because of the pervasive externalities to which they give rise.

Micro-behaviour and macro-performance

How do network activities translate into the macro-performance of economies?

The discussion in the previous section implied that to the extent that the worth of contacts is reflected in wages and salaries, social capital is a component of human capital. It should be noted though that in poor countries, where labour markets can malfunction badly, or can even be non-existent, attributing returns to the various factors of production is especially problematic. But even if we were to leave that problem aside, we know that networks give rise to externalities. This makes the translation from micro-behaviour to macro-performance an especially difficult subject.

To illustrate, consider a simple formulation of economy-wide production possibilities. Let individuals be indexed by j ($j = 1, 2, \ldots$). Let K denote the economy's stock of physical capital and L_j the labour-hours put in by person j. I do not specify the prevailing system of property rights to physical capital, nor do I describe labour relations, because to do so would be to beg the questions being discussed here. But it is as well to keep in mind that

in a well-developed market economy K would be dispersed private property, in others K would be in great measure publicly owned, in yet others much would be communally owned, and so forth. It is also worth remembering that in market economies labour is wage based, that in subsistence economies 'family labour' best approximates the character of labour relations, and that labour cooperatives are not unknown in certain parts of the world; and so on.

Let h_j be the human capital of person j (years of schooling, health). His or her effective labour input is then $h_j L_j$. h_j is what one may call 'traditional human capital' (we leave aside the networks to which j belongs). Human capital is embodied in workers. Given the economy's knowledge base and institutions (the latter I take here to be the engagements brought about by the interpersonal networks), human capital in conjunction with physical capital produces an all-purpose output, Y, which we may call gross national product (GNP). Each of the aggregate indices requires for its construction prices for the multitude of components that make up the aggregate. In industrial market economies, the required prices are typically market prices. When externalities are pervasive, the construction of such indices poses special problems. Let us therefore assume away problems of aggregation by imagining the economy to possess a single good, Y. Problems nevertheless remain in measuring the pathways that link micro-behaviour to macro-performance. Let us study them.

Total factor productivity

Write $H = {}_j(h_j L_j)$. H is aggregate human capital. Now suppose that output possibilities are given by the relationship

$$Y = AF(K, H), \qquad (A > 0), \qquad (1)$$

where F is the economy's aggregate production function. F is non-negative and is assumed to be an increasing function of both K and H.

In eq. (1) A is *total factor productivity*. It is a combined index of institutional capabilities (including the prevailing system of property rights) and publicly shared knowledge. A macroeconomy characterized by the production function F would produce more, other things being equal, if A were larger (that is, if publicly shared knowledge was greater or institutional capabilities higher). Of course, the economy would also produce more, other things being equal, if K or h_j or L_j were larger. In short, technological possibilities for transforming the services of physical and human capital into output, when embedded in the prevailing institutional structure of the economy, account for eq. (1).

Consider now a scenario where civic cooperation increases in the community: the economy moves from a bad equilibrium system of mutual beliefs to a good one. The increase would make possible a more efficient allocation of resources in production. A question arises: would the increase in cooperation appear as a heightened value of A, or would it appear as an increase in H, or as increases in both?

The answer lies in the extent to which network externalities are like public goods. If the externalities are confined to small groups (that is, small groups are capable of undertaking cooperative actions on their own – with little effect on others – and take such actions in the good equilibrium), the improvements in question would be reflected mainly through the h_j's of those in the groups engaged in increased cooperation. On the other hand, if the externalities are economy-wide (as in the case of an increase in quasi-voluntary compliance in the economy as a whole owing to an altered set of beliefs, even about members of society one does not personally know), the improvements would be reflected mainly through A. Either way, the directional changes in macroperformance (though not the magnitude of the changes) would be the same. Other things being equal, an increase in A or in some of the h_j's (brought about by whichever of the mechanisms we have considered) would mean an increase in GNP, an increase in wages, salaries and profits, and possibly an increase in investment in both physical and human capital. The latter would result in faster rate of growth in output and consumption, and, if a constant proportion of income were spent on health, a more rapid improvement in health as well.

Interpreting cross-section findings

It will be useful to connect the macroeconomic account to the findings from less aggregated data. In his analysis of statistics from the 20 administrative regions of Italy, Putnam (1993) found civic tradition to be a strong predictor of contemporary economic indicators. He showed that indices of civic engagement in the early years of the 20th century were highly correlated with employment, income and infant survival in the early 1970s. Putnam also found that regional differences in civic engagement can be traced back several centuries and that, controlling for civic traditions, indices of industrialization and public health have no impact on current civic engagement. As he put it, the causal link appears to be from civics to economics, not the other way round. How do his findings square with the formulation in eq. (1)?

The same sort of question can be asked of even less aggregated data. Narayan and Pritchett (1999) have analysed statistics on household expenditure and social engagements in a sample of some 50 villages in Tanzania, to discover that households in villages where there is greater participation in village-level social organizations on average enjoy greater income per head. The authors have also provided statistical reasons for concluding that greater communitarian engagements result in higher household expenditure rather than the other way round.

To analyse these findings in terms of our macroeconomic formulation, consider two autarkic communities, labelled by i (=1, 2). I simplify by assuming that members of a community are identical. Denote the human capital per person in community i by h_i. By h_i I now mean not only the traditional forms of human capital (health and education), but also network capital. Let L_i denote the number of hours worked by someone in i, by N_i the size of i's population, and by K_i the total stock of physical assets in i. Aggregate output, Y_i, is

$$Y_i = A_i F(K_i, N_i h_i L_i). \qquad (2)$$

Improvements in civic cooperation are reflected in increases in A, or h, or both. It follows that if civic cooperation were greater among people in community 1 than in community 2, we would have $A_1 > A_2$, or $h_1 > h_2$, or both. Imagine now that the two communities have the same population size, possess identical amounts of physical capital, and work the same number of hours. GNP in community 1 would be greater than GNP in community 2 (that is, $Y_1 > Y_2$). More generally, an observer would discover that, controlling for differences in K and L, there is a positive association between a community's cooperative culture (be it total factor productivity, A_i, or human capital, h_i) and its mean household income (Y_i/N). This is one way to interpret the finding reported in Narayan and Pritchett (1999).

Consider now a different thought-experiment. Imagine that in year 1900 the two communities had been identical in all respects but for their cooperative culture, of which community 1 had more (that is, in 1900, $A_1 > A_2$, or $h_1 > h_2$, or both). Imagine next that, since 1900, both A_i and h_i have remained constant. Suppose next that people in both places have followed a simple saving rule: a constant fraction s_K (>0) of aggregate output has been invested each year in accumulating physical capital. (For the moment I imagine that net investment in human capital in both communities is nil.) In order to make the comparison between the communities simple, imagine finally that the communities have remained identical in their demographic features. It is then obvious that in year 1970 community 1 would be richer than community 2 in terms of output, wages and salaries, profits, consumption and wealth. This is one way to interpret Putnam's (1993) findings.

Notice that we have not had to invoke possible *increases* in total factor productivity (A_i) or human capital (h_i) to explain why a cooperative culture is beneficial. Total factor productivity and human capital have done all the work in our analysis of the empirical finding: we have not had to invoke secular improvements in them to explain why a more cooperative society would be expected to perform better economically.

As the communities in our thought experiment are both autarkic, there is no flow of physical capital from one to the other. This is an economic distortion for the combined communities: the rates of return on investment in physical capital in the two places remain unequal. The source of the distortion is the enclave nature of the two communities, occasioned in our example by an absence of markets linking them. There would be gains to be enjoyed if physical capital could flow from community 2 to community 1.

Autarky is an extreme assumption, but it is not a misleading assumption. What the model points to is that, to the extent that social capital is exclusive, it inhibits the flow of resources, in this case a movement of physical capital from one place to the other. Put another way, if markets fail to function well, capital does not move from community 2 to community 1 to the extent it ideally should. When social networks within each community block the growth of markets, their presence inhibits economic progress.

Dark matters

Two potential weaknesses of resource allocation mechanisms built on social capital are easy enough to identify.

Exclusivity. Networks are exclusive, not inclusive. This means that 'anonymity', the hallmark of competitive markets, is absent from the operations of networks. When market enthusiasts proclaim that one person's money is as good as any other person's in the market place, it is anonymity they invoke. In allocation mechanisms governed by networks, however, 'names' matter. Transactions are personalized. This implies inefficiencies: resources are unable to move to their most productive uses.

Inequalities. The benefits of cooperation are frequently captured by the more powerful within the network. McKean (1992), among others, has discovered that the local elite (usually wealthier households) capture a disproportionate share of the benefits of common-property resources, such as coastal fisheries and forest products. Her finding is consistent with the possibility that all who cooperate benefit.

Exploitation within networks

The reason why social capital continues to radiate a warm glow in the literature is that the subject has been motivated by examples of the Prisoner's Dilemma. However, one-period games involving the use of common property resources don't give rise to the Prisoner's Dilemma (Dasgupta, 2005). Consider an indefinitely repeated game among N players, in which (*a*) the stage game possesses a unique Nash equilibrium and (*b*) the 'minmax' payoffs of the players are lower than their respective payoffs in the Nash equilibrium. As is well known (Fudenberg and Maskin, 1986), if the players discount their future payoffs at a low enough rate, there are social norms that can sustain an outcome where the time-average of the per-period payoff to a player is *less than* the payoff at the unique Nash equilibrium. That player would be worse off in a long-term relationship with the others than if the players were not in a long-term

relationship. The social norm sustaining that outcome would be *exploitative* of the player.

Inequality is not the same as exploitation, which is why to demonstrate exploitation in an empirically satisfactory way will prove to be very hard: any such demonstration would involve comparison of an observable state of affairs with a counterfactual. However, some stark examples are suggestive. In Indian villages, access to local common-property resources is often restricted to the privileged (for example, caste Hindus), who are also among the more prosperous landowners. The outcasts (euphemistically called members of 'schedule castes') are among the poorest of the poor. Stark inequities exist, too, in patron–client relationships in agrarian societies, which make it very likely that the 'client' is worse off in consequence of that relationship than without it. Ogilvie (2003) has unearthed striking differences between the life chances of women in 17th-century Germany (embedded in dense networks) and the life chances of women in 17th-century England (not so embedded in dense networks): English women were better off.

Morals

Social capital is an aggregate of interpersonal networks. From the economic point of view, belonging to a network helps a person to coordinate his strategies with others. We should not prejudge the character of the strategies on which members of a network coordinate. As with any other form of capital asset, social capital can be put to good use or bad.

PARTHA DASGUPTA

See also **common property resources; cooperation; repeated games; social networks, economic relevance of; social norms.**

Bibliography

Arrow, K.J. 2000. Observations on social capital. In *Social Capital: A Multifaceted Perspective*, ed. P. Dasgupta and I. Serageldin. Washington, DC: World Bank.

Baland, J.-M. and Platteau, J.-P. 1996. *Halting Degradation of Natural Resources: Is There a Role for Rural Communities?* Oxford: Clarendon Press.

Besley, T., Coate, S. and Loury, G. 1992. The economics of rotating savings and credit associations. *American Economic Review* 82, 782–810.

Bowles, S. and Gintis, H. 2002. Social capital and community governance. *Economic Journal* 112, 419–36.

Bromley, D.W. 1992. *Making the Commons Work: Theory, Practice and Policy.* San Francisco: Institute for Contemporary Studies.

Burt, R.S. 1992. *Structural Holes: The Social Structure of Competition.* Cambridge, MA: Harvard University Press.

Coleman, J.S. 1988. Social capital in the creation of human capital. *American Journal of Sociology* 94, 95–120.

Dasgupta, P. 2000. Economic progress and the idea of social capital. In *Social Capital: A Multifacted Perspective*, ed. P. Dasgupta and I. Serageldin. Washington, DC: World Bank.

Dasgupta, P. 2005. Common property resources: economic analytics. *Economic and Political Weekly* 40, 1610–22.

Dasgupta, P. and Heal, G. 1979. *Economic Theory and Exhaustible Resources.* Cambridge: Cambridge University Press.

Dasgupta, P. and Mäler, K.-G. 1991. The environment and emerging development issues. *Proceedings of the Annual World Bank Conference on Development Economics, 1990.* Washington, DC: World Bank.

Fudenberg, D. and Maskin, E. 1986. The folk theorem in repeated games with discounting or with incomplete information. *Econometrica* 54, 533–56.

Grootaert, C. and van Bastelaer, T. 2002. *The Role of Social Capital in Development: An Empirical Assessment.* Cambridge: Cambridge University Press.

Jodha, N.S. 1986. Common property resources and the rural poor. *Economic and Political Weekly* 21, 1169–81.

McKean, M. 1992. Success on the commons: a comparative examination of institutions for common property resource management. *Journal of Theoretical Politics* 4, 256–68.

Narayan, D. and Pritchett, L. 1999. Cents and sociability: household income and social capital in rural Tanzania. *Economic Development and Cultural Change* 47, 871–89.

Ogilvie, S. 2003. *A Bitter Living: Women, Markets, and Social Capital in Early Modern Germany.* Oxford: Oxford University Press.

Ostrom, E. 1990. *Governing the Commons: The Evolution of Institutions for Collective Action.* Cambridge: Cambridge University Press.

Putnam, R.D., Leonardi, R. and Nanetti, R.Y. 1993. *Making Democracy Work: Civic Traditions in Modern Italy.* Princeton, NJ: Princeton University Press.

Putnam, R.D. 2000. *Bowling Alone: The Collapse and Revival of American Community.* New York: Simon and Schuster.

Seabright, P. 1997. Is cooperation habit-forming? In *The Environment and Emerging Development Issues*, vol. 2, ed. P. Dasgupta and K.-G. Mäler. Oxford: Clarendon Press.

Udry, C. 1990. Credit markets in northern Nigeria: credit as insurance in a rural economy. *World Bank Economic Review* 4, 251–69.

social choice

Social choice theory, pioneered in its modern form by Arrow (1951), is concerned with the relation between individuals and the society. In particular, it deals with the aggregation of individual interests, or judgements, or well-beings, into some aggregate notion of social welfare, social judgement or social choice. It should be obvious that the aggregation exercise can take very different forms depending on exactly what is being aggregated (e.g., the personal interests of different people, or their moral or political judgements) and what is to be derived on that basis (e.g., a measure of social welfare, or public decisions regarding what is to be done or what outcomes are to be accepted). The formal similarities between these exercises in the analytical format of aggregation should not make us overlook the diversities in the nature of the exercises performed (see Sen, 1977a, 1986). In fact, the axioms chosen for different exercises are often quite divergent, and the general conception of aggregation in social choice theory permits such variation.

1. Welfare economics and social choice

Although the origins of social choice theory – in one form or another – can be traced back at least two hundred years (Borda, 1781; Condorcet, 1785; Bentham, 1789), the formal theory of social choice was initiated by Kenneth Arrow (1951) less than four decades ago. Arrow drew on some existing notions of welfare economics. One concept of a *social welfare function* had been introduced by Bergson (1938). This was defined in a very general form indeed: as a real-value function W(.), determining social welfare, 'the value of which is understood to depend on all the variables that might be considered as affecting welfare' (p. 417). Such a social welfare function – swf for short – might be thought to be a real-valued function defined on X, the set of alternative social states. It is a bit more permissive to see a Bergson social welfare function as an ordering R of X (more permissive because not all orderings can be numerically represented).

Various uses to which a swf can be put in welfare economics were investigated, particularly by Samuelson (1947). His exercises made use of several criteria that a *swf* may be required to satisfy, including the Pareto criterion, demanding that unanimous individual preference over any pair of states should yield the corresponding social preference over that pair.

None of the conditions that Samuelson imposed on a *swf* for his exercises required any general specification of how the social ordering might change if *different* sets (strictly, *n*-tuples) of individual orderings were considered. If any *n*-tuple of individual preference orderings is called a 'profile', then Samuelson's exercises – and those considered by Bergson – were all 'single-profile' problems, without additional requirements of *inter-profile* consistency.

Arrow (1951) defined a social welfare function – henceforth *SWF* (to be distinguished from a Bergson–Samuelson *swf*) – as a functional relation specifying one social ordering R for any given *n*-tuple of individual orderings (Ri), with one ordering Ri for each person i: R = f(Ri)).

Note that if a Bergson–Samuelson swf is defined as a social ordering R (rather than as a real-valued function W(.)), then an Arrow SWF is a function the value of which would be a Bergson–Samuelson swf. Arrow's exercise, in this sense, is concerned with the way of arriving at a Bergson–Samuelson swf.

Arrow proceeded to impose a few conditions that any reasonable SWF could be expected to satisfy. His 'impossibility theorem' (more formally called 'the General Possibility Theorem') shows that no SWF can satisfy all these conditions together. One of the conditions deals specifically with the multiple-profile characteristics of a *SWF*, viz., the independence of irrelevant alternatives (condition I). This requires that the chosen alternatives from any subset of social states must remain unaltered as long as the individual preferences over this subset remain unaltered, even though the individual preferences may have been revised over other subsets. Another condition is a weak version of the Pareto principle (condition P) which requires that unanimous *strict* preference over a pair must be reflected in the same strict preference for the society. Another requirement is that of unrestricted domain (condition U), which demands that the domain of the *SWF* must include all logically possible *n*-tuples of individual orderings, that is, the *SWF* should be able to specify a social ordering R no matter what the individual orderings happen to be. Finally, there is a condition of non-dictatorship (condition D), which demands that there is no individual such that if he or she prefers any *x* to any *y*, then *x* is socially preferred to *y*, no matter what the other individuals prefer.

One version of the 'impossibility theorem' of Arrow establishes that, if the set of individuals is finite and the number of distinct social states is at least three, then there is no social welfare function (SWF) satisfying conditions U, I, P, and D.

This result has been the starting point of much of modern social choice theory. Even though the focus has somewhat shifted in recent years from impossibility results to other issues, there is no question at all that Arrow's formulation of the social choice problem in presenting his 'impossibility theorem' laid the foundations of social choice theory as it has evolved.

Two *interpretational* issues may be sorted out first before formal social choice theory is considered for a general examination. The first issue concerns the interpretation of 'social preference'. As has already been remarked, the nature of the social choice exercise can vary in many different ways, and one source of variation is the nature of the end-point that is sought (in particular the interpretation of R). Consider the relation of strict social preference *xPy*. It can be given different interpretations depending on the nature of the exercise. For example, *xPy* can stand for the judgement that society is better off in state *x* than in state *y*. Such a judgement can be the view of a particular individual (in his or her capacity as an aggregating judge), or the mechanical outcome of some institutional process of aggregating judgements (e.g., the result of a voting procedure). Or, alternatively, *xPy* can stand for the statement that, in the choice exactly over the pair *(x, y)*, *x* alone must be chosen. A further alternative is to interpret *xPy* as the requirement that *y* must not be chosen from any set which contains *x* (whether or not it contains any other alternative). These and other interpretations give different views of 'social preference', and careful attention has to be paid to the nature of the exercise depending on the interpretation given. Although Arrow's 'impossibility theorem' and similar results apply to *all* the interpretations (and here there is a genuine economy in the general axiomatic method), extensive variations in the relevance of the results to different types of problems must be recognized.

Second, a different source of variation relates to the interpretation of the individual preference orderings. The individual ordering can stand for the ranking of personal well-being, and if so, the exercise is one of *well-being aggregation*. An example may be found in arriving at overall judgements of the well-being of the community based on rankings of individual well-beings. To take a very different type of example, in making a committee decision, the different judgements of the members of the committee may be aggregated together in an overall judgement or an overall decision, and that exercise is one of *judgement aggregation*. This is not to deny that the judgements of members of the committee may, in fact, be influenced by their individual interests, but the nature of the exercise is primarily that of aggregating the possibly divergent judgements of the members of a committee to arrive at an over-all committee view. In some other exercises, for example, in electing a candidate or a member of Parliament or a Mayor, the individual votes may well reflect a clear-cut mixture of individual interests and political beliefs, so that the exercise may have features of interest aggregation as well as judgement aggregation. Once again, it is worth emphasizing that while the formal results such as Arrow's 'impossibility theorem' apply to each of the interpretations, the exact substantive content of the result would depend on the particular interpretation chosen.

The specific context of Arrow's exercise was that of supplementing the work of Bergson and Samuelson in deriving social welfare functions for welfare-economic studies. If the individual orderings are interpreted as utility rankings of individuals, and social preferences interpreted as a judgement of social welfare, the Arrow theorem asserted that there is no way of combining individual utility orderings into an overall social welfare judgement satisfying the four specified conditions. The result can be easily translated into a choice-theoretic framework by adopting a choice-based notion of 'social preference', e.g., the 'base' relation or the 'revealed preference' relation of social choice. On this interpretation, it would appear that there is no way of arriving at a social choice procedure specifying what is to be chosen (over pairs, or over larger subsets), satisfying the appropriately interpreted (i.e., in terms of choice) conditions specified by Arrow (see Blair, Bordes, Kelly and Suzumura, 1976; and Sen, 1977a, 1982).

This is, of course, a negative result. A great deal of social choice theory, at least in the early stage, consisted of trying to deal with this result, suggesting different interpretations, different extensions, different ways of 'resolution', and other responses to the 'impossibility' identified by Arrow.

The main lines of response to Arrow's result will be examined presently. It is, however, worth emphasizing that the 'impossibility theorem' must not be seen as primarily a 'negative' achievement. The axiomatic method, as used here, can take a set of axioms which look reasonable enough and then derive some joint implications of these axioms. If the implications are unacceptable, the axioms can be re-examined. Interpreted thus, the axiomatic method is a procedure for assessing a set of principles reflected in the axiom structure, and it persistently invites attention to the content and acceptability of the axioms chosen.

Arrow's impossibility result brought out the unviability of the welfare-economic structure that had emerged in the discussion preceding the birth of modern social choice theory. After the rejection of 'interpersonal comparisons' of well-being (on this see Robbins, 1932, 1938), it was increasingly accepted that social choices or social judgements would have to be based on individual utility orderings without interpersonal comparisons. The four axioms chosen by Arrow make a good deal of sense in that context, and had indeed been used – formally or informally – in the pre-existing literature. What Arrow's theorem demonstrates is the unviability of that structure. The primary impact of Arrow's initial result was to demand that the entire question of the basis of social welfare judgement be re-examined. While this is, in one sense, a negative result, in another sense it opened up various ways of reformulating the social choice problem as a result of the demonstrated unviability of the pre-existing approach. The later literature in social choice theory bears testimony to the fact that many of these ways have been found to be both feasible *and* useful. Several of these approaches will be examined later on in this note, but the positive contribution of the negative impossibility result presented by Arrow has to be kept in view to see these advances in their appropriate perspective.

2. Variations and extensions of Arrow's impossibility result

The literature of social choice theory contains a large number of theorems that take the form of presenting variations of the type of impossibility identified by Arrow. In fact, Arrow himself has presented several distinct versions. The one presented in 1951 contained a formulational error, which was identified and corrected by Blau (1957). A later version, which was the one cited in the last section, is presented in Arrow (1963). Various other variations can be found in the literature, modifying one condition or another, and presenting impossibility results based on conditions that are more demanding in

some respects and less demanding in others (see particularly Blau, 1957; Murakami, 1968; Pattanaik, 1971; Fishburn, 1973, 1974; Hansson, 1976; Brams, 1976; Plott, 1976; Kelly, 1978; Monjardet, 1979; Roberts, 1980b; Chichilnisky, 1982; McManus, 1982; Suzumura, 1983; Hurley, 1985; Nitzan and Paroush, 1985, among many others). Each of Arrow's conditions has been modified in one way or another in these different variants.

One particular variation, which is both illuminating and simple, relates to results presented by Wilson (1972) and Binmore (1976). This shows that given unrestricted domain and independence, all permissible social welfare functions will either have social rankings 'imposed' irrespective of individual preferences, or have a dictator, or have a 'reverse dictator' (a person such that whenever he prefers x to y, society prefers y to x). Arrow's impossibility theorem can be seen as a corollary of this when the Pareto principle is also demanded, since Pareto will eliminate both 'imposition' and 'reverse dictatorship', leaving dictatorship as the only remaining possibility.

One line of variation that has been very extensively investigated is that of weakening the demand of 'collective rationality', i.e., relaxing the requirement that social choice must be based on a social *ordering* (complete, reflexive and transitive). The 'range' of the social welfare function is supposed to include only social orderings, and the proposed relaxation weakens that demand. It can be shown that if the transitivity of only *strict* social preference is demanded (without also demanding the transitivity of social indifference), then all of Arrow's conditions can be simultaneously satisfied and there is no impossibility (see Sen, 1969, 1970; see also Schick, 1969).

This condition of transitivity of strict preference, formally called 'quasi-transitivity', when imposed on social preferences, for a social welfare function satisfying unrestricted domain, independence and the Pareto principle, has the effect of confining social choice procedures to 'oligarchies' (this result was first presented in an unpublished paper by Gibbard, and reported in Sen, 1970). An oligarchic group consists of a set of individuals such that if any one of them prefers any x to any y, then x must be taken to be socially preferred to or indifferent to y, and if all the individuals in that group unanimously prefer x to y, then x must be taken to be socially strictly preferred to y. One extreme case of oligarchy is that of a one-person oligarchy, which corresponds to Arrow's dictatorship. The other extreme makes the oligarchy group include every individual in the community. In this latter case, the fact that all of them taken together happen to be decisive is not remarkable (it follows in fact immediately from the Pareto principle). But it also gives every member of the community 'veto' power in the sense that whenever anyone prefers any x to any y, this precludes the possibility of y being socially strictly preferred to x, and this has the effect of producing lots of social indifferences all around (see Sen, 1969).

This 'veto' result can be obtained even without demanding quasi-transitivity of social preference, by supplementing the weaker demand of 'acyclicity' (i.e. the absence of strict preference cycles) with some other conditions, as has been investigated by Mas-Colell and Sonnenschein, 1972; Schwartz, 1972, 1986; Guha, 1972; Brown, 1974, 1975; Blau, 1976; Blau and Deb, 1977; Monjardet, 1979, and others. Partial 'veto' results have been established with still weaker conditions (see Blair and Pollak, 1982; Kelsey, 1984).

On a somewhat different line of investigation, it has been possible to somewhat weaken the condition of full transitivity of social preference and still retain exactly the impossibility identified by Arrow, i.e., dictatorship following from conditions U, I and P. This is easily done by replacing the requirement of ordering by that of having 'semi-orders', but it can be relaxed further (see Blair and Pollak, 1979; Blau, 1979).

These investigations of relaxation of collective rationality have not been confined only to the weakening of transitivity of social preference. It is possible to drop the requirement of completeness of social preference, permitting the possibility that many pairs of states x and y may be not socially rankable vis-à-vis each other, and still the impossibility result may survive if the Arrow conditions are correspondingly redefined to cover this case with sufficient richness of social ranking, in line with Arrow's original motivation (see Barthelemy, 1983; Weymark, 1983).

Yet another line of investigation consists of relaxing the requirement that social choice must be 'binary' in nature, in the sense of the choice function being representable by a binary relation (whether or not that binary relation R is called social preference). Some positive possibility results were obtained by Schwartz (1970, 1972), Fishburn (1973), Plott (1973), Bordes (1976), and Campbell (1976), by demanding consistency conditions on choice functions that are weaker than the requirement of binary choice.

One way of doing this is to convert preference cycles into indifference classes. For example, take the case of the so-called 'paradox of voting' in which person 1 prefers x to y, and y to z, person 2 prefers y to z, and z to x, and person 3 prefers z to x, and x to y. In this case the majority rule yields x being socially preferred to y, y being socially preferred to z, and z being socially preferred to x, producing a strict preference cycle, with no alternative that is not beaten by another alternative. If, in this case, all the three alternatives are declared socially indifferent, by converting the cycle into an indifference class, then much of Arrow's requirements can be retained. However, one type of consistency will certainly be violated by this formulation of social choice, to wit, relating social choice over the pair (x, y) to that over the triple (x, y, z). Due to the majority preference for x over y, and the demand of the 'independence' condition (I) that individual preferences only over (x, y) be considered when

choosing over this pair only, x must be chosen and y rejected in the choice over the pair (x, y). But in the choice over the triple (x, y, z), even state y can be selected as a member of the indifference class, when the majority cycle is converted into indifference. The choosability of y from the larger set (x, y, z), and its non-choosability from the smaller set (x, y) contained in the larger set, does violate a standard condition of consistency of choice, variously called Property α or the 'Chernoff condition', or standard 'contraction consistency'. In the absence of this consistency, the choice function cannot possibly be represented in a binary form, i.e., through a binary relation R such that the choices correspond to the R-maximal elements (with R being derived from the internal properties of choice, e.g., xRy when x is chosen in the presence of y). But this condition (Property α) is, in fact, much weaker than the requirement that the choice function be binary.

Since binariness may not in itself be a compelling requirement, the plausibility of this line of resolution of Arrow's impossibility depends on the value of the consistency conditions that these 'solutions' may actually satisfy. By imposing some relatively appealing consistency conditions, it can be shown that the dictatorship result of Arrow, and the other related results regarding oligarchy, veto power, etc., derived in the binary framework can reappear easily enough in non-binary choice as well (see Blair, Bordes, Kelly and Suzumura, 1976; Sen, 1977a). It can also be pointed out that even when the social choice procedures do not satisfy binariness in the sense of being *representable* by a binary relation, there would, of course, be binary relations that are *generated* by the choice function. For example, the 'revealed preference' relation (xRy if x is chosen in the presence of y) will be defined by any choice function, since the choice of any alternative (say, x) from any set containing another alternative (say, y), will yield the deduction xRy. The issue of binariness arises when it is further demanded that is what is chosen from each subset consists exactly of the R-maximal elements of that set, according to that binary relation R. It can be shown that binariness in this form demands much the same thing whether we concentrate on the 'revealed preference' relation, or the 'base relation' (the latter being defined as: xRy if and only if x is chosen specifically from the pair x, y). Binariness according to the 'revealed preference' relation is equivalent to that according to the 'base' relation (see Herzberger, 1973).

Although the demands of binariness provide one way of re-establishing Arrow's 'impossibility' results, a different way is not to demand binariness at all, but to translate all of Arrow's demands to one specific binary relation generated by social choice, e.g., the 'revealed preference', or the 'base' relation. The Arrow theorem will hold exactly in the same way for each such interpretation of R, provided the Arrow conditions are correspondingly reinterpreted. In this sense binariness is not a central issue in the inescapability of the 'impossibility' result of

Arrow (on this see Sen, 1977a, 1982; on related matters, see Grether and Plott, 1982; Suzumura, 1983; and Matsumoto, 1985).

One general conclusion that seems to emerge from these investigations of relaxation of 'collective rationality' properties is the durability and robustness of Arrow's 'impossibility' result. The tension between different types of principles seems to survive various ways of relaxing these principles, and the particular 'impossibility theorem' of Arrow is a centre piece of a much broader picture. Demands on consistency of social choice can be dramatically changed without the 'impossibility' features disappearing.

3. Domain restrictions

When presenting his impossibility result, Arrow had suggested the possibility that a resolution might be found in terms of restricting the domain of the social welfare function (no longer requiring that it works no matter what the individual preferences happen to be). It is, of course, clear that there are many preference combinations for which such procedures as the method of majority decision will yield perfectly consistent social choice. Arrow (1951) himself had explored a particular type of restriction of individual preferences called 'single-peaked preferences' (earlier discussed by Black, 1948). This corresponds to the case in which the alternatives can be so arranged on a line that everyone's intensity of preference has one peak only, i.e., the preference drops monotonically as we move from left to right, or rises monotonically, or it rises to a peak and then falls. Arrow showed that if individual preferences are single-peaked and the number of voters is odd, then majority decision will yield transitive social preference.

The positive possibility result for single-peaked preferences can be generalized in many different ways. It can be shown that individual preferences being single-peaked in every triple of alternatives is equivalent to the condition that in every triple there is one state such that no one regards it to be 'worst'. It can be shown that a similar agreement on some alternative being regarded as not 'best' would do, and so would an agreement on some alternative being not 'medium'. Altogether, this sufficiency condition is called 'value restriction', and the particular type of agreement (whether 'not best', or 'not worst', or 'not medium') may vary from triple to triple (see Sen, 1966). Also the requirement of oddness of the number of voters can be eliminated if the demand is not for full transitivity of social preference, but only the absence of preference cycles and the existence of a majority winner (Sen, 1969). In this general line of investigation, necessary and sufficient conditions for transitivity as well as acyclicity of majority decisions (i.e., for the existence of a majority winner) have been identified by Inada (1969, 1970) and Sen and Pattanaik (1969). The former requirement is called 'extremal

restriction'. The relationships among these and other related conditions are discussed in Inada (1969), Sen (1970), Pattanaik (1971), Fishburn (1973), Salles (1976), Slutsky (1977), Kelly (1978), Monjardet (1979), Blair and Muller (1983), Larsson (1983), Suzumura (1983), Dummett (1984), Arrow and Raynaud (1986), Jain (1986) among many others.

On a different line of analysis, domain conditions can be specified not only in terms of general qualitative correspondence of individual preferences, but also in terms of number-specific requirements on the distribution of voters over the different preferences (see particularly Plott, 1967; Tullock, 1967, 1969; Saposnik, 1975; Slutsky, 1977; Gaertner and Heinecke, 1978; Grandmont, 1978; Dummett, 1984).

These domain conditions all deal specifically with the method of majority decision, but the problem can be investigated more generally. Domain conditions for other voting rules have been investigated (see, for example, Pattanaik, 1971). More recently, the necessary and sufficient domain conditions for the existence of any social welfare function satisfying all of Arrow's other conditions (whether or not based on counting majority) have been investigated (see Kalai and Muller, 1977, and Maskin, 1976; see also Dasgupta, Hammond and Maskin, 1979; Kalai and Ritz, 1980, and Chichilnisky and Heal, 1983).

These domain restrictions are indeed very demanding, and counterexamples can be found without any loss of plausibility in terms of real-life situations (see particularly Kramer, 1973). But if these restrictions are not fulfilled, then there is no general 'solution' to be found in opting for the majority rule, or some other rule like that. Indeed, it can be shown for the majority rule that the cycles that may be generated may well be extremely extensive, yielding 'total cycles' involving all social states (see Schofield, 1978; McKelvey, 1979). This line of investigation too, like the one on collective rationality (discussed in the last section), yields rather discouraging results. No general solution of impossibility theorems of the type presented by Arrow can be easily found by opting for a rule like the majority decision, hoping that the domain conditions will be somehow satisfied.

In many economic decisions, it is quite straightforward to see that these conditions will indeed be all violated. However, when the number of alternatives happen to be small, and when there is complex balancing of conflicting considerations, as in many political contexts (elections, committee decisions over rival proposals, etc.), there might possibly be some room for optimism. If cycles or other types of intransitivities turn out to be rather rare in these cases, then the approach of domain restriction may well offer some help. In contrast, in welfare-economic problems, that hope is very limited.

Indeed, if we take such a simple social-welfare problem as the division of a given cake between three or more individuals, with each person voting according to his or her own share of the cake, it can be easily shown that there will indeed be majority cycles. But it is worth noting in this context that the method of majority decision is not particularly appropriate for such economic problems anyway. Any distribution of the given cake can be improved by choosing one of the persons (even the poorest one) and dividing a part of his or her share for the benefit of all others, thereby producing an 'improvement' according to the majority rule. Indeed, we can go on 'improving' the distribution in this way, following the majority ranking procedure, making the worst-off individual more and more worse off all the time. As a criterion for welfare-economic judgement, majority rule is, in fact, a non-starter. The recognition of this fact makes it less tragic that majority cycles will tend to arise easily enough in many economic problems involving distributional variations. The majority rule would not have offered any 'real solution' to the task of making social welfare judgements in this type of economic problems even if it had been fully consistent. It is more in the context of political decisions involving a few diverse alternatives (rather than welfare-economic judgements in general) that majority rule and related decision procedures have some prima facie plausibility. It is, thus, of some interest that it is in the context of these problems that the domain conditions investigated by the social choice literature are of direct relevance and offer some hope.

4. Manipulability and implementation

A different type of problem for voting procedures arises from the possibility of 'manipulation' of the decision mechanism by the voters voting 'dishonestly'. A voting procedure is 'manipulable' when it is in the interest of some voter for some set of individual preferences to vote differently from his or her sincere preference.

The ubiquity of the possibility of manipulation had been conjectured for a long time, but it was established only recently in a remarkable theorem first presented by Gibbard (1973), and then by Satterthwaite (1975). A similar result, and a pointer to positive possibility if the conditions are relaxed, was presented by Pattanaik (1973.) The Gibbard–Satterthwaite manipulation theorem establishes that every non-dictatorial voting scheme with at least three distinct outcomes must be manipulable.

Gibbard established this theorem as a corollary of another one dealing with 'game forms' in general, of which voting schemes happen to be special cases. A game form does not restrict the strategies to be chosen by the individuals to the orderings of social states (i.e., to 'ballots'), and each person's strategy set can be any set of signals. Gibbard established that no non-dictatorial game form with at least three possible outcomes can be 'straightforward' (a concept first used by Farquharson, 1956), in the sense that each person would have a

dominant strategy (i.e., a best strategy with respect to his ordering of the outcomes, irrespective of what the strategies of others might be). Thus for every non-dictatorial game form of this type, there is at least one person who does not have a dominant strategy for some preference ordering of outcomes. From this the manipulability theorem follows immediately. If a voting scheme were *non*-manipulable, then everyone would have had a dominant strategy, viz., recording his or her *true* preference irrespective of what others do. Since the existence of dominant strategies is disestablished, so is the existence of *honest* dominant strategies.

Various variations of this discouraging result and some avenues of escape have been investigated in the literature, which is quite vast (but excellent discussions can be found in Barbera, 1977; Pattanaik, 1978; Laffont, 1979; Peleg, 1984; Brams and Fishburn, 1983; Moulin, 1983; and Jain, 1986).

The focus on 'honest' revelation of preferences has gradually given way to discussions of equilibrium and of implementation (for an early pointer in this direction, see Dummett and Farquharson, 1961). If the object of the exercise is effectiveness in the sense of getting an appropriate outcome (rather than seeking honesty as such), then the thing to investigate is indeed the existence of an effective mechanism rather than a 'strategy-proof' one. If, for example, a non-strategy-proof mechanism yields an equilibrium of dishonest behaviour that produces the same outcome as honest revelation of preferences would, then that mechanism could well be regarded as successful in terms of effectiveness.

The shift in attention towards equilibrium and implementation has opened up new lines of investigation, which are being explored (see particularly Dutta and Pattanaik, 1978; Dasgupta, Hammond and Maskin, 1979; Sengupta and Dutta, 1979; Peleg, 1984; Moulin, 1983). The implementation literature also links up with more standard problems of public economics, in which it has received attention in a somewhat different but related form (see, particularly, Groves and Ledyard, 1977; Green and Laffont, 1979; Laffont, 1979).

5. Information: utility, compensations and fairness

The alleged impossibility of interpersonal comparisons of utility was entirely accepted in the early works on social choice theory. Arrow's (1951) format gave no room to interpersonally comparable utility information, and indeed took utility information in the form of non-comparable ordinal utility rankings. This was entirely in line with the dominant position of welfare economics at that time. Even though there were formats for interpersonal comparisons of utility suggested in some contributions to welfare economics (see particularly Vickrey, 1945, and Harsanyi, 1955), these suggestions were not followed up in the formal social-choice-theoretic literature until much later.

There had been earlier attempts to by-pass the need for utility comparisons by using the notion of compensation tests (e.g., whether the gainers *could* compensate the losers), and this had led to the identification of problems of internal consistency as well as of cogency (see Kaldor, 1939; Hicks, 1939; Scitovsky, 1941; Little, 1950; Samuelson, 1950; Baumol, 1952; Gorman, 1953; Graaff, 1957). The problem of cogency is perhaps deeper, in some ways, than that of consistency. To make sure that gainers have gained so much that they can compensate the loser does, of course, have some immediate plausibility as a requirement. However, the relevance of the compensation tests suffers from the following limitation. If compensations are not paid, then it is not clear in what way the situation can be taken to be an improvement (since those who have lost may well be a great deal poorer, needier or more deserving – whatever our criteria for such judgments might be – than the gainers). And if compensations *are* in fact paid, then *after* the compensation what we observe is a Pareto improvement, so that no compensation tests are in fact needed. Thus, the compensation approach suffers from having to face a choice between being unconvincing or being irrelevant.

Another approach that by-passes the need for interpersonal comparisons proper is that of 'fairness', presented first by Foley (1967). Here a person's advantage is judged by comparing his bundle of goods with those enjoyed by others, and a situation is called 'equitable' if no individual prefers the bundle of goods enjoyed by another person to his own. If an allocation is both Pareto optimal and equitable then it is called 'fair'. (There is some non-uniformity of language in the literature, and sometimes the term 'fair' has been defined simply as 'equitable', e.g., in Feldman and Kirman, 1974 and Pazner and Schmeidler, 1974.) This approach has been pursued by a number of authors (such as Kolm, 1969; Schmeidler and Vind, 1972; Varian, 1974, 1975; Goldman and Sussangkarn, 1978; Archibald and Donaldson, 1979; Crawford, 1979; Crawford and Heller, 1979; Svensson, 1980; Champsaur and Laroque, 1982; Suzumura, 1983, and others). There are interesting problems of the existence of fair allocations and of the consistency of fairness with other principles.

It should be noted that the comparisons involved in the calculus of 'fairness' are not interpersonal ones, but in fact comparisons of different positions that the same individual might occupy (e.g., having commodity bundles), as it is evaluated by the given person. The criterion of 'non-envy' does clearly have some appeal, even though it can be argued that our deprivations may be related not only to other people's commodity bundles but also to non-commodity features of their advantage. For example, a person with a disability may well prefer to be in somebody else's position without that disability, but that is not the same thing as envying that other person's commodity bundle. As such, it could be argued, that the informational base of the fairness calculus is fundamentally limited.

Another difference between the 'fairness' approach and the standard social-choice-theoretic procedures relates to the more limited aim of the former. As Varian (1974) puts it, the fairness criterion in fact limits itself to answering the question as to whether there is a 'good' allocation (pp. 64–5). It is certainly true that social choice theory has been abundantly more ambitious, perhaps unrealistically so. On the other hand, it can be argued that even the features of 'goodness' identified by the approach of fairness (e.g., equitability with efficiency) may often fail to be satisfied by any feasible allocation at all, so that the question of ranking the 'non-good' allocations is not really avoidable. In addition, it can be argued that insofar as the foundation of the 'fairness' approach is based only on comparing the commodity bundles of different persons without going further into the relative advantages enjoyed by the persons (taking everything into account), the criterion of 'goodness' used in the 'fairness' literature is itself rather a limited one. It is perhaps for these reasons that the use of interpersonal comparisons of well-being in social choice theory (in the literature on social welfare functionals, to be discussed presently) has tended to aim at going a great deal further than the 'fairness' literature was programmed to achieve.

6. Social welfare functionals and interpersonal comparisons

The empirical problem of obtaining information on interpersonal comparisons of utility has to be distinguished from the formal problem of accommodating such information within the structure of social choice theory. The format of social welfare functions used by Arrow, and the related formats of collected choice rules (involving such various forms as social decision functions, social choice functions, etc.), make no provision for any utility information finer than that of non-comparable individual orderings. One way of extending that framework is to permit the use of more utility information, through what have been called 'social welfare functionals' (SWFL): $R = F(\{U_i\})$. For each set (strictly, n-tuple) of utility functions $U_i,...,U_n$ (one function per person), the social welfare functional F determines exactly one social ordering R. However, since utility functions can be nominally varied through alternative presentations without involving any 'real' change (e.g., doubling all the utility numbers), any social welfare functional has to be combined with some 'invariance' requirement. If two utility n-tubles (U_i) and (U_i^*) are judged to be informationally equivalent, differing from each other only in representation, then $F\{U_i\}) = F(\{U_i^*\})$. The assumed structure of measurability and interpersonal comparability of utilities can be incorporated through specifying these invariance requirements (see Sen, 1970, 1977b; d'Aspremont and Gevers, 1977; Roberts, 1980a; Blackorby, Donaldson and Weymark, 1984).

Arrow's social welfare function is a special case of a social welfare functional with the invariance requirement corresponding to ordinal non-comparability (i.e., if one *n*-tuple of utility functions is replaced by another obtained from the first by taking positive, monotonic transformations of each utility function – not necessarily the same for all – then the social ordering R determined by the first *n*-tuple will also be yielded by the second). It is obvious that Arrow's 'impossibility theorem' can be translated in the format of social welfare functionals with *ordinal non-comparability*. More interestingly, this result can be generalized to the case of *cardinal non-comparability* also. When individual utilities can be cardinally measured but not in any way interpersonally compared, the same impossibility result continues to hold (see Sen, 1970). On the other hand, introducing interpersonal comparability without cardinality (i.e., using ordinal comparability only) does resolve the Arrow dilemma, and various possible SWFLs exist fulfilling all of Arrow's conditions in this case. An example is provided by Rawls's maximin rule (or the lexicographic version of it), defining these exercises in terms of utility comparison, rather than that of indices of 'primary goods', as in Rawls's own framework.

Richer utility information can be systematically used to admit various social choice procedures not admissible in Arrow's framework. The use of various axioms to characterize particular rules utilizing richer utility information can be found in an influential and important contribution by Suppes (1966). In the recent years the more formal frameworks of social choice theory (in particular, that of SWFLs) have been extensively used to derive axiomatically a number of standard aggregation procedures, such as the Rawlsian lexicographic maximin, utilitarianism, and some others (see particularly Hammond, 1976, 1977, 1979; Strasnick, 1976; d'Aspremont and Gevers, 1977; Arrow, 1977; Sen, 1977b; Deschamps and Gevers, 1978, 1979; Maskin, 1978; Gevers, 1979; Roberts, 1980a; Blackorby, Donaldson and Weymark, 1984; d'Aspremont, 1985). While these results are formal and do not address the question of the empirical content of interpersonal comparisons of utility (though this too is discussed by Hammond, 1977), the axiom structures have been related to various empirical insights thrown up by the substantive literature on interpersonal comparisons.

One format that has also been investigated relates to the intermediate possibility of making *partial* interpersonal comparisons of utilities. Various formal structures of partial comparability and partial cardinality have, in fact, been investigated in the social-choice-theoretic literature (see Sen, 1970; Blackorby, 1975; Fine, 1975; Basu, 1979; Bezembinder and van Acker, 1979). This is a less ambitious approach, admitting that not all types of interpersonal comparisons may be possible, and such comparability may be at best partial, with many undecided cases. Nevertheless *some* definite results

can be obtained even on the basis of the incomplete structures.

Various other informational frameworks involving richer utility data can be and have been investigated, and some of them lend themselves to fruitful social-choice-theoretic use. One of the structures that need further investigation is the problem of combining *n*-tuples of 'extended orderings' reflecting each person's interpersonal comparisons. These are ordinal structures, but instead of there being one interpersonal comparison covering all the individuals in the different possible positions, this starts with the set of interpersonal comparisons made by different individuals (one 'extended ordering' per person), and addresses the problem of aggregation in that framework. Some interesting results in this area have been obtained (see Hammond, 1976; Kelly, 1978; Suzumura, 1983; Gaertner, 1983). The task, however, is rather a difficult one, since the information to marshall is extremely extensive, and progress in this area has tended to be rather slow. On the other hand, since social choice theory has to be concerned with the problem of combining different persons' possibly divergent views, that 'extended' problem certainly has a good deal of relevance and potential importance.

7. Liberty and rights

The informational limitations of the early social-choice-theoretic structures have led to responses in the later literature not only in the form of enriching the utility information (by the use of such structures as social welfare functionals, SWFL), but also that of making more systematic use of *non-utility* information. One of the areas that has been investigated in this context is that of rights in general and of liberty in particular. Liberty can be an important consideration in matters of social choice, but it cannot be adequately captured in terms of utility information, however rich it might be. If it is asserted that a person should be free to do what he or she likes in certain purely personal matters, that assertion is based on the non-utility characteristics of the 'personal nature' of these choices, and not primarily on utility considerations. As John Stuart Mill (1859) had argued, even if others might be offended by someone's personal behaviour in such matters as religious practice, it would not be appropriate to count the disutility of the offended in the same way as the utility of the person whose freedom of religious practice is under consideration. Various notions of 'protected spheres', 'personal domains', etc., have been formalized in the social-choice-theoretic literature in specifying domains of personal liberty.

One of the results obtained in this field that has led to a great deal of controversy concerns the conflict between the Pareto principle and certain minimal conditions of liberty when imposed on a social choice framework with unrestricted (or a fairly wide) domain. The 'impossibility of the Paretian liberal', presented in Sen (1970), has led to a variety of responses, including extensions, disputations, and suggestions of different ways of 'resolving' the conflict (see Ng, 1971; Batra and Pattanaik, 1972; Gibbard, 1974; Blau, 1975; Seidl, 1975; Campbell, 1976; Kelly, 1976, 1978; Aldrich, 1977; Breyer, 1977; Ferejohn, 1978; Karni, 1978; Suzumura, 1978, 1983; Mueller, 1979; Barnes, 1980; Bernholz, 1980; Breyer and Gardner, 1980; Breyer and Gigliotti, 1980; Fountain, 1980; Gardner, 1980; McLean, 1980; Weale, 1980; Baigent, 1981; Gaertner and Krüger, 1981; Gärdenfors, 1981; Hammond, 1981; Schwartz, 1970, 1972, 1986; Sugden, 1981, 1985; Austen-Smith, 1982; Levi, 1982; Krüger and Gaertner, 1983; Basu, 1984; Kelsey, 1985; Wriglesworth, 1985; Coughlin, 1986; Elster and Hylland, 1986; Gaertner, 1986; Riley, 1986; Webster, 1986, among others). The literature is vast, and covers issues of political compatibility, moral cogency and strategic consistency; it has been critically surveyed and assessed by Suzumura (1983) and Wriglesworth (1985). Various alternative formulations of liberty, in terms of social judgments, social decisions and social institutions can be shown to yield corresponding impossibility results (see Sen, 1983).

It is not really surprising that conditions of liberty or rights which make essential use of non-utility information may clash with exclusively utility-based principles, such as the Pareto principle. Non-utility considerations cannot be immovable objects if utility considerations, even in a rather limited context (as in the Pareto principle), are made into an irresistible force. One role of this type of impossibility result lies in pointing to the possibility that utility data may not be informationally adequate for social judgement or social choice, even when the utility information comes in the most articulate and complete form. Other lessons have also been suggested, and each interpretation has also been substantively disputed.

While impossibility results like this have received good deal of attention, relatively little effort has so far been spent on investigating the positive implications of various theories of rights, liberties and freedom, in the general area of social choice. The need for caution in formulating the demands of liberty because of problems of internal consistency has in fact been investigated. But the more general question of developing a fruitful and positive theory of rights and liberty within the general structure of social choice theory has not yet been much investigated.

8. Independence and neutrality

The independence of irrelevant alternatives, used by Arrow, plays a major part in the social choice formats in the Arrovian tradition. It is also crucial for Arrow's impossibility theorem. The nature, implications and acceptability of the independence condition have been subjected to a good deal of critical examination in the literature (see particularly Gärdenfors, 1973; Hansson, 1973; Ray, 1973; Fine and Fine, 1974; Fishburn, 1974; Mayston, 1974; Young, 1974a, 1974b; Binmore, 1976; Kelly, 1978; Pattanaik, 1978; Moulin, 1983; Suzumura, 1983; Peleg, 1984; Hurley, 1985; Schwartz, 1986).

One of the objections that was originally raised about the relevance of Arrow's impossibility theorem related to the acceptability of the independence condition. Some authors (in particular Little, 1950 and Samuelson, 1967) argued that seeking inter-profile consistency in any form (including Arrow's 'independence' condition) is largely gratuitous. It was also argued that traditional welfare economics had never sought such a condition, and because of the crucial use of condition I, 'Arrow's work has no relevance to the traditional theory of welfare economics, which culminates in the Bergson–Samuelson formulations' (Little, 1950, pp. 423–5). 'For Bergson,' argued Samuelson (1967), 'one and only one of the ... possible patterns of individuals' orderings is needed' (pp. 48–9), and the question of inter-profile consistency does not arise.

In response to this line of objection, several 'single-profile impossibility theorems' in the spirit of Arrow's original theorem have been derived and discussed (see particularly Parks, 1976; Kemp and Ng, 1976; Pollak, 1979; Roberts, 1980b; Rubinstein, 1981; Hurley, 1985). These results depend on dropping inter-profile consistency in favour of rather strong intra-profile requirements, typically including some condition of single-profile neutrality, requiring that whatever combination of individual orderings be decisive for establishing xRy should be sufficient for establishing aRb if each individual ranking over (x, y) is the same as that over (a, b) in that given profile. The nature of the alternatives – whether x and y, or a and b – is, thus, not to make any difference, in relating individual preferences over particular pairs to social preference over those pairs, for any given profile of individual preferences.

These results are interesting, but it must be noted that the requirements on which they are based (e.g., of single-profile neutrality) are rather strong. Also the dictatorial result that follows from the other conditions is that of single-profile dictatorship, which might not be thought to be as objectionable as the existence of one inter-profile dictator who wins for every possible preference profile (as in Arrow's theorem).

No matter what one thinks of these single-profile impossibility results, it can certainly be argued that the original objection raised by Little and Samuelson about the relevance of inter-profile conditions for social choice theory is hard to sustain. Given the motivation underlying demands for consistency in the relation between individual preferences and social choice, it is not at all clear why such consistency requirements should be thought to be applicable only for a given profile and not between different profiles of individual preferences (no matter how close these profiles are in relevant respects).

It could, of course, be argued that utility orderings (or preferences) are not an adequate informational base anyway for social choice and if that position were taken, then the very idea of a social welfare function would have to be rejected in favour of some richer informational formulation, such as a social welfare functional SWFL. If, on the other hand, the motivation underlying the use of a social welfare function is accepted, and it is agreed that for a given preference n-tuple (i.e., a given profile), there is only one social ordering, then it is not clear why it would be thought to be perfectly okay that social preferences might change over a given pair when there is a change of individual preferences over some pair of alternatives quite unconnected with this particular one. The need for some interprofile consistency is hard to deny altogether. It could, of course, be argued that Arrow's particular inter-profile condition is not the appropriate one to use for inter-profile consistency, but that would not be an objection to inter-profile conditions as such, only to the particular formulation of Arrow's condition I. It should also be noted that there are other inter-profile conditions that can be used in order to generate impossibility results like Arrow's, without any use of condition I (see in particular Chichilnisky, 1982).

If Arrow's condition I is dropped, a number of alternative possibilities do, in fact, open up for social choice procedures. For one thing, 'positional' information can be used to rank alternative social states and to arrive at social choice. In fact, in an early contribution to social choice theory, Borda (1781) had used a decision procedure that violates condition I in arriving at overall rankings based on rank-order weights. This method – often called the Borda rule – is a special case of a general class of 'positional' rules. The general properties of 'positional' rules have been fruitfully investigated by Gärdenfors (1973) and Fine and Fine (1974), among others. The Borda ruling in particular has also received attention, and various particular rules have been investigated, critically examined and axiomatized (see Young, 1974a; Fishburn and Gehrlein, 1976; Hansson and Sahlquist, 1976; Gardner, 1977; Farkas and Nitzan, 1979; and Nitzan and Rubinstein, 1981, among others).

Positional rules take note of the fact that an alternative x preferred to another alternative y may be proximate to each other in a person's preference ordering without any other alternative in between, or may be separated by the existence of one or more other alternatives intermediate between the two. The rationale of positional rules relates to attaching importance to the placing of intermediate alternatives in individual preferences, which can be taken as suggesting that the gap between the two must be, other things given, larger. This argument is not entirely convincing. Many intermediate alternatives can be placed in a small interval, while large intervals may happen to be empty because of the contingent fact that there happens to be no other alternative that fits in just there. On the other hand, if information is thought to be

extremely hard to get in social choice (a view that was certainly taken by Borda, 1781), then it is not entirely unreasonable to attach some significance to the fact that the placing of intermediate alternatives might be indicative of something. With some implicit assumption of uniformity of distribution of alternatives over the preference line (or some other suitable belief), the positional rules may have some clear rationale, and the Borda rule in particular might be particulary handy and useful.

It is possible to use positional information also in the context of richer informational base, e.g., when interpersonal comparisons of utilities are permissible. Indeed 'interpersonal positional rules' may have some distinct advantage both (1) over rules that make non-positional use of interpersonally comparable individual orderings, and (2) over non-comparable positional rules. Such interpersonal positional rules may also be demonstrably more reasonable, in some contexts, than voting procedures like the majority rule which use neither interpersonal comparisons nor positional information (on this see Sen, 1977b; Gaertner, 1983).

9. Concluding remarks

In understanding the literature of social choice theory it is important to bear in mind that while there are considerable analytical similarities between different problems tackled in this vast literature, the interpretations of the results and of their implications must take note of the particular nature of each of the substantively different problems. The axiomatic method, which has been so extensively used in the literature, offers enormous scope for efficient economy, but that economy will be self-defeating if the substantive differences are not carefully taken into account in interpreting exactly the content of the theorems derived. For example, the classic 'impossibility' result of Arrow may impose informational constraints that are much more reasonable in aggregating political preferences of different individuals over a small set of alternative proposals (or candidates) than in arriving at aggregative judgements of social justice taking note of conflicting individual interests over possible distributions of commodity vectors.

There is sometimes a temptation to see social choice theory as providing a particular 'method' of dealing with problems of aggregation. There is some truth in this diagnosis, in the sense that the discipline of axiomatic procedures has some exacting demands. On the other hand, the axioms can vary a great deal, and the interpretation of the axioms also will vary with the nature of the problems considered. The monolithic view of something called 'the social-choice-theoretic approach', which is often referred to both by those who wish to use it and those who wish to criticize it, may be deeply misleading. For some arguments on different sides on this question, see Elster and Hylland (1986).

There are, in fact, two different ways of seeing social choice theory. First, it is a field, and in this field there is scope for having different approaches. There are many problems of interpersonal aggregation, and in the broader sense, social choice theory is a field in which such aggregation – of different types – is studied. Second, social choice theory also provides a method of analysis, in which the insistence on the explicitness of axioms and on the clarity of assumptions imposes exacting formulational demands. Indeed, some of the more notable achievements of social choice theory have come from this insistence on explicitness and clarity (e.g., Arrow's own demonstration of the impossibility of combining a set of assumptions that were being implicitly invoked in the literature of the welfare economics of that period, including eschewing interpersonal comparisons of utility). While the second interpretation is a narrower one than the first, it is nevertheless broad enough to permit different types of axioms to be used, and different political, economic and social beliefs to be incorporated in the axiom structure. Neither interpretation would give any cogency to the search for 'the social-choice-theoretic approach'.

One reason why social choice theory has received as much attention as it has in the last few decades relates to the importance of the field with which that theory has been concerned (and which characterizes that theory in the *broader* sense). Another reason has been the fruitfulness of making implicit ideas explicit, and of following their implications consistently and clearly. As a methodological discipline, social choice theory has contributed a great deal to clarifying problems that had been obscure earlier. While insistence on clarity at all costs has also some limitations (sometimes the narrowness of the axiom structure used in social choice theory has indeed been seen as a limitation), social choice theory has undoubtedly been a creative tradition among other methodological traditions that can be used to analyse economic, social and political problems involving group aggregation. The vast literature surveyed in this article can be ultimately judged by what has been achieved in terms of clarifying the obscure and illuminating the unclear. Perhaps the successes have been rather mixed, but that fact is not surprising.

AMARTYA SEN

See also **Arrow's theorem; constitutions, economic approach to; public choice; social choice (new developments); social welfare function.**

Bibliography

Aldrich, J. 1977. The dilemma of a Paretian liberal: Some consequences of Sen's theorem. *Public Choice* 30, 1–21.

Archibald, G.C. and Donaldson, D. 1979. Notes on economic equality. *Journal of Public Economics* 12, 205–14.

Arrow, K.J. 1951. *Social Choice and Individual Values*. New York: Wiley.

Arrow, K.J. 1963. *Social Choice and Individual Values*. 2nd edn, New York: Wiley.

Arrow, K.J. 1977. Extended sympathy and the possibility of social choice. *American Economic Review* 67, 219–25.

Arrow, K.J. and Raynaud, H. 1986. *Social Choice and Multicriterion Decision-Making*. Cambridge, Mass.: MIT Press.

Austen-Smith, D. 1982. Restricted Pareto and rights. *Journal of Economic Theory* 26, 89–99.

Baigent, N. 1981. Decomposition of minimal libertarianism. *Economic Letters* 7, 29–32.

Barbera, S. 1977. Manipulation of social decision functions. *Journal of Economic Theory* 15, 226–78.

Barnes, J. 1980. Freedom, rationality and paradox. *Canadian Journal of Philosophy* 10.

Barthelemy, J.P. 1983. *Arrow's theorem: Unusual domains and extended codomains*. In Pattanaik and Salles (1983).

Basu, K. 1979. *Revealed Preference of Governments*. Cambridge: Cambridge University Press.

Basu, K. 1984. The right to give up rights. *Economica* 51, 413–22.

Batra, R.M. and Pattanaik, P.K. 1972. On some suggestions for having non-binary social choice functions. *Theory and Decision* 3, 1–11.

Baumol, W.J. 1952. *Welfare Economics and the Theory of the State*. Cambridge, Mass.: Harvard University Press.

Bentham, J. 1789. *An Introduction to the Principles of Morals and Legislation*. London: Payne. Reprinted, Oxford: Clarendon Press, 1907.

Bergson, A. 1938. A reformulation of certain aspects of welfare economics. *Quarterly Journal of Economics* 52, 310–34.

Bernholz, P. 1980. A general social dilemma: Profitable exchange and intransitive group preference. *Zeitschrift für Nationalökonomie* 40, 1–23.

Bezembinder, Th. and van Acker, P. 1979. A note on Sen's partial comparability model. Department of Psychology, Katholieke Universiteit, Nijmegen, The Netherlands.

Binmore, K. 1976. Social choice and parties. *Review of Economic Studies* 43, 459–64.

Black, D. 1948. On the rationale of group decision making. *Journal of Political Economy* 56, 23–34.

Black, D. 1958. *The Theory of Committees and Elections*. London: Cambridge University Press.

Blackorby, C. 1975. Degrees of cardinality and aggregate partial orderings. *Econometrica* 43, 845–52.

Blackorby, C., Donaldson, D. and Weymark, J.A. 1984. Social choice with interpersonal utility comparisons. *International Economic Review* 25, 327–56.

Blair, D.H., Bordes, G., Kelly, J.S. and Suzumura, K. 1976. Impossibility theorems without collective rationality. *Journal of Economic Theory* 13, 361–79.

Blair, D.H. and Muller, E. 1983. Essential aggregation procedures on restricted domains of preferences. *Journal of Economic Theory* 30, 34–53.

Blair, D.H. and Pollak, R.A. 1979. Collective rationality and dictatorship: the scope of the Arrow theorem. *Journal of Economic Theory* 21, 186–94.

Blair, D.H. and Pollak, R.A. 1982. Acyclic collective choice rules. *Econometrica* 50, 931–43.

Blau, J.H. 1957. The existence of a social welfare function. *Econometrica* 25, 302–13.

Blau, J.H. 1975. Liberal values and independence. *Review of Economic Studies* 42, 413–20.

Blau, J.H. 1976. Neutrality, monotonicity and the right of veto: a comment. *Econometrica* 44, 603.

Blau, J.H. 1979. Semiorders and collective choice. *Journal of Economic Theory* 21, 195–206.

Blau, J.H. and Deb, R. 1977. Social decision functions and veto. *Econometrica* 45, 871–9.

Bonner, J. 1986. *Politics, Economics and Welfare*. Brighton: Wheatsheaf.

Borda, J.C. 1781. Memoire sur les élections au scrutin. *Mémoires de l'Académie Royale des Sciences*; English translation by A. de Grazia, *Isis* 44 (1953).

Bordes, G. 1976. Consistency, rationality and collective choice. *Review of Economic Studies* 43, 447–57.

Bose, A. 1975. *Marxian and Post-Marxian Political Economy*. Harmondsworth: Penguin Books.

Brams, S.J. 1975. *Game Theory and Politics*. New York: Free Press.

Brams, S.J. 1976. *Paradoxes in Politics*. New York: Free Press.

Brams, S.J. and Fishburn, P.C. 1983. *Approval Voting*. Boston: Birkhäuser.

Breyer, F. 1977. The liberal paradox, decisiveness over issues, and domain restrictions. *Zeitschrift für Nationalökonomie* 37, 45–60.

Breyer, F. and Gardner, R. 1980. Liberal paradox, game equilibrium and Gibbard optimum. *Public Choice* 35, 469–81.

Breyer, F. and Gigliotti, G.A. 1980. Empathy and respect for the right of others. *Zeitschrift für Nationalökonomie* 40, 59–64.

Brown, D.J. 1974. An approximate solution to Arrow's problem. *Journal of Economic Theory* 9, 375–83.

Brown, D.J. 1975. Aggregation of preferences. *Journal of Economics* 89, 456–69.

Buchanan, J.M. and Tullock, G. 1962. *The Calculus of Consent*. Ann Arbor: University of Michigan Press.

Camacho, A. 1974. Societies and social decision functions. In Leinfellner and Kohler (1974).

Campbell, D.E. 1976. Democratic preference functions. *Journal of Economic Theory* 12, 259–72.

Campbell, D.E. 1980. Algorithms for social choice functions. *Review of Economic Studies* 47, 617–27.

Champsaur, P. and Laroque, G. 1982. Strategic behavior in decentralized planning procedures. *Econometrica* 50, 325–44.

Chichilnisky, G. 1982. Social aggregation rules and continuity. *Quarterly Journal of Economics* 96, 337–52.

Chichilnisky, G. and Heal, G. 1983. Necessary and sufficient conditions for a resolution of the social choice paradox. *Journal of Economic Theory* 31, 68–87.

Condorcet, M. de. 1785. *Essai sur l'Application de l'Analyse à la Probabilité des Décisions Rendues à le Pluralité des Voix*. Paris.

Coughlin, P.J. 1986. Rights and the private Pareto principle. *Economica* 53.

Crawford, V.P. 1979. A procedure for generating Pareto-efficient egalitarian-equivalent allocations. *Econometrica* 47, 49–60.

Crawford, V.P. and Heller, W.P. 1979. Fair division with indivisible commodities. *Journal of Economic Theory* 21, 10–27.

Dasgupta, P., Hammond, P. and Maskin, E. 1979. The implementation of social choice rules: some general results on incentive compatibility. *Review of Economic Studies* 46, 185–216.

Dasgupta, P. and Heal, G. 1979. *Economic Theory and Exhaustible Resources*. London: James Nisbet, and Cambridge: Cambridge University Press.

d'Aspremont, C. and Gevers, L. 1977. Equity and informational basis of collective choice. *Review of Economic Studies* 46, 199–210.

d'Aspremont, D. 1985. Axioms for social welfare orderings. In Hurwicz, Schmeidler and Sonnenschein (1985).

Davis, O.A., De Groot, M.H. and Hinich, M.J. 1972. Social preference orderings and majority rule. *Econometrica* 40, 147–57.

Deb, R. 1976. On Schwartz's rule. *Journal of Economic Theory* 16, 103–10.

Deschamps and Gevers, L. 1978. Leximin and utilitarian rules: a joint characterisation. *Journal of Economic Theory* 17, 143–163.

Deschamps and Gevers, L. 1979. Separability, risk-taking and social welfare judgements. In Laffont (1979).

Dummett, M. 1984. *Voting Procedures*. Oxford: Clarendon Press.

Dummett, M. and Farquharson, R. 1961. Stability in voting. *Econometrica* 29, 133–43.

Dutta, B. 1980. On the possibility of consistent voting procedures. *Review of Economic Studies* 47, 603–16.

Dutta, B. and Pattanaik, P.K. 1978. On nicely consistent voting systems. *Econometrica* 46, 163–70.

Elster, J. and Hylland, A., eds. 1986. *Foundations of Social Choice Theory*. Cambridge: Cambridge University Press.

Farkas, D. and Nitzan, S. 1979. The Borda rule and Pareto stability: a comment. *Econometrica* 47, 1305–6.

Farquharson, R. 1956. Straightforwardness in voting paradoxes. *Oxford Economic Papers* 8, 80–9.

Farrell, M.J. 1976. Liberalism in the theory of social choice. *Review of Economic Studies* 43, 3–10.

Feldman, A. and Kirman, A. 1974. Fairness and envy. *American Economic Review* 64, 995–1005.

Ferejohn, J.A. 1978. The distribution of rights in society. In Gottinger and Leinfellner (1978).

Fine, B.J. 1975. A note on interpersonal aggregation and partial comparability. *Econometrica* 43, 173–4.

Fine, B.J. and Fine, K. 1974. Social choice and individual ranking. *Review of Economic Studies* 41, 303–22, 459–75.

Fishburn, P.C. 1973. *The Theory of Social Choice.* Princeton: Princeton University Press.

Fishburn, P.C. 1974. On collective rationality and a generalized impossibility theorem. *Review of Economic Studies* 41, 445–59.

Fishburn, P.C. and Gehrlein, W.V. 1976. Borda's rule, positional voting, and Condorcet's simple majority principle. *Public Choice* 28, 79–88.

Foley, D. 1967. Resource allocation in the public sector. *Yale Economic Essays* 7, 73–6.

Fountain, J. 1980. Bowley's analysis of bilateral monopoly and Sen's liberal paradox in collective choice theory: a note. *Quarterly Journal of Economics* 95.

Gaertner, W. 1983. Equity- and inequity-type Borda rules. *Mathematical Social Sciences* 4, 137–54.

Gaertner, W. 1986. Pareto, independent rights exercising and strategic behaviour. *Journal of Economics: Zeitschrift für Nationalökonomie* 46.

Gaertner, W. and Heinecke, A. 1978. Cyclically mixed preferences: A necessary and sufficient condition for transitivity of the social preference relation. In Gottinger and Leinfellner (1978).

Gaertner, W. and Krüger, L. 1981. Self-supporting preferences and individual rights: the possibility of Paretian libertarianism. *Economica* 48, 17–28.

Gaertner, W. and Krüger, L. 1983. Alternative libertarian claims and Sen's paradox. *Theory and Decision* 15, 211–30.

Gardenfors, P. 1973. Positional voting functions. *Theory and Decision* 4, 1–24.

Gärdenfors, P. 1981. Rights, games and social choice. *Nous* 15.

Gardner, R. 1977. The Borda game. *Public Choice* 30, 43–50.

Gardner, R. 1980. The strategic inconsistency of Paretian liberalism. *Public Choice* 35, 241–52.

Gevers, L. 1979. On interpersonal comparability and social welfare orderings. *Econometrica* 47, 75–90.

Gibbard, A. 1973. Manipulation of voting schemes: a general result. *Econometrica* 41, 587–601.

Gibbard, A. 1974. A Pareto-consistent libertarian claim. *Journal of Economic Theory* 7, 338–410.

Goldman, S.M. and Sussangkarn, C. 1978. On the concept of fairness. *Journal of Economic Theory* 19, 210–16.

Gorman, W.M. 1953. Community preference fields. *Econometrica* 21, 63–80.

Gottinger, H.W. and Leinfellner, W. 1978. *Decision Theory and Social Ethics: Issues in Social Choice.* Dordrecht: Reidel.

Graaff, J. der. 1957. *Theoretical Welfare Economics.* Cambridge: Cambridge University Press.

Grandmont, J.M. 1978. Intermediate preferences and majority rule. *Econometrica* 46, 317–30.

Green, J. and Laffont, J.-J. 1979. *Incentives in Public Decision Making.* Amsterdam: North-Holland.

Grether, D.M. and Plott, C.R. 1982. Nonbinary social choice: an impossibility theorem. *Review of Economic Studies* 49, 143–9.

Groves, T. and Ledyard, J. 1977. Optimal allocation of public goods: a solution to the 'free rider' problem. *Econometrica* 45, 783–810.

Guha, A.S. 1972. Neutrality, monotonicity and the right of veto. *Econometrica* 40, 821–6.

Hammond, P.J. 1976. Equity, Arrow's conditions and Rawls' difference principle. *Econometrica* 44, 793–804.

Hammond, P.J. 1977. Dual interpersonal comparisons of utility and the welfare economics of income distribution. *Journal of Public Economics* 6, 51–71.

Hammond, P.J. 1979. Equity in two person situation: some consequences. *Econometrica* 47, 1127–36.

Hammond, P.J. 1981. Liberalism, independent rights, and the Pareto principle. In L.J. Cohen et. al., *Logic, Methodology and Philosophy of Sciences*, Amsterdam: North-Holland.

Hansson, B. 1973. The independence condition in the theory of social choice. *Theory and Decision* 4.

Hansson, B. 1976. The existence of group preferences. *Public Choice* 28, 89–98.

Hansson, B. and Sahlquist, H. 1976. A proof technique for social choice with variable electorate. *Journal of Economic Theory* 13, 193–200.

Harsanyi, J.C. 1955. Cardinal welfare, individualistic ethics, and interpersonal comparisons of utility. *Journal of Political Economy* 63, 309–21.

Harsanyi, J.C. 1979. Bayesian decision theory, rule utilitarianism and Arrow's impossibility theorem. *Theory and Decision* 11, 289–318.

Herzberger, H.G. 1973. Ordinal preference and rational choice. *Econometrica* 41, 187.

Hicks, J.R. 1939. *Value and Capital.* Oxford: Clarendon Press.

Hurley, S. 1985. Supervenience and the possibility of coherence. *Mind* 94, 501–26.

Hurwicz, L., Schmeidler, D. and Sonnenschein, H., eds. 1985. *Social Goals and Social Organization: Essays in Memory of Elisha Pazner.* Cambridge: Cambridge University Press.

Inada, K. 1984. On the economic welfare function. *Econometrica* 32, 316–38.

Inada, K. 1964. A note on the simple majority decision rule. *Econometrica* 37, 490–506.

Inada, K. 1969. On the simple majority decision rule. *Econometrica* 32, 525–31.

Inada, K. 1970. Majority rule and rationality. *Journal of Economic Theory* 2, 27–40.

Jain, S. 1986. Special majority rules: a necessary and sufficient condition for quasi–transitivity with quasi–transitive individual preferences. *Social Choice and Welfare* 3, 99–106.

Kalai, E. and Muller, E. 1977. Characterization of domains admitting non-dictatorial social welfare functions and nonmanipulable voting procedures. *Journal of Economic Theory* 16, 457–69.

Kalai, E. and Ritz, Z. 1980. Characterization of private alternative domains admitting Arrow social welfare functions. *Journal of Economic Theory* 22, 23–36.

Kaldor, N. 1939. Welfare propositions in economics. *Economic Journal* 49, 549–52.

Karni, E. 1978. Collective rationality, unanimity and liberal ethics. *Review of Economic Studies* 45, 571–4.

Kelly, J.S. 1976. The impossibility of a just liberal. *Economica* 43, 67–75.

Kelly, J.S. 1978. *Arrow Impossibility Theorems*. New York: Academic Press.

Kelsey, D. 1984. Acyclic choice without the Pareto principle. *Review of Economic Studies* 51, 693–9.

Kelsey, D. 1985. The liberal paradox: a generalization. *Social Choice and Welfare* 1, 245–50.

Kemp, M.C. and Ng, Y.K. 1976. On the existence of social welfare functions, social orderings and social decision functions. *Economica* 43, 59–66.

Kolm, S.Ch. 1969. The optimum production of social justice. In *Public Economics*, ed. J. Margolis and H. Guitton. London: Macmillan.

Kramer, G.H. 1973. On a class of equilibrium conditions for majority rule. *Econometrica* 41, 285–97.

Kramer, G.H. 1978. A dynamic model of political equilibrium. *Journal of Economic Theory* 16, 310–34.

Krüger, L. and Gaertner, W. 1983. Alternative libertarian claims and Sen's paradox. *Theory and Decision* 15.

Laffont, J.J., ed. 1979. *Aggregation and Revelation of Preferences*. Amsterdam: North-Holland.

Larsson, B. 1983. *Basic Properties of the Majority Rule*. Lund: University of Lund.

Leinfellner, W. and Kohler, E., eds. 1974. *Developments in the Methodology of Social Sciences*. Dordrecht: Reidel.

Levi, I. 1982. Liberty and welfare. In Sen and Williams (1982).

Little, I.M.D. 1950. *A Critique of Welfare Economics*. Oxford: Clarendon Press. 2nd edn, 1957.

Machina, M. and Parks, R. 1981. On path independent randomized choices. *Econometrica* 49, 1345–7.

Mas-Colell, A. and Sonnenschein, H.F. 1972. General possibility theorems for group decisions. *Review of Economic Studies* 39, 185–92.

Maskin, E. 1976. Social welfare functions on restricted domain. Mimeo.

Maskin, E. 1978. A theorem on utilitarianism. *Review of Economic Studies* 45, 93–6.

Matsumoto, Y. 1985. Non-binary social choice: revealed preferential interpretation. *Economica* 52, 185–94.

Mayston, D.J. 1974. *The Idea of Social Choice*. London: Macmillan.

McKelvey, R.D. 1979. General conditions for global intransitivities in formal voting models. *Econometrica* 47, 1085–112.

McLean, I.S. 1980. Liberty, equality and the Pareto principle: a comment on Weale. *Analysis* 40.

McManus, M. 1982. Some properties of topological social choice functions. *Review of Economic Studies* 49, 447–60.

Mill, J.S. 1859. *On Liberty*. Reprinted, Harmondsworth: Penguin Books, 1974.

Monjardet, B. 1979. Duality in the theory of social choice. In Laffont (1979).

Moulin, H. 1983. *The Strategy of Social Choice*. Amsterdam: North-Holland.

Mueller, D.C. 1979. *Public Choice*. Cambridge: Cambridge University Press.

Murakami, Y. 1968. *Logic and Social Choice*. New York: Dover.

Myerson, R.B. 1983. Utilitarianism, egalitarianism, and the timing effect in social choice problems. *Econometrica* 49, 883–97.

Ng, Y.K. 1971. The possibility of a Paretian liberal: impossibility theorems and cardinal utility. *Journal of Political Economy* 79, 1397–402.

Ng, Y.K. 1979. *Welfare Economics*. London: Macmillan.

Nitzan, S. and Paroush, J. 1985. *Collective Decision Making: An Economic Outlook*. Cambridge: Cambridge University Press.

Nitzan, S. and Rubinstein, A. 1981. A further characterization of the Borda Ranking Methods. *Public Choice* 36, 153–8.

Nozick, R. 1974. *Anarchy, State and Utopia*. Oxford: Blackwell.

Parks, R.P. 1976. An impossibility theorem for fixed preferences: a dictatorial Bergson–Samuelson social welfare function. *Review of Economic Studies* 43, 447–450.

Pattanaik, P.K. 1971. *Voting and Collective Choice*. Cambridge: Cambridge University Press.

Pattanaik, P.K. 1973. On the stability of sincere voting situations. *Journal of Economic Theory* 6, 558–74.

Pattanaik, P.K. 1978. *Strategy and Group Choice*. Amsterdam: North-Holland.

Pattanaik, P.K. and Salles, M., eds. 1983. *Social Choice and Welfare*. Amsterdam: North-Holland.

Pazner, E.A. and Schmeidler, D. 1974. A difficulty in the concept of fairness. *Review of Economic Studies* 41, 441–3.

Peleg, B. 1978. Consistent voting systems. *Econometrica* 46, 153–62.

Peleg, B. 1984. *Game Theoretic Analysis of Voting in Committees*. Cambridge: Cambridge University Press.

Plott, C.R. 1967. A notion of equilibrium and its possibility under majority rule. *American Economic Review* 57, 788–806.

Plott, C.R. 1973. Path independence, rationality and social choice. *Econometrica* 41, 1075–91.

Plott, C.R. 1976. Axiomatic social choice theory: an overview and interpretation. *American Journal of Political Science* 20, 511–96.

Pollak, R.A. 1979. Bergson-Samuelson social welfare functions and the theory of social choice. *Quarterly Journal of Economics* 93, 73–90.

Ray, P. 1973. Independence of irrelevant alternatives. *Econometrica* 41, 987–91.

Riley, J. 1986. *Liberal Utilitarianism: Social Choice Theory and J.S. Mill's Philosophy*. Cambridge: Cambridge University Press.

Robbins, L. 1932. *An Essay on the Nature and Significance of Economic Science*. London: Macmillan.

Robbins, L. 1938. Interpersonal companions of utility. *Economic Journal* 48, 635–41.

Roberts, K.W.S. 1980a. Interpersonal comparability and social choice theory. *Review of Economic Studies* 47, 421–39.

Roberts, K.W.S. 1980b. Social choice theory: the single and multiple-profile approaches. *Review of Economic Studies* 47, 441–50.

Rubinstein, A. 1981. The single profile analogues to multiple profile theorems: mathematical logic's approach. Murray Hill: Bell Laboratories.

Salles, M. 1976. Characterization of transitive individual preferences for quasi-transitive collective preferences under simple games. *International Economic Review* 17, 308–18.

Samuelson, P.A. 1947. *Foundations of Economic Analysis*. Cambridge, Mass.: Harvard University Press.

Samuelson, P.A. 1950. Evaluation of real national income. *Oxford Economic Papers* 2, 1–19.

Samuelson, P.A. 1967. Arrow's mathematical politics. In *Human Values and Economic Policy*, ed. S. Hook. New York: New York University Press.

Saposnik, R. 1975. On transitivity of the social preference relation under simple majority rule. *Journal of Economic Theory* 10, 1–7.

Satterthwaite, M. 1975. Strategy-proofness and Arrow's conditions: existence and correspondence theorems for voting procedures and social welfare functions. *Journal of Economic Theory* 10, 187–217.

Schick, F. 1969. Arrow's proof and the logic of preference. *Journal of Philosophy* 36, 127–44.

Schmeidler, D. and Sonnenschein, H. 1978. Two proofs of the Gibbard–Satterthwaite theorem on the possibility of a strategyproof social choice function. In Gottinger and Leinfellner (1978).

Schmeidler, D. and Vind, K. 1972. Fair net trades. *Econometrica* 40, 637–42.

Schofield, N. 1978. Instability of simple dynamic games. *Review of Economic Studies* 40, 575–94.

Schofield, N. 1983. Generic instability of majority rule. *Review of Economic Studies* 50, 695–705.

Schwartz, T. 1970. On the possibility of rational policy evaluation. *Theory and Decision* 1, 89–106.

Schwartz, T. 1972. Rationality and the myth of the maximum. *Nous* 6, 97–117.

Schwartz, T. 1986. *The Logic of Collective Choice*. New York: Columbia University Press.

Scitovsky, T. 1941. A note on welfare propositions in economics. *Review of Economic Studies* 9, 77–88.

Seidl, C. 1975. On liberal values. *Zeitschrift für Nationalökonomie* 35, 257–92.

Sen, A.K. 1966. A possibility theorem on majority decisions. *Econometrica* 34, 75–9.

Sen, A.K. 1969. Quasi-transivity, rational choice and collective decisions. *Review of Economic Studies* 36, 381–93.

Sen, A.K. 1970. *Collective Choice and Social Welfare*. San Francisco: Holden-Day; Edinburgh: Oliver & Boyd. Republished, Amsterdam: North-Holland.

Sen, A.K. 1977a. Social choice theory: a re-examination. *Econometrica* 45, 58–89.

Sen, A.K. 1977b. On weights and measures: informational constraints in social welfare analysis. *Econometrica* 45, 1539–72.

Sen, A.K. 1982. *Choice, Welfare and Measurement*. Oxford: Blackwell, and Cambridge, Mass.: MIT Press.

Sen, A.K. 1983. Liberty and social choice. *Journal of Philosophy* 80, 5–28.

Sen, A.K. 1986. Social choice theory. In *Handbook of Mathematical Economics* Vol III, ed. K.J. Arrow and M. Intriligator. Amsterdam: North-Holland.

Sen, A.K. and Pattanaik, P.K. 1969. Necessary and sufficient conditions for rational choice under majority decision. *Journal of Economic Theory* 1, 178–202.

Sen, A.K. and Williams, B. 1982. *Utilitarianism and Beyond*. Cambridge: Cambridge University Press.

Sengupta, M. 1980. The knowledge assumption in the theory of strategic voting. *Econometrica* 49, 1301–4.

Sengupta, M. and Dutta, B. 1979. A condition for Nash stability under binary and democratic group decision functions. *Theory and Decision* 10, 293–310.

Slutsky, S. 1977. A characterization of societies with consistent majority decision. *Review of Economic Studies* 44, 211–26.

Strasnick, S. 1976. Social choice theory and the derivation of Rawls' difference principle. *Journal of Philosophy* 73, 85–99.

Sugden, R. 1981. *The Political Economy of Public Choice*. Oxford: Martin Robertson.

Sugden, R. 1985. Liberty, preference and choice. *Economics and Philosophy* 1, 213–30.

Suppes, P. 1966. Some formal models of grading principles. *Synthese* 6, 284–306.

Suzumura, K. 1978. On the consistency of libertarian claims. *Review of Economic Studies* 45, 329–42. A correction, 46, (1979), 743.

Suzumura, K. 1980. Liberal paradox and the voluntary exchange of rights-exercising. *Journal of Economic Theory* 22, 407–42.

Suzumura, K. 1983. *Rational Choice, Collective Decisions and Social Welfare*. Cambridge: Cambridge University Press.

Svensson, L.G. 1980. Equity among generations. *Econometrica* 48, 1251–6.

Tullock, G. 1967. The general irrelevance of the general possibility theorem. *Quarterly Journal of Economics* 81, 256–70.

Tullock, G. 1969. *Toward a Mathematics of Politics*. Ann Arbor: University of Michigan Press.

Varian, H. 1974. Equity, envy and efficiency. *Journal of Economic Theory* 9, 63–91.

Varian, H. 1975. Distributive justice, welfare economics and the theory of fairness. *Philosophy and Public Affairs* 4, 223–47.

Vickrey, W. 1945. Measuring marginal utility by reactions to risk. *Econometrica* 13, 319–33.

Vickrey, W. 1960. Utility, strategy and social decision rules. *Quarterly Journal of Economics* 75, 507–25.

Ward, B. 1965. Majority voting and alternative forms of public enterprise. In *The Public Economy of Urban Communities*, ed. J. Margolis. Baltimore: Johns Hopkins Press.

Weale, A. 1980. The impossibility of a liberal egalitarianism. *Analysis* 40.

Webster, N. 1986. Liberals and information. *Theory and Decision* 20, 41–52.

Weymark, J.A. 1983. Arrow's theorem with social quasi-orderings. *Public Choice* 42, 235–46.

Wilson, R.B. 1972. Social choice theory without the Pareto principle. *Journal of Economic Theory* 5, 478–86.

Wilson, R.B. 1975. On the theory of aggregation. *Journal of Economic Theory* 10, 89–99.

Wriglesworth, J. 1985. *Libertarian Conflicts in Social Choice*. Cambridge: Cambridge University Press.

Young, H.P. 1974a. An axiomatization of the Borda's rule. *Journal of Economic Theory* 9, 43–52.

Young, H.P. 1974b. A note on preference aggregation. *Econometrica* 42, 1129–31.

social choice (new developments)

With the exception of the research on single-peaked preferences and their multidimensional generalizations, for the most part the early literature on social choice theory dealt with abstract sets of alternatives and domains of preferences and feasible sets that exhibited little structure. Since the late 1970s there has been a dramatic shift of focus, with structured sets of alternatives and restricted domains coming to the fore. In particular, a great deal of attention has been directed towards the kinds of concrete problems that arise in economics, with alternatives being allocations of goods and preferences and feasible sets satisfying the sorts of restrictions encountered in economic models. In this article, we provide an overview of some of the recent contributions to four topics in normative social choice theory in which economic modelling has played a prominent role: Arrovian social choice theory on economic domains, variable-population social choice, strategy-proof social choice, and axiomatic models of resource allocation. Structured environments have also been considered in positive social choice theory, notably in the political economy literature. See Austen-Smith and Banks (2005) for an introduction to this literature. Other areas of social choice

theory have also been active in recent years: see Arrow, Sen and Suzumura (2002; 2008) for recent surveys of these topics.

Arrovian social choice on economic domains

Arrow's theorem (see Arrow, 1963) is concerned with the aggregation of profiles of individual preference orderings into a social ordering of a set of alternatives X. Let $\mathscr{R}$ denote the set of all orderings of X. In Arrow's theorem, there is a finite set of individuals $N=\{1, \ldots, n\}$ with $n \geq 2$, each of whom has a weak preference ordering R_i on X. An (*Arrovian*) *social welfare* function f assigns a social ordering $R = f(\mathbf{R})$ of X to each profile $\mathbf{R}=(R_1, \ldots, R_n)$ of individual preference orderings in some domain $\mathscr{D}$ of profiles. Arrow's theorem demonstrates that it is impossible for a social welfare function to satisfy independence of irrelevant alternatives, henceforth IIA (the social ranking of a pair of alternatives depends only on the individual rankings of these alternatives), weak Pareto (if everyone strictly prefers one alternative to a second, then so does society), and nondictatorship (nobody's strict preferences are always respected) if the domain is unrestricted ($\mathscr{D} = \mathscr{R}^n$) and $|X| \geq 3$.

Arrow's theorem is not directly applicable to economic problems. In economic problems, both the social alternatives and the individual preferences exhibit considerable structure and, therefore, a social welfare function only needs to be defined on a restricted domain of preference profiles. For a comprehensive survey of the literature on Arrovian social choice on economic domains, see Le Breton and Weymark (2008).

When X is a subset of the real line $\mathbb{R}$, a preference R_i is *single-peaked* if there is a unique best alternative $\pi(R_i)$ in X, the *peak*, and alternatives on the same side of the peak are worse the further away from the peak they are. Let $\mathscr{S}$ denote the set of all single-peaked preferences on X. If the alternatives in X are different levels of a single public good, it is natural to expect individual preferences to be single-peaked. Black (1948) has shown that ranking pairs of alternatives by majority rule produces a social ordering if the individuals have single-peaked preferences when n is odd. More generally, it follows from results in Moulin (1980) that on $\mathscr{S}^n$, any *generalized median social welfare* function satisfies all the Arrow axioms except his domain assumption with nondictatorship strengthened to anonymity (permuting preferences leaves the social ordering invariant). These functions are defined by first *fixing* $n-1$ single-peaked preferences (which can be interpreted as being the preferences of phantom voters) and then, for any profile of single-peaked preferences in $\mathscr{S}^n$ of the n real individuals, applying majority rule to the resulting profile of $2n-1$ preferences, both real and phantom. Note that the total number of preferences in one of these profiles is odd, so Black's theorem applies. Each specification of the preferences of the phantom voters defines a distinct generalized median social welfare function.

Ehlers and Storcken (2002) have characterized all the social welfare functions on this domain that satisfy IIA and weak Pareto.

A domain $\mathscr{D}$ of preference profiles is *Arrow inconsistent* if no social welfare function satisfying Arrow's three non-domain axioms exists on $\mathscr{D}$. In a seminal article, Kalai, Muller and Satterthwaite (1979) identified a sufficient condition for $\mathscr{D}$ to be Arrow inconsistent when $\mathscr{D}$ is the Cartesian product of individual preference domains $\mathscr{D}_i$. A set of alternatives is *free* if preference profiles are unrestricted on this set. A domain is *saturating* if (i) there are at least two free pairs, (ii) any two free pairs of alternatives can be connected to each other by means of a series of overlapping free triples, and (iii) any other pair of alternatives is *trivial* in the sense that there is only one way in which any individual ranks these alternatives. When each of the individual preference domains $\mathscr{D}_i$ is the same, saturating preference domains are Arrow inconsistent. Because a free pair is part of a free triple when the domain is saturating, Arrow's theorem implies that there is a dictator on this pair when IIA and weak Pareto are satisfied. The same person must be a dictator on all free pairs because adjacent free triples in the connection procedure have two alternatives in common. On trival pairs, by weak Pareto, everyone is a dictator. This method of showing that a domain is Arrow inconsistent is known as the *local approach*.

Kalai, Muller and Satterthwaite (1979) have also shown that, when $X = \mathbb{R}_+^m$, interpreted as the set of all allocations of m divisible public goods, the domain of all profiles of *classical public goods preferences* (that is, continuous, strictly monotonic, and convex preferences) is saturating and, hence, is Arrow inconsistent when $m \geq 2$. Other examples of saturating domains include the set of all expected utility preferences on the set of lotteries on three or more certain outcomes (Le Breton, 1986) and the set of Euclidean spatial preferences on $\mathbb{R}_+^m$ or $\mathbb{R}^m$, that is, preferences for which there is a global best alternative and alternatives are ranked by the negative of their distance from this alternative (Le Breton and Weymark, 2002). The Arrow inconsistency of the spatial preference domain was originally shown by Border (1984) using a different proof strategy.

When alternatives are allocations of private goods and individuals only care about their own consumption, Bordes and Le Breton (1989) have identified a strengthening of the concept of a saturating domain that implies that the domain is Arrow inconsistent. If X consists of all the allocations of two or more divisible private goods in which everyone is guaranteed to receive a positive amount of some good, then the domain satisfies this condition if individuals can have any *classical private goods preference*, that is, a preference that is continuous, strictly monotonic, and convex over own consumption (see also Maskin, 1976; Border, 1983).

The examples considered so far all have the feature that the set of alternatives has a Cartesian structure. If X

incorporates feasiblity constraints, this is not the case. Using a modification of the local approach, Bordes, Campbell and Le Breton (1995) have shown that the domain of classical private goods preferences is Arrow inconsistent if the set of alternatives is the set of feasible allocations with positive consumptions of all goods for an exchange economy with two or more divisible private goods. Bordes and Le Breton (1990) have also adapted the local approach to analyse Arrow consistency in assignment, matching and pairing problems. In an *assignment problem*, one of n indivisible objects is assigned to each of the n individuals. In a *matching problem*, there are two groups of n individuals with each person from one group matched to one person from the other group. In a *pairing problem*, an even number n of individuals is grouped in pairs. If the preference domains in these problems are such that individuals only care about which individual or good they are matched, paired or assigned to, but are otherwise unrestricted, then the domain is Arrow inconsistent when $n \geq 4$.

The preceding discussion suggests that economic domain restrictions do not provide a satisfactory way of circumventing Arrow's social welfare function impossibility theorem when the set of alternatives is not one-dimensional. This conclusion is reinforced by the results in Redekop (1995) that show that, in order for a subset of a domain of Arrow-inconsistent economic preferences to be Arrow consistent, the sub-domain must be topologically small. Roughly speaking, this requirement severely limits the amount of preference diversity that can be present in the domain.

Arrow's theorem can also be formulated in terms of a social choice correspondence. For each preference profile **R** in its preference domain $\mathscr{D}$, a *social choice correspondence* C specifies the socially optimal alternatives $C(A, \mathbf{R})$ in each agenda A (feasible subset of X) in its agenda domain $\mathscr{A}$. In its choice-theoretic formulation, the Arrow axioms are Arrow's choice axiom (for a fixed preference profile, if agenda A is a subset of agenda B, then the set of alternatives chosen in A consists of the restriction to A of the set of alternatives chosen from B when this restriction is nonempty), independence of infeasible alternatives (the alternatives chosen from an agenda only depend on the preferences for alternatives in this agenda), Pareto optimality (only Pareto optimal alternatives are chosen), and nondictatorship (the chosen alternatives are not always a subset of one individual's best feasible alternatives). Arrow's theorem shows that these conditions are inconsistent if the preference domain is unrestricted and the agenda domain consists of all the finite subsets of X. When the agenda domain is closed under finite unions (as is the case in the choice-theoretic version of Arrow's theorem), Arrow's choice axiom is necessary and sufficient for the chosen alternatives in each admissible agenda to be generated by maximizing a profile-dependent social ordering of X (see Hansson, 1968).

In some economic applications, the ability to restrict the agenda domain, not just the preference domain, has weakened the constraints on the admissible social choice correspondences sufficiently for the Arrovian axioms to be consistent. This observation was first made by Bailey (1979), who noted that the set of feasible allocations in an exchange economy does not contain a finite number of alternatives, and so does not satisfy Arrow's agenda domain assumption. While the example Bailey used to show the consistency of the Arrow axioms is problematic, as Donaldson and Weymark (1988) have shown, if each agenda in the agenda domain is the set of feasible allocations for an exchange economy with divisible private goods (different aggregate endowments yield different agendas) and if each profile in the preference domain is a profile of classical private goods preferences for which no individual is indifferent between a consumption bundle with strictly positive components and one that has zero consumption of some good, then the Arrow axioms are consistent. For example, the *equal division Walrasian social choice correspondence* satisfies these axioms. For each exchange economy, this correspondence selects the set of Walrasian (competitive) equilibrium allocations using an equal division of the aggregate endowment as each individual's endowment vector.

When production is possible, an agenda is the set of feasible allocations given the aggregate resource endowment and the production technologies. Possible restrictions on agendas include compactness, comprehensiveness (that is, they satisfy free disposal), and convexity. When there are only public goods, Le Breton and Weymark (2002) have shown that the Arrow axioms are consistent if the preference domain includes only Euclidean spatial preferences on $\mathbb{R}^m_+$ with $m \geq 2$ and the agenda domain includes only compact sets with nonempty interiors. With these domain assumptions, a social choice correspondence satisfying the Arrow axioms can be constructed by fixing a utility representation for each preference and then using an individualistic Bergson–Samuelson social welfare function to choose the best alternatives from each agenda for each preference profile.

In both of these examples, one of the choice-theoretic versions of the Arrow axioms is vacuous. In the exchange economy example, it is Arrow's choice axiom, whereas in the spatial example, it is independence of infeasible alternatives. For a public goods economy with at least two divisible goods, none of the Arrow axioms is vacuous if the agenda domain includes only compact comprehensive sets with nonempty interiors and the preference domain includes only classical public goods preferences. By means of an example, Donaldson and Weymark (1988) have shown that the Arrow axioms are consistent with these domain assumptions. However, their example exhibits dictatorial features and it is not known if the axioms are still consistent if nondictatorship is replaced with anonymity (permuting preferences for a

given agenda does not change the set of chosen alternatives). Donaldson and Weymark have also established a private goods version of this possibility theorem.

Arrovian impossibility results have also been obtained with the social choice correspondence framework using a strengthened form of independence of infeasible alternatives, due to Donaldson and Weymark (1988), called independence of Pareto-irrelevant alternatives. This condition requires the chosen alternatives from each agenda to depend only on the preferences over the Pareto optimal alternatives. For example, for public goods economies, Duggan (1996) has shown that this strengthened independence condition and the other Arrow axioms are inconsistent if $X = \mathbb{R}^m_+$ with $m \geq 3$, the agenda domain consists of all the compact, comprehensive and convex subsets of X, and the preference domain is the set of all profiles of classical public goods preferences for which individual preferences are strictly convex in own consumption.

In contrast with the local approach used to analyse social welfare functions, no unifying methodology has been developed to investigate the consistency of Arrow's choice-theoretic axioms, with the consequence that little is yet known about where the boundary between possibility and impossibility for social choice correspondences lies.

Variable-population social choice

The Arrovian framework is based on ordinal preferences that are interpersonally noncomparable and, hence, any social decision rule that makes use of interpersonal utility comparisons, such as classical utilitarianism, is ruled out from the outset. Sen (1974) has argued that this informational poverty plays a fundamental role in precipitating Arrovian impossibilities, and has proposed a generalization of the concept of an Arrovian social welfare function called a social welfare functional to allow for interpersonal utility comparisons. Each individual i is assumed to have a utility function U_i on the set of alternatives X_i in which he is alive and a *social welfare functional* maps each admissible profile of individual utility functions into a social ordering of the set of all alternatives X. In fixed-population social choice, $X_i = X$ for all i. There is an extensive literature that has investigated the implications for the functional form of these functionals of combining different assumptions concerning the measurability and interpersonal comparability of utility with various normative criteria, including analogues of the Arrovian axioms, when there is a fixed population: see Bossert and Weymark (2004) for a survey. In this section, we provide an introduction to the main issues that arise in selecting appropriate social objective functions when the population size is not fixed. A detailed treatment of this topic and further references may be found in Blackorby, Bossert and Donaldson (2005).

Since the 1980s, population ethics has established itself as an important branch of moral philosophy. Parfit (1984) has been particularly influential in bringing this issue to the attention of philosophers and, more generally, to scholars in various disciplines interested in applied ethics. An up-to-date account of variable-population issues in moral philosophy is given by Broome (2004). Although there are many economic applications of variable-population social choice, such as the design of aid packages (that may have population consequences) for developing countries, the choice of budgets devoted to prenatal care, and policies affecting the intergenerational allocation of resources, the economics literature, with few exceptions, did not initially pay much attention to this topic. Much of the recent interest in these issues can be traced to the influential article by Blackorby and Donaldson (1984), who extended the welfarist model of social choice to allow for a variable population.

In this setting, each alternative $x \in X$ is a complete description of the relevant state of affairs including the size and composition of the population. Furthermore, alternatives are interpreted as full histories of the world, from the remote past to the distant future. Thus, the set of those alive in x contains everyone who has ever lived in this alternative and not merely those who are alive in a given period. This assumption is important to avoid counter-intuitive conclusions regarding the termination of lives. As a consequence, ending someone's life does not change population size; it affects the lifetime and, possibly, the lifetime utility of the person in question.

For each $x \in X$, $u_i = U_i(x)$ is the lifetime well-being (utility) of any individual i alive in x and $U(x)$ is the vector of utilities of these individuals. The standard convention is to assign a lifetime utility level of zero to a *neutral* life. A life, taken as a whole, is a neutral life from the viewpoint of the individual leading it if it is as good as a life without any experiences (a state of permanent unconsciousness). Note that it is not necessary to invoke states of non-existence of an individual in order to define the notion of neutrality. In particular, it is not claimed that an individual can gain or lose by being brought into existence. Therefore, an existing person's life is worth living if the individual's lifetime utility is positive.

Welfarism is the principle that the only features of an alternative that are socially relevant are the utilities of the individuals alive in this alternative. Welfarism implies that the social ordering of X for any profile of individual utility functions in the domain of the social welfare functional can be determined by a single *social welfare ordering* of all possible vectors of individual utilities $\mathscr{U} = \cup_{n \in \mathbb{N}} \mathbb{R}^n$, where $\mathbb{N}$ is the set of positive integers; that is, if a social welfare functional is welfarist, there exists an ordering R on $\mathscr{U}$ such that alternative $x \in X$ is at least as good as alternative $y \in X$ for the profile of utility functions U if and only if $U(x)RU(y)$. The set of individuals

alive in x and y need not be the same. Thus, given welfarism, the problem of variable-population social evaluation can be reduced to the problem of establishing a social welfare ordering R on the set $\mathscr{U}$ of all utility vectors (of varying dimension). If there are $n \in \mathbb{N}$ individuals alive in an alternative, without loss of generality they can be labelled $1, \ldots, n$ provided that the standard anonymity property is satisfied. A representation of the restriction of R to fixed-population comparisons is an individualistic Bergson–Samuelson social welfare function.

The most commonly discussed examples of variable-population social welfare orderings are extensions of utilitarianism. According to *average utilitarianism* (AU) (resp. *classical utilitarianism* (CU)), average (resp. total) utilities are used as the criterion to compare any two utility vectors. Formally, for all $n, m \in \mathbb{N}$, all $u \in \mathbb{R}^n$, and all $v \in \mathbb{R}^m$, $u R_{\text{AU}} v$ if and only if $\frac{1}{n} \sum_{i=1}^{n} u_i \geq \frac{1}{m} \sum_{i=1}^{m} v_i$ (resp. $u R_{\text{CU}} v$ if and only if $\sum_{i=1}^{n} u_i \geq \sum_{i=1}^{m} v_i$). Clearly, fixed-population comparisons are the same according to R_{AU} and R_{CU}, but this is not necessarily the case if n and m differ.

Average utilitarianism is rejected by most contributors to this area. Its fundamental problem is that the value of adding a person, *ceteris paribus*, depends on the utilities of those alive. This has rather unfortunate consequences. Suppose, for example, that everyone is extremely well-off in an alternative and we consider the addition of an individual who, if brought into existence, would have a lifetime utility just slightly below the average of the existing population and no one else's utility is affected. According to AU, this person should not be brought into existence. The following example is even more disturbing. Consider a society in which everyone is extremely miserable by all standards (and well below neutrality). AU recommends the *ceteris paribus* addition of anyone with a lifetime utility slightly above the average, even if this utility level is well below neutrality.

Classical utilitarianism suffers from what Parfit (1984) calls the *repugnant conclusion*. A variable-population social welfare ordering R implies the repugnant conclusion if, for any population size n, for any positive level of utility ξ (no matter how high), and for any level of utility $\varepsilon \in (0, \xi)$ (no matter how close to zero), there exists a population size $m > n$ such that a population with n people in which everyone has a lifetime utility of ξ is considered inferior to a population of m individuals each of whom has a lifetime utility of ε; that is, for any situation in which everyone alive has an arbitrarily high level of well-being, there is always a situation of mass poverty (with everyone arbitrarily close to neutrality) that is considered superior.

In order to avoid the repugnant conclusion and, at the same time, the counter-intuitive implications of average utilitarianism, Blackorby and Donaldson (1984) have proposed *critical-level utilitarianism* (CLU) with a positive critical level as an alternative criterion. CLU employs a parameter $\alpha \in \mathbb{R}$ (the critical level) and is defined by letting, for all $n, m \in \mathbb{N}$, all $u \in \mathbb{R}^n$, and all $v \in \mathbb{R}^m$, $u R_{\mathrm{CLU}} v$ if and only if $\sum_{i=1}^{n} [u_i - \alpha] \geq \sum_{i=1}^{m} [v_i - \alpha]$. The special case corresponding to $\alpha = 0$ is CU. The parameter α has an intuitive interpretation: it is the level of utility that, if experienced by an additional person, makes the alternative resulting from the *ceteris paribus* addition of such a person to any given society as good as the original. Because the critical level is constant, the problems of AU alluded to above are avoided. If, moreover, α is positive, the repugnant conclusion is avoided because there is a positive difference between the critical level and the level of utility representing neutrality.

In addition to providing a thorough analysis of critical-level utilitarianism and its main alternatives, Blackorby, Bossert and Donaldson (2005) have discussed several extensions of the basic model. For example, the critical-level utilitarian orderings can be generalized by considering transformed utilities rather than the utilities themselves. If the transformation is chosen to be strictly concave, the corresponding social ordering represents inequality aversion in utilities. Furthermore, they have considered orderings that use non-welfare information such as birth dates and lifetimes in addition to lifetime utilities, as well as variants that incorporate uncertainty. Moreover, they have analysed variable-population choice problems and applications.

Strategy-proof social choice

A *social choice* function g chooses one alternative from the set of alternatives X for each preference profile in the domain $\mathscr{D}$. If it is only known that the true profile is in $\mathscr{D}$, in order to implement the desired choice $g(\mathbf{R})$ when the profile is $\mathbf{R}$, individuals must have an incentive to truthfully report their preferences. *Strategy-proofness* is the requirement that nobody can obtain a preferred outcome by reporting a false preference regardless of what the preferences of the other individuals are. Strategy-proofness places severe constraints on the kinds of social choice functions that can be considered and, on some domains, conflicts with other social desiderata (for introductions to recent developments in strategy-proof social choice theory, see Sprumont, 1995; Barberà, 2001).

The classic result on strategy-proofness is the Gibbard (1973)–Satterthwaite (1975) theorem, which shows that no social choice function g can satisfy both nondictatorship and Pareto optimality when $\mathscr{D} = \mathscr{R}^n$ if $|X| \geq 3$. The same conclusion follows if Pareto optimality is replaced with *unanimity*, the requirement that an alternative is chosen if everybody agrees that it is uniquely best. Either of these conditions implies that the range $\mathrm{rg}(g)$ of g is all of X when the domain is unrestricted. A variant of the Gibbard–Satterthwaite theorem states that on an unrestricted domain, if $|\mathrm{rg}(g)| \geq 3$, then strategy-proofness implies that someone must be a

dictator on $\mathrm{rg}(g)$ (that is, g always chooses one of this person's best alternatives on $\mathrm{rg}(g)$).

More positive results are obtained if it is known that preferences are single-peaked. Moulin (1980) has shown that if $X \subseteq \mathbb{R}$, $\mathscr{D} = \mathscr{S}^n$, and the social choice function g only depends on the peaks of the individual preferences, then g satisfies strategy-proofness if and only if it is a minmax social choice function and it satisfies strategy-proofness, Pareto optimality and anonymity if and only if it is a generalized median social choice function. A *minmax* social choice function g is defined by specifying an alternative x_S in the closure of X for each coalition of individuals with $x_T \leq x_S$ if $S \subseteq T$ and setting

$$g(\mathbf{R}) = \min_{S \subseteq N} \left\{ \max_{i \in S} \{ \pi(R_i), x_S \} \right\}, \quad \forall \mathbf{R} \in \mathscr{S}^n.$$

For each $\mathbf{R} \in \mathscr{S}^n$, a *generalized median social choice* function chooses the median of the actual individual preference peaks and the fixed peaks of $n-1$ phantom voters. These functions are minmax rules in which the alternatives x_S are the same for coalitions of the same size. Barberà, Gul and Stacchetti (1993) have provided an alternative characterization of minmax rules in terms of winning coalitions that has proved to be quite useful. If, as is the case with minmax rules, the chosen alternative for each profile depends only on each person's most-preferred alternative(s) on the range, the social choice function satisfies the *tops-only* property. On the domain $\mathscr{S}^n$, Barberà and Jackson (1994) have shown that the tops-only property assumed by Moulin (1980) is implied by strategy-proofness if either $\mathrm{rg}(g)$ is an interval or g satisfies Pareto optimality.

The original strategies used to prove the Gibbard–Satterthwaite theorem cannot be adopted to analyse strategy-proofness when preferences are continuous. The problem is that these proofs alter profiles by moving two alternatives to the top two positions in a person's preference, but this is not possible with continuous preferences if X is a connected set, as there can be no second-ranked alternative. This difficulty was overcome by Barberà and Peleg (1990) who established a version of the Gibbard–Satterthwaite theorem for continuous preferences on a metric space of alternatives using the option set methodology introduced by Laffond (1980), Satterthwaite and Sonnenschein (1981), and Barberà (1983). An *option set* identifies the set of outcomes that are feasible given the preferences of a subgroup of individuals for some admissible reported preferences of the rest of the population. For example, when there is a dictator d, the option set generated by R_d consists of the best alternatives on the range for this preference and the option set generated by any other person's preference is the whole range. The option set methodology proceeds by identifying the structure imposed on option sets by the properties that one wants the social choice function to satisfy.

In order for a social choice function to be strategy-proof, it must ignore most of the information about individual preferences. On many domains in which all admissible preferences have unique best alternatives on the range, strategy-proofness implies the tops-only property provided that the range of the social choice function satisfies some regularity condition. Weymark (2004) has proposed a proof strategy for establishing the tops-only property that avoids the model specificity of earlier proofs.

A social choice function defined on a domain of profiles of separable preferences on a product set of alternatives is *decomposable* if the value chosen for a component depends only on the individual marginal preferences for that component. The first decomposability results were established by Border and Jordan (1983) who, for example, showed that, for the domain of all profiles of separable quadratic preferences on a multidimensional Euclidean space, a social choice function satisfies strategy-proofness and unanimity if and only if it decomposes into strategy-proof, unanimous social choice functions on each component. Furthermore, these one-dimensional mechanisms can be any member of Moulin's class of minmax social choice functions. Since the development of the option set methodology, decomposability results for strategy-proof social choice functions have been established for a number of other domains of separable preferences. For example, Barberà, Gul and Stacchetti (1993) have shown that, if X is a discrete product set in a Euclidean space and individuals can have any separable preference that satisfies a multidimensional analogue of single-peakedness, then the conclusions of Border and Jordan's theorem hold if the range of the social choice function is all of X. Whether strategy-proofness and auxiliary conditions such as unanimity imply decomposability depends on how much preference variability is present in the domain. Establishing a decomposability theorem typically involves first showing that the tops-only property is satisfied, as in Barberà, Gul and Stacchetti (1993). Much of the literature on this issue has been synthesized and extended by Le Breton and Sen (1995; 1999).

If X is a product set, but only a subset Z of X is feasible, decomposability results are still possible, but not every combination of the corresponding one-dimensional social choice functions is admissible. For example, using the model in Barberà, Gul and Stacchetti (1993) with the best alternative for each preference required to be in Z, Barberà, Massó and Neme (1997) have shown that any social choice function g that is strategy-proof and whose range is Z is decomposable into one-dimensional minmax rules on each component, but, in order for a combination of such minmax rules to always produce a feasible outcome, g must satisfy a rather complicated condition called the *intersection property*.

In the preceding discussion, everyone has the same set of admissible preferences, and so it is possible that they might agree on what is best. When there are private goods and individuals care only about their own consumption, one generally expects there to be distributional conflicts. We illustrate the implications of strategy-proofness with private goods in two problems: the allotment problem and the exchange of divisible goods.

In an *allotment problem*, there is a fixed amount Ω of a divisible good to allocate. If individuals care only about own consumption, each person's preference is defined on $X = [0, \Omega]$. If these preferences are single-peaked, a prominent solution to this problem is the *uniform rule* (see Benassy, 1982) which, for each admissible profile $\mathbf{R} \in \mathscr{S}^n$, chooses the unique allocation $x = (x_1, \ldots, x_n) \in X^n$ (x_i is person i's allocation) for which (i) if $\Omega \leq \sum_{i=1}^{n} \pi(R_i)$, there exists $\lambda \in \mathbb{R}_+$ such that, for all $i \in N$, $x_i = \min\{\pi(R_i), \lambda\}$ and (ii) if $\Omega \geq \sum_{i=1}^{n} \pi(R_i)$, there exists $\lambda \in \mathbb{R}_+$ such that, for all $i \in N$, $x_i = \max\{\pi(R_i), \lambda\}$. Sprumont (1991) has shown that, if the domain is the set of all profiles of continuous single-peaked preferences on X, then a social choice function satisfies strategy-proofness, Pareto optimality and private-goods anonymity (permuting preferences results in the same permutation of the individual allocations) if and only if it is the uniform rule. Sprumont's article also includes the first explicit theorem about the tops-only property in the strategy-proofness literature.

When X is the set of allocations of an exchange economy with two or more divisible private goods, the general conclusion is that strategy-proofness and Pareto optimality conflict with other desirable properties for a social choice function on a sufficiently rich domain of classical private goods preference profiles. If the aggregate endowment is privately owned and participation in the collective choice procedure is voluntary, the social choice function must satisfy individually rationality; that is, each person is guaranteed a consumption bundle weakly preferred to his endowment. Hurwicz (1972) has shown that strategy-proofness, Pareto optimality and individual rationality are inconsistent for two-person, two-good exchange economies on such a preference domain. This impossibility theorem has only recently been extended to the general n-person, m-good case by Serizawa (2002).

With monotonic preferences, a dictator in an exchange economy always receives the whole endowment. For the domain of classical private goods preference profiles, Zhou (1991) has shown that strategy-proofness, Pareto optimality and nondictatorship are inconsistent when there are at least two goods, but only two individuals. When there are at least three individuals, Satterthwaite and Sonnenschein (1981) have shown by example how to construct Pareto optimal, strategy-proof, nondictatorial social choice functions for this domain. In their example, someone is *bossy* (that is, there is an individual who can change the consumption bundle of someone else by reporting a different preference without affecting his or her own consumption bundle) and, for each profile, one of two individuals receives all of the endowment. It is generally agreed that bossy mechanisms are unsatisfactory. Serizawa and Weymark (2003) have shown that any social choice function that satisfies strategy-proofness and Pareto optimality cannot guarantee everyone a consumption bundle bounded away from the origin on a rich domain of classical private goods preferences.

Given that any strategy-proof and Pareto optimal social choice function g must fail even minimal distributional desiderata on such domains, Barberà and Jackson (1995) have explored the implications of abandoning Pareto optimality. For private ownership exchange economies with classical private goods preferences, they have shown that if g is strategy-proof, non-bossy, and satisfies some other auxiliary conditions, then trade must be restricted to occur in a limited set of fixed proportions with possibly upper limits on the amounts that can be exchanged. In the case of two goods and two individuals, if g satisfies strategy-proofness and individual rationality, there are only two such proportions, and the choice procedure resembles the fixed-price trading rules studied by Benassy (1982) with different buying and selling prices for each good.

Axiomatic models of resource allocation

The literature on axiomatic models of resource allocation has a close affinity to the literature on Arrovian social choice on economic domains. As is the case with Arrovian social choice when there are multiple agendas, the research on axiomatic models of resource allocation investigates the implications of normative criteria (axioms) when both individual preferences and the set of feasible agendas satisfy the kinds of restrictions found in economic models. What distinguishes this literature is the set of axioms considered, many of which rely on the special structure provided by economic models for their definition. In this section, we present a very selective introduction to the models and axioms considered in this literature and describe a few of the theorems that have been obtained. For a comprehensive survey of this literature, see Thomson (2008).

In an *allocation problem*, there is an aggregate social endowment $\Omega \in \mathbb{R}_{++}^m$ of m private goods that are to be allocated among $n \geq 2$ individuals based on their preferences. In the canonical allocation problem, $m \geq 2$ and all goods are divisible. An *economy* is then described by a pair $E = (\mathbf{R}, \Omega)$, where $\mathbf{R}$ is a profile of classical private goods preferences. Let $\mathscr{E}$ denote the set of all such economies. Given the endowment Ω, the corresponding agenda $A(\Omega)$ is the set of feasible allocations $x = (x_1, \ldots, x_n)$ that exhaust Ω, where $x_i \in \mathbb{R}_+^m$ is person i's consumption bundle.

A *solution* is a mapping that selects a subset of the feasible allocations for each economy in $\mathscr{E}$. Note that a solution φ can be identified with a social choice correspondence C by setting $C(A(\Omega), \mathbf{R}) = \varphi(E)$ for all $E \in \mathscr{E}$. A solution satisfies *efficiency* if it always chooses Pareto optimal allocations and it satisfies *no envy* if, at any selected allocation, nobody strictly prefers anyone else's allocated consumption bundle. No envy, which was independently introduced by Tinbergen (1953), Foley (1967), and Kolm (1972), is the fundamental fairness condition considered in this literature. An example of a solution satisfying both efficiency and no envy for this class of economies is the *equal division Walrasian solution* φ^W, which is defined from the equal division Walrasian social choice correspondence in the manner described above.

The literature on fair allocation has expanded the scope of the canonical model in several respects. For example, economies with varying populations or with production have been considered. In addition, variations of this model have been explored, for example, by allowing for indivisibilities or, when there is only one good, preference restrictions such as single-peakedness.

An alternative set-up with public goods has also been examined. The existence of solutions satisfying efficiency and no envy in public goods environments is a more complex matter than in the private goods case, largely because arguments involving pure exchange cannot be invoked when there are public goods. Furthermore, the technology that permits us to transform private goods into public goods is also important. With some additional assumptions, however, efficiency and no envy can be satisfied (see Diamantaras, 1991, for example).

Prominent among the new axioms that have been introduced and used in characterizations of existing and new solutions is *consistency*, which is discussed in detail in Thomson (1990). In order to define consistency, the notion of a solution must be extended to include economies with different numbers of individuals. Let x be an allocation that is selected by such a solution for an $(n+1)$-person economy. Now suppose that person k leaves the economy with the consumption x_k. Define a *reduced* n-person economy by removing k and subtracting x_k from the total endowment. Consistency demands that the allocation $(x_1, \ldots, x_{k-1}, x_{k+1}, \ldots, x_{n+1})$ is selected in the reduced economy.

Other important properties include monotonicity conditions with respect to the quantities of the resources available, with respect to the technology, or with respect to the population. A solution φ satisfies *resource monotonicity* if, whenever the social endowment expands, no one becomes worse off in any chosen allocation. In private goods models with production, an axiom similar in spirit to resource monotonicity is *technology monotonicity*. It requires that if the only difference between two economies is that the technology of one dominates that of the other, then everyone should be at least as well-off in any allocation chosen for the former economy as in any allocation chosen for the latter. *Population monotonicity* is a solidarity axiom. As is the case for consistency, it applies in models with variable population. Suppose that the population is expanded, but the total endowment is unchanged. Population monotonicity demands that the burden imposed on the existing population by the presence of the additional individuals is shared by all its members; no one who is present before the population expansion is better off as a consequence of the population augmentation.

If there is only one divisible good, each economy $E=(\mathbf{R}, \Omega)$ defines an allotment problem, as in the preceding section. When all preference profiles are single-peaked, the *uniform solution* simply applies the uniform rule for the allotment problem to each economy in the domain. In addition to the characterization of this solution presented in the preceding section, there have been axiomatizations of the uniform solution using no envy, consistency, and variants of either resource monotonicity or population monotonicity, among other axioms (see Thomson, 2008).

If some of the goods to be allocated are indivisible, much of the theory developed in the context of perfectly divisible goods still applies. Due to the specific nature of the problem of allocating indivisible objects, some interesting additional results can be obtained. As an illustration, consider an assignment problem in which n indivisible objects are to be allocated to n individuals and there is also a perfectly divisible good ('money') that can be consumed in any amount, positive or negative: see Thomson (2008) for references to contributions that permit the number of goods and individuals to differ. A commodity bundle for person i is now a pair $(t_i, j) \in \mathbb{R} \times N$, where t_i (resp. j) is the amount of money (resp. object) allocated to i. It is assumed that i's preference R_i on $\mathbb{R} \times N$ is strictly monotonic in money and that money can be used to compensate for the receipt of a less desirable good in the sense that, for all $t_i \in \mathbb{R}$ and all j, $k \in N$, there exists $s_i \in \mathbb{R}$ such that i is indifferent between (s_i, k) and (t_i, j). An economy now consists of a preference profile $\mathbf{R}$ with the properties introduced above, an aggregate endowment of money $T \in \mathbb{R}$, and the n indivisible objects. Because the set of objects is fixed, an economy can be characterized by a pair $E = (\mathbf{R}, T)$. A feasible allocation for E is a pair (t, ρ), where $t \in \mathbb{R}^n$ is a vector of *balanced* monetary allocations (that is, $\sum_{i=1}^n t_i = T$) and $\rho: N \to N$ is a permutation with $\rho(i)$ specifying the object allocated to person i.

Solutions and the efficiency and no-envy axioms are defined in the usual manner. If the monetary allocations are restricted to be non-negative, it is clear that solutions satisfying no envy may not exist. For example, if $T = 0$ and everyone regards the same object as being uniquely best regardless of the amount of money received, whoever is allocated this object is envied by everyone else because no monetary compensations are possible. Sufficient conditions for the existence of solutions satisfying no envy with non-negative monetary allocations are discussed in Thomson (2008). As is to be expected, these conditions ensure that there is a sufficient amount of money available to carry out the requisite compensation payments.

In the case of perfectly divisible goods, we have noted that the equal division Walrasian solution φ^W satisfies efficiency and no envy. Interestingly, in the indivisible good model considered here, the allocations generated by an adaptation of this Walrasian solution to the present framework are the *only* allocations satisfying no envy (see Svensson, 1983). Moreover, efficiency *follows* as a consequence of no envy. In this model, for an economy E, if everyone is provided with the same endowment $t_0 \in \mathbb{R}_+$ of money, a *Walrasian equilibrium* is a feasible allocation (t, ρ) and a price $p_k \in \mathbb{R}_+$ for each good $k \in N$ such that the bundle $(t_i, \rho(i))$ is weakly preferred by individual i among all bundles that have values no more than t_0. The solution φ^W is then defined by letting $\varphi^W(E)$ be the set of Walrasian equilibrium allocations that can be obtained in this way.

A number of fairness principles besides no envy have been considered in the literature (see Fleurbaey and Maniquet, 2008; Thomson, 2008). Particularly notable among them is *egalitarian equivalence*, which is due to Pazner and Schmeidler (1978). In the canonical allocation problem, egalitarian equivalence requires that, for each economy $E = (\mathbf{R}, \Omega)$, each selected allocation x has the property that there exists a consumption bundle $z_0 \in \mathbb{R}_+^m$ that everyone is indifferent to. Pazner and Schmeidler (1978) have shown that on the domain of economies $\mathcal{E}$ for this problem, solutions exist that satisfy both egalitarian equivalence and Pareto optimality. However, egalitarian equivalence need not satisfy independence of infeasible alternatives, as the egalitarian allocation $(z_0, \ldots, z_0)$ associated with x need not be feasible.

There is now an extensive literature that employs the framework and many of the axioms described in this section to re-examine the foundations of egalitarian theories. If individuals are held responsible in part for the outcomes they receive, conditional versions of egalitarianism demand that individual differences caused by factors beyond the individuals' control should be compensated for, whereas inequities that can be attributed to choices for which an individual is responsible do not attract that kind of equalization. Variants of this theory have been advocated by, for example, Roemer (1993): see Fleurbaey and Maniquet (2008) for a detailed survey of this literature.

Concluding remarks

As noted above, the response of Sen (1974) to Arrovian social welfare function impossibilities was to abandon the ordinal non-comparability of individual utilities built into the Arrow framework. However, he maintained the spirit of IIA by assuming that the social ranking of any two alternatives should depend only on the individual utilities obtained with them. This independence assumption is the cornerstone of the welfarist approach employed in the literatures on social choice with interpersonal utility comparisons and on variable-population social choice.

A different resolution of the Arrovian dilemma has been described and defended in Fleurbaey (2007). Rather than abandoning ordinal non-comparability, Arrow's IIA assumption is relaxed so as to (i) to allow the social ranking of a pair of alternatives to depend on how these alternatives are ranked relative to some other alternatives and (ii) to incorporate some principle of fairness. This proposal has been explored in a series of articles by Fleurbaey and various co-authors. For example, Fleurbaey, Suzumura and Tadenuma (2005) have shown that, when the alternatives are the set of all allocations of $m \geq 2$ divisible private goods and the domain of the social welfare function is the set of profiles of classical private goods preferences, then weak Pareto and a private goods version of anonymity are compatible with an independence condition that incorporates fairness considerations of the sort embodied in egalitarian equivalence. Independence conditions such as this or ones based on envy-freeness employ non-local information about preferences, including information about alternatives that may not be feasible if resource constraints are taken into account. This line of research provides a bridge between the literatures on Arrovian social choice on economic domains and axiomatic models of resource allocation discussed above by employing some form of independence condition (as in the former) while at the same time requiring social decisions to be fair (as in many of the latter contributions).

WALTER BOSSERT AND JOHN A. WEYMARK

See also **Arrow's theorem; mechanism design; mechanism design (new developments); social choice; strategic voting; strategy-proof allocation mechanisms; voting paradoxes.**

Bibliography

Arrow, K.J. 1963. *Social Choice and Individual Values*, 2nd edn. New York: Wiley.

Arrow, K.J., Sen, A.K. and Suzumura, K. 2002. *Handbook of Social Choice and Welfare*, vol. 1. Amsterdam: North-Holland.

Arrow, K.J., Sen, A.K. and Suzumura, K. 2008. *Handbook of Social Choice and Welfare*, vol. 2. Amsterdam: North-Holland.

Austen-Smith, D. and Banks, J.S. 2005. *Positive Political Theory II: Strategy and Structure*. Ann Arbor: University of Michigan Press.

Bailey, M.J. 1979. The possibility of rational social choice in an economy. *Journal of Political Economy* 87, 37–56.

Barberà, S. 1983. Strategy-proofness and pivotal voters: a direct proof of the Gibbard–Satterthwaite theorem. *International Economic Review* 24, 413–17.

Barberà, S. 2001. An introduction to strategy-proof social choice functions. *Social Choice and Welfare* 18, 619–53.

Barberà, S., Gul, F. and Stacchetti, E. 1993. Generalized median voting schemes and committees. *Journal of Economic Theory* 61, 262–89.

Barberà, S. and Jackson, M. 1994. A characterization of strategy-proof social choice functions for economies with pure public goods. *Social Choice and Welfare* 11, 241–52.

Barberà, S. and Jackson, M.O. 1995. Strategy-proof exchange. *Econometrica* 63, 51–87.

Barberà, S., Massó, J. and Neme, A. 1997. Voting with constraints. *Journal of Economic Theory* 76, 298–321.

Barberà, S. and Peleg, B. 1990. Strategy-proof voting schemes with continuous preferences. *Social Choice and Welfare* 7, 31–8.

Benassy, J.-P. 1982. *The Economics of Market Disequilibrium*. New York: Academic Press.

Black, D. 1948. On the rationale of group decision-making. *Journal of Political Economy* 56, 23–34.

Blackorby, C., Bossert, W. and Donaldson, D. 2005. *Population Issues in Social Choice Theory, Welfare Economics, and Ethics*. Cambridge: Cambridge University Press.

Blackorby, C. and Donaldson, D. 1984. Social criteria for evaluating population change. *Journal of Public Economics* 25, 13–33.

Border, K.C. 1983. Social welfare functions for economic environments with and without the Pareto principle. *Journal of Economic Theory* 29, 205–16.

Border, K.C. 1984. An impossibility theorem for spatial models. *Public Choice* 43, 293–305.

Border, K.C. and Jordan, J.S. 1983. Straightforward elections, unanimity and phantom voters. *Review of Economic Studies* 50, 153–70.

Bordes, G., Campbell, D.E. and Le Breton, M. 1995. Arrow's theorem for economic domains and Edgeworth hyperboxes. *International Economic Review* 36, 441–54.

Bordes, G. and Le Breton, M. 1989. Arrovian theorems with private alternatives domains and selfish individuals. *Journal of Economic Theory* 47, 257–81.

Bordes, G. and Le Breton, M. 1990. Arrovian theorems for economic domains: assignments, matchings and pairings. *Social Choice and Welfare* 7, 193–208.

Bossert, W. and Weymark, J.A. 2004. Utility in social choice. In *Handbook of Utility Theory*, vol. 2, ed. S. Barberà, P.J. Hammond and C. Seidl. Boston: Kluwer.

Broome, J. 2004. *Weighing Lives*. Oxford: Oxford University Press.

Diamantaras, D. 1991. Envy-free and efficient allocations in large public good economies. *Economics Letters* 36, 227–32.

Donaldson, D. and Weymark, J.A. 1988. Social choice in economic environments. *Journal of Economic Theory* 46, 291–308.

Duggan, J. 1996. Arrow's theorem in public goods environments with convex technologies. *Journal of Economic Theory* 68, 303–18.

Ehlers, L. and Storcken, T. 2002. Arrow's theorem in spatial environments. Cahier No. 2002–03, Département de sciences économiques, Université de Montréal.

Fleurbaey, M. 2007. Social choice and just institutions: new perspectives. *Economics and Philosophy* 23, 15–43.

Fleurbaey, M. and Maniquet, F. 2008. Compensation and responsibility. In Arrow, Sen and Suzumura (2008).

Fleurbaey, M., Suzumura, K. and Tadenuma, K. 2005. Arrovian aggregation in economic environments: how much should we know about indifference surfaces? *Journal of Economic Theory* 124, 22–44.

Foley, D.F. 1967. Resource allocation and the public sector. *Yale Economic Essays* 7, 45–98.

Gibbard, A. 1973. Manipulation of voting schemes: a general result. *Econometrica* 41, 587–601.

Hansson, B. 1968. Choice structures and preference relations. *Synthese* 18, 443–58.

Hurwicz, L. 1972. On informationally decentralized systems. In *Decision and Organization: A Volume in Honor of Jacob Marschak*, ed. C.B. McGuire and R. Radner. Amsterdam: North-Holland.

Kalai, E., Muller, E. and Satterthwaite, M.A. 1979. Social welfare functions when preferences are convex, strictly monotonic, and continuous. *Public Choice* 34, 87–97.

Kolm, S.-C. 1972. *Justice et Equité*. Paris: CNRS.

Laffond, G. 1980. Révelation des preferences et utilités unimodales. Doctoral dissertation, Conservatoire National des Arts et Méticrs.

Le Breton, M. 1986. Essais sur les fondements de l'analyse économique de l'inégalité. Doctoral dissertation, Université de Rennes 1.

Le Breton, M. and Sen, A. 1995. Strategyproofness and decomposability: weak orderings. Discussion Papers in Economics No. 95-04, Indian Statistical Institute, Delhi Centre.

Le Breton, M. and Sen, A. 1999. Separable preferences, strategyproofness, and decomposability. *Econometrica* 67, 605–28.

Le Breton, M. and Weymark, J.A. 2002. Social choice with analytic preferences. *Social Choice and Welfare* 19, 637–57.

Le Breton, M. and Weymark, J.A. 2008. Arrovian social choice theory on economic domains. In Arrow, Sen and Suzumura (2008).

Maskin, E.S. 1976. Social welfare functions for economics. Unpublished manuscript, Darwin College, Cambridge University.

Moulin, H. 1980. On strategy-proofness and single peakedness. *Public Choice* 35, 437–55.

Parfit, D. 1984. *Reasons and Persons*. Oxford: Oxford University Press.

Pazner, E.A. and Schmeidler, D. 1978. Egalitarian equivalent allocations: a new concept of economic equity. *Quarterly Journal of Economics* 92, 671–87.

Redekop, J. 1995. Arrow theorems in economic environments. In *Social Choice, Welfare, and Ethics*, ed.

W.A. Barnett, H. Moulin, M. Salles and N.J. Schofield. Cambridge: Cambridge University Press.

Roemer, J.E. 1993. A pragmatic theory of responsibility for the egalitarian planner. *Philosophy and Public Affairs* 22, 146–66.

Satterthwaite, M.A. 1975. Strategy-proofness and Arrow's conditions: existence and correspondence theorems for voting procedures and social welfare functions. *Journal of Economic Theory* 10, 187–217.

Satterthwaite, M.A. and Sonnenschein, H. 1981. Strategy-proof allocation mechanisms at differentiable points. *Review of Economic Studies* 48, 587–97.

Sen, A. 1974. Informational bases of alternative welfare approaches: aggregation and income distribution. *Journal of Public Economics* 3, 387–403.

Serizawa, S. 2002. Inefficiency of strategy-proof rules for pure exchange economies. *Journal of Economic Theory* 106, 219–41.

Serizawa, S. and Weymark, J.A. 2003. Efficient strategy-proof exchange and minimum consumption guarantees. *Journal of Economic Theory* 109, 246–63.

Sprumont, Y. 1991. The division problem with single-peaked preferences: a characterization of the uniform rule. *Econometrica* 59, 509–19.

Sprumont, Y. 1995. Strategyproof collective choice in economic and political environments. *Canadian Journal of Economics* 28, 68–107.

Svensson, L.-G. 1983. Large indivisibilities: an analysis with respect to price equilibrium and fairness. *Econometrica* 51, 939–54.

Thomson, W. 1990. The consistency principle. In *Game Theory and Applications*, ed. T. Ichiishi, A. Neyman and Y. Tauman. New York: Academic Press.

Thomson, W. 2008. Fair allocation rules. In Arrow, Sen and Suzumura (2008).

Tinbergen, J. 1953. *Redelijke Inkomensverdeling*, 2nd edn. Haarlem: N.V. DeGulden Pers.

Weymark, J.A. 2004. Strategy-proofness and the tops-only property. Working Paper No. 04-W09, Department of Economics, Vanderbilt University.

Zhou, L. 1991. Inefficiency of strategy-proof allocation mechanisms in pure exchange economies. *Social Choice and Welfare* 8, 247–54.

social contract

Social contract theory, as I shall understand it here, is a theory about how the moral assessment of actions, practices, institutions, laws, constitutions, or related items is based – directly or indirectly – on the consent – actual or hypothetical – of the members of society.

Although social contract theories could be formulated to require merely that a majority of the members society consent to the relevant item, such formulations are not particularly plausible and generally have not been advocated. The standard requirement is that *all* members consent to the item, and I shall assume this in what follows.

Social contract theory is also sometimes understood to be (*a*) an empirical theory about how government actually arose, or (*b*) a metaethical theory about the general nature of morality and moral reasons. Here, however, we shall focus on contractarianism as a substantive theory of justice.

The origins of social contract theory can be traced back to the discussion of justice (in the voice of Glaucon) by Plato (c. 430–347 BCE) in *The Republic* (1961), but systematic development of the theory really started in the 17th century, with the beginnings of the modern state and the challenging of the alleged divine right of kings and aristocrats to rule. The first comprehensive statement of social contract theory came from Thomas Hobbes (1588–1679) in his *Leviathan* (1651), in which he offered a social contract justification for almost unlimited powers of the state. Other important historical figures associated with social contract theory include John Locke (1632–1704) (1690), Jean-Jacques Rousseau (1712–78) (1762), and Immanuel Kant (1724–1804) (1785). Since the publication of John Rawls's *A Theory of Justice* (1971), there has been a significant renewal of interest – among economists, political scientists, and philosophers – in social contract ethical and political theory.

Social contract theory can be used to address several quite different topics. It might be formulated (*a*) as a full theory of *individual morality* (ethics), (*b*) as a theory of *what duties we morally owe each other* (which does not include any impersonal duties), (*c*) as a *theory of political authority* in the sense of the conditions under which we have a duty to obey the dictates of others, (*d*) as a theory of *legitimacy* in the sense of the conditions under which others are not permitted to forcibly interfere with our actions (even if wrong), or (*e*) as a theory of the moral permissibility or *justice of political institutions*. For simplicity, I shall focus on social contract theory as a theory of the justice of political institutions – although most remarks apply to most other versions.

There are two broad kinds of social contract theory: *actual* contract (rights-based) theory and *hypothetical* contract theory (contractarianism or contractualism). I consider each in turn.

Actual social contract theory

Actual social contract theory holds that a set of political institutions (for example) is just if and only if it has been directly consented to by those governed or conforms to the requirements of a constitution to which all have consented. The best-known actual social contract theory is that of Locke (1690) – although it has also been interpreted as a hypothetical contract theory (see below).

In order for consent to have moral force, the agent must be rationally competent and the consent must be free (not coerced) and suitably informed (for example,

not based on fraud). Establishing exactly what is required for consent to be valid is a very important topic. For excellent discussion, see Simmons (1993).

Actual consent can be explicit (as in 'I hereby consent') or implicit (as when one allows a friend to enter one's house without explicitly granting permission). Identifying the exact conditions under which an action other than explicit consent is nonetheless a case of (implicit) consent is an important topic insightfully analysed by Simmons (1993). Given that explicit consent is extremely rare, most actual social contract theories (plausibly) allow both kind of consent to have moral force.

Actual social contract theories presuppose that individuals have certain (choice-protecting) natural rights of self-governance (a kind of natural freedom). Any restriction of this natural freedom is deemed unjust unless the individual has consented to it. Given the insecurity of life, health, and possessions in the absence of government, it typically makes sense for people to give up some of their natural freedom and submit to the authority of government – on the condition that the political authority thus established does a reasonably good job protecting people's pre-political rights. When they do so, actual social contract theory holds that the resulting political institutions are just.

Actual social contract theory faces three important challenges. One is whether individuals have the natural rights that the theory postulates. A second is that, for almost all existing societies, there never was a relevant universal agreement on political institutions – which entails, given the theory, that no set of political institutions is just. A third is that, even if there was a relevant universal agreement in some past generation, it's not clear why this would be relevant for the present generation (the members of which have not consensually given up their natural rights). (For further discussion, see Simmons 1993.)

Hypothetical social contract theory

Given the problem of securing universal actual consent to political institutions (or any set of rules), most contemporary social contract theories appeal to hypothetical consent. They hold that political institutions (for example) are just if and only if (*a*) they would be universally agreed to under specified conditions, or (*b*) they conform to a political constitution that would be agreed to under specified conditions. For brevity, call these theories *contractarian*.

Contractarian theories differ in their specification of the circumstances under which the relevant hypothetical agreement takes place. These conditions include the motivations and beliefs of the parties to the agreement as well as the non-agreement point (the outcome if no agreement is made).

Hobbesian approaches provide realistic, morally neutral, specifications of the circumstances, and attempt thereby to reduce morality to individual or collective rationality. They specify that individuals are mainly interested in promoting their own wellbeing and assume that agents have reasonably full knowledge of their situation. The non-agreement point (state of nature) is a state of war of all against all in which there is neither government nor morally constrained behaviour. Hobbes (1651), Buchanan (1975), and Gauthier (1986) are all in this tradition. (See also Hampton, 1986; Kavka, 1986; and Binmore, 1994; 1998.) A main objection to the Hobbesian approach is that it yields an impoverished conception of morality: individuals are protected by morality only to the extent that cooperation with them is useful to others. Because babies, young children and severely disabled people offer others no benefits from cooperation, knowledgeable predominantly self-interested agents would not enter into agreements with them. Such vulnerable individuals might be protected by the terms of the hypothetical agreement, but that would be only to the extent that others happen to care about them (for example, parents for their children). Absent any such contingent concern, such individuals are deemed merely resources to be exploited.

Lockean forms of contractarianism are similar to Hobbesian ones except that the non-agreement point (state of nature) is not pre-moral. (As indicated above, Locke, 1690, has been read both as an actual social contract theory and as a hypothetical social contract theory.) Although there is no government, individuals have certain natural rights, and generally respect the rights of others. Although the non-agreement point is not the dire Hobbesian state of war of all against all, the absence of a generally accepted adjudication and enforcement agency often leads to feuds when someone believes that her rights have been violated without adequate rectification. In addition, the absence of government makes it difficult to provide various public goods (such as roads and national defence). The result is that it would be rational for all to agree to give up some of their natural rights and to establish a state.

Kantian forms of contractarianism (sometimes called 'contractualism') view the social contract device as representing reasonable reciprocity between moral equals who are self-legislating free and equal members of the kingdom of ends. Harsanyi (1953; 1955) and Rawls (1971), for example, allow that individuals are primarily concerned with promoting their own good, but they impose a *veil of ignorance* that blocks parties from having any knowledge of their capacities, positions, or desires. One objection to this approach is that it eliminates all individual differences and thus reduces the agreement to the choice of one individual behind the veil of ignorance. Scanlon (1998), by contrast, allows individuals full knowledge of their situations, but stipulates that the parties are motivated by a desire to reach a fair and reasonable agreement (as opposed to simply promoting their own good). (Habermas, 1993, is similar in spirit.) So understood,

contractarianism takes ability to justify our conduct to each other as central to morality. Of course, the notion of a fair and reasonable agreement is a moral notion and does an enormous amount of work here.

I shall not here attempt to adjudicate between Hobbesian, Lockean and Kantian contractarianism. Instead, I shall focus on general challenges to contractarianism.

Contractarianism is sometimes charged with ignoring the interests of beings – such as animals, infants and fetuses – that are not able to communicate linguistically, make commitments, and so on. Although some Hobbesian contractarian theories do ignore these interests, this is not an essential part of contractarianism. Some theories (such as Scanlon's) take these interests into account by allowing that trustees representing the interests of such beings are parties to the agreement.

A more fundamental criticism of contractarianism, often raised by Marxists, feminists and communitarians, is that it is *individualistic*. Contractarianism is indeed *normatively individualistic*, which is to say that it claims that the ultimate right-making features are features of individual people (viz. their consent), not irreducible features of collectivities. It does not, however, assume *ontogenetic* (or *developmental*) *individualism*, the view that denies that individual people are shaped and formed by the social context in which they find themselves. Nor does it assume *ontological individualism*, the view that individual persons are ontologically prior to society. Nor is contractarianism committed to the view that people are (inevitably or contingently) *egoistic* or *materialistic* in their desires (for example, caring only about the bundle of material goods that they control). Many contractarian theorists have made such assumptions, but such assumptions are not essential to contractarianism.

The main challenge to contractarianism questions the claim that *hypothetical* consent has any normative force. We generally deem it wrong for someone to take one's car without one's permission, even if one would have consented had one been asked. Moreover, in those cases where hypothetical consent seems to have some moral force, it may be because it is an indicator of what is in that person's *interests*. For example, my hypothetical consent for you to move my car away from the fire is an indication that it is in my *interest* that you do so. It may be that the appeal to my interests is what is doing the moral work here and that hypothetical consent does no real work.

Consent under the right conditions is clearly normatively significant. Whether actual or hypothetical consent can do all the work required of it for contractarian theories is a matter of ongoing debate.

PETER VALLENTYNE

See also **Buchanan, James M.; collective rationality; contract theory; Harsanyi, John C.; Hume, David; implicit contracts; justice; justice (new perspectives); market failure; Pareto** efficiency; Pareto principle and competing principles; social norms; tragedy of the commons.

Bibliography

Binmore, K. 1994. *Game Theory and the Social Contract. Volume 1: Playing Fair.* Cambridge, MA: MIT Press.

Binmore, K. 1998. *Game Theory and the Social Contract. Volume 2: Just Playing.* Cambridge, MA: MIT Press.

Buchanan, J. 1975. *The Limits of Liberty: Between Anarchy and Leviathan.* Chicago: University of Chicago Press.

Gauthier, D. 1986. *Morals by Agreement.* London: Oxford University Press.

Habermas, J. 1993. *Justification and Application: Remarks on Discourse Ethics,* trans. C. Cronin. Cambridge, MA: MIT Press.

Hampton, J. 1986. *Hobbes and the Social Contract Tradition.* Cambridge, MA: Cambridge University Press.

Harsanyi, J. 1953. Cardinal utility in welfare economics and the theory of risk taking. *Journal of Political Economy* 61, 434–5. Reprinted in J. Harsanyi, *Essays on Ethics, Social Behavior, and Scientific Explanation.* Dordrecht: Reidel, 1976.

Harsanyi, J. 1955. Cardinal welfare, individualistic ethics, and interpersonal comparisons of utility. *Journal of Political Economy* 63, 309–21. Reprinted in J. Harsanyi, *Essays on Ethics, Social Behavior, and Scientific Explanation.* Dordrecht: Reidel, 1976.

Hobbes, T. 1651. *Leviathan.* New York: Penguin Books, 1968.

Kant, I. 1785. *Groundwork of the Metaphysics of Morals.* New York: Harper Torchbooks, 1958.

Kavka, G. 1986. *Hobbesian Moral and Political Theory.* Princeton: Princeton University Press.

Locke, J. 1690. *The Second Treatise of Government.* Indianapolis: Bobbs-Merrill, 1952.

Plato. 1961. *The Republic.* In *The Collected Dialogues of Plato,* ed. E. Hamilton and H. Cairns. Princeton: Princeton University Press.

Rawls, J. 1971. *A Theory of Justice.* Cambridge, MA: Belknap Press of Harvard University Press.

Rousseau, J.J. 1762. *Contrat Social.* In *Political Writings,* ed. C.E. Vaughan. Cambridge: Cambridge University Press, 1915.

Scanlon, T.M. 1998. *What We Owe to Each Other.* Cambridge, MA: Belknap Press of Harvard University Press.

Simmons, A.J. 1993. *On the Edge of Anarchy.* Princeton: Princeton University Press.

social democracy

Social democracy refers to a political theory, a social movement or a society that aims to achieve the egalitarian objectives of socialism while remaining committed to the values and institutions of liberal democracy.

Born in an era of sharp ideological polarities and intense social conflicts, social democracy has often been seen as a pragmatic compromise between capitalism and socialism. As Leszek Kołakowski has put it: 'The trouble with the social-democratic idea is that it does not stock or sell any of the exciting ideological commodities which totalitarian movements – communist, fascist, or leftist – offer dream-hungry youth.' Instead of an 'ultimate solution for all human misfortune' or a 'prescription for the total salvation of mankind', said Kołakowski, social democracy offers merely 'an obstinate will to erode by inches the conditions which produce avoidable suffering, oppression, hunger, wars, racial and national hatred, insatiable greed and vindictive envy' (Kołakowski, 1982, p. 11). For much of the 20th century, the comparative modesty of this ambition led many to underestimate social democracy. Both the Marxist Left and the free market Right disparaged the purportedly unimaginative parties and trade unions that were the principal carriers of social democratic ideas, and scorned the vacillating compromises they saw as inherent in a social democratic analysis of politics. But if we review the historical development of social democracy, it becomes clear that such interpretations misjudge its character and significance. In fact, social democracy deserves to be recognized as one of the 20th century's most creative and durable influences on the politics and economics of the advanced industrialized nations.

Formation

What would later be called 'social democracy' first emerged in the late 19th century in the labour movements of north-west Europe. Early non-European outposts were also established in Australia and New Zealand around the same time. In nations such as Britain, France, Germany and Sweden, advocates of the interests of the working class inhabited polities that were characterized by rapid industrialization and the slow, inconsistent emergence of liberal constitutionalism and democratic citizenship. These circumstances created a complex structure of constraints and opportunities for labour movements that differed from those in southern or eastern Europe. In this relatively liberal environment, the politicized elements of the working class could build powerful political parties and trade unions to represent and protect their interests, and ultimately, or so they hoped, use democratic means to abolish the profound poverty and social oppression that working-class leaders saw as the ineluctable consequences of industrialization. The leaders and theorists of these movements, figures such as Keir Hardie, Jean Jaurès, Eduard Bernstein and Hjalmar Branting, were influenced by a variety of ideological traditions, most obviously Marxism, but also progressive liberalism, republicanism and 'utopian' socialism (for one account of such non-Marxist influences see Stedman Jones, 2004). They drew upon all of these intellectual currents as they began to sketch the outlines of a social democratic political theory. While important first approximations of this 'revisionist' socialism were articulated in the late 19th century by the Fabian Society and the Independent Labour Party in Britain, and by the republican socialists led by Jaurès in France (see Tanner, 1997; Berman, 2006, pp. 28–35), the frankest and most influential theoretical case for social democracy in this period was made by Bernstein, who explicitly confronted the forces of Marxist orthodoxy led by Karl Kautsky within the German Sozialdemokratische Partei Deutschlands (SPD). Bernstein's ideas, most fully expressed in his 1899 book *The Preconditions of Socialism*, laid the foundations for subsequent social democratic thinking by directly contesting two doctrines central to the self-understanding of Marxism during this period: historical materialism and the class struggle. Bernstein argued that capitalism was not doomed to collapse of its own accord, as a result of inevitable internal crises and the immiseration of the mass of the population. On the contrary, he thought that capitalism had in fact shown itself to be a flexible and adaptable economic system, capable of sustaining itself for the foreseeable future. While there was therefore no inevitability to the collapse of capitalism, continued Bernstein, it was certainly possible for significant modifications to be made to its structure through political action. In a democratic society, progressive social reforms could significantly improve the position of the working class. In this sense, Bernstein provocatively added, 'what is usually termed "the final goal of socialism" … is nothing to me, the movement is everything' (Bernstein, 1898, pp. 168–9). Bernstein was equally sceptical of the doctrine of class struggle. In contrast to the widespread Marxist assumption that socialism required the working class to monopolize political power, he stressed that cross-class alliances would be necessary for socialists to enter government, and that socialism was in any case best seen as addressed to the people as a whole rather than as an ideology tethered to only one social group. In these senses, Bernstein presented social democracy as the 'legitimate heir' of liberalism, with its aim being 'the development and protection of the free personality' (Bernstein, 1899, p. 147).

Bernstein's revisionism was extremely controversial, but his strategic vision appeared increasingly relevant to party leaders after the First World War, as socialist parties began to mobilize greater political support and found themselves on the cusp of power in many nations. The Russian revolution had now established a clear distinction between two different forms of socialist struggle, the reformist and the revolutionary. In response, the socialist parties of north-west Europe (and of Australia and New Zealand) were increasingly drawn towards reformism in practice, if not always in theory. Before the Second World War, however, the experience of such parties in government was for the most part short-lived and ineffective. In Britain, France and Germany, notionally socialist parties

endured traumatic periods in government. Faced by economic crisis and ultimately depression, these parties had few intellectual resources to draw upon as they found themselves fighting capitalist crisis armed only with socialist rhetoric. At the 1931 congress of the SPD, the trade union leader Fritz Tarnow aptly summarized the dilemma: 'Are we standing at the sickbed of capitalism not only as doctors who want to heal the patient, but also as prospective heirs who can't wait for the end and would gladly help the process along with a little poison?' (quoted in Berman, 2006, p. 110).

However, one party did manage to advance beyond this dilemma, and in effect forged the path that would later be followed by other socialist parties after 1945: the Swedish Socialdemokratiska Arbetarepartiet (SAP). In office more or less continuously from 1932, the SAP's achievement under their leader Per Albin Hansson was twofold. First, in political terms, the Swedish social democrats created a durable and popular political identity that encompassed the industrial working class but reached beyond this social constituency. In particular, the Swedish social democrats forged a cross-class alliance with the agrarian party, creating a coalition of workers and farmers that guaranteed the SAP's hold on office. Second, the social democratic-led governments of this period introduced a range of policies that would later be seen as virtually definitive of social democratic policy-making, although they were not generally regarded as such at this time. Two broad policy agendas were pursued in Sweden: first, the use of counter-cyclical measures to engineer an upturn in the economy. Influenced by the early (pre-*General Theory*) writings of Keynes and the so-called 'Stockholm School' of Swedish economists, the social democratic finance minister Ernst Wigforss undertook active state intervention such as public works to fight depression, in effect developing a form of 'proto-Keynesianism'. Similar remedies had been proposed within the Labour Party in Britain and the German SPD, but only in Sweden were senior social democratic politicians sufficiently open to these ideas and in a position to implement them. Although the Swedish economy's subsequent recovery was impressive, the role of the government's expansionary polices in this recovery is debatable; their importance as a political symbol is less so. In any case, as Erik Lundberg has argued, perhaps the real merit of the SAP-led government's economic programme 'consisted fundamentally in the avoidance of backward or unintelligent policy measures', something that certainly cannot be said of every government of this period (Lundberg, 1985, p. 9). Second, and complementary to this 'proto-Keynesianism', the Swedish social democrats introduced a range of social welfare measures, including unemployment insurance, a housing programme, and enhanced state pensions (for further discussion of this government, see Sassoon, 1996, pp. 42–6; Tilton, 1990, pp. 39–69; Berman, 2006, pp. 152–76).

By the outbreak of the Second World War, the parameters of social democratic ideology had been established. First, social democrats were committed to parliamentary democracy rather than violent insurrection or direct democracy. This not only meant that social democrats saw peaceful, constitutional methods as the best means of reforming capitalism, but also that they saw a system of parliamentary representation as the most plausible form of democratic government and the mass party as the best vehicle for aggregating and advancing their political objectives. These democratic commitments meant that in the early 20th century social democrats often led the struggle to expand the franchise to all men and women. Second, social democrats tailored their electoral appeals to the 'people' as a whole and not simply to one social class. From its inception, social democracy has been understood by its advocates as aiming at the construction of cross-class coalitions. A form of 'social patriotism' has dominated social democratic political discourse, which presented economic redistribution as synonymous with the national interest. As Per Albin Hansson famously argued in 1928, the Swedish social democrats sought to establish Sweden as a 'people's home' (*folkhemmet*) where 'no one looks down upon anyone else … and the stronger do not suppress and plunder the weaker' (quoted in Tilton, 1990, p. 127). Third, social democrats believed that it was above all through legislation and government policy that this vision of an egalitarian society would be realized (for further discussion of these three points, see Esping-Andersen, 1985, pp. 4–11; Przeworski, 1985). Animating all three of these basic social democratic assumptions was a distinctive political theory that sought to combine a strong commitment to individual freedom and democracy with the recognition that freedom and democratic participation can only be accessed by all citizens in circumstances of relative material equality (see Jackson, 2007). Nonetheless, there remained significant disagreement before the Second World War about precisely which policies could best advance these political ideals. Many on the Left ultimately believed that some form of collective ownership of capital was the most coherent route to greater equality. The policy instruments characteristically associated with social democracy, and pioneered in Sweden, had yet to be firmly established in the minds and hearts of social democrats themselves.

Golden age

The period from 1945 to the early 1970s is often referred to as the 'golden age' of social democracy. One difficulty with this characterization is that social democratic parties were not always in government when many of the reforms identified as 'social democratic' were actually implemented. Although the electoral prospects of social democratic parties were considerably brighter than before the war, this was not a period in which social democratic parties established electoral hegemony (except in Scandinavia). A second difficulty with describing this period as a golden age for social democracy is that it could be countered that these years were above all a period of extraordinary success and dynamism for capitalism. The golden age, it might be argued, was in fact a long post-war capitalist boom, which brought sustained economic growth and rising living standards for all. Social democracy could therefore be said to be basically irrelevant to this more fundamental economic trend. Both of these caveats need to be taken seriously, but in my view they do not decisively undermine the case for identifying this era as a broadly social democratic one.

Three crucial post-war developments in the political economy of the advanced industrialized nations have rightly been seen as illustrating the powerful influence of social democracy in this period: the establishment of 'social citizenship' in Western Europe; the emergence of full employment as a legitimate, and in some cases pre-eminent, objective of government policy; and the increased role given to trade unions in economic policymaking. Post-war social conditions bred much greater popular support for these initiatives than had existed pre-war: the radicalizing impact of a 'people's war' against fascism, coupled with deep-seated hostility towards the maladies of unregulated capitalism manifested in the 1930s, meant that both the public and political elites were receptive to new policy frameworks of a broadly social democratic kind. The most influential authors of such frameworks, William Beveridge and John Maynard Keynes, identified themselves as progressive liberals (although it should be noted that in the 1940s, Beveridge was much more sympathetic to socialism than is often recognized: see Harris, 1997, pp. 428–43, 480). However, while the ideas of Beveridge and Keynes significantly modified earlier social democratic thinking, there was no doubt that their proposals meshed with basic social democratic aspirations, enhanced the economic credibility of social democracy as a model for public policy, and offered a practical way forward for parties of the Left previously committed to an imprecise socialist rhetoric. Conservative parties did their best to keep pace with the perceptible march leftwards of public opinion and expert advice, but as they did so they inevitably abandoned their own ideological terrain. The ideological common sense of the age was now based around progressive taxation, public spending and state intervention in the economy.

Social democrats had long emphasized the need to reshape the pattern of resource distribution thrown up by the market so that all citizens could exercise the rights associated with democratic citizenship and be 'decommodified', that is have access to an income based on their needs, rather than on their success or otherwise in the labour market (Esping-Andersen, 1990, pp. 21–3). The most advanced welfare states of the immediate post-war period – in Britain and Sweden – were created by social democratic governments. The 1945–51 British Labour government led by Clement Attlee enacted William Beveridge's vision of the welfare state: unemployment insurance, pensions and family allowances all largely followed Beveridge's lead, while Aneurin Bevan's universal and tax-funded National Health Service built on and in some respects deepened Beveridge's ideas. By 1950, the British sociologist T.H. Marshall could write that in Britain social rights had been added to the civil and political rights of each citizen. 'Social citizenship' was now a reality (Marshall, 1950). The post-war SAP governments in Sweden were slower off the mark than the Labour Party in undertaking welfare measures, but by the 1950s the celebrated universalism of the Swedish welfare state had begun to emerge. Most importantly, an expanded pension scheme introduced in 1959, the ATP, gave the middle class a stake in the welfare state. As in the 1930s, the SAP were shrewd at building cross-class alliances, this time breaking with the farmers to create a wage-earners' coalition between blue- and white-collar employees, organized around universal, but also earnings-related, pension rights (Esping-Andersen, 1985, pp. 108–10; Eley, 2002, pp. 318–9).

Of course, the welfare state was not exclusively a social democratic initiative: other ideological traditions and political actors played a role in its creation, notably the Christian Democratic parties of Austria, Germany and Italy. But social democracy possesses a distinctive vision of welfare, which has dominated the post-war political trajectory of certain nations (notably in Scandinavia) and which has also been influential in other national welfare systems (for example, Britain). The social democratic welfare state is universal in scope and egalitarian in its impact, rather than targeted as a residual measure at those on low incomes. And, even where these social democratic features are absent from a nation's social policy, it has often been the electoral threat from the Left that has prompted the Right to undertake its own social reforms.

The second policy development of this period that can fairly be classified as social democratic is the emergence of full (male) employment as a central political goal. The 'right to work', an old leftist slogan, was obviously connected to the 'social rights' discussed above. The post-war vision of social citizenship was premised on the assumption that every able-bodied male citizen would be in paid work, both as a matter of social justice, and for the more pragmatic reason that income tax and social insurance payments were needed to fund the new welfare state. The advantages of full employment from a social democratic perspective were clear. As the economist Thomas Balogh wrote, full employment 'removes the need for servility, and thus alters the way of life, the relationship between the classes. It changes the balance of forces in the economy' (quoted in Scharpf, 1991, p. 16). In comparison to the labour movement's impotence under the mass unemployment of the 1930s, the more or less full employment of the golden age strengthened the hand of labour in conflicts with capital, significantly increasing the bargaining power of wage-earners.

Prior to the golden age, social democrats had usually maintained that only a substantial socialization of the economy would enable governments to secure full employment. The post-war influence of Keynes was decisive in shifting social democratic thinking on this question, and in enabling social democrats to formulate a radical but economically credible policy agenda. Keynes was thought to have demonstrated that state regulation and intervention could in fact stabilize capitalism, since full employment and steady economic growth could be maintained by an expansionary fiscal policy aimed at sustaining economic demand in periods of economic downturn. This conviction was apparently vindicated by the economic boom of the post-war years, although it is doubtful whether some of the policies conventionally labelled as 'Keynesian' played a significant role in sustaining this period of economic stability. Discretionary government intervention and deficit spending did not play a large role in the political economy of the social democratic golden age. Indeed, leading social democratic nations such as Norway and Sweden had the largest budget surpluses in the Organisation for Economic Co-operation and Development (OECD) in the 1960s. More significant was the growth of public spending: the rise of the welfare state was itself a measure that helped to boost aggregate demand and stabilize the business cycle (Glyn, 1995, p. 42).

Cordial relations with the trade union movement were critical to the success of Keynesian policymaking, since under conditions of full employment some form of wage restraint was necessary to avoid inflationary pay settlements. Trade unions now emerged as central players in post-war politics and corporatist wage-bargaining was a key feature of the new economic order. Once again, Sweden was in the vanguard of this development. Two trade union economists, Gösta Rehn and Rudolf Meidner, developed the so-called 'Rehn–Meidner model', which incorporated an active labour market policy and solidaristic wage bargaining into a comprehensive economic strategy designed to maintain growth and employment, lower inflation and narrow income inequality. The Rehn–Meidner model was adopted by the SAP in the late 1950s and was an important influence on the policy of successive social democratic governments in Sweden. Under these governments, the principle of 'equal pay for equal work' was implemented, so that, once the fair rate for a certain job had been agreed by employers and the unions, less efficient firms could no longer enjoy the subsidy of paying their employees lower wages than those undertaking the same job elsewhere in the economy. Poorly performing firms now had an incentive to improve their efficiency or they would go out of business; the active labour market policy was designed to find work for those who found themselves unemployed as a result. Social democratic objectives were therefore harnessed to an attempt to improve economic efficiency and engineer a more dynamic economy (see Turvey, 1952; Tilton, 1990, pp. 189–214; Sassoon, 1996, pp. 203–6). Although no other social democratic party developed an economic strategy of comparable sophistication to the Rehn–Meidner model, the greater involvement of trade unions in economic policy-formation, and the widespread negotiation of agreements over pay and conditions between employers and trade unions, can be seen as a further example of the influence of social democracy after the Second World War.

It is unlikely that, shorn of these three distinctively social democratic elements, post-war capitalism could have secured the same progressive economic outcomes visible in the golden age: steady economic growth, low unemployment, low inflation, rising wages and narrowing income inequality. This radical shift in the character of capitalism inevitably prompted lively debate among social democrats about their political objectives. While many social democrats had earlier assumed that their ultimate objective (however distant) was the creation of a new order called 'socialism' (and thus the abolition of capitalism), it now seemed that many of their aspirations could in fact be realized in the context of a reformed capitalism. Reflecting on this point, and on the electoral failures of some social democratic parties in the 1950s and early 1960s, reformist leaders began to articulate a new social democratic revisionism, initiating fresh theoretical controversy with the Left of their parties. The SPD ostentatiously settled its account with Marxism in 1959, with its adoption of the Bad Godesberg Programme. This new statement of SPD priorities dispensed with the socialist end goal that had previously been a constitutive feature of party rhetoric and accepted that what had previously been seen as short-term goals were now exhaustive of social democratic objectives: full employment, a just distribution of wealth, and consistent economic growth. Likewise, in the British Labour Party, revisionists such as the party leader Hugh Gaitskell and his ally Anthony Crosland set out the case for a democratic socialism that discriminated more carefully between ends and means. They defined socialist ends as certain durable ethical principles, in particular the pursuit of greater economic equality, rather than as the attainment of a particular model of economic organization. Equality, argued the British revisionists, could in fact be realized through a variety of means, including some socialization of industry or collective capital ownership, but also through a strong welfare state, the regulation of the labour market, Keynesian economic management, and the progressive taxation of income and wealth. A compelling intellectual case for this was set out in Crosland's 1956 book *The Future of Socialism*, which has remained an essential reference point for debates about social democratic strategy in Britain ever since.

By the late 1960s, revisionist social democracy had therefore acquired a sharper theoretical definition and had some significant policy achievements to its credit. A revival in the electoral popularity of both Willy Brandt's SPD and Harold Wilson's Labour Party saw both parties in government as the 1960s drew to a close. But by then the golden age was also nearing its end.

Crisis and adaptation

While social democracy was certainly an indispensable ingredient of the social capitalism that permeated western Europe after the Second World War, it would be foolish to neglect the importance of capitalism itself to this distinctive phase of economic history. Strong economic growth and rising productivity had provided the necessary economic conditions for expansive social reforms; by the mid-1970s, these conditions were no longer in place. Instead, social democrats were placed under severe pressure by a threefold crisis.

First, a global economic slowdown was triggered by various factors, including a general decline in the growth of labour productivity and increases in the price of raw materials after the OPEC oil price rises of 1973 and 1979. Growing industrial conflict between unions and employers added to the gloomy economic outlook. The result was a period of rising unemployment and inflation, and low levels of economic growth, across all industrialized nations.

Second, the 1970s saw the beginning of a marked change in the class structure of industrialized economies and a feminization of their workforces. The core constituency of social democratic parties – workers in manufacturing industries – declined as a proportion of the labour force, while there was a substantial increase in the proportion of employees working in service industries. At the same time, there was a steady rise in female employment rates, particularly in the service sector. The male breadwinner model that had been taken for granted in earlier social democratic thinking was now outdated, while the shift to service-sector employment, or so it was claimed, fractured class solidarity and thus the electoral base of social democracy. The growth of service sector employment also created problems for social democratic economic policy: increased employment in private sector services was associated with greater wage inequality, which raised the prospect of a trade-off between the two social democratic goals of full employment and greater income equality. By the 1990s, one social democratic route to resolving this trade-off, namely, increasing employment in public sector services (the policy pursued in Sweden in the 1970s and 1980s) was said to be ruled out by the budgetary constraints imposed on governments by increasingly tax-resistant electorates and financial markets nervous of budget deficits (Iversen and Wren, 1998).

Third, in response to the previous two developments, the intellectual basis of social democracy was subjected to a concerted and ingenious attack. A 'New Right' emerged, which was quick to attribute the blame for the economic downturn of the 1970s to ham-fisted Keynesian intervention in the market, over-powerful trade unions enforcing inflationary pay settlements and wasteful public expenditure on welfare programmes, since the latter undermined individual responsibility and required efficiency-inhibiting levels of direct taxation. Although primarily influential in English-speaking countries, the 'neoliberal' prescriptions of austere counter-inflationary measures, welfare state retrenchment, privatization and deregulation set the tone for the discussion of economic policy throughout the industrialized nations in the 1980s and 1990s. The policymaking autonomy of national governments was in any case reduced in this period by the (neoliberal-inspired) relaxation of capital controls across the OECD in the 1980s and the associated growth in capital mobility across national boundaries.

These developments led many to claim that the prospects for social democracy within one country were bleak, a point apparently underlined by the electoral dominance of conservative parties in some nations during the 1980s and 1990s, in particular Britain and Germany. There is no doubt that social democrats were defeated on a number of important issues in this period and were forced to adapt their programmatic objectives to take account of the changed economic and political context. But did this necessarily signal the end of the basic social democratic aspirations discussed earlier? There are three reasons for thinking this latter conclusion is too strong. First, although clearly a period of social democratic retreat, social democratic parties remained electorally viable. Second, some established social democratic institutions remained resilient in the face of neoliberal attack. Third, a new revisionism had developed by the end of the 20th century, which sketched out a plausible, if modest, agenda for the pursuit of greater economic equality and full employment in the revivified global capitalism of the new millennium. This survey of social democracy will conclude by elaborating on these three points.

First, then, although frequently remembered as a period of electoral failure for the Left, in fact the 1970s, 1980s and 1990s also saw some significant social democratic victories. France was led by its first socialist head of state, François Mitterrand, from 1981 to 1995; in newly democratized Spain the socialist party under Felipe González remained in power for over a decade after 1982; in Australia Bob Hawke and then Paul Keating presided over the longest ever period of Labor government, from 1983 to 1996. The SAP largely maintained its grip on power in Sweden (albeit with periods out of office between 1976–82 and 1991–4), while the Austrian social democrats were in office throughout the 1970s and

1980s, either as a single-party government or in coalition. By the 1990s, after long periods in opposition, the British Labour Party under Tony Blair and the SPD under Gerhard Schröder had returned to power. This evidence undermines the crude sociological determinism that associates the electoral fortunes of social democratic parties with the proportion of manual workers in the labour force. In fact there has never been an obvious correlation between social democratic party support and the size of the working class. Social democratic parties have had low levels of support when there has been a high proportion of manual workers in the labour force, for example the SPD in the 1950s and 1960s, and have had high levels of support when that proportion has been in decline, for example the SAP after the 1970s (Kitschelt, 1994, p. 41). Of course, electoral success is by itself an insufficient measure of social democratic resilience. All the governments of the Left just mentioned have faced criticism for conceding too much ground to the Right, and some have been accused of implementing the dictates of neoliberal capitalist restructuring at the expense of pursuing a distinctively social democratic course. How valid are these claims?

Here we reach the second point mentioned earlier: although a perceptible swing to the Right occurred in public policy in this period, sometimes under nominally social democratic governments, this backlash was never far-reaching enough to undo the most deeply entrenched of the social democratic institutions established after 1945. Although the welfare state and progressive taxation of income were placed under considerable pressure, they remained at the heart of the public understanding of social justice in most industrialized nations with significant labour movements. Even in Britain, where the Thatcher government carried out the most thorough attack on the achievements of the golden age, Bevan's National Health Service remained intact, and quite substantial redistribution continued to take place through the tax and benefit system. As the 21st century began, 12.3 per cent of the UK population were living in poverty, but the tax and benefit system still reduced by 61 per cent the number of British citizens left in poverty by the market (Glyn, 2006, p. 171). Between 1980 and 2001 social spending actually increased as a share of GDP across the OECD. Although the rate of increase was slower than in previous decades, in this period social spending in northern European countries, those most influenced by social democracy, increased more than in 'liberal' economies (Glyn, 2006, pp. 165–6). While the scale of the free-market triumph should therefore be kept in proportion, equally the problems that confronted social democracy as traditionally understood should not be underestimated. In particular, after the 1970s the attainment of full employment became a much more elusive and apparently intractable political goal. Some social democratic governments, for example in Austria and Sweden, had initially had some success in responding to the downturn in the 1970s with a classical Keynesian reflation, but in general this period saw an increase in unemployment rates, and greater scepticism among both policymaking elites and public opinion about the capacity of governments to maintain full employment, or at least about whether it was possible to do so while simultaneously narrowing inequality.

Third, this more pessimistic political climate prompted a fresh attempt to revise party programmes in order to adapt social democratic politics to this new 'post-industrial' era. The resulting 'post-industrial social democracy' can be distilled into three elements (see Vandenbroucke, 2001; White, 1999). First, the acceptance of certain constraints on the policymaking autonomy of national governments and of some strict assumptions about how governments can maintain economic growth and stability. In particular, social democratic politicians ruled out the use of large budget deficits and significant tax increases to fund social democratic goals. Although this was sometimes presented as the result of immutable economic changes (for example 'globalization'), it is perhaps more accurate to see this commitment as a political one. After the tumultuous events of the 1970s, social democratic parties were obliged to impress on both financial markets and the electorate their competence as economic managers and to signal that there would be no return to the high inflation and sluggish economic growth that marked the end of the golden age. While this commitment undoubtedly set strict limits on the activities of social democratic governments, it nonetheless still left some latitude for egalitarian policy interventions (remember, for example, that budget deficits were never a core feature of social democratic policymaking in the golden age). The second element of this new revisionism therefore stressed the importance of supply-side measures to improve both economic efficiency and social justice. Social democrats became increasingly interested in supplementing *ex post* redistribution through the welfare system with an attempt to equalize the *ex ante* distribution of financial assets and human capital. In particular, investment in education and training, and active labour market programmes, were seen as key instruments that would both boost employment rates and improve productivity. Similarly, social democratic policymakers expressed some interest in ensuring a more equitable distribution of financial assets through universal 'stakeholder' grants (Paxton, White and Maxwell, 2006). Third, the reduction of poverty and inequality through the welfare state and labour market regulation remained a social democratic priority, but with a distinctive focus on two issues: first, ensuring that the welfare state contributes to the reduction of unemployment by improving the incentives for low-paid workers to enter, and remain in, the labour market; and second, ensuring that the welfare state adequately supports family life and female participation in the workforce. The increased use of in-work benefits such as earned income tax credits was proposed to address the first of these issues, while the expansion of welfare systems to include universal nursery provision (an objective already accomplished in Scandinavia) was proposed to assist with the second. Although these aspirations were more modest than the equivalent programmes of social democratic parties in the 'golden age', they were nonetheless recognizably social democratic, and in crucial respects departed from neoliberalism. In particular, the revisionist emphasis on increasing employment through supply-side measures and generous in-work benefits offered an answer to the claim that in the 'post-industrial' economy there must be an inevitable trade-off between employment and equality. However, there was undoubtedly a tension at the heart of this revisionism: these new methods of reducing inequality and unemployment still required significant social spending, but the acceptance of tax resistance and budgetary restraint limited the resources available to social democratic governments. Critics were correct to point out that a social democracy cannot be obtained on the cheap.

At the beginning of the 21st century those countries most deeply influenced by social democracy, in essence the Scandinavians, still differed significantly from those in which social democracy had penetrated less thoroughly. In Sweden, 6.4 per cent of the population lived in poverty; in Britain, the equivalent figure was 12.3 per cent, and in the United States 17 per cent (Glyn, 2006, p. 171). As Kołakowski has suggested, social democracy 'has invented no miraculous devices to bring about the perfect unity of men or universal brotherhood' (Kołakowski, 1982, p. 11). But when we review the history of social democracy during the 20th century it is clear that it has some non-negligible achievements to its credit. In particular, social democrats have greatly improved the position of the disadvantaged in the industrialized nations, and have ensured that the interests of the poor were represented in political systems formerly monopolized by the rich.

BEN JACKSON

See also **Bernstein, Eduard; Beveridge, William Henry; Fabian economics; Keynesianism; socialism; welfare state.**

Bibliography

Berman, S. 2006. *The Primacy of Politics*. Cambridge: Cambridge University Press.

Bernstein, E. 1898. The theory of collapse and colonial policy. In *Marxism and Social Democracy: The Revisionist Debate 1896–1998*, ed. H. Tudor and J.M. Tudor. Cambridge: Cambridge University Press, 1988.

Bernstein, E. 1899. *The Preconditions of Socialism*. Cambridge: Cambridge University Press, 1993.

Crosland, C.A.R. 1956. *The Future of Socialism*. London: Jonathan Cape.

Eley, G. 2002. *Forging Democracy: The History of the Left in Europe*. Oxford: Oxford University Press.

Esping-Andersen, G. 1985. *Politics against Markets: The Social Democratic Road to Power*. Princeton: Princeton University Press.

Esping-Andersen, G. 1990. *The Three Worlds of Welfare Capitalism*. Princeton: Princeton University Press.

Glyn, A. 1995. Social democracy and full employment. *New Left Review* 211, 33–55.

Glyn, A. 2006. *Capitalism Unleashed*. Oxford: Oxford University Press.

Harris, J. 1997. *William Beveridge: A Biography*. Oxford: Oxford University Press.

Iversen, T. and Wren, A. 1998. Equality, employment and budgetary restraint: the trilemma of the service economy. *World Politics* 50, 507–46.

Jackson, B. 2007. *Equality and the British Left*. Manchester: Manchester University Press.

Kitschelt, H. 1994. *The Transformation of European Social Democracy*. Cambridge: Cambridge University Press.

Kołakowski, L. 1982. What is living (and what is dead) in the social-democratic idea? *Encounter* 58, 11–17.

Lundberg, E. 1985. The rise and fall of the Swedish model. *Journal of Economic Literature* 23, 1–36.

Marshall, T.H. 1950. *Citizenship and Social Class*. Cambridge: Cambridge University Press.

Paxton, W., White, S. and Maxwell, D. 2006. *The Citizen's Stake*. Bristol: Policy Press.

Pierson, C. 2001. *Hard Choices? Social Democracy in the 21st Century*. Cambridge: Polity Press.

Przeworski, A. 1985. *Capitalism and Social Democracy*. Cambridge: Cambridge University Press.

Sassoon, D. 1996. *One Hundred Years of Socialism*. London: Fontana.

Scharpf, F. 1991. *Crisis and Choice in European Social Democracy*. Ithaca: Cornell University Press.

Stedman Jones, G. 2004. *An End to Poverty? A Historical Debate*. London: Profile.

Tanner, D. 1997. The development of British socialism 1900–18. In *An Age of Transition: British Politics 1880–1914*, ed. E.H.H. Green. Edinburgh: Edinburgh University Press.

Tilton, T. 1990. *The Political Theory of Swedish Social Democracy*. Oxford: Oxford University Press.

Turvey, R. 1952. *Wages Policy under Full Employment*. London: William Hodge.

Vandenbroucke, F. 2001. European social democracy and the third way: convergence, divisions, and shared questions. In *New Labour: The Progressive Future?* ed. S. White. Basingstoke: Palgrave.

White, S. 1999. Social liberalism, stakeholder socialism and post-industrial social democracy. *Renewal* 7, 29–38.

social discount rate

Discount rates are required to evaluate the future costs and benefits of economic policies. It is widely recognized, for instance, that a dollar today is worth more than a dollar five years from now. But how much more?

All cost–benefit analyses with a temporal element must choose a rate or rates of discount. Discount rates are especially important for evaluating global warming, the loss of biodiversity, hazardous waste disposal, and related environmental issues. Infrastructure investments, such as roads, bridges, and tunnels, often last for generations. In these cases many costs and benefits come in the relatively distant future.

Consider Table 1, which compares present and future benefits using various positive rates of discount:

Just as economists use discount rates, so do they require a rate of compounding when measuring compensation for past injustices. For instance, if we are to make restitution for some previous loss, we must decide how to convert past dollar amounts into a present sum.

Economists (for example, Lind et al., 1982) typically suggest that a discount rate for dollars should have two components: the productivity of capital, and the social rate of time preference. The productivity of capital refers to our ability to turn present resources into a larger future value. If, for instance, we have a dollar today we can invest it and reap greater value in the future, at least on average. This is one reason why current resources are worth more than future resources. The social rate of time preference refers to our impatience. Even if we cannot invest capital productively, many of us would rather consume today than one year from now. This provides another reason why current resources might be of greater value than future resources.

Observed market rates of interest will not reflect these values with complete accuracy. First, governments typically tax the return to capital. The social return on capital is therefore greater than the private return. Second, private capital markets face some degree of transactions costs, credit rationing, and bureaucratic regulation. Once again, the social return to capital is likely higher than the private return. Third, government taxation and borrowing (to fund projects) will involve deadweight loss and crowding out. All of these variables should be reflected in any proper measure of the cost of capital, and for these

Table 1 *Estimated number of future benefits equal to one present benefit based on different discount rates*

Years in the future	1%	3%	5%	10%
30	1.3	2.4	4.3	17.4
50	1.6	4.3	11.4	117.3
100	2.7	19.2	131.5	13,780.6
500	144.7	2,621,877.2	39,323,261,827	4.96×10^{20}

reasons we should not simply take market interest rates as given.

A related debate concerns whether governments should take risk into account when using a social discount rate. Kenneth Arrow (1971) argued that the government should use a riskless rate. A riskless rate, of course, will be much lower. Short-term US Treasury securities often yield no more than a 1 per cent real rate of return; the average rate of return on private capital in the United States can run as high as 10 or 15 per cent.

According to Arrow's reasoning, government faces little or no financial risk from bad policies, given that it can spread costs across a very large number of taxpayers. Nonetheless, this argument has been criticized from at least two directions. First, private corporations can spread their risks across a large number of diversified shareholders. Second, Arrow's argument focuses too much on the purely financial side of risk. Financial losses on a project are, taken alone, mere transfers. The relevant risks include possible variation in the value of the project for its intended beneficiaries. When one measures these risks, the absolute size of the government, or the number of taxpayers, is of little concern.

Changing discount factors over time may lower the effective social discount rate (Weitzman, 2001). In this case the lowest discount rate will have a highest contribution to an expected value calculation made from the vantage point of today. Richard Posner (2004, pp. 153–4) summarizes the argument:

> Suppose there's an equal chance that the applicable interest rate throughout this and future centuries will be either 1 percent or 5 percent. The present value of $1 in 100 years is 36.9 cents if the interest rate used to compute the present value is 1 percent but only .76 cents (a shade over three-quarters of a cent) if it is 5 percent. Now consider the 101st year and remember the assumption that the two alternative discount rates are equally probable. If the interest rate used to discount the future to the present value is 1 percent, then the present value of $1 at the end of that year will have shrunk from 36.9 cents to 36.6 cents. If instead the interest rate used is 5 percent, the present value of .76 cents will have shrunk to about .75 cents. This means that the *average* present value of $1 at the end of the 101st year will be 18.68 cents, implying an average discount rate of less than 2 percent, rather than 3 percent. The reason is that the more rapid decline in value under the higher discount rate (5 percent) reduces its influence on present value.

In other words, when there is uncertainty about future discount rates, the lower rates have a greater relative weight the further we look into the future.

In recent times 'gamma discounting' has become a popular approach. Weitzman (2001, p. 260) notes, 'even if every individual believes in a constant discount rate, the wide spread of opinion on what it should be makes the effective social discount rate decline significantly over time.' This is for the same reason as discussed above. More generally, many individuals will argue that the present is more important than 30 years from now, but that after some point further differences in time should cease to matter. Perhaps what happens 300 years from now is not much less important than what happens 200 years from now. This view has found significant support in polls of both ordinary and distinguished economists (Weitzman, 2001).

Most of the debate has focused on discounting dollars; but the discounting of utility is a separate issue. Most likely, the discount rate on utility should be lower than the discount rate on dollars. Utility is not 'productive' over time as is invested capital; we are therefore left with only the rate of time preference as a discount factor. And for inter-generational policies, the rate of social time preference is arguably zero. Before individuals are born, they do not experience a disutility of waiting (Cowen and Parfit, 1992).

Another question is how much discounting should apply to very large changes in individual welfare. Recall that cost–benefit analysis is best suited to analysing small changes at the margin, where market prices (adjusted for risk, taxes, and transactions costs) measure values. So it is reasonable to argue that a dollar today is worth more than a dollar 20 years from now. But what if we are discounting lives across hundreds of years? Is a single life today really worth more than the entire survival of the human race 500 years from now? It is quite easy to generate such a conclusion, for reasonable parameter values, through the uncritical application of discounting. In these cases it appears that a more explicit ethical judgement should override the cost–benefit approach.

In sum, the debate over discount rates has seen continual progress. Most economists now agree that there is no single correct rate of social discount for all problems. Instead, the proper discount rate depends on the problem under consideration, the time horizon, the assumptions about risk, and the magnitude of the associated costs and benefits. A social discount rate is a very useful tool, and it can nicely complement our broader normative judgements. But it does not remove the need to keep the relevant ethical issues in mind.

TYLER COWEN

See also **risk; term structure of interest rates; time consistency of monetary and fiscal policy.**

Bibliography

Arrow, K. 1971. *Essays in the Theory of Risk-Bearing.* Amsterdam: North-Holland.

Cowen, T. and Parfit, D. 1992. Against the social discount rate. In *Justice Across the Generations: Philosophy, Politics, and Society, Sixth Series*, ed. P. Laslett and J. Fishkin. New Haven: Yale University Press.

Lind, R. et al. 1982. *Discounting for Time and Risk in Energy Policy*. Baltimore, MD: Johns Hopkins University Press.

Posner, R. 2004. *Catastrophe: Risk and Response*. Oxford: Oxford University Press.

Weitzman, M. 2001. Gamma discounting. *American Economic Review* 91, 260–71.

social insurance

Individuals face uncertainty about their labour income. Chronic medical conditions can arise that reduce or remove their ability to work. Alternatively, unemployment spells can occur, temporarily reducing their earnings. The individual burden of these risks can be eased by private insurance and by social insurance schemes. A social insurance scheme is a government transfer programme whereby individuals who claim a condition or state that reduces their labour income, such as disability or unemployment, obtain a transfer from the government for the duration of this state. Social insurance can also be carried out through income taxation.

The design of optimal social insurance is a classic problem in economics. This problem is interesting because neither private insurance or public arrangements can fully distinguish between low income by choice and low income by necessity. For example, the diagnosis of medical factors is often subjective, and symptoms for certain conditions, such as back pain or recurrent migraines, are hard to verify. If full insurance for chronic medical conditions were available, individuals could exploit this by reporting fake symptoms to claim insurance benefits and stop working. Similarly, an unemployed worker's ability to find new employment depends on the effort she exerts in the job search. If full unemployment insurance is available, an unemployed worker would have no incentive to search for a new job if her search effort cannot be monitored. Private information on earning ability implies that full insurance might defeat itself by removing the incentive to work. Hence, the essential trade-off in the design of optimal social insurance schemes is the one between risk sharing and incentives.

The trade-off between insurance and incentives

Consider the following simple social insurance problem (the set-up for this example is adapted from Diamond and Mirrlees, 1978). Individuals live for two periods. In the first period of their lives, they are endowed with one unit of consumption and they consume. In the second period, they consume and they may work if they are able to. The probability of being able to work is π, a number between zero and 1. Work is publicly observable and the probability distribution of ability is known, but ability is private information. Preferences over consumption in each period are represented by $u(c)$, where $c \geq 0$ is consumption and the function u is strictly increasing and strictly concave. The utility cost of working if able is γ, a number strictly greater than zero. An individual produces one unit of output if she works, zero otherwise.

The government wishes to induce able agents to work in period two and to maximize their lifetime expected utility, given by:

$$u(c_1) + \pi u(c_2^w) + (1 - \pi) u(c_2^n) - \pi \gamma,$$

where c_1 is consumption in period one and c_2^w and c_2^n are consumption conditional on work and no work in period two.

There is no private insurance, but the government is able to pool risk by distributing consumption, conditional on work in period two, so that the expected value of lifetime consumption equals the expected value of lifetime work for any individual. Individuals cannot save but the government has access to a storage technology with return $R > 0$. Hence, an individual's lifetime consumption profile must satisfy the constraint:

$$Rc_1 + \pi c_2^w + (1 - \pi) c_2^n \leq R + \pi. \tag{1}$$

Since the government cannot observe ability, the consumption profile must also satisfy an incentive compatibility constraint:

$$u(c_2^w) - \gamma \geq u(c_2^n), \tag{2}$$

to ensure that able individuals will work.

The optimal social insurance scheme satisfies the following two inequalities when the incentive compatibility constraint (2) is binding:

$$\frac{u'(c_2^w)}{u'(c_2^n)} < 1, \tag{3}$$

$$\frac{u'(c_1)}{Eu'(c_2)} < R, \tag{4}$$

where $Eu'(c_2)$ is the expected marginal utility of consumption in period two.

It immediately follows from (3) that full insurance is not possible and $c_2^w > c_2^n$. Hence, private information implies that the first-best optimum is not attainable. Equation (4) is derived from the government's Euler equation and uncovers another dimension of the incentive problem. It implies that the consumption path associated with the optimal insurance scheme displays a wedge between the marginal intertemporal rate of substitution and the intertemporal rate of transformation, R. This intertemporal wedge indicates that the marginal cost of transferring consumption to period two is greater than the value of forgone consumption at time one. The additional cost stems from the need to maintain incentive compatibility. This requires consumption to be conditional on work and therefore stochastic in the second period. Since utility from consumption is strictly concave, a given increase in expected utility requires more resources when consumption is stochastic rather than deterministic.

The intertemporal wedge implies that individuals would prefer to reduce their consumption in the first period. Thus, in general, the optimal intertemporal consumption profile cannot be obtained without preventing access to the capital market or imposing a tax on savings. For example, Golosov and Tsyvinski (2006) study optimal disability insurance in competitive equilibrium and show that the intertemporal wedge requires disability benefits to be asset-tested.

Private information on individuals' ability to work gives rise to adverse selection in the social insurance problem. The resulting partial insurance and the presence of an intertemporal wedge (4) hold generally with adverse selection (Golosov, Kocherlakota and Tsyvinski, 2003) and also characterize social insurance under moral hazard (Rogerson, 1985), as in the unemployment example described in the introduction. Moreover, (4) exemplifies a general feature of social insurance with private information, that is, that the government would want to control trade in related commodities (Atkinson and Stiglitz, 1976; Varian, 1980).

Long-run inequality

Social insurance models with infinite horizons generate normative implications for long-run consumption inequality. The key step in deriving these implications is to formulate the social insurance problem recursively. The pioneering work of Green (1987) provides an early example of a recursive solution to a dynamic social insurance problem with infinite horizon and constant discounting. The main insight is to restate the optimal social insurance scheme as a contract between the government (the principal) and individuals (the agents). Such a contract retains memory of the history of outcomes and assigns current transfers and promises of future transfers based on that history and the current outcome. Despite their large dimensionality, such histories can be summarized by agents' *promised value*. (These results can be found in Spear and Srivastava, 1987; Thomas and Worrall, 1990; Abreu, Pierce and Stacchetti, 1990; Phelan and Townsend, 1991.) This one-dimensional object, which corresponds to an agent's expected lifetime utility at a point in time, encodes an agent's history and permits a recursive formulation of the problem. Incentive compatibility implies that current transfers and promises of future transfers will depend on promised utility and on the current outcome. The government's promises of future transfers can be represented as *continuation values* that determine agents' promised utility in the subsequent period.

For example, in an infinite horizon version of the earlier disability insurance example, the individual history for some period $t > 1$ is the sequence of work outcomes from period 1 to period $t - 1$. This can be summarized by an individual's expected lifetime utility at the beginning of time t, her promised value. The optimal consumption allocation at time t and the continuation value will depend on the promised value and on the current work outcome. The continuation value corresponds to an individual's expected lifetime utility in the subsequent period. Hence, individuals' future consumption depends on current work and on the past history of work.

The history dependence of the consumption path resulting from an optimal insurance scheme implies that the trade-off between insurance and incentives shapes the evolution of the consumption distribution. Specifically, incentive compatibility implies that not only current transfers but also continuation values will be conditional on current outcomes. An immediate consequence is that the degree of consumption inequality tends to continually increase. An additional implication can be derived from the intertemporal wedge, which implies that individuals will face a downward-sloping path of promised lifetime utility under the optimal social insurance scheme.

Taken together, these results give rise to an extreme conclusion. Consumption inequality should grow without bound, with all individuals in the population converging to their minimum promised lifetime utility, except for a vanishing fraction converging to their bliss point. This *immizeration* property is robust. It obtains in partial (Green, 1987; Thomas and Worrall, 1990) and general (Atkeson and Lucas, 1992) equilibrium, under weak assumptions on preferences (Phelan, 1998), and holds in adverse selection and, in somewhat weaker form (Pavoni, 2004), in moral hazard environments.

The radical implications for consumption inequality generated by optimal social insurance models have prompted research on alternative normative criteria for the government's problem, based on an intergenerational interpretation of the infinite horizon framework. In standard models of social insurance, future generations are considered only indirectly, via the altruism of the earlier ones. Phelan (2006) proposes a government objective with equal weight on all future generations and shows that this implies a finite amount of inequality in the limit. Farhi and Werning (2005) explore a class of social insurance allocations that take into account individuals currently alive, as well as future generations, by assigning the latter a vanishingly small weight in the government's objective. They find that long-run inequality remains bounded and all individuals avoid misery.

The recursive principal–agent approach to social insurance problems underlies most macroeconomic applications. Two recently prominent areas of interest are optimal income taxation with unobserved skills and optimal unemployment insurance with hidden effort.

Optimal income taxation

Income taxes are an important instrument for social insurance. Hence, the basic trade-off between insurance and incentives also underlies the design of optimal tax systems. This intuition drives Mirrlees's (1971) seminal study of optimal income taxes. The main assumption is that labour income is observable but it depends on individual effort and skills, which are private information. Taxes are restricted to depend on labour income only, but, conditional on this, the government is relatively unconstrained. Lump-sum taxes and arbitrarily progressive or regressive tax schemes can all be part of the armoury. Mirrlees studies a static economy and finds that optimal marginal income taxes are low and slightly declining in income. Diamond (1998) and Saez (2001) find that marginal income taxes are high and sharply increasing in income at low income levels. Diamond and Saez's results can be interpreted as a prescription for a rapid phase-out of social benefits for low-income individuals, which is consistent with the US system. The properties of the optimal marginal income tax are very sensitive on the assumed properties of the skill distribution, which explains the difference in findings.

Albanesi and Sleet (2006) and Kocherlakota (2005) apply Mirrlees's approach to dynamic economies and derive implications for optimal capital and labour income taxes. The optimal taxes depend on the agents' history. This is achieved by conditioning the tax payments on the entire history of labour income, or, equivalently, on outstanding wealth when skill shocks are i.i.d. The properties of optimal capital income taxes stem from the intertemporal wedge associated with the optimal consumption path. As noted earlier, this wedge implies that individuals would like to save more for two reasons: their lifetime consumption path is downward-sloping and they face consumption risk, given that the optimal scheme provides only partial insurance. Capital income must then be taxed to prevent this excessive saving. The optimal marginal asset tax, however, has a very specific form: it is decreasing in labour income. Hence, it is stochastic and negatively correlated with consumption. Excessive saving is discouraged by making after-tax returns on assets correlate positively with labour income, and thus reducing the hedging value of holding assets. Albanesi and Sleet (2006) also study the properties of marginal labour income taxes and find that they should be high at low income and decreasing in wealth. Kocherlakota (2006) provides an extensive review of these findings. Albanesi (2006) studies optimal capital income taxes in economies with moral hazard and idiosyncratic capital returns. She shows that the intertemporal wedge can be negative in this class of economies. In this case, the optimal marginal capital income taxes are increasing in income.

Unemployment insurance

Hopenhayn and Nicolini (1997) consider the design of optimal unemployment insurance under moral hazard. The probability of finding a new job depends on the unemployed worker's search effort, which is private information. (Shimer and Werning, 2005, analyse optimal unemployment insurance in a search model with adverse selection.) The unemployed worker is risk averse and cannot borrow or save. Upon finding a new job, the worker will be employed for ever at a fixed wage. The optimal unemployment insurance scheme is self-financing and comprises two elements: a sequence of unemployment benefits and a lump-sum tax levied when the worker finds new employment. The presence of moral hazard implies that the replacement ratio – that is, the fraction of previous salary transferred to the worker in the form of unemployment benefits – must be strictly smaller than one when the worker is searching for a new job. Hence, the optimal scheme provides only partial insurance against unemployment.

The main results are that unemployment benefits should be decreasing over the course of the unemployment spell and the employment tax is increasing with the length of the unemployment spell, under mild conditions on preferences. The decreasing benefits result – first derived in the seminal paper of Shavell and Weiss (1979) – is a manifestation of the intertemporal wedge in this setting. The intertemporal wedge implies that promised utility should be decreasing as long as the worker is unemployed, which requires unemployment benefits to decrease over time. The employment tax result stems from consumption smoothing, since continuation utility rises discretely when a worker becomes employed. By taxing a newly employed worker, the government optimally smooths this jump in consumption. A worker's promised utility declines further for longer unemployment spells, hence, the employment tax is increasing with the length of unemployment.

Pavoni (2004) enriches Hopenhayn and Nicolini's model by introducing a realistic feature: a worker's human capital depreciates over the unemployment spell. The optimal unemployment insurance programme displays two novel features. First, if human capital depreciates rapidly enough during unemployment, benefits are bounded below by a minimal assistance level. Second, the optimal employment tax should decrease with the length of the unemployment spell. These new findings are a consequence of the fact that a worker's wage upon employment depends positively on her human capital. As human capital depreciates over the length of the unemployment spell, expected utility upon employment declines. This reduces and eventually eliminates the government's need to decrease benefits over time to induce a decline in promised lifetime utility over the unemployment spell. Similarly, it reduces the value of the employment tax required to smooth consumption in the transition from unemployment to employment.

More recently, Pavoni and Violante (2007) study the optimal design of welfare-to-work programmes. These programmes are a mix of government expenditures on passive policies, such as unemployment insurance and social assistance, and active policies, such as job-search monitoring, training and wage taxes or subsidies, targeted to the unemployed. Most governments in fact use a combination of passive and active policies in dealing with unemployment. There are several novel findings. First, the optimal welfare-to-work programme endogenously generates a permanent policy of last resort, which resembles a social assistance programme. The unemployed worker is given a constant lifetime benefit and is not active in job search or training. Second, optimal unemployment benefits are generally decreasing or constant during unemployment, but they must increase after a successful spell of training. These findings result from the fact that utility conditional on employment decreases with the length of the unemployment spell and increases after job training. The central assumption is once again human capital depreciation over the unemployment spell. Indeed, it is shown that human capital depreciation is necessary for policy transition to be part of an optimal welfare-to-work programme. Pavoni and Violante find that, by providing more insurance to skilled workers and more incentives to unskilled workers, the optimal welfare-to-work programme delivers significant welfare gains with respect to the existing US system.

Concluding remarks

Most studies of optimal social insurance exclude the presence of private insurance contracts. Does the government have a special role in the provision of insurance with private information? The seminal work of Prescott and Townsend (1984) demonstrates that the first welfare theorem holds for a large class of economies with private information. Golosov and Tsyvinski (2007) study an adverse selection economy and allow for private insurance provision. They show that, if all trades are observable and individuals can sign binding contracts with private insurers *ex ante*, the only effect of public provision of social insurance is crowding out of private insurance. On the other hand, if certain trades are not observable, privately provided insurance is not Pareto optimal and government policies can increase welfare. Bisin and Guatioli (2004) examine a moral hazard economy and show that, if agents' contractual relationships with competing insurance providers cannot be monitored, the competitive equilibrium allocation is not Pareto optimal.

The observability of trades allows insurance providers to restrict participation in additional contractual relationships with other agents. Absent this, individuals can undo the incentive effects of an insurance contract by purchasing additional insurance or engaging in trades that provide a limited amount of self-insurance. These results suggest that, if exclusivity of private insurance contracts cannot be enforced, government provision of insurance plays a critical role.

STEFANIA ALBANESI

See also **optimal taxation; social insurance and public policy; unemployment insurance.**

I wish to thank Claudia Olivetti, Nicola Pavoni and Bruce Preston for very helpful comments and discussion.

Bibliography

Abreu, D., Pierce, D. and Stacchetti, E. 1990. Toward a theory of discounted repeated games with imperfect monitoring. *Econometrica* 58, 1041–63.

Albanesi, S. 2006. Optimal taxation of risky productive capital with private information. Working Paper No. 12419. Cambridge, MA: NBER.

Albanesi, S. and Sleet, C. 2006. Dynamic optimal taxation with private information. *Review of Economic Studies* 73, 1–30.

Atkeson, A. and Lucas, R. 1992. On efficient distribution with private information. *Review of Economic Studies* 59, 427–53.

Atkinson, A. and Stiglitz, J. 1976. The design of the tax structure: direct versus indirect taxation. *Journal of Public Economics* 6, 55–75.

Bisin, A. and Guatioli, D. 2004. Moral hazard and non-exclusive contracts. *Rand Journal of Economics* 35, 306–28.

Diamond, P. 1998. Optimal taxation: an example with a U–shaped pattern of optimal marginal tax rates. *American Economic Review* 88, 83–95.

Diamond, P. and Mirrlees, J. 1978. A model of social insurance with variable retirement. *Journal of Public Economics* 10, 295–336.

Farhi, E. and Werning, I. 2005. Inequality, social discounting and progressive estate taxation. Working Paper No. 11408. Cambridge, MA: NBER.

Golosov, M., Kocherlakota, N. and Tsyvinski, A. 2003. Optimal indirect and capital taxation. *Review of Economic Studies* 70, 569–87.

Golosov, M. and Tsyvinski, A. 2006. Designing optimal disability insurance: a case for asset testing. *Journal of Political Economy* 114, 257–79.

Golosov, M. and Tsyvinski, A. 2007. Optimal taxation with endogenous insurance markets. *Quarterly Journal of Economics* 122, 487–534.

Green, E. 1987. Lending and the smoothing of uninsurable income. In *Contractual Arrangements for Intertemporal Trade*, ed. E. Prescott and N. Wallace. Minneapolis: University of Minnesota Press.

Hopenhayn, H. and Nicolini, J. 1997. Optimal unemployment insurance. *Journal of Political Economy* 105, 412–38.

Kocherlakota, N. 2005. Zero expected wealth taxes: a Mirrlees approach to dynamic optimal taxation. *Econometrica* 73, 1587–621.

Kocherlakota, N. 2006. Advances in dynamic optimal taxation. In *Advances in Economics and Econometrics: Theory and Applications, Ninth World Congress*, vol. 1, ed. R. Blundell, W. Newey and T. Persson. Cambridge: Cambridge University Press.

Mirrlees, J. 1971. An exploration in the theory of optimum income taxation. *Review of Economic Studies* 38, 175–208.

Pavoni, N. 2004. Optimal unemployment insurance, with human capital depreciation, and duration dependence. Mimeo, University College London.

Pavoni, N. and Violante, G. 2007. Optimal welfare-to-work programs. *Review of Economic Studies* 74, 283–318.

Phelan, C. 1998. On the long run implications of repeated moral hazard. *Journal of Economic Theory* 79, 174–91.

Phelan, C. 2006. Opportunity and social mobility. *Review of Economic Studies* 73, 487–504.

Phelan, C. and Townsend, R. 1991. Computing multi-period, information constrained optima. *Review of Economic Studies* 58, 853–81.

Prescott, E. and Townsend, R. 1984. Pareto optima and competitive equilibria with adverse selection and moral hazard. *Econometrica* 52, 21–45.

Rogerson, W. 1985. Repeated moral hazard. *Econometrica* 53, 69–76.

Saez, E. 2001. Using elasticities to derive optimal income tax rates. *Review of Economic Studies* 68, 205–29.

Shavell, S. and Weiss, L. 1979. The optimal payment of unemployment insurance benefits over time. *Journal of Political Economy* 87, 1347–62.

Shimer, R. and Werning, I. 2005. Liquidity and insurance for the unemployed. Working Paper No. 11689. Cambridge, MA: NBER.

Spear, S. and Srivastava, S. 1987. On repeated moral hazard with discounting. *Review of Economic Studies* 54, 599–617.

Thomas, J. and Worrall, T. 1990. Income fluctuations and asymmetric information: an example of a repeated principal agent problem. *Journal of Economic Theory* 51, 367–90.

Varian, H. 1980. Redistributive taxation as social insurance. *Journal of Public Economics* 14, 49–68.

social insurance and public policy

Social insurance expenditures are the largest and fastest-growing component of government expenditures in the developed world. In the United States, for example, only about four per cent of federal government spending in 1953 was devoted to social insurance; by 2003, this had grown eleven-fold to 44 per cent of federal spending (Gruber, 2005). There has been a corresponding growth in research into the behavioural impacts of a wide variety of social insurance programmes. This article reviews that research and its implications for policy, and loosely follows the structure of Gruber (2005), Chapter 12.

At its core, the design of social insurance reflects a trade-off between insurance and incentives. As laid out nicely by Baily (1978), the optimal level of social insurance benefits sets equal the consumption smoothing gains from a social insurance programme to the moral hazard costs of that programme. Programmes that serve largely to displace existing efficient forms of consumption smoothing are less valuable than programmes that greatly increase the ability to smooth consumption around adverse events. Programmes that induce large distortions by insuring adverse behaviour are less valuable than programmes that target those who are truly adversely impacted and not changing their behaviour to qualify. The research on social insurance within economics has largely focused on the latter of these issues, the moral hazard induced by social insurance programmes. More recently there has been more attention paid to the former issue, with an eye towards optimal programme design.

Institutional features of social insurance programmes

Social insurance programmes are distinguished from other types of government spending by four characteristics. Workers participate by 'buying' insurance through payroll taxes or mandatory contributions by themselves or their employer. These contributions make them eligible to receive benefits if some measurable event occurs, such as disability or on-the-job injury. These benefits are conditioned only on making contributions and on the occurrence of the adverse event. And they are typically not means-tested: benefits do not depend on the level of one's current income or assets.

The most common social insurance programme the world over is public pensions through programmes such as Social Security in the United States; in the United States, this is the single largest government expenditure (see Gruber and Wise, 1999, for an overview of social security programmes in developed nations). The other major source of social insurance around the world is social health insurance, such as through programmes like Medicare in the United States, which provides universal health insurance coverage to the elderly, or National Health Insurance in Canada, which provides universal health insurance coverage to the entire population; see Cutler (2002) for an overview of health insurance programmes in developed nations.

Why have social insurance?

In the canonical expected utility model, with insurance available at actuarially fair prices, optimizing consumers will choose to fully insure themselves against idiosyncratic risk. Yet, in markets without social insurance, we often observe much less than full insurance. This motivates the desire for government interventions through programmes such as social insurance.

The major theoretical justification for social insurance is *adverse selection* in private insurance markets. As noted in the classic analysis of Akerlof (1970), asymmetric information between buyers and sellers can lead to market failure, whereby trades that would be beneficial to both insured and insurer are not made. Private insurers may be wary that the individuals demanding insurance from them represent the highest risks in the population, leading the insurer to be unwilling to offer insurance at a price that corresponds to the risk of the average person. As shown in the equally classic Rothschild and Stiglitz (1976) model, this can lead to a breakdown of the insurance market, or at best to a situation where high-risk individuals buy full insurance at high prices, while low-risk individuals are only partially insured at low prices. While there is controversy over the extent of adverse selection in some insurance markets, this is clearly an enormous problem in markets such as those for annuities or health insurance. (The issue of adverse selection in health insurance markets is discussed in Cutler and Zeckhauser, 2000, who summarize in particular the very compelling Cutler and Reber, 1998, article on this issue. Finkelstein and Poterba, 2004, present evidence for adverse selection in annuities markets.)

While market failure is the most appealing justification to economists for social insurance, other rationales may be much more important to policymakers. One such rationale is administrative efficiency. While the government may be less efficient at producing goods than the private sector, it is clearly more efficient at providing insurance; for example, while health insurance administrative costs are 12 per cent of premiums in the private US health insurance market, they are 1.3 per cent of premiums under Canadian National Health Insurance (Gruber, 2005). Even more important in practice is paternalism. Politicians fear that individuals simply will not insure themselves appropriately against sizable risks, so that social insurance is required to ensure proper protection.

The benefits of social insurance

The arguments presented above suggest a number of reasons why private insurance markets may not make it possible for a risk-averse individual to satisfy his desire for consumption smoothing. Yet they do not suggest that consumption smoothing is completely unavailable, because individuals may have private means to smooth consumption: their own savings, labour supply of family members, borrowing from friends, and so on. The justification for social insurance depends on the extent to which social insurance is more efficient than a

consumers' own private consumption smoothing mechanisms. If social insurance is simply displacing, or 'crowding out', equally efficient self-insurance, then there is no gain to the government intervention (but there may be costs, discussed below). The extent to which individuals can self-insure against adverse events will obviously vary with the predictability of the risk (for example, disability may be much less predictable than retirement), the size of the risk (for example, the total income loss from disability or retirement may be much larger than that from unemployment), and the availability of other forms of consumption smoothing (for example, own savings or spousal labour supply).

Tests for the extent of self-insurance against risk have typically proceeded by examining whether individuals can smooth their consumption across adverse events without social insurance, or, relatedly, the extent to which social insurance programmes simply crowd out other forms of self-insurance rather than smoothing consumption. For example, Gruber (1997) and Browning and Crossley (2001) find there is relatively modest consumption smoothing benefits to unemployment insurance during jobless spells. On the other hand, available evidence suggests that retirement income support only partially crowds out private retirement savings, so that there is a real impact on consumption in retirement from programmes such as Social Security. (The evidence on Social Security and savings is described in Gruber, 2005, ch. 13. International evidence on Social Security and consumption is presented in Gruber and Wise, 2008.) There is also evidence of a crowd-out effect of public health insurance in the United States, both from the general provision of this insurance and the asset limitations that accompany qualification for these programmes (Gruber and Yelowitz, 1999). In developing countries, a very large literature finds that families are typically quite good at insuring themselves against modest risks, but not against large risks like disability (Case, 1995; Gertler and Gruber, 2002).

The limitations of these consumption-based tests, however, are highlighted by Chetty and Looney (2006). As they point out, individuals may be using very inefficient means of smoothing consumption, so that even if social insurance doesn't increase consumption smoothing it raises social efficiency. A further difficulty arises because even inefficient consumption smoothing mechanisms (such as savings for idiosyncratic and large risks, which are better insured by pooling across individuals) may have social benefits (if there is too little savings for other reasons such as capital taxation). Clearly, more work is needed to fully evaluate the benefits side of the social insurance equation.

The costs of social insurance

The costs side of social insurance is moral hazard, the adverse behaviour that is encouraged by insuring against an adverse event. As with adverse selection, moral hazard arises due to asymmetric information in insurance markets. By trying to insure against an adverse event (true injury), the insurer may encourage individuals to pretend that an adverse event has happened to them when it actually hasn't. For example, a common type of social insurance in developed nations is insurance against on-the-job injuries (the workers' compensation programme in the United States), which provides partial income replacement for those injured at work. Since work-based injury is often difficult to verify, however (for example for back strains or mental stress), workers may claim to be injured on the job simply to take a partially paid vacation. The rise in leisure induced by the combination of financial incentives and imperfect observability is an example of moral hazard.

Moral hazard can arise along many dimensions. In examining the effects of social insurance, three types of moral hazard play a particularly important role. The first is reduced precaution against entering the adverse state; for example, because individuals have medical insurance that covers illness, they may reduce preventive activities to protect their health. The second is increased expenditures when in the adverse state; for example, because individuals have medical insurance, they may use more medical care than they otherwise would, or because individuals have workers' compensation they don't work hard to rehabilitate their injury. Finally, there may be supplier responses to insurance against the adverse state; for example, physicians may provide too much care to those with health insurance or firms may not be careful enough in protecting workers with injury insurance.

An enormous literature has arisen in public finance to assess the importance of moral hazard in social insurance programmes. The pioneer of this literature was Martin Feldstein, whose work in the early 1970s emphasized the labour market distortions caused by Social Security, unemployment insurance, and other programmes. A large literature has followed up on these issues, as review comprehensively by Krueger and Meyer (2002). In this section, I briefly review the key conclusions from this literature.

Much of the work in this area has been focused on the effects of retirement income security programmes on retirement behaviour: by insuring the elderly against income loss from retirement, these programmes may induce retirement. The literature within countries typically concludes that these programmes are an important, but not the dominant, reason for earlier retirement. Cross-country comparisons such as Gruber and Wise (1999), however, suggest that in the long run these programmes may play the dominant role in determining retirement behaviour.

A second area of much focus has been the impact of unemployment insurance: higher unemployment insurance benefits have been shown to significantly increase unemployment durations (for example, Meyer, 1989), while there is no evidence that the longer durations are resulting in more effective job search. A more limited set of studies has uncovered even larger evidence of responsiveness of injury and illness durations to the financial incentives embedded in workers' compensation and related programmes (for example, Krueger, 1991; Pettersson-Lidbom and Thoursie, 2006). The evidence for programmes that compensate for permanent disability suggests more modest effects on disability rates (for example, Gruber, 2000), which is consistent with the better ability to monitor long-run disability than short-run inability to work.

Unemployment and, to some extent, work-related injury reflect not only the decisions of employees but those of employers also. In most countries, the taxes that finance these insurance programmes are not experience-rated; that is, employers pay tax rates that do not depend on the utilization of the system by their workers. In the United States, there is experience rating, but it is only partial. A small but important literature suggests that this imperfect experience rating is a large source of adverse selection, as workers and firms combine to increase government-subsidized leisure through, for example, temporary layoffs (Topel, 1983; Anderson and Meyer, 2000).

There is also a large literature on moral hazard in medical expenditure programmes. There is clear evidence from the seminal Rand Health Insurance Experiment in the United States that the insured use care excessively, as those randomized into less generous insurance plans used less health care with no adverse impact on health (Newhouse and the Insurance Experiment Group, 1993). At the same time, a number of studies find that having no health insurance is associated with adverse health outcomes (for example, Lurie et al., 1984; Currie and Gruber, 1996a; 1996b). These findings suggest a 'medical effectiveness' curve which relates spending to health improvements: the curve is steep for initial health care use, but flattens out for the excessive care used by those with all health expenditures covered.

As with other social insurance programmes, there are also important supply-side moral hazard issues in health care. A number of studies document the importance of reimbursement structures for the treatment of patients by health providers, finding in particular that prospective reimbursement systems that ensure that providers bear part of the risk for excessive treatment lead to less utilization without adverse health impacts (Newhouse, 1996).

Optimal social insurance

These sets of findings have important implications for the optimal design of social insurance programmes. First, social insurance should be only partial. For example, replacement rates under labour market-based programmes (the extent to which social insurance benefits replace earnings before the adverse event) should be less than full, much less so in several cases. The fact that consumption smoothing is not full, while moral hazard is important, suggests in the framework of Baily (1978) less than full replacement. There is too little work on most of these programmes to state the optimal replacement rates with confidence, but there are clear directions for reform. The enormous moral hazard associated with Workers' Compensation in the United States or with social security in many European nations, for example, suggests that there may be welfare gains from reducing benefit levels.

Another important example is health care, where the optimal insurance policy is one in which individuals bear a large share of medical costs within some affordable range, and are only fully insured when costs become unaffordable. This structure is optimal because coverage for all medical expenditures (what is often called 'first dollar coverage', since all dollars of medical spending, starting with the first dollar, are covered) has little consumption smoothing benefit, but a large moral hazard cost. The consumption smoothing benefit from first dollar coverage is small because there is little utility gain to risk-averse individuals from insuring a small risk. At the same time, this first dollar coverage has substantial moral hazard cost because it encourages individuals to overuse the medical system, demanding care for which the social costs exceed the social benefits. An example of an optimal insurance plan would be Feldstein's (1973) 'major risk insurance' plan, in which individuals would face a high co-payment (such as 50 per cent) on all services until they spent a sizable share of their income (such as ten per cent) on medical care, beyond which there would be no more co-payments.

The second lesson is that there should be more supply-side risk bearing in insured markets. For example, more tightly experience rating employers for the costs of unemployed or injured workers could significantly reduce overuse of these social insurance systems. And more risk bearing by providers in the medical system has been shown to reduce utilization without adversely impacting health.

Finally, these findings also open the question of moving from the traditional 'defined benefit' approach to social insurance to a 'defined contribution' approach where individuals are mandated to provide for their own protection against adverse events. This model has been seen most forcefully in the debate over US Social Security privatization, whereby the traditional programme under which today's workers save collectively for today's retirees would be replaced by one where individuals saved for their own retirement. Partial privatization is in use in a number of countries, but is a source of contentious debate in the United States (see Gruber, 2005, ch. 13, for a brief review of the arguments on both sides of this debate).

This debate has not generally proceeded, however, to other social insurance programmes. For example,

Feldstein and Altman (1998) have suggested a programme under which individuals save in their own unemployment insurance accounts for jobless spells. Using such mandatory savings accounts to finance short-run risks can potentially provide the consumption smoothing benefits of unemployment insurance while reducing the moral hazard costs of the programme. This is an approach worth consideration in other social insurance contexts, where the essential question becomes the value of the redistribution that is lost by such 'self-insurance' approaches.

JONATHAN GRUBER

See also **adverse selection; moral hazard; pensions; social insurance; Social Security in the United States.**

Bibliography

Akerlof, G.A. 1970. The market for lemons: quality uncertainty and the market mechanism. *Quarterly Journal of Economic* 84, 488–500.

Anderson, P.M. and Meyer, B.D. 2000. The effects of the unemployment insurance payroll tax on wages, employment, claims and denials. *Journal of Public Economics* 78, 81–106.

Baily, M. 1978. Some aspects of optimal unemployment insurance. *Journal of Public Economics* 10, 379–402.

Browning, M. and Crossley, T. 2001. Unemployment insurance levels and consumption changes. *Journal of Public Economics* 80, 1–23.

Case, A. 1995. Symposium on consumption smoothing in developing countries. *Journal of Economic Perspectives* 9(3), 81–82.

Chetty, R. and Looney, A. 2006. Consumption smoothing and the welfare consequences of social insurance in developing economies. *Journal of Public Economics* 90, 2351–6.

Currie, J. and Gruber, J. 1996a. Saving babies: the efficacy and cost of recent changes in the Medicaid eligibility of pregnant women. *Journal of Political Economy* 104, 1263–96.

Currie, J. and Gruber, J. 1996b. Health insurance eligibility, utilization of medical care, and child health. *Quarterly Journal of Economics* 111, 431–66.

Cutler, D. 2002. Equality, efficiency and market fundamentals: the dynamics of international medical care reform. *Journal of Economic Literature* 40, 881–906.

Cutler, D.M. and Reber, S.J. 1998. Paying for health insurance: the trade-off between competition and adverse selection. *Quarterly Journal of Economics* 113, 433–66.

Cutler, D. and Zeckhauser, R. 2000. The anatomy of health insurance. In *Handbook of Health Economics*, vol. IA, ed. A. Culyer and J. Newhouse. Amsterdam: North-Holland.

Feldstein, M.S. 1973. The welfare loss of excess health insurance. *Journal of Political Economy* 81, 251–80.

Feldstein, M.S. and Altman, D. 1998. Unemployment insurance savings accounts. Working Paper No. 6860. Cambridge, MA: NBER.

Finkelstein, A. and Poterba, J. 2004. Adverse selection in insurance markets: policyholder evidence from the U.K. annuity market. *Journal of Political Economy* 112, 183–208.

Gertler, P. and Gruber, J. 2002. Insuring consumption against illness. *American Economic Review* 92, 51–70.

Gruber, J. 1997. The consumption smoothing benefits of unemployment insurance. *American Economic Review* 87, 192–205.

Gruber, J. 2000. Disability insurance benefits and labor supply. *Journal of Political Economy* 108, 1162–83.

Gruber, J. 2005. *Public Finance and Public Policy*, 1st edn. New York: Worth Publishers.

Gruber, J. and Wise, D.A. 1999. Introduction and summary. In *Social Security and Retirement around the World*, ed. J. Gruber and D.A. Wise. Chicago: University of Chicago Press.

Gruber, J. and Wise, D.A. 2008. *Social Security and Well-Being Around the World*. Chicago: University of Chicago Press.

Gruber, J. and Yelowitz, A. 1999. Public health insurance and private savings. *Journal of Political Economy* 107, 1249–74.

Krueger, A.B. 1991. Workers' compensation insurance and the duration of workplace injuries. Mimeo, Princeton University.

Krueger, A.B. and Meyer, B.D. 2002. Labor supply effects of social insurance. In *Handbook of Public Economics*, vol. 4, ed. A.J. Auerbach and M. Feldstein. Amsterdam: North-Holland.

Lurie, N., Ward, N.B., Shapiro, M.F. and Brook, R.H. 1984. Termination from Medi-Cal – does it affect health? *New England Journal of Medicine* 311, 480–4.

Meyer, B.D. 1989. A quasi-experimental approach to the effects of unemployment insurance. Working Paper No. 3159. Cambridge, MA: NBER.

Newhouse, J. 1996. Reimbursing health plans and health providers: selection vs. efficiency in production. *Journal of Economic Literature* 34, 1236–63.

Newhouse, J.P. and the Insurance Experiment Group. 1993. *Free For All? Lessons from the RAND Health Insurance Experiment*. Cambridge, MA: Harvard University Press.

Pettersson-Lidbom, P. and Thoursie, P.S. 2006. Temporary disability insurance and labor supply: evidence from a natural experiment. Mimeo, Stockholm University.

Rothschild, M. and Stiglitz, J. 1976. Equilibrium in competitive insurance markets: an essay on the economics of imperfect information. *Quarterly Journal of Economics* 90, 629–49.

Topel, R.H. 1983. On layoffs and unemployment insurance. *American Economic Review* 73, 541–59.

social interactions (empirics)

The empirical economics literature on social interactions addresses the significance of the social context in economic decisions. Decisions of individuals who share a social milieu are likely to be interdependent. Recognizing the nature of such interdependence in a variety of conventional and unconventional settings and measuring empirically the role of social interactions poses complex econometric questions. Their resolution may be critical for a multitude of phenomena in economic and social life and of matters of public policy. Questions like why some countries are Catholic and others Protestant, why crime rates vary so much across cities in the same country, why fads exist and survive, and why there is residential segregation and neighbourhood tipping are all in principle issues that may be examined as social interactions phenomena.

The social context enters in a variety of ways. One is that individuals care not only about their own purely private outcomes – for example, the kinds of cars they drive or the education they acquire – but also about outcomes of others, such as the kinds of cars or the education of their friends. This type of interpersonal effect is known as *endogenous* social effect (or interaction), because it depends on *decisions* of others in the same social milieu. Individuals may also care about personal characteristics of others, that is, whether they are young or old, black or white, rich or poor, trendy or conventional, and so on, and about other attributes of the social milieu that may not be properly characterized as deliberate decisions of others. The latter is known as exogenous social or *contextual* effect. In addition, individuals in the same or similar social settings tend to act similarly because they share common unobservable factors. Such an interaction pattern is known as *correlated* effects. This terminology is due to Manski (1993).

Emergence of social interdependencies is natural if individuals share a common resource or space in a way that is not paid but still generates constraints on individual action. This is also known as pecuniary externalities. Individuals who try to form expectations about future outcomes of current decisions, like occupational choice, may rely on lessons from the actions of others and therefore end up mimicking their behaviour. Endogenous social interactions are a case of real externalities, a pervasive feature of economic behaviour.

Theorizing in this area must lie in the interface of economics, sociology and psychology, and often is imprecise. Terms like social interactions, neighbourhood effects, social capital and peer effects are often used as synonyms, although they may have different connotations. Empirical distinctions between endogenous, contextual and correlated effects are critical for policy analysis because of the 'social multiplier,' as we see further below.

Joint dependence among individuals' decisions *and* characteristics within a social milieu is complicated further by the fact that in many interesting circumstances individuals in effect choose the social context. For example, individuals choose their friends and their neighbourhoods and thus their neighbourhood effects as well. Such choices involve information that is in part unobservable to the analyst, and therefore require making inferences among the possible factors which contribute to decisions (Brock and Durlauf, 2001; Moffitt, 2001). The present article focuses on highlighting the significance of key empirical findings and owes a lot to Durlauf (2004), the most comprehensive review to date that examines the methodological basis, statistical reliability and conceptual and empirical breadth of the neighbourhood effects literature.

Empirical framework

Let individual i's outcome ω_i, a scalar, be a linear function of a vector of observable individual characteristics, X_i, of a vector of contextual effects, $Y_{n(i)}$, which describe i's neighbourhood $n(i)$, and of the expected value of the ω_j's of the members of neighbourhood $n(i)$, $j \in n(i)$. It is straightforward to incorporate social interactions into economic models in a manner that is fully compatible with economic reasoning, that is, by positing that individuals maximize a utility function subject to constraints and obtain a behavioural equation such as:

$$\omega_i = k + cX_i + dY_{n(i)} + Jm_{n(i)} + \varepsilon_i, \quad (1)$$

where ε_i is a random error and k a constant. Ignore for the moment the fact that individual i may have deliberately chosen neighborhood $n(i)$. The assumption that the expectation of ε_I in (1) is zero, conditional on individual characteristics, on contextual effects and on the event that i is a member of neighbourhood $n(i)$, allows us to focus on the estimation of the model. The critical next step for translating theoretical models into empirical applications is to assume *social equilibrium* and that individuals hold *rational expectations* over $m_{n(i)}$. That is, individuals' expectations are confirmed in that they are exactly equal to what the model predicts on average. So, taking the expectation of ω_i and setting it equal to $m_{n(i)}$ allows us to solve for $m_{n(i)}$. Substituting back into (1) yields a *reduced form equation*, an expression for individual i's outcome in terms of all observables:

$$\omega_i = \frac{k}{1-J} + cX_i + \frac{J}{1-J}cX_{n(i)} + \frac{d}{1-J}Y_{n(i)} + \varepsilon_i. \quad (2)$$

This simple linear model obscures the richness that nonlinear social interactions models make possible, like multiplicity of equilibria (Brock and Durlauf, 2001). Yet it does facilitate studying other aspects. For example, it does confirm that endogenous social effects generate feedbacks which magnify the effects of neighbourhood characteristics. That is, the effect of a unit increase in

$Y_{n(i)}$ is $\frac{d}{1-J}$, and not just d, as one would expect from (1). It also confirms why it is tempting for empirical researchers to study individual outcomes as functions of all observables. Following the pioneering work of Datcher (1982), a great variety of individual outcomes have been studied in the context of different neighbourhoods and typically significant effects have been found. Deriving causal results requires suitable data.

Manski (1993) emphasized that the practice of including neighbourhood averages of individual effects as contextual effects, $Y_{n(i)} = X_{n(i)}$, may cause failure of identification of endogenous as distinct from exogenous interactions, that is, to estimate J separately from d. That is, if the neighbourhood attributes *coincide* with the neighbourhood averages of its inhabitants' characteristics, or $Y_{n(i)} = X_{n(i)}$, then regressing individual outcomes on neighbourhood averages of individual characteristics as contextual effects allows us to estimate a function of the parameters of interest, $\frac{Jc+d}{1-J}$, the coefficient of $X_{n(i)}$ in a regression according to (2). A statistically significant estimate of this coefficient implies that at least one type of social interaction is present, either J or d or both are non-zero. This is known as Manski's *reflection problem*, which is specific to *linear* models: the equilibrium value of the outcome $m_{n(i)}$ is linearly related to the neighbourhood attributes, and therefore its effect on individual outcomes may not be distinguishable from their 'reflection'.

Complicating the basic model in natural ways, as by assuming correlated effects – like group members sharing group-specific unobservable effects, that is, the performance of students in the same class is affected by the quality of their teacher, which is unobservable, over and above peer effects from classmates – introduces additional difficulties with identification, even if individuals are randomly assigned to groups. Brock and Durlauf (2007) provide an exhaustive analysis of the various possibilities in binary choice situations with unobserved group effects. When more than two choices are available, there may be additional possibilities for identifying choice-specific effects by working with subsets of choices (Brock and Durlauf, 2005). Graham (2006), discussed further below, offers a promising approach for continuous outcomes when individuals are randomly assigned to groups, but is not focusing on the distinction between exogenous and endogenous interactions. Yet more possibilities appear when panel data (that is, repeated observations over time on the same decision-making units) are available. If contextual effects take time to make their impact on the endogenous social effect, the linear dependence is broken and the lack of identification – Manski's reflection problem – is mitigated. Sometimes, and depending upon the nature of the data as well, it may be impossible, especially in linear models, to identify social interactions in the presence of unobserved group effects. Moving from linear models to binary and other non-linear choice models improves identification even with cross-section data (Brock and Durlauf, 2007).

If it is plausible to exclude some neighbourhood averages of individual covariates, then identification may be possible. Also, if nonlinearities are inherent in the basic model specification, identification again may be possible, even in the case where the contextual effects coincide with neighbourhood averages of individual characteristics. Nonlinearities may eliminate the reflection problem. A noteworthy case in point here is Drewianka (2003), who studies two-sided matching in the marriage market and finds that it allows identification of endogenous *and* exogenous social interactions. The logic of the model requires that the two sides of the market contain an additional source of variation: the greater the number of potential marriage partners, the higher is the probability that a match will occur. There is an inherent multiplier effect at work here. One's prospects of finding a marriage partner depends on the rate at which other people match up, an endogenous social effect. Drewianka's results show that a ten per cent increase in the fraction of the population that is unmarried causes the marriage rate of never-married men to fall by ten per cent and that of never-married women by seven per cent.

An interesting consequence of endogenous social interactions is the amplification of differences in average neighbourhood behaviour across neighbourhoods. In fact, Glaeser, Sacerdote and Scheinkman (2003) use directly such patterns in the data to estimate a *social multiplier*. This is defined for a change in a particular fundamental determinant of an outcome as the ratio of a total effect, which includes a direct effect on an individual outcome plus the sum total of the indirect effects through the feedback from the effects on others in the social group, to the direct effect. It is easy to see it as the ratio of the 'group level' coefficient, the coefficient of $Y_{n(i)}$ in eq. (2), to the 'individual level' coefficient, the coefficient of $Y_{n(i)}$ in eq. (1): $\frac{d}{1-J}\frac{1}{d} = \frac{1}{1-J}$. It follows that a social multiplier greater than 1 implies endogenous social interactions, $0 < J < 1$. This approach must deal, in practice, with dependence across decisions of individuals belonging to the same group, which is implied by non-random sorting in terms of unobservables. It is particularly useful in delivering ranges of estimates for the endogenous social effect and when individual data are hard to obtain.

This is the case with crime data. Glaeser, Sacerdote and Scheinkman (1996) motivate their study of crime and social interactions by the extraordinary variation in incidence of crime across US metropolitan areas over and above differences in fundamentals. If social interactions are present, variations in observed outcomes are larger than would be expected from variations in underlying fundamentals. Glaeser, Sacerdote and Scheinkman (2003) regress actual crime rates against predicted crime rates, which are formed by multiplying percentages of US individuals in each of eight age categories by the

crime rate of persons in that category. They perform such regressions at the level of county and state cross-sectionally and for the entire United States over time. Their results imply large social multipliers, which increase with the level of aggregation exactly as their basic theory would predict.

It is possible to modify this basic model in order to study several other areas involving economic decisions akin to social interactions. For example, diffusion of innovations, herding and adoption of norms or other institutions by a population involve ideas that are conceptually related to social interactions. Transmission of job-related information is of particular relevance (see Ioannides and Loury, 2004). Also, J, the endogenous social effect, may be negative, as in the case of land development, which is conceivably due to congestion.

Identification of social interactions using observational data on 'natural experiments'

Several researchers have sought to identify social interactions by exploiting uniquely suitable features of observational data, which are often referred to as 'natural experiments'. For example, consider outcomes for children from families with several children who share the common influence of unobservable family factors, such as parental values and competence, taste for education and time spent with children, and other unobservables that affect the upbringing of household members living in close proximity. They also share the variation in neighbourhood effects that is produced by families' residential moves. By using observations on several children from the same family who are separated in age by at least three years, Aaronson (1998) controls for family-specific characteristics. This obviates the need to control for the impact of self-selection in terms of unobservable neighbourhood characteristics. Aaronson uses data from the Panel Study of Income Dynamics and finds large and statistically significant contextual neighbourhood effects, but his models exclude endogenous social effects. His results are robust to changes in estimation techniques and in sample and variable definitions, but are sensitive to the formulation of neighbourhood characteristic proxy. Incomplete specification of family characteristics is an important concern, and its consequences for the robustness of estimated relationships are aptly demonstrated by Ginther, Haveman and Wolfe (2000).

Grinblatt, Keloharju and Ikaheimo (2004) use data for *all* residents of two large Finnish provinces – amounting to millions of observations – and establish that automobile purchase decisions by close residential neighbours influence one another. The measured endogenous neighbourhood effects are strongest among individuals belonging to the same 'social class' (especially if they belong to lower-income classes), or when the cars they purchase are of the same make or even the same model. These findings militate in favour of information sharing rather than 'keeping up with the Joneses'. We note that excluding neighbourhood averages of demographics as contextual effects is reasonably plausible in this case: there is no reason why the average *age* of my neighbours should affect directly my taste in *cars*. Conceptually related is the study of Aizer and Currie (2004), who use data from more than 3.5 million birth certificates from California to examine 'information sharing effects' in the utilization of publicly funded prenatal care. They conclude that it is not information sharing, but is, instead, differences in the behaviour of institutions that explain the established correlations between neighbourhood and ethnic group membership in prenatal care use.

Luttmer (2005) uses data from the US National Survey of Families and Households, augmented with census data from the Public Use Microdata Areas, and examines how self-reported well-being varies with own and neighbours' incomes and with other characteristics. He interprets his findings as direct evidence that people have preferences regarding their neighbours' incomes. That is, after an individual's own income is controlled for, higher earnings of neighbours are associated with lower levels of self-reported happiness on a variety of measures.

Sacerdote (2001) exploits the fact that at Dartmouth College freshman-year room-mates and dorm-mates are randomly assigned, thus producing a natural quasi-experimental setting for studying peer effects. Sacerdote posits that an individual's grade point average is a function of an individual's own academic ability prior to college entrance, of social habits, and of the academic ability and grade point average of his room-mates. Sacerdote finds that peers have an impact on each others' grade point average and on decisions to join social groups such as fraternities. He does not, however, find residential peer effects in other major college decisions, such as choice of college major. He finds peer effects in grade point average at the individual room level – you keep up with your room-mates! – whereas peer effects in fraternity membership occur at both the room level and the entire dorm level – dorms are conformist! These data provide strong evidence for the existence of peer effects in student outcomes, even among highly selected college students who may be otherwise quite homogeneous albeit in close proximity to one another. Peer effects are smaller the more directly a decision is related to labour market activities.

Peer effects in classrooms and schools

Social interactions in classrooms – peer effects – are particularly interesting in understanding schooling as an economic activity and its consequences for inequality of social outcomes. Whether students benefit from classmates with different characteristics and academic performance and whether the effect is different depending upon whether one's classroom peers are more or less able are important for education policy and the actual

functioning of schools. In other words, deciding whether or not students should be 'tracked' – that is, administratively segregated in terms of different characteristics – is the sort of policy question which rests on understanding peer effects quantitatively.

Hoxby (2001) posits a relationship between individual academic achievement by a male student in a particular school and grade as the sum of what the mean achievement among males would have been in the absence of peer effects, of a term that is proportional to the percentage of females in the classroom, plus an error. She extends such a relationship to the case of several racial groups, which is particularly appropriate for the Texas Schools Project data that she uses. Her identification strategy involves exploring the panel structure of the data under the plausible assumption that there is natural idiosyncratic variation across successive cohorts in terms of gender, race and other individual attributes. Hoxby finds that students are affected by the achievement levels of their peers: an exogenous one-point increase in peers' reading scores raises a student's own score between 0.14 and 0.4 points. Peer effects are stronger intra-race, and there is evidence of contextual effects: both male and female students perform better in classrooms that are more female despite the fact that females' math performance is about the same as that of males.

The role of gender is corroborated by research by Arcidiacono and Nicholson (2005), who use data on the universe of students admitted to US medical schools for a particular year. One positive peer effect in US medical schools that they find pertains to female students, who benefit from attending medical schools that have other female students with relatively high scores on the verbal reasoning section of the Medical College Admission Test.

Of particular interest have been studies of the impact of school racial integration in the US on student performance. Consider Boston's Metropolitan Council for Educational Opportunities (METCO) programme, a voluntary desegregation programme. The programme allows mainly black inner-city kids from Boston public schools to commute to mainly white suburban communities in the Boston area that accommodate them in their public schools. Angrist and Lang (2004) show that, although the receiving districts, which tend to have higher mean academic performance, experience a mean decrease due to the programme, the effects are merely 'compositional', and there is little evidence of statistically significant effects of METCO students on their non-METCO classmates. Analysis with micro data from a particular receiving district (Brookline, Massachusetts) generally confirms this finding, but also produces some evidence of negative effects on *minority* students in the receiving district. METCO is a noteworthy social experiment, which was initiated by civil rights activists seeking to bring about de facto desegregation of schools. Lack of evidence of negative peer effects is particularly useful for informing desegregation policy. Still, there is self-selection in the participants on both sides.

Estimation of social interactions in experimental settings

Experimental data used by social interactions studies come from two types of deliberate experiments: field and laboratory experiments. A well-known field experiment is Project STAR, an experimental programme in the US state of Tennessee that randomly assigned entering kindergarten students into three different class sizes and then randomly assigned teachers to them. A recent study that utilizes Project STAR data is Graham (2006). He seeks to estimate a relationship like (1) by measuring 'excess' variance patterns across groups of exogenously given, but varying, sizes of classrooms that are associated with randomly assigned students and teachers in the presence of correlated effects in the form of unobservable group effects. Graham compares excess variance across small and large classrooms and finds social multipliers between 1.07 and 2.31, and 1.05 to 3.07, for math and reading achievement, respectively. Studies of this type aim at distinguishing excess between-classroom variance that is due to social interactions from that due to group-level heterogeneity.

Duflo and Saez (2003), using experimental data, study how social interactions among employees of a large US university may influence participation in a tax-deferred retirement account plan. The experiment more than tripled the attendance rate of those who received a small monetary reward for participating, doubled that of those not thus 'treated' but who belonged to the same departments as the treated, and significantly increased participation in the target programme by individuals from treated departments, and did so by almost as much as those who did not receive direct encouragement. While clearly the effect of social interactions may coexist with differential treatment and motivational reward effects, social interactions are also relevant for the effects of treatment on attendance and of attendance on participation. The authors conclude that the role of social interactions in amplifying the effect of treatment is unambiguous, in spite of the fact that they cannot distinguish unambiguously between the three different effects.

Moving to Opportunity (MTO) is a set of large randomized field experiments that were conducted by the US Department of Housing and Urban Development in several large US cities. The experiments offered poor households (who were chosen by lottery from among residents of high-poverty public housing projects) housing vouchers and logistical assistance through non-governmental organizations for the purpose of relocating to precisely defined 'better' neighborhoods. Several studies based on data from these experiments show that outcomes after relocation improved for children,

primarily for females, in terms of education, risky behaviour and physical health, but the effects on male youth were adverse. Regarding outcomes for adults, such as economic self-sufficiency or physical health, the picture is more mixed. Kling, Ludwig and Katz (2005) find that four to seven years after relocation families (primarily female-headed ones with children) lived in safer neighbourhoods that had lower poverty rates than those of a control group that were not offered vouchers. Unfortunately, there is serious controversy over how to interpret these findings in the context of policy design for large-scale policy interventions (Sobel, 2006).

As for laboratory experiments, a notable study is by Ichino and Falk (2006). The experiment involves workers in pairs stuffing envelopes, with control being provided by subjects working alone in a room. These authors find that standard deviations of output are significantly smaller within pairs than between pairs, that is individuals keep up with their neighbours. They also find that social interactions raise productivity: average output per person is greater when subjects work in pairs. They also show that social interactions are asymmetric: low-productivity workers are more sensitive to the behaviour of high-productivity workers. Their setting does reduce some of the noise associated with 'natural' experiments but does not allow for contextual effects.

Identification of social interactions with self-selection to groups and sorting

The presence of non-random sorting on unobservables is a major challenge for the econometric identification of social interactions models. The critical role of local public finance in education in the United States has been studied extensively as a link between sorting into residential communities and socio-economic outcomes. Brock and Durlauf (2001) turned adversity into advantage by recognizing that self-selection itself, that is that individuals choose their neighbourhoods making $n(i)$ in eq. (1) endogenous, may provide additional evidence on identification. That is, if it is possible to estimate a neighbourhood selection rule, then correction for selection bias via the mean estimated bias, the so-called Heckman correction term, introduces an additional regressor in the right-hand side of (1) whose neighbourhood average is not a causal effect. Ioannides and Zabel (2002) implement this method successfully using micro data for a sample of households and their ten closest residential neighbours from the American Housing Survey and contextual information for the census tracts in which these individuals reside.

Endogeneity of the average of one's neighbours housing demands, an endogenous social effect, is instrumented by treating housing demands by a group of close neighbours as a simultaneous system of equations. By choosing neighbourhoods, census tracts in this application, individuals choose desirable social interactions. Ioannides and Zabel work with an otherwise standard housing demand model and find a very significant and large endogenous social effect along with very significant contextual effects in the form of unobservable group effects. Several other studies have sought to use instrumental variables to account for self-selection. Still, the identification of valid instruments is often quite hard and requires deep understanding of the actual setting.

Conclusions

Social interactions are ubiquitous. Interest in estimating their effects is expanding rapidly in numerous areas of economics and is motivating important methodological advances. For econometricians, key challenges include social interactions effects on market outcomes coexisting with feedbacks from the characteristics of individual market participants via their impacts on prices, consequences of self-selection and the attendant role of the presence of individual and group unobservables. Fundamentally, and in the light of ever-improving data availability, social interactions empirics will rely increasingly critically on careful theorizing that involves precise definitions of social interactions and justifies stochastic specification, possibly by calling on psychology and sociology to define appropriate boundaries, and must facilitate use of data from different sources. The likely payoff is enormous: better understanding of social forces in the modern economy, with individuals sharing information while self-selecting into social groups and living and working in close proximity to one another, as in firms and cities, the hallmark of modern economic life, and informed design of policy interventions.

YANNIS M. IOANNIDES

See also **educational finance; natural experiments and quasi-natural experiments; neighbours and neighbourhoods; psychology of social networks; social interactions (theory); social multipliers; Tiebout hypothesis.**

Bibliography

Aaronson, D. 1998. Using sibling data to estimate the impact of neighborhoods on children's educational outcomes. *Journal of Human Resources* 23, 915–46.

Aizer, A. and Currie, J. 2004. Networks or neighborhoods? Interpreting correlations in the use of publicly funded maternity care in California. *Journal of Public Economics* 88, 2573–85.

Angrist, J. and Lang, K. 2004. Does school integration generate peer effects? Evidence from Boston's METCO Program. *American Economic Review* 94, 1613–34.

Arcidiacono, P. and Nicholson, S. 2005. Peer effects in medical school. *Journal of Public Economics* 89, 327–50.

Brock, W. and Durlauf, S. 2001. Interaction-based models. In *Handbook of Econometrics*, vol. 5, ed. J. Heckman and E. Leamer. Amsterdam North-Holland.

Brock, W. and Durlauf, S. 2005. A multinomial choice model with social interactions. In *The Economy as an Evolving Complex System III*, ed. L. Blume and S. Durlauf. Oxford and New York: Oxford University Press.

Brock, W. and Durlauf, S. 2007. Identification of binary choice models with social interactions. *Journal of Econometrics* 140, 52–75.

Datcher, L. 1982. Effects of community and family background on achievement. *Review of Economics and Statistics* 64, 32–41.

Drewianka, S. 2003. Estimating social effects in matching markets: externalities in spousal search. *Review of Economics and Statistics* 85, 408–23.

Duflo, E. and Saez, E. 2003. The role of information and social interactions in retirement plan decisions: evidence from a randomized experiment. *Quarterly Journal of Economics*, 118, 815–42.

Durlauf, S. 2004. Neighbourhood effects. In *Handbook of Urban and Regional Economics, Volume 4: Cities and Geography*, ed. J. Henderson and J.-F. Thisse. Amsterdam: North-Holland.

Ginther, D., Haveman, R. and Wolfe, B. 2000. Neighborhood attributes as determinants of children's outcomes: how robust are the relationships? *Journal of Human Resources* 35, 603–42.

Glaeser, E., Sacerdote, B. and Scheinkman, J. 1996. Crime and social interactions. *Quarterly Journal of the Economics* 112, 508–48.

Glaeser, E., Sacerdote, B. and Scheinkman, J. 2003. The social multiplier. *Journal of the European Economic Association* 1, 345–53.

Graham, B. 2006. Identifying social interactions through conditional variance restrictions. Working paper, Department of Economics, University of California, Berkeley.

Grinblatt, M., Keloharju, M. and Ikaheimo, S. 2004. Interpersonal effects in consumption: evidence from the automobile purchases of neighbors. Working Paper No. 10226. Cambridge, MA: NBER.

Hoxby, C. 2001. Peer effects in the classroom: learning from gender and race variation. Working Paper No. 7867. Cambridge, MA: NBER.

Ichino, A. and Falk, A. 2006. Clean evidence on peer effects. *Journal of Labor Economics* 24, 39–57.

Ioannides, Y. and Loury, L. 2004. Job information networks, neighborhood effects, and inequality. *Journal of Economic Literature* 42, 1056–93.

Ioannides, Y. and Zabel, J. 2002. Interactions, neighborhood selection, and housing demand. Working paper, Department of Economics, Tufts University.

Kling, J., Ludwig, J. and Katz, L. 2005. Neighborhood effects on crime for female and male youth: evidence from a randomized housing experiment. *Quarterly Journal of the Economics* 120, 87–130.

Luttmer, E. 2005. Neighbors as negatives: relative earnings and well-being. *Quarterly Journal of Economics* 120, 963–1002.

Manski, C. 1993. Identification of endogenous social effects: the reflection problem. *Review of Economic Studies* 60, 531–42.

Moffitt, R. 2001. Policy interventions, low-level equilibria and social interactions. In *Social Dynamics*, ed. S. Durlauf and H. Peyton Young. Cambridge, MA: MIT Press.

Sacerdote, B. 2001. Peer Effects with random assignment: results for Dartmouth roommates. *Quarterly Journal of Economics* 116, 681–704.

Sobel, Ml. 2006. Spatial concentration and social stratification: does the clustering of disadvantage 'beget' bad outcomes?' In *Poverty Traps*, ed. S. Bowles, S. Durlauf and K. Hoff. Princeton, NJ: Princeton University Press.

social interactions (theory)

Social interactions refer to particular forms of externalities, in which the actions of a reference group affect an individual's preferences. The reference group depends on the context and is typically an individual's family, neighbours, friends or peers. Social interactions are sometimes called non-market interactions to emphasize the fact that these interactions are not regulated by the price mechanism.

Veblen's (1934) analysis of conspicuous consumption – that is, consumption that signals wealth – is perhaps the first contribution to the economic literature on social interactions. Duesenberry (1949) and Leibenstein (1950) are also among the earliest contributors. Although Veblen's *Theory of the Leisure Class* has had a remarkable impact in the social sciences, Schelling's (1971; 1972) pioneering formal analysis of the influence of social groups in behaviour was particularly important for later developments in economics.

Models of social interaction seem particularly adapted to solving a pervasive problem in the social sciences, namely, the observation of large differences in outcomes in the absence of commensurate differences in fundamentals. Many models of social interactions exhibit *strategic complementarities*, which occur when the marginal utility to one person of undertaking an action increases with the average amount of the action undertaken by his peers. Consequently, a change in fundamentals has a direct effect on behaviour and an indirect effect of the same sign. Each person's actions change, not only because of the direct change in fundamentals, but also because of the change in the behaviour of his or her peers. The result of all these indirect effects is the *social multiplier*.

When this social multiplier is large, we expect to see a large variation of aggregate endogenous variables relative to the variability of fundamentals, which seem to characterize phenomena as diverse as stock market crashes, religious differences, and differences in crime rates. In fact, if social interactions are large enough, multiple equilibria can occur – that is, one may observe different outcomes from exactly the same fundamentals. The existence of multiple equilibria also helps us to understand high levels of variance of aggregates.

Social interaction models have implications for the sorting of people and activities across space. As Schelling (1971) demonstrated, when individuals can choose locations, the presence of these interactions may result in segregation across space, even in situations where the typical individual would be content to live in an integrated neighbourhood, provided his group does not form too small a minority. Cities exist because of agglomeration economies which are likely to come from non-market complementarities. In dynamic settings, social interactions can produce s-shaped curves which help to explain the observed time series patterns of phenomena as disparate as telephone adoption and women in the workplace.

Closely related topics include social learning, where agents learn from observing choices by other agents (for example, Arthur, 1989; Bickhchandani, Hirshleifer and Welch, 1992), and local interaction games (for example, Ellison, 1993; Morris. 2000).

Schelling's critical mass model

In Chapter 3 of his *Micromotives and Macrobehavior* (1978), Schelling discusses a *critical mass model* where he supposes that there is an activity which some individuals will always undertake, others will undertake only if a large enough fraction of the population is engaged in the action, and still others may never undertake. Formally, agents are parameterized by an $x \in [0, 1]$, and can choose between undertaking an action or not. The gain in utility for an agent of undertaking the action is given by $u(x, t)$, where t is the fraction of the population engaging in the action. Schelling assumes that $u(x, t)$ decreases with x, that is, agents can be inversely ordered by their gains from undertaking the action. He also assumes that $u(x, t)$ increases with t, that is, the gain is larger if a larger fraction of the population is engaged in the action. This assumption is exactly what was later named 'strategic complementarity'. Each agent x takes t as given and chooses to undertake the action if and only if $u(x, t) \geq 0$. An equilibrium is a fraction t^* such that $u(t^*, t^*) = 0$. Clearly, for such a t^* every agent $x \leq t^*$ will undertake the action while, if $x > t^*$, agent x would refrain. Schelling constructs a numerical example where multiple equilibria arise and noticed that even when uniqueness prevails such models display a 'multiplier effect'. In his example, the presence of a smaller number of individuals that would undertake the activity unconditionally would have a more than proportionate effect on the equilibrium level of the activity. Granovetter (1978) proposes a very similar model to analyse riots and other collective actions. He noted that as parameter changes some equilibria may disappear, leading to drastic changes in the equilibrium outcomes.

Versions of the critical mass model set out in Schelling (1978) have later been used to study a myriad of economic questions, often with a more detailed micro-economic foundation to justify strategic complementarities. Examples include income inequality (Loury, 1977; Durlauf, 1996), social customs (Akerlof, 1980), the big push in industrialization (Murphy, Shleifer and Vishny, 1989), crime (Sah, 1991), education (Benabou, 1993), savings and consumption norms (Lindbeck, 1997), the transmission of culture (Bisin and Verdier, 2000), and the timing of desertion by soldiers (de Paula, 2005). A continuous action version of the same model, where an agent's utility depends on the average action of the population, is used by Cooper and John (1988) to model macroeconomic coordination failures. Much of this work has ignored market responses to the presence of social interactions. Among the exceptions are Becker and Murphy (2000), who produce a systematic analysis of the effect of prices on market behaviour when social interactions are present, and Pesendorfer (1995), who examines how a monopolist would exploit the presence of non-market interactions.

Models inspired in statistical physics

Schelling's (1971) paper sets out a model where individuals occupy discrete points on the line or plane and interact locally. However, most of the developments that followed use the simpler critical mass model. Follmer (1974) was the first to use explicitly a random field model (also known as an interactive particle system) imported from statistical physics to model social interactions. In these models one typically postulates an individual's interdependence and analyses the equilibrium behaviour that emerges. Typical questions concern the existence and multiplicity of equilibria that are consistent with the postulated individual behaviour, and the sensitivity of these equilibria to parameters. Follmer models an economy in which the preferences of an individual depend on the preference of his peers, and shows that randomness in individual preferences may affect the aggregate, even as the number of agents grows to infinity – a failure of the law of large numbers. Blume (1993) and Brock (1993) recognize the connection between models of discrete choice with interaction effects and some random field models.

Glaeser, Sacerdote and Scheinkman (1996) observe that crime rates across large American cities seem to vary too much to be explained by the usual socio-economic variables. They construct a theoretical model connecting the structure of social interactions among individuals with the variation of aggregate behaviour across space, providing a framework for investigating the importance of social interactions. They set up a simple model of local interactions inspired by the voter model in the literature on interacting particle systems (for example, Ligget, 1985) and show that the model is able to generate the large observed variance across aggregates from small amounts of variability in the fundamentals. A simple, one-sided version of their model works as follows. Individuals occupy discrete points on a circle and choose between two actions $\{0, 1\}$. With probability π, the individual chooses action 1 with probability p and action 0 with probability $1 - p$. With probability $(1 - \pi)$ he imitates his predecessor's action. The parameter $(1 - \pi)$ can be thought of as measure of the intensity of social interactions. In a population of n individuals, if we write a^i for the action of agent i then

$$\frac{1}{\sqrt{n}} \sum_{i=1}^{n} (a^i - p) \to N\left(0, \frac{2 - \pi}{\pi} p(1 - p)\right).$$

Although the limit average action is always p, the (limit) variance of the normalized average action across groups is a function of π. As π converges to zero, this variance becomes arbitrarily large. Social interactions increase the variance of the crime rate across population groups. In a similar vein, Topa (2001) examines the spatial distribution of unemployment with the aid of contact processes (for example, Ligget, 1985), another class of random field models, and shows that social interactions help explain the variation in unemployment rates among Chicago census tracts.

Brock and Durlauf (2001) develop a model that is very much in the spirit of Schelling's critical mass framework. Individuals choose between two actions, and the payoff an individual experiences when taking an action depends on a baseline utility that is common across individuals, on an idiosyncratic preference parameter, and on the distance between his action and the average expected action in the population. By making specific assumptions on the probability distribution of the idiosyncratic preferences, Brock and Durlauf obtain a joint probability measure over choices that is related to that of the mean-field version of the Curie–Weiss model of statistical mechanics. They then show that the model may have one or three equilibria, depending on the values of some parameters. Multiple equilibria are more likely to appear when the baseline utility of the two actions is not very different or when the desire for conformity is strong. Durlauf (1997) and Ioannides (1997; 2006) consider generalizations of the Brock–Durlauf framework with a richer interaction structure that accommodates local interactions. Horst and Scheinkman (2006), who do not use explicitly the language of random fields, also consider infinite systems with arbitrary interaction structures.

Most of this literature is static, but dynamics models, usually involving myopic agents, have been developed, for example, by Blume (1993), Blume and Durlauf (1999), Brock and Durlauf (2001), and Young (1993;

1998), who is especially interested in the evolution of social norms and customs.

The social multiplier

The *social multiplier* measures the ratio of the effect on the average action caused by a change in a parameter to the effect on the average action that would occur if individual agents ignored the change in actions of their peers. This social multiplier can also be thought of as a ratio $\frac{\Delta_P}{\Delta_I}$ where Δ_I is the average response of an individual action to an exogenous parameter (that affects only that person) and Δ_P is the (per capita) response of the peer group to a change in the same parameter that affects the entire peer group. Unless an equilibrium selection mechanism is present, the social multiplier is well defined only if the equilibrium average action is unique, but models that exhibit large social multipliers can explain large differences in outcomes across populations with small differences in exogenous variables. If agents have idiosyncratic random preferences, this same multiplier amplifies the differences in realizations of these preferences across samples. In models with continuous actions but otherwise fairly general interaction structures, Glaeser and Scheinkman (2003) and Horst and Scheinkman (2006) show that *moderate social interactions*, a condition that limits the effect of the actions by peers on the optimal choice of an individual, is sufficient for uniqueness of equilibrium. They also show that if, in addition, strategic complementarities are present then the social multiplier exceeds one. Typically, the forces that lead to multiple equilibria also lead to large social multipliers. For instance, in Brock and Durlauf's (2001) model, in the region where uniqueness prevails, the social multiplier is bigger when the desire to conform is stronger and when the fraction of agents that are close to being indifferent between the two possible actions is larger (see Glaeser and Scheinkman, 2003).

Choice of peer group

In several models (for example, Gabszewicz and Thisse, 1996; Benabou, 1993; Glaeser, Sacerdote and Scheinkman, 1996; Mobius, 2000) the peer group that concerns an agent is formed by geographical neighbours. Mobius (1999) shows that, in a context that generalizes Schelling's (1972) tipping model, the persistence of segregation depends on the particular form of the near-neighbour relationship.

Kirman (1983), Kirman, Oddou and Weber (1986), and Ioannides (1990) use random graph theory to treat the peer group relationship as random. This approach is particularly useful in deriving properties of the probable peer groups as a function of the original probability of connections. Another literature deals with individual incentives for the formation of networks (for example, Boorman, 1975; Jackson and Wolinsky, 1996; Bala and

Goyal, 2000). (A related problem is the formation of coalition in games; for example, Myerson, 1991.) Benabou (1993) and Glaeser and Scheinkman (2001) use Tiebout's equilibrium approach (see for example Bewley, 1981) to model peer group choice.

Empirical issues

Several statistical problems arise in estimating social interactions effects. It is often difficult for a researcher to identify correctly the peer groups. Another problem is that, ideally, one should distinguish between three effects in understanding group behaviour: correlation of individual characteristics, influences of group characteristics on individuals, and the influence of group behaviour on individual behaviour (Manski, 1993). Although the last two effects could both be merged into a social interactions effect, the correlation across individual error terms could remain a problem.

This problem does not arise in randomized experiments that allocate persons into different groups. Katz, Kling and Liebman (2001) and Ludwig, Hirschfeld and Duncan (2001) use data generated by the Moving to Opportunity experiment to provide evidence for the existence of neighbourhood spillovers on juvenile crime. Sacerdote (2001) exploits variation in peer groups generated by the random assignment of freshman roommates at Dartmouth and finds evidence of peer effects on academic effort, grade point average, and fraternity membership.

In the absence of randomized experiments, Case and Katz (1991) use peer group background characteristics as instruments for peer group outcomes, which in certain cases yield valid estimates of social interactions. They find some evidence that peer behaviour influences self-reported juvenile crime. However, as Manski (1993) stresses in the presence of correlations among unobservables, which is particularly likely to arise with sorting, the instrumental variables estimator may overstate social interactions.

Brock and Durlauf (2001) discuss structural identification for their model. In Brock and Durlauf (2000) they provide estimators for the parameters of the model and account for the endogeneity of peer groups. This structural approach leads to a natural behavioural interpretation of the parameters, but it requires individual-level observations and it may suffer from mis-specification.

A less structural approach, proposed by Glaeser, Sacerdote and Scheinkman (1996), is to use the variances of group average outcomes to identify social interactions. Using this methodology they found evidence that social interactions can help explain the large differences in community crime rates. This approach has been formalized further by Graham (2004). Another possibility is to use the logic of the multiplier. The relationship between exogenous variables and outcomes for individuals is compared with the relationship between exogenous variables and outcomes for groups. The ratio is a measure of the size of social interactions. Glaeser, Sacerdote and Scheinkman (2003) apply this method and find evidence of interactions in social group membership in the Dartmouth room-mates data and in crime, and of human capital spillovers at the state and the Public Use Microsample Area (PUMA) level. Yet another alternative is to identify the presence of interactions based on spatial clustering in the behavioural data. (see Topa, 2001; Conley and Topa, 2002). Finally, results in De Paula (2005) suggest that dynamic models might be easier to identify.

JOSÉ A. SCHEINKMAN

See also **ergodicity and nonergodicity in economics; neighbours and neighbourhoods; social capital; social interactions (empirics); social multipliers; statistical mechanics.**

Research supported by the National Science Foundation through grant SES 0350770. I thank Alberto Bisin, Aureo de Paula, Ed Glaeser, and Yannis Ioannides for comments.

Bibliography

Akerlof, G. 1980. A theory of social customs of which unemployment may be one consequence. *Quarterly Journal of Economics* 94, 749–75.

Arthur, W. 1989. Increasing returns, competing technologies and lock-in by historical small events: the dynamics of allocation under increasing returns to scale. *Economic Journal* 99, 116–31.

Bala, V. and Goyal, S. 2000. A non-cooperative model of network formation. *Econometrica* 68, 1181–229.

Becker, G. and Murphy, K. 2001. *Social Markets: Market Behavior in a Social Environment.* Cambridge, MA: Belknap-Harvard University Press.

Benabou, R. 1993. Workings of a city: location, education, and production. *Quarterly Journal of Economics* 107, 619–52.

Bewley, T. 1981. A critique of Tiebout's theory of local public expenditures. *Econometrica* 49, 713–40.

Bickhchandani, S., Hirshleifer, D. and Welch, I. 1992. A theory of fads, fashion, custom, and cultural exchange as information cascades. *Journal of Political Economy* 100, 992–1026.

Bisin, A. and Verdier, T. 2000. Beyond the melting pot: cultural transmission, marriage, and the evolution of ethnic and religious traits. *Quarterly Journal of Economics* 115, 955–88.

Blume, L. 1993. The statistical mechanics of strategic interaction. *Games and Economic Behavior* 5, 387–424.

Blume, L. and Durlauf, S. 1999. Equilibrium concepts for models with social interactions. Mimeo. Cornell University.

Boorman, S. 1975. A combinatorial optimization model for transmission of job information through contact networks. *Bell Journal of Economics* 6, 216–49.

Brock, W. 1993. Pathways to randomness in the economy: emergent nonlinearity and chaos in economics and finance. *Estudios Economicos* 81, 3–55.

Brock, W. and Durlauf, S. 2000. Interactions-based models. In *Handbook of Econometrics*, ed. J. Heckman and E. Learnerx Amsterdam: North-Holland.

Brock, W. and Durlauf, S. 2001. Discrete choice with social interactions. *Review of Economic Studies* 68, 235–60.

Case, A. and Katz, L. 1991. The company you keep: the effects of family and neighborhood on disadvantaged families. Working Paper No. 3705. Cambridge, MA: NBER.

Conley, T. and Topa, G. 2002. Socio-economic distance and spatial patterns in unemployment. *Journal of Applied Econometrics* 17, 303–27.

Cooper, R. and John, A. 1988. Coordinating coordination failures in Keynesian models. *Quarterly Journal of Economics* 103, 441–64.

de Paula, A. 2005. Inference in a synchronization game with social interactions. Mimeo. Department of Economics, Princeton University.

Duesenberry, J. 1949. *Income, Saving, and the Theory of Consumer Behavior.* Cambridge, MA: Harvard University Press.

Durlauf, S. 1993. Nonergodic economic growth. *Review of Economic Studies* 60, 349–66.

Durlauf, S. 1996. A theory of persistent income inequality. *Journal of Economic Growth* 1, 75–93.

Durlauf, S. 1997. Statistical mechanics approaches to socio-economic behavior. In *The Economy as an Evolving, Complex System II*, ed. W. Arthur, S. Durlauf and D. Lane. Reading, MA: Addison-Wesley.

Ellison, G. 1993. Learning, local interaction and coordination. *Econometrica* 61, 1047–72.

Follmer, H. 1974. Random economies with many interacting agents. *Journal of Mathematical Economics* 1, 51–62.

Gabszewicz, J. and Thisse, J.-F. 1986. Spatial competition and the location of firms. In *Location Theory*, ed. R. Arnott; *Fundamentals of Pure and Applied Economics 5*, ed. J. Lesourne and H. Sonnenschein. Amsterdam: Harwood Academic.

Glaeser, E., Sacerdote, B. and Scheinkman, J. 1996. Crime and social interactions. *Quarterly Journal of Economics* 111, 507–48.

Glaeser, E., Sacerdote, B. and Scheinkman, J. 2003. The social multiplier. *Journal of the European Economic Association* 1, 345–53.

Glaeser, E. and Scheinkman, J. 2001. Measuring social interactions. In *Social Dynamics*, ed. S. Durlauf and P. Young. Cambridge, MA: MIT Press.

Glaeser, E. and Scheinkman, J. 2003. Non-market interactions. In *Advances in Economics and Econometrics: Theory and Applications*, ed. M. Dewatripont, L. Hansen and S. Turnovsky. Cambridge: Cambridge University Press.

Graham, B. 2004. Identifying social interactions through excess variance contrasts. Mimeo. Harvard University.

Granovetter, M. 1978. Threshold models of collective behavior. *American Journal of Sociology* 83, 1420–43.

Horst, U. and Scheinkman, J. 2006. Equilibria in systems of social interactions. *Journal of Economic Theory* (forthcoming).

Ioannides, Y. 1990. Trading uncertainty and market structure. *International Economic Review* 31, 619–38.

Ioannides, Y. 1997. The evolution of trading structures. In *The Economy as an Evolving, Complex System II*, ed. W. Arthur, S. Durlauf and D. Lane. Reading, MA: Addison-Wesley.

Ioannides, Y. 2006. Topologies of social interactions. *Economic Theory* 28, 559–84.

Jackson, M. and Wolinsky, A. 1996. A strategic model of economic and social networks. *Journal of Economic Theory* 71, 44–74.

Katz, L., Kling, A. and Liebman, J. 2001. Moving to opportunity in Boston: early results of a randomized mobility experiment. *Quarterly Journal of Economics* 116, 607–54.

Kirman, A. 1983. Communication in markets: a suggested approach. *Economic Letters* 12, 1–5.

Kirman, A., Oddou, C. and Weber, S. 1986. Stochastic communication and coalition formation. *Econometrica* 54, 129–38.

Leibenstein, H. 1950. Bandwagon, snob, and Veblen effects in the theory of consumers' demand. *Quarterly Journal of Economics* 64, 183–207.

Ligget, T. 1985. *Interacting Particle Systems*. New York: Springer Verlag.

Lindbeck, A. 1997. Incentives and social norms in household behavior. *American Economic Review* 87, 370–77.

Loury, G. 1977. A dynamic theory of racial income differences. In *Women, Minorities and Employment Discrimination*, ed. P. Wallace and A. La Mond. Lexington, MA: Lexington Books.

Ludwig, J., Hirschfeld, P. and Duncan, G. 2001. Urban poverty and juvenile crime: evidence from a randomized housing-mobility experiment. *Quarterly Journal of Economics* 116, 665–79.

Mailath, G., Samuelson, L. and Shaked, A. 2001. Evolution and endogenous interactions. In *The Evolution of Economic Diversity*, ed. U. Pagano and A. Nicita. London: Routledge.

Manski, C. 1993. Identification of endogenous social effects: the reflection problem. *Review of Economic Studies* 60, 531–42.

Morris, S. 2000. Contagion. *Review of Economic Studies* 67, 57–78.

Mobius, M. 2000. The formation of ghettos as a local interaction phenomenon. Working Paper. Massachusetts Institute of Technology.

Murphy, K., Shleifer, A. and Vishny, R. 1989. Industrialization and the Big Push. *Journal of Political Economy* 97, 1003–26.

Myerson, R. 1991. *Game Theory: Analysis of Conflict*. Cambridge, MA: Harvard University Press.

Pesendorfer, W. 1995. Design innovation and fashion cycles. *American Economic Review* 85, 771–92.

Sacerdote, B. 2001. Peer effects with random assignment: results for Dartmouth roommates. *Quarterly Journal of Economics* 116, 681–704.

Sah, R. 1991. Social osmosis and crime. *Journal of Political Economy* 99, 1272–95.

Schelling, T. 1971. Dynamic models of segregation. *Journal of Mathematical Sociology* 1, 143–86.

Schelling, T. 1972. A process of residential segregation: neighborhood tipping. In *Racial Discrimination in Economic Life*, ed. A. Pascal. Lexington, MA: Lexington Books.

Schelling, T. 1978. *Micromotives and Macrobehavior*. New York: Norton.

Topa, G. 2001. Social interactions, local spillovers and unemployment. *Review of Economic Studies* 68, 261–95.

Veblen, T. 1934. *The Theory of the Leisure Class: An Economic Study of Institutions*. New York: Modern Library.

Young, H. 1993. The evolution of conventions. *Econometrica* 61, 57–84.

Young, H. 1998. *Individual Strategy and Social Structure: An Evolutionary Theory of Institutions*. Princeton, NJ: Princeton University Press.

social learning

Since the 1950s, people living in developing countries have gained access to technologies, such as high-yielding agricultural seed varieties and modern medicine, that have the potential to dramatically alter the quality of their lives. Although the adoption of these technologies has increased wealth and lowered mortality in many parts of the world, their uptake has been uneven. Entire communities sometimes stubbornly oppose the use of modern medicine or contraceptives. And while high-yielding crop varieties might have spread widely, it took as long as two decades in some cases for this readily available technology to be adopted.

The traditional explanation for the observed differences in the response to new opportunities, across and within countries, is based on heterogeneity in the population (Griliches, 1957; Mansfield, 1968). An alternative explanation, which has grown in popularity in recent years, is based on the idea that individuals are often uncertain about the returns from a new technology. For example, farmers might not know the (expected) yield that will be obtained from a new and uncertain technology and young mothers might be concerned about the side effects from a new contraceptive. In these circumstances, a neighbour's decision to use a new technology indicates that she must have received a favourable signal about its quality and her subsequent experience with it serves as an additional source of information. Because information must flow sequentially from one neighbour to the next, social learning provides a natural explanation for the gradual diffusion of new technology even in a homogeneous population. Social learning can also explain the wide variation in the response to external interventions across otherwise identical communities, simply as a consequence of the randomness in the information signals that they received.

Early contributions by Banerjee (1992) and Bikhchandani, Hirshleifer, and Welch (1992) gave rise to an enormous theoretical literature on social learning (see Bramoulle and Kranton, 2004, for an excellent summary). This article, however, is concerned with a smaller empirical literature on social learning and economic development that has emerged in recent years.

The adoption of new agricultural technology

Following Munshi (2004), consider a simple model of agricultural investment in which there are two technologies: a new high yielding variety (HYV) and a traditional variety. The yield from the uncertain HYV technology for grower i in period t is specified as

$$y_{it} = y(Z_i) + \eta_{it} \qquad (1)$$

where $y(Z_i)$ is the yield under normal growing conditions and Z_i is a vector of soil characteristics and prices. η_{it} is a mean-zero serially independent disturbance term with variance λ_i^2, measuring deviations from normal growing conditions that cannot be observed by the grower.

When $y(Z_i)$ is uncertain, the optimal acreage allocated to HYV by the risk-averse grower is increasing in his best estimate of the expected yield $\hat{y}_{it}$ and decreasing in the variance of that estimate σ_{it}^2, as well as λ_i^2. Because the grower's expected utility is declining in σ_{it}, he will utilize all the information about $y(Z_i)$ that is available to him to arrive at his best estimate $\hat{y}_{it}$.

At the beginning of each period the grower receives an unbiased information signal about the value of his expected yield. He combines this signal with his prior to compute his best estimate of his expected yield, which in turn determines the acreage that he allocates to HYV. Subsequently, he observes his neighbours' acreage decisions, which reveal the signals that they received, as well as all the yields realized in the village. This social information is used to update the grower's prior for the next period. Under the assumption that the expected yield is constant across growers, the acreage function is additively separable in the expected yield, and that the individual information signals are normally distributed, Munshi derives the grower's acreage A_{it} as a linear function of his lagged acreage A_{it-1}, lagged (mean) acreage in the village $\overline{A}_{t-1}$, and lagged yields $\overline{y}_{t-1}$:

$$A_{it} = \pi_0 + \pi_1 A_{it-1} + \pi_2 \overline{A}_{t-1} + \pi_3 \overline{y}_{t-1} + \varepsilon_{it}. \qquad (2)$$

When information is pooled efficiently within the village, A_{it-1} contains all the information about the

expected yield y that was available at the beginning of period $t-1$, specifically the entire history of information signals and yield realizations up to that time. Conditional on A_{it-1} and fixed individual characteristics subsumed in ε_{it}, $\overline{A}_{t-1}$ represents the new information that was received by the village in period $t-1$ through the exogenous signals. $\overline{y}_{t-1}$, in turn, represents the information that was obtained from the yield realizations in that period.

As Manski (1993) notes, neighbours' past decisions will be correlated with the grower's current decision if any determinant of that decision is correlated across neighbours and over time. The prospects for the identification of social learning improve, however, when we focus on the relationship between the current decision and lagged yield realizations. The information signal received by the grower in period t, u_{it}, determines his acreage decision A_{it} (through his expected yield estimate), and so is subsumed in ε_{it}. Growers are never systematically misinformed, $E(u_{it})=y$, and so $\overline{y}_{t-1}$ and ε_{it} will be correlated as well. One solution to this problem is to difference $\overline{y}_{t-2}$ from $\overline{y}_{t-1}$, leaving us with $\overline{\eta}_{t-1} - \overline{\eta}_{t-2}$ from eq. (1). The acreage response to *yield shocks* in the village would then identify the presence of social learning.

By specifying yield to be the sum of a constant term $y(Z_i)$ and an idiosyncratic shock η_{it}, we implicitly assume that input markets function smoothly and that input and output prices do not change over time. In practice, changes in the yield from period $t-2$ to $t-1$ could reflect changes in prices or access to scarce resources (such as credit) that are unobserved by the econometrician but directly determine the grower's period-t acreage decision. A spurious yield (shock) effect could in that case be obtained.

To provide additional support for the presence of social learning, Munshi exploits differences in the diffusion of information across crops. Although we have assumed that the expected yield is the same across growers, it will more generally depend on the grower's characteristics. The grower could condition for differences between his own and his neighbours' observed characteristics when learning from them, but the prospects for social learning decline once we allow for the possibility that some of these characteristics will be unobserved, or imperfectly observed. Take the case where all the neighbours' characteristics are unobserved by the grower. He could rely on his own information signals and yield realizations, ignoring information from his neighbours, to obtain a consistent but inefficient estimate of his expected yield. Alternatively, he could continue to learn from his neighbours' decisions and yields, but some bias will then inevitably appear. The testable prediction is that the grower will choose individual learning if the population is heterogeneous and the yield with the new technology is sufficiently sensitive to unobserved characteristics; otherwise he will prefer to learn from his neighbours.

Munshi tests this prediction with farm-level data on wheat and rice cultivation over a three-year period at the onset of the Indian Green Revolution. Rice-growing areas in India are characterized by wide variation in soil characteristics. The early rice varieties were also sensitive to soil characteristics such as salinity, as well as to managerial inputs, which are difficult to observe. As predicted, HYV acreage responds to lagged yield shocks in the village with wheat but not rice. If the view that rice growers were informationally disadvantaged is correct, then these growers should have compensated for their lack of information by experimenting on their own land. Munshi shows that rice adopters allocate more land to HYV than wheat adopters, despite the fact that their farms are smaller and the likelihood of HYV adoption is significantly higher for wheat.

While Munshi assumes that the yield $y(Z_i)$ is exogenous and uncertain, Foster and Rosenzweig (1995) assume that the grower's objective is to learn his optimal (profit-maximizing) input use Z_i. The point of departure for their work is the target-input model of Jovanovic and Nyarko (1996), but we will see that the signal extraction aspect and, hence, the basic structure of the learning process remains the same across these different models of learning. With a slight change of notation, Foster and Rosenzweig assume that the grower attempts to learn the optimal or target input use on his land θ^*,

$$\tilde{\theta}_{ijt} = \theta^* + u_{ijt} \tag{3}$$

where $\tilde{\theta}_{ijt}$ is the optimal input use on plot i for farmer j in period t and u_{ijt} is an i.i.d. random variable with a (known) variance σ_u^2. Notice the similarity with eq. (1), where the grower's objective was to learn the value of the (expected) yield $y(Z_i)$. Previously he collected information on $y(Z_i)$ from various sources to finally arrive at the optimal acreage. In the current set-up, the grower collects information on θ^* from various sources to arrive at his profit-maximizing input level.

Summing the profit over all plots and taking expectations, the grower's expected profit can be expressed as

$$\pi_{jt} = \left[\eta_h - \frac{\eta_{ha}}{A_j} \frac{H_{jt}}{2} - \sigma_{\theta jt}^2 - \sigma_u^2 \right] H_{jt} + \eta_a A_j, \tag{4}$$

where η_a is the yield from traditional varieties, η_h is the yield from HYV on the plot most suitable for the HYV technology, and η_{ha} represents the loss from using land less suitable for HYV. A_j is the total number of plots and H_{jt} is the number of plots allocated to HYV. The term in square brackets above represents the yield from HYV, which is declining in two sources of error: σ_u^2 and $\sigma_{\theta jt}^2$. σ_u^2 corresponds to λ^2 in Munshi's model and cannot be avoided. However, $\sigma_{\theta jt}^2$, which is associated with suboptimal input use and corresponds to σ_{it}^2 in Munshi's

framework, will go to zero as the grower learns the value of θ^* from his own and his neighbours' experiences.

Munshi's test of social learning is based on the relationship between the grower's current HYV acreage and his neighbours' lagged HYV yields. In contrast, Foster and Rosenzweig derive implications for the relationship between the grower's profit (yield) with HYV and cumulative experience with the new technology corresponding to eq. (4):

$$\pi_{jt} = (\eta_h + \beta_{ot} S_{jt} + \beta_{vt} \bar{S}_{-jt}) H_{jt} + \eta_a A_{jt} + \xi_{jt}$$

where the term in parentheses represents the profit (yield) from HYV, which is increasing in the cumulative experience with the new technology on own land S_{jt} and neighbours' land S_{-jt}. The potential sources of spurious correlation that arise in Munshi's analysis evidently apply here as well; time-varying changes in growing conditions or access to scarce resources would affect current HYV yield as well as S_{jt}, $\bar{S}_{-jt}$. Once again it is possible to appeal to the restrictions from the theory to provide additional support for the presence of social learning. While Munshi shows that the effect of neighbours' past decisions and experiences on the grower's current decision will vary across crops, depending on growing conditions and the technology, Foster and Rosenzweig derive predictions for changes in the pattern of learning over time. Foster and Rosenzweig's learning model generates the predictions that (i) β_{ot}, β_{vt} will be declining over time, and (ii) $\frac{\beta_{ot}}{\beta_{vt}} = \frac{\beta_{ot+1}}{\beta_{vt+1}}$. These predictions are successfully tested, consistent with the presence of social learning.

The fertility transition

In Munshi and Myaux's (2006) simple model of the fertility transition no one regulates fertility prior to the inception of a family planning programme. While this remains a potential equilibrium, a new equilibrium in which a sufficient fraction of the community uses convenient modern contraceptives to regulate fertility could also emerge. The object of interest in this case might not be the performance of the new contraceptive (or the side effects associated with its use) but the nature of the future social norm. Individuals gradually learn about the equilibrium that will ultimately prevail in their community as they interact sequentially with each other over time. These changes in beliefs can be mapped into changes in actions: the probability that the individual chooses modern contraceptives in period t is determined by her decision in period $t-1$ and the probability that she interacted with a user in that period. With random interactions within the community, this last probability is in turn measured by the proportion of users in the community in period $t-1$.

As with the identification of social learning in agriculture, estimation of the individual decision-rule described above is complicated by the fact that lagged contraceptive prevalence in the community could proxy for any unobserved determinant of the contraception decision that is correlated across individuals and over time. However, the model of the fertility transition as a process of changing social norms places additional restrictions on the relationship between the individual's contraception decision and lagged contraceptive prevalence in the local area. In particular, social effects should be restricted to the narrow social group within which norms restricting fertility were traditionally enforced. Contraceptive prevalence outside that social group should have no effect on the individual's contraception decision.

Munshi and Myaux test these predictions using data from rural Bangladesh. In their research setting, the traditional norm was characterized by early and universal marriage, followed by immediate and continuous childbearing. Religious authority provided legitimacy and enforced the rules that sustained this equilibrium. Changes in social norms should then have occurred independently within religious groups (Hindus and Muslims) within the village. As predicted, lagged contraceptive prevalence within the individual's religious group within the village has a strong effect on her contraception decision whereas cross-religion effects are entirely absent, both for Hindus and for Muslims.

Education

A recent paper (Yamauchi, 2007) studies social learning and investment in education with the same three-year farm panel at the onset of the Indian Green Revolution that was used by Foster and Rosenzweig (1995) and Munshi (2004). Schooling levels among the growers in the sample were determined long before the unexpected availability of the new HYV technology and so the returns to schooling can be estimated directly at the level of the village using realized incomes. A positive relationship between schooling enrolment among the children and the returns to schooling in the previous generation is then seen to be indicative of social learning.

The usual problem when testing this prediction is that returns to education in the previous generation could be correlated with returns in the current generation, which directly determine school enrolment but are unobserved by the econometrician. Yamauchi consequently derives testable predictions that provide additional support for the presence of social learning at the level of the village. He shows formally that social learning will be faster when the income variance is lower and when there is *greater* heterogeneity in educational attainment in the village. This last prediction is not inconsistent with Munshi's observation that social learning will be slower in heterogeneous populations where neighbours' characteristics are unobserved or imperfectly observed. Schooling is an easily observed characteristic and Yamauchi's insight is that more variance in this characteristic leads to more

precise estimates of the returns to schooling. Matching these predictions, schooling enrolment among the children is increasing in the returns to schooling in the village in the previous generation and, more importantly, is increasing in the interaction of the returns and the variance in educational attainment in the village.

Conclusion

How important is social learning in the development process? Foster and Rosenzweig report results from a simulation exercise based on the estimated parameters from their learning model, which compares profits without learning, with learning from own experience, and with learning from own and neighbours' experience. Profitability from the new HYV was lower than profitability from the traditional variety to begin with, but HYV profits exceed traditional profits after four years of experience without learning. With social learning, this point is reached one year earlier. Similarly, Yamauchi's simulations indicate that an increase in schooling inequality within the village could increase enrolment levels by nearly ten per cent. Access to social information appears to be readily available in many practical applications, particularly since there is little cost to the individual from providing information to his neighbours. Thus, the value of interventions that provide the seed for the subsequent spread of such information could be quite high. Understanding how best to design such interventions would seem to be an important area for future research.

KAIVAN MUNSHI

See also **agricultural economics; education in developing countries; fertility in developing countries; learning and information aggregation in networks; religion and economic development; social norms; technical change; technology.**

Bibliography

Banerjee, A.V. 1992. A simple model of herd behavior. *Quarterly Journal of Economics* 117, 797–817.

Bikhchandani, S., Hirshleifer, D. and Welch, I. 1992. A theory of fads, fashion, custom, and cultural change as informational cascades. *Journal of Political Economy* 100, 992–1026.

Bramoulle, J. and Kranton, R. 2004. Public goods in social networks: how networks can shape social learning and innovation. Working paper, University of Maryland.

Foster, A.D. and Rosenzweig, M.R. 1995. Learning by doing and learning from others: human capital and technical change in agriculture. *Journal of Political Economy* 103, 1176–209.

Griliches, Z. 1957. Hybrid corn: an exploration in the economics of technological change. *Econometrica* 25, 501–22.

Jovanovic, B. and Nyarko, Y. 1996. Learning by doing and the choice of technology. *Econometrica* 64, 1299–310.

Mansfield, E. 1968. *The Economics of Technological Change.* New York: Norton.

Manski, C.F. 1993. Identification of endogenous social effects: the reflection problem. *Review of Economic Studies* 60, 531–42.

Munshi, K. 2004. Social learning in a heterogeneous population: technology diffusion in the Indian Green Revolution. *Journal of Development Economics* 73, 185–213.

Munshi, K. and Myaux, J. 2006. Social norms and the fertility transition. *Journal of Development Economics* 60, 1–38.

Yamauchi, F. 2007. Social learning, neighborhood effects, and investment in human capital: evidence from Green-Revolution India. *Journal of Development Economics* 83, 37–62.

social multipliers

A *social multiplier* arises when choices are subject to social interactions that lead to strategic complementarity, whereby an increase in the level of an action among others in a relevant social group leads to an increase in the same action at the individual level. While multiplier effects may arise in diverse contexts, including that of inter-firm behaviour (see, for example, Cooper and John, 1988), the specific notion of a social multiplier refers to the effects of choice interdependence among individuals in social contexts such as families, neighbourhoods, schools, professional associations, or friendship networks, within which individual choices are thought to be influenced by the choices of others directly, rather than via the effect of others' choices on market variables. Such influences are also referred to as *endogenous* (social) effects. Under such interactions, a change in fundamentals exerts a direct effect on individual action, with the actions of others held fixed, and an indirect effect in the same direction representing the complementary response to the fundamentals-induced change in others' actions. Around a stable equilibrium, the total effect (direct plus indirect) of a change in fundamentals on the average equilibrium action exceeds (in magnitude) the average partial (that is, direct) effect of the change. The social multiplier can be defined as the ratio of the former effect to the latter effect, and therefore exceeds 1 (around a stable equilibrium) under the assumption of strategic complementarity. (The effects of changes in fundamentals around an unstable equilibrium are discussed below.) Social multipliers have been invoked to explain large differences in behaviour across groups that are not readily explained by variation in group characteristics, as well as improbably large swings over time in social practices.

To illustrate formally, consider a choice model in which each agent in a large but finite group (size N) chooses the level of an action, $x_i \geq 0$, according to a reaction function, $x_i = f_i(p, \theta_i, \bar{x})$, where p is a common exogenous parameter such as the unit price of the action, θ_i is an individual factor such as inherent taste for the action or income (presumed exogenous), and $\bar{x} \equiv \frac{1}{N}\sum_i x_i$ is the (presumed or expected) average action level in the population. (The individual is included in the population average on the assumption that his influence on average action is negligible; this assumption is relaxed below.) The model is said to involve *global* social interactions in the sense that individual action is complementary to the average action in the relevant population, rather than depending only on the actions of some subset of the population. Assume without loss of generality that $\partial x_i/\partial p \leq 0$ and $\partial x_i/\partial \theta_i \geq 0$; strategic complementarity implies $\partial x_i/\partial \bar{x} > 0$. Equilibrium is a set of actions, x_i^e, for $i = 1, 2, 3, \ldots, N$, such that $x_i^e = f_i(p, \theta_i, \bar{x}^e)$ for each i, and $\bar{x}^e = \frac{1}{N}\sum_i x_i^e$.

If we evaluate the effects of a price change around a stable equilibrium, the social multiplier is defined by

$$ M \equiv \frac{\frac{d\bar{x}^e}{dp}}{\frac{1}{N}\sum_i \frac{\partial x_i}{\partial p}}. $$

(The existence of equilibrium, stable or otherwise, is not guaranteed under the minimal conditions given here, but we assume existence for simplicity. See Glaeser, Sacerdote and Scheinkman, 2003, for discussion of existence conditions.) It can be shown (Becker and Murphy, 2000) that the multiplier can be expressed as

$$ \frac{1}{1 - \frac{1}{N}\sum_i \frac{\partial x_i}{\partial \bar{x}}}. $$

The last expression shows how the strength of the social interactions, as indicated by the term $\frac{1}{N}\sum_i \frac{\partial x_i}{\partial \bar{x}}$, or γ for short, influences the magnitude of the multiplier. For example, if $\gamma = .6$, then the multiplier (M) equals 2.5. This means that the equilibrium effect on the average action of an exogenous change in price is two and a half times as great (in absolute magnitude, and in the same direction) as the average partial effect of the price change on individual action. When social interactions are 'moderate' in the sense that $\gamma < 1$ around the given equilibrium (Glaeser and Scheinkman, 2003), the value of the multiplier exceeds 1 (provided also $\gamma > 0$) and the model exhibits multiplier effects in the intended sense. The consequences of strong social interactions, that is, $\gamma \geq 1$, are discussed below.

In the same model, variation in the vector of individual factors, $\{\theta_1, \theta_2, \ldots, \theta_N\}$, of the population also generates a social multiplier. With some additional assumptions we can define the social multiplier for a change in the average value of θ_i, denoted $\bar{\theta}$. Assuming that the partial effects, $\partial x_i/\partial \theta_i$, are constant and identical across individuals, the effect on the average equilibrium action of a marginal change in $\bar{\theta}$ does not depend on the composition of the changes in the underlying θ_i values, and so is well defined. The social multiplier is now given by the ratio:

$$\frac{d\bar{x}^e}{\frac{d\bar{\theta}}{\frac{\partial x_i}{\partial \theta_i}}}.$$

Again it can be shown that this ratio equals $\frac{1}{1-\gamma}$, where $\gamma = \frac{1}{N}\sum_i \frac{\partial x_i}{\partial \bar{x}}$, so the magnitude of the multiplier around a given equilibrium does not depend on the underlying independent variable (p or $\bar{\theta}$). (If the population is small in the sense that a change in any individual x_i exerts a non-negligible effect on $\bar{x}$, then a change in any single θ_i also sets off multiplier effects. See, for example, Cooper and John, 1988; and Glaeser, Sacerdote and Scheinkman, 2003. This analysis may be seen as applying to small group dynamics such as classrooms.)

A critical implication of social multipliers for policy analysis is that failure to consider multipliers may lead to significant underestimation of the aggregate effects of interventions when these are based on estimates of the effect of the intervention at the individual level. Conversely, elasticities measured at the aggregate level will overestimate the average individual response to a change in a variable affecting that individual alone.

Behaviour of the multiplier

Given the formulation that $M = \frac{1}{1-\gamma}$, we see that the multiplier increases in γ over the open interval $[0,1)$, diverging in the limit as γ approaches 1. However, in the neighbourhood of a stable equilibrium, it can be shown that $\gamma < 1$, guaranteeing a finite social multiplier. This latter condition says that, for the average individual, a change in average peer action leads to a less than commensurate change in individual action, *ceteris paribus*.

If social interactions are not moderate (that is, $\gamma \geq 1$) in the neighbourhood of equilibrium, the equilibrium is unstable. In addition, when $\gamma \geq 1$ over some or all the domain of mean peer behaviour, multiple equilibria or non-existence of equilibrium may obtain. Comparative statics around unstable equilibria are possible (provided $\gamma \neq 1$), but in such cases the aggregate equilibrium response to a parameter takes the opposite direction from the partial individual response: for example, $\gamma = 1.5$ yields $M = -2$. Consequently, multiplier effects – defined in the sense that the aggregate equilibrium response reinforces the partial individual response – do not apply. Becker and Murphy (2000) define the social multiplier as the value of γ itself and so do not restrict social multipliers to effects around stable equilibria. However, they too discount the relevance of comparative statics around unstable equilibria, instead invoking instability to help explain phenomena such as fads, which are characterized by explosive growth in popularity followed by precipitous declines in same.

Applications

The lesson of social multipliers is that relatively small differences in fundamentals can lead to large differences in outcomes. This insight has been used to help explain large differences in crime rates across cities (Glaeser, Sacerdote and Scheinkman, 1996), in unemployment rates across census tracts (Topa, 2001), in medical treatment rates across hospital market regions (Bell, 2002; and Burke, Fournier and Prasad, 2006), and in stock market participation across social groups (Hong, Kubik and Stein, 2004), among other patterns not readily explained on the basis of the relevant fundamentals. Conventional wisdom suggests that peer effects may be an important determinant of educational outcomes, and this is the subject of much active research, but conclusive findings on the significance of social multipliers on academic achievement have not yet emerged, given the difficulties of empirical estimation.

Social multiplier effects may accelerate shifts in social norms over time initiated by technological change. Goldin and Katz (2002) link the birth control pill, via direct effects as well as indirect multiplier effects, to the dramatic increases in women's career investment and age of first marriage in the 1970s. Contraception has also been linked to the large increases in out-of-wedlock births since the 1960s (Akerlof, Yellen and Katz, 1996), where the technology's direct impact eventually served, via social interactions, to erode the associated social stigma against out-of-wedlock births. More recently, it has been argued that social multiplier effects have magnified the impact of technological change on obesity rates in the United States in recent decades and led to a larger value of the social norm for body size (Burke and Heiland, 2007).

Extensions

Analysis of social interactions under discrete choice has been described extensively in Brock and Durlauf (2001) and Blume and Durlauf (2003). Utility is given a random component in these models, so that individual behaviour varies probabilistically with exogenous factors, and in general mean behaviour does not vary linearly with mean group characteristics. The magnitude of the social multiplier, as well as the existence and number of equilibria, again depends on the intensity of social influence on individual choice, although the analytic representation of the multiplier is less straightforward than in the linear, continuous choice model (see Glaeser and Scheinkman, 2003). This discrete model framework has been applied extensively in the analysis of qualitative, non-market choices such as dropping out of school and engaging in criminal behaviour, choices that are intuitively viewed as particularly susceptible to social influences. The model has the further advantage of being readily amenable to empirical analysis, as the structural choice equation maps directly into a logistic (or, with modification, probit) specification.

Social multipliers also arise in models of *local* social interactions, in which individual action is influenced only by the actions of others in some 'local' subset of the population as defined within a particular spatial model. For example, individuals could be situated at points on a circle and the influential peer group defined as the neighbours to the immediate left and the immediate right. Under this structure, and on the assumption of symmetric and linearly separable reaction functions under continuous choice, the social multiplier (as defined above) equals $\frac{1}{1-\lambda}$, where λ represents the marginal effect of average action among the (local) peer group on individual action. The magnitude of the multiplier therefore reveals the strength of the social interactions but not necessarily their topology. Models on lines, lattices, tori, and other spaces are also possible, as are numerous other specifications of the reference group. Multipliers typically arise, but the specific results derived here do not hold in general. See, for example, Ioannides (2006).

Mean group characteristics, $\bar{\theta}$, may be a direct source of externalities on individual outcomes. In the presence of such an externality, dubbed an *exogenous* or *contextual* effect (Manski, 1993), variation in some mean group characteristic may induce an effect on the mean outcome that exceeds the effect that would be predicted on the basis of the influence of the characteristic at the individual level. For example, if girls tend to have higher test scores than boys, and if an increase in the classroom proportion of girls reduces disruption and so enables higher test scores for all, the total effect on average test scores of an increase in the proportion of girls will exceed the averaged individual gender effects. (However, if girls have lower test scores on average, the two effects of gender composition will oppose each other.) Therefore, even in the absence of *endogenous* effects, for example even if an increase in mean peer achievement has no effect on individual achievement, exogenous interactions may create the appearance of a multiplier effect. While Manski adopts the stance that social multipliers occur only in the presence of endogenous social interactions, on the grounds that contextual effects are merely additive, Glaeser, Sacerdote and Scheinkman (2003) endorse a broader definition of multipliers that includes the contributions of both exogenous and endogenous social interactions. Although the two types of interactions are analytically distinct, it may be impossible to separate their respective effects empirically. Manski (1993) was the first to note this identification problem. Subsequent progress in distinguishing between exogenous and endogenous interactions in the linear model has been made by Brock and Durlauf (2001) and Cohen-Cole (2006).

The canonical derivation of social multipliers relies on a static decision model. However, recent research shows that results may not extend to dynamic decision contexts. In a life-cycle consumption model with social interactions, Binder and Pesaran (2001) show that the emergence of social multipliers depends on the distribution of patience parameters in the economy. Bisin, Horst and Ozgur (2006) show that multiplier effects may disappear under forward-looking behaviour and that the magnitude of social multipliers (in both static and dynamic settings) depends on the amount of information people have about other people's types. These findings indicate that the empirical relevance of social multipliers is likely to depend not just on the salience of social influences but also on the nature of intertemporal decision-making, both of these factors being the subject of significant ongoing debate within economics.

Measurement

The researcher who seeks to measure social multipliers encounters a number of formidable identification problems inherent in the empirical analysis of social interactions. One of the most pervasive challenges concerns the non-random selection of individuals into social groups, resulting in a spurious correlation of actions within groups based on similarities in underlying traits, some of which inevitably go unobserved (an example of *correlated effects*, in Manski's terminology). Under such conditions, naive estimation strategies may produce the appearance of a social multiplier where none exists. Even when all relevant sources of correlated effects can be controlled for, it may be possible to identify only the combined effects of endogenous and exogenous social interactions. The attempt to overcome these problems constitutes an area of active research in econometrics. Since identification conditions depend on the nature of the structural model, qualitative research into the empirical context of interest can serve as an important complement to quantitative analysis.

In a model in which individual action depends linearly on mean peer-group action and linear in all other factors, the definition of the social multiplier as a ratio of marginal effects suggests a regression-based estimation strategy. As explicated in Glaeser, Sacerdote and Scheinkman (2003) and Graham and Hahn (2005), the method involves estimating the respective reduced form equations for mean group behaviour and individual behaviour and taking the ratio of the estimated coefficients on mean group characteristics and individual characteristics, using cross-sectional data on groups defined, for example, by geographic boundaries. Under random assignment into groups, this ratio measures the social multiplier at the given level of aggregation. A multiplier value significantly greater than 1 is consistent with the presence of strategic complementarities deriving from endogenous social interactions. However, if exogenous effects of mean group characteristics cannot be ruled out, the measured multiplier captures both endogenous and exogenous effects due to the collinearity of mean characteristics and mean outcomes. In discrete-choice social interactions models, this collinearity does not hold, however, so endogenous and exogenous effects can be identified separately provided social effects in general can be identified (see Brock and Durlauf, 2001; 2007).

Using variants of this ratio method, Glaeser, Sacerdote and Scheinkman (2003) estimate social multipliers on criminal activity in the United States ranging from 1.72 at the county level to 2.8 at the state level and 8.16 at the national level. As the authors acknowledge, the estimates must be viewed sceptically, given that they do not control for sorting on unobservable demographic factors. Although some models predict that multipliers should increase with the level of aggregation, biases due to unobserved factors may also increase with the level of aggregation.

Graham (2006) employs an alternative measurement of social multipliers, also in the linear context, based on 'excess variance contrasts' that controls for some types of group-level heterogeneity, in his example the effects of teacher inputs on student achievement. The method relies on variation in average class size across sets of classrooms, where a Tennessee policy ensured random assignment of students and teachers to classrooms on the basis of size. Graham obtains point estimates of 1.76 and 1.97 for the social multipliers on maths and reading achievement in kindergarten, respectively, and shows that his estimation method will detect (true) multiplier effects in many cases in which standard regression methods would not. However, the method requires random assignment into groups on the basis of size, and does not separate exogenous from endogenous effects.

Although nonlinearities in the structural model may alleviate the problem of isolating endogenous social multipliers from exogenous effects, they do not alleviate the problems caused by non-random group selection. Although selection biases cannot be eliminated with complete confidence, they can be mitigated in a number of ways. In addition to including various fixed effects as data permit, researchers may exploit information about the selection process itself, as in Ioannides and Zabel (2003), or by eliminating selection altogether, as in experimental research (see, for example, Ichino and Falk, 2006) and in research that exploits natural experiments (Hoxby and Weingarth, 2006). In each of these last three examples, the evidence points to significant social multipliers (on housing demand, worker output, and school achievement, respectively). But even in cases of random group assignment results may be questioned: recent evidence based on a simulated interaction environment shows that an estimated social multiplier, when measured among randomly assigned groups, may be biased downward relative to its true value (Arcidiacono et al., 2005).

MARY. A. BURKE

See also **consumption externalities; interdependent preferences; social interactions (empirics); social interactions (theory); social learning; social norms; social networks, economic relevance of.**

Bibliography

Akerlof, G., Yellen, J.L. and Katz, M. 1996. An analysis of out-of-wedlock childbearing in the United States. *Quarterly Journal of Economics* 111, 277–317.

Arcidiacono, P., Foster, G., Goodpaster, N. and Kinsler, J. 2005. Estimating spillovers in the classroom with panel data. Mimeo, Duke University.

Becker, G. and Murphy, K.M. 2000. *Social Economics: Market Behavior in a Social Environment.* Cambridge, MA: Belknap-Harvard University Press.

Bell, A.M. 2002. Locally interdependent preferences in a general equilibrium environment. *Journal of Economic Behavior and Organization* 47, 309–33.

Binder, M. and Pesaran, M.H. 2001. Life-cycle consumption under social interactions. *Journal of Economic Dynamics and Control* 25, 35–83.

Bisin, A., Horst, U. and Ozgur, O. 2006. Rational expectations equilibria of economies with local interactions. *Journal of Economic Theory* 127, 74–116.

Blume, L.E. and Durlauf, S. 2003. The interactions-based approach to socioeconomic behavior. In *Social Dynamics*, ed. S. Durlauf and H.P. Young. Cambridge, MA: MIT Press.

Brock, W. and Durlauf, S. 2001. Interactions based models. In *Handbook of Econometrics*, vol. 5, ed. J. Heckman and E. Leamer. Amsterdam: North-Holland.

Brock, W. and Durlauf, S. 2007. Identification of binary choice models with social interactions. *Journal of Econometrics* 140, 52–75.

Burke, M.A., Fournier, G. and Prasad, K. 2006. The emergence of local norms in networks. *Complexity* 11, 65–83.

Burke, M.A. and Heiland, F. 2007. Social dynamics of obesity. *Economic Inquiry* 45, 571–591.

Cohen-Cole, E. 2006. Multiple groups identification in the linear-in-means mode. *Economics Letters* 92, 157–62.

Cooper, R. and John, A. 1988. Coordinating coordination failures in Keynesian models. *Quarterly Journal of Economic Research* 103, 441–63.

Glaeser, E., Sacerdote, B. and Scheinkman, J. 1996. Crime and social interactions. *Quarterly Journal of Economics* 111, 507–48.

Glaeser, E., Sacerdote, B. and Scheinkman, J. 2003. The social multiplier. *Journal of the European Economic Association* 1, 345–53.

Glaeser, E. and Scheinkman, J. 2003. Non-market interactions. In *Advances in Economics and Econometrics: Theory and Applications, Eighth World Congress*, eds. M. Dewatripont, L.P. Hansen and S. Turnovsky. Cambridge: Cambridge University Press.

Goldin, C. and Katz, L. 2002. The power of the pill: oral contraceptives and women's career and marriage decisions. *Journal of Political Economy* 110, 730–70.

Graham, B. 2006. Identifying social interactions through conditional variance restrictions. Working paper, Department of Economics, University of California, Berkeley.

Graham, B. and Hahn, J. 2005. Identification and estimation of the linear-in-means model of social interactions. *Economics Letters* 88, 1–6.

Hong, H., Kubik, J. and Stein, J. 2004. Social interaction and stock-market participation. *Journal of Finance* 59, 137–63.

Hoxby, C. and Weingarth, G. 2006. Taking race out of the equation: school reassignment and the structure of peer effects. Mimeo, Harvard University.

Ichino, A. and Falk, A. 2006. Clean evidence on peer effects. *Journal of Labor Economics* 24, 39–57.

Ioannides, Y. 2006. Topologies of social interactions. *Economic Theory* 28, 559–84.

Ioannides, Y. and Zabel, J.E. 2003. Neighborhood effects and housing demand. *Journal of Applied Econometrics* 18, 563–84.

Manski, C. 1993. Identification of endogenous social effects: the reflection problem. *Review of Economic Studies* 60, 531–42.

Topa, G. 2001. Social interactions, local spillovers, and unemployment. *Review of Economic Studies* 68, 261–95.

social networks, economic relevance of

Classical foundations of social network theory

Sociology's founding theorists often made use of network metaphors, since the discipline is defined by the study of relations; sociology is the study of positions, rather than persons. Durkheim (1984), for example, focused on the primary role played by the division of labour in the idealized transition from traditional to modern society. In traditional society, labour is divided mostly within rather than between households, so outside the household, economic, political, and social relations are based on similarity ('mechanical solidarity'). Modern society is characterized by a complex division of labour outside the household, resulting in relations based on complementary differences and dependence ('organic solidarity'). While Durkheim believed that organic solidarity provided much greater potential for growth and development, he also noted that such systems are more vulnerable to breakdowns in connectivity.

Of the classical social theorists, Georg Simmel made the most explicit use of network foundation in his work. What defines *being social* for Simmel is the super-individual quality of a collective (Simmel, 1908, p. 123). A *group* takes on this unique characteristic when its existence cannot be linked to the loss of particular members. When there are only two people involved, there is only one tie, and the group can be dissolved with the loss of either person; so for Simmel, the minimal social group is three persons, and also the first opportunity for structural variation in the ties – a 'two-path' or a completed triangle. In his work on *tertius gaudens* (the 'third who

enjoys'), Simmel explores the unique returns to those who occupy an exclusive bridging role between two others. He argues that power and control are a function of how *others* are connected to each other, rather than an individual attribute:

> The power *tertius gaudens* must expend in order to attain his advantageous position does not have to be great in comparison with the power of each of the two parties, since the quantity of his power is determined exclusively by the strength which each of them has relative to each other. (Simmel, 1908)

Simmel's work is the foundation for much current research on power and positions in networks (Blau, 1964; Cook et al., 1983; Burt, 1992).

Key concepts in modern social network analysis

Modern social network analysis focuses on explaining where networks come from and how networks affect outcomes. There are multiple summary measures for network structure that span multiple units of analysis. For nodes, the most basic measure is 'degree', the number of ties each node has to others in the network, which aggregates up to the population degree distribution. For dyads, measures can either characterize properties of the dyads themselves (for example, duration and type of relation) or the properties of the two nodes (for example, 'homophily', a form of correlation between nodal attributes). Higher-order configurations include triads and other cycles, the size and distribution of connected sets, path lengths and distributions, and many more (see Wasserman and Faust, 1994). Overall network attributes are typically functions of the lower-order properties, and include measures like density, centralization, core-periphery structure and clustering. At each level, researchers ask either 'what underlying processes produce this network structure?' or 'what impact does this structure have on some outcome?' The endogeneity of some of the implied processes leads to interesting and complicated dynamics, as well as causal ambiguity.

Where do networks come from?

Networks emerge as the result of relationship formation and dissolution. These processes are influenced by several factors: contextual effects (for example, population composition, institutional mediation), individual propensities (for example, expansiveness and attractiveness), dyadic factors (for example, homophily or mutuality), and explicit social rules about relational configurations (such as incest prohibitions or social balance rules for friendship).

Real-world social networks almost always display more clustering and heterogeneity than similar-volume Erdos–Reyni random networks; where ties are distributed uniform random across all possible pairs. Why? Part of the answer is propinquity: individual behaviour is organized by social contexts that filter the pool of available partners (Feld, 1981). Preferences for similarity combine with attribute heterogeneity to cluster ties within groups (Blau, 1977). Heterogeneities in individual propensities to send or receive ties also contribute to departures from a random graph. More active persons and groups will account for a disproportionate share of relations, creating hubs or clusters of activity. Recent 'scale free' network research has this simple 'preferential attachment' mechanism at its core (Barabasi and Albert, 1999).

Most social network research examines processes that operate above the individual level, focusing on the generative mechanisms explicitly governed by dyadic, triadic, and higher order relational norms. One of the first dyadic models proposed for directed relations was mutuality, which captures the commonly observed propensity for out-ties to be reciprocated. Another common dyadic model is 'homophily', the preference for similarity in social relations (Blau, 1977). Friendship pairs tend to be similar with respect to socioeconomic status, gender, race, and delinquent behaviour (Cohen, 1977; McPherson, Smith-Lovin and Brashears, 2006) and romantic ties are strongly assortative by age, education and race (Mare, 1991; Morris, 1995). The general term given to this dyadic process is selective mixing – which may be assortative, disassortative or idiosyncratic.

The most famous model for triads is social balance theory. At its simplest, balance theory predicts that a friend of a friend will also be a friend, as enmity among one's friends leads to strain and is avoided (Davis, 1963; Holland and Leinhardt, 1971). The beauty of balance models is that they provide a clear (if endogenous) link between individual action and network structure. Consider the example below in Figure 1, which maps the potential change space for a single unbalanced triad. The initial triad is unbalanced, since actor a's friends (+) are

enemies (−) with each other. This pattern is expected to create dissonance and thus each actor has a motive to change relations, any one of which would result in one of the three balanced outcomes. Each change, however, may alter the balance of other triads containing these nodes, resulting in a chain reaction of tie formation and dissolution over time.

Many triad and higher-order models for generating networks have been explored. Johnsen (1986) generalizes a set of simple balance-like rules to describe complex group-level structures. Moody (1999) demonstrates that a stable macro-structure cannot be assumed from the underlying choice patterns, since triad transitions that create balance from one actor's position will create imbalance for another. Chase (1980) demonstrates that perfect linear hierarchies require transitive triadic dominance relations, and that empirically (at least among chickens) relation formation rules are more important than individual attributes in predicting domination. Bearman, Moody and Stovel (2004) show that a simple rule prohibiting four cycles in heterosexual networks (the smallest possible cycle, equivalent to swapping partners) can generate romantic networks very similar to those observed among adolescents.

Finally, relational rules may also be based on the *timing* of ties. The most well-known example is the norm of serial monogamy in romantic relations. Monogamous networks are completely disconnected at any time point – they are nothing more than collections of isolated dyads. By contrast, if partnerships can be maintained concurrently, larger connected sets can form, and a giant component emerges quickly, even in the absence of nodes with many partners. This example has been well studied because it has implications for the spread of infection (Moody, 2000; Morris and Kretzschmar, 1997; Morris, Goodreau and Moody, 2008).

What do networks do?

The relations that comprise a network define a set of positions. The positions provide individuals with opportunities for access (a person to call for help) and exposure to diffusion (such as a virus through sexual contact or information about a job). The overall structure determines the population dynamics of social exchange and the evolutionary stability of the system.

At the individual level, opportunities for access to resources follow from one's position in the network. The set of direct contacts – called the egocentric or local network – is typically characterized by number, heterogeneity, and strength (Marsden, 1987). Recent work on social capital is in this tradition, building on the strength of ties, questions of trust, and the mix of types of people one is connected to (Lin, 2002). The effect of indirect ties – the partners of one's partners, and beyond – has produced some of the best-known network research, starting with Lee's (1969) *The Search for an Abortionist* and

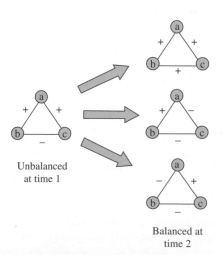

Unbalanced at time 1

Balanced at time 2

Figure 1   An example of a balance transition

continuing with Granovetter's (1973) classic 'The Strength of Weak Ties' on job search. Since the people one is very close to know what one knows, the most profitable information sources come from those connected to other social worlds.

While we may be influenced by our local network partners, their opinions are similarly shaped, so, as with balance theory, the *system* of beliefs (its coherence, polarization and dynamics) co-evolves with network ties. Studies of diffusion over networks have focused on a range of different types of processes, from the spread of information, norms, and innovation, to the spread of infection (Coleman, Katz, and Menzel, 1966; Rogers, 1962; Morris, 2004). Speed, pervasiveness and stability of spread are determined by the variations in the transmission network: density, the length and number of paths, clustering and relationship timing (Pool and Kochen, 1978; Watts and Strogatz, 1998; Moody, 2000; Morris and Kretzschmar, 1997).

The systemic perspective on endogeneous network processes leads naturally to a full game theoretic approach: a set of rules that govern interaction at the micro level which aggregate up to produce complex dynamics at the population level. Topics studied this way include the predominance of matrilateral cross-cousin marriage in classificatory kinship systems (White, 1963), the pervasive spread of unpopular ideas (Centola, Willer and Macy, 2005), and the conditions for the persistence of altruism (Bowles, Choi and Hapfensitz, 2003).

Social networks and economic outcomes

The starting point for most network approaches to economic sociology is Granovetter's (1985) paper on social embeddedness. Granovetter draws a middle ground for economic action between purely autonomous and economically rational actors ('undersocialized action') and deeply embedded and largely scripted normative behaviour ('oversocialized action'). He argues that 'personal relations and structures' generate trust, discourage malfeasance and are subject to intentional construction. Here, as always in the network approach, there is both a local and a global dynamic:

> 'Embeddedness' refers to the fact that economic action and outcomes, like all social action and outcomes, are affected by actors dyadic (pairwise) relations *and* by the structure of the overall network of relations. As a shorthand, I will refer to these as the relational and the structural aspects of embeddedness. (Granovetter, 1992, p. 33)

At the dyadic level, the focus has traditionally been on contact strength, with questions turning on issues of trust and exchange in local networks (Uzzi, 1999). At the structural level, the focus is on how larger sets of nodes are mutually reconnected, providing contexts where

mutual obligations can be reinforced and behaviour monitored (Moody and White, 2003).

The role of networks in shaping economic behaviour is particularly clear in settings where markets are not well developed or where capital resources are low. In China's emerging market economy, for example, firms will often trade with well-trusted associates even if they can find the goods cheaper elsewhere (Keister, 2001). Similar patterns are also found in Western commodity trading markets, however, suggesting there is a benefit to trading with known partners even in well-regulated settings (Baker, 1984; Uzzi, 1996; 1999).

Ron Burt (1992) builds on both Granovetter and Simmel in his research, linking profit opportunities to 'structural holes' – the absence of a tie between two persons who have something to exchange. As in Granovetter, information is more likely to accrue to those who connect such disconnected worlds. As in Simmel, those in a 'middle-man' position can play each of the actors they are connected to off each other, gaining a control advantage in the network. Structural holes are essentially arbitrage opportunities. As such, in a dynamic setting there are incentives for holes to be closed, which can either shorten the window of profitability, or generate action to protect the advantage.

Conclusion

The history of research on social networks in sociology is broad and deep, ranging from the earliest foundation of the discipline to nearly every empirical area under current research. The general theme of this work is to relate individual network positions or network structures either to generative social processes (when explaining network formation) or to substantive outcomes (when using networks to predict behaviour). For economic questions, a network embeddedness approach allows one to identify social bounds to strategic action based on connections with others.

Much of the current research on social networks is computationally intensive. Given the endogeneity inherent in network processes, statistical models cannot assume independence, and dynamic models are not analytically tractable. To analyse network formation, a new class of statistical models is being developed to help distinguish between different mechanisms that lead to similar structural features (Morris, 2003; Handcock et al., 2003; Snijders, 2001; Snijders et al., 2006). The new computational models allow us to examine a wide range of questions without the need for traditional simplifying assumptions. This set of tools promises to put the 'social' back into social science research methods.

JAMES MOODY AND MARTINA MORRIS

See also **assortative matching; economic sociology; economy as a complex system; graph theory; mathematics of networks; network formation.**

Bibliography

Baker, W. 1984. The social structure of a national securities market. *American Journal of Sociology* 89, 775–811.

Barabasi, A.-L. and Albert, R. 1999. Emergence of scaling in random networks. *Science* 286, 509–12.

Bearman, P., Moody, J. and Stovel, K. 2004. Chains of affection: the structure of adolescent romantic networks. *American Journal of Sociology* 110, 44–91.

Blau, P.M. 1964. *Exchange and Power in Social Life*. New York: Wiley.

Blau, P.M. 1977. *Inequality and Heterogeneity: A Primitive Theory of Social Structure*. New York: The Free Press.

Blau, P.M. and Schwartz, J.E. 1984. *Crosscutting Social Circles: Testing a Macrostructural Theory of Intergroup Relations*. Orlando: Academic Press.

Bowles, S., Choi, J.K. and Hopfensitz, A. 2003. The co-evolution of individual behaviors and social institutions. *Journal of Theoretical Biology* 223, 135–47.

Burt, R.S. 1992. *Structural Holes: The Social Structure of Competition*. Cambridge, MA: Harvard University Press.

Centola, D., Willer, R. and Macy, M. 2005. The emperor's dilemma: a computational model of self-enforcing norms. *American Journal of Sociology* 110, 1009–40.

Chase, I.D. 1980. Social process and hierarchy formation in small groups: a comparative perspective. *American Sociological Review* 45, 905–24.

Cohen, J.M. 1977. Sources of peer group homogeneity. *Sociology of Education* 50, 227–41.

Coleman, J.S., Katz, E. and Menzel, H. 1966. The diffusion of an innovation among physicians. *Sociometry* 20, 253–70.

Cook, K.S., Emerson, R.M., Gillmore, M.R. and Yamagishi, T. 1983. The distribution of power in exchange networks: theory and experimental evidence. *American Journal of Sociology* 89, 275–305.

Davis, J.A. 1963. Structural balance, mechanical solidarity, and interpersonal relations. *American Journal of Sociology* 68, 444–62.

Durkheim, E. 1984. *The Division of Labor in Society*, tr. W.D. Halls. New York: The Free Press.

Feld, S.L. 1981. The focused organization of social ties. *American Journal of Sociology* 86, 1015–35.

Granovetter, M. 1973. The strength of weak ties. *American Journal of Sociology* 81, 1287–303.

Granovetter, M. 1985. Economic action and social structure: the problem of embeddedness. *American Journal of Sociology* 91, 481–510.

Granovetter, M. 1992. Problems of explanation in economic sociology. In *Networks and Organizations: Structure, Form and Action*, ed. N. Nohria and R.G. Eccles. Boston, MA: Harvard Business School Press.

Handcock, M.S., Hunter, D.R., Butts, C.T., Goodreau, S.M. and Morris, M. 2003. *Statnet: An R Package for the Statistical Modeling of Social Networks*. Funding support from NIH grants R01DA012831 and R01HD041877. Available at http://www.csde.washington.edu/statnet, accessed 22 March 2007.

Holland, P. and Leinhardt, J. 1971. Transitivity in structural models of small groups. *Comparative Groups Studies* 2, 107–24.

Johnsen, E.C. 1985. Network macrostructure models for the Davis–Leinhardt set of empirical sociomatrices. *Social Networks* 7, 203–24.

Johnsen, E.C. 1986. Structure and process: agreement models for friendship formation. *Social Networks* 8, 257–306.

Keister, L.A. 2001. Exchange structures in transition: a longitudinal study of lending and trade relations in Chinese business groups. *American Sociological Review* 66, 336–60.

Lee, N.H. 1969. *The Search for an Abortionist*. Chicago: University of Chicago Press.

Lin, N. 2002. *Social Capital: A Theory of Social Structure and Action*. Cambridge: Cambridge University Press.

Mare, R.D. 1991. Five decades of educational assortative mating. *American Sociological Review* 56, 15–32.

Marsden, P.V. 1987. Core discussion networks of Americans. *American Sociological Review* 52, 122–31.

McPherson, M., Smith-Lovin, L. and Brashears, M.E. 2006. Social isolation in America: changes in core discussion networks over two decades. *American Sociological Review* 71, 353–75.

Moody, J. 1999. *The Structure of Adolescent Social Relations: Modeling Friendship in Dynamic Social Settings*. Chapel Hill: University of North Carolina.

Moody, J. 2000. The importance of relationship timing for diffusion. *Social Forces* 81, 25–56.

Moody, J. and White, D.R. 2003. Social cohesion and embeddedness: a hierarchical conception of social groups. *American Sociological Review* 68, 103–27.

Morris, M. 1993. Epidemiology and social networks: modeling structured diffusion. *Sociological Methods and Research* 22, 99–126.

Morris, M. 1995. Data driven network models for the spread of infectious disease. In *Epidemic Models: Their Structure and Relation to Data*, ed. D. Mollison. Cambridge: Cambridge University Press.

Morris, M. 2003. Local rules and global properties: modeling the emergence of network structure. In *Dynamic Social Network Modeling and Analysis*, ed. R. Breiger, K. Carley and P. Pattison. Washington, DC: National Academy Press.

Morris, M. 2004. *Network Epidemiology: A Handbook for Survey Design and Data Collection*. International Studies in Demography. Oxford: Oxford University Press.

Morris, M., Goodreau, S. and Moody, J. 2008. Sexual networks, concurrency, and STD/HIV. In *Sexually Transmitted Diseases*, 4th edn. ed. K.K. Holmes, P.F. Sparling, W.E. Stamm, P. Piot, J.N. Wasserheit, L. Corey, M. Cohen and H. Watts. New York: McGraw-Hill.

Morris, M. and Kretzschmar, M. 1997. Concurrent partnerships and the spread of HIV. *AIDS* 11, 641–48.

Myers, D.J. 2000. The diffusion of collective violence: infectiousness, susceptibility, and mass media networks. *American Journal of Sociology* 106, 173–208.

Pool, I.D.S. and Kochen, M. 1978. Contacts and influence. *Social Networks* 1, 5–51.

Rogers, E.M. 1962. *Diffusion of Innovations*. New York: The Free Press.

Simmel, G. 1908. In *The Sociology of Georg Simmel*, ed. K.H. Wolf. New York: The Free Press, 1950.

Snijders, T.A.B. 2001. The statistical evaluation of social network dynamics. *Sociological Methodology* 31, 361–95.

Snijders, T.A.B, Pattison, P.E., Robins, G.L. and Handcock, M.S. 2006. New specifications for exponential random graph models. *Sociological Methodology* 36, 99–153.

Uzzi, B. 1996. The sources and consequences of embeddedness for the economic performance of organizations: the network effect. *American Sociological Review* 61, 674–98.

Uzzi, B. 1999. Embeddedness in the making of financial capital: how social relations and networks benefit firms seeking financing. *American Sociological Review* 64, 481–505.

Valente, T.W. 1995. *Network Models of the Diffusion of Innovations*. Cresskill, NJ: Hampton Press.

Wasserman, S. and Faust, K. 1994. *Social Network Analysis*. Cambridge: Cambridge University Press.

Watts, D.J. and Strogatz, S.H. 1998. Collective dynamics of 'small-world' networks. *Nature* 393, 440–2.

White, H.C. 1963. *Anatomy of Kinship: Mathematical Models for Structures of Cumulated Roles*. Englewood Cliffs, NJ: Prentice-Hall.

White, H.C. 2002. *Markets from Networks: Socioeconomic Models of Production*. Princeton: Princeton University Press.

Zuckerman, E.W. 2003. On networks and markets. *Journal of Economic Literature* 46, 545–65.

social networks in labour markets

The use of social networks is widespread both in employers' recruiting and in workers' job-seeking. Social contacts help workers to find jobs, and employers to find employees. Indeed, social contacts convey rich and reliable information, which they spread widely and fast throughout the labour market. They thus constitute cost-effective search channels that both enrich the information available to both firms and workers, and enhance its quality.

Formal versus informal information sources

The study of social networks in labour markets highlights the nature of labour market transactions as very different from trading in goods, and reflects the importance of idiosyncrasies. The role of job market search and its dealing with frictions goes at least as far back as Stigler (1962). Everyday experience indicates that access to information is heavily influenced by social structure. Individuals use connections with others, such as friends and social and professional acquaintances, to maintain information networks. Rees (1966) first drew attention to differences among workers in their use of the variety of available informational outlets. In this context, *formal* sources of information include state and private employment agencies, newspaper advertisements, union hiring halls, school and college placement services and, more recently, the internet (Kuhn and Skuterud, 2000). *Informal* sources include referrals from employees and other employers, direct inquiries by job seekers and indirect ones through social connections. A recent literature in economics has developed about the details of social interactions that affect the job search process. This literature is complemented by the more extensive sociological analysis of networks. Several sociological works, including notably Granovetter (1974) and Boorman (1975), have been very influential within the economics literature. This article explores the salience within both theoretical and empirical economics research of a social networks approach in the study of labour markets.

Stylized facts

Several stylized facts about labour market networks have been established by empirical work on job information networks (Ioannides and Loury, 2004). The *first* stylized fact is that there is widespread use of friends, relatives, and other acquaintances in job search, and it has increased over time. The *second* stylized fact about job information networks is that the use of friends and relatives in job search often varies by location and by demographic characteristics. Differences in using informal contacts by age, race and ethnicity show conflicting patterns that suggest that important subtleties associated with the operation of social networks are at work. This is confirmed by international comparative evidence. Pellizzari (2004) explores the empirical evidence for the countries of the European Union as of 2003, using the European Community Household Panel, and compares with the United States, using the National Longitudinal Survey of Youth (NLSY). Pellizzari documents large cross-country and cross-industry variation in the wage differentials between jobs found through formal methods and those found through informal ones. Across countries and industries, premiums and penalties are equally frequent. Such differences may be attributed to different recruitment strategies by firms and to different institutional and social practices which may compound the impact of differences in the industrial compositions of economies. The *third* stylized fact about job information networks is that job search through friends and relatives is generally productive. Both employed and unemployed workers who used friends to search for jobs received more offers per contact and accepted more offers per contact than did workers who used other sources of information about job openings. The *fourth* stylized fact about job information networks is that part of the variation in the productivity of job search through networks by demographic group simply reflects differences in usage. In particular, US data suggest that almost one-fifth of the total difference in the probability of gaining employment between black and white youth resulted from racial differences in the use of social contacts.

Recent theoretical treatments

We crudely distinguish two mechanisms through which social contacts impact on the functioning of the labour market. First, referrals relay information across the two sides of the labour market, firms and workers. Second, workers' connections disseminate job information within the supply side of the labour market through word-of-mouth communication.

Hires mediated by referrals reduce employer uncertainty about prospective workers' productivity, for a number of reasons (Montgomery, 1991). One is that incumbent workers are likely to refer their trusted acquaintances and help them be better informed about their prospective employers. A second reason is that the long-term nature of the relationship between incumbent employees and their employers provides the latter with superior information on the incumbents' productivity-related traits. It is thus not surprising that evidence shows that referral bonuses bring high returns to firms. Yet excessive reliance on referrals deprives firms and individuals who happen to be outside the social networks of firms' workers of mutually beneficial matches.

Recent findings have improved our understanding of the supply-side effects of social networks (Calvó-Armengol and Jackson, 2004; 2006). In their models, workers rely both on own search effort and on information exchange with their social circles to find jobs. Information passing across acquaintances can display a variety of real-life features; for example, when connections differ in terms of intensities, information recipients can be ranked so as to reflect these relational preferences. Calvó-Armengol and Jackson's models are the first to explain several important stylized facts about labour markets, which are hard to explain altogether without an explicit social network model. We turn to those next.

First, information passed from employed individuals to their unemployed acquaintances makes it more likely that these acquaintances will become employed. This generates a positive correlation between employment and wages of networked individuals *within* and *across* periods. Such positive long-run correlation arises despite the short-run rival nature of job information in the following sense: indirect contacts who are two links away in a network are potential competitors for any job held by any common friend. *Second*, duration dependence and persistence in unemployment, both of which are well documented, can be understood as social effects: the longer an individual is unemployed the more likely it is that her social environment is associated with unfavourable future unemployment prospects. This explanation for duration dependence complements more common ones, such as unobserved heterogeneity. This effect resembles an externality and is also responsible for stickiness in aggregate employment dynamics. The closer the economy is to very high employment (or unemployment), the harder it is to leave that state. For similar reasons, parts of the economy can experience a boom while simultaneously other parts of the economy are experiencing a bust.

These are implications of *exogenous* information networks. With an endogenous network that results from agents' participation decisions, the model's predictions are the following. *Third*, the likelihood of dropping out of the labour force is higher for an individual whose social contacts have poor employment experience, or for an individual with few acquaintances. *Fourth*, small differences in initial conditions of different individuals and of network structure can lead to large differences in drop-out rates. Indeed, when an individual drops out, the prospects worsen for all those who remain, and this generates spillover effects in others' decisions to participate or to drop out. Differences in collective employment histories combine with differences in network structure to produce sustained inequality of wages and drop-out rates that feed on each other. So history matters and is responsible for producing persistent income inequality for reasons that are very different from those due to inequalities in human capital investments. Because spillover effects work in reverse, selective and targeted (rather than separate) interventions in the labour market that provide incentives for individuals not to drop out are likely to have amplified effects.

Current empirical treatments

Empirical research has yet to employ fully formal network concepts. It relies typically on concepts of association because of geographic or cultural proximity. There is evidence of persistent correlations in patterns of unemployment in US cities. Socio-economic characteristics, and in particular ethnic and occupational distance, seem to explain a substantial component of the spatial dependence in unemployment. Topa (2001) and Conley and Topa (2002) argue that social interactions can indeed explain the spatial correlation patterns present in the data. Weinberg, Reagan and Yankow (2004) show that one standard deviation improvement in neighbourhood social characteristics and in job proximity raises individuals' hours worked by six per cent and four per cent on average, respectively. Such social interactions have non-linear effects. The greatest impact is in the worst neighbourhoods. Being in a disadvantaged neighbourhood is more important than the labour activity of one's neighbours per se. Bayer, Ross and Topa (2005) document that people who live close to each other, defined as being in the same census block – a US census block encompasses

3,500 to 5,000 residents of a contiguous geographical area – also tend to work together, that is, in the same census block. Using data from Dartmouth College (where roommates are assigned randomly), Marmaros and Sacerdote (2002) find large positive correlations between getting help from fraternity or sorority contacts and obtaining prestigious, high-paying jobs. Still, other research points to self-selection as the likely origin of such effects: Oreopoulos (2003) finds that, when neighbourhoods are not selected, neighbourhood quality plays *little* role in determining a youth's eventual earnings, likelihood of unemployment, and welfare participation, while correlations among outcomes for siblings are much higher.

As richer network data become available, further empirical tests of the implications of labour market networks should be developed, which ultimately may call for more elaborate network modelling tools in labour economics. Such research deserves attention.

ANTONI CALVÓ-ARMENGOL AND YANNIS M. IOANNIDES

See also **mathematics of networks; network formation; social interactions (empirics); social interactions (theory); psychology of social networks.**

Bibliography

Bayer, P., Ross, S. and Topa, G. 2005. Place of work and place of residence: informal hiring networks and labor market outcomes. Working Paper No. 11019. Cambridge, MA: NBER.

Boorman, S. 1975. A combinatorial optimization model for transmission of job information through contact networks. *Bell Journal of Economics and Management Science* 6, 216–49.

Calvó-Armengol, A. and Jackson, M. 2004. The effects of social networks on employment and inequality. *American Economic Review* 94, 426–54.

Calvó-Armengol, A. and Jackson, M. 2006. Networks in labor markets: wages and employment dynamics and inequality. *Journal of Economic Theory* (forthcoming).

Conley, T. and Topa, G. 2002. Socio-economic distance and spatial patterns in unemployment. *Journal of Applied Econometrics* 17, 303–27.

Granovetter, M. 1974. *Getting a Job: A Study of Contacts and Careers*, 1st edn. Cambridge, MA: Harvard University Press.

Ioannides, Y. and Loury, L. 2004. Job information networks, neighborhood effects, and inequality. *Journal of Economic Literature* 42, 1056–93.

Kuhn, P. and Skuterud, M. 2000. Job search methods: internet versus traditional. *Monthly Labor Review* 123(10), 3–11.

Marmaros, D. and Sacerdote, B. 2002. Peer and social networks in job search. *European Economic Review* 46, 870–9.

Montgomery, J. 1991. Social networks and labor market outcomes: towards an economic analysis. *American Economic Review* 81, 1408–18.

Oreopoulos, P. 2003. The long-run consequences of living in a poor neighbourhood. *Quarterly Journal of Economics* 118, 1533–75.

Pellizzari, M. 2004. Do friends and relatives really help in getting a good job? Discussion Paper No. 623. Centre for Economic Performance, London School of Economics.

Rees, A. 1966. Information networks in labor markets. *American Economic Review, Papers and Proceedings* 56, 559–66.

Stigler, G. 1962. Information in the labor market. *Journal of Political Economy* 70(5), 94–105.

Topa, G. 2001. Social interactions, local spillovers and unemployment. *Review of Economic Studies* 68, 261–95.

Weinberg, B., Reagan, P. and Yankow, J. 2004. Do neighborhoods affect work behavior? Evidence from the NLSY79. *Journal of Labor Economics* 24, 891–924.

social norms

Social norms are customary rules of behaviour that coordinate our interactions with others. Once a particular way of doing things becomes established as a rule, it continues in force because we prefer to conform to the rule given the expectation that others are going to conform (Schelling, 1960; Lewis, 1969). This definition covers simple rules that are self-enforcing at a primary level, such as which hand to extend in greeting or which side of the road to drive on, and more complex rules that trigger sanctions against those who deviate from a first-order rule. (We express outrage if someone cuts in front of someone else in a queue.) The former are sometimes called *conventions* and the latter *norms* (Sugden, 1986; Coleman, 1990; Bicchieri, 2006), but in fact there are numerous gradations and levels of response to norm violation that make this dichotomy problematic. Hence I shall use the term 'norm' in its inclusive sense in what follows.

David Hume (1739) was the first to call attention to the central role that norms play in the construction of social order. Norms define property rights, that is, who is entitled to what. They determine what commodities are accepted as money. They shape our sense of obligation to family and community. They determine the meanings we attach to words. Indeed it is hard to think of a form of interaction that is not governed to some degree by social norms. (For book-length treatments of the subject see Lewis, 1969; Ullman-Margalit, 1977; Sugden, 1986; Young, 1998a; Posner, 2000; Hechter and Opp, 2001; Bicchieri, 2006.)

Norms and equilibria

Norms can be represented as equilibria of suitably defined games; indeed, Hume's analysis of norms can be viewed as one of the earliest examples of game-theoretic reasoning. Nevertheless, not every equilibrium of a game is a norm.

First, the term generally applies only to games with *multiple* equilibria. People can queue for service *or* they can push. They can use gold coins *or* glass beads as money. They can drive on the left *or* on the right.

Second, even if a game has multiple equilibria, they do not necessarily qualify as norms. To illustrate the distinction, consider two individuals who get to divide a dollar provided they can agree on how to divide it. Each makes a demand, and if the demands sum to at most one the demands are met; otherwise they get nothing. This is a coordination game and it has many equilibria. For example, if one person demands 43 cents and the other demands 57 cents, the demands are in equilibrium: no one can gain from a unilateral deviation. But this is not a norm; it is an idiosyncratic equilibrium for these two individuals. Fifty–fifty division, by contrast, is a norm because it is usual and customary in games of this kind, and everyone knows it.

Norm enforcement

Broadly speaking there are three different mechanisms by which norms are held in place. Some are sustained by a pure coordination motive. If it is the norm to drive on the left, I adhere to the norm in order to avoid accidents. If gold is the commonly accepted currency, it would be a waste of time to try to conduct my business with glass beads. These are 'social' phenomena, because they are held in place by shared expectations about the appropriate solution to a given coordination problem, but there is no need for social enforcement.

Other norms are sustained by the threat of social disapproval or punishment for norm violations (Sugden, 1986; Coleman, 1990). If queuing is the norm, I will be censured if I try to push my way to the front. If duelling is the proper response to an insult, I will lose status in the community if I do not challenge the one who insulted me. If I am expected to avenge the murder of my brother and fail to carry it out, I may be ostracized by other family members. (Exactly why third parties bother to express disapproval or carry out punishments is a matter of debate, but the evidence suggests that they sometimes do so even at considerable personal cost – Fehr, Fischbacher, and Gächter, 2002.)

A third enforcement mechanism arises through the internalization of norms of proper conduct. If it is the norm not to litter, I will avoid littering even in situations where no one can see me. If I eat a meal in a foreign city and fail to tip the waiter, I need not fear the consequences because there is no continuing relationship; nevertheless, I may think the worse of myself for having done it. More generally, norms often take on the character of virtuous or right action (Hume, 1739), and departures from a norm can trigger emotions of shame or guilt even when third-party enforcement is absent (Coleman, 1990; Elster, 1989; 1999). This fact is especially useful in large-scale societies, where it may be difficult to monitor others' compliance with equilibrium behaviour.

Norms and efficiency

It remains to be explained why a dictionary on economics, in contrast to sociology or law, should bother with an entry on social norms. What economic purpose do they serve? The answer is that norms coordinate expectations, and thereby reduce transaction costs in interactions that possess multiple equilibria (Wärneryd, 1994).

This point seems clear enough intuitively; it can also be demonstrated experimentally (Roth, 1985). Consider a game in which two players can divide a pile of chips in any way they like, but if they fail to agree on a division within a specified period of time they forfeit all of them. When all the chips can be cashed in for the same amount of money, the norm is to divide the chips equally; moreover, the great majority of players do in fact coordinate in this manner. Notice, however, that *any* division of the chips, not just 50–50, can constitute an equilibrium of the one-shot game if both players expect that it will be played. Now consider a variant in which the players get to cash in the chips for different amounts of money (which is publicly known). In this case there are two potential focal solutions – divide the chips equally and divide the money-value of the chips equally – but there is no norm to steer the players' expectations towards one or the other. As a result, the frequency of disagreement rises substantially.

More generally, a norm has economic value if it creates a uniquely salient or focal solution to a coordination problem, thus reducing the risk of coordination failure. In this sense norms are a form of social capital (Coleman, 1987). This does not mean, however, that norms are invariably welfare-enhancing; indeed some norms would appear not to have *any* direct welfare implications.

Consider norms of etiquette, such as the fine points of table manners. The welfare consequences are so trivial that it is hard to see why anyone bothers with them. No one is harmed, for example, if I wear a hat to dinner or eat peas with my fingers. The fact is, however, that such indiscretions may do serious harm to my reputation. In particular, they signal that I am a person who does not care about social norms, which may lead others to doubt my reliability in more important interactions (Posner, 2000). Complex social rituals allow people to signal their sensitivity to *norms in general*; they also provide a training ground for learning to follow norms, and for disciplining those who fail to do so.

Even when norms do have direct welfare implications, one cannot conclude that societies will opt for efficient norms. It is doubtful, for example, that norms of retribution are efficient, or that pushing is superior to queuing. Yet these are the operative norms in quite a few cases. The problem is that norms are not 'chosen'; they arise from historical accident and the accumulation of precedent. Once expectations converge on an inefficient norm, it can be very difficult to dislodge.

Over the longer run, it is conceivable that societies could somehow extricate themselves from inefficient outcomes. One way that this could happen is that societies with superior norms simply displace societies with inferior norms, through growth, conquest, or migration. Another possibility is that societies with inferior norms imitate the practices of more successful ones (Robson and Vega-Redondo, 1996; Boyd and Richerson, 2002). Yet a third possibility is that norm change comes from within, the result of gradual and almost imperceptible changes in expectations that 'tip' the society into a new way of doing things without anyone intending it. I discuss this possibility in more detail below.

Excess uniformity

I have argued that some norms may be inferior from the standpoint of welfare, yet stay in place for long periods of time. Others may have little direct effect on welfare but serve an important signalling function. Another way in which norms can affect economic welfare is by imposing excess uniformity on behaviour. People might be better off if they adapted their actions to their particular circumstances. To illustrate, consider a situation in which a principal and an agent are bargaining over the terms of a contract. To be concrete, think of a landlord bargaining with a prospective tenant over the terms for renting a plot of land. In theory, the optimal contract will depend on a variety of factors that may be idiosyncratic to the contracting parties, including information asymmetries, monitoring costs, and attitudes towards risk (Cheung, 1969). Yet, in practice, contracts often exhibit a high degree of uniformity, and employ 'usual and customary' terms that mask idiosyncratic differences (Bardhan, 1984). In late 20th-century Illinois agriculture, for example, more than 50 per cent of the contracts specified fixed shares between tenant and landlord, and of these more than 90 per cent specified the shares 1/2–1/2, 2/5–3/5, or 1/3–2/3 for landlord and tenant respectively (Young and Burke, 2001).

The logic of 'usual and customary' contractual terms is that they create a focal solution in a situation that has many possible solutions, thereby reducing transaction costs. Such norms are not unique to agricultural contracts: for example, building contractors and architects get customary markups over cost, franchisees pay standard percentages to their parent companies, real estate agents receive customary commissions on house sales, and so forth. While such norms may reduce transaction costs, however, the uniformity imposed by the norm may prevent the contracting parties from fully wringing out all of the potential gains in their particular circumstances. Thus, in evaluating the efficiency of a norm, one must consider both the savings in transaction costs and the costs imposed by excess uniformity.

The evolution of norms

If a norm merely represents one equilibrium out of many, how does society settle on a particular one starting from out-of-equilibrium conditions? We may distinguish three ways in which norms become established and change over time: top-down influences, including official edicts and role models; bottom-up influences in which local customs and practices coalesce into norms; and lateral influences in which established norms from one type of interaction are transferred to related types of interactions. The law, for example, operates partly from the top down: statutes and judicial rulings identify norms of acceptable behaviour in people's relations with others. At the same time, the boundary between acceptable and unacceptable behaviour is constantly in flux due to variations in the way that individual cases are resolved by individual courts (a bottom-up effect). And precedents in one domain can be transferred by analogy to other domains (a lateral effect). An example of the latter is the extension of laws regarding persons to those involving corporations. (For more on the interplay between norms and the law see Ellickson, 1991; Posner, 2000).

As these examples suggest, the evolution of norms is a complex process that involves the interplay of many different forces. One may nevertheless gain insight into the process by examining how small variations in behaviour at the individual level can trigger major norm shifts at the societal level. Consider a symmetric, two-person coordination game G that is played by pairs of agents drawn from a large population. Assume for simplicity that the total number of agents is even, and that in each period everyone is paired with someone else through a random matching process. Each matched pair chooses actions simultaneously and receives the corresponding payoffs in G. Assume that each agent makes a 'trembled best response' to the distribution of choices in the previous period. Specifically, suppose that for some $0 < \varepsilon$, $\lambda < 1$, each agent chooses a best response with probability $\lambda(1 - \varepsilon)$, trembles with probability $\lambda \varepsilon$ (in which case he chooses an action uniformly at random), and chooses the same action as before with probability $1 - \lambda$ due to inertia (Kandori, Mailath and Rob, 1993; Young, 1993a).

This type of learning process has rather striking implications for the social norms that are most likely to emerge and remain in place for long periods of time. To illustrate, consider a competition between alternative forms of money. Suppose there are m different commodities, indexed $1 \leq k \leq m$. Assume that, in a given pairwise interaction, each player's payoff is a_k if both adopt the kth form of money, whereas their payoffs are zero if they adopt different forms. It can be shown that this trembled best-response process selects the efficient equilibrium with high probability, that is, when ε is small the probability is high that, in the long run, almost everyone in the population will be using the form of money with the highest payoff (Kandori and Rob, 1995).

Unfortunately, efficiency may fail when the game does not have the structure of a pure coordination game. Suppose, for example, that a potential form of money generates payoffs in two ways: as a medium of exchange and as a form of adornment. In each pairwise interaction, let $a_k > 0$ be the payoff in each period from using k as a medium of exchange (on the assumption that the other player also uses k), and let $b_k > 0$ be the payoff from using it instead as jewellery. When there are just two commodities one obtains the symmetric payoff matrix

$$a_1 + b_1, a_1 + b_1 \qquad b_1, b_2$$
$$b_2, b_1 \qquad a_2 + b_2, a_2 + b_2$$

Assume that this is a coordination game: $a_1 + b_1 > b_2$ and $a_2 + b_2 > b_1$. Commodity k is *efficient* if it maximizes $a_k + b_k$. However, the evolutionary process defined above selects the *risk-dominant* commodity, that is, the commodity k that maximizes $a_k + 2b_k$ (Young, 1998b). The latter criterion gives twice as much weight to the payoffs from adornment (which do not require coordination) as to the payoffs that arise from using the same medium of exchange (which do require coordination). Moreover, this conclusion holds under a wide range of assumptions about the nature of the trembled best-response process (Blume, 2003).

Evolution and fairness norms

The evolutionary framework outlined above has implications not only for the efficiency of norms but for their distributive properties as well. Consider a situation in which a principal and an agent must agree on the form of contract that will govern their relationship. Different types of contracts will have different distributive implications, some favouring the principal, others favouring the agent. Let us restrict attention to just those contracts that leave both parties better off than they would be under their outside options. To illustrate, suppose that just three contract forms are available: A, B, C. Assume a one-shot, take-it-or-leave-it bargaining process in which a given contract is adopted if and only if both parties simultaneously agree to it; otherwise they fall back on their outside options. Without loss of generality one can assume that the outside options have zero utility for both players. Consider the following example:

		Agents		
		A	B	C
	A	5, 1	0, 0	0, 0
Principals	B	0, 0	3, 3	0, 0
	C	0, 0	0, 0	1, 5

Contract A favours the principal, C favours the agent, and B is a compromise between A and C. (Of course, in reality there may be many more contract forms, but this

does not change the analysis in any fundamental way.) Consider an evolutionary process like the one for money conventions, but with two distinct populations – one of principals, the other of agents – that are randomly matched in each period. Assume, as in the previous model, that they play trembled best responses with inertia, where 'best response' is defined relative to the frequency of play of the opposite population in the previous period.

A *contractual norm* is a situation in which the same contract is agreed to by everyone. In this example all three norms are efficient: none of them Pareto dominates another. It can be shown, however, that the evolutionary process favours exactly one of them, namely, the compromise contract B. More generally, in evolutionary processes of this sort there tends to be a selection bias toward outcomes that represent a compromise for the two parties, and against extreme outcomes that lie near the boundary of the payoff-possibility set (Young, 1998b). This suggests that *norms of fairness* may result from evolutionary forces, an idea that is explored by Binmore (1994; 2005) and Young (1998a).

General implications

Although evolutionary accounts of norm formation vary in their details, they have several qualitative implications that hold quite generally. One is that different societies often employ different norms for solving the same type of coordination problem. This follows from the fact that norms represent alternative equilibria that can become established through different sequences of chance events. This is known as the *local conformity/global diversity* effect (Young, 1998a). It has been documented in a variety of settings, including agricultural contracting (Young and Burke, 2001), and the manner in which subjects divide payoffs in experimental situations (Henrich et al., 2004).

A second implication is that, due to stochastic perturbations, norms occasionally shift, and these shifts tend to be quite rapid compared with the long periods of stasis when a given norm is in place. This is the *tipping* or *punctuated equilibrium* effect (Young, 1998a).

A third implication is that some norms are inherently more stable or durable than others: once established they tend to remain in place for long periods of time even when buffeted by stochastic shocks. These *stochastically stable norms* depend on the payoff structure of the underlying game, and also on the nature of the stochastic perturbations (Foster and Young, 1990; Young, 1993a; Kandori, Mailath and Rob, 1993; Samuelson, 1997). Irrespective of these details, the important point is that some norms are remarkably resilient under changing circumstances. Due to their longevity, such norms may come to be seen as right and necessary, though in fact they are the product of chance and contingency, and are sustained simply because they coordinate people's expectations about how to interact with one another.

H. PEYTON YOUNG

See also **Hume, David; social capital.**

Bibliography

Bardhan, P. 1984. *Land, Labor, and Rural Poverty: Essays in Development Economics.* New York: Columbia University Press.

Bicchieri, C. 2006. *The Grammar of Society: The Nature and Dynamics of Social Norms.* New York: Cambridge University Press.

Binmore, K. 1994. *Game Theory and the Social Contract, Vol. 1: Playing Fair.* Cambridge, MA: MIT Press.

Binmore, K. 2005. *Natural Justice.* Oxford: Oxford University Press.

Blume, L. 2003. How noise matters. *Games and Economic Behavior* 44, 251–71.

Boyd, R. and Richerson, P. 2002. Group beneficial norms can spread rapidly in a structured population. *Journal of Theoretical Biology* 215, 287–96.

Cheung, S. 1969. *The Theory of Share Tenancy.* Chicago: University of Chicago Press.

Coleman, J. 1987. Norms as social capital. In *Economic Imperialism: The Economic Approach Applied Outside the Field of Economics*, ed. G. Radnitzky and P. Bernholz. New York: Paragon House.

Coleman, J. 1990. *Foundations of Social Theory.* Cambridge, MA: Harvard University Press.

Ellickson, R. 1991. *Order Without Law: How Neighbors Settle Disputes.* Cambridge, MA: Harvard University Press.

Elster, J. 1989. Social norms and economic theory. *Journal of Economic Perspectives* 3(4), 99–117.

Elster, J. 1999. *Alchemies of the Mind.* Cambridge, UK: Cambridge University Press.

Fehr, E., Fischbacher, U. and Gächter, S. 2002. Strong reciprocity, human cooperation, and the enforcement of social norms. *Human Nature* 13, 1–25.

Foster, D. and Young, H.P. 1990. Stochastic evolutionary game dynamics. *Theoretical Population Biology* 38, 219–32.

Harsanyi, J. and Selten, R. 1988. *A General Theory of Equilibrium Selection in Games.* Cambridge, MA: MIT Press.

Hechter, M. and Opp, K.-D., eds. 2001. *Social Norms.* New York: Russell Sage Foundation.

Henrich, J., Boyd, R., Bowles, S., Fehr, E. and Gintis, H. (eds.) 2004. *Foundations of Human Sociality: Economic Experiments and Ethnographic Evidence in 15 Small-Scale Societies.* Oxford: Oxford University Press.

Hume, D. 1739. *A Treatise of Human Nature.* Oxford: Oxford University Press, 1978.

Kandori, M., Mailath, G. and Rob, R. 1993. Learning, mutation, and long-run equilibria in games. *Econometrica* 61, 29–56.

Kandori, M. and Rob, Rl. 1995. Evolution of equilibria in the long run: a general theory and applications. *Journal of Economic Theory* 65, 29–56.

Lewis, D. 1969. *Convention: A Philosophical Study.* Cambridge, MA: Harvard University Press.

Posner, E. 2000. *Law and Social Norms.* Cambridge, MA: Harvard University Press.

Robson, A. and Vega-Redondo, F. 1996. Efficient equilibrium selection in evolutionary games with random matching. *Journal of Economic Theory* 70, 65–92.

Roth, A. 1985. Toward a focal point theory of bargaining. In *Game-Theoretic Models of Bargaining*, ed. A. Roth. Cambridge, UK and New York: Cambridge University Press.

Samuelson, L. 1997. *Evolutionary Games and Equilibrium Selection.* Cambridge, MA: MIT Press.

Schelling, T.C. 1960. *The Strategy of Conflict.* Cambridge, MA: Harvard University Press.

Sugden, R. 1986. *The Economics of Rights, Cooperation and Welfare.* Oxford: Basil Blackwell.

Ullmann-Margalit, E. 1977. *The Emergence of Norms.* Oxford: Oxford University Press.

Wärneryd, K. 1994. Transaction cost, institutions, and evolution. *Journal of Economic Behavior and Organization* 25, 219–39.

Young, H.P. 1993a. The evolution of conventions. *Econometrica* 61, 57–84.

Young, H.P. 1993b. An evolutionary model of bargaining. *Journal of Economic Theory* 59, 145–68.

Young, H.P. 1998a. *Individual Strategy and Social Structure.* Princeton, NJ: Princeton University Press.

Young, H.P. 1998b. Conventional contracts. *Review of Economic Studies* 65, 773–92.

Young, H.P. and Burke, M. 2001. Competition and custom in economic contracts: a case study of Illinois agriculture. *American Economic Review* 91, 559–73.

social preferences

For the longest time economists reacted allergically to preference formulations that allowed for anything but material self-interest (cf. Binmore, Shaked and Sutton, 1985). The reaction was well founded: by adding elements to the agent's utility function, potentially one allows economic theory to explain everything and, therefore, nothing. Any behaviour can be explained by assuming it is preferred. However, this strong position has sometimes made economics seem out of touch with the world economists try to explain. Even economists care about the outcomes achieved by others, in addition to their own outcomes. Moreover, they also care about how those outcomes are achieved. Only in 1982, however, was the weakness of taking material self-interest for granted demonstrated by Werner Güth and his co-authors, who showed that economic theory failed in the simplest of decision settings (Güth, Schmittberger and Schwarze,

1982), the ultimatum game. In this game a first mover offers a share of a monetary 'pie' to a second mover who either accepts the proposal, in which case it is divided as proposed, or rejects the proposal, in which case both players earn nothing. Since then this game has become the workhorse of experimenters intent on exploring carefully the extent to which people behave in ways that are contrary to their material self-interest.

While it is interesting to document the fact that people consider the outcomes of others when they make choices in experimental games, there are at least two other particularly compelling aspects of the research that has developed since the 1980s. First, these deviations from self-interest can be replicated, and have been, both inside and outside the laboratory. Replication suggests that these behaviours are not just errors or flukes, and therefore, although self-interest is a convenient modelling assumption, it should not be used as the basis for policy formulation. Second, this research illustrates that there is a difference between theory failing because of a false assumption and its failing because of flawed logic. Research shows that people do use economic reasoning, but that they, or most of them, are not narrowly self-interested.

The original results of the ultimatum game provided the impetus for a large body of research. Initially, some researchers were convinced that the explanation was not a concern for others but simple error (for example, Binmore, Shaked and Sutton, 1985). However, this explanation was soon swept aside by volumes of evidence from a variety of games that suggested that the payoffs of other players entered into the strategic choices of experimental participants (see the reviews of Bowles, 2004; or Sobel, 2005). Despite all this research, a precise definition of social preference has not been settled upon. In most cases, 'social preference' is defined loosely as *a concern for the payoffs allocated to other relevant reference agents in addition to the concern for one's own payoff*. (A largely separate branch of research has focused on altruism and warm glow motives for giving to others, especially in the context of public goods provision. This work is discussed elsewhere in the dictionary.)

Within the standard outcome-oriented definition, research has focused on identifying the more pro-social preferences for altruism and inequality aversion while considerably less attention has been given to their opposites, spite and eminence. The evidence from the hundreds of ultimatum games conducted since 1982 suggests that, on the second-mover side of the game, few people are willing to accept the low offers associated with the subgame perfect equilibrium prediction. In fact, offers of less than 20 per cent of the pie are routinely rejected, and as offers increase they are more likely to be accepted (Camerer, 2003). Turning down positive offers is clearly against one's material self-interest, but it is consistent with aversion to unequal payoffs (inequality aversion). As the stakes increase, the probability of a rejection falls,

but even when the pie is as large as three months expenditures the rejection rate is not zero (Cameron, 1999).

Interpreting the motivation of the first mover in the ultimatum game is not as straightforward, though. One hypothesis is that proposers offer half the pie because they are inequality averse. We cannot, however, distinguish this reasoning from that of completely selfish, but astute, proposers who anticipate that low offers will be rejected and offer half because they know it will be accepted. The dictator game evolved to identify the motives of first movers (Forsythe et al., 1994). The dictator game is played just like the ultimatum game except for one very important design change: second movers are passive recipients of whatever they are allocated. In other words, they cannot reject offers. If the enlightened self-interest hypothesis is correct, we would expect to see first movers allocating nothing in the dictator game. This is not the case. Although allocations in the dictator game are susceptible to changes in the presentation of the game (Hoffman et al., 1994; Eckel and Grossman, 1996), it is common for people to allocate positive amounts. In fact, it is common for the behaviour of non-student participants in the two games to be indistinguishable (Carpenter, Burks and Verhoogen, 2005) suggesting that many people prefer equal outcomes.

There is some question as to whether the simple outcome-oriented definition of social preference is sufficient. An example illustrates why. Instead of offers being generated by other participants, imagine second movers in the ultimatum game being assigned offers randomly by a computer programme. If inequality aversion is a sufficient description of the motivations of participants, this change should have no impact on behaviour. However, it does: responders are much less likely to reject computer-generated offers than offers that come from real proposers (Blount, 1995). This indicates that people are also interested in the process and intentions that generate outcomes. The definition of social preference should perhaps be expanded accordingly to *a concern for the payoffs allocated to other relevant reference agents and the intentions that led to this payoff profile in addition to the concern for one's own payoff*.

Expanding the definition of social preference to include a process component allows us to also classify reciprocity – treating only kind acts with kindness – as a social preference. Pure reciprocity, however, is more elusive than inequality aversion because one needs to show that outcomes and intentions matter. Only a few experiments have been conducted to show that intentions matter, but the results are compelling. For example, imagine two binary choice versions of the ultimatum game (Falk, Fehr and Fischbacher, 2003). In game A, the proposer can decide between claiming the lion's share of a ten-dollar pie (8, 2) and sharing the pie equally (5, 5). In game B, the first option is the same (8, 2) but the second is even worse for the second mover because the proposer demands the whole pie (10, 0). Inequality aversion predicts that the (8, 2) offer will be rejected at the same rate in the two games because the other offer is irrelevant – the decision-maker should focus only on the outcome presented. Reciprocity, on the other hand, suggests that one would be much less likely to reject (8, 2) in game B because it is the kinder of the two offers. Indeed, people are almost five times more likely to reject the (8, 2) offer in game A. An alternative approach is to compare the response of participants to different outcome allocations after another participant has made a kind or unkind act to the response when there is no initial move by another participant (Charness and Rabin, 2002). Reciprocity is identified by the subtraction of the first outcomes and intentions experiment from the second baseline inequality-aversion experiment.

In the trust (or investment) game, a first mover decides how much to send to a second mover. Any amount sent is multiplied by $k > 1$ before it reaches the second-mover. The second mover then decides how much to send back. Because of the multiplication, sending money is socially efficient yet a first mover should send money only if she trusts the second mover to send back at least enough to cover the investment. The standard interpretation is that the first mover must expect the second mover to be motivated by reciprocity before it makes sense to invest in the partnership (Berg, Dickaut and McCabe, 1995). However, one can just as easily invoke inequality aversion to explain the fact that people tend to send back more when they receive more (Cox, 2004). The same problem exists with the related experiments developed to test for the notion of gift exchange in the labour market context (for example, Fehr and Schmidt, 1999).

Other, more indirect, evidence for reciprocity and the more nuanced definition of social preference comes from the experimental literature on voluntary contributions to public goods. In these settings participants are given an endowment and asked to decide how much to contribute to a 'group project'. The incentives are of a social dilemma; contributing nothing is a dominant strategy but contributing everything is socially efficient. Playing the public goods game in strategic form asks participants to decide how much they want to contribute conditional on the contributions of others. Half the participants are conditionally cooperative in that they generate contribution schedules that are increasing in the contributions of others (Fischbacher, Gächter and Fehr, 2001). The fact that people condition their contributions according to those of others suggests that intentions and reciprocity matter.

To identify reciprocity separately from inequality aversion one may employ a design in which the two forces pull in different directions. Imagine that one can punish free riders in the public goods game: a participant can impose a penalty p at a cost c. In most cases people punish despite it being dominant to free ride on the

punishment done by others (that is, punishment is just a second-order public good), and this tends to stabilize contributions (Fehr and Gächter, 2000). However, in most cases $p > c$, which means cooperators reduce the inequality between themselves and the free rider by punishing. To isolate the role of, in this case negative, reciprocity one can allow $p < c$, which actually increases the inequality. Although they do it less often, people punish when the sanction delivered is lower than the cost, and this is a nice demonstration of reciprocity (Carpenter, 2007).

Several attempts have been made to organize the evidence on social preferences into parsimonious, but flexible, utility functions. One of the most successful outcome-oriented approaches is the Fehr and Schmidt (1999) specification, perhaps because it is relatively easy to work with. Here the utility of player i increases in her own payoff, x_i, but decreases in any difference between her payoff and the payoffs of other relevant players. For two-player games this is just:

$$u_i(x_i, x_j) = \begin{cases} x_i - \alpha_i(x_j - x_i) & \text{if } x_i < x_j \\ x_i - \beta_i(x_i - x_j) & \text{if } x_i \geq x_j \end{cases}$$

where α_i is player i's degree of inferiority aversion and β_i is her degree of superiority aversion. It is natural to expect $\alpha_i > \beta_i$.

While this utility function is a good first approximation because it has been shown to be consistent with much of the experimental data (if one is willing to make assumptions about the distribution of α's and β's in the population) it is limited in two ways. First, as one can see in Figure 1, the predictions can be coarse. It is not hard to graph the indifference curves associated with the Fehr–Schmidt specification, but if one superimposes a budget constraint on the indifference mapping there are just two predictions: keep it all or give away half unless the constraint has exactly the same slope as the indifference curve, in which case any amount between nothing and half is possible.

The fact that intentions play no role is a second problem faced by all the outcome-oriented approaches. A trade-off does, however, exist because incorporating intentions makes the specifications considerably harder

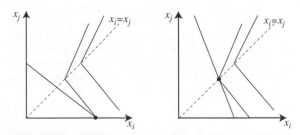

Figure 1

to work with. The outcome- and process-oriented specifications evolved from the notion of *psychological games*, which posits that utility will depend on both outcomes and beliefs (Geanakoplos, Pearce and Stacchetti, 1989). Beliefs are important because emotional responses are often triggered by expectations about how one should be treated. Perhaps the specification that is easiest to work with is the Charness–Rabin utility function, which incorporates a term θq to capture reciprocal motivations:

$$u_i(x_i, x_j) = (\rho r + \sigma s + \theta q)x_j + (1 - \rho r - \sigma s - \theta q)x_i.$$

The parameters r and s indicate which of the two players has the advantage ($r = 1$ if $x_i > x_j$, $s = 1$ if $x_j > x_i$ and $r = s = 0$ otherwise) and the parameters ρ and σ represent outcome-oriented preferences (Charness and Rabin, 2002). To recover the Fehr–Schmidt specification we simply assume $\sigma < 0 < \rho < 1$ and $\theta = 0$. Reciprocity and intentions are at work if $\theta > 0$ because we set $q = -1$ if player j has misbehaved and $q = 0$ otherwise.

Why should economists care about social preferences? By ignoring social preferences, economists have incompletely characterized many important interactions (Fehr and Fischbacher, 2002). Because many people are motivated by notions of fairness and reciprocity, social preferences can hinder the dynamics of competition that are assumed to drive equilibria, especially in the context of labour markets. For example, wages may never fall to the competitive equilibrium level because bosses understand that workers are reciprocally motivated. By lowering the wage, the boss also lowers morale and productivity (Bewley, 1999). Likewise, the economic theory of collective action is only narrowly applicable because it fails to realize that most people are predisposed to cooperate, but hate being taken advantage of (Andreoni, 1988). Designing incentives is ultimately more challenging when one accounts for the heterogeneity of social motivations identified in economic experiments.

Future research on social preferences is likely to extend in a number of interesting directions. Experimenters have begun to move from the laboratory to the field to identify the preferences of more representative samples and to investigate the external validity of these preferences (that is, what important behaviours and outcomes do social preferences correlate with?). Within the laboratory it will be interesting to better isolate the role of outcomes versus the role of intentions, to examine the co-evolution of preferences and institutions, and to examine the difference between social preferences and social norms. Is it the case, for example, that norms dictate how one should treat others regardless of whether the prescribed behaviour is consistent with one's underlying preferences?

JEFFREY CARPENTER

See also **altruism in experiments; economic man; public goods experiments; trust in experiments.**

Bibliography

Andreoni, J. 1988. Why free ride? Strategies and learning in public good experiments. *Journal of Public Economics* 37, 291–304.

Berg, J., Dickaut, J. and McCabe, K. 1995. Trust, reciprocity and social history. *Games and Economic Behavior* 10, 122–42.

Bewley, T. 1999. *Why Wages Don't Fall During a Recession.* Cambridge, MA: Harvard University Press.

Binmore, K., Shaked, A. and Sutton, J. 1985. Testing noncooperative bargaining theory: a preliminary study. *American Economic Review* 75, 1178–80.

Blount, S. 1995. When social outcomes aren't fair: the effect of causal attribution on preferences. *Organizational Behavior & Human Decision Processes* 62, 131–44.

Bowles, S. 2004. *Microeconomics: Behavior, Institutions and Evolution.* Princeton: Princeton University Press.

Camerer, C. 2003. *Behavioral Game Theory: Experiments on Strategic Interaction.* Princeton: Princeton University Press.

Cameron, L. 1999. Raising the stakes in the ultimatum game: experimental evidence from Indonesia. *Economic Inquiry* 37, 47–59.

Carpenter, J. 2007. The demand for punishment. *Journal of Economic Behavior & Organization* 62, 522–42.

Carpenter, J., Burks, S. and Verhoogen, E. 2005. Comparing students to workers: the effects of social framing on behavior in distribution games. In *Field Experiments in Economics. Research in Experimental Economics*, ed. J. Carpenter, G. Harrison and J. List. Greenwich, CT and London: JAI/Elsevier.

Charness, G. and Rabin, M. 2002. Understanding social preferences with simple tests. *Quarterly Journal of Economics* 117, 817–70.

Cox, J.C. 2004. How to identify trust and reciprocity. *Games and Economic Behavior* 46, 260–81.

Eckel, C. and Grossman, P. 1996. Altruism in anonymous dictator games. *Games and Economic Behavior* 16, 181–91.

Falk, A., Fehr, E. and Fischbacher, U. 2003. On the nature of fair behavior. *Economic Inquiry* 41, 20–6.

Fehr, E. and Fischbacher, U. 2002. Why social preferences matter – the impact of non-selfish motives on competition, cooperation and incentives. *Economic Journal* 112, 1–33.

Fehr, E. and Gächter, S. 2000. Cooperation and punishment in public goods experiments. *American Economic Review* 90, 980–94.

Fehr, E. and Schmidt, K. 1999. A theory of fairness, competition, and cooperation. *Quarterly Journal of Economics* 114, 769–816.

Fischbacher, U., Gächter, S. and Fehr, E. 2001. Are people conditionally cooperative? Evidence from a public goods experiment. *Economic Letters* 71, 397–404.

Forsythe, R., Horowitz, J., Savin, N.E. and Sefton, M. 1994. Fairness in simple bargaining experiments. *Games and Economic Behavior* 6, 347–69.

Geanakoplos, J., Pearce, D. and Stacchetti, E. 1989. Psychological games and sequential rationality. *Games and Economic Behavior* 1, 60–79.

Güth, W., Schmittberger, R. and Schwarze, B. 1982. An experimental analysis of ultimatum bargaining. *Journal of Economic Behavior and Organization* 3, 367–88.

Hoffman, E., McCabe, K., Shachat, J. and Smith, V. 1994. Preferences, property rights, and anonymity in bargaining games. *Games and Economic Behavior* 7, 346–80.

Sobel, J. 2005. Interdependent preferences and reciprocity. *Journal of Economic Literature* 43, 392–436.

Social Security in the United States

Social Security is a US federal government programme that provides retirement, survivors, and disability benefits for eligible workers and their dependents. It is largely a pay-as-you-go (PAYG) tax-financed system; current workers are taxed and these revenues are used to finance the old age, survivors (widows, widowers, surviving divorced spouses, surviving children, and parents of deceased qualified workers), and disability insurance payments. Health insurance for the elderly is administered separately in another federal programme called Medicare. Retirement, survivor and disability benefits are financed by Old Age, Survivors, and Disability Insurance (OASDI) taxes, and Medicare is financed by Health Insurance (HI) taxes. In addition to the Social Security Disability Insurance, there is Supplemental Security Income that pays benefits based on financial need. Health insurance for the poor, regardless of age, is called Medicaid, and this is financed jointly from general federal tax revenues and the states.

Eligibility and coverage

As of 2007, nearly 150 million US workers and their dependents are covered by this compulsory system. (Among the few workers who are exempted from Social Security are federal civilian workers hired prior to 1984, some state and local government workers, and household, agricultural, and self-employed workers whose earnings are too low.)

There are about 40 million retirees and survivors drawing benefits from the system. Eligibility for full retirement benefits typically requires working for ten years, although early retirement at age 62 is possible with reduced retirement benefits. The normal retirement age is currently 66, set to rise to 67 by the time the 1960 cohort retires with full benefits in 2027. Retirement benefits are about 40 per cent of lifetime average earnings of a median wage earner. High-wage earners get less than this average replacement rate and low-wage earners get more than 40 per cent. Up to an additional one-half of the retired worker's full benefits may be paid to a spouse if

the spouse has not worked or has low earnings. Under certain conditions, a divorced spouse can also get benefits if the marriage lasted at least ten years. If there are children eligible for Social Security, each receives up to 50 per cent of full benefits, with a maximum of 150–80 per cent of a worker's own benefit payments as a family limit. (For details and the separate Disability Insurance program, see http://www.ssa.gov.)

The tax rate for OASDI in 2007 is 12.4 per cent, paid equally by the employer and the employee (or entirely if the individual is self-employed). There is no tax on earnings over $90,000 in 2005 dollars, with this maximum rising every year by the national average wage index. (The HI tax rate is 2.9 per cent on all earnings.)

A private retirement fund is typically invested in a portfolio of stocks, commercial paper, and government bonds. Social Security is largely a PAYG system with current taxpayers paying for current retirees' benefits. Since the mid-1980s, Social Security has collected more in taxes than it pays in benefits. This surplus is loaned to the US Treasury in return for special-issue Treasury bonds that are used to finance other federal government expenditures. This Trust Fund stood at $1.8 trillion at the end of 2006. Although this sum seems quite large, it is actually quite small compared with the annual OASDI benefits which amounted to $461 billion in 2006. (This sum does not include Disability Insurance payments, which amounted to $91 billion in 2006.)

A short history of social security

In 1889, Germany became the first country to provide old-age insurance on a large scale. Designed by the Chancellor, Otto von Bismarck, the German system required mandatory participation, taxed the employer and the employee, and used government taxes to provide retirement and disability benefits. The retirement age was 70, though it was lowered to 65 in 1916.

In the United States, the German system was viewed as a model. In 1935 The Social Security Act, which covered workers in commerce and industry, was signed by President Roosevelt. The original proposal and draft legislation, the Economic Security Act, was a three-tiered social pension programme: (*a*) old-age welfare payments, (*b*) mandatory, contributory old-age retirement programme, which evolved into what is now called Social Security, and (*c*) voluntary annuity sales through which the federal government would sell certificates to workers who could, upon reaching the retirement age, convert them into monthly annuities to supplement their basic compulsory retirement benefits. However, the Congress rejected this third proposal by President Roosevelt, which was the earliest attempt to allow for a private saving programme for retirement with mandatory annuitization at retirement.

Since 1935, many changes were made to the original 1935 act. In 1937, workers were required to pay two per cent of their payroll to support the Social Security system. In 1939, the programme started to cover dependents and survivors. In 1950, coverage was extended beyond commerce and industry, the tax rate was raised to three per cent and benefits were also raised. In 1956, the tax rate was raised to four per cent and disability insurance was introduced. Early retirement for women was permitted in 1956 and for men in 1961, when payroll taxes were raised to six per cent. In 1972, cost-of-living-adjustments (COLAs) were introduced, automatically indexing benefit levels to inflation, and payroll taxes were raised to 9.2 per cent. In 1977, the tax rate jumped to 9.9 per cent.

In 1983 the National Commission on Social Security Reform was formed to suggest ways to deal with the actuarial imbalance of the system. The commission proposed (a) a phased-out increase in the retirement age from 65 to 67, (b) an increase in the self-employment tax, (c) partial taxation of benefits to upper income retirees, and (d) expanding the coverage to include federal civilian and non-profit organization employees. The tax rate was raised to 10.8 per cent. The payroll tax rate was raised to 11.4 per cent in 1985 and to 12.4 per cent in 1993. In 1996 the Social Security Trustees (1996) reported that the Social Security system would start to run deficits in 2012, and the trust funds would be exhausted by 2029. They predicted that, in order to keep the benefit levels unchanged, the tax rate would have to rise to 18 per cent.

The simple economics of social security

A simple textbook treatment will be presented here. For a more rigorous treatment of social security and overlapping generations models, the reader should see the seminal papers by Samuelson (1958) and Diamond (1965). For a recent and thorough coverage, see Ljungqvist and Sargent (2005).

Consider a simple overlapping generations model in which individuals live for two periods. Their earnings are y when young, and 0 when old. They face a social security tax $\tau > 0$ when young and receive a retirement benefit b when old. They earn interest on their saving, s at the rate r. There is no uncertainty. The economy is assumed to have perfectly functioning credit markets that allow individuals to borrow any amount at the market rate of interest r. Population grows at the rate $n > 0$, and productivity (and hence real income) grows at the rate $g > 0$. The lifetime utility function of the individuals is given by

$$\log(c_1) + \beta \log(c_2) \tag{1}$$

where c_1 and c_2 are consumption when young and when old, respectively, and $\beta > 0$ is the subjective discount factor. The budget constraints faced by the individuals are given as

$$c_1 + s = y, \tag{2}$$

$$c_2 = (1+r)s + b. \tag{3}$$

We can write the lifetime budget constraint facing the agents in present value form

$$c_1 + \frac{c_2}{(1+r)} = y + \frac{b}{(1+r)}. \tag{4}$$

The social security system is unfunded, which requires that taxes collected from the young equal benefits given out to the retirees at each period:

$$b = (1+g)(1+n)\tau y. \tag{5}$$

The Lagrangian for the individuals' optimization problem is given by

$$L = \log(c_1) + \beta \log(c_2) + \lambda \left[y + \frac{b}{(1+r)} - c_1 - \frac{c_2}{(1+r)} \right], \tag{6}$$

where $\lambda > 0$ is the Lagrange multiplier.

The first-order conditions for maximizing the Lagrangian with respect to the choice variables c_1 and c_2 (and λ) yield optimal lifetime consumption quantities as

$$c_1 = \frac{(1+r)(1-\tau)y + b}{(1+r)(1+\beta)}, \quad \text{and} \tag{7}$$

$$c_2 = \frac{\beta(1+r)(1-\tau)y + \beta b}{(1+\beta)}. \tag{8}$$

The second-order conditions are satisfied given our choices of functional forms and parameter restrictions.

Using eqs (2), (3) and (8), we obtain the individual's private saving under social security as

$$s^{ss} = \frac{\beta(1-\tau)y - b/(1+r)}{(1+\beta)}. \tag{9}$$

In the absence of an unfunded social security system, that is, with $\tau = 0$ and $b = 0$ saving in this 'laissez-faire' world is given by

$$s^{lf} = \frac{\beta y}{(1+\beta)}. \tag{10}$$

Equations (9) and (10) show that the introduction of an unfunded social security system reduces private saving. The productive capital stock in an economy is determined to a large extent by private saving, and therefore an unfunded social security system results in a lower capital stock and per capita consumption. (If factor prices were determined by the capital–labour ratio, then the general equilibrium effect would be to mitigate the reduction in private saving as the return to capital would rise in response to the decrease in private saving. However, the direction of the change in saving is still the same, and, in quantitative versions of the life cycle model, the magnitude of this effect is quite large. Also, if labour were endogenous, one could show that a public pension system with no linkage between contributions and benefits distorts labour supply and causes it to decline, increasing the welfare cost of social security.)

What is the impact on the individual's lifetime welfare? To address this issue, we need to compare the lifetime consumption values with and without social security. Substituting eq. (5) into eqs (7) and (8) yields

$$c_1^{ss} = \frac{(1+r)y - \tau y[(1+r) - (1+g)(1+n)]}{(1+r)(1+\beta)},$$
(11)

$$c_2^{ss} = \frac{\beta(1+r)y - \beta\tau y[(1+r) - (1+g)(1+n)]}{(1+r)(1+\beta)},$$
(12)

where c_1^{ss} and c_2^{ss} denote young- and old-age consumption under social security. Imposing $\tau = 0$ and $b = 0$, and using c_1^{lf} and c_2^{lf} to denote young- and old-age consumption under laissez-faire, eqs (11) and (12) become

$$c_1^{lf} = \frac{(1+r)y}{(1+r)(1+\beta)},$$
(13)

$$c_2^{lf} = \frac{\beta(1+r)y}{(1+r)(1+\beta)}.$$
(14)

Given our assumptions on the parameters,

$$c_1^{lf} > c_1^{ss} \text{ and } c_2^{lf} > c_2^{ss} \text{ if and only if}$$
$$[(1+r) - (1+g)(1+n)] > 0.$$
(15)

In other words, unfunded social security yields lower young- and old-age consumption to the individual if the rate of return on capital, $1+r$ exceeds the rate of growth of the economy, $(1+g)(1+n)$. Historically, the annual growth rate of the US economy has been just under three per cent on average, whereas the return on capital has exceeded this amount significantly, by about two to five percentage points, depending on alternative definitions of the capital stock. (The condition presented in eq. (15) is known as the condition for dynamic efficiency. If the reverse were true, then a reduction in saving would reduce the dynamic inefficiency – or capital overaccumulation – in the economy and improve welfare by raising consumption in both periods.) In fact, reform proposals in the early 21st century to partially privatize social security aim to exploit this return differential in order to build up private retirement funds for old-age consumption.

Pros and cons of unfunded social security

As the previous section demonstrates, the most important welfare cost of an unfunded social security system is that it discourages private saving and the accumulation of capital. There are other social costs associated with a PAYG system, as well as several social benefits. In this section, these costs and benefits will be listed and the empirical evidence about their quantitative importance will be evaluated.

Welfare costs

1. Unfunded social security discourages saving. Young workers with high marginal propensities to save are taxed, and these resources are given to retirees, with low marginal propensities to save.
2. It discourages work effort since the payroll tax paid by the worker has less than perfect linkage, if any, with the retirement benefit that the worker will receive. The labour supply in the economy is adversely affected and so is the demand for labour as firms are reluctant to create new jobs.
3. It distorts the retirement decision and encourages early retirement.
4. It imposes a hardship on workers and companies that are facing borrowing and credit constraints. Removing the high payroll taxes devoted to this mandatory retirement system would bring welcome relief to a large number of households and firms.

Welfare benefits

1. Unfunded social security provides longevity insurance. If an individual lives longer than expected, an annuity that provides for old-age consumption is welfare-enhancing. (Yaari, 1965, showed that in a simple two-period overlapping generations model rational individuals without an altruistic motive would find it optimal to annuitize all their wealth. However, the annuity markets in the United States are thin. Indeed, Friedman and Warshawsky, 1990, and Mitchell et al., 1999, document that private annuities are unattractively priced in the United States and that the adverse selection problem likely contributes to the low volume of these contracts. See also Feldstein, 2005, for an excellent survey of issues surrounding social security.)
2. It provides insurance against (privately uninsurable) income shocks over the life cycle.
3. It serves as a partial substitute for a commitment device for individuals with time-inconsistent preferences.
4. It insulates the individuals to a large extent against aggregate shocks such as stock market risks.

Empirical evidence

Auerbach and Kotlikoff (1987) presented one of the earliest attempts at using a large-scale overlapping generations framework to address fiscal policy effects in a calibrated, general equilibrium setting. In their model, social security did not have any potential benefit and it imposed a deadweight loss on the society. Hubbard and Judd (1987) introduced the longevity insurance aspect of social security, but still the negative impact of social security on saving outweighed this potential benefit. İmrohoroğlu, İmrohoroğlu and Joines (1995; 1999) consider variations of the earlier quantitative models but they arrive at the same conclusion.

Diamond (1977) and Feldstein (1985) argue that a fraction of the population might lack the foresight to save for retirement. However, İmrohoroğlu, İmrohoroğlu and Joines (2003) use Strotz's (1956) time-inconsistent preferences to evaluate the welfare benefit of an unfunded social security system in a quantitative model that includes most of the costs and benefits of social security listed in the previous section, and find that social security is a poor commitment device and that individual's welfare is not increased unless for relatively high (and implausible) degrees of short-term discount rates.

Despite the widespread quantitative evidence on the inefficiency of social security, the institution seems resistant to reform. One reason may be due to potentially large transitional costs of privatizing the system. For example, Conesa and Krueger (1999) find large transitional costs, with a majority of the currently alive population suffering welfare losses, and thus blocking any reform proposal. Fuster, İmrohoroğlu and İmrohoroğlu (2007), on the other hand, argue that these transitional costs are easily paid for by a growing economy with strong bequest motives and flexible labour markets. Cooley and Soares (1999) and Boldrin and Rustichini (2000) study the political-equilibrium considerations that allow the introduction and maintenance of an unfunded social security system. Krueger and Kuebler (2006) introduce aggregate uncertainty in the form of 'investment risk' but still find that the capital crowding-out effect dominates and social security reduces welfare.

The future of social security

Notwithstanding the economic burden of PAYG systems, as the previous section summarizes, the future holds much worse predictions for OECD countries. Without exception, all social security programmes are facing financing difficulties, due to significant increases in longevity and the decline in fertility. Most European countries and Japan are ageing very rapidly. The United States is not far behind. The ratio of the number of retirees to the number of workers, the so-called dependency ratio, was near 20 per cent in 2000, but is expected to rise to almost 50 per cent in 2060, meaning only two workers per retiree.

To get an idea of how much of a strain the ageing of the US population puts on the current pension system, consider the following two findings from De Nardi,

İmrohoroğlu and Sargent (1999). First, responding to a proposal to index retirement age to the increases in longevity so that the dependency ratio is held constant at the current level of 20 per cent, they calculate that the retirement age has to eventually rise to 76. Second, maintaining retirement benefits at their current levels when the population is ageing (and the costs of medical services are rising) requires an increase in the Social Security payroll tax rate from its current level of about 10 per cent to an eventual 40 per cent. If a reform is not undertaken, even these high tax rates will not be enough to pay for pensions in the future.

SELAHATTIN İMROHOROĞLU

See also **overlapping generations model of general equilibrium; retirement; social insurance; social insurance and public policy.**

Bibliography

Auerbach, A. and Kotlikoff, L. 1987. *Dynamic Fiscal Policy.* Cambridge: Cambridge University Press.

Boldrin, M. and Rustichini, A. 2000. Political equilibria with social security. *Review of Economic Dynamics* 3, 41–78.

Conesa, J. and Krueger, D. 1999. Social Security reform with heterogeneous agents. *Review of Economic Dynamics* 2, 757–95.

Cooley, T. and Soares, J. 1999. A positive theory of social security based on reputation. *Journal of Political Economy* 107, 135–60.

De Nardi, M. İmrohoroğlu, S. and Sargent, T. 1999. Projected U.S. demographics and social security. *Review of Economic Dynamics* 2, 575–615.

Diamond, P. 1965. National debt in a neoclassical debt model. *Journal of Political Economy* 55, 1126–50.

Diamond, P. 1977. A framework for social security analysis. *Journal of Public Economics* 8, 275–98.

Feldstein, M. 1985. The optimal level of Social Security benefits. *Quarterly Journal of Economics* 100, 303–20.

Feldstein, M. 2005. Rethinking social insurance. *American Economic Review* 95, 1–24.

Friedman, B. and Warshawsky, M. 1990. The cost of annuities: implications for saving behavior and bequests. *Quarterly Journal of Economics* 105, 135–54.

Fuster, L., İmrohoroğlu, A. and İmrohoroğlu, S. 2007. Elimination of social security in a dynastic framework. *Review of Economic Studies* 74, 113–45.

Hubbard, G. and Judd, K. 1987. Social security and individual welfare. *American Economic Review* 77, 630–46.

İmrohoroğlu, A., İmrohoroğlu, S. and Joines, D. 1995. A life cycle analysis of social security. *Economic Theory* 6, 83–114.

İmrohoroğlu, A., İmrohoroğlu, S. and Joines, D. 1999. Social security in an overlapping generations economy with land. *Review of Economic Dynamics* 2, 638–65.

İmrohoroğlu, A., İmrohoroğlu, S. and Joines, D. 2003. Time inconsistent preferences and social security. *Quarterly Journal of Economics* 118, 745–84.

Krueger, D. and Kubler, F. 2006. Pareto improving social security reform when financial markets are incomplete. *American Economic Review* 96, 737–55.

Ljungqvist, L. and Sargent, T.J. 2005. *Recursive Macroeconomic Theory*. Cambridge: MIT Press.

Mitchell, O., Poterba, J.M., Warshawsky, M.J. and Brown, J.R. 1999. New evidence on the money's worth of individual annuities. *American Economic Review* 89, 1299–318.

Samuelson, P. 1958. An exact consumption loan model of interest with or without the social contrivance of money. *Journal of Political Economy* 66, 467–82.

Social Security Trustees. 1996. *1996 Annual Report of the Board of Trustees of the Federal Old-Age and Survivors Insurance and Disability Insurance Trust Funds*. Washington, DC: U.S. Government Printing Office. Online. Available at http://www.ssa.gov/history/reports/trust/1996, accessed 8 June 2007.

Strotz, R. 1956. Myopia and inconsistency in dynamic utility maximization. *Review of Economic Studies* 23, 165–80.

Yaari, M. 1965. Uncertain lifetime, life insurance, and the theory of the consumer. *Review of Economic Studies* 32, 137–50.

social status, economics and

People are social animals who care about their standing in society and about what other members of society think about them. Stated differently, people care about their 'prestige' or the 'respect' that they are accorded by individuals with whom they interact. Many people would gladly pay large sums of money for a knighthood. Similarly, the value of the Nobel Prize lies not entirely in the monetary prize itself; hence, we would not be surprised to find that many people might actually be willing to pay to obtain this prize. These observations on human nature are far from new. Hobbes, for example, wrote that: 'Men are continuously in competition for honour and dignity' (cited in Hirschman, 1973, p. 634).

While the focus of traditional economics has been on the monetary rewards that are exchanged through a market mechanism, sociologists have stressed social status and other social rewards as important motivations for human behaviour. The term 'social status' was first introduced by Max Weber as 'an effective claim to social esteem in terms of negative or positive privileges' (Weber, 1922, p. 305).

One of the important factors that distinguishes societies and determines their success is the form of the incentives they provide to their members. Generally speaking, there are three broad types of such incentives: (*a*) private monetary rewards, such as wages and profits;

(*b*) social rewards including status and prestige; and (*c*) rules, laws and regulations that enforce certain types of behaviour while penalizing others. Societies may differ in the mix of incentives and rules that they employ. This mixture has a significant effect on economic performance. Social status is thus part of the incentive structure provided to individuals in every society. These incentives affect an individual's choice of actions, occupation, education level and so forth. They thus are of significant economic importance (see also the review in Weiss and Fershtman, 1998).

If prizes, knighthoods and other status symbols were up for sale like any other commodity, their value would be deflated. The value of a medal, award or title will be small if it were obtainable by many people. Thus, the value of status symbols depends on the allocation rule that determines who is eligible to receive a particular symbol, who is excluded from eligibility, and the number (and certainly the identity) of its recipients. This property distinguishes social status from economic rewards. Giving a medal to one individual (may) reduce the medal's value for another individual. Social status may thus be viewed as the ranking of individuals, or groups of individuals, in society. This ranking may be based on personal attributes, actions, occupations or group affiliations. Yet, by definition, if someone climbs up in rank, someone else climbs down.

Ranking matters only if people agree on how ranking is established. For social status to matter, a society must generally agree about the relative position of its members. The crucial feature of social status is that it 'rests on collective judgement, or rather a consensus of opinion within a group. No one person can by himself confer status on another, and if a man's social position were assessed differently by everybody he met, he would have no social status at all' (Marshall, 1977, p. 198). An interesting – and relatively unexplored – issue is the role of social status in a multicultural society where every group maintains its own ranking, each of which may be affected by different characteristics. Fershtman, Hvide and Weiss (2006) have shown that the gains made from (social) trade in a culturally diverse society can be translated into higher output and wages.

The categories comprising social status are diverse. We can distinguish between 'status group' and 'individual status'. In the first category, originally conceived by Weber, social status is obtained by affiliation with a group, be it a social class, profession, club, and so on. Members of the status group share a similar status. In the second category, social status is obtained through individual attributes or actions. One should also distinguish between status that is acquired through specific actions or group affiliation and status that has been inherited. People may have social status simply by being born into an aristocratic class. The specific structure for gaining and maintaining status thus plays an important role in determining its effect on the economy.

One of the major problems in incorporating social status into economic models is that it is not directly observed. How do we measure status? How do we identify the ranking that determines who is perceived as important and who is not? And how do we quantify this ranking? Another task necessary for modelling is identification of the variables that affect social status. What determines social status in different societies? Conducting surveys by asking people to rank occupations according to their 'prestige' or to state which of the individual's attributes affects his or her social status has been the accepted method for responding to these questions. Treiman (1977), for example, found that the ranking of occupational status is stable over time and similar in different societies. Moreover, status rank has been found to be systematically dependent on occupational attributes; that is, occupations requiring high levels of education and providing high income also confer high social status.

Social status and consumption

Individuals may use consumption choices to signal that they have properties that affect their social status. The most familiar form of such signalling is 'conspicuous consumption', a signal relevant whenever relative wealth is a factor in determining social status. Yet the quest for social status may affect other consumption choices as well. For instance, individuals may buy and display books or go to the theatre to signal the level of education they have obtained. They may join clubs, buy a house or hire a maid if such actions signal their desired status.

The concept of 'conspicuous consumption' was first introduced by Veblen (1899), who argued that individuals often consume highly attention-getting goods and services in order to signal their wealth and thereby achieve greater social status. 'In order to hold the esteem of men it is not sufficient merely to possess wealth or power. The wealth and power must be put in evidence, for esteem is awarded only on evidence' (Veblen, 1899, p. 36). The extreme form of such behaviour is known as the 'Veblen effect', witnessed whenever individuals are willing to pay higher prices for functionally equivalent goods (for a discussion, see also Leibenstein, 1950, and Frank, 1985a, 1985b). The Veblen effect may indeed be empirically significant in some luxury good markets (see Creedy and Slottje, 1991; Heffetz, 2004).

Veblen distinguished between (a) 'invidious comparison' – whenever an individual from a higher class consumes conspicuously to distinguish himself from an individual from a lower class, and (b) 'pecuniary emulation' – whenever an individual from a lower class consumes conspicuously to imitate a member of the upper class. Bagwell and Bernheim (1996) used a signalling model to investigate the conditions under which the Veblen effect may result from the desire to signal wealth (see also Ireland, 1994).

Conspicuous consumption may lead to excessive consumption and suboptimal saving but this conclusion depends on the specific details and timing of the 'conspicuous consumption race'. As Corneo and Jeanne (1996) have shown, if the signalling is typically done late in the life cycle, conspicuous consumption may actually encourage saving.

Letting social status be determined by relative wealth may help to explain some of the puzzles we observe in human behaviour. The empirical evidence indicates that saving continues in old age and hardly declines with wealth. These observations imply that saving behaviour cannot be explained solely by consumption motives. Status concerns derived from relative wealth may provide some explanation of this phenomenon.

It is important to note that the striving for social status is not the only explanation for conspicuous consumption. Such behaviour may also signal professional success or ability. Simply think about a situation in which you need to choose a lawyer without any knowledge about the candidates' ability. Often a lawyer's dress, the car she drives or how her office is decorated may affect your choice whenever these items may signal ability or success.

Status and the labour market

Some professions enjoy higher status than others. For example, in most countries, being a physician yields a higher status than being a butcher. In such cases, simply belonging to a profession, and not the individual's characteristics, is rewarded by the relevant professional status. Consequently, individuals choose their occupations not just according to the wages they will be paid but also according to the status associated with that occupation. However, at equilibrium, wages are also affected by the quest for status. Adam Smith, in *The Wealth of Nations* (1776), was the first to state the argument of compensating wage differences by pointing out that: 'Honour makes a great part of the rewards of all honourable professions. …The most detestable of all employment, that of a public executioner, is, in proportion to the quantity of work done, better paid than any common trade whatever' (Book I, ch. X). Still, no empirical evidence has been found supporting the phenomenon of high status being associated with low wages, because to do so one would need to control for ability, which is not directly observable.

Status concerns may also explain some degree of wage rigidity. Unemployed individuals are often reluctant to accept temporary but low-paid jobs because doing so implies a loss of status (see Blinder, 1988). On the other hand, immigrants are less reluctant to apply for low-status jobs; they tend to stress wages over occupational status partly because their reference group is new immigrants and not the wider society.

The workplace itself is a venue for social interaction. Relative wages may be important in forming 'local' status at the firm level. The willingness of workers to exert effort may be affected by social rewards. When workers are concerned with their 'local' status, wage inequality across firms will tend to exceed wage inequality within firms. Within each firm, productivity differentials will exceed wage differentials as some reward is derived from higher status (see Frank 1984a, 1984b; for empirical and experimental evidence for the relevance of wage comparisons, see Clark and Oswald, 1996, and Zizzo and Oswald, 2001).

Social status and growth

The great variability in growth across different economies is a major puzzle for economists. While most of the literature offers an economic explanation for this phenomenon, others claim that some of the variation can be attributed to cultural factors. Social status affects growth primarily by affecting individuals' choices of occupation, investment and education. For example, it has been argued that contempt for entrepreneurs and the high status of the idle gentleman in 19th-century England were the main causes for its economic decline during that period (see Wiener, 1981).

A common feature of recent growth models is the existence of externalities associated with human capital or certain occupations. Each worker, when choosing his level of schooling or occupation, ignores the impact of his choice on overall economic performance. Whenever social status is attached to such activities or occupations, it can be perceived as a corrective mechanism. Baumol (1990) emphasizes the role of social status or social prestige associated with 'non productive' (rent-seeking) activities versus 'productive' activities. The implications are simple: a status structure that awards higher status to 'productive' activities is conducive to growth. Fershtman and Weiss (1993) used a simple general equilibrium model in which wages and social status are determined endogenously to show that changes in the demand for status, triggered by changes in preferences, may affect growth rates.

But status has a collective nature and may be determined endogenously by the type of people who choose each occupation or profession. The drive for status may thus be counter-productive and induce an inefficient allocation of talent among the different occupations. A large emphasis on status may encourage the 'wrong' individuals – those with low ability and great wealth – to choose a 'productive' occupation or acquire schooling, thereby forcing workers with high ability but little wealth to leave growth-enhancing occupations. This crowding-out effect may, by itself, discourage growth (see Fershtman, Murphy and Weiss, 1996).

Social status as a corrective mechanism

It has long been recognized that activities that generate externalities but cannot be priced are not efficiently regulated by private rewards. It was Arrow (1971) who initially suggested the role of social rewards as a mechanism designed to resolve the inefficiencies arising from externalities (see Elster, 1989, for a critical view for this approach). According to Arrow, an individual who chooses an action or occupation that produces positive externalities is appreciated by other members of the society and obtains social status, whereas an individual who produces negative externalities is treated with contempt (or a negative social status). The use of such a social mechanism is appealing as it implies that the problem of market inefficiencies due to externalities can be resolved or diminished. On the other hand, the use of social rewards is limited in itself and as a corrective mechanism. As mentioned previously, a profusion of medals reduces their value – a property that limits the scope of their use.

Social status and the evolution of preferences

While most economists are sympathetic to the idea that the concern for social status is an important aspect of human decision-making, they remain reluctant to incorporate this variable into mainstream economic modelling. The ruling paradigm of *homo economicus* is that of an individual whose utility depends on his consumption bundle and who makes employment decisions based on the wages to be received for performing a particular job. For sociologists, the dominant paradigm is that of people who, as social animals, wish to maximize their standing in society.

The reluctance to incorporate status concerns into the utility function is rooted in the assumption that models including this variable often allow too broad a range of behaviour and thus ultimately display little predictive power (for more on this view see Postlewaite, 1998). The debate centres on whether status concerns are a 'direct effect', reflecting the fact the people are (also) social animals, or an 'indirect effect', meaning that people care about social status because status affects the goods and services that they and their children will consume (for an illustration of the indirect approach, see Cole, Mailath and Postlewaite, 1992).

Incorporating preferences for status as 'hard-wired' into the utility function raises the question of why people (or other animals) have ingrained preferences for social status. One approach to dealing with the issue, common in economics, is not to deal with the formation of preferences. Social biologists have adopted a different approach. They argue that feelings and social concerns are hard-wired in human actors; moreover, concerns that increase fitness tend to be more common. Fershtman and Weiss (1998a; 1998b) applied an evolutionary approach and showed that caring about social status can be part of stable evolutionary preferences.

CHAIM FERSHTMAN

See also **social norms.**

Bibliography

Arrow, K. 1971. Political and economic evaluation of social effects and externalities. In *Frontier of Quantitative Economics*, ed. M. Intriligator. Amsterdam: North-Holland.

Bagwell-Simon, L. and Bernheim, D. 1996. Veblen effects in a theory of conspicuous consumption. *American Economic Review* 86, 349–73.

Baumol, W. 1990. Entrepreneurship: productive, unproductive and destructive. *Journal of Political Economy* 98, 893–921.

Blinder, A. 1988. The challenge of high unemployment. *American Economic Review* 78, 1–15.

Clark, A. and Oswald, A. 1996. Satisfaction and comparison income. *Journal of Public Economics* 61, 359–81.

Cole, H., Mailath, G. and Postlewaite, A. 1992. Social norms, saving behaviour and growth. *Journal of Political Economy* 100, 1092–125.

Corneo, G. and Jeanne, O. 1996. Social organization, status and saving behaviour. *Journal of Public Economics* 70, 37–52.

Creedy, J. and Slottje, D. 1991. Conspicuous consumption in Australia. Working Paper No. 307. Department of Economics, University of Melbourne.

Elster, J. 1989. Social norms and economic theory. *Journal of Economic Perspectives* 3(4), 99–117.

Fershtman, C., Hvide, H. and Weiss, Y. 2006. Cultural diversity, status concerns and the organization of work. *Research in Labor Economics: The Economics of Immigration and Social Diversity*, ed. S. Polachek., C. Chiswick and H. Rapoport. Amsterdam: Elsevier, JAI Press.

Fershtman, C., Murphy, K. and Weiss, Y. 1996. Social Status, Education and Growth. *Journal of Political Economy* 104, 108–32.

Fershtman, C. and Weiss, Y. 1993. Social status, culture and economic performance. *Economic Journal* 103, 946–59.

Fershtman, C. and Weiss, Y. 1998a. Why do we care what others think about us? In *Economics, Values and Organization*, ed. A. Ben Ner and L. Putterman. Cambridge: Cambridge University Press.

Fershtman, C. and Weiss, Y. 1998b. Social rewards, externalities and stable preferences. *Journal of Public Economics* 70, 53–73.

Frank, R. 1984a. Interdependent preferences and the competitive wage structure. *Rand Journal of Economics* 15, 510–20.

Frank, R. 1984b. Are workers paid their marginal products? *American Economic Review* 74, 549–70.

Frank, R. 1985a. *Choosing the Right Pond: Human Behavior and the Quest for Status*. London and New York: Oxford University Press.

Frank, R. 1985b. The demand for unobservable and other nonpositional goods. *American Economic Review* 75, 101–16.

Heffetz, O. 2004. Conspicuous consumption and the visibility of consumers' expenditures. Working paper. Center for Health and Wellbeing, Princeton University.

Hirschman, A. 1973. An alternative explanation of contemporary harriedness. *Quarterly Journal of Economics* 87, 634–7.

Ireland, N. 1994. On limiting the market for status signals. *Journal of Public Economics* 53, 91–110.

Leibenstein, H. 1950. Band wagon, snob, and Veblen effects in the theory of consumers demand. *Quarterly Journal of Economics* 622, 183–207.

Marshall, T. 1977. *Class, Citizenship and Social Development: Essays*. Chicago: University of Chicago Press.

Postlewaite, A. 1998. The Social Basis of Interdependent Preferences. *European Economic Review* 42, 779–800.

Smith, A. 1776. *The Wealth of Nations*, Reprinted, New York: Modern Library, 1934.

Treiman, D. 1977. *Occupational prestige in comparative perspective*. New York: Academic Press.

Veblen, T. 1899. *The theory of the leisure class: An economic study of institutions*. London: Unwin Books, 1994.

Weber, M. 1922. *Economy and Society*. Berkeley: University of California Press, 1978.

Wiener, M. 1981. *English Culture and the Decline of the Industrial Spirit, 1850– 1980*. Cambridge: Cambridge University Press.

Weiss, Y. and Fershtman, C. 1998. Social status and economic performance: a survey. *European Economic Review* 42, 801–20.

Zizzo, D. and Oswald, A. 2001. Are people willing to reduce others' income? *Annales d'économie et de statistice*, July/December, 39–65.

social welfare function

The concept of a social welfare function is central in welfare economics, the branch of economics that explores the implications of various ethical criteria for deciding what promotes social welfare, what public policies the society should choose, and so on. It was first introduced by Bergson (1938), and was subsequently elaborated by Samuelson (1947). A related, though not identical, notion with the same name was introduced by Arrow (1951); as we shall see below, Arrow's interpretation was different from Bergson's earlier concept. The technical literature on the social welfare function is large. The following discussion will, however, focus mainly on conceptual issues relating to the social welfare function.

The basic notation

Consider a given society consisting of individuals 1, 2, ..., n, with X a given set of alternative social states. For our purpose, a social state can be thought of as a complete description of all 'relevant' aspects of the state of affairs prevailing in society, but one can also use more limited interpretations of a social state. An element of X can be thought of as a vector, each component of the vector representing some particular feature of the state of society. The elements of X will be denoted by x, y, z, and so forth. Each individual i is assumed to have a preference ordering ('at least as good as') defined over X (an ordering over X is defined to be a binary relation T over X such that T is reflexive (for all x in X, xTx), connected (for all distinct x and y in X, xTy or yTx), and transitive (for all x, y, and z in X, if xTy and yTz, then xTz)). Such a preference ordering will be denoted by R_i, R_i', and so on. An n-tuple of preference orderings over X, with the preference ordering of each individual figuring exactly once in the n-tuple, will be called a preference profile. For all social states x and y, and for every individual i, [xP_iy if and only if xR_iy and not yR_ix] and [xI_iy if and only if xR_iy and yR_ix]. Intuitively, P_i denotes the strict preference relation for individual i and I_i denotes the indifference relation for individual i. R, R', and so forth will denote social weak preference relations ('socially at least as good as') over X. Thus, xRy will denote that x is at least as good as y for the society. Given a social weak preference relation R, one can define P ('socially better than') and I ('socially indifferent to') in terms of R in the same way as P_i and I_i are defined in terms of R_i.

The Bergson–Samuelson social welfare function

The Bergson–Samuelson social welfare function (SWF (BS)) is a function W that specifies exactly one real number, $W(x)$, for each social state x in X. The intended interpretation is this: for all x and y in X, $W(x) \geq W(y)$ denotes that x offers at least as high a level of social welfare as y, and similarly for $W(x) > W(y)$ and $W(x) = W(y)$. $W(x)$, $W(y)$, and so on are thus ordinal indices of social welfare attached to social states. What determines the form of the function? For example, what determines whether x should be assigned a higher number than y or y should be assigned a higher number than x? This, as both Bergson (1938) and Samuelson (1947) pointed out, will depend on the value judgements that we use. It is not surprising that very little can be done with the mathematical notion of an SWF (BS) at this level of generality. Even then, the abstract notion, by itself, makes one thing clear: if one wants to say anything specific about social welfare, one must introduce explicit value judgements.

Specific conclusions emerge from the Bergson–Samuelson social welfare function as one introduces additional value judgements that restrict the form of the function. Thus, assuming that each individual has an (ordinal) utility function U^i defined over X, Samuelson considers an 'individualistic' form for SWF (BS) where the social welfare indices attached to social states depend exclusively on the individual utilities, so that one can write the SWF (BS) as

$$W = F(U^1, \ldots, U^n) \quad \ldots \quad (1)$$

Assuming further that social states are simply allocations (a specification of the quantity of each commodity figuring in each consumer's consumption bundle and the quantity of each commodity figuring in each producer's production plan), F is increasing in U^1, U^2, ..., and U^n, and each consumer's utility depends only on her consumption bundle, Samuelson (1947, pp. 229–49) derived conditions for the maximization of social welfare subject to the relevant resource constraints and technological constraints of the society.

Arrow's social welfare function

Arrow (1951) introduced a somewhat different concept of a social welfare function (SWF (A)). An SWF (A) is a functional rule G, which, for every possible preference profile, $(R_1, \ldots, R_n)$, belonging to a non-empty class of preference profiles, defines exactly one ordering R over X. Thus, we write

$$R = G(R_1, \ldots, R_n). \quad \ldots \quad (2)$$

The intuitive interpretation of R figuring in this definition is that it represents the society's weak preference relation over social states, R being constrained to be an ordering. Thus, an SWF (A) gives us a (unique) social ordering of all social states once the individuals' orderings over the social states are given.

What is the relationship between the notion of an SWF (BS) and that of an SWF (A)? Suppose an SWF (BS) has the form given by (1), and suppose the utility functions U^i ($i=1,2,\ldots,n$) are ordinal so that they have no more significance beyond just the orderings that they, respectively, represent (recall that the Bergson–Samuelson social welfare indices are also ordinal). Then noting that, for every profile of real valued utility functions over X, we have Bergson–Samuelson social welfare indices for social states, such that the ordering implied by the social welfare indices does not change with a change in the profile of utility functions so long as the orderings implied by the utility functions do not change (see Samuelson, 1947, p. 228), we would have a unique social ordering over X for every profile of individual orderings over X and, hence, an Arrow-type social welfare function. Thus, underlying an SWF (BS) as given in (1), there is always an SWF (A). (The converse is not necessarily true since the social ordering R of Arrow may not be representable by a real valued function over X.)

Arrow's impossibility theorem

If Arrow did not impose any restrictions on an SWF (A), then the definition, by itself, would be of no more substantive interest than just the definition of an SWF (BS) without any specific assumptions about the properties of the SWF (BS). Arrow, however, proceeded to introduce specific restrictions regarding the form of an SWF (A). He postulated that an SWF (A), G, must satisfy the following properties.

Universal Domain (UD): the domain of G must include every possible preference profile.

Weak Pareto Criterion (WP): for every profile of individual orderings, $(R_1, \ldots, R_n)$, in the domain of G and for all x and y in X, if xP_iy for every individual i, then xPy.

The weak Pareto criterion, which is a weak version of the familiar Pareto principle, just says that, if every individual strictly prefers social state x to social state y, then the social ordering must rank x higher than y.

Independence of Irrelevant Alternatives (IIA): Consider any two profiles of preference orderings, $(R_1, \ldots, R_n)$ and $(R'_1, \ldots, R'_n)$, in the domain of G and any two social states x and y. If, for every individual i, $[xR_iy$ if and only if $xR'_iy]$ and $[yR_ix$ if and only if $yR'_ix]$, then $[xRy$ if and only if $xR'y]$ and $[yRx$ if and only if $yR'x]$ where the social orderings R and R' correspond to the preference profiles $(R_1, \ldots, R_n)$ and $(R'_1, \ldots, R'_n)$, respectively.

IIA requires that, if the individual orderings change but everybody's ranking of a pair of social states remains unchanged, then the social ranking of those two social states must remain unchanged though the social ranking over other pairs of social states may change.

Non-dictatorship (ND): there does not exist any individual k such that for all social states x and y and for every profile of individual orderings $(R_1, \ldots, R_n)$ in the domain of G, if xP_ky, then xPy.

ND just says that there should not be any individual such that, whenever she strictly prefers any social state x to any other social state y, x must rank higher than y in the social ordering, irrespective of other people's preferences.

Arrow's (1951) famous impossibility theorem tells us that, if there are at least three social states in X, then there does not exist any SWF (A) that simultaneously satisfies WP, IIA, and ND. The result has the flavour of a paradox since, prima facie, the properties postulated by Arrow for his social welfare function seem plausible. It may be useful to consider two examples to illustrate how Arrow's theorem 'works'. Consider first the simple majority rule (SMR) which says that, for every preference profile, $(R_1, \ldots, R_n)$, and for all x and y in X, xRy if and only if the number of individuals who consider x to be at least as good as y is greater than or equal to the number of individuals who consider y to be at least as good as x. While the SMR satisfies WP, IIA, and ND, it does not yield a social ordering for every preference profile. Thus, if we have three individuals, 1, 2, and 3, and three alternatives, x, y, and z, then, for the preference profile such that $(xP_1y$ & yP_1z & $xP_1z)$, $(yP_2z$ & zP_2x & $yP_2x)$, and $(zP_3x$ & xP_3y & $zP_3y)$, the SMR gives us $(xPy$ & yPz & $zPx)$ which is not an ordering (this, in fact is the well

known 'voting paradox'). Let us take a second example, the Borda rule, which for every preference profile specifies the social ordering as follows. On the assumption that X has m elements, if an individual places a social state x in the first position in his preference ordering, then x gets m points from him; if an individual places x in the second position in his preference ordering, then x gets $m - 1$ points from him; and so on. (In stating the Borda rule, I have ignored the case where an individual may be indifferent between two social states. For a complete specification of the Borda rule, however, the assignment of points in such cases needs to be specified.) A social state a is considered to be socially at least as good as a social state b under the Borda rule if and only if the sum of all the points received by a from all individuals is greater than or equal to the corresponding sum for b. The Borda rule satisfies all the conditions of Arrow excepting IIA. To see that it violates IIA, let us consider the case where we have two individuals (1 and 2), three alternatives (x, y, and z), and two preference profiles, (R_1, R_2) and (R'_1, R'_2) as follows: xP_1y & yP_1z & xP_1z; zP_2y & yP_2x & zP_2x; xP'_1y & yP'_1z & xP'_1z; and zP'_2x & xP'_2y & zP'_2y. Given the profile (R_1, R_2), each of x and z receives a total of four points, and, hence, we have xIz, but, given the preference profile (R'_1, R'_2), x receives a total of five points while z receives a total of four points, and hence, we have $xP'z$. This violates IIA since the social ranking of x and z changes when we go from (R_1, R_2) to (R'_1, R'_2), though the ranking of x and z has remained the same for each individual.

An impossibility theorem such as Arrow's (1951) compels us to think what has gone 'wrong', which of the requirements are unreasonable, and which of the restrictions need to be discarded or modified to provide a way out of the impasse. In this brief article, I shall not explore these questions, which have been discussed in great detail in the large literature that followed Arrow (1951). Instead, I turn to some basic issues about how one is to interpret the notion of the social welfare function itself.

Alternative intuitive interpretations

Some important questions have been raised by a number of economists (see, for example, Bergson, 1954; Little, 1952; and Sen, 1977a) about the interpretation of a social ranking of social states, such as the social ordering yielded by Arrow's social welfare function and the ordering implied by the welfare indices given to us by a Bergson–Samuelson social welfare function.

It has been claimed by both Little (1952) and Bergson (1954) that the social ordering R figuring in Arrow's definition of a social welfare function is the result of an aggregation procedure or 'constitution' that aggregates a given profile of individual preference orderings reflecting the individuals' judgements or opinions. In contrast, as Bergson pointed out, the welfare indices that come from his social welfare function were intended to reflect a given individual's personal value judgements about what was good for the society (in a somewhat similar fashion, Sen, 1977a, makes a distinction between committee decision and social welfare judgements). Arrow (1963) agreed that he did intend his social welfare function to be a constitution or a rule for aggregating people's opinions, but he claimed that such aggregation was, indeed, the central issue of welfare economics.

An example may be helpful in clarifying the distinction. Suppose someone, say, individual i, says that a complete ban on smoking in all public places is better than a prohibition of smoking only in a few designated public places. Suppose, when asked to give the reason why he thinks so, he gives us as the reason the fact that 99 per cent of the population in the society have the opinion that a complete ban on smoking will be better for society. This may be a good enough reason if i's original statement is a statement about how society should rank the two policies for the purpose of social action, given the existing opinions or judgements of the people. It is, however, possible that individual i made his original statement as his personal judgement about what would promote society's welfare. In that case, we would find his response to the request for justification a little strange: we would feel that he should give 'independent reasons' for his statement rather than referring to the judgements of other people. Typically, in aggregating people's opinions or judgements through a 'constitution' or a committee decision procedure, we do not look into the basis of people's opinions; we take them as given and simply try to find out what will be a reasonable way of reconciling different opinions. In contrast, in forming our personal social welfare judgements, typically we do not aggregate individuals' opinions or judgements (in fact, we often question the basis of other people's judgements when they do not conform to our ethical beliefs), though we do take into account other people's well-being. Also, in our personal judgements about social welfare, we often compare the welfare losses or gains of one person with those of another, while, in aggregating opinions or judgements, we rarely take into account the strength with which one individual favours x over y with the strength with which another person favours y over x (see Sen, 1977a, p. 159).

Arrow (1963) sees the basic purpose of welfare economics as that of analysing procedures for aggregating individual opinions so as to arrive at social decisions. Therefore, he interprets the social ordering that results from such aggregation as the basis of social action, the aggregation procedure being his social welfare function. Nevertheless, as he pointed out, ordering R in the definition of an SWF (A) could also be interpreted as reflecting the social welfare judgement of a given individual, say, i. In that case, the SWF (A) would reflect the rule (s) by which i derived his personal social welfare judgement regarding social states, given the preferences of the individuals in the society. If, however, we adopt this interpretation of the SWF (A), then it will be appropriate to interpret the preference orderings that constitute the arguments of the SWF (A) as reflecting the individuals' welfares rather than their value judgements or opinions since it is not clear why a person would use other people's judgements to form his own social welfare judgement.

Both the interpretations of a social welfare function would seem to be important for welfare economics conceived in a broad fashion. In some ways, the analysis of personal judgements about social welfare and the aggregation of the opinions of the individuals in society correspond to two distinct phases that can often be discerned in a democratic process. The first stage is the stage of deliberation where people engage in ethical debates about each other's personal social welfare judgements. The second stage is the stage of voting or aggregation of people's judgements, where people's opinions or judgements, as they emerge from the debates and deliberations of the first stage, are taken as given, and attention is focused on arriving at a 'reasonable compromise' on the basis of these judgements (see Little, 1952; and Pattanaik, 2005).

The informational basis of social welfare judgements

Arrow's analytical structure does not permit us to consider certain types of information, which, intuitively, we often regard as important for forming our social welfare judgements. I note two such informational constraints.

Cardinal utility and interpersonal comparisons of utilities: the SWF (A) defines social ordering as a function of the individuals' preference orderings over X. Thus, social ordering does not use any cardinal feature of individual welfare, and interpersonal comparisons either of the levels of individual welfare or of individual welfare gains and losses does not play any role in the determination of Arrow's social ordering. The same is also true of the 'individualistic' Bergson–Samuelson social welfare function (see (1)), given Samuelson's assumption that the individual utility functions are all ordinal. Such complete eschewal of cardinal notions of individual welfare and all interpersonal comparisons of individual welfares goes counter to our intuition when one interprets the Arrow's social ordering as reflecting someone's social welfare judgement rather than simply as the result of aggregating opinions through a 'constitution'. As I have noted earlier, in forming our social welfare judgements, we typically take into account the welfare of individuals. In doing so, we also often resort to interpersonal comparisons of individuals' welfare levels or changes in their welfares. The SWF (A) would not allow us to do this. For example, consider two allocations $x = (98, 2)$ and $y = (97, 3)$ of 100 units of some desired resource between two individuals, 1 and 2. Suppose someone wants to say that a move from x to y, involving a redistribution in favour of 2 will improve social welfare because, at x, 1 has a higher level of utility than 2, and, even after redistribution, 1's welfare continues to be higher than 2's welfare. This justification cannot be given in the Arrow framework since the framework does not permit such interpersonal comparison of welfare levels. Nor can the person justify his social welfare comparison of x and y in the framework by saying that individual 1's utility loss from the switch from x to y is outweighed by 2's gain of utility since Arrow's framework permits neither cardinal individual utilities nor interpersonal comparisons of utility differences. In the literature that followed Arrow (1951), a series of important contributions (see, for example, Harsanyi, 1955; Sen, 1970b, 1977b, 1979; d'Aspremont and Gevers, 1977; and Gevers, 1979) have explored social welfare judgements based on much richer utility information incorporating cardinal and interpersonally comparable individual utilities, and have demonstrated that Arrow-type impossibility results often lose their bite in this expanded analytical structure.

Non-utility information: Sen (1977b) demonstrated that, though the definition of an SWF (A), by itself, does not rule out the possibility of using non-utility information, such as the information contained in the description of social states, in making social welfare judgements, a somewhat stronger version of WP, together with IIA, does rule out that possibility. The stronger version requires that, for all social states x and y and for every preference profile $(R_1, \ldots, R_n)$, [if $x\, I_i\, y$ for all individuals i, then $x\, I\, y$], and, further, [if xR_iy for all individuals i and xP_iy for some individual i, then xPy].

Individual rights based on the notion that an individual should be able to make free choices in affairs relating to his or her private life is an important example of an ethical value based on non-utility information. The concept of an individual's private life, which John Stuart Mill (1859) considered so important, cannot, however, be articulated in terms of individual utilities alone. While i's religion may cause just as much disutility for his neighbours as his playing loud music in early hours of the morning, Mill (1859) would have considered i's religion, but not his playing loud music in early hours of the morning, to be an aspect of i's private life. Sen (1970a) investigated the implications of granting individuals the right to make free choices with respect to their private affairs irrespective of how others feel about their choices. In his celebrated result on the impossibility of the Paretian liberal, Sen (1970a) demonstrated that respect for such individual rights clashes sharply with WP, even if one discards IIA and replaces the Arrow requirement that the social weak preference relation R be an ordering by the much weaker requirement that R be reflexive and connected and P be acyclic (P is said to be acyclic if and only if there do not exist $x_1, x_2, \ldots, x_n$ in X such that x_1Px_2 & x_2Px_3 & ... & $x_{n-1}Px_n$ & x_nPx_1). While Sen departed radically from the Arrow format by introducing individual rights that have non-utility information as their basis, he still retained one basic feature of the Arrow format. His analysis was in terms of a social weak preference

relation specified by a function of individual orderings over the social states. Sen's formulation of an individual's right to make free choices in his own private life was introduced as a restriction on this function, the restriction being contingent on the individual's preferences over social states that differed only with respect to some features of his private life. An alternative version of Sen's theorem uses the notion of social choice rather than social preference. The point under consideration applies to that version as well. Given any set of feasible social states, social choice from that set is still a function of individual preference orderings over social states, and the rights of an individual are formulated as restrictions on social choices, the restrictions being contingent on the individual's preferences over certain social states.

Several subsequent writers (see, for example, Sugden, 1985; and Gaertner, Pattanaik, and Suzumura, 1992), who argued for a formulation of individual rights in terms of game forms, had to abandon Sen's format altogether, given their conception of an individual's right as the individual's freedom to choose any of the actions or strategies permissible under the right rather than in terms of constraints imposed on social weak preferences (or social choice) by the individual's preferences over certain types of social states. (See, however, Pattanaik and Suzumura, 1996, for an attempt to put the problem of social choice of a rights structure, viewed as a game form, back in the framework of an Arrow-type social welfare function.)

To conclude, it will perhaps be fair to say that, while the concept of a social welfare function has been a powerful analytical tool for investigating implications of value judgements relating to social welfare, the individualistic version (see (1)) of the Bergson–Samuelson formulation, as well as Arrow's formulation of the concept, had certain limitations and that some important developments in welfare economics have their origin in attempts to overcome those limitations.

PRASANTA K. PATTANAIK

See also **Arrow, Kenneth Joseph; Arrow's theorem; Bergson, Abram.**

Bibliography

Arrow, K.J. 1951. *Social Choice and Individual Values*. New York: Wiley.

Arrow, K.J. 1963. *Social Choice and Individual Values*, 2nd edn. New Haven: Yale University Press.

Bergson, A. 1938. A reformulation of certain aspects of welfare economics. *Quarterly Journal of Economics* 52, 310–34.

Bergson, A. 1954. On the concept of social welfare. *Quarterly Journal of Economics* 68, 232–52.

d' Aspremont, C. and Gevers, L. 1977. Equity and informational basis of collective choice. *Review of Economic Studies* 46, 199–210.

Gaertner, W., Pattanaik, P.K. and Suzumura, K. 1992. Individual rights revisited. *Economica* 59, 161–77.

Gevers, L. 1979. On interpersonal comparability and social welfare orderings. *Econometrica* 47, 75–90.

Harsanyi, J.C. 1955. Cardinal welfare, individualistic ethics, and interpersonal comparisons of utility. *Journal of Political Economy* 63, 77–94.

Little, I.M.D. 1952. Social choice and individual values. *Journal of Political Economy* 60, 422–32.

Mill, J.S. 1859. *On Liberty*. New York: Liberal Press, 1956.

Pattanaik, P.K. 2005. Little and Bergson on Arrow's concept of social welfare. *Social Choice and Welfare* 25, 369–79.

Pattanaik, P.K. and Suzumura, K. 1996. Individual rights and social evaluation: a conceptual framework. *Oxford Economic Papers* 48, 194–212.

Samuelson, P.A. 1947. *Foundations of Economic Analysis*. Cambridge, MA: Harvard University Press.

Sen, A.K. 1970a. The impossibility of a Paretian liberal. *Journal of Political Economy* 78, 152–7.

Sen, A.K. 1970b. Interpersonal aggregation and partial comparability. *Econometrica* 38, 393–409.

Sen, A.K. 1977a. Social choice theory: a re-examination. *Econometrica* 45, 53–89.

Sen, A.K. 1977b. On weights and measures: informational constraints in social welfare analysis. *Econometrica* 45, 1539–72.

Sen, A.K. 1979. Interpersonal comparisons of welfare. In *Economic and Human Welfare: Essays in Honour of Tiber Scitovsky,* ed. M. Boskin. New York: Academic Press.

Sugden, R. 1985. Liberty, preference, and choice. *Economics and Philosophy* 1, 213–29.

socialism

It is said that the word 'socialism' was first used by Pierre Leroux, a supporter of Saint-Simon, in 1832, and was quickly taken up by Robert Owen. The word has meant many different things to different people. It has been used as a synonym for communism, i.e. as a bright vision of a future in which there are neither rich nor poor, neither exploiters nor exploited, in which, to use an expression borrowed from Charles Taylor, 'generic man is harmoniously united in the face of nature'. It is by definition the solution of most if not all economic problems, the end of 'alienation'. As such it has religious overtones: Man was at one time in harmony with society, and will become so once again. For others, these utopian-sounding aims are either meaningless or a vague ideal, the higher stage of communism. 'Socialism' is, so to speak, here on earth, and can be seen (according both to Soviet doctrine and to right-wing critics) in the 'really existing socialism' of countries in the Soviet sphere, who claim to be on the way towards a communist future. Still others criticize this 'really existing socialism' from the left, declaring it not to be socialism at all; their criteria for what constitutes socialism are not always very

clear, some using the marxist vision as their point of departure, others laying stress on the lack of democracy, the hierarchical nature of society, and other departures from what, in their view, ought to be. The term 'socialism' is also used, or misused, to describe the aims and programme of the British Labour party, or the state of affairs actually achieved under a series of social-democratic governments in Sweden. The term at one time had an appeal to moderates. Thus the moderate-reforming party of the Third Republic in France chose to call itself Radical-Socialist, though its leaders, such as Edouard Herriot, had no aims which could qualify as socialist. Then, at the extreme right of the political spectrum, Hitler's party was self-described as national-*socialist*.

So one should proceed at an early stage to a definition, or rather to exclusions. Not Hitler, obviously. Nor Herriot either. If one were to adopt a definition which corresponds with Marx's vision of socialism (of which much more below), there is the evident danger of adopting an impossibly rigid criterion by which to judge any real-world society: thus, whatever reasons there may be to criticize or condemn today's USSR, it would be rather pointless to 'accuse' it of not having ensured the withering away of the state, or not having 'surmounted' (*aufgehoben*) the division of labour. Let us provisionally accept the following as a definition of socialism: a society may be seen to be a socialist one if the major part of the means of production of goods and services are not in private hands, but are in some sense socially owned and operated, by state, socialized or cooperative enterprises. 'The major part' is enough. Just as any non-dogmatic socialist would accept that most 'capitalist' countries contain sizeable state and cooperative sectors but still deserve the label 'capitalist'. This leaves three big questions unanswered:

(1) What are the relationships between management and workforce *within* the enterprise?
(2) How do the production units interrelate? (i.e. by plan, by contractual or market relations, or some combination of both).
(3) If the state or other public bodies own and operate any part of the economy, who controls the state, and how. One remembers the remark attributed to Engels, that if state ownership is the criterion of socialism, the first socialist institution was the regimental tailor.

If the world 'socialist' was coined in 1832, the idea of socialism long preceded it. Among the first to put forward principles which contain strong socialist elements was Gerard Winstanley, representing the Levellers of Cromwell's time. They believed in equality, wished property to be held in common, opposed concentrations of private wealth. During the French revolution Babeuf denounced inequalities of wealth and advocated the overthrow of the government, which he saw as representing property-owners. Robert Owen could be described as a paternalist, in that he believed in good treatment of his employees (as can be seen even today in the housing he built for his workers in New Lanark), but he also envisaged what would now be called producers' cooperatives. As essentially a practical man, he can be distinguished from those 'utopian socialists' who, before Marx, have painted a series of pictures of imaginary socialist-type societies. Leszek Kolakowski (1981) analyses the ideas of men like Fourier, Saint-Simon, Proudhon, and notes certain elements of similarity with those of Marx, and also some essential differences. They have in common, inter alia, a hate for the 'bourgeois' order, a society based upon greed, profit, the mercantile spirit. The French revolution substituted plutocracy for aristocracy. Unlike Marx, they did not consider this to be a progressive stage in the history of mankind, but, like Marx, they stressed the ugly features of capitalist industrialism and wished to do away with it, substituting a new harmony, cooperation, the reassertion of the true rights of Man. They rejected Adam Smith's basic idea that common good is generally attained through the competitive profit-making process. As, for instance, was asserted by Saint-Simon, the basic cause of human misery is free competition and the anarchy of the market. The so-called utopians varied in their approach to the issue of equality: thus in Fourier's 'phalansteries' the means of production were held in common, children were to be brought up together, the family would dissolve, there would be provision of subsistence for all, but Fourier would encourage individual enrichment through work (though not the inheritance of riches or unearned incomes). Some advocated violent revolution to achieve their objectives, others hated violence and hoped to persuade their fellow-citizens to adopt freely the ideas of the good and just society of their imagination.

As will be argued later, Marx differed from his predecessors not because he conceived of a realistic alternative to capitalism: there was much that was utopian in his ideas too. However, *firstly* he did not go into detail as to how a future society would function; nothing in Marx is similar to such notions as phalansteries, or radiant cities of 1800 persons with 810 different human characteristics, or the idea that dirty work that needs doing will be done by boys, who, as everyone knows, like dirt; Marx favoured the emancipation of women, but he did not follow Fourier in drawing up a 'table des termes de l'alternat amoureux'.

Secondly, and more important, he provided a set of powerfully argued historical reasons as to why the desired state of affairs must come to pass. As Engels said at his graveside: 'Just as Darwin discovered the law of development of organic nature, so Marx discovered the law of development of human history.' The class struggle, the growth of monopoly capitalism, the proletarianization of the petty bourgeoisie (peasants, shopkeepers, small businessmen of all kinds), the growing misery of the masses, the growth of class-consciousness, the logic and consequences of large-scale industry, the belief that, having spectacularly developed the forms of production, the bourgeois-capitalist relations of production act as fetters on the further development of productive forces, all these things will lead inexorably towards socialism. Ever-deepening crises, the falling rate of profit, the refusal of the poverty-stricken masses to accept their lot, i.e. the accumulation of capitalist contradictions, will bring the system down. The proletariat, having overthrown the bourgeoisie, would inaugurate the classless society. In the marxist tradition there are various interpretations of the relative importance of historic necessity (i.e. inevitableness, a march towards a predestined goal) and voluntariness (deliberate human action designed to achieve the goal). These two principles coexist uneasily, and they can be seen as mutually inconsistent, but they can be reconciled. To take two examples, it *is* meaningful to assert that, should a professional soccer team play a school side, the professionals would 'inevitably' win. The same would be (was) true of a conflict between the Germans and say the Luxemburg army. However, the outcome requires human action, on the part of the footballers and the German soldiers respectively.

This calls for two kinds of comments. One relates to the interpretation of history, the other to the utopian elements of so-called scientific socialism.

It hardly needs stressing that capitalism has not evolved in the manner foreseen by Marx. He himself stressed, in a famous passage, that no mode of production passes from the historical scene before its productive potential is exhausted. He believed that capitalism was reaching exhaustion already when he was writing *Das Kapital*. Over a hundred years later it is still not exhausted, and ever-new technological revolutions, while certainly presenting new problems and dilemmas of which we shall speak, continue to enlarge the productive potential of capitalist society. It is also clear that the concept of 'proletarianization' was wide of the mark. Yes, great concentrations of 'monopoly-capital' do exist, but so do very large numbers of small businesses and a far larger number of 'professionals' of all sorts and grades who are, or consider themselves to be, middle class. This fact has given rise to much debate among marxists, typified by the argument between Poulantzas and Erik Wright (see for example, Wright, 1979). We need not go into this argument, which turns on who could or could not be considered to be working class. The political and social fact remains that a large and growing proportion of the citizenry of developed countries do not own the means of production and are emphatically not class-conscious proletarians.

Furthermore, the development of the forces of production has made possible a substantial improvement in the living standards even of those who in any definition are workers. Clearly, they do not have 'nothing to lose but their chains'. It is neither original nor amusing to say that men who have 'nothing to lose' except a three-bedroomed house, a car, a video-tape machine and a holiday in Spain are not very likely to be revolutionaries, or indeed particularly interested in socialism. It is true none the less. Marx himself, and some of his followers, when willing to recognize that living standards could rise, insisted that this does not remove the essential antagonism between labour and capital, the existence of exploitation and alienation. In a sense this is so, though one must avoid an oversimplified zero-sum-game approach; situations arise in which both profits and wages can rise together, as they have done in successful capitalist countries in the twenty-five years that followed the last war. Nor is there any necessary correlation between the depth of human misery and the spirit of revolt. None the less, the lack of support for the socialist alternative in developed countries cannot be treated as merely a temporary aberration. It is also true that revolutions, whatever their merits or necessities, impose grave hardship upon people, notably the masses. The association of the word 'socialism' with revolution is therefore an important reason for many 'proletarians' *not* to support the socialist idea, at least in developed countries. 'Underdeveloped socialism' is a different question, to be tackled later.

Now to the utopian nature of Marx's 'scientific socialism'. The key points to make are:

1. *Abundance*. Here Marx reflects the optimism of his century, yet natural resources are not inexhaustible. Human needs and wants increase – as indeed Marx himself recognized. Conservationist and ecological socialism can be strongly defended, but this is precisely because resources (even the air we breathe, the water we drink) are finite. It is not the case that the problem of production has been 'solved', and that socialists will not require to take seriously the question of the allocation of scarce resources. I define 'abundance' as a sufficiency for all reasonable requirements at zero price.

2. *The non-acquisitive 'new man'*. His (and her) appearance surely presupposes abundance. Marx himself was perfectly clear that a share-poverty 'socialism' would reproduce 'the old rubbish'. Men do not become good by being so persuaded, or by reading good books. If there is enough for everyone, then there is no need to strive to keep things for oneself, one's family, one's locality, one's institution. If there is scarcity, therefore opportunity cost, therefore a situation in which there are mutually exclusive alternatives, then conflict on priorities of resource allocation is inevitable. This does not in fact require any assumption about individual egoism. Even unselfish persons tend to identify the needs they know with the common good. Indeed, in a complex modern society there is no generally accepted and objectively based criterion as to what 'the common good' is. Nor can any individual apprehend the multitude of alternative uses potentially available for the resources he or she desires, either for him/her self or for the given township, library, orchestra, football team, industry or whatever.

3. *The political assumptions*. These are linked with (1) and (2), above. The state withers away, not only because it is assumed that its 'essential' repressive functions are not needed when no ruling class imposes its will on the masses, but also because, to re-cite Charles Taylor, Marx assumed a 'generic man harmoniously united in the face of nature'. Consequently there would be no need for legal institutions, coercive powers, police, indeed any politics as we know them. Civil society and individuals will have merged, the task of the 'administration of things' would not be undertaken by political institutions, would be merely technical. There *is* no marxist *political* theory of socialism.

4. *The economic assumptions*.

(a) *Value theory and economic calculation*. The suppression of the market, of commodity production, of money, seems to involve the 'withering away' of the law of value. What is to replace it? Presumably it will continue to be important to use resources economically to provide the goods and services desired by society. How are calculations to be made? On this Marx is almost totally silent. Engels, in *Anti-Dühring*, speaks of assessing use-values and relating them to the labour-time required to provide for them. This runs at once into several rather evident problems. First is the theoretical one that Marx most emphatically (at the very beginning of the first volume of *Das Kapital*) asserted that different use-values were not comparable, so could not be added up or subtracted. A pen, a cup, a book, a skirt, a light-bulb (to take a few examples at random) satisfy different needs. The one thing they have in common, apart from satisfying various needs, is that they are the products of labour. How, in any case, are Engels's use-values to be computed, by whom, on the basis of what criteria? In a book wholly devoted to marxian use-value (*valeur d'usage*), G. Roland (1985) goes at length into the basically unsatisfactory treatment by Marx of use-value, due apparently to his anxiety to distance himself from subjective value theory. This has created some awkward problems for Soviet pricing theory, or at the very least does nothing to help. The dogmatists insist that Marxian labour-values ought to underlie Soviet prices, or alternatively that these be modified into the equivalence of 'prices of production', but both of these share the characteristic of being based on effort, on cost. This not only fails to give due weight to utility (or user preferences), but also runs into yet another problem, or rather two interlinked problems; measuring labour inputs, and the failure to take into account other scarcities. A few brief remarks are appropriate on each of these points. Can one actually identify the labour content, including the labour embodied in machine and materials, and the 'share' in joint overheads, of hundreds of thousands or even millions of different goods and services? This is a hugely difficult if not impossible task, even if one calculated only in hours of labour. But then what of skilled labour? How is it to be 'reduced' to simple labour? Marx does not handle this 'reduction' satisfactorily in discussing value in capitalist society, and in the end one is left with actual wage ratios as the only usable criterion, which is unhelpfully circular. And then can one treat labour as the only scarce factor? What of land, oil, timber, what of time (not labour-time, but, say, delay in construction)? Novozhilov remarked that the most modern equipment would be scarce even under full communism unless it was assumed that technical progress ceases. Space forbids further remarks about other deficiencies of the labour-theory inherited from Marx. (Thus demand or price must affect labour-content if there are economies or diseconomies of scale, or if relative prices influence choice of techniques.) And if the purist retorts that Marxian value theory is not supposed to apply to socialist economies at all, then he or she must be asked: 'What is your alternative?'. This has (so far) usually taken the form of some surrogate labour-theory (such as hours of human effort), with all the deficiencies of such an approach.

(b) 'Simplicity'. The lack of interest – until comparatively recently – of Marx and marxists in the question of economic calculation under socialism is explicable by a grave misunderstanding, i.e. by the belief that the complexities of modern industrial society are a consequence of commodity production and 'commodity fetishism', which conceal relations which, as Marx said, were inherently 'clear and transparent'. 'Everything will be quite simple without this so-called value', said Engels. Planning under socialism 'will be child's play', said Bebel. 'To organize the entire economy on the lines of the postal service, … under the leadership of the armed proletariat, this is immediate [*sic*] task' (Lenin, in 1917), and so on. But evidently in a modern industrial society with hundreds of millions of people, hundreds of thousands of productive units, millions of products and services (if disaggregated down to specific items, there *are* millions), it is a hugely complex task to discover exactly who needs what, and to identify the most effective means of providing for needs, especially if one bears in mind that any output requires the acquisition (or allocation) of dozens or more of inputs. Barone (in his path-breaking 'Ministry of production in a collectivist state') pointed this out in 1908, but failed to get a hearing from the socialists of his time. It is nonsense to talk of labour under socialism being 'directly social', in the sense of being

applied with advance knowledge of needs – contrasting with *ex post* validation through the market under capitalism. This can only be so if perfect knowledge and foresight were assumed, and the need to test *ex post* for possible error assumed to be unnecessary. All socialists (rightly!) reject theories which assume perfect foresight, perfect markets, perfect competition, when put forward by neoclassical model-builders. So, apart from problems of value theory, there is the sheer complexity of marketless, quantitative planning, the formidable obstacles in the way of identifying requirements and providing for their satisfaction.

(c) *Political-social implications.* Lest the above be seen as 'merely technical', and so remediable by computers, the objective requirements of marketless planning in a complex industrial economy are centralizing (who but the centre can identify need and ensure the allocation of means of production?), hierarchical, bureaucratic, and concentrate immense power over both people and things in the hands of the state apparatus. The importance of political democracy is undeniable, but the officials (who else?) who plan the output and allocation of sheet steel, sulphuric acid and flour are taking decisions unconnected with democratic voting – save in the sense that such voting should affect broad priorities. There were moments when Marx, Engels, Lenin, showed that they understood the inevitability of hierarchy: thus Lenin saw the socialist economy as a sort of 'single office, a single factory', with 'a single will linking all the sub-units together' to ensure the parts of the economy fitted together 'like clockwork' (Lenin, 1962, pp. 157). But whereas clockwork functions automatically (i.e. is not unlike the 'hidden hand', or maybe the hidden pendulum), in a marketless economy the parts have to be moved by human beings charged with the purpose. The contrast between 'the administration of man' and 'the administration of things' (a phrase borrowed by Marx from Saint-Simon) is a false contrast: I am quite unable to 'administer' this piece of paper, but I can persuade a secretary to type it, a postman to deliver it to the publisher, and (hopefully!) the publisher decides to tell the printers to print it! All Soviet experience underlines the political and social consequences of the high concentration of hierarchically organized economic power.

5. *Division of labour and 'alienation'.* There is, and must surely be, a division of labour between productive units (those that produce sulphuric acid, steel or hairdressing services are unlikely also to be making hats, computer software or music). Marx's notion of a universal man, who fishes, looks after sheep and writes literary criticism, without being a professional fisherman, shepherd or critic, makes no sense, other

than in the (sensible but weaker) form of aiming at a greater degree of job interchangeability. Thus the author of these lines was once a soldier, then a bureaucrat, then a university teacher, but could not be all of these at once. The vertical division of labour (e.g. between management and those managed) could also be modified by some system of rotation or election, but management is also a skill, and human intelligence is not of itself a guarantee of tolerable administrative ability: we all know of good specialists who could not (and would not wish to) administer anything well. One is then struck by the inherent unreality of such books as by I. Mészáros (1972). Mészáros fully and correctly sets out Marx's view, and he does state that 'the political road to the supersession of alienation and reification' is a long one and success is not guaranteed. But he still sees the 'transcendence of alienation' as a meaningful goal, as if separation of Man from his product, his subordination to outside forces, the division of labour, can be overcome through the elimination of private ownership. And Kolakowski (1981, p. 172) is surely right when he notes that for Marx 'the fundamental premise of alienation is already present as soon as goods become commodities', and that 'the division of labour leads necessarily to commerce'. So alienation appears to be the inescapable consequence of an inescapable division of labour, so how can it be *aufgehoben*? Private ownership represents a particular manifestation of 'outside' control, and it is an important part of any socialist programme to give to labour a greater influence over the work process. But what can one make of Bettelheim (1968), when he criticizes Yugoslav-type self-management enterprises for what is surely the wrong reason: that they are controlled not by the workforce but by the market. It ought to be clear that production is for use, and that *what* is produced ought in the last analysis to conform to user needs, i.e. to be controlled by a force outside the production unit itself. This could be the market, in which bargaining takes place between producer and user. It could be a planning agency, who informs the production unit what is should be doing. *Tertium non datur*.

6. *Labour, wages, 'the proleteriat'*. Several distinct points need to be made.

(a) *The end of the wages system*. This is not what real workers want. Money wages give freedom of choice, including the choice of hiring the services of each other (to repair the roof, baby-mind, drive to work or whatever). Marx's idea of tokens denominated in hours of labour ('which are not money and do not circulate') makes very little sense, and not surprisingly has not been applied. If goods are distributed free, this usually limits consumer choice: you take what you are *given*.

(b) *Labour direction* is the sole known alternative to material incentives or other forms of inequality. This was understood by Kautsky, Trotsky and Bukharin, when they discussed this question. The term 'labour market' has an opprobrious sound, reminiscent perhaps of a slave market. Yet workers are freer, have greater choice, more possibility to bargain, than under direction of labour, necessarily exercised by officials with power over persons.

(c) *The proletariat as redemptor humanis* is essentially a religious concept, unrelated to the qualities and desires of the real working class. Eloquent words on this subject have been written by Andre Goroz: 'No empirical observation or actual experience of struggle can lead to the discovery of the historic mission of the proletariat, which, according to Marx, is the constituent of its class being' (Gorz, 1980, p. 22). Rudolf Bahro wrote that 'the proletariat, the collective subject of general emancipation, remains a philosophical hypothesis in which is concentrated the utopian element of marxism', and he added, rightly, that 'the immediate objectives of subordinate classes and strata are always conservative' (*sind immer Konservativ*) (1977, p. 174). But if one accepts these and other similar arguments, it follows that, as Lenin said, the working class left to itself will limit itself to 'trade union' types of demands, and so it is the task of the revolutionary intelligentsia to provide the revolutionary theory. This in turn leads to what has been called 'substitutionism', i.e. a party dominated at the top by non-workers, which in its turn dominates society, an outcome prophesied by Bakunin well over a hundred years ago. It is clearly not the case that, to cite Marx's letter to Weydemeyer in 1852, 'the class struggle leads necessarily to the dictatorship of the proletariat', which 'is but a transition to the withering away of classes' (letter dated 5 March 1852, Marx, 1962, p. 427).

Marxists may now be impatiently protesting that the above analysis is a vision of full communism, that no one, certainly not Marx himself, expected this to be realized quickly, or even certainly. The much-used words 'socialisme ou barbarie' show a recognition that barbarism can be an outcome if the socialist idea fails. Trotsky spoke often of a 'transitional epoch' during which money, markets, commodity production, are indeed indispensable. Soviet discussions refer to the indeterminate length of time required to move from 'socialism' (i.e. Soviet reality, which they define as socialism) to full communism. For example, a book devoted to the subject and published for the fiftieth anniversary of the revolution duly lists the characteristics of communism (abundance; from each according to his ability, to each according to needs; the elimination of commodity money relations, and so on), but goes on to stress that communism must be preceded by the lower 'socialist' phase, and that to try to overleap that phase is 'a harmful utopia' (Gatovski, et al (eds), 1967, p. 9, p. 43).

Marx himself used 'socialism' and 'communism' almost as interchangeable terms. Whether 'really existing socialism' should be seen as a transitional society or as socialist is to some extent just a terminological question. In either case it is supposed to be evolving towards fully fledged socialism or communism. But does it? Should it? What are the signs by which such an evolution can be identified?

Bettelheim has good evidence for his view that, for Marx and Engels, when the workers acquire the means of production, 'there will be in socialist society, even at the beginning, no commodities, no value, no money, and consequently no prices and no wages' (Bettelheim, 1968, p. 32). Equally strongly, the French critic Cornelius Castoriadis roundly asserts that 'Marx knew nothing of transitional societies infinitely contained within each other like Russian dolls or Chinese boxes, which Trotskyists later invented' (Castoriadis, 1979, p. 299). Marx did specifically say, in the *Grundrisse*, that 'nothing is more absurd than to imagine that the associated producers' would choose to interrelate via commodity production, exchange, markets. We have already noted that, in his *Critique of the Gotha Programme*, Marx envisaged an immediate conversion of wages into tokens denominated in hours of labour, not despite but *because* society will still bear the stigmata of pre-socialist attitudes. In the 1920s in the Soviet Union it seemed obvious to the party comrades that reducing the area of market relations was in some sense the equivalent of an advance towards socialism. Indeed those who forced 25 million peasant households to join so-called collective farms thought that this was part of the class struggle, though the effect was to turn independent 'petty-bourgeois' households, who did to a considerable degree control their own means of production and their product, into something akin to a new sort of state serfdom. If socialism is to do with the liberation of the 'direct producers', then surely this was a march in the wrong direction.

Similarly, can we say that the Hungarian or Soviet reformers of today are wrong in advocating an extension of 'commodity-money relations'? And if the point is made that such a judgment would be premature, but that communism is still an aim to pursue when circumstances are propitious, it is legitimate to ask: what circumstances can be imagined in which communism/socialism in Marx's sense *could* come about? No wonder the Soviet orthodoxy of today is to speak of 'mature socialism' as a long-term stage, with communism seen as a remote objective of no short-term operational significance.

There were also socialist alternatives to Marx, during and since his lifetime. William Morris combined some ideas derived from Marx with ethical socialism and devotion to arts and crafts. Others further developed Christian socialism of various kinds, and indeed much could be made of the contrast between Christian ideals and the mercantile spirit, the 'dark satanic mills' ('and we will build Jerusalem in England's green and pleasant land'). The British Labour party in its origins and for many decades afterwards was heavily influenced by Christian beliefs, especially those based on Methodist and other nonconformist creeds (thereby attracting some contemptuous remarks from Lenin). The Fabian society (Shaw, the Webbs and others) by contrast, preached non-religious (and non-violent) socialism, opposed extremes of inequality, and advocated industrial democracy. However, though they too influenced the Labour party, the Society remained a small intellectual group, with a tendency to believe that an elite (themselves), or even a strong dictator, would show the way. It is perhaps no accident that both Shaw and the Webbs lived to express an admiration for Stalin – even though they themselves would recoil from cruelty and killing. Mention must also be made of G.D.H. Cole and 'guild socialism', with decentralized decision-making by producers' associations.

On the continent, social-democracy nominally retained its allegiance to Marxism. However, already in 1899 Edward Bernstein advocated a non-revolutionary revision of many of Marx's theories. While the leaders of German social-democracy, men like Bebel and Kautsky, rejected Bernstein's 'revisionism', it was in fact rooted in the considerable improvement of the workers' living standards, the weakening of revolutionary spirit. In the end, while retaining marxism as their nominal creed, German and other continental social-democrats (notably the 'Austro-marxists', such as Otto Bauer) adopted a non-revolutionary position which differed little from Bernstein's and became a party of moderate reform within capitalist society.

In Russia, side by side with the growth of marxism (initially preached by men such as Plekhanov and Zieber) there arose other and non-marxist socialist currents, sometimes labelled 'populist'. They believed that a Russian road to some form of socialism could be found, perhaps based on traditional communal institutions, which would enable capitalism to be by-passed. These ideas came from men such as Mikhailovsky and Vorontsov. As we shall see, Marx himself did not reject this possibility. There were also some influential anarchist socialists, owing inspiration to Bakunin, of which Prince Peter Kropotkin was a colourful example.

Since 1945 European social-democrats have tended to abandon their already tenuous allegiance to Marx and Marxism, and it may be hard to discern the extent of commitment to socialism of any sort in the programme and policies of the German and French parties. By contrast, the recent evolution of the Italian communist party has put it close to a social-democratic, evolutionist position. Opinions vary within the British and the Scandinavian Labour parties. Further change may well depend greatly on what happens to contemporary capitalism.

Of course the future may reserve surprises for us all. While material resources may be finite, the scientific-technical revolution may enable us to economise labour on a big scale. The resulting high level of unemployment may be a chronic disease. True, by freeing factory and office labour, we could, in a more rational society, greatly enlarge labour-intensive forms of providing a higher quality of life. But precisely this is opposed, and successfully so, by the New Right, by the 'Chicago' ideology, which is vehemently against public expenditures. Yet we may already be reaching a stage in which the *profitable* (privately profitable) use of labour can cover only a portion of those available for work. A possible reading of Marx places emphasis on equating the realm of freedom with freedom *from* work (i.e. from necessity), with a much shorter working week, and Gorz too sees freedom as a situation where one can undertake handicrafts and other hobbies. This would be a paradoxical reversal of the view that the one scarce factor of production is labour, since then it would be the abundant factor, the problem being how to share it out. This would not be the era of abundance. To cite an example, fish could be caught by modern trawlers using fewer fishermen, but dangers of over-fishing would compel a strict limitation on numbers caught. This brings one back to the idea of an environment-preserving, ecologically conscious, employment-sharing socialism as an attractive alternative to capitalism. But this was not Marx's alternative.

A case for socialism can be made, not only along the 'ecological' lines mentioned above. In the developed world, massive resources are devoted to persuading people to buy trivia, to keep up with the Joneses. Unemployment is a scourge which is a threat to public order. External diseconomies (and external economies too) frequently cause the pursuit of private micro-profit to conflict with more general interest. The 'quality of life' may not be readily quantifiable, but several economists (for instance Kuznets, Tobin) have noted that conventional measures of economic growth by GNP can conceal real losses, or indeed count real *costs* of urban living as a net addition to welfare. The inequalities of income and property-ownership have all too often no visible connection with the contribution to society or to production of the individuals concerned. Schumpeter (cited in Brus, 1980) rightly pointed out that no social system 'can function which is based exclusively on free contracts … and in which everyone is guided only by personal short-term interest'. Furthermore, fanatics of the New Right are engaged in reducing essential public services, disintegrating where possible the welfare state, cutting back public transport, pursuing dogmatic monetarism, in the naive belief that primitive laissez-faire is the best of all possible worlds. It may turn out that that the grave-diggers of capitalism will be those ultra-'liberal' ideologists who fail to understand how modern capitalism really works, that the so-called imperfections (price and wage stickiness, administered prices, oligopoly and so on) are

preconditions for the functioning of the system. On the assumption of perfect competition, perfect markets, perfect foresight, there is no role for the entrepreneur, no reason for firms to exist, and logically enough profits tend to zero in equilibrium. The idea that rational investment decisions are possible when we face so many inflationary uncertainties (what will the rate of the dollar, or the rate of interest, be in a year's time?) is somewhat far-fetched, to put it mildly, and inconsistent with meaningful 'rational expectations'. The belief that all markets clear, that unemployment is 'voluntary leisure preference', curable by freeing the labour market, will sound very odd to future generations.

No socialist should deny the need for economic calculations. With no price mechanism it is not possible to calculate or compare cost, or to measure the intensity of wants. Microdemand cannot be derived from voting or from clamour, nor should there be 'dictatorship over needs', to cite the title of a critique of East European socialism (Feher et al., 1983). There really is no alternative to allowing choice, i.e. to 'voting' with money. Choice necessarily involves competition between actual and potential suppliers. Yet the limitations of the price mechanism also require to be clearly seen. As the Hungarian economist Janos Kornai (1971) has pointed out, major decisions are not and cannot be taken on the basis of price information alone. The currently fashionable 'methodological individualism' goes far to deny the very existence of the general interest, distinct from that of individuals composing the society, confining 'public goods' to defence and lighthouses. (Yet it is not even true that the interests of a firm are only the sum total of that of the individuals composing it!)

Socialism as an idea lays stress on the general interest, but has not always avoided overstressing this at the expense of the individuals, for otherwise the dangers of totalitarianism (albeit of a paternalist kind) may loom ahead. The notion that Man is at the mercy of blind forces he cannot control, or of mighty and remote corporations (faceless, *sociétés anonymes,* or worse still, inhuman computers) sets up a search for a 'socialist' alternative, more human, fairer, and not necessarily less 'efficient' in terms of human welfare. Acquisitiveness and competitiveness may be unavoidable, must indeed be utilized, but do not require to be encouraged. Individualist profit-seeking as the dominant purpose in life, can be regarded by socialists as inhuman and ultimately destructive of society. A greater – not exclusive, but a greater – emphasis on caring for others may be a precondition for survival. More directly destructive would be nuclear war. There was a long-standing attachment of the idea of socialism with that of peace. This can be less confidently argued today, alas (when Chinese and Vietnamese soldiers shot at each other, could they both be 'socialist'?). Experience does show that states aiming to be socialist can commit aggressive acts, and accumulate immense stores of destructive weapons. None the less, the autonomous role of the arms lobby and of hate-propaganda may be particularly associated with militant capitalism.

Socialist ideas in the Third World raise some specific problems. While it is dangerous to generalize about so heterogenous a group of countries, in many of them the logic and spirit of capitalism is rejected. There too, to re-quote Bahro, ordinary people are 'immer konservativ', and it is capitalism which is new, which threatens traditional ties and attitudes. The effect may or may not be to provide mass support for socialist slogans: we have had such phenomena as Khomeini and Moslem fundamentalism by way of reaction. But socialist ideas do attract many, in places as far apart and as different as Chile, India, Egypt, Zimbabwe. Of course many blunders have been committed in the name of pursuing socialist policies, not least in relations with the peasantry. But there are many examples which demonstrate that there are countries where free-market capitalism, far from being associated with free and democratic institutions, requires repressive police-state measures. Pinochet's Chile is but one such example.

The relationship between socialism and economic development is a subject in itself, on which volumes could be written. It has often been pointed out that, paradoxically, marxist-inspired revolutions have occurred in relatively backward countries. Indeed, the Russian Empire in 1917 was in no sense 'ripe' for socialism. The preconditions were absent, and the Mensheviks considered themselves to be orthodox marxists when they denounced the Bolsheviks for trying to overleap the pre-destined historical stages. Lenin, on the contrary, believed that it was possible, indeed essential, to seize power when opportunity offered and *then* to create the preconditions, with (he hoped) the help of revolutions in developed industrial countries. Some of the less agreeable features of the Soviet system can be ascribed to isolation in a hostile world, or to 'socialism in one country', though it would be wrong, in the light of later experience (such as the evolution of the relations between the USSR and China) to regard this one factor as decisive. But, true enough, backward countries seeking to introduce 'socialism' introduce backward 'socialism'. It becomes an industrializing ideology, mobilizing the masses and imposing sacrifices for the goal of modernization, of industrialization, with a substantial admixture of nationalism. Whatever may have been their conscious aim, a strong case can be made for the proposition that Lenin and Mao re-established their respective empires, after a period of breakdown and disintegration, which in China's case lasted almost a century.

Marx's attitude to the socialist transformation of backward countries was by no means clear-cut. While his basic model did point to a socialist revolution occurring in highly industrialized capitalist countries, his correspondence with Vera Zasulich showed that he had great difficulty in applying his ideas to Russia. Theodor Shanin has edited a lively and (in the best sense of the word) provocative volume (Shanin, 1984), which does show Marx's perplexity, his partial recognition that there would perhaps be a road which by-passes capitalism. This was far from the view of Russian marxists, and the correspondence with Zasulich remained unpublished until 1924. However, on other occasions Marx took a different view, as when he regarded British rule in India as progressive, in the sense of introducing capitalist relations into a traditionalist society.

Any analysis of 'really existing socialism' would have to take account of the major role of nationalism, though this at least would have astonished Marx. It influenced Soviet internal and foreign policies, it surely played a key role in the split between Russia and China, it may be seen in the treatment of the Hungarian minority by the Romanians. The Soviet author Vasili Grossman, in his major novel *Life and Fate* (1985), put into the mouth of one of his characters the thought that the battle of Stalingrad completed the process of transforming Bolshevism into National-Bolshevism (needless to say, the book was not published in the Soviet Union). We are very far from the idea that 'the workers have no fatherland', and the proper translation of the Soviet official doctrine of 'proletarian internationalism' is 'acceptance of the leadership of Moscow on all important questions'.

There is one aspect of 'backward socialism' which has profound political and social significance. In the USSR, in China, and in many Third World countries, the peasantry formed a large part of the population and there was a sizeable petty bourgeoisie. Far from having exhausted its potentialities, the 'marketization' of the economy was still in its early stages. In Marx's model the bulk of the petty bourgeoisie has been eliminated by monopoly capital. But in these countries, in the name of the class struggle, it was destroyed by coercive state policies, i.e. by police measure. Indeed, the police has to be ever-watchful in case the banned private activities are reborn. This is one reason, among others, for there being socialist police states, which have only a remote connection with Marx's 'dictatorship of the proletariat'.

Much could be said about socialist analyses of under-development, and such names as André Gunder Frank, Samir Amin and Arrighi Emmanuel come to mind. How far was underdevelopment due to capitalism and to links with the world capitalist market? Do socialist remedies require a break with that market? Is the poverty of the Third World due to 'unequal exchange' and exploitation? Is there any operational meaning in the so-called transfer of values? Thus if (say) Zaire buys a machine from the United States and a precisely similar machine at the same price from India, are 'values transferred' in the one case and not in the other?

If Amin is to be believed, such a deal would actually impoverish Zaire if the purchase is from the United States, for presumably the machine would contain much less labour than the similar machine bought from India, or than whatever Zaire exports to America in exchange for it. Yet frankly this is nonsense. Which by no means excludes the possibility, or even the likelihood, of unequal gains from trade.

It is of interest, in the light of some socialist theories of development, to compare the experience of various countries which follow widely different models. In doing so it is evidently important not to select countries which suit a prearranged roman à thèse. Thus Cuba's record on literacy, health, the poor, compares favorably with (say) Guatemala, its economic performance is outshone by South Korea and Singapore, but it would be far-fetched to imagine that Cuba under another Batista would have equalled such countries as these; many factors are involved other than the economic system. More to the point would be to compare South Korea with North Korea: same people, same historical experience until 1945. In this instance South Korea undoubtedly out-performs the North. In Africa the free-market orientated Côte d'Ivoire has done better, even for its poor, than those of its neighbours who have opted for socialist-type solutions, but again, some African countries have achieved an appalling mess for reasons very far removed from socialism: Ghana and Uganda can serve as examples.

Those who assign to capitalism, or the links with the world market, the responsibility for income inequality, unemployment, regional underdevelopment, etc. should be made to study China. China also illustrates the correctness of the idea advanced by Arthur Lewis: the general level of wages in a given country depends not on the relative productivity of specific workers: thus an Indian or Chinese driver of a five-ton truck is probably as 'productive' as his American or British equivalent. It is determined by what he called opportunity-cost, notably (in predominantly peasant countries) the very low productivity and rewards available in agriculture. Thus wages in Shanghai, even in the modern industrial sector, are very low indeed. Were China a capitalist country, this would be the effect of the enormous 'reserve army of labour' constituted by 800 million peasants, whose income is much lower than that of Shanghai workers. In China it is a matter of public policy that urban wages be not too excessively far above the levels in rural areas. The effect is not dissimilar.

True enough, any comparison between China and India must note the great inequalities of income in India, and also the fact that the lowest strata of the poor in India are very poor indeed, compared with China. However, as was pointed out by Amartya Sen, India since independence has found it politically indispensable to avoid mass famine, while China suffered acutely from the politically imposed effects of the Great Leap Forward: millions died.

Nor should one ignore the big regional disparities in China, or the very considerable inequalities which existed even before Deng's reform policy was adopted. Also Yugoslavia's regional inequalities persist. Of course in both these instances there are historical and geographic explanations. All that can be said is that these matters resist speedy solutions under all systems.

To return to the developed world, the Soviet model has come to serve as a negative factor, and Western socialists, and indeed Eurocommunists, have tried to distance themselves from it.

The negative influence of the Soviet example is partly due to the revelations about the Stalin terror and Gulag. But, paradoxically, it was the Stalin period which, with all its horrors, did show a high degree of dynamism, high growth rates, evoking some enthusiasm and commitment from many Soviet citizens as well as foreign observers. It was brutal, it was crude, but they were forcing through a huge industrialization programme, preparing for war, fighting it, eventually winning it. There is, unfortunately, some Stalin-nostalgia in the Soviet Union today, analysed vividly by the emigré Viktor Zaslavsky (1982). 'Really existing socialism' has become grey, dull, undramatic, inefficient, more than a little corrupt. The ruling stratum under Stalin was young and faced sizeable risks of purge and execution. People could find little to enthuse about under the Brezhnev gerontocracy; the privileged abused their privileges without fear of punishment, shortages and poor quality contrasted with official claims of successes. Of course, under Stalin, things were in fact much worse. There were indeed horrors, but they were little understood outside the Soviet Union. (Thus the brutalities of collectivization and the famine that followed it were fairly successfully concealed from view.) The result was that the Soviet Union and the 'socialism' it represented became for a time a pole of attraction for millions. 'I have seen the future, and it works', 'Soviet communism – a new civilization', to cite two contemporary judgements. Today the Soviet model no longer impresses or convinces. It is not in chaos, it is not about to fall apart, but it is no beacon, can inspire nobody either in or out of the Soviet Union. And this despite the fact that much has gone wrong in the capitalist West. We will see if the new generation of leaders can restore the lost dynamism.

A few left-wing intellectuals transferred their allegiance to Mao. As was the case with some Western admirers of Stalin's Russia of the Thirties, this allegiance or admiration was based on misunderstanding, on ignorance. The 'Maoists' simply did not know about the real Great Leap Forward and its millions of victims, or just what the 'Great Proletarian Cultural Revolution' was really about. The post-Mao reaction brought them to their senses. The Yugoslav self-management model too has had its admirers, and indeed its principles are attractive, and will be looked at below. However, grave economic problems have hit Yugoslavia. By no means all of them are connected with the self-management model, but the fact remains that the negative aspects now tend to predominate in observers' minds. Then there was Poland. The 'Solidarnosc' story, in the present context, is one which not only highlights governmental economic ineptitude, but more important, makes spectacular nonsense of the communist claims to represent the workers, or to be the advance-guard of the proletariat.

So, to summarize, socialism is not, at present, a politically attractive slogan, and this despite the quite vigorous efforts of the New Right to destroy 'consensus-capitalism'. Worse, the immediate political programme of (for instance) Labour's left in Great Britain may be a sure recipe for trouble, reminiscent of the tragic errors of the Allende regime in Chile (which I had the sad experience of witnessing: price control, import controls, large wage increases, the disruption of the normal functioning of the market with no coherent idea of how to replace it).

Democratic socialism, however defined, can come only if the majority of the people are convinced that the old order has outlived itself, that major changes in a socialist direction are urgently needed. In a percipient analysis, S.C. Kolm has noted a repeated tendency: a left-wing government is elected, and its economic policies begin to hurt those middle strata (or middle-class, or left-centre parties) whose votes brought this government to power. The result is a rightward shift of opinion, and either the loss of the parliamentary majority (as in France, in 1937–8, for example) or a successful right-wing coup, as in Chile. Some draw far-reaching conclusions about there not being any democratic road to socialism (although, for example, in Chile there was no left-wing majority in Congress, Allende having been elected on a 'reformist' programme and with some support from left-wing Christian Democrats). Whatever may be the actual or anticipated resistance of the powers-that-be, one can only repeat that democratic socialism requires the support over a prolonged period of the democratic majority – and right now this is not available – except for Swedish-style welfare state social-democracy (which has again won an election in Sweden on a welfare-state programme).

Perhaps Sweden is in fact the model we should study, if what we seek is a programme which a moderate, non-revolutionary, democratic-socialist party ought to 'sell' to the electorate. Yes, it is a high-tax solution, but one which the electorate, at least in Sweden, can be persuaded to prefer to any Swedish translation of Thatcherism. In my book on *Feasible Socialism* (Nove, 1983), I rejected the notion that Sweden is a socialist republic ('and not only because it is a monarchy'), and of course there is a large 'capitalist' sector. But there is no serious current of opinion in Sweden which would support a policy of nationalizing the privately owned enterprises, or other drastic changes of existing arrangements. So if this is in fact the practical policy recipe of moderate-socialism or social-democratic parties in Western Europe, then this might be seen as a medium-term objective. Leaving the term 'socialist' as a distant perspective, just as the official Soviet propaganda now views full communism. Just as the Soviet government does not tell people that they actually intend at any particular date to abolish wages

and prices, so a Western socialist party should not be committed to 'the introduction of socialism' as a policy for today. But there should be a longer-term objective. What objective?

For reasons already examined at length, it cannot be the socialism/communism foretold by Marx. Then what can it be? Let us examine this subject, bearing in mind the three points made earlier: what relationship between management and workforce; how do productive units interrelate; and what sort of state can be envisaged – bearing in mind that a state there would and must be, with important functions to perform.

So let us look at self-management. Why has its Yugoslav version lost much of its attractiveness? As already suggested, some of the reasons have little to do with the self-management model as such: centrifugal tendencies in a multi-national state with a relatively weak central authority; unwise policies on interest rates (which have been negative in real terms) and on foreign exchange; lack of any effective control over bank credits, to cite some examples. However, certain lessons can none the less be drawn.

One is that self-management is not necessarily desired by the workforce, in the sense that many wish to spend long hours sitting in committee-rooms or studying the firm's accounts. However, the formal responsibility of management to the workforce is an important principle, as is the right of participation, which can be exercised when something goes wrong or feelings run high.

A second point relates to the lack of interest of much of the workforce in the longer term. This is a consequence of that fact that the capital assets do not belong to them, and when they leave they have no saleable asset to dispose of. Their only interest is in the income they can earn. This inclines them to a short-term view, to a desire to increase current income rather than invest in the future. One effect is to increase inflationary pressure.

Thirdly, neither the workforce nor the management has any real responsibility for investment decisions, past or present. Suppose they prove disastrous, who is to blame? If indeed the initial investment decision (to set up the firm) was mistaken, and it was taken before there could be a workers' council or the election (appointment) of a manager, why should management or labour be penalized? This is one aspect of a wider problem: that of how to cope with failure under socialism (other than be assuming that it will not occur!).

Fourthly, by making the workforce's incomes dependent on the given enterprise's financial results (subject, to be sure, to a legal minimum), one ensures unequal pay for equal work, and thus a chronic source of tension and discontent. Thus suppose citizen A and citizen B both drive five-ton lorries from Zagreb to Split, but A works for a more successful enterprise than B; they may well receive very different pay. The resultant pressure for higher pay in the financially less successful enterprises is yet another source of inflationary pressure.

Fifthly, Yugoslavia suffers from unemployment. Yet material incentives based upon dividing net revenues among the existing labour force builds in a reluctance to employ extra labour, whenever such employment would diminish the sum represented by net (distributable) revenue per head. In choosing between investment variants, there is for the same reason a tendency to choose the more capital-intensive variant, in comparison with the profit-orientated capitalist or the 'plan-fulfilling' Soviet manager.

For what should be obvious reasons, self-management requires a market. The self-managed units decide what to produce by reference to market criteria, and purchase their inputs by freely negotiating contracts with suppliers. Charles Bettelheim was quite right when he wrote that 'commodity production' (i.e. for exchange) must exist so long as units of production are autonomous and not wholly integrated into the plan. Yet he criticizes Yugoslav-type self-management: the workers do not really control their means of production and the product – the market does. This presupposes the existence of some unrealizable alternative, in which what is done and the acquisition of means to do it are controlled by no outside force at all. Yet needs have to be conveyed somehow, if not through negotiating contracts then via instructions from a superior authority.

Another significant moral to draw from Yugoslav experience relates to regional questions. In a country which, for historical and geographical reasons, has a relatively highly developed north and a backward south, measures to correct these disparities have had little success. Experience elsewhere shows that such matters defy solution in very different systems (for instance, compare Italy's *mezzogiorno*, or the megalopolis problem in such countries as Mexico and Brazil). However, the combination of autonomous 'self-managed' units and centrifugal forces, with the centre in a relatively weak position, tends to perpetuate or even reinforce regional inequalities. Indeed – and Soviet experience with *sovnarkhozy* (regional economic councils) points in the same direction – one might conclude that regional power over enterprises is very likely to result in irrationalities. The reason is clear: a local authority has information about the needs of its locality and, unless prevented, will tend to give them priority to the detriment of other localities, with duplication of investments as yet another undesirable consequence. In other words, if one were to imagine a modern industrial society with complex interregional links, there are two possible logical solutions: central control or enterprise autonomy (the 'enterprise' could, in some circumstances, be large or even, in such cases as electricity supply, a centrally controlled monopoly). If power over resources were given to an authority covering one area, it would divert resources for its own purposes, with potentially disruptive effects.

Finally, one must refer to the very considerable literature, of which Ward's fascinating excursion into 'Illyria' is the original example (Ward, 1958), which appears to prove that self-managed enterprises, in which the workforce's income depends on that enterprise's net revenue, are of their nature inefficient. Some of the conclusions are irrelevant to the real world. Thus Ward's model shows that it would 'pay' the firm to reduce output if prices rose, but this would only be so under the assumption of so-called 'perfect competition', in which such considerations as real competition do not enter. For example, in real competition one is concerned not to lose customers to one's competitors, who might not be regained if prices fall, as in future they might. Nor are self-managed enterprises likely to dismiss fellow-workers without some extremely strong reasons. None the less, as already noted, they may choose labour-saving, capital-intensive investment variants even when unemployment is a major social problem. It may be necessary (and it surely is possible) to devise fiscal means to counteract this tendency. As for efficiency, this depends (inter alia) on the attitude of the workforce. Would the sense of participation increase commitment and loyalty, and so the quality of the work effort? These considerations seldom figure in economic analysis (with Albert O. Hirschman an honourable exception). Some unimaginative model-builders would doubtless also conclude that the reluctance of Japanese firms to shed labour is 'inefficient', yet any loss can be counterbalanced by the sense of 'belonging' that goes with security of employment. A recent study of Israeli *kibbutzim* noted that one finds no resistance there to labour-saving innovations, which can be encountered in private firms, because such innovations do not threaten loss of jobs.

There are lessons to be learnt from the experience of the Mondragon cooperatives in northern Spain. Unlike the Yugoslav enterprises, they pay wages, so that there is an identifiable profit. They also ensure that the workforce has shares in the business (if necessary lending them the money to acquire them), and this also gives them a longer-term stake in its prosperity. It is, however, worth recalling that the Mondragon enterprises function in an area of strong local loyalties, just as the *kibbutz* members are committed volunteers. The outcome may be different with different human material.

Socialists must be aware that there are bound to be problems connected with property ownership and long-term responsibility, involving also risk-taking and the consequences of failure. Where uncertainty exists – i.e. in any conceivable situation – there must be the possibility of failure. A capitalist can go bankrupt, but what of 'socialist bankruptcy'? One cannot 'solve' this question simply by assuming either perfect foresight or perfect planning. The existence of genuine autonomy of decision-making is surely an aim desirable in itself, and freedom necessarily involves both uncertainty and freedom to err, to act in ways not necessarily consistent with the general interest or the national plan.

What, then, could a 'feasible socialism' be like? Should the word be redefined? Surely a non-utopian definition of socialist values should be counterposed to the crude laissez-faire ideology of the New Right. Some of the traditional slogans associated with socialism have become deservedly unpopular. There are good reasons to associate nationalization with bureaucracy, satisfying neither the workforce nor the customers. It is in a review in *Radical Philosophy* (Spring 1985) that one can read: 'A regime devoted to equality in its literal sense would have to be authoritarian, ready to crush inequalities whenever they reasserted themselves, as they inevitably and constantly would.' The New Right's view of 'liberty' may be distasteful, but one must recognize that the aims of equality and freedom can conflict with one another. Socialism cannot be happy with a purely acquisitive society. Indeed such a society would fall apart, for why should civil servants, judges, police officers, not be crude income-maximizers, i.e. behave as most doctors seem to do in America? Yet acquisitiveness is not a value to be disparaged, the vast majority of citizens do have material aspirations. Thus a conscientious doctor does his best for his patients, even if they cannot pay an economic fee, but he or she is also not averse to acquire a country cottage and go on holiday to Greece. Furthermore, at least since the days of Adam Smith it has been rightly noted that there are worse ambitions than making money: the men who, in the process of competing for power, sent their comrades to be shot in cellars were not seeking to maximize profits. What is to be sought is a balance between (enlightened) self-interest and a sense of social responsibility. Inevitably this differs as between individuals.

Individuals also differ greatly in what might be called 'producers' preferences'. Some like to be independent innovators, others prefer routine. Some gladly take responsibility, others prefer to avoid it. Some opt for life in a commune or *kibbutz*, others would be very unhappy there. While Marx's vision of a universal Man is a fantasy, it is not at all a fantasy to provide both for variety and for the opportunity to change one's specialization if the spirit so moves one. A socialism based on one economic model might be a sort of procrustean bed for a sizeable part of the population. (Imagine, for example, *compulsory* communal living!) Hence it seems desirable to redefine 'socialism' as a mixed economy: enterprises large and small, many if not most self-managed or cooperative, with some private enterprises too. If the private sector does not play a dominant role, its existence should be consistent with a sensibly defined socialism; otherwise its suppression would be the constant task of a 'socialist' police (unless, of course, it proves not to be needed, in which case 'privateers' no more require to be banned than to outlaw private water-carriers when everyone has tap water). A major objective would be not only to ensure variety of choice of occupations, but also work for all, when unemployment is in danger of becoming a major social curse. Only in ideological textbooks of economics do labour markets automatically clear. One must anticipate the need to take job-creating action. One must also anticipate that freedom to organize involves freedom to form not only political parties but also interest groups which will press for additional resources. Since money will undoubtedly continue to exist, it would be possible to issue too much of it in the face of pressures, so inflation (and some species of monetarism) will not just go away. Freedom of choice implies both a market and competition, both in consumers' goods and producers' goods and services, though there must also be some large-scale natural monopolies (such as electricity, water, public transport), where responsibility of management to the users is as important as its responsibility to its workforce.

Mises, Hayek, and later on also Friedman, have argued that efficiency in resource allocation is impossible under socialism. At a formal level they were answered by Lange, Lerner, Dickinson, but there were and are major practical obstacles in realizing their socialist models, which are anchored (as are so many of the neoclassicals') in static equilibrium assumptions, and it is unclear why either the central planning board or the managers in Lange's model should act out their parts in the prescribed manner. It should be admitted that the absence of (or severe limits on) a real capital market can cause inefficiencies, that rewards for risk-taking and innovation may well sit uneasily with social or state ownership of capital assets. Nor is this all. Kornai, in his Dublin lecture (Kornai, 1985) pointed to contradictions between the requirements of efficiency and socialist ethics. But the world is full of contradictions, and one usually arrives at some species of compromise; 'maximization' in terms of just one objective function can seldom be encountered in really existing societies (a fully fledged and devoted 'profit maximizer' would probably suffer a nervous breakdown, if not already dead of cardiac arrest). Mises and company are right to insist that economically meaningful prices are needed, wrong to assert that socialist prices cannot be meaningful (though today's Soviet prices are indeed irrational, reflecting neither use-value nor relative scarcity). But it must be emphasized how far the contemporary Western system is from the free-market model of the textbooks. Thus in his challenging 'Profits without production', Seymour Mellman notes and deplores the narrow concentration on short-term profits, by executives who have no long-term commitment to their corporation (on average they move to another one within five years or so). Current uncertainties about prices, interest rates, inflation, are hardly conducive to 'rational' long-term investment decisions. Too often critics of socialist economics (with its imperfections) implicitly compare it with a Chicago utopia, which is in its own way as unreal as a marxist one. Perfect markets and perfect plans are equally utopian.

But in the end much will depend on the ability of contemporary capitalism to surmount its many problems, not least that of mass unemployment and ecological decline (acid rain, deforestation, over-fishing, etc.). The masses will not opt for a different system unless faced with the bankruptcy of the existing one. To repeat, it was Marx who wrote that no mode of production passes from the scene unless and until its productive potential is exhausted. With Soviet-type socialism seen as obsolete, in contradiction with the forces of production, it offers no alternative model. A great deal remains to be done to revive socialism as an aim worthy of effort and sacrifice.

ALEC NOVE

See also **command economy; economic calculation in socialist countries; Engels, Friedrich; Marx, Karl Heinrich; material balances; planning; socialism (new perspectives).**

Bibliography

Bahro, R. 1977. *Die Alternative: zur Kritik des real existierenden Sozialismus.* Cologne: Europäische Verlaganstalt.

Bettelheim, C. 1968. *La transition vers l'économie socialiste.* Paris: F. Maspero.

Brus, W. 1980. Political system and economic efficiency – the East European context. *Journal of Comparative Economics* 4(1), 40–55.

Castoridis, C. 1979. *Les carrefours du labyrinthe.* Paris: Editions du Scuil. Trans. K. Soper and M.H. Ryle as *Crossroads in the Labyrinth*, Brighton: Harvester, 1984.

Feher, F., Heiler, A. and Markus, G. 1983. *Dictatorship Over Needs.* Oxford: Basil Blackwell.

Gatovsky, L., ed. 1967. *Zakonomernosti i puti sozdaniia materialnotekhnicheskoi bazy kommunizma* (On legality and the means of creating a material-technical base for communism). Moscow: Akademii Nauk SSSR.

Goroz, A. 1980. *Adieux au prolétariat.* Paris: Editions Galilée.

Grossman, V. 1985. *Life and Fate.* Trans. from the Russian by R. Chandler, London: Collins Harvill.

Kolakowski, L. 1976. *Main Currents of Marxism: its rise, growth and dissolution.* Vol. 1, trans. from the Polish by P.S. Falla, Oxford: Clarendon Pres, 1978.

Kornai, J. 1971. *Anti-Equilibrium. On economic systems theory and tasks of research.* Amsterdam and London: North-Holland.

Kornai, J. 1985. The dilemmas of the socialist economy. Geary Lecture, Dublin, 1979. In *Contradictions and Dilemmas*, ed. J. Kornai, Corvina: Kner Printing House.

Kuznets, S. 1941. *National Income and its Composition, 1919–1938.* New York: National Bureau of Economic Research.

Lenin, V.I. 1962. *Sochineniia* (Works). 5th edn, Vol. 36, Moscow.

Marx, K. 1962. *Marx–Engels Works* (Russian). Vol. XXVIII, Moscow.

Mészáros, I. 1972. *Marx's Theory of Alienation.* 3rd edn, London: Merlin Press.

Nove, A. 1983. *The Economics of Feasible Socialism.* London: Allen & Unwin.

Roland, G. 1985. *Valeur d'usage chez Karl Marx.* Brussels.

Schumpeter, J. 1976. *Capitalism, Socialism and Democracy.* 5th edn, London: Allen & Unwin.

Shanin, T. 1984. *Late Marx and the Russian Road.* London: Routledge & Kegan Paul.

Ward, B. 1958. The firm in Illyria: market syndicalism. *American Economic Review* 48, September, 566–89.

Wright, E.O. 1979. *Class Structure and Income Determination.* New York: Academic Press.

Zaslavsky, V. 1982. *The Neo-Stalinist State: class, ethnicity, and consensus in Soviet society.* New York: Sharpe; Brighton: Harvester.

socialism (new perspectives)

Marxian theory

In the Marxian theory of historical materialism, the ruling class in each mode of production has its special method for extracting the economic surplus from the direct producers; that method follows from the characteristic property relations under the mode. Under the slave mode, the surplus produced by slaves is forcibly appropriated by the slave owner; under feudalism, the lord extracts surplus serf labour through the corvée and various forms of taxation. Capitalism, Marx argued, was the first mode in which surplus extraction was not obviously coercive: no capitalist owns his workers or forcibly takes their product. Indeed, under capitalism workers and capitalists form contracts in which labour power is exchanged for a wage. The capitalist keeps the product of the worker's labour.

Indeed, Marx wished to explain capitalist surplus extraction as a process that would emerge under competitive contracting in which workers and capitalists bargain; in the end, competitive markets set the terms of labour exchange. (As Makowski and Ostroy, 1993, have written, prices are what appear after the dust of the competitive brawl has cleared. It is incorrect to think of prices as *directing* trade; rather, bargaining among many pairs of individuals reaches an equilibrium summarized by a price.)

Why is it that capitalists end up getting the better part of the deal – that is, why do they end up with the surplus while the worker ends up with his wage, which in the Marxian view was only enough for him to subsist upon? The answer lies not in the fact that the capitalist is more clever or has the police on his side; it is that capital is scarce relative to the available supply of labour, and workers must bid for the right to use that scarce capital, which provides them with a wage. Were labour scarce, then capital would have to bid for labour, and profits would be bid down to a minimal level at which capitalists were indifferent between continuing to own capital and becoming workers. *Why* capitalism seems to have been characterized, throughout its history, as a situation of capital scarcity is not fully understood. Marx argued that capitalists as a class, perhaps represented by the state, undertook strategies to guarantee a 'reserve army of the unemployed' in order to maintain the imbalance. Indeed, the proletarianization of the agricultural periphery is an important process by which labour abundance has been maintained until the present (see Rosa Luxemburg, 1913). Keynes and Schumpeter envisaged a time when capital would cease to be scarce, bringing about the euthanasia of the capitalist class.

Thus, the fundamental source of the accumulation of wealth in the hands of a small class, through profits created in production, is the fact that abundant workers must bid for the 'privilege' of using their labour power on privately owned productive assets that increase its productivity immensely. This provides them with a wage greater than they could have earned in the non-capitalist sector (back on the family farm, so to speak, or selling apples from a street cart), and also produces an additional amount that, according to the labour–capitalist bargain, belongs to the capitalist. Capitalists consume a part of this surplus product, and invest the rest in other profit-making activities.

Some writers have argued that capitalism *is* a system that extracts the surplus from workers coercively; they point to the struggles between workers and bosses at the point of production. It is, I believe, important to point out that capitalist accumulation could transpire, in principle, if capitalists were competitive and if coercion at the point of production of the worker by the capitalist and his agents did not occur. That coercion, upon which many have focused as a central evil of capitalism, exists only because labour contracts are incomplete and not costlessly enforceable. Imagine that the worker and capitalist could contract about every eventuality that might occur during production. If, in addition, the contract were costlessly enforceable (imagine an omnipotent arbitrator who is to hand to deal with any disagreement), then there would be no petty coercion at the point of production: capitalists would not try to speed up assembly lines, force workers to work overtime, cheat them of their wages, discipline them in demeaning ways, and so on. I believe that Marx thought that the essence of capitalism was the accumulation of capital *even* under such conditions. That actual capitalism is not perfectly competitive, that contracts are incomplete, and that capitalists and workers will haggle over who is to do what when a situation not described in the contract comes up, is something which makes capitalism more unpleasant than the ideal type would be, but is not of its essence.

Marx (1867) believed that the property relations of each mode of production would last only so long as they succeeded in inducing production in an efficient way. 'The water mill gives us the feudal lord, the steam engine,

the industrial capitalist.' He believed that eventually productive forces would develop to such an extent that the capitalist mode of extracting economic surplus would no longer be effective. Indeed, the next stage in economic history, Marx conjectured, would be socialism, a period in which the means of production were collectively owned and the economic surplus thereby became the property of the workers.

We must define exploitation in the Marxian sense. Marx said that workers were exploited because the labour required to produce the goods they could purchase with their wages – including the labour needed to reproduce the capital stock used up in that production – was smaller in quantity that the labour expended by workers for which they received those wage goods. The 'surplus' labour, the difference between these two quantities, became embodied in goods which, according to the contract, are owned by capitalists and which they sell for profit. Why does the worker put up with this situation? Because he has no access to the means of production; the surplus labour he supplies is, so to speak, the rent he pays the capitalist for access to those means.

Exploitation is defined as the fact that workers labour for more hours than are 'embodied' in the goods they receive as the real wage. Note that, although Marx insisted the wage was one of subsistence, this is entirely unnecessary for the argument. All that must be the case, for exploitation to exist, is that the hours of labour embodied in goods which wages purchase are fewer than the hours worked by workers.

Marx viewed socialism as the system that would end capitalist exploitation in this sense. There are, however, at least prime facie, several ways of ending exploitation short of collectivizing ownership in the means of production, and hence in collectivizing the product they produce above what workers receive as wages. One is syndicalism, in which groups of workers own their factories collectively; another is people's capitalism, a system in which firms are privately owned by citizens, each of whom owns a small share of all firms. Syndicalism would quickly generate a system with highly unequal ownership of productive assets, so some groups of workers would 'exploit' others through trade, not to speak of hiring contract labourers. Designing a people's capitalism in which Marxian exploitation was eliminated is possible in the abstract, but it would be difficult to implement in actuality. One should note that the distribution of shares in firms to citizens that would abolish exploitation could not be an equal one. Consider the situation of a person who does not work out of choice (a 'surfer') but collects dividends: he would be exploiting others, in the Marxian sense, because the amount of labour embodied in the goods he can purchase with his income is greater than the labour he expends. Indeed, for exploitation to be abolished (in the sense that the labour accounts balance) those who choose not to work should receive zero shares of the capital stock. It is in fact possible to design a system of share ownership so that, when individuals choose their labour supply to maximize their preferences over labour and income, the income they receive from wages *and their dividends* is precisely enough to purchase goods embodying *exactly the labour they expended* and, in addition, the allocation of labour and goods is Pareto-efficient. This arrangement, called the proportional solution by Roemer and Silvestre (1993), solves an interesting intellectual problem, but it has little importance as a way of solving the problem of capitalist exploitation because of the difficulty of actually computing the shares of firms citizens should receive, when information about preferences is asymmetric. (The proportional solution, however, may be used by a small community – for example, of fishermen – who collectively own a resource, such as a lake, and wish to exploit it efficiently, avoiding congestion and overuse.) Moreover, as we shall see below, it is not necessarily ethically desirable if workers are significantly heterogeneous in skill.

Socialism, then, became identified with collectivization of the means of production. Workers would produce more goods than they consumed (nobody claims that investment should be zero under socialism), but the existence of a surplus product would not constitute exploitation because it would be owned by all. This presumably meant that the state, which represented the working class, would decide upon its use. Whether this would be the case because workers obtain the suffrage and vote in a party to represent them, or because a party proclaiming itself to represent the working class takes power by non-democratic means, is another question.

A terminological point is in order. Some advocates of socialism define it as a system in which everyone reaches his full potential, racism and sexism vanish, and citizens view each other as brothers. This is a mistake. To be true to the theory of historical materialism, socialism should be defined as a nexus of property relations that eliminates capitalist exploitation. Whether such a system possesses other nice characteristics in consequence is a scientific question, one that cannot be settled by definition.

A special word must be said about equality. If workers are highly heterogeneous in skills, eliminating capitalist exploitation does not eliminate inequality in incomes. Nevertheless, there has been a tradition of viewing socialism as a system of quite equal incomes. This is partly due to the level of abstraction of Marx's thinking, in which he often viewed capitalism as characterized by a mass of homogeneous workers struggling against a small elite of homogeneous capitalists. It is, however, also due to the belief that many of the inequalities in workers' skills come from unequal opportunities fostered by capitalism and, were capitalism to be eliminated, workers would therefore *become* more equal in skills. I believe this view of what socialist transformation, in the sense of Marx, would accomplish is too optimistic – on which more below, when we return to the conception of socialism as equality.

Social democracy

In the event, the world saw two major kinds of socialist experiment: one, initiated by the Bolshevik revolution, was brought to power by a communist party which ruled undemocratically, and shunned the use of markets, which, it feared, would bring with them the old capitalist mentality, whereby traders tried to accumulate capital and hence to exploit others. The other was social democracy, in which parties representing workers won state power through democratic means, and attempted to tax profits for the purpose of investment and augmenting workers' consumption (the so-called social wage). The social democratic path did not as a principle abolish private ownership of capital assets, although some firms were nationalized.

In principle, both of these techniques could abolish the kind of exploitation associated with capitalism. If communist parties were perfect agents of their collective principal, the working masses, they could set the rate of accumulation at that level desired by workers (there is a problem here of how to aggregate workers' disparate preferences over that rate), and then invest the surplus in the way that would best meet the interests of workers (another preference aggregation problem). And under social democracy, private capital could be taxed at a rate sufficiently high that, although rates of exploitation would not be zero, they would be small. To keep capital from fleeing to more profitable venues, under that situation, workers would have to be sufficiently skilled so that, even under such a regime, capitalists' profits would be sufficiently great. Thus, guaranteeing highly skilled labour would appear to be a part of the social-democratic formula if capital is freely mobile.

With regard to equalizing the distribution of income, both the Soviet-type economies (the Soviet Union and eastern Europe) and the Nordic social democracies did an excellent job. (Indeed, at least in the Soviet Union, it is arguable that skilled workers contributed more labour, in efficiency units, then they received back in goods.) The major difference was that the Soviet economies equalized incomes at low levels, while the social democracies did so at high levels. To what was due the failure of the Soviet economies? We still do not have a completely satisfying explanation, but it seems as if their abrogation of markets was an important factor.

Although the Soviet-type economies used markets from time to time, beginning with Lenin's introduction of the New Economic Policy in the 1920s, they were never allowed to operate with the kind of freedom that would have fostered technological innovation, and by the 1960s it was the lack of innovation that was largely responsible for the low level of living standards. (Of course, when the state acted to concentrate talent in one sector, such as the space industry, it was able to achieve impressive results, but the Soviet economy never succeeded in fostering innovation across the board.) These problems were foreseen much earlier, however, in the debate around market socialism that began in the 1930s, with Oskar Lange's argument that markets could in large part replace central planning in a socialist economy. Lange (1936) proposed that central planners announce to industrial managers prices for their inputs and outputs, and require the managers to report back with the amounts of inputs they would demand and outputs they would produce at those prices, if they were to equate the prices to marginal costs (a necessary condition for Pareto efficiency). The planners would then sum up, observe the discrepancies in the supply and demand of each commodity, announce a second set of prices, raising those for goods in excess demand and lowering those for goods in excess supply, and go through the exercise again, hoping to eliminate the imbalances. Lange believed this process would converge rapidly to an equilibrium; then the planners would post the equilibrium prices and instruct firms to produce accordingly.

Lange also suggested that each household receive a certain fraction of the firms' profits, perhaps allocated according to family size.

Lange did not deal properly with the demands of consumers. But, even if we assume that these could be incorporated into the scheme, what is the point of his kind of planning? Why not simply let the market run autonomously? Lange had no convincing answer: he did say that the central planning bureau (CPB) would be able to achieve equilibrium much faster than the market, avoiding the disequilibrium phase that he considered to be socially costly. (Today, this seems to be a quaint view, given the millions of commodities that are produced in a complex economy. Indeed, economic theory still has no full explanation of how the market 'finds' the equilibrium, and there are theorems that the kind of 'tâtonnement' Lange proposed would, with high probability, not converge.) Perhaps Lange thought that the CPB would control economic activity through setting interest rates of various kinds, thus directing firms to invest in the directions the planners desired.

Friedrich Hayek (1940), however, offered a critique of another type. He wrote that it was an illusion to believe that managers could respond with their input demands, facing prices announced by planners, because they *did not know* their production functions, and therefore could not compute marginal costs. Capitalist firm managers, he said, learn how much they can produce with given inputs by the discipline of competition. It is the competitive brawl that teaches managers how to cut costs and produce efficiently, and to suppose that managers would know how to do so in the sterilized situation envisaged by Lange was wrong. Indeed, how would the CPB deal with innovation, with new commodities? The secret of real markets, Hayek argued, was that they provide incentives and a mechanism for people (entrepreneurs) with local information about needs and production possibilities to realize their ideas. Thus fixing the set of

managers *ex ante* was already dooming the system to conservatism and inefficiency.

It is interesting to note that Hayek never wrote that socialist managers would be opportunistic or self-serving – that they would lie to the CPB in order to influence their allotment of inputs. Hayek postulated that managers were 'loyal and capable'. This is in sharp contrast to the critique of socialist management that emerged after 1970 among Western capitalist economists, when the principal–agent problem was formulated, and 'shirking' and opportunism became central issues.

Markets, incentives and coordination

Indeed, this raises a critical question about the failure of centrally planned socialism: was it due to *lack of incentives* or to *lack of coordination*? Markets perform two functions: they provide incentives for workers and entrepreneurs to improve their skills and discover new commodities so as to increase their income, but they also coordinate economic activity. It may not be simple theoretically to distinguish precisely these two functions, but they are clearly different. Matching of workers to firms, for example, occurs in large part by observing wage offers; firms shop for inputs by observing price offers. Of course, the system does not work perfectly, but there is doubtless a strong element of coordination engendered by a competitive price system. (Price systems do not coordinate some things properly, such as control of externalities and the supplies of public goods, and therein lies a major liberal justification of state intervention.)

The history of the Soviet economy is replete with stories of poor incentives *and* poor coordination: we do not have a complete account of the relative importance of these two failures in the lacklustre performance of centrally planned economies in their late period. One also reads, however, of how hard Soviet workers worked, and how ingenious they were at making do with poor inputs (see, for instance, Burawoy and Lukacs, 1985). I believe it is important to answer the question posed above, for upon it may rest the possibility of a future for socialism.

Suppose markets are needed mainly to generate incentives to work hard, to form skills, to invent, and so on. This implies that it will be difficult to use markets *and* to redistribute income in a relatively equal manner through taxation. After all, if workers form skills in order to increase their incomes, but then their incomes are taxed away, why form skills? But suppose that markets are needed mainly to coordinate economic activity: then in principle wage income (which would adjust competitively to reflect marginal value products) could be taxed to produce an income distribution of equality without harming production. In the second case, workers would form skills and innovate because they enjoyed doing so, or felt valued for their social contributions.

I suspect the coordination problem is relatively more important in the failure of centrally planned economies, and the incentive problem relatively less important, than most currently believe. Many economists, especially, assume that the opportunist kind of behaviour so prevalent in the theory of *homo oeconomicus* is a deep aspect of human nature, and therefore that it must have been rampant in the Soviet Union.

Markets are essential in any complex economy, at least for coordination and perhaps for incentives. But, as we have seen in the Nordic countries, tremendous accomplishments with respect to income distribution can be achieved with taxation and 'wage solidarity'. One might say that the future of socialism lies in emulating the Nordic social democracies. They may, however, not be easy to emulate, as the solidarity of their citizenries may be due to their homogeneity – linguistic, religious, and ethnic. Perhaps welfare states of that magnitude cannot be achieved in highly heterogeneous societies.

A future for socialism may still, then, require an alternative to conventional private ownership of firms with significant redistribution through taxation, because the solidarity necessary for democratic approval of that degree of redistribution may not evolve in large heterogeneous societies. If firms are not to be privately owned, as they are in the Nordic model, then a central question concerns the way that accountability of firm management is achieved. There is a principal–agent problem between the firm manager (agent) and the shareholder–citizens (principal). How do the latter keep the manager from running off with the profits and even the assets of the firm? The classical solution is that firm ownership must be highly concentrated, so that a small number of shareholders stand to gain huge amounts by carefully monitoring the management. In this view, distributing shares of firms equally to all citizens would destroy management accountability, resulting in unbridled corruption and inefficiency.

Recently a second theory has been proposed: that the guarantor of firm accountability is the corporate raider. When the raider sees the price of a firm's stock fall, because the firm is not performing well (perhaps due to management corruption or lack of imagination), he will buy a majority of shares and reorganize the firm to be efficient, thus increasing its stock price and providing him with a large capital gain. Here, too, if credit markets are imperfect, we need wealthy individuals to keep firms running well.

If these two mechanisms of accountability exhaust the possibilities, then market economies in which firm profits are distributed in a relatively equal manner to citizens are impossible. An apparent alternative, however, exists in Germany and Japan, where firms are monitored by boards of directors consisting largely of officers of banks that have a relationship to the firm. It is beyond my scope to describe this mechanism here: suffice to say it provides an alternative to relying upon hugely wealthy individuals for guaranteeing firm accountability. If market socialism has a future, it may well be with this kind of arrangement:

firms will be monitored by bankers from the public sector, whose reputations and careers depend upon doing a good job, or they may be monitored by other stakeholders of the firm.

Another alternative (proposed by Roemer, 1994), with no present real-world examples, is a system in which firm ownership is distributed to citizens in an initially equal way but ownership rights are circumscribed. An owner will collect dividends from the firms in her portfolio, and even trade equity shares on a stock market, but she may not liquidate her equity holdings for cash. This would be accomplished by denominating corporate shares in a special unit of account. The values of shares in that unit would oscillate according to supply and demand, reflecting traders' views about the future profitability of firms, as in a standard stock market. At death, a citizen's portfolio would escheat to the Treasury, and young adults, at the age of 21, would each receive their endowment of shares. Some inequality in the values of shareholdings would emerge as a consequence of differential luck and skill in the stock market during a lifetime, but that equality would not be passed on to descendants. In other words, this scheme is a method by which the nation's profit income could be distributed in a relatively equal manner to citizens, while the virtues of a stock market, with respect to the valuation of shares, and the disciplining of management are retained.

There are surely possibilities for undermining the intentions of such a system. If there are also individuals (such as foreigners) who are allowed to invest in these firms, then possibilities arise for citizens to capitalize their holdings, to cash out their shares. Old citizens will want firms in which they hold shares to sell their assets and pay out the entire value of the firm as dividends. Whether regulation could make the system workable is an open question.

Finally, there is the possibility of state ownership of firms. We do not as yet have a definitive experiment to test whether state ownership can work, for the Soviet-type experiments also involved *lack of democracy*; it is logically possible that democratic accountability could keep state-owned firms running efficiently. There are, however, problems here as well: politicians, to whom state-owned firm managers ultimately report, have their own interests that do not always coincide with those of the public. The electoral mechanism is probably too crude a tool to force politicians to monitor firms in the public interest. (Indeed, state-owned firms often pay their workers too much, to garner their political support.) I conjecture that non-state ownership of firms will be significant in any future socialist experiment.

We return finally to the relationship of socialism to equality. Do socialists believe that an economy which implements 'from each according to his ability, to each according to his work', which by definition eliminates Marxian exploitation, is desirable? Most socialists probably desire more equality than this, at least in societies where workers are highly heterogeneous in skills. Thus, socialists have come to be, and perhaps always were, more egalitarian than the Marxian definition would imply. Popular usage suggests that socialism should be defined as a regime of income equality, a departure from the Marxian tradition.

The proposals we have discussed above are all concerned with the allocation of profit income. But is the allocation of profits so important with regard to equalizing the distribution of income? In contemporary advanced economies, profits (including interest and rents) comprise at most one-quarter of national income; even if this part were distributed in an egalitarian manner to all households, and remained of the same size, the distribution of income would still, in most advanced countries, be quite unequal. Should, then, the difference between socialism, popularly conceived, and capitalism lie mainly in the distribution of wage income or of the role of redistributive taxation of labour income?

Rather than trying to define at what Gini coefficient a society becomes socialist, one can be satisfied with ordering regimes in the world with respect to their degree of socialism. The central instruments for socialist implementation then become, as well as the redistribution of profit income, intensive investment in education, with a bias towards rectifying the disadvantages children suffer due to being raised by poorly educated parents, in order to equalize market-determined labour incomes, and redistribution of labour income through taxation. The channel of intensive investment in education of the disadvantaged is important because the provision of skills has value to persons for reasons other than the instrumental one of providing income: education renders life more meaningful and fruitful. But if the education channel alone turns out to be too costly, or too ineffective, to engender the changes in income distribution which are desirable, then other methods must be used as well. The issue of *feasible* socialism, therefore, will hinge upon the package of reforms that are effective and can be realized through democratic means.

Feasible socialism: immigration and unemployment

How politically feasible is socialism – that is, to what degree can we expect democracies to implement the reforms that move societies further along the socialist scale? Here the most hopeful historical evidence is provided by the Nordic and north European countries. Two problems seem to be paramount for the continuation of the socialist trajectory in these economies: those of immigration and unemployment.

The welfare states of the north European countries, as mentioned earlier, evolved during the period when their populations were largely homogeneous, along ethnic, linguistic, and religious dimensions. Homogeneity may be a necessary condition for the democratic implementation of significant redistribution if the welfare state is motivated by either a purely redistributive or an insurance function. For, with respect to the insurance function, it is not in the interest of highly educated and high-wage natives in, let us say, Denmark, to pool their risks with poorly educated, low-wage immigrants. And with respect to the purely redistributive function, ethnic, linguistic and religious heterogeneity reduce solidarity, to put it mildly, which must be the motivation of purely redistributive taxation.

Unemployment is a problem not only for the deleterious welfare effects its victims suffer but because it is a severe form of economic inefficiency. If 'socialist' countries have high unemployment levels, and 'capitalist' countries low levels, eventually the inefficiency of the former may well reduce per capita income significantly below that of the latter, and populations of the socialist countries will begin to find the higher incomes offered, on average, by the capitalist regimes an attractive alternative. If we assume that, in the coming century, the United States (and, let us say, China) continue to offer low-unemployment, low-taxation, high-growth regimes, but with relatively little redistribution, then democratic polities in Europe and the rest of the world may be reluctant to move further along the socialist spectrum. This, of course, assumes that there is some sacrifice in economic growth entailed by redistributive institutions, a point that I have not defended here but have taken for granted, and which may be incorrect. Indeed, a growing literature asserts that equality increases productivity (see Bardhan and Bowles, 2000).

It is perfectly natural for fertility rates to fall when social insurance replaces the family as the source of income in old age: and smaller families, probably more than anything, entail the liberation of women. (They are also, of course, an effect of that liberation.) But European fertility rates now necessitate either a significant flow of immigrants from poorer countries, or a sharp decline in per capita incomes in Europe for retired workers, or an increase in the length of working life (which itself would exacerbate the unemployment problem). So lower fertility renders the progress towards socialism more complex at least, if not infeasible.

Consequently, the issue of multiculturalism becomes a key intellectual problem for socialists. What degree of integration or assimilation of immigrants is necessary for democratic European polities to be willing and interested to continue and perhaps expand their welfare states? (Recall, we refer here not simply to the redistributive motive but to the risk-bearing motive of natives wishing to pool risks with immigrants.) We do not, I think, yet know the answer. And will this degree of assimilation, whatever it turns out to be, be acceptable to poor Southerners or Easterners who are contemplating migration to the North or West?

Socialism, in the sense of equality of incomes, with a democratic implementation requires either a self-interested insurance motive or a selfless solidaristic motive among the majority of voter–citizens. We can hope that, as national populations come to experience more equality, they would come to have a deeper preference for it: socialists, at least, believe that solidaristic preferences can intensify with the experience of equality because equality is a public good, a fact that will be appreciated when it is experienced. (Indeed, we have not, in this article, discussed the *negative* externalities that socialists believe accompany a regime with a highly concentrated ownership of private firms, in which corporate and even state policy is set to further the interests of only the wealthiest sliver of society.) But the initial transitions along this path, taken by relatively self-centred voters, must come from the insurance motive. Here, then, is an important problem for progress towards socialism in our time.

JOHN E. ROEMER

See also **labour's share of income; Marx, Karl Heinrich; redistribution of income and wealth; risk sharing; social democracy; socialism; surplus.**

Bibliography

Bardhan, P. and Bowles, S. 2000. Wealth inequality, wealth constraints and economic performance. In *Handbook of Income Distribution*, ed. A. Atkinson and F. Bourguignon. Amsterdam: Elsevier Science Press.

Burawoy, M. and Lukacs, J. 1985. Mythologies of work: A comparison of firms in state socialism and advanced capitalism. *American Sociological Review* 50, 723–37.

Hayek, F.A. 1940. Socialist calculation: the competitive solution. *Economica* 7, 125–49.

Lange, O. 1936. On the economic theory of socialism. In *On the Economic Theory of Socialism*, ed. B. Lippincott. Minneapolis: University of Minnesota Press, 1956.

Luxemburg, R. 1913. *The Accumulation of Capital*. New York: Monthly Review Press, 1972.

Makowski, L. and Ostroy, J. 1993. General equilibrium and market socialism: clarifying the logic of competitive markets. In *Market Socialism: The Current Debate*, ed. P. Bardhan and J. Roemer. New York: Oxford University Press.

Marx, K. 1867. *Capital*, vol. 1. Basingstoke: Penguin Classics, 1992.

Roemer, J. 1994. *A Future for Socialism*. Cambridge, MA: Harvard University Press.

Roemer, J. and Silvestre, J. 1993. The proportional solution for economies with both private and public ownership. *Journal of Economic Theory* 59, 426–44.

socialist calculation debate

The calculation debate centres on two issues. First, is it possible for a socialist society, with common ownership of the means of production, to use planning to replicate the performance of a capitalist society, with private ownership of these means of production? This is the 'replication' thesis. Second, is it possible to do 'better' than replication? This is the 'improvement' thesis. Both theses were widely accepted by economists until shortly before the planned economies passed into history with the demise of the Soviet Union (Lavoie, 1985).

We state the problem in terms of possibility in order to emphasize an ambiguity in the debate. What do we conclude when we observe a centrally planned economy that is unlike any known capitalist economy? Has the replication thesis failed? Not necessarily. Confronting such a situation, Drewnowski (1961) combined the replication thesis with a (planner's) revealed preference axiom to argue for the 'improvement' thesis. Some 36 years later Drewnowski (1997) was candid about the profession's failure in foreseeing the collapse of the planned economies:

> an event which will forever dominate the history of the second half of the 20th century, the downfall of communism, or, more correctly, the disintegration of Soviet-type political and economics systems. The striking feature of that great upheaval was that it came as a virtual surprise ... the shock of the collapse of seemingly invincible systems was unique in history. (Drewnowski, 1997, p. 919)

Though it was not always recognized as such at the time when the debate was open, in hindsight it is clear that the calculation debate was about the interactions among models, motives and incentives. The question was whether a model that is a 'true' description of a capitalist society might be the basis for a planned economy that satisfies the replication thesis. But to answer in the affirmative, we need to specify the motives of agents and the incentives they confront. If we agree that the model describes what we want to attain, then it can serve as the basis of the public interest. We can stipulate that public interest has some motivational force for everyone and then ask whether the incentives in a planned economy are such that planners will seek the public interest. Again, with hindsight, we can answer that they are not.

As a guide to how the calculation debate played out, we rely on the model-theoretic insights of Walter Eucken. A philosophical economist in the liberal tradition who survived great danger in Germany during the Hitler era (Gerber, 1994), Eucken was brought by F.A. Hayek to the founding meeting of the Mont Pèlerin Society in 1947 as evidence of a surviving liberal tradition and so a hope for a self-governing Germany. In 1948, Eucken's associates Ludwig Erhart (Gerber, 1994, p. 31) who as director of the Office of Economic Administration used the power entrusted in him by the American occupiers to end the centrally administered economy (Mendershausen, 1949).

Eucken (1948) asked whether the same model could be used to describe both a centrally administered and a decentralized market economy. He noted that J.S. Mill argued that two models are required, one for each institutional setting. There is no reason to believe, Mill claimed, that the characters of the people in different institutional settings will be the same over time, Mill (1844, VI, p. x, § 3). Socialism, which may not be feasible now, might become so in the future. Therefore, there is no reason to believe that the same model can encompass both systems. Eucken supported the two-model view, not by appealing to differences in the characters of the agents over time, but by appealing to the economist's understanding at a given moment in time. We will return to this claim below.

Von Mises on Mill

The calculation debate was launched in 1920 by a brief article in which von Mises challenged socialist economists to apply their arguments against market economies to their own systems. On the assumption that the collectivist state would allow private ownership of consumer goods, and thus private exchange, von Mises saw no particular problem with the allocation of consumer goods in such a setting (von Mises, 1920, pp. 90–2). The problem comes with collective ownership without exchange, since there will then be no prices. If prices do not reflect opportunity costs, no 'rational' economic system is possible.

Von Mises's 1920 paper cites a small number of socialists, none of whom he respected. That situation changed in *Socialism: An Economic and Sociological Analysis* (German editions 1922 and 1932; English edition 1936), in which he draws the reader's attention to John Stuart Mill's thoughts on socialism:

> The writer who has occupied himself most thoroughly with this problem is John Stuart Mill. All subsequent arguments are derived from his. ... They have provided for decades one of the main props of the socialist idea, and have contributed more to its popularity than the hate-inspired and frequently contradictory arguments of socialist agitators. (Von Mises, 1936, pp. 154–5)

It is here that von Mises confronts the second model of economics, the model of socialism that Mill made for people of the future, somewhat like us only a little more sensitive to the opinion of others.

Mill explained how a successful communal system must rely upon a different motivational package from that which characterizes market economies. The problem of the commons was central to classical political economy. T.R. Malthus's 1798 *Essay on Population* had addressed William Godwin's proposal of replacing private property with commons. Malthus asked why individuals would renounce or defer marriage and child-bearing if someone else were to support them in case of difficulty. For Mill, a workable commons would require a great deal of motivation from the desire for approbation. For a system of equality to hold, the commoners must know that they cannot support each other's children

without mass misery. They must understand the self-restraint that characterizes a market economy and, without any incentive other than a desire for praise and an aversion for blame, act on the basis of that understanding (Mill, 1848, II, p. 1, § 3). When embodied in the public understanding, the model must be self-motivating. We will return to the question of self-motivating models below.

Von Mises (1936) surmised that Mill failed to link material rewards with effort in a market economy because he wrote before the advent of marginal productivity theory (1936, p. 155). Von Mises acknowledged the problem of workers who are not paid by the piece – the partial basis of Mill's defence of socialism – and suggests this is the fault of the worker:

> Doubtless the individual working for a time wage has no interest in doing more than will keep his job. But if he can do more, if his knowledge, capability and strength permit, he seeks for a post where more is wanted and where he can thus increase his income. It may be that he fails to do this out of laziness, but that is not the fault of the system. (Von Mises, 1936, p. 155)

To deal with public opinion motivation, von Mises denies Mill's hypothesis that human nature might alter and claims instead that motivation will be the same under both settings:

> It is not impossible that under Socialism the public spirit will be so general that disinterested devotion to the common welfare will take the place of self-seeking. Here Mill lapses into the dreams of the Utopians and conceives it possible that public opinion will be powerful enough to incite the individual to increased zeal for labour, that ambition and self-conceit will be effective motives, and so on.

> It need only be said that unfortunately we have no reason to assume that human nature will be any different under Socialism from what it is now. And nothing goes to prove that rewards in the shape of distinctions, material gifts, or even the honourable recognition of fellow citizens, will induce the workers to do more than the formal execution of the tasks allotted to them. Nothing can completely replace the motive to overcome the irksomeness of labour which is given by the opportunity to obtain the full value of that labour. (Von Mises, 1936, p. 157)

Von Mises continues to suggest that 20th-century socialists object to this argument by pointing to heroic acts as counter-examples. He confesses that he does not understand heroes, people who are ready to die for their principles, driven purely by their 'union of will and deed' (1936, p. 158). He acknowledges that the fate of civilization itself depends on the heroic (1936, pp. 157–8) – but for the purposes of economic analysis, he is content to deal with the ordinary. Although von Mises was disinclined to pursue the heroic as a motivational force under socialism, there is an important sense in which the Soviet Union seems to have attempted to function on the basis of the heroic: the *Stakhanovite* (someone who joyfully over-fulfils the production quota), a Soviet-era concept of consequence (Kotkin, 1995, pp. 207–15), deserves attention.

Mill, too, was concerned with the character of the majority in any institutional setting; hence, to learn about social transformation one needed to apply the inverse deductive method, Mill (1844, VI, p. x, § 4). The idea that people might develop over time was ruled out of consideration by Hayek (1935a) in his summary of the history of the discussion:

> John Stuart Mill, in his autobiography, numbered himself among the socialists, because he believed that certain ideas would be realized in the distant future. In this connection Cairnes pointed out that true socialism does not consist of a body of ideas which can only be realized if human nature and the conditions of human life are radically transformed. Socialism subsists in the recommendation of certain modes of action and in the utilization of the authority of the state for particular purposes. This appears to me also as the correct view. (Hayek, 1935a, p. 47)

The replication thesis

Fred Taylor's 1928 Presidential Address to the American Economic Association sets the stage for market socialism. He starts with a model of the private market economy:

> First, on the basis of a vast complex of institutions, customs and laws, the citizen adopts a line of conduct which provides him with a money income of greater or less volume. Secondly, that citizen comes on the market with said income demanding from those persons who have voluntarily assumed the rôle of producers, whatever commodities, he, the citizen chooses. Thirdly, the producers promptly submit to the dictation of the citizen in this matter, providing always that said citizen brings along with his demand entire readiness to pay for each commodity a price equal to the cost of producing that commodity. (Taylor, 1929, p. 1)

The 'correct general procedure' for market socialism is to maintain consumer sovereignty. Taylor thus argues that the planners must use the market model in the new institutional setting:

> (1) the state would assure to the citizen a given money income and (2) the state would authorize the citizen to spend that income as he chose in buying commodities produced by the state – a procedure which would virtually authorize the citizen to dictate just what commodities the economy authorities of the state should produce. (Taylor, 1929, p. 1)

To implement this under socialism, Taylor proposes 'trial and error'. The idea is that the planners start

somewhere, and then if managers have an algorithm which implements the market model of capitalism, they adjust factor valuations on the basis of shortage or surpluses (Taylor, 1929, p. 8). Equilibrium follows. In H.D. Dickinson's judgement, Taylor's statement of the replication thesis withstood later criticism:

> Taylor has the distinction of being the first writer to point out the way to answer Professor Mises's attack on socialists. Moreover, he answers in anticipation the later criticisms of Professors Hayek and Robbins, that had not been published when his paper first appeared. (Dickinson, 1938, p. 532)

The improvement thesis

Although market socialism literature took off from Taylor, Barone had earlier described an iterative solution to the problem of maximizing consumer surplus, for example, 'maximum collective welfare' (Barone, 1908, pp. 270–2), and he discussed the problem of how to determine productive coefficients experimentally (1908, pp. 287–8). Further, with his argument that one can do better than replicate the capitalist solution in the case of multiple prices, he announced the improvement thesis:

> Hence, when the first area is larger than the second it is possible that multiple prices may be consistent with increase welfare for the community. And as such a proceeding is more possible when production is socialized, this is in reality a sound argument in defence of socialized production, in certain cases, when such conditions are proved to exist. (Barone, 1908, p. 283)

As the literature developed, the Taylor solution concept was refined and extended by a host of authorities. Responding to von Mises and developing Taylor's ideas without knowledge of Barone, H.D. Dickinson (1933, p. 29) added that socialist markets would eliminate or reduce the divergence between private and social costs. For Dickinson, the improvement thesis was made in the welfare metric of orthodox economics characterized by consumer sovereignty (Hutt, 1940).

Maurice Dobb (1933), who had previously supported market socialism entailing consumer sovereignty, abandoned this position in 1933 (1933, p. 591). Instead, Dobb (1933) asked whether the state might actually create tastes (1933, p. 532) and why the rules of equating at the margin would apply to a state which can create technical progress (1933, pp. 596–7). There followed one of the most interesting and important contributions to the debate, Abba Lerner's defence of market socialism. Lerner mentions half a dozen mistakes in Dickinson's algorithm and counters that those in power do not wish to reconcile markets with socialism:

> The cautious guardians of socialism are for retaining the superior strategic fastnesses of simple faith. The heresy must be eradicated. Against the 'Dickinson'

thesis is the raised the antithetical slogan: 'The categories of capitalist economy are inapplicable to the socialist society.' (Dobb, 1933, p. 52)

Hayek entered the debate at this point with his edited collection *Collectivist Economic Planning*, which contained translations of Barone's and von Mises's original papers, as well as Hayek's summary of the discussion and his restatement of the issues in terms of the dynamics of the problem. Here the questions posed were: supposing that a solution were found to a static problem, what incentives does the socialist management have to solve the problem correctly? Why would the model have motivational force? (Hayek, 1935b)

Hayek also noted the ambiguity in the notion of 'impossible' in his discussion of Taylor and Dickinson. In his account, their papers

> were directed to show that on the assumption of complete knowledge of all the relevant data, the values and the quantities of the different commodities to be produced might be determined by the application of the apparatus which theoretical economics explains the formation of prices and the direction of production in a competitive system. Now it must be admitted that this is not an impossibility in the sense that it is logically contradictory. But to argue that a determination of prices by such a procedure being logically conceivable in any way invalidates the contention that it is not a possible solution, only proves that the real nature of the problem has not been perceived. (Hayek, 1935b, pp. 207–8)

The great step in algorithm development was then taken by Oskar Lange (1936), with comments and corrections from Lerner (1936). The market algorithm now improved upon the competitive equilibrium by solving for a social optimal in which the same consumer preferences pertained as in the capitalist economy. Market socialism became the Lange–Lerner model.

Hayek's (1940) review of the market socialism advanced by Taylor, Dickinson, and Lange questioned the details of the algorithm, pointing out many gaps in the explanations. Hayek's puzzle as to whether an algorithm based on a model of price-taking competitive equilibrium was appropriate would later blossom into perhaps the most influential single article in the debate, his 1945 'Use of Knowledge in Society' paper. The problem for the capitalist economy was reformulated as a division of knowledge. Prices reflect the decentralized knowledge possessed by all the participants in a market economy. Markets were now viewed in terms of information aggregation. Hayek also notes the importance of the socialist market economists, singling out Lange and Lerner for particular praise, and suggesting that the political aspect of the debate is now at end, leaving only methodological issues to resolve.

The entire focus of the calculation debate was challenged again in 1961 when Drewnowski revived

Dobb's position and asked why a socialist economy would wish to mimic or modestly improve upon a market. Why would not the planned economy find a political optimum as distinct from the consumer sovereignty constrained optimum? Indeed, if one looked at existing socialism in the Soviet Union, very little attention was paid to consumer wishes.

Proving the algorithm correct

The line of defence on the replication thesis consisted in taking a model of a private market economy and turning it into an algorithm that might be given to various functionaries of a planned economy. The defences differed on the basis of the model itself, or the algorithm, and in the latter instance the issue of possibility came into play. The question that emerged next was whether the solution of the model and algorithm would be the same. Specifically, would the agents in the model of capitalism make the same choices as those in the planned economy if, in fact, the incentives facing them differed under the two institutional settings?

The debate now moved from theoretical issues of information generation to understanding behaviour in socialist economies in the light of the incentives facing the socialist agents. In 1936, Durbin discussed the responsibility of economists to get things right, arguing that the economist certifies the model's correctness, but not the algorithm's correctness. There is no guarantee that it will be executed as ordered. That is someone else's responsibility:

> "The calculations will not be made." "The mobile resources will be unwilling to move." "The production units that ought to expand will refuse to do so." All these criticisms may or may not be true. They may or may not be the real problems of policy. But they are not problems that the professor of economic theory is competent to discuss. They are problems of social behaviour. They can only be resolved, if they can be resolved at all, by a comprehensive sociological and principally psychological analysis. (Durbin, 1936, p. 678)

Later, reviewing the Taylor and Lange contributions, Dickinson points out the weakness in the algorithm. He focused on whether agents in the socialist algorithm would behave the same way as agents in the capitalist model:

> Mr. Lange ... speaks of rules that the Central Planning Board would have to impose on the managers of socialist enterprise. But what guarantee would there be that these rules would, in fact, be observed? Pure equilibrium theory makes it possible to deduce the rules that socialist managers ought to follow: to determine whether, or under what circumstances, socialists managers are likely to follow them requires data that only realistic social studies can provide. (Dickinson, 1938, p. 533)

Is market socialism incentive compatible?

For capitalism and socialism to be described by one model, we require common technology, common preferences and common incentives. At the simplest level we require all three to have a price system. Could a strongly planned economy actually move to the pseudo-planning of market socialism? Eucken (1948) asked whether Lange's solution is incentive compatible in a centrally administered state:

> Would it not perhaps have been possible to graft prices on to the controlling mechanism of the centrally administered economy in the following way? The central administration would have distributed consumption goods by rationing, as well as fixing prices. With regard to consumption goods, demand and supply would have been equated by rationing. But with regard to the factors of production, there would have been no rationing. Entrepreneurs would have applied for these to the state authorities. The factors would have been priced, and then these prices adjusted according to the extent of demand. By this adjustment of prices would not demand and supply have been possible? In this way, the German authorities would have been proceeding in accordance with proposals outlined by, for example, O. Lange. Wouldn't it have been possible to follow this proposal? (Eucken, 1948, p. 93)

Not surprisingly, given what we know today, his answer was an emphatic 'no':

> This method of control was out of the question for the central administration, for it would have meant to some extent letting the control of the means of production – in this case leather or iron – out of its hands.

Competition means the end of central authority:

> Competition can be used to improve efficiency, but as a mechanism of direction for an important section of the economy it cannot be applied without abdication of the central authority. (Eucken, 1948, p. 94)

Eucken claimed that the outcomes, as well as the language used to describe outcomes in the two institutional settings, will differ. The critical term is 'unemployment'. In an institution-free, Post-Keynesian economic model, one could draw a single production possibility curve for any society. Without unemployment, all would attain the frontier (for example, Samuelson, 1948, p. 20). Eucken notes that Keynesian unemployment, which is both privately and socially costly, would not be a problem in a centrally administered economy:

> every worker can be taken on regardless of costs. In an exchange economy, workers are dismissed because there exists a measure of scarcity with regard to single units … Workers are dismissed if the return resulting from their employment does not cover the costs. … even it is estimated that the costs of employing several thousand

workers on road construction are not covered, the central administration does not have to cut the work short. In these conditions full employment is always attainable. (Eucken, 1948, p. 179)

But without prices to allocate resources, we revert to barter. Here is Eucken and Meyer's (1948) description of the barter economy that will emerge in the centrally administered economy:

> In dirty, overcrowded, and unlighted trains they travel into the farm areas which are a hundred miles away in order to get from the farmers some food in exchange for part of their city rations, part of their wages in kind, if they receive such, or simply for other possessions which they still retain. They collect beechnuts from the forest and leftover ears of grain and potatoes on the harvested fields. In their free time or during their vacations they cut peat, collect wood in the forests, cultivate a small vegetable plot, or search for rabbit food. Housewives spend uncounted hours in lines before stores, in lines before distribution offices for ration coupons, and in lines before various other government offices. (Eucken and Meyer, 1948, pp. 56–7)

The economic consequence of barter is 'misery' and risks 'death by starvation':

> From the economic point of view, such extra work is senseless waste. From the point of view of the individual German, however, it is exceedingly important because it saves him from ultimate misery and frequently even from death by starvation. (1948, pp. 56–7)

What puts society in this interior of the production possibility set? The 'unemployment' problem in capitalism maps to the 'barter problem' in planned economies because a move from planning to markets is not in the interests of those who hold state power.

The consequences of the calculation debate

The major principles textbooks of the post-Second World War era (Elzinga, 1992) drew from the planned economies' lower consumption–income ratio the inference that they were growing faster than market economies. In doing so, they supposed that 'unemployment' had the same meaning in both systems (Samuelson, 1970, p. 3; Lipsey and Steiner, 1975, p. 899; McConnell, 1963, p. 751). As a consequence, his insight was temporarily obscured. The realization that planning entailed barter led to the abolition of the German centrally administered economy, but this insight was lost until the Soviet economies were near collapse. Then it became all too obvious that the disequilibrating prices and pervasive shortages were the result of planners' pursuit of private self-interest as opposed to the public good (Levy, 1990; Shleifer and Vishny, 1992).

Learning from the Soviet failure

The surprising collapse of the Soviet economies occasioned a useful series of discussions that attempted to understand both the failure of the Soviet economies and the failure of Western economists. It is appropriate that von Mises's and Hayek's works were re-read with considerable care by those on all sides of the debate.

The heroic in the Soviet economy

As noted above, von Mises rejected the possibility of socialism motivated by public opinion. The fact that heroic motivation is a puzzle for him may reflect the fact that economists have lost contact with the classical economists' discussion of sympathetic motivation and the desire for approbation (Peart and Levy, 2005). For Mill, the strength of sympathetic motivation measures a civilization. The willingness of Americans to die because of their obligation towards the enslaved indicated to him the superiority of American civilization (Peart and Levy, 2005). For Mill, the desire for approbation is sufficiently strong that anti-social behaviour is attenuated because an individual wishes to avoid the disapproval that follows when he or she imposes a cost on society. Socialism on a small scale is then unproblematic to the extent that a socialist community has tight consensus concerning costly behaviour. If one disagrees with such proscriptions, one can leave that society. Mill's dissent from large-scale socialism arose from his concern that it would be inconsistent with diversity of opinion: there would be no place to go to.

Scholars have asked why the Soviet political authorities ignored the advice of expert engineers to insist upon the heroic. A belief in the efficacy of heroic motivation might explain the political appeal of the gigantic Soviet engineering failures such as the Great Dnieper Dam, the Steel City of Magnitogorsk and the White Sea Canal (Graham, 1993; Kotkin, 1995). When his engineers asked the authorities to consider what was feasible, Stalin is quoted as having responded: 'There are no fortresses that Bolsheviks cannot storm' (Graham, 1993, p. 42).

In his study of Soviet engineering failures Graham puzzles over why the political leadership gave up the feasible and risked the certainty of the ordinary for a chance at the heroic. Such a choice makes no sense if one is thinking of the incentives of a market economy with a democratic government. But if socialism itself is justified by risking the seemingly impossible, then the political calculus seems inescapable.

> Had all the costs, social and economic, of the Dnieper dam been considered more carefully, and had the benefits of a single enormous hydroelectric power plant been weighted against those of several small ones, including thermal power plants, a different decision probably would have been made. These alternatives, now quite obviously more desirable, were outlined by Russian engineers during the early planning stages. The final decision to go ahead with the giant dam was based not on technical and social analysis but on ideological and political pressure. Stalin and the top leaders of the Community Party wanted the largest power plan ever built in order to impress the world and the Soviet population with their success and that of the Communist social order. (Graham, 1993, p. 52)

If industrial development is viewed as if it required wartime heroics then, on the basis of American experience of the Civil War, one might well expect a mix of volunteers and conscripts. And this, of course, was observed:

> One of the characteristics of industrialization under Stalin was the coexistence of volunteer and forced labour, of heroic self-sacrifice and violent coercion. (Graham, 1993, p. 59)

Heroic motivation also provides some insight into the speed of the Soviet collapse. If people chose socialism over ordinary capitalism with the hope or expectation of the heroic, then, when the 'heroic' was revealed as a sham, there was no reason to believe that the public opinion that supported the system would hold. Events like the explosion at Chernobyl – 'a product of the standard Soviet industrialization policy that emphasized gigantic projects over smaller ones' (Graham, 1993, p. 90) – could no longer be hidden.

The 'impossibility thesis'?

At the beginning of this article we distinguished between the 'replication' thesis and the 'improvement' thesis to identify the source of ambiguity in the question 'Is socialism possible?' Let us return to that theme in the context of a very simple question. Was it possible for the Soviet system to create projects which could have been deemed efficient *ex post* by the standard calculus? Let us be precise and define the heroic as something valued for its own sake. Soviet engineering for instrumental, non-heroic projects, for example military hardware, certainly passes any market test for efficiency. But when the political incentives were such as to view the project as an end it itself then, even when the *ex post* efficient projects were feasible from an engineering point of view, they were not selected by the political authorities. Thus, although efficient projects were possible, in the sense that anything feasible is 'possible', they were not observed and never existed. The debate between Caplan (2004) and Boettke and Leeson (2005) over the 'impossibility' of socialism struggles with this ambiguity, with the question of whether the Soviet Union failed because of the 'impossibility' of calculation or because of the perverse incentives of its rulers.

There is another way in which the 'impossibility' of socialism is ambiguous. Von Mises's argument asserts that a socialist economy cannot replicate the performance of a market economy. From this, can we infer that socialist economies could not exist with the consent of the participants? Surely not. There may well be benefits from socialism that over-compensate the material losses. This would seem an odd reading of von Mises if one supposes, as both sides of the Caplan–Boettke and Leeson debate do, that von Mises is engaged in an argument with Marx and his followers. It is not so odd if he is arguing with Mill's socialism.

Is socialism robust?

The attractions of socialism to disinterested scholars are evident in Roemer's efforts to re-imagine a socialism that can be defended after the fall of the Soviet Union. In a remarkable display of historical rethinking, Roemer works through the Lange–Hayek argument giving Hayek the better of each question (Roemer, 1994, p. 30). This is not just on technical economic grounds.

> Hayek claims that Lange never justifies disallowing market determination of industrial prices in his proposal along with consume goods' prices and wages. Indeed, Lange actually does but the justification seems weak if not wrong. He says that disequilibrium in industrial prices is very costly in the economy, since these prices determine the prices of all other goods, and that the CPB [Central Planning Board] can find the equilibrium faster than the market can. I am puzzled by this. It seems that perhaps Lange feared that, were he to allow the market to determine all prices in his model (except the interest rate), he would be giving up too much. As Hayek notes … the Lange proposal already makes great concessions to those who opposed pervasive planning, perhaps Lange believed it would not have been politically wise to go further. (Roemer, 1994, pp. 30–1)

Roemer identifies the problem with the Soviet Union as the incentives of the principal–agent sort (Roemer, 1994, pp. 35–9) and he argues that this can be attenuated by allowing democratic competition (Roemer, 1994, pp. 39–40). A model of the political process is proposed in which equilibrium brings with it a weighted average of the utilities of the poor and the rich (Roemer, 1994, p. 65). To deal with investment decisions, Roemer (1994) proposes the creation of institutions akin to the Japanese *keiretsu* (a group of interrelated companies, both in terms of products and owners, which is centred around a bank) sufficiently independent of the state so as not to allow politician–manager interactions (1994, p. 74).

Roemer's model of the political process is not the only one available. A long line of formal voting models worry about the results of changing the sequence of votes. If an agenda-setter can control the voting sequence he is able to force the results (for example, McKelvey, 1979). Faith in the performance of *keiretsu* may have declined somewhat from the early 1990s. Perhaps we should give some thought as to what can go wrong with a model of socialism as compared with a model of capitalism (Levy, 2002).

If modellers are putting forward proposals out of sympathy for extra-scientific goals (Peart and Levy, 2005), then the arguments about model uncertainty in the context of least favourable priors take on a new urgency (Brock, Durlauf and West, 2007).

Eucken's argument that moving from a planned economy to market socialism is not incentive compatible does not, of course, preclude moving from a capitalist economy to market socialism. The question is whether that is where society would stay. If market socialism were to be judged a failure, would we return to market capitalism or move to less liberal institutions (Levy, Peart and Farrant, 2005)?

DAVID M. LEVY AND SANDRA J. PEART

See also **Hayek, Friedrich August von; Lange, Oskar Ryszard; Lerner, Abba Ptachya; Mises, Ludwig Edler von; planning; socialism.**

Bibliography

Barone, E. 1908. The ministry of production in the collectivist state. In *Collectivist Economic Planning*, ed. F.A. Hayek. London: Routledge, 1935.

Boettke, P. and Leeson, P. 2005. Still impossible after all these years. *Critical Review* 17, 155–70.

Brock, W.A., Durlauf, S.N. and West, K.D. 2007. Model uncertainty and policy evaluation: some theory and empirics. *Journal of Econometrics* 136, 629–64.

Caplan, B. 2004. Is socialism really 'impossible'? *Critical Review* 16, 33–52.

Dickinson, H.D. 1933. Price formation in a socialist community. *Economic Journal* 43, 237–50.

Dickinson, H.D. 1938. On the economic theory of socialism. *Economic Journal* 48, 531–33.

Dobb, M.H. 1933. Economic theory and the problem of a socialist economy. *Economic Journal* 43, 588–98.

Drewnowski, J. 1961. The economic theory of socialism: a suggestion for reconsideration. *Journal of Political Economy* 69, 341–54.

Drewnowski, J. 1997. Review of the roundtable talks and the breakdown of communism. *Europe–Asia Studies* 49, 919–20.

Durbin, E.F.M. 1936. Economic calculus in a planned economy. *Economic Journal* 46, 676–90.

Elzinga, K.G. 1992. The eleven principles of economics. *Southern Economic Journal* 58, 861–79.

Eucken, W. 1948. On the theory of the centrally administered economy: an analysis of the German experiment tr. T.W. Hutchison. *Economica* 15, 79–100 and 173–93.

Eucken, W. and Meyer, F.W. 1948. The economic situation in Germany. *Annals of the American Academy of Political and Social Science* 260, 53–63.

Gerber, D.J. 1994. Constitutionalizing the economy: German neo-liberalism, competition law and the 'new' Europe. *American Journal of Comparative Law* 42, 25–84.

Graham, L.R. 1993. *The Ghost of the Executed Engineer: Technology and the Fall of the Soviet Union*. Cambridge, MA: Harvard University Press.

Hayek, F.A. 1935a. The nature and the history of the problem. In *Collectivist Economic Planning*, ed. F.A. Hayek. London: Routledge.

Hayek, F.A. 1935b. The present state of the debate. In *Collectivist Economic Planning*, ed. F.A. Hayek. London: Routledge.

Hayek, F.A. 1940. Socialist calculation: the competitive solution. *Economica* 7, 125–49.

Hayek, F.A. 1945. The use of knowledge in society. *American Economic Review* 35, 519–30.

Hutt, W.H. 1940. Economic institutions and the new socialism. *Economica* 7, 419–34.

Kotkin, S. 1995. *Magnetic Mountain: Stalinism as a Civilization*. Berkeley: University of California Press.

Lange, O. 1936. On the economic theory of socialism: part one. *Review of Economic Studies* 4, 53–71.

Lavoie, D. 1985. *Rivalry and Central Planning: The Socialist Calculation Debate Reconsidered*. Cambridge: Cambridge University Press.

Lerner, A.P. 1936. A note on socialist economies. *Review of Economic Studies* 4, 72–6.

Levy, D.M. 1990. The bias in centrally planned prices. *Public Choice* 67, 213–36.

Levy, D.M. 2002. Robust institutions. *Review of Austrian Economics* 15, 131–42.

Levy, D.M., Peart, S.J. and Farrant, A. 2005. The spatial politics of the road to serfdom. *European Journal of Political Economy* 21, 982–99.

Lipsey, R.G. and Steiner, P.O. 1975. *Economics*, 4th edn. New York: Harper & Row.

Malthus, T.R. 1798. *An Essay on the Principle of Population as It Affects the Future Improvement of Society, with Remarks on the Speculations of Mr. Godwin, M. Condorcet, and Other Writers*. London: J. Johnson.

McConnell, C. 1963. *Economics: Principles, Problems, and Policies*, 2nd edn. New York: McGraw-Hill.

McKelvey, R.D. 1979. General conditions for global intransitivities in formal voting models. *Econometrica* 47, 1085–113.

Mendershausen, H. 1949. Prices, money and the distribution of goods in postwar Germany. *American Economic Review* 39, 646–72.

Mill, J.S. 1844. A System of Logic. In *The Collected Works of John Stuart Mill*, vol. 8, ed. J.M. Robson. Toronto: University of Toronto Press, 1981.

Mill, J.S. 1848. The Principles of Political Economy. In *The Collected Works of John Stuart Mill*, vol. 3, ed. J. Robson. Toronto: University of Toronto Press, 1965.

Mises, L.E. von. 1920. Economic calculation in the socialist commonwealth, tr. S. Adler. In *Collectivist Economic Planning*, ed. F.A. Hayek. London: Routledge, 1935.

Mises, L.E. von. 1936. *Socialism: An Economic and Sociological Analysis*, tr. J. Kahane. Indianapolis: Liberty Classics, 1981.

Peart, S.J. and Levy, D.M. 2005. *The "Vanity of the Philosopher": From Equality to Hierarchy in Post-Classical Economics*. Ann Arbor: University of Michigan Press.

Roemer, J.E. 1994. *A Future for Socialism*. Cambridge: Harvard University Press.

Samuelson, P.A. 1948. *Economics: An Introductory Analysis*. New York: McGraw-Hill.

Samuelson, P.A. 1970. *Economics*, 8th edn. New York: McGraw-Hill.

Shleifer, A. and Vishny, R. 1992. Pervasive shortages under socialism. *Rand Journal of Economics* 23, 237–46.

Taylor, F.M. 1929. The guidance of production in a socialist state. *American Economic Review* 19, 1–8.

soft budget constraint

The concept of 'soft budget constraint' was coined by János Kornai in his famous book *Economics of Shortage* (1980). In that book, Kornai developed a comprehensive theory of the socialist economy (completed in 1992 by *The Socialist System*). The starting point was that in the socialist economy firms had soft budget constraints as opposed to the hard budget constraints firms face under capitalism. 'The classical capitalist firm had a hard budget constraint. If it is insolvent, it will sooner or later become bankrupt. ... As opposed to this, the budget constraint of the traditional socialist firm is soft. If it works with a loss, that does not yet lead to real bankruptcy, i.e. ceasing operations' (1980, p. 29). Kornai used the concepts of mainstream economic theory to contrast the situation of the capitalist firm and that of the socialist firm. Interestingly, in standard microeconomic theory only households have a budget constraint, not firms. The latter do not face any financial constraint and are only maximizing profits. Also, a budget constraint is usually supposed to be hard; otherwise, it is not a constraint and would never bind. The concept of a soft budget constraint thus seemed intriguing. It reflected indeed a very important observation about the environment that the managers of socialist firms were facing as compared with the environment faced by the managers of capitalist firms. It immediately implied differences in the incentives managers were facing in the two economic systems, and one could understand that these different incentives had economy-wide consequences.

The basic observation that socialist firms had soft budget constraints and could count on continued financial support in case they were making losses led Kornai to develop a general theory of shortage in the socialist economy. Since the budget constraints of firms under socialism are not binding, one understands that they must necessarily meet a resource constraint, that is, experience shortage. This is, however, only an equilibrium implication. Shortage, or hitting the resource constraint, was thus achieved via the effect of soft budget constraints on the demand behaviour of firms.

The soft budget constraint and the socialist firm

Soft budget constraints had indeed a major effect on the demand behaviour of socialist firms. Demand for labour, input and capital tended to be in general higher than what it would be if firms had hard budget constraints. Moreover, demand was hardly responsive to price variations. Shortage tended to aggravate this phenomenon as it led to a hoarding motive in demand that tended to further aggravate shortages. Shortages were thus ubiquitous. Kornai developed a comprehensive theory of firm behaviour under shortage based on the premise of soft budget constraints. As of 2007 this is the most complete theory of the socialist economy that has been published, and most probably it will never be superseded. It is impossible in this short article even to summarize the exhaustive analysis Kornai made of firm behaviour and its general consequences for the economy. The interested reader is strongly encouraged to read the original analysis.

Like many general theories, Kornai's theory has led to extensions, refinements and also criticisms. Several papers (for example, Kornai and Weibull, 1983; Goldfeld and Quandt, 1988; 1990; 1993; Magee and Quandt, 1994) have examined formally the link between soft budget constraints and the supply and demand behaviour of socialist firms. Gomulka (1985) emphasized that households had hard budget constraints under socialism. In that case, the price system should have been able to eliminate shortages for consumer goods. Firms were, however, not responsive to signals from consumer goods markets under socialism since they were themselves resource-constrained on their supply side.

Why did firms in the socialist economy have soft budget constraints? Kornai related this to paternalism: 'Paternalism is the direct explanation for the softening of the budget constraint' (1980, p. 568). While *Economics of Shortage* was written while Kornai was still living under the socialist regime, this explanation remained partial for obvious reasons of self-censorship. In *The Socialist System*, he came back to the question at more length and explicitly defined causal mechanisms from the undivided power of the Communist Party to the dominance of state ownership, the preponderance of bureaucratic coordination to the environment of firms, including softness of budget constraints.

The soft budget constraint as a general incentive problem

While debates on soft budget constraints, their causes and consequences, were mostly confined to economists studying the socialist economy, it was recognized that the soft budget constraint syndrome had broader applications than in the context of socialist firms. Bail-outs of large firms and banks occur repeatedly under capitalism. Kornai's concept of the soft budget constraint was widely popularized by a now famous article by Dewatripont and Maskin (1995) in which the soft budget constraint problem was formalized within the context of contract theory. Dewatripont and Maskin formulated the soft budget constraint as a rather general dynamic commitment problem within the framework of contract theory. Specifically, a principal can decide to invest in a project but would like to commit to not refinancing the project if its return is low. However, if he or she cannot commit not to refinance the firm, there are conditions under which they end up bailing out the firm in order to cut its losses. One would thus observe projects that are on the whole unprofitable but that nevertheless get refinanced: in other words they are loss-making firms with soft budget constraints. The reason for this dynamic commitment problem is that the initial funds spent on the firm are sunk costs and only the net return to refinancing is taken into account in the bail-out decision. Dewatripont and Maskin thus showed the soft budget constraint to be a very general incentive problem, not necessarily one to be related only to the context of socialism or government ownership.

The Dewatripont–Maskin model considers the following adverse selection problem. The government faces a population of firms, each needing one unit of funds in initial period 1 in order to start its project. A proportion α of these projects are of the 'good, quick' type: after one period, the project is successfully completed, and generates a gross (discounted) financial return $R_g > 1$. Moreover, the manager of the firm (possibly also workers) obtains a positive net (discounted) private benefit E_g. In contrast, there is a proportion $(1 - \alpha)$ of bad and slow projects which generate no financial return after one period. If terminated at that stage, managers in the firm obtain a private benefit E_t. Instead, if refinanced, each project generates after two periods a gross (discounted) financial return π_b^* and a net (discounted) private benefit E_b. Initially, α is common knowledge but individual types are private information. A simple result easily follows: if $1 < \pi_b^* < 2$ and $E_b > 0$, refinancing bad projects is sequentially optimal for the government, and bad entrepreneurs who expect to be refinanced apply for initial financing. The government would, however, be better off if it were able to commit not to refinance bad projects, since it would thereby deter managers with bad projects from applying for initial financing, provided $E_t < 0$.

Termination is here, by assumption, a disciplining device that allows the uninformed investor (creditor) to turn away bad types and only finance good ones. The problem is that termination is not sequentially rational if π_b^* is>1: once the first unit has been sunk into a bad project, its net continuation value is positive so that, in the absence of commitment, the soft budget constraint syndrome arises. In this set-up, because irreversibility of investment is such a general economic feature, the challenge for theory is more to explain why hard budget constraints prevail rather than why budget constraints are soft in the first place.

Dewatripont and Maskin considered the soft budget constraint problem within a banking set-up. They found that centralized banking was more prone to the soft budget constraint problem than decentralized banking. In particular, to the extent that decentralized banking makes refinancing more costly because of incentive problems between multiple investors, it tends to favour hard budget constraints (see also Povel, 2004; Huang and Xu, 1999). Indeed, when refinancing is more costly, investors are less willing to bail out and more willing to terminate projects, thereby changing the incentives of enterprise managers.

While the dynamic commitment aspect of the Dewatripont–Maskin model of the soft budget constraint coincides with Kornai's formulation, differences appear when it comes to explain the *cause* of the soft budget constraint problem. Indeed, Kornai emphasized the paternalism of government as the cause of soft budget constraints for the socialist firms. Paternalism plays no role in the Dewatripont–Maskin model. The main driving force is the sunk cost of investment combined with asymmetric information and inability to commit. These apparent differences in the explanation of soft budget constraints are, however, related to the fact that both explanations are located at different levels of abstraction and also related to different concepts of scientific explanation. Indeed, Dewatripont and Maskin develop a logical explanation of soft budget constraints deriving sufficient conditions for soft budget constraints. This logical explanation can then be applied to different empirical and institutional set-ups such as the one they discuss, namely, the comparison between centralized and decentralized banking. From that point of view, one can claim that paternalism is neither a necessary nor a sufficient condition for soft budget constraints. However, it would be misguided to take this as a criticism of Kornai's explanation, which is situated at a different level. His concept of soft budget constraint is an abstraction destined to capture an empirical regularity rooted in the behaviour of socialist firms. The explanation of soft budget constraints by paternalism is an explanation based on empirical plausibility rooted in the institutional context of the socialist economy. Because these are explanations at different levels (pure logic on one hand, a concept grasping an empirical regularity on the other hand), they should not be seen as contradictory and mutually exclusive. Kornai's explanation gives us crucial insights into the mechanisms of the socialist economy and the transition process, while Dewatripont and Maskin give us logical conditions that may be applied to different institutional contexts.

The Dewatripont–Maskin model has made the soft budget constraint concept an integral part of mainstream economic theory. It has also been recognized as a general incentive problem that has played a role in diverse banking crises that have occurred since the 1980s, including the 1997 east Asian crisis.

Models of soft budget constraints under socialism

Soft budget constraint models following the seminal model of Dewatripont and Maskin have been developed in a wide variety of contexts.

First, beyond the models cited above that look mostly at the consequences of soft budget constraints, models were developed to understand various aspects of the socialist economy. For example, Qian (1994) developed a model showing that in the context of the socialist economy, shortages were a good way of alleviating some of the negative consequences of the soft budget constraint. Indeed, he showed that price caps that lead to shortages could be beneficial in terms of somewhat reducing soft budget constraints. Indeed, shortages reduce the likelihood that bad projects will be refinanced. This therefore reduces the incentive to submit poor projects. *A contrario*, with flexible prices and soft budget constraints, enterprises will bid for scarce resources, leading to price inflation and crowding out of consumers who face hard budget constraints. This explains why in many socialist economies, partial price liberalization as in the Soviet Union under President Gorbachev or in Poland, Hungary and China, and even advanced price liberalization as in the former Yugoslavia, led to strong inflationary pressures. Other models by Wang (1991) and Debande and Friebel (2004) showed similarly that reforms giving more autonomy to socialist enterprises, by relaxing the monitoring of enterprise activities, tended to exacerbate the soft budget constraint phenomenon.

Qian and Xu (1998) developed a model to explain how soft budget constraints had a negative effect on innovation under socialism. This is an important theme since the inferiority of the socialist economy to capitalism in generating innovation is one of the main reasons for its demise. Knowing that firms had soft budget constraints, central planners were very cautious in approving investments with innovations. Indeed, the return to innovations is uncertain and it is important to be able to stop innovations that turn out to be disappointing. The market economy is able to do that quite well because firms have hard budget constraints; but this was not the case in the socialist economy. The caution of central planners in approving investments was relatively greater in areas where innovation was riskier and science was relatively new, precisely in the areas where returns to innovation were greater. This was for example the case for the computer industry where socialist firms were considerably behind capitalist firms. In industries where science was older and where risk was likely to be smaller, the centrally planned economy fared better. This was the case for the aerospace industry.

Dewatripont and Roland (1997) developed a model to analyse the links between the soft budget constraint and another incentive problem that was present in the socialist economy: the ratchet effect. The term 'ratchet effect' was coined by Berliner (1952) in his analysis of management behaviour in Soviet-style firms. In such firms, managers were given what appeared to be strong incentives to fulfil their production plans. Indeed, they had inducements to *overfulfil* the plans: each percentage point over the target was rewarded by additional bonuses. Nevertheless, managers tended to pass up the opportunity for these bonuses and instead were conservative in their plan overfulfilment, rarely exceeding two per cent over target. Berliner's explanation for this conservatism was that managers feared that next year's target would be 'ratcheted up' (that is, made more demanding) if they exceeded this year's goal. By producing at 110 per cent instead of 102 per cent, their bonus would be higher today, but so would their target tomorrow. Dewatripont and Roland show that the ratchet effect and the soft budget constraint could be interrelated in the sense that the need to bail out weaker firms gave central planners a stronger incentive to ratchet up the plans of the better-performing firms.

Models of soft budget constraints under transition

The issue of soft budget constraints, which was initially fundamentally neglected in the early transition from Communism, especially by the advocates of the Washington consensus and the big-bang approach to transition (apart from some mention here and there), became an increasingly important issue in trying to understand the restructuring process of firms and of banks. First, the transfer of ownership from the state to the private sector changed the incentives to bail out firms but did not automatically eliminate soft budget constraints. Indeed, by using the Dewatripont–Maskin model, it is easy to understand (see for example Kornai, Maskin and Roland, 2003) that privatization, by shifting the financing of firms from the government to the private sector, and in particular to private banks, changes the motive for bailing out firms from *ex post* care for employment and the welfare of those working inside firms to *ex post* profit maximization. However, this only reduces the extent of soft budget constraints since the logic of Dewatripont–Maskin does not require state ownership to generate soft budget constraints. What is important is the sunk cost nature of investment that would lead an investor to refinance *ex post* a firm even though it would want to commit not to do so *ex ante*.

Complementary to the changes in ownership, there are other institutional changes during the transition process that would tend to harden budget constraints. Several models were developed along those lines. Segal (1998) examined the role of demonopolization and the development of competition in the hardening of budget constraints. The argument is that demonopolization and competition both increase entry of firms and reduce their profits. The reduction in profitability reduces the incentive to bail out but the increase in entry reduces the likelihood of any single firm of being bailed out. The argument requires a certain amount of excess capacity of

firms in the market at any moment in time which is consistent with competition between firms.

Berglöf and Roland (1998) advanced a slightly different argument about the entry of new firms. If newly entering firms have a sufficiently high expected return in a competitive market economy, then the expected return to lending to those firms might be higher than the expected return to bailing out existing firms. This would then yield hard budget constraints as firms with poor projects would know that they were unlikely to be bailed out. However, for the argument to be valid, the expected return of the newly entering firms should be significantly higher than that of older firms. Indeed, with an expected return that is only marginally higher than for old firms, soft budget constraints will still be present. The reason is that the competition for funds is tilted against the newly entering firms because initial investments in older projects are sunk costs. Therefore, the comparison between the return to bailing out existing firms and financing entering firms does not count the initial funding of existing firms, so giving the latter an advantage in the competition for funds. The upshot is that, if the rhythm of innovation in the economy is not sufficiently strong, that is, if newly entering firms do not have an expected profitability that is significantly higher than existing firms, then soft budget constraints may persist.

Yet another argument related to firm restructuring was made by Perotti (1993) and Coricelli and Miles-Ferretti (1993). This is basically an argument about the negative externalities of soft budget constraints. Here, the hardening of budget constraints depends on the existing firms' links with their chains of suppliers and clients. The idea is that banks may be reluctant not to bail out firms if closing the operations of those firms creates financial difficulties for the more healthy firms who do business with the weaker firms. The softness of banks may in turn dull the incentives of firms to restructure. From that point of view, it is harder to restructure an enterprise operating in a business network where budget constraints are very soft. Firms setting up new business networks in an environment with hard budget constraints will then face hard budget constraints themselves.

Another key institutional variable affecting the hardness of budget constraints is the degree of decentralization of banking. This is what the original Dewatripont–Maskin model was about. The general idea is that under decentralized banking renegotiating initial loan contracts is more difficult because it involves more inefficiencies and it thus more costly than under centralized banking. The particular mechanism in the model is the following: if the initial bank is liquidity-constrained and refinancing must involve another bank, the initial bank has less incentive to monitor the firm to ensure a higher return to refinancing than would be the case under centralized banking. Indeed, it must share the returns to monitoring with the other bank, which is assumed to have less experience with the existing project and thus is unable to monitor the use of its money. Because the bank that monitors does not reap the full returns from monitoring, it has less incentive to monitor. Other mechanisms would deliver the same result. Povel (1995) examined a model in which a project is financed from the outset by two banks. Suppose that an agreement on a restructuring plan is necessary to refinance a poor project and that each bank's assessment of the continuation value of the project is private information. The asymmetric information between banks can give rise to a delay in their negotiating an acceptable restructuring plan. However, if the value of the project declines over time, this delay may render refinancing unprofitable. Huang and Xu (1999) studied a related model in which two banks (investors) agree to lend jointly to a project but have conflicting interests concerning how to organize the project should it be refinanced. This conflict may make it too costly to reach an agreement on a strategy after bail-out. Huang and Xu apply this argument to illuminating the east Asian crisis of the late 1990s. They note that the Korean *jaebeols* were subject to centralized financing and suffered from lack of financial discipline and soft budget constraints. By contrast, Taiwan's economy was characterized by dispersed financial institutions and decentralized banking. In the event, Taiwan suffered much less from the crisis than Korea (even though it, too, was attacked by speculators).

Decentralization of government, under federalism, may also have the effect of hardening budget constraints under certain circumstances, as argued by Qian and Roland (1998) on the basis of the transition experience in China. Decentralization of government was an important feature of Chinese reforms. Among other features, it led local governments to compete with each by investing in infrastructure investment other than to attract capital in order to boost growth in their province or region. This form of fiscal competition leads to overinvestment in infrastructure because the return of infrastructure investment to a province is higher than the return to society as a whole, as local governments do not internalize the effects of attracting capital away from other provinces. A positive side effect of this fiscal competition, however, can be that local governments prefer now to put their money in infrastructure investment rather than in bailing out loss-making enterprises. While decentralization of government can harden the budget constraints of enterprises that are controlled by local government, it can lead to soft budget constraints for local governments. Indeed, local governments can always structure the composition of their expenditures in such a way as to coax additional funds from the central government. Say that local governments are responsible for hospitals: they can strategically underfund hospitals and spend their budget on other items so as to obtain additional funds from the central government to improve hospital services. Zhuravskaya (2000) found evidence in the case of Russia that is consistent with such a story, using a data-set from 35 large cities in 29 regions of the Russian Federation between 1992 and 1997. She found that any increase in own revenues by local governments tends to be offset by a decrease of nearly one to one of shared revenues, which is consistent with soft budget constraints. Moreover, weaker fiscal incentives (a negative correlation between shared revenues and own revenues) tend, everything else equal, to decrease spending on education and health. There is also a negative impact on the quality of health and education, as measured by infant mortality and evening school attendance by children due to crowded schools. This is consistent with the idea that local governments have an incentive to distort their expenditures by neglecting health and education so as to try to attract more grants.

The soft budget constraint phenomenon has been analysed not only in the context of enterprises or local governments. It has been studied in the context of banks. Mitchell (1998), for example, built a model of bank passivity where banks fail to liquidate bad projects because they themselves expect to be bailed out by government. Similarly, Berglöf and Roland (1995) show that banks may strategically exploit the government's concerns for jobs in loss-making firms to extract government subsidies to refinance firms, in effect forcing the government to pay part of the refinancing of firms. Policies of bank monitoring and recapitalization may help to eliminate soft budget constraints of banks, but policies of recapitalization face incentive problems of their own (see Aghion, Bolton and Fries, 1999).

Other visions of soft budget constraints

Interpretations of the soft budget constraint different from the Dewatripont–Maskin model are also present in the economics profession. Boycko, Shleifer and Vishny (1996), for example, do not identify soft budget constraints as a dynamic commitment problem where the principal would prefer not to commit to bailing out but ends up doing so. They see soft budget constraints as a deliberate choice of a governmental body to subsidize an enterprise in order to prevent management from laying off workers. According to this interpretation, soft budget constraints are not an inefficiency but rather the consequence of government preferences. This interpretation has had some following partly because the Dewatripont–Maskin view of soft budget constraints has often been associated too narrowly with bank–enterprise relationships and because, in reality, soft budget constraints occur more often in government–firm or government–bank relationships.

More recently, Robinson and Torvik (2006) have proposed a political economy theory of the soft budget constraint in a framework where politicians cannot commit to electoral promises. In this context, soft budget constraints, while resulting from an inability to commit to not bailing out a project, can be seen as commitment devices to redistribute transfers to particular constituencies of voters and to secure re-election and create an incumbency advantage. In other words, it emerges as a kind of political patronage.

Empirical work on soft budget constraints

It is very difficult to do empirical work measuring soft budget constraints as *ex post* bail-outs that were not desired *ex ante*. Seen in this light, soft budget constraints must be distinguished from other forms of funding that do not have these features. Moreover, the behaviour of firms is based on expectations of bail-out that are not easy to measure either. It is therefore not surprising that much of the empirical work surrounding soft budget constraints has been about using proxies that can find some justification but are necessarily quite noisy. The empirical work has often been quite detached from the theory. Various studies have looked at tax arrears as a proxy for soft budget constraints (Claessens and Peters, 1997; Schaffer, 1998; and Coricelli and Djankov, 2001). Others have looked at open subsidies (Djankov and Nenova, 2000; Grigorian, 2000) or tax concessions (Alfandari, Fan and Freinkman, 1996; Brown and Earle, 2000; and Shleifer and Treisman, 2000). Other studies have looked at payment arrears of firms and the tendencies of state-owned banks to lend to distressed enterprises (Cull and Xu, 2000; Gao and Schaffer, 1998; Coricelli and Djankov, 2001; Claessens and Djankov, 1998; and Schaffer, 1998).

There have nevertheless been a few serious attempts to study soft budget constraints empirically in line with existing theoretical models. Petterson-Liblom and Dahlberg (2005), for example, have made such an attempt and tried to measure the effect of soft budget constraints on the borrowing behaviour of Swedish municipal governments between 1974 and 1992. They find that, if a municipality expects to be bailed out with certainty, as opposed to a situation where it is not bailed out, it will increase its debt level by 30 per cent. Bail-out expectations obviously cannot be measured directly. However, instead they use observed bail-out as a noisy measure of bail-out expectations and use an instrumental variable approach to get consistent measurement using as instrument observed bail-outs in neighbouring municipalities.

A comprehensive survey of the literature on soft budget constraints can be found in Kornai, Maskin and Roland (2003).

GÉRARD ROLAND

See also **adverse selection; agency problems; privatization; socialism; state capture and corruption in transition economies.**

Bibliography

Aghion, P., Bolton, P. and Fries, S. 1999. Optimal design of bank bailouts: the case of transition economies. *Journal of Institutional & Theoretical Economics* 155, 51–70.

Alfandari, G., Fan, Q. and Freinkman, L. 1996. Government financial transfers to industrial enterprises and restructuring. In *Enterprise Restructuring and Economic Policy in Russia*, ed. S. Commander, Q. Fan and M. Schaffer. Washington, DC: Economic Development Institute of the World Bank.

Berglöf, E. and Roland, G. 1995. Bank restructuring and soft budget constraints in financial transition. *Journal of the Japanese and International Economies* 9, 354–75.

Berglöf, E. and Roland, G. 1998. Soft budget constraints and banking in transition economies. *Journal of Comparative Economics* 26, 18–40.

Berliner, J. 1952. The informal organization of the Soviet firm. *Quarterly Journal of Economics* 66, 342–65.

Boycko, M., Shleifer, A. and Vishny, R. 1996. A theory of privatization. *Economic Journal* 106, 309–19.

Brown, D. and Earle, J. 2000. Privatization and enterprise restructuring in Russia: new evidence from panel data on industrial enterprises. Working Paper No. 1. Moscow: Russian European Center for Economic Policy.

Claessens, S. and Djankov, S. 1998. Politicians and firms in seven central and eastern European countries. Policy Research Working Paper No. 1954. Washington, DC: World Bank.

Claessens, S. and Kyle Peters, Jr. R. 1997. State enterprise performance and soft budget constraints: the case of Bulgaria. *Economics of Transition* 5, 305–22.

Coricelli, F. and Djankov, S. 2001. Hardened budgets and enterprise restructuring: theory and an application to Romania. *Journal of Comparative Economics* 29, 749–63.

Coricelli, F. and Milesi-Ferretti, M.G. 1993. On the credibility of 'big bang' programs: a note on wage claims and soft budget constraints in economies in transition. *European Economic Review* 37, 387–95.

Cull, R. and Xu, L.C. 2000. Bureaucrats, state banks, and the efficiency of credit allocation: the experience of Chinese state-owned enterprises. *Journal of Comparative Economics* 28, 1–31.

Debande, O. and Friebel, G. 2004. A positive theory of 'give-away' privatization. *International Journal of Industrial Organization* 22, 1309–25.

Deviatov, A. and Ickes, B.W. 2005. Reputation and the soft budget constraint. Mimeo, New Economic School, Moscow.

Dewatripont, M. and Maskin, E. 1995. Credit and efficiency in centralized and decentralized economies. *Review of Economic Studies* 62, 541–55.

Dewatripont, M. and Roland, G. 1997. Transition as a process of large scale institutional change. In *Advances in Economic Theory*, vol. 2, ed. D. Kreps and K. Wallis. Cambridge: Cambridge University Press.

Dewatripont, M. and Roland, G. 2000. Soft budget constraints, transition and financial systems. *Journal of Institutional and Theoretical Economics* 156, 245–60.

Djankov, S. and Nenova, T. 2000. Why did privatization fail in Kazakhstan? Working paper. Washington, DC: World Bank.

Gao, S. and Schaffer, M.E. 1998. Financial discipline in the enterprise sector in transition countries: how does China compare? Discussion Paper No. 98/01. Edinburgh: Centre for Economic Reform and Transformation, Heriott-Watt University.

Goldfeld, S.M. and Quandt, R.E. 1988. Budget constraints, bailouts and the firm under central planning. *Journal of Comparative Economics* 12, 502–20.

Goldfeld, S.M. and Quandt, R.E. 1990. Output targets, the soft budget constraint and the firm under central planning. *Journal of Economic Behavior and Organization* 14, 205–22.

Goldfeld, S.M. and Quandt, R.E. 1993. Uncertainty, bailouts, and the Kornai effect. *Economics Letters* 41, 113–19.

Gomulka, S. 1985. Kornai's soft budget constraint and the shortage phenemenon: a criticism and restatement. *Economics of Planning* 19(1), 1–11.

Grigorian, D.A. 2000. Ownership and performance of Lithuanian enterprises. Policy Research Working Paper No. 2343. Washington, DC: World Bank.

Huang, H. and Xu, C. 1999. Financial institutions and the financial crisis in east Asia. *European Economic Review* 43, 903–14.

Kornai, J. 1980. *Economics of Shortage*. Amsterdam: North-Holland.

Kornai, J. 1992. *The Socialist System. The Political Economy of Communism*. Princeton: Princeton University Press; Oxford: Oxford University Press.

Kornai, J., Maskin, E. and Roland, G. 2003. Understanding the soft budget constraint. *Journal of Economic Literature* 41, 1095–136.

Kornai, J. and Weibull, J.W. 1983. Paternalism, buyers' and sellers' market. *Mathematical Social Sciences* 6, 153–69.

Magee, K.L. and Quandt, R.E. 1994. The Kornai effect with partial bailouts and taxes. *Economics of Planning* 27, 27–38.

Mitchell, J. 1998. Strategic creditor passivity, regulation and bank bailouts. Discussion Paper No. 1780. London: CEPR.

Perotti, E. 1993. Bank lending in transition economies. *Journal of Banking and Finance* 17, 1021–32.

Petterson-Lidblom, P. and Dalhberg, M. 2005. An empirical approach for estimating the causal effect of soft budget constraints on economic outcomes. Mimeo, Institute for International Economic Studies.

Povel, P. 2004. Multiple banking as a commitment not to rescue. *Research in Finance* 21, 175–99.

Qian, Y. 1994. A theory of shortage in socialist economies based on the soft budget constraint. *American Economic Review* 84, 145–56.

Qian, Y. and Roland, G. 1998. Federalism and soft budget constraint. *American Economic Review* 88, 1143–62.

Qian, Y. and Xu, C. 1998. Innovation and bureaucracy under soft and hard budget constraints. *Review of Economic Studies* 65, 151–64.

Robinson, J.A. and Torvik, R. 2006. A political economy theory of the soft budget constraint. Working Paper No. 12133. Cambridge, MA: NBER.

Schaffer, M.E. 1998. Do firms in transition have soft budget constraints? A reconsideration of concepts and evidence. *Journal of Comparative Economics* 26, 80–103.

Segal, I.R. 1998. Monopoly and soft budget constraint. *Rand Journal of Economics* 29, 596–609.

Shleifer, A. and Treisman, D. 2000. *Without a Map: Political Tactics and Economic Reform in Russia*. Cambridge, MA: MIT Press.

Wang Y. 1991. Economic reform, fixed capital investment expansion and inflation: a behaviour model based on the Chinese experience. *China Economic Review* 2, 3–27.

Zhuravskaya, E. 2000. Incentives to provide local public goods: fiscal federalism, Russian style. *Journal of Public Economics* 76, 337–68.

Solow, Robert (born 1924)

Robert Merton Solow was born in Brooklyn, New York, in August 1924, so he was just five years old when the Great Depression began in the United States and, depending on how it is dated, about 16 years old when it ended. The timing is significant. He has characterized himself as part of 'the generation of economists that was moved to study economics by the feeling that we desperately needed to understand the Depression' (1990, p. 183). The search for answers about what makes the macroeconomy tick, and why it occasionally goes badly off track, has been the touchstone of his illustrious career.

Solow was educated in the New York City public schools at a time when that was a great place to be educated – a time and place that produced, among others, Kenneth Arrow, William Baumol and Robert Fogel. (That's just in economics; the number of distinguished scientists is legendary.) Graduating at age 16, he became the first person in his family to attend college when he enrolled at Harvard in 1940. But with the Second World War raging, Harvard did not hold his attention for long, and he volunteered for the US Army in 1942, serving for three years in North Africa and Italy – an experience which, he has written, 'formed my character' (1988).

Once back at Harvard, and now married to the love of his life, Barbara, Solow decided to specialize in economics and was fortunate to have Wassily Leontief assigned as his tutor. This happenstance probably also changed his life – and that of the economics profession – forever; for while economics at Harvard may not have been very inspiring at the time, nor particularly modern, nor at all mathematical, Leontief was a shining exception. He took the brilliant young student from Brooklyn under his wings and, six years later, Solow won the first of what were to be many accolades: the coveted David A. Wells Prize for the best Harvard Ph.D. dissertation. The thesis, which Solow never published because 'I thought I could do it better' (1988), was a highly original application of the then-novel theory of Markov processes to model the size distribution of wage income in the United States.

(I actually read this fine piece of work, in typescript, while writing my own dissertation on the size distribution of income at MIT about 20 years later.)

Solow's first and, in a real sense, only job was at MIT, where he was an assistant professor and then an associate professor of statistics in what was then the Department of Economics and Social Science from 1950 to 1958. It was during that period that he wrote his two classic papers on the theory and empirics of economic growth (described below), which must have made it easy for MIT to promote him to professor of *economics* in 1958. Fifteen years later he was designated an Institute Professor, a post he held until his (*de jure*, not *de facto*) retirement from MIT in 1995. In 1979 he served as president of the American Economic Association. In 1987 he was awarded the Nobel Prize in Economic Science. And in 1999 he was honoured with the National Medal of Science – one of only seven economists ever to have won that distinction. Since 2000 Solow has divided his time between MIT and the Russell Sage Foundation in New York, where he is the foundation's only permanent fellow.

In terms of research *publication* dates, the years 1956–61 in Solow's career were probably as outstanding a quinquennium as any economist has ever had. (The research production dates, one guesses, were probably two years or so earlier.) His two most famous papers, of course, came in 1956 and 1957: 'A contribution to the theory of economic growth', *Quarterly Journal of Economics* (1956), and 'Technical change and the aggregate production function', *Review of Economics and Statistics* (1957). These two papers, which introduced what came to be called 'the Solow model' and 'the Solow residual', respectively, are still, almost 50 years later, among the most widely cited papers in all of economics, despite the fact that they are so much a part of the corpus of modern economics that citations are often omitted. At this point the Nobel Prize, had there been one at the time, was in the bag. And indeed, when his Nobel award was announced, these were the two papers principally cited.

The two landmark growth papers were followed by the classic book *Linear Programming and Economic Analysis* (with Paul Samuelson and Robert Dorfman, 1958), which included, among other things, the first turnpike theorem. Next, in a 1960 paper titled 'Investment and technical progress', Solow introduced the important – and probably realistic – idea that new technology might have to be 'embodied' in new capital. (After all, a vintage 2000 computer does not develop new and more powerful chips just because Intel invents them.) This theoretical innovation opened the door to a new and rich (though more complicated) class of 'vintage' models, which then proliferated in the literature. Finally, in collaboration with Arrow, Hollis Chenery, and B.S. Minhas, he invented the constant-elasticity-of-substitution (CES) production function in their paper 'Capital-labor substitution and economic efficiency' (1961). This clever

functional form has proven to be both a theoretical and an empirical workhorse.

This remarkable burst of intellectual activity and creativity clearly established Robert M. Solow, by then aged 37, as *the* major figure in what was then the subfield of economic theory that was attracting the most attention (growth theory). And it won him the American Economic Association's John Bates Clark Medal, which is awarded every second year to the most outstanding economist under the age of 40, in 1961. But along the way he also found time to collaborate with Samuelson in bringing the Phillips curve to America in their, 'Analytical aspects of anti-inflation policy', *American Economic Review*, 1960. Whew!

The Solow model

Now back to 1956. Why is the Solow model so important in the history of economic thought? Prior to Solow's 'Contribution' and the contemporaneous paper by Trevor Swan (1956), growth theory was stuck in an awkward position. Roy Harrod (1939) and Evsey Domar (1946), noting the long-run (relative) fixity of the capital-output ratio, had posited a fixed-proportions technology:

$$Y_t = fK_t \tag{1a}$$

$$Y_t = gL_t, \tag{1b}$$

where f and g, two constants, are the reciprocals of the capital-output and labour-output ratios. Equating investment ($I_t - \frac{dK_t}{dt}$) and saving (S_t), and assuming that saving is proportional to income yields:

$$\frac{dK_t}{dt} = S_t = sY_t, \tag{2}$$

and appending an exogenously growing labour force,

$$L_t = L_0 e^{nt}, \tag{3}$$

leads to the simple Harrod–Domar growth model. The model is easily solved for a so-called balanced growth path in which output, capital, and labour are all growing at the same rate, so that ratios like $y = \frac{Y}{L}$ and $k = \frac{K}{L}$ are constant. By (3), that common growth rate must be n – or $n+\mu$, where μ is the rate of technical progress, in the presence of exogenous technological progress. (In that case, L must be measured in efficiency units.) By (2), the growth rate of capital is $\left(\frac{1}{K}\right)\left(\frac{dK}{dt}\right) = s\left(\frac{Y}{K}\right)$ which, by (1a) is just sf. So the fundamental Harrod–Domar equation equates the so-called natural rate of growth, n, to the so-called 'warranted' rate of growth, sf:

$$sf = n. \tag{4}$$

Output in an economy with a labour force growing at rate n can therefore grow at that same rate *indefinitely* as long as sf is equal to n. But therein lies the problem that bothered Solow. The product sf is a number, which is equal to n only by coincidence. What if they are unequal? If $sf > n$, K will grow faster than L *forever*; if $sf < n$, it will grow slower – also forever. The economy therefore seems to be poised on a knife-edge between explosive growth of y and k and implosive contraction. Only if sf happens to be equal to n is it capable of supporting steady growth.

To Solow, the child of the Depression, this was a most unsatisfactory state of affairs in modelling, because real economies do not behave that way. The Great Depression was such a noteworthy event precisely because it was so *unusual*. History teaches us that capitalist economies do not often either implode or explode. Something approximating steady growth is much more normal. Could all this really be the result of coincidence? Solow thought not.

Furthermore, he had long been interested in production theory and viewed factor proportions – at least at the aggregate level – as more variable than fixed. So he proposed making just one change in the Harrod–Domar model: replacing the fixed-coefficients technology (1) by a more conventional neoclassical production function:

$$Y = F(K, L) \tag{5}$$

which, under constant returns to scale, can be written:

$$\frac{Y}{L} = F(\tfrac{K}{L}, 1) \text{ or}$$
$$y = f(k). \tag{6}$$

Now the solution looks the same as in the Harrod–Domar model because saving (= investment) per head is $sf(k)$, which is constant along a steady-state growth path (defined as a path with constant $k = \frac{K}{L}$). The growth rate of the capital stock is therefore $\frac{\dot{K}}{K} = \frac{sf(k)}{k}$, and equating this to the growth rate of labour (n) leads to the famous Solow equation:

$$sf(k) = nk. \tag{7}$$

This resembles the Harrod–Domar equation, (4), and in fact, if $f(\cdot)$ is a linear function, the two are identical. But by making $f(\cdot)$ concave, Solow eliminated the bothersome knife-edge. Now no coincidence of constants is necessary to allow steady growth; instead, k will adjust to guarantee it, as illustrated by the famous Solow diagram (see Figure 1).

Many *complications* can be and have been added to the basic Solow model – some of them present already in Solow's original 1956 paper. But the two most essential and important lessons are extremely robust. First, the economy's growth path is most likely stable, not perched on a precarious knife edge. Consider the growth rate of $k = \frac{K}{L}$ which, in steady-state equilibrium, is zero.

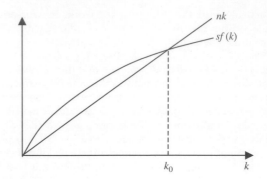

Figure 1

Out of equilibrium, it is:

$$\left(\frac{1}{k}\right)\frac{dk}{dt} = \left(\frac{1}{K}\right)\frac{dK}{dt} - \left(\frac{1}{L}\right)\frac{dL}{dt}$$
$$= \frac{sf(k)}{k} - n, \text{ so that}$$
$$\frac{dk}{dt} = sf(k) - nk. \tag{8}$$

Thus k will be *increasing* whenever $sf(k) > nk$ and *decreasing* whenever $sf(k) < nk$. A glance at Figure 1 (or taking the derivative of equation (8) with respect to k) shows that the dynamic model is *stable* – the reverse of the Harrod–Domar conclusion – as long as $sf'(k) < n$ (see again Figure 1.) Thus this new wrinkle achieved Solow's modelling objective: the economy normally stays on track.

But the Solow model accomplished more. Contrary to what had been common intuition until 1956 (and is still common intuition among those not schooled in the Solow model), the model shows that *the economy's growth rate* (n in this case, or $n + \mu$ with exogenous technical progress) *is independent of its saving rate* (s). How can that be true when, in the model, saving is the same as investment? The answer, which was surprising at the time but is now a commonplace, is that a society's propensity to save affects its *level* of output, as shown by equation (7), but not its steady-state *growth rate*.

Of course, this lesson can be learned too well. Figure 2 shows what happens if a Solow economy manages to *increase* its saving rate, from s_0 to s_1. The steady-state level of capital per head will eventually rise from k_0 to k_1, raising output per head from $f(k_0)$ to $f(k_1)$. Thus societies that save more will be richer, just as expected. But Solow pointed out that they *will not* grow faster in steady state; in fact, the growth rates are identical in the two equilibria shown in Figure 2.

However, we should remember that the transition from one steady state to another can take a very long time (see Atkinson, 1969). During the adjustment period from point A to point B, which may last for decades, the

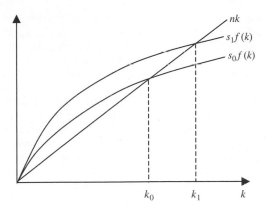

Figure 2

high-saving society will indeed grow faster than the low-saving society. It's just that diminishing returns to capital eventually catch up to it, bringing the growth rate back down to n. Furthermore, what later came to be called 'endogenous growth theory' argued that capital deepening (which often meant knowledge capital or human capital) can boost the *growth rate* forever, if returns are not diminishing.

The Solow residual

Solow's famous 1957 paper picked up on the diminishing returns point. It had two main purposes: to verify that, as an empirical matter, returns to capital are in fact diminishing; and to estimate the relative contributions of technological progress and capital deepening (rising k) to growth.

He began by adding what we now call 'Hicks-neutral' technical progress to the production function (5) to get:

$$Y = A_t F(K_t, L_t). \qquad (9)$$

Under constant returns and assuming marginal productivity pricing of factor inputs, Euler's theorem implies that:

$$\left(\frac{1}{Y_t}\right)\frac{dY_t}{dt} = \left(\frac{1}{A_t}\right)\frac{dA_t}{dt} + \beta\left(\frac{1}{K_t}\right)\frac{dK_t}{dt}$$
$$+ (1-\beta)\left(\frac{1}{L_t}\right)\frac{dL_t}{dt},$$

where β is capital's factor share. Subtracting the growth rate of labour from both sides leads to the 'other' famous Solow equation:

$$\left(\frac{1}{y}\right)\frac{dy}{dt} = \left(\frac{1}{A}\right)\frac{dA}{dt} + \beta\left(\frac{1}{k}\right)\frac{dk}{dt}, \qquad (10)$$

where $\left(\frac{1}{A}\right)\frac{dA}{dt}$ is what came to be called *the Solow residual*, that is, the portion of the growth of labour productivity that is not accounted for by capital deepening.

As an empirical matter, Solow (and those who have followed in his tradition) solved (10) for the residual:

$$\left(\frac{1}{A}\right)\frac{dA}{dt} = \left(\frac{1}{y}\right)\frac{dy}{dt} - \beta\left(\frac{1}{k}\right)\frac{dk}{dt}. \qquad (11)$$

Since everything on the right-hand side of (11) is directly measurable, this equation can be used to compute a time series on the growth rate of A. Doing so for the years 1909–1949 led Solow to the conclusion that only about one-eighth of the advance of $y = \frac{Y}{L}$ over those 40 years could be attributed to capital deepening, leaving about seven-eighths to 'the residual'. This surprising (at the time) conclusion has, of course, changed numerically over the years as Solow's technique was refined (by, for example, Edward Denison, 1962 and others) and replicated on newer data. But the qualitative finding that technological change is far more important than capital deepening has held up extremely well.

What about the shape of the aggregate production function (9)? Does it really display the curvature implied by diminishing returns? By using his synthetic time series on A_t, it was a simple matter of arithmetic to compute a time series on $f(k_t) = \frac{y_t}{A_t}$ (which is *not* constant in the real world) and then to inspect the shape of the $f(k)$ function. Eyeballing the scatter plot gave Solow 'an inescapable impression of curvature, of persistent but not violent diminishing returns' (1957, p. 318). And his regression estimates of a variety of nonlinear functional forms, including the Cobb–Douglas, rejected linearity. (In the Cobb-Douglas case, his original estimate of β was 0.35, not far from current estimates.)

One footnote to the history of economic thought is of interest here. Solow's empirical work included the years of the Great Depression and the Second World War, in contrast to the modern proclivity to omit those years. He noted that the data points for 1943–49 fell noticeably above the smooth estimated $f(k)$ function, however. Solow commented on that fact, hazarded a few guesses as to why the war years might have been abnormal, but ended up confessing that 'I leave this a mystery'. Honesty pays. Shortly thereafter Warren Hogan (1958) found a computational error in Solow's original paper and showed that, once the error was corrected, those seven data points were no longer aberrant.

Other work

I will be much briefer about some of Solow's other major works.

In a famous paper presented at the December 1959 meetings of the American Economic Association, Samuelson and Solow (1960) not only brought the Phillips curve to America but noted how what we now call a wage-Phillips curve can be transformed into a price-Phillips curve, and offered some reasons why that curve might not represent a stable menu of policy choices.

Shortly thereafter, Solow joined the 'New Frontier' as a staff member of President Kennedy's original Council of Economic Advisers in 1961–62 and continued as a consultant through the Kennedy-Johnson years. That was quite a staff, including the likes of Kenneth Arrow and Arthur Okun. As Solow put it years later, 'What I cherish about that experience was the conscious effort to use macroeconomic theory to interpret the world and, in a small way, change it' (1990, p. 192).

In 1973 Solow and I co-authored 'Does fiscal policy matter?' (1973), a widely discussed contribution to the then-raging monetarist-Keynesian debate. The article established several counterintuitive implications that stem from the so-called government budget constraint, one version of which is:

$$\frac{dM}{dt} + P_b \frac{dB}{dt} = G + iB - (Y + iB), \qquad (12)$$

where M is the money supply, B is the number of government bonds (paying interest rate i and selling for price P_b), G is government (non-interest) expenditure, and $T(\,\cdot\,)$ is the tax function. (While Blinder and Solow, 1973, used a fixed-price model, Tobin and Buiter, 1976, soon established parallel results for a full-employment model with a variable price level). The right-hand side of equation (12) is the budget deficit, and the left-hand side is the value of the money and/or bonds that must be issued to cover it. In steady-state equilibrium, neither M nor B is changing; so the budget must be balanced:

$$G + iB = T(Y + iB). \qquad (13)$$

Blinder and Solow used a dynamic IS-LM model with wealth effects to prove the following surprising results:

- The model is always stable under money financing of deficits; but it can be either stable or unstable under bond financing of deficits, depending on parameter values.
- *If* the model is stable under debt financing, then bond-financed deficits are actually *more* expansionary in the long run than money-financed deficits, in stark contrast to standard teaching in macroeconomics. (Remember, monetarists were arguing at the time that bond-financed deficit spending had no effects on aggregate demand.)
- An open-market purchase that creates more money is expansionary in the short run (for the usual reasons), but will be cancelled out in the long run by the government budget constraint.

While the first result is purely technical, the other two have interesting intuitions behind them. Other things equal, more wealth, whether in the form of M or B, leads to higher output. (This assumes that government bonds are net wealth to the private sector.) In the short run, money creation is more expansionary than bond creation for the usual reasons. But for that very reason the budget deficit closes more slowly (and therefore with more cumulative wealth creation) under bond financing, leading ultimately to a larger multiplier (provided the system is dynamically stable), which is the second result.

A similar dynamic adjustment explains the third result, which anticipated parts of the message of both Sargent and Wallace's 'Unpleasant monetarist arithmetic' (1981) and the so-called fiscal theory of the price level. Start from an initial steady-state equilibrium with a balanced budget. An expansionary open-market operation, which raises Y, brings in more tax revenue, thereby creating a budget surplus. Under pure money financing $\frac{dB}{dt} = 0$, so (12) implies that the surplus leads to money destruction. And (13) assures us that, with unchanged fiscal policy, the process of money destruction must continue until all of the new money has been withdrawn. Thus, in a real sense, the government budget constraint renders conventional monetary policy (an open-market operation) *impossible* in the long run.

In 1981 Solow teamed with Ian McDonald to produce a widely cited model of wage bargaining. The paper offered a fresh approach to a question that dates back to the earliest days of the Keynesian period: why, over business cycles, does employment fluctuate so much while real wages fluctuate so little? So-called implicit contract theories tend to predict that both employment (E) and real wages (w) will be stabilized; but empirical observation shows clearly that E is quite cyclical. McDonald and Solow used bargaining theory to explain the phenomenon of highly variable E combined with relatively stable w. Their arguments are lengthy and technical, and defy brief summarization. Suffice it to say that each of several specifications of how efficient bargaining might be conducted points toward this conclusion.

No summary of Robert M. Solow's contributions can be limited to reviewing his research, for Solow has also been a 'good citizen' *par excellence*. He has given his name, his wisdom, and his precious time to all manner of worthy causes and organizations, ranging from the Institute for Advanced Study to the Sierra Club. These include the National Academy of Sciences, the Manpower Demonstration Research Corporation, the Center for Advanced Study in the Behavioral Sciences, the Woods Hole Oceanographic Institution, the National Science Board, and the German Marshall Fund. There are many more.

Perhaps even more important are his remarkable achievements as a teacher of economics, particularly Ph.D. students. During the 1950s and early 1960s Samuelson, Solow, and a few of their friends created what came to be considered the world's pre-eminent economics department at MIT, an institution that had no great tradition in the subject. Generations of America's and the world's best graduate students came to study at MIT largely because Samuelson and Solow were there. While there are no objective measures of such things, Solow's clarity, wit, and mastery of a variety of subjects surely made him one of the finest teachers of economics who ever lived.

Perhaps for this reason, he was also the dissertation adviser of choice for scores of MIT's most promising graduate students over a period of time spanning 45 years. The list of Solow dissertation students, particularly in the 1960s, reads like an all-star team. In the two years 1966 and 1967 alone (based on completion dates), he supervised the Ph.D. dissertations of (in alphabetical order) George Akerlof, Robert Gordon, Robert Hall, William Nordhaus, Eytan Sheshinski, Joseph Stiglitz, and Martin Weitzman. Extending the time span back to 1956 and forward to 1971 brings in (now in chronological order) Alain Enthoven, Ronald Jones, Ronald Findlay, Peter Diamond, Ray Fair, Avinash Dixit, Jeremy Siegel, and the present author. More recently, Olivier Blanchard and David Romer were added to the list. It is a remarkable collection, for both its quality and its intellectual diversity. And probably every person on that list came to idolize Solow.

For most leading scholars, a list of students even half that extensive and distinguished would probably be considered their most enduring legacy. But in Robert M. Solow's case the research accomplishments are so fundamental, and have been and will be so enduring, that they eclipse even his amazing accomplishments as a teacher.

ALAN S. BLINDER

See also **endogenous growth theory; fiscal theory of the price level; Phillips curve (new views).**

Selected works

1956. A contribution to the theory of economic growth. *Quarterly Journal of Economics* 70, 65–94.

1957. Technical change and the aggregate production function. *Review of Economics and Statistics* 39, 312–20.

1958. (With R. Dorfman and P. Samuelson.) *Linear Programming and Economic Analysis*. New York: McGraw-Hill.

1960. Investment and technical progress. Proceedings of the First Stanford Symposium. In *Mathematical Methods in the Social Sciences, 1959*, ed. K. Arrow, S. Karlin and P. Suppes. Stanford, CA: Stanford University Press.

1960. (With P. Samuelson.) Analytical aspects of anti-inflation policy. Papers and Proceedings of the Seventy-second Annual Meeting of the American Economic Association. *American Economic Review* 50, 177–94.

1961. (With K. Arrow, H. Chenery, and B. Minhas.) Capital-labor substitution and economic efficiency. *Review of Economics and Statistics* 43, 225–50.

1973. (With A. Blinder.) Does fiscal policy matter? *Journal of Public Economics* 2, 319–37.

1981. (With I. McDonald.) Wage bargaining and employment. *American Economic Review* 71, 896–908.

1988. Autobiography. *Les Prix Nobel: The Nobel Prizes 1987*, ed. W. Odelberg. Stockholm: Nobel Foundation. Online.

Available at http://nobelprize.org/economics/laureates/1987/solow-autobio.html, accessed 17 June 2005.

1990. My evolution as an economist. *Lives of the Laureates: Ten Nobel Economists*, 2nd edn., ed. W. Breit and R. Spencer. Cambridge, MA: MIT Press.

Bibliography

Atkinson, A. 1969. The timescale of economic models: how long is the long run? *Review of Economic Studies* 36, 137–52.

Denison, E. 1962. *The Sources of Economic Growth in the United States and the Alternatives Before Us*. Supplementary Paper 13. Washington, DC: Committee for Economic Development.

Domar, E. 1946. Capital expansion, rate of growth and employment. *Econometrica* 14, 137–47.

Harrod, R. 1939. An essay in dynamic theory. *Economic Journal* 49(193), 14–33.

Hogan, W. 1958. Technical progress and production functions. *Review of Economics and Statistics* 40, 407–11.

Sargent, T. and Wallace, N. 1981. Some unpleasant monetarist arithmetic. *Federal Reserve Bank of Minneapolis Quarterly Review* 5(3), 1–17.

Swan, T. 1956. Economic growth and capital accumulation. *Economic Record* 32, 334–61.

Tobin, J. and Buiter, W. 1976. Long-run effects of fiscal and monetary policy on aggregate demand. In *Monetarism: Studies in Monetary Economics*, vol. 1, ed. J. Stein. Amsterdam: North-Holland.

Sombart, Werner (1863–1941)

Sombart was born in Ermsleben (Germany), the son of a well-to-do National Liberal member of the Prussian Diet. He studied economics, history, philosophy and law in Berlin, Pisa and Rome. In 1888 he received his Ph.D. from the University of Berlin and became an officer of the Bremen chamber of commerce. Two years later he was appointed extraordinary professor of political economy at the University of Breslau; in 1906 he became full professor at the Handelshochschule in Berlin and in 1917 transferred to the University of Berlin.

Sombart started his career as a left-wing advocate of social reform, influenced by Marxian theory. This was the reason why for a long time he could hold only second-rate positions within the German university system.

His dissertation on the economic and social conditions of the Roman campagna (1888) was brilliant and much less controversial than his later works.

An important work was his description of the German economy in the 19th century (1903). His outstanding study on the historical genesis of modern capitalism from its medieval origins to modern times (1902) may be considered Sombart's magnum opus. The really important edition was the second, published between 1919 and

1927, which differed completely from the first. Its three volumes treated three stages of capitalist development: early capitalism (Frühkapitalismus), high capitalism (Hochkapitalismus) – beginning with the industrial revolution in the 1760s – and late capitalism (Spätkapitalismus), starting with the First World War. The scope of this study was extremely broad. The reader is confronted with an amazing richness of facts. However, the data that Sombart presented were mostly second-hand and contained many speculative notions. Still, this work of 'unsubstantial brilliance' was 'highly stimulating even in its errors' (Schumpeter).

In the course of his research on *Der moderne Kapitalismus* Sombart published several special studies on the psychology and spirit of capitalism, on the Jews, on war and on luxury. In an outstanding study of the bourgeois individual in general and the entrepreneur in particular he analysed the emergence of a certain economically oriented mentality and psychology, the 'Wirtschaftsgeist' (economic spirit), for the development of capitalism (1913a).

In a voluminous work on the role of the Jews in economic life (1911) he described the Jews much along the lines of Max Weber's analysis of Puritanism, as the most dynamic part of the population, introducing 'capitalist spirit' into commerce and industry. At this time, Sombart was *not* anti-Semitic: on the contrary, he perceived the Jews' contribution to the rise of capitalism positively, regarded them as one of the most valuable 'species' of mankind (1912, p. 56) and was in favour of Zionism as the national renaissance of the Jewish people. However, the foundations for his later anti-Semitic turn were laid: his description of the characteristics of the Jews treating them as a 'species' was full of prejudices and exaggerations and included a discussion of their 'race' and 'blood' peculiarities.

Sombart further analysed the development of luxury consumption, which he connected in a very original way with the erotic, and its economic importance as a creator of new markets and industries (1913b, vol. 1). In a similar way Sombart treated war as a creator of new markets due to the services and goods required by the military (1913b, vol. 2). Two years later, during the First World War, Sombart wrote a chauvinistic, strongly anti-English book, which glorified war and militarism (1915).

The best way to observe his political views is to follow his discussion of Marx: Sombart was never a Marxist. But when the third volume of *Capital* appeared, he praised Marx as an outstanding thinker and described his theory in a very positive way (1894), which in turn was warmly welcomed by Engels. Also, in the first edition of his famous work on socialism and social movements (1896), he discussed Marx from a sympathetic point of view. Its tenth edition, now titled *Der proletarische Sozialismus (Marxismus)* (1924), was violently anti-socialist and full of hatred and personal insults against Marx. However, this did not hinder Sombart from stating three years later

that his *Der moderne Kapitalismus* was written in the Marxian spirit and that he regarded it in a certain way as the conclusion of Marx's work (1927, p. xix).

What had occurred was an evolution in Sombart's assessment of capitalist development. While he still appreciated Marx as a historian of capitalism, he now disliked the latter's optimistic view of the future, his regard for capitalism as the creator of the better world to come (1927, p. xx). Marx had been an admirer of technical progress and of the historical forces which fostered it. Sombart no longer believed that industrial development was automatically beneficial and he realized its destructive potential. He contrasted the uniformity and ugliness of modern civilization with the cultural variety of the pre-industrial past.

This enmity towards industrial development and what he called the 'economic age' was the reason why Sombart sought to ally himself with right-wing anti-capitalism and temporarily turned to fascism. When the Nazis came to power he made a contribution towards a programme of German (National) Socialism (1934). Contrary to proletarian socialism, which accepted the industrial society and only intended to redistribute its surplus, Sombart perceived German socialism as rejecting the industrial age (1934, pp. 160–8). Beside the corporative state, the Führerprinzip (leader principle), a state interventionist regulation of the German economy, autarky and a partial re-agrarianization of Germany, he advocated state planning and, as a key idea, control of technological development based on what would now be called technology assessment (1934, pp. 263–7). However, his proposals were not welcomed by the Nazis, who, as Sombart was later to realize, intended to use the most advanced technologies available in order to win political hegemony.

It is not easy to regard Sombart as member of an economic school. He first rejected and then accepted the deductive method. Although he claimed that *Der moderne Kapitalismus* bridged the gap between the abstract-theoretical and the empirical-historical method (1927, p. xvi), there is no doubt that this book lacked an analytical framework and, though not without theory, was mainly a historical description of a large mass of facts. Sombart even 'out-Schmollered Schmoller' (Schumpeter) and had to be regarded as belonging to the younger, or, as Schumpeter called it, the 'youngest' historical school. Sombart himself, however, would certainly not have admitted that. He distinguished between three kinds of political economy (1930): the 'richtende Nationalökonomie' (judging economics), which was intended to decide what was right and wrong, and whose scientific character Sombart denied; the 'ordnende Nationalökonomie' (ordering or systematizing economics) that tried to apply quantitative exact methods, and, finally, the 'verstehende Nationalökonomie' (understanding or interpretative economics), which should be a 'Geistwissenschaft' (science of social mind) and was both theoretical and historical and tried to grasp the motives of economic life. It goes without saying that Sombart regarded his own work as being in the tradition of the latter.

Sombart extended his claims to being a theorist beyond economics. He also perceived sociology as a 'Geistwissenschaft' and developed a new type of sociology, which he called 'Noo-Soziologie', which was supposed to be a general theory of culture (1936).

He remains one of the most brilliant and interesting personalities of the German economics profession, being a gifted writer with profound historical insights. While many works of the Historical School are boring collections of facts, the best of Sombart's books are sparkling and still make fascinating reading.

B. SCHEFOLD

See also **Historical School, German.**

Selected works

1888. Über Pacht- und Lohnverhältnisse in der römischen Campagna. Ph.D. dissertation, Berlin.
1894. Zur Kritik des ökonomischen Systems von Karl Marx. *Archiv für soziale Gesetzgebung und Statistik* 7, 555–94.
1896. Sozialismus und soziale Bewegungen im 19. *Jahrhundert.* Jena: Gustav Fischer. Trans. as *Socialism and the Social Movement*, London: Dent; New York, Dutton, 1909. 10th edn published as *Der Proletarische Sozialismus* (*Marxismus*), 1924.
1902. *Der moderne Kapitalismus: Historisch-systematische Darstellung des gesamteuropäischen Wirtschaftslebens von seinen Anfängen bis zur Gegenwart.* 2 vols, Leipzig. 2nd edn, 3 vols, Munich and Leipzig: Duncker & Humblot, 1916–27.
1903. *Die deutsche Volkswirtschaft im neunzehnten Jahrhundert.* Berlin: Georg Bondi.
1909. *Das Lebenswerk von Karl Marx.* Jena: Gustav Fischer.
1911. *Die Juden und das Wirtschaftsleben.* Leipzig: Duncker & Humblot. Trans. as *The Jews and Modern Capitalism*, London: T.F. Unwin, 1913 .
1912. *Die Zukunft der Juden.* Leipzig: Duncker & Humblot.
1913a. *Der Bourgeois: Zur Geistesgeschichte des modernen Wirtschaftsmenschen.* Munich and Leipzig: Duncker & Humblot. Trans. and ed. M. Epstein as *The Quintessence of Capitalism: A Study of the History and Psychology of the Modern Business Man*, New York: Dutton, 1915.
1913b. *Studien zur Entwicklungsgeschichte des modernen Kapitalismus.* Munich and Leipzig: Duncker & Humblot. Vol. 1: *Luxus und Kapitalismus.* Trans. into English as a report under the auspices of the Works Progress Administration and the Department of Social Science, Columbia University, Project number 465–97–3–81. Vol. 2: *Krieg und Kapitalismus.*
1915. *Händler und Helden. Patriotische Besinnungen.* Munich and Leipzig: Duncker & Humblot.
1930. *Die drei Nationalökonomien.* Munich and Leipzig: Duncker & Humblot.

1934. *Deutscher Sozialismus.* Berlin: Buchholz & Weisswange. Trans and ed. K.F. Geiser as *A New Social Philosophy*, Princeton: Princeton University Press; Oxford: Oxford University Press, 1937.
1936. *Soziologie: Was sie ist und was sie sein soll.* Berlin: Sitzungsberichte der Preussischen Akademie der Wissenschaften.

Bibliography

Crosser, P.K. 1941. Werner Sombart's philosophy of National-Socialism. *Journal of Social Philosophy* 6, 263–70.
Mitchell, W.C. 1929. Sombart's Hochkapitalismus. *Quarterly Journal of Economics* 43 February, 303–23.

Sonnenfels, Joseph von (1733–1817)

Sonnenfels was born of Jewish parents who shortly afterwards converted to Catholicism; the family moved in 1744 from Moravia to Vienna, where the father taught oriental languages. Joseph first served in the army from 1749 to 1754, when he began to study law and literature at Vienna University. A prominent member of the Enlightenment literati, he was in late 1763 appointed to the newly founded chair in 'Police and Cameralistic Sciences' at the University of Vienna. Until his death he was prominent in constitutional reform, also engaging in a campaign for the abolition of torture and of usury. The textbook which he wrote for his own teaching, the *Grundsätze der Polizei Handlungs- und Finanzwissenschaft* (1765, 1769, 1776), remained the official text in the Austrian Empire until 1848, running to eight editions and several abbreviated teaching editions.

The *Grundsätze* devotes a volume each to police, commerce and finance. The leading idea running through all three is the importance of a large population gainfully employed for the general welfare of the state. Coupled with this is a conception of the accumulation of wealth as the 'multiplication of means of subsistence', governed however by the necessity of maintaining equilibrium in the society, which is the task of police. The functioning of police is therefore tied less directly to economic welfare than is the case with Justi, to whom Sonnenfels makes reference. The treatment of commerce owes a great deal to Forbonnais' *Elémens du commerce*, who also laid emphasis on the advantages of a large population and the need for proportion within it. But while Sonnenfels takes much from Forbonnais, he remains more concerned with general political order than the economic structure of an advancing society.

K. TRIBE

Selected works

1765, 1769, 1776. *Grundsätze der Polizei, Handlungs- und Finanzwissenschaft.* 3 vols. Vienna: Camensina. 8th and final edn, 1818–22.

South Sea bubble

The South Sea Company was founded in 1711, in the expectation that peace between Spain and England after the end of the War of the Spanish Succession would produce profitable trading opportunities with the 'South Seas' (that is, Spanish America). The company's trading activity remained intermittent and unprofitable throughout the 1710s. In 1719, a new scheme was launched – the conversion of government debt into equity of the South Sea Company. Debt-holders of the 1710 lottery loan were offered the option to convert their holdings into company shares. The government agreed to make interest payments to the company instead of to debt-holders. As old (and illiquid) loans were swapped for liquid company shares, debt-holders gained. The government negotiated a lower rate of interest, and the South Sea Company made a modest profit. The 1719 equity-for-debt swap is generally seen as Pareto-improving.

The 1720 conversion scheme differed in important ways. Key elements included (*a*) the absence of a fixed conversion ratio – higher prices of South Sea stock meant that more debt could be bought with each share, (*b*) issuance of new stock on instalment, with only a small down payment required, (*c*) massive lending against shares, and (*d*) a high degree of corruption in the awarding of the contract. The South Sea conversion also shared important characteristics with John Law's Mississippi scheme in France, which produced a similar run-up (and crash) of prices half a year earlier.

Both the Bank of England and the South Sea Company competed for the contract to convert government bonds into equity. After bribes to MPs, ministers, and members of the court (of about £1.3 million), the South Sea Company won the right to perform the conversion in March 1720. By this time, the price of its shares had increased to 255, from 128 at the beginning of the year. The share prices of other companies moved up and down in parallel with South Sea stock, but less sharply (see Figure 1). The company proceeded to issue fresh shares in four subscriptions, and offered to convert debt into shares on (modestly) generous terms. By late June, prices had risen to 765, and forward prices during the summer rose as high as 950. When regular trading resumed, prices began to weaken, but the fourth subscription was still strongly oversubscribed. In September, prices fell quickly. By the year end they had almost declined to their January level.

Interpretations

Since Mackay's classic *Extraordinary Popular Delusions and the Madness of Crowds*, the South Sea bubble has often been cited as a prime example of irrational investor behaviour. In contrast, Peter Garber (2000) argued that the share prices increased in line with 'changing view(s) of market fundamentals'. If the scheme had succeeded in improving economic conditions in England as a whole (as John Law's logic would predict; Verde, 2004), the

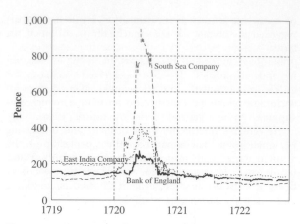

Figure 1   Share price of major English companies, 1718–22. *Source*: Neal (1990); data from ICPSR, Study No. 1008.

firm's large capital base might have allowed it to pursue profitable ventures. Yet most of these remained vague, and the company had no track record of successfully making money from anything other than financial transactions. It is doubtful whether future profits could ever have been high enough to justify the company's market capitalization in the summer of 1720. Even Garber accepts that prices above 400 are hard to square with reasonable expectations of future profits. Easy credit, investor preferences for lottery-like payoffs (as a result of shares being sold with only a small down payment), and restricted free float (caused by company lending against its own shares) may have contributed to the start of the bubble.

Recent work has focused on the reasons why the bubble, once under way, could have expanded greatly. Dale (2004) argues that apparent mispricing of subscription receipts proves investor irrationality, while others have argued that the gap can be explained by the option-like nature of receipts. Temin and Voth (2004) examined the trades of a goldsmith bank, Hoare's, which made large profits buying and selling South Sea stock in 1720. They argue that the bank was aware of the overpricing, but invested in South Sea stock regardless. Predictability of investor sentiment made it rational to 'ride' the bubble, and to sell out with a profit as soon as it began to deflate. This strategy is similar to hedge fund behaviour on Nasdaq in the late 1990s (Brunnermeier and Nagel, 2004). If other large investors faced similar incentives, the lack of a coordinated early attack becomes easier to understand. The role of market microstructure imperfections was probably limited, as opportunities to sell short were abundant. However, the nature of the settlement process and the artificial reductions of free float engineered by the company may have contributed to the bubble.

Consequences

The rise and crash of share prices in 1720 had few direct economic consequences. As prices declined, former debt-holders demanded compensation. Parliament investigated the scheme in which it had played such an important role. Directors had most of their assets expropriated. In contrast to the resolution of the Mississippi bubble in France, those who had tendered government bonds for company shares received partial compensation in the form of fresh government debt. The political consequences were possibly more formidable than the immediate economic repercussions. Leading politicians who had taken bribes, such as the Chancellor of the Exchequer, John Aislabie, were forced out of office and incarcerated. Robert Walpole, sometimes referred to as England's first prime minister, distinguished himself both through his opposition to the scheme and competent handling of its fallout. He succeeded Aislabie at the Exchequer and remained in power until 1742.

The collapse of the South Sea bubble is sometimes seen as a factor behind the Bubble Act. This appears to be erroneous, as the Act was passed before the bubble deflated (Carswell, 1993). Its passage and rigorous enforcement after the summer of 1720 probably owed more to the company's efforts to support its own sagging share price. Because of the Act, new equity issues became very rare for almost a century. The Act was repealed only in 1825.

HANS-JOACHIM VOTH

See also **bubbles; Law, John.**

Bibliography

Brunnermeier, M. and Nagel, S. 2004. Hedge funds and the technology bubble. *Journal of Finance* 59, 2013–40.
Carswell, J. 1993. *The South Sea Bubble.* Stroud: Sutton.
Dale, R. 2004. *The First Crash.* Princeton: Princeton University Press.
Garber, P. 2000. *Famous First Bubbles: The Fundamentals of Early Manias.* Cambridge, MA: MIT Press.
Mackay, C. 1995. *Extraordinary Popular Delusions and the Madness of Crowds.* San Francisco: Essential Library.
Neal, L. 1990. *The Rise of Financial Capitalism: International Capital Markets in the Age of Reason.* New York: Cambridge University Press.
Temin, P. and Voth, H.J. 2004. Riding the South Sea bubble. *American Economic Review* 94, 1654–68.
Verde, F. 2004. Government equity and money: John Law's system in 1720 France. Working paper, Federal Reserve Bank of Chicago.

sovereign debt

A straightforward definition of sovereign debt is simply that it is debt issued by a sovereign state. Sovereignty conveys supreme legal authority within the geographical boundaries of the nation, giving national authorities autonomy over the regulation of economic activity inside the country through legislation, administration, and judicial enforcement. This means that foreign governments cannot interfere with economic activities within sovereign boundaries and cannot enforce contractual relationships without the cooperation of national authorities. Sovereignty denies creditors the right to reach inside national borders to confiscate assets or attach sources of revenue, public or private, to satisfy outstanding debts.

The importance of sovereignty for international capital markets is evidenced by the frequency of sovereign debt crises and defaults in history. Over the last century, debt crises have been associated with default on international debt by governments in emerging market economies, but between the 16th and 19th centuries, sovereign default was similarly frequent in Continental Europe (Reinhart, Rogoff and Savastiano, 2003, survey the history of default). Although international debt is at the centre of recent crises, the economics of sovereign debt apply to debt issued in either international or domestic markets held by either foreign or domestic residents.

Repayment incentives

The first concern of the analytical literature on sovereign debt is what motivates borrowers to repay. When the terms of a debt contract cannot be enforced directly by a court, a sovereign debtor can be expected to honour its debts only if it faces adverse consequences for default. The literature focuses on the possibility of sanctions for debt repudiation or default which include the disruption of international trading opportunities by foreign governments or private market participants. For example, a repudiating debtor could face a trade embargo, suspension of trade preferences or loss of access to international capital markets. Each is a case of reduced access to international trade, whether trade takes place contemporaneously or across time.

A consequence of the observation that indirect sanctions provide the incentive to repay debt is that lending to sovereigns is constrained by the willingness of the debtor to repay rather than by its ability to repay. The concept of willingness to pay was first applied to sovereign debt in an analytical model by Eaton and Gersovitz (1981) following its use in earlier writings (for example, by Wallich, 1943). Eaton and Gersovitz emphasize that repayments are made out of the enlightened self-interest of the borrower and that debt limits are determined by the expected present value of the cost to the debtor of sanctions that will be imposed if debt is repudiated. They demonstrate two versions based on punishments given by disruptions of commodity trade and credit embargoes, respectively. Some immediate implications of the Eaton and Gersovitz model are that credit is rationed in equilibrium and that more severe sanctions increase loan supply.

It is helpful to compare lending sustained by the threat of indirect sanctions to conventional domestic lending with collateral. The use of collateral relies on a legal system to enforce debtors' property rights in collateral assets and to transfer the ownership of these assets to creditors in specific events. Under a collateralized loan, the debtor repays to avoid losing collateral, and a solvent debtor should do so only if the value of the collateral exceeds the required repayments. The value of loan collateral determines a non-sovereign debtor's willingness to pay. In a time-consistent equilibrium, lenders should not lend more than the discounted present value of the assets that secure the loan.

Sovereign borrowing is analogous to legally enforced borrowing with collateral if the imposition of sanctions subsequent to default is exogenous. In the first version of the Eaton and Gersovitz (1981) model, static trade sanctions are imposed exogenously. One insight of the paper is that indirect penalties such as the disruption of commodity trade require the cooperation of third parties. Lenders, public or private, can count on repayment only if sovereign nations impose barriers to a recalcitrant debtor's trade.

Eaton and Gersovitz consider how repayment incentives might arise within the credit relationship itself using the threat of a credit embargo in response to debt repudiation. Lending and repayment are derived as perfect equilibrium outcomes in an extensive-form game in which a risk-averse borrower with random income seeks to smooth consumption over time. The borrower repays if the net gain from future credit access is at least as large as the utility cost of the current repayment. The gains from credit access depend on the amount that creditors will lend which in turn depends on the expected future repayments by the borrower. Perfect equilibrium is sustained in this model by the threat of permanent loan autarky. The strategies of creditors are restricted so that this punishment is imposed after a borrower defaults in any amount. Therefore, borrowers either repay in full or repudiate their entire debt.

The restriction to a simple choice between full repayment and debt repudiation is one drawback of the basic model. If the gains from international trade or credit market access are stochastic, then allowing state-contingent repayments can raise welfare for borrowers or lenders. Grossman and van Huyck (1988) argue that default and debt renegotiation may result from an implicit state-contingent contract guided by a standard loan contract with fixed repayments. Several authors model debt renegotiation as the outcome of an implicit state-contingent contract with stochastic gains from trade.

Bulow and Rogoff (1989a) analyse a repeated model of bargaining over repayments supported by the threat of trade sanctions. By making an initial loan, a creditor acquires the right to invoke a trade embargo if the debtor fails to repay. Each period the creditor and debtor bargain over the amount paid by the debtor to avoid an embargo. This provides a foundation for the exogenous penalty in the first version of the Eaton and Gersovitz

model. The repayment is a price paid by the debtor to purchase the right to impose sanctions from the creditor for one period at a time. The enforcement of property rights is essential to this model. The sovereign initially has the right to free trade which it sells for the price of the loan to the creditor. The creditor then rents access to free trade back to the sovereign debtor. Thus, the Bulow and Rogoff model parallels lending with collateral in that it requires the enforcement of property rights by disinterested parties.

The possibility that lending to sovereigns might be enforced by market participants themselves by denying credit market access to defaulting borrowers is addressed by several authors. Bulow and Rogoff (1989b) question the force of reputations-based punishments for sovereign default and argue that official imposition of sanctions or enforcement of creditor rights is necessary. In particular, they argue that permanent loan autarky as proposed by Eaton and Gersovitz in the consumption-smoothing model of debt does not deter sovereign default when debtors can accumulate foreign assets. They assume that a sovereign borrower can purchase a foreign interest-bearing deposit which it can draw against in default. Under this assumption, a sovereign borrower raises its welfare by defaulting and saving the amount it would have repaid in a foreign deposit. In this model, international capital flows are sustainable in perfect equilibrium, but only if the sovereign lends rather than borrows. An interpretation of Bulow and Rogoff is that reputations alone cannot sustain sovereign borrowing.

This result received a fair amount of attention because it implies that sovereign borrowing cannot be self-enforcing and that earning a reputation for default is insufficient to deter opportunistic debtor behaviour. The special assumption of the paper is that foreign financial obligations to a government are enforceable while its financial obligations are not. The foreign creditors can commit to make future payments to the sovereign even when it is not in their self-interest to do so as events unfold. Contrary to the interpretation of Bulow and Rogoff, the capacity to commit requires exogenous enforcement by another sovereign state.

Kletzer and Wright (2000) model capital flows for smoothing a sovereign borrower's consumption in the absence of any exogenous means of contract enforcement and show that asymmetric commitment is essential for the conclusion of Bulow and Rogoff (1989b). The paper demonstrates self-enforcing equilibrium in an economy that treats both sides of the market as sovereign supported by punishments proof to renegotiation. Permanent credit embargoes are not credible threats because the gains from risk-sharing that motivate lending in the first place give incentives for forgiveness and new lending. Kletzer and Wright prove that sovereign borrowing is sustainable with free entry by new counterparties, as envisioned by Bulow and Rogoff, so that reputations alone are sufficient to enforce repayments.

The general result is that credit markets are possible under the anarchy that characterizes international relations between sovereigns.

Kletzer and Wright show that the equilibrium is implemented by one-period loan contracts with state-contingent repayments that are not negative. State-contingent repayments are interpreted as the outcome of an implicit contract guided by non-contingent loan contracts following Grossman and van Huyck (1988) and Bulow and Rogoff (1989a). Non-commitment is exactly the assumption that any net payment (a new loan) made by a lender to the sovereign is voluntary. Introducing one-sided commitment allows conventional insurance, while the absence of commitment is consistent with renegotiation of conventional loans. Kletzer and Wright show that the opportunistic acceptance of defaulting borrower's deposits can only eliminate sovereign borrowing if international insurance contracts are exogenously enforced by other sovereigns.

A different approach to demonstrating the self-enforcement of sovereign borrowing through reputations is taken by Cole and Kehoe (1998). In a model with asymmetric information, actions reveal information about the borrower to counterparties in other financial and trading relationships. The borrower then cares about reputational spillovers to these other relationships (this possibility is noted by Bulow and Rogoff, 1989b). They demonstrate equilibrium repayment in the presence of one-sided enforcement of the sovereign's own foreign loans.

Sovereigns may enforce contractual obligations of domestic borrowers to foreigners because they recognize national benefits from doing so. Sovereign states set the rules inside national borders so that enforcement of private obligations is at the discretion of the sovereign. Until the 1950s, sovereign immunity applied with respect to assets held abroad. A commercial creditor could not pursue legal remedies against a sovereign debtor without its consent in either the United States or UK. By the mid-1970s both countries adopted more restrictive legal theories of sovereign immunity that allow private parties to seek remedy in home courts if the dispute concerns commercial transactions (Brownlie, 2003). Contemporary emerging market bond issues are overwhelmingly issued under the governing law of a few financial centres (primarily, New York and London) and incorporate waivers of sovereign immunity, thus voiding one-sided commitment opportunities.

Insights about the credibility of sanctions can be gained from the historical experience of sovereign default. Although trade sanctions were imposed on some sovereign defaulters before the 1930s, creditor nations appear reluctant to interfere with the international trade of countries in default (see, for example, Eichengreen and Portes, 1989). Ozler (1993), Lindert and Morton (1989) and others find that historically default had limited impact on loan terms for sovereign debtors, suggesting that credit market sanctions may be weak. Short-lived punishment or exclusion from borrowing while existing debts are renegotiated and settled, however, is consistent with credible punishment. Esteves (2005) studies debt renegotiation between 1870 and 1913 and finds that capital market access was disrupted during renegotiation but regained after resolution was reached. Many recent financial crises and debt restructurings conform to this pattern.

Debt restructuring and current issues

Debt renegotiation and restructuring is central to achieving gains from international risk sharing with limited financial instruments. The protracted and costly process of sovereign default resolution and debt restructuring remains the subject of policy concern and debate (for a review, see Eichengreen, 2003). The difficulty of achieving the acquiescence of a large number of creditors to a sovereign debt restructuring agreement has been repeatedly demonstrated for more than a century.

Tirole (2002) identifies the interaction between problems of common agency in lending and debtor sovereignty with inefficiencies in international capital markets that produce crises. He observes that debtor sovereignty impedes solutions adopted in domestic financial markets to the externalities between various creditors due to common agency in lending. With legal enforcement of contracts, bond and loan covenants can establish property rights across heterogeneous creditors in the event of default so that individual creditors internalize the impact of their actions on other creditors. In corporate debt markets, the power of the sovereign can be used to mitigate moral hazard on the part of both the debtor and its creditors, in contrast to the case of sovereign debt.

One of the problems associated with common agency in sovereign lending is the difficulty of achieving collective action among creditors for resolving default. In default or bankruptcy, an important role of court proceedings is to aggregate various debts and resolve conflicting claims between creditors to reach completion of financial restructuring. For a sovereign debtor, creditors who accept reduced repayments in default increase the value of debt held by creditors who do not participate. Free-riding of this type is related to the public goods nature of the capacity to impose sanctions. Any debt rollover, reduction or exchange can be prone to hold out creditors prolonging and contributing to crises. A minority of creditors can also interfere with a debt restructuring by exercising their legal rights under the governing law under which bonds were issued.

The collective action problem for sovereign debt renegotiation (and, hence, lending to sovereigns) has been addressed in various ways over time. Prominent approaches include the formation of bondholder committees, representation of creditors by banks, and official sector intervention. In recent years, two approaches, neither new, have received considerable attention. One of these concerns contractual innovation. An example of the issue is the ability of individual creditors to block debt swaps under unanimous consent clauses required for US corporate borrowing. Collective action clauses allowing less than unanimous consent to restructurings were included in major sovereign debt issues in the United States after active encouragement. An analysis of these clauses is given by Eichengreen, Kletzer and Mody (2004), and the possibility of voiding benefits for hold-outs under unanimous action clauses is discussed by Buchheit and Gulati (2000). The other approach is statutory innovation which may challenge debtor sovereignty. Although such proposals have been advanced over the years, the prospects for an international debt restructuring mechanism are diminished for now in favour the pursuit of market-based innovation.

KENNETH M. KLETZER

Bibliography

Brownlie, I. 2003. *Principles of Public International Law*, 6th edn. New York: Oxford University Press.

Buchheit, L. and Gulati, G. 2000. Exit consents in sovereign bond exchanges. *UCLA Law Review* 48, 59–84.

Bulow, J. and Rogoff, K. 1989a. A constant recontracting model of sovereign debt. *Journal of Political Economy* 97, 155–78.

Bulow, J. and Rogoff, K. 1989b. LDC debt: is to forgive to forget? *American Economic Review* 79, 43–50.

Cole, H. and Kehoe, P. 1998. Models of sovereign debt: partial versus general reputations. *International Economic Review* 39, 55–70.

Eaton, J. and Gersovitz, M. 1981. Debt with potential repudiation: theory and estimation. *Review of Economic Studies* 48, 289–309.

Eichengreen, B. 2003. Restructuring sovereign debt. *Journal of Economic Perspectives* 17(4), 75–98.

Eichengreen, B., Kletzer, K. and Mody, A. 2004. Crisis resolution: next steps. *Brookings Trade Forum 2003*, 279–337.

Eichengreen, B. and Portes, R. 1989. Setting defaults in the era of bond finance. *World Bank Economic Review* 3, 211–39.

Esteves, R. 2005. Quis custodiet quem? Sovereign debt and bondholders' protection before 1914. Mimeo, Department of Economics, University of California, Berkeley.

Grossman, H. and van Huyck, J. 1988. Sovereign debt as a contingent claim: excusable default, repudiation, and reputation. *American Economic Review* 78, 1088–97.

Kletzer, K. and Wright, B. 2000. Sovereign debt as intertemporal barter. *American Economic Review* 90, 621–39.

Lindert, P. and Morton, P. 1989. How sovereign debt has worked. In *Developing Country Debt and the World Economy*, ed. J. Sachs. Chicago: University of Chicago Press.

Ozler, S. 1993. Have commercial banks ignored history? *American Economic Review* 83, 608–20.

Reinhart, C., Rogoff, K. and Savastano, M. 2003. Debt intolerance. *Brookings Papers on Economic Activity* 2003(1), 1–74.

Tirole, J. 2002. *Financial Crises, Liquidity and the International Monetary System*. Princeton, NJ: Princeton University Press.

Wallich, H. 1943. The future of Latin American dollar bonds. *American Economic Review* 33, 324–35.

Soviet and Russian research, development and innovation

Understanding the functioning and performance of the Soviet R&D system and the innovation-averse character of the planned economy has posed a challenge to economists and science policy specialists since the 1950s. It was the 'sputnik' shock of 1957 that triggered serious analysis of the Soviet R&D system. Initial research (De Witt, 1961; Korol, 1965) focused on its scale in terms of personnel and on its basic organizational and behavioural characteristics, a landmark being the 1969 Organisation for Economic Co-operation and Development (OECD) report on science policy in the USSR (Zaleski et al., 1969). This report highlighted the organizational separation of research from production, the dominant role not only in basic research but also in much applied work of the USSR Academy of Sciences, which played a central role in the overall science policy of the country and occupied a position of prestige in society, and the relatively modest role in R&D of the higher educational sector. Later works contributed to a better understanding of the Soviet research system, its social context and the influence on science and technology of the Communist Party (for example, Kruse-Vaucienne and Logsdon, 1979; Lubrano and Solomon, 1980; and Fortescue, 1986).

By this time there was an appreciation that, notwithstanding the scale of the R&D system, somewhat overstated by Soviet official statistics, overall innovative performance was poor. A pioneering attempt to analyse the economics of innovation was that of Joseph Berliner (1976). For Berliner, Soviet industry was characterized by an all-pervasive lack of responsibility, risk aversion and weak incentives for successful innovation, coupled with mild penalties for failure in the absence of competition. Attempted economic reforms had achieved little to improve the situation, and Berliner was not optimistic that any major structural change would boost innovative performance. The outcome of inadequate innovation was the focus of work undertaken by researchers at the University of Birmingham (Amann, Cooper and Davies, 1977). This demonstrated that from the mid-1950s to the mid-1970s in many sectors of the economy there had been no diminution in the technological gap between the Soviet Union and leading industrial economies. It was found that the rate of diffusion of major new technologies tended to be slower than the rates typical of advanced market economies. Further research into industrial innovation (Amann and Cooper, 1982) built on the findings of Berliner, but with more attention to the specific historical conditions in which particular industrial sectors developed. While the Soviet economic system was by its nature averse to innovation, some sectors showed better performance than others, above all the defence industries, where high-level political intervention and priority resource allocation to some extent overcame the inertia of dysfunctional economic institutions. That these institutions and their implications for the development of science and technology had deep roots was shown by the pioneering work of Bailes (1978) and Lewis (1979). Further insights into poor performance in the adoption of innovations was later provided by the application of a principal–agent framework, notably by Dearden, Ickes and Samuelson (1990).

In the late 1960s and 1970s, recognizing the limits of the domestic civil R&D and innovation systems, the Soviet authorities attempted to boost technological progress by importing the latest technology from the West, with the automobile industry as a pioneer (Holliday, 1979). The results were disappointing. As shown by Hanson (1981), the Soviet system was unable to reap the productivity gains potentially available and was incapable of reproducing the technology itself to diffuse the achievements more widely.

Soviet R&D was heavily militarized. At various times the authorities attempted to harness the resources and skills of the military sector to boost performance in the civil economy. However, 'spin-off' from the defence sector was modest; and when a policy of 'conversion' was pursued in the final years of the Soviet regime under General Secretary Gorbachev the practical results were modest because of organizational barriers, inadequate incentives and a marked quality gap between the military and civilian sectors.

In terms of personnel and spending, the USSR had a research system of significant size. However, appreciation of its true scale and the productivity of Soviet science was complicated by data problems. According to Soviet official statistics, in 1990 expenditure on science was five per cent of national income. However, this figure was inflated by double counting. Researchers of the Centre for Science Research and Statistics (CSRS), Moscow, concluded that for Russia in 1990 the actual share was two per cent of GDP. Similarly, official data overstated the number of R&D personnel, partly because employment was not expressed in terms of full-time equivalents.

In the USSR the innovation process was always understood, implicitly by government officials and often explicitly by economists, as a linear process; that is, new products and processes were developed on the basis of ideas and inventions originating in basic and applied research, and then development, after which they were 'introduced' into the sphere of production and then diffused more widely. Only in the very final years did some analysts become aware of the work of Chris Freeman and other Western science policy specialists who challenged the linear model and argued for a richer understanding involving feedback relationships. However, in present-day Russia the linear model is still widely met and increased spending on science often presented as the principal means of promoting innovation.

Since the collapse of the USSR, there has been no attempt by either Russian or other scholars to revisit the Soviet R&D and innovation system in a comprehensive manner. However, the work of Harrison (2003; 2005) has provided some additional insights into the system during the Stalin era, with a focus on principal–agent problems and incentives. The analysis undertaken before the end of the system drew upon the theoretical advances of Janos Kornai (1980; 1992) only to a limited extent (for example, Hanson and Pavitt, 1987), in particular Kornai's analysis of the role of soft budget constraints. The pioneering work of the late Yurii Yaremenko, perhaps the most innovative of all economists working in the USSR in its final 20 years, was also disregarded (in particular, Yaremenko, 1981). Yaremenko analysed the Soviet economy in terms of a hierarchical, multilevel system, with each level having access to resources, human and material, differentiated by quality. The qualitative heterogeneity of resources became firmly institutionalized. An individual enterprise, or entire branch of production, could rise up the hierarchy and secure access to higher-quality resources only by a policy decision of the political–economic authorities. Lower-level sectors habitually deprived of high-quality resources compensated for their lack by resort to larger quantities of lower-quality inputs. At the upper levels, occupied by defence industries and some priority civil sectors, higher-quality resources permitted the use and development of more advanced technologies, but innovations possible in these privileged conditions were unsuited to diffusion to lower levels of the economy lacking an appropriate resource environment for their successful application.

The collapse of Communism at the end of 1991 had a profound impact on Russian R&D and innovation. Positive factors included the end of ideological controls and censorship, as well as a new freedom for scientists to travel abroad. But for most of the 1990s negative factors predominated: funding collapsed, the number of researchers contracted sharply, many institutes of the Academy of Sciences and industry developed non-research activities in order to survive, and many talented younger scientists found work abroad (Glaziev and Schneider, 1993; Schneider, 1994; Gokhberg, Peck and Gács, 1997). A comparison of the situation in Russia in 2004 with that of 1990, after 15 years of transformation, shows the following principal features: R&D expenditure as a percentage of GDP was 1.35 compared with 2.03, although it reached a low point of 0.74 in 1992; in real terms R&D spending was 42 per cent of the 1990 level; and the number of researchers had fallen to 401,425 compared with 992,571, the decline moderating but still under way. In 2004 almost 50 per cent of all researchers were 50 years old or more, compared with 35 per cent in 1994 (CSRS and Roskomstat data). A landmark was the OECD (1994) review of Russian science, technology and innovation policies, which remains, with its background report, an important source on the system and problems of transition. This argued that post-Communist economic transformation in the prevailing circumstances of Russia would inevitably entail a substantial contraction of the scale of the R&D effort and set out measures designed to convert the Soviet-era legacy into a coherent market-oriented National Innovation System.

Research undertaken by Russian and Western economists and science policy specialists reveals that, notwithstanding reform measures, the Russian R&D system as of 2006 still retains many Soviet characteristics (Dezhina and Saltykov, 2005; Radosevic, 2003). There is still organizational fragmentation, with the majority of R&D organizations being remote from the business sector; a large proportion of research organizations remain in state ownership; budget spending predominates, with only a modest contribution from the private sector; higher education plays a limited role in R&D, the Academy system still absorbs a large share of total expenditure; and the military share of R&D remains substantial, almost as large as in Soviet times. The inertia is such that it cannot be said that Russia possesses a National Innovation System, understood as a coherent set of interrelated institutions promoting innovation as a natural outcome of their day-to-day functioning. This situation has arisen in part because strong vested interests within the Academy of Sciences and industry have pursued their own survival strategies, resisting the government's reform initiatives.

Innovation activity in Russia's market economy remains relatively weak. Explanatory factors include weak competition, inadequate managerial skills, underdeveloped technological capabilities at the company level, weak demand for new products and processes, inadequate finance, with little long-term bank lending and a lack of venture capital, and modest foreign direct investment in manufacturing (OECD, 2006; World Bank, 2006). Russian government initiatives to improve the situation include the creation of special economic zones and the formation of venture capital funds. The domestic intellectual property rights regime has been improved but enforcement remains weak. For Russia, promoting a more effective R&D system and improved innovative performance is becoming a policy priority in the face of competition from other dynamic emerging economies, in particular China and India (Cooper, 2006). Notwithstanding relative strength in human capital and the possession of a large research system in terms of

personnel, the Soviet past still hampers Russia in the field of research, development and innovation, and for economists this represents an interesting case of the costs of institutional inertia.

JULIAN COOPER

See also **command economy; emerging markets; planning; technical change.**

Bibliography

Amann, R., Cooper, J.M. and Davies, R.W. 1977. *The Technological Level of Soviet Industry.* New Haven and London: Yale University Press.

Amann, R. and Cooper, J.M. 1982. *Industrial Innovation in the Soviet Union.* New Haven, CT, and London: Yale University Press.

Amann, R. and Cooper, J.M. 1986. *Technical Progress and Soviet Economic Development.* Oxford: Basil Blackwell.

Bailes, K.E. 1978. *Technology and Society under Lenin and Stalin.* Princeton: Princeton University Press.

Berliner, J.S. 1976. *The Innovation Decision in Soviet Industry.* Cambridge, MA: MIT Press.

Cooper, J. 2006. Of BRICs and brains: comparing Russia with China, India, and other populous emerging economies. *Eurasian Geography and Economics* 47, 255–84.

De Witt, N. 1961. *Education and Professional Employment in the USSR.* Washington, DC: National Science Foundation.

Dearden, J., Ickes, B.W. and Samuelson, L. 1990. To innovate or not to innovate: incentives and innovation in hierarchies. *American Economic Review* 80, 1105–24.

Dezhina, I.G. and Saltykov, B.G. 2005. The national innovation system in the making and the development of small business in Russia. *Studies on Russian Economic Development* 16, 184–90.

Fortescue, S. 1986. *The Communist Party and Soviet Science.* Basingstoke: Macmillan.

Glaziev, S. and Schneider, C.M. 1993. *Research and Development Management in the Transition to a Market Economy.* Laxenburg: IIASA.

Gokhberg, L., Peck, M.J. and Gács, J. 1997. *Russian Applied Research and Development: Its Problems and Promise.* Laxenburg: IIASA.

Hanson, P. 1981. *Trade and Technology in Soviet-Western Relations.* London and Basingstoke: Macmillan.

Hanson, P. and Pavitt, K. 1987. *The Comparative Economics of Research, Development and Innovation in East and West: A Survey.* Chur, Switzerland: Harwood Academic Publishers.

Harrison, M. 2003. The political economy of a soviet military R&D failure: steam power for aviation, 1932 to 1939. *Journal of Economic History* 63, 178–212.

Harrison, M. 2005. A Soviet quasi-market for inventions: jet propulsion, 1932–1946. *Research in Economic History* 23, 1–59.

Holliday, G.D. 1979. *Technology Transfer to the USSR 1928–1937 and 1966–1975: The Role of Western Technology in Soviet Economic Development*. Boulder, CO: Westview Press.

Kontorovich, V. 1994. The future of the Soviet science. *Research Policy* 23, 113–21.

Kornai, J. 1980. *The Economics of Shortage*. Amsterdam: North-Holland.

Kornai, J. 1992. *The Socialist System: The Political Economy of Communism*. Oxford: Clarendon Press.

Korol, A. 1965. *Soviet Research and Development: Its Organization, Personnel and Funds*. Cambridge, MA: MIT Press.

Kruse-Vaucienne, U.M. and Logsdon, J.M. 1979. *Science and Technology in the Soviet Union. A Profile*. Washington, DC: Graduate Program in Science, Technology and Public Policy, George Washington University.

Lewis, R. 1979. *Science and Industrialisation in the USSR*. London and Basingstoke: Macmillan.

Lubrano, L.I. and Solomon, S.G. 1980. *The Social Context of Soviet Science*. Boulder, CO: Westview Press.

OECD (Organisation for Economic Co-operation and Development). 1994. *Science, Technology and Innovation Policies. Russian Federation*. Vol. 1: *Evaluation Report*; vol. 2, *Background Report*. Paris: OECD.

OECD. 2006. *OECD Economic Surveys. Russian Federation*. Paris: OECD.

Radosevic, S. 2003. Patterns of preservation, restructuring and survival: science and technology policy in Russia in post-Soviet Era. *Research Policy* 32, 1105–24.

Schneider, C.M. 1994. *Research and Development Management: From the Soviet Union to Russia*. Laxenburg: Physica-Verlag; Heidelberg: IIASA.

World Bank, Russia. 2006. *Russian Economic Report No. 13*. Moscow.

Yaremenko, Y.V. 1981. *Strukturnye izmeneniya v sotsialisticheskoi ekonomike*. Moscow: Mysl'.

Zaleski, E., Kozlowski, J.P., Wienert, H., Davies, R.W., Berry, M.J. and Amann, R. 1969. *Science Policy in the USSR*. Paris: OECD.

Soviet economic reform

Lively organizational churning took place in the Soviet economy behind the façade of ideologically mandated institutional immobility. A command economy was introduced in 1918, gave way in 1921 to a market socialism-type system known as the New Economic Policy (NEP), and was re-established in the late 1920s and early 1930s. In the remaining 60 years of its existence it was subject to three types of organizational change: the reordering of the command hierarchy; the adjustment of the border between the command core of the economy and the peripheral sectors where the market was allowed to operate; and the introduction of market-inspired institutions into the command core itself, which we call 'economic reform' and which is the subject of this article.

The Soviet economy was organized like a pyramid, with party and government leadership at the top, enterprises at the bottom, and ministries for particular sectors in the middle. Its organizational chart was regularly redrawn as ministries and enterprises were merged or split, enterprises were transferred from one ministry to another, and functions were redistributed among staff and line departments. These changes usually came with less fanfare than economic reforms and attracted relatively little interest from researchers, and their effects on economic performance have not been documented.

Administrative allocation of resources in the core of the Soviet economy (industry, transport and communications, construction, distribution of producer goods, foreign trade, part of agriculture) coexisted with market or quasi-market allocation at the interface between these sectors and households, as well as among the households. The boundary between market and command economy was repeatedly adjusted, as when rationing of consumer goods alternated with their sale at fixed prices in state stores. Institutional changes in this category had immediate perceptible effects on consumer well-being, as attested to by contemporaries.

Output commands and input quotas in physical terms never covered all the activities of enterprises. Because of the strains on the planners' information-processing capacity, some commands were formulated in aggregate terms, and some production activities were left out of the central plan. Actual physical details of aggregate commands, as well as decisions on the provision of some goods and services, were left for the producers and users to work out. Also, the command allocation of resources was shadowed by monetary flows. Enterprises paid each other for supplies at centrally fixed prices, kept accounts in monetary terms and received plan targets for cost, total value of output, and other monetary magnitudes.

In the classical Soviet system, enterprise managers' discretion in production decisions was limited to low priority goods. Money circulating in the command sector was meant to passively track planned direction of resources, playing merely the accounting role. Financial targets ranked low in the enterprise plans. The thrust of economic reforms was to make these subordinate features of the economy, which served as props for the command mechanism, at least as important as the latter. Enterprise discretion, the influence of users on suppliers, and reliance on monetary plan targets were all intended to increase. Economic reforms attracted far greater attention than the other organizational changes because they arguably represented a retreat from the Stalin-era ideological orthodoxy and held out hope for systemic change, fuelling, for example, theorizing on the convergence of socialism and capitalism (Ellman, 1980, p. 200).

The prototypical reform

The first Soviet economic reform was announced in 1965. It replaced the main indicator of performance, namely, gross value of output, with sales, so as to bring the users' evaluation to bear on the producers' performance. To unshackle the initiative of the enterprises, the number of output targets for specific products in physical terms was reduced, and some other targets, including cost reduction, labour productivity, the number of employees, and the average wage were abolished. The volume of profit and its ratio to the enterprise's assets became binding targets. Enterprises were allowed to undertake small-scale investments on their own, and for this purpose to create a development fund financed out of profit, depreciation charges, and proceeds from selling unneeded equipment.

Pecuniary incentives built into the economic mechanism, working automatically, were to play an equal, if not greater, role than specific commands from superiors in guiding enterprises. A substantial bonus fund and a fund for residential construction and social development were created to replace the tiny fund that previously served these purposes. Inspired by the Western experience of profit sharing, reformers made enterprise profit the source of financing for the funds. While previously the evaluation of performance and rewards and punishments were all tied to the degree of plan fulfillment, the size of the funds was tied to the actual growth of profit or sales and to the ratio of profit to assets through a formula. The coefficients of the formula were to be fixed for a number of years, rather than changed by the ministry at its discretion every plan period.

Another element of stable rules of the game for the enterprises was to be provided by firm five-year plan targets. Before, these were changed annually, if not more often, at the discretion of the supervisory body. Enterprises always paid for labour and intermediate inputs, but got their plant and equipment free as a result of centralized investment decisions. The reform introduced a capital charge, a fixed percentage of the value of fixed and working capital to be remitted to the state budget. The intentions to supplant input rationing by wholesale trade in intermediate products and to allow enterprises themselves to determine their wage bills were announced.

Reform was introduced gradually and by 1970 applied to almost all of industry (Schroeder, 1971, p. 38). However, in the process many reform measures were reversed or modified in ways subverting their initial intent. By the early 1970s, commands for production of specific goods had returned to the pre-reform level of detail, and targets for labour productivity and cost reduction were brought back (Iasin, 1989, p. 100). Funds for decentralized investment were first restricted and then used as just another source for financing centrally planned projects (Dyker, 1981, p. 139). Rules for incentive fund formation and for awarding bonuses were changed repeatedly, making the latter dependent on meeting a number of plan targets, rather than on the growth of profit (Ellman, 1984,

pp. 87–9). Trade in intermediate goods and operational five-year plans for the enterprises never got off the ground or were realized only on a small scale.

Because of the eventual reversal of so many reform provisions, arguments about their effects are limited to the period of 1966–70, when they were still being implemented. According to the official Soviet data, industrial output continued to grow through this period at the same rate as in the previous five years. The recalculation by the US Central Intelligence Agency (CIA) shows a marginal slowdown in industrial growth (US Congress, 1982, pp. 191–2), and some of the estimates using the CIA data detect a pickup in total factor productivity (TFP) growth or a slowdown in its decline in 1966–70. To be able to attribute it to the effects of reform, one would need a thorough growth accounting exercise, which has not been undertaken for lack of data. Thus, changes in capacity utilization in industry help explain part of the variation in TFP in the 1970s. Capacity utilization in the late 1960s was increasing for reasons unrelated to the reform, though the extent of the increase, and hence its impact, is unknown (Kontorovich, 1990, pp. 44, 46).

There may have been one-shot gains as enterprises strove to qualify for the new incentive rules by assuming more ambitious targets, selling unneeded equipment, and drawing down inventories (Schroeder, 1971, p. 44). Over the longer period, qualitative evidence on the micro level shows the persistence of enterprise behaviour patterns which the reform intended to modify – hiding resources from the planners, meeting plan targets in ways that sacrifice the customers' needs and waste resources, and neglecting the tasks that the planners had not specifically ordered. The reform also had several unintended ill effects on performance. Greater discretion in choosing product mix and enhanced interest in value of output measured in administrative, cost-plus prices led enterprises to shift product mix in the direction of more intermediate input-intensive items, irrespective of customers' needs (Kushnirsky, 1982, pp. 16–17; Iasin, 1989, p. 99). The relationship between the dynamics of output in physical and in value terms shifted, and planners, who did not acknowledge the shift in their calculations, were led to make more wasteful allocation decisions (Khanin, 1991, pp. 41–9).

The treadmill of reform

The expression 'treadmill of reform', coined by Schroeder (1979), refers to the post-1965 Soviet attempts at reform followed by retreats. The treadmill appears even more daunting if one considers Soviet history and the collective experience of command economies. A prominent official in charge of heavy industry briefly attempted to do away with rationing of producer goods within his sector as early as 1932, with disastrous results (Davies, 1984, pp. 213–14).

In the late 1940s, as an element of Communist rule, a command economy was imposed on eight east European countries, some of which were the first to try economic reform after Stalin's death in 1953. Most of the elements of the Soviet 1965 reform had been formulated and tried by its east European neighbours in the previous decade, with similar results. In Hungary in 1954, the number of commands issued to the enterprises was reduced, and the intention was announced to promote decentralization through the use of incentives and delivery contracts. The number of commands soon crept back up. In 1956, half a dozen enterprises were experimenting with greater independence and use of profit as the success indicator (Kornai, 1959, pp. 218–21, 234). New incentives based on profit sharing were introduced in the late 1950s, and a capital charge in 1964. These measures proved, in the official estimation, to be ineffective (Bauer, 1987, p. 135; Berend, 1990, pp. 108–9).

Reforms in Poland starting in 1956 included the reduction in the number of commands, provisions for decentralized investment by the enterprises, and creation of a substantial bonus fund, both to be financed out of profit. These changes were rolled back in 1959–60 (Montias, 1962, pp. 294–307, 320–4). In Czechoslovakia in 1958, profit was made the main success indicator for the enterprises, with the latter also given the right to decide on some investment projects and finance them internally. These reforms were reversed in 1960–2 (Myant, 1989, pp. 82–4; Teichova, 1988, p. 148).

By the time the USSR embarked on its first economic reform in 1965, these and other east European countries were already on their second round, to be followed by further retreats, modifications, and new attempts too numerous to list here. Only Hungary escaped from the treadmill, having established a market socialist economy after 1968. The Czechoslovak reform of 1967 and the Polish reform of 1982 had similar aspirations, but the former was aborted after the Soviet invasion of 1968 and the latter was unsuccessful.

Decrees on 'improvement of the economic mechanism' adopted in 1979 aimed to correct some of the ill effects and reaffirm some of the neglected planks of the previous attempt. To curb the enterprises' manipulation of product mix in favour of more material-intensive products, the main target was changed from sales to a variant of net product, and bonus payments were made conditional on fulfilling delivery obligations in terms of volume, assortment, quality and timeliness. The use of fixed rates, rather than ministry discretion, in planning wages, profit distribution, and other financial and physical targets for the enterprises was to be broadened. The intention to make five-year plans operational and to make ministries self-financed was reiterated. An internal investigation of the effects of this reform found that it had no positive impact on performance due to its piecemeal nature (Ellman and Kontorovich, 1998, p. 109).

Though reforms are discussed as discrete events dated by the year of their announcement, each one was implemented in stages over a number of years, with amendments, modifications, and reversals to follow. This, combined with the redrawing of the organizational chart, such as the merger of enterprises into larger associations initiated in 1973, assured that Soviet industry was in the process of almost continuous administrative upheaval. The tempo of reorganization reached a fever pitch in the system's final years.

In 1984 a 'large scale experiment', again attempting to cut the number of commands to the enterprises, and extending and strengthening some of the provisions of the 1979 decree, was introduced in five industrial ministries. By 1987 it covered the whole of industry, but in the summer of that year a 'radical reform' was announced, to be implemented in 1988.

Its most radical provision was the election of the enterprise managers by their employees. Plan targets for total value of output, profit, and most other aspects of enterprise performance became mere recommendations. Obligatory targets, rechristened as state orders, were kept for output of major products and commissioning of production capacity. They were supposed to cover only a small share of the enterprises' output. Producers were free to dispose of the output over and above the state orders however they saw fit, though most prices remained fixed. All investment by enterprises was to be financed from their own revenue. Uniform stable long-term rates were to determine the distribution of profit among payments into the state budget and contributions to the investment, bonus, and social development funds, with enterprises retaining a greater share than before. The ministries were forbidden to issue any commands not on the government-approved list, a practice common during the previous reform attempts. Now the enterprises could go to court to overturn such commands and demand compensation for damages they caused.

The effects of the last reform on overall economic performance were drowned out by those of the collapsing political system and of severely misaligned fiscal, monetary, and investment policies. The reform increased financial resources at the enterprises' disposal, and these were used to boost employee remuneration and decentralized investment, aggravating already significant shortages of consumer and investment goods. Faced with compulsory targets for only part of their output, enterprises reduced production of non-compulsory items and shifted their product mix towards higher-priced items. This disrupted the flow of inputs and made other enterprises unable to meet their obligatory output targets, contributing to an output contraction in many sectors in 1989 and economy-wide in 1990–91 (Krueger, 1993).

The elections of managers were repealed in 1990. The rates governing enterprise profit sharing were manipulated to rein in the growth of employee remuneration. In order to combat supply disruptions, centralized distribution of non-centrally planned products was authorized in mid-1988. While the official objective of curtailing the share of state orders in the output was regularly reiterated, by the end of 1991 only 14 percent of producer goods distribution officially went through non-administrative channels.

Amended and revised, the 1987 reform defined the formal economic mechanism of the Soviet economy until its very end. Western experts were both urging Hungarian-type market socialism on the USSR and predicting its imminent adoption (Schroeder, 1979, p. 340). While proposals for such a transformation were floated in the country's final years, they were never realized.

Reasons for the failure of reforms

The evidence of failure consists of the record of reform decrees modified and overturned; proclaimed intentions never realized in practice; the treadmill, with its repeated attempts to implement the same changes over and over again; and qualitative evidence that reform planks actually implemented failed to effect intended changes in enterprise behaviour, and often had negative unintended consequences.

Politics

Most Western research focused on the Soviet reform of 1965, and the prevailing interpretation attributed its failure to self-interested sabotage by ministry and other middle-level officials (Schroeder, 1979, p. 324). Resistance from the state and party bureaucracy was also seen as a formidable obstacle to the next serious reform. Such an explanation was consistent with the view of Soviet politics that became predominant in Western academia in the 1970s. The totalitarian model, with the top ruler being the sole unchallenged political actor, had been abandoned in favour of a pluralist model with politics propelled by interest-group rivalry.

The 'bureaucratic resistance' explanation of reform failure was incomplete at best, for the formal repeal of reform decrees could be done only by the top political authority. It implied the capitulation of the rulers in the face of interest-group pressure. But this would be highly unlikely given the record of Soviet economic history. In 1921, Lenin thought up and introduced his market-socialist NEP, to the dismay of much of the party, which nevertheless did not mount effective resistance. In 1957, Khrushchev abolished ministries, to the unhappiness of officialdom, but without resistance (Berliner, 1983, p. 383). Gorbachev dismantled or emasculated every economic bureaucracy in the space of a few years. Sectoral ministries were merged, had their staffs cut, and lost their power over the enterprises. The planners themselves carried out the abolition of the central planning when ordered to do so. The Central Committee and the regional party committees gave up their economic functions in an equally disciplined fashion.

The military-industrial complex, the sector alleged to have the most political power, was ordered to convert to civilian production and complied (Ellman and Kontorovich, 1998).

A more plausible political explanation of the failure of 1965 reform is that the top political echelon was divided, with the General Secretary being less than enthusiastic about it from the beginning (Baibakov, 1998, p. 171). Fear of political destabilization in the wake of the Czechoslovak ferment of 1968 is thought to have further soured the rulers on the reform and brought it to an end.

Systemic incompatibility of reforms

Political explanations imply that the reforms as designed were effective, and their failure was due to faulty implementation. Yet the provisions of the Soviet 1965 reform that were faithfully implemented did not achieve their objectives and produced harmful side effects, some of which are described above. The uniform ineffectiveness and the unintended negative consequences of reforms implemented in different countries over a long period of time suggest flaws in their conceptual foundation, namely, the implanting of market-like institutions into a command economy.

Economic reforms were an exercise in imitation, their intuitive plausibility deriving from the appeal to the workings of competitive markets. Trying to make money, prices, and profit that already existed in the command economy perform some of their market economy functions, the reformers fell into a linguistic trap, committing the equivocation fallacy. Enterprise profit from centrally ordered transactions, calculated in terms of administratively set prices in an economy with passive money, had, despite the name, different properties from profit in a competitive market, and could not perform the latter's functions. There is no reason why the pursuit of this profit would lead enterprises to efficient decisions, or why the stress on the volume of sales would make them sensitive to the wishes of their customers.

While market-type institutions in the command economy context lack the efficiency properties of their market economy namesakes, they also interfere with the vital function of central planning. A modern industrial economy, with its narrow specialization, depends for survival on the regular flow of intermediate goods and services from producers to users. In the absence of market, the task of balancing, or assuring a tolerably smooth flow of inputs, falls on the economy's administrators. The complexity of the problem (the large number of products and producers, and continuous shocks to supply) overwhelms the planners' information-processing capabilities. If balancing is done poorly enough, the economy may come to a halt. Because of the difficulty and importance of balancing, it received the lion's share of attention of planners and ministry officials, and economic institutions were moulded so as to make their task easier (Grossman, 1963). Despite all that, the degree of balancing remained a grave problem in the functioning of the Soviet economy.

Most of the economic reform measures, if implemented, endangered the regular flow of supplies (Kontorovich, 1988, pp. 312–13). A reduction in the number of output commands and an increase in the level of their aggregation meant abandoning the balancing of some products, or making it less precise. Tying incentives to growth rather than to plan fulfilment deprived the officials of one of the tools of ensuring that certain quantities of goods were produced and shipped to users at designated intervals (Litwack, 1990). Abiding by stable long-term profit-sharing coefficients and other norms required the planners to give up their discretion to adaptively adjust the plan as its imbalances were being revealed (Litwack, 1991, p. 262). Developing long-range plans with the degree of detail that would make them operational defied the planners' capabilities, already strained by constructing a balanced annual plan. Enterprise investment funds, when effective, diverted inputs from the uses designated by the planners (Dyker, 1981, pp. 136–9; Ellman, 1984, pp. 95–6).

The recognition of systemic incompatibility sheds a different light on the role of bureaucratic resistance. Reforms forced the middle-level economic officials, who were responsible for balancing, to surrender many of the tools for accomplishing the task on which they were being evaluated. Faced with contradictory demands, they procrastinated, violated reform decrees, and exerted informal pressure on the enterprises in order to perform the function that was vital for the economy. To the degree that they succeeded, their actions protected the economy from the harm that systemically incompatible change would have inflicted on it (Kontorovich, 1988, pp. 313–14; Litwack, 1991, p. 264). The formal reversal of reform decrees occurred when the top officials recognized their ill effects.

Reasons for the treadmill

Repeated introduction and abandonment of a policy or an organizational change may result from efficient adaptation to changing circumstances, the shifting outcomes of interest group struggles, or changing ideas or ideology. However, all of these explanations are predicated on the effectiveness of recurring change (Siegmund, 1997, pp. 372–4). The treadmill of reform can be explained, in the spirit of North (2005), by considering the rulers' beliefs about the economy. The evidence, centring on the last and most aggressively reformist General Secretary, suggests that they were not aware that the reforms were unworkable. Memoirs of Gorbachev's aides and advisors, as well as his own pronouncements, show that he came to power with the intent of improving the workings of the system and with a belief in its ability to be reformed. This reflected a persistent strand in the Soviet thinking, which viewed the economy as highly malleable, as manifested in the overwhelmingly normative slant of Soviet economics and in much of economic policy.

Gorbachev (1995, p. 117) attributed the failure of the 1965 reform to bureaucratic resistance at the intermediate level of hierarchy, and blamed saboteurs for everything that went wrong with his own reforms. Even after having smashed the system and all its institutes, Gorbachev would insist that 'Sectoral bureaucrats could eat alive anyone, including the Chairman of the Council of Ministers and even the General Secretary' (Abalkin, Medvedev and Kotkovskii, 1995, p. 123). If sabotage was seen as the reason for the previous failure, then trying again, and harder, made sense. The experts on whom the rulers relied for the elaboration of the content of reforms shared this outlook. Some of the best Soviet economists argued that the 1965 reform had a strong positive effect on the economy. They viewed bureaucratic resistance as the main obstacle to a reform's success, as in Zaslavskaia's sensational 'Novosibirsk paper' from 1983 (Hanson, 1992, p. 59). The 1987 reform, incorporating the planks that had been tried and proven unsuccessful over the preceding 30 years, embodied the best the Soviet economists had to offer at the time.

The belief in reformability of the command economy was maintained against long odds. The Soviet rulers had rich experience in supervising large, complex, long-term projects for the creation of weapons systems. With the best domestic experts in charge, every available bit of foreign scientific information was utilized, prototypes were tested experimentally, and failed designs were set aside, while successful ones were developed, resulting in a formidable arsenal. The preparation of reforms outwardly followed the pattern that proved successful in the technical fields: a more or less public discussion among the economists, managers, and officials; blueprints prepared by committees including the representatives of these groups; and experimental trials preceding broad implementation.

Soviet reformers could also draw on foreign expertise, specifically, the east Europeans' evaluation of their own countries' experiences, as well as the work of Western Sovietologists. Thus, Kornai (1959, p. 225) argued that central planning is a coherent system, and piecemeal changes to it, such as less detailed plans or more discretion given to the enterprises, are bound to hurt performance. Granick (1959, p. 123–4) noted that even the minimal use of money and quasi-market institutions in the classical Soviet economic model conflicted with central planning. Grossman (1963, p. 119) demonstrated that the 'command principle' could not coexist with the market mechanism in a stable way, and correctly predicted the fate of the 1965 reform as it was being initiated (Grossman, 1966, p. 54). The Western analysis of reform failures was based on information from Soviet publications.

One would expect the Soviet economists to learn about the futility of reforms from the experiments, or from the experience of their implementation, or from east European and Western publications, and eventually to convey this conclusion to the rulers (though Western analyses from the 1970s and 1980s, stressing bureaucratic resistance, could be read as confirming Soviet domestic misconceptions). This is what happened to the east European economists, who started from the same level of understanding as their Soviet colleagues in the 1950s (Kornai, 1959, pp. xiii–xxv). The Hungarian 1968 market reform and the aborted Czechoslovak one in 1967 were preceded by public discussions among economists with a level of sophistication that was not achieved in the USSR until after 1987.

The different pace of learning appears to reflect tighter political constraints in the USSR than in some of its east European satellites. Unlike in the development of new weapons, the rulers had preferences not just over the results of economic reforms, but also over their methods. At various times, they considered particular reform measures to be unacceptable for reasons of political stability and ideological propriety. This not only forestalled public and professional discussions but also created disincentives for individual research. Trying to learn from the results of countries with laxer political constraints simply did not pay in terms of professional advancement. Ideas that were not allowed to be articulated just did not occur to people, at least not to those who might ever be asked for advice. Conversely, the reforms which the rulers did accept could not be undermined by negative experimental results. Economic experiments were ordered by the political authorities and conducted by lower-ranking officials. The latter understood the experiment to embody the official policy and did what they could to make it a success. Choosing the better-performing enterprises to participate in the experiment or affording special treatment to the participants usually assured the result.

In the late 1980s, as political and ideological constraints on acceptable economic advice were being removed, economists' learning greatly accelerated, and proposals for a transition to a market economy were publicly announced. The demise of the command economy, weakened as it was by the final round of economic reforms, came not through the implementation of these radical proposals but as a result of the disintegration of the underlying political system.

VLADIMIR KONTOROVICH

See also **command economy; socialism; Soviet growth record; Soviet Union, economics in.**

Bibliography

Abalkin, L.I., Medvedev, V.A. and Kotkovskii, L. Ia., eds. 1995. *Istoricheskie sud'by radikal'noi reformy.* Prague: Laguna.
Baibakov, N.K. 1998. *Ot Stalina do Ieltsina.* Moscow: GazOilPress.
Bauer, T. 1987. Reforming or perfectioning the economic mechanism. *European Economic Review* 31, 132–8.
Berend, I.T. 1990. *The Hungarian Economic Reforms 1953–1988.* Cambridge: Cambridge University Press.
Berliner, J.S. 1983. Planning and management. In *The Soviet Economy: Toward the Year 2000,* ed. A. Bergson and H.S. Levine. London: George Allen and Unwin.
Davies, R.W. 1984. The socialist market: a debate in Soviet industry, 1932–33. *Slavic Review* 43, 201–23.
Dyker, D.A. 1981. Decentralization and the command principle – some lessons from Soviet experience. *Journal of Comparative Economics* 5, 121–48.
Ellman, M. 1980. Against convergence. *Cambridge Journal of Economics* 4, 199–210.
Ellman, M. 1984. *Collectivization, Convergence and Capitalism: Political Economy in a Divided World.* London: Academic Press.
Ellman, M. and Kontorovich, V., eds. 1998. *The Destruction of the Soviet Economic System: An Insiders' History.* Armonk, NY: M. E. Sharpe.
Gorbachev, M.S. 1995. *Zhizn' i reformy.* Book 1. Moscow: Novosti.
Granick, D. 1959. An organizational model of Soviet industrial planning. *Journal of Political Economy* 67, 109–30.
Grossman, G. 1963. Notes for a theory of the command economy. *Soviet Studies* 15, 101–23.
Grossman, G. 1966. Economic reforms: a balance sheet. *Problems of Communism* 15(6), 43–55.
Hanson, P. 1992. *From Stagnation to Catastroika. Commentaries on the Soviet Economy, 1983–1991.* The Washington Papers No. 155. Washington, DC: CSIS.
Iasin, Ie.G. 1989. *Khoziaistvennaia sistema i radikal'naia reforma.* Moscow: Ekonomika.
Khanin, G.I. 1991. *Dinamika ekonomicheskogo razvitiia SSSR.* Novosibirsk: Nauka.
Kontorovich, V. 1988. Lessons of the 1965 Soviet Economic Reform. *Soviet Studies* 40, 308–16.
Kontorovich, V. 1990. Utilization of fixed capital and Soviet industrial growth. *Economics of Planning* 23, 37–50.
Kornai, J. 1959. *Overcentralization in Economic Administration.* New York: Oxford University Press. 1994.
Krueger, G. 1993. *Goszakazy* and the Soviet Economic Collapse. *Comparative Economic Studies* 15, 1–18.
Kushnirsky, F.I. 1982. *Soviet Economic Planning, 1965–1980.* Boulder, CO: Westview Press.
Litwack, J.M. 1990. Ratcheting and economic reform in the USSR. *Journal of Comparative Economics* 14, 254–68.
Litwack, J.M. 1991. Discretionary behaviour and soviet economic reform. *Soviet Studies* 43, 255–79.
Montias, J.M. 1962. *Central Planning in Poland.* New Haven, CT: Yale University Press.
Myant, M. 1988. *The Czechoslovak Economy 1948–1988.* Cambridge: Cambridge University Press.
North, D.C. 2005. *Understanding the Process of Economic Change.* Princeton, NJ: Princeton University Press.

Schroeder, G.E. 1971. Soviet economic reform at an impasse. *Problems of Communism* 20(4), 36–46.

Schroeder, G.E. 1979. The Soviet economy on a treadmill of 'reforms'. In US Congress, Joint Economic Committee, *Soviet Economy in a Time of Change*, vol. 1. Washington, DC: Government Printing Office.

Siegmund, U. 1997. Are there nationalization–privatization cycles? *Economic Systems* 21, 370–4.

Teichova, A. 1988. *The Czechoslovak Economy 1918–1980*. London: Routledge.

US Congress, Joint Economic Committee. 1982. *USSR: Measures of Economic Growth and Development, 1950–1980*. Washington, DC: Government Printing Office, 1982.

Soviet growth record

The end of 1991 saw the collapse of the Communist regime, which was created 74 years earlier following the October Revolution of 1917, and with it the disintegration of the Soviet Union into 14 newly created independent states. In this way one of the most heroic social experiments of modern times came to a close. That experiment failed to provide a sustainable alternative social order to that offered by the capitalist model of different mixes of market mechanisms and government interventions, and of different variants of democratic and free political regimes. The Communist alternative, as presented by the Soviet model, was composed, on the economic side, of central planning and public ownership of most means of production, and of an authoritarian regime on the political and social sides; and both built on the foundation of the political economy teaching of Karl Marx as modified by Vladimir Lenin and his fellow-Bolsheviks. This Communist alternative claimed to be more equitable and just, and at the same time also more efficient, than the capitalist alternative. Specifically, it claimed to achieve faster economic growth, and thereby the potential to catch up with and overtake the capitalist market economies.

The end of the Communist system, indeed its demise from within, provides proof that the system as a whole was not sustainable, and very probably that economic factors had a role to play. This, however, was not the accepted view by everybody during the Soviet period. Throughout that era, more so during its earlier years, the Soviet system presented in the eyes of many an economic challenge (in addition to a military one) to the developed West, the threat of 'taking over and surpassing', and an alternative, possibly superior growth strategy to developing countries. This economic challenge, including its military implications, aroused great interest and thereby generated academic and intelligence input into the study of the Communist economic model and development strategy and the resulting outcomes in terms of economic growth, efficiency and equity. This input became all the more difficult, for both Soviet and Western scholars, because of the veil of secrecy that was imposed by the Soviet authorities on all sorts of information and statistics, and the distinct, though in many cases obscure methodology underlying the economic information and data that were published. Often the published data were biased in order to present a rosier picture than the Communist reality.

A survey of Western attempts to estimate the Soviet growth and efficiency record and to evaluate and explain it was prepared by this author during the last years of the Soviet Union (Ofer, 1987). It included, among others, mainstream estimates as well as some of the main debates and disagreements over these estimates and the methodologies used to compile them. Even though some of the figures presented turned out to be somewhat biased in favour of the Soviet record (in the light of data and analysis that became available in later years), the main observations and conclusions regarding the Soviet experience seem to have survived the test of time: among other things, they showed that, following a period of relatively rapid growth and successful industrialization and modernization, the pace of growth slowed down over time up to a virtual stop; that, overall, Soviet growth lagged behind that of many Western and other countries; that Soviet growth was achieved with a greater use of labour and investment resources per unit of output, and a lower contribution to improved productivity, that is, at lower efficiency levels and also with much greater human sacrifice, economic and otherwise; and that the Second World War, the Cold War and imperial ambitions came at a price in terms of lower growth and standards of living. Furthermore, the record showed that Soviet growth was achieved also at high costs of mortgaging future resources and capabilities, thereby making future growth and the transition, at least for a while, even more difficult. There was some degree of greater economic equity among the majority of the people, but it was bought at the cost of lower standard of living and the denial of freedom.

In this article I re-examine the Soviet growth record in the light of new information that became available since the collapse of the Soviet Union, including insights gained from the collapse itself, which was not predicted by most commentators, and from experience gained to date during the transition period. As will be seen, a fair amount of new research and thinking has emerged since the collapse. In most cases the data and analysis in Ofer (1987) and, even more important, the long list of references there will serve as a kind of benchmark for further discussion and, to save space, discussed only briefly here.

The growth record revisited: the basic figures

The Communist revolution took place in 1917, but only in mid-1928 was the Communist economic system and growth strategy put into place. Collectivization of agriculture was combined with a rapid industrialization drive, all organized under a command system of nationalization and central planning. The first decade following the revolution was devoted to taking control of the economy and government, reconstruction following the devastation caused by the First World War, revolution and civil war, and to experimentation and debates over the appropriate economic strategy, integrated with the struggle for political control. It is generally established that by 1928 the GDP in the Soviet Union regained approximately the level of 1913, while GDP per-capita (GDPPC) lagged by at least ten per cent behind. This followed a sharp decline due to war and revolution, reaching a low by 1920 and 1921. Most estimates for the Soviet period start in 1928. The growth record therefore captures performance during the era of the 'Soviet model'.

Economic growth and productivity can be measured in a number of ways. For growth we will concentrate on the levels and rates of the growth of GDP per capita. Total GDP and various segments thereof will be used when appropriate. The levels and growth rates of productivity will be represented mostly by 'total factor productivity' (TFP), that is, output per unit of combined inputs, usually labour and capital. In what follows we use as a base mostly data assembled and estimated by Angus Maddison (1995; 2001) derived from what he considered (and we concur) the best and most reliable sources. For the Soviet Union Maddison relied mostly on the work of Abram Bergson (1961), who put together the most comprehensive body of Soviet national accounting using the same methodology as the CIA (Maddison, 1995, pp. 141–2). In both cases there were more recent adjustments, as explained below. Maddison's monumental work *Monitoring the World* (1995) includes the additional advantage that it provides comparisons of the growth record of many countries, through a uniform methodology and units of accounts. Estimates of variables not included in Maddison's work are presented following a check of their basic consistency with the lead estimates. Remaining disagreements and the underlying alternative estimates will be presented and discussed.

The rate of growth of GDP per capita over the entire 1928–90 period was 2.6 per cent per annum, a fivefold increase. This growth record puts the Soviet Union in the group of fast developers and separates it from a large number of countries, mostly in the Third World, that failed to take off or that started much later. Over that period the Soviet Union transformed itself from a predominantly agricultural economy, but with a considerable industrial base, into a modern industrial and urban economy. Soviet growth also achieved a modest degree of catching up with the US economy and other developed countries. The relative level of GDPPC of the Soviet Union grew from nearly 21 per cent of the US level in 1928 to almost 30 per cent in 1990. Somewhat higher

ratios for 1990 are estimated by others. However, similar or better achievements are shown for many advanced countries as well as for a number of developing countries, especially in east Asia (Maddison 1995, Tables 1–4, p. 25). At the same time many countries, mostly of the Third World, remained far behind.

The average growth rate shown above is made up of significantly higher rates up to 1970, of between three and four per cent a year (if the First World War is disregarded) followed by steeply declining growth trend thereafter, down to half of one per cent per year during 1985–90. Some decline in growth rates and sustained growth, following a faster take-off, is considered normal. The Soviet record raises the question of whether growth along the Soviet model was sustainable in the future. Likewise, it is observed that, after reaching the highest relative level of GDPPC of 37.5 per cent of that of the United States in 1970, up from 21 per cent in 1928, the Soviet economy retreated below that level, even below most other developed countries. It is this break in the trend, from catching up to retreat, which started at an early stage that sounds the alarm bell over the merit of the Soviet model and its long-term sustainability.

The trends in the rates of growth of total GDP are similar, with relatively high rates up to around 1970 and sharply declining ones thereafter. The difference between the trends in GDP and GDPPC represents a trend of declining population growth, indeed a very sharp 'demographic transition', and declining fertility on top of the heavy losses associated with the collectivization drive and famine, other atrocities by Stalin, and then with the Second World War (see more on this below).

The figures on growth mentioned above are somewhat lower than those calculated by the CIA and accepted by most scholars during the period before the fall of the Soviet regime. The downward adjustments were made, including by the CIA itself (Noren and Kurtzweg, 1993), on the basis of more information that came out from the Soviet Union during the 1980s, of criticism of the then existing estimates by émigré and Soviet scholars, most notably Igor Birman (1989), and on the basis of parallel efforts by others, most notably the International and the European Comparative Projects (Summers and Heston, 1991; Heston, Summers and Aten, 2002; United Nations, 1999 and earlier years) projects of estimating and comparing the national accounts of many countries and their structure, based on the purchasing power parity methodology. The CIA revisions were mostly a result of downgrading the quality of Soviet goods and services and of the productivity of providers of public services below previous estimates, and from higher estimates of hidden inflation rates (especially of military production). Based on these downward revisions, affecting mostly estimates since about 1970 during the last two decades, and the adding of 1985–90 to the series, the 1928–90 annual rates of growth were reduced by about half a percentage point, from 4.2 to 3.7 per cent for total GDP and from 3.0 to 2.6

per cent for GDPPC. The adjusted estimates also revealed an even steeper decline following the break near 1970. The new estimates also reduced the relative standing of the Soviet economy by 1990 and before: the Soviet Union/United States ratio for GDPPC for 1990, estimated previously by the CIA at 0.43 per cent, was adjusted down 0.30–0.34 (the higher figure is from United Nations, 1999, ECP96). While these are quite significant adjustments, they do not change the basic picture of the overall growth record or the declining trend. As we shall see, the factors responsible for these trends also remain pretty much the same.

During the years just before and immediately following the fall, an avalanche of estimates reduced the relative standing of the Soviet economy and its growth record much below the adjustments cited above. These were supported by the economic crisis in the Soviet Union, Russia and the former Soviet states, before and then following the collapse, and included sharp declines in output during the early transition period. Some claimed that the observed fall in measured output after the collapse reflected, among other things, an artificial over-estimation of the level of output before the collapse (see, for example, Åslund, 2002, pp. 21–39). Some of the arguments supporting this claim were incorporated into the adjusted estimates mentioned above. However, in some cases such observations were based on a failure to distinguish between the level of output and consumer welfare. First, most of the fall in estimated GDP was due to sharply reduced investment and military expenditure and only a smaller decline in consumption. Second, the price liberalization and the creation of markets reduced shortages and queues, and in this way improved consumer welfare. Finally, the production and consumption preferences of the Russian society shifted from that of the Communist leadership to another, determined by the people, who valued consumption much more and investment and military spending much less. In the eyes of the people, the former GDP was evaluated so much lower. It took a few years until more balanced views on the size of the Soviet GDP returned, and estimates of overall Soviet growth, including the downward adjustments, were generally accepted (Hanson, 2003, p. 249; see also Åslund, 2002, p. 34).

This is, however, not yet the case regarding Soviet growth during 1928–37 (or 1940). Here the gaps among scholars seem to have widened rather than converged since the fall of the Soviet Union. Two Russian economists, Vasili Selyunin and Grigori Khanin, produced an estimate of Soviet growth, with much lower estimates for 1928–40 (Harrison, 1993). These estimates reinforced similar estimates for the same period by Naum Jasny, estimates that were the focus of an older dispute, unearthed in a recent article by Howard Wilhelm (2003). Per contra, a recent book by Robert Allen (2003) challenges the entire established view of Soviet growth, based mostly on an upward revision of rates of growth, of GDP and of consumption, also during 1928–37 (Allen, 2003, and Wilhelm, 2003, Appendix A, pp. 212–22). Maddison's figures for that period are taken from Moorstein and Powell (1966, p. 361) which are somewhat higher than the final estimate by Bergson (1961, p. 48).

The estimates by Bergson and by Moorsteen and Powell are calculated with 1937 prices as weights. Given the sharp structural changes, and that of the accompanying relative prices, the choice of price weights, according to the notorious theory on 'price index relativity', makes a great difference. In particular, the use of 1928 prices as weights produces significantly faster growth during 1928–37. Allen claims, correctly, that in such cases a geometric average of the two measures or a similar 'compromise' is more appropriate. This argument was well known to Bergson and all other scholars working on Soviet growth. Bergson justified his decision to stick to 1937 prices by claiming that the 1928 prices were not free-market prices and extremely distorted. This observation is supported also by Hunter and Szyrmer (1992, ch. 3 and pp. 305–11), who substituted in their calculations for 1928 a set of 'equilibrium calculated prices'. When these prices were used as weights the rate of growth of GDPPC during 1928–37 turned out to be similar to that based on 1937 prices (see more on Allen's position below). Let us point out in conclusion that any upward adjustment for 1928–40 will have to come at the expense of future growth, of both GDPPC and consumption, thus making the declining trend even steeper. Indeed, despite the above, Allen seems to agree with the overall estimates for GDPPC presented by Maddison (Allen, 2003, pp. 220–2).

A few words have to be added on the changing industrial structure of the Soviet Union. The general patterns of change were similar to those associated with modern economic growth: there was a decline in the share of agriculture, a marked increase in the share of manufacturing and related industries, construction and transportation, combined together as the M sector, and some rise in the share of services – all in terms of shares of the labour force and of GDP. The intensity of the changes was marked during 1928–37 and much less during the period since 1970. The increase in services was slower than normal in market economies, explained by the absence of markets, the view that services were 'nonproductive', and the policy of requiring households to supply many services during their non-working hours. The decline in the shares of agriculture was somewhat slower than 'normal', especially of the labour share, reflecting low labour and general productivity, due to the inefficiency of collectivization and low investment. Regarding manufacturing, and the M sector, there was the bias in favour of producer goods at the expense of consumer-good industries. This bias was reflected in the emphasis on investment and defence at the expense of consumption. The Soviet economy, as well as those of other Communist countries, was characterized as 'over-industrialized' in comparison with market economies at similar levels of development.

The growth strategy: extensive and capital-led growth

Growth strategy

What was the growth strategy of the Soviet Union, and how successful was its implementation? What explains the initial take-off and acceleration, and the eventual decline? How do the Soviet growth mechanisms and patterns compare with those in other countries in nature and in effectiveness?

Following intensive and prolonged debates during the 1920s, the Soviet leadership opted for a strategy of rapid growth and industrialization based on high rates of investment, mostly in heavy industry and the producer-goods industries – sector 'A', according to the Marxian jargon. It rejected the other option of industrializa-tion along with more 'balanced growth' between the various branches of the economy (Erlich, 1960; Hunter and Szyrmer, 1992). The chosen strategy followed the Marxian doctrine of 'expanded reproduction', as articu-lated at the time by a growth model developed by G.A. Fel'dman (Allen, 2003, pp. 53–60). It demonstrated that an initial high rate of investment can provide for rapid growth and, after some delay, also an overall higher level of consumption. On the ground this strategy was trans-lated into maximization of capital investment through forced savings and maximum mobilization of the labour force, including of women, while consumption levels were kept at the minimum sustainable level. Both became possible only through the power of the authoritarian regime and the 'command economy', not to mention the merciless rule of and atrocities committed by Stalin.

The rates of gross investment climbed over time to over 30 per cent of GDP. Such rates allowed the capital stock to grow much faster than in most other countries, and contributed the lion's share of total growth. The high share of investment left a mere 55 per cent of total output to household consumption, a share that declined further towards the end of the regime to below 50 per cent of GDP, 15–20 per cent below common levels in market economies. The most extreme version of this growth model was manifested during 1928–40, when the industrialization effort drew millions of people from agriculture and rural areas to monumental construction sites in old and newly created cities, and when con-sumption levels failed to increase, according to the accepted view, or even declined or increased modestly according to dissenters on both sides. There is no doubt, however, that the accompanying human losses and suffering during that period kept personal welfare at extremely low levels. The Second World War added to the losses and suffering, and only during the early 1950s did private consumption levels start to rise to a moderate

level. By 1990, consumption per capita in the Soviet Union stood at just a quarter of that in the USA as compared with about a third for GDPPC (investment per capita stood at 55 per cent of that in the USA). While over the Soviet period there was some catching up towards the US level in terms of GDPPC, there was probably very little in terms of private consumption. It can be concluded that, while the first part of the Soviet growth strategy, emphasizing investment, was fully implemented, the subsequent rise in consumption was very modest, and failed to arrive at the 'promised land'. A major reason for this, as is demonstrated below, is the failure of the Soviet system to generate enough produc-tivity growth through technological and other efficiency improvements.

An increasing share of the remaining GDP was devoted to defence and to 'public consumption' on edu-cation and health care, among other areas. The increase in defence spending, eventually up to about 15 per cent of GDP, is explained by the perceived outside threats, first from Germany and then from the West, and by the ambition of the leadership to build an empire and become a world power (see more below). Investments in human capital through education and health services were extremely important for the industrialization drive. However, they also improved the welfare of the popu-lation, typically to a higher degree than in other countries with similar levels of economic development. Especially noteworthy here is the encouragement of women to enrol in professional and higher education.

Extensive growth

Economic growth in general is generated mainly by two sources, inputs, or factors of production, namely, capital and labour, and by increasing productivity in the use of inputs, getting more output per unit of input. Among the factors increasing productivity are technological innova-tions, improvements in the organization of production, increasing returns to scale, greater effort, and structural shifts from less to more productive activities (such as from agriculture to manufacturing). The common way to account for the contribution of each factor to total growth is through a 'Cobb–Douglas' production function, which assumes constant returns to scale. It is estimated as output produced by capital and labour, and at a per capita level. The residual growth left after the contributions of all inputs are accounted for is 'total factor productivity' (TFP). Inputs are combined according to their relative weights in the production process.

The Soviet growth model and record described above are commonly characterized as involving 'extensive' growth' or as 'inputs-led' or even 'capital-led growth, driven mostly by increasing contributions of the main inputs and much less by greater productivity. The alter-native, 'modern' economic growth as defined by among others Simon Kuznets (1966) is characterized in most developed countries as 'intensive', whereby productivity growth accounts for a larger share of total growth and the bulk of per capita growth.

But although the mobilization of inputs was fundamental to the Soviet growth strategy, there was no strategic intention to neglect 'productivity growth'; on the contrary, much was invested in R&D, in quality manpower, material resources and institutional support. Technological innovation and productivity growth were targets of multiple incentives. Yet the outcomes were rather disappointing. While during 1928–40 TFP contributed 1.7 per cent a year to GDP growth, nearly half the per capita growth, this declined to about 0.5 per cent during the 1950s and 1960s and then moved to negative territory for the rest of the period, thereby contributing significantly to the decline in the rates of growth. The higher contribution of TFP during the early period is explained by the dissemination of readily available technologies, and by major structural changes and initial industrialization, advantages that were exhausted later.

The poor outcomes in productivity during the entire post-Second World War period are explained by attributes of the system that limit its capability to generate new technologies and other improvements in productivity, by the priority allotted to the military sector, and by systemic barriers to the diffusion of innovations and general TFP measures across the production sector. Other TFP failures resulted from the interaction between the input-driven growth and the systemic nature of central planning. A study by Joseph Berliner (1976) contains a detailed analysis of the technological weaknesses of the Soviet system. There are two major clusters of factors, strongly intrinsic to central planning, that are responsible for the low rate of TFP: first, the rigidity of the system and the high cost of flexibility and change; and second, the strong reliance on quantity and quantitative rather than qualitative achievements and incentives.

The hierarchical, top-down and command character of central planning discourages initiatives from below and suppresses competition: both are believed to be essential generators of innovation. Central planning is also a rigid system that minimizes flexibility in production and supply networks. The high costs of reliable information flows across long hierarchical command ladders provide a great advantage to routine, and to inertia in regard to change, to 'planning from the achieved level' as against adopting flexible and innovating plans, and to stable supply networks as against shifting ones – which are needed when new materials or new markets are to be preferred. All are barriers to innovation and dissemination.

Control under central planning is so much simpler – indeed only possible – on the basis of quantitative performance measures of output, inputs, supply flows, than on the basis of qualitative performance measures, better or new products, new production materials and processes, and greater labour efficiency. For these reasons, incentives under central planning reward first of all quantitative performance. Production plans are tight, in order to eliminate hidden production reserves and to encourage growth, and incentives to managers are mostly based on plan fulfilment. This also helps to assure smoother supply flows among producers. It comes at the expense of quality, of productivity and of setting aside time and inputs for improvements and innovation. Tight plans create shortages and a seller's market that severely limits the power of buyers to control for quality. There are incentives to innovation, but they are mostly dominated by those for quantitative plan fulfilment. The emphasis on quantity also explains the almost complete absence of exit of inefficient enterprises with obsolete equipment. Their contribution to the fulfilment of branch plans cannot be dispensed with. They are kept alive through the mechanism, termed by Janos Kornai, of the 'soft budget constraint', the pumping of additional resources into enterprises that create bottlenecks in the flow of output. On balance, enterprise managers refrain from the introduction of improvements and innovations, whose dissemination across the economy is thereby severely restricted. The extensive nature of the growth strategy as described above also contributed to the bias in favour of quantity over qualities and of inertia over change.

The process of innovation and technological change suffered also from the autarkic nature of the economy, which limited the free flow of advanced technology and production processes from the West. Huge effort and resources were invested in 'reverse engineering' of Western technology, and in other cases in 'reinventing the wheel', both costly and wasteful alternatives. Civilian innovation also suffered from the high priority accorded to military R&D and to military production in general. Military R&D took the lion's share, by some estimates up to two-thirds or more, of the entire Soviet R&D effort, and on top of this only very limited spillover of military innovations was allowed to benefit civilian production. Furthermore, in order to assure the prompt fulfilment of military production plans and to secure its proper quality, an entire body of priorities was granted to the military at the expense of the civilian sector: in quality of human resources and material inputs, in quality control, in price discrimination and in supply. The superior military and space technological achievements of the Soviet Union became possible partly by the imposition of a heavy burden on the civilian sector, and an even heavier burden on civilian R&D and on Soviet TFP.

Part of the relatively high TFP during the 1950s and the decline thereafter can be explained by the process of reconstruction following the devastation caused by the Second World War. However, another part of the TFP decline over time is to be blamed on some of the factors listed above as negatively affecting TFP, such as the increasing difficulty over time in incorporating new and more complex technologies. Another important factor is the increasing complexity of the economy, imposing a heavier burden on the planning process, on the flows of information and on the supply networks. It was expected that at some point the computer revolution may have eased those difficulties, but this did not really happen, partly due to the slow development of information technology (IT) in the Soviet Union, but probably more to the growing complexity of the economy such that even much more advanced IT would not be able to simulate.

In addition to external autarky the Soviet economy also suffered from what may be called 'internal autarky', the segmentation of the economy along vertical lines corresponding to the planning hierarchy, with very limited horizontal networking. This structure was amenable to the development and dissemination of sector-specific technologies, but was an obstacle to economy-wide ones. This is one explanation of the slow development of IT of all types, including in computing, and other advanced general-purpose technologies. The internal segmentation of the economy also delayed the dissemination of new technologies across branch lines. This is why central planning failed to seize an advantage over the market economy, where the dissemination of innovations is delayed by patent protection. Finally, declining TFP may also have been caused by the trend of deteriorating discipline and work motivation and enforcement during the post-Stalinist era.

After the death of Stalin the Soviet Union became engaged in endless attempts at reform, directed mostly towards partial decentralization, in order to increase flexibility and reduce the cost of change. Most such attempts failed and were abandoned, and gave way to recentralization. The more radical, though still partial, reform measures initiated by President Gorbachev during 1985–91 increased the level of disorganization of the economy and contributed to the disintegration of the centrally planned system.

The above discussion of the low and declining level of TFP fits well the Cobb–Douglas production setting, in the sense that the listed factors affected the entire production process in a neutral way. There were, however, two attempts to link the low productivity performance of the economy to the high rates of capital investment and the nature of this investment. The more notable study, by Martin Weitzman (1970), estimated a constant elasticity of substitution (CES) production function for the Soviet economy and found that, while the level of productivity growth was quite respectable and constant over time, the elasticity of substitution (ES) between capital and labour was very low. Later estimates repeated this finding for the entire post-war period (Easterly and Fischer, 1995). Low and declining ES is consistent with low and declining capital productivity. Low ES focuses on the increasing technological and organizational difficulties in keeping up with the fast trend of capital deepening that resulted from the growth strategy, especially when the rate of growth of the labour force declined over time (see below). This resulted in a fast decline in the marginal productivity of capital. While the substitution of capital

for labour is normal along the path of modernization and industrialization, the level and pace of the needed substitution, dictated by the high rate of investment, were too demanding for Soviet planners and innovators. Soviet sources reported thousands of fully equipped work stations with no workers to operate them. Low ES is therefore one important argument supporting the non-sustainability in the long run of capital-led growth. But the sources of low ES seem to be very similar to the explanations for low TFP in general listed above.

An interesting interpretation of the same phenomenon is offered by Vladimir Popov (2007). His argument is that it is much easier for the central planners to build and install new enterprises and plants than to re-innovate and replace production lines in existing ones, as was shown above. Accordingly, high rates of growth of output were achieved during the 1950s and early 1960s, when a new generation of enterprises was built. These rates declined later when the production lines aged and the authorities encountered organizational and technical difficulties in replacing them. The outcome was continued use of old equipment for long periods of time, well beyond its normal age in market economies, with very high maintenance and labour costs and long stoppages. The reluctance to retire obsolete enterprises resulted also from the pressures to meet tight production plans, as mentioned above (see Ickes and Ryterman, 1997). In later years, continued investments in new plants, albeit at lower rates, faced extreme labour shortages. This stage arrived during the late 1970s and 1980s, and contributed further to declining marginal utility of capital and to low ES.

The low and declining rates of TFP growth, or for that matter of any 'residual' beyond the contribution of capital and labour, made the Soviet growth pattern more and more 'extensive', that is, dependent almost exclusively on increasing amounts of labour and capital. At the same time, there are limits to the potential growth of both inputs, and therefore 'extensive' growth is bound to decline and eventually stagnate (Bergson, 1973). In order to keep the rate of growth of capital at high levels, under conditions of no or little growth of TFP, and with normally lower growth of labour inputs (see below), the share of investment in GDP must increase constantly, first reducing the share of consumption, and later also its absolute level. In addition to the negative effects on incentives and morale, this eventually becomes also politically unsustainable, even under an authoritarian regime. This danger forced the planners during the 1980s to reduce the rates of growth of material capital, from the previous eight to nine per cent per year to six per cent and below.

The increase in labour inputs was also coming to a virtual halt because of two reinforcing trends: first, the decline in the rate of growth of the population, and second, reaching practically the maximum rate of labour force participation. The first trend is part and parcel of the modernization process, but in the Soviet Union it accelerated beyond the 'normal' pace due to heavy material pressures on the population – the depressed standard of living, the meagre provision of housing and of household durables and services, and the pressure on women to participate fully in the labour force. On top of this, most women were forced to spend much time during after-work hours on household chores. As a result the rate of population growth declined from 1.8 per cent a year during the 1950s down to just 0.9 per cent in the 1980s, when much of this remaining growth was concentrated in the Muslim areas. As we shall see, the Soviet Union underwent a too rapid 'demographic transformation' that contributed to the decline in growth rates. The proposition suggested by Allen of a danger of population explosion in the Soviet Union under an alternative economic scenario is totally unfounded (Allen, 2003, pp. 111–32).

The rate of growth of labour inputs, however measured, declined during the post-Second World War period from more than 1.5 per cent annually, faster than the population's, down to 0.1 per cent, far below the population's, during 1985–90. During the last decades of the Soviet period the population reached the rates of growth of highly developed countries. The same is true of the rates of labour force participation, especially of women; indeed, on this score the Soviet rates surpassed those in some developed countries. As a result, as time went by the 'extensive' model turned more and more into 'capital-led'. The negative effects on output per capita of the saturation of labour participation rates are clear. The excess decline in the rate of growth of population negatively affects GDPPC through an increased per capita burden of public services such as defence and public administration.

Extensive, or 'capital-led' growth, Soviet-style, is therefore first of all a dead end. Second, any given level of GDPPC is achieved with more labour (including during off-work hours) and capital, which leaves the population a smaller share for consumption as compensation, than in market economies at similar levels of development.

Soviet growth strategy: catching up and haste

As mentioned above, the strategy of maximum growth was also motivated by the desire and drive to catch up with the West and surpass it. The rationale was provided by Stalin in 1931: 'We are fifty or a hundred years behind the advanced countries. We must make good the distance in ten years. Either we do it or they crush us' (cited by Berliner, 1976, p. 161). This goal was repeated time and again by other Soviet leaders. The growth strategy described above was designed to accomplish just that; but the sense of urgency added one more element to it, namely, 'virtuous haste' as termed by Gregory Grossman (1983). Haste is defined here as actions taken in order to bring about higher rates of growth in the near future at the expense of future growth. The economic merits of

such an action, or strategy, depend on the rate of time preference of the decision-maker. The more impatient he is, the greater is the room for more worthwhile haste. In what follows we list actions taken by the Soviet leadership that seem to testify to a very high rate of time preference. Alternatively, such actions can result from miscalculations regarding the real costs of their actions in terms of future growth. It seems that both played a role and that miscalculation was also important. Haste can also be looked at as an act of borrowing higher growth rates than available in the present. Since the Soviet Union could hardly borrow abroad, it had to mortgage its own resources and future growth. One can therefore learn about the degree of miscalculation by comparing the rate of interest that the system was ready to pay (its rate of time preference) with the rate it actually paid. This is equivalent to comparing the integral below an intended curve depicting growth rates over the period with the curve of actual growth.

A high rate of time preference seems to contradict the essence of the growth strategy, described above, of high rates of investment and readiness to postpone consumption into the future, which signify low rates of time preference. One way to reconcile this apparent contradiction is to note that patience was imposed on the people rather than on the leadership. The people would probably have preferred a 'normal' maximization of utility over time. The objective function of the leadership reflects high time preference, the maximization of present growth even at the expense of future growth. Therefore, compared with an objective function that maximizes the welfare of the people, the policy of haste will be depicted as a steeper down-sloping trend of growth rates, cutting the 'optimal' curve from above.

Soviet growth was notorious for over-depleting natural resources and in causing costly damage to the environment. When the rates of oil extraction started to decelerate during the 1970s, pressures to keep oil flowing resulted in over-pumping and the flooding of wells. Overuse of fertilizers and land in order to grow more cotton depleted large areas of arable land and of water sources in Uzbekistan. Lake Baikal and other water and land resources were contaminated by overuse and by industrial and nuclear waste; air pollution and other environmental issues were disregarded. Relevant here are also the acceleration of the demographic transition and the early mobilization of labour at the cost of labour shortages later. The high and increasing volume of investment, combined with pressures to complete projects fast, was responsible for lower quality, for a thinner technological content, and for mistaken decisions when taken in a hurry, all with negative effects on future efficiency and growth. Investment is treated by growth theory as a major vehicle for the introduction of new technologies and of processes of 'endogenous growth'. Haste prevented this from taking place in the Soviet Union despite high investment rates. The haste to construct many new projects simultaneously was also responsible for the permanently large and increasing stock of incomplete projects, further reducing the levels of capital productivity.

Rapid industrialization required vast investments in infrastructure, in urbanization, utilities, and in transport and communication networks. Haste demanded that these should be limited to the minimum level necessary or below, saving investment resources for 'productive' projects with shorter payback periods. This was also true of infrastructure services inside enterprises, where the peripheral activities serving production lagged behind. Over time, and especially towards the end of the regime, the level of maintenance of infrastructure services deteriorated. Over the second half of the period the quality of education and health services also deteriorated, a result of conservatism and inertia and little contact with the outside world, as well as mounting resource constraints.

In sum, haste contributed to the trend of declining rates of growth, over and above the factors mentioned earlier. The deficit in growth during the later period can be thought of as payments of interest and payments against the still growing stock of debt. The accumulated debt was endowed to the new regime in the forms of obsolete production capacities, over-depleted natural resources, a large environmental deficit, run-down physical and service infrastructure, a debt that would have to be repaid as a precondition of and barrier to the resumption of growth. It definitely contributed to the initial decline in output and other difficulties during the transition.

More generally, 'haste' is part and parcel of the selected growth strategy, and of the economic system and political regime. One justification for such a choice, discussed in the economics literature, but also advanced by the Soviet Communist leaders, is that it provided the (only?) method, or at least an efficient and quick one, of economic take-off, the break away from the vicious circle of the low-development trap. We now know that the selection of the Soviet model of growth had to take into account a future model shift. While it is clear that the choice made by the Soviet leaders was motivated, at least partly, by internal and external power considerations, it is worthwhile examining the merit of such a choice from a purely economic point of view. As mentioned above, the choice of an economic growth strategy was hotly debated in the Soviet Union during the 1920s (Erlich, 1960).

Two-stage development strategies and the cost of switching

In 1991 Russia joined, *ex post* and with no initial intention, a group of countries with two-stage development strategies: a first stage for the purpose of take-off, an initial modernization and industrialization drive, and a second stage, starting with the transition, for joining the

'normal' route to 'modern economic growth', consisting of a more balanced growth via market mechanisms and private property, moderate government intervention, and a democratic and open society (Rodrik, 2005). The argument justifying a special initial strategy, is that a successful take-off requires more drastic policy steps, more forceful institutions and more determined and intensive government intervention. The economic payoff of such a strategy differs across countries. However, usually little attention is paid to the costs of switching to the second stage of normal growth. These costs are higher the wider the differences are between the economic, social and political tools used for the take-off effort and those required for 'modern economic growth'. Therefore, even if the first stage is considered successful, the overall evaluation and the calculation of net benefits have to take into account these costs too. The costs of switching are made of up three elements: the costs of delaying the shift beyond its appropriate timing, caused by vested interests of the regime in power and fear of the unknown future; the actual costs of switching and transition when they come; and the size of the accumulated debt from 'haste', which is an intrinsic part of any take-off. The delay and the transition periods are longer, and all three cost elements higher, the wider the difference is between the economic and institutional set-up of the take-off and the post-switching system. One clear advantage of a uniform strategy of 'balanced growth' from the start is that economic modernization and growth go hand in hand with the evolutionary development of the appropriate institutions, so that virtually no radical switching is required.

The case of the Soviet Union is one where the above described gap between the two stages is near the maximum thinkable; so accordingly also are the expected costs. The central political and economic control and the suppression of freedoms of all kinds made it easier to take strategic decisions and to impose a heavier burden on the population. But these came at the costs of destroying and blocking the development of social and political institutions needed eventually for the second stage.

As mentioned above, to this day there are some disagreements as to the merit of the Soviet take-off strategy, standing alone. Most observers see some benefits in, even an economic justification for, the big push forward and the early Soviet industrialization drive. To be sure, nobody condones Stalin's atrocities, and few if any are ready to accept the human costs associated with that period, whatever the achievements. But, assuming that similar, even better results could have been achieved with a milder variant of the same model, one can appreciate some of the economic outcomes – relatively fast growth and radical structural change. Even if one takes into account the destruction caused by the Second World War and the exaggerated concentration on the Cold War and on military build-up, the Soviet economy did succeed by

1970 in accomplishing a degree of catching up and, even more than that, in putting the Soviet Union on the road towards economic modernization. Robert Allen (2003), writing after the collapse of the Soviet regime, goes beyond the above and argues that the take-off strategy saved the Soviet Union from the dismal fate of many developing countries, and that no alternative strategy could have accomplished such a feat. At the same time, the only serious attempt at a counterfactual calculation produced a feasible and better alternative, along a route to more balanced growth, and, what is significant, a strategy that would have reduced considerably the costs of switching (Hunter and Szyrmer, 1992).

The figures presented in the first part of this article suggest a degree of support for the Soviet growth and modernization record during the regime's first three decades or so, but only when no account is taken of the oceans of human suffering inflicted on the population by Stalin and, somewhat less so, by the other leaders. At the same time, the calculation so far, including the cost of shifting, demonstrates without any doubt that, even on the basis of pure economic considerations alone, the Soviet growth project was probably not worthwhile.

The Soviet system was established as a long-term and sustained alternative economic and social system, with no intention whatsoever to switch later. Yet, during the 1950s and early 1960s, the Soviet leaders became worried about the low efficiency of the economy and lagging technology, and the prospect of declining growth rates. For a time there were quite open discussions over the proper future course, including various proposals to move to a more decentralized system, in the direction of what was termed 'market socialism'. The discussions and proposals were limited to the economic mechanisms and did not include any ideas regarding changing the political regime. Yet, in retrospect, this seems to have been the appropriate time for switching to a new system altogether. *Per contra*, the discussions culminated in 1964 in a package of the so-called Kosygin reforms. It was a watered down variant of previous proposals. Only mild attempts were made at implementation, and within a few years the Soviet Union had sunk into the 'stagnation era' under Brezhnev and his followers. The actual delay in the shift is fully understandable, given the strong regime and economic system and strategy introduced in 1928. The totalitarian Communist regime, backed by a strong ideological paradigm and an even stronger military backing, could not have been expected to replace itself only on the basis of rational and valid arguments. Would President Gorbachev have acted as he did had he known the final consequences?

The change came 15–20 years later. During 1970–90, the period of delay, the average rate of growth of GDPPC stood at just above one per cent per year or at near zero if 1991 is added in, as compared with 3.4 per cent during 1928–70. It was followed, starting in 1990 or 1991, by the transitional output decline of nearly 40 per cent, which was halted and reversed only from 1999. This decline is partly a manifestation of the difficulties in making the extreme shift of institutions, formal and informal. Over the entire period of delay plus early transition – that is, from 1970 to 1998 – GDPPC *declined* at an average annual rate of 1.3 per cent.

The entire period related to the take-off lasted, therefore, 70 years, from 1928 to 1998, rather than about 30, its segment of rapid growth. The rate of growth associated with this entire period stands at a mere 1.5 per cent per year, less than half the rate during the period of fast growth. This is not a great achievement by comparison with other countries at similar levels of development, or even developed countries. It is even more dismal when the higher sacrifices in terms of labour inputs and lower consumption levels are factored in.

One can think of other dates than 1998 to signal the end of the transition period: one is when Russia resumes its pre-fall GDPPC level of 1990 or 1991, possibly around 2008; yet another date could be when Russia regained the highest level of GDPPC of 37.5 per cent of the US level, reached back in 1970. An optimistic date, based on much guessing, could come between five and ten years after 2008. If true, then the delay plus transition periods will last much longer than the period of rapid growth itself, reducing the hypothetical growth rates still further.

Concluding note

The Soviet Union initiated and implemented a heroic social experiment that included a rapid and fairly successful, albeit rather distorted, process of modernization and growth. It culminated more then 70 years later in a dead end that required a difficult 'transition' in order to join the main road to modern economic growth. *Ex ante* it was not intended to be an experiment, nor a temporary phenomenon. There are many lessons to draw: one is that, even on the basis of economic accounting alone, and when the costs of transition are included, it is highly doubtful whether the experiment, judged as a take-off strategy, paid off. It clearly failed as a sustained alternative. Yet the counterfactual analysis, based on the experience of other countries that did better, is always conditioned on the issue of feasibility: could an alternative strategy, as indeed was advocated for the Soviet Union during the 1920s, have pulled the Soviet Union into take-off and sustained growth?

GUR OFER

See also **command economy; Soviet economic reform; Soviet Union, economics in; Soviet and Russian research, development and innovation; transition and institutions.**

Bibliography

Allen, R.C. 2003. *Farm to Factory: A Reinterpretation of Soviet Industrial Revolution*. Princeton and Oxford: Princeton University Press.

Åslund, A. 2002. *Building Capitalism: The Transformation of the Former Soviet Bloc.* Cambridge: Cambridge University Press.

Bergson, A. 1961. *The Real National Income of Soviet Russia since 1928.* Cambridge, MA: Harvard University Press.

Bergson, A. 1973. Toward a new growth model. *Problems of Communism* 22, 1–9.

Bergson, A. 1997. How big was the Soviet GDP? *Comparative Economic Studies* 39, 1–14.

Bergson, A. and Levine, H.S. 1983. *The Soviet Economy: Toward the Year 2000.* London and Boston: G. Allen & Unwin.

Berliner, J. 1976. *The Innovation Decision in Soviet Industry.* Cambridge, MA: MIT Press.

Birman, I. 1989. *Personal Consumption in the USSR and the USA.* London: Macmillan.

Easterly, W. and Fischer, S. 1995. The Soviet economic decline. *World Bank Economic Review* 9, 341–72.

Erlich, A. 1960. *The Soviet Industrialization Debate, 1924–1928.* Cambridge: Cambridge University Press.

Grossman, G. 1983. The economics of virtuous haste: a view of Soviet industrialization and institutions. In *Marxism, Central Planning, and the Soviet Economy: Economic Essays in Honor of Alexander Ehrlich*, ed. P. Desai. Cambridge, MA: MIT Press.

Hanson, P. 2003. *The Rise and Fall of the Soviet Economy.* London: Longman.

Harrison, M. 1993. Soviet economic growth since 1928: the alternative statistics of G.I. Khanin. *Europe–Asia Studies* 45, 141–67.

Heston, A., Summers, R. and Aten, B. 2002. *Penn World Table Version 6.1* (ICPb). Center for International Comparisons at the University of Pennsylvania (CICUP). Online. Available at http://pwt.econ.upenn.edu/php_site/pwt61_form.php, accessed 8 February 2007.

Hunter, H. and Szyrmer, J.M. 1992. *Faulty Foundations: Soviet Economic Policies 1928–40.* Princeton: Princeton University Press.

Ickes, B.W. and Ryterman, R. 1997. Entry without exit: economic selection under socialism. Working paper, Penn State University.

Kuznets, S. 1966. *Modern Economic Growth. Rate, Structure, and Spread.* New Haven and London: Yale University Press.

Maddison, A. 1995. *Monitoring the World Economy.* Paris: OECD.

Maddison, A. 2001. *The World Economy: A Millennial Perspective.* Paris: OECD.

Moorstein, R. and Powell, R.P. 1966. *The Soviet Capital Stock 1928–1962.* Homewood, IL: Richard D. Irwin.

Noren, J. and Kurtzweg, L. 1993. The Soviet economy unravels: 1985–91. In US Congress, Joint Economic Committee. *The Former Soviet Union in Transition*, vol. 1. Washington, DC.

Ofer, G. 1987. Soviet economic growth: 1928–1985. *Journal of Economic Literature* 25, 1767–1933.

Popov, V. 2007. Life cycle of the centrally planned economy: why Soviet growth rates peaked in the 1950s. In *Transition and Beyond: A Tribute to Mario Nuti*, ed. S. Estrin, G. Kolodko and M. Uvalic. Basingstoke: Palgrave Macmillan.

Rodrik, D. 2005. Growth strategies. In *Handbook of Economic Growth*, ed. P. Aghion and S. Durlauf. Amsterdam: North-Holland.

Summers, R. and Heston, A. 1991. The Penn World Table (Mark 5): an expanded set of international comparisons, 1950–1988. *Quarterly Journal of Economics* 106, 327–68.

United Nations. 1999. *International Comparison of Gross Domestic Product in Europe 1996* (ECP96). Geneva: UN Statistical Commission and Economic Commission for Europe. (Also ECP80, 85, 90, 93).

Weitzman, M. 1970. Soviet postwar economic growth and capital–labor substitution. *American Economic Review* 60, 676–92.

Wilhelm, J.H. 2003. The failure of the American Sovietological economics profession. *Europe–Asia Studies* 55, 59–74.

Soviet Union, economics in

Economics in the Soviet Union progressed through a number of distinct stages of development that were linked to the evolution of the Soviet system itself. It was also replete with multi-level contradictions. Actively building upon the foundations laid down by the British classical economists, Karl Marx's prognosis of capitalist collapse had envisaged a new role for economic ideas as the blueprint for the creation of a socialist utopia; even so, official Soviet doctrine came to view Western 'bourgeois' economics with unreserved hostility. After 1929 much of Soviet economics declined dramatically to become a subservient puppet of its Communist Party masters, yet before 1929 the Russian contribution to mainstream economic theory had bloomed to a new level of international respectability. Even under Joseph Stalin's gaze the works of some Western economists were translated into Russian – J.M. Keynes's *General Theory* was published in the USSR in 1948 – but other Western authors who were critical of the Soviet system were banned. And just as the USSR collapsed at the end of the 1980s and central planning was finally laid to rest, a number of Western economists were actively planning exactly how to transform Soviet-type economies into their market-based opposites. To summarize the results of the Soviet experiment for economics widely interpreted, it was an essential failure coupled with an episodic success.

1917–1929

The first problem encountered after 1917 was whether 'economics' as a discipline would be required at all, given that planning was supposed to be a science of collectively organized production. Did economic laws as natural regulators apply only to commodity production, or would new economic laws be developed in socialism? N.I. Bukharin's *Economics of the Transition Period* (1920) exemplified the liquidationist call for an end to all monetary accounting, anticipating that the naturalization of economic thinking would occur. This approach was however soon discarded when the New Economic Policy (NEP) was introduced in 1921. Bolshevik leaders then enthusiastically embraced the principles of 'sound money' and market exchange. For example V.I. Lenin declared that what was needed was 'less politics and more economics' and advised that communists should 'learn to trade'. But against Bukharin's proposal for an industrial democracy, Lenin still declared that politics should take precedence over economics and that the proletarian dictatorship must prevail. This contradictory attitude remained an undercurrent in all official Soviet proclamations on economics until 1991.

Despite such inconsistencies there is no doubt that the 1920s were the decade in which Russian and Ukrainian economists contributed the most to international developments in economic theory. Immediately recognizable names like N.D. Kondratiev, E.E. Slutsky and A.V. Chayanov were only the tip of an iceberg in terms of the contributions made by Soviet-based economists to fields like business-cycle analysis, agricultural economics, monetary theory and the economics of planning. Kondratiev and Slutsky were actively involved in cross-country collaborations through personal connections with Western economists like Wesley Mitchell and through their membership of international bodies such as the Econometric Society. Moreover, if the influence of Russian-born émigré economists is also considered, then the international impact of economics originating in the Soviet Union before Stalin's rise to dominance is difficult to deny.

To take the relevant fields of economics in turn: with respect to business-cycle analysis, Kondratiev and his colleagues in the Moscow Conjuncture Institute built upon the pre-revolutionary contributions of M.I. Tugan-Baranovsky, and added various extra dimensions of analysis including Mitchell-style empiricism, greater statistical sophistication (such as Fourier analysis) and a more direct interest in contemporary policy concerns (the effects of peasant taxation). Kondratiev (1922) began the decade with a detailed study of the Russian grain market, using differing levels of farm marketability to explain its collapse during the war and revolution, before embarking upon a more general analysis of the world economy (1925) using a three-cycle schema of long, medium and short cycles that was later employed by Joseph Schumpeter. The notion of 50-year-long cycles generated by the periodic creation of basic capital goods quickly achieved some international notoriety. Other members of the Conjuncture Institute focused on topics such as

scientific discovery as causation (T.I. Rainov), the methodology of cycle analysis (N.S. Chetverikov), seasonal fluctuations (Ya. P. Gerchuk) and the theory of economic prognosis (A.L. Vainshtein). Outside the Conjuncture Institute, S.A. Pervushin (1928) analysed cyclical movements as being composed of various shifts in price relations, and suggested that in Russia from 1890 to 1913 domestic factors such as the harvest were increasing in importance as causative influences.

Even before 1917, agriculture had been a key focus for Russian economists; after 1917, this interest blossomed into a multitude of different approaches. As director of the Institute for Agricultural Economics, Chayanov (1925) focused on the structure and optimal size of farms, and the motivating drive of peasants. He argued that peasants should not be modelled as *homo economicus*, but rather a labour-consumption balance was more appropriate to them, in which both the monetary and the non-monetary needs of the family were evaluated against the drudgery of labour performed. L.N. Litoshenko, by contrast, viewed peasant proprietors as driven solely by acquisitive motives, and supported rural class differentiation as a means of improving agricultural techniques. E.A. Preobrazhensky (1926) identified the agrarian sector of the economy as a source of funds for state-induced industrialization that should be 'pumped over' into the industrial sector by price policy, as part of a process called 'primitive socialist accumulation'. Against this idea Bukharin encouraged peasants to 'enrich themselves' as a means of fostering growth in both agriculture and industry. And Kondratiev provided detailed forecasts of international grain markets using the latest insights of cycle theory in order to facilitate Russian agricultural exports as a means of financing industrial development (Barnett, 1998).

Monetary theory experienced an unexpected boost in the early Soviet context through the need to establish a stable currency in civil war conditions. Debates over the role of money in a socialist economy occurred from 1917 onwards, with contributors from the left and the right clashing over fundamentals. Tugan-Baranovsky advocated a system of paper money in which metallic reserves would not circulate, while Marxists like S.G. Strumilin proposed various non-monetary accounting schemes such as labour time vouchers and energy units. Preobrazhensky celebrated the profligate issue of paper currency as a method of political confrontation, but L.N. Yurovsky's idea (1925) for a parallel gold-backed currency to counter the resultant hyperinflation eventually won out after 1921. Keynes's design for a currency board for north Russia in 1918 was quickly overtaken by events. These issues were part of a larger debate over the significance of war communism (1918–1920), a system that A. A. Bogdanov characterized as being driven by acute shortages, rather than being the result of the rational application of socialist economics.

Most obviously of all, the economics of planning attracted a huge amount of effort from many Soviet economists, especially in relation to industrialization strategy. In 1917 Tugan-Baranovsky outlined a method of planning using marginalist techniques, but this soon became doctrinal heresy. The first detailed Soviet effort was the GOELRO electrification plan of 1920 that employed an engineering approach, but as NEP progressed a more sophisticated planning methodology developed. Debate centred on questions such as the weight of extrapolation from current trends (genetic planning) against desired ultimate goals (teleological planning), the nature of the interrelation between state industry and private agriculture, and the role of Quesnay-type balances in the planning process (Barnett, 2005).

For example, building on the agricultural census developed by pre-revolutionary *zemstvo* (local government) statisticians, P.I. Popov pioneered the preparation of an economy-wide balance for 1923/24 from within the Central Statistical Administration. As part of the genetic approach to planning developed inside the State Planning Agency, V.G. Groman and V.A. Bazarov (1925) uncovered various empirical regularities that operated in the NEP economy, such as the law of market saturation, using models adapted from the natural sciences. From within the Conjuncture Institute, N.N. Shaposhnikov applied the net present value principle to the planning of capital investment projects, and Kondratiev prepared a detailed plan for agriculture and forestry for 1924–8 that was based on an indicative 'perspectives' approach. By the end of the 1920s, both yearly 'control figures' and five-year imperative plans had been developed, the latter containing hundreds of pages of figures that plotted the progress of all branches of Soviet industry as centralized directives.

Slutsky's economics-related work in the 1920s was in a theoretical league of its own. His first important contribution was to the mathematical modelling of currency emission, where he compared various complex formulae with the reality of Soviet monetary policy after 1917, in order to calculate the income that the state received from emission. He then turned his attention to the praxeological foundations of economics and provided a set of axioms for describing the parameters of an economic system. And in the same year as he published the groundbreaking paper suggesting that cyclical processes in the economy might be modelled as the summation of independent chance causes (1927), he also wrote a critique of Bohm-Bawerk's conception of value. An important feature of Slutsky's random cycles was periodic disarrangement and consequently regime change, which served to distinguish them from approximately regular business cycles. The 1927 article was the result of many years of work on the theory of stochastic processes that eventually produced a new conception of the stochastic limit. After the closure of the Conjuncture

Institute in 1930 Slutsky was not arrested, and he continued to pursue related topics such as the use of the extrapolation method in relation to random processes. Slutsky's mathematical work was also employed in the set theoretic approach to probability theory developed by A.N. Kolmogorov.

In various unconnected fields, in 1924 A.A. Konyus drew upon the contributions of Irving Fisher to develop a sophisticated cost of living index that lay between the Paasche and Laspeyres indices, and in 1926 he suggested the idea that consumer preferences could be represented in terms of both prices and income. In 1925 Strumilin assessed the economic benefits of education by evaluating the length of study undertaken against the increased skills obtained, suggesting that education should be taken to the point where marginal cost equalled marginal revenue. And in 1928 G.A. Fel'dman developed a two-sector model of economic growth in which the ratio of the capital stock of the producer and consumer goods sectors was related to projected growth rates, predicting that (for a stable growth path) investment must be divided between the two sectors in identical proportion to the stock of capital. This model had direct implications for Soviet planners.

Finally, the emigration of economists like Jacob Marschak, Simon Kuznets, Evgeny Domar and W.W. Leontief after 1917 transferred some of the existing themes of Russian economics to the USA. Leontief's input-output approach (1941), Marschak's Walrasian market socialism, Domar's growth theory and Kuznets' work on secular trend all owed an important debt to their Russian origins, although such work was transformed by the American context. But the 1920s were unquestionably successful in terms of producing many influential developments of relevance outside of the immediate Soviet context.

1929–1953

After 1929 the storm clouds of Stalinism poured down upon Soviet economists with uncompromising force. Key figures like Bukharin and Kondratiev were sentenced to long periods in jail, and important centres like the Conjuncture Institute were closed: in response Kondratiev wrote an account of the methodology of economic statics and dynamics. Lesser members of the various economics groupings like Vainshtein and Konyus were dispersed. What took the place of the pioneering Soviet economics of the 1920s was a polarization to ideological and technical extremes.

Vacuous general statements about the current direction of Soviet policy, such as Stalin's 'law of the harmonious development of the national economy', sat alongside the minute detail of the planning of every branch of Soviet industry. This polarization facilitated the Orwellian discrepancy between officially declared aims (economic equality for all) and actual government policies (a system of slave labour camps). The central question of deciding overall plan targets was resolved politically in the Politburo (with assistance from Gosplan and the Commissariats), and the advisory function of economists evaporated. Soviet economic discourse consequently became an instrument of its political masters. However, there remained limited scope for dissent in tangential fields such as mathematics. For example, L.V. Kantorovich initiated his conception of optimal planning in 1939, in response to the problem of distributing the manufacture of parts to available machine tools so as to produce the maximum number of sets of components, or a method of machine loading to obtain the highest productivity. To generalize this idea, an optimal plan was one in which the proposed product assortment was optimally distributed amongst firms at the lowest possible cost of production. Shadow prices or 'objectively determined valuations' were to be used in this process. Kantorovich won the Nobel Prize (jointly with Tjalling Koopmans) in 1975 for this work on linear programming.

Further afield, émigré economists like Boris Brutzkus and S.N. Prokopovich contributed to the debates over the nature of the Soviet system in the 1930s. Brutzkus (1935) took an Austrian-type approach that focused on the centralizing and information-gathering issues, while Prokopovich (1924) criticized the reliability of official Soviet statistics and the state control of property. The socialist calculation debate that viewed the problem either as a computational issue relating to solving sets of equations (Oscar Lange) or as erroneously assuming perfect knowledge (F.A. Hayek) failed to fully engage with the reality of Soviet planning, where bureaucratic and interest group factors were dominant. The Institutionalist response was quite different, with Thorstein Veblen (1921) analysing the Soviet success in Russia in terms of long-prevalent national traditions of economic organization, and John Commons viewing the USSR as the outcome of collective action. Keynes (1925) characterized Bolshevism as business in subordination to religion.

In terms of contextual influences, it is difficult to exaggerate the effect that the Second World War had on Soviet society. The Soviet economy had been placed on a war footing since the mid-1930s, and centralized accounting of material production was seen as crucial to military success. Hence the Soviet system was the direct result of the needs of a war economy, including features such as consumer rationing and strict hierarchical control of industrial production. When the war was finally over, planners found it difficult to adapt to a consumer-led approach. Consequently a semi-underground second economy developed in the USSR that was built upon informal connections, yet was tolerated because of its role in facilitating plan fulfilment. This was theorized by Aron Katsenelinboigen (1977) as a network of coloured markets with differing degrees of non-legality, black being the most extreme. This aspect of the

Soviet economy encouraged criminality and had serious consequences for the market reforms of the late 1980s.

1953–1985

After Stalin's death in 1953, a thaw began in Soviet intellectual life that had positive consequences for Soviet economics. Previously repressed economists like Vainshtein resurfaced, alongside the coming of age of a new generation of Soviet economists who demonstrated a greater mathematical sophistication than their predecessors. Key members of the mathematical school included N.Ya. Petrakov, V.S. Nemchinov and V.V. Novoshilov, and their organizational base was the Central Economic Mathematical Institute of the Academy of Sciences. Alongside the concern for mathematics went an interest in cybernetics, in Kantorovich-type optimality problems and in economic reforms in general.

The outcome of this new infusion was the development of the concept of a system of optimally functioning economy (SOFE), in which the idea of an optimal plan was taken a number of steps further by adding the notions that plans should be calculated using optimal prices and with concern for resource scarcity and incentive rewards. It also included the idea of an economy as a hierarchical structure in which decision-making should occur at various levels appropriate to the specific task being considered, such as the national economy, a given industry or an individual enterprise, rather than always at the apex of the pyramid. Another important centre of this period was the Institute of Economics and Organization of Industrial Production in Siberia, where in 1965 Abel Aganbegyan stirred controversy by presenting a very negative evaluation of the Soviet economy as antiquated and undemocratic. Foreign commentators also highlighted problems endemic to the Soviet system such as gigantomania (large units were easier to plan), investment cycles linked to production bottlenecks, and soft budget constraints (Nove, 1990).

The Khrushchev era witnessed the first major attempts at economic reform since Stalin, initiated by E.G. Liberman's campaign (1962) to allow enterprises the capacity to plan their own production programmes. While some reforms were implemented in 1965 that attempted to improve industrial management, the impetus for change diminished after 1968, when Soviet tanks were sent into Czechoslovakia. In the 1970s Soviet economics continued to develop along various mathematical, ideological and empiricist paths, including the suggestion of oxymorons such as 'planned markets', while the Soviet economy entered a period of relative decline. Burdened by the need to maintain nuclear parity with the West and by imperial overstretch (Afghanistan), its superpower status was increasingly difficult to sustain. Answers about how to solve these problems were not always forthcoming from within official Soviet economics, and, if they were, they were rarely immediately heeded. For example, it took Aganbegyan's reformist economics nearly 20 years to be taken seriously by Soviet leaders.

However, in less contentious areas worthwhile contributions came in the post-Stalin period from Vainshtein on measuring national wealth and D. I. Oparin on multi-sector accounting. The standard of empirically focused economic investigation without obvious ideological significance was usually high, and it was often easy to separate out the pseudo-Marxian framework from the valuable detailed content. However, in areas with direct policy significance the quality and reliability of the economics declined dramatically. For example, official statistics on growth rates in the USSR after 1928, perhaps the most contentious subject of all for Soviet economists, certainly suffered from ideological interference, but even Western economists who tackled this topic objectively (such as Kuznets) came to somewhat ambiguous conclusions. In general Soviet economic theory between 1929 and 1985 was thoroughly dominated by political concerns, but this did not necessarily invalidate every aspect of the all of the work done in this period.

1985–1991

With Mikhail Gorbachev's accession to power, the twin tracks of cultural openness (*glasnost*) and economic restructuring (*perestroika*) were opened. This was presented as a return to a NEP-style mixed economy, with 'socialist markets' and entrepreneurial cooperation being hailed as solutions. This opened the floodgates to the reprinting of material by repressed economists like Bukharin and Litoshenko, and to the open discussion in Soviet economics journals of topics like the neoclassical theory of production and distribution. Soon after this, Western authors like Hayek and von Mises were being translated. In 1989, a government programme was issued acknowledging that 'the market' must take precedence over the plan. Some of the key economists of the Gorbachev period were Aganbegyan, Stanislav Shatalin and Leonid Abalkin, who were part of an old guard that had always advocated reform. They were, however, quickly overtaken in the audacity of their transition programmes by a younger generation of economists such as Grigory Yavlinsky and Yegor Gaidar. Various concepts were applied to explain the dilemmas facing the Soviet economy in this period, such as monetary overhang (savings without prospect of being spent), market Stalinism (state decrees pronouncing market activity) and then spontaneous privatization (impromptu transfer of state property).

A key point to recognize is that, in the USSR at the end of the 1980s and after being suppressed for so long, pro-market economics was a 'revolutionary' set of ideas that gave Russian advocates a sense of liberation from decades of stale dogma, although some had made the intellectual change much earlier. The main focus of the early proponents of market reform was mass privatization of state-owned property and the liberalization of state-controlled prices, as contained in Yavlinsky's 500-days programme of 1990. Macroeconomic stabilization policies (such as balancing the state budget) took third place. This shock-therapy approach was influenced by Western economists like Jeffrey Sachs, and also by eastern European theorists such as Janos Kornai. Advice on upholding the appropriate sequence of economic reforms, as articulated by Ronald McKinnon (1991) based on previous experience in Latin America and Asia, was not fully heeded in the 'transition fever' of the time. Many Western economists were actively involved in propagating market-friendly ideas and policies even in the early Gorbachev period, and they often overwhelmed any lingering loyalty to Marxist economics through the greater sophistication and more rigorous technical framework of their work.

It should also be acknowledged that the particular conception of 'the market' espoused by many Russian economists at this time was somewhat one-sided, an amalgam of favourable assessments such as Frank Knight's conception of entrepreneurial capacity, early Austrian capital theory and Milton Friedman-style monetarism, complemented by Kornai's account (1980) of the Soviet shortage economy. The Keynesian tradition was difficult to discern. In discarding the socialist heritage, the baby (market critiques) had been thrown out with the bath water (central planning). In hindsight, many market advocates appeared rather naive about the consequences of releasing the genie of unfettered self-interest from the Soviet bottle. The question of whether 'the market' as a universal general mechanism (Adam Smith's 'invisible hand') actually existed, as against the idea of many different types of markets as institutions that were influenced by local and national conventions, grew in significance as the reforms progressed. Perhaps one reason for the sidestepping of Keynes's work was that it had been openly discussed even under Stalin, and hence was seen by some as tainted.

In terms of results, the consequences of the post-Soviet transition are fundamentally contested, with some economists celebrating the creation of a successful market economy in Russia (Anders Aslund) while others protested against oligarch-controlled mafia capitalism (Boris Kagarlitsky). What is unquestionable is that the initial rush to privatization and liberalization has been tempered by more recent concern for the development of stable legal institutions and a mature business culture. If banking panics and financial crashes are a sign of a functioning market system, then in the summer of 1998 Russia experienced a classic example, sparked by the government defaulting on its debts.

Conclusion

With the collapse of the USSR at the end of the 1980s, and China's conversion to 'capitalism with a socialist mask', the West has undoubtedly triumphed in the battle of comparative economic systems. However, it did so at significant ideological cost. Firstly, state intervention in the economy had to be embraced by many Western governments after 1936. Secondly, the goals of full employment, a welfare state and (more recently) fairness in international exchange have become accepted policies in many developed countries, even if they are not often achieved. And thirdly, Western economics was transformed after 1945, with the mainstream integration of ideas such as public goods, monopolistic competition, cooperative games, social cost and even status goods, that would make it unrecognizable to someone like Bukharin versed in the Austrian approach of 1914, and even more so to Marx, who had never even recognized the 'marginal revolution' of the 1870s. The refusal of Soviet economists to officially engage with new developments in Western economics was thus understandable, since had they had done so they would have realized the simplistic caricature that had been foisted onto them.

Furthermore, the existence of the USSR altered to some extent the purpose of mainstream economics as it was conducted in the West. Before 1917, economists served the state mainly as academic theorists and as advisers on specific topics, for example on monetary or fiscal policy. After 1945, with the onset of the Cold War, the role of some economists widened to providing more general advice on economic development issues. Moreover, the responsibility of a few Western economists increased even further than this. For example W. W. Rostow, famous for his stages theory of economic growth, was a development economist with a detailed knowledge of British trade cycles. Summoned to serve in the Kennedy administration he became a key advocate of increased US involvement in Vietnam, arguing for more American troops on the ground and heightened bombing of the North, in order to contain the expansion of communism in South-East Asia. That it was a professionally trained economic historian giving advice to US presidents on geopolitics was indicative.

To conclude, the USSR collapsed spectacularly, and its official economic ideology was derisory even when first issued, yet many Russian/Soviet economists produced work of lasting value that was influential at the time of its first issue, and is still being referred to today. All things considered, the Soviet contribution was (in spite of Stalin's efforts) perhaps the most influential experimental failure in the development of economic ideas thus far.

VINCENT BARNETT

See also **economic calculation in socialist countries; socialism; Soviet economic reform; Soviet growth record; Stalinism, political economy of.**

Bibliography

Aganbegyan, A. 1988. *The Challenge: Economics of Perestroika*. London: Hutchinson.

Aslund, A. 1995. *How Russia Became a Market Economy*. Washington, DC: Brookings Institution.

Barnett, V. 1998. *Kondratiev and the Dynamics of Economic Development*. London: Macmillan.

Barnett, V. 2005. *A History of Russian Economic Thought*. London: Routledge.

Bazarov, V. 1925. On 'recovery processes' in general. In *Foundations of Soviet Strategy for Economic Growth*, ed. N. Spulber. Indiana: Indiana University Press, 1964.

Bergson, A. and Kuznets, S. 1963. *Economic Trends in the Soviet Union*. Cambridge, MA: Harvard University Press.

Brus, W. and Laski, K. 1989. *From Marx to the Market*. Oxford: Clarendon.

Brutzkus, B. 1935. *Economic Planning in Soviet Russia*. London: Routledge.

Bukharin, N. 1920. *The Politics and Economics of the Transition Period*. London: Routledge and Kegan Paul, 1979.

Chayanov, A. 1925. *The Theory of Peasant Economy*. Homewood: Irwin, 1966.

Ellman, M. 1973. *Planning Problems in the USSR*. Cambridge: Cambridge University Press.

Fel'dman, G. 1928. On the theory of growth rates of national income. In *Foundations of Soviet Strategy for Economic Growth*, ed. N. Spulber. Indiana: Indiana University Press, 1964.

Groman, V. 1925. On certain regularities empirically observable in our economy. In *Foundations of Soviet Strategy for Economic Growth*, ed. N. Spulber. Indiana: Indiana University Press, 1964.

Kagarlitsky, B. 1995. *Restoration in Russia*. London: Verso.

Kantorovich, L. 1965. *The Best Use of Economic Resources*. Oxford: Pergamon.

Katsenelinboigen, A. 1977. Coloured markets in the Soviet Union. *Soviet Studies* 29, 62–85.

Keynes, J. M. 1925. *A Short View of Russia*. London: Hogarth Press.

Kondratiev, N. 1922. Regulation of the grain market and the supply of grain. In *The Works of Nikolai Kondratiev*, ed. N. Makasheva, W. Samuels, and V. Barnett. London: Pickering and Chatto, 1998.

Kondratiev, N. 1925. The long waves in economic life. *Review of Economic Statistics* 17 (1935), 105–15.

Konyus, A. 1924. The problem of the true index of the cost of living. *Econometrica* 7 (1939), 10–29.

Kornai, J. 1980. *Economics of Shortage*. Amsterdam: North Holland.

Leontief, W. 1941. *The Structure of American Economy*. New York: Oxford University Press.

Liberman, E. 1962. The plan, profits and bonuses. In *Socialist Economics*, ed. A. Nove and M. Nuti. Harmondsworth: Penguin, 1972.

McKinnon, R. 1991. *The Order of Economic Liberalization*. Baltimore, MD: Johns Hopkins University Press.

Nove, A. 1990. *Studies in Economics and Russia*. London: Macmillan.

Pervushin, S. 1928. Cyclical fluctuations in agriculture and industry in Russia. *Quarterly Journal of Economics* 42, 564–92.

Preobrazhensky, E. 1926. *The New Economics*. Oxford: Clarendon, 1965.

Prokopovich, S. 1924. *The Economic Condition of Soviet Russia*. London: King and Son.

Slutsky, E. 1927. The summation of random causes as the source of cyclic processes. *Econometrica* 5 (1937), 105–45.

Sutela, P. 1991. *Economic Thought and Economic Reform in the Soviet Union*. Cambridge: Cambridge University Press.

Veblen, T. 1921. *The Engineers and the Price System*. New York: Viking.

Yurovsky, L. 1925. *Currency Problems and Policy of the Soviet Union*. London: Parsons.

Spain, economics in

The development of economic thought in Spain should be understood as occurring in a country which, due to the delay in the development of its economy and scientific thought, became a recipient of economic ideas from more advanced centres. Spain went from empire to decadence, and awareness of this led people to search for strategies which would remedy the situation. The delay in the reception of foreign theories was determined by strong institutional obstacles, though these did not prevent theories from becoming known. This process of reception and redevelopment of foreign ideas was not always linear, nor was it uniform in time. The Spanish case is of interest when explaining and analysing and the effectiveness of different theories once they cross the borders of the country in which they have been developed.

The conquest of America and the world of scholasticism

The first traces of economic thought in the Iberian Peninsula are found in the age of the Arab civilization. A writer and politician of Tunisian origin, between the 14th and 15th centuries, Ibn-Jaldún, wrote *The Muqaddima*, a treatise on the science of civilization, in which he added an explanation of the birth and decline of dynasties, and which contains the first global vision of economics. *The Muqaddima* also contains an analysis of the cyclical nature of economics, as well as an embryonic model of development.

Later, the conquest of America set a new stage for the prosperous Hispanic monarchy. The transport of precious metals from the Indies and the increase in commercial exchange sparked off a revolution in prices which accompanied the commercial revolution. Population increase and the cultivation of new lands brought about a decline in productivity and, together with a lack of technical innovation, the problems of pauperism and the monetary and financial problems associated with unprecedented inflation, which originated in the discovery of the New World, led a group of theologians and members of the colonial administration to set forth their opinions on the situation of a country which had begun a long process of decline.

The first diagnoses came from a group of doctors, naturalists, politicians, philosophers and theologians, scholastic latecomers, members of the so-called School of Salamanca, whose founder was Francisco de Vitoria (1483–1546). These writers were to be found at the prestigious University of Salamanca, though they had been educated at the most important European universities. They were witnesses to the problems which originated during the conquest of America, and though the trade practices of wealthy Spaniards returning to Spain from Latin America were always ahead of theological doctrine, the scholars attempted to reconcile economics and morality by using natural law based on probabilism and casuistry. This led them to apply Thomistic philosophy to resolve practical affairs and to explain economic problems which had arisen in the world of the Counter-Reformation before the scenario which had opened with the discovery of America. They used the manuals of confessors and wrote a set of works in which they tackled economic problems with the utmost insight, amongst which stood out a subjective theory of value, the quantity theory of money, the theory of exchange, the general doctrine of interest and the workings of the market, as well as questions of fiscal policy in relation to distributive justice.

The cleric Martín de Azpilcueta early on formulated the quantity theory of money in his *Comentario resolutorio de cambios* of 1556, and, through considering the variations in exchange rates, came close to the purchasing power parity theory. His successor, Tomás de Mercado, author of *Suma de Tratos y Contratos* of 1571, managed to integrate quantity theory into a theory of prices. The Spanish scholastics built a bridge between medieval scholasticism and modern philosophy. They were contemporary with the writers of the first mercantile system but, unlike them, they were privileged by having had a philosophic-cum-systematic training of Aristotelian–Thomistic origin, which enabled them to focus on economic questions not only from a moral viewpoint but also from a more analytical one.

The ideas of the School of Salamanca were disseminated throughout the rest of Europe and the American colonies from the universities of Coimbra, Alcalá, Mexico, Lima, Rome and Paris. Together with Azpilcueta and Mercado, their principal members included clerics such as Domingo de Soto, Luís de Molina and Juan de Lugo. Scholastic doctrines at the end of the 16th century coexisted with those of the first mercantilist writers, in some cases resulting in an authentic symbiosis.

The age of mercantilism

A group of writers, royal advisers, social reformers and Spanish political economists swelled the ranks of the bullionists, among whom the first of the so-called mercantilists were often found. However, Schumpeter discovered among them a few economics writers whose ideas were distinct from primitive bullionism, and whom he considered to be authors of a quasi-systems, such as Luís Ortíz, who wrote a Memoir in 1558, the year following the first bankruptcy of Phillip II of the House of Austria. Aware of Spain's economic decline, Ortíz investigated a set of remedies which constituted a programme of industrial development.

Ortíz attributed the problems of the Spanish economy to internal price rises in Castile, to the inability of national industry to meet demand from the American continent, and to social disdain towards the skilled classes. The remedies consisted of prohibiting exports of raw materials from Spain and preventing imports of foreign goods. In addition, he investigated Spain's political and social institutions in relation to the workings of the economic system.

With an optimistic vision based on the potentialities of the Spanish economy, some Spanish mercantilists saw the fundamental problem not in the limitations of nature itself, but in social and institutional factors. This led to their proposing the elimination of idleness, the abolition of laws which considered manual labour contemptible, limitations on of luxuries for the rich, and changes in a taxation system whose burden fell unjustly on farm workers, aggravating depopulation. Some, such as Ortíz, saw market unification as necessary, the nobility and the Church being the main obstacle.

The expulsion of the 'Moriscos' at the beginning of the 17th century undermined the demographic base, already weakened by epidemics in the previous century; this was combined with a reduction in shipments of precious metals from the Indies. The disastrous policies of the King's advisors only added to the problems caused by the territorial and political dispersal of the Spanish monarchy. Some writers, such as Sancho de Moncada, in his work *Restauración política en España* of 1619, summed up the efforts made by the best writers of the time. He was concerned with quantifying the crisis and identifying the variables which brought it about and the links between them.

As a whole, the mercantilists at the beginning of the 17th century concluded that the root of all evil was international trade, since it enabled foreigners to extract from Spain undeveloped raw materials, as well as silver and gold. The remedy consisted in prohibiting manufactured foreign imports, as well as prohibiting the export of raw materials from Spain. This would stimulate the domestic market and internal spending, benefiting the people by increasing the monarchy's revenue. Some mercantilists, such as Martínez de Mata, insisted on increasing production when establishing a relation between cost and the production function.

Not all voices clamoured for prohibition; they explained the crisis in terms of the enrichment of foreigners and the unfair commerce which Spain maintained with them; some, exceptionally, such as Alberto Struzzi, who published a *Diálogo sobre el comercio destos Reinos de Castilla* in 1624, and, later, Diego José Dormer, author of several politico–historical discourses in 1684, defended economic plans which would reactivate commercial relations with foreign countries, with a different strategy based on improving the domestic competitiveness of Spain's economic sectors.

A second mercantilist phase began in Spain with the economic recovery at the end of the 17th century, during the reigns of the last monarch of the House of Austria, Charles II, and the first Bourbon, Philip V. The political economists of the period, amongst whom Jerónimo de Uztáriz and Bernardo de Ulloa were outstanding, were already more seasoned writers. The former, the most important of the Spanish mercantilists, wrote the *Theórica y práctica de Comercio y de Marina* in 1724, a work which aroused the interest of Adam Smith, and which was translated into several European languages. Uztáriz obtained quantitative data on the Spanish economy, on the basis of which he proposed the lowering of production costs in order to lower domestic prices as a solution. He also proposed a free trade zone with America.

In general, the 'mature' Spanish mercantilists concerned themselves with questions such as political unification, the elimination of internal customs and excessive indirect taxes, which accumulated in a chain effect and raised final sale prices. They equally supported privileged commercial companies and the existence of a strong navy, which could be used not only for military purposes but also for commercial ones. Their fundamental aim was to encourage industry, in terms of which they continued to defend the protection of domestic industry from foreign competition, in this way relegating agriculture, though only in order of preference. Foreign countries were no longer considered enemies, but were now an example to be followed, especially the Netherlands, France and the United Kingdom.

A characteristic of the Spanish mercantilists in this last period was their belief in the need to establish a more complex analysis of underdevelopment, with reference to the disadvantages of international trade (with consequent problems for the balance of trade) as well as the problems stemming from low labour productivity, which had its origin in the socio–political and cultural structure of the period. To break out of this vicious cycle of underdevelopment, they proposed that industry should be the key to development and consumption the driving force of industry.

The Enlightenment and economic reform

The long expansive cycle that began at the end of the 17th century had brought about a marked population increase, crop expansion, growth in industrial production and greater development of internal and external trade. Spain, however, was still a feudal society in which the land ownership system had not changed and, despite crop development, agrarian production techniques had not substantially changed.

Although Spanish mercantilism had been characterized as a 'hardy perennial', the age of Enlightenment in Spain started under the reign of the House of Bourbon of Charles III (1759–88), in which mercantilist stances in their most extreme version were abandoned, bringing about internal market liberalization and a set of reforms which began to liberate the economy.

For the economic writers of the period, the problem was still the same as in the past: to remove obstacles to economic growth, but with a clear awareness that this should be combined with certain institutional and legal changes, provided that they did not affect the absolute power of the monarch in any way. Despite the survival of mercantilist ideas, they now had new analytical tools at their disposal, thanks especially to economic ideas received from abroad, such as those of the new British political arithmeticians and of the French Physiocrats and agrarian reformers, together with those of the so-called Gournay group, and later, those of Adam Smith himself, which would circulate uninterruptedly until shortly before the end of the century, when the multiple wars with France and the United Kingdom largely hindered the circulation of ideas.

Some politicians who enjoyed great power and who were close to the monarch, such as the Count of Campomanes, author of The Discurso sobre el fomento de la industria popular (1774), amongst other works, and a genuine inspirer of this first transitional phase between mercantilism and enlightenment, drew up a renewal programme whose central element was economic reform. His modernizing ideas were founded on a gradual transformation consisting of (a) liberalizing colonial commerce, (b) abolishing the price-regulated 'tax' on cereals, (c) stimulating their commerce, and (d) putting an end to the increase in the considerable unproductive properties of the Church, and to the excessive interventionism of guilds. All in all, he designed a coherent system of development based on domestic economic freedom, together with protection against foreign trade, which granted a key role to the promotion of agriculture based on the figure of the independent farm worker, the expansion of 'popular industry' and occupational increase.

The Enlightenment saw the creation of the Sociedades Económicas de Amigos del País, organizations which encouraged the study of regional economy, of agricultural techniques and of economic science in general. In these societies, after the Italian model, the first professorships in civil economy and commerce were created, giving rise to the beginning of the institutionalization of political economy in Spain at the end of the 18th century. These societies disseminated economic ideas developed in other European countries, such as those of the British political arithmeticians of the 17th century and those of the French thinkers, mercantilists, advocates of agrarian reform, and physiocrats, namely, Cantillon, Melon, Forbonnais, Mirabeau, Montesquieu, Turgot, and, later on, Condillac, Necker and Adam Smith.

The Spanish economics writers, of whom the most prominent were Pablo de Olavide, Enrique Ramos, Nicolás de Arriquíbar, Bernardo Danvila and Francesco Roma i Rossell, were acquainted with the ideas of physiocracy and were especially attracted to their advocacy of agrarian reform, even if the theoretical and analytical core on which they were based, as well as ideas of exclusive agricultural productivity and the single tax, were not accepted as a whole. In this sense, many of the first enlightened Spaniards were more in tune with the ideas developed by the so-called Gournay group, which better reconciled the importance of agriculture, industry and commerce, and in which indigenous mercantilism was in its turn very pronounced.

It was also in this period that knowledge of the Wealth of Nations began to spread. However, it was not translated until the 1790s. Not all viewpoints were in favour of liberal agrarianism; others opted for industrial development, the influence of German sources being obvious, especially in the territories of the Crown of Aragon, by which the ideas of the Baron von Bielfeld and von Justi led to more industrialist formulations and justified greater participation by the state and by the nobility in the development of economic activities. Enlightened Spaniards of this first period found themselves at a crossroads between tradition and innovation, and although the influence of mercantilism remained, they created a favourable environment so that in the following years the enlightenment movement would manifest itself with greater strength, giving impetus to the modernization of Spain.

A second stage of enlightenment coincided with the French Revolution. At that time Gaspar Melchor de Jovellanos was commissioned by the Real Sociedad Económica Matritense to write the most important work on economics in 18th century Spain, the Informe de Ley agraria of 1795, based on the concept of the economy as the main 'government science'. He designed a programme of pragmatic, gradual and moderate liberal reforms, which included the redistribution of land, freedom from leasing, and imposing limits on property inheritance. He also proposed abolishing the privileges of the powerful stock-breeding organization, the 'Mesta', and the liberalization of domestic commerce, which, however, did not include freedom to export or with revision of the taxation system. All of this was contemplated with a programme of public investment and special attention to education, since it conferred decisive importance on human capital. Jovellanos, very well-acquainted with the works of European economists such as Cantillon, Galiani, Mirabeau, Turgot and Necker, among many others, was one of the Spanish economists who read Smith's work with discernment, as a result of which the idea of self-interest as the driving force of economic activity would be consistently present in his work, as it would be for many Spanish economists of the time.

In short, it may be said that in Spain the Enlightenment occurred later than in other countries and had different overtones arising not only from Spain's political and institutional situation but also from other factors such as its comparative economic backwardness. This did not present an obstacle to the enlightened Spaniards when disseminating the main economic ideas synthesized in the United Kingdom, France, Italy and in the Germanic countries; in any case, the canonical plan for the dissemination of economic ideas through the mercantilism–physiocracy–liberal economy chain did not appear to be suitable for Spain, where, along with a more persistent neo-mercantilism, fundamentally of an Italian and French nature, the weak presence of physiocracy in its theoretical aspects could be detected, and also a strong influence from the British and French agrarian reformers, who were widely dispersed throughout the peninsula.

The enlightened Spaniards applied the ideas they received to the solution of the problems of their age. In this sense it may be considered to have been an active reception, stimulated both by the economic societies in which the Economy was studied and which promoted the translation of the principal texts of the economists mentioned above, and by economic publications, which multiplied during the second half of the century. These disseminated ideas, making the last decade of the century an especially fertile one, in which, despite heightened censorship because of the French Revolution, the principal works of the European economists saw the light of day.

Classical economics in the liberal age

The Court of the Spanish Inquisition had already denounced Lorenzo Normante, the first lecturer in political economy, who had begun his classes in the Aragonese Economics Society under the influence of Melon, Genovesi and Condillac, for considering economic ideas which were not in accordance with dictates of the Catholic religion. The same trial dealt with the *Wealth of Nations*, which was censored by the Inquisition in 1792, being accused of tolerance and naturalism.

Despite this, Smith's influence had already been felt by some Spanish economists, who generally made an adapted interpretation of his work. Government collaboration and the suppression of his identity allowed a compendium of his work to be translated by the Marquis of Condorcet in the same year, and only two years later the first complete edition saw the light of day, with a few adjustments here and there to avoid the rigours of the censor. *The Wealth of Nations* was translated by one of the

Spanish economists on whom the influence of Smith was clear, the diplomat José Alonso Ortíz. Smith's work quickly became a reference point for the teaching of economics, even though it must be emphasized that its late translation and the prevailing doctrinal influences in Spain simultaneously brought several interpretations of Smith's work into being, especially those of an agrarian nature, which coincided with an avalanche of influences which converged at the beginning of the 19th century.

The introduction of Smith's work was accompanied by the first versions of the mathematical economics of N.-F. Canard, of the neo-physiocratic doctrines of Germain Garnier, or the partially physiocratic ones of Jean Herrenschwand, and those of the other advocates of agrarian reform of a non-physiocratic origin, both British and French. Together with these, the doctrines of the French economist J.B. Say had special relevance. His *Traité d'Économie Politique* was translated early in Spain (1804–7) and used as an official text to teach economics, when Spanish economics took its first step towards institutionalization in universities at the beginning of the 19th century.

Nevertheless, the influence of Smith lasted throughout the first decades of the 19th century; however, by that time the influence of the French economist was decisive. This strong influence (eight editions in Spanish of the *Traité* alone) can be explained by its support for keeping the same groups in power and its more pragmatic nature, which favoured its reception, together with greater accessibility of the French language in Spain, and because it was not subject to the vicissitudes of censorship. The fact is that it inspired all university texts for the first three decades of the 19th century. Smith's ideas were used by his followers to defend non-Malthusian interpretations with respect to population and an industrial development model, modified with a defence of customs prohibition. Some authors, such as Gonzalo de Luna, made an interpretation of the ideas of Say and Smith in a neo-mercantilist vein, which defended industrial development models without altering the political bases of the old regime.

Only those who were in exile for political reasons when the absolute monarch, Fernando VII, returned escaped such influences. They lived in the United Kingdom and became acquainted with the ideas of the classical British authors, especially Ricardo, McCulloch and James Mill. Amongst them stood out José Canga Argüelles, the Chancellor of the Exchequer in the brief liberal period which began in 1820 and author of *Elementos de la Ciencia de la Hacienda* (1825), and the most important Spanish economist of the 19th century, Álvaro Flórez Estrada, the main introducer in Spain of the ideas of Ricardo and James Mill, and in whose works the influence of Sismondi and Richard Jones could also be appreciated. In Flórez Estrada's *Curso de Economía*, published for the first time during his exile in London in 1828, a limited influence of Ricardo's ideas (the *Principles* were not translated until the 20th century) could be seen. Also disseminated with the ideas of Say and Ricardo, though with less intensity, were those of Malthus (almost exclusively his ideas on population), Sismondi (his ideas on the agrarian development model) and Bentham, Condillac and Gaetano Filangieri (though their influence was not solely in the sphere of economics).

Say's influence dwindled from the 1840s, allowing greater doctrinal plurality. The influence of Say's disciples left a clear mark on the main Spanish economists of the time, such as Eusebio María del Valle, Andrés Borrego and Manuel Colmeiro. It was a period characterized by eclecticism, in which distancing from the ideas of the classical school is associated on one hand with criticism of deductive methodology, and on the other with recognizing the negative effects of the development of capitalism on the lowest social classes. This favoured reception of Sismondi's ideas and those of the social Christianity of Alban de Villeneuve-Bargemont, so influential in France. All this included a distancing from free trade and favouring industrial protection.

Richard Cobden's journey to Spain in 1844, in his crusade for the free trade league, united with reception of the ideas from the French liberal economists, directed most Spanish economists towards free trade, a trend supported by such texts as *Cinco proposiciones sobre los males que causa la Ley de Aranceles a la nación*, published in 1837 by the liberal economist Pablo Pebrer, and the writings of another exile, José Joaquín de Mora, which sparked off the lengthy so-called protectionism–free trade debate, which continued until the beginning of the protectionist era with the Restoration in 1874. The decisive influence of the ideas of Frederic Bastiat and his *Harmonies Économiques* established an age of economic optimism and defence of liberal ideas in their French version, an influence which separated Spanish economists from knowledge of the most relevant developments in the classical British economy, minimizing the influence of economists like John Stuart Mill, an economist of more advanced and complex interpretation, especially after the events of 1848. Perhaps it was the sparseness of the theoretical analysis in Bastiat's work which favoured its dissemination, together with his outspoken opposition to the situation of social confrontation, which by that time was beginning to spread from France, fuelled by criticism of the republican groups and of incipient socialism in its utopian aspects.

The debate was situated in the territory of common-sense economic discourse, which brought about pragmatic empiricism and led to the pronounced stagnation of economic science. In the last three decades of the century even marginalist ideas were unknown, perhaps due to a lack of mathematical preparation of the economists, most of them university professors in law schools, or perhaps due to their lack of dedication to investigation. Neither did the historicism of the German Historical School of Schmoller make its mark, and the influence of the Historical School was reduced to methodological relativism which legitimized public intervention, both commercially and socially. Neither did incipient theoretical Marxism make its mark; Bakunin's anarchism was introduced first, and when this happened it was initially separated from the workers' movement, and with a level of popularization in which the ideas of Marx arrived clearly transformed.

Undoubtedly, Spanish economists lost the thread of economic science in the second half of the 19th century; the main criticisms of classical thought came from their own ranks, where the protectionist change of tack that took place in the 1870s later combined with concern for social problems. This led to a defence of liberalism tempered by historicist influence which legitimized protectionism and social reform when not influenced by fundamentalist pseudo-religious postures which would later be taken up in corporatist thought. Likewise, the German Philosopher Krause's ideas, received through the influence of the philosophy of H. Ahrens, contributed to elaborating the first labour legislation of a paternalist nature in defence of the working class, which in these authors, amongst whom stand out G. de Azcárate, A. Álvarez Buylla and J.M. Piernas Hurtado, was compatible with free exchange and economic liberalism. In this breeding ground, tinged with a spirit of regeneration and extra-scientific approaches, the populist ideas of Henry George had a great impact, his *Progress and Poverty* having a tardy but enthusiastic reception in Spain.

The economic modernization of Spain

In the last few years of the 19th century and in the first few of the 20th century, the educational reform project carried out by the Institución Libre de Enseñanza allowed Spanish undergraduates to travel to foreign universities and become acquainted with doctrines developed beyond Spanish borders, especially in Germany. This was how the renewal of economic studies began, after its long stagnation in the previous century.

Antonio Flores de Lemus, the principle Spanish economist of the first half of the 20th century, was educated at the turn of the century in the lecture halls of German universities under the tuition of Schmoller and Wagner, and, on returning to Spain, became the great disseminator of a neo-historicist and realist stream in economic science which dominated Spanish science until the outbreak of the civil war in 1936. A university department chair in the service of the Inland Revenue, he did research on official economic statistics. The reformation of the Inland Revenue and monetary problems were some of the questions in which his intervention was decisive, such as the period between 1927 and 1929, when the possible return to the gold standard was being considered. Some of his successors, such as Luís Olariaga, an expert in monetary matters, Francisco Bernis, an Inland Revenue scholar, economist and statistician,

Olegario Fernández Baños, or Germán Bernácer, who developed a macroeconomic model with Keynesian connotations, distanced themselves from the neo-historicist stance in order to embrace more recent developments in economic science.

At the same time the influence of corporatist doctrines was equally making itself felt, based on *fin de siècle* conservative Christian thought, and an outward admiration for the Fascist movement and the corporatist doctrines of the Italians which quickly took root in the most conservative sectors of the Spanish intelligentsia. The Spanish corporatist model inspired economic policy during the dictatorship of Primo de Rivera (1923–30) and Franco's fledgling dictatorship, where its influence united with economic planning, on which the economic policy in the first years of Franco's autarky was based.

In any case, corporatism in Spain had neither the significance nor the duration which it enjoyed in Portugal, and had no level of analysis, being reduced to plans which were ostensibly superior to capitalism and socialism, but in which the economy was completely subordinated to political ends.

On the other hand, the marginalist revolution was almost unknown in Spain during the 19th century and for a good part of the 20th, with the exception of some teachers in the engineering schools. Later, thanks to the creation of the Faculty of Political and Economic Sciences at the University of Madrid (1943), neoclassical microeconomics began to spread in university lecture halls. Equally influential was the arrival in Spain in 1941 of Heinrich Von Stackelberg, an economist of German origin with a mathematical education. He was mainly responsible for training future university lecturers in economic theory, participating in courses at the Faculty of Economic Sciences, created only three years before his death. The delay in the reception of neoclassical thought is attributable to the economists' poor mathematical preparation and perhaps to the neo-historicist focus which prevailed in Spain in the first decades of the 20th century.

The reception of Keynesianism in Spain

Knowledge of the work of Keynes was hindered by the fact that the *General Theory* appeared only months before the outbreak of the Spanish civil war, though previously the worldwide dissemination of *The Economic Consequences of the Peace* had allowed Keynesian ideas to be known through the press, and Keynes himself had visited Spain in 1930. In the intense debate held on monetary questions in Spain in the period 1927–30, no influence can be detected of the ideas contained in the *Tract on Monetary Reform*. With the benefit of hindsight, during Keynes's visit, agreement can be detected between Keynes's ideas and those of the Spanish economists who had rejected the revaluation of the peseta because of its repercussions on production and employment,

especially with the *Dictamen sobre la implantación del patron oro* drawn up in 1929. In the years leading up to the Spanish civil war, Keynesian ideas were interpreted in the context of attempts to escape from the depression through economic policy which did not conform with the orthodox line: the promotion of public works at the cost of a budget deficit, the abandonment of the gold standard and the granting of credit facilities, in the framework of moderate protectionism.

The later reception of the *General Theory* was complicated, not only because of the civil war but also because the tradition of a more neo-historicist slant had prevailed in the academic world until then.

There was first 'underground' dissemination among a depleted group of Spanish economists during the civil war and the first few years of the Second World War. Later, however, in the 1940s, there was a wide dissemination of Keynesian literature within the context of wider discussion of post-war reconstruction policies and the framework necessary for stability. There was a varied reception to Keynesian thought by some economists such as Manuel de Torres, Emilio de Figueroa, or Joan Sardá, who in some cases considered the primary formulations of the neoclassical synthesis, while others combined the ideas of the *General Theory* with some contributions from Austrian and Scandinavian economists.

Another group of experienced economists, very knowledgeable of Keynesian literature, such as Germán Bernácer and Luís Olariaga, harboured reservations about Keynesian ideas, especially on an analytical level.

On the economic and political level, the interpretation is more complex due to the political situation during General Franco's long dictatorship. In this context, Keynesian proposals were used by the regime itself and by some of its principal advisors, such as Higinio Paris Eguilaz, to encourage employment policies and economic stability. This meant misuse on the part of an official authority which used a 'bastardized' Keynesianism identified with systematic interventionism, resulting in a confusing clash between Keynesianism and corporatist dirigisme, in which Franco's 'autarky' was propped up by an interventionist recipe of price and wage controls, rationing, credit manipulation, and an import substitution policy, rounded off by a system of licences and official authorizations. This model distanced itself completely from Keynesianism since, in reality, its economic strategy relied upon a revision of the assisted capitalist model which had begun in the time of the dictatorship of Primo de Rivera (1923–30), of clearly corporatist influence, and was now being updated in the light of pre-war national socialist experiences.

Together with these, there was a group of expert economists which used Keynesianism for establishing the basis of a more rational economy, and it was precisely from this group that the main criticism came of the autarkic regime, of its disastrous and suffocating interventionism and of its self-destructive effects. Within this new generation of economists educated at the new Facultad de Ciencias Económicas there were also subtle differences, among those who embraced the postures of the Ordo group and the school of Freiburg (liberals linked to research institutes, tolerated, but critical of Franco's regime), as well as some university lecturers who defended mixed viewpoints in which well-received Hayekian liberalism was tempered from viewpoints which appealed to a 'liberating' intervention along the lines of Röpke or Eucken.

The 1950s saw a considerable expansion of Keynesianism through debate by several disciples of Professor Manuel de Torres, such as Enrique Fuentes Quintana, Manuel Varela and Emilio de Figueroa on the applicability in Spain of the ideas contained in the *General Theory*; the debate was useful for advancing Keynesian ideas on unemployment towards more relevant problems which ran deep in Spanish economic backwardness, such as the limitations of Spanish industry and the consequences of very restricted international trade. It also tested awareness of the need to maintain applicable economic policy measures, with better knowledge of economic macro-magnitude principles, which led to the reactivation of the first studies in national accounting and the first estimations on the national revenue and its provincial distribution. In 1954, studies for the drafting of the first input–output table for the Spanish economy began under the direction of Professors Manuel de Torres and Valentín Andrés Álvarez who, with the assistance of the Italian economist Vera Cao-Pinna and the Instituto per la Cogiuntura di Roma, allowed the first table to be drawn up in 1958, with data referring to the Spanish economy of 1954.

Resistance to this expansion of Keynesianism was less than in other countries; abandonment of the neo-historicist line helped, of a realistic nature which had dominated the teaching of economics up to the Spanish civil war, along with the fact that, unlike in Portugal, economic corporatism did not attain an influence beyond the first moments of Franco's regime.

The clearest influence on Keynesian economic policy was the Stabilization Plan (1959), which helped to bring about a thawing in Franco's autarky, which by then had become untenable. A paradox existed in that, in reality, this late application of the Keynesian programme consisted of a 'cooling off' and restriction on the economy, with a view to an opening up of the Spanish economy to foreign markets. Later on, a more mature assimilation of Keynesian ideas came about, the main focus of which was the Bank of Spain's Research Department, where an econometric model was developed for the Spanish economy on a Keynesian base. Later, recognition of the effectiveness of monetary policy to combat inflation and macroeconomic imbalance contributed in an indirect way to the favourable penetration of the ideas of Milton Friedman, which were almost unknown in Spain until the 1980s.

Concluding remarks

Unlike in other countries, the academic institutionalization of economic studies, begun in the 19th century, did not gel into the creation of specific centres for the study of economic science, which until then had been taught in law schools, and later in schools of commerce at non-graduate level.

From the beginning of the 20th century there were several attempts to create a specific centre, which were not successful until 1943, with the creation of the School of Political, Economic and Commercial Sciences at the University of Madrid. This was the main factor which led to the consolidation of the academic study of economics in Spain, and to the fact that economic theory, first microeconomics and later macroeconomics, became known in the Spanish academic world. From these same lecture halls later in the 1960s Keynesian ideas spread, in their diverse formulations, initially the neoclassical synthesis, and later on in the 1960s and 1970s Keynesian macroeconomics in its truest interpretation and its later developments, as with the monetarist theories and those of the new macroeconomics.

Assimilation of economic theory and the training of economists constituted an essential step in that the policies of stabilization and the opening up to foreign markets, first within Franco's regime in a phase in which Spain, once the autarky had been exhausted and abandoned, began to be incorporated into the main international economic organizations and later on during the democratic transition. These policies allowed change from a closed economic model to an open economy, from which integration into the international economic panorama became possible.

ALFONSO SÁNCHEZ HORMIGO

See also **Bernácer, Germán; British classical economics; cameralism; Catholic economic thought; France, economics in (before 1870); Gournay, Jacques Claude Marie Vincent, Marquis de; heterodox economics; Historical School, German; Italy, economics in; mercantilism; physiocracy; Portugal, economics in; religion, economics of; scholastic economics; Stackelberg, Heinrich von.**

Bibliography

Almenar, S. 2005. Chair, tribune and seat: Spanish economists in parliament (1844–1923). An exploration. In *Economists in Parliament in the Liberal Age (1848– 1920)*, ed. M. Augello and M. Guidi. Aldershot: Ashgate.

Almenar, S. and Llombart, V. 2001. Spanish societies, academies and economic debating societies. In *The Spread of Political Economy and the Professionalisation of Economists*, ed. M. Augello and M. Guidi. London and New York: Routledge.

Almenar, S. and Llombart, V. 2007. *History of Spanish Economic Thought*. New York and London: Routledge.

Almenar, S. and Sánchez Hormigo, A. 2004. Bibliografía de historia del pensamiento económico en España. In *Economía y economistas españoles*, vol. 9, ed. E. Fuentes Quintana. Barcelona: Galaxia Gutenberg-Funcas.

Azpilcueta, Martín de. 1556. *Comentario resolutorio de cambios*, ed. J.M. Pérez Prendes and L. Pereña. Madrid: CSIC, 1965.

Campomanes, Conde de. 1774–1775. *Discurso sobre el fomento de la industria popular y Discurso sobre la educación popular de los artesanos y su fomento*, ed. J. Reeder. Madrid: Instituto de Estudios Fiscales, 1975.

Cardoso, J.L. and Lluch, E. 1999. Las teorías económicas contempladas a través de una óptica nacional. In *Economía y economistas españoles*, vol. 1, ed. E. Fuentes Quintana. Barcelona: Galaxia Gutenberg-Funcas.

Colmeiro, M. 1861. *Biblioteca de economistas españoles de los siglos XVI, XVII y XVIII*, ed. L. Perdices y J. Reeder. Madrid: IEF, Fundación ICO y Real Academia de Ciencias Morales y Políticas, 2005.

Estapé, F. 1990. *Introducción al pensamiento económico. Una perspectiva española*. Madrid: Espasa Calpe.

Flórez Estrada, A. 1828. *Curso de Economía Política*, ed. S. Almenar, 2 vols. Madrid: Instituto de Estudios Fiscales, 1980.

Fuentes Quintana, E., ed. 1999–2004. *Economía y economistas españoles*, 9 vols. Barcelona: Galaxia Gutenberg-Funcas.

Grice-Hutchinson, M. 1993. *Economic thought in Spain: Selected Essays*. Aldershot: Edward Elgar.

Jovellanos, G.M. de. 1795. Informe en el Expediente de Ley Agraria. In *G.M. de Jovellanos, Escritos económicos*, ed. V. Llombart. Madrid: I.E.F., Fundación ICO, Real Academia de Ciencias Morales y Políticas, 2000.

Llombart, V. 2004. Traducciones españolas de economía política: 1700–1812: catálogo bibliográfico y una nueva perspectiva. *Cyber Review of Modern Historiography*, No. 9. Firenze: Firenze University Press.

Lluch, E. 1973. *El pensament econòmic a Catalunya (1760–1840). Els orígens ideològics del proteccionisme i la pressa de consciència de la burgesia catalana*. Barcelona: Edicions 62.

Lluch, E. 1980. Sobre la historia nacional del pensamiento económico. In *A. Flórez Estrada, Curso de Economía*, ed. S. Almenar. Madrid: I.E.F.

Lluch, E. 1999. Las historias nacionales del pensamiento económico y España. In *Economía y economistas españoles*, vol. 1, ed. E. Fuentes Quintana. Barcelona: Galaxia Gutenberg-Funcas.

Martín Rodríguez, M. 1989. La institucionalización de los estudios de Economía Política en la Universidad Española (1784–1857). In *Marqués de Valle Santoro, Elementos de Economía Política con aplicación particular a España*. Madrid: I.E.F.

Ortiz, L.E. 1558. *Memorial del contador Luis de Ortiz a Felipe II*, ed. J. Larraz. Madrid: Instituto de España, 1970.

Perdices, L. and Reeder, J. 2003. *Diccionario de Pensamiento Económico en España (1500–2000)*. Madrid: Síntesis, Fundación ICO.

Perrotta, C. 1993. Early Spanish mercantilism: the first analysis of underdevelopment. In *Mercantilism: The Shaping of an Economic Language*, ed. L. Magnusson. London: Routledge.

Schwartz, P. 2000. The Wealth of Nations censured. Early translations in Spain. In *Contributions to the History of Economic Thought in. Essays in Honour of R.D. Collison Black*, ed. E. Murphy y R. Prendergast. London: Routledge.

Smith, R.S. 1968. English economic thought in Spain, 1776–1848. In *The Transfer of Ideas: Historical Essays*, ed. C.D. Godwin et al. Durham, NC: South Atlantic Quarterly.

Smith, R.S. 1971. Spanish mercantilism: a hardy perennial. *Southern Economic Journal* 38, 1–11.

Uztáriz, Jerónimo de. 1724. *Theórica y práctica de comercio y de marina*. Madrid: Aguilar, 1968.

Velarde, J. 1974. *Introducción a la historia del pensamiento económico español en el siglo XX*. Madrid: Editora Nacional.

spatial econometrics

Spatial econometrics is concerned with models for dependent observations indexed by points in a metric space or nodes in a graph. The key idea is that a set of locations can characterize the joint dependence between their corresponding observations. Locations provide a structure analogous to that provided by the time index in time series models. For example, near observations may be highly correlated but, as distance between observations grows, they approach independence. However, while time series are ordered in a single dimension, spatial processes are almost always indexed in more than one dimension and not ordered. Even small increases in the dimension of the indexing space permit large increases in the allowable patterns of interdependence between observations. The primary benefit of this modelling strategy is that complicated patterns of interdependence across sets of observations can be parsimoniously described in terms of relatively simple and estimable functions of objects like the distances between them.

The fundamental ingredients in any spatial model are the index space and locations for the observations. In contrast to the typical time series situation where calendar observation times are natural indices and immediately available, the researcher will often need to decide upon an index space and acquire measurements of locations/distances. The role of measured locations/distances is to characterize the interdependence between economic agents' variables, particularly those that are unobservable – for example, regression error terms. The appropriate index space depends on the economic application, and its choice is inherently a judgement call by the researcher. Fortunately, the economics of the application often provide considerable guidance and the index space/metric(s) can be tailored to promote a good fit between the economic model and the empirical work. For example, when local spillovers or competition are the central economic features, obvious candidate metrics are measures of transaction/travel costs limiting the range of the spillovers or competition. If productivity measurement were the focus, distances between observed firms or sectors could be based upon economic mechanisms that might generate co-movement in productivity – for example, measures of similarity between production technologies. Index spaces are not limited to the physical space or times inhabited by the agents and can be as abstract as required by the economics of the application.

Locations/distances are almost never perfectly measured, and this puts a premium on empirical methods that are robust to their mismeasurement. Even if the ideal metric were physical distance, usually agents' physical locations are imprecise, known only within an area – for example, census tract or county. At best this will result in imprecise distance information between agents, and if inter-agent distances are approximated with measurements based on these areas, such as distance between centroids, errors result. Moreover, in the great majority of applications the ideal metric is *not* physical distance and must be either estimated or approximated, inevitably resulting in some amount of measurement error.

There are two main approaches to modelling a spatial data generation process (DGP). The first is to model explicitly a population residing in an underlying metric space and the process of drawing an observed sample from this population. The second is to model the data-set of observed agents' outcomes as being determined by a system of simultaneous equations. In the remainder of this article, I discuss each of these approaches in turn for the simplest case of cross-sectional data. It is important to note, however, that the methods in the following section – covariance and generalized method of moments (GMM) estimation, spatial correlation robust inference – can be directly applied to panel or repeated cross-section data by simply including time as one of the components in the spatial index (*s* defined below). Most if not all cluster/group effect models can be considered a special case of spatial models with a binary metric indicating common group/cluster membership. See Wooldridge (2003) for an excellent review of these models. I do not discuss them here because their associated empirical techniques and sampling schemes do not translate well to more general spatial models. I conclude with a brief discussion of areas of econometrics where links to spatial econometrics are perhaps underappreciated.

1. Models for samples from a population

This section discusses spatial econometric models that view the data as being a sample from some arbitrarily large population (see, for example, Conley, 1999, for a more formal treatment). The population of individuals is assumed to reside in some metric space, typically $\Re^k$ or an integer lattice, with each individual i located at a point s_i.

The basic model of dependence characterizes dependence between agents' random variables via their locations. The data are assumed to be weakly dependent (perhaps after de-trending). (Andrews, 2005, is an important exception that explicitly considers strong cross sectional dependence arising from common shocks.) If two agents' locations s_i and s_j are close, then their random variables ϕ_{s_i} and ϕ_{s_j} may be highly dependent. As the distance between s_i and s_j grows large, ϕ_{s_i} and ϕ_{s_j} become essentially independent. Notions of weak dependence can be formalized in essentially the same manner as for time series, for example, with mixing coefficients. Under regularity conditions limiting the strength of dependence, laws of large numbers and central limit results can be obtained for properly normalized averages of ϕ_s. See, for example, Takahata (1983) or Bolthausen (1982). These approximations almost always use what is called an increasing domain approach to limits, with the corresponding thought experiment being that, as the sample size grows, an envelope containing the locations would be growing without bound.

When one works within this framework, it is often useful to approach an empirical problem in two steps. First, decide upon a (small) set of metrics based on the economics of the application, and then consider statistical modelling of dependence as a function of the metrics. It is much easier to conduct statistical modelling given a metric than to try to simultaneously vary both the model specification and the metric itself.

Statistics that describe spatial correlation patterns are simple to construct. Any statistic relating co-variation of ϕ_{s_i} and ϕ_{s_j} to some measure of their proximity could be used to characterize patterns in dependence. Classic references are Moran (1950) and Geary (1954), and the text by Cliff and Ord (1981) contains a good treatment. One useful approach is based on nonparametric estimation of a covariance function (see for example Conley and Topa, 2002, or Conley and Ligon, 2002). The ϕ_s process is covariance stationary if its expectation is the same at all locations and $cov(\phi_s, \phi_{s+h})$ depends only on the relative displacement h. For high-dimensional h, it is useful to consider a special case called isotropy where covariances depend only on the length of h; covariance depends upon distance but not direction. Take an isotropic covariance stationary ϕ_s with expectation zero for simplicity. Its covariance function f can be expressed in a regression equation involving distances $d_{i,j}$:

$$E(\phi_{s_i}\phi_{s_j}|s_i, s_j) = f(d_{i,j}). \qquad (1)$$

The function f in eq. (1) can be estimated parametrically or, as is particularly useful in preliminary data analysis, via a nonparametric regression of $\phi_{s_i}\phi_{s_j}$ on $d_{i,j}$. Investigation of correlation patterns when there is more than one candidate metric can by done by simply letting f be a function of more than one distance measure.

In cases where ϕ_s is not isotropic or non-stationary, f can still be interpretable as a measure of average co-movement. If the process is covariance stationary but not isotropic, an estimate at a given distance d_0, call it $\hat{f}(d_0)$, will converge to a weighted average of $cov(\phi_s, \phi_{s+h})$ for displacements h that have length d_0. The relative weights of different directions h will depend on their frequency of sampling. An analogous interpretation of f is available when ϕ_s is non-stationary, $cov(\phi_s, \phi_{s+h})$ depends on s, but still weakly dependent with averages of $cov(\phi_s, \phi_{s+h})$ across s remaining convergent. In this case, $\hat{f}(d_0)$ will converge to a weighted average of $cov(\phi_s, \phi_{s+h})$ across those h with length d_0 and across all s. Typically, this is still a valuable measure of co-movement. If non-stationarity is suspected, it is also very useful to construct localized versions of measures of spatial correlation. Localized f estimates for subregions of the locations can easily be constructed by just confining the observations used to estimate (1); see Anselin (1995) for extensive treatment of localized versions of Moran (1950) and Geary (1954) measures of spatial correlation.

Estimates of f can also be viewed directly as test statistics for the null hypothesis of independence. Under the null hypothesis of independence, the sampling distribution of an f estimator can be approximated and compared to the realized value of f estimates to test the hypothesis of independence. Such tests for independence remain valid even with measurement errors in distances (see Conley and Ligon, 2002).

Parameter estimation via moment conditions
In most econometric applications, the parameters of interest can be estimated using GMM. GMM estimation with weakly spatially dependent data is straightforward, and the spatial dependence is relevant for inference and efficiency (see Conley, 1999). Consider instrumental variables (IV) estimation in the linear model with outcome y_{s_i}, regressors x_{s_i}, and instruments z_{s_i}

$$y_{s_i} = x'_{s_i}\beta + u_{s_i}$$
$$\text{and} \tag{2}$$
$$Ez_{s_i}u_{s_i} = 0$$

The IV estimator is identified by the moment condition (2): that the instruments are not correlated with the error term. Since this is a moment condition with respect to the marginal distribution of the data across agents, it is valid with or without spatial dependence. The familiar solution remains: $\beta = \left(Ez_{s_i}x'_{s_i}\right)^{-1}Ez_{s_i}y_{s_i}$. Consistent estimates of β can be obtained using sample averages to approximate these expectations since a law of large numbers applies to weakly dependent spatial data. Thus,

the usual IV estimator, $\hat{\beta}_N = (\frac{1}{N}\sum_{i=1}^{N}z_{s_i}x'_{s_i})^{-1}\frac{1}{N}\sum_{i=1}^{N}z_{s_i}y_{s_i}$, remains consistent with weak spatial dependence. It is of course feasible to construct $\hat{\beta}_N$ without any knowledge of locations/distances, so it is trivially robust to measurement error in them. The impact of such spatial dependence is only upon inference, getting correct standard errors or testing.

This logic carries over to any GMM estimator of a parameter θ_0 that is identified from a moment condition involving a (potentially nonlinear) function g:

$$Eg(\phi_s; \theta_0) = 0.$$

The majority of econometric models with nonlinearity or limited dependent variables can be estimated via some choice for g. Under mild regularity conditions, θ_0 can be consistently estimated by minimum distance methods using $\frac{1}{N}\sum_{i=1}^{N}g(\phi_{s_i}; \cdot)$ to approximate $Eg(\phi_s; \cdot)$. A GMM estimator is the argument minimizing the criterion function, $J_N(\theta)$, which takes the same form as with time series or independent data:

$$J_N(\theta) = \left[\frac{1}{N}\sum_{i=1}^{N}g(\phi_{s_i}; \theta)\right]'\Omega\left[\frac{1}{N}\sum_{i=1}^{N}g(\phi_{s_i}; \theta)\right],$$

where Ω is some positive definite matrix. Just as for the time series case (Hansen, 1982), an efficient GMM estimator can be obtained by taking Ω to be a consistent estimator of the limiting variance-covariance matrix of $\frac{1}{\sqrt{N}}\sum_{i=1}^{N}g(\phi_{s_i}; \theta_0)$, whose form depends on the spatial covariance structure of the data. One such covariance matrix estimator is described in the following subsection.

Inference
The usual approach to inference using large sample approximations can be employed with weakly spatially dependent data. Returning to the IV model, the typical approximation for the distribution for $\hat{\beta}_N$ is based on the expression:

$$\sqrt{N}(\hat{\beta}_N - \beta) = \left(\frac{1}{N}\sum_{i=1}^{N}z_{s_i}x'_{s_i}\right)^{-1}\left[\frac{1}{\sqrt{N}}\sum_{i=1}^{N}z_{s_i}u_{s_i}\right]. \tag{3}$$

Under regularity conditions, the first term in the product converges to the matrix $Ez_{s_i}x'_{s_i}$. The second term in brackets has a limiting normal distribution:

$$\frac{1}{\sqrt{N}}\sum_{i=1}^{N}z_{s_i}u_{s_i} \Rightarrow N(0, V) \tag{4}$$

where V is the limiting variance-covariance matrix of

$$\frac{1}{\sqrt{N}}\sum_{i=1}^{N}z_{s_i}u_{s_i}.$$

V contains terms of the form $Ez_{s_i}u_{s_i}z'_{s_i}u_{s_i}$ and cross-covariance terms, $Ez_{s_i}u_{s_i}z'_{s_j}u_{s_j}$, that will be non-zero for at least some i,j pairs. With weak dependence, the covariance between variables indexed i and j will eventually vanish as the distance between s_i and s_j grows.

In some cases, V has a nice form. For example, suppose locations were on an integer lattice, Z^k; samples consist of all integer coordinates in a region (assumed to grow as $N \to \infty$); and variables are covariance stationary. In this case, V can be expressed as an infinite sum of a covariance function:

$$V = \sum_{h \in Z^k} Cov(z_s u_s, z_{s+h} u_{s+h}). \qquad (5)$$

With integer locations on the line, this expression coincides with its analog for covariance stationary time series.

With a consistent estimate of V, call it $\hat{V}_N$, the approximate distribution implied by (3) and (4) can be used for inference:

$$\sqrt{N}(\hat{\beta}_N - \beta) \overset{Approx}{\sim} N\left(0, \left(\frac{1}{N}\sum_{i=1}^{N} z_{s_i} x'_{s_i}\right)^{-1} \right.$$
$$\left. \times \hat{V}_N \left(\frac{1}{N}\sum_{i=1}^{N} z_{s_i} x'_{s_i}\right)^{-1'}\right).$$

There are of course many ways V could be estimated. If it were assumed to have a parametric form, for example, by parameterizing the covariance function in (5), then consistent estimates could be obtained by GMM. Perhaps the most popular approach has been nonparametric estimation of V following Conley (1996; 1999). This approach is analogous to time series heteroskedasticity and autocovariance (HAC) consistent covariance matrix estimation, and can be viewed as a smoothed periodogram spectral density estimator. (See Priestley, 1981, for a discussion of the vast literature on spectral methods in time series and some extensions to spatial processes. Spectral methods for spatial processes date back to at least the 1950s; for example, Whittle, 1954; Bartlett, 1955; Grenander and Rosenblatt, 1957; Priestley, 1964). With the use of residuals $\hat{u}_{s_i}$ to approximate u_{s_i}, V can be estimated as a weighted sum of cross products $z_{s_i}\hat{u}_{s_i}z'_{s_j}\hat{u}_{s_j}$:

$$\hat{V}_N = \frac{1}{N}\sum_{i=1}^{N}\sum_{j=1}^{N} K_N(s_i, s_j) \cdot z_{s_i}\hat{u}_{s_i}z'_{s_j}\hat{u}_{s_j}.$$

$K_N(\cdot, \cdot)$ is a kernel used to weight pairs of observations, with close observations receiving a weight near 1 and those far apart receiving weights near zero. $K_N(s_i,s_j)$ is commonly specified to be uniform kernel that is 1 if s_i and s_j are within a cut-off distance and zero otherwise. (This indicator function K_N is not guaranteed to provide positive definite (PD) covariance matrix estimates; however, this is very rarely a problem in practice. PD estimates can be insured by an alternate choice of kernel; see Conley, 1999.) $\hat{V}_N$ will be consistent if as $N \to \infty$, $K_N(s, s+h) \to 1$ for any given displacement h, but slowly enough so that the variance of $\hat{V}_N$ collapses to zero.

In practice, this estimator will require a decision about the exact form of $K_N(\cdot, \cdot)$. With a uniform kernel, this is just an operational definition of which observations are near and which are far. A conservative distinction between near and far observations can be made even with multiple candidate metrics by assigning a far classification only when all metrics agree. There is no need for the data to be covariance stationary, nor is the specific sampling framework here necessary. Analogous HAC methods can be applied to weakly dependent but nonstationary data, including that generated by simultaneous equations DGPs like those discussed in the following section 2 (see Pinkse, Slade and Brett, 2002; Kelejian and Prucha, 2007).

The main reason nonparametric estimators like $\hat{V}_N$ are often preferred to parametric models for V is their robustness to measurement errors in locations/distances. Parametric V estimators are generally inconsistent with such errors, while $\hat{V}_N$ remains consistent under mild conditions. $\hat{V}_N$ can be consistent with spatially correlated and even endogenous errors; a sufficient condition is simply that they be bounded (Conley, 1999). With location/distance errors, the weight assigned to pair i,j can be altered relative to the weight $K_N(\cdot, \cdot)$ would assign with exact locations. But $\hat{V}_N$ remains consistent, because the altered weights will still satisfy the necessary conditions for consistency of $\hat{V}_N$: the weight on observations at any true displacement will still converge to 1, slowly enough. Even when working with parametric models of V, $\hat{V}_N$ remains of interest since the discrepancy between it and a parametric V estimator can provide a useful joint test for the absence of location/distance errors and proper parametric specification (Conley and Molinari, 2007).

More important than $\hat{V}_N$ remaining consistent is its robustness in practice to moderate amounts of location error. Consider the impact of introducing location error for $\hat{V}_N$ defined with a kernel $K_N(s_i,s_j)$ equal to 1 if s_i and s_j are within L_N units, and zero otherwise. If the magnitude of measurement error is moderate relative to L_N, then the weights on most pairs of points would be unchanged if erroneously measured locations were used in place of true locations. Changes in weights occur only for those points whose true distance is near enough to the cut-off L_N that location errors result in measured and true distances being on opposite sides of L_N. With moderate amounts of location error, these pairs of observations with true distance near L_N will usually not be a large portion of the sample, so $\hat{V}_N$ will tend to be close to its value with true locations. Similar results obtain for other kernels as weights arising from moderately mismeasured locations remain close to those received with true locations (see Conley and Molinari, 2007).

2. Population simultaneous equation models

The second approach to modelling spatial data is with a simultaneous equations model, most directly interpretable as a model for a *population* of N agents. This approach explicitly specifies a joint model for the population, in contrast to typical models in Section 1, where the joint determination of outcomes in the population is not explicitly treated. These simultaneous equation models are directly applicable to situations where the entire population of agents is observed, like all US states or counties or even all firms in an industry. Typical applications include studies of games being played among these agents or of spillovers across agents; see, for example, Case, Hines and Rosen (1993) and Pinkse, Slade and Brett (2002).

The most common type of model is a simultaneous spatial autoregression (SAR). Its simplest formulation for an $N \times 1$ outcome vector Y_N is:

$$Y_N = \rho W_N Y_N + \varepsilon_N, \qquad (6)$$

with scalar parameter ρ and IID shocks ε_N (typically Gaussian). The $N \times N$ matrix W_N is commonly referred to as a 'spatial weights' matrix and assumed known. W_N has zero main-diagonal elements, and its off-diagonal elements reflect some notion of interaction. Typical W_N contain (i,j) elements that are non-zero only if locations i and j are adjacent on a graph or elements inversely related to distances between locations. W_N is usually row-standardized so that its rows sum to 1. The parameter space is restricted so that $(I - \rho W_N)^{-1}$ exists and the model has reduced form:

$$Y_N = (I - \rho W_N)^{-1} \varepsilon_N.$$

Thus Y_N is a linear combination of the ε_N IID shocks. Though SAR models are finite (usually) irregular lattice models, their origins date to at least the infinite regular lattice models of Whittle (1954). Textbook treatment of SARs can be found in Anselin (1988).

Typical specifications for W_N imply a great deal of heterogeneity across observations. Variances will typically differ across the elements of Y_N by construction unless $\rho = 0$. Unconditional heteroskedasticity is thus coupled with spatial dependence. Covariances between pairs of agents will differ in patterns that are of course determined by W_N but will depend on the entire structure of this matrix and will not generally follow a simple pattern in terms of some metric. For example, with W_N defined based upon a graph, covariance between agents i and j will not be a function of their graph distance, though it can be characterized in terms of properties of the graph (Martellosio, 2004). A given graph will 'hard-wire' patterns in correlations across agents. For example Wall (2004) notes, with model (6) for US states with W_N based on adjacency, that Missouri and Tennessee are

constrained to be the least spatially correlated states, while relative correlations between other pairs of states change depending on ρ. Even with a more flexible parameterization – for example, specifying the elements of W_N to be flexible functions of distance, as in Pinkse, Slade and Brett (2002) – there is still a tendency for heterogeneity in the implied joint distribution to be difficult to anticipate. While this complicates their use as statistical models, as discussed below, it is in my view likely to be a desirable property in a structural model. For example, if the model's joint distribution is to be taken seriously as capturing equilibrium outcomes for N asymmetric agents playing a game, then one would expect 'hard-wired' heterogeneity depending on the exact structure of the game.

Though the population of agents is observed, large sample approximations taking limits as $N \to \infty$ are still potentially useful. However, the requisite limit theorems technically differ from those referenced in Section 1. Since the DGP is changing as N grows, triangular array limit results are required. Consistency and distribution results for Gaussian maximum likelihood estimators (MLEs) with spatial dependence have existed at least since Mardia and Marshall (1984). An extensive set of SAR limiting distribution results is obtained by Lee (2004a) for likelihood-based estimators under a variety of conditions upon 'spatial weights' matrices like W_N. Quite useful limit theorem results can also be found in Kelejian and Prucha (2001). Correct specification of W_N is essential for these results, as SAR estimators will generally be inconsistent when there is measurement error in locations/distances used to specify this matrix (the same holds true for other parametric models of dependence structure).

A great deal of the literature has focused on computational issues involving MLEs. Non-trivial W_N matrices make computation of normalizing constants challenging. Substantial progress has been made in techniques for computing MLEs by exploiting sparseness or specific structure of 'spatial weights' matrices and re-parameterization to facilitate computation (see Pace and Barry, 1997; Barry and Pace, 1999; LeSage and Pace, 2007). These numerical techniques allow likelihood-based inference for even very large data-sets in certain applications or specifications. It is also feasible, of course, to estimate SAR parameters without computing MLEs, by using only a subset of the implications of the model to obtain method of moments estimates (see Kelejian and Prucha, 1999, and Lee, 2007, and subsequent work by these authors). This literature has been successful in addressing most computational issues with SAR models.

The key remaining difficulties in using SAR models are in terms of model specification and interpretation. Even for the simplest SAR model (6), it is hard to characterize implications of different ρ without explicitly calculating their implied joint distributions. The parameter ρ is not a simple correlation coefficient; in general it is not comparable across different specifications for W_N. In my experience, explicit calculations of descriptive measures of the implied joint distributions for many different ρ are required to understand whether varying this parameter will trace out a useful path through the space of joint distributions.

Unless one has access to virtually complete data on a population, SAR models are very difficult to properly specify as structural models. To take an optimistic case, suppose model (6) with Gaussian ε applied to a population of N agents, but a subset of agents were sampled. The likelihood of such a sample is well-defined, and in principle its form could be found by integrating out all the unobserved variables. But this calculation requires the exact form of W_N, which depends on all the unobserved agents, a full structure which will rarely be observed if only a small fraction of the agents are sampled. Proper specification of W_N is perhaps feasible only if the vast majority of the population is sampled – for example, if only a few states or counties are missing.

Even with complete data on a population, SARs are difficult to specify because they are inherently fragile. Changing a single element of W_N will in general influence the entire joint distribution of Y and it is difficult to intuitively understand the impact of a given change in W_N. Increasing flexibility by parameterizing W_N by taking its elements to be a series expansion in distance(s), as in Pinkse, Slade and Brett (2002), is of limited help. There remains only an indirect link between the series expansion and the implied joint distribution. It is hard to see how much additional flexibility in, for example, allowed covariance structure is gained by adding another term in the expansion.

I think these difficulties should be considered a consequence of modelling a large-dimensional system of structural simultaneous equations rather than SAR-specific problems. It seems likely to be difficult to anticipate changes in equilibrium outcomes resulting from changes in individual agents' decision rules or best-response functions in any modelling framework. In my view, SARs remain a useful first step towards the goal of constructing good large-dimensional structural simultaneous equation models.

Of course SAR models need not be intended as structural models; they can be viewed, for instance, as tools to incorporate spatial dependence into forecasting models. A mis-specified but parsimonious model might still forecast well. However, the cumbersome relation between specification of 'spatial weights' and the implied joint distribution makes it hard to fashion parsimonious SAR models. This seems ample reason to avoid their use in forecasting. Directly specifying measures of dependence like covariances as a parsimonious function of distance appears far easier, even if the true DGP were an SAR.

3. Links between spatial econometrics and other areas

Work on interactions-based models has much in common with simultaneous equations-style spatial models (see Brock and Durlauf, 2001, for an extensive review). In these models, the behaviour of individuals is influenced by the characteristics and/or behaviour of others. Insofar as the relevant set of 'others' can be described in a spatial framework, they can be thought of as spatial econometric models. Much of this work is theory, taking the approach of specifying conditional probability measures to capture individuals' behaviours and then deriving the implied properties of the compatible joint distribution(s). Empirical work with these models has just begun and will share many of the same challenges described above; some can even be cast directly as SARs (see Lee, 2004b).

Spatial models are potentially very useful in modelling high-dimensional vector time series. Limited degrees of freedom with typical length samples require substantial restrictions upon the DGP to make progress. The potential of spatial models to capture complicated interdependence with a small number of parameters (given auxiliary location/distance information) makes them well suited for use in characterizing a variety of restrictions upon high-dimensional vector DGPs. Good examples of the benefits of spatial approaches to this type of time series modelling are Chen and Conley (2001), Giacomini and Granger (2004), and Bester (2005a; 2005b).

TIMOTHY G. CONLEY

See also **generalized method of moments estimation; heteroskedasticity and autocorrelation corrections; social interactions (empirics); spectral analysis; stratified and cluster sampling; statistical mechanics.**

Bibliography

Andrews, D. 2005. Cross-section regression with common shocks. *Econometrica* 73, 1551–85.

Anselin, L. 1988. *Spatial Econometrics: Methods and Models.* Boston: Kluwer Academic Publishers.

Anselin, L. 1995. Local indicators of spatial association. *Geographical Analysis* 27, 93–115.

Barry, R. and Pace, R. 1999. A Monte Carlo estimator of the log determinant of large sparse matrices. *Linear Algebra and its Applications* 289, 41–54.

Bartlett, M. 1955. *An Introduction to Stochastic Processes.* Cambridge: Cambridge University Press.

Bester, C. 2005a. Random field and affine models for interest rates: an empirical comparison. Working paper, University of Chicago.

Bester, C. 2005b. Bond and option pricing in random field models. Working paper, University of Chicago.

Bolthausen, E. 1982. On the central limit theorem for stationary mixing random fields. *Annals of Probability* 10, 1047–50.

Brock, W. and Durlauf, S. 2001. Interactions-based models. In *Handbook of Econometrics* 5, ed. J. Heckman and Leamer. Amsterdam: North-Holland.

Case, A., Hines, J. and Rosen, H. 1993. Budget spillovers and fiscal policy interdependence: evidence from the states. *Journal of Public Economics* 52, 285–307.

Chen, X. and Conley, T. 2001. A new semiparametric spatial model for panel time series. *Journal of Econometrics* 105, 59–83.

Cliff, A. and Ord, J. 1981. *Spatial Processes*. London: Pion Limited.

Conley, T. 1996. Econometric modeling of cross-sectional dependence. Ph.D. thesis, University of Chicago.

Conley, T. 1999. GMM estimation with cross sectional dependence. *Journal of Econometrics* 92, 1–45.

Conley, T. and Ligon, E. 2002. Economic distance, spillovers, and cross country comparisons. *Journal of Economic Growth* 7, 157–87.

Conley, T. and Molinari, F. 2007. Spatial correlation robust inference with errors in location or distance. *Journal of Econometrics* 140(1), 76–96.

Conley, T. and Topa, G. 2002. Socio-economic distance and spatial patterns in unemployment. *Journal of Applied Econometrics* 17, 303–27.

Geary, R. 1954. The contiguity ratio and statistical mapping. *Incorporated Statistician* 5, 115–45.

Giacomini, F. and Granger, C. 2004. Aggregation of space–time processes. *Journal of Econometrics* 118, 7–26.

Grenander, U. and Rosenblatt, M. 1957. Some problems in estimating the spectrum of a time series. *Proceedings of the Third Berkeley Symposium on Mathematical Statistics and Probability* 7, 77–93.

Hansen, L. 1982. Large sample properties of generalized method of moments estimators. *Econometrica* 50, 1029–54.

Kelejian, H. and Prucha, I. 1999. A Generalized moments estimator for the autoregressive parameter in a spatial model. *International Economic Review* 40, 509–33.

Kelejian, H. and Prucha, I. 2001. On the asymptotic distribution of the Moran I test statistic with applications. *Journal of Econometrics* 104, 219–57.

Kelejian, H. and Prucha, I. 2007. HAC estimation in a spatial framework. *Journal of Econometrics* 140(1), 131–54.

Lee, L. 2004a. Asymptotic distributions of quasi-maximum likelihood estimators for spatial autoregressive models. *Econometrica* 72, 1899–925.

Lee, L. 2004b. Identification and estimation of spatial econometric models with group interactions, contextual factors and fixed effects. Working paper, Ohio State University.

Lee, L. 2007. GMM and 2SLS estimation of mixed regressive, spatial autoregressive models *Journal of Econometrics* 140(1), 155–89.

LeSage, J. and Pace, R. 2007. A matrix exponential spatial specification. *Journal of Econometrics* 140(1), 190–214.

Mardia, K. and Marshall, R. 1984. Maximum likelihood estimation of models for residual covariance in spatial regression. *Biometrika* 71, 135–46.

Martellosio, F. 2004. The correlation structure of spatial autoregressions. Working paper, University of Southampton.

Moran, P. 1950. Notes on continuous stochastic phenomena. *Biometrika* 37, 17–23.

Pace, R. and Barry, R. 1997. Quick computation of regressions with a spatially autoregressive dependent variable. *Geographical Analysis* 29, 232–47.

Pinkse, J., Slade, M. and Brett, C. 2002. Spatial price competition: a semiparametric approach. *Econometrica* 70, 1111–53.

Priestley, M. 1964. Analysis of two-dimensional processes with discontinous spectra. *Biometrika* 51, 195–217.

Priestley, M. 1981. *Spectral Analysis and Time Series*, 2 vols. New York: Academic Press.

Takahata, H. 1983. On the rates in the central limit theorem for weakly dependent random fields. *Zeitschrift fur Wahrscheinlichkeitstheorie und verwandte Gebiete* 64, 445–56.

Wall, M. 2004. A close look at the spatial structure implied by the CAR and SAR models. *Journal of Statistical Planning and Inference* 121, 311–24.

Whittle, P. 1954. On stationary processes on the plane. *Biometrika* 2, 434–49.

Wooldridge, J. 2003. Cluster-sample methods in applied econometrics. *American Economic Review* 93, 133–8.

spatial economics

What is spatial economics? In a nutshell, spatial economics is concerned with the allocation of (scarce) resources over space and the location of economic activity. Depending on how this definition is read, the realm of spatial economics may be either extremely broad or rather narrow. On the one hand, economic activity has to take place somewhere so that spatial economics may be concerned with anything that economics is concerned about. On the other hand, location analysis focuses mostly on one economic question, namely, location choice. This is only one decision among a large number of economic decisions.

Which boundaries for spatial economics?

In practice, we can distinguish three sets of questions for which the importance of the spatial dimension is very different. Consider first the core questions of spatial economics. For example, why are there cities? Why do some regions prosper while others do not? Why do we observe residential segregation? Why do firms from the same industry cluster? These are intrinsically 'spatial' questions, that is, questions in which the spatial dimension plays a dominant role. For instance, it would be difficult to speak meaningfully about the existence and growth of cities without some explicit consideration of space. Then there is a second group of issues concerned with, *inter alia*, the analysis of technological spillovers, the determinants of trade flows, or even the functioning of social networks. These issues all have a spatial dimension but its importance remains to be determined. Put differently, these are 'contested' issues between spatial economists and other economists. Finally, there is an extremely broad range of economic questions for which the spatial dimension is likely to be less important. For example, what are the drivers of investment? How important are firing costs to explain unemployment? What are the returns to education? To answer these questions, the main role of space is to provide possibly a major source of variation for empirical research. If, for instance, different regions of a country have different education systems with, say, different age limits for compulsory schooling, this variation can be used to produce some meaningful estimates of the returns to education. Even if we take this third set of questions to lie outside spatial economics, it nonetheless remains the case that spatial economics is concerned with a broad and heterogeneous set of questions, which involve very different spatial scales (from the very small to the very big) with imprecise boundaries. It is quite possible that this breadth and heterogeneity has hindered the development of the field. This is also what makes it interesting.

The centrality of spatial frictions

To see what makes spatial economics specific, it is useful to reformulate the question about its definition in the following way. Is spatial economics only about adding a spatial dimension? In the *Theory of Value*, Debreu (1959) answers affirmatively. A commodity is defined by all its characteristics including its location: the same good traded in different locations must be treated as different commodities. This 'answer' runs into serious problems, as pointed most clearly by Starrett (1974). Consider the extreme case of homogenous space where firms face the same convex production set, and consumer preferences are the same (and locally not satiated). Transporting commodities between locations is costly. Then the *spatial impossibility theorem* states that, with a finite number of locations, consumers, and firms, no equilibrium involves transportation. The intuition behind this result is straightforward: since economic activities are perfectly divisible and agents have no objective reason to distinguish between locations, each location operates in autarchy to save on transport costs. To avoid this very counterfactual result (no trade), one of the assumptions behind the spatial impossibility theorem needs to be relaxed. If one takes transport costs as an unavoidable fact of life, one must assume either some non-homogeneity of space or some non-convexity of production sets.

As shown by a first branch of trade theory, it is possible to develop a framework for spatial economics that builds only on local productivity differences. This approach was pioneered by Ricardo (1821), who developed a theory of land use based on relative fertility. This approach was later generalized to consider exogenous technological differences for all types of goods. A second branch of trade theory builds instead on differences in factor endowments over space. This is the so-called Hecksher–Ohlin theory of trade. The Ricardian and Hecksher–Ohlin approaches have led to sophisticated theories of location and trade that rely on the existence of (exogenous) 'comparative advantages' across locations. As shown by a large body of theoretical work in international trade, these approaches can be readily incorporated in the Arrow–Debreu framework. Although these approaches are central to the sister discipline of international trade, they played a much less important role in the development of spatial economic theory.

The pioneers

Instead, spatial economics has focused on the existence of non-convexities in the presence of transport costs. A key reason for this focus is that, although comparative advantage constitutes an appealing explanation for understanding trade flows at the world level, it provides at best a partial explanation for the location patterns of industries within countries, and it is at pains to explain major concentrations of population in large metropolitan areas. Instead, non-convexities in production or consumption seem to hold more promise for providing convincing answers to the core questions of spatial economics. The easiest way to model these non-convexities is to assume some indivisibility in a partial equilibrium framework. This type of work was pioneered by von Thünen (1826). In his model, a competitive farming sector bids for some homogenous land. The key non-convexity is that the output must be sold at a central market. With costly transport costs, farmers are willing to bid for land up to the point where the rent at a given distance from the market is equal to the gross revenue from the output minus the cost of non-land factors and minus transport costs. With a competitive land market, land goes to the highest bidder and the equilibrium typically involves concentric rings of specific land use around the central market.

To understand land use patterns in cities, Alonso (1964) developed an approach that was based on similar principles. His model again assumes a homogenous space, but replaces von Thünen's market by a central business district to which residents must commute at a cost to find work. This very parsimonious microeconomic model manages to replicate key stylized facts about land use and land prices within cities using rigorous microeconomic modelling. This is a showcase for the power of microeconomic approaches. It has spawned a large literature, which first extended the basic model in a number of directions and then went on to model multi-centric cities (for further details, see Fujita, 1989).

Independently of the 'Thünen tradition' that relies on an exogenous focal point for trade or production, another tradition was developed following Weber's (1909) work. Weber deals with the location problem of an indivisible and competitive plant that faces transport costs in order to ship its inputs from their sources and its outputs to their markets. With the use of essentially linear-programming techniques, the optimal location (which minimizes total transport costs) can be derived. Like Alonso's monocentric model, Weber's approach has been extended in many directions to consider, among others, more flexible production functions and the optimal location of public facilities.

Hotelling (1929) also explored the location problem faced by producers but went in a very different direction. His fundamental insight is that, because of indivisibilities, there will not be infinitely many producers at each point so that Weber's price-taking assumption is not tenable. With a small number of producers the location decision will involve more than minimizing transport costs since location also affects the competitive process. To make his point, he assumes evenly distributed consumers over a finite segment, each consuming one unit of a homogenous good. The market is served by two firms that need to choose a location and each customer patronizes the firm that minimizes the sum of the 'mill' price and shipping costs. At a first stage, firms choose a location and then compete in price. This deceptively simple game has received a lot of attention. Firms face a fundamental trade-off between central locations, which allow them to capture a larger share of the market, and more peripheral locations, which allow them to mitigate the intensity of competition. The resolution of this trade-off depends on the fine details of the assumptions being made (and particularly how an increase in the distance affects the price setting power of producers). Difficulties with existence of equilibrium have also turned out to be an important issue in this literature.

Modern approaches to spatial economic theory

Non-convexities in production lie at the heart of spatial economics. The literature discussed so far treats them as exogenous. It was not long before the literature started to worry about what these non-convexities were about. Nowhere was this worry stronger than in the 'new urban economics' literature, where Alonso's assumption of an exogenously given central business district quickly started to look very ad hoc. To understand central business districts or, more generally, why economic activity agglomerates, spatial economics had to provide microeconomic foundations for (local) increasing returns. Being able to generate increasing returns from plausible assumptions without leading to a degenerate market structure (for example, a monopoly firm for the entire economy)

was a fundamental challenge for spatial economics. This was also true for many other fields such as industrial organization and international trade, where increasing returns were also needed to explain key stylized facts. Spatial economists were fortunate because they could draw on the insights provided by Adam Smith (1776) and Alfred Marshall (1890). Although Smith's argument about the division of labour being limited by the extent of the market pre-dates Marshall's *Principles* by more than a century, Marshall's 'magic trilogy' proved much more influential. Following Marshall, local increasing returns could arise because of knowledge spillovers, linkages between input suppliers and final producers, and thick local labour market interactions. What the modern literature on the microfoundations of increasing returns has achieved is a formalization of these insights (see Duranton and Puga, 2004, for an extensive review of this literature). Three main mechanisms can be used to generate local increasing returns: sharing, matching, and learning. Sharing mechanisms show how small non-convexities like small fixed costs paid by heterogeneous producers can be spread across larger quantities as market size increases and thus yield aggregate increasing returns. Matching mechanisms explore how larger markets might improve the probability and quality of matching. Finally, learning mechanisms explore the benefits of local size for the creation and diffusion of knowledge.

The second major problem faced by spatial economics is that many fundamental issues having to do with regional and urban development call for general equilibrium modelling. For instance, some cities can afford to specialize because they can trade with other cities. Hence, looking at one isolated city in the tradition of Thünen and Alonso may not be enough for some purposes. Similarly, the agglomeration of economic activity in core regions may occur because firms find larger markets there and because consumers find cheaper and more diverse supplies. These two forces are mutually reinforcing. This is the famous circular and cumulative causation mechanisms first emphasized by Myrdal (1957).

To model spatial economies, two main approaches came to dominate the intellectual landscape. The first follows the work of Henderson (1974) and is know as the 'urban systems' approach. In this type of framework, cities arise endogenously as the result of a trade-off between agglomeration economies and urban crowding. Both types of forces are modelled with the use of various microeconomic foundations. Cities can also trade with one another and the workers decide where to work. This strand of literature has been successful in replicating many stylized facts about urban systems, from the tendency of many cities to specialize while others diversify to their role in the innovation process.

Following the work of Krugman (1991), the 'new economic geography' is the second main general equilibrium approach in spatial economics. This approach puts trade costs (rather than commuting costs in urban systems) at the heart of the agglomeration–dispersion trade-off. Agglomeration in the larger market is beneficial for firms because it gives them better access to consumers. Following this, workers also want to be in the larger market in order to be able to buy goods without having to pay inter-regional trade costs. Krugman's model is based on Dixit and Stiglitz's (1977) model of product differentiation and offers a formalization of Myrdal's circular and cumulative causation. It goes beyond that because agglomeration is not always an equilibrium outcome. This is because, under agglomeration, most goods sold in the periphery need to be shipped from the core and thus prices there may be quite high. In turn, this can make it profitable for firms to locate in the periphery. When trade costs are high, the even dispersion of manufacturing is indeed the unique equilibrium in Krugman's model. On the other hand, when trade costs are low, serving the residual demand in the periphery can be achieved at a low cost and agglomeration occurs. This strand of literature has grown exponentially since 1990, culminating with Fujita, Krugman, and Venables's (1999) book.

The difficulties of spatial empirical work

What about the evidence, then? Ultimately, it is observation that should allow us to judge of the relevance of our theories. To discuss very briefly what the issues for empirical work are, it is useful to retain the partial versus general equilibrium distinction made above. A typical 'partial equilibrium' question that has attracted much attention over the years is that of location choices of new firms. To look at the determinants of location choice, one would like to somehow explain location choices in terms of a bunch of possible determinants. This is a difficult exercise, for several reasons. The first one has to do with the nature of the problem. Location choices are not continuous. Because they are discrete, firms decide to locate 'somewhere' rather than be spread continuously. Put differently, new firms choose between discrete alternatives so that one has to use discrete choice methods, which are more complex to implement than standard regression approaches. Then there is a whole range of possible determinants for location choices. This makes this type of exercise very data-intensive and particularly prone to missing variables biases. It is also likely that location decisions are made not only on the basis of the characteristics of the spatial units where firms locate, but are also influenced by what happens in neighbouring units. More generally, it is likely to be the case that different determinants of location matter at different spatial scales. To take these concerns into account, spatial econometrics has developed a set of tools. Spatial econometrics resembles standard time-series analysis in that it takes the values of the explanatory variables of the neighbouring spatial units (as well as their error term) into account. The fundamental complication is that

spatial dependence can 'go both ways', unlike time dependence in time series. An alternative would be to ignore spatial units altogether and work directly on continuous space. Although there have been some developments in that direction (see Cressie, 1993, for a review), data limitations confine this type of approach to a small range of problems.

When one looks at more general equilibrium issues, many of the difficulties of partial equilibrium analysis are still there while other concerns also become prominent. Take the analysis of regional disparities as an example. A first issue is that such general equilibrium problems have several endogenous variables. In our example regional population and income are likely to be simultaneously determined. To analyse this type of question, two polar approaches (and everything in between) may be adopted. A more descriptive approach consists in focusing on one particular variable, say, local income, and trying to explain its spatial variation using a range of potential factors as indicated by theoretical models. Many of these factors such as the local population are then likely to be endogenous. This requires finding appropriate instruments for such endogenous variables (since unfortunately natural experiments are even scarcer in this field than elsewhere in economics). In some cases, good instruments may not be available. In contrast to descriptive analysis, structural approaches require writing down a particular model and deriving a set of equations that can then be estimated. The main problem faced by this type of approach is that many possible models are likely to have some explanatory power. To return to our example, regional disparities in a country are likely to be caused by the factors highlighted by the urban systems approach (local external effects and so forth) and those highlighted by the new economic geography (trade costs and pecuniary externalities), as well as factor endowments, institutions, and so on. The list of plausible determinants for regional disparities is long and it is very problematic to impose strong priors regarding the validity of one specific model. For many issues in spatial economics, model selection is in fact a huge concern (see Sutton, 2000, for more). Despite these difficulties, it is fair to say that much has been learnt about cities and regions since the mid-1970s (see Rosenthal and Strange, 2004, and Head and Mayer, 2004, for recent reviews).

The road ahead (?)

What current challenges does spatial economics face? On the theoretical front, three main problems remain open. The first is to provide a unified general equilibrium approach to spatial economics and end the often uneasy coexistence between urban systems and the new economic geography. Despite some attempts, as of 2005 there is no such unified framework, and providing one will be difficult. The main obstacles are about modelling. General equilibrium models of spatial economics entail making detailed assumptions about the spatial structure, the production structure, and the mobility of people, goods and ideas, all this under increasing returns. In such cases, nonlinearities occur everywhere and analytical solutions are the exception rather than the rule. Despite this, a general but tractable model of cities and regions is probably worth fighting for. A second key challenge regards the microfoundations of trade costs. Trade costs play a fundamental role in many models but their microeconomic foundations have received only scant attention. This will probably involve looking beyond transport costs and open the black box of the multiplicity of transactions costs associated with trade between different parties. A third major challenge regards the development of a 'theory of proximity' (for lack of a better name). Such theory would provide some answers as to why direct interactions between economic agents matter and how. Non-market interactions will no doubt loom large in any theory of proximity.

On the empirical front, a first key challenge is to develop new tools for spatial analysis. With very detailed data becoming available, new tools are needed. Ideally, all the data work should be done in continuous space to avoid border biases and arbitrary spatial units. We are still a long way from being able to do so. The second main challenge for empirical work is of a very different nature. Applied work has over the years managed to produce a reasonable set of estimates regarding a range of issues such as the intensity of local externalities or the determinants of urban growth, among others. No doubt, further progress is necessary and will occur but the main challenge is now to understand the mechanisms behind the elasticities or the decompositions that have been produced. For instance, the elasticity of local productivity to the density of economic activity is now well circumscribed between two and five per cent. We ignore nearly everything about the relative importance of the possible mechanisms behind such numbers. Finally, being able to distinguish between theories – for instance, between factor endowments, urban systems and new geography to explain regional patterns of economic activity – is also a fundamental task where research has barely begun to make progress.

To conclude, one may want to raise the issue of the position of this field within economics and its relationship with other areas of investigation. It is fair to say that, with the advent of Alonso's modern urban economics and that of strategic models of location, spatial economics traded its breadth of knowledge against some depth on much smaller subset of questions. Since the mid-1970s, spatial economics has managed to broaden again its focus by remaining open to outside influences. For instance, the new economic geography finds its roots in international trade theory, while much modern empirical work is heavily influenced by modern applied labour economics and industrial organization. For spatial economics, there is scope for further expansion. Over the years, housing and real estate economics have become fairly detached from the rest of spatial economics and the time may be ripe for new encounters and new cross-fertilizations. A similar statement also holds for local public economics. Finally, outside economics, the part of geography that deals with economic issues, 'economic geography', has a focus that considerably overlaps with spatial economics. The relationship between the two disciplines has been fraught with difficulties. On the one hand many geographers react very negatively to the renewed interest by economists in spatial issues. On the other hand, economists tend to ignore the work done by economic geographers. Despite these difficulties, geographers may learn something from the economists' more rigorous approach while the greater breadth of geographers may offer a great source of inspiration for economists.

GILLES DURANTON

See also **central place theory; GIS data in economics; housing supply; international trade theory; location theory; monocentric versus polycentric models in urban economics; neighbours and neighbourhoods; new economic geography; spatial econometrics; systems of cities; urban agglomeration; urban economics; urban growth; urban housing demand; urban production externalities; urban transportation economics.**

Bibliography

Alonso, W. 1964. *Location and Land Use*. Cambridge, MA: Cambridge University Press.

Cressie, N. 1993. *Statistics for Spatial Data*. New York: John Wiley.

Debreu, G. 1959. *Theory of Value*. New Haven: Yale University Press.

Dixit, A. and Stiglitz, J. 1977. Monopolistic competition and optimum product diversity. *American Economic Review* 67, 297–308.

Duranton, G. and Puga, D. 2004. Micro-foundations of urban agglomeration economies. In *Handbook of Regional and Urban Economics*, vol. 4, ed. V. Henderson and J.-F. Thisse. Amsterdam: North-Holland.

Fujita, M. 1989. *Urban Economic Theory: Land Use and City Size*. Cambridge: Cambridge University Press.

Fujita, M., Krugman, P. and Venables, A. 1999. *The Spatial Economy: Cities, Regions, and International Trade*. Cambridge, MA: MIT Press.

Head, K. and Mayer, T. 2004. The empirics of agglomeration and trade. In *Handbook of Regional and Urban Economics*, vol. 4, ed. V. Henderson and J.-F. Thisse. Amsterdam: North-Holland.

Henderson, J. 1974. The sizes and types of cities. *American Economic Review* 64, 640–56.

Hotelling, H. 1929. Stability in competition. *Economic Journal* 39, 41–57.

Krugman, P. 1991. Increasing returns and economic geography. *Journal of Political Economy* 99, 484–99.

Marshall, A. 1890. *Principles of Economics*. London: Macmillan.

Myrdal, G. 1957. *Economic Theory and Under-developed Regions*. London: Duckworth.

Rosenthal, S. and Strange, W. 2004. Evidence on the nature and sources of agglomeration economies. In *Handbook of Regional and Urban Economics*, vol. 4, ed. V. Henderson and J.-F. Thisse. Amsterdam: North-Holland.

Ricardo, D. 1821. *The Principles of Political Economy*, 3rd edn. Homewood, IL: Irwin, 1963.

Smith, A. 1776. *An Inquiry into the Nature and Causes of the Wealth of Nations*. London: Printed for W. Strahan and T. Cadell.

Starrett, D. 1974. Principles of optimal location in a large homogeneous area. *Journal of Economic Theory* 9, 418–48.

Sutton, J. 2000. *Marshall's Tendencies: What Can Economists Know?* Cambridge. MA: MIT Press.

von Thünen, J. 1826. *Isolated State: An English edition of Der isolierte Staat*, trans. C. Wartenberg and P. Hall. Oxford: Pergamon Press, 1966.

Weber, A. 1909. *Alfred Weber's Theory of the Location of Industries*, trans. C. Friedrich. Chicago: University of Chicago Press, 1929.

spatial market integration

Markets aggregate demand and supply across actors distributed in space. At the international level, monetary policy, exchange rate adjustment and the distribution of the gains from trade depend fundamentally on how well prices equilibrate across countries, as vast literatures on the law of one price and purchasing power parity emphasize (Froot and Rogoff, 1995; Anderson and van Wincoop, 2004). At the national level, well-functioning markets ensure that macro-level economic policies (for example, with respect to exchange rates, trade, and fiscal or monetary policy) change the incentives and constraints faced by micro-level decision-makers. Macroeconomic policy commonly becomes ineffective without strong market transmission across space of the signals sent by central governments. Similarly, well-functioning markets underpin growth stimuli originating in micro-level phenomena. For example, without good access to distant markets that can absorb excess local supply, firms' adoption of improved production technologies will tend to cause producer prices to drop, erasing the gains from technological change and thereby dampening incentives for firms to adopt new technologies that can stimulate economic growth. Poorly integrated markets thereby choke off the prospective gains from technological change. Markets also play a fundamental role in managing risk associated with demand and supply shocks in that well-integrated markets facilitate adjustment in net export flows across space, thereby reducing price variability faced by consumers and producers. Finally, the spatial extent of markets has profound implications for antitrust policy (Stigler and Sherwin, 1985).

The micro-level realities of markets in much of the world, however, involve poor communications and transport infrastructure, limited rule of law, and restricted access to commercial finance, all of which can sharply limit the degree to which markets function as effectively as textbook models typically assume. A long-standing empirical literature documents considerable commodity price variability across space, especially in developing countries, with various empirical tests of market integration suggesting significant and puzzling forgone arbitrage opportunities (Fackler and Goodwin, 2001). The international trade literature similarly finds substantial and sometimes persistent deviations from the law of one price and from purchasing power parity, even among advanced market economies (Froot and Rogoff, 1995; Anderson and van Wincoop, 2004). These results raise important questions about the nature of spatial market integration in actual economies.

The concept of spatial market integration

Although contemporary economics rests fundamentally upon the concept of markets, the discipline struggles with the important and practical challenges of clearly defining a market empirically and of establishing whether markets are efficient in allocating scarce goods and services (Barrett, 2001). Much of the problem revolves around the concept of 'market integration' one employs and the empirical evidence thereby needed to demonstrate that condition. In macroeconomics and international economics, a common conceptualization of market integration focuses on 'tradability', the notion that a good is traded between two economies or that market intermediaries are indifferent between exporting from one nation to another and not doing so. Tradability signals the transfer of excess demand from one market to another, as captured in actual or potential physical flows. Positive trade flows are sufficient to demonstrate spatial market integration under the tradability standard. But prices need not be equilibrated across markets. Spatial market integration conceptualized as tradability is therefore consistent with Pareto-inefficient distributions.

For this reason, the primary approach one finds in the spatial market integration literature focuses instead on the notion of competitive equilibrium and Pareto efficiency manifest in zero marginal profits to arbitrage. At the heart of most analyses of market integration lies the Enke–Samuleson–Takayama–Judge (ESTJ) spatial equilibrium model (Enke, 1951; Samuelson, 1952; Takayama and Judge, 1971), in which the dispersion of prices in two locations for an otherwise identical good is bounded from above by the cost of arbitrage between the markets when trade volumes are unrestricted and bounded from below when trade volumes reach some ceiling value (for example, associated with a trade quota). More precisely, in ESTJ spatial equilibrium

$$p^0 = p^1 + \tau^{10} \text{ if } q^{10} \in (0, q^{10}*)$$
$$\leq p^1 + \tau^{10} \text{ if } q^{10} = 0$$
$$\geq p^1 + \tau^{10} \text{ if } q^{10} = q^{10}*$$

where p^0 and p^1 are the prices in two spatially distinct markets, 0 and 1, respectively, τ^{10} is the cost of moving the good from market 1 to market 0, q^{10} is the physical volume of trade between the two markets and $q^{10}*$ is a maximal permitted trade volume between the two markets (for example, due to a trade quota). These equilibrium conditions imply both firm-level profit maximization and long-run competitive equilibrium at market level. The strict equality reflects the form of competitive equilibrium assumed under the law of one price. If trade occurs and is unrestricted, the marginal trader earns zero profits and prices in the two markets co-move perfectly. The theory, however, implies multiple competitive equilibria. The first weak inequality reflects a segmented equilibrium in which no trade occurs. Prices can be uncorrelated within the price band created by the costs of inter-market arbitrage. The latter weak inequality reflects binding trade quotas that may yield positive marginal quasi-rents to arbitrage.

Note that trade is neither necessary nor sufficient for the attainment of ESTJ competitive equilibria. Hence the difference between tradability-based and efficiency-based conceptualizations of market integration. In the prevailing view, spatial market integration occurs when the ESTJ equilibrium condition holds, irrespective of whether trade occurs.

Empirical estimation methods

The empirical challenge of measuring spatial market integration arises because the ESTJ equilibrium condition involves four variables – prices, transactions costs, trade volumes and trade volume quotas – yet few studies employ more than price data. Spatial price analysis studies typically test for co-movement in time series of prices measured simultaneously at different places. But even with proper controls for autocorrelation or non-stationarity, such studies inevitably impose great structure on the nature of market relationships: for example, linear pricing, continuous unidirectional tradability, and stationary transactions costs series. Tests of the hypothesis of market efficiency thereby become indistinguishable from tests of the veracity of the assumptions that underpin model specification. Simple linear time series tests of market integration-cum-equilibrium using co-integration, error correction or Granger causality models have therefore drawn considerable criticism (Barrett, 1996; Baulch, 1997; Fackler and Goodwin, 2001).

More recent innovations use mixture distribution estimation methods in an attempt to integrate price data

with transactions costs, trade volume data, or both, while relaxing some of the strong assumptions that underpin conventional time series methods of testing for market integration. Baulch's (1997) parity bounds model (PBM) that integrates price and transactions costs series is perhaps the best known of these methods. Barrett and Li (2002) extended the PBM to incorporate trade data. These methods have their shortcomings too, however. They rely on inherently arbitrary distributional assumptions in estimation and typically ignore the time series properties of the data, not permitting analysis of the dynamics of inter-temporal adjustment to short-run deviations from long-run equilibrium and potentially important distinctions between short-run and long-run integration (Ravallion, 1986).

A fragile empirical foundation for guiding policy

Even satisfaction of the ESTJ spatial equilibrium condition does not imply welfare maximization unless the costs of commerce and the quasi-rents associated with binding trade quotas are minimized. In order for markets to fulfil the promise they offer for risk management, efficient distribution of production according to comparative advantage, clear transmission of policy signals, and maintenance of micro-level incentives to innovate, there should be neither segmented competitive equilibria nor effective trade quotas. When the costs of commerce are high or trade restrictions bind, it can be difficult to draw out clear implications for policy even from empirical analyses that take seriously the implications of ESTJ spatial equilibrium. Given limited data, in particular a paucity of data on transactions costs and trade volumes, and the intrinsic limitations of existing empirical methods, economists still have only a fragile empirical foundation for reaching clear, strong judgements about spatial market integration as a guide for corporate or government policy.

CHRISTOPHER B. BARRETT

See also **agricultural markets in developing countries; purchasing power parity; spatial economics.**

Bibliography

Anderson, J. and van Wincoop, E. 2004. Trade costs. *Journal of Economic Literature* 42, 691–751.

Barrett, C. 1996. Market analysis methods: are our enriched toolkits well suited to enlivened markets? *American Journal of Agricultural Economics* 78, 825–9.

Barrett, C. 2001. Measuring integration and efficiency in international agricultural markets. *Review of Agricultural Economics* 23, 19–32.

Barrett, C. and Li, J. 2002. Distinguishing between equilibrium and integration in spatial price analysis. *American Journal of Agricultural Economics* 84, 292–307.

Baulch, B. 1997. Transfer costs, spatial arbitrage, and testing for food market integration. *American Journal of Agricultural Economics* 79, 477–87.

Enke, S. 1951. Equilibrium among spatially separated markets: solution by electrical analogue. *Econometrica* 19, 40–7.

Fackler, P. and Goodwin, B. 2001. Spatial price analysis. In *Handbook of Agricultural Economics*, vol. 1B, ed. B. Gardner and G. Rausser. Amsterdam: Elsevier.

Froot, K. and Rogoff, K. 1995. Perspectives on PPP and long-run real exchange rates. In *Handbook of International Economics*, vol. 3, ed. G. Grossman and K. Rogoff. Amsterdam: Elsevier.

Ravallion, M. 1986. Testing market integration. *American Journal of Agricultural Economics* 68, 102–9.

Samuelson, P. 1952. Spatial price equilibrium and linear programming. *American Economic Review* 42, 283–303.

Stigler, G. and Sherwin, R. 1985. The extent of the market. *Journal of Law and Economics* 28, 555–85.

Takayama, T. and Judge, G. 1971. *Spatial and Temporal Price Allocation Models*. Amsterdam: North-Holland.

spatial mismatch hypothesis

First formulated by Kain (1968), the spatial mismatch hypothesis states that, residing in urban segregated areas distant from and poorly connected to major centres of employment growth, black workers face strong geographic barriers to finding and keeping well-paid jobs. In the US context, where jobs have been decentralized and blacks have stayed in the central parts of cities, the main conclusion of the spatial mismatch hypothesis is that distance to jobs is the main cause of high unemployment rates and low earnings among blacks. The spatial mismatch literature has focused on race under the presumption that (inner-city) blacks are not residing close to (suburban) jobs, either because they are discriminated against in the (suburban) housing market or because they want to live near members of their own race. Most of this literature has focused on black workers, and it is only recently that the analysis has been extended to other minority workers, especially Hispanics.

Since Kain's study, hundreds of others have been conducted trying to test the spatial mismatch hypothesis (see, in particular, the literature survey by Ihlanfeldt and Sjoquist, 1998). The usual approach is to relate a measure of labour-market outcomes, typically employment or earnings, to another measure of job access, typically some index that captures the distance between residences and centres of employment. Some control variables (typically human capital variables) are also included.

The main econometric problem with this test is that residential location is endogenous, since families are not randomly assigned residential locations but instead choose them (Ihlanfeldt, 2005). Thus, self-selection rather than distance to jobs may explain why black workers have adverse labour market outcomes. This problem has been dealt with mainly by exploiting inter-city variations in black residential centralization to estimate the effect of job access on black employment (Weinberg, 2000). Another way is to focus the analysis on youth who still reside with their parents, since residential location is decided by parents for their children (Raphael, 1998). Given the limits of these approaches, the general conclusions for youth workers are: (*a*) poor job access indeed worsens labour-market outcomes, (*b*) black and Hispanic workers have worse access to jobs than white workers, and (*c*) racial differences in job access can explain between one-third and one-half of racial differences in employment.

The theoretical foundations of these empirical results, however, remain unclear. If researchers do agree on the causes (housing discrimination and/or social interactions) and on the consequences of the spatial mismatch hypothesis (higher unemployment rates and lower wages for black workers), the economic mechanisms and thus the policy implications are difficult to identify.

A first theoretical view (Brueckner and Zenou, 2003) is to argue that suburban housing discrimination skews black workers towards the central business district (CBD) and thus keeps black residences remote from the suburbs. Since black workers who work in the suburban business district (SBD) have more costly commutes, few of them will accept SBD jobs, which makes the black CBD labour pool larger than the SBD pool. Under either a minimum wage or an efficiency wage model, this enlargement of the CBD pool leads to a high unemployment rate among CBD workers.

Another theory (Zenou, 2002) has proposed that distance has a negative impact on workers' productivity. Indeed, because of the lack of good public transportation in large US metropolitan areas, especially from the central city to the suburbs, blacks have relatively low productivity at suburban jobs because they arrive late to work due to the unreliability of the mass transit system, which causes them to frequently miss transfers. If this is true, then firms may draw a red line beyond which they will not hire workers. So, if housing discrimination against blacks forces them to live far away from jobs, then firms will be reluctant to hire black workers because they have relatively lower productivity than whites.

YVES ZENOU

See also **housing policy in the United States; poverty traps; residential segregation; unemployment; urban economics; wage inequality, changes in.**

Bibliography

Brueckner, J. and Zenou, Y. 2003. Space and unemployment: the labour-market effects of spatial mismatch. *Journal of Labor Economics* 21, 242–66.

Ihlanfeldt, K. 2005. A primer on spatial mismatch within urban labor markets. In *A Companion to Urban*

Economics, ed. R. Arnott and D. McMillen. Boston: Blackwell Publishing.

Ihlanfeldt, K. and Sjoquist, D. 1998. The spatial mismatch hypothesis: a review of recent studies and their implications for welfare reform. *Housing Policy Debate* 9, 849–92.

Kain, J. 1968. Housing segregation, Negro employment, and metropolitan decentralization. *Quarterly Journal of Economics* 82, 32–59.

Raphael, S. 1998. The spatial mismatch hypothesis and black youth joblessness: evidence from the San Francisco Bay area. *Journal of Urban Economics* 43, 79–111.

Weinberg, B. 2000. Black residential centralization and the spatial mismatch hypothesis. *Journal of Urban Economics* 48, 110–34.

Zenou, Y. 2002. How do firms redline workers? *Journal of Urban Economics* 52, 391–408.

specie-flow mechanism

The 'specie-flow mechanism' is an analytic version of automatic, or market, adjustment of the balance of international payments. In competitive markets with specie-standard institutions, behaviour will lead to national price levels and income flows consistent with equilibrium in the international accounts, commonly interpreted in this context to mean zero trade balances.

The classic exposition of the mechanism, for the better part of two centuries all but universally accepted, at least as a first approximation, was provided by David Hume in a 1752 essay, 'Of the Balance of Trade'. While it is appropriate to associate the essence of the model with Hume, all the ingredients of Hume's argument had long been available. There were even notable prior attempts to fit the analytic pieces into a self-contained model. Further, even if we give to Hume all the considerable credit due to his systematic, compact statement, his version is not the whole of the specie-flow mechanism; and the specie-flow mechanism is not the whole analysis of balance of payments adjustment.

Hume's presentation is a simple application of the quantity theory of money in a setting of international trade and its financing. With a pure 100 per cent reserve gold standard, and beginning with balance in the international accounts, a decrease in the money stock of country A results in a directly proportionate fall in its price level, which is also a decrease relative to the initially unaffected price levels of other countries; as country A's price level falls, consumer response, in Hume's account, will reduce A's imports and increase its exports; when the exchange rate is bid to the gold point, the export trade balance will be financed by gold inflow, which will raise prices in A and lower prices abroad until the international price differentials and net trade flows are eliminated. The line of causation runs from changes in money to changes in prices to changes in net trade flows to international movements of gold that eliminate the earlier price differentials and thereby correct the trade imbalance and stop the shipment of gold. In equilibrium, the distribution of gold among countries (and regions within countries) yields national (and regional) price levels consistent with zero trade balances.

This theory of trade equilibration links with the Ricardian theory of production specialization. In a comparative advantage model of two countries, two commodities, and labour input, country A has absolute advantages of different degrees in both goods. To have two-way trade, the wage rate of country A must be greater than that abroad, within the wage-ratio range specified by the proportions of A's productive superiority in the two goods. Gold will flow until the international wage ratio yields domestic prices that equate total import and export values.

The conclusion that trade imbalances, and thus gold flows, cannot long obtain was in fundamental contrast to the mercantilistic emphasis on persistent promotion of an export balance and indefinite accumulation of gold. Still, the mercantilists decidedly associated gold inflows with export surpluses of goods and services; a good many writers had posited a direct relation between the money stock and the price level; similarly, it had been indicated that relative national price level changes would affect trade flows. However, while we should bow to such predecessors of Hume as Isaac Gervaise (1720) and Richard Cantillon (1734) and perhaps nod to Gerard de Malynes (1601) for attempts to construct adjustment models, Hume put the elements together with unmatched elegance and awareness of implication – and influence.

Hume's version was specifically a *price*-specie-flow mechanism, with the prices being national price levels (and exchange rates). Even as a price mechanism, the model has problems.

While it is reasonable to presume that price levels will move in the same directions (even if not in the same proportions) as the huge changes in the money stock envisioned by Hume, there remain questions of the impact on import and export expenditures. Vertical demand schedules in country A for imports and in other countries for A's exports would leave the physical amounts of imports and exports unresponsive to price changes. If, following Hume, we upset the initial equilibrium by a large decrease in money and thus in prices in country A, foreign expenditure on A's goods will fall proportionally with the fall in A's prices. The import balance of A will be financed with gold outflow, resulting in a further fall in A's prices and export value and an increase in prices abroad and in A's import expenditure. The gold flow, rather than correcting the trade flow, will increase the import trade balance of A when demand elasticities are zero (or sufficiently small). The import and export demand (and supply) elasticity conditions required for price (including exchange rate) changes to be equilibrating – conditions which are empirically realistic – came much later to be summarized in the 'Marshall–Lerner condition'. Under the most unfavourable circumstances of infinite supply elasticities and initially balanced trade, all that is required for stability is that the arithmetic sum of the elasticities of foreign demand for A's exports and of A's demand for imports be greater than unity.

Aside from the nicety of specifying elasticity conditions for stability, is it appropriate to couch the model in terms of diverging national price levels or of changes in a country's import prices compared to its export prices? Suppose country A has a commodity export balance, resulting perhaps from a shift in international demands reflecting changed preferences in favour of A's goods or imposition of a tariff by A or a foreign crop failure. As gold flows in, A's expenditures expand and prices are expected to rise. Prices of A's *domestic* goods (which do not enter foreign trade) do rise; but prices of *internationally* traded goods are affected little, if at all, for the increase in A's demand for such goods is countered by decrease in demand for them in gold-losing countries. Consumers in A, facing the domestic–international price divergence, shift to now relatively cheapened international goods (imports and A-exportables) from more expensive domestic goods, thus increasing import volume and value and also absorption of exportables. Producers in A shift out of international goods into domestic, thus reducing exports and expanding imports. Corresponding, but opposite, substitutions and shifts are diffused among the other countries. These respective domestic adjustments in consumption and production would continue until the gold flow ceases and the trade imbalance is corrected.

Substantial modern empirical research, however, is more supportive of Hume's changes in the terms of trade or of transitory divergences in relative prices of traded and non-traded goods than of the assumed invariant applicability of the equilibrium 'law of one price' commonly adopted in the modern 'monetary approach' to the balance of payments.

When gold flows into country A, portfolio equilibria of individuals and firms are upset, with cash balances now in excess. People try to spend away redundant balances. Expenditure rises and money income becomes larger. With greater income, demands for goods – including foreign goods – increase: at any given commodity price, quantity demanded has become larger. Import quantities and values rise. Changes in money give rise abroad to opposite portfolio adjustments and changes of income, thereby decreasing A's exports. In all this, there are some changes (upward in A and downward abroad) in prices of domestic goods and production factors, but the adjustment process entails income changes as well as price changes.

Some such role of changes in money income and demand schedules was noted – in different contexts and with different degrees of clarity and emphasis – by many writers in the 19th and early 20th centuries. But single-minded emphasis on income, with little or no explicit role for the money stock and prices, came only with application to balance of payments adjustment of the national income theory of J.M. Keynes. However, such application – with its regalia of marginal propensities and secondary, supplemental repercussions of multipliers – is not contingent on, or uniquely associated with, an international gold standard. Further, neglect of money in the foreign-trade multiplier analysis is a grievous omission. Equilibrium in the income model is characterized by equating of the flows of income leakages (saving, tax payments, imports) and income injections (investment, government expenditure, exports). But such equality of total leakages and injections permits a continuing trade imbalance. And a trade imbalance financed by a gold flow – or accompanied by money change generally – leads to further change in income; that is, income had not reached a genuine equilibrium.

The actual world, even with the classical gold standard in the generation prior to the Second World War, has not conformed well in institutions and processes with the construct of Hume. A world generally of irredeemable paper money and universally of demand deposits along with fractional-reserve banking and discretionary money policy – a world including the International Monetary Fund arrangement of indefinitely pegged exchange rates – has relied on selected adjustment procedures more than on automatic adjustment mechanisms. So Hume's model in its own terms is inadequate and in important empirical respects is even inappropriate. But it provided analytical coherency and expositional emphasis in an early stage of a discussion which continues to evolve.

WILLIAM R. ALLEN

Bibliography

Blaug, M. 1985. *Economic Theory in Retrospect*. Cambridge: Cambridge University Press.

Darby, M. and Lothian, J. 1983. *The International Transmission of Inflation*. Chicago: University of Chicago Press.

Fausten, D. 1979. The Humean origin of the contemporary monetary approach to the balance of payments. *Quarterly Journal of Economics* 93, 655–73.

Rotwein, E., ed. 1970. *David Hume: Writings on Economics*. Madison: University of Wisconsin Press.

Yeager, L. 1976. *International Monetary Relations: Theory, History, and Policy*. 2nd edn. New York: Harper & Row.

specification problems in econometrics

Specification problems in econometrics arise because economic theory often identifies a generally agreed upon framework (such as market determination of price and volume) but leaves up to the individual analyst the translation of the framework into a fully defined empirical model. With virtually no guidance from theory, a data analyst is expected to choose a list of relevant variables, the functional form, the separation of variables into endogenous and exogenous, the dynamics, and the error distributions. Substantial doubt about these assumptions is a characteristic of the analysis of non-experimental data and much experimental data as well. If this uncertainty is left unattended, it can cause serious doubt about the corresponding inferences.

To emphasize the distinction between the general framework and an instance of the framework on which the data analysis rests, the specific set of assumptions used to draw inferences from a data set is called a 'specification'. The treatment of doubt about the specification is called 'specification analysis'. The research strategy of trying many different specifications is called a 'specification search'.

Estimation, sensitivity analysis and simplification searches: three treatments of specification ambiguity

When an inference is suspected to depend crucially on a doubtful assumption, two kinds of actions can be taken to alleviate the consequent doubt about the inferences. Both require a list of alternative assumptions. The first approach is *statistical estimation*, which uses the data to select from the list of alternative assumptions and then makes suitable adjustments to the inferences to allow for doubt about the assumptions. (I am including under the heading 'estimation' the kind of two-valued estimation problem that economists usually call hypothesis testing.) The second approach is a *sensitivity analysis* that uses the alternative assumptions one at a time, thereby demonstrating either that all the alternatives lead to essentially the same inferences or that minor changes in the assumptions make major changes in the inferences. For example, a doubtful variable can simply be included in the equation (estimation), or two different equations can be estimated, one with and one without the doubtful variable (sensitivity analysis).

The borderline between the techniques of estimation and sensitivity analysis is not always clear since a specification search can be either a method of estimation of a general model or a method of studying the sensitivity of an inference to choice of model. Stepwise regression, for example, which involves the sequential deletion of 'insignificant' variables and insertion of 'significant' variables is best thought to be a method of estimation of a general model rather than a study of the sensitivity of estimates to choice of variables, since no attempt is generally made to communicate how the results change as different subsets of variables are included.

A fundamental distinction between estimation and sensitivity analysis is whether the logic is two-valued or three-valued. The logic of traditional econometric estimation is two-valued: either one takes the action or one does not. A sensitivity analysis offers a third possible conclusion: the data cannot be relied upon to make the decision.

If the data are very informative, estimation is the preferred approach. But parameter spaces can always be enlarged beyond the point where data can be helpful in distinguishing alternatives. When abbreviated parameter spaces appear to be used, there usually lurks behind the scene a much larger space of assumptions that ought to be explored. If this larger space has been explored through a pre-testing procedure and if the data are sufficiently informative in indicating that estimation is the preferred approach, then adjustments to the inferences are in order to account for the pre-testing bias and model uncertainty. If the data are not adequately informative about the parameters of the larger space, we need to have answers to the sensitivity question whether ambiguity about the best method of estimation implies consequential ambiguity about the inferences. A data analysis should therefore combine estimation with sensitivity analysis, and only those inferences that are clearly favoured by the data or are sturdy enough to withstand minor changes in the assumptions should be retained.

Estimation and sensitivity analyses are two phases of a data analysis. Simplification is a third. The intent of simplification is to find a simple model that works well for a class of decisions. A specification search can be used for simplification, as well as for estimation and sensitivity analysis. Confusion among these three kinds of searches ought to be eliminated since the rules for a search and measures of success properly depend on the intent.

Criticism and revision: a deeper problem

These first three kinds of specification searches fall within the reach of traditional econometric theory, both frequentist and Bayesian. But neither theory of inference can deal with another reason for exploring more than one model: a data anomaly that forces the analyst to alter the model and thus to carry out an action that was wholly unanticipated and unplanned. This is not allowed within traditional theories of inference, which presume that responses to data are fully planned and completely committed before the data are observed. From a frequentist standpoint, a fully committed plan is needed to determine sampling properties. From a Bayesian perspective, subjective probabilities fully determine the responses to the data, and, if elicitation of probabilities is allowed to be determined after the data are explored, then we double count the data evidence – once to form the 'priors' and again to update the priors.

In settings in which the theory and the method of measurement are clear, responses can be conveniently planned in advance. In practice, however, most data analysts have very low levels of commitment to whatever plans they may have formulated before reviewing the data. Even when planning is extensive, most analysts reserve the right to alter the plans if the data are judged 'unusual'. A review of the planned responses to the data after the data are actually observed can be called *criticism*, the function of which is either to detect deficiencies in the original family of models that ought to be remedied by enhancements of the parameter space or to detect inaccuracies in the original approximation of prior information. When either the model or the prior information is revised, the planned responses are discarded in favour of what at the time seem to be better responses.

The form that criticism should take is not clear cut. Much of what appears to be criticism is in fact a step in a process of estimation, since the enhancement of the model is completely predictable. An example of an estimation method masquerading as criticism is a t-test to determine if a specific variable should be added to the regression. In this case the response to the data is planned in advance and undergoes no revision once the data are observed.

Criticism and the prospect of the revision of planned responses create a crippling dilemma for both classical and Bayesian inference. According to classical inference, the choice of procedure for analysing the data should be based entirely on sampling properties, but these are impossible to compute unless the response to every conceivable data set is planned and fully committed. A Bayesian has problems whether or not a criticism is successful. When a criticism is successful, that is to say when the family of models is enhanced or the prior distribution is altered in response to anomalies in the data, there is a severe double counting problem if estimation then proceeds as if the model and prior distribution were not data-instigated. Even if criticism is not successful, the prospect of successful criticism makes the inferences from the data weaker than conventionally reported because the commitment to the model and the prior is weaker than is admitted.

Estimation: choice of variables for linear regression
Specification problems are not limited to, but are often discussed within, the context of the linear regression model, probably because the most common problem facing analysers of economic data is doubt about the exact list of explanatory variables. The first results in this literature addressed the effect of excluding variables that belong in the equation, and including variables that do not. Let y represent the dependent variable, x an included explanatory variable, and z a doubtful explanatory variable. If we assume that:

$$E(y|x, z) = \alpha + x\beta + z\theta.$$
$$\text{Var}(y|x) = \sigma^2,$$
$$E(z|x) = c + xr$$
$$\text{Var}(z|x) = s^2$$

where $\alpha, \beta, \theta, c, r, \sigma^2$ and s^2 are unknown parameters, and y, x, and z are observable vectors, then β can be estimated with z included in the equation:

$$b_{.z} = (x'M_z x)^{-1}(x'M_z y),$$

or with z excluded:

$$b = (x'x)^{-1}x'y,$$

where

$$M_z = I - z(z'z)^{-1}z'.$$

The first two moments of these estimators are straightforwardly computed:

$$E(b_{.z}) = \beta, \qquad E(b) = \beta + r\theta$$
$$\text{Bias}(b_{.z}) = 0 \qquad \text{Bias}(b) = r\theta$$
$$\text{Var}(b_{.z}) = \sigma^2(x'M_z x)^{-1} \quad \text{Var}(b) = \sigma^2(x'x)^{-1}$$

where

$$\text{Bias}(b) = E(b) - \beta.$$

A quick algebraic calculation reveals that $\text{Var}(b_{.z}) \geq \text{Var}(b)$. These moments form two basic results in 'specification analysis' made popular by Theil (1957): (*a*) if a relevant variable is excluded ($\theta \neq 0$), the estimator is biased by an amount that is the product of the coefficient of the excluded variable times the regression coefficient from the regression of the included on the excluded variable; (*b*) if an irrelevant variable is included ($\theta = 0$), the estimator remains unbiased, but has an inflated variance.

This bias result can be useful when the variable z is unobservable and information is available on the probable values of θ and r, since then the bias in b can be corrected. But if both x and z are observable, these results are not useful by themselves because they do not unambiguously select between the estimators, one doing well in terms of bias but the other doing well in terms of variance. The choice will obviously depend on information about the value of $r\theta$ since a small value of $r\theta$ implies a small value for the bias of b. The choice will also depend on the loss function that determines the trade-off between bias and variance.

For mathematical convenience, the loss function is usually assumed to be quadratic:

$$L(\beta^*, \beta) = (\beta^* - \beta)^2,$$

where β^* is an estimator of β. The expected value of this loss function is known as the mean squared error of the estimator, which can be written as the variance plus the square of the bias:

$$\text{MSE}(b_{.z}, \beta) - \text{MSE}(b, \beta) = r(\text{Var}(\theta_{.x}^*) - \theta\theta')r'$$

where $\theta^*_{.x}$ is the least squares estimator of θ controlling for x, and where the notation allows x and z to represent collections of variables as well as singlets. Through inspection of this formula we can derive the fundamental result in this literature. The estimator based on the restricted model is better in the mean squared error sense than the unrestricted estimator if and only if θ is small enough that $\mathrm{Var}(\theta^*_{.x}) - \theta\theta'$ is positive definite. If θ is a scalar, this condition can be described as 'a true t less than one', $\theta^2/\mathrm{Var}(\theta^*_{.x}) < 1$.

This result is also of limited use since its answer to the question 'Which estimator is better?' is another question, 'How big is θ?' A clever suggestion is to let the data provide the answer, and to omit z if its estimated t value is less than one. Unfortunately, because the estimated t is not exactly equal to the true t, this two-step procedure does not yield an estimator that guarantees a lower mean squared error than unconstrained least squares. Thus the question remains: how big is θ? For more discussion consult Judge and Bock (1983).

Since the choice of estimator depends on what we already know about the parameter, we need to find a way to include in the analysis some explicit dependence on the prior state of knowledge. A Bayesian analysis allows the construction of estimators that make explicit use of prior information about θ. It is convenient to assume that the information about θ takes the form of a preliminary data set in which the estimate of θ is zero (or some other number, if you prefer). Then the Bayes estimate of β is a weighted average of the constrained and unconstrained estimators:

$$b_B = \left(v^{-1} + v'^{-1}\right)^{-1}\left(v^{-1}b_{.z} + v'^{-1}b\right)$$

where v' is the prior variance for h, and v is the sampling variance for $\theta^*_{.x}$. Instinct might suggest that this compromise between the two estimators would depend on the sampling variances of $b_{.z}$ and b, but the correct weights are inversely proportional to prior variance and the sample variance for θ.

A card-carrying Bayesian regards this to be the solution to the problem. Others will have a different reaction. What the Bayesian has done is only to enlarge the family of estimators. The two extremes are still possible since we may have $v'=0$ or $v' \to \infty$ but in addition there are the intermediate values $v'>0$. Thus the Bayesian answer to the question is another question: 'What is the value of v'?'

Sensitivity analysis

At this point we have to switch from the estimation mode to the sensitivity mode, since precise values of v' will be hard to come by on a purely a priori basis and since the data usually will be of little help in selecting v' with great accuracy. A sensitivity analysis can be done from a classical point of view simply by contrasting the two extreme estimates, $b_{.z}$ and b, corresponding to the extreme values

of v'. A Bayesian approach allows a much richer set of sensitivity studies. A mathematically convenient analysis begins with a hypothetical value for v', say v'_0, which is selected to represent as accurately as possible the prior information that may be available. A neighbourhood around this point is selected to reflect the accuracy with which v'_0 can be chosen. For example, v' might be restricted to lie in the interval

$$v'_0/(1+c) < v' < v'_0(1+c),$$

where c measures the accuracy of v'_0. The corresponding interval of Bayes's estimates b_B is

$$1/\left(v'_0(1+c) + v\right) < (b_B - b_{.z})/v(b - b_{.z})$$
$$< (1+c)/\left(v'_0 + v(1+c)\right),$$

where it is assumed that $(b - b_{.z})>0$. If this interval is large for small values of c, then the estimate is very sensitive to the definition of the prior information. For example, suppose that interest focuses on the sign of β. Issues of standard errors aside, if $b_{.z}$ and b are the same sign, then the inference can be said to be sturdy since no value of c can change the sign of the estimate b_B. But if $b>0>b_{.z}$, then the values of c in excess of the following will cause the interval of estimates to overlap the origin:

$$c^* = \max\left[u - 1, u^{-1} - 1\right],$$

where $u = -\left(v/v'_0\right)(b/b_{.z})$ Thus if u is close to one, the inference is fragile. This occurs if differences in the absolute size of the two estimates are offset by differences in the variances applicable to the coefficient of the doubtful variable. Measures like these can be found in Leamer (1978; 1982; 1983b).

Robustness

When a set of acceptable assumptions does not map into a specific decision, the inference is said to be fragile. A decision can then sensibly be based on a minimax criterion that selects a 'robust' procedure that works well regardless of the assumption. The literature on 'robustness' such as that reviewed by Krasker, Kuh and Welsch (1983) has concentrated on issues relating to the choice of sampling distribution, but could be extended to choice of prior distribution.

Simplification, proxy selection and data selection

A specification search involving the estimation of many different models can be a method of estimation or a method of sensitivity analysis. Simplification searches are also common, the goal of which is to find a simple quantitative facsimile that can be used as a decision-making tool. For example, a model with a high R^2 can be expected to provide accurate forecasts in a stable environment, whether or not the coefficients can be given a causal interpretation. In particular, if two explanatory variables are highly correlated, then one can be excluded from the equation without greatly reducing the overall fit since the included variable will take over the role of the excluded variable. No causal significance necessarily attaches to the coefficient of the retained variable.

A specification search can also be used to select the best from a set of alternative proxy variables, or to select a data subset. These problems can be dealt with by enlarging the parameter space to allow for multiple proxy variables or unusual data points. Once the space is properly enlarged, the problems that remain are exactly the same as the problems encountered when the parameters are coefficients in a linear regression, namely estimation and sensitivity analysis.

Data-instigated models

The subjects of estimation, sensitivity analysis and simplification deal with concerns that arise during the planning phase of a statistical analysis when alternative responses to hypothetical data-sets are under consideration. A distinctly different kind of specification search occurs when anomalies in the actual data suggest a revision in a planned response, for example, the inclusion of a variable that was not originally identified. This is implicitly disallowed by formal statistical theories that presuppose the existence of a response to the data that is planned and fully committed. I like to refer to a search for anomalies as 'Sherlock Holmes inference', since, when asked who might have committed the crime, Holmes replied, 'No data yet … It is a capital mistake to theorize before you have all the evidence. It biases the judgements.' This contrasts with the typical advice of theoretical econometricans: 'No theory yet. It is a capital mistake to look at the data before you have identified all the theories.'

Holmes is properly concerned that an excessive degree of theorizing will make it psychologically difficult to see anomalies in the data that might, if recognized, point sharply to a theory that was not originally identified. On the other hand, the econometrician is worried that data evidence may be double counted, once in the Holmesian mode to instigate models that seem favoured by the data and again in the estimation mode to select the instigated models over original models. Holmes is properly unconcerned about the double counting problem, since he has the ultimate extra bit of data: the confession. We do not have the luxury of running additional experiments and the closest that we can come to the Holmesian procedure is to set aside a part of the data set in hopes of squeezing a confession after we have finished identifying a set of models with a Holmesian analysis of the first part of the data. Unfortunately, our data-sets never do confess, and the ambiguity of the inferences that is clearly present after the Holmesian phase lingers on with very little attenuation after the estimation phase. Thus we are forced to find a solution to the Holmesian conundrum of

how properly to characterize the data evidence when models are instigated by the data, that is to say, how to avoid the double counting problem. Clearly, what is required is some kind of penalty that discourages but does not preclude Holmesian discoveries. Leamer (1978) proposes one penalty that rests on the assumption that Holmesian analysis mimics the solution to a formal pre-simplification problem in which models are explicitly simplified before the data are observed in order to avoid observation and processing costs that are associated with the larger model. Anomalies in the data-set can then suggest a revision of this decision. Of course, real Holmesian analysis cannot actually solve this sequential decision problem since in order to solve it one has to identify the complete structure that is simplified before observation. But we can nonetheless act as if we were solving this problem, since by doing so we can compute a very sensible kind of penalty for Holmesian discoveries (Leamer, 1978).

Criticism

Criticism refers to a search for data anomalies that might force a revision of the model. Criticism and data-instigated models are not as frequent as they may appear. As remarked before, much of what is said to be criticism is only a step in a method of estimation, and many models that seem to be data-instigated are in fact explicitly identified in advance of the data analysis. For example, forward stepwise regression, which adds statistically significant variables to a regression equation, cannot be said to be producing data-instigated models because the set of alternative models is explicitly identified before the data analysis commences and the response to the data is fully planned in advance. Stepwise regression is thus only a method of estimation of a general model. Likewise, various diagnostic tests that lead necessarily to a particular enhancement of the model, such as a Durbin–Watson test for first-order autocorrelation and Ramsey's (1969) so-called specification error test for a special kind of nonlinearity, select but do not instigate a model.

'Goodness of fit' tests that do not have explicit alternatives are sometimes used to criticize a model. However, the Holmesian question is not whether the data appear to be anomalous with respect to a given model but rather whether there is a plausible alternative that makes the data appear less anomalous and, most importantly, more understandable. Goodness of fit tests have nothing to do with understanding. They measure statistical properties that may or may not be meaningful in the context in which the data are being studied. In large samples, all models have large goodness of fit statistics, and the size of the statistic is no guarantee, or even a strong suggestion, that there exists a plausible alternative model that is substantially better than the one being used.

Unexpected parameter estimates are probably the most effective criticisms of a model. A Durbin–Watson statistic that indicates a substantial amount of autocorrelation can be used legitimately to signal the existence of left-out variables in settings in which there is strong prior information that the residuals are white noise. Apart from unexpected estimates, graphical displays and the study of influential data points may stimulate thinking about the inadequacies in a model.

EDWARD E. LEAMER

See also **Bayesian econometrics; econometrics; statistical inference.**

Bibliography

Box, G.E.P. 1980. Sampling and Bayes' inference in scientific modelling and robustness. *Journal of the Royal Statistical Society* 143, 383–430.
Judge, G.G. and Bock, M.E. 1983. Biased estimation. In *Handbook of Econometrics*, vol. 1, ed. Z. Griliches and M. Intriligator. Amsterdam: North-Holland.
Krasker, W.S., Kuh, E. and Welsch, R.E. 1983. Estimation for dirty data and flawed models. In *Handbook of Econometrics*, vol. 1, ed. Z. Griliches and M. Intriligator. Amsterdam: North-Holland.
Leamer, E.E. 1978. *Specification Searches*. New York: Wiley.
Leamer, E.E. 1982. Sets of posterior means with bounded variance priors. *Econometrica* 50, 725–36.
Leamer, E.E. 1983a. Let's take the con out of econometrics. *American Economic Review* 73, 31–43.
Leamer, E.E. 1983b. Model choice and specification analysis. In *Handbook of Econometrics*, vol. 1, ed. Z. Griliches and M. Intriligator. Amsterdam: North-Holland.
Ramsey, J.B. 1969. Tests for specification errors in classical linear least squares regression analysis. *Journal of the Royal Statistical Society* 32, 350–71.
Theil, H. 1957. Specification errors and the estimation of economic relationships. *Review of the International Statistical Institute* 25, 41–51.

spectral analysis

Spectral analysis is a statistical approach for analysing stationary time series data in which the series is decomposed into cyclical or periodic components indexed by the frequency of repetition. Spectral analysis falls within the frequency domain approach to time series analysis. This is in contrast with the time domain approach in which a time series is characterized by its correlation structure over time. While spectral analysis provides a different interpretation of a time series from time domain approaches, the two approaches are directly linked to each other.

Statistical spectral analysis tools were first developed in the middle of the 20th century in the mathematical statistics and engineering literatures, and many of the important early contributions are discussed in the classic textbook treatment by Priestley (1981). The label 'spectral' has been adopted because of the close link to the physics of light. While the analogy with the physics of light is fairly useless in economics, economists recognized by the 1960s that spectral analysis is a useful empirical tool for understanding the cyclical nature of many time series, and it provides a powerful theoretical framework for developing econometric methodology: for example, the theoretical underpinnings of Granger causality (Granger, 1969) are based in spectral analysis. Since the 1960s, spectral analysis tools have become standard parts of the time series econometrics toolkit, and have influenced a broad range of areas within econometrics. A comprehensive list of references would be long but some notable examples are: band spectral regression (Engle, 1974), generalized method of moments (GMM) (Hansen, 1982), heteroskedasticity autocorrelation (HAC) covariance matrix estimation and inference (Newey and West, 1987; Andrews, 1991; Kiefer and Vogelsang, 2005), unit root testing (Phillips and Perron, 1988), cointegration (Stock and Watson, 1988; Phillips and Hansen, 1990), semiparametric methods (Robinson, 1991), structural identification of empirical macroeconomics models (Blanchard and Quah, 1989; King et al., 1991), testing for serial correlation (Hong, 1996), measures of persistence (Cochrane, 1988), measures of fit for calibrated macro models (Watson, 1993), estimation of long memory models (Geweke and Porter-Hudak, 1983).

Let y_t, $t=1,2, \ldots$ denote a second-order stationary time series with mean $\mu=E(y_t)$ and autocovariance function $\gamma_j = cov(y_t, y_{t-j})$. Most empirical economists find it natural to characterize relationships between random variables in terms of correlation structure, and γ_j conveniently summarizes the statistical structure of y_t. Autocovariances are fundamental population moments of a time series not directly connected to any specific modelling choice. In contrast, the idea of decomposing y_t into cyclical components may appear to impose restrictions on y_t; but an important result, the spectral representation theorem, indicates that nearly any stationary time series can be represented in terms of cyclical components. By using notation from Hamilton (1994), nearly any stationary (discrete-time) time series can be represented as

$$y_t = \mu + \int_0^\pi [\alpha(\omega)\cos(\omega t) + \delta(\omega)\sin(\omega t)] \, d\omega,$$

$$(1)$$

where ω denotes frequency and $\alpha(\omega)$ and $\delta(\omega)$ are mean zero random processes such that for any frequencies $0 < \omega_1 < \omega_2 < \omega_3 < \omega_4 < \pi$,

$$cov\left(\int_{\omega_1}^{\omega_2} \alpha(\omega) \, d\omega, \int_{\omega_3}^{\omega_4} \alpha(\omega) \, d\omega \right) = 0,$$

$$cov\left(\int_{\omega_1}^{\omega_2} \delta(\omega) \, d\omega, \int_{\omega_3}^{\omega_4} \delta(\omega) \, d\omega \right) = 0,$$

and for any frequencies $0 < \omega_1 < \omega_2 < \pi$ and $0 < \omega_3 < \omega_4 < \pi$,

$$cov\left(\int_{\omega_1}^{\omega_2} \alpha(\omega)\, d\omega, \int_{\omega_3}^{\omega_4} \delta(\omega)\, d\omega \right) = 0.$$

It is fundamental that a stationary time series can be decomposed into (random) cyclical (cosine and sine) components.

A useful way of interpreting (1) is to measure how the cyclical components contribute to the variation in y_t. Similar to the way in which the area under a density of a random variable determines the probability of a range of values, the area under the spectral density of y_t measures the contribution to the variance of y_t from the cyclical components for a range of frequencies.

Let $f(\omega)$ denote the spectral density of y_t where $\omega \in [-\pi, \pi]$. It can be shown that $f(\omega) \geq 0$ and that $f(-\omega) = f(\omega)$. A fundamental property of $f(\omega)$ is that $var(y_t) \equiv \gamma_0 = \int_{-\pi}^{\pi} f(\omega)d\omega$ and the contribution to the variance of y_t from components with frequencies $\omega \in (\omega_1, \omega_2)$ where $-\pi < \omega_1 < \omega_2 < \pi$ is given by $\int_{\omega_1}^{\omega_2} f(\omega)\, d\omega$. Therefore, loosely speaking, frequencies for which $f(\omega)$ takes on large values correspond to cyclical components that make relatively large contributions to the variation in y_t. For those more comfortable thinking in terms of the time domain cycle length, it is easy to convert frequency to cycle length. Suppose $f(\omega_1)$ corresponds to a peak (global or local) of $f(\omega)$; then components with frequencies close to ω_1 make important contributions to the variation in y_t. Consider $\cos(\omega_1 t)$ and $\sin(\omega_1 t)$ and rewrite them as $\cos(2\pi\frac{\omega_1 t}{2\pi})$ and $\sin(2\pi\frac{\omega_1 t}{2\pi})$. Recall that the cosine and sine functions are periodic with period 2π in their argument. Therefore, $\cos(2\pi\frac{\omega_1 t}{2\pi})$ and $\sin(2\pi\frac{\omega_1 t}{2\pi})$ repeat whenever $\frac{\omega_1 t}{2\pi}$ is an integer. Setting $\frac{\omega_1 t}{2\pi} = 1$ indicates that the functions repeat every $t = \frac{2\pi}{\omega_1}$ time periods. The quantity $\frac{2\pi}{\omega_1}$ is called the period corresponding to frequency ω_1.

For a concrete example consider a monthly time series where $f(\omega)$ has a peak at $\omega = \frac{\pi}{6}$. Thus cycles with period $2\pi/(\pi/6)=12$ months (annual cycles) are important for variation of y_t. Suppose quarterly cycles (period is three months) are also important, then $f(\omega)$ will have a peak at $\omega = \frac{2\pi}{3}$. The highest frequency for which we can learn about y_t is $\omega = \pi$ or two-period cycles because cycles that last fewer than two periods do not have data observed within the cycle. This practical limitation on the frequency is called the 'alias effect'. The length of the sample size, T, also limits what can be learned about long cycles. For cycles that last T time periods, that is, frequency $\omega = \frac{2\pi}{T}$, we observe data for exactly one cycle. For cycles longer than T time periods the data does contain information about those cycles, but the information is incomplete because only part of the cycle is observed. It is difficult to learn about very low-frequency components from the data, and in particular it is difficult to learn about $f(0)$.

What does $f(\omega)$ look like? The spectral representation theorem implicitly defines the integral of $f(\omega)$ but not $f(\omega)$ itself. Because the variance of y_t is the area under $f(\omega)$, there is a direct link between $f(\omega)$ and γ_j. Suppose $\sum_{j=-\infty}^{\infty} |\gamma_j| < \infty$, then the spectral density can be expressed as

$$f(\omega) = \frac{1}{2\pi} \sum_{j=-\infty}^{\infty} \cos(\omega j)\gamma_j$$

$$= \frac{1}{2\pi}\left[\gamma_0 + 2\sum_{j=1}^{\infty} \cos(\omega j)\gamma_j \right] \qquad (2)$$

where the last expression uses $\cos(0) = 1$, $\cos(-\omega j) = \cos(\omega j)$ and $\gamma_{-j} = \gamma_j$. It straightforward to show that a converse relationship holds

$$\gamma_j = \int_{-\pi}^{\pi} f(\omega)\cos(\omega j)\, d\omega.$$

This dual relationship between $f(\omega)$ and γ_j makes spectral analysis a powerful analytical tool beyond the direct interpretation of the spectral density in assessing importance of cyclical components.

For the class of stationary autoregressive moving average (ARMA) models $f(\omega)$ takes on a simple form. Let L denote the lag operator, $Ly_t = y_{t-1}$, and define lag polynomials $\phi(L) = 1 - \phi_1 L - \phi_2 L^2 - \ldots - \phi_p L^p$ and $\theta(L) = 1 + \theta_1 L + \theta_2 L^2 + \ldots + \theta_q L^q$. Suppose y_t is a stationary $ARMA(p, q)$ process given by $\phi(L)(y_t - \mu) = \theta(L)\varepsilon_t$ where ε_t is a mean zero uncorrelated time series (white noise process) with $var(\varepsilon_t) = \sigma_\varepsilon^2$. Then

$$f(\omega) = \frac{\sigma_\varepsilon^2 \theta(e^{i\omega})\theta(e^{-i\omega})}{2\pi\phi(e^{i\omega})\phi(e^{-i\omega})} \qquad (3)$$

where $i = \sqrt{-1}$. If the lag polynomials can be factored as $\phi(L) = (1 - \lambda_1 L)(1 - \lambda_2 L)\ldots(1 - \lambda_p L)$ and $\theta(L) = (1 - \delta_1 L)(1 - \delta_2 L)\cdots(1 - \delta_q L)$, then $f(\omega)$ can be written as

$$f(\omega) = \frac{\sigma_\varepsilon^2 \prod_{j=1}^{q}(1 + \delta_j^2 - 2\delta_j\cos(\omega))}{2\pi \prod_{j=1}^{p}(1 + \lambda_j^2 - 2\lambda_j\cos(\omega))}.$$

In the $AR(1)$ case $f(\omega) = \frac{\sigma_\varepsilon^2}{2\pi}\left(1 + \phi_1^2 - 2\phi_1\cos(\omega)\right)^{-1}$. If $\phi_1 > 0$ (typical for many macroeconomic and finance time series), $f(\omega)$ has a single peak at $\omega=0$. As ω increases, $f(\omega)$ steadily declines. As ϕ_1 approaches one, the peak at $\omega=0$ increases and sharpens. Thus, variation of autoregressive time series with strong persistence is driven primarily by short frequency/long cycle components. At the other extreme, when the time series is uncorrelated ($\phi_1 = 0$), the spectral density is constant/flat for all ω, so cyclical components contribute equally at all frequencies to the variation in y_t. An uncorrelated series is called a white noise process because of the analogue to white light which is comprised equally of all visible frequencies of light (all colours).

The special case of $f(0)$ is important for inference in time series models because the asymptotic variance of many time series estimators depends on $f(0)$. For example, consider $\bar{y} = T^{-1}\sum_{t=1}^{T} y_t$, the natural estimator of μ. A simple calculation gives

$$var(\bar{y}) = T^{-1}\left(\gamma_0 + 2\sum_{j=1}^{T-1}\left(1 - \frac{j}{T}\right)\gamma_j\right).$$

If $\sum_{j=-\infty}^{\infty}|\gamma_j| < \infty$, then $\lim_{T\to\infty} Tvar(\bar{y}) = \gamma_0 + 2\sum_{j=1}^{\infty}\gamma_j = 2\pi f(0)$. Therefore, the asymptotic variance of a sample average, often called the long-run variance, is proportional to the spectral density at frequency zero. Inference about the population mean would require a standard error, that is, an estimate of $f(0)$. The link between asymptotic variances and zero frequency spectral densities extends to estimation of linear regression parameters and nonlinear estimation obtained using GMM. The estimation of asymptotic variance matrices that are proportional to a zero frequency spectral density is commonly known as HAC covariance matrix estimation.

Estimates of the spectral density can be obtained either parametrically or nonparametrically. For the case of ARMA models, parametric estimators are straightforward in principle and involve replacing the lag polynomial coefficients in (3) with estimators. Although estimation methods for ARMA models are well established, there are numerical and identification issues that can complicate matters when an MA component is included, especially in the case of vector time series. In contrast, pure AR models are easy to estimate (including the vector case) and, in principle, AR models can well approximate a stationary time series with suitable choice of the lag order. For these reasons, autoregressive spectral density estimators are the most widely used parametric estimators (see, for example, Berk, 1974; Perron and Ng, 1998; den Haan and Levin, 1997). One important practical challenge of implementing autoregressive spectral density estimators is the choice of autoregressive lag order. Advice in the literature on choice of lag order often depends on the intended use of the spectral density estimator.

Nonparametric estimators of the spectral density are appealing at the conceptual level because they do not depend on specific parameterization of the model. In principle nonparametric estimators are flexible enough to provide good estimators for a very wide range of stationary time series. In practice, though, implementation of nonparametric estimators can be a delicate matter, and large sample sizes are required for accuracy. Notwithstanding the practical challenges, nonparametric spectral density estimators are widely used in econometrics

primarily because of the central role they play in HAC covariance matrix literature due to the influential contributions by Newey and West (1987) and Andrews (1991). These so-called Newey–West or Newey–West–Andrews standard errors are routinely used in practice; yet many empirical researchers are unaware of the direct link to nonparametric spectral density estimation.

Nonparametric estimators are obtained by estimation of (2) using sample autocovariances

$$\hat{\gamma}_j = T^{-1}\sum_{t=j+1}^{T}(y_t - \bar{y})(y_{t-j} - \bar{y}).$$

The challenge is that $f(\omega)$ depends on an infinite number of autocovariances of which only a finite number can be estimated. The highest-order autocovariance that be estimated is γ_{T-1}, but it is estimated badly because there is only one observation, $(y_T - \bar{y})(y_1 - \bar{y})$. Plugging the $\hat{\gamma}_j$ into (2) gives the estimator

$$I_T(\omega) = \frac{1}{2\pi}\left[\hat{\gamma}_0 + 2\sum_{j=1}^{T-1}\cos(\omega j)\hat{\gamma}_j\right],$$

which is the periodogram. Like $f(\omega)$ the periodogram is non-negative. For $\omega \neq 0$ the periodogram is asymptotically unbiased but its variance does not shrink as the sample size grows, so it is not a consistent estimator. At frequency zero, the situation is even more problematic because simple algebra can used to show that $I_T(0) = 0$. This result holds as long as $\hat{\gamma}_j$ is computed using a quantity that sums to zero (like $y_t - \bar{y}$). Therefore, $I_T(0)$ is useless for estimating $f(0)$. Fortunately, the periodogram can be modified to obtain better estimates of $f(\omega)$. Consider

$$\hat{f}(\omega) = \frac{1}{2\pi}\left[\hat{\gamma}_0 + 2\sum_{j=1}^{T-1}k\left(\frac{j}{M}\right)\cos(\omega j)\hat{\gamma}_j\right]$$

where $k(x)$ is a weighting function or kernel such that $k(0)=1$, $k(-x)=k(x)$, $\int_{-\infty}^{\infty}k(x)^2 dx < \infty$ and $|k(x)| \leq 1$. The number M is called the bandwidth or, for some $k(x)$ functions, the truncation lag. The kernel downweights the higher order $\hat{\gamma}_j$, and the bandwidth controls the speed at which downweighting occurs. A recent paper by Phillips, Sun and Jin (2006) achieves downweighting by exponentiating the kernel, for example by using $k(j/T)^\rho$, where ρ controls the degree of downweighting.

While a large number of kernel functions have been proposed and analysed since the 1940s, two have become widely used in econometrics: the Bartlett kernel and the quadratic spectral (QS) or Bartlett–Priestley kernel. These kernels are in the class of kernels that guarantee $\hat{f}(\omega) \geq 0$. The Bartlett kernel is

$$k(x) = \begin{cases} 1 - |x| & \text{for} \quad |x| \leq 1 \\ 0 & \text{for} \quad |x| > 1 \end{cases},$$

and it puts linearly declining weights on $\hat{\gamma}_j$ up to lag $M-1$ and weight zero on higher lags so that M plays the role of a truncation lag. Consistency of zero frequency Bartlett kernel estimators was established by Newey and West (1987) in a very general setting. The QS kernel is

$$k(x) = \frac{25}{12\pi^2 x^2}\left(\frac{\sin(6\pi x/5)}{6\pi x/5} - \cos(6\pi x/5)\right),$$

and it does not truncate; weight is placed on all $\hat{\gamma}_j$. The weights decline in magnitude as j increases but some weights can be negative. Andrews (1991) showed in a general setting that the QS kernel minimizes the approximate mean square error of $\hat{f}(0)$ for a particular class of kernels.

The idea of downweighting $\hat{\gamma}_j$ is natural and is not merely a technical trick. For any stationary time series, $\lim_{j\to\infty}|\gamma_j| = 0$, therefore downweighting $\hat{\gamma}_j$, or replacing it with zero when j is large, is similar to replacing an unbiased, high variance estimator with a biased, small variance estimator. If γ_j shrinks quickly as j increases, the bias induced by downweighting is small. For $\hat{f}(\omega)$ to be a consistent estimator $M\to\infty$ and $M/T\to 0$ as $T\to\infty$. These conditions suggest that while downweighting is required it cannot be too severe. Unfortunately, these conditions do not restrict the value of M that can be used for a given sample because, given T, any value of M can be embedded in a rule that satisfies these conditions. For example, suppose $T=100$. Then the bandwidth rules $M = 10\sqrt{T}$ and $M = 0.2\sqrt{T}$ satisfy the conditions for consistency but yield very different bandwidths of $M=100$ and $M=2$.

The finite sample properties of $\hat{f}(\omega)$ are complicated and depend on M and $k(x)$. Formulas for the exact bias and variance of $\hat{f}(\omega)$ have been worked out by Neave (1971) when $\omega\neq 0$, and by Ng and Perron (1996) for $\omega=0$. Because the exact formulas are complicated, approximations are often used. The variance can be approximated by

$$\frac{T}{M}var(\hat{f}(\omega)) \approx \begin{cases} V & \text{for} \quad 0<\omega<\pi \\ 2V & \text{for} \quad \omega = 0, \pi \end{cases}$$

$$(4)$$

where $V = f(\omega)^2\int_{-\infty}^{\infty}k(x)^2 dx$. An approximation for the bias was derived by Parzen (1957) and it depends on the behaviour of $k(x)$ around $x=0$. For the Bartlett and QS kernels the approximate bias formulas are $-\frac{1}{M}\sum_{j=-\infty}^{\infty}|j|\gamma_j$ and $-\frac{18}{125}\pi^2\frac{1}{M^2}\sum_{j=-\infty}^{\infty}j^2\gamma_j$ respectively. Under suitable regularity conditions, an asymptotic normality result holds:

$$\sqrt{\frac{T}{M}}\left(\hat{f}(\omega) - E(\hat{f}(\omega))\right)\to^d N(0, V).$$

According to these approximations, the variance is proportional to $f(\omega)^2$ and increases as M increases

whereas the bias depends on additional nuisance parameters but decreases as M increases. These well known results are discussed at length in Priestley (1981) and are the source of commonly held intuition that says that, as M increases, bias decreases but variance increases. This intuition is usually valid but only holds for $\hat{f}(0)$ when M is small. As M increases the relationship between bias/variance and M is more complicated, as discussed by Ng and Perron (1996). Recall that if no downweighting of $\hat{\gamma}_j$ is used, then $\hat{f}(0)$ becomes $I_T(0) = 0$. Obviously, this estimator has a large bias and zero variance. As M increases, less downweighting is used and, once M is large enough, the bias/variance relationship flips with bias increasing and variance decreasing in M.

An asymptotic approximation that can capture this more complex bias/variance bandwidth relationship for $\hat{f}(0)$ can be obtained using fixed-b asymptotics. Suppose $\hat{f}(0)$ is embedded into a sequence of random variables under the assumption that $b=M/T$ is a fixed constant with $b\in(0,1)$. Neave (1970) first used this approach to derive an alternate asymptotic variance formula for $\hat{f}(\omega)$. Let $B(r)$ denote a standard Brownian bridge, that is, $B(r) = W(r)-rW(1)$ where $W(r)$ is a standard Wiener process, and let $\Rightarrow$ denote weak convergence. Under suitable regularity conditions Kiefer and Vogelsang (2005) show that $\hat{f}(0) \Rightarrow f(0)Q(b)$, where

$$Q(b) = \frac{2}{b}\left(\int_0^1 B(r)^2 dr - \int_0^{1-b}B(r+b)B(r)\, dr\right)$$

for the Bartlett kernel and

$$Q(b) = -\frac{1}{b^2}\int_0^1\int_0^1 k''\left(\frac{r-s}{b}\right)B(r)B(s)\, drds$$

for the QS kernel with analogous results for $\omega \neq 0$ obtained by Hashimzade and Vogelsang (2007). Phillips, Sun and Jin (2006) obtain similar results for exponentiated kernels. The fixed-b asymptotic result approximates $\hat{f}(0)$ by the random variable $Q(b)$ which is similar to a chi-square random variable. When $\hat{f}(0)$ is used to construct standard errors of an estimator like $\bar{y}$, fixed-b asymptotics provides an approximation for $t = (\bar{y} - \mu)/\sqrt{2\pi\hat{f}(0)/T}$ of the form $t \Rightarrow W(1)/\sqrt{Q(b)}$. This limiting random variable is invariant to $f(0)$ but depends on the random variable $Q(b)$. Because $Q(b)$ depends on M (through $b=M/T$) and $k(x)$, the fixed-b approximation captures much (but not all) of the randomness in $\hat{f}(0)$. In contrast, the standard approach appeals to a consistency result for $\hat{f}(0)$ to justify approximating $\hat{f}(0)$ by $f(0)$ and t is approximated by a $N(0,1)$ random variable that does not depend on M or $k(x)$. Theoretical work by Jansson (2004) and Phillips, Sun and Jin (2005) has established that the fixed-b approximation for t in the case of Gaussian data is more accurate than the standard normal approximation. Some results for the non-Gaussian case have been obtained by Goncalves and Vogelsang (2006).

The fixed-b approximation provides approximations of the bias and variance of $\hat{f}(0)$ that are polynomials in b. For the Bartlett kernel Hashimzade, Kiefer and Vogelsang (2005) show that $E(\hat{f}(0) - f(0)) \approx f(0)(-b + \frac{1}{3}b^2)$ and

$$var(\hat{f}(0)) \approx var(f(0)Q(b))$$
$$= f(0)^2 \left(\frac{4}{3}b - \frac{7}{3}b^2 + \frac{14}{15}b^3 + \frac{2}{9}b^4 \right.$$
$$\left. - \frac{1}{15b^2}(2b-1)^5 1(b > \frac{1}{2}) \right),$$
$$(5)$$

where $1(b > \frac{1}{2}) = 1$ if $b > \frac{1}{2}$ and 0 otherwise. Because $b = M/T$, the bias and variance of $\hat{f}(0)$ are approximated by high order polynomials in M/T. The leading term in the variance exactly matches the standard variance formula (4). Because of the higher order terms, the fixed-b variance is more closely related to the exact variance. A plot of the variance polynomial would show that as M increases, variance is initially increasing but once M becomes large enough, variance decreases in M. The fixed-b bias can be combined with the Parzen bias to give

$$E(\hat{f}(0) - f(0)) \approx -\frac{1}{M} \sum_{j=-\infty}^{\infty} |j| \gamma_j$$
$$+ f(0) \left(-\frac{M}{T} + \frac{1}{3} \left(\frac{M}{T} \right)^2 \right).$$
$$(6)$$

This combined formula better approximates the behaviour of the exact bias. As M increases, the first term shrinks, but the second and third terms increase in magnitude. Depending on the relative magnitudes of $\sum_{j=-\infty}^{\infty} |j| \gamma_j$ and $f(0)$, bias will be decreasing in M when M is small, but as M increases further bias becomes increasing in M. It is interesting to note that $1/M$ and M/T terms in (6) match the terms in the type of bias approximations used by Velasco and Robinson (2001) in third order Edgeworth calculations.

Bandwidth choices that minimize approximate mean square error (MSE) of $\hat{f}(0)$ were used by Andrews (1991) and Newey and West (1994) where the bias and variance were approximated using only the leading terms in (5) and (6). A simple, closed form, solution is obtained for M that depends on $\sum_{j=-\infty}^{\infty} |j| \gamma_j$ and $f(0)$. Andrews (1991) recommends plugging in parametric estimators of these unknown quantities, whereas Newey and West (1994) recommend using nonparametric estimators. Including the higher order terms provided by (5) and (6) would allow a higher order approximation to the MSE. Given the polynomial structure of (5) and (6) with respect to M, the first order condition to this optimization problem is a high order polynomial in M

with coefficients that depend $\sum_{j=-\infty}^{\infty} |j| \gamma_j$ and $f(0)$. Given plug-in estimates, obtaining the value of M that minimizes the approximate MSE would amount to numerically finding the root of a polynomial, which is not difficult. Such an analysis does not appear to exist in the econometrics literature.

While the focus of this article has been on the spectral analysis of a univariate time series, extending the concepts, notation, and estimation methods to the case of a vector of time series is straightforward. A vector of time series can be characterized by what is called the spectral density matrix. The diagonal elements of this matrix are the individual spectral densities. The off-diagonal elements are called the cross-spectral densities. The cross-spectral densities in general can be complex valued functions even when the data is real valued. The cross-spectral densities capture correlation between series and co-movements of series can be characterized in terms of cross-amplitude, phase and coherency, which are real valued functions. Many of the ideas and concepts in the original Granger (1969) causality paper were expressed in terms of cross-spectral densities.

TIMOTHY J. VOGELSANG

See also **generalized method of moments estimation; heteroskedasticity and autocorrelation corrections; serial correlation and serial dependence.**

Bibliography

Andrews, D.W.K. 1991. Heteroskedasticity and autocorrelation consistent covariance matrix estimation. *Econometrica* 59, 817–54.

Berk, K. 1974. Consistent autoregressive spectral estimates. *Annals of Statistics* 2, 489–502.

Blanchard, O. and Quah, D. 1989. The dynamic effects of aggregate demand and supply disturbances. *American Economic Review* 79, 655–73.

Cochrane, J.H. 1988. How big is the random walk in GNP? *Journal of Political Economy* 96, 893–920.

den Haan, W.J. and Levin, A. 1997. A practicioner's guide to robust covariance matrix estimation. In *Handbook of Statistics: Robust Inference*, vol. 15, ed. G. Maddala and C. Rao. New York: Elsevier.

Engle, R. 1974. Band spectrum regression. *International Economic Review* 15, 1–11.

Geweke, J. and Porter-Hudak, S. 1983. The estimation and application of long memory time series. *Journal of Time Series Analysis* 4, 221–38.

Goncalves, S. and Vogelsang, T.J. 2006. Block bootstrap puzzles in HAC robust testing: the sophistication of the naive bootstrap. Working paper, Department of Economics, Michigan State University.

Granger, C. 1969. Investigating causal relations by econometric models and cross-spectral methods. *Econometrica* 37, 424–38.

Hamilton, J.D. 1994. *Time Series Analysis*. Princeton, NJ: Princeton University Press.

Hansen, L.P. 1982. Large sample properties of generalized method of moments estimators. *Econometrica* 50, 1029–54.

Hashimzade, N., Kiefer, N.M. and Vogelsang, T.J. 2005. Moments of HAC robust covariance matrix estimators under fixed-*b* asymptotics. Working paper, Department of Economics, Cornell University.

Hashimzade, N. and Vogelsang, T.J. 2007. Fixed-*b* asymptotic approximation of the sampling behavior of nonparametric spectral density estimators. *Journal of Time Series Analysis*.

Hong, Y. 1996. Consistent testing for serial correlation of unknown form. *Econometrica* 64, 837–64.

Jansson, M. 2004. The error rejection probability of simple autocorrelation robust tests. *Econometrica* 72, 937–46.

Kiefer, N.M. and Vogelsang, T.J. 2005. A new asymptotic theory for heteroskedasticity autocorrelation robust tests. *Econometric Theory* 21, 1130–64.

King, R., Plosser, C., Stock, J. and Watson, M. 1991. Stochastic trends and economic fluctuations. *American Economic Review* 81, 819–40.

Neave, H.R. 1970. An improved formula for the asymptotic variance of spectrum estimates. *Annals of Mathematical Statistics* 41, 70–7.

Neave, H.R. 1971. The exact error in spectrum estimates. *Annals of Mathematical Statistics* 42, 901–75.

Newey, W.K. and West, K.D. 1987. A simple, positive semi-definite, heteroskedasticity and autocorrelation consistent covariance matrix. *Econometrica* 55, 703–8.

Newey, W.K. and West, K.D. 1994. Automatic lag selection in covariance estimation. *Review of Economic Studies* 61, 631–54.

Ng, S. and Perron, P. 1996. The exact error in estimating the spectral density at the origin. *Journal of Time Series Analysis* 17, 379–408.

Parzen, E. 1957. On consistent estimates of the spectrum of a stationary time series. *Annals of Mathematical Statistics* 28, 329–48.

Perron, P. and Ng, S. 1998. An autoregressive spectral density estimator at frequency zero for nonstationarity tests. *Econometric Theory* 14, 560–603.

Phillips, P.C.B. and Hansen, B.E. 1990. Statistical inference in instrumental variables regression with i(1) processes. *Review of Economic Studies* 57, 99–125.

Phillips, P.C.B. and Perron, P. 1988. Testing for a unit root in time series regression. *Biometrika* 75, 335–346.

Phillips, P.C.B., Sun, Y. and Jin, S. 2005. Optimal bandwidth selection in heteroskedasticity-autocorrelation robust testing. Working Paper No. 2005-12, Department of Economics, UCSD.

Phillips, P.C.B., Sun, Y. and Jin, S. 2006. Spectral density estimation and robust hypothesis testing using steep origin kernels without truncation. *International Economic Review* 47, 837–94.

Priestley, M.B. 1981. *Spectral Analysis and Time Series*, vol. 1. New York: Academic Press.

Robinson, P. 1991. Automatic frequency domain inference on semiparametric and nonparametric models. *Econometrica* 59, 1329–63.

Stock, J. and Watson, M. 1988. Testing for common trends. *Journal of the American Statistical Association* 83, 1097–107.

Velasco, C. and Robinson, P.M. 2001. Edgeworth expansions for spectral density estimates and studentized sample mean. *Econometric Theory* 17, 497–539.

Watson, M.W. 1993. Measures of fit for calibrated models. *Journal of Political Economy* 101, 1011–41.

speculative bubbles

We maintain that a speculative bubble exists if the market price of an asset differs from its fundamental value – the expected present value of the stream of future dividends attached to the asset. In an economy with a finite sequence of trading dates, the fundamental theorem of asset pricing (see Dybvig and Ross, 1987) guarantees that the equilibrium market price of any asset equals its fundamental value. But in some economies with an infinite sequence of trading dates, this result does not hold, and speculative bubbles may arise. An investor might buy an asset at a price higher than its fundamental value if she expects to sell it later on at a higher price – Harrison and Kreps (1978) call this process 'speculative behaviour'. In general equilibrium models, however, agents take prices as given and trade assets to transfer income across time and states. These models do not contemplate 'speculative behaviour' as it is usually understood. Therefore, the term 'speculative bubble' may seem inappropriate in some theoretical frameworks. Santos and Woodford (1997) talk broadly about 'asset pricing bubbles'.

There have been famous historical examples of sudden asset price increases followed by an abrupt fall as the Dutch 'tulipmania' (1634–7), the 'Mississippi bubble' (1719–20) and the 'South Sea bubble' (1720). Kindleberger (1978) argues that these are examples of bubbles, whereas Garber (2000) provides market-fundamental explanations for these episodes. More recently, we have seen sharp changes in stock and housing markets. The Japanese stock and land prices experienced a sharp rise in the late 1980s and a dramatic fall in the early 1990s. During the 'technology bubble', the Nasdaq Composite Index rose by more than 300 per cent between August 1996 and March 2000, and then fell sharply, reaching the August 1996 level in October 2002. This pattern has been especially intense for the Internet-related sector (Ofek and Richardson, 2003).

There is a vast literature following the variance-bound tests proposed by LeRoy and Porter (1981) and Shiller (1981) that finds significant excess volatility of stock prices (see Gilles and LeRoy, 1991, for a survey). The violation of these variance bounds suggests that asset prices are not determined by fundamental values (see

Flood and Hodrick, 1990, and Cochrane, 1992 for a discussion). Various tests have been proposed to detect the presence of rational bubbles in asset prices (see Camerer, 1989, and Cuthbertson, 1996, for a survey). But these tests have important shortcomings. Estimating the fundamental values of an asset is usually a complex task. Hence, rejections of the null hypothesis could be due to an incorrect specification of the fundamental value and not necessarily to the existence of a bubble (Flood and Hodrick, 1990). Even in the most famous apparent bubble episodes, some authors have provided a fundamentalist explanation (see, for example, Donaldson and Kamstra, 1996; Pástor and Veronesi, 2006). To avoid the uncertainty associated with the specification of the fundamental value, Diba and Grossman (1988a) develop a test to detect bubbles based on the investigation of the stationary properties of asset prices and dividends. The main drawback of this test, as Evans (1991) shows, is its limited power to detect periodically collapsing bubbles. Given the severe problems in establishing empirically the existence of bubbles, it is of great importance to understand the theoretical conditions under which bubbles may exist.

If all traders are rational, a backward induction argument precludes the existence of bubbles for assets traded at a finite sequence of dates. More specifically, assume that the economy ends at time T, and there is an asset that provides a dividend of d_T at time T. Then the price of the asset at $T-1$ must be equal to the present value of d_T. By backward induction a bubble cannot exist at any point in time t less that T. Hence, a rational bubble begins on the first date of trading. Moreover, in present value terms the size of the bubble must be constant. (This is usually called the martingale property of bubbles.) Diba and Grossman (1988b) argue that negative rational bubbles cannot exist because it would imply that investors expect that the price of the asset will become negative at a finite future date. Tirole (1982) concludes that, in an economy with a finite number of infinitely lived traders, any asset must be valued according to its market fundamental. However, Tirole (1985) shows that under certain circumstances a deterministic overlapping generations economy allows for the existence of bubbles. In infinite-horizon optimization economies, bubbles are not compatible with the transversality condition: the present value of optimal asset holdings must converge to zero. But by definition the discounted price of the asset will converge to the size of the bubble. Hence, either the asset is in zero net supply or the size of the bubble is equal to zero.

Santos and Woodford (1997) explore the existence of asset pricing bubbles in an infinite-horizon competitive framework, allowing for potentially incomplete markets, arbitrary borrowing limits and incomplete participation of agents (this framework considers jointly economies with a finite number of infinitely lived households and overlapping generations economies). They show that the price of any asset in positive net supply must be equal to its fundamental value, provided that the present value of aggregate wealth is finite. This latter condition is satisfied empirically (see Abel et al., 1989) since in industrialized economies the aggregate share of income that goes to capital is greater than the investment rate. Loewenstein and Willard (2000) extend these results to a finite horizon economy where assets are negotiated continuously. Some key conditions underlying the negative results of Santos and Woodford (1997) are rational expectations, symmetric information and competitive behaviour.

This analysis has important implications for monetary theory because it precludes the existence of valued fiat money as a store of wealth in a broad class of economies. Santos (2006) extends these results to an economy with liquidity constraints and proves that these constraints must be binding infinitely often for all agents in the economy. Hence, in his simple model the aggregate value of the money supply must be equal to the value of aggregate output infinitely often. This is in the spirit of the quantity theory of money. On a related matter, the absence of rational bubbles guarantees that the initial real value of public debt is equal to the present value of future net public revenues. This is a necessary condition to establish the validity of the 'fiscal theory of the price level' (Sims, 1994; Woodford, 1995).

The presence of bubbles has also been explored in theoretical frameworks with asymmetric information or boundedly rational agents. Allen, Morris and Postlewaite (1993) find necessary conditions for the existence of bubbles in a model with asymmetric information and a finite sequence of trading dates, and provide examples satisfying these conditions. The existence of a bubble is possible because there is private information which is not common knowledge (all agents know that all agents know, and so on, ad infinitum) that the stock price will fall. Everybody realizes that the stock is overpriced but each agent expects to sell at a higher price before the true value becomes publicly known.

Bubbles may appear in the presence of agency problems associated with short-run optimization behaviour. Allen and Gorton (1993) show that for some compensation schemes a manager may purchase a stock with some prospect of capital gains although with certainty the price will fall below its current level at some point in the future. Allen and Gale (2000) develop a model in which intermediation by the banking sector leads to an agency problem that results in asset bubbles. Investors borrow from banks to buy a risky asset, and they can default in the case of low payoffs. Hence risky assets are more attractive, and therefore investors bid up asset prices.

The behavioural finance literature (see Barberis and Thaler, 2003; Shleifer, 2000 for a survey) often assumes that some agents – called noise traders – are not fully rational. In models in which noise traders and rational agents coexist, the price of an asset can deviate from its fundamental value if rational agents are limited in their capacity to eliminate the mispricing. Shleifer (2000) describes bubbles as the interaction between a significant number of positive feedback investors (who buy securities when prices rise and sell when prices fall), and rational arbitrageurs who anticipate the bursting of bubbles. In this framework, rational arbitragers buy initially after a good-news event to increase the price of the asset and to stimulate the demand of the positive feedback traders; later, they undo their position before the bubble explodes. Abreu and Brunnermeier (2003) develop a model in which noise traders coexist with rational arbitrageurs who become aware of the existence of a bubble sequentially. These rational arbitrageurs would like to exit the market just before the bubble bursts, because before bursting the asset displays high capital gains. The bubble can explode for exogenous reasons, or endogenously when a sufficient number of arbitrageurs decide to abandon the market. In this setting some news could facilitate synchronization and, as a consequence, the bursting of the bubble. Scheinkman and Xiong (2003) develop a model in which overconfidence generates disagreements among agents regarding asset fundamentals. They show that the price of an asset can be above its fundamental value.

In summary, asset prices seem rather volatile – more than their fundamental values. By definition this implies the existence of speculative bubbles. Most empirical exercises to detect the presence of bubbles seem inconclusive. The conditions under which general equilibrium models generate bubbles seem rather fragile, since optimizing agents are unwilling to accumulate arbitrary amounts of wealth. Most recent work has explored the existence of bubbles in economies with limited rationality, asymmetric information and strategic behaviour. The main challenge for these approaches is to explain the mechanisms that lead agents to hold overpriced assets. Specifically, if agents accumulate those assets for arbitrary reasons, then these exercises will not be very enlightening.

MIGUEL A. IRAOLA, MANUEL S. SANTOS

See also **arbitrage pricing theory; excess volatility tests; noise traders; present value.**

Bibliography

Abel, A.B., Mankiw, N.G., Summers, L.H. and Zeckhauser, R.J. 1989. Assessing dynamic efficiency: theory and evidence. *Review of economic studies* 56, 1–20.

Abreu, D. and Brunnermeier, M.K. 2003. Bubbles and crashes. *Econometrica* 71, 173–204.

Allen, F. and Gale, D. 2000. Bubbles and crises. *Economic Journal* 110, 236–55.

Allen, F. and Gorton, G. 1993. Churning bubbles. *Review of Economic Studies* 60, 813–36.

Allen, F. Morris, S. and Postlewaite, A. 1993. Finite bubbles with short sale constraints and asymmetric information. *Journal of Economic Theory* 61, 206–29.

Barberis, N. and Thaler, R. 2003. A survey of behavioral finance. In *Handbook of the Economics of Finance*, vol. 1, ed. G.M. Constantinides, M. Harris and R.M. Stulz. Amsterdam: North-Holland.

Camerer, C. 1989. Bubbles and fads in asset prices. *Journal of Economic Surveys* 3, 3–41.

Cuthbertson, K. 1996. *Quantitative Financial Economics: Stocks, Bonds and Foreign Exchange.* Chichester: Wiley & Sons Ltd.

Cochrane, J. 1992. Explaining the variance of price dividend ratios. *Review of Financial Studies* 5, 243–80.

Diba, B.T. and Grossman, H.I. 1988a. Explosive rational bubbles in stock prices. *American Economic Review* 78, 520–30.

Diba, B.T. and Grossman, H.I. 1988b. The theory of rational bubbles in stock prices. *Economic Journal* 98, 746–54.

Donaldson, R.G. and Kamstra, M. 1996. A new dividend forecasting procedure that rejects bubbles in asset prices: the case of 1929's stock crash. *Review of Financial Studies* 9, 333–83.

Dybvig, P.H. and Ross, S.A. 1987. Arbitrage. In *The New Palgrave: A Dictionary of Economics*, vol. 1, ed. J. Eatwell, M. Milgate and P. Newman. London: Macmillan.

Evans, G.W. 1991. Pitfalls in testing for explosive bubbles in asset prices. *American Economic Review* 81, 922–30.

Flood, R.P. and Hodrick, R.J. 1990. On testing for speculative bubbles. *Journal of Economic Perspectives* 4(2), 85–101.

Garber, P.M. 2000. *Famous First Bubbles.* Cambridge, MA: MIT Press.

Gilles, C. and LeRoy, S. 1991. Econometric aspects of the variance-bounds tests: a survey. *Review of Financial Studies* 4, 753–91.

Harrison, J.M. and Kreps, D.M. 1978. Speculative investor behavior in a stock market with heterogeneous expectations. *Quarterly Journal of Economics* 92, 323–36.

Kindleberger, C.P. 1978. *Manias, Panics and Crashes: A History of Financial Crises.* New York: Basic Books.

LeRoy, S.F. and Porter, R.D. 1981. The present value relation: test based on implied variance bounds. *Econometrica* 49, 555–74.

Loewenstein, M. and Willard, G.A. 2000. Rational equilibrium asset-pricing bubbles in continuous trading markets. *Journal of Economic Theory* 91, 17–58.

Ofek, E. and Richardson, M. 2003. DotCom mania: the rise and fall of internet stock prices. *Journal of Finance* 58, 1113–37.

Pástor, L. and Veronesi, P. 2006. Was there a Nasdaq bubble in the late 1990s? *Journal of Financial Economics* 81, 61–100.

Santos, M.S. 2006. The value of money in a dynamic equilibrium model. *Economic Theory* 27, 39–58.

Santos, M.S. and Woodford, M. 1997. Rational asset pricing bubbles. *Econometrica* 65, 16–57.

Scheinkman, J.A. and Xiong, W. 2003. Overconfidence and speculative bubbles. *Journal of Political Economy* 11, 1183–219.

Shiller, R.J. 1981. Do stock prices move too much to be justified by subsequent changes in dividends? *American Economic Review* 71, 421–36.

Shleifer, A. 2000. *Inefficient Markets: An Introduction to Behavioral Finance.* Oxford: Oxford University Press.

Sims, C.A. 1994. A simple model of the determination of the price level and the interaction of monetary and fiscal policy. *Economic Theory* 4, 381–99.

Tirole, J. 1982. On the possibility of speculation under rational expectations. *Econometrica* 50, 1163–81.

Tirole, J. 1985. Asset bubbles and overlapping generations. *Econometrica* 53, 1499–528.

Woodford, M. 1995. Price level determinacy without control of a monetary aggregate. *Carnegie-Rochester Conference Series on Public Policy* 43, 1–46.

Spence, A. Michael (born 1943)

Andrew Michael Spence was awarded the Nobel Prize in 2001 for his work on information economics. His pioneering doctoral dissertation on signalling formed the basis of an enormous subsequent literature in the economics of information. Spence's insight is that individuals can take actions that provide information to others, even though the actions themselves have no effect on productivity or on that which is desired by the buyer.

In Akerlof's classic 1970 paper on lemons 'The Market for "Lemons": Quality Uncertainty and the Market Mechanism', a market characterized by adverse selection is described where goods that are put on the used market are not a random selection of all goods. Spence asked whether there were actions that could be taken that might identify one type of seller from another in a market for lemons. For example, sellers who knew that their goods were particularly high-quality might offer warranties, whereas those who were concerned about the quality of their products would refrain from doing so.

This simple insight stimulated Spence to create an entire new way of thinking about information in markets. His dissertation, which, published in 1974, became one of the most influential of the second half of the 20th century, is a work that is well ahead of its time, incorporating complexity of equilibrium concepts, game-theoretic strategies and notions of multiplicity that were not to be fleshed out until a number of years later.

Education as signalling

Spence's classic paper 'Job Market Signaling' (1973a) considers whether education might be used merely as a signal of worker quality rather than as a tool to enhance productivity. There are two assumptions behind it. First, information is asymmetric. Workers know their own productivity, but firms do not. Second, there is a negative correlation between the cost of acquiring education and worker productivity. That is, the individuals who are most productive in the labour market are also those who find it cheapest to acquire education. The easiest way to think of this is that the most able people can pass an exam with fewer hours of study than the least able people.

Spence shows that an equilibrium exists where high-ability types acquire more schooling than low-ability types, even though schooling has no inherent effect on productivity. The simplest analysis is shown in Figure 1.

Suppose that there are two groups, the high-ability group with costs shown by C_{II} and a low ability group shown C_I. Suppose further that the employers make the assumption that individuals who have y^* or more education are high-ability, whereas those with $<y^*$ education are of low ability. The low-ability group has a cost of schooling equal to y, whereas the high-ability group has a cost of schooling equal to $y/2$. The productivity of high-quality workers is 2, whereas the productivity of low-quality workers is 1. Spence shows that the employers' expectations will be validated in equilibrium.

First, consider a high-ability group. If workers are deemed to be high-ability, then their output and wage is 2. If workers are deemed to be low-ability, then their output and wage is 1. Low-productivity individuals, whose cost is given by C_I, find it best to accept the low wage and to invest in zero units of schooling, as shown by the point at the origin in the diagram labelled group I. But because high-ability individuals have sufficiently low costs, it is profitable for them to acquire y^* of schooling because $2(-y^*/2) > 1$.

The assumptions behind this model lead immediately to predictions on which signals may survive in the market. In order for a signal to be effective, it must be correlated with productivity, and also have different costs across high-ability and low-ability groups. So, for example, skill in origami is rarely suggested as a signal in labour markets because origami prowess is unlikely to be highly correlated with productivity in most jobs.

The basic insight that comes out of the signalling structure can be extended. In the job market signalling

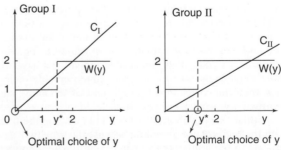

Figure 1

paper itself, Spence lays out an early version of the statistical discrimination argument, where male and female workers may end up with different levels of education, but the signal that a given level of education conveys in the male population may be quite different from that conveyed by the same level of education in the female population. This results from multiple equilibria in the basic signalling model, a point that Spence makes clear even in his very early analysis.

When Spence presented his Nobel lecture in 2001 (Spence, 2002), he revisited the signalling analysis and discovered that differential costs were unnecessary. Indeed, it is possible to have costs that are the same across two groups but benefits that differ, such as the case in the labour market setting where some workers choose to be measured and are paid piece rates and other workers choose not to be measured and are paid straight salaries. Those workers who choose to be measured are of higher ability (Lazear, 1986).

Additionally, a large literature on advertising traces its roots to the Spence signalling analysis. Certain firms choose to bear the cost of advertising because the value to them is higher than the value to other firms. For example, a firm with a very good product benefits more by having information about its product available to the public than a firm with a bad product. While both types of products may benefit in the short run from signalling through advertising, the firm with the better product can enjoy longer-run gains, and therefore higher return to the information conveyed by advertising.

In an extension 'Time and Communication in Economic and Social Interaction' (1973b) Spence considers how willingness to spend time acts as a signal. When a political leader attends an event at which he has no role, he signals that he values the organization that has sponsored the event. Indeed, the fact that he has no role makes the signal stronger. Another example involves pricing sporting events at less than the market-clearing level, while using time spent in a queue as an allocative mechanism. Although the wasted time creates deadweight loss, it may be that time is a better signal of interest in the sporting event than is the money that an individual is willing to spend on the tickets. To the extent that the quality of the event depends on the active participation of the fans, it may be optimal to use time in addition to money to allocate tickets to potential buyers.

Time operates differently from other costs in the traditional signalling model that applied in the job market. The high cost of time actually may preclude using education as a signal. If the most able individuals have the most valuable use of time in other (unobservable) activities, then the correlation between cost of schooling and productivity may be reversed. The highly productive individuals in the labour market would also be the ones with the high cost of schooling because the time that they allocate to schooling is so valuable. In this case schooling would not be an effective signal. Spence works out

situations where the schooling may be an adverse signal of productivity in his 2001 Nobel address.

Spence notes in his early signalling paper that signalling creates a divergence between social and private value. While individuals may have an incentive to signal and firms may have an incentive to pay for that signal, the signal in the purest case reflects wasted resources. In the simplest version, where information has no sorting or allocative role, individuals who acquire education receive higher earnings, and those who do not receive lower earnings. The market equilibrium has higher inequality of income than one without signalling, but no higher output in the labour market. Because signalling is in itself unproductive, resources devoted to schooling are wasteful. A law that said 'firms are precluded from paying on the basis of education' would be welfare enhancing.

Spence after signalling

Spence's interest in the economics of information has motivated him to consider issues that arise in product markets. Specifically, he questions whether the information that is present in a market provides appropriate signals to producers and consumers, or whether those signals would lead to inefficient outcomes under certain circumstances. Information, he reasons, is an important component of markets that provides incentives and alters behaviour. One of the first questions that Spence considers is whether the information that is transmitted by the market induces firms to produce the correct quality of a good. Spence (1975) shows that, as a general matter, the market produces the wrong quality, although not necessarily a quality that is too high or too low. Price is determined by the marginal consumer, but the rents associated with the particular quality of a good depend on the area under the demand curve up to the quantity produced, and not just the value that the marginal consumer places on it. If the marginal consumer's valuation of higher quality is lower than the average consumer's, then the market underproduces quality. If the marginal consumer's valuation is higher than the average consumer's, then the market overproduces quality. (A related idea is explored in Spence and Owen, 1977.)

Spence's forays into quality and product choice have led him to consider more general questions relating to the interaction between information and industry structure. In 'Cost Reduction, Competition, and Industry Performance' (1981a), Spence examines the trade-off between a duplication of R&D expenditures and having the competitive industry that makes the market statically efficient. If there is only one firm, then that firm will charge a monopoly price. However, if there is a multiplicity of firms, then each firm must do its own R&D when R&D is appropriable. The output is closer to the competitive equilibrium with more firms than with fewer. The optimal degree of competition trades off replication of R&D expenditures against competitive outcomes. R&D in this world creates natural monopoly-like characteristics for the industry. Just as it is inefficient to duplicate electrical lines that go from power plants to individual houses, it is also inefficient to force every firm to do its own R&D. One solution discussed is to have an R&D consortium with appropriate subsidies. The firms in the consortium can share the information generated, and the subsidies may provide enough incentives to undertake efficient R&D activity.

A related idea is explored in the 'The Learning Curve and Competition' (1981b). The learning curve makes costs a function of past output. The more firms there are in an industry, the less output that any single firm produces. Each firm learns more slowly and costs are higher when there is a large number of firms than when there is a small number of firms. But the advantage of a large number of firms is that output is closer to the competitive level. Unfortunately, the competitive level is at a higher cost than if output were produced by one firm alone. This trade-off is analysed and a market structure is suggested.

Michael Spence as a person

An American by birth, Spence was born in New Jersey, when his father, a Canadian, was spending time on a project in Washington during the Second World War. His mother, also an American by birth, was raised in Winnipeg. The Spence family moved to Toronto from rural Canada when Mike was nine years old. Mike lived in Toronto until he went to Princeton in 1962. Having grown up in Canada, Mike acquired important skills, in particular the ability to play hockey. An excellent athlete, Spence played hockey for four years while at Princeton. He graduated summa cum laude and went on to be a Rhodes Scholar, attending Oxford University, where he studied mathematics. After getting his training at Oxford, he enrolled for a Ph.D. at Harvard. His most important mentors were Kenneth Arrow, Thomas Schelling and Richard Zeckhauser. In 1972 he completed his landmark thesis on signalling and was awarded a Ph.D. Mike taught at the Harvard's J. F. Kennedy School of Government for a couple of years. In 1973 he moved to Stanford University's economics department, but was lured back to Harvard in 1976. After a few more years doing research, Mike became the Dean of the Faculty of Arts and Sciences in 1984, and stayed in that position until he moved back to Stanford in 1990, when he took over as Dean of the Graduate School of Business. He retired from that position in 1999.

Mike's output in family life has been as high-quality as his output in research. He has three children. Graham, born in 1979, graduated from Princeton; Catherine, born in 1982, graduated from Columbia; and Marya, born in 1985, enrolled at Harvard in 2003. Michael Spence, a great friend and exceptional individual, excels at everything that he attempts.

Mike's hobbies include windsurfing and motorcycle riding.

EDWARD LAZEAR

See also **Akerlof, George Arthur; signalling and screening; Stiglitz, Joseph E.**

Selected works

1973a. Job market signaling. *Quarterly Journal of Economics* 87, 355–74.
1973b. Time and communication in economic and social interaction. *Quarterly Journal of Economics* 87, 651–60.
1974. *Market Signaling: Informational Transfer in Hiring and Related Processes*. Cambridge, MA: Harvard University Press.
1975. Monopoly quality and regulation. *Bell Journal of Economics* 6, 417–29.
1977. (With B. Owen.) Television programming, monopolistic competition, and welfare. *Quarterly Journal of Economics* 91, 103–26.
1981a. Cost reduction, competition, and industry performance. *Econometrica* 52, 101–22.
1981b. The learning curve and competition. *Bell Journal of Economics* 12, 49–70.
2002. Signaling in retrospect and the informational structure of markets. *Les Prix Nobel 2001*. Stockholm: The Nobel Foundation. Online. Available at http://nobelprize.org/economics/laureates/2001/spence-lecture.pdf, accessed 10 June 2006.

Bibliography

Akerlof, G. 1970. The market for 'lemons': quality uncertainty and the market mechanism. *Quarterly Journal of Economics* 84, 488–500.
Lazear, E. 1986. Salaries and piece rates. *Journal of Business* 59, 405–31.

Spencer, Herbert (1820–1903)

Though largely ignored today, Herbert Spencer was one of the most influential scientists and philosophers of the late 19th century. He was born at Derby to a family of Dissenters, and educated at home. As a young man he worked on the London & Birmingham Railway, acquiring considerable practical knowledge of civil engineering and, through his observation of railway cuts, expertise in geology. He had no university training, but read extremely widely in an array of fields.

He was an early and enthusiastic partisan of Darwin and of evolutionist ideas. In 1860 he published a prospectus for a 'system of synthetic philosophy', a general compendium of knowledge, which was to occupy him for much of the rest of his life. He set out to survey, from the 'evolutionary point of view', the fields of biology, psychology, sociology and ethics, publishing in turn *First*

Principles (1862); *Principles of Biology* (2 vols, 1864–7); *Principles of Sociology* (3 vols, 1876–96) and *Principles of Ethics* (2 vols, 1879–93).

His most important work bearing on social policy was the polemical *The Man Versus the State* (1884). Spencer was a highly vocal champion of social Darwinism, applying the principle he termed 'survival of the fittest' to a broad variety of struggles, including economic competition. He conceived society by analogy to an organism, arguing that it developed according to immanent processes of growth, and hence that the positive actions and interventions of politicians were likely to be harmful or superfluous.

M. DONNELLY

Selected works

1862. *First Principles*. London.

1864–7. *The Principles of Biology*, 2 vols. London: Williams & Norgate.

1870–72. *The Principles of Psychology*, 2 vols. 2nd edn. London: Williams & Norgate. (First published 1855.)

1876–96. *Principles of Sociology*, 3 vols. London: Williams & Norgate.

1879–93. *The Principles of Ethics*, 2 vols. London: Williams & Norgate.

1884. *The Man Versus the State*. Indianapolis: Liberty Classics, 1982.

Bibliographic addendum

Offer, J., ed. 2000. *Herbert Spencer: Critical Assessments*. London: Routledge, 2000.

Weinstein, D. 1998. *Equal Freedom and Utility: Herbert Spencer's Liberal Utilitarianism*. Cambridge: Cambridge University Press.

Spiethoff, Arthur August Kaspar (1873–1957)

Spiethoff was born on 13 May 1873 in Düsseldorf and died on 4 April 1957 in Tübingen. He was a student of Adolph Wagner and research assistant to Gustav Schmoller. In 1908 he was appointed professor at Prague University and from 1918 held a chair of political economy at Bonn University until his retirement in 1939. He was the long-time editor of *Schmollers Jahrbuch* and (with Edgar Salin) of *Hand- und Lehrbücher aus dem Gebiet der Sozialwissenschaften*.

Spiethoff is best known for his path-breaking research into business cycles, as well as for his studies on methodology, culminating in his concept of 'economic style' (*Wirtschaftsstil*). In his methodological studies Spiethoff, strongly influenced by the German Historical School, sought for a solution to the antinomy of history and theory: the quest for generalizing statements about an ever-changing reality. He stressed the distinction between

two methods of inquiry: pure economic theory (brought to perfection by Quesnay, Ricardo, Thünen, Menger, Jevons and Pareto) and 'observational' (*anschauliche*) or 'economic Gestalt theory' (in the tradition of the mercantilists, List, Sombart and Schmoller). Pure economic theory, whether or not it exclusively deals with timeless phenomena, such as those common to all forms of economic life, abstracts and isolates arbitrarily, depending on the particular purpose in view. 'Observational theory', on the other hand, takes its time-conditioned data from the real world and abstracts only from their historical uniqueness to isolate the regular and essential features. It thus yields an 'explanatory description' – that is, an effigy, or replica, of reality – 'purged of historical accidents' (Spiethoff, 1953b, p. 76). With its findings derived from time-conditioned data, 'economic Gestalt theory' is a 'historical' theory, the validity and applicability of its generalizations dependent on the existence and dominance of a certain 'economic style', representing uniformities of economic life in a certain historical epoch (for example, the 'economic styles' of medieval town economy, of free market capitalism or of interventionism). Spiethoff's ultimate aim was an all-embracing general economic theory that would include as many different 'historical' theories as there are 'economic styles', together with the pure theory of timeless phenomena.

The foremost field of application of Spiethoff's methodological approach has always been his research on business cycles. In his writings (see Schweitzer, 1941, for a comprehensive bibliography), starting from the work of Clément Juglar, he emphasized three points: first, the necessity not to focus exclusively on crisis or overproduction, but instead to visualize the phenomenon of cyclical fluctuations as an entity; second, the strategic role to be ascribed to capital investment in the explanation of business cycles; and third, the fact that booms and depressions should not be considered as merely an accidental and insignificant concomitant of economic activity but must be understood as the essential form of capitalist life itself. This basic perception made him one of the founding fathers of modern business cycle research.

In keeping with this notion of 'time-conditioned' theory, Spiethoff considered his findings to be valid only for a certain 'economic style', representing an age marked by the prevalence of a highly developed capitalist economy and a free-market system. This era lasted from 1820 to 1913, with the capitalist economy not yet fully developed in earlier periods, while increasingly becoming subject to manipulation, planning and management in later times. Spiethoff's striving for 'historical' generalization took the form of distilling a 'typical cycle' to give account of the recurrent and essential features of all historically known business cycles. This 'typical cycle', now generally accepted, consists of three 'cyclical stages' (upswing, crisis and downswing), two of which may be subdivided into five 'cyclical phases': the downswing comprises the recession phase with investment declining and a 'first revival' during which the decline in investment is halted, while the upswing includes the 'second revival' with rapidly increasing investment, the boom phase characterized by rising interest rates and, finally, 'capital scarcity' with declining investment paving the way for the next downturn.

With his observation of 'cyclical periods', during which years of either boom or depression preponderate, Spiethoff in fact anticipated what later would become known as the Kondratieff cycle or 'long wave'.

INGO BARENS

Selected works

1933. Overproduction. In *Encyclopaedia of the Social Sciences*, vol. 11. New York: Macmillan.

1952. The 'historical' character of economic theories. *Journal of Economic History* 12, 131–9.

1953a. Pure theory and economic Gestalt theory: ideal types and real types. In *Enterprise and Secular Change: Readings in Economic History*, ed. F.C. Lane and J.C. Riemersma. London: George Allen & Unwin.

1953b. Business cycles. *International Economic Papers* 3, 75–171. Originally published in German as 'Krisen', in *Handwörterbuch der Staatswissenschaften*, vol. 6, Jena: G. Fischer, 1923.

1955. *Die wirtschaftlichen Wechsellagen: Aufschwung, Stockung, Krise*. 2 vols. Zurich: Polygraph.

Bibliography

Clausing, G., ed. 1933. *Der Stand und die nächste Zukunft der Konjunkturforschung. Festschrift für Arthur Spiethoff*. Berlin: Duncker & Humblot.

Clausing, G. 1958. Arthur Spiethoffs wissenschaftliches Lebenswerk. *Schmollers Jahrbuch für Gesetzgebung, Verwaltung und Volkswirtschaft* 78, 257–90.

Lane, F.C. 1956. Some heirs of Gustav von Schmoller. In *Architects and Craftsmen in History: Festschrift für Abbott Payson Usher*, ed. J.T. Lombie. Tübingen: Mohr.

Lane, F.C. and Riemersma, J.C. 1958. Introduction to Arthur Spiethoff. In *Enterprise and Secular Change: Readings in Economic History*, ed. F.C. Lane and J.C. Riemersma. London: George Allen & Unwin.

Schweitzer, A. 1941. *Spiethoff's Theory of the Business Cycle*, vol. 8. Laromie: University of Wyoming Publications (contains an extensive bibliography).

spline functions

In the everyday use of the word, a 'spline' is a flexible strip of material used by draftsmen in the same manner as French curves to draw a smooth curve between specified points. The mathematical *spline function* is similar to the draftsman's spline. It has roots in the aircraft, automobile and shipbuilding industries. Formally, a spline function is a piecewise continuous function with a specified degree of continuity imposed on its derivatives. Usually the pieces are polynomials. The abscissa values, which define the segments, are referred to as 'knots', and the set of knots is referred to as the 'mesh'.

The terminology and impetus for most contemporary work on spline functions can be traced to the seminal work of I.J. Schoenberg (1946), although the basic idea can be found in the writings of E.T. Whittaker (1923) and, in Schoenberg's (1946, p. 68) own modest opinion, in the earlier work of Laplace. Today the literature on spline functions comprises an integral part of modern approximation theory. Useful monographs covering splines are De Boor (2001), Eubank (1988), Green and Silverman (1994), Poirier (1976), Schumaker (1981), and Wahba (1990). The many important contributions of Grace Wahba in the 1970s and 1980s (for example, Kimeldorf and Wahba, 1970; Wahba, 1978; 1983) united the approximation theory and the emerging statistics literatures involving spline functions.

Given a degree d and a knot vector $t = [t_1, t_2, \ldots, t_K]'$, where $t_1 < t_2 < \ldots < t_K$, the collection of polynomial splines having s continuous derivatives forms a linear space. For example, the collection of *linear* splines with knot sequence t is spanned by the functions

$$1, x, (x - t_1)_+, \ldots, (x - t_K)_+.$$

where $(\cdot)_+ = \max(\cdot, 0)$. This set is called the *truncated power basis* of the space. In general, the basis for a spline space of degree d and smoothness s is made up of monomials up to degree d together with terms of the form $(x - t_k)_+^{s+j}$, where $1 \leq j \leq d - s$. For example, cubic splines have $d - 3$ and $s - 2$ so that the basis has elements

$$1, x, x^2, x^3, (x - t_1)_+^3, \ldots, (x - t_K)_+^3.$$

Unfortunately, these truncated power functions have poor numerical properties. For example, in linear regression problems the condition of the design matrix deteriorates rapidly as the number of knots increases. A popular alternative representation is the so-called *B-spline basis* (see De Boor. 2001). These functions are constructed to have support only on a few neighbouring intervals defined by the knots.

The importance of spline functions in approximation theory is explained by the following *best approximation property*. Consider the data points $(x_i, y_i)(i = 1, 2, \ldots, n)$ and suppose without loss of generality that $0 < x_1 < x_2 < x_3 < \ldots < x_n < 1$. Given $\lambda > 0$, consider the optimization problem

$$\min_{f(\cdot)} \sum_{i=1}^{n} [y_i - f(x_i)]^2 + \lambda \int_0^1 [D^m f(x)]^2 dx,$$

(1)

where D^m denotes the differentiation operator of degree m, $f(\cdot)$ is a function defined on $[0, 1]$ such that $D^j f, j \leq m - 1$, is absolutely continuous, and $D^m f$ is in the set of measurable square integrable functions on $[0, 1]$. The first term in (1) comprises the familiar least squares measure of fit and the second term comprises a measure of the smoothness in $f(\cdot)$. The parameter λ measures the trade-off between fit and smoothness. The solution to (1) is a polynomial *smoothing spline* of degree $2m - 1$ with knots at all the abscissa data points. As $\lambda \rightarrow 0$, the solution is referred to as an *interpolating spline* and it fits the data exactly. The choice of λ is crucial and the method of cross-validation is a popular method for choosing λ (see for example Craven and Wahba, 1979, or Green and Silverman, 1994, pp. 30–8). The most popular choice for m is $m = 2$ yielding a *natural cubic spline* as the solution to (1). The adjective 'natural' implies that the second derivative equals zero at the endpoints.

To interpret the first term in (1) as the log-likelihood of normal linear regression model, smoothing splines can be viewed as the outcome of penalized (reflected in the second term in (1)) maximum likelihood estimation. A Bayesian interpretation of smoothing splines, provided by Kimeldorf and Wahba (1970) and expanded by Silverman (1985) and Wahba (1978; 1983), views (1) as a log-posterior density with $\exp[-\frac{1}{2}\lambda \int [D^m f(x)]^2 dx]$ serving as a prior density over the space of all smooth functions.

In econometrics spline functions are most often employed to parametrize a regression function. For example, splines were the functional form chosen to parametrize the treatment in the first major social experiment in economics: the New Jersey Income-Maintenance Experiment. Such *regression splines* usually include only a few knots and not necessarily at the design points. This usage may simply reflect the flexibility and good approximation properties of splines, or the attempt to capture *structural change*. For example, a researcher may believe the relationship between two variables y and x is locally a polynomial, but that at precise points in terms of x the relationship 'changes', not in a discontinuous fashion in level but rather continuously in derivative of order $2m - 1$. Common choices for such x variables are time, age, education or income, to name a few, with a nearly unlimited number of choices of candidates for y variables.

In statistics spline functions are used in isotonic regression, histogram smoothing, density estimation, interpolation of distribution functions for which there is no closed-form analytic representation, and nonparametric regression. In the latter case spline smoothing corresponds approximately to smoothing by a kernel method with bandwidth depending on the local density of design points.

While spline functions have proved to be valuable approximation tools, they also arise naturally in their own right in economics. Income tax functions with increasing marginal tax rates constitute a *linear spline*, as do familiar 'kinked' demand curves and 'kinked' budget sets. *Quadratic splines* serve as useful ways of generating asymmetric loss functions for use in decision theory. In distributed lag analysis, spline functions have been used as natural generalizations of Almon polynomial lags. *Periodic cubic splines* have proved useful in seasonal adjustment and in analysis of electricity load curves. Spline functions in these applications are attractive partly because, given the knots, the spline can be expressed as linear functions of unknown parameters, hence facilitating statistical estimation.

Knots play different roles in the approximation theory and structural change literatures. In the former they are largely nuisance parameters, and, apart from parsimony considerations, the number and location of the knots are of no particular importance other than that they serve to define a smooth best-fitting curve. When viewed as *change points*, however, the knots become parameters of interest. In applications involving structural change, the number of potential knots is small, and their location reflects subject-matter considerations. For example, when fitting a time trend with a regression spline, the knots may reflect the effect on the dependent variable of a specific event of interest – for example, a war. A prior distribution can then be specified over the interval bounded by the start and end of the war.

Estimation of the number and location of the knots is hindered by numerical and statistic complications. The knots enter spline functions nonlinearly, and there are typically numerous local minima in the residual sum-of-squares surface. Many of these local minima correspond to knot vectors with *replicate knots*, that is, knots which pile up on top of each other, signalling that further discontinuities in the derivatives of the function are required. When knot locations are set free, knots move to areas where the function is less smooth. If, in addition, the *number* of knots is unknown, the difficulties multiply. For example, under the null hypothesis that adjacent intervals are identical, the location of the unnecessary knot is unidentified.

Different solutions have emerged to the problem of unknown location and number of knots. Some introduce a large number of potential knots from which a subset is to be selected (for example, Halpern, 1973; Friedman and Silverman, 1989; Smith and Kohn, 1996). The problem then becomes one of variable selection where each knot corresponds to a column of a design matrix from which a 'significant' subset is to be determined. In some Bayesian nonparametric regression studies (for example, DiMatteo, Genovese and Kass, 2001; Smith, Wong and Kohn, 1998; Denison, Mallick and Smith, 1998), knot locations are treated as parameters and given prior distributions. Additional constraints are usually imposed to keep knots some minimum distance apart. The definitive treatment of the problem of unknown location and number of knots has not yet emerged.

Early applications of splines to *multivariate* problems (see Green and Silverman, 1994, ch. 7) involved tensor product spaces that of necessity depended on the choice of coordinate system. An example is the two-dimensional *thin plate spline* of Wahba (1990) which simulates how a thin metal plate would behave if forced through some control points. This is similar to the one-dimensional draftsman's spline. The tensor product structure of these spaces implicitly defines the domain of an unknown function to be a hyperrectangle, and this can restrict the ability to capture important features in the data that are not oriented along one of the major axes. There is a considerable literature on constructing and representing smooth, piecewise polynomial surfaces over meshes in many variables. In particular, much has been written about the case in which the underlying partition consists of triangles or high-dimensional simplices. Because of their invariance to affine transformations, *barycentric coordinates* (that is, coordinates expressed as weighted combinations of the vertices of the triangle) are used to construct spline spaces over such meshes. The *triogram* methodology of Hansen, Kooperberg and Sardy (1998) employs continuous, piecewise linear (planar) bivariate splines defined over adaptively selected triangulations in the plane. Analogous to stepwise knot addition and deletion in a univariate spline space, the underlying triangulation is constructed adaptively by adding and deleting vertices.

DALE J. POIRIER

See also **statistical inference; statistics and economics; structural change; two-stage least squares and the *k*-class estimator.**

Bibliography

Craven, P. and Wahba, G. 1979. Smoothing noisy data with spline functions: estimating the correct degree of smoothing by the method of cross-validation. *Numerische Mathematik* 31, 377–403.

De Boor, C. 2001. *A Practical Guide to Splines*. New York: Springer-Verlag.

Denison, D., Mallick, B. and Smith, A. 1998. Automatic Bayesian curve fitting. *Journal of the Royal Statistical Society: Series B* 60, 333–50.

DiMatteo, I., Genovese, C. and Kass, R. 2001. Bayesian curve-fitting with free knot splines. *Biometrika* 88, 1055–71.

Eubank, R. 1988. *Spline Smoothing and Nonparametric Regression*. New York: Marcel-Dekker.

Friedman, J. and Silverman, B. 1989. Flexible parsimonious smoothing and additive modeling. *Technometrics* 31, 3–21.

Green, P. and Silverman, B. 1994. *Nonparametric Regression and Generalized Linear Models*. London: Chapman and Hall.

Halpern, E. 1973. Bayesian spline regression when the number of knots is unknown. *Journal of the Royal Statistical Society: Series B* 35, 347–60.

Hansen, M., Kooperberg, C. and Sardy, S. 1998. Triogram models. *Journal of the American Statistical Association* 93, 101–19.

Kimeldorf, G. and Wahba, G. 1970. A correspondence between Bayesian estimation on stochastic processes and smoothing by splines. *Annals of Mathematical Statistics* 41, 495–502.

Poirier, D. 1976. *The Econometrics of Structural Change with Special Emphasis on Spline Functions*. Amsterdam: North-Holland.

Schoenberg, I. 1946. Contributions to the problem of approximation of equidistant data by analytic functions: Parts I and II. *Quarterly Journal of Applied Mathematics* 4, 45–99; 112–41.

Schumaker, L. 1981. *Spline Functions: Basic Theory*. New York: Wiley.

Silverman, B. 1985. Some aspects of the spline smoothing approach to nonparametric regression curve fitting (with discussion). *Journal of the Royal Statistical Society: Series B* 47, 1–52.

Smith, M. and Kohn, R. 1996. Nonparametric regression using Bayesian variable selection. *Journal of Econometrics* 75, 317–43.

Smith, M., Wong, C. and Kohn, R. 1998. Additive nonparametric regression with autocorrelated errors. *Journal of the Royal Statistical Society: Series B* 60, 311–31.

Wahba, G. 1978. Improper priors, spline smoothing and the problem of guarding against model errors in regression. *Journal of the Royal Statistical Society: Series B* 40, 364–72.

Wahba, G. 1983. Bayesian 'confidence intervals' for the cross-validated smoothing spline. *Journal of the Royal Statistical Society: Series B* 45, 133–50.

Wahba, G. 1990. *Spline Models for Observational Data*. Philadelphia: Society for Industrial and Applied Mathematics.

Whittaker, E. 1923. On a new method of graduation. *Proceedings of the Edinburgh Mathematical Society* 41, 63–75.

spontaneous order

I am convinced that if it were the result of deliberate human design… this mechanism would have been acclaimed as one of the greatest triumphs of the human mind.
—F.A. Hayek (1945, p. 87)

A spontaneous order is a recognizable pattern that is produced by a process that is neither directed by deliberate design nor created for a specific purpose, though it may produce useful results. The economy is a spontaneous order of dizzying magnitude, with billions of individuals coordinating their plans in order to satisfy their individual desires, with no overarching direction. One of most important characteristics of a spontaneous order is that it consists of patterns that are recognizable with human reason despite the fact that they are not the result of human design. In the market, the distribution and allocation of resources according to some principle or predictable relationship, such as the fact that prices correspond to opportunity costs or that the price of a good equals the firm's marginal cost, is an example of such a pattern. These systematic and predictable patterns, while obviously recognizable with human reason, are not the result of any person or group actively trying to achieve this result.

The origin of money or a medium of exchange can be explained purely by self-interested actions on the part of individuals who had no intention of creating it. 'As each economizing individual becomes increasingly more aware of his economic interest, he is led, by this interest, without any agreement, without legislative compulsion, and even without regard to the public interest, to give his commodities in exchange for other, more saleable, commodities even if he does not need them for an immediate consumption purpose' (Menger, 1871, p. 260). As more people begin realizing the gain to be had from collecting goods that others want, a medium of exchange spontaneously arises even though no person was intending this result, only their own gain. In this example, it is easy to see that emergence of a medium of exchange was 'of human action but not of human design' (Hayek, 1967).

Hayek (1948, p. 78) defines the economic problem as 'how to secure the best use of resources known to any member of society, for ends whose relative importance only these individuals know'. In the market, the spontaneous order that serves this function is the systematic elimination of 'mutual ignorance on the part of potential market participants' (Kirzner, 1992, p. 44). The mechanism that produces this result is the price system. The price system permits the spread and use of dispersed knowledge of time and place so the actions of individuals are coordinated. The self-interested actions of entrepreneurs searching for profit opportunities and competing for potential trading partners is what insures that prices embody the relevant information that people need in order to adjust their plans to the constantly changing circumstances of the market. Wealth creation throughout the market system is driven by the continuous discovery of new opportunities to trade, and the ceaseless desire to innovate in a manner that will either lower the costs associated with previous trades or open the opportunity for new trades to be made.

This result, while spectacular, is not an end to be actively sought. That the spontaneous order of the market process is means-driven implies that the information that is generated through this process is not available by any other means. We cannot know the utility function of even one other individual (perhaps not even our own), much less that of every individual in society. The only way to generate the allocation of resources that best provides for the varying ends of individuals is to allow the market process to function. That is why we stress that the process that generates a spontaneous order is ends-independent. Although it performs an extremely useful function in the coordination of knowledge and action, this coordination is not the result of action aimed at coordination, but the result of action aimed at gaining profit.

The fact that the generation of a spontaneous order relies not on purpose but on the elements following particular rules of behaviour is significant in that such behavioural rules need to be self-enforcing. In the market process, the behaviour that produces the order is profit-seeking, and as such is self-enforcing as long as people prefer more to less. However, there is an additional criterion that must be met for the order to emerge: '…people must also observe some conventional rules, that is, rules which do not simply follow from their desires and their insight into relations of cause and effect, but which are normative and tell them what they ought or ought not to do' (Hayek 1973, p. 45). There must be an accepted and well-defined sphere over which the individual has control. The importance of clearly defined and enforced property rights cannot be overstated in terms of its impact on the emergence and maintenance of a spontaneous market order.

Features of a spontaneous order

The unique characteristics of a spontaneous order imply certain features that are in contrast to planned or designed orders. Since the order is a result of a process that depends only on the individual elements following certain rules, the complexity and scale of a spontaneous order can be much greater than that of a planned order. The limitations of planned orders are readily apparent when we first consider the necessity of channelling everything through one mind. What is amazing is the possible complexity of the spontaneous order. Leonard Reed's 'I, Pencil' (1958) is among the best expositions of the degree of complexity that the market process achieves, and, since he wrote of the amazingly complicated task of producing pencils, it is safe to say that the process has become even more complex when one considers such products as automobiles, airplanes and computers. (See Adam Smith, 1776, pp. 15–16, for a similar examination of the common woollen coat worn by a day labourer and the multitude of exchanges that must be coordinated to produce even this homely product. See also Milton and Rose Friedman, 1980, pp. 1–29, for a classic discussion of the implications of the 'I, Pencil' story for our understanding of the market order.) This ability to maintain orders of such complexity is the source of the continued productive and technological growth that has characterized modern history. The importance of this feature of the spontaneous order lies in the astounding ability not just to develop such a complex order but, more importantly, to maintain it though circumstances are changing at almost every instant.

A related feature that spontaneous orders have in common is their abstract nature. Because the order is not the result of deliberate action, the individuals within it do not need to understand or even be aware of the process at work. This feature allows for a level of abstraction that is beyond the understanding of the human mind. The evidence for this in nature is abundant. Although the frontier of scientific knowledge is constantly being pushed out, it seems that there is no limit to what we still have to learn about the workings of the universe. This is true also in the realm of social sciences, but to an even greater extent. Since there is no one mind that must understand the implications of every action, there is almost no limit to experimentation in the market. The profit incentive provides for the generation of endless variation in attempts to better satisfy consumer desires. The abstract quality of the market process is that the goal, satisfying consumers, is never clearly defined. It is the market test, the 'systematic plan changes generated by the flow of market information released by market participation' that directs activity toward this end, not a comprehensive plan (Kirzner, 1973, p. 10).

Finally, spontaneous orders are dependent on the structure within which the elements are acting. (An excellent discussion of this point can be found in Vernon Smith's 2002 Nobel Prize address, where he contrasts 'ecological rationality' with 'constructivist rationality'; see V. Smith, 2003.) A purposeful design is dependent on the conception of the designer and his control over the elements. The specific state of a spontaneous order is determined not only by the characteristics of the elements but also by the structure that determines motivation. In the market the entrepreneurial profit incentive is a characteristic of the process. However, this characteristic produces coordination only if the easiest way to get a profit is to provide buyers and sellers with opportunities to better their trading positions. If, instead, the best way to profit is to manipulate the political process in order to confiscate others' wealth, then the order that emerges will not be recognizable as the coordination of knowledge and action.

Conclusion

Social norms, language, money and the price system are all examples of spontaneously emerging institutions in the social realm which, while certainly products of purposive human action, are nevertheless not of direct human design. The role of the political process is to define the structure within which the filter processes and the equilibrium processes of the 'invisible hand' operate. Understanding this concept means admitting a limit to the social states that are achievable. It means realizing that 'most of these social states that are romantically imagined to be possible are inconsistent with the motivational postulate of economics, with human nature as it exhibits its uniformities' (Buchanan, 1997, p. 37). Instead the political process must be confined to creating 'conditions in which an orderly arrangement can establish and ever renew itself' (Hayek, 1960, p. 161). That the market process is a spontaneous order means that our span of control is limited, yes, but there is no need to despair! For recognizing the spontaneous order of the market economy as an ongoing entrepreneurial process of discoveries for mutual benefit and wealth creation also means that we can achieve states of civilization that could not be imagined were it necessary that we designed them.

PETER BOETTKE AND JENNIFER DIRMEYER

See also **invisible hand.**

Bibliography

Buchanan, J. 1997. There is science of economics. *The Collected Works of James M. Buchanan*, vol. 12. Indianapolis: Liberty Press.

Friedman, M. and Friedman, R. 1980. *Free to Choose.* New York: Harcourt Brace Jovanovich.

Hayek, F.A. 1960. *The Constitution of Liberty.* Chicago: University of Chicago Press.

Hayek, F.A. 1948. The use of knowledge in society. In *Individualism and Economic Order*. Chicago: University of Chicago Press.

Hayek, F.A. 1967. The results of human action but not of human design. In *Studies in Philosophy, Politics and Economics*. London: Routledge & Kegan Paul.

Hayek, F.A. 1973. *Law Legislation and Liberty, Volume 1: Rules and Order.* Chicago: University of Chicago Press.

Kirzner, I. 1973. *Competition and Entrepreneurship*. Chicago: University of Chicago Press.

Kirzner, I. 1992. *The Meaning of the Market Process.* New York: Routledge.

Menger, C. 1871. *Principles of Economics*. New York: New York University Press, 1981.

Read, L. 1958. I, pencil: my family tree as told to Leonard E. Read. *The Freeman: Monthly Magazine of the Foundation for Economic Education*, December. Reprinted in *The Freeman* 46(5) (1996).

Smith, A. 1776. *An Inquiry into the Nature and Causes of the Wealth of Nations*. Chicago: University of Chicago Press, 1976.

Smith, V. 2003. Constructivist and ecological rationality in economics. *American Economic Review* 93, 465–508.

sports, economics of

Why study the economics of sport? In purely financial terms it is not a very important part of the economy. In 1998, out of the $8 trillion US economy, the commercial sports industry generated only $17.7 billion of revenue – a sizeable sum to be sure, but only 0.2 per cent of US GDP and one-quarter the size of the automotive repair industry ($69.6 billion) in that year (U.S. Bureau of the Census, 1999, Table 6.1). We have no field of study on the economics of car repair. We advance five reasons for special interest in sport.

First, the cultural significance of sport is huge. Either by choice or because of economic constraints, organizers have extracted only a portion of the consumer surplus generated by sporting events;

Second, professional sports as business activities are unique in that the service is (sporting) competition itself, and one supplier cannot create the product without the cooperation of his or her (sporting) rival.

Third, investing public money in building facilities to host sports teams and sporting events such as the Olympics is widely advocated by politicians and boosters, and using cost–benefit methods to evaluate economic impacts is an important contribution of economists;

Fourth, the sports industry is also a marvellous laboratory for testing and applying economic theory: it is Frederick Winslow Taylor's fantasy, where practically every aspect of worker (player) performance is observed and most are measured.

Fifth, when properly handled, sport can be one of the most effective ways of introducing students to fundamental economic concepts.

Sport and culture

Games and sports are a universal cultural phenomenon. Many animal species engage in what humans call play, and perhaps it is the lengthy adolescence characteristic of humans which has facilitated such an astonishing range of game playing. All ancient civilizations had their sports, and athletic contests are characteristic of even the least developed societies. In most civilizations sport has been connected with religion, as was true, for example, of the ancient Olympic Games. Modern sports, however, are characterized by secular rules. They also tend to be more formalized than traditional games, encouraging the gathering of statistical records for the purposes of evaluation and comparison.

Almost all of the popular modern sports were formalized somewhere between 1840 and 1900: for example, baseball (1846), soccer (1848), Australian football (1859), boxing (1865), cycling (1867), rugby union (1871), tennis (1874), American football (1874), ice hockey (1875), basketball (1891), rugby league (1895), motor sport (1895) and the modern Olympic Games (1896). Many historians have noted the coincidence between the creation of the rigid, scientific sports, and the advent of rigid industrialized societies in the United Kingdom and United States during the 19th century. The only major modern sports whose organization pre-dates this era are golf, cricket and horse racing, all of which had established rules and clubs from the mid-18th century.

The fact that almost all the sports which today dominate the world originated either in the United Kingdom or the United States also reflects Anglo-Saxon economic dominance. Soccer was spread at the end of the 19th century, when British expatriates or foreign students studying in Britain taught local elites the virtues of British culture and British games. American economic hegemony has given further impetus either to sports that were invented in Britain and taken up by Americans (such as golf and tennis) or to indigenous inventions such as basketball.

The cultural significance of sport has made most countries reluctant to see it exploited for business purposes. Even in the United States, the National Collegiate Athletic Association (NCAA) insists in retaining the myth of amateurism within the confines of the billion-dollar college sports industry. The myth of amateurism prevailed until recent years in a wide variety of international sports. Increasing commercialization is still opposed today in world soccer, and even in baseball the fans object to the idea of corporate sponsorship emblazoned on the sacred team shirt or on the bases.

This reluctance to mix business with sport explains in part how sport can attain such cultural centrality without achieving an equivalent economic status. Another reason is the difficulty involved in capturing rents. Each match in a championship is a mini-soap opera, involving high drama and the unexpected, and much of the pleasure derives from discussion among fans both before and after the event, but the protagonists have few opportunities to secure property rights over these external effects. Even the news media that pore ceaselessly over the latest sporting gossip pays little for the privilege. Another explanation may be the elastic supply of sporting contests (even if demand is frequently inelastic). Sporting competition is generally cheap to supply. Thus, even if the fans complain bitterly about the prices they pay, only a small fraction of personal disposable income is devoted to sport. Nonetheless, sports consumption is also a luxury good, and total expenditure rises rapidly as national income increases.

The business of professional sport

The starting point for the economic analysis of professional sport is the uncertainty of outcome and competitive balance. Almost from their inception, professional sports leagues have sought to restrain economic competition in the name of creating a more attractive contest. Thus, baseball's National League, the world's first professional sports league, was created in the United States in 1876 and introduced the reserve clause, a mechanism for tying each player in perpetuity to his club in 1879. From the 1880s, the clubs in the league argued that this restraint was necessary to ensure that all the best players did not end up playing for the biggest teams. Were this to happen, they reasoned, the championship would become predictable, fans would lose interest and the competition would collapse. In economic terms, they were arguing that a non-cooperative equilibrium would lead to a less balanced distribution of results than would be selected by a planner interested in maximizing interest in the sport. Given the frequent use of this argument by sports leagues to defend antitrust challenges levelled against economic restraints in the sporting labour and product markets, it might reasonably be termed the 'competitive balance defence'.

Economic analysis of the competitive balance defence began with Simon Rottenberg's 1956 article in the *Journal of Political Economy*. Rottenberg argued that the reserve clause would have no effect on the distribution of playing talent in a league, since players would always migrate to teams where their economic return (or marginal revenue product) was highest, regardless of whether the club or the player owned the right to the revenue stream. This is clearly an articulation of the Coase theorem, despite being published four years before Coase's 'On the problem of social cost'. Moreover, references to empirical tests of the Coase theorem commonly present sports leagues as an outstanding example. Much of the sports literature is an elaboration of these issues. El-Hodiri and Quirk (1971) developed the first theoretical model of revenue sharing, and used this to advance the proposition that revenue sharing would not affect the distribution of talent in a league, also for Coasean reasons. Much of the early literature involved examining a sport league's objective function, which in North America typically has been considered to be characterized by profit maximization (see, for example, Jones, 1969), while Sloane (1971) drew attention to the not-for-profit structure and culture that obtained in British sports, notably football, and in the rest of the world (for a detailed comparison of the American and European sports business models see Szymanski and Zimbalist, 2005). Since the 1990s there has been renewed interest in the theoretical modelling of league structures, focusing on the North American model (see, for example, Fort and Quirk, 1995; Vrooman 1995) and the European model (Késenne, 2000; Hoehn and Szymanski, 1999). Szymanski (2003) advances a synthesis based on the contest or tournament framework, and there is a good deal of current research in this vein.

Much of this analysis has been applied in the context of antitrust either in the United States or in the European Union. The courts, in line with most economists, consider leagues as cartels that impose restraints on economic competition in labour and product markets. In North America, the competitive balance defence has been applied in cases such as *NCAA v. Board of Regents of the University of Oklahoma*, 468 U.S. 85 (1984), which argued that collective selling of broadcast rights was necessary to maintain competitive balance. In this case, as in several others, the defence failed, not because the courts did not accept that leagues were special cases where ancillary restraints could be justified, but because the restraints

were excessive given their stated aims (see, for example, Flynn and Gilbert, 2001). As a result, most labour market restraints in sports have been built into collective bargaining agreements that are exempt from antitrust, while broadcasting rights generally enjoy the exemption granted by Congress in the 1962 Sports Broadcasting Act. In Europe, where equivalent exemptions do not exist, labour market restraints akin to the reserve clause have been ruled illegal (by the European Court of Justice in the 1995 *Bosman* case), while collective selling of broadcast rights has been treated differently in EU member states, some outlawing collective selling (Italy and Spain) and others permitting it (Germany and the UK).

Economic impact of sports teams and facilities

Top-level professional sports teams have an immense cultural impact on their communities but very little, if any, positive economic impact. All independent empirical research concurs on this point (see, for example, Siegfried and Zimbalist, 2000). There are three principal reasons why no positive economic development effect should be expected from a new team or facility. First, most of the money spent at a facility is replacement for spending at other entertainment activities elsewhere in the metropolitan area. Second, leakages of spending at a ballpark or arena out of the local economy tend to be much greater than at other locally owned entertainment venues, thereby depressing sports expenditure multipliers. Third, public subsidies for facility construction and maintenance create a negative budgetary impact, engendering either higher taxes or lower services, thereby dragging down economic expansion. Nonetheless, the existence of externalities, public good benefits and consumer surplus (consumer demand for sporting contests is generally assumed to be inelastic) may justify some level of public support for facility construction.

Sport as laboratory

Sports involve forms of competitive behaviour that are readily observed and measured. Often we know the precise output and productivity of agents, and their compensation packages are frequently a matter of public record. Hence, it is not surprising that labour economists have looked to the sports field for natural experiments of theories such as incentive design, teamwork, and discrimination.

Gerald Scully (1974) pioneered attempts to empirically estimate a player's marginal revenue product (MRP) in baseball, and much of the subsequent research on labour market performance has followed his innovation. He also applied his method, which involved estimating individual contributions to winning through measures of batting performance, and then the value of winning for team revenues, to compare the relationship between wages and MRPs for players from different races, thus providing a test of discrimination. Disputes over Scully's MRP methodology has kept a lively debate going in the literature (see, for example, Zimbalist, 1992).

Another example of the use of sports data to test labour market theories is Ehrenberg and Bognanno (1990), who used scores of professional golfers in different tournaments to test the incentive effects of different prize structures. Other important studies include the use of trading in baseball cards to test for discrimination on the part of fans (Nardinelli and Simon, 1990) and the effect of strikes on demand in baseball (Schmidt and Berri, 2004).

Sports in the classroom

Sports economics has become extremely popular in the classroom, often because many students find it easiest to appreciate economic reasoning in the context of an activity with which they have a certain degree of familiarity. As the discussion of the competitive balance defence suggests, using sport to teach economics can be risky given that sports are often a special case. However, using sport to study concepts such as demand or labour market incentives can be illuminating.

For example, the case of spectator demand can be used to develop a number of key economic concepts. Since the marginal cost to an owner of an additional fan attending any given game is approximately zero, profit and revenue maximization at the ballpark are congruent. This implies that a profit-maximizing strategy for an owner is to set ticket prices so that the price elasticity of demand equals 1. Much of the early empirical literature, however, yielded estimates of inelastic demand at existing prices. This literature suffered from various deficiencies: the use of weighted average, rather than individual, ticket prices; the absence of important control variables; and the failure to specify the model properly. Possibly significant control variables include the probability of home team victory, the uncertainty of outcome, the quality of the visiting team and importance of the contest in league competition, the weather, the age of the stadium, the number of star players on the teams, the distribution and level of local income, and whether the game is televised. Modelling should be affected by the existence of a stadium capacity constraint, the positive dynamic (demonstration) effect of having a full or nearly full stadium on future attendance, and the simultaneity effect whereby higher attendance increases the home field advantage, which improves home team performance and increases attendance. Finally, the profit-maximizing owner will seek to maximize not just gate revenues but all stadium revenues (ticket sales, net concessions, catering, signage, memorabilia, and parking). Many of these considerations will cause ticket prices to appear to be set below the point of unitary elasticity when, in fact, they are set at revenue-maximizing levels.

Conclusion

The economics of sport is a relatively new and rapidly growing field. While the literature is gaining sophistication and relevance, there is still much ground to be covered. Nonetheless, there are some important areas of research that have not been covered properly here, such as research in college athletics in the United States or studies of team production in sports.

In our view the most pressing area for economic research remains the competitive balance defence, both in terms of theoretical foundations and in terms of empirical evidence.

STEFAN SZYMANSKI AND ANDREW ZIMBALIST

Bibliography

Coase, R. 1960. The problem of social cost. *Journal of Law and Economics* 3, 1–44.

Ehrenberg, R. and Bognanno, M. 1990. Do tournaments have incentive effects? *Journal of Political Economy* 98, 1307–24.

El-Hodiri, M. and Quirk, J. 1971. An economic model of a professional sports league. *Journal of Political Economy* 79, 1302–19.

Flynn, M. and Gilbert, R. 2001. An analysis of professional sports leagues as joint ventures. *Economic Journal* 111, F27–46.

Fort, R. and Quirk, J. 1995. Cross subsidization, incentives and outcomes in professional team sports leagues. *Journal of Economic Literature* 33, 1265–99.

Hoehn, T. and Szymanski, S. 1999. The Americanization of European football. *Economic Policy* 28, 205–40.

Jones, J. 1969. The economics of the National Hockey League. *Canadian Journal of Economics* 2, 1–20.

Késenne, S. 2000. Revenue sharing and competitive balance in professional team sports. *Journal of Sports Economics* 1, 56–65.

Nardinelli, C. and Simon, C. 1990. Customer racial discrimination in the market for memorabilia: the case of baseball. *Quarterly Journal of Economics* 105, 575–96.

Neale, W. 1964. The peculiar economics of professional sports. *Quarterly Journal of Economics* 78, 1–14.

Rottenberg, S. 1956. The baseball players' market. *Journal of Political Economy* 64, 242–58.

Scully, G. 1974. Pay and performance in major league baseball. *American Economic Review* 64, 915–30.

Schmidt, M. and Berri, D. 2004. The impact of labor strikes on consumer demand: an application to professional sports. *American Economic Review* 94, 344–57.

Siegfried, J. and Zimbalist, A. 2000. The economics of sports facilities and their communities. *Journal of Economic Perspectives* 14(3), 95–114.

Sloane, P. 1971. The economics of professional football: the football club as a utility maximiser. *Scottish Journal of Political Economy* 17(2), 121–46.

Szymanski, S. 2003. The economic design of sporting contests. *Journal of Economic Literature* 41, 1137–87.

Szymanski, S. and Zimbalist, A. 2005. *National Pastime: How Americans Play Baseball and the Rest of the World Plays Soccer*. Washington, DC: Brookings Institution.

U.S. Bureau of the Census. 1999. *Service Annual Survey: Current Business Reports* Washington, DC: U.S. Department of Commerce. Online. Available at http://www.census.gov/prod/2000pubs/bs98.pdf, accessed 12 August 2005.

Vrooman, J. 1995. A general theory of professional sports leagues. *Southern Economic Journal* 61, 971–90.

Zimbalist, A. 1992. Salaries and performance in major league baseball. In *Diamonds Are Forever: The Business of Baseball*, ed. P. Sommers. Washington, DC: Brookings Institution.

spurious regressions

For the first three-quarters of the 20th century the main workhorse of applied econometrics was the basic regression

$$Y_t = a + bX_t + e_t. \tag{1}$$

Here the variables are indicated as being measured over time, but could be over a cross-section; and the equation was estimated by ordinary least squares (OLS). In practice more than one explanatory variable x would be likely to be used, but the form (1) is sufficient for this discussion. Various statistics can be used to describe the quality of the regression, including R^2, t-statistics for β, and Durbin–Watson statistic d which relates to any autocorrelation in the residuals. A good fitting model should have $|t|$ near 2, R^2 quite near one, and d near 2.

In standard situations the regression using OLS works well, and researchers used it with confidence. But there were several indications that in special cases the method could produce misleading results. In particular, when the individual series have strong autocorrelations, it had been realized by the early 1970s by time series analysis that the situation may not be so simple; that apparent relationships may often be observed by using standard interpretations of such regressions. Because a relationship appears to be found between independent series, they have been called 'spurious'. Note that, if $b = 0$, then e_t must have the same time series properties as Y_t, that is, it will be strongly autocorrelated, and so the assumptions of the classical OLS regression will not be obeyed. The possibility of getting incorrect results from regressions was originally pointed out by Yule (1926) in a much cited paper that discussed 'nonsense correlations'. Kendall (1954) also pointed out that a pair of independent autoregressive series of order one could have a high apparent correlation between them; and so if they were put into a regression a spurious relationship could be obtained.

The magnitude of the problem was found from a number of simulations. The first simulation on the topic was by Granger and Newbold (1974), who generated pairs of independent random walks, from (1) with $a = b = 1$. Each series had 50 terms and 100 repetitions were used. If the regression is run, using series that are temporarily uncorrelated, one would expect that roughly 95 per cent of values of $|t|$ on b would be less than 2. This original simulation using random walks found $|t| \leq 2$ on only 23 occasions; out of the 100, $|t|$ was between 2 and 4 on 24 occasions, between 4 and 7 on 34 occasions, and over 7 on the other 19 occasions.

The reaction to these results was to reassess many of the previously obtained empirical results in applied time series econometrics, which undoubtedly involved highly autocorrelated series but had not previously been concerned by this fact. Just having a high R^2 value and an apparently significant value of b was no longer sufficient for a regression to be satisfactory or its interpretations relevant. The immediate questions were how one could easily detect a spurious regression and then correct for it. Granger and Newbold (1974) concentrated on the value of the Durbin–Watson statistic: if the value is too low, it suggests that the regressions results cannot be trusted. Remedial methods such as using a Cochrane–Orcutt technique to correct autocorrelations in the residuals, or differencing the series used in a regression, were inclined to introduce further difficulties and could not be recommended. The problem arises because the equation is mis-specified; the proper reaction to having a possible spurious relationship is to add lagged dependent and independent variables until the errors appear to be white noise, according to the Durbin–Watson statistic. A random walk is an example of a I(1) process, that is, a process that needs to be differenced to become stationary. Such processes seem to be common in parts of econometrics, especially in macroeconomics and finance. One approach that is widely recommended is to test whether X_t, Y_t are I(1) and, if so, to difference before one performs the regression. There are many tests available; a popular one is due to Dickey and Fuller (1979).

A theoretical investigation of the basic unit root, ordinary least squares, spurious regression case was undertaken by Phillips (1986). He considered the asymptotic properties of the coefficients and statistics of eq. (1), $\hat{a}, \hat{b}$, the t-statistic for b, R^2 and the Durbin–Watson statistics $\hat{\rho}$. To do this he introduced the link between normed sums of functions of unit root processes and integrals of Weiner processes. For example, if a sample X_t of size T is generated from a driftless random walk, then

$$T^{-2} \sum_{1}^{T} X_t^2 \to \sigma_\varepsilon^2 \int_0^1 W^2(t)\, dt,$$

where σ_ε^2 is the variance of the shock, and $W(t)$ is a Weiner process. As a Weiner process is a continuous time random process on the real line [0,1], the various sums involved are converging and can thus be replaced by integrals of a stochastic process. This transformation makes the mathematics of the investigation much easier, once one becomes familiar with the new tools. Phillips is able to show that

- the distributions of the t-statistics for $\hat{a}$ and $\hat{b}$ from (1) diverge as t becomes large, so there is no asymptotically correct critical values for these conventional tests;
- $\hat{b}$ converges to some random variable whose value changes from sample to sample;
- Durbin–Watson statistics tend to zero; and
- R^2 does not tend to zero but to some random variable.

What is particularly interesting is not only that do these theoretical results completely explain the simulations but also that the theory deals with asymptotics, $T \to \infty$, whereas the original simulations had only $T = 50$. It seems that spurious regression occurs at all sample sizes.

Haldrup (1994) has extended Phillips's result to the case for two independent I(2) variables and obtained similar results. (An I(2) variable is one that needs differencing twice to get to stationarity, or, here, difference once to get to random walks.) Marmol (1998) has further extended these results to fractionally integrated I(d) processes. Durlauf and Phillips (1988) regress I(1) process on deterministic polynomials in time, thus polynomial trends, and found spurious relationships.

Although spurious regressions in econometrics are usually associated with I(1) processes, which were explored in Phillips's well-known theory and in the best known simulations, what is less appreciated is that the problem can also occur, although less clearly, with stationary processes.

Table 1 shows simulation results from independent series generated by two first order autoregressive models with coefficients a_1 and a_2 where $0 \leqslant a_1 = a_2 \leqslant 1$ and with inputs e_{xt}, e_{yt} both Gaussian white noise series, using regression 1 estimated using OLS with sample sizes varying between 100 and 10,000.

It is seen that sample size has little impact on the percentage of spurious regressions found (apparent significance of the b coefficient in (1)). Fluctuations down columns do not change significantly with the number of

iterations used. Thus, the spurious regression problem is not a small sample property. It is also seen to be a serious problem with pairs of autoregressive series which are not unit root processes. If $a = 0.75$, for example, then 30 per cent of regressions will give spurious implications. Further results are available in the original paper but will not be reported in detail. The Gaussian error assumption can be replaced by other distributions with little or no change in the simulation results, except for an exceptional distribution such as the Cauchy. Spurious regressions also occur if $a_1 \neq a_2$, although less frequently, and particularly if the smaller of the two a values is at least 0.5 in magnitude.

The obvious implications of these results is that applied econometricians should not worry about spurious regressions *only* when dealing with I(1), unit root, processes. Thus, a strategy of first testing whether a series contains a unit root before entering into a regression is not relevant. The results suggest that many more simple regressions need to be interpreted with care when the series involved are strongly serially correlated. Again, the correct response is to move to a better specification, using lags of all variables.

Concerns about spurious regressions produced interest in tests for unit roots, of which there are now many; and empirical works with time series will usually test between I(1) or I(0), or may sometimes consider more complicated alternatives. If series are found to be I(1), simple regressions have been replaced with considerations of cointegration and construction of error-correction models.

A recent survey of studies of spurious relationships is Pilatowska (2004).

CLIVE W. J. GRANGER

See also **Durbin-Watson statistic.**

Table 1 *Regression between independent AR(1) series*

Sample series	$a = 0$	$a = 0.25$	$a = 0.5$	$a = 0.75$	$a = 0.9$	$a = 1.0$
100	4.9	6.8	13.0	29.9	51.9	89.1
500	5.5	7.5	16.1	31.6	51.1	93.7
2,000	5.6	7.1	13.6	29.1	52.9	96.2
10,000	4.1	6.4	12.3	30.5	52.0	98.3

$a_1 = a_2 = a$ percentage of $|t| > 2$
Source: Granger, Hyung and Jeon (2001).

Bibliography

Baltagi, B. 2001. *A Companion to Theoretical Econometrics*. Oxford: Blackwell.

Dickey, D. and Fuller, W. 1979. Distribution of the estimates for autoregressive time series with a unit root. *Journal of the American Statistical Association* 74, 427–31.

Durlauf, S. and Phillips, P. 1988. Trends versus random walks in time series analysis. *Econometrica* 56, 1333–54.

Granger, C. 2001. Spurious regression in econometrics. In *A Companion to Theoretical Econometrics*, ed. B. Baltagi. Oxford: Blackwell.

Granger, C., Hyung, N. and Jeon, Y. 2001. Spurious regression with stationary series. *Applied Economics* 33, 899–904.

Granger, C. and Newbold, P. 1974. Spurious regressions in econometrics. *Journal of Econometrics* 2, 111–20.

Haldrup, N. 1994. The asymptotics of single-equation cointegration regressions with I1 and I2 variables. *Journal of Econometrics* 63, 153–81.

Kendall, M. 1954. *Exercises in Theoretical Statistics*. London: Griffin.

Marmol, F. 1998. Spurious regression theory with non-stationary fractionally integrated processes. *Journal of Econometrics* 84, 233–50.

Phillips, P. 1986. Understanding spurious regressions in econometrics. *Journal of Econometrics* 33, 311–40.

Phillips, P. 1988. New tools for understanding spurious regressions. *Econometrica* 66, 1299–325.

Pilatowska, M. 2004. Realization of the congruence postulate as a method of avoiding the effects of a spurious relationship. In *Dynamic Economic Models*, vol. 6, ed. Z. Zieliński. Torun: Nicolaus Copernicus University.

Yule, G. 1926. Why do we sometimes get nonsense correlations between time-series? A study in sampling and the nature of time-series. *Journal of the Royal Statistical Society* 89, 1–64.

Sraffa, Piero (1898–1983)

In the history of economics, Piero Sraffa is an enigma. His reputation as a major economic theorist rests on but three works: the *Economic Journal* article of 1926, the Introduction to his edition of Ricardo's *Principles* (the first of the 11 volumes of the complete *Works and Correspondence* of *David Ricardo*, which established Sraffa as the finest scholar to have edited a major work in the literature of economics), and the 99 pages of *Production of Commodities by Means of Commodities*, a sparse, terse collection of logical propositions, the significance of which is a matter of often heated debate.

A reclusive figure, of great personal warmth and puckish humour, Sraffa spent most of his life in Cambridge. Yet his influence extended far beyond academic economics.

Throughout the 1930s he spent every Thursday afternoon and evening, during term, with Ludwig Wittgenstein. It was Sraffa who forced Wittgenstein to accept that the theory of language advanced in the *Tractatus Logico-Philosophicus* was logically inadequate, paving the way for the recognition of the social content of signs and language presented in *Philosophical Investigations*. In the preface of the latter, Wittgenstein acknowledged the importance of Sraffa's criticism of his arguments over many years, and added 'I am indebted to *this* stimulus for the most consequential ideas of this book'. He later commented that after discussion with Sraffa he felt 'like a tree stripped of all its branches', but that a consequence of this drastic pruning was healthier growth.

Of perhaps wider significance was Sraffa's close friendship with Antonio Gramsci. After Gramsci's imprisonment, Sraffa led the effort to ameliorate the harsh conditions in which he was held, and to secure his release. The essential contents of Gramsci's letters from prison, written to his sister-in-law Tatiana, were channelled through Sraffa in Cambridge to the exiled Italian Communist Party. And it was Sraffa who ensured, by establishing an account with unlimited credit at a Milanese bookshop, that Gramsci was supplied with the materials he needed to work on his *Prison Notebooks*.

The philosophical debates with Wittgenstein, and his political and intellectual commitment to socialism, point to important elements in Sraffa's intellectual make-up. His economics, always rigorous, became in the 1930s increasingly formal, to the extent that the search for logically precise and unambiguous expression inhibited his writing. His socialism demanded an economics that was concrete; that, however abstract, was appropriate to the interpretation of real economic institutions and phenomena.

The compelling empiricism of Marxian socialist thought, and the rejection of the use of subjective concepts, are themes running throughout Sraffa's economics. Economics should be constructed from variables and relationships which are, at least in principle, observable and measurable. The classical analysis of value and distribution is constructed on just such 'empirical' foundations, whilst neoclassical theory, based as it is on unverifiable hypotheses concerning individual choice, evidently is not.

But although Sraffa's contribution to economic theory may have been motivated by these methodological concerns, its substance involves the logic of theoretical argument, in particular the demonstration of the logical consistency of the classical analysis of value and distribution, and, as a corollary, the logical deficiencies of neoclassical theory.

Life and works

Piero Sraffa was born in Turin on 5 August 1898. His father was Angelo Sraffa, a professor of commercial law who later became Chancellor of the Bocconi University in Milan. The Piazza Sraffa in Milan is named for Angelo, not Piero. Piero's mother was Irma Tivoli.

Sraffa was educated at the Liceo d'Azeglio in Turin, where he was greatly influenced by Umberto Cosmo, who introduced him to socialist ideas and, in 1919, to Antonio Gramsci. Sraffa's studies at the University of Turin, from 1916 to 1920, were interrupted by military service, which he spent as both a ski instructor and an engineer, blowing up bridges to stem the Austrian advance. He attended relatively few lectures. Nonetheless his honours thesis, 'Monetary Inflation in Italy During and After the War', was considered by his supervisor, Luigi Einaudi, to be quite brilliant. It was published in 1920.

After graduation, Sraffa worked for a few weeks in a bank to learn some banking 'from the inside'. He then went to the London School of Economics (1921–2) where he attended lectures by Cannan, Foxwell and Gregory.

During his stay in London, Sraffa visited Cambridge, bearing a letter of introduction to Keynes from Mary

Berenson, a friend of Sraffa's family who, ten years earlier, had entertained Keynes and other young Cambridge graduates in her villa near Florence. Keynes was at the time engaged in the debate on the reconstruction of the international monetary system, and had agreed to be editor of a weekly supplement to the *Manchester Guardian*, dealing with the monetary and financial problems of Europe. Keynes asked Sraffa to contribute an article on the Italian banking system, which was at the time experiencing a severe crisis. The article proved to be too long for the newspaper, and was published in the *Economic Journal* instead, a shorter version appearing in the *Manchester Guardian*.

These two articles were also published in Italian, and caused the Fascist regime considerable irritation. Now back in Italy, Sraffa was accused by Mussolini of 'banking defeatism' and 'sabotage of Italian finance'. Keynes invited Sraffa to return to England until things had calmed down. However, on 23 January 1923 Sraffa was detained at Dover, and after being questioned for three hours was informed that he had been refused permission to land by order of the Home Secretary. The reason for the denial of entry is still not clear. Keynes secured the removal of Sraffa's name from the list of 'undesirables' in 1924.

Fortunately the situation in Italy was less severe than had been expected, though Sraffa resigned the job he had obtained as Director of the Bureau of Labour Statistics of the Province of Milan. The Bureau had been established by the socialist provincial administration in 1922, and was experiencing difficulties with the Fascist government. A few months later he was appointed to a lectureship in Political Economy and Public Finance at the University of Perugia.

The preparation of his lectures at Perugia stimulated him to write 'Sulle relazioni fra costo e quantita prodotta' (1925). As a result of this article Sraffa was appointed to a professorship in Political Economy at the University of Cagliari, a post he held *in absentia* to the end of his life, donating his salary to the support of the library. Edgeworth's high opinion of the article led to an invitation to submit a version to the *Economic Journal* (1926). This, in turn, led to Sraffa's being offered the lectureship in Cambridge, which he took up in October 1927.

Before leaving Italy, Sraffa had pursued his interest in monetary problems by translating Keynes's *Tract on Monetary Reform* into Italian, and writing several short reviews of books on money and banking for the *Giornale degli Economisti* (1925b; 1926b; 1926c; 1927b).

In 1919, Sraffa had joined the Socialist Students' Group at the University of Turin and had participated actively in the political life developing around *Ordine Nuovo*, the magazine founded in 1919 by Gramsci, Tasca, Terracini and Togliatti, the group who were to play the crucial role in the split from the Socialist Party at the Congress of Livorno in 1921 and the foundation of the Italian Communist Party.

Whilst in London Sraffa wrote three articles for *Ordine Nuovo* on the condition of the working class in England and the role of trade unions. In 1924 he published an open letter to Gramsci in *Ordine Nuovo* criticizing the Communist Party for its dogmatic refusal to contemplate an alliance with other democratic groups against Fascism. A few years later Gramsci, now imprisoned, accepted Sraffa's argument.

Sraffa also opposed the orthodox Party line in two letters to *Stato Operaio* in 1927. In a discussion of the devaluation of the Italian lira, he criticized the prevalent view that policy decisions are always mechanically and '*directly* dictated by the *immediate* interests of the banks and the big industrialists' (1927a, p. 180). He advanced instead the view that political bodies such as the Fascist Party have their own interests which can enter into the dynamics of the decision process.

In October 1927 Sraffa began his lectures in Cambridge, presenting courses on the theory of value and on the relationship between banks and industry in continental Europe. He was to lecture for only three years, finding the very process increasingly difficult. Joan Robinson, who attended the lectures on her return from India, recalled them vividly, not least because Sraffa liked to develop a dialogue with his class – a procedure unknown in Cambridge. In 1930 Sraffa was appointed Marshall Librarian, and also placed in charge of graduate studies. He gave up lecturing for good.

Shortly after arriving in Cambridge, Sraffa had shown Keynes the set of propositions (derived from Marx's reproduction schemes) which were to grow into *Production of Commodities*. But this work was somewhat overwhelmed both by the intense debate in Cambridge surrounding Keynes's *Treatise on Money* and, later, *The General Theory* (it was Sraffa who organized the famous 'circus' which discussed the *Treatise* in 1931), and by Sraffa assuming in 1930 the editorship of the Royal Economic Society edition of *The Works and Correspondence of David Ricardo*. Sraffa's work on the theory of value, and his interest in monetary theory, coalesced in a critical review of Hayek's *Prices and Production* (Sraffa, 1932a; 1932b).

Sraffa's participation in political debate was necessarily limited after his arrival in Cambridge. He maintained contact with the leadership of the Italian Communist Party in Paris, and in 1927 wrote to the *Manchester Guardian* denouncing the ill treatment of the imprisoned Gramsci. He visited Gramsci and attempted, to no avail, to use the influence of his uncle, an eminent judge, to secure Gramsci's release. Following Gramsci's death in 1937, it was Sraffa who conveyed to Togliatti Gramsci's wishes concerning the editing of the *Quaderni dal Carcere*.

In another service to a friend, Sraffa travelled to Austria following the *Anschluss* to inform Wittgenstein's family that Ludwig had renounced his Austrian citizenship.

In 1939 Sraffa was elected to a Fellowship of Trinity College (he had previously held dining rights at King's), a post he took up shortly after the outbreak of war. When Italy entered the war in June 1940, Sraffa was interned as an enemy alien on the Isle of Man. Keynes managed to extricate him by the end of the summer. Sraffa never gave up his Italian citizenship.

By the late 1940s, the publication of the edition of Ricardo had been long delayed. This was partly due to the reorganization required when in 1943, after six volumes were already in the press, Ricardo's letters to Mill and, amongst other writings, the papers on Absolute and Exchangeable Value were discovered. But delay was also caused by the difficulty Sraffa was having in writing the introductions to the volumes, particularly the introduction to the *Principles*. The second problem was solved after 1948 with the assistance of Maurice Dobb. Dobb and Sraffa would discuss each paragraph in detail. Dobb would write it up. Sraffa would revise what Dobb had written. Dobb would rewrite. And so on, until the job was done. The first four volumes of the *Works and Correspondence of David Ricardo* were published in 1951, the next five volumes in 1952, a bibliographical miscellany formed the tenth volume published in 1955, and, after a number of false starts by others, a general index, compiled by Sraffa himself, was published as the eleventh volume of the set in 1971.

The edition is widely acknowledged to be a scholarly masterpiece. George Stigler (1953) commented,

> Ricardo was a fortunate man. He lived in a period – then drawing to a close – when an untutored genius could still remake economic science … . And now, 130 years after his death, he is as fortunate as ever: he has been befriended by Sraffa – who has been befriended by Dobb.
>
> Keynes told us, in 1933, that Sraffa, 'from whom nothing is hid', would give us the full works of Ricardo within the year. The truth of the first part of the statement had as its cost the falsification of the second, and it has been a splendid bargain. For Sraffa's *Ricardo* is a work of rare scholarship. The meticulous care, the constant good sense, and the erudition make this a permanent model for such work; and the host of new materials seem to suggest that Providence meets half-way the deserving scholar

The *Ricardo* completed, Sraffa could return to the work he had done on the theory of value and distribution in the 1930s and early 1940s. Old notes were reassembled, including a proof of the Perron–Frobenius theorem on non-negative square matrices, which the Trinity mathematician Besicovitch (an analyst with no prior knowledge of the theorem) had provided on a postcard delivered on Christmas Day 1944. The result of assembling these old notes was *Production of Commodities by Means of Commodities* (1960).

This book was greeted with almost universal puzzlement. It seemed to present, in odd, formal terms, propositions which had become familiar with the development of linear models in the 1940s and 1950s. Had Sraffa's brilliant insights of 30 years before been overtaken by events in mainstream theory? Of the earlier reviewers only Dobb (1961), Meek (1961) and Newman (1962) grasped the fact that this was a work of profound significance, with implications for the logical foundations of both classical and neoclassical theories of value and distribution.

Sraffa spent the rest of his life in Cambridge though up to 1973 he visited Italy in every vacation, staying in his apartment in Rapallo, and going to Rome to attend meetings of the economic section of the Academia dei Lincei. Other than economics and politics, Sraffa's great interest was the collection of books. He assembled a magnificent collection of economics books and pamphlets, including a first edition of *Kapital* inscribed by Marx himself (which he later presented to the Istituto Gramsci) and a copy of the *Wealth of Nations* containing Adam Smith's bookplate. He shared this enthusiasm with Keynes. Together they discovered, identified, and wrote an introduction to an edition of David Hume's *An Abstract of a Treatise on Human Nature* (1938). When he died, on 3 September 1983, Sraffa left his collection to Trinity College.

Contributions to economic theory

The early years

Sraffa's dissertation (1920) dealt with central *practical* issues occupying writers on monetary matters at the end of the First World War – the causes and consequences of inflation, the stabilization of internal prices and exchange rates within an unstable international monetary system, the argument for restoring the gold standard and revaluing the currency to the pre-war gold parity.

Sraffa argued that since the abandonment of the gold standard at the beginning of the war had been followed by halving of the purchasing power of gold, then a return to the gold standard would require a rise in the value of the metal, forcing countries which fixed parity either to devalue or to bear the consequences of deflation. Since, in these circumstances, the monetary authorities could not achieve stability of both prices and the rate of exchange, it was better to opt for the former. There is no law which forced the authorities to stabilize the currency at the pre-war level. The *normal* value of the currency is completely 'conventional', that is, it can be at *any* level that common opinion expects it to be (Sraffa, 1920, p. 42). Sraffa, in opposition to Einaudi, and in common with the position which Cassel, Hawtrey and Keynes were to urge on the Genoa Economic Conference in April 1922, favoured a 'managed currency'.

Sraffa's 'practical' case embodied an implicit theoretical argument. He stressed the role of the state, moulded by the pressure of the major economic classes, in determining the distribution of income. Monetary policy was thus considered in terms of its impact on the real wage. Sraffa, like Keynes in *A Tract on Monetary Reform*, accepted a version of the quantity theory. But whereas Keynes's views on the determination of the distribution of income were essentially neoclassical, Sraffa's position was more akin to that of the classical economists and Marx. Keynes argued that social forces and monetary factors have only temporary effects on the distribution of income, influencing only the disequilibrium real wage rate. Unless, that is, they were able to effect the real factors which determined the equilibrium distribution (Keynes, 1923, p. 27). Sraffa, however, argued that monetary policies, and hence inflation and deflation, are aspects of the social conflicts that directly regulate the equilibrium or normal real wage rate (Sraffa, 1920, pp. 25, 40–2).

A similar view of the role of economic institutions in social conflict was spelt out in the *Economic Journal* and *Manchester Guardian* articles. The articles focused on the financial needs of newly developing Italian industry, and on the evolution of the links between industry and the banks. His study of the formation of large groups or 'concentrations' of financial and industrial power is similar to that in Hilferding's *Finance Capital*. He stressed the enormous economic and political power which such groups can acquire and outlined the way in which conflicts within the groups might affect economic policy. He also revealed the accounting tricks which had been used by two major banks to disguise their financial difficulties, and showed how the authorities had evaded legal restrictions in order to favour some major financial groups (Sraffa, 1922b, p. 676). No wonder Mussolini was so upset.

Laws of returns

Sraffa's early writings, although imbued with a certain critical radicalism, provided no hint of the theoretical tour de force that was to come. It is true that he had emphasized the role of social classes and institutions in the normal (as opposed to disequilibrium) operation of the economy, but in Sraffa's examination of the neoclassical (predominantly Marshallian) theory of cost (1925; 1926) these concerns with the 'objective' characteristics of economic activity were transformed into a penetrating critique of the logical foundations of the theory of the equilibrium of the competitive firm and of the supply curve.

Sraffa's starting point in the *Annali di Economia* was a distinction between an analysis in which the relationship between cost and quantity produced was determined by 'objective' factors, such as the ordering of different qualities of land, and a relationship which was based on 'subjective' factors, namely the marginal disutility which accompanies the offer of increased quantities of factor services. The former relation is, as Wicksteed had argued (1914), essentially descriptive; it is the latter which is, in

neoclassical theory, analytic. The supply curve is simply the demand curve 'reversed' (Wicksteed, 1914).

In the *Economic Journal* Sraffa made the same point in a rather different way. In classical economics, he argued, the 'laws of returns' did not derive from a unified analysis of cost. Quite the contrary. The discussion of increasing returns was associated with the analysis of accumulation, most notably Adam Smith's examination of the relationship between the extent of the market and the division of labour. Diminishing returns, on the other hand, were the distinctive component of the theory of rent. The suggested symmetry of increasing and diminishing returns is a quite different construction, characteristic of the neoclassical supply curve.

But if cost is subjective disutility, how does a phenomenon which is defined by personal psyche manifest its influence in the determination of the equilibrium of the competitive firm, and in the determination of the supply curve of the industry? The apparent similarity between the determination of equilibrium in individual choices, and the equilibrium of the firm, is quite spurious. For whereas all the determinants of individual choice – preferences and endowments – are peculiar to the individual, the determinants of the equilibrium in production in a competitive economy – the technical conditions of production and the supply of factor services – are external to the firm.

Sraffa then demonstrated that neither increasing *nor* diminishing returns are compatible with the assumption of perfect competition in the determination of the supply curve of an industry, except in the peculiar case in which economies or diseconomies of scale are external to the firm but internal to the industry.

Diminishing returns are incompatible with perfect competition, since the presumption of price taking precludes any impact of the output of individual firms, or, in Marshall's *ceteris paribus* world, the output of individual industries, on prices, unless it is assumed that endowments are fixed in individual firms, or are peculiar to individual industries.

Increasing returns are also incompatible with assumption of perfect competition – other than those which are external to the firm, but internal to the industry.

Only the assumption of constant returns to scale is compatible with the assumption of perfect competition:

> In normal cases the cost of production of commodities produced competitively ... must be regarded as constant in respect of small variations in the quantity produced. And so, as a simple way of approaching the problem of competitive value, the old and now obsolete theory which makes it dependent on the cost of production alone appears to hold its ground as the best available. (Sraffa, 1926a, pp. 540–1)

There were two ways out of the conundrum, either to adopt the general equilibrium reasoning which Sraffa had deployed so effectively against the notion of the supply curve, or to abandon the assumption of perfect competition. Marshall's theory must be abandoned (Sraffa, 1930a; 1930b).

The first course was ruled out on the grounds that examination of 'the conditions of simultaneous equilibrium in numerous industries', though a well-known approach, is far too complex; 'the present state of our knowledge ... does not permit of even much simpler schemata being applied to the study of real conditions' (1926a).

The second course recognizes both the 'everyday experience ... that a very large number of undertakings – and the majority of those which produce manufactured consumers' goods – work under conditions of individual diminishing costs', and that

> the chief obstacle against which they have to contend when they want gradually to increase their production does not lie in the cost of production ... but in the difficulty of selling the larger quantity of goods without reducing the price, or without having to face increased marketing expenses. This ... is only an aspect of the usual descending demand curve, with the difference that instead of concerning the whole of a commodity, whatever its origin, it relates only to the goods produced by a particular firm ... (1926a).

Sraffa's second option launched the Cambridge analysis of imperfect competition, first in Richard Kahn's fellowship dissertation (1931) at King's, then in Joan Robinson's *Economics of Imperfect Competition*.

Apart from his contribution in the *Economic Journal* symposium on increasing returns, Sraffa did not participate further in the debate on the Marshallian theory of cost. The reasons are not hard to seek.

First, imperfect competition theory, instead of providing a new, more concrete approach to the analysis of value and distribution, was simply absorbed into neoclassical theory. The fact that imperfectly competitive models do not provide a foundation for a theory of value, seemed to enhance the status of partial equilibrium analysis, rather than hasten its rejection; with the competitive theory of value still holding sway at the level of general equilibrium (a neat rationale is provided by Hicks, 1946, pp. 83–4). The survival of the 'U' shaped cost curve as an analytical tool, constructed from the presumption of increasing, then diminishing returns, is in no small part attributable to the longevity provided by models of the imperfectly competitive firm. Nonetheless, the appearance of the 'U' shaped curve in models of the competitive firm, more than 60 years after Sraffa clearly demonstrated the illegitimacy of the construction, is an indication of just how intellectually disreputable theoretical economics can be.

Second, Sraffa's implicit identification of classical and Marxian theory with the notion that competitive value is 'dependent on the cost of production' is clearly wrong, as examination of neoclassical models which take account

of 'simultaneous equilibrium in numerous industries' readily demonstrates. Sraffa had deployed general equilibrium reasoning to demolish the theory of the competitive firm and the industry supply curve. Further criticism of neoclassical theory would require consideration of general equilibrium models of value and distribution. And a constructive rehabilitation of classical theory would require a general analysis too. It would require, that is, an analysis of 'the process of diffusion of profits throughout the various stages of production and of the process of forming a normal level of profits throughout all the industries of a country' – the problem Sraffa acknowledged was 'beyond the scope of this article' (1926a, p. 550).

Monetary theory

There was no sign of Sraffa's emerging critique of neoclassical theory in his review of Hayek's *Prices and Production* (1932a; 1932b). Instead, the review displays some similarities between Sraffa's position and that held by Keynes soon after the publication of the *Treatise*.

Sraffa argued that Hayek had failed to identify the essential properties of money by neglecting the fact that

> money is not only a medium of exchange, but also a store of value and the standard in terms of which debts, and other legal obligations, habits, opinions, conventions, in short all kinds of relations between men, are more or less rigidly fixed. (Sraffa, 1932a, p. 43)

The absence of any conception of wage agreements and debts fixed in money terms prevented Hayek from analysing correctly the effects on the distribution of income of a general fall or rise in prices. Since money had been thoroughly 'neutralized' it could not effect the distribution of income or the rate of accumulation. Hence Hayek could characterize 'forced saving' as a disequilibrium phenomenon, with no permanent effects.

Sraffa argued that this conclusion was contrary to a 'common sense' view of the economy. During a period of inflation

> one class has, for a time, robbed another class of a part of their incomes; and has saved the plunder. When the robbery comes to an end, it is clear that their victims cannot possibly consume the capital which is now well out of their reach. (Sraffa, 1932a, p. 48; see also 1932b, p. 249)

Sraffa's view that class conflict determines the *normal* real wage, and that monetary policy may be part of that conflict, is an echo of Sraffa's earlier position, yet nothing is said on the theory of distribution as such.

The most enduring construction in the article is Sraffa's invention of the concept of the own rate of return. Sraffa utilized the idea to elucidate the concept of equilibrium underpinning much of Hayek's discussion. In particular he demonstrates that whilst in disequilibrium there may be as many 'natural' rates of interest (that is, own rates of return) as there are commodities, competition will tend to equalize these natural rates just as competition eliminates any divergence between market prices and normal prices – indeed these are two aspects of the same process.

Yet Sraffa does not consider how the equilibrium rate of interest is determined. Nor does he criticize Hayek's association of the rate of interest with the length of the production process.

Ricardo

Sraffa's edition of *The Works and Correspondence of David Ricardo* proved to be more than a great scholarly achievement. For in his introduction to *The Principles of Political Economy and Taxation* Sraffa presented an entirely new interpretation of Ricardo's theory of value and distribution. Sraffa's interpretation established a new, theoretically consistent version of the surplus approach to the analysis of distribution in the *Essay on Profits*. Further, he demonstrated that this approach was sustained in the *Principles* by Ricardo's use of the labour theory of value, and that, contrary to the accepted view of Ricardo's analysis presented by Jacob Hollander (1904), Ricardo did not retreat from his use of the labour theory of value in successive versions of the *Principles*.

In the *Essay on Profits* Ricardo stated that it is the rate of profit in agriculture which determines the rate of profit in the economy as a whole. Sraffa argued that

> The rational foundation of the principle of the determining role of the profits of agriculture, which is never explicitly stated by Ricardo, is that in agriculture the same commodity, namely corn, forms both the capital (conceived as composed of the subsistence necessary for workers) and the product; so that the determination of profit by the difference between total product and capital advanced, and also the determination of the ratio of this profit to the capital, is done directly between quantities of corn without any question of valuation. (Sraffa, 1951, p. xxxi)

The beautiful simplicity of this interpretation – the rate of profit being determined in the agricultural sector as a ratio of quantities of corn, and in the other sectors as a ratio of values, with the price ratio between corn and other commodities adjusting so as to equalize the rate of profit (corn, being the wage good, is part of the capital in all sectors) – suggested itself in the preparation of *Production of Commodities by Means of Commodities*. If there is but one 'basic' commodity in the economy (that is, but one commodity which enters directly or indirectly the production of all others), then not only must all the inputs to that commodity consist of itself, but the general rate of profit must be determined by the ratio of surplus of the commodity produced to its means of production.

This powerful result, in which the rate of profit is clearly determined as the ratio of surplus to means of production was sustained in the *Principles*, where it was generalized to incorporate the fact that surplus and

means of production will consist of heterogeneous 'bundles' of commodities. The homogeneity necessary to find the ratio of surplus to means of production was achieved by evaluating the two bundles in terms of the labour embodied directly and indirectly in their production.

As is well known, this generalization foundered on the fact that commodities do not exchange at their labour values, and hence that the ratio, evaluated in terms of labour values, does not measure the rate of profit. As Sraffa pointed out, Ricardo's persistent struggle with this difficulty was expressed as the fact that prices might change due to a change in distribution when labour values (dependent upon conditions of production) were unchanged. Hence he sought an 'invariable standard of value' which would tie movements in price to movements in labour values alone:

> Ricardo was not interested for its own sake in the problem of why two commodities produced by the same quantities of labour are not of the same exchangeable value. He was concerned with it only in so far as thereby relative values are affected by changes in wages. The two points of view of difference and of change are closely linked together; yet the search for an invariable measure of value, which is so much at the centre of Ricardo's system, arises exclusively from the second and would have no counterpart in an investigation of the first. (Sraffa, 1951, p. xlix)

Sraffa was able to demonstrate that Ricardo had continued the search for an invariable standard to the end of his life, sustaining thereby the exposition of his theory of distribution in terms of the labour theory of value. The conclusive proof of Sraffa's argument was found in Ricardo's papers on *Absolute and Exchangeable Value* discovered together with other papers and letters in 1943, their existence having been previously unknown.

Sraffa's interpretation of Ricardo had a considerable impact at the time of its publication, not least because there was great interest in the analysis of growth at the time. The analysis of distribution plays a central role in the classical theory of growth (accumulation by the capitalists is determined by their share of the product). The problems in the theory of the rate of profit posed by the neoclassical analysis of growth, and in some versions of Keynesian growth theory, excited interest in Ricardo's approach.

However, the real importance of Sraffa's new interpretation was for the understanding of Marx's analysis of value and distribution, which is based on Ricardo's theory, and for the general rehabilitation of the surplus approach to value and distribution, which had, for so long, been regarded as logically deficient.

Production of Commodities by Means of Commodities

The subtitle of Sraffa's book (1960) is *Prelude to a Critique of Economic Theory*, and in the Preface he suggested that 'If the foundation holds, the critique may be attempted later, either by the writer or by someone younger and better equipped for the task' (Sraffa, 1960, p. vi), echoing Ricardo's comment in the Preface to his *Principles* (1817, p. 6).

Production of Commodities is a peculiarly sparse book. The argument has been pared to the absolute minimum to sustain the propositions which Sraffa wishes to advance. Yet the precision and logical elegance of the argument are 'the work of an artist working in the medium of economic theory' (Newman, 1962).

The theoretical essence of the book may be distilled from the argument of Part I in which Sraffa deals with single-product industries and circulating capital. There Sraffa demonstrates that the approach to the analysis of value and distribution adopted by Ricardo and by Marx is logically consistent. Taking as data the size and composition of output, the conditions of reproduction, and the real wage it may be shown that (1) in an economy which is capable only of reproducing itself, relative prices are determined by the conditions of production; and (2) that, in an economy which is capable of producing a physical surplus over and above the needs of reproduction, relative prices are determined by the conditions of production of basic commodities, and the manner in which the surplus is distributed. If, in the latter case, the surplus is distributed as a rate of profit, then the data determine relative prices and that rate of profit. The economically meaningful solution – that with non-negative prices – is unique. (The prices of non-basics depend upon their own conditions of production and the prices of basics, but the prices of basics are not affected by the prices of non-basics.) These propositions had already been advanced by Dmitriev (1898), though they were not well known.

Sraffa then drops the assumption that the real wage is given. The degree of freedom thus introduced into the analysis is expressed in the locus of the rates of profit associated with any particular values of the wage (in terms of the *numéraire*). For any given value of the wage there is a unique rate of profit (and associated prices), and vice versa. There is a maximum wage, when the rate of profit is equal to zero; and a maximum rate of profit when the wage is equal to zero. Closure of the model requires either that the real wage be given (that is, determined outside the determination of the rate of profit and normal prices) as in classical theory; or that the rate of profit be given.

Sraffa's suggestion that the rate of profit is 'susceptible of being determined from outside the system of production, in particular by the level of the money rates of interest' (1960, p. 33), is essentially symmetrical with the classical approach in which the real wage is 'given'. For the classical economist the real wage is determined by social and historical forces, circumstances which may be analysed quite separately from the determination of relative prices and the rate of profit. Likewise, it may be argued that the money rate of interest, to which, in a competitive economy, the rate of profit must conform, is determined by the normal operations of monetary institutions, especially the state. This position is reminiscent of Sraffa's earlier work in monetary theory, and of Keynes's remark that 'the rate of interest is a highly conventional, rather than a highly psychological, phenomenon' (Keynes, 1936, p. 203).

While the 'data' of Sraffa's analysis of value and distribution are identical with the data of the analyses advanced by Ricardo and by Marx (other than in his not taking the real wage as given), and hence his results are a validation of their arguments, his method of solution is different. Whereas Ricardo and Marx sought to determine the rate of profit as a ratio of aggregates, Sraffa solves for the rate of profit and prices simultaneously. Indeed, his argument demonstrates the necessity of doing so. Yet in his construction of the standard commodity, Sraffa seeks to recreate the clarity of the classical derivation of the rate of profit from the ratio between surplus and means of production.

Sraffa first constructs from the given conditions of production a 'standard system', an hypothetical economy in which the composition of means of production and net product (wages and profits) are the same. If the wage is expressed as a proportion of standard net product, w, then the proportion of net product accruing to profits is $(1-w)$. If the ratio of total net product to surplus is R – a ratio which may be evaluated because the composition of inputs and net output is the same – then the rate of profit will be equal to $R(1-w)$.

The standard system is therefore a direct descendant of the agricultural sector in the *Essay on Profits*, the rate of profit is expressed as a ratio between two physical quantities.

Sraffa then demonstrates that if the standard net product is adopted as *numéraire*, and hence as the measure in terms of which the wage is expressed, then the rate of profit will be equal to $R(1-w)$, exactly as in the standard system in which the relationship between the wage and the rate of profit is expressed in purely physical terms.

The purpose of this construction is, Sraffa tells us, to 'give transparency to a system and render visible what was hidden' (1960, p. 23). The rate of profit is seen to be determined by the magnitude of surplus. Yet the use of the standard commodity must be distinguished from Ricardo's use of the 'corn sector', or the use of the labour theory of value by Ricardo and Marx. In Sraffa's case, the rate of profit is *determined* by the solution of the simultaneous equations, the standard commodity is a purely auxiliary construction. In the case of Ricardo and Marx, the rate of profit is determined (albeit imperfectly) by calculating ratio of surplus to means of production by means of the labour theory of value.

It cannot be said that the standard commodity is entirely successful as a means of rendering visible what might otherwise be hid. It is, perhaps, too complex, lacking the simple force of the labour theory. It has the virtue, however, of being analytically correct.

Considerable puzzlement was engendered by Sraffa's statement in the Preface of his book that 'The investigation is concerned exclusively with such properties of an economic system as do not depend on changes in the scale of production ...' (1960, p. v). The absence of any reference to demand led unsuspecting readers to equate his results with the non-substitution theorem, and hence with the assumption of constant returns to scale. However, a careful reading of Sraffa's analysis reveals that no knowledge of any relationship between *changes* in outputs and *changes* in inputs, or between price and quantity is *necessary* for the solution of the equations, and hence for the determination of the rate of profit and prices (given the wage). This contrasts with neoclassical theory, in which the determination of prices is dependent upon knowledge of functional relationships between supply and demand. If, in Sraffa's analysis, quantities should change, then any consequential change in conditions of production will result in changes in prices.

In Part II of his book, Sraffa extends his analysis to multi-product industries and fixed capital, and to the analysis of economies with more than one non-reproducible input. As might be expected, the analysis is considerably more complex, and in some cases the results less clear-cut (the solution of the system may not, for example, be unique, and the definition of basics and non-basics is more abstract than is the case with single-product industries). Yet the basic structure of classical analysis is preserved – the prices, the rate of profit, and other distributive variables (say, land rents), are determined by the conditions of production, given the wage.

Sraffa's analysis is a triumphant restatement of the classical analysis of value and distribution. It is therefore somewhat misleading to refer to a 'Sraffa-based critique of Marx', a phrase which implies, perhaps unintentionally, that Sraffa has developed a method of analysis which is conceptually different from that advanced by his predecessors in the theory of surplus value. This is not the case.

The label 'neo-Ricardian' which is often attached to Sraffa's work (a term which he himself vehemently rejected) is also unfortunate, implying as it does that the argument of *Production of Commodities* is in some way a solution to problems posed by Ricardo, but not encountered by Marx or by other surplus theorists. The confusion may have derived from simplistically identifying Sraffa with Ricardo, given his edition of Ricardo's works, or from a confusion of the standard commodity with Ricardo's invariable standard of value (even though the latter cannot exist). It may also derive from a fear that any weakening of 'commitment' to the labour theory of value implies a rejection of surplus theory. This is to confuse a tool which is used to solve an analytical problem, with the data of the problem. The labour theory

of value is not a datum in surplus theory (if it were, Quesnay would not be a surplus theorist), it is a means of demonstrating that the rate of profit is determined by the magnitude of surplus (less rent).

Almost as a by-product of his examination of the fundamentals of classical theory, Sraffa produced a decisive critique of the neoclassical theory of the rate of profit (and hence of the neoclassical theory of long-run normal prices). An examination of the relationship between the changes in the distribution of income and the consequent changes in relative prices leads to the conclusion that such changes 'cannot be reconciled with *any* notion of capital as a measurable quantity independent of distribution and prices' (1960, p. 38). In Part III of *Production of Commodities* Sraffa extended his examination of changes in distribution and prices to the case in which changes in distribution lead to changes in the technique of production. He demonstrates that as distribution is varied switches between methods of production, according to which is cheapest, do not follow any particular pattern. Indeed, a technique which is cheapest when the rate of profit is low, may be superseded by another technique at a higher rate of profit, and at a yet higher rate of profit the first technique may again prove to be cheapest and so supersede the second technique. In other words, competitive choice of technique does not result in any particular ordering of techniques. Most notably, the capital intensity of production is *not* an inverse function of the rate of profit, as is implied by the concept of the marginal productivity of capital.

The discussion of 'reswitching' (see Symposium, 1966; Garegnani, 1970) following Levhari's failed attempt (1965) to demonstrate that Sraffa's result was confined to decomposable systems, blossomed into a general critique of the logical foundations of the neoclassical theory of the rate of profit. The conclusion of the debate may be stated as:

> it is not possible, using the data of neoclassical theory – the preferences of individuals, the technology, and the size and distribution of the endowment – to determine the normal long-run rate of profit and the associated prices.

Neoclassical models of competitive value which are consistent in their own terms (say, the model presented in Debreu's *Theory of Value*, 1959) do not determine a long-run equilibrium, in which stocks of produced means of production are adjusted to the demand for them and, in consequence, there is a uniform rate of profit. The definition of equilibrium used in such models is different from the traditional long-run equilibrium (Garegnani, 1976; Milgate, 1979). If the model in Debreu (1959) were constrained to yield a uniform rate of profit, it would be over-determined to degree $k-1$, where k is the number of reproducible means of production. Paradoxically, this latter result is derivable from Hahn's (1982) attempt to refute the above conclusion.

Conclusion

Piero Sraffa's consistently critical approach to the neoclassical theory of value and distribution was motivated by a distaste for 'subjective' models, but was conducted in purely logical terms, at least in his later works.

His admiration for the 'objective' structure of classical theory, and for the 'openness' of that structure which permits the incorporation of concrete institutional factors into the formal analysis, led him to attempt to establish that theory on logically more rigorous grounds than had hitherto been available.

The critical debate set off by *Production of Commodities* has been somewhat blunted by the change in the notion of equilibrium used in general equilibrium theory (the implications of this change for the operational content of economic theory, that is, for the relationship between the theory and the competitive market economy it purports to analyse, have not as yet been satisfactorily analysed).

The well-known propensity of economists to ignore uncomfortable results has also led to the critique being viewed as an esoteric debate in capital theory, with little general significance. This view is clearly wrong. Any critique of the neoclassical theory of value and distribution is a critique of the entire corpus of neoclassical analysis, for the theory of price formation is central to all neoclassical results. (See, for example, Eatwell and Milgate, 1984, on the relevance of these results for the theory of output and employment.)

But it was the revival of the classical (and Marxian) approach to value and distribution, with all the consequences that has for the study of employment, accumulation, technical change, and so on, which was Sraffa's central concern. *Production of Commodities* was designed to lay the groundwork for that revival.

JOHN EATWELL AND CARLO PANICO

See also **capital theory (paradoxes); reswitching of technique; surplus.**

Selected works

1920. *L' Inflazione Monetaria in Italia*. Milan: Primiata Scuola Tipografica Salesiana.
1921a. Open shop drive. L' Ordine Nuovo, 5 July, 3.
1921b. Industriali e governo inglese contr i lavoratori. *L' Ordine Nuovo*, 24 July, 3.
1921c. Labour leaders. *L' Ordine Nuovo*, 4 August, 1–2.
1922a. The bank crisis in Italy. *Economic Journal* 32, 179–97.
1922b. Italian banking to-day. The Manchester Guardian Commercial: Reconstruction in Europe, Supplement, 7 December, 675–6.
1924a. Problemi di oggi e di domani. *L' Ordine Nuovo*, 1–15 April, 4.
1924b. Obituary – Maffeo Panteloni, *Economic Journal* 34, 648–53.

1925a. Sulle relazioni tra costi e quantità prodotta. *Annali di Economia* II, 277–328.
1925b. A short review of Hastings, H.B., *Cost and Profit Their: Relation to Business Cycles*. Boston: Houghton Mifflin, 1923. Giornale degli Economisti 66, 389–90.
1926a. The laws of return under competitive conditions. *Economic Journal* 36, 535–50.
1926b. A short review of Lehfeldt, R.A., *Money*. London: Oxford University Press, 1926. Giornale degli Economisti 67, 230.
1926c. A short review of Segre, M., Le banche nell'ultimo decennio, con particolare riguardo al loro sviluppo patologico nel dopo guerra, Milano: La Stampa Commercial, 1926. Giornale degli Economisti 67, 230.
1927a. Il vero significato della 'quoto 90', two letters to A Tasca with a reply. *Stato Operaio* 1 (November–December 1927), 1089–95. Reprinted in *Capitalismo Italiano del novecento*, Bari: Laterza, 1972.
1927b. A short review of Phillips, H.W., *Modern Foreign Exchange and Foreign Banking*. London: Macdonald and Evans, 1926. Giornale degli Economisti 68, 610.
1927c. The methods of Fascism. The case of Antonio Gramsci. *Manchester Guardian,* 24 October.
1930a. Symposium on 'Increasing Returns and the Representative Firm'. A criticism. *Economic Journal* 50, 89–92.
1930b. Symposium on 'Increasing Returns and the Representative Firm'. Rejoinder. *Economic Journal*, 50, 93.
1930c. An alleged correction of Ricardo. *Quarterly Journal of Economics* 44, 539–44.
1932a. Dr. Hayek on money and capital. *Economic Journal* 42, 42–53.
1932b. Money and capital: a rejoinder. *Economic Journal* 42, 249–51.
1938. *D. Hume: An Abstract of a Treatise on Human Nature (1740)*. Cambridge: Cambridge University Press. (Introduction with J.M. Keynes.)
1951–73. *The Works and Correspondence of David Ricardo*. 11 vols. Cambridge: Cambridge University Press.
1960. *Production of Commodities by Means of Commodities*. Cambridge: Cambridge University Press.
1962. Production of Commodities. A comment. *Economic Journal* 72, 477–9.

Bibliography

Debreu, G. 1959. *Theory of Value: An Axiomatic Analysis of Economic Equilibrium*. New York: Wiley.
Dmitriev, V.K. 1898. The theory of value of David Ricardo: an attempt at a rigorous analysis. In *Economic Essays on Value, Competition, and Utility*, trans D. Fry, ed. D.M. Nuti. Cambridge: Cambridge University Press, 1974.
Dobb, M. 1961. Review of Production of Commodities by Means of Commodities. De Economist. Reprinted In *A*

Critique of Economic Theory. ed. E.K. Hunt and J.G. Schwartz. Harmondsworth: Penguin, 1972.

Eatwell, J. 1975. Mr Sraffa's standard commodity and the rate of exploitation. *Quarterly Journal of Economics* 89, 543–55.

Eatwell, J. and Milgate, M., eds. 1984. *Keynes's Economics and the Theory of Value and Distribution.* London: Duckworth.

Garegnani, P. 1970. Heterogeneous capital, the production function and the theory of distribution. *Review of Economic Studies* 37, 407–36.

Garegnani, P. 1976. On a change in the notion of equilibrium in recent work on value and distribution. In *Essays in Modern Capital Theory*, ed. M. Brown, K. Sato and P. Zarembka. Amsterdam: North Holland.

Garegnani, P. 1978. Sraffa's revival of Marxist economic theory. *New Left Review* No. 112, 71–5.

Hahn, F. 1982. The neo-Ricardians. *Cambridge Journal of Economics* 6, 353–74.

Hicks, J. 1946. *Value and Capital.* 2nd edn. Oxford: Clarendon Press.

Hollander, J. 1904. The development of Ricardo's theory of value. *Quarterly Journal of Economics* 18, 455–91.

Keynes, J.M. 1923. *A Tract on Monetary Reform.* London: Macmillan.

Keynes, J.M. 1936. *The General Theory of Employment, Interest and Money.* London: Macmillan.

Levhari, D. 1965. A non-substitution theorem and switching of techniques. *Quarterly Journal of Economics* 79, 98–105.

Meek, R.L. 1961. Mr Sraffa's rehabilitation of classical economics. *Scottish Journal of Political Economy* 8, 119–36.

Milgate, M. 1979. On the origin of the notion of 'intertemporal equilibrium'. *Economica* 46, 1–10.

Newman, P. 1962. Production of Commodities by Means of Commodities. *Schweizerische Zeitschrift fur Volkswirtschaft und Statistik* 98, 58–75.

Pasinetti, L.L. 1985. In memoria di Piero Sraffa: economista italiano a Cambridge. *Economica politica* 2, 315–32.

Ricardo, D. 1817. *On the Principles of Political Economy and Taxation.* London: Murray.

Roncaglia, A. 1978. *Sraffa and the Theory of Prices.* London: Wiley.

Roncaglia, A. 1983. Piero Sraffa: una bibliografia ragionata. *Studi Economici* 21, 137–61.

Roncaglia, A. 1984. Sraffa e le banche. *Rivista Milanese di Economia* 10, April–June, 104–12.

Stigler, G.J. 1953. Sraffa's Ricardo. *American Economic Review* 43, 586–99.

Symposium. 1966. Symposium on paradoxes in capital theory. *Quarterly Journal of Economics* 80, November.

Wicksteed, P. 1914. The scope and method of political economy in the light of the 'marginal' theory of value and distribution. *Economic Journal* 24, 1–23.

Sraffian economics

Piero Sraffa, born in 1898 at Turin as the son of a well-off professor of law, lived from the twenties to his death in 1983 the quiet bachelor life of a don at King's and Trinity Colleges, Cambridge. Though his published and unpublished works are few, Sraffa has four claims to fame in the science of economics and the history of ideas.

(i) His 1926 article, 'The Laws of Returns Under Competitive Conditions', was a seminal progenitor of the monopolistic competition revolution. It alone could have justified a lifetime appointment.

(ii) An intimate of Keynes and Wittgenstein, Sraffa is said to have speeded Wittgenstein on his second philosophical road to Damascus by a rail station query, 'What then is the meaning of this [Sicilian] gesture?' The young Sraffa provided books and pin money to the marxist Antonio Gramsci jailed by Mussolini, and he remained quietly interested in leftist matters. Sraffa was an organizer of the famous 1931–5 Cambridge 'Circus' which included Joan and Austin Robinson, Roy Harrod, James Meade and many others. Using Richard Kahn as messenger Gabriel, Maynard Keynes derived much benefit for his nascent *General Theory* from their brilliant group. Except for the chapter 17 discussion of *own* rates of interest, where Keynes must have benefited from Sraffa's 1932 polemic with Friedrich Hayek, there are few signs of a Sraffian interest in the macro-economics of effective demand. Sharing with Keynes an antiquarian's preoccupation with rare books, Sraffa and Keynes jointly discovered, identified, and edited the valuable *Abstract of A Treatise on Human Nature* that David Hume had published anonymously as a puff for his great initial work on philosophy.

(iii) Sraffa's editing of *The Works and Correspondence of David Ricardo*, a lone-wolf effort over a quarter of a century (aided much toward the end by Maurice Dobb) is one of the great scholarly achievements of all time, ranking in its perfections with the team efforts of the editors of Horace Walpole and James Boswell.

(iv) Finally, in the seventh decade of his life, Piero Sraffa published a classic in capital theory, *The Production of Commodities by Means of Commodities* (1960). As with Mozart if not Mendelssohn, Sraffa's death leaves posterity wistful that his full potential never came into print: what would we not give the good fairies, if somewhere in the attic of a country house there should be discovered a manuscript presenting Sraffa's planned critique of marginalism?

A fresh survey of Sraffa requires the reverse of a chronological order. First comes his 1960 book, which has spawned an extensive literature but still needs – if the technology is to be adequately handled – to have Sraffa's special equalities embedded in the general inequalities–equalities of the 1937 von Neumann model. The essentially completed Sraffa–Leontief circulating capital model provides by itself a prism with which to diffract the paradigms of Marx, Ricardo, and various brands of neoclassicism; and, self-reflexively, it can serve to help judge Ricardo's editor.

The unity in Sraffa's scientific vision, from before 1926 until death at age 85, then becomes visible.

Truly general time phasing

The polemics of Böhm-Bawerk, Knight, and other capital theorists are illuminated, and seen through, by the 1937 von Neumann general model once that is made explicitly *time-phased* and open-ended, to allow (a) for *primary* factors (labour, land, ...) not necessarily producible within the system, and (b) for net consumptions of outputs not necessarily ploughed back into the system for self-propelled growth.

We can summarize the n outputs produced by J activities, where each jth activity uses as inputs a vector of M primary factors and a vector of commodity-inputs, while producing a vector of joint products:

$$q_i^{t+1} = \sum_{j=1}^{J} b_{ij} \min \left[Q_{1j}^t / a_{1j}, \ldots, Q_{nj}^t / a_{nj}; \right.$$

$$\left. L_{1j}^t / l_{1j}, \cdots, L_{Mj}^t / l_{Mj} \right] \qquad (1)$$

$$= Q_{i1}^{t+1} + \cdots + Q_{iJ}^{t+1} + C_i^{t+1},$$

$$i = 1, \ldots, n \gtreqless J, \qquad Q_{ij}^t \geqq 0, \qquad C_i^{t+1} \geq 0,$$

$$L_{mj}^t \geq 0$$

$$(2)$$

$$a = [a_{ij}] = \begin{bmatrix} a_{11} & \cdots & a_{1J} \\ \vdots & & \vdots \\ a_{n1} & \cdots & a_{nJ} \end{bmatrix} \geqq 0;$$

$$b = [b_{ij}] = \begin{bmatrix} b_{11} & \cdots & b_{1J} \\ \vdots & & \vdots \\ b_{n1} & \cdots & b_{nJ} \end{bmatrix} \geq 0$$

$$l = [l_{mj}] = \begin{bmatrix} l_{11} & \cdots & l_{1J} \\ \vdots & & \vdots \\ l_{M1} & \cdots & l_{MJ} \end{bmatrix} \geqq 0; \quad J \gtreqless M \gtreqless n$$

$$(3)$$

The a's are input/output coefficients; l's are Walrasian factor coefficients; b's are joint-product proportions (as with 1 bu. wool and 2 pounds mutton per sheep). We use standard notation: for a matrix or vector z, $z > 0$ means all its elements are positive; $z = 0$ means all elements zero; $z \geqq 0$ means no element negative, with all possibly zero; $z \geq$ means no element negative, but at least one being strictly positive; z' is the transpose of z: thus, $[1 \cdots 1]'$ is a column vector of ones. I is the identity matrix, with 1's in the main diagonal and zero elsewhere:

$$I = [\delta_{ij}], \quad \text{as in } I = [1], \quad I = \begin{bmatrix} 1 & 0 \\ 0 & 1 \end{bmatrix}, \ldots, \text{etc.}$$

For systems in a *stationary* state, regardless of the variable we have $z^{t+1} = z^t = z$. An important case is where supplies of primary factors are specified to be positive constants:

$$\sum_{1}^{J} L_{mj} = L_m, \qquad m = 1, \ldots, M.$$

Remark: the circulating capital models occupying most pages in Sraffa (1960) are a special case of (3) where each column of (b_{ij}) can be written to contain a single 'one' with the remaining elements zero. In general, any or all a_{ij}'s could be assumed by a mathematician to be zero; but, to be interesting, every b must have at least one positive element in each of its rows and its columns, so that every good is produced somewhere and every activity produces at least one good. An activity might not require any *direct* primary input; but, if we are not to be in the Land of Cockaigne, where lollipops grow freely on trees and don't even require picking, indirectly or directly every good must require some primary factor(s).

General Hawkins–Simon conditions

For a circulating-capital system to be 'productive', in the sense of providing positive net consumptions, a must satisfy simple Hawkins–Simon conditions, such as that powers of a, a^k go to zero as $k \to \infty$. When b involves joint products, matters are more complex and the literature needs the equivalent of the following two constructive tests:

Axiom 1 (No Land of Cockaigne). The following standard linear programming problem must have a solution of zero:

Subject to

$$lx \leqq 0, \qquad x \geqq 0$$
$$[b - a] x \geq 0,$$
$$\max_x [1 \cdots 1] [b - a] x = Z^* = 0$$

Axiom 2 (Generalized Hawkins–Simon). The following standard LP problem will have a positive solution if, and only if, the system is 'productive'.

Subject to

$$lx \leqq [L_1 \cdots L_M] > 0, \qquad x \geqq 0$$
$$[b - a] x \geqq C \geqq 0$$
$$C \geqq (c \cdots c)' \geqq C$$
$$\max_{x, C} c = c^* > 0$$

Example: Suppose, as in the joint-production passages of Sraffa (1960), that $(b - a)$ is $n \times n$, with $J = n$. It will then be *sufficient* that $b - a$ have row sums positive for the productiveness axiom to be satisfied. (Contrary to what seems to be suggested by the mathematician C. F. Manara (1979), these row-sum conditions are *not* necessary—as $b - a$ with diagonal elements near to 1's and off-diagonals of -10 and -0.01 demonstrates.)

Competitive pricing relations

In the absence of any uncertainty, or restrictions on entry and knowledge, perfect competition will be led in (1) by Darwinian arbitrage to equality of the profit rate in all processes positively operated. In matrix terms, with P a non-negative row vector of n goods prices, W a non-negative row vector of M primary-factor prices, and r the common profit or interest rate, we have the general dynamic equalities–inequalities:

$$P^{t+1}b \leqq (P^t a + W^t l)(1 + r^t),$$
$$\left(P^t, P^{t+1}, W^t, r^t\right) \geq 0 \qquad (4)$$

$$P^{t+1}q^{t+1} = \left[P^t ax^{t+1} + W^t lx^{t+1}\right](1 + r^t)$$
$$= \left[P^t aq^t + W^t L^t\right](1 + r^t) \qquad (4')$$

where x^{t+1} is the column-vector of the J 'intensities' at which the respective activities are carried on:

$$0 \leqq x_j^{t+1} \leqq L_{mj}^j/l_{mj}, \qquad m = 1, \ldots, M; \qquad J = 1, \ldots, J$$
$$\leqq Q_{ij}/a_{ij}, \qquad i = 1, \ldots, n$$
$$\geqq q_i^{t+1}/b_{ij} \qquad (4'')$$

The convention is understood in (4) and (1) that when a denominator vanishes, the term in which it appears is ignorable. In (4) P's and W's are measured in any numeraire and equality of r's in all processes used does not imply equality of different goods' *own* rates of interest when P^{t+1} is not proportional to P^t.

Wherever a strong inequality holds in (4), that activity must cease to operate, in accordance with the perfect-competition *duality* conditions:

$$0 = x_j^{t+1}\left[\sum_{i=1}^n P_i^{t+1} b_{ij} - \sum_{i=1}^n P_i^t a_{ij}(1 + r^t)\right.$$
$$\left. - \sum_{m=1}^M W_m^t l_{mj}(1 + r^t)\right]$$
$$j = 1, \ldots, J \qquad (4''')$$

In stationary equilibrium, $(P^t, W^t, r^t) = (P, W, r)$ and all time scripts can be omitted. All *own* rates of interest are then the same.

Then (4) becomes

$$Pb \leqq (Pa + Wl)(1 + r) \qquad (5a)$$

$$Pbx = (Pax + Wlx)(1 + r) \qquad (5b)$$

These equality-of-profit-rate conditions are not merely competitive arbitrage conditions. Taken together with the steady-state version of (1)–(2), namely with

$$(b - a)x \geqq C \geq 0, \qquad lx \leqq L > 0 \qquad (5c)$$

(5a)–(5b) are the necessary and sufficient conditions for *inter-temporal* production efficiency (or inter-temporal Pareto efficiency on the allocation side). Unless there exists, at the observed steady state, an existent $(P, W,$ uniform $r)$, the society must be planning wastefully. So aside from Marx's 1867 innovation being a backward step in positivistic realism, it was a bad (avoidable) blunder from the standpoint of social planning and efficiency – a point never glimpsed by Marx or his admiring editor Engels. Because Sraffa's brief text never grapples with the *inter-temporal* relations implied by *his* steady states, his readers are left unaware of this technocratic property of dualistic competitive pricing.

Sraffa (1960) considers for the most part very special cases of (5), and of (1)–(2), cases for which n happens to equal J with $b - a$ square and of rank n and with all equalities holding in (4). Save for the brief chapter on land, he mostly works with labour as the only primary factor and seems to presuppose that $[l_{11} \cdots l_{1n}]$ $[b - a(1 + r)]^{-1}$ happens to be positive in an open neighbourhood above r equal to zero. (*Remark*: in many technologies, even when J greatly exceeds n, the number of activities operated at a *positive* level will be equal to n; this endogenous fact should not obscure the more general truth, that changes in demand will generically alter the choice of n viable activities – so that, as will be seen, von Neumann's rectangular matrices cannot be sidestepped.)

Properties of competitive equilibrium

The following twenty properties of Sraffian steady-state systems are straightforwardly verifiable:

1. When goods require more than one primary input, as for example labour and land, a shift in demand away from a land-intensive good and toward a labour-intensive good, will at each ruling profit rate raise the price ratio of the latter relative to the former; and it will tend to raise labour's distributive share at the expense of land's. *The labour theory of value*, even in the absence of the complication of time-phasing and interest, *thus generally fails and the theory of distribution cannot be separated from the complications of value theory* (of supply–demand pricing theory).

2. Even when labour is the only primary factor, time-phasing means that Smith's bipartite formula of

wage-*plus*-interest is indeed a necessary statement of price and of national income. Save in singular cases where all goods happen to have exactly the same percentage of direct-wage cost to total cost – what Marx called the case of 'equal organic compositions of capital' – a change in the interest rate must alter relative commodity prices – again vitiating the simple labour theory of value.

3. Suppose there are no joint products, all raw materials being used up in a single employment. It follows from the last paragraph that competitive prices in time-phased systems must generally at positive interest rates differ from the 'marked-up values' of Marx's *Capital* (1967, Volume I, ch. IX) which replace a uniform industry-by-industry rate of profit by a uniform rate of surplus value: the 1867 marked up values mark up the direct wage costs only, with allegedly no mark up earned on raw-materials outlays for produced goods (for 'constant' capitals). Ratios of these peculiar 'marked-up values' agree with ratios of zero-profit-rate prices, and do so no matter how great the capitalists' surplus! So, in consequence of paragraph numbered 2 above, the 1867 Marxian constructs do systematically depart from realistic competitive prices and need to be 'transformed' into correct Sraffian competitive prices *by abandoning them* – as L. von Bortkiewicz had demonstrated in 1907.

4. Staying with the assumption of no joint products, we can deduce the existence of a *factor-price trade-off frontier*: at any specified Sraffian interest rate, there is defined a convex trade-off frontier between the maximal real wage rate of labour and real rent rate of land (where *any* good is the numeraire for measuring such real factor prices); a rise in the interest rate must shift inward the convex contour relating real factor prices (but equal upward increments of r can induce inward shifts in the frontier that both accelerate and decelerate). Though Sraffa does not use this name for the frontier, for the case where labour is the sole primary factor, he recognizes the properties of this basic frontier.

5. The amount of net consumptions produced in the stationary state will be maximal when the competitive system chooses those golden-rule techniques that are supported by a zero rate of interest. (If all primary factors grow at a Harrod natural rate of $[1+g]t$, the maximal per capita net consumptions will be realized only when competition mandates use of the golden-rule techniques supportable by an interest rate of g: $1 + r = 1 + g$.) This paragraph states something different from the earlier remarks around equations (5) concerning inter-temporal Pareto efficiency.

6. At any interest rate r, prevailing in Sraffa's time-phased system, the competitively viable techniques observed can be verified from (5) to achieve inter-temporal Pareto optimality. By contrast, consider the 1867 Marxian techniques that maximize the rate of surplus value for a specified vector of relative primary-factor prices. Not only are these techniques unrealistic describers

of the positive facts about the laws of motion of capitalism. In addition they would achieve a Planner's nightmare, in general producing permanently in the steady state less of goods than the system is capable of and involving more of society's scarce primary factors than is technologically needed. When Sraffa's readers begin to worry about consumption preferences over different time periods, they will additionally have to revert back from 1867 Marxian values to (5)'s dualistic competitive pricing to restore *inter-temporal* Pareto efficiency of consumption.

How demand-tastes affect pricing and distribution

The above half-dozen rules of a Sraffian system are valid either for *his* von Neumann technologies or for *all* versions of neoclassical technologies (involving convexity and first-degree homogeneity). Although Sraffa reserved judgement for half a century on whether he wanted to assume constant returns to scale, experiments with returns laws that depart from that property will be found to rob his algebra of *any* interesting economic applications, as the paucity of results on this point in the literature of the last quarter of a century attests. (Thus, specify for the coal and iron example of the opening page of Sraffa (1960) that in the iron industry doubling inputs quadruples output, while in the wheat industry tripling inputs doubles output. Then none of the pricing relations have other than empty definitional content!) The Perron input/output matrices don't define *existent* prices of production.

Further properties of steady-state competitive systems are the following.

7. Prior to Sraffa (1960), the Leontief literature had established and generalized the 1949 Non-Substitution Theorem. When labour is the only primary factor and there are no joint products, at any observed profit rate the competitive prices at which all goods are positively produced must be independent of the composition of steady-state demand: even when alternative techniques could be reasonably substituted, changes in demand can *never mandate their new use*. Also, with labour the only primary factor, a shift in demand toward goods that involve a high relative fraction of direct-wage cost must be at the observed profit rate raise the wage-profit share. As already indicated, when labour must cooperate with land, shifts in demand toward or away from goods that are relatively labour-rather-than-land-intensive must at the observed competitive profit rate raise or lower their relative prices and presumptively alter the distribution of income between workers and landlords. The Ricardian dream of ridding distribution theory from the complications of (consumer-demand) value theory is seen to be, in general, a pipe dream – as Ricardo realized in his occasional lapses into good sense (as for example, when recognizing that Napoleonic-war shift of demand toward labour-intensive soldiering would raise the wage share

prior to a repopulating of the countryside). At a given interest rate, in the absence of technical innovation and non-labour primary factors, a rise in the money wage rate cannot force a permanent substitution of machines for labour since machines' steady-state costs then rise proportionally with the wage rate.

Some joint-product phenomena

The following properties of joint-product systems are also common to Sraffian technologies of the von Neumann type and to *all* versions of neoclassical technologies.

8. When joint products are admitted – surely the realistic case – the classical economists' hope to deduce steady-state price ratios from technology and supply alone is *generally* frustrated. When one species of sheep produces wool and mutton in joint proportions – or when one round trip of a ship supplies east and west transportation jointly – each alteration in tastes and demands for the joint products alters their steady-state price ratio and does so even under perfect Arrow–Debreu certainty.

Sraffa's favourite case of joint products would involve as many independent activities used as there are goods: $J = n$. For a simple example, consider 2 sheep species, each producing 2 products: say, species 1 produces 2 of wool and 1 of mutton, while species 2 produces 1 of wool and 2 of mutton; for simplicity let both sheep require the same inputs and cost, which might as well be labour only.

So long as consumer's demand involves relative expenditures on the two goods not too unequal – no good ever attracting more than two-thirds of consumers' dollars – Sraffa is correct in expecting cost-technology alone to determine competitive pricing of $P_{\text{mutton}}/P_{\text{wool}} = 1.0$. But, as soon as people want to spend more than two-thirds of their incomes on mutton, Sraffa loses his equality of number of goods and number of positively used activities: only the meat-intensive species is competitively viable; consumer demand functions are then price determining.

Let us add a third sheep species, producible like the others but yielding 1.75 of wool and 1.75 of mutton. Now $J = 3 > 2 = n$. When people singularly spend exactly half their incomes on wool and mutton, only the species 3 is viable and $P_m/P_w = 1.0$ as set by demand. When people spend a bit more on mutton than wool, less than $J = 3$ activities are positively viable, namely, $2 = n < J = 3$, as species 3 and species 2 alone survive. For a range of demands, $P_m/P_w = 1/3$, a numerical value that can be calculated from Sraffa's 2 cost-prices of production relations. However, as the demand for mutton relative to that of wool runs the gamut of possible ratios, we reach the limit of each Sraffian horizontal step and must traverse the staircase's vertical risers with market price being consumer-demand determined. In realistic cases, J is large relative to n; and there are many different n-by-n

square sub-matrices that consumer-demand functions will endogenously select out of the rectangular n-by-J matrix of technology.

In sum, globally demand generally helps determine relative prices of Sraffa's joint products between zero and infinity, doing so along a staircase's vertical risers and (locally horizontal) step segments.

This general rule of global dependence on demand of joint-products' price ratios does admit of an exceptional case where an aspect of a Non-Substitution theorem does obtain. Let us, so to speak, introduce so weak a degree of jointness of production that we are still in a close neighbourhood of indecomposable non-joint-production. This will occur when a positive *net* amount of each good is available only from a single process, and where each process does produce one such positive amount net. Under this stipulation the square matrix $(b_{ij} - a_{ij})$, after feasible renumbering of the goods and of the processes, will be specified to have positive elements in its diagonal and negative off-diagonal elements. Its general Hawkins–Simon conditions will also require that the inverse $[b - a]^{-1}$ exists and is positive. Under these strong stipulations, as is well known, the same Non-Substitution theorem that holds for circulating-capital and exponential-depreciation models will hold for joint production.

Otherwise, however, demand conditions can in general have an essential effect on relative prices; and can do so even when we grant Sraffa special indulgences (such as equality of J and n, non-singularity of $b - a$, and labour the only primary factor). Thus, often, even when $(l_{1j})(b - a)^{-1}$ is positive, some elements of $(b - a)^{-1}$ will be negative – with the result that for some ratios of consumptions, $(b - a)^{-1}C$ will not define positive feasible gross outputs, and price ratios will have to be influenceable by demand-tastes.

Sraffian artifacts: standard commodity baskets

The following properties are special and hold only for singular subsets of von Neumann technologies or for singular subsets of neoclassical technologies.

9. Postulate no joint products, no primary factors other than labour, no alternative techniques for producing goods, and that all goods considered are basics in the sense that a is indecomposable so that every good requires for its production something directly or indirectly of every good as input. Then, Sraffa (1960) deduces the existence of a market basket of goods or *standard commodity*, with unique positive weights $(Q_1^*, \ldots, Q_n^*)'$, such that the *real wage expressed in terms of the standard commodity* and paid to the workers postfactum (at the *end* of the period when they work at the beginning of the period) is a declining *linear* relation:

$$w = 1 - (r/r^*) \qquad (6a)$$

where r^* is the maximum positive profit rate the system can pay, with the column vector (Q_i^*) uniquely definable by the eigenvector relation:

$$a(1 + r^*)[Q_i^*] = [Q_i^*] > 0$$
$$[l_{11} \cdots l_{1n}] [I - a]^{-1} Q^* = 1 \qquad (6b)$$

Some real prices rise faster with the profit rate than the standard commodity's price does; some rise less fast. When the real *standard*-basket wage rate is half its zero-profit-rate level, the post-factum wage share of the basket's cost is also one half; and similarly for any other fraction.

Sraffa, for reasons not easy to understand, thought that (6)'s truth somehow provided Ricardo with a defence for his labour theory of value. Even in the restricted case where (6) does validly obtain, one perceives no successful resurrection of Ricardo's desired labour theory of value or absolute standard of value that is provided by Sraffa's demonstration of the existence of this standard commodity: price ratios are still not equal to zero-profit-rate price ratios as the crude labour theory of value wants them to be.

10. It is by now understood that a Sraffian Standard Commodity often fails to exist, for a variety of reasons:

(i) In circulating-capital and exponentially depreciating models, as soon as competition mandates a switch in technology when interest rates or relative primary factor prices change, there will then generally not be any market-basket weights that entail a linear tradeoff frontier between the real wage and the profit rate.

(ii) In these same models, realistic decomposabilities can negate the existence of positive weights that yield linearity.

(iii) In still other cases, as shown by Takahiro Miyao in 1977, for various vectors of direct-labour coefficients, there can be an infinity of positive weights that produce the same normalized linearity as the Sraffian standard.

(iv) In joint-product cases, as C. F. Manara (1979) instanced. $(b - a[1+r])$ may have only complex eigenvalues and eigenvectors. Again, no Sraffian standard validity obtains. (For such Manara cases, at some feasible interest rates certain processes cease to be competitively viable; so Ricardo can never find his middling composite, which is neither too time-consuming nor too time-economizing, to provide him with the chimera of an absolute standard of value for making comparisons across time and space.)

(v) There exist many joint-product cases in which Sraffa's eigenvector of $(b - a[1+r])$ is real, but involves some negative elements. Playing the game of defining market baskets with negative weights cannot, by some analogy with foreign-trade exports and imports or debtor-creditor ownings and owings, make economic sense useful to a Ricardian critical of Adam Smith or of J. B. Clark.

(vi) There are many joint-product cases in which Sraffa's eigenvector for $(b - a[1+r])$ is all positive. But still there may be no standard market basket that yields a *linear* factor-price frontier validity applicable over the *whole* interval of feasible profit rates. (On one side of some critical r, all of Sraffa's $J(=n)$ processes are competitively viable. On the other side of that r, less than n processes can earn the competitive profit rate. The *true* market cost of Sraffa's nominated market basket there ceases to obey the linear law that allegedly Ricardo's value theory could benefit from.)

(vii) Instead of abandoning the hunt for a chimera, some Sraffians sought comfort in the belief that the important joint-product cases are not those of the wool-mutton type but rather are those of the new-machine-old-machine type or are of the permanently-durable-land type; and nursed a hope that for these important cases, the 'pathologies' of non-existent standard commodities might be absent.

If anyone ever believed this, it was an illusion. Under (i) above, we saw land–labour models in which induced changes in a_{ij} coefficients take place, which induce violation of the Sraffian desired linearity. A locus linear piecewise is not linear.

Also, a model where the only jointness is of the durable-good type may well not admit of a standard commodity defined over the whole interval of feasible profit rates. The following counterexample settles the issue of possible non-existence:

$$b = \begin{bmatrix} 1 & 1 \\ 7/8 & 0 \end{bmatrix}, \quad a = \begin{bmatrix} 7/8 & 0 \\ 0 & 1/2 \end{bmatrix},$$
$$[l_{11} l_{12}] = \begin{bmatrix} 1 & 9 \end{bmatrix}$$

$$(7)$$

Here is its story. Good 1 can be consumed directly or be used as a new machine to produce, in cooperation with labour, Good 1 itself along with the joint product of an old machine. The used machine, called Good 2, can be used with labour to produce Good 1 and worthless scrap. (Later, we'll recognize that Good 2 might be an object of final utility for its own sake, distinct from Good 1's marginal and total utility.)

This example's numbers involve the need in Process 2 for (relatively) much labour and little of the old machine to produce 1 of Good 1. Process 1, by contrast, produces 1 of Good 1 with relatively little labour and a fair amount of new machines: for Process 1, labour with 7/8 of a new machine, produces 1 of a new machine and 7/8 of an old machine; for Process 2, 9 of direct labour, with 1/2 of an old machine, produces 1 of a new machine and scrap.

Calculation shows that competition must then work out as follows:

At very low profit rates, the used machine is so consuming of high-wage direct labour that it (and Process 2) cannot be used in production under viable competition.

The price ratio P_2/P_1 will then be set completely by utility tastes for Goods 1 and 2: with enough yen for Good 2, P_2/P_1 can be anything from zero to infinity at low profit rates.

Between the critical profit rates of 1.59%, and 109.67% both processes could be used competitively. If both are useable throughout that interval, then over that interval (and only it) Sraffa will get his desired linear standard commodity; but it must fail him at the lower interval of profit rates. Worse still, even in the higher interval where algebra does yield him a linear function, economics can veto the relevance of Sraffa's standard function: as soon as people have strong enough marginal utility for Good 2, *all* of it is bid away from production uses! Process 2 becomes competitively unviable and Sraffian cost-prices lose relevance; demand becomes decisive, frustrating the primitive classicist's yearning to determine prices from supply considerations alone. The 1960 purported defence of Ricardo's absolute standard has collapsed.

Moral

The Walrasian paradigms are in general unavoidable in the most unrestricted von Neumann paradigm. What began as a classicist-inspired critique of 'marginalism' ends up as a demonstration that the classical-economics paradigm does need to be broadened into the post-1870 mainstream economics – even when smooth Clarkian marginal products are scrupulously avoided! Where a critique succeeds, and is needed, is in exposing how special are the one-sector, homogeneous-scalar-capital paradigms of Clark.

The genuine Fisher, von Neumann, Arrow–Debreu structure of time-phased general equilibrium stands confirmed by the Sraffa–Leontief probings.

1960 light on Clarkian oversimplifications

Heroic works clarifying faults of neoclassical parables have been done by Joan Robinson, Luigi Pasinetti, Pierangelo Garegnani, Bertram Schefold, J. S. Metcalfe and I. Steedman, C. F. Manara, V. K. Dimitriev, and many others.

11. Before she knew of Sraffa's 1960 model, Joan Robinson had usefully debunked the notion that some aggregate Platonic Kapital, $K = \sum P_j k_j$, enters into real-world production functions and defines (aggregate) *marginal net product* of Kapital, $\partial[K(t + 1) - K(t)]/\partial K(t)$, to give the real-world interest or profit rate. Sraffa's model showed once and for all the falsity of the following neoclassical apologetics:

> Roundabout ['mechanized'] methods are productive and the interest rate measures the incremental social product obtainable by extending the degree of round-aboutness through the effective action of *saving* (by the exercise of painful 'waiting' and 'abstinence').

Sraffa showed this: as soon as there are more capital goods than 1, it is impossible to say of every pair of techniques which one is the more roundabout, time-intensive, or mechanized. Increases in the interest rate above zero can first raise various P_2/P_1 ratios and then later lower them. So above a critical high interest rate, competition may revert back to a technique that had been viable only at very low interest rates.

> Sraffian *reswitching* thus implies: lowering the interest rate may result in lower steady-state consumption levels, prior to its ultimately raising them to the maximal golden-rule level.

Oskar Lange once wrote, sardonically but seriously, that Ludwig von Mises, the enemy of socialism, deserved a statue in the socialist Hall of Fame for compelling Abba Lerner and Lange (to say nothing of earlier Pareto, Barone and Fred Taylor) to work out how socialists might devise efficient decentralizing-pricing planning algorithms. One can insist, seriously, that a neoclassical Hall of Fame deserves Sraffa's statue.

12. Actually all of Sraffa's findings about the impossibility of defining 'more roundaboutness' and 'less roundaboutness' apply precisely to smoothly differentiable *vector*-capital Clarkian models. Neoclassicists should reproach themselves, and thank Robinson and Sraffa, for belated recognition that the locus of (stationary-state consumption, interest rate) need not be a one-way trade-off. This recognition is achievable quite apart from the dramatic case of *double-switching*.

Example. Let $[L, K_1, K_2, C_1, C_2, \dot{K}_1 = dK_1/dt, \dot{K}_2]$ depict, for a 2-sector neoclassical economy in which marginal products are *not* illusory, its labour, its 2 heterogeneous capital good's stocks, its 2 consumptions, and its net investments. Its production function and steady-state profit-rate are given by

$$L = F\big[K_1, K_2; C_1 + \dot{K}_1, C_2 + \dot{K}_2\big] \quad (8a)$$

$$r = -(\partial F/\partial K_i)/(\partial F/\partial \dot{K}_i), \quad i = 1, 2 \tag{8b}$$

where F is a first-degree-homogeneous, smooth, convex function. Set L by convention at a plateau of unity, and for simplicity specify C_2/C_1 always to be unity; and, as a condition of stationarity, make all $\dot{K}_i$ vanish. Then the above pair of relations permit us to determine the level of consumption as a function of the profit rate: $C_1 = f(r)$. What neoclassicals insufficiently realized before 1960, away from $r = 0$ there is no necessity for C_1 to fall as r rises. This valid Sraffa–Robinson point does not score against 'marginalism'. Their razor cuts as deeply against the discrete-activity technology!

'Marginalism' per se is not what deserves the razor of Sraffa's critique. As already said, whatever Clarkian marginalism can display for good or ill is already capable of being displayed by a von Neumann discrete-activities technology.

Space permits only a few further mentions of paradoxical neoclassical phenomena that can result from positive profit rates in time-phased models. A critique to reveal them, as shown by several valuable works of Metcalfe and Steedman, cuts as much within von Neumann-activities models as within smooth marginalist models of the marginal-product type.

13. The Non-Substitution Theorem, which makes relative costs and prices, at each interest rate, independent of the composition of final demands when only one primary factor is present, is common to discrete and marginalist models. When more than one primary factor is present, the Non-Substitution Theorem becomes a Substitution Theorem – again, generally both for discrete and marginalist models.

14. Both for discrete and marginalist no-joint-product models, for each interest rate there is a convex trade-off between various real factor returns (expressed in terms of any good as numeraire). For both models, it is not necessarily true that the trade-off relation between the interest rate itself and real factor returns is convex. (Point 14 has a considerable overlap with Point 4.)

15. For both models, the substitutions of technique that win out under competition are Pareto efficient. For both models, any technique observed to be viable under competition is a golden-rule technique for that Harrod-growth rate which is equal to the observed interest rate!

16. For both models, when the real wage rises relative to the real land-rent (independently of the good used as numeraire), any mandated technical substitution must (if anything) involve a shift to lower embodied-dated-labour contents of each good and to higher embodied-dated-land contents. This is as true for a positive interest rate as for a zero one, since we deal exclusively with golden-rule technologies. (The difference, when r is zero, is that there is then no need for *dating* of the direct-and-indirect labour and land contents.)

17. As shown by Metcalfe and Steedman, Pasinetti, Samuelson, and others, in both models one must guard against a tempting fallacy.

In both models, when r is zero, the change in technique induced by a rise in the Wage/Rent ratio must economize on the actual Labour/Land ratios observed to be used in the various industries in the stationary state. Something like this remains true for very low interest rates. (For economy of space, explanatory qualifications are omitted.) However, for large enough r, what are called embodied-dated-labour contents and embodied-dated-land contents can be substantially different from synchronized-stationary-state labour and land actually observed to be used. So beware of believing that observed $(\Delta W_m)(\Delta L_m)$ must be negative when r is positive. This has implications

for the correct Heckscher–Ohlin and Stolper–Samuelson trade theorems under time-phasing.

18. Related to the above is the following observation. At $r = 0$, the competitive system is out on its true *stationary-state production-possibility-frontier*. At positive r, although the system moves to a new *steady*-state (not stationary-state!) golden-rule set of techniques, what it can produce and does produce in the stationary state is generally not out on the stationary-state production-possibility frontier – but rather is inside that frontier. As demand changes, what we observe inside that frontier need not trace out well-behaved concave loci.

Some Marxians, bemused by notions of 'unequal exchange', think that the above phenomena are Pareto inefficient: the system and the world get less product than is producible in the zero-interest-rate golden-rule state. Yes, of course. But that is unavoidable in any scenario where the economic system could only obtain the produced inputs needed for Schumpeter's golden-rule zero-interest rate utopia by doing current 'waiting' and sacrificing of current consumptions. The curse of the poor societies is their poverty – even when *intertemporal* Pareto efficiency is always obtaining. Again, all this is as true of *discrete* as of *marginalist* technologies.

Sraffian refutation of Marxism

On the basis of this elaborate description of findings about Sraffian time-phased systems, one can apply the results to appraise and correct (1) neoclassical economics, (2) Marxian economics, (3) Ricardian and classical economics. Its thrust on neoclassical economics was sampled in the previous section.

20. Sraffian economics, as earlier passages make clear, devastatingly repudiates that central part of Marx's economics, Capital, Volume I (1867) which proposed a new paradigm involving an equal 'rate of surplus value' by industries or departments. Sraffa and Darwinian arbitrage require an *equal* 'rate of profit' by industries or departments. Under exploitative capitalism, Marx misidentifies what is out there to be observed in the competitive market, for the reason that Marx has the capitalists garnering too little in the industries using much capital goods and garnering too much in those using relatively little capital goods. It is a gratuitous error, and a sterile one, since Marx's paradigm does not help predict the laws of motion of competitive capitalism, or help understand the average magnitude of the profit level around which Marx's errors spread. Equal profit rates are not a capitalist shibboleth; dual to the primal variables of the von Neumann technology is an equalized profit rate.

Ian Steedman, a scholar sympathetic to the *Weltanschauung* of Marx and of economic reform, has documented the Sraffian rejection of 1867 Marxian rate of surplus value in his book, *Marx After Sraffa* (1981). So here it need only be said: in the end, from the posthumous publication of *Capital*, volume III (1894), one can recognize that the 'transformation' from Marx's 1867 [marked-up] 'values' to bourgeois 'prices' involves abandoning the 1867 relations and returning to the pre-Marx and post-Marx cost-profit relations of the Sraffian model.

One cannot quite leave it at that. Karl Marx does deserve a statue in the Sraffian Hall of Fame. Not, of course, for his sterile detour into the rate of surplus value. As documented by Samuelson in the 1974 Lloyd Metzler *Festschrift*, Marx was the first scholar after Quesnay to grapple explicitly with input/output capital models. Implicitly, Marx was the first to use (a_{ij}) coefficients. Moreover, in his Tableaus of Steady Reproduction and of Expanded Reproduction, *Capital*, volume II (1885), Karl Marx was the first to present coherent row-and-column arrays of steady-state equilibrium. This is Marx's imperishable contribution to analytical economics, and it is impervious to the deflating of hyperbole concerning Marx as allegedly a great *Mathematical* economist.

1926 reconsidered from 1960

There were two main parts to Sraffa's celebrated 1926 paper. The first part, which today we realize was by far the more important one, dealt with the phenomena of *increasing returns to scale*. So long as demand is insufficiently large to take a firm beyond such an initial phase of increasing returns or decreasing costs, the firm cannot find a maximum-profit equilibrium while still remaining a perfect competitor. Cournot knew this in 1838, and it was not new doctrine even then. The turn of the century literature on trusts and industrial organization, in America and Germany, was never in doubt on this. By 1926 the whole issue would have been old hat save for Alfred Marshall's tergiversations on the compatibility of *increasing* (internal, statical and reversible) *returns* and perfection of competition. Marshall, at the time of his death in 1924, was at the height of his prestige, with the greatest capacity for good and for potential confusion. Therefore, it was important for Piero Sraffa to restate elegantly that real-life firms, with localized and segmented markets, were in imperfect-competition equilibrium with falling marginal production costs that were offset by selling and transport costs and by price declines inducible by expansion of their own outputs and sales. Independently, E.H. Chamberlin and J.M. Clark were saying much the same thing at the time in America. But the ruling establishment of Cambridge, understandably, could learn something best from one of its own publishing in the *Economic Journal*. The 1960 book has no relation to this part of Sraffa's early work.

Within the mid-1920s, Sraffa's other thesis generated a disproportionate amount of interest. Not only was the familiar downward-sloping Marshallian supply curve to be ruled out as incompatible with perfection of competition; the young Sraffa was newly arguing that upward-sloping supply curves were also of vacuous importance for Marshallian partial equilibrium. All that Sraffa left his reader, then, was a horizontal, *constant-cost* competitive supply curve.

This is plain wrong. Sraffa's 1960 book demonstrates that, when primary factors other than a single homogeneous labour exist, rightward shifting Marshallian and Walrasian demand curves will generally trace *rising* price intersections on the relevant supply curves. Joan Robinson's famous 1941 *Economic Journal* article on rising supply price was the first East Anglian recognition of the formal comparative statics of general equilibrium. I doubt that she or Piero ever noticed the incompatibility with 1926 Sraffa; or the incompatibility of Heckscher–Ohlin and Stolper–Samuelson in the foreign-trade literature with Sraffa's thesis of *constant* costs and implied *linear* production-possibility frontiers. The pre- and post-Ricardian classical literature is full of examples of wine grown on special vineyard lands: for generations students have followed Viner's 1931 example of calling this 'the pure Ricardian case' – while the modern teachers of the young know this as the Jones–Samuelson–Haberler specific-factors model.

Students of rhetoric should be interested to analyse the elements of style that enable the erudite author of a faulty thesis to persuade himself and several generations of thinkers of its truth and importance. Because Sraffa wrote so little, and wrote so rarely on the mainstream topics of contemporary scholarship, his skills as a writer have perhaps been insufficiently noticed.

Knowing what we now know – that it is a mistake to believe that constant-cost cases exhaust the categories of admissible competitive price – we are in a position to study the young Sraffa's extra-scientific motivations. Why does a sophisticated intelligence make *this* mistake? Just as inside a fat man is a thin man trying to get out, so outside the Sraffa of the post-World War I heyday of Walras–Marshall neoclassicism, there was already in 1925 an atavistic classical economist trying to get back in. This is why one can say that, from before 1925 to after 1960, there is a discernible consistency in Piero Sraffa's thought and ideology. An objective reader will want to be alert to this tendency.

1960 verdicts on 1817 Ricardianism

Sraffa's models, we have by now seen, tellingly reject the following Ricardian stereotypes:

(*a*) Prices are determinable by the labour theory of value.
(*b*) Land and rent can be ignored by concentrating on external-marginal land with zero rent (or on internal margins in every positive-rent acre).
(*c*) The complication for pricing of time and the interest rate can somehow be avoided or ameliorated by defining an intermediate standard commodity or workbasket, which is less time consuming and interest-rate-inflated than the most time consuming

goods (old wine and tall trees) and is more time consuming and interest-rate inflated than the most directly produced goods (shrimp picked up by labour on the seashore).

(*d*) One can correctly understand the distribution of income among workers, landowners, and capitalists independently of the complications of demand theory (consumers' demand functions, marginal utility, revealed preferences, etc.).

(*e*) It is superficial to base goods' and factors' pricings on mere supply and demand.

(*f*) Adam Smith committed some grave errors in decomposing price and national income eclectically into wage-plus-rent-plus-profit components, and in enunciating his notion of 'labour command'.

Enough has been given of current misunderstandings related to Ricardo for the present exegesis. What does a close rereading of Smith and Ricardo reveal under the light of the post-1960 analysis?

Most of both scholars' actual inferences about the real world are compatible with the post-1960 findings. Smith's scorecard is certainly not inferior to Ricardo's, even after we make allowances for the latter's tendency to proclaim as being universal what is only likely.

Ricardo began to write on microeconomics because he thought he discerned basic logical flaws in Smith's system. Just as Steedman showed, in the Ronald Meek *Festschrift*, that Marx's criticisms of Ricardo could not stand up to modern examination, Robert L. Bishop (1985) has shown that Ricardo's criticisms of Smith similarly cannot stand up.

Both Smith and Ricardo tried to compare, by one scalar parameter, the diverse price vector of two times and places: $[P_i/P_1, P_i/W_1, W_m/W_1; r]$ for China and Scotland, or for the Englands of 1780 and 1688. We know that just cannot be done: then, now, or ever. Smith's 'labour command' notions merely proceeded from the prosaic observation that people always have about two-thirds of a 24-hour day available to them: $\sum P_j C_j/W$ is the hours people have to work for what they consume; because per capita C's tend to rise only with longterm productivity, the above ratio tends to decline only slowly as people enjoy more leisure.

Ricardo never supplanted this imperfect measure by a better one, for all his palaver about absolute standards of value. Worse: gratuitously he attributed to Adam Smith *unwarranted* deviations from the labour theory of value by virtue of Smith's labour command passages. Actually, there is no valid connection. Smith only introduces deviations between what $[P_i/P_j, P_i/W]$ are and what they would be under a labour theory of value when these deviations are warranted by (1) scarcities of needed lands and natural resources, and (2) timephasing of production that involves produced goods as inputs and which takes place when the competitive market displays positive interest and profit rates.

For all of Ricardo's scolding of Smith as a recanter from the labour theory of value, Ricardo up to his dying month admitted the cogency of the point (2) – that goods of the same content but involving manifestly different time involvements and relative profits would have prices that systematically fail to be proportional to their embodied labour contents. And on point (1), as we have already seen, Smith was right; and Ricardo was wrong in his belief that he could get rid of the complication of rent by use of external (or internal) margins. George Stigler's Pickwickian 1958 defence of 'Ricardo's 93% labour theory of value' on the positivistic grounds that his labour theory of value allegedly averages out in practice to errors of only about 7 per cent makes one wonder what scientists would think of a defence of Lamarck as being 49.99% right; or of the stone-fire-air-and water paradigm as offering a theory of matter that is at least 0.001% as accurate as Mendelev's 93-element periodic table. For some comparisons a 7% error becomes a 70% or a 700% error.

Ricardo does seem to make two major advances on Smith. Paradoxically for Sraffa's hero, *Ricardo made a giant step beyond* Smith toward marginalism. His ubiquitous numerical examples presuppose almost a continuum of alternative doses of labour-and-produced-goods applied to the same acre(s) of land. Inside Ricardo there is a von Thünen and a J.B. Clark striving to be born!

A second Ricardian advance is not so important or clear-cut. You must read Smith closely to perceive his understanding that the rent of inelastically supplied land is price determined rather than price-determining. In Ricardo the point is made crystal clear; in the chapter on land in Sraffa (1960), you must read with sophistication to perceive the point.

Also, Ricardo stresses, indeed overstresses, the point that the profit rate would not have to decline – as saving brings into existence new capital stocks and enlarged populations – if new lands were ever available in unlimited and redundant supply. A good point, even if readers of his expositions might be forgiven for misunderstanding him to imply that the only reason for a drop in interest and profit rates is a rise in rent: actually, as Smith knew, a persistent excess in the growth rate of capitals relative to population must in many realistic technologies raise the real wage and lower profit rates *even in the presence of superabundant lands*.

The post-1960 Sraffian analysts who grade Ricardo's blue books must often mark down his submissions. It is time therefore to study how Ricardo's 1951 editor dealt with these issues.

1960 light on the 1951 editor of Ricardo

The history of humane letters involves only history. Samuel Johnson's mistakes may be more interesting than his correct observations. To the antiquarian, antiquarianism is all there is to the history of the humanities.

The history of scientific thought is a two-fold matter. We are interested in Newton's alchemy and biblical prophecy because we are interested in Newton the man and scientist. At the same time his stepsister's theology is likely to elicit a yawn from even the most besotted antiquarian. How Newton discerned that a homogeneous sphere of non-zero radius attracts as if all its mass were at its center point, that is part of the history of cumulative science. Say that this attitude involves an element of Whig history if you will, but remember that working scientists have some contempt for those historians and philosophers of science who regard efforts in the past that failed as being on a par with those that succeeded, success being measurable by latest-day scientific juries who want to utilize hindsight and ex post knowledge.

Economics is in between belles lettres and cold science. Serious economists below the age of 60 will judge Sraffa's edition of Ricardo both for its antiquarian and its scientific interests and insights. How then will they judge it?

From an antiquarian view the work is a jewel of perfection. Reviewers' enthusiasm has been unbounded. By luck and Sraffa's energetic skills, virtually every scrap written by David Ricardo has been made available to the interested reader. This is a boon to scholars who lack the slightest interest in the history of thought for its own sake: Baconian scientific observation of Ricardo's economics has now been made possible by Sraffa's labours.

Editorial emendations have also been done in the new edition with skill and brevity. You might almost say that the editor has for the most part stayed chastely out of the act, letting David and his friends speak their pieces without an accompaniment of Greek Chorus expressions of approval or disapproval.

From the scientific viewpoint, and now a minority viewpoint is being expressed here, there is something anticlimactic about the great Sraffa edition of Ricardo. It is not just that we see, as if imprisoned in amber, the backward and forward gropings of a scholar who from his 1814 entrance into microeconomics until his death in 1823 makes almost no progress in resolving his self-created ambiguities and problematics. Somehow one had hoped that the whole picture would be a prettier scientific picture, so that the editor's Herculean framings would be for a more worthwhile object.

There is, however, no point in lamenting that Ricardo was only what he was. It is the 'road not taken' by the editor that occasions a twinge of regret. From the scientist's rather than the antiquarians' viewpoint, we appreciate from an editor and commentator what Jacob Viner gave economists in his magnificent 1937 *Studies in the Theory of International Trade* and what Eli Heckscher supplied in his *Mercantilism*. It is what Clifford Truesdell's lengthy introductions to the collected works of Euler provide, and what Abraham Pais succeeds in bringing off in his 1984 survey of the scientific physics of Albert Einstein. Admittedly old Edwin Cannan carried to excess his patronizing reviews of past economic giants, not only faulting them for their sins in failing to believe what Cannan believed in 1928 but also managing to convict them of the crime of not being so smart as himself. Surely, there is a golden mean somewhere between Cannan's dominating the act and Sraffa's avoiding getting into it?

Fortunately, in his Introduction to Ricardo's *Principles* (written late in the day, with the help of Maurice Dobb), Piero Sraffa does let himself go a little bit. Thus, he conjectures that Ricardo, in a lost 1814 manuscript or letter or conversation, may have worked out a model in which the profit rate is determined within agriculture, as a ratio of (so to speak) corn to corn; and, Sraffa all but says, in such a model distribution theory is successfully emancipated from value theory. Unlike Viner and Cannan, who can be very hard indeed on the guinea pigs they are judging, one reads Sraffa in his Introduction as being quite indulgent of Ricardo. When he quotes Ricardo as purporting to get rid of the complication of rent by concentrating on the external margin, Sraffa never seems tempted to add that this is a *non sequitur*. When Ricardo tries to overdifferentiate his product from Smith's, Sraffa never writes: 'Of course, when Smith made the emergence of positive interest cause a divergence of price from labour contents, he was doing what Ricardo often admits must be done – namely, formulating a two-factor rather than a one-factor model of pricing.' The critique of mainstream twentieth century that Sraffa never lived to articulate was evidently festering inside the editor of Ricardo during the 1930–1951 period and serving to soften his critical judgments.

Salutations

Did any scholar have so great an impact on economic science as Piero Sraffa did in so few writings? One doubts it. And there cannot be many scholars in any field whose greatest works were published exclusively in their second half century of life.

Piero Sraffa was much respected and much loved. With each passing year, economists perceive new grounds for admiring his genius.

PAUL A. SAMUELSON

See also **Sraffian economics (new developments)**.

Bibliography

Bishop, R. 1985. Competitive value when only labor is scarce. *Quarterly Journal of Economics* 100, November, 1257–92.

Bortkiewicz, L. von. 1907. On the correction of Marx's fundamental theoretical construction in the third volume of *Capital*. Trans. in *Karl Marx and the Close of his System*, ed. P.M. Sweezy, New York: August M. Kelley, 1949.

Garegnani, P. 1970. Heterogeneous capital, the production function and the theory of distribution. *Review of Economic Studies* 37, 407–36.

Garegnani, P. 1984. Value and distribution in the classical economists and Marx. *Oxford Economic Papers* 36(2), 291–325.

Manara, C.F. 1979. Sraffa's model for the joint production of commodities by means of commodities. In *Essays on the Theory of Joint Production*, ed., L.L. Pasinetti, New York: Columbia University Press.

Marx, K. 1867, 1885, 1894. *Capital*, Vols. I, II, III. English translations in C.H. Kerr, ed. Chicago: Chicago University Press, or other editions.

Metcalfe, J.S. and Steedman, I. 1972. Reswitching and primary input use. *Economic Journal* 82, March, 140–57.

Miyao, T. 1977. A generalization of Sraffa's standard commodity and its complete characterization. *International Economic Review* 18, February, 151–62.

Neumann, J. von. 1945. A model of general economic equilibrium. In *Review of Economic Studies* 13, 1–9. Trans. of an article in *Ergebnisse eines mathematischen Kolloquiums*, Vol. 8, ed. K. Menger, Leipzig, 1937.

Pasinetti, L.L. 1977. *Lectures on the Theory of Production.* New York: Columbia University Press.

Robinson, J. 1956. *The Accumulation of Capital.* London: Macmillan.

Samuelson, P.A. 1975. Marx as mathematical economist: steady-state and exponential growth equilibrium. In *Trade, Stability, and Macroeconomic, Essays in Honor of Lloyd A. Metzler*, ed. G. Horwich and P. Samuelson, New York: Academic Press, 269–307. Also ch. 225 in *The Collected Scientific Papers of Paul A. Samuelson*, Vol. III, Cambridge, Mass.: MIT Press, 1972.

Samuelson, P.A. 1975. Trade pattern reversals in time-phased Ricardian systems and intertemporal efficiency. *Journal of International Economics* 5, November, 309–63. Also ch. 251 in *The Collected Scientific Papers of Paul A. Samuelson*, Vol. IV, Cambridge, Mass.: MIT Press, 1977.

Samuelson, P.A. 1983. Thünen at two hundred. *Journal of Economic Literature* 21, December, 1468–88.

Schefold, B. 1971. *Theorie der Kuppelproduktion (Mr. Sraffa on Joint Production).* Privately printed, Basel.

Seton, F. 1957. The transformation problem. *Review of Economic Studies* 25, 149–60.

Sraffa, P. 1926. The laws of returns under competitive conditions. *Economic Journal* 36, December, 535–50.

Sraffa, P., ed. 1951–73. *The Works and Correspondence of David Ricardo.* Cambridge: Cambridge University Press, for the Royal Economic Society; with editorial collaboration by M.H. Dobb.

Sraffa, P. 1960. *Production of Commodities by Means of Commodities: Prelude to a Critique of Economic Theory.* Cambridge: Cambridge University Press.

Steedman, I. 1977. *Marx After Sraffa.* London: New Left Books. Reprinted, London: Verso, 1981.

Steedman, I. 1982. Marx on Ricardo. In *Classical and Marxian Political Economy: Essays in Honour of Ronald L. Meek*, ed. I. Bradley and M. Howard, London: Macmillan.

Stigler, G.J. 1958. Ricardo and the 93% labor theory of value. *American Economic Review* 48, June, 357–67.

Sraffian economics (new developments)

1 Introduction

In an earlier era, Sraffians took aim at the neoclassical assertion that the demand for and supply of labour and other resources determine factor incomes. In stalking the big game, a smaller question served as bait: is there a homogenous substance, aggregate capital, whose marginal product determines the return to capital? After a brief episode of disagreement in the 1960s, the neoclassical side conceded that in models with even a minimal disaggregation of capital goods the answer to the smaller question was 'no'. Despite this bloodletting, the chase petered out. When the hunter and hunted ran out of formal modeling disagreements, a settlement was drawn up.

The agreement stipulated that there are two neoclassical theories. The first is an aggregative model that tells the familiar parables of Solow growth theory: increases in savings raise the ratio of the value of capital to labour employed, the rate of interest falls as production becomes more capital intensive, and so on. But once a multiplicity of capital goods is introduced these parables no longer hold true. General equilibrium theory, on the other hand, places no limit on the number of capital or consumption goods and still gives a coherent, determinate account of markets and price determination and hence of the distribution of factor incomes.

While both parties could agree to this decree, they took different views as to who walked away with the more valuable share of the community property. By the time of the split in the early 1970s, general equilibrium had already been singled out as the jewel of microeconomic theory. The results that had to be jettisoned were confined to the steady-state effects of capital accumulation, leaving the prize results of general equilibrium theory – the existence of equilibrium and its welfare properties – untouched. Moreover, a multiplicity of consumption goods will by itself (without multiple capital goods) imply that the distribution of income cannot be determined by *the* marginal products of capital and labour. The disavowal by high theory of aggregate capital therefore seemed a small loss. Hahn (1982), which marked the end of engagement with Sraffian dissent, argued not only that general equilibrium theory left the line of neoclassical succession intact but also that all legitimate Sraffian results could be obtained from suitably specialized general equilibrium models. The resilience of general equilibrium theory to the Sraffian assault had the curious effect that aggregative empirical work could proceed unaffected by the Sraffa episode. Even a literature such as growth accounting for which the Sraffa critique was pertinent showed no influence – it took off just as the Sraffa attack reached its height. Since general equilibrium theory viewed all forms of aggregation as suspect, overlooking Sraffian concerns about capital aggregation seemed to be one of the ordinary compromises that applied work demands.

The Sraffian reaction was more complicated. Some held that modern general equilibrium theory, although internally consistent, fails because it does not explain how relative prices converge through time to long-run values, in this view the proper goal of economic science (Garegnani, 1976; Eatwell, 1982; Kurz and Salvadori, 1995). Another strand simply promoted Sraffian and classical economics more generally as a distinct type of economic theory (for example, Harcourt, 1974; Marglin, 1984). Sraffians take as their starting place a wage or wage-share of output that is determined by non-economic forces, for example by political power or by social consensus. The general equilibrium model in contrast explains factor prices as the outcome of the endogenous play of supply and demand. Different assumptions, different theories: let the evidence decide.

An agreement to disagree should leave all parties dissatisfied. If Sraffian economics and general equilibrium theory are merely two contenders, each with its own starting place about the causal forces that move factor prices, to be adjudicated by empirical test, then all the wrangling in the 1960s was for nought. For if wages are determined by political power, say, rather than supply and demand, then political power could remain the prime determinant even if capital always aggregated perfectly or in models where prices do not converge to long-run values. What makes the Sraffa–neoclassical debate significant is its critical dimension, the Sraffian arguments that the forces of competition cannot pin down the distribution of income, that any supply-and-demand theory is riddled with internal flaws. The neoclassical side of the debate has contributed its share of confusion: the mere existence of competitive equilibria does not speak to the adequacy of a supply-and-demand account of factor incomes. When translated into the language of general equilibrium, the Sraffian complaints presumably concern the determinacy and stability of equilibrium, not existence or optimality. Yet Hahn (1982), for example, treats determinacy only casually and leaves stability unaddressed. Fortunately, some decades of delay after the noisy 1960s and 1970s, the literature on Sraffa has turned to these points.

The Sraffian complaints about supply-and-demand theories of price determination can be spelled out in two ways. The first appears in Sraffa's *Production of Commodities by Means of Commodities* (1960): the laws of competitive markets do not fully determine factor prices or the distribution of income. Competition requires that the same rate of return is earned in every sector; when laid out as a system of equations, this requirement leaves one more variable than equation, thus revealing a single dimension of indeterminacy, or, as Hahn (1982) put it, a 'missing equation'. Hahn and other neoclassical economists responded that the missing equation would be found as soon as supply-equals-demand equalities are incorporated into Sraffa's model. Sraffians vacillate on

market clearing for factors of production: the land market has to clear, but the labour market does not. This asymmetry drives the single dimension of indeterminacy. As we will see, if the same market-clearing conditions that Sraffians impose on land markets to quash extra dimensions of indeterminacy are applied to the labour market then even the standard single dimension of indeterminacy can disappear. This conclusion would seem to undercut Sraffa's book: if indeterminacy stems solely from failing to require the labour market to clear then Sraffians hardly need an elaborate model to press their point. In essentially any setting, the deletion of labour-market clearing will turn the wage into a free variable and hence leave the distribution of income indeterminate. On this score at least, there would be no need to object to the aggregate neoclassical production function.

But the story is not so simple: the neoclassical presumption that the full gamut of market-clearing conditions necessarily brings determinacy is not correct. Although the ingredients have to be recombined, the Sraffian tradition takes just the right modeling steps that lead factor prices in general equilibrium models to be indeterminate. Sraffians have long insisted that linear activities provide a more faithful representation of technology than the differentiable production functions that dominate practical work in neoclassical economics. Although linear activities by themselves do not generate indeterminacy, they can when endowments of capital goods are governed by rational savings decisions rather than by chance. The Sraffian view of the economy as an ongoing cycle of reproduction thus paves the way for factor-price indeterminacy. On the other hand, the particular way Sraffa and his followers have spelled out their long-run view of the economy, by requiring that relative prices be constant through time, undermines their 'missing equation' criticism: linear activities models with constant relative prices have determinate factor prices. And the aggregation of capital has no bearing on the matter: the determinacy of a supply-and-demand theory of factor incomes depends on how many activities operate compared with the number of scarce factors, not on the number of capital goods.

The second completion of the Sraffa critique focuses on the potential for the value of capital per worker to behave badly, for example to increase in response to a rise in the interest rate. Although it might seem that this possibility could by itself lead to instability, this turns out not to be the case. Instability can arise in general equilibrium but it stems from the demand side of the model, not the failure of capital goods to aggregate.

Little of the Sraffian–neoclassical settlement therefore withstands scrutiny. While a couple of assertions in Solow growth theory about steady states hinge on whether the economy has a single sector and whether capital aggregates, the operation of competitive markets does not. The neoclassical confidence that the general equilibrium model answers all Sraffian challenges is equally misplaced: the Sraffian indeterminacy thesis can be reclaimed. As for neoclassical growth theory, its main message that the return to saving diminishes as savings increase can be re-expressed to avoid the limitations of single-sector models. But here too the Sraffian tradition points the way to important corrections. The characteristic neoclassical equality between an economy's interest rate and its marginal rate of transformation is an artifact of differentiable technologies. With linear activities this equality need not obtain, although for the failure of the neoclassical maxim to be robust, we must follow the neoclassical program, rejected by some Sraffians, of letting utility functions determine consumption.

This article's focus on determinacy and stability aims to complement Paul Samuelson's SRAFFIAN ECONOMICS, which draws the lessons to be learned from Sraffa regarding aggregative parables.

2 The single dimension of Sraffian indeterminacy

We set a benchmark model of linear activities. Let there be N goods, one type of labour, L types of land, and finitely many activities. Each activity i when operated at the unit level requires an investment one period in advance of $a_i = (a_{1i}, \ldots, a_{Ni}) \geq 0$ units of the N material inputs and then a contemporaneous application of $\ell_i \geq 0$ units of labour and $\Lambda_i = (\Lambda_{1i}, \ldots, \Lambda_{Li}) \geq 0$ units of the L land types to produce the outputs $b_i = (b_{1i}, \ldots, b_{Ni}) \geq 0$. The level at which activity i is operated is given by y_i. We assume to begin that the prices $p = (p_1, \ldots, p_N) \geq 0$ of the material goods purchased as inputs equal the prices of the same material goods when sold as outputs one period later. Profit maximization requires that no activity makes positive economic profits and that any activity i in use $(y_i > 0)$ makes zero economic profits. Let r be the intertemporal interest rate, $w \geq 0$ be the wage, and $\rho = (\rho_1, \ldots, \rho_L) \geq 0$ be the rental rates on land. So profit maximization dictates, for each activity i,

$$p_1 b_{1i} + \cdots + p_N b_{Ni} \leq (1+r)(p_1 a_{1i} \\ + \cdots + p_N a_{Ni}) + w\ell_i + \rho_1 \Lambda_{1i} \\ + \cdots + \rho_L \Lambda_{Li} \qquad (2.1)$$

and that equality holds for any i such that $y_i > 0$.

The Sraffa literature equivocates on market clearing for resources. While land types are required to have a 0 rental rate when in excess supply, the situation for labour is often left unspecified. Let e_{Λ_k} denote the supply of type k land and e_ℓ denote the supply of labour. We then have the market-clearing conditions

$$\sum_i \Lambda_{ki} y_i \leq e_{\Lambda_k} \\ \sum_i \Lambda_{ki} y_i < e_{\Lambda_k} \Rightarrow \rho_k = 0, \qquad (2.2)$$

for land type $k = 1, \ldots, L$. The analogous conditions for labour are

$$\sum_i \ell_i y_i \leq e_\ell, \qquad \sum_i \ell_i y_i < e_\ell \Rightarrow w = 0.$$

Letting y denote a vector activity levels, an *equilibrium* is a $(p, w, 1 + r, y)$ that satisfies (2.1) and (2.2). When we impose labour-market clearing, we say so explicitly. In the background lurk additional market-clearing conditions for produced goods, which we introduce in Section 3.

Since we are interested only in relative prices, we can normalize prices by choosing one of the goods or a bundle of goods as numéraire. We set

$$p_1 + \cdots + p_N = 1. \qquad (2.3)$$

In the basic Sraffa model, the focus of the first generation of literature, each activity produces only one good and uses no land, and only one activity is available to produce each good j and is therefore given the index j. So in this case we can rescale each (b_i, a_i, ℓ_i) so that b_{ii} (the sole non-zero coordinate of b_i) equals 1. Assuming that each activity is in use, then (2.1) gives us the classical Sraffa equalities:

$$p_i = (1 + r)(p_1 a_{1i} + \cdots + p_N a_{Ni}) + w\ell_i, \\ i = 1, \ldots, N. \qquad (2.4)$$

The simplest version of Sraffa's missing equation or single dimension of indeterminacy thesis amounts to the observation that (2.4) and the normalization (2.3) comprise $N + 1$ equations but contain $N + 2$ price variables $(p, w, 1 + r)$. A complete argument that any $(p, w, 1 + r) \gg 0$ that satisfies (2.3) and (2.4) is locally contained in a one-dimensional set of points that solve the same equalities might seem to require a rank condition, but the linearity and homogeneity of the model make this unnecessary (see the *Note on the dimension of indeterminacy of the basic Sraffa model* at the end of the text). Since typically we can parameterize the solutions to (2.3) and (2.4) by w or r, the distribution of income is indeed indeterminate. Competition, Sraffians suggest, does not pin down a division of social wealth between capital and labour.

The indeterminacy of the above model has little significance from the neoclassical point of view: the only alarming possibility would be if market-clearing equalities for some reason could not close the model. The Sraffian literature does not engage this argument, however, and instead takes either w or r to be exogenous, set by political factors or by a macroeconomic determination of the interest rate. The rationale for this practice is perpetually unclear: is it that market-clearing equalities cannot fill the indeterminacy gap or does some principle trump the laws of supply and demand?

We now document the efforts of Sraffa and his followers to maintain precisely a single dimension of indeterminacy. 'Single dimension of indeterminacy' is not standard terminology; a more conventional but equivalent description would be that Sraffians aim to show that (2.1) and (2.3) locally determine prices once w or r has been set, and hence that, given w or r and given an equilibrium, a small exogenous change in demand will leave prices unaffected. Indeed, the view that prices in the long-run are affected only by technology and the distribution of income, not the composition of demand, has long been a top item of the Sraffian theoretical agenda.

Single-dimensional indeterminacy faces three threats – rent, joint production, and the choice of technique – the same topics that dominate the second half of Sraffa's book and the second generation of the Sraffian literature. Although we will see that the arguments available against extra dimensions of indeterminacy can sometimes be turned against the presence of any indeterminacy at all, our position is that zero, one, and more than one are all plausible equilibrium possibilities for the dimension of indeterminacy. It may seem in some of our exhibits that demand and market-clearing should eradicate any indeterminacy, but Section 3 will show that they are compatible.

Exhibit A: land and rent

If we add an additional non-produced factor with a positive price to a Sraffa model the number of endogenous price variables will increase. If no further activities are drawn into production, the dimension of indeterminacy will normally go up.

For an elementary example, let there be one produced good ($N = 1$), one type of land ($L = 1$) with endowment $e_\Lambda > 0$, and multiple activities. Let the rental rate of the single land type be ρ and again normalize activities so that when activity i is operated at the unit level one unit of output is produced. Suppose activity i with $(a_{1i}, \ell_i, \Lambda_i) \gg 0$ is in use and $y_m = 0$ for $m \neq i$. Then (2.1) reduces to

$$1 = (1 + r)a_{1i} + w\ell_i + \rho\Lambda_i, \qquad (2.5)$$

$$1 \leq (1 + r)a_{1m} + w\ell_m + \rho\Lambda_m, \quad m \neq i, \qquad (2.6)$$

where we have substituted in our normalization $p_1 = 1$. To ensure that (2.2) does not constrain ρ to equal 0, the land must be fully employed: $\Lambda_i y_i = e_\Lambda$. With ρ as an additional free variable, if $(\bar{w}, 1 + \bar{r}, \bar{\rho}) \gg 0$ satisfies (2.5) and (2.6) and each inequality in (2.6) is strict, an additional dimension of indeterminacy obtains: if we independently vary w and $1 + r$ a small amount then ρ can be adjusted so that (2.5) and (2.6) remain satisfied.

Another type of equilibrium occurs when two or more activities are in use. If two activities i and j are in use, then (2.5) is replaced by two equalities,

$$1 = (1 + r)a_{1i} + w\ell_i + \rho\Lambda_i, \\ 1 = (1 + r)a_{1j} + w\ell_j + \rho\Lambda_j. \qquad (2.7)$$

Condition (2.6) now holds for $m \neq i, j$, and full employment for land is given by $\Lambda_i y_i + \Lambda_j y_j = e_\Lambda$. Evidently the argument for an additional dimension of indeterminacy now fails; we cannot independently vary w and $1 + r$ and expect to satisfy both equalities in (2.7) using the single free variable ρ.

The Sraffa literature largely focuses on the second type of equilibrium with a single dimension of indeterminacy. Is this the more likely type? To make the best case, notice that in the first type of equilibrium the produced input must be accumulated in an amount that leads the stock of land to be fully utilized using only activity i – otherwise (2.2) would require $\rho = 0$. If e_c is the stock of the produced input accumulated each period, then to fully employ both e_c and the entire land supply e_Λ using only activity i there must be an activity level $y_i \geq 0$ such that $a_{1i}y_i = e_c$ and $\Lambda_i y_i = e_\Lambda$. Since e_c must therefore equal $e_\Lambda a_{1i} / \Lambda_i$, perhaps one could conclude that the accumulation of this exact amount is unlikely to occur. But consider how the shape of the production possibilities set changes as e_c changes. If we fix w arbitrarily (and suppose implicitly there is no constraint on labour supply), then, outside of exceptional values for w and barring flukes in the input usage coefficients, at most two activities can have the least cost per unit of output and be employed by profit-maximizing firms. If exactly two activities are in use, then the economy can raise its usage of the produced input and increase output by switching the mix of the two activities towards whichever activity economizes on the use of land and uses the produced input intensively – the activity j with the higher a_{1j}/Λ_j ratio. This remixing delivers a linear increase in current output as e_c rises. Since increases in e_c must come from the previous period, the production possibilities frontier (PPF) for the current and previous period's consumption is also linear at points where two activities are in use. Once remixing has been exhausted (a 'switch point'), a new activity with a higher a_{1j}/Λ_j must be adopted if more capital is to be used to produce more current output. At the switch point, the first type of equilibrium occurs where one activity is in use and the PPF exhibits a kink (non-differentiability). But optimizing agents may well choose to save a quantity of the produced input so as to end up at a kink in the PPF. See Figure 1, which pictures the tangency between a kink in a PPF and a smooth indifference curve, where consumption at other periods is fixed at optimal levels. (In a multi-agent model, one may interpret the indifference curve as the boundary of the set of consumptions that can Pareto improve on the optimum.) Such one-activity-in-use

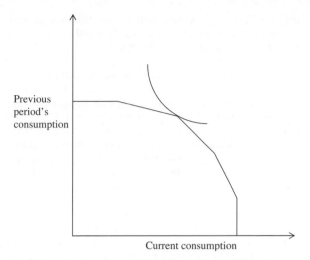

Figure 1   Production possibility frontier and indifference curve

optima are robust to perturbations of the model. Hence, contrary to Sraffian practice, the first type of equilibrium with the extra dimension of indeterminacy should not be excluded.

We will see in Exhibit C that if nevertheless we dismissed the first type of equilibrium as unlikely then we would be compelled also to dismiss equilibria with the traditional Sraffian single dimension of indeterminacy. The view that a single dimension of indeterminacy obtains across modeling environments therefore cannot be upheld.

The Sraffian theory of rent often considers cases where a given type of land is used in only one sector out of many. A different rationale is then available for concluding that a single dimension of indeterminacy will be the norm. Let $N > 1$ and suppose some good j is produced by single-output activities that use various types of land in addition to labour and material inputs. We consider the simplest scenario, extensive rent, where each type of land is used by only one activity (Sraffa, 1960, ch. 11; Quadrio-Curzio, 1980; for more general models, see Salvadori, 1986, and Bidard, 2004, ch. 17). We assign each land type the same index as the activity that uses it. Renormalizing again so that activities produce one unit of output when run at the unit level, the profit-maximization conditions for the production of j appear as

$$p_j \leq (1+r)(p_1 a_{1i} + \cdots + p_N a_{Ni}) + w\ell_i + \rho_i \Lambda_{ii},$$

$$(2.8)$$

for each activity i that produces j and where equality holds if $y_i > 0$. The market-clearing requirements (2.2) for land types are assumed to hold. Let the remainder of the economy's sectors satisfy the assumptions of the basic Sraffa model: for each good $k \neq j$, one single-output activity that uses no land is available and in operation. To avoid the distraction of feedback from changes in p_j into

j's cost of production, we assume that j does not directly or indirectly enter into the production of any other good. We pick some good besides j as numéraire and fix w. Finally, letting c_i denote the non-land cost of production $(1+r)(p_1 a_{1i} + \cdots + p_N a_{Ni}) + w\ell_i$ for an activity i that produces j, we assume that w and technology coefficients are such that no ties occur among the c_i: if i and m are distinct activities that produce j then $c_i \neq c_m$.

As in the previous $N = 1$ example, the presence of an extra dimension of indeterminacy will depend on the extent of production. But since $N > 1$ we may view the extent of the production of j as a consequence of the demand for j rather than of different levels of accumulation. As increases in demand progressively raise p_j the economy will first use the type i land for which c_i is lowest. When this land type is exhausted p_j must rise further, until the type i for which c_i is second lowest earns the rate of return r, at which point production can expand further. And so on. The supply 'function' thus consists of steps where the 'horizontals' indicate that some type i is partly but not fully utilized and the 'verticals' that a set of land types is fully utilized but that p_j is not yet high enough for the lowest cost of the remaining types to be drawn into production.

On the horizontals the standard one dimension of indeterminacy obtains. If l types of land are used to produce j, the single zero-profit equality in (2.4) for p_j is replaced by l equalities from (2.8). But the additional $l - 1$ equalities are matched by $l - 1$ additional endogenous rental rates – the one land type that is partly utilized is constrained to have a 0 rental rate. Hence, since (2.4) and (2.3) generate a single dimension of indeterminacy, so do the horizontal equilibria. On the verticals an additional dimension of indeterminacy appears: with l types of land in use, $l - 1$ additional zero-profit equalities are again present, but now, since the last type of land to be brought into production is no longer constrained to have a 0 rental rate, there are l additional endogenous factor prices.

The Sraffa literature concentrates on the horizontals rather than the verticals, in line with the Sraffian tradition of taking demand to be exogenous. If the demand for j were completely inelastic – unresponsive to price – then it would be an unlikely accident if this inelastic demand happened to coincide with one of the verticals of the supply function. The horizontals have the added advantage for Sraffians that any small shift in demand will leave the equilibrium at the same step and hence have no effect on any price. From the neoclassical point of view, completely inelastic demand seems far-fetched and does not obtain even when agents have Leontiev utilities. An inelastic demand argument for the horizontal equilibria also sometimes has no bite. When $N = 1$ there is no division of demand into separate outputs; only an inelastic accumulation of the produced input can then allow escape from an extra dimension of indeterminacy.

But in the $N > 1$ case and if we grant an elastic demand function for j, the additional indeterminacy of the vertical equilibria is hardly a reason for worry. More land is brought into cultivation because of demand-led increases in p_j; at a vertical equilibrium the demand for j therefore can pin down p_j and determine each rental rate. While the vertical equilibria dash the Sraffian hope to show that demand is locally irrelevant for price determination, they have no broader significance. The ease with which demand disposes of additional dimensions of indeterminacy underscores the pressing need for demand functions in the Sraffa model; without explicit demands, we will never be able to check whether any apparent case of indeterminacy is the genuine article.

Outside of our attention to extra dimensions of indeterminacy, the above account of rent stays close to Sraffa (1960). Sraffa pays heed to the supply-and-demand restrictions on rental rates: he complies with the rule that factors in excess supply must have a zero price and argues that when the scale of production expands the price of a scarce factor used in production should increase. While Sraffa applies these principles only to land, they pertain to labour as well, as we will see in Exhibit C.

Exhibit B: joint production
The simplest case of joint production occurs when $N = 2$ and there is no land. Profit maximization then requires

$$p_1 b_{1i} + p_2 b_{2i} \leq (1 + r)(p_1 a_{1i} + p_2 a_{2i}) + w\ell_i,$$
$$(2.9)$$

for each activity i and with equality if $y_i > 0$. We again consider two types of equilibria. In the first type, one activity i is in use and each of the unused activities satisfies (2.9) with strict inequality. Then just one equality constrains the four prices $(p_1, p_2, w, 1 + r)$, and, so, given the normalization (2.3), there are two dimensions of indeterminacy. In the second type of equilibrium, two activities are in use, in which case just the standard single dimension of Sraffian indeterminacy obtains.

Dual to the dimensions of indeterminacy in the two types of equilibria are the dimensions of possible net productions of goods. In first type, the net production in a steady state must lie in the one-dimensional cone

$$\{((b_{1i} - a_{1i})y_i, (b_{2i} - a_{2i})y_i) : y_i \geq 0\}$$

while in the second type, with activities i and j active, net production lies in the two-dimensional cone

$$\{((b_{1i} - a_{1i})y_i + (b_{1j} - a_{1j})y_j, (b_{2i} - a_{2i})y_i$$
$$+ (b_{2j} - a_{2j})y_j) : (y_i, y_j) \geq 0\}.$$

The first type of equilibrium might therefore seem implausible: if labour is inelastically supplied, then this supply determines y_i and hence pins down exactly one vector of net outputs in the one-dimensional cone. But

this restriction does not undermine the one-activity equilibria; there is ample room for the relative price p_1/p_2 to equilibrate demand to the fixed supply. While the one-activity equilibria therefore cannot be dismissed as pathological, it would seem, as in the extensive rent example in Exhibit A, that the additional indeterminacy will vanish once we introduce explicit market-clearing conditions. See the 'sheep' example in SRAFFIAN ECONOMICS for an illustration of how an economy can move from one type of equilibrium to the other as demand shifts. Many Sraffians contend that the two-activity equilibria – called 'square' since the number of activities in use equals the number of goods – are more likely (Steedman, 1976; Schefold, 1978a; 1978b; 1988; Lippi, 1979). One interesting rationale for this view, Schefold (1990), argues that, if agents always consume goods in fixed proportions, as with Leontiev utilities, then price adjustment will not be able to bring demand in line with supply in the one-dimensional cone. But if the fixed proportions of consumption vary from person to person, then a change in p_1/p_2 will have a differential effect on the scale of consumption of dissimilar agents. The ratio of demands for the two outputs will then vary with p_1/p_2 and equilibration can occur (Salvadori, 1982; 1990; Bidard, 1997; and see also the Samuelson–Schefold interchange in Bharadwaj and Schefold, 1990, and for an overview of the extensive literature Salvadori and Steedman, 1988).

Exhibit C: choice of activities
So far, we have considered how an extra dimension of indeterminacy can arise. When a choice of activities is available to produce one or more goods even the standard single dimension of Sraffian indeterminacy can disappear. Let each activity produce just one good. We suppose there is no land and consider equilibria – $(p, w, 1 + r, y)$ that satisfy (2.1) and (2.3) – where each good is produced. If we ignore whether the labour market clears, the standard single dimension of Sraffian indeterminacy will obtain and we may index equilibrium prices $(p, w, 1 + r)$ by w. For most values of w, and if we bar flukes of the production coefficients, the resulting equilibrium $(p, w, 1 + r)$ will permit exactly N activities to satisfy their 0-economic profit conditions, that is, earn exactly the rate of return r. If an additional $N + 1$st activity were required to satisfy its 0-profit condition, then the prices $(p, w, 1 + r)$ that satisfy (2.1) and (2.3) will be pinned down. Hence, with a menu of only finitely many activities available, there are only finitely many w at which $N + 1$ zero-profit equalities could be satisfied at an equilibrium obeying (2.1) and (2.3) – these are the 'switch points' at which the economy moves from one set of N cost-minimizing activities to another. Since there are only finitely many switch points, the required values for w might seem like flukes. But once we impose market clearing for labour, equilibrium can demand a switch point and Sraffian indeterminacy then disappears.

For an example, let $N = 1$ and thus by normalization $p_1 = 1$. Suppose two activities are available with input coefficients (a_{11}, ℓ_1) and (a_{12}, ℓ_2). If both activities are in use, then the profit-maximization condition (2.1) and the normalization (2.3) reduce to the two equalities

$$1 = (1 + r)a_{11} + w\ell_1, \qquad 1 = (1 + r)a_{12} + w\ell_2.$$
$$(2.10)$$

Flukes in coefficients aside, (2.10) will determine a unique $(w, 1 + r)$. In contrast, if only one activity is in use and the idle activity makes strictly negative profits (its inequality in (2.1) is strict), then the standard single dimension of Sraffian indeterminacy obtains. But now consider market clearing, which we impose on both labour and the material input. If labour is inelastically supplied in the quantity e_ℓ and if e_c is the amount of the material input accumulated each period, then full employment of the input and labour requires

$$a_{11}y_1 + a_{12}y_2 = e_c, \qquad \ell_1y_1 + \ell_2y_2 = e_\ell,$$
$$(2.11)$$

where if $y_i > 0$ then activity i satisfies its zero-profit condition with equality. Evidently equilibria with both activities in use can be robust (they are not accidents of the parameters). For typical values of the model's parameters – e_c, e_ℓ, and the a_{ij} coefficients – (2.10) and (2.11) will have a unique solution $(\bar{w}, 1 + \bar{r}, \bar{y})$ and any small variation in the parameters will through a small adjustment of $(w, 1 + r, y)$ lead to a new unique solution. So if we begin with a model that has a two-activity equilibrium $(\bar{w}, 1 + \bar{r}, \bar{y})$ that is strictly positive in each coordinate then as the model is perturbed a two-activity equilibrium will continue to be present. Marginal products for e_c and e_ℓ are also well-defined at these equilibria and equal $(w, 1 + r)$. We have taken the savings/accumulation level e_c to be exogenous, but we could let e_c be a function of the prices $(w, 1 + r)$ without affecting the robustness of the two-activity equilibria. Indeed, a two-activity equilibrium could well be the unique equilibrium – as when, for example, the accumulation level e_c is increasing in r. (That is, if $(1 + r, w)$ and $(1 + r', w')$ both satisfy (2.10) and $r > r'$ then e_c is strictly larger with $(1 + r, w)$ than with $(1 + r', w')$.) The robustness argument in no way hinges on there being a single material good. With two goods and three activities, the analogues to (2.10) to (2.11) would each consist of three variables and equations, and again indeterminacy would disappear.

The Sraffian dismissal of cases where $N + 1$ activities are in use rarely receives explicit defence. The rationale presumably is that the labour market is not required to clear, in which case there is no reason to suppose that w should equal one of the unusual values where $N + 1$ activities all earn the same rate of return.

The indeterminacy-reducing effect of factor-market clearing has already appeared. In the $N = 1$ example in Exhibit A we saw the dimension of indeterminacy drop from 2 to 1 when two activities are in use rather than one. Indeed, that example and the present example are essentially the same: r and the rental rate on land were the endogenous price variables in Exhibit A whereas r and the wage are the endogenous price variables here (w also appears in Exhibit A but we treated w as a parameter and ignored labour-market clearing). And just as in Exhibit A, the one-activity-in-use equilibrium requires a special configuration for (e_c, e_ℓ): when one activity i is in use and $(w, 1 + r) \gg 0$ we must have $e_c = e_\ell a_{1i}/\ell_i$. Similar conclusions hold when $N > 1$ (see Section 3).

Sraffians cannot have it both ways: if the case against an extra dimension of indeterminacy in the presence of land – that the special resource configurations are unlikely – is compelling, then consistency would seem to demand rejection of the single dimension of indeterminacy in the classical Sraffa setting. Of course, one may argue instead that labour unlike land is traded in a market that does not clear. But then the indeterminacy of the wage becomes an assumption rather than a conclusion: in any model where the labour market does not clear, the wage can be treated as a free parameter. We will expand on this point in the next section.

Gathering our exhibits together, we can summarize concisely what determines the extent of indeterminacy. Counting labour as an input, the dimension of indeterminacy equals the difference between the number of positively priced (hence fully utilized) inputs and the number of activities in use (see FACTOR PRICES IN GENERAL EQUILIBRIUM). In basic Sraffian indeterminacy, $N + 1$ inputs are used by N activities: hence 1 dimension of indeterminacy. In the $N = 1$ extensive-rent example with 1 activity in use, 3 inputs are used by 1 activity: 2 dimensions of indeterminacy. In the extensive-rent example where l types of land are in use and all l are fully utilized, $N + 1 + l$ inputs have a positive price and are used by $N + l - 1$ activities: again 2 dimensions of indeterminacy. In the joint production example with 2 goods, 3 inputs have a positive price: 2 dimensions of indeterminacy occur when 1 activity is in use and 1 dimension of indeterminacy occurs when 2 activities are in use. Finally, in our choice-of-activities example indeterminacy disappears when 2 positively priced inputs are used by 2 activities.

As we will now see, a general equilibrium account of factor-price indeterminacy also reports a dimension of indeterminacy equal to the difference between the number of scarce inputs and the numbers of activities in use.

3 Sraffian indeterminacy with explicit market clearing

The forces of supply and demand have been nipping at the heels of Sraffian indeterminacy: the labour market clearing requirement in Exhibit C sometimes eliminates indeterminacy, and our informal appeals to output demand functions appeared to eclipse the extra indeterminacy that arose in Exhibits A and B. So perhaps a careful inclusion of market clearing for all goods will snuff all the indeterminacy out. This turns out not to be the case.

We include labour among the markets that are required to clear. The only formal feature of labour in Sraffian models that distinguishes it from land is that homogenous labour is used in every sector whereas a specific type of land need not be. (In reality, different varieties of labour are used in different industries and some type of land is used in every industry.) So we treat labour (and stocks of other inputs) as we have previously modelled land: for a positive price to rule, demand must equal supply. Labour markets are distinctive of course – labour can require an efficiency wage to induce effort, wage contracts can serve as decades-long insurance contracts, workers can be in unions, and so forth – and perhaps these special traits lead labour markets not to clear. But if we simply exempt labour from market clearing by fiat, one purpose of the Sraffa model is undermined. Wage indeterminacy will obtain whenever the labour market does not have to clear – whether or not capital aggregates, relative prices are constant through time, or linear activities describe technology. Models that allow the labour market not to clear in effect *assume* that markets do not pin down the distribution of income; they do not demonstrate that principle.

We now distinguish explicitly between material goods when they are inputs at an earlier point in time and outputs at a later point. Two periods will be enough; material inputs will be supplied inelastically at time 1, and labour supplied and output sold at period 2. Relative prices will no longer be restricted to remain proportional through time. Relative prices that vary from period to period run counter to Sraffian tradition but are indispensable: if indeterminacy is to survive in the presence of market clearing, additional free price variables are imperative.

The prices of the N material inputs supplied at time 1 will be denoted $p^1 = (p_1^1, \ldots, p_N^1)$ while the prices of the goods sold at time 2 will be $p^2 = (p_1^2, \ldots, p_N^2)$. As in the basic Sraffa model, suppose just N activities are available, one for each produced good, and let y denote the activity levels. Output demand is given by the demand function

$$x(p^1, p^2, w, 1 + r) = (x_1(p^1, p^2, w, 1 + r), \ldots,$$
$$x_N(p^1, p^2, w, 1 + r)),$$

and the exogenous supply of labour and material inputs is given by e_ℓ and the N-vector e. Together these ingredients must obey Walras' law:

$$\frac{1}{1+r} p^2 \cdot x(p^1, p^2, w, 1 + r) = \frac{1}{1+r} w e_\ell + p^1 \cdot e.$$

An equilibrium at which $(p^1, p^2, w, 1 + r, y) \gg 0$ satisfies

$$p_i^2 = (1 + r)(p_1^1 a_{1i} + \cdots + p_N^1 a_{Ni}) + w\ell_i,$$
$$i = 1, \ldots, N, \tag{3.1}$$

$$x_i(p^1, p^2, w, 1 + r) = y_i, \qquad i = 1, \ldots, N, \tag{3.2}$$

$$a_{j1} y_1 + \cdots + a_{jN} y_N = e_j, \qquad j = 1, \ldots, N, \tag{3.3}$$

$$\ell_1 y_1 + \cdots + \ell_N y_N = e_\ell, \tag{3.4}$$

$$p_1^1 + \cdots + p_N^1 = 1, \tag{3.5}$$

$$p_1^2 + \cdots + p_N^2 = 1. \tag{3.6}$$

There are two normalizations, (3.5) and (3.6), since the model uses an interest rate r rather than present value prices. The explicit inclusion of demand ensures that the model takes into account the indeterminacy-reducing effect of demand that we saw in Exhibits A and B.

The equilibria described by the above equalities typically will exhibit indeterminacy. If we fix y at an equilibrium value, then the market-clearing equalities for the material inputs and labour, (3.3) and (3.4), will remain satisfied as $(p^1, p^2, w, 1 + r)$ varies. Of the remaining $2N + 2$ equalities, one of the remaining market-clearing or zero-profit equalities is redundant due to Walras' law. But since the remaining $2N + 1$ equalities have the $2N + 2$ endogenous variables $(p^1, p^2, w, 1 + r)$, one dimension of indeterminacy will typically obtain. The qualification 'typically' is necessary because a rank condition must hold in order to prove indeterminacy via the implicit function theorem (Mandler, 1999).

Several points give the above reasoning a Sraffian flavour. First, just as in Sraffa's book, aggregate quantities remain fixed and hence the indeterminacy operates on prices alone (though as the prices consistent with the fixed aggregate quantities change, individual incomes and individual consumption vary). Of course, unlike Sraffa, we know that markets clear at the fixed aggregate quantities. Second, linear activities are essential. With a differentiable technology a given vector of aggregate quantities would be incompatible with multiple equilibrium price vectors (Mandler, 1997). Third, it is no accident that the present supply-and-demand model and the basic Sraffian model both display a single dimension of indeterminacy. The degree of indeterminacy is the same because first, with inelastic input supply the market-clearing conditions for inputs do not restrict prices, and second, the $N - 1$ independent market-clearing equalities for output are exactly counterbalanced by the $N - 1$ new second-period prices (we lose one price due to the added normalization (3.6)). This match between added equilibrium conditions and added prices variables means that if we open the door to the variations considered in the previous section – inelastically supplied land, joint production, choice of activities – the dimensions of indeterminacy of the two approaches will still coincide. In both models, the dimension will equal the difference between the number of inelastically supplied first-period factors that have a positive price and the number of activities in use.

Indeterminacy therefore need not always obtain when the Sraffa price equations are embedded in a supply-and-demand model. If more than one activity per produced good is available, then equilibria may have $N + 1$ rather than N activities in use. This possibility should come as no surprise since the $N = 1$ no-indeterminacy example in Exhibit C had two activities in use. Indeed if $N = 1$ and we introduced choice among activities, the present supply-and-demand model would be exactly the model in Exhibit C (the market-clearing equality omitted in Exhibit C is superfluous by Walras' law and (3.5)–(3.6) imply $p^1 = p^2$). Despite their neglect in the Sraffian literature, equilibria where the number of activities in use exceeds the number of produced goods are perfectly plausible if all factor markets are required to clear.

To complete the case for the compatibility of Sraffian indeterminacy and market clearing, we must deal with a famous counterargument that with overwhelming probability any equilibrium will have at least as many activities in use as positively priced factors (Mas-Colell, 1975; Kehoe, 1980) – we faced a similar argument in Exhibit A. Conditions (3.3) and (3.4) consist of $N + 1$ equalities in the N unknowns y; hence for almost every endowment (e, e_ℓ) there will exist no y that obeys these equalities. Consequently for these generic endowments there will be no equilibria satisfying (3.1)–(3.6). If only these N activities are available, then at the generic endowments one of the material inputs or labour will be in excess supply and have a 0 price. Hence for generic endowments the number of positively priced factors will not exceed the number of activities in use. But the seemingly unusual endowments (e, e_ℓ) at which (3.3) and (3.4) do have a solution can arise systematically (see FACTOR PRICES IN GENERAL EQUILIBRIUM and Mandler, 1995). The material endowments e are the outcome of past savings–investment decisions; agents will not knowingly accumulate so much of a material input that it ends up in excess supply. Even when resources can be used productively no matter how great their supply, the endowments where $N + 1$ resources are used by N activities can still arise – those endowments appear at kinks on the production possibilities frontier (for example, see Figure 1 where the material input is accumulated just to the point where the available land is fully utilized using only one activity). This view of capital as a set of accumulated

goods rather than a random endowment of nature fits well with Sraffa's view of production as circular.

We have had to do some damage to the Sraffian tradition to embed its indeterminacy claims in a market-clearing model. Inelastic factor supply finds no echo in the Sraffa literature. Not all factors have to be supplied inelastically for indeterminacy to obtain – it would be enough if some subset of k factors with positive prices were supplied inelastically and were used by fewer than k activities – but some must be.

More heretical from the Sraffian point of view, we have had to let relative prices vary across time periods. Time-varying prices allow output prices to clear the output markets without constraining input prices. The $N = 1$ case obscures this feature of the indeterminacy since then there are no relative output prices. But when $N > 1$ relative prices will be constant through time in a market-clearing model only if the economy is in a steady state, and steady states typically are determinate. For example, an overlapping generations model, even with a linear activity analysis technology, has locally unique steady states (Mandler, 1999). Indeed steady states will be determinate in virtually any model where markets clear and saving responds to the rate of return (including Marxian models where investment is increasing in the profit rate). There are two reasons. First, in any given period of a steady state, the prices of that period's given stocks of producible factors are constrained to equal the prices of the same factors being produced for the next period; the indeterminacy arguments we gave earlier therefore cannot be applied. Second, for factors that are not produced, endowment levels then *should* be seen as random parameters, and it would therefore be a fluke if some set of k such factors were used by fewer than k activities without one of them being in excess supply.

4 Sraffian instability

One may also read the Sraffian critique of neoclassical economics as arguing that the failure of capital to aggregate can lead the savings–investment market to be unstable. The case for instability relies on 'reverse capital deepening', where the ratio of the value of an economy's capital goods relative to the number of workers employed increases as a function of the interest rate. Consider a constant-relative-price Sraffa model with one or more activities available to produce each good, no joint production and hence no fixed capital, and where the economy is in a steady state. Set some consumption good to be the numéraire. Then, if we fix the economy's vector of outputs per worker, the ratio of the value of capital to labour is well-defined at any given r. An increase in r can affect the value-of-capital to labour ratio in two ways. First, for any produced good j, a 'real effect' can change which activity produces j at lowest cost, which in turn will alter the vector of capital goods per worker used in the production of j. Second, even if the activities in use remain the same (and with the composition of output still fixed), a change in r will affect the relative prices of capital goods. This 'price effect' will typically change the value of the capital goods each worker uses. So although the real effect of an increased interest rate might lead to a decrease in the quantity of each capital good used per worker, the price effect can cause the value of capital per worker to rise; hence reverse capital deepening can occur. This possibility can appear in an economy with one consumption good and just one capital good. An increase in the interest rate will then lower the physical capital to labour ratio, but the price of the capital good can increase by enough that the ratio of the value of capital to labour rises (Bloise and Reichlin, 2009). Since this chain of events can happen with a single capital good, it is, strictly speaking, misleading to identify reverse capital deepening with the failure of capital goods to aggregate.

Here is the instability scenario. To ensure that Sraffian 'long-run' prices rule, we must assume that before and after a shock the economy is in a steady state, where relative prices are constant. If for simplicity there is no fixed capital, the steady state assumption implies that the value of capital per worker equals investment per worker. Consequently, with reverse capital deepening, investment per worker can exceed savings per worker when the interest rate is above its equilibrium value, thus pushing the interest rate higher, further away from equilibrium. Similarly, investment per worker can fall short of savings per worker when the interest rate is below equilibrium, pushing the interest rate lower.

The difficulty with this reasoning is that there is no market where long-run or steady-state savings and investment meet. When a shock to savings or investment occurs, the only instability that could undermine the economy must appear in markets at some specific set of dates – presumably the markets concurrent with the shock. Those markets, however, can equilibrate only at the out-of-long-run-equilibrium prices that obtain when the economy with the pre-shock endowments of resources begins its transition to a new post-shock long run. Long-run prices are therefore of dubious pertinence to the stability issue.

Garegnani (2000) and Schefold (2005) have argued that Sraffian instability surfaces in market-clearing general equilibrium models as a failure of tâtonnements to converge to equilibrium prices. This innovation in the Sraffian agenda has cleared away the cobwebs from the well-rehearsed interchange where Sraffians grouse that capital goods do not aggregate and Walrasians reply that equilibria in the Arrow–Debreu model exist.

The traditional model of a tâtonnement does not apply to an economy with linear activities: whenever positive economic profits can be earned, firms will want to expand without bound, and hence excess demands and tâtonnement price adjustments will be ill-defined. The most detailed and specific proposal to embed Sraffian

instability in a general equilibrium model, Schefold (2005), steps around this problem by assuming that output prices are always set so that no activity makes positive profits, and letting the tâtonnement operate only on factor prices, which are the primary object of interest. Given a vector of factor prices, one may calculate the prices for outputs that minimize costs and then the consumer demand for outputs that result from these factor and output prices. The profit-maximizing decisions of firms, assuming they produce these output levels, then determine factor demands, and the difference between factor demand and factor supply leads to a tâtonnement price adjustment. Even in this setting, factor demand can be multi-valued since there can be many cost-minimizing factor combinations that produce any given output vector. To tackle this problem, one may define the tâtonnement with a differential inclusion rather than a differential equation (Mandler, 2005, and for differential inclusions generally see Aubin and Cellina, 1984).

Goods are distinguished by the date at which they appear and are of two types, *factors* which are not produced and *outputs* which can be produced. Factors can include the initial period's stock of capital goods as well as various types and dates of labour, land, and raw materials. Technology is given by a matrix of linear activities A where each activity (a column of A) produces only one output but may use any of the non-produced factors and produced goods as inputs. We follow the standard sign convention where positive entries in A denote outputs and negative entries indicate inputs, index goods so that outputs come first and factors second, and assume that positive quantities of all outputs can be produced simultaneously. To permit intermediate goods, an output can have negative as well as positive entries in A. Let A_o denote the output rows of A and let A_f denote the factor rows of A. For an arbitrary vector of factor prices p_f, we may find the competitive output prices $p_o(p_f)$ by solving the cost minimization problem $\min_y - p_f A_f y$ subject to $A_o y \geq (1, \ldots, 1)$, $y \geq 0$, and setting $p_o(p_f)$ equal to the Lagrange multipliers at a solution to this problem. With output prices set in this way, consumers' output and factor excess demands become functions of p_f alone. Let $x_o(p_f)$ and $x_f(p_f)$ denote these functions, which we assume obey Walras' law. The demand for factors by firms is a x_f in the set

$$X_f(p_f) = \{x_f : x_f \text{ maximizes } (p_o(p_f), p_f) \cdot (x_o(p_f), x_f)$$
$$\text{subject to } (x_o(p_f), x_f) = Ay, y \geq 0\}.$$

As we mentioned, $X_f(p_f)$ may have multiple elements.

An equilibrium is a p_f such that $x_f(p_f) \in X_f(p_f)$. A *factor tâtonnement* is then a function $p_f(t)$, differentiable almost everywhere, such that when differentiable there is a $x_f \in X_f(p_f(t))$ with

$$\dot{p}_f(t) = x_f(p_f(t)) - x_f.$$

Due to the sign convention governing A and since $x_f(p_f)$ is an excess demand, $x_f(p_f)$ and x_f will both usually be negative.

Since the factors can appear at any date, the model can cover a classical Sraffa economy with just labour and produced goods as inputs, so long as the economy is finite-lived: the factors subject to the tâtonnement would consist of the initial period's stocks of capital goods and labour at all dates, while all capital goods that appear after the initial period would be classified as outputs.

While our initial story of reverse-capital-deepening instability was driven by responses of the value of capital to the interest rate, comparable stories are possible that refer just to the non-produced factors that appear in the above factor tâtonnement. Suppose in the classical Sraffa setting that some bundle of activities is cost-minimizing at both low and high r's and some other set of activities is cost-minimizing at intermediate r's (this is called 'reswitching'). And suppose further that a steady state with a small labour supply will use one of these sets of activities and a steady state with a large labour supply will use the other set. Then, if the economy initially is in a steady state at either a high or intermediate r and has the small labour supply, an exogenous shift to the large labour supply would lower r in a new steady state and hence raise w, hardly the intuitive price response to a supply increase. This tale compares steady states and tracks the movement of the wage through time, whereas all prices in a factor tâtonnement respond simultaneously to disequilibrium. Schefold (2005) nevertheless suggests that reswitching can lead a factor tâtonnement to be unstable.

Evaluation of this claim faces an immediate difficulty: no matter how well-behaved firms' factor demands are, consumer behaviour, which here appears as the factor supply function $x_f(p_f)$, can by itself lead to instability. To block this path, let us assume that demand obeys the weak axiom, the traditional tool used in general equilibrium theory to tame an exchange economy's demand function and ensure tâtonnement stability. In the present setting the weak axiom states that, for any p_f and p_f', $p_o(p_f') \cdot x_o(p_f) + p_f' \cdot x_f(p_f) \leq 0$ and $(x_o(p_f'), x_f(p_f')) \neq (x_o(p_f), x_f(p_f))$ imply $p_o(p_f) \cdot x_o(p_f') + p_f \cdot x_f(p_f') > 0$. Just as in an exchange economy, the weak axiom implies that a factor tâtonnement is stable (Mandler, 2005). Thus, no matter how many potential capital theory paradoxes are packed into the technology, if price adjustments are guided by excess demand and demand obeys the weak axiom, stability obtains. In fact, if the weak axiom is satisfied then in a factor tâtonnement the distance between the out-of-equilibrium prices that the auctioneer calls out and any equilibrium price vector will decline monotonically.

A tâtonnement is a highly artificial model of how an economy responds to disequilibrium: price adjustments are governed by 'notional' consumer demands, which

cannot be satisfied at nonequilibrium prices, rather than by rationed or constrained demands that could be. On top of this problem, an intertemporal tâtonnement requires the prices of goods that appear at different time periods to adjust simultaneously. Perhaps in a more realistic setting the paradoxes of capital theory will turn out to be a distinct source of instability – but the case remains to be made.

5 Back to growth theory

Sraffa saw the economy as embedded in time, with endowments of produced inputs determined by the accumulation of capital, and he modelled technology using the plausible primitive of linear activities, not production functions packaged with suspicious differentiability assumptions designed to make factor returns determinate. These points add up to an effective criticism of a supply-and-demand theory of factor pricing, but the details of the argument need to be rearranged. The impossibility of capital aggregation plays no role.

But does the Sraffian stress on capital aggregation at least serve as an effective criticism of growth theory and the parables of the Solow model? On the surface, the fact that in a comparison of steady states the value of capital per worker or consumption per worker can increase with the interest rate may appear to undermine boilerplate neoclassical maxims on how to allocate resources through time. For example, it might seem that increases in savings could raise interest rates and lower future consumption. Unfortunately, as in the analysis of stability, the Sraffian focus on steady-state comparisons and on the value of capital per worker misleads. The move from one steady state to another involves the adjustment of myriad individual consumption and savings decisions at multiple points in time: consequently the impact of a change in savings today on steady-state consumption can diverge from the impact on consumption at a specific future date with all other consumption levels held constant. And when capital goods and consumption goods are separate commodities, changes in the relative prices of capital goods can break the linkage between increases in savings – sacrifices of present consumption – and increases in the value of capital; thus an increase in savings that lowers r *and* the value of capital per worker is not remarkable.

Following Solow (1963), define an economy's rate of return between the present and some future date t as the return in consumption at t as present-day consumption is sacrificed. If there is one consumption good per period, the gross rate of return between the present and t is the ratio of the gain in consumption at t relative to the quantity of today's consumption forgone, holding consumption at all other dates fixed. No reference to the value of capital is involved. With this definition, the familiar neoclassical maxims reappear: with linear activities and free disposal production possibilities sets are convex and a sacrifice of consumption today must lead to a weak increase in consumption at t (holding all other consumption levels fixed) and the rate of return between today and t must weakly diminish in the quantity of consumption sacrificed. Comparisons of steady states can also be cleansed of references to the value of capital, but here a less than impressive set of claims is available. If again there is one consumption good per period and we avoid settings with infinitely many agents, such as the overlapping generations model, then any increase in steady-state consumption per worker entails an increase of the amount of at least one of the capital goods used per worker, and the move to a steady state with higher consumption per worker requires a sacrifice of consumption at some set of dates prior to arrival at the new steady state. The no-free-lunch moral of neoclassical growth theory rears its head.

But beneath these broad conclusions lies a vein of caveats, so far unmined by the Sraffian movement. When technology is described by linear activities, the rate of return can differ depending on whether it is defined by decreases or increases in present consumption. The reason is that the production possibilities set may well exhibit a kink at precisely the consumption levels chosen either by private agents in a market economy or by a benevolent planner who maximizes a sum of utilities (see Figure 1, but read the axes as present consumption and date t consumption). In the market-economy case, an exact match between the market interest rate r and the rate of return can then fail to obtain, though r must lie between the lower bound given by the rate of return for small increases in present consumption and the upper bound given by the rate of return for small decreases. In the planning case, the mismatch is between agents' intertemporal rates of substitution and the technological rate of return. Neoclassical growth theory has largely ignored these discrepancies, perhaps because of the blinders of its long reliance on differentiable production functions. But as in static factor pricing, linear activities in a growth setting open up a conceptual gap between prices and material rates of return: r no longer has to align with the physical return to sacrifices in consumption.

The presence of a kink in a production possibilities set will hinge on the number of activities in use, just as with factor-price indeterminacy. An irony crops up here: it is only in an optimal growth exercise that maximizes agents' utilities that an allocation at which the production possibilities set is kinked normally would be selected. Consider a planner with access to linear activities and stocks of resources at dates from the present (period 1) to the distance future (period T) that can be used to produce a sequence of consumption levels (again one consumption good per period). Resources are inelastically supplied and an arbitrary number of intermediate capital goods is permitted. Any consumption sequence $\bar{x} = (\bar{x}_1, \ldots, \bar{x}_T)$ that is on the economy's PPF must satisfy the property that, for any $t = 1, \ldots, T$, $\bar{x}_t$ solves the problem of maximizing x_t subject to $(x_t, (\bar{x}_i)_{i \neq t})$ being in

the production possibilities set. Pick one such problem with $t > 1$ where, if we view $\bar{x}_1$ as a parameter, the solution $x_t(\bar{x}_1)$ is non-constant. Consider those activities in use at some initial solution whose usage levels change as $\bar{x}_1$ changes. If no subset of k of these activities utilizes or produces more than k goods at the initial optimum, then (barring flukes of the production coefficients) the function $x_t(\bar{x}_1)$ will be differentiable at the initial $\bar{x}_1$: the PPF is smooth. (The good x_t should not be included in the count of the number of goods utilized or produced.) In the remaining cases, the initial $\bar{x}_1$ is a point where $x_t(\bar{x}_1)$ is not differentiable and the PPF is kinked; here inputs are accumulated just to the point where some set of k activities uses or produces more than k goods. Since almost every $\bar{x}$ on the PPF does not sit at a kink, a planner who selects a consumption stream arbitrarily could safely ignore the nondifferentiable points and declare that given the selection $\bar{x}$ the forces of technology alone determine the marginal rate of intertemporal transformation between any two time periods. If, furthermore, this planner decentralized the economy's investment decisions to private entrepreneurs the planner would have to choose this rate as the market rate of interest. Curiously, the Sraffian hostility to utility maximization and substitution in consumption (see Exhibits A and B) also leads to the conclusion that a $\bar{x}$ at a PPF kink is an unusual event; hence the Sraffian view comes to the aid of the neoclassical identification of interest rates with rates of technological transformation. On the other hand a planner who maximizes a sum of agent utilities could well choose a consumption stream at a kink (see Exhibit A and Figure 1). While the Sraffian emphasis on linear activities serves as a welcome corrective to the neoclassical habit of assuming that any function or surface is differentiable, in the end it is utility functions that prevent a linear activities growth model from providing a purely technological determination of the interest rate.

As we have seen, this lesson goes beyond growth theory. Market economies also gravitate to kinks on PPFs. Consequently, even when an economy has a single consumption good per period, which allows consumption output to be modelled with an aggregate production function, that production function may well not be differentiable when evaluated at the factor endowments that arise in equilibrium. Empirical work that relies on a differentiability assumption – for example, the classical growth-accounting estimates of total factor productivity (Solow, 1957; Kendrick, 1961; Denison, 1962) – is therefore subject to coherent Sraffian criticism.

6 Conclusion

The Sraffian insistence on linear activities casts critical light on the instinctive neoclassical habit of assuming that interest rates and marginal rates of transformation must be equal and that production functions must be differentiable. Another more abstract Sraffian principle proves just as illuminating: economic activity is ongoing, not a one-time exchange among agents with disparate endowments and preferences. As we have seen, it is when the endowments of capital goods are determined by rational accumulation rather than by chance that factor-price indeterminacy can appear.

The Sraffian view of equilibrium, which revives earlier classical ideas, fits well with subsequent mainstream developments. Modern macroeconomics, both new Keynesian and new classical, has rejected models where the government's actions, such as an aggregate demand stimulus, systematically surprise agents; instead government actions are governed by a distribution that agents know. The new understanding of expectations is not driven by a belief that agents are never surprised or never hold an incorrect model of the economy but in order to pinpoint results that are immune to invalidation as agents learn and adapt to their environment, a precept close to the Sraffian view of equilibria as ongoing. The Sraffian perspective has wide application. For example, while the production of capital endowments by past equilibrium activity can lead to factor-price indeterminacy, a similar dependence of the present on past equilibrium decisions can eliminate some disturbing features of other brands of indeterminacy (Mandler, 2002). In overlapping-generations indeterminacy, market clearing is compatible with agents at the beginning of economic time unanimously anticipating any future price path that lies in a multidimensional set. But if one sees an equilibrium as ongoing, rather than commencing anew each period, the indeterminacy problem disappears after an equilibrium gets under way. The agents in an economy that has already followed an anticipated price path for a number of periods and that experiences no shock will continue to hold their previously formed expectations, and given those expectations equilibrium prices at each period will be locally unique.

MICHAEL MANDLER

See also **capital theory (paradoxes); determinacy and indeterminacy of equilibria; factor prices in general equilibrium; general equilibrium; neo-Ricardian economics; Sraffian economics.**

Note on the dimension of indeterminacy of the basic Sraffa model

We may rewrite (2.4) as $p(I - (1 + r)A) = w\ell$, where A is the matrix whose ith column is a_i and $\ell = (\ell_1, \ldots, \ell_N)$. Due to the homogeneity of (2.4) in (p, w), we can replace (2.3) with $w=1$ without changing the relative prices $\frac{1}{\|p\|}p$ in any solution $(p, w, 1 + r)$ to (2.3) and (2.4) or changing the dimension of the set of solutions. If at a solution $(\bar{p}, \bar{w}, 1 + \bar{r}) \gg 0$ to $p(I - (1 + r)A) = \ell$, $I - (1 + \bar{r})A$ has rank N, then, for r near $\bar{r}$, $p = \ell(I - (1 + r)A)^{-1}$ solves $p(I - (1 + r)A) = \ell$. Hence locally there is a one-dimensional set of solutions.

If $I - (1 + \bar{r})A$ has rank $<N$, then $\{p : p(I - (1 + \bar{r})A) = \ell\}$ has dimension ≥ 1, and so locally the solution set contains a set of dimension ≥ 1.

Bibliography

Aubin, J. and Cellina, A. 1984. *Differential Inclusions*. Berlin: Springer-Verlag.

Bharadwaj, K. and Schefold, B., eds. 1990. *Essays on Piero Sraffa*. London: Unwin Hyman.

Bidard, C. 1997. Pure joint production. *Cambridge Journal of Economics* 21, 685–701.

Bidard, C. 2004. *Prices, Reproduction, and Scarcity*. Cambridge: Cambridge University Press.

Bloise, G. and Reichlin, P. 2009. An obtrusive remark on capital and comparative statics. *Metroeconomica*.

Denison, E. 1962. *The Sources of Economic Growth in the United States*. Washington, DC: Committee for Economic Development.

Eatwell, J. 1982. Competition. In *Essays in Classical and Marxian Political Economy*, ed. I. Bradley and M. Howard. London: Macmillan.

Garegnani, P. 1976. On a change in the notion of equilibrium in recent work on value and distribution. In *Essays on Modern Capital Theory*, ed. M. Brown, K. Sato and P. Zarembka. Amsterdam: North-Holland.

Garegnani, P. 2000. Savings, investment, and capital in a system of general intertemporal equilibrium. In *Critical Essays on Piero Sraffa's Legacy in Economics*, ed. H. Kurz. Cambridge: Cambridge University Press.

Hahn, F. 1982. The neo-Ricardians. *Cambridge Journal of Economics* 6, 353–74.

Harcourt, G. 1974. The Cambridge controversies: the afterglow. In *Contemporary Issues in Economics*, ed. M. Parkin and A. Nobay. Manchester: Manchester University Press.

Kehoe, T. 1980. An index theorem for general equilibrium models with production. *Econometrica* 48, 1211–32.

Kendrick, J. 1961. *Productivity Trends in the United States*. Princeton: Princeton University Press.

Kurz, H. and Salvadori, N. 1995. *Theory of Production: A Long-period Analysis*. Cambridge: Cambridge University Press.

Lippi, M. 1979. *I prezzi di produzione*. Bologna: Il Mulino.

Mandler, M. 1995. Sequential indeterminacy in production economies. *Journal of Economic Theory* 66, 406–36.

Mandler, M. 1997. Sequential regularity in smooth production economies. *Journal of Mathematical Economics* 27, 487–504.

Mandler, M. 1999. Sraffian indeterminacy in general equilibrium. *Review of Economic Studies* 66, 693–711.

Mandler, M. 2002. Classical and neoclassical indeterminacy in one-shot versus ongoing equilibria. *Metroeconomica* 53, 203–22.

Mandler, M. 2005. Well-behaved production economies. *Metroeconomica* 56, 477–94.

Marglin, S. 1984. *Growth, Distribution, and Prices.* Cambridge, MA: Harvard University Press.

Mas-Colell, A. 1975. On the continuity of equilibrium prices in constant-returns production economies. *Journal of Mathematical Economics* 2, 21–33.

Quadrio-Curzio, A. 1980. Rent, income distribution, and orders of efficiency and rentability. In *Essays on the Theory of Joint Production*, ed. L. Pasinetti. London: Macmillan.

Salvadori, N. 1982. Existence of cost-minimizing systems within the Sraffa framework. *Zeitschrift für Nationalökonomie* 42, 281–98.

Salvadori, N. 1986. Land and choice of techniques within the Sraffa framework. *Australian Economic Papers* 25, 94–105.

Salvadori, N. 1990. Comment. In *Essays on Piero Sraffa*, ed. K. Bharadwaj and B. Schefold. London: Unwin Hyman.

Salvadori, N. and Steedman, I. 1988. Joint production analysis in a Sraffian framework. *Bulletin of Economic Research* 40, 165–95.

Schefold, B. 1978a. Multiple product techniques with properties of single product systems. *Zeitschrift für Nationalökonomie* 38, 29–53.

Schefold, B. 1978b. On counting equations. *Zeitschrift für Nationalökonomie* 38, 253–85.

Schefold, B. 1988. The dominant technique in joint production systems. *Cambridge Journal of Economics* 12, 97–123.

Schefold, B. 1990. On changes in the composition in output. In *Essays on Piero Sraffa*, ed. K. Bharadwaj and B. Schefold. London: Unwin Hyman.

Schefold, B. 1997. *Normal Prices, Technical Change, and Accumulation.* London: Macmillan.

Schefold, B. 2005. Reswitching as a cause of instability of intertemporal equilibrium. *Metroeconomica* 56, 438–76.

Solow, R. 1957. Technical changes and the aggregate production function. *Review of Economics and Statistics* 39, 312–20.

Solow, R. 1963. *Capital Theory and the Rate of Return.* Amsterdam: North-Holland.

Sraffa, P. 1960. *Production of Commodities by Means of Commodities.* Cambridge: Cambridge University Press.

Steedman, I. 1976. Positive profits with negative surplus value: a reply to Wolfstetter. *Economic Journal* 86, 873–6.

s-S models

The s-S model is the canonical model of inaction arising from costs of adjustment. Individuals do not always react to changes in their environment. Consumers rarely buy a new house or car after every fluctuation in their permanent income. Firms often leave prices fixed for months even though information is arriving at a much greater frequency. Fixed costs of adjustment provide a natural explanation of this inertial behaviour. If an agent faces a fixed cost to taking some action and if the loss to non-adjustment is small in the neighbourhood of the optimal choice, then it will pay to leave things be until the benefits of adjustment exceed the costs. Bar-Ilan and Blinder (1996) call this the 'optimality of usually doing nothing'.

The term s-S derives from inventory theory. In Arrow, Harris and Marschak's (1951) seminal paper, a firm allows its inventory holdings to decline below a level s before placing an order that replenishes inventories to a level S. Subsequently, the term s-S has come to denote an entire class of models of discrete and infrequent adjustment in which the optimal strategy is characterized by a set of triggers and targets. s-S models have been applied to explain inertia in a variety of microeconomic settings, including money demand, cash management, pricing, durable goods, and investment. In macroeconomics, the principal application has been to provide microfoundations for price stickiness and thereby the real effects of money. This is the menu cost model of price stickiness.

Microeconomics: the basic idea

The hallmark of the s-S policy is the combination of inaction and discrete adjustment. The basic idea can be simply illustrated in a static setting. Consider an agent who must choose x to minimize some twice differentiable, concave payoff function $\pi(x)$. The agent is endowed with a value x_0. The wrinkle is that there is a fixed cost k to changing x from its initial value. The optimal policy which balances the costs and benefits of adjustment is illustrated in Figure 1. If x_0 is less than S_L or greater than S_H, the benefit of adjusting to S^* outweighs the fixed cost of adjustment k. If x_0 is between S_L and S_H, inaction is optimal, and consequently $[S_L, S_H]$ is referred to as the range of inaction. The points S_L and S_H are, respectively, the upper and lower adjustment triggers. S^* is the adjustment target.

The combination of inaction and discrete adjustment is a direct result of the fixed cost of adjustment k. If instead the cost of adjustment were some twice differentiable convex function $c(x - x_0)$ with $c(0) = c'(0) = 0$, then there would be no inaction. Since the marginal cost of adjustment is zero at x_0, it would always be optimal to move closer to S^*.

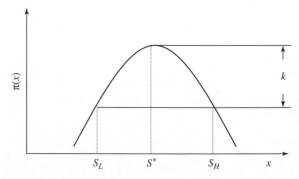

Figure 1   An s-S policy

One of the immediate lessons of Figure 1 is that the range of inaction $[S_L, S_H]$ may be large even if the fixed cost k is small. This is because the payoff function is normally very flat in the neighbourhood of S^*. Formally, if k is small we may apply Taylor's theorem and approximate $\pi(x)$ in the neighbourhood of S^* by a quadratic $\pi(S^*) + \frac{1}{2}\pi''(S^*)(x - S^*)^2$. Recall S^* is the optimal choice so the first-order term is zero. S_L and S_H are the points at which this payoff function is equal to $\pi(S^*) - k$. Solving for S_L and S_H yields

$$|S_L - S^*| = |S_H - S^*| = \sqrt{\frac{k}{-\frac{1}{2}\pi''(S^*)}}.$$

The size of the range of inaction is increasing in the adjustment cost and decreasing in the concavity of the loss function. Here second order adjustment costs are sufficient to generate a first-order range of inaction (Mankiw, 1985).

One-sided rules

Adding exogenous forces that act on x makes the problem dynamic. If the increments to x in the absence of adjustment are Markov, then the optimal trigger-target strategy will be stationary. The literature considers two cases. The first is monotonic drift in x. This leads to 'one-sided' s-S rules. Agents let x drift until it passes a trigger s and then reset it to some value S. One-sided rules arise naturally in inventory theory (Arrow, Harris and Marschak, 1951), money demand (Baumol, 1952) and pricing under inflation (Sheshinski and Weiss, 1977). Scarf (1959) provides a general proof of the optimality of one-sided s-S policies, requiring only that π is concave.

The canonical model is due to Sheshinski and Weiss. In their model, time is continuous and indexed by t. They interpret $x = \ln p_i - \ln p$ as the log difference between a firm's price p_i and a price index p. Payoffs therefore depend on the firm's relative price p_i/p. They assume that the price index grows at rate g and that the firm discounts future payoffs at a rate ρ. Let V denote the profits of a firm that has just paid the fixed cost of adjustment and must choose a new price. The firm's optimization problem becomes:

$$V = \max_{\ln S, T} \frac{\int_0^T \pi(\ln S - gt)e^{-\rho t}dt - e^{-\rho T}k}{1 - e^{-\rho T}}.$$

Here $\ln S$ is natural logarithm of the target price S, and T is the time between successive price adjustments. The adjustment trigger $s = Se^{-gT}$. The first order conditions are:

$$\int_0^T \pi'(\ln S - gt)e^{-\rho t}dt = 0, \quad \text{and}$$

$$\pi(\ln s) = \rho(V - k).$$

The first equation says that the present value of marginal losses over the cycle should be set to zero. If the firm did not discount profits and the loss function were quadratic, this would mean that the trigger S and target s would be symmetrically placed about the static optimal price. With discounting the firm backloads losses; S is closer to the optimum than s. The second equation says that, at the trigger, the instantaneous profit is equal to the cost of delaying the next cycle.

Most of the comparative statics of this model are similar to those of the static model outlined above. An increase in the cost of adjustment k, an increase in the rate of inflation g, or a reduction in the concavity of π causes the target S to rise and the trigger s to fall. One somewhat surprising result is that an increase in g has an ambiguous effect on the frequency of adjustment. For given s-S bands an increase in g increases the frequency of adjustment. The widening of the bands, however, works in the opposite direction.

Two-sided rules

If we relax the assumption that the shocks to x are monotonic, then the optimal policy will involve two triggers, one above and one below the target. Such two-sided rules have been used to study cash management (Miller and Orr, 1966), pricing (Barro, 1972), durable consumption (Grossman and Laroque, 1990) and investment (Abel and Eberly, 1994).

As an example of a two-sided model, take the static example from the beginning of this section and assume that x follows a Brownian motion in the absence of adjustment. In this dynamic model, the agent would not necessarily adjust to the static optimum; the shape of the profit function and the drift of the shock may cause the agent to pick a target on one side or the other. Even so, many of the comparative statics of this example are similar to the static model. The range of inaction tends to widen if the profit function becomes less concave or the adjustment cost rises.

One entirely new effect is that of the variance of the shock to x on the adjustment policy. If x follows a driftless Brownian motion, then an increase in the variance of x leads to wider bands and to more frequent adjustment. The intuition is simple. If the agent keeps the bands fixed, then the increase in the variance of x makes adjustment more frequent. The costs of adjustment rise relative the benefits. Wider bands are optimal. If the agent widens the bands in proportion to the standard deviation of the shock, the frequency of adjustment is unchanged but the benefits to adjustment rise. More frequent adjustment is optimal.

The widening of the bands in this case reflects an option value to waiting. s-S adjustment is like exercising an option: the greater is uncertainty, the more this option must be in the money. Dixit (1991) shows that, given this option value, fourth-order adjustment costs lead to a first-order range of inaction.

Macroeconomic models

The work of Blinder (1981) and Caplin (1985) inspired interest in the aggregate implications of s-S policies. Blinder provided several examples that illustrate the difficulty of aggregating s-S policies and the potential complexity of the resulting dynamics. Caplin demonstrated that the s-S model of inventories can explain the commonly observed finding in the inventory literature that production is more volatile than sales, an observation that is at odds with the production smoothing model of inventories (the idea is that the discreteness of orders in the s-S model adds to the variability in demand). In recent years, a large body of research has examined the relationship between microeconomic frictions and aggregate dynamics. Two themes run through this literature. The first is a need to surmount difficult modelling issues brought on by the heterogeneity inherent in discrete adjustment. The second is that aggregate dynamics may look very different from the microeconomic dynamics.

The curse of dimensionality

The main difficulty encountered in constructing aggregate models of with s-S behaviour is that the cross-sectional distribution of agents' deviations from their optimum becomes a state variable. This distribution determines how many agents are near their triggers and hence the amount of adjustment that will take place in the near future. To handle the high dimension of this distribution, current models follow one of three approaches. Some, like Caplin and Spulber (1987), Caplin and Leahy (1991, 1997), Danziger (1999) and Gertler and Leahy (2006) make distributional assumptions that reduce the number of state variables. Others, like Blinder (1981), Bertola and Caballero (1990), and Caballero and Engel (1991; 1999), ignore equilibrium considerations and construct aggregates by integrating across actions of isolated individuals facing possibly correlated shocks. A third group, Dotsey, King and Wolman (1999), Willis (2002), and Caplin and Leahy (2006), solve stochastic general equilibrium models, employing approximations that reduce dimensionality.

The Caplin–Spulber model

The Caplin–Spulber model illustrates a few basic lessons in s-S aggregation. Three log-linear relationships form the backbone of the model: the aggregate price index p is the average of individual prices p_i; output y equals real balances $m - p$; and a firm's optimal price p^* rises with the price level and with real balances, $p^* - p = \alpha(m - p)$. Caplin and Spulber close the model with three assumptions: m is continuous and monotonically increasing; firms follow one-sided s-S pricing policies (when $p_i - p^*$ falls to s, it is raised to S) and the initial distribution of relative prices is uniform between s and S.

The surprising result is that money is neutral in this setting in spite of the fact that all prices adjusted infrequently. The reason is simple. Output can only change if the distribution of prices shifts relative to the money supply. As the money supply increases, firms' prices tend to fall relative to p^*. The few firms that hit s adjust to S. In this way, firms rearrange themselves within the interval (s,S). Given the uniformity assumption, the distribution of $p_i - p^*$ is unchanged. The price index moves with the money supply because a small number of firms change their prices by a large amount.

The Caplin–Spulber model makes two important points. First, discrete adjustment at the microeconomic level does not necessarily imply discrete adjustment at the macroeconomic level. Heterogeneity in adjustment times tends to smooth macroeconomic dynamics. Second, the aggregate implications of s-S dynamics depend on the evolution of the distribution of idiosyncratic deviations from the optimum. In this case, the distribution does not change, and the s-S model looks exactly like its frictionless counterpart.

The Caplin–Leahy model

Caplin and Leahy (1991) make three simple amendments to the model of Caplin and Spulber: they assume that money is non-monotonic and follows a Brownian motion, that firms follow symmetric, two-sided pricing policies, and that the initial distribution of prices is uniform with a range equal to the distance between the trigger and the target. In this setting, they show that the distribution retains this shape. It rises and falls within the range of inaction like an elevator.

The Caplin–Leahy model provides an example of an economy in which the distribution of idiosyncratic deviations changes over time. The model also illustrates a new phenomenon: the state-dependence of the effects of aggregate shocks. When the distribution of prices is interior to the range of inaction, shocks to the money supply shift the distribution and output changes. When the distribution of prices reaches the edge of the range of inaction, further movements in the money supply cause price adjustment. The distribution of deviations rearranges itself as in the Caplin–Spulber model and money is neutral.

The Bertola–Caballero–Engel model

In a series of papers, Bertola, Caballero and Engel develop a very flexible framework for modelling aggregate and distributional dynamics with idiosyncratic shocks. Caballero and Engel (1991) consider prices and one-sided adjustment, and Bertola and Caballero (1990) consider durables and the two-sided adjustment. In this discussion, I will focus on durables and two-sided adjustment.

The model takes the neoclassical model without adjustment costs to be its benchmark. This neoclassical model would predict that an individual would like to hold a quantity $x_i^*(z_i)$ where z_i is a vector of individual characteristics. They then postulate that individuals follow s-S policies: if $x_i - x_i^* \notin (S_L, S_H)$ then the individual adjusts his holdings so that $x_i - x_i^* = S^*$. They assume that innovations in x_i^* are i.i.d. across agents i and over time, and are characterized by a stochastic matrix P_t, where P_t may depend on the aggregate state of the economy at date t. Since P_t has a finite number of states, the $x_i - x_i^*$ take on a finite number of values; denote these deviations by $\hat{x}_i$. The stochastic matrix P_t and the s-S adjustment policy, induce transitions on the $\hat{x}_i$. Denote these transitions by the stochastic matrix $\hat{P}_t$.

This model provides a simple accounting of aggregate dynamics. Let X_t denote the aggregate holdings of durables and let the column vector f_t denote the cross-sectional density of the $\hat{x}_i$. Then $X_i = \int x_i^* di + f_t' \hat{x}$ and $f_{t+1} = f_t \hat{P}_{t+1}$. In this view, s-S dynamics provide a theory of the error term in the neoclassical model. Deviations from the neoclassical model are associated with fluctuations in the density f.

The model allows one to consider the relative roles of idiosyncratic and common shocks. Shocks that are common across firms tend to shift f as they do in the model of Caplin and Leahy. Idiosyncratic shocks, however, tend to mute the effects of aggregate shocks. If only idiosyncratic shocks are present the error $f_t' \hat{x}$ settles down to an ergodic density of $\hat{P}$. If idiosyncratic shocks dominate, the microeconomic frictions are lost in the aggregate like in the model of Caplin and Spulber.

Caballero (1993) applies this framework to show that an aggregate s-S model may look like a partial adjustment model. A shock causes some agents to adjust and leaves others near the adjustment trigger. The result is that subsequent idiosyncratic shocks lead to further adjustment.

Like Caplin and Spulber and Bertola and Caballero, Golosov and Lucas (2004) emphasize the fact that the firms that adjust in an s-S model are those with the greatest desire to adjust. This differentiates s-S models from models in which the time between price adjustments is fixed such as the Taylor or Calvo models of price adjustment.

Equilibrium interactions

The simplicity of the Bertola–Caballero–Engel model comes from the absence of equilibrium interactions. The question arises whether the introduction of equilibrium interactions would enhance or limit the differences between the s-S model and its frictionless counterpart. As usual, the answer is 'it depends'.

On the one hand, Thomas (2002) shows that endogenous prices may blunt s-S dynamics. If an unusually large number of people are about to purchase a car, the price of cars should rise, thereby dissuading some agents from making purchases. Endogenous prices movements

therefore tend to smooth any deviations of the s-S model from the frictionless neoclassical benchmark. Caplin and Leahy (2004) argue that in a one-sided model the aggregate dynamics become observationally equivalent to a model without adjustment costs.

On the other hand, Ball and Romer (1990) argue that in the presence of strategic complementarities, non-adjustment by some agents can encourage non-adjustment by others. For example, if due to increasing returns the profitability of investment by any single firm is increasing in aggregate investment, then each firm may forgo investment because others have decided to forgo investment. Strategic complementarities may therefore cause the range of inaction to widen, allowing for greater differences from the frictionless benchmark.

s-S models remain an active area of research. At the microeconomic level, recent work has shown how s-S frictions may trap information leading to interesting boom–bust cycles (Caplin and Leahy, 2006) and how adverse selection may amplify s-S frictions (House and Leahy, 2004). At the macroeconomic level, work progresses on the importance of fixed costs in explaining price inertia (Golosov and Lucas, 2003; Gertler and Leahy, 2006; Midrigan, 2006).

JOHN LEAHY

See also **adjustment costs; Arrow, Kenneth Joseph; microfoundations; monetary business cycle models (sticky prices and wages); options; Phillips curve; stochastic optimal control.**

Bibliography

Abel, A. and Eberly, E. 1994. A unified model of investment under uncertainty. *American Economic Review* 84, 1369–84.

Arrow, K., Harris, T. and Marschak, J. 1951. Optimal inventory policy. *Econometrica* 19, 205–72.

Ball, L. and Romer, D. 1990. Real rigidities and the non-neutrality of money. *Review of Economic Studies* 57, 183–203.

Baumol, W. 1952. The transactions demand for cash: an inventory theoretic approach. *Quarterly Journal of Economics* 66, 545–56.

Bar-Ilan, A. and Blinder, A. 1996. Consumer durables: evidence on the optimality of usually doing nothing. *Journal of Money, Credit and Banking* 24, 258–72.

Barro, R. 1972. A theory of monopolistic price adjustment. *Review of Economic Studies* 39, 17–26.

Bertola, G. and Caballero, R. 1990. Kinked adjustment costs and aggregate dynamics. In *NBER Macroeconomics Annual 1990*, ed. O. Blanchard and S. Fischer. Cambridge, MA: MIT Press.

Blinder, A. 1981. Retail inventory behavior and business fluctuations. *Brookings Papers on Economic Activity* 1981(2), 443–505.

Caballero, R. 1993. Durable goods: an explanation for their slow adjustment. *Journal of Political Economy* 101, 351–84.

Caballero, R. and Engel, E. 1991. Dynamic (S, s) economies. *Econometrica* 59, 1659–86.

Caballero, R. and Engel, E. 1999. Explaining investment dynamics in US manufacturing: a generalized (S, s) approach. *Econometrica* 67, 741–82.

Caplin, A. 1985. The variability of aggregate demand with (S, s) inventory policies. *Econometrica* 53, 1395–410.

Caplin, A. and Spulber, D. 1987. Menu costs and the neutrality of money. *Quarterly Journal of Economics* 102, 703–25.

Caplin, A. and Leahy, J. 1991. State-dependent pricing and the dynamics of money and output. *Quarterly Journal of Economics* 106, 683–708.

Caplin, A. and Leahy, J. 1994. Business as usual, market crashes and wisdom after the fact. *American Economic Review* 84, 548–65.

Caplin, A. and Leahy, J. 1997. Aggregation and optimization with state-dependent pricing. *Econometrica* 65, 601–23.

Caplin, A. and Leahy, J. 2004. On the relationship between representative agent models and (S, s) models. In *Productivity, East Asia Seminar on Economics 13*, ed. T. Ito and A. Rose. Chicago: University of Chicago Press.

Caplin, A. and Leahy, J. 2006. Equilibrium in a durable goods market with lumpy adjustment. *Journal of Economic Theory* 128, 187–213.

Danziger, L. 1999. A dynamic economy with costly price adjustments. *American Economic Review* 89, 878–901.

Dixit, A. 1991. Analytic approximations in models of hysteresis. *Review of Economic Studies* 58, 141–52.

Dotsey, M., King, R. and Wolman, A. 1999. State dependent pricing and the general equilibrium dynamics of money and output. *Quarterly Journal of Economics* 104, 655–90.

Gertler, M. and Leahy, J. 2006. A Phillips curve with an Ss foundation. Working Paper No. 11971. Cambridge, MA: NBER.

Grossman, S. and Laroque, G. 1990. Asset pricing and optimal portfolio choice in the presence of illiquid durable consumption goods. *Econometrica* 58, 25–51.

Golosov, M. and Lucas, R. 2003. Menu costs and Phillips curves. Working Paper No. 10187. Cambridge, MA: NBER.

House, C. and Leahy, J. 2004. An Ss model with adverse selection. *Journal of Political Economy* 112, 581–614.

Midrigan, V. 2006. Menu costs, multiproduct firms and aggregate fluctuations. Working paper, Ohio State University.

Mankiw, N. 1985. Small menu costs and large business cycles: a macroeconomic model of monopoly. *Quarterly Journal of Economics* 100, 529–39.

Miller, M. and Orr, D. 1966. A model of demand for money by firms. *Quarterly Journal of Economics* 80, 413–35.

Scarf, H. 1959. The optimality of (S, s) policies in the dynamic inventory problem. In *Mathematical Methods in Social Sciences*, ed. K. Arrow, S. Karlin and P. Suppes. Stanford: Stanford University Press.

Sheshinski, E. and Weiss, Y. 1977. Inflation and the cost of price adjustment. *Review of Economic Studies* 44, 287–303.

Thomas, J. 2002. Is lumpy investment relevant for the business cycle? *Journal of Political Economy* 110, 508–34.

Willis, J. 2002. General equilibrium of a monetary model with state-dependent pricing. Working paper, Federal Reserve Bank of Kansas City.

stability and growth pact

Proposed by Germany in 1995, backed by France, and created by the Treaty of Amsterdam in 1997, the Stability and Growth Pact (SGP) is the European Union's (EU) answer to concerns about fiscal unsustainability. It consisted initially of three regulations: (*a*) on the strengthening of the surveillance and coordination of economic policies (Council of the European Union, 1997b), (*b*) on speeding up and clarifying the implementation of the excessive deficit procedure (Council of the European Union, 1997c, and (*c*) on the SGP (Council of the European Union, 1997a). The SGP extended fiscal discipline into the Economic and Monetary Union (EMU) after 1 January 1999 in the manner foreseen by the Treaty of Maastricht (1992) for the convergence period 1993–8: budget deficit and public debt must be, respectively, no more than three per cent of GDP and 60 per cent of GDP.

When, in May 1998 the European Union was deciding which of the EU–15 countries would enter into the EMU in light of the Treaty of Maastricht criteria, some countries, including Germany, would have failed to qualify under the debt criterion. Since then, the debt criterion has been interpreted in terms of trend rather than level in the Treaty of Maastricht, and as a consequence in the Treaty of Amsterdam. The nature of the SGP thus changed within a year of its ratification and a year before its implementation. *De facto* the rule was then to abide by the deficit criterion, leaving the debt criterion to be interpreted in trend. The change in the interpretation of the Treaty of Maastricht resulted in a change in the interpretation of the Treaty of Amsterdam, weakening its original double objective.

What happens when a country does not abide by the three per cent rule? If it is a first violation, the country will have to make a non-interest-bearing deposit with the European Commission. The amount of this deposit comprises a fixed component equal to 0.2 per cent of GDP, and a variable component linked to the size of the deficit. Each following year the Council may decide to intensify the sanctions by requiring an additional deposit, though the annual amount of deposits may not exceed the upper limit of 0.5 per cent of GDP. A deposit is converted into a fine if, in the view of the Council, the excessive deficit has not been corrected after two years. After three consecutive years of violation, the country will see its three deposits become a fine. While the three per cent limit might seem tight, the probability that a country will fail to abide by the Pact for three years in a row, and thus be fined, was originally perceived as low. However, revised numbers from Eurostat indicate that Greece has always been above the three per cent deficit ceiling since its entry into the EMU in 2001. Additionally, Portugal's deficit was greater than three per cent in 2001, 2004 and 2005, Germany's deficits exceeded three per cent from 2002 to 2005, France's deficits exceeded three per cent from 2002 to 2004, and Italy, the UK and the Netherlands breached the three per cent rule in 2004. Further, the deposit rules were not enforced. Not only did Portugal not have to make a deposit, but France and Germany were not fined. The credibility of the SGP was dramatically weakened for the second time.

On 20 March 2005 the European Union decided to try to improve the credibility of the Pact: the Council adopted a report entitled 'Improving the Implementation of the Stability and Growth Pact' (Council of the European Union, 2005). The report was endorsed by the European Council in its conclusions of 22 March, and is now an integral part of the SGP. On 27 June 2005 two additional regulations amended Regulations 1466/97 and 1467/97 (Council of the European Union, 1997b; 1997c). The European Council unanimously agreed to introduce some flexibility into the SGP, creating a *de facto* SGP II. This flexibility was particularly introduced via the concept of 'relevant factors', which are country-specific. Examples of relevant factors are: (*a*) budgetary efforts towards increasing or maintaining at a high level financial contributions to foster international solidarity and to achieve European policy goals (notably the unification of Europe if this has a detrimental effect on the growth and fiscal burden of a Member State), (*b*) structural reforms (for example, pensions, social security), (*c*) policies supporting R&D, and (*d*) medium-term budgetary efforts (consolidating during good economic times, a reduction in debt levels, and an increase in public investment). The European Council is the final judge of the relevance of a given factor.

In the meantime, the European Union introduced a 'code of conduct' to counterbalance the new flexibility that had been introduced. This code of conduct established a country-specific Medium Term budgetary Objective (MTO) that serves as the actual deficit target around which some flexibility is allowed so long as the country abides by the three per cent reference value. To better understand the effects of these amendments, the background economics literature is reviewed, as well as the institutional design of the original SGP and its amended version.

The rationales for a supra-national fiscal rule

One of the major goals of the SGP was to make fiscal discipline a permanent feature of the EMU. Safeguarding sound government finances was regarded as a means to strengthen the conditions for price stability in the Eurozone. However, it was also recognized that the loss by individual countries of the exchange rate instrument in the EMU must be offset by automatic fiscal stabilizers at the national level to help economies adjust to asymmetric shocks. The three per cent deficit limit was calculated under the assumption that the long run average nominal GDP growth rate is five per cent, whereas the long-run inflation rate is two per cent.

Beetsma (2001) provided a summary of the different arguments in favour of a fiscal rule. Arguments in support of a Europe-wide fiscal rule are of three types: benefits to domestic governments; benefits to other governments; and collective benefits.

Benefits to domestic governments

The main benefit to domestic governments, namely, public finance sustainability, has been studied by several researchers, including Amador (2000), Ballabriga and Martinez-Mongay (2005), Bohn (1995), Mongelli (1999), Nielsen (1992), and Perotti, Strauch and von Hagen (1998). The SGP aims at ensuring the sustainability of EU public finances, and hence is supposed to prevent governments hampering growth through unsustainable fiscal policies. For illustration purposes, it should be noticed that the primary balance as a percentage of GDP is close to zero or even positive (a surplus) for most of the euro area members. Thus, what pushes countries like France, and Germany above the three per cent deficit ceiling seems to be, primarily, interest payments on debt.

Does Europe really need a fiscal rule to prevent unsustainable domestic public finances? If the answer is 'yes', it is because financial markets do not work properly. If government bond yields include risk premia, increasing indebtedness may cause bond yields to rise, thus raising the cost of borrowing and imposing discipline on governments. Market discipline of this kind may be especially relevant and important in the EMU, in which governments of the Member States can issue debt but do not have the possibility of monetizing and inflating away excessive debt. Spreads between European bonds have narrowed considerably since 1991 for the Eurozone members. Bernoth, von Hagen and Schuknecht (2004) explain that, for Deutschmark/euro denominated bonds, EMU membership reduces the linear effect of debt on default risk premia. Accordingly, EMU members enjoy a lower risk premium than before, but this benefit declines with the size of public debt compared with Germany's. This is consistent with the view that markets anticipate fiscal support for EMU countries in financial distress unless these countries had previously been highly undisciplined. Thus, the disciplinary function of credit markets still exists. If the domestic benefit provided by the SGP to governments is not striking, does the SGP discipline governments in their relationships with others, or ease the coordination of fiscal policies between governments?

Benefits to other governments

A first issue is the likelihood of free-riding behaviours: the fear that governments would run higher deficits in a monetary union. Indeed, under the uncovered interest parity assumption, the understanding was that countries would run deficits that would be financed partly by the Eurozone through an overall rise in the Eurozone bonds interest rate. Undeniably, a country not belonging to a monetary union and running a high deficit will have to face a rise in its domestic interest rate. In a monetary union and integrated capital market, a country running a high deficit will face a much lower rise in its interest rate since it is now equalized at the European level. This may create the incentive for some countries to free-ride on others. If every country behaves this way, on the one hand the overall interest rate rises, and on the other hand the Eurozone faces a greater risk of public finance unsustainability. In the extreme scenario, Eichengreen and Wyplosz (1998) argue that the financial turmoil triggered by a default on the debt of any member country would have significant cross-border effects. In this context, focusing his discussion of free-riding and the SGP on the effects of centralized monetary policy combined with decentralized fiscal policy, Uhlig (2002) regards the SGP as necessary in preventing free-riding in the form of excessively high deficits.

The second issue is moral hazard, which differs from free-riding to the extent that it is 'post-contractual opportunism'. In other words, once countries belong to the EMU, countries' loss functions change: without the SGP, governments could weigh more the use of fiscal policies to increase the likelihood of a re-election rather than keeping their public finances in line with the Maastricht guidelines. Dixit (2001) and Dixit and Lambertini (2001) demonstrate that fiscal discretion leads to equilibrium levels of output and inflation very different from Pareto-optimal choices. The SGP should thus prevent countries from changing their attitudes once within the Eurozone.

Collective benefits

Collective benefits of the SGP are at least threefold: the coordination of domestic fiscal policies; the policy-mix argument; and the reinforcement of the ECB's credibility. First, there is the question of fiscal coordination among member countries. A lack of coordination could lead to asymmetric economic shocks on both the aggregate demand and aggregate supply in every country due to large differences in fiscal policies. The coordination of fiscal policies is intended to eliminate those large differences in fiscal policies across countries, and thus implicitly create a Europe-like fiscal policy. The coordination argument is different from the policy-mix argument in the sense that it addresses only the coordination of fiscal policies, whereas

the policy-mix argument addresses the question of the coordination of the European monetary policy to a European-like fiscal policy.

Second, Beetsma (2001) and Issing (2002) analyse the policy-mix argument and claim that the advent of a central monetary authority was important in establishing the correct mix of fiscal and monetary policy within the Eurozone. Article 99(3) of the Treaty establishing the European Community stipulates: 'In order to ensure closer coordination of economic policies and sustained convergence of the economic performances of the Member States, the Council shall, on the basis of reports submitted by the Commission, monitor economic developments in each of the Member States' (Treaty establishing the European Community, 2002). Beetsma and Uhlig (1999) build a model of centralized monetary policymaking and decentralized fiscal policymaking and find that a monetary union combined with an appropriately designed fiscal rule will be strictly preferred to fiscal autonomy, as there are benefits to coordination of a Europe-wide monetary policy with a Europe-wide fiscal policy.

The third reason is the maintenance of the credibility of the European Central Bank (ECB) through insuring its primacy as a monetary authority. As noted by Buti and van den Noord (2004, p. 6), the EMU is, '[commonly] seen as a regime of monetary leadership where fiscal policy is to support the central bank in its task to keep inflation in check'. This power is drawn from the following European Council resolution which accompanies the Pact: '[it] is also necessary to ensure that national budgetary policies support stability oriented monetary policies' (Buti and van den Noord, 2004, p. 6). Around the time that the Maastricht Treaty was drafted, Beetsma and Bovenberg (1999) showed that the European budgetary situation could undermine the credibility of the future European Central Bank. If a country's fiscal situation becomes unsustainable, other countries might be forced to bail out the insolvent national government. Alternatively, the European Central Bank could be forced to monetize national debts, and thereby create additional inflation in the EU although this would be forbidden in theory by the statutes of the ECB. In this regard, the SGP is a secondary safety device.

Institutional design

Formally, the SGP consists of three elements: a political commitment, a preventive element, and a dissuasive element.

The political commitment

Peer support and peer pressure are an integral part of the Stability and Growth Pact: the Council and the Commission are expected to motivate countries to adhere to the pact, and make public their positions and decisions at all appropriate stages of SGP procedure. The idea is to make the SGP more transparent. Member States may also establish a committee of experts to advise them on the main macroeconomic projections, a notion that has roots in the economic literature (Wyplosz, 2005). With this aim, Council Regulation 1466/97 reinforces the multilateral examination of budget positions and the coordination of economic policies.

The SGP foresees the submission of all Member States to stability and convergence programmes. Stability and convergence programmes must present information on the adjustment path and the expected path of the general government debt ratio, as well as the main assumptions made about expected economic development. New to SGP II and in line with the recommendations of the literature, structural reforms are encouraged by the possibility of taking them into account on the path towards adjustment.

The preventive arm of the SGP

The preventive arm of the Pact was, for the first time, given real substance with the implementation of the Medium Term budgetary Objective (MTO). In the 1997 version of the SGP, the MTO was the same for every country: a close-to-balance or surplus budget. Since 2005 the MTO has been given a new definition and is part of a broader new addition to the SGP: the code of conduct. Member States have to define a specific MTO in cyclically adjusted terms. Thus, cycles are now taken into consideration. As recommended by the literature, country specificities must be taken into account. This new device means that surpluses from periods of economic growth are required to be used for debt and deficit reduction.

The goals of the MTO are threefold. The first is to provide a margin with respect to the three per cent of GDP deficit ceiling. This margin is calculated by taking into account the past output volatility and budgetary sensitivity to output fluctuations of each Member State. The second goal is fiscal sustainability, for instance, taking into account the economic and budgetary impact of ageing populations. Influenced by the economic literature (Blanchard and Giavazzi, 2004; Buti, Eijffinger and Franco, 2003; Fatas, 2005), the third goal is to take into account the need for public investment and represents the structural side of the SGP. The MTOs are revised every four years or whenever a major reform is implemented.

The Council also has the leeway to issue an 'early warning' to Member States before an excessive deficit has occurred. Articles 6(2) and 10(2) of Council Regulation 1466/97 state that

> In the event that the Council identifies significant divergence of the budgetary position from the medium-term budgetary objective, or the adjustment path towards it, it shall, with a view to giving early warning in order to prevent the occurrence of an excessive deficit, address, in accordance with Article 103 (4) a

recommendation to the Member State concerned to take the necessary adjustment measures.

The dissuasive element

If a country breaches the three per cent value for three consecutive years, it is considered to be in violation of the SGP. In order to dissuade countries from excessive deficits, Council Regulation 1467/97 establishes the Excessive Deficit Procedure (EDP). When the council decides that an excessive deficit exists, it makes recommendations to the Member State and establishes a deadline of six months (raised from four) for corrective policies to be implemented. If a Member State fails to implement the policies based on the Council's decisions, the Council imposes sanctions (deposits, and then fines), which are levied within ten months of the first report of an excessive deficit. A country cannot avoid the deposits, and ultimately the fine, unless the Council decides to abrogate some or all of the sanctions. Abrogation depends on the significance of the progress made by the participating Member State concerned in correcting the excessive deficit, if the breach has resulted from an unusual event or a major economic decline (that is, an annual decline in real GDP of at least two per cent), or if the country's deficit is due to 'relevant factors'. Any fines already imposed are not reimbursable. Interest on the deposits lodged with the Commission, and the yield from fines, are distributed among Member States without an excessive deficit, in proportion to their share in the total GNP of eligible Member States.

Assessing the Pact

The economic literature and the Pact

Before the creation of the SGP, the economic literature reflected on the need of a fiscal rule for Europe. In 1997 this fiscal rule materialized into the SGP. Since then, the literature has addressed criticisms to the scientific justifications of the specific design of the SGP, as well as the proposed alternatives. Most of the justifications covered by the literature are inscribed in the following notions: sustainability of public finances; free-riding and moral hazard; coordination and policy mix; and finally the credibility of the ECB.

As for sustainability, the literature is divided with respect to the actual effects of the SGP. Prior to the SGP, financial markets played an active role in disciplining governments, and this still holds true (Bernoth, von Hagen and Schuknecht, 2004). Moreover, some authors challenge the arbitrariness of the definition of sustainability implicit in the SGP (the 3 per cent–60 per cent rule). On the one hand, since budget composition is different across countries, De Grauwe (2003) argues that countries should be able to choose their own debt target instead of the 60 per cent ceiling, and as a consequence have different deficit targets. On the other hand, Coeure and Pisani-Ferry (2005) call for a better concept of sustainability, including, for example, pension regimes. This is something that has been addressed in the amended version of the SGP through the notion of 'relevant factors', one of them being the change in pension expenditures.

As for free-riding, the SGP may prevent it (Warin and Wolff, 2005), but it seems not to prevent moral hazard. Indeed, although included in the definition of the Pact, the dissuasive arm seems to malfunction due to moral hazard behaviours: De Haan, Berger and Jansen (2003) explain that this is, likely, one of the reasons why some countries – Germany and France, for instance – decide to put more emphasis on solving their internal troubles by relaxing their fiscal policies instead of strictly abiding by the letter of the SGP. The SGP should prevent countries from changing their attitudes once within the Eurozone, but a recent literature on the political budget cycle (PBC) and the SGP explains that incumbent governments within the Eurozone display an inclination to raise public expenditure, and thus deficits, before an election, although they have to abide by the SGP (Mink and de Haan, 2005). Buti and van den Noord (2003) analyse the fiscal policies over the 1999–2002 period and find some evidence of expansionary fiscal policies motivated by near-term elections. This result is confirmed by von Hagen (2003) who concludes that there is evidence that fiscal policies were used during the period 1998–2002 before elections. Those results mean that the SGP does not seem to prevent moral hazard behaviours.

As for coordination, the SGP may not represent the optimal means of dealing with this problem. Eichengreen (1990), Cohen (1990), and Branson (1990) study the necessity for a federal budget to augment national budgets. MacDougall (1977), De Grauwe (1990), Italianer and Vanheukelen (1993), and Bryson (1994) analyse the need for a centralized budget as a way of establishing automatic stabilizers with income transfers from better-off to worse-off countries.

As for the ECB's credibility, every EMU member enjoys a lower risk premium than before the creation of the euro, which can be explained by many reasons, such as the liquidity of the market and the improved credibility for the central bank in charge of the European monetary policy. Did the SGP play a role? Since Germany and France did not abide by the SGP for some years, it is difficult to grant this benefit exclusively to the SGP.

In this context, the SGP has had mixed results. The amended version of 2005 embodies some of the changes called for by the economic literature. The lack of flexibility was one of the main criticisms. For instance, Cooper and Kempf (2000) call for some flexibility at the fiscal level, as the ECB lacks the tools necessary for stabilization in the presence of country-specific shocks. The new definition of the MTO based on country specificities seems to go into this direction, although this is misleading. Indeed, the MTO in SGP I was ignored by

the countries which considered only the three per cent of GDP reference value. Its new definition, based on country specificities, is now at the core of the assessment of countries' fiscal policies. In other words, it seems that for national policies what matters is no longer three per cent but their specific medium-term objectives, by definition lower than three per cent. In this regard, the new preventive arm seems to be tighter than the original SGP, and hence less flexible.

Flexibility refers to the idea of an optimal fiscal policy even though it can be above the three per cent deficit limit. In order to target this optimal fiscal policy and prevent at the same time the existence of a political budget cycle, authors such as Wyplosz (2005), Beetsma and Debrun (2005), Annet, Decressin and Deppler (2005), and Marinheiro (2005) argue for the strong version of institutional reform – the creation of an independent Fiscal Policy Committee (FPC), and a reconfiguring of the debt targets so that they are established, country by country, on a basis of the starting position. This would not automatically mean the end of a fiscal rule in Europe, but it would mean the end of the SGP. However, SGP II seems to go in this direction. Indeed, it allows countries to decide whether they need an independent committee to scrutinize their domestic fiscal policy. In fact, this independent committee is not the same as the FPC: the FPC could decide an optimal fiscal policy above the three per cent deficit limit, which is not the case with a national independent committee.

Another solution to introduce some flexibility without renegotiating the Treaty of Amsterdam, as well as giving some weight back to the debt criterion, is proposed by Pisani-Ferry (2002): allowing countries to opt out of the Excessive Deficit Procedure based on the deficit, and abide by the 60 per cent of GDP debt criterion. In this spirit, a country can have a deficit greater than three per cent as long as its debt is below 60 per cent. But before we consider other amendments, what is the future of the Pact from an institutional perspective?

The future of the Pact

The changes produced by SGP II to SGP I are twofold and concern both the preventive and dissuasive arms of the Pact. First, there is new definition of the medium-term objective (preventive arm). However, it is acceptable if countries do not abide by the medium-term objective as long as they do not go over the reference value of three per cent: regulation 1466/97 explains,

> [in] order to enhance the growth-oriented nature of the Pact, major structural reforms which have direct long-term cost-saving effects, including by raising potential growth, and therefore a verifiable impact on the long-term sustainability of public finances, should be taken into account when defining the adjustment path to the medium-term budgetary objective for countries that have not yet reached this objective and in allowing a

temporary deviation from this objective for countries that have already reached it.

This amendment does not relax the reference value of three per cent, but relaxes the constraint imposed by the definition of the medium-term budgetary objective. The amendment adds: 'in order not to hamper structural reforms that unequivocally improve the long-term sustainability of public finances, special attention should be paid to pension reforms introducing a multi-pillar system that includes a mandatory, fully funded pillar, because these reforms entail a short-term deterioration of public finances during the implementation period.'

The second change was to the dissuasive arm and deals with exceptional circumstances. The dissuasive arm is looser than it was. The European Commission is asked to prepare a report in case of a breach of the deficit reference value by a Member State. If the breach is not justified by an economic downturn (a recession of at least two per cent of GDP) or an exceptional external event, countries have to make deposits to the European Commission that will be transformed into fines in the third consecutive year if a country could not abide by the reference value for three years in a row. The amended regulation loosens the constraint by introducing the notion of relevant factors. Moreover, before asking for deposits when a country breaches the deficit reference value for the first or second time, the Commission should look at the medium-term economic position of a country, at relevant factors, and at the overall quality of public finances.

In the long run, the most important question in assessing the effects of SGP II versus SGP I is to know whether the preventive arm – tighter than under SGP I – will outweigh the loosening of the dissuasive arm. The answer is in the hands of national governments. In retrospect, the SGP does not seem to provide an ideal answer to the branches of the literature studying the potential need for a fiscal rule. This is not surprising, since the SGP is, by its nature, as much a politically designed rule extending the Treaty of Maastricht as an economically designed one.

THIERRY WARIN

See also **European Central Bank; fiscal federalism.**

Bibliography

Amador, J. 2000. Fiscal Policy and budget deficit stability in a continuous time stochastic economy. Working Paper No. 384. Universidade Nova de Lisboa.

Annet, A., Decressin, J. and Deppler, M. 2005. Reforming the stability and growth pact. Policy Discussion Paper 05/02, IMF.

Ballabriga, F.C. and Martinez-Mongay, C. 2005. Sustainability of EU public finances. Economic Papers No. 225, European Commission.

Beetsma, R. 2001. Does EMU need a stability pact? In *The Stability and Growth Pact – The Architecture of Fiscal Policy in EMU*, ed. A. Brunila, M. Buti and D. Franco. Basingstoke: Palgrave.

Beetsma, R. and Bovenberg, A. 1999. The interaction of fiscal and monetary policy in a monetary union: balancing credibility and flexibility. In *The Economics of Globalization: Policy Perspectives from Public Economics*, ed. A. Razin and E. Sadka. Cambridge: Cambridge University Press, 1999.

Beetsma, R. and Debrun, X. 2005. Implementing the stability and growth pact: enforcement and procedural flexibility. Working Paper No. 433, European Central Bank.

Beetsma, R. and Uhlig, H. 1999. An analysis of the stability and growth pact. *Economic Journal* 109, 546–71.

Bernoth, K., von Hagen, J. and Schuknecht, L. 2004. Sovereign risk premia in the European government bond market. Working Paper No. 369, European Central Bank.

Blanchard, O. and Giavazzi, F. 2004. Reforms that can be done: improving the SGP through a proper accounting of public investment. Discussion Paper No. 4220, CEPR.

Bohn, H. 1995. The sustainability of budget deficits in a stochastic economy. *Journal of Money, Credit and Banking* 27, 257–71.

Branson, W. 1990. Intégration des marchés financiers et croissance durant la transition vers l'UEM. In *Vers l'Union économique et monétaire européenne*, ed. Ministère de l'Economie. Paris: La Documentation Française.

Bryson, J. 1994. Fiscal Policy coordination and flexibility under European monetary union: implications for macroeconomic stabilization. *Journal of Policy Modeling* 16, 541–57.

Buti, M., Eijffinger, S. and Franco, D. 2003. Revisiting the stability and growth pact: grand design or internal adjustment? Economic Papers No. 180, European Commission.

Buti, M. and van den Noord, P. 2003. Discretionary fiscal policy and elections: the experience of the early years of EMU. Economics Department Working Paper No. 351, OECD.

Buti, M. and van den Noord, P. 2004. Fiscal Policy in EMU: rules, discretion and political incentives. Economic Papers No. 206, European Commission.

Coeure, B. and Pisani-Ferry, J. 2005. Fiscal Policy in EMU: towards a sustainability and growth pact. *Oxford Review of Economic Policy* 21, 598–617.

Cohen, D. 1990. Union monétaire, solvabilité des Etats et politiques budgétaires. In *Vers l'Union économique et monétaire européenne*, ed. Ministère de l'Economie. Paris: La Documentation Française.

Council of the European Union. 1997a. Resolution of the European council on the stability and growth pact. Amsterdam, 17 June. *Official Journal* C 236, 02/08/1997, 1–2.

Council of the European Union. 1997b. Council regulation (EC) No 1466/97 of 7 July 1997 on the strengthening of the surveillance of budgetary positions and the surveillance and coordination of economic policies. *Official Journal* L 209, 02/08/1997, 1–5.

Council of the European Union. 1997c. Council regulation (EC) No 1467/97 of 7 July 1997 on speeding up and clarifying the implementation of the excessive deficit procedure. *Official Journal* L 209 , 02/08/1997, 6–11.

Council of the European Union. 2005. Council report to the European council on improving the implementation of the stability and growth pact. 21 March. Online. Available at http://www.eu2005.lu/en/actualites/documents_travail/2005/03/21stab/stab.pdf, accessed 30 August 2006.

Cooper, R. and Kempf, H. 2000. Designing stabilisation policy in a monetary union. Working Paper No. 7607. Cambridge, MA: NBER.

De Grauwe, P. 1990. La discipline budgétaire dans les unions monétaires. In *Vers l'Union économique et monétaire européenne*, ed. Ministère de l'Economie. Paris: La Documentation Française.

De Grauwe, P. 2003. *Economics of Monetary Union*, 5th edn. Oxford: Oxford University Press.

De Haan, J., Berger, H. and Jansen, D.-J. 2003. The end of the stability and growth pact? Working Paper No. 1093, CESifo.

Dixit, A. 2001. Games of monetary and fiscal interactions in the EMU. *European Economic Review* 45, 589–613.

Dixit, A. and Lambertini, L. 2001. Monetary–fiscal policy interactions and commitment versus discretion in a monetary union. *European Economic Review* 45, 977–87.

Eichengreen, B. 1990. Is Europe an optimal currency area? Discussion Paper No. 478, CEPR.

Eichengreen, B. and Wyplosz, C. 1998. The stability pact: more than a minor nuisance. *Economic Policy* 26, 67–113.

Fatas, A. 2005. Is there a case for sophisticated balanced-budget rules? Economics Department Working Papers No. 466, OECD.

Issing, O. 2002. On macroeconomic policy co-ordination in EMU. *Journal of Common Market Studies* 40, 345–58.

Italianer, A. and Vanheukelen, M. 1993. Proposals for community stabilization mechanisms: some historical applications. *European Economy* 5, 493–510.

MacDougall, D. 1977. *Report of the Study Group on the Role of Public Finance in European Integration*. Brussels: Commission of the European Communities.

Marinheiro, C. 2005. Has the stability and growth pact stabilized? Evidence from a panel of 12 European countries and some implications for the reform of the Pact. Working Paper No. 1411, CESifo.

Mink, M. and de Haan, J. 2005. Has the stability and growth pact impeded political budget cycles in the European union? Working Paper No. 1532, CESifo.

Mongelli, F. 1999. The effects of the European economic and monetary union (EMU) on fiscal sustainability. *Open Economies Review* 10, 31–61.

Nielsen, S. 1992. A note on the sustainability of primary budget deficits. *Journal of Macroeconomics* 14, 745–54.

Perotti, R., Strauch, R. and von Hagen, J. 1998. Sustainability of public finances. CEPR and ZEI.

Pisani-Ferry, J. 2002. Fiscal discipline and policy coordination in the Eurozone: assessment and proposals. In *Budgetary Policy in E(M)U: Design and Challenges*, Ministerie van Financiën. The Hague.

Treaty establishing the European community. 2002. *Official Journal* C 325, 24 December.

Uhlig, H. 2002. One money, but many fiscal policies in Europe: what are the consequences? Discussion Paper No. 3296. London: CEPR.

von Hagen, J. 2003. Fiscal discipline and growth in Euroland: experiences with the stability and growth pact. *Wirtschaftspolitische Blätter* 50, 163–83.

Warin, T. and Wolff, L. 2005. Europe's deficit free riders: a panel data analysis. *European Political Economy Review* 3, 5–17.

Wyplosz, C. 2005. Fiscal Policy: institutions versus rules. *National Institute Economic Review* 191, 64–78.

stable population theory

Demographers are centrally concerned with the relationship between a population's age structure and its rates of mortality, fertility and immigration. Alfred Lotka (Sharpe and Lotka, 1911; Lotka, 1939) devised the fundamental mathematical form of this relationship for a population with no migration, in the event that age-specific mortality and fertility do not change over time.

Renewal equation

Lotka dealt primarily with females; we shall say more about males shortly. Mortality is described by an instantaneous mortality rate $\mu(a)$ at each age a, and determines the probability $l(a)$ of surviving from birth to age a, as $l(a) = \exp\left(-\int_0^a ds\, \mu(s)\right)$. Demographers refer to $l(a)$ as survivorship. Fertility is described by the rate $m(a)$ of female births per female of age a. The goal of the theory is to track changes in population number and age composition over time. Suppose that we know the numbers of females at all ages at some initial time that we denote as $t = 0$. Writing $B(t)$ for the rate at which females are born to the total population at any later time $t > 0$, Lotka derived what is called the 'renewal equation',

$$B(t) = \int_0^t ds\, l(s)\, m(s)\, B(t - s) + h(t).$$

To understand this equation, note first that the number of females $n(s,t)$ at age s at time t must be simply the number of survivors, $l(s)\, B(t - s)$, of the $B(t - s)$ females born at time $t - s$. Thus, the first term on the right side of the equation sums births to females whose ages range from zero to t, and who were born at any time between zero and t. The second term $h(t)$ represents births to all females who were alive at the initial time $t = 0$, and whose ages are therefore greater than t. As time passes, the females whose children are counted in $h(t)$ get older and eventually die. Thus after a long time passes (that is, for large values of t) only the first term on the right of the renewal equation remains. Lotka found that the solutions to this simpler equation are of a particular form that describes what is called a stable population.

Classical stable theory

A *stable population* has an unchanging age structure and a constant exponential growth rate r; both structure and growth rate are determined by vital rates (mortality, fertility). In a stable population, births at time $t - a$ must differ from births at time t by an exponential factor in the growth rate, $B(t - a) = e^{-ra} B(t)$. Also, a stable population obeys the renewal equation at long times t when the term $h(t)$ is zero. Inserting the above exponential relationship into the renewal equation shows that the growth rate r satisfies what is called Lotka's *characteristic equation*,

$$1 = \int_0^\infty da \, l(a) \, m(a) \, e^{-ra}.$$

In a stable population, the number of females at age s at time t is $n(s, t) = l(s) B(t - s) = l(s) e^{-rs} B(t)$. Hence the proportion of a stable population that is between ages a and $a + da$ in a stable population is $u(a) = C \, l(a) \, e^{-ra}$, with

$$C = \frac{1}{\int_0^\infty da \, l(a) \, e^{-ra}}.$$

The per-capita birth rate (also called the crude birth rate) in a stable population is $b = \int_0^\infty da \, m(a) \, u(a)$, and the per-capita death rate (also called the crude death rate) in a stable population is $d = \int_0^\infty da \, \mu(a) \, u(a)$. The stable growth rate is $r = (b - d)$ (as may be shown by using the relationship $\mu(a) = -(1/l(a)) \, (dl(a)/da)$ in d and integrating by parts).

Given unchanging vital rates and any initial population in the full renewal equation, Lotka showed that the population obtained by solving the renewal equation must eventually approach a stable population whose growth rate is determined by the characteristic equation. This result explains the adjective stable in the term, stable age distribution. Lotka's original proof was put on a secure mathematical footing by Feller (1941; 1971); such stability is characteristic of many populations that undergo mortality and renewal (including, for example, light bulbs or laptop computers that die at a rate that depends on their age and must therefore be replaced at some rate). The property of stability of the age distribution when mortality and fertility are constant in time is known as 'demographic strong ergodicity'.

In Lotka's classical theory of demography, male births are accounted for by noting that the human sex ratio at birth is remarkably constant over time and place, close to 1.05 male births for every female birth, except in cases where deliberate preference for one or the other sex leads to an excess mortality of the less favoured sex. Thus the numbers of males born are computed simply as a constant multiple of the number of females born. Male age structure will obviously become stable along with the female age structure, but male mortality and thus survivorship are usually different from female.

In practice, the renewal equation, in which time is a continuous variable, is often replaced by a discrete-time version in which time advances in discrete units. In that case the age composition of a population is represented by a vector of population numbers in successive discrete age classes, and the renewal equation is replaced by a matrix recursion. Leslie (1945) formulated and analysed the properties of the discrete equation, which has much the same properties as the continuous version. Coale (1972), working with time as a continuous variable, and Keyfitz (1977), working with time as a discrete or a continuous variable, provide authoritative discussions of the mathematics and application of stable population theory.

Applying classical stable theory

Stable theory and the characteristic equation yield powerful, fundamental insights into the relationship between mortality, fertility, population growth rate and age structure. We mention a few important examples here. The fact that a stable population's age structure is proportional to $l(a)e^{-ra}$ shows that a population's age pyramid is steeper for faster growing populations and shallower for low-mortality populations. Economists are often interested in the proportions of a population at young (usually under 20 yrs), working (20 to 65 yrs) and old (over 65 yrs) ages, and in the dependency ratio (the sum of young and old divided by the number working). With mortality fixed, the dependency ratio for a stable population is high when the growth rate r is either large and negative (fewer young, many old) or large and positive (many young, fewer old), and there is an optimal growth rate at which dependency ratio is minimized. With growth rate fixed, the dependency ratio increases when mortality declines because more people survive to reach old age. These properties of stable populations carry over qualitatively to real populations, illuminating the effects of changes in mortality and fertility on population composition. In particular many industrialized countries in the 21st century have small positive or even negative growth rates as well as low mortality; today's populations thus have a larger fraction of older individuals and a smaller fraction of young than the higher-fertility and lower-mortality populations of the 19th and early 20th centuries. Changes in population age structure play an important role in the theory of economic demography, especially in understanding the role of transfers between different age segments of a population (Lee, 1994).

In stable theory, the contribution of fertility to population growth rate is described by the net replacement rate,

$$\text{NRR} = \int_0^\infty da\, l(a)\, m(a),$$

which is the expected lifetime reproduction of a female. A stable population grows ($r > 0$), declines ($r < 0$) or is stationary depending on whether the NRR is greater than, less than, or equal to one. For any given pattern of mortality the condition NRR = 1 defines replacement fertility, the fertility rate at which $r = 0$. The generation time of a stable population is the average number of years over which a cohort of mothers produces its daughters, and is defined as

$$T = \int_0^\infty da\, a\, l(a)\, m(a)\, e^{-ra}.$$

Generation times in contemporary populations are set primarily by the age pattern of reproduction. Historically generation times have ranged from 20 to 25 years for human populations, being on the lower end in high-fertility populations that have a high growth rate.

Stable population theory has found wide application in ecology and evolutionary biology, as discussed extensively by Caswell (2001). In the biological context, the stable population growth rate r is often used to measure the fitness (in a Darwinian sense) of a particular biological combination of fertility and mortality. Another useful characterization of fertility and mortality patterns is given by the reproductive value $v(a)$ of an individual of age a in a stable population (Fisher, 1930),

$$v(a) = \int_a^\infty db\, \frac{l(b)}{l(a)}\, m(b)\, e^{-r(b-a)}.$$

Clearly $v(a)$ is the discounted present value of an individual's future reproduction after age a, using the discount rate r. The characteristic equation tells us that $v(0) = 1$, so the reproductive value measures the future contribution of an individual of age a relative to a newborn. In evolutionary theory, and in some questions in economic demography, we are interested in the effect on population growth rate of a change in either mortality or fertility at a particular age. An increase in fertility at age a changes r by an amount proportional to the reproductive value $v(a)$, whereas an increase in mortality at age a changes r by an amount proportional to the product $l(a)e^{-ra}\, v(a)$.

An elegant and important concept derived from stable theory is that of population momentum (Keyfitz, 1971). Suppose that a stable population is growing at some rate $r > 0$ and that the population makes an instantaneous transition to replacement fertility. Stable theory implies that with this reduced fertility the population will eventually become stable and stationary with growth rate $r = 0$. Keyfitz asked: what is the ratio of the size of the final stationary population to the size of the population just before the fertility transition occurs? The answer is that the ratio equals *population momentum*. Write $v_0(a)$ for the reproductive value in the final stationary population, T_0 for the generation time in the final stationary population, and $u(a)$ for the stable structure of the initially growing population. Then Keyfitz's population momentum equals $(1/T_0) \int_0^\infty da\, v_o(a)\, u(a)$. In the real world, fertility transitions take time and so actual momentum is generally larger than Keyfitz's momentum, as shown by Li and Tuljapurkar (1999).

Stable theory has also been extended to include the effects of migration, with much attention focused on the effects of immigration into low-fertility populations. Arthur and Espenshade (1988) and Feichtinger and Steinmann (1992) analyse the case where a stream of immigrants of known age distribution and total number is added annually to a population with below-replacement fertility. This case is relevant to a number of industrialized countries. Over time, the population's age structure will again converge to a stable age structure that is determined jointly by the age structure of the immigrant flow, the vital rates of the resident population, and the rate at which immigrants' vital rates converge to those of the residents. Feichtinger and Steinmann point out that the general theory of stable populations with immigration is closely connected with the theory of manpower systems and other social processes; for a review of the latter see Vassiliou (1997).

The concepts and mathematics of stable theory are relevant to the dynamics of populations that are structured by variables other than age. In demography, we may be interested in the joint distribution of age and parity (a female's parity is the number of offspring that she has had), or of age and health status. In biological applications, we may be concerned with a stage variable (for example, size or developmental state) rather than with age: this is especially the case for plants, insects and other organisms in which age is not directly observable in nature, or in which vital rates depend directly on size or state rather than on age. In such cases, Lotka's renewal arguments can nonetheless be applied, except that we must keep track of the distribution of individuals according to both their age and their stage. Vital rates now include mortality and fertility as function of age and stage, as well as rates at which individuals move between stages during the course of life. It should be obvious that such stage-based demography is relevant to economic analyses of the human life cycle in which we are interested in the transitions made by individuals between stages such as marriage, divorce and employment. The stable theory of stage-based demography parallels the theory we describe here, and an account of the biological theory is given by Caswell (2001).

Stable theory has been an essential component of successful efforts by demographers to create widely applicable models of vital rates, as nicely discussed by Preston, Heuveline and Guillot (2000). The authors also describe an extension of stable theory, in which a population's different age segments change at different rates over time. Thus, if the growth rate of individuals aged a at time t is $r(a, t)$, the population density of individuals at age $a + b$ at a later time $t + b$ is given by

$$n(a + b, t + b) = n(a, t) \frac{l(a + b)}{l(a)}$$
$$\times \exp\left\{ \int_0^b ds\, r(a + s, t + s) \right\}.$$

This expression allows us to use observations at different times on population structures (such as censuses or surveys) to estimate the growth rates of particular cohorts. Since populations in the real world are rarely stable, this approach is often useful.

Demographic weak ergodicity

Ansley Coale (1957) pointed out that mortality and fertility rates in practice will vary with time, violating the assumptions of strong demographic ergodicity and calling into question the use of classical stable theory. He argued, however, that human populations should forget their more remote history, in the sense that today's population composition should be most strongly influenced by the recent rather than the distant past. Lopez (1961; 1967) provided a mathematical framework for this argument by defining *demographic weak ergodicity*: if mortality and fertility rates change with time in some arbitrary (but demographically sensible) way, and two populations with distinct initial age structures are subject to the same sequence of changing rates, then as time goes by the age structures of these two populations will become proportional. Lopez's original proof has been strengthened and extended in more recent work, as discussed for example by Tuljapurkar (1982). An important consequence of demographic weak ergodicity is that we are justified in focusing attention on relatively recent changes in mortality and fertility as being the key to current and future age structures.

Stochastic stable theory

The work of Coale and Lopez provided the impetus for a powerful generalization of stable theory to cases when mortality and fertility rates change with time in a stochastic fashion. Analysis of historical mortality and fertility rates for any real population reveals that these rates change with time. Some changes are slow and secular, as has been true for the 20th century decline in mortality and for the transition to low fertility in some parts of the world. But many changes in rates occur over time intervals of a generation or less and are often quickly reversed: 20th century examples include fertility swings in industrialized countries, and short-term (decadal or faster) variability in age-specific mortality rates. Cohen (1976) formulated a mathematical description of such variability for a discrete-time population model in which fertility and mortality rates vary over time in response to some underlying stochastic process. (Cohen's original model assumed a finite-state Markov process but his results have been extended to many other stochastic processes.) The vital rates (mortality, fertility) vary over time but have a stationary probability distribution. A central assumption on the variability in rates is that any initial population structure (for example, with only children present, or only adults present) will eventually lead to a population in which every age class is represented. Given this condition, Cohen showed that, if the same random sequence of vital rates is applied to two populations with different initial age structures, the two populations will change subsequently so that their age structures become proportional over time. Thus the stochastic sequence of mortality and fertility rates generates a stationary stochastic sequence of population age structures that is maintained over time. This property is called *demographic stochastic weak ergodicity*. Let $X(t)$ denote the time sequence of vital rates and $Y(t)$ the time sequence of population structures. Then there is a joint stationary probability distribution of these two quantities. The stationary stochastic sequence of age structures is a *time-varying stable population*.

The property of stochastic weak ergodicity allows demographers to focus attention on the time-varying stable population. We note here some useful properties of this kind of stable population; see Tuljapurkar (1990) and Caswell (2001) for further discussion. The growth rate r of a classical stable population is replaced here by a long-run stochastic growth rate that satisfies a stochastic analog of the Lotka characteristic equation. The age structure itself can be described in terms of its moments, means, variances, covariances and autocorrelations. Variability over time in population structure reflects both the time-averages of mortality and fertility and the variances and covariances of these vital rates over time. Finally, stochastic theory allows us to characterize population trajectories in terms of probabilities of future events; for example, we can compute the probability that a dependency ratio will become unusually high or low over some specified time interval. Stochastic stable theory has been useful in demographic applications, especially to forecasting and fiscal problems (see below), and in a variety of ecological applications.

Population forecasts

The age structures of human populations can be effectively described, analysed, and even projected with reasonable accuracy using classical one-sex renewal theory. In many situations demographers are called upon to make extremely long-term forecasts of population number and composition; for example, the United States Social Security Administration's trustees require a 75-year forecast. In most forecasts made by institutions such as census bureaus or the United Nations, experts first make a set of alternative projections of mortality, fertility and immigration, and then generate projections along these alternative futures. These alternative futures are referred to as scenarios. In most cases, the vital rates in these projected scenarios settle to constant values after some initial period of time, often over a generation length or so. Correspondingly, the long-run forecast populations always approach classical stable populations that correspond to the alternative long-run rates.

Stochastic stable population theory generalizes this forecasting approach considerably by projecting vital rates as non-stationary stochastic processes. In most cases, the stochastic variability in the rates is fixed or changes very slowly, and there is a long-run slow secular change in the time-average vital rates. With vital rates projected in this way it is possible to generate stochastic population forecasts. At long times, these forecasted population structures usually approach stable time-varying populations. A major advantage of using stochastic stable theory is that probabilistic projections can be made of population structures, dependency ratios, and associated quantities of policy interest (Lee and Tuljapurkar, 2000).

SHRIPAD TULJAPURKAR

See also **economic demography; population ageing; population dynamics.**

Bibliography

Arthur, W.B. and Espenshade, T.J. 1988. Immigration policy and immigrants' ages. *Population and Development Review* 14, 315–26.

Caswell, H. 2001. *Matrix Population Models: Construction, Analysis and Interpretation*, 2nd edn. Sunderland, MA: Sinauer.

Coale, A.J. 1972. *The Growth and Structure of Human Populations: A Mathematical Investigation*. Princeton: Princeton University Press.

Cohen, J.E. 1976. Ergodicity of age structure in populations with Markovian vital rates, I: countable states. *Journal of the American Statistical Association* 71, 335–9.

Feichtinger, G. and Steinmann, G. 1992. Immigration into a population with fertility below replacement level – the case of Germany. *Population Studies* 46, 275–84.

Feller, W. 1941. On the integral equation of renewal theory. *Annals of Mathematical Statistics* 12, 243–67.

Feller, W. 1971. *An Introduction to Probability Theory and its Applications*, vol. 2, 2nd edn. New York: Wiley.

Fisher, R.A. 1930. *The Genetical Theory of Natural Selection*. Oxford: Clarendon Press.

Keyfitz, N.C. 1971. On the momentum of population growth. *Demography* 8, 71–80.

Keyfitz, N.C. 1977. *Introduction to the Mathematics of Population – With Revisions*. New York: Addison-Wesley.

Lee, R.D. 1994. The formal demography of population aging, transfers, and the economic life cycle. In *The Demography of Aging*, ed. L. Martin and S. Preston. Washington, DC: National Academy Press.

Lee, R. and Tuljapurkar, S. 2000. Population forecasting for fiscal planning: issues and innovations. In *Demography and Fiscal Policy*, ed. A. Auerbach and R. Lee. Cambridge: Cambridge University Press.

Leslie, P.H. 1945. On the use of matrices in certain population mathematics. *Biometrika* 33, 183–212.

Li, N. and Tuljapurkar, S. 1999. Population momentum. *Population Studies* 53, 255–62.

Lopez, A. 1961. *Problems in Stable Population Theory*. Princeton: Office of Population Research.

Lopez, A. 1967. Asymptotic properties of a human age distribution under a continuous net maternity function. *Demography* 4, 680–7.

Lotka, A.J. 1939. A contribution to the theory of self-renewing aggregates, with special reference to industrial replacement. *Annals of Mathematical Statistics* 10, 1–25.

Preston, S.H., Heuveline, P. and Guillot, M. 2000. *Demography: Measuring and Modeling Population Processes*. Malden, MA: Blackwell.

Sharpe, F.R. and Lotka, A.J. 1911. A problem in age-distribution. *Philosophical Magazine* 21, 435–8.

Tuljapurkar, S. 1982. Population dynamics in variable environments. IV: Weak ergodicity in the Lotka equation. *Journal of Mathematical Biology* 14, 221–30.

Tuljapurkar, S. 1990. *Population Dynamics in Variable Environments*. Berlin: Springer Verlag.

Vassiliou, P.-C.G. 1997. The evolution of the theory of non-homogeneous Markov systems. *Applied Stochastic Models and Data Analysis* 13, 159–76.

Stackelberg, Heinrich von (1905–1946)

Heinrich von Stackelberg was born on 31 October 1905, in Kudinowo, near Moscow, where his father was the director of a factory. The homeland of the family was the Baltic state of Estonia, although his mother was born in Argentina of Spanish descent. The family escaped the Russian revolution, retiring first to Yalta in the Crimea, and afterwards to Germany. They initially settled in Ratibor, Silesia but moved to Cologne in 1923. He completed his high school education in Cologne, studied economics at the University of Cologne, obtaining his 'Diplomvolkswirt' (master of economics) in 1927, 'Dr. rer.pol.' in 1930, and his habilitation in 1935.

He began his scientific career in 1928 as an assistant professor at the University of Cologne (1928–35). From 1935 until 1941 he was 'Dozent' and 'ausserordentlicher Professor' (associate professor) at the University of Berlin, and from 1941 until 1944 full professor at the University of Bonn. During World War II he was for some time drafted to military service. In 1944 and 1945 he held a guest professorship at the University of Madrid. He died at the early age of 41 in Madrid on 12 October 1946.

Stackelberg was the most gifted theoretical economist in Germany during his time. His habilitation thesis *Marktform und Gleichgewicht* (1934) has had a lasting influence on price theory. 'Stackelberg asymmetric duopoly' is known all over the world. His contributions to Austrian capital theory are the basis for all modern extensions of this theory. His textbook *Grundzüge der theoretischen Volkswirtschaftslehre* (1943) was the first 'modern' introduction to economics in the sense that it is based on a coherent theory of household and firm behaviour. Moreover, Stackelberg contributed to several other fields: cost theory, exchange rate theory, saving theory and others. In Germany he was one of the few leading economists who introduced mathematics into economics and took up the Anglo-Saxon approach in price and cost theory (Edgeworth, Marshall, Hicks, Harrod, Chamberlin and others).

The difficulty of oligopoly theory consists in the fact that the oligopolists are in a game theoretic situation which, in general, cannot be put into the form of a pure maximum problem. Stackelberg's seminal idea was that this can nevertheless be done if – in the case of a duopoly – one firm takes a 'dependent' position (i.e. takes the actual price or production of the other firm as given) and the other an 'independent' one (i.e. knows this behaviour and fixes its price or production accordingly so that it maximizes its profits or other utility indices). If both firms wish to be in the 'dependent' position, a Cournot-type equilibrium results; on the other hand if both firms wish to be in the 'independent' position, a contradiction arises since each firm assumes a behaviour of the other which is incompatible with its actual behaviour. If they nevertheless fix their prices (or production) at that level, a 'Bowley'-type oligopoly solution, as Stackelberg calls it, would emerge. Since it is unclear which position the firms will take, Stackelberg considered the oligopoly as a market form without equilibrium. *Marktform und Gleichgewicht* (1934) is comparable with Chamberlin's *The Theory of Monopolistic Competition* (1933) and Joan Robinson's *The Economics of Imperfect Competition* (1933), but goes further in the analysis and in mathematical rigour.

Stackelberg accepted Austrian capital theory, which emphasizes the time structure of production ('zeitlicher Aufbau der Produktion'). The main drawback of this theory is that one of its basic concepts, namely the average gestation period, could not be well defined and measured for a modern interdependent economy. In 'Kapital und Zins in der stationären Verkehrswirtschaft' (1941) Stackelberg suggests the following solution. In a simple economy where the original factor input takes place in period O and the product ripens by nature (such as in the production of wood), the subsistence fund S, the yearly income (= harvest) Y, the interest factor $q=1+r$, where r = rate of interest, and the gestation period T satisfy the relation $S = Y/q^T$. Stackelberg defines an economy as equivalent to this simple economy, if they correspond with respect to S, Y and r, where S is identified with labour income L. Thus the average gestation period may be calculated by $T = (\log Y - \text{Log } L)/\log q$.

In the article 'Beitrag zur Theorie des individuellen Sparens' (1939), Stackelberg deals with the problem: why does a household save? What are the effects of interest rate expectations on household saving? He took up the conceptual framework of Hicks and Allen (1934) and applied it to the allocation of expenditures in the time space. He derived Böhm-Bawerk's law of under-evaluation of future commodities from the law of declining marginal rates of substitution and showed how the optimal allocation of expenditure in time depends on it.

Stackelberg was a neoclassical economist. In his opinion, Keynes really added nothing new to available economic knowledge. He also considered Keynes's interest rate theory as a special case of Böhm-Bawerk's theory of exchange of present against future commodities (see 'Zins und Liquidität', 1947).

Stackelberg kept intimate relations with that group of German economists (Walter Eucken, Erwin v. Beckerath and others) who during the war prepared the transition of the German economy to a free enterprise system. In spite of his untimely death, his influence especially on economic theory in Germany was most important in the sense that he initiated the reorientation of German economic thinking to the Anglo-Saxon approach. His very original contributions to economic theory have had a lasting effect.

WILHELM KRELLE

Selected works

1932. *Grundlagen einer reinen Kostentheorie*. Vienna: Julius Springer.

1934. *Marktform und Gleichgewicht*. Vienna and Berlin: Julius Springer.

1938a. Probleme der unvollkommenen Konkurrenz. *Weltwirtschaftliches Archiv*, 95–141.

1938b. Arbeitszeit und Volkswirtschaft. *Jahrbuch des Arbeitswissenschaftlichen Instituts*, Berlin, 61–86.

1938c. Das Brechungsgesetz des Verkehrs. *Jahrbüchr für Nationalökonomie und Statistik*, 680–96.

1939a. Beitrag zur Theorie des individuellen Sparens. *Zeitschrift für Nationalökonomie*, 167–200.

1939b. Theorie der Vertriebspolitik und Qualitätsvariation. *Schmollers Jahrbuch für Gesetzgebung, Verwaltung und Volkswirtschaft im Deutschen Reich*, 43–85.

1940. Die Grundlagen der Nationalökonomie, Bemerkungen zu dem gleichnamigen Buch von Walter Eucken. *Weltwirtschaftliches Archiv* 41, 245–85.

1941. Kapital und Zins in der stationären Verkehrswirtschaft. *Zeitschrift für Nationalökonomie*, 25–61.

1941a. Elemente einer dynamischen Theorie des Kapitals. *Archiv für mathematische Wirtschafts-und Sozialforschung*, 8–29, 70–93.

1943. *Grundzüge der theoretischen Volkswirtschaftslehre*. Stuttgart and Berlin: Kohlhammer.

1944. Theorie des Wechselkurses bei vollständiger Konkurrenz. *Jahrbücher für Nationalökonomie und Statistik*, 1–65.

1947. Zins und Liquidität. Eine Auseinandersetzung mit Keynes. *Schweizerische Zeitschrift fü Volkswirschaft und Statistik*, 311–28.

1951. *Grundlagen der theoretischen Volkswirtschaftslehre*. Tübingen and Zurich: J.C.B. Mohr and Polygraphischer Verlag.

Bibliography

Chamberlin, E. 1933. *The Theory of Monopolistic Competition*. Cambridge, Mass: Harvard University Press.

Hicks, J.R. and Allen, R.G.D. 1934. A reconsideration of the theory of value. *Economica*, NS 1, 196–219.

Robinson, J. 1933. *The Economics of Imperfect Competition*. London: Macmillan.

Stakhanovism

Stakhanovism was a movement begun in the Soviet Union in 1935 to increase labour productivity by the popularization of work techniques reputedly initiated by workers themselves.

On 31 August 1935, Aleksei Grigorevich Stakhanov, a 30-year-old miner in the Donets Basin, cut 102 tons of coal during his six-hour shift. This amount represented 14 times his quota, and within a few days was hailed by *Pravda* as a world record. Anxious to celebrate and reward individuals' achievements in production that could serve as stimuli to other workers, the Soviet Union's Communist Party launched the Stakhanovite movement, or Stakhanovism. The title 'Stakhanovite', conferred on workers and peasants who set production records or otherwise demonstrated mastery of their assigned tasks, quickly superseded that of 'shock worker' (*udarnik*). Day by day throughout the autumn of 1935, the campaign intensified, culminating in an All-Union Conference of Stakhanovites in industry and transportation which met in the Kremlin in late November. Outstanding Stakhanovites mounted the podium to recount how, defying their quotas and often the scepticism of fellow workers and bosses, they applied new techniques of production to achieve stupendous results for which they were rewarded with wages that reached dizzying heights. Stalin captured the upbeat mood of the conference when, by way of explaining how such records were possible only in the 'land of socialism', he uttered the phrase, 'Life has become better, and happier too.' Widely disseminated, and even set to song, Stalin's words served as the motto of the movement.

The year 1936 was declared a Stakhanovite year. Competitions among workers during designated Stakhanovite months, ten-day periods (*dekady*), and shifts spread the movement everywhere in the Soviet Union. Not a single place of work was without its Stakhanovites. Even the Gulag got into the act. The highest proportion of Stakhanovites could be found in the extractive industries, the energy sector, and railway transportation, where upwards of 40 per cent of all workers were so designated by August 1936. Young male workers who had passed technical training courses, were classified as at least semi-skilled and had an average of three to five years experience were most likely to be represented among Stakhanovites. However, what was characteristic in industry was not the case in agriculture, where the most prominent Stakhanovites were women. They included Pasha Angelina – brigade leader of the first all-female tractor brigade – and Maria Demchenko, the sugar-beet cultivator. The enormous publicity surrounding these and other female collective- and state-farm workers suggests a continuation of the party's efforts, begun during collectivization, to forge an alliance with rural women against the previously dominant patriarchy of peasant families. Lending support to this interpretation is the rather frequent mention in rural female Stakhanovites' testimonials of having been orphans and of overcoming the resistance of unenlightened husbands.

Stakhanovism encompassed lessons about not only how to work but how to live. In addition to providing a model for success on the shop floor, in the mine, or in the field, it conjured up images of the good life. Many of the qualities Stakhanovites were supposed to exhibit at work – cleanliness, neatness, punctuality, preparedness, and a keenness for learning – were applicable at home, too. These qualities were associated with *kulturnost* ('culturedness'), the acquisition of which marked an individual as a New Soviet Man or Woman. Advertisements for perfume in journals intended for Stakhanovites, articles about Stakhanovites on shopping sprees, photographs of Stakhanovites sharing their happiness with their families, newsreels showing them moving into comfortable apartments and driving new automobiles presented to them as gifts, all symbolized *kulturnost*. Wives of male Stakhanovites had an important part to play in the movement as helpmates preparing nutritious meals, keeping their apartments clean and comfortable, and otherwise creating a cultured environment in the home so that their husbands were well rested and eager to work with great energy. It was also important to demonstrate that Stakhanovites were admired by their workmates and considered worthy of holding public office.

Stakhanovites, however, were not necessarily popular. Even before the raising of output norms in early 1936, workers who had not been favoured with the best conditions and consequently struggled to fulfil their norms expressed resentment of Stakhanovites by verbally and even physically abusing them. Foremen and engineers, only too well aware that 'recordmania' and the provision of special conditions for Stakhanovites created disruptions in production and bottlenecks in supplies, also on occasion 'sabotaged' the movement. At least that was the accusation made against many who often served as scapegoats for the failure of Stakhanovism to fulfil its promise of unleashing the productive forces of the country. Thus, at least in an indirect way, Stakhanovism fed the Terror of 1936–8.

In the course of those years, quite a few Stakhanovites received special educational training followed by promotion into the ranks of management; others were sent on tours of worksites where they demonstrated their skills; many became deputies in the Supreme Soviets of the USSR and its constituent republics. Eventually, Stakhanovism was routinized, becoming merely another task for party and trade union committees to carry out. It continued into the war and even enjoyed something of a revival in the post-war years when it was exported to eastern Europe. Stakhanov himself served in a number of honorific administrative positions in the coal-mining industry before descending into alcoholism and disability. When he died in 1977 he was all but forgotten, although the eastern Ukrainian town of Kadievka (Lugansk oblast) where he had worked and set his record had its name changed to Stakhanov. The 50th anniversary of Stakhanovism in 1985 was observed in the Soviet Union with an outpouring of popular and scholarly literature, museum displays, lectures, and special exhibitions, all of which were intended to inspire the 'Stakhanovites of the 1980s'. It is hard to imagine Stakhanovism inspiring anything in the post-Soviet era except perhaps ridicule.

LEWIS H. SIEGELBAUM

See also **labour discipline; Soviet Union, economics in; Soviet economic reform; Stalinism, political economy of.**

Bibliography

Buckley, M. 1999. Was rural Stakhanovism a movement? *Europe-Asia Studies* 51, 299–314.

Co-operative Publishing Society of Foreign Workers in the U.S.S.R. 1936. *Labour in the Land of Socialism: Stakhanovites in Conference.* Moscow.

Davies, R. and Khlevnyuk, O. 2002. Stakhanovism and the Soviet economy. *Europe-Asia Studies* 54, 867–904.

Siegelbaum, L. 1988. *Stakhanovism and the Politics of Productivity in the USSR, 1935–1941.* New York: Cambridge University Press.

Stalinism, political economy of

'Stalinism' refers to a political–economic system of state ownership and administrative-resource allocation directed by a monopoly political party (in Stalin's case, the Politburo of the Central Committee of the Communist Party, or Stalin himself), which combines economic incentives and 'repression' to achieve its economic and political goals (Gregory, 2004). Stalinism was first created in the Soviet Union as Stalin gained totalitarian authority and has subsequently been practised in modified forms by dictators in China (Mao), Cuba (Castro), North Korea (Kim Jong Il), Iraq (Saddam Hussein) and Cambodia (Pol Pot). 'Early' Stalinism dates from the expulsion from ruling circles of Stalin's last significant opposition in 1929 and 1930, when Stalin still had to build majorities within the Politburo (Khlevnyuk, 1996). 'High Stalinism' dates from late 1934 until his death in March 1953, when Stalin's decisions could no longer be challenged and his top officials carried out orders rather than participated in collective decisions. By the mid-1930s, the Politburo had ceased to meet regularly, and the party and state were run by informal groups appointed directly by Stalin. Stalinism is distinguished from the 'administrative-command' system of the post-Stalin era by the latter's less extreme use of repression and by its collective leadership.

The economic policy of Stalinism was marked by extremely high rates of capital accumulation in heavy industry and defence, with lesser priority for consumer goods and services. Nevertheless, there were investment cycles both in Stalin's Russia and throughout the Soviet empire, even though investment was dictated politically rather than by market forces. Although some interpret these investment cycles as temporary bouts of moderation, Stalin deliberately reduced investment when he feared that low real wages would harm work effort or, worse, lead to uncontainable civil unrest. Stalin used his secret police to gauge the mood of workers and peasants for this purpose.

Although Stalin's predecessor, V.I. Lenin, established the institutions of terror, such as a special secret police for political enemies and an arbitrary system of 'socialist legality', it was Stalin who initiated 'mass operations' against his enemies. In January 1930, he ordered the arrest, deportation, and execution of kulaks (wealthier peasants and virtually any perceived regime opponent). In this 'dekulakization' campaign, more than two million peasants were deported to 'special settlements' or to the 'corrective labour camps' of the Gulag. The term 'Gulag' denotes the Main Administration of Camps under the jurisdiction of the interior ministry. Stalin used the assassination of the Leningrad party boss in December of 1934 to purge the party and state leadership of remaining political enemies. Their executions were pronounced at the Moscow show trials, the first being the 1936 trial of G. Zinoviev and L. Kamenev, expelled Politburo members and opponents of Stalin. The purge of the party elite broadened into 'mass operations', or the Great Terror, in

July 1937 initiated by telegrams from Stalin and operational plans by NKVD head, N. Ezhov, which set execution and imprisonment quotas for 65 regions. The NKVD was the Peoples' Commissariat for Internal Affairs, which was charged with carrying out terror operations. The Great Terror was tightly controlled by Stalin from start to finish. Although the original quotas called for 70,000 executions, almost 700,000 executions took place between July 1937 and November 1938. Although various theories exist as to why Stalin ordered these mass executions, it appears that he wished to create a new generation of party leaders to replace the 'Old Bolsheviks' who had lost their revolutionary fervour, and also wished to rid the Soviet Union of 'socially harmful' classes, who could have formed fifth columns during the Second World War.

After the Great Terror, Stalin used 'lesser terror', whereby he imposed criminal penalties on huge numbers of ordinary people for workplace violations, such as theft of state or collective property, no matter how petty. He continued, however, 'greater terror' with massive deportations of national groups in the 'mobilization years' of 1940–6. Although many expected a relaxation of terror at the end of the war, in fact the Gulag's population more than doubled. Stalin's lesser terror decrees remained in effect until his death in March of 1953, although many of them had fallen into disuse.

Stalin as dictator

From approximately 1932 until his death Stalin was a true dictator: he had his way on every matter and was not afraid to abuse and humiliate his associates (Rees, 2004; Gorlizki and Khlevnyuk, 2004). As Khlevnyuk (2001a, p. 325, emphasis added), concluded, 'Stalin himself was not merely a symbol of the regime but the leading figure who made the principal decisions and initiated *all* state actions of *any* significance.' After 1930, Stalin increasingly bypassed formal procedures as reflected in the declining frequency of Politburo meetings (Rees and Watson, 1997, p. 12) and the use of ad hoc subcommittees that he personally appointed (Wheatcroft 2004, p. 91). Stalin continued to bind his associates into complicity by requiring each Politburo member to approve his decisions once he had made them (Gorlizki and Khlevnyuk, 2004).

Despite his dominance, Stalin faced massive principal–agent problems with his associates. His correspondence is full of concern about 'paper fulfilment' and of angry calls for monitoring fulfilment and increased responsibility for designated officials (Gregory, 2004, pp. 165, 266).

The erosion of collective rule is consistent with Hayek's (1944) insight that the rise of a sole dictator is inevitable in such an environment. Stalin's ascendancy is explained by the need for a tie-breaker as the Politburo members quarrelled, but, more importantly, by the fact that Stalin was more ambitious, brutal and controlled than his rivals. There were no 'moderates' or 'extremists'

within the Politburo after 1930. Stalin did not have to confront major ideological or policy differences. The divisions that did exist were on lines of narrow self-interest based on departmental position.

Stalin made top-level appointments personally, was deeply suspicious of professional administrators and technocrats, and trusted only a few old Bolsheviks. Stalin was particularly concerned about rent seeking by those within his narrow circle who represented branch or territorial interests, 'who cause us to deceive each other' (Khlevnyuk et al., 2001, p. 80), and 'who turn our Bolshevik party into a conglomerate of branch groups' (Rees and Watson, 1997, p. 16).

Planning under the Stalinist system was quite different from its textbook description. The state planning agency, Gosplan, prepared only highly aggregated plans, stating: 'Gosplan is not a supply organization and cannot take responsibility either for centralized specification of orders by product type or by customer or the regional distribution of products' (cited in Gregory, 2004, p. 139). Gosplan refused to plan actual transactions, labelling them 'syndicate work' (Belova and Gregory, 2002, p. 271). Gosplan only reluctantly represented the state in inter-ministerial conflicts, claiming that 'we are simply not equipped to deal with such matters' (Belova and Gregory, 2002, p. 271). In short, after its 1929 purge and subsequent politicization, Gosplan limited its exposure by doing as little as possible. The ultimate power to direct resources belonged to the dictator. Stalin did not want a planning board with immense powers or numerous staff. He did not trust information from those accountable for results, a phenomenon Wintrobe (1998) labelled the 'dictator's curse'. Truth-telling was the specialized task of Gosplan and other agencies such as the NKVD, which became Stalin's solution to the wider principal–agent problem (Belova and Gregory, 2002, pp. 269–73).

Hayek's (1944, p. 82) assertion that a totalitarian system 'cannot tie itself down in advance to general and formal rules that prevent arbitrariness... It must constantly decide questions which cannot be answered by formal principles only' was true of Stalinist practice. There were few formal rules; the rules that existed were subject to override. Fresh guidelines were issued to plan each new year or quarter, rather than general planning rules being carried over. Ministries operated without corporate-governance charters (Gregory and Markevich, 2002, pp. 793–4). 'Administrative' enforcement was encouraged through appeals to vertical superiors (Belova, 2001). Planning procedures were complicated, contradictory, and confusing (Markevich, 2003). Enterprise usually received a few output and assortment assignments midway through the plan period, while secondary targets for costs and productivity were worked out retrospectively for reporting purposes. All plans, labelled 'draft' or 'preliminary', were no more than informal agreements which could be changed subsequently by virtually any superior.

The 'correcting' and 'finalizing' of plans was a never-ending process; the 'final' plan remained always on the horizon (Markevich, 2003).

Resources were allocated in the course of the 'battle for the plan' during which superiors were barraged by requests to intervene. Intervention was generally arbitrary but was based on some implicit rules of thumb, such as the priority of heavy industry and the military: 'All orders for the Ministry of Defense must be fulfilled exactly according to the schedule not allowing any delays' (Gregory, 2004, pp. 160–1). Plan interventions created havoc for producers. The most important industrial leader of the 1930s expressed his frustration as follows: 'They give us every day decree upon decree; each one is stronger and without foundation' (Khlevnyuk, 1993, p. 32).

Stalin himself favoured such ad hoc administrative allocation: 'Only bureaucrats can think that planning work ends with the creation of the plan. *The creation of the plan is only the beginning*. The real direction of the plan develops only after putting the plan together' (Stalin, 1937, p. 413; emphasis added). Ad hoc allocation could not have been better designed for the exercise of political influence. Everything was tentative and subject to arbitrary change by someone higher up in the chain of command. Savvy politicians like Stalin would have been able to weigh the political benefits of satisfying an influential regional or industrial leader.

Stalin's unwillingness to bind himself in advance to rules cascaded down through the political system, preventing the emergence of a 'law-governed' economy. The commitment failure outlined by Kornai, Maskin and Roland (2003) – the predatory principle forcing agents to break formal rules so as to exploit them – seems to fit the Stalin model. In the Soviet case producers could break rules citing the threat to production from the rule, while superiors reserved the right to punish hapless scapegoats for breaking the same rules.

It was the power to live outside formal rules that sentenced Stalin and his Politburo to lives of toil, drudgery and tedium (Gregory, 2004, pp. 68–72). Threats of resignation and pleas for lengthy vacations were commonplace. A representative Politburo meeting, held on 5 March 1932, had 69 participants and 171 points on its agenda (Khlevnyuk et al., 1995, p. 232). The greatest burden fell on Stalin who in 1934, a typical year, spent 1,700 hours in official meetings, the equivalent of more than 200 eight-hour days (Khlevnyuk, 1996, pp. 190–1). Virtually every communication requested a decision from him.

Accumulation and consumption
Dictators could aim for economic growth (Olson, 1993; Glaeser et al., 2004) or for self-enrichment, or share rents to build loyalty and political power (Wintrobe, 1998). Stalin was clearly obsessed with accumulation (hardly a surprise given Marx's emphasis on accumulation), which was captured in the growth models of Preobrazhenskii and Fel'dman in the early Soviet period (Erlich, 1960; Spulber, 1964). At the core of Stalin's strategy to 'build socialism' were massive programmes for the hydroelectric dams, machinery complexes, vehicle works, blast furnaces, railways, and canals that were included on its itemized 'title lists' of approved projects.

Although Politburo meetings in the 1930s left few formal minutes, records reveal that the Politburo consistently set the *nominal* investment budget, grain collections, and foreign exchange, three variables related to investment (Gregory, 2004, chs 4 and 5). The investment budget allotted 'investment rubles' to industrial and regional agencies for construction and machinery, although no one appeared to know the real investment that resulted. Grain collections were designed to contribute to a budget surplus through the excess of state sale prices over purchase prices. Stalin personally directed foreign exchange to foreign capital goods rather than the luxury goods sometimes demanded by the Bolshevik elite.

If Stalin's goal was indeed to maximize investment, two facts are, at first glance, confusing. First, Stalin was extremely concerned about consumption. In Stalin's words, the 'provisioning of workers' was one of 'the most contested issues' before the Politburo, and trade was 'the most complicated ministry' (Gregory, 2004, pp. 93–4). Stalin personally ordered consumer goods to cities where labour productivity was declining (Gregory, 2004, ch. 4). The Politburo decided retail trade plans, prices, assortment, and even the opening of new stores. The second confusing fact is that Stalin reduced the nominal investment budget on two occasions, in 1933 and 1937 (Davies, 2001). Such evidence can be interpreted either as unstable dictatorial preference or as a consistent rule of thumb to decide the volume of investment. Stalin's capacity for calculation, patience, and self-control suggest the stable preferences approach is correct.

Figure 1 illustrates the model that Stalin and his Politburo used to set investment and consumption. The figure captures the Marxian concept of the surplus product, the gap between output and consumption, as the outcome of a distributive struggle. The model has theoretical precursors in Schrettl (1982; 1984), and is set out more fully by Gregory (2004, ch. 4). It belongs to a general class of models in which a ruler's freedom of action is circumscribed by social 'tolerance limits' (Kornai, 1980, pp. 211–14) or a revolution or disorder constraint (Acemoglu and Robinson, 2000). If the Politburo increased investment too much, it risked provoking the workers to provide less effort or even rebel.

In the Stalinist system, the demand for labour was always enough for full employment and all able-bodied persons were required by law to work (Granick, 1987); thus, employment, N, was fixed exogenously; individual

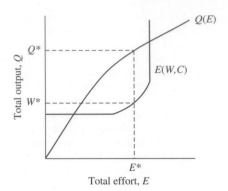

Figure 1   The investment maximization model

effort, e, was variable, so total effort, $E = e \cdot \overline{N}$, was variable although employment was not. Total output, Q, depended on total effort, E. Total effort varied with the real wage, w, as follows. The aggregate wage bill W, the consumer goods received by workers, is measured along the vertical axis in the same units as output, and is proportional to the real wage given that employment is fixed, that is, $W = w \cdot \overline{N}$. There is a reservation wage, analogous to a tolerance limit or disorder constraint, below which effort is zero; there is also a 'fair' wage at which effort is maximized. This effort curve bears some resemblance to Akerlof (1984) as applied at the micro-level. As the economy moves from the fair wage to the reservation wage, effort declines; at the lower limit unrest threatens to boil over into strikes and rebellion. Thus the effort curve intersects the horizontal axis at the reservation wage and becomes vertical at the fair wage. Effort also depends on the level of repression or coercion, C, discussed below. To maximize effort, the dictator would pay the fair wage and get the maximum output, but this would not maximize the surplus. To maximize the surplus, $Q - W$, he would choose the intermediate wage, effort, and output levels denoted W^*, E^*, and Q^*.

An effort curve of this shape makes the consequences of plan mistakes asymmetric: paying the workers too little could be much worse than paying them too much. Paying workers too little not only cuts effort but also risks outright confrontation with the state. An investment-maximizing dictator must tread a fine line between the pursuit of investment and the triggering of serious disorder.

Stalin managed worker morale and effort in two ways. When investment and consumption were about right in the aggregate, plan mistakes were taken care of by reallocating consumer goods to those left short. In the case of aggregate mistakes, when too much investment threatened to provoke the workers, investment was cut back, such as in 1933 and 1937. In this sense, Stalin's behaviour was stable and consistent, given the constraints that he perceived.

The fair wage was set by a mass psychology that was unpredictable and hard to manipulate. If workers concluded from the propaganda of economic successes that they were being cheated, the fair wage would rise, forcing the Politburo to cut investment back. Stalin used the vast informant network of the NKVD to monitor protests, strikes, anti-Soviet statements, and factory-wall graffiti, and eavesdrop to gauge mass opinion (Berelovich, 2000). Stalin had obvious political motives to do this, but within our framework wages and fairness lay at the cross-hairs of politics and economics.

Figure 1 suggests other options. Stalin and his closest subordinates could seek to manipulate the effort curve by offering ideological rewards in place of material payoffs. The attempt to transform *homo economicus* into *homo sovieticus* led, however, to a vicious circle of wage equalization and declining productivity (Kuromiya, 1988; Davies, 1996). The Stakhanovite movement, the most publicized mobilization campaign of the 1930s, was driven by progressive piece rates that permitted participating workers to drive up their incomes by overfulfilling norms (Davies and Khlevnyuk, 2002). Stalin abandoned it because it tended to raise fair-wage aspirations among non-participating workers, and also threatened inflation (Gregory, 2004, pp. 104–6). Stalin also saw targeted rationing as a way to force accumulation without a loss of effort of high-priority workers: 'He who does not work *on industrialization* shall not eat' (cited in Gregory, 2004, p. 98; emphasis added). Finely targeted rationing, however, required a massive bureaucracy and proved to be a blunt instrument (Davies and Khlevnyuk, 1999).

Coercion and accumulation

The dictator could also use coercion to cause workers to lower their reservation wage without reducing effort. As long as coercion displaces the effort curve in Figure 1 downward while leaving the production curve undisturbed, the surplus is increased and coercion is 'successful'. Stalin conducted three notable experiments with coercion to foster accumulation: the forced collectivization of the peasantry, the criminalization of workplace behaviour, and the use of forced labour.

Politically, collectivization aimed to impose Soviet power in the countryside and eliminate the stratum of richer peasants. Economically, collectivization was to give the state agricultural products at low state-dictated delivery prices, which could be sold domestically and abroad for a profit. In a word, collectivization's aim was to lower peasant living standards while controlling their effort administratively. Collectivization was triggered by the peasants' perceived unwillingness to contribute sufficiently to investment-led industrialization (Wheatcroft and Davies, 2002; Davies and Wheatcroft, 2004). The collective farms enabled Moscow to replace local decision making with central plans on sown acreage and obligatory deliveries. Acreage expanded but yields collapsed, while the share delivered to the state increased. Excessive procurements, bad weather, and plan errors combined to strip the countryside of grain; first the livestock were slaughtered, then the farmers themselves starved, threatened by severe punishments including death for theft of agricultural products. Davies and Wheatcroft (2004) dispel Conquest's (1987) notion that Stalin manufactured the famine of 1932/33 to kill class enemies; rather they show an inept leadership subsequently trying to ameliorate the effects of its own bungling.

In 1932/33, Stalin intentionally directed food to those able to work in the fields and denied it to those already hospitalized by hunger (Davies and Wheatcroft, 2004, ch. 13). Ellman (2000) has also applied the entitlement theory of Sen (1983) to the 1946/7 famine and shows that the role of the state was essentially negative: it again selected those who died by denying them entitlements. Thus, concentrating grain stocks in the hands of the Soviet state actually increased the number of deaths. The 5.5 million–6.5 million famine deaths in 1932/33 far exceeded deaths in pre-revolutionary famines (Davies and Wheatcroft, 2004, pp. 402–3).

A variety of studies, including Millar (1974), conclude that collectivization did not increase the 'agricultural surplus' defined as the gap between the value of output produced in agriculture and output consumed in agriculture, due to the need to shift investment resources into agriculture to make up for the loss of animal draft power slaughtered during collectivization.

As the 1940s began, Stalin redirected coercion from specific class enemies to the entire public-sector work force. A battery of intimidating laws criminalized workplace violations which had previously been punished by administrative sanctions within the enterprise. The law of 26 June 1940 (Kozlov, 2004, vol. 1) made absenteeism, defined as any 20 minutes' unauthorized absence or even idling on the job, a criminal offence, punishable by up to six months' corrective labour with a 25 per cent reduction in pay. Repeat offences counted as unauthorized quitting, punishable by two to four months' imprisonment. Enterprise managers were made criminally liable for failure to report worker violations. In August 1940 the minimum sentence for petty theft at work and 'hooliganism' was set at one year's imprisonment. The notorious decree of June 1947 raised the minimum sentence for any theft of state or socialized property to five and seven years imprisonment. These punitive laws remained on the books until Stalin died.

A report prepared as background for Khrushchev's secret de-Stalinization speech of February 1956 (Kozlov, 2004, vol. 1, statistical appendix) shows that from 1940 through June 1955 a total of 35.8 million persons were sentenced for criminal offences. With repeat offenders not allowed for, this would represent about one-third of the adult population of roughly 100 million. Of the 35.8

million, 15.1 million were imprisoned and a quarter of a million were executed. These laws placed cumulative totals of millions of people in prisons and camps outside of the 'civilian' workforce. The goal of lesser terror was to increase the effort for the same real wage by criminally punishing work indiscipline. Unlike collectivization, there have been no decisive evaluations of the success or failure of these draconian policies. What is known is that they fell into misuse and were formally removed after Stalin's death. If they had been successful, they would have been retained.

The Gulag

Collectivization, the Great Terror, the repression of state employees in the 1940s, and the arrests of 'national contingents' created huge flows into the Gulag, the interior ministry's chief administration of labour camps, created in 1930 to manage camps which at their peak housed more than 2.5 million inmates. Similar numbers of deportees were confined to labour settlements in the remote interior. The forced labourers were engaged in forestry, mining and construction, where they made up substantial shares of employment, but never more than about three per cent of the total workforce including farm workers (Khlevnyuk, 2001; 2003). The cumulative total of persons sentenced to the Gulag in the course of its existence, probably in excess of 20 million, remains the subject of debate. The Gulag's own central catalogs are inconsistent (Kozlov, 2004, vol. 2); it appears that even the Gulag did not know the correct number.

Political repression strategy (collectivization, terror, war) rather than economics dictated the Gulag's development, but, once created, the Gulag represented a tempting source of cheap labour. The Gulag's consistent economic raison d'être was to explore and colonize regions that were resource-rich but inhospitable, since forced labour could be ordered around the country at will (Khlevnyuk, 2003). Subsistence wages combined with the enforcement of effort through close supervision were supposed to promote the low-cost accumulation depicted in Figure 2.

The primacy of politics over economics is shown by the fact that the NKVD and its successor the MVD (the Ministry for Internal Affairs) did not lobby for expansion. The NKVD projected a shrinking number of inmates for the third five-year plan (1938–42), just as the first victims of Stalin's Great Terror began to flood in (Gregory, 2003, p. 4). In the late 1940s Gulag officials proposed to release all but the most dangerous prisoners from camps (Tikhonov, 2003), but this was unacceptable to Stalin. In 1953, within three months of Stalin's death, MVD chief Lavrenty Beriia released one and a half million prisoners, 60 per cent of the Gulag's inmates according to a plan prepared five years earlier. The MVD was increasingly alarmed by the Gulag's economic and social costs. Its economic costs were reflected in growing

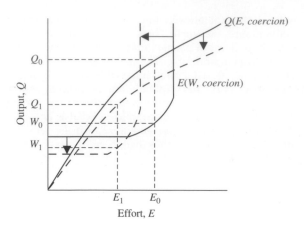

Figure 2   'Failed' coercion?

financial deficits; the social costs were measured by high rates of recidivism. Although the camps were supposed to segregate hardened criminals from youth offenders, the camps were mixing bowls and the high turnover spread the culture and mores of camp life throughout society.

Hopes for huge profits from the Gulag were quickly dashed. In the Far Eastern camps (Nordlander, 2003) early optimism about huge surpluses in gold mining was replaced by pessimism as output per inmate fell precipitously. The fact the White Sea–Baltic Canal (Morukov, 2003) was finished on time and on budget stimulated high expectations, until its major construction flaws became apparent. The Gulag leaders underestimated the risks of building Noril'sk (Borodkin and Ertz, 2003) and exposed the illusions that inmates could be coerced into supplying effort without economic rewards (Ertz, 2003).

By the post-war years, Gulag officials had concluded that the camps were operating at a loss. Labour productivity was much lower than that of free workers, while guarding detainees was very expensive; in 1950 there was one guard to ten inmates, leading to the widespread practice of 'unguarded' prison contingents. Prisoners formed protective networks and actually operated a number of camps (Heinzen, 2004). The arsenal of punishments was not sufficient to motivate prisoners, and trade-offs were complicated: prisoners placed on reduced rations for failing to meet work quotas could not work effectively. The most effective incentive systems, such as early release for exemplary work, deprived the Gulag of its best workers. Material incentives played an ever larger role in motivating penal labour (Borodkin and Ertz, 2003; 2005; Ertz, 2005). In the last years of the Gulag, prisoners were paid civilian wages (albeit at lower scales) and the distinctions between penal and free labour became blurred.

Evaluating repression

Effective coercion requires that penalties be accurately assessed and targeted, and that the agents of repression be

well informed about offenders and the costs of their crimes. The relationship between true effort and punishment was 'noisy', and oppressive law could do little more than ensure that workers or inmates were physically at work and did not steal too much. Agricultural controllers could order the collective farms to sow more land but could not assess whether the land was being farmed efficiently (Davies and Wheatcroft, 2004). In industry, attempts to pin 'normal' effort down to objective technological criteria proved fruitless (Davies and Khlevnyuk, 2002, p. 877; Filtzer, 2002, pp. 232–41).

The investigation of low effort could yield an error of Type I that punished the innocent, or a Type II error that acquitted the guilty. Type I errors are reflected in the high rates of penalization that condemned hard workers along with neer-do-wells, drunks and thieves. Virtually every worker became liable to prosecution for some offence, including one-time and accidental violations. Rational managers might wish to prosecute only problem workers and repeat offenders, but the laws penalized even petty offences, and managers who failed to report offences were threatened with the same. As a result, the innocent were bundled along with the guilty in extraordinarily large numbers.

In March 1937, Nikolai Ezhov, Stalin's chief instrument of the Great Terror, told his officials to expect 'some innocent victims … Better that ten innocent people should suffer than one spy get away. When you chop wood, chips fly' (Montefiore, 2003, p. 194). The attitude of Vyacheslav Molotov, Stalin's prime minister (interviewed by Chuev, 1991, p. 416) was, 'never mind if extra heads fall', even when one of those 'heads' was that of his own wife. Type II errors were evidenced by the fact that, although penalization rates were very high, offending rates were even higher (Filtzer, 2002, pp. 167–8). A judicial system that was supposed to 'make the chips fly' somehow failed to chop the wood. The combination of severe penalization and low conviction probability for the guilty is consistent with high-cost policing and justice administration (Becker, 1968, p. 184); the high rate of conviction of innocents, however, is a cost of the dictator's efforts to achieve a lower rate of offending than society was willing to tolerate (Djankov et al., 2003). Although the Gulag did not generate an *internal* surplus, it could have displaced the effort curve of civilian workers if they expected the Gulag wage as their punishment for low effort. But if workers expect Type I errors, they will be punished regardless of effort; if they expect to benefit from Type II errors, they can shirk without fear.

Error rates were not exogenous. They were fashioned by the counteractions of those threatened, who could take steps to reduce their risks. Workers and managers diverted effort from production into mutual insurance: since the threat was shared among them, they could agree to cover up each other's shortcomings. Post-war managers tolerated lateness and absence to maintain goodwill, and underreported such violations, while pursuing quitters who undermined morale and the factory's capacity to fulfil the plan (Filtzer, 2002). The rural police and courts pooled risks with the rural community in sheltering the young offenders who had deserted factories or technical schools (Kozlov, 2004, vol. I). Mutual insurance tended to cut the individual risk of punishment. Regional party officials defied even the most powerful central organizations to protect their own (Harris, 1999, pp. 156–63). High-level patrons could protect the most egregious embezzlers (Belova, 2001).

Figure 2 explains the phenomenon of 'failed coercion'. The threat of punishment raised the 'noise' of the system as managers and regional officials reported false results and formed mutual protection networks, shifting the output–effort curve down. The leftward movement of the effort curve captures the diversion of activity from production to the avoidance of repression. Failed coercion reduces the surplus and lowers output.

Faced with widespread enforcement failures at lower levels, Stalin forced the legal system, local party offices, and the militia to increase arrest and conviction rates or suffer penalties themselves. The most common method of forcing repression was to distribute quotas by region and profession to officials at lower levels (Kozlov, 2004, vol. 1). In the Great Terror of 1937/38 local officials had to work feverishly to achieve a set number of confessions per day (Vatlin, 2004). To fulfil such plans the police officials imputed individual guilt from increasingly trivial differences in behaviour. Whether or not these measures reduced the Type II errors, they seem likely to have encouraged false denunciation and confession and so added to the errors of Type I.

After Stalin

Stalin's successors reverted to collective leadership and toned down repression. They also inherited an economy in secular decline. Thus, the story of the post-Stalin leadership is that of unsuccessful attempts to reform the Stalinist system without altering its basic characteristics, other than mass repression.

PAUL R. GREGORY

See also **Akerlof, George Arthur; inflation; Preobrazhensky, Evgenii Alexeyevich; rationing; rent seeking; Soviet Union, economics in; Stakhanovism.**

Bibliography

Acemoglu, D. and Robinson, J. 2000. Political losers as barriers to economic development. *American Economic Review* 90(2), Papers and Proceedings, 126–30.
Akerlof, G. 1984. Gift exchange and efficiency wages: four views. *American Economic Review* 74(2), Papers and Proceedings, 79–83.
Becker, G. 1968. Crime and punishment: an economic approach. *Journal of Political Economy* 76, 169–217.

Belova, E. 2001. Economic crime and punishment. In *Behind the Façade of Stalin's Command Economy: Evidence from the State and Party Archives*, ed. P. Gregory. Stanford, CA: Hoover Institution Press.
Belova, E. and Gregory, P. 2002. Dictator, loyal and opportunistic agents: the Soviet archives on creating the Soviet economic system. *Public Choice* 113, 265–86.
Berelovich [Berelowitch], A. 2000. *Sovetskaia Derevnia Glazami VChK– OGPU– NKVD, 2: 1923–1929.* Moscow: Rosspen.
Borodkin, L. and Ertz, S. 2003. Coercion versus motivation: forced labor in Norilsk. In *The Economics of Forced Labor: The Soviet Gulag*, ed. P. Gregory and V. Lazarev. Stanford, CA: Hoover Institution Press.
Borodkin, L. and Ertz, S. 2005. Forced labor and the need for motivation: wages and bonuses in the Stalinist camp system. *Comparative Economic Studies* 47, 418–35.
Chuev, F. 1991. *Sto sorok besed s Molotovym.* Moscow: Terra.
Conquest, R. 1987. *The Harvest of Sorrow: Soviet Collectivization and the Terror-Famine.* Oxford: Oxford University Press.
Davies, R. 1996. *The Industrialisation of Soviet Russia, 4: Crisis and Progress in the Soviet Economy, 1931–1933.* Basingstoke: Macmillan.
Davies, R. 2001. Why was there a Soviet investment cycle in 1933–7? PERSA Working Paper No. 16. Department of Economics, University of Warwick. Online. Available at http://www2.warwick.ac.uk/fac/soc/economics/staff/faculty/harrison/archive/persa/016fulltext.pdf, accessed 12 September 2006.
Davies, R. and Khlevnyuk, O. 1999. The end of rationing in the Soviet Union, 1934–1935. *Europe–Asia Studies* 51, 557–610.
Davies, R. and Khlevnyuk, O. 2002. Stakhanovism and the Soviet economy. *Europe–Asia Studies* 54, 897–903.
Davies, R. and Wheatcroft, S. 2004. *The Industrialisation of Soviet Russia, 5: The Years of Hunger: Soviet Agriculture, 1931–1933.* Basingstoke: Palgrave.
Djankov, S., Glaeser, E., La Porta, R., Lopez-de-Silanes, F. and Shleifer, A. 2003. The new comparative economics. *Journal of Comparative Economics* 31, 595–619.
Erlich, A. 1960. *The Soviet Industrialization Debate, 1924–1928.* Cambridge, MA: Harvard University Press.
Ellman, M. 2000. The 1947 Soviet famine and the entitlement approach to famines. *Cambridge Journal of Economics* 24, 603–30.
Ertz, S. 2003. Building Norilsk. In *The Economics of Forced Labor: The Soviet Gulag*, ed. P. Gregory and V. Lazarev. Stanford, CA: Hoover Institution Press.
Ertz, S. 2005. Trading effort for freedom: workday credits in the Soviet camp system. *Comparative Economic Studies* 47, 476–90.
Filtzer, D. 2002. *Soviet Workers and Late Stalinism: Labour and the Restoration of the Stalinist System after World War II.* Cambridge: Cambridge University Press.

Glaeser, E., La Porta, R., Lopez-de-Silane, F. and Shleifer, A. 2004. Do institutions cause growth? *Journal of Economic Growth* 9, 271–303.

Gorlizki, Y. and Khlevnyuk, O. 2004. *Cold Peace: Stalin and the Soviet Ruling Circle, 1945–1953*. New York: Oxford University Press.

Granick, D. 1987. *Job Rights in the Soviet Union: Their Consequences*. New York: Cambridge University Press.

Gregory, P. 2003. An introduction to the economics of the Gulag. In *The Economics of Forced Labor: The Soviet Gulag*, ed. P. Gregory and V. Lazarev. Stanford, CA: Hoover Institution Press.

Gregory, P. 2004. *The Political Economy of Stalinism: Evidence From the Soviet Secret Archives*. New York: Cambridge.

Gregory, P. and Markevich, A. 2002. Creating Soviet industry: the house that Stalin built. *Slavic Review* 61, 4787–814.

Harris, J. 1999. *The Great Urals: Regionalism and the Evolution of the Soviet System*. Ithaca, NY: Cornell University Press.

Hayek, F. 1944. *The Road to Serfdom*. Chicago: University of Chicago Press.

Heinzen, J. 2004. Corruption in the Gulag: dilemmas of officials and prisoners. *Comparative Economic Studies* 47, 2456–75.

Khlevnyuk, O. 1993. *Stalin i Ordzhonikidze: konflikty v Politburo v 30–e gody*. Moscow: Rossiia molodaia.

Khlevnyuk, O. 1996. *Politburo. Mekhanizmy politcheskoi vlasti v 1930-e gody*. Moscow: ROSSPEN.

Khlevnyuk, O. 2001a. Stalinism and the Stalin period after the 'Archival Revolution'. *Kritika* 2, 319–28.

Khlevnyuk, O. 2001b. The economy of the Gulag. In *Behind the Façade of Stalin's Command Economy: Evidence from the State and Party Archives*, ed. P. Gregory. Stanford, CA: Hoover Institution Press.

Khlevnyuk, O. 2003. The economy of the OGPU, NKVD, and MVD of the USSR, 1930–1953: the scale, structure, and trends of development. In *The Economics of Forced Labor: The Soviet Gulag*, ed. P. Gregory and V. Lazarev. Stanford, CA: Hoover Institution Press.

Khlevnyuk, O. and Devis, R.U. [Davies, R.W.], Kosheleva, L., Ris, E. [Rees, E.] and Rogovaia, L., eds. 2001. *Stalin i Kaganovich. Perepiska. 1931–1936 gg*. Moscow: ROSSPEN.

Khlevnyuk, O., Kvashonkin, A., Kosheleva, L. and Rogovaia, L. eds. 1995. *Stalinskoe Politbiuro v 30-e gody. Sbornik dokumentov*. Moscow: AIRO-XX.

Kornai, J. 1980. *The Economics of Shortage*, 2 vols. Amsterdam: North-Holland.

Kornai, J., Maskin, E. and Roland, G. 2003. Understanding the soft budget constraint. *Journal of Economic Literature* 41, 1095–136.

Kozlov, V., ed. 2004. *Istoriia Stalinskogo Gulaga*, 6 vols. Moscow: ROSSPEN.

Kuromiya, H. 1988. *Stalin's Industrial Revolution: Politics and Workers, 1928–1932*. Cambridge: Cambridge University Press.

Markevich, A. 2003. Was the Soviet Economy planned? Planning in the People's Commissariats in the 1930s. PERSA Working Paper No. 9. Department of Economics, University of Warwick.

Millar, J. 1974. Mass collectivization and the contribution of Soviet agriculture to the first Five-Year Plan: a review article. *Slavic Review* 33, 4750–66.

Montefiore, S. 2003. *Stalin: The Court of The Red Tsar*. London: Weidenfeld and Nicolson.

Morukov, M. 2003. The White Sea–Baltic Canal. In *The Economics of Forced Labor: The Soviet Gulag*, ed. P. Gregory and V. Lazarev. Stanford, CA: Hoover Institution Press.

Nordlander, D. 2003. Magadan and the economic history of Dalstroi in the 1930s. In *The Economics of Forced Labor: The Soviet Gulag*, ed. P. Gregory and V. Lazarev. Stanford, CA: Hoover Institution Press.

Olson, M. 1993. Dictatorship, democracy, and development. *American Political Science Review* 87, 3567–76.

Rees, E., ed. 2004. *The Nature of Stalin's Dictatorship: The Politburo, 1924–1953*. Basingstoke: Palgrave.

Rees, E. and Watson, D. 1997. Politburo and Sovnarkom. In *Decision-Making in the Stalinist Command Economy, 1932–37*, ed. E. Rees. Basingstoke and London: Macmillan.

Schrettl, W. 1982. *Consumption, Effort, and Growth in Soviet-Type Economies: A Theoretical Analysis*. Ann Arbor, MI: University Microfilms International.

Schrettl, W. 1984. Anspruchsdenken, Leistungsbereitschaft und Wirtschaftszyklen. In *Wachstumsverlangsamung und Konjunkturzyklen in unterschiedlichen Wirtschaftssystemen*, ed. A. Schuller. Berlin: Duncker and Humblot.

Sen, A. 1983. *Poverty and Famines: An Essay on Entitlement and Deprivation*. Oxford: Oxford University Press.

Spulber, N. 1964. *Soviet Strategy for Economic Growth*. Bloomington, IN: Indiana University Press.

Stalin, I. 1937. *Voprosy Leninizma*, 10th edn. Moscow: Gospolitizdat.

Tikhonov, A. 2003. The end of the Gulag. In *The Economics of Forced Labor: The Soviet Gulag*, ed. P. Gregory and V. Lazarev. Stanford, CA: Hoover Institution Press.

Vatlin, A.Iu. 2004. *Terror raionnogo masshtaba: 'massovye operatsii' NKVD v Kuntsevskom raione Moskovskoi oblasti. 1937–1938 gg*. Moscow: ROSSPEN.

Wheatcroft, S. 2004. From Team Stalin to degenerate tyranny. In *The Nature of Stalin's Dictatorship: The Politburo, 1924–1953*, ed. E. Rees. Basingstoke: Palgrave.

Wheatcroft, S. and Davies, R. 2002. The Soviet famine of 1932–3 and the crisis in agriculture. In *Challenging Traditional Views of Russian History*, ed. S. Wheatcroft. Basingstoke: Palgrave.

Wintrobe, R. 1998. *The Political Economy of Dictatorship*. Cambridge: Cambridge University Press.

standard commodity. *See* **Sraffian economics.**

standards of living (historical trends)

Methodology
Research specialists agree that the standard of living has many elements, such as material goods and services, health, socio-economic fluidity, education, inequality, and political and religious freedom. Opinions differ on the precise measures to be used within each category and on the weights that should be attached to each. Health, for example, is measurable by length of life, morbidity (illness or disability) and physical fitness. Conceivably, one might attempt comparisons using all feasible measures, but this is expensive and time-consuming, and in any event good measures within categories are often highly correlated.

Weighting is a contentious issue in any attempt to summarize the standard of living, or otherwise compress diverse measures into a single number. Economists and other social scientists know that tastes are individualistic and diverse but they recognize general tendencies. Here I consider the available historical information on material standards and health.

Material aspects
Measure
The most widely used measure of the material standard of living is Gross Domestic Product (GDP) per capita, adjusted for changes in the price level (inflation or deflation). This measure reflects only economic activities that flow through markets, omitting productive endeavours unrecorded in market exchanges, such a preparing meals at home or maintenance done by the homeowner. It ignores work effort required to produce income and does not consider conditions surrounding the work environment, which might affect health and safety. Crime, pollution and congestion, which many people consider important issues affecting their quality of life, are also excluded from GDP. Moreover, technological change, relative prices and tastes affect the course of GDP and the products and services that it includes, which creates what economists call an 'index number' problem that is not readily solvable. Nevertheless most economists believe that real GDP per capita does summarize or otherwise quantify important aspects of the average availability of goods and services.

Time trends
Table 1 shows the course of the material standard of living in the United States from 1820 to 1998. Over this period of 178 years real GDP per capita increased by 21.7 times, or an average of 1.73 per cent per year. Although

Table 1   *GDP per capita in the United States, 1820–1998*

Year	GDP per capita (1990 international dollars)	Annual growth rate (%) from previous period
1820	1,257	
1870	2,445	1.34
1913	5,301	1.82
1950	9,561	1.61
1973	16,689	2.45
1990	23,214	1.94
1998	27,331	2.04

Source: Maddison (2001, tables A-1c and A-1d).

the evidence available to estimate GDP directly is meagre, this rate of increase was probably many times higher than experienced during the colonial period. This conclusion is justified by considering the implications of extrapolating the level observed in 1820 ($1,257) backward in time at the growth rate measured since 1820 (1.73 per cent). Under this supposition, real per capita GDP would have doubled every 40 years (halved every 40 years going backward in time) and so by the mid-1700s there would have been insufficient income to support life. Because the cheapest diet able to sustain good health would have cost nearly $500 per year, the tentative assumption of modern economic growth contradicts what actually happened. Moreover, historical evidence suggests that important ingredients of modern economic growth, such as technological change and human and physical capital, accumulated relatively slowly during the colonial period.

Cycles
Although real GDP per capita is given for only seven dates in Table 1, it is apparent that economic progress has been uneven over time. If annual or quarterly data were given, it would show that business cycles have been a major feature of the economic landscape since industrialization began in the 1820s. By far the worst downturn in US history occurred during the Great Depression of the 1930s, when real per capita GDP declined by approximately one-third and the unemployment rate reached 25 per cent.

Regions
The aggregate numbers also disguise regional differences in the standard of living. In 1840 personal income per capita was twice as high in the Northeast as in the North Central States. Regional divergence increased after the Civil War when the South Atlantic became the nation's poorest region, attaining only one-third of the living standards in the Northeast. Regional convergence occurred in the 20th century, and industrialization in the South significantly improved the region's economic standing after the Second World War.

Health

Life expectancy

Two measures of health are widely used in economic history: life expectancy at birth (or average length of life) and average height, which measures nutritional conditions during the growing years. Table 2 shows that life expectancy has approximately doubled since the mid-19th century, reaching 76.7 years in 1998. If depressions and recessions have adversely affected the material standard of living, epidemics have been a major cause of sudden declines in health in the past. Fluctuations during the 19th century are evident from the table, but as a rule growth rates in health have been considerably less volatile than those for GDP, particularly during the 20th century.

Childhood mortality greatly affects life expectancy, which was low in the mid-1800s substantially because mortality rates were very high for this age group. For example, roughly one child in five born alive in 1850 did not survive to age one, but today the infant mortality rate is less than one per cent. The period since 1850 has witnessed a significant shift in deaths from early childhood to old age. At the same time, the major causes of death have shifted from infectious diseases originating with germs or micro-organisms to degenerative processes that are affected by lifestyle choices such as diet, smoking and exercise.

Time trends

The largest gains were concentrated in the first half of the 20th century, when life expectancy increased from 47.8 years in 1900 to 68.2 years in 1950. Factors behind the growing longevity include the ascent of the germ theory of disease, programmes of public health and personal hygiene, better medical technology, higher incomes, better diets, more education, and the emergence of health insurance.

Explanations

Numerous important medical developments contributed to improving health. The research of Pasteur and Koch was particularly influential in leading to acceptance of the germ theory in the late 1800s. Prior to their work, many diseases were thought to have arisen from miasmas or vapours created by rotting vegetation. Thus, swamps were accurately viewed as unhealthy, but not because they were home to mosquitoes and malaria. The germ theory gave public health measures a sound scientific basis, and shortly thereafter cities began cost-effective measures to remove garbage, purify water supplies, and process sewage. The notion that 'cleanliness was next to Godliness' also emerged in the home, where bathing and the washing of clothes, dishes and floors became routine.

The discovery of Salvarsan in 1910 was the first use of an antibiotic (for syphilis), which meant that the drug was effective in altering the course of a disease. This was an important medical event, but broad-spectrum antibiotics were not available until the middle of the century. The most famous of these early drugs was penicillin, which was not manufactured in large quantities until the 1940s. Much of the gain in life expectancy was attained before chemotherapy and a host of other medical technologies were widely available. A cornerstone of improving health from the late 1800s to the middle of the 20th century was therefore prevention of disease by reducing exposure to pathogens. Also important were improvements in immune systems created by better diets and by vaccination against diseases such as smallpox and diphtheria.

Heights

Since the early 1980s, historians have increasingly used average heights to assess health aspects of the standard of living. Average height is a good proxy for the nutritional status of a population because height at a particular age reflects an individual's history of *net* nutrition, or diet minus claims on the diet made by work (or physical activity) and disease. The growth of poorly nourished children may cease, and repeated bouts of biological stress – whether from food deprivation, hard work, or disease – often leads to stunting or a reduction in adult height. The average heights of children and of adults in countries around the world are highly correlated with their life expectancy at birth and with the log of the per capita GDP in the country where they live.

Applications

This interpretation for average heights has led to their use in the study of the health of slaves, health inequality, living standards during industrialization, and trends in mortality. The first important results in the 'new anthropometric history' dealt with the nutrition and

Table 2   *Life expectancy at birth in the United States, 1850–1998*

Year	Life expectancy
1850	38.3
1860	41.8
1870	44.0
1880	39.4
1890	45.2
1900	47.8
1910	53.1
1920	54.1
1930	59.7
1940	62.9
1950	68.2
1960	69.7
1970	70.8
1980	73.7
1990	75.4
1998	76.7

Source: Haines (2006).

health of Americans slaves as determined from stature recorded for identification purposes on slave manifests required in the coastwise slave trade. The subject of slave health has been a contentious issue among historians, in part because vital statistics and nutrition information were never systematically collected for slaves (or for the vast majority of the American population in the mid-19th century, for that matter). Yet the height data showed that children were astonishingly small and malnourished while working slaves were remarkably well fed. Adolescent slaves grew rapidly as teenagers and were reasonably well-off in nutritional aspects of health.

Time trends

Table 3 shows the time pattern in height of native-born American men obtained in historical periods from military muster rolls, and for men and women in recent decades from the National Health and Nutrition Examination Surveys. This historical trend is notable for the tall stature during the colonial period, the mid-19th

Table 3   *Average height of native-born US men and women by year of birth, 1710–1970*

Year	Centimeters		Inches	
	Men	Women	Men	Women
1710	171.5		67.5	
1720	171.8		67.6	
1730	172.1		67.8	
1740	172.1		67.8	
1750	172.2		67.8	
1760	172.3		67.8	
1770	172.8		68.0	
1780	173.2		68.2	
1790	172.9		68.1	
1800	172.9		68.1	
1810	173.0		68.1	
1820	172.9		68.1	
1830	173.5		68.3	
1840	172.2		67.8	
1850	171.1		67.4	
1860	170.6		67.2	
1870	171.2		67.4	
1880	169.5		66.7	
1890	169.1		66.6	
1900	170.0		66.9	
1910	172.1		67.8	
1920	173.1		68.1	
1930	175.8	162.6	69.2	64.0
1940	176.7	163.1	69.6	64.2
1950	177.3	163.1	69.8	64.2
1960	177.9	164.2	70.0	64.6
1970	177.4	163.6	69.8	64.4

Source: Steckel (2006) and sources therein.

century decline, and the surge in heights of the 20th century. Comparisons of average heights from military organizations in Europe show that Americans were taller by two to three inches. Behind this achievement were a relatively good diet, little exposure to epidemic disease, and relative equality in the distribution of wealth. Americans could choose their foods from the best of European and Western Hemisphere plants and animals, and this dietary diversity combined with favourable weather meant that Americans never had to contend with harvest failures. Thus, even the poor were reasonably well fed in colonial America.

Cycles and explanations

Loss of stature began in the second quarter of the 19th century when the transportation revolution of canals, steamboats and railways brought people into greater contact with diseases. The rise of public schools meant that children were newly exposed to major diseases such as whooping cough, diphtheria, and scarlet fever. Food prices also rose during the 1830s and growing inequality in the distribution of income or wealth accompanied industrialization. Business depressions, which were most hazardous for the health of those who were already poor, also emerged with industrialization. The Civil War of the 1860s and its troop movements further spread disease and disrupted food production and distribution. A large volume of immigration also brought new varieties of disease to the United States at a time when urbanization brought a growing proportion of the population into closer contact with contagious diseases. Estimates of life expectancy among adults at ages 20, 30 and 50, which was assembled from family histories, also declined in the middle of the 19th century.

In the 20th century, heights grew most rapidly for those born between 1910 and 1950, an era when public health and personal hygiene took vigorous hold, incomes rose rapidly and there was reduced congestion in housing. The latter part of the era also witnessed a larger share of income or wealth going to the lower portion of the distribution, implying that the incomes of the less well-off were rising relatively rapidly. Note that most of the rise in heights occurred before modern antibiotics were available, which means that disease prevention was a more significant cause of improving health than the ability to alter its course after onset. The growing control that humans have exercised over their environment, particularly increased food supply and reduced exposure to disease, may be leading to biological (but not genetic) evolution of humans with more durable vital organ systems, larger body size, and later onset of chronic diseases.

Recent stagnation

Between the middle of the 20th century and the present, however, the average heights of American men have stagnated, increasing by only a small fraction of an inch over the since the 1950s. Table 3 refers to the native born, so recent increases in immigration cannot account for the stagnation. In the absence of other information, one might be tempted to suppose that environmental conditions for growth are so good that most Americans have simply reached their genetic potential for growth. Unlike in the United States, heights and life expectancy have continued to grow in Europe, which has the same genetic stock as that from which most Americans descend. By the 1970s several American health indicators had fallen behind those in Norway, Sweden, the Netherlands and Denmark. While American heights were essentially flat after the 1970s, heights continued to grow significantly in Europe. The Dutch men are now the tallest, averaging six feet, about two inches more than American men. Lagging heights leads to questions about the adequacy of health care and lifestyle choices in America. As discussed below, it is doubtful that lack of resource commitment to health care is the problem because America invests a greater share of GDP than the Netherlands. Greater inequality and less access to health care could be important factors in the difference. But access to health care alone, whether due to low income or lack of insurance coverage, may not be the only issues – health insurance coverage must be used regularly and wisely. In this regard, Dutch mothers are known for regular pre- and post-natal checkups, which are important for early childhood health.

Note that significant differences in health and the quality of life follow from these height patterns. The comparisons are not part of an odd contest that emphasizes height, nor is 'big' per se assumed to be beautiful. Instead, we know that, on average, stunted growth has functional implications for longevity, cognitive development and work capacity. Children who fail to grow adequately are often sick, suffer learning impairments and have a lower quality of life. Growth failure in childhood has a long reach into adulthood because individuals whose growth has been stunted are at greater risk of death from heart disease, diabetes and some types of cancer. Therefore it is important to know why Americans are falling behind.

International perspective
Per capita GDP comparisons
Table 4 places American economic performance in perspective relative to other countries. In 1820 the United States was fifth in world ranking, roughly 30 per cent below the leaders (United Kingdom and the Netherlands), but still two to three times better off than the poorest sections of the globe. It is notable that in 1820 the richest country (the Netherlands at $1,821) was approximately 4.4 times better off than the poorest (Africa at $418), but by 1950 the ratio of richest to

Table 4 *GDP per capita by country and year, 1820–1998*

| Country | 1990 international dollars | | | | | | |
	1820	1870	1913	1950	1973	1998	Ratio 1998 to 1820
Austria	1,218	1,863	3,465	3,706	11,235	18,905	15.5
Belgium	1,319	2,697	4,220	5,462	12,170	19,442	14.7
Denmark	1,274	2,003	3,912	6,946	13,945	22,123	17.4
Finland	781	1,140	2,111	4,253	11,085	18,324	23.5
France	1,230	1,876	3,485	5,270	13,123	19,558	15.9
Germany	1,058	1,821	3,648	3,881	11,966	17,799	16.8
Italy	1,117	1,499	2,564	3,502	10,643	17,759	15.9
Netherlands	1,821	2,753	4,049	5,996	13,082	20,224	11.1
Norway	1,104	1,432	2,501	5,463	11,246	23,660	21.4
Sweden	1,198	1,664	3,096	6,738	13,493	18,685	15.6
Switzerland	1,280	2,202	4,266	9,064	18,204	21,367	16.7
United Kingdom	1,707	3,191	4,921	6,907	12,022	18,714	11.0
Portugal	963	997	1,244	2,069	7,343	12,929	13.4
Spain	1,063	1,376	2,255	2,397	8,739	14,227	13.4
United States	1,257	2,445	5,301	9,561	16,689	27,331	21.7
Mexico	759	674	1,732	2,365	4,845	6,655	8.8
Japan	669	737	1,387	1,926	11,439	20,413	30.5
China	600	530	552	439	839	3,117	5.2
India	533	533	673	619	853	1,746	3.3
Africa	418	444	585	852	1,365	1,368	3.3
World	667	867	1,510	2,114	4,104	5,709	8.6
Ratio of richest to poorest	4.4	7.2	8.9	20.6	21.7	20.0	

Source: Maddison (2001, table B-21).

poorest had widened to 21.8 ($9,561 in the United States versus $439 in China), which is roughly the level it is today (in 1998, it was $27,331 in the United States versus $1,368 in Africa). These calculations understate the growing disparity in the material standard of living because several African countries today fall significantly below the average, whereas it is unlikely that they did so in 1820 because GDP for the continent as a whole was close to the level of subsistence.

Per capita GDP growth

It is clear that the poorer countries are better off today than they were in 1820 (by 3.3 times in both Africa and India). At the simplest level, the explanation is that the countries that are now rich grew much faster after 1820. The last column of Table 4 shows that Japan realized the most spectacular gain, climbing from approximately the world average in 1820 to the fifth richest today, with more than a thirtyfold increase in real per capita GDP. All countries that are rich today had rapid increases in their material standard of living, realizing more than tenfold increases since 1820. The underlying reasons for this diversity of economic success is a central question in the field of economic history.

Life expectancy

Table 5 shows that disparities in life expectancy have been much less than those in per capita GDP. In 1820 all countries were bunched in the range of 21 to 41 years, with Germany at the top and India at the bottom, giving a ratio of less than two to one. It is doubtful that any country or region has had a life expectancy below 20 years for long periods of time because death rates would have exceeded any plausible upper limit for birth rates,

leading to population implosion. The 20th century witnessed a compression in life expectancies across countries, with the ratio of levels in 1999 being 1.56 (81 in Japan versus 52 in Africa). Japan has also been a spectacular performer in health, increasing life expectancy from 34 years in 1820 to 81 years in 1999. Among poor unhealthy countries, health aspects of the standard of living have improved more rapidly than the material standard of living relative to the world average. Because many public health measures are cheap and effective, it has been easier to extend life than it has been to promote material prosperity, which has numerous complicated causes.

Height comparisons

Figure 1 compares stature in the United States and the United Kingdom. Americans were very tall by global standards in the early 19th century as a result of their rich and varied diets, low population density, and relative equality of wealth. Unlike other countries that have been studied (France, the Netherlands, Sweden, Germany, Japan and Australia), both the United States and the UK suffered significant height declines during industrialization (as defined primarily by the achievement of modern economic growth) in the 19th century. Note, however, that the amount and timing of the height decline in the UK has been the subject of a lively debate. See for example the February 1993 issue of the *Economic History Review* for papers by Roderick Floud, Kenneth Wachter and John Komlos; only the Floud–Wachter figures are given here.

One may speculate that the timing of the declines shown in Figure 1 is probably more coincidental than emblematic of linkage among similar causal factors across the two countries. While it is possible that growing trade and commerce spread disease, as in the United States, it is more likely that a major culprit in the UK was rapid urbanization and associated increased in exposure to diseases. This conclusion is reached by noting that urban-born men were substantially shorter than the rural-born, and between the periods of 1800–30 and

Table 5 *Life expectancy at birth by country and year, 1920–1999*

Country	1820	1900	1950	1999
France	37	47	65	78
Germany	41	47	67	77
Italy	30	43	66	78
Netherlands	32	52	72	78
Spain	28	35	62	78
Sweden	39	56	70	79
United Kingdom	40	50	69	77
United States	39	47	68	77
Japan	34	44	61	81
Russia	28	32	65	67
Brazil	27	36	45	67
Mexico	n.a.	33	50	72
China	n.a.	24	41	71
India	21	24	32	60
Africa	23	24	38	52
World	26	31	49	66

Source: Maddison (2001, table 1-5a).

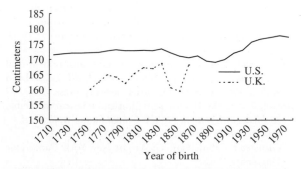

Figure 1 Average height of soldiers in Britain and of native-born American soldiers, 1710–1970. *Sources*: Steckel (2006, Fig. 12) and Floud, Wachter and Gregory (1990, Table 4.8).

1830–70 the share of the British population living in urban areas leaped from 38.7 per cent to 54.1 per cent.

RICHARD H. STECKEL

See also **anthropometric history; cliometrics.**

Bibliography

Easterlin, R.A. 1971. Regional income trends, 1840–1950. In *The Reinterpretation of American Economic History*, ed. R.W. Fogel and S.L. Engerman. New York: Harper and Row.

Engerman, S.L. 1997. The standard of living debate in international perspective: measures and indicators. In *Health and Welfare during Industrialization*, ed. R.H. Steckel and R. Floud. Chicago: University of Chicago Press.

Floud, R., Wachter, K.W. and Gregory, A.S. 1990. *Height, Health, and History: Nutritional Status in the United Kingdom, 1750–1980*. Cambridge: Cambridge University Press.

Haines, M. 2006. Vital statistics. In *Historical Statistics of the United States: Millennial Edition*, ed. S. Carter et al. New York: Cambridge University Press.

Komlos, J. 1998. Shrinking in a growing economy? The mystery of physical stature during the industrial revolution. *Journal of Economic History* 58, 779–802.

Maddison, A. 2001. *The World Economy: A Millennial Perspective*. Paris: OECD.

Meeker, E. 1980. Medicine and public health. In *Encyclopedia of American Economic History*, ed. G. Porter. New York: Scribner.

Pope, C.L. Adult 1992 mortality in America before 1900: a view from family histories. In *Strategic Factors in Nineteenth Century American Economic History: A Volume to Honor Robert W. Fogel*, ed. C. Goldin and H. Rockoff. Chicago: University of Chicago Press.

Porter, R., ed. 1996. *The Cambridge Illustrated History of Medicine*. Cambridge: Cambridge University Press.

Steckel, R.H. 1986. A peculiar population: the nutrition, health, and mortality of American slaves from childhood to maturity. *Journal of Economic History* 46, 721–41.

Steckel, R.H. 1995. Stature and the standard of living. *Journal of Economic Literature* 33, 1903–40.

Steckel, R.H. 1998. Strategic ideas in the rise of the new anthropometric history and their implications for interdisciplinary research. *Journal of Economic History* 58, 803–21.

Steckel, R.H. 2000. Industrialization and health in historical perspective. In *Poverty, Inequality and Health*, ed. D. Leon and G. Walt. Oxford: Oxford University Press.

Steckel, R.H. 2006. Health, nutrition and physical wellbeing. In *Historical Statistics of the United States: Millennial Edition*, ed. S. Carter, S. Gartner, M. Haines, A. Olmstead, R. Sutch and G. Wright. New York: Cambridge University Press.

Steckel, R.H. and Floud, R., eds. 1997. *Health and Welfare during Industrialization*. Chicago: University of Chicago Press.

state capture and corruption in transition economies

In an ideal democratic society, citizens determine economic policy by a direct voting procedure or by selecting appropriate candidates at polls. In the real world, economic policy is affected either directly by self-interested elected officials or bureaucrats, or indirectly by special interests such as industrial lobbyists or even large individual enterprises. The actual channels of the (primarily negative) influence of special interests on economic policy are called 'state capture'. In most contexts, state capture necessarily involves 'corruption', that is, abuse of public office for private gain.

In any single instance of state capture, there are winners and losers. Typically, winners are politically important or simply large firms or whole industries and bureaucrats that receive favours from those firms. On the losing side, there are small, politically unimportant businesses, and, ultimately, the public interest and consumer welfare. For transition economies, the overall growth rate of the enterprise sector is estimated to be ten per cent lower in a capture state than in a state where the effect is less pronounced (Hellman, Jones and Kaufmann, 2003).

The principal characteristics of transition economies, as related to the issue of state capture, are an excessively concentrated industrial structure inherited from the planned economy, a high level of discretion of public officials with respect to economic policy, weak institutions of political control such as party system or independent media, suboptimal allocation of authority between government layers, and low labour mobility, both horizontal and vertical. In short, all of these factors might be traced to the command economy legacy, one-party political system, and legal system subordinated to political authorities, which were characteristic for most now-in-transition countries in the mid-1980s.

In transition economies, political channels that transmit information and, if necessary, money from firms and other special interests to politicians are not only less efficient but also less institutionalized. Furthermore, conceptualization of a specific exchange, both by parties to the transaction and by a student of transition, might vary widely. For example, what is considered as a fully legal lobbying activity or campaign contribution in an OECD country might be thought of as a bribe or even outright extortion in some other economy.

Theory and transition specifics

A bureaucrat's ability to extract bribes, the manifest form of corruption, is primarily determined by her ability to create rents by shaping the playing field for business (Rose-Ackerman, 1999; Shleifer and Vishny, 1993). In a perfect analogy with a monopolist's behaviour, if a certain kind of business activity requires obtaining a licence from a regulating body, the bureaucrat who is able to collect kickbacks for granting a licence has incentives to keep the number of licences issued less than socially optimal, thus increasing the 'price' of those licences that are actually issued. Furthermore, this provides incentives for bureaucrats to create new rent-seeking opportunities by introducing as much licensing and regulation as possible. A monopolistic bureaucrat sets a lower bribe level than a chain of successive monopolies, but the total volume of bribes is obviously higher, and so the impact on social welfare of centralized versus decentralized corruption is ambiguous.

Typically, special interests, through the channels of state capture, seek protection from competition such as barriers to entry to the local market. At the international level, protection usually takes the form of either import quotas or tariff protection. At the local level, it takes the form of licensing, small-scale regulation, and preferential treatment such as tax exemptions granted to individual firms. State capture, as a stable, long-standing relationship between existing businesses and incumbent politicians, has adverse effects on business and politics. In business, state capture prevents innovation by new or potential entrants, and, as a result, by the incumbent firms as well, and thus constrains market development. In politics, it decreases the chances of challengers to mount an aggressive campaign against an incumbent, which in turn reduces the electoral accountability of incumbents.

It is possible to further refine the concept of state capture. Hellman, Jones, and Kaufmann (2003), based on their econometric analysis of a large sample of firms in transition economies, argue that is helpful to distinguish state capture from influence by special interests and administrative corruption, with each of the three forms of business involvement in politics having distinct causes and consequences. In this refined definition, *state capture* is the process by which firms shape rules of the game through semi-institutionalized bribes paid to public officials and politicians; *influence* is the same process without direct transfers (this category corresponds to costly 'informational lobbying' à la Grossman and Helpman, 2001); and *administrative corruption* encompasses all 'petty' forms of bribery related to law enforcement and regulation.

Political power and economic power

At the beginning of the transition, most of the political institutions that help to mitigate agency problems in the developed world such as independent courts and media, grass-roots political parties, and an institutionalized civil society were virtually non-existent. Inadequate provision of basic public goods such as property rights protection and enforcement of contracts, which are crucial for economic development, forced economic agents to seek alternative tools, for example by supplementing their productive investment with investment in private

protection. For large businesses, state capture was a potentially powerful tool. In this view, bribes to bureaucrats or other forms of privately financed purchase of a public good becomes a strategy of an economic agent to increase efficiency and predictability in his business relations.

A specific feature of many transition economies, as compared with developing countries with a similar level of GDP per capita, is that they have been left with the remnants of the command economy with its highly centralized industry. The effect was more pronounced in industrially developed countries such as Russia or Slovakia, and less in countries such as Vietnam or Albania.

The existence of enterprises with very large employment levels produced a specific form of state capture. The managers of large enterprises, regardless of whether they were profitable, have a large menu of political instruments in their hands from which to choose. In particular, they could either rely on ties inherited from the times of the plan, or operate in a newly formed web of quasi-market exchanges including various forms of barter (Gaddy and Ickes, 2002). In Russian practice, quasi-market exchanges have often relied on government power to set individual tax rates and energy tariffs for enterprises. In the Hellman, Jones, and Kaufmann (2003) classification this often amounts to influence, not state capture, as government intervention is caused not by direct bribery, but by a complex chain of exchanges made possible by agents' participation in the same social network. The downside of this phenomenon is that inefficient enterprises are not driven out of the market, and at the same time provide political pressure to deter entry of new enterprises.

Proponents of 'big bang' reforms argued that establishing the rule of law, including institutions of property rights protection, requires the creation of a 'grass-roots' demand for the rule of law. In practice, it meant that the former state property should first go into the hands of private owners, and only then would those owners become natural proponents of a system of property rights protection. With the high degree of concentration present in transition economies, and with undeveloped factor markets, this forecast (from private ownership to demand for the rule of law to property rights) has not been borne out. The rich – the beneficiaries of early privatization – have obvious incentives to protecting their property privately, for example via state capture. Consequently, they do not have incentives to lobby for the establishment of well-working state institutions such as independent courts or efficient bureaucrats. Instead, they seek to increase their political influence and modify the existing state institutions so that resources and wealth continue to be redistributed in their favour (Polishchuk and Savvateev, 2004; Sonin, 2003). (This is an example of how the Coase theorem is not valid when there are wealth effects.)

In Russia, the country that entered transition after the longest period of Communist rule, and was the most industrialized of the transition economies, a decade after economic reforms were launched most of the productive assets were concentrated in the hands of a few individuals, the so-called oligarchs (see OLIGARCHS). In part, it is an efficiency requirement that, when property rights are poorly protected, control rights over assets become concentrated: the larger a single owner's share is, the greater her incentives are to pursue improvements, such as in corporate governance. (Recent cross-country studies show that the worse the general protection and enforcement of property rights are, the greater is the concentration of control rights.) On the other hand, wealth inequality very often imposes heavy costs on the economy, primarily because it produces widespread inequality of opportunity. Still, the main problem is that the oligarchs who rely on state capture for protection of their property rights and enforcement of contracts do not form a natural constituency for the rule of law. This effect is especially strong when economic inequality is accompanied by underdeveloped democratic institutions, which is typically the case in formerly Communist countries.

Decentralization and interjurisdictional mobility

A critical check on the extent of state capture comes from institutions of federalism. The traditional approach to fiscal federalism focused on externalities in the provision of public goods that arise from preference heterogeneity in different jurisdictions. In the 1990s, a new approach emerged emphasizing the accountability of government officials (agency problems) at both the central and the local levels (for example, Qian and Weingast, 1997; see also Bardhan and Mookherjee, 2006). All former command economies started transition with an overly centralized government structure, and faced the problem of optimal reallocation of economic and political authority.

The effect of decentralization on state capture is twofold. First, more authority allocated to the local level makes the agency problem at this level less prevalent, thus increasing accountability and reducing corruption. On the other hand, special interests might have much more influence over local government bodies; therefore, shifting authority downward might make state capture both more desirable and easier to achieve. In Bardhan and Mookherjee (2006) a reallocation of authority towards local government has a number of consequences. First, the amount of bribes collected by the central government decreases; second, local governments become captured by local special interests. Since local capture might be more easily supported by a social network, the total amount of bribes in the economy declines. However, since corruption at the central level is more money-based, the special interests are less entrenched than at the local level, and thus local capture brings more

economic inefficiency. Ultimately, decentralization, while reducing bribe-based corruption measures, reduces economic efficiency.

Starting with Tiebout (1956), interjurisdictional mobility has been considered a major constraint on the local governments' power to abuse their prerogatives. When mobility is high and subjects can relocate from a jurisdiction with a predatory or hostile government, the local government's monopoly power over laws, regulation, and their execution is compromised. (See Slinko, Yakovlev and Zhuravskaya, 2005, for unique evidence on political capture at the local level.) Since the capacity to extract bribes is increasing in the bureaucrats' power to manipulate regulation, high mobility – of both firms and individuals – would force local governments to compete in providing a business-friendly environment. For a local government, the incentives to devote resources to, for example, fighting corruption, increase with the resources devoted by neighbouring governments.

The endless transition

In most transition countries, the transition from the command economy started in the late 1980s and early 1990s. It might be argued that most of the economic problems they face are no longer those of transition, but those of economic development.

Since the transition began, fear of the *Leviathan state* has been swiftly replaced by the fear of the *capture state*, where large and powerful businesses tilt the playing field through bribery, media ownership, huge campaign contributions, and direct participation in politics. Lurid stories told about some extreme forms of capture, such as the use of corrupt secret service and police officers against business competitors, have crowded out images of millions perishing in forced-labour camps or dying in numerous famines caused by Communist economic management. There is a false sense of symmetry between the state manipulating its subjects and subjects manipulating the state: while the former has proved to be dangerous on a large scale, the second is a mere obstacle to economic development. There are several ways to deal with this obstacle. A backward-looking way is through restoration of the repressive capacity of the state. Another is through the development of political and civic institutions that may put a check on elected officials and bureaucrats, and through the institutionalization of business influence on politics in an efficient way.

KONSTANTIN SONIN

See also **oligarchs; political institutions, economic approaches to; transition and institutions.**

Bibliography

Bardhan, P. and Mookherjee, D. 2000. Capture and governance at local and national levels. *American Economic Review* 90, 135–9.

Bardhan, P. and Mookherjee, D. 2006. Decentralization, corruption and government accountability: an overview. In *Handbook of Economic Corruption*, ed. S. Rose-Ackerman. Northampton, MA: Edward Elgar.

Gaddy, C. and Ickes, B. 2002. *Russia's Virtual Economy*. Washington, DC: Brookings Institution Press.

Grossman, G. and Helpman, E. 2001. *Special Interest Politics*. Cambridge, MA: MIT Press.

Hellman, J., Jones, G. and Kaufmann, D. 2003. Seize the state, seize the day: state capture and influence in transition economies. *Journal of Comparative Economics* 31, 751–73.

Persson, T. and Tabellini, G. 2000. *Political Economics*. Cambridge, MA: MIT Press.

Polishchuk, L. and Savvateev, A. 2004. Spontaneous (non)emergence of property rights. *Economics of Transition* 12, 103–27.

Qian, Y. and Weingast, B. 1997. Federalism as a commitment to preserving market incentives. *Journal of Economic Perspectives* 11(4), 83–92.

Roland, G. 2000. *Transition and Economics: Politics, Markets, and Firms*. Cambridge, MA: MIT Press.

Rose-Ackerman, S. 1999. *Corruption and Government: Causes, Consequences, and Reform*. Cambridge: Cambridge University Press.

Shleifer, A. and Vishny, R. 1993. Corruption. *Quarterly Journal of Economics* 108, 601–17.

Slinko, I., Yakovlev, E. and Zhuravskaya, E. 2005. Laws for sale: evidence from Russia. *American Law and Economics Review* 7, 284–318.

Sonin, K. 2003. Why the rich may favor poor protection of property right. *Journal of Comparative Economics* 31, 715–31.

Tiebout, C. 1956. A pure theory of local expenditures. *Journal of Political Economy* 64, 416–24.

state-dependent preferences

Theories of individual decision making under uncertainty pertain to situations in which a choice of a course of action, by itself, does not determine the outcome. To formulate these theories Savage (1954) introduced what has become the standard analytical framework. It consists of three sets: a set S, of *states of the world* (or *states*, for short); an arbitrary set C, of *consequences*; and the set F, of all the functions from the set of states to the set of consequences. Elements of F, referred to as 'acts', represent courses of action, consequences describe anything that may happen to a person, and states are the resolutions of uncertainty, that is, 'a description of the world so complete that, if true and known, the consequences of every action would be known' (Arrow, 1971, p. 45). Decision makers are characterized by preference relations, $\succsim$, on F. With few exceptions, preference relations are taken to be complete (that is, for all f and g in F, either $f \succsim g$ or $g \succsim f$) and transitive binary relations on F. The symbols $f \succsim g$ have the interpretation 'the course of

action f is preferred or indifferent to the course of action g'. The strict preference relation, $\succ$, and the indifference relation, $\sim$, are the asymmetric and symmetric parts of $\succsim$, respectively.

To speak loosely, a preference relation is state-dependent when the prevailing state of nature is itself of direct concern to the decision maker. For example, taking out a health insurance policy is choosing an action whose consequences – the indemnities – depend on the realization of the decision maker's state of health. In this example, the state is the decision maker's state of health. It affects the decision maker's well-being directly, and indirectly, through the payoff prescribed by the health insurance policy. The preference relation may display ordinal state dependence, in which case the underlying state may affect the decision maker's preferences by altering his ordinal ranking of the consequences; or cardinal state dependence, by altering his risk attitudes; or both.

To define state dependence formally, it is convenient to adopt the model of Anscombe and Aumann (1963). In this model the state space is finite, and the consequences are lotteries, that is, probability distributions that assign strictly positive probability to a finite number of outcomes. Denote by $L(X)$ the set of lotteries on an arbitrary set, X, of possible outcomes. Given a preference relation $\succsim$ on F; a state s; and f, f', g, g' in F, define a preference relation on F conditional on s, $\succsim_s$, by $f \succsim_s f'$ if $g \succsim g'$ whenever $f(s) = g(s)$, $f'(s) = g'(s)$ and $g(s') = g'(s')$ for all $s' \in S - \{s\}$. Because acts are functions, $f(s)$ is defined uniquely. Thus $\succsim_s$ defines a preference relation on $L(X)$ conditional on s. This induced preference relation is also denoted by $\succsim_s$.

A state $s \in S$ is said to be *null* if $f \succsim_s f'$ for all $f, f' \in F$, otherwise it is *non-null*.

Definition. A preference relation $\succsim$ on F is *state dependent* if $\succsim_s \neq \succsim_{s'}$ for some non-null s and s' in S.

Because consequences are lotteries, if a preference relation $\succsim$ on F displays state dependence, then $\succsim_s$ and $\succsim_{s'}$ must differ on the ranking of some lotteries in $L(X)$. This may be due to distinct attitudes toward risk and/or distinct ordering of outcomes, that is, degenerate lotteries that assign the given outcomes probability one. Circumstances in which the dependence of the decision maker's preferences on the state constitutes an indispensable feature of the decision problem include the choice of health insurance coverage (see Arrow, 1974; Karni, 1985); the choice of flight insurance coverage (see Eisner and Strotz, 1961); the choice of optimal consumption and life insurance plans in the face of uncertain life span (see Yaari, 1965; Karni and Zilcha 1985); and the provision of collective protection (see Cook and Graham, 1977).

Subjective expected utility representations

Preferences among acts are a matter of personal judgement, presumably combining the decision maker's valuation of the consequences and his or her beliefs regarding the likely realization of alternative events (that is, subsets of the state space). Subjective expected utility theory pertains to preference relations whose structures allow the decision maker's valuations of the consequences to be expressed numerically, by a utility function; his or her beliefs to be quantified by a (subjective) probability measure on the set of states; and the acts to be evaluated by the expectations of the utility of the corresponding consequences with respect to the subjective probability. In other words, the theory depicts the decision makers' choice among alternative acts as expected utility maximizing behaviour.

By the classic von Neumann–Morgenstern expected utility theorem, $\succsim$ on F satisfies the axioms of expected utility theory (that is, $\succsim$ is a complete and transitive binary relation satisfying the Archimedean and independence axioms) if and only if there exist real-valued functions $w(\cdot, s)$ on X, $s \in S$, such that for all $f, g \in F$,

$$f \succsim g \Leftrightarrow \sum_{s \in S} \sum_{x \in X} w(x, s) f(x, s)$$
$$\geq \sum_{s \in S} \sum_{x \in X} w(x, s) g(x, s). \tag{1}$$

Furthermore, the functions $\{w(\cdot, s) \mid s \in S\}$ are unique up to cardinal unit comparable transformation. (That is, if some other utility functions $\{w'(\cdot, s) \mid s \in S\}$ represent $\succsim$ on F, in the sense of equation (1), there exit $b > 0$ and real numbers $a(s)$, one for each state, such that $w'(s) = bw(s) + a(s)$ for all $s \in S$.)

The function $w(\cdot, \cdot)$ captures the decision maker's valuation of the outcomes and his or her beliefs about the likely realization of the states. The axioms of expected utility theory do not imply a unique decomposition of w into subjective probability distribution on S and utility on outcome-state pairs. Indeed, let $p(s)$, $s \in S$, be any list of positive numbers that sum up to 1, where $p(s) = 0$ if and only if s is null. Define $u(x, s) = w(x, s)/p(s)$ for all non-null $s \in S$ and all $x \in X$, and $u(x, s) = \bar{u}$ if s is null. Then, by equation (1), $\succsim$ is represented by $\sum_{s \in S} p(s) \sum_{x \in X} u(x, s) f(x, s)$. The question is as follows: are there additional conditions that would imply a *unique* decomposition of $w(x,s)$ into a product of utility representing the (possibly state-dependent) valuation of the outcomes and probabilities representing beliefs that govern the decision maker's choice among acts?

Anscombe and Aumann (1963) show that a preference relation is non-trivial (that is, $f \succ g$ for some $f, g \in F$) satisfying the axioms of expected utility theory and state independence, that is, $\succsim_s = \succsim_{s'}$, for all non-null $s, s' \in S$ if and only if there exist a real-valued function, u, on X, and a subjective probability distribution, π, on S such that, for all $f, g \in F$,

$$f \succsim g \Leftrightarrow \sum_{s \in S} \pi(s) \sum_{x \in X} u(x) f(x, s)$$
$$\geq \sum_{s \in S} \pi(s) \sum_{x \in X} u(x) g(x, s). \tag{2}$$

Moreover, u is unique up to positive linear transformation, and π is unique satisfying $\pi(s) = 0$ if and only if s is null.

The subjective expected utility representation (2) separates risk attitudes, represented by the utility function, from beliefs, represented by the subjective probabilities. However, the uniqueness of the probabilities depends crucially on the premise that constant acts are constant utility acts. This premise is not implied by the axioms. In particular, *state-independent preferences do not imply state-independent utility function*. To see why, let γ be a strictly positive real-valued function on S and $\Gamma = \sum_{s \in S} \gamma(s) \pi(s)$. Define $\hat{u}(x, s) = u(x)/\gamma(s)$, for all $x \in X$ and $s \in S$, and let $\hat{\pi}(s) = \gamma(s) \pi(s)/\Gamma$, for all $s \in S$. Then, by equation (2) and the uniqueness of u, for all $f, g \in F$,

$$f \succsim g \Leftrightarrow \sum_{s \in S} \hat{\pi}(s) \sum_{x \in X} \hat{u}(x, s) f(x, s)$$
$$\geq \sum_{s \in S} \hat{\pi}(s) \sum_{x \in X} \hat{u}(x, s) g(x, s). \tag{3}$$

Thus, the utility–probability pair $(\hat{u}, \hat{\pi})$ induces a subjective expected utility representation of $\succsim$ that is equivalent to the one induced by the pair (u, π). There are infinitely many distinct utility-probability pairs that represent the same preference relation in the sense of equation (2). Moreover, because π and $\hat{\pi}$ are distinct, even if beliefs exist a priori and are coherent enough to allow their representation by probabilities, it is not evident which of the infinitely many probability distributions consistent with $\succsim$ actually represents the decision makers beliefs. But if the probabilities that figure in the representation are meaningless, there seems to be no compelling reason to prefer the expected utility representation (2) over the more general additive representation (1). On the contrary, because the additive representation does not require that the preferences be state independent, it is applicable to the analysis of problems, such as the demand for health and life insurance, in which the assumption of state-independent preferences is clearly inadequate.

Hypothetical preferences and subjective expected utility representations of state-dependent preferences

An alternative analytical framework, introduced by Karni and Schmeidler (1981), postulates the existence of a preference relation on hypothetical lotteries, whose prizes are outcome-state pairs. This preference relation is assumed to satisfy the axioms of expected utility and to be consistent with the actual preference relation on acts. Because the hypothetical lotteries imply distinct, hence incompatible, marginal distributions on the state space, preferences among such lotteries are *introspective* and may be expressed verbally only as hypothetical choices.

Decision makers are supposed to be able to conceive of such hypothetical lotteries and to invoke, for the purpose of their evaluation, the same mental processes that govern their actual decisions.

To express these ideas formally, denote by $L(X \times S)$ the set of all probability distributions on $X \times S$ that assign strictly positive probabilities to a finite number of outcome-state pairs. A lottery $\ell \in L(X \times S)$ is said to be *non-degenerate* if $\sum_{y \in X} \ell(y, s) > 0$ for all $s \in S$. Denote by $\hat{\succsim}$ an introspective preference relation on $L(X \times S)$. For each $s \in S$, define the conditional introspective preferences, $\hat{\succsim}_s$, on $L(X \times S)$ analogously to the definition of $\succsim_s$ (that is, $\ell \hat{\succsim}_s \ell'$ if and only if $\ell \hat{\succsim} \ell'$, for all $\ell, \ell' \in L(X \times S)$ such that $\ell(\cdot, s') = \ell'(\cdot, s')$ for all $s' \in S - \{s\}$).

To speak loosely, the introspective preference relation $\hat{\succsim}$ and the actual preference relation $\succsim$ are consistent if they are induced by the same utilities. Formally, define a mapping, H, from $L(X \times S)$ to F as follows: For each non-degenerate $\ell \in L(X \times S)$, let $H(\ell)(x, s) = \ell(x, s) / \sum_{y \in X} \ell(y, s)$, for all $(x, s) \in X \times S$. A state s is said to be *obviously null* if $f \succsim f'$ for all f and f' in F and there are $\ell, \ell' \in L(X \times S)$ such that $\ell \hat{\succ}_s \ell'$. A state s is *obviously non-null* if $f \succ_s g$ for some f and g. A state s is *essential* if there are ℓ and ℓ' in $L(X \times S)$ such that $\ell \hat{\succ}_s \ell'$.

Strong consistency: For all $s \in S$ and non-degenerate ℓ and ℓ' in $L(X \times S)$, $H(\ell) \succ_s H(\ell')$ implies $\ell \hat{\succ}_s \ell'$, and if s is obviously non-null, then $\ell \hat{\succ}_s \ell'$ implies $H(\ell) \succ_s H(\ell')$.

Theorem (Karni and Schmeidler, 1981): Let $\succsim$ be a non-trivial binary relation on F and $\hat{\succsim}$ a binary relation on $L(X \times S)$. Then each of the two relations satisfies the axioms of expected utility and jointly they satisfy strong consistency if and only if there exist a real-valued function, u, on $X \times S$ and a probability distribution, π, on S such that, for all f and g in F,

$$f \succsim g \Leftrightarrow \sum_{s \in S} \pi(s) \sum_{x \in X} u(x, s) f(x, s)$$
$$\geq \sum_{s \in S} \pi(s) \sum_{x \in X} u(x, s) g(x, s) \tag{4}$$

and, for all ℓ and ℓ' in $L(X \times S)$,

$$\ell \hat{\succsim} \ell' \Leftrightarrow \sum_{s \in S} \sum_{x \in X} u(x, s) \ell(x, s)$$
$$\geq \sum_{s \in S} \sum_{x \in X} u(x, s) \ell'(x, s). \tag{5}$$

Moreover, the function u is unique up to cardinal unit comparable transformation, the probability π restricted to the event of all essential states is unique, and for s obviously null $\pi(s) = 0$ and for s obviously non-null $\pi(s) > 0$.

The subjective expected utility representation in (4) applies whether the preference relation, $\succsim$, is state-dependent or state-independent. Furthermore, as Karni and Mongin (2000) observed, because the utility function is identified using hypothetical lotteries, the probability

measure π in the representation theorem above quantifies the decision maker's beliefs. A similar result in a somewhat different framework is proved in Karni (2003); a probabilistically sophisticated version of this approach appears in Grant and Karni (2004).

A weaker version of this result, based on restricting the consistency condition to a subset of hypothetical lotteries that have the same marginal distribution on S, due to Karni, Schmeidler and Vind (1983), yields a subjective expected utility representation with state-dependent preferences. Wakker (1987) has extended the theory of Karni, Schmeidler and Vind (1983) to include the case in which the set of consequences is a connected topological space. However, the arbitrary choice of the subset of hypothetical lotteries renders the probabilities in these works arbitrary.

Other theories that yield subjective expected utility representations invoke preferences on conditional acts (that is, preference relations over the set of acts conditional on events). Fishburn (1973) and Karni (2007) advanced such theories assuming consequence sets that have distinct structures. Skiadas (1997) proposed a non-expected utility model, based on hypothetical preferences, that yield a representation with state-dependent preferences. In this model, acts and states are primitive concepts, and preferences are defined on act-event pairs. For any such pair the consequences (utilities) represent the decision maker's expression of his holistic valuation of the act. The decision maker is not supposed to be aware of whether or not the given event occurred; hence, his evaluation of the act reflects, in part, his anticipated feelings, such as disappointment aversion.

Moral hazard and state-dependent preferences

Drèze (1961; 1987) and Karni (2006) present distinct theories of individual decision making under uncertainty with moral hazard and state-dependent preferences. Both assume that decision makers can exercise some control over the likely realization of events.

Drèze does not specify the means by which this control is exercised, relying instead on their manifestation in the decision maker's choice behaviour. In particular, departing from Anscombe and Aumann's (1963) 'reversal of order' assumption, Drèze assumes that decision makers strictly prefer that the uncertainty of the lottery payoff be resolved before that of the acts, presumably to allow them to exploit this information by taking action to affect the likely realization of the underlying states. Drèze obtains a unique separation of state-dependent utilities from a set of probability distributions over the set of states of nature. Choice is represented as expected utility maximizing behaviour where the expected utility associated with any given act is itself the maximal expected utility with respect to the probabilities in the set.

Karni (2005b) replaces the state space with a set of effects – phenomena on which decision makers can place bets and whose realization they can influence by their actions. In Karni's theory the choice set consists of action-bet pairs. Actions affect the decision maker's well-being directly (for example, actions may correspond to levels of effort) and indirectly (through their impact on the decision maker's beliefs); bets are functions from effects to monetary payoffs. Karni gives necessary and sufficient conditions for the existence of subjective expected utility representations with unique, action-dependent, subjective probabilities; effect-dependent utility functions representing the evaluation of wealth; and a distinct function that captures the direct impact of the choice of action on the decision maker's well-being.

Attitudes toward risk

As with state-independent preferences, the economic analysis of many decision problems involving state-dependent preferences requires measures of risk aversion. Such measures are developed in Karni (1985).

EDI KARNI

See also **expected utility hypothesis; Savage's subjective expected utility model; uncertainty.**

Bibliography

Anscombe, F. and Aumann, R. 1963. A definition of subjective probability. *Annals of Mathematical Statistics* 43, 199–205.
Arrow, K. 1971. *Essays in the Theory of Risk Bearing.* Chicago: Markham Publishing Co.
Arrow, K. 1974. Optimal insurance and generalized deductibles. *Scandinavian Actuarial Journal*, 1–42.
Cook, P. and Graham, D. 1977. The demand for insurance and protection: the case of irreplaceable commodities. *Quarterly Journal of Economics* 91, 143–56.
Drèze, J. 1961. Les fondements logiques de l'utilité cardinale et de la probabilité subjective. *La Décision*. Colloques Internationaux de CNRS.
Drèze, J. 1987. Decision theory with moral hazard and state-dependent preferences. In *Essays on Economic Decisions under Uncertainty*. Cambridge: Cambridge University Press.
Eisner, R. and Strotz, R. 1961. Flight insurance and the theory of choice. *Journal of Political Economy* 69, 355–68.
Fishburn, P. 1973. A mixture-set axiomatization of conditional subjective expected utility. *Econometrica* 41, 1–25.
Grant, S. and Karni, E. 2004. A theory of quantifiable beliefs. *Journal of Mathematical Economics* 40, 515–46.
Karni, E. 1985. *Decision Making under Uncertainty: The Case of State-Dependent Preferences*. Cambridge, MA: Harvard University Press.
Karni, E. 2003. On the representation of beliefs by probabilities. *Journal of Risk and Uncertainty* 26, 17–38.
Karni, E. 2006. Subjective expected utility theory without states of the world. *Journal of Mathematical Economics* 42(3), 167–88.
Karni, E. 2007. Foundations of Bayesian theory. *Journal of Economic Theory* 132(1), 168–88.
Karni, E. and Mongin, P. 2000. On the determination of subjective probability by choice. *Management Science* 46, 233–48.
Karni, E. and Schmeidler, D. 1981. An expected utility theory for state-dependent preferences. Working Paper No. 48–80. Foerder Institute for Economic Research, Tel Aviv University.
Karni, E., Schmeidler, D. and Vind, K. 1983. On state-dependent preferences and subjective probabilities. *Econometrica* 51, 1021–31.
Karni, E. and Zilcha, I. 1985. Uncertain lifetime, risk aversion and life insurance. *Scandinavian Actuarial Journal*, 109–23.
Savage, L. 1954. *The Foundations of Statistics*. New York: John Wiley.
Skiadas, C. 1997. Subjective probability under additive aggregation of conditional preferences. *Journal of Economic Theory* 76, 242–71.
Wakker, P. 1987. Subjective probabilities for state-dependent continuous utility. *Mathematical Social Sciences* 14, 289–98.
Yaari, M. 1965. Uncertain lifetime, life insurance and the theory of the consumer. *Review of Economic Studies* 32, 137–50.

state space models

State space models is a rather loose term given to time series models, usually formulated in terms of unobserved components, that make use of the state space form for their statistical treatment.

At the simplest level, structural time series models (STMs) are set up in terms of components such as trends and cycles that have a direct interpretation. Signal extraction, or smoothing, provides a description of these features that is model-based and hence avoids the ad hoc nature of procedures such as moving averages and the Hodrick–Prescott filter. While smoothing uses all the observations in the sample, filtering yields estimates at a given point in time that are constructed only from observations available at that time. For key time series, filtered estimates form the basis for 'nowcasting' in that they give an indication of the current state of the economy and the direction in which it is moving. They also provide the starting point for forecasts of future observations and components.

Local level model

A simple model with permanent and transitory components illustrates the basic ideas of filtering and smoothing. Suppose that the observations consist of a random walk component plus a random irregular term, that is

$$y_t = \mu_t + \varepsilon_t, \quad \varepsilon_t \sim NID(0, \sigma_\varepsilon^2), \quad t = 1, \ldots, T \tag{1}$$

$$\mu_t = \mu_{t-1} + \eta_t, \qquad \eta_t \sim NID(0, \sigma_\eta^2), \tag{2}$$

where the irregular and level disturbances, ε_t and η_t respectively, are mutually independent and the notation $NID(0, \sigma^2)$ denotes normally and independently distributed with mean zero and variance σ^2. The signal–noise ratio, $q = \sigma_\eta^2 / \sigma_\varepsilon^2$, plays the key role in determining how observations should be weighted for prediction and signal extraction. In a large sample, filtering is equivalent to a simple exponentially weighted moving average; the higher q is, the more past observations are discounted. When q is zero, the level is constant and all observations have the same weight. The reduced form of the model has the first differences following a first-order moving average process, that is

$$\Delta y_t = \xi_t + \theta \xi_{t-1}, \qquad \xi_t \sim NID(0, \sigma_\xi^2)$$

where $\theta = [(q^2 + 4q)^{1/2} - 2 - q]/2$. This produces the same forecasts. However, the structural form in (1) also yields nowcasts and smoothed estimates of the level, μ_t, throughout the series. In the middle of a large sample, the smoothed estimates are approximately equal to

$$\frac{1 + \theta}{1 - \theta} \sum_j (-\theta)^{|j|} y_{t+j}, \qquad q > 0, \tag{3}$$

while the filtered estimates are

$$(1 + \theta) \sum_{j \geq 0} (-\theta)^j y_{t-j}, \qquad q > 0. \tag{4}$$

At the end of the sample this estimate also yields the forecast of future levels and future observations. Note that, although the above expressions are useful for displaying the weighting of the observations, finite sample computation is best done by a simple forward recursion for filtering and a subsequent backward one for smoothing.

State space form
The state space form (SSF) is a simple device whereby a dynamic model is written in terms of just two equations. The model in (1) and (2) is a special case. The general linear SSF applies to a multivariate time series, y_t, containing N elements. These observable variables are related to an $m \times 1$ vector, α_t, known as the *state vector*, through a *measurement equation*

$$y_t = Z_t \alpha_t + d_t + \varepsilon_t, \qquad t = 1, \ldots, T \tag{5}$$

where Z_t is an $N \times m$ matrix, d_t is an $N \times 1$ vector and ε_t is an $N \times 1$ vector of serially uncorrelated disturbances with mean zero and covariance matrix H_t. In general the elements of α_t are not observable. However, they are assumed to be generated by a first-order Markov process,

$$\alpha_t = T_t \alpha_{t-1} + c_t + R_t \eta_t, \qquad t = 1, \ldots, T \tag{6}$$

where T_t is an $m \times m$ matrix, c_t is an $m \times 1$ vector, R_t is an $m \times g$ matrix and η_t is a $g \times 1$ vector of serially uncorrelated disturbances with mean zero and covariance matrix, Q_t. Equation (6) is the *transition equation*. The specification of the system is completed by assuming that the initial state vector, α_0, has a mean of a_0 and a covariance matrix P_0, and that the disturbances ε_t and η_t are uncorrelated with the initial state. The disturbances are often assumed to be uncorrelated with each other in all time periods, though this assumption may be relaxed to allow contemporaneous correlation, the consequence being a slight complication in some of the filtering formulae.

The definition of α_t for any particular statistical model is determined by construction. Its elements may or may not be identifiable with components that have a substantive interpretation, for example as a trend or a seasonal. From the technical point of view, the aim of the state space formulation is to set up α_t in such a way that it contains all the relevant information on the system at time t and that it does so by having as small a number of elements as possible. The SSF is not, in general, unique.

The Kalman filter (KF) is a recursive procedure for computing the optimal estimator of the state vector at time t, based on the observations up to and including y_t. In a Gaussian model, the disturbances ε_t and η_t, and the initial state, are all normally distributed. Because a normal distribution is characterized by its first two moments, the Kalman filter can be interpreted as updating the mean and covariance matrix of the conditional distribution of the state vector as new observations become available. The conditional mean minimizes the mean square error and when viewed as a rule for all realizations it is the minimum mean square error estimator (MMSE). Since the conditional covariance matrix does not depend on the observations, it is the unconditional MSE matrix of the MMSE. When the normality assumption is dropped, the KF is still optimal in the sense that it minimizes the mean square error within the class of all linear estimators. Given initial conditions, a_0 and P_0, the Kalman filter delivers the optimal estimator of the state vector as each new observation becomes available. When all T observations have been processed, it yields the optimal estimator of the current state vector based on the full information set. When the initial conditions cannot be specified a diffuse prior is often placed on the initial state. This amounts to setting $P_0 = \kappa I$, and letting the scalar κ go to infinity. Stable algorithms for handling diffuse priors are set out in Durbin and Koopman (2001).

Prediction is carried out straightforwardly by running the KF without updating. Mean square errors of the forecasts are produced at the same time. Smoothing is carried out by a backward filter initialized with the estimates delivered by the KF at time T. The aim is to compute the optimal estimator of the state vector at time t using information made available after time t as well as before. Efficient smoothing algorithms are described in Durbin and Koopman (2001, pp. 70–3). The weights are implicit, but Koopman and Harvey (2003) give an algorithm for computing and displaying them at any point in time. The state space smoother is far more general than the classic Wiener–Kolomogorov (WK) filter. The WK filter computes weights explicitly and for simple models it is possible to obtain expressions for the estimator in the middle of a doubly infinite sample without too much difficulty. Formula (3) is a case in point. However, the WK filter is limited to time-invariant models and even here it has no computational advantages over the state space fixed-interval smoothing algorithm. In the second edition of his celebrated text describing the WK filter, Whittle (1984, p. xi) writes 'In its preoccupation with the stationary case and generating function methods, the 1963 text essentially missed the fruitful concept of state structure. This … has now come to dominate the subject.'

The system matrices $\mathbf{Z}_t$, $\mathbf{H}_t$, $\mathbf{T}_t$, $\mathbf{R}_t$ and $\mathbf{Q}_t$ may depend on a set of unknown parameters, and one of the main statistical tasks will often be the estimation of these parameters. Thus in the random walk plus noise model, (1), the parameters σ_η^2 and σ_ε^2 will usually be unknown. As a by-product, the KF produces a vector of prediction errors or *innovations* and in a Gaussian model these can be used to construct a likelihood function that can be maximized numerically with respect to the unknown parameters.

Since the state vector is a vector of random variables, a Bayesian interpretation of the Kalman filter as a way of updating a Gaussian prior distribution on the state to give a posterior is quite natural. The mechanics of filtering, smoothing and prediction are the same irrespective of whether the overall framework is Bayesian or classical. Smoothing gives the mean and variance of the state, conditional on all the observations. For the classical statistician, the conditional mean is the MMSE, while for the Bayesian it minimizes the expected loss for a symmetric loss function. With a quadratic loss function, the expected loss is given by the conditional variance. The real differences between classical and Bayesian treatments arise when the parameters are unknown. In a Bayesian framework, the hyperparameters, as they are often called, are random variables. The development of simulation techniques based on Markov chain Monte Carlo (MCMC) has now made a full Bayesian treatment a feasible proposition. This means that it is possible to simulate a distribution for the state that takes account of hyperparameter uncertainty.

Applications

The use of unobserved components opens up a new range of possibilities for economic modelling. Furthermore, it provides insights and a unified approach to many other problems. The examples below give a flavour.

The local linear trend model generalizes (1) by the introduction of a stochastic slope, β_t, which itself follows a random walk. Thus

$$\mu_t = \mu_{t-1} + \beta_{t-1} + \eta_t, \qquad \eta_t \sim NID(0, \sigma_\eta^2),$$
$$\beta_t = \beta_{t-1} + \zeta_t, \quad \zeta_t \sim NID(0, \sigma_\zeta^2),$$
$$(7)$$

where the irregular, level and slope disturbances, ε_t, η_t and ζ_t, respectively, are mutually independent. If both variances σ_η^2 and σ_ζ^2 are zero, the trend is deterministic. When only σ_ζ^2 is zero, the slope is fixed and the trend reduces to a random walk with drift. Allowing σ_ζ^2 to be positive but setting σ_η^2 to zero gives an integrated random walk trend, which when estimated tends to be relatively smooth. Signal extraction of the trend by setting the signal–noise ratio, $q = \sigma_\zeta^2/\sigma_\varepsilon^2$, to 1/1600 gives the Hodrick–Prescott filter for quarterly data. Adding a cyclical component to (1) provides a vehicle for detrending based on a model the parameters of which can be estimated from the data.

Orphanides and van Norden (2002) have recently stressed the importance of tracking the output gap in *real time*. Given the parameter estimates, real time estimation of components such as the output gap is just an exercise in filtering. However, as new observations become available the estimate of the gap at a particular point in time can be improved by smoothing. Harvey, Trimbur and van Dijk (2007) adopt a Bayesian approach which has the advantage of giving the full distribution of the output gap. Statistics such as the probability that the output gap is increasing are readily calculated. Following Kuttner (1994), Harvey, Trimbur and van Dijk (2007) also construct an unobserved components model relating the output gap to inflation in what is effectively a Phillips curve relationship.

A number of authors, beginning with Sargent (1989), have estimated the structural parameters of dynamic stochastic general equilibrium (DSGE) models using state space methods. The linear rational expectations model is first solved for the reduced-form state equation in its predetermined variables. Once this has been done, the model is put in state space form and the parameters are estimated by maximum likelihood. Alternatively a Bayesian approach can be adopted; see Smets and Wouter (2003, p. 1138).

Data irregularities

Some of the most striking benefits of the structural approach to time series modelling become apparent only when we start to consider more complex problems. In particular, the SSF offers considerable flexibility with regard to dealing with data irregularities, such as missing observations and observations at mixed frequencies. Missing observations are easily handled in the SSF simply by omitting the updating equations while retaining the prediction equations. Filtering and smoothing then go through automatically and the likelihood function is constructed using prediction errors corresponding to actual observations. With flow variables, such as income, the issue is one of temporal aggregation. This may be dealt with by the introduction of a cumulator variable into the state. The study by Harvey and Chung (2000) on the measurement of British unemployment provides an illustration of how mixed frequencies are handled and how using an auxiliary series can improve the efficiency of nowcasting and forecasting a target series. The challenge was how to obtain timely estimates of the underlying change in unemployment. Estimates of the numbers of unemployed according to the International Labor Organization (ILO) definition are given by the Labour Force Survey (LFS), which consists of a rotating sample of approximately 60,000 households. These estimates have been published on a quarterly basis since the spring of 1992, but from 1984 to 1991 estimates were available for the spring quarter only. Another measure of unemployment, based on administrative sources, is the number of people claiming unemployment benefit. This measure, known as the claimant count, is available monthly, with very little delay and is an exact figure. It does not provide a figure corresponding to the ILO definition, but it moves roughly in the same way as the LFS figure. The first problem is how to extract the best estimate of the underlying monthly change in a series which is subject to sampling error and which may not have been recorded every month. The second is how to use a related series to improve this estimate. These two issues are of general importance, for example in the measurement of the underlying rate of inflation or the way in which monthly figures on industrial production might be used to produce more timely estimates of national income. State space methods deal with the mixed frequencies in the target series, with the rather complicated error structure coming from the rotating sample (see Pfeffermann, 1991) and with the different frequency of the auxiliary series.

Continuous time

Continuous time STMs observed at discrete intervals can easily be put in SSF (see Harvey, 1989, ch. 9). An important case is the continuous time version of (7) where the smoothed trend is a cubic spline. Setting up such a model for a cubic spline enables the smoothness parameter to be estimated by maximum likelihood and the fact that irregularly spaced data may be handled means that it can be used to fit a nonlinear function to cross-sectional data. The model can easily be extended, for example to include other components, and it can be compared with

alternative models using standard statistical criteria (see Kohn, Ansley and Wong, 1992).

Nonlinear and non-Gaussian models

Some of the most exciting recent developments in time series have been in nonlinear and non-Gaussian models. For example, it is possible to fit STMs with heavy-tailed distributions on the disturbances, thereby making them robust with respect to outliers and structural breaks. Similarly, non-Gaussian models, designed to deal with count data and qualitative observations, can be set up with stochastic components. In the general formulation of a state space model, the distribution of the observations is specified conditional on the current state and past observations, that is

$$p(\mathbf{y}_t|\boldsymbol{\alpha}_t, \mathbf{Y}_{t-1}), \qquad t = 1, \dots, T \qquad (8)$$

where $\mathbf{Y}_{t-1} = \{\mathbf{y}_{t-1}, \mathbf{y}_{t-2}, \dots\}$. Similarly the distribution of the current state is specified conditional on the previous state and observations so that

$$p(\boldsymbol{\alpha}_t|\boldsymbol{\alpha}_{t-1}, \mathbf{Y}_{t-1}). \qquad (9)$$

The initial distribution of the state is given as $p(\boldsymbol{\alpha}_0)$. In a linear Gaussian model the conditional distribution in (8) and (9) are characterized by their first two moments and so they are specified by the measurement and transition equations. The Kalman filter updates the mean and covariance matrix of the state. In more general models, computer-intensive methods, using techniques such as importance sampling, have to be applied (see Durbin and Koopman, 2001). Within a Bayesian framework, methods are normally based on MCMC. Particle filtering is often used for signal extraction; see the review in Harvey and de Rossi (2006).

The use of state space methods highlights a fundamental distinction in time series models between those motivated by description and those set up to deal directly with forecasting. This is epitomized by the contrast between STMs on the one hand and autoregressions and autoregressive-integrated-moving average (ARIMA) models on the other. In a linear Gaussian world, the reduced form of an STM is an ARIMA model and questions regarding the merits of STMs for forecasting revolve round the gains, or losses, from the implied restrictions on the reduced form and the guidance, or lack of it, given to the selection of a suitable model (see the discussion in Harvey, 2006). Once nonlinearity and non-Gaussianity enter the picture, the two approaches can be very different. Models motivated solely by forecasting tend to be set up in terms of a distribution for the current observations conditional on past observations rather than in terms of components. For example, changing variance can be captured by a model from the generalized autoregressive conditional heteroscedasticity (GARCH) class, where conditional variance is a function of past observations, as opposed to a stochastic volatility (SV) model in which the variance is a dynamic unobserved component. The readings in Shephard (2005) describe SV models and discuss the use of computationally intensive methods for estimating them.

The realization that the statistical treatment of a wide range of dynamic models can be dealt with directly in a unified framework is important. For engineers, using state space methods is a natural way to proceed. For many economists, brought up with regression and autoregression, state space is an alien concept. This is changing. State space methods are now becoming an important part of the toolkit of econometricians and economists.

ANDREW HARVEY

See also **data filters; Kalman and particle filtering; prediction formulas.**

Bibliography

Durbin, J. and Koopman, S. 2001. *Time Series Analysis by State Space Methods*. Oxford: Oxford University Press.

Harvey, A. 1989. *Forecasting, Structural Time Series Models and Kalman Filter*. Cambridge: Cambridge University Press.

Harvey, A. 2006. Forecasting with unobserved components time series models. In *Handbook of Economic Forecasting*, vol. 1, ed. G. Elliot, C. Granger and A. Timmermann. Amsterdam: North-Holland.

Harvey, A. and Chung, C.-H. 2000. Estimating the underlying change in unemployment in the UK (with discussion). *Journal of the Royal Statistical Society, Series A* 163, 303–39.

Harvey, A. and de Rossi, G. 2006. Signal extraction. In *Palgrave Handbook of Econometrics*, vol. 1, ed. K. Patterson and T. Mills. Basingstoke: Palgrave Macmillan.

Harvey, A., Trimbur, T. and van Dijk, H. 2007. Trends and cycles in economic time series: a Bayesian approach. *Journal of Econometrics* 140(2), 618–49.

Kohn, R., Ansley, C. and Wong, C.-H. 1992. Nonparametric spline regression with autoregressive moving average errors. *Biometrika* 79, 335–46.

Koopman, S. and Harvey, A. 2003. Computing observation weights for signal extraction and filtering. *Journal of Economic Dynamics and Control* 27, 1317–33.

Kuttner, K. 1994. Estimating potential output as a latent variable. *Journal of Business and Economic Statistics* 12, 361–8.

Orphanides, A. and van Norden, S. 2002. The unreliability of output gap estimates in real-time. *Review of Economics and Statistics* 84, 569–83.

Pfeffermann, D. 1991. Estimation and seasonal adjustment of population means using data from repeated surveys. *Journal of Business and Economic Statistics* 9, 163–75.

Sargent, T. 1989. Two models of measurements and the investment accelerator. *Journal of Political Economy* 97, 251–87.

Smets, F. and Wouter, R. 2003. An estimated dynamic stochastic general equilibrium model of the euro area. *Journal of the European Economic Association* 1, 1123–75.

Shephard, N. 2005. *Stochastic Volatility*. Oxford: Oxford University Press.

Whittle, P. 1984. *Prediction and Regulation*, 2nd edn. Blackwell: Oxford.

statistical decision theory

Decision theory is the science of making optimal decisions in the face of uncertainty. Statistical decision theory is concerned with the making of decisions when in the presence of statistical knowledge (data) which sheds light on some of the uncertainties involved in the decision problem. The generality of these definitions is such that decision theory (we drop the qualifier 'statistical' for convenience) formally encompasses an enormous range of problems and disciplines. Any attempt at a general review of decision theory is thus doomed; all that can be done is to present a description of some of the underlying ideas.

Decision theory operates by breaking a problem down into specific components, which can be mathematically or probabilistically modelled and combined with a suitable optimality principle to determine the best decision. Section 1 describes the most useful breakdown of a decision problem – that into actions, a utility function, prior information and data. Section 2 considers the most important optimality principle for reaching a decision – the Bayes principle. The frequentist approach to decision theory is discussed in Section 3, with the minimax principle mentioned as a special case. Section 4 compares the various approaches.

The history of decision theory is difficult to pin down, because virtually any historical mathematically formulated decision problem could be called an example of decision theory. Also, it can be difficult to distinguish between true decision theory and formally related mathematical devices such as least squares estimation. Decision theory became clearly formulated as a science through work of John von Neumann and Oscar Morgenstern, culminating in their book *Theory of Games and Economic Behavior* (1944), and Abraham Wald, culminating in his book *Statistical Decision Functions* (1950). (The books do discuss some of the earlier history of decision theory.) General introductions to decision theory can be found, at an advanced level, in Blackwell and Girshick (1954) and Savage (1954); at an intermediate level in Raiffa and Schlaifer (1961), Ferguson (1967), De Groot (1970), Berger (1985), and French and Rios Insua (2000); and at a basic level in Raiffa (1968), Lindley (1985), and Winkler (1972).

1 Elements of a decision problem

In a decision problem, the most basic concept is that of an action a. The set of all possible actions that can be taken will be denoted by A. Any decision problem will typically involve an unknown quantity or quantities; this unknown element will be denoted by θ.

Example 1 A company receives a shipment of parts from a supplier, and must decide whether to accept the shipment or to reject the shipment (and return it to the supplier as unsatisfactory). The two possible actions being contemplated are:

a_1: accept the shipment, a_2: reject the shipment.

Thus $A = \{a_1, a_2\}$. The uncertain quantity which is crucial to a correct decision is:

θ = the proportion of defective parts in the shipment.

Clearly action a_1 is desirable when θ is small enough, while a_2 is desirable otherwise.

The key idea in decision theory is to attempt a quantification of the gain or loss in taking possible actions. Since the gain or loss will usually depend upon θ as well as the action a taken, it is typically represented as a function of both. In economics this function is generally called the *utility function*, following the work of Frank Ramsey in the 1920s, and is denoted by $U(\theta, a)$. It is to be understood as the gain achieved if action a is taken and θ obtains. (The scale for measuring 'gain' will be discussed later.) In the statistical literature it is customary to talk in terms of loss instead of gain, with typical notation $L(\theta, a)$ for the loss function. Loss is just negative gain, so defining $L(\theta, a) = -U(\theta, a)$ results in effective equivalence between the two formulations (whatever maximizes utility will minimize loss).

Example 1 (continued) The company determines its utility function to be given by:

$$U(\theta, a_1) = 1 - 10\theta, \quad U(\theta, a_2) = -0.1.$$

To understand how these might be developed, note that if a_2 is chosen the shipment will be returned to the supplier and a new shipment sent out. This new shipment must then be processed, all of which takes time and money. The overall cost of this eventuality is determined to be 0.1 (on the scale being used). The associated utility is -0.1 (a loss is a negative gain). Note that this cost is fixed: that is, it does not depend on θ.

When a_1 is chosen, quite different considerations arise. The parts will be utilized with, say, gain of 1 if none is defective. Each defective part will cause a reduction in income by a certain amount, however, so that the true overall gain will be 1 reduced by a linear function of the proportion of defectives. $U(\theta, a_1)$ is precisely of this form. The various constants in $U(\theta, a_1)$ and $U(\theta, a_2)$ are chosen to reflect the various importance of the associated costs.

The scale chosen for a utility function turns out to be essentially unimportant, so that any convenient choice can be made. If the gain or loss is monetary, a suitable monetary unit often can provide a natural scale. Note, however, that utility functions can be defined for any type of gain or loss, not just monetary. Thus, in example 1, the use of defective parts could lead to faulty final products from the company, and affect the overall quality image or prestige of the company. Such considerations are not easily stated in monetary terms, yet can be important to include in the overall construction of the utility function. (For more general discussion of the construction of utility functions, see Berger, 1985.)

The other important component of a decision problem is the information available about θ. This information will often arise from several sources, substantially complicating the job of mathematical modelling. We content ourselves here with consideration of the standard statistical scenario where there are available (a) data, X, from a statistical experiment relating to θ; and (b) background or *prior* information about θ, to be denoted by $\pi(\theta)$. Note that either of these components could be absent.

The data, X, is typically modelled as arising from some *probability density* $p_\theta(X)$. This, of course, is to be interpreted as the probability (or probability density) of the particular data value when θ obtains.

Example 1 (continued) It is typically too expensive (or impossible) to test all parts in a shipment for defects, so that a statistical sampling plan is employed instead. This generally consists of selecting, say, n random parts from the shipment, and testing only these for defects. If X is used to denote the number of defective parts found in the tested sample, and if n is fairly small compared with the total shipment size, then it is well known that $p_\theta(X)$ is approximately the binomial density:

$$p_\theta(X) = \frac{n!}{X!(n-X)!} \theta^X (1 - \theta)^{n-X}.$$

The *prior* information about θ is typically also described by a probability density $\pi(\theta)$. This density is the probability (or mass) given to each possible value of θ in the light of beliefs as to which values of θ are most likely.

Example 1 (continued) The company has been receiving a steady stream of shipments from this supplier and has recorded estimates of the proportion of defectives for each shipment. The records show that 30 per cent of the shipments had θ between 0.0 and 0.025, 22 per cent of the shipments had θ between 0.025 and 0.05, 15 per cent had θ between 0.05 and 0.075, 11 per cent had θ between 0.075 and 0.10, 13 per cent had θ between 0.10 and 0.15, and the remaining 9 per cent had θ bigger than 0.15. Treating the varying θ as random, a probability density which provides a good fit to these percentages is the beta (1,14) density given (for $0 \leq \theta \leq 1$) by:

$$\pi(\theta) = 14(1 - \theta)^{13}.$$

(For example, the probability that a random θ from this density is between 0.0 and 0.025 can be calculated to be 0.30, agreeing exactly with the observed 30 per cent.) It is very reasonable to treat θ for the current shipment as a random variable from this density, which we will thus take as the prior density.

2 Bayesian decision theory

When θ is known, it is a trivial matter to find the optimal action: simply maximize the gain by maximizing $U(\theta, a)$ over a. When θ is unknown, the natural generalization is to first 'average' $U(\theta, a)$ over θ, and then maximize over a. The correct method of 'averaging over θ' is to determine the overall probability density of θ, to be denoted $\pi^*(\theta)$ (and to be described shortly), and then consider the *Bayesian expected* utility:

$$U^*(a) = E^{\pi^*}[U(\theta, a)]$$
$$= \int U(\theta, a)\pi^*(\theta)\mathrm{d}\theta.$$

(This last expression assumes that θ is a continuous variable taking values in an interval of numbers. If it can assume only one of a discrete set of values, then this integral should be replaced by a sum over the possible values.) Maximizing $U^*(a)$ over a will yield the optimal *Bayes action*, to be denoted by a^*.

Example 1 (continued) Initially, assume that no data, X, are available from a sampling inspection of the current shipment. Then the only information about θ is that contained in the prior $\pi(\theta)$; $\pi^*(\theta)$ will thus be identified with $\pi(\theta) = 14(1 - \theta)^{13}$. Calculation yields:

$$U^*(a_1) = \int_0^1 (1 - 10\theta)\,14(1 - \theta)^{13}\mathrm{d}\theta = 0.33,$$

$$U^*(a_2) = \int_0^1 (-0.1)\,14(1 - \theta)^{13}\mathrm{d}\theta = -0.1.$$

Since $U^*(a_1) > U^*(a_2)$, the Bayes action is a_1, to accept the shipment.

When data, X, are available, in addition to the prior information, the overall probability density π^* for θ must combine the two sources of information. This is done by *Bayes's theorem* (from Bayes, 1763), which gives the overall density, usually called the *posterior density*, as:

$$\pi^*(\theta) = p_\theta(X) \cdot \pi(\theta)/m(X),$$

where:

$$m(X) = \int p_\theta(X)\pi(\theta)\mathrm{d}\theta$$

(or a summation over θ if θ assumes only a discrete set of values), and $p_\theta(X)$ is the probability density for the experiment with the observed values of the data X inserted.

Example 1 (continued) Suppose a sample of $n = 20$ items is tested, out of which $X = 3$ defectives are observed. Calculation gives that the posterior density of θ is:

$$\pi^*(\theta) = p_\theta(3) \cdot \pi(\theta)/m(3)$$
$$= \left[\frac{20!}{3!17!}\theta^3(1 - \theta)^{17}\right] \cdot \left[14(1 - \theta)^{13}\right]/m(3)$$
$$= (185, 504)\theta^3(1 - \theta)^{30},$$

which can be recognized as the beta (4,31) density. This density describes the location of θ in the light of all available information. The Bayesian expected utilities of a_1 and a_2 are thus:

$$U^*(a_1) = \int_0^1 (1 - 10\theta)\pi^*(\theta)\mathrm{d}\theta$$
$$= \int_0^1 (1 - 10\theta)(185, 504)\theta^3(1 - \theta)^{30}\mathrm{d}\theta$$
$$= -0.14,$$

and

$$U^*(a_2) = \int_0^1 (-0.1)\pi^*(\theta)\mathrm{d}\theta = -0.1.$$

Clearly a_2 now has the largest expected utility and should be the action chosen; in other words, the lot of parts should be rejected.

3 Frequentist decision theory

An alternative approach to statistical decision theory arises from taking a 'long run' perspective. The idea is to imagine repeating the decision problem a large number of times, and to develop a decision strategy which will be optimal in terms of some long-run criterion. This is called the *frequentist* approach, and is essentially due to Neyman, Pearson and Wald (see Neyman and Pearson, 1933; Neyman, 1977; Wald, 1950).

To formalize the above idea, let $d(X)$ denote a decision strategy or decision rule. The notation reflects the fact that we are imaging repetitions of the decision problem which will yield possibly different data X, and must therefore specify the action to be taken for any possible X. The utility of using $d(X)$ when θ obtains is thus

$U(\theta, d(X))$. The statistical literature almost exclusively works with loss functions instead of utility functions; for consistency with this literature we will thus use the loss function $L(\theta, d) = -U(\theta, d)$. (Of course, we want to minimize loss.)

The first step in a frequentist evaluation is to compute the *risk function* (expected loss over X) of d, given by:

$$R(\theta, d) = E_\theta\{L(\theta, d(X))\}$$
$$= \int L(\theta, d(X))p_\theta(X)dX.$$

(Again, this integral should be a summation if X is discrete valued.) For a fixed θ this risk indicates how well $d(X)$ would perform if utilized repeatedly for data arising from the probability density $p_\theta(X)$. For various common choices of L this yields familiar statistical quantities. For instance, when L is 0 or 1, according to whether or not a correct decision is made in a two action hypothesis testing problem, the risk becomes the 'probabilities of type I or type II errors'. When L is 0 or 1, according to whether or not an interval $d(X)$ is contains θ, the risk is 1 minus the 'coverage probability function' for the confidence procedure $d(X)$. When $d(X)$ is an estimate of θ and $L(\theta, d) = (\theta - d)^2$, the risk is the 'mean squared error' commonly considered in many econometric studies. (If the estimator $d(X)$ is unbiased, then this mean squared error is also the variance function for d.)

Example 2 Example 1, involving acceptance or rejection of the shipment, is somewhat too complicated to handle here from the frequentist perspective; we thus consider the simpler problem of merely estimating θ (the proportion of defective parts in the shipment). Assume that loss in estimation is measured by *squared error*; that is:

$$L(\theta, d(X)) = [\theta - d(X)]^2.$$

A natural estimate of θ, based on X (the number of defectives from a sample of size n), is the sample proportion of defectives $d_1(X) = X/n$. For this decision rule (or *estimator*), the risk function when X has the binomial distribution discussed earlier (so that X takes only the discrete values $0, 1, 2, \ldots n$) is given by:

$$R(\theta, d) = \sum_{X=0}^{n} \left(\theta - \frac{X}{n}\right)^2 p_\theta(X) = \theta(1 - \theta)/n.$$

The second step of a frequentist analysis is to select some criterion for defining optimal risk functions (and hence optimal decision rules). One of the most common criteria is the *minimax principle*, which is based on consideration of the maximum possible risk:

$$R^*(d) = \max_\theta R(\theta, d).$$

This indicates the worst possible performance of $d(X)$ in repeated use, and hence has some appeal as a criterion based on a cautious attitude. Using this criterion, an optimal decision rule is, of course, defined as one which minimizes $R^*(d)$, and is called a *minimax decision rule*.

Example 2 (continued) It is easy to see that:

$$R^*(d_1) = \max_\theta R(\theta, d_1) = \max_\theta \frac{\theta(1 - \theta)}{n} = \frac{1}{4n}.$$

However, d_1 is not the minimax decision rule. Indeed, the minimax decision rule turns out to be:

$$d_2(X) = \left(X + \sqrt{n}/2\right)/\left(n + \sqrt{n}\right),$$

which has $R^*(d_2) = 1/[4(1 + \sqrt{n})^2]$ (compare with Berger, 1985, p. 354). The minimax criterion here is essentially the same as the minimax criterion in game theory. Indeed, the frequentist decision problem can be considered to be a zero-sum two-person game with the statistician as player II (choosing $d(X)$), an inimical 'nature' as player I (choosing θ), and payoff (to player I) of $R(\theta, d)$. (Of course, it is rather unnatural to assume that nature is inimical in its choice of θ.) (For further discussion of this relationship, see Berger, 1985, ch. 5.)

Minimax optimality is but one of several criteria that are used in frequentist decision theory. Another common criterion is the invariance principle, which calls for finding the best decision rule in the class of rules which are 'invariant' under certain mathematical transformations of the decision problem. (See Berger, 1985, ch. 6, for discussion.)

There also exist very general and elegant theorems which characterize the class of acceptable decision rules. The formal term used is 'admissible': a decision rule, d, is *admissible* if there is no decision rule, d^*, with $R(\theta, d^*) \leq R(\theta, d)$, the inequality being strict for some θ. If such a d^* exists, then d is said to be *inadmissible*, and one has obvious cause to question its use. Very common decision rules, such as the least squares estimator in three or more dimensional normal estimation problems (with sum of squares error loss), can turn out rather astonishingly to be inadmissible, so this avenue of investigation has had a substantial impact on decision theory. A general discussion, with references, can be found in Berger (1985).

4 Comparison of approaches

For solving a real decision problem, there is little doubt that the Bayesian approach is best. It incorporates all the available information (including the prior information, $\pi(\theta)$, which the frequentist approach ignores), and it tends to be easier than the frequentist approach by an order of magnitude. Maximizing $U^*(a)$ over all actions is generally much easier than minimizing something like $R^*(d)$ over all decision rules; the point is that, in some sense, the frequentist approach needlessly complicates the issue by forcing consideration of the right thing to do for each possible X, while the Bayesian worries only about what to do for the actual data X that are observed. There are also fundamental axiomatic developments (see, Ramsey, 1931; Savage, 1954; and Fishburn, 1981, for a general review) which show that only the Bayesian approach is consistent with plausible axioms of rational behaviour. Basically, the arguments are that situations can be constructed in which the follower of any non-Bayesian approach, say the minimax analyst, will be assured of inferior results.

Sometimes, however, decision theory is used as a formal framework for investigating the performance of statistical procedures, and then the situation is less clear. In Example 2, for instance, we used decision theory mainly as a method to formulate rigorously the problem of estimating a binomial proportion θ. If one is developing a statistical rule, $d(X)$, to be used for binomial estimation problems in general, then its repeated performance for varying X is certainly of interest. Furthermore, so the argument goes, prior information may be unavailable or inaccessible in problems where routine statistical analyses (such as estimating a binomial proportion θ) are to be performed, precluding use of the Bayesian approach.

The Bayesian reply to these arguments is that (*a*) optimal performance for each X alone will guarantee good performance in repeated use, negating the need to consider frequentist measures explicitly; and (*b*) even when prior information is unavailable or cannot be used, a Bayesian analysis can still be performed with so called objective prior densities: see Bernardo and Smith (1994), and Berger (2006).

Example 2 (continued) If no prior information about θ is available, one might well say that choosing $\pi(\theta) = 1$ reflects this lack of knowledge about θ. A Bayesian analysis (calculating the posterior density and choosing the action with smallest Bayesian expected squared error loss) yields, as the optimal estimate for θ when X is observed:

$$d_3(X) = (X + 1)/(n + 2).$$

This estimate is considerably more attractive than, say, the minimax rule $d_2(X)$ (see Berger, 1985, p. 375).

In practical applications of decision theory, it is the Bayesian approach which is dominant, yet the frequentist approach retains considerable appeal among theoreticians. A general consensus on the controversy appears quite remote at this time. This author sides with the Bayesian approach in the above debate, while recognizing that there are some situations in which the frequentist approach might be useful. For an extensive discussion of these issues, see Berger (1985); Chernoff and Moses (1959) provide an insightful introduction.

JAMES O. BERGER

See also **Bayesian statistics; decision theory in econometrics; game theory; utility.**

Bibliography

Bayes, T. 1763. An essay towards solving a problem in the doctrine of chances. *Philosophical Transactions of the Royal Society* 53, 370–418.

Berger, J. 1985. *Statistical Decision Theory and Bayesian Analysis*. New York: Springer-Verlag.

Berger, J. 2006. The case for objective Bayesian analysis. *Bayesian Analysis* 1, 385–402.

Bernardo, J. and Smith, A. 1994. *Bayesian Theory*. New York: John Wiley.

Blackwell, D. and Girshick, M.A. 1954. *Theory of Games and Statistical Decisions*. New York: Wiley.

Chernoff, H. and Moses, L. 1959. *Elementary Decision Theory*. New York: John Wiley.

De Groot, M.H. 1970. *Optimal Statistical Decisions*. New York: McGraw-Hill.

Ferguson, T.S. 1967. *Mathematical Statistics: A Decision Theoretic Approach*. New York: Academic Press.

Fishburn, P.C. 1981. Subjective expected utility: a review of normative theories. *Theory and Decision* 13, 139–99.

French, S. and Rios Insua, D. 2000. *Statistical Decision Theory*. London: Arnold.

Lindley, D.V. 1985. *Making Decisions*. New York: Wiley.

Neyman, J. 1977. Frequentist probability and frequentist statistics. *Synthese* 36, 97–131.

Neyman, J. and Pearson, E.S. 1933. On the problem of the most efficient tests of statistical hypotheses. *Philosophical Transactions of the Royal Society* 231, 289–337.

Raiffa, H. 1968. *Decision Analysis: Introductory Lectures on Choices under Uncertainty*. Reading, MA: Addison-Wesley.

Raiffa, H. and Schlaifer, R. 1961. *Applied Statistical Decision Theory*. Boston: Division of Research, Graduate School of Business Administration, Harvard University.

Ramsey, F.P. 1931. Truth and probability. In *The Foundations of Mathematics and Other Logical Essays*. London: Kegan, Paul, Trench and Trubner. Reprinted in *Studies in Subjective Probability*, ed. H. Kyburg and H. Smokler, New York: Wiley, 1964.

Savage, L.J. 1954. *The Foundations of Statistics*. New York: Wiley.

von Neumann, J. and Morgenstern, O. 1944. *Theory of Games and Economic Behavior*. Princeton: Princeton University Press.

Wald, A. 1950. *Statistical Decision Functions*. New York: Wiley.

Winkler, R.L. 1972. *An Introduction to Bayesian Inference and Decision*. New York: Holt, Rinehart & Winston.

statistical inference

Deduction is the process whereby we pass from a general statement to a particular case: the reverse procedure, from the particular to the general, is variously called induction, or inference. Statistical inference is ordinarily understood to involve repetition or averaging, as when an inference is made about a population on the basis of a sample drawn from it. Economic facts are typically established by means of statistical inference. Economists construct a model of the world and deduce from it implications for the real world. These are checked against the available data, leading to some degree of support for the model. Statistical inference is concerned with how this support should be calculated.

Statistical inference incorporates a parameter θ which describes the model. In the simplest cases θ is a real number but in many models it is a set of numbers or even a function. The other basic element is the data x being observations made on the actual economic system. So θ corresponds to the general element and x to the particular. The model describes how the data follow from the parameter value. This is usually in the form of a probability distribution $p(x|\theta)$: the probability of x, given the value of θ. The problem of statistical inference is to make some statement about θ given the value of x. A simple example is provided by a model that says one variable y has linear regression on another z, the regression line having equation $y = \alpha + \beta z$ and the parameter being the pair $(\alpha, \beta) = \theta$. Data may then be collected for several pairs (y_i, z_i), $i = 1, 2, \ldots, n$ and an inference made about θ. The probability specification will ordinarily be that, for any z, y is normally distributed about $\alpha + \beta z$ with constant variance σ^2. If σ^2 is unknown then it will need to be included with α and β in θ.

Two types of inference statement are ordinarily made about θ: estimation and testing. The main distinction being that in testing some values of θ are singled out for special consideration, whereas in estimation all values of θ are treated equally. In the regression example, the hypothesis may be made that z does not affect y in the sense that $\beta = 0$. It would then be usual to test the hypothesis $\beta = 0$. In estimation, on the other hand, $\beta = 0$ plays no special role and the reasonable values of β on the basis of x are required. Estimation takes two forms, point and interval. In the former θ is estimated by a single number, the point estimate; or in the multidimensional case by a set of numbers. In the latter an interval, or region, of values of θ which are reasonably supported by the data is given. In the regression example

$$b = \sum (y_i - y_\cdot)(z_i - z_\cdot) \Big/ \sum (z_i - z_\cdot)^2$$

is the least-squares point estimate of β, $y_\cdot$ and $z_\cdot$ being the means of the y- and z- values respectively. An interval estimate would be of the form $b \pm ts$, where s is the standard deviation evaluated from the data and t is the value obtained from Student's t-distribution. Point estimates are usually inadequate because they do not include any expression of the uncertainty that exists about the parameter: interval estimates are much to be preferred

and usually, as in the regression case, start with the point estimate b and construct the interval about it. Interval estimates and tests are often related by the fact that the interval contains those parameter values which would not be judged significant were a test of that value to be carried out.

There is no general agreement on how statistical inference should be performed though, in some common situations, there is good agreement about the numerical results. It is possible to recognize three main schools named after Fisher; Neyman, Pearson and Wald (NPW); and Bayes.

The Fisherian school is the least formalized and is the one most favoured by scientists, especially those on the biological side, in medicine and agriculture. Because of its lack of a strict mathematical structure it is the hardest to describe succinctly, yet, because of this it is often the easiest to use. The name is entirely apposite since it is essentially the creation of one man, R.A. Fisher (1925, 1935). Estimation is based on the log-likelihood function $L(\theta) = \log p(x|\theta)$. Here $p(x|\theta)$, the probability of data x given parameter θ, is considered as a function of θ for the observed values of the data, now considered as fixed. A point estimate of θ is provided by the maximum likelihood value $\hat{\theta}$, that maximizes, over θ, $L(\theta)$. The precision of $\hat{\theta}$ can be found using minus the second derivative of $L(\theta)$ at $\hat{\theta}$. An interval estimate is then of the form $\hat{\theta} \pm s$, where s depends on the measure of precision. Extensions to the multi-dimensional dimensional case are readily available and, although cases are known where the method is unsatisfactory, it often works extremely well and is deservedly popular. In the case of normal means, maximum likelihood and least-squares estimates agree. A Fisherian test of the hypothesis that θ is equal to a specified value θ_0 is found by constructing a statistic $t(x)$ from the data x and calculating the probability, were $\theta = \theta_0$, of getting the value of $t(x)$ observed, or more extreme. This probability is called the significance level: the smaller it is, the more doubt is cast on θ having the value θ_0. The best-known example is the F-test for the equality of means in an analysis of variance. It is typical of the Fisherian approach that few rules are available for the choice of the statistic $t(x)$. His genius was enough to produce reasonable answers in important cases. Often $t(x)$ is based on a point estimate of θ.

In some ways NPW is a formalized version of Fisher's approach. It has been much developed in the United States, though even there much applied work is Fisherian and it is the theoreticians who espouse NPW. There are many good expositions: for example, Lehmann (1959, 1983). Statistical inferences are thought of as decisions about θ and the merit of a decision is expressed in terms of a loss function measuring how bad the decision is when the true value is θ. If $t(x)$ is a point estimate of the real parameter θ, squared error $\{t(x) - \theta\}^2$ is the loss function ordinarily used, the loss diminishing the nearer the estimate is to the true value. In testing, the decisions are to reject or to accept the null value θ_0 being tested. The simplest loss function is zero for a correct decision and some constant, positive value for each incorrect one. The probability of rejection of $\theta = \theta_0$ when in fact it is true is typically the significance level in Fisher's approach. Having the concept of a decision and a loss function, it becomes possible to ask the question, what is the best decision (estimate or test)? The criterion used to answer this is the expected loss, the expectation being over the data values according to the probability specification $p(x|\theta)$. Thus, for point estimate $t(x)$, the expected loss is $\int \{t(x) - \theta\}^2 p(x|\theta) \mathrm{d}x$. The problem then is to choose $t(x)$ to minimize this function. There is a substantial difficulty in that this expected loss depends on θ, which is unknown. Consequently additional criteria have to be used in order to select the optimum decision. For example, the decisions may be restrained in some way, as when a point estimate is restricted to be unbiased. A basic result is that the only sensible decisions are those which arise from the following procedure. Select a probability distribution $p(\theta)$ for θ and minimize the expected loss obtained by averaging over both x and θ – in the point estimation case, $\int \int \{t(x) - \theta\}^2 p(x|\theta) p(\theta) \mathrm{d}x \mathrm{d}\theta$. This expectation being a number, the minimization is usually possible without ambiguity. However, the choice of $p(\theta)$ remains to be made. It is important to notice that in NPW theory the distribution of θ is merely introduced as a device for producing a reasonable decision (the technical term is 'admissible') and is not necessarily held to express opinions about θ.

The third system of inference is named, quite inappropriately, after the discoverer of Bayes's,' theorem. Laplace was the first significant user. Inference is a passage from the special x to the general θ on the basis of a model $p(x|\theta)$ going in the opposite direction, from θ to x. In the Bayesian view, inference is similarly accomplished by a probability distribution $p(\theta|x)$ of θ, given x. The two distributions are related by Bayes's,' theorem, $p(\theta|x) \propto p(x|\theta) p(\theta)$, where $p(\theta)$ is a distribution for θ. NPW and Bayes are similar in their introduction of probabilities for θ. A basic difference is that the Bayesian approach recognizes $p(\theta)$ as a statement of belief about θ, and not, as does NPW, just as a technical device. With this strong statement about $p(\theta)$ both x and θ have probabilities attached and the full force of the probability calculus can be employed: in particular, to make the inference $p(\theta|x)$. Now the inference is couched, not in terms of estimates or tests, but by means of a probability distribution. If $p(\theta|x)$ is centred around $t(x)$, say as its mean, then $t(x)$ may be conveniently thought of as a point estimate of θ. If θ_0 is of special interest $p(\theta_0|x)$ may be used as a test of the hypothesis that $\theta = \theta_0$. But the full inference is the distribution $p(\theta|x)$. Consequently, once the big step of introducing $p(\theta)$ has been made, the inference problem is solved by use of the probability calculus: no other considerations are needed. For example, typically θ is multi-dimensional $\theta = (\theta_1, \theta_2, \ldots, \theta_m)$ and

only a few parameters are of interest, the remainder are called nuisance parameters. If only θ_1 matters, inferences about it are easily made by the marginal distribution $p(\theta_1|x)$ found by integrating out the nuisance parameters from $p(\theta|x)$. The regression example above for slope β (α and σ^2 being nuisance) provides an illustration.

Until World War I, Bayesian and non-Bayesian views had alternated in popularity, but the work of R.A. Fisher was so influential that it led to an almost complete suppression of the Bayesian view, which was reinforced by the work of Neyman, Pearson and Wald. Savage (1954) renewed interest in the Bayesian approach by providing it with its axiomatic structure, following Ramsey (1931) whose original ideas had lain unappreciated. Savage was much influenced by the work of de Finetti (his most accessible work is 1974/5) who provided a new view of probability that has had considerable impact upon subsequent thinking. Today the three disciplines lie uneasily together.

The Bayesian approach is the most formalized of the three inferential methods because everything is expressed within the single framework of the probability calculus, which is itself very well formalized. It has been relatively little used largely because of the perceived difficulty of assigning a distribution to θ. An important property of this method is that it is easily extended to include decision-making. As with NPW theory, a class of decisions d is introduced together with a loss function $l(d, \theta)$ expressing the loss in selecting d when θ obtains, and choosing that decision d that minimizes the expected (over θ) loss $\int l(d, \theta) p(\theta|x) \mathrm{d}\theta$, using the inference $p(\theta|x)$. (This is in contrast to the NPW approach, using the expectation over x.)

There are two basic differences between the Bayesian paradigm and the other two. These concern the logical structure, and the likelihood principle. Both the Fisherian and NPW paradigms tackle an inference problem by thinking of several, apparently sensible procedures, investigating their properties and choosing that procedure which overall has the best properties. Fisher's work on maximum likelihood and its demonstrated superiority to the method of moments provides an example. In neither of these approaches are there general procedures: for example, there is no way known of constructing an interval estimate. Within NPW, Wald did introduce the minimax principle but it is generally unsatisfactory in the inference context and has not been used in practice. Against this, the techniques that are available, like maximum likelihood and analysis of variance, are easy to use and interpret (though the interpretation is often wrong: see equations (1) and (2) below). The lack of a formal structure has enabled statisticians to extemporize and come up with valuable concepts and techniques that are of substantial practical value though sometimes with weak justifications. The Bayesian paradigm proceeds differently. It begins by laying down reasonable, elementary properties to be demanded of an inference and then, by deduction, discovers which procedures have these properties. In that sense it is the complete opposite of the Fisherian and NPW views that start with the procedures. It is the method used in other branches of mathematics where the basic properties provide the axioms for the subsequent, logical development. Though there are important variants, all the axiom systems proposed lead to the result that the only inference procedures satisfying them are those that use probability: that the only sensible inference for θ, given x, is a probability statement about θ, given x. The Bayesian position is therefore a deduction from simple requirements about our inferences. NPW comes near to recognizing this in its technical introduction of $p(\theta)$. The Fisherian view never addresses the problem.

The second difference between the Bayesian and other views involves the likelihood principle. The model provides $p(x|\theta)$ which, for fixed θ, is a probability for x. Considered as a function of θ for fixed x, it is called the likelihood for θ (given x). It was an important contribution of Fisher's to emphasize the distinction between the probability and likelihood aspects, and to show us, for example, in the maximum likelihood estimate, the importance of the likelihood function. However, Fisher did not consider the likelihood to be the only tool for inference. In a significance test, based on a statistic $t(x)$, he used the significance level, which is an integral over values of x giving more extreme values to t than that observed, for the tested value θ_0. Clearly this cannot be calculated from the likelihood function which holds x fixed and varies θ. NPW uses the expected (over x) loss and therefore does not use the likelihood function. On the other hand, the only feature of the data used in a Bayesian procedure is the likelihood, supplementing it with the distribution for θ. The likelihood principle says that if two data sets, x and y, have the same likelihood, then the inferences from x and y should be the same. Most statistical procedures in common use today violate the principle, but Bayesian procedures do not. The latter part of that statement is clearly true from Bayes's,' theorem which, in order to calculate the inference, uses only the likelihood. Here is an example of its violation when an unbiased estimate is used.

Given θ, x is a random sample from a population in which each value is either 1 or 0 with probabilities θ and $1-\theta$. In one case the sample is selected to be of size n and r of the values are found to be 1. In the second case, r is chosen and the population sampled until r 1's have been observed, the total sample being of size n. In each case the likelihood is $\theta^r (1-\theta)^{n-r}$ and so, by the likelihood principle, the inferences should be the same. However, in the first case the unbiased estimate of θ based on (r, n) is the familiar r/n: in the second case it is $(r-1)/(n-1)$. Significance tests of $\theta = \frac{1}{2}$, say, are different in the two cases because 'more extreme' in one case means more extreme values of r for fixed n, and in the other more extreme values of n for fixed r. There are many impressive

arguments in favour of the likelihood principle, even outside the Bayesian paradigm, yet it is not accepted by most statisticians and almost all inferential procedures used today violate it: maximum likelihood estimation is the obvious exception.

There is another interesting consequence of the axiomatic, Bayesian approach leading to the probabilistic form of inference, and that is that any non-probabilistic inference will somewhere violate the basic properties set out in the axioms. Indeed, it is true that every non-Bayesian procedure has a counter-example where it behaves in an absurd fashion. In illustration let $(l(x), u(x))$ be a confidence interval for θ at level α based on data x. The precise meaning of this is that

$$p(l(x) < \theta < u(x)|\theta) = \alpha, \quad \text{for all } \theta. \qquad (1)$$

Notice that this is a probability statement about x, given θ, based on $p(x|\theta)$. In words, the probability that the random interval $(l(x), u(x))$ contains θ is α, for all given θ. It is easy to produce examples for x in which the interval is the whole line; $l(x) = -\infty$, $u(x) = \infty$, and $\alpha = 0.95$. Here we are 95 per cent confident that θ is real. This is absurd in the case of the observed x, although it is true that for 95 per cent of x's the statement will be true. Contrast this with the Bayesian statement that

$$p(l(x) < \theta < u(x)|x) = \alpha, \quad \text{for all } x, \qquad (2)$$

based on $p(\theta|x)$. This is about θ, given x: in words, the probability is α that θ lies between $l(x)$ and $u(x)$. Clearly, with $\alpha < 1$, it could never happen that the interval is the whole real line.

A key ingredient in any form of inference is clearly probability, whose laws are well understood. But there is considerable dispute over the interpretation of probability: disputes which have practical consequences. There are two broad groups: subjective and frequentist views. In the subjective view, a probability is an expression of the subject's belief. Thus (2) expresses a belief that θ lies between the numbers $l(x)$ and $u(x)$. In the frequentist view, probabilities are related to observed frequencies. Thus (1) says that the frequency with which the interval contains θ is α. The latter are objective, in the sense that the frequencies can be objectively observed by all subjects. The great majority of statisticians today adopt the frequency view, claiming an objectivity for their methods. Most Bayesians hold to the subjective approach, claiming that economists have to express beliefs about the system they are discussing. It is undoubtably true that many users of statistics think of the frequency statements, like (1), as belief statements, like (2). It had been thought that the two views were opposites but de Finetti showed that the frequentist view of probability is a special case of the subjective view, namely when the data are believed to be exchangeable. The values $x_1, x_2, \ldots, x_n$ are exchangeable if their probability distribution is invariant under permutation of the x's. A random sample from a population would ordinarily be judged to possess this invariance. The case mentioned earlier where each x_i is either 1 or 0, with probabilities θ and $1 - \theta$ respectively, is the standard example. Here θ is a frequency probability, or chance, about which there are beliefs $p(\theta)$ changed by the data x to new beliefs $p(\theta|x)$.

Resistance to the Bayesian approach and subjective probability has centred around the genuine difficulty of assessing beliefs, especially when there is little knowledge of the parameter. Rather than face the formidable, and perhaps impossible, task of measuring belief, statisticians have concentrated on frequentist methods, sometimes ignoring their defects. A related difficulty with the subjective approach is the lack of objectivity in the sense that two subjects may, on the basis of the same data, have different beliefs. The Bayesian response is that this reflects reality and if each economist were to express all his beliefs probabilistically, we would have a clearer appreciation of the situation; and, in any case, different beliefs come together with increasing amounts of data. This is why observational studies are so important. Economics is predominantly frequentist but does have a substantial school, particularly in econometrics, of the Bayesian persuasion. The close connection between that view and decision-making makes it more attractive to the economist than to the laboratory scientist who sees himself as acquiring knowledge, not making decisions.

Inference that is statistical, involving repetition, is naturally allied to the frequency view: whereas inference, in general, has no frequency basis. But de Finetti's observation connecting exchangeability (which is essentially a finite, frequentist property) with subjectivity shows that the Bayesian view embraces both statistical and non-statistical inference. Consequently, the subjective, Bayesian paradigm has enormous potential, encompassing almost all problems of passing from the special to the general. The guilt (corresponding to θ) of a defendant in a law court on the basis of evidence x is an example. The likelihood principle says that the only relevant features are the probabilities of the evidence on the assumptions of innocence, and of guilt. Whether this potential will be realized depends in large part on overcoming the practical difficulties of assessment of beliefs.

Statistical inference depends on a probability specification $p(x|\theta)$ for data x, given parameter θ. If NPW it uses, in addition, a loss function: if Bayesian it introduces an additional probability specification for θ, $p(\theta)$. An important topic studies how the inference is affected by changes in any of the specifications. The inference is said to be robust if the change has little effect on it. For example, it is usual to choose $p(x|\theta)$ to be normal, largely because this assumption is relatively easy to handle and leads to many, simple and powerful answers. We might ask what happens if the normal is replaced by the very similar Student's t-distribution with its rather longer tails. For the mean μ of the normal, the sample mean is,

by any standard, an excellent point estimate of μ: a trimmed mean, in which a few extreme observations are discarded, is not quite as good but is still reasonable. With the t-distribution however, the situation is reversed and the trimmed mean behaves better than the sample mean. The former is more robust. Of recent years a lot of work has been put into the study of robust inference procedures to replace less robust ones like least squares.

The scientist who is able to plan his experiment, either in the laboratory or in the field, has a much simpler inference problem than the economist who, almost entirely, has to rely on data that have arisen naturally instead of being planned. The planned experiment can take cognizance of factors additional to those the scientist is directly interested in. This can be done either by explicitly including them in the experiment, or by a suitable randomization procedure that has a high chance of eliminating any unwanted effects. The economist is usually denied both opportunities, though sometimes extra factors can be included. The inference procedure should therefore recognize uncertainties that the laboratory experiment has eliminated. This is not always done and inference in economics remains less satisfactory than in other sciences. The concept of causation is harder to understand in economics. In the regression of y on z above, it is easy to think of z causing changes in y: but it may be that changes in y and z are both caused by related fluctuations in a third variable w. The attempt by econometricians to avoid this difficulty by including many variables has led to complexities of interpretation due to the high dimensionality of the problem.

Statistical inference is ordinarily thought of as a passage from data x to parameter θ but there is another form in which the inference is from past data x to future observations y, with no explicit reference to a parameter. This is often called prediction, and the obvious application is to time series with $x = (x_1, x_2, \ldots, x_n)$, x_i being the value of some quantity at time t_i and $y = x_{n+1}$; so that the quantity has been observed up to time t_n and it is required to predict its value at t_{n+1}. The usual way to proceed is to model the time series in some parametric form involving θ and to infer the value of θ on the basis of x. The model will specify $p(y|x, \theta)$ and one possibility is to predict y using $p(y|x, \hat{\theta})$, where $\hat{\theta}$ is a point estimate of θ. In the Bayesian view $p(y|x)$ is directly available for prediction, where $p(y|x) = \int p(y|x, \theta)p(\theta|x)\mathrm{d}\theta$, using standard probability calculations. It is arguable that all practical inference problems are of this type and that the model, and θ, are only introduced as a means of solving them.

Many statistical procedures are complicated and require extensive computations: in some cases, one does not even know how to find a procedure. One possibility is to use approximation techniques and find a procedure which loses only little information in comparison with the optimum method and yet is simple. Asymptotic methods often provide such approximations. Data often consist of a random sample, or of a time series $(x_1, x_2, \ldots, x_n)$ involving n observations. It is often possible to study the limiting behaviour as n increases without limit. For example, with random samples, the maximum likelihood estimate θ is asymptotically normally distributed with mean equal to the true value and variance σ^2/n, with σ^2 calculable in terms of the second derivative of the loglikelihood. Although this is only true as n goes to infinity, it can be used to produce a 95 per cent confidence interval for θ of the form $\theta \pm 1.96\sigma/n^{1/2}$. This is then an approximation, for large n, to the exact interval. Asymptotic methods have been very successful though it is often difficult to know how fast the limit is approached and whether a particular n is large or not. Stirling's asymptotic formula is remarkably accurate for n as low as 3. Some asymptotic results are not realized until n is well into the thousands.

The present position in statistical inference is historically interesting. The bulk of practitioners use well-established methods like least squares, analysis of variance, maximum likelihood and significance tests: all broadly within the Fisherian school and chosen for their proven usefulness rather than their logical coherence. If asked about their rigorous justification most of these people would refer to ideas of the NPW type: least-squares estimates are best, linear unbiased; F-tests have high power and maximum likelihood values are asymptotically optimal. Yet these justifications are far from satisfactory: the only logically coherent system is the Bayesian one which disagrees with the NPW notions, largely because of their violation of the likelihood principle. The practitioner is most reluctant to adopt this logical approach because of its apparent impracticality. The impracticality is largely an illusion and current work is energetically overcoming it. So the next few decades should be interesting as the various theories get amended and one emerges triumphant, or some new ideas avoid the contradictions. Whatever happens, inference will surely remain one of the most important of subjects, simply because of the ubiquity of inference problems in all aspects of human endeavour.

D.V. LINDLEY

See also **maximum likelihood.**

Bibliography

De Finetti, B. 1974–5. *Theory of Probability: a Critical Introductory Treatment.* 2 vols, London: Wiley. Translation from the 1970 Italian original by A. Machi and A. Smith.

Fisher, R.A. 1925. *Statistical Methods for Research Workers.* Edinburgh: Oliver & Boyd.

Fisher, R.A. 1935. *The Design of Experiments.* Edinburgh: Oliver & Boyd.

Lehmann, E.L. 1959. *Testing Statistical Hypotheses.* New York: Wiley.

Lehmann, E.L. 1983. *The Theory of Point Estimation.* New York: Wiley.

Ramsey, F.P. 1931. *The Foundations of Mathematics and Other Logical Essays.* London: Kegan, Paul, Trench, Trubner.

Savage, L.J. 1954. *The Foundations of Statistics.* New York: Wiley.

statistical mechanics

Statistical mechanics is a branch of physics which studies the aggregate behaviour of large populations of objects, typically atoms. A canonical question in statistical mechanics is how magnets can appear in nature. A magnet is a piece of iron with the property that atoms tend on average to be spinning up or down; the greater the lopsidedness the stronger the magnet. (Spin is binary.) While one explanation would be that there is simply a tendency for individual atoms to spin one way rather than another, the remarkable finding in the physics literature is that interdependences in spin probabilities between the atoms can, when strong enough, themselves be a source of magnetization. Classic structures of this type include the Ising and Currie–Weiss models (cf. Ellis, 1985).

Economists, of course, have no interest in the physics of such systems. On the other hand, the mathematics of statistical mechanics has proven to be useful for a number of modelling contexts. As illustrated by the magnetism example, statistical mechanics models provide a language for modelling interacting populations. The mathematical models of statistical mechanics are sometimes called 'interacting particle systems' or 'random fields', where the latter term refers to interdependent populations with arbitrary index sets, as opposed to a variables indexed by time. It is the mathematics of statistical mechanics models that economists have found valuable in studying the evolution of populations.

Statistical mechanics models are useful to economists as these methods provide a framework for linking micro-economic specifications to macroeconomic outcomes. A key feature of a statistical mechanical system is that, even though the individual elements may be unpredictable, order appears at an aggregate level. At one level this is an unsurprising property; laws of large numbers provide a similar linkage. However, in statistical mechanics models properties can emerge at an aggregate level that are not describable at the individual level. Magnetism is one example of this as it is a feature of a system, not an individual element; the existence of aggregate properties without individual analogues is sometimes known as 'emergence'. Emergent properties are in fact why statistical mechanics models appear to be such an intriguing set of tools for economists since they suggest that there may be aspects of macroeconomic outcomes that are not reducible to the microeconomic specification from which they derive. As such, emergence is a way to make progress on understanding aggregate behaviour in the presence of heterogeneous agents. This is especially important in light of results by Hugo Sonnenschein and others that show that the Arrow–Debreu type general equilibrium framework does not, by itself, impose many restrictions on which data can be observed. (See AGGREGATE DEMAND THEORY and AGGREGATION (THEORY) for the basic results and different efforts to overcome this lack of empirical content to general equilibrium theory.) In order to produce empirical implications, statistical mechanics models impose stronger (and in many ways different) restrictions on individual interrelationships than are found in Arrow–Debreu models, so in this sense are clearly less general. What is interesting is that the aggregate properties of statistical mechanics models often do not depend on details of the interaction structure, a property known as 'universality'.

The general structure of statistical mechanics models may be understood as follows. Consider a population of elements ω_i, where i is an element of some arbitrary index set I. Let $\underset{\sim}{\omega}$ denote vector all elements in the population and $\underset{\sim I-i}{\omega}$ denote all the elements of the population other than i. Concretely, each ω_i may be thought of as an individual choice. A statistical mechanics model is specified by the set of probability measures

$$\mu(\underset{\sim}{\omega} | \underset{\sim I-i}{\omega}) \qquad (1)$$

for all i. These probability measures describe how each element of a system behaves given the behaviour of other elements. Following our example, (1) can be interpreted as describing the probabilities of a given person's choice given the choices of others. The objective of the analysis of the system is to understand the joint probability measures for the entire system,

$$\mu(\underset{\sim}{\omega}), \qquad (2)$$

that are compatible with the conditional probability measures (1). Thus, the goal of the exercise is to understand the probability measure for the population of choices given the conditional decision structure for each choice. Stated this way, one can see how statistical mechanics models are conceptually similar to various game-theory models, an idea found in Blume (1993), who uses statistical mechanics methods to study the convergence properties of populations in which individual agents interact with their neighbours via a sequence of coordination games.

This formulation of statistical mechanics models, with conditional probability measures representing the micro-level description of the system, and associated joint probability measures the macro-level or equilibrium description of the system, also illustrates an important difference between physics and economics reasoning. For the physicist, treating conditional probability measures as

primitive objects in modelling is natural. One does not ask 'why' one atom's behaviour reacts to other atoms. In contrast, conditional probabilities are not natural modelling primitives to an economist. The microeconomic foundations of a model (that is, the specification of preferences, technology, beliefs and possibly institutional framework) produce conditional probabilities as descriptions of how individuals behave in the environment. Hence, one does not start by taking as a given a conditional probability description that imposes the requirement that the likelihood that an individual student drops out of high school is an increasing function of the drop-out decisions of other students. Rather, one specifies a decision problem for the student in which peer influence matters, possibly through a direct desire to conform or via information communicated by the decisions of others. This decision problem will have a probabilistic structure in which the individual outcome ω_i depends on others in the population, just as in the standard statistical mechanics case, but the form of this dependence is derivative from the specification of the decision problem. A defect of a number of economic models using statistical mechanics is the tendency to follow the physics literature and treat (1) as an appropriate way to formulate microfoundations.

Dynamic versions of statistical mechanics models are usually modelled in continuous time. One considers the process $\omega_i(t)$ and, unlike the atemporal case, probabilities are assigned to at each point in time to the probability of a change in the current value. Operationally, this means that for sufficiently small δ

$$\mu(\omega_i(t+\delta) | \omega_i(t+\delta) \neq \omega_i(t))$$
$$= f(\underset{\sim i-I}{\omega}(t), \omega_i(t))\delta + o(\delta). \qquad (3)$$

What this means is that at each t there is a small probability that $\omega_i(t)$ will change value; such a change is known as a flip when the support of $\omega_i(t)$ is binary. This probability is modelled as depending on the current value of element i as well as on the current (time t) configuration of the rest of the population. Since time is continuous whereas the index set is countable, the probability that two elements change at the same time is 0 when the change probabilities are independent. Systems of this type lead to the question of the existence and nature of invariant or limiting probability measures for the population, that is, the study of

$$\lim_{t \Rightarrow \infty} \mu(\underset{\sim}{\omega}(t) | \underset{\sim}{\omega}(0)). \qquad (4)$$

Discrete time systems can of course be defined analogously; for such systems a typical element is $\omega_{i,t}$. In such cases, it perhaps most natural to assume that changes in the individual elements of the system are simultaneous.

The conditional probability structure described by (1) can lead to very complicated calculations for the joint probabilities (2). In the interests of analytical tractability, physicists have developed a set of methods referred to as mean field analyses. These methods typically involve replacing the conditioning elements in (1) with their expected values, that is

$$\mu(\omega_i | E(\underset{\sim}{\omega}_{I-i})). \tag{5}$$

A range of results exist on how mean field approximation relate to the original probabilities models they approximate. From the perspective of economic reasoning, mean field approximations have a substantive economic interpretation as they implicitly mean that agents make decisions based on their beliefs about the behaviours of others rather than the behaviours themselves. Brock and Durlauf (2001a) develop an environment in which the equilibrium set of choices, modelled as an expectational form of a noncooperative (that is, Nash) equilibrium, turns out to the mean field approximation of a model that is interpretable as a social planning equilibrium in which the planner determines all choices, accounting for complementarities in payoffs across individuals.

Properties

Existence

The first question one naturally asks for environments of the type described concerns the existence of a joint or invariant probability measure over the population of elements in which conditional probabilities for the behaviours of the elements have been specified. Existence results of this type differ from classic results such as the Kolmogorov extension theorem in that they concern the relationship between conditional probabilities and joint probabilities, rather than the relationship (as occurs in the Kolmogorov case) between joint probabilities measured on finite sets of elements versus an infinite collection that represents the union of the various elements. Liggett (1985; 1991) provides a comprehensive survey of results. These results are quite technical but do not, in my judgement, require conditions that appear to be reasonable from the perspective of socio-economic systems, at least in the sense that they do not seem to have any interesting behavioural content.

Uniqueness or multiplicity

The existence of a joint or invariant measure says nothing about how many such measures exist. When there are multiple measures compatible with the conditional probabilities, the system is said to be nonergodic. Notice that for the dynamical models the uniqueness question involves the dependence of the invariant measure on the initial configuration on $\omega(0)$ or $\underset{\sim}{\omega}_0$. Heuristically, for atemporal models, nonergodicity is thus the probabilistic analog to multiple equilibria, whereas for temporal models nonergodicity is the probabilistic analog to multiple steady states.

One of the fascinating features of statistical mechanics models is their capacity to exhibit nonergodicity in nontrivial cases. Specifically, nonergodicity can occur when the various direct and indirect connections between individuals in a population create sufficient aggregate interdependence across agents. As such statistical mechanics models use richer interactions structure than appear, for example in conventional time series models. To see this, consider a Markov chain $\Pr(\omega_t = 1 | \omega_{t-1} = 1) \neq 1$ and $\Pr(\omega_t = 0 | \omega_{t-1} = 0) \neq 1$, then $\lim_{j \Rightarrow \infty} \Pr(\omega_{t+j} | \omega_0)$ will not depend on ω_0. However, suppose that $I = Q$, that is, the index set is the set of integers so that we are considering the evolution of a countable collection of elements. Suppose further that the system has a local Markov property of the form $\Pr(\omega_{i,t} | \underset{\sim}{\omega}_{t-1}) = \Pr(\omega_{i,t} | \omega_{i-1,t-1} \omega_{i,t-1} \omega_{i+1,t-1})$; in words, the behaviour of a particular $\omega_{i,t}$ depends on its value at $t-1$ as well as its 'nearest neighbours'. In this case, it is possible that $\lim_{j \Rightarrow \infty} \Pr(\omega_{i,t+j} | \underset{\sim}{\omega}_0)$ does depend on $\underset{\sim}{\omega}_0$ even though no conditional probability $\Pr(\omega_{i,t} | \omega_{i-1,t-1} \omega_{i,t-1} \omega_{i+1,t-1})$ equals 1. The reason for this is that, in the case of an evolving set of interacting Markov processes, there are many indirect connections. For example, the realization of $\omega_{i-2,t-2}$ will affect $\omega_{i,t}$ because of its effect on $\omega_{i-1,t-1}$; no analogous property exists when there is a single element at each point in time. In fact, the number of elements at time $t-k$ that affect $\omega_{i,t}$ is, in this example, growing in k. This does not mean that such a system necessarily has multiple invariant measures, merely that it can when there is sufficient sensitivity of $\Pr(\omega_{i,t} | \underset{\sim}{\omega}_{t-1}) = \Pr(\omega_{i,t} | \omega_{i-1,t}, \omega_{i,t-1}, \omega_{i+1,t-1})$ to the realizations of $\omega_{i-1,t-1} \omega_{i,t-1}$ and $\omega_{i+1,t-1}$. For many statistical mechanics models, this dependence can be reduced to a single parameter. For example, a dynamic version of the Ising model may be written $\Pr(\omega_{i,t} | \underset{\sim}{\omega}_{t-1}) \propto e^{J(\omega_{i-1,t-1} + \omega_{i,t-1} + \omega_{i+1,t-1})}$; so J fully characterizes the degree of dependence in the system. In statistical mechanics models, one often finds threshold effects, that is, when J is below some $J < \bar{J}$, the system exhibits a unique invariant measure whereas, if $J > \bar{J}$, multiple measures exist.

Applications

The earliest uses of statistical mechanics models in economics appear to be Follmer (1974) and Allen (1982). Follmer analyses the question of when idiosyncratic preference shocks affect aggregate prices. He models these shocks as binary and shows that, if the shocks obey nearest neighbour-type interdependence, then it is possible for the shocks to affect the aggregate price level. This occurs specifically if the interdependences are strong enough to produce multiple invariant measures, that is, the law of large numbers breaks down for the shocks. Allen (1982) applies statistical mechanics ideas to the diffusion of technical change.

Statistical mechanics ideas were largely dormant until the early 1990s, when a number of researchers independently began using the tools, often in very different contexts. One area where this work has proven valuable is game theory. Blume (1993; 1995) employed statistical mechanics methods to understand the role of different interactions structures in evolutionary game theory. Brock (1993) provides a wide-ranging exploration of the relationship between various types of statistical mechanics models and particular socio-economic environments, with particular attention to the difference between environments with and without a social planner.

Other authors have applied statistical mechanics to particular substantive contexts. Durlauf (1993) employs a discrete time model to study economic growth. For this work, the motivation was twofold: first, to formalize the idea that local spillover effects can create a development trap and, second, to identify how leading sectors can expand and thereby lead to a take-off to sustained industrialization and growth. Bak et al. (1993) analyse a model in which industrial demand linkages can cause idiosyncratic shocks to produce aggregate fluctuations. Kelly (1994) shows how these models can explain how shocks of different sizes lead to very different macroeconomic consequences. Other applications include financial market fluctuations (Horst, 2005), information transmission (Kosfeld, 2005), technical change (Auerswald et al., 2000) and trade networks and unemployment (Oomes, 2003).

Current theoretical research using statistical mechanics models has attempted to extend their use to more general specifications than have appeared in the physics and mathematics literatures. Brock and Durlauf (2006) extend various properties of statistical mechanics models to contexts where agents face more than two choices; the choices are not ordered so this approach creates links between statistical mechanics models and multinomial choice models in the various social sciences. Other authors have focused on continuous choice spaces (for example, Bisin, Horst and Özgür, 2006; Horst and Scheinkman, 2006). Ioannides (2006) considers how a range of alternative interaction structures affect aggregate outcomes. Bisin, Horst and Özgür, (2006) consider issues of self-consistent beliefs for different local interaction structures. Horst and Scheinkman (2006) consider the relationship between conditional probabilities of the form (1) and the underlying decision problems of agents, thus facilitating better microfoundations when these methods are used. These various directions seem promising in allowing statistical mechanics methods to describe richer socio-economic environments.

Researchers have also begun to bring statistical mechanics models to empirical work. The econometric analyses of statistical mechanics models of Brock and Durlauf (2001a; 2001b; 2006; 2007), Conley and Topa (2003) and Topa (2001) have begun to be studied. As initially discussed in Manski (1993), complicated identification problems exist in uncovering behavioural

interdependences even when individual-level data are available. One important message from Brock and Durlauf (2001a; 2001b) is that the nonlinearities that are embedded in the probability structure of statistical mechanics models are important in overcoming what Manski has called the reflection problem. Further, Brock and Durlauf (2007) shows how the presence of multiple equilibria can be used to uncover behavioural interdependences in the presence of group level unobservables.

Once one moves to econometric applications, it is essential to allow for richer forms of individual heterogeneity than are found in the various theoretical models. Indeed, most theoretical models assume that individual agents are described by the same conditional probability measure; one exception is Glaeser, Sacerdote and Scheinkman (1996). At this point, essentially nothing is known about the properties of statistical mechanics models in which empirically salient forms of heterogeneity have been introduced. For this reason, I believe that advances in the use of statistical mechanics methods in economic theory and econometrics will prove to be strongly complementary.

Additional reading

Thompson (1988) is a standard physics textbook on statistical mechanics. Badii and Politi (1997) provide a useful discussion of statistical mechanics that segues from physical to statistical and computational models. Liggett (1985; 1991) are magisterial mathematical treatments of the probability structures that underlie statistical mechanics models as I have described them. Kinderman and Snell (1980) is an informal but enjoyable treatment and useful for building intuition. In the statistical mechanics literature, models where there is heterogeneity in the interaction weights linking individual elements are known as 'spin glasses'; see Fischer and Hertz (1991) for a readable treatment. Durlauf (1997) develops a statistical mechanics framework that nests a number of models that have appeared in economics, particularly those associated with complex systems; related perspectives are found in Ioannides (1997).

STEVEN N. DURLAUF

See also **economy as a complex system; emergence; ergodicity and nonergodicity in economics; Markov processes; social interactions (empirics); social interactions (theory); spatial econometrics.**

Bibliography

Allen, B. 1982. Some stochastic processes of interdependent demand and technological diffusion exhibiting externalities among adopters. *International Economic Review* 23, 595–608.
Auerswald, P., Kauffman, S., Lobo, J. and Shell, K. 2000. The production recipes approach to technological

innovation: an application to learning by doing. *Journal of Economic Dynamics and Control* 24, 389–450.

Badii, R. and Politi, A. 1997. *Complexity: Hierarchical Structures and Scaling in Physics*. New York: Cambridge University Press.

Bak, P., Chen, K., Scheinkman, J. and Woodford, M. 1993. Aggregate fluctuations from independent sectoral shocks: self-organized criticality in a model of production and inventory dynamics. *Ricerche Economiche* 47, 3–30.

Bisin, A., Horst, U. and Özgür, O. 2006. Rational expectations equilibria of economies with social interactions. *Journal of Economic Theory* 127, 74–116.

Blume, L. 1993. The statistical mechanics of strategic interaction. *Games and Economic Behavior* 5, 387–424.

Blume, L. 1995. The statistical mechanics of best-response strategy revision. *Games and Economic Behavior* 11, 111–45.

Brock, W. 1993. Pathways to randomness in the economy: emergent nonlinearity and chaos in economics and finance. *Estudios Economicos* 8, 3–55.

Brock, W. and Durlauf, S. 2001a. Discrete choice with social interactions. *Review of Economic Studies* 68, 235–60.

Brock, W. and Durlauf, S. 2001b. Interactions-based models. In *Handbook of Econometrics*, vol. 5, ed. J. Heckman and E. Leamer. Amsterdam: North-Holland.

Brock, W. and Durlauf, S. 2006. A multinomial choice model with social interactions. In *The Economy as an Evolving Complex System III*, ed. L. Blume and S. Durlauf. New York: Oxford University Press.

Brock, W. and Durlauf, S. 2007. Identification of binary choice models with social interactions. *Journal of Econometrics* 140, 52–75.

Conley, T. and Topa, G. 2003. Identification of local-interactions models with imperfect location data. *Journal of Applied Econometrics* 18, 605–18.

Durlauf, S. 1993. Nonergodic economic growth. *Review of Economic Studies* 60, 349–66.

Durlauf, S. 1997. Statistical mechanics approaches to socioeconomic behavior. In *The Economy as an Evolving Complex System II*, ed. W. Arthur, S. Durlauf and D. Lane. Reading, MA: Addison-Wesley.

Ellis, R. 1985. *Entropy, Large Deviations, and Statistical Mechanics*. New York: Springer-Verlag.

Fischer, K. and Hertz, J. 1991. *Spin Glasses*. New York: Cambridge University Press.

Follmer, H. 1974. Random economies with many interacting agents. *Journal of Mathematical Economics* 1, 51–62.

Glaeser, E., Sacerdote, B. and Scheinkman, J. 1996. Crime and social interactions. *Quarterly Journal of Economics* 111, 507–48.

Horst, U. 2005. Financial price fluctuations in a stock market model with many interacting agents. *Economic Theory* 25, 917–832.

Horst, U. and Scheinkman, J. 2006. Equilibria in systems of social interactions. *Journal of Economic Theory* 130, 44–77.

Ioannides, Y. 1997. The evolution of trading structures. In *The Economy as an Evolving Complex System II*, ed. W. Arthur, S. Durlauf and D. Lane. Reading, MA: Addison-Wesley.

Ioannides, Y. 2006. Topologies of social interactions. *Economic Theory* 28, 559–84.

Kelly, M. 1994. Big shocks versus small shocks in a dynamic stochastic economy with many interacting agents. *Journal of Economic Dynamics and Control* 18, 397–410.

Kindermann, R. and Snell, J. 1980. *Markov Random Fields and Their Applications*. Providence, RI: American Mathematical Society. Online. Available at http://www.ams.org/online_bks/conm1/conm1-whole.pdf, accessed 13 September 2006.

Kosfeld, M. 2005. Rumours and markets. *Journal of Mathematical Economics* 41, 646–64.

Liggett, T. 1985. *Interacting Particle Systems*. New York: Springer-Verlag.

Liggett, R. 1991. *Stochastic Interacting Systems: Contact, Voter, and Exclusion Models*. New York: Springer-Verlag.

Manski, C. 1993. Identification of endogenous social effects: the reflection problem. *Review of Economic Studies* 60, 531–42.

Oomes, N. 2003. Local trade networks and spatially persistent unemployment. *Journal of Economic Dynamics and Control* 27, 2115–49.

Thompson, C. 1988. *Classical Equilibrium Statistical Mechanics*. New York: Oxford University Press.

Topa, G. 2001. Social interactions, local spillovers, and unemployment. *Review of Economic Studies* 68, 261–95.

statistics and economics

The close interrelationship between economics and statistics, going back to their common roots in 'Political Arithmetic', played a crucial role in availing the development of both disciplines during their practical knowledge (pre-academic) period. Political economy was first separated from political arithmetic and became an academic discipline – the first social science – at the end of the 18th century, partly as a result of political arithmetic losing credibility. Statistics emerged as a 'cleansed' version of political arithmetic, focusing on the collection and tabulation of data, and continued to develop within different disciplines including political economy, astronomy, geodesy, demography, medicine and biology; however, it did not become a separate academic discipline until the early 1900s.

During the 19th century the development of statistics was institutionally nurtured and actively supported by the more empirically oriented political economists such as Thomas Malthus who helped to create section F of the Royal Society, called 'Economic Science and Statistics', and subsequently to found the Statistical Society of London. The teaching of statistics was introduced into the university curriculum in the 1890s, primarily in economics departments (see Walker, 1929).

The close relationship between economics and statistics was strained in the first half of the 20th century, as the descriptive statistics tradition, associated with Karl Pearson, was being transformed into modern (frequentist) statistical inference in the hands of Fisher (1922, 1925, 1935a, 1956) and Neyman and Pearson (1933), and Neyman (1935, 1950, 1952). During the second half of the 20th century this relationship eventually settled into a form of uneasy coexistence. At the dawn of the 21st century there is a need to bring the two disciplines closer together by implementing certain methodological lessons overlooked during the development of modern statistics.

1 The 17th century: political arithmetic, the promising beginnings

If one defines statistics broadly as 'the subject matter of collecting, displaying and analysing data', the roots of the subject are traditionally traced back to John Graunt's (1620–74) *Natural and Political Observations upon the Bills of Mortality*, published in 1662 (see Hald, 1990; Stigler, 1986), the first systematic study of demographic data on birth and death records in English cities. Graunt detected surprising regularities stretching back over several decades in a number of numerical aggregates, such as the male/female ratio, fertility rates, death rates by age and location, infant mortality rates, incidence of new diseases and epidemics, and so on. On the basis of these apparent regularities, Graunt proceeded to draw certain tentative inferences and discuss their implications for important public policy issues. Hald summarized the impact of this path-breaking book as follows:

> Graunt's book had immense influence. Bills of mortality similar to the London bills were introduced in other cities, for example, Paris in 1667. Graunt's methods of statistical analysis were adopted by Petty, King and Davenant in England; Vauban in France; by Struyck in the Netherlands; and somewhat later by Sussmilch in Germany. Ultimately, these endeavours led to the establishment of governmental statistical offices. Graunt's investigation on the stability of the sex ratio was continued by Arthuthnott and Nicolas Bernoulli. (Hald, 1990, p. 103)

Graunt's book had close affinities in both content and objectives to several works by his close friend William Petty (1623–87) on 'Political Arithmetick' published during the 1670s and 1680s; Graunt and Petty are considered joint founders of the 'political arithmetic' tradition (Redman, 1997). The fact that Graunt had no academic credentials and published only the single book led to some speculation in the 1690s, which has persisted to this day, that Petty was the real author of *The Bills of Mortality*. The current prevailing view (see Greenwood, 1948; Kreager, 1988) is that Petty's potential influence on Graunt's book is marginal at best. Stone aptly summarizes this view as follows:

> Graunt was the author of the book associated with his name. More than likely, he discussed it with his friend; Petty may have encouraged him to write it, contributed certain passages, helped obtaining the Bills for the county parish … at Romsey, the church in which Petty's baptism is recorded and in which he is buried; he may even have suggested the means of interpolating the numbers of survivors between childhood and old age. But all this does not amount to joint let alone sole authorship. (Stone, 1997, p. 224)

Hull (1899), one of Petty's earliest biographers and publisher of his works, made a strong case against Petty being the author of the 'Bills of Mortality' by comparing his methodological approach to that of Graunt:

> Graunt exhibits a patience in investigation, a care in checking his results in every possible way, a reserve in making inferences, and a caution about mistaking calculation for enumeration, which do not characterize Petty's work to a like degree.
>
> The spirit of their work is often different when no question of calculation enters. Petty sometimes appears to be seeking figures that will support a conclusion which he has already reached; Graunt uses his numerical data as a basis for conclusions, declining to go beyond them. He is thus a more careful statistician than Petty, but he is not an economist at all. (Hull, 1899, pp. xlix and lxxv)

Both Graunt and Petty used limited data to draw conclusions and make predictions about the broader populations, exposing themselves to severe criticisms as to the appropriateness and reliability of such inferences. For instance, using data on christenings and burials in a single county parish in London, they would conjure up estimates of the population of London (which included more than 130 parishes), and then on the basis of those estimates, and certain contestable assumptions concerning mortality and fertility rates, proceed to project estimates of the population of the whole of England. The essential difference between their approaches is that Graunt put enough emphasis on discussing the possible *sources of error* in the collection and compilation of his data, as well as in his assumptions, enabling the reader to assess the reliability (at least qualitatively) of his inferences. Petty, in contrast, was more prone to err on the side of political expediency by drawing inferences that would appeal to the political powers of his time (see Stone, 1997).

Graunt and Petty considered statistical analysis a way to draw *inductive inferences* from observational data, analogous to performing experiments in the physical sciences (see Hull, 1899, p. lxv). Political arithmetic stressed the importance of a new method of quantitative measurement – 'the art of reasoning by figures upon things relating to the government' – and was instrumental in the development of both statistics and economics (see Redman, 1997, p. 143). The timing of this emphasis on quantitative measurement and the collecting of data was not coincidental. The *empiricist* turn pioneered by Francis Bacon (1561–1626) had a crucial impact on intellectual circles such as the London Philosophical Society and the British Association, with which Graunt and Petty were associated – these circles included Robert Boyle, John Wallis, John Wilkins, Samuel Hartlib, Christopher Wren and Isaac Newton. As summarized by Letwin:

> The scientific method erected by Bacon rested on two main pillars: natural history, that is, the collection of all possible facts about nature, and induction, a careful logical movement from those facts of nature to the laws of nature. (Letwin, 1965, p. 131)

Graunt and Petty were also influenced by philosopher John Locke (1632–1704), through personal contact. Locke was the founder the British empiricist tradition, which continued with George Berkeley (1685–1753) and David Hume (1711–76). Indeed, all three philosophers wrote extensively on political economy as it relates to empirical economic phenomena, and Locke is credited with the first use of the most important example of analytical thinking in economics, the demand-supply reasoning in determining price (see Routh, 1975).

Graunt's and Petty's successors in the political arithmetic tradition, Gregory King (1648–1712) and Charles Davenant (1656–1714) continued to emphasize the importance of collecting data as the only objective way to frame and assess sound economic policies. Their efforts extended the pioneering results of Grant and Petty and provided an improved basis for some of the original predictions (such as the population of England), but they did not provide any new methodological insights into the analysis of the statistical regularities originally enunciated by Graunt. The enhanced data collection led to discussions of how certain economic variables should be measured over time, and a new literature on index numbers was pioneered by William Fleetwood (1656–1723). The roots of national income accounting, which eventually led to the current standardized macro-data time series, can be traced back to the efforts of these early pioneers in political arithmetic (see Stone, 1997).

According to Hald:

> His [Graunt's] life table was given a probabilistic interpretation by the brothers Huygens; improved life tables were constructed by de Witt in the Netherlands and by Halley in England and used for the computation of life annuities. The life table became a basic tool in medical statistics, demography, and actuarial science. (Hald, 1990, p. 1034)

The improved life tables, with proper probabilistic underpinnings, were to break away from the main political arithmetic and become part of a statistical/probabilistic tradition that would develop independently in Europe in the next two centuries, giving rise to a new literature on life tables and insurance mathematics (see Hald, 1990).

A methodological digression. This was a crucial methodological development for data analysis because it was the first attempt to provide probabilistic underpinnings to Graunt's statistical regularities. Unfortunately, the introduction of probability in the life tables was of limited scope and had no impact on the broader development of political arithmetic, which was growing during the 18th century without any concerns for any probabilistic underpinnings. Without such underpinnings, however, one cannot distinguish between real regularities and artifacts.

2 The 18th century: the demise of political arithmetic

At the dawn of the 18th century political arithmetic promised a way to provide an objective basis for more reliable framing and assessment of economic and social policies. As described by Petty, the method of political arithmetic replaces the use of 'comparative superlative words, and intellectual arguments' with 'number, weight, or measure; to use only arguments of sense; and to consider only such causes as have visible foundations in nature, leaving those that depend on the mutable minds, opinions, appetites, and passions of particular men, to the consideration of others' (Hull, 1899, p. 244).

English political institutions, including the House of Commons, the House of Lords and the monarchy, took full advantage of the newly established methods of political arithmetic and encouraged, as well as financed, the collection of new data as needed to consider specific questions of policy (see Hoppit, 1996). Putting these methods to the (almost exclusive) service of policy framing by politicians carried with it a crucial danger for major abuse. An inherent problem for social scientists in general has always been to distinguish between inferences relying on sound scientific considerations and those motivated by political or social preferences and leanings.

The combination of (*a*) the absence of sound probabilistic foundations that would enable one to distinguish between real regularities and artefacts, and (*b*) the inbuilt motivation to abuse data in an attempt to make a case for one's favourite policies, led inevitably to extravagant and unwarranted speculations, predictions and claims. These indulgences eventually resulted in the methods of political arithmetic losing credibility. The extent of the damage was such that Greenwood, in reviewing 'Medical Statistics from Graunt to Farr', argued:

> One may fairly say on the evidence here summarized that the eighteenth-century political arithmeticians of England made no advance whatever upon the position reached by Graunt, Petty and King. They were

second-rate imitators of men of genius. (Greenwood, 1948, p. 49)

An important component of the evidence provided by Greenwood was the 'population controversy', which often involved idle speculation in predicting the population of England. This speculation began with Graunt with a lot of cautionary notes attached, but it continued into the 18th century with much less concern about the possible errors that could vitiate such inferences. The discussions were from two opposing schools of thought: the *pessimists*, who claimed that the population was decreasing, and the *optimists*, who argued the opposite; their conflicting arguments were based on the same bills of mortality popularized by Graunt. Neither side had reliable evidence for its predictions because the data provided no sound basis for reliable inference. All predictions involved highly conjectural assumptions of fertility and mortality rates, the average number of people living in each house, and so on. The acrimonious arguments between the two sides revealed the purely speculative foundations of all such claims and contributed significantly to the eventual demise of political economy (see Glass, 1973, for a detailed review).

The above quotation from Greenwood might be considered today as an exaggeration, but it describes accurately the prevailing perception at the end of the 18th century. An unfortunate consequence of disparaging the methods of political arithmetic was the widely held interpretation that it provided decisive evidence for the ineffectiveness of Bacon's *inductive method*. Indeed, one can argue that this cause was instrumental in the timing of the emergence of *political economy* at the end of the 18th century, as the first social science to break away from political arithmetic. Adam Smith (1723–90) declared: 'I have no great faith in political arithmetick' (1776, p. 534). James Steuart (1712–80) was even more critical:

> Instead of appealing to political arithmetic as a check on the conclusions of political economy, it would often be more reasonable to have recourse to political economy as a check on the extravagances of political arithmetic. (quoted by Redman, 1997)

During the late 18th century, political economy defined itself by contrasting its methods with those of political arithmetic, arguing that it did not rely only on tables and figures in conjunction with idle speculation, but was concerned with the theoretical issues, causes and explanations underlying the process that generated such data. Political economists contrasted their primarily deductive methods to the discredited inductive methods utilized by political arithmeticians. As argued by Hilts:

> Of importance to the history of statistics in England was the fact that the political economists were fully conscious of their deductive proclivities and saw political economy as methodologically distinct from the inductive science of statistics. (Hilts, 1978, p. 23)

At this point it should be emphasized that the terms induction and deduction had different connotations during the 18th century, and care should be taken when interpreting some of the claims of that period (see Redman, 1997). Despite the criticisms by leading political economists of the inductive method, broadly understood as using the data as a basis of inference, the tradition of collecting, compiling and charting data as well as drawing inferences concerning broad tendencies on such a basis, continued to grow throughout the 18th and 19th centuries, and was influential in the development of political economy. Some political economists such as Thomas Malthus (1766–1834) and John McCulloch (1789–1864) continued to rely on the British empiricist tradition of using data as a basis of inference, but were at great pains to separate themselves from the 18th century's discredited political arithmetic tradition. Indeed, the leading political economists of that period, including Adam Smith and David Ricardo (1772–1823), used historical data extensively in support of their theories, conclusions and policy recommendations developed by deductive arguments (see Backhouse, 2002a).

At the close of the 18th century, the only bright methodological advance in the withering tradition of political arithmetic was provided by William Playfair's (1759–1823) *The Commercial and Political Atlas*, published in 1786. This book elevated the analysis of tabulated data to a more sophisticated level by introducing the power of graphical techniques in displaying and analysing data. Playfair introduced several innovating techniques such as hachure, shading, colour coding, and grids with major and minor divisions of both axes to render the statistical regularities in the data even more transparent. In a certain sense, the graphical techniques introduced by Playfair made certain empirical regularities more transparent and rendered certain conclusions easier to draw. The graphs in this book represent economic time series, measuring primarily English trade (imports/exports) with other countries during the 18th century. Indeed, Playfair's writings were mainly on political economy; his first book, *Regulation of the Interest of Money*, was published in 1785 (see Harrison, 2004).

In what follows the developments in probability theory will be discussed only when they pertain to the probabilistic underpinnings of statistical analysis; for a more detailed and balanced discussion see Hald (1990; 1998; 2007). The probabilistic underpinnings literature on probability developed independently from political arithmetic in England, and there was no interaction between the two until the mid-19th century.

Viewed from today's vantage point, the primary problem with Grant's inferences based on data pertaining to a single parish in London, was how 'representative' the data were for the population of London as a whole, which included more than 130 other parishes. This problem was formalized much later in terms of whether the data can be realistically viewed as a 'random sample' from the population of London. Defining what a random sample is, however, requires probability theory, which was not adequately understood until the late 19th century (see Peirce, 1878).

Jacob Bernoulli. The first important result relating to the probabilistic underpinnings of statistical regularities was Jacob Bernoulli's (1654–1705) Law of Large Numbers (LLN), published posthumously in 1713 by his nephew Nicolas Bernoulli (1687–1759). Bernoulli's theorem showed that *under certain circumstances*, the relative frequency of the occurrence of a certain event A, say $\bar{X} = \frac{1}{n}\sum_{i=1}^{n} X_i = \frac{m}{n}$ (m occurrences of $\{X_i=1\}$ and $n-m$ occurrences of $\{X_i=0\}$ in n trials) provides an estimate of the probability $\mathbb{P}(A) = p$ whose accuracy increases as n goes to infinity. In modern terminology $\bar{X}$ constitutes a consistent estimator of p. Bernoulli went on to use this result in an attempt to provide an interval estimator of the form: p is in $(\bar{X} \pm \varepsilon)$ for some $\varepsilon > 0$, but his estimator was rather crude (see Hald, 1990).

A methodological digression. The *circumstances* assumed by Bernoulli were specified in terms of the trials being *independent and identically distributed* (IID). It turned out that the same probabilistic assumption defines the notion of a *random sample* mentioned in relation to the probabilistic underpinnings concerning Graunt's statistical regularities, though the two literatures were developing independently. The role of these probabilistic underpinnings was not made explicit, however, until the early 1920s (see Section 4.2). Indeed, the role of the IID assumptions is often misunderstood to this day. For instance, Hilts argues:

> Mathematically the theorem stated [LLN], in very simplified language, that an event which occurs with a certain probability, appears with a frequency approaching that probability as the number of observations is increased. (Hilts, 1973, p. 209)

Strictly speaking, the LLN says nothing of a sort, because, unless the trials are IID, the result does *not* follow. This insight was clearly articulated by Uspensky:

> It should, however, be borne in mind that little, if any, value can be attached to the practical applications of Bernoulli's theorem, unless the conditions presupposed in this theorem are at least approximately fulfilled: independence of trials and constant probability of an event for every trial. (Uspensky, 1937, p. 104)

Laplace. The first successful attempt to integrate data analysis with the probabilistic underpinnings should be credited to Pierre-Simon Laplace (1749–1827), a famous French mathematician and astronomer, and Thomas Bayes (1702–61), a British mathematician and Presbyterian minister. In papers published in 1764 and 1765 (see Hald, 2007) respectively, they proposed the first inverse probability (posterior-based) interval for p for the form p of the form 'p is in $(\bar{x} \pm [\varepsilon|\bar{x}])$' for some $\varepsilon > 0$, by assuming a prior distribution $p \sim U(0,1)$ that is p is a

uniformly distributed random variable (see Hacking, 1975). This gave rise to the *inverse probability* approach (known today as the Bayesian approach) to statistical inference, which was to dominate statistical induction until the 1920s, before the Fisherian revolution. In 1812 Laplace (see Hald, 2007) also provided the first frequentist interval estimator of p of the form p is in $(\bar{X} \pm \varepsilon)$ for some $\varepsilon > 0$. The difference between this result and a similar result by Bernoulli is that Laplace used a more accurate approximation based on *convergence in distribution* as the basis of his result; the first *central limit theorem* supplying an asymptotic approximation of the binomial by the Normal distribution (see Hald, 1990).

3 The 19th century: political economy and statistics

The demise of political arithmetic by the early 19th century was instrumental in contributing to the creation of two separate fields: political economy and statistics. Political economy was created to provide more reasoned explanations for the causes and contributing factors giving rise to economic phenomena. Statistics was demarcated by the narrowing down of the scope of political arithmetic in an attempt to cleanse it from the unwarranted speculation that undermined its credibility during the 18th century.

3.1 The Statistical Society of London

Given their common roots, the first institution created to foster the development of the field of statistics, the Statistical Society of London, was created in 1834 with the active participation of several political economists, including Thomas Malthus and Richard Jones (1790–1855), who, together with John Drinkwater (1801–51), Henry Hallam (1777–1859) and Charles Babbage (1791–1871), were to found the Society after some prompting from Quetelet, who visited England in 1833. Other political economists who played very active roles in the early stages of the Society included Thomas Tooke (1774–1858), John R. McCulloch (1789–1864) and Nassau Senior (1790–1864). The first council included notable personalities such as Earl FitzWilliam (1748–1833), William Whewell (1794–1866), G.R. Porter (1792–1852) and Samuel Jones-Loyd (1796–1883).

In an attempt to protect themselves from the disrepute on speculation based on data brought about by political arithmeticians, the new society was founded upon the explicit promise to put the emphasis, not on inference, but upon the collection and tabulation of data of relevance to the state. The founding document stated:

> The Statistical Society of London has been established for the purposes of procuring, arranging, and publishing Facts calculated to illustrate the condition and prospects of the Society. (*Journal of the Statistical Society of London*, 1834, p. 1)

The seal on the cover of the *Journal of the Statistical Society of London* (*JSSL*) was a wheatsheaf around which was written '*aliis exterendum*' ('to be threshed by others'). That is, the aim of the society is to painstakingly gather the facts and let others draw whatever conclusions might be warranted:

> The Statistical Society will consider it to be the first and most essential rule of its conduct to exclude carefully all Opinions from its transactions and publications – to confine its attention rigorously to facts – and, as far as it may be found possible, to facts which can be stated numerically and arranged in tables. (*JSSL*, 1834, pp. 1–2)

Of particular interest is the way the statement of the aims of the society separated statistics from political economy:

> The Science of Statistics differs from Political Economy because although it has the same end in view, it does not discuss causes, nor reason upon probable effects; it seeks only to collect, arrange, and compare, that class of facts which alone can form the basis of correct conclusions with respect to social and political government. (*JSSL*, 1834, p. 2)

The overwhelming majority of the published papers in the *JSSL* were in the political arithmetic tradition of Graunt, relating primarily to economic, medical and demographic data, with two major improvements: ameliorated methods for the collection and tabulation of data giving rise to more accurate and reliable data, and more careful reasoning being used to yield less questionable inferences. This is particularly true for data relating to life tables and mortality rates associated with epidemics. The best examples of such an output are given by William Farr (1807–83), who is considered to be the founder of medical statistics because his analysis of such data contributed to medical advances and crucial changes in policies concerning public health (see Greenwood, 1948). For a more extensive discussion of the methodological and institutional developments associated with data collection and tabulation in England and France see Schweber (2006) and Desrosières (1998).

By the 1850s it became apparent that the early founding declaration of the society to publish papers that stay away from 'Opinions' – drawing conclusions on the basis of data – was unrealistic, unattainable and unjustifiable in the minds of the members of the society. Despite this initial promise, slowly but surely *JSSL* publications began to go beyond the mere reporting and tabulation of data relating to economic, political, demographic, medical, moral and intellectual issues, including poverty figures and education statistics. The motto '*aliis exterendum*' was removed in 1857 from their seal to reflect the new vision of the society (see RSS, 1934).

3.2 The probabilistic underpinnings in the 19th century

During the early 19th century, a completely separate tradition in statistical analysis of data was being developed in Europe (mainly in France and Germany) in the fields of *astronomy* and *geodesy*. This literature was developing completely independently of political arithmetic, but by the 1840s the two traditions had merged in the hands of Adolphe Quetelet (1796–1874): see Porter (1986).

Legendre and Gauss. In the early 19th century the analysis of astronomical and geodesic data by Adrien-Marie Legendre (1752–1833), Carl Friedrich Gauss (1777–1855) and Laplace introduced curve-fitting as a method to summarize the information in data (see Farebrother, 1999). In modern notation the simplest form of curve-fitting can be expressed in the form of a linear model $\mathbf{y} = \mathbf{X}\boldsymbol{\beta} + \boldsymbol{\varepsilon}$, where $\mathbf{y} := (y_1, y_2, \ldots, y_n)$ and $\mathbf{X} := (\mathbf{x}_1, \mathbf{x}_2, \ldots, x_n)$ denote a vector and a matrix of observations, respectively, $\boldsymbol{\beta} := (\beta_1, \beta_2, \ldots, \beta_m)$ a vector of unknown parameters and $\boldsymbol{\varepsilon} := (\varepsilon_1, \varepsilon_2, \ldots, \varepsilon_n)$ a vector of errors. Legendre (1805) is credited with inventing least squares as a mathematical approximation method, by proposing the minimization of $\ell(\boldsymbol{\beta}) = (\mathbf{y} - \mathbf{X}\boldsymbol{\beta})^{\top}(\mathbf{y} - \mathbf{X}\boldsymbol{\beta})$ as a way to estimate $\boldsymbol{\beta}$. Gauss (1809) should be credited with providing the probabilistic underpinnings for this estimation problem by transforming the mathematical approximation error into a generic statistical error:

$$\varepsilon_k(x_k, m) = \varepsilon_k \sim \text{NIID}(0, \sigma^2), k = 1, 2, \ldots, n, \ldots,$$

(1)

where $\text{NIID}(0, \sigma^2)$ stands for 'Normal, Independent and Identically Distributed with mean 0 and variance σ^2'. Laplace provided the first justification of the Normality assumption based on the central limit theorem in 1812 (see Hald, 2007). What makes Gauss's contribution all-important from today's vantage point is that the probabilistic assumptions in (1) provide the framework that enables one to assess the reliability of inference. Ironically, Gauss's embedding of the mathematical approximation problem into a statistical model is rarely appreciated as the major contribution that it is (see Spanos, 2008). Instead, what Gauss is widely credited with is the celebrated Gauss-Markov theorem (see Section 4.6).

Quetelet. The 'law of error' was elevated to a most important method in analysing social phenomena by Adolphe Quetelet (1796–1874), a Belgian astronomer and polymath, in the 1840s. His statistical analysis of data differed in that his methods were integrated with the probabilistic underpinnings that were lacking in the analysis of political arithmeticians; his probabilistic perspective was primarily influenced by the work of Joseph Fourier (1768–1830), a French mathematician and physicist. Quetelet's most important contribution was to explicate Graunt's regularities in terms of the notion of probabilistic (chance) regularity which combined the unpredictability at the individual level with the abiding regularity at the aggregate level. By fitting the Normal curve over the histogram of a great variety of social data, his objective was to eliminate 'accidental' influences and determine the average physical and intellectual features of a human population, including normal and abnormal behaviour. His *modus operandi* was the notion of the 'average man' (see Desrosières, 1998). The 'average man' began as a simple way of summarizing the systematic characteristic of a population, but in some of Quetelet's later work, 'average man' is presented as an ideal type, and any deviations from this ideal were interpreted as *errors* of nature.

A methodological digression. In addition to the substantive issues raised by his approach to 'social physics' (see Cournot, 1843), the methodological underpinnings of Quetelet's statistical analysis were rather weak. When the Normal curve is fitted over a histogram of data $\mathbf{x}:=(x_1,x_2,\ldots,x_n)$ in an attempt to summarize the statistical regularities, one implicitly assumes that data $\mathbf{x}$ constitutes a realization of an IID process $\{X_k, k=1,2,\ldots,n,\ldots\}$ (see Spanos, 1999); these are highly questionable assumptions for most of the data used in Quetelet (1942). The concern to evaluate the precision (reliability) of an inference, introduced earlier by Laplace and Gauss, was absent from Quetelet's work. Hence, his analysis of statistical regularities did not give rise to any more reliable inferences than those of the political arithmetic tradition a century earlier; the necessity to assess the validity of the premises (NIID assumptions) for inductive inference was not clearly understood at the time.

3.3 The 'mathematical' turn

In the last quarter of the 19th century there was a concerted effort to render both statistics and economics more rigorous by introducing the language of mathematics into both disciplines. In statistics this effort was spearheaded by Edgeworth, Galton and Pearson and in economics by Edgeworth, Jevons, Walras and Irving Fisher. The mathematical turn of this period was motivated by the strong desire to emulate the physical sciences and introduce quantification into these fields, which involved both calculus and probability theory (see Backhouse, 2002a; 2002b).

Galton. Quetelet's use of the Normal curve to analyse social data had a powerful influence on Francis Galton (1822–1911), who provided a different interpretation to the 'law of error'. Galton (1869) interpreted the variation around the mean, not as errors from the ideal type, but as the very essence of nature's variability. Using this variability he introduced the notions of regression and correlation in the 1890s as a way to determine relationships between different data series $\{(x_k, y_k), k=1, 2, \ldots, n\}$. Regression and correlation opened the door for providing statistical explanations which revolutionized statistical modelling in the biological and social sciences (see Porter, 1986). Retrospectively, Galton was the founder of the biometrics tradition, which had a great influence on the development of statistics in the 20th

century in the hands of Karl Pearson (1857–1936) and Udny Yule (1871–1951).

Pearson significantly extended the summarization of data in the form of smoothing histograms, by introducing a whole family of new frequency curves – known today as the Pearson family – to supplement the Normal curve, and applied these techniques extensively to biological data, with notable success. He also provided clear probabilistic underpinnings for Galton's regression and correlation methods. *Yule* (1897) established a crucial link between the Legendre–Gauss least-squares and the linear regression model, by showing that least-squares can be used to estimate the parameters of linear regression, bringing together two seemingly unrelated literatures (see Spanos, 1999). This was an important breakthrough that, unfortunately, also introduced a confusion between two different perspectives on empirical modelling: curve-fitting as a mathematical approximation method, and the probabilistic perspective where regression is viewed as a purely probabilistic concept defined in terms of the first moment of a conditional distribution (see Stigler, 1986; Spanos, 2008). Yule published a highly influential textbook in statistics in 1911 in which he successfully blended the biometric with the 'economic statistics' tradition.

Edgeworth. Of particular interest for the fundamental interaction between statistics and economics is the case of Francis Edgeworth (1845–1926), primarily an economist. His mathematical self-training enabled him to provide a bridge between the theory of errors tradition going back to Gauss and Laplace, the biometric tradition of Galton and Pearson, and the more traditional economic statistics of the 19th century focusing on economic time series data and index numbers. His direct influence on statistics, however, was rather limited because the style and mathematical level of his writings were too demanding for the statisticians of the late 19th century. Bowley (1928), 'at the request of the Council prepared a summary of his mathematical work which may have served to make his achievement known to a wider circle' (see RSS, 1934, p. 238). Edgeworth contributed crucially to the mathematization of economics and the theory of index numbers (see Backhouse, 2002b).

William Stanley Jevons (1835–82), English economist and logician. In his book *The Theory of Political Economy* (1871), he used calculus to formulate the marginal utility theory of value, and the notion of partial equilibrium, which provided the foundation for the marginalist revolution in economics (see Backhouse, 2002a).

Léon Walras (1834–1910) was a French mathematical economist, one of the protagonists in the marginalist revolution and the innovator of general equilibrium theory. His perspective on the use of mathematics in economics was greatly influenced by *Augustin Cournot* (1801–77), a French philosopher, mathematician and economist. Cournot is credited with the notion of functional relationships among economic variables, which led him to the supply and demand curves (see Backhouse, 2002a).

In the United States the process of mathematization began somewhat later with *Irving Fisher* (1867–1947), who followed in the footsteps of Walras, Jevons and Edgeworth in introducing mathematics into economics and making significant contributions to the theory of index numbers (see Backhouse, 2002b).

These early pioneers in the mathematization of economics shared a vision of using statistics to provide pertinent empirical foundations to economics (see Moore, 1908). Fisher described this goal as a life-long ambition:

> I have valued statistics as an instrument to help fulfill one of the great ambitions of my life, namely, to do what I could toward making economics into a genuine science. (Fisher, 1947, p. 74)

The same vision was clearly articulated much earlier by Jevons:

> The deductive science of Economics must be verified and rendered useful by the purely empirical science of statistics. (Jevons, 1871, p. 12)

Indeed, Neville Keynes attributed to *statistics* a much greater role in the quantification of economics than hitherto:

> The functions of statistics in economic enquiries are: ... descriptive, ... to suggest empirical laws, which may or may not be capable of subsequent deductive explanation, ... to supplement deductive reasoning by checking its results, and submitting them to the test of experience, ... the elucidation and interpretation of particular concrete phenomena, ... enabling the deductive economist to test and, where necessary, modify his premises, ... measure the force exerted by disturbing agencies. (Keynes, 1890, pp. 342–46)

At the dawn of 20th century, pioneers such as Moore (1908; 1911), who aspired to help in securing empirical foundations for economics, had several advantages – for example, the institutionalization of the collection and compilation of economic data via the establishment of government statistical offices, the systematic development of index numbers, and so on. The mathematization of economics provided them with economic models amenable to empirical enquiry (see Backhouse, 2002a; 2002b). In addition, at the end of the 19th century there were several developments in statistical methods, including least-square curve-fitting, regression, correlation, periodogram analysis and trend modelling that seemed tailor-made for analysing economic data (see Mills, 1924; Stigler, 1954; Heckman, 1992; Hendry and Morgan, 1995).

4 The 20th century: a strained relationship

To enliven the discussion of the tension created in the 1920s between economic statistics and statistical inference, the account below refers to the confrontation between the two protagonists who represented the different perspectives, Bowley and Fisher.

4.1 Economic statistics as against statistical inference

The early 20th century statistics scene was dominated by Karl Pearson (1857–1936) and his research in biology at the Galton Laboratory established in 1904. Pearson's research at this laboratory consolidated the biometrics tradition, whose primary outlet was the in-house journal *Biometrika*. Pearson established the department of 'Applied Statistics' at University College in 1911, which, at the time, was the only place one could study for a degree in statistics (see Walker, 1958).

Arthur Bowley (1869–1957) was a typical successful 'economic statistician' of the early 20th century who authored one of the earliest textbooks in statistics, *Elements of Statistics* (1901), while a part-time lecturer at the London School of Economics. Bowley understood statistics as comprising two different but interrelated components, the *arithmetic* and the *mathematical*. The former was concerned with statistical techniques as they relate to measurement, compilation, interpolation, tabulation and plotting of data, as well as the construction of index numbers; this constitutes *Part I – General Elementary Methods*, and comprises the first 258 pages of Bowley (1902). The mathematical dimension (*Part II – The Application of the Theory of Probability to Statistics*, the last 74 pages of Bowley, 1902) was concerned with the use of probability theory in minimizing and evaluating the errors associated with particular inferences. The last 12 pages of Bowley (1902) are devoted to a discussion of 'regression and correlation' as expounded by Pearson (1896) and Yule (1897).

Bowley (1906) illustrated what he meant by 'errors' using the 'probable error' for the arithmetic average $\bar{x}_n := \frac{1}{n}\sum_{k=1}^{n}x_k$ of the data $(x_1, x_2, \ldots, x_n)$ as:

$$\bar{x} \pm SD(\bar{x}), \qquad (2)$$

with $SD(\bar{x})$ denoting the standard deviation of $\bar{x}$. Taking the Normal distribution as an example he argued that the claim in (2) can be interpreted as saying that 'the chance that a given observation should be within this distance of the true average is 2:1' (1906, p. 549). This interpretation can be best understood as based on a Bayesian credible interval evaluation, instead of a frequentist confidence interval developed in the 1930s.

From this perspective, Bowley interpreted the work of Pearson and Edgeworth as concerned with providing different ways to evaluate these 'probable errors' (for example, $SD(\bar{x})$) using either a fitted frequency curve or an asymptotic approximation, respectively (see Bowley, 1906, p. 550). It is interesting to note that in the 5th edition of Bowley's statistics book, published in 1926, *Part II* increased threefold to 210 pages, but contains no reference to Fisher's work, which, at the time, was well on its way to transform Karl Pearson's descriptive statistics into modern statistical inference.

Ronald Fisher (1893–1962) pioneered a recasting of statistics (1915; 1921; 1922), moving away from the Edgeworth–Pearson reliance on large sample approximations based on inverse probability (Bayesian) methods, and focusing on finite sample frequentist inference relying on sampling distributions. This recasting was initially inspired by Gossett's (1908) derivation of the student's t distribution for a given sample size *n*. Fisher made this recasting explicit in his 1921 paper by severely criticizing the inverse probability (Bayesian) approach and articulating a more complete picture of his approach to statistical inference in his 1922 classic paper.

In the early 1920s Bowley was a professor of statistics (second in fame only to Karl Pearson) at the London School of Economics (LSE), known primarily for his contributions in the area of survey sampling, and Fisher was a young statistician at Rothamstead Agricultural Station trying to make sense of a 200-year accumulation of experimental data. Bowley was aware of Fisher's early work: we know that in 1924 Bowley requested and promptly received Fisher's offprints for the LSE library (see Box, 1978, p. 171). Moreover, by some accident of faith, the two were neighbours at Harpenden, interacting socially as bridge companions (see Box, 1978, p. 85). Indeed, Bowley encouraged Fisher to publish his correction of Pearson's (1900) evaluation of degrees of freedom associated with his goodness-of-fit test (see Fisher, 1922a).

The next academic encounter between the professor and the young aspiring statistician was in 1929 when Fisher applied for an academic position in Social Biology at the LSE, but was turned down in favour of Lancelot Hogben (see Box, 1978, p. 202). Fisher's first academic position was at University College as Professor of Eugenics, in 1933. The tension between their different perspectives on statistics became public in their first showdown at Fisher's presentation to the Royal Statistical Society in 18 December 1934 entitled 'The Logic of Inductive Inference', where he attempted to explain his published work on recasting the problem of statistical induction since his 1922 paper. Bowley was appointed to move the traditional vote of thanks and open the discussion, and after some begrudging thanks for Fisher's 'contributions to statistics in general' – by then Fisher's 1925 book had made him famous – he went on to disparage his new approach to statistical inference based on the likelihood function by describing it as abstruse, arbitrary and misleading. His comments were predominantly sarcastic and discourteous, and went as far as to accuse Fisher of giving insufficient credit to Edgeworth (see Fisher, 1935, pp. 55–7). The litany of churlish comments and currish remarks continued with the rest of the old guard: Isserlis, Irwin and the philosopher Wolf (1935, pp. 57–64), who was brought in by Bowley to undermine Fisher's philosophical discussion on induction. Jeffreys

complained about Fisher's criticisms of the Bayesian approach (1935, pp. 70–2). To Fisher's support came Egon Pearson, Neyman and, to a lesser extent, Bartlett. Pearson (1935, pp. 64–5) argued that:

> When these ideas [on statistical induction] were fully understood … it would be realized that statistical science owed a very great deal to the stimulus Professor Fisher had provided in many directions. (Pearson, 1935, pp. 64–5)

Neyman was equally supportive, praising Fisher's path-breaking contributions, and explaining Bowley's reaction to Fisher's critical review of the traditional view of statistics as understandable attachment to old ideas (1935, p. 73).

Fisher, in his reply to Bowley and the old guard, was equally contemptuous:

> The acerbity, to use no stronger term, with which the customary vote of thanks has been moved and seconded … does not, I confess, surprise me. From the fact that thirteen years have elapsed between the publication, by the Royal Society, of my first rough outline of the developments, which are the subject of to-day's discussion, and the occurrence of that discussion itself, it is a fair inference that some at least of the Society's authorities on matters theoretical viewed these developments with disfavour, and admitted with reluctance. … However true it may be that Professor Bowley is left very much where he was, the quotations show at least that Dr. Neyman and myself have not been left in his company. … For the rest, I find that Professor Bowley is offended with me for 'introducing misleading ideas'. He does not, however, find it necessary to demonstrate that any such idea is, in fact, misleading. It must be inferred that my real crime, in the eyes of his academic eminence, must be that of 'introducing ideas'. (Fisher, 1935, pp. 76–82)

Fisher's reference to 'his academic eminence', although containing a dose of sarcasm, it was not totally out of place. Bowley became a member of the Council of the Royal Statistical Society as early as 1898, served as its Vice-President in 1907–8 and again in 1912–14, and President in 1938–40. He was awarded the society's highest honour, the Guy Medal in gold, in 1935; he received the Guy in silver as early as 1895. In contrast, Fisher had no academic position until 1933, and even that came with the humiliating stipulation that he would *not* teach statistics from his new position as Professor of Eugenics at University College (see Box, 1978, p. 258).

Fisher made it clear that he associated the 'old guard' in statistics with Bowley-type economic statistics:

> Statistical methods are essential to social studies, and it is principally by the aid of such methods that these studies may be raised to the rank of sciences. This particular dependence of social studies upon statistical methods has led to the unfortunate misapprehension

that statistics is to be regarded as a branch of economics, whereas in truth methods adequate to the treatment
of economic data, in so far as these exist, have mostly
been developed in the study of biology and the other
sciences. (Fisher, 1925, p. 2)

The unbridgeable gap between Bowley and the 'old
guard' on one side, and Fisher, Neyman and Pearson on
the other, was apparent six months earlier when Bowley
was assigned the same role for Neyman's first presentation. Despite the fact that Neyman began his presentation
by praising Bowley for his earlier contributions to survey
sampling methods, he grouped him with Fisher and
accused him of the same abstruseness:

> I am not certain whether to ask for an explanation or to
> cast a doubt. It is suggested in the paper that the work
> is difficult to follow and I may be one of those who
> have been misled by it. I can only say I have read it
> at the time it appeared and since, and I have read
> Dr Neyman's elucidation of it yesterday with great care.
> I am referring to Dr Neyman's confidence limits. I am
> not at all sure that the 'confidence' is not a 'confidence
> trick'. (Neyman, 1934, pp. 608–9)

His 'confidence trick' remark is not very surprising in
view of Bowley's own interpretation of (2) in inverse
probabilistic (Bayesian) terms. Predictably, Egon Pearson
and Fisher came to Neyman's rescue from the rebukes of
old guard.

Retrospectively, Bowley's charge of abstruseness,
levelled at both Fisher and Neyman, might best be
explained in terms of David Hume's (1711–76) 'tongue
in cheek' comment two centuries earlier:

> The greater part of *mankind* may be divided into two
> classes; that of *shallow* thinkers, who fall short of the
> truth; and that of *abstruse* thinkers, who go beyond it.
> The latter class are by far the most rare; and I may add,
> by far the most useful and valuable. They suggest hints,
> at least, and start difficulties, which they want, perhaps,
> skill to pursue; but which may produce fine discoveries,
> when handled by men who have a more just way of
> thinking. ... All people of *shallow* thought are apt to
> decry even those of *solid* understanding, as *abstruse*
> thinkers, and methaphysicians, and refiners; and never
> will allow any thing to be just which is beyond their
> own weak conceptions. (Hume, 1987, pp. 253–4)

In summary, the pioneering work of Fisher, Egon
Pearson and Neyman, was largely ignored by the Royal
Statistical Society (RSS) establishment until the early
1930s. By 1933 it was difficult to ignore their contributions, published primarily in other journals, and the
'establishment' of the RSS decided to display its tolerance
to their work by creating 'the Industrial and Agricultural
Research Section', under the auspices of which both
papers by Neyman and Fisher were presented in 1934 and
1935 respectively. In their centennial volume published
in 1934, the RSS acknowledged the development of

'mathematical statistics', referring to Galton, Edgeworth,
Karl Pearson, Yule and Bowley as the main pioneers, and
listed the most important contributions in this sub-field
which appeared in its *Journal* during the period 1909–33,
but the three important papers by Fisher (1922a;
1922b; 1924) are conspicuously absent from that list.
The list itself is dominated by contributions in vital,
commercial, financial and labour statistics (see RSS, 1934,
pp. 208–23). There is only one reference to Egon Pearson,
for his 1933 paper 'Control and Standardization of Quality of Manufactured Products' – the very paper used as
self-justification by the RSS in creating the new section. It
is interesting to note that by the late 1920s the revolutionary nature of Fisher's new approach to statistics was
clearly recognized by many. Tippet (1931) was one of the
earliest textbook attempts to blend the earlier results on
regression and correlation within Fisher's new approach.
In the United States, Hotelling (1930) articulated a most
elucidating perspective on Fisher's approach.

4.2 The Fisher–Neyman–Pearson approach

The main methods of the Fisher–Neyman–Pearson
(F–N–P) approach to statistical inference, point estimation, hypothesis testing and interval estimation, were
in place by the late 1930s. The first complete textbook
discussion of this approach, properly integrated with
its probabilistic underpinnings, was given by Cramer
(1946). The methodological discussions concerning the
form of inductive reasoning underlying the new frequentist approach, however, were to linger on until the
1960s and beyond; see the exchange between Fisher
(1955), Pearson (1955) and Neyman (1956).

One of the most crucial insights of the F–N–P
approach to statistical inference, which set it apart from
previous approaches to statistics, was the explicit specification of the premises of statistical induction in terms
of the notion of a *statistical model*:

> The postulate of randomness thus resolves itself into
> the question, 'Of what population is this a random
> sample?' which must frequently be asked by every
> practical statistician. (Fisher, 1922, p. 313)

He defined the initial choice of the statistical model in
the context of which the data will be interpreted as a
'representative sample' as the problem of *specification*,
emphasizing the fact that: 'the adequacy of our choice
may be tested *posteriori*' (1922, p. 314). Indeed, the first
three tests discussed in Fisher (1925, pp. 78–94) are
misspecification (M-S) tests for the Normality, Independence and Identically Distributed assumptions. Fisher
(1922; 1925; 1935), and later Neyman (1938/1952; 1950),
emphasized the importance of both model specification
and validation vis-à-vis the data:

> Guessing and then verifying the 'chance mechanism',
> the repeated operations of which produces the observed
> frequencies. (Neyman 1977, p. 99)

Pearson (1931a, 1931b) was among the first to discuss the implications of non-Normality as well as develop M-S tests for it; see Lehmann (1999) for the early concern about the consequences of misspecification in the 1920s.

The F–N–P discernments concerning statistical model *specification*, *M-S testing*, and *respecification* can be summarized in the form of what might be called the F–N–P *perspective* (articulated in Spanos, 2006a) which can be summarized as follows:

1. Every statistical (inductive) inference is based on certain *premises*, in the form of (*a*) a *statistical model* $\mathcal{M}$ parameterizing the probabilistic structure of an observable stochastic process $\{Z_t, t \in \mathbb{N}\}$, and (*b*) a set of data $\mathbf{Z}:=(\mathbf{z}_1,\ldots,\mathbf{z}_n)$, viewed as a 'typical realization' of this process.
2. A statistical model is specified in terms of a complete and internally consistent set of probabilistic assumptions concerning the underlying stochastic process $\{Z_t, t \in \mathbb{N}\}$. For example, the Normal/linear regression model is specified in terms of assumptions [1]–[5] (Table 1) concerning the observable process $\{(y_t|\mathbf{X}_t = \mathbf{x}_t), t \in \mathbb{N}\}$, and not the errors.
3. *Statistical adequacy.* Securing the validity of assumptions [1]–[5] vis-à-vis the data in question is necessary for establishing 'statistical regularities' and ensuring the reliability of inference (see Spanos, 2006a; 2006b; 2006c).

The importance of the F–N–P perspective stems from the fact that the *statistical model* enables one:

(i) to assess the validity (statistical adequacy) of the premises for inductive inference – by testing the assumptions using misspecification tests; and
(ii) to provide relevant error probabilities for appraising the reliability of the associated inference (see Spanos, 2006a).

It is well known that the *reliability* of any inference procedure depends crucially on the validity of the pre-specified *statistical model* vis-à-vis the data in question. The optimality of these procedures is defined by their capacity to give rise to valid inferences (*trustworthiness*), which is calibrated in terms of the associated error probabilities – how often these procedures lead to erroneous inferences (see Mayo, 1996). In the case of confidence interval estimation the calibration is usually gauged in terms of minimizing *the coverage error probability*: the probability that the interval does *not* contain the true value of the unknown parameter(s). In the case of hypothesis testing the calibration is ascertained in terms of minimizing *the type II error probability* – the probability of accepting the null hypothesis when false, for a given *type I error probability* (see Cox and Hinkley, 1974). It is also known, but often insufficiently appreciated, that when any of the model assumptions are invalid, the *reliability of inference* is called into question (see Pearson, 1931a; Bartlett, 1935, for early discussions). Departures from the model assumptions will give rise to a discrepancy between the *nominal* error probabilities (valid premises), and the *actual* error probabilities (misspecified premises), giving rise to unreliable inferences (see Spanos and McGuirk, 2001, Spanos, 2005).

Although the nature of the F–N–P statistical induction became clear by the late 1930s, the form of the underlying *inductive reasoning* was clouded by a disagreement between the two protagonists (see Mayo, 2005). Fisher argued for 'inductive inference' spearheaded by his significance testing (see Fisher, 1955; 1956), and Neyman argued for 'inductive behaviour' based on Neyman–Pearson testing (see Neyman, 1956; Lehmann, 1993; Cox, 2006). Neither account, however, gave satisfactory answers to the question 'when do data $\mathbf{Z}$ provide *evidence* for (or against) a hypothesis or a claim H?' The *pre-data* error–probabilistic account of inference seemed inadequate for a *post-data* evaluation of the inference reached to provide a clear evidential interpretation of the results (see Hacking, 1965).

The F–N–P paradigm, in addition to (*a*) the pre-data as against post-data error probabilities, still grapples with some additional philosophical/methodological issues including (*b*) the fallacies of acceptance and rejection (for example statistical as against substantive significance), (*c*) double use of data, (*d*) statistical model selection (specification) as against model validation, (*e*) structural as against statistical models. These and other methodological issues have been extensively debated in other social sciences such as psychology and sociology (see Morrison and Henkel, 1970; Lieberman, 1971; Godambe and Sprott, 1971), but largely ignored in economics until recently.

Mayo (1996) argued convincingly that some of these chronic methodological issues and problems can be addressed by supplementing the Neyman–Pearson approach to testing (see Pearson, 1966) with a post-data assessment of inference based on *severe testing reasoning*. This extended frequentist approach to inference, called *the error-statistical approach*, has been used by Mayo (1991) to address (*c*), by Mayo and Spanos (2006) to address the fallacies of acceptance and rejection, and by Spanos (2006b; 2007) to deal with the issues (*d*) and (*e*), respectively.

Table 1   *The Normal/linear regression model*

Statistical GM:	$y_t = \beta_0 + \boldsymbol{\beta}_1^\top \mathbf{x}_t + u_t, \ t \in \mathbb{N}$,
[1] Normality:	$(y_t|\mathbf{X}_t = \mathbf{x}_t) \sim \mathsf{N}(.,.)$,
[2] Linearity:	$E(y_t|\mathbf{X}_t = \mathbf{x}_t) = \beta_0 + \boldsymbol{\beta}_1^\top \mathbf{x}_t$, linear in $\mathbf{x}_t$,
[3] Homoskedasticity:	$Var(y_t|\mathbf{X}_t = \mathbf{x}_t) = \sigma^2$, free of $\mathbf{x}_t$,
[4] Independence:	$\{(y_t|\mathbf{X}_t = \mathbf{x}_t), \ t \in \mathbb{N}\}$ an independent process,
[5] t-invariance:	$\boldsymbol{\theta} := (\beta_0, \boldsymbol{\beta}_1, \sigma^2)$ do not change with t.

4.3 Economic statistics in the early 20th century

In the 1930s applied economists were more keyed to Bowley's traditional view of economic statistics than to F–N–P statistical inference perspective. Indeed, Bowley was elected president (the first from Britain) of the Econometric Society for 1938–9. The more economics-oriented 'statistics textbooks' written in the 1920s and 1930s, including Bowley (1920/1926/1937), Mills (1924/1938), Ezekiel (1930), Davis and Nelson (1935) and Secrist (1930), largely ignored the new statistical inference paradigm. Their perspective was primarily one of 'descriptive statistics', supplemented with the Pearson–Yule curve-fitting perspective on correlation and regression, and certain additional focus on the analysis of time series data, including index numbers (see Persons, 1925).

Economic statistics, as exemplified in Mills (1924), provided the framework for the work at the National Bureau of Economic Research (NBER), of which Mills was a staff member. The empirical work on business cycles by Burns and Mitchell (1946) represents an excellent use of descriptive statistics in conjunction with graphical methods, as understood at the time. Their detailed, carefully crafted and painstaking statistical analysis of business cycles, however, suffers from the same crucial weakness as all descriptive statistics: the premises for inductive inference (the underlying statistical model) is not explicitly specified, and as a result one cannot assess the reliability of inferences based on such statistics. For instance, without clearly specified probabilistic premises one can easily misidentify temporal dependence type cycles with regular business cycles (see Spanos, 1999).

The conventional wisdom at the time is summarized by Mills (1924) in the form of a distinction between 'statistical description vs. statistical induction'. In statistical description measures such as the sample mean $\bar{x} = \frac{1}{n}\sum_{k=1}^{n} x_k$, the sample variance $s_x^2 = \frac{1}{n-1}\sum_{k=1}^{n}(x_k - \bar{x}_n)^2$, the correlation $r = \frac{\sum_{k=1}^{n}(x_k - \bar{x}_n)(y_k - \bar{y}_n)}{\sqrt{\left[\sum_{k=1}^{n}(x_k - \bar{x}_n)^2\right]\left[\sum_{k=1}^{n}(y_k - \bar{y}_n)^2\right]}}$, and so on, 'provide just a summary for the data in hand' and 'may be used to perfect confidence, as accurate descriptions of the given characteristics' (1924, p. 549). However, when the results are to be extended *beyond* the data in hand – statistical induction – their validity depends on certain inherent a priori assumptions such as (a) the 'uniformity' for the *population* and (b) the 'representativeness' of the *sample* (1924, pp. 550–2).

A methodological digression. Unfortunately, Mills's misleading argument concerning descriptive statistics lingers on even today. The reality is that there are *appropriate* and *inappropriate* summaries of the data, which depend on the inherent probabilistic structure of the data. For instance, if data $\{(x_k, y_k), k = 1,\ldots,n\}$ are *trending*, like most economic time series, the summary statistics $(\bar{x}, s_x^2, \bar{y}, s_y^2, r)$ represent artefacts – highly misleading

descriptions of the features of the data in hand. When viewed in the context of a probabilistic framework, $(\bar{x}, s_x^2, \bar{y}, s_y^2, r)$ are unreliable estimators of $E(X_k)$, $E(Y_k)$, $Var(X_k)$, $Var(Y_k)$, $Corr(X_k, Y_k)$; they provide reliable and precise estimates only when certain probabilistic assumptions concerning the underlying the vector process $\{(X_k, Y_k) \ k \in \mathbb{N}\}$, such as independent and identically distributed (IID), are valid for the data in hand. Any departures from these premises require one to qualify the reliability and precision of these estimates. In an important sense one of Fisher's lasting contribution to statistics was to (a) make the IID assumptions explicit as part of the problem of specification, by formalizing Mills's a priori 'uniformity' and 'representativeness' assumptions, and (b) render them empirically testable. It is important to note that ignoring statistical adequacy is a very different criticism of Burns and Mitchell than that of Koopmans (1947); see below.

The paper by Yule (1926), entitled 'Why do we sometimes get nonsense correlations between time series'?, provided a widely discussed wakeup call in economics, because it raised serious doubts about the appropriateness of the linear regression model when the data $\{(x_k, y_k), k = 1,\ldots,n\}$ constitute time series, by pointing out the risk of getting *spurious* results. As commented in Spanos (1989b), the source of the spurious (nonsense) correlation problem is *statistical inadequacy* (see Section 4.8 below). Yule's (1927) autoregressive (AR(p)) and Slutsky's (1927) moving average (MA(q)) models can be viewed as attempts to specify statistical models to capture the temporal dependence in time series data.

Stochastic processes. The AR(p) and MA(q) models were given proper probabilistic underpinnings by Wold (1938) using the newly developed theory of stochastic processes by Kolmogorov and Khitchin in the early 1930s (see Doob, 1953). This was a crucial and timely development in probability theory which extended significantly the intended scope of the F–N–P approach beyond the original IID frame-up, by introducing several dependence and heterogeneity concepts, such as Markov dependence, stationarity and ergodicity (see Spanos, 1999, ch. 8).

4.4 The Econometric Society and the Cowles Commission

The vision statement of the Econometric Society founded in 1930 read:

> Its main object shall be to promote studies that aim at a unification of the theoretical-quantitative and the empirical-quantitative approach to economic problems. (Frisch, 1933, p. 106)

The impression among quantitatively oriented economists in the early 1930s was that the F–N–P sampling theory methods were inextricably bound up with agricultural experimentation. It was generally believed that these methods are relevant only for analysing 'random samples' of experimental data, as Frisch argued:

> In problems of the kind encountered when the data are the result of *experiments* which the investigator can control, the sampling theory may render very valuable services. Witness the eminent works of R.A. Fisher and Wishart on problems of agricultural experimentation. (Frisch, 1934, p. 6)

In place of the statisticians' linear regression Frisch proposed his errors-in-variables scheme, which treated all observable variables symmetrically by decomposing them into a latent systematic (deterministic) component and a white-noise error with economic theory providing relationships among the systematic components. Fisher's reaction to Frisch's scheme was that economists were perpetuating a major *confusion* between 'statistical' regression coefficients and 'coefficients in abstract economic laws' (see Bennett, 1990, p. 305).

Tinbergen's (1939) empirical modelling efforts were in the spirit of the Pearson–Yule curve-fitting tradition, which paid little attention to the validity of the premises of inference. In reviewing this work Keynes (1939) destructively criticized the use of regression in econometrics and raised numerous substantive and statistical problems, but not the reliability of inference problem (see Spanos, 2006a).

The first attempt to bring together Frisch's errors-in-variables scheme with Fisher's linear regression model was made by Koopmans (1939), which had no success. Koopmans' primary influence on econometrics was as a leading figure in the Cowles Commission in Chicago in the 1940s (see Heckman, 1992).

The first successful attempt to bring the F–N–P methods into econometrics modelling was made by Haavelmo (1944), who argued convincingly against the prevailing view that sampling methods are only applicable to random samples of experimental data (see Spanos, 1989a). Contrary to this view, the F–N–P perspective provides the proper framework for modelling time series data which exhibit both dependence and heterogeneity:

> For no tool developed in the theory of statistics has any meaning ... without being referred to some stochastic scheme. (Haavelmo, 1944, p. iii)
> ... economists might get more useful and reliable information (and also fewer spurious results) out of their data by adopting more clearly formulated probabilistic models. (1944, p. 114)

The part of Haavelmo's monograph that had the biggest impact on the development of econometrics was, however, the technical 'solution' to the simultaneity problem that was formalized and extended by the Cowles Commission in the form of the simultaneous equations model (SEM): see Koopmans, 1950. Despite the introduction of frequentist methods of inference by the Cowles Commission, the theory-driven specification of

the *structural model*:

$$\mathbf{\Gamma}^\top \mathbf{y}_k = \mathbf{\Delta}^\top \mathbf{x}_k + \varepsilon_k, \quad \varepsilon_k \sim N(0, \mathbf{\Omega}), \quad k \in \mathbb{N},$$
$$(3)$$

(using the traditional notation, see Spanos, 1986), leaves any inferences concerning the structural parameters ($\mathbf{\Gamma}$, $\mathbf{\Delta}$, $\mathbf{\Omega}$) highly susceptible to the unreliability of inference problem.

Methodological digression. The unreliability of inference arises primarily because it is often insufficiently appreciated that the statistical reliability of such inference depends crucially on the *statistical adequacy* of the (implicit) *reduced form model*:

$$\mathbf{y}_k = \mathbf{B}^\top \mathbf{x}_k + \mathbf{u}_k, \quad \mathbf{u}_k \sim N(\mathbf{0}, \mathbf{\Sigma}), \quad k \in \mathbb{N}.$$
$$(4)$$

That is, unless (4), viewed as multivariate linear regression model (assumptions [1]–[5] in Table 1 in vector form), is statistically adequate ([1]–[5] are valid for the data in question), any inference based on (3) is likely to be unreliable. Note that *identification* refers to being able to define the structural parameters ($\mathbf{\Gamma}$, $\mathbf{\Delta}$, $\mathbf{\Omega}$) uniquely in terms of the statistical parameters ($\mathbf{B}, \mathbf{\Sigma}$). In practice (4) is not even estimated explicitly, let alone have its assumptions [1]–[5] tested thoroughly before drawing any inferences concerning ($\mathbf{\Gamma}$, $\mathbf{\Delta}$, $\mathbf{\Omega}$) (see Spanos, 1986; 1990). A more expedient way one that highlights the reliability issue, is to view (3) as a structural model which is embedded into the statistical model (4), giving rise to a special type of substantive information restrictions. Hence, the theory-dominated perspective of the Cowles Commission, despite the importance of the technical innovations introduced in dealing with simultaneity, has (inadvertently) undermined the problem of statistical adequacy in empirical modelling (see Spanos, 2006a). As argued by Heckman:

> The Haavelmo–Cowles way of doing business – to postulate a class of models *in advance of looking at the data* and to consider identification problems within the prescribed class – denies one commonly used process of inductive inference that leads to empirical discovery. ... The Haavelmo program as interpreted by the Cowles Commission scholars refocused econometrics away from the act of empirical discovery and toward a sterile program of hypothesis testing and rigid imposition of a priori theory onto the data. (Heckman, 1992, pp. 883–4)

Koopmans (1947), in reviewing Burns and Mitchell (1946), criticized their focusing on the purely empirical nature of their results without any guidance from economic theory. He pronounced their empirical findings as representing the 'Kepler stage' of data analysis, in contrast to the 'Newton stage', where the original empirical regularities were given a structural (theoretical) interpretation using the law of universal gravitation (LUG). What Koopmans (1947) neglected to point out is that it was not the theory that guided Kepler to the regularities, but the statistical regularities exhibited by the data. Indeed, Kepler established these regularities 60 years before Newton was inspired by them to come up with his LUG. The Cowles Commission approach, which Koopmans misleadingly associates with the Newton stage, was equally (if not more) vulnerable to the reliability of inference problem. There is no reason to believe that the reduced form (4) implied by the structural form (3), which was specified in complete ignorance of the probabilistic structure of the data, will constitute a statistically adequate model. The specification of statistical models relying exclusively on substantive information is not conducive to reliable/precise inferences. The crucial difference between Kepler's empirical results and those in Burns and Mitchell (1946) and Klein (1950) – based largely on Koopmans's preferred approach – is that Kepler's constitute real statistical regularities in the sense that his estimated model of elliptical motion, viewed retrospectively in the context of the linear regression model (Table 1), is statistically adequate; assumptions [1]–[5] are valid for his original data (see Spanos, 2008).

4.5 Textbook econometrics: the Gauss–Markov perspective

The textbook approach to econometrics was largely shaped in the early 1960s by two very successful textbooks by Johnston (1963) and Goldberger (1964) by viewing the SEM as an extension/modification of the classical linear model. These textbooks demarcated the intended scope of econometrics to be the 'quantification of theoretical relationships', and reverted back to the 'curve-fitting' perspective of the Legendre–Gauss 19th century tradition, instead of adopting the F–N–P perspective (see Spanos, 1995; 2007).

The cornerstone of textbook econometrics is the so-called *Gauss–Markov theorem*, which is based on the linear model:

$$\mathbf{y} = \mathbf{X}\boldsymbol{\beta} + \varepsilon, \quad E(\varepsilon) = \mathbf{0}, \quad E(\varepsilon\varepsilon^\top) = \sigma^2 \mathbf{I}_n,$$

$$(5)$$

where $\mathbf{I}_n$ is the identity matrix. In the context of (5), Gauss in 1823 (see Hald, 2007) proved that the least squares estimator $\hat{\boldsymbol{\beta}}_{LS} = (\mathbf{X}^\top\mathbf{X})^{-1}\mathbf{X}^\top\mathbf{y}$ has minimum variance within the class of *linear* and *unbiased* estimators of $\boldsymbol{\beta}$. For the sake of historical accuracy it is important to point out that Markov had nothing to do with this theorem (see Neyman, 1952, p. 228). This theorem, and the perspective it exemplifies, provide the central axis around which textbook econometrics revolves (see Greene, 2003).

A methodological digression. Spanos (1986) challenged the traditional interpretation that the Gauss–Markov theorem provides a formal justification for least squares

via the optimality of the estimators it gives rise to, arguing that the results of this theorem provide a poor basis for *reliable* and *precise* inference. This is primarily because the Gauss–Markov theorem yields the mean and variance of $\hat{\boldsymbol{\beta}}_{LS}$ but *not* its sampling distribution, that is $\hat{\boldsymbol{\beta}}_{LS} \overset{?}{\sim} D(\boldsymbol{\beta}, \sigma^2(\mathbf{X}^\top\mathbf{X})^{-1})$. Hence, even the simplest forms of inference, like testing $H_0 : \boldsymbol{\beta} = \mathbf{0}$ would require one to use either inequalities like Chebyshev's to approximate the relevant error probabilities (Spanos, 1999, pp. 550–2), or invoke asymptotic approximations; neither method would, in general, give rise to reliable and precise inferences (Spanos, 2006a, pp. 46–7).

The Gauss–Markov 'curve-fitting' perspective promotes 'saving the theory' by attributing the stochastic structure to the error term and favouring broad premises (weak assumptions) in an attempt to protect the inference from the perils of misspecification. This move, however, relegates the essentialness of ensuring the reliability and precision of inference. Weak assumptions, such as the Gauss–Markov assumptions in (5), do not guarantee reliable inferences, but they usually give rise to much less precise inferences than specific premises comprising assumptions such as [1]–[5] (Table 1): Spanos, 2006a. As perceptively noted by Heckman:

> In many influential circles, ambiguity disguised as simplicity or 'robustness' is a virtue. The less said about what is implicitly assumed about a statistical model generating data, the less many economists seem to think is being assumed. The new credo is to let sleeping dogs lie. (Heckman, 1992, p. 882)

In addition, the 'error-fixing' strategies of the textbook approach, designed to deal with departures from the linearity, homoskedasticity, no-autocorrelation assumptions, do not usually address the reliability of inference problem (Spanos and McGuirk, 2001).

Some of the important technical developments in both econometrics and statistics since the 1980s, such as the *generalized method of moments* (see Hansen, 1982), as well as certain *nonparametric* (see Pagan and Ullah, 1999) and *semiparametric* methods (see Horowitz, 1998), are motivated by this Gauss–Markov perspective. These methods, although very useful for a number of different aspects of empirical modelling, do not provide the answer to statistical misspecification, and often compromise the reliability/precision of substantive inferences (see Spanos, 1999, pp. 553–5).

4.6 Demarcating the boundaries of modern statistics

As argued above, the F–N–P perspective has been largely ignored in empirical modelling in economics, despite the wholesale adoption of Fisher's estimation and the Neyman–Pearson testing methods. One of the primary obstacles has been the problem of blending the substantive subject matter and statistical information and their roles in empirical modelling. Many aspects of empirical modelling, in both the physical and social sciences, implicate both sources of information in a variety of functions, and others involve one or the other, more or less separately. For instance, the development of *structural* (theoretical) *models* is primarily based on *substantive information*; that activity, by its very nature, cannot be separated from the disciplines in question, but where does this leaves statistics? It renders the problem of demarcating its boundaries as a separate discipline extremely difficult (see Lehmann, 1990; Cox, 1990).

A methodological digression. Spanos (2006c) argued that the lessons learned in blending the substantive and statistical information in econometric modelling can help delineate the boundaries of statistics as a separate discipline. Certain aspects of empirical modelling, which focus on *statistical information* and are concerned with the nature and use of statistical models, can form a body of knowledge that is shared by all applied fields. *Statistical model specification, the use of graphical techniques* (going back to Playfair), *misspecification (M-S) testing and respecification,* together with the relevant inference procedures, constitute aspects of statistical modelling that can be developed *generically* without requiring any information concerning 'what substantive variables the data **Z** quantify or represent'. All these aspects of empirical modelling belong to the *realm of statistics* and can be developed generically without any reference to substantive subject matter information. This, in a sense, will broaden the scope of modern statistics because the current literature and textbooks pay little attention to some of these aspects of modelling (see Cox and Hinkley, 1974).

The statistical and substantive information can be amalgamated, without compromising their integrity, by embedding structural models into adequate statistical models, which would provide the premises for statistical inference. That is, the substantive restrictions need to be thoroughly tested and accepted in the context of the statistical model in order for the resulting empirical model to enjoy both statistical and substantive meaning (see Spanos, 2006b; 2007).

4.7 The Box–Jenkins turn in statistics

An important development in statistics that had a lasting effect on econometrics and created a tension with textbook econometrics, was the publication of Box and Jenkins (1970). Building on the work of Wold (1938), they proposed a new statistical perspective on time series modelling which placed it within the F–N–P modelling framework where the premises of inference is specified by a *statistical model*. In addition to transforming descriptive time series analysis into statistical inference proper, the Box–Jenkins approach introduced several noteworthy innovations into empirical modelling that influenced empirical modelling in economics.

(i) Modelling begins with a family of *statistical models* in the form of the ARIMA(p,d,q):

$$y_t^* = \alpha_0 + \sum_{k=1}^{p} \alpha_k y_{t-k}^*$$
$$+ \sum_{\ell=1}^{q} \beta_\ell \varepsilon_{t-\ell} + \varepsilon_t, \quad \varepsilon_t \sim \text{NIID}(0, \sigma^2),$$
$$t \in \mathbb{N}, \tag{6}$$

where $y_t^* := \Delta^d y_t$, that was thought to capture adequately the temporal dependence and heterogeneity (including seasonality) in time series data.

(ii) Statistical modelling was viewed as an *iterative process* that involves several stages, *identification, estimation, diagnostic checking,* and *prediction.*

(iii) *Diagnostic checks,* based on the residuals from the fitted model, offered a way to detect model inadequacies with a view to improve the original model.

(iv) *Exploratory data analysis* (EDA) was legitimized as providing an effective way to select (identify) a model within the ARIMA(p,d,q) family.

(v) The deliberate choice of a more *general specification* in order to put the model 'in jeopardy' (see Box and Jenkins, 1970, p. 286) is exploited in assessing the adequacy of a selected model.

The Box–Jenkins approach constituted a major departure from the rigid textbook approach, where the model is assumed to be specified by economic theory in advance of any data. Indeed, the predictive success of the ARIMA(p,d,q) models in the 1970s exposed the statistical inadequacy of traditional econometric models, sending the message that econometric models could ignore the temporal dependence and heterogeneity of times series data at their peril (see Granger and Newbold, 1986).

The weaknesses of traditional econometric modelling techniques brought out by the Box–Jenkins modelling motivated several criticisms from within econometrics, including those by Hendry (1977) and Sims (1980), that led to the autoregressive distributed lag (ADL(p,q)) and the vector autoregressive (VAR(p)) family of models, respectively. The LSE tradition (see Hendry, 1993), embraced and extended the Box–Jenkins innovations (i)–(v), rendering the general-to-specific approach the backbone of its empirical modelling methodology (see Hendry, 1995).

4.8 Unit roots and cointegration

The Box–Jenkins ARIMA(p,d,q) modelling approach raised the question 'how does one decide on the value of $d \geq 0$ in $\Delta^d y_t$, that is appropriate to induce stationarity?' It turned out that the value of d is related to the number of unit roots in the $AR(m)$ representation, $y_t = \gamma_0 + \sum_{k=1}^{m} \gamma_k y_{t-k} + u_t$, of the underlying stochastic process $\{y_t, t \in \mathbb{N}\}$. Efforts to answer this question led to the unit root 'revolution', initiated by Dickey and Fuller (1979) in the statistics literature. This had an immediate impact on the econometrics literature, which generalized and extended the initial results in a number of different directions (see Phillips and Durlauf, 1986; Phillips, 1987). This literature eventually led to further important developments, which brought out a special relationship (cointegration) among unit root processes and error-correction models (see Engle and Granger, 1987; Johansen, 1991; Hendry, 1995).

A methodological digression. The (non-standard) sampling distribution results associated with unit roots were used by Phillips (1986) to shed light on the chronic problem of spurious regression raised by Yule (1926). This problem was revisited by Granger and Newbold (1974) using simulations of time series data $\{(x_t, y_t), t=1,\ldots,n\}$ generated by two *uncorrelated* Normal unit root processes:

$$y_t = y_{t-1} + \varepsilon_{1t}, \quad x_t = x_{t-1} + \varepsilon_{2t},$$

$$E(\varepsilon_{1t}) = 0, \quad E(\varepsilon_{2t}) = 0, \quad E(\varepsilon_{1t}^2) = \sigma_{11},$$

$$E(\varepsilon_{2t}^2) = \sigma_{22}, \quad E(\varepsilon_{1t}\varepsilon_{2t}) = 0.$$

Their results demonstrated that when these data were used to estimate the linear regression model, $y_t = \beta_0 + \beta_1 x_t + u_t$, the inferences based on the estimated model were completely unreliable. In particular, they noted a huge discrepancy between the *nominal* (α=.05) and *actual* ($\hat{\alpha} = .76$) error probabilities when testing the hypothesis β_1=0.

In a very influential paper, Phillips (1986) explained this by deriving analytically the (non-standard) sampling distributions of the least-squares estimators $(\hat{\beta}_0, \hat{\beta}_1)$ under the above unit root scheme, showing how different they were from the assumed distributions. What was not sufficiently appreciated was that the discrepancy between the nominal and actual error probabilities is a classic symptom of unreliable inferences emanating from a statistically misspecified model, that is *misspecification*, due to ignoring the temporal dependence/heterogeneity in the data, is the real source of spurious regression. One would encounter similar unreliabilities when the data exhibit deterministic trends or/and Markov dependence, or/and non-Normalities (see Spanos and McGuirk, 2001, Spanos, 2005). Deriving the sampling distributions under all scenarios of possible misspecifications is impractical (there is an infinity of such scenarios), and does *not* address the unreliability of inference issue. What is needed is to respecify the original model to account for the disregarded information that gave rise to the detected departures. For instance, for the above Granger and Newbold data, if one were to estimate the dynamic linear regression model:

$$y_t = \alpha_0 + \alpha_1 x_t + \alpha_2 x_{t-1} + \alpha_3 y_{t-1} + \varepsilon_t, \quad t \in \mathbb{N},$$

the above noted unreliabilities would disappear (see Spanos, 2001).

4.9 Recent developments in microeconometrics

Arguably, some of the most important developments in econometrics since 1980 have taken place in an area broadly described as microeconometrics (see Manski and MacFadden, 1981; Heckman and Singer, 1984; Cameron and Trivedi, 2005) for a recent textbook survey. This area includes *discrete and limited dependent* and *duration models* for *cross-section data*, as well as *panel data* models. The roots of these statistical models go back to the statistical literature on the probit/logit and analysis of variance models (see Agresti, 2002, ch. 16), but they have been generalized, extended and adapted for economic data.

A welcome facet of microeconometrics is the specification of statistical models that often takes into consideration the probabilistic structure of the data (see Heckman, 2001). Unfortunately, this move does not often go far enough in securing statistical adequacy. This becomes apparent when one asks, 'what are the probabilistic assumptions providing a complete specification for the probit/logit, duration and the fixed and random effect models?' Without such complete specifications, one would not even know what potential errors to probe for to secure statistical adequacy.

While these developments in microeconometrics are of great importance, their potential value has been offset by the insufficient attention paid to the task of ensuring reliability and precision of inference. Their statistical results are still largely dominated by the Gauss–Markov perspective, in the sense that:

(i)   the probabilistic structure of the models in question is specified, almost exclusively, in terms of unobservable *error terms*,
(ii) the error probabilistic assumptions are often vague and incomplete, and invariably involve non-testable orthogonality conditions,
(iii) the statistical analysis focuses primarily on constructing consistent and asymptotically Normal estimators, and
(iv) respecification is often confined to 'error-fixing'.

In view of (i)–(iv), even questions of ensuring statistical adequacy cannot be posed unequivocally for these statistical models.

Spanos (2006a; 2006d) proposed complete specifications for these statistical models in terms of probabilistic assumptions relating to the observable stochastic processes involved, but there is a long way to go to develop adequate misspecification testing and the respecification results needed to ensure the reliability and precision of inference when applying these statistical models to actual data.

5 Conclusion

The demise of political arithmetic by the end of the 18th century, due to the unreliability of the inferences its methods gave rise to, contains important lessons for both economics and statistics. Petty's attitude of 'seeking figures that will support a conclusion already reached by other means' lingers on in applied econometrics more than three centuries later. The problem then was that, in addition to the quality and the accuracy of data, the probabilistic underpinnings of establishing statistical regularities were completely lacking. Fisher's recasting of statistical induction has changed that, and it is now known that the explicit specification of the *statistical model* enables one to (*a*) assess the validity of the premises for inductive inference, and (*b*) provide relevant error probabilities for assessing the reliability of ensuing inferences. It has taken several decades to understand how one can assess the model assumptions vis-à-vis the observed data using *misspecification tests* (see Spanos, 1999), but one hopes it will take less time before modellers understand the necessity to implement such tests with the required care and thoroughness to ensure the *reliability* of the resulting statistical inferences (see Spanos, 2006a).

The Box–Jenkins modelling approach exposed the inattention to statistical adequacy in traditional econometric modelling and strengthened the call for adopting the F–N–P perspective. This will bring modern statistical inference closer to econometrics to the benefit of both disciplines. Careful implementation of this perspective will certainly improve the reliability of empirical evidence in economics and other applied disciplines. Moreover, the *ab initio* separation of the statistical and substantive information can help demarcate and extend the intended scope of statistics. The error-statistical extension/ modification of frequentist statistics (Mayo, 1996) can address some of the inveterate problems concerning inductive reasoning and broaden the intended scope of statistical inference in these disciplines by enabling one to consider questions of substantive adequacy, shedding light on causality issues, omitted variables and confounding effects (see Spanos, 2006b).

ARIS SPANOS

See also **Bowley, Arthur Lyon; Davenant, Charles; Edgeworth, Francis Ysidro; Fisher, Ronald Aylmer; King, Gregory; Petty, William.**

Bibliography

Agresti, A. 2002. *Categorical Data Analysis*, 2nd edn. New York: Wiley.

Backhouse, R.E. 2002a. *The Ordinary Business of Life: A History of Economics from the Ancient to the Twenty-First Century*. Princeton, NJ: Princeton University Press.

Backhouse, R.E. 2002b. *The Penguin History of Economics*. London: Penguin Books.

Bartlett, M.S. 1935. Some aspects of the time-correlation problem in regard to tests of significance. *Journal of the Royal Statistical Society* 98, 536–43.

Bennett, J.H. 1990. *Statistical Inference and Analysis: Selected correspondence of R.A. Fisher*. Oxford: Clarendon Press.

Bernoulli, J. 1713. *Ars Conjectandi*. Basilea: Thurnisius, tr. E. D. Sylla, Baltimore: Johns Hopkins University Press, 2006.

Bowley, A.L. 1902, 1920, 1926, 1937. *Elements of Statistics*, 2nd, 4th, 5th and 6th edns. London: Staples Press.

Bowley, A.L. 1906. Address to the economic science and statistics. *Journal of the Royal Statistical Society* 69, 540–58.

Bowley, A.L. 1928. *F.Y. Edgeworth's Contributions to Mathematical Statistics*. Clifton: Augustus M. Kelley.

Box, G.E.P. and Jenkins, G.M. 1970. *Time Series Analysis: Forecasting and Control*. San Francisco: Holden-Day.

Box, J.F. 1978. *R.A. Fisher: The Life of a Scientist*. New York: Wiley.

Burns, A.F. and Mitchell, W.C. 1946. *Measuring Business Cycles*. New York: NBER.

Cameron, A.C. and Trivedi, P.K. 2005. *Microeconometrics: Methods and Applications*. Cambridge: Cambridge University Press.

Cournot, A. 1843. *Exposition de la théorie des chances et des probabilités*. Paris: Hachette.

Cox, D.R. 1990. Role of models in statistical analysis. *Statistical Science* 5, 169–74.

Cox, D.R. 2006. *Principles of Statistical Inference*. Cambridge: Cambridge University Press.

Cox, D.R. and Hinkley, D.V. 1974. *Theoretical Statistics*. London: Chapman & Hall.

Cramer, H. 1946. *Mathematical Methods of Statistics*. Princeton, NJ: Princeton University Press.

Davis, H.T. and Nelson, W.F.C. 1935. *Elements of Statistics*. Indiana: Principia Press.

Desrosières, A. 1998. *The Politics of Large Numbers: A History of Statistical Reasoning*. Cambridge, MA: Harvard University Press.

Dickey, D.A. and Fuller, W.A. 1979. Distribution of the estimators for autoregressive time series with a unit root. *Journal of the American Statistical Association* 74, 427–31.

Doob, J.L. 1953. *Stochastic Processes*. New York: Wiley.

Engle, R.F. and Granger, C.W.J. 1987. Co-integration and error-correction: representation, estimation and testing. *Econometrica* 55, 251–76.

Ezekiel, M. 1930. *Methods of Correlation Analysis*, 2nd edn. New York: Wiley.

Farebrother, R.W. 1999. *Fitting Linear Relationships: A History of the Calculus of Observations 1750–1900*. New York: Springer.

Fisher, I. 1947. Response of Irving Fisher. *Journal of the American Statistical Association* 42, 4–5.

Fisher, R.A. 1915. Frequency distribution of the values of the correlation coefficient in samples from an indefinitely large population. *Biometrika* 10, 507–21.

Fisher, R.A. 1921. On the 'probable error' of a coefficient deduced from a small sample. *Metron* 1, 2–32.

Fisher, R.A. 1922. On the mathematical foundations of theoretical statistics. *Philosophical Transactions of the Royal Society, A* 222, 309–68.

Fisher, R.A. 1922a. On the interpretation of χ^2 from contingency tables, and the calculation of p. *Journal of the Royal Statistical Society* 85, 87–94.

Fisher, R.A. 1922b. The goodness of fit of regression formulae and the distribution of regression coefficients. *Journal of the Royal Statistical Society* 85, 597–612.

Fisher, R.A. 1924. The conditions under which χ^2 measures the discrepancy between observation and hypothesis. *Journal of the Royal Statistical Society* 87, 442–50.

Fisher, R.A. 1925. *Statistical Methods for Research Workers*. Edinburgh: Oliver & Boyd.

Fisher, R.A. 1935. The logic of inductive inference. *Journal of the Royal Statistical Society* 98, 39–54, 55–82.

Fisher, R.A. 1935a. *The Design of Experiments*. Edinburgh: Oliver & Boyd.

Fisher, R.A. 1955. Statistical methods and scientific induction. *Journal of the Royal Statistical Society, B* 17, 69–78.

Fisher, R.A. 1956. *Statistical Methods and Scientific Inference*. Edinburgh: Oliver & Boyd.

Frisch, R. 1933. Editorial. *Econometrica* 1, 1–4.

Frisch, R. 1934. *Statistical Confluence Analysis by Means of Complete Regression Schemes*. Oslo: Universitetets Okonomiske Institutt.

Galton, F. 1869. *Hereditary Genius: An Inquiry into its Laws and Consequences*. London: Macmillan.

Gauss, C.F. 1809. *Theoria Motus Corporum Coelestium in Sectionibus Conicis Solem Ambientium*. Hamburg: F. Perthes & I.H. Besser.

Glass, D.V. 1973. *Numbering the People*. Farnborough: Saxton House.

Godambe, V.P. and Sprott, D.A. 1971. *Foundations of Statistical Inference: A Symposium*. Toronto: Holt, Rinehart and Winston.

Goldberger, A.S. 1964. *Econometric Theory*. New York: Wiley.

Gossett, W.S. 1908. The probable error of the mean. *Biometrika* 6, 1–25.

Granger, C.W.J. and Newbold, P. 1974. Spurious regressions in econometrics. *Journal of Econometrics* 2, 111–20.

Granger, C.W.J. and Newbold, P. 1986. *Forecasting Economic Time Series*, 2nd edn. London: Academic Press.

Graunt, J. 1662. *Natural and Political Observations upon the Bills of Mortality*. London: John Martyn. Repr. in Hull (1899).

Greene, W.H. 2003. *Econometric Analysis*, 5th edn. Englewood Cliffs, NJ: Prentice-Hall.

Greenwood, M. 1948. *Medical Statistics from Graunt to Farr*. Cambridge: Cambridge University Press.

Haavelmo, T. 1944. The probability approach to econometrics. *Econometrica* 12(Suppl), 1–115.

Hacking, I. 1965. *Logic of Statistical Inference*. Cambridge: Cambridge University Press.

Hacking, I. 1975. *The Emergence of Probability*. Cambridge: Cambridge University Press.

Hald, A. 1990. *History of Probability and Statistics and Their Applications before 1750*. New York: Wiley.

Hald, A. 1998. *A History of Mathematical Statistics from 1750 to 1930*. New York: Wiley.

Hald, A. 2007. *A History of Parametric Statistical Inference from Bernoulli to Fisher*. New York: Springer.

Hansen, L.P. 1982. Large sample properties of generalized method of moments estimators. *Econometrica* 97, 93–115.

Harrison, B. 2004. *Oxford Dictionary of National Biography*. Oxford: Oxford University Press.

Heckman, J.J. 1992. Haavelmo and the birth of modern econometrics: a review of the history of econometric ideas by Mary Morgan. *Journal of Economic Literature* 30, 876–86.

Heckman, J.J. 2001. Micro data, heterogeneity, and the evaluation of public policy: Nobel lecture. *Journal of Political Economy* 109, 673–748.

Heckman, J.J. and Singer, B. 1984. Econometric duration analysis. *Journal of Econometrics* 24, 63–132.

Hendry, D.F. 1977. On the time series approach to econometric model building. In *New Methods in Business Cycle Research*, ed. C.A. Sims. Minnesota: Federal Reserve Bank of Minneapolis.

Hendry, D.F. 1993. *Econometrics: Alchemy or Science?* Oxford: Blackwell.

Hendry, D.F. 1995. *Dynamic Econometrics*. Oxford: Oxford University Press.

Hendry, D.F. and Morgan, M.S. 1995. *The Foundations of Econometric Analysis*. Cambridge: Cambridge University Press.

Hilts, V.L. 1973. Statistics and social science. In *Foundations of Scientific Method: the Nineteenth Century*, ed. R.N. Giere and R.S. Westfall. Bloomington, IN: Indiana University Press.

Hilts, V.L. 1978. Aliss exterendum, or, the origins of the Statistical Society of London. *Isis* 69, 21–43.

Hotelling, H. 1930. British statistics and statisticians today. *Journal of the American Statistical Association* 25, 186–90.

Hoppit, J. 1996. Political arithmetic in eighteenth-century England. *Economic History Review* 49, 516–40.

Horowitz, J.L. 1998. *Semiparametric Methods in Econometrics*. New York: Springer.

Hull, H.C. 1899. *The Economic Writings of Sir William Petty*. Cambridge: Cambridge University Press.

Hume, D. 1987. *Essays, Moral, Political and Literary*, ed. E.F. Miller Indianapolis, ID: Liberty Fund.

JSSL (*Journal of the Statistical Society of London*). 1834. Prospects of the objects and plan of operation of the statistical society of London. Reprinted in *Journal of the Statistical Society of London*, 1869, 385–7.

Jevons, W.S. 1871. *The Theory of Political Economy*. New York, NY: Augustus M. Kelley, 1911.

Johansen, S. 1991. Estimation and hypothesis testing of cointegrating vector of Gaussian vector autoregressive models. *Econometrica* 59, 1551–80.

Johnston, J. 1963. *Econometric Methods*. New York: McGraw-Hill.

Keynes, J.M. and Tinbergen, J. 1939–40. Professor's Tinbergen's method. *Economic Journal* 49, 558–68; A Reply, by J. Tinbergen, and Comment by Keynes, 50, 141–56.

Keynes, J.N. 1890/1917. *The Scope and Method of Political Economy*. Fairfield, NJ: Augustus M. Kelley.

Klein, L.R. 1950. *Economic Fluctuations in the United States 1921–1941*. Cowles Commission for Research in Economics, Monograph No. 11. New York: Wiley.

Kolmogorov, A.N. 1933. *Foundations of the theory of Probability*, 2nd edn. New York: Chelsea.

Koopmans, T.C. 1939. *Linear Regression Analysis of Economic Time Series*. Netherlands Economic Institute, Publication No. 20. Haarlem: F. Bohn.

Koopmans, T.C. 1947. Measurement without theory. *Review of Economics and Statistics* 29, 161–72.

Koopmans, T.C. 1950. *Statistical Inference in Dynamic Economic Models*, Cowles Commission Monograph, No. 10. New York: Wiley.

Kreager, P. 1988. New light on Graunt. *Population Studies* 42, 129–40.

Legendre, A.M. 1805. *Nouvelles Méthodes pour la détermination des orbites des comètes*. Paris: Courcier.

Lehmann, E.L. 1990. Model specification: the views of Fisher and Neyman, and later developments. *Statistical Science* 5, 160–8.

Lehmann, E.L. 1993. The Fisher and Neyman–Pearson theories of testing hypotheses: one theory or two? *Journal of the American Statistical Association* 88, 1242–9.

Lehmann, E.L. 1999. 'Student' and small-sample theory. *Statistical Science* 14, 418–26.

Letwin, W. 1965. *The Origins of Scientific Economics*. New York: Anchor Books.

Lieberman, B. 1971. *Contemporary Problems in Statistics: A Book of Readings for the Behavioral Sciences*. Oxford: Oxford University Press.

Manski, C.F. and MacFadden, D. 1981. *Structural Analysis of Discrete Data with Econometric Applications*. Cambridge, MA: MIT Press.

Mayo, D.G. 1991. Novel evidence and severe tests. *Philosophy of Science* 58, 523–52.

Mayo, D.G. 1996. *Error and the Growth of Experimental Knowledge*. Chicago: University of Chicago Press.

Mayo, D.G. 2005. Philosophy of statistics. In *Philosophy of Science: An Encyclopedia*, ed. S. Sarkar and J. Pfeifer. London: Routledge.

Mayo, D.G. and Spanos, A. 2004. Methodology in practice: statistical misspecification testing. *Philosophy of Science* 71, 1007–25.

Mayo, D.G. and Spanos, A. 2006. Severe testing as a basic concept in a Neyman–Pearson philosophy of induction. *British Journal for the Philosophy of Science* 57, 323–57.

Mills, F.C. 1924. *Statistical Methods*. New York: Henry Holt, 1938.

Moore, H.L. 1908. The statistical complement of pure economics. *Quarterly Journal of Economics* 23, 1–33.

Moore, H.L. 1911. *The Law of Wages*. New York: Macmillan.

Morgan, M.S. 1990. *The History of Econometric Ideas*. Cambridge: Cambridge University Press.

Morrison, D.E. and Henkel, R.E. 1970. *The Significance Test Controversy: A Reader*. Chicago: Aldine.

Neyman, J. 1934. On the two different aspects of the representative method: the method of stratified sampling and the method of purposive selection. *Journal of the Royal Statistical Society* 97, 558–625.

Neyman, J. 1935. On the problem of confidence intervals. *Annals of Mathematical Statistics* 6, 111–6.

Neyman, J. 1950. *First Course in Probability and Statistics*. New York: Henry Holt.

Neyman, J. 1952. *Lectures and Conferences on Mathematical Statistics and Probability*, 2nd edn. Washington, DC: U.S. Department of Agriculture.

Neyman, J. 1956. Note on an article by Sir Ronald Fisher. *Journal of the Royal Statistical Society B* 18, 288–94.

Neyman, J. 1977. Frequentist probability and frequentist statistics. *Synthèse* 36, 97–131.

Neyman, J. and Pearson, E.S. 1933. On the problem of the most efficient tests of statistical hypotheses. *Philosophical Transactions of the Royal Society of London*. Series A 231, 289–337.

Pagan, A.R. and Ullah, A. 1999. *Nonparametric Econometrics*. Cambridge: Cambridge University Press.

Pearson, E.S. 1931a. The analysis of variance in cases of non-normal variation. *Biometrika* 23, 114–33.

Pearson, E.S. 1931b. Note on tests for normality. *Biometrika* 22, 423–24.

Pearson, E.S. 1955. Statistical concepts in the relation to reality. *Journal of the Royal Statistical Society*. Series B 17, 204–7.

Pearson, E.S. 1966. The Neyman–Pearson story: 1926–34. In *Research Papers in Statistics: Festschrift for J. Neyman*, ed. F.N. David. New York: Wiley.

Pearson, K. 1896. Mathematical contributions to the theory of evolution. III. Regression, heredity, and panmixia. *Philosophical Transactions of the Royal Society of London*. Series A 187, 253–318.

Pearson, K. 1900. On the criterion that a given system of deviations from the probable in the case of a correlated system of variables in such that it can be reasonably supposed to have arisen from random sampling. *Philosophical Magazine* 50, 157–175.

Peirce, C.S. 1878. The probability of induction. *Popular Science Monthly* 12, 705–18.

Persons, W.M. 1925. Statistics and economic theory. *Review of Economics and Statistics* 7, 179–97.

Phillips, P.C.B. 1986. Understanding spurious regression in econometrics. *Journal of Econometrics* 33, 311–40.

Phillips, P.C.B. 1987. Time series regressions with a unit root. *Econometrica* 55, 227–301.

Phillips, P.C.B. and Durlauf, S.N. 1986. Multiple time series regression with integrated processes. *Review of Economic Studies* 53, 473–95.

Playfair, W. 1786. *The Commercial and Political Atlas*. London: T. Burton, 1801.

Porter, T.M. 1986. *The Rise of Statistical Thinking 1820–1900*. Princeton, NJ: Princeton University Press.

Quetelet, A. 1942. *A Treatise on Man and the Development of his Faculties*. Edinburgh: Chambers.

Redman, D.A. 1997. *The Rise of Political Economy as a Science*. Cambridge, MA: MIT Press.

Routh, G. 1975. *The Origins of Economic Ideas*. London: Macmillan.

RSS (Royal Statistical Society). 1934. *Annals of the Royal Statistical Society 1834–1934*. London: The Royal Statistical Society.

Schweber, L. 2006. *Disciplining Statistics: Demographic and Vital Statistics in France and England, 1830–1885*. Durham, NC: Duke University Press.

Secrist, H. 1930. *An Introduction to Statistical Methods*, 2nd edn. New York: Macmillan.

Sims, C.A. 1980. Macroeconomics and reality. *Econometrica* 48, 1–48.

Slutsky, E. 1927. The summation of random causes as the source of cyclic processes. In Russian; English tr. in *Econometrica* 5, 1937.

Smith, A. 1776. *An Inquiry into the Nature and Causes of the Wealth of Nations*, ed. R.H. Campell, A.S. Skinner and W.B. Todd. Oxford: Clarendon Press.

Spanos, A. 1986. *Statistical Foundations of Econometric Modelling*. Cambridge: Cambridge University Press.

Spanos, A. 1989a. On re-reading Haavelmo: a retrospective view of econometric modeling. *Econometric Theory* 5, 405–29.

Spanos, A. 1989b. Early empirical findings on the consumption function, stylized facts or fiction: a retrospective view. *Oxford Economic Papers* 41, 150–69.

Spanos, A. 1990. The simultaneous equations model revisited: statistical adequacy and identification. *Journal of Econometrics* 44, 87–108.

Spanos, A. 1995. On theory testing in econometrics: modeling with nonexperimental data. *Journal of Econometrics* 67, 189–226.

Spanos, A. 1999. *Probability Theory and Statistical Inference: Econometric Modeling with Observational Data*. Cambridge: Cambridge University Press.

Spanos, A. 2001. Time series and dynamic models. In *A Companion to Theoretical Econometrics*, ed. B. Baltagi. Oxford: Blackwell.

Spanos, A. 2005. Misspecification, Robustness and the Reliability of Inference: the simple t-test in the presence of Markov dependence. Working Paper, Virginia Polytechnic Institute and State University.

Spanos, A. 2006a. Econometrics in retrospect and prospect. In *New Palgrave Handbook of Econometrics*, vol. 1, ed. T.C. Mills and K. Patterson. London: Macmillan.

Spanos, A. 2006b. Revisiting the omitted variables argument: substantive vs. statistical adequacy. *Journal of Economic Methodology* 13, 179–218.

Spanos, A. 2006c. Where do statistical models come from? Revisiting the problem of specification. In *Optimality: The Second Erich L. Lehmann Symposium*, ed. J. Rojo. Lecture Notes-Monograph Series, vol. 49. Beachwood, OH: Institute of Mathematical Statistics.

Spanos, A. 2006d. Revisiting the statistical foundations of panel data models. Working paper, Virginia Polytechnic Institute and State University.

Spanos, A. 2008. Curve-fitting, the reliability of inductive inference and the error-statistical approach. *Philosophy of Science*.

Spanos, A. and McGuirk, A. 2001. The model specification problem from a probabilistic reduction perspective. *Journal of the American Agricultural Association* 83, 1168–76.

Stigler, G.J. 1954. The early history of the empirical studies of consumer behavior. *Journal of Political Economy* 62, 95–113.

Stigler, S.M. 1986. *The History of Statistics: the Measurement of Uncertainty before 1900*. Cambridge, MA: Harvard University Press.

Stone, J.R.N. 1954. *The Measurement of Consumers' Expenditure and Behaviour in the United Kingdom, 1920–1938*. Cambridge: Cambridge University Press.

Stone, R. 1997. *Some British Empiricists in the Social Sciences 1650–1900*. Cambridge: Cambridge University Press.

Tinbergen, J. 1939. *Statistical Testing of Business Cycle Research*, 2 vols. Geneva: League of Nations.

Tippet, L.H.C. 1931. *The Methods of Statistics*. London: Williams & Norgate.

Uspensky, J.V. 1937. *Introduction to Mathematical Probability*. New York: McGraw-Hill.

Walker, H.M. 1929. *Studies in the History of Statistical Method*. Baltimore: Williams & Wilkins.

Walker, H.M. 1958. The contributions of Karl Pearson. *Journal of the American Statistical Association* 53, 11–22.

Wold, H.O. 1938. *A Study in the Analysis of Stationary Time Series*. Uppsala: Almquist & Wicksell.

Yule, G.U. 1897. On the theory of correlation. *Journal of the Royal Statistical Society* 60, 812–54.

Yule, G.U. 1911. *An Introduction to the Theory of Statistics*. London: Griffin & Co.

Yule, G.U. 1926. Why do we sometimes get nonsense correlations between time series? – a study in sampling and the nature of time series. *Journal of the Royal Statistical Society* 89, 1–64.

Yule, G.U. 1927. On a method of investigating periodicities in disturbed series, with special reference to Wolfer's sunspot numbers. *Philosophical Transactions of the Royal Society of London*. Series A 226, 267–98.

status and economics

The study of status by economists is not new, although some find it surprising since status initially seems to be a purely social phenomenon. A person's status is a ranking in a hierarchy that is socially recognized. People may participate in a number of status hierarchies defined by the different social groups of which they are members. Status may be defined and recognized narrowly, as when it is based on skills or accomplishments, or widely, when it is valued and recognized by an entire society. People within a society may differ in their assessment of the importance of a given status characteristic, so consistent rankings may be impossible. Research on status has focused on several questions. Why do individuals value status? What kinds of status-seeking behaviours do we observe? Is status-seeking a good thing?

Status-seeking behaviour

Adam Smith (1759) noted that individuals are willing to expend effort to attain status, a practice that Veblen (1926), who coined the term 'conspicuous consumption', viewed as wasteful. Social psychologists study the subject, and have developed *status characteristics theory*. Status-seeking acts vary by culture, and include purchasing status goods or houses in the 'right' neighbourhood and seeking education that is not 'useful' (Veblen, 1926), as well as contributing money to charity (Getzner, 2000) and planning weddings that strain the family's financial resources (Bloch, Vijayendra and Desai, 2004). In these examples status-seeking is clearly 'costly' to the individual. Assessing the full impact of status also requires evaluating several possible indirect consequences of individuals' status-seeking decisions.

Bernheim (1994) models behaviour that is a costly signal about an individual's social desirability. Agents possess unobservable characteristics and tastes for associating with others with whom they are similar. Individuals seek the esteem of others, where esteem is based on public perceptions of their types. The result is the development of social norms whereby individuals choose similar actions in order to increase their popularity. In this case, the underlying characteristics may have cultural value, for example, whether an individual is 'selfish' or 'generous'. Sub-populations within a society might value these characteristics differently, however, resulting in different sets of social norms. This means that this type of status characteristic can only be ranked within its cultural context.

In other cases, the status characteristic may have value for its own sake in addition to its status value. Here returns to status are not based solely on an absolute level of status, but also on one's relative status ranking. Thus status-seeking becomes a contest where individuals are willing to incur costs in order to win. Positional externalities are inherent in the status contest since actions that increase one individual's relative status decrease another's. The resulting positional arms races generally reduce social welfare and probably account for much of Veblen's distaste for status-seeking behaviour.

Earnings or wealth are straightforward examples of this type of positional externality. As every department chair knows, people care not only about their absolute earnings but also about relative earnings. Bolton (1991) models bargaining with agents who care about their absolute and relative earnings and finds that this model organizes some experimental data better than a pure self-interest model. In the laboratory, subjects behave as if they care about relative earnings. Concern about relative earnings may explain why risk-averse individuals may make decisions that, on the surface, seem more consistent with risk-seeking behaviour (Robson, 1992.)

Fershtman, Murphy and Weiss (1996) argue that status-seeking can affect economic growth. They model social status as a reward for education-seeking in economic growth-enhancing occupations. If individuals differ only by ability or income, then awarding status to growth-enhancing occupations will increase growth. However, if individuals differ by both wealth and learning ability, the 'wrong' agents may acquire education; inefficiency arises because low-talent/high-income workers invest in education to obtain higher status occupations. This reduces wages in the high-status occupations and crowds out high-talent/low-income workers. In addition to the inefficient distribution of talent in the economy, status-seeking then reduces economic growth.

Endowed status

While the status characteristics discussed above result from decisions an individual makes and may be productivity related, this is not necessarily the case. Other status characteristics, such as race, are inherited. In other cases, individuals may have some, but not complete, control, for example, for physical attractiveness. Nevertheless, both experimental and empirical work documents the effect of these types of status on economic decision-making.

Artificial status. Even artificial status may act like a type of power in economic interactions. Ball et al. (2001) report results from experiments designed to control for productivity-related attributes and to isolate the effect of status. Status is awarded by the experimenters in order to ensure that all subjects recognize the relevant status characteristics. Half the participants are publicly awarded gold stars based on the sum of their answers to a trivia quiz, and then trade in experimental markets where the traders on one side of the market have stars, and those on the other side do not. They find that earnings average 15 per cent higher for the high-status traders, and that the pattern of higher earnings persists even when status is randomly awarded. Thye (2000) finds a similar result in a bargaining game. Kumru and Vesterlund (2005) use the same procedure to induce status differences, in a sequential voluntary contributions game. They find that low-status agents tend to mimic the decisions of high-status agents, which creates an incentive for high-status agents to raise their own contributions. This suggests that status may play a role in increasing welfare in games where the socially optimal outcome is not the equilibrium in a pure self-interest model of behaviour.

Beauty. Economists' theory of incomplete information explains why costly status credentials that signal productivity may be linked to higher income. This occurs, for example, when job candidates with college degrees are preferred over those with no college education for jobs requiring no higher education-related skills. A systematic link between earnings and status characteristics that seem entirely unrelated to productivity, such as beauty, is more puzzling. For example, Hamermesh and Biddle's (1994) find a 15 per cent wage premium to more beautiful workers; this is consistent with status characteristics theory, but probably not with the marginal productivity theory of wages. Equally puzzling is Hamermesh and Parker's (2005) finding that more attractive faculty members earn higher teaching evaluations. In a trust-game experiment, Wilson and Eckel (2006) find that people are more likely to trust attractive people, although they are actually no more trustworthy than less attractive people. Other examples come from Solnick and Schweitzer (1999) on ultimatum games and Mulford et al. (1998) and Kahn, Hottes and Davis (1971) in Prisoner's Dilemma games.

In an experimental study of employment, Mobius and Rosenblatt (2006) are able to decompose the positive effect of beauty into three mechanisms. More attractive people are more confident, and confidence increases productivity. If one holds confidence fixed, however, more attractive people are wrongly considered more able by employers, a stereotype that may result in hiring mistakes. Beautiful workers also may have better communication and social skills, which may be beneficial when tasks require working with others. The result concerning confidence may explain the higher evaluations of attractive teachers.

Gender. In almost all societies men hold higher status than women. In the United States, for example, women earn less money than men, even after attempting to control for productivity related attributes such as education (Altonji and Blank, 1999; Darity and Mason, 1998.) Experimental tasks can allow researchers to look at environments where productivity does not affect outcomes in any way. Earnings differentials persist even in these environments. Women also may face discrimination in hiring. Goldin and Rouse (2000) collect hiring data for symphony orchestra musicians where in some cases musicians were auditioned from behind a screen. They find that women were more likely to be hired or advanced to the next round of auditions in the 'blind' condition and less likely than men to be advanced in the 'non-blind' condition.

Differential treatment by gender also exists in experimental bargaining games. Solnick (2001) and Eckel and Grossman (2001) conducted ultimatum bargaining games where subjects knew their opponents' gender. Both studies find that offers made by men and women are about the same, while offers made to women are significantly lower than those made to men. Field experiments also find differences in bargaining by gender; women are initially offered higher prices in the market for cars (Ayers and Siegelman, 1995) and trading cards (List, 2004).

Race. Race is a form of status, and also follows the pattern of low status being related to low earnings. Even after controlling for productivity related attributes, African-Americans still earn less than whites (Altonji and Blank, 1999; Darity and Mason, 1998). Economic research on the motivations for race-related wage discrimination dates back to Becker (1957). Arrow (1972) argues that status in working environments is related to discrimination in yet another way. Conferring high status on some individuals can compensate for 'nearness' to individuals with whom an agent prefers not to associate. This hypothesis explains the result that highly educated African-Americans face a relatively larger wage gap. In this case education, another status characteristic, works against their earnings potential.

Any status characteristic needs to be observable in order to affect others' behaviour. Since African-Americans have less education than whites on average, Arrow argues that using education as a criterion in employment screening can be an effective means of discrimination. Visual cues may form another basis for discrimination. Historically, skin colour was used as a signal of social status – since people who worked outside tended to tan, so lighter skin was a signal of wealth and high status. Arrow (1972) describes skin colour as a 'cheap' signal about race. Darity and Mason (1998) survey a number of studies on race and skin colour in the United States and find that even within races, darker skin colour is related to lower earnings. This suggests that race-related status hierarchies are not bilateral. Eckel and Wilson (2004) show subjects pictures of their opponents prior to playing a trust game. A different sample of subjects from the same subject pool evaluated the pictures for a number of characteristics. They find that trust and reciprocity are significantly related to skin colour as well as perceived 'friendliness' and 'reliability' of the recipients. Field experiments also find differences in bargaining by race; minorities are initially offered higher prices in the market for cars (Ayers and Siegelman, 1995) and trading cards (List, 2004).

Names. Names are a signal of status in societies where there is segmentation based on race or ethnicity. Bertrand and Mullianathan (2003) find that employers in the United States are less likely to call the sender of a résumé identified with an African-American sounding name than they are the sender of an identical resume with a white-sounding name. Fryer and Levitt (2004) calculate an index of 'distinctively black'-sounding names, but once they control for other socio-economic variables, names are not associated with economic disadvantage. Fershtman and Gneezy (2001) conduct laboratory experiments in Israel that pair subjects with partners who have names that denote a distinctive Ashkenazic or Eastern Jewish heritage. They find that subjects with an Ashkenazic partner are three times more likely to make an efficient transfer in a trust game, although similar amounts are transferred in dictator games, regardless of the ethnic heritage of the recipient. This suggests that while people of Ashkenazic heritage are of higher status, more highly educated, and, in the experiment, more trusted, they are not more worthy when it comes to income distributions in an experiment.

Discussion

Amidst all the evidence that status affects economic decision-making, the puzzle is: why? Discrimination research often focuses on two possible explanations, namely, animus and statistical discrimination. While animus seems a consistent explanation for preferring individuals endowed with high status, it is less compelling when considering status that can be obtained. Statistical discrimination is consistent with Bernheim's (1994) model of social norms; however, Goldin and Rouse's (2000) result that women are hired more frequently than men in a setting where gender is unknown, and less frequently when it is known, suggests that status may be an inaccurate proxy for a person's hidden attribute of interest.

Status may serve a role in helping to prevent coordination failures in games with multiple plausible outcomes. If all individuals use the strategy 'defer to high-status individuals' in a game such as the 'battle of the sexes', status can serve to largely eliminate dominated outcomes. Status may also provide a means of avoiding low payoff equilibrium outcomes in Prisoner's Dilemma type games. Kumru and Vesterlund's (2005) result that low-status individuals follow the behaviour of high-status leaders in public goods games suggests a positive role for status in guiding behaviour.

Evolutionary arguments provide the most plausible explanation for the existence of status-related behaviour. Tastes that increase the likelihood that individuals survive and produce offspring are most likely to persist in a population. Among social animals the highest-ranking individual is generally the largest with size being linked to sexual maturity, strength and health. These high-status individuals generally enjoy preferential access to food and mates and produce a disproportionate number of society's offspring. Status hierarchies can also enhance the transmission of useful information if the prestige of high-status, successful individuals attracts close observation by others (Henrich and Gil-White, 2001). To the extent that status characteristics are passed along to offspring, this strategy produces a stronger 'next generation' than one without status hierarchies. This argument suggests that current tastes may result from what was a successful survival strategy for our distant ancestors. On the other hand, in most parts of the world humans have access to food and medical care that reduce the importance of status preference as a survival strategy. Given the inefficiencies that status-seeking produces, therefore, these tastes may no longer be optimal.

SHERYL BALL

See also **behavioural game theory; experimental economics; psychological games; women's work and wages.**

Bibliography

Altonji, J.G. and Blank, R.M. 1999. Race and gender in the labor market. In *Handbook of Labor Economics*, vol. 3C, ed. O. Ashenfelter and D. Card. Oxford: Elsevier.

Arrow, K. 1972. *Models of Job Discrimination. Racial Discrimination in Economic Life*. Lexington, MA: Lexington Books.

Ayers, I. and Siegelman, P. 1995. Race and gender discrimination in bargaining for a new car. *American Economic Review* 85, 304–21.

Ball, S., Eckel, C., Grossman, P. and Zame, W. 2001. Status in markets. *Quarterly Journal of Economics* 116, 161–88.

Becker, G. 1957. *The Economics of Discrimination*. Chicago: University of Chicago Press.

Bernheim, B.D. 1994. A theory of conformity. *Journal of Political Economy* 102, 841.

Bertrand, M. and Mullainathan, S. 2004. Are Emily and Greg more employable than Lakisha and Jamal? A field experiment on labor market discrimination. *American Economic Review* 94, 991–1013.

Bloch, F., Rao, V. and Desai, S. 2004. Wedding celebrations as conspicuous consumption: signaling social status in rural India. *Journal of Human Resources* 39, 675–95.

Bolton, G.E. 1991. A comparative model of bargaining: theory and evidence. *American Economic Review* 81, 1096–136.

Darity, W.A., Jr. and Mason, P.L. 1998. Evidence on discrimination in employment: codes of color, codes of gender. *Journal of Economic Perspectives* 12(2), 63–90.

Eckel, C. and Grossman, P. 2001. Chivalry and solidarity in ultimatum games. *Economic Inquiry* 39, 171–88.

Eckel, C. and Wilson, R.K. 2004. Initiating trust: the conditional effects of skin color on trust among strangers. Working paper, Rice University.

Fershtman, C. and Gneezy, U. 2001. Trust and discrimination in a segmented society: an experimental approach. *Quarterly Journal of Economics* 116, 351–77.

Fershtman, C., Murphy, K.M. and Weiss, Y. 1996. Social status, education, and growth. *Journal of Political Economy* 104, 108–32.

Fershtman, C. and Weiss, Y. 1993. Social status, culture and economic performance. *Economic Journal* 103, 946–59.

Fryer, R.G. and Levitt, S.D. 2004. The causes and consequences of distinctively black names. *Quarterly Journal of Economics* 119, 767–805.

Getzner, M. 2000. Hypothetical and real economic commitments, and social status, in valuing a species protection programme. *Journal of Environmental Planning and Management* 43, 541–59.

Goldin, C. and Rouse, C. 2000. Orchestrating impartiality: the impact of 'blind' auditions on female musicians. *American Economic Review* 90, 715–41.

Hamermesh, D.S and Biddle, J.E. 1994. Beauty and the labor market. *American Economic Review* 84, 1174–94.

Hamermesh, D.S. and Parker, A. 2005. Beauty in the classroom: instructors' pulchritude and putative pedagogical productivity. *Economics of Education Review* 24, 369–76.

Henrich, J. and Gil-White, F. 2001. The evolution of prestige: freely conferred deference as a mechanism for enhancing cultural transmission. *Evolution and Human Behavior* 22, 165–96.

Kahn, A., Hottes, J. and Davis, W.L. 1971. Cooperation and optimal responding in the prisoner's dilemma game: effects of sex and physical attractiveness. *Journal of Personality and Social Psychology* 17, 267–79.

Kumru, C. and Vesterlund, L. 2005. The effects of status on voluntary contribution. Working paper, University of Pittsburgh.

List, J.A. 2004. The nature and extent of discrimination in the marketplace: evidence from the field. *Quarterly Journal of Economics* 119, 49–89.

Mobius, M.M. and Rosenblat, T.S. 2006. Why beauty matters. *American Economic Review* 96, 222–35.

Mulford, M., Orbell, J., Shatto, C. and Stockard, J. 1998. Physical attractiveness, opportunity and success in everyday exchange. *American Journal of Sociology* 103, 1565–93.

Robson, A.J. 1992. Status, the distribution of wealth, private and social attitudes to risk. *Econometrica* 60, 837–57.

Solnick, S. 2001. Gender differences in the ultimatum game. *Economic Inquiry* 39, 189–200.

Solnick, S.J. and Schweitzer, M.E. 1999. The influence of physical attractiveness and gender on ultimatum game decisions. *Organizational Behavior and Human Decision Processes* 79, 199–215.

Smith, A. 1759. *The Theory of Moral Sentiments*, ed. D.D. Raphael and A.L. MacFie. Indianapolis, IN: Liberty Press/Liberty Classics, 1982.

Thye, S.R. 2000. A status value theory of power in exchange relations. *American Sociological Review* 65, 407–32.

Van Huyck, J. and Battalio, R. 2002. Prudence, justice, benevolence, and sex: evidence from similar bargaining games. *Journal of Economic Theory* 104, 227–46.

Veblen, T. 1926. *A Theory of the Leisure Class: An Economic Study of Institutions.* New York, NY: Vanguard Press.

Wilson, R.K. and Eckel, C.C. 2006. Judging a book by its cover: beauty and expectations in the trust game. *Political Research Quarterly* 59, 189–202.

Steindl, Josef (1912–1993)

Steindl was born in Vienna on 14 April 1912. He studied economics in Vienna and received his Ph.D. working under Richard Strigl. He worked in the Austrian Institute for Economic Research (AIER) from 1935 to 1938, the year of his emigration to England. He was a lecturer at Balliol College, Oxford, from 1938 to 1941, and then a research worker at the Oxford Institute of Statistics. He worked there with Michal Kalecki, who left a lasting mark on his theoretical work. He returned to Austria in 1950. He was barred from teaching at the University of Vienna for ideological reasons and resumed his job at AIER, where he worked until his retirement in 1978. In 1970, however, the University of Vienna bestowed upon him a honorary professorship. He was visiting professor at Stanford University in 1974/5.

Steindl dealt with the economic problems of the size of firms (1945) and of the distribution of firms according to size (1965a). He explained the pattern of size distribution of firms by means of random processes (birth and death processes). Other fields of interest were education (1967) and technology (for example, 1980).

The research which may prove longest lasting is his work on the development and the present phase of capitalist economies. His main work (1952) deals with the tendency to stagnation of the mature capitalist economy. His point of departure was that oligopoly leads to increased profit margins and consequently to a fall in effective demand. The ensuing decline in the degree of capacity utilization causes *ceteris paribus* a lower level of investment and a decline in the rate of growth in mature capitalist economies. The slowing down of capital growth reduces further the utilization of capacity and leads to a cumulative process of declining growth. Steindl thus treats the utilization parameter differently from Kalecki, for whom it is a purely passive variable. Another difference consists in the explanation of the growth trend of the capitalist economy without having recourse to exogenous factors like innovations.

Maturity and Stagnation in American Capitalism was largely ignored during a period of high employment and intensive growth. Only when the old weaknesses of unemployment and stagnation reappeared did the book arouse wider interest and prove its lasting significance. The evolution of his ideas is shown in the introduction to (1976) and in the penetrating analysis both of present economic trends (for example, 1979; 1985a; 1985b) and of the present state of economics (1984).

K. LASKI

Selected works

1945. *Small and Big Business: Economic Problems of the Size of Firms*. Oxford: Basil Blackwell.

1952. *Maturity and Stagnation in American Capitalism*. Oxford: Basil Blackwell. Republished, New York: Monthly Review Press, 1976.

1965a. *Random Processes and the Growth of Firms: A Study of the Pareto Law*. London: Griffin.

1965b. The role of manpower requirement in the educational planning experience of the Austrian EIP. In *Manpower Forecasting in Educational Planning*. Paris: OECD.

1979. Stagnation theory and stagnation policy. *Cambridge Journal of Economics* 3, 1–14.

1980. Technical progress and evolution. In *Research, Development and Technological Innovation*, ed. D. Sahal. Lexington, MA: Lexington Books.

1981. Ideas and concepts of long run growth. *Banca Nazionale del Lavoro Quarterly Review* 136(March), 35–48.

1982. The role of household savings in the modern economy. *Banca Nazionale del Lavoro Quarterly Review* 140(March), 69–88.

1984. Reflections on the present state of economics. *Banca Nazionale del Lavoro Quarterly Review* 148(March), 3–14.

1985a. Distribution and accumulation. *Political Economy: Studies in the Surplus Approach* 1(1).

1985b. Structural problems in the present crisis. *Banca Nazionale del Lavoro Quarterly Review* 154(September), 223–32.

Steuart, Sir James (1713–1780)

Biographical

The Steuart family owned two estates, Goodtrees, which is near Edinburgh, and Coltness on the outskirts of Glasgow. Goodtrees was the seat of Sir James Steuart, the second baronet, Solicitor-General and a member of the Union Parliament. Sir James married Anne Dalrymple, the eldest daughter of the Lord President of the Court of Session, by whom he had five children of whom James was the only son. James was born on 10 October 1713, presumably at Goodtrees.

James attended the Parish School at North Berwick, proceeding in due course to Edinburgh University where he studied, inter alia, constitutional and Scots law. Thereafter James made the expected progression and passed the Bar examinations in 1735 at the age of 22.

Steuart became the third baronet in 1717 on the death of his father but did not spend time either enjoying his new status or his standing as an advocate. Rather he embarked upon a Foreign Tour (1735–40). It was during this period that he lost his remarkable mother, an event which may have affected his future fate.

Steuart travelled with a fellow advocate, Carnegie of Boysack, and the pair initially went to Holland where they pursued further study. But in due course they travelled through France, settling for a period in Avignon. Avignon was at this time a Papal Territory and a haven for those Scots who had been 'out' in the Jacobite Rebellion of 1715. It was here that Steuart met the Duke of Ormond, a fervent supporter of the Cause who in turn directed Steuart's steps to Madrid where he met the Earl Marischall, another of the architects of the ill-fated '15. It may have been the influence of these two men that directed Steuart's steps to Rome in the late 1730s. Steuart seems to have been captivated by the Old Pretender and his staff (there is very little mention of Prince Charles) and in a way which was to have a profound influence upon his future.

Steuart met Lord Elcho in Lyons, en route home, and it may be that he persuaded the future commander of the Prince's Life Guards to join the movement. In any event Steuart was active on behalf of the Jacobites after his return to Scotland in 1740 and it was because of this that he was sent to France as ambassador in 1745, following the success at Prestonpans. But after the battle of Culloden in April 1746 Steuart entered a long period of exile and to begin with maintained an active link with the Party. But the early 1750s saw a withdrawal from the Jacobite interest and Steuart, together with Lord Elcho, eventually settled in Angoulême where they lived in some style, with the support of Elcho's mother (Wemyss, 2003).

Steuart was bored, however, and it was probably significant that the exiled Parlement of Paris came to the locality in 1753. It was here that Steuart met Mercier de la Rivière, the latter-day Physiocrat so much admired by Adam Smith, with whom Steuart formed a long and lasting friendship. When the Parlement returned to Paris in 1754 Steuart followed where he was entertained by Mercier de Rivière and probably introduced to Montesquieu and Mirabeau.

The scientific opportunities were considerable, but in fact Steuart left Paris and France in 1755 to avoid compromising his position further in the event of hostilities with Britain. He left Paris in short before the dissemination of the *Tableau économique*. The first two books of the *Principles* were completed in the isolation of Tübingen (Germany) by August 1759. His work on the *Policy of Grain* and the *Dissertation on the German Coin* belong to this period.

The Steuart family left Tübingen in 1761 following Lord Barrington's successful attempt to have Steuart's son, also James, appointed as a coronet in the British Dragoons. They travelled west to Rotterdam and Antwerp before settling temporarily in the Spa. It was here that Steuart was arrested by the French authorities and subsequently imprisoned. The arrest was thought to be due to Steuart's close knowledge of the weakness of the French economy, although another gloss has been put upon the event by Paul Chamley. Chamley indicated that Steuart had been caught in the possession of plans for the invasion of Santo Domingo (Haiti): plans which had been prepared by Mercier de la Rivière, 'who had a personal pecuniary interest in an English invasion of the island and may also have realised that it would do his friend Steuart no harm in the eyes of London if he were arrested by the French' (Chamley, 1965, pp. 44–6; Skinner, in Steuart, 1998, vol. 1, pp. xlv–xlvi).

Steuart returned to England in 1763, the year of peace with France, under the mistaken belief that the British government had acted upon his behalf. Steuart did not in fact receive a pardon for past misdemeanours until 1771. But in the meantime he enjoyed the protection of Lord Barrington, sometime Secretary at War, whom he had first met on the Foreign Tour.

After the frenetic period when he brought the *Principles* (1767) to completion, he pursued, or rather continued to pursue, work of an academic nature as the *Works* amply confirm. He also found time to write a series of letters on the American War (Raynor and Skinner, 1994) which are chiefly interesting for his suspicion of military victory and his advocacy of free trade with the Colonies whatever the outcome. These letters were written between 1775 and 1778.

Steuart was apparently a good neighbour, actively interested in the economic affairs of the locality (Lanarkshire) and in the politics of the region. Steuart died on 26 November 1780. He was interested in the family vault at Cambusnethan (Lanarkshire) which is now sadly in ruins. Coltness has been demolished apart from some remnants of the original stable block.

Steuart married Lady Frances Wemyss (Lord Elcho's sister) on 25 October 1743 and their son, also James, was born the following year. Sir James Steuart-Denham had a distinguished military career. He served mainly in Ireland and on his death in 1839 was the Senior General in the British army, notable for his reform of cavalry tactics. He married Alicia Blacker of Carrick but there were no children. (The name 'Denham' was added in 1773 following the transfer of the estate of Westshields to the third baronet on the death of Archibald Denham) (Skinner, 2006, p. 73).

The *Principles*: methodology

It should be noted that one of the most important features of Sir James Steuart's career was his extensive knowledge of the Continent. The Foreign Tour (1735–40) and exile as a result of his association with the Jacobites meant that by the end of the Seven Years' War Sir James had spent almost half of his life in Europe. In this time he mastered four languages (French, German, Spanish and Italian), a fact which may help to explain Joseph Schumpeter's judgement that 'there is something un-English (which is not merely Scottish) about his views and his mode of presentation' (1954, p. 176n).

In the course of his travels Steuart visited a remarkable number of places which included Antwerp, Avignon, Brussels, Cadiz, Frankfurt, Leyden, Liège, Madrid, Paris, Rome, Rotterdam, Tübingen, Utrecht, Venice and Verona. He seems, moreover, consistently to have pursued experiences which were out of the common way. For example, when he settled at Angoulême he took advantage of his situation to visit Lyons and the surrounding country. During his residence in Tubingen, he undertook a tour of the schools in the Duchy of Württemburg. Earlier he had spent no less than 15 months in Spain where he was much struck by the irrigation schemes in Valencia, Mercia and Granada, the mosque in Cordoba and the painful consequences of the famine in Andalusia in the spring of 1737. In fact very little seems to have been lost on him and it is remarkable how often specific impressions found their way into the main body of the *Principles*. In his major book Steuart noted the economic consequences of the Seven Years' War in Germany, the state of agriculture in Picardy, the arrangement of the kitchen gardens round Padua, and the problem of depopulation in the cities of the Austrian Netherlands.

Steuart drew attention to the difficulties under which he laboured in the preface to the *Principles* precisely because he thought they would be of interest to the reader. He pointed out that the 'composition' was the 'successive labour of many years spent in travelling' (1966, p. 304) during which he had examined different countries 'constantly, with an eye to my own subject':

> I have attempted to draw information from every one with whom I have been acquainted: this however I found to be very difficult until I had attained to some previous knowledge of my subject. Such difficulties confirmed to me the justness of Lord Bacon's remark, that he who can draw information by forming proper questions, is already possessed of half the science. (1966, pp. 5–6)

Steuart wrote very much in the style of a man finding his way through a new field. This, added to the fact that nearly eight years separate the first and last books, presented obvious problems; problems of which Steuart was always conscious but which he viewed with very mixed feelings:

> Had I been master of my subject on setting out, the arrangement of the whole would have been rendered more concise; but had this been the case, I should never had been able to go through the painful deduction which forms the whole train of my reasoning and upon which … the conviction it carries along with it in a great measure depends. (1966, p. 7)

Steuart sought to establish a system of thought whose content met the requirements of Newtonian methodology. The leading feature of Steuart's method is objective empiricism. He was thus entirely in accord with his friend Hume (Skinner, 2005) but like Hume, he recognized that the mere collection of facts was not of itself sufficient. The first step on the route to knowledge is the collection and description of facts; the second, the statement of certain 'principles' reached through a process of induction.

Steuart also recognized that the scientist can only advance by concerning 'himself' with *cause* and *consequence*, that is, by thinking deductively. He solved the problem of how to combine the two techniques by using induction to establish his basic hypotheses, or 'principles', and deduction for what Hasbach described as the 'clarification of phenomena' (1891).

Steuart was quite clear as to the techniques of reasoning to be employed. The rules were simple, if difficult to obey: observation, induction, deduction, verification. There remained the question of the technique to be followed in building up a body of knowledge and here Steuart's answer was equally clear.

The scientist should begin with the simple (and thus apparently abstract) case and gradually take account of more and more complex (and thus 'realistic') cases. The first objective must be clarity and Steuart thus recognized that the attainment of the second, relevance, can only come about through the use of the abstraction in the early stages of study. He argued that in building up a body of knowledge:

> Every branch of it must, in setting out, be treated with simplicity and all combinations not absolutely necessary must be banished from the theory. (1966, p. 227)

But, since the object is relevance and since the 'more extensive any theory be made, the more it will be useful', it follows that as we proceed 'combinations will crowd in and every one of these must be attended to' (1966, p. 227). Steuart always employed this technique in dealing with a body of knowledge; that is, he gradually builds up his argument in a series of steps which progressively increase in complexity.

At the same time, Steuart recognized that theoretical edifices constructed in this manner present the economist with particular difficulties which arise from the nature of the subject matter itself. In Steuart's view, the economist or social scientist can only show 'how consequences *may* follow from one another; to foretell what *must* follow is exceedingly difficult if not impossible' (1966, p. 365). While we can and must establish general principles these do not provide rules of behaviour which must always hold good. Steuart thus concluded (somewhat ironically, in the context of a critique of Hume's quantity theory) that:

> I think I have discovered that in this, as in every other part of political economy, there is hardly such a thing as a general rule to be laid down' (1966, p. 339)

Given the need for a systematic statement of particular principles, established in accordance with the discipline of an appropriate methodology, there remained the problem of establishing a useful 'method' in respect of the *organization* of the discourse as a whole.

> The thing to be done is to fall upon a distinct method … by contriving a chain of ideas, which may be directed towards every part of the plan, and which at the same time, may be made to arise methodically from one another. (1966, p. 28)

Here again, Steuart followed Hume's lead.

The 'plan' is contained in the first two books and is based upon a theory of economic development. Steuart's dominant theme was to be change and growth, and it is this which gives his work cohesion.

The historical perspective

Steuart opened his account with 'society in the cradle' before going on to trace the origins of, and the process of transition between, the various stages of the progress of man.

In this context, Steuart made use of a theory of stages, now recognized as a piece of apparatus which was central to the work of the Scottish Historical School. He cites, for example, the Tartars and Indians as relatively primitive socio-economic types of organization (1966, p. 56) while concentrating primarily on the third and fourth stages – the stages of agriculture and commerce. In the former case, Steuart observed that those who lacked the means of subsistence could acquire it only through becoming dependent on those who owned it; in the latter, he noted that the situation was radically different in that all goods and services command a price. He concluded, in passages of quite striking clarity:

> I deduce the origin of the great subordination under the feudal government, from the necessary dependence of the lower classes for their subsistence. They consumed the produce of the land, as the price of their subordination, not as the reward of their industry in making it produce.

He continued:

> I deduce modern liberty from the independence of the same classes, by the introduction of industry, and circulation of an adequate equivalent for every service. (1966, pp. 208–9).

Steuart also observed that 'an opulent, bold and spirited people, having the fund of the Prince's wealth in their own hands, have it also in their own power, when it becomes strongly their inclination, to shake off his authority' (1966, p. 216).

The alteration in the distribution of power which was reflected in the changing balance between proprietor and merchant led Steuart to the conclusion that 'industry must give wealth and wealth *will* give power' (1966, p. 213). As an earnest of this position, he drew attention (significantly in his Notes on Hume's *History*) to the reduced position of the Crown at the end of the reign of Elizabeth: a revolution which appears 'quite natural when we set before us the causes which occasioned it. Wealth must give power; and industry, in a country of luxury, will throw it into the hands of the commons' (1966, p. 213n).

It was perhaps for this reason that Steuart's French translator, Senovert (1789), advised his readers that of the advantages to be gained from a reading of the *Principles*, 'Le premier sera de convaincre, sans doute, que la révolution qui s'opère sous nos yeux était dans l'ordre des choses nécessaires' (1966, p. 24n). Senovert, in short, was convinced of the inevitability of the Revolution and believed that the *Principles* confirmed the point.

Economic analysis

To economists he has always been Sir James Steuart, because that is how he appears on the title page of his 1767 book. This is subtitled, *An Essay on the Science of Domestic Policy in Free Nations, in which are particularly considered, Population, Agriculture, Trade Industry, Money, Coin, Interest, Circulation, Banks, Exchange, Public Credit and Taxes*. It offers a detailed, comprehensive and often original account of the application of economic argument to this enormous range of questions.

The population theory with which the book opens anticipates much that Malthus went on to say, and Marx even suggested in the first volume of *Capital* that, 'admirers of Malthus do not even know that the first edition of the latter's work on population contains, except in the purely declamatory part, very little but excerpts from Steuart' (Marx, 1867, p. 333).

His analysis of the balance of payments has also been much admired. He went considerably further than Hume by incorporating a detailed analysis of the capital account, and this led him to the conclusion (among several where he differs from Hume) that a country with a persistent capital account deficit will be unable to find an equilibrium price level at which specie flows cease.

Steuart's travels on the Continent during his 18 years of exile from 1745 to 1763 acquainted him with monetary developments in Paris and Amsterdam, and this enriched his theoretical and empirical chapters on money and banking. But it is his analysis of economic policy which has attracted much modern attention. The contrast between his analysis and Smith's in *The Wealth of Nations* published just nine years later is especially marked. Steuart's years of exile had given him a detailed knowledge of economic and financial policy on the Continent, and in particular in France, Germany and Holland, and he advocated a degree of state intervention into every aspect of economic life, which contrasted sharply with the principles that Smith enunciated. Skinner (1981) has suggested that it was precisely Steuart's long years of residence on the Continent that

led him to evolve a 'system' which was so much more dirigiste than that of his great Scottish contemporary.

In his book, Steuart offers extensive and detailed advice to an idealized statesman, who is assumed to possess unlimited knowledge and whose 'inclinations are always to be virtuous and benevolent' (1966, p. 333). Steuart believed that markets do not always clear, and this was especially the case with the labour market, where there was always liable to be an imbalance between 'demand' and the supply of 'work'. Manufacturers, merchants and workers sought to 'consolidate' any high living standards they temporarily achieved into permanently higher incomes, and they often achieved this by restricting competition. Once prices and wages were consolidated at high levels, employment necessarily suffered as soon as foreign manufacturers began to produce more cheaply. With these assumptions about the behaviour of workers and entrepreneurs, and the impotence or non-existence of corrective market forces, there was an extensive range of policies through which state intervention could be expected to increase wealth, welfare and employment.

As soon as domestic production became overpriced, imports would undermine domestic employment and the creation of wealth, and Steuart therefore proposed that 'a branch of trade should be cut off' where the Statesman shall find,

> upon examining the whole chain of consequences, … the nation's wealth not at all increased, nor her trade encourages, in proportion to the damage at first incurred by the importation. (1966, p. 293)

In addition to protecting industry against imports, Steuart advocated export subsidies, because he saw the alternative to, for instance, subsidizing exports of fish by £250,000 so that what cost £1,000,000 could be sold overseas for £750,000, as the total loss of £750,000 of potential domestic output. Without the subsidy,

> those employed in the fishery will starve; … the fish taken will either remain upon hand, or be sold by the proprietors at a great loss; they will be undone, and the nation for the future will lose the acquisition of £750,000 a year. (1966, pp. 256–7)

Steuart was also concerned that as industry and population grew, the price of subsistence would rise as the population forced farming onto inferior land, where 'the progress of agriculture demands an additional expence'. In order to 'preserve the intrinsic value of goods at the same standard as formerly; [the Statesman] must assist agriculture with his purse, in order that exportation may not be discouraged' (1966, p. 200).

As well as seeking to avert the influence of agricultural diminishing returns by subsidizing agriculture in order to keep export costs down, Steuart actually proposed the setting up of a 'policy of grain' in 'the Common Markets of England', where the government would buy up all the grain that farmers were prepared to produce at 'the minimum price expedient for the farmers', and sell all that could be marketed at 'the maximum price expedient for the wage-earners', and store any excess in state granaries. Steuart actually drafted this anticipation of the European Economic Community's agricultural policies of the 1970s and the 1980s in 1759 while he was in exile in Tübingen.

Steuart also anticipated post-Second World War industrial policies, for he argued that a Statesman should not hesitate to intervene directly in the finance and management of any new undertaking where he saw economic potential, and should

> inquire into the capacity of those at the head of it; order their projects to be laid before him; and when he finds them reasonable and well planned, he ought to take unforeseen losses upon himself … the more care and expence he is at in setting the undertaking on foot, the more he has a right to direct the prosecution of it towards the general good. (1966, p. 391)

Steuart was a powerful advocate of public works to create employment whenever there was an excess supply of labour. The government should always finance the employment of 'the deserving and the poor', and they should be employed to extend a nation's social and economic infrastructure rather than for unproductive purposes:

> If a thousand pounds are bestowed upon making a firework, a number of people are thereby employed, and gain a temporary livelihood. If the same sum is bestowed for making a canal for watering the fields of a province, a like number of people may reap the same benefit, and hitherto accounts stand even; but the firework played off, what remains, but the smoke and stink of the powder? Whereas the consequence of the canal is a perpetual fertility to a formerly barren soil. (1767, vol. 1, p. 519)

All these interventionist policies needed to be financed, and Steuart actually welcomed the high taxation this would entail. He argued that taxes redistribute income and wealth and create employment, for they

> advance the public good, by drawing from the rich, a fund sufficient to employ both the *deserving* and the *poor* in the service of the state. (1767, vol. 1, pp. 512–13)

They also increase the power and prestige of the Statesman, for

> By taxes the Statesman is enriched, and by means of this wealth, he is enabled to keep his subjects in awe, and to preserve his dignity and consideration. (1966, p. 304)

Economists who believe in the efficacy of market forces have often been concerned that high taxation may have adverse supply side effects, but Steuart actually believed that taxation would often have *favourable* supply side effects. High taxes

> may discourage idleness; and idleness will not be totally rooted out, until people be forced, in one way or other, to give up superfluity and days of recreation … When the hands employed are not diligent, the best expedient is to raise the price of their subsistence by taxing it. (1966, pp. 691–5)

Steuart was aware that this analysis of the social and economical benefits from high taxation would not be popular with his contemporaries, and that 'the politics of my closet is very different from those of the century in which I live', but he comforted himself with the thought that 'reason is reason', and that in another century these startling opinions would be acknowledge as correct (1767, vol. 1, p. 514).

Steuart's industrial policies amount (as Eltis, 1986, has suggested) to the setting up of a corporate state with a social contract between producers who are protected against foreign competition and whose employment is guaranteed, and the state to whom they pay high taxes. Some of these are then returned to inefficient producers, while the rest furthers the state's social and political objectives.

In addition to welcoming high taxation as a tool for the finance of industrial policies, Steuart was an advocate of state banks which would issue paper money. By making money less scarce, he believed that they would reduce interest rates and so benefit industry and commerce. He argued that John Law's Mississippi Scheme could have been successful in France with only a few minor modifications in the manner it was set up and administered, and that this could have established the long-term rate of interest at two per cent in France.

The many kinds of government expenditure Steuart so strongly advocated could also be financed through borrowing, and here again Steuart was ahead of his time. He believed that in the limit, whatever a government could raise from taxation could be devoted to the payment of interest on public debt so that at a five per cent rate of interest, governments could borrow 20 times their tax revenues:

> If no check be put on the augmentation of public debts, if they be allowed constantly to accumulate, and if the spirit of a nation can patiently submit to the natural consequences of such a plan, it must end in this, that all property, that is income, will be swallowed up in taxes; and these will be transferred to the creditors.

But even in that state of affairs where all property income is paid as interest to those who have lent to the government does not represent the limit of the state's power to borrow. It can go on to tax the recipients of debt interest and so provide the wherewithal to finance still further borrowing, for these taxes 'may be mortgaged again to a new set of men, who will retain the denomination of creditors' (1767, vol. 2, pp. 633–4). Some may doubt that governments can at the same time continue to borrow, and defraud those from whom they borrowed in the past by taxing away their interest so that this provides the finance for still further borrowing. Won't there be a refusal to go on lending to such governments? No, opines Steuart, because

> The prospect of a second revolution of the same kind with the first would be very distant; and in matters of credit, which are constantly exposed to risk, such events being beyond the reach of calculation, are never taken into any man's account who has money to lend. (1966, p. 647)

Hence Steuart was perceptive enough to appreciate that sovereigns (and sovereign governments) can continually defraud their creditors, while new lenders will still queue up to be defrauded because the prospect of this will be so distant and problematical that it has a negligible influence on the immediate willingness to lend.

Steuart's book was well received at first, but Smith, who believed that economies would make full use of their labour and capital in the complete absence of government-inspired employment policies, and at the same time wholly distrusted the omniscience and benevolence of governments, greatly weakened Steuart's reputation as a serious economist by totally ignoring the existence of his book in *The Wealth of Nations*. Four years before its publication, he wrote that 'Without once mentioning [Steuart's book], I flatter myself, that every false principle in it, will meet with a clear and distinct confutation in mine' (1977, p. 164).

In the 19th century Marx gave Steuart his due, and there are 13 references to him in the first volume of *Capital*. Several 19th-century German economists have compared Steuart's historical and institutional approach to political economy favourably with Smith's deductive methodology, but most accolades to the richness and originality of Steuart's contribution only emerged after the Keynesian revolution.

His monetary and employment theory have been much praised, most comprehensively by Vickers (1959), though Hutchison (1978) and Schumpeter (1954) have also recognized his Keynesian anticipations. Steuart's monetary theory has much more in common with Keynes than the mere proposition that sufficient monetary expansion will reduce interest rates to two per cent. In Steuart's argument, money expenditure is not closely linked to the money supply, for idle balances will often be freely held, and the price level depends upon

> demand and competition … Let the specie of a country … be augmented or diminished, in ever so great a proportion, commodities will still rise and fall according to the principles of demand and competition … Let the quantity of coin be ever so much increased, it is

the desire of spending it alone which will raise prices. (1966, pp. 344–5)

But Steuart's monetary and employment theory describe only one element of his thought which has anticipated modern developments. S.R. Sen, the distinguished Indian economic planner who published an important book on Steuart in 1957, commended him as 'the first Economic Adviser to the Government of India', praised his case for detailed intervention into every aspect of economic life and suggested that 'it would not be any great exaggeration to say that A.P. Lerner's chapter on functional finance seems almost a paraphrase of Steuart' (Sen, 1957, p. 122). Twenty years later, Akhtar (1979), of the New York Federal Reserve Bank restated Steuart in 30 equations, and compared his growth theory favourably with Smith's.

The classical counter-revolution of the 1980s has, of course, challenged the case for detailed state intervention which became so fashionable after the Keynesian revolution, and Steuart's dirigisme has been criticized by Anderson and Tollison (1984). It will be evident that there has been a more extensive response to the interventionist political economy of Sir James Steuart in the 20th century than there was in his own time.

Reception

Contemporary reaction was mixed. Hume is said to have been critical of the 'form and style' of the book while James Boswell considered the work to be 'irregular and fanciful' (Skinner, 1966, p. xlvi). Hugh Blair wrote to Hume that 'Sir James' Book is the most ponderous piece of lumber that I have ever looked into' (NLSms. 23153). One contemporary review was cautious, noting that:

> We have no idea of a statesman having any connection with the affair, and we believe that the superiority which England has at present over all the world, in point of commerce, is owing to her excluding statesman from the executive part of all commercial concerns. (*Critical Review* 23 (1767), p. 412)

The *Monthly Review* (36, p. 464) went so far as to accuse Steuart of imbibing prejudices abroad 'by no means consistent with the present state of England and the genius of Englishmen'. Steuart replied:

> Can it be supposed, that during an absence of near twenty years, I should in my studies have all the while been modelling my speculations of English notions? If, from this work I have any merit at all, it is by divesting myself of English notions, so far as to be able to expose in a fair light the sentiments and policy of foreign nations, relatively to their own situation. (1966, pp. 4–5. This passage occurs in the second edition of the *Principles*, published in the *Works*)

But if Steuart did not fare well among at least some of the key figures of the Enlightenment (and later!) the situation was rather different upon the Continent and elsewhere. During the 1780s the text was twice translated into German while there was a French version in 1789. Kobayashi (in Steuart, 1998) has suggested that Steuart's model of 'primitive accumulation' may help to explain the popularity of his work in contemporary Ireland and Germany. Keith Tribe (1988, p. 133) on the other hand, noted that 'until the final decade of the eighteenth century Sir James Steuart's *Inquiry* was better known and more frequently cited than Smith's *Wealth of Nations*'.

But perhaps the most intriguing link is with North America. The pirated Dublin edition of 1770 was circulated widely in the Colonies and attracted the attention of Alexander Hamilton who was naturally concerned about the economic prospects of the infant republic. Hamilton rejected Smith's 'fuzzy philosophy' in favour of a policy of protection as a means of counterbalancing the competitive advantages of the British economy in the years following the Peace of Paris (1783). This perspective seems to have been widely shared, and is essentially a variant of Steuart's stage of 'infant trade'.

ORIGINAL ARTICLE BY WALTER ELTIS, REVISED AND EXPANDED BY
ANDREW SKINNER

Selected works

1767. *An Inquiry into the Principles of Political Oeconomy; being an Essay on the Science of Domestic Policy in Free Nations*, 2 vols. London.

1805. *Works, Political, Metaphysical and Chronological*, 6 vols. London.

The *Works* includes a revised edition of the 1767 edition of the Principles.

1966. *Principles*, 2 vols, ed. A.S. Skinner. Edinburgh: Oliver & Boyd for the Scottish Economic Society.

1998. *Principles*, 4 vols (variorum), ed. A.S. Skinner with N. Kobayashi and H. Mizuta. London: Pickering and Chatto.

Biography

Chamley (1963; 1965); Skinner (in Steuart, 1966; 1998). See also the article on Steuart in the *Oxford Dictionary of National Biography* (2004).

Bibliography

Akhtar, M.A. 1978. Steuart on growth. *Scottish Journal of Political Economy* 26, 57–74.

Akhtar, A. 1979. An analytical outline of Sir James Steuart's macroeconomic model. *Oxford Economic Papers* 31, 283–302.

Anderson, G.M. and Tollison, R.B. 1984. Sir James Steuart as the apotheosis of mercantilism and his relation to Adam Smith. *Southern Economic Journal* 51, 456–68.

Chamley, P. 1963. *Economie Politique et Philosophie chez Steuart et Hegel*. Paris: Librairie Dalloz.

Chamley, P. 1965. *Documents Relatifs à Sir James Steuart*. Paris: Librairie Dalloz.

Coltness Collections, The. 1842. Edinburgh: Maitland Club.

Davie, E.G. 1967. Anglophone and Anglophile. *Scottish Journal of Political Economy* 14, 291–304.

Eagly, R.V. 1961. Sir James Steuart and the aspiration effect. *Economica* 28, 53–61.

Eltis, W. 1986. Sir James Steuart's corporate state. In *Ideas in Economics*, ed. R.D.C. Black. London: Macmillan.

Eltis, W. 1987. Steuart, Sir James. In *The New Palgrave: A Dictionary of Economics*, vol. 4, ed. J. Eatwell, M. Milgate and P. Newman. London: Macmillan.

Grossman, H. 1943. The evolutionist revolt against classical political economy. *Journal of Political Economy* 51, 506–22.

Hasbach, W. 1891. *Untersuchungen über Adam Smith*. Leipzig: Duncker & Humblot.

Hirschman, A.O. 1977. *The Passions and the Interests: Political Arguments for Capitalism before its Triumph*. Princeton: Princeton University Press.

Hont, I. 1983. The rich country–poor country debate in Scottish political economy. In *Wealth and Virtue: The Shaping of Political Economy in the Scottish Enlightenment*, ed. I. Hont and M. Ignatieff. Cambridge: Cambridge University Press.

Hutchison, T. 1978. *On Revolutions and Progress in Human Economic Knowledge*. Cambridge: Cambridge University Press.

Hutchison, T. 1988. *Before Adam Smith*. Oxford: Basil Blackwell.

Johnston, E.A.G. 1937. *Predecessors of Adam Smith*. New York: Kelley, 1960.

Jones, P., ed. 1988. *Philosophy and Science in the Age of Enlightenment*. Edinburgh: John Donald.

King, J.E. 1988. *Economic Exiles*. London: Macmillan.

Low, J.M. 1952. An eighteenth century controversy in the theory of economic progress. *Manchester School of Economic and Social Studies* 20, 311–20.

Marx, K. 1867. *Capital*. Moscow: Progress Publishers for Lawrence & Wishart, 1974.

Meek, R.L. 1967. The rehabilitation of Sir James Steuart. In *Economics and Ideology and other Essays*. London: Chapman and Hall.

Perelman, M. 1983. Classical political economy and primitive accumulation. *History of Political Economy* 15, 451–94.

Raynor, D. and Skinner, A.S. 1994. Sir James Steuart: nine letters on the American Conflict, 1775–1778. *William and Mary Quarterly* 51, 775–6.

Schumpeter, J.A. 1954. *History of Economic Analysis*. New York: Oxford University Press.

Sen, S.R. 1957. *The Economics of Sir James Steuart*. London: Bell.

Skinner, A.S. 1981. Sir James Steuart: author of a system. *Scottish Journal of Political Economy* 38, 20–42.

Skinner, A.S. 2005. David Hume and James Steuart. In *The Reception of David Hume in Europe*, ed. P. Jones. London: Thoemmes.

Skinner, A.S. 2006. Sir James Steuart, *Principles of Political Economy*. In *A History of Scottish Economic Thought*, ed. A and S. Dow. London: Routledge.

Tortajada, R. 1999. *The Economics of James Steuart*. London: Routledge.

Tribe, K.P. 1988. *Governing Economy: The Reformation of German Economic Discourse*. Cambridge: Cambridge University Press.

Vickers, D. 1959. *Studies in the Theory of Money 1690–1776*. Philadelphia: Chilton.

Vickers, D. 1979. Sir James Steuart. *Journal of Economic Literature* 8, 1190–5.

Wemyss, A. 2003. *Elcho of the '45*, ed. J.S. Gibson. Edinburgh: Saltire Society.

Yang, H.S. 1994. *The Political Economy of Trade and Growth: An Analytical Interpretation of Sir James Steuart's Inquiry*. Cheltenham: Edward Elgar.

Stewart, Dugald (1753–1828)

Stewart was the most important early commentator on Adam Smith's work. He was born in Edinburgh in 1753 and died there in 1828. He was the brilliant and well-connected son of an Edinburgh professor and was destined for an academic career from the earliest age. Educated at Edinburgh and Glasgow Universities, Stewart was taught by Adam Ferguson and Thomas Reid and became a close acquaintance of Adam Smith. He was appointed to the Edinburgh Chair of Moral Philosophy on Ferguson's retirement in 1785 and held it until 1810, when ill-health forced his retirement. A charismatic and influential teacher, his vast erudition and synthetic skill was shaped by an acute sensitivity to the ideological responsibilities of the pedagogue. He was a prolific writer whose contemporary reputation was built on the first volume of his *Elements of the Philosophy of the Human Mind* (vol. 1, 1792; vol. 2, 1815; vol. 3, 1826) and its companion text book *Outlines of Moral Philosophy* (1793). These works circulated widely in the universities of Britain, America and the continent in the early 19th century and did much to establish Scottish Common Sense Philosophy as the most influential vehicle of elite education in the age of the American, French and Industrial Revolutions. Stewart's collected works were published posthumously in 1854–60.

Stewart's *Account of the Life and Writings of Adam Smith LL.D* (1793) was frequently republished, often as an introduction to Smith's works. He discussed the *Wealth of Nations* in relation to the *Theory of Moral Sentiments* and both in relation to Smith's abortive plan for publishing a theory of jurisprudence. At Edinburgh he lectured on the principles of government and political economy in 1800–8 to an influential group of students who were to do much to form Whig and Tory opinion in the early 19th century. These lectures were intended for publication but the manuscript was accidentally destroyed and never rewritten. Their substance can, however, by inferred from a posthumous text which was compiled from his notes and published with his collected works.

Stewart was the first academic to detach the study of political economy from that of the theory of government and to treat each as a distinct branch of political science and it is in this methodological innovation rather than for any particular economic theory that his importance for political economy lies. His lectures were addressed to those 'who study Political Economy with a view to the improvement of the theory of legislation' (Stewart, vol. 9, p. 255). He defined political economy as the sum of 'all those speculations which have for their object the happiness and improvement of Political Society'. This, not 'the mistaken notions concerning Political Liberty which have been so widely disseminated in Europe by the writing of Mr Locke', was the only proper foundation on which a true science of government could be raised (Stewart, vol. 8, pp. 10, 23). In general, Stewart's lectures offered an intelligent, critical presentation of the arguments of the *Wealth of Nations* illuminated by occasional information about Smith's last thoughts, by a sustained and sympathetic reappraisal of the Physiocrats and by a persistent preoccupation with perfectibility, progress, and the gradual improvement of the British Constitution.

NICHOLAS PHILLIPSON

Selected works

1854–60. *The Collected Works of Dugald Stewart*, 10 vols, ed. Sir W. Hamilton. Edinburgh: Thomas Constable & Co.

Bibliography

Collini, S., Winch, D. and Burrow, J. 1893. *That Noble Science of Politics: A Study in Nineteenth-century Intellectual History*. Cambridge: Cambridge University Press.

Phillipson, N.T. 1983. Dugald Stewart and the pursuit of virtue in Scottish university education; Dugald Stewart and Scottish moral philosophy in the Enlightenment. In *Universities, Society and the Future*, ed. N. Phillipson. Edinburgh: Edinburgh University Press.

sticky wages and staggered wage setting

Nominal wages are regarded as 'sticky' if they fail to adjust to the level that would prevail in an equilibrium with costless wage adjustment and full information. Modern analysis of sticky wages, including their allocative effects and policy implications, owes an enormous debt to Keynes (1936). Keynes believed that nominal wages were likely to adjust less rapidly than prices in response to nominal shocks. He inferred that real wages would move counter-cyclically, so that a monetary contraction would push up the real wage and reduce employment. Keynes applied his theory to explain the collapse of employment during the Great Depression, and also argued that wage stickiness had major implications for the choice of a monetary regime. But while Keynes's work is justly celebrated, it is important to recognize that the belief that nominal wage rigidities play a central role in the monetary transmission channel was held by some pre-eminent classical economists of the early 19th century, such as Henry Thornton.

The staggered wage contracts model: background and development

In the wake of the Keynesian revolution, considerable interest developed in constructing quantitative models that incorporated the hallmarks of Keynes's framework, including sticky wages. A shortcoming of these early models was in their characterization of expectations as adaptive. By the mid-1970s, the first models incorporating sticky wages into a general equilibrium framework with rational expectations appeared in seminal work by Fisher (1977) and Gray (1976). These authors assumed that wages were set a fixed number of periods in advance, with the predetermined wage set so that the labour market was expected to clear at the 'maturity' of the contract. While a considerable innovation, the 'Fischer–Gray' contract formulation effectively constrained the real effects of monetary shocks to be no longer than the duration of the longest contract.

Taylor (1980) introduced staggered wage contracts as a mechanism for allowing monetary shocks to exert real effects lasting beyond the length of the longest contract, a feature he termed the 'contract multiplier'. Taylor's wage-contracting model was meant to be consistent with several empirical observations about wage-setting: (a) wages are typically set in nominal terms and remain unchanged for sustained periods; (b) wage-setting tends to be asynchronous across different groups of workers; and (c) workers appear to take the wages set by other workers into account when adjusting their own wage, as well as aggregate demand. (See Taylor, 1999, for a comprehensive survey of staggered contracting models.)

Specifically, Taylor divided workers into N cohorts, and assumed that each cohort was constrained to adjust its contract wage at fixed intervals once every N periods. To illustrate the key features accounting for a contract multiplier, it is helpful to consider the special case in which wage contracts last two periods. In this case, Taylor's model can be expressed as three equations:

$$x_T = (1/2)^*(w_T + E_T w_{T+1})$$
$$+ (1/2)^* g(y_T + E_T y_{T+1}) \tag{1}$$

$$w_T = (1/2)^*(x_T + x_{T-1}) \tag{2}$$

$$m_T = y_T + w_T \tag{3}$$

where variables are in logs, and E_T denotes the conditional expectation operator. The first equation determines the contract wage x_T, which is the fixed wage each member of the cohort currently readjusting its wage receives over the life of the contract. The contract wage is specified as depending on average economy-wide wages expected over the contract life, and on current and future output (with the parameter g determining the sensitivity). The second equation expresses the average economy-wide wage w_T as a simple average of the contract wages still in effect. Finally, the model includes a simple quantity theory relation between the exogenous money stock m_T and nominal demand (the price level is a constant markup over the average wage). If we substitute (2) into (1), it is evident that contract wages have both a forward- and a backward-looking component, with the latter playing a crucial role in allowing the model to generate persistence.

Solving Taylor's model for the contract wage yields:

$$x_T = h^* x_{T-1} + (1 - h)^* m_T$$
$$h = (1 - g^\wedge(1/2))^\wedge 2 / (1 - g) \tag{4}$$

where the money supply is assumed to follow a random walk. If g is sufficiently small, the composite parameter h is close to unity, implying that a monetary shock has a small initial effect on the contract wage (and on average wages or prices), and thus exerts a large and persistent effect on output; by contrast, h approaches zero as g approaches unity, consistent with monetary shocks exerting large immediate effects on wages and prices, but little effect on output. Importantly, since g is assumed to be a free parameter, Taylor's model can rationalize an arbitrarily high degree of output persistence even if contracts last only two periods.

Taylor's staggered contracts model represented a major innovation in so far as it seemed to provide an empirically realistic model of monetary transmission within a rational expectations framework. However, while the staggered contracts framework became widely utilized for generating persistent responses to monetary shocks, the assumption that nominal wage stickiness was the primary source of monetary non-neutrality proved less durable. Thus, when Calvo (1983) developed an alternative staggered contracts framework that departed from Taylor's by specifying contract durations to be random, he assumed that prices rather than wages were sticky. This shift towards specifying sticky prices as the source of monetary non-neutrality was motivated by empirical evidence that appeared inconsistent with counter-cyclical real wages, and persisted until the late 1990s.

Some critiques, and the model's evolution

The staggered contracts model of Taylor and Calvo has evolved in response to several critiques. First, Fuhrer and Moore (1995) criticized its inability to account for inflation persistence or for the output costs of disinflation. In particular, while one-time changes in money could have persistent real effects, these authors showed that permanent changes in money growth had fleeting effects on inflation and output. Two alternative approaches have emerged in response to this critique. One route has effectively embedded additional persistence into the contracting structure by assuming that some agents have adaptive expectations (as in Roberts, 1997) or that workers not receiving a signal to change their wage (or price) follow a mechanical indexation scheme (Christiano, Eichenbaum and Evans, 2005). An alternative approach adopted by Ball (1995) and Erceg and Levin (2003) retains rational expectations, but assumes that agents learn gradually about shocks. Either approach may account for inflation inertia and prolonged output losses due to disinflation; however, in the latter inflation persistence is not intrinsic as in the indexation schemes, but instead depends on features that determine the speed of learning (including policymaker credibility and transparency).

Chari, Kehoe and McGratten (2000) challenged the ability of Taylor-style contracts to generate a contract multiplier in a version with explicit micro-foundations. These authors argued that the key parameter 'g' in eq. (1) should not be regarded as a free parameter, since in their model it was determined by structural parameters characterizing tastes and technology; moreover, no reasonable calibration could account for a low enough value of 'g' to deliver a sizeable contract multiplier. This challenge spawned an expansive literature showing how it was possible to account for a sizeable contract multiplier in a more richly specified micro-founded model. While Chari, Kehoe and McGratten included only price rigidities, Erceg (1997) and Huang and Liu (2002) demonstrated that a *combination* of wage and price rigidities could help generate a substantial contract multiplier in a micro-founded setting in which workers acted as monopolistic competitors in the labour market and set wages in a staggered fashion; an extensive literature (including these authors) has shown that persistence may also be enhanced through various real rigidities.

Finally, Caplin and Spulber (1987) introduced state-dependent pricing into the staggered contracts framework, allowing agents to choose when to adjust their contracts rather than constraining them to adjust at exogenous intervals ('time-dependent' pricing). Their paper suggested that this innovation had pivotal implications for the monetary transmission mechanism: money could be neutral under some conditions. However, in subsequent analysis in a state-dependent framework with maximizing agents, Dotsey, King and Wolman (1999) found that the dynamic responses to a monetary policy shock were qualitatively similar to those of the standard Taylor model with time-dependent contracts.

In light of these critiques, the recent literature has tended to incorporate both wage and price rigidities within a micro-founded staggered contracts framework that allows for some form of inflation inertia. Christiano, Eichenbaum and Evans (2005) showed that such a model provides a good quantitative characterization of the responses of macro variables to a monetary policy shock. Importantly, the nearly acyclical empirical response of the real wage to a monetary policy shock has also helped renew support for incorporating sticky wages (as well as sticky prices) into macro models; with sticky prices alone, the real wage would be strongly pro-cyclical. Finally, interest in sticky wages has also been buttressed by the welfare analysis of Erceg, Henderson and Levin (2000). These authors showed that a combination of wage and price rigidities may account for the policymaker's apparent trade-off between stabilizing inflation and the output gap, and has important normative implications for the design of policy rules.

A significant shortcoming of recent models is that their empirical support comes primarily from aggregate data. Moreover, current micro-founded sticky wage models appear deficient in their characterization of worker–firm attachments, and because they fail to take account of the sizeable costs of renegotiating labour contracts. Models that are developed to address such limitations may well have substantially different normative implications from current models, even if their dynamic properties remain similar.

CHRISTOPHER J. ERCEG

See also **monetary transmission mechanism.**

Bibliography

Ball, L. 1995. Disinflation and imperfect credibility. *Journal of Monetary Economics* 35, 5–24.

Calvo, G. 1983. Staggered prices in a utility-maximizing framework. *Journal of Monetary Economics* 12, 383–98.

Caplin, A. and Spulber, D. 1987. Menu costs and the neutrality of money. *Quarterly Journal of Economics* 102, 703–25.

Chari, V., Kehoe, P. and McGratten, E. 2000. Sticky price models of the business cycle: can the contract multiplier solve the persistence problem? *Econometrica* 69, 1151–79.

Christiano, L., Eichenbaum, M. and Evans, C. 2005. Nominal wage rigidities and the dynamic effects of shocks to monetary policy. *Journal of Political Economy* 113, 1–45.

Dotsey, M., King, R. and Wolman, A. 1999. State-dependent pricing and the general equilibrium dynamics of money and output. *Quarterly Journal of Economics* 104, 655–90.

Erceg, C. 1997. Nominal rigidities and the propagation of monetary disturbances. International Finance Discussion Paper 590. Washington, DC: Board of Governors of the Federal Reserve System.

Erceg, C., Henderson, D. and Levin, A. 2000. Optimal monetary policy with staggered wage and price contracts. *Journal of Monetary Economics* 46, 281–313.

Erceg, C. and Levin, A. 2003. Imperfect credibility and inflation persistence. *Journal of Monetary Economics* 50, 915–44.

Fisher, S. 1977. Long-term contracts, rational expectations, and the optimal money supply rule. *Journal of Political Economy* 85, 191–205.

Fuhrer, J. and Moore, G. 1995. Inflation persistence. *Quarterly Journal of Economics* 110, 127–59.

Gray, Joanna 1976. Wage indexation: a macroeconomic approach. *Journal of Monetary Economics* 2, 221–35.

Huang, Kevin and Liu, Z. 2002. Staggered price-setting, staggered wage-setting, and business cycle persistence. *Journal of Monetary Economics* 46, 405–33.

Humphrey, T. 2004. Classical deflation theory. *Federal Bank of Richmond Economic Quarterly* 90, 11–32.

Keynes, J. 1936. *The General Theory of Interest, Employment, and Money.* London: Macmillan.

Roberts, J. 1997. Is inflation sticky? *Journal of Monetary Economics* 39, 173–96.

Taylor, J. 1980. Aggregate dynamics and staggered contracts. *Journal of Political Economy* 88, 1–24.

Taylor, J. 1999. Staggered wage and price setting in macroeconomics. In *Handbook of Macroeconomics*, ed. J. Taylor and M. Woodford. Amsterdam: North-Holland.

Stigler, George Joseph (1911–1991)

George Stigler begins his autobiographical statement for the 1982 Nobel Prize in Economic Sciences (Stigler, 1983a) with these words:

> I was born in Renton, a suburb of Seattle, Washington, in 1911. I was the only child of Joseph and Elizabeth Stigler, who had separately migrated to the United States at the end of the 19th century, my father from Bavaria and my mother from what was then Austria-Hungary (and her mother was in fact Hungarian). I attended schools in Seattle through the University of Washington, from which I was graduated in 1931. I spent the next year at Northwestern University.
>
> My main graduate training was received at the University of Chicago from which I received the Ph.D. in 1938. The University of Chicago then had three economists – each remarkable in his own way – under whose influence I came. Frank H. Knight was a powerful, sceptical philosopher, at that time vigorously debating Austrian capital theory but gradually losing interest in the details of economic theory. Jacob Viner was the logical disciplinarian, and equally the omniscient student of the history of economics. Henry Simons was the passionate spokesman for a rational, decentralized organization of the economy. I was equally influenced by two fellow students, W. Allen Wallis and Milton Friedman.

His statement ends this way:

> I met my wife, Margaret L. Mack, at the University of Chicago. We were married in 1936. She died in 1970. I have three sons, Stephen (a statistician), David (a lawyer), and Joseph (a social worker). We are a close-knit family, and each summer we gather at a cottage on the Muskoka Lakes in Canada.

The Nobel Committee adds, 'George J. Stigler died on December 1, 1991'.

What is not mentioned in the texts between these paragraphs are the honours. The presidency of the American Economic Association in 1964, of the History of Economics Society in 1977, and of the Mont Pelerin Society in 1977 along with the Nobel Prize in 1982 testify to a unique career. Specialist accounts focus on his contributions to the study of regulation (Peltzman, 1993), industrial organization (Demsetz, 1993) and the economics of science (Diamond, 2005). Biographical sketches by those who knew him well (Becker, 1993; Friedman, 1993; Wallis, 1993), as well as his 1988 *Memoirs,* attempt to link his life and work. Following his death, McCann and Perlman (1993) assessed Stigler's career; Longawa (1993) provided the bibliography. More than a decade later, we attempt to capture the permanent challenge of Stigler's work here.

Economists today, trained in a straightforward mathematical discipline, find it difficult to appreciate Stigler's combination of technical competence, scholarly erudition (Becker, 1993; Rosen, 1993; Rosenberg, 1993), and mordant wit (Friedland, 1993). The erudition is a serious problem. A student who has read neither Aristotle nor Bernard Mandeville may fail to appreciate the change of attitude between an article by Stigler that opened with a passage from Aristotle's *Ethics* (1943, p. 355) and his identification of revealed preference with the philosophy advanced in Mandeville's *Fable of the Bees* (1966, p. 68). Moreover, there is what Friedland calls 'shyness' and Friedman refers to as 'sensitivity'. Stigler's 1988 autobiographical account spells out his contributions to the ongoing scientific discussion, but it hides virtuosity behind a veil of modesty.

Virtuosity and modesty

We begin with a contribution that is not discussed in Stigler's autobiography but which might have made the career of an ordinary scholar, what is known in the operations research literature as 'Stigler's diet problem'. Given a scientific consensus on nutrients, what is the least-cost diet? We start with 9 nutrients and 77 foods. Recent accounts of Stigler's treatment remain appreciative:

> Stigler used trial and error, and mathematical insight and agility to solve his (9×77) set of inequalities. Based on cost and nutrient content, he was able to 'weed' the original 77 foods down to 15 as the

eliminated foods were dominated by those in the list of 15. ... Stigler's diet for 1939 data cost $39.93 per year. ... Stigler's 1939 diet problem was the first 'large-scale' problem that was solved using the simplex method (Dantzig 1963 [pp. 551–67]). In 1947, nine clerks, using hand-operated desk calculators, pivoted away for 120 clerk-days and found the linear programming minimum cost of $39.69. Stigler knew what he was doing! (Garille and Gass, 2001, p. 2)

Authorities preface their judgement of the technical virtuosity of the performance with this judgement of the oddity of it all:

> Stigler's diet problem is a prime example of an OR model that faithfully describes the real-world situation but whose solution validity is close to zero. As Stigler (1945, p. 312) cautioned: 'No one recommends these diets for anyone, let alone everyone.' (Garille and Gass, 2001, p. 2)

What was the effort about, then? Stigler explains when he compares his solution to those of 'competent dieticians' which cost two or three times as much as his:

> The dieticians take account of the palatability of goods, variety of diet, prestige of various foods, and other cultural facets of consumption. ... the particular judgments of the dieticians as to minimum palatability, variety, and prestige are at present highly personal and non-scientific, and should not be presented in the guise of being parts of a scientifically-determined budget. The second reason is that these cultural judgments, while they appear modest enough to government employees and even to college professors, can never be valid in such a general form. No one can now say with any certainty what the cultural requirements of a particular person may be ... If the dieticians persist in presenting minimum diets, they should at least report separately the physical and cultural components of these diets. (1945, p. 314)

So, for Stigler, claims about the goals of individuals that had neither scientific nor philosophical weight were embedded in the dieticians' solutions. At a minimum, he insisted, these claims should be made transparent.

Later editions of Stigler's textbook discuss his contribution to linear programming:

> This method of isolating products is intimately related to a method known as linear programming, and the 'shadow prices' of that method are the implicit alternative cost of inputs. (1966, p. 119)

To which he adds the footnote, 'See almost any other book on economics' and a reference to an article by Paul Samuelson on Frank Knight's theorem on linear programming.

We do not learn that Stigler's data and solution were the test case for George Dantzig to prove the worth of his

simplex. Instead, we see played out Stigler's distaste for the economist who overemphasizes his own originality, a view reiterated in the essays on J.S. Mill's originality. There, too, Stigler endorsed Mill's stance of impartiality between ideas he created and those he adopted from others (1955; 1982, pp. 96–7).

We turn next to the problem that closes Stigler's diet paper – how we, as scientists, know the goals of individuals outside some very narrow physical sense.

Knight's discipline

Stigler's writings with an autobiographical component invariably stress the importance of Frank H. Knight as exemplar. In a self-assessment that is viewed sceptically by McCann and Perlman (1993), Stigler wrote 'A more improbable Moses, if Knight would ever forgive the analogy, could not be designed' (1988, p. 17). He stressed the impact of Knight's personal integrity, his life as a scholar who renounced money and fame as impediments to truth seeking (1988, p. 18). Is it then something of a surprise that the opening chapter in *Memories of an Unregulated Economist* asks 'Are economists good people?' Here, what worries Stigler most seems to have been the possibility that scholarship might be compromised by monetary rewards, specifically in the case of consulting. He thinks this is not so in his own case, but he recognizes that scholarship might also be influenced by sympathy for the client's case (1988, p. 133). Indeed, the idea that a statistician might be tempted by sympathy for the client is a concern in statistical consulting (Vardeman and Morris, 2003). Such ethical issues are rarely considered in economics.

Stigler was Knight's discipline for a long time. In his 1943 *Economica* review of *Theory of Competitive Price* (the partially completed 1946 *Theory of Price*), Lachmann pronounced the book to be a coherent version of Knight's distribution theory in which classical productive inputs are replaced by anonymous factors and cost is opportunity forgone. This view has triumphed so completely, outside perhaps of neo-Ricardian theorizing, that the radicalism of the Knight–Stigler position at the time is now largely forgotten.

In 1943 Stigler put forward a Knightian challenge to new welfare economics, defending classical economic policy against the new orthodoxy. First, he offered the objection:

> Consider theft; our present policy toward this means of livelihood probably has adverse effects on the national income. Prevention of theft and punishment of thieves involves substantial expenditures for policemen, courts, jails, locks, insurance salesman, and the like. By compensating successful thieves for the amounts they would otherwise steal, we save these resources and hence secure a net gain. (If this policy leads to an undue increase in declarations of intent to steal, the retired successful thieves – who, after all, have special talents in

this direction – may be persuaded to assume the police functions.) (1943, p. 356)

Since it would 'outrage our moral sensibilities to pay voluntary tribute to thieves', something must be wrong. Stigler sketched the alternative:

> The familiar admonition not to argue over differences in tastes [*de gustibus non est disputandum*] leads not only to dull conversations but to bad sociology. It is one thing to recognize that we cannot *prove*, by the usual tests of adequacy of proof, the superiority of honesty over deceit or the desirability of a more equal income distribution. But it is quite another thing to conclude that therefore ends of good policy are beyond the realm of scientific discussion.
>
> For surely the primary requisite of a working social system is a consensus on ends. The individual members of the society must agree upon the major ends which that society is to seek. (1943, p. 357)

The 1943 paper is cited in the first edition of the full *Theory of Price* as defending a Knightian consensus view of the law (Knight 1947, p. 62):

> it is the fundamental tenet of those who believe in free discussion that matters of fact and logic can (eventually) be agreed upon by competent men of good will, that matters of taste cannot be ... (1946, pp. 15–16)

Consensus in deep goals was as critical to policy as consensus about theory and fact was to science.

The defence of new welfare economics was brief and devastating (Samuelson, 1943). Here is K.J. Arrow's judgment:

> Professor Stigler has made it a burden of reproach to the new welfare economics that it does not take into account the consensus on ends. It is not clear from his discussion whether he regards the agreed-on ends as being obvious from introspection or casual observation ... or as requiring special inquiry; his comments seem rather to incline in the former direction, in which case he lays himself open to Professor Samuelson's request for immediate enlightenment on various economic issues. (1951, pp. 83–4)

We find much of the Knightian spirit also in the 1946 minimum wage law article. Stigler did not reprint this article in the *Citizen and the State*, where he described it mischievously as an example of how the misguided economist, lacking a market failure to correct, 'lamented the intervention of the state' (1975, p. x). The reader can predict the judgement about the topic but perhaps not the remedy. Given that minimum wage laws fail to ameliorate poverty, what should we do? The shared goal of equal treatment drives Stigler's results in a surprising direction:

> One principle is fundamental in the amelioration of poverty; those who are equally in need should be

helped equally. If this principle is to be achieved, there must be an objective criterion of need; equality can never be achieved when many cases are judged (by many people) 'on their merits' ... It is the corollary of this position that assistance should not be based upon occupation. The poor farmer, the poor shopkeeper, and the poor miner are on an equal footing. (1946, pp. 364–5).

Stigler endorsed a negative income tax in this context:

> There is a great attractiveness in the proposal that we extend the personal income tax to the lowest income brackets with negative rates in these brackets. Such a scheme could achieve equality of treatment with what appears to be a (large) minimum of administrative machinery. (1946, p. 365)

Stigler's belief at the time that greater equality is a shared goal might explain his sharp reaction to 'suggestions' that he and Friedman tone down their egalitarianism in their joint study *Roofs and Ceilings* (Hammond and Hammond, 2006). The full history of 'natural experiments' in economics has yet to be written. When it is, the study of how the San Francisco housing market responded to the great earthquake (Friedman and Stigler, 1946) will take pride of place (Rockoff, 1991).

The work on regulation cited in the Nobel award falls between his Knightian and his Mandevillean periods. Neither electricity (Stigler and Friedland, 1962) nor securities regulation (Stigler, 1964) attained the articulated goals of regulation policy.

Mandeville's disciple

Stigler renounced the procedure of imputing goals from articulated speech when his papers on regulation were collected:

> It seems unfruitful, I am now persuaded, to conclude from the studies of the effects of various policies that those policies which did not achieve their announced goals, or had perverse effects (as with a minimum wage law), are simply mistakes of the society. A policy adopted and followed for long time, or followed by many difference states, could not usefully be described as a mistake: eventually its real effects would become known to interested groups. To say that such policies are mistaken is to say that one cannot explain them. (1976, p. x)

Ten years before, in the 1966 edition of *The Theory of Price*, Stigler tells the student about the 'penetrating' Bernard Mandeville, the philosopher of revealed preference. This is the passage he quotes:

> I don't call things Pleasures which Men say are best, but such as they seem to be most pleased with; ... John never cuts any Pudding, but just enough that you can't say he took none; this little Bit, after much chomping

and chewing you see goes down with him like chopp'd Hay; after that he falls upon the Beef with a voracious Appetite, and crams himself up to his Throat. Is it not provoking to hear John cry every Day that Pudding is all his Delight, and that he don't value the Beef of a Farthing? (1966, p. 68)

Instead of supposing policy goals are revealed in discussion, he now follows Mandeville in imputing goals from consequences (Stigler, 1971). So, regulatory capture is not a 'failure' of regulation, but is instead the very *point* of regulation.

An intriguing variation on Mandeville is presented in Stigler and Becker (1977). The title, 'De Gustibus Non Est Disputandum', teases Knight and the early Stigler. Stigler and Becker posit that individuals are homogeneous with respect to unobservable ends of life revealed in the material world by means of physical goods and other inputs. Attracting much attention as the first of a sequence of papers on rational addiction, the paper perhaps ought to be seen as an attempt to recover common goals in choice. Following the supposition that goals are better revealed in mute choices than articulated speech, in the Tanner Lectures (Stigler, 1982), Stigler now defends the sort of productivity ethics once denounced by Knight.

Information as commodity

In Stigler's judgement (and that of Becker), the 1961 article 'Economics of Information' was the most important of the contributions listed in the citation for the Nobel award. The problem is deceptively straightforward: how long will the rational consumer search for a lower price? From the claim that it pays to search more for higher-priced commodities, Stigler obtains the result that, the higher the price of a commodity, the lower is its percentage deviation from the central tendency. The importance of characterizing a competitive equilibrium in terms of a statistical distribution of prices instead of a point mass from 'one price in the market' is hard to overstate. This approach explains Stigler's scepticism regarding models of oligopoly that involve price movements in one direction but not another (1947; 1978). Stigler's insistence on the veil of language also helps explain why he maintains that we recover prices at which transactions were made even if they differ from reported prices (Stigler and Kindahl, 1970).

The implications that flow from Stigler's notion that information is a commodity extend to his theory of oligopoly (Stigler, 1964). Here, Stigler takes joint profit maximization as the default and explains oligopoly on the basis of the probability of detecting collusion. Now that the economics literature has rediscovered Adam Smith's principle of sympathetic behaviour as an explanation of group formation and cooperative behaviour (Sally, 2001), Stigler's argument that cooperation is the

default in small groups reveals great prescience. Perhaps most unsettling to economists is the additional implication regarding the endogeneity of economic advice. For, if information is a commodity and economists provide information to clients, then we, too, are inside the economic process. This endogeneity of economic advice raises causality questions that are central to 'Do Economists Matter?' (Stigler, 1976). The issue here is whether our opinions are the cause or the effect? Now this is an identification problem!

Science as consensus
As we read the record, the Knightian emphasis on consensus underscores Stigler's view of scientific practice. For Stigler, as for his teachers, the past of economics was part of what has been called the extended present. Indeed, Rosen (1993) and Rosenberg (1993) explain, perhaps more clearly than did Stigler himself, how Stigler drew inspiration from the past. What distinguishes him from Knight is the view that we might justifiably exclude views that are sufficiently far from the centre:

> There is merit in excluding the lunatic from discourse. If a man tells me he is Napoleon, or that matter, Josephine, discussion would serve no purpose. ... Occasionally the lone dissenter with the absurd view will prove to be right – a Galileo with a better scheme of the universe, a Babbage with a workable computer – but if we give each lunatic a full, meticulous hearing, we should be wasting vast time and effort. So long as we do not suppress the peaceful lunatic, we leave open the possibility that he may convince others that he is right.
>
> ... The larger the group [with a common outlook], the more certain we can be that it is not *insane* in the sense of being divorced from apparent fact and plausible reasoning. (Stigler, 1975, pp. 3–4)

This view explains Stigler's reading of the Sraffian account of classical economics and the challenge offered by Samuel Hollander (Hollander, 1979; Stigler, 1990; Hollander, 1990). In the case of the Chicago reaction to Ronald Coase's 'mistake' in the discussion of externalities, the outcome differed. Stigler recounts how the response was to invite Coase to dinner to talk about the 'mistake' which, by the end of the evening, became the 'Coase theorem' (Stigler, 1988, pp. 75–8).

Stigler's confidence in the relationship between the size of the community asserting a factual claim and the probability that the claim is correct depends upon the independence of inquiry. Yet once we believe that common acceptance warrants belief, we violate the independence of acceptance upon which the probability claim rests. If we accept a result *because* it is widely accepted, can we really think the result is even more firmly established for the next 'researcher'? This is, perhaps, the most significant weakness of Stigler's generation, the neglect of what in retrospect turn out to be game-theoretic issues.

And here Stigler is particularly stubborn (Demsetz, 1993). One technical point at which game theory is missing occurs in Stigler's assertion (1966, pp. 94–5) that the institutional framework fails to affect the equilibrium in the case of auctions. Stigler's (1952) work on T.R. Malthus fails to appreciate Malthus's argument that William Godwin's proposal for communism would create a large-number prisoner's dilemma. Godwin himself came to understand Malthus's prisoner's dilemma argument (Godwin, 1801, p. 74).

If there is systematic error in one aspect of our understanding of the past, then our demand for coherence may well impose, as a general equilibrium condition, errors in all those aspects that are connected. So one error about the past leads to others. As evidence of how the Malthus error has cascaded, consider Stigler on the most famous characterization of economics: '[Malthus's] pessimism was the source for the characterization of economics as "the dismal science"' (1988, p. 5). A great deal of scholarship has now demonstrated that this is precisely wrong (Persky, 1990; Levy, 2001; Peart and Levy, 2005).

Chicago School

Stigler's Nobel lecture (1983b) is an important defence of the progressive nature of economics. To make the progressive case, one needs to know how the discipline changes over time. In an important sense, the Chicago School doctrine of motivational homogeneity (Stigler and Becker, 1977) is a return to the classical economic doctrine of analytical egalitarianism against contending schools of thought that had focused instead on racial, ethnic or class differences. These latter accounts prevailed in early 20th-century economic analysis, and presupposed unequal economic competence (Peart and Levy, 2005). In early neoclassical economics, inferiority was adduced from claims about positive time preference (Peart, 2000). Stigler never succumbed to this temptation (Stigler, 1941, pp. 212–19; Stigler and Becker, 1977).

The Chicago defence of the egalitarian roots of classical economics in the face of racist claims by eugenicists and other 'progressive' 'experts' (Peart and Levy, 2005), suggests that, fundamentally, economics has hardly progressed in its foundational elements. We find the same defense of homogenous capacity in Adam Smith arguing against Plato, in J.S. Mill against Carlyle, and in Chicago. But perhaps this ought not to surprise economists. Whatever is important for policy will be contested. The answer to Samuelson's request, which Arrow echoed, then perhaps lies in considering whether 'common ends', policy goals, are recommended by those who believe themselves able to rule others or by those who believe themselves to be essentially the same as others. In his defence of equal capacity of economic agents, George Stigler will be remembered as second to none in the 20th century.

DAVID M. LEVY AND SANDRA J. PEART

See also **Arrow, Kenneth Joseph; Becker, Gary S.; Chicago School; egalitarianism; Knight, Frank Hyneman; Malthus, Thomas Robert; Mandeville, Bernard; Mill, John Stuart; revealed preference theory; Ricardo, David; Smith, Adam; Sraffa, Piero; welfare economics.**

Selected works

1941. *Production and Distribution Theories: The Formative Period*. New York: Macmillan.
1943. The new welfare economics. *American Economic Review* 33, 355–9.
1945. The cost of subsistence. *Journal of Farm Economics* 27, 303–14.
1946. The economics of minimum wage legislation. *American Economic Review* 36, 358–65.
1946. *The Theory of Price*. New York: Macmillan.
1946. (With M. Friedman.) *Roofs or Ceilings? The Current Housing Problem*. Irvington-on-Hudson, NY: Foundation for Economic Education.
1947. The kinked oligopoly demand curve and rigid prices. *Journal of Political Economy* 55, 432–49.
1952. The Ricardian theory of value and distribution. *Journal of Political Economy* 60, 187–207.
1955. The nature and role of originality in scientific progress. *Economica* n.s. 22, 293–302.
1961. The economics of information. *Journal of Political Economy* 69, 213–25.
1962. (With C. Friedland.) What can regulators regulate? The case of electricity. *Journal of Law and Economics* 5, 1–16.
1964. A theory of oligopoly. *Journal of Political Economy* 72, 44–61.
1966. *The Theory of Price*. New York: Macmillan.
1970. (With J. Kindahl.) *The Behavior of Industrial Prices*. New York: NBER.
1971. Smith's travels on the ship of state. *History of Political Economy* 3, 265–77.
1975. *The Citizen and the State: Essays on Regulation*. Chicago: University of Chicago Press.
1976. Do economists matter? *Southern Economic Journal* 42, 347–54.
1977. (With G. Becker.) De gustibus non est disputandum. *American Economic Review* 67, 76–90.
1978. The literature of economics: the case of the kinked oligopoly demand curve. *Economic Inquiry* 16, 185–204.
1982. *The Economists as Preacher, and Other Essays*. Chicago: University of Chicago Press.
1983a. Autobiography. In *Les Prix Nobel. The Nobel Prizes 1982*, ed. W. Odelberg. Stockholm: Nobel Foundation. Online. Available at http://nobelprize.org/nobel_prizes/economics/laureates/1982/stigler-bio.html, accessed 13 September 2006.
1983b. Nobel Lecture: The process and progress of economics. *Journal of Political Economy* 91, 529–45.
1988. *Memoirs of an Unregulated Economist*. New York: Basic Books.

1990. Ricardo or Hollander?*Oxford Economic Papers* n.s. 42, 765–8.

Bibliography

Arrow, K. 1951. *Social Choice and Individual Values*, 2nd edn. New York: John Wiley & Sons.

Becker, G. 1993. George Joseph Stigler: January 17, 1911–December 1, 1991. *Journal of Political Economy* 101, 761–7.

Dantzig, G. 1963. *Linear Programming and Extensions*. Princeton: Princeton University Press.

Demsetz, H., 1993. George J. Stigler: midcentury neoclassicalist with a passion to quantify. *Journal of Political Economy* 101, 793–808.

Diamond, A., Jr. 2005. Measurement, incentives and constraints in Stigler's economics of science. *European Journal of the History of Economic Thought* 12, 635–61.

Friedland, C. 1993. On Stigler and Stiglerisms. *Journal of Political Economy* 101, 780–3.

Friedman, M. 1993. George Stigler: a personal reminiscence. *Journal of Political Economy* 101, 768–73.

Garille, S. and Gass, S. 2001. Stigler's diet problem revisited. *Operations Research* 49, 1–13.

Godwin, W. 1801. *Thoughts Occasioned by the Perusal of Dr. Parr's Spital Sermon*. London: G. G. & J. Robinson.

Hammond, J. and Hammond, C. 2006. *Making Chicago Price Theory: Friedman–Stigler Correspondence, 1945–1958*. London: Routledge.

Hollander, S. 1979. *The Economics of David Ricardo*. Toronto: University of Toronto Press.

Hollander, S. 1990. A reply to Professor Stigler and Dr. Peach. *Oxford Economic Papers* n.s. 42, 769–71.

Knight, Frank H. 1947. *Freedom and Reform*. New York: Harper & Bros.

Lachmann, L. 1943. *The Theory of Competitive Price*. *Economica* n.s. 10, 264–5.

Levy, D. 2001. *How the Dismal Science Got Its Name: Classical Economics and the Ur-Text of Racial Politics*. Ann Arbor: University of Michigan Press.

Longawa, V. 1993. George J. Stigler: A bibliography. *Journal of Political Economy* 101, 849–62.

McCann, C. Jr. and Perlman, M. 1993. On thinking about George Stigler. *Economic Journal* 103, 994–1014.

Peart, S. 2000. Irrationality and intertemporal choice in early neoclassical thought. *Canadian Journal of Economics* 33, 175–88.

Peart, S. and Levy, D. 2005. *The 'Vanity of the Philosopher': From Equality to Hierarchy in Post-Classical Economics*. Ann Arbor: University of Michigan Press.

Peltzman, S. 1993. George Stigler's contribution to the economic analysis of regulation. *Journal of Political Economy* 101, 818–32.

Persky, J. 1990. Retrospectives: a dismal romantic. *Journal of Economic Perspectives* 4(4), 165–72.

Rockoff, H. 1991. History and economics. *Social Science History* 15, 239–64.

Rosen, S., 1993. George J. Stigler and the industrial organization of economic thought. *Journal of Political Economy* 101, 809–17.

Rosenberg, N. 1993. George Stigler: Adam Smith's best friend. *Journal of Political Economy* 101, 833–48.

Sally, D. 2001. On sympathy and games. *Journal of Economic Behavior & Organization* 44, 1–30.

Samuelson, P. 1943. Further commentary on welfare economics. *American Economic Review* 33, 604–7.

Vardeman, S. and Morris, M. 2003. Statistics and ethics: some advice for young statisticians. *American Statistician* 57, 21–6.

Wallis, W. 1993. George J. Stigler: In Memoriam. *Journal of Political Economy* 101, 774–9.

Stiglitz, Joseph E. (born 1943)

Joseph E. Stiglitz helped to create the economics of information, which analyses equilibrium in markets in which there are asymmetries of information among the market participants. For that work, he received the Nobel Prize in Economics in 2001, jointly with George A. Akerlof and A. Michael Spence. Stiglitz's work demonstrated the many and sometimes subtle ways in which markets can fail to lead to efficient outcomes. His work elucidated a broad set of phenomena that had largely been ignored before 1970 because they were outside the limits of the standard paradigm: incentive contracts, bankruptcy, quantity rationing, financial structure, equilibrium price distributions, innovation, and dysfunctional institutions. This work contributed to a paradigm shift in economics. In the new paradigm, the price system only imperfectly solves the information problem of scarcity because of the many other information problems that arise in the economy. Stiglitz has also proved central theorems in many fields: development economics, finance, trade theory, public economics, and industrial organization.

The broad plan from which much of Stiglitz's work originates had two central goals. The first was to show that many of the implications of the standard neoclassical model do not remain valid once the assumption of perfect information is dropped. His famous paper on adverse selection (Rothschild and Stiglitz, 1976) opens with these words:

> Economic theorists traditionally banish discussions of information to footnotes. Serious consideration of costs of communication, imperfect knowledge, and the like would, it is believed, complicate without informing. ... [T]his comforting myth is false. Some of the most important conclusions of economic theory are not robust to considerations of imperfect information. (1976, p. 629)

The second goal was to provide a better theoretical understanding of the workings of the economic system as a whole. As Stiglitz (2007) explains,

> By the time I had finished my graduate studies, I had realized that the model of the economic system that was being taught – and that was at the center of policy analyses – was *not* a model of a modern capitalist economy. It was little more than a fancy version of a primitive agriculture exchange/production economy, slightly updated to include manufacturing – so long as there were diminishing returns. There was but a short distance between Ricardo and Walras, and between Walras and Samuelson. ...
>
> ...Capital was nothing more than seed that was harvested but not consumed...
>
> ...technology was stagnant (or at most exogenous).

A critical assumption of the standard neoclassical model is that there is a price for each quality of good in the market and for each action one would wish to contract for. Buyers have no problem ascertaining quality, and firms produce the quality that they have agreed to produce. Firms do not need to motivate their workers. Lenders do not worry about borrowers repaying. Owners do not worry about managers taking the right actions. Stiglitz in his lectures to students in the 1980s gave the example of a stylist cutting hair: in the standard model, there would be a price for each hair that he cut.

A real-life experiment that helped economists evaluate the standard neoclassical theory were the experiences with market socialism in Eastern Europe, in which government owned the firms but there was a manager of each firm whose job it was to maximize profits of that firm, facing market prices. Stiglitz (1994) argues that if the neoclassical model were an accurate characterization of a market economy, then market socialism would have been successful. Because of the importance of incentive problems and of non-price institutions (such as banks) within the economy, the inefficiencies that arose under market socialism were not accidents, but rather the inevitable consequence of (a) the limitations of the information contained in prices, and (b) the gap between the set of goods for which markets can practically exist and the much broader set of present and future goods and actions on which welfare depends.

Two ideas inform much of Stiglitz's work.

1. The 'control/information' system of market economies embraces far more than the price system of the neoclassical model. The exchange problem is intertwined with the process of selection over hidden characteristics, the provision of incentives for hidden behaviours and for innovation, and the coordination of choices over institutions.

2. Competitive equilibrium in economies with imperfect information and missing markets is not, in general, Pareto efficient. Market outcomes can be improved on

by government intervention, e.g., taxes and subsidies. A simple illustration is that if the care that the insured take to avoid an accident is not observable to the insurance company, then commodities like fire extinguishers that decrease insured agents' losses should be subsidized, while commodities like alcohol that increase their losses should be taxed (Greenwald and Stiglitz, 1986, p. 247).

Until Arrow's work on medical care (Arrow, 1963), the only reasons for a missing market that had been well explored were environmental externalities and the inability, or undesirability, to exclude from use (the problem of public goods). Stiglitz's contributions would help to radically broaden the understanding of the sources of externalities to include information externalities, group reputation effects, agglomeration effects, knowledge spillovers, and pecuniary externalities (see, for example, Greenwald and Stiglitz, 1986 on pecuniary externalities and Hoff, 2001 on coordination failures). In the process, Stiglitz's work would help to change the profession's understanding of capitalism, although the policy recommendations of economists did not change as much as Stiglitz had hoped.

Citations are an objective, if imperfect, measure of influence. Kim, Morse and Zingales (2006) compiled a list of the 146 articles published in economics journals from 1970 to 2002 that had received by June 2006 more than 500 citations from the ISI Web of Science/Social Science Citation Index. Six of Stiglitz's papers appear on this list (no other author has more). In descending order of number of citations, the papers are Stiglitz and Weiss (1981), Rothschild and Stiglitz (1970), Dixit and Stiglitz (1977), Shapiro and Stiglitz (1984), Rothschild and Stiglitz (1976), and Grossman and Stiglitz (1980). This article will place each of these papers in the context of Stiglitz's research programme.

Biographical data

Stiglitz's early experiences shaped his lifelong professional interests in understanding how an economy handles risk, and in bringing economic theory to bear on real-world problems. Stiglitz was born in Gary, Indiana, a city marred, in his words, by 'huge inequality, poverty, and discrimination' (Stiglitz, 2007). He was the middle of three children. After the failure of an earlier business, his father became an independent insurance agent. One part of his job was to find new insurers for firms whose businesses had burned down and whose insurance policies had been cancelled. Stiglitz's mother worked in the family insurance business when Stiglitz was young and later taught elementary school in a low-income inner city neighbourhood of Gary. After retiring from elementary education, she worked in adult remedial education, where she encountered some of the same students whom she had taught as children in inner-city schools.

Stiglitz's genius was recognized early. In high school, he was assigned independent study in lieu of some of the regular classes, which he had outstripped. (His father apparently took this the wrong way: he was concerned that something might be wrong with the other children in the school.) Stiglitz followed his older brother to Amherst College and graduated in 1964. There he studied economics, physics, history and mathematics, and was president of his senior class. In that position, he was a maverick: he tried unsuccessfully to stop the college from funding the training of Amherst sports teams in Bermuda during college vacations, and organized an exchange programme between Amherst and a segregated college in the US South.

He obtained a Ph.D. from the Massachusetts Institute of Technology (MIT) in 1967. Stiglitz (2007) writes that particularly important influences at MIT on his later work were his statistics teacher, Harold Freeman, who taught the recently developed theory of subjective probability, and Ken Arrow, with whom he took a class as a second-year graduate student. At that time, Arrow was writing the final formalization of the Arrow–Debreu paradigm. Realizing the model's limitations, Arrow was also beginning a research agenda into the consequences of imperfect information.

Stiglitz joined the faculty of MIT in 1966, spent time between 1969 and 1971 at the Institute for Development Studies at the University of Nairobi under a Rockefeller Foundation grant, and then moved from university to university: Yale (1967–74), St Catherine's College, Oxford (Visiting Fellow, 1973–74), Stanford (1974–76), All Souls' College, Oxford (1976–79), the Institute of Advanced Studies at Princeton (1978–79), Princeton University (1979–88), Stanford (1988–2000), and Columbia University (2000 to date).

Stiglitz received the John Bates Clark Medal in 1979, awarded biennially by the American Economic Association for the most distinguished work by an economist under the age of 40. In 1987, Stiglitz became founding editor of the *Journal of Economic Perspectives*.

In 1993 Stiglitz joined President Clinton's Council of Economic Advisors, which he chaired in 1995–97. In 1997, Stiglitz was appointed to the position of Chief Economist of the World Bank. The East Asian financial crisis occurred in 1997–99, and Stiglitz argued publicly against the policies of the International Monetary Fund (IMF) towards the crisis. Disagreements arose both about the consequences of the policies (given the uncertainties about both the structure of the economy and future events), and about welfare judgments of the acceptable trade-offs between competing goals. Stiglitz's positions led to conflict with other officials in Washington. In November 1999, he resigned from the World Bank and returned to academia. At Columbia, he co-founded the Initiative for Policy Dialogue, which studies policy issues and provides training to policymakers from developing countries.

The economics of uncertainty

The economics of uncertainty is concerned with the principles that an individual uses in evaluating a random distribution of returns. Applications extend from how individuals allocate their portfolios between safe assets and risky assets, to how farmers allocate their land among different crops. The work in the 1960s by James Tobin and others equated an increase in risk with an increase in variance. Rothschild and Stiglitz (1970) set forth an alternative definition of an increase in risk. Comparing income distributions with the same mean, they proposed a definition that corresponded to a preference ordering among every expected-utility-maximizer with a concave utility function. This ordering was not the same as a ranking based on increases in variance. In a companion paper, Rothschild and Stiglitz (1971) demonstrated the usefulness of their definition in deriving comparative static results. They showed that such results depended on a simple criterion: the concavity or convexity of a 'first-order condition' characterizing the individual's optimal decision with respect to the random variable that was the source of risk. That work unified work in an area that until then had been in great confusion.

The economics of information

In the 1960s, James Mirrlees (1971) began working on the problem of how a government could design an optimal tax schedule, taking into account that government can observe individuals' incomes, but not their ability and effort. Given this asymmetry of information, the analytical problem is to distribute a given tax burden according to differences in ability to pay. Stiglitz recognized the similarities between this problem and the problems that arise in *markets* with asymmetric information. For example, insurance companies and banks want to design a menu of offers that will maximize their profits, taking into account that they do not know each individual's risk of accident or bankruptcy and the care that an individual expends to avoid the insured for event. Employers want to design labour contracts to maximize productivity, taking into account that they can observe only imperfectly workers' ability and effort. Together with a small group of pioneers in the 1970s, and influenced in particular by Akerlof (1970) and Spence (1974), Stiglitz devised models in which these kinds of problems could be analysed. The work on hidden characteristics (adverse selection) and incentive problems (moral hazard) came to be the core of the economics of information.

Hidden characteristics

A selection problem arises whenever there is imperfect information about the characteristics of the items being transacted, and different sides of the transaction know different things (so information is *asymmetric*). This problem is pervasive. Rothschild and Stiglitz (1976) constructed a celebrated model of the insurance market in which individuals differ only in terms of their privately known risk type, and the insurance firms know the overall distribution of risks in the population. The model uses the canonical textbook apparatus of consumer choice – budget lines and indifference curves – but produces very surprising (and counter-Walrasian) results.

To illustrate Stiglitz's *modus operandi* – to begin with highly simplified models of particular markets that allow him to identify a general principle – I describe the Rothschild and Stiglitz (1976) model in detail.

Consider an individual whose situation is described by his income if he is lucky enough to avoid an accident (W_{NA}) and his income if an accident occurs (W_A). His initial endowment point E is (W, $W{-}d$), where d represents the damages incurred in the accident. An individual purchases insurance in order to alter his pattern of income across these two states of nature.

Begin with the benchmark case in which insurance companies know an individual's probability of accident. Given this probability, let $\bar{W}$ denote his expected income. Then competitive equilibrium would be at a point A, illustrated in Figure 1, where the insurance company breaks even and the individual's indifference curve, denoted $\bar{U}^L$, is tangent to the budget line. Since risk-averse individuals who are offered a break-even

price for insurance will choose full insurance, the equilibrium is along the 45 degree line. The line that goes through the endowment point E and the point A is the locus of contracts at which an insurance company breaks even ('the fair-odds line').

Next, bring in asymmetric information. Suppose that individuals are of two types that differ in their accident probability, and each individual knows his own type. Given the differences in risks, the high-risk individual has lower expected wealth (denoted $\widehat{W}$ in Figure 2) than the low-risk individual (denoted $\bar{W}$). If the insurance company could observe who was low-risk and who was high-risk, then, on the same reasoning as above, the equilibrium contracts would be at A and B in Figure 2. A higher accident probability gives rise to a flatter indifference curve (the two types' indifference curves satisfy the 'single crossing property'), and also to more costly insurance (a flatter fair-odds line). If, however, the insurance firms do not know the characteristics of individuals, then clearly they cannot offer contracts A and B. For in that case all individuals would claim they were the low-risk type and choose the contract A, and the insurance companies would not break even. Offers that survive the competitive process cannot specify a price at which customers choose to buy all the insurance they want, because the high-risk individuals would always purchase more insurance at that price than the low-risk individuals, and the insurance firms would not break even. Competitive offers of contracts instead consist of both a price and a quantity.

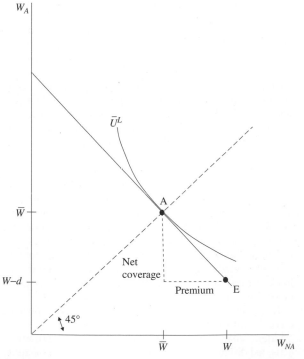

Figure 1

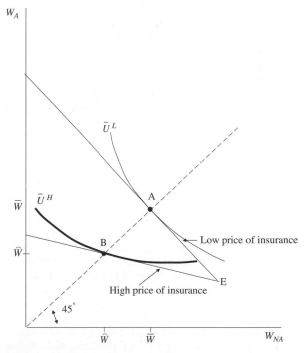

Figure 2

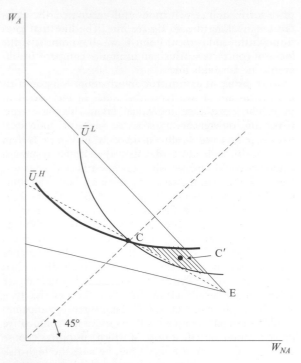

Figure 3

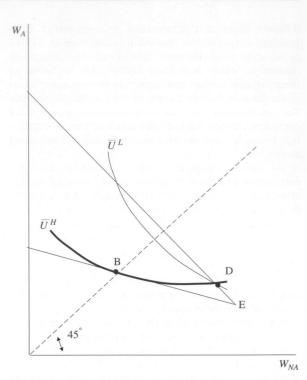

Figure 4

Consider then the price and quantity offer at point C in Figure 3, which would break even if all individuals purchased it (a 'pooling' contract). Rothschild and Stiglitz demonstrate that this *cannot* be equilibrium. Any contract such as C′ in the shaded area in Figure 3 – with slightly *less* insurance coverage than C but at a *lower* price per dollar of coverage – would attract only the low-risk individuals. Given that, the contract C generates losses and would be withdrawn.

The only possible equilibrium, which is illustrated in Figure 4, is one in which the market distinguishes types by offering a contract B with complete coverage (which will be chosen by the high-risk individuals), and a contract D with partial insurance coverage (which the low-risk individuals prefer to full insurance at B and which the high-risk individuals do not prefer to full insurance at B). In this case, the market 'solves' the screening problem, but at the cost of foreclosing otherwise feasible and desirable exchanges.

If, however, there are so few high-risk types that low-risk individuals are strictly better off at a pooling contract, then there can be no competitive equilibrium at all! The pooling contract with full insurance breaks the candidate separating equilibrium, but a contract at a slightly lower price and lower quantity of insurance cream skims the low-risk individuals. Thus the contract with full insurance cannot break even and is withdrawn.

With these intuitive graphs, Rothschild and Stiglitz demonstrate the non-robustness to considerations of imperfect information of two central results of the neo-classical model – that equilibrium is characterized by supply equals demand, and that equilibrium always exists. In papers published over the next two decades, Stiglitz demonstrated how market responses to the screening problem could explain puzzles in equity, credit, and labour markets. For instance, in equity markets, when insiders in a firm have more information than outsiders, the controlling insiders' willingness to issue equity conveys a signal that says that on average the shares are overpriced. The market responds by lowering the price. This discourages firms from issuing new shares and provides an explanation for the fact that firms have limited access to equity (Greenwald, Stiglitz and Weiss, 1984).

In credit markets, if prospective borrowers have more information than lenders about the riskiness of their investments, there are situations in which a lender will set his interest rate below the market-clearing rate. He will not wish to raise his interest rate to what the market will bear if an increase in the rate would lead the lowest-risk borrowers to drop out of the market and reduce the lender's expected return. In this case, credit rationing will occur, as demonstrated in Stiglitz and Weiss (1981).

In many markets, individuals can, at a cost, provide credible information about their characteristics. Then 'hidden characteristics' become public information. This led Stiglitz to examine the incentives to acquire and transmit information. A central point is that the *private* returns to the provision of information differ from the *social* returns; thus the level of information that is public in a signalling equilibrium has, in general, no efficiency properties. To see this, consider the following simple example from Stiglitz (1975).

Suppose that there are two ability types whose productivity is A^H and A^L (the more able can do in one hour what the less able take A^H/A^L hours to do). Suppose a fraction of the population p is high ability and a fraction $1-p$ is low ability. Ability is private information but, at a cost C, an individual can reveal it (as would occur, for instance, if there is a credential that only a high-ability type is able to obtain but that certifies a skill unrelated to productivity). Then there exist two equilibria – a *separating* equilibrium and a *pooling* equilibrium – if

$$A^L < A^H - C < pA^H + [1-p]A^L.$$

If all other high-ability types screen, then the first inequality implies that the remaining high-ability individual has an incentive to screen. In doing do, he earns more than his alternative wage in the screening equilibrium, A^L. This establishes that a separating equilibrium exists.

However, if no other high-ability types screen, then both types are paid their average productivity, and the second inequality implies that the remaining high-ability individual has no incentive to screen. This establishes that a pooling equilibrium exists, as well, and that it Pareto dominates the separating equilibrium. In the separating equilibrium, each worker fails to take into account the effect of his decision to screen on the wage of unscreened individuals. Individual decisions create a diffuse externality.

The fact that the investor in information must obtain a positive expected private return from his information-gathering activities led Stiglitz to a fundamental result in finance. There had long been a theory, the *efficient markets theory*, which states that the observation of prices in capital markets suffices to reveal all relevant private information. Grossman and Stiglitz (1980) showed that the theory was incorrect: if information is costly and markets are competitive, then there must be an 'equilibrium degree of disequilibrium' – persistent discrepancies between prices and 'fundamental values' that provide incentives for individuals to obtain information. In capital markets, prices serve two functions: besides being used in the conventional way to clear markets, they also convey information. When individuals invest in information and thereby learn that the return to a security is going to be high (or low), they bid its price up (or down), and thus the price system makes that information publicly available. But if all information were publicly conveyed, there would be no incentives for individuals to invest in information.

Differential information can be a source of pure economic rents. Stiglitz argued that firms exploit that fact by creating 'noise'. Knowing that it is costly for customers

to search, Salop and Stiglitz (1977) showed that stores can exploit that by varying their prices to extract rents from customers with high search costs. The market equilibrium prices serve to discriminate (imperfectly) among individuals with different search costs. These results overturned a standard theory, *the law of the single price* (a given commodity is sold at the same price in all stores). In a similar vein, Edlin and Stiglitz (1995) argued that managers will have an incentive to enhance asymmetries of information between them and rival managers and boards of directors, and thereby limit the scope for takeovers.

Hidden actions and agency theory

In the standard neoclassical model, there are no conflicts of interest between economic actors. *Principal–agent* theory introduces conflicts of interest by specifying actions that cannot be observed. In this theory, a principal, who delegates a task to an agent, designs a contract that makes payment depend on observable circumstances (for example, revenues) that are correlated with the desired, but unobservable, actions of the agent. Stephen Ross, James Mirrlees, and Joe Stiglitz contemporaneously developed principal–agent theory.

Stiglitz's initial contribution was stimulated by his observations in Kenya during parts of 1969–71. He analysed a puzzle that had been recognized at least since Alfred Marshall – the apparent inefficiency of the institution of sharecropping, which assigns the tenant only a share of the marginal return to his effort. Stiglitz (1974) showed that sharecropping could be advantageous to tenants and landlords because of the *savings in monitoring costs* compared to a wage system with costly monitoring, the *increases in output* compared with a wage system with imperfect monitoring, and the *reduction in risk borne by tenants* compared with a system where workers pay fixed land rents but do not have access to risk markets.

Four insights in this paper and Stiglitz's other work in principal–agent theory have been important for further developments in economics.

1. It is sometimes useful to take the *transaction*, rather than the *market*, as the unit of analysis.
2. There is a trade-off between incentives and insurance if the principal has greater ability to bear risk than the agent. Since the first-order effect of distorting incentives is zero, the provision of a small level of insurance is in this case always welfare-increasing.
3. The distribution of wealth influences the extent of agency problems, in both rich and poor countries.
4. Agency problems are pervasive in a complex modern economy. Agency theory has contributed to new theories of public finance, of corporate governance, and of positive political economy.

However, in many situations, incentive contracts cannot be written because an individual's individual contribution to output is not well observed. In that case, pure economic rents can play an important role in providing incentives, as in Shapiro and Stiglitz (1983). But integrating the idea of rents into a model of a competitive economy initially posed a puzzle. If price exceeds marginal cost, why doesn't competition lead to price-cutting?

An antecedent to Shapiro and Stiglitz (1984) was the model in Shapiro (1983) of rents as an incentive to quality that is unobservable at the time of purchase. In that model, firms develop a reputation for quality by the goods that they produce. The prospect of the loss of rents to a firm that 'milks' its reputation by selling at a high price less than the promised quality induces the firm to live up to consumers' expectations. Competition does not lead to price-cutting because consumers come to learn that, if the price is too low, firms do not have an incentive to maintain their reputation, and therefore the offer of high-quality goods at a low price is not credible.

Shapiro and Stiglitz (1984) extends to a labour market the idea of rents as an incentive device for difficult-to-monitor effort. Workers in this model are identical (so there is no selection issue). Firms observe at random intervals whether a worker is working or shirking. To elicit effort, each firm would want to offer a higher wage than other firms so that, if it finds a worker shirking, he suffers a cost when he is fired. But if it benefited one firm to raise its wage, it would benefit all firms. This might seem like the dilemma where, if every spectator in the stadium stands up to get a better view, no one sees any better. But, when all employers raise their wages, those actions have a real effect on the economy: unemployment emerges since the higher wage rations firms' demand for workers. Now a worker who is fired cannot immediately find another job. This makes job loss costly. Unemployment creates an incentive to work on the job rather than shirk, and so competitive equilibrium will be characterized by unemployment and pure labour rents.

Macroeconomics

Informational imperfections limit the scope of equity and credit markets, as well as insurance and labour markets. In a series of papers with Bruce Greenwald and Andrew Weiss (for example, Greenwald, Stiglitz and Weiss, 1984), Stiglitz drew out the implications for the fluctuations in output and employment that have characterized capitalism throughout its history. The central argument was this. Limitations in the scope of equity markets in the presence of significant bankruptcy costs lead firms to behave in a risk-averse manner. They pay attention to own risk, while traditional theory suggests that the only risk that firms should care about is the correlation with the stock market. Higher levels of investment or production entail increased debt, and, as debt is increased, the bankruptcy probability is increased. Firms will therefore produce and invest only up to the point where expected marginal returns equal expected marginal

bankruptcy costs. This has four implications that contrast with what would occur with perfect markets:

1. *Amplification of small shocks.* Changes in the net worth of the firm or in the riskiness of the environment affect the production and investment decisions of the firm (in contrast to the standard theory). For a highly leveraged firm, small changes in demand can result in large changes in output and employment. Thus, disturbances to the economy tend to be amplified.
2. *Persistence.* If for some reason net worth is reduced at a given time, production falls in subsequent periods. Only gradually will production be restored to normal, as net worth builds up again.
3. *Risk-averse banks.* Banks are a specialized kind of firm whose production activity is making loans. A reduction in the net worth of banks and an increase in the riskiness of their environment will lead them to contract their output – that is, to make fewer loans, which has multiplier effects throughout the economy.
4. *Worsening of the applicant pool for loans during a recession.* For any given bankruptcy cost, there is a critical net worth such that, below that net worth, firms act in a risk-loving manner, and, above that net worth, in a risk-averse manner. If the economy moves into a recession and firms find their net worth decreases, good (that is, risk-averse) firms reduce their loan applications, bad (that is, risk-loving) firms increase their loan applications, so that there is an increasing proportion of bad (that is, low net worth) applicants. These effects may be so strong as to lead to a situation where banks make no loans at all!

During Stiglitz's tenure as Chief Economist of the World Bank, the contrast between his perspective on macroeconomics and the perspective based on well-functioning markets came to a head. These are two starkly different ways of looking at the world. If there are well-functioning markets, then opening up capital markets will lead to efficient outcomes. This view was identified with the US Treasury Department and the IMF in the 1990s. During the 1997–99 East Asian financial crisis, a condition of IMF financial support was that the East Asian economies adopt contractionary fiscal and monetary policies. The contractionary monetary policy would raise interest rates and, at some point, reverse private capital outflows and restore the ability of the East Asian countries to repay their foreign debts. It was argued that the basic reason for the East Asian financial crisis was lack of transparency, or corruption, in the business practices of these economies, which frightened away foreign investors.

In contrast, Furman and Stiglitz (1998) argued that a lack of transparency did not cause the crisis (although it aggravated the effects of the downturn once it began). They argued that small developing countries are financially fragile. There are pervasive externalities in banks' and firms' decisions to obtain short-term loans from abroad. Each bank and each firm takes the risk environment as given, and yet the aggregate set of decisions determines the risk of a financial crisis. This meant that some limits on free capital markets were appropriate in developing countries. Moreover, Furman and Stiglitz argued that policies that increased interest rates in the East Asian economies would greatly erode the net worth of debtors, and the erosion of their net worth would lead to a recession that could not easily be reversed.

Stiglitz's views ultimately were influential. However, the openness of his conflict with the IMF and US Treasury frayed his relationships with many people in Washington and hastened his departure from the World Bank.

Development economics

Whereas macroeconomics remains split between different schools with contrasting views on the importance of market imperfections, the centrality of market imperfections in the field of development economics is not questioned. Before the development of the economics of information (and also the development of game theoretic models of political economy), economists lacked a broad framework for understanding of the sources of the imperfections in markets. Economists who tried to design policies to fit developing country markets generally assumed rigidities in markets, but did not explain them by reference to a choice-based perspective. Abhijit Banerjee (2001, p. 465) has characterized development economics in this era as the 'ugly duckling' of economics: 'It was full of strange assumptions and contrary logic, and all the other [fields of] economics made fun of it.'

Stiglitz's work in development economics played a major role in transforming the field. His models were important in establishing (*a*) that positive feedback mechanisms can give rise to multiple equilibria and underdevelopment traps; (*b*) that, because the causes of market failures and constraints on growth vary greatly from setting to setting, analysis has to be done case by case; and (*c*) that non-market institutions need have no efficiency properties. Important applications of these ideas are below.

Trade-off between diversity of goods and scale economies. In a path-breaking model, Dixit and Stiglitz (1977) posed a question seemingly unrelated to development. This paper addressed the question: will a market solution yield the socially optimum kinds and quantities of commodities if there are multiple possible varieties of goods, each produced by a single firm with increasing returns to scale in production? The desire by consumers for diversity meant that there would be many firms, but not necessarily the optimal number.

Dixit and Stiglitz used a modelling assumption that turned out to be very useful analytically. By assuming a continuum of goods, their set-up lets the modeller respect the discrete nature of many location decisions and yet analyse the model in terms of the behaviour of continuous variables like the share of manufacturing in a particular region.

This model became a building block in models in the new fields of endogenous growth theory and economic geography. To understand the flavour of this work, consider an economy with three sectors: a low-technology sector, an advanced sector, and an intermediate sector that produces an array of non-traded, i.e., domestic, goods, modelled as Dixit–Stiglitz commodities, which are inputs into the advanced sector. An expansion of the advanced sector increases the demand for non-traded inputs, and so lowers their average costs and increases the available variety. With a greater variety of intermediate inputs, production in the advanced sector is more efficient. It can thus be the case that, when many other firms enter the advanced sector, it pays the remaining firms in the traditional sector to do so; but, when all other firms remain in the traditional, low-technology sector, it pays the remaining firm to do so, too. A low-level equilibrium can therefore be sustained even when the economy is fully open to international trade.

Breakdown of the Washington consensus. The standard neoclassical model predicts that growth is inevitable in capital-poor market economies: over time, all economies will converge in per capita income. This model led a generation of economists to a simple set of policy prescriptions that would set the preconditions for growth: maintain macroeconomic stability (since high inflation interferes with the workings of the price system), limit government ownership of enterprise, and deregulate ('stabilize, privatize, and liberalize'). This so-called Washington Consensus has broken down, in part because it has become clear that there are no sure-fire formulas for success, and in part because of the evolution of economic theory away from the perfect markets paradigm. The three central developments in economic theory have been the economics of information, game theory, and institutional economics. In the new economic theory, development is no longer seen primarily as a process of capital accumulation, but instead as a process of organizational change. Evidence of the breakdown of the Washington consensus is that a recent World Bank volume that reviews economic growth in developing countries in the 1990s states that 'The central message of this volume is that there is no unique universal set of rules [to promote growth]…[W]e need to get away from formulae and the search for elusive "best practices".' (World Bank, 2005, p. xiii).

Dysfunctional institutions. In the past, many scholars have made the argument that institutions that emerge out of individual actions are necessarily optimal: they are there because their benefits outweigh the costs. Stiglitz's work on sharecropping (Stiglitz 1974) exemplifies that approach. However, as Stiglitz has often remarked, that analysis is partial equilibrium. That analysis studies the optimal contract while holding fixed everything else in the economy.

In many cases, however, contracts that individuals enter into impose externalities on other agents. There may be no forces that ensure the Pareto efficiency of the set of contracts that individuals adopt. For instance, when insurance is provided through family and friends as well as through the market, the informal insurance will *raise* welfare if it provides (sufficient) peer monitoring (Arnott and Stiglitz, 1990), but, otherwise, such insurance will *lower* welfare because the additional insurance exacerbates moral hazard and so raises the cost of market insurance – an effect that no individual internalizes. The analysis of the inefficiency of contracting choices generalizes widely, for example, to technological change (Acemoglu, 1997), to neighbourhood formation (Hoff and Sen, 2005), and to institutional change, as I discuss next.

One of the most important economic transformations in modern history began with the collapse of Communism in Eastern Europe and the former Soviet Union.... The transition process in the 1990s entailed the creation of a new set of economic and political institutions and, in most countries, produced an unexpectedly deep and prolonged depression. In Russia and many other transition countries, the rapid transfer of state enterprises to private hands ('Big Bang privatization') did not lead to a political demand for institutional reforms needed to govern private property, as many economists had expected. Hoff and Stiglitz (2004; 2007) investigate the influence of economic policies on the demand for a rule of law. They show that Big Bang privatization can create powerful incentives to strip assets and to delay the establishment of a rule of law. This may result in a long period of economic decline. The cause is that no individual takes account of the effect of his economic choices on long-run institutional change.

Evaluation

The high level of idealization in much of Stiglitz's formal work, and the surprising (or at least counter-Walrasian) results often obtained, have led his harshest critics to see in his work a predilection for the intriguing exception rather than the general rule: granted that market failures occur, how much do they really matter? From a staunch admirer, Avinash Dixit, one hears the statement that a paper by Stiglitz begins with the phrase, 'Assume there are two types'. However, the statement that 'there are two types' (or 'two actions') in Stiglitz's papers of the 1970s and 1980s marked a radical departure from the standard model, which implicitly assumes that in each market there is only one type (or one action); that is, that information in the market is symmetric. This modest relaxation of the perfect information assumption reveals that symmetric information is essential to the results of the standard neoclassical paradigm.

Stiglitz's work demonstrates that the standard theory misconstrues many of the virtues of the market. The standard theory exaggerates the role of prices in conveying information about scarcity, and fails to take account of the difficulties of making the price system work. At the same time, the standard theory fails to recognize some central virtues of the market – its ability to address problems of selection, incentives, information gathering, and innovation – because the standard paradigm is silent about all these problems.

Compared to the state of economics in 1970, mainstream theory can now accommodate a far broader ranger of phenomena. But there is no unified single framework, as there was in the Walrasian paradigm. Instead, there is a fragmented collection of disparate models. Distinct models are capable of explaining the same phenomena but are difficult to distinguish empirically.

A further problem that remains for future work is that multiple forms of private information exist within any sector. With multiple incentive problems, it is necessary to consider the distortion among incentives that results when incentives for more easily observed actions are created at the expense of less easily observed actions.

In these ways, Stiglitz's theoretical work has contributed to the resurgence of empirical work in economics. Kim, Morse and Zingales (2006) document a reversal in the previous 30 years in the importance of theoretical work, which dominated the profession in the 1970s and 1980s, and gave way to the primacy of empirical work in the early 1990s. Much of that empirical work is a response to a body of theory that established that neither markets nor governments work perfectly. Stiglitz's demonstration that imperfect information undermines the results of the standard neoclassical model has shifted not only models of thought in economics, but also the relative importance of different sources of knowledge about the economy.

KARLA HOFF

See also **adverse selection; credit rationing; moral hazard; principal and agent (i); principal and agent (ii); selection bias and self-selection; Washington Consensus.**

Selected works

1970. (With M. Rothschild.) Increasing risk: I. A definition. *Journal of Economic Theory* 2, 225–43.

1971. (With M. Rothschild.) Increasing risk: II. Its economic consequences. *Journal of Economic Theory* 3, 66–84.

1974. Incentives and risk sharing in sharecropping. *Review of Economic Studies* 41, 219–55.

1975. The theory of 'screening', education, and the distribution of income. *American Economic Review* 65, 283–300.

1976. (With M. Rothschild.) Equilibrium in competitive insurance markets: an essay on the economics of imperfect information. *Quarterly Journal of Economics* 90, 630–49.

1977. (With S. Salop.) Bargains and ripoffs: a model of monopolistically competitive price dispersion. *Review of Economic Studies* 44, 493–510.

1977. (With A. Dixit.) Monopolistic competition and optimal product diversity. *American Economic Review* 67, 297–308.

1980. (With S. Grossman.) On the impossibility of informationally efficient markets. *American Economic Review* 70, 393–408.

1981. (With A. Weiss.) Credit rationing in markets with imperfect information. *American Economic Review* 71, 393–410.

1984. (With C. Shapiro.) Equilibrium unemployment as a worker discipline device. *American Economic Review* 74, 433–44.

1984. (With B. Greenwald and A. Weiss.) Informational imperfections in capital markets and macroeconomic fluctuations. *American Economic Review* 74, 194–99.

1986. (With B. Greenwald.) Externalities in economies with imperfect information and incomplete markets. *Quarterly Journal of Economics* 101, 229–64.

1991. (With R. Arnott.) Moral hazard and non-market institutions: dysfunctional crowding out or peer monitoring. *American Economic Review* 81, 179–90.

1993. (With B. Greenwald.) Financial market imperfections and business cycles. *Quarterly Journal of Economics* 108, 77–114.

1994. *Whither Socialism*. Cambridge, MA: MIT Press.

1995. (With A. Edlin.) Discouraging rivals: managerial rent-seeking and economic inefficiencies. *American Economic Review* 85, 1301–12.

1998. (With J. Furman.) Economic crises: evidence and insights from East Asia. *Brookings Papers on Economic Activity* 1998(2), 1–114.

2004. (With K. Hoff.) After the Big Bang: obstacles to the emergence of the rule of law in post-Communist societies. *American Economic Review* 94, 753–63.

2007. (with K. Hoff.) Exiting a Lawless State. Manuscript.

2007. *Selected Scientific Papers of Joseph E. Stiglitz*, vol. 1. Oxford: Oxford University Press.

Bibliography

Acemoglu, D. 1997. Training and innovation in an imperfect labour market. *Review of Economic Studies* 64, 445–64.

Akerlof, G. 1970. The market for 'lemons': quality uncertainty and the market mechanism. *Quarterly Journal of Economics* 84, 488–500.

Arrow, K. 1963. Uncertainty and the welfare economics of medical care. *American Economic Review* 53, 941–69.

Banerjee, A. 2001. Comment on K. Hoff and J.E. Stiglitz, 'Modern Economic Theory and Development'. In *The Future of Development Economics in Perspective*, ed. G. Meier and J.E. Stiglitz. Oxford: Oxford University Press.

Hoff, K. 2001. Beyond Rosenstein-Rodan: the modern theory of coordination problems in development. *Annual World Bank Conference on Development Economics 2000*, 145–176.

Hoff, K. and Sen, A. 2005. Homeownership, community interactions, and segregation. *American Economic Review* 95, 1167–89.

Kim, E.H., Morse, A. and Zingales, L. 2006. What has mattered to economics since 1970? *Journal of Economic Perspective* 20(4), 189–202.

Mirrlees, J. 1971. An exploration in the theory of optimum income taxation. *Review of Economic Studies* 38, 175–208.

Shapiro, C. 1983. Premiums for high quality products as returns to reputations. *Quarterly Journal of Economics* 98, 659–80.

Spence, A.M. 1974. *Market Signaling: The Information Structure of Hiring and Related Processes*. Boston, MA: Harvard University Press.

World Bank. 2005. *Economic Growth in the 1990s: Learning from a Decade of Reform*. Washington, DC: World Bank.

stochastic adaptive dynamics

Stochastic adaptive dynamics require analytical methods and solution concepts that differ in important ways from those used to study deterministic processes. Consider, for example, the notion of asymptotic stability: in a deterministic dynamical system, a state is locally asymptotically stable if any sufficiently small deviation from the original state is self-correcting. We can think of this as a first step toward analysing the effect of stochastic shocks; that is, a state is locally asymptotically stable if, after the impact of a small, one-time shock, the process evolves back to its original state.

This idea is not entirely satisfactory, however, because it treats shocks as if they were isolated events. Economic systems are usually composed of large numbers of interacting agents whose behaviour is *constantly* being buffeted by perturbations from various sources. These *persistent* shocks have substantially different effects from *one-time* shocks; in particular, persistent shocks can accumulate and tip the process out of the basin of attraction of an asymptotically stable state. Thus, in a stochastic setting, conventional notions of dynamic stability – including *evolutionarily stable strategies* – are inadequate to characterize the long-run behaviour of the process. Here we shall outline an alternative approach that is based on the theory of large deviations in Markov processes (Freidlin and Wentzell, 1984; Foster and Young, 1990; Young, 1993a).

Types of stochastic perturbations

Before introducing formal definitions, let us consider the various kinds of stochastic shocks to which a system of interacting agents may be exposed. First, there is the interaction process itself whereby agents randomly encounter other agents in the population. Second, the agents' behaviour will be *intentionally* stochastic if they are employing mixed strategies. Third, their behaviour may be *unintentionally* stochastic if their payoffs are subject to unobserved utility shocks. Fourth, mutation processes may cause one type of agent to change spontaneously into another type. Fifth, in and out-migration can introduce new behaviours into the population or extinguish existing ones. Sixth, the system may be hit by aggregate shocks that change the *distribution* of behaviours. This list is by no means exhaustive, but it does convey some sense of the range of stochastic influences that arise quite naturally in economic (and biological) contexts.

Stochastic stability

The early literature on evolutionary game dynamics tended to sidestep stochastic issues by appealing to the law of large numbers. The reasoning is that, when a population is large, random influences at the individual level will tend to average out, and the aggregate state variables will evolve according to the *expected* (hence deterministic) direction of motion. While this approximation may be reasonable in the short and medium run, however, it can be quite misleading when extrapolated over longer periods of time. The difficulty is that, even when the stochastic shocks have very small probability, their accumulation can have dramatic long-run effects that push the process far away from its deterministic trajectory.

The key to analysing such processes is to observe that, when the aggregate stochastic effects are 'small' and the resulting process is ergodic, the long-run distribution will often be concentrated on a very small subset of states – possibly, in fact, on a *single* state. This leads to the idea of *stochastic stability*, a solution concept first proposed for general stochastic dynamical systems by Foster and Young (1990, p. 221): 'the stochastically stable set (SSS) is the set of states S such that, in the long run, it is nearly certain that the system lies within every open set containing S as the noise tends slowly to zero.' The analytical technique for computing these states relies on the theory of large deviations first developed for continuous-time processes by Freidlin and Wentzell (1984), and subsequently extended to general finite-state Markov chains by Young (1993a). It is in the latter form that the theory is usually applied in economic contexts.

An illustrative example

The following simple model illustrates the basic ideas. Consider a population of n agents who are playing the 'Stag Hunt' game:

	A	B
A	10, 10	0, 7
B	7, 0	7, 7

The *state* of the process at time t is the current number of agents playing A, which we shall denote by $z_t \in Z = \{0, 1, 2, \ldots, n\}$. Time is discrete. At the start of period $t+1$, one agent is chosen at random. Strategy A is a best response if $z_t \leq .7n$ and B is a best response if $z_t \geq .7n$. (We assume that the player includes herself in assessing the current distribution, which simplifies the computations.) With high probability, say $1-\varepsilon$, the agent chooses a best response to the current distribution of strategies; while with probability ε she chooses A or B at random (each with probability $\varepsilon/2$).

We can interpret such a departure from best response behaviour in various ways: it might be a form of experimentation, it might be a behavioural 'mutation', or it might simply be a form of ignorance – the agent may not know the current state. Whatever the explanation, the result is a *perturbed best response process* in which individuals choose (myopic) best responses to the current state with high probability and depart from best response behaviour with low probability.

This process is particularly easy to visualize because it is one-dimensional: the states can be viewed as points on a line, and in each period the process moves to the left by one step, to the right by one step, or stays put. Figure 1 illustrates the situation when the population consists of ten players.

The transitions indicated by solid arrows have high probability and represent the direction of best response, that is, the main flow of the process. The dashed arrows go against the flow and have low probability, which is the same order of magnitude as ε. (The process can also loop by staying in a given state with positive probability; these loops are omitted from the figure to avoid clutter.)

In this example the transition probabilities are easy to compute. Consider any state z to the left of the critical value $z^* = 7$. The process moves right if and only if one more agent plays A. This occurs if and only if an agent currently playing B is drawn (an event with probability $1 - z/10$) and this agent mistakenly chooses A (an event with probability $\varepsilon/2$). In other words, if $z < 7$ the probability of moving right is $R_z = (1 - z/10)(\varepsilon/2)$. Similarly, the probability of moving left is $L_z = (z/10)(1 - \varepsilon/2)$. The key point is that the right transitions have much smaller probability than the left transitions when ε is small. Exactly the reverse is true for those states $z > 7$. In this case the probability of moving right is $R_z = (1 - z/10)(1 - \varepsilon/2)$, whereas the probability of moving left is $L_z = (z/10)(\varepsilon/2)$. (At $z = 7$ the process

Figure 1   Transitions for the perturbed best response process in the Stag Hunt game and a population of ten agents. *Note:* Each vertex represents the number of agents playing action A at a given time. Solid arrows are transitions with high-probability, dotted arrows are transitions with low (order-ε) probability.

moves left with probability .15, moves right with probability .35, and stays put with probability .50.)

Computing the long-run distribution

Since this finite-state Markov chain is irreducible (each state is reachable from every other via a finite number of transitions), the process has a unique long-run distribution. That is, with probability 1, the relative frequency of being in any given state z equals some number μ_z *independently of the initial state*. Since the process is one-dimensional, the equations defining μ are particularly transparent, namely, it can be shown that for every $z < n$, $\mu_z R_z = \mu_{z+1} L_{z+1}$. This is known as the *detailed balance condition*. It has a simple interpretation: in the long run, the process transits from $z+1$ to z as often as it transits from z to $z+1$.

The solution in this case is very simple. Given any state z, consider the directed tree T_z consisting of all right transitions from states to the left of z and all left transitions from states to the right of z. This is called a *z-tree* (see Figure 2).

An elementary result in Markov chain theory says that, for one-dimensional chains, the long-run probability of being in state z is proportional to the product of the probabilities on the edges of T_z:

$$\mu_z \propto \prod_{y<z} R_y \prod_{y>z} L_y. \tag{1}$$

This is a special case of the *Markov chain tree theorem*, which expresses the stationary distribution of any finite chain in terms of the probabilities of its z-trees. (Versions of this result go back at least to Kirchhoff's work in the 1840s; see Haken, 1978, s. 4.8. Freidlin and Wentzell, 1984, use it to study large deviations in continuous-time Wiener processes.)

Formula (1) allows us to compute the order-of-magnitude probability of each state without worrying about its exact magnitude. Figure 2 shows, for example, that μ_3, the long-run probability of state $z = 3$, must be proportional to ε^6, because the 3-tree has six dotted arrows, each of which has probability of order ε. Using this method we can easily compute the relative probabilities of each state.

Stochastic stability and equilibrium selection

This example illustrates a general property of adaptive processes with small persistent shocks. That is, the persistent shocks act as a *selection mechanism*, and the *selection strength increases the less likely the shocks are*. The reason is that the long-run distribution depends on the probability of escaping from various states, and the

| 0 | 1 | 2 | 3 | 4 | 5 | 6 | 7 | 8 | 9 | 10 |

Figure 2   The unique 3-tree

critical escape probabilities are *exponential* in ε. Figure 1 shows, for example, that the probability of all-B (the left endpoint) is larger by a factor of $1/\varepsilon$ than the probability of any other state, and it is larger by a factor of $1/\varepsilon^4$ than the probability of all-A (the right endpoint). It follows that, as ε approaches zero, the long-run distribution of the process is concentrated entirely on the all-B state. It is the unique stochastically stable state.

While stochastic stability is defined in terms of the limit as the perturbation probabilities go to zero, sharp selection can in fact occur when the probabilities are quite large. To illustrate, suppose that we take $\varepsilon = .20$ in the above example. This defines a very noisy adjustment process, but in fact the long-run distribution is still strongly biased in favour of the all-B state. It can be shown, in fact, that the all-B state is nearly 50 times as probable as the all-A state. (See Young, 1998b, ch. 4, for a general analysis of stochastic selection bias in one-dimensional evolutionary models.)

A noteworthy feature of this example is that the stochastically stable state (all-B) does not correspond to the Pareto optimal equilibrium of the game, but rather to the risk dominant equilibrium (Harsanyi and Selten, 1988). The connection between stochastic stability and risk dominance was first pointed out by Kandori, Mailath and Rob (1993). Essentially their result says that, in any symmetric 2×2 game with a uniform mutation process, the risk dominant equilibrium is stochastically stable provided the population is sufficiently large. The logic of this connection can be seen in the above example. In the pure best response process ($\varepsilon = 0$) there are two absorbing states: all-B and all-A. The basin of attraction of all-B is the set of states to the left of the critical point, while the basin of attraction of the all-A is the set of states to the right of the critical point. The left basin is bigger than the right basin. To go from the left endpoint into the opposite basin therefore requires more 'uphill' motion than to go the other way around. In any symmetric 2×2 coordination game the risk dominant equilibrium is the one with the widest basin, hence it is stochastically stable under uniform stochastic shocks of the above type.

How general is this result? It depends in part on the nature of the shocks. On the one hand, if we change the probabilities of left and right transitions in an arbitrary way, then we can force any given state – including non-equilibrium states – to have the highest long-run probability; indeed this follows readily from formula (1). (See Bergin and Lipman, 1996.) On the other hand, there are many natural perturbations that do lead to the risk dominant equilibrium in 2×2 games. Consider the following class of perturbed best response dynamics. In state z, let $\Delta(z)$ be the expected payoff from playing A against the population minus the payoff from playing B against the population. Assume that in state z the probability of choosing A divided by the probability of choosing B is well-approximated by a function of form $e^{h(\Delta(z))/\beta}$ where $h(\Delta)$ is non-decreasing in Δ, strictly increasing at $\Delta=0$, and skew-symmetric ($h(\Delta)=-h(-\Delta)$). The positive scalar β is a measure of the noise level. In this set-up, a state is *stochastically stable* if its long-run probability is bounded away from zero as $\beta \rightarrow 0$. Subject to some minor additional regularity assumptions, it can be shown that, in any symmetric 2×2 coordination game, if the population is large enough, the unique stochastically stable state is the one in which everyone plays the risk-dominant equilibrium (Blume, 2003).

Unfortunately, the connection between risk dominance and stochastic stability breaks down – even for uniform mutation rates – in games with more than two strategies per player (Young, 1993a). The difficulty stems from the fact that comparing 'basin sizes' works only in special situations. To determine the stochastically stable states in more general settings requires finding the path of least resistance – the path of greatest probability – from every absorbing set to every other absorbing set, and then constructing a rooted tree from these critical paths (Young, 1993a). (An absorbing set is a minimal set of states from which the *unperturbed* process cannot escape.) What makes the one-dimensional situation so special is that there are only two absorbing sets – the left endpoint and the right endpoint – and there is a unique directed path going from left to right and another unique path going from right to left. (For other situations in which the analysis can be simplified, see Ellison, 2000; Kandori and Rob, 1995.)

There are many games of economic importance in which this theory has powerful implications for equilibrium selection. In the non-cooperative Nash bargaining model, for example, the Nash bargaining solution is essentially the unique stochastically stable outcome (Young, 1993b). Different assumptions about the one-shot bargaining process lead instead to the selection of the Kalai–Smorodinsky solution (Young, 1998a; for further variations see Binmore, Samuelson and Young, 2003). In a standard oligopoly framework, marginal cost pricing turns out to be the stochastically stable solution (Vega-Redondo, 1997).

Speed of adjustment

One criticism that has been levelled at this approach is that it may take an exceedingly long time for the evolutionary process to reach the stochastically stable states when it starts from somewhere else. The difficulty is that, when the shocks have very small probability, it takes a long time (in expectation) before enough of them accumulate to tip the process into the stochastically stable state(s). While this is correct in principle, the waiting time can be very sensitive to various modelling details. First, it depends on the size and probability of the shocks themselves. As we have already noted, the shocks need not be small for sharp selection to occur, in which case the waiting time need not be long either. (In the above example we found that an error rate of 20 per cent still

selects the all-B state with high probability.) Second, the expected waiting time depends crucially on the *topology* of interaction. In the above example we assumed that each agent reacts to the distribution of actions in the whole population. If instead we suppose that people respond only to actions of those in their immediate geographic (or social) neighbourhood, the time to reach the stochastically stable state is greatly reduced (Ellison, 1993; Young, 1998b, ch. 6). Third, the waiting time is reduced if the stochastic perturbations are not independent, either because the agents act in a coordinated fashion, or because the utility shocks among agents are statistically correlated (Young, 1998b, ch. 9; Bowles, 2004).

Path dependence

The results discussed above rely on the assumption that the adaptive process is ergodic, that is, its long-run behaviour is almost surely independent of the initial state. Ergodicity holds if, for example, the number of states is finite, the transition probabilities are time-homogeneous, and there is a positive probability of transiting from any state to any other state within a finite number of periods. One way in which these conditions may fail is that the weight of history grows indefinitely. Consider, for example, a two-person game G together with a population of potential row players and another population of potential column players. Assume that an initial history of plays is given. In each period, one row player and one column player are drawn at random, and each of them chooses an ε-trembled best reply to the opposite population's previous actions (alternatively, to a random sample of fixed size drawn from the opponent's previous actions). This is a stochastic form of fictitious play (Fudenberg and Kreps, 1993; Kaniovski and Young, 1995). The *proportion* of agents playing each action evolves according to a stochastic difference equation in which the magnitude of the stochastic term decreases over time; in particular it decreases at the rate $1/t$.

This type of process is not ergodic. It can be shown, in fact, that the long-run proportions converge almost surely either to a neighbourhood of all-A or to a neighbourhood of all-B, where the relative probabilities of these two events depend on the initial state (Kaniovski and Young, 1995). Processes of this type require substantially different techniques of analysis from the ergodic processes discussed earlier; see in particular Arthur, Ermoliev and Kaniovski (1987), Benaim and Hirsch (1999) and Hofbauer and Sandholm (2002).

Summary

The introduction of persistent random shocks into models with large numbers of interacting agents can be handled using methods from stochastic dynamical systems theory; moreover, there is virtually no limit on the dimensionality of the systems that can be analysed using these techniques. Such processes can exhibit path dependence if the weight of history is allowed to grow indefinitely. If instead past actions fade away or are forgotten, the presence of persistent random shocks makes the process ergodic and its long-run behaviour is often easier to analyse. An important feature of such ergodic models is that some equilibrium states are much more likely to occur in the long run than others, and this holds independently of the initial state. The length of time that it takes to reach such states from out-of-equilibrium conditions depends on key structural properties of the model, including the size and frequency of the stochastic shocks, the extent to which they are correlated among agents, and the network topology governing agents' interactions with one another.

H. PEYTON YOUNG

See also **agent-based models; evolutionary economics.**

Bibliography

Arthur, W.B., Ermoliev, Y. and Kaniovski, Y. 1987. Adaptive growth processes modelled by urn schemes. *Kybernetika* 6, 49–57; English trans in *Cybernetics* 23, 779–89.

Benaim, M. and Hirsch, M.W. 1999. Mixed equilibria and dynamical systems arising from fictitious play in perturbed games. *Games and Economic Behavior* 29, 36–72.

Bergin, J. and Lipman, B.L. 1996. Evolution with state-dependent mutations. *Econometrica* 64, 943–56.

Binmore, K., Samuelson, L. and Young, H.P. 2003. Equilibrium selection in bargaining models. *Games and Economic Behavior* 45, 296–328.

Blume, L.E. 1995. The statistical mechanics of best-response strategy revision. *Games and Economic Behavior* 11, 111–45.

Blume, L.E. 2003. How noise matters. *Games and Economic Behavior* 44, 251–71.

Bowles, S. 2004. *Microeconomics: Behavior, Institutions, and Evolution.* Princeton, NJ: Princeton University Press.

Ellison, G. 1993. Learning, local interaction, and coordination. *Econometrica* 61, 1047–71.

Ellison, G. 2000. Basins of attraction, long run equilibria, and the speed of step-by-step evolution. *Review of Economic Studies* 67, 17–45.

Foster, D. and Young, H.P. 1990. Stochastic evolutionary game dynamics. *Theoretical Population Biology* 38, 219–32.

Freidlin, M. and Wentzell, A. 1984. *Random Perturbations of Dynamical Systems.* Berlin: Springer-Verlag.

Fudenberg, D. and Kreps, D. 1993. Learning mixed equilibria. *Games and Economic Behavior* 5, 320–67.

Haken, H. 1978. *Synergetics.* Berlin: Springer-Verlag.

Harsanyi, J. and Selten, R. 1988. *A General Theory of Equilibrium Selection in Games.* Cambridge MA: MIT Press.

Hofbauer, J. and Sandholm, W. 2002. On the global convergence of stochastic fictitious play. *Econometrica* 70, 2265–94.

Kandori, M., Mailath, G. and Rob, R. 1993. Learning, mutation, and long-run equilibria in games. *Econometrica* 61, 29–56.

Kandori, M. and Rob, R. 1995. Evolution of equilibria in the long run: a general theory and applications. *Journal of Economic Theory* 65, 383–414.

Kaniovski, Y. and Young, H.P. 1995. Learning dynamics in games with stochastic perturbations. *Games and Economic Behavior* 11, 330–63.

Karlin, S. and Taylor, H.M. 1975. *A First Course in Stochastic Processes.* New York: Academic Press.

Vega-Redondo, F. 1997. The evolution of Walrasian behaviour. *Econometrica* 65, 375–84.

Young, H.P. 1993a. The evolution of conventions. *Econometrica* 61, 57–84.

Young, H.P. 1993b. An evolutionary model of bargaining. *Journal of Economic Theory* 59, 145–68.

Young, H.P. 1998a. Conventional contracts. *Review of Economic Studies* 65, 776–92.

Young, H.P. 1998b. *Individual Strategy and Social Structure.* Princeton. NJ: Princeton University Press.

stochastic dominance

In order to determine whether a relation of stochastic dominance holds between two distributions, the distributions are first characterized by their *cumulative distribution functions* (CDFs). For a given set of incomes, the value of the CDF at income y is the proportion of incomes in the set that are no greater than y. In the context of a random variable Y, the value of the CDF of the distribution of Y at y is the probability that Y should be no greater than y.

Suppose that we consider two distributions A and B, characterized respectively by CDFs F_A and F_B. Then distribution B dominates distribution A stochastically at first order if, for any argument y, $F_A(y) \geq F_B(y)$. This definition often looks as though it is the wrong way round, but a moment's reflection shows that it is correct as stated. If y denotes an income level, then the inequality in the definition means that the proportion of individuals in distribution A with incomes no greater than y is no smaller than the proportion of such individuals in B. In other words, there is at least as high a proportion of poor people in A as in B, if poverty means an income smaller than y. If B dominates A at first order, then, whatever poverty line we may choose, there is always more poverty in A than in B, which is why we say that A is the dominated distribution.

Higher orders of stochastic dominance can also be defined. To this end, we define repeated integrals of the CDF of each distribution. Formally, we define a sequence of functions by the recursive definition

$$D^1(y) = F(y), \quad D^{s+1}(y) = \int_0^y D^s(z)\, dz,$$

$$\text{for } s = 1, 2, 3, \ldots.$$

Thus the function D^1 is the CDF of the distribution under study, $D^2(y)$ is the integral of D^1 from 0 to y, $D^3(y)$ is the integral of D^2 from 0 to y, and so on. By definition, distribution B dominates A at order s if $D_A^s(y) \geq D_B^s(y)$ for all arguments y. The lower limit of 0 is used for clarity of exposition; in general it is the lowest income in the pooled distributions. The definition makes it clear that first-order dominance implies dominance at all higher orders, and more generally that dominance at order s implies dominance at all orders higher than s. Since the implications go in only one direction, it follows that higher-order dominance is a weaker condition than lower-order dominance. I will give a more detailed interpretation of the functions D^s shortly in the context of poverty indices.

In theoretical arguments, it is sometimes desirable to distinguish weak from strong stochastic dominance. The above definitions are of weak dominance. For strong dominance, it is required that the inequality should be strict for at least one value of the argument y. In empirical investigations, the distinction is of no interest, since no statistical test can detect a significant difference between weak and strong inequalities.

Some applications make use of the concept of *restricted stochastic dominance*. This means that the relevant inequality is required to hold over some restricted range of the argument y rather than for all possible values. In empirical work, it is often only restricted dominance that can usefully be studied, since with continuous distributions there are usually too few data in the tails of the distributions for statistically significant conclusions to be drawn. Again, for measures of poverty, it is only dominance over the range of incomes up to the poverty line that is of interest.

Relation between stochastic dominance and welfare

When studying either income inequality or poverty, one is automatically in a normative context. Most modern studies make explicit or implicit use of a *social welfare function* (SWF). In a paper by Blackorby and Donaldson (1980) (henceforth BD), various ethically desirable criteria are developed and the sorts of SWF that respect these criteria are characterized.

One of these criteria is the *anonymity* of individuals. If we take all the worldly goods of a rich man and give them to a poor man, and then give the few worldly goods of the poor man to the rich man, then social welfare should be unchanged. Formally, a SWF that respects this requirement is symmetric with respect to its arguments, which are the incomes of the members of society.

Another requirement is the *Pareto principle*. According to it, we should rank situation B better than situation A if at least one individual is better off in B than in A, and no one is worse off. In order for a SWF to respect the Pareto principle, it must be increasing in all its arguments.

Another requirement, for measures of poverty only, is that a poverty index should not depend at all on the incomes of the non-poor. BD show that this implies a separability condition on the SWF. If in addition we require that the poverty index should be defined for arbitrary poverty lines, then the separability condition becomes a requirement of additive separability. The SWF can therefore be written as

$$W(y_1, \ldots, y_N) = \sum_{i=1}^{N} u(y_i), \qquad (1)$$

where the 'utility' function u is increasing in its argument. Alternatively, the SWF can be any increasing transform of the function W. In all cases, the SWF is symmetric and increasing in its arguments, and so satisfies BD's ethical criteria.

It can be seen that first-order stochastic dominance of A by B means that B has higher social welfare than A for all SWFs of the form (1). This can be shown by a simple integration by parts, under the assumption that the function u is differentiable. In fact, this dominance is also a necessary condition for B to have higher welfare than A for all SWFs of the form (1). It follows from the above argument that, if we use first-order stochastic dominance as a criterion for ranking distributions, then we need not restrict attention to a specific SWF, since any SWF of the form (1) gives the same ranking if one distribution dominates the other at first order.

A more restricted class of SWFs is given by functions of the form (1) where we impose the additional restriction that the second derivative of u is negative. It turns out that all the SWFs of this more restricted class give a unanimous ranking of two distributions if one dominates the other at second order. This sort of result can be extended to higher orders of dominance. As the dominance condition becomes progressively weaker, the class of SWFs that give unanimous rankings becomes progressively smaller, subject to more and more restrictive conditions on the function u and its derivatives.

Relation between stochastic dominance and poverty

The so-called *headcount ratio* is sometimes used as a measure of the amount of poverty in a given income distribution. This ratio is the proportion of individuals in the distribution with incomes below, or equal to, the poverty line. If this line is denoted by z, then the headcount ratio is the value of the CDF at z. If we have two populations, A and B, characterized by two CDFs, F_A and

F_B, then, for poverty line z, the headcount ratio is higher in A than in B if and only if $F_A(z) > F_B(z)$. If the inequality $F_A(y) > F_B(y)$ holds for all values of y up to z, then we have restricted first-order stochastic dominance up to z.

Corresponding to any income y less than z, we define the *poverty gap* as $z - y$. When we restrict attention to the welfare of people with incomes less than z, it is convenient to use a function π that measures the disutility of the poverty gap, rather than a function u of the sort used in a SWF. Thus we have a class of *poverty indices*, defined as follows:

$$\Pi(z) = \int_0^z \pi(z - y) \, dF(y),$$

If $\pi' > 0$, which means that the disutility of poverty increases with the poverty gap, it can be shown that, for all poverty indices of the above form, there is more poverty in A than in B if B dominates A at first order over the range of incomes less than or equal to z.

Earlier, a sequence of functions D^s was introduced, these functions being repeated integrals of the CDF. A useful explicit representation of these functions is given by the formula

$$D^s(z) = \frac{1}{(s - 1)!} \int_0^z (z - y)^{s-1} dF(y). \quad (2)$$

The formula clearly holds for $s = 1$, if we remember that $0! = 1$. It is not hard to show by induction that it also holds for integers greater than 1.

For $s = 2$, the formula becomes

$$D^2(z) = \int_0^z (z - y) \, dF(y),$$

from which we see that, for given z, $D^2(z)$ is the average poverty gap for poverty line z. If, for all $z \in [z_-, z_+]$, $D_A^2(z) > D_B^2(z)$, then it follows that the average poverty gap is greater in A than in B for all poverty lines in the interval $[z_-, z_+]$. But this condition is just restricted second-order stochastic dominance of A by B over that interval.

As with welfare functions, this result can be extended. By progressively restricting the admissible class of poverty indices, in particular by imposing signs on the derivatives of π, it can be seen that all poverty indices in these more restricted classes unanimously see more poverty in A than in B if there is a progressively higher order of stochastic dominance; see Davidson and Duclos (2000) for more details. An essential reference on poverty measurement is Atkinson (1987), in which the axiomatic approach of BD is extended to poverty measurement. See also three papers by Foster and Shorrocks (1988a; 1988b; 1988c).

Relation between stochastic dominance and inequality

If a richer person in distribution A transfers some income to a poorer person in such a way that the richer person stays richer after the transfer, the post-transfer distribution B stochastically dominates A at second order. The Pigou–Dalton principle of transfers says that 'Robin-Hood' transfers of the sort described should improve welfare. But it is easy to see that distribution B does not dominate A at first order, and indeed this is right and proper according to the Pareto principle, since the richer person is worse off after the transfer.

This example shows that, when we discuss inequality, we are not talking about the same thing as welfare. Any reasonable measure of inequality must declare that there is no inequality if everyone has the same income, even if everyone is in abject poverty.

The classical tool for studying inequality is the *Lorenz curve*. For any proportion p between zero and one, the ordinate of the corresponding point on the Lorenz curve for a given income distribution is the proportion of total income that accrues to the first $100p$ per cent of people when they are sorted in order of increasing income. By construction, the Lorenz curve fits into the unit square, lies below the 45-degree line that is the diagonal of that square, and is (weakly) convex. Figure 1 displays a typical Lorenz curve.

A distribution B is said to *Lorenz dominate* another distribution A if the Lorenz curve of B lies everywhere above that of A. We then say that there is less inequality in B than in A. But this comparison of A and B is not a welfare comparison, and, in particular, does not allow a comparison of poverty. This defect is remedied by the concept of *generalized Lorenz dominance*, based on the generalized Lorenz curve introduced by Shorrocks

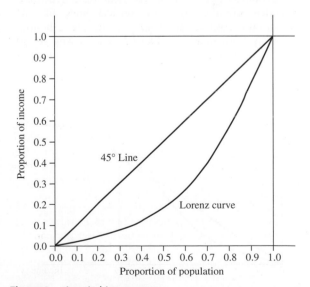

Figure 1   A typical Lorenz curve

(1983). The ordinates of this curve are the Lorenz ordinates multiplied by the average income of the distribution. It turns out that generalized Lorenz dominance is the same thing as second-order stochastic dominance. Either one of these concepts implicitly mixes notions of welfare and inequality, as shown by the fact that the function u in a SWF of form (1) that respects second-order dominance has a negative second derivative, which implies diminishing marginal (social) utility of income. The discussion of the previous section shows that higher-order dominance criteria put more and more weight on the welfare of the poorest members of society.

Graphical representation and quantiles

Consider the setup in Figure 2, where the CDFs of two distributions A and B are plotted. The functions D^2 used for second-order dominance comparisons can be evaluated for a given argument, like z_1 in the figure, as the areas beneath the CDFs, by the usual geometric interpretation of the Riemann integral. We see that distribution B dominates A at second order because, although the CDFs cross, the areas between them are such that the condition for second-order dominance is always satisfied. Thus the vertical line MN marks off a large positive area between the graphs of the two CDFs up to the point at which they cross, and thereafter a small negative area bounded on the right by MN.

For generalized Lorenz dominance, it can be shown that what must be non-negative everywhere is the area between the two curves, bounded not by a vertical line like MN but rather by a horizontal line like KL. This area is the difference between the areas under two *quantile functions*, a quantile function being by definition the inverse of the CDF. Although it is tedious to demonstrate it algebraically, it is intuitively clear that, if the areas bounded on the right by vertical lines like MN are always positive, then so are the areas bounded above by horizontal lines like KL. This is why generalized Lorenz dominance and second-order stochastic dominance are equivalent conditions. The whole theory

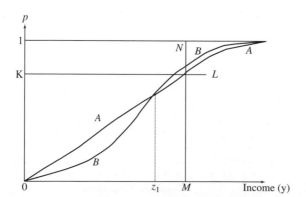

Figure 2   Generalized Lorenz and second-order dominance

of stochastic dominance can be developed using quantiles rather than incomes; this is called a *p-approach*. Such approaches are used to advantage in Jenkins and Lambert (1997; 1998), Shorrocks (1998), and also Spencer and Fisher (1992).

Another thing that emerges clearly from Figure 2 is that the threshold income z_1 up to which first-order stochastic dominance holds is always smaller than the threshold z_2 up to which we have second-order dominance. In the figure, we have second-order dominance everywhere, and so we can set z_2 equal to the highest income in either distribution. More generally, we can define a threshold z_s as the greatest income up to which we have dominance at order s. The z_s constitute an increasing sequence.

A result shown in Davidson and Duclos (2000) is that, if the distribution B dominates A at first-order over a range $[0,z]$, with $z > 0$, then, no matter what happens for incomes above z, there is always *some* order s such that B dominates A at order s over the full range of the two distributions, provided only that that range is finite.

Estimation and inference

Suppose that we have a random sample of N independent observations y_i, $i = 1, \ldots, N$, from a population. Then it follows from (2) that a natural estimator of $D^s(z)$ (for a non-stochastic z) is

$$\hat{D}^s(z) = \frac{1}{(s-1)!} \int_0^z (z-y)^{s-1} \, d\hat{F}(y)$$

$$= \frac{1}{n(s-1)!} \sum_{i=1}^{N} (z-y_i)^{s-1} \mathrm{I}(y_i \leq z),$$

$$(3)$$

where $\hat{F}$ denotes the empirical distribution function of the sample and $\mathrm{I}(.)$ is an indicator function equal to 1 when its argument is true and 0 otherwise. For $s = 1$, the formula (3) estimates the population CDF by the empirical distribution function. For arbitrary s, it has the convenient property of being a sum of independent and identically distributed (IID) variables, which makes it easy to show that (3) is consistent and asymptotically normal. The asymptotic variance is also easy to estimate in a distribution-free manner, by which is meant that no parametric assumptions need be made about the distributions under study.

When two distributions are compared for stochastic dominance, two kinds of situations typically arise. The first is when there are two independent populations, with random samples from each. The other arises when we have independent paired drawings from the same population. For instance, one variable could be before-tax income, and the other after-tax income for the same individual. Explicit expressions for the asymptotic variance of the difference between the estimates of $D^s(z)$

for the case of independent samples were given as early as 1989 in an unpublished thesis (Chow, 1989). The sampling distribution of a related estimator for poverty indices and independent samples is found in Kakwani (1993), Bishop, Chow and Zheng (1995) and Rongve (1997). For a different approach to inference on stochastic dominance, see Anderson (1996). A comprehensive approach to inference on stochastic dominance is found in Davidson and Duclos (2000).

The approach proposed by McFadden (1989) is based on the supremum of the difference between the estimates (3) for two independent populations. For $s = 1$, this turns out to be a variant of the Kolmogorov–Smirnov test, with known properties. For higher values of s, although it is easy to compute the statistic, its asymptotic properties under the null are not analytically tractable. However, simulation-based methods can surmount this difficulty; see Barrett and Donald (2003).

A somewhat vexed question in testing for dominance is whether to test the null hypothesis of dominance or that of non-dominance. The latter has the advantage that, if the null is rejected, all that remains is dominance. More generally, the former approach rejects the null of dominance only when there is clear evidence against it, and the latter accepts the alternative of dominance only when there is clear evidence in its favour. The former approach is more common in the literature; see for instance Richmond (1982), Beach and Richmond (1985), Wolak (1989), and Bishop, Formby and Thistle (1992). The latter is discussed in an unpublished paper (Howes, 1993) and in Kaur, Prakasa Rao and Singh (1994).

RUSSELL DAVIDSON

See also **Dalton, Edward Hugh John Neale; ethics and economics; Gini ratio; income mobility; inequality (measurement); Pigou, Arthur Cecil; poverty; social welfare function; tax incidence.**

Bibliography

Anderson, G. 1996. Nonparametric tests of stochastic dominance in income distributions. *Econometrica* 64, 1183–93.

Atkinson, A. 1987. On the measurement of poverty, *Econometrica* 55, 749–64.

Barrett, G. and Donald, S. 2003. Consistent tests for stochastic dominance. *Econometrica* 71, 71–103.

Beach, C. and Richmond, J. 1985. Joint confidence intervals for income shares and Lorenz curves. *International Economic Review* 26, 439–50.

Bishop, J., Chow, K. and Zheng, B. 1995. Statistical inference and decomposable poverty measures. *Bulletin of Economic Research* 47, 329–40.

Bishop, J. Formby, J. and Thistle, P. 1992. Convergence of the South and non-South income distributions, 1969–1979. *American Economic Review* 82, 262–72.

Blackorby, C. and Donaldson, D. 1980. Ethical indices for the measurement of poverty. *Econometrica* 48, 1053–62.

Chow, K. 1989. Statistical inference for stochastic dominance: a distribution free approach. Ph.D. thesis, University of Alabama.

Davidson, R. and Duclos, J.-Y. 2000. Statistical inference for stochastic dominance and for the measurement of poverty and inequality. *Econometrica* 68, 1435–64.

Foster, J. and Shorrocks, A. 1988a. Poverty orderings. *Econometrica* 56, 173–7.

Foster, J. and Shorrocks, A. 1988b. Poverty orderings and welfare dominance. *Social Choice and Welfare* 5, 179–98.

Foster, J. and Shorrocks, A. 1988c. Inequality and poverty orderings. *European Economic Review* 32, 654–62.

Howes, S. 1993. Asymptotic properties of four fundamental curves of distributional analysis. Unpublished paper, STICERD, London School of Economics.

Jenkins, S. and Lambert, P. 1997. Three 'I's of poverty curves, with an analysis of UK poverty trends. *Oxford Economic Papers* 49, 317–27.

Jenkins, S. and Lambert, P. 1998. Three 'I's of poverty curves and poverty dominance: TIPs for poverty analysis. *Research on Economic Inequality* 8, 39–56.

Kakwani, N. 1993. Statistical inference in the measurement of poverty. *Review of Economics and Statistics* 75, 632–39.

Kaur, A., Prakasa Rao, B. and Singh, H. 1994. Testing for second-order stochastic dominance of two distributions. *Econometric Theory* 10, 849–66.

McFadden, D. 1989. Testing for stochastic dominance. In *Studies in the Economics of Uncertainty*, ed. T. Fomby and T. Seo. New York: Springer-Verlag.

Richmond, J. 1982. A general method for constructing simultaneous confidence intervals. *Journal of the American Statistical Association* 77, 455–60.

Rongve, I. 1997. Statistical inference for poverty indices with fixed poverty lines. *Applied Economics* 29, 387–92.

Shorrocks, A. 1983. Ranking income distributions. *Economica* 50, 3–17.

Shorrocks, A. 1998. Deprivation profiles and deprivation indices. In *The Distribution of Household Welfare and Household Production*, ed. S. Jenkins et al. Cambridge: Cambridge University Press.

Spencer, B. and Fisher, S. 1992. On comparing distributions of poverty gaps. *Sankhya: The Indian Journal of Statistics Series B* 54, 114–26.

Wolak, F. 1989. Testing Inequality constraints in linear econometric models, *Journal of Econometrics* 41, 205–35.

stochastic frontier models

The stochastic frontier model was first proposed by Aigner, Lovell and Schmidt (1977) and Meeusen and van den Broeck (1977) in the context of production function estimation. The model extends the classical production function estimation by allowing for the presence of technical inefficiency. The idea is that, although the production technology is common knowledge to a group of producers, efficiency in using that technology in the production process may vary by producers, with the degree of efficiency depending possibly on factors such as experience, management skills, and so on. Given the technology, fully efficient producers may realize the full potential of the technology and obtain the maximum possible output for given inputs, while less efficient producers see their output fall short of the maximum possible level. Therefore, the underlying technology defines a *frontier* of production, and actual outputs observed in the data may fall below the frontier because of the presence of technical inefficiency.

A stochastic production frontier model can be specified as

$$\ln y_i = \ln y_i^* - u_i, \quad u_i \geq 0, \quad (1)$$

$$\ln y_i^* = f(x_i; \beta) + v_i, \quad (2)$$

where y_i is the observed output of producer i, y_i^* is the potential output which is subject to a zero-mean random error v_i, x_i and β are vectors of inputs and the corresponding coefficients, respectively, and $u_i \geq 0$ is the effect of technical inefficiency. Equation (2) defines the *stochastic frontier* of the production function; it is stochastic because of v_i. Given that $u_i \geq 0$, observed log of output ($\ln y_i$) is bounded below the frontier. The value of $100 * u_i$ is the percentage by which output can be increased using the same inputs if production is fully efficient. The model without u_i reduces to the classical specification of a production function.

A popular empirical strategy in estimating the above model is to impose distributional assumptions on u_i and v_i, from which a likelihood function can be derived and estimated. For instance, one may assume that

$$v_i \sim N(0, \sigma^2), \quad (3)$$

$$u_i \sim N^+(\mu, \sigma_u^2), \quad (4)$$

where $N^+(\cdot)$ indicates the positive truncation of a normal distribution. The positive truncation gives non-negative values of u_i and hence ensures that firms are constrained by the technology frontier. By making μ and/or σ_u^2 functions of observables (such as ages and years of schooling), one can model the determinants of inefficiency.

The distribution assumption of (4) encompasses many of the models in the literature as special cases. For instance, the half-normal distribution of u_i proposed by Aigner, Lovell and Schmidt (1977) is obtained by restricting $\mu = 0$ and σ_u^2 to be a constant. The half-normal density has a mode at 0 which implies that the majority of the producers are clustered near full efficiency level. The assumption may be unnecessarily restrictive,

in particular for industries in which certain degree of inefficiency is expected for the producers. The assumption is relaxed by having $\mu \neq 0$ to allow the mode to depart from 0. Since limited theory is available in guiding the choice of u_i's distribution, various distribution assumptions are explored in the literature for their flexibility in shaping the distribution (for example, the Gamma distribution of Greene, 1980) and/or for checking the robustness of estimation results.

It is often of great empirical interest to estimate the degree of inefficiency (u_i) for each producer (observation). The observation-level estimates are obtained using the estimator $E(u_i | v_i - u_i)$ proposed by Jondrow et al. (1982). The value of $100 \times E(u_i | v_i - u_i)$ gives the percentage by which output is increased if production is fully efficient. Similarly, an efficiency index is estimated using $E(\exp(-u_i) | v_i - u_i)$ (Battese and Coelli, 1988). The estimated value gives the actual output as a share of potential output, and the value is bounded between 0 and 1. A likelihood ratio test of the null hypothesis that u_i equals 0 can be performed to test for the presence of inefficiency. It amounts to testing the model against its OLS counterpart (the model without u_i). The distribution of the test statistic, however, is non-standard, because the value of 0 is on the boundary of u_i's support. Alternatively, given that an obvious difference between v_i and $v_i - u_i$ is the skewness of the latter, Schmidt and Lin (1984) suggest a simple test based on the sample skewness of the OLS residuals. If $v_i - u_i$ is the correct specification, the residuals would skew to the left and the null hypothesis of a normal error would be rejected.

If panel data is available, the model may be written as (for the ease of illustration, assume that the deterministic part of the frontier function is linear):

$$\ln y_{it} = \alpha + x_{it}' \beta + v_{it} - u_i, \quad u_i \geq 0, \quad (5)$$

where α is a constant. One may impose distributional assumptions on v_{it} and u_i to derive the likelihood function of the model. Alternatively, a distribution-free approach suggested by Schmidt and Sickles (1984) is available. In this approach, one defines $\alpha_i = \alpha - u_i$ and assumes that the u_i is an individual-specific parameter. With the definition of α_i substituted into (5), the model is estimated by standard fixed-effect panel estimators which yield consistent estimates of α_i for a large T. Since $\alpha_i = \alpha - u_i$ and $u_i \geq 0$, one then recovers the estimated values of α and u_i using the normalization equations of $\hat{\alpha} = max\{\hat{\alpha}_i\}$ and $\hat{u}_i = \hat{\alpha} - \hat{\alpha}_i$. This normalization procedure amounts to counting the most efficient firm in the sample as 100 per cent efficient.

By duality, technical inefficiency in the production also leads to a higher cost of production. The estimated cost of technical inefficiency often has important policy implications, and estimation can be done using a stochastic cost frontier model in a cost minimization framework. The model specification is:

$$\ln C_i = \ln C_i^* + u_i', \quad u_i' \geq 0, \tag{6}$$

$$\ln C_i^* = g(w_i, y_i; \gamma) + v_i', \tag{7}$$

where C_i is the observed cost of producer i, C_i^* is the efficient level of cost which is subject to a zero-mean random error v_i', w_i is the vector of input prices, γ is the vector of coefficients, and $u_i' \geq 0$ is the effect of inefficiency on the cost of production. Equation (7) defines the stochastic cost frontier, and the observed cost lies above the frontier. The value of $100 * u_i'$ measures the extra cost as a percentage of the minimum cost. Econometric analysis of (6) and (7) is similar to that of the production function model. A notable difference is that the cost model's OLS residuals skew to the right if inefficiency presents in the data.

An advantage of a cost function approach over a production function approach is that the issue of *allocative inefficiency* can be addressed in addition to the technical inefficiency. Allocative inefficiency refers to the use of improper input combinations, that is, the marginal rate of technical substitution between inputs departs from the input price ratio. The improper input mix increases the cost of production, and the effect is not the same as technical inefficiency. Because the analysis of allocative inefficiency requires information of input prices, it is usually carried out in a cost minimization framework. To jointly estimate both technical and allocative inefficiency, Schmidt and Lovell (1979) provide the solution technique for a cost system in which the production technology is Cobb-Douglas. Kumbhakar (1997) presents a theoretical solution for a model with a translog cost function, and the difficulty in the empirical implementation of this model is discussed and resolved in Kumbhakar and Wang (2006) and Kumbhakar and Tsionas (2005).

Although the stochastic frontier model is most often applied to the estimation of production and cost functions, an increasing body of research has adopted the methodology to other fields of study. Hofler and Murphy (1992) apply this estimation approach to labour market search models. Due to the costs of search, observed wages tend to fall below the maximum offers that are available in the market; this shortfall is analogous to a technical inefficiency. Another application is found in the study of financing constraints on investment, where Wang (2003) models the frictionless level of investment as the frontier, and actual investment falls below the frontier because of financing constraints. This approach allows Wang to quantify the effect of financing constraints on investment (represent by u_i), which is infeasible with the conventional linear-regression approach. In an application to economic growth, Kumbhakar and Wang (2005) employ the stochastic frontier approach and model growth convergence as countries' movements towards the world production frontier. A country may fall short of producing the maximum possible output because of technical inefficiency, and the phenomenon of technological catch-up is observed if the country moves towards the world production frontier over time. By making u_i a function of time and other macro variables, Kumbhakar and Wang test and confirm the convergence hypothesis.

The stochastic frontier model also finds applications in finance. For example, a long-standing issue in the finance literature is whether the initial public offering (IPO) underpricing – the phenomenon whereby the initial offer price of an IPO is below the closing day's bid price – is deliberate on the firm's part or not. Hunt-McCool, Koh and Francis (1996) adopt the stochastic frontier model to investigate the issue, in which u_i measures the difference between the maximum predicted offer price and the actual offer price. The advantage of the stochastic frontier model in this application is that it can be used to measure the level of deliberate underpricing in the pre-market without using aftermarket information.

Kumbhakar and Lovell (2000) offer an excellent review of the existing models in the stochastic frontier literature. The more recent developments in the literature aim to make the model more flexible. For instance, correlations between v_i and u_i are made possible through copula functions. If time series or panel data are available, then it is possible to make u_t or u_{it} serially correlated. Semiparametric and nonparametric estimation methods are also adopted to estimate the frontier of the production function (for example, $f(x_i; \beta)$) and the frontier of the cost function (for example, $g(w_i, y_i; \gamma)$) so that they are not restricted to specific functional forms.

HUNG-JEN WANG

See also **X-efficiency.**

Bibliography

Aigner, D., Lovell, C.A.K. and Schmidt, P. 1977. Formulation and estimation of stochastic frontier production function models. *Journal of Econometrics* 6, 21–37.

Battese, G.E. and Coelli, T.J. 1988. Prediction of firm-level technical efficiencies with a generalized frontier production function and panel data. *Journal of Econometrics* 38, 387–99.

Greene, W.H. 1980. On the estimation of a flexible frontier production model. *Journal of Econometrics* 13, 101–15.

Hofler, R.A. and Murphy, K.J. 1992. Underpaid and overworked: measuring the effect of imperfect information on wages. *Economic Inquiry* 30, 511–29.

Hunt-McCool, J., Koh, S.C. and Francis, B.B. 1996. Testing for deliberate underpricing in the IPO premarket: a stochastic frontier approach. *Review of Financial Studies* 9, 1251–69.

Jondrow, J., Lovell, C.A.K., Materov, I.S. and Schmidt, P. 1982. On the estimation of technical inefficiency in the stochastic frontier production function model. *Journal of Econometrics* 19, 233–8.

Kumbhakar, S.C. 1997. Modeling allocative inefficiency in a translog cost function and cost share equations: an exact relationship. *Journal of Econometrics* 76, 351–6.

Kumbhakar, S.C. and Lovell, C.A.K. 2000. *Stochastic Frontier Analysis*. Cambridge, UK and New York: Cambridge University Press.

Kumbhakar, S.C. and Tsionas, E.G. 2005. The joint measurement of technical and allocative inefficiencies: an application of Bayesian inference in nonlinear random-effects models. *Journal of American Statistical Association* 100, 736–47.

Kumbhakar, S.C. and Wang, H.-J. 2005. Estimation of growth convergence using a stochastic production frontier approach. *Economics Letters* 3, 300–5.

Kumbhakar, S.C. and Wang, H.-J. 2006. Estimation of technical and allocative inefficiency: a primal system approach. *Journal of Econometrics* 134, 419–40.

Meeusen, W. and van den Broeck, J. 1977. Technical efficiency and dimension of the firm: some results on the use of frontier production functions. *Empirical Economics* 2, 109–22.

Schmidt, P. and Lin, T.F. 1984. Simple tests of alternative specifications in stochastic frontier models. *Journal of Econometrics* 24, 349–61.

Schmidt, P. and Lovell, C.A.K. 1979. Estimating technical and allocative inefficiency relative to stochastic production and cost frontiers. *Journal of Econometrics* 9, 343–66.

Schmidt, P. and Sickles, R.C. 1984. Production frontiers and panel data. *Journal of Business and Economic Statistics* 2, 367–74.

Wang, H.-J. 2003. A stochastic frontier analysis of financing constraints on investment: the case of financial liberalization in Taiwan. *Journal of Business and Economic Statistics* 3, 406–19.

stochastic optimal control

In the long history of mathematics, stochastic optimal control is a rather recent development. Using Bellman's principle of optimality along with measure-theoretic and functional-analytic methods, several mathematicians such as H. Kushner, W. Fleming, R. Rishel, W.M. Wonham and J.M. Bismut, among many others, made important contributions to this new area of mathematical research during the 1960s and early 1970s. For a complete mathematical exposition of the continuous time case, see Fleming and Rishel (1975), and for the discrete time case, see Bertsekas and Shreve (1978).

The assimilation of the mathematical methods of stochastic optimal control by economists was very rapid. Several economic papers started to appear in the early 1970s, among which we mention Merton (1971) on consumption and portfolio rules using continuous time methodology and Brock and Mirman (1972) on optimal economic growth under uncertainty using discrete time techniques. Since then, stochastic optimal control methods have been applied in most major areas of economics such as price theory, macroeconomics, monetary economics and financial economics. Chang (2004) offers a rigorous mathematical exposition of stochastic control methods and numerous examples from economics.

In this article we (a) state the stochastic optimal control problem, (b) explain how it differs from deterministic optimal control and why that difference is crucial in economic problems, (c) present intuitively the methodology of optimal stochastic control and, finally, (d) give an illustration from optimal stochastic economic growth.

Consider the problem:

$$J[k(t), t, \infty] = \max E_t \int_t^\infty e^{-\rho s} u[k(s), v(s)] \mathrm{d}s$$

$$(1)$$

subject to the conditions

$$\mathrm{d}k(t) = T[k(t), v(t)]\mathrm{d}t + \sigma[k(t), v(t)]\mathrm{d}Z(t),$$
$$k(t) \text{given.}$$

$$(2)$$

Here $v = v(t) = v(t, \omega)$ is the control random variable, $k = k(t) = k(t, \omega)$ is the state random variable, $\rho \geq 0$ is the discount on future utility, u denotes a utility function, T is the drift component of technology, σ is the diffusion component, $\mathrm{d}Z$ is a Wiener process and E_t denotes expectation conditioned on $k(t)$ and $v(t)$.

We note immediately that (1) and (2) generalize the deterministic optimal control by incorporating *uncertainty*. The modelling of economic uncertainty is achieved by allowing both the control and state variables to be random and more importantly by postulating that condition (2) is described by a stochastic differential equation of the Itō type.

In the problem described by (1) and (2), if $\sigma(k, v)=0$ and if k and v are assumed to be real variables instead of random, then (1) and (2) reduce to the special case of deterministic optimal control. Thus, the stochastic optimal control problem differs from the deterministic optimal control in the sense that the former generalizes the latter, or equivalently, in the sense that the latter is a special case of the former. This is crucial mathematical difference.

For the economist, the generalization achieved from stochastic optimal control means that the analysis of dynamic economic models becomes more realistic. The economic theorist who uses stochastic optimal control in positive or in welfare economics, in free market or centrally planned economies allows for randomness. Measurement errors, omission of important variables, non-exact relationships, incomplete theories and other methodological complexities are modelled in stochastic optimal control by allowing the control and state variables to be random, and also, by incorporating *pure randomness* through the white noise factor d$Z(t)$. The random variable d$Z(t)$ describes increments in the Wiener process $\{Z(t), t \geq 0\}$ that are independent and normally distributed with mean, E[d$Z(t)$]=0 and variance Var[d$Z(t)$]=dt.

In particular, eq. (2) is a significant economic generalization of the analogous equation in deterministic control. The reader may recall that in deterministic control the constraint is given by $\dot{k} \equiv dk(t)/dt = T[k(t), v(t)]$. Because d$k(t)$ in (2) is a random variable we can compute its mean and variance. They are given by

$$E[\,dk(t)\,] = T[\,k(t), v(t)\,];$$
$$Var[\,dk(t)\,] = \sigma^2[\,k(t), v(t)\,]dt.$$

Thus (2) is a meaningful generalization of its counterpart in deterministic control because it involves means, standard deviations and pure randomness in capturing the complexities of economic reality. A comprehensive analysis of Itō equations, such as (2), is given in Malliaris and Brock (1982).

The problem in (1) and (2) is a stochastic analogue of the deterministic one studied in Arrow and Kurz (1970, pp. 27–51). A standard technique for our problem, as in the case of Arrow and Kurz, is Bellman's (1957, p. 83) 'Principle of Optimality' according to which 'an optimal policy has the property that, whatever the initial state and control are, the remaining decisions must constitute an optimal policy with regard to the state resulting from the first decision'. The problem in eqs (1) and (2) is studied here for the undiscounted, finite horizon case, that is, for $\rho = 0$ and $N < \infty$.

Using Bellman's technique for dynamic programming, eqs (1) and (2) can be analysed as follows:

$J[k(t), t, N]$

$= \max E_t \int_t^N u(k, v) ds$

$= \max E_t \int_t^{t+\Delta t} u(k, v) ds + \max E_{t+\Delta t} \int_{t+\Delta t}^N u(k, v) ds$

$= \max E_t \int_t^{t+\Delta t} u(k, v) ds + J[k(t + \Delta t), t + \Delta t, N]$

$= \max E_t \left\{ \int_t^{t+\Delta t} u(k, v) ds + J[k(t + \Delta t), t + \Delta t, N] \right\}$

$= \max E_t \left\{ u[k(t), v(t)]\Delta t + J[k(t), t, N] \right.$

$\left. + J_k \Delta k + J_t \Delta t + \frac{1}{2} J_{kk} (\Delta k)^2 \right.$

$\left. + J_{kt} (\Delta k)(\Delta t) + \frac{1}{2} J_{tt} (\Delta t)^2 + o(\Delta t) \right\}. \quad (3)$

Observe that Taylor's theorem is used to obtain (3) and therefore it is assumed that J has continuous partial derivatives of all orders less than three in some open set containing the line segment connecting the two points $[k(t), t]$ and $[k(t + \Delta t), t + \Delta t]$. Let (2) be approximated and write

$$\Delta k = T(k, v)\Delta t + \sigma(k, v)\Delta Z + o(\Delta t). \quad (4)$$

Insert (4) into (3) and use the multiplication rules

$$(\Delta Z) \times (\Delta t) = 0, \quad (\Delta t) \times (\Delta t) = 0 \text{ and}$$
$$(\Delta Z) \times (\Delta Z) = \Delta t$$

to get

$$0 = \max E_t \left[u(k, v)\Delta t + \left(J_k T + J_t + \frac{1}{2} J_{kk} \sigma^2 \right)\Delta t \right.$$
$$\left. + J_k \sigma \Delta Z + o(\Delta t) \right].$$

$$(5)$$

For notational convenience let

$$\Delta J = \left[J_t + J_k T + \tfrac{1}{2} J_{kk} \sigma^2 \right]\Delta t + J_k \sigma \Delta Z.$$

$$(6)$$

Using (6), eq. (5) becomes

$$0 = \max E_t[u(k, v)\Delta t + \Delta J + o(\Delta t)]. \quad (7)$$

This is a partial differential equation with boundary condition $[(\partial J)/\partial K][k(N), N, N] = 0$. Pass E_t through the parentheses of (7) and, after dividing both sides by Δt, let $\Delta t \to 0$ to conclude

$$0 = \max \left[u(k, v) + J_t + J_k T(k, v) + \frac{1}{2} J_{kk} \sigma^2(k, v) \right].$$

$$(8)$$

This last equation is usually written as

$$-J_t = \max \left[u(k, v) + J_k T(k, v) + \frac{1}{2} J_{kk} \sigma^2(k, v) \right]$$

$$(9)$$

and is known as the Hamilton–Jacobi–Bellman equation of stochastic control theory.

Next, we define the costate variable $p(t)$ as

$$p(t) = J_k[k(t), t, N]$$

from which it follows that its partial derivative with respect to k is

$$p_k = \partial p/\partial k = J_{kk}. \quad (10)$$

Therefore, we may rewrite (9) as

$$-J_t = \max H(k, v, p, \partial p/\partial k). \qquad (11)$$

where H is the functional notation of the expression inside the brackets of (9). Assume next that a function v exists that solves the maximization problem of (11) and denote such a function by

$$v^0 = v^0(k, p, \partial p/\partial k). \qquad (12)$$

Note that v^0 is a function of $k(t)$ and t alone, along the optimum path, because J_k is a function of $k(t)$ and t alone. In the applied control literature, and more specifically in economic applications, v^0 is called a *policy function*. Assuming then that a policy function v^0 exists, (11) may be rewritten as

$$-J_1 = \max H(k, v, p, \partial p/\partial k)$$
$$= H[k, v^0(k, p, \partial p/\partial k), p, \partial p/\partial k]$$
$$= H^0(k, p, \partial p/\partial k). \qquad (13)$$

This last equation is again a functional notation of the right-hand side expression of (9) under the assumption of the existence of an optimum control v^0, that is,

$$H^0(k, p, \partial p/\partial k) = u(k, v^0) + pT(k, v^0)$$
$$+ \frac{1}{2}\frac{\partial p}{\partial k}\sigma^2(k, v^0). \qquad (14)$$

Equipped with the above analysis we can now state

Proposition 1 (Pontryagin stochastic maximum principle). Suppose that $k(t)$ and $v^0(t)$ solve for $t \in [0, N]$ the problem:

$$\max E_0 \int_0^N u(k, v)dt$$

subject to the conditions

$$dk = T(k, v)dt + \sigma(k, v)dZ, \quad k(t)\text{given}.$$

Then, there exists a costate variable $p(t)$ such that for each t, $t \in [0, N]$:

(1) v^0 maximizes $H(k, v, p, \partial p/\partial k)$ where

$$H(k, v, p, \partial p/\partial k) = u(k, v) + pT(k, v) + \frac{1}{2}\frac{\partial p}{\partial k}\sigma^2;$$

(2) the costate function $p(t)$ satisfies the stochastic differential equation

$$dp = -H_k^0 dt + \sigma(k, v^0)J_{kk}dZ; \text{ and}$$

(3) the transversality condition holds

$$p[k(N), N] = \frac{\partial J}{\partial k}[k(N), N, N] = 0,$$
$$p(N)k(N) = 0.$$

Finally, we briefly illustrate the stochastic optimal control technique to the stochastic Ramsey problem studied in Merton (1975). The problem is to find an optimal saving policy s^0 to

$$\text{maximize } E_0 \int_0^T u(c)dt \qquad (15)$$

subject to

$$dk = \left[sf(k) - (n - \sigma^2)k\right]dt - \sigma k dZ \qquad (16)$$

and $k(t) \geq 0$ for each t. Here, u is a strictly concave, von Neumann–Morgenstern utility function of per capita consumption c for the representative consumer and $f(k)$ is a well-behaved production function. Note that $c=(1 - s)f(k)$ and that eq. (16) generalizes Solow's equation of neoclassical economic growth. Uncertainty enters (16) via randomness in the rate of growth of the labour force. Let

$$J[k(t), t, T] = \max_s E_t \int_t^T u[(1 - s)f(k)]dt.$$

The Hamilton–Jacobi–Bellman equation is given by

$$0 = \max\bigg\{ u[(1 - s)f(k)] + J_t$$
$$+ J_k\left[sf(k) - (n - \sigma^2)k\right] + \frac{1}{2}J_{kk}\sigma^2 k^2 \bigg\} \qquad (17)$$

which yields

$$\frac{du}{dc}\left[(1 - s^0)f(k)\right] = J_k. \qquad (18)$$

To solve for s^0, in principle, one solves (18) for s^0 as a function of k, $T - t$ and J_k, and then substitutes this solution into (17) which becomes a *partial differential equation for J*. Once (17) is solved, then its solution is substituted back into (18) to determine s^0 as a function of k and $T - t$. The nonlinearity of the Hamilton–Jacobi–Bellman equation causes difficulties in finding a closed form solution for the optimal saving function. However, if we let $\sigma=0$ in (17), one obtains the classical Ramsey rule of the certainty case.

From the fact that numerous economic questions involve uncertainty and can be formulated as stochastic optimal control problems, one may conclude that economic interest is likely to be lively in this area for some time to come.

A.G. MALLIARIS

See also **dynamic programming; Markov processes; Wiener process.**

Bibliography

Arrow, K.J. and Kurz, M. 1970. *Public Investment, the Rate of Return, and Optimal Fiscal Policy.* Baltimore, MD: Johns Hopkins Press.

Bellman, R. 1957. *Dynamic Programming.* Princeton: Princeton University Press.

Bertsekas, D.P. and Shreve, S.E. 1978. *Stochastic Optimal Control: The Discrete Time Case.* New York: Academic Press.

Brock, W.A. and Mirman, L. 1972. Optimal economic growth and uncertainty: the discounted case. *Journal of Economic Theory* 4, 479–513.

Chang, F.R. 2004. *Stochastic Optimization in Continuous Time.* New York: Cambridge University Press.

Fleming, W.H. and Rishel, R.W. 1975. *Deterministic and Stochastic Optimal Control.* New York: Springer.

Malliaris, A.G. and Brock, W.A. 1982. *Stochastic Methods in Economics and Finance.* Amsterdam: North-Holland.

Merton, R.C. 1971. Optimal consumption and portfolio rules in a continuous-time model. *Journal of Economic Theory* 3, 373–413.

Merton, R.C. 1975. An asymptotic theory of growth under uncertainty. *Review of Economic Studies* 42, 375–93.

stochastic volatility models

Stochastic volatility (SV) is the main concept used in the fields of financial economics and mathematical finance to deal with the endemic time-varying volatility and codependence found in financial markets. Such dependence has been known for a long time; early commentators include Mandelbrot (1963) and Officer (1973). It was also clear to the founding fathers of modern continuous time finance that homogeneity was an unrealistic if convenient simplification; for example, Black and Scholes (1972, p. 416) wrote, '... there is evidence of non-stationarity in the variance. More work must be done to predict variances using the information available.' Heterogeneity has deep implications for the theory and practice of financial economics and econometrics. In particular, asset pricing theory is dominated by the idea that higher rewards may be expected when we face higher risks, but these risks change through time in complicated ways. Some of the changes in the level of risk can be modelled stochastically, where the level of volatility and degree of codependence between assets is allowed to change over time. Such models allow us to explain, for example, empirically observed departures from Black–Scholes–Merton prices for options and understand why we should expect to see occasional dramatic moves in financial markets.

The outline of this article is as follows. In the first section I trace the origins of SV and provide links with the basic models used today in the literature. In the second section I briefly discuss some of the innovations in the second generation of SV models. In the third section I briefly discuss the literature on conducting inference for SV models. In the fourth section I talk about the use of SV to price options. In the fifth section I consider the connection of SV with realized volatility. An extensive review of this literature is given in Shephard (2005).

The origin of SV models

The origins of SV are messy. I will give five accounts, which attribute the subject to different sets of people. Clark (1973) introduced Bochner's (1949) time-changed Brownian motion (BM) into financial economics. He wrote down a model for the log-price M as

$$M_t = W_{\tau_t}, \quad t \geq 0, \tag{1}$$

where W is Brownian motion (BM), t is continuous time, τ is a time change and $W \perp\!\!\!\perp \tau$, where $\perp\!\!\!\perp$ denotes independence. The definition of a time-change is a non-negative process with non-decreasing sample paths, although Clark also assumed τ has independent increments. Then $M_t | \tau_t \sim N(0, \tau_t)$. Further, so long (for each t) as $E\sqrt{\tau_t} < \infty$, then M is a martingale (written $M \in \mathcal{M}$) for this is necessary and sufficient to ensure that $E|M_t| < \infty$. More generally, if (for each t) $\tau_t < \infty$, then M is a local martingale (written $M \in \mathcal{M}_{loc}$). Hence Clark was solely modelling the instantly risky component of the log of an asset price, written Y, which in modern semimartingale (written $Y \in \mathcal{SM}$) notation we would write as

$$Y = A + M.$$

The increments of A can be thought of as the instantly available reward component of the asset price, which compensates the investor for being exposed to the risky increments of M. The A process is assumed to be of finite variation (written $A \in \mathcal{FV}$).

To the best of my understanding, the first published direct volatility clustering SV paper is that by Taylor (1982). His discrete time model of daily returns, computed as the difference of log-prices

$$y_i = Y_i - Y_{i-1}, \quad i = 1, 2, \ldots,$$

where I have assumed that $t = 1$, represents one day to simplify the exposition. He modelled the risky part of returns, $m_i = M_i - M_{i-1}$, as a product process

$$m_i = \sigma_i \varepsilon_i. \tag{2}$$

Taylor assumed ε has a mean of zero and unit variance, while σ is some non-negative process, finishing the model by assuming $\varepsilon \perp\!\!\!\perp \sigma$. Taylor modelled ε as an autoregression and

$$\sigma_i = \exp(h_i/2),$$

where h is a non-zero mean Gaussian linear process. The leading example of this is the first order autoregression

$$h_{i+1} = \mu + \phi(h_i - \mu) + \eta_i, \quad \eta_i \sim NID(0, \sigma_\eta^2).$$
$$(3)$$

In the modern SV literature the model for ε is typically simplified to an i.i.d. process, for we deal with the predictability of asset prices through the A process rather than via M. This is now often called the log-normal SV model in the case where ε is also assumed to be Gaussian. In general, M is always a local martingale.

A key feature of SV, which is not discussed by Taylor, is that it can deal with leverage effects. Leverage effects are associated with the work of Black (1976) and Nelson (1991), and can be implemented in discrete time SV models by negatively correlating the Gaussian ε_i and η_i. This still implies that $M \in \mathcal{M}_{loc}$, but allows the direction of returns to influence future movements in the volatility process, with falls in prices associated with rises in subsequent volatility.

Taylor's discussion of the product process was predated by a decade in the (until recently) unpublished Rosenberg (1972). Rosenberg introduces product processes, empirically demonstrating that time-varying volatility is partially forecastable and so breaks with the earlier work by Clark. He suggests an understanding of aggregational Gaussianity of returns over increasing time intervals and pre-dates a variety of econometric methods for analysing heteroskedasticity.

In continuous time the product process is the standard SV model

$$M_t = \int_0^t \sigma_s dW_s, \qquad (4)$$

where the non-negative spot volatility σ is assumed to have càdlàg sample paths (which means it can possess jumps). The squared volatility process is often called the spot variance.

The first use of continuous-time SV models in financial economics was, to my knowledge, by Johnson (1979), who studied the pricing of options using time-changing volatility models in continuous time (see also Johnson and Shanno, 1987; Wiggins, 1987). The best-known paper in this area is Hull and White (1987). Each of these authors desired to generalize the Black and Scholes (1973) approach to option pricing models to deal with volatility clustering. In the Hull and White approach, σ^2 follows the solution to the univariate SDE

$$d\sigma^2 = \alpha(\sigma^2)dt + \omega(\sigma^2)dB,$$

where B is a second Brownian motion and $\omega(.)$ is a non-negative deterministic function.

The probability literature has demonstrated that SV models and their time-changed BM relatives are fundamental. This theoretical development will be the fifth strand of literature that I think of as representing the origins of modern stochastic volatility research. Suppose we simply assume that $M \in \mathcal{M}_{loc}^c$, a process with continuous local martingale sample paths. Then the celebrated Dambis–Dubins–Schwartz theorem shows that M can be written as a time-changed Brownian motion. Further, the time-change is the quadratic variation (QV) process

$$[M]_t = \text{p} - \lim_{n\to\infty} \sum_{j=1}^n (M_{t_j} - M_{t_{j-1}})(M_{t_j} - M_{t_{j-1}})',$$
$$(5)$$

for any sequence of partitions $t_0 = 0 < t_1 < \ldots < t_n = t$ with $\sup_j\{t_j - t_{j-1}\} \to 0$ for $n \to \infty$. What is more, as M has continuous sample paths, so must $[M]$. Under the stronger condition that $[M]$ is absolutely continuous, then M can be written as a stochastic volatility process. This latter result, which is called the martingale representation theorem, is due to Doob (1953). Taken together, this implies that time-changed BMs are canonical in continuous sample path price processes, and SV models are special cases of this class. A consequence of the fact that, for continuous sample path time-change BM, $[M] = \tau$, is that in the SV case

$$[M]_t = \int_0^t \sigma_s^2 ds.$$

The SV framework has an elegant multivariate generalization. In particular, write a p-dimensional price process M as (4) but where σ is a matrix process whose elements are all càdlàg, W is a multivariate BM process. Further $[M]_t = \int_0^t \sigma_s \sigma_s' ds$.

Second-generation model building

Univariate models
General observations
In initial diffusion-based models the volatility was Markovian with continuous sample paths. Research in the late 1990s and early 2000s has shown that more complicated volatility dynamics are needed to model either options data or high frequency return data. Leading extensions to the model are to allow jumps into the volatility SDE (for example, Barndorff-Nielsen and Shephard, 2001; Eraker, Johannes and Polson, 2003) or to model the volatility process as a function of a number of separate stochastic processes or factors (for example, Chernov et al., 2003; Barndorff-Nielsen and Shephard, 2001).

Long memory

In the SV literature considerable progress has been made on working with both discrete and continuous time long-memory SV. This involves specifying a long-memory model for σ in discrete or continuous time.

Breidt, Crato and de Lima (1998) and Harvey (1998) looked at discrete time models where the log of the volatility was modelled as a fractionally integrated process. In continuous time there is work on modelling the log of volatility as fractionally integrated Brownian motion by Comte and Renault (1998). More recent work, which is econometrically easier to deal with, is the square-root model driven by fractionally integrated BM introduced in an influential paper by Comte, Coutin and Renault (2003) and the infinite superposition of non-negative OU processes introduced by Barndorff-Nielsen (2001).

Jumps

In detailed empirical work a number of researchers have supplemented standard SV models by adding jumps to the price process or to the volatility dynamics. Bates (1996) was particularly important as it showed the need to include jumps in addition to SV, at least when volatility is Markovian. Eraker, Johannes and Polson (2003) deals with the efficient inference of these types of models. A radical departure in SV models was put forward by Barndorff-Nielsen and Shephard (2001), who suggested building volatility models out of pure jump processes called non-Gaussian OU processes. Closed form option pricing based on this structure is studied briefly in Barndorff-Nielsen and Shephard (2001) and in detail by Nicolato and Venardos (2003). All these non-Gaussian OU processes are special cases of the affine class advocated by Duffie, Pan and Singleton (2000) and Duffie, Filipovic and Schachermayer (2003).

Multivariate models

Diebold and Nerlove (1989) introduced volatility clustering into traditional factor models, which are used in many areas of asset pricing. In continuous time their type of model has the interpretation

$$M_t = \sum_{j=1}^{J} \int \beta_{(j)s} dF_{(j)s} + G_t,$$

where the factors $F_{(1)}, F_{(2)}, \ldots, F_{(J)}$ are independent univariate SV models and G is correlated multivariate BM. Some of the related papers on the econometrics of this topic include King, Sentana and Wadhwani (1994) and Fiorentini, Sentana and Shephard (2004), who all fit this kind of model. These papers assume that the factor loading vectors are constant through time.

A more limited multivariate discrete time model was put forward by Harvey, Ruiz and Shephard (1994) who

allowed $M_t = C \int_0^t \sigma_s dW_s$, where σ is a diagonal matrix process and C is a fixed matrix of constants with a unit leading diagonal. This means that the risky part of prices is simply a rotation of a p-dimensional vector of independent univariate SV processes.

Inference based on return data

Moment-based inference

The task is to carry out inference on $\theta = (\theta_1, \ldots, \theta_K)'$, the parameters of the SV model based on a sequence of returns $y = (y_1, \ldots, y_T)'$. Taylor (1982) and Melino and Turnbull (1990) calibrated their models using the method of moments. Systematic studies, using a GMM approach, of which moments to heavily weight in SV models was given in Andersen and Sørensen (1996), Genon-Catalot, Jeantheau and Larédo (2000), Sørensen (2000) and Hoffmann (2002).

A difficulty with using moment-based estimators for continuous time SV models is that it is not straightforward to compute the moments y. In the case of no leverage, general results for the second order properties of y and their squares were given in Barndorff-Nielsen and Shephard (2001). Some quite general results under leverage are also given in Meddahi (2001).

In the discrete time log-normal SV models the approach advocated by Harvey, Ruiz and Shephard (1994) has been influential. Their approach was to remove the predictable part of the returns, so we think of $Y = M$ again, and work with $\log y_i^2 = h_i + \log \varepsilon_i^2$. If the volatility has short memory, then this form of the model can be handled using the Kalman filter, while long-memory models are often dealt with in the frequency domain. Either way, this delivers a Gaussian quasi-likelihood which can be used to estimate the parameters of the model. The linearized model is non-Gaussian due to the long left-hand tail of $\log \varepsilon_i^2$ which generates outliers when ε_i is small.

Simulation-based inference

In the 1990s a number of econometricians started to use simulation-based inference to tackle SV models. To discuss these methods it will be convenient to focus on the simplest discrete time log-normal SV model given by (2) and (3).

MCMC allows us to simulate from $\theta, h|y$, where $h = (h_1, \ldots, h_T)'$. Discarding the h draws yields samples from $\theta|y$. Summarizing yields fully efficient parametric inference. In an influential paper, Jacquier, Polson and Rossi (1994) implemented an MCMC algorithm for this problem. A subsequent paper by Kim, Shephard and Chib (1998) gave quite an extensive discussion of various MCMC algorithms. This is a subtle issue and makes a very large difference to the computational efficiency of the methods (see, for example, Jacquier, Polson and Rossi, 2004; Yu, 2005).

Kim, Shephard and Chib (1998) introduced the first filter using a so-called particle filter. As well as being of substantial scientific interest for decision making, the advantage of a filtering method is that it allows us to compute marginal likelihoods for model comparison and one-step-ahead predictions for model testing.

Although MCMC-based papers are mostly couched in discrete time, a key advantage of the general approach is that it can be adapted to deal with continuous time models by the idea of augmentation. This was fully worked out in Elerian, Chib and Shephard (2001), Eraker (2001) and Roberts and Stramer (2001).

A more novel non-likelihood approach was introduced by Smith (1993) and later developed by Gourieroux, Monfort and Renault (1993) and Gallant and Tauchen (1996) into what is now called indirect inference or the efficient method of moments. Here I briefly give a stylized version of this approach.

Suppose there is an auxiliary model for the returns (for example, GARCH) whose density, $g(y; \psi)$, is easy to compute and, for simplicity of exposition, has $\dim(\psi) = \dim(\theta)$. Then compute its MLE, which we write as $\hat{\psi}$. We assume this is a regular problem so that $\partial \log g(y; \hat{\psi})/\partial \psi = 0$ recalling that y is the observed return vector. Simulate a very long process from the SV model using parameters θ, which we denote by y^+, and evaluate the score using not the data but this simulation. This produces

$$\left. \frac{\partial \log g(y^+; \psi)}{\partial \psi} \right|_{\psi = \hat{\psi}}, \quad y^+ \sim f(y; \theta).$$

Then move θ around until the score is again zero, but now under the simulation. Write the point where this happens as $\tilde{\theta}$. It is called the indirect inference estimator.

Options

Models

SV models provide a basis for realistic modelling of option prices. We recall the central role played by Johnson and Shanno (1987) and Wiggins (1987). The best-known paper in this area is by Hull and White (1987), who looked at a diffusion volatility model with leverage effects. They assumed that volatility risk was unrewarded and priced their options either by approximation or by simulation. Hull and White (1987) indicated that SV models could produce smiles and skews in option prices, which are frequency observed in market data. The skew is particularly important in practice, and Renault and Touzi (1996) proved that can be achieved in SV models via leverage effects.

The first analytic option pricing formulae were developed by Stein and Stein (1991) and Heston (1993). The only other closed form solution I know of is the one based on the Barndorff-Nielsen and Shephard (2001) class of non-Gaussian OU SV models. Nicolato and

Venardos (2003) provided a detailed study of such option pricing solutions; see also the textbook exposition in Cont and Tankov (2004, ch. 15). Slightly harder computationally to deal with is the more general affine class of models highlighted by Duffie, Filipovic and Schachermayer (2003).

Econometrics of SV option pricing

In theory, option prices themselves should provide rich information for estimating and testing volatility models. I discuss the econometrics of options in the context of the stochastic discount factor (SDF) approach, which has a long history in financial economics and is emphasized in, for example, Cochrane (2001) and Garcia, Ghysels and Renault (2006). For simplicity I assume interest rates are constant. We start with the standard Black–Scholes (BS) problem, which will take a little time to recall, before being able to rapidly deal with the SV extension. We model

$$d \log Y = (r + p - \sigma^2/2)dt + \sigma dW,$$
$$d \log \widetilde{M} = hdt + bdW,$$

where $\widetilde{M}$ is the SDF process and r the riskless short rate, and σ, h, b and p, the risk premium, are assumed constant for the moment.

We price all contingent payoffs $g(Y_T)$ as $C_t = E(\widetilde{M}_T/\widetilde{M}_t)g(Y_T)|\mathcal{F}_t)$, the expected discounted value of the claim where $T > t$. For this model to make financial sense we require that $\widetilde{M}_t Y_t$ and $\widetilde{M}_t \exp(tr)$ are local martingales, which is enough to mean that adding other independent BMs to the $\log \widetilde{M}$ process makes no difference to C or Y, the observables. These two constraints imply, respectively, $p + b\sigma = 0$ and $h = -r - b^2/2$. This means that (C^{BS}, Y) is driven by a single W.

When we move to the standard SV model we can remove this degeneracy. The functional form for the SV Y process is unchanged, but we now allow

$$d \log \widetilde{M} = hdt + adB + bdW, \quad d\sigma^2 = \alpha dt + \omega dB,$$

where we assume that $B \perp\!\!\!\perp W$ to simplify the exposition. The SV structure means that p will have to change through time in response to the moving σ^2. B is again redundant in the SDF (but not in the volatility) so the usual SDF conditions again imply $h = -r - \frac{1}{2}a^2$ and $p + b\sigma = 0$. This implies that the move to the SV case has little impact, except that the sample path of $\sigma^2 \perp\!\!\!\perp W$. So the generalized BS (GBS) price is

$$C_t^{GBS}(\sigma_t^2) = E\left(\frac{\widetilde{M}_T}{\widetilde{M}_t} g(Y_T)|\mathcal{F}_t \right)$$
$$= E\left\{ C_t^{BS}\left(\frac{1}{T-t}\int_t^T \sigma_u^2 du \right) |\sigma_t^2, Y_t \right\}.$$

Now C^{GBS} is a function of both Y_t and σ_t^2, which means that (C^{GBS}, Y) is not degenerate. From an econometric viewpoint this is an important step, meaning inference on options is just the problem of making inference on a complicated bivariate diffusion process. When we allow leverage back into the model, the analysis becomes slightly more complicated algebraically.

In some recent work econometricians have been trying to use data from underlying assets and option markets to jointly model the dynamics of (C^{GBS}, Y). The advantage of this joint estimation is that we can pool information across data types and estimate all relevant effects which influence Y, σ^2 and $\tilde{M}$. Relevant papers include Chernov and Ghysels (2000), Pastorello, Patilea and Renault (2003), Das and Sundaram (1999) and Bates (2000).

Realized volatility

The advent of very informative high-frequency data has prompted econometricians to study estimators of the increments of the quadratic variation (QV) process and then to use this estimate to project QV into the future in order to predict future levels of volatility. The literature on this starts with independent, concurrent papers by Andersen and Bollerslev (1998), Barndorff-Nielsen and Shephard (2001) and Comte and Renault (1998). Some of this work echoes earlier important contributions from, for example, Rosenberg (1972) and Merton (1980).

A simple estimator of $[Y]$ is the *realized QV process*

$$[Y_\delta]_t = \sum_{j=1}^{\lfloor t/\delta \rfloor} \left(Y_{\delta j} - Y_{\delta(j-1)}\right)\left(Y_{\delta j} - Y_{\delta(j-1)}\right)',$$

thus as $\delta \downarrow 0$ so $[Y_\delta]_t \xrightarrow{P} [Y]_t$. If $A \in \mathcal{F}V^c$, then $[Y] = [M]$, while if we additionally assume that M is SV then $[Y_\delta]_t \xrightarrow{P} \int_0^t \sigma_s \sigma_s' ds$.

In practice it makes sense to look at the increments of the QV process. Suppose we are interested in analysing daily return data, but in addition have higher-frequency data measured at the time interval δ. The i-th daily realized QV is defined as

$$V(Y_\delta)_i = \sum_{j=1}^{\lfloor 1/\delta \rfloor} \left(Y_{i+\delta j} \quad Y_{i+\delta(j-1)}\right)$$
$$\times \left(Y_{i+\delta j} - Y_{i+\delta(j-1)}\right)' \xrightarrow{P} V(Y)_i$$
$$= [Y]_i - [Y]_{i-1},$$

the i-th daily QV. The diagonal elements of $V(Y_\delta)_i$ are called realized variances and their square roots are called realized volatilities.

Andersen et al. (2001) have shown that to forecast the volatility of future asset returns a key input should be predictions of future daily QV. Recall, from Ito's formula, that, if $Y \in \mathcal{SM}^c$ and $M \in \mathcal{M}$, then writing F_t as the filtration generated by the continuous history of Y up to time t then

$$E(y_i y_i' | \mathcal{F}_{i-1}) \simeq E(V(Y)_i | \mathcal{F}_{i-1}).$$

A review of some of this material is given by Barndorff-Nielsen and Shephard (2006a).

A difficulty with this line of argument is that the QV theory tells us only that $V(Y_\delta)_i \xrightarrow{p} V(Y)_i$; it gives no impression of the size of $V(Y_\delta)_i - V(Y)_i$. Jacod (1994) and Barndorff-Nielsen and Shephard (2002) have strengthened the consistency result to provide a univariate central limit theory

$$\frac{\delta^{-1/2}([Y_\delta]_t - [Y]_t)}{\sqrt{2 \int_0^t \sigma_s^4 ds}} \xrightarrow{d} N(0,1),$$

while giving a method for consistently estimating the integrated quarticity $\int_0^t \sigma_s^4 ds$ by using high-frequency data. This analysis was generalized to the multivariate case by Barndorff-Nielsen and (Shephard 2004a). This type of analysis greatly simplifies parametric estimation of SV models, for we can now have estimates of the volatility quantities SV models directly parameterize. Barndorff-Nielsen and Shephard (2002), Bollerslev and Zhou (2002) and Phillips and Yu (2005) study this topic from different perspectives.

Recently there has been interest in studying the impact of market microstructure effects on the estimates of realized covariation. This causes the estimator of the QV to become biased. Leading papers on this topic are Zhou (1996), Fang (1996), Bandi and Russell (2003), Hansen and Lunde (2006) and Zhang, Mykland and Aït-Sahalia (2005). Further, one can estimate the QV of the continuous component of prices in the presence of jumps using the so-called realized bipower variation process. This was introduced by Barndorff-Nielsen and Shephard (2004b; 2006b).

NEIL SHEPHARD

See also **capital asset pricing model; options; options (new perspectives); realized volatility.**

My research is supported by the Economic and Social Science Research Council (UK) through the grant 'High frequency financial econometrics based upon power variation'.

Bibliography

Andersen, T. and Bollerslev, T. 1998. Answering the skeptics: yes, standard volatility models do provide accurate forecasts. *International Economic Review* 39, 885–905.

Andersen, T., Bollerslev, T., Diebold, F. and Labys, P. 2001. The distribution of exchange rate volatility. *Journal of the American Statistical Association* 96, 42–55. Correction published in vol. 98 (2003), p. 501.

Andersen, T. G. and Sørensen, B. 1996. GMM estimation of a stochastic volatility model: a Monte Carlo study. *Journal of Business and Economic Statistics* 14, 328–52.

Bandi, F. and Russell, J. 2003. Microstructure noise, realized volatility, and optimal sampling. Mimeo. Graduate School of Business, University of Chicago.

Barndorff-Nielsen, O. 2001. Superposition of Ornstein–Uhlenbeck type processes. *Theory of Probability and its Applications* 45, 175–94.

Barndorff-Nielsen, O. and Shephard, N. 2001. Non-Gaussian Ornstein–Uhlenbeck-based models and some of their uses in financial economics (with discussion). *Journal of the Royal Statistical Society, Series B* 63, 167–241.

Barndorff-Nielsen, O. and Shephard, N. 2002. Econometric analysis of realised volatility and its use in estimating stochastic volatility models. *Journal of the Royal Statistical Society, Series B* 64, 253–80.

Barndorff-Nielsen, O. and Shephard, N. 2004a. Econometric analysis of realised covariation: high frequency covariance, regression and correlation in financial economics. *Econometrica* 72, 885–925.

Barndorff-Nielsen, O. and Shephard, N. 2004b. Power and bipower variation with stochastic volatility and jumps (with discussion). *Journal of Financial Econometrics* 2, 1–48.

Barndorff-Nielsen, O. and Shephard, N. 2006a. Variation, jumps, and high frequency data in financial econometrics. In *Advances in Economics and Econometrics: Theory and Applications*, vol. 1, ed. R. Blundell, P. Torsten and W. Newey. Cambridge: Cambridge University Press.

Barndorff-Nielsen, O. and Shephard, N. 2006b. Econometrics of testing for jumps in financial economics using bipower variation. *Journal of Financial Econometrics* 4, 1–30.

Bates, D. 1996. Jumps and stochastic volatility: exchange rate processes implicit in deutsche mark options. *Review of Financial Studies* 9, 69–107.

Bates, D. 2000. Post-'97 crash fears in the S-&P 500 futures option market. *Journal of Econometrics* 94, 181–238.

Black, F. 1976. Studies of stock price volatility changes. *Proceedings of the Business and Economic Statistics Section, American Statistical Association*, 177–81.

Black, F. and Scholes, M. 1972. The valuation of option contracts and a test of market efficiency. *Journal of Finance* 27, 399–418.

Black, F. and Scholes, M. 1973. The pricing of options and corporate liabilities. *Journal of Political Economy* 81, 637–54.

Bochner, S. 1949. Diffusion equation and stochastic processes. *Proceedings of the National Academy of Science of the United States of America* 85, 369–70.

Bollerslev, T. and Zhou, H. 2002. Estimating stochastic volatility diffusion using conditional moments of integrated volatility. *Journal of Econometrics* 109, 33–65.

Breidt, F., Crato, N. and de Lima, P. 1998. On the detection and estimation of long memory in stochastic volatility. *Journal of Econometrics* 83, 325–48.

Chernov, M. and Ghysels, E. 2000. A study towards a unified approach to the joint estimation of objective and risk neutral measures for the purpose of options valuation. *Journal of Financial Economics* 56, 407–58.

Chernov, M., Gallant, A., Ghysels, E. and Tauchen, G. 2003. Alternative models of stock price dynamics. *Journal of Econometrics* 116, 225–57.

Clark, P. 1973. A subordinated stochastic process model with fixed variance for speculative prices. *Econometrica* 41, 135–56.

Cochrane, J. 2001. *Asset Pricing*. Princeton: Princeton University Press.

Comte, F., Coutin, L. and Renault, E. 2003. Affine fractional stochastic volatility models. Mimeo. University of Montreal.

Comte, F. and Renault, E. 1998. Long memory in continuous-time stochastic volatility models. *Mathematical Finance* 8, 291–323.

Cont, R. and Tankov, P. 2004. *Financial Modelling with Jump Processes*. London: Chapman and Hall.

Das, S. and Sundaram, R. 1999. Of smiles and smirks: a term structure perspective. *Journal of Financial and Quantitative Analysis* 34, 211–40.

Diebold, F. and Nerlove, M. 1989. The dynamics of exchange rate volatility: a multivariate latent factor ARCH model. *Journal of Applied Econometrics* 4, 1–21.

Doob, J. 1953. *Stochastic Processes*. New York: John Wiley and Sons.

Duffie, D., Filipovic, D. and Schachermayer, W. 2003. Affine processes and applications in finance. *Annals of Applied Probability* 13, 984–1053.

Duffie, D., Pan, J. and Singleton, K. 2000. Transform analysis and asset pricing for affine jump-diffusions. *Econometrica* 68, 1343–76.

Elerian, O., Chib, S. and Shephard, N. 2001. Likelihood inference for discretely observed non-linear diffusions. *Econometrica* 69, 959–93.

Eraker, B. 2001. Markov chain Monte Carlo analysis of diffusion models with application to finance. *Journal of Business and Economic Statistics* 19, 177–91.

Eraker, B., Johannes, M. and Polson, N. 2003. The impact of jumps in returns and volatility. *Journal of Finance* 53, 1269–300.

Fang, Y. 1996. Volatility modeling and estimation of high-frequency data with Gaussian noise. Ph.D. thesis. Sloan School of Management, MIT.

Fiorentini, G., Sentana, E. and Shephard, N. 2004. Likelihood-based estimation of latent generalised ARCH structures. *Econometrica* 12, 1481–517.

Gallant, A. and Tauchen, G. 1996. Which moments to match. *Econometric Theory* 12, 657–81.

Garcia, R., Ghysels, E. and Renault, E. 2006. The econometrics of option pricing. In *Handbook of Financial*

Econometrics, ed. Y. Aït-Sahalia and L. Hansen. Amsterdam: North-Holland.

Genon-Catalot, V., Jeantheau, T. and Larédo, C. 2000. Stochastic volatility as hidden Markov models and statistical applications. *Bernoulli* 6, 1051–79.

Gourieroux, C., Monfort, A. and Renault, E. 1993. Indirect inference. *Journal of Applied Econometrics* 6, S85–S118.

Hansen, P. and Lunde, A. 2006. Realized variance and market microstructure noise (with discussion). *Journal of Business and Economic Statistics* 24, 127–61.

Harvey, A. 1998. Long memory in stochastic volatility. In *Forecasting Volatility in Financial Markets*, ed. J. Knight and S. Satchell. Oxford: Butterworth-Heinemann.

Harvey, A., Ruiz, E. and Shephard, N. 1994. Multivariate stochastic variance models. *Review of Economic Studies* 61, 247–64.

Heston, S. 1993. A closed-form solution for options with stochastic volatility, with applications to bond and currency options. *Review of Financial Studies* 6, 327–43.

Hoffmann, M. 2002. Rate of convergence for parametric estimation in stochastic volatility models. *Stochastic Processes and their Application* 97, 147–70.

Hull, J. and White, A. 1987. The pricing of options on assets with stochastic volatilities. *Journal of Finance* 42, 281–300.

Jacod, J. 1994. Limit of random measures associated with the increments of a Brownian semimartingale. Preprint No. 120. Laboratoire de Probabilitiés, Université Pierre et Marie Curie, Paris.

Jacquier, E., Polson, N. and Rossi, P. 1994. Bayesian analysis of stochastic volatility models (with discussion). *Journal of Business and Economic Statistics* 12, 371–417.

Jacquier, E., Polson, N. and Rossi, P. 2004. Stochastic volatility models: univariate and multivariate extensions. *Journal of Econometrics* 122, 185–212.

Johnson, H. 1979. Option pricing when the variance rate is changing. Working paper. University of California, Los Angeles.

Johnson, H. and Shanno, D. 1987. Option pricing when the variance is changing. *Journal of Financial and Quantitative Analysis* 22, 143–51.

Kim, S., Shephard, N. and Chib, S. 1998. Stochastic volatility: likelihood inference and comparison with ARCH models. *Review of Economic Studies* 65, 361–93.

King, M., Sentana, E. and Wadhwani, S. 1994. Volatility and links between national stock markets. *Econometrica* 62, 901–33.

Mandelbrot, B. 1963. The variation of certain speculative prices. *Journal of Business* 36, 394–419.

Meddahi, N. 2001. An eigenfunction approach for volatility modeling. Cahiers de recherche No. 2001–29. Department of Economics, University of Montreal.

Melino, A. and Turnbull, S. 1990. Pricing foreign currency options with stochastic volatility. *Journal of Econometrics* 45, 239–65.

Merton, R. 1980. On estimating the expected return on the market: an exploratory investigation. *Journal of Financial Economics* 8, 323–61.

Nelson, D. 1991. Conditional heteroskedasticity in asset pricing: a new approach. *Econometrica* 59, 347–70.

Nicolato, E. and Venardos, E. 2003. Option pricing in stochastic volatility models of the Ornstein–Uhlenbeck type. *Mathematical Finance* 13, 445–66.

Officer, R. 1973. The variability of the market factor of the New York stock exchange. *Journal of Business* 46, 434–53.

Pastorello, S., Patilea, V. and Renault, E. 2003. Iterative and recursive estimation in structural non-adaptive models. *Journal of Business and Economic Statistics* 21, 449–509.

Phillips, P. and Yu, J. 2005. A two-stage realized volatility approach to the estimation for diffusion processes from discrete observations. Discussion Paper No. 1523. Cowles Foundation, Yale University.

Renault, E. and Touzi, N. 1996. Option hedging and implied volatilities in a stochastic volatility model. *Mathematical Finance* 6, 279–302.

Roberts, G. and Stramer, O. 2001. On inference for nonlinear diffusion models using the Hastings–Metropolis algorithms. *Biometrika* 88, 603–21.

Rosenberg, B. 1972. The behaviour of random variables with nonstationary variance and the distribution of security prices. Working paper 11, Graduate School of Business Administration, University of California, Berkeley. Reprinted in N. Shephard (2005).

Shephard, N. 2005. *Stochastic Volatility: Selected Readings*. Oxford: Oxford University Press.

Smith, A. 1993. Estimating nonlinear time series models using simulated vector autoregressions. *Journal of Applied Econometrics* 8, S63–S84.

Sørensen, M. 2000. Prediction based estimating equations. *Econometrics Journal* 3, 123–47.

Stein, E. and Stein, J. 1991. Stock price distributions with stochastic volatility: an analytic approach. *Review of Financial Studies* 4, 727–52.

Taylor, S. 1982. Financial returns modelled by the product of two stochastic processes – a study of daily sugar prices 1961–79. In *Time Series Analysis: Theory and Practice*, vol. 1, ed. O. Anderson. Amsterdam: North-Holland.

Wiggins, J. 1987. Option values under stochastic volatilities. *Journal of Financial economics* 19, 351–72.

Yu, J. 2005. On leverage in a stochastic volatility model. *Journal of Econometrics* 127, 165–78.

Zhang, L., Mykland, P. and Aït-Sahalia, Y. 2005. A tale of two time scales: determining integrated volatility with noisy high-frequency data. *Journal of the American Statistical Association* 100, 1394–411.

Zhou, B. 1996. High-frequency data and volatility in foreign-exchange rates. *Journal of Business and Economic Statistics* 14, 45–52.